Phylogeny of the Living World—Bacteria

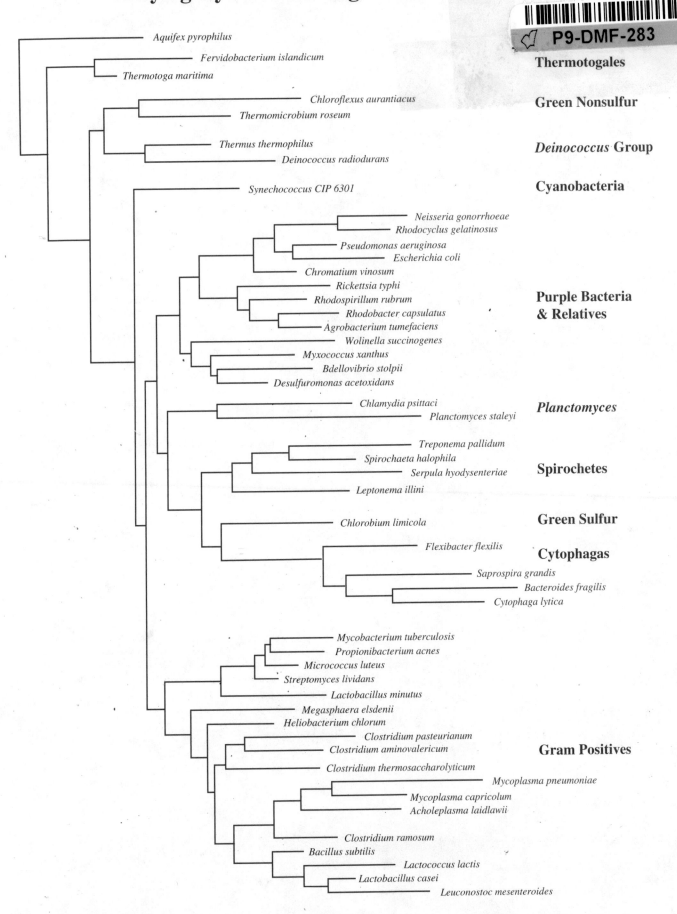

PHYLOGENETIC TREE OF BACTERIA. This tree is derived from 16S ribosomal RNA sequences. Eleven major groups of Bacteria can be defined as indicated. Compare with Figure 16.1. *Data of Carl R. Woese.*

Brock
Biology
of Microorganisms

MICHAEL T. MADIGAN dedicates this book to the memory of Andy and Marcy (Yorg and Pedder or "The Feeders")—the two best friends he ever had.

JOHN M. MARTINKO dedicates this book to his wife Judy and to his family. Thanks for your unending patience, good humor, and support during the seemingly unending process of getting this finished!

JACK PARKER dedicates this book to Justine, D'Arcy, and Grant Parker: three of the best and brightest.

About the Authors

MICHAEL T. MADIGAN (in photo with Andy) received a bachelor's degree in biology and education from Wisconsin State University at Stevens Point in 1971 and M.S. and Ph.D. degrees in 1974 and 1976, respectively, from the University of Wisconsin, Madison, Department of Bacteriology. His graduate work involved study of the biology of hot spring photosynthetic bacteria under the direction of Thomas D. Brock. Following three years of postdoctoral training in the Department of Microbiology, Indiana University, where he worked on photosynthetic bacteria with Howard Gest, he moved to Southern Illinois University at Carbondale, where he is now Professor of Microbiology. He has been a coauthor of *Biology of Microorganisms* since the fourth edition (1984) and teaches courses in introductory microbiology and bacterial diversity. In 1988 he was selected as the outstanding teacher in the College of Science, and in 1993 its outstanding researcher. His research has dealt almost exclusively with anoxygenic phototrophic bacteria, especially those representatives that inhabit extreme environments, and has focused on the nitrogen-fixing and major photosynthetic properties of these organisms. He has published nearly 75 research papers and has coedited a major treatise on photosynthetic bacteria. His nonscientific interests include reading, hiking, tree planting, and caring for his dogs and horses. He lives on a quiet lake about five miles from the SIU campus with his wife, Nancy, two dogs, Willie and Plum, and King, Peggy, and Silas (horses).

JOHN M. MARTINKO attended The Cleveland State University and majored in biology with a chemistry minor. As an undergraduate student he participated in a unique cooperative education program gaining research experience in several microbiology and immunology laboratories. He then worked for two years at Case Western Reserve University as a Laboratory Manager, continuing his cooperative education research on the structure and serology of *Streptococcus pyogenes* cell wall antigens and on the epidemiology of streptococcal infections. He next went to the State University of New York (SUNY) at Buffalo where he did research on antibody specificity for his M.A. and Ph.D. (1978) in Microbiology. As a postdoctoral fellow, he worked at Albert Einstein College of Medicine in New York on the structure of major histocompatibility complex proteins. Since 1981, he has been in the Department of Microbiology at Southern Illinois University at Carbondale where he is currently the Chair and an Associate Professor. His research interests include the structure, genetics, and evolution of histocompatibility complex proteins and T-cell receptors, the molecular evolution of blood group antigens, and the effects of stress on the immune response. His teaching interests include undergraduate and graduate courses in immunology and a team-taught general microbiology course, where he is responsible for immunology, host defense, and infectious diseases. He lives with his wife, Judy, a junior high school science teacher, and their daughters, Martha and Helen, in Carbondale where he coaches his daughters' soccer teams.

JACK PARKER received his bachelor's degree in biology and also received his doctoral degree in a biology program (Ph.D., Purdue University, 1973). However, his research project dealt with bacterial physiology and he completed his Ph.D. research while in the microbiology department at the University of Michigan. Following this he spent four years studying bacterial genetics at York University in Toronto, Ontario. He has taught courses in bacterial genetics, general genetics, molecular biology, and molecular genetics, and has participated in courses in introductory microbiology, medical microbiology, and virology primarily at Southern Illinois University at Carbondale, where he is now a Professor in the Department of Microbiology and Dean of the College of Science. His research has been in the broad area of molecular genetics and gene expression and for the last 15 years has been focused most specifically on studies of how cells control the accuracy of protein synthesis. He is the author of approximately 50 research papers. His home is on the edge of the Shawnee National Forest in deep southern Illinois where he lives with his wife, Beth, and three children, Justine, D'Arcy, and Grant.

Brock

Biology of Microorganisms

EIGHTH EDITION

Michael T. Madigan
Southern Illinois University–Carbondale

John M. Martinko
Southern Illinois University–Carbondale

Jack Parker
Southern Illinois University–Carbondale

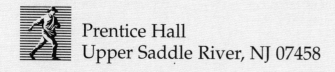

Prentice Hall
Upper Saddle River, NJ 07458

Library of Congress Cataloging-in-Publication Data

MADIGAN, MICHAEL T.
    Brock biology of microorganisms / Michael T. Madigan, John M.
Martinko, Jack Parker.—8th ed.
        p.   cm.
    Prev. ed. cataloged under: Biology of Microorganisms.
    Includes bibliographical references and index.
    ISBN 0-13-520875-0 (alk. paper)
    1. Microbiology.   I. Martinko, John M. II. Parker, Jack
QR41.2.B77 1996   96-8585
576—dc20   CIP

Editions of *Biology of Microorganisms*

First Edition, 1970, Thomas D. Brock
Second Edition, 1974, Thomas D. Brock
Third Edition, 1979, Thomas D. Brock
Fourth Edition, 1984, Thomas D. Brock,
    David W. Smith, and Michael T. Madigan
Fifth Edition, 1988, Thomas D. Brock and
    Michael T. Madigan
Sixth Edition, 1991, Thomas D. Brock and Michael T. Madigan
Seventh Edition, 1994, Thomas D. Brock, Michael T. Madigan,
    John M. Martinko, and Jack Parker
Eighth Edition, 1997, Michael T. Madigan,
    John M. Martinko, and Jack Parker

Acquisitions Editor: David Kendric Brake
Development Editor: Laura J. Edwards
Production Editor: Debra A. Wechsler
Editor in Chief: Paul F. Corey
Editorial Director: Tim Bozik
Director, Production & Manufacturing: David W. Riccardi
Executive Managing Editor, Production: Kathleen Schiaparelli
Assistant Managing Editor, Production: Shari Toron
Marketing Manager: Kelly McDonald
Creative Director: Paula Maylahn
Art Director: Heather Scott
Art Manager: Gus Vibal
Interior Designer: Sheree Goodman
Cover Designer: Joseph Sengotta
Copy Editor: Carol J. Dean
Proofreader: Stephanie Hiebert
Indexer: Barbara Littlewood
Manufacturing Manager: Trudy Pisciotti
Photo Editor: Lorinda Morris-Nantz
Cover Research: Karen Branson
Illustrators: Academy ArtWorks, Inc.
Art Editors: Warren J. Ramezzana, Grace Hazeldine
Cover Photograph: "Emerald Pool"—color from algae and bacteria
                     at Black Sand Basin, Yellowstone National
                     Park. Alan L. Detrick/Photo Researchers, Inc.

Printed in the United States of America
10 9 8 7 6 5 4 3 2 1

ISBN 0-13-520875-0

Prentice-Hall International (UK) Limited, *London*
Prentice-Hall of Australia Pty. Limited, *Sydney*
Prentice-Hall Canada Inc., *Toronto*
Prentice-Hall Hispanoamericana, S.A., *Mexico*
Prentice-Hall of India Private Limited, *New Delhi*
Prentice-Hall of Japan, Inc., *Tokyo*
Simon & Schuster Asia Pte, Ltd., *Singapore*
Editora Prentice-Hall do Brasil, Ltda., *Rio de Janeiro*

Overview

Brock BIOLOGY OF MICROORGANISMS, Eighth Edition

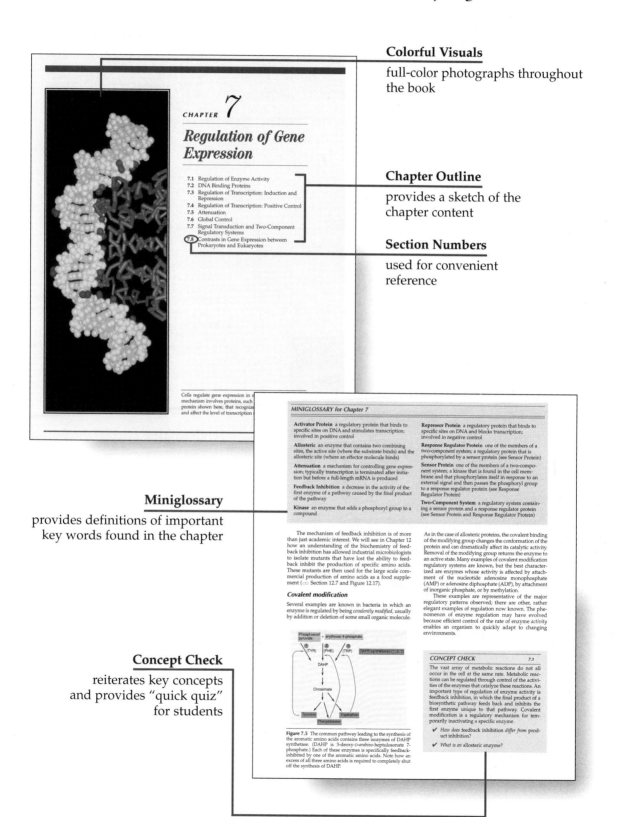

Colorful Visuals

full-color photographs throughout the book

CHAPTER *7*

Regulation of Gene Expression

7.1 Regulation of Enzyme Activity
7.2 DNA Binding Proteins
7.3 Regulation of Transcription: Induction and Repression
7.4 Regulation of Transcription: Positive Control
7.5 Attenuation
7.6 Global Control
7.7 Signal Transduction and Two-Component Regulatory Systems
7.8 Contrasts in Gene Expression between Prokaryotes and Eukaryotes

Chapter Outline

provides a sketch of the chapter content

Section Numbers

used for convenient reference

Cells regulate gene expression in mechanism involves proteins, such protein shown here, that recognize and affect the level of transcription

---

MINIGLOSSARY for Chapter 7

Activator Protein a regulatory protein that binds to specific sites on DNA and stimulates transcription; involved in positive control

Allosteric an enzyme that contains two combining sites, the active site (where the substrate binds) and the allosteric site (where an effector molecule binds)

Attenuation a mechanism for controlling gene expression; typically transcription is terminated after initiation but before a full-length mRNA is produced

Feedback Inhibition a decrease in the activity of the first enzyme of a pathway caused by the final product of the pathway

Kinase an enzyme that adds a phosphoryl group to a compound

Repressor Protein a regulatory protein that binds to specific sites on DNA and blocks transcription; involved in negative control

Response Regulator Protein one of the members of a two-component system; a regulatory protein that is phosphorylated by a sensor protein (see Sensor Protein)

Sensor Protein one of the members of a two-component system; a kinase that is found in the cell membrane and that phosphorylates itself in response to an external signal and then passes the phosphoryl group to a response regulator protein (see Response Regulator Protein)

Two-Component System a regulatory system containing a sensor protein and a response regulator protein (see Sensor Protein and Response Regulator Protein)

Miniglossary

provides definitions of important key words found in the chapter

The mechanism of feedback inhibition is of more than just academic interest. We will see in Chapter 12 how an understanding of the biochemistry of feedback inhibition has allowed industrial microbiologists to isolate mutants that have lost the ability to feedback inhibit the production of specific amino acids. These mutants are then used for the large scale commercial production of amino acids as a food supplement (∞ Section 12.7 and Figure 12.17).

Covalent modification

Several examples are known in bacteria in which an enzyme is regulated by being *covalently modified*, usually by addition or deletion of some small organic molecule.

As in the case of allosteric proteins, the covalent binding of the modifying group changes the conformation of the protein and can dramatically affect its catalytic activity. Removal of the modifying group returns the enzyme to an active state. Many examples of covalent modification regulatory systems are known, but the best characterized are enzymes whose activity is affected by attachment of the nucleotide adenosine monophosphate (AMP) or adenosine diphosphate (ADP), by attachment of inorganic phosphate, or by methylation.

These examples are representative of the major regulatory patterns observed; there are other, rather elegant examples of regulation now known. The phenomenon of enzyme regulation may have evolved because efficient control of the rate of enzyme *activity* enables an organism to quickly adapt to changing environments.

Concept Check

reiterates key concepts and provides "quick quiz" for students

Figure 7.5 The common pathway leading to the synthesis of the aromatic amino acids contains three isozymes of DAHP synthetase. (DAHP is 3-deoxy-D-*arabino*-heptulosonate 7-phosphate.) Each of these enzymes is specifically feedback-inhibited by one of the aromatic amino acids. Note how an excess of all three amino acids is required to completely shut off the synthesis of DAHP.

CONCEPT CHECK   *7.1*

The vast array of metabolic reactions do not all occur in the cell at the same rate. Metabolic reactions can be regulated through control of the activities of the enzymes that catalyze these reactions. An important type of regulation of enzyme activity is feedback inhibition, in which the final product of a biosynthetic pathway feeds back and inhibits the first enzyme unique to that pathway. Covalent modification is a regulatory mechanism for temporarily inactivating a specific enzyme.

✔ *How does feedback inhibition differ from product inhibition?*

✔ *What is an allosteric enzyme?*

Full-color Diagrams

color coded to help
the student learn

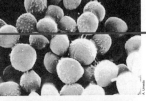

Figure 7.13 Positive control of enzyme induction. (a) In the absence of an inducer, neither the activator protein nor the RNA polymerase can bind to the DNA. (b) An inducer molecule binds to the activator protein, which in turn binds to the activator binding site. This allows RNA polymerase to bind to the promoter and begin transcription.

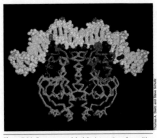

Figure 7.14 Computer model of the interaction of a positive regulatory protein with DNA. This figure shows the cyclic AMP binding protein, a regulatory protein involved in the control of several operons (see Section 7.6). The α-carbon backbone of this protein is shown in blue and purple. The protein is shown binding to a DNA double helix, which is shown in yellow and light blue. Note that binding of this protein to DNA has caused the DNA to be bent by almost 90°. Reprinted with permission from Science 253:1001–1007 (1991), © AAAS.

Section Numbers

used for assignment
and cross reference

regulation may have evolved because efficient control of the rate of enzyme activity. However, working together, these mechanisms result in an efficient regulation of cell metabolism so energy is not wasted carrying out unnecessary reactions.

Control systems that vary the level of *expression* of particular genes are the main subject of this chapter. However, the actual number of different regulatory mechanisms is vast, and most genes seem to be regulated by more than one. We begin by briefly discussing the processes involved in regulating the *activity* of preformed enzymes before considering how the synthesis of enzymes is controlled.

7.1

Regulation of Enzyme Activity

There are a large number of mechanisms of posttranslational regulation. In some cases an enzyme is synthesized as part of a larger inactive precursor protein, and the enzyme must be activated by removing a portion of the precursor protein. Another mechanism is to reduce the level of activity by actually degrading the enzyme molecules. However, we discuss here a reversible and temporary form of regulation involving less drastic changes to the enzyme molecule.

Tables

summarize important
information for
easy reference

TABLE 7.1 A few of the global control systems known in *Escherichia coli*[a]

System	Signal	Primary activity of regulatory protein	Number of genes regulated
Aerobic respiration	Presence of O_2	Repressor	20
Anaerobic respiration	Lack of O_2	Activator	20+
Catabolite repression	Cyclic AMP concentration	Activator	300+
Heat shock	Temperature	Alternative sigma	17
Nitrogen utilization	NH_3 limitation	Activator/alternative sigma	12+
Oxidative stress	Oxidizing agent	Activator	12+
SOS response	Damaged DNA	Repressor	17

[a]For many of the global control systems, regulation is complex. A single regulatory protein can play more than one role. For instance, the regulatory protein for anaerobic respiration, FNR, is an activator protein for many promoters but a repressor for others. Regulation can also be indirect or the activator for the nitrogen utilization system activates promoters recognized by an alternative proteins involved are members of two-component systems (see Section 7.7). Many genes are for a discussion of the SOS response, see Section 9.3.)

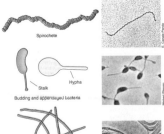

Coccus

Rod

Spirillum

Spirochete

Stalk

Hypha

Budding and appendaged bacteria

Filamentous

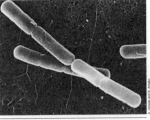

(a)

(b)

(c)

Figure 3.10 Spherical and rod-shaped bacteria, which associate in characteristically different ways as viewed by scanning electron microscopy. (a) *Streptococcus*, a chain-forming organism that divides only in one plane. (b) *Staphylococcus*, a coccus that divides in more than one plane. Cells of both organisms are about 1 μm in diameter. (c) Chains of the rod-shaped bacterium *Bacillus*. Cells are about 0.8 μm in diame-

Figure 3.9 Representative cell shapes (morphology) in prokaryotes. Next to each drawing is a phase photomicrograph showing an example of that morphology. Organisms are coccus, *Thiocapsa roseopersicina* (diameter of a single cell = 1.5 μm); rod, *Desulfuromonas acetoxidans* (diameter = 1 μm); spirillum, *Rhodospirillum rubrum* (diameter = 1 μm); spirochete, *Spirochaeta stenostrepta* (diameter = 0.25 μm); budding and appendaged, *Rhodomicrobium vannielii* (diameter = 1.2 μm); filamentous, *Chloroflexus aurantiacus* (diameter = 0.8 μm).

which is a complex but highly organized process. Two identical daughter cells result from the division of one parent cell; each daughter cell receives a nucleus with an identical set of chromosomes.

Eukaryotic cells also contain distinct structures called **organelles**, within which important cellular functions occur (Figure 3.11). Organelles are absent from prokaryotes, although major physiological processes that take place in organelles, such as respiration and photosyn-

Outstanding Micrographs

keyed to colorful line art

WORKING WITH NUCLEIC ACIDS: THE TOOLS

Our knowledge of molecular biology and genetics has depended on the development of adequate research tools. Advances in knowledge of how nucleic acids work have generally been tied to the development of new methods. We discuss here some of these methods.

1. Extraction and purification of DNA The first requirement is a sample of DNA free of other cellular

2. Detecting the presence of DNA There are several methods for detecting the presence of DNA in a solution. One of the most widely used is by its absorption of ultraviolet radiation. DNA strongly absorbs ultraviolet radiation at a wavelength of 260 nm. The absorption is due to the purine and pyrimidine bases. As seen in the figure, double-stranded DNA absorbs less strongly than single-stranded DNA. This is because

become positioned in the gradient at positions corresponding to their densities. If ethidium bromide has been added, observation of the centrifuge tube with ultraviolet radiation after the centrifugation reveals bands of DNA by fluorescence (see later). This method is called the *buoyant density method* and permits both determination of density and separation of molecules of differing density.

4. Gel electrophoresis One of

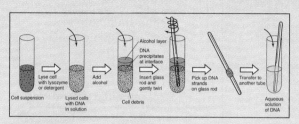

Cell suspension — Lyse cell with lysozyme or detergent — Lysed cells with DNA in solution — Add alcohol — Cell debris — Alcohol layer / DNA precipitates at interface / Insert glass rod and gently twirl — Pick up DNA strands on glass rod — Transfer to another tube — Aqueous solution of DNA

chemicals. The steps in the purification of DNA are shown here. The aqueous solution in the final step is treated with RNAse to remove RNA. Proteins are then removed by use of denaturing solvents (usually phenol). By repeating the purification steps a number of times, a solution can be obtained that is virtually free of any components other than DNA.

Note that the solution of DNA obtained never consists of native DNA molecules of the length found in the cell. The purification process causes the DNA to be broken into fragments of various (random) lengths. If the DNA has been handled gently during purification, the lengths of the fragments will be about one-hundredth of the length of the whole chromosome.

the interaction between the bases on the opposite strands of the double-stranded DNA (hydrogen bonding) reduces the ultraviolet absorbance.

3. Density-gradient centrifugation of DNA DNA molecules vary in density, depending on their exact chemical composition. DNA molecules with a higher content of guanine plus cytosine (GC) are denser than molecules with low GC. The density of DNA can be determined by centrifugation at very high speed in a *gradient* of cesium chloride (CsCl). The DNA solution is added to a solution of CsCl and centrifuged at high speed for several hours until equilibrium is reached. The CsCl forms a density gradient from the top to bottom of the tube, and DNA molecules form bands at appropriate densities. At equilibrium, the DNA molecules

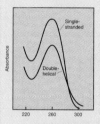

Single-stranded / Double-helical

Absorbance vs Wavelength (nm): 220, 260, 300

the most widespread methods of

ANTISENSE NUCLEIC ACID

Regulation of the synthesis of proteins often involves transcriptional control. Less often, genes are controlled at the level of translation. Most control networks, whether they are transcriptional or translational, utilize regulatory proteins. However, it is now clear that in some cases it is a regulatory RNA, not a regulatory *protein*, that is involved. One type of regulatory RNA, called **antisense RNA**, is known to be used in the regulation of several different bacterial genes. Antisense RNA acts by forming base pairs with a complementary, or sense, strand of RNA. When the sense RNA is mRNA, the resulting double-stranded structure can prevent translation. Antisense RNA can be synthesized from the same gene as the sense RNA by having a second promoter oriented in the direction opposite that of the first or by having a second gene with the promoter at the other end. Antisense RNA does not have to be used only to regulate

the synthesis of a protein. In some plasmids, it controls the initiation of DNA synthesis.

Antisense nucleic acids can be specifically designed and synthesized by scientists in the laboratory and delivered directly to cells. These short (15–25 nucleotides) synthetic chains (oligonucleotides) are usually made of DNA rather than RNA. Their sequence can be made to allow them to bind to a specific mRNA and prevent translation (or allow the molecule to be recognized by nucleases). Antisense nucleic acid can also bind to the DNA in the nucleus and prevent transcription. The latter is possible because some DNA can form a *triple* helix! The "extra" strand (the oligonucleotide) forms specific interactions with those parts of the bases that are in the major groove of a normal double helix to give *triplex DNA*. (Not all DNA sequences can form triple helices, at least not without the aid of special enzymes.)

Synthetic antisense oligonucleotides can be designed to be extremely specific, whether they bind to a message or to the regulatory region of a gene. The specificity arises because a sequence of only 20 bases should occur no more often than once in 10^{12} bases of "random" DNA. Therefore, it is unlikely that the antisense RNA would bind to anything other than its known target in any cell. This specificity might allow antisense nucleic acids to become an important new type of antibiotic, and this possibility is being pursued by a number of pharmaceutical companies. Antisense nucleic acids could be designed to be used against specific viruses or to regulate particular genes in either disease-causing (pathogenic) organisms or human tumor cells. The possible utility of these molecules is just one example of how understanding the structure of a gene may have very practical applications (⊂ Chapter 10). ∎

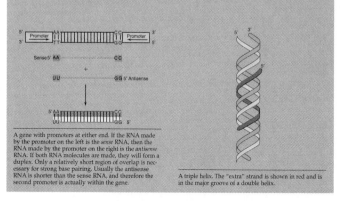

A gene with promoters at either end. If the RNA made by the promoter on the left is the *sense* RNA, then the RNA made by the promoter on the right is the *antisense* RNA. If both RNA molecules are made, they will form a duplex. Only a relatively short region of overlap is necessary for strong base pairing. Usually the antisense RNA is shorter than the sense RNA, and therefore the second promoter is actually within the gene.

A triple helix. The "extra" strand is shown in red and is in the major groove of a double helix.

Material for Review

REVIEW QUESTIONS

1. If an enzyme can be effectively inhibited by feedback inhibition, why would cells also have mechanisms to regulate its synthesis?

2. Describe why a protein that binds to a specific sequence of double-stranded DNA is unlikely to bind to the same sequence if the DNA is single-stranded.

3. Describe the regulation of two different operons, one having an effector that is an *inducer* and the other having an effector that is a *corepressor*.

4. The maltose regulon is inducible and is regulated by an activator protein. The lactose operon is inducible but is regulated by a repressor protein. Explain how induction can be brought about by either positive control (activator protein) or negative control (repressor protein).

5. In most cases operators are very close to the promoters they control, while activator binding sites can be some distance away. Explain why this should be so.

6. Describe how transcriptional attenuation works. What is actually being "attenuated"? Why hasn't the type of attenuation that controls several different amino acid biosynthetic pathways in *Escherichia coli* also been found in eukaryotes?

7. Describe the mechanism by which catabolic activator protein (CAP), the regulatory protein for catabolite repression, functions using the lactose operon as an example. For this operon the CAP protein is not a repressor. Describe the regulatory region of a gene for which the CAP protein *is* a repressor. (*Hint:* Think about your answer to Question 5.)

8. What are the two components that give the name to signal transduction regulation in prokaryotes? What is the function of each of the components?

9. One of the members of a two-component system is typically located in the cell membrane. What reason can you think of why this might be so?

10. Many genes are under multiple control systems. In the lactose operon, there is lactose-specific regulation and regulation by a global control system. Describe how each of the controls on the lactose operon actually functions. Why do you think both systems are necessary?

APPLICATION QUESTIONS

1. The amino acids isoleucine and valine share a common pathway for most steps in their biosynthesis. In *Escherichia coli* the first common step can be subject to feedback inhibition by valine but not by isoleucine. In most strains, though, the addition of valine does not cause isoleucine deprivation. However, in other strains it does (that is, adding valine causes isoleucine starvation and the cells stop growing). What explanation can you give for the difference between the normal "valine-resistant" strains and those whose growth is sensitive to valine?

2. What would happen to regulation from a promoter under negative control if the region where the regulatory protein binds were deleted? What if the promoter were under positive control?

3. Promoters from *Escherichia coli* under positive control are not close matches to the DNA consensus sequence for *E. coli* (∞ Section 6.6). Why?

4. Interestingly, the attenuation control of some of the pyrimidine biosynthetic pathway genes in *Escherichia coli* actually involves coupled transcription and translation. Can you describe a mechanism whereby the cell could somehow make use of translation to help it measure the level of pyrimidine nucleotides?

SUPPLEMENTARY READINGS

Alberts, B., D. Bray, J. Lewis, M. Raff, K. Roberts, and J. D. Watson. 1994. *Molecular Biology of the Cell,* 3rd edition. Garland, New York. This standard reference text for cellular molecular biology covers most areas of the regulation of gene expression. There is a heavy emphasis on eukaryotes.

Hoch, J.A., and T. J. Silhavy. 1995. *Two-Component Signal Transduction.* American Society for Microbiology, Washington, DC. This volume gives detailed information on a number of different two-component regulatory systems in Bacteria.

Lodish, H., D. Baltimore, A. Berk, S. L. Zipursky, P. Matsudaira, and J. Darnell. 1995. *Molecular Cell Biology,* 3rd edition. Scientific American Books, New York. This is another large reference text covering most areas of molecular biology and gene regulation, with a strong emphasis on eukaryotes.

Neidhardt, F. C., R. Curtiss III, J. L. Ingraham, E. C. C. Lin, K. B. Low, B. Magasanik, W. Reznikoff, M. Riley, M. Schaechter, and H. E. Umbarger (eds.). 1996. *Escherichia coli and Salmonella: Cellular and Molecular Biology,* 2nd edition. American Society for Microbiology, Washington, DC. This compendium contains information on the regulation of many specific pathways and on important cellwide systems in these important organisms.

Stryer, L. 1995. *Biochemistry,* 3rd edition. W. H. Freeman. New York. An excellent biochemistry text with considerable information on the regulation of enzyme activity and synthesis.

 On–line resources for this chapter are on the World Wide Web at http://www.prenhall.com/~brock (click on the table of contents li[...]

APPENDIX 2 Advanced Mathematics of Microbial Growth and Chemostat Operation

Exponential Growth

Although analyses of microbial growth can be done graphically or algebraically, as discussed in Chapter 5, for some purposes it is necessary to use a differential equation to express the quantitative relationships of growth:

$$\frac{dX}{dt} = \mu X \qquad (1)$$

where X may be cell number or some specific cellular component such as protein and μ is the *instantaneous growth-rate constant.* When this equation is integrated, we obtain a form that reflects the activities of a typical microbial population in batch culture:

$$\ln X = \ln X_0 + \mu(t) \qquad (2)$$

where ln refers to the natural logarithm (logarithm base e), X_0 is cell number at time 0, X is cell number at time t, and t is elapsed time during which growth is measured. This equation fits experimental data from the exponential phase of bacterial cultures very well, such as those in Figure 5.2. Taking the antilogarithm of each side gives

$$X = X_0 e^{\mu t} \qquad (3)$$

This equation is useful because it allows prediction of population density at a future time from the present value and μ, the growth-rate constant. As discussed in Section 5.2, an important constant parameter for an exponentially growing population is the doubling (or generation) time. Doubling of the population has occurred when $X/X_0 = 2$. Rearranging and substituting this value into Equation (3) gives

$$2 = e^{\mu(t_{gen})} \qquad (4)$$

Taking the natural logarithm of each side and rearranging gives

$$\mu = \frac{\ln 2}{t_{gen}} = \frac{0.693}{t_{gen}} \qquad (5)$$

The generation time, t_{gen}, may be used to define another growth parameter, k, as follows:

$$k = \frac{1}{t_{gen}} \qquad (6)$$

where k is the growth-rate constant for a batch culture. Combining Equations (5) and (6) shows that the two growth-rate constants, μ and k, are related:

$$\mu = 0.693k$$

It is important to understand that μ and k are both reflections of the same growth process of an exponentially increasing population. The difference between them may be seen in their derivation: μ is the *instantaneous rate constant,* and k is an *average* value for the population over a finite period of time. (See Table A2.1 for calculation of k values from experimental results.) This distinction is more than a mathematical point. As was emphasized in Chapter 5, microbial growth studies must deal with *population* phenomena, not the activities of individual cells. The constant k reflects this averaging assumption. However, the constant μ, being instantaneous, is a closer approximation of the rate at which individual activities are occurring. Further, the instantaneous constant μ allows us to consider bacterial growth dynamics in a theoretical framework separate from the traditional batch culture.

Mathematical relationships of chemostats

An especially important application of the instantaneous growth-rate constant, μ, is the chemostat, a culture device (∞ Section 5.5) in which population size and growth rate may be maintained at constant values of the experimenter's choosing over a wide range of values. As we saw in Section 5.5, the rate of bacterial growth is a function of nutrient concentration. This function represents a saturation process that may be described by the equation

TABLE A2.1 Calculation of k values

Bacterial population densities are expressed in scientific notation with powers of 10, and so Equation (3) may be converted to terms of logarithm base 10 and k substituted for the instantaneous constant μ:

$$k = \frac{\log_{10} X_t - \log_{10} X_0}{0.301 t}$$

Example 1:
$X_0 = 1000 \; (= 10^3) \; \log_{10} \text{ of } 1000 = 3$
$X_t = 100{,}000 \; (= 10^5) \; \log_{10} \text{ of } 100{,}000 = 5$
$t = 4 \text{ hr}$
$k = \dfrac{5 - 3}{(0.301)4} = \dfrac{2}{1.204}$
$k = 1.66 \text{ doublings/hr}$
$t_{gen} = 0.60 \text{ hr (36 min) for population to double}$

Example 2:
$X_0 = 1000 \; (= 10^3) \; \log_{10} \text{ of } 1000 = 3$
$X_t = 100{,}000{,}000 \; (= 10^8) \; \log_{10} \text{ of } 10^8 = 8$
$t = 120 \text{ hr}$
$k = \dfrac{8 - 3}{(0.301)120} = \dfrac{5}{36.12}$
$k = 0.138 \text{ doubling/hr}$
$t_{gen} = 7.2 \text{ hr (430 min) for population to double}$

Brief Contents

Contents

CHAPTER *18*

Eukarya: Eukaryotic Microorganisms   **769**

CHAPTER *19*

Host–Parasite Relationships   **785**

CHAPTER *20*

Concepts of Immunology   **813**

CHAPTER *21*

Clinical and Diagnostic Microbiology and Immunology   **865**

CHAPTER *22*

Epidemiology and Public Health Microbiology   **902**

CHAPTER *23*

Major Microbial Diseases   **929**

APPENDIX *1*

Energy Calculations in Microbial Bioenergetics   *A-1*

APPENDIX *2*

Advanced Mathematics of Microbial Growth and Chemostat Operation   *A-5*

APPENDIX *3*

Bergey's Classification of Prokaryotes   *A-7*

Glossary   *G-1*

Index   *I-1*

Preface

A new Golden Age of Microbiology is upon us! An entire bacterial genome can be sequenced in a few months' time, our understanding of the molecular bases of microbial diseases has blossomed, we have developed keen insight into the inner workings and evolutionary history of microorganisms, and career opportunities in the biomedical sciences are unprecedented. What a great time to learn microbiology! We therefore take great pride in introducing students to the field of microbiology as it exists today through our book, *Brock Biology of Microorganisms 8/e.*

Like previous editions of *Biology of Microorganisms, Brock Biology of Microorganisms 8/e (BBOM 8/e)* clearly presents the essential concepts of microbiology and illustrates them with the latest information available from the scientific literature. However, our main goal remains the same as it has always been: to explain the basic science and applications of microbiology at a level that the introductory student can readily appreciate and master but in a way that generates excitement about this fundamental area of biology.

With this new edition, the title of our book has changed, as Thomas D. Brock—the architect of the previous seven editions of *Biology of Microorganisms*—has passed the torch to his co-authors. However, the reader can be assured that we are firmly committed to maintaining the outstanding reputation this book has enjoyed for over a quarter of a century. It is thus fitting and proper that the name *Brock* be incorporated into the title because although he is no longer a co-author, *Brock Biology of Microorganisms 8/e* continues to emphasize strong, well-documented science, a philosophy that made Thomas Brock one of the premier scientists and educators of the last half of this century. The current co-authors of *BBOM 8/e* have integrated this philosophy into their own coverage of the three major areas of microbiology today: general/organismal (Madigan), molecular/genetic (Parker), and medical/immunological (Martinko).

Changes from the 7th Edition

Users of previous editions of *Biology of Microorganisms* will note some organizational changes in this book. The medical/immunology block of chapters, previously placed in the middle of the book, has been moved to the end. From the examination of over 75 lecture outlines for introductory microbiology courses and from discussions with several users of the book, the authors were convinced that this move made good pedagogical sense and was consistent with the mainstream of introductory microbiology courses. As a result of this change, the chapters on metabolic diversity, ecology, evolution, and the major microbial groups have been moved forward and now closely follow the introductory and genetics chapters. We have also moved the chapter on microbial growth forward to immediately follow the chapter on nutrition and metabolism. This is the point in introductory microbiology courses where the concepts of exponential growth are usually taught and frequently illustrated with laboratory experiments; thus, the new edition places the growth material where it should be most useful for tying together lecture concepts with laboratory practice. The sequence of chapters in *BBOM 8/e* also closely follows the guidelines established by the Education Division of the American Society for Microbiology (ASM) for teaching introductory microbiology courses.

Two new chapters appear in *Brock Biology of Microorganisms 8/e.* The first, Chapter 7, is entitled "Regulation of Gene Expression." This chapter, whose foundation was previously present in the chapter entitled "Macromolecules and Molecular Genetics," focuses on how genes are controlled in prokaryotes. This material has been strengthened by the addition of new information on "two-component regulatory systems" (using chemotaxis as a molecular model) and other examples of the exciting area of bacterial gene regulation. The second new chapter, Chapter 11, is entitled "Microbial Control Agents." In this chapter we have consolidated discussion of the basic science and mode of action of antibiotics and chemotherapeutic agents, as well as other antimicrobial agents. Instructors who wish to cover this material as a unit will now find it logically organized and in one place. The chapter on industrial microbiology also contains some material on antibiotics, but the focus here is on *industrial production* rather than on the characteristics of the agents themselves.

Besides these organizational changes and new chapters, *Brock Biology of Microorganism 8/e* has undergone extensive revision in every chapter. The genetics revolution of the past 20 years has spawned new and powerful research tools that have impacted virtually every area of microbiology; as a result, bacterial genetics now offers solutions to applied problems in medicine, agriculture, and the environment. All these areas are covered here, stressing real-life examples that join the theoretical with the practical and make the material come alive.

Despite these changes, however, users of previous editions of this book can rest assured that *Brock Biology of Microorganisms 8/e* retains its emphasis on the microorganisms themselves. Although important research tools, microorganisms are significant in their own right and their basic biology and ecological activities remain the focus of this book. In addition, we have made every effort to produce an up-to-date book that maintains the authority, clarity, and breadth of coverage that instructors expect from *Biology of Microorganisms*. We trust that users will agree and we welcome comments from students and instructors alike on any aspect of our book.

Pedagogical Aids

A variety of teaching and learning aids are built into *Brock Biology of Microorganisms 8/e*, some of them new to this edition.

Art and Photographs   Virtually every piece of full color art has seen revision and several new pieces of art have been added to keep pace with the rapid developments occurring in the field of microbiology and to assist instructors in teaching the essential concepts of science. The color coding of macromolecules in all pieces of art remains the same as in previous editions: the two strands of DNA are in different shades of green, RNA is orange, and proteins are brown. From Chapter 1 through Chapter 23 this consistency in the use of color will help reinforce essential concepts and give students helpful visual feedback. As is a tradition with *Biology of Microorganisms*, this new edition contains superb photos obtained by the authors directly from researchers. Students and instructors alike will also appreciate how the electron micrographs in *BBOM 8/e* look just the way a scientist would see them in a research journal—as *black and white* photographs. The authors have resisted the practice of some microbiology textbooks to use false color to "enhance" electron micrographs; ironically, although admittedly adding color, such practice only decreases resolution. Thus, in *BBOM 8/e* the reader will be treated to electron micrographs that are crisp, clear, and as scientifically accurate as those published in the primary microbiological literature. Many of the light micrographs are in color, of course, because they were taken with color film to capture the natural colors of many microorganisms and microbial habitats.

Concept Checks and Concept Links   The "In Brief" segments, first introduced in the 6th edition, have been blended into a new learning aid we call **Concept Checks.** A concept check consists of a brief overview of the material in the previous section followed by several short questions designed to ensure that students have gotten the "take-home" message from what they have just read. Concept Checks can be thought of as "speed bumps" along the road of text, figures, and tables in each chapter, and are intended to either reinforce what has just been learned or to signal that a key point has been missed.

In addition to Concept Checks, **Concept Links** are new to this edition. Concept Links, indicated by the chain link icon (∞), alert the reader to ties between what is being read and related material found elsewhere in the book. The section numbering system in *BBOM 8/e*, employed since the first edition and a unique organizational feature among microbiology textbooks, accompany Concept Links to direct the reader to the related material.

Study Questions, Supplementary Readings and Miniglossaries   As usual, relevant and challenging **Study Questions,** many of them new to this edition, can be found at the end of each chapter. However, two levels of questions are now included. *Review Questions* emphasize factual material—the "database" necessary for an understanding of the concepts—while *Application Questions* require students to apply what they know, synthesize information, and solve a problem. The authors hope that the application questions will help build critical thinking skills in beginning microbiology students as they assimilate the basic information and master key microbiological concepts. As usual, the most recent **Supplementary Readings** will be found at the end of each chapter for students and instructors who wish to go beyond the textbook for a more detailed treatment of the material.

We maintain the popular **Miniglossaries** in the new edition. The old sports adage "You can't follow the *players* without a *program*," has a counterpart in microbiology: "You can't follow the *concepts* without the *language*." Thus, early in each chapter the key terms necessary to understand the ensuing material are succinctly defined and gathered together in one place for quick reference while reading the chapter. The authors feel that regular use of the miniglossaries will quickly build vocabulary and help reinforce critical concepts.

Boxes   Boxes have become a popular means of focusing on particular issues in both textbooks and in the popular press. In *Brock Biology of Microorganisms 8/e* we have written a number of **new boxes** and have organized all of the boxes along three major themes: "Learning from the Past" (enrichment boxes placing a microbiological concept in historical context), "Techniques and Applications" (boxes that describe a particular method or application of a method central to the field of microbiology), and "A Focus On" (boxes that take an in-depth look at a particular aspect or issue in microbiology of both scientific and general interest). In addition, we have made considerable efforts to illustrate our boxes, either with photos or with art (or with both in some cases), in order for the reader to better visualize the points discussed.

All in all, we feel that the combination of learning aids and enrichment materials woven into *Brock Biology of Microorganisms 8/e* will make for a strong learning experience but will at the same time support, rather than detract from the main message—text, figures, and tables—that make up the core of the book.

Chapter Highlights in *Brock Biology of Microorganisms 8/e* include:

Chapter 1—Introduction: Overview of Microbiology and Cell Biology builds on the approach of previous editions as the introduction to the basic biology of the cell. New material on culturing methods and Koch's postulates along with a new box on the development of solid culture media, enrich this chapter.

Chapter 2—Cell Chemistry maintains its important early position in the book as the chemical primer every student needs to master. The emphasis remains on understanding the *fundamental chemistry of macromolecules* and vivid illustrations help solidify this essential information. A new illustrated box on Pasteur and stereoisomerism enrich this chapter with historical perspective.

Chapter 3—Cell Biology emphasizes the structure of the prokaryotic cell with a completely rewritten section on the bacterial nucleoid and significant new material on cell wall structure. A new box discussing the possibility that bacterial endospores could survive for millions of years will be sure to stir controversy in both students and instructors alike.

Chapter 4—Nutrition and Metabolism has been reorganized to begin with an overview of basic nutrition and culture media from which it flows into concepts of enzymes, energetics, and metabolism. A new section on fatty acid biosynthesis completes the previous discussion of anabolic processes.

Chapter 5—Microbial Growth focuses on the concepts of exponential growth and population growth and flows from here into a discussion of environmental effects on microbial growth. Applied aspects (for example, control of microbial growth) have been moved to a new chapter (Chapter 11). New material on methods for measuring microbial growth bring the laboratory and the classroom closer together.

Chapter 6—Macromolecules and Molecular Genetics is shortened from the previous edition to maintain the focus on basic macromolecular syntheses and the key proteins and structures required to carry them out. Vivid and pedagogical use of color will guide the student through the central concepts of molecular biology as they occur in prokaryotes, contrasting them when appropriate with molecular processes in eukaryotes.

Chapter 7—Regulation of Gene Expression is a totally new chapter that blends discussion of both classical regulatory phenomena like induction and repression with new material on global control mechanisms. The hot topic of "signal transduction" involving two-component regulatory systems is introduced here using bacterial chemotaxis as a model of this complex regulatory process.

Chapter 8—Viruses has been streamlined somewhat to retain the focus on the essentials of viral replication. This chapter ends with an expanded section on viroids and prions—virus-like particles that are not viruses—comparing and contrasting these interesting genetic elements with viruses and emphasizing their unique replication features and pathogenic properties.

Chapter 9—Microbial Genetics builds on the material in Chapters 6 through 8 to create a modern picture of genetics in prokaryotes. Coverage of both classical bacterial genetics and the molecular phenom-

ena behind it is well supported by new art and illustrated boxes depicting the historical development of the field of bacterial genetics. New material on the bacterial genetic map and genomic sequencing reflect the enormous strides made in these areas in recent years. This chapter closes by contrasting prokaryotic and eukaryotic genetics using the highly studied *Saccharomyces cerevisiae* (baker's yeast) as a model eukaryote.

Chapter 10—Genetic Engineering and Biotechnology remains an up-to-the-minute chapter, discussing all the basic tools and methods of recombinant DNA technology and their latest applications in the field of biotechnology. The chapter has been greatly updated and strengthened with the addition of dramatic color photos illustrating new genetically engineered systems. But as always, the coverage in this chapter emphasizes the *basic science* behind the applications, tying it to the principles of molecular biology and genetics developed in Chapters 6–9.

Chapter 11—Microbial Growth Control is a new chapter consolidating coverage of material on chemical and physical agents used to prevent microbial growth. Discussion of major antibiotics of clinical significance highlight this chapter with the emphasis remaining on the structure and mode of action of antimicrobial agents and the basic principles behind microbial growth control.

Chapter 12—Industrial Microbiology focuses on large-scale microbial fermentations with antibiotic production remaining the driving force. However, several new pieces of art have been added to better illustrate the industrial production of important chemicals other than antibiotics, such as vitamins, amino acids and citric acid, and several new color photos of "industrial microbiology in action" will help students better grasp the variety and scale of industrial microbial processes. The "Home Brew" box, always a hit with students, is now illustrated with color photos to show the major steps in the small-scale brewing of beers and ales. Material on sewage and wastewater treatment round out this chapter, reminding students and instructors alike that this essential part of everyday life is a prime example of the large-scale use of microorganisms.

Chapter 13—Metabolic Diversity has been heavily revised to bring out the latest information on the nearly limitless ways in which microorganisms obtain the energy needed for growth. New material on syntrophic relationships among chemoorganotrophic bacteria and iron oxidation by phototrophic bacteria (an example of a totally new *concept* in microbiology), bring this chapter up to the minute.

Chapter 14—Microbial Ecology begins with new discussion of basic ecological principles underlying the activities of cells, cell populations, and communities of microorganisms in nature. New material has been added in virtually every section of this chapter and new boxes like "Microbial Life Deep Underground" dramatically emphasize the old adage in microbiology that "bugs are everywhere".

Chapter 15—Microbial Evolution, Systematics, and Taxonomy has been shortened somewhat from the seventh edition by deletion of the detailed phylogenetic pictures of each major domain of life; the latter material has been updated and integrated as an introduction to each of the organisms chapters (Chapters 16–18). A more detailed treatment of classical bacterial taxonomy has been added to this chapter to better tie together phylogenetic and more traditional approaches of bacterial classification. New material on the "RNA world" and early life forms reflect the rapid advancements in our understanding of the origin of life.

Chapter 16–18—these "organisms" chapters remain the core of the traditional microbial diversity theme of *Biology of Microorganisms*. Each chapter (16–Prokaryotic Diversity: Bacteria; 17–Prokaryotic Diversity: Archaea; and 18–Eukarya: Eukaryotic Microorganisms) begins with a phylogenetic overview and then proceeds to a detailed description of the major microbial groups. Every section of these three chapters has seen updating as a result of the great strides made in understanding microbial diversity in recent years.

Chapter 19—Host–Parasite Relationships maintains the long tradition in this book of covering microbial interactions with humans. The focus remains on the struggle between the host and the parasite and the major weapons both possess to maintain health or to induce disease. To streamline this chapter coverage of cells involved in nonspecific immunity has been moved to Chapter 20 as part of an introduction to the immune system.

Chapter 20—Concepts of Immunology has been heavily updated to incorporate the latest concepts in probably the fastest moving field in all of biology. The focus here is on the *basic science* behind immunology; much of the material on applied and clinical microbiology has been moved to the next chapter. However, a new box on "Catalytic Antibodies" ties basic concepts to future applications and emphasizes how basic science frequently spawns exciting applications.

Chapter 21—Clinical and Diagnostic Microbiology and Immunology consolidates the microbiology and immunology applications of clinical significance. Several highly innovative clinical diagnostic methods have emerged from advances in microbiology and immunology and are described in detail in this chapter. Because these methods are rapidly changing the way hospital microbiology is done, this chapter will surely be important for courses with a health professions emphasis. Look for a number of new color photos here that show how clinical microbiology is done today.

Chapter 22—Epidemiology and Public Health Microbiology sets the stage for consideration of specific diseases in the following chapter by outlining the principles of disease transmission and the concepts of public health. An exciting new section on emerging and resurgent infectious diseases brings this chapter up-to-the-minute, describing how these diseases are so abruptly and successfully transmitted when conditions are right. New tables on epidemic diseases, immunization statistics, reportable diseases, resurgent and emerging diseases, and virulence factors help tie the concepts of disease transmission to real-life human situations and also lend reference value to this chapter.

Chapter 23—Major Microbiology Diseases maintains broad coverage of microbial diseases using an ecological approach of grouping individual diseases by their mode of transmission. As you would expect in such a chapter, every section has received updating, especially the coverage of AIDS. This chapter comes alive with excellent color photos and disease statistics that show human disease symptoms and reflect the worldwide prevalence and trends of these diseases.

Supplements

Several Supplements accompany this textbook. These include a **Student Study Guide,** which highlights key topics and contains a wealth of additional objective and subjective review questions. For the instructor, a valuable **Instructor's Manual** and **Test Item File** is available which describes various ways to structure an introductory microbiology course using *BBOM 8/e* as text, and gives answers to all the Study Questions. In addition, a set of **250 Color Transparencies** (containing over 350 individual pieces of art) will greatly assist instructors in organizing and presenting class lectures. In addition, a **CD-ROM** containing all art and tables from *Brock Biology of Microorganisms 8/e* is available to instructors for instant access to these materials in the classroom.

The **Contemporary View** program sponsored jointly by Prentice Hall and *The New York Times*, and initiated with the sixth edition of *Biology of Microorganisms*, complements the new edition as well. Through this program, core subject matter from the text is supplemented by articles describing microbiology in "real-life" situations pared from the pages of *The New York Times*. Contemporary View will strengthen the connection between the classroom and the real world, as students see how important an understanding of microbiology is to activities in their everyday lives.

Acknowledgments

This book is not the product of just the authors. Many scientists gave valuable reviews of an earlier draft or on specific sections in the draft and others made special efforts to provide color photographs taken directly from their laboratory research. We are extremely grateful for their efforts. These include:

Laurie Achenbach, *Southern Illinois University*
Robert Andrews, *Iowa State University*
Ester Angert, *Harvard University*

Judy Armitage, *Oxford University, England*
Jeanette Baker, *Monsanto Company*
Carl E. Bauer, *Indiana University*
John A. Breznak, *Michigan State University*
Clare Bunce, *Cold Spring Harbor Archives*
Bryon Burch, *The Beverage People, Santa Rosa, California*
Mary Burke, *Oregon State University*
Richard W. Castenholz, *University of Oregon*
David P. Clark, *Southern Illinois University*
Mary Lynne Perille Collins, *University of Wisconsin, Milwaukee*
Morris Cooper, *Southern Illinois University*
Stephen Cooper, *University of Michigan*
Phillip R. Cunningham, *Wayne State University*
Michael Dalbey, *University of California, Santa Cruz*
Lawrence Dreyfeus, *University of Missouri, Kansas City*
Judith L. Edmiston, *University of Texas, Austin*
Stephen Edmondson, *Southern Illinois University*
George Feher, *University of California, San Diego*
Martin Flajnik, *University of Miami*
Susan French, *Busch Creative Services, Anheuser Busch Company*
Paul Golnick, *State University of New York (SUNY), Buffalo*
Wendy Gorman, *Northern Arizona University*
Gerhard Gottschalk, *University of Göttingen, Germany*
John Haddock, *Southern Illinois University*
Barbara Hemmingsen, *San Diego State University*
Marcus Hüttel, *Max Planck Institute for Marine Microbiology, Germany*
Sally Jackson, *Baylor University*
Ronald Jenke, *Minneapolis Community College*
Sam Kaplan, *University of Texas Medical Center, Houston*
Allan Konopka, *Purdue University*
Susan Koval, *University of Western Ontario, Canada*
Stephen Kowalczykowski, *University of California, Davis*
William Lorowitz, *Pittsburgh State University*
Bonnie Maidak, *University of Illinois, Urbana*
Caleb Makukutu, *Kingwood College*
Renee D. Mastrocco, *Rockefeller University Archives*
David Maxwell, *Auheuser Busch Co., St. Louis*
Kevin McBride, *Calgene Company*
Peter McConnachie, *Southern Illinois University School of Medicine*
Mark D. Moore, *University of Texas Medical Center*
Dino Moras, *Université Louis Pasteur*
Edward Moticka, *Southern Illinois University*
William Muhlach, *Southern Illinois University*
Suzanne Nailor, *American Sterilizer Company*
Frederick C. Neidhardt, *University of Michigan*
Arthur Nonomura, *Hampshire Chemical Co.*
Kaori Ohki, *Tokai University, Japan*

John M. Olson, *University of Massachusetts*
Jörg Overmann, *University of Oldenburg, Germany*
Aharon Oren, *Hebrew University, Jerusalem*
Norman Pace, *Indiana University*
Stephen R. Padgette, *Monsanto Company*
Elizabeth Parker, *Carbondale, Illinois*
Nina Parker, *Minot State University Science Division*
Bobbie Pettriess, *Wichita State University*
Norbert Pfennig, *University of Konstanz, Germany*
William Picking, *St. Louis University*
John N. Reeve, *Ohio State University*
Udo Reischel, *University of Regensburg, Germany*
Monica Riley, *Marine Biological Laboratory, Woods Hole, MA*
Gary P. Roberts, *University of Wisconsin*
Kenneth E. Rudd, *National Institutes of Health*
Gordon Sauer, *University of Kansas*
Georg E. Schulz, *University of Freiburg, Germany*
Garriet Smith, *University of South Carolina, Aiken*
Jolynn F. Smith, *Southern Illinois University*
Mitchell Sogin, *Marine Biological Laboratory, Woods Hole, MA*
Barton Spear, *Spear Studio, Erie, Pennsylvania*
Gary Stacey, *University of Tennessee*
James T. Staley, *University of Washington*
Karl O. Stetter, *University of Regensburg, Germany*
Todd Stevens, *DOE/Batelle, Richland, Washington*
Mary Stunkard, *River Falls Area Hospital, River Falls, WI*
Mark Tamplin, *University of Florida*
Tom Terry, *University of Connecticut*
Jacques Vasse, *Laboratoire de Biologie Moléculaire des Relations Plantes-Microorganismes, CNRS-INRA, Castanet-Tolosan, France*
Judith Vogt, *Fort Hays University*
David M. Ward, *Montana State University*
Pamela Weathers, *Polytechnic Institute*
Fritz Widdel, *Max Planck Institute for Marine Microbiology, Germany*
Russell G. Wilkinson, *University of Melbourne, Australia*
Carl R. Woese, *University of Illinois*
Aideen Young, *Albert Einstein College of Medicine*
Davide Zannoni, *University of Bologna, Italy*
Stephen H. Zinder, *Cornell University*

Finally, the authors are grateful to all the people at Prentice Hall who have contributed in significant ways to this edition, including in particular David K. Brake and Laura J. Edwards (editorial) and Debra Wechsler (production). We also wish to acknowledge the contributions of Carol J. Dean and Stephanie Hiebert to copy editing, Barbara Littlewood (Madison, WI), who composed the index, and Toni Gower, Southern Illinois University, for her expert word processing skills.

Michael T. Madigan
John M. Martinko
Jack Parker

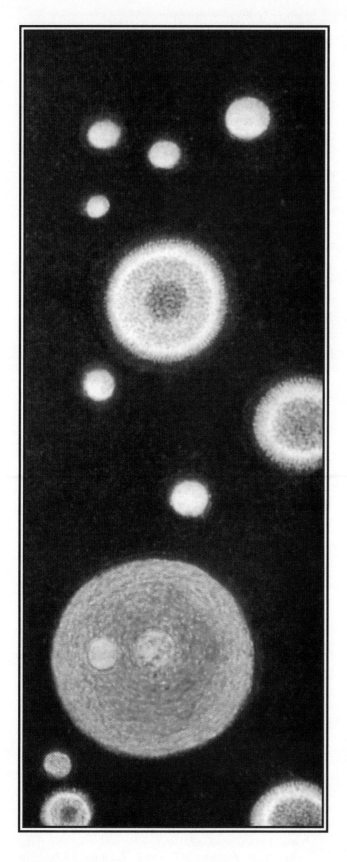

CHAPTER 1

Introduction: An Overview of Microbiology and Cell Biology

A photograph of bacterial colonies taken by Walter Hesse, a contemporary of the famous early microbiologist Robert Koch, over 100 years ago. See the box, Solid Media and Development of the Petri Plate.

MINIGLOSSARY for Chapter 1

Archaea a group of phylogenetically related prokaryotes distinct from Bacteria

Bacteria a group of phylogenetically related prokaryotes distinct from Archaea

Cell the fundamental unit of living matter

Cytoplasm the fluid portion of a cell, bounded by the cell membrane but excluding the nucleus (if present)

DNA deoxyribonucleic acid, the hereditary material of cells and some viruses

Ecology the study of organisms in their natural environments

Entropy a measure of the degree of disorder in a system; entropy always increases in a closed system

Enzymes protein catalysts that function to speed up chemical reactions

Eukarya all eukaryotic organisms

Eukaryote a cell possessing a membrane-enclosed nucleus and usually other organelles

Evolution change in a line of descent over time leading to the production of new species or varieties within a species

Genetics heredity and variation of living organisms

Metabolism all biochemical reactions in a cell

Microorganism a microscopic organism consisting of a single cell or cell cluster, including the viruses

Mutation an inheritable change in the nucleotide (base) sequence of the DNA of an organism

Prokaryote a cell lacking a true nucleus

Pure Culture a culture containing a single kind of microorganism

RNA ribonucleic acid, involved in protein synthesis as messenger RNA, transfer RNA, and ribosomal RNA

Spontaneous Generation the hypothesis that living organisms can originate from nonliving matter

Sterile absence of all living organisms and viruses

Transcription the synthesis of RNA using a DNA template

Translation the synthesis of proteins using the genetic information in RNA as a template

---

Microbiology is the study of microorganisms, a large, diverse group of microscopic organisms that exist as single cells or cell clusters; it also includes viruses, which are microscopic but not cellular. Microbial cells are thus distinct from the cells of animals and plants, which are unable to live alone in nature and can exist only as parts of multicellular organisms (Figure 1.1). A single microbial cell is generally able to carry out its life processes of growth, energy generation, and reproduction independently of other cells, either of the same kind or of a different kind.

What, then, is microbiology all about? We can list several aspects of microbiology here: (1) It is about living cells and how they work. (2) It is about microorganisms, an important class of cells capable of free-living (independent) existence. It is especially about bacteria, a large group of structurally simple cells of enormous basic and practical significance. (3) It is about microbial diversity and evolution, about how different kinds of microorganisms arose and why. (4) It is about what microorganisms do in the world at large, in human society, in our own bodies, and in the bodies of animals and plants. (5) It is about the central role that microbiology plays as a basic biological science and how an understanding of microorganisms helps in understanding the biology of higher organisms, including humans.

Why study microbiology? Microbiology, one of the most important of the biological sciences, is studied for two major reasons:

1. As a *basic biological science*, microbiology provides some of the most accessible research tools for probing the nature of life processes. Our most sophisticated understanding of the chemical and physical principles behind living processes has arisen from studies using microorganisms. This is in large part because microbial cells share many biochemical properties with cells of multicellular organisms. And this, coupled with the fact that microbial cells can reach extremely high cell densities in culture and are readily amenable to genetic study, makes them excellent models for understanding cell function in plants and animals. To illustrate this, one only need consider the fact that the discovery that DNA is the genetic material of cellular organisms emerged from a study of genetic transfer in bacteria (∞ Origins of Bacterial Genetics, Chapter 9) to realize that biology owes much to the microorganism.

2. As an *applied biological science*, microbiology deals with many important practical problems in medicine, agriculture, and industry. Some of the most important diseases of humans, animals, and plants are caused by microorganisms. Microorganisms

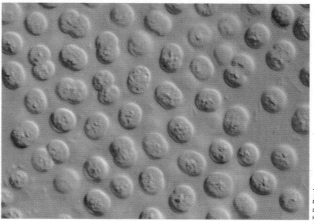

Figure 1.1 Living organisms are composed of cells. (a) Plants and (b) animals are composed of many cells; they are called *multicellular*. A single plant or animal cell cannot have an independent existence; each of its cells is dependent on the other. (c) Microorganisms are free-living cells. A single microbial cell can have an independent existence. Shown is a photomicrograph of photosynthetic microorganisms called *cyanobacteria*. Each cell is about 3 μm in diameter.

play major roles in soil fertility and animal production. Many large-scale industrial processes are microbially based, which has led to the development of a whole new discipline, *biotechnology*.

In this book, both the basic and applied aspects of microbiology are covered in an integrated fashion. We discuss the experimental basis of microbiology, the general principles of cell structure and function, the classification and diversity of microorganisms, biochemical processes in cells, and the genetic basis of microbial growth and evolution. From an applied viewpoint, we discuss disease processes in humans that are caused by microorganisms, the nature of the immune response, the roles of microorganisms in food and agriculture, and industrial and biotechnological processes employing microorganisms.

The material in this textbook serves as a foundation for advanced work in microbiology. It also serves as a basis for further studies in cell biology, biochemistry, molecular biology, and genetics. Although a student may begin to study microbiology primarily because of an interest in its applied problems, the basic concepts

learned will serve as a foundation for advanced study in many areas of contemporary biology. A firm grasp of microbiological principles will also serve as a basis for understanding biological processes in higher organisms, including humans.

1.1
Microorganisms as Cells

The **cell** is the fundamental unit of all living matter. A single cell is an entity, isolated from other cells by a cell membrane (and perhaps a cell wall) and containing within it a variety of chemical materials and subcellular structures (Figure 1.2). The **cell membrane** is the *barrier* that separates the inside of the cell from the outside. Inside the cell membrane are the various structures and chemicals that make it possible for the cell to function. Key structures are the **nucleus** or **nucleoid,** where the *information* needed to make more cells is stored, and the **cytoplasm,** where the *machinery* for cell growth and function is present.

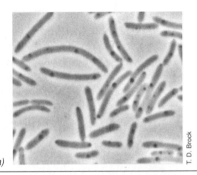

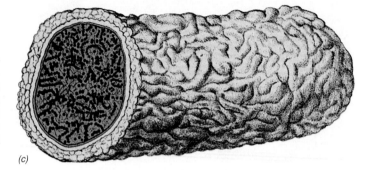

Figure 1.2 Cells. (a) Photomicrograph of rod-shaped bacterial cells as seen in the light microscope; a single cell is about 1 μm in diameter. (b) Longitudinal section through a bacterial cell as viewed with an electron microscope. The two lighter areas represent the nucleoid. (c) An artist's interpretation in three dimensions of a gram-negative bacterial cell, showing the characteristic wrinkled outer structure.

All cells contain certain types of complex chemical components: **proteins, nucleic acids, lipids,** and **polysaccharides.** Collectively, these are called *macromolecules.* Because these chemical components are common throughout the living world, it is thought that all cells have descended from a single common ancestor, the *universal ancestor.* Through billions of years of evolution, the tremendous diversity of cell types that exist today has arisen.

Although each kind of cell has a definite structure and size, a cell is a dynamic unit, constantly undergoing change and replacing its parts. Even when it is not growing, a cell is continually taking materials from its environment and working them into its own fabric. At the same time, it continuously discards waste products into its environment. A cell is thus an *open system,* forever changing yet generally remaining the same.

The hallmarks of a cell

A living cell is a complex chemical system. What are the characteristics that set living cells apart from nonliving chemical systems? We list five major characteristics here (Figure 1.3).

1. **Self-feeding or nutrition.** Cells take up chemicals from the environment, transform these chemicals from one form to another, release energy, and eliminate waste products.

2. **Self-replication or growth.** Cells are capable of directing their own synthesis. As a result of nutritional processes, a cell grows and divides, forming two cells, each nearly identical to the original cell.

3. **Differentiation.** Many cells can undergo changes in form or function, a process called *differentiation.* When a cell differentiates, certain substances or structures that were not formed previously are now formed, or substances or structures that had been formed previously are no longer formed. Cell differentiation is often part of a cellular life cycle in

which cells form specialized structures involved in sexual reproduction, dispersal, or survival when confronted with unfavorable conditions.

4. **Chemical signaling.** Cells respond to chemical and physical stimuli in their environment, and in the case of motile cells, can actually move toward or away from the environmental stimulus, a process called *taxis.* In addition, cells can often *interact* or *communicate* with other cells, generally by means of **chemical signals.** Multicellular organisms, such as plants and animals, are composed of many different cell types that have arisen as a result of differentiation from single cells and are arranged to form tissues and organs characteristic of a particular organism. In multicellular organisms, complex interactions between these different cell types lead to the behavior and function of these cells. One of the striking things about the cells of multicellular organisms is that they are incapable of independent existence in nature and exist only as part of a whole plant or animal. This interdependence of the cells of higher organisms is one of the hallmarks of multicellular life. In the microbial world chemical communication occurs, although it is less highly developed and is limited to just certain groups.

5. **Evolution.** Unlike inanimate structures, unicellular and multicellular organisms *evolve.* This means that hereditary changes (which occur at low but regular rates in all cells) can influence the overall fitness of the cell or higher organism in a positive or negative way. The result of evolution is selection for those organisms best suited to life in a particular environment.

The improbable cell

A fundamental law of physics is that the universe is moving constantly toward a condition of greater disorder (increasing entropy), with molecules and atoms becoming

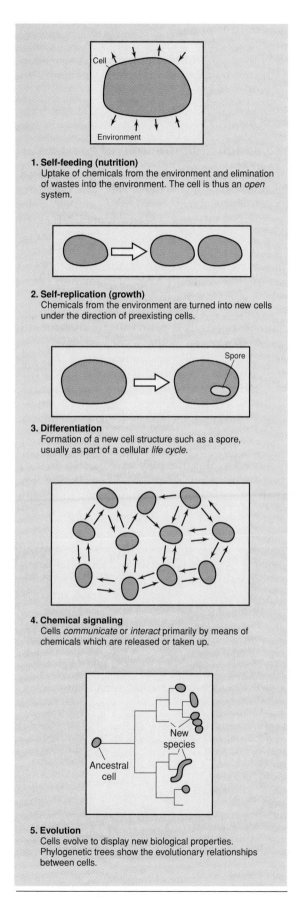

1. Self-feeding (nutrition)
Uptake of chemicals from the environment and elimination of wastes into the environment. The cell is thus an *open* system.

2. Self-replication (growth)
Chemicals from the environment are turned into new cells under the direction of preexisting cells.

3. Differentiation
Formation of a new cell structure such as a spore, usually as part of a cellular *life cycle*.

4. Chemical signaling
Cells *communicate* or *interact* primarily by means of chemicals which are released or taken up.

5. Evolution
Cells evolve to display new biological properties. Phylogenetic trees show the evolutionary relationships between cells.

Figure 1.3 The hallmarks of cellular life.

arranged in a random manner. Life then, seems like a miracle since the living cell is anything but random, being a highly ordered, exceedingly improbable (nonrandom) structure. More careful analysis of cells, however, shows that there is no divergence between biology and physics. How can this be? First, we note that the idea of randomness implies that a system is in *equilibrium* with its surroundings. A living cell, on the other hand, is definitely *not* in equilibrium with its surroundings. We say that the living cell is a *nonequilibrium* system. How is it possible for a cell to maintain this nonequilibrium condition?

As we have seen, a living cell is actually an *open system*, a system in which energy is taken in from the surroundings and used to maintain cell structure. A cell carries out energy transformations, and some of this energy is used to maintain the structure of the cell itself. Thus, for a cell to function as a cell, its structure must be maintained. Indeed, *the basis of cell function is cell structure*. The importance of structure as a foundation of life is further emphasized when we recall that cells are *self-replicating systems*. Cell reproduces cell. Therefore, making a living cell is a matter of making the right structure.

The nonrandom nature of a living cell is shown most dramatically by an analysis of its chemical composition and a comparison of that chemical composition with the chemical composition of the earth. The average chemical composition of a living cell is quite different from the average chemical composition of the earth. The cell therefore is not a random assortment of chemical elements found on Earth but instead a *selective* chemical system composed primarily of C, H, O, N, S, and P, the major elements of life (Figure 1.4). This further emphasizes the special or nonrandom nature of a living cell.

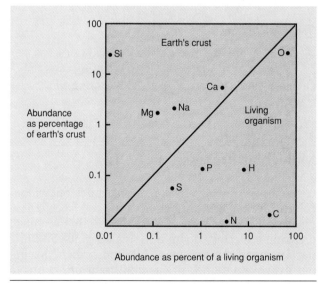

Figure 1.4 Chemical differences between a living organism and the earth. Note that the key elements C, H, O, N, P, and S are much more abundant in living organisms than in nonliving matter. Thus, living organisms *concentrate* these elements from the environment.

Where did the first cells come from? In some way the first cell must have come from a noncell, something before the cell, a procellular structure. Although formation of the first cell was an improbable event that may have taken several hundred million (if not over a billion) years to occur, once the first cell arose, a series of highly probable events followed including growth and division to form populations of cells from which evolutionary forces could begin to select for improvements and diversification.

Astronomers tell us that there are probably vast numbers of planets in the universe with Earthlike conditions on which life might arise or might have arisen. If life is an almost inevitable event when the proper physical and chemical conditions are available, then we can anticipate that there are other planets in the universe with living organisms similar to those found on Earth. The precise assemblage of organisms would be different, of course, because so much of evolution depends upon accidents of history; thus the time stage and evolution of these other planets would likely be earlier or later than those on Earth. But if we were able to sample some of these other planets, it is likely we would not find any surprises; as we will see, major chemical and physical principles govern the structure and function of all cells.

CONCEPT CHECK   *1.1*

Microbiology is the study of microorganisms, those organisms that exist in nature as single cells. Microbiology is a broad biological discipline with strong roots in both basic and applied science. Living cells are open systems that show several principal characteristics including nutrition, growth, differentiation, chemical signaling, and evolution.

✔ *Why might the study of a genetic phenomenon be easier and more productive in a microorganism than in a large animal?*

✔ *How can cells exist as highly ordered systems in a universe in which entropy is constantly increasing?*

✔ *Which elements are the major elements of life?*

1.2

Molecular Processes in Cells

Cells can be studied as *machines* that carry out chemical transformations. In this view, the cell is a *chemical machine* that converts energy from one form to another, breaking molecules down into smaller units, building up larger molecules from smaller ones, and carrying out many other kinds of chemical transformations (a molecule is defined as two or more atoms chemically bonded to one another). The term **metabolism** is used to refer to the collective series of chemical processes that occur in living organisms, both biosynthetic and degradative. Cells can also be studied as *coding devices*

analogous to computers, possessors of information that is either passed on to offspring or is processed into another form. The term **genetics** is used to describe the heredity and variation of living organisms and to describe the mechanisms by which these are brought about. Let us now look at these two cellular functions in more detail.

Machine function: The role of enzymes

If a cell is a chemical machine, what are the driving forces that make it function? The components of the cell's chemical machine are **enzymes,** protein molecules capable of catalyzing specific chemical reactions. Whether or not a cell can carry out a particular chemical reaction depends on the presence in the cell of a particular enzyme that catalyzes the reaction. Specificity of enzymes is generally very high, and so even very closely related chemical reactions in the cell are catalyzed by separate enzymes. The specificity of an enzyme is determined primarily by its structure. As proteins, enzymes are polymers of long chains of amino acids (there are 21 different amino acids) that are connected in specific and highly precise ways. It is the amino acid *sequence* of an enzyme that determines its structure as well as its catalytic specificity. A single protein molecule can often have 300 or more amino acid residues. The long chain of amino acids becomes folded into a specific configuration, leading to the formation of various domains that play specific roles in the function of the protein (Figure 1.5). The machine function of a cell (metabolism) is ultimately determined by the amounts and types of the various enzymes it contains.

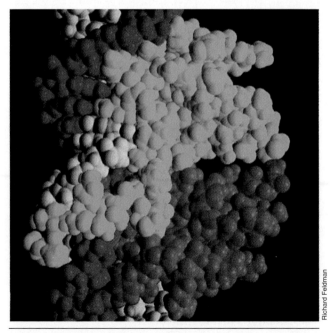

Figure 1.5 The structure of a protein as shown in a computer-generated model. Different domains (regions) of the protein are shown in different colors.

Coding function: DNA, RNA, protein

It is when we examine how the amino acid sequence of a protein is determined that we consider cells as coding devices. How is the cell able to arrange amino acids into precise sequences, each constituting a separate kind of protein? To understand this, we must consider the cell as a device for storing information and converting it to the appropriate form. In this context, the cell can be viewed as a repository of protein sequence information stored in a coded manner.

This code, called the *genetic code,* is stored in the sequence of nucleotides (each of which contains a nitrogen base, adenine, guanine, cytosine, or thymine) in the hereditary molecule **deoxyribonucleic acid (DNA).** DNA is present in the cell as two long molecules, which are intertwined to form a helix, the famous DNA double helix (Figure 1.6). A **gene** is a segment of DNA that encodes a specific protein.

DNA undergoes two major cellular transformations. First, all DNA molecules in a cell undergo *replication* before cell division, allowing each new cell to receive a complete set of genetic instructions (Figure 1.6). Second, the genetic code written in the DNA must be *transcribed* and *translated:* the DNA base sequence dictates the specific amino acid sequence of the protein. This process is carried out by a special, highly complex translation machinery in the cell. The translation apparatus is at the very core of cell function.

The translation system

In human language, it is primarily a convenience for text to be translated from one language to another. In the language of the cell, however, translation is essential if the cell is to function. This is so because the genetic information in the DNA molecule is only a repository; it is incapable of being interpreted directly. A protein is made using only the translation system as an intermediate. The translation apparatus is thus the central and most basic attribute of the cell.

Because of the central role of the translation system, it is difficult to change it. In fact, the translation system, although differing in certain details, is essentially the same in all kinds of organisms. It is very likely that in the origin of life the translation system was one of the first things to arise and, once formed, was retained in essentially the same form throughout the long history of life on Earth. Interestingly, the translation apparatus cannot use DNA directly, as the information in DNA must first be transcribed into ribonucleic acid (RNA). This may also be a reflection of the origin of life on Earth in that many biologists believe that there was a period in evolution before DNA existed in which RNA was the repository of genetic information (∞ Section 15.3).

Although we will describe the steps in protein synthesis in some detail in Chapter 6, at the moment we note that the production of proteins requires two processes (Figure 1.6):

1. **Transcription,** the formation of **messenger ribonucleic acid** molecules, which contain a complementary copy of the genetic information stored in DNA.

2. **Translation,** a process of linking amino acids together to make proteins, occurs on **ribosomes,** structures composed of another form of RNA and several proteins, on which the translation process actually occurs.

In the translation process messenger RNA, which contains the genetic instructions for the proteins to be made, combines with ribosomes, and through the action of several other factors to be described in detail later, eventually yields a protein that folds to perform a specific function in the cell (Figure 1.6). Transcription of different genes leads to the formation of different proteins, and we will see that the complement of proteins produced by a given organism is characteristic of that organism and is also subject to change by growth conditions and other factors.

CONCEPT CHECK   *1.2*

Cells are chemical machines that convert energy from one form to another. The chemical reactions of cells are carried out by specialized proteins called enzymes. Cells are also coding devices, translating information in the genes (DNA base sequences) into the structure of proteins (amino acid sequences).

✔ *What are the products of transcription and translation?*

✔ *Why must DNA replicate before cell division?*

1.3

Growth, Mutation, and Evolution

The connection between the two attributes of a cell, its machine function and its coding function, is expressed through the process of *cell growth.* A living cell grows in size and then divides and forms two cells. In the orderly growth process that results in two cells being formed from a single cell, all the constituents of the cell double in amount. Growth in size requires the functioning of the chemical machinery of the cell to supply energy and precursors for biosynthesis of macromolecules. But each of the two cells must contain *all* the genetic information necessary for the formation of more cells, and so during cell growth and division there must be a duplication of DNA. And it is important not just that the DNA content doubles when one cell turns into two but also that the *precise sequence* of bases in the DNA is copied (see Figure 1.6). The fidelity of this copying function is very high, and so the two progeny of a single cell are identical in DNA base sequence to the parent. We say that the two progeny are *genetically identical* to the parent (top part of Figure 1.7).

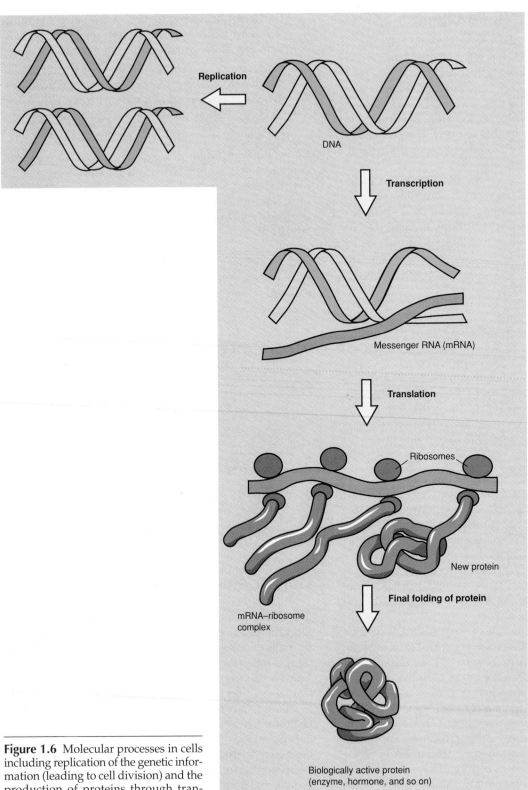

Replication

DNA

Transcription

Messenger RNA (mRNA)

Translation

Ribosomes

New protein

Final folding of protein

mRNA–ribosome
complex

Biologically active protein
(enzyme, hormone, and so on)

Figure 1.6 Molecular processes in cells including replication of the genetic information (leading to cell division) and the production of proteins through transcription and translation.

Mutations

However, mistakes in copying DNA do occur occasionally, and so one of the two offspring cells may not always be identical to the parent (bottom part of Figure 1.7). The mistakes that are made in copying DNA are called **mutations.** A mutation is a permanent change in the sequence of nucleotides in the DNA molecule that is passed on to one or more offspring. In some cases, a mutation has no detectable effect, but in many cases it results in the for-

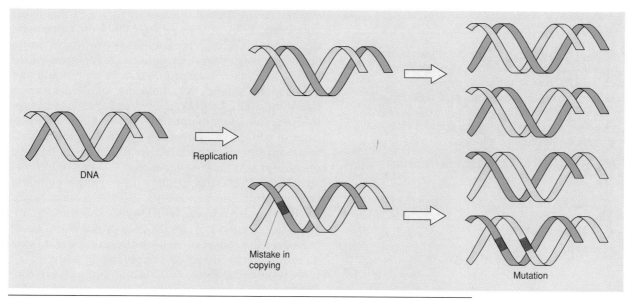

Figure 1.7 Fidelity of DNA replication and the phenomenon of mutation. The top portion of the figure shows normal replication events; the bottom portion, the introduction of a mutation.

mation of a malfunctioning protein (or no protein at all), and so the cell with the mutation is defective, possibly either dying or deteriorating. Such cells eventually disappear from the population and are no longer seen. Thus, mutations are generally harmful. However, in rare cases a mutation may result in the formation of an improved protein, for example, an enzyme better able to function than that in the parent cell. The cell containing this altered enzyme thus has a *selective advantage* and, as a result of further cell divisions, may eventually replace the parent type. The survival value of a mutation usually depends on the environment in which the organism is living. In some environments, a mutation may be advantageous, and in other cases it may be harmful.

Mutations and evolution

The process of natural variation through mutation and the consequent effects on the cell that we have just discussed is called **natural selection,** a phenomenon that is at the basis of the process of **evolution.** Although Charles Darwin first proposed the theory of evolution from observations on multicellular organisms, Darwinian evolution is demonstrated most dramatically and effectively at the microbial level. Not only has microbial evolution provided some of the most critical support for the theory of evolution, it has also led to some very important practical consequences. For instance, disease-causing microorganisms have arisen that are resistant to certain antibiotics, and therefore the diseases caused by these microorganisms cannot be treated with these antibiotics. We see this as a dramatic and very practical result of microbial evolution. Many other important consequences of natural selection will be discussed later in this book. We note here most importantly that the vast diversity of microorganisms, as well as the diversity of higher organisms, is due to the process of evolution through mutation and natural selection.

Genetic potential of the cell

As we have noted, the DNA coding sequence that specifies the amino acid sequence of a specific protein molecule is called a *gene.* How many proteins are there in a cell? A simple bacterial cell, the organism *Escherichia coli,* has about 1900 different *kinds* of proteins under any given set of conditions and contains a total of about 2.4 million individual protein molecules. Some proteins in the *E. coli* cell are present in very large numbers, even greater than 100,000 molecules per cell, whereas other proteins are present in the cell only in a very small number of molecules. Thus, the cell has some means of controlling the *expression* (transcription and translation) of its genes so that not all genes are expressed to the same extent or at the same time.

How many genes does a cell have? If we determine the amount of DNA per cell, we will find that our *E. coli* cell has about 4.7 million base pairs of DNA (4700 kilobases). If a single gene averages 1.1 kilobases in length, that means that the cell has about 4200 genes if all the DNA encodes protein. However, we know that some of the DNA does not encode proteins and we thus estimate that the *E. coli* bacterial cell has more than 3000 total *coding* genes, of which only about 1000–2000 are expressed at any particular time. Eukaryotic cells can have many more genes than this, as a human cell, for example, contains about 1000 times the DNA of an *E. coli* cell.

CONCEPT CHECK   *1.3*

Cell division requires that all the constituents of the cell, including the genes, be duplicated. A single bacterial cell can have more than 4000 genes. Occasional mistakes are made in the duplication of genes, resulting in mutations. Evolution occurs because mutants that are better adapted to the environment than the parent are favored and are selected.

   ✔ *How is evolution affected by the coding function of cells?*

   ✔ *How many total nucleotides are present in the DNA of the bacterium* Escherichia coli? *(Remember, DNA is double-stranded.)*

1.4

Cell Structure

What is the structure of a cell? All cells have a barrier separating inside from outside that is called the **cytoplasmic (cell) membrane** (Figure 1.8). It is through the cell membrane that all nutrients and other substances of vital importance to the cell pass in, and it is through this membrane that waste materials and other cell products pass out. When the membrane is damaged, the interior contents of the cell leak out and the cell usually dies. As we shall see, some drugs and other chemical agents damage the cytoplasmic membrane and in this way bring about the destruction of cells.

The cytoplasmic membrane is a very thin, highly flexible layer and is structurally weak. By itself, it usually cannot hold the cell together, and an additional stronger layer, called the **cell wall,** is usually necessary (Figure 1.8). The wall is a relatively rigid layer that is present outside the membrane and protects the membrane and strengthens the cell. Plant cells and most microorganisms have such rigid cell walls. Animal cells, however, do not have walls; these cells have developed other means of support and protection.

Within a cell, and bounded by the cytoplasmic membrane, is a complicated mixture of substances and structures called the **cytoplasm.** These materials and structures, bathed in water, carry out the functions of the cell. The major components of the cytoplasm, other than water, include macromolecules, ribosomes, small organic molecules (mainly precursors of macromolecules), and various inorganic ions.

Prokaryotic and eukaryotic cells

Upon careful study of the internal structures of cells, two basic types have been recognized: **prokaryote** and **eukaryote** (Figure 1.8*a* and *b*). These two types of cells are structurally very different. A major structural difference between prokaryotes and eukaryotes, other than size, is the arrangement of DNA within the cell.

Eukaryotes contain a nucleus enclosed by a nuclear membrane that contains several DNA molecules and undergoes division by the well-known process of **mitosis.** By contrast, the prokaryotic nuclear region, called the *nucleoid,* is not surrounded by a membrane and consists of a single DNA molecule whose division is nonmitotic. Unlike prokaryotes, eukaryotic cells typically contain other membrane-enclosed internal structures in addition to the nucleus, such as mitochondria and chloroplasts (the latter in photosynthetic cells only), and also have a *cytoskeleton,* a series of internal structures that structurally support the cell and help organize and move its internal components. **Bacteria** and **Archaea,** two major evolutionary lineages (see Section 1.5), are the only prokaryotes. There are several groups of eukaryotic microorganisms, including **algae, fungi,** and **protozoa.** In addition, all multicellular life forms (plants and animals) are constructed of eukaryotic cells.

Microorganisms in general are very small. A rod-shaped prokaryote is typically about 1–5 micrometers (μm) long and thus is completely invisible to the naked eye. To illustrate how small a bacterium is, consider that 500 bacteria 1 μm in length could be placed end to end across the period at the end of this sentence.

Viruses are not cells. Viruses lack many of the attributes of cells, of which the most important is that they are not dynamic open systems. A single virus particle is a static structure, quite stable and unable to change or replace its parts. Only when it is associated with a cell does a virus become able to replicate and acquire some of the attributes of a living system. Thus, unlike cells, viruses have no metabolism of their own. Although viruses have genetic information (either DNA or RNA), they lack the translation apparatus and use the cell's machinery for protein synthesis. Viruses are known to infect various organisms including microorganisms. Some size comparisons of viruses and cells are shown in Figure 1.9. Although many viruses cause disease in the organisms they infect, virus infection does not always lead to disease. We will see in Chapter 8 that

CONCEPT CHECK   *1.4*

All cells have a critical barrier, the cytoplasmic membrane, that separates the cytoplasm from its surroundings. The most important structure inside the cell is the nucleus or nucleoid, the site of the genetic information, DNA. Cells can be divided into two large groups depending on the organization of their DNA and several other properties. In eukaryotes, the nucleus is enclosed by a nuclear membrane, whereas in prokaryotes no such membrane exists.

   ✔ *What are the major structural differences between prokaryotic and eukaryotic cells?*

   ✔ *Do viruses contain nucleic acid? If so, what kinds?*

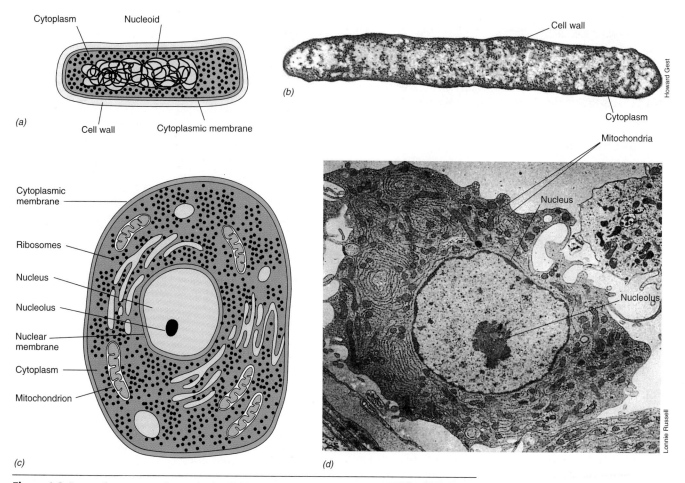

Figure 1.8 Internal structure of microbial cells. (a) Diagram of a prokaryote. (b) Electron micrograph of a prokaryote. The cell is about 1 μm in diameter. (c) Diagram of a eukaryote. (d) Electron micrograph of a eukaryote (animal cell). The cell is about 25 μm in diameter.

besides causing disease, viruses can have other profound effects on cells.

1.5
Evolutionary Relationships among Living Organisms

Although cells can be differentiated structurally as being either prokaryotic or eukaryotic, cell structure does not necessarily imply *evolutionary* relationship. However, phylogenetic (evolutionary) relationships among microorganisms can now be determined, and Chapter 15 describes the methods used to discern such relationships. These methods rely on nucleic acid sequence comparisons, in particular, the sequences of *ribosomal RNA*—structural RNA of the ribosome, the key cell structure involved in translation (see Section 1.2). Indeed, one of the great recent discoveries in biology is that changes in the sequence of nucleotides in ribosomal RNA (ultimately dictated by mutations in the DNA that encodes ribosomal RNA) can be used as a measure of evolutionary relationships among cells.

From studies on ribosomal RNA sequences, three evolutionarily distinct cellular lineages can be defined, two of which are prokaryotic in structure and one of which is eukaryotic. The groups have been given the names **Bacteria, Archaea,** and **Eukarya** (Figure 1.10).* However, despite the fact that both are structurally prokaryotic at the molecular level, Bacteria and Archaea

---

*To avoid the perception that all prokaryotes are closely related (which indeed they are not), the terms *Bacteria* and *Archaea* are used to describe two of the three taxonomic domains (the highest of biological taxons) (see Chapter 15 and Figure 1.11) of cellular organisms. The third group of organisms is the *Eukarya* (all of which are eukaryotic cells). However, the word *bacteria* is so firmly entrenched in the science of microbiology that it is unlikely ever to lose its synonymity with the word *prokaryote*. This fact thus creates a semantic dilemma. Are all bacteria (prokaryotes) still Bacteria (a phylogenetic unit)? No. In this book, the word *Bacteria* (with a capital B) is reserved for describing organisms in a *phylogenetic sense*, that is, as members of the domain Bacteria. In contrast, the word *bacteria* (lowercase b) appears many times in this book and is used to refer to prokaryotes in general, without reference to phylogeny. Because a detailed discussion of the Archaea is reserved for Chapter 17, most references to bacteria in the chapters preceding Chapter 17 are to Bacteria. However, no phylogenetic implication should be attached to the words *bacteria* and *bacterium* (lowercase b) no matter where they are used in this book.

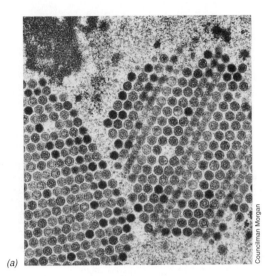

(a)

Councilman Morgan

Typical animal cell

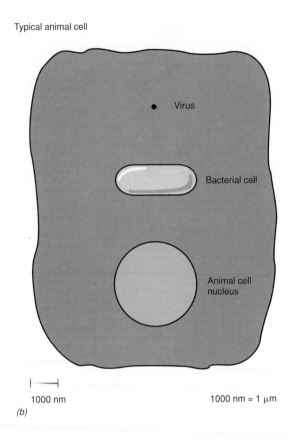

Virus

Bacterial cell

Animal cell nucleus

1000 nm

1000 nm = 1 μm

(b)

Figure 1.9 Virus size and structure. (a) Particles of adenovirus, a virus that causes respiratory infections in humans. A single virus particle is about 100 nm in diameter. (b) The size of a virus in comparison to a bacterial and animal cell.

are as evolutionarily distinct from one another as either group is from the Eukarya. All three groups are thought to have diverged from a common ancestral organism, the "universal ancestor," early in the history of life on Earth (Figure 1.11). Because the cells of higher animals and plants are all eukaryotic, it follows that eukaryotic microorganisms were the ancestors of multicellular organisms, whereas Bacteria and Archaea represent evolutionary branches that never evolved past the microbial stage (Figure 1.12). The organelles of eukaryotic cells, the mitochondria and the chloroplast, are phylogenetically related to Bacteria and became part of the eukaryotic cell eons ago in a process called *endosymbiosis*, discussed in detail in Chapter 15.

Classification

In addition to understanding and appreciating the phylogenetic origins of cellular organisms, it is important for a variety of reasons to be able to identify and classify microorganisms. For example, rapid identification of a particular human disease-causing microorganism (pathogen) is usually essential to knowing how to treat an infected person. We discuss many methods for identifying microorganisms in Chapter 21, Clinical and Diagnostic Microbiology and Immunology. Several cri-

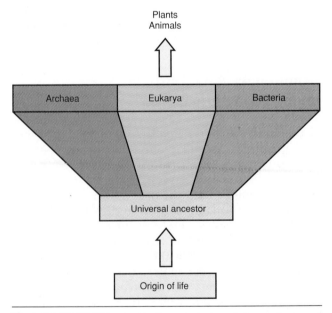

Plants
Animals

| Archaea | Eukarya | Bacteria |

Universal ancestor

Origin of life

Figure 1.11 Cellular evolution as deduced from ribosomal RNA sequencing.

	Prokaryotic	Eukaryotic
Macroorganisms	None known	Eukarya: Animals Plants
Microorganisms	Archaea	Eukarya: Algae Fungi Protozoa
	Bacteria	

Figure 1.10 The major types of cellular life forms.

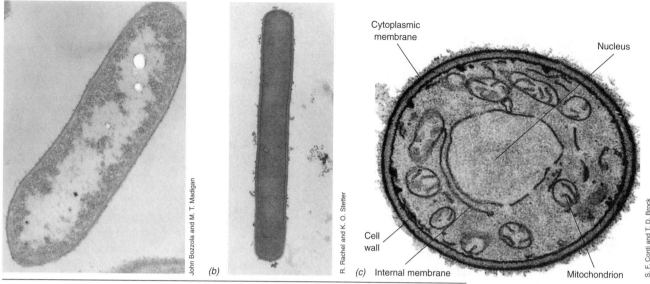

Figure 1.12 Electron micrographs of sections of a cell from each of the domains of living organisms. (a) *Heliobacterium modesticaldum* (Bacteria); the cell measures 1 × 3 μm. (b) *Methanopyrus kandleri* (Archaea); the cell measures 0.5 × 4 μm. (c) *Saccharomyces cerevisiae* (Eukarya); the cell measures about 8 μm in diameter.

teria have been used to characterize microorganisms, and at present both phylogenetic standing and other cellular characteristics are used for classification purposes. After close study of the structure and function of a microorganism, including its genetics, metabolism, behavior, and other key properties, it is usually possible to recognize a set of characteristics unique to a given organism. Once an organism has been defined by such a set of characteristics, it can be given a name.

Microbiologists use the binomial system of nomenclature first developed by Linnaeus for plants and animals. The *genus* name is applied to a number of related organisms; each different type of organism within the genus has a *species* name. Genus and species names are always used together to describe a specific type of organism, whether it be a single cell or a group of such cells. (In writing, the genus and species names are either underlined or printed in italics. For example, the bacterium *Escherichia coli*, or *E. coli* for short, has a genus name, *Escherichia*, and a species name, *coli*.)

Microbial diversity

An understanding of microbial diversity requires an appreciation of the evolutionary roots of cells. Because evolution has shaped all life on Earth, the structural and functional diversity we see in cells represents successful evolutionary events that, through the process of natural selection, have conferred survival value (fitness) on the microorganisms extant today. Microbial diversity can be seen in terms of variations in cell size and morphology, metabolic strategies, motility, cell division, developmental biology, adaptation to environ-

mental extremes, and many other structural and functional aspects of the cell. We describe the major groups of Bacteria, Archaea, and Eukarya in Chapters 16, 17, and 18, respectively, and consider there many of the aspects that make microbial life so amazingly diverse. Here we give only a quick overview of things to come.

Several evolutionary branches occur within the **Bacteria,** including all disease-causing (pathogenic) prokaryotes and most of the bacteria commonly found in soil, water, animal digestive tracts, and many other environments. Some of these organisms contain pigments that allow them to use light as an energy source in a process called *phototrophy*, others rely on organic chemicals as energy sources, and some can even use inorganic chemicals as fuel to drive cellular processes. Most of the prokaryotes encountered in everyday life are phylogenetically Bacteria, and some have evolved special structures such as spores for aiding survival. Oxic environments (those containing O_2) as well as various anoxic habitats are inhabited by various species of Bacteria.

With the prokaryotes called **Archaea,** in contrast, we see a much different picture. Most Archaea are anaerobes, cells incapable of living in air. Many also thrive under unusual growth conditions, inhabiting what humans would consider extreme environments: hot springs (to temperatures *above* the boiling point of water), extremely salty bodies of water, and highly acidic or alkaline soils and water. Indeed, certain species of Archaea currently define the limits of biological tolerance to physiochemical extremes. Certain Archaea also show unusual biochemical features, such as the methanogens, which are prokaryotes that produce methane (natural gas) as an integral part of their energy metabolism.

Among the **Eukarya** that are microorganisms are the algae, fungi, and protozoa. *Algae* contain chlorophyll, a green pigment that serves as a light-gathering molecule, making it possible for them to carry out phototrophy. Algae are common in aquatic habitats and can also be found in soil. *Fungi*—molds, yeasts, and mushrooms—lack chlorophyll and obtain their energy from organic compounds in soil and water. Fungi are thought to play a major role in the breakdown of dead organic matter in these and other environments. *Protozoa* are colorless, motile Eukarya that obtain food by ingesting other organisms or organic particles. Protozoa lack the cell walls of algae and fungi and in this respect resemble animal cells. Many protozoa are free-living microorganisms, but several cause disease in humans and other animals.

CONCEPT CHECK   1.5

Living organisms can be divided into prokaryotes and eukaryotes based on cell structure, but from an evolutionary viewpoint, living organisms form three major groups: Bacteria, Archaea, and Eukarya. Structural and functional diversity among microorganisms is substantial.

✔ *Which of the cells shown in Figure 1.12 are prokaryotic? Which are eukaryotic? What does cell structure say about evolutionary relationships?*

1.6
Populations, Communities, and Ecosystems

Up to now, we have been discussing cells as if they lived in isolation in their environments. Nothing could be further from the truth. Cells live in nature in association with other cells, in assemblages that we call **populations.** Such populations are composed of groups of related cells, generally derived by successive cell divisions from a single parent cell. The location in the environment where a population lives is called the **habitat** of that population.

In nature, populations of cells rarely live alone. Rather, they live in association with other populations of cells in assemblages called **communities** (Figure 1.13). Frequently, populations in communities interact, either in beneficial ways or in harmful ways. If two populations interact in a beneficial way, these populations will then maintain themselves better when together than when separate. In such cases we speak of the cooperative nature of the populations. In other cases, two populations living in the same habitat may interact in a way that is harmful to one of the populations. If such harmful interaction occurs, the population that is harmed will be reduced in number, or even replaced. If the effect is severe enough, the population may be completely eliminated.

The effect of organisms on their habitats

The living organisms in a habitat also interact with the physical and chemical environment of that habitat. Habitats differ markedly in their physical and chemical characteristics, and a habitat that is favorable for the growth of one organism may be harmful for another organism. Thus, the microbial community that we see in any given habitat is determined to a great extent by the physical and chemical characteristics of that environment.

In addition, organisms can modify the physical and chemical properties of their environment. Organisms carrying out metabolic processes remove nutrients from the environment and use them to build new cells. At the same time, organisms excrete waste products of their metabolism into the environment. Therefore, as time progresses, the environment is gradually changed through life processes. We speak collectively of the living organisms together with the physical and chemical constituents of their environment as an **ecosystem.**

Since single microbial cells are too small to be seen with the unaided eye, our knowledge of microorganisms in nature begins with studies using the microscope. Examination of natural materials such as soil, mud, water, spoiling food, human bodily excretions, and other living and dead material under the microscope reveals that such materials teem with vast numbers of microbial cells. Although such tiny cells seem inconsequential, we can state that microorganisms are *small but powerful.* Even though a single cell usually cannot, by itself, cause a perceivable effect, on a habitat, this single cell may be capable of multiplying rapidly and producing a massive number of progeny that may cause a major effect on the habitat. One of the most important and impressive attributes of microorganisms is *rapid growth* (multiplication), as single cells do not remain single very long and can quickly develop into *populations* that can significantly change the chemical properties of a habitat. Thus, although microorganisms may seem to occupy inconsequential niches in nature, they are extremely important components of ecosystems. Microorganisms also have very serious effects on higher plants and animals. In later chapters, after we have learned some of the details of microbial structure and function, we will again discuss the ways in which microorganisms affect animals, plants, and the whole global ecosystem.

CONCEPT CHECK   1.6

In general, microorganisms exist in nature in populations that interact with other populations in microbial communities. The activities of microbial communities can greatly affect the chemical and physical properties of their habitats.

✔ *What is a microbial habitat?*

✔ *How do microorganisms change the chemical and physical properties of their habitats?*

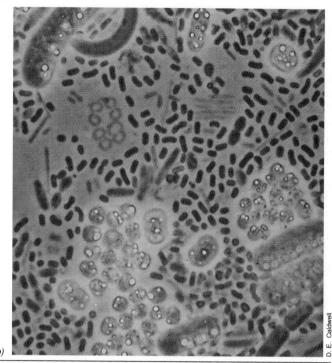

(a)

(b)

T. D. Brock

D. E. Caldwell

Figure 1.13 Examples of microbial communities. (a) A dense community of algae and cyanobacteria in the surface waters of a nutrient-rich lake (Lake Mendota, Wisconsin). (b) Photomicrograph of a bacterial community that developed in the depths of a small lake (Wintergreen Lake, Michigan), showing cells of various sizes.

1.7
Laboratory Culture of Microorganisms

Although we can get an idea of what microorganisms look like from a microscopic study of a natural habitat, we can best study their characteristics by obtaining them in a pure culture. A **pure culture** is a culture consisting of only one kind of microorganism. In order to obtain a pure culture, we must be able to grow the organism in the laboratory. This requires that we provide it with the proper nutrients and environmental conditions so that it can grow. It is also essential that we keep other organisms from entering the culture. Such unwanted organisms, called **contaminants,** are ubiquitous, and microbiological technique revolves around the avoidance of contaminants. Once we have isolated a pure culture, we can then proceed to study its biochemistry, physiology, genetics, and other characteristics.

Culture media

What are the conditions required for microbial growth? Microorganisms are cultured in water to which appropriate nutrients have been added. The aqueous solution containing such necessary nutrients is called a **culture medium.** The nutrients present in the culture medium provide the microbial cell with the ingredients required for the cell to produce more cells like itself. Besides an energy source, which can be an organic or an inorganic chemical, or light, a culture medium must have a source of carbon, nitrogen, and several other necessary nutrients to be described in detail in Chapter 4. Culture media can be prepared for use in either a liquid state or a gel (semisolid) state. A liquid culture medium is converted to the semisolid state by the addition of a gelling agent, usually agar. Culture media containing agar are dispensed in flat, covered dishes called **Petri dishes,** where microbial cells can grow and form visible masses called *colonies* (Figure 1.14). The history of the development of solid culture media and its impact on microbiology are discussed in the box, Solid Media and Development of the Petri Plate.

Aseptic technique

Before actually proceeding to the culture of microorganisms, we must first consider how to exclude contaminants. Microorganisms are everywhere. Because of their small size, they are easily dispersed in the air and on surfaces. Therefore, we must **sterilize** the culture medium soon after its preparation to eliminate microorganisms already contaminating it; this is usually done by heat. However, it is equally important to take precautions during the subsequent *handling* of a sterile culture medium to exclude from it all but the desired organisms. Thus, other materials that come into contact with a sterile culture medium must themselves be sterile.

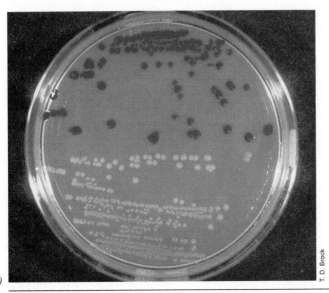

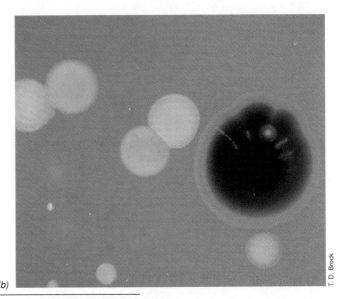

Figure 1.14 Colonies formed on an agar-containing medium on a Petri plate. (a) Red colonies of the bacterium *Serratia marcescens* (top half of plate) streaked alongside colonies of a mutant derivative unable to make the red pigment (bottom half of plate). (b) Mixed colonies of yellow and purple pigmented bacteria.

The technique used in the prevention of contamination during manipulations of cultures and sterile culture media is called **aseptic technique.** Its mastery is required for success in the microbiology laboratory, and it is one of the first methods learned by the novice microbiologist. Airborne contaminants are the most common problem because the air always contains dust particles that generally have a community of microorganisms on them. When containers are opened, they must be handled in such a way that contaminant-laden air does not enter (Figures 1.15 and 1.16). Aseptic transfer of a culture from one tube of medium to another is usually accomplished with an inoculating loop or needle that has been sterilized by incineration in a flame (Figure 1.15). Cultures in which growth has taken place can also be transferred to the surface of agar plates (Figure 1.16), where colonies develop from the growth and division of single cells. Picking and restreaking from an isolated colony is a major method of obtaining pure cultures from microbial communities containing many different organisms.

1.8

The Impact of Microorganisms on Human Affairs

One goal of the microbiologist is to understand how microorganisms work, and through this understanding to devise ways in which benefits may be increased and damages curtailed. Microbiologists have been eminently successful in achieving these goals, and microbiology has played a major role in the advancement of human health and welfare. An overview of the impact of microorganisms on human affairs is shown in Figure 1.17.

Microorganisms as disease agents

One measure of the microbiologist's success is shown by the statistics in Figure 1.18 (page 19), which compare the present causes of death in the United States to those at the beginning of the twentieth century. At the beginning of the twentieth century, the major causes of death were infectious diseases; currently, such diseases are of only minor importance. Control of infectious disease has come as a result of our comprehensive understanding of disease processes as well as improved sanitary practices and the discovery and use of antimicrobial agents. As we will see later in this chapter, microbiology as a science had its beginnings in these studies of disease.

CONCEPT CHECK   *1.7*

Microorganisms can be grown in the laboratory in culture media containing the essential nutrients they require. Successful cultivation of pure cultures of microorganisms can be done only if aseptic technique is practiced.

✔ *What is meant by the word* sterile? *What would happen if freshly prepared culture media were not sterilized?*

✔ *Why is aseptic technique necessary for successful cultivation of pure cultures in the laboratory?*

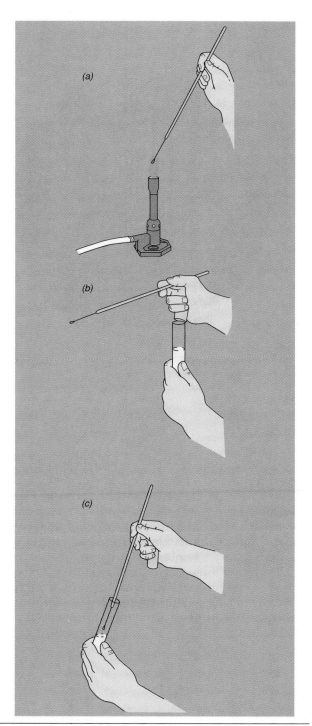

Figure 1.15 Aseptic transfer. (a) Loop is heated until red-hot and cooled in air briefly. (b) Tube is uncapped. (c) Sample is removed and tube is recapped. Sample is transferred to a sterile tube. Loop is reheated before being taken out of service.

However, although we now live in a world where many pathogenic microorganisms are under control, for the individual dying slowly of acquired immune deficiency syndrome (AIDS), the cancer patient whose immune system has been devastated as a result of treatment with an anticancer drug, or the individual infected with a multiple-drug-resistant pathogen, microorganisms can still be the major threat to survival. Although such tragic situations barely appear in our health statistics, they are of no less concern. Further, microbial diseases still constitute the major causes of death in many of the developing countries of the world. Although eradication of smallpox from the world has been a stunning triumph for medical science, millions still die yearly from such pervasive illnesses as malaria, tuberculosis, cholera, African sleeping sickness, and severe diarrheal diseases.

Thus, microorganisms are still serious threats to human existence. But on the other hand we must emphasize that most microorganisms are *not* harmful to humans. In fact, most microorganisms cause no harm at all and instead are actually *beneficial*, carrying out processes that are of immense value to human society. Even in the health care industry, microorganisms play beneficial roles. For instance, the pharmaceutical industry is a multibillion dollar industry built in part on the large-scale production of antibiotics by microorganisms. A number of other major pharmaceutical products are also derived, at least in part, from the activities of microorganisms.

Microorganisms and agriculture

Our whole system of *agriculture* depends in many important ways on microbial activities. A number of major crops are members of a plant group called the **legumes,** which live in close association with special bacteria that form structures called *nodules* on their roots. In these root nodules, atmospheric nitrogen (N_2) is converted to fixed nitrogen compounds that the plants can use for growth. In this way, the activities of the root nodule bacteria reduce the need for costly plant fertilizer. Also of major agricultural importance are the microorganisms that are essential for the digestive process in ruminant animals such as cattle and sheep. These important farm animals have a special digestive organ called the **rumen** in which microorganisms carry out the digestive process. Without these microorganisms, cattle and sheep production would be virtually impossible. Microorganisms also play key roles in the cycling of important nutrients in plant nutrition, particularly carbon, nitrogen, and sulfur. Microbial activities in soil and water convert these elements to forms that are readily accessible to plants. In addition to benefits to agriculture, microorganisms also have harmful effects. Animal and plant diseases due to microorganisms have major economic impact.

Microorganisms and the food industry

Once food crops and animals are produced, they must be delivered in wholesome form to consumers. Microorganisms play important roles in the *food industry.* We note first that because of food spoilage vast amounts of

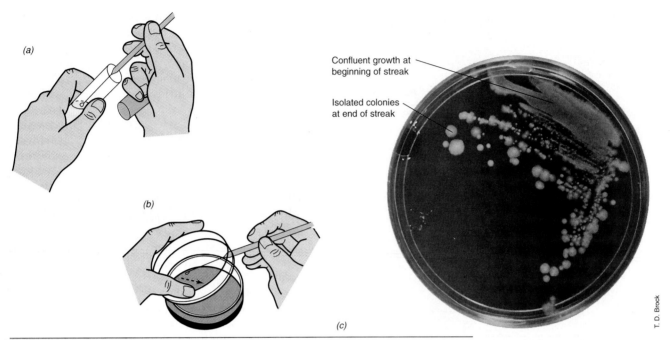

Figure 1.16 Method of making a streak plate to obtain pure cultures. (a) Loop is sterilized, and then a loopful of inoculum is removed from tube. (b) Streak is made over a sterile agar plate, spreading out the organisms. Following the initial streak, subsequent streaks are made at angles to it, the loop being resterilized between streaks. (c) Appearance of the streaked plate after incubation. Note the presence of isolated colonies. It is from such well-isolated colonies that pure cultures can usually be obtained.

money are wasted every year. The canning, frozen-food, and dried-food industries exist to prepare foods in such ways that they will not undergo microbial spoilage.

However, not all microorganisms have harmful effects on foods. Dairy products manufactured, at least in part, via microbial activity include cheese, yogurt, and buttermilk, all products of major economic value. Sauerkraut, pickles, and some sausages also owe their existence to microbial activity. Baked goods are made using yeast. Even more pervasive in our society are alcoholic beverages, also based on the activities of yeast.

All these applications of microorganisms in food and agriculture are of ancient origin, but microbiology has not rested on the past. Consider, for instance, the microorganism's contribution to a carbonated soft drink. The major sugar in many soft drinks is *fructose*, produced from cornstarch via microbial activity. In diet soft drinks, the artificial sweetener *aspartame* is a combination of two amino acids, both produced microbiologically. Finally, the *citric acid* added to many soft drinks to give them tang and bite is produced in a large-scale industrial process using a fungus. The microbial pro-

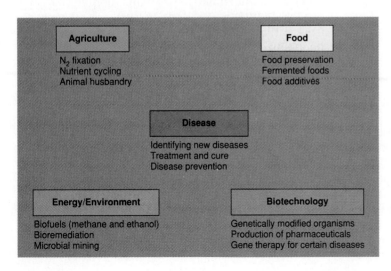

Figure 1.17 The impact of microorganisms on human affairs. Although many people think of microorganisms in the context of infectious diseases, few microorganisms actually cause disease. Microorganisms affect many aspects of our lives in addition to playing a role as disease agents.

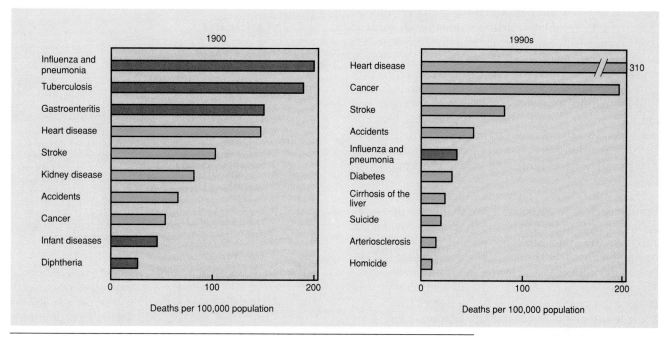

Figure 1.18 Death rates for the 10 leading causes of death in the United States: 1900 and 1990s. Infectious diseases were the leading causes of death in 1900, whereas today they are much less important. Microbial diseases are shown in red, nonmicrobial diseases in green. Data from the United States National Center for Health Statistics.

duction of many of these products will be discussed in Chapter 12.

Microorganisms, energy, and the environment

Our complex industrial society is energy-driven, and here also microorganisms play major roles. Much natural gas (methane) is a product of bacterial action, arising from the activities of methanogenic bacteria. It is harvested worldwide in vast amounts as a primary fuel. A few other mineral and energy products are also the result of microbial activity, but of even greater interest is the relationship of microorganisms to the petroleum industry. Crude oil is subject to vigorous microbial attack, and drilling, recovery, and storage of crude oil all have to be done under conditions that minimize microbial damage.

Because human activity will result in the complete consumption of available fossil fuels during the next century, we must seek new ways to supply the energy needs of society. In the future, microorganisms may provide major alternative energy sources. Phototrophic microorganisms can harvest light energy for the production of **biomass,** energy stored in living organisms. Microbial biomass and existing waste materials such as domestic refuse, surplus grain, and animal wastes, can be converted to "biofuels," such as methane and ethanol, by other microorganisms.

Microorganisms can also be used to help clean up pollution created by human activities, a process called

bioremediation. Various organisms have now been isolated from nature that consume spilled oil, solvents, and other environmentally toxic pollutants, either directly at the site of the spill or later on after the toxic materials have pervaded soils or entered the groundwater. The great diversity of microorganisms noted earlier contains vast genetic resources for solutions to cleaning up the environment, and much research in this area is taking place at present. The field of biotechnology has contributed to this effort by developing methods for genetically modifying natural organisms to make them even better bioremediation agents.

Microorganisms and the future

The preceding recital of the benefits of microorganisms is only the beginning. One of the most exciting new areas of microbiology is **biotechnology.** In the broad sense, biotechnology entails the use of microorganisms in large-scale industrial processes, but by *biotechnology* today we usually mean the application of genetic procedures, generally to create novel microorganisms capable of synthesizing specific products of high commercial value. Biotechnology is highly dependent on **genetic engineering,** the discipline that concerns the artificial manipulation of genes and their products.

Genes from human sources, for instance, can be broken into pieces, modified, and added to or subtracted from, using microorganisms and their enzymes as precise and sophisticated molecular tools. It is even possible to make completely artificial genes using genetic engineering techniques. Once the desired gene has been selected or created, it can be inserted into a microorganism where it can be made to reproduce and make the desired gene product. For instance, human insulin, a hormone found in abnormally low amounts in people with the disease diabetes, has now been produced microbiologically with the human insulin gene engineered into a microorganism. We discuss genetic engineering and biotechnology in detail in Chapter 10.

The overwhelming influence of microorganisms in human society is clear. We have many reasons to be aware of microorganisms and their activities (Figure 1.17). As the eminent French scientist Louis Pasteur, one of the founders of microbiology, expressed it: "The role of the infinitely small in nature is infinitely large." Therefore, before we begin a detailed study of microbiology, let us consider briefly the contributions that Pasteur and others have made to the foundation of the science of microbiology.

CONCEPT CHECK   *1.8*

Microorganisms can be both beneficial and harmful to humans. Although we tend to emphasize the harmful microorganisms (infectious disease agents), many more microorganisms are beneficial than harmful.

✔ *In what ways are microorganisms important in the food and agricultural industries?*

✔ *What is biotechnology and how might it improve the lives of humans?*

1.9

A Brief History of Microbiology

Although the existence of creatures too small to be seen with the eye had long been suspected, their discovery was linked to the invention of the microscope. Robert Hooke described the fruiting structures of molds (eukaryotic cells) in 1664 (Figure 1.19), but the first per-

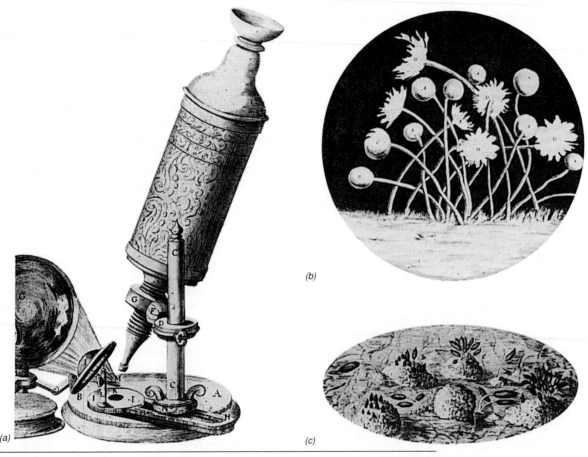

(b)

(a)   *(c)*

Figure 1.19 (a) The microscope used by Robert Hooke. (b,c) Two drawings by Robert Hooke which represent one of the first microscopic descriptions of microorganisms. (b) A blue mold growing on the surface of leather; the round structures contain spores of the mold. (c) A mold growing on the surface of an aging, deteriorating rose leaf.

son to see microorganisms in any detail was the Dutch amateur microscope builder Antoni van Leeuwenhoek, who used simple microscopes of his own construction (Figure 1.20*a*). Leeuwenhoek's microscopes were extremely crude by today's standards, but by careful manipulation and focusing he was able to see organisms as small as prokaryotes. He reported his observations in a series of letters to the Royal Society of London, which published them in English translation. Drawings of some of Leeuwenhoek's "wee animalcules," as he referred to them, are shown in Figure 1.20*b*. His observations were confirmed by other workers, but progress in understanding the nature and importance of these tiny organisms came slowly. Only in the nineteenth century did improved microscopes become available and widely distributed. During its history, the science of microbiology has traditionally taken the greatest steps forward when better microscopes have been developed, for these have enabled scientists to penetrate ever deeper into the mysteries of the cell.

Microbiology as a science did not develop until the latter part of the nineteenth century. This long delay occurred because, in addition to microscopy, certain basic techniques for the study of microorganisms needed to be devised. In the nineteenth century, investigation of two perplexing questions led to the development of these techniques and laid the foundation of microbiological science: (1) Does spontaneous generation occur? (2) What is the nature of contagious disease? By the end of the nineteenth century both questions had been answered, and the science of microbiology was firmly established as a distinct and growing field.

Pasteur and the downfall of spontaneous generation

The basic idea of spontaneous generation can easily be understood. If food is allowed to stand for some time, it putrefies. When the putrefied material is examined microscopically, it is found to be teeming with bacteria. Where do these bacteria come from, since they are not seen in fresh food? Some people said they developed from seeds or germs that had entered the food from the air, whereas others said that they arose spontaneously from nonliving materials.

Spontaneous generation would mean that life could arise from something nonliving, and many people could not imagine something so complex as a living cell arising spontaneously from nonliving materials. The most powerful opponent of spontaneous generation was the French chemist Louis Pasteur, whose work on this problem was the most exacting and convincing. Pasteur first showed that structures were present in air that resembled closely the microorganisms seen in putrefying materials. He did this by passing air through guncotton filters, the fibers of which stopped solid particles. After the guncotton was dissolved in a mixture of alcohol and ether, the particles that it had trapped fell to the bottom of the liquid and were examined on a microscope slide. Pasteur found that in ordinary air there constantly exists a variety of microbial cells and that they could not be distinguished from the organisms found in much larger numbers in putrefying materials. Pasteur concluded that the organisms found in putrefying materials originated from microorganisms present in the air. He postulated that these cells are constantly being deposited on all objects. If this conclusion was correct, it would mean that if food were treated to destroy all the living organisms contaminating it, then it should not putrefy.

Pasteur used heat to eliminate contaminants since it had already been established that heat effectively kills living organisms. In fact, other workers had shown that

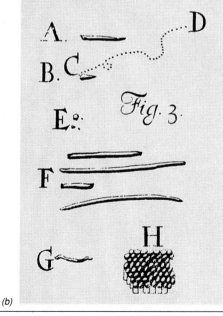

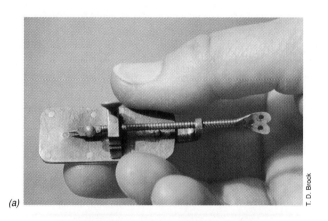

Figure 1.20 (a) Photograph of a replica of Leeuwenhoek's microscope. (b) Leeuwenhoek's drawings of bacteria, published in 1684. Even from these crude drawings we can recognize several morphological types of common bacteria. A, C, F, and G, rod-shaped; E, spherical or coccus-shaped; H, cocci packets.

when a nutrient solution was sealed in a glass flask and heated to boiling, it never putrefied. The proponents of spontaneous generation criticized such experiments by declaring that fresh air was necessary for spontaneous generation and that the air itself inside the sealed flask was affected in some way by heating so that it could no longer support spontaneous generation. Pasteur skirted this objection simply and brilliantly by constructing a swan-necked flask, now called a *Pasteur flask* (Figure 1.21). In such a flask putrefying materials could be heated to boiling; after the flask was cooled, air could reenter but the bends in the neck prevented particulate matter, bacteria, or other microorganisms from getting into the main body of the flask. Material sterilized in such a flask did not putrefy, and no microorganisms ever appeared as long as the neck of the flask did not

contact the sterile liquid. However, if the flask was tipped to allow the sterile liquid to contact the neck of the flask, putrefaction occurred and the liquid soon teemed with microorganisms. This simple experiment effectively settled the controversy surrounding the theory of spontaneous generation.

Killing all the bacteria or other microorganisms in or on objects is a process we now call **sterilization,** and the procedures that Pasteur and others used were eventually refined and carried over into microbiological research (see Section 1.7). Disproving the theory of spontaneous generation thus led to the development of effective sterilization procedures, without which microbiology as a science could not have developed. Food science also owes a debt to Pasteur, as his principles are applied in the canning and preservation of many foods.

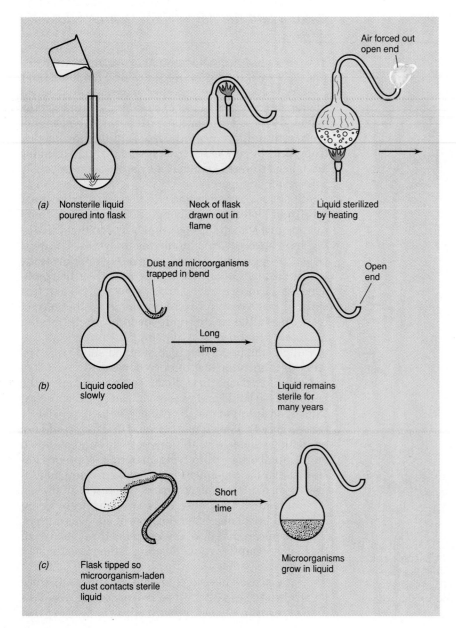

(a) Nonsterile liquid poured into flask

Neck of flask drawn out in flame

Liquid sterilized by heating

Air forced out open end

Dust and microorganisms trapped in bend

(b) Liquid cooled slowly

Long time

Liquid remains sterile for many years

Open end

(c) Flask tipped so microorganism-laden dust contacts sterile liquid

Short time

Microorganisms grow in liquid

Figure 1.21 Pasteur's experiment with the swan-necked flask. (a) Sterilizing the contents of the flask. (b) If the flask remains upright, no microbial growth occurs. (c) If microorganisms trapped in the neck reach the sterile liquid, they grow rapidly.

Endospores

Although Pasteur was successful in sterilizing materials with simple boiling, some workers found that boiling was insufficient. We now know that this failure resulted from the presence in these materials of bacteria that formed unusually heat-resistant structures called *endospores.* Initial work on the endospore was carried out by two men: John Tyndall in England, and Ferdinand Cohn in Germany. Both scientists observed that some preparations, such as the fruit juice solutions used by Pasteur, were relatively easy to sterilize, requiring only 5 minutes (min) of boiling, whereas others were not sterilized by much longer periods of boiling, sometimes even hours. Notably difficult to sterilize were hay infusions. In addition, once hay had been brought into the laboratory, even the sugar solutions could no longer be reliably sterilized by hours of boiling. Cohn performed detailed microscopic observations and discovered endospores inside cells of old cultures of species of *Bacillus.* Cohn and another German scientist, Robert Koch, applied this observation to the study of disease, as discussed later. Bacterial endospores are the most heat-resistant living structures known, and most sterilization methods are designed to kill these spores.

Pasteur went on to many other triumphs in microbiology and medicine. Chief among these was his development of vaccines for the diseases anthrax, fowl cholera, and rabies during a very productive period from 1880 to 1890. These medical and veterinary breakthroughs were not only highly significant in their own right, but helped solidify the concept of the germ theory of disease whose principles were being developed at this time by a contemporary of Pasteur, Robert Koch. We review this important period in the history of microbiology now.

Koch and the germ theory of disease

Proof that microorganisms could cause disease provided the greatest impetus for the development of the science of microbiology. Indeed, even in the sixteenth century it was thought that something could be transmitted from a diseased person to a well person to induce in the latter the disease of the former. Many diseases seemed to spread through populations and were called *contagious;* the unknown agent that did the spreading was called the *contagion.* After the discovery of microorganisms, it was more or less widely held that these organisms were responsible for contagious diseases, but proof was lacking. Discoveries by Ignaz Semmelweis and Joseph Lister provided some evidence for the importance of microorganisms in causing human diseases, but it was not until the work of Robert Koch, a physician, that the *germ theory of disease* was clearly conceptualized and given experimental support.

In his early work, published in 1876, Koch studied *anthrax,* a disease of cattle which sometimes also

occurs in humans. Anthrax is caused by a spore-forming bacterium now called *Bacillus anthracis,* and the blood of an animal infected with anthrax teems with cells of this large bacterium. Koch used the mouse as an experimental animal in his anthrax studies, but because there were no commercially available sources of white laboratory mice at that time, he used the common gray house mouse collected from a nearby horse barn.

Koch established by careful microscopy that the bacteria were always present in the blood of an animal that was succumbing to the disease. However, mere association of the bacterium with the disease did not prove that it actually *caused* the disease; it might instead be a *result* of the disease. Therefore, Koch demonstrated that it was possible to take a small amount of blood from a diseased mouse and inject it into a second mouse, which subsequently became diseased and died. He could then take blood from this second animal, inject it into another, and again obtain the characteristic disease symptoms. By repeating this process as often as 20 times, successively transferring small amounts of blood containing bacteria from one animal to another, he proved that the bacteria did indeed cause anthrax: the twentieth animal died just as rapidly as the first, and in each case Koch could demonstrate by microscopy that the blood of the dying animal contained large numbers of the spore-forming bacterium.

However, Koch carried this experiment further. He found that the bacteria could also be cultivated in nutrient fluids outside the animal body and that even after many transfers in culture the bacteria could still cause the disease when reinoculated into an animal. Bacteria from a diseased animal and bacteria in culture both induced the same disease symptoms upon injection. On the basis of these and other experiments Koch formulated the following criteria, now called **Koch's postulates,** for proving that a specific type of microorganism causes a specific disease:

1. The organism should be constantly present in animals suffering from the disease and should not be present in healthy individuals.

2. The organism must be cultivated in a pure culture away from the animal body.

3. Such a culture, when inoculated into susceptible animals, should initiate the characteristic disease symptoms.

4. The organism should be reisolated from these experimental animals and cultured again in the laboratory, after which it should still be the same as the original organism.

Koch's postulates are summarized in Figure 1.22. Koch's postulates not only supplied a means of demonstrating that specific organisms cause specific diseases but also provided a tremendous spur for the develop-

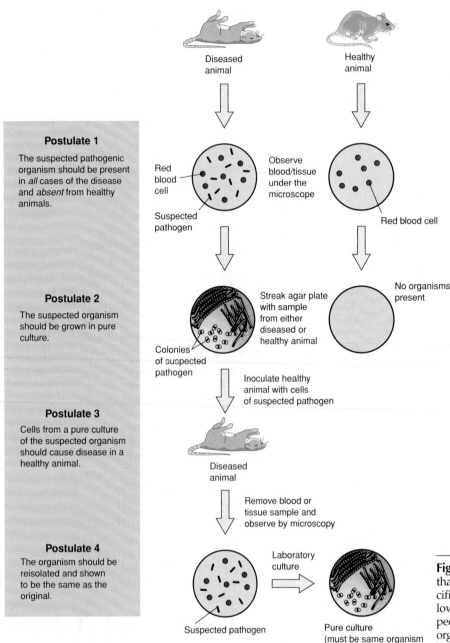

Postulate 1

The suspected pathogenic organism should be present in *all* cases of the disease and *absent* from healthy animals.

Postulate 2

The suspected organism should be grown in pure culture.

Postulate 3

Cells from a pure culture of the suspected organism should cause disease in a healthy animal.

Postulate 4

The organism should be reisolated and shown to be the same as the original.

Diseased animal

Healthy animal

Red blood cell

Suspected pathogen

Observe blood/tissue under the microscope

Red blood cell

No organisms present

Colonies of suspected pathogen

Streak agar plate with sample from either diseased or healthy animal

Inoculate healthy animal with cells of suspected pathogen

Diseased animal

Remove blood or tissue sample and observe by microscopy

Suspected pathogen

Laboratory culture

Pure culture (must be same organism as before)

Figure 1.22 Koch's postulates for proving that a specific microorganism causes a specific disease. Note how it is essential that following isolation of a pure culture of the suspected pathogen, a laboratory culture of the organism should both initiate the disease and be recovered from the diseased animal.

ment of the science of microbiology by stressing the importance of laboratory culture. Using Koch's postulates as a guide, investigators following Koch revealed the causes of many important diseases of humans and other animals. These discoveries led to the development of successful treatments for the prevention and cure of many infectious diseases, thereby greatly improving the scientific basis of clinical medicine.

Koch and pure cultures

As we have noted, in order to successfully study the activities of a microorganism, such as a microorganism that causes a disease, one must be sure that it alone is present in a culture. That is, the culture must be *pure* (see Section 1.7). With objects as small as microorganisms, ascertaining purity is not easy, for even a very tiny sample of blood or animal fluid may contain several kinds of organisms that may all grow together in culture. Koch realized the importance of pure cultures. He developed several ingenious methods of obtaining them, of which the most useful is that involving the isolation of single *colonies* on solid media (see the box). Koch observed that when a solid nutrient surface, such as a potato slice, was exposed to air and then incubated, bacterial colonies developed, each having a characteristic shape and color. He inferred that each colony had arisen from a single bacterial cell that had fallen on the

SOLID MEDIA AND DEVELOPMENT OF THE PETRI PLATE

Robert Koch was the first to grow bacteria on solid culture media. Koch initially employed gelatin as a solidifying agent for the various nutrient fluids he used to culture pathogenic bacteria and developed a method for preparing horizontal slabs of solid media which were kept free of contamination by covering them with a bell jar.

Nutrient gelatin was a marvelous culture medium for the isolation and study of various bacteria, but it had several drawbacks, the most important being that gelatin does not remain solid at body temperature (37°C), the optimum temperature for growth of most human pathogens. Thus, a more versatile solidifying agent was needed, and this turned out to be agar.

Agar is a polysaccharide derived from red algae. It was used widely in the nineteenth century, especially in tropical countries, as a gelling agent. The first use of agar as a solidifying agent for bacteriological culture media was made by Walter Hesse, an associate of Koch. The actual suggestion that agar be used instead of gelatin was made by Hesse's wife, Fannie. Fannie Hesse had used agar in the preparation of fruit jellies, and when it was tried as a solidifying agent in nutrient media, it was immediately found to be superior to gelatin. Hesse wrote to Koch about this discovery, and Koch quickly adapted agar to his own studies, including his classic studies on the isolation of the bacterium *Mycobacterium tuberculosis*, the cause of the disease tuberculosis (see text).

In 1887 Richard Petri published a brief paper describing a modification of Koch's flat plate technique. Petri's enhancement, which turned out to be amazingly useful, was the development of the special double-sided dishes that bear his name. The advantage of Petri dishes was apparent; they could be easily stacked and sterilized separately from the medium, and, following the addition of molten medium to the smaller of the two double dishes, the larger dish could be used as a cover to prevent contamination. Colonies that formed on the surface of the agar in the Petri dish remained fully exposed to air and could easily be manipulated for further study. The original idea of Petri has not been improved on to this day, as the Petri dish, made either of reusable glass and sterilized by dry heat or of disposable plastic and sterilized by ethylene oxide (a gaseous sterilant), is the mainstay of the microbiology laboratory.

Finally, it should also be noted that Koch was keenly aware of the implications his pure culture methods had for the study of microbial systematics. Koch observed that different colonial forms (differing in color, colony morphology, size, and the like) developed on solid media exposed to a contaminated object, and that these colonial forms bred true and could be distinguished from one another by their colony characteristics. Cells from different colonies also differed microscopically and often in their temperature or nutrient requirements as well. Koch realized that these differences among microorganisms met all the requirements that taxonomists had established for the classification of larger organisms, such as plant and animal species. In Koch's own words: *"All bacteria which maintain the characteristics which differentiate one from another when they are cultured on the same medium and under the same conditions, should be designated as species, varieties, forms, or other suitable designation."* Thus, Koch's discovery of solid culture media and his emphasis on pure culture microbiology reached far beyond the realm of medical bacteriology; his discoveries supplied critically needed tools for development of the field of bacterial taxonomy, genetics, and several related disciplines. ∎

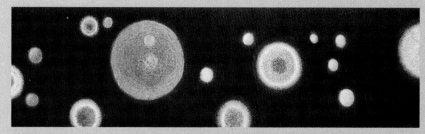

A photograph of colonies formed on agar taken by Walter Hesse, an associate of Robert Koch. The colonies include those of fungi (molds) and bacteria and were obtained during studies Hesse initiated on the microbiological content of air in Berlin, Germany, in 1882. From Hesse, W. 1884. "Ueber quantitative Bestimmung der in der luft enthaltenen Mikroorganismen," in Struck (ed.), *Mittheilungen aus dem Kaiserlichen Gesundheitsamte.* August Hirschwald.

surface, found suitable nutrients, and begun to multiply. Because the solid surface prevented the bacteria from moving around, all the offspring of the initial cell had remained together, and when a large enough number of organisms was present, the mass of cells became visible to the naked eye. He assumed that colonies with different shapes and colors were derived from different kinds of microorganisms. When the cells of a single colony were spread out on a fresh surface, many colonies developed, each with the same shape and color as the original.

Koch realized that this discovery provided a simple way of obtaining pure cultures: he found that if mixed cultures were spread on solid nutrient surfaces, the individual cells were so far apart that the colonies they produced did not mingle. Many organisms could not grow on potato slices, and so he devised semisolid media in which gelatin and later *agar* (see the box) was used as a solidifying agent. Agar is the major agent used for solidifying culture media today.

It is important to realize that Koch's postulates have relevance beyond simply identifying organisms that cause specific diseases. The essential point here is that from the study of pure cultures, one can show that *specific organisms have specific effects*. This principle that different organisms have unique biological activities was important in establishing microbiology as an independent biological science (see the box). Because of these major contributions of Koch, by the beginning of the twentieth century the disciplines of bacteriology and microbiology were on a firm footing.

Koch and tuberculosis

The greatest accomplishment of Robert Koch's work in medical bacteriology was with tuberculosis. At the time Koch began this work (1881), one-seventh of all reported deaths of human beings were caused by tuberculosis. Even at this time there was strong evidence that tuberculosis was a contagious disease, but the suspected causal organism had never been seen, either in diseased tissues or in culture. Koch's aim from the beginning of his work on tuberculosis was to demonstrate the causal agent of tuberculosis, and to this end he employed all the methods he had so carefully developed previously: microscopy, staining of tissues, pure culture isolation, and animal inoculation.

As is now well-known, *Mycobacterium tuberculosis,* the "tubercle bacillus," is very difficult to stain because of large amounts of lipid present on its outer surface. But Koch devised a staining procedure for *M. tuberculosis* in tissue samples using alkaline methylene blue in conjunction with a second stain (bismarck brown) that stained only the tissue (Koch's method was the forerunner of the Ziehl–Nielsen stain used today for staining acid-fast bacteria like *M. tuberculosis,* see Section 16.31). Using his newly developed staining method, Koch observed bright-blue, rod-shaped cells of *M. tuberculosis* in tuberculous tissues, the latter staining a light brown (Figure 1.23). However, from his previous work on anthrax, Koch realized that simply identifying an organism associated with tuberculosis was not enough; he must *culture* the organism in order to prove that it was the specific cause of tuberculosis.

Producing cultures of *Mycobacterium tuberculosis* was not easy, but eventually Koch was successful in obtaining colonies of this organism on coagulated blood serum. Later he used agar, which had just been introduced as a solidifying agent (see the box). Under the best conditions *M. tuberculosis* grows slowly in culture, but Koch's persistence and patience eventually led to pure

cultures of this organism from a variety of human and animal sources. From here it was relatively easy to obtain definitive proof that the organism he had isolated was the true cause of the disease tuberculosis. Guinea pigs can be readily infected with *M. tuberculosis* and eventually succumb to systemic tuberculosis. Koch showed that diseased guinea pigs contained masses of *M. tuberculosis* cells in their tissues and that pure cultures obtained from such animals transmitted the disease to uninfected animals. Thus, Koch successfully satisfied all four criteria of his famous postulates (Figure 1.22), and the cause of tuberculosis was understood.

Koch announced to the world his discovery of the tubercle bacillus in a famous lecture given in Berlin in 1882. The news of Koch's discovery spread quickly to England and the United States (the *New York Times* called it "one of the great scientific discoveries of the age"), and his conclusions as to the etiology of tuberculosis were quickly accepted. For his work on tuberculosis, including not only discovery of the causative agent but also development of methods for its specific staining (Figure 1.23) and preparation of a substance called *tuberculin,* useful in the diagnosis of tuberculosis, Koch received the 1905 Nobel Prize for Physiology or Medicine just 5 years before his death.

Koch had many other triumphs in microbiology besides discovering the causes of anthrax and tuberculosis. These included the discovery and isolation of the causative agent of cholera, *Vibrio cholerae,* the discovery of the importance of water filtration in the control of cholera, the development of the concept of disease "carriers," and the publication of the first photomicrographs of bacteria. Clearly, Robert Koch's contributions to the development of modern microbiology were monumental.

A summary of some of the most important discoveries in the field of microbiology, from the time of van Leeuwenhoek up into the twentieth century, are summarized in Table 1.1.

Developments of microbiology in the twentieth century

In the twentieth century, the field of microbiology has developed rapidly in two separate directions—applied and basic. On the applied side, the practical advances made by Koch led to extensive developments in *medical microbiology* and *immunology* in the early part of the century, with the discovery of many new bacterial pathogens (∞Discoverers of the Main Bacterial Pathogens, Chapter 23) and the working out of the principles by which these pathogens infect the body and are in turn resisted by the body's defenses. Other early practical advances were in the field of *agricultural microbiology,* which led to an understanding of microbial processes in the soil that are beneficial or harm-

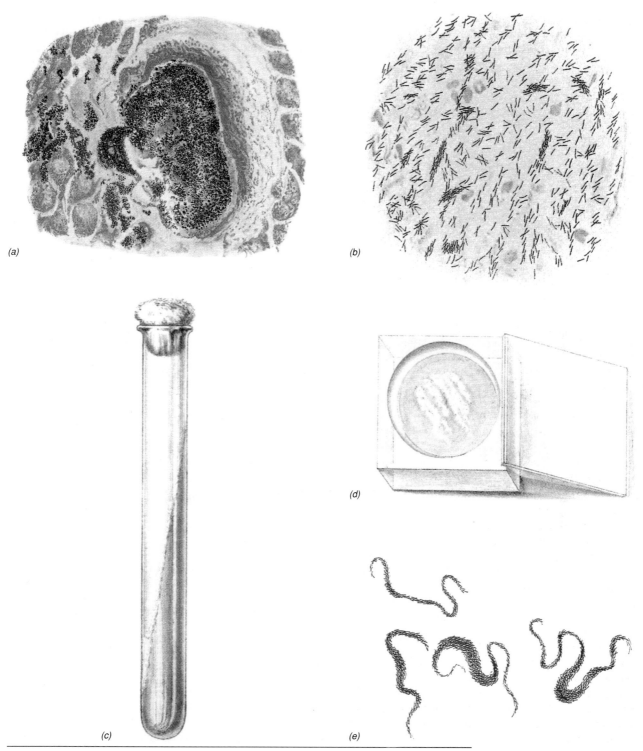

Figure 1.23 Robert Koch's drawings of cells of *Mycobacterium tuberculosis* in tissues and in laboratory culture. (a) Section through a tubercle from lung tissue. Cells of *M. tuberculosis* stain blue, whereas the lung tissue stains brown. (b) Cells of *M. tuberculosis* in a sputum sample of a tuberculous patient. (c,d,e) Growth of *M. tuberculosis* in pure culture: (c) Growth on a slant culture of coagulated blood serum. (d) Growth on a glass plate of a similar medium inside a glass box (with lid open). (e) A colony of *M. tuberculosis* cells taken from the plate in (d) and observed microscopically at 700×; cells appear as long "cordlike" forms (compare with Figure 16.101*b*). Original drawings appeared in Koch, R. 1884. "Die Aetiologie der Tuberkulose." *Mittheilungen aus dem Kaiserlichen Gesundheitsamte* 2:1–88.

TABLE 1.1	Three Hundred Years of Microbiology: Some Key Papers in Microbiology, 1684–1983 [a]	

Year	Investigator(s)	Discovery
1684	Antoni van Leeuwenhoek	Discovery of bacteria
1798	Edward Jenner	Smallpox vaccination
1857	Louis Pasteur	Microbiology of the lactic acid fermentation
1860	Louis Pasteur	Role of yeast in alcoholic fermentation
1864	Louis Pasteur	Settled spontaneous generation controversy
1867	Robert Lister	Antiseptic principles in surgery
1881	Robert Koch	Methods for study of bacteria in pure culture
1882	Robert Koch	Discovery of cause of tuberculosis
1884	Robert Koch	Koch's postulates
1884	Robert Koch	Discovery of cause of cholera
1884	Christian Gram	Gram-staining method
1889	Sergei Winogradsky	Concept of chemolithotrophy
1889	Martinus Beijerinck	Concept of a virus
1890	Sergei Winogradsky	Autotrophic growth of chemolithotrophs
1901	Martinus Beijerinck	Enrichment culture method
1901	Karl Landsteiner	Human blood groups
1908	Paul Ehrlich	Chemotherapeutic agents
1928	Frederick Griffith	Discovery of pneumococcus transformation
1929	Alexander Fleming	Discovery of penicillin
1944	Oswald Avery, Colin Macleod, Maclyn McCarty	Explanation of Griffith's work—DNA is genetic material
1944	Selman Waksman	Discovery of streptomycin
1953	James Watson, Francis Crick	Structure of DNA
1959	Arthur Pardee, Francois Jacob, Jacques Monod	Gene regulation by a repressor protein
1959	Rodney Porter	Immunoglobulin structure
1959	F. Macfarlane Burnet	Clonal selection theory
1960	Francois Jacob, David Perrin, Carmon Sanchez, Jacques Monod	Concept of an operon
1975	Georges Kohler, Cesar Milstein	Monoclonal antibodies
1976	Susumu Tonegawa	Rearrangement of immunoglobulin genes
1977	Fred Sanger, Steven Niklen, Alan Coulson	Methods for sequencing DNA
1983	Luc Montagnier	Discovery of HIV, the cause of AIDS

[a] Major reference sources here include Brock, T. D. (1961), *Milestones in Microbiology*, Prentice Hall, Englewood Cliffs, NJ; Brock, T. D. (1990), *The Emergence of Bacterial Genetics*, Cold Spring Harbor Press, Cold Spring Harbor, NY. *Year* refers to the year in which the discovery was published.

ful to plant growth. Later in the twentieth century, such studies on soil microbiology led to the discovery of important uses of microorganisms, such as in the formation of *antibiotics* and *industrial chemicals*. This led, especially after World War II, to the field of *industrial microbiology*.

Finally, soil microbiology has also provided an important foundation for studies on microbial processes in water bodies such as lakes, rivers, and oceans, studies classified under the field of *aquatic microbiology*. One branch of aquatic microbiology deals with the development of processes for providing safe water for human society. The handling of human wastes, espe-

cially domestic sewage, has required the development of large-scale engineering processes for sewage treatment, most of which are microbial. Thus, the field of *sanitary microbiology* has developed, which is of importance not only to biologists but also to engineers whose responsibility is the design of these large-scale processes. To provide safe drinking water, procedures for eliminating harmful bacteria from water supplies have been developed, which are classified under the field of *drinking water microbiology*. By the late twentieth century, all the subdisciplines of applied microbiology mentioned here have coalesced into a field called *microbial ecology*, which will be discussed in Chapter 14.

In addition to the *applied* aspects of microbiology that have provided such important advances for human society, extensive developments have occurred in our understanding of the *basic* principles of microbial function. In the early part of the twentieth century, the most important developments in basic microbiology involved the discovery of new kinds of bacteria and their proper classification (*bacterial taxonomy*). Bacterial classification required a study of the nutrients that bacteria consume and the products that they make, studies comprising part of the field of *bacterial physiology*. One part of physiology that became of major importance as the twentieth century progressed involved the study of the physical and chemical structure of bacteria, studies included in the field of *bacterial cytology* (the word *cytology* refers to the study of the cell). Another major development from physiology was the study of bacterial enzymes and the chemical reactions that they carry out, a field called *bacterial biochemistry.*

Another very important area of basic research has involved the study of heredity and the variation that bacteria undergo during their growth and development, studies that fall under the discipline of *bacterial genetics.* Although some ideas of bacterial variation were known early in the twentieth century, it was not until the discovery of genetic exchange in bacteria in about 1950 that bacterial genetics really became a major field of study. Bacterial genetics, biochemistry, and physiology developed mainly during the 1950s, leading by the early 1960s to an advanced understanding of DNA, RNA, and protein synthesis. The field of *molecular biology* arose to a great extent from these bacterial studies.

Another important development in the twentieth century involved the study of viruses. Although disease-causing viruses were first discovered at the end of the nineteenth century, it was not until the middle of the twentieth century that the true nature of viruses was determined. Much of this work involved the study of viruses that infect bacteria, called *bacteriophages.* An important development was the realization that virus infection was analogous to genetic transfer, and the relationships between viruses and other genetic elements was worked out primarily from research on bacteriophages.

By the 1970s, our knowledge of the basic processes of bacterial physiology, biochemistry, and genetics had advanced to such a great extent that it was possible to manipulate the genetic material of cells experimentally using bacteria as tools. It also became possible to introduce genetic material (DNA) from foreign sources into bacteria and control its replication and characteristics. This led to development of the field of *biotechnology.* Although biotechnology originally arose from basic studies, its use in promoting human welfare required application of the principles of physiology and industrial microbiology, a good example of how basic and applied research advance together. Also at about this same time, nucleic acid sequencing was worked out and used as a tool to discern phylogenetic relationships among prokaryotes, which led to revolutionary new concepts in the field of biological classification and to the first true understanding of the evolutionary history of microorganisms.

We have briefly outlined the practical significance of microbiology and examined the historical development of the science. We now proceed to a careful discussion of the basic properties and activities of microorganisms. Only after a solid understanding of basic microbiology has been established is it possible to discuss the applications intelligently. Thus, the material in this book moves from the basic to the applied, and from the simple to the more complex. We focus now on the essentials of cell chemistry necessary for a modern understanding of microorganisms.

CONCEPT CHECK   *1.9*

Louis Pasteur's work on spontaneous generation led to the development of methods for control of the growth of microorganisms. Robert Koch developed a set of criteria that provided an experimental framework for the study of infectious microorganisms, and developed the first methods for the growth of pure cultures of microorganisms. In the twentieth century, basic and applied aspects of microbiology have worked hand in hand to yield a number of important practical advances and a revolution in molecular biology.

✔ *How did Pasteur's famous experiment defeat the theory of spontaneous generation?*

✔ *How can Koch's postulates prove cause and effect in a disease?*

✔ *Who was the first person to use solid culture media in microbiology? What advantages do solid media offer for the culture of microorganisms?*

Material for Review

REVIEW QUESTIONS

1. List five key properties associated with the living state. Which of these are properties of *all* cells? Which are properties of only *some types* of cells?

2. Cells can be thought of as both machines and coding devices. Explain how these two attributes of a cell differ.

3. How many genes does a typical prokaryotic cell have? Do eukaryotes have a smaller or a larger number of genes than prokaryotes?

4. Compare and contrast the prokaryotic and the eukaryotic cell. List the properties that are common to both types. List the properties that are different in these two cell types.

5. Are viruses cells? List three ways in which viruses differ from cells.

6. What is needed for translation to occur in a cell? What is the product of the translational process?

7. In what ways are Bacteria and Archaea similar? In what ways are they different? How can one distinguish a cell of the Bacteria from a cell of the Archaea?

8. What is an ecosystem? Do microorganisms live in pure cultures in an ecosystem? What effects can microorganisms have on their ecosystem?

9. What is a pure culture and how can one be obtained? Why was knowledge of how to obtain a pure culture important for development of the science of microbiology?

10. How would you convince a friend that microorganisms are much more than just agents of disease?

11. Explain the principle behind the use of the Pasteur flask in studies on spontaneous generation.

12. Explain why the invention of solid culture media was of great importance to the development of microbiology as a science.

APPLICATION QUESTIONS

1. A "yeast cake" consists of billions of cells, and a human consists of billions of cells. However, it is obvious that a yeast cake differs from a human. Briefly discuss this fundamental difference and explain its significance.

2. Discuss why the words *transcription* and *translation* are appropriate ones to describe the production by cells of RNA and protein, respectively.

3. Pasteur's experiments on spontaneous generation were of enormous importance for the advance of microbiology, having an impact on the methodology of microbiology, ideas on the origin of life, and the preservation of food, to name just a few. Explain briefly how the impact of his experiments was felt on each of the topics listed.

4. Describe the various lines of proof Robert Koch used to definitively associate the bacterium *Mycobacterium tuberculosis* with the disease tuberculosis. How would his proof have been flawed if any of the tools he developed for studying bacterial diseases had not been available for his study of tuberculosis?

SUPPLEMENTARY READINGS

Brock, T. D. 1975. *Milestones in Microbiology.* American Society for Microbiology, Washington, DC. (reprint of 1961 Prentice Hall edition). The key papers of Pasteur, Koch, and others are translated, edited, and annotated for the beginning student.

Brock, T. D. 1988. *Robert Koch: A Life in Medicine and Bacteriology.* Science Tech, Madison, WI. The standard biography of the discoverer of the germ theory of disease.

Brock, T. D. (ed.). 1989. *Microbes and Infectious Diseases.* Scientific American Books, New York. Historical and modern ideas about the role of microorganisms as disease-causing agents.

Brock, T. D. 1990. *The Emergence of Bacterial Genetics.* Cold Spring Harbor Press, Cold Spring Harbor, NY. Traces the path of development of bacterial genetics from the late nineteenth century through the era of gene splicing.

Bulloch, W. 1935. *The History of Bacteriology.* Oxford University Press, London. The standard history of bacteriology; emphasis on medical aspects.

Campbell, N. A. 1995. *Biology,* 4th edition. Benjamin-Cummings, Redwood City, CA. The best introductory biology textbook for students of biology.

Dobell, C. (ed. and trans.). *Antoni van Leeuwenhoek and His "Little Animals."* Constable, London (1960, Dover Publications, New York). An introduction to Leeuwenhoek's life, his work, and his times.

Dubos, R. 1988. *Pasteur and Modern Science.* Science Tech, Madison, WI. A brief, well-illustrated biography of Louis Pasteur.

Lederberg, J. (ed.). 1992. *Encyclopedia of Microbiology.* Academic Press, San Diego, CA. A four-volume, 2100-page treatise on all aspects of microbiology.

Postgate, J. 1992. *Microbes and Man,* 3rd edition. Cambridge University Press, Cambridge, England. A short treatment of microorganisms and how they affect the lives of humans.

Raven, P. H., and **G. B. Johnson.** 1996. *Biology,* 6th edition. Times Mirror/Mosby College Publishing, St. Louis, MO. An excellent modern treatment of biological principles.

 On~line resources for this chapter are on the World Wide Web at:
http://www.prenhall.com/~brock (click on the table of contents link and then select Chapter 1).

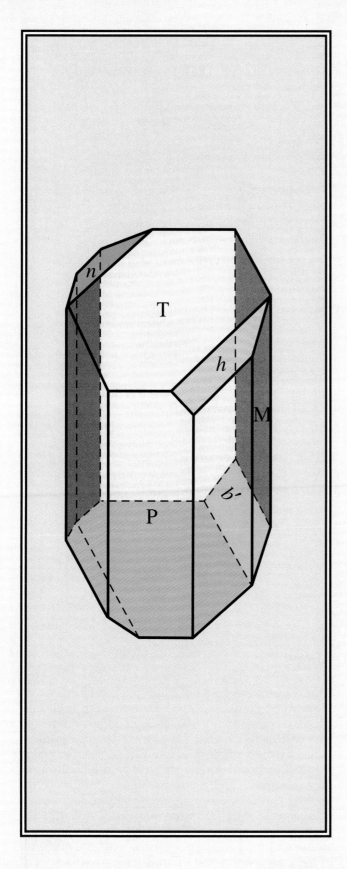

Cell Chemistry

Louis Pasteur, one of the fathers of the science of microbiology, began his career as a chemist and studied, among other things, the chiral properties of chemical crystals such as those of tartaric acid shown here. From this work Pasteur developed a deep-seated belief that the chemical reactions carried out by living organisms were inherently asymmetric and this led to some of his greatest discoveries in microbiology. See the box, Chemical Isomers and Living Organisms.

MINIGLOSSARY for Chapter 2

Atomic Weight the total mass of protons and neutrons in an atom

Covalent Bond a chemical bond in which electrons are shared between two atoms

Denaturation destruction of the folding properties of a protein leading (usually) to loss of biological activity

Glycosidic Bond a type of covalent bond that links sugar units together in a polysaccharide

Hydrogen Bond a weak chemical bond between a hydrogen atom and a second, more electronegative element, usually an oxygen or nitrogen atom

Isotope one form of an element in which the number of neutrons varies from that in other forms of the same element

Lipid glycerol bonded to fatty acids and other groups such as phosphate by an ester or ether linkage

Macromolecule polymer of covalently linked monomeric units

Molecule two or more atoms chemically bonded to one another

Nonpolar possessing hydrophobic (water-repelling) characteristics and not easily dissolved in water

Nucleotide a monomer of a nucleic acid containing a nitrogen base (adenine, guanine, cytosine, thymine, or uracil), a molecule of phosphate, and a sugar, either ribose (in RNA) or deoxyribose (in DNA)

Peptide Bond a type of covalent bond joining amino acids in a polypeptide

Phosphodiester Bond a type of covalent bond linking nucleotides together in a polynucleotide

Polar possessing hydrophilic characteristics and generally water-soluble

Polymer a chemical compound formed by polymerization and consisting of repeating units called monomers

Polynucleotide a polymer of nucleotides bonded to one another by phosphodiester bonds

Polypeptide a polymer of amino acids bonded to one another by peptide bonds

Polysaccharide a polymer of sugar units bonded to one another by glycosidic bonds

Primary Structure in an informational macromolecule, such as a polypeptide, the precise sequence of monomeric units

Protein a polypeptide or group of polypeptides that form a molecule of specific biological function

Quaternary Structure in proteins, the number and arrangement of individual polypeptides in the final protein molecule

Radioisotope an isotope of an element that undergoes spontaneous decay with the release of particles

Secondary Structure the initial pattern of folding of a polypeptide or a polynucleotide, usually dictated by opportunities for hydrogen bonding

Stereoisomer one form of a molecule that is the mirror image of another form of the same molecule

Tertiary Structure the final folded structure of a polypeptide that has previously attained secondary structure

---

o understand microbiology, one must have some understanding of the chemical processes taking place within cells. As pointed out in Chapter 1, living cells are governed by the same chemical and physical principles that dictate the properties of nonliving matter. However, as self-replicating entities, cells contain a variety of molecules not usually found in inanimate structures. The chemical nature of these molecules is the subject of this chapter. We begin by introducing chemical principles of atoms, molecules, and atomic bonding and proceed through a discussion of the biochemistry of macromolecules, the building blocks of the cell. The student may find it useful to refer to material in this chapter from time to time while mastering the principles of cellular metabolism (Chapter 4), growth (Chapter 5), and molecular biology, virology, and genetics (Chapters 6 through 9).

2.1

Atoms

Atoms are the basic units of matter and can be defined as the smallest particles of an element that can enter into chemical combinations. The atoms of different elements have characteristic chemical and physical properties determined by the number of constituent particles of which they are composed. Atoms consist of particles called **protons, electrons,** and **neutrons.** Protons are *positively* charged, and electrons are *negatively* charged; neutrons, as their name implies, are *uncharged* particles. About 20 different elements are found in cells. However only 6 of these are present in abundance: hydrogen, carbon, nitrogen, oxygen, phosphorus, and sulfur. The atomic structures of these 6 important elements are shown in Table 2.1.

| | | | | | | Electrons in | Simplified |
| TABLE 2.1 | Atomic structure of the major elements of life | | | | | | |

Element	Atomic number	Atomic weight	Protons	Neutrons	Electrons	Electrons in outer shell	Simplified atomic structure[a]
Hydrogen	1	1	1	0	1	1	H•
Carbon	6	12	6	6	6	4	•C̈•
Nitrogen	7	14	7	7	7	5	•N̈•
Oxygen	8	16	8	8	8	6	:Ö:
Phosphorus	15	31	15	16	15	5	•P̈•
Sulfur	16	32	16	16	16	6	:S̈:

[a]Only the outer shell electrons are depicted in the simplified structure.

The **nucleus** of an atom contains protons and neutrons and constitutes the majority of the atomic mass. Each proton carries an electric charge of +1, and the number of protons in the atoms of an element corresponds to the **atomic number** of that element. The **atomic weight** of an element is the total mass of protons and neutrons in the atoms of that element. Hence, the atomic number of carbon is 6 and its atomic weight is 12 (Table 2.1). Although the number of protons in the atoms of a given element is constant, the number of *neutrons* may vary, giving rise to **isotopes** of a given element. For example, atoms of the element sulfur always contain 16 protons but exist in nature in three isotopic forms. Sulfur-32 (^{32}S), the most abundant form, contains 16 protons and 16 neutrons, sulfur-34 (^{34}S), a less abundant sulfur isotope, contains 16 protons and 18 neutrons, and sulfur-35 (^{35}S), a radioactive isotope or **radioisotope,** contains 16 protons and 19 neutrons. Unlike that of ^{32}S and ^{34}S, the atomic nucleus of ^{35}S is highly unstable and spontaneously disintegrates, releasing an electron (beta particle). However, despite having different atomic weights, all isotopes of sulfur are chemically alike because they have the same number of protons and the same number of electrons in their outer shell (Table 2.1).

Electrons

Electrons are negatively charged particles of negligible mass and are present in atoms in numbers equal to protons. Electrons orbit the nucleus in *shells,* zones of electron density at increasing distances from the nucleus of the atom. The chemical properties of elements are for the most part determined by the number of electrons in the outermost shell of their atoms (see Table 2.1). The outermost shell of most elements of atomic number 8 and above contains between one and eight electrons. The potential of an element for combination with other elements to form molecules is to a major degree governed by whether the outer electron shell is full or partially depleted. For the biologically relevant elements, the most stable electron configuration in the outer shell of atoms is eight (the "octet rule" of general chemistry), except for the first shell, which holds only two electrons. Table 2.1 shows that the outer shells of atoms of the major bioelements, hydrogen, carbon, nitrogen, oxygen, phosphorus, and sulfur, are all incomplete; these elements thus readily combine with other atoms to yield molecules in which all atoms achieve stable outer electron shells.

CONCEPT CHECK   2.1

The uniqueness of each type of chemical element is determined by the number and distribution of protons and neutrons in the nucleus and of electrons in the outer shells of its atoms.

✔ *Define the word* atom. *Of what is an atom composed?*

✔ *Why does the element carbon have an atomic number of 6 but an atomic weight of 12?*

✔ *How do* isotopes *of a given element differ?*

2.2
Molecules and Chemical Bonding

When atoms combine with one another by chemically bonding, the resulting combination is referred to as a **molecule.** Several types of chemical bonds are known, including ionic, covalent, and hydrogen bonds. **Ionic bonds** are bonds in which electrons are *not* equally shared between two atoms, resulting in charged molecules. For example, in NaCl the atom of chlorine is much more electronegative (electron-attracting) than that of sodium and formally removes the electron from the outer shell of sodium to form, in effect, Na^+Cl^-. The electrical attraction between the two atoms holds the molecule together. Ionic bonds are biologically important in nucleic acid–protein interactions and a few other molecular interactions. However, ionic bonds are not significant in the bonding of the important bioelements listed in Table 2.1 and thus will not be further discussed here. Biologically relevant molecules are held together by *covalent* and *hydrogen* bonds.

Covalent bonds

Bonds in which electrons are *shared* between two atoms are referred to as **covalent bonds.** Consider the element hydrogen. A hydrogen atom contains a single electron and readily combines with itself to form a molecule of hydrogen gas, H_2 (Figure 2.1a). The H_2 molecule consists of two hydrogen atoms, each containing two shared electrons, the maximum allowed in the initial electron shell surrounding any nucleus. The atoms in the H_2 molecule are held together by covalent bonds. By comparison, carbon has an outer electron shell containing four electrons and can accept four additional electrons in this shell. Hence, a carbon atom readily combines with four hydrogen atoms, leading to the formation of methane (Figure 2.1b), a simple *organic* (carbon-containing) compound. By sharing electrons in this fashion, the carbon atom and the hydrogen atoms achieve stable outer electron shells: eight electrons in the outer shell for carbon, and two each for the hydrogens. Likewise, when water forms from the elements hydrogen and oxygen (Figure 2.1c), the outer electron shells in both elements reach the stable configuration.

If more than one electron pair is shared between two atoms, double or even triple bonds are formed (Figure 2.2). Double bonds occur most frequently as $C=O$ or $C=C$ bonds and are found in a variety of important biological molecules. For triple bonds, three electron pairs are shared between two atoms. Triple bonds are rare in biological molecules, but microorganisms occasionally metabolize triply bonded compounds, the best example being N_2 (Figure 2.2), which is reduced in nitrogen fixation (∞ Section 13.27).

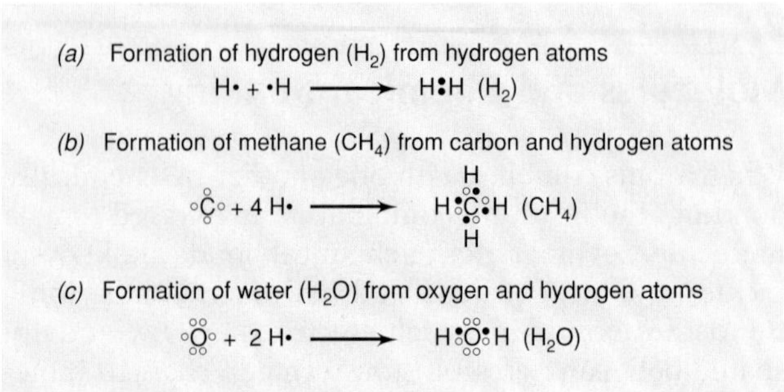

Hydrogen bonds

A second type of bond that is extremely important in biological systems is the **hydrogen bond.** Hydrogen bonds form between hydrogen atoms and more electronegative elements like oxygen and nitrogen. Hydrogen bonds are weak bonds. However, when many hydrogen bonds are formed within and between macromolecules, the overall stability of the molecule is greatly increased.

Water molecules readily undergo hydrogen bonding (Figure 2.3). Since an oxygen atom is relatively electronegative and a hydrogen atom is not, the covalent bond between oxygen and hydrogen is one in which the shared electrons in the outer shells orbit nearer the oxygen nucleus than the hydrogen nucleus. This creates a *slight* electric charge separation, oxygen slightly negative, hydrogen slightly positive; this is indicated by the sign δ, as in $\delta+$ or $\delta-$, to indicate partial charge separation (Figure 2.3). The $\delta+$ of the hydrogen atom attracts the oxygen of a second water molecule, in effect creating a positively charged "bridge" between the two electronegative oxygen atoms of adjacent water molecules. The most common hydrogen bonds in macromolecules are those that involve $O^{\delta-}\cdots H^{\delta+}-O^{\delta-}$, $O^{\delta-}-H^{\delta+}\cdots N$,

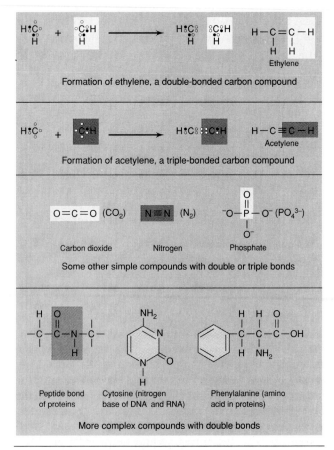

Formation of ethylene, a double-bonded carbon compound

Formation of acetylene, a triple-bonded carbon compound

$O=C=O$ (CO_2) $N\equiv N$ (N_2)   Phosphate (PO_4^{3-})

Carbon dioxide   Nitrogen   Phosphate

Some other simple compounds with double or triple bonds

Peptide bond of proteins   Cytosine (nitrogen base of DNA and RNA)   Phenylalanine (amino acid in proteins)

More complex compounds with double bonds

Figure 2.2 Covalent bonding of some biologically important molecules containing double or triple bonds.

(a) Formation of hydrogen (H_2) from hydrogen atoms

$H\cdot + \cdot H \longrightarrow H:H$ (H_2)

(b) Formation of methane (CH_4) from carbon and hydrogen atoms

$C + 4 H\cdot \longrightarrow H:C:H$ (CH_4)

(c) Formation of water (H_2O) from oxygen and hydrogen atoms

$O + 2 H\cdot \longrightarrow H:O:H$ (H_2O)

Figure 2.1 Formation of typical single covalently bonded molecules.

and $N^{\delta-}\cdots H^{\delta+}\!-\!N^{\delta-}$ interactions in proteins and nucleic acids (Figure 2.3). We will see later that hydrogen bonds play major roles in the biological properties of macromolecules, especially in the higher order structure of proteins and nucleic acids.

Other atomic interactions

Molecules are involved in other types of atomic interactions not referred to as formal chemical bonds. **Van der Waals forces** are nonspecific *attractive* forces that occur when the distance between two atoms is reduced to the range of 3–4 angstroms (Å). Van der Waals forces occur because of momentary charge asymmetries around atoms due to electron movement. If atoms become closer than 3–4 Å, however, overlap of the atom's electron shells causes repulsive forces to occur. Van der Waals forces can play a significant role in the binding of substrates to enzymes (∞ Section 4.5) and in protein–nucleic acid interactions.

    Hydrophobic interactions can also be important in the properties of biologically relevant molecules. Hydrophobic interactions occur because nonpolar (water-repelling) molecules tend to cluster together in an aqueous environment. Because of this, nonpolar portions of a macromolecule tend to associate, as do polar portions of a macromolecule (but for the opposite reason). Hydrophobic interactions are a major consideration in the folding of macromolecules such as proteins and play an important role in the binding of substrates to enzymes. Hydrophobic interactions are also often important in the noncovalent association of certain proteins, for example, as in a protein that consists of several polypeptide subunits (see Section 2.8).

Bond strengths and the importance of carbon

The relative strength of chemical bonds can be measured by the energy required to dissociate them. Table 2.2 lists the bond energies of a number of biologically relevant chemical bonds. Note that double and triple bonds are much stronger than single bonds and that hydrogen bonds are by comparison very weak. Hydrogen bonds form and dissociate in the cell spontaneously and rapidly, whereas covalent bonds are formed and broken only by specific chemical reactions brought about by enzymes (∞ Section 4.5). Noncovalent interactions such as van der Waals forces and hydrophobic bonds are also very weak. However, like hydrogen bonds, these interactions become significant when large numbers of them develop either within or between various molecules.

    All cellular structures contain an abundance of carbon in chemical combination with other elements. Carbon is able to combine not only with many other ele-

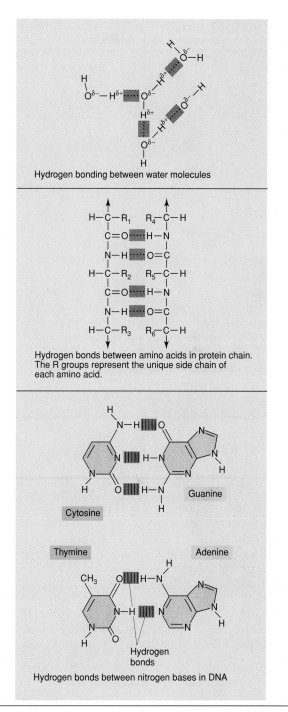

Figure 2.3 Hydrogen bonding of water and some biologically relevant molecules.

ments but also with itself, thus forming larger chemical structures of considerable diversity and complexity. An enormous number of different carbon compounds have been identified in various biological systems. For example, many biological compounds contain a carbon atom bonded to an oxygen atom (Table 2.3). Differences in the carbon–oxygen bond (single or double) and the

TABLE 2.2	Energies of some covalent bonds and noncovalent interactions

Covalent bonds	Bond energy (kJ/mol)[a]
Single bonds	
H—H	436
C—H	411
C—O	369
C—N	294
C—S	260
Double bonds	
C=C	616
C=O	704
O=O	402
Triple bonds	
C≡C	805
N≡N	955
Noncovalent interactions	**Energy**
Hydrogen bonds	4.2–8.4
Hydrophobic interactions	4.2–8.4
van der Waals attractions	4.2–8.4

[a]Bond energies are given as the amount of heat needed to break the bonds.

nature of the atoms surrounding the carbon–oxygen bond dictate the chemical properties of the molecule. Thus, carboxylic acids, for example, are chemically distinct from alcohols and each plays a specific role in cellular biochemistry (Table 2.3).

CONCEPT CHECK   2.2

Hydrogen bonds and other types of weak bonds play important roles in the way biomolecules function. Carbon is at the core of cell structure.

✔ *How does a* covalent bond *differ from a* hydrogen bond?

✔ *Is an oxygen atom or a hydrogen atom a molecule? Why or why not? Is water a molecule?*

✔ *In water, which atoms show covalent bonding and which show hydrogen bonding?*

2.3

Water as a Biological Solvent

The chemistry of life occurs in water. In fact, microbial cells are 70–90% water by weight, and all chemical reactions that occur in the cytoplasm of a cell take place in this aqueous environment. Water is an ideal biological solvent. Although pure water is electrically neu-

tral, having an equal number of electrons and protons (see Figure 2.1), water molecules contain two elements, oxygen and hydrogen, of quite different electronegativity, and bonding between these elements leads to charge asymmetry (Figure 2.3) and the property of *polarity.* The polar properties of water make it an excellent solvent because many biologically important molecules are themselves polar and thus readily dissolve in water. As we will see in Chapter 3, dissolved substances are continually passed into and out of the cell through transport activities of the cytoplasmic membrane.

The polar properties of water also allow for ready hydrogen bonding, which we have seen is important in the overall structure of macromolecules like proteins and nucleic acids (Figure 2.3). Water forms three-dimensional networks with other molecules including macromolecules and, by so doing, spatially positions these molecules for potential interactions. But in addition to hydrogen bonding, the polar nature of water makes it highly *cohesive,* meaning that water molecules tend to have a high affinity for one another and form chemically ordered arrangements in which hydrogen bonds are constantly forming, breaking, and re-forming. The cohesive nature of water is responsible for many of its biologically important properties, such as high surface tension and high specific heat. Also, the fact that water expands on freezing to yield a less dense solid form (ice) has profound effects on life in temperate and polar aquatic environments. In a lake, for example, ice on the surface insulates the water beneath the ice and prevents its freezing, thus allowing aquatic organisms to live under the ice.

The high polarity of water is also beneficial to the cell because it tends to force *nonpolar* substances to aggregate and remain together. We will see a practical example of this in the next chapter when we study membrane structure and function. Membranes contain nonpolar substances such as lipids, which aggregate and function as a barrier to the flow of polar molecules in and out of the cell.

Life originated in water, and anywhere on Earth where liquid water exists, microorganisms are likely to be found. (We will see some spectacular examples of this in Chapters 5 and 17.) We now consider the major substances dissolved in the water of a cell.

CONCEPT CHECK   2.3

Life could not exist without liquid water; the polar properties of water make it an excellent solvent for biological systems.

✔ *List three reasons why it is essential that cytoplasm be an aqueous environment.*

✔ *What is the most abundant molecule in a cell?*

TABLE 2.3	Carbon–oxygen and other compounds of biochemical importance	

Chemical species	Structure	Biological importance
Carboxylic acid	$\overset{\displaystyle O}{\overset{\displaystyle \|}{-C-OH}}$	Organic, amino, and fatty acids
Aldehyde	$\overset{\displaystyle O}{\overset{\displaystyle \|}{-C-H}}$	Functional group of reducing sugars such as glucose
Alcohol	$\overset{\displaystyle H}{\overset{\displaystyle \|}{\underset{\displaystyle \|}{\underset{\displaystyle H}{-C-OH}}}}$	Lipids, carbohydrates
Keto	$\overset{\displaystyle O}{\overset{\displaystyle \|}{-C-}}$	Pyruvate, citric acid cycle intermediates
Ester	$\overset{\displaystyle H \quad\;\; O}{\overset{\displaystyle \| \quad\;\; \|}{\underset{\displaystyle \|}{\underset{\displaystyle H}{-C-O-C-}}}}$	Lipids of Bacteria and Eukarya, amino acid attachment to tRNAs
Phosphate ester	$\overset{\displaystyle O^- \quad\;\;}{\overset{\displaystyle \| \quad\;\; \|}{\underset{\displaystyle \| \quad\;\; \|}{\underset{\displaystyle O \quad\;\;}{^-O-P-O-C-}}}}$	Nucleic acids, DNA and RNA
Thioester	$\overset{\displaystyle O}{\overset{\displaystyle \|}{R_1-C-S-R_2}}$	Energy metabolism, biosynthesis of fatty acids
Ether	$\overset{\displaystyle H \quad\;\; H}{\overset{\displaystyle \| \quad\;\; \|}{\underset{\displaystyle \| \quad\;\; \|}{\underset{\displaystyle H \quad\;\; H}{-C-O-C-}}}}$	Lipids of Archaea, sphingolipids
Acid anhydride	$\overset{\displaystyle O \quad\;\; O^-}{\overset{\displaystyle \| \quad\;\; \|}{\underset{\displaystyle \|}{\underset{\displaystyle O_-}{R-C-O-P=O}}}}$	Energy metabolism, for example, acetyl phosphate
Phosphoanhydride	$\overset{\displaystyle O^- \quad\;\; O^-}{\overset{\displaystyle \| \quad\;\; \|}{\underset{\displaystyle \| \quad\;\; \|}{\underset{\displaystyle O \quad\;\; O}{^-O-P-O-P-O^-}}}}$	Energy metabolism, for example, ATP

2.4

Small Molecules: Monomers

The main chemical components of cells are structures called **macromolecules.** Macromolecules are built up of individual building blocks that are connected in spe-cific ways. A single building block is called a **monomer,** and the macromolecule is called a **polymer;** polymers are composed of similar or repeating units. There are only a few types of polymers important in cell biochemistry, and each is made of a characteristic set of monomers.

Monomers are generally small organic compounds that are grouped in classes according to their chemical properties. There are four classes of monomers to be considered here: *sugars*, the monomeric constituents of **polysaccharides**; *fatty acids*, the monomeric units of **lipids**; *nucleotides*, the basic units of the **nucleic acids** (DNA and RNA); and *amino acids*, the monomeric constituents of **proteins**. The four classes of macromolecules can be subdivided into "informational" and "noninformational" types. Nucleic acids and proteins are considered informational macromolecules because the *sequence* of monomeric units within them is highly specific and carries biological information and the means to process this information. Lipids and polysaccharides, on the other hand, are not informational because the sequence of monomers in these polymers is frequently highly repetitive and the sequence itself is generally of less functional importance. However, it should be appreciated that both the nature *and* the sequence of the monomeric units of any macromolecule are important in distinguishing it chemically from related macromolecules.

CONCEPT CHECK   **2.4**

Macromolecules are polymers of monomers. Important cellular macromolecules are nucleic acids, proteins, polysaccharides, and lipids.

✔ *Which macromolecules are informational?*

2.5

Carbohydrates and Polysaccharides

Carbohydrates (sugars) are organic compounds containing carbon, hydrogen, and oxygen in a ratio of 1:2:1. The structural formula for glucose, the most abundant of all sugars, is $C_6H_{12}O_6$ (Figure 2.4). The most biologically relevant carbohydrates are those containing 4, 5, 6, and 7 carbon atoms (designated as C_4, C_5, C_6, and C_7). C_5 sugars (pentoses) are of special significance because of their role as structural backbones of nucleic acids. Likewise, C_6 sugars (hexoses) are the monomeric constituents of cell wall polymers and energy reserves. Figure 2.4 shows the structural formulas of a few common sugars. Derivatives of simple carbohydrates can be formed by replacing one or more of the hydroxyl groups by other chemical species. For example, the important bacterial cell wall polymer **peptidoglycan** (∞ Section 3.5) contains the glucose derivatives *N*-acetylglucosamine and *N*-acetylmuramic acid (Figure 2.5). Besides sugar derivatives, sugars having the same *structural* formula can still differ in their *stereoisomeric* properties (see Section 2.9). Hence, a large number of different sugars are potentially available to the cell for the construction of polysaccharides.

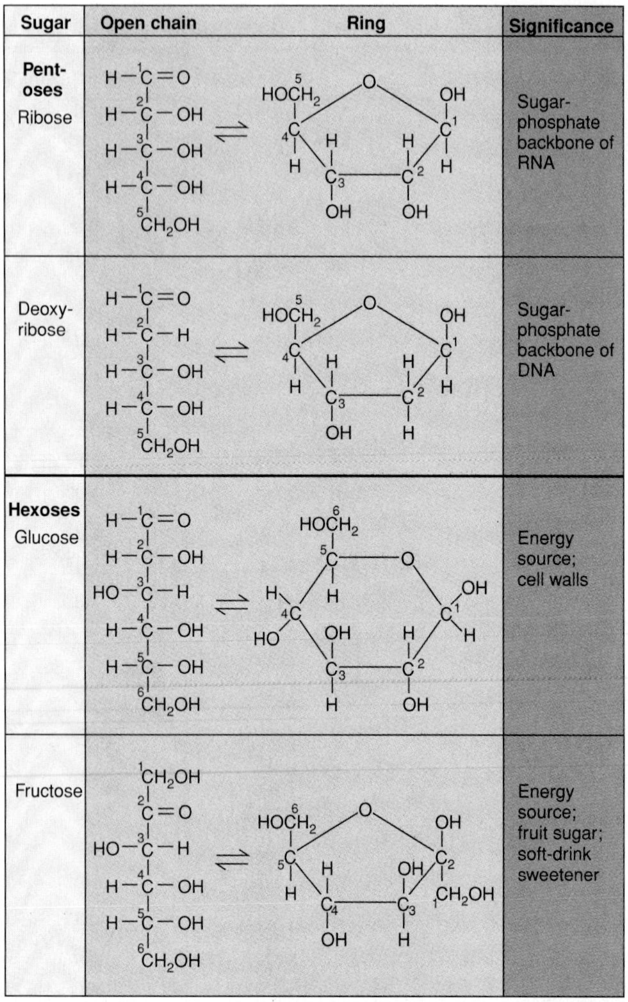

Figure 2.4 Structural formulas of a few common sugars. The formulas can be represented in two alternate ways, open chain and ring. The open chain is easier to visualize, but the ring form is the commonly used structure.

Polysaccharides and the glycosidic bond

Polysaccharides are high-molecular-weight carbohydrates containing many (sometimes hundreds or even thousands) monomeric units connected to one another by covalent bonds referred to as **glycosidic bonds** (Figure 2.6). If two sugar units (monosaccharides) are joined by a glycosidic linkage, the resulting molecule is called a *disaccharide*. The addition of one more monosaccharide yields a *trisaccharide*, and several more an *oligosaccharide*; an extremely long chain of monosaccharides in glycosidic linkage is called a **polysaccharide.**

The glycosidic bond can exist in two different orientations, referred to as alpha (α) and beta (β) (Figure 2.6*a*). Polysaccharides with a repeating structure composed of glucose units linked between carbons 1 and 4 in the *alpha* orientation (for example, glycogen and starch, Figure 2.6*b*) function as important carbon and

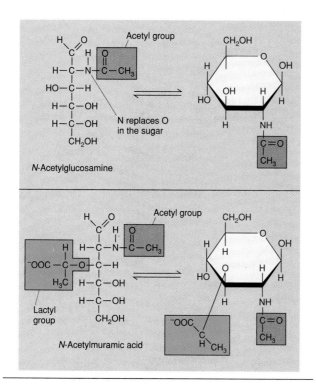

Figure 2.5 Sugar derivatives found in the cell walls of most Bacteria. Note that the parent structure is *glucose* in both cases. Note that like sugars themselves, sugar derivatives can be drawn in either open-chain or ring form.

energy reserves in bacteria, plants, and animals. Alternatively, glucose units joined by *beta*-1,4 linkages are present in cellulose (Figure 2.6*b*), a stiff plant and algal cell wall component functionally unrelated to glycogen or starch. Thus, even though starch and cellulose are both composed solely of glucose units, their functional properties are entirely different because of the different configurations, α and β, of their glycosidic bonds. Polysaccharides can also combine with other classes of macromolecules, such as protein and lipid, to form **glycoproteins** and **glycolipids**. These compounds play important roles in cell membranes as cell surface receptor molecules. The compounds reside on the external surfaces of the membrane where they are in contact with the environment. Glycolipids also constitute a major portion of the cell wall of gram-negative bacteria and as such

---

CONCEPT CHECK 2.5

Sugars (carbohydrates) combine into long polymers called polysaccharides. Polysaccharides can also contain other molecules such as protein or lipid, forming complex polysaccharides.

✔ *How can glycogen and cellulose differ so much in their physical properties when they both consist of 100% glucose?*

---

impart a number of unique surface properties to these organisms (∞Section 3.6).

2.6
Fatty Acids and Lipids

Fatty acids are the main constituents of **lipids.** Fatty acids have interesting chemical properties because they contain both highly hydrophobic (water-repelling) and highly hydrophilic (water-soluble) regions. Palmitate,* for example (Figure 2.7), is a 16-carbon fatty acid composed of a chain of 15 saturated (fully hydrogenated) carbon atoms and a single carboxylic acid group. Other common fatty acids are stearic (C_{18} saturated) and oleic (C_{18} monounsaturated) acids (Figure 2.7).

Triglycerides and complex lipids

Simple lipids (fats) consist of fatty acids bonded to the C_3 alcohol *glycerol* (Figure 2.7). Simple lipids are also referred to as **triglycerides** because three fatty acids are linked to the glycerol molecule.

Complex lipids are simple lipids that contain additional elements such as phosphate, nitrogen, or sulfur, or small hydrophilic carbon compounds such as sugars, ethanolamine, serine, or choline (Figure 2.7). **Phospholipids** are a very important class of complex lipids, as they play a major structural role in the cytoplasmic membrane (∞Section 3.3).

The chemical properties of lipids make them ideal structural components of membranes. Because they are *amphipathic,* that is, show properties of hydrophobicity and hydrophilicity, lipids aggregate in membranes with the hydrophilic portions toward the external or internal (cytoplasmic) environment, while maintaining their hydrophobic portions away from the aqueous milieu (∞Section 3.3). Such structures are ideal permeability barriers because of the inability of water-soluble sub-

---

CONCEPT CHECK 2.6

Lipids contain both hydrophobic and hydrophilic units; their chemical properties make them ideal structural components for cell membranes.

✔ *What part of a fatty acid molecule is hydrophobic? Hydrophilic?*

✔ *How does a phospholipid differ from a triglyceride?*

✔ *Draw the chemical structure of butyrate, a C_4 fully saturated fatty acid.*

---

*Fatty acids can exist in both protonated (RCOOH) and unprotonated (RCOO⁻) forms, depending on pH. Because at pH 7 fatty acids are generally unprotonated, this is indicated by adding the suffix *-ate* to the root term for the fatty acid. Thus, palmitic acid is $C_{15}H_{31}COOH$, and palmitate is $C_{15}H_{31}COO^-$.

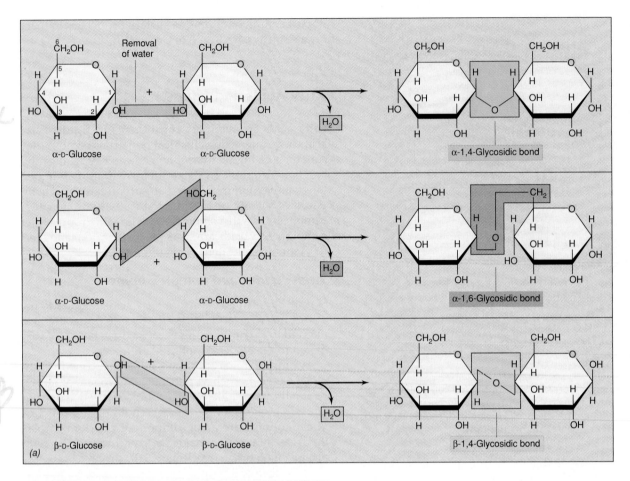

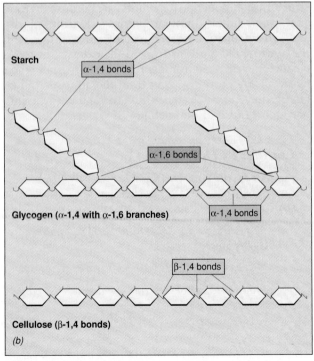

Figure 2.6 Polysaccharides. (a) Formation of polysaccharides from monosaccharides. The removal of a molecule of H_2O in each case is an example of a dehydration reaction. (b) General structures of some common polysaccharides. Note color coding to (a).

stances to flow through the hydrophobic fatty acid portion of the lipids. Indeed, the major function of the cytoplasmic membrane is to serve as a barrier to the diffusion of substances into or out of the cell.

2.7

Nucleotides and Nucleic Acids

The nucleic acids, deoxyribonucleic acid (DNA), and ribonucleic acid, (RNA), are polymers of monomers called **nucleotides.** DNA and RNA are thus both **polynucleotides.** DNA carries the genetic blueprint for the cell, and RNA acts as an intermediary molecule to convert the blueprint into defined amino acid sequences in proteins (∞Section 1.2). Despite these important cellular functions, nucleic acids are composed of only a relatively few simple building blocks. Each nucleotide is composed of three separate units: a five-carbon sugar, either ribose (in RNA) or deoxyribose (in DNA), a nitrogen base, and a molecule of phosphate, PO_4^{3-}. Figure 2.8 shows schematic drawings of single nucleotides of DNA and RNA.

Nucleotides

The nitrogen bases of nucleic acids belong to either of two chemical classes. *Purine* bases, **adenine** and **guanine,** contain two fused carbon–nitrogen rings, whereas

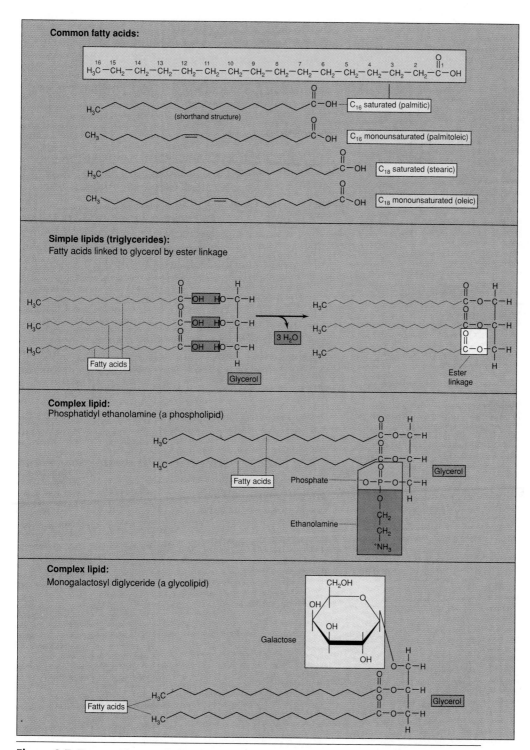

Figure 2.7 Fatty acids, simple lipids (fats), and complex lipids. Simple lipids are formed by a dehydration reaction between fatty acids and glycerol.

pyrimidine bases, **thymine, cytosine,** and **uracil,** contain a single six-membered carbon–nitrogen ring (Figure 2.9). Guanine, adenine, and cytosine are found in both DNA and RNA; thymine is present (with minor exceptions) only in DNA, and uracil is present only in RNA. In a nucleotide, a base is attached to a pentose sugar by a glycosidic linkage between carbon atom 1 of the sugar and a nitrogen atom of the base, either the nitrogen

atom labeled atom 1 (pyrimidine base) or nitrogen atom 9 (purine base). Without a phosphate, a base bonded to its sugar is referred to as a **nucleoside.** Nucleotides are thus nucleosides containing one or more phosphates (Figure 2.10).

Nucleotides play other roles in the cell besides their major role as constituents of nucleic acids. Nucleotides, especially adenosine triphosphate (ATP) (Figure 2.10),

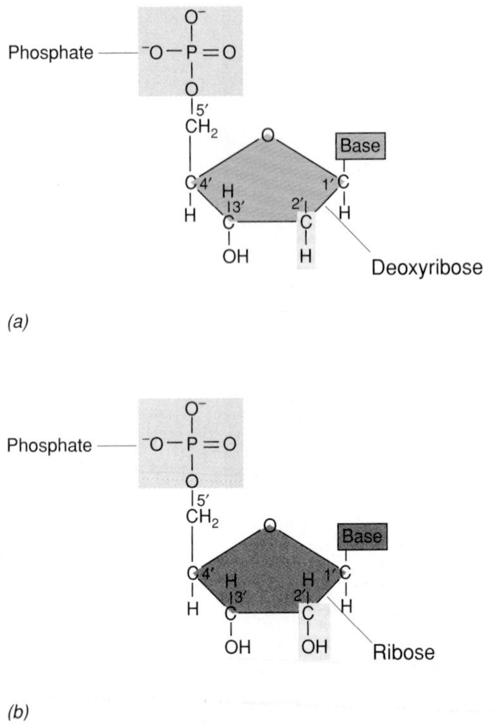

Phosphate

(a)

Phosphate

(b)

Figure 2.8 Nucleotides. (a) Of DNA. (b) Of RNA. Note the numbering system employed and the chemical differences on carbon atom 2′ between deoxyribose and ribose. Both deoxyribonucleotides (in DNA) and ribonucleotides (in RNA) contain a 5′-phosphate.

function as carriers of chemical energy and can release sufficient energy during the hydrolytic cleavage of a phosphate bond to drive energy-requiring reactions in the cell (∞ Section 4.8). Other nucleotides or nucleotide derivatives function in oxidation–reduction reactions in the cell (∞ Section 4.6) as carriers of sugars in the

biosynthesis of polysaccharides (∞ Section 4.18) and as regulatory molecules inhibiting or stimulating the activities of certain enzymes or metabolic events. However, here we are discussing the role of a nucleotide as a building block of nucleic acid, the major *informational* function of nucleotides.

Nucleic acids

Nucleic acids are long polymers in which nucleotides are covalently bonded to one another in a defined sequence, forming structures called **polynucleotides.** The backbone of the nucleic acid is a polymer in which sugar and phosphate molecules alternate (Figure 2.11). Hence, when we refer to a specific *sequence* of nucleotides in a nucleic acid, we are really referring to the variable portions of the nucleotide—the nitrogen bases; the sugar–phosphate backbone is always the same.

The precise sequence of nucleotides in a DNA or RNA molecule is referred to as the molecule's *primary structure.* As we have discussed, the sequence of bases in a DNA or RNA molecule is *informational*, representing the genetic information necessary to reproduce an identical copy of the organism. We will see later that the replication of DNA and the production of RNA are highly complex processes (∞ Chapter 6) and that a virtually error-free mechanism is necessary to ensure the faithful transfer of genetic traits from one generation to another.

In chemical terms nucleic acids are composed of nucleotides covalently attached to one another via phosphate from carbon 3 [referred to as the 3′ (3 prime) carbon] of one sugar to carbon 5 (5′) of the adjacent sugar[†]

[†]Because the ring structure in the base is numbered also, the prime numbering system is used to refer to positions of carbon atoms on the sugars.

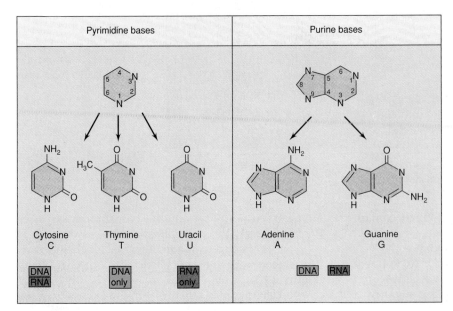

Figure 2.9 Structure of bases of DNA and RNA. The letters C, T, U, A, and G are used to designate the individual bases. Note the numbering system of the rings.

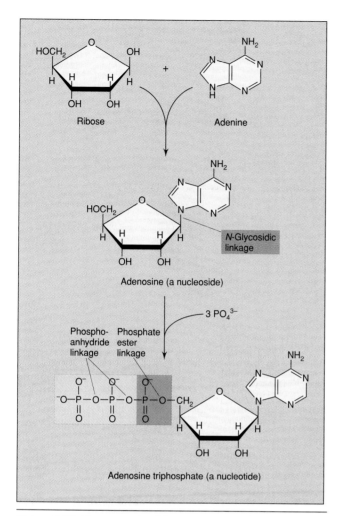

Figure 2.10 Components of the important nucleoside triphosphate, adenosine triphosphate. The energy of hydrolysis of a phosphoanhydride is greater than that of a phosphate ester and will have significance in Chapter 4 (∞ Section 4.8).

(Figure 2.11*a*). The phosphate linkage is chemically a **phosphodiester** since a single phosphate is connected by ester linkage to two separate sugars.

DNA

In cells, DNA is present in double-stranded form. Each cellular chromosome contains two strands of DNA, each strand containing several million nucleotides linked by phosphodiester bonds. The strands themselves associate with one another by hydrogen bonds that form between the nucleotides of one strand and the nucleotides of the other. When positioned adjacent to one another, purine and pyrimidine bases can undergo hydrogen bonding (see Figure 2.3). Chemically, the most stable hydrogen bonding configuration occurs when guanine (G) forms hydrogen bonds with cytosine (C), and adenine (A) forms hydrogen bonds with thymine (T) (see Figure 2.3). Specific base pairing, A with T and G with C, means that the two strands of DNA are *com-*

plementary in base sequence; wherever a G is found in one strand, a C is found in the other, and wherever a T is present in one strand, its complementary strand has an A (Figure 2.11*b*). The molar amounts of guanine and cytosine are therefore *identical* in double-stranded DNA from any source; likewise, the molar amounts of adenine and thymine are the same. It should also be noted that although DNA is generally double-stranded, some viruses contain only single-stranded DNA (∞ Section 8.1).

RNA

With the exception of certain viruses that contain double-stranded RNA, all ribonucleic acids are *single-stranded* molecules. However, RNA molecules can fold back upon themselves in regions where complementary base pairing can occur to form a variety of highly folded structures. The pattern of folding observed in RNA is referred to as its *secondary structure* (Figure 2.11*b*).

RNA plays three crucial roles in the cell. **Messenger RNA** (mRNA) contains the genetic information of DNA in a single-stranded molecule *complementary* in base sequence to a portion of the base sequence of DNA. **Transfer RNA** (tRNA) molecules are the "adaptor" molecules in protein synthesis. The tRNA molecule effectively adapts the genetic information from the language of nucleotides to the language of amino acids, the building blocks of proteins. **Ribosomal RNA** (rRNA) molecules, of which several distinct types are known, are important structural and catalytic components of the ribosome, the protein-synthesizing system of the cell. These various RNA molecules are discussed in detail in Chapter 6.

CONCEPT CHECK   2.7

The informational content of a nucleic acid is determined by the sequence of nitrogen bases along the polynucleotide chain. Both RNA and DNA are informational macromolecules. RNA can often fold into various configurations to obtain secondary structure.

✔ *What is a* nucleotide?

✔ *How many carbon atoms are present in a pyrimidine base? In a purine base?*

✔ *How does a* nucleo*side differ from a* nucleo*tide?*

2.8

Amino Acids and Proteins

Amino acids are the monomeric units of proteins. Most amino acids consist of only carbon, hydrogen, oxygen, and nitrogen, but 2 of the 21 common amino acids found

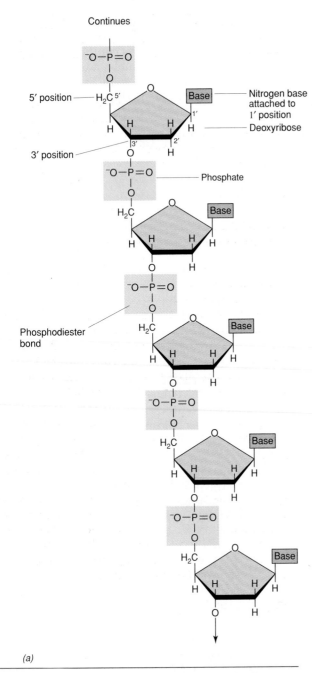

(a)

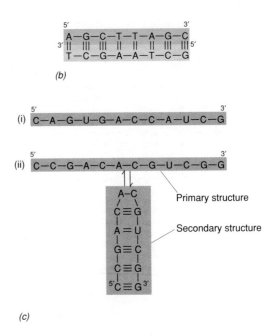

(b)

(i) 5′ C—A—G—U—G—A—C—C—A—U—C—G 3′

(ii) 5′ C—C—G—A—C—A—C—G—U—C—G—G 3′

Primary structure

Secondary structure

(c)

Figure 2.11 Structure of part of a DNA chain. (a) The nitrogen bases can be adenine, guanine, cytosine, or thymine. In RNA, an OH group is present on the 2′ carbon of the pentose sugar and uracil replaces thymine. (b) Simplified structure of DNA in which only the nitrogen bases are shown. Note how the two strands are complementary in base sequence (A═T; G≡C) and bonded by hydrogen bonds. Note also how the two strands of DNA are shown in two different shades of green; this convention is used throughout this book. (c) RNA: (i) A sequence showing only primary structure; (ii) A sequence that allows for secondary structure. RNA is shown in orange throughout this book.

in cells also contain sulfur and one contains selenium. All amino acids contain two important functional groups, a *carboxylic acid* group (—COOH) and an *amino* group (—NH$_2$) (Figure 2.12). These groups are functionally important because covalent bonds, between the carbon of the carboxyl group of one amino acid and the nitrogen of the amino group of a second amino acid (with elimination of a molecule of water), form the **peptide bond,** a type of covalent bond characteristic of proteins (see Figure 2.13).

All amino acids conform to the general structure shown in Figure 2.12. Amino acids differ in the nature of the side group (abbreviated R in Figure 2.12) attached to the α-carbon. The α-carbon is the carbon atom *immediately adjacent* to the carboxylic acid group. The side chains on the alpha carbon vary considerably, from as simple as a hydrogen atom in the amino acid glycine, to aromatic ringed structures in amino acids such as phenylalanine (Figure 2.12). The chemical properties of an amino acid are to a major degree governed by the nature of the side chain, and thus amino acids that show similar chemical properties can be grouped into amino acid "families" as shown in Figure 2.12. For example, the side chain may itself contain a carboxylic acid group, such as in aspartic acid or glutamic acid, rendering the amino acid acidic. Alternatively, several amino acids contain nonpolar hydrophobic side chains and are grouped together as nonpolar amino acids. The amino acid cysteine contains a sulfhydryl group (—SH), which is frequently important in connecting one chain of amino acids to another by *disulfide linkage* (R—S—S—R).

The large number of chemically distinct amino acids that exist makes it possible for cells to produce an enormous number of different proteins with widely different biochemical properties. For example, proteins that

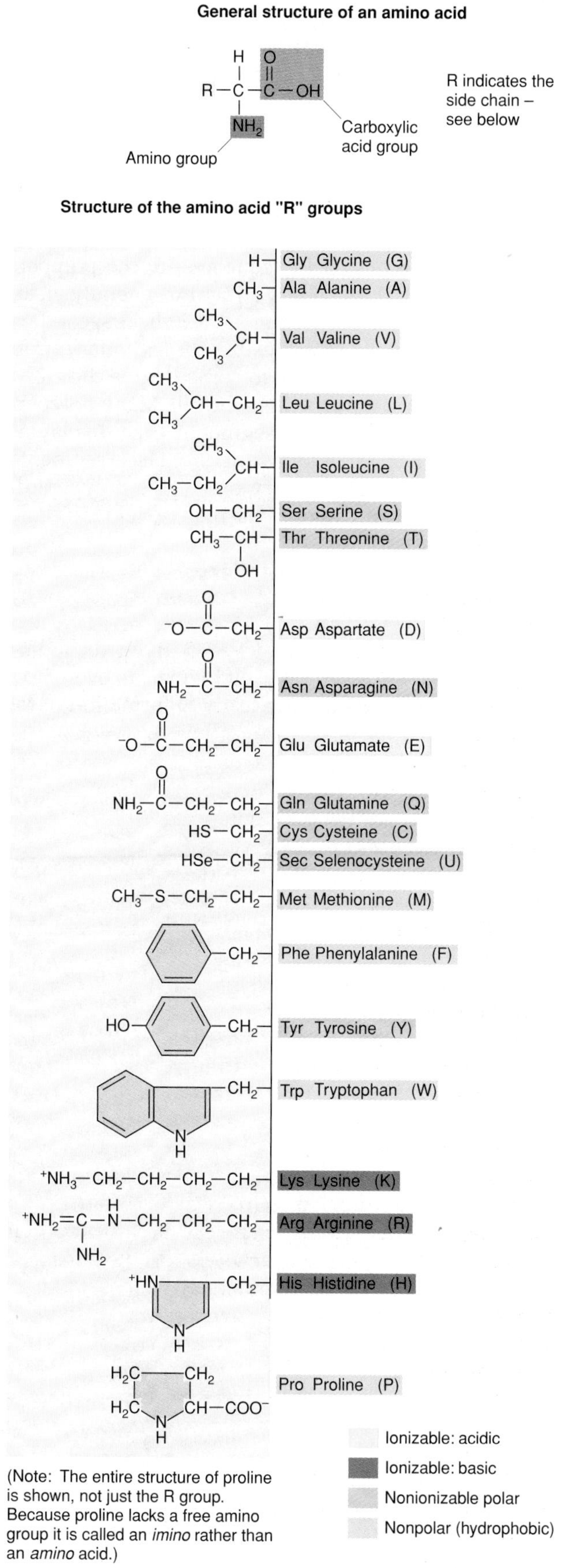

General structure of an amino acid

R indicates the side chain — see below

Amino group

Carboxylic acid group

Structure of the amino acid "R" groups

H—	Gly Glycine	(G)
CH₃—	Ala Alanine	(A)
	Val Valine	(V)
	Leu Leucine	(L)
	Ile Isoleucine	(I)
OH—CH₂—	Ser Serine	(S)
	Thr Threonine	(T)
	Asp Aspartate	(D)
	Asn Asparagine	(N)
	Glu Glutamate	(E)
	Gln Glutamine	(Q)
HS—CH₂—	Cys Cysteine	(C)
HSe—CH₂—	Sec Selenocysteine	(U)
CH₃—S—CH₂—CH₂—	Met Methionine	(M)
—CH₂—	Phe Phenylalanine	(F)
HO——CH₂—	Tyr Tyrosine	(Y)
—CH₂—	Trp Tryptophan	(W)
⁺NH₃—CH₂—CH₂—CH₂—CH₂—	Lys Lysine	(K)
⁺NH₂=C—N—CH₂—CH₂—CH₂—	Arg Arginine	(R)
⁺HN——CH₂—	His Histidine	(H)
	Pro Proline	(P)

(Note: The entire structure of proline is shown, not just the R group. Because proline lacks a free amino group it is called an *imino* rather than an *amino* acid.)

Ionizable: acidic

Ionizable: basic

Nonionizable polar

Nonpolar (hydrophobic)

Figure 2.12 Structure of the 21 common amino acids. The three-letter codes for the amino acids are to the left of the names, and the one-letter codes are in parentheses to the right of the names.

are in direct contact with highly hydrophobic regions of the cell, such as proteins embedded in the lipid-rich cytoplasmic membrane, generally contain higher overall proportions of hydrophobic amino acids (or contain regions extremely rich in hydrophobic amino acids) than proteins that function in the aqueous environment of the cytoplasm.

Structure of proteins: Primary and secondary structure

Proteins play key roles in cell function. Two major classes of proteins are *catalytic* proteins (enzymes) and *structural* proteins. Enzymes serve as catalysts for the wide variety of chemical reactions that occur in cells (∞ Chapter 4). Structural proteins are those that become integral parts of the structures of cells in membranes, walls, and cytoplasmic components. In essence, a cell is what it is because of the kinds of proteins it contains. Therefore, an understanding of protein structure is essential for an understanding of cell function.

Proteins are polymers of various lengths containing defined sequences of amino acids covalently bonded by peptide bonds (Figure 2.13). Two amino acids bonded together constitute a *dipeptide,* three amino acids a *tripeptide,* and so on. Many amino acids covalently linked via peptide bonds constitute a **polypeptide,** and proteins consist of one or more polypeptides. The number of amino acids varies from one protein to another. Proteins with as few as 15 and as many as 10,000 amino acids are known. Since proteins may vary in their composition, sequence, and number of amino acids, it is easy to see that enormous variation in protein structure (and thus function) is possible.

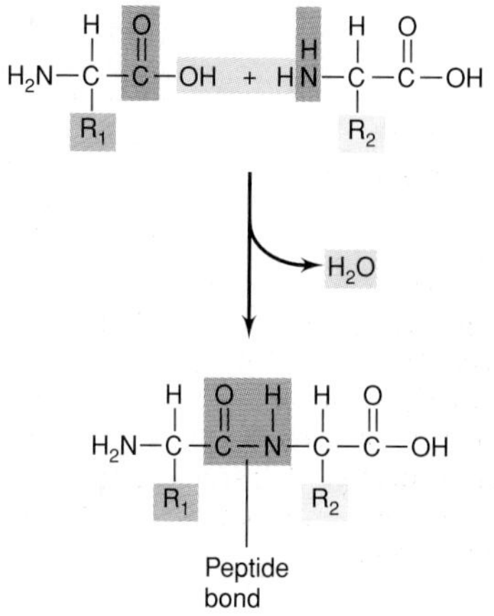

Peptide bond

Figure 2.13 Peptide bond formation. R_1 and R_2 refer to the variable portion (side chain) of the amino acid (see Figure 2.12).

All proteins are folded molecules and show complex arrangements of structure. The linear array of amino acids is referred to as the **primary structure** of the polypeptide; primary structure thus gives a complete description of all covalent bonds present in the molecule. In many ways the primary structure of a polypeptide can be considered the most important, because a given primary structure allows only certain types of higher order structure to occur. The juxtaposition of α-carbon R groups dictated by the primary structure forces the polypeptide to twist and fold in a specific way. This process leads to formation of the **secondary structure** of the protein (Figure 2.14). Hydrogen bonds, the weak noncovalent linkages discussed earlier (see Section 2.2), play important roles in the type of secondary structure that a protein attains.

A typical secondary structure for many polypeptides is the α-*helix* (Figure 2.14a). Imagine a linear polypeptide wound around a cylinder. Under these conditions oxygen and nitrogen atoms from different amino acids become positioned close enough together in the twisted structure to allow hydrogen bonding to occur. This opportunity for H bonding (and the inherent stability associated with it) helps direct many polypeptides to take on an α-helix secondary structure (Figure 2.14a).

Many polypeptides conform to a different type of secondary structure referred to as the β-*sheet*. In the β-sheet, the chain of amino acids in the polypeptide folds back and forth upon itself instead of forming a helix; this type of folding exposes hydrogen atoms that can undergo extensive hydrogen bonding (Figure 2.14b). Some polypeptides contain both regions of α-helix and regions of β-sheet secondary structure, the type of folding being determined by the available opportunities for hydrogen bonding and hydrophobic interactions (recall that these will ultimately be dictated by the primary structure—the amino acid sequence—of the polypeptide). Since β-sheet secondary structure generally yields a rather rigid structure, whereas α-helical secondary structures are usually more flexible, the secondary structure of a given polypeptide to some degree dictates a functional role for the protein in the cell. Many polypeptides fold into two or more segments, each displaying α-helix or β-sheet secondary structure. These segments, referred to as *domains,* are regions of the polypeptide that have specific functions in the final protein molecule.

Structure of proteins: Tertiary and quaternary structure

Once a polypeptide has achieved a given secondary structure it folds back upon itself to form an even more stable molecule. This folding leads to formation of the

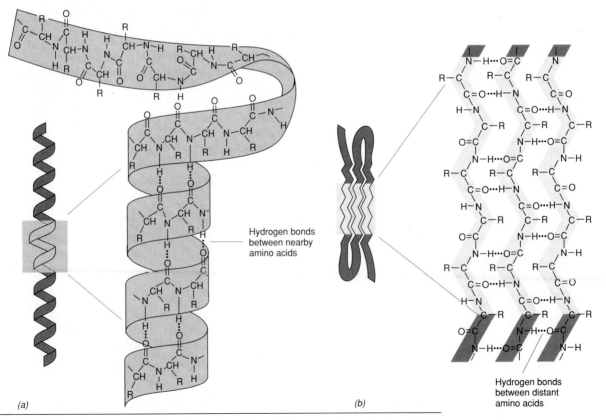

Hydrogen bonds between nearby amino acids

Hydrogen bonds between distant amino acids

(a)

(b)

Figure 2.14 Secondary structure of polypeptides. (a) α-Helix secondary structure. Note that hydrogen bonding does not involve the R groups but instead occurs between the atoms involved in the peptide bonds. (b) β-Sheet secondary structure.

tertiary structure of the protein. Like secondary structure, tertiary structure of a protein is ultimately determined by primary structure, but tertiary structure is also governed to some extent by the secondary structure of the molecule. As a result of the formation of secondary structure, the side chain of each amino acid in the polypeptide is positioned in a specific way. If additional hydrogen bonds, covalent bonds, hydrophobic interactions, or other atomic interactions are able to form, the polypeptide will fold to accommodate them and attain a unique three-dimensional shape (Figure 2.15).

Frequently a polypeptide folds in such a way that adjacent sulfhydryl (—SH) groups of cysteine residues are exposed (Figure 2.15b). These free —SH groups can join covalently to form a disulfide (—S—S—) bridge between the two amino acids. If the two cysteine residues are located in different polypeptide chains of a protein, the disulfide bond physically links the two molecules (Figure 2.15b). In addition, a single polypeptide can spontaneously fold and bond to itself covalently if two cysteine residues form a disulfide linkage within the molecule. The tertiary folding of the polypeptide ultimately forms exposed regions or grooves in the molecule (Figures 2.15 and 2.16) which may be of importance in binding other molecules (for example, the binding of a substrate to an enzyme) (∞ Section 4.5).

If a protein consists of more than one polypeptide, and many proteins do, the arrangement of polypeptide subunits in space to form the final protein molecule is referred to as the **quaternary structure** of the protein (Figure 2.16). It should be remembered that in proteins showing quaternary structure, each subunit of the final protein itself contains primary, secondary, and tertiary structure. Some proteins displaying quaternary structure contain many identical subunits; others contain several nonidentical subunits, whereas still others may contain more than one identical subunit and a second nonidentical subunit in the final protein molecule. The subunits of multisubunit proteins are held together either by noncovalent interactions (hydrogen bonding, van der Waals forces, or hydrophobic interactions) or by covalent linkages, generally intersubunit disulfide bonds.

Denaturation of proteins

When proteins are exposed to extremes of heat or pH, or to certain chemicals or metals that affect their folding properties, they are said to undergo **denaturation** (Figure 2.17). In general, the biological properties of a protein are lost when it is denatured. When proteins are denatured, peptide bonds are generally unaffected, and the sequence of amino acids (primary structure) in the polypeptide remains unchanged. However, denaturation causes the polypeptide chain to unfold, destroying the higher order structure of the molecule, in particular hydrogen bonds. The denatured polypeptide retains its primary structure because it is held together by covalent peptide bonds. Depending on the severity of the denaturing conditions, refolding of the polypeptide may occur after removal of the denaturant (Figure 2.17). However, the fact that denaturation is generally associated with loss of biological activity of the protein

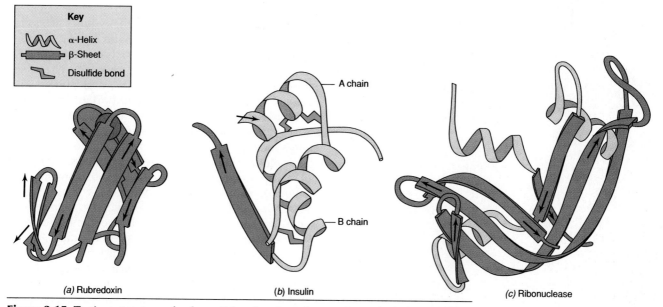

Figure 2.15 Tertiary structure of polypeptides showing where regions of α-helix or β-sheet secondary structure might be located. (a) Rubredoxin; the red sphere is a molecule of heme. (b) Insulin, a protein containing two polypeptide chains; note how the B chain contains both α-helix and β-sheet secondary structure and how disulfide linkages (—S—S—) may help in dictating folding patterns (tertiary structure). (c) Ribonuclease, a large protein with several regions of α-helix and β-sheet.

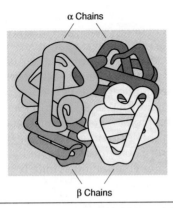

Figure 2.16 Quaternary structure of hemoglobin, a protein containing four polypeptide subunits. There are two *kinds* of polypeptide in hemoglobin, α chains (shown in blue and red) and β chains (shown in orange and yellow). Separate colors are used to distinguish the four chains.

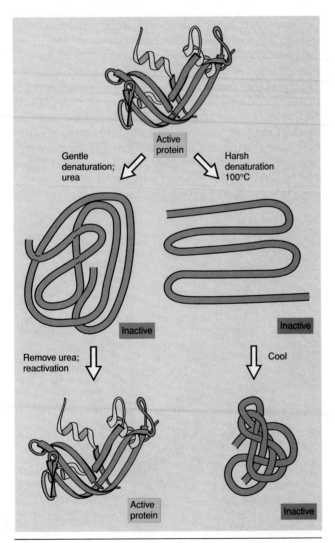

Figure 2.17 Denaturation of a protein using ribonuclease (whose structure was discussed in Figure 2.15c) as an example. Note how harsh denaturation generally yields a permanently destroyed molecule.

clearly shows that biological activity is not inherent in the primary structure of proteins but instead is a result of the unique folding of the molecule as ultimately directed by primary structure. Folding of a polypeptide therefore accomplishes two things: (1) the polypeptide obtains a unique shape that is compatible with a *specific* biological function, and (2) the folding process converts the molecule to its most chemically stable form.

CONCEPT CHECK   **2.8**

The building blocks of proteins are amino acids, which combine to form polypeptides. The primary structure of a protein is determined by its amino acid sequence, but it is the folding (higher order structure) of the polypeptide that determines how the protein functions in the cell.

✔ *Draw the structure of the amino acid alanine and label the α-carbon.*

✔ *Why is aspartic acid considered an* acidic *amino acid and tryptophan a* nonpolar *amino acid?*

✔ *Define the terms* primary, secondary, *and* tertiary *with respect to protein structure.*

2.9

Stereoisomerism

It was mentioned in reference to carbohydrates in Section 2.5 that two molecules may have the same molecular formula but exist in different structural forms. These related but not identical molecules are referred to as **isomers** (Figure 2.18) (see the box, Chemical Isomers and Living Organisms). Isomers are important in biology, especially in the chemistry of sugars. For example, *Escherichia coli* grows well on glucose as a carbon and energy source but does not grow on the closely related sugar allose (Figure 2.18a), presumably because it cannot incorporate or metabolize this hexose isomer. Many isomers of common sugars are found as constituents of the cell walls of Bacteria and Archaea (∞Section 3.5). Sugar isomers are also known that contain the same molecular and structural formulas except that one is a "mirror image" of the other, just as the left hand is a mirror image of the right. In carbohydrate chemistry two identical sugars that are mirror images of one another are called **stereoisomers** or **enantiomers** and have been given the designations D and L (Figure 2.18b). D Sugars predominate in biological systems.

Stereoisomerism is also important in protein chemistry. Like sugars, amino acids can exist as D or L stereoisomers. However, in the case of protein, life has

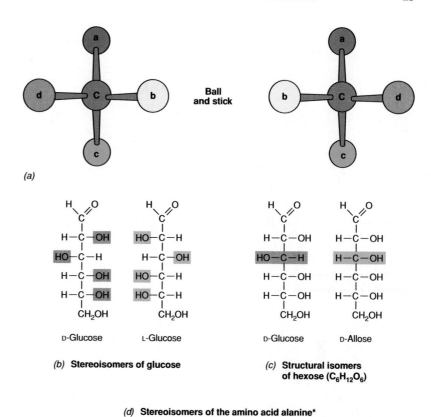

Figure 2.18 Stereoisomers and structural isomers. (a) A ball-and-stick model showing mirror images (stereoisomers). (b) Stereoisomers of glucose (note the mirror-image structure). (c) Structural isomers (diastereoisomers) of glucose (note the *non*-mirror-image nature of the structures). (d) Stereoisomers of the amino acid alanine.

evolved to use the L form rather than the D (Figure 2.18d) (see the box). D-Amino acids are found occasionally in nature, most commonly in the cell wall polymer peptidoglycan (∞Section 3.5) and in certain peptide antibiotics (∞Section 11.8). Prokaryotes are equipped to handle the conversion of D-amino acids to the L form by way of enzymes that specifically catalyze this transformation. Cells contain enzymes called **racemases** whose function is to convert the unusual form (L sugar or D-amino acid) to the readily metabolizable form (D sugar or L-amino acid).

CONCEPT CHECK   2.9

Stereoisomers are mirror-image molecules. With rare exception, amino acids in cells are L stereoisomers, whereas sugars are generally D stereoisomers.

✔ *Look at Figure 2.18. Why are D-glucose and D-allose diastereoisomers whereas D-glucose and L-glucose are stereoisomers?*

✔ *How can cells interconvert the D and L forms of various molecules?*

CHEMICAL ISOMERS AND LIVING ORGANISMS

Louis Pasteur was trained as a chemist, and his first research work was on the chemistry of stereoisomers, specifically, on the optical properties of crystals of tartaric acid (see the accompanying figure), published in 1848. Because stereoisomers are mirror images (that is, they show handedness), they are also *chiral*, meaning that they are optically active: crystals or solutions of one stereoisomer rotate polarized light in one direction, whereas the other stereoisomer rotates light in the opposite direction. Pasteur also noticed that racemic mixtures of tartaric acid were often contaminated with fungi, but that as the organisms grew on the tartaric acid, they consumed only one of the two stereoisomeric forms. When this occurred, an optically inactive solution of tartaric acid became optically active! From these studies Pasteur recognized the significance of asymmetry in living systems, and his interest in this problem led him from chemistry into biology, a field that would occupy him for the rest of his life.

For Pasteur, the fact that living organisms discriminated between stereoisomers was of profound philosophical significance. He saw life processes as inherently asymmetric, whereas nonliving chemical processes were not asymmetric. Pasteur's vision of the asymmetry of life has been well confirmed over the nearly 150 years since he did his work. In fact, when Pasteur began to study spontaneous generation (Section 1.9), he used his knowledge of the asymmetry of living organisms to bolster his confidence that spontaneous generation could not exist. In addition, Pasteur's early conclusions that fermentations such as lactic acid and alcoholic fermentations are due to the activities of living organisms were based in part on his discovery that chemical reactions in living organisms, in contrast to abiotic systems, yield highly asymmetric products.

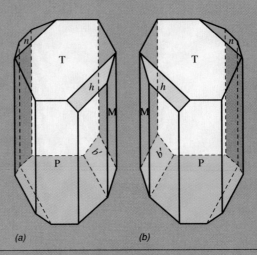

(a)   (b)

Pasteur's drawings of tartaric acid ($C_4H_6O_6$) crystals, used to illustrate his famous paper on optical activity. (a) Left-handed crystal. (b) Right-handed crystal. Note that the two crystals are mirror images. The letters on the faces of the crystals were Pasteur's way of labeling the mirror image faces of the two crystals. Color has been added here to show the mirror image faces more clearly.

Why is it that the amino acids of proteins in living organisms are of only the L configuration, whereas when the same amino acids are made chemically, they consist of mixtures of D and L forms? We know that this is because chemical reactions in living organisms are carried out by asymmetric catalysts, **enzymes.** Asymmetry begets asymmetry, and because enzymes are asymmetric, they introduce asymmetry into the molecules they produce. (It should be noted that D-amino acids are found in a few biological molecules, most notably bacterial cell walls, but are absent from cellular proteins.)

The preceding statement begs the question of how the *first* asymmetry arose in the living world. Was it by chance or by design? Is there somewhere in the universe a planet with living organisms whose proteins are comprised of only D-amino acids? A planet, perhaps, where everything got started along the opposite asymmetry to that of life on Earth and in which life is now trapped forever in this opposite asymmetry? Possibly. We will return to the whole question of evolution and the origin of life in Chapter 15.

Isomers such as the L-amino acids or the L-tartrate crystals shown here, gradually, with time, racemize spontaneously, resulting in a mixture of D and L forms. Organic material derived from dead organisms, occurring as part of fossils, gradually becomes a racemic mixture. Over geological time periods, an equal amount of D- and L-amino acids will be left. The amino acids of fossils consist of variable proportions of L- and D-amino acids, the L/D ratio depending upon how long the organism has been dead. And, from knowledge of the rate at which amino acids spontaneously racemize, the L/D ratio of a fossil can actually be used as a way of obtaining its approximate age.

Thus, we see that chemistry and biology are closely intertwined. Living organisms can carry out chemical reactions that yield specific stereoisomers, and it was Pasteur's grasp of this phenomenon that helped him push his beliefs about the nature of microbial activities forward in the face of considerable criticism from other, less enlightened scientists of the day. ∎

Material for Review

REVIEW QUESTIONS

1. What is the numerical relationship of protons to neutrons and electrons in atoms? Which of these particles carries an electric charge?

2. Which elements are the major elements found in living organisms? Why are oxygen and hydrogen particularly abundant in living organisms?

3. Define the word *molecule*. How many atoms are in a molecule of hydrogen gas? In a molecule of glucose?

4. Refer to the structure of the nitrogen base *cytosine* shown in Figure 2.2. Draw this structure and then label the positions of all single bonds and double bonds in the cytosine molecule.

5. Compare and contrast the words *monomer* and *polymer*. Give three examples of biologically important polymers and list the monomers of which they are composed.

6. RNA and DNA are similar types of macromolecules but show distinct differences as well. List three ways in which RNA differs chemically or physically from DNA. What is the cellular function of DNA and RNA?

7. Why are *amino acids* so named?

8. Write a general structure for an amino acid. What is the importance of the R group to final protein structure? Why does the amino acid *cysteine* have special significance for protein structure?

9. Chemically, what type of reaction between two amino acids leads to formation of the peptide bond? (You may wish to review Figure 2.13 before answering.)

10. Draw the peptide bond. Now redraw this structure in atomic form (see Figure 2.2) showing the arrangement of electrons within the atoms of the peptide bond.

APPLICATION QUESTIONS

1. Observe the following nucleotide sequences of RNA: (a) GUCAAAGAC, (b) ACGAUAACC. Can either of these RNA molecules have secondary structure? If so, draw the potential secondary structure(s).

2. A few soluble (cytoplasmic) proteins contain a high content of hydrophobic amino acids. How would you predict these proteins would fold as to their tertiary structure and why?

3. Cells of the genus *Halobacterium*, an archaean that lives in very salty environments, contain over 5 molar (*M*) potassium (K^+). Because of this high K^+ content, many cytoplasmic proteins of *Halobacterium* cells are enriched in two specific amino acids that are present in much higher proportions in *Halobacterium* proteins than in functionally similar proteins from *Escherichia coli* (which has only very low levels of K^+ in its cytoplasm). Which amino acids are enriched in *Halobacterium* proteins and why? *Hint:* Which amino acids could best neutralize the positive charges due to K^+?

4. It is often the case that proteins that show α-helix secondary structure are more flexible than proteins showing β-sheet secondary structure. Discuss why this could be the case.

5. When an egg is placed in a beaker of boiling water, changes in the egg occur almost immediately. Describe what happens and why the contents of a boiled egg look so different from the contents of a fresh egg.

6. In light of your answer to the preceding question, explain how it can be that certain prokaryotes, called *hyperthermophiles*, thrive (and indeed grow optimally) in boiling hot springs. How must the proteins of hyperthermophiles differ from proteins in the egg?

7. Only one of the 21 common amino acids has no D or L form (that is, does not show chirality) and therefore has no stereoisomeric forms. Which amino acid is it and why?

SUPPLEMENTARY READINGS

Alberts, B., D. Bray, J. Lewis, M. Raff, K. Roberts, and **J. D. Watson.** 1994. *Molecular Biology of the Cell,* 3rd edition. Garland, New York. An excellent textbook of cell biology with emphasis on the molecular aspects of cells. Chapters 1–3 give a detailed account of cell chemistry and macromolecular structure. Highly recommended for advanced reading.

Stryer, L. 1995. *Biochemistry,* 3rd edition. W. H. Freeman, San Francisco. One of the best general biochemistry texts with excellent artwork making the material come alive. Highly recommended.

Voet, D., and **J. D. Voet.** 1995. *Biochemistry,* 2nd edition. John Wiley, New York. An excellent biochemistry text with strong coverage of informational macromolecules.

Watson, J. D., N. H. Hopkins, J. W. Roberts, J. A. Steitz, and **A. M. Weiner.** 1987. *Molecular Biology of the Gene,* 4th edition. Benjamin-Cummings, Redwood City, CA. Chapters 4 and 5 consider aspects of cell chemistry, in particular nucleic acid structure and functions. Becoming dated but still useful.

 On~line resources for this chapter are on the World Wide Web at: http://www.prenhall.com/~brock (click on the table of contents link and then select Chapter 2).

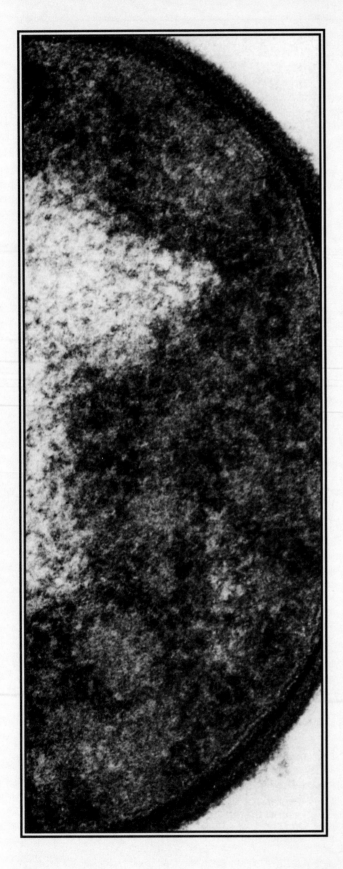

CHAPTER *3*

Cell Biology

The electron microscope allows microbiologists to study the internal structure of cells such as the gram-positive bacterium, *Bacillus subtilis*, shown here in a transmission electron micrograph of a sectioned cell.

MINIGLOSSARY for Chapter 3

Active Transport an energy-dependent transport process in which a substance is transported across the cytoplasmic membrane at the expense of ATP or a membrane potential

Chemotaxis movement of an organism toward (*positive*) or away from (*negative*) a chemical gradient

Chloroplast the chlorophyll-containing photosynthetic organelle of eukaryotic photosynthetic organisms

Chromosome a DNA molecule, usually circular in prokaryotes and linear in eukaryotes, carrying genes essential to cellular function

Cytoplasmic Membrane the permeability barrier of the cell, separating the cytoplasm from the environment

Endospore a highly heat-resistant, thick-walled, differentiated cell produced by certain gram-positive Bacteria

Eukaryote a cell containing a membrane-enclosed nucleus and usually other organelles

Flagellum a long, thin cellular appendage capable of rotation in prokaryotic cells and responsible for swimming motility

Gas Vesicles gas-filled cytoplasmic structures bounded by protein and conferring buoyancy on cells

Gram-Negative a prokaryotic cell whose cell wall contains relatively little peptidoglycan but contains an outer membrane composed of lipopolysaccharide, lipoprotein, and other complex macromolecules

Gram-Positive a prokaryotic cell whose cell wall consists chiefly of peptidoglycan and lacks the outer membrane of gram-negative cells

Group Translocation an energy-dependent transport process in which the substance transported is chemically modified during the transport process

Lipopolysaccharide (LPS) lipid in combination with polysaccharide and protein forming the major portion of the cell wall in gram-negative Bacteria

Magnetosomes particles of magnetite (Fe_3O_4) organized into nonunit membrane-enclosed structures in the cytoplasm of magnetotactic Bacteria

Mitochondrion (mitochondria) an organelle found in most eukaryotic cells in which respiration and energy generation occurs

Nucleoid an aggregated state of the circular chromosome of prokaryotic cells

Organelle a unit membrane-enclosed structure found in the cytoplasm of eukaryotic cells

Peptidoglycan a polysaccharide composed of alternating repeats of acetylglucosamine and acetylmuramic acid with the latter in adjacent layers cross-linked by short peptides

Periplasm a gellike region between the outer surface of the cytoplasmic membrane and the inner surface of the lipopolysaccharide layer of gram-negative Bacteria

Phototaxis movement of an organism toward light

Poly-β-hydroxybutyrate (PHB) a common storage material of prokaryotic cells consisting of a polymer of β-hydroxybutyrate or another β-alkanoic acid

Prokaryote a cell that lacks a membrane-enclosed nucleus and that usually has a single circular DNA molecule as its chromosome

Protoplast an osmotically protected cell whose cell wall has been removed

Ribosome small particles composed of RNAs and proteins that function in protein synthesis

In this chapter we present the principles of structure and function relationships in microbial cells. We emphasize the biology of the prokaryotic cell, but we compare and contrast prokaryotes with eukaryotes and discuss in detail a few eukaryotic cell structures.

Cells, like houses, are built by connecting simple building blocks in various ways to create more complex structures. We discussed the chemical nature of cellular building blocks in Chapter 2 and emphasized how these simple structures can be polymerized to form macromolecules. Cells are basically well-defined assemblages of macromolecules. Despite great diversity in the chemical composition of the macromolecules found in different cells, from a structural perspective, all cells solve a number of biological problems in common ways. For example, most cells employ a lipid bilayer as the structural foundation of the cytoplasmic membrane. Although the bilayer may vary in its exact chemical composition from species to species, all cytoplasmic membranes have the same basic structure and all serve as permeability barriers. Also, ribosomes, the structures on which proteins are synthesized in all organisms, differ somewhat in chemical detail from species to species but not in cellular function. Thus, *unity* exists in many structure and function relationships in all cells.

Because cells are microscopic, we begin this chapter with a discussion of microscopes and microscopy. The microscope is a major tool of the microbiologist. Historically it was the microscope that first revealed the secrets of cell structure, and even today it remains a powerful tool in cell biology studies.

3.1

Microscopes and Microscopy

Microscopic examination of microorganisms makes use of either the **light microscope** or the **electron microscope.** For most routine work, the light microscope is used, whereas for special research purposes, especially in studies on internal cell structure, the electron microscope is used in addition to the light microscope. All microscopes employ the principle that specific lenses magnify the image of a cell such that details of its structure are most apparent. In addition to magnification, however, is *resolution*, the ability to distinguish two adjacent points as separate. Although magnification can be increased virtually without limit, resolution cannot; resolution is dictated by the physical properties of light. It is thus resolution and not magnification that ultimately defines the limits of what we are able to see with a microscope. We begin our discussion with the light microscope, for which the limits of resolution are about 0.2 μm [200 nanometers (nm)], and then proceed to describe the electron microscope, for which resolution is improved over that of the light microscope by about 1000-fold.

The compound light microscope

The **light microscope** has been of crucial importance for the development of microbiology as a science and remains a basic tool of routine microbiological research. Several types of light microscopes are commonly used in microbiology: *bright-field, phase contrast, dark-field,* and *fluorescence.* The **bright-field microscope** is most commonly used in elementary biology and microbiology courses and consists of two series of lenses (objective lens and ocular lens), which function together to resolve the image (Figure 3.1). With this microscope, specimens are visualized because of the differences in contrast that exist between them and the surrounding medium. Contrast differences arise because cells absorb or scatter light in varying degrees. Many bacterial cells are difficult to see well with the bright-field microscope because of their lack of contrast with the surrounding medium. Pigmented organisms may be an exception, however, because the color of the organism adds contrast, thus improving visualization of the cells (Figure 3.2).

Magnification and resolution

The total magnification of a compound microscope is the *product* of the magnification of its objective and ocular lenses (Figure 3.1*b*). Magnifications of about 1500× are

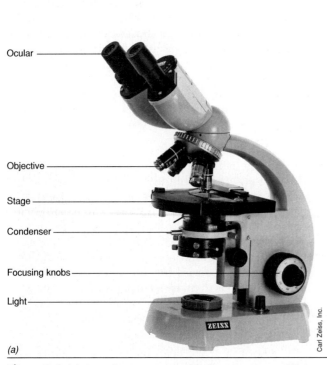

(a)

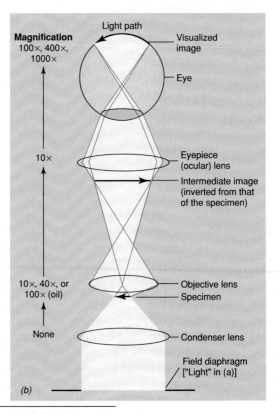

(b)

Figure 3.1 (a) A modern compound light microscope. Various key parts of the microscope are labeled. (b) Path of light through a compound light microscope. Besides 10×, eyepieces (oculars) are available in 15–30×.

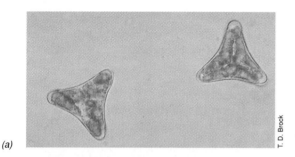

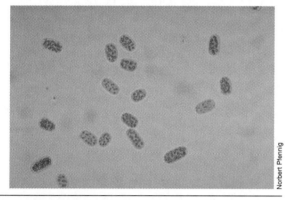

Figure 3.2 Photomicrographs of pigmented microorganisms by bright-field microscopy. (a) A green alga (eukaryote). (b) A purple phototrophic bacterium (prokaryote). The algal cells are about 15 μm in diameter, and the bacterial cells are about 5 μm in diameter.

near the upper limit obtainable with a compound light microscope. This limit is set because of a property of a lens called **resolution.** Resolving power is a function of the wavelength of light used and an innate property of the objective lens known as its *numerical aperture* (a measure of light-gathering ability). In general, there is a correspondence between the magnification of a lens and its numerical aperture: lenses with higher magnification usually have higher numerical apertures. The diameter of the smallest resolvable object is equal to 0.5λ/numer-

ical aperture, where λ is the wavelength of light used. Based on this formula, resolution is greatest when blue light is used to illuminate a specimen and the objective that is used has a very high numerical aperture.

The highest resolution possible in a compound light microscope is about 0.2 μm. This means that two objects closer together than 0.2 μm are not resolvable as distinct and separate. Most microscopes used in microbiology have oculars that magnify 10–15× and objectives of 10–100× (Figure 3.1*b*); at 1000×, objects 0.2 μm in diameter can just be resolved. With the 100× objective, and with certain other objectives of very high numerical aperture, a high grade optical oil is used between the specimen and the objective. Lenses on which oil is to be used are called *oil-immersion lenses.* Oil is used with these lenses because it has a higher numerical aperture than air (which has a numerical aperture of 1) and also a higher refractive index, and this greatly improves resolution.

Phase contrast, dark-field, and fluorescence microscopy

The **phase contrast microscope** was developed to improve contrast differences between cells and the surrounding medium, making it possible to see cells without staining them (Figure 3.3). Phase contrast microscopy is based on the principle that cells differ in refractive index from their surroundings and hence bend some of the light rays that pass through them. Light passing through a specimen of refractive index different from that of the surrounding medium is retarded. This effect is amplified by a special ring in the objective lens of a phase contrast microscope, leading to the formation of a dark image on a light background (Figure 3.3*b*). The phase contrast microscope is widely employed in research applications because it can be used to observe wet-mount (living) preparations. Staining, on the other hand, although a widely used procedure in light microscopy (discussed later), generally kills cells and can distort their features.

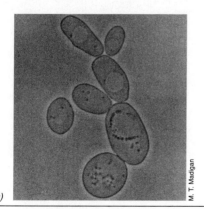

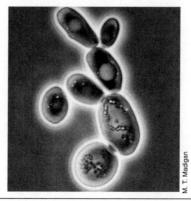

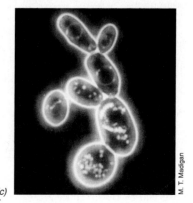

Figure 3.3 Photomicrographs of the same field of cells of the baker's yeast, *Saccharomyces cerevisiae,* taken by different types of light microscopy. (a) Bright-field. (b) Phase contrast. (c) Dark-field.

The **dark-field microscope** is a light microscope in which the lighting system has been modified to reach the specimen from the sides only. The only light reaching the lens is light scattered by the specimen, and thus the specimen appears light on a dark background (Figure 3.3c). Resolution by dark-field microscopy is quite high, and objects can frequently be resolved by dark-field that are not resolvable in bright-field or phase contrast microscopes. Dark-field microscopy is also an excellent way to observe the motility of microorganisms, as bundles of flagella are often resolvable with this technique (see Figure 3.48a).

The **fluorescence microscope** is used to visualize specimens that *fluoresce*, that is, emit light of one color when light of another color shines upon them. Fluorescence occurs either because of the presence within cells of naturally fluorescent substances such as chlorophyll or other fluorescing components (*autofluorescence*) (see Figure 3.4a) or because the cells have been treated with a fluorescent dye (Figure 3.4b). Fluorescence microscopy is widely used in clinical diagnostic microbiology and also in microbial ecology (∞ Chapters 14 and 21).

Staining: Increasing contrast for bright-field microscopy

Dyes can be used to stain cells and increase their contrast so that they can be more easily seen in the bright-field microscope. Dyes are organic compounds, and each class of dye has an affinity for specific cellular materials. Many dyes commonly used in microbiology are positively charged (cationic) and combine strongly with negatively charged cellular constituents such as nucleic acids and acidic polysaccharides. Examples of cationic dyes include *methylene blue, crystal violet,*

and *safranin*. Because cell surfaces are generally negatively charged, these dyes combine with structures on the surfaces of cells and hence are excellent general-purpose stains.

The simplest staining procedures are done with dried preparations (Figure 3.5). A slide containing a dried suspension of microorganisms is flooded for a minute or two with a dilute solution of a dye, rinsed several times in water, and blotted dry. It is usual to observe dried stained preparations of bacteria with a high power (oil-immersion) lens (Figure 3.5).

Differential stains are so named because they are used in procedures that do not stain all kinds of cells equally. An important differential staining procedure widely used in bacteriology is the *Gram stain* (Figure 3.6a). On the basis of their reaction to the Gram stain, bacteria can be divided into two major groups: **gram-positive** and **gram-negative.** After Gram staining, gram-positive bacteria appear purple and gram-negative bacteria appear red (Figure 3.6b). This difference in reaction to the Gram stain arises because of differences in the cell wall structure of gram-positive and gram-negative cells (as discussed later in this chapter). The Gram stain is one of the most useful staining procedures in the bacteriological laboratory; it is almost essential in identifying an unknown bacterium to determine first whether it is gram-positive or gram-negative.

The electron microscope

Electron microscopes are widely used for studying the detailed structure of cells. To study the internal structure of cells, a **transmission electron microscope (TEM)** is essential. In the TEM, electrons are used instead of light rays and electromagnets function as lenses, the

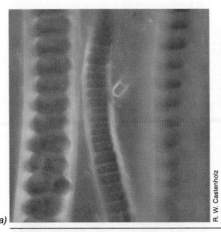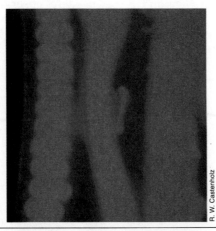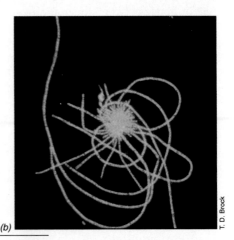

(a)   R. W. Castenholz   *(b)*   R. W. Castenholz   T. D. Brock

Figure 3.4 Photomicrographs of various microorganisms as visualized by fluorescence microscopy. (a) Cyanobacteria. Left, cells observed by bright-field microscopy. Right, same cells observed by fluorescence after shining light of 546 nm on them. The red color is due to autofluorescence of chlorophyll and other pigments. (b) Cells of the filamentous bacterium *Leucothrix mucor* stained with the fluorescent dye, acridine orange, which fluoresces green.

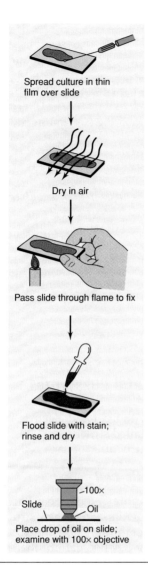

Figure 3.5 Staining cells for microscopic observation.

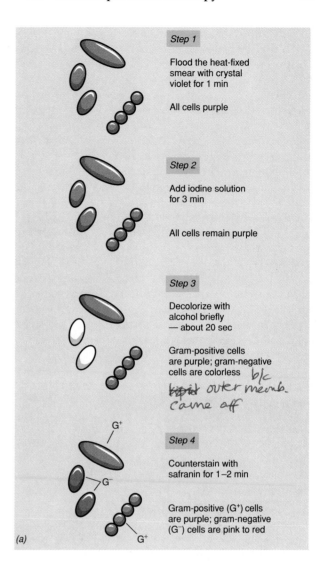

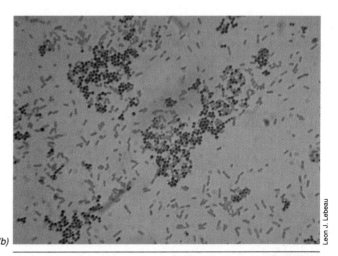

Figure 3.6 The Gram stain. (a) Steps in the Gram stain procedure. (b) Photomicrograph of Bacteria that are gram-positive (blue-purple) and gram-negative (pink-red). The species are *Staphylococcus aureus* and *Escherichia coli,* respectively.

whole system operating in a high vacuum (Figure 3.7). The resolving power of the electron microscope is much greater than that of the light microscope, and thus the electron microscope enables one to see many structures of even molecular size, such as proteins and nucleic acids (see Figure 3.42). However, electron beams do not penetrate very well, and if one is interested in seeing internal cell structure, even a single cell is too thick to be viewed directly. Consequently, special techniques of *thin sectioning* are needed to prepare specimens for the electron microscope. A single bacterial cell, for instance, is cut into many very thin slices, which are then examined individually with the electron microscope (see Figure 3.8a). To obtain sufficient contrast, the preparations are treated with a special electron microscope stain such as osmic acid, permanganate, uranium, lanthanum, or lead. Because these substances are composed of atoms

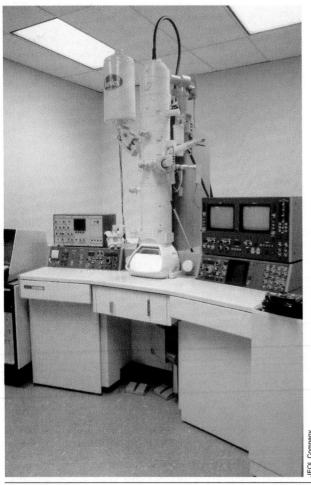

JEOL Company

Figure 3.7 A modern electron microscope. This instrument encompasses both transmission and scanning electron microscope functions.

of high atomic weight, they scatter electrons well and thus improve contrast (see Figure 3.8*a*).

If only the *external* features of an organism need be observed, thin sections are not necessary and intact cells or cell components can be observed directly by TEM with a technique called *negative staining* (see, for example, Figure 3.63). Alternatively, one can use the **scanning electron microscope (SEM)** (Figures 3.7 and 3.8*b*). With this tool, the specimen is coated with a thin film of a heavy metal such as gold. An electron beam from the SEM is then directed down on the specimen and scans back and forth across it. Electrons scattered by the metal are collected, and they activate a viewing screen to produce an image (see Figures 3.8*b* and 3.10). In the SEM, even fairly large specimens can be observed, and the depth of field is extremely good. A wide range of magnification can be obtained with the SEM, from as low as 15× up to about 100,000×, but only the *surface* of an object can be visualized. All electron microscopes are fitted with cameras to allow a photograph, called an *electron micrograph*, to be taken.

3.2

Overview of Cell Structure: Prokaryotes and Eukaryotes

Careful study with light and electron microscopes has revealed the detailed structure of cells, and we discuss here the major structures of each cell type.

The prokaryotic cell

A typical prokaryotic cell of either Bacteria or Archaea generally has the following major structures: cell wall, cytoplasmic membrane, ribosomes, inclusions, and nucleoid (Figure 3.8*a*). What are these structures and what are their functions?

The **cytoplasmic membrane** is the critical permeability barrier, separating the inside from the outside of the cell. The **cell wall** is a rigid structure outside the cytoplasmic membrane which provides support and protection from osmotic lysis. **Ribosomes** are small particles composed of protein and ribonucleic acid (RNA), which are visible in the transmission electron microscope. A single prokaryotic cell may have as many as 10,000 ribosomes. Ribosomes are part of the translation apparatus, and synthesis of cell proteins takes place on these structures. (We will discuss ribosomes, proteins, and RNA in detail in Chapter 6.) Prokaryotes occasionally contain **inclusions** consisting of storage material made up of compounds of carbon, nitrogen, sulfur, or phosphorus. Such inclusions can be formed when these nutrients are in excess in the environment and function as repositories of these nutrients when limitations occur.

The nuclear region of the prokaryotic cell differs quite significantly from that of the eukaryotic cell. Prokaryotic cells do not possess a true nucleus, the function of the nucleus being carried out by a single mole-

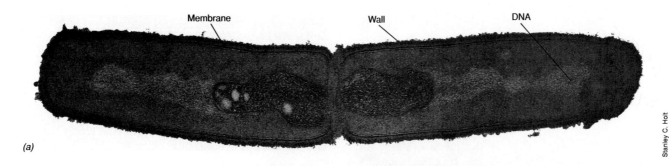

Membrane   Wall   DNA

(a)

Stanley C. Holt

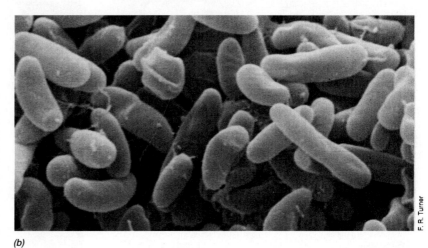

(b)

F. R. Turner

Figure 3.8 Electron micrographs of bacterial cells taken with (a) transmission and (b) scanning electron microscopes. (a) Thin section of a typical gram-positive bacterium, *Bacillus subtilis*. The cell has just divided, and two membrane-containing structures are attached to the cross-wall. Note the light region in the middle which is DNA. The cell is about 0.8 μm in diameter. (b) Cells of the phototrophic bacterium *Rhodospirillum sodomense*. A single cell is about 0.75 μm wide.

cule of deoxyribonucleic acid (DNA). DNA is present in a more-or-less free state within the prokaryotic cell but is often seen in electron micrographs (Figure 3.8*a*) in an aggregated form referred to as the **nucleoid.** In analogy to the eukaryote, the DNA molecule of the prokaryote is called a **chromosome.**

Many, but not all, bacteria are able to move. Movement of a prokaryotic cell is usually by means of a structure called a **flagellum** (plural, **flagella**). Each flagellum consists of a single, coiled tube of protein. The *rotation* of flagella propels the cell through liquids. Bacterial flagella can be seen with the light microscope if special staining techniques are used and are readily visible with the electron microscope (see Figure 3.46).

Morphology of prokaryotes

The *shape* of a cell is referred to as its *morphology*. Several distinct shapes of bacteria can be recognized and have been given different names. Schematic examples of some of these bacterial shapes along with phase photomicrographs are shown in Figure 3.9. A bacterium that is spherical or ovoid in morphology is called a **coccus** (plural, **cocci**). A bacterium with a cylindrical shape is called a **rod.** Some rods are curved, frequently forming spiral-shaped patterns and are then called **spirilla.**

In many prokaryotes, the cells remain together in groups or clusters after division, and the arrangements

in these groups are often characteristic of different organisms. For instance, cocci or rods may occur in long chains (Figure 3.10*a*). Some cocci form thin sheets of cells, whereas others occur in three-dimensional cubes or irregular cubelike clusters (Figure 3.10*b*). Several groups of bacteria are immediately recognizable by their unusual shapes. Examples include **spirochetes,** which are tightly coiled bacteria, **appendaged bacteria,** which possess extensions of their cells as long tubes or stalks, and **filamentous bacteria,** which form long, thin cells or chains of cells (Figure 3.9). It should be noted that the morphologies of prokaryotic cells shown in Figure 3.9 and 3.10 are *representative* ones; many variations of each of these basic morphological types have been found in newly isolated organisms.

The eukaryotic cell

Eukaryotic cells are larger and more complex in structure than prokaryotic cells, and a key difference is that eukaryotes contain *true nuclei*. The **nucleus** is a special membrane-enclosed structure within which DNA is located (Figure 3.11). The DNA in the nucleus is organized into **chromosomes,** structures that remain essentially invisible except at the time of cell division. Before cell division occurs, the chromosomes are duplicated and then condense, become thicker, and undergo division as the nucleus divides. The process of nuclear division in eukaryotes is called **mitosis** (see Figure 3.73),

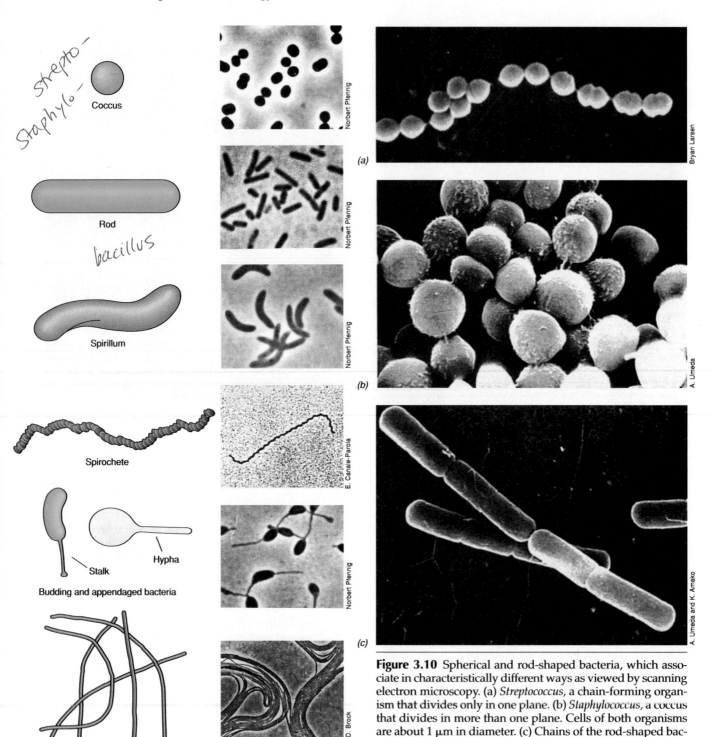

Figure 3.9 Representative cell shapes (morphology) in prokaryotes. Next to each drawing is a phase photomicrograph showing an example of that morphology. Organisms are coccus, *Thiocapsa roseopersicina* (diameter of a single cell = 1.5 μm); rod, *Desulfuromonas acetoxidans* (diameter = 1 μm); spirillum, *Rhodospirillum rubrum* (diameter = 1 μm); spirochete, *Spirochaeta stenostrepta* (diameter = 0.25 μm); budding and appendaged, *Rhodomicrobium vannielii* (diameter = 1.2 μm); filamentous, *Chloroflexus aurantiacus* (diameter = 0.8 μm).

Figure 3.10 Spherical and rod-shaped bacteria, which associate in characteristically different ways as viewed by scanning electron microscopy. (a) *Streptococcus*, a chain-forming organism that divides only in one plane. (b) *Staphylococcus*, a coccus that divides in more than one plane. Cells of both organisms are about 1 μm in diameter. (c) Chains of the rod-shaped bacterium *Bacillus*. Cells are about 0.8 μm in diameter.

which is a complex but highly organized process. Two identical daughter cells result from the division of one parent cell; each daughter cell receives a nucleus with an identical set of chromosomes.

Eukaryotic cells also contain distinct structures called **organelles,** within which important cellular functions occur (Figure 3.11). Organelles are absent from prokaryotes, although major physiological processes that take place in organelles, such as respiration and photosyn-

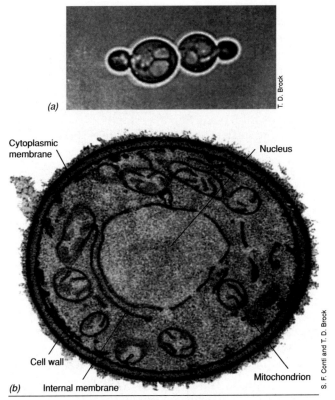

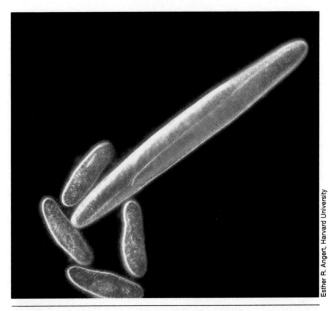

(a)

Cytoplasmic
membrane

Nucleus

Cell wall

(b)   Internal membrane

Mitochondrion

Figure 3.11 The yeast cell, a typical eukaryotic microorganism. (a) Photomicrograph of a yeast cell in the process of budding. (b) Electron micrograph of a thin section through a yeast cell, showing various structures. A single cell is about 8 μm in diameter.

Figure 3.12 Photomicrograph of a giant prokaryote, the surgeonfish symbiont *Epulopiscium fishelsoni*. The rod-shaped *E. fishelsoni* cell in this field is about 600 μm (0.6 mm) long and is shown with four cells of the protozoan (eukaryote) *Paramecium*, each of which measures about 150 μm in length. *E. fishelsoni* is phylogenetically related to *Clostridium* species.

thesis, may still occur in prokaryotic cells. One kind of organelle found in most eukaryotes is the **mitochondrion** (plural, **mitochondria**) (Figure 3.11; see also Figure 3.75). Mitochondria are the organelles within which the energy-generating functions of the cell occur. The energy produced in mitochondria is then used throughout the cell.

Algae are eukaryotic microorganisms that carry out the process of *photosynthesis*. In these organisms, as well as in green plants, an additional type of organelle is found: the **chloroplast.** The chloroplast is green and is the site where chlorophyll is localized and where the light-gathering functions involved in photosynthesis occur (see Figures 3.2*a* and 3.76).

The size of microbial cells and the significance of being small

Prokaryotes vary in size from cells as small as 0.1–0.2 μm in width to those more than 50 μm in diameter; a few very large prokaryotes, such as the surgeonfish symbiont *Epulopiscium fishelsoni* (Figure 3.12), are up to 50 μm in diameter and can be more than 0.5 millimeters (mm) in length (Figures 3.12 and 3.13). However, the dimensions of an average rod-shaped prokaryote, the bacterium *Escherichia coli*, for example, are about 1 × 3 μm (Figure 3.13). Typical eukaryotic cells may be 2 μm to more than

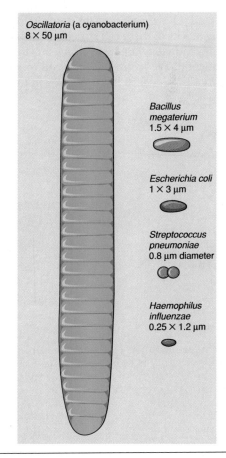

Oscillatoria (a cyanobacterium)
8 × 50 μm

Bacillus megaterium
1.5 × 4 μm

Escherichia coli
1 × 3 μm

Streptococcus pneumoniae
0.8 μm diameter

Haemophilus influenzae
0.25 × 1.2 μm

Figure 3.13 Comparison of sizes of a variety of prokaryotes.

200 μm in diameter. Thus, prokaryotes are very small cells compared to eukaryotes, and the small size of prokaryotes affects a number of their biological properties. For example, the rate at which nutrients and waste products pass into and out of a cell, a factor that can greatly affect cellular metabolic rates and growth rates, is in general *inversely* proportional to cell size. This is because transport rates are to some degree a function of the amount of *membrane surface area* available, and relative to cell volume, small cells have more surface available than do large cells. This point can be seen most readily in the case of a sphere, in which the *volume* is a function of the cube of the radius ($V = \frac{4}{3}\pi r^3$), whereas the *surface area* is a function of the square of the radius ($SA = 4\pi r^2$). The surface-to-volume ratio of a sphere can thus be expressed as $3/r$ (Figure 3.14). A cell with a smaller r value therefore has a *higher* ratio of surface area to volume than a larger cell and thus can have a more efficient exchange with its surroundings than a large cell. This advantage of the small cell typically allows for more rapid growth rates and larger populations of prokaryotic cells than eukaryotic cells in most microbial habitats. This in turn affects an organism's ecology in that high numbers of rapidly metabolizing cells can cause major physiochemical changes in an ecosystem over a relatively short period of time. We will pick this theme up again when we consider microorganisms in their natural habitats in Chapters 14 and 19.

We proceed now to consider several cell structures in more detail. Our objective is to describe the chemical building blocks of these cell structures and to explain how they are connected in a way that leads to a defined cellular function. We begin with the cell membrane, a structure that is critical to life processes in the cell.

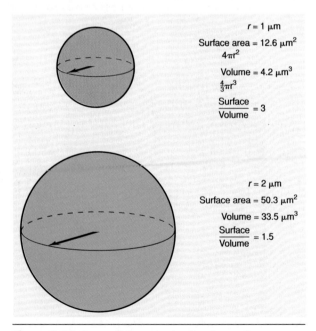

Figure 3.14 As a cell *increases* in size, its surface area-to-volume ratio *decreases*.

CONCEPT CHECK　　　3.2

Although prokaryotes and eukaryotes are distinguished by nuclear structure, other important differences exist between these two cell types. Prokaryotes are smaller in size than eukaryotes, and eukaryotes contain a membrane-enclosed nucleus and organelles within which many important functions are carried out. The small size of prokaryotic cells affects their physiology, growth rate, and ecology.

✔ List three morphological types of prokaryotes.

✔ What is a flagellum and what does it do?

✔ What physical property of cells increases as cells become smaller?

3.3
Cytoplasmic Membrane: Structure

The **cytoplasmic membrane** is a thin structure that completely surrounds the cell. Only 8 nm thick, this vital structure is the critical barrier separating the inside of the cell (the cytoplasm) from its environment. If the membrane is broken, the integrity of the cell is destroyed, the internal contents leak into the environment, and the cell dies. The cytoplasmic membrane is also a *highly selective barrier*, enabling a cell to concentrate specific metabolites and excrete waste materials.

Chemical composition of membranes

The general structure of most biological membranes is a **phospholipid bilayer** (Figure 3.15). As discussed in Section 2.6, phospholipids contain both highly hydrophobic (fatty acid) and relatively hydrophilic (glycerol) moieties and can exist in many different chemical forms as a result of variation in the nature of the fatty acids or phosphate-containing groups attached to the glycerol backbone. As phospholipids aggregate in an aqueous solution, they tend to form bilayer structures spontaneously—the fatty acids point inward toward each other in a hydrophobic environment, and the hydrophilic portions remain exposed to the aqueous external environment (Figure 3.15); the bilayer character of membranes probably represents the most stable arrangement of lipid molecules in an aqueous environment.

Thin sections of the cytoplasmic membrane can be visualized with the electron microscope; a representative example is seen in Figure 3.16a. To prepare the membrane for electron microscopy, cells must first be treated with osmic acid or some other electron-dense material that combines with hydrophilic components of the membrane (Figure 3.16b). By careful

TABLE 3.1	Comparative permeability of membranes to various molecules

Substance	Rate of permeability[a]
Water	100
Glycerol	0.1
Tryptophan	0.001
Glucose	0.001
Chloride ion (Cl⁻)	0.000001
Potassium ion (K⁺)	0.0000001
Sodium ion (Na⁺)	0.00000001

[a]Relative scale—permeability with respect to permeability of water, given as 100.

the membrane in the *same* direction (Figure 3.21). **Antiporters** transport one substance across the membrane in one direction while transporting the second substance in the *opposite* direction (Figure 3.21).

The necessity for carrier-mediated transport mechanisms in microorganisms can readily be appreciated. If diffusion were the only type of transport mechanism available, cells would not be able to acquire the proper concentrations of solutes. In diffusion processes, both the rate of uptake and the intracellular level are proportional to the external concentration. Active transport mechanisms overcome this problem by enabling the cell to accumulate solutes *against* a concentration gra-

dient. As shown in Figure 3.22, carrier-mediated transport shows a saturation effect: if the concentration of substrate in the medium is high enough to saturate the carrier, which is frequently the case even at quite low substrate concentrations, the rate of uptake (and often the internal level as well) becomes maximal.

One characteristic of carrier-mediated transport processes is the *highly specific nature* of the transport event. The binding and carrying of a substance across the membrane resembles an enzyme reaction (∞ Section 4.5). Certain carrier proteins react only with a single kind of molecule, but many show affinities for a chemical *class* of molecules. For instance, there are carriers that transport certain, usually related, amino acids and others that transport a variety of related sugars. This economy in uptake reduces the need for separate transport proteins for every single amino acid or every single sugar the cell needs to transport.

The action and energy requirements of transport proteins

Membrane transport proteins are generally integral proteins, with portions of the protein being exposed to both the cytoplasm and the external environment. Arranged in this fashion, it is possible for solutes bound on the external surface of the cell to be carried through the membrane by a conformational change in the transport protein (Figure 3.23).

Most transport processes are linked to the expenditure of energy and result in a much higher concentration of the transported molecules inside than outside the cell. If a solute is transported by an energy-dependent process, then energy can be used to pump the solute *against* the concentration gradient. Energy can be derived from either high energy phosphate compounds, such as adenosine triphosphate (ATP) (∞ Section 4.8) or by the

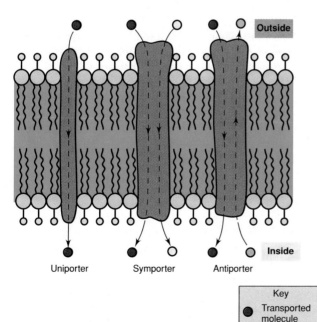

Figure 3.21 Schematic drawing showing the operation of various types of transport proteins.

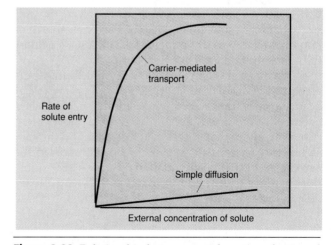

Figure 3.22 Relationship between uptake rate and external concentration in passive uptake and active transport. Note that in the carrier-mediated process the uptake rate shows saturation at relatively low external concentrations.

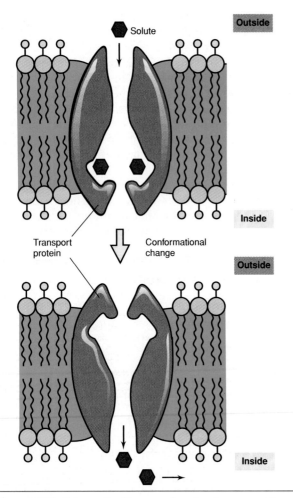

Figure 3.23 Schematic diagram showing how a conformational change in a transmembrane protein could serve to drive the transport event.

dissipation of a gradient of protons or sodium ions across the membrane. The ion gradients are themselves established during energy-releasing reactions in the cell (∞ Chapters 4 and 13) and can be used as a source of potential energy to drive the uptake of solutes against the concentration gradient (see later).

Two major mechanisms of energy-linked transport are known. **Group translocation** is the process whereby a substance is transported while simultaneously being chemically modified, generally by phosphorylation. Alternatively, in the process of **active transport** the substance can accumulate to a high concentration in the cytoplasm in chemically unaltered form. Active transport requires energy and is linked to the energy available in ion gradients or ATP.

Group translocation

Group translocation is a transport process in which the substance is *chemically altered* in the course of passage across the membrane. Since the product that appears inside the cell is chemically different from the external substrate, no actual concentration gradient of the external solute per se is produced across the membrane. The best studied cases of group translocation involve transport of the sugars glucose, mannose, fructose, *N*-acetylglucosamine, and β-glucosides, which are *phosphorylated* during transport by the **phosphotransferase system.**

The phosphotransferase system in the bacterium *Escherichia coli* is composed of 24 proteins, at least 4 of which are necessary to transport a given sugar. The proteins in the phosphotransferase system are themselves alternately phosphorylated and dephosphorylated in a cascading fashion until a transmembrane transport protein called Enzyme II_c receives the phosphate group and phosphorylates the sugar in the actual transport process (Figure 3.24). The high energy phosphate bond that supplies the necessary energy for the phosphotransferase system comes from a key metabolic intermediate called *phosphoenol pyruvate.* A small protein called HPr, the enzyme that phosphorylates it (Enzyme I), and Enzyme II_a are cytoplasmic proteins, whereas Enzymes II_b and II_c are membrane proteins (Figure 3.24). HPr and Enzyme I are nonspecific components of the phosphotransferase system and participate in all phosphotransferase reactions, whereas specific Enzymes II exist for the uptake of each individual sugar.

Other substances transported by group translocation include purines, pyrimidines, and fatty acids. However, many substances, including several sugars, are not taken up by the phosphotransferase system but instead are accumulated by the process of active transport (Table 3.2).

Concerning the energy requirements of the phosphotransferase system, it should be noted that although one high energy phosphate bond (one ATP equivalent) is consumed in the process of transporting the glucose molecule (Figure 3.24), the phosphorylation of glucose is the first step in its intracellular metabolism anyway (∞ Section 4.10). Group translocation systems in general transport molecules in such a way that an early intermediate of the respective biochemical pathway is usually generated. In the case of glucose phosphorylation by the phosphotransferase system, the uptake of glucose is thus essentially energetically neutral.

Active transport

Active transport is an energy-dependent pumping system in which the substance being transported combines with a membrane-bound carrier, which then releases the *chemically unchanged* substance inside the cell. Since the substance is not altered during the transport process, if it is not consumed in cell reactions, its concentration inside may reach many times the external concentration. Substances transported by active transport include some sugars, most amino acids and organic acids, and a number of inorganic ions such as sulfate, phosphate, and

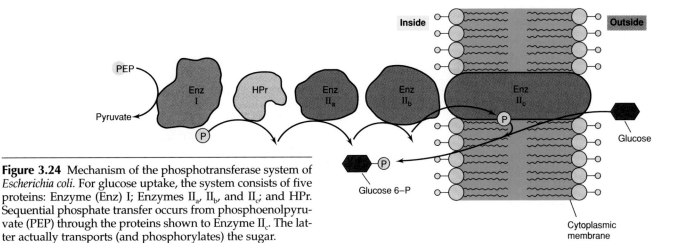

Figure 3.24 Mechanism of the phosphotransferase system of *Escherichia coli*. For glucose uptake, the system consists of five proteins: Enzyme (Enz) I; Enzymes II_a, II_b, and II_c; and HPr. Sequential phosphate transfer occurs from phosphoenolpyruvate (PEP) through the proteins shown to Enzyme II_c. The latter actually transports (and phosphorylates) the sugar.

potassium. Glucose is taken up by active transport processes in some bacteria and by the phosphotransferase system in others (Table 3.2).

As in any other pump, active transport requires that work be performed. In bacteria, the energy for driving the pump comes from ATP in the case of some transporters, or more commonly from the separation of hydrogen ions (protons) across the membrane, called the *proton motive force* (∞ Section 4.13). Energy released from the breakdown of organic or inorganic compounds, or from the energy of light, is used to establish a separation of protons across the membrane, with the proton concentration highest outside the cell and lowest inside (∞ Section 4.13). This results in an energized membrane as depicted in Figure 3.25. It is the electrochemical potential residing in the proton motive force that drives the uptake of nutrients by active transport (Figure 3.25). Each membrane carrier involved in active transport has specific sites for both its substrate (for example, glucose or potassium) and a proton (or protons). As the substrate is taken up, protons move across the membrane and the proton motive force is diminished (Figure 3.25).

The proton motive force is the energy link between membrane transporters and the metabolic machinery, making it possible for the carriers to "pump" nutrients inward. Cations, such as K⁺, may be actively transported into the cytoplasm by uniporters in response to

the proton motive force because the interior of the cell is negative when the membrane is energized (Figure 3.25). Uptake of anions occurs together with that of protons by symporters, and so it is effectively the undissociated acid that enters the cell (Figure 3.25). Excess sodium (Na⁺) within the cell can be pumped out by a sodium–proton antiporter, maintaining the net electric charge across the membrane (Figure 3.25). Transport of uncharged molecules such as sugars or amino acids can also be linked to the electric charge differences: the symporter transports both the substrate and one or more protons. Proton pumps linked to transport are key constituents of all prokaryotic membranes and are also present in the inner membranes of mitochondria and chloroplasts.

Substances taken up by active transport but not linked to dissipation of a proton gradient use the energy of ATP to drive the transport reaction. For example, in *Escherichia coli*, lactose is actively transported at the expense of a proton motive force, whereas the related disaccharide maltose is actively transported at the expense of ATP.

TABLE 3.2	Mechanism of glucose uptake by various bacteria
Phosphotransferase system	**Active transport**
Escherichia coli	*Pseudomonas aeruginosa*
Bacillus subtilis	*Azotobacter vinelandii*
Clostridium pasteurianum	*Micrococcus luteus*
Staphylococcus aureus	*Mycobacterium smegmatis*

CONCEPT CHECK   *3.4*

Because the cytoplasmic membrane is a tight barrier to diffusion, specific transport events function to bring nutrients into the cell. Most transport reactions require energy. In group translocation, the compound transported is chemically modified, while in active transport it is not. The proton motive force is the energy source for most active transport events.

✔ *Why are transport proteins necessary?*

✔ *How does a* symporter *differ from an* antiporter?

✔ *Why can it be said that during group translocation a concentration gradient is not established but that during active transport it is?*

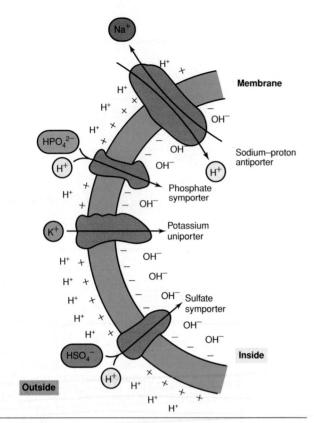

Figure 3.25 Use of ion separation in the proton motive force, in this case the separation of protons from hydroxyl ions across the membrane, to transport inorganic ions by specific transport proteins. Note that there is a separation both of protons and of electrical charge.

3.5

The Cell Wall of Prokaryotes: Peptidoglycan and Related Molecules

Because of the concentration of dissolved solutes inside the bacterial cell, a considerable turgor pressure develops, estimated at 2 atmospheres (atm) in a bacterium like *Escherichia coli*; this is roughly the same as the pressure in an automobile tire. To withstand these pressures, bacteria contain **cell walls,** which also function to give shape and rigidity to the cell. The prokaryotic cell wall is difficult to visualize well with the light microscope but can be readily seen in thin sections of cells with the electron microscope.

Bacteria can be divided into two major groups, called **gram-positive** and **gram-negative.** The original distinction between gram-positive and gram-negative was based on a special staining procedure, the *Gram stain* (see Section 3.1), but differences in cell wall structure are at the base of these differences in the Gram staining reaction. Gram-positive and gram-negative cells differ markedly in the appearance of their cell walls, as is shown in Figure 3.26. The gram-negative cell wall is a multilayered structure and quite complex, whereas the gram-positive cell wall consists of primarily a single type of molecule and is often much thicker. Close examination of Figure 3.26 shows that there is also a significant textural difference between the surfaces of gram-positive and gram-negative Bacteria, as revealed by the scanning electron microscope.

The focus of this section is on the polysaccharide component of the cell walls of prokaryotes, both Bacteria and Archaea. These include, in particular, peptidoglycan, but also a variety of related and unrelated polysaccharides found in Archaea. In Section 3.6 we describe the special wall components found in gram-negative Bacteria.

Peptidoglycan

In the cell walls of Bacteria there is one rigid layer that is primarily responsible for the strength of the wall. In most Bacteria, additional layers are present outside this rigid layer. The rigid layer of both gram-negative and gram-positive Bacteria is very similar in chemical composition. Called **peptidoglycan** (or **murein**), this layer is a thin sheet composed of two sugar derivatives, *N-acetylglucosamine* and *N-acetylmuramic acid*, and a small group of amino acids consisting of L-alanine, D-alanine, D-glutamic acid, and either lysine or diaminopimelic acid (DAP) (Figure 3.27). These constituents are connected to form a repeating structure, the *glycan tetrapeptide* (Figure 3.28).

The basic structure of peptidoglycan is a thin sheet in which the glycan chains formed by the sugars are connected by *peptide cross-links* formed by the amino acids. The glycosidic bonds connecting the sugars in the glycan chains are very strong, but these chains alone cannot provide rigidity in all directions. The full strength of the peptidoglycan structure is realized only when these chains are cross-linked by amino aids. This cross-linking occurs to characteristically different extents in different Bacteria, with greater rigidity coming from more complete cross-linking. In gram-negative Bacteria, cross-linkage usually occurs by direct peptide linkage of the amino group of diaminopimelic acid to the carboxyl group of the terminal D-alanine (Figure 3.29a). In gram-positive Bacteria, cross-linkage is usually by a peptide interbridge, the kinds and numbers of cross-linking amino acids varying from organism to organism. In *Staphylococcus aureus*, the best-studied gram-positive organism, each interbridge peptide consists of five molecules of the amino acid glycine connected by peptide bonds (Figure 3.29b). The overall structure of a peptidoglycan molecule is shown in Figure 3.29c.

In gram-positive Bacteria, as much as 90% of the cell wall consists of peptidoglycan, although another kind of constituent, teichoic acid (see discussion later) is usually present in small amounts. And, although some bacteria are thought to have only a single layer of peptidoglycan surrounding the cell, many Bacteria, especially gram-positive Bacteria, have several (up to about 25) peptidoglycan layers. In gram-negative Bacteria only

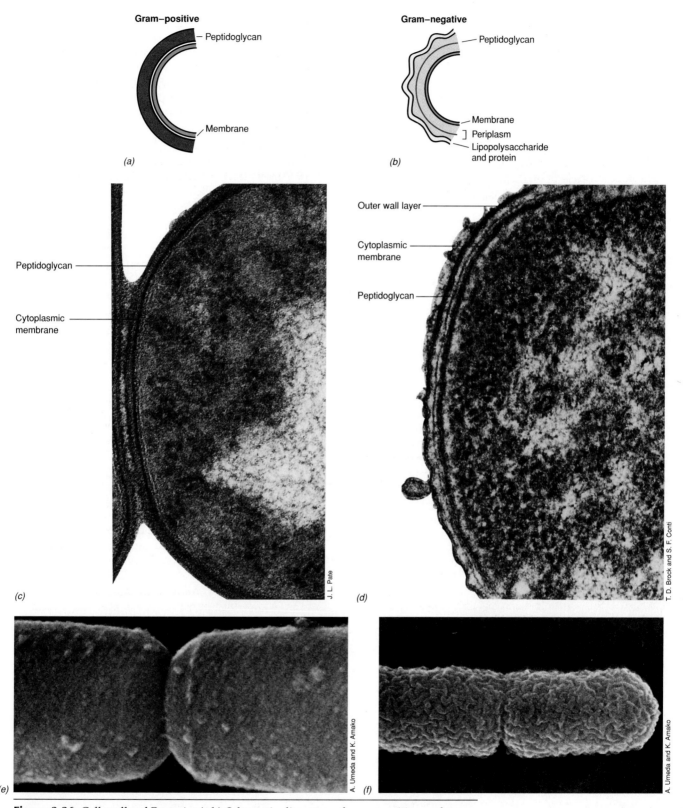

Figure 3.26 Cell walls of Bacteria. (a,b) Schematic diagrams of gram-positive and gram-negative cell walls. (c) Electron micrograph showing the cell wall of a gram-positive bacterium, *Arthrobacter crystallopoietes.* (d) Gram-negative bacterium, *Leucothrix mucor.* (e,f) Scanning electron micrographs of gram-positive (*Bacillus subtilis*) and gram-negative (*Escherichia coli*) Bacteria. Note the surface texture in the cells shown in (e) and (f). A single cell of *B. subtilis* or *E. coli* is about 1 μm in diameter.

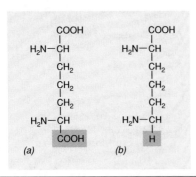

Figure 3.27 (a) Diaminopimelic acid. (b) Lysine.

about 10% of the wall is peptidoglycan, the majority of the wall consisting of a complex layer as discussed in Section 3.6. However, the shape of both gram-positive and gram-negative cells is thought to be determined by the lengths of the peptidoglycan chains and by the manner and extent of cross-linking of the chains.

Diversity in peptidoglycan

Peptidoglycan is present only in Bacteria; the sugar N-acetylmuramic acid and the amino acid diaminopimelic acid are never found in the cell walls of Archaea or Eukarya. However, not all Bacteria have DAP in their peptidoglycan. This amino acid is present in all gram-negative Bacteria and in some gram-positive species, but most gram-positive cocci have lysine instead of DAP, and a few other gram-positive Bacteria have other amino acids. Another unusual feature of the bacterial cell wall is the presence of two amino acids that have the D configuration, D-alanine and D-glutamic acid. As we saw

in Chapter 2, in proteins amino acids are always of the L configuration.

Several generalizations regarding peptidoglycan structure can be made. The glycan portion is uniform, with only the sugars N-acetylglucosamine and N-acetylmuramic acid being present, and these sugars are always connected in β-1,4 linkage. The tetrapeptide of the repeating unit shows major variation only in one amino acid, the lysine–diaminopimelic acid alternation. However, the D-glutamic acid at position 2 can be hydroxylated in some organisms, whereas substitutions occur in amino acids at positions 1 and 3 in a few others.

More than 100 different peptidoglycan types are known, and the greatest variation among them occurs in the interbridge. Any of the amino acids present in the tetrapeptide can also occur in the interbridge, but in addition, a number of other amino acids can be found there, such as glycine, threonine, serine, and aspartic acid. However, certain amino aids are never found in the interbridge: branched-chain amino acids, aromatic amino acids, sulfur-containing amino acids, and histidine, arginine, and proline. Thus, it can be stated that although the precise chemistry of peptidoglycan can vary, the structural makeup of peptidoglycan is the same in all forms of the molecule: glucosamine and muramic acid form the backbone, and the muramic acid molecules are cross-linked with amino acids.

Teichoic acids and a summary of the gram-positive wall

Gram-positive Bacteria frequently have acidic polysaccharides attached to their cell wall called **teichoic acids** (from the Greek word *teichos*, meaning "wall"). The

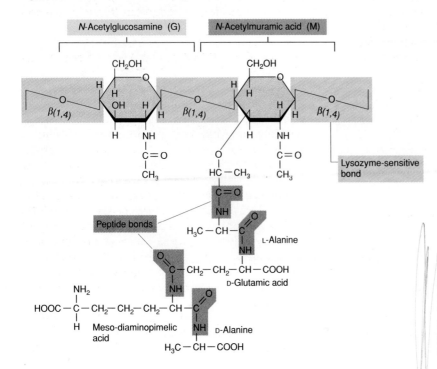

Figure 3.28 Structure of one of the repeating units of the peptidoglycan cell wall structure, the glycan tetrapeptide. The structure given is that found in *Escherichia coli* and most other gram-negative Bacteria. In some Bacteria, other amino acids are found.

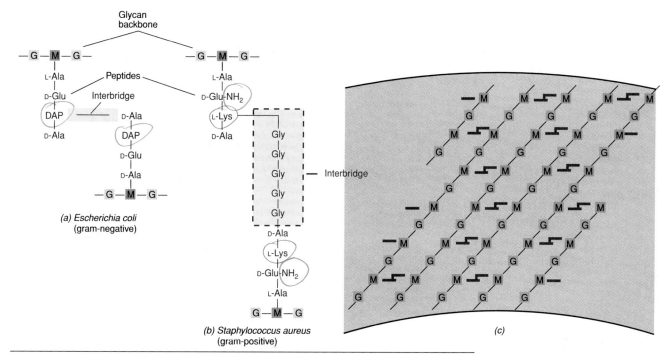

Figure 3.29 Manner in which the peptide and glycan units are connected in formation of the peptidoglycan sheet. (a) Direct interbridge in gram-negative Bacteria. (b) Glycine interbridge in *Staphylococcus aureus* (gram-positive). (c) Overall structure of peptidoglycan. The diagram depicts several ribbons of peptidoglycan cross-linked to one another. To visualize an entire single layer of peptidoglycan, imagine these cross-linked ribbons extending around a cylinder or sphere representing the cell as shown. G, *N*-Acetylglucosamine; M, *N*-acetylmuramic acid; bold lines in (c) indicate peptide cross-links.

term *teichoic acids* includes all wall, membrane, or capsular polymers containing glycerophosphate or ribitol phosphate residues. These polyalcohols are connected by phosphate esters and usually have other sugars and D-alanine attached (Figure 3.30*a*). Because they are negatively charged, teichoic acids are partially responsible for the negative charge of the cell surface as a whole and may function to effect passage of ions through the cell wall. Certain glycerol-containing acids are bound to membrane lipids of gram-positive Bacteria; because these teichoic acids are intimately associated with lipid, they have been called *lipoteichoic acids*.

Figure 3.30*b* summarizes the structure of the cell wall of gram-positive Bacteria and shows how teichoic acids and lipoteichoic acids are arranged in the overall wall structure.

Pseudopeptidoglycan and other cell walls of Archaea

Certain methanogenic Archaea, prokaryotes whose metabolism is linked to the production of natural gas (methane), contain cell walls constructed of a polysaccharide very similar to that of peptidoglycan. This material is called *pseudopeptidoglycan* (Figure 3.31*b*). The backbone of pseudopeptidoglycan is composed of alternating repeats of *N*-acetylglucosamine and *N*-acetyltalosaminuronic acid

(the latter replaces the *N*-acetylmuramic acid of peptidoglycan) (compare Figures 3.28 and 3.31*b*). The backbone of pseudopeptidoglycan also varies from peptidoglycan in that the glycosidic bonds are of the 1,3 type instead of the 1,4 type found in peptidoglycan (Figures 3.28 and 3.31*b*).

Cell walls of other Archaea lack both peptidoglycan and pseudopeptidoglycan and consist of polysaccharide, glycoprotein, or protein. *Methanosarcina* species, also methanogenic Archaea, contain thick polysaccharide walls composed of glucose, glucuronic acid, galactosamine, and acetate. Extremely halophilic (salt loving) Archaea such as *Halococcus* produce walls similar to that of *Methanosarcina* but which also contain an abundance of sulfate (SO_4^{2-}) residues. A structure for the *Halococcus* cell wall is shown in Figure 3.31*c*.

The most common wall type among Archaea is the paracrystalline surface layer (S-layer) (see Section 3.11) consisting of protein or glycoprotein, generally of hexagonal symmetry (Figure 3.31*a*). S-layers have been found among species of all groups of Archaea, the extreme halophiles, the methanogens, and the hyperthermophiles. *Methanospirillum* (∞Figure 17.5*c*) and *Methanothrix* (∞Figure 17.7*d*), two methanogenic Archaea, have extremely complex cell envelopes. These organisms grow as long chains of cells separated from one another by a dense "spacer" region, and this

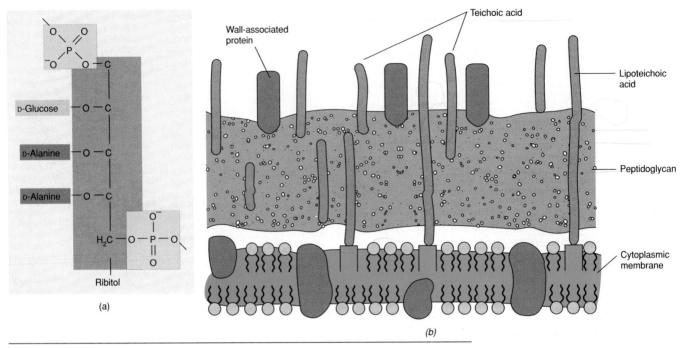

Figure 3.30 Teichoic acids and the overall structure of the gram-positive cell wall. (a) Structure of the ribitol teichoic acid of *Bacillus subtilis*. The teichoic acid is a polymer of the repeating ribitol units shown here. (b) Summary diagram of the gram-positive cell wall.

entire structure is then enclosed within an S-layer sheath.

We thus see in species of Archaea a great variety of cell wall types, varying from molecules that closely resemble peptidoglycan to cell walls totally lacking a polysaccharide component. But with one exception, *Thermoplasma* (∞ Section 17.4), all Archaea contain a cell wall of some sort, and as in Bacteria, the archaeal cell wall functions to prevent osmotic lysis and to define cell shape. In addition, because they lack peptidoglycan in their cell walls, all Archaea are naturally resistant to the action of lysozyme, an enzyme that destroys this molecule (see below).

Protoplast formation

Peptidoglycan, the signature molecule of Bacteria, can be destroyed by certain agents. One such agent is the enzyme **lysozyme,** a protein that breaks the 1,4-glycosidic bonds between *N*-acetylglucosamine and *N*-acetylmuramic acid in peptidoglycan (Figure 3.28), thereby weakening the wall. Water then enters the cell and the cell swells and eventually bursts, a process called **lysis** (Figure 3.32*a*). Lysozyme is found in animal secretions including tears, saliva, and other body fluids and presumably functions as a major line of defense against infection by Bacteria.

If the proper concentration of a solute that does not penetrate the cell, such as sucrose, is added to the medium, the solute concentration outside the cell balances that inside. Under these conditions, lysozyme still digests peptidoglycan, but lysis does not occur, and instead a **protoplast** is formed (Figure 3.32*b*). If such

sucrose-stabilized protoplasts are placed in water, lysis occurs immediately. The word *spheroplast* is often used as a synonym for protoplast, although the two words have slightly different meanings: protoplasts are generally free of residual cell wall material, whereas spheroplasts usually contain pieces of wall material attached to the otherwise membrane-enclosed structure.

Although most prokaryotes cannot survive without their cell walls, a group of Bacteria are able to do so; these are the mycoplasmas, a group that cause certain infectious diseases (∞ Section 16.27). Mycoplasmas are essentially free-living protoplasts and are able to survive without cell walls either because they have unusually tough mem-

---

CONCEPT CHECK   3.5

The cell walls of Bacteria contain a polysaccharide called peptidoglycan. This material consists of strands of alternating repeats of *N*-acetylglucosamine and *N*-acetylmuramic acid, with the latter cross-linked between strands by short peptides. Archaea lack peptidoglycan but contain walls made of other polysaccharides or of protein. The enzyme lysozyme destroys peptidoglycan, leading to cell lysis.

✔ *List the monomeric components of peptidoglycan.*

✔ *Why is peptidoglycan such a strong macromolecule?*

✔ *How does pseudopeptidoglycan resemble peptidoglycan? How do the two molecules differ?*

✔ *How is a protoplast generated?*

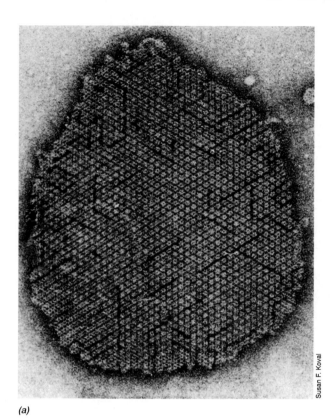

(a)

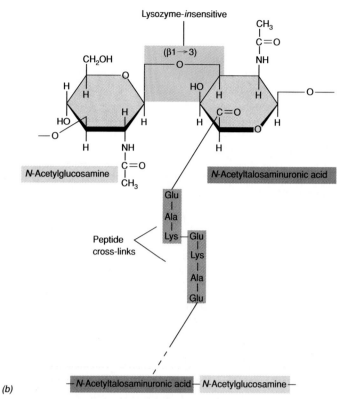

(b)

(c)

Figure 3.31 The S-layer and other cell wall types in Archaea. (a) Transmission electron micrograph of a portion of an S-layer showing the paracrystalline nature of this cell wall layer. Shown is the S-layer from the prokaryote *Aquaspirillum serpens* (a member of the Bacteria); this S-layer displays hexagonal symmetry as do many of the S-layers found in Archaea. (b) Structure of pseudopeptidoglycan, the cell wall polymer of *Methanobacterium* species. Note the resemblance to the structure of peptidoglycan shown in Figure 3.28. (c) Cell wall structure of *Halococcus*, an extremely halophilic archaean. The wall consists of a repeating three-part pattern. Note the abundance of sulfate groups. UA, uronic acid; Glu, glucose; Gal, galactose; GluNac, *N*-acetylglucosamine; GalNAc, *N*-acetylgalactosamine; gly, glycine; GulNUA, *N*-acetylgulosaminuronic acid; Man, mannose.

branes or because they live in osmotically protected habitats, such as the animal body. Certain mycoplasmas have sterols in their cell membranes, which lends strength and rigidity to this structure. *Thermoplasma* (mentioned previously) is the only known archaean to lack a cell wall.

3.6

The Outer Membrane of Gram-Negative Bacteria

Besides, peptidoglycan, gram-negative Bacteria contain an additional wall layer made of **lipopolysaccharide.** This layer is effectively a second lipid bilayer, but it is not constructed solely of phospholipid, as is the cytoplasmic membrane; instead it contains polysaccharide and protein. The lipid and polysaccharide are intimately linked in the outer layer to form specific lipopolysaccharide structures. Because of the presence of lipopolysaccharide, the outer layer is frequently called the **lipopolysaccharide layer,** or simply **LPS.**

Chemistry of the LPS layer

Although complex, the chemical structures of some LPS layers are now understood. As seen in Figure 3.33, the polysaccharide consists of two portions, the core polysaccharide and the O-polysaccharide. In *Salmonella,*

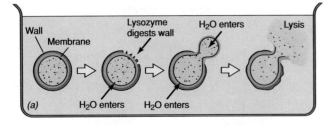

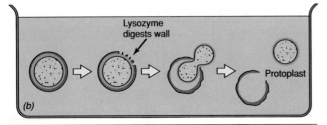

Figure 3.32 Protoplasts. (a) In dilute solution breakdown of the wall releases the protoplast, but it immediately lyses because the cytoplasmic membrane is very weak. (b) In a solution containing a high concentration of a solute such as sucrose, water does not enter the protoplast and it remains stable.

where it has been best studied, the **core polysaccharide** consists of ketodeoxyoctonate (KDO), seven-carbon sugars (heptoses), glucose, galactose, and N-acetylglucosamine. Connected to the core is the *O*-polysaccharide, which usually contains galactose, glucose, rhamnose, and mannose (all six-carbon sugars) as well as one or more unusual dideoxy sugars such as abequose, colitose, paratose, or tyvelose. These sugars are connected in four- or five-membered sequences, which often are branched. When the sugar sequences are repeated, the long *O*-polysaccharide is formed.

The relationship of the *O*-polysaccharide to the rest of the LPS layer is shown in Figure 3.34. The lipid portion of the lipopolysaccharide, referred to as **lipid A** (Figure 3.33) is not a glycerol lipid, but instead the fatty acids are connected by ester amine linkage to a disaccharide composed of N-acetylglucosamine phosphate (Figure 3.33). The disaccharide is attached to the core *O*-polysaccharides through KDO (Figure 3.33). Fatty acids commonly found in lipid A include caproic, lauric, myristic, palmitic, and stearic acids. In the outer membrane, the LPS associates with various proteins to form the *outer* half of the unit membrane structure. A **lipoprotein** complex is found on the *inner* side of the outer membrane of a number of gram-negative Bacteria (Figure 3.34). This lipoprotein is a small (~7200-molecular-weight) protein that serves as an anchor between the outer membrane and peptidoglycan. In the *outer* leaf of the outer membrane, LPS replaces phospholipids; the latter are found predominantly in the inner leaf (Figure 3.34).

Endotoxin

One important biological property of the outer membrane layer of many gram-negative Bacteria is that it is frequently *toxic* to animals. Gram-negative Bacteria that are pathogenic for humans and other mammals include members of the genera *Salmonella*, *Shigella*, and *Escherichia*, among others. The toxic property of the outer membrane layer of these bacteria is responsible for some of the symptoms of infection that they bring about. The toxic properties are associated with part of the lipopolysaccharide layer, in particular, lipid A, of these organisms. The term *endotoxin* is used to refer to this toxic component of LPS, as we will discuss in Section 19.10. However, interestingly, LPS from several nonpathogenic bacteria can also show endotoxin activity. Thus, the organism itself need not be pathogenic to contain toxic cell wall components.

Porins and the periplasm

Unlike the cytoplasmic membrane, the outer membrane of gram-negative Bacteria is relatively permeable to small molecules even though it is basically a lipid

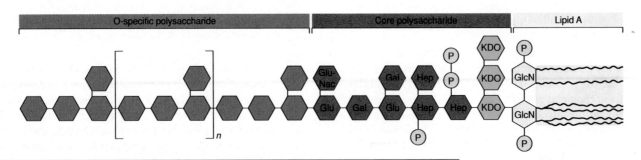

Figure 3.33 Structure of the lipopolysaccharide of gram-negative Bacteria. The precise chemistry of lipid A and the polysaccharide components vary among species of gram-negative Bacteria, but the sequence of major components (lipid A–KDO–core–O-specific) is generally uniform. The O-specific polysaccharide varies enormously among species. KDO, ketodeoxyoctonate; Hep, heptose; Glu, glucose; Gal, galactose; GluNac, N-acetylglucosamine; GlcN, glucosamine. The lipid A portion of the LPS layer can be toxic to animals and comprises the *endotoxin complex* (∞ Section 19.10).

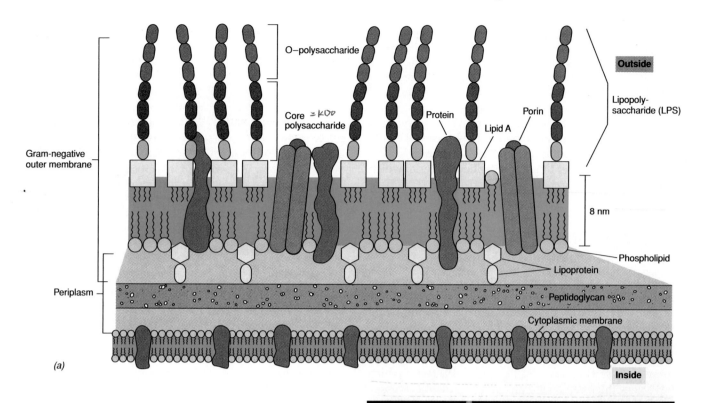

O–polysaccharide

Core = KDO
polysaccharide

Protein

Porin

Lipid A

Outside

Lipopoly-
saccharide (LPS)

Gram-negative
outer membrane

8 nm

Phospholipid

Lipoprotein

Periplasm

Peptidoglycan

Cytoplasmic membrane

(a)

Inside

(b)

Georg E. Schulz

Figure 3.34 Bacterial LPS layer. (a) Arrangement of lipopolysaccharide, lipid A, phospholipid, porins, and lipoprotein in gram-negative Bacteria. See Figure 3.33 for details of the structure of the LPS layer. (b) Molecular model of porin proteins. Note the three pores present, one formed from each of the proteins forming a porin molecule. The view is perpendicular to the plane of the membrane. Model based on X-ray diffraction studies of *Rhodobacter blastica* porin.

bilayer. How, then, can molecules move across the outer membrane? Proteins called **porins** are present in the outer membrane of gram-negative Bacteria, and these proteins function as channels for the entrance and exit of hydrophilic low-molecular-weight substances (Figure 3.34). Several porins have now been identified, and both specific and nonspecific classes are known. *Nonspecific porins* form water-filled channels through which small substances of any type can pass. By contrast, some porins are highly specific because they contain a specific binding site for one or more substances. The largest porins allow entry of substances up to about 5000 molecular weight.

Structural studies have shown that most porins are proteins containing three identical subunits. Porins are transmembrane proteins (Figure 3.34a) and associate to form small membrane holes about 1 nm in diameter

(Figure 3.34b). Apparently a mechanism exists for opening and closing the pores because resistance to certain antibiotics is related to porin structure. Presumably, such pores can be closed to prevent antibiotic uptake.

The outer membrane is thus relatively permeable to small hydrophobic molecules. However, the outer membrane is *not* permeable to enzymes or other large molecules. In fact, one of the major functions of the outer layer may be to keep certain enzymes, which are present outside the cytoplasmic membrane, from diffusing away from the cell. These enzymes are present in a region called the **periplasm** (see Figures 3.34 and 3.35). In *Escherichia coli* this space between the outer surface of the cytoplasmic membrane and the inner surface of the LPS-containing outer membrane occupies a distance of about 12–15 nm and is gellike in consistency, presumably because of the abundance of periplasmic proteins found there (Figure 3.35). The periplasm of gram-negative Bacteria generally contains three types of proteins: *hydrolytic enzymes*, which function in the initial degradation of food molecules, *binding proteins*, which begin the process of transporting substrates, and *chemoreceptors*, which are proteins involved in the chemotaxis response (see Sections 3.10 and 7.7). Periplasmic binding proteins function to bind a substance and take it to the membrane-bound carrier. Binding proteins of this type seem to be absent in gram-positive Bacteria, which also lack a lipopolysaccharide layer and a defined periplasmic space.

Relationship of cell wall structure to the Gram stain

Are the structural differences between the cell walls of gram-positive and gram-negative Bacteria responsible in any way for the Gram stain reaction? In the Gram stain (see Section 3.1), an insoluble crystal violet–iodine complex is formed inside the cell, and this complex is extracted by alcohol from gram-*negative* but not from gram-*positive* Bacteria. Gram-positive Bacteria, which have very thick cell walls consisting of several layers of peptidoglycan, become dehydrated by the alcohol. This causes the pores in the walls to close, preventing the insoluble crystal violet–iodine complex from escaping. In gram-negative Bacteria, alcohol readily penetrates the lipid-rich outer layer, and the thin peptidoglycan layer also does not prevent solvent passage, thus, the crystal violet-iodine complex is easily removed. However, the Gram reaction is not related directly to cell wall chemistry since yeasts, which have a thick cell wall but one of an entirely different chemical composition, also stain gram-positive. Thus, it is not the chemical constituents but the *physical structure* of the wall that is responsible for a gram-positive reaction.

Now that we have a picture of the basic structure of the cell walls of Bacteria and Archaea, we consider how the wall is synthesized, focusing on the peptidoglycan layer from which most of the information on this subject is available. Because final synthesis of this crucial cellular molecule occurs *outside* the cytoplasmic membrane, careful control and coordination of peptidoglycan synthesis is necessary for cellular integrity to remain. We see how this happens now.

CONCEPT CHECK   3.6

In addition to peptidoglycan, gram-negative Bacteria contain an outer membrane consisting of lipopolysaccharide, protein, and lipoprotein. Proteins called porins allow for permeability across the outer membrane, and a space called the periplasm is present, which contains various proteins involved in important cellular functions.

✔ *What components constitute the LPS layer of gram-negative Bacteria?*

✔ *What is the function of porins and where are they located in a gram-negative cell wall?*

✔ *Why does alcohol readily decolorize gram-negative bacteria?*

3.7
Cell Wall Synthesis and Cell Division

When a cell enlarges during the division process, new cell wall synthesis must take place, and this new wall material must be added in some way to the preexisting wall without loss of structural integrity. This process occurs as shown in Figure 3.36a. Small openings in the macromolecular structure of the wall are created by enzymes called **autolysins**, similar in function to lysozyme, which are produced within the cell. New wall material is then added across the openings (Figure 3.36). The junction between new and old peptidoglycan forms a ridge on the cell surface of gram-positive Bacteria (Figure 3.36b), analogous to a scar. Thus, it is essential that new peptidoglycan be spliced onto preexisting peptidoglycan *before* severing bonds within the latter to ensure that cell turgor pressure does not burst the cell wall at a splice point. If this does not take place, a process of spontaneous lysis called **autolysis** can occur.

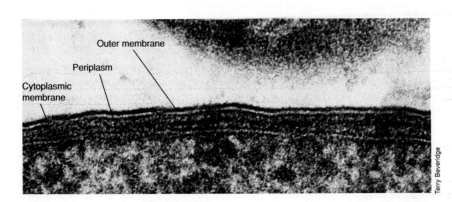

Outer membrane
Periplasm
Cytoplasmic membrane

Figure 3.35 High magnification thin section of the cell envelope of *Escherichia coli* showing the periplasmic gel bounded by the outer membrane and the cytoplasmic membrane. The large, dark particles in the cytoplasm are ribosomes. The thickness of the cytoplasmic membrane is 7.5 nm.

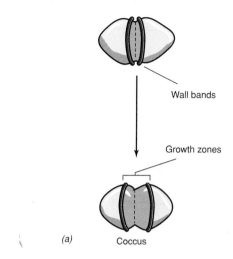

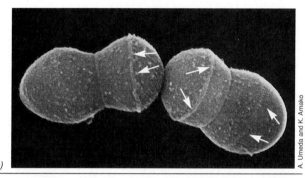

Figure 3.36 Cell wall synthesis in gram-positive Bacteria. (a) Localization of new cell wall synthesis during cell division. In cocci, new cell wall synthesis is localized at only one point. (b) Scanning electron micrograph of *Streptococcus hemolyticus* showing division bridges (arrows). A single cell is about 1 μm in diameter.

Biosynthesis of peptidoglycan

The peptidoglycan layer can be thought of as a stress-bearing fabric, much like a sheet of rubber. Synthesis of new peptidoglycan during cell growth involves controlled cutting by autolysins of bonds connecting small areas of preexisting peptidoglycan, with the simultaneous insertion of new pieces of peptidoglycan, much as a patch is inserted into a piece of woven fabric. As this process continues during cell division, cell volume increases until a cross-wall (septum) forms and the cell divides into two cells (Figure 3.36).

Two carrier molecules participate in peptidoglycan synthesis, uridine diphosphate and a lipid carrier. The lipid carrier, called **bactoprenol,** is a C_{55} isoprenoid alcohol that is connected via phosphodiester linkage to N-acetylmuramic acid to which a pentapeptide is attached (Figure 3.37). The second amino sugar of peptidoglycan, N-acetylglucosamine, is then added followed by addition of the pentaglycine bridge.

The assembly of polymers such as peptidoglycan *outside* the cytoplasmic membrane presents special problems

Figure 3.37 Bactoprenol (undecaprenolphosphate), the lipid carrier of the cell wall peptidoglycan building blocks.

of transport and control. Bactoprenol is involved in transport of peptidoglycan building blocks across the membrane, where the disaccharide pentapeptide is then inserted into a growing point of the cell wall (Figure 3.38). The function of bactoprenol is to render sugar intermediates sufficiently hydrophobic so that they will pass through the hydrophobic cytoplasmic membrane. The lipid carrier inserts the disaccharide pentapeptide complex into the glycan backbone and then moves back inside the cell to pick up another peptidoglycan precursor unit (Figure 3.38).

The final step in cell wall synthesis is formation of the peptide cross-links between adjacent glycan chains. The formation of peptide cross-links involves an unusual type of peptide bond formation, called **transpeptidation,** which is also noteworthy because it is the reaction inhibited by the antibiotic *penicillin*. This cross-linking reaction involves peptide formation with one of several different amino acids, depending on the organism involved. In gram-negative Bacteria, such as *Escherichia coli*, the cross-linking is between diaminopimelic acid (DAP) on one peptide and D-alanine on an adjacent peptide (Figure 3.38b). Initially, there are *two* D-alanine groups at the end of the peptidoglycan precursor, but one D-alanine group is split off during the transpeptidation reaction. The peptide bond between the two molecules of D-alanine serves to *activate* the subterminal D-alanine, thereby favoring its reaction with the DAP (Figure 3.38). This reaction occurs outside the cytoplasmic membrane, where energy is not available, and the transpeptidation reaction replaces the requirement of energy input. In *Staphylococcus aureus* the transpeptidation reaction occurs via the pentaglycine bridge (see Figure 3.29).

Inhibition of transpeptidation by penicillin thus leads to the formation of a weakened peptidoglycan. Further damage to the cell, resulting in lysis and death, occurs as autolysins continue to act, but because new peptidoglycan cross-links cannot occur, the cell wall becomes progressively weaker and osmotic lysis takes place. However, because peptidoglycan *synthesis* must be occurring for penicillin to act, penicillin-induced lysis occurs only in *growing* cells. In nongrowing cells,

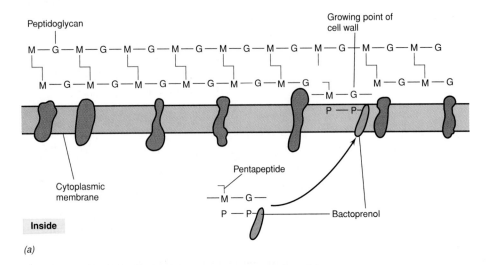

(a)

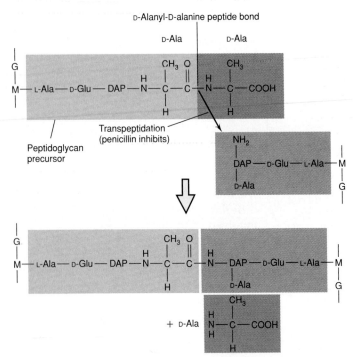

(b)

Figure 3.38 Peptidoglycan synthesis. (a) Transport of peptidoglycan precursors across the cytoplasmic membrane to the growing point of the cell wall. (b) The transpeptidation reaction that leads to the final cross-linking of two peptidoglycan chains. Penicillin inhibits this reaction.

CONCEPT CHECK   3.7

New cell wall is synthesized during bacterial growth by inserting new glycan units into preexisting wall material. A long-chain alcohol called bactoprenol facilitates transport of new glycan units through the cytoplasmic membrane to become part of the growing cell wall.

✔ *What are* autolysins *and why are they necessary?*

✔ *What is the function of bactoprenol?*

✔ *What is* transpeptidation *and why is it important?*

---

the action of autolysins does not occur, and so breakdown of the cell wall peptidoglycan is prevented. It is fascinating to consider that one of the key developments in human medicine, the discovery of penicillin, is linked to a specific biochemical reaction involved in cell wall synthesis, transpeptidation.

3.8
Arrangement of DNA in Prokaryotes

Now that we have developed an understanding of the structure and function of the cytoplasmic membrane and cell wall, we proceed to consider the structure of the genetic material, DNA, that directs the biosynthesis of everything that makes up a cell. We have seen that the cell's genetic information is carried by double-stranded DNA molecules formed of complementary antiparallel strands of polynucleotides (∞ Sections 1.2 and 2.7). As we shall see (in Section 3.14), the cellular DNA in eukaryotes is complexed with several proteins in structures called **chromosomes**. The cellular DNA in prokaryotes is also typically called a chromosome, or a **prokaryotic chromosome**, but except in certain Archaea has far less protein associated with it, being closer to a naked DNA molecule. However, there are some proteins associated with this molecule, and they seem to be important in maintaining structure. There is no nucleus in prokaryotic cells, and the bacterial chromosome is not separated from the rest of the cell by a membrane. In addition to the chromosome, one or more small circular DNA molecules, called *plasmids*, may be present in prokaryotic cells (∞ Sections 6.3 and 8.24).

Supercoiling and chromosome structure

The chromosome of prokaryotes is typically a covalently closed *circular* molecule, although some Bacteria are known that have linear chromosomal DNA (∞Section 6.3). The total amount of DNA in the chromosome of a bacterium such as *Escherichia coli* is about 4700 kilobase pairs. Not surprisingly, this is considerably less than that of eukaryotic cells, but it is greater than that of viruses or organelles (Figure 3.39). Although prokaryotic DNA is not confined to a nucleus as it is in eukaryotes, it does tend to aggregate as a distinct structure within the cell and is visible when observed with the electron microscope (Figure 3.40). The term **nucleoid** is used to describe aggregated DNA in the prokaryotic cell (Figure 3.40) and under special staining conditions the nucleoid can actually be observed in cells examined with just the light microscope (Figure 3.41). Ribosomes are

(a)
(b)

Figure 3.40 The bacterial nucleoid. (a) Transmission electron micrograph of a thin section of *Escherichia coli*. (b) Same as (a), but with the nucleoid colored.

absent in the region of the cell where the bacterial nucleoid is situated, probably because nucleoid DNA exists in a gellike form that tends to exclude particulate matter.

If gently lysed, DNA can be released from prokaryotic cells (Figure 3.42), and the extensive folding and twisting necessary to store the DNA in the cell becomes readily apparent. The amount of twisting and folding can be appreciated when it is considered that the 4.7 *million* base pairs in the genome of *Escherichia coli*, if opened and linearized, would be about 1 *mm* in length, yet the *E. coli* cell is only about 2–3 μm long! To package this much DNA into the cell requires that the DNA be **supercoiled** (Figure 3.43; ∞ Section 6.2). Supercoiled DNA takes on a considerably more compact shape than its

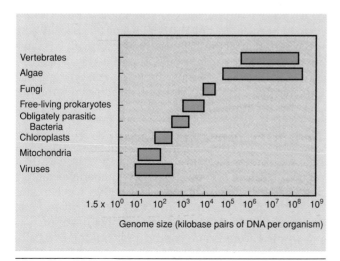

Figure 3.39 Range of genome sizes in various groups of organisms.

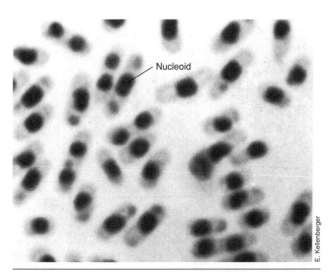

Figure 3.41 Light photomicrograph of cells of *Escherichia coli* treated in such a way as to make the nucleoid visible.

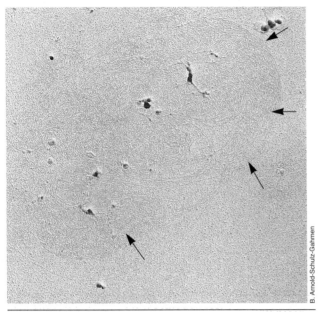

B. Arnold-Schulz-Gahmen

Figure 3.42 Electron micrograph of an isolated nucleoid from a cell of *Escherichia coli*. Cells were gently lysed to allow the highly compacted nucleoid to emerge intact. Arrows point to edge of strands.

freely circularized counterpart. However, the bacterial chromosome is not a simple supercoil as shown in Figure 3.43. There are over 50 **supercoiled domains** in the *E. coli* chromosome, and these domains are stabilized by association with structural proteins (Figure 3.44). This structure allows the very long DNA molecule to be folded and twisted to fit into the cell. In some of the Archaea the chromosomal DNA is extensively complexed with proteins, to an extent very similar to that found in the eukaryotic chromosome (see Section 3.14).

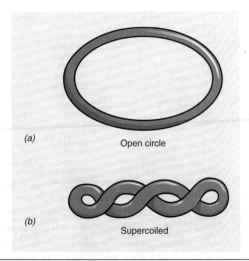

Figure 3.43 The bacterial chromosome. (a) Open circular form. (b) Supercoiled form. Note that in either case the DNA is present in a covalently closed form typical of most prokaryotes.

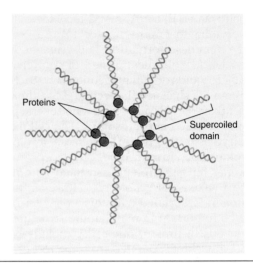

Figure 3.44 The double-stranded DNA in the chromosome of the bacterium *Escherichia coli* contains over 50 supercoiled domains, each of which is stabilized by binding to specific proteins.

Chromosomal copy number

Rapidly growing prokaryotic cells usually contain multiple copies or partially completed copies (generally two to four) of the bacterial chromosome, and only when cell growth has ceased does the chromosome number approach one per cell. The reason for this is that rapidly growing cells can actually divide faster than the DNA replication machinery can make new copies of the bacterial chromosome. Thus, to ensure that a complete copy of the bacterial chromosome is ready for each daughter cell at cell division, new rounds of DNA synthesis are initiated before the old round is completed. This leads to multiple copies or partial copies in rapidly growing cells.

However, organisms like prokaryotes that reproduce asexually are typically **haploid** in genetic complement; that is, the cell's minimum genome contains a single copy of each chromosome. Because most prokaryotes seem to have only a single chromosome (∞ Section 6.3), this means that all the cell's essential information is present on this single chromosome. Typically this means one copy of each gene. As we just discussed, fast-growing bacteria may have multiple copies of their chromosome, but these are all duplicates.

Genetic exchange in prokaryotes

In spite of their haploid nature and the fact that their reproduction is asexual, extensive genetic exchange processes exist in prokaryotic cells, but the mechanisms are quite distinct from the process in eukaryotes. First, the process is quite fragmentary, almost never involving whole chromosome complements of the two cells. Second, the DNA is transferred in only one direction, from a donor to a recipient. Third, the mechanisms by which DNA transfer occurs are specialized. Three distinct types of mechanisms for DNA transfer have been recognized: (1) *conjugation*, in which DNA transfer occurs as a result of cell-to-cell contact (the process that most closely resembles sex in eukaryotes); (2) *transduction*, in which DNA transfer is mediated by viruses; and (3) *transformation*, in which free DNA is involved. In transformation, the donor cell generally lyses, releasing DNA into the medium, and some of this free (naked) DNA is taken up by recipient cells. All three mechanisms of gene transfer have been shown to occur in certain Archaea as well as Bacteria. We discuss the details of these various DNA transfer processes in Chapter 9.

CONCEPT CHECK 3.8

The DNA of the typical prokaryotic cell exists in a very long, single, circular molecule, called the prokaryotic chromosome, which is present in the cell in a highly aggregated state called the nucleoid. The nucleoid is not surrounded by a membrane and is present free in the cytoplasm. It is in a highly supercoiled form.

✔ *What is a prokaryotic chromosome and how many chromosomes does a prokaryote have?*

✔ *What is a nucleoid?*

3.9
Flagella and Motility

Many prokaryotes are motile, and this ability to move independently is usually due to a special structure, the **flagellum** (plural, **flagella**) (Figure 3.45). Certain bacterial cells can move along solid surfaces by *gliding* (∞ Section 16.13), and certain aquatic microorganisms can regulate their position in a water column by gas-filled structures called gas vesicles (see Sections 3.12 and 16.2).

However, the majority of motile prokaryotes move by means of flagella. Motility allows the cell to reach different regions of its microenvironment. In the struggle for survival, movement to a new location may mean the difference between survival and death of the cell. But, as in any physical process, cell movement is closely tied to an energy expenditure. We begin now with a detailed consideration of flagellar motility in prokaryotes.

Bacterial flagella

Bacterial flagella are long, thin appendages free at one end and attached to the cell at the other end. They are so thin (about 20 nm) that a single flagellum can never be seen directly with the light microscope but only after staining with special flagella stains that increase their diameter (Figure 3.45; see also Figure 3.47). Flagella are also readily seen with the electron microscope (Figure 3.46).

Flagella are arranged differently on different bacteria. In **polar flagellation** the flagella are attached at one or both ends of the cell. Occasionally a tuft (group) of flagella may arise at one end of the cell, an arrangement called *lophotrichous* (*lopho* means "tuft"; *trichous* means "hair") (Figure 3.47). Tufts of flagella of this type can be seen in living cells by dark-field microscopy (see Section 3.1 and Figure 3.48a), and using a laser source with dark-field microscopy, even a single flagellum can be observed; in both of these methods the flagella appear light and are attached to light-colored cells against a dark background. In extremely large prokaryotes, tufts of flagella can also be observed by phase contrast microscopy (Figure 3.48b). In **peritrichous flagellation** the flagella are inserted at many places around the cell surface (*peri* means "around"). The type of flagellation, polar or peritrichous, is often used as a characteristic in the classification of bacteria (see Figure 3.47).

Flagellar structure

Flagella are not straight but helically shaped; when flattened, they show a constant distance between two adjacent curves, called the *wavelength*, and this wavelength is

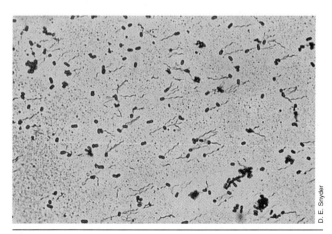

Figure 3.45 Photomicrograph of cells of the bacterium *Proteus mirabilis* stained to show flagella.

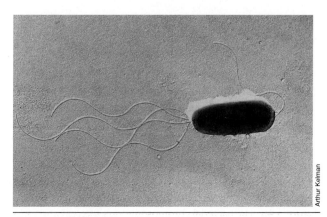

Figure 3.46 Electron micrograph of a bacterial cell, showing flagella. The cell is approximately 3 μm long.

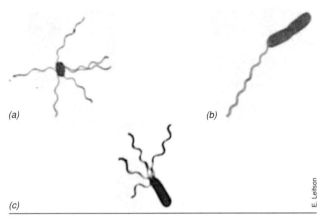

(a)

(b)

(c)

E. Leifson

Figure 3.47 Light photomicrographs of prokaryotes containing different flagellar arrangements. Cells are stained with Leifson flagella stain. (a) Peritrichous. (b) Polar. (c) Lophotrichous.

constant for a given organism (Figures 3.46–3.48). The filament of bacterial flagella is composed of subunits of a protein called **flagellin.** The shape and wavelength of the flagellum are in part determined by the structure of the flagellin protein and also to some extent by the direction of rotation of the filament. The basic flagellar structure to be described here varies little among prokaryotes, as the basic architecture and mechanism of flagellar action are very similar in species of both Bacteria and Archaea. Even the structure of flagellin is highly conserved among related groups of bacteria, further suggesting that flagellar motility has deep evolutionary roots.

The base of the flagellum is different in structure from that of the filament (Figure 3.49). There is a wider

region at the base of the flagellum called the *hook.* The hook consists of a single type of protein and functions to connect the filament to the motor portion of the flagellum. This motor, called the *basal body*, is anchored in the cytoplasmic membrane and cell wall. The basal body consists of a small central rod that passes through a system of rings. In gram-negative Bacteria, an outer ring is anchored in the lipopolysaccharide layer and another in the peptidoglycan layer of the cell wall, and an inner ring is located within the cytoplasmic membrane (Figure 3.49). In gram-positive Bacteria, which lack the outer lipopolysaccharide layer, only the inner pair of rings is present. Surrounding the inner ring and anchored in the cytoplasmic membrane are a pair of proteins called *Mot* (Figure 3.49). These proteins actually drive the flagellar motor causing rotation of the filament. A final set of proteins, called the *Fli* proteins (Figure 3.49) function as the motor switch, reversing rotation of the flagella in response to intracellular signals.

Several genes are required for flagellar synthesis and subsequent motility. In *Escherichia coli* and *Salmonella typhimurium*, where studies have been most extensive, over 40 genes, called *fla*, *fli*, and *flg*, are necessary for motility. These genes have several functions, including encoding structural proteins of the flagellar apparatus, export of flagellar components through the membrane to the outside of the cell, and regulation of the many biochemical events surrounding the synthesis of new flagella. Control of flagella synthesis is tightly regulated in the cell both by metabolic factors and by signals emerging from the cell division cycle.

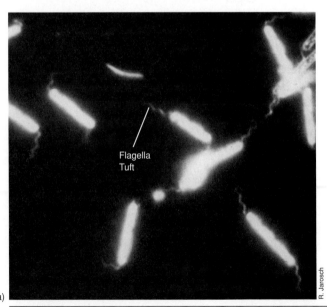

Flagella
Tuft

R. Jarosch

(a)

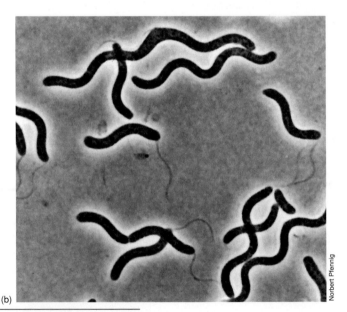

Norbert Pfennig

(b)

Figure 3.48 Bacterial flagella as observed in living cells. (a) Dark-field photomicrograph of a group of large rod-shaped bacteria with flagellar tufts at each pole. A single cell is about 2 μm wide. (b) Phase contrast photomicrograph of the large phototrophic purple bacterium *Rhodospirillum photometricum.* A single cell measures about 3 × 30 μm.

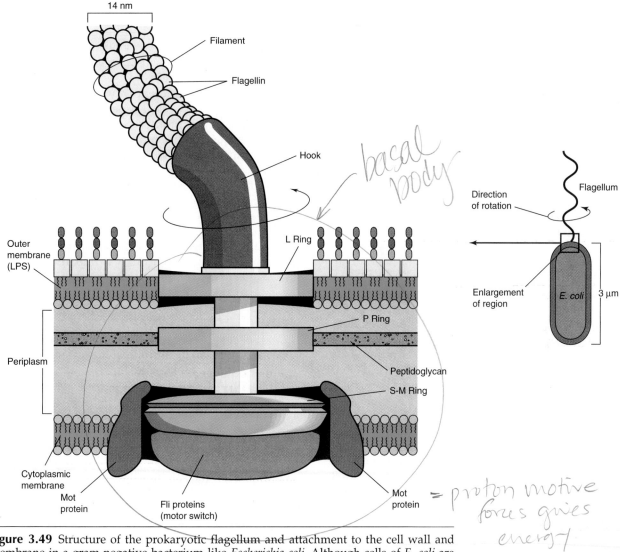

Figure 3.49 Structure of the prokaryotic flagellum and attachment to the cell wall and membrane in a gram-negative bacterium like *Escherichia coli*. Although cells of *E. coli* are peritrichously flagellated, for simplicity, only a single flagellum is shown. The L ring is embedded in the LPS layer, and the P ring in peptidoglycan. The S-M ring is embedded in the cytoplasmic membrane. The Mot proteins function as the flagellar motor, whereas the Fli proteins function as the motor switch. See text for further details.

Flagellar movement

How is motion imparted to the flagellum? Each individual flagellum is actually a semirigid structure that does not flex but, as mentioned previously, moves by rotation, in the manner of a propeller. Visual evidence of this can be obtained by observing the behavior of motile cells tethered by their flagella to microscope slides. Such cells rotate around the point of attachment at rates of revolution consistent with those inferred for flagellar movement in free-swimming cells.

The rotary motion of the flagellum is imparted from the basal body, which functions as a motor. The energy required for rotation of the flagellum comes from the proton motive force (see Sections 3.4 and 4.13).

Proton movement across the membrane through the Mot complex (Figure 3.49) drives rotation of the flagellum, and calculations have shown that about 1000 protons must be translocated per single rotation of the flagellum.

Flagella do not rotate at a constant speed but instead can increase or decrease their rotational speed in relation to the strength of the proton motive force. Flagella rotation can move bacteria through liquid media at speeds of up to 60 cell lengths/second (sec). Although this is only about 0.00017 kilometer/hour (km/hr), when comparing this speed with that of higher organisms in terms of the number of lengths moved per second, it is extremely fast. The fastest animal, the cheetah, moves at a maximum rate of about 110 km/hr, but this represents only about 25 body lengths/sec. Thus, when size is accounted for, prokaryotic cells swimming at 50–60 lengths/sec are actually moving much faster than larger organisms.

The motions of polar and lophotrichous organisms are different from those of peritrichous organisms. Peritrichously flagellated organisms generally move in a straight line in a slow, stately fashion. Polar organisms, on the other hand, move more rapidly, spinning around and dashing from place to place. The different behavior of flagella on polar and peritrichous organisms is illustrated in Figure 3.50.

Flagellar growth

The individual flagellum grows not from the base, as does an animal hair, but from the tip. Flagellin molecules formed in the cell pass up through the hollow core of the flagellum and add on at the terminal end. The synthesis of a flagellum from its flagellin protein molecules occurs by a process called *self-assembly:* all the information for the final structure of the flagellum resides in the protein subunits themselves. Growth of the flagellum occurs more-or-less continuously, although the rate of growth slows as the filament elongates. However, if a portion of the tip is broken off, it is regenerated.

When a cell divides, the two daughter cells must acquire a full complement of flagella. In polarly flagellated organisms, the process of cell division probably occurs as shown in Figure 3.51, the new flagellum forming from the oldest pole. In a monopolarly flagellated cell such as *Caulobacter*, where this process has been studied in detail, the two poles of the cell probably differ in some way, so the flagellum is formed at one pole and not at the other. In peritrichously flagellated organisms, at cell division preexisting flagella are distributed equally between the two daughter cells and new flagella are synthesized and fill in the gaps.

Motility in eukaryotes

Many eukaryotic cells are motile, and two types of organelles of motility are recognized: flagella and cilia (Figure 3.52). **Flagella,** as just discussed, are long, filamentous structures, but in eukaryotes flagella are constructed in a much different way than in prokaryotes and they move in a whiplike manner instead of rotating like the flagella in prokaryotes. Eukaryotic flagella are much larger than those of prokaryotes and are composed of protein structures called **microtubules.** Movement is imparted to the flagellum by coordinated sliding of the several microtubules present within each flagellum. Energy for microtubule sliding is supplied by ATP.

Cilia are similar to eukaryotic flagella in fine structure but differ in being shorter and more numerous (Figure 3.52). Cilia function much like oars in a rowboat; these rigid structures beat in synchrony to impart rapid movement to the ciliated cell. In microorganisms, cilia are found

---

CONCEPT CHECK   3.9

Motility in microorganisms is commonly associated with flagella. In prokaryotes the flagellum is a complex structure made of several proteins, most of which are anchored in the cell wall and membrane. The flagellum filament, which is made of a single kind of protein, rotates at the expense of a proton gradient, which drives the flagellar motor. Flagella in eukaryotes differ in structure and function from those of prokaryotes, moving in a whiplike motion imparted by sliding microtubules.

✔ *How does* polar flagellation *differ from* peritrichous flagellation?

✔ *What is* flagellin *and where is it found?*

✔ *How does a bacterial flagellum move a cell forward?*

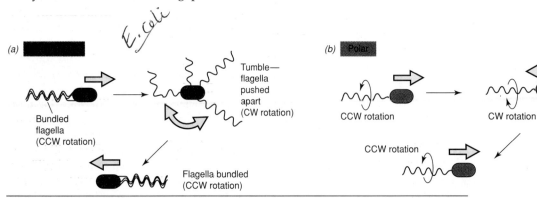

Figure 3.50 Manner of movement in polarly and peritrichously flagellated prokaryotes. (a) Peritrichous: forward motion is imparted by all flagella rotating counterclockwise (CCW) in a bundle. Clockwise (CW) rotation causes the cell to tumble, and then a return to counterclockwise rotation leads the cell off in a new direction. (b) Polar: cells change direction by reversing flagellar rotation (thus pulling instead of pushing the cell) and then return to pushing. The large yellow arrows show the direction the cell is traveling.

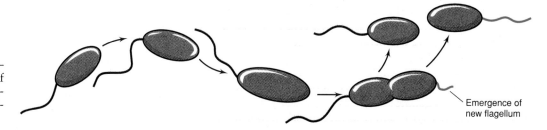

Figure 3.51 Synthesis of flagella during cell division of a polarly flagellated bacterium.

Emergence of new flagellum

primarily in one group of protozoa called the *ciliates;* we discuss the biology of these organisms in Chapter 18.

3.10
Bacterial Behavior: Chemotaxis, Phototaxis, and Other Taxes

Although not all prokaryotes are motile, many are, and it is reasonable to assume that motility confers a selective advantage on cells under certain environmental conditions. Prokaryotes encounter *gradients* of physical and chemical agents in nature, and the motility machinery in the cell is designed to respond in a positive or negative way to these gradients by directing movement of the cell either toward or away from the signal molecule, respectively. Such directed movements are called *taxes*, and a variety of such responses occur in microorganisms. **Chemotaxis,** a response to chemicals, and **phototaxis,** a response to light, are two well-known taxes, and we focus on these here.

Chemotaxis

To understand chemotaxis we can focus on the behavior of a single bacterial cell faced with a chemical gradient of an attractant (Figure 3.53). Unlike larger organisms, prokaryotes are too small to sense a gradient along their body length. They must instead, while moving, compare the chemical or physical state of their environment with that sensed a few seconds before. In other words, bacteria respond to the *temporal* (rather than *spatial*) *gradient* of signal molecules as they swim along.

In the absence of a gradient cells move in a random fashion that includes **runs,** where the cell is swimming

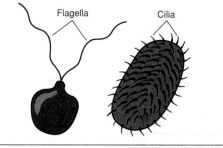

Figure 3.52 Cilia and flagella, organs of motility in eukaryotic cells.

Flagella   Cilia

forward in a smooth fashion, and **tumbles,** when the cell stops and jiggles about (Figure 3.53*a*). Following a tumble, the direction of the next run is random (Figure 3.53*a*). Thus, by means of runs and tumbles the organism moves about randomly in its environment but does not really go anywhere. However, if a gradient of a chemical attractant is present, these random movements become biased. As the organism senses higher concentrations of the attractant (through periodic sampling of the concentration of the chemical in its environment), runs become longer and tumbles less frequent. The net result of this behavior is that the organism moves up the concentration gradient of the attractant (Figure 3.53*b*). If the organism is sensing a repellant, the same general mechanism applies, although in this case it is the *decrease* in concentration of the repellant (rather than the *increase* in concentration of an attractant) that promotes runs. Forward movement in a run occurs when the flagellar motor is rotating counterclockwise. When the flagella rotate clockwise, the bundle pushes apart, forward motion ceases, and the cells tumble (Figure 3.50).

Bacterial chemotaxis can be most easily demonstrated by immersing a small glass capillary containing an attractant in a suspension of motile bacteria that does not contain the attractant. From the tip of the capillary, a gradient is set up into the surrounding medium, with the concentration of chemical gradually decreasing with distance from the tip (Figure 3.54). If the capillary contains an attractant, the bacteria will move toward the capillary, forming a swarm around the open tip (Figure 3.54); subsequently many of the motile bacteria will move into the capillary. Some bacteria will move into the capillary even if it contains a solution of the same composition as the medium because of random movements. But if an attractant is present, the concentration of bacteria within the capillary can be many times higher than the external concentration. On the other hand, if the capillary contains a repellant, the concentration of bacteria within the capillary will be considerably less than the concentration outside (Figure 3.54). Using this simple method, it is possible to rapidly screen chemicals for their ability to act as attractants or repellants for a given bacterium.

How do bacteria use temporal changes in chemical concentrations to control flagellar rotation? This is a complex story and involves several regulatory events at the genetic and biochemical levels. We therefore reserve

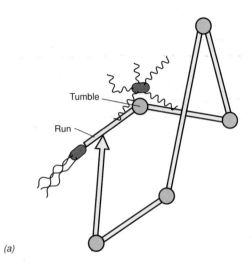

(a)

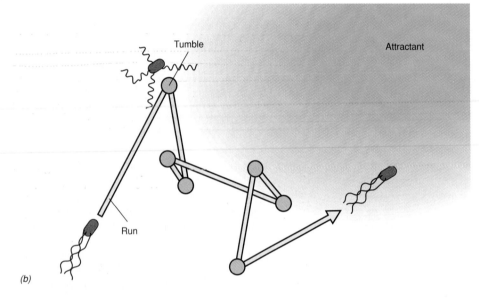

(b)

Figure 3.53 Chemotaxis. (a) In the absence of a chemical attractant the cell swims randomly in runs, changing direction during tumbles. (b) In the presence of an attractant runs become biased, and the cell moves up the gradient of the attractant.

detailed discussion of the mechanism of chemotaxis for Chapter 7 (∞Section 7.7) where this topic can be better understood in a biochemical-genetic context. However, suffice it for now to say that the molecular mechanism of chemotaxis involves sensory proteins in the membrane called *chemoreceptors* that sense the chemical gradient with time and interact with cytoplasmic proteins to affect flagellar motor direction. Thus, because the direction of flagellar rotation governs whether the cell runs or tumbles, chemotaxis can be thought of as a chemically driven *sensory response system* affecting flagellar function.

Phototaxis

Many phototrophic (photosynthetic) microorganisms move toward light, a process called *phototaxis*. The advantage of phototaxis is that it allows a phototrophic organism to orient itself for the most efficient photosynthesis. This can be shown if a light spectrum is spread

across a microscope slide on which there are motile phototrophic bacteria; the bacteria accumulate at the wavelengths at which their photosynthetic pigments absorb (Figure 3.55) (∞Section 13.3 for a discussion of photosynthetic pigments).

Two different taxes are observed in phototrophic prokaryotes. One, called *scotophobotaxis*, can be observed only microscopically and occurs when a phototrophic bacterium happens to swim outside the illuminated field of view of the microscope into darkness. This event signals the cell to tumble, reverse direction, and once again swim in a run, thus reentering the light. The fact that phototrophic cells show scotophobic behavior and accumulate in regions of the spectrum in which their pigments absorb (Figure 3.55a) strongly suggests that the scotophobotactic response is somehow triggered by changes in energy generation (ATP synthesis or the state of the proton motive force). However, the actual mechanism of scotophobotaxis is yet to be elucidated.

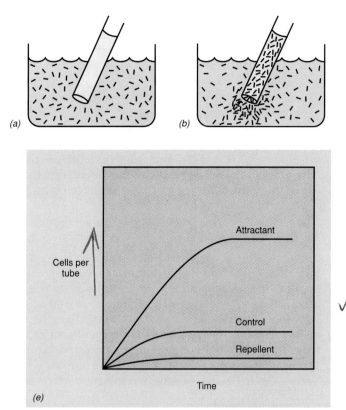

(e)

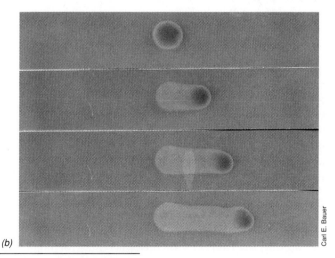

Figure 3.54 Capillary technique for studying chemotaxis in bacteria. (a) Insertion of capillary into a bacterial suspension. (b) Accumulation of bacteria in a capillary containing an attractant. (c) Control capillary contains a salt solution that is neither an attractant nor a repellent. Cell concentration inside the capillary becomes the same as that outside. (d) Repulsion of bacteria by a repellant. (e) Time course showing cell numbers in capillaries containing various chemicals.

However, in addition to scotophobotaxis, phototrophic microorganisms can carry out true phototaxis, a directed movement up a light gradient toward an increasing intensity of light. This can be thought of as analogous to positive chemotaxis except that in this case the attractant is not a chemical but instead is light. In some species, such as the highly motile phototrophic organism *Rhodospirillum centenum*, *entire colonies* of cells show phototaxis and move in unison toward the light (Figure 3.55*b*). Although the molecular details of pho-

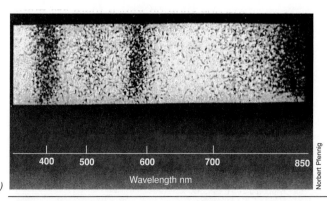

Figure 3.55 Phototaxis. (a) Scotophobic accumulation of the phototrophic bacterium *Thiospirillum jenense* at light wavelengths at which its pigments absorb. A light spectrum was displayed on a microscope slide containing a dense suspension of the bacteria; after a period of time, the bacteria had accumulated selectively and the photomicrograph was taken. The wavelengths at which accumulations occur are those at which bacteriochlorophyll *a* absorbs (compare with Figure 13.4*b*). (b) Phototaxis of an entire colony of the purple phototrophic bacterium *Rhodospirillum centenum* over a 2-hr time course (time 0 at top). These strongly phototactic cells move in unison toward the light source on the right.

totaxis are not yet known, there is good evidence that several parts of the regulatory system that govern chemotaxis are also involved in phototaxis. These include in particular cytoplasmic proteins (Che proteins) that control the direction of rotation of the flagella (∞Section 7.7). This evidence has emerged from the study of mutants of phototrophic bacteria defective in phototaxis; such mutants frequently have defective chemotaxis systems as well. A *photoreceptor,* analogous to a chemoreceptor but able to sense a gradient of *light* instead of chemicals, is responsible for orchestrating the phototaxis response. It is hypothesized that the photoreceptor can in some way interact with the proteins that affect flagella rotation to maintain the cell in a run if it is swimming toward an increasing intensity of light.

Other taxes

Other bacterial taxes, such as movement toward or away from oxygen or toward or away from conditions of high ionic strength, are also beginning to be understood in molecular terms now that some of the general principles of sensory response systems have been elucidated, primarily from work on chemotaxis. In most of these cases a common mechanism applies: cells periodically sample their environment and process this information through a signal transduction pathway (∞Section 7.7) that leads to control of the direction of flagellar rotation. These can be considered simple behavioral responses, and a rationale for elucidating the molecular mechanisms of bacterial taxes is to gain a better understanding of similar responses, such as nerve transmission, in higher organisms. Thus, from a behavioral point of view, motile prokaryotes are well attuned to the chemical and physical state of their environment and as such can move toward or away from various stimuli presumably as a means of remaining competitively successful.

CONCEPT CHECK **3.10**

Motile bacteria can respond to chemical and physical gradients in their environment, and the processes of chemotaxis and phototaxis are good examples of this. Random movement of a prokaryotic cell can be biased either toward or away from a chemical (or toward light in a phototrophic microorganism) by controlling the degree to which runs or tumbles occur. The latter are controlled by the direction of rotation of the flagellum, which in turn is controlled by a network of sensory and response proteins.

✔ *Define the word* chemotaxis.

✔ *What causes a* run *versus a* tumble?

✔ *How does* scotophobotaxis *differ from* phototaxis?

3.11
Bacterial Cell Surface Structures and Cell Inclusions

Prokaryotes can produce a variety of structures that are attached to or in some way protrude from the cell surface, and also several different types of internal cell structures. We survey these structures here, but it should be understood that not all bacteria will contain all these structures and many bacteria may contain none of them. Thus, these are *optional* structures produced by some kinds of prokaryotes but not others.

Fimbriae and pili

Fimbriae and pili are structurally similar to flagella but are not involved in motility. **Fimbriae** are considerably shorter than flagella and are more numerous (Figure 3.56) but, like flagella, consist of protein. Not all organisms have fimbriae, and the ability to produce them is an inherited trait. The functions of fimbriae are not known for certain in all cases, but there is some evidence that they enable organisms to stick to surfaces including animal tissues in the case of some pathogenic bacteria or to form pellicles or scums on the surfaces of liquids.

Pili are similar structurally to fimbriae but are generally longer, and only one or a few pili are present on the surface. Pili can be visualized under the electron microscope because they serve as specific receptors for

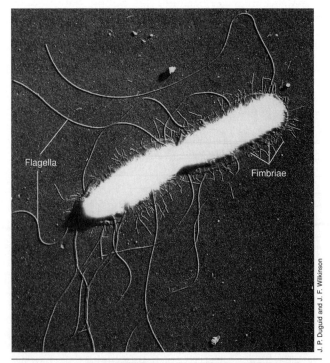

Figure 3.56 Electron micrograph of a dividing cell of *Salmonella typhi,* showing flagella and fimbriae. A single cell is about 0.9 μm in diameter.

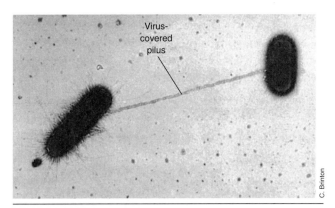

Virus-covered pilus

C. Brinton

Figure 3.57 The presence of pili on an *Escherichia coli* cell is revealed by the use of viruses that specifically adhere to the pilus. The cell is about 0.8 μm in diameter.

certain types of virus particles, and when coated with virus they can be easily seen (Figure 3.57). There is strong evidence that pili are involved in the process of conjugation in prokaryotes, as will be discussed in Section 9.9. Pili are also involved in attachment to human tissues by some pathogenic bacteria.

Paracrystalline surface layers (S-layers)

Many prokaryotes contain a cell surface layer composed of a two-dimensional array of protein. These layers are called *S-layers*. S-layers have been detected in representatives of virtually every phylogenetic grouping of Bacteria and are nearly universal among Archaea. In some species of Archaea the S-layer is also the cell wall (see Section 3.5). S-layers have a crystalline appearance and show various symmetries, such as hexagonal, tetragonal, or trimeric, depending upon the number and structure of the protein or glycoprotein subunits of which they are composed (see Figure 3.31a to view an electron micrograph of an S-layer). By virtue of their wide phylogenetic distribution, S-layers can obviously associate with a variety of cell wall structures, including the LPS layer of gram-negative Bacteria (see Section 3.6), the peptidoglycan layers of gram-positive Bacteria, and even directly with the cell membrane in certain Archaea (see Section 3.5).

The major function of S-layers is unknown. However, as the interface between the cell and its environment it is likely that in cells that produce them the S-layer at least functions as an external permeability barrier, allowing the passage of low-molecular-weight substances while excluding large molecules. Evidence also exists that in pathogenic (disease-causing) bacteria that contain S-layers, this structure may confer protection on the bacterium against certain host defense mechanisms.

Capsules and slime layers: The glycocalyx

Many prokaryotic organisms secrete on their surfaces slimy or gummy materials (Figure 3.58). A variety of these structures consist of polysaccharide, and a few consist of protein. The terms **capsule** and **slime layer** are sometimes used to describe polysacchride layers, but the more general term **glycocalyx** is also applied. Glycocalyx is defined as the polysaccharide-containing material lying outside the cell. The glycocalyx varies in different organisms but usually contains glycoproteins and a large number of different polysaccharides, including polyalcohols and amino sugars. The glycocalyx may be thick or thin, rigid or flexible, depending on its chemical nature in a specific organism. The rigid layers are organized in a tight matrix that excludes particles, such as india ink; this form is referred to as a *capsule*. If the glycocalyx is more easily deformed, it will not exclude particles and is more difficult to see; this arrangement is referred to as a *slime layer*.

Glycocalyx layers have several functions in bacteria. Outer polysaccharide layers play an important role

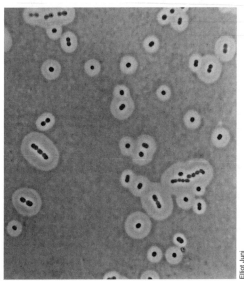

Elliot Juni

(a)

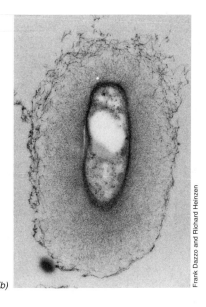

Frank Dazzo and Richard Heinzen

(b)

Figure 3.58 Bacterial capsules. (a) Demonstration of the presence of a capsule in an *Acinetobacter* species by negative staining with india ink observed by phase contrast microscopy. The india ink does not penetrate the capsule, and so it is revealed in outline as a light structure on a dark background. (b) Electron micrograph of a thin section of a *Rhizobium trifolii* cell stained with ruthenium red to reveal the capsule. The diameter of the cell proper (not including the capsule) is about 0.7 μm.

in the *attachment* of certain pathogenic microorganisms to their hosts. As we will see in Section 19.6, pathogenic microorganisms that enter the animal body by specific routes usually do so because of binding reactions that occur between outer cell surface components (such as the glycocalyx) and specific host tissues. The glycocalyx plays other roles as well. There is some evidence that encapsulated bacteria are more difficult for phagocytic cells of the immune system (∞Section 20.3) to recognize and subsequently destroy. In addition, because outer polysaccharide layers probably bind a significant amount of water, there is reason to believe a glycocalyx layer plays some role in resistance to desiccation.

Carbon storage polymers

Granules or other inclusions are often seen within cells. Their nature differs in different organisms, but they almost always function in the storage of energy or as a reservoir of structural building blocks. Inclusions can often be seen directly with the phase contrast microscope, but their contrast can usually be increased by using dyes. Inclusions often show up very well with the electron microscope (Figure 3.59). Most cellular inclusions are bounded by a thin *nonunit* membrane consisting of lipid separating the inclusion from the cytoplasm proper.

In prokaryotic organisms, one of the most common inclusion bodies consists of **poly-β-hydroxybutyric acid (PHB),** a lipidlike compound that is formed from β-hydroxybutyric acid units (Figure 3.59a). The monomers of this acid are connected by ester linkages, forming the long PHB polymer, and these polymers aggregate into granules. The length of the monomer in the polymer can vary considerably, from as short as C_4 to as long as C_{18} in certain organisms. Thus, the collective term *poly-β-hydroxyalkanoate* (PHA) has been coined to describe this whole class of carbon/energy storage polymers. A wide variety of prokaryotes, including representatives of both the Bacteria and the Archaea, produce PHAs, however, Eukarya do not naturally produce PHAs.

Another storage product formed by prokaryotes is **glycogen,** which is a starchlike polymer of glucose subunits (we discussed the chemistry of glycogen in Section 2.5). Glycogen granules are usually smaller than PHB granules and can be seen only with the electron microscope, but the presence of glycogen in a cell can be detected in the light microscope because the cell appears a red-brown color when treated with dilute iodine because of a glycogen–iodine reaction. Like PHB and other PHAs, glycogen is a storage depot for carbon and energy.

Other storage materials and inclusions

Many microorganisms accumulate large reserves of inorganic phosphate in the form of granules of **polyphosphate.** These granules are stained by many basic dyes; one of these dyes, toluidine blue, becomes reddish violet in color when combined with polyphosphate. This phenomenon is called *metachromasy* (color change), and granules that stain in this manner are often called **metachromatic granules.**

A variety of prokaryotes are capable of oxidizing reduced sulfur compounds such as hydrogen sulfide, elemental sulfur, and thiosulfate. These oxidations are linked to either reactions of energy metabolism (∞Sections 13.8 and 13.10) or biosynthesis (∞Section 13.7), but in both instances **elemental sulfur** frequently accumulates inside the cell in large, readily visible granules (Figure 3.60). The granules of elemental sulfur remain as long as a source of reduced sulfur is still present. However, as the reduced sulfur source becomes limiting, the sulfur in the granules is oxidized, usually to sulfate, and the granules slowly disappear as this reaction proceeds.

Magnetosomes are intracellular crystal particles of the iron mineral magnetite, Fe_3O_4 (Figure 3.61). Magnetosomes impart a permanent magnetic dipole to a cell, allowing it to respond to a magnetic field. Bacteria that produce

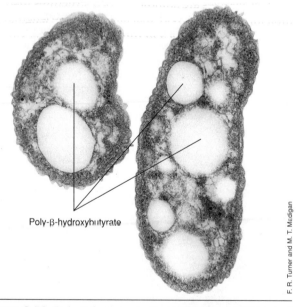

(a)

(b)

Poly-β-hydroxybutyrate

F. R. Turner and M. T. Madigan

Figure 3.59 Poly-β-hydroxybutyrate (PHB). (a) Chemical structure of PHB, a common poly-β-hydroxyalkanoate. A monomeric unit is shaded. Other alkanoate polymers are made by substituting longer-chain hydrocarbons for the —CH₃ group on the β carbon. (b) Electron micrograph of a thin section of cells of the phototrophic bacterium *Rhodospirillum sodomense* containing granules of PHB.

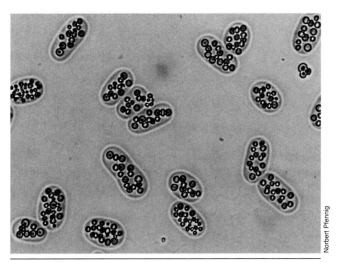

Figure 3.60 Bright-field photomicrograph of cells of the purple sulfur bacterium *Chromatium buderi*. Note the sulfur globules inside the cell. A single cell measures about 4 × 7 μm.

magnetosomes exhibit *magnetotaxis*, the process of orienting and migrating along geomagnetic field lines (∞ Section 16.11 for more on magnetotactic bacteria). Although the combining form *-taxis* is used in the word *magnetotaxis*, there is no evidence that magnetotactic bacteria employ the sensory systems of chemotactic or phototactic bacteria (see Sections 3.10 and 7.7). Instead, the alignment of magnetosomes in the cell simply imparts magnetic properties to it which then orient the cell in a particular direction in its environment.

Magnetosomes are surrounded by a membrane containing phospholipids, proteins, and glycoproteins. Magnetosome membrane proteins probably play a role in precipitating Fe^{3+} (brought into the cell in soluble form by chelating agents) as Fe_3O_4 in the developing magnetosome. The morphology of magnetosomes appears to be species-specific, varying in shape from square (Figure 3.61) to rectangular to spike-shaped in certain bacteria.

Magnetosomes have been described from a variety of different primarily aquatic bacteria and have even been found in some algae (eukaryotes). Algal magnetosomes presumably function to make the algal cells magnetotactic as in bacteria. Measurements of the magnetic moment of magnetotactic algal cells indicates that the magnetic force within these cells is much greater than that of magnetotactic bacteria, which is consistent with the larger size of the algal cells.

3.12

Gas Vesicles

A number of prokaryotic organisms that live a floating existence in lakes and the sea produce **gas vesicles,** which confer buoyancy on the cells. Gas vesicles are a means of motility, allowing cells to float up and down in a water column in response to environmental factors. The most dramatic instances of flotation due to gas vesicles are seen in cyanobacteria that form massive accumulations (blooms) in lakes (Figure 3.62). Gas-vesiculate cells rise to the surface of the lake and are blown by winds into dense masses. Gas vesicles are also present in certain purple and green phototrophic bacteria (∞ Section 16.1) and in some nonphototrophic bacteria that live in lakes and ponds. Some Archaea also contain gas vesicles.

Gas vesicles are spindle-shaped structures made of protein, hollow but rigid, that are of variable lengths and diameters. Gas vesicles in different organisms vary in length from about 300 to 700 nm and in width from 60

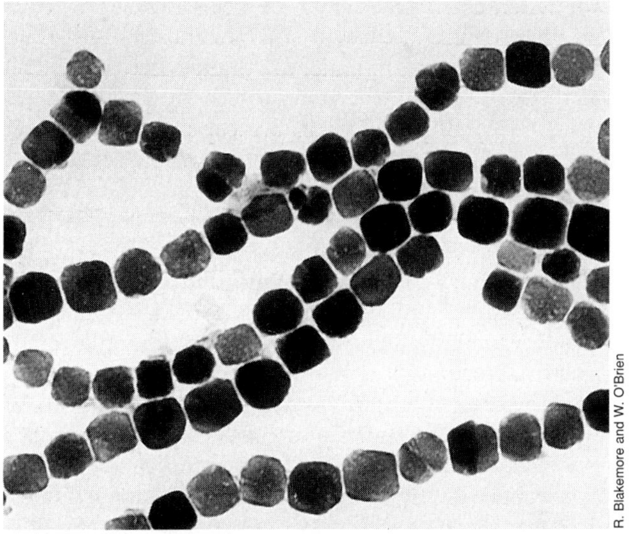

Figure 3.61 Magnetosomes. Magnetic particles of Fe_3O_4 isolated from the magnetotactic bacterium *Aquaspirillum magnetotacticum*. Each particle is about 50 nm in length (∞ Figure 16.49).

Figure 3.62 Flotation of cyanobacteria from a bloom on a nutrient-rich lake, caused by the presence of gas vesicles.

to 110 nm, but the vesicles of any given organism are more or less of constant size. They are present in the cytoplasm and may number from a few to hundreds per cell. The gas vesicle membrane is composed only of protein, is about 2 nm thick, and is impermeable to water and solutes but permeable to most gases; thus, gas vesicles exist as gas-filled structures surrounded by the constituents of the cytoplasm (Figure 3.63).

The rigidity of the gas vesicle membrane is essential for the structure to resist the pressures exerted on it from without; it is probably for this reason that it is composed of a protein able to form a rigid membrane rather than of lipid, which would form a fluid , highly mobile membrane. However, even the gas vesicle membrane cannot resist high hydrostatic pressure and can be collapsed, leading to a loss of buoyancy. Once collapsed, gas vesicles cannot be reinflated. The presence of gas vesicles can be determined by either light or electron microscopy (Figure 3.64), but their identity is never certain unless they disappear when the cells are subjected to high hydrostatic pressure.

Molecular structure of gas vesicles

Gas vesicles contain only two different types of protein (Figure 3.65). The major gas vesicle protein, called GvpA, is a small, highly hydrophobic protein. GvpA is the shell protein and makes up 97% of the total protein of the gas vesicle. The second protein, called GvpC, is a larger protein but is present in much smaller amounts; the function of GvpC protein is to strengthen the shell of the gas vesicle (Figure 3.65). Gas vesicles are constructed of several copies of the GvpA protein aligned as parallel ribs forming a watertight surface. GvpA protein folds as a β-sheet and thus gives considerable rigidity to the overall vesicle structure (Figure 3.65). Studies have shown that most of the hydrophobic amino acids of GvpA protein face the gas surface side of the vesi-

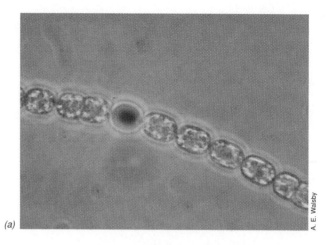

(a)

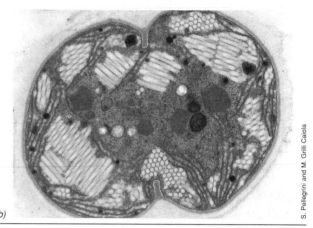

(b)

Figure 3.64 Gas vesicles of the cyanobacteria *Anabaena* and *Microcystis*. (a) *Anabaena flos-aquae*. The cell in the center (a heterocyst) lacks gas vesicles. In the other cells, the vesicles group together as phase-bright objects that scatter light. (b) Transmission electron micrograph of the cyanobacterium *Microcystis*. Gas vesicles are arranged in bundles, here observable in both longitudinal and cross section.

cles, whereas more hydrophilic amino acids face the cytoplasm. The ribs of GvpA protein are strengthened by GvpC protein, which acts as a cross-linker, binding several GvpA ribs together (Figure 3.65). Studies of genes coding for the gas vesicle proteins GvpA and GvpC from taxonomically diverse gas-vesiculate bacteria have shown a remarkable degree of DNA sequence homology, suggesting great evolutionary conservation in the structure of these proteins. Thus, although the final shape of the gas vesicle can vary in different organisms from long and thin to short and fat (compare Figures 3.63 and 3.64), these differences are not due to major differences in the primary structure (that is, amino acid sequence) of gas vesicle proteins in each case but instead are a function of how the proteins are arranged to form the intact vesicle.

The composition of the gas inside a gas vesicle is the same as that of the gas in which the organism is suspended, and gas is present in the vesicle at about 1 atm

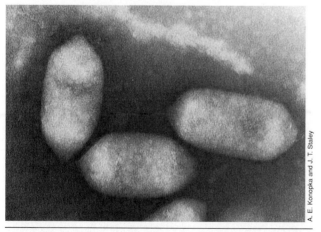

Figure 3.63 Electron micrographs of gas vesicles purified from the bacterium *Microcyclus aquaticus* and examined in negatively stained preparations. A single gas vesicle is about 100 nm in diameter.

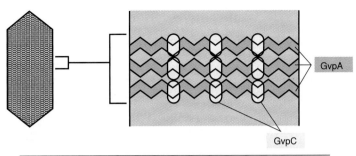

Figure 3.65 Model of how the two proteins that make up the gas vesicle, Gvp A and GvpC, interact to form a water-tight but gas-permeable structure. GvpA makes up the rib and is a rigid β-sheet. GvpC is the cross-linker and is of an α-helix structure.

pressure. Because the gas vesicle attains a density of about 5–20% of that of the cell proper, intact gas vesicles decrease the density of the cell, thereby increasing its buoyancy. Aquatic phototrophic organisms in particular benefit from this motility strategy because it allows them to adjust their position rapidly in a water column to regions where the light intensity for photosynthesis is optimal.

3.13
Endospores

Certain species of Bacteria produce special structures called **endospores** within their cells (Figure 3.66). Endospores are differentiated cells that are very resistant to heat and cannot be destroyed easily, even by harsh chemicals. Endospore-forming bacteria are found most commonly in the soil, and virtually any sample of soil has some endospores present.

The discovery of bacterial endospores was of immense importance to microbiology because the knowledge of such remarkably heat-resistant forms was essential for the development of adequate methods of sterilization, not only of culture media but also of foods and other perishable products (∞Section 1.6). Although many organisms other than bacteria form spores, the bacterial endospore is unique in its degree of heat resistance. Endospores are also resistant to other harmful agents such as drying, radiation, acids, and chemical disinfectants, and can remain dormant for extremely long periods of time. The life cycle of a spore-forming organism is illustrated in Figure 3.67.

Endospore structure

Endospores (so called because the spore is formed *within* the cell) are readily seen under the light microscope as strongly refractile bodies (see Figure 3.66). Spores are very impermeable to dyes, so occasionally they are seen as unstained regions within cells that have been stained with basic dyes such as methylene blue. To stain spores specifically, special spore-staining procedures must be used. The structure of the spore as seen with the electron microscope is vastly different from that of the vegetative cell, as shown in Figure 3.68. The structure of the spore is much more complex than that of the vegetative cell in that it has many layers. The outermost layer is the **exosporium,** a thin, delicate covering made of protein. Within this are the **spore coats,** composed of layers of protein. Below the spore coat is the **cortex,** which consists of loosely cross-linked peptidoglycan, and inside the cortex is the **core** or **spore protoplast,** which contains the usual cell wall (core wall), cytoplasmic membrane, cytoplasm, nucleoid, and so on. Thus the spore differs structurally from the vegetative cell primarily in the kinds of structures found outside the core wall.

One chemical substance that is characteristic of endospores but not present in vegetative cells is **dipicolinic acid** (Figure 3.69). This substance has been found in all endospores examined and is located in the core. Spores are also high in calcium ions, most of which are combined with dipicolinic acid. The calcium–dipicolinic acid complex of the core represents about 10% of the dry weight of the endospore.

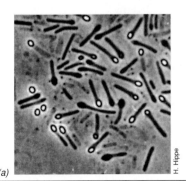
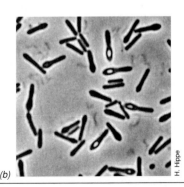
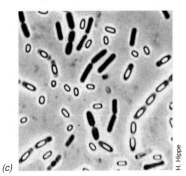

Figure 3.66 The bacterial endospore. Phase contrast photomicrographs illustrating several types of endospore morphologies and intracellular locations. (a) Terminal. (b) Subterminal. (c) Central.

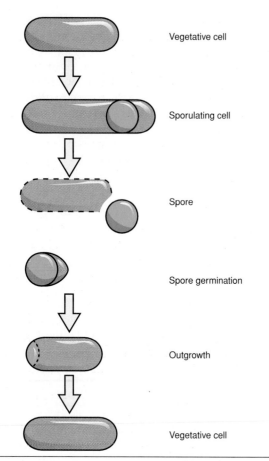

Figure 3.67 Life cycle of an endospore-forming bacterium.

Properties of the endospore core

The core of a mature endospore differs greatly from the vegetative cell from which it was formed. Besides having an abundant calcium dipicolinate (Figure 3.69) content, the core is in a partially dehydrated state. The core of a mature endospore contains only 10–30% of the water content of the vegetative cell, and thus the consistency of the core cytoplasm is that of a gel. Dehydration of the core greatly increases the heat resistance of the endospore but has also been shown to confer resistance to chemicals, such as hydrogen peroxide (H_2O_2), and causes enzymes remaining in the core to become inactive.

In addition to the low water content of the spore, the pH of the core cytoplasm is about one unit lower than that of the vegetative cell and contains high levels of core-specific proteins called *small acid-soluble spore proteins* (SASPs). These are made during the sporulation process and have at least two functions. SASPs bind tightly to DNA in the core and protect it from potential damage from ultraviolet radiation, dessication, and dry heat. However, in addition, SASPs function as a carbon and energy source for the outgrowth of a new vegetative cell from the endospore, a process called *germination* (discussed later).

Endospore formation

During endospore formation, a vegetative cell is converted to a nongrowing, heat-resistant structure—the endospore. As previously described and as summa-

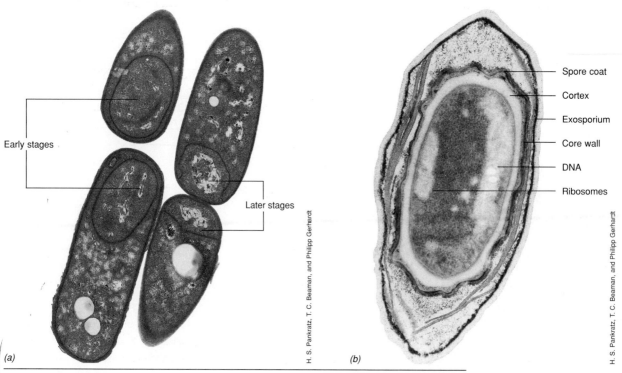

Figure 3.68 Electron microscopy of the bacterial spore. (a) Formation of spores within mother cells (sporangia) of *Bacillus megaterium.* A single cell is about 1.5 μm in diameter. (b) Mature free spore.

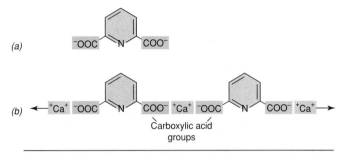

Figure 3.69 Dipicolinic acid (DPA). (a) Structure of DPA. (b) How Ca^{2+} cross-links two DPA molecules to form a complex.

rized in Table 3.3, the differences between the endospore and the vegetative cell are profound. Sporulation involves a very complex series of events in *cellular differentiation*. Bacterial sporulation does not occur when cells are dividing exponentially but only when growth ceases owing to the exhaustion of an essential nutrient. Thus, cells of *Bacillus*, a typical endospore-forming bacterium, cease vegetative growth and begin sporulation when a key nutrient such as the carbon or nitrogen source becomes limiting.

Many genetically directed changes in the cell underlie the conversion from vegetative growth to sporulation. The structural changes occurring in sporulating cells of *Bacillus* are shown in Figure 3.70, and the process can be divided into several stages. In *Bacillus subtilis*, where detailed studies of sporulation have been done, the entire sporulation process takes about 8 hr. Genetic studies of mutants of *Bacillus*, each blocked at one of the various stages of sporulation shown in Figure 3.70, have

shown that as many as 200 genes are involved in the sporulation process. Sporulation requires that the synthesis of some proteins involved in vegetative cell functions cease and that specific spore proteins be made. This is accomplished by activation of a variety of spore-specific genes including *spo*, *ssp* (which encodes SASPs), and many other genes in response to an environmental trigger to sporulate. The proteins encoded by these genes catalyze the series of events leading from a moist, metabolizing vegetative cell to a dry, metabolically inert but extremely resistant endospore (Table 3.3 and Figure 3.70).

Germination

An endospore is able to remain dormant for many years (decades, centuries? See the box, How Long Can an Endospore Survive?), but it can convert back to a vegetative cell relatively rapidly. This process involves three steps: activation, germination, and outgrowth (Figure 3.71). *Activation* is accomplished by heating freshly formed endospores for several minutes at a sublethal but elevated temperature or by storing spore suspensions for weeks or months at 4°C or room temperature. Activated spores are conditioned to germinate when placed in the presence of specific nutrients. *Germination*, usually a rapid process (on the order of several minutes), involves loss of refractility of the spore, increased ability to be stained by dyes, and loss of resistance to heat and chemicals. Loss from the spores of calcium dipicolinate and cortex components occurs during this stage, and the SASPs are degraded. The next stage, *outgrowth*, involves visible swelling due to water uptake and synthesis of

TABLE 3.3	Differences between endospores and vegetative cells	
Characteristic	**Vegetative cell**	**Endospore**
Structure	Typical gram-positive cell	Thick spore cortex Spore coat Exosporium
Microscopic appearance	Nonrefractile	Refractile
Calcium content	Low	High
Dipicolinic acid	Absent	Present
Enzymatic activity	High	Low
Metabolism (O_2 uptake)	High	Low or absent
Macromolecular synthesis	Present	Absent
mRNA	Present	Low or absent
Heat resistance	Low	High
Radiation resistance	Low	High
Resistance to chemicals (for example, H_2O_2) and acids	Low	High
Stainability by dyes	Stainable	Stainable only with special methods
Action of lysozyme	Sensitive	Resistant
Water content	High, 80–90%	Low, 10–25% in core
Small acid-soluble proteins (product of *ssp* genes)	Absent	Present
Cytoplasmic pH	About pH 7	About pH 5.5–6.0 (in core)

A FOCUS ON . . .

HOW LONG CAN AN ENDOSPORE SURVIVE?

In this chapter we have discussed the dormancy and resistance properties of bacterial endospores and have pointed out that endospores can survive for long periods in a dormant state. But how long is long?

Published evidence for endospore longevity has shown that endospores can remain viable (that is, capable of germination into vegetative cells) for at least several decades and probably for much longer than that. A suspension of spores of the bacterium *Clostridium aceticum* (see photograph) prepared in 1947 was placed in growth medium in 1981, 34 years later, and in less than 12 hr growth commenced, leading to a robust culture. *Clostridium aceticum* was originally isolated by the Dutchman K. T. Wieringa in 1940 but was thought to have been lost until this vial of *C. aceticum* spores was found in a storage room at the University of California at Berkeley and revived.[a]

Other more extreme examples of endospore longevity have been documented. Bacteria of the genus *Thermoactinomyces* are thermophilic endospore formers that are widespread in nature in soil, plant litter, and fermenting plant material. Microbiological examination of a Roman archaeological site in the United Kingdom that was dated to over 2000 years ago yielded significant numbers of viable *Thermoactinomyces* spores in various pieces of debris. Additionally, *Thermoactinomyces* spores were recovered in fractions of sediment cores from a Minnesota lake known to be over 7000 years old. Although contamination is always a possibility in such studies, samples in both of these cases were processed in such a way as to virtually rule out contamination with "recent" spores.[b]

What factors could limit the age of an endospore? Cosmic radiation has been considered a major factor because it can introduce mutations in DNA.[b] It has been hypothesized

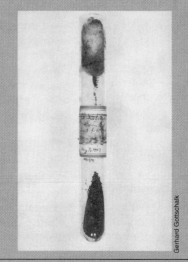

Photograph of a test tube containing spores of the bacterium *Clostridium aceticum* prepared on May 7, 1947. After remaining dormant for over 30 years, the spores were suspended in a culture medium after which growth occurred within 12 hr.

that over periods of thousands of years, the cumulative effects of cosmic radiation could introduce so many mutations into the genome of an organism that even highly radiation-resistant structures such as endospores would succumb to the genetic damage. However, extrapolations from actual experimental assessments of the effect of natural radiation on endospores suggest that if suspensions of endospores were partially shielded from cosmic radiation, for example, by being embedded in layers of organic matter, they could retain viability over periods as great as *several hundred thousand years* and perhaps even longer. Amazing, but is this the upper limit?

In 1995 a group of scientists reported the revival of bacterial spores they claimed were 25–40 million years old.[c] The spores were allegedly preserved in the gut of an extinct bee trapped in amber of known geological age. The presence of endospore-forming bacteria in these bees was previ-

ously suspected from electron microscopic studies of the insect gut which showed endospore-like structures and because *Bacillus*-like DNA was recovered from the insect. DNA stored under the proper conditions is known to be quite resistant to decay, but could actual *viable cells* have survived this long? Although this was originally considered highly unlikely, the results of growth experiments showed otherwise. Samples of bee tissue incubated in a sterile culture medium quickly yielded endospore-forming bacteria. Rigorous precautions were taken to demonstrate that the endospore-forming bacterium revived from the amber-encased bee was not a modern-day contaminant. Moreover, comparisons of the nucleotide sequence in a specific gene obtained from DNA from the insect's gut and from DNA prepared from the revived bacterium showed that the bacterium isolated was the likely source of the DNA in the insect gut.

If this claim of almost unbelievable endospore longevity is supported by repetition of the results in independent laboratories (and such confirmation is crucial for verifying such a highly controversial finding), then endospores stored under the proper conditions can remain viable indefinitely. This is a remarkable testimony to the endospore, a structure that undoubtedly evolved to help cells remain viable for relatively short periods but that turned out to be such a well-designed structure that dormancy for hundreds of thousands, if not millions, of years, may be possible. ∎

[a]Braun, M., F. Mayer, and G. Gottschalk. 1981. *Clostridium aceticum* (Wieringa), a microorganism producing acetic acid from molecular hydrogen and carbon dioxide. *Arch. Microbiol.* 128:288–293.

[b]Gest, H., and J. Mandelstam. 1987. Longevity of microorganisms in natural environments. *Microbiol. Sci.* 4:69–71.

[c]Cano, R. J., and M. K. Borucki. 1995. Revival and identification of bacterial spores in 25- to 40-million-year-old Dominican amber. *Science* 268:1060–1064.

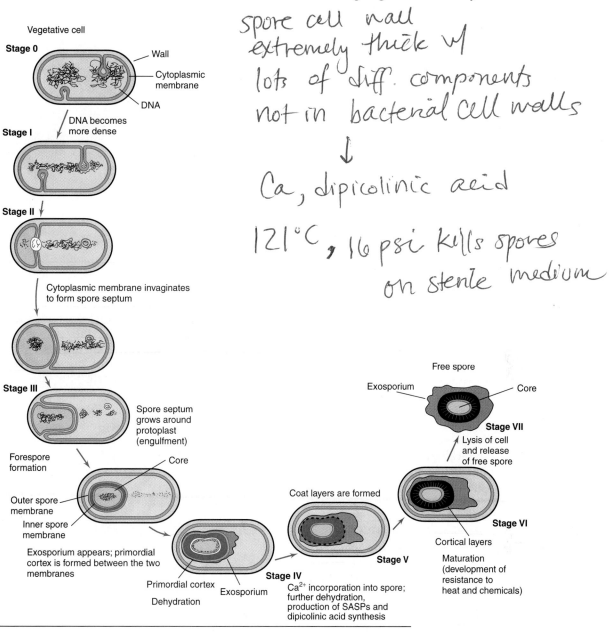

*spore cell wall
extremely thick w/
lots of diff. components
not in bacterial cell walls*
↓
Ca, dipicolinic acid

*121°C, 16 psi kills spores
on sterile medium*

Figure 3.70 Stages in endospore formation. The stages listed (0 through VII) are those most clearly distinguishable microscopically and are used in studies on the kinetics of the sporulation process. *bacillus*

CONCEPT CHECK   3.13

The endospore is a highly resistant differentiated bacterial cell produced by certain types of gram-positive Bacteria. Spore formation leads to a nearly dehydrated spore core that contains essential macromolecules and a variety of substances such as calcium dipicolinate and small acid-soluble proteins, absent from vegetative cells. Spores can remain dormant indefinitely but germinate quickly when the appropriate trigger is applied.

✔ *What is* dipicolinic acid *and where is it found?*

✔ *What are* SASPs *and what is their function?*

✔ *What happens when an endospore germinates?*

---

new RNA, proteins, and DNA. The cell emerges from the broken spore coat and eventually begins to divide (Figure 3.71). The cell remains in vegetative growth until environmental signals that trigger sporulation are once again sensed.

3.14

The Nucleus: Defining Organelle of Eukaryotes

We learned earlier that cells can be prokaryotic (Archaea and Bacteria) or eukaryotic (Eukarya) (∞ Sections 1.4 and 1.5). The root word *karyon* is Greek for "nut" or "kernel" and refers to the **nucleus**. Cells with a nucleus

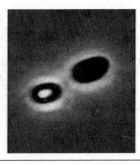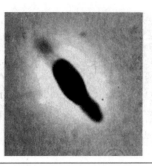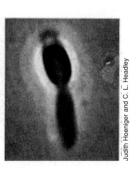

Figure 3.71 Germination: Conversion of the endospore to a vegetative cell; photomicrographs showing the sequence of events.

are eukaryotes. This membrane-enclosed organelle contains the eukaryotic cell's DNA. In many eukaryotic cells the nucleus is a large organelle many micrometers in diameter, easily visible with the light microscope even without staining. In smaller eukaryotes, however, special staining procedures often are required to see the nucleus.

Two key steps in biological information flow, DNA replication and RNA synthesis (transcription), occur in the nucleus, whereas the process of protein synthesis (translation) occurs in the cytoplasm. We will learn much more about these processes and the organization of eukaryotic genes in Chapter 6. However, it is important to mention here that the DNA in the nucleus is found in **chromosomes.** Further, the typical eukaryotic cell contains far more DNA than a prokaryotic cell (see Figure 3.39). Some of this is related to the fact that eukaryotes have more genes, but in many eukaryotes there is a tremendous amount of DNA that serves no known coding function. Some is also related to the fact that in eukaryotes genes are commonly split into **exons** (coding regions) and **introns** (noncoding regions). RNA synthesized from such a gene must be *processed* to remove the noncoding regions before it can be used (∞ Section 6.7). This processing also occurs in the nucleus. Thus, the nucleus is both a storehouse and a processing factory for genetic information. In prokaryotes, which lack a membrane-enclosed nucleus, the processes of transcription and translation are tightly coupled (∞ Section 7.5), and the extensive RNA processing typical of eukaryotes does not occur.

Nuclear structure

The nuclear membrane consists of a pair of parallel unit membranes separated by a space of variable thickness. The inner membrane is usually a simple sac, but the outer membrane is in many places continuous with the cytoplasmic membrane. The dual-membrane arrangement does, however, facilitate functional specificity because the inner and outer membranes specialize in interactions with the nucleoplasm and cytoplasm, respectively. The nuclear membrane contains many pores (Figure 3.72), which are formed from holes in both unit membranes at places where the inner and outer membranes are joined. The pores are about 9 nm

wide and permit facile passage in and out of the nucleus of macromolecules of up to about 60,000 molecular weight. However, certain proteins of greater than 60,000 molecular weight, such as DNA and RNA polymerases (which are up to 200,000 in molecular weight) are also able to pass into the nucleus from their sites of synthesis in the cytoplasm. Whether they pass through nuclear pores or are threaded through the lipid bilayer by specific transport proteins is not known. A typical animal cell nucleus contains 3000–4000 nuclear pores.

A structure often seen within the nucleus is the **nucleolus,** an area rich in RNA that is the site of ribosomal RNA synthesis. Ribosomal proteins synthesized in the cytoplasm are transported into the nucleolus and used along with ribisomal RNA to assemble the small and large subunits of the eukaryotic ribosome (∞ Section 6.9). They are then exported to the cytoplasm where they function in protein synthesis.

Chromosomes and DNA

As we have discussed (see Section 3.8), the DNA of most prokaryotes is contained primarily in a single molecule of free DNA, whereas in eukaryotes DNA is present in more complex structures, the *chromosomes.* **Chromosome** means "colored body," for chromosomes were first seen as structures colored by certain stains. Many chromosome stains involve dyes that react strongly with basic (that is, positively charged) proteins called **histones,** which in eukaryotes often are attached to the DNA.

During cell division, the nucleus divides following a doubling of the chromosome number, a process called **mitosis** (see Figure 3.73), yielding two cells, each with a full complement of chromosomes. Histones and small protein tubes called microtubules play important roles in the mitotic process. Histones are spaced along the DNA double helix at regular intervals, the DNA itself being wound around each histone molecule. The packing forms a discrete structure called a **nucleosome** (Figure 3.74). Nucleosomes aggregate and form a fibrous material called **chromatin.** Chromatin itself can be compacted by folding and looping to eventually form the intact chromosome. Because DNA is negatively charged (owing to the large number of phosphate groups present), there

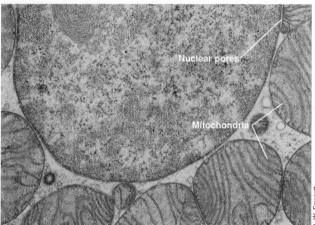

Figure 3.72 The nucleus and nuclear pores. (a) Electron micrograph of a yeast cell by the freeze-etch technique, showing a surface view of the nucleus. The cell is about 8 μm wide. (b) Thin section of mouse adipose tissue showing a portion of the nucleus and several mitochondria. The nucleus is about 2 μm wide. Note the pores in the nuclear membrane in both (a) and (b).

chromosome. For instance, the total amount of DNA per yeast cell is only three times that in *Escherichia coli*, but yeast has 16 chromosomes, and so the average yeast DNA molecule is much shorter than the *E. coli* chromosome. In higher organisms, however, the length of the DNA molecule in a single chromosome is many times greater than that in the prokaryotic chromosome if it were opened and linearized.

The DNA content per nucleus varies from species to species in much the same way as does nuclear size. In addition, the chromosome number also varies greatly, from just a few to many hundreds. Another variable feature is *genome size,* which is the actual amount of DNA per cell. The genome size of various eukaryotes was compared with that of viruses and prokaryotes in Figure 3.39 and will be discussed again in Chapter 6 (∞ Table 6.2).

Organisms that reproduce sexually are typically **diploid** in genetic complement, and the cell's minimum genome contains one copy of each chromosome *from each parent.* Therefore, there typically are at least *two* copies of each gene present, and these may not be identical. During sexual reproduction **gametes** are formed that are *haploid*; that is, they contain *one* copy of each chromosome (see Section 3.8). These are analogous to the sperm and egg of multicellular eukaryotes. The gametes fuse to form a single cell called a **zygote,** and the nucleus of the zygote usually results from fusion of the nuclei of the two gametes. The zygote nucleus thus has twice the chromosome complement of the gametes. Some eukaryotes (yeast is a good example) can replicate asexually as haploid cells (∞ Section 9.12) and thus have both haploid and diploid forms.

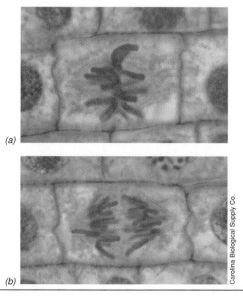

Figure 3.73 Mitosis, as seen in the light microscope. These are onion root tip cells that have been stained to reveal nucleic acid and chromosomes. (a) Metaphase. Chromosomes are paired in the center of the cell. (b) Anaphase. Chromosomes are separating.

is a strong tendency for various parts of the molecule to repel each other. Histones neutralize some of these negative charges, permitting contraction of the chromosomes. Microtubules function to form the *spindle apparatus,* which is the actual structure that moves chromosomes to the two poles of the dividing cell (Figure 3.73).

In each chromosome, the DNA is a single *linear* molecule to which histones (and other proteins) are attached. In yeast (and probably many other microorganisms), the length of the DNA in a single chromosome is actually shorter than that of a linearized prokaryotic

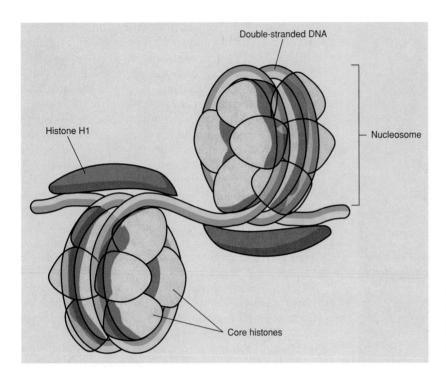

Double-stranded DNA

Histone H1

Nucleosome

Core histones

Figure 3.74 Packaging of DNA around a core of histone proteins to form a nucleosome. Nucleosomes are arranged along the DNA strand somewhat like beads on a string. This arrangement is typical of DNA in eukaryotic cells.

CONCEPT CHECK   3.14

The DNA of the eukaryotic cell is present in a number of separate linear molecules, each associated with histone proteins (forming the nucleosome) and formed into structures called chromosomes. The chromosomes are present in a membrane-enclosed structure called the nucleus. Each species of organism has a specific number of chromosomes, with the haploid number existing in sperm or eggs and the diploid number in the zygote.

✔ *In the eukaryotic cell what steps in biological information flow occur in the nucleus?*

✔ *What are the differences between the chromosome found in eukaryotic cells and the prokaryotic chromosome?*

3.15
Organelles: Mitochondria and Chloroplasts

Eukaryotic cells have a number of important functions localized in discrete bodies called **organelles.** Besides the nucleus, the two most important organelles are *mitochondria*, in which energy metabolism is carried out, and *chloroplasts*, in which the process of photosynthesis is carried out in plants and algae. We discuss these two organelles in more detail here.

Mitochondria

In eukaryotic cells the processes of respiration and oxidative phosphorylation (a mechanism of ATP formation) (∞ Section 4.12) are localized in membrane–enclosed structures, the **mitochondria** (singular, **mitochondrion**). Mitochondria are of prokaryotic size and can be rod-shaped or nearly spherical (Figure 3.75; see also Figure 3.11). A typical animal cell such as a liver cell can contain 1000 mitochondria, but the number per cell depends somewhat on the cell type and size; a yeast cell may have as few as two mitochondria per cell. The mitochondrial membrane, which lacks sterols, is much less rigid than a cell's cytoplasmic membrane. Mitochondria therefore show a considerable plasticity, which makes their shape as seen in electron micrographs highly variable (Figure 3.75).

The mitochondrial membrane is constructed in a manner similar to other unit membranes: a bilayer of phospholipid with embedded proteins. However, unlike the cytoplasmic membrane, the outer mitochondrial membrane is rather permeable. In this structure, protein channels are present that allow passage of any molecule of molecular weight less than approximately 10,000. It is for this reason that ATP, produced within the mitochondrion, can move to the cytoplasm where it is used in energy-requiring reactions. In addition to the outer membrane, mitochondria possess a system of folded inner membranes called *cristae*. These inner membranes, formed by invagination of the outer membrane, are the site of enzymes involved in respiration and ATP production and of specific transport proteins that regulate the passage of metabolites into and out of

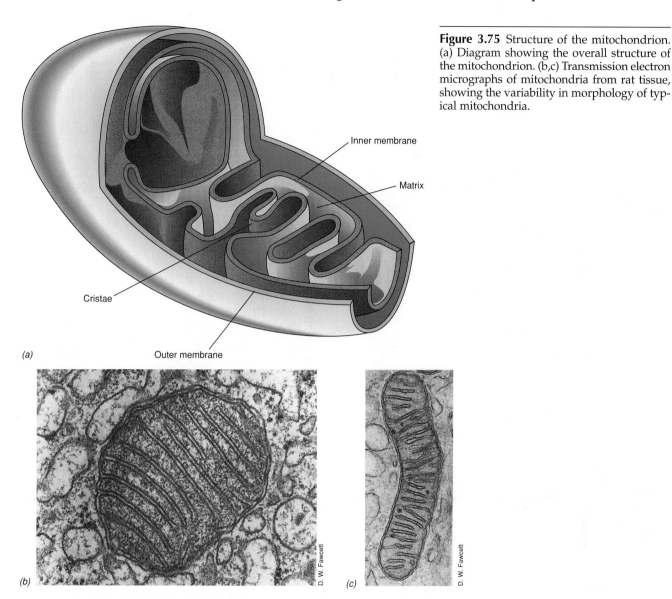

Figure 3.75 Structure of the mitochondrion. (a) Diagram showing the overall structure of the mitochondrion. (b,c) Transmission electron micrographs of mitochondria from rat tissue, showing the variability in morphology of typical mitochondria.

Inner membrane

Matrix

Cristae

(a)   Outer membrane

(b)   D. W. Fawcett

(c)   D. W. Fawcett

the *matrix* of the mitochondrion (Figure 3.75). The matrix contains a number of enzymes involved in the oxidation of organic compounds, in particular, enzymes of the citric acid cycle (∞ Section 4.14). The mitochondrion can thus be viewed as the energy storehouse of the cell.

Chloroplasts

Chloroplasts are chlorophyll-containing organelles found in all eukaryotic organisms able to carry out photosynthesis. Chloroplasts of many algae are relatively large and hence are readily visible with the light microscope (Figure 3.76). The size, shape, and number of chloroplasts vary markedly but, unlike mitochondria, they are generally much larger than bacteria.

Like mitochondria, chloroplasts have a very permeable outer membrane, a much less permeable inner membrane, and an intermembrane space. The inner membrane surrounds the lumen of the chloroplast, called the *stroma*, but is not folded into cristae like the inner membrane of the mitochondrion. Instead chlorophyll, photosynthesis-specific proteins, the photosynthetic electron transport chain, and all other components needed for photosynthesis are located in a series of flattened membrane discs called **thylakoids** (Figure 3.77). The thylakoid membrane is highly impermeable to ions and other metabolites because its function is to establish the proton motive force necessary for ATP synthesis (∞ Sections 4.13 and 13.3). In green algae and green plants, thylakoids are usually associated in stacks of discrete structural units called *grana* (∞ Figure 13.5).

The chloroplast stroma contains large amounts of the enzyme ribulose bisphosphate carboxylase, called *RubisCO* for short. This enzyme is the key enzyme of the Calvin cycle, the series of reactions by which most photosynthetic organisms convert CO_2 to organic form

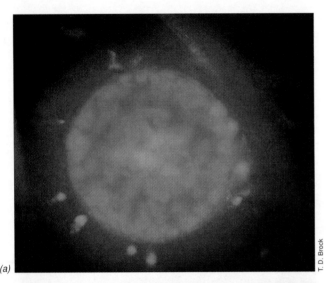

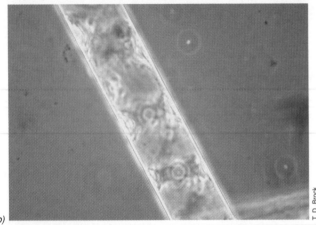

Figure 3.76 Photomicrographs of algal cells showing the presence of chloroplasts. (a) Fluorescence photomicrograph of the diatom *Stephanodiscus*. The chlorophyll in the chloroplasts absorbs light and fluoresces red. (b) Phase contrast photomicrograph of *Spirogyra* showing the characteristic spiral chloroplasts.

(∞Section 13.7). RubisCO makes up over 50% of the total chloroplast protein and produces phosphoglyceric acid, a key compound in the biosynthesis of glucose (∞Sections 4.18 and 13.7). The permeability of the outer chloroplast membrane allows glucose and ATP produced during photosynthesis to diffuse into the cytoplasm where they can be used to build new cell material.

Relationships of organelles to bacteria

On the basis of their relative autonomy and morphological resemblance to bacteria, it was suggested long ago that mitochondria and chloroplasts are descendents of ancient prokaryotic organisms. This theory of *endosymbiosis* (*endo* means "within") says that eukaryotes arose from the engulfment of a prokaryotic cell by

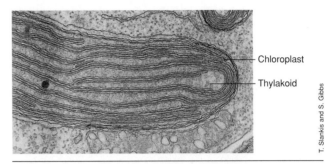

Figure 3.77 Electron micrograph showing a chloroplast of the golden brown alga *Ochromonas danica*. Note the thylakoids.

a larger cell (∞ Section 15.4). Several pieces of evidence support this scenario:

1. *Mitochondria and chloroplasts contain DNA*. Although most of their functions are encoded by nuclear DNA, a few organellar components are encoded within the organellar genome, most notably ribosomal RNAs, transfer RNAs, and certain proteins of the respiratory chain. Mitochondrial and choloroplast DNA exist in *covalently closed circular form*, as it does in prokaryotes (see Section 3.8), but is generally present in more than one copy. Mitochondrial DNA can be visualized in cells by special staining methods (Figure 3.78).

2. *Mitochondria and chloroplasts contain their own ribosomes*. Ribosomes, the cell's protein synthesis "factories" (∞ Section 6.9), exist in either a large form [80 Svedberg units (S)] typical of the cytoplasm of eukaryotic cells or a smaller form (70S) unique to prokaryotes. Mitochondrial and chloroplast ribosomes are 70S in size, the same as those of prokaryotes.

3. *Antibiotic specificity*. Many of the antibotics that kill or inhibit Bacteria by specifically interfering with 70S ribosome function, for example, streptomycin, also inhibit protein synthesis in mitochondria and chloroplasts.

4. *Phylogeny*. Phylogenetic studies using comparative ribosomal RNA sequencing methods (∞Section 15.6) have shown convincingly that the chloroplast and mitochondrion are related to Bacteria. These studies have clearly shown that the modern eukaryotic cell arose from an association of two organisms. The mitochondrion and chloroplast are descendents of different groups of Bacteria because their ribosomal RNA sequences closely match those of certain species of Bacteria. The same techniques show that the cytoplasmic component of eukaryotes evolved totally independently.

Presumably organelles evolved (following endosymbiotic events) by the progressive loss of more and more

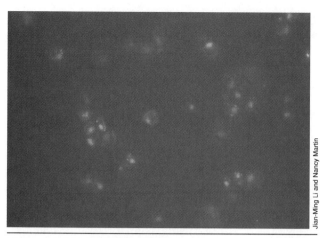

Jian-Ming Li and Nancy Martin

Figure 3.78 Cells of the yeast *Saccharomyces cerevisiae* stained to show mitochondrial DNA. Each mitochondrion has two to four circular chromosomes which stain blue with the fluorescent dye used. See also Figure 3.11.

CONCEPT CHECK   **3.15**

Two key organelles of eukaryotes are the chloroplast, involved in photosynthesis, and the mitochondrion, involved in respiration. It is likely that these organelles were originally Bacteria that established permanent residence inside another cell (endosymbiosis). Ample molecular genetic evidence supports this contention.

✔ *How is DNA arranged in organelles?*

✔ *Why does streptomycin affect organelles?*

---

of their genetic independence, eventually becoming functionally specialized and dependent on their cytoplasmic host cell. The result is the chloroplasts and mitochondria we see today.

3.16
Comparisons of the Prokaryotic and Eukaryotic Cell

At this stage is might be useful to draw some comparisons between the prokaryotic and eukaryotic cell. It should be clear by now that there are profound differences in the internal structure of these two cell types. One important distinction is that eukaryotes have many types of cellular functions segregated into membrane-containing structures (organelles). We discussed mitochondria and chloroplasts earlier, and Table 3.4 lists a number of other membranous structures found in eukaryotes.

Table 3.5 compares the prokaryotic and eukaryotic cell in several categories and emphasizes the great *structural* differences between prokaryotes and eukaryotes. However, we should recall here at the end of our journey through cell biology that all cells contain common types of molecules: proteins, nucleic acids, polysaccharides, and lipids, and that all use many of the same kinds of metabolic machinery. Chemical differences in the building blocks and variations in the assembly of macromolecules to form cells lead to the structural and functional diversity we see in living organisms today.

We now turn to the important aspects of nutrition and energy metabolism. How does the cell create structurally complex molecules that are eventually assembled to make new cells? We consider these problems and how cells solve them in the next chapter.

TABLE 3.4	Membrane-containing structures in eukaryotes	
Stucture	**Characteristics**	**Function**
Mitochondria	Prokaryotic in size, complex internal membrane arrays	Energy generation: respiration
Chloroplasts	Green, chlorophyll-containing, many shapes, often quite large	Photosynthesis
Endoplasmic reticulum	Not a distinct organelle, but extensive array of internal membranes; sites of ribosomes	Protein synthesis
Golgi bodies	Membrane aggregates of distinct structure	Secretion of enzymes and other macromolecules
Vacuoles	Round, membrane-enclosed bodies of low density	Food digestion (food vacuoles); waste product excretion (contractile vacuoles)
Lysosomes	Submicroscopic membrane-enclosed particles	Contain and release digestive enzymes
Peroxisomes	Submicroscopic membrane-enclosed particles	Photorespiration in plants
Glyoxysomes	Submicroscopic membrane-enclosed particles	Enzymes of glyoxylate cycle
Nucleus	Large, generally centrally located	Contains genetic material (genomic DNA)

TABLE 3.5	Comparison of the prokaryotic and eukaryotic cell	

Properties	Prokaryote	Eukaryote
Phylogenetic groups	Bacteria, Archaea	Eukarya: Algae, fungi, protozoa, plants, animals
Nuclear structure and function:		
Nuclear membrane	Absent	Present
Nucleolus	Absent	Present
DNA	Single molecule generally covalently closed and circular, not complexed with histones (other DNA in plasmids)	Linear, present in several chromosomes, usually complexed with histones
Division	No mitosis	Mitosis; mitotic apparatus with microtubular spindle
Sexual reproduction	Fragmentary process, unidirectional; no meiosis; usually only portions of genetic complement reassorted	Regular process; meiosis; reassortment of whole chromosome complement
Introns in genes	Rare	Common
Cytoplasmic structure and organization:		
Cytoplasmic membrane	Usually lacks sterols; hopanoids may be present	Sterols usually present; hopanoids absent
Internal membranes	Relatively simple; limited to specific groups	Complex; endoplasmic reticulum; Golgi apparatus
Ribosomes	70S in size	80S, except for ribosomes of mitochondria and chloroplasts, which are 70S
Membranous organelles	Absent	Several present
Respiratory system	Part of cytoplasmic membrane; mitochondria absent	In mitochondria
Photosynthetic pigments	In internal membranes or chlorosomes; chloroplasts absent	In chloroplasts
Cell walls	Present (in most), composed of peptidoglycan (Bacteria), other polysaccharides, protein, glycoprotein (Archaea)	Present in plants, algae, fungi, usually polysaccharide; absent in animals, most protozoa
Endospores	Present (in some), very heat-resistant	Absent
Gas vesicles	Present (in some)	Absent
Forms of motility:		
Flagellar movement	Flagella composed of a single type of protein arranged in a fiber and anchored into the cell wall and membrane; flagella rotate	Flagella or cilia; composed of microtubules; do not rotate
Nonflagellar movement	Gliding motility; gas vesicle–mediated	Cytoplasmic streaming and ameboid movement; gliding motility
Cytoskeleton containing microtubules	Absent	Present; microtubules are present in flagella, cilia, basal bodies, mitotic spindle apparatus, centrioles
Size	Generally small, usually <2 μm in diameter	Usually larger, 2 to >100 μm in diameter

Material for Review

REVIEW QUESTIONS

1. What is the function of staining in light microscopy? Why are cationic dyes used for general staining purposes?

2. What are the major morphologies of prokaryotes? Draw cells for each morphology you list.

3. Describe in a single sentence the manner in which a unit membrane is formed from phospholipid molecules.

4. Explain in a single sentence why ionized molecules do not readily pass through the membrane barrier of a cell. How *do* such molecules get through the cytoplasmic membrane?

5. Describe a major chemical difference between membranes of Bacteria and of Archaea.

6. List three kinds of membrane proteins and give a short explanation of the function of each.

7. Water molecules penetrate cytoplasmic membranes fairly readily, but hydrogen ions (protons) do not, even though a proton is smaller than a water molecule. Why?

8. Compare and contrast *group translocation* and *active transport*. For each of these processes, include a discussion of specificity, energy requirement, and transport against a concentration gradient.

9. Why is the bacterial cell wall rigid layer called *peptidoglycan?* What are the chemical reasons for the rigidity that is conferred on the cell wall by the peptidoglycan structure?

10. Since a single peptidoglycan molecule is very thin, explain in chemical terms how the very *thick* peptidoglycan-containing cell wall of gram-positive Bacteria is formed.

11. List several functions for the outer wall layer in gram-negative Bacteria.

12. What is the bacterial periplasm? What types of Bacteria have a periplasm and of what significance is the periplasmic space?

13. Write a clear explanation (two or three sentences) of why sucrose is able to stabilize bacterial cells from lysis by lysozyme.

14. Describe the structure and function of a bacterial flagellum. What is the energy source for the flagellum?

15. In a few sentences, write an explanation for how a motile bacterium is able to sense the direction of an attractant and move toward it.

16. What types of cytoplasmic inclusions are formed by prokaryotes? How does an inclusion of poly-β-hydroxybutyric acid (PHB) differ from a magnetosome in composition and metabolic role?

17. Both lysozyme and penicillin bring about bacterial cell lysis but by different mechanisms. Describe the mechanism by which each of these agents causes cell lysis.

18. In a few sentences, indicate how the bacterial endospore differs from the vegetative cell in structure, chemical composition, and ability to resist extreme environmental conditions.

19. The discovery of the bacterial endospore was of great practical importance. Why?

20. How does the eukaryotic nucleus differ from the prokaryotic nucleoid? In what ways are these two structures similar?

21. Set up a table following the format of Table 3.5, with the second and third columns blank, and then fill in the blanks. As you do so, think back to the figures in this chapter that illustrate the properties being considered.

APPLICATION QUESTIONS

1. Calculate the size of the smallest resolvable object if 600-nm light is used to observe a specimen with a 100× oil-immersion lens having a numerical aperture of 1.32? How could resolution be improved using this same lens?

2. Calculate the surface-to-volume ratio of a spherical cell 15 μm in diameter and a cell 2 μm in diameter. What are the consequences of these differences in surface-to-volume ratio for cell function?

3. Imagine a planet where life evolved in a totally nonaqueous environment. Cells on this planet contain highly hydrophobic cytoplasm and live in water-free environments. Predict and draw the structure of the type of cytoplasmic membranes organisms on this planet would have and discuss why such a membrane would be best suited to these organisms.

4. From what you know about the nature of the bacterial cell wall and membrane, explain why a rod-shaped bacterial cell becomes a spherical structure when its wall is removed under conditions such that cell lysis cannot occur.

5. Assume you are given two cultures, one of a species of gram-negative Bacteria and one of a species of Archaea. Other than by sequencing ribosomal RNA, discuss at least five different ways you could tell which culture is which.

SUPPLEMENTARY READINGS

Alberts, B., D. Bray, J. Lewis, M. Raff, K. Roberts, and **J. D. Watson.** 1994. *The Molecular Biology of the Cell,* 3rd edition. Garland, New York. A substantial textbook with extensive coverage of cell structure, with emphasis on the eukaryotic cell.

Campbell, N. A. 1995. *Biology,* 4th edition. Benjamin-Cummings, Redwood City, CA. An excellent general biology textbook with good coverage of basic cell structure and function.

Cole, J. A., C. Dow, and **S. Mohan** (eds.). 1992. *Prokaryotic Structure and Function: A New Perspective.* Soc. Gen Microbiol. Symp. 47. Cambridge University Press, New York. Chapters, written by experts, on the structure, function, and regulation of synthesis of bacterial cell components.

Cooper, S. 1991. *Bacterial Growth and Division.* Academic Press, San Diego, CA. A consideration of replication of the bacterial chromosome in relation to cell division processes. Also covers major processes of cell division in eukaryotic cells.

Neidhardt, F. C., J. L. Ingraham, and **M. Schaechter.** 1990. *Physiology of the Bacterial Cell—A Molecular Approach.* Sinauer Associates, Sunderland, MA. A very readable textbook of bacterial physiology with an emphasis on molecular aspects of cell structure and function, growth, and genetics.

Riley, M. (ed.). 1990. *The Bacterial Chromosome.* American Society for Microbiology, Washington, DC. A detailed treatment of the structure and processing of bacterial DNA.

Schlegel, H. G. 1993. *General Microbiology,* 7th edition. Cambridge University Press, New York. An excellent overview of cell structure and many other aspects of microbiology for the beginning student.

On~line resources for this chapter are on the World Wide Web at:
http://www.prenhall.com/~brock (click on the <u>table of contents</u> link and then select Chapter 3).

Nutrition and Metabolism

Enzymes are the cell's biocatalysts, and the biochemical reactions a cell can carry out are a function of its complement of enzymes. Here is shown a molecular model of the enzyme lysozyme, which catalyzes the breakage of glycosidic bonds in peptidoglycan and can thus kill bacterial cells.

MINIGLOSSARY for Chapter 4

Activation Energy the energy required to bring substrates to the reactive state

Aerobe a microorganism able to use O_2 in respiration

Anabolism the sum total of all biosynthetic reactions in the cell

Aseptic Technique methods for maintaining sterile culture media and other sterile objects free from microbial contamination during manipulations

Autotroph an organism capable of biosynthesizing all cell material from CO_2 as the sole carbon source

Catabolism biochemical reactions leading to the production of usable energy (usually ATP) by the cell

Chemolithotroph an organism that uses inorganic chemicals as energy sources (electron donors)

Chemoorganotroph an organism that uses organic chemicals as energy sources (electron donors)

Citric Acid Cycle a cyclic series of reactions resulting in the conversion of acetate to two CO_2

Coenzyme a small nonprotein molecule that participates in a catalytic reaction as part of an enzyme

Complex Medium a culture medium composed of digests of chemically undefined substances such as yeast and meat extracts

Culture Medium an aqueous solution of various nutrients suitable for the growth of microorganisms

Defined Medium a culture medium whose precise chemical composition is known

Electron Acceptor a substance that can accept electrons from some other substance, thereby becoming reduced in the process

Electron Donor a substance that can donate electrons to some electron acceptor, thereby becoming oxidized in the process

Enzyme a protein that has the ability to speed up (catalyze) a specific chemical reaction

Fermentation anaerobic catabolism in which an organic compound serves as both an electron donor and an electron acceptor and in which ATP is produced by substrate-level phosphorylation

Free Energy (G) energy available to do work

Heterotroph an organism requiring organic compounds as a carbon source

Oxidative Phosphorylation the production of ATP at the expense of a proton motive force formed by electron transport

Phototroph an organism capable of using light as an energy source

Proton Motive Force an energized state of the membrane resulting from the separation of charge and the elements of water (H^+ versus OH^-) across the membrane

Pure Culture a microbial culture containing a single kind of microorganism

Reduction Potential the inherent tendency (measured in volts) of a compound to donate electrons

Respiration the process in which a compound is oxidized with O_2 or an O_2 substitute functioning as the terminal electron acceptor, usually accompanied by ATP production by oxidative phosphorylation

Siderophores iron chelators that can bind iron present at very low concentrations

Sterile absence of all living organisms and viruses

Substrate-Level Phosphorylation production of ATP by the direct transfer of a high energy phosphate molecule from a phosphorylated organic compound to ADP

A key feature of a living system is the ability to direct chemical reactions and organize molecules into specific structures. The ultimate expression of this organization is self-replication (growth). The term **metabolism** is used to refer to all the chemical processes taking place within a cell. In this chapter we focus on metabolism and in the next chapter consider the actual process of cell growth.

Microbial cells are built of chemical substances of a wide variety of types, and when a cell grows, all these chemical constituents increase in amount. The basic chemical elements of a cell come from outside the cell, from the environment, but these chemical elements are transformed by the cell into the characteristic constituents of which the cell is composed. The chemicals from the environment of which a cell is built are called **nutrients.** Nutrients are taken up into the cell and are changed into cell constituents. This process by which a cell is built up from the simple nutrients obtained from its environment is called **anabolism.** Because anabolism results in the biochemical synthesis of new cell material, it is also called **biosynthesis.**

Biosynthesis is an *energy-requiring process*, and each cell must thus have a means of obtaining energy. Cells also need energy for other functions such as cell movement (motility) and transport of nutrients (∞ Sections 3.4 and 3.9). Like other nutrients, the energy source is also obtained from the environment, and two kinds of energy sources are used: *light* and *chemicals*. Although a number of organisms obtain their energy from light, most microorganisms obtain energy from the oxidation of chemical compounds. Chemicals used as energy sources are broken down into simpler constituents, and as this breakdown occurs, energy is released. The process by which chemicals are broken down and energy released is called **catabolism.** Catabolic reactions are the focus of this chapter.

4.1

Overview of Metabolism

A simplified view of cell metabolism is shown in Figure 4.1, which depicts how catabolic degradative reactions supply energy needed for cell functions and how anabolic reactions bring about the synthesis of cell components from nutrients. Note that in anabolism, nutrients from the environment are converted to cell *components,* whereas in catabolism, energy sources from the environment are converted to *waste products.*

Energy classes of microorganisms

It is conventional to group microorganisms in metabolic classes depending on the sources of *energy* they use. All the terms used to describe these classes employ the combining form *troph,* derived from a Greek word meaning "to feed." Thus, organisms that use *light* as an energy source are called **phototrophs** (*photo* is from the Greek word for "light"), and organisms that use *chemicals* as energy sources are called **chemotrophs.** Most of the organisms we deal with in microbiology use *organic* compounds as energy sources and thus are types of chemotrophs called **chemoorganotrophs.** Organisms able to use *in*organic chemicals as energy sources are called **chemolithotrophs.**

A knowledge of cell metabolism is essential for understanding the biochemistry of microbial growth. Also, a knowledge of metabolism aids in developing laboratory procedures for culturing microorganisms and in developing suitable procedures for preventing the growth of unwanted microorganisms. Because many of the important practical consequences of microbial growth, such as infectious disease or the production of useful products, are linked to microbial metabolism, a knowledge of microbial nutrition and metabolism is also of great use in medical and industrial microbiology. We begin with an overview of nutrition before considering metabolism.

CONCEPT CHECK   *4.1*

Metabolism involves two classes of chemical transformations, building up (biosynthetic) processes, called anabolism, and breaking down processes, called catabolism. Two kinds of energy sources can be used by cells—light and chemicals.

✔ *What class of metabolic reactions in the cell are energy-yielding reactions?*

✔ *Why are* chemoorganotrophs *so named?*

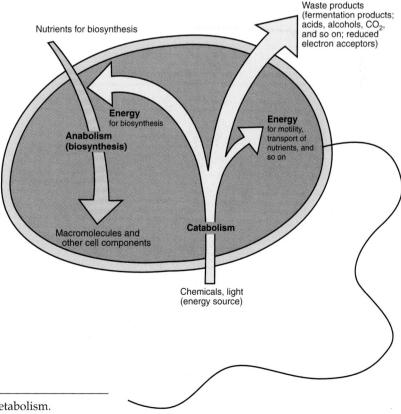

Figure 4.1 A simplified view of cell metabolism.

4.2

Microbial Nutrition

Chemical composition of a cell

Cells contain large amounts of small molecules as well as macromolecules. The cell may obtain most of the small molecules it needs from the environment in preformed condition or it may biosynthesize them from simpler molecules. Macromolecules, by contrast, are always synthesized in the cell. Although there are many naturally occurring elements, virtually the whole mass of a cell consists of substances composed of only four types of atoms: *carbon, oxygen, hydrogen,* and *nitrogen* (∞Figure 1.4). These four elements make up the backbone of macromolecules as well as all small organic molecules. A number of other elements are less abundant than C, O, H, and N but are nevertheless just as important to overall cell metabolism. These include phosphorus, potassium, calcium, magnesium, sulfur, iron, zinc, manganese, copper, molybdenum, cobalt, and a few other elements, depending on the organism.

Water accounts for about 90% of the wet weight of a cell, and macromolecules the bulk of the dry weight of a cell (Table 4.1). And of the four classes of macromolecules present in a cell, *proteins* are the most abundant by weight. Inorganic ions and monomers constitute the balance of the cell dry weight (Table 4.1). We now consider how nutrients are obtained by the cell from its environment.

Carbon and nitrogen

Nutrients can be divided into two classes: (1) *macronutrients,* which are required in large amounts (Table 4.2), and (2) *micronutrients,* which are required in only small amounts (see Table 4.3). We begin with a discussion of the major macronutrients carbon and nitrogen.

Most prokaryotes require an organic compound of some sort as their source of **carbon.** Nutritional studies have shown that many bacteria can assimilate various organic carbon compounds and use them to make new cell material. Amino acids, fatty acids, organic acids, sugars, nitrogen bases, aromatic compounds, and countless other organic compounds have been shown to be used by one bacterium or another. On a dry weight basis, a typical cell is about 50% carbon and carbon is the major element in all classes of macromolecules.

After carbon, the next most abundant element in the cell is **nitrogen.** A typical bacterial cell is about 12% nitrogen (by dry weight), and nitrogen is a major constituent of proteins, nucleic acids, and several other constituents in the cell. Nitrogen can be found in nature in both organic and inorganic forms (Table 4.2). However, the bulk of available nitrogen in nature is in *inorganic* form, either as ammonia (NH_3), nitrate (NO_3^-), or N_2. Most bacteria are capable of using ammonia as the sole nitrogen source, and many can also use nitrate. However, nitrogen gas (N_2) can be a nitrogen source for certain bacteria, the *nitrogen-fixing bacteria,* and we discuss the properties of these organisms in detail later (∞Sections 13.27, 14.14, 14.23, and 16.16).

TABLE 4.1	Chemical composition of a prokaryotic cell[a]		
Molecule	**Percent of dry weight[b]**	**Molecules per cell**	**Different kinds**
Total macromolecules	96	24,610,000	~2500
Protein	55	2,350,000	~1850
Polysaccharide	5	4,300	2[c]
Lipid	9.1	22,000,000	4[d]
DNA	3.1	2.1	1
RNA	20.5	255,500	~660
Total monomers	3.5		~350
Amino acids and precursors	0.5		~100
Sugars and precursors	2		~50
Nucleotides and precursors	0.5		~200
Inorganic ions	1		18
Total	100%		

[a]Data from Neidhardt, F. C., et al. (eds.). 1996. Escherichia coli and Salmonella typhimurium—*Cellular and Molecular Biology,* 2nd edition. American Society for Microbiology, Washington, DC.

[b]Dry weight of an actively growing cell of *E. coli* $\cong 2.8 \times 10^{-13}$ g.

[c]Assuming peptidoglycan and glycogen to be the major polysaccharides present.

[d]There are several classes of phospholipids, each of which exists in many kinds because of variability in fatty acid composition between species and because of different growth conditions.

TABLE 4.2 Macronutrients in nature and in culture media

Element	Usual form of nutrient found in the environment	Chemical form supplied in culture media
Carbon (C)	CO_2, organic compounds	Glucose, malate, acetate, pyruvate, hundreds of other compounds, or complex mixtures (yeast extract, peptone, and so on)
Hydrogen (H)	H_2O, organic compounds	H_2O, organic compounds
Oxygen (O)	H_2O, O_2, organic compounds	H_2O, O_2, organic compounds
Nitrogen (N)	NH_3, NO_3^-, N_2, organic nitrogen compounds	*Inorganic:* NH_4Cl, $(NH_4)_2 SO_4$, KNO_3, N_2 *Organic:* Amino acids, nitrogen bases of nucleotides, many other N-containing organic compounds
Phosphorus (P)	PO_4^{3-}	KH_2PO_4, Na_2HPO_4
Sulfur (S)	H_2S, SO_4^{2-}, organic S compounds, metal sulfides (FeS, CuS, ZnS, NiS, and so on)	Na_2SO_4, $Na_2S_2O_3$, Na_2S, cysteine, or other organic sulfur compounds
Potassium (K)	K^+ in solution or as various K salts	KCl, KH_2PO_4
Magnesium (Mg)	Mg^{2+} in solution or as various Mg salts	$MgCl_2$, $MgSO_4$
Sodium (Na)	Na^+ in solution or as NaCl or other Na salts	NaCl
Calcium (Ca)	Ca^{2+} in solution or as $CaSO_4$ or other Ca salts	$CaCl_2$
Iron (Fe)	Fe^{2+} or Fe^{3+} in solution or as FeS, $Fe(OH)_3$, or many other Fe salts	$FeCl_3$, $FeSO_4$, various chelated iron solutions (Fe^{3+} EDTA, Fe^{3+} citrate, and so on)

Other macronutrients:
P, S, K, Mg, Ca, Na, Fe

Phosphorus occurs in nature in the form of organic and inorganic phosphates and is required by the cell primarily for synthesis of nucleic acids and phospholipids. **Sulfur** is required because of its structural role in the amino acids cysteine and methionine (∞ Section 2.8) and because it is present in a number of vitamins, such as thiamine, biotin, and lipoic acid, as well as in coenzyme A. Sulfur undergoes a number of chemical transformations in nature carried out exclusively by microorganisms (∞ Section 14.15) and is available to organisms in a variety of forms. Most cell sulfur originates from inorganic sources, either sulfate (SO_4^{2-}) or sulfide (HS^-) (Table 4.2).

Potassium is required by all organisms. A variety of enzymes, including some of those involved in protein synthesis, specifically require potassium. **Magnesium** functions to stabilize ribosomes, cell membranes, and nucleic acids and is also required for the activity of many enzymes. **Calcium** (which is not an essential nutrient for the growth of many microorganisms) helps stabilize the bacterial cell wall and plays a key role in the heat stability of endospores (∞ Section 3.13). **Sodium** is required by some but not all organisms, and its need often reflects the habitat of the organism. For example, seawater has a high sodium content and marine microorganisms generally require sodium for growth, whereas closely related freshwater species are usually able to grow in the absence of sodium.

Although sometimes considered a micronutrient, **iron** is required by cells in larger amounts than other trace metals and is thus best considered a macronutrient. Iron plays a major role in cellular respiration, being a key component of the cytochromes and iron–sulfur proteins involved in electron transport (see Section 4.12 and Table 4.3). However, because most inorganic iron salts are highly insoluble, many organisms produce specific iron-binding agents called **siderophores,** which solubilize iron salts and transport iron into the cell. One major group of siderophores consists of derivatives of hydroxamic acid, which chelate ferric (Fe^{3+}) iron very strongly (Figure 4.2a). Once the iron–hydroxamate complex has passed into the cell, the iron is released and the hydroxamate can exit the cell and be utilized again for iron transport. In some bacteria, the iron-binding compounds are not hydroxamates but instead are phenolic acids. Enteric bacteria such as *Escherichia coli* and *Salmonella typhimurium* produce complex phenolic siderophores called **enterobactins.** These siderophores are derivatives of the aromatic compound catechol and have an extremely high binding affinity for iron. The structure of *E. coli* enterobactin is shown in Figure 4.2b. As we will see in Chapter 19, availability of iron has important consequences in the ability of many harmful (pathogenic) bacteria to grow in the body.

TABLE 4.3	Micronutrients (trace elements) needed by living organisms[a]

Element	Cellular function
Chromium (Cr)	Required by mammals for glucose metabolism; no known microbial requirement
Cobalt (Co)	Vitamin B_{12}; transcarboxylase (propionic acid bacteria)
Copper (Cu)	Certain proteins, notably those involved in respiration, for example, cytochrome c oxidase; or in photosynthesis, for example, plastocyanin; some superoxide dismutases
Manganese (Mn)	Activator of many enzymes; present in certain superoxide dismutases and in the water-splitting enzyme of photosystem II in oxygenic phototrophs
Molybdenum (Mo)	Present in various flavin-containing enzymes; also in molybdenum nitrogenase, nitrate reductase, sulfite oxidase, DMSO-TMAO reductases, some formate dehydrogenases, oxotransferases
Nickel (Ni)	Most hydrogenases; coenzyme F_{430} of methanogens; carbon monoxide dehydrogenase; urease
Selenium (Se)	Formate dehydrogenase; some hydrogenases; the amino acid selenocysteine
Tungsten (W)	Some formate dehydrogenases; oxotransferases of hyperthermophiles (for example, aldehyde: ferredoxin oxidoreductase of *Pyrococcus furiosus*)
Vanadium (V)	Vanadium nitrogenase; bromoperoxidase
Zinc (Zn)	Present in the enzymes carbonic anhydrase, alcohol dehydrogenase, RNA and DNA polymerases, and many DNA-binding proteins
Iron (Fe)[b]	Cytochromes, catalases, peroxidases, iron–sulfur proteins (for example, ferredoxin), oxygenases, all nitrogenases

[a]Not every micronutrient listed is required by all cells; some metals listed are found in enzymes present in only specific microorganisms.
[b]Needed in greater amounts than other metals—not generally considered a trace element.

Micronutrients (trace elements)

Although required in just tiny amounts, micronutrients are nevertheless just as critical to cell function as are macronutrients. Micronutrients are metals, many of which play a structural role in various enzymes, the cells' catalysts. Table 4.3 summarizes the major micronutrients of living systems and gives examples of enzymes in which each plays a role.

Because the requirement for trace elements is so small, for the laboratory culture of microorganisms it is frequently unnecessary to add trace elements to the culture medium. However, if a culture medium contains highly purified chemicals dissolved in high purity distilled water, a trace element deficiency can occur. In such cases a small amount of a solution of trace metals (Table 4.3) is added to the medium to make available the necessary metals.

Growth factors

Growth factors are *organic* compounds that, like micronutrients, are required in very small amounts and only by some cells. Growth factors include vitamins, amino acids, purines, and pyrimidines. Although most microorganisms are able to synthesize all these compounds, certain others require one or more of them preformed from the environment.

Vitamins are the most commonly needed growth factors. Most vitamins function as parts of coenzymes (see, for instance, Figures 4.8, 4.14, and 4.22), and these are summarized in Table 4.4. Many microorganisms are able to synthesize all the components of their coenzymes, but some are unable to do so and must be provided with certain parts of these coenzymes in the form of vitamins. Lactic acid bacteria, which include

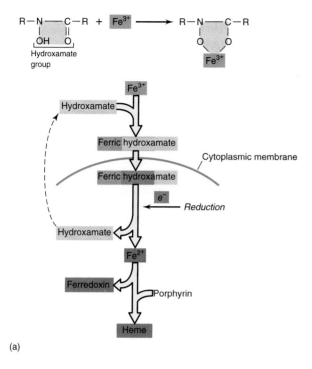

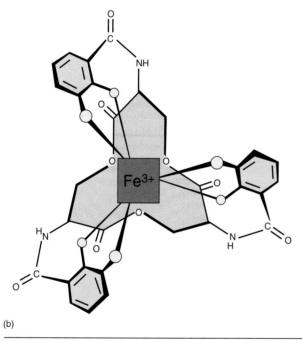

Figure 4.2 Iron-chelating agents produced by microorganisms, (a) Hydroxamate. Iron is bound as Fe^{3+} and released inside the cell as Fe^{2+}. (b) Ferric enterobactin of *Escherichia coli*. The oxygen atoms of each catechol molecule are shown in yellow.

the genera *Streptococcus, Lactobacillus, Leuconostoc,* and others (∞ Section 16.25), are renowned for their complex vitamin requirements, which are even more extensive than those of humans (see Table 4.5)! The vitamins most commonly required by microorganisms are thi-

TABLE 4.4	Vitamins and their functions
Vitamin	**Function**
p-Aminobenzoic acid	Precursor of folic acid
Folic acid	One-carbon metabolism; methyl group transfer
Biotin	Fatty acid biosynthesis; β-decarboxylations; CO_2 fixation
Cobalamin (B_{12})	Reduction of and transfer of single carbon fragments; synthesis of deoxyribose
Lipoic acid	Transfer of acyl groups in decarboxylation of pyruvate and α-ketoglutarate
Nicotinic acid (niacin)	Precursor of NAD^+; electron transfer in oxidation–reduction reactions
Pantothenic acid	Precursor of coenzyme A; activation of acetyl and other acyl derivatives
Riboflavin	Precursor of FMN, FAD in flavoproteins involved in electron transport
Thiamine (B_1)	α-Decarboxylations; transketolase
Vitamins B_6 (pyridoxal-pyridoxamine group)	Amino acid and keto acid transformations
Vitamin K group; quinones	Electron transport; synthesis of sphingolipids
Hydroxamates	Iron-binding compounds; solubilization of iron and transport into cell
Coenzyme M (CoM)	Required by certain methanogens; plays a role in methanogenesis

CONCEPT CHECK   *4.2*

The hundreds of chemical compounds present inside a living cell are formed from starting materials available as nutrients from the environment. Elements required in fairly large amounts are called macronutrients, while metals and organic compounds needed in very small amounts are called micronutrients and growth factors, respectively.

✔ *On a dry weight basis, what is the most abundant class of macromolecules in the cell?*

✔ *What two classes of macromolecules contain the bulk of the nitrogen in a cell?*

✔ *Why is an element like Co^{2+} considered a micronutrient whereas an element like C is considered a macronutrient?*

amine (vitamin B$_1$), biotin, pyridoxine (vitamin B$_6$), and cobalamin (vitamin B$_{12}$).

4.3

Culture Media

We can summarize our discussion of microbial nutrition by examining the chemical composition of several culture media for growing microorganisms in the laboratory (Table 4.5). Two broad classes of culture media are used in microbiology: **chemically defined** and **undefined (complex)**. Chemically defined media are prepared by adding precise amounts of highly purified inorganic or organic chemicals to distilled water. Therefore, the *exact chemical composition* of a defined medium is known. In many cases, however, knowledge of the exact composition of a medium is not critical. In these instances complex media may suffice or for various reasons may even be advantageous. Complex media employ digests of casein (milk protein), beef, soybeans, yeast cells, or any of a number of other highly nutritious (yet chemically undefined) substances. Such digests are available commercially in powdered form and can be weighed out rapidly and dissolved in distilled water to give a medium. However, a major concession in using a complex medium is loss of control of its precise nutrient composition.

Nutritional requirements and biosynthetic capacity

Table 4.5 shows three recipes for culture media, two defined and one complex. The complex medium is easiest to prepare and supports good growth of either organism shown in the table, the enteric bacterium *Escherichia coli* or the lactic acid bacterium *Leuconostoc mesenteroides*, an extremely fastidious (nutritionally demanding) bacterium. The simple defined medium supports excellent growth of *E. coli* but not of *L. mesenteroides*; growth of the latter organism in defined medium requires the addition of several organic nutrients and growth factors not needed by *E. coli* (Table 4.5). With this in mind, which organism, *E. coli* or *L. mesenteroides*, has a greater *biosynthetic* capacity? Obviously, *E. coli*, since its ability to grow on a simple defined culture medium means that it has the ability to synthesize *all* its organic cellular constituents from a single carbon compound, in this case glucose (Table 4.5). By contrast, *L. mesenteroides* has multiple growth factor requirements indicative of limited biosynthetic capacity. The complex nutritional needs of *L. mesenteroides* can be satisfied by either preparing a defined medium as shown in Table 4.5 (although in this case it could take quite a bit of time to do so) or by using a complex medium (Table 4.5), which by contrast can usually be quickly prepared.

TABLE 4.5 Examples of culture media for microorganisms with simple and demanding nutritional requirements		
Defined culture medium for *Escherichia coli*	**Defined culture medium for** *Leuconostoc mesenteroides*	**Complex culture medium for** *either E. coli* or *L. mesenteroides*
K$_2$HPO$_4$ 7 g	K$_2$HPO$_4$ 0.6 g	Glucose 15 g
KH$_2$PO$_4$ 2 g	KH$_2$PO$_4$ 0.6 g	Yeast extract 5 g
		Peptone 5 g
(NH$_4$)$_2$SO$_4$ 1 g	NH$_4$Cl 3 g	KH$_2$PO$_4$ 2 g
MgSO$_4$ 0.1 g	MgSO$_4$ 0.1 g	Distilled water 1000 ml
CaCl$_2$ 0.02 g	Glucose 25 g	pH 7
Glucose 4–10 g	Sodium acetate 20 g	
Trace elements (Fe, Co, Mn, Zn, Cu, Ni, Mo) 2–10 µg each	Amino acids (alanine, arginine, asparagine, aspartate, cysteine, glutamate, glutamine, glycine, histidine, isoleucine, leucine, lysine, methionine, phenylalanine, proline, serine, threonine, tryptophan, tyrosine, valine) 100–200 µg of each	
Distilled water 1000 ml pH 7		
	Purines and pyrimidines (adenine, guanine, uracil, xanthine) 10 mg of each	
	Vitamins (biotin, folate, nicotinic acid, pyridoxal, pyridoxamine, pyridoxine, riboflavin, thiamine, pantothenate, *p*-aminobenzoic acid) 0.01–1 mg of each	
	Trace elements (see first column) 2–10 µg each	
	Distilled water 1000 ml	
	pH 7	

It is important to understand when examining recipes for culture media such as those shown in Table 4.5 that *different microorganisms can have vastly different nutritional requirements*. Thus, for successful culture of a given microorganism it is necessary to understand its nutritional requirements and then supply it with its essential nutrients in the proper form and proportions in a culture medium. If care is taken in preparing culture media, it is usually quite easy to culture microorganisms in the laboratory.

With this foundation in nutritional principles, we turn our attention to the cell's use of nutrients to drive energy-yielding reactions—catabolism.

CONCEPT CHECK   **4.3**

Culture media supply the nutritional needs of microorganisms and can be either chemically defined or undefined (complex).

✔ *Why is the routine culture of* Leuconostoc mesenteroides *easier in a complex medium than in a chemically defined medium?*

✔ *In which medium, simple defined or complex (shown in Table 4.5), do you think* Escherichia coli *would grow faster? Why?*

4.4

Energetics

Energy is defined as the ability to do work. In this chapter, we discuss how living organisms use chemical energy. **Chemical energy** is the energy released when organic or inorganic compounds are oxidized. The first law of thermodynamics tells us that energy can be converted from one form to another but can be neither created nor destroyed. Because its many forms are interconvertible, energy is most conveniently expressed by a single energy unit. In biology the most commonly used energy units are the kilocalorie (kcal) and the kilojoule (kJ). A kilocalorie is defined as the quantity of heat energy necessary to raise the temperature of 1 kilogram (kg) of water 1°C. One kilocalorie is equivalent to 4.184 kJ. Because the kJ is widely used in microbial energetics, we will use this convention throughout this book.

Free energy

Chemical reactions are accompanied by *changes* in energy. The amount of energy involved in a chemical reaction is expressed in terms of the gain or loss of energy during the reaction. There are two expressions of the amount of energy released during a chemical reaction, abbreviated H and G. H, for *enthalpy,*

expresses the total amount of energy released during a chemical reaction. However, some of the energy released is not available to do useful work but instead is lost as heat energy. G, **free energy,** is used to express the energy released *that is available to do useful work.* The change in free energy during a reaction is expressed as $\Delta G^{0\prime}$, where the symbol Δ should be read "change in." Reactant and product concentrations and pH (when H^+ is a reactant or product) affect the observed free-energy changes. The superscripts ^{0} and $^\prime$ mean that a given free-energy value was obtained under "standard" conditions: pH 7, 25°C, all reactants and products initially at 1 M concentration*. If in the reaction

$$A + B \rightarrow C + D$$

the $\Delta G^{0\prime}$ is *negative,* then free energy is released and the reaction as written occurs spontaneously; such reactions are called **exergonic.** If, on the other hand, $\Delta G^{0\prime}$ is *positive,* the reaction does not occur spontaneously, but instead the reverse reaction (to the left) occurs spontaneously; such reactions are called **endergonic.**

In an exergonic reaction, the reaction proceeds until the concentration of products builds up, and then the reverse reaction, the conversion of products back to reactants, increases. An equilibrium is eventually reached in which the forward and reverse reactions are exactly balanced. This balanced condition does not mean that reactant and product occur in equal concentrations. The concentration of products and reactants at equilibrium is related to the free energy of the reaction. If the reaction proceeds with a *large* negative $\Delta G^{0\prime}$, then the equilibrium is far toward the products and very little of the reactants remains. In contrast, if the reaction proceeds with a *small* negative $\Delta G^{0\prime}$, then at equilibrium there are nearly equal amounts of products and reactants. By determining the concentrations of products and reactants at equilibrium, it is possible to calculate the free-energy yield of any reaction (see Appendix 1 for further details).

Free energy of formation and calculating $\Delta G^{0\prime}$

In addition to speaking of the free energy *yield* of reactions, it is also necessary to talk about the free energy *of* individual substances. This is the *free energy of formation*, the energy yielded or energy required for the *formation* of a given molecule from its constituent elements. By con-

---

*Although standard conditions are rarely an accurate reflection of actual conditions in a microbial habitat, especially as regards the concentration of reactants and products (see Appendix 1), in most cases standard free energy values are satisfactory for use in bioenergetic calculations.

vention, the free energy of formation (G^0_f) of the elements (for instance, C, H_2, N_2) is zero. If the formation of a *compound* from elements proceeds exergonically, then the free energy of formation of the compound is negative (energy is released), whereas if the reaction is endergonic (energy is required), then the free energy of formation of the compound is positive. A few examples of free energies of formation are given in Table 4.6. For most compounds G^0_f is *negative*, reflecting the fact that compounds tend to form spontaneously from elements. Again, the relative probabilities of different reactions (formations in this case) can be derived from comparison of the respective energies of formation. Thus, we see that glucose [G^0_f, -917.22 kJ/mole (mol)] is more likely to form from carbon, hydrogen, and oxygen than is methane (G^0_f, -50.75 kJ/mol) to form from carbon and hydrogen. The positive G^0_f for nitrous oxide ($+104.18$ kJ/mol) tells us that this molecule does not form spontaneously but rather decomposes to nitrogen and oxygen. The free energies of formation of a variety of compounds of microbiological interest are given in Appendix 1.

Using free energies of formation, it is possible to calculate the *change* in free energy occurring in a given reaction. For a simple reaction such as A + B → C + D, $\Delta G^{0\prime}$ is calculated by subtracting the *sum* of the free energies of formation of the reactants (in this case A and B) from that of the products (C and D). Thus,

$\Delta G^{0\prime}$ of A + B → C + D

$$= G^0_f[C + D] - G^0_f[A + B]$$

The saying "products minus reactants" summarizes the necessary steps for calculating changes in free energy during chemical reactions. However, it is necessary to balance the reaction chemically before free-energy calculations can be made. Appendix 1 details the steps involved in calculating free energies for any hypothetical reaction.

TABLE 4.6	Free energy of formation for a few compounds of biological interest
Compound	**Free energy of formation**[a]
H_2O	-237.17
CO_2	-394.4
H_2	0
O_2	0
NH_4^+	-79.37
N_2O	$+104.18$
Acetate	-369.4
Glucose	-917.22
CH_4	-50.75
CH_3OH	-175.4

[a]The free-energy values (G^0_f) are in *kJ/mol*.

CONCEPT CHECK   **4.4**

The chemical reactions of the cell are accompanied by changes in energy, expressed in kJ. A chemical reaction can occur with the release of free energy, in which case it is called exergonic, or with the consumption of free energy, in which case it is called endergonic.

✔ *What is free energy?*

✔ *In general, are catabolic reactions exergonic or endergonic?*

✔ *Using the data in Table 4.6, calculate $\Delta G^{0\prime}$ for the reaction $CH_4 + \frac{1}{2}O_2 \rightarrow CH_3OH$.*

4.5
Catalysis and Enzymes

Free-energy calculations tell us only what conditions prevail when the reaction or system is at equilibrium; they do not tell us how long it will take for equilibrium to be reached. The formation of water from gaseous oxygen and hydrogen is a good example. The energetics of this reaction is quite favorable: $H_2 + \frac{1}{2}O_2 \rightarrow H_2O$, $\Delta G^{0\prime} = -237$ kJ. However, if we were to simply mix O_2 and H_2 together, no measurable formation of water would occur within our lifetime. The explanation is that the rearrangement of oxygen and hydrogen atoms to form water requires that the chemical bonds of the reactants be broken first. The breaking of bonds requires energy, and this energy is referred to as **activation energy**. Activation energy is the amount of energy (in kJ) required to bring all molecules in a chemical reaction to the reactive state. For a reaction that proceeds with a net release of free energy (that is, an exergonic reaction), the situation is as diagrammed in Figure 4.3.

Enzymes

The idea of activation energy leads us to the concept of catalysis. A **catalyst** is a substance that serves to *lower* the activation energy of a reaction. A catalyst serves to *increase* the rate of reaction even though it itself is not changed. It is important to note that catalysts do not affect the energetics or the equilibrium of a reaction; catalysts affect only the *speed* at which reactions proceed.

Most reactions in living organisms would not occur at appreciable rates without catalysis. The catalysts of biological reactions are proteins called **enzymes.** Enzymes are highly specific in the reactions that they catalyze. That is, each enzyme catalyzes only a *single type* of chemical reaction, or in the case of certain enzymes, a class of closely related reactions. This specificity is related to the precise three-dimensional structure of the

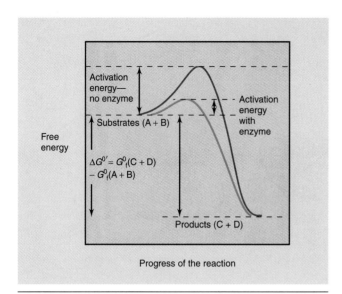

Figure 4.3 Progress of a hypothetical reaction: A + B → C + D and the concept of activation energy. Chemical reactions may not proceed spontaneously even though energy would be released, because the reactants must first be activated. Once activation has occurred, the reaction then proceeds spontaneously. Catalysts such as enzymes lower the required activation energy.

enzyme molecule. In an enzyme-catalyzed reaction, the enzyme temporarily combines with the reactant, which is termed a **substrate** (S) of the enzyme, forming an **enzyme–substrate complex.** Then, as the reaction proceeds, the **product** (P) is released and the enzyme (E) is returned to its original state:

$$E + S \rightleftharpoons E\text{---}S \rightleftharpoons E + P$$

The enzyme is generally much larger than the substrate(s), and the combination of enzyme and substrate(s) usually depends on weak bonds, such as hydrogen bonds, van der Waals forces, and hydrophobic interactions (∞ Section 2.8) to join the enzyme to the substrate. The small portion of the enzyme to which substrates bind is referred to as the **active site** of the enzyme.

Enzyme catalysis

The catalytic power of enzymes is impressive. Enzymes typically increase the rate of chemical reactions about 10^8 to 10^{20} times the rate that would occur spontaneously. To catalyze a specific reaction, an enzyme must do two things: (1) bind the correct substrate, and (2) position the substrate relative to the catalytically active groups at the enzyme's active site. Binding of substrate to enzyme produces the enzyme–substrate complex (Figure 4.4). This serves to align reactive groups and places strain on specific bonds in the substrate(s). The result of enzyme–substrate complex formation is a

reduction in the activation energy required to make the reaction proceed (Figure 4.3) with the conversion of substrate(s) to product(s). These steps are summarized diagrammatically in Figure 4.4 for the glycolytic enzyme *fructose bisphosphate aldolase* (see Section 4.10).

Note that the reaction depicted in Figure 4.3 is exergonic because the free energy of formation of the substrate is *greater* than that of the product. Enzymes can also catalyze endergonic reactions, converting energy-poor substrates to energy-rich products. In this case, not only must an activation energy barrier be overcome, but sufficient free energy must also be put *into* the system to raise the energy level of the substrates to that of the products. Although theoretically all enzymes are reversible in their action, in practice, enzymes catalyzing highly exergonic or highly endergonic reactions are essentially unidirectional. If a particularly exergonic reaction needs to be reversed during cellular metabolism, a distinctly different enzyme is frequently involved in the reaction.

Structure of enzymes

As we have discussed, enzymes are proteins, polymers of subunits called amino acids (∞ Section 2.8). Each enzyme has a specific three-dimensional shape. The linear array of amino acids (primary structure) folds and twists into a specific configuration to achieve secondary and tertiary structure. The precise conformation of an enzyme may be seen more easily in a computer-generated space-filling model (Figure 4.5). In this example of the enzyme *lysozyme*, the large cleft is the site where the substrate binds (the active site).

Being proteins, enzymes are also subject to the effects of physical and chemical variables, most notably temperature and pH. For example, enzymes from organisms capable of growth at high temperatures (thermophiles and hyperthermophiles) (∞ Sections 5.8 and 17.3) are generally quite heat-stable, while those from nonthermophiles are usually rapidly inactivated by heat. Such differences arise from variations in amino acid *sequence* in different enzymes. Recall that differences in amino acid sequence dictate differences in *folding* of the protein. A specifically folded protein thus assumes specific binding and physical properties.

Many enzymes contain small nonprotein molecules that participate in the catalytic function but are not considered substrates in the usual sense. These small enzyme-associated molecules are divided into two categories on the basis of the nature of their association with the enzyme, *prosthetic groups* and *coenzymes.* **Prosthetic groups** are bound very tightly to their enzymes, usually permanently. The heme group present in cytochromes is an example of a prosthetic group, and cytochromes will be described in detail later in this chapter. **Coenzymes** are bound rather loosely to

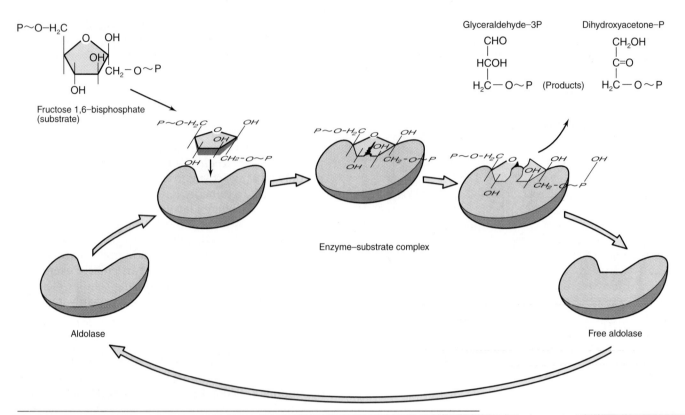

Figure 4.4 The catalytic cycle of an enzyme as depicted for the enzyme fructose bisphosphate aldolase. This enzyme catalyzes the following reaction: fructose 1,6-bisphosphate $\rightleftharpoons$ glyceraldehyde 3-phosphate + dihydroxyacetone phosphate in glycolysis (see Figure 4.12). Following binding of fructose 1,6-bisphosphate in the formation of the enzyme–substrate complex, the conformation of the enzyme is altered, placing strain on certain bonds of the substrate, which break and yield the two products.

Figure 4.5 Computer-generated space-filling model of the enzyme lysozyme. The substrate binding site (active site) is in the large cleft on the left side of the model. See Section 3.5 for a discussion of the mode of action of lysozyme.

enzymes, and a single coenzyme molecule may associate with a number of different enzymes at different times during growth. Coenzymes serve as intermediate carriers of small molecules from one enzyme to another. Most coenzymes are derivatives of vitamins (see Table 4.4).

Enzymes are named either for the substrate they bind or for the chemical reaction they catalyze, by addition of the combining form *-ase*. Thus cellul*ase* is an enzyme that attacks cellulose, glucose oxid*ase* is an enzyme that catalyzes the oxidation of glucose, and ribonucle*ase* is an

CONCEPT CHECK   ***4.5***

The reactants in a chemical reaction must first be activated before the reaction can take place, and this requires the use of a catalyst. Enzymes are catalytic proteins that are highly specific in the reactions they catalyze, and this specificity resides in the folding pattern of the polypeptide(s) in the protein.

✔ *What is the function of a catalyst?*

✔ *What class of macromolecules are enzymes?*

✔ *Where on an enzyme does its substrate bind?*

enzyme that decomposes ribonucleic acid. A more formal nomenclature system employing a specific numbering system is used to classify enzymes more precisely.

4.6

Oxidation–Reduction

The utilization of chemical energy in living organisms involves **oxidation–reduction** (also called **redox**) reactions. Chemically, an oxidation is defined as the *removal* of an electron or electrons from a substance. A reduction is defined as the *addition* of an electron (or electrons) to a substance. In biochemistry, oxidations and reductions frequently involve the transfer of not just electrons but whole hydrogen atoms also. A hydrogen atom (H) consists of an electron plus a proton. When the electron is removed, the hydrogen atom becomes a *proton* (or hydrogen ion, H^+). We will on occasion need to distinguish between oxidation–reduction reactions involving electrons only or hydrogen atoms only, but reserve this distinction for the appropriate time (see Sections 4.12 and 4.13).

Electron donors and acceptors

Oxidation–reduction reactions involve electrons being donated by an electron donor and being accepted by an electron acceptor. For example, hydrogen gas, H_2, can release electrons and hydrogen ions (protons) and become oxidized:

$$H_2 \rightarrow 2\,e^- + 2\,H^+$$

However, electrons cannot exist alone in solution; they must be part of atoms or molecules. The equation as drawn thus gives us chemical information but does not itself represent a real reaction. The above reaction is only a *half reaction*, a term that implies the need for a second half reaction. This is because for any *oxidation* to occur, a subsequent *reduction* must also occur. For example, the oxidation of H_2 could be coupled to the reduction of many different substances including O_2 in a second reaction:

$$\tfrac{1}{2} O_2 + 2\,e^- + 2\,H^+ \rightarrow H_2O$$

This half reaction, which is a reduction, when coupled to the oxidation of H_2 above, yields the following overall balanced reaction:

$$H_2 + \tfrac{1}{2} O_2 \rightarrow H_2O$$

In reactions of this type, we will refer to the substance *oxidized,* in this case H_2, as the **electron donor,** and the substance *reduced,* in this case O_2, as the **electron acceptor** (Figure 4.6). The key to understanding biological

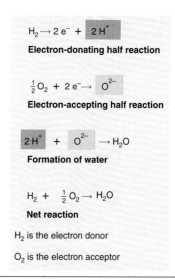

$$H_2 \rightarrow 2\,e^- + \boxed{2\,H^+}$$

Electron-donating half reaction

$$\tfrac{1}{2} O_2 + 2\,e^- \rightarrow \boxed{O^{2-}}$$

Electron-accepting half reaction

$$\boxed{2\,H^+} + \boxed{O^{2-}} \rightarrow H_2O$$

Formation of water

$$H_2 + \tfrac{1}{2} O_2 \rightarrow H_2O$$

Net reaction

H_2 is the electron donor

O_2 is the electron acceptor

Figure 4.6 Example of an oxidation–reduction reaction: The formation of H_2O from H_2 and O_2.

oxidations and reductions is to keep straight the proper half reactions—there must always be one reaction involving an electron *donor* and another reaction involving an electron *acceptor.*

Reduction potentials

Substances vary in their tendency to give up electrons and become oxidized or to accept electrons and become reduced. This tendency is expressed as the **reduction potential** (E_0') of the substance. This potential is measured electrically in reference to a standard substance, H_2. By convention, reduction potentials are expressed for half reactions written as *reductions.* Thus, oxidized form + $e^- \rightarrow$ reduced form. If protons are involved in the reaction, as is often the case, then the reduction potential is to some extent influenced by the hydrogen ion concentration (pH). By convention in biology, reduction potentials are given for neutrality (pH 7) because the cytoplasm of the cell is neutral or nearly so. Using these conventions, at pH 7 the reduction potential (E_0') of

$$\tfrac{1}{2} O_2 + 2\,H^+ + 2\,e^- \rightarrow H_2O$$

is +0.816 volts (V), and that of

$$2\,H^+ + 2\,e^- \rightarrow H_2$$

is −0.421 V.

Oxidation–reduction couples and complete O–R complete reactions

Most molecules can be either electron donors or electron acceptors under different circumstances, depending on what other substances they react with. The same atom on each side of the arrow in the half reactions can be thought of as representing an oxidation–reduction (O–R) couple, such as $2 H^+/H_2$ or $\frac{1}{2} O_2/H_2O$. When writing an O–R couple, the *oxidized* form is always placed on the left.

In constructing complete oxidation–reduction reactions from their constituent half reactions, it is simplest to remember that the reduced substance of an O–R couple whose reduction potential is more negative *donates* electrons to the oxidized substance of an O–R couple whose potential is more positive. Thus, in the couple $2 H^+/H_2$, which has a potential of -0.42 V, H_2 has a great tendency to *donate* electrons. On the other hand, in the couple $\frac{1}{2} O_2/H_2O$, which has a potential of $+0.82$ V, H_2O has a very slight tendency to donate electrons, but O_2 has a great tendency to *accept* electrons. It follows then that in a reaction of H_2 and O_2, H_2 serves as the electron *donor* and becomes oxidized, and O_2 serves as the electron *acceptor* and becomes reduced (Figure 4.6). Even though by chemical convention both half reactions are written as reductions, in an actual O–R reaction one of the two half reactions must be written as an oxidation and therefore proceeds in the reverse direction. Thus, note that in the reaction shown in Figure 4.6, the oxidation of H_2 to $2 H^+ + 2 e^-$ is

reversed from the formal half reaction, written as a reduction.

The electron tower

A convenient way of viewing electron transfer in biological systems is to imagine a vertical tower (Figure 4.7). The tower represents the range of reduction potentials for O–R couples from the most negative at the top to the most positive at the bottom. The reduced substance in the pair at the top of the tower has the *greatest* amount of potential energy (roughly the energy that it took to lift the substance to the top), whereas the oxidized substance in the couple at the bottom of the tower has the *greatest* tendency to accept electrons.

As electrons from the electron donor at the top of the tower fall, they can be "caught" by acceptors at various levels. The difference in electric potential between two substances is expressed as $\Delta E_0'$. The farther the electrons drop from a donor before they are caught by an acceptor, the greater the amount of energy released; that is, $\Delta E_0'$ *is proportional to* $\Delta G^{0'}$ (Figure 4.7). O_2, at the bottom of the tower, is the most favorable electron acceptor used by organisms. In the middle of the tower, O–R

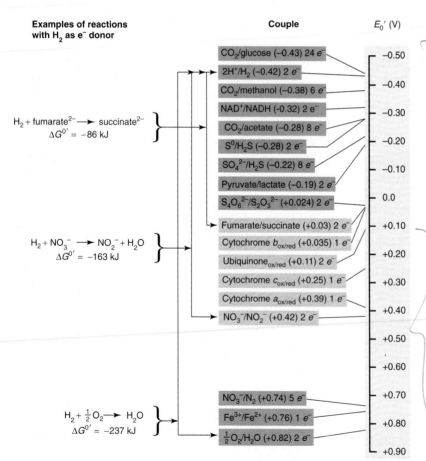

Figure 4.7 The electron tower. O–R couples are arranged from the strongest reductants (negative reduction potentials) at the top to the strongest oxidants (positive reduction potentials) at the bottom. As electrons are donated from the top of the tower, they can be "caught" by acceptors at various levels. The farther the electrons fall before they are caught, the greater the difference in reduction potential between electron donor and electron acceptor and the more energy is released. As an example of this, on the left is shown the differences in energy released when a single electron donor, H_2, reacts with any of three different electron acceptors, fumarate, nitrate, and oxygen.

couples can act as either electron donors or acceptors. For instance, the $2 H^+/H_2$ couple has a reduction potential of -0.42 V. The fumarate–succinate couple has a potential of $+0.02$ V. Hence, the oxidation of hydrogen (the electron donor) can be coupled to the reduction of fumarate (the electron acceptor):

$$H_2 + fumarate^{2-} \rightarrow succinate^{2-}$$

On the other hand, the oxidation of succinate to fumarate can be coupled to the reduction of NO_3^- or $\frac{1}{2} O_2$:

$$Succinate^{2-} + NO_3^- \rightarrow fumarate^{2-} + NO_2^- + H_2O$$

$$Succinate^{2-} + \tfrac{1}{2} O_2 \rightarrow fumarate^{2-} + H_2O$$

Hence, under conditions where oxygen is absent (called *anoxic*) in the presence of H_2, fumarate can be an electron acceptor (producing succinate), and under other conditions (for example, anoxic in the presence of NO_3^-, or aerobic) succinate can be an electron donor (producing fumarate). Indeed, all the transformations involving fumarate and succinate described here are carried out by various microorganisms under certain nutritional and environmental conditions.

In catabolism the electron donor is often referred to as an **energy source.** Many potential electron donors exist in nature (∞ Chapters 13 and 14), but for now it is essential to understand that it is the *complete* oxidation–reduction reaction that actually releases energy. As discussed in the context of the electron tower, the amount of energy released in an O–R reaction depends on the nature of *both* the electron donor and the electron acceptor: the greater the difference between reduction potentials of the two half reactions, the more energy there will be released when they react (Figure 4.7) (see Appendix 1).

CONCEPT CHECK   4.6

Oxidation–reduction reactions, which are involved in the energy-yielding reactions of cells, involve the transfer of electrons from one reactant to another. The energy source, which is the electron donor, gives up one or more electrons, which are transferred to an electron acceptor. The tendency of a compound to accept or release electrons is expressed quantitatively by its reduction potential.

✔ *In the reaction $H_2 + \frac{1}{2} O_2 \rightarrow$ what is the electron* donor *and what is the electron* acceptor?

✔ *What is the E_0' of the $2 H^+/H_2$ couple?*

✔ *Why is NO_3^- a better electron acceptor than fumarate?*

4.7
Electron Carriers

In the cell, the transfer of electrons in an oxidation–reduction reaction from donor to acceptor involves one or more intermediates referred to as **carriers.** When such carriers are used, we refer to the initial donor as the **primary electron donor** and to the final acceptor as the **terminal electron acceptor.** The net energy change of the complete reaction sequence is determined by the *difference* in reduction potentials between the primary donor and the terminal acceptor. The transfer of electrons through the intermediates involves a series of oxidation–reduction reactions, but the energy change from these individual steps must add up to the value obtained by considering only the starting and ending compounds.

The intermediate electron carriers may be divided into two general classes: those freely diffusible and those firmly attached to enzymes in the cytoplasmic membrane. The fixed carriers function in membrane-associated electron transport reactions and are discussed in Section 4.12. Freely diffusible carriers include the coenzymes nicotinamide-adenine dinucleotide (NAD^+) and NAD-phosphate ($NADP^+$) (Figure 4.8). NAD^+ and $NADP^+$ are *hydrogen atom* carriers and always transfer two hydrogen atoms to the next carrier in the chain. Such hydrogen atom transfer is referred to as a dehydrogenation.[†]

The reduction potential of the $NAD^+/NADH$ (or $NADP^+/NADPH$) couple is -0.32 V, which places it fairly high on the electron tower; that is, NADH (or NADPH) is a good electron *donor.* However, although the NAD^+ and $NADP^+$ couples have the same reduction potentials, they generally function in different capacities in the cell. $NAD^+/NADH$ is directly involved in energy-generating (catabolic) reactions, whereas $NADP^+/NADPH$ is involved primarily in biosynthetic (anabolic) reactions.

Coenzymes increase the diversity of O–R reactions by making it possible for chemically dissimilar molecules to interact as initial electron donor and ultimate electron acceptor, the coenzyme acting as intermediary. As we have discussed, most biological reactions are catalyzed by specific enzymes that can react with only a limited range of substrates. Oxidation–reduction reactions may be considered to proceed in three stages: removal of electrons from the primary donor, transfer of electrons through one or a series of electron carriers, and addition of electrons to the terminal acceptor. Each step in the reaction is catalyzed by a different enzyme, each of which

---

[†]Strictly speaking NAD^+ or $NADP^+$ carries two electrons and one proton, the second H^+ being released to solution. Therefore, $NAD^+ + 2 e^- + 2 H^+$ actually yields $NADH + H^+$. However, for simplicity, we write $NADH + H^+$ as NADH.

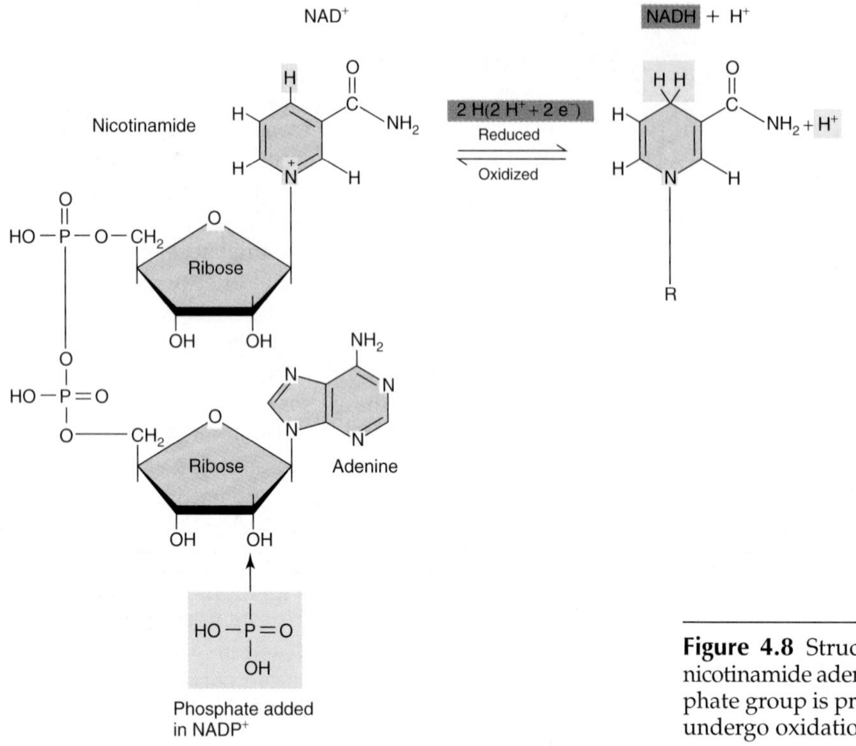

Figure 4.8 Structure of the oxidation–reduction coenzyme nicotinamide adenine dinucleotide (NAD$^+$). In NADP$^+$, a phosphate group is present, as indicated. Both NAD$^+$ and NADP$^+$ undergo oxidation–reduction as shown.

binds to its substrate and to its specific coenzyme. Figure 4.9 is a schematic diagram showing the functioning of the coenzyme NAD$^+$ in a two-part reaction. Note that after a coenzyme has performed its chemical function in one reaction, it can diffuse through the cytoplasm until it collides with another enzyme that requires the coenzyme in that form. Following conversion of the coenzyme back to its original form, the whole process can be repeated (Figure 4.9).

CONCEPT CHECK   4.7

The transfer of electrons from donor to acceptor in a cell involves the participation of one or more electron carriers. Some electron carriers are membrane-bound, whereas others are freely diffusible, transferring electrons from one place to another in the cell.

✔ *What is the difference between an* electron *and a* hydrogen atom?

✔ *Is NADH a better electron donor than H$_2$? Why or why not?*

4.8

High Energy Compounds

Energy released as a result of oxidation–reduction reactions must be conserved for cell functions. In living organisms, chemical energy released in O–R reactions is most commonly transferred to a variety of phosphate compounds in the form of **high energy phosphate bonds;** these compounds then function as the energy source to drive energy-requiring reactions in the cell.

In phosphorylated compounds, phosphate groups are attached via oxygen atoms by *ester* or *anhydride* bonds, as illustrated in Figure 4.10. However, not all phosphate bonds are high energy bonds. As a means of expressing the energy of phosphate bonds, the free energy released when water is added and the phosphate is hydrolyzed can be given. As seen in Figure 4.10, the $\Delta G^{0\prime}$ of hydrolysis of the phosphate bond in glucose 6-phosphate is only -13.8 kJ/mol, whereas the $\Delta G^{0\prime}$ of hydrolysis of the phosphate bond in phosphoenolpyruvate is -51.6 kJ/mol, almost four times that of glucose 6-phosphate. Thus, phosphoenolpyruvate, a phosphoanhydride, is considered a *high energy compound* and glucose 6-phosphate, a phosphate ester, is not.

Adenosine triphosphate (ATP)

The most important high energy phosphate compound in living organisms is adenosine triphosphate (ATP). ATP consists of the ribonucleoside adenosine, to which three phosphate molecules are bonded in series (Figure 4.10). ATP serves as the prime energy carrier in living organisms, being generated during exergonic reactions and being used to drive endergonic reactions. From the structure of ATP (Figure 4.10) it can be seen that two of the phosphate bonds of ATP are phosphoanhydrides and have high free energies of hydrolysis.

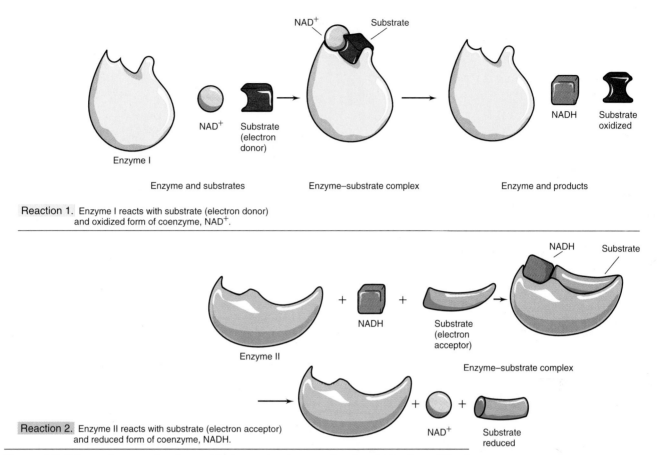

Reaction 1. Enzyme I reacts with substrate (electron donor) and oxidized form of coenzyme, NAD$^+$.

Reaction 2. Enzyme II reacts with substrate (electron acceptor) and reduced form of coenzyme, NADH.

Figure 4.9 Schematic example of an oxidation–reduction reaction involving the oxidized and reduced forms of the coenzyme nicotinamide-adenine dinucleotide, NAD$^+$ and NADH.

It should be emphasized that although we express the energy of high energy phosphate bonds in terms of the free energy of hydrolysis, in actuality it is undesirable for these bonds to hydrolyze in cells in the absence of a second reaction that can utilize the energy released because the free energy of hydrolysis would then be lost to the cell as heat. The free energy of high energy phosphate bonds is generally used to drive biosynthetic reactions and other aspects of cell function through carefully regulated processes in which the energy released from ATP hydrolysis is coupled to energy-requiring reactions.

Coenzyme A

In addition to high energy phosphate compounds, certain other high energy compounds are produced in the cell and can conserve the energy released in exergonic reactions. These include derivatives of coenzyme A (for example, acetyl-CoA; see structure in Figure 4.22). Coenzyme A derivatives contain *sulfo*anhydride (thioester) instead of *phospho*anhydride (Figure 4.10) bonds and yield sufficient free energy on hydrolysis to drive the synthesis of a high energy phosphate bond (see

Table 2.3 for the chemical structure of thioesters and phosphoanhydrides). For example, in the reaction: acetyl-S-CoA + H$_2$O + ADP + P → acetate$^-$ + HS-CoA + ATP + H$^+$, the energy released from the hydrolysis of coenzyme A is conserved in the synthesis of ATP. Coenzyme A derivatives (acetyl-CoA is just one of many) are especially important to the energetics of anaerobic microorganisms, in particular those whose energy metabolism involves fermentation (see Sections 4.10, 13.19, and 13.20); we will thus return to the importance of these compounds in Chapter 13.

CONCEPT CHECK 4.8

The energy released in oxidation–reduction reactions is not wasted but is conserved by the formation of certain biochemical compounds that contain high energy phosphate or sulfur bonds. One of the most common high energy phosphate compounds is ATP, which serves as a prime energy carrier in the cell.

✔ *Why are ATP and ADP considered high energy phosphate compounds whereas AMP is not?*

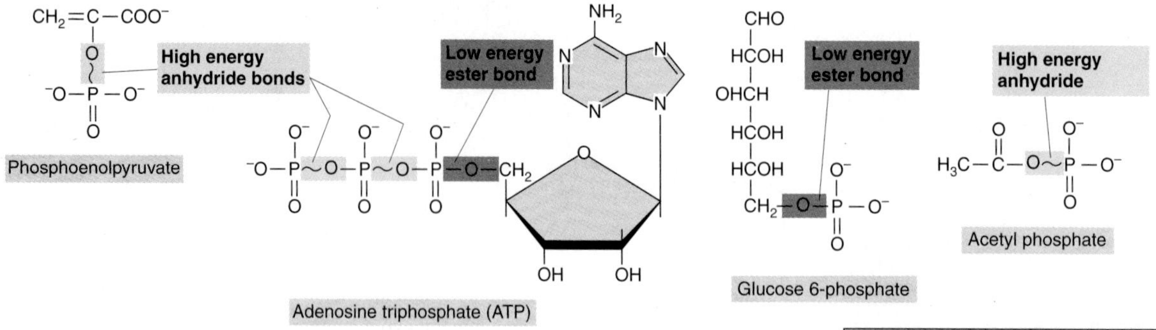

Figure 4.10 High energy phosphate bonds. The table shows the free energy of hydrolysis of some of the key phosphate esters and anhydrides, indicating that some of the phosphate ester bonds are of higher energy than others. Structures of four of the compounds are given to indicate the position of low energy and high energy bonds. ATP contains three phosphates, but only two of them are high energy (shown in blue). ADP contains two phosphates of which only one is high energy. AMP does not contain a high energy phosphate bond.

Compound	$G^{0'}$ kJ/mol
High energy	
Phosphoenolpyruvate	−51.6
1,3-Bisphosphoglycerate	−52.0
Acetyl phosphate	−44.8
ATP	−31.8
ADP	−31.8
Low energy	
AMP	−14.2
Glucose 6-phosphate	−13.8

4.9

Fermentation and Substrate-Level Phosphorylation

In the following pages we will be concerned with the mechanisms by which ATP is synthesized as a result of oxidation–reduction reactions involving *organic* compounds. The pathways for the oxidation of organic compounds and conservation of energy in ATP can be divided into two major groups: (1) **fermentation,** in which the O–R process occurs in the *absence* of any added terminal electron acceptors; and (2) **respiration,** in which molecular oxygen or some other oxidant serves as the terminal electron acceptor.

In the absence of externally supplied electron acceptors, many organisms perform *internally balanced* oxidation–reduction reactions of organic compounds with the release of energy, a process called **fermentation.** There are many different types of fermentations (∞ Chapter 13), but under fermentative conditions only *partial* oxidation of the carbon atoms of the organic compound occurs and therefore only a small amount of the potential energy available is released. The oxidation in a fermentation is coupled to the subsequent reduction of an organic compound generated from catabolism of the initial fermentable substrate; thus, no externally supplied electron acceptor is required (Figure 4.11).

ATP is produced in fermentations by a process called **substrate-level phosphorylation.** In substrate-level phosphorylation, ATP is synthesized during specific enzymatic steps in the catabolism of the organic compound. This is in contrast to **oxidative** (or **electron transport**)

phosphorylation (discussed later), where ATP is produced via membrane-mediated events not connected directly to the metabolism of specific substrates.

An example of fermentation is the catabolism of glucose by a lactic acid bacterium:

$$\text{Glucose} \rightarrow 2 \text{ lactate}^- + 2 \text{ H}^+$$

$$(C_6H_{12}O_6 \rightarrow 2 C_3H_4O_3^- + 2 H^+)$$

Note that this is a balanced reaction and that the products, lactate plus protons, have the same proportion of hydrogen and oxygen atoms as glucose. Likewise, a similar situation exists in the catabolism of glucose by yeast in the absence of oxygen:

$$\text{Glucose} \rightarrow 2 \text{ ethanol} + 2 \text{ carbon dioxide}$$

$$(C_6H_{12}O_6 \rightarrow 2 C_2H_6O + 2 CO_2)$$

Note that in this reaction some of the carbon atoms end up in CO_2, a more *oxidized* form than the carbon atoms in the starting molecule, glucose, whereas other carbon atoms end up in ethanol, which is more *reduced* (that is, it has more hydrogens and electrons per carbon atom) than glucose. In each fermentation, oxidation–reduction is internally balanced.

The energy released in the fermentation of glucose to ethanol or lactate (−235 and −198 kJ/mol, respectively) is conserved by substrate-level phosphorylations in the form of high energy phosphate bonds in ATP, with a net production of *two* such bonds in each case. We now discuss the details of the fermentation of glucose and the manner in which some of the energy released is conserved in high energy phosphate bonds.

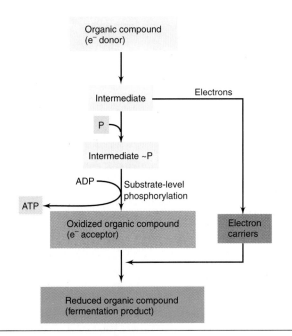

Figure 4.11 Carbon and electron flow in fermentation. Note that there is no externally supplied electron acceptor and that ATP is produced by donation of a high energy phosphate group from a phosphorylated intermediate directly to ADP.

CONCEPT CHECK   4.9

Some compounds can serve as electron donors in the absence of an externally added electron acceptor in a process called fermentation. A fermentation is an internally balanced oxidation–reduction reaction in which some atoms of the energy source become more reduced whereas others become more oxidized, and energy is produced by substrate-level phosphorylation.

✔ *Define the word* fermentation.

✔ *How does fermentation most clearly differ from respiration?*

4.10
Glycolysis: The Embden–Meyerhof Pathway

A common biochemical pathway for the fermentation of glucose is **glycolysis,** also named the **Embden–Meyerhof pathway** for its major discoverers. Glycolysis can be divided into three major stages, each involving a series of individually catalyzed enzymatic reactions (Figure 4.12).

Stage I of glycolysis is a series of preparatory rearrangements, reactions that do not involve oxidation–reduction and do not release energy but that lead to the production from glucose of two molecules of the key intermediate, *glyceraldehyde 3-phosphate.* In Stage II, oxidation–reduction occurs, high phosphate bond energy is produced in the form of ATP, and two molecules of pyruvate are formed. In Stage III, a second oxidation–reduction reaction occurs and fermentation products (for example, ethanol and CO_2, or lactic acid) are formed (Figure 4.12).

Stages I and II: Preparatory and redox reactions

In Stage I, glucose is phosphorylated by ATP yielding glucose 6-phosphate; phosphorylation reactions of this type often occur preliminary to oxidation. The initial phosphorylation of glucose *activates* the molecule for the subsequent reactions. Glucose 6-phosphate is converted to an isomeric form, fructose 6-phosphate, and a second phosphorylation leads to the production of *fructose 1,6-bisphosphate,* which is a key intermediate product of glycolysis. The enzyme *aldolase* catalyzes the splitting of fructose 1,6-bisphosphate into two three-carbon molecules, glyceraldehyde 3-phosphate and its isomer, dihydroxyacetone phosphate (see also Figure 4.4).[‡] Note that thus far, there have been no oxidation–reduction reactions and that all the reactions, including the consumption of ATP, proceed without electron transfers.

The oxidation reaction of glycolysis occurs in Stage II during the conversion of glyceraldehyde 3-phosphate to 1,3-bisphosphoglyceric acid. In this reaction (which occurs twice, once for each molecule of glyceraldehyde 3-phosphate), an enzyme whose coenzyme is NAD^+ accepts two hydrogen atoms and NAD^+ is converted to NADH; the enzyme catalyzing this reaction is called *glyceraldehyde-3-phosphate dehydrogenase.* Simultaneously, each glyceraldehyde-3-P molecule is phosphorylated by the addition of a molecule of inorganic phosphate. This reaction, in which inorganic phosphate is converted to organic form, sets the stage for energy conservation by substrate-level phosphorylation; ATP formation is made possible because each of the phosphates on a molecule of 1,3-bisphosphoglyceric acid represents a high energy phosphate bond (see Figure 4.10). The synthesis of ATP occurs when each molecule of 1,3-bisphosphoglyceric acid is converted to 3-phosphoglyceric acid, and later on in the pathway, when each molecule of phosphoenolpyruvate is converted to pyruvate (Figure 4.12).

In glycolysis, *two* ATP molecules are consumed in the two phosphorylations of glucose, and *four* ATP molecules are synthesized (two from each 1,3-bisphospho-

---

[‡]There is an enzyme that catalyzes the interconversion of dihydroxyacetone phosphate and glyceraldehyde 3-phosphate. For simplicity, we consider here only glyceraldehyde 3-phosphate since it is the compound that is further metabolized.

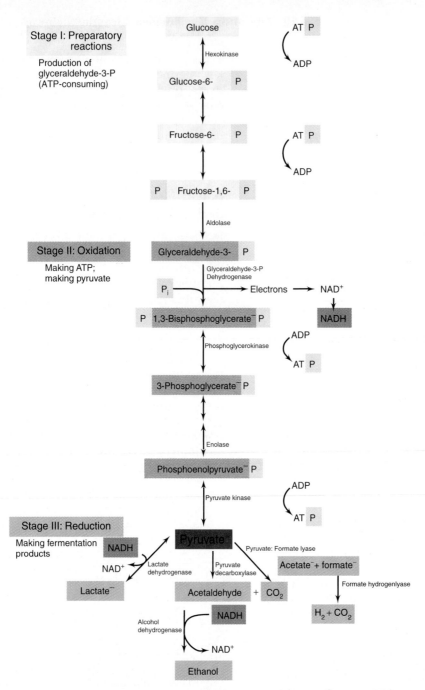

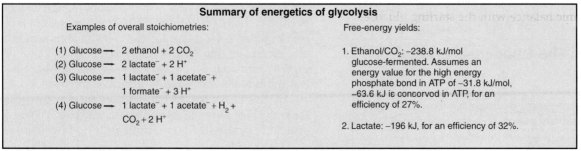

Figure 4.12 Embden–Meyerhof pathway (glycolysis), the sequence of enzymatic reactions in the conversion of glucose to pyruvate and then to fermentation products (enzymes are shown in small type). Note that when fructose bisphosphate is split, *two* molecules of glyceraldehyde 3-P are formed, although for simplicity, only one molecule is shown in the figure. Also note that pyruvate is the central "hub" of glycolysis—all fermentation products are made from pyruvate and just a few common examples are given. Compare the color coding of this figure with that of Figure 4.11.

glyceric acid converted to pyruvate). Thus, the *net gain* to the organism is two molecules of ATP per molecule of glucose fermented.

Stage III: Production of fermentation products

During the formation of two molecules of 1,3-bisphosphoglyceric acid, two molecules of NAD^+ are reduced to NADH (see Figure 4.12). However, a cell contains only a small amount of NAD^+, and if all of it were converted to NADH, the oxidation of glucose would stop; the continued oxidation of glyceraldehyde 3-phosphate can proceed only if there is a molecule of NAD^+ present to accept released electrons. This "roadblock" is overcome in fermentation by the oxidation of NADH back to NAD^+ through reactions involving the reduction of pyruvate to any of a variety of **fermentation products.** In the case of yeast, pyruvate is reduced to ethanol with the release of CO_2. In lactic acid bacteria (or in muscle tissue rendered anoxic by vigorous exercise) pyruvate is reduced to lactate (see lower part of Figure 4.12). Many routes of pyruvate reduction in various fermentative prokaryotes are known (∞Chapter 13), but the net result is the same; NADH must be returned to the oxidized form, NAD^+, for the energy-yielding reactions of fermentation to continue. As a diffusible coenzyme, NADH can move away from glyceraldehyde-3-phosphate dehydrogenase, attach to an enzyme that reduces pyruvate to lactic acid (lactate dehydrogenase), and diffuse away once again following conversion to NAD^+ to repeat the cycle all over again (see Figure 4.9 for details of this mechanism).

In any energy-yielding process, oxidation must balance reduction and there must be an electron acceptor for each electron removed. In this case, the *reduction* of NAD^+ at one enzymatic step in glycolysis is balanced with its *oxidation* at another. The final product(s) must also be in oxidation–reduction balance with the starting substrate, glucose. Hence, the products discussed here, ethanol plus CO_2 or lactate plus protons, are in electrical and atomic balance with the starting glucose.

Glucose fermentation: Net and practical results

The ultimate result of glycolysis is the consumption of glucose, the net synthesis of two ATPs, and the production of fermentation products. For the organism the crucial product is ATP, which is used in a wide variety of energy-requiring reactions, and fermentation products are merely waste products. However, the latter substances are hardly considered waste products by the distiller, the brewer, or the baker (see the box, The Products of Yeast Fermentation). Thus, fermentation is more than just an energy-yielding process. It is a means

TECHNIQUES & APPLICATIONS

THE PRODUCTS OF YEAST FERMENTATION

The aerobic and anaerobic processes of energy generation may seem dull and prosaic, but they are at the basis of one of the most striking discoveries of the human race, alcoholic fermentation. Although many prokaryotes form alcohol, it is a simple eukaryote that is most commonly exploited, the yeast *Saccharomyces cerevisiae*. Found in various sugar-rich environments such as fruit juices and nectar, yeasts have the ability to carry out the two opposing modes of metabolism discussed in this chapter, fermentation and respiration. When oxygen is present, yeasts grow efficiently on the sugar substrate, making yeast cells and CO_2. However, when O_2 is absent, yeasts switch to an anaerobic metabolism, resulting in a reduced cell yield but significant amounts of alcohol. Every home wine maker or brewer is an amateur microbiologist, perhaps without even realizing it (∞ Home Brew, Chapter 12). When grapes are squeezed to make juice, small numbers of yeast cells present on the grapes in the vineyard are transferred to the must. During the first several days of the wine-making process, these yeast cells grow by respiration but consume O_2, making the juice anoxic. As soon as the oxygen is depleted, fermentation can begin and the process of alcohol formation takes over. This switch from aerobic to anaerobic metabolism is crucial, and special care must be taken to make sure air is kept out of the fermenting vessel. Lots of things can go wrong in wine making. The wrong yeast might get started, insufficient sugar in the juice may result in too low an alcohol concentration, and spoilage can occur from growth of bad yeasts or bacteria.

Wine is only one of many products made with yeast. Others include beer and distilled spirits such as whisky, vodka, and gin. In distilled spirits, the ethanol, produced in relatively low amounts (10–15% by volume) by the yeast, is concentrated by distilling to make a beverage containing 40–60% alcohol. Even alcohol for motor fuel is made with yeast in parts of the world where sugar is plentiful but petroleum is in short supply (such as Brazil). Yeast also serves as the leavening agent in bread, although here it is not the alcohol that is important but CO_2, the other product of the alcohol fermentation. We discuss yeast and yeast products in some detail in Chapter 12.

We can thus appreciate how the lowly yeast cell, forced to carry out a fermentative lifestyle because the oxygen it needs for respiration is absent, has impacted the lives of humans. Besides being "waste products" of the glycolytic pathway in yeast, ethanol and CO_2 are the key ingredients in the products of the alcoholic beverage and baking industries, respectively. ∎

of producing natural products useful to humans. We discuss industrial production of fermentation products in more detail in Chapter 12.

4.11

Respiration

We have just discussed the catabolism of glucose as it occurs in the *absence* of external electron acceptors. A relatively small amount of energy is released in fermentation and only a few ATP molecules synthesized. This small energy release may be understood in terms of the formal principles of oxidation–reduction reactions. Fermentation processes yield little energy for two reasons: (1) the carbon atoms in the starting compound are only partially oxidized, and (2) the difference in reduction potentials between the primary electron donor and terminal electron acceptor (that is, the vertical distance on the electron tower) is small. However, if O_2 or some other *external* terminal acceptor is present, all the substrate molecules can be oxidized completely to CO_2 and a far *higher* yield of ATP is theoretically possible. The process by which a compound is oxidized using O_2 as external electron acceptor is called **aerobic respiration** (Figure 4.13).

The greater energy release during respiration occurs because respiring cells surmount the two limitations just listed for fermentation: (1) the carbon atoms in the starting compound can be completely oxidized to CO_2; and (2) the terminal electron acceptor has a relatively positive reduction potential, leading to a large net difference in potentials between primary donor and terminal acceptor and therefore the synthesis of much ATP.

Our discussion of respiration deals with the biochemical mechanisms involved in both the carbon and electron transformations: (1) the biochemical pathways

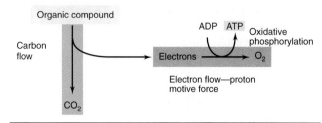

Figure 4.13 Carbon and electron flow in aerobic respiration. Note that an externally supplied electron acceptor (O_2) is present and that ATP can now be produced by oxidative phosphorylation.

involved in the transformation of organic carbon to CO_2 and (2) the way electrons are transferred from the organic compound to the terminal electron acceptor, driving ATP synthesis (Figure 4.13). We begin with a discussion of electron flow.

4.12

Electron Transport Systems

Electron transport systems are composed of *membrane-associated* electron carriers. These systems have two basic functions: (1) to accept electrons from an electron donor and transfer them to an electron acceptor, and (2) to conserve some of the energy released during electron transfer for synthesis of ATP.

Several types of oxidation–reduction enzymes are involved in electron transport: (1) NADH dehydrogenases, which transfer hydrogen atoms from NADH; (2) riboflavin-containing electron carriers, generally called flavoproteins [which contain flavin mononucleotide (FMN) or flavin-adenine dinucleotide (FAD)]; (3) iron–sulfur proteins; and (4) cytochromes, which are proteins containing an iron–porphyrin ring called *heme*. In addition, one class of *nonprotein* electron carriers is known, lipid-soluble quinones, sometimes called coenzymes Q. Quinones can diffuse freely through the membrane, generally transferring electrons from iron–sulfur proteins to cytochromes. We now consider each of these classes of electron transport components in more detail.

NADH dehydrogenases are proteins bound to the inside surface of the cell membrane. They accept hydrogen atoms from NADH (Figure 4.8), generated in various cellular reactions, and pass the hydrogen atoms to flavoproteins.

Flavoproteins are proteins containing a derivative of riboflavin (Figure 4.14); the flavin portion, which is bound to a protein, is a prosthetic group that is alternately reduced as it accepts hydrogen atoms and oxidized when electrons are passed on. Note that flavoproteins *accept* hydrogen atoms and *donate* electrons; we will consider what happens to the two protons later. Two flavins are commonly observed in cells, flavin mononucleotide and flavin-adenine dinucleotide, in which FMN is bonded to ribose and adenine through a second phosphate. Riboflavin, also called vitamin B_2, is a required growth factor for some organisms (see Section 4.2).

The **cytochromes** are proteins with iron-containing porphyrin ring prosthetic groups (heme) attached to them (Figure 4.15). They undergo oxidation and reduction through loss or gain of a *single electron* by the iron atom at the center of the cytochrome:

$$Cytochrome–Fe^{2+} \rightleftharpoons cytochrome–Fe^{3+} + e^-$$

Several classes of cytochromes are known, differing in their reduction potentials. One cytochrome can transfer electrons to another that has a more positive reduction

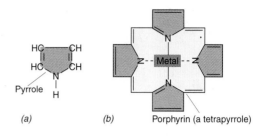

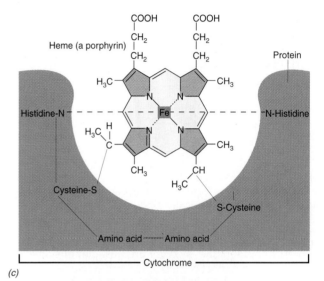

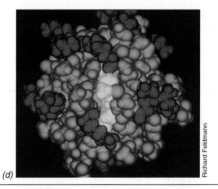

Figure 4.15 Cytochrome and its structure. (a) Structure of the pyrrole ring. (b) Four pyrrole rings are condensed, leading to formation of the porphyrin ring. Various metals can be incorporated into the porphyrin ring system. (c) In cytochromes, the porphyrin ring is covalently linked via disulfide bridges to cysteine molecules in the protein. Note the presence of iron in the center of the ring. (d) Computer-generated model of cytochrome *c*. The protein completely surrounds the porphyrin ring (light color) in the center. Cytochromes carry electrons only, not entire hydrogen atoms.

Figure 4.14 Flavin mononucleotide (FMN) (riboflavin phosphate, a hydrogen atom carrier). The site of oxidation–reduction is the same in FMN and flavin-adenine dinucleotide (FAD).

potential and can itself accept electrons from a cytochrome or quinone molecule with a less positive reduction potential. The different cytochromes are designated by letters, such as cytochrome *a*, cytochrome *b*, cytochrome *c*. The cytochromes of one organism may differ slightly from

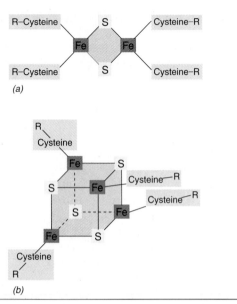

(a)

(b)

Figure 4.16 Arrangement of the iron–sulfur centers of nonheme iron–sulfur proteins. (a) Fe_2S_2 center. (b) Fe_4S_4 center. The cysteine linkages are from the protein portion of the molecule. Iron–sulfur proteins typically carry electrons only.

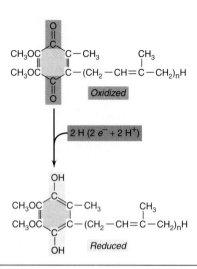

Figure 4.17 Structure of oxidized and reduced forms of coenzyme Q, a quinone. The five-carbon unit in the side chain (an isoprenoid) occurs in a number of multiples. In bacteria, the most common number is $n = 6$; in higher organisms, $n = 10$. Note that oxidized quinone requires two hydrogen atoms (2 H) to become fully reduced. An intermediate form, the semiquinone (one H more reduced than oxidized quinone), is formed during the reduction of a quinone (see Figure 4.19).

those of another, and so there are designations such as cytochrome a_1, cytochrome a_2, cytochrome aa_3, and so on. Occasionally, cytochromes form tight complexes with other cytochromes or with iron–sulfur proteins. An example of such a complex is the cytochrome bc_1 complex, which contains two different b-type cytochromes and one c-type cytochrome. It plays an important role in energy metabolism (see next section).

In addition to the cytochromes, where iron is bound to heme, several **nonheme iron–sulfur proteins** are associated with the electron transport chain in various places. Several arrangements of sulfur and iron have been found in different nonheme iron–sulfur proteins, but the Fe_2S_2 and Fe_4S_4 clusters are the most common (Figure 4.16). The iron atoms are bonded to free sulfur and to the protein via sulfur atoms from cysteine residues (Figure 4.16). *Ferredoxin*, a common iron–sulfur protein in biological systems, has an Fe_2S_2 configuration. The reduction potentials of iron–sulfur proteins vary over a wide range depending on the number of iron atoms and sulfur atoms present and how the iron centers are attached to protein. Thus different iron–sulfur proteins can serve at different points in the electron transport process. Like cytochromes, iron–sulfur proteins carry *electrons only*, not hydrogen atoms.

The **quinones** (Figure 4.17) are highly hydrophobic molecules involved in electron transport. Some quinones found in bacteria are related to vitamin K, a growth factor for higher animals. Like flavoproteins, quinones serve as *hydrogen atom* acceptors and *electron* donors.

CONCEPT CHECK   **4.12**

Electron transport systems consist of a series of membrane-associated electron carriers that function in an integrated fashion to carry electrons from the electron donor to an electron acceptor such as oxygen. Except for the quinones, which are lipid-soluble, nonproteinaceous molecules, and some water-soluble c-type cytochromes, the carriers are bound to membrane proteins and are arranged in the membrane in the order of their reduction potentials, from most electronegative to most electropositive.

✔ *How is iron present in cytochromes compared with a protein like ferredoxin?*

✔ *In what major way do quinones differ from other electron carriers in the membrane?*

4.13

Energy Conservation from Electron Transport

The overall process of electron transport in the electron transport chain is shown in Figure 4.18. During electron transport, ATP is produced by the process of oxidative phosphorylation. The production of ATP is linked directly to the establishment of a **proton motive force** across the membrane, electron transport reactions serving to establish this energized state of the membrane. We now consider the details of this process.

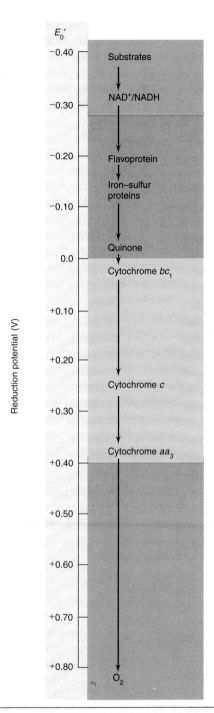

E_0'

Reduction potential (V)

-0.40 — Substrates

-0.30 — NAD⁺/NADH

-0.20 — Flavoprotein

        Iron–sulfur proteins

-0.10 —

0.0 — Quinone

      Cytochrome bc_1

+0.10 —

+0.20 —

       Cytochrome c

+0.30 —

       Cytochrome aa_3

+0.40 —

+0.50 —

+0.60 —

+0.70 —

+0.80 — O_2

Figure 4.18 One example of an electron transport system, leading to the transfer of electrons from substrate to O_2. This particular sequence is typical of the electron transport chain of the mitochondrion and some Bacteria (for example, *Paracoccus denitrificans*). The chain in *Escherichia coli* lacks cytochromes *c* and *a*, and instead electrons go directly from cytochrome *b* to cytochrome *o* or *d*, which acts as a terminal oxidase. By breaking up the complete oxidation into a series of discrete steps, energy conservation is possible through proton motive force formation leading to ATP synthesis. Compare colors here with those in Figure 4.7.

The proton motive force: Chemiosmosis

To understand the manner in which electron transport is linked to ATP synthesis, we must first discuss the manner in which the electron transport system is oriented in the cell membrane. The overall structure of the membrane was outlined in Section 3.3 (∞ Figure 3.17). It was shown there that proteins are embedded in the lipid bilayer of membranes and that the orientation of proteins in the membrane is such that most have access to both the outside and the inside of the cell (trans-membrane proteins).

The electron transport carriers discussed earlier are oriented in the membrane in such a way that a *separation* of protons from electrons occurs during the transport process. Hydrogen atoms, removed from hydrogen atom carriers such as NADH, are separated into electrons and protons, the electrons being transported through the chain by specific carriers and the protons being extruded outside the cell into the environment (in gram-negative prokaryotes protons are extruded to the periplasm), resulting in a slight acidification of the external milieu (Figure 4.19). At the end of the electron transport chain, the electrons are passed to the final electron acceptor (in the case of aerobic respiration, this is O_2) and reduce it.

When O_2 is reduced to H_2O, it requires H^+ from the cytoplasm to complete the reaction, and these protons originate from the dissociation of water into H^+ and OH^-; $H_2O \rightarrow H^+ + OH^-$. The use of H^+ in the reduction of O_2 to H_2O and the extrusion of H^+ cause a net accumulation of OH^- on the *inside* of the membrane. Despite their small size, because they are charged, neither H^+ nor OH^- freely passes through the membrane, and so equilibrium cannot be spontaneously restored. Thus, although electron transport to O_2 can be thought of as producing water, what is actually produced are the *elements* of water, H^+ and OH^-, which accumulate on opposite sides of the membrane. The net result is the generation of a *pH gradient* and an *electrochemical* potential across the membrane, with the *inside* of the cytoplasm electrically negative and alkaline, and the *outside* of the membrane electrically positive and acidic. This pH gradient and electrochemical potential cause the membrane to be energized (much like a battery), and this electrical energy can be used by the cell.

In the same way that the energized state of a battery is expressed as its electromotive force (in volts), the energized state of a membrane is expressed as the **proton motive force** (also in volts). The energized state of the membrane induced as a result of electron transport processes can be used directly to do useful work such as ion transport (∞ Section 3.4) or flagellar rotation (∞ Section 3.9), or it can be used to drive the formation of high energy phosphate bonds in ATP, as will be described later. The idea of a proton gradient driving ATP synthesis was first proposed as the *chemiosmotic the-*

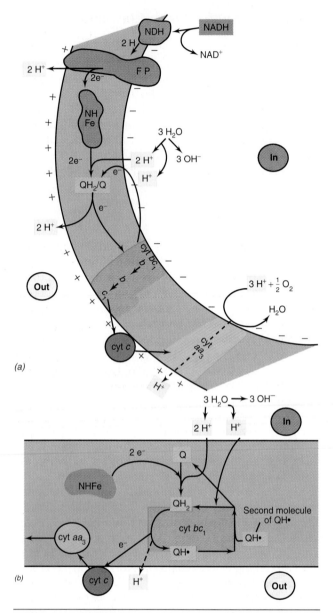

(a)

(b)

Figure 4.19 Generation of a proton motive force. (a) Orientation of electron carriers in the bacterial membrane, showing the manner in which electron transport can lead to charge separation and generation of a proton gradient. The + and − charges originate from the protons (H^+) or hydroxyl ions (OH^-) accumulating on opposite sides of the membrane. NDH, NADH dehydrogenase; FP, flavoprotein; NHFe, nonheme iron–sulfur protein; Q, coenzyme Q; cyt, cytochromes. IN and OUT refer to the cytoplasm and the environment (the periplasm of gram-negative bacteria), respectively. (b) The Q cycle. Electrons shuttle between QH_2 and the cytochrome bc_1 complex. For each molecule of QH_2 oxidized to QH· (semiquinone), one electron is passed to cyt c and one proton is passed to the outside. The bc_1 complex can also disproportionate two QH· to QH_2 plus Q, with the uptake of a proton from the cytoplasm. This mechanism serves to increase the number of protons pumped at the Q-bc_1 site of the electron transport chain.

ory in 1961 by the English scientist Peter Mitchell; Mitchell later received the Nobel Prize for this important contribution.

Generation of the proton motive force

The key steps in proton motive force formation involve the activities of the flavin enzymes, quinones, and the cytochrome bc_1 complex (Figure 4.19). The series of oxidation–reduction reactions occurring during electron transport may be analyzed by examining each pair of carriers sequentially (Figure 4.19a). Following the donation of two hydrogen atoms from NADH to FAD, two H^+ are extruded when FADH donates two electrons (only) to an iron–sulfur protein. Two protons are taken up from the dissociation of water in the cytoplasm when the nonheme iron protein reduces coenzyme Q. Coenzyme Q passes electrons one at a time to the cytochrome bc_1 complex. The cytochrome bc_1 complex contains several proteins and is present in the electron transport chain of most organisms able to respire. It also plays a role in photosynthetic electron flow (∞ Sections 13.4 and 13.5). The major function of the cytochrome bc_1 complex is to transfer electrons from quinones to cytochrome c linked to the translocation of protons across the membrane. The cytochrome bc_1 complex is oriented in the membrane in such a way that protons are discharged to the environment when electrons are transferred to an acceptor, resulting in the accumulation of OH^- in the cytoplasm and protons on the outer surface of the membrane (Figure 4.19). How does this process actually occur?

Reduced coenzyme Q (QH_2) donates one electron to the bc_1 complex, extruding a proton and converting QH_2 to QH·, the semiquinone form of coenzyme Q. The semiquinone can be reduced to QH_2 by one of the b-type cytochromes in the bc_1 complex, along with the uptake of a proton. For every two QH· molecules that enter the complex, one is reduced to QH_2 while the other is oxidized to Q. This "Q cycle" (Figure 4.19b), acts to increase the number of protons extruded across the membrane at the Q-bc_1 site. Electrons travel from the bc_1 complex to cytochrome c and cytochrome a, the latter of which serves in conjunction with the terminal oxidase (Figure 4.19b). Finally, the reduction of $\frac{1}{2}O_2$ to H_2O occurs and the electron transport reactions are completed (Figure 4.19).

The electron transport scheme shown in Figure 4.19 is just one of many different carrier sequences observed in different organisms. However, the important feature of all of them is the generation of a *proton gradient*, acidic outside and alkaline inside. The gradient results in a proton motive force that actually drives the synthesis of ATP.

The proton motive force and ATP formation

How is the proton motive force used to synthesize ATP? An important component of this process is a membrane-bound enzyme complex called *ATP synthase*, or *ATPase* for short, which contains two main parts, a multisubunit headpiece present on the inside of the membrane and a proton-conducting tailpiece that spans the membrane (Figure 4.20). This enzyme catalyzes a reversible reaction between ATP and ADP + P_i (inorganic phosphate) as shown in Figure 4.20. Operating in one direction, this enzyme catalyzes the formation of ATP by allowing the controlled reentry of protons across the energized membrane. Just as the formation of the proton gradient was energy-*driven,* the controlled dissipation of the proton motive force is energy-*releasing,* and some of the energy is conserved in the synthesis of ATP in a process called **oxidative phosphorylation.** The mechanism by which proton reentry into the cell drives ATP synthesis is not entirely clear, but the process is thought to involve a conformational change in ATPase proteins catalyzed by proton translocation and the return of the ATPase to its native conformation, releasing energy that is coupled to the synthesis of ATP. The β subunits of the F_1 portion of ATPase are the actual catalytic sites of ATP synthesis. As protons enter from the outside, the β subunits rotate relative to F_0, and this catalyzes the reaction ADP + P

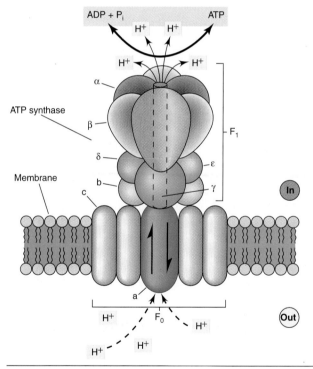

Figure 4.20 Structure and function of the membrane-bound ATP synthase (ATPase), which functions as a proton channel between the cytoplasm and the cell exterior (environment). (a) Structure. F_1 consists of five polypeptides, α (three copies), β (three copies), γ, δ, and ε. This is the catalytic protein responsible for the interconversion of ATP and ADP + P_i. F_0 is integrated in the membrane and consists of three polypeptides, a, b (two copies), and c (four copies). It is responsible for channeling protons across the membrane. As protons enter, the dissipation of the proton motive force drives ATP synthesis from ADP + P_i. The reverse reaction, ATP→ ADP + P_i, drives the extrusion of protons to the cell exterior. Thus, ATP synthase is reversible in its action.

→ ATP (Figure 4.20). Measurements of the numbers of protons consumed by an ATPase per ATP produced yield a value of 4.

ATPase can also catalyze the reverse reaction, that is, the hydrolysis of ATP and the extrusion of 4 H^+ to the outer portion of the membrane (Figure 4.20). This results in the conversion of phosphate bond energy to the energy of a proton motive force. Thus, high energy phosphate bonds and the proton motive force can be looked at as *different forms of cell energy.* Because a membrane potential is used by the cell to drive a number of different reactions, transport and motility being chief among them, ATPases are present even in organisms that do not carry out oxidative phosphorylation, such as lactic acid bacteria.

The operation of the electron transport chain and creation of a proton motive force can be studied with various chemicals that affect these processes. Two classes of chemicals are known: **inhibitors** and **uncouplers.** Inhibitors block electron flow and thus ATP synthesis. Examples include carbon monoxide (CO), which prevents the reduction of O_2 to H_2O by aa_3-type oxidases, and cyanide (CN^-), hydrogen sulfide (H_2S), or azide (N_3^-), which bind tightly to cytochromes and block electron transport. In contrast, uncouplers inhibit ATP synthesis *without* affecting the ATPase. All these agents, such as dinitrophenol and dicumarol, are lipid-soluble substances that increase membrane permeability, thereby promoting the leakage of protons across the membrane. The latter results in dissipation of the proton motive force and hence inhibition of ATP synthesis.

CONCEPT CHECK   *4.13*

When electrons are transported through a membrane-integrated electron transport system, protons are extruded to the outside of the membrane forming the proton motive force. Key electron carriers include flavins, quinones, the cytochrome bc_1 complex, and other cytochromes, depending on the organism. The cell utilizes the proton motive force via ATPases to yield ATP.

✔ *How do electron transport reactions generate the proton motive force?*

✔ *What structure in the cell converts the proton motive force to ATP?*

4.14

Carbon Flow: The Citric Acid Cycle

We now consider the metabolic aspects of carbon flow in respiration. The early steps in the respiration of glucose involve the same biochemical steps as those of glycolysis (see Figure 4.12). As we noted, a key intermediate in glycolysis is pyruvate. Whereas in fermentation pyruvate is converted to fermentation products, in respiration pyruvate is oxidized fully to CO_2. One major pathway by which pyruvate is completely oxidized to CO_2 is called the **citric acid cycle** (CAC) as outlined in Figure 4.21.

Pyruvate is first decarboxylated, leading to the production of one molecule of NADH and an acetyl molecule coupled to coenzyme A (acetyl-CoA) (Figure 4.22). The acetyl group of acetyl-CoA combines with the four-carbon compound oxalacetate, leading to the formation of citric acid, a six-carbon organic acid, the energy of the high energy acetyl-CoA bond being used to drive this synthesis (Figure 4.22). Dehydration, decarboxylation, and oxidation reactions follow, and two additional CO_2 molecules are released. Ultimately, oxalacetate is regenerated and can function again as an acetyl acceptor, thus completing the cycle.

CO_2 release and fuel for electron transport

For each pyruvate molecule oxidized through the cycle, three CO_2 molecules are released (Figure 4.21), one during the formation of acetyl-CoA, one by the decarboxylation of isocitrate, and one by the decarboxylation of α-ketoglutarate. As in fermentation, the electrons released during the oxidation of intermediates in the CAC are transferred to enzymes containing the coenzyme NAD^+ or FAD. However, respiration differs from fermentation in the manner in which NADH and FADH are oxidized. In respiration, the electrons from NADH, instead of being used to reduce an intermediate such as pyruvate, are transferred to oxygen or other terminal electron acceptors through the action of the *electron transport system* described in Section 4.13. Thus, unlike the situation in fermentation, the presence of an electron acceptor in respiration allows for the complete oxidation of glucose to CO_2 with a much greater yield of energy.

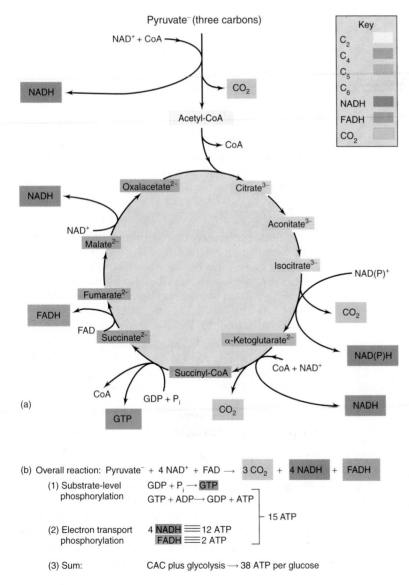

Figure 4.21 The citric acid cycle (CAC). The three-carbon compound pyruvate is oxidized to CO_2, with the electrons used to make NADH and FADH. The reoxidation of NADH and FADH in the electron transport chain leads to the generation of a proton motive force and thus to the synthesis of ATP. The CAC begins when the two-carbon compound acetyl-CoA condenses with the four-carbon compound oxalacetate to form the six-carbon compound citrate. Through a series of oxidations and transformations, this six-carbon compound is ultimately converted back to the four-carbon compound oxalacetate, which then begins another cycle with addition of the next molecule of acetyl-CoA. (b) The overall balance sheet of fuel (NADH/FADH) for the electron transport chain and CO_2 generated in the CAC.

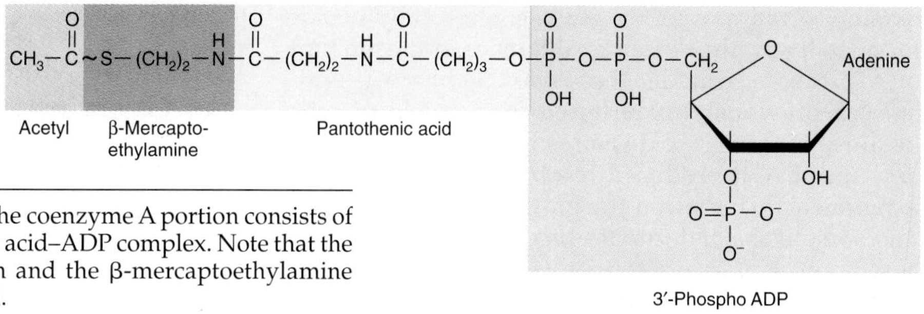

Figure 4.22 Structure of acetyl-CoA. The coenzyme A portion consists of the β-mercaptoethylamine–pantothenic acid–ADP complex. Note that the C—S bond between the acetyl portion and the β-mercaptoethylamine portion is a high energy thioester bond.

Biosynthesis and the citric acid cycle

It should also be mentioned that besides playing a key role in catabolic reactions, the citric acid cycle is important to the cell for biosynthetic reasons as well. This is because the cycle is composed of a number of key carbon skeletons that can be drawn off for biosynthetic purposes when needed. Particularly important in this regard are the intermediates α-ketoglutarate and oxalacetate, which are the precursors of a number of amino acids (see Section 4.19), and succinyl-CoA, needed to form the porphyrin ring of the cytochromes, chlorophyll, and several other tetrapyrrole compounds. Oxalacetate is also important because it can be converted to phosphoenolpyruvate, a precursor of glucose (see Section 4.18). In addition to these, acetyl-CoA provides the starting material for fatty acid biosynthesis (see Section 4.21). Thus, it can be stated that the citric acid cycle plays two major roles in the cell: bioenergetic and biosynthetic.

CONCEPT CHECK   *4.14*

Respiration involves the complete oxidation of an organic electron donor to CO_2 and H_2O with much greater energy release than during fermentation. The citric acid cycle plays a major role in the respiration of organic compounds.

   ✔ *How many molecules of CO_2 and pairs of hydrogen atoms are released per acetate consumed in the citric acid cycle?*

4.15

Balance Sheet of Aerobic Respiration and Energy Storage

The net result of reactions of the citric acid cycle is the complete oxidation of pyruvic acid to three molecules of CO_2 with the production of four molecules of NADH and one molecule of FADH. As we have seen, the NADH and FADH molecules can be reoxidized through the electron transport system; the yield of ATP is up to 3 ATP molecules per molecule of NADH and 2 ATPs per molecule of FADH. In addition, the oxidation of α-ketoglutarate to

succinate involves a substrate-level phosphorylation, producing guanosine triphosphate (GTP), which can be converted to ATP. Thus, a total of 15 ATP molecules can be synthesized for each turn of the cycle. Since the oxidation of glucose yields 2 molecules of pyruvic acid, a total of 30 molecules of ATP can be synthesized in the citric acid cycle. Also, when oxygen is available, the 2 NADH molecules produced during glycolysis can be reoxidized by the electron transport system, yielding 6 more molecules of ATP. Finally, 2 molecules of ATP are produced by substrate-level phosphorylation during the conversion of glucose to pyruvic acid. Thus, aerobes can form up to 38 ATP molecules from 1 glucose molecule in contrast to the 2 molecules of ATP produced fermentatively.

If we assume that the high energy phosphate bond of ATP has an energy of about 32 kJ/mol, then 1216 kJ (38 × 32) of energy can be converted to high energy phosphate bonds in ATP by the complete oxidation of glucose to CO_2 and H_2O. Since free-energy calculations show that the total amount of energy available from the complete oxidation of glucose by oxygen is 2822 kJ/mol, aerobic respiration is about 43% efficient, the rest of the energy being lost as heat.

If an organism were 100% efficient, all the energy yield of a biochemical reaction would be conserved in the form of high energy bonds or a membrane potential. However, organisms are not 100% efficient, and part of the energy is not conserved but is lost as heat. Thus we must distinguish between energy *release*—the total energy yield of a reaction—and energy *conservation*—the energy available to the organism to do work. Interestingly, although the *yield* of ATP from fermentation is very low, the *efficiency* of fermentations can be reasonably high. The lactic fermentation [glucose → 2 lactate], for example, releases 198 kJ and drives the synthesis of 2 ATPs, for an efficiency of 32%. Thus, fermentations release relatively *low amounts* of energy, but the conversion process is thermodynamically reasonably efficient.

ATP and cell yield

The amount of ATP produced by an organism has a direct effect on cell yield. The yield of cells obtained in a culture, that is, the biomass produced, can vary greatly with the composition of the growth medium and the environmental conditions under which growth occurs. For example, if one measures the grams of yeast cells

basics

obtained from growth on glucose aerobically versus fermentatively, the dramatic difference in the amount of ATP produced under these two conditions (as calculated previously) is reflected in the yield of cells obtained. That cell yield is directly proportional to the amount of ATP produced has been confirmed from experimental studies on the growth yields of various microorganisms and implies that the energy costs for assembly of macromolecules (which are the major energy costs for a growing microbial cell) (∞ Chapter 6) are much the same for all microorganisms.

Energy storage

ATP is present in relatively low concentration in the cell, about 2 millimolar (mM) in an actively growing cell, and thus ATP plays only a *catalytic* role during growth of the cell; ATP is continuously being broken down to drive biosynthetic reactions and resynthesized at the expense of catabolic reactions. For energy *storage*, most microorganisms produce insoluble polymers that can later be oxidized for the production of ATP. Examples include the polyglucose polymers starch and glycogen, the lipid polymer poly-β-hydroxybutyrate and other polyhydroxyalkanoates, and elemental sulfur, stored by many sulfur chemolithotrophs. These polymers are deposited within the cell as large granules that can often be seen with the light or electron microscope (∞ Sections 3.11 and 16.5). In the absence of an external energy source, the cell may oxidize these polymers and thus be able to make new cell material or simply maintain itself even when nutrients are temporarily unavailable in the environment.

Polymer formation is important to the cell for two reasons. First, potential energy is stored in a stable form, and second, insoluble polymers have little effect on the internal osmotic pressure of cells. If the same number of units were present as monomers in the cell, the high solute concentration would increase cellular osmotic pressure, resulting in an influx of water and possible lysis. Thus, storage polymers make possible the storage of energy in a readily accessible form that does not interfere with other cellular processes.

CONCEPT CHECK 4.15

Although only two high energy phosphate bonds are conserved as ATP during fermentation of 1 glucose molecule to 2 lactate or 2 ethanol plus CO_2, as many as 38 ATPs can be formed during aerobic respiration. Storage polymers are common in prokaryotes and function as carbon and/or energy reserves.

✔ *Considering just the reactions of glycolysis, why can 8 ATPs be made under aerobic conditions but only 2 ATPs anaerobically by fermentation?*

✔ *Of what advantage to the cell are storage polymers?*

4.16

Alternate Modes of Energy Generation: An Overview

This chapter deals only with energy generation by either fermentation or respiration of organic compounds, but microorganisms have many other possibilities for obtaining energy. These alternate modes of energy metabolism form the theme of Chapter 13 but are summarized here to provide an overview of catabolic processes.

Anaerobic respiration, chemolithotrophy, and phototrophy

One alternate mode of energy generation is a variation on respiration in which electron acceptors *other than* oxygen are used. Because of the analogy to aerobic respiration, these processes are called **anaerobic respiration.** Electron acceptors used in anaerobic respiration include nitrate (NO_3^-), ferric iron (Fe^{3+}), sulfate (SO_4^{2-}), carbonate (CO_3^{2-}), and even certain organic compounds. Because of their positions on the electron tower (none of these acceptors have as electropositive an E_0' as the O_2/H_2O couple) (see Figure 4.7), less energy is released when these electron acceptors are used instead of oxygen. However, the utilization of these alternate electron acceptors permits microorganisms to respire in environments where oxygen is absent. The contrasts between aerobic and anaerobic respiration are presented in Figure 4.23.

A second mode of energy generation involves the use of *inorganic* rather than organic chemicals. Organisms able to use inorganic chemicals as energy sources are a type of chemotroph called **chemolithotrophs** (literally, *rock-eating*). Examples of inorganic energy sources include hydrogen sulfide (H_2S), hydrogen gas (H_2), and ammonia (NH_3). Chemolithotrophic metabolism usually involves aerobic respiratory processes such as those described in this chapter but using an inorganic energy source rather than an organic one (Figure 4.23). Chemolithotrophs have electron transport components like chemoorganotrophs and form a proton motive force. However, one important distinction between chemolithotrophs and chemoorganotrophs is in their sources of *carbon* for biosynthesis. Chemoorganotrophs can generally use compounds such as glucose as carbon sources as well as energy sources, but chemolithotrophs cannot use their inorganic energy compounds as sources of carbon. Most chemolithotrophs utilize carbon dioxide as a carbon source and are, hence, **autotrophs.** Organisms that use organic compounds as carbon sources are called **heterotrophs.**

A large number of microorganisms, as well as higher plants, are *phototrophic*, using light as an energy source in the process of photosynthesis. We call such organisms **phototrophs** (literally, *light-eating*). The mechanisms by which light is used as an energy source are unique and complex, but the underlying result is the generation of

(a) **Aerobic respiration**

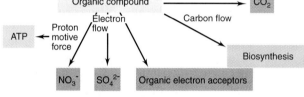

(b) **Anaerobic respiration**

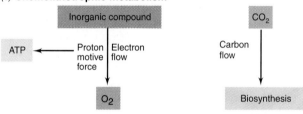

(c) **Chemolithotrophic metabolism**

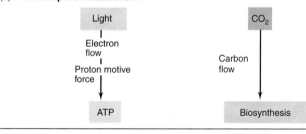

(d) **Phototrophic metabolism**

Figure 4.23 Energetics and carbon flow in (a) aerobic respiration, (b) anaerobic respiration, (c) chemolithotrophic metabolism, and (d) phototrophic metabolism. Note the importance of electron transport leading to proton motive force formation in each case.

a proton motive force that can be used in the synthesis of ATP. Most phototrophs use energy conserved in ATP for the assimilation of carbon dioxide as the carbon source for biosynthesis. Such phototrophs are also, hence, autotrophs. However, as we will see in Chapter 13, photosynthesis in microorganisms has some special features and complications. There are, for instance, two types of photosynthesis in microorganisms, one form similar to that of higher plants, and a unique type of photosynthesis found only in certain prokaryotes.

Importance of the proton motive force to alternate bioenergetic strategies

One overall conclusion to be drawn at this point is that in terms of energy metabolism, microorganisms show an amazing diversity of bioenergetic strategies. Thousands of organic compounds, many reduced inorganic compounds, and light can be used by one or another microorganism as an energy source. However, with the exception of most fermentations, where substrate-level phosphorylations prevail, metabolic diversity in respiration and photosynthesis revolves around *different ways of generating a proton motive force.* Thus, regardless of whether electrons come from the oxidation of organic or inorganic chemicals or from phototrophic processes, they all traverse a membrane-bound electron transport chain and in so doing generate a proton motive force (Figure 4.23); energy conservation in all cases occurs through function of the membrane-bound ATPase (Figure 4.20). In Chapter 13 we will examine some of the details of the different bioenergetic strategies that result in generation of the proton motive force.

CONCEPT CHECK   4.16

Electron acceptors other than oxygen can function as terminal electron acceptors for energy generation. Because oxygen is absent, this is called anaerobic respiration. Chemolithotrophs use inorganic compounds as sources of energy, whereas phototrophs use light as an energy source. The proton motive force is involved in all forms of respiration and photosynthesis.

✔ *In terms of their energy source(s), how do chemoorganotrophs differ from chemolithotrophs?*

✔ *What is the carbon source for autotrophic organisms?*

4.17

Biosynthetic Pathways: Anabolism

We have just discussed **catabolism,** the biochemical processes by which microorganisms obtain energy from organic compounds. We now consider the other major reaction series, **anabolism,** the biochemical processes by which microorganisms build up the vast array of chemical substances of which they are composed (Figure 4.1). The chemical composition of the cell was outlined in Table 4.1. Note that the dominant constituents (other than water) are macromolecules. Macromolecules, as we know (∞ Sections 2.5–2.8), are made by the polymerization of monomers. We now summarize the key biosynthetic reactions that form these monomers. Unlike catabolic pathways, which can be extremely diverse, anabolic pathways tend to be more uniform across species. Thus, the broad principles stressed here apply to a wide variety of different organisms.

Energy for anabolism is provided by ATP or the proton motive force (Figure 4.24). Energy conserved during catabolic reactions in ATP and the proton gradient is consumed during biosynthesis (Figure 4.24). In Sections 4.10 and 4.14 we described the intermediates in the oxidation of glucose in both glycolysis and the citric acid cycle; in those sections our main concern was not the ultimate fate of these intermediates but how their transformations lead to the formation of ATP. Now we are concerned with how these intermediates are used in biosynthesis. Some of the same intermediates that are formed during catabolism are used during anabolism. These key intermediates, summarized in Figure 4.25, are actually very few in number.

CONCEPT CHECK   4.17

Anabolism, also called biosynthesis, is the process by which the cell builds up the vast array of chemical substances that are needed for growth. The ATP produced during catabolism is consumed during anabolism.

✔ *Why can ATP and the proton motive force be considered equivalent and interchangeable forms of energy?*

✔ *What major biochemical pathways generate important biosynthetic intermediates?*

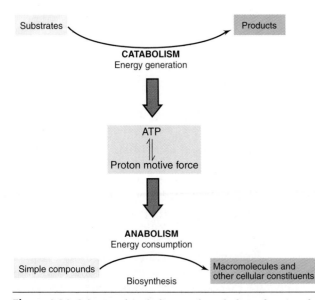

Figure 4.24 Scheme of anabolism and catabolism showing the key role of ATP and the proton motive force in integrating the processes.

4.18

Sugar Metabolism

A small but significant amount of a cell is composed of polysaccharide (see Table 4.1 for the chemical composition of a cell). Polysaccharides are key constituents of the cell walls of many organisms, and in Bacteria the peptidoglycan cell wall has a polysaccharide backbone. In addition, cells often store energy in the form of the polysaccharides **glycogen** and **starch.** The sugars in polysaccharides are primarily six-carbon sugars called *hexoses.* The most common hexose in polysaccharides is **glucose,** which, as we have seen, is also an excellent energy source for many microorganisms. In addition to hexoses, two five-carbon sugars, called *pentoses,* are key constituents of nucleic acids, **ribose** in RNA and **deoxyribose** in DNA. It is thus convenient to separate our discussion into two parts dealing separately with hexoses and pentoses.

Hexoses

Six-carbon sugars needed for biosynthesis either can be obtained from the environment or can be synthesized within the cell from nonsugar starting materials. A summary of the main pathways is given in Figure 4.26. It can be seen that two key intermediates of hexose metabolism are **glucose 6-phosphate** and **uridine diphosphoglucose (UDPG),** a nucleotide diphosphate sugar. We already have discussed the role of glucose 6-phosphate in glycolysis (see Figure 4.12). Besides its involvement in energy metabolism, glucose 6-phosphate can also feed into the pathways for polysaccharide synthesis, in which case it is converted to UDPG. UDPG is an activated form of glucose that is synthesized from uridine triphosphate (UTP) and glucose 1-phosphate as outlined in Figure 4.26. UDPG is the starting material not only for the synthesis of glucose-containing polysaccharides but also for the synthesis of other nucleoside diphosphate sugars needed in biosynthesis. Thus, glucose 6-phosphate is the central intermediate in glucose *catabolism,* and UDPG is the central intermediate in glucose *anabolism.*

When a microorganism has a hexose available as an energy and carbon source, this hexose becomes the precursor of the hexoses needed for biosynthesis. However, many microorganisms are able to grow using nonhexose carbon sources. In such situations, the hexose needed for biosynthesis must be synthesized within the cell by a process called **gluconeogenesis.** Gluconeogenesis is the creation of new glucose molecules from noncarbohydrate precursors. The starting material for gluconeogenesis is phosphoenolpyruvate (PEP), one of the key intermediates in glycolysis. As we saw in Section 4.10 (see Figure 4.12), PEP is formed in the lower part of the glycolytic pathway, and by *reversal* of this pathway (but using a different enzyme than that used in glycolysis), glucose 6-phosphate can be formed.

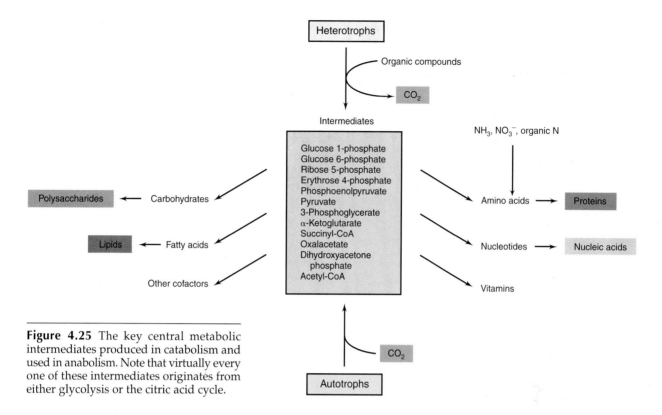

Figure 4.25 The key central metabolic intermediates produced in catabolism and used in anabolism. Note that virtually every one of these intermediates originates from either glycolysis or the citric acid cycle.

Where does the PEP needed for gluconeogenesis come from? There are a number of ways in which PEP can be formed, but a major one is by decarboxylation of oxalacetate, which itself is a key intermediate in the citric acid cycle (see Figures 4.12 and 4.21).

Pentoses

In most instances, five-carbon sugars are formed by the removal of one carbon atom from a hexose (Figure 4.26). Several different pathways to accomplish this are

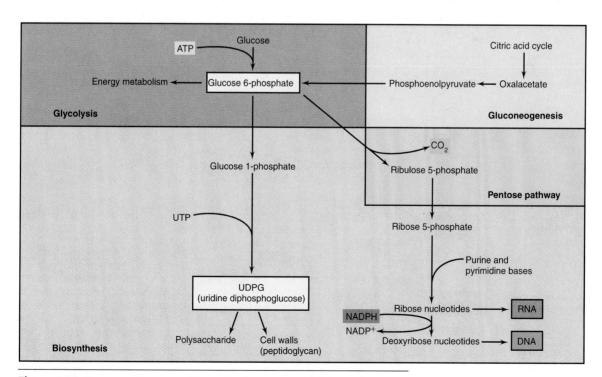

Figure 4.26 Hexose and pentose biosynthesis: summary of the main routes.

known. One of the common pathways is the oxidative decarboxylation of glucose 6-phosphate, yielding CO_2 and the five-carbon intermediate ribulose 5-phosphate. Ribulose 5-phosphate is converted to two other pentose sugars, one of which is ribose 5-phosphate, the pentose needed for nucleotide synthesis. Once a ribose nucleotide is formed, it can feed directly into RNA synthesis. For DNA synthesis, deoxyribose is needed and is synthesized by enzymatic reduction (using NADPH) of the 2' oxygen on the ribose of a ribose nucleotide. This important reduction in the cell is carried out by the enzyme *ribonucleotide reductase*. These various reactions are summarized in Figure 4.26.

4.19
Amino Acid Biosynthesis

As we have noted, there are 21 common amino acids in proteins. Organisms that cannot obtain some or all amino acids preformed from the environment must synthesize them from other sources. The structures of the amino acids were discussed in Section 2.8 and illustrated in Figure 2.12. There are two aspects of amino acid biosynthesis: (1) synthesis of the **carbon skeleton** of each amino acid, and (2) incorporation of the **amino group.**

Amino acids can be grouped into *families* based on the precursor used for the synthesis of the carbon skeleton, and these families are summarized in Figure 4.27. Note that the precursor molecules are ones that we have encountered before in glycolysis and the citric acid cycle and were summarized in Figure 4.25.

Amination

The attachment of the amino group to a carbon skeleton to form an amino acid can be done either at the very end, after the carbon skeleton is completely synthesized,

---

CONCEPT CHECK   4.18

Although sugars are important energy sources for most cells, they are also needed for the biosynthesis of cell walls, nucleic acids, and so on. If sugars are not available as a nutrient from the environment, the cell must synthesize them from other sources. Key sugars in the cell contain six carbons (hexoses) or five carbons (pentoses).

✔ *What is the starting substance for the synthesis of a glucose polymer like glycogen?*

✔ *How is deoxyribose synthesized?*

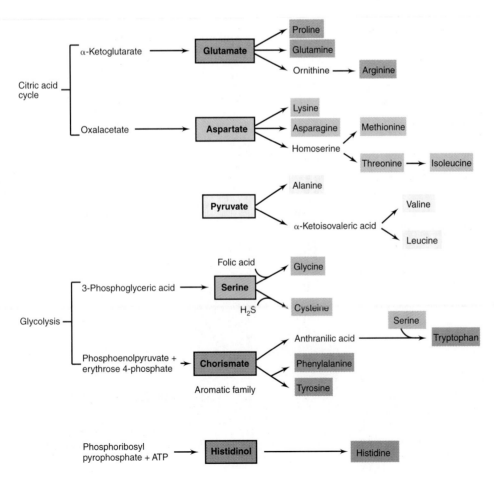

Figure 4.27 Amino acid families. The amino acids on the right are derived from the starting materials on the left. The compounds in boxes are the parent molecules of each amino acid family. Note how many amino acids originate from carbon skeletons from either glycolysis or the citric acid cycle. Except for chorismate and histidinol, non-amino acid intermediates are not shown in color.

Material for Review

REVIEW QUESTIONS

1. Define the terms *chemoorganotroph, chemolithotroph, phototroph, autotroph,* and *heterotroph.*

2. Examine the data in Table 4.1. Why is it accurate to say that a bacterial cell consists primarily of macromolecules? What class of macromolecules constitutes the largest weight in the cell? What class of macromolecules is present in the cell in the largest number of molecules? From what you know about energy, where would you say that the largest amount of energy in the cell is stored?

3. What are siderophores and why are they necessary?

4. Why would the following medium *not* be considered a chemically defined medium: glucose, 5 grams (g); NH_4Cl, 1 g; KH_2PO_4, 1 g; $MgSO_4$, 0.3 g; yeast extract, 5 g; distilled water, 1 liter (l).

5. Describe how would you calculate $\Delta G^{0\prime}$ for the reaction: glucose $+ 6 O_2 \rightarrow 6 CO_2 + 6 H_2O$. If you were told that this reaction is highly *exergonic,* what would be the sign (negative or positive) of the $\Delta G^{0\prime}$ you would expect for this reaction?

6. Distinguish between $\Delta G^{0\prime}$ and G^0_f.

7. Why are enzymes needed by the cell?

8. Describe a rationale for why an enzyme from the bacterium *Escherichia coli* loses its catalytic ability after being boiled.

9. Describe the difference between a *coenzyme* and a *prosthetic group.*

10. The following is a series of coupled electron donors and electron acceptors. Using just the data given in Figure 4.7, order this series from most energy-yielding to least energy-yielding. H_2/Fe^{3+}, H_2S/O_2, methanol/NO_3^- (producing NO_2^-), H_2/O_2, Fe^{2+}/O_2, NO_2^-/Fe^{3+}, H_2S/NO_3^-.

11. What is an electron carrier? Give three examples of electron carriers and indicate their oxidized and reduced forms.

12. Is the reaction glucose 6-phosphate $+ ADP \rightarrow$ glucose $+ ATP$ exergonic or endergonic (refer to Figure 4.10 to help answer this question).

13. Where in glycolysis is NADH *produced*? Where is NADH *consumed*?

14. Iron plays an important role within the cell in energy-generating processes. Give three examples in which iron plays a role as an electron carrier. How is iron provided as a nutrient in culture media?

15. What is meant by the term *proton motive force* and why is this concept so important in biology?

16. The chemicals dinitrophenol and cyanide both act as cellular poisons but in quite different ways. Compare and contrast the modes of action of these two chemicals.

17. Give in simplified form the equation for the oxidation–reduction reaction that takes place in the cytochrome molecule. Can you suggest a role for the portions of the cytochrome molecule that are *not* involved in oxidation–reduction?

18. Knowing the function of the electron transport chain, can you imagine an organism that could live if it completely lacked the components (for example, cytochromes) needed for an electron transport chain? (*Hint:* Focus your answer on the mechanism of ATPase.)

19. Work through the energy balance sheets for fermentation and respiration and account for all sites of ATP synthesis. Organisms can obtain nearly 20 times more ATP when growing aerobically on glucose than anaerobically. Write one sentence that accounts for this difference.

20. Why can it be said that the citric acid cycle plays two major roles in the cell?

21. What are the similarities and differences in aerobic respiration in *Escherichia coli*, a chemoorganotroph, and *Thiobacillus thioparus*, a chemolithotroph?

22. A substance such as glucose can function in an organism both as an energy source and as a building block of cell material. What are the names of the two types of metabolic processes involved in these two disparate functions? What is the fate of the carbon atoms of the glucose molecules that are used in energy generation? List three groups of carbon compounds derived from glucose that are building blocks of cell material.

23. After an organism *synthesizes* glycogen, it can later break down and *utilize* the glycogen. However, the biochemical pathways for glycogen synthesis and glycogen utilization are distinct. Describe these two distinct types of pathways. Can you see any advantage in terms of cell function if separate pathways are used for these two processes?

24. Figure 4.25 indicates that there are only a few intermediate compounds that serve as the starting points for anabolism. For each of the amino acid families, list the intermediates from Figure 4.25 that are involved in the biosynthetic reactions.

25. Purine and pyrimidine metabolism can be approached in the same way that amino acid metabolism was just approached in Question 24. List the comparable intermediates in purine and pyrimidine metabolism.

26. Describe the process by which a fatty acid such as palmitate (C_{16} saturated) is synthesized in a cell.

APPLICATION QUESTIONS

1. Design a defined culture medium for an organism that can grow aerobically on acetate as a carbon and energy source. Make sure all the nutrient needs of the organism are accounted for and in the correct relative proportions.

2. *Desulfovibrio* can grow anaerobically with H_2 as electron donor and SO_4^{2-} as electron acceptor (which is reduced to H_2S). Based on this information and the data in Table A1.2 (Appendix 1), indicate which of the following components *could not* exist in the electron transport chain of this organism and why: cytochrome *c*, ubiquinone, cytochrome c_3, cytochrome *a*, ferredoxin.

3. Again, using the data in Table A1.2, predict the sequence of electron carriers in an organism growing aerobically and producing the following electron carriers: ubiquinone, cytochrome *a*, cytochrome *b*, NADH, cytochrome *c*, FAD.

4. Why is it essential for the theory of chemiosmosis to assume that membranes are impermeable to the elements of H_2O, H^+, and OH^-, but that membranes can be permeable to H_2O itself?

5. Explain the following observation: cells of *Escherichia coli* fermenting glucose grow faster when NO_3^- is supplied to the culture (NO_2^- is produced), and then grow even faster (and stop producing NO_2^-) when the culture is highly aerated.

6. Applying what you know about glycolysis, the citric acid cycle, and gluconeogenesis, explain why a mutant of *Escherichia coli* containing a defective malate dehydrogenase (that is, cannot carry out the reaction malate^{2-} + $NAD^+ \rightarrow$ oxaloacetate^{2-} + NADH + H^+) (see Figure 4.21) is able to grow on glucose but not on acetate, whereas nonmutated (wild-type) *E. coli* can be grown on either substrate.

SUPPLEMENTARY READINGS

Atlas, R. M. 1993. *Handbook of Microbiological Media.* CRC Press, Boca Raton, FL. A comprehensive listing of media for the growth of bacteria and some tips on preparing culture media.

Dawes, I. W., and **I. W. Sutherland.** 1992. *Microbial Physiology,* 2nd edition. Blackwell Scientific, London. A short but fairly detailed account of bacterial physiology.

Gottschalk, G. 1986. *Bacterial Metabolism,* 2nd edition. Springer-Verlag, New York. A college-level text dealing exclusively with bacterial topics. Although a bit dated, it contains an in-depth treatment of anoxic metabolism.

Harris, D. A. 1995. *Bioenergetics at a Glance.* Blackwell Science, Cambridge, MA. A brief, highly visual treatment of the key concepts of bioenergetics. Contains good detail on chemiosmotic events.

Moat, A. G., and **J. W. Foster.** 1995. *Microbial Physiology,* 3rd edition. John Wiley, New York. A very good general-purpose textbook of microbial physiology.

Neidhart, F. C., R. Curtiss III, J. L. Ingraham, E. C. C. Lin, K. B. Low, B. Magasanik, W. Reznikoff, M. Riley, M.

Schaechter, and **H. E. Umbarger.** 1996. Escherichia coli *and* Salmonella: *Cellular and Molecular Biology,* 2nd edition. American Society for Microbiology, Washington, DC. A massive treatment of all aspects of the physiology, biochemistry, and genetics of *E. coli* and *Salmonella* species.

Neidhardt, F. C., J. L. Ingraham, and **M. Schaechter.** 1990. *Physiology of the Bacterial Cell—A Molecular Approach.* Sinauer Associates, Sunderland, MA. A very readable account of bacterial physiology.

Schlegel, H. G., and **B. Bowien.** 1989. *Autotrophic Bacteria.* Science Tech, Madison, WI. The definitive book on phototrophic and chemolithotrophic metabolism.

Stryer, L. 1995. *Biochemistry,* 4th edition. W. H. Freeman, San Francisco. Excellent general coverage of basic energetics, enzymes, and other aspects of biochemistry.

White, D. 1995. *The Physiology and Biochemistry of Prokaryotes.* Oxford University Press, New York. A college-level textbook of microbial physiology and biochemistry emphasizing the experimental science behind the principles. Highly recommended.

On~line resources for this chapter are on the World Wide Web at: http://www.prenhall.com/~brock (click on the table of contents link and then select Chapter 4).

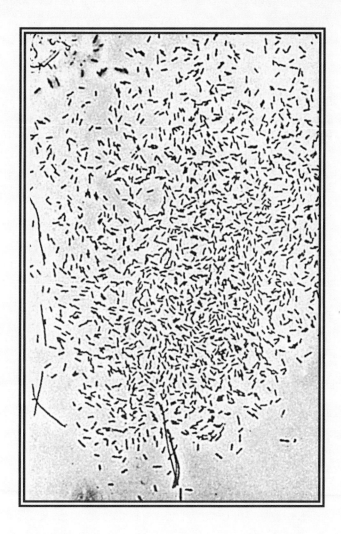

Microbial Growth

Single-celled microorganisms, such as these bacterial cells growing on a microscope slide immersed in a boiling hot spring, grow exponentially; a single cell elongates and divides to form two cells, each of which is itself capable of growth. In this way, large numbers of cells can develop in a very short period of time.

MINIGLOSSARY for Chapter 5

Acidophile an organism that grows best at low pH

Aerobe an organism that can use O_2 in respiration; some require O_2 for growth

Alkaliphile an organism that grows best at high pH

Anaerobe an organism that cannot use O_2 in respiration and whose growth may be inhibited by O_2

Batch Culture a closed-system microbial culture of fixed volume

Chemostat a device that allows for the continuous culture of microorganisms in which both growth rate and cell number can be controlled independently

Compatible Solute a molecule that is accumulated in the cytoplasm for adjustment of water activity but that does not inhibit biochemical processes

Exponential Growth growth of a microorganism where the cell number doubles within a fixed time period

Facultative with respect to O_2, an organism that can grow in either its presence or absence

Generation Time the time required for a population of microbial cells to double

Growth an increase in cell number

Halophile a microorganism that requires NaCl for growth

Hyperthermophile a microorganism that has a growth temperature optimum of 80°C or greater

Lag Phase a period preceding the exponential growth phase when cells may be metabolizing but are not yet growing

Lysis loss of cellular integrity with release of cytoplasmic constituents

Mesophile an organism that grows best at temperatures between 20 and 45°C

Microaerophile an aerobic organism that can grow only when oxygen tensions are reduced from that in air

Psychrophile an organism with a growth temperature optimum of 15°C or lower and a maximum growth temperature below 20°C

Psychrotolerant an organism capable of growth at low temperatures but whose growth temperature optimum is above 20°C

Stationary Phase the period immediately following exponential growth when the growth rate of the population falls to zero

Thermophile an organism whose growth temperature optimum lies between 45 and 80°C

Viable capable of reproducing

Xerophile an organism that is able to live, or that lives best, in very dry environments

---

We have thus far discussed cell chemistry (Chapter 2), cell biology (Chapter 3), and the general principles of microbial nutrition and metabolism (Chapter 4). Before we begin our study of the biosynthesis of macromolecules and the genetics of microorganisms (Chapter 6), we consider here microbial growth. In microbiology, the word **growth** is defined as *an increase in the number of cells.* Growth is an essential component of microbial function, as any given cell has a finite life span in nature and the species is maintained only as a result of continued growth of the population. However, in addition to understanding the basic science of microbial growth, many practical situations call for the *control* of microbial growth. Knowledge of how microbial populations can rapidly expand is quite useful in designing methods to control microbial growth; we will study these methods in Chapter 11.

In the present chapter we first discuss the general principles of microbial growth and describe the growth cycle of microbial populations. Because so much is known about bacterial growth, we focus our discussion on prokaryotes. However, it should be kept in mind that the general principles of exponential growth apply to all unicellular microorganisms, whether prokaryotic or eukaryotic. After consideration of means to measure growth and a special tool, the chemostat, for manipulating growth of a microbial population, we focus on the major environmental factors influencing microbial growth: temperature, pH, water potential, and oxygen. We will then be ready to take our knowledge of microbial growth and place it in the context of the fundamental molecular biological processes that occur as a cell prepares to divide and become two cells.

5.1

Overview of Cell Growth

The bacterial cell is essentially a synthetic machine that is able to duplicate itself. The synthetic processes of bacterial cell growth involve as many as 2000 chemical reactions of a wide variety of types. Some of these reactions involve energy transformations. Other reactions involve biosynthesis of small molecules—the building blocks of macromolecules—as well as the various cofactors and coenzymes needed for enzymatic reactions. However, the main reactions of cell synthesis are *polymerization reactions*, the processes by which polymers (macromolecules) are made from monomers. The major reactions of macromolecular synthesis will be discussed in Chapter 6: DNA synthesis, RNA synthesis, and protein synthesis. Once polymers are made, the stage is set for the final events of cell growth: assembly of macromolecules and formation of cellular structures such as the cell wall, cytoplasmic membrane, flagella, ribosomes, inclusion bodies, enzyme complexes, and so on.

Binary fission

In most prokaryotes, growth of an individual cell continues until the cell divides into two new cells, a process called *binary fission* (*binary* to express the fact that *two* cells have arisen from one cell). In a growing culture of a rod-shaped bacterium such as *Escherichia coli*, for example, cells are observed to elongate to approximately twice the length of an average cell and then form a partition that eventually separates the cell into two daughter cells (Figure 5.1). This partition is referred to as a *sep-*

tum and is a result of the inward growth of the cytoplasmic membrane and cell wall from opposing directions until the two daughter cells are pinched off (Figure 5.1). During the growth cycle all cellular constituents increase in number such that each daughter cell receives a complete chromosome and sufficient copies of all other macromolecules, monomers, and inorganic ions to exist as an independent cell. Partitioning of the replicated DNA molecule between the two daughter cells depends on the DNA remaining attached to membranes during division, with septum formation leading to separation of chromosome copies, one going to each daughter cell (Figure 5.1).

The time required for a complete growth cycle in bacteria is highly variable and is dependent on a number of factors, both nutritional and genetic. Under the best nutritional conditions the bacterium *Escherichia coli* can complete the cycle in about 20 min; a few bacteria can grow even faster than this, but many grow much slower. The control of cell division is a complex process and appears to be intimately tied to chromosomal replication events.

CONCEPT CHECK   **5.1**

Microbial growth involves an increase in the *number* of cells rather than in the size of individual cells. Growth of most microorganisms occurs by binary fission. Cell division and chromosome replication are usually coordinately regulated.

✔ *Define* microbial growth.

5.2

Population Growth

As we have mentioned, *growth* is defined as an increase in the *number* of microbial cells in a population, which can also be measured as an increase in microbial *mass*. **Growth rate** is the change in cell number or cell mass *per unit time*. During this cell division cycle, all the structural components of the cell double. The interval for the formation of two cells from one is called a **generation**, and the time required for this to occur is called the **generation time**. The generation time is thus the time required for the cell population to double. Because of this, the generation time is also sometimes called the *doubling time*. Note that during a single generation, both the cell number and cell mass double. Generation times vary widely among organisms. Many bacteria have generation times of 1–3 hr, but a few very rapidly growing organisms are known that divide in as little as 10 min, and others have generation times of several hours or even days.

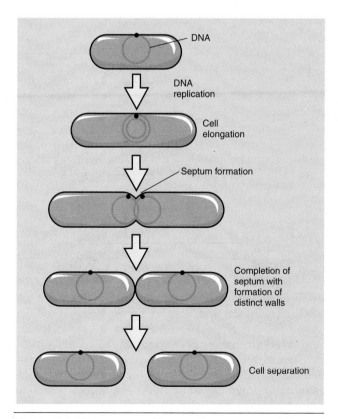

Figure 5.1 The process of binary fission in a rod-shaped prokaryote. For simplicity, the nucleoid is depicted as a single circle in green.

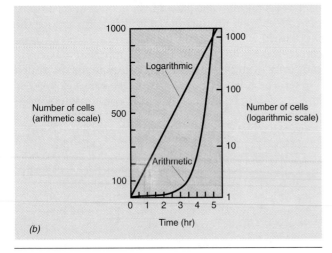

Time (hr)	Total number of cells
0	1
0.5	2
1	4
1.5	8
2	16
2.5	32
3	64
3.5	128
4	256
4.5	512
5	1,024
5.5	2,048
6	4,096
.	.
.	.
10	1,048,576

(a)

(b)

Figure 5.2 The rate of growth of a microbial culture. (a) Data for a population that doubles every 30 min. (b) Data plotted on an arithmetic (left ordinate) and a logarithmic (right ordinate) scale.

Exponential growth

A growth experiment beginning with a single cell having a doubling time of 30 min is presented in Figure 5.2. This pattern of population increase, where the number of cells *doubles* during each unit time period, is referred to as **exponential growth.** When the cell number from such an experiment is graphed on arithmetic coordinates as a function of elapsed time, one obtains a curve with a constantly increasing slope (Figure 5.2*b*). However, deriving growth rate information from such curves is difficult. The number of cells on a logarithmic ($\log_{10}$) scale is presented in Figure 5.2*b* in a graph in which cell number is plotted logarithmically and time is plotted arithmetically (a *semilogarithmic* graph), resulting in a straight line. This straight-line function is an immediate indicator that the cells are growing exponentially. Semilogarithmic graphs are also convenient and simple to use for estimating generation times from a set of results. The doubling time may be read directly from the graph (Figure 5.3; see also Figure 5.8).

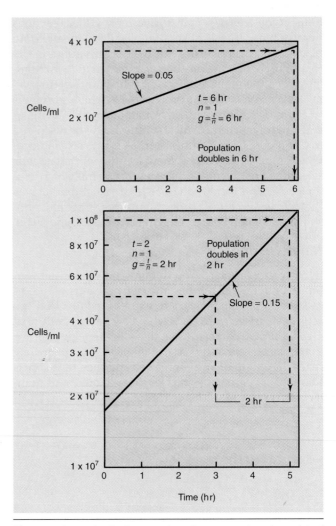

Figure 5.3 Method of estimating the generation times (g) of exponentially growing populations with generation times of 6 and 2 hr, respectively, from data plotted on semilogarithmic graphs. The slope of each line is $0.301/g$. All numbers are expressed in scientific notation; that is, 10,000,000 is 1×10^7, 60,000,000 is 6×10^7, and so on.

One of the characteristics of exponential growth is that the *rate* of increase in cell number is slow initially but increases at an ever faster rate. This results, in the later stages, in an explosive increase in cell numbers. For example, in the experiment in Figure 5.2, the *rate* of cell production in the first 30 min of growth is one cell per 30 min. However, between 4 and 4.5 hr of growth, the rate of cell production is considerably faster at 256 cells per 30 min (Figure 5.2). A practical implication of exponential growth is that when a nonsterile product such as milk is allowed to stand under conditions such that microbial growth can occur, a few hours during the early stages of exponential growth are not detrimental, whereas standing *for the same length of time* during the later stages is disastrous.

Calculating generation times

For many purposes in microbiology, it is useful to know the generation time of a microbial population during exponential growth, and such information can be obtained from a mathematical analysis of cell number data such as those in Figure 5.2. The increase in cell number that occurs in an exponentially growing bacterial culture is a simple geometric progression of the number 2. As two cells double (to become four cells), we can express this as $2^1 \rightarrow 2^2$. As four cells become eight, we express this as $2^2 \rightarrow 2^3$, and so on. Because of this geometric progression, there is a direct relationship between the number of cells present in a culture initially and the number present after a period of exponential growth:

$$N = N_0 2^n$$

where N = final cell number, N_0 = initial cell number, and n = *number of generations* that have occurred during the period of exponential growth. The generation time g of the cell population is calculated as t/n, where t is simply the hours or minutes of exponential growth. Thus, from a knowledge of the initial and final cell numbers in an exponentially growing cell population, it is possible to calculate n, and from n and knowledge of t, the generation time g.

To express the equation $N = N_0 2^n$ in terms of n, the following transformations are necessary:

$$N = N_0 2^n$$

$$\log N = \log N_0 + n \log 2$$

$$\log N - \log N_0 = n \log 2$$

$$n = \frac{\log N - \log N_0}{\log 2} = \frac{\log N - \log N_0}{0.301}$$

With n now expressed in terms of readily measurable quantities, N and N_0, generation times can be calculated. As an example of how to perform a calculation, we use actual data from the lower graph in Figure 5.3. The generation time of 2 hr, which in this case was determined directly from the graph, can also be derived from the facts that $N = 10^8$, $N_0 = 5 \times 10^7$, and $t = 2$. Thus,

$$n = \frac{\log 10^8 - \log (5 \times 10^7)}{0.301} = \frac{8 - 7.69}{0.301} = 1$$

Thus, the generation time $t/n = 2/1 = 2$ hr. The generation time g can also be calculated from the slope of the line obtained in the semilogarithmic plot of exponential growth, as slope = $0.301/g$.

Armed with knowledge of n and t, one can calculate g for different microorganisms growing exponentially under different culture conditions. Generation times in microbiological research are often useful as indications of the physiological state of a cell population and are frequently used to test the negative or positive effect of some treatment on the bacterial culture. Moreover, from knowledge of the generation time of a bacterium and the number of cells originally present (see Section 5.4 for how cell numbers are determined), it is possible to obtain a population of cells of known number by simply incubating the culture for a specific period of time. Thus, cells in the "midexponential" phase of growth (often desirable for study of enzymes and other components isolated from the cells) can easily be obtained by knowledge of the culture generation time under a given set of conditions.

CONCEPT CHECK   **5.2**

Microbial populations show a characteristic type of growth pattern called exponential growth, which is best seen by plotting the number of cells at various time periods on a semilogarithmic graph. From knowledge of the initial and final cell numbers and the time of exponential growth, the generation time of the cell population can be calculated directly.

✔ *Distinguish between the terms* growth rate *and* generation time.

✔ *Why does exponential growth lead to large cell populations in so short a period of time?*

✔ *What is a* semilogarithmic *plot?*

5.3
Growth Cycle of Populations

The data presented in Figure 5.2 reflect only part of the growth cycle of a microbial population, the part called *exponential growth*. In an enclosed vessel, referred to as a *batch culture*, a typical *growth curve* for a population of cells is obtained as illustrated in Figure 5.4. This growth curve can be divided into several distinct phases called the **lag phase, exponential phase, stationary phase,** and **death phase.**

Lag phase

When a microbial population is inoculated into a fresh medium, growth usually does not begin immediately but only after a period of time called the *lag phase*, which may be brief or extended depending on the history of the culture and growth conditions. If an exponentially growing culture is inoculated into the same medium under the same conditions of growth, a lag is not seen and exponential growth begins immediately. However, if the inoculum is taken from an old (stationary phase,

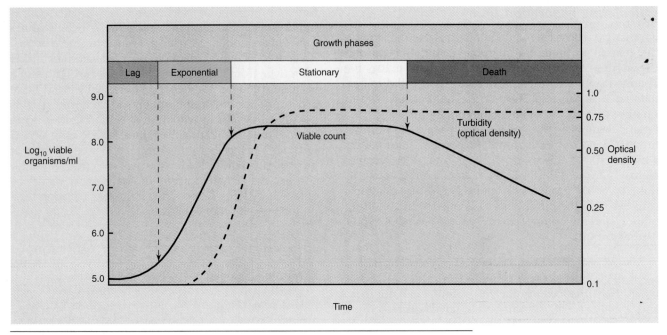

Figure 5.4 Typical growth curve for a bacterial population. See Section 5.4 for a description of the counting methods employed.

which is discussed later) culture and inoculated into the same medium, a lag usually occurs even if all the cells in the inoculum are *viable*, that is, able to reproduce. This is because the cells are usually depleted of various essential constituents and time is required for their resynthesis. A lag also ensues when the inoculum consists of cells that have been damaged (but not killed) by treatment with heat, radiation, or toxic chemicals because of the time required for the cells to repair the damage.

A lag is also observed when a population is transferred from a rich culture medium to a poorer one. This happens because for growth to occur in a particular culture medium the cells must have a complete complement of enzymes for synthesis of the essential metabolites not present in that medium. On transfer to a new medium, time is required for synthesis of the new enzymes.

Exponential phase

The *exponential phase* of growth has already been discussed. As noted, it is a consequence of the fact that each cell divides to form two cells, each of which also divides to form two more cells, and so on. Most unicellular microorganisms grow exponentially, but rates of exponential growth vary greatly. The rate of exponential growth is influenced by environmental conditions (temperature, composition of the culture medium) as well as by genetic characteristics of the organism itself. In general, prokaryotes grow faster than eukaryotic microorganisms, and small eukaryotes grow faster than large ones.

Stationary phase

In a batch culture (see Section 5.5), exponential growth cannot occur indefinitely. One can calculate that a single bacterium with a generation time of 20 min would, if it continued to grow exponentially for 48 hr, produce a population that weighed about 4000 times the weight of the earth! This is particularly impressive because a single bacterial cell weighs only about one-trillionth (10^{-12}) of a gram. Obviously, something must happen to limit growth of the population long before this time. What generally happens is that either an essential nutrient of the culture medium is used up or some waste product of the organism builds up in the medium to an inhibitory level and exponential growth ceases. The population has reached the **stationary phase.**

In the stationary phase there is no net increase or decrease in cell number. However, although no growth usually occurs in the stationary phase, many cell functions may continue, including energy metabolism and some biosynthetic processes. In some organisms, slow growth may even occur during the stationary phase; some cells in the population grow, whereas others die, the two processes balancing out so that no net increase or decrease in cell number occurs (this is a phenomenon called *cryptic growth*).

Studies with *Escherichia coli* have identified several genes that are necessary for the survival of cells that have entered the stationary phase. Some of these genes, called *sur* (for *survival*) genes, have as yet unknown functions, but mutations in *sur* genes lead to rapid cell death as cells enter the stationary phase. Because in nature it is likely that many bacterial cells are in a nongrowing or very slow growing state (∞ Chapter 14), it is not surprising that several genes have evolved to deal with conditions in which the cells in a population have reached the stationary phase or for other reasons are growing at extremely low rates.

Death phase

If incubation continues after a population reaches the stationary phase, the cells may remain alive and continue to metabolize, but they may also die. If the latter occurs, the population is said to be in the *death phase*. In some cases death is accompanied by actual cell **lysis.** Note in Figure 5.4 how the death phase of the growth cycle is also an exponential function; however, in most cases, the rate of cell death is much slower than that of exponential growth.

Before we conclude this section it should be emphasized that the phases of the bacterial growth curve shown in Figure 5.4 are reflections of the events in a *population* of cells, not in individual cells. The terms *lag phase, exponential phase, stationary phase,* and *death phase* do not apply to individual cells but only to *populations* of cells.

Now let us turn our attention to methods for determining cell numbers in microbial cultures.

CONCEPT CHECK   **5.3**

A microbial population generally shows a characteristic growth pattern when inoculated into a fresh culture medium. There is an initial lag phase, and then growth commences in an exponential fashion. As essential nutrients are depleted, or toxic products build up, growth ceases and the population enters the stationary phase. If incubation continues, cells may begin to die and the population is said to be in the death phase.

✔ *To what phase of the growth curve do the mathematics of growth apply?*

✔ *When does a lag phase usually* not *occur?*

✔ *Why do cells enter the stationary phase?*

5.4

Measurement of Growth

Population growth is measured by following changes in the number of cells or weight of cell mass. There are several methods for counting cell numbers or estimating cell mass, suited to different organisms or different problems.

Total cell count

The number of cells in a population can be measured by counting a sample under the microscope, a method called the **direct microscopic count.** Two kinds of direct microscopic counts are done, either on samples dried on slides or on samples in liquid. With liquid samples, special *counting chambers* must be used. In such a counting chamber, a grid is marked on the surface of the glass slide, with squares of known small area (Figure 5.5). Over each square on the grid is a volume of known size, very small but precisely measured. The number of cells per unit area of grid can be counted under the microscope, giving a measure of the number of cells

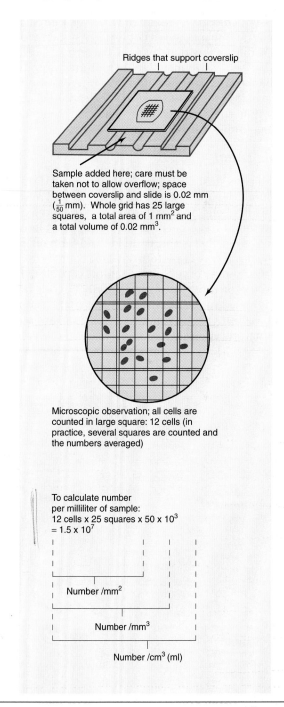

Ridges that support coverslip

Sample added here; care must be taken not to allow overflow; space between coverslip and slide is 0.02 mm ($\frac{1}{50}$ mm). Whole grid has 25 large squares, a total area of 1 mm^2 and a total volume of 0.02 mm^3.

Microscopic observation; all cells are counted in large square: 12 cells (in practice, several squares are counted and the numbers averaged)

To calculate number per milliliter of sample: 12 cells x 25 squares x 50 x 10^3 = 1.5 x 10^7

Number /mm^2

Number /mm^3

Number /cm^3 (ml)

Figure 5.5 Direct microscopic counting procedure using the Petroff–Hausser counting chamber.

per small chamber volume. Converting this value to the number of cells per milliliter of suspension is easily done by multiplying by a conversion factor based on the volume of the chamber sample (Figure 5.5).

Direct microscopic counting is a quick way of estimating microbial cell number. However, it has certain limitations: (1) Dead cells are not distinguished from living cells. (2) Small cells are difficult to see under the microscope, and some cells are probably missed. (3) Precision is difficult to achieve. (4) A phase contrast microscope is required when the sample is not stained. (5) The method is not usually suitable for cell suspensions of low density. With bacteria, if a cell suspension has less than 10^6 cells/milliliter (ml), few if any bacteria will be seen in the microscope field. However, dilute suspensions may be counted if a sample is first concentrated and then resuspended in a small volume.

Viable count

In the method just described, both living and dead cells are counted. In many cases we are interested in counting only live cells, and for this purpose *viable* cell counting methods have been developed. A viable cell is defined as one that is able to divide and form offspring, and the usual way to perform a viable count is to determine the number of cells in the sample capable of forming *colonies* on a suitable agar medium. For this reason, the viable count is often called the **plate count**, or **colony count.** The assumption made in this type of counting procedure is that *each viable cell can yield one colony.*

There are two ways of performing a plate count: the spread plate method and the pour plate method (Figure 5.6). With the **spread plate method,** a volume of an appropriately diluted culture usually no greater than 0.1 ml is spread over the surface of an agar plate using a sterile glass spreader. The plate is then incubated until the colonies appear, and the number of colonies is counted. It is important that the surface of the plate be dry, so that the liquid that is spread soaks in. Volumes greater than 0.1 ml are rarely used because the excess liquid does not soak in and may cause the colonies to coalesce as they form, making them difficult to count. In the **pour plate method** (Figure 5.6), a known volume (usually 0.1–1.0 ml) of culture is pipetted into a sterile Petri plate; melted agar medium is then added and mixed well by gently swirling the plate on the table top. Because the sample is mixed with the molten agar medium, a larger volume can be used than with the spread plate; however, with the pour plate method the organism to be counted must be able to briefly withstand the temperature of melted agar, 45°C.

Dilutions

With both the spread plate and pour plate methods, it is important that the number of colonies developing on the plates not be too large because on crowded plates some cells may not form colonies and some colonies may fuse, leading to erroneous measurements. It is also essential that the number of colonies not be too small, or the statistical significance of the calculated count will be low. The usual practice, which is the most valid sta-

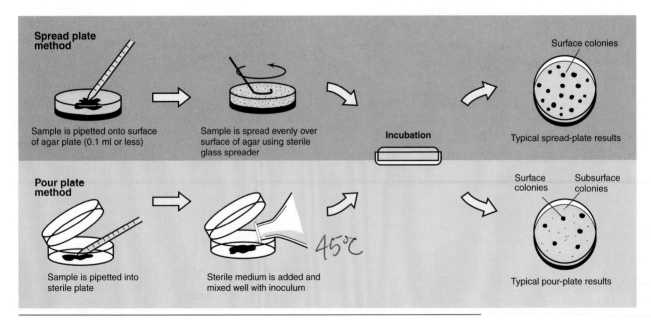

Figure 5.6 Two methods of performing a viable count (plate count). In either case the sample must usually be diluted before plating.

tistically, is to count colonies only on plates that have between 30 and 300 colonies.

To obtain the appropriate colony number, the sample to be counted must almost always be *diluted*. Since one rarely knows the approximate viable count ahead of time, it is usually necessary to make more than one dilution. Several 10-fold dilutions of the sample are commonly used (Figure 5.7). To make a 10-fold (10^{-1}) dilution, one can mix 0.5 ml of sample with 4.5 ml of diluent, or 1.0 ml sample with 9.0 ml diluent. If a 100-fold (10^{-2}) dilution is needed, 0.05 ml can be mixed with 4.95 ml diluent, or 0.1 ml with 9.9 ml diluent. Alternatively, a 10^{-2} dilution can be made by making two successive 10-fold dilutions. In most cases, such *serial dilutions* are needed to reach the final dilution desired. Thus, if a 10^{-6} ($1/10^6$) dilution is needed, it can be achieved by making three successive 10^{-2} ($1/10^2$) dilutions or six successive 10^{-1} dilutions (Figure 5.7).

Sources of error in plate counting

The number of colonies obtained in a viable count depends not only on the inoculum size but also on the suitability of the culture medium and the incubation conditions used; it also depends on the length of incubation. The cells deposited on the plate will not all develop into colonies at the same rate, and if a short incubation time is used, less than the maximum number of colonies will be obtained. Furthermore, the size of colonies often varies. If some tiny colonies develop, they may be missed during the counting. It is usual to determine the incubation conditions (medium, temperature, time) that will give the maximum number of colonies of a given organism and then use these conditions throughout. Viable counts can be subject to large error, and if accurate counts

are desired, great care must be taken and replicate plates of key dilutions must be prepared. Note that two or more cells in a clump form only a single colony, and so a viable count may be erroneously low. To more clearly state the result, viable counts are often expressed as the number of *colony-forming units* obtained rather than as the number of *viable cells* (since a colony-forming unit may contain one or more cells).

Despite the difficulties associated with viable counting, the procedure gives the best information on the number of viable cells and so is widely used. In food, dairy, medical, and aquatic microbiology, viable counts are employed routinely. The method has the virtue of high sensitivity: samples containing very few cells can be counted, thus permitting sensitive detection of microbial contamination of products or materials. Moreover, the use of highly selective culture media and growth conditions (∞ Section 21.2) in viable counting procedures allows for the counting of only particular cell types in a mixed population of microorganisms.

Measurements of cell mass and turbidity

In some cases it is necessary to estimate the *mass* of cells present in a culture rather than the actual cell *number*. However, because cell mass is proportional to cell number, a determination of one parameter can be used to estimate the other. Net cell mass can be measured by concentrating (by centrifugation, for example) a known volume of culture and weighing the pellet obtained. The usual procedure is to determine the dry weight of cells following drying of the pelleted cells at 90–110°C overnight. The dry mass of bacterial cells is usually 10–20% of the wet mass.

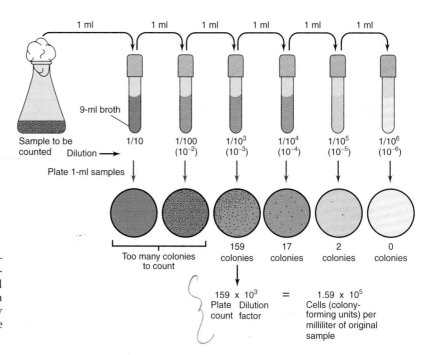

Figure 5.7 Procedure for viable count using serial dilutions of the sample. The sterile liquid used for making dilutions can simply be water, but a balanced salt solution or growth medium may yield a higher recovery. The dilution factor is the reciprocal of the dilution.

A more rapid and quite useful method of obtaining an estimate of cell number or mass is by use of *turbidity measurements*. A cell suspension looks cloudy (turbid) to the eye because cells scatter light passing through the suspension. The more cells present, the more light scattered and hence the more turbid the suspension. Turbidity can be measured with a *photometer* or a *spectrophotometer*, devices that pass light through a cell suspension and detect the amount of unscattered light that emerges (Figure 5.8). The major difference between these two instruments is that a photometer employs a simple filter (usually red, green, or blue) to generate incident light of relatively broad wavelength, whereas a spectrophotometer employs a prism or diffraction grating to generate incident light in a narrow band of wavelengths to impinge on the sample (Figure 5.8a). However, both devices measure only *unscattered* light, and readings are recorded in photometer units (for example, "Klett units" for the Klett-Summerson photometer) (see Figure 5.8b) or optical density units (OD) for a spectrophotometer.

For unicellular organisms, photometer units or OD are proportional (within certain limits) to cell mass and thus also to cell number; therefore, turbidity readings can be used as a substitute for other counting methods. However, before using turbidity as an estimate of cell number or mass, a *standard curve* must first be prepared for each organism to be studied, relating some direct measurement of cell number (microscopic or viable count) or mass (dry weight) to the *indirect* measurement obtained from turbidity (Figure 5.8c). Such a curve can contain data for both cell number and cell mass, allowing for an estimate of both parameters from a single turbidity reading (Figure 5.8c).

At high concentrations of cells, light scattered away from the detecting unit by one cell can be rescattered back by another (and thus appear to the photocell as if it had never been scattered), and when this occurs, the one-to-one correspondence between cell number and turbidity loses linearity (Figure 5.8c). Nevertheless, within limits turbidity measurements can be reasonably accurate and have the virtue of being quick and easy to perform. In addition, turbidity measurements can usually be made without destroying or significantly disturbing the sample. For these reasons turbidity measurements are widely employed to follow the growth rate of microbial cultures; the same sample can be checked repeatedly, and the measurements plotted on a semilogarithmic plot versus time (Figure 5.8b) and used to calculate the generation time of the growing culture.

5.5
Continuous Culture

Our discussion of population growth thus far has been confined to batch cultures, growth occurring in a fixed volume of a culture medium that is continually being altered by the actions of the growing organisms until it is no longer suitable for growth. In the early stages of exponential growth in batch cultures, conditions may remain relatively constant, but in later stages when cell numbers become quite large, drastic changes in the chemical composition of the culture medium usually occur. For many studies, it is desirable to keep cultures in constant environments for long periods, and this is done by employing *continuous cultures*. A continuous culture is essentially a flow system of constant volume to which medium is added continuously and from which continuous removal of any overflow can occur. Once such a system is in equilibrium, cell number and nutrient status remain *constant*, and the system is said to be in **steady state.**

The chemostat

The most common type of continuous culture device used is a **chemostat** (Figure 5.9), which permits control of both the population density and the growth rate of the culture. Two elements are used in the control of a chemostat—the *dilution rate* and the *concentration of a limiting nutrient,* such as a carbon or nitrogen source. In a batch culture nutrient concentration can affect both the growth rate and the growth yield of a microorganism (Figure 5.10). At very low concentrations of a given nutrient, the growth rate is reduced, probably because the nutrient cannot be transported into the cell fast enough to satisfy metabolic demand, whereas at moderate or higher levels of nutrient, growth *rates* may not be affected while cell *yield* continues to increase (Figure 5.10). In contrast to a batch culture, in a chemostat, growth rate and growth yield can be controlled independently of each other, the former by adjusting the dilution rate and the latter by varying the concentration of a nutrient present in a limiting amount.

CONCEPT CHECK   5.4

Growth is measured by the change in number of cells with time. Cell counts done microscopically measure the total number of cells in a population, whereas viable cell counts (plate counts) measure only the living population. Measurements of cell mass or turbidity are indirect but very useful measures of cell growth.

✔ *Why is a* viable count *more sensitive than a* microscopic count?

✔ *What is the major assumption made in relating plate count results to cell number?*

✔ *Describe how you would dilute a bacterial culture by 10⁻⁷.*

✔ *Of what use is a spectrophotometer in the study of microbial growth?*

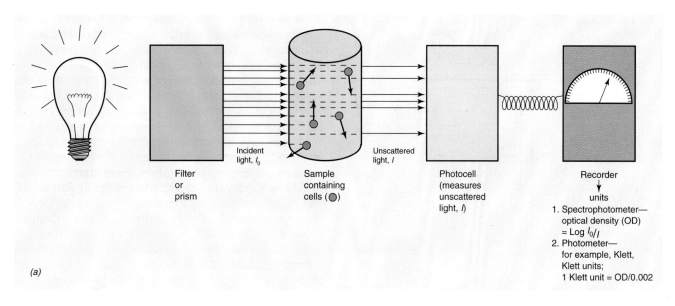

(a)

Filter
or
prism

Incident
light, I_0

Sample
containing
cells ($\bullet$)

Unscattered
light, I

Photocell
(measures
unscattered
light, I)

Recorder
↓
units

1. Spectrophotometer—
   optical density (OD)
   = Log I_0/I
2. Photometer—
   for example, Klett,
   Klett units;
   1 Klett unit = OD/0.002

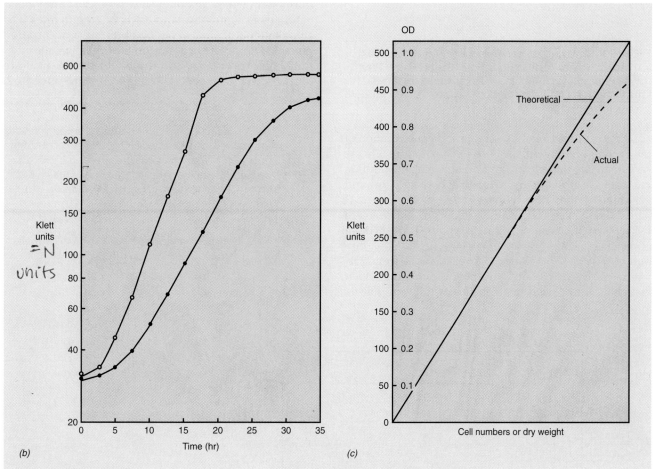

(b)

(c)

Cell numbers or dry weight

Theoretical

Actual

Figure 5.8 Turbidity measurements of microbial growth. (a) Measurements of turbidity are made in a spectrophotometer or photometer. The photocell measures incident light unscattered by cells in suspension and gives readings in optical density or photometer units. (b) Typical growth curve data obtained in Klett units for two organisms growing at different growth rates. For practice, calculate the generation time (g) of the two cultures using the formula $n = \frac{\log N - \log N_0}{0.301}$ where N and N_0 are two different Klett values taken between a time interval t. (c) Relationship between cell number or dry weight and turbidity readings. Note that the one-to-one correspondence between these relationships breaks down at high turbidities.

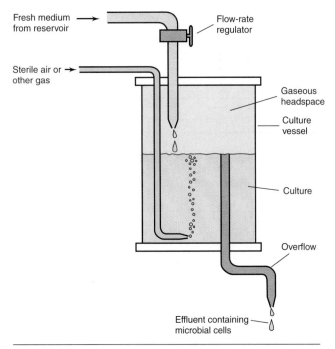

Figure 5.9 Schematic for a continuous culture device (chemostat). In such a device, the population density is controlled by the concentration of limiting nutrient in the reservoir, and the growth rate is controlled by the flow rate (see Figure 5.11). Both parameters can be set by the experimenter.

Effects of varying dilution rate and concentration of the growth-limiting nutrient are given in Figure 5.11. As seen, there are rather wide limits over which the dilution rate controls growth rate, although at both very low and very high dilution rates the steady state breaks down. At high dilution rates, the organism cannot grow fast enough to keep up with its dilution, and the culture is washed out of the chemostat. At the other extreme, at very low dilution rates, a large fraction of the cells may die from starvation because the limiting nutrient is not being added fast enough to permit maintenance of cell metabolism.

There is probably a minimum amount of energy necessary to maintain cell structure and integrity, called **maintenance energy,** and nutrients used for maintenance energy are not available for biosynthesis and cell growth. Thus, at very low dilution rates steady-state conditions are not maintained, and the population slowly washes out. Note also in Figure 5.11 that because the limiting nutrient is quickly assimilated, its concentration in the chemostat vessel itself is virtually zero until very high dilution rates are obtained and the culture begins to wash out.

The *cell density* (cells/ml) in the chemostat is controlled by the level of the limiting nutrient, just as cell yield was controlled in a batch culture (Figure 5.10). If the concentration of this nutrient in the incoming medium is raised, with the dilution rate remaining constant, the cell density will increase although growth rate will remain the same and the steady-state concentration of the nutrient in the culture vessel will still be virtually zero. Thus, by adjusting dilution rate and nutrient level, the experimenter can obtain at will a variety of population densities growing at a variety of growth rates. The actual shape of the curve for bacterial concentration given in Figure 5.11 depends on the organism, the environmental conditions, and the limiting nutrient used. Further discussion of chemostats occurs in Appendix 2.

CONCEPT CHECK   5.5

Continuous culture devices (chemostats) are a means of maintaining cell populations in exponential growth for long periods. In a chemostat, the rate at which the culture is diluted governs the growth rate and the population size is governed by the concentration of the growth-limiting nutrient entering the vessel.

✔ *How do microorganisms in a* chemostat *differ from microorganisms in a* batch culture?

✔ *How is growth rate controlled in a chemostat?*

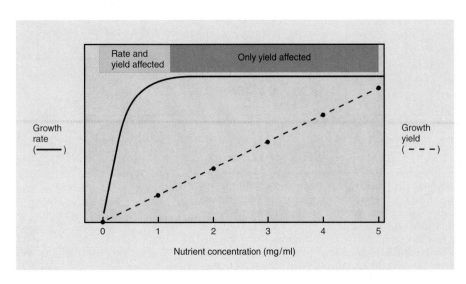

Figure 5.10 Relationship among nutrient concentration, growth rate (solid line), and growth yield (dashed line) in a batch culture (closed system). At low nutrient concentrations both growth rate and growth yield are affected.

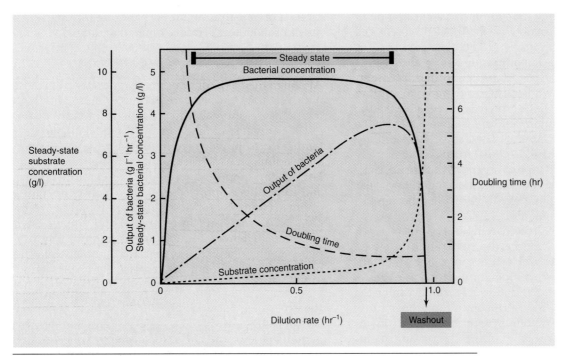

Figure 5.11 Steady-state relationships in the chemostat. The dilution rate is determined from the flow rate and the volume of the culture vessel. Thus, with a vessel of 1000 ml and a flow rate through the vessel of 500 ml/hr, the dilution rate would be 0.5 hr^{-1}. Note that at high dilution rates, growth cannot balance dilution, the population washes out, and the substrate concentration rises to the concentration in the medium reservoir (since there are no bacteria to use the inflowing substrate). However, throughout most of the range of dilution rates shown, the population density remains constant and the substrate concentration remains at a very low value (that is, steady state). Note that although the population density remains constant, the growth rate (doubling time) varies over a wide range. Thus, the experimenter can obtain populations with widely varying growth rates without affecting population density. See Appendix 2 for further details of chemostat operation.

5.6
Effect of Environmental Factors on Growth

Up to now we have described growth of microorganisms under essentially ideal laboratory conditions. However, the activities of microorganisms are greatly affected by the chemical and physical conditions of their environments. Understanding environmental influences helps us to explain the distribution of microorganisms in nature and makes it possible for us to devise methods for controlling microbial activities and destroying undesirable organisms. Not all organisms respond equally to a given environmental factor. In fact, an environmental condition may be harmful to one organism and actually beneficial to another. Organisms can tolerate some adverse conditions under which they cannot grow, and hence we must distinguish between the effects of environmental conditions on the *viability* of an organism and on growth, differentiation, and reproduction.

Regardless of whether organisms are interacting with other organisms in natural communities or with each other in pure culture in the laboratory, the environment can significantly affect their ability to carry out metabolic reactions and grow. Many environmental factors could be considered in this connection, however, four main factors have been identified that clearly play major roles in controlling microbial growth: temperature, pH, water availability, and oxygen. We consider each of these factors in detail here.

5.7
Effect of Temperature on Microbial Growth

Temperature is one of the most important environmental factors influencing the growth and survival of organisms. It can affect living organisms in either of two opposing ways. As the temperature rises, chemical and enzymatic reactions in the cell proceed at more rapid rates and growth becomes faster. However, *above* a certain temperature, proteins, nucleic acids, and other cellular components may be irreversibly damaged. Thus, as the temperature is increased within a given range, growth and metabolic function increase up to a point where inactivation reactions set in. Above this point, cell functions fall sharply to zero. Thus, we find that for every organism there is a **minimum temperature** below which growth no longer occurs, an **optimum temperature** at which growth is most rapid, and a **maximum temperature** above which growth is not possible (Figure 5.12). The optimum temperature is always nearer the *maximum* than the minimum. These three temperatures, often called the **cardinal temperatures,** are generally characteristic of each type of organism but are not completely fixed, as they can be modified slightly by other factors of the environment—in particular, the composition of the growth medium.

The maximum growth temperature of a given organism most likely reflects the inactivation discussed previously. However, the factors controlling an organism's *minimum* growth temperature are not as clear. As mentioned earlier (∞ Section 3.4), the cytoplasmic membrane must be in a fluid state for proper functioning. Perhaps the minimum temperature of an organism results from "freezing" of the cytoplasmic membrane so it no longer functions properly in nutrient transport or proton gradient formation. This explanation is supported by experiments in which the minimum temperature for an organism is altered to some extent by adjustments in membrane lipid composition (see Section 5.8). It is also observed that the cardinal temperatures of different microorganisms differ widely; some organisms have temperature optima as low as 5–10°C, and some higher than 100°C. The temperature range throughout which growth occurs is even wider than this, from below freezing to greater than boiling (the archaean *Pyrodictium brockii* has a temperature maximum of 110°C, and a related organism can grow at up to 113°C!). However, no single organism can grow over this whole temperature range, and the usual range for a given organism is about 30°, although some have a much broader temperature range than others.

Temperature classes of organisms

Although there is a continuum of organisms, from those with very low temperature optima to those with high temperature optima, it is possible to broadly distinguish *four groups* of microorganisms in relation to their temperature optima: **psychrophiles,** with low temperature optima, **mesophiles,** with midrange temperature optima, **thermophiles,** with high temperature optima, and **hyperthermophiles,** with very high temperature optima (Figure 5.13). Mesophiles are found in warm-blooded animals and in terrestrial and aquatic environments in temperate and tropical latitudes. Psychrophiles and thermophiles are found in unusually cold and unusually hot environments, respectively. Hyperthermophiles are found in extremely hot habitats such as hot springs, geysers, and deep-sea hydrothermal vents (see Sections 5.8 and 14.9).

In *Escherichia coli*, a typical mesophile, a detailed study of growth as a function of temperature has precisely defined the cardinal temperatures. The optimum temperature of *E. coli* in a rich complex medium is 39°C, the maximum is 48°C, and the minimum is 8°C. These values are subject to slight strain differences, and in general, the maximum and minimum temperatures supporting growth of an organism are higher and lower, respectively, when tested in complex rather than defined media.

CONCEPT CHECK   5.7

Temperature is a major environmental factor controlling microbial growth. Various microorganisms differ greatly in their temperature requirements for growth.

✔ *What are the approximate cardinal temperatures for Escherichia coli? To what temperature class does it belong?*

✔ *How does a* hyperthermophile *differ from a* psychrophile?

5.8

Microbial Growth at Temperature Extremes

Because humans live and work on the surface of the earth where temperatures are generally moderate, it is natural to consider very hot and very cold environments as being "extreme." And they are extreme for human habitation because humans would die quickly

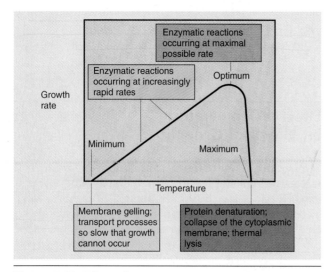

Figure 5.12 Effect of temperature on growth rate and the molecular consequences for the cell.

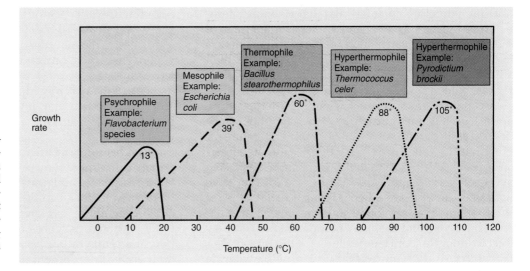

Figure 5.13 Relation of temperature to growth rates of a typical psychrophile, a typical mesophile, a typical thermophile, and two different hyperthermophiles. The temperature optima of the example organisms are shown on the graph.

if immersed in boiling or freezing water. However, the natural habitats of many microorganisms can be either extremely hot or extremely cold, and the organisms that live there have evolved to grow optimally under these conditions. We consider such organisms here.

Cold environments and psychrophiles

Much of the earth's surface experiences fairly low temperatures. The oceans, which make up over half of the earth's surface, have an average temperature of 5°C, and the depths of the open oceans have constant temperatures of about 1–3°C. Vast land areas of the Arctic and Antarctic are permanently frozen or are unfrozen for only a few weeks in summer (Figure 5.14a). These cold environments are rarely sterile, and some microorganisms can be found alive and growing at any low temperature at which liquid water still exists. Even in many frozen materials there are usually microscopic pockets of liquid water present where microorganisms can grow. It is important to distinguish between environments that are cold *throughout* the year and those that are cold *only* in winter. The latter, characteristic of con-

(a)

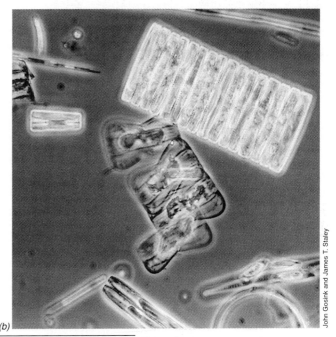

(b)

Figure 5.14 Microorganisms from Antarctic sea ice. (a) A core of permanently frozen seawater from McMurdo Sound, Antarctica. Note the dense coloration due to pigmented microorganisms, and the boot included to show scale. (b) Phase contrast micrograph of phototrophic microorganisms from the core shown in (a). Most organisms are either diatoms or green algae (both eukaryotic microorganisms).

tinental temperate climates, may have summer temperatures as high as 40°C and winter temperatures of −20°C or colder. Such highly variable environments are much less favorable for cold-adapted organisms than are the constantly cold environments found in polar regions, at high altitudes, and in the depths of the oceans.

As noted earlier, organisms with low temperature optima are called **psychrophiles**. A psychrophile can be defined as an organism with an optimal temperature for growth of 15°C or lower, a maximum growth temperature below 20°C, and a minimal temperature for growth at 0°C or lower. Organisms that grow at 0°C but have optima of 20–40°C are called *psychrotolerant.*

Psychrophiles are found in environments that are constantly cold, and they may be rapidly killed even by brief warming to room temperature. For this reason, their laboratory study requires that great care be taken to ensure that they never warm up during sampling, transport to the laboratory, plating, pipetting, or other manipulations. Some of the best-studied psychrophiles have been algae that grow in dense masses within and under the ice in polar regions (Figure 5.14*b*). Psychrophilic algae are also often seen on the surfaces of snowfields and glaciers in such large numbers that they impart a distinctive red or green coloration to the surface (Figure 5.15*a*). The most common snow alga is *Chlamydomonas nivalis;* its brilliant red spores are responsible for the red

color (Figure 5.15*b*). The alga probably grows within the snow as a green-pigmented vegetative cell and then sporulates; as the snow dissipates by melting, erosion, and vaporization, the spores become concentrated on the surface. Snow algae are most commonly seen on melting permanent snowfields in midsummer to late summer and are especially common in sunny, dry areas, probably because in more rainy areas they are washed away from the snowfields.

Psychrotolerant microorganisms are much more widely distributed than psychrophiles and can be isolated from soils and water in temperate climates as well as from meat, milk and other dairy products, cider, vegetables, and fruit stored under refrigeration (4°C). As noted, psychrotolerant microorganisms grow best at a temperature between 20 and 40°C. Because temperate environments warm up in summer, it is understandable that they cannot support the heat-sensitive psychrophiles, the warming essentially providing a selective force favoring psychrotolerant species and excluding psychrophilic forms. It should be emphasized that although psychrotolerant microorganisms do grow at 0°C, they do not grow very well, and one must often wait several weeks before visible growth is seen in culture media. Various genera of Bacteria, fungi, algae, and protozoa have members that are psychrotolerant.

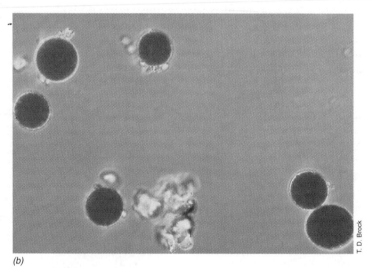

(b)

(a)

Figure 5.15 Snow algae. (a) Snow bank in the Sierra Nevada, California, with red coloration caused by the presence of snow algae. Pink snow such as this is common on summer snow banks at high altitudes throughout the world. (b) Photomicrograph of red-pigmented spores of the snow alga *Chlamydomonas nivalis.*

Molecular adaptations to psychrophily

Psychrophiles produce enzymes that function optimally in the cold and that are often denatured or otherwise inactivated at even very moderate temperatures. Another feature of psychrophiles is that compared to mesophiles, active transport occurs well at low temperature, an indication that the cytoplasmic membranes of psychrophiles are constructed in such a way that low temperatures do not inhibit membrane phenomena. Studies on the composition of cytoplasmic membranes from psychrophiles have shown them to contain a higher content of *unsaturated* fatty acids (∞ Section 4.21), which help to maintain a semifluid state of the membrane at low temperatures (membranes composed of predominantly saturated fatty acids would become waxy and nonfunctional at low temperatures). The lipids of some psychrophilic bacteria also contain *polyunsaturated* fatty acids and long-chain hydrocarbons with multiple double bonds. In the latter connection, a hydrocarbon with nine double bonds ($C_{31:9}$) has been identified from the lipids of several Antarctic bacteria.

Freezing

Despite the ability of some organisms to grow at low temperatures, there is a lower limit below which reproduction is impossible. Pure water freezes at $0°C$ and seawater at $-2.5°C$, but freezing is not continuous and microscopic pockets of water continue to exist at much lower temperatures. Although freezing prevents microbial growth, it does not always cause microbial death. In addition, the medium in which the cells are suspended considerably affects sensitivity to freezing. Water-miscible liquids such as glycerol and dimethylsulfoxide (DMSO), when added at about 10% (final concentration) to the suspending medium, penetrate the cells and protect by reducing the severity of dehydration effects and preventing ice crystal formation. In fact, the addition of such agents, called *cryoprotectants*, is a common way of *preserving* microbial cultures at very low temperatures (usually -70 to $-196°C$).

High temperature environments and thermophiles and hyperthermophiles

Organisms whose growth temperature optimum is above $45°C$ are called **thermophiles,** and those whose optimum is above $80°C$ are called **hyperthermophiles.** Temperatures as high as these are found in nature only in certain restricted areas. For example, soils subject to full sunlight are often heated to temperatures above $50°C$ at midday, and some soils may become warmed to even $70°C$, although a few centimeters under the surface the temperature is much lower. Fermenting materials such as compost piles and silage usually reach temperatures of $60–65°C$. However, the most extensive and extreme high temperature environments are found in nature in association with volcanic phenomena.

Many hot springs have temperatures near boiling, and steam vents (fumaroles) may reach $150–500°C$. Hydrothermal vents in the bottom of the ocean have temperatures of $350°C$ or greater (see Section 14.10 for a discussion of hydrothermal vents). Hot springs occur throughout the world but are especially concentrated in the western United States, New Zealand, Iceland, Japan, the Mediterranean region, Indonesia, Central America, and central Africa. The area with the largest single concentration of hot springs in the world is Yellowstone National Park, Wyoming. Although some springs vary in temperature, others are very constant, not varying more than $1–2°C$ over many years.

Many hot springs are at the boiling point for the altitude ($92–93°C$ at Yellowstone, $99–100°C$ at locations where the springs are close to sea level). As the water overflows the edges of the spring and flows away from the source, it gradually cools, setting up a *thermal gradient*. Along this gradient, various microorganisms grow (Figure 5.16 and see the cover of this book), with different species growing in the different temperature ranges. By studying the species distribution along such thermal gradients and by examining hot springs and other thermal habitats at different temperatures around the world, it is possible to determine the upper temperature limits for each kind of organism (Table 5.1). From this information we conclude that (1) prokaryotic organisms in general are able to grow at temperatures higher than those at which eukaryotes can grow; (2) the most thermophilic of all prokaryotes are certain species of Archaea; and (3) nonphototrophic organisms are able to grow at higher temperatures than can phototrophic forms. However, it should be emphasized that not all organisms from a group are able to grow near the upper limits for that group. Usually only a relatively few species or genera are able to function successfully near the upper temperature limit.

Molecular adaptations to thermophily

How can thermophiles and hyperthermophiles thrive at high temperatures? First, their enzymes and other proteins are much more stable to heat than are those of mesophiles, and these macromolecules actually function *optimally* at high temperatures. How is heat stability achieved? Studies of thermophilic enzymes have shown that they often differ very little in amino acid sequence from an enzyme that catalyzes the same reaction in a mesophile. It appears that a critical amino acid substitution in one or a few locations in the enzyme allows it to fold in a different way and thereby withstand the denaturing effects of heat.

Heat stability of proteins from hyperthermophiles is also improved as a result of the increased number of *salt bridges* (bridging of charges on amino acids by Na^+ or other cations) present and the densely packed highly hydrophobic interiors of the proteins, which naturally

Figure 5.16 Growth of thermophilic cyanobacteria in hot springs in Yellowstone National Park. (a) Aerial photograph of a very large boiling spring, Grand Prismatic Spring. The orange color in the outflow channel is due to the rich carotenoid pigments of bacteria and cyanobacteria. (b) Characteristic V-shaped pattern formed by cyanobacteria at the upper temperature for photosynthetic life, 70–74°C, in the thermal gradient formed from a boiling hot spring. The pattern develops because the water cools more rapidly at the edges than in the center of the channel. The spring flows from the back of the picture toward the foreground.

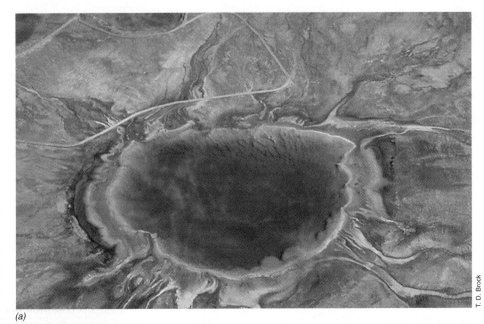

(a)

(b)

have membrane lipids rich in *saturated* fatty acids, thus allowing the membranes to remain stable and functional at high temperatures. Saturated fatty acids form much stronger hydrophobic bonds than do unsaturated fatty acids, which accounts for the membrane stability. Hyperthermophiles, virtually all of which are Archaea, do not contain fatty acids in the lipids of their membranes but instead have hydrocarbons of various lengths composed of repeating units of the five-carbon compound phytane (∞ Figure 3.20a) bonded by ether link-

| TABLE 5.1 | Presently known upper temperature limits for growth of living organisms |

Group	Upper temperature limits (°C)
Animals	
Fish and other aquatic vertebrates	38
Insects	45–50
Ostracods (crustaceans)	49–50
Plants	
Vascular plants	45
Mosses	50
Eukaryotic microorganisms	
Protozoa	56
Algae	55–60
Fungi	60–62
Prokaryotes	
Bacteria	
Cyanobacteria	70–74
Anoxygenic phototrophic bacteria	70–73
Chemoorganotrophic bacteria	90
Archaea	
Hyperthermophilic methanogens	110
Sulfur-dependent hyperthermophiles	113

resist unfolding in the aqueous milieu. In addition to enzymes and other proteins in the cell, the protein-synthesizing machinery (that is, ribosomes and other constituents) of thermophiles and hyperthermophiles, as well as such structures as the cytoplasmic membrane, are likewise heat-stable. We mentioned earlier that psychrophiles have membrane lipids rich in *unsaturated* fatty acids, thus making the membranes fluid and functional at low temperatures. Conversely, thermophiles

age to glycerolphosphate. We discussed the details of the unique membrane architecture of hyperthermophilic Archaea in Chapter 3 (∞ Section 3.3) and will consider other aspects of heat stability in hyperthermophiles in Section 17.5.

Why are eukaryotes absent from environments with temperatures above about 60°C (Table 5.1)? This most likely involves the stability of organellar membranes, which must remain fairly porous to permit passage of large molecules like ATP and RNA. It is likely that porous membranes such as these would be more temperature-labile than the typical lipid bilayers of prokaryotes (or lipid monolayers of some hyperthermophiles) (∞ Section 3.3). Thus, above 60°C, the organelles of eukaryotes cannot survive and the only life forms observed are prokaryotes.

Ecology and biotechnological applications of thermophiles and hyperthermophiles

In most boiling hot springs (Figure 5.17), a variety of hyperthermophiles are usually present. The growth of such organisms can be studied by immersing microscope slides into the spring and retrieving them after a few days. Microscopic examination of the slides reveals colonies of prokaryotes (Figure 5.17b) that have devel-

oped from single bacterial cells that attached to and grew on the glass surface.

Ecological studies of organisms living in boiling springs have shown that growth rates are fairly rapid, and doubling times of as short as 1 hr have been recorded. Both aerobic and anaerobic species have been found, and many morphological and physiological types exist (Table 5.2 and ∞ Section 17.3). Phylogenetic studies using ribosomal RNA sequencing (∞ Section 15.6) have shown that an abundant diversity of hyperthermophiles exists. Some hyperthermophiles show growth temperature optima greater than 100°C and thus are grown in the laboratory in pressurized vessels to prevent boiling.

Thermophilic prokaryotes with growth temperature optima below 80°C have been widely found in artificial thermal environments (Table 5.2). The hot water heater, domestic or industrial, usually has a temperature of 55–80°C and is a favorable habitat for the growth of thermophilic prokaryotes. Organisms resembling *Thermus aquaticus*, a common hot spring organism (see the back cover of this book), have been isolated from hot water at many installations. Electric power plants, hot industrial process water, and other artificial thermal sources probably also provide sites where thermophiles can grow.

Thermophilic and hyperthermophilic microorganisms are of interest for more than just basic biological rea-

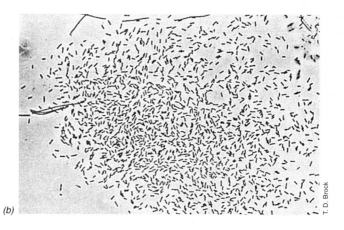

Figure 5.17 Bacterial growth in boiling water. (a) A small boiling spring in Yellowstone National Park, Boulder Spring. This spring is superheated, having a temperature 1–2°C above the boiling point. The mineral deposits around the spring consist mainly of silica. (b) Photomicrograph of a bacterial microcolony that developed on a microscope slide immersed in a boiling spring such as that shown in (a).

(a)

(b)

T. D. Brock

TABLE 5.2	Major groups of thermophilic and hyperthermophilic prokaryotes	

Genus	Temperature range (°C)
Bacteria	
Phototrophic Bacteria	
Cyanobacteria	55–70
	(One strain, 74)
Purple bacteria	45–60
Green bacteria	40–73
Gram-positive Bacteria	
Bacillus	50–70
Clostridium	50–75
Lactic acid bacteria	50–65
Actinomycetes	55–75
Other Bacteria	
Thiobacillus	50–60
Spirochete	54
Desulfotomaculum	37–55
Gram-negative aerobes	50–75
Gram-negative anaerobes	50–75
Thermotoga/Aquifex	55–95
Thermus	60–80
Archaea	
Methanogens	45–110
Sulfur-dependent hyperthermophiles	60–113
Thermoplasma	37–60

sons. These organisms offer some major advantages for industrial and biotechnological processes, many of which run more rapidly and efficiently at high temperatures. Enzymes from thermophiles and hyperthermophiles are capable of catalyzing biochemical reactions at high temperatures (∞ Section 12.9 and Figure 12.19) and are generally more stable than enzymes from mesophiles, thus prolonging the shelf life of enzyme preparations. A practical example of a heat-stable enzyme of great applied importance is the DNA polymerase isolated from the thermophile *Thermus aquaticus. Taq polymerase*, as this enzyme

is known, has been used to automate the repetitive steps involved in the *polymerase chain reaction* (PCR) technique, a method of amplifying specific DNA sequences (∞ Section 10.9). Because *Taq* polymerase does not denature at the high temperatures needed to melt DNA in the PCR method, it is possible to perform several repetitive melting and polymerization steps without having to add fresh DNA polymerase, as was the case when enzymes from mesophilic microorganisms were used. Several other uses of heat-stable enzymes (∞ Section 12.9 and Figure 12.19) and other thermal stable products are known and are being developed for industrial applications.

5.9
Acidity and Alkalinity (pH)

Acidity or alkalinity of a solution is expressed by its **pH** on a scale on which neutrality is pH 7 (Figure 5.18). Those pH values that are less than 7 are said to be *acidic*, and those greater than 7 are *alkaline* (or *basic*). It is important to remember that pH is a *logarithmic function;* a change of 1 pH unit represents a *10-fold* change in hydrogen ion concentration. Thus, vinegar (pH near 2) and household ammonia (pH near 11) differ in hydrogen ion concentration by a billionfold.

pH and microbial growth

Each organism has a pH range within which growth is possible and usually has a well-defined pH optimum. Most natural environments have pH values between 5 and 9, and organisms with optima in this range are most common. Only a few species can grow at pH values of less than 2 or greater than 10. Organisms that live at low pH are called **acidophiles.** Fungi as a group tend to be more acid-tolerant than bacteria. Many fungi grow optimally at pH 5 or below, and a few grow well at pH values as low as 2. Several bacteria are also acidophilic. In fact, some of these bacteria are *obligate* acidophiles, unable to grow at all at neutral pH. Obligately acidophilic bacteria include several species of *Thiobacillus* (∞ Section 16.5) and several genera of Archaea, including *Sulfolobus* and *Thermoplasma* (∞ Sections 17.3 and 17.4).

Thiobacillus species, such as *T. ferroxidans,* and *Sulfolobus* exhibit an interesting property related to their acidophilic nature: they oxidize sulfide minerals and produce sulfuric acid. We discuss the role of these organisms in mining processes in Section 14.17. Probably the most critical factor for obligate acidophily is the cytoplasmic membrane. When the pH is raised to neutrality, the cytoplasmic membrane of obligately acidophilic bacteria actually dissolves and the cells lyse, suggesting that high concentrations of hydrogen ions actually are required for membrane stability.

A few organisms have high pH optima for growth, sometimes as high as pH 10–11, and are known as **alkaliphiles.** Alkaliphilic microorganisms are usually found

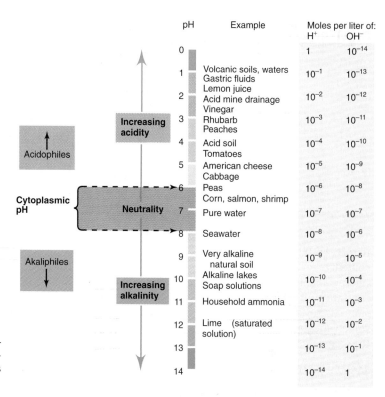

Figure 5.18 The pH scale. Note that although microorganisms can live at very low or very high pH, the cell's internal pH remains near neutrality.

in highly basic habitats such as soda lakes and high carbonate soils. Most alkaliphilic prokaryotes studied have been aerobic nonmarine bacteria, and many are *Bacillus* species. Some extremely alkaliphilic bacteria are also halophilic (salt-loving), and most of these are Archaea (∞ Section 17.1). Some alkaliphiles have found industrial uses because they produce hydrolytic enzymes, such as proteases, which function well at alkaline pH, and are used as supplements for household detergents (∞ Section 12.9).

Finally, concerning pH and microbial growth, it should be emphasized that despite the requirements of a particular organism for a specific pH for growth, the optimal growth pH represents the pH of the *extracellular* environment only; the *intracellular* pH must remain near neutrality in order to prevent destruction of acid- or alkali-labile macromolecules in the cell. In extreme acidophiles or extreme alkaliphiles the intracellular pH may vary by 1–1.5 pH units from neutrality, but for the majority of microorganisms, whose pH optimum for growth is between pH 6 and 8 (referred to as **neutrophiles**), the cytoplasm remains neutral or very nearly so (Figure 5.18).

Buffers

In a batch culture the pH can change during growth as the result of metabolic reactions that consume or produce acidic or basic substances. Thus, chemicals called *buffers* are frequently added to microbial culture media to keep the pH relatively constant. Such pH buffers generally work over only a narrow pH range; hence different buffers must be used to buffer at different pH values. For near-neutral pH ranges (pH 6–7.5), phosphate, usually supplied

as KH_2PO_4, is an excellent buffer. Many other buffers for use in microbial growth media or for the assay of enzymes extracted from microbial cells are available, and the best buffering system for one organism or enzyme may be considerably different from that of another. Thus, the optimal buffer for use in a particular situation must usually be determined empirically, although for assaying enzymes *in vitro*, a certain buffer that works well in an assay for the enzyme from one organism will usually work well for assaying this same enzyme from other organisms.

CONCEPT CHECK 5.9

The acidity or alkalinity of an environment can greatly affect microbial growth. Some organisms have evolved to grow best at low or high pH, but most organisms grow best between pH 6 and 8. The internal pH of a cell must stay near neutrality even though the external pH is highly acidic or basic.

✔ *What is the increase in concentration of protons (H⁺) when going from pH 7 to pH 4?*

✔ *What are* buffers *and why are they needed?*

5.10
Water Availability

As we saw in Chapter 2, water is the solvent of life. All organisms require water, and water availability is an important factor affecting the growth of microorganisms in

TABLE 5.3   **Water activity of several substances**

Water activity, a_w	Material	Some organisms growing at stated water activity
1.000	Pure water	*Caulobacter, Spirillum*
0.995	Human blood	*Streptococcus, Escherichia*
0.980	Seawater	*Pseudomonas, Vibrio*
0.950	Bread	Most gram-positive rods
0.900	Maple syrup, ham	Gram-positive cocci
0.850	Salami	*Saccharomyces rouxii* (yeast)
0.800	Fruit cake, jams	*Saccharomyces bailii, Penicillium* (fungus)
0.750	Salt lake, salt fish	*Halobacterium, Halococcus*
0.700	Cereals, candy, dried fruit	*Xeromyces bisporus* and other xerophilic fungi

nature. Water availability not only depends on the water content of an environment, that is, how moist or dry a solid microbial habitat may be, but is also a function of the concentration of solutes such as salts, sugars, or other substances that are dissolved in water. This is because dissolved substances have an affinity for water, which makes the water associated with solutes unavailable to organisms.

Water activity and osmosis

Water availability is generally expressed in physical terms such as **water activity**. Water activity, abbreviated a_w, is a ratio of the vapor pressure of the air in equilibrium with a substance or solution to the vapor pressure at the same temperature of pure water. Thus values of a_w vary between 0 and 1 and some representative values are given in Table 5.3. Water activities in agricultural soils generally range between 0.90 and 1.00.

Water diffuses from a region of high water concentration (low solute concentration) to a region of lower water concentration (higher solute concentration) in the process of *osmosis*. In most cases, the cytoplasm of a cell has a higher solute concentration than the environment, so water tends to diffuse into the cell and the cell is said to be in *positive water balance*. However, when a cell is in an environment of low water activity, there is a tendency for water to flow out of the cell. Thus, when a cell is immersed in a solution of low water activity, such as a salt or sugar solution, it *loses* water to the environment and plasmolyses. The salt or sugar solution can, in effect, be considered analogous to a *dry* environment.

In nature, osmotic effects are of interest mainly in habitats with high concentrations of salts. Seawater contains about 3% sodium chloride (NaCl) plus small amounts of many other minerals and elements. Microorganisms found in the sea usually have a specific requirement for the sodium ion in addition to growing optimally at the water activity of seawater (Figure 5.19). Such organisms are called **halophiles**. The growth of halophiles requires at least some NaCl, but the optimum varies with the organism; thus the terms *mild halophile* and *moderate halophile* are used to describe halophiles with low (1–6%) and moderate (6–15%) NaCl requirements, respectively (Figure 5.19).

Most microorganisms are unable to cope with environments of very low water activity and either die or become dehydrated and dormant under such conditions. **Halotolerant** organisms can tolerate some reduction in the a_w of their environment but generally grow best in the absence of the added solute (Figure 5.19). By contrast, some organisms thrive at very low water activity, and these organisms are of interest not only from the standpoint of their adaptation to life under these conditions but also from an applied standpoint such as that of the food industry, where solutes such as salt and sucrose are commonly used as food additives to inhibit microbial growth. Organisms capable of growth in very salty environments are called **extreme halophiles** (Figure 5.19); extreme halophiles generally require 15–30% NaCl, depending on the species, for optimum growth (∞ Section 17.1). Organisms able to live in environments high in sugar are called **osmophiles**, and those able to grow in very dry environments (made dry by lack of water) are called **xerophiles**. Examples of these various organisms are given in Table 5.3.

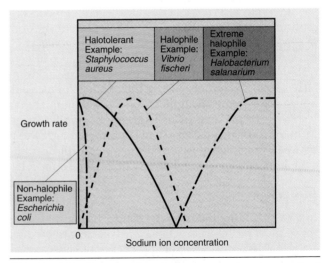

Figure 5.19 Effect of sodium ion concentration on growth of microorganisms of different salt tolerances. The optimum NaCl concentration for marine microorganisms such as *V. fischeri* is about 3%; for extreme halophiles, it is between 15 and 30%, depending on the organism.

Compatible solutes

How do organisms grow under conditions of low water activity? When an organism grows in a medium with a low water activity, it can obtain water from its environment only by increasing its *internal* solute concentration. An increase in internal solute concentration can be accomplished by either pumping inorganic ions into the cell from the environment or synthesizing or concentrating an organic solute. Organisms are known that employ either of these mechanisms, and several examples are given in Table 5.4.

The solute used inside the cell for adjustment of cytoplasmic water activity must be noninhibitory to biochemical processes within the cell; such compounds are called **compatible solutes.** Several different compatible solutes are known in microorganisms (Table 5.4 and Figure 5.20). These substances are all highly water-soluble sugars or sugar alcohols, other alcohols, amino acids or their derivatives, or in the case of extremely halophilic Archaea and a very few extremely halophilic Bacteria, potassium ions (K^+) (Table 5.4). Compatible solutes are either synthesized by the microorganisms directly or in some cases (such as glycine betaine or K^+) accumulated from the environment. The concentration of compatible solutes in the cell is a function of the level of external solutes, and in each organism the maximal amount of compatible solute(s) made or that can be accumulated is a genetically directed characteristic; this results in different organisms tolerating different ranges of water potential (Tables 5.3 and 5.4). Thus, nonhalotolerant, halotolerant, halophilic, and extremely halophilic microorganisms are to some extent defined by their genetic capacity to produce or accumulate compatible solutes.

Gram-positive cocci of the genus *Staphylococcus* are notoriously halotolerant bacteria (in fact, a common isolation procedure for them is to use media containing 7.5%

NaCl), and these organisms use the amino acid *proline* as a compatible solute. *Betaine* is a derivative of the amino acid glycine in which the protons on the amino group are replaced by three methyl groups; this leaves a permanent positive change on the N atom, which increases its solubility. Glycine betaine is widely distributed as a compatible solute, especially among halophilic Bacteria and cyanobacteria (Table 5.5). Extremely halophilic species of the phototrophic bacterial genus *Ectothiorhodospira*, most of which inhabit salt lakes and soda lakes (∞ Section 17.1), produce a novel compatible solute called *ectoine*, which is a derivative of the cyclic amino acid proline (Figure 5.20). Ectoine is also produced by nonphototrophic extreme halophiles when grown in the absence of glycine betaine. A variety of glycosides are produced by marine algae, but with rare exception they accumulate only in low amounts because the cells are not very halophilic. Xerophilic yeasts and the halophilic green alga *Dunnaliella salina*, which lives in extremely salty lakes like Great Salt Lake, produce mainly *glycerol* as a compatible solute. Other examples of compatible solutes are listed in Table 5.5, and structures are shown in Figure 5.20.

CONCEPT CHECK   **5.10**

Water can become unavailable to an organism when the dissolved solute concentration in its environment increases. To counteract this situation organisms produce or accumulate intracellular compatible solutes that function to maintain the cell in positive water balance. Some microorganisms have evolved to grow best at reduced water potential, and some even require high levels of salts in order to grow.

✔ *What is the a_w of pure water?*

✔ *What is a* compatible solute *and why is it needed?*

TABLE 5.4   **Compatible solutes of microorganisms**

Organism	Major solute(s) accumulated	Minimum a_w for growth
Bacteria, nonphototrophic	Glycine betaine, proline (mainly gram-positive), glutamate (mainly gram-negative)	0.97–0.90
Freshwater cyanobacteria	Sucrose, trehalose	0.98
Marine cyanobacteria	α-Glucosylglycerol	0.92
Marine algae	Mannitol, various glycosides, proline, dimethylsulfonium propionate	0.92
Salt lake cyanobacteria	Glycine betaine	0.90–0.75
Halophilic anoxygenic phototrophic Bacteria (*Ectothiorhodospira/Halorhodospira* and *Rhodospirillum* species)	Glycine betaine, ectoine, trehalose	0.90–0.75
Extremely halophilic Archaea (for example, *Halobacterium*) and some Bacteria (for example, *Haloanaerobium*)	KCl	0.75
Dunaliella (halophilic green alga)	Glycerol	0.75
Xerophilic yeasts	Glycerol	0.83–0.62
Xerophilic filamentous fungi	Glycerol	0.72–0.61

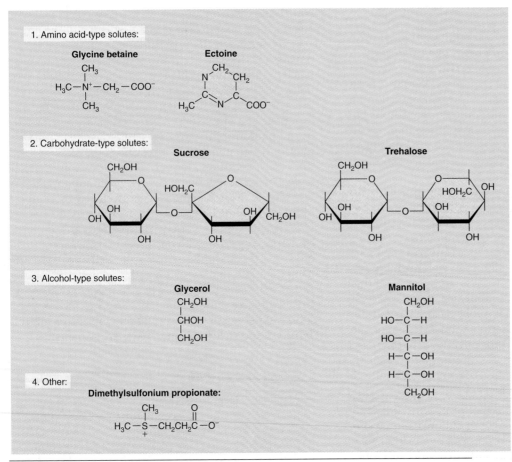

Figure 5.20 Structures of some common compatible solutes in microorganisms. The structures of glutamate and proline, other common solutes, were shown in Figure 2.12. The formal name of ectoine is 1,4,5,6-tetrahydro-2-methyl-4-pyrimidine carboxylate. Note that all the compounds shown here are very water-soluble.

5.11

Oxygen

Microorganisms vary in their need for, or tolerance of, oxygen. In fact, microorganisms can be divided into several groups depending on the effect of oxygen, as outlined in Table 5.5. **Aerobes** are species capable of growth at full oxygen tensions (air is 21% O_2), and many can even tolerate elevated concentrations of oxygen (hyperbaric oxygen). **Microaerophiles**, by contrast, are aerobes that can use O_2 only when it is present at levels reduced from that in air, usually because of their limited capacity to respire or because they contain some oxygen-sensitive molecule such as an oxygen-labile enzyme. Organisms that lack a respiratory system cannot use oxygen as a terminal electron acceptor. Such organisms are called **anaerobes,** but there are two kinds of anaerobes: **aerotolerant anaerobes,** which can tolerate oxygen and grow in its presence even though they cannot use it, and **obligate** (or **strict**) **anaerobes,** which are killed by oxygen (Table

5.5). The reason obligate anaerobes are killed by oxygen is probably because they are unable to detoxify some of the products of oxygen metabolism (see later discussion). When oxygen is reduced, several toxic products such as hydrogen peroxide (H_2O_2), superoxide (O_2^-), and hydroxyl radical (OH·) are formed. Many obligate anaerobes are rich in flavin enzymes (∞ Section 4.12), which can react spontaneously with O_2 to yield these toxic products. By contrast, aerobes have enzymes that decompose toxic oxygen products, whereas anaerobes seem to lack all or some of these enzymes.

Microbial culture and the effects of oxygen

For the growth of many aerobes, it is necessary to provide extensive aeration. This is because O_2 is only poorly soluble in water and the O_2 used up by the organisms during growth is not replaced fast enough by diffusion from the air. Forced aeration of cultures is therefore frequently desirable and can be achieved either by vigorously shaking the flask or tube on a shaker or by bubbling sterilized

TABLE 5.5	Oxygen relationships of microorganisms			
Group	**Relationship to O_2**	**Type of Metabolism**	**Example**	**Habitat[a]**
Aerobes				
Obligate	Required	Aerobic respiration	*Micrococcus luteus*	Skin, dust
Facultative	Not required, but growth better with O_2	Aerobic, anaerobic respiration, fermentation	*Escherichia coli*	Mammalian large intestine
Microaerophilic	Required but at levels lower than atmospheric	Aerobic respiration	*Spirillum volutans*	Lake water
Anaerobes				
Aerotolerant	Not required, and growth no better when O_2 present	Fermentation	*Streptococcus pyogenes*	Upper respiratory tract
Obligate	Harmful or lethal	Fermentation or anaerobic respiration	*Methanobacterium formicicum*	Sewage sludge digestors, anoxic lake sediments

[a]Listed are typical habitats of the example organism.

air into the medium through a fine glass tube or porous glass disc. Usually aerobes grow much better with forced aeration than when O_2 is provided by simple diffusion.

For anaerobic culture, the problem is to *exclude*, not provide, oxygen. And because oxygen is ubiquitous in the air, special methods are needed to culture anaerobic microorganisms. Obligate anaerobes vary in their sensitivity to oxygen, and a number of procedures are available for reducing the O_2 content of cultures—some simple and suitable mainly for less sensitive organisms, others more complex but necessary for growth of strict anaerobes.

Bottles or tubes filled completely to the top with culture medium and provided with tightly fitting stoppers provide anoxic conditions for organisms not too sensitive to small amounts of oxygen. It is also possible to add a chemical called a *reducing agent* that reacts with oxygen and reduces it to H_2O. A good example is *thioglycolate*, which is added to a medium, called *thioglycolate broth*, commonly used to test an organism's requirements for O_2 (Figure 5.21). After thioglycolate reacts with oxygen throughout the tube, oxygen can penetrate only near the top of the tube where the medium contacts air. Obligate aerobes grow only at the top of such tubes. Facultative organisms grow throughout the tube but best near the top. Microaerophiles grow near the top but not right at the top (Figure 5.21). Anaerobes grow only near the bottom of the tube, where oxygen cannot penetrate (Figure 5.21). A redox indicator dye called *resazurin* is added to the medium because the dye changes color in the presence of oxygen and thereby indicates the degree of penetration of oxygen into the medium (Figure 5.21).

To remove all traces of O_2 for the culture of anaerobes, it is possible to place an O_2-consuming gas in a jar holding the tubes or plates. One of the simplest devices for this is an *anaerobic jar*, a heavy-walled jar with a gastight seal within which tubes, plates, or other containers to be incubated are placed. The air in the jar is

replaced with a mixture of H_2 and CO_2, and in the presence of a chemical catalyst the traces of O_2 left in the vessel or culture medium are consumed, thus leading to

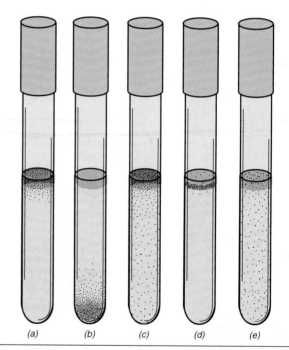

(a)   (b)   (c)   (d)   (e)

Figure 5.21 Aerobic, anaerobic, facultative, microaerophilic, and aerotolerant anaerobe growth, as revealed by the position of microbial colonies within tubes of a culture medium. A small amount of agar has been added to keep the liquid from becoming disturbed and the redox dye, resazurin, which is pink when oxidized and colorless when reduced, is added as a redox indicator. (a) Oxygen penetrates only a short distance into the tube, so obligate aerobes grow only at the surface. (b) Anaerobes, being sensitive to oxygen, grow only away from the surface. (c) Facultative aerobes are able to grow in either the presence or the absence of oxygen and thus grow throughout the tube. (d) Microaerophiles grow away from the most oxic zone. (e) Aerotolerant anaerobes grow throughout the tube.

anoxic conditions (see Section 21.1 and Figure 21.6 for a discussion of the use of the anaerobic jar).

For strict anaerobes, such as methanogenic bacteria, it is necessary not only to carefully remove all traces of O_2 but also to carry out all manipulations of cultures in an anoxic atmosphere, as these organisms can be killed by even a brief exposure to O_2. In these cases, a culture medium is first boiled to render it oxygen-free, and then a reducing agent such as H_2S is added and the mixture is sealed under an oxygen-free gas. All manipulations are carried out under a tiny jet of oxygen-free hydrogen or nitrogen gas that is directed into the culture vessel when it is open, thus driving out any O_2 that might enter. For extensive research on anaerobes, special boxes fitted with gloves, called *anaerobic glove boxes*, permit work with open cultures in completely anoxic atmospheres (Figure 5.22).

Toxic forms of oxygen

Oxygen is a powerful oxidant and an excellent electron acceptor for respiration (∞ Section 4.11). Oxygen in its normal ground state is referred to as **triplet oxygen.** However, one major form of toxic oxygen is called **singlet oxygen,** a higher energy form of oxygen in which outer shell electrons surrounding the nucleus become highly reactive and are able to carry out a variety of spontaneous and undesirable oxidations within the cell. Singlet oxygen is produced both photochemically and biochemically, the latter through the action of various peroxidase enzymes. Organisms that frequently encounter singlet oxygen, such as airborne bacteria and phototrophic microorganisms, contain pigments called **carotenoids,** which function to convert singlet oxygen to nontoxic forms.

Other highly toxic forms of oxygen include **superoxide anion** (O_2^-), **hydrogen peroxide** (H_2O_2), and **hydroxyl radical (OH·),** all of which are produced as inadvertent by-products during the reduction of O_2 to H_2O in respiration (Figure 5.23). Flavoproteins, quinones, thiols, and iron–sulfur proteins (∞ Section 4.12) can also carry out the reduction of O_2 to O_2^-. Superoxide is highly reactive and can oxidize virtually any organic compound in the cell, including macromolecules. Peroxides such as H_2O_2 can damage cell components but are generally not as toxic to the cell as superoxide or hydroxyl radical. The latter is the most reactive of all toxic oxygen species and can instantly oxidize any organic substance in the cell. However, the hydroxyl radical is only a transient species in most cells because the major source of OH· is ionizing radiation, to which most cells are not commonly exposed. Small amounts of OH· can also be produced from H_2O_2 (Figure 5.23), but when peroxides do not accumulate in the cell (because of the action of catalase, which is discussed later), this source of hydroxyl radical is virtually eliminated. We will see later that various toxic oxygen species can be produced by certain immune cells in the animal body and used to kill microbial invaders (∞ Section 20.3).

Enzymes that destroy toxic oxygen

With such an array of toxic oxygen derivatives, it is perhaps not surprising that organisms have evolved enzymes that destroy certain oxygen products (Figure 5.24). The most common enzyme in this category is **catalase,** which attacks hydrogen peroxide; the activity of catalase is illustrated in Figures 5.24*a* and 5.25. Another enzyme that acts on hydrogen peroxide is **peroxidase** (Figure 5.24*b*), which differs from catalase in requiring a reductant, usually NADH, producing H_2O_2 as a product. Superoxide is destroyed by the enzyme **superoxide dismutase** (SOD) (Figure 5.24*c*), which com-

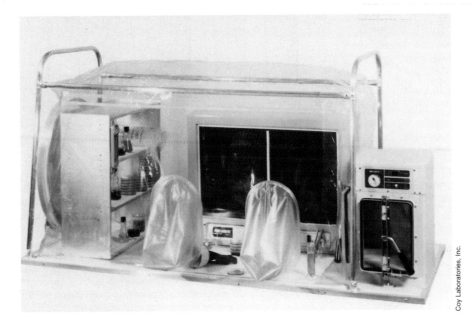

Figure 5.22 Anaerobic glove bag for manipulating and incubating cultures under anoxic conditions. The airlock on the right, which can be evacuated and filled with oxygen-free gas, serves as a port for adding and removing materials to and from the glove bag.

Coy Laboratories, Inc.

$$O_2 + e^- \rightarrow O_2^- \quad \text{Superoxide}$$
$$O_2^- + e^- + 2\,H^+ \rightarrow H_2O_2 \quad \text{Hydrogen peroxide}$$
$$H_2O_2 + e^- + H^+ \rightarrow H_2O + OH\cdot \quad \text{Hydroxyl radical}$$
$$OH\cdot + e^- + H^+ \rightarrow H_2O \quad \text{Water}$$

$$\text{Overall:}\quad O_2 + 4\,e^- + 4\,H^+ \rightarrow 2\,H_2O$$

Figure 5.23 Four-electron reduction of O_2 to water by step-wise addition of electrons. All the intermediates formed are reactive and toxic to cells.

bines two molecules of superoxide to form one molecule of hydrogen peroxide and one molecule of oxygen. Superoxide dismutase and catalase working together can thus bring about the conversion of superoxide back to oxygen (Figure 5.24).

Aerobes and facultative aerobes generally contain both superoxide dismutase and catalase, although a few obligate aerobes lack catalase. Superoxide dismutase is indispensable to aerobic cells, and the low levels (or complete absence) of this enzyme in obligate anaerobes may be the major reason why oxygen is toxic to them. Some aerotolerant anaerobes, such as lactic acid bacteria, also lack superoxide dismutase, but they are somehow able to use protein-free Mn^{2+} complexes to carry out the dismutation of O_2^- to H_2O_2 and O_2. Such a reaction may have functioned as a primitive form of superoxide dismutase in ancient organisms. This is supported by the fact that all known SODs contain a metal cofactor, usually Mn^{2+}, but also Fe^{2+} or Cu^{2+} plus Zn^{2+}, at the enzyme's active site. SODs have been isolated from several species of Bacteria, Archaea, and Eukarya.

Anaerobic microorganisms

Considering our discussion of toxic forms of oxygen and the fact that anaerobes frequently lack the means to defend against them, the student may get the impres-

(a) Catalase:
$$H_2O_2 + H_2O_2 \rightarrow 2\,H_2O + O_2$$

(b) Peroxidase:
$$H_2O_2 + NADH + H^+ \rightarrow 2\,H_2O + NAD^+$$

(c) Superoxide dismutase:
$$O_2^- + O_2^- + 2\,H^+ \rightarrow H_2O_2 + O_2$$

(d) Superoxide dismutase/catalase in combination:
$$4\,O_2^- + 4\,H^+ \rightarrow 2\,H_2O + 3\,O_2$$

Figure 5.24 Enzymes acting on toxic oxygen species. (a) Catalases and (b) peroxidases are generally porphyrin-containing proteins, although some flavoproteins may act in this manner. (c) Superoxide dismutases are metal-containing proteins, the metals being copper and zinc, manganese, or iron. (d) Combined reaction of superoxide dismutase and catalase.

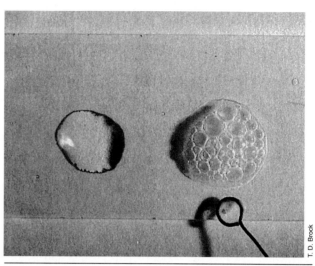

Figure 5.25 Method for testing a microbial culture for the presence of catalase. A heavy loopful of cells from an agar culture was mixed on a slide with a drop of 30% hydrogen peroxide. The immediate appearance of bubbles is indicative of the presence of catalase. The bubbles are O_2 produced by the reaction $H_2O_2 + H_2O_2 \rightarrow 2\,H_2O + O_2$.

sion that anaerobic organisms are quite rare. However, nothing could be further from the truth. Anoxic environments abound on Earth and include muds and other sediments of lakes, rivers, and oceans; bogs and marshes; waterlogged soils; canned foods; intestinal tracts of animals; the oral cavity of animals, especially around the teeth; certain sewage treatment systems; deep underground areas such as oil pockets; and some underground waters (see Chapter 14 for descriptions of some of these habitats). In most of these habitats, anoxic conditions and the accompanying low reduction potentials are due to the activities of organisms, mainly bacteria, that consume oxygen during respiration and produce highly reducing substances such as H_2 and H_2S. If no replacement oxygen is available, the habitat becomes anoxic.

So far as is known, obligate anaerobiosis occurs in three groups of microorganisms: a wide variety of prokaryotes, a few fungi, and a few protozoa. One of the best-known groups of obligately anaerobic Bacteria belongs to the genus *Clostridium*, a group of gram-positive spore-forming rods. Clostridia are widespread in soil, lake sediments, and intestinal tracts and are often responsible for spoilage of canned foods (∞ Sections 11.6 and 16.26). Other obligately anaerobic bacteria are found among the methanogens and many other archaeans, the sulfate-reducing and homoacetogenic bacteria (∞ Chapters 16 and 17) and many of the bacteria that inhabit the animal gut (∞ Section 19.4). Among obligate anaerobes, however, the sensitivity to oxygen varies greatly; some organisms are able to tolerate traces of oxygen, whereas others are not.

Growth and molecular biology

We have now completed our study of microbial growth and the major environmental factors that control it, and we are ready to examine the molecular biological processes that occur during microbial growth: DNA replication, transcription, and translation.

These fundamental molecular processes must be coordinated in such a way that at the time of cell division each daughter cell receives an equal share of newly synthesized cell materials and contains at least one copy of every essential protein and nucleic acid needed to exist as an independent entity. These molecular processes occur in similar ways in all organisms. Thus, a psychrophilic organism inhabiting Antarctic sea ice (Figure 5.14) and a hyperthermophile growing in boiling water (Figure 5.17) employ similar mechanisms in processing genetic information, but the actual genes in the two organisms are very different; these genetic differences are at the heart of microbial diversity, to be considered in Chapters 16–18. A detailed understanding of the coding functions of a cell and how genetics influences cellular evolution is necessary to bring the diversity of microbial life we have seen in this chapter into clear focus.

CONCEPT CHECK   5.11

Aerobes require oxygen to live, whereas anaerobes do not and may even be killed by O_2. Several toxic forms of oxygen can be formed in the cell, but enzymes are present that can neutralize most of them. Special methods may be necessary to grow strictly aerobic or anaerobic bacteria.

- ✔ *What is the chemical structure of superoxide anion?*
- ✔ *How does a reducing agent work?*
- ✔ *How does superoxide dismutase protect a cell?*

Material for Review

REVIEW QUESTIONS

1. What is the difference between the growth rate of an organism and its generation time? Express both functions in terms of g, t, and n.

2. Why is it useful to plot growth data from a growing microbial culture?

3. Describe the growth cycle of a population of bacterial cells from the time this population is first inoculated into fresh medium. How can the growth pattern differ when it is measured by total count or by viable count?

4. Describe one direct and one indirect method by which microbial growth can be measured. Make sure that the methods you choose agree with your definition.

5. Describe briefly the process by which a single cell develops into a visible colony on an agar plate. With this explanation as a background, describe the principle behind the viable count method.

6. In a chemostat, if the dilution rate is increased, will the growth rate increase, decrease, or stay the same? If the concentration of the limiting substrate in the medium reservoir is increased, will the growth rate increase, decrease, or stay the same? Will the population density increase, decrease, or stay the same?

7. Examine the graph describing the relationship between growth rate and temperature (Figure 5.12). Give an explanation, in biochemical terms, of why the optimum temperature for an organism is closer to its maximum than its minimum.

8. Would you expect to find a psychrophilic microorganism alive in a hot spring? Why? It is frequently possible to isolate thermophilic microorganisms from cold-water environments. Give an explanation of how this can be.

9. Concerning the pH of the environment and of the cell, in what ways are acidophiles and alkaliphiles different? In what ways are they similar?

10. Write an explanation in molecular terms for how a halophile is able to make water molecules flow *into* a cell.

11. List three *chemical classes* of compatible solutes produced by various microorganisms. List at least two things they all have in common.

12. Contrast an aerotolerant and an obligate anaerobe in terms of sensitivity and ability to grow in the presence of oxygen (O_2). How does an aerotolerant anaerobe differ from a microaerophile?

13. Compare and contrast the enzymes *catalase* and *superoxide dismutase* from the following points of view: substrates, oxygen products, organisms containing them, role in oxygen tolerance of the cell.

APPLICATION QUESTIONS

1. Starting with four bacterial cells per milliliter in a rich nutrient medium, with a 1-hr lag phase and a 20-min generation time, how many cells will there be in 1 l of this culture after 1 hr? After 2 hr? After 2 hr if one of the initial four cells was dead?

2. Calculate the generation time in a growth experiment in which a medium was inoculated with 5×10^6 cells/ml of *Escherichia coli* cells and, following a 1-hr lag, grew exponentially for 5 hr, after which the population was 5.4×10^9 cells/ml.

3. Return to Chapter 2 and locate a figure that best describes what happens to a cell of a mesophile like *Escherichia coli* when placed in a culture medium at 75°C. Contrast this with a figure from Chapter 5 that best describes what would happen if cells of *Pyrodictium brockii* were placed

under the same conditions. Describe why neither organism would grow.

4. From what you know concerning growth at reduced water potential and the phenomenon of compatible solutes, describe what would happen if you took a cell of the extreme halophile *Halobacterium salinarum* from its growth medium (containing 25% NaCl) and placed it in distilled water. Also, predict what a cell of *Escherichia coli*, a typical nonhalophile, would do if you did the reverse experiment (distilled water to 25% NaCl).

5. From what you know about the structure of the cytoplasmic membrane, describe why toxic forms of oxygen such as superoxide anion (O_2^-) could do severe damage to a cell by oxidizing phospholipids (the latter are a major

SUPPLEMENTARY READINGS

Brock, T. D. (ed.) 1986. *Thermophiles: General, Molecular, and Applied Microbiology.* John Wiley, New York. A good treatment of thermophilic bacteria, with emphasis on hyperthermophiles.

Koch, A. L. 1995. *Bacterial Growth and Form.* Chapman and Hall, New York. An in-depth but fun to read consideration of several topics relating to bacterial growth, in particular osmotic effects, cell wall structure, and cell wall synthesis.

Kristjansson, J. K. (ed.) 1992. *Thermophilic Bacteria.* CRC Press, Boca Raton, FL. An overview of different types of thermophilic and hyperthermophilic prokaryotes.

Lederberg, J. (ed.) 1992. *Encyclopedia of Microbiology.* Academic Press, San Diego. Many sections of this treatise on microbiology deal with topics related to bacterial growth.

Monod, J. 1949. The growth of bacterial cultures. *Annu. Rev. Microbiol.* 3:371–394. The classic review of the bacterial growth curve.

Neidhardt, F. C., R. Curtiss III, J. L. Ingraham, E. C. C. Lin, K. B. Low, B. Magasanik, W. Reznikoff, M. Riley, M. Schaechter, and **H. E. Umbarger** (eds.) 1996. Escherichia coli *and* Salmonella: *Cellular and Molecular Biology,* 2nd edition. American Society for Microbiology, Washington DC. An advanced-level reference containing the most complete source of growth information on these bacteria.

Neidhardt, F. C., J. L. Ingraham, and **M. Schaechter.** 1990. *Physiology of the Bacterial Cell.* Sinauer Associates, Sunderland, MA. A textbook of bacterial physiology that deals with all aspects of bacterial growth.

White, D. 1995. *The Physiology and Biochemistry of Prokaryotes,* Oxford University Press, New York. An advanced undergraduate- or graduate-level text covering microbial growth and the physiology that supports it. Includes many references to the primary literature. Highly recommended.

On~line resources for this chapter are on the World Wide Web at: http://www.prenhall.com/~brock (click on the table of contents link and then select Chapter 5).

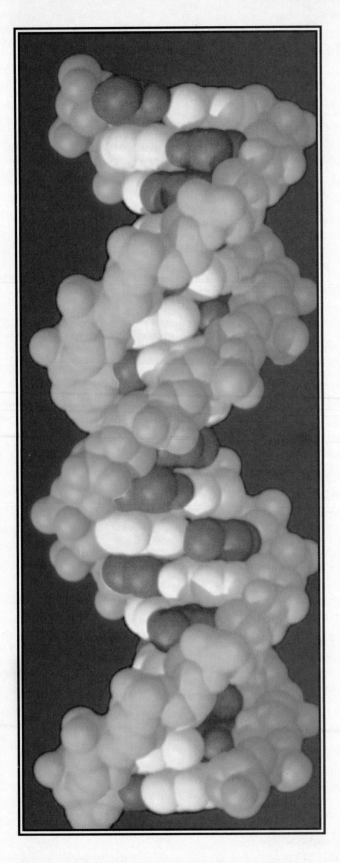

Macromolecules and Molecular Genetics

Using microorganisms, scientists were able to elucidate the fundamental processes of biological information flow that occur in all living organisms. Shown here is a computer model of DNA depicting the overall alignment of the double helix.

MINIGLOSSARY for Chapter 6

Anticodon a sequence of three bases in a tRNA molecule that base-pairs with a codon during protein synthesis

Aminoacyl-tRNA Synthetases a group of enzymes each one of which catalyzes the attachment of the correct amino acid to a tRNA

Antiparallel in reference to double-stranded DNA, one strand runs $5' \rightarrow 3'$ and other $3' \rightarrow 5'$

Chromosome a genetic element, usually circular in prokaryotes and linear in eukaryotes, carrying genes essential to cellular function

Codon a sequence of three bases in mRNA that encodes an amino acid

DNA Polymerase an enzyme that synthesizes a new strand of DNA in the $5' \rightarrow 3'$ direction using an antiparallel DNA strand as a template

Exon the coding DNA sequences in a split gene (contrast with introns)

Gene a segment of DNA specifying a protein (via mRNA), a tRNA, or an rRNA

Genome the total complement of genes contained in a cell or virus

Hybridization formation of a duplex nucleic acid molecule with strands derived from different sources by complementary base pairing

Intron the intervening noncoding DNA sequences in a split gene (contrast with exon)

Messenger RNA (mRNA) an RNA molecule that contains the genetic information necessary to encode a particular protein

Molecular Chaperones a group of proteins that help other proteins fold or refold from a partially denatured state

Operon a cluster of genes whose expression is controlled by a single operator

Primary Transcript an unprocessed RNA molecule that is the direct product of transcription

Primer a molecule (usually a polynucleotide) to which DNA polymerase can attach the first nucleotide during DNA replication

Promoter a site on DNA to which RNA polymerase can bind and begin transcription

Replication synthesis of DNA using DNA as a template

Restriction Enzyme an enzyme that recognizes and makes double-stranded breaks at specific DNA sequences

Ribosomal RNA (rRNA) types of RNA found in the ribosome; some participate actively in the process of protein synthesis

Ribosome a cytoplasmic particle composed of ribosomal RNA and protein, which is a central part of the protein-synthesizing machinery of the cell

Ribozyme an RNA molecule that can catalyze chemical reactions

RNA Polymerase an enzyme that synthesizes RNA in the $5' \rightarrow 3'$ direction using an antiparallel DNA strand as a template

RNA Processing the conversion of a precursor RNA to its mature form

Semiconservative Replication DNA synthesis yielding new double helices, each consisting of one parental and one progeny strand

Transcription the synthesis of RNA using a DNA template

Transfer RNA (tRNA) an adaptor molecule used in translation that has specificity for both a particular amino acid and for one or more codons

Translation the synthesis of protein using the genetic information in messenger RNA as a template

We now begin a study of the flow of information in microorganisms that will extend over the next five chapters. As we noted in Chapter 1, two hallmarks of life are *energy transformation* and *information flow*. In Chapter 4 we dealt with the problem of energy transformation: *metabolism*. Now we deal with the problem of information flow: *genetics*. **Genetics** is the discipline that deals with the mechanisms by which traits are passed from one organism to another and how they are expressed.

In the present chapter we discuss how genetic information is organized and expressed in cells, and in the next chapter we deal with how this expression is regulated.

Subsequent chapters concern the genetic information of some noncellular forms, the transfer of genetic information, and how genetic information can be manipulated.

The study of genetics at the molecular level is central to an understanding of the variability of organisms and the evolution of species. Since biological information flow is the basis of cellular function, genetics is also a major research tool in attempts to understand the molecular mechanisms by which cells function. Genetics and biochemistry work together in the continuing quest to discern the ultimate basis of life.

Genetics also provides us with approaches to the construction of new organisms of potential use in human affairs. As such, it has provided us with some of the most important advances in agriculture, medicine, and industry. An understanding of genetic mechanisms makes it possible for researchers to manipulate species and construct new organisms. It also makes it possible for scientists and physicians to develop means for controlling the important infectious diseases of humankind. We will have much to say about the application of genetics to human affairs in subsequent chapters.

6.1

Macromolecules and Genetic Information

All the processes that take place in the cell involve molecules. Many of the molecules involved in the steps of genetic information flow are very large; they are called **macromolecules.** However, long before we knew what these molecules were, or even what steps in the flow might be, it was clear that there was a functional unit of genetic information. This unit has come to be called the **gene.**

What is a gene and what is its function?

A somewhat oversimplified definition of a gene is an entity that specifies the structure of a single polypeptide or *protein* chain. We discussed the chemistry of proteins in Chapter 2 and noted that they consist of one or more polypeptide chains. A polypeptide is composed of a series of amino acids connected in peptide linkage. There are usually 20 different amino acids present in proteins, and a single protein molecule typically has several hundred amino acid residues (∞ Section 2.8). The gene is the element of information that specifies the *sequence* of amino acids of the protein. Genes are stored information, whereas proteins are the cell's functional entities.

In all cells the genes themselves are composed of *deoxyribonucleic acid* (DNA). The information in the gene is present as the sequences of bases in the DNA. Like protein, DNA is a macromolecule. Interestingly, the information stored in the DNA specifies the sequence of a protein only through the intermediary of another macromolecule, *ribonucleic acid* (RNA). RNA can serve either as a true informational intermediate (a messenger) or in some cases as a more active part of the cell's machinery. Because all three of these molecules, DNA, RNA, and protein, contain biological information in their sequences, they are often called **informational macromolecules** (∞ Section 2.4).

In DNA and RNA the information is encoded in the *base sequence* of the purine and pyrimidine bases of the polynucleotide chain. When we discuss the information content of a nucleic acid, we thus speak of the *coding* properties of this material. The amino acid sequence of the polypeptide is *coded* by the sequence of purine and pyrimidine bases within the nucleic acid,

with *three* bases encoding a single amino acid. We will discuss this coding function in detail in this chapter.

The steps in information flow

When a cell divides and forms two cells, all types of informational macromolecules are synthesized. The molecular processes underlying genetic information flow can be divided into three stages, which are described briefly here (Figure 6.1).

1. *Replication.* The DNA molecule, which serves as the cell's genetic material, is a **double helix** of two long chains (∞ Section 2.7). During replication, DNA, containing the master genetic blueprint, duplicates. The products of DNA replication are two double helices, the two strands thus becoming four strands.

2. *Transcription.* DNA does not function directly in protein synthesis but through an RNA intermediate. The transfer of the information to RNA is called **transcription,** and the RNA molecule carrying the information to encode a protein is called **messenger RNA** (mRNA). In most cases, at any particular location on the chromosome, only one strand of the DNA is transcribed, and the information of this strand is then contained in the mRNA. Some regions of DNA that are transcribed do not encode proteins but rather contain information for other types of RNA, such as **transfer RNA** (tRNA) and **ribosomal RNA** (rRNA). Therefore, we must expand our definition of a gene to include a region of DNA that encodes one of these types of RNA. As we shall see, these other types of RNA molecules also have important functions in the cell.

3. *Translation.* The specific sequence of amino acids in each protein is directed by a specific sequence of bases in the mRNA (which was transcribed from the DNA). This information in the nucleic acids is present as a **genetic code.** It takes *three* bases on the mRNA to encode a single amino acid, and each triplet of bases is called a **codon.** There is a direct linear correspondence between the base sequence of a gene and the amino acid sequence of a polypeptide (Figure 6.1). The genetic code is actually translated into protein by means of the protein-synthesizing system. This system consists of **ribosomes** (which are themselves made up of proteins and rRNA), transfer RNA, and a number of enzymes. The ribosomes are the structures to which messenger RNA attaches. Transfer RNA is the key link between codon and amino acid. There are one or more separate tRNA molecules corresponding to each amino acid, and the tRNA has a triplet of three bases, the **anticodon,** which is *complementary* to the codon of the messenger RNA. An enzyme brings about the attachment of the correct amino acid to the correct tRNA.

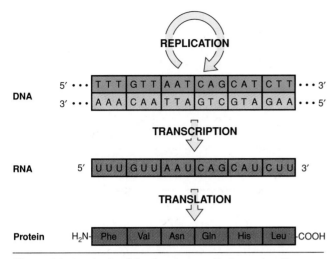

Figure 6.1 Synthesis of the three types of informational macromolecules. Note that in any particular region, only one of the two strands of the DNA double helix is transcribed.

In the processes of replication, transcription, and translation, the information in the sequence of the nucleic acid may specify either the sequence of another nucleic acid or of a protein. However, the transfer of sequence information from nucleic acid to protein is unidirectional; the sequence of a protein **does not** specify the sequence of a nucleic acid. This one-way transfer of genetic information from nucleic acid to protein is sometimes referred to as the *central dogma of molecular biology* because it holds for all life forms on our planet.

The three transfer steps shown in Figure 6.1 are those used in all cells. In Chapter 7 we will learn that information transfer in the reproduction of some viruses can involve two other types of transfer. One is *RNA replication*, where RNA is used as a template for RNA synthesis. The other is *reverse transcription*, where the use of sequence information in RNA specifies a sequence in DNA. Note that in both cases the information transfer is from nucleic acid to nucleic acid, and therefore neither violates the central dogma.

Prokaryotic and eukaryotic genetics

The three information transfer steps, replication, transcription, and translation, are used in all cells. However, as we shall see, there are some differences in the mechanisms of these processes in prokaryotes and in eukaryotes. In part this is due to differences in the organization of the genetic information and to the fact that eukaryotes have a nucleus.

We emphasized in Chapter 3 the basic differences in the organization of DNA in prokaryotes and eukaryotes. In summary, the typical bacterial genome consists of a single, covalently closed *circular* molecule of DNA distributed in the cytoplasm of the cell, and the eukaryotic genome consists of several *linear* pieces of DNA present in individual chromosomes in the cell nucleus.

In prokaryotes there is no membrane separating the chromosome from the cytoplasm (∞ Section 3.14), and therefore the production of messenger RNA from the DNA template can be closely linked to translation of the messenger RNA by the ribosomes. In eukaryotes the chromosomes are located inside the nucleus and the ribosomes in the cytoplasm, so transcription and translation are spatially separated. Transcription occurs inside the nucleus, and the RNA molecules must then be transported into the cytoplasm. In addition protein-encoding genes of eukaryotes are frequently split into two or more coding regions, with noncoding regions separating coding regions. The coding sequences are called **exons,** and the intervening noncoding regions, **introns.** Both intron and exon regions are transcribed into the **primary transcript,** or **pre-mRNA,** and the functional mRNA is subsequently formed by enzymatic removal of noncoding regions. A summary contrasting genetic phenomena in prokaryotes and eukaryotes is given in Figure 6.2.

CONCEPT CHECK   **6.1**

The three key processes of macromolecular synthesis are DNA replication; transcription, the synthesis of RNA from a DNA template; and translation, the synthesis of proteins using messenger RNA as template. Although the basic processes are the same in both prokaryotes and eukaryotes, the organization of the genetic information is more complex in eukaryotes because many genes have distinct coding regions (exons) and noncoding regions (introns).

✔ *What three informational macromolecules are involved in genetic information flow?*

✔ *In all cells there are three processes involved in genetic information flow. What are they?*

6.2
DNA Structure

We dealt with the general structure of nucleic acids in Chapter 2. In the next few sections of this chapter we shall discuss various subjects related to DNA, the nucleic acid that is the genetic material of all cells: (1) details of DNA structure necessary for an understanding of molecular genetics, (2) the types of genetic elements containing DNA that are found in cells, (3) methods for studying DNA experimentally, and (4) the mechanism of DNA replication. With this information as a basis, we will then be able to turn to a discussion of how the information of DNA is transcribed into RNA.

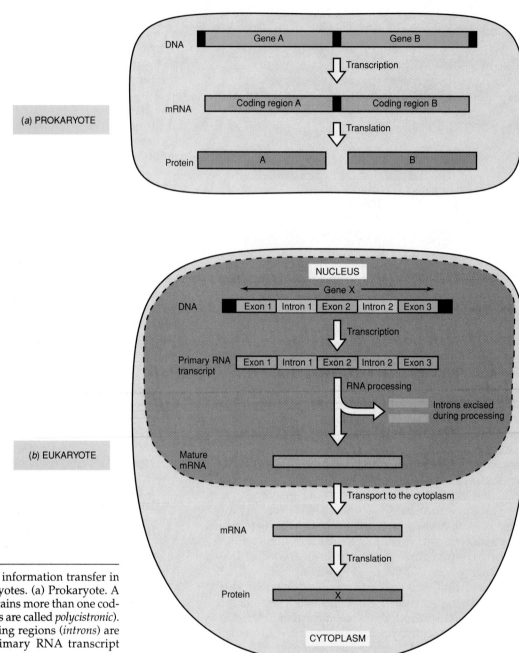

Figure 6.2 Contrast of information transfer in prokaryotes and eukaryotes. (a) Prokaryote. A single mRNA often contains more than one coding region (such mRNAs are called *polycistronic*). (b) Eukaryote. Noncoding regions (*introns*) are removed from the primary RNA transcript before translation.

As we have noted, only four different nucleic acid bases are found in DNA: adenine (A), guanine (G), cytosine (C), and thymine (T). The genetic information for all cellular processes is stored in DNA in the *sequence* of bases along the polynucleotide chain. As already shown in Figure 2.11, the backbone of the DNA chain consists of alternating units of phosphate and the sugar *deoxyribose;* connected to each sugar is one of the nucleic acid *bases*. Note especially the numbering system for the positions of sugar and base; the phosphate connecting two sugars spans from the 3′-carbon of one sugar to the 5′-carbon of the adjacent sugar. This numbering system is frequently used in discussing DNA replication

and should be kept in mind. The phosphate linkage in DNA is a phospho*diester* because a single phosphate is connected by ester linkage to two separate sugars. At one end of the DNA molecule the sugar has a phosphate on the 5′-hydroxyl, whereas at the other end the sugar has a free hydroxyl at the 3′-position.

The biochemistry of DNA replication is shown in Figure 6.3. As seen, the precursor of the new unit added is a deoxyribonucleoside *tri*phosphate. Replication of DNA proceeds by insertion of a new nucleoside triphosphate at the free 3′- (hydroxyl) end, with the subsequent loss of two phosphates (generating a deoxyribonucleoside *mono*phosphate); thus, DNA synthesis *always* pro-

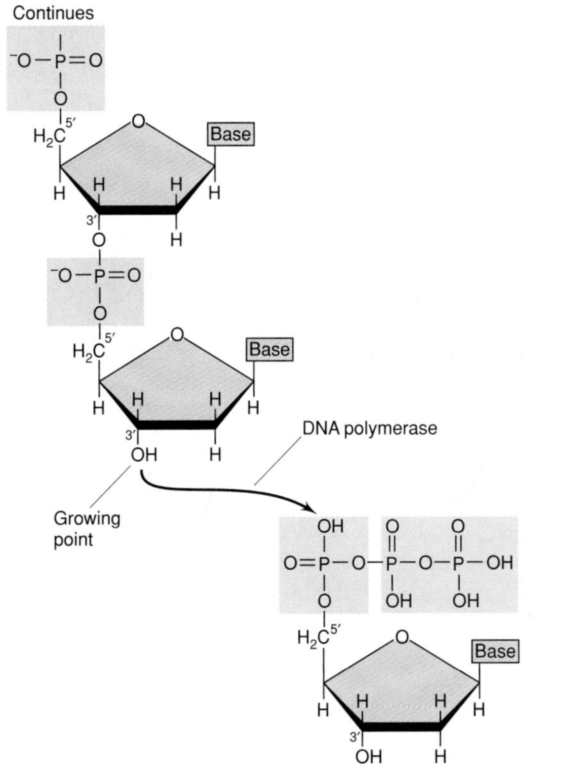

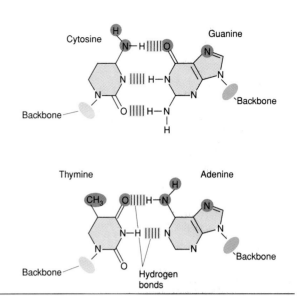

Figure 6.4 Specific pairing between adenine (A) and thymine (T) and between guanine (G) and cytosine (C) via hydrogen bonds. These two base pairs are the base pairs typically found in double-stranded DNA. Atoms that are found in the major groove of the double helix and that interact with proteins are highlighted in red. The deoxyribose phosphate backbones of the two strands of DNA are also indicated.

Figure 6.3 Structure of the DNA chain and mechanism of growth by addition from a deoxyribonucleoside triphosphate at the 3'-end of the chain. Growth always proceeds from the 5'-phosphate to the 3'-hydroxyl end. The enzyme DNA polymerase catalyzes the addition reaction. The four deoxyribonucleotides that serve as precursors are deoxythymidine triphosphate (dTTP), deoxyadenosine triphosphate (dATP), deoxyguanosine triphosphate (dGTP), and deoxycytidine triphosphate (dCTP). The two terminal phosphates of the triphosphate are split off as pyrophosphate (PP_i). Thus, two high energy phosphate bonds are consumed on the addition of a single nucleotide.

ceeds toward the 3'-end of the molecule ($5' \rightarrow 3'$). As we will see, this requirement that DNA synthesis always proceeds $5' \rightarrow 3'$ has important consequences in the replication of double-stranded DNA for both cells and viruses.

DNA as a double helix

As we shall discuss in Chapter 8, the chromosomes of some viruses are single-stranded. However, in all **cellular organism chromosomes**, DNA does not exist as a single-stranded polynucleotide but as two polynucleotide strands that are not identical in base sequence but instead are **complementary**. The complementarity of DNA arises because of the specific pairing of the purine and pyrimidine bases: adenine always pairs with thymine, and guanine always pairs with cytosine (Figure 6.4). The two strands in the resulting **double-stranded** molecule are arranged in an *antiparallel* fashion (see

Figure 6.5). This means the two strands are in a "head-to-toe" arrangement. In Figure 6.5 the strand on the left is arranged 5' to 3' top to bottom, whereas the other strand is 5' to 3' bottom to top. However, the two strands are not arranged side by side in the DNA molecule but instead are wrapped around each other in a helix, the **double helix** (Figure 6.6). In this double helix, DNA has two distinct grooves, the *major groove* and the *minor groove*. There are many important proteins that interact specifically with DNA (as we shall see in Chapter 7). In general, these proteins interact predominantly with the major groove, where there is a considerable amount of space. Because of the regularity of the double helix, some atoms of the bases are always exposed in the major groove (and some in the minor groove). Atoms in the major groove that are known to be important in interactions with proteins are shown in Figure 6.4.

The size of a DNA molecule can be expressed in terms of its *molecular weight*, but because a single nucleotide has a molecular weight of about 330, and because DNA molecules are many nucleotides long, the molecular weight mounts up rapidly. (The nucleic acid in even small viruses, for instance, may have a molecular weight in the millions; the DNA in cells in the billions.) A more convenient way of expressing the sizes of DNA molecules is in terms of the *number of thousands* of nucleotide bases per molecule. Thus, a DNA molecule with 1000 bases contains 1 *kilobase* of DNA. If the DNA is a double helix, then one speaks of *kilobase pairs*. Thus, a double helix 5000 bases in length would have a length of 5 kilobase pairs. The bacterium *Escherichia*

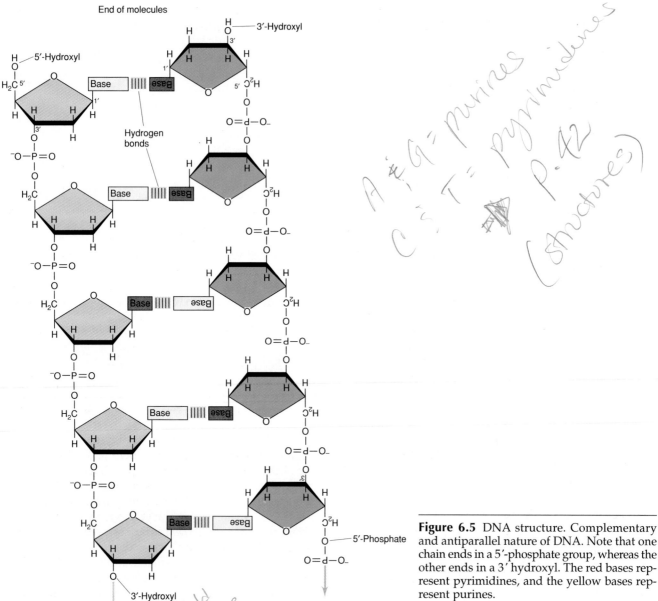

Figure 6.5 DNA structure. Complementary and antiparallel nature of DNA. Note that one chain ends in a 5'-phosphate group, whereas the other ends in a 3' hydroxyl. The red bases represent pyrimidines, and the yellow bases represent purines.

coli has about 4700 kilobase pairs of DNA in its chromosome. Each base pair takes up 3.4 Å [3.4 × 10⁻¹⁰ meter (m)] in length along the helix, and each turn of the helix contains 10 base pairs. Therefore, 1 kilobase of DNA has 100 turns of the helix and is 0.3 μm long. Calculations such as this can be very interesting, as we shall soon see.

Supercoiled DNA

Large DNA molecules, representing hundreds or millions of base pairs, are often represented in figures very simply as rods or, as in the case of the bacterial chromosome, uniform circles. It would be simplest to imagine that double-stranded DNA molecules have exactly the "correct" number of turns that one would predict by knowing the number of base pairs. Such a DNA molecule is said to be *relaxed*. However, consideration of the

length of a simple, relaxed double helix and the size of microbial cells and viruses indicate that there must be some higher order structure, because such a molecule could not be packed into a cell. For instance, if we calculate the length of DNA in the *Escherichia coli* chromosome, we will find it to be more than 1 mm, 1000 times longer than the *E. coli* cell itself! Such packaging problems also occur in viral DNA and the DNA of eukaryotic cells. How is it possible to pack so much DNA into such a little space? The solution: *supercoiling.*

Supercoiling is a state in which double-stranded DNA molecules are further twisted. Figure 6.7 shows a diagram of how this could happen in a circular DNA duplex. Supercoiling puts the DNA molecule under torsion. (Take a rubber band and twist it about itself. This twisting generates a tightly coiled structure that is under

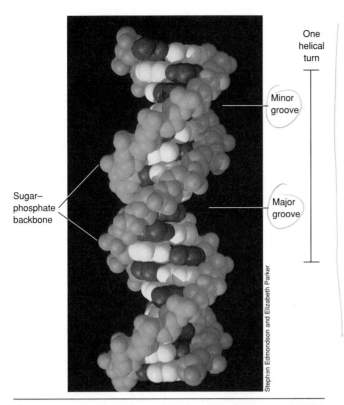

Figure 6.6 A computer model of a short segment of DNA showing the overall arrangement of the double helix. One of the sugar–phosphate backbones is shown in blue and the other in green. The pyrimidine bases are shown in red and the purines in yellow. Note the locations of the major and minor grooves. The model was produced using software from the Computer Graphics Laboratory, University of California at San Francisco.

considerable torsion. This torsion is held, however, only if the circular structure is maintained. Cut the twisted rubber band and see what happens!) DNA can be supercoiled in either a *positive* or *negative* direction. **Negative supercoiling** occurs when the DNA is twisted about its axis in the *opposite* direction from that of the right-handed double helix. It is in this form that supercoiled DNA is predominantly found in nature.

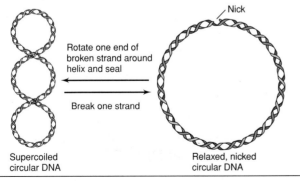

Figure 6.7 Supercoiled circular DNA and relaxed, nicked circular DNA interconversions. A nick is a break in a phosphodiester bond of one strand.

How is supercoiling brought about? In eukaryotic chromosomes, the formation of the nucleosome (∞ Section 3.14) introduces negative supercoils. In Bacteria and Archaea, there is a special enzyme called **DNA gyrase,** which introduces negative supercoils. The process can be thought to occur in several stages. First, the circular DNA molecule is twisted, then a break occurs where the two chains come together, and then the broken double helix is resealed on the opposite side of the intact strand (Figure 6.8). DNA gyrase is a *topoisomerase,* specifically a topoisomerase II. Note the derivation of the name *topoisomerase. Topology* is the branch of mathematics that deals with the properties of geometric figures that are unaltered when the figures are twisted or contorted. We are dealing here with the topology of DNA, and a topoisomerase is an enzyme that affects this topology. Of some interest is the fact that two antibiotics that act on Bacteria, *nalidixic acid* and *novobiocin,* inhibit the action of DNA gyrase. Novobiocin is also effective against several species of Archaea, where it also seems to inhibit DNA gyrase.

There is another enzyme that is able to *remove* supercoiling in DNA. This enzyme, called *topoisomerase I,* introduces a single-strand break in the DNA and causes the passage of one single strand of the double helix through the other. As was shown in Figure 6.7, a break in the backbone (a **nick**) of either strand allows the DNA to return to the relaxed state. This is true whether the supercoiling is positive or negative. Such enzymes are found in both prokaryotes and eukaryotes. Linear DNA, as in eukaryotic chromosomes, is prevented from returning to the relaxed state by the proteins bound to it. To prevent the entire bacterial chromosome from becoming relaxed every time a nick is made, the chromosome contains approximately 50 *supercoiled domains.* A nick in the DNA in one of these domains does not relax the DNA in the others. It is unclear what holds the DNA in these domains, but it is likely to involve proteins.

Through the action of these topoisomerases, the DNA molecule can be alternately coiled and relaxed. Because coiling is necessary for packing the DNA into the confines of a cell and relaxing is necessary so DNA can be replicated, these two complementary processes clearly play an important role in the behavior of DNA in the cell. In most prokaryotes, the actual level of negative supercoiling is the result of a balance between the activity of DNA gyrase and topoisomerase I. In addition, however, supercoiling is known to affect gene expression. Certain genes are more actively transcribed when DNA is supercoiled, whereas transcription of other genes is inhibited by excessive supercoiling.

A few prokaryotes contain an enzyme called *reverse gyrase.* This topoisomerase is capable of introducing *positive* supercoils in DNA. The organisms that contain this enzyme are among those that grow at extremely high temperatures (∞ Section 5.8). Some DNA in these organisms seems to be "relaxed," that is, with neither

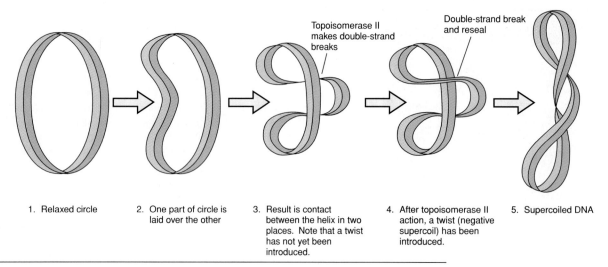

1. Relaxed circle

2. One part of circle is laid over the other

3. Result is contact between the helix in two places. Note that a twist has not been introduced.

Topoisomerase II makes double-strand breaks

4. After topoisomerase II action, a twist (negative supercoil) has been introduced.

Double-strand break and reseal

5. Supercoiled DNA

Figure 6.8 Introduction of supercoiling in a circular DNA by action of topoisomerase II, which makes double-strand breaks.

positive nor negative supercoils. This may be a matter of a balance between reverse gyrase and the activity of histonelike proteins. Interestingly, one organism, the hyperthermophilic archaean *Methanothermus fervidus*, has nucleosome-like structures in which the DNA seems to be positively supercoiled.

Positive supercoiling, brought about by reverse gyrase or other means, is a kind of "overwinding" and might play an important role in protecting the DNA from being denatured. However, in order to be of use, the information in DNA must be accessible to the cell's machinery. Therefore, in all cells, the structure of DNA is likely to be quite dynamic.

CONCEPT CHECK   **6.2a**

DNA is generally arranged as a double-stranded molecule that assumes a helical configuration. The very long DNA molecule is able to be packaged into the cell because it is supercoiled. In most prokaryotes this supercoiling is brought about by enzymes called topoisomerases.

✔ *What are the ends of DNA strands called, and how do the ends differ?*

✔ *Explain what* antiparallel *means in regard to the structure of double-stranded DNA.*

Other important features of DNA structure

In regions of the chromosome that encode proteins, the sequence of the DNA is dictated in large measure by the amino acid sequence of the encoded proteins and the nature of the genetic code (see Section 6.9). However, there are frequently base sequences in DNA that are

present not because of their coding properties but because they influence the *secondary structure* of DNA, or the way in which DNA interacts with proteins.

Long DNA molecules are quite flexible, but stretches of DNA less than 100 base pairs are much more rigid. Some short segments of DNA can be bent by proteins that interact with them. However, certain sequences themselves result in bends in the DNA. The sequences of this **bent DNA** often involve several runs of five or six adenines (in the same strand), each separated by four or five bases; an example of such a sequence is shown in Figure 6.9. In addition, some sequences contribute to the ability of DNA to bend when certain proteins interact with the DNA. DNA bending seems to be commonly involved in the regulation of gene expression, as we shall discuss in Chapter 7.

Short, repeated sequences are often found in DNA molecules. Many proteins have been found that interact with regions of DNA containing repeated sequences (∞ Chapter 7) but that are repeated in inverse orientation. This type of repeat is called an **inverted repeat.** Inverted repeats give the DNA sequence a twofold symmetry. As shown in Figure 6.10, nearby inverted repeats could theoretically lead to the formation of **stem-loop** (cruciform) structures in DNA. (Note that the stem of a stem-loop is a short double helix with normal base pairing and antiparallel strands.)

Figure 6.9 Double-stranded DNA with runs of five or six A's can form a bent structure.

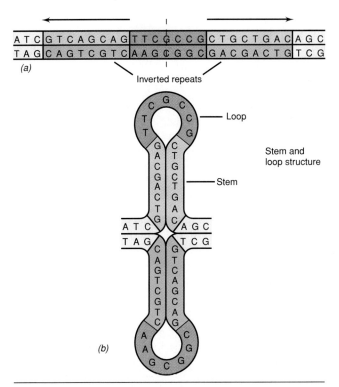

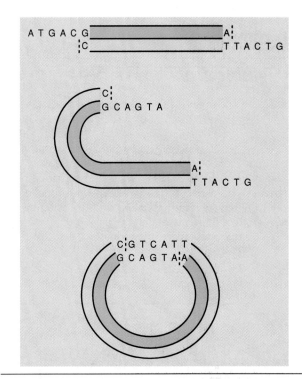

Figure 6.10 Inverted repeats and the formation of a stem-loop structure. (a) Nearby inverted repeats in DNA. The arrows indicate the symmetry around the imaginary axis (dashed line). (b) Formation of stem-loop structures (cruciform structures) by pairing of complementary bases on the *same* strand.

Figure 6.11 Linear DNA with complementary single-stranded ends ("sticky ends") can cyclize by base pairing of the complementary ends.

By examining Figure 6.10 it can be seen that either strand from this region could form a stem-loop. Therefore, some inverted repeats found in DNA that is transcribed may be important because of the stem-loops found in the RNA. Such *secondary structure* formed by base pairing within a single strand of nucleic acid is very important in both transfer RNA (see Section 6.8) and ribosomal RNA (∞ Section 15.5 and Figure 15.8c).

The *ends* of linear DNA molecules can also have interesting sequences that lead to changes in structure. Some DNA molecules have single-stranded regions at each end that are complementary. This leads to the possibility that the two ends can find each other and associate by complementary base pairing, as illustrated in Figure 6.11 for the formation of a circle. DNA with single-strand complementary sequences at the ends is said to have "sticky ends." Some linear DNA molecules have **hairpin** structures at each end. A hairpin is like a stem-loop but with almost no loop. Hairpins can be formed from a single-stranded region at the end of a molecule that contains an inverted repeat, as illustrated in Figure 6.12. We shall discuss other important sequences found at the ends of the linear DNA in eukaryotic chromosomes in Section 6.5.

Because of the enormous length of the DNA in a cell, it can almost never be handled experimentally as a com-

plete unit. The mere manipulation of DNA in the test tube leads to its fragmentation into molecules of smaller size. The need to study the size and shape of DNA is evident. Some of the tools for studying DNA are presented in the box, Working with Nucleic Acids: The Tools. As described, a useful technique for studying the sizes of DNA molecules is electrophoresis. As noted, the nucleic acid molecules migrate through the pores of the gel at rates depending on their molecular weight or molecular shape. Small or compact molecules migrate more rapidly than large or loose molecules. In one figure (see the box), a number of DNA fragments have been separated out in the gel.

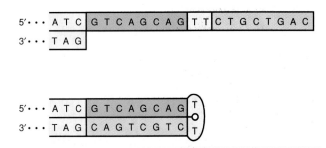

Figure 6.12 A hairpin structure at one end of a linear DNA molecule. If the linear DNA had been completely double-stranded, the sequences shown in green would have been inverted repeats.

TECHNIQUES & APPLICATIONS

WORKING WITH NUCLEIC ACIDS: THE TOOLS

Our knowledge of molecular biology and genetics has depended on the development of adequate research tools. Advances in knowledge of how nucleic acids work have generally been tied to the development of new methods. We discuss here some of these methods.

1. Extraction and purification of DNA The first requirement is a sample of DNA free of other cellular

2. Detecting the presence of DNA There are several methods for detecting the presence of DNA in a solution. One of the most widely used is by its absorption of ultraviolet radiation. DNA strongly absorbs ultraviolet radiation at a wavelength of 260 nm. The absorption is due to the purine and pyrimidine bases. As seen in the figure, double-stranded DNA absorbs less strongly than single-stranded DNA. This is because

become positioned in the gradient at positions corresponding to their densities. If ethidium bromide has been added, observation of the centrifuge tube with ultraviolet radiation after the centrifugation reveals bands of DNA by fluorescence (see later). This method is called the *buoyant density method* and permits both determination of density and separation of molecules of differing density.

4. Gel electrophoresis One of

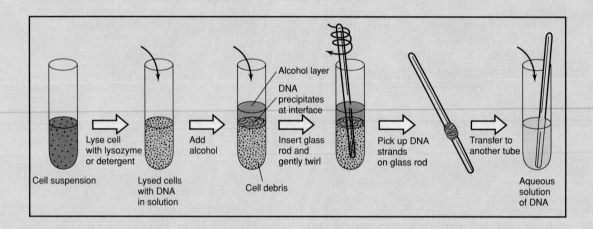

chemicals. The steps in the purification of DNA are shown here. The aqueous solution in the final step is treated with RNAse to remove RNA. Proteins are then removed by use of denaturing solvents (usually phenol). By repeating the purification steps a number of times, a solution can be obtained that is virtually free of any components other than DNA.

Note that the solution of DNA obtained never consists of native DNA molecules of the length found in the cell. The purification process causes the DNA to be broken into fragments of various (random) lengths. If the DNA has been handled gently during purification, the lengths of the fragments will be about one-hundredth of the length of the whole chromosome.

the interaction between the bases on the opposite strands of the double-stranded DNA (hydrogen bonding) reduces the ultraviolet absorbance.

3. Density-gradient centrifugation of DNA DNA molecules vary in density, depending on their exact chemical composition. DNA molecules with a higher content of guanine plus cytosine (GC) are denser than molecules with low GC. The density of DNA can be determined by centrifugation at very high speed in a *gradient* of cesium chloride (CsCl). The DNA solution is added to a solution of CsCl and centrifuged at high speed for several hours until equilibrium is reached. The CsCl forms a density gradient from the top to bottom of the tube, and DNA molecules form bands at appropriate densities. At equilibrium, the DNA molecules

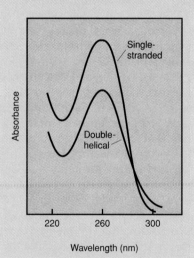

the most widespread methods of studying nucleic acids is gel electrophoresis. Introduction of electrophoresis methods has revolutionized research on molecular genetics. *Electrophoresis* is the procedure by which charged molecules are allowed to migrate in an electric field, the rate of migration being determined by the size of the molecules and their electric charge. In gel electrophoresis, the nucleic acid is suspended in a gel, usually made of polyacrylamide or agarose. The gel is a complex network of fibrils, and the *pore size* of the gel can be controlled by the way in which the gel is prepared. The nucleic acid molecules migrate through the pores of the gel at rates dependent on their molecular weight and molecular shape. Small molecules or compact molecules migrate more rapidly than large or loose molecules. After a defined period of time of migration (usually a few hours), the locations of the DNA molecules in the gel are assessed by making the DNA molecules fluorescent and observing the gel with ultraviolet radiation.

Shown here is a photograph of an electrophoresis apparatus. The horizontal frame, made of lucite plastic, holds the gel. The gel is submerged in buffer that makes an electrical connection to the power supply (shown in foreground). The gel is stained after electrophoresis with ethidium bromide and observed by use of ultraviolet radiation. In each lane, a mixture of DNA fragments was applied. A computerized scanner can be used to locate the positions of the DNA bands.

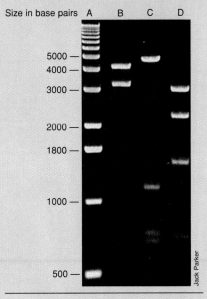

A typical gel

Separation of DNA fragments by these relatively simple means has proven extremely useful. However, large molecules (greater than 40,000 base pairs) are not separated from one another. New electrophoretic techniques have been developed that allow separation of large fragments. One of these methods, called *pulse field gel electrophoresis* or **PFGE,** involves sending short pulses of electricity to an array of electrodes surrounding the agarose gel. PFGE and related techniques are very valuable for analyzing DNA molecules the size of those found in a small eukaryotic chromosome.

5. Detecting DNA by fluorescence When nucleic acids are treated with dyes that are fluorescent and are able to combine firmly with the nucleic acid chain, the nucleic acid is rendered fluorescent. The dye *ethidium bromide* is widely used to render DNA fluorescent because it combines tightly within the DNA molecule. Ethidium bromide interacts with double-stranded DNA. If the DNA is then observed with an ultraviolet source, it will fluoresce.

6. Labeling nucleic acids Radioactivity is widely used in nucleic acid research because radioactivity can be detected in extremely tiny amounts. Radioactive nucleic acids can be detected either directly with a scintillation counter or indirectly via their effect on photographic film (autora-

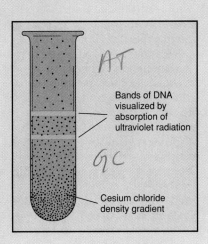

Bands of DNA visualized by absorption of ultraviolet radiation

Cesium chloride density gradient

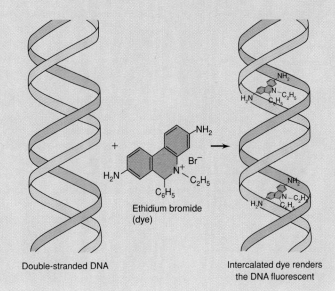

Double-stranded DNA

Ethidium bromide (dye)

Intercalated dye renders the DNA fluorescent

diography). Autoradiography of radioactive nucleic acids is one of the most widely used techniques in molecular genetics because it can be applied to the detection of nucleic acid molecules during gel electrophoresis.

A nucleic acid can be made radioactive by incorporation of radioactive phosphate during nucleic acid synthesis. If radioactive phosphate is added to a culture while nucleic acid synthesis is taking place, the newly synthesized nucleic acid becomes radioactively labeled.

$$^{32}PO_4 \rightarrow \ ^{32}P\text{-labeled nucleotides}$$
$$\rightarrow \ ^{32}P\text{-labeled nucleic acid}$$

Alternatively, end-labeling of purified DNA that contains a free hydroxyl group at the 5′-position can be done, using radioactive ATP labeled in the third phosphate. The enzyme polynucleotide kinase specifically removes the third phosphate from ATP and attaches it to the free hydroxyl group at the 5′-end of the molecule. End-labeling is an extremely useful technique as it permits labeling of preformed molecules. By tracing the radioactivity through subsequent chemical steps, the end of the molecule can be followed.

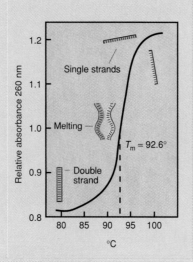

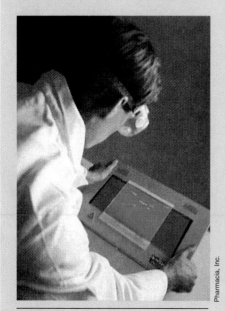

Using fluorescence to locate nucleic acid bands on a gel

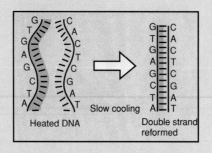

$$^{32}P\text{-P-P-adenosine}$$
$$+ \ HO\text{-deoxyribose-DNA} \rightarrow$$
$$^{32}P\text{-O-deoxyribose-DNA} + ADP$$

New methods of labeling nucleic acids have been developed that make use of nonradioactive chemicals that can be incorporated into DNA and can be detected by a variety of reagents that give either a colored product or even emit light (which can be detected with X-ray film). Some of these methods have nearly the sensitivity of a radioactive label but are more convenient to use and do not generate radioactive waste.

7. Denaturing nucleic acids As we have noted, the strands of a double helix can be separated by heating, a process generally called *melting*. As shown in part 2 above, double-stranded molecules show lower ultraviolet absorbance than single-stranded molecules. Therefore, if the ultraviolet absorbance of a nucleic acid solution is measured while it is being heated, the

increase in absorbance when the double-stranded molecules are converted to single-stranded molecules will show the temperature at which strand separation occurs. The figure shows the change in absorbance at 260 nm when a solution containing double-stranded DNA is gradually heated. The midpoint of the transition, called T_m, is a function of the GC content of the DNA. If the heated DNA is allowed to cool slowly, the double-stranded native DNA may reform.

Strands can also be separated at room temperature by changing the ionic conditions of a solution of DNA. Because temperature is not involved, the process is usually referred to as *denaturation*. However, the end result is the same. And just as with strands separated by heat, if ionic conditions are slowly returned to normal, then the double helix can reform.

8. Nucleic acid hybridization *Hybridization* is the artificial construction of a double-stranded nucleic acid by complementary base pairing of two single-stranded nucleic acids. When a DNA solution that has been heated (see earlier discussion) is allowed to cool slowly, many of the complementary strands reassociate and the original double-stranded complex reforms, a process called *reannealing*. The reannealing occurs only if the base sequences of the two strands are complementary. Thus, nucleic acid hybridization permits the formation of artificial double-stranded hybrids of DNA, RNA, or DNA:RNA. Nucleic acid hybridization provides a powerful tool for studying the genetic relatedness between nucleic acids. It also permits the detection of pieces of nucleic acid that are complementary to a single-stranded molecule of known sequence. Such a single-stranded molecule of known sequence is called a **probe.** For instance, a radioactive nucleic acid probe can be used to *locate*, in an unknown mixture, a nucleic acid sequence complementary to the probe. Detection of nucleic acid hybridization is usually done with membrane filters constructed of nitrocellulose. Single-stranded DNA is first bound to the filter, and then the probe is added.

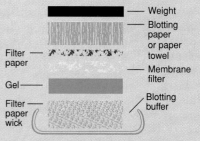

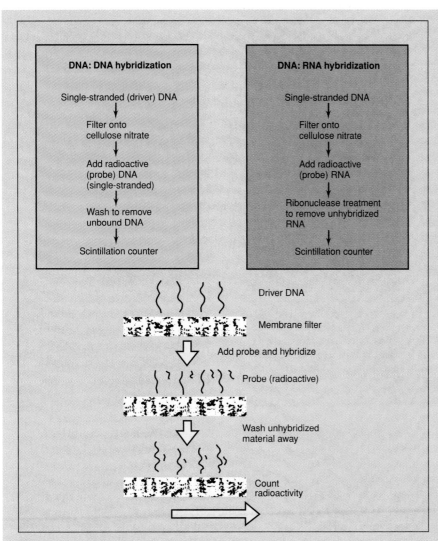

DNA: DNA hybridization

Single-stranded (driver) DNA
↓
Filter onto cellulose nitrate
↓
Add radioactive (probe) DNA (single-stranded)
↓
Wash to remove unbound DNA
↓
Scintillation counter

DNA: RNA hybridization

Single-stranded DNA
↓
Filter onto cellulose nitrate
↓
Add radioactive (probe) RNA
↓
Ribonuclease treatment to remove unhybridized RNA
↓
Scintillation counter

Driver DNA

Membrane filter

Add probe and hybridize

Probe (radioactive)

Wash unhybridized material away

Count radioactivity

available for determining the sequences of DNA molecules. Appropriate treatments are used to generate DNA fragments that end at the four bases and that are radioactive. Then the fragments are subjected to electrophoresis so molecules with one nucleotide difference in length are separated on the gel. This electrophoresis procedure involves *four* separate lanes, one for fragments ending at each of the four bases of the DNA, adenine, guanine, cytosine, and thymine. The positions of these fragments are located by autoradiography, and from a knowledge of which base is represented by each lane, the sequence of the DNA can be read off.

Two different procedures have been developed to accomplish such determinations, called the *Maxam–Gilbert* and the *Sanger dideoxy* procedures. In the Maxam–Gilbert procedure chemicals are used that break

Probe that does not base-pair with the DNA on the filter is then washed off. If necessary, hybridization conditions can be manipulated to favor the formation of DNA:DNA or DNA:RNA hybrids. Nucleic acid hybridization is a powerful tool in genetics.

Hybridization can also be done after gel electrophoresis. The nucleic acid molecules are transferred by blotting from the gel to a sheet of membrane filter material, and the probe is then added to the filter. The procedure when DNA is in the gel and RNA or DNA is the probe is often called a *Southern blot procedure*, named for the scientist E. M. Southern, who first developed it. When RNA is in the gel and DNA or RNA is the probe, the procedure is called a *Northern blot*. A

Western blot (sometimes called an immunoblot) involves protein–antibody binding rather than nucleic acids; ∞ Section 21.9. The figure also shows the use of a nucleic acid probe to search for complementary sequences in a mixture. The DNA fragments have been spread out by gel electrophoresis and then transferred to the membrane filter. The RNA probe, which is radioactively labeled, is allowed to reanneal to the DNA on the filter and its position is determined by autoradiography.

9. Determining the sequence of DNA   Although the base sequences of both DNA and RNA can be determined, it turns out for chemical reasons that it is easier to sequence DNA. Even automated machines are now

Laying the membrane filter on the gel

the DNA preferentially at each of the four nucleotide bases under conditions in which only one break per chain is made. (Thus, four separate test tubes are prepared.)

In the Sanger dideoxy procedure the sequence is actually determined by making a *copy* of the single-stranded DNA, using the enzyme *DNA polymerase*. This enzyme uses deoxyribonucleoside triphosphates as substrates and adds them to a *primer*. In the incubation mixtures (four separate test tubes) are small amounts of each of the dideoxy analogs of the deoxyribonucleoside triphosphates. Because the dideoxy sugar lacks the 3'-hydroxyl, continued lengthening of the chain cannot occur. The dideoxy analog thus acts as a *specific chain-termination reagent*. Fragments of variable length are obtained, depending on the incubation conditions. The nucleic acid fragments formed are radioactive from using either a radioactive primer or a radioactive deoxynucleoside triphosphate in the reactions. Electrophoresis of these fragments is then carried out, and the positions of the radioactive bands are determined by autoradiography. By aligning the four dideoxynucleotide lanes and noting the vertical position of each fragment relative to its neighbor, the sequence of the DNA copy can be read directly from the gel (see figure below).

A major advantage of the Sanger method is that it can be used to

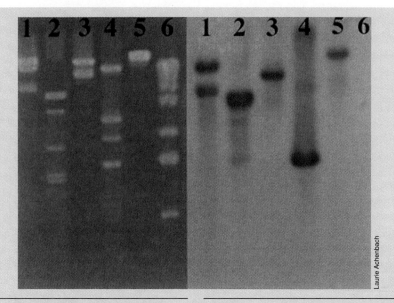

Laurie Achenbach

Agarose gel electrophoresis of DNA molecules. Purified molecules of DNA from several different plasmids were treated with restriction enzymes and then subjected to electrophoresis.

Southern blot of the DNA gel shown to the left. After blotting, hybridization with a radioactively labeled probe was carried out. The positions of the bands have been detected by X-ray autoradiography. Note that only some of the DNA fragments have sequences complementary to the labeled probe. Lane 6 contained DNA used as a size marker and none of the bands hybridized to the probe.

sequence RNA as well as DNA. To sequence RNA, a single-stranded DNA copy is made (using the RNA as the template) by the enzyme reverse transcriptase. By making the single-stranded DNA in the presence of dideoxynucleotides, various-sized

DNA fragments are generated suitable for Sanger-type sequencing. from the sequence of the DNA, the RNA sequence is deduced by base-pairing rules. The Sanger method has been instrumental in rapidly sequencing ribosomal RNAs for use

Normal deoxynucleotide

Dideoxy analog   Missing OH

DNA chain   Direction of chain growth

No free 3'-OH, replication will stop at this point

in studies on microbial evolution (∞ Chapter 18).

For determining the DNA sequence of a long molecule, such as a whole gene, it is necessary to proceed in stages. First, the DNA is broken into small overlapping fragments and the sequence of each fragment determined. Using the overlaps as a guide, the sequence of the whole molecule can be deduced.

The demands of projects that involve sequencing entire genomes (∞ Sections 9.11 and 10.15) have led to the development of automated DNA sequencing systems. With such systems the sequencing reactions are still based on the dideoxy methodology, but fluorescent dye–labeled primers (or bases) are used so that bands can be easily detected. The products are separated by automated electrophoresis and the bands detected by fluorescence spectroscopy. In one procedure each of the four different reactions utilizes a different fluorescent label so that all four reactions can be run on a single lane. The results are analyzed by computer and a sequence printed out with each of the four bases being color coded (see figure below). ■

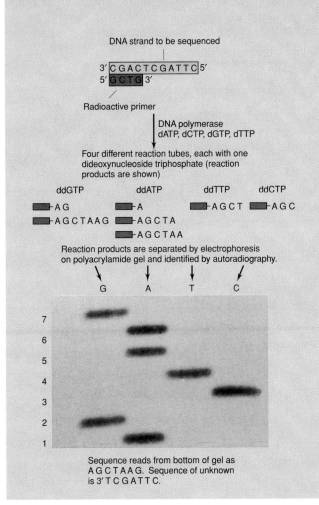

Sequence reads from bottom of gel as A G C T A A G. Sequence of unknown is 3′ T C G A T T C.

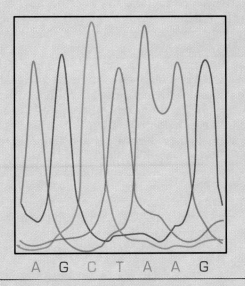

Results of sequencing the same DNA as on the photograph to the left, but using an automated sequencer and fluorescent labels.

The effect of temperature on DNA structure

Although the hydrogen bonds between the base pairs are individually very weak (∞ Section 2.2), there are many such bonds holding together the strands of a typical double-stranded DNA molecule. There may be millions or even billions of such bonds, depending on the number of base pairs in the molecule. Remember that each adenine–thymine base pair has two such bonds, and each guanine–cytosine base pair has three. This makes GC pairs stronger than AT pairs.

When isolated from cells and kept at temperatures near room temperature and at physiological salt concentrations, DNA remains in a double-stranded form. However, if the temperature is raised, the hydrogen bonds will break, but the covalent bonds holding a chain together will not, and so the DNA strands will separate. This process is generally called *melting* and, as shown in the box, can be measured experimentally. DNA with high numbers of GC pairs melts at a higher temperature than a similar-sized molecule with more AT pairs. (The fact that DNA with even high amounts of GC melts below 100°C is why the structure of DNA in the chromosomes of organisms that live at extremely high temperatures is of such interest.) If the heated DNA is allowed to cool slowly, the double-stranded native DNA can reform. Interestingly, such a process can be used not simply to reform native DNA but also to form *hybrid* molecules whose two strands come from separate sources.

Hybridization of nucleic acids

Hybridization is the artificial construction of a double-stranded nucleic acid by complementary base pairing of two single-stranded nucleic acids. The procedure for constructing nucleic acid hybrids is shown in the box. Since adenine pairs with uracil as well as with thymine, both DNA:DNA and DNA:RNA hybrids can be made. There must be a high degree of complementarity between two single-stranded nucleic acid molecules if they are to form a stable hybrid. In the most common use of hybridization, one of the molecules is used as a radioactive *probe* to detect a specific nucleic acid sequence, and formation of hybrids is detected by observing the formation of double-stranded molecules containing radioactivity.

One of the most common uses of hybridization is to detect DNA sequences that are complementary to mRNA molecules. Detection of DNA:RNA hybridization is usually done with membrane filters made of nitrocellulose or nylon. The DNA is first made single-stranded (melted or denatured) and immobilized on the filter. The membrane is then treated to prevent any nonspecific nucleic acid binding. Following this, a radioactive RNA probe is added. This RNA is able to remain attached to the membrane only if it can base-pair with the immobilized DNA. After appropriate incubation, the unhybridized RNA is washed out and the radioactivity still bound to the filter is measured (see the box).

CONCEPT CHECK   6.2b

Some sequences in double-stranded DNA are inverted repeats. Some of these repeats may be recognition sites for proteins, whereas others are places where secondary structure occurs in the DNA or an RNA copy of it. The strands of a double-helical DNA molecule can be separated experimentally by heat in a process called melting. Two complementary single strands can hybridize to form a stable double-stranded molecule. RNA can also hybridize with single-stranded DNA.

✔ *What is an* inverted repeat?

✔ *Why can some nucleic acids hybridize?*

6.3
Genetic Elements

As we have seen, the genetic material of all cells is double-stranded DNA (∞ Section 2.7). Before proceeding to describe how cells replicate their DNA, we consider what kind of structures must be replicated. Structures containing genetic material can be called *genetic elements*. The **genome** is the total complement of genes in a cell or virus. Although the main genetic element is the *chromosome*, other genetic elements are found and play important roles in gene function in both prokaryotes and eukaryotes (Table 6.1). Two key properties of genetic elements are (1) their ability to self-replicate, and (2) their genetic coding properties.

The chromosome

In Section 3.8 we discussed the fact that a typical prokaryote has a single chromosome containing all (or most) of the genes found inside the cell. Eukaryotes have multiple chromosomes as a part of their genome. Also, the typical prokaryotic chromosome is a circular DNA molecule, whereas the DNA in all known eukaryotic chromosomes is linear. In Table 6.2 the number, size, and configuration of chromosomes in a few known microorganisms, both prokaryotic and eukaryotic, are given. Note that the chromosome of the bacterium *Borrelia burgdorferi*, the

---

TABLE 6.1   Kinds of genetic elements

Element	Description
Prokaryote	
Chromosome	Extremely long, usually circular, double-stranded DNA molecule
Plasmid	Typically a relatively short, usually circular, double-stranded DNA molecule
Virus	Single- or double-stranded DNA or RNA molecule
Transposable element	Double-stranded DNA molecule always found within another DNA molecule
Eukaryote	
Chromosome	Extremely long, linear, double-stranded DNA molecule
Plasmid	Typically a relatively short circular or linear double-stranded DNA molecule
Mitochondrion or chloroplast	Intermediate-length DNA molecules, usually circular
Virus	Single- or double-stranded DNA or RNA molecules
Transposable element	Double-stranded DNA molecule always found within another DNA molecule

TABLE 6.2	Sizes, shapes, and numbers of chromosomes in microorganisms			
			Chromosome	
Organism	**Comments**	**Size (base pairs)[a]**	**Number**	**Geometry**
Bacteria				
Mycoplasma pneumoniae	Causes pneumonia (∞ Chapter 23)	7.8×10^5	1	⬡
Borrelia burgdorferi	Causes Lyme disease (∞ Chapter 23)	9.5×10^5	1	⌇⌇⌇⌇
Haemophilus influenzae	Gram-negative, can cause disease (∞ Chapter 23)	$1,830,137^b$	1	◯
Rhodobacter sphaeroides	Gram-negative, phototrophic	4.0×10^6	2	◯
Escherichia coli	Gram-negative, genetic model	4.7×10^6	1	◯
Bacillus subtilis	Gram-positive, genetic model	5.0×10^6	1	◯
Myxococcus xanthus	Complex developmental cycle (∞ Chapter 16)	9.5×10^6	1	◯
Archaea				
Methanococcus voltae	Methanogen (∞ Chapters 5 and 17)	1.9×10^6	1	◯
Thermococcus celer	Grows at high temperature (∞ Chapters 5 and 17)	1.9×10^6	1	◯
Haloferax mediterranei	Grows in high salt (∞ Chapters 5 and 17)	2.9×10^6	1	◯
Sulfolobus acidocaldarius	Grows at high temperature and high acidity (∞ Chapters 5 and 17)	3.0×10^6	1	◯
Eukarya[c]				
Giardia lamblia	Flagellated protozoan that causes acute gastroenteritis (∞ Chapters 18 and 23)	1.2×10^7	4	⌇⌇⌇⌇
Saccharomyces cerevisiae	Yeast, widely used in science and industry (∞ Chapters 9 and 18)	$12,057,500^d$	16	⌇⌇⌇⌇
Dictyostelium discoideum	Cellular slime mold, developmental model (∞ Chapter 18)	5.0×10^7	7	⌇⌇⌇⌇
Tetrahymena thermophila	Ciliated protozoan (∞ Chapter 18)	2.1×10^8	5	⌇⌇⌇⌇

[a]In the case of the Eukarya the genome sizes and chromosome number are for the haploid form.

[b]*Haemophilus influenzae* Rd was the first cellular organism to have its genome entirely sequenced.

[c]All the organisms listed are single-celled.

[d]*Saccharomyces cerevisiae* was the first eukaryote to have its genome completely sequenced. The number given here does not include the mitochondrial genome and some repeated sequences.

causative agent of Lyme disease (∞ Section 23.10) is linear. The ends of this chromosome may have hairpin repeats like those shown in Figure 6.12. Although very uncommon, linear chromosomes are now known to exist in a few other Bacteria. The bacterium *Streptomyces lividans* also has a linear chromosome, but it has proteins covalently bound to its ends. We will discuss the importance of such proteins in Section 6.5.

The few examples of prokaryotes given in Table 6.2 are not random choices. They include Archaea as well as Bacteria, and examples of among the smallest and largest known prokaryotic chromosomes. We shall discuss the apparent fact that *Rhodobacter sphaeroides* has two chromosomes later.

Only a few examples of eukaryotic microorganisms are also given in Table 6.2, but all their chromosomes have linear DNA and multiple chromosomes even in the haploid state. Both these conditions are the rule in eukaryotic organisms, whatever their size. Higher eukaryotes do have much more DNA. The haploid human genome has only a few more chromosomes than yeast but has over 200 times more DNA.

The eukaryotic chromosome is a linear DNA molecule containing genes, which has a special DNA

sequence called a *telomere* at each end and a *centromere* somewhere between the telomeres. Centromeres are important for partitioning the chromosomes during cell division. As we shall see, telomeres play an important role in the replication of these molecules (see Section 6.5). The number of chromosomes is constant within a species but varies widely among species. The overall structure of a eukaryotic chromosome can be visualized as a series of compact, highly folded units called **nucleosomes,** each containing about 200 DNA base pairs, separated by linkers of less extensively complexed DNA (∞ Section 3.14 and Figure 3.74). Within each nucleosome, the DNA molecule is wound around a cluster of histone molecules organized in a precise, repeatable pattern. One function of the nucleosome structure is to permit packing of the long DNA molecules within the cell. For instance, if all the DNA of the chromosomes of a human cell were stretched end to end, the length would be close to 2 m, yet this DNA is packed into 46 chromosomes in a total length of only 200 µm.

The chromosomal organization in eukaryotes involves two features not generally found in prokaryotes:

1. *Split genes.* Many eukaryotic protein-encoding genes are split or interrupted by noncoding DNA sequences inserted between the sequences that actually code for a single polypeptide (see Figure 6.2). These noncoding intervening sequences are called **introns,** and the coding sequences are called **exons.** The number of introns per gene is variable and ranges from none to more than 50. During transcription, both introns and exons are copied, and the intron sequences are subsequently cut out and removed when the messenger RNA is processed into its final form.

2. *Repetitive sequences.* Eukaryotes generally contain much more DNA per genome than is needed to encode all the proteins required for cell function. Eukaryotic DNA can be divided into several classes. **Single-copy DNA** contains the coding sequences for the main proteins of the cell (and the associated intron DNA). **Moderately repetitive DNA,** found in a few to relatively large numbers of copies, codes for some major macromolecules of the cell: histones, immunoglobulins (involved in immune mechanisms, as discussed in Chapter 20), ribosomal RNA, and transfer RNA. **Highly repetitive (satellite) DNA** is found in a very large number of copies. In humans, about 20–30% of the DNA is found in repetitive sequences, and one 300-base-pair sequence is repeated approximately 300,000 times. The function of most highly repetitive DNA is unknown.

However, as is true of so much of biology, except for the central dogma there seem to be few categorical statements. Introns have also been found in the protein-encoding genes of prokaryotes, but they are much less common and are removed from the transcript by a different process. Repetitive DNA sequences are also present in prokaryotes. These include low-copy-number repeated sequences, such as genes that encode rRNA, and even a few short but more highly repetitive sequences. For instance, there is a 38-base-pair sequence found in *Escherichia coli* that may be repeated 500–1000 times, making up almost 1% of the genome. However, the two generalizations are still correct: eukaryotic genomes are characterized by having split genes and repetitive sequences.

Nonchromosomal genetic elements: Viruses and plasmids

A number of genetic elements that are *not* cellular chromosomes have been recognized. Some nonchromosomal genetic elements that we will discuss briefly here include viruses, plasmids, the genomes of mitochondria and chloroplasts, and transposable elements.

Viruses are genetic elements, either DNA or RNA, that control their own replication and transfer from cell to cell. The viral genome is also referred to as a chromosome, but it is distinct from the cellular chromosomes. Both linear and circular viral chromosomes are known. Viruses are of special interest because they are often (but not always) responsible for disease states. We discuss viruses in Chapter 8, and virus diseases in Chapter 23.

Plasmids are typically small genetic elements that exist and replicate separately from the chromosome. The great majority of plasmids are double-stranded DNA, and although most plasmids are circular, some are linear. Plasmids differ from viruses in two ways: (1) they do not cause cellular damage (generally they are beneficial), and (2) viruses have extracellular forms, whereas plasmids do not. Although plasmids have been recognized in only a few eukaryotes, they have been found in most prokaryotic species. We discuss plasmids in Chapter 9. Some plasmids find wide use in gene manipulation and genetic engineering, as outlined in Chapter 10.

Many prokaryotes seem to contain one or more plasmids in addition to their chromosome. Some plasmids contain genes whose protein products can confer important properties on the host cell, such as resistance to antibiotics. Many plasmids are rather small (a few kilobase pairs), but some are quite large (several megabases). None are as large as the chromosome, however. From this information, you might think that in prokaryotes a chromosome is simply defined as the largest genetic element in the cell. Although the definition of what constitutes a chromosome in Bacteria is still controversial, it is coming to mean a genetic element that contains genes whose products are involved in essential metabolic steps *under all growth conditions.* Such genes are sometimes referred to as *housekeeping genes.* For instance, a gene encoding DNA gyrase is always required by a cell, whereas a gene that enables a bacterium to be resistant to an antibiotic is required only under certain conditions (the presence of the antibiotic). It has not yet been conclusively demonstrated that any prokaryote has more than one chromosome by this definition, but it is possible that some do. Such proof requires evidence that each "chromosome" contains single-copy genes that are essential. However, there are several Bacteria that may well have more than one chromosome, including *Rhodobacter sphaeroides* (see Table 6.2). This may also be true of the spirochete *Borrelia burgdorferi* (see Table 6.2), which has a complex genome containing a large, linear chromosome and several circular and linear plasmids.

Nonchromosomal genetic elements: Organelles and transposable elements

Mitochondria and **chloroplasts** contain nonchromosomal genetic elements and are found in eukaryotes. As we discussed in Section 3.15, the mitochondrion is the site of respiratory enzymes and plays a major role in energy generation in most eukaryotes. The chloroplast is a green, chlorophyll-containing structure that is the site of phototrophic ATP formation. From a genetic viewpoint, mitochondria and chloroplasts can be viewed as independently replicating genetic elements. However, these organelles are much more complex than plasmids and viruses because they contain not only DNA but also a complete machinery for protein synthesis, including 70S ribosomes, transfer RNA, and all the other components necessary for translation and formation of functional proteins. One intriguing feature of mitochondria and chloroplasts is that, despite the fact that they contain many genes and a complete translation system, their existence is not independent of the chromosomes because most proteins in them are coded not by organelle DNA but by chromosomal DNA. We discuss the genetics of mitochondria in Section 9.14.

Transposable elements are pieces of DNA having the ability to move from one site on a chromosome to another. Transposable elements are found in prokaryotes and eukaryotes and play important roles in genetic variation. There are three types of transposable elements: insertion sequences, transposons, and some special viruses. *Insertion sequences* are the simplest type and carry no genetic information other than that required for them to move into new locations. *Transposons* are larger and contain other genes. We discuss both of these types in more detail in Chapter 9. In Chapter 8 we discuss a virus, Mu, that is also a transposable element. The unique feature of transposable elements is that *they all replicate as part of some other molecule of DNA.* In spite of this, some of these elements are clearly "self-replicating" in the sense that they control their own replication and as such fit our definition of a genetic element. Others might simply be considered "jumping genes."

CONCEPT CHECK   *6.3*

In addition to the chromosomes, a number of other genetic elements exist in cells. Plasmids are DNA molecules that exist separately from the chromosome of the cell. Mitochondria and chloroplasts contain their own DNA genomes. Viruses are genetic elements, either DNA or RNA, that control their own replication. Transposable elements exist as a part of other genetic elements.

✔ *What is a* genome?

✔ *What genetic material is found in all cellular chromosomes?*

✔ *What is the difference between the number of chromosomes in prokaryotes and eukaryotes?*

6.4
Restriction and Modification of DNA

Organisms are occasionally faced with the problem of coping with foreign DNA, generally derived from viruses, that may derange cellular metabolism or initiate processes leading to cell death. In a unicellular microorganism this problem must be dealt with without the aid of some of the complex host defense mechanisms available to higher organisms (∞ Chapters 19 and 20). However, many prokaryotes have an extremely effective mechanism of dealing with foreign DNA, enzymatic destruction.

The enzymes involved in the destruction of foreign DNA are called **restriction endonucleases** and are remarkably specific in their action, an essential property if destruction of cellular DNA is to be avoided. Restriction endonucleases combine with DNA only at sites with specific sequences of bases. Clearly such enzymes could not exist unless the DNA of the host organism were protected from attack. In order to protect itself from its own restriction enzymes, a cell has enzymes that chemically **modify** the specific sequences on its own DNA so that these sequences are not attacked. Therefore, restriction enzymes are one-half of a **restriction-modification system.**

Restriction enzymes

There are several different kinds of restriction enzymes. The most common seem not only to recognize specific sequences of bases but also to make double-stranded breaks *within* these sequences. Many of these sequences exhibit twofold symmetry around a given point. Thus, one restriction endonuclease from *Escherichia coli*, called *Eco*RI, has the following recognition sequence:

$$5'\ldots G\!\downarrow\!A\!-\!A\!-\!T\!-\!T\!-\!C\!-\ldots 3'$$

$$3'\ldots C\!-\!T\!-\!T\!-\!A\!-\!A\!\underset{\uparrow}{-}G\!-\ldots 5'$$

The cleavage sites are indicated by arrows, and the axis of symmetry by a dashed line. Note that the two strands have the same sequence if one is read from the left and the other from the right (or, in terms of polynucleotide strands, if both are read $5' \rightarrow 3'$ or both read $3' \rightarrow 5'$). Such a structure is called a **palindrome.** (A palindrome is a sequence of characters that reads the same when read from either right or left—for instance, *Sex at noon taxes* and *Able was I ere I saw Elba.* The term *palindrome* is derived from the Greek meaning "to run back again.")

Many restriction enzymes are composed of two identical polypeptide subunits, each of which recognizes and cuts the sequence on a single strand. Since the sequences recognized by restriction enzymes are relatively short, and frequently palindromic, such enzymes always make *double*-stranded breaks, and such double-stranded breaks are not subject to correction by repair enzymes. This ensures that an invading nucleic acid will be destroyed.

Almost all known restriction enzymes are from prokaryotes. The recognition sequences and cutting sites for a few restriction enzymes are given in Table 6.3. Note that the recognition sites for the enzymes listed are 4, 5, 6, and 8 base pairs. In a "random" DNA molecule, one would expect any 4-base-pair sequence to occur approximately once every 256 base pairs based on the probability of $\frac{1}{4} \times \frac{1}{4} \times \frac{1}{4} \times \frac{1}{4}$ (assuming each base pair is equally probable in the DNA). Therefore, such an enzyme would cut a large DNA molecule into many specific fragments. A specific 6-base-pair sequence should appear every 4096 base pairs in random DNA, and an 8-base-pair sequence should appear only once about every 1 million base pairs (a megabase pair). The *Escherichia coli* chromosome, which is about 4.7 megabase pairs, is cut 21 times by the enzyme *Not*I, which recognizes an 8-base-pair sequence indicating that in this chromosome the *Not*I recognition sequence is used somewhat more often than one might have predicted.

Many restriction enzymes are now known, and more are being sought. The reason for this is not just that they are very interesting enzymes. They are also of great importance in DNA research. Note that the enzyme *Eco*RI can cut *any* double-stranded DNA that has its recognition sequence and cuts only at that sequence. This enables scientists to cut large DNA molecules into smaller fragments. Such fragments with defined termini, created as a result of the action of specific restriction enzymes, are amenable to determination of nucleotide sequences, thus permitting the working out of the complete sequence of DNA molecules (see the box).

Another use of certain restriction enzymes is that they permit the conversion of DNA molecules into fragments that can be joined by DNA ligase (see Section 6.5). This enables laboratory researchers to clone DNA, as will be discussed in Chapter 10.

Modification: Protection from restriction

An integral part of the cell's restriction-modification system is the modifying enzyme, which chemically **modifies** the specific sequences on its *own* DNA so these sequences are not attacked by the cell's own restriction enzymes. Such modification generally involves *methylation* of specific bases within the recognition sequence so the restriction nuclease can no longer act. Thus, for each restriction enzyme there must also be a modification enzyme, the two enzymes being closely associated. For example, the sequence recognized by the *Eco*RII restriction enzyme (also see Table 6.3) is

$$C–C–A–G–G$$

$$G–G–T–C–C$$

and modification of this sequence results in methylation of two cytosines:

$$\overset{\text{m}}{C}–C–A–G–G$$

$$G–G–T–C–\underset{\text{m}}{C}$$

Note that a given nucleotide sequence can be a substrate for *either* a restriction enzyme *or* a modification enzyme but not both. This is because modification makes the sequence unreactive with the restriction enzyme, and action of the restriction enzyme destroys the recognition site of the modification enzyme.

Restriction enzymes are such important tools in modern molecular genetic research that they have become widely available commercially. A number of companies purify and market restriction enzymes with

---

TABLE 6.3   Recognition sequences of a few restriction endonucleases

Organism	Enzyme designation	Recognition sequence[a]
Bacillus subtilis	*Bsu*RI	GG↓C̆C
Brevibacterium albidum	*Bal*I	TGG↓C̆CA
Escherichia coli	*Eco*RI	G↓AǍTTC
Escherichia coli	*Eco*RII	↓C̆CAGG (not a palindromic sequence)
Haemophilus haemolyticus	*Hha*I	GC̆G↓C
Haemophilus influenzae	*Hind*II	GTPy↓PuAC̆
Haemophilus influenzae	*Hind*III	A↓AGCTT
Nocardia otitidis-caviarum	*Not*I	GC↓GGC̆CGC
Thermus aquaticus	*Taq*I	T↓CGǍ

[a]Arrows indicate the sites of enzymatic attack. Asterisks indicate the site of methylation (modification). G, guanine; C, cytosine; A, adenine; T, thymine; Pu, any purine; Py, any pyrimidine. Only the 5′ → 3′ sequence is shown.

a variety of specificities. If the DNA sequence of a particular region of a molecule is known, a research worker can generally obtain a restriction enzyme that can cut in this region.

Different restriction enzyme systems in one host

If an incoming viral genome is modified by the host, then the restriction enzyme will ignore it. The virus will replicate and release progeny (all with their DNA modified) into the environment. If these progeny viruses infect a new cell of the same type, none of them will be susceptible to the restriction enzyme. At least, none will if all members of the host species have the same restriction-modification system. However, they do not. For instance, different *strains* of *Escherichia coli* have different restriction-modification systems. Therefore, a virus propagated on one of them will be restricted if it infects another. In addition, restriction systems have been discovered where the restriction enzyme recognizes only DNA that has a modification of a type seen only in other cells. These enzymes then specifically restrict DNA previously modified in a "foreign" host. In a typical strain of *E. coli* there are at least four different restriction systems operating. One is a "classical" restriction-modification system that restricts incoming unmodified DNA, and the other three restrict DNA with different foreign modifications. Some strains of *E. coli* have even more restriction-modification systems operating. In Chapter 8 we will see how selective pressures operating on a virus and its host may lead to the accumulation of a variety of such defense mechanisms.

Restriction enzyme analysis of DNA

As noted, a DNA molecule can be cut at a specific location by a given restriction enzyme. Because the base sequences recognized by most restriction enzymes are four to six nucleotides long, there will generally be only a limited number of such sequences in a piece of DNA. After cleaving the DNA (Figure 6.13*a*), the fragments can be separated by agarose gel electrophoresis, as shown in Figure 6.13*b*. The distance migrated by any band of DNA in such a gel can be determined by calibrating the electrophoresis system with DNA molecules of known size. By judicious use of several restriction enzymes of different specificities, and by use of overlapping fragments, it is possible to construct a **restriction enzyme map** in which the positions cut by each of the several restriction enzymes can be designated (Figure 6.13*c*).

Several procedures are now available for determining the base sequences of DNA molecules. In fact, automated machines are available for sequencing DNA. Details are presented in the box. By successively determining the sequences of small overlapping fragments of DNA, it is possible to determine the sequences of

very large pieces of DNA. The sequences are now known for thousands of genes, as well as for the complete genome of many viruses (∞ Chapter 8), several prokaryotes, and the eukaryotic microorganism *Saccharomyces cerevisiae* (∞ Chapter 9).

CONCEPT CHECK 6.4

Restriction enzymes are cellular enzymes that recognize specific short base sequences in DNA and make two single-stranded breaks at locations within the recognition sites. A restriction enzyme does not affect the cell that produces it because its own DNA is methylated at the recognition site by a modification enzyme specific for that site.

✔ *Of what use are restriction enzymes to organisms in nature?*

✔ *Of what use are restriction enzymes in the laboratory?*

6.5
DNA Replication

The problem of DNA replication can be simply put: the nucleotide base sequence residing in each long molecule of the DNA double helix must be precisely duplicated to form a copy of the original molecule. The cell has solved this seemingly complex problem in an elegant fashion: by means of *complementary base pairing*. As we have discussed (see Figure 6.4), adenine pairs specifically with thymine and guanine pairs with cytosine. If the DNA double helix is opened up, a new strand can be synthesized as the complement of each of the parental strands. As shown in Figure 6.14, replication is **semiconservative,** the two resulting double helices consisting of one progeny and one parental strand.

Templates and primers

The DNA molecule that is copied to form a complement is called a *template*. A template is a preformed pattern that is copied, but the *new* DNA molecule is not covalently connected to the *old* DNA molecule.

The chemistry of DNA, the nature of its precursors, and the activities of the enzymes involved in replication place some important restrictions on the manner in which this new strand is synthesized. The precursor of each new nucleotide in the chain is a nucleoside 5′-*triphosphate*, of which the two terminal phosphates are removed and the internal phosphate is attached covalently to deoxyribose of the growing chain (see Figure 6.3). The addition of the nucleotide to the growing chain requires the presence of a free hydroxyl group, and such a free hydroxyl group is available only at the

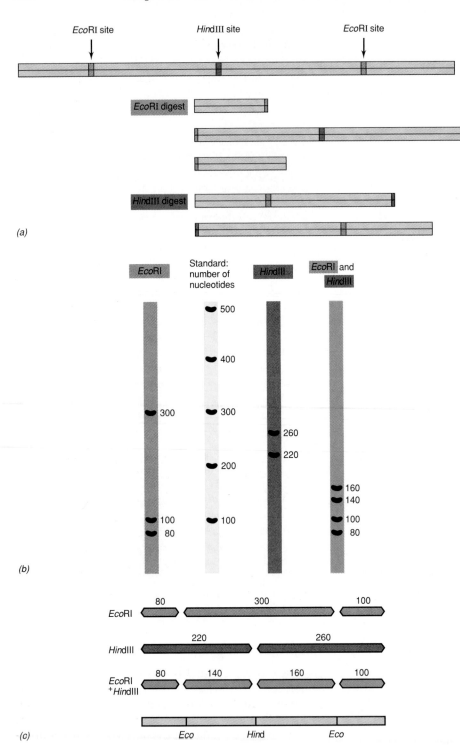

Figure 6.13 Restriction enzyme analysis of DNA. (a) A DNA molecule with two *Eco*RI and one *Hind*III restriction sites. (b) Results of electrophoresis of digests with each of the enzymes separately and a "double digest" with both enzymes. The lane with the standards allows one to determine the size of each fragment (and of the entire molecule, which in this case is 480 base pairs). Note that comparison of the results of the single and double digests indicates that the 300-base-pair fragment generated by *Eco*RI digestion must have the *Hind*III site within it. (c) By comparing the fragments generated, it is possible to deduce this map. If you ignore the original figure given in (a) and make a map just using the data in (b), you can also obtain another map that is identical to this but rotated 180°. Both maps are correct unless you have more information that defines which end of the molecule is left and which is right.

3′-end of the molecule. This chemical restriction leads to an important law that is at the basis of many facets of DNA replication: *DNA replication always proceeds from the end with the 5′-phosphate to the 3′-hydroxyl end, the 5′-phosphate of the incoming nucleotide being attached to the 3′-hydroxyl of the previously added nucleotide.*

The enzymes that catalyze addition of the nucleotides are called **DNA polymerases.** *All* DNA polymerases synthesize new DNA in the 5′ → 3′ direction. *However, no* *known DNA polymerase can begin a new chain. All these enzymes can only add a nucleotide onto a preexisting 3′-OH group.* Therefore, for a *new* chain to be started, there must be a **primer,** a site at which the DNA polymerase can attach the first nucleotide. In most cases this primer is a short stretch of *RNA*.

When the double helix is opened up at the beginning of replication, an RNA-polymerizing enzyme acts first, resulting in formation of this RNA primer. A spe-

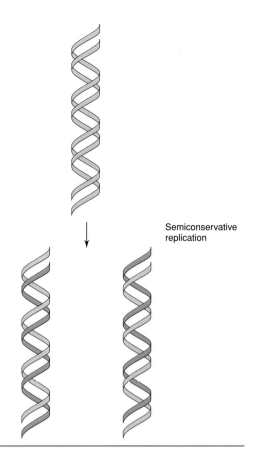

Semiconservative
replication

Figure 6.14 DNA replication is a semiconservative process. In both prokaryotes and eukaryotes the process is always semiconservative. Note that the new double helices each contain one new and one old strand.

cific RNA-polymerizing enzyme, called *primase*, participates in primer synthesis by laying down a short stretch of RNA. At the growing end of this RNA primer is a 3'-OH group to which DNA polymerase can add the first deoxyribonucleotide. Once priming has begun, continued extension of the molecule occurs as *DNA* rather than RNA. Thus, the newly synthesized molecule has a structure like that shown in Figure 6.15. The primer must eventually be removed, as we shall see.

To understand the complete replication of a double-stranded DNA molecule, it is easiest to choose an actual example and see how it is replicated. Most of the information on the mechanism of DNA replication has been obtained from the bacterium *Escherichia coli*, and the following discussion deals primarily with this organism.

Initiation of DNA synthesis

As is the case for most prokaryotes, the chromosome of *Escherichia coli* is a circular DNA molecule. Also like most Bacteria, there is a single location on this chro-

mosome where DNA synthesis is initiated, the so-called **origin of replication.** The origin of replication consists of a specific sequence of about 300 bases that is recognized by specific initiation proteins. At the origin of replication, the DNA double helix is opened up and the initiation of DNA replication occurs on the two single strands. As replication proceeds, the site of replication, called the **replication fork,** moves down the DNA.

Replication is frequently bidirectional from the origin of replication, as shown in Figure 6.16, and therefore there are *two* replication forks moving in opposite directions. In circular DNA, bidirectional replication leads to the formation of characteristic structures called **theta structures** (Figure 6.16). Most large DNA molecules, whether from prokaryotes or eukaryotes, have bidirectional replication from fixed origins. A single eukaryotic chromosome has many origins. This is not simply because the DNA is longer because, as we have seen, this is not always the case (∞ Section 3.14). It may reflect the fact that DNA polymerases from eukaryotes do not replicate as fast as the prokaryotic enzymes. DNA replication is carefully regulated, and the site where this regulation takes place is the origin.

Leading and lagging strands

There are three different DNA polymerases in *Escherichia coli*, called DNA polymerases I, II, and III. It is DNA polymerase III that is the primary enzyme of replication at the replicating forks. However, several other enzymes are also involved. The details of events at the replication fork are illustrated in Figure 6.17. At the replication fork, the DNA double helix is unwound and a small single-stranded region is formed by the action of specific proteins called *helicases*. Helicases are ATP-dependent enzymes that hydrolyze ATP as they move down the helix in advance of the replicating fork. The single-

Figure 6.15 Structure of the RNA–DNA combination that results at the initiation of DNA synthesis.

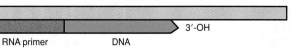

Old 3'
New PPP-5'

DNA
5'
RNA primer   DNA   3'-OH

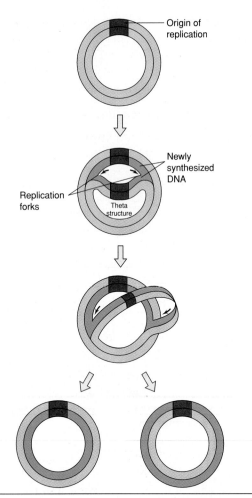

Figure 6.16 In circular DNA, bidirectional replication from an origin leads to the formation of replication intermediates resembling the Greek letter theta (θ).

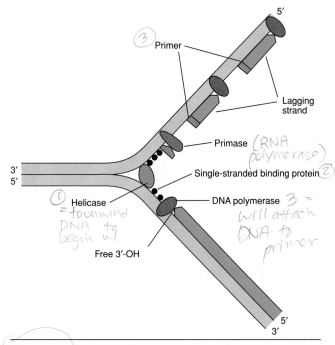

[handwritten annotations on figure: "① = to unwind DNA to begin w/", "Primase (RNA polymerase)", "② ", "③ = will attach DNA to primer"]

Figure 6.17 Events at the DNA replication fork.

stranded region generated is complexed with a special protein, the *single-strand binding protein*, which stabilizes the single-stranded DNA, preventing the formation of intrastrand hydrogen bonds.

Figure 6.17 reveals an important difference between replication of the two strands, which arises from the fact that DNA replication always proceeds from 5′-phosphate to 3′-hydroxyl (always adding a *new* nucleotide to the 3′-OH of the growing chain). On the strand growing from the 5′-phosphate to the 3′-hydroxyl, called the **leading strand,** DNA synthesis can occur *continuously* because there is always a free 3′-OH at the replication fork to which a new nucleotide can be added. But on the opposite strand, called the **lagging strand,** DNA synthesis must occur *discontinuously* (because there is no 3′-OH at the replication fork to which a new nucleotide can attach). Where is the 3′-OH on this strand? At the *opposite* end, *away* from the growing point. Therefore, on the lagging strand, a small (11-base) RNA primer must be synthesized by primase to provide free 3′-OH groups. After synthesizing the primer, primase is replaced by the enzyme DNA polymerase III. Then deoxyribonucleotides

are added until DNA polymerase III reaches the previously synthesized DNA.

At this point, DNA polymerase III stops and is released from the DNA. The next enzyme that is involved, *DNA polymerase I,* has more than one activity. It can clearly synthesize DNA. However, at the same time it is adding nucleotides on to the 3′-OH, it has an *exonuclease* activity that removes the RNA primer from in front of it (Figure 6.18). When the primer has been removed and replaced with DNA, DNA polymerase I is then released. The last phosphodiester bond is made by an enzyme called *DNA ligase.* (This enzyme can seal any nicks made in DNAs that have a 5′-phosphate and 3′-OH and along with DNA polymerase I is also involved in DNA repair.)

Each short stretch of DNA made by DNA polymerase III on the lagging strand is called an *Okazaki fragment* and is about 1000 bases long. Each of these must be primed individually. By contrast, the leading strand is primed only once, at the origin. Because of bidirectionality there are two replicating forks going in opposite directions (see Figure 6.16), which means that at the origin there are two leading strands and two lagging strands started (Figure 6.19).

As can be seen from the preceding discussion, the *Escherichia coli* enzyme called DNA polymerase II does not seem to be involved in chromosomal replication. Interestingly, however, this enzyme is closely related to the major chromosomal DNA polymerase activities of both the Archaea and the Eukarya.

While DNA synthesis is continuing at the replication fork, changes in the coiling of the DNA are occurring, modified by unwinding enzymes and topoisomerases (see Section 6.2). Unwinding is obviously an essential

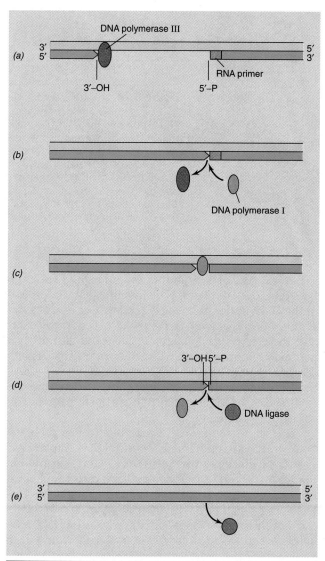

Figure 6.18 Sealing two fragments on the lagging strand. (a) DNA polymerase III is synthesizing DNA in the 5′ → 3′ direction toward the RNA primer of a previously synthesized fragment on the lagging strand. (b) On reaching the fragment, DNA polymerase I replaces III. (c) DNA polymerase I continues synthesizing DNA while removing the RNA primer from the previous fragment. (d) DNA ligase replaces DNA polymerase I after the primer has been removed. (e) DNA ligase seals the two fragments together.

feature of DNA replication, and because supercoiled DNA is under strain, it unwinds more easily than DNA that is not supercoiled. Thus, by regulating the degree of supercoiling, topoisomerases regulate the process of replication (and also transcription, as discussed later).

Fidelity of DNA replication: Proofreading

Errors in DNA replication introduce mutations. Mutation rates in living organisms are remarkably low, between 10^{-8} and 10^{-11} errors per base pair inserted. Part of the reason for this accuracy is that DNA polymerase actually gets *two* chances to incorporate the correct base at a given site. The first chance occurs when complementary bases are inserted by base-pairing rules, A with T and G with C, using the template strand as the pattern. The second chance occurs because of a second enzymatic activity, referred to as **proofreading,** associated with DNA polymerase III (Figure 6.20). In addition to inserting nucleotides in the replicating strand, DNA polymerase also contains a 3′ → 5′ *exonuclease* activity that can remove a misinserted nucleotide and replace it with the correct nucleotide. Proofreading activity is summoned if an incorrect base has been inserted because misinsertion creates unstable base pairing. This proofreading activity gives the polymerase activity a second chance to insert the correct base (Figure 6.20). (Note that the proofreading exonuclease activity is the *opposite* of the 5′ → 3′ exonuclease activity of DNA polymerase I used to remove the primer from "in front" of the polymerase.)

Exonuclease proofreading occurs in prokaryotes, eukaryotes, and viral DNA replication systems. In addition to exonucleolytic proofreading capabilities, prokaryotes and eukaryotes contain *endonucleolytic* proteins capable of removing a misinserted nucleotide long after DNA polymerase has passed the point of the error (∞ Sections 9.1 and 9.2). The combination of endonucleolytic and exonucleolytic (proofreading) activities ensures nearly error-free replication of the extremely long DNA sequences that make up genomic DNA.

We have mentioned a number of enzymes and other proteins that combine or act on DNA. A summary of some of these enzymes is given in Table 6.4.

Replicating linear genetic elements

We used a circular DNA molecule in discussing the steps in DNA replication. Circular DNA molecules are common; most prokaryotic chromosomes are circular,

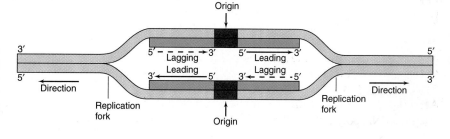

Figure 6.19 At an origin of replication that directs bidirectional replication, two replication forks must start. Therefore, two leading strands must be primed, one in each direction.

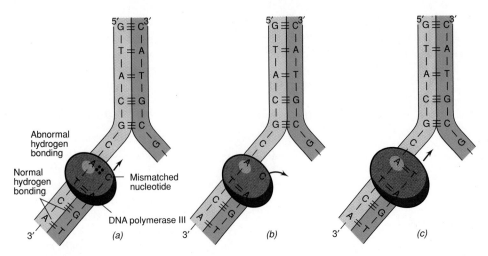

Figure 6.20 Proofreading by the $3' \rightarrow 5'$ exonuclease activity of DNA polymerase III. (a) A mismatch in base pairing at the terminal base pair causes the polymerase to pause briefly. This is a signal for the proofreading activity (b) to excise the mismatched nucleotide, after which the correct base is incorporated (c) by polymerase activity.

CONCEPT CHECK   6.5b

DNA synthesis begins at a unique location called the origin of replication. The double helix is unwound by helicase and is stabilized by single-stranded binding protein. Extension of the DNA occurs continuously on the leading strand but discontinuously on the lagging strand. Most errors in base pairing are corrected by proofreading functions associated with the action of DNA polymerase.

✔ *What is the* leading strand?

✔ *What is the* lagging strand?

as are most plasmids and some viruses. Almost all the steps in replication are identical whether the chromosome is linear or circular. However, there is one problem with replication of linear genetic elements that cir-

cular ones do not have, and that problem is at the extreme 5'-end of each strand. To understand the problem, refer back to Figure 6.15. Imagine that the left end of the DNA in this diagram is actually one end of a linear chromosome. Even if the RNA primer is very short and there is a special enzyme to remove it, no DNA polymerase can replace it with DNA since *all* DNA polymerases require a primer. Therefore, if nothing is done, the DNA molecule will become shorter each time it is replicated. Genetic elements that are linear have clearly solved this problem!

In fact, there are many solutions to this problem. Some viruses having linear chromosomes actually circularize themselves by their sticky ends, as shown in Figure 6.11. Some other viruses have direct repeats at each end of their chromosomes. A recombination process (a joining together of different DNA molecules) uses the repeats to join several partially replicated DNA molecules together into a very large molecule from which

TABLE 6.4   Enzymes affecting DNA

Enzyme	Action	Function in the cell
Restriction endonuclease	Cuts DNA at specific base sequences	Destroys foreign DNA
DNA ligase	Links DNA molecules	Completes replication process
DNA polymerase I	Attaches nucleotides to the growing DNA molecule, removes RNA primers	Fills gaps in DNA, primarily for DNA repair, and removes primers
DNA polymerase III	Attaches nucleotides to the growing DNA molecule, proofreads each inserted nucleotide	Replicates DNA
DNA gyrase (topoisomerase II)	Increases the twisting pattern of DNA, promoting supercoiling	Maintains compact structure of DNA
DNA helicase	Binds to DNA near replicating fork	Promotes DNA strand separation
DNase	Degrades DNA to nucleotides	Destroys DNA
DNA methylase	Places methyl groups on DNA bases, thus inhibiting restriction endonuclease action	Modifies cellular DNA so it is not affected by its own restriction endonuclease
Primase	Makes short RNA chains using a DNA template	Needed to make primer to be used by DNA polymerase
Topoisomerase I	Relaxes supercoiled DNA	Helps maintain the proper level of supercoiling

perfect copies are cut by endonucleases (∞ Section 8.12). Several types of viruses and many linear plasmids solve the problem of replicating linear DNA by using not an *RNA* primer but rather a *protein* primer. Although all DNA polymerases must add each nucleotide to a free —OH group, some DNA polymerases can add the first base onto an —OH group found on specific proteins that bind to the ends of these linear chromosomes (Figure 6.21). These proteins are encoded by the plasmid or virus, and they function to recognize the ends of the chromosomes. These protein primers are not removed, so these particular types of plasmids and viruses have proteins covalently attached to the 5'-ends of their DNA. This may also be the means by which some linear chromosomes of Bacteria, such as those of *Streptomyces lividans,* are replicated.

None of these methods of replicating linear DNA are used to complete the ends of eukaryotic chromosomes (telomeres). Telomeres of eukaryotic chromosomes contain repetitive DNA: a short sequence (often six base pairs) tandemly repeated from 20 to several hundred times (Figure 6.22*a*). The sequences from different eukaryotes are closely related, and one strand always has several guanines. This guanine-rich sequence can be

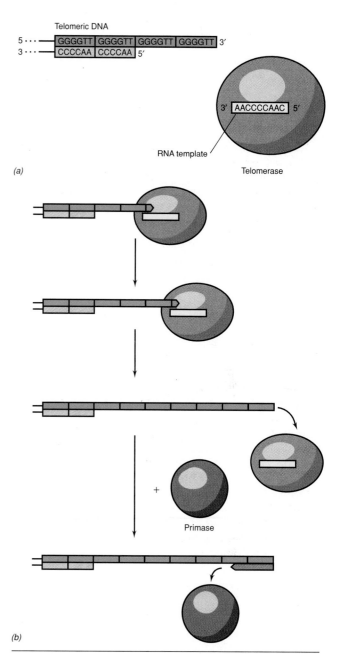

(a)

(b)

Figure 6.22 Model for the action of the telomerase at one end of a eukaryotic chromosome. (a) A diagram of the sequence of the end of the DNA in a telomere, with four of the guanine-rich repeats, and the enzyme telomerase, which contains a short RNA template. (b) Steps in elongation of the guanine-rich strand catalyzed by telomerase. After telomerase finishes, the lagging strand can be primed with an RNA primer by primase. The next step (not shown) would be completion of the lagging strand by DNA polymerase and ligase.

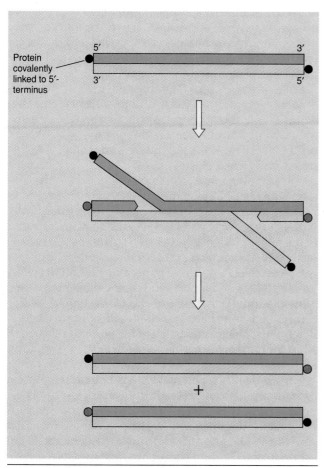

Figure 6.21 Replication of linear DNA using protein primers. The new strands of DNA are primed by proteins that stay covalently attached to the 5'-ends.

added onto the 3'-end of a DNA molecule by an interesting enzyme called **telomerase** (see Figure 6.22). Telomerases add onto the 3'-ends of linear DNA. *They do not need a DNA template because they contain a small RNA template as a cofactor.* These enzymes can work repetitively to make a long extension. Once this extension is long enough, the other strand can be primed with an

RNA primer in the normal fashion. The telomeres do not need to be a precise number of repeats long, just long enough to ensure that no genetic information becomes lost during DNA replication.

CONCEPT CHECK   **6.5c**

The ends of linear genetic elements present a problem to the replication machinery that circular genetic elements do not. Some prokaryotic linear elements solve this problem using a protein primer. Eukaryotes solve the problem using a special enzyme called telomerase to extend one strand of the DNA.

✔ *What is a* protein primer?

✔ *What is* telomerase?

6.6

Transcription

Ribonucleic acid (RNA) plays a number of important roles in the expression of genetic information in the cell. Three major types of RNA have been recognized: **messenger RNA (mRNA), transfer RNA (tRNA), and ribosomal RNA (rRNA).** These are all products of *transcription* of the information in an organism's DNA. There are three key differences between the chemistry of RNA and that of DNA: (1) RNA has the sugar *ribose* instead of *deoxyribose;* (2) RNA has the base *uracil* instead of the base *thymine;* and (3) except in certain viruses, RNA is not double-stranded. A change from *deoxyribose* to *ribose* affects some of the chemical properties of a nucleic acid, and enzymes that affect DNA in general have no effect on RNA, and vice versa. The change from *thymine* to *uracil* does not affect base pairing, as the two nucleotide bases pair with adenine equally well.

It should be emphasized that RNA acts at two levels, genetic and functional. At the *genetic* level, RNA can carry the genetic information from DNA (mRNA) (or in the case of RNA viruses, play a direct genetic function). At the *functional* level, RNA acts as a macromolecule in its own right, serving a functional and structural role in ribosomes (rRNA) or an amino acid transfer role in protein synthesis (tRNA). Some RNA even has catalytic (enzymatic) activity. In this section we focus our discussion on how RNA is made.

Overview of transcription

The transcription of genetic information from DNA to RNA is carried out through the action of the enzyme **RNA polymerase,** which catalyzes the formation of phosphodiester bonds between ribonucleotides. RNA polymerase requires the presence of DNA, which acts as a template. The precursors of RNA are the ribonucleoside triphosphates ATP, GTP, UTP, and CTP. The

chemistry of RNA synthesis is much like the chemistry of DNA synthesis (see Figure 6.3). During elongation of an RNA chain, the nucleotides are added to the 3'-OH of the ribose of the preceding nucleotide, which are polymerized with the release of the two high energy phosphate bonds. Thus, in RNA synthesis (as in DNA synthesis), the overall direction of chain growth is from the 5'-end to the 3'-end and the *template* strand is antiparallel. Unlike DNA polymerase, however, *RNA polymerase can start chains* (the initial nucleotide in an RNA chain then retains all three phosphates). The first base in the RNA is almost always a purine, either adenine or guanine.

In most cases, the DNA template for RNA polymerase is a double-stranded DNA molecule, but only *one* of the two strands is transcribed for any given gene. The enzyme RNA polymerase differs markedly among Bacteria, Archaea, and Eukarya. The following discussion deals only with RNA polymerase from Bacteria, which has the simplest structure (and about which the most is known). Later in this section we will discuss the types of RNA polymerases in different organisms.

All RNA polymerases from Bacteria studied are complex enzymes with closely related subunit structures. The enzyme from *Escherichia coli* has four different types of protein subunits, designated β, β', α, and σ (sigma), with α appearing in two copies. The subunits interact to form the active enzyme, but the sigma factor is not as tightly bound as the others and easily dissociates, leading to the formation of what is called the *core enzyme* ($\alpha_2\beta\beta'$). The core enzyme alone can catalyze the formation of RNA, and the role of sigma is in *recognition* of the appropriate site on the DNA for the initiation of RNA synthesis. The process of RNA synthesis involving RNA polymerase and sigma is illustrated in Figure 6.23.

RNA polymerase is a large protein and forms contacts with the DNA over many bases simultaneously. As noted (see Section 6.2), proteins can interact specifically with DNA because parts of the base pairs are exposed in the major groove. In order to *start* an RNA chain correctly, RNA polymerase must first recognize the proper region on the DNA. These particular sites on the DNA where RNA polymerase binds are called **promoters.** Note that only *one* strand of the DNA double helix is transcribed at a time. Which strand is transcribed is determined by the orientation of the promoter sequence. RNA polymerase travels away from the promoter region, synthesizing RNA as it moves.

Once the RNA polymerase has bound, the process of transcription can proceed. In this process, the DNA double helix at the promoter is *opened up* by the RNA polymerase (Figure 6.23). As the polymerase moves, it causes the DNA to unwind in short segments, transcription of these segments occurs, and the DNA double helix closes up again. As a result of this transient unwinding, the bases of the template strand are *exposed* and then can be copied into the RNA complement. Thus, the pro-

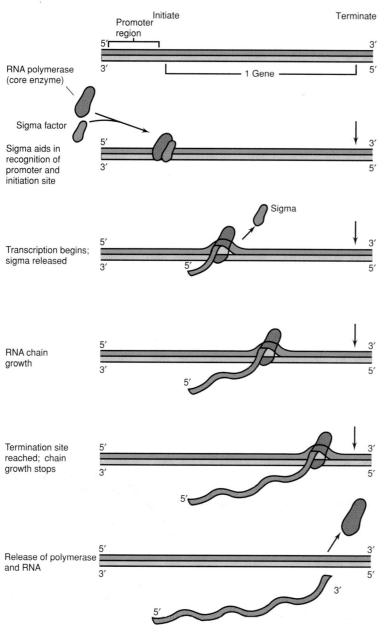

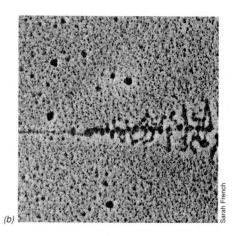

(b)

Sarah French

(a)

Figure 6.23 Transcription. (a) Steps in messenger RNA synthesis. The initiation and termination sites are specific nucleotide sequences on the DNA. RNA polymerase moves down the DNA chain, causing temporary opening of the double helix and transcription of one of the DNA strands. When a termination site is reached, chain growth stops and the mRNA and polymerase are released. (b) Electron micrograph of transcription occurring along a gene on the *Escherichia coli* chromosome. The region of active transcription represents about two kilobase pairs of DNA. Transcription proceeds from left to right.

moter *points* the RNA polymerase in one or the other direction. When a region of DNA has two nearby promoters pointing in opposite directions, then transcription from one of the promoters occurs in one direction (on one of the strands) and transcription from the other occurs in the opposite direction (on the other strand).

Once a small portion of RNA has been formed, the sigma factor dissociates; most of the elongation is therefore carried out by the core enzyme alone (Figure 6.23). Thus, sigma is involved in the formation of only the initial RNA polymerase–DNA complex. As the newly synthesized RNA dissociates from the DNA, the opened DNA closes into the original double helix. Transcription also stops at specific regions called **transcription terminators.**

Therefore, unlike replication, which involves copying an entire genome, transcription usually involves much smaller units of DNA, often a single gene. This allows the cell to transcribe different genes at very different frequencies. As we shall see in Chapter 7, regulation of the transcription of specific genes can be a very efficient mechanism of controlling gene expression.

Promoters

As we have noted, the promoter plays a key role in the initiation of RNA synthesis. Promoters are specific DNA sequences where RNA polymerase enzymes attach. The sequences of a large number of promoters from a variety of organisms have been determined. Figure 6.24 shows the sequence of a few promoters from *Escherichia coli*. It is the sigma factor, as part of the RNA polymerase, that recognizes these promoters.

A single organism can have several different sigma factors, and these can recognize different promoter sequences. All the sequences in Figure 6.24 are recognized by the same sigma factor, the major sigma factor in *E. coli*. If you examine the sequences, you will see that they are not identical. However, two sequences within the promoter region *are highly conserved* between promoters, and it is these that are recognized by sigma. Both sequences precede (are *upstream* of) the site where transcription starts. One is a region 10 bases before the start of transcription, the −10 region (called the *Pribnow box*). Notice that although each promoter is slightly different, many bases are the same. When comparing the −10 regions of all the promoters recognized by this sigma to determine which base occurs most often at each position, one arrives at the *consensus sequence* TATAAT. In our example, each promoter has from three to five matches for these bases. The second region of conserved sequence is about 35 bases from the start of transcription. The consensus sequence in the −35 region is TTGACA. Once again, most of the sequences are not *exactly* the same as the consensus sequence.

Note that in Figure 6.24 the sequence of only one strand is given. By convention among geneticists the strand shown is the one oriented with its 5'-end upstream (therefore, it is *not* the strand used as the template by RNA polymerase). Showing only the sequence of one strand is simply "shorthand" to save the space of writing the other strand. It is essential, though, to remember that promoters are double-stranded, as is the region to be transcribed.

Other sigma factors in other organisms are sometimes much more specific; very little leeway is allowed in the critical bases that are recognized. In *E. coli*, promoters that are most like the consensus are usually more effective in binding RNA polymerase. The more effective promoters are called *strong promoters* and are of considerable value in genetic engineering, as will be discussed in Chapter 10.

Transcription terminators

As important as initiation of transcription is *termination* of transcription. **Termination** of RNA synthesis occurs at specific base sequences on the DNA. A common termination sequence on the DNA is one containing an inverted repeat with a central nonrepeating segment (see Section 6.2 and Figure 6.10 for an explanation of inverted repeats). When such a DNA sequence is transcribed, the RNA can form a stem-loop structure by intrastrand base pairing (Figure 6.25). When such stem-loop structures *in the RNA* are followed by runs of uridines, they are effective transcription terminators. Other termination sites are regions where a GC-rich sequence is followed by an AT-rich sequence. Such kinds

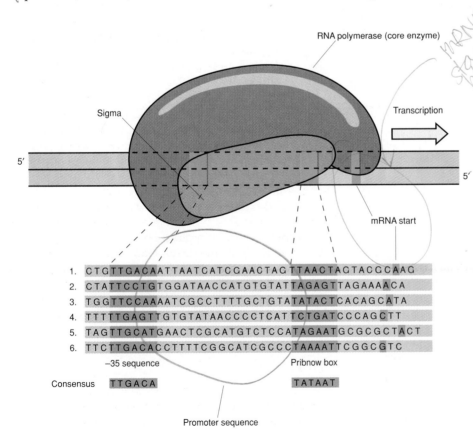

Figure 6.24 The interaction of RNA polymerase with the promoter. Shown below the diagram are six different promoter sequences identified in *Escherichia coli*. The contacts of the RNA polymerase with the −35 sequence and the Pribnow box are shown. Transcription begins at a unique base just downstream from the Pribnow box. Below the actual sequences at the −35 and Pribnow box regions are consensus sequences derived from comparing many promoters.

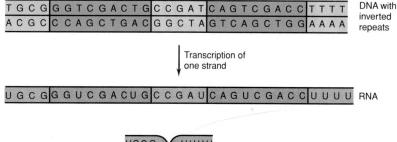

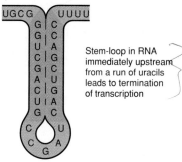

Figure 6.25 Inverted repeats in transcribed DNA lead to formation of a stem-loop structure in the RNA, which can result in termination of transcription.

of structures lead to termination without addition of any extra factors and are sometimes termed *intrinsic terminators.*

Other types of terminator sequences have been discovered that require protein factors in addition to RNA polymerase in order to function. In *Escherichia coli* one type of transcription terminator requires a protein called *Rho*. Rho does not bind to RNA polymerase or to DNA but binds tightly to RNA and moves down the chain toward the RNA polymerase–DNA complex. Once RNA polymerase has paused at a *Rho-dependent termination site,* Rho can then cause the RNA and polymerase to leave the DNA, thus terminating transcription. Other proteins involved in transcription termination are, like Rho, RNA-binding proteins. In all cases the sequences involved in termination operate at the level of RNA. However, remember that RNA is transcribed from DNA, and so transcription termination is ultimately determined by *specific nucleotide sequences on the DNA.*

CONCEPT CHECK   **6.6a**

The three major types of RNA are messenger RNA (mRNA), transfer RNA (tRNA), and ribosomal RNA (rRNA). The transcription of RNA from the DNA involves the enzyme RNA polymerase, which adds bases onto 3′-ends of growing chains. However, unlike DNA polymerase, RNA polymerase can start a chain. RNA polymerase recognizes a specific start site on the DNA called the promoter. RNA synthesis stops at a transcription terminator.

✔ *What is a* promoter?

✔ *What is a* transcription terminator?

RNA polymerases in Eukarya and Archaea

Our overview of transcription dealt with fundamental principles of this process in all organisms, but to this point our discussion of the details has concerned only Bacteria. We mentioned that the RNA polymerase from Bacteria was the simplest. What about these enzymes from other types of organisms?

Eukaryotic organisms have three different types of RNA polymerase in their nuclei, each responsible for the synthesis of a particular type of RNA. All three enzymes in eukaryotes are relatively complex, with many different subunits. **RNA polymerase I** *synthesizes most types of rRNA;* **RNA polymerase II** *synthesizes all the mRNA,* and **RNA polymerase III** *synthesizes tRNA* (and one type of rRNA). The reason for this specificity is that each type of RNA polymerase recognizes only those promoters that occur with the particular class of gene. In Bacteria the promoter for a gene encoding a protein could well be identical to a promoter for a gene encoding a tRNA. That does not happen in eukaryotes. As is the case in Bacteria, the great majority of genes in eukaryotes encode proteins, and therefore the great majority of genes are transcribed by RNA polymerase II.

These RNA polymerases need to interact specifically with promoter sequences; however, this interaction seems to be considerably more complex than the case in Bacteria, and many other proteins are involved. Figure 6.26 shows a representation of an RNA polymerase II interacting with a typical promoter. This particular promoter contains a *TATA box,* a conserved sequence resembling in some respects the Pribnow box in Bacteria (see Figure 6.24), and an initiator element near the transcription start site. One or both of these sequences are usually present in a promoter recognized by polymerase II, the eukaryotic polymerase most like

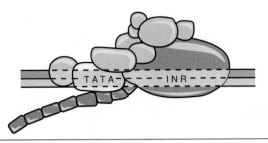

Figure 6.26 The interaction of eukaryotic RNA polymerase II with a promoter. The polymerase itself (shown in brown) is positioned at the initiator element (INR) of the promoter. A TATA box binding protein (shown in yellow) is shown bound at the TATA box. The polymerase has a repetitive sequence at one end (shown as a tail-like structure) that can be phosphorylated. The other proteins shown in blue are a few of the very large number of accessory factors required for initiation.

that from Bacteria. A variety of other sequences and other proteins are typically involved in initiating transcription from these promoters. In genes transcribed by RNA polymerase II there seems to be a connection between transcription termination and a processing event that takes place at the 3'-end of the mRNA (see Section 6.7), but not much is known about this process.

Like the Bacteria, the Archaea have a single RNA polymerase, but it is more closely related to the RNA polymerase II of eukaryotes than it is to the polymerase of Bacteria (∞ Figure 15.14). The promoter sequences from a variety of archaeans have been compared, and most contain a conserved AT-rich sequence of 6–8 base pairs about 18–27 base pairs upstream of the transcription start site (Figure 6.27). Once again this sequence is similar to that of the Pribnow box found in bacterial promoters (Figure 6.24) but is more similar in sequence, and at about the same position as the *TATA box* element found in the promoters recognized by eukaryotic RNA polymerase II. In fact, the archaeal sequence is also called a TATA box. Less is known about the transcription termination signals in the Archaea, but in some genes it seems clear that inverted repeats followed by an AT-rich sequence are involved, sequences very similar to those found in many bacterial transcription terminators. However, such sequences are not found in other archaeal genes. One other type of possible transcription terminator contains no inverted repeats, but rather the nucleotide sequence contains repeated stretches with runs of T's.

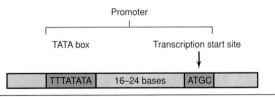

Promoter

TATA box   Transcription start site

| TTTATATA | 16–24 bases | ATGC |

Figure 6.27 The sequence elements of a typical promoter from the Archaea.

Specific inhibitors of RNA polymerase action

A number of antibiotics and synthetic chemicals have been shown to specifically inhibit RNA synthesis. For example, a group of antibiotics called *rifamycins* inhibits by attacking the β subunit of the RNA polymerase enzyme. Rifamycin has marked specificity for prokaryotes but also inhibits RNA synthesis in chloroplasts and mitochondria of some eukaryotes. Rifamycin has been an especially useful tool in studying nucleic acid synthesis in virus-infected cells. A group of antibiotics called *streptovaricins* is related to rifamycins in structure and function. *Streptolydigin* is an antibiotic that also inhibits transcription by binding to the β subunit of RNA polymerase, but it binds to a different site than rifamycin. Another chemical, *amanitin*, inhibits RNA synthesis in eukaryotes without affecting prokaryotes. Amanitin specifically inhibits RNA polymerase II, the one that synthesizes mRNA.

Actinomycin inhibits RNA synthesis by combining with DNA and blocking elongation. Actinomycin binds most strongly to DNA at guanine-cytosine base pairs, fitting into the major groove on the double strand where RNA is synthesized.

Messenger RNA and operons

The RNA carrying the information that is translated into a protein is called **messenger RNA (mRNA).** Most mRNA, in both prokaryotes and eukaryotes, is unstable and is degraded by cellular nucleases. This is in contrast to rRNA and tRNA, which are sometimes referred to as stable RNA. In prokaryotes, a single mRNA molecule often codes for more than one protein (see Figure 6.2). In *prokaryotic* (both Bacteria and Archaea) genetic elements, genes coding for related enzymes are often clustered together. In these situations the RNA polymerase proceeds down the chain and transcribes the whole series of genes into a single long mRNA molecule. An mRNA coding for such a group of genes is called a **polycistronic mRNA** (∞ Section 9.5). Subsequently, when this polycistronic mRNA participates in protein synthesis (see Section 6.9), several polypeptides coded by a single mRNA can be synthesized at one time.

We will discuss regulation of mRNA synthesis in Chapter 7, but introduce here the concept of the operon. An **operon** is a complete unit of gene expression, generally involving genes coding for several polypeptides on a polycistronic mRNA or genes coding for ribosomal RNA. In some cases, the transcription of the mRNA for an operon is under the control of a specific region of the DNA, the **operator,** which is adjacent to the coding region of the first gene in the operon. As we shall see in Chapter 7, the operator functions by being able to bind certain regulatory proteins.

For the most part, polycistronic mRNA does not exist in eukaryotes. However, this is because of differences in *translation,* not in transcription. We will discuss these differences later in this chapter after we deal with additional steps often required to convert a transcript from a eukaryotic protein-encoding gene into usable mRNA.

6.7

RNA Processing and Ribozymes

As we discussed (see Section 6.6) transcription can produce several types of RNA: messenger RNA, transfer RNA, and ribosomal RNA. In prokaryotes, a transcript of a protein-encoding gene is typically the actual mRNA and is used directly to make protein. However, transcripts of other types of genes generally need to be *processed* to reach final form. In eukaryotes, all transcripts must be processed before being used. The conversion of a *precursor* RNA to a *mature* RNA is called **RNA processing.** In prokaryotes and eukaryotes, tRNAs and rRNAs are made initially as long precursor molecules, which are then cut to make the final mature RNAs. In eukaryotes, and much less commonly in prokaryotes, mRNA is also the result of processing a pre-mRNA. As discussed in Section 6.3, the genes of eukaryotes are often split, with noncoding intervening sequences, *introns,* separating the coding regions, *exons.* The *primary transcript* from such a gene must be extensively processed to remove the noncoding regions before the translation process can be initiated. Only a few introns have been discovered in protein-encoding genes in prokaryotes and in certain bacteriophage. The processing step by which introns are removed and exons are joined is called **splicing.**

However, although introns are found in several types of genes in both prokaryotes and eukaryotes, the splicing machinery that removes introns from eukaryotic pre-mRNA is unique. The process involves a complex containing several different ribonucleoproteins (each contains both a small RNA and several proteins) called a **spliceosome.** The spliceosome is a highly com-

plex structure capable of removing introns and joining adjacent exons to form a mature mRNA. Figure 6.28 is a diagram of the two-step reaction by which an intron found in eukaryotic pre-mRNA is removed. Note that there are some conserved bases at the splice junctions and that the intron is removed as a lariat structure. These removed introns are degraded by the cell. In higher eukaryotes there are often many introns in a single gene, and so it is clearly important not only that they be removed but they be removed in the correct order. Some introns (particularly those found in tRNA genes and in genes of the mitochondria or chloroplasts) are removed by a different process involving just proteins. Several introns, including all of those that are found in Bacteria and bacteriophage, are *self-splicing* (ribozymes; see next subsection).

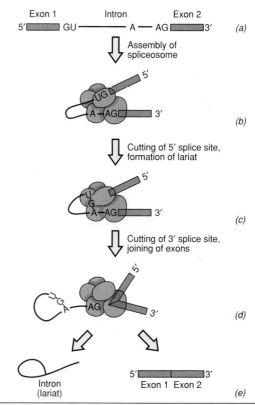

Figure 6.28 Removal of an intron from the transcript of a eukaryotic protein-encoding gene. (a) The pre-mRNA with a single intron. The sequence GU is conserved at the 5′ splice site and AG at the 3′ splice site. There is also an interior A which serves as a branch point. (b) Several small ribonucleoprotein particles (shown in brown) assemble on the RNA to form a spliceosome. Each of these particles contains distinct small RNA molecules that are involved in the splicing mechanism. (c) The 5′ splice site has been cut with the simultaneous formation of a branch point. (d) The 3′ splice site has been cut, while the two exons were joined. Note that overall two phosphodiester bonds were broken but two others were formed.

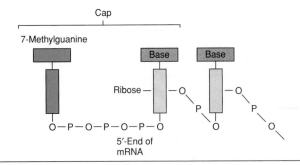

Figure 6.29 Structure of the cap added to the 5'-end of eukaryotic mRNA.

Two other unique steps occur in the processing of eukaryotic mRNA. Both steps take place in the nucleus before transport of the mature mRNA into the cytoplasm. The first step is called **capping** and actually occurs before transcription is complete. Capping consists of adding a methylated guanine nucleotide at the 5'-phosphate end, called the *cap* (Figure 6.29). Also, a number of the bases in mRNA are methylated after transcription, and the 2'-hydroxyl group of the ribose is occasionally methylated.

The remaining processing step consists of trimming the 3'-end of the pre-mRNA and adding a *poly-A tail*. This step is called **tailing** and, as we discussed (see Section 6.6), may occur in conjunction with termination. All three steps leading to the formation of eukaryotic mRNA are shown in Figure 6.30.

We thus see that the synthesis of a mature, functional RNA is a complex and dynamic process that involves considerably more than the simple transcription of a DNA template.

Ribozymes

We emphasized in Chapter 4 the role of proteins as biochemical catalysts. Many important cellular processes involve ribonucleoproteins, complexes including both RNA and protein. The role of RNA in such complexes was *assumed* to be structural (a place for the proteins to bind) or involved in base pairing with other nucleic acids. As we have seen, there is a short RNA molecule in the enzyme *telomerase* that functions as a template. However, it has now been shown that certain types of RNA can act as *enzymes* as well. Catalytic RNAs, referred to as **ribozymes,** are involved in a number of important cellular reactions. RNA enzymes work like protein enzymes in that an "active site" exists that binds the substrate and catalyzes formation of a product. Ribozymes have been discovered in both prokaryotes and eukaryotes, and in organelles, and others have now been synthesized in laboratories. Studies show that some very short RNAs, as few as 19 bases, can function as ribozymes.

Most ribozymes are **self-splicing introns.** They are *RNA-splicing enzymes* that remove themselves from an RNA molecule while joining adjacent exons together. In one well-studied case of splicing in a ribosomal RNA in *Tetrahymena* (a protozoan), a 413-nucleotide *intron* acts as a ribozyme and splices itself out of a longer precursor rRNA, joining two adjacent exons to form the final rRNA (Figure 6.31). The intron ribozyme acts as a sequence-specific endoribonuclease and, once removed from the precursor RNA, circularizes with the further removal of a short oligonucleotide fragment (Figure 6.31). This particular type of self-splicing intron (a *group I* self-splicing intron) is widespread in nature and is the only type known in Bacteria and bacteriophage.

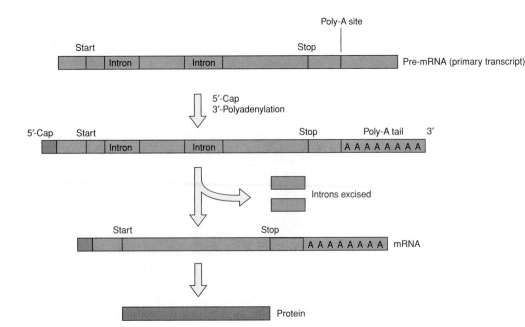

Figure 6.30 An overview of the processing of the pre-mRNA into mature mRNA in eukaryotes. The processing steps including adding a cap at the 5'-end, removing the introns, and clipping of the 3'-end of the transcript while adding a poly-A tail. All these steps are carried out in the nucleus. The location of the start and stop codons to be used during translation are also indicated.

Absolute proof that these ribozymal transformations occur in the absence of specific protein has come from experiments in which the gene for the entire precursor rRNA from *Tetrahymena* has been transferred to *Escherichia coli* where the segment can be transcribed. This transcribed segment carries out the splicing reaction in the complete absence of *Tetrahymena* proteins. There is evidence that proteins play some role in the splicing reaction of members of *group II self-splicing* introns. In this type of intron the splicing reaction itself is quite like that used by the spliceosome (see Figure 6.28), but no accessory RNAs are required.

Self-splicing introns differ from most protein enzymes in that they normally can act only once. However, there is another ribozyme, RNase P, that can act repeatedly on many different substrate molecules because it does not digest itself in the reaction. RNase P is a ribonucleoprotein, but the small RNA (377 nucleotides in *E. coli*) is the catalytic component, not the protein. As is the case for proteins with enzymatic activity, all ribozymes must be folded into the proper structure for activity. In some cases, this structure might be supplied by the secondary structure of the RNA itself. In others, like RNase P, specific proteins may help keep the RNA in the active conformation. RNase P functions in the cell to modify primary transcripts coding for transfer RNAs (see next section).

The discovery of ribozymes has caused a reevaluation of other cellular processes that involve RNA. For instance, it is now clear that ribosomal RNA plays an active role in protein synthesis, apparently catalyzing the formation of the peptide bonds that link amino acids together in a protein (see Section 6.9). But clearly most enzymes are protein. Why do ribozymes exist? It has been proposed that they are the vestigial remains of a simpler form of life, "RNA life," which may have predated the era of proteins as the cell's major catalysts. We discuss this concept in more detail in Chapter 15.

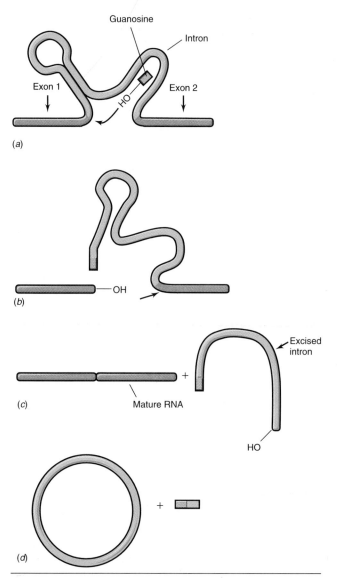

Figure 6.31 Self-splicing ribozymal intron of the protozoan *Tetrahymena*. There is considerable secondary structure in such molecules, which is critical for the splicing reaction. (a) A ribosomal RNA precursor contains a 413-nucleotide intron. (b) Following the addition of the nucleoside guanosine, the intron splices itself out and joins the two exons. (c) The intron is spliced out. (d) The intron circularizes with the loss of a 15-nucleotide fragment.

CONCEPT CHECK   **6.7**

RNA molecules are often modified after transcription, an operation called RNA processing. Three distinct processing steps can occur to eukaryotic pre-mRNAs: splicing, capping, and tailing. Introns found in some other transcripts are self-splicing, and the RNA itself catalyzes the reaction. RNA molecules with catalytic activity are called ribozymes and play an important role in the cell.

✔ *What is* splicing?

✔ *What is a* ribozyme?

6.8

Transfer RNA

The translation of the RNA message into the amino acid sequence of protein is brought about through the action of transfer RNA (tRNA) (Figure 6.32). Transfer RNA is an adaptor molecule having two specificities, one for a codon on mRNA, the other for an amino acid. The transfer RNA and its specific amino acid are brought together by means of specific enzymes that ensure that a particular tRNA receives its correct amino acid. These

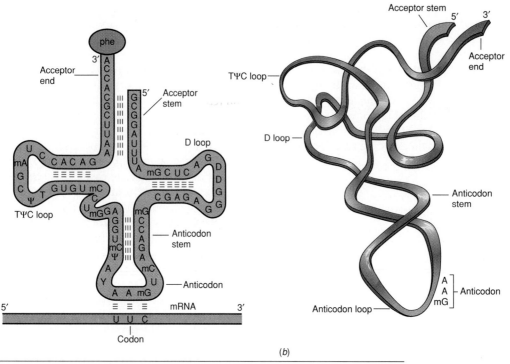

Figure 6.32 Structure of a transfer RNA, yeast phenylalanine tRNA. (a) The conventional cloverleaf structure. The amino acid is attached to the ribose of the terminal A at the acceptor end. A, adenine; C, cytosine; U, uracil; G, guanine; ψ, pseudouracil; D, dihydrouracil; m, methyl; Y, a modified purine. (b) In actuality the molecule folds so that the D loop and TψC loops are close together and associate by hydrophobic interactions.

enzymes, called **amino acid activating enzymes** or **aminoacyl-tRNA synthetases,** have the important function of recognizing *both* the amino acid *and* the specific tRNA for that amino acid.

Structure of tRNA

The detailed structure of tRNA is now well understood. There are about 60 different specific tRNAs in bacterial cells and 100–110 in mammalian cells. Transfer RNA molecules are short, single-stranded molecules with lengths (among different tRNAs) of 73–93 nucleotides. When compared, it has been found that certain bases and secondary structures are constant for all tRNAs, and there are other parts that are variable. Transfer RNA molecules also contain some purine and pyrimidine bases differing slightly from the normal bases found in RNA in that they are chemically modified, often methylated. Some of these unusual bases are pseudouridine, inosine, dihydrouridine, ribothymidine, methyl guanosine, dimethyl guanosine, and methyl inosine. These modifications are added to the bases after transcription. Base modifications as well as other types of processing (see Section 6.7) are necessary to make a functional tRNA from the transcript of a tRNA-encoding gene. Although the molecular structure of tRNA is single-stranded, there are extensive double-stranded regions within the molecule as a result of internal base pairing when the molecule folds back on itself.

The structure of tRNA is generally drawn in cloverleaf fashion, as in Figure 6.32a. Some regions of secondary structures are given names having to do either with the bases most often found there (the TψC loop and the D loop) or with specific functions (anticodon loop and acceptor end). The three-dimensional structure of a tRNA is more clearly shown in Figure 6.32b. Note that bases that appear widely separated in the cloverleaf model are actually close together when viewed in three dimensions. This means that some of the bases in the "loops" are actually paired.

One of the variable parts of the tRNA molecule contains the **anticodon,** the site recognizing the codon on the mRNA. The anticodon is found in the *anticodon loop,* shown in Figure 6.32. There are just *three* nucleotides in the anticodon loop that are specifically involved in the recognition process and that base-pair with the codon (see Section 6.10). Other portions of the tRNA interact with the ribosome (both rRNA and protein), other protein factors, and the activating enzyme. At the 3'-end of the chain, where a sequence of nucleotides projects from the rest of the molecule, the sequence of these three nucleotides is always the same, cytosine-cytosine-adenine (CCA), and it is to the ribose sugar of the terminal A that the amino acid is covalently attached via an ester linkage. From this acceptor portion of the tRNA, the amino acid is transferred to the growing polypeptide chain on the ribosome by a mechanism that will be described in the next section.

Recognition, activation, and charging of tRNA

Recognition of the correct tRNA by an aminoacyl-tRNA synthetase involves specific contacts between key regions of the nucleic acid and particular amino acids of its respective synthetase (Figure 6.33). As might be expected because of the unique sequence in this region, the *anticodon* of the tRNA is important in recognition by the synthetase. However, other contact sites between the tRNA and the synthetase are also important. Studies of tRNA binding to aminoacyl-tRNA synthetases in which specific bases in the tRNA have been changed by genetic mutation have shown that only a small number of key nucleotides in a tRNA besides the anticodon region are involved in recognition; the key recognition nucleotides are often part of the acceptor stem of the tRNA molecule (see Figure 6.32). In a few cases, recognition of a tRNA by its cognate synthetase is totally independent of the anticodon region. It should be emphasized at this point that the fidelity of this recognition process is crucial, for if the wrong amino acid is attached to the tRNA, it may be inserted in the improper place in the polypeptide, leading to the synthesis of a faulty protein.

The specific chemical reaction between amino acid and tRNA catalyzed by the aminoacyl-tRNA synthetase first involves *activation* of the amino acid by reaction with ATP:

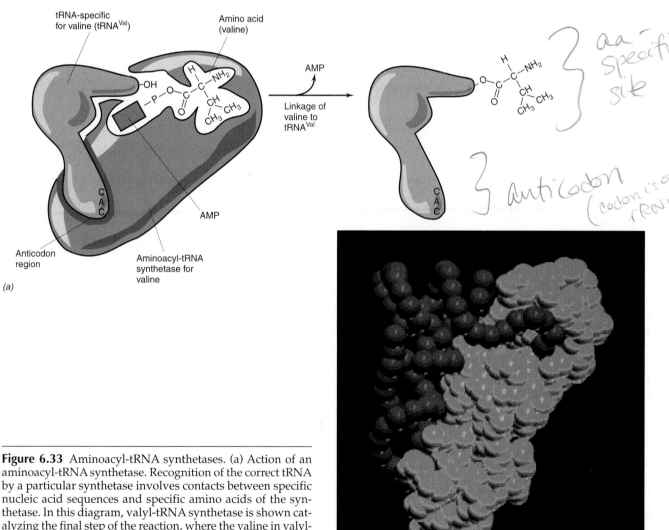

(a)

(b)

Figure 6.33 Aminoacyl-tRNA synthetases. (a) Action of an aminoacyl-tRNA synthetase. Recognition of the correct tRNA by a particular synthetase involves contacts between specific nucleic acid sequences and specific amino acids of the synthetase. In this diagram, valyl-tRNA synthetase is shown catalyzing the final step of the reaction, where the valine in valyl-AMP is transferred to tRNA. (b) A computer model showing the interaction of glutaminyl-tRNA synthetase (blue) with its tRNA (red). Reprinted with permission from *Science 252:* 1682–1689 (1991) © AAAS.

Dino Moras

$$\text{Amino acid} + \text{ATP} \rightleftharpoons \text{aminoacyl-AMP} + \text{P–P}$$

The aminoacyl-AMP intermediate formed normally remains bound to the enzyme until collision with the appropriate tRNA molecule, and the activated amino acid is then transferred to the tRNA to form a *charged* tRNA:

$$\text{Aminoacyl-AMP} + \text{tRNA} \rightleftharpoons \text{aminoacyl-tRNA} + \text{AMP}$$

The pyrophosphate (P–P) formed in the first reaction is split by a pyrophosphatase, forming two molecules of inorganic phosphate. Since ATP is used and AMP is formed, a total of *two* high energy phosphate bonds are required for the activation of an amino acid and charging a tRNA. Once activation and charging have occurred, the aminoacyl-tRNA (AA-tRNA) leaves the synthetase and is brought to the ribosome by a protein factor. The mechanism of protein synthesis is discussed in the next section.

CONCEPT CHECK 6.8

One or more transfer RNAs exist for each amino acid found in protein. Enzymes called aminoacyl-tRNA synthetases function to attach an amino acid to a tRNA. Once the correct amino acid is attached to its tRNA, further specificity resides only in the codon-anticodon interaction.

✔ *What is the function of a tRNA?*

✔ *What is the function of an aminoacyl-tRNA synthetase?*

6.9

Translation: The Process of Protein Synthesis

It is the amino acid *sequence* that determines the structure (and ultimately the function) of the final active protein. The key objective of protein synthesis is thus placement of the proper amino acid at the proper place in the polypeptide chain. This is the role of the protein-synthesizing machinery of the cell.

Steps in protein synthesis

Ribosomes are the site of protein synthesis. Each ribosome is constructed of two subunits. In prokaryotes, the ribosome subunits are of 30S (Svedberg units) and 50S, yielding intact 70S ribosomes.* Each subunit is itself a ribonucleoprotein complex made up of specific

---

*The numbers 30S, 50S, and 70S refer to the sedimentation coefficients of ribosome subunits or intact ribosomes when subjected to centrifugal force in an ultracentrifuge.

TABLE 6.5	**Ribosome structure**[a]	
Property	**Prokaryote**	**Eukaryote**
Overall size	70S	80S
Small subunit	30S	40S
Number of proteins	~21	~30
RNA size (number of bases)	16S (1500)	18S (2300)
Large subunit	50S	60S
Number of proteins	~34	~50
RNA size (number of bases)	23S (2900)	28S (4200)
	5S (120)	5.8S (160)
		5S (120)

[a]Ribosomes of mitochondria and chloroplasts of eukaryotes are similar to prokaryotic ribosomes (∞ Section 9.5).

ribosomal RNAs and ribosomal proteins. The 30S subunit contains 16S rRNA and about 21 proteins, while the 50S subunit contains 5S and 23S rRNA and about 34 proteins (Table 6.5 and Figure 6.34a). In *Escherichia coli*, there are 53 different ribosomal proteins, most present at one copy per ribosome. The actual synthesis of a protein involves a complex cycle in which the various ribosomal components play specific roles.

Although a continuous process, protein synthesis can be thought of as occurring in a number of discrete steps: **initiation, elongation, termination-release,** and **polypeptide folding.** The first two steps are outlined in Figure 6.34b. In addition to mRNA, tRNA, and ribosomes, the process involves a number of proteins designated initiation, elongation, and termination factors; guanosine triphosphate provides energy for the process.

Initiation of protein synthesis

In prokaryotes initiation always begins with a free 30S ribosome subunit, and an **initiation complex** forms consisting of a 30S ribosome subunit, mRNA, formylmethionine tRNA, and initiation factors. Guanosine triphosphate is required for this step. To this initiation complex a 50S ribosome subunit is added to make the active 70S ribosome. At the end of the translational process, the released ribosome separates again into 30S and 50S subunits. Just preceding the initiation codon on the mRNA is a sequence of from three to nine nucleotides (the so-called **Shine–Dalgarno sequence**) that is involved in the binding of the mRNA to the ribosome. This ribosome binding site at the 5'-end of the mRNA is complementary to the 3'-end of the 16S RNA of the ribosome, and it is thought that base pairing ensures effective formation of the ribosome–mRNA complex.

The presence of the Shine–Dalgarno site on the mRNA and its specific interaction with 16S rRNA allow prokaryotic ribosomes to use polycistronic mRNA

(handwritten annotations:) in 50S subunit

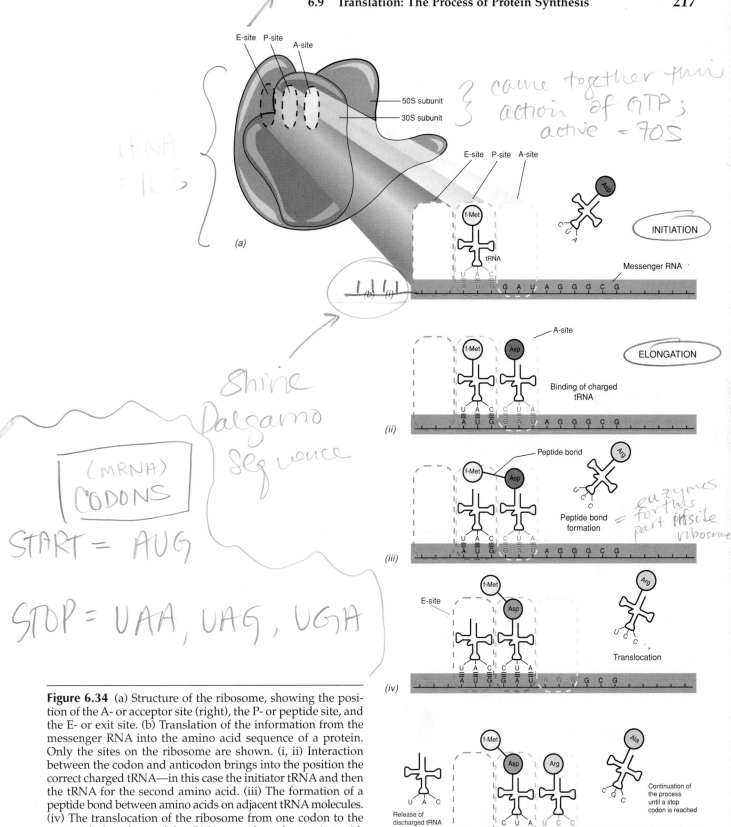

(handwritten annotations around figure:)
came together thru action of GTP; active = 70S

tRNA = ICS

Shine Dalgarno Sequence

(MRNA) CODONS

START = AUG

STOP = UAA, UAG, UGA

enzymes for this part inside ribosome

Figure 6.34 (a) Structure of the ribosome, showing the position of the A- or acceptor site (right), the P- or peptide site, and the E- or exit site. (b) Translation of the information from the messenger RNA into the amino acid sequence of a protein. Only the sites on the ribosome are shown. (i, ii) Interaction between the codon and anticodon brings into the position the correct charged tRNA—in this case the initiator tRNA and then the tRNA for the second amino acid. (iii) The formation of a peptide bond between amino acids on adjacent tRNA molecules. (iv) The translocation of the ribosome from one codon to the next with the release of the tRNA, now free of an amino acid, from the E-site. (v) The next charged tRNA binds to the A-site.

because the ribosome can find each initiation site within a message (see Section 6.6). Eukaryotic ribosomes typically recognize an mRNA by its 5′-cap and initiate only at the first possible initiation codon. Therefore, they normally cannot translate polycistronic mRNA.

Initiation always begins with a special initiator aminoacyl-tRNA binding to the **start codon,** AUG. In Bacteria this is **formylmethionine** tRNA. Subsequently, the formyl group at the N-terminal end of the polypeptide is removed; the terminal amino acid of the completed protein is hence methionine. In Eukarya and Archaea, initiation begins with methionine instead of formylmethionine. (Although all proteins are *initiated* with a methionine, this amino acid is often removed by a specific protease after translation.)

Elongation and termination

The mRNA is threaded through the ribosome primarily bound to the 30S subunit. The ribosome contains other sites where the tRNAs interact. Two of these sites are located primarily on the 50S subunit, and they are termed the P-site and the A-site (see Figure 6.34b). The A-site, the **acceptor** site, is the site where the new AA-tRNA first attaches. The P-site, the **peptide** site, is the site where the growing peptide is held by a tRNA. During peptide bond formation, the peptide moves to the tRNA at the A-site as a new peptide bond is formed. Several soluble (nonribosomal) elongation factors are required for **elongation,** as well as additional molecules of GTP (to simplify Figure 6.34b, the elongation factors are omitted and only a portion of the ribosome is shown). The tRNA that holds the peptide must now be moved (translocated) from the A-site to the P-site, thus opening up the A-site for another AA-tRNA.

Translocation requires a specific elongation factor and one molecule of GTP per each tRNA translocated. At each translocation step the message is advanced three nucleotides, exposing a new codon at the ribosome A-site. It had been thought that translocation caused the empty tRNA to be released from the ribosome. However, it now appears that translocation pushes this empty tRNA to a third site, called the E-site. It is from this **exit** site that the tRNA is actually released from the ribosome.

The precision of the translocation step is critical to the accuracy of protein synthesis. The ribosome must move exactly three bases (one codon) at each step.

When several ribosomes are simultaneously translating a single message, the complex is called a **polysome** (Figure 6.35). Polysomes increase the speed and efficiency of mRNA translation, and because each ribosome acts independently of the others, each ribosome in a polysome complex can make a complete polypeptide (Figure 6.35). Note in Figure 6.35 how ribosomes closest to the 5′-end (the beginning) of the mRNA molecule have short polypeptides attached to them because only a few codons have been read, while ribosomes closest to the 3′-end of the message have nearly finished polypeptides.

It was long thought that all proteins folded spontaneously into their active form while they were being synthesized (Figure 6.35). However, we now know this is not the case. Many proteins require the assistance of other proteins called **molecular chaperones** for proper folding or for assembly into larger complexes. The chaperones themselves do not become part of the assembled proteins. These proteins seem to be both extremely widespread, and their sequences highly conserved. The activity of a type of molecular chaperone called a *chaperonin* is shown in Figure 6.36. The unfolded or improperly folded protein enters the molecular chaperone where it is folded and then released. Energy for the folding comes from ATP. Other cellular chaperones are involved in carrying the unfolded protein to the chaperonin. In addition to folding newly synthesized proteins, chaperones also can refold proteins that have partially denatured in the cell. Such protein denaturation can occur because the organism has temporarily experienced high temperatures in its environment.

The **termination** of protein synthesis occurs when a codon is reached that does not specify an AA-tRNA. There are three codons of this type, and they are called **stop,** or **nonsense, codons** (see Section 6.10); they serve as the stopping points for protein synthesis. No tRNA binds to a stop codon, but instead proteins called *release factors* read the chain-terminating signal and serve to cleave the attached polypeptide from the terminal tRNA. Following this, the ribosome dissociates, and the subunits are then free to form new initiation complexes.

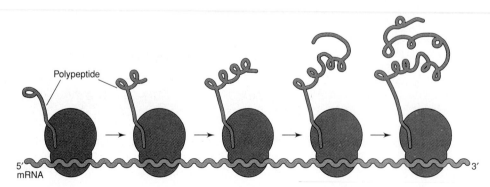

Polypeptide

5′
mRNA
3′

Figure 6.35 Translation by several ribosomes on a single messenger RNA (polysome). Note how the ribosomes nearest the 5′-end of the message are at an earlier stage in the translation process.

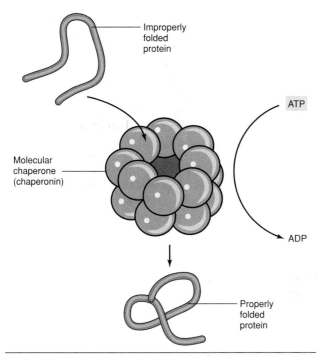

Figure 6.36 The action of a molecular chaperone. An improperly folded protein is taken into the barrel-like structure of the chaperonin. ATP is used to supply the energy required to fold the protein properly.

Role of ribosomal RNA in protein synthesis

Ribosomes are composed of a series of proteins and ribosomal RNAs (see Table 6.5). Two decades ago it was assumed that ribosomal proteins were probably the functional components of the ribosome, whereas the role of the ribosomal RNAs was largely structural, that is, serving as a support for ribosomal proteins. However, it is now clear that rRNA plays a critical *functional* role in all stages of protein synthesis, from initiation to termination. The role of the many proteins present in the ribosome, although less clear, may be as facilitators of RNA function by stabilizing or positioning the key functional sequences in the various ribosomal RNAs.

In prokaryotes, it is clear that 16S rRNA is involved in initiation through base pairing between the ribosome binding sequence (the Shine–Dalgarno sequence) upstream of the start codon on the mRNA and a complementary sequence in 16S rRNA. There is strong evidence that mRNA and rRNA interactions also occur during elongation.

The 23S rRNA seems to play a role in translocation, and the elongation factors are known to interact with 23S rRNA. The 16S rRNA is also involved in termination, possibly interacting with the mRNA or through interactions with the release factors.

Strong evidence also exists for a role for rRNA in ribosome subunit association, as well as for tRNA positioning in the decoding (A- or P-, see Figure 6.34b) sites

on the ribosome and even in catalyzing peptide bond formation. Charged tRNAs that enter the ribosome recognize the correct codon by codon:anticodon base pairing, but they are also physically attached to the ribosome by interactions of the anticodon stem-loop of the tRNA with specific locations within 16S rRNA. The acceptor end of the tRNA interacts with the 23S rRNA.

Finally, the peptidyl transferase reaction (the actual formation of peptide bonds that occurs on the 50S subunit of the ribosome) is associated with 23S rRNA. It seems likely that the reaction itself is catalyzed by ribozymal activity (see Section 6.7).

Ribosomal RNA thus plays a major role in translation. Ribosomal function is clearly dependent on the major RNA species present.

Secretory proteins

Many proteins are used *outside* the cell and must somehow get from the site of synthesis on ribosomes through the cytoplasmic membrane. In prokaryotes, periplasmic enzymes and extracellular enzymes are secretory proteins. In eukaryotes there are a large number of membrane-enclosed organelles, for example, mitochondria, into which proteins must be transported, as well as proteins that must be secreted outside the cell.

How is it possible for a cell to selectively transfer some proteins across a membrane while leaving most proteins in place in the cytoplasm? Proteins that must be transported through membranes are synthesized with an extra N-terminal peptide sequence, about 15–20 amino acids in length, which is called the **signal sequence.** In this signal sequence, hydrophobic amino acids predominate and permit the enzyme to be threaded through the hydrophobic lipid membrane. In many cases, the ribosomes that synthesize secretory proteins are bound directly to the cytoplasmic membrane, so the protein is formed and passes through the membrane simultaneously. Once the protein has been secreted, the signal sequence is removed by a peptidase enzyme, an example of the process of **posttranslational modification.**

The study of protein secretion has important practical implications for genetic engineering (∞ Chapter 10). If bacteria are genetically engineered to serve as agents for the production of foreign proteins, it is desirable to manipulate the signal sequence in order to arrange for the desired protein to be excreted so it can be readily isolated and purified.

Effect of antibiotics on protein synthesis

A large number of antibiotics inhibit protein synthesis by interacting with the ribosome. These interactions are quite specific, and many have been shown to involve rRNA. Several of these antibiotics are medically useful, and several are also effective research tools because they are specific for different steps in protein synthesis. For instance, *streptomycin* inhibits initiation, whereas *puromycin, chlo-*

ramphenicol, cycloheximide, and tetracycline inhibit elongation. Even when two antibiotics inhibit the same overall step in protein synthesis, the mechanisms of the inhibition can be quite different. Puromycin binds to the A-site on the ribosome, and the growing peptide is transferred to it instead of to an AA-tRNA. The puromycin–peptide is then released from the ribosome, halting elongation. Chloramphenicol inhibits elongation by blocking formation of the peptide bond.

Many antibiotics specifically inhibit ribosomes of organisms from only one or two of the phylogenetic domains. Of the antibiotics just listed, chloramphenicol and streptomycin are specific for the ribosomes of Bacteria and cycloheximide for ribosomes of Eukarya. Since mitochondria and chloroplasts have ribosomes of the prokaryotic type, it is of interest that antibiotics inhibiting protein synthesis in Bacteria also generally inhibit protein synthesis in mitochondria and chloroplasts. This is one piece of evidence supporting the hypothesis that mitochondria and chloroplasts were originally derived by intracellular infection of a eukaryotic cell by prokaryotes (∞ Sections 3.15 and 15.4). The fact that an antibiotic inhibits eukaryotic ribosomes as well as those from Bacteria does not necessarily mean it cannot be used in medicine. Although tetracycline also inhibits eukaryotic ribosomes, it apparently does not do so at the concentrations that result from administering it during antibacterial therapy.

One interesting inhibitor of protein synthesis is diphtheria toxin, an agent involved in the pathogenesis of the disease diphtheria. As discussed in Section 19.8, diphtheria toxin inactivates an elongation factor and hence is a powerful inhibitor of protein synthesis in eukaryotes and Archaea. The overall pattern of antibiotic sensitivity of the Archaea is unlike that of either the Bacteria or the Eukarya (∞ Section 15.8).

CONCEPT CHECK   6.9

The ribosome plays a key role in the translation process, bringing together mRNA and amino acid–charged tRNAs. There are three sites on the ribosome: the acceptor site, where the charged tRNA first combines; the peptide site, where the growing polypeptide chain is held; and an exit site. During each step of amino acid addition, the message advances three nucleotides (one codon) and the tRNA moves from the acceptor to the peptide site. Termination of protein synthesis occurs when a stop codon, which does not code for any amino acid, is reached. A major target of antibiotic action is the ribosome.

✔ What are the components of a ribosome?

✔ What functional roles does rRNA play in protein synthesis?

6.10
The Genetic Code

As noted, a triplet of three bases encodes a specific amino acid. It is conventional to present the genetic code as mRNA rather than as DNA because it is with mRNA that the translation process occurs. The 64 possible codons of mRNA are presented in Table 6.6. Note that in addition to the codons specifying the various amino acids, there are also special codons for starting (AUG) and for stopping (UAA, UAG, UGA) translation.

Perhaps the most interesting feature of the genetic code is that most amino acids are encoded by several different but related base triplets. This means that in most cases there is no one-to-one correspondence between the amino acid and the codon—knowing the amino acid at a given location does not mean that the codon at that location is automatically known.[†] The property of a code in which there is no one-to-one correspondence between word and code is called **degeneracy** (a term derived from the technical field of cryptography).

What is the significance of degeneracy? Degeneracy means that either (1) a single tRNA molecule can pair with more than one codon or (2) there is more than one tRNA for each amino acid. Actually, both situations are true. For some amino acids there are more than one tRNA molecule (there are, for instance, six different tRNA molecules in *Escherichia coli* that carry the amino acid leucine). However, there is not a tRNA corresponding to every possible anticodon. In some cases, tRNA molecules form *standard* base pairs at only the first two positions of the codon, tolerating unusual base pairs at the third position. This apparent mismatch phenomenon, called **wobble**, is illustrated in Figure 6.37. The pairing between G and U is allowed at the wobble position.

Start and stop codons

As seen in Table 6.6, a few triplets do not correspond to any amino acid. These triplets (UAA, UAG, UGA) are the **nonsense,** or **stop, codons,** and they signal the termination of translation of the gene coding for a specific protein (see Section 6.9).

What is the mechanism by which the proper *starting point* for translation is found? The message is read by reading the **start codon, AUG,** which at the beginning of the message, codes for the amino acid N-formylmethionine (or methionine in Eukarya and Archaea). The importance of having a well-defined starting point is readily understood if we consider that with a triplet code it is absolutely essential that translation begin at the correct location because if it does not, the whole

---

[†]The reverse is true, however. Knowing the DNA codon, one can specify the amino acid in the protein (assuming the proper reading frame is known). This permits the determination of amino acid sequences from DNA base sequences.

| TABLE 6.6 | The genetic code as expressed by triplet base sequences of mRNA[a] | | | | | | |

Codon	Amino acid	Codon	Amino acid	Codon	Amino acid	Codon	Amino acid
UUU	Phenylalanine	UCU	Serine	UAU	Tyrosine	UGU	Cysteine
UUC	Phenylalanine	UCC	Serine	UAC	Tyrosine	UGC	Cysteine
UUA	Leucine	UCA	Serine	UAA	None (stop signal)	UGA	None (stop signal)
UUG	Leucine	UCG	Serine	UAG	None (stop signal)	UGG	Tryptophan
CUU	Leucine	CCU	Proline	CAU	Histidine	CGU	Arginine
CUC	Leucine	CCC	Proline	CAC	Histidine	CGC	Arginine
CUA	Leucine	CCA	Proline	CAA	Glutamine	CGA	Arginine
CUG	Leucine	CCG	Proline	CAG	Glutamine	CGG	Arginine
AUU	Isoleucine	ACU	Threonine	AAU	Asparagine	AGU	Serine
AUC	Isoleucine	ACC	Threonine	AAC	Asparagine	AGC	Serine
AUA	Isoleucine	ACA	Threonine	AAA	Lysine	AGA	Arginine
AUG (start)[b]	Methionine	ACG	Threonine	AAG	Lysine	AGG	Arginine
GUU	Valine	GCU	Alanine	GAU	Aspartic acid	GGU	Glycine
GUC	Valine	GCC	Alanine	GAC	Aspartic acid	GGC	Glycine
GUA	Valine	GCA	Alanine	GAA	Glutamic acid	GGA	Glycine
GUG	Valine	GCG	Alanine	GAG	Glutamic acid	GGG	Glycine

[a]The boxes of codons are colored according to the scheme: ☐ ionizable: acidic, ■ ionizable: basic, ☐ nonionizable polar, and ☐ nonpolar (∞ Figure 2.12). The nucleotide on the left is at the 5′-end of the triplet.
[b]AUG encodes *N*-formylmethionine at the beginning of mRNAs of Bacteria.

reading frame will be shifted and an entirely different protein (or no protein at all) will be formed (see later).

As we discussed in Section 6.9, ribosomes from Bacteria recognize a specific AUG as a start codon with the aid of an upstream sequence on the mRNA called the Shine–Dalgarno sequence. This extra help at initiation explains why a few messages from Bacteria actually use other codons, such as GUG, for a start codon. However, even these unusual start codons specify *N*-formylmethionine.

Open reading frames

We have shown that a protein is encoded by a specific segment of DNA, but that for the protein to be synthesized, the DNA segment must first be transcribed into mRNA. The transcription of DNA into mRNA

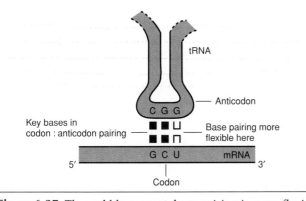

Figure 6.37 The wobble concept: base pairing is more flexible for the third base of the codon.

requires the presence of a promoter just upstream of the transcription start site. How does an experimenter know that the transcribed mRNA encodes a protein? One approach is to examine the base sequence of the mRNA to see whether there is a *start codon*, such as AUG, present near the beginning of the sequence. If such a start codon is followed by a long base sequence before a *stop codon* in the same reading frame is present, then it is likely that this mRNA codes for a protein. Such a base sequence is called an **open reading frame (ORF)** because the start codon sets the proper reading frame for translation and a stop codon in the same frame ends translation. A computer can be programmed to scan long base sequences in DNA databases to look for open reading frames. The search for ORFs is very useful in genetic engineering (∞ Chapter 10) when one has isolated and sequenced an unknown piece of DNA and is not certain whether or not it encodes protein. The computer search for ORFs permits the researcher to locate genes that were previously unsuspected.

In a sense, the term *open reading frame* is equivalent to the term *gene*. If a piece of DNA lacks a detectable open reading frame, it is unlikely that this piece of DNA codes for protein. In eukaryotic cells especially, a large amount of noncoding DNA is present.

Other genetic codes

When the genetic code was cracked during the 1960s, all the prokaryotes and eukaryotes examined were found to use the same code. When the mRNA for mammalian hemoglobin was given to the *Escherichia coli* protein-synthesizing machinery (such as ribosomes and tRNA), mammalian hemoglobin was synthesized. Therefore, the genetic code appeared to be a **universal code** in that the exact same code was used by all living systems. Once techniques became available to sequence DNA easily, genes from many organisms began to be sequenced. However, comparison of these DNA sequences to the amino acid sequence of the proteins they encode has led to a few unexpected surprises. One of these, the discovery of introns in eukaryotic genes, was very surprising but did not change our ideas about the genetic code. However, it has also been discovered that some organelles and some cells use genetic codes that are slight variations of the "universal" genetic code (see the box, Selenocysteine: The Twenty-first Amino Acid).

The original findings of these alternative codes were in the genomes of mitochondria. So far as is known, only the mitochondria of plants use the universal code without change. The other organelles in plants, the chloroplasts, also use this standard code. The mitochondria of all other eukaryotes use codes with one or a few slight differences. A few of these variations are shown in Table 6.7. Note that there is *not* simply a mitochondrial code, although there are a few common themes, such as the general use of UGA as a tryptophan codon. It is also clear that these alternate codes are very closely related to the universal code and are almost certainly derived from it evolutionarily. Several examples of cells are now known whose chromosomes also use slightly different codes, and a few examples of these are also given in Table 6.7. Note that all these alternative chromosomal genetic codes have different assignments for what are normally stop codons. These organisms simply have fewer nonsense codons because one or two are now read as sense.

If every codon has an assignment, you might imagine that it is very difficult to change the genetic code in an organism. For instance, the change of AUA from an isoleucine codon to a methionine codon means that every protein that once had an isoleucine encoded by AUA now has a methionine at this position. Such a protein may not function normally. This may not be a severe problem if *codon usage* is not random. After the genetic code had been worked out by biochemists and before any genes had actually been sequenced, it was assumed that the degenerate codons for an amino acid would be used at an equal frequency. This is another assumption that DNA sequencing has shown to be incorrect! Codon usage is highly biased, and this bias changes from organism to organism. In *Escherichia coli*, for instance, only about 1 out of 20 isoleucine residues is encoded by an AUA, the other 19 being encoded by AUU and AUC. It is thought that one of the steps that can lead to codon reassignment is that the codon becomes rarely used in a genome. This is easier to achieve in mitochondria because they have very small genomes (∞ Section 9.15).

Mistranslation

Another problem in the translation of the genetic code is that errors sometimes occur. This means that a codon may be "read" improperly and the wrong amino acid inserted. Amino acids whose codons differ by only a single base, for example, phenylalanine (UUU) and leucine (UUA), are most likely to be mistranslated. In rare instances leucine may be added to the growing polypeptide instead of phenylalanine even when the codon is UUU. In the normal cell these rare errors occur in only a small number of all the protein molecules and hence have no detrimental effect. For example, experimental measurements of mistranslation have shown that only 1 in 10^3–10^4 codons is misread. However, certain antibiotics that act on ribosomes, such as streptomycin and neomycin, increase translation errors to such an extent that many protein molecules in the cell are abnormal and the cell can no longer function properly.

Other types of translational errors can also occur, such as a ribosome shifting into the wrong reading frame or erroneously reading a stop codon as a sense codon. Amazingly, certain genetic elements seem to have evolved to take advantage of such translational "errors" to make essential proteins (∞ Section 8.22).

| TABLE 6.7 | Variations in the genetic code[a] |

Codon	Universal code	Other codes in cellular chromosomes			Other mitochondrial codes		
		Mycoplasma	*Paramecium*	*Euplotes*	Yeast	Protozoa	Mammals
UGA	Stop	Tryptophan	Stop	Cysteine	Tryptophan	Tryptophan	Tryptophan
UAA/UAG	Stop	Stop	Glutamine	Stop	Stop	Stop	Stop
AUA	Isoleucine	Isoleucine	Isoleucine	Isoleucine	Methionine	Methionine	Methionine
CUA	Leucine	Leucine	Leucine	Leucine	Threonine	Leucine	Leucine
AGA/AGG	Arginine	Arginine	Arginine	Arginine	Arginine	Arginine	Stop

[a]The universal genetic code is used in the chromosomes of most cells, chloroplasts, plant mitochondria, and their viruses and plasmids. A few organisms use slightly different codes in their chromosomes (in the nucleus). The examples of these other nuclear codes are from *Mycoplasma* (Bacteria) and two different ciliated protozoa (Eukarya). All nonplant mitochondria use variations of the universal code, whereas plant mitochondria use the universal code. The examples here are only a few of the different types known.

A FOCUS ON . . .

SELENOCYSTEINE: THE TWENTY-FIRST AMINO ACID

The genetic code has codons for 20 amino acids that are assembled into proteins during translation. However, many proteins contain other amino acids. In fact, there are well over 100 different amino acids found in at least a few proteins. Until recently, it was thought that these "extra" amino acids were made by modifying one of the standard amino acids *after* it was incorporated into protein, a process called *posttranslational modification.* However, it is now clear that one of these extra amino acids is put into protein by the translational machinery itself. This one exception is *selenocysteine.*

Selenocysteine has the same structure as cysteine, but it has a selenium atom rather than a sulfur atom. It was known for some time that a few proteins contain this unusual amino acid. For example, *Escherichia coli* makes two different formate dehydrogenase enzymes and both contain a single selenocysteine residue. When the gene encoding one of these enzymes was sequenced, it was found that the codon that corresponded to the selenocysteine was a UGA. UGA is normally an efficient stop codon in *E. coli,* but it has now been demonstrated that it can be translated directly as selenocysteine in certain mRNA molecules, not only in *E. coli* but also in other prokaryotes and in eukaryotes, including humans. Therefore, selenocysteine is the twenty-first amino acid known to be encoded by the genetic code.

How can a codon sometimes be a stop codon and sometimes a sense codon *in the same chromosome?* The answer apparently lies in the *context* of the codon, the sequence of the bases surrounding the UGA codon and in their secondary structure. In certain contexts, the translational machinery interprets UGA as "selenocysteine." In all other contexts, UGA means "stop translation." Selenocysteine has its own tRNA (as do all the standard amino acids) and also has a special protein factor that brings only this tRNA to the ribosome.

Selenocysteine is even more readily oxidized than cysteine. Therefore, enzymes that contain this amino acid must be protected from oxygen. It has been proposed that UGA might once have been a normal sense codon, calling only for selenocysteine, but that the increase in oxygen in our environment following the evolution of photosynthesis (∞ Chapter 15) selected for proteins that contain cysteine (whose codons are UGU and UGC). This allowed the coding assignment of UGA to be altered except in a few special cases. ∎

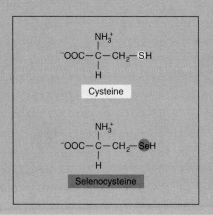

Cysteine

Selenocysteine

Overlapping genes

Although the evidence is strong that the nucleotide sequence specifying one product is separate and distinct from the sequence specifying another product, studies on the small bacterial virus φX174 have shown that this virus has insufficient genetic information to encode all the proteins necessary for its reproduction, but that genetic economy is introduced by using the same piece of DNA for the coding of more than one product. It is a process made possible by reading of the same nucleotide sequence in *two different reading frames,* beginning at different sites. There are now known a number of interesting patterns of *overlapping genes,* most of which occur in viruses (∞ Sections 8.8 and 8.9).

In this chapter we have seen that biological information flows from DNA, the cell's genetic material, to RNA, and finally to protein. The mechanisms of the three main steps in information transfer, replication, transcription, and translation are quite similar in prokaryotes and eukaryotes. The information in a gene can be accurately replicated so that when a cell divides the progeny has the same genetic information. The information in the gene can also be transcribed into RNA, and for genes that encode proteins, translated into proteins. Although some RNA has catalytic activity, it is the proteins that are the components of the cell that carry out the thousands of reactions making up the cell's metabolism. We now turn our attention to how cells regulate whether a particular gene or set of genes is expressed.

CONCEPT CHECK   6.10

The genetic code is expressed in terms of RNA, and a single amino acid may be encoded by several different but related codons. In addition to the nonsense, or stop, codons, there is also a specific start codon that signals the location where the translation process should begin.

✔ *What is meant by a* degenerate *code?*

✔ *What is the* "universal" *genetic code?*

Material for Review

REVIEW QUESTIONS

1. Describe the *central dogma* of molecular biology.

2. Genes were discovered before their chemical nature was known. Define a gene without mentioning its chemical nature. Of what is a gene composed?

3. Inverted repeats can give rise to stem-loops. Show this by giving the sequence of double-stranded DNA containing an inverted repeat and show how the transcript from this region can form a stem-loop.

4. Is the sequence 5′-GCACGGCACG-3′ referred to as an inverted repeat? Explain your answer.

5. DNA molecules that are AT-rich separate into two strands more easily when the temperature is raised than do DNA molecules that are GC-rich. Write an explanation for this observation based on the properties of AT and GC base pairing.

6. What are restriction enzymes? What is the prime function of a restriction enzyme in the cell that produces it (that is, why do cells have restriction enzymes)? How is it that the restriction enzyme in a cell does not cause degradation of that cell's DNA?

7. Nucleic acid hybridization is at the basis of many modern genetic techniques. Write a short explanation for each of the following statements:

   a. The strength of the DNA hybrid is greater if two long than if two short DNA molecules are involved.

   b. Even if the base sequences are not *exactly* complementary, hybridization can still occur between two relatively long DNA molecules, but this is less likely to occur if the molecules are short.

8. A structure commonly seen in circular DNA during repli-cation is called a *theta structure.* Draw a diagram of the replication process and show how a theta structure could arise.

9. Why are errors in DNA replication so rare? What additional enzyme activity (other than polymerization) is associated with DNA polymerase III and how does it serve to reduce errors?

10. Describe the mechanism by which the two enzymes DNA polymerase and DNA ligase function together to effect DNA synthesis.

11. What is a topoisomerase and what is its function? Contrast the actions of topoisomerases I and II.

12. What are ribozymes and what types of biochemical reactions are they generally associated with?

13. There are three processing steps in producing most eukaryotic mRNA but not prokaryotic mRNA. Write a short description of each of these three steps.

14. What are aminoacyl-tRNA synthetases and what types of reactions do they carry out? Approximately how many different types of these enzymes are present in the cell? How does a synthetase recognize its correct substrates?

15. Do genes for tRNAs have promoters? Do they have start codons? Explain.

16. The start and stop sites for mRNA synthesis (on the DNA) are different from the start and stop sites for protein synthesis (on the mRNA). Explain.

17. Imagine you have isolated a new antibiotic that inhibits the translocation process in protein synthesis. Explain what would happen, in terms of mRNA synthesis, polypeptide synthesis, and cell growth, if such an antibiotic were added to a growing culture.

APPLICATION QUESTIONS

1. If the central dogma pertains to all organisms on this planet, what might it tell us about evolution?

2. Give a chemical explanation of why DNA isolated from members of the genus *Myxococcus* has a higher melting temperature than DNA isolated from members of the genus *Cytophaga*.

3. Typical restriction enzymes are part of a restriction-modification system. Explain why they *have* to be. Restriction enzymes that recognize modified DNA are not part of a restriction-modification system. Explain why they *cannot* be.

4. The genome of the bacterium *Neisseria gonorrhoeae* consists of a single double-stranded DNA molecule that contains 2220 kilobase pairs. Calculate the length of this DNA molecule in centimeters. If 85% of this DNA molecule is made up of the open reading frames of genes encoding proteins and the average protein is 300 amino acids long, how many protein-encoding genes does *Neisseria* have? What kind of information do you think might be present in the other 15% of the DNA?

5. Circular DNA molecules, such as those of most bacterial chromosomes, circumvent one problem encountered in replication of linear DNA. What is this problem? Also, having a circular chromosome results in a new problem: the two daughter chromosomes are interlocked after replication is completed. Is there any type of enzyme discussed in this chapter that might help separate these molecules so they can be partitioned?

6. Two methods for determining the sequence of a DNA molecule are the Maxam–Gilbert and the Sanger methods. In what ways do these two methods differ? In what ways are they similar?

7. Digestion of a short piece of *linear* DNA 1500 base pairs long with the restriction enzyme *Taq*I resulted in fragments of 200, 500, and 800 base pairs. When the same DNA was digested with the restriction enzyme *Hind*III, fragments of 300 and 1200 base pairs were obtained. Simultaneous digestion with both enzymes gave fragments of 100, 200, 500, and 700 base pairs. From these data,

prepare a restriction map of the 1500-base-pair piece of DNA. (Of what help was it to know that this piece of DNA was linear?)

8. A classic experiment to show that DNA replication was semiconservative was performed by Mathew Meselson and Franklin Stahl. These workers used the heavy isotope of nitrogen, N^{15}, to label DNA during the replication process. DNA labeled with N^{15} can be separated from regular DNA by ultracentrifugation. In the Meselson–Stahl experiment, cells whose DNA was fully labeled with N^{15} were transferred to a medium containing regular (light) N^{14}. As replication proceeded, DNA containing light nitrogen was obtained and could be separated in the ultracentrifuge. Three kinds of molecules can be anticipated: both strands heavy (initial parent), half heavy–half light, and both strands light. Describe the anticipated result of this experiment after the first round of replication. Label each of the two strands obtained as to whether it will be heavy or light. Describe the anticipated results after a further round of replication.

9. What would be the consequence (in terms of both mRNA and protein synthesis) if the *promoter* region for the gene encoding an enzyme such as β-galactosidase were deleted from the DNA? If the *base sequence* of the promoter were changed so that the binding of RNA polymerase was weaker?

10. Many bacterial mRNA molecules are polycistronic, each mRNA coding for more than one protein. Imagine an mRNA that codes for two proteins with an intervening noncoding region between the two coding regions. From your understanding of how the translation process works, explain why the end result would be two separate proteins rather than one mixed (hybrid) protein.

11. Explain why a *nonsense* codon functions as a stopping point for protein synthesis. Why doesn't a nonsense codon also function as a stopping point for mRNA synthesis?

12. What would be the consequence (in terms of protein synthesis) if one base of the anticodon of a specific tRNA were changed to another base? Why would it matter which base was changed? Why would it matter what it was changed to? Why might it matter which amino acid was charged to the normal tRNA?

13. If you look at the sequence of a short stretch of bases from the middle of a particular mRNA, you should be able to identify three different reading frames (and determine the sequence of the protein encoded by each). However, if this mRNA were in a cell, each ribosome that translates it would use only one of those reading frames. How do ribosomes determine which reading frame is correct?

14. What would be the result (in terms of protein synthesis) if RNA polymerase initiated *transcription* one base upstream of its normal starting point? Why? What would be the result (in terms of protein synthesis) if *translation* began one base downstream of its normal starting point? Why?

15. In the bacterium *Salmonella typhimurium* glutamyl-phosphate reductase and glutamate kinase, two of the enzymes involved in the synthesis of proline, are apparently translated from a single polycistronic message. Draw a diagram of the region of the chromosome containing the two genes encoding these enzymes. Show the correct relative position(s) of that portion of the DNA that contains or encodes all of the following: promoter(s), Shine–Dalgarno sequence(s), start and stop codons, and transcription terminator(s). Do you believe these genes probably contain introns? Why or why not?

16. If the genes you diagrammed in answering Question 15 were actually two genes encoding two different tRNAs, how would your diagram be different?

17. If the genes you diagrammed in answering Question 15 were from a eukaryote, your diagram would be quite different. In addition, there might be other important sequences present. Change your sketch accordingly and explain all the differences.

SUPPLEMENTARY READINGS

Alberts, B., D. Bray, J. Lewis, M. Raff, K. Roberts, and J. D. Watson. 1994. *Molecular Biology of the Cell,* 3rd edition. Garland, New York. This book has become a standard reference book for cellular molecular biology. Although the emphasis on eukaryotes is strong, prokaryotes are not slighted.

Freifelder, D., and **G. M. Malacinski.** 1993. *Essentials of Molecular Biology,* 2nd edition. Jones and Bartlett, Boston. An excellent introductory textbook, emphasizing the fundamental principles of gene structure and expression.

Hill, W. E., A. Dahlberg, R. A. Garrett, P. B. Moore, D. Schlessinger, and **R. Warner.** 1990. *The Ribosome: Structure, Function, and Evolution.* American Society for Microbiology, Washington, DC. A compendium of current research on the biology of the ribosome.

Kornberg, A., and **T. A. Baker.** 1992. *DNA Replication,* 2nd edition. W. H. Freeman, New York. An advanced textbook covering the biochemistry, molecular biology, and genetics of DNA replication in prokaryotes and eukaryotes.

Lodish, H., D. Baltimore, A. Berk, S. L. Zipursky, P. Matsudaira, and **J. Darnell.** 1995. *Molecular Cell Biology,* 3rd edition. Scientific American Books, New York. One of the "big" cell biology textbooks, with strong emphasis on macromolecular processes in eukaryotes.

Stryer, L. 1995. *Biochemistry,* 3rd edition. W. H. Freeman, New York. A biochemistry text with excellent coverage of molecular biology.

Voet, D., and **J. D. Voet.** 1995. *Biochemistry,* 2nd edition. John Wiley, New York. A biochemistry text with considerable material on molecular biology.

Watson, J. D., N. H. Hopkins, J. W. Roberts, J. A. Steitz, and **A. M. Weiner.** 1987. *Molecular Biology of the Gene,* 4th edition. Benjamin-Cummings, Menlo Park, CA. This book has been the standard textbook on macromolecules and DNA for over 20 years.

Wolfe, S. L. 1995. *An Introduction to Cell and Molecular Biology.* Wadsworth, Belmont, CA. An excellent introductory text covering the molecular biology of prokaryotes and eukaryotes.

On~line resources for this chapter are on the World Wide Web at: http://www.prenhall.com/~brock (click on the <u>table of contents</u> link and then select Chapter 6).

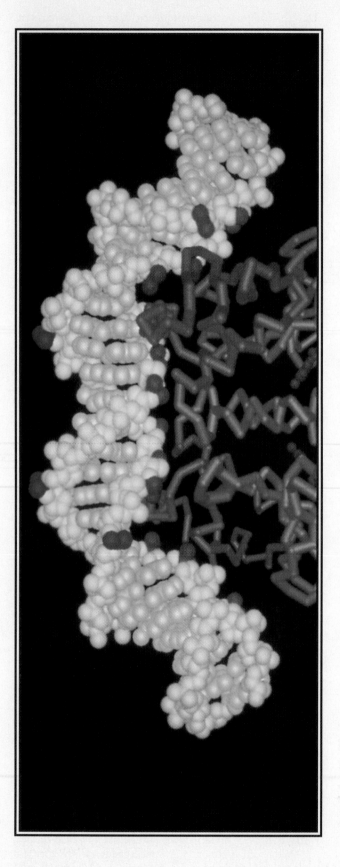

Regulation of Gene Expression

Cells regulate gene expression in many ways. One common mechanism involves proteins, such as the cyclic AMP binding protein shown here, that recognize specific DNA sequences and affect the level of transcription from the gene.

MINIGLOSSARY for Chapter 7

Activator Protein a regulatory protein that binds to specific sites on DNA and stimulates transcription; involved in positive control

Allosteric an enzyme that contains two combining sites, the active site (where the substrate binds) and the allosteric site (where an effector molecule binds)

Attenuation a mechanism for controlling gene expression; typically transcription is terminated after initiation but before a full-length mRNA is produced

Feedback Inhibition a decrease in the activity of the first enzyme of a pathway caused by the final product of the pathway

Kinase an enzyme that adds a phosphoryl group to a compound

Repressor Protein a regulatory protein that binds to specific sites on DNA and blocks transcription; involved in negative control

Response Regulator Protein one of the members of a two-component system; a regulatory protein that is phosphorylated by a sensor protein (see Sensor Protein)

Sensor Protein one of the members of a two-component system; a kinase that is found in the cell membrane and that phosphorylates itself in response to an external signal and then passes the phosphoryl group to a response regulator protein (see Response Regulator Protein)

Two-Component System a regulatory system containing a sensor protein and a response regulator protein (see Sensor Protein and Response Regulator Protein)

I n the previous chapter we saw how the information stored as a sequence of bases in a region of DNA (a gene) can be transcribed into RNA and then translated into a sequence of amino acids to yield a specific protein. Most proteins are enzymes (∞ Section 4.5) and carry out the reactions responsible for the cell's metabolism: the reactions that allow it to process nutrients, to build new cellular material, to grow, and to divide. In Chapter 4 we discussed a few of the key enzymatic reactions occurring during anabolism and catabolism, in Chapter 5 the process of growth, and in Chapter 6 the major reactions in the synthesis of macromolecules. In this chapter we consider the major processes of cellular control, how the numerous chemical reactions in the cell are orchestrated in an efficient manner.

Hundreds of different enzymatic reactions occur simultaneously during a single cycle of cell growth. However, not all these reactions occur to the same extent. Some compounds are needed in large amounts, and the reactions that lead to them must occur frequently. Alternatively, other compounds are needed in only small amounts, and therefore the reactions that lead to them need not occur as often. For maximum utilization of available resources, cells need to *control* the level of expression of the genetic information.

Although the cell may need different proteins in different amounts, an individual protein might be needed in roughly the same amount under most growth conditions. For instance, cells typically need relatively little *DNA primase* (∞ Section 6.5) but require high levels of *single-stranded binding protein* (∞ Section 6.5), even though both proteins are necessary for synthesizing DNA. From what we have discussed about the nature of promoters (∞ Section 6.6), you might pre-

dict that the gene encoding single-stranded binding protein has a promoter with a sequence similar to the *consensus sequence* and that the promoter for the primase gene does not; such an arrangement would allow the proper amounts of each of these proteins.

However, far more common is the situation in which a particular reaction needs to occur frequently under some conditions but not under others. For instance, one might expect there to be more glutamate dehydrogenase activity in a cell growing on ammonia than in a cell growing on another nitrogen source (∞ Section 4.19). Additionally, under many growth conditions a cell might not need to carry out a particular reaction at all. For instance, enzymes required for the breakdown of the sugar lactose are useful to the cell only if lactose is present in its environment. Most microorganisms have the genetic information to encode many more different kinds of proteins than are actually present in the cell under any particular condition (∞ Section 1.3). Thus, the need to regulate biochemical reactions in response to changing growth conditions, or as part of a developmental process, is clear. How does this type of regulation occur?

There are two major modes of regulation in the cell. One controls the *activity* of preexisting enzyme and one controls the *amount* (or even the complete presence or absence) of an enzyme (Figure 7.1). Regulation of the activity of an enzyme obviously happens *after* the protein has been synthesized (that is, posttranslationally). By contrast, regulation of the amount of enzyme synthesized can occur at the level of transcription (how much messenger RNA [mRNA] is made) or at the level of translation (whether or not the mRNA is translated to make the protein). Regulation of the synthesis of an enzyme is a coarser level of control than regulating

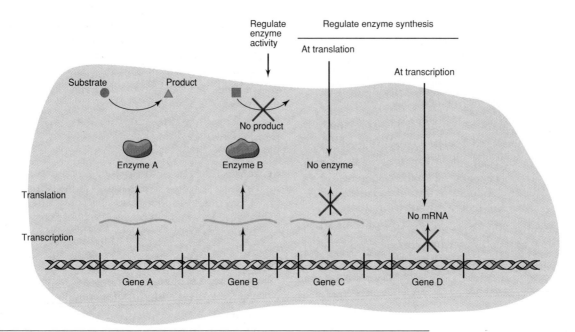

Figure 7.1 An overview of the mechanisms that can be used in regulation. The product of gene A is enzyme A, which is synthesized constitutively and carries out its reaction. Enzyme B is also synthesized constitutively but its activity can be inhibited. The synthesis of the product of gene C can be prevented by control at the level of translation. The synthesis of the product of gene D can be prevented by control at the level of transcription.

activity. However, working together, these mechanisms result in an efficient regulation of cell metabolism so energy is not wasted carrying out unnecessary reactions.

Control systems that vary the level of *expression* of particular genes are the main subject of this chapter. However, the actual number of different regulatory mechanisms is vast, and most genes seem to be regulated by more than one. We begin by briefly discussing the processes involved in regulating the *activity* of preformed enzymes before considering how the synthesis of enzymes is controlled.

7.1
Regulation of Enzyme Activity

There are a large number of mechanisms of posttranslational regulation. In some cases an enzyme is synthesized as part of a larger inactive precursor protein, and the enzyme must be activated by removing a portion of the precursor protein. Another mechanism is to reduce the level of activity by actually degrading the enzyme molecules. However, we discuss here a reversible and temporary form of regulation involving less drastic changes to the enzyme molecule.

Product inhibition

A simple mechanism by which an enzymatic reaction may be regulated is a process called *product inhibition.* As we have seen (∞ Section 4.5), an enzyme combines with its *substrate,* the reaction occurs, and the *product* is

released. Because enzymatic reactions are generally *equilibrium reactions*, as product builds up, the reaction can occur in the *reverse* direction, from product to substrate. Product inhibition occurs primarily if the product of the enzymatic reaction is not used in subsequent reactions. But if the product is itself the substrate for another enzyme, then it will be continually removed and product inhibition will not occur. Note that in product inhibition it is the enzyme that *formed* the product that is also *inhibited* by the product.

Feedback inhibition

A major mechanism for the control of enzymatic activity involves the phenomenon of **feedback inhibition.** Feedback inhibition is seen primarily in the regulation of entire biosynthetic pathways, such as the pathway involved in the synthesis of an amino acid or purine. As we have seen, such pathways involve many enzymatic steps, and the final product, the amino acid or nucleotide, is many steps removed from the starting substrate (∞ Section 4.19 and Figure 4.27). Yet, this final product is able to feed back to the first step in the pathway and regulate its own biosynthesis. How?

In feedback inhibition the amino acid or other end product of the biosynthetic pathway inhibits the activity of the *first* enzyme in this pathway. Thus, as the end product builds up in the cell, its further synthesis is inhibited. If the end product is used up, however, synthesis can resume (Figure 7.2).

How is it possible for the end product to inhibit the activity of an enzyme that acts on a substrate quite

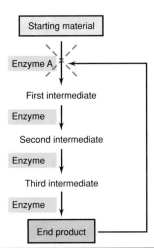

Figure 7.2 Feedback inhibition of enzyme activity. The activity of the first enzyme of the pathway is inhibited by the end product, thus controlling production of end product.

unrelated to it? This occurs because of a property of the inhibited enzyme known as **allostery.** An allosteric enzyme has two important binding sites, the *active* site, where the substrate binds, and the *allosteric* site, where the inhibitor (sometimes called an "effector") binds reversibly. When an inhibitor binds, generally noncovalently, at the allosteric site, the conformation of the enzyme molecule changes so that the substrate no longer binds efficiently at the active site (Figure 7.3). When the concentration of the inhibitor falls, equilibrium favors dissociation of the inhibitor from the allosteric site, returning the active site to its catalytic shape. Allosteric enzymes are very common in both anabolic and catabolic pathways and are especially important in branched pathways. For example, the amino acids proline and arginine are both synthesized from glutamic acid. Figure 7.4 shows that these two amino acids can control the first enzyme unique to their own synthesis without affecting the other so that a surplus of proline, for example, does not cause the organism to be starved for arginine.

In addition, some biosynthetic pathways are regulated by the use of **isozymes** (short for isofunctional enzymes: *iso* means "same" or "constant"). These enzymes catalyze the same reaction but are subject to different regulatory control. An example is synthesis of the aromatic amino acids (Figure 7.5; ∞ Figure 4.27). Three different isozymes catalyze the first reaction in

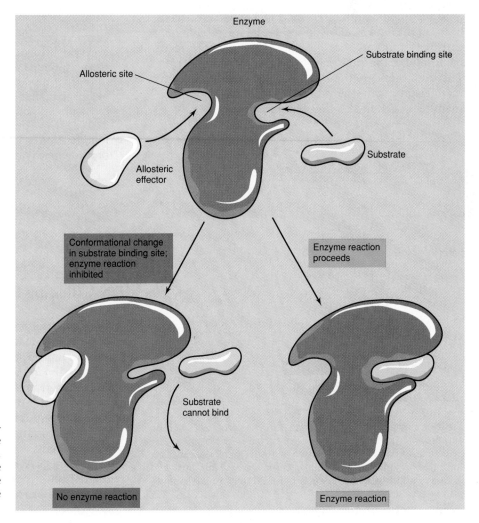

Figure 7.3 Mechanism of enzyme inhibition by an allosteric effector. When the effector combines with the allosteric site, the conformation of the enzyme is altered so that the substrate can no longer bind.

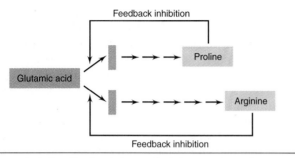

Figure 7.4 Feedback inhibition (solid arrows) in a branched biosynthetic pathway. A key intermediate in each pathway is shown in pink. Compare with Figure 12.17.

this pathway, and each enzyme is regulated independently by each of the three different end product amino acids. Unlike the earlier examples of feedback inhibition where inhibitors completely stopped an enzyme activity, in this case the total amount of the initial enzyme activity is diminished in a stepwise fashion and falls to zero only when all three products are present in excess.

The mechanism of feedback inhibition is of more than just academic interest. We will see in Chapter 12 how an understanding of the biochemistry of feedback inhibition has allowed industrial microbiologists to isolate mutants that have lost the ability to feedback inhibit the production of specific amino acids. These mutants are then used for the large scale commercial production of amino acids as a food supplement (∞ Section 12.7 and Figure 12.17).

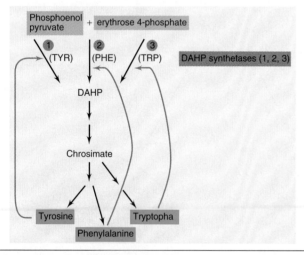

Figure 7.5 The common pathway leading to the synthesis of the aromatic amino acids contains three isozymes of DAHP synthetase. (DAHP is 3-deoxy-D-*arabino*-heptulosonate 7-phosphate.) Each of these enzymes is specifically feedback-inhibited by one of the aromatic amino acids. Note how an excess of all three amino acids is required to completely shut off the synthesis of DAHP.

Covalent modification

Several examples are known in bacteria in which an enzyme is regulated by being *covalently modified*, usually by addition or deletion of some small organic molecule. As in the case of allosteric proteins, the covalent binding of the modifying group changes the conformation of the protein and can dramatically affect its catalytic activity. Removal of the modifying group returns the enzyme to an active state. Many examples of covalent modification regulatory systems are known, but the best characterized are enzymes whose activity is affected by attachment of the nucleotide adenosine monophosphate (AMP) or adenosine diphosphate (ADP), by attachment of inorganic phosphate, or by methylation.

These examples are representative of the major regulatory patterns observed; there are other, rather elegant examples of regulation now known. The phenomenon of enzyme regulation may have evolved because efficient control of the rate of enzyme *activity* enables an organism to quickly adapt to changing environments.

CONCEPT CHECK   7.1

The vast array of metabolic reactions do not all occur in the cell at the same rate. Metabolic reactions can be regulated through control of the activities of the enzymes that catalyze these reactions. An important type of regulation of enzyme activity is feedback inhibition, in which the final product of a biosynthetic pathway feeds back and inhibits the first enzyme unique to that pathway. Covalent modification is a regulatory mechanism for temporarily inactivating a specific enzyme.

✔ *How does* feedback inhibition *differ from* product inhibition?

✔ *What is an* allosteric enzyme?

7.2

DNA Binding Proteins

Small molecules are often involved directly in the regulation of protein activity. For instance, in the example given in Figure 7.4, the amino acids proline and arginine bind directly to the enzyme and inhibit it. The situation with regard to regulating enzyme *synthesis* is quite different. Although small molecules are often involved in regulating transcription, they rarely do so directly. Instead they typically influence the binding of certain proteins, called *regulatory proteins*, to specific sites on the DNA, and it is these proteins that actually regulate transcription. In this section we shall discuss a few general properties of proteins that bind to DNA.

Interaction of proteins with nucleic acids

Protein–nucleic acid interactions are central to replication, transcription, and translation, as well as to the regulation of these processes. Two general kinds of protein–nucleic acid interactions are noted: nonspecific and specific, depending on whether the protein attaches *anywhere* along the nucleic acid or whether the interaction is sequence-specific. As an example of proteins that do *not* interact in a sequence-specific fashion, we mention the **histones**, proteins that are extremely important in the structure of the eukaryotic chromosome (∞ Section 3.14), although less significant in prokaryotes. Histones are relatively small proteins that have a high proportion of positively charged amino acids (arginine, lysine, histidine). DNA, as we have noted, is a polynucleotide and has a high proportion of negatively charged phosphate groups, making it a negatively charged molecule. These phosphate groups are on the outside of the DNA double helix. Histones, because of their positive charge, combine strongly and relatively nonspecifically with the negatively charged DNA. In the eukaryotic cell there is generally enough histone so that all the phosphate groups of the DNA are covered. Association of histones with DNA leads to the formation of nucleosomes, the unit particles of the eukaryotic chromosome (∞ Section 3.14). Even these relatively nonspecific interactions can affect gene expression. If the DNA is covered with histones, other proteins such as RNA polymerase will not be able to

bind and transcription cannot take place. DNA replication removes histones and allows other proteins to bind, and apparently certain other proteins can disrupt the condensed structure of DNA. However, loss of histones need not automatically lead to transcription but may simply leave the gene capable of being activated by other factors.

There are also a number of proteins that interact with DNA in a *sequence-specific* manner. These interactions occur by association of the amino acid side chains of the proteins with the bases as well as with the phosphate and sugar molecules of the DNA. The major groove in DNA, because of its size, is an important site of protein binding. In Figure 6.4 several of the atoms of the base pairs found in the major groove and known to interact with proteins are identified. In order to achieve *specificity* in such interactions, the protein must interact simultaneously with more than one nucleic acid base, frequently several. We have already described a structure in DNA called an *inverted repeat* (∞ Figure 6.10). Such inverted repeats are frequently the locations at which protein molecules combine specifically with DNA (Figure 7.6). Note that this interaction does not involve the formation of cruciform structures in the DNA. Proteins that interact specifically with DNA are frequently *dimers*, composed of two identical polypeptide chains. On each polypeptide chain is a region, called a *domain*, that interacts specifically with a region of DNA in the major groove. A consideration of this type of interaction provides an

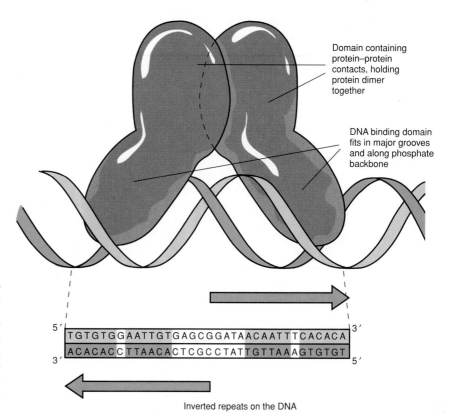

Domain containing protein–protein contacts, holding protein dimer together

DNA binding domain fits in major grooves and along phosphate backbone

Figure 7.6 A protein dimer combines specifically with *two sites* on the DNA. The specific DNA sequences that interact with the protein are *inverted repeats*. The nucleotide sequence of the operator gene of the lactose operon is shown and the inverted repeats, which are sites at which the *lac* repressor makes contact with the DNA, are shown in shaded boxes.

5' TGTGTGGAATTGTGAGCGGATAACAATTTCACACA 3'
3' ACACACCTTAACACTCGCCTATTGTTAAAGTGTGT 5'

Inverted repeats on the DNA

explanation for the fact that such proteins interact with inverted repeats: in this way, *each* of the polypeptides of the protein dimer combines with each of the DNA strands (Figure 7.6). Because the protein recognizes *contact points* associated with specific base pairs, its binding is sequence-specific.

Structure of DNA binding proteins

Studies of the structure of several DNA binding proteins from both prokaryotes and eukaryotes have revealed a few types of common protein substructures that are apparently critical for proper binding of many of these proteins to DNA. One of these is termed the *helix-turn-helix motif* (Figure 7.7). The helix-turn-helix consists of a stretch of amino acids that form an α-helix secondary structure (the so-called recognition helix), which is joined to a short stretch of three amino acids, the first of which is usually a glycine that functions to "turn" the protein (Figure 7.7*a*). The other end of the "turn" is connected to a second helix, which stabilizes the first by interacting hydrophobically with it. Recognition of specific DNA sequences occurs by a combination of noncovalent interactions including hydrogen bonds and van der Waals contacts (∞ Section 2.2) between the protein and base pairs on the DNA. Many different DNA binding proteins from Bacteria show the helix-turn-helix structure, including many repressor proteins such as the bacteriophage lambda repressor (Figure 7.7*b*) and the *lac* and *trp* repressors of *Escherichia coli* (see Section 7.3).

Two other types of protein substructures are commonly found in DNA binding proteins. One of these, the *zinc finger*, is frequently found in eukaryotic regulatory proteins that bind to DNA. The zinc finger is a substructure of protein that, as its name implies, binds a zinc ion (Figure 7.8*a*). It seems most likely that part of the "finger" of amino acids that is created forms an α-helix and this interacts with the DNA in the major groove. There are typically at least two such fingers on the protein involved in binding. The other protein substructure commonly found in DNA binding proteins is the *leucine zipper*. This substructure is formed by the side chains on leucine residues spaced every seven amino acids, and it somewhat resembles a zipper. Unlike the helix-turn-helix and the zinc finger, the leucine zipper does not seem to interact with DNA itself but serves to hold two other α-helices in the correct position to bind DNA (Figure 7.8*b*).

Once a protein combines at a specific site on the DNA, a number of outcomes can occur. In some cases, the protein is an enzyme that carries out some specific action on the DNA, such as RNA polymerase, which makes RNA using DNA as the template. However, in other cases the protein that binds can *block* transcription or can *activate* it. We shall discuss how such regulation is brought about in the next few sections of this chapter.

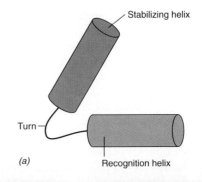

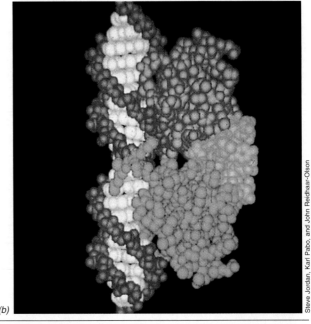

Figure 7.7 The helix-turn-helix structure of some DNA binding proteins. (a) A simple model of the helix-turn-helix elements. (b) A computer model of the bacteriophage lambda repressor, a typical helix-turn-helix protein, bound to its operator gene. One subunit of the dimeric repressor is shown in dark blue and the other in dark green. The light colors represent regions on the dimers involved in subunit interactions. Reprinted with permission from *Science* 241:53–57 (1988), © AAAS.

Steve Jordan, Karl Pabo, and John Reidhaar-Olson

CONCEPT CHECK   7.2

Certain proteins can bind to DNA because of specific interactions between certain regions of the proteins and specific regions of the DNA molecule. In some cases the interactions are not sequence-specific, but in other cases they are. Proteins that bind to nucleic acid may be enzymes that use nucleic acid as substrates, or they may be regulatory proteins that affect how genes function.

✔ *Why are some interactions specific to certain DNA sequences?*

✔ *What is a protein domain?*

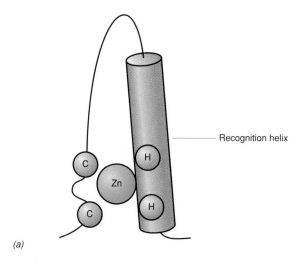

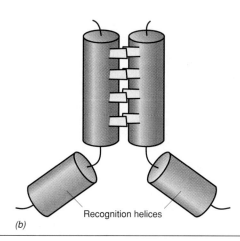

Figure 7.8 Simple models of protein substructures found in eukaryotic DNA binding proteins. α-Helices are represented by cylinders. Recognition helices are the domains involved in DNA-binding. (a) The zinc finger structure. The amino acids holding the Zn^{2+} ion always include at least two cysteine residues (C) with the other residues being histidine (H). (b) The leucine zipper structure. The leucine residues (shown in yellow) are always spaced exactly every seven amino acids. The interaction of the leucine side chains helps hold the two helices together.

7.3
Regulation of Transcription: Induction and Repression

In Section 7.1 we considered how cells regulate enzyme *activity*. We now begin a discussion on how cells regulate enzyme *synthesis*. The regulation of activity is typically very rapid (seconds or less), while the regulation of enzyme synthesis is a relatively slow process (a few to several minutes). If a new enzyme needs to be synthesized, it will take some time before that enzyme is present in the cell in sufficient amounts to affect metabolism. Alternatively, if synthesis of an enzyme is stopped, it may be a considerable amount of time before the existing enzyme is diluted out sufficiently to no longer affect metabolism. Several different mechanisms for controlling enzyme synthesis are known in bacteria, and all of them are greatly influenced by the *environment* in which the organism is growing, in particular by the presence or absence of specific small molecules. These molecules can interact with specific proteins to control transcription or, more rarely, translation. We begin our discussion by describing repression and induction, simple forms of regulation that govern gene expression at the level of *transcription*.

Enzyme repression

Often the enzymes catalyzing the synthesis of a specific product are not synthesized if this product is present in the medium. For example, the enzymes involved in formation of the amino acid arginine are synthesized only when arginine is *not* present in the culture medium; external arginine *represses* the synthesis of these enzymes. As can be seen in Figure 7.9, if arginine is added to a culture growing exponentially in a medium devoid of arginine, growth continues at the previous rate, but the formation of the enzymes involved in arginine synthesis stops. Note that this is a *specific* effect, as the syntheses of all other enzymes in the cell continue at the same rates as previously.

Enzyme repression is a very widespread phenomenon in bacteria—it occurs as a means of controlling the synthesis of a wide variety of enzymes involved in the biosynthesis of amino acids, purines, and pyrimidines. In almost all cases it is the final product of a particular biosynthetic pathway that represses the enzymes of this pathway. In these cases repression is quite specific, and the process usually has no effect on the synthesis of enzymes other than those involved in the specific biosynthetic pathway. The value to the organism of enzyme repression is obvious because it effectively ensures that the organism does not waste energy synthesizing unneeded enzymes.

Enzyme induction

A phenomenon complementary to repression is *enzyme induction*, the synthesis of an enzyme only when its substrate is present. Figure 7.10 shows this process in the case of the enzyme β-galactosidase, which is involved in utilization of the sugar lactose. If lactose is absent from the medium the enzyme is not synthesized, but synthesis begins almost immediately after lactose is added. Enzymes involved in the catabolism of carbon and energy sources are often inducible. Again, one can see the value to the organism of such a mechanism, as it provides a means whereby the organism does not synthesize an enzyme until it is needed.

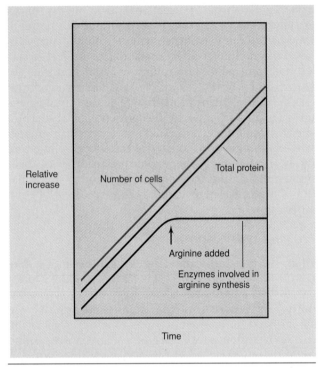

Figure 7.9 Repression of enzymes involved in arginine synthesis by addition of arginine to the medium. Note that the rate of total protein synthesis remains unchanged.

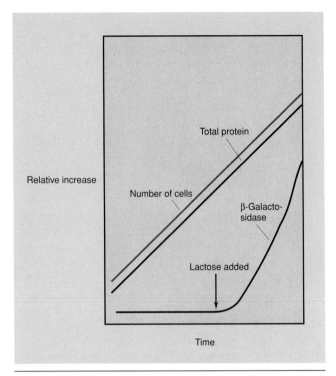

Figure 7.10 Induction of the enzyme β-galactosidase on the addition of lactose to the medium. Note that the rate of total protein synthesis remains unchanged.

The substance that initiates enzyme induction is called an **inducer**, and a substance that represses production is called a **corepressor**; these substances, which are always small molecules, are often collectively called **effectors**. Not all inducers and corepressors are substrates or end products of the enzymes involved. For example, *analogs* of these substances may induce or repress even though they are not substrates of the enzyme. Isopropylthiogalactoside (IPTG), for instance, is an inducer of β-galactosidase even though it cannot be hydrolyzed by the enzyme. In nature, however, inducers and corepressors are probably normal cell metabolites.

Mechanism of induction and repression

Enzyme repression or induction acts at the level of transcription; enzyme synthesis is controlled by initiating or terminating mRNA production for a particular enzyme or group of enzymes. How can inducers and corepressors affect transcription in such a specific manner? They do this indirectly by combining with specific regulatory proteins which then in turn affect mRNA synthesis. In the case of a repressible enzyme, the corepressor (for example, arginine) combines with a specific **repressor protein**, the arginine repressor, that is present in the cell (Figure 7.11). The repressor protein is an allosteric protein (see Sections 7.1 and 7.2), its conformation being altered

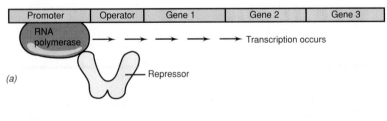

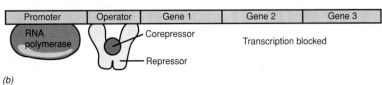

Figure 7.11 The process of enzyme repression. (a) Transcription of the operon occurs because the repressor is unable to bind to the operator. (b) After a corepressor (small molecule) binds to the repressor, the repressor binds to the operator and blocks transcription; mRNA and the proteins it codes for are not made.

when the corepressor combines with it. This altered repressor protein can then combine with a specific region of the DNA near the promoter of the gene, the **operator region**. This region gave its name to the **operon**, which as we have discussed (∞ Section 6.6) is a cluster of genes whose expression is under the control of a single operator. All the genes in an operon are transcribed as a single unit. The operator is adjacent to the promoter where synthesis of mRNA is initiated. If the repressor binds to the operator, the synthesis of mRNA is blocked and the protein or proteins specified by this mRNA cannot be synthesized. If the mRNA is polycistronic, *all* the proteins encoded by this mRNA will be repressed.

Enzyme induction can also be controlled by a *repressor*. In this case, the situation previously described is reversed. The specific repressor protein is active in the *absence* of the inducer, completely blocking the synthesis of mRNA, but when the inducer is added, it combines with the repressor protein and inactivates it. Inhibition of mRNA synthesis being overcome, the enzyme or enzymes can be made (Figure 7.12). All systems involving repressors have the same underlying mechanism, *inhibition* of the synthesis of mRNA by the action of specific repressor proteins that are themselves under the control of specific small-molecule inducers and repressors. Because the repressor's role is inhibitory, regulation involving repressors is often referred to as **negative control.**

It should be emphasized that not all enzymes of the cell are controlled by simple induction or repression and that the synthesis of some enzymes is not strongly controlled at all. Enzymes whose level of synthesis is about the same under all growth conditions are called *constitutive enzymes.* Constitutive enzymes are generally key cellular enzymes required for growth under all nutritional conditions and are thus synthesized continuously in the growing cell. However, enzymes not under the control of repressor systems need not be constitutively synthesized. There are many mechanisms known by which enzyme synthesis is regulated.

CONCEPT CHECK   *7.3*

The amount of an enzyme in the cell can be controlled by increasing (induction) or decreasing (repression) the amount of mRNA that encodes the enzyme. This transcriptional regulation involves regulatory proteins that bind to DNA and to small molecules called effectors. For one type of transcriptional regulation, the regulatory protein is called a repressor and it functions by inhibiting mRNA synthesis.

✔ *How does a repressor inhibit the synthesis of a specific mRNA?*

✔ *What is an operon?*

7.4
Regulation of Transcription: Positive Control

Repression constitutes a kind of regulation called **negative control.** The controlling element—the repressor protein—brings about the *repression* of mRNA synthesis. Even though the repressor has a negative role, a system using a repressor can control enzyme induction, as we saw with β-galactosidase. However, another type of control has also been recognized that is called **positive control.** In positive control, a regulator protein *promotes* the binding of RNA polymerase, thus acting to *increase* mRNA synthesis. We will now consider a system that involves positive regulation, the regulation of maltose catabolism in *Escherichia coli.*

The maltose regulon

The enzymes for the utilization of the sugar maltose in *Escherichia coli* are synthesized only after the addition of maltose to the medium. The pattern of induction of these enzymes follows that shown for β-galactosidase

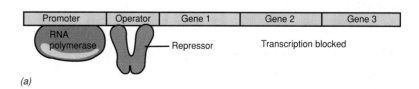

(a)

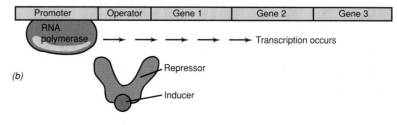

(b)

Figure 7.12 The process of enzyme induction using a repressor. (a) A repressor protein binds to the operator region and blocks the action of RNA polymerase. (b) An inducer molecule binds to the repressor and inactivates it. Transcription by RNA polymerase occurs and an mRNA for that operon is formed.

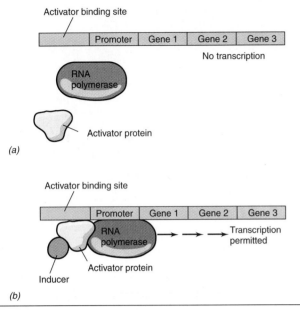

(a)

(b)

Figure 7.13 Positive control of enzyme induction. (a) In the absence of an inducer, neither the activator protein nor the RNA polymerase can bind to the DNA. (b) An inducer molecule binds to the activator protein, which in turn binds to the activator binding site. This allows RNA polymerase to bind to the promoter and begin transcription.

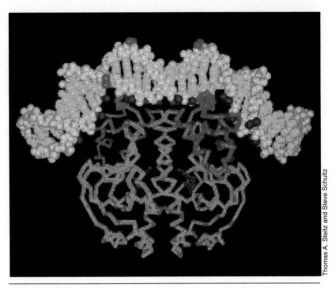

Figure 7.14 Computer model of the interaction of a positive regulatory protein with DNA. This figure shows the cyclic AMP binding protein, a regulatory protein involved in the control of several operons (see Section 7.6). The α-carbon backbone of this protein is shown in blue and purple. The protein is shown binding to a DNA double helix, which is shown in yellow and light blue. Note that binding of this protein to DNA has caused the DNA to be bent by almost 90°. Reprinted with permission from *Science* 253:1001–1007 (1991), © AAAS.

in Figure 7.10, but in this case it is maltose, not lactose, that is the inducer. The synthesis of the enzymes for maltose utilization is controlled at the level of transcription, but by an **activator protein,** not by a repressor. The *maltose activator protein* cannot bind to the DNA unless the protein first binds maltose, the effector. When the activator protein does bind to DNA, it allows RNA polymerase to begin transcription (Figure 7.13). Activators, like repressors, recognize specific sequences on the DNA. The sequence that serves as the binding site of the activator is not called an operator but an *activator binding site*. Nonetheless, the genes controlled by this activator binding site *are* called an operon.

In negative control, the repressor binds to the operator and blocks transcription. How does an activator protein work? Positively controlled promoters have nucleotide sequences that are not close matches to the consensus sequence (∞ Figure 6.24). Even with the correct sigma factor, the RNA polymerase has difficulty recognizing these promoters. The activator protein, when bound to DNA, helps the RNA polymerase either recognize the promoter or begin transcription. The activator protein may cause a change in the structure of the DNA, perhaps by bending it (Figure 7.14), allowing the RNA polymerase to make the correct contacts with the DNA. The activator protein may also interact directly with the RNA polymerase. This can happen either when the activator

binding site is close to the promoter (Figure 7.15*a*) or when it is several hundred base pairs away from the promoter (Figure 7.15*b*).

The genes needed for maltose utilization are spread out in several operons, each of which has an activator binding site to which the maltose activator protein can bind. Therefore, the maltose activator protein actually controls more than one operon. When more than one operon is under the primary control of the same regulatory protein, these operons are collectively known as a **regulon.** Therefore, the enzymes for maltose utilization are encoded by the *maltose regulon*. Regulons are also known for operons under negative control. The arginine biosynthetic enzymes (mentioned in Section 7.3) are encoded by the *arginine regulon* whose operons are all under the control of the arginine repressor protein.

Many genes in *Escherichia coli* have promoters under positive control, and many have promoters under negative control. However, there are other types of regulation known. In addition, many genes (perhaps most genes) either have a promoter with multiple types of control or have more than one promoter, each with its own control system! We next discuss a type of regulation (attenuation) found in prokaryotes where regulation of transcription is typically coupled to translation. Then we turn our attention to global control networks and describe how cells can regulate many genes in response to particular environmental conditions.

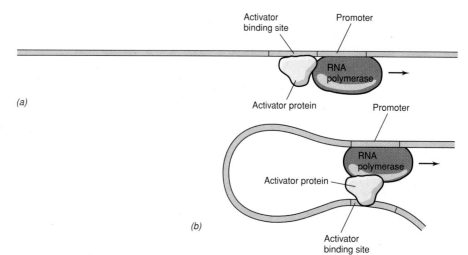

(a)

(b)

Figure 7.15 Some activator proteins interact with RNA polymerase. (a) The activator binding site is near the promoter. (b) The activator binding site is several hundred base pairs from the promoter. In this case, the DNA must be looped to allow the activator and the RNA polymerase to contact.

CONCEPT CHECK   **7.4**

Positive regulators of transcription are called activator proteins. They bind to activator binding sites on the DNA and stimulate transcription by RNA polymerase. Activator protein activity, like repressor protein activity, is modified by effectors. For positive control of enzyme induction, the effector promotes the binding of the activator protein and thus stimulates mRNA synthesis.

✔ *Compare and contrast the activities of an activator protein and a repressor protein.*

✔ *Distinguish between an operon and a regulon.*

7.5

Attenuation

Another element of control, called **attenuation,** has been recognized in several operons. The word *attenuation* means "to lessen in amount." In the previous two sections we described regulating transcription at *initiation*; that is, repressors block the synthesis of RNA whereas activator proteins encourage synthesis. In transcription attenuation the control occurs *after* initiation of RNA synthesis but before its completion. That is, the number of *completed* transcripts from a gene or an operon is reduced, even though the number of initiated transcripts is not. Most of the first examples of attenuation involved regulating genes controlling the biosynthesis of certain amino acids in gram-negative Bacteria. The first such system to be described was the *tryptophan operon* in *Escherichia coli,* and we focus on it here.

Attenuation and the tryptophan operon

The tryptophan operon contains structural genes for five proteins of the tryptophan biosynthetic pathway, plus the promoter and regulatory sequences at the beginning of the operon (Figure 7.16). Like many operons, the tryptophan operon has more than one type of regulation. One type is repression, and one of the regulatory sequences is an operator to which the tryptophan repressor can bind. In addition to promoter and operator regions, there is a sequence called the **leader sequence,** which codes for a polypeptide that contains tandem tryptophan codons near its terminus and functions as an **attenuator** (Figure 7.16). If tryptophan is plentiful in the cell, the leader peptide will be synthesized. On the other hand, if tryptophan is in short supply, the tryptophan-rich leader peptide will *not* be synthesized. The striking fact is that synthesis of the leader peptide results in *termination* of transcription of the tryptophan structural genes, whereas if synthesis of the leader peptide is blocked by tryptophan deficiency, transcription of the tryptophan structural gene can occur.

How does *translation* of the leader peptide regulate *transcription* of the tryptophan genes downstream? This can be explained by considering that these two processes in prokaryotic cells are occurring virtually simultaneously (Figure 7.17). Thus, while *transcription* of downstream DNA sequences is still proceeding, *translation* of sequences already transcribed has begun. Apparently, as the mRNA is released from the DNA, the ribosome binds to it and translation begins. Attenuation occurs (RNA polymerase stops transcription) because a portion of the newly formed mRNA folds into a double-stranded loop that signals cessation of RNA polymerase action (∞ Figure 6.25). The stem-loop structures formed by mRNA are brought about because two stretches of nucleotide bases near each other are complementary and can thus base-pair.

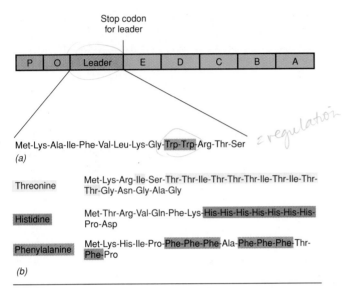

Met-Lys-Ala-Ile-Phe-Val-Leu-Lys-Gly-Trp-Trp-Arg-Thr-Ser

(a)

Threonine	Met-Lys-Arg-Ile-Ser-Thr-Thr-Ile-Thr-Thr-Thr-Ile-Thr-Ile-Thr-Thr-Gly-Asn-Gly-Ala-Gly
Histidine	Met-Thr-Arg-Val-Gln-Phe-Lys-His-His-His-His-His-His-His-Pro-Asp
Phenylalanine	Met-Lys-His-Ile-Pro-Phe-Phe-Phe-Ala-Phe-Phe-Phe-Thr-Phe-Pro

(b)

Figure 7.16 Structure of the tryptophan operon and of tryptophan and other leader peptides in *Escherichia coli*. (a) Arrangement of the tryptophan operon. Note that the leader encodes a short peptide containing two tryptophan residues near its terminus. (b) Amino acid sequence of leader peptides synthesized in some other amino acid biosynthetic operons. Because isoleucine is made from threonine, it is an important constituent of the threonine leader peptide.

If tryptophan is plentiful, the ribosome will translate the leader sequence until it comes to the stop codon. The remainder of the leader RNA can then assume a stem-loop, a *transcription pause site*, which is followed by a uracil-rich sequence that actually causes termination. However, if tryptophan is in short supply, the ribosome pauses at a tryptophan codon; the presence of the stalled ribosome at this position allows an alternative stem-loop to form (sites 2 and 3 in Figure 7.17). This stem-loop is *not* a termination signal, and it effectively prevents the terminator (sites 3 and 4 in Figure 7.17) from forming. RNA polymerase then moves past the nonfolded termination site and begins transcription of the tryptophan structural genes. Thus, we see that in attenuation there is a highly integrated system in which transcription and translation interact, with the rate of transcription being influenced by the rate of translation.

Thus, in the tryptophan biosynthetic pathway, two distinct mechanisms for the regulation of transcription exist, repression and attenuation. Repression is a mechanism that has large effects on the rate of enzyme synthesis, whereas attenuation brings about a finer control. Working together, these two mechanisms precisely regulate the synthesis of tryptophan biosynthetic enzymes, and hence the biosynthesis of tryptophan. Attenuation has also been shown to occur in *Escherichia coli* in the biosynthetic pathways for histidine, threo-

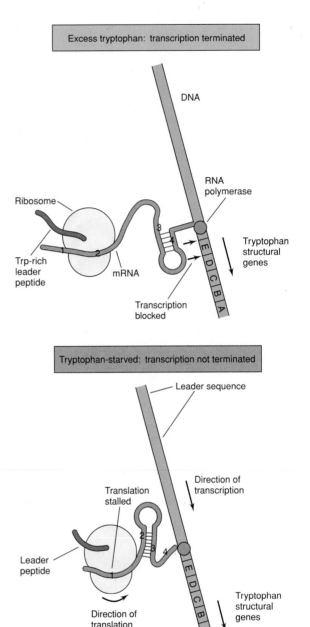

Figure 7.17 Control of transcription of tryptophan operon structural genes by attenuation in *Escherichia coli*. The leader peptide is coded by regions 1 and 2 of the mRNA. Two regions of the growing mRNA chain are able to form double-stranded loops, 2:3 and 3:4. Under conditions of excess tryptophan, the ribosome translates the complete leader peptide, and so region 2 cannot pair with region 3. Regions 3 and 4 then pair to form a loop that blocks RNA polymerase. If translation is stalled because of tryptophan starvation, loop formation via 2:3 pairing occurs, loop 3:4 does not form, and transcription proceeds past the leader sequence.

nine-isoleucine, phenylalanine, and several other amino acids and essential metabolites as well. As shown in Figure 7.16b, the leader peptide for each amino acid biosynthetic operon is rich in that particular amino acid. The *his* operon is dramatic in this regard because its leader contains *seven* histidines in a row near the end of the peptide (Figure 7.16*b*).

Other attenuation mechanisms

Gram-positive Bacteria, such as *Bacillus,* also use attenuation to regulate certain amino acid biosynthetic operons. As in gram-negative Bacteria, the mechanisms also involve attenuation of transcription and alternative secondary structures, which in one configuration lead to termination. However, the mechanisms are translation-independent and rather than a translating ribosome, an RNA binding protein is involved. This protein binds as a result of interacting with an *effector,* in some cases a tRNA, but in other cases the amino acid itself. In the *Bacillus subtilis* tryptophan operon the protein is called the *trp attenuation protein,* and in the presence of the amino acid tryptophan binds to the leader and favors transcription termination. If tryptophan is limiting, the protein does not bind and transcription proceeds.

Many cases are now known where attenuation involves genes unrelated to amino acid biosynthesis, and the mechanisms obviously do not involve measuring the amount of amino acid. In *Escherichia coli* some of the operons involved in pyrimidine biosynthesis are regulated by attenuation, and the same is true for the pyrimidine biosynthetic genes of *Bacillus.* However, the mechanisms are quite different from each other, although each monitors the level of pyrimidine nucleotides in the cell. In *E. coli* a translated mRNA leader is involved, but in *Bacillus* there is no coupling of transcription and translation. Instead, an RNA binding protein controls the alternative structures of the mRNA.

Finally, a type of regulation called *translational attenuation* is also known. In these cases the translation of the leader peptide prevents the *translation* of the next gene on the polycistronic mRNA. The mechanism apparently involves accessibility of the Shine–Dalgarno sequence of the regulated gene (∞ Section 6.9). Translational attenuation is known to regulate expression of several antibiotic resistance genes in gram-positive bacteria.

7.6
Global Control

Often an organism needs to regulate many different genes simultaneously in response to a change in its environment. For instance, when the bacterium *Escherichia coli* is starved for phosphate, over 80 different genes are transcribed in response, bringing about the synthesis of new proteins. These proteins play roles in adapting the bacterium to a phosphate-deficient environment. There are several such sets of genes in *E. coli* whose products are required to respond to particular conditions (Table 7.1). Because these control mechanisms operate on a wide cellular basis, they are referred to as *global control systems* and may include one or more regulons. Some scientists use the term *stimulon* to refer to a group of genes that becomes active in response to an environmental signal. Other scientists use the term *modulon* to emphasize that global control often modulates (adjusts) other controls on the same genes.

In addition to allowing an organism to respond to a signal by activating a network of genes, global regulation can be used to prevent some genes from responding unnecessarily. For instance, Sections 7.3 and 7.4 covered how the enzymes for lactose or maltose utilization can be induced by adding either lactose or maltose to the growth medium. However, it would be wasteful to induce these enzymes if the cells were already growing on a carbon source that they could use more efficiently. In fact, one of the global regulatory networks, **catabolite repression,** prevents this problem.

Catabolite repression

In catabolite repression the synthesis of a variety of unrelated enzymes, primarily catabolic, are inhibited when cells are grown in a medium that contains an energy source such as glucose. Catabolite repression has been called the **glucose effect** because glucose was the first substance shown to initiate it, although in some organisms glucose does not cause this form of enzyme repression. Catabolite repression occurs when the organism is offered a catabolizable energy source in the presence of a more readily catabolizable energy source, such as glucose.

One consequence of catabolite repression is that it can lead to so-called **diauxic growth** if the two energy sources are present in the medium at the same time and if the enzyme needed for utilization of one of the energy sources is subject to catabolite repression. In diauxic growth, the organism grows first on one energy source and there is then a temporary cessation before growth is

---

CONCEPT CHECK   7.5

Attenuation is a mechanism whereby gene expression (typically at the level of transcription) is controlled after initiation of RNA synthesis. Most attenuation mechanisms involve a coupling of transcription and translation and can therefore occur only in prokaryotes.

✔ *Why can control systems involving coupled transcription and translation occur only in* prokaryotes?

✔ *Explain how the formation of one stem-loop in the RNA can block the formation of another.*

TABLE 7.1   A few of the global control systems known in *Escherichia coli*[a]

System	Signal	Primary activity of regulatory protein	Number of genes regulated
Aerobic respiration	Presence of O_2	Repressor	20
Anaerobic respiration	Lack of O_2	Activator	20+
Catabolite repression	Cyclic AMP concentration	Activator	300+
Heat shock	Temperature	Alternative sigma	17
Nitrogen utilization	NH_3 limitation	Activator/alternative sigma	12+
Oxidative stress	Oxidizing agent	Activator	12+
SOS response	Damaged DNA	Repressor	17

[a]For many of the global control systems, regulation is complex. A single regulatory protein can play more than one role. For instance, the regulatory protein for anaerobic respiration, FNR, is an activator protein for many promoters but a repressor for others. Regulation can also be indirect or require more than one regulatory protein. Note that the activator for the nitrogen utilization system activates promoters recognized by an alternative sigma factor. In addition, some of the regulatory proteins involved are members of two-component systems (see Section 7.7). Many genes are regulated by more than one global control system. (For a discussion of the SOS response, see Section 9.3.)

resumed on the other energy source. This phenomenon is illustrated in Figure 7.18 for growth on a mixture of glucose and lactose. The enzyme β-galactosidase, which is responsible for utilization of lactose, is inducible, but its synthesis is also subject to catabolite repression. Thus, as long as glucose is present in the medium, β-galactosidase is not synthesized; the organism grows only on the glucose and leaves the lactose untouched. When the glucose is exhausted, catabolite repression is abolished. After a lag, β-galactosidase is synthesized and growth on lactose can occur. Notice that Figure 7.18 shows that the cells grow more rapidly on glucose. Thus, catabolite repression ensures that the cells use the *best* carbon source first.

How does catabolite repression work? Catabolite repression involves control of transcription by an acti-

vator protein (see Section 7.4). In the case of catabolite-repressible enzymes, binding of RNA polymerase occurs only if another protein, called **catabolite activator protein (CAP)**, has bound first. An allosteric protein, CAP binds to DNA only if it has first bound a small molecule called *cyclic adenosine monophosphate* or **cyclic AMP** (see Figure 7.14). Cyclic AMP (Figure 7.19) has been shown to be a key element in a variety of control systems, not only in bacteria but in higher organisms also. Cyclic AMP is synthesized from ATP by an enzyme called *adenylate cyclase*, and glucose inhibits the synthesis of cyclic AMP and stimulates its transport out of the cell. When glucose is transported into the cell, the cyclic AMP level in the cell is lowered, and binding of RNA polymerase to the promoter does not occur. Thus, catabolite repression is really a result of a deficiency of cyclic AMP and can be overcome by adding this compound to the medium.

Although this may sound like a simple positive regulatory system (as in Figure 7.13), each of the operons that CAP controls is *also* under control of a specific regulatory protein. Therefore, catabolite repression modulates several unrelated regulatory systems and thus is an example of global control. As long as glucose is present, catabolite repression prevents expression of all other catabolic operons under this global controlling element. The complete regulatory region of the *lactose*

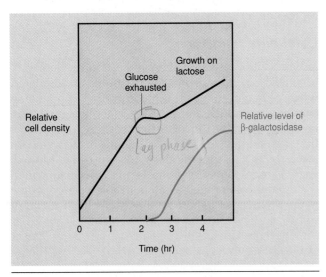

Figure 7.18 Diauxic growth on a mixture of glucose and lactose. Glucose represses the synthesis of β-galactosidase. After glucose is exhausted, a lag occurs until β-galactosidase is synthesized, and then growth can resume on lactose.

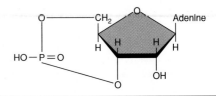

Figure 7.19 Cyclic adenosine monophosphate (cyclic AMP, cAMP) is produced from ATP by the enzyme adenylate cyclase.

operon is shown in Figure 7.20. For transcription to occur, two requirements must be met: (1) the level of cyclic AMP must be high enough so that the CAP protein binds to the CAP binding site, and (2) there must be an inducer such as lactose present so that the lactose repressor does not block transcription by binding to the operator.

Cyclic AMP has a number of regulatory roles in eukaryotes that do not involve catabolite repression and is also an extracellular signal for the aggregation process in certain cellular slime molds (∞ Section 18.3).

Other global control networks

Genes belonging to global control systems do not all use a simple combination of repressors or activators to achieve regulation. Several involve *alternative sigma factors*, including some shown in Table 7.1, and in these cases, regulation is brought about by changing either the amount or the activity of these factors. Most genes in *Escherichia coli* require the sigma factor referred to as σ^{70} (the superscript 70 indicates the size of this protein, 70 kilodaltons) for transcription and have promoters like those shown in Figure 6.24. The genes that are induced by an increase in temperature (heat shock) have promoters with a quite different sequence, and RNA polymerase requires a different sigma factor (σ^{32}) to recognize them. It is the *amount* of this alternative sigma factor in the cell that regulates the *heat shock response*, and the amount of σ^{32} itself is controlled not by transcription but by the stability of the factor.

Most proteins are very stable; once made, they continue to perform their functions and are passed along at cell division. However, some proteins are unstable. They are recognized by enzymes in the cell called **proteases** and are rapidly degraded. In *E. coli*, σ^{32} is normally degraded within a minute or two after it is synthesized. However, when cells experience heat shock, this degradation process is inhibited. This means there will be more σ^{32}, and therefore it can direct more RNA polymerase to more heat shock promoters.

Global regulatory systems must have a common way to transmit a signal from the environment to the gene(s). We have seen how this is accomplished with catabolite repression, but in the case of the heat shock response, how does a bacterium know what the temperature is? This mechanism seems to involve one of the *heat shock proteins*, a protein called DnaK. DnaK is essential for the normal growth of *E. coli* at any temperature, but the amount that is synthesized is increased by heat shock (recall that genes under the control of a global control system are also usually regulated in other ways). The protein DnaK is a **chaperonin**, one of a group of proteins called *molecular chaperones* (∞ Section 6.9). DnaK somehow helps other proteins fold properly and is also involved in the

degradation of σ^{32}. Possibly when the temperature is increased, the activity of DnaK is directed more toward folding proteins and less is available for the degradation of σ^{32}. Certainly an increase in temperature could influence formation of the correct secondary and tertiary structures of proteins or even cause them to denature slightly (∞ Section 2.8). This would result in an increased level of σ^{32}, and the genes encoding the heat shock proteins would be transcribed. However, since the amount of DnaK increases as part of the heat shock response, it eventually builds up and σ^{32} is degraded again, bringing the cell back to its normal state.

Many genes belong to more than one global control system and therefore can have several overlapping regulatory systems. In some global control systems, more than one regulatory protein might be involved in regulating a single operon. An example is control of the *lac* operon by both the lactose repressor and the catabolite activator protein. However, in this case each regulatory protein essentially operates independently of the other. In the next section we discuss "two-component" regulatory systems, regulatory mechanisms in which at least two proteins are involved in generating a response to a single signal and which seem to be widespread in nature.

CONCEPT CHECK   7.6

Cells often need to regulate many genes in response to a single environmental signal. Such cellwide regulatory responses are termed global control. In many cases the genes involved may also be under control of other regulatory circuits. Catabolite repression is an example of global control, and it serves to help cells make the most efficient use of carbon sources.

✔ *Explain how catabolite repression can involve an activator protein.*

✔ *Why might it not be efficient to have all the genes responding to an environmental signal be in a single operon?*

7.7

Signal Transduction and Two-Component Regulatory Systems

Bacteria regulate cell metabolism in response to a wide variety of environmental fluctuations, including temperature changes, changes in pH and oxygen availability, changes in the availability of nutrients, and even changes in the number of cells present. Therefore, there

CAP binding site

Promoter

−35 sequence

```
T A A T G T G A G T T A G C T C A C T C A T T A G G C A C C C C A G G C T T T A C A T T T A T G C T T C C G G C T
A T T A C A C T C A A T C G A G T G A G T A A T C C G T G G G G T C C G A A A T G T A A A T A C G A A G G C C G A
```

Figure 7.20 The genetic elements involved in regulation of the lactose operon. The first gene in this operon, *lacZ*, encodes the enzyme β-galactosidase, which breaks down lactose. The operon contains two other genes that are also involved in lactose metabolism. Notice that the two halves of the operator (where the repressor would bind) are almost perfect inverted repeats. There are also inverted repeats in the CAP binding site although these are less perfect. Also shown are the transcriptional start site and the −35 sequence and the Pribnow box, which are part of the promoter (∞ Figure 6.24). In addition, the location of the base pairs encoding the Shine–Dalgarno sequence and the start codon are also given. These two sequences would function on the mRNA (∞ Section 6.9).

must be mechanisms by which bacteria receive signals from the environment and transmit them to the specific target to be regulated. We have seen in preceding sections that some signals can be small molecules that enter the cell (often by specific uptake mechanisms) and act as *effectors*. For instance, in the case of the maltose regulon (see Section 7.4) the sugar maltose binds to the maltose activator protein, causing the protein to bind to specific DNA sequences and activate transcription. However, in many cases the external signal is not transmitted directly to the regulatory protein. Instead, a signal is first detected by a sensor and then transmitted in a changed form to the rest of the regulatory machinery, a process called **signal transduction.**

Sensor kinases and response regulators

Many of the regulatory systems by which cells sense and then respond to environmental signals are called **two-component systems.** Such systems are characterized by having two different proteins: (1) a specific **sensor protein** located in the cell membrane, and (2) a partner **response regulator protein.** The *sensor protein* has **kinase** activity and is often referred to as a *sensor kinase*. A *kinase* is an enzyme that phosphorylates compounds. Sensor kinases detect a signal from the environment on their outer surface and in response phosphorylate themselves (autophosphorylation) at a specific histidine residue on their cytoplasmic surface (see Figure 7.21). This phosphoryl group is then transmitted to another protein inside the cell, the *response regulator*. The response regulator is typically a DNA binding protein that regulates transcription. In Figure 7.21 the phosphorylated response regulator is acting as a repressor protein, while the unphosphorylated response regulator does not bind to DNA.

The mechanism used by the response regulator to control transcription depends on the system being described. In *Escherichia coli* the osmolarity of the environment controls which of two proteins, OmpC or OmpF, is synthesized as part of the outer membrane. The response regulator of this system is OmpR. When OmpR is phosphorylated, it acts as an *activator* of transcription of the *ompC* gene and a *repressor* of transcription of the *ompF* gene.

Two-component systems are now known to regulate a large number of genes in many different bacteria. A few examples include nitrogen assimilation in *Escherichia coli*, nitrogen fixation in *Klebsiella* and *Rhizobium*, and sporulation in *Bacillus* (which has a very complex regulatory system). In *E. coli* alone it is estimated that at least 50 different two-component systems operate. Two-component systems closely related to those in bacteria have also been found in lower eukaryotes, such as the yeast *Saccharomyces cerevisiae*. Higher eukaryotes also use phosphorylation as a mechanism of signal transduction in order to respond to environmental changes.

In order to complete a regulatory circuit, there must be a way to terminate the signal. Typically, this involves a *phosphatase*, an enzyme that can remove the phosphoryl group from the response regulator protein. In some cases this reaction is carried out by the sensor kinase itself, while in other systems there is a third protein that carries out this reaction. Therefore, there are "two-component" systems with three components! Actually some systems have even more components, as the signal may be processed through several steps. However, in all cases two-component systems have a sensor kinase and a response regulator.

Not all response regulators regulate transcription, as some actually regulate cell behavior. We have previously discussed the fact that bacteria can move toward or away from particular chemicals, a process referred

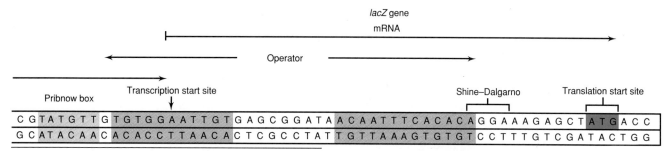

lacZ gene
mRNA

Operator

Pribnow box

Transcription start site

Shine–Dalgarno

Translation start site

```
C G T A T G T T G T G T G G A A T T G T G A G C G G A T A A C A A T T T C A C A C A G G A A A G A G C T A T G A C C
G C A T A C A A C A C A C C T T A A C A C T C G C C T A T T G T T A A A G T G T G T C C T T T G T C G A T A C T G G
```

Figure 7.20 (continued)

to as *chemotaxis* (∞ Section 3.10). We noted that bacteria are too small to actually sense *spatial* gradients of a chemical, but rather they respond to *temporal* gradients. That is, they can sense the change in concentration of a chemical outside the cell *over time.* Bacteria use a two-component system to sense the temporal changes in chemical concentration and regulate flagellar motion.

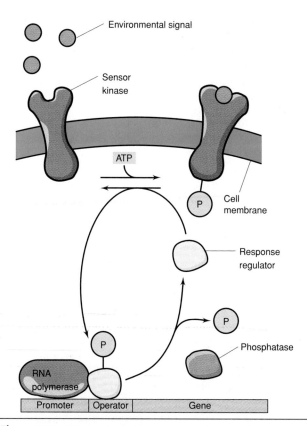

Figure 7.21 The control of gene expression by a two-component system. The main components of the system include a *sensor kinase* in the cell membrane that phosphorylates itself in response to an environmental signal. The phosphoryl group is then transferred to the other main component, a *response regulator.* In the system diagrammed in this figure the phosphorylated response regulator serves as a repressor. There must also be a phosphatase in the system to cycle the response regulator.

Mechanism of chemotaxis

The mechanism of chemotaxis is quite complex and involves a variety of different proteins. A number of *sensory proteins* are in the cell membrane, and these sense the presence of attractants and repellants. These proteins allow the cell to sense whether, over time, the concentration of the substance increases or decreases as the cell moves. The cell thus responds to the *change* in concentration rather than the *absolute* concentration of the chemical stimulus. The sensory proteins are called *methyl-accepting chemotaxis proteins* (**MCPs**), or *receptor-transducer proteins,* or simply **transducers.** In *Escherichia coli,* four different MCPs have been identified, and each is a transmembrane protein (Figure 7.22). Each MCP can sense a variety of compounds. For example, the *Tar* transducer of *E. coli* can sense the attractants aspartate and maltose as well as repellants such as the heavy metals cobalt and nickel.

MCPs bind attractants or repellants directly, or in some cases indirectly, through interactions with periplasmic binding proteins. Binding of an attractant or repellant sets in play a series of interactions with cytoplasmic proteins that eventually affects flagellar rotation. If rotation of the flagellum is *counterclockwise,* the cell will continue to move in a run. If the flagellum rotates *clockwise,* however, the cell will tumble (∞ Section 3.10).

The current model for flagellar control shows that the transducers are in contact with the cytoplasmic proteins CheW and CheA (Figure 7.22). CheA is the *sensor kinase* in this two-component system. When a transducer has bound a chemical, it changes conformation and (with CheW) causes a change in the autophosphorylation of CheA (forming CheA-P). *Attractants* decrease the rate of autophosphorylation, whereas *repellants increase* this rate. Phosphorylated CheA (CheA-P) then phosphorylates CheY (forming CheY-P), a *response regulator.* CheY-P interacts with the flagellar motor to induce clockwise flagellar rotation and tumbling (the motor switch itself consists of proteins encoded by *fla* genes).

CheA-P can also phosphorylate CheB, another response regulator, but this is a much slower reaction

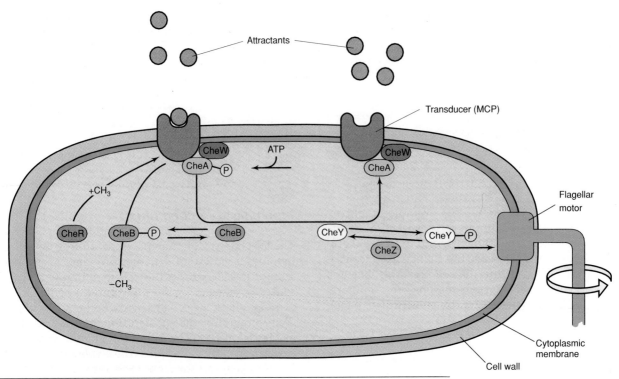

Figure 7.22 Interactions of transducers, chemotaxis (Che) proteins, and the flagellar motor in bacterial chemotaxis. The transducer (MCP) forms a complex with the *sensor kinase* CheA and the coupling protein CheW. This combination results in a signal-regulated autophosphorylation of CheA to CheA-P. CheA-P can then phosphorylate the *response regulators* CheB and CheY. Phosphorylated CheY (CheY-P) interacts directly with the flagellar motor switch. CheZ dephosphorylates CheY-P. CheR continually adds methyl groups to the transducer. CheB-P (but not CheB) removes them. The degree of methylation of the transducers controls their ability to respond to attractants and repellants and leads to adaptation. The structure of the flagellar motor was shown in Figure 3.49.

than the phosphorylation of CheY. We shall discuss the activity of CheB-P later. Thus, CheY is the central protein in the system because it serves as the *response regulator* for chemotaxis, governing the direction of rotation of the flagellum. When CheY is phosphorylated, the flagellar motor switches from a counterclockwise to a clockwise rotation, causing the cell to tumble. If unphosphorylated, CheY cannot bind, the flagellar motor continues counterclockwise rotation, and the cell undergoes a run. Another protein, CheZ, dephosphorylates CheY, returning it to a form promoting runs instead of tumbles. Because repellants increase the level of CheY-P, they lead to tumbling, whereas attractants lead to a lower level of CheY-P and smooth swimming.

Note that the system described can *signal* that a chemical has been bound and regulate flagellar rotation but seems to be unable to note a change with the passage of time. There is a second component to chemotaxis, and this is **adaptation.**

As their name implies, MCPs can be methylated. There is a cytoplasmic protein, CheR, that continually adds methyl groups to the MCPs at a slow rate using S-adenosylmethionine as a methyl donor. The phos-

phorylated form of the response regulator CheB is a demethylase that removes methyl groups from the MCPs. The level of methylation of the MCPs affects their conformation and controls adaptation to a sensory signal. It allows resetting of the signaling state of the receptor even though the concentration of the chemical remains unchanged.

If the level of an attractant remains high, the level of phosphorylation of CheA (and, therefore, of CheY and CheB) will remain low, the cell will swim smoothly, and the level of methylation of the MCPs will increase (because CheB-P is not present to demethylate). However, the MCPs no longer respond to the attractant when they are fully methylated. Therefore, even though the level of attractant might remain high, the level of CheA-P (and CheB-P) increases and the cell begins to tumble. However, now the MCPs can be demethylated by CheB-P, and when this happens, the receptors can once again respond to attractants. The situation is the opposite with regard to repellants (fully methylated MCPs respond best to repellants).

The control of chemotaxis is obviously quite complicated and involves a number of regulatory switches.

Unlike the case in many other two-component systems, in chemotaxis the signal transduction system regulates the activity of the gene products, not their synthesis. Signal transduction is an important regulatory mechanism in both prokaryotes and eukaryotes.

7.8

Contrasts in Gene Expression between Prokaryotes and Eukaryotes

We have discussed only a few of the mechanisms by which cells can control the activity of a protein and also only a few major mechanisms that can regulate the synthesis of a protein. Most of the mechanisms we considered for regulating synthesis operate at the level of transcription, and all involve regulatory proteins. Interestingly, regulatory RNA also exists (see the box, Antisense Nucleic Acid).

Although many of the major regulatory patterns are shared between prokaryotes and eukaryotes, there are many differences. Because of the lack of compartmentation in prokaryotes, the processes of transcription and translation are *coupled*. Also, the messenger RNA of prokaryotes is frequently polycistronic, with more than one protein being translated from the same message.

In eukaryotes, on the other hand, transcription and translation take place in separate compartments in the cell and the integration of these processes seen in prokaryotes is lacking. How about induction and repression in eukaryotes? Although many eukaryotes do exhibit repression, there is no good evidence for the kind of negative control so commonly found in prokaryotes. However, positive control mechanisms are common in eukaryotes. If operons exist in eukaryotes, they involve the control of only single enzymes, rather than the multienzyme control systems so commonly seen in prokaryotes, and there is no evidence for polycistronic RNA molecules in eukaryotes except in a few viruses. These are translated inefficiently. However, eukaryotes occasionally and viruses commonly make single large proteins (polyproteins) from a single monocistronic mRNA, and these proteins are then cleaved into several smaller active proteins. Posttranslational cleavage is rare in prokaryotes. Eukaryotes also have regulation involving splicing of mRNA, regulation that does not exist in prokaryotes.

CONCEPT CHECK   7.7

Signal transduction systems transmit environmental signals to the cell. In prokaryotes signal transduction typically involves two-component systems, which include a sensor protein located in the membrane and a cytoplasmic response regulator protein. The sensor protein is a kinase, and the activity of the response regulator depends on its state of phosphorylation. Most two-component systems regulate transcription, but that regulating bacterial chemotaxis operates at the level of protein activity.

✔ *What are* kinases *and what is their role in two-component regulatory systems?*

✔ *Could a response regulator be an activator or a repressor?*

ANTISENSE NUCLEIC ACID

Regulation of the synthesis of proteins often involves transcriptional control. Less often, genes are controlled at the level of translation. Most control networks, whether they are transcriptional or translational, utilize regulatory proteins. However, it is now clear that in some cases it is a regulatory *RNA*, not a regulatory *protein*, that is involved. One type of regulatory RNA, called **antisense RNA**, is known to be used in the regulation of several different bacterial genes. Antisense RNA acts by forming base pairs with a complementary, or sense, strand of RNA. When the sense RNA is mRNA, the resulting double-stranded structure can prevent translation. Antisense RNA can be synthesized from the same gene as the sense RNA by having a second promoter oriented in the direction opposite that of the first or by having a second gene with the promoter at the other end. Antisense RNA does not have to be used only to regulate the synthesis of a protein. In some plasmids, it controls the initiation of DNA synthesis.

Antisense nucleic acids can be specifically designed and synthesized by scientists in the laboratory and delivered directly to cells. These short (15–25 nucleotides) synthetic chains (oligonucleotides) are usually made of DNA rather than RNA. Their sequence can be made to allow them to bind to a specific mRNA and prevent translation (or allow the molecule to be recognized by nucleases). Antisense nucleic acid can also bind to the DNA in the nucleus and prevent transcription. The latter is possible because some DNA can form a *triple* helix! The "extra" strand (the oligonucleotide) forms specific interactions with those parts of the bases that are in the major groove of a normal double helix to give **triplex DNA.** (Not all DNA sequences can form triple helices, at least not without the aid of special enzymes.)

Synthetic antisense oligonucleotides can be designed to be extremely specific, whether they bind to a message or to the regulatory region of a gene. The specificity arises because a sequence of only 20 bases should occur no more often than once in 10^{12} bases of "random" DNA. Therefore, it is unlikely that the antisense RNA would bind to anything other than its known target in any cell. This specificity might allow antisense nucleic acids to become an important new type of antibiotic, and this possibility is being pursued by a number of pharmaceutical companies. Antisense nucleic acids could be designed to be used against specific viruses or to regulate particular genes in either disease-causing (pathogenic) organisms or human tumor cells. The possible utility of these molecules is just one example of how understanding the structure of a gene may have very practical applications (Chapter 10).   ■

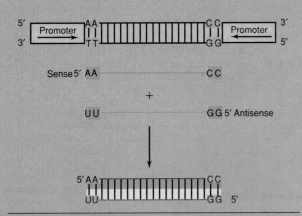

A gene with promoters at either end. If the RNA made by the promoter on the left is the *sense* RNA, then the RNA made by the promoter on the right is the *antisense* RNA. If both RNA molecules are made, they will form a duplex. Only a relatively short region of overlap is necessary for strong base pairing. Usually the antisense RNA is shorter than the sense RNA, and therefore the second promoter is actually within the gene.

A triple helix. The "extra" strand is shown in red and is in the major groove of a double helix.

Material for Review

REVIEW QUESTIONS

1. If an enzyme can be effectively inhibited by feedback inhibition, why would cells also have mechanisms to regulate its synthesis?

2. Describe why a protein that binds to a specific sequence of double-stranded DNA is unlikely to bind to the same sequence if the DNA is single-stranded.

3. Describe the regulation of two different operons, one having an effector that is an *inducer* and the other having an effector that is a *corepressor.*

4. The maltose regulon is inducible and is regulated by an activator protein. The lactose operon is inducible but is regulated by a repressor protein. Explain how induction can be brought about by either positive control (activator protein) or negative control (repressor protein).

5. In most cases operators are very close to the promoters they control, while activator binding sites can be some distance away. Explain why this should be so.

6. Describe how transcriptional attenuation works. What is actually being "attenuated"? Why hasn't the type of attenuation that controls several different amino acid biosynthetic pathways in *Escherichia coli* also been found in eukaryotes?

7. Describe the mechanism by which catabolic activator protein (CAP), the regulatory protein for catabolite repression, functions using the lactose operon as an example. For this operon the CAP protein is not a repressor. Describe the regulatory region of a gene for which the CAP protein *is* a repressor. (*Hint:* Think about your answer to Question 5.)

8. What are the two components that give the name to signal transduction regulation in prokaryotes? What is the function of each of the components?

9. One of the members of a two-component system is typically located in the cell membrane. What reason can you think of why this might be so?

10. Many genes are under multiple control systems. In the lactose operon, there is lactose-specific regulation and regulation by a global control system. Describe how each of the controls on the lactose operon actually functions. Why do you think both systems are necessary?

APPLICATION QUESTIONS

1. The amino acids isoleucine and valine share a common pathway for most steps in their biosynthesis. In *Escherichia coli* the first common step can be subject to feedback inhibition by valine but not by isoleucine. In most strains, though, the addition of valine does not cause isoleucine deprivation. However, in other strains it does (that is, adding valine causes isoleucine starvation and the cells stop growing). What explanation can you give for the difference between the normal "valine-resistant" strains and those whose growth is sensitive to valine?

2. What would happen to regulation from a promoter under negative control if the region where the regulatory protein binds were deleted? What if the promoter were under positive control?

3. Promoters from *Escherichia coli* under positive control are not close matches to the DNA consensus sequence for *E. coli* (∞ Section 6.6). Why?

4. Interestingly, the attenuation control of some of the pyrimidine biosynthetic pathway genes in *Escherichia coli* actually involves coupled transcription and translation. Can you describe a mechanism whereby the cell could somehow make use of translation to help it measure the level of pyrimidine nucleotides?

SUPPLEMENTARY READINGS

Alberts, B., D. Bray, J. Lewis, M. Raff, K. Roberts, and **J. D. Watson**. 1994. *Molecular Biology of the Cell*, 3rd edition. Garland, New York. This standard reference text for cellular molecular biology covers most areas of the regulation of gene expression. There is a heavy emphasis on eukaryotes.

Hoch, J.A., and **T. J. Silhavy**. 1995. *Two-Component Signal Transduction*. American Society for Microbiology, Washington, DC. This volume gives detailed information on a number of different two-component regulatory systems in Bacteria.

Lodish, H., D. Baltimore, A. Berk, S. L. Zipursky, P. Matsudaira, and **J. Darnell**. 1995. *Molecular Cell Biology*, 3rd edition. Scientific American Books, New York. This is another large reference text covering most areas of molecular biology and gene regulation, with a strong emphasis on eukaryotes.

Neidhardt, F. C., R. Curtiss III, J. L. Ingraham, E. C. C. Lin, K. B. Low, B. Magasanik, W. Reznikoff, M. Riley, M. Schaechter, and **H. E. Umbarger** (eds.). 1996. Escherichia coli *and* Salmonella: *Cellular and Molecular Biology*, 2nd edition. American Society for Microbiology, Washington, DC. This compendium contains information on the regulation of many specific pathways and on many important cellwide systems in these important organisms.

Stryer, L. 1995. *Biochemistry*, 3rd edition. W. H. Freeman. New York. An excellent biochemistry text with considerable information on the regulation of enzyme activity and synthesis.

 On~line resources for this chapter are on the World Wide Web at: http://www.prenhall.com/~brock (click on the <u>table of contents</u> link and then select Chapter 7).

CHAPTER *8*

Viruses

Viruses can multiply only within cells. However, all viruses, such as the poliovirus shown here, have extracellular forms by which they can be transmitted from cell to cell and organism to organism.

MINIGLOSSARY for Chapter 8

Bacteriophage a virus that infects prokaryotic cells

Lysogen a bacterium containing a prophage

Minus (Negative)-Strand Nucleic Acid an RNA or DNA strand that has the opposite sense of (is complementary to) the mRNA of a virus

Oncogene a gene whose expression causes formation of a tumor

Plaque a zone of lysis or cell inhibition caused by virus infection of a lawn of sensitive cells

Plus (Positive)-Strand Nucleic Acid an RNA or DNA strand that has the same sense as the mRNA of a virus

Prion an infectious agent whose extracellular form may contain no nucleic acid

Provirus (Prophage) the genome of a temperate virus when it is replicating with, and usually integrated into, the host chromosome

Retrovirus a virus whose RNA genome has a DNA intermediate as part of its replication cycle

Reverse Transcription the process of copying information found in RNA into DNA

Temperate Virus a virus whose genome is able to replicate along with that of its host and not cause cell death in a state called lysogeny

Transformation a process by which a normal cell becomes a cancer cell (but see alternative usage in Chapter 9)

Virion the complete virus particle; the nucleic acid surrounded by a protein coat and in some cases other material

Virulent Virus a virus that lyses or kills the host cell after infection; a nontemperate virus

Virus a genetic element containing either RNA or DNA that replicates in cells but is characterized by having an extracellular state

---

Viruses are genetic elements that can replicate independently of a cell's chromosomes but not independently of cells themselves (∞ Section 6.3). In order to multiply, viruses must enlist a cell in which they can replicate. Such a cell is called a **host.** Viruses are characterized by also having an extracellular state.

Viruses are not the only type of genetic element that takes advantage of the metabolic machinery encoded by the cell's own chromosomes (∞ Section 6.3). Like these other elements, viruses can confer important new properties on their host cell. These properties will be inherited when the host cell divides if each new cell also inherits the viral genome. These changes are often not harmful and may even be beneficial. However, viruses, unlike genetic elements such as plasmids (∞ Sections 6.3 and 9.8), have an extracellular form that enables them to easily transmit themselves from one host to another. This extracellular form has enabled some viruses to replicate themselves in a host in a way that is destructive to the host cell. This destructive replication accounts for the fact that some viruses are agents of disease. In many cases, whether a virus causes disease or hereditary change depends on the host cell and on the environmental conditions.

In this chapter we shall discuss some of the ways in which viruses can redirect the metabolism of the host cell in order to replicate. This chapter is divided into three parts. The first part deals with basic concepts of virus structure and function. The second part deals with the nature and manner of multiplication of the bacterial viruses (bacteriophages). The third part deals with im-portant groups of animal viruses. In both the second and third parts we shall describe some of the basic molecular biology of virus multiplication. These discussions expand on the concepts of macromolecular synthesis and gene regulation we covered in Chapters 6 and 7.

Scientists have studied and continue to study viruses for what they can tell us about the genetics and biochemistry of cellular metabolism and, in the case of some viruses, the development of disease. However, as we shall see in Chapters 9 and 10, viruses are also important tools for the microbial geneticist and the genetic engineer.

8.1
General Properties of Viruses

Viruses have both an extracellular and an intracellular state. In the **extracellular** state, a virus is a submicroscopic particle containing nucleic acid surrounded by protein and occasionally containing other macromolecular components. In this extracellular state, the **virus particle,** also called the **virion,** is metabolically inert and does not carry out respiratory or biosynthetic functions. The virion is the structure by which the **virus genome** is carried from the cell in which it has been produced to another cell where the viral nucleic acid can be introduced. Once in the new cell, the **intracellular state** is initiated. In the intracellular state, **virus replication** occurs: the virus genome is produced, and the components that make up the virus coat are synthesized. When a virus genome is introduced into a host cell and reproduces, the process is called **infec-**

tion. A cell that a virus can infect and in which it can replicate is called a **host.** Viral genomes are very limited in size, and they encode primarily those functions that they cannot adapt from their hosts. Therefore, during replication inside a cell, there is a heavy dependence on host cell structural and metabolic components. The virus redirects preexisting host machinery and metabolic functions necessary for virus replication.

As we have seen (∞ Section 6.1), all cells have double-stranded deoxyribonucleic acid (DNA) as their genetic material. However, viruses can have either DNA or ribonucleic acid (RNA) as their genetic material, and it can be either single-stranded or double-stranded. Viruses are sometimes divided into two types based on whether they have DNA or RNA as their genetic material, and *all* viruses contain one or the other in the virion. However, there is a third group of viruses that use *both* DNA and RNA as their genetic material but at different stages of their reproductive cycle (Figure 8.1). The latter include the retroviruses, which contain an RNA genome in the virion but replicate through a DNA intermediate, and the human hepatitis B virus, which contains DNA in the virion but has an RNA intermediate in replication. These classes can be further subdivided according to whether the nucleic acid in the virion is single- or double-stranded (Figure 8.1). In spite of the diversity of genome structure, viruses obey the *central dogma* (∞ Section 6.1): all genetic information flows from nucleic acid to protein. In addition, all viruses use the cell's translational machinery, and so no matter what the genome structure of the virus, messenger RNA (mRNA) must be generated that can be translated on the host's ribosomes.

Viruses can also be classified on the basis of the hosts they infect. Thus, we have animal viruses, plant viruses, and bacterial viruses. Bacterial viruses, sometimes called *bacteriophages* (or *phage* for short, from the Greek *phagein* meaning "to eat"), have been studied

---

THE NAME "VIRUS"

The word **virus** originally referred to any poisonous emanation, such as the venom of a snake, and later came to be used more specifically for the causative agent of any infectious disease. Pasteur often referred to bacteria that caused infectious diseases as viruses. By the end of the nineteenth century, a large number of bacteria had been isolated and shown to be causal agents of specific infectious diseases, but there were some diseases for which a bacterial cause had not been shown. One of these was foot-and-mouth disease, a serious skin disease of animals. In 1898, Friedrich Loeffler and Paul Frosch presented the first evidence that the cause of foot-and-mouth disease was an agent so small that it could pass through filters that could hold back all known bacteria. That the agent was not an ordinary toxin could be shown by the fact that it was active at very low dilution and could be transmitted in filtered material from animal to animal. Loeffler and Frosch concluded "that the activity of the filtrate is not due to the presence in it of a soluble substance, but due to the presence of a causal

agent capable of reproducing. This agent must then be obviously so small that the pores of a filter which will hold back the smallest bacterium will still allow it to pass. . . . If it is confirmed by further studies . . . that the action of the filtrate . . . is actually due to the presence of such a minute living being, this brings up the thought that the causal agents of a large number of other infectious diseases . . . which up to now have been sought in vain, may also belong to this smallest group of organisms."

A year later, the Dutch microbiologist Martinus Beijerinck published his work on tobacco mosaic disease, a crippling leaf disease of tobaccos and tomatoes. In 1892, D. Ivanowsky of Russia had first shown that the causal agent of tobacco mosaic disease was filterable, but Beijerinck went much further and provided strong evidence that although the causal agent was filterable, it had many of the properties of a living organism. He called the agent a *Contagium vivum fluidum*, a living germ that is soluble. He postulated that the agent must be incorporated into the living protoplasm of the cell in order to reproduce, and that its reproduction must be brought

about with the reproduction of the cell. This postulate comes very close to our current understanding of how viruses reproduce. Beijerinck also noted that there were other plant diseases for which causal agents had not been isolated, and these might also be caused by filterable agents. Soon a number of other filterable agents were shown to be the causes of both plant and animal diseases. Such agents came to be called **filterable viruses,** but as further work on these agents was carried out, the word "filterable" was gradually dropped. Today, the original meaning of "virus" has been forgotten, and the word is now used to refer to the kinds of agents discussed in this chapter. Bacterial viruses were first discovered by the British scientist F. W. Twort in 1915, and independently by the French scientist F. d'Herelle in 1917, who called them *bacteriophages* (from the combining form *phago* meaning "to eat"). Although bacteriophages are viruses, the name "phage" is still widely used to refer to this particular class of filterable infectious agents. ■

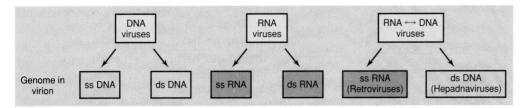

Figure 8.1 Viral genomes. The genomes of viruses can be composed of either DNA or RNA, and some use both as their genomic material at different stages in their life cycle. However, only one type of nucleic acid is found in the virion of any particular type of virus. This can be single-stranded (ss), double-stranded (ds), or in the case of the hepadnaviruses, partially double-stranded.

primarily as convenient model systems for research on the molecular biology and genetics of virus reproduction. Many of the basic concepts of virology were first worked out with bacterial viruses and subsequently applied to viruses of higher organisms. Because of their frequent medical importance, *animal viruses* have been extensively studied. The two groups of animal viruses most studied are those infecting insects and those infecting warm-blooded animals. *Plant viruses* are often important in agriculture but have been less studied than animal viruses. In this chapter, we discuss the structure, replication, and genetics of viruses infecting bacteria and warm-blooded animals.

CONCEPT CHECK   8.1

A virion is the extracellular form of a virus and contains either an RNA or a DNA genome. The virus genome is introduced into a new host cell by infection. The virus redirects the host metabolism in order to replicate.

✔ *How does a virus differ from a plasmid?*

✔ *How does a virion differ from a cell?*

8.2
Nature of the Virion

Virions vary widely in size and shape. Viruses are smaller than cells, ranging in size from 0.02 to 0.3 μm. A common unit of measure for viruses is the *nanometer*, which is 1000 times smaller than 1 μm and 1 million times smaller than 1 mm. Smallpox virus, one of the largest viruses, is about 200 nm in diameter; poliovirus, one of the smallest, is only 28 nm in diameter.

As we have stated, some viruses contain RNA, others DNA, and the nucleic acid can be either double- or single-stranded, depending on the virus. Viral *genomes* are also smaller than those of cells. Most bacterial genomes are between 1000 and 9000 kilobase pairs of DNA, with the smallest known being about 590 kilobase pairs. (Interestingly, the bacteria with the smallest genomes are, like viruses, parasites that replicate in other cells; ∞ Sections 16.22 and 16.23.) However, one of the largest known viral genomes, that of vaccinia, is only 190 kilobase pairs. Some viruses have genomes so small they contain less than five genes. The sizes of the genomes of a few representative types of viruses are given in Table 8.1. As can be seen in the table, the

---

TABLE 8.1 Some types of viral genomes[a]

Virus	Host	Type of nucleic acid in virion	Structure	Number of molecules	Size
H-1 parvovirus	Animals	Single-stranded DNA	Linear	1	5,176 bases
φX174	Bacteria	Single-stranded DNA	Circular	1	5,386 bases
Simian virus 40 (SV40)	Animals	Double-stranded DNA	Circular	1	5,224 base pairs
Poliovirus	Animals	Single-stranded RNA	Linear	1	7,433 bases
Cauliflower mosaic virus	Plants	Double-stranded DNA	Circular	1	8,025 base pairs
Cowpea mosaic virus	Plants	Single-stranded RNA	Linear	2 different	9,370 bases (total)
Reovirus type 3	Animals	Double-stranded RNA	Linear	10 different	23,549 base pairs (total)
Bacteriophage λ	Bacteria	Double-stranded DNA	Linear	1	48,514 base pairs
Herpes simplex virus type I	Animals	Double-stranded DNA	Linear	1	152,260 base pairs

[a]The sizes of the viral genomes chosen for this table are known accurately because they have been sequenced. However, this accuracy can be misleading because only a particular strain or isolate of a virus was sequenced. Therefore, the sequence and exact number of bases for other isolates may be slightly different. No attempt has been made to choose the largest and smallest viruses known, but rather to give a fairly representative sampling of the sizes and structures of the genomes of viruses containing both single- and double-stranded RNA and DNA.

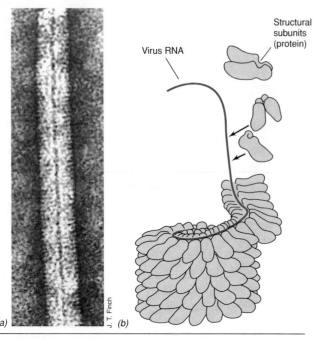

J. T. Finch

Figure 8.2 An example of the arrangement of virus nucleic acid and protein coat in a simple virus, tobacco mosaic virus. (a) Electron micrograph at high resolution of a portion of the virus particle. (b) Assembly of the tobacco mosaic virion. The RNA assumes a helical configuration surrounded by the protein capsid. The center of the particle is hollow.

genome of some viruses, such as reovirus, is not present in a single molecule but is segmented into more than one molecule.

The structures of virions (virus particles) are quite diverse, varying widely in size, shape, and chemical composition. The nucleic acid of the virion is always located within the particle, surrounded by a protein coat called the *capsid*. The terms *coat, shell,* and *capsid* are often used interchangeably to refer to this outer layer. The protein coat is always formed of a number of individual protein molecules, called *structural subunits*, which are arranged in a precise and highly repetitive pattern around the nucleic acid (Figure 8.2). The small genome size of most viruses restricts the number of different viral proteins. A few viruses have only a single kind of protein in their capsid, but most viruses have several chemically distinct kinds of structural subunits that are themselves associated in specific ways to form larger assemblies called *morphological units* or capsomers. It is the morphological unit that is seen with the electron microscope.

The information for proper aggregation of the structural subunits into capsomers is contained within the structure of the proteins themselves, and the overall process of assembly is thus called **self-assembly.** For many viruses, this self-assembly process is assisted by *molecular chaperones,* proteins that assist in folding and assembly but that themselves are not a part of the final

structure (∞ Section 6.9). A single virion generally has a large number of morphological units.

The complete complex of nucleic acid and protein, packaged in the virus particle, is called the virus **nucleocapsid.** Although the virus structure just described is frequently the total structure of a virus particle, a number of viruses have more complex structures. These viruses are *enveloped* viruses in which the nucleocapsid is enclosed in a membrane (Figure 8.3). (Viruses without membranes are sometimes called *naked* viruses.) *Virus membranes* are generally lipid bilayer membranes (∞ Section 3.3), but associated with these membranes are often *virus-specific* proteins. Inside the virion are often one or more virus-specific *enzymes*. Such enzymes usually play a role during the infection and replication process, as we will discuss later in this chapter.

Virus symmetry

The nucleocapsids of viruses are constructed in highly symmetric ways. Symmetry refers to the way in which the protein morphological units are arranged in the virus shell. When a symmetric structure is rotated around an axis, the same form is seen again after a certain number of degrees of rotation. Two kinds of symmetry are recognized in viruses, which correspond to the two primary shapes, rod and spherical. Rod-shaped viruses have helical symmetry, and spherical viruses have icosahedral symmetry. In all cases, the characteristic structure of the virus is determined by the structure of the protein subunits of which it is constructed.

A typical virus with **helical symmetry** is the tobacco mosaic virus (TMV) illustrated in Figure 8.2. It is an RNA virus in which the 2130 identical protein subunits (each 158 amino acids in length) are arranged in a helix. In TMV, the helix has $16\frac{1}{2}$ subunits per turn, and the overall dimensions of the virion are 18×300 nm. The lengths of helical viruses are determined by the length of the nucleic acid, but the width of the helical virus particle is determined by the size and packaging of the protein subunits.

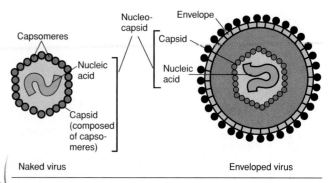

Figure 8.3 Comparison of naked and enveloped virus, two basic types of virus particles.

An **icosahedron** is a symmetric structure roughly spherical in shape that has 20 faces. Icosahedral symmetry is the most efficient arrangement for subunits in a closed shell because it uses the smallest number of units to build a shell. The simplest arrangement of morphological units is 3 per face, for a total of 60 units per virus particle. The 3 units at each face can be either identical or different. Most viruses have more nucleic acid than can be packed into a shell made of just 60 morphological units. The next possible structure that permits close packing contains 180 units, and many viruses have shells with this configuration. Other known configurations involve 240 units and 420 units.

To help understand icosahedral symmetry, a model can be made following the instructions given in Figure 8.4. When discussing symmetry, one speaks of *axes of rotation*. A flat triangle shape, for instance, has one threefold axis of symmetry because there are three possible rotations that will lead to the exact configuration seen originally. Three-dimensional objects such as viruses can have more than one axis of symmetry. An icosahedron, for instance, has three different axes of symmetry, twofold, threefold, and fivefold (see Figure 8.4). When a rod is placed through the twofold axis of symmetry (one of the edges) in the model, the model can be turned once around this axis (one-half of the way or 180°) to

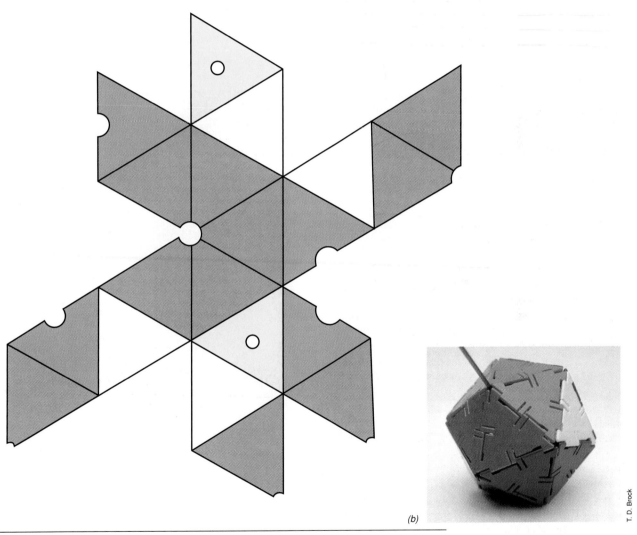

(a)

(b)

T. D. Brock

Figure 8.4 Demonstration of icosahedral symmetry. (a) A pattern that can be used to make a model of an icosahedral virus. Make a photocopy of this figure, cut it out, and fold at every line. Then tape the adjoining faces together to make a three-dimensional structure. (A copy machine that enlarges will provide a copy that can be more easily folded.) When the structure is folded, the axes of symmetry are evident. Cut the paper at these axes, as marked, before folding. A metal rod or wire can then be inserted completely through the center of the model to study each axis. (b) Photograph of a model showing a rod through the axis of fivefold symmetry.

obtain the same configuration again. When the rod is placed through one of the threefold axes of symmetry (one of the faces), the model can be turned three times, and if the rod is placed through one of the fivefold axes of symmetry (one of the vertices) the model can be turned five times. An electron micrograph of a typical icosahedral virus is shown in Figure 8.5a.

Enveloped viruses

Many viruses have complex membranous structures surrounding the nucleocapsid (Figure 8.5b). Enveloped viruses are common in the animal world (for example, influenza virus), but some enveloped bacterial viruses are also known. The virus envelope consists of a lipid bilayer with proteins, usually glycoproteins, embedded in it. Although the glycoproteins of the virus membrane are encoded by the virus, the lipids are derived from the membranes of the host cell; proteins of the host cell membrane are somehow excluded. The symmetry of enveloped viruses is expressed not in terms of the virion as a whole but in terms of the nucleocapsid present inside the virus membrane.

What is the function of the membrane in a virus particle? We will discuss this in detail later but note that because of its location in the virion, the membrane is the structural component of the virus particle that interacts first with the cell. The specificity of virus infection, and some aspects of virus penetration, are controlled in part by characteristics of virus membranes.

Complex viruses

Some virions are even more complex, being composed of several separate parts with separate shapes and symmetries. The most complicated viruses in terms of struc-

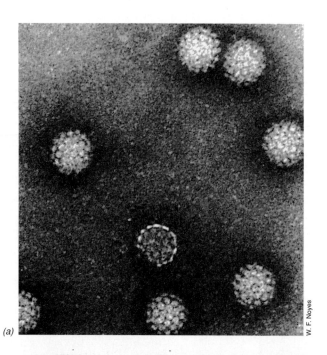

(a)

W. F. Noyes

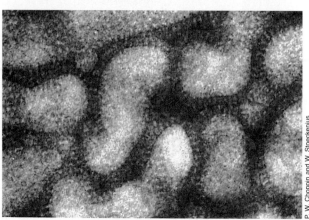

(b)

P. W. Choppin and W. Stoeckenius

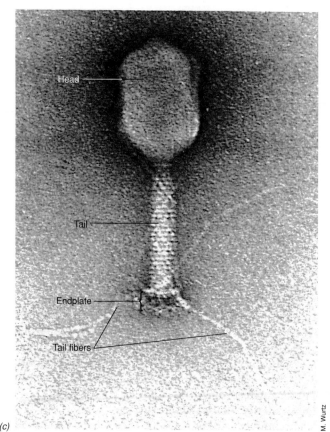

(c)

Head

Tail

Endplate

Tail fibers

M. Wurtz

Figure 8.5 Electron micrographs of various virus particles. (a) Human wart virus, a virus with icosahedral symmetry. The individual particles are about 55 nm in diameter. (b) Influenza virus, an enveloped virus. The individual particles are about 80 nm in diameter. (c) Bacterial virus (bacteriophage) T4 of *Escherichia coli*. Note the complex structure. The tail components are involved in attachment of the virion to the host and infection of the nucleic acid. The head is about 85 nm in diameter.

ture are some of the bacterial viruses, which possess not only icosahedral heads but also helical tails (Figure 8.5c). In some bacterial viruses, such as the T4 virus of *Escherichia coli,* the tail itself is a complex structure. For instance, T4 has almost 20 separate proteins in the tail, and the T4 head has several more proteins. In such complex viruses, assembly is also complex. For instance, in T4 the complete tail is formed as a subassembly, and then the tail is added to the DNA-containing head. Finally, tail fibers formed from another protein are added to make the mature, infectious virus particle (see the discussion of T4 assembly in Section 8.11).

Enzymes in virions

We have stated that virions do not carry out metabolic processes. Outside a host cell, a virion is metabolically inert. However, some virions do contain enzymes that play roles in the infection process. For instance, many viruses contain their own nucleic acid polymerases that transcribe the viral nucleic acid into messenger RNA once the infection process has begun. The retroviruses are RNA viruses that replicate inside the cell as DNA intermediates. These viruses possess an enzyme, an RNA-dependent DNA polymerase called *reverse transcriptase,* that transcribes the information in the incoming RNA into a DNA intermediate. A number of viruses contain enzymes that aid in entering cells or in release of the virus from the host cells in the final stages of the infection process. One group of such enzymes, called *neuraminadases,* breaks down glycosidic bonds of glycoproteins and glycolipids of the connective tissue of animal cells, thus aiding in the liberation of the virus. Virions infecting some bacteria possess an enzyme, *lysozyme* (∞ Section 3.5), that makes a small hole in the bacterial cell wall that allows the viral nucleic acid to

enter. The same enzyme is produced in large amounts in the later stages of infection, causing lysis of the host cell and release of the virions. We will discuss some of these enzymes in more detail later.

8.3
The Virus Host

Because viruses, like plasmids, replicate only inside living cells, research on viruses requires use of appropriate hosts. For the study of bacterial viruses, pure cultures are used either in liquid or on semisolid (agar) media. Because most bacteria are so easy to culture, it is quite easy to study bacterial viruses, and this is why such detailed knowledge of bacterial virus multiplication is available.

With animal viruses, the initial host may be a whole animal that is susceptible to the virus, but for research purposes it is desirable to have a more manageable host. Many animal viruses can be cultivated in *tissue* or *cell cultures,* and the use of such cultures has enormously facilitated research on animal viruses.

Cell cultures

A cell culture is obtained by promoting growth of cells taken from an organ of the experimental animal. Cell cultures are generally obtained by aseptically removing pieces of the tissue in question, dissociating the cells by treatment with an enzyme that breaks apart the intercellular cement, and spreading the resulting suspension out on the bottom of a flat surface, such as a bottle or a Petri dish. The cells generally produce glycoprotein-like materials that permit them to adhere to glass surfaces. The thin layer of cells adhering to the glass or plastic dish, called a *monolayer,* is then overlaid with a suitable culture medium and the culture incubated. The culture media used for cell cultures are generally quite complex, employing a number of amino acids and vitamins, salts, glucose, and a bicarbonate buffer system. To obtain best growth, addition of a small amount of blood serum is usually necessary, and several antibiotics are generally added to prevent bacterial contamination.

Some cell cultures prepared in this way grow indefinitely and can be established as *permanent cell lines.* Such cell cultures are most convenient for virus research because cell material is continuously available for research purposes. In other cases, indefinite growth does not occur, but the culture may remain alive for a number of days. Such cultures, called *primary cell cultures,* may still be useful for virus research, although new cultures will have to be prepared from fresh sources from time to time.

In some cases, cell culture monolayers cannot be obtained, but whole organs, or pieces of organs, can be cultured. Such **organ cultures** may still be useful in

---

CONCEPT CHECK   **8.2**

In the virion of the naked virus, only nucleic acid (DNA or RNA) and protein are present, with the nucleic acid on the inside; the whole unit is called the nucleocapsid. Enveloped viruses have one or more lipoprotein layers surrounding the nucleocapsid. The nucleocapsid is arranged in a symmetric fashion, with a precise number and arrangement of structural subunits surrounding the virus nucleic acid. Although viruses are metabolically inert, in some viruses, one or more enzymes are present within the virion. Such enzymes play a role in the initial stages of the infection process.

✔ *What is the difference between a* naked virus *and an* enveloped virus?

✔ *Where are the genes located that encode virus-specific proteins?*

virus research because they permit growth of viruses under more-or-less controlled laboratory conditions.

8.4
Quantification of Viruses

In order to obtain any significant understanding of the nature of viruses and virus replication, it is necessary to be able to *quantify* the number of virus particles. Virions are almost always too small to be seen under the light microscope. Although they can be observed under the electron microscope, the use of this instrument is cumbersome for routine study. In general, viruses are quantified by measuring their effects on the host cells that they infect. It is common to speak of a *virus infectious unit*, which is the smallest unit that causes a detectable effect when placed with a susceptible host. By determining the number of infectious units per volume of fluid, a measure of virus quantity can be obtained. We discuss here several approaches to assessment of the virus infectious unit.

Plaque assay

When a virus particle initiates an infection on a layer or lawn of host cells growing spread out on a flat surface, a zone of *lysis* or *growth inhibition* may occur that results in a clear area in the lawn of growing host cells. This clearing is called a **plaque,** and it is assumed that each plaque has originated from replication events that began with one virion.

Plaques are essentially "windows" in the lawn of confluent cell growth. With bacterial viruses, plaques may be obtained when virus particles are mixed into a thin layer of host bacteria that is spread out as an agar overlay on the surface of an agar medium (Figure 8.6a). During incubation of the culture, the bacteria grow and form a turbid layer that is visible to the naked eye. However, wherever a successful virus infection has been initiated, lysis of the cells occurs, resulting in the formation of a clear zone called a *plaque* (Figure 8.6b).

The plaque procedure also permits the isolation of pure virus strains because if a plaque has arisen from a single virion, all the virions in this plaque are probably genetically identical. Some of the virions from this plaque can be picked and inoculated into a fresh bacterial culture to establish a pure virus line. The development of the plaque assay technique was as important for the advance of virology as Koch's development of solid media (∞ Section 1.9) for bacteriology.

Plaques may be obtained for animal viruses by using animal cell culture systems as hosts. A monolayer of cultured animal cells is prepared on a plate or flat bottle, and the virus suspension overlaid. Plaques are revealed by zones of destruction of the animal cells (Figure 8.7).

In some cases, the virus may not actually destroy the cells but may cause changes in morphology or growth rate that can be recognized. For instance, tumor viruses may not destroy cells but may cause the cells to grow faster than uninfected cells, a phenomenon called *transformation*. As we have noted, the general arrangement of cells in a tissue culture is a monolayer. This is because growth generally ceases when the cells, as a result of growth, come in contact with each other (a phenomenon known as *contact inhibition*). Transformed cells have altered growth requirements and continue to grow, piling up to form a small *focus of growth* (called a *focus of infection* when the transformation has been brought about by virus infection). By counting foci of infection, a quantitative measure of virus may be obtained.

Efficiency of plating

One important concept in quantitative virology involves the idea of *efficiency of plating*. Counts made by plaque assay are always lower than counts made with the electron microscope. The efficiency with which virions infect host cells is rarely 100% and may often be considerably less. This does not mean that virions that have not caused infection are inactive, although this is sometimes the case. It may merely mean that under the conditions used, successful infection with these particles has not occurred. Although with bacterial viruses, efficiency of plating is often higher than 50%, with many animal viruses it may be very low, 0.1 or 1%. Why virus particles vary in infectivity is not well understood. In some cases it is possible that the conditions used for quantification are not optimal. Because it is technically difficult to count virions with the electron microscope, it is difficult to assess the actual efficiency of plating, but the concept is important in both research and medical practice. Because the efficiency of plating is rarely close to 100%, when the plaque method is used to quantify virus, it is accurate to express the concentration (called the *titer*) of the virus suspension not as the absolute number of virion units but as the number of *plaque-forming units*.

Animal infectivity methods

Some viruses do not cause recognizable effects in cell cultures but cause death in the whole animal. In such cases, quantification can be done only by some sort of titration in infected animals. The general procedure is to carry out a serial dilution of the unknown sample, generally at 10-fold dilutions, and to inject samples of each dilution into numbers of sensitive animals. After a suitable incubation period, the fraction of dead and live animals at each dilution is tabulated and an *end point dilution* is calculated. This is the dilution at which, for example, *half* of the injected animals die. Although such serial dilution methods are much more cumbersome and much less accurate than cell culture methods, they may be essential for the study of certain types of viruses.

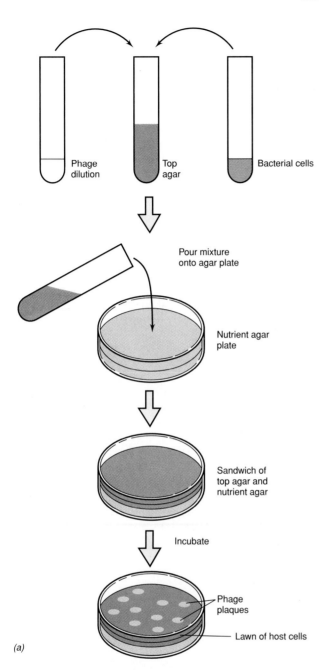

(a)

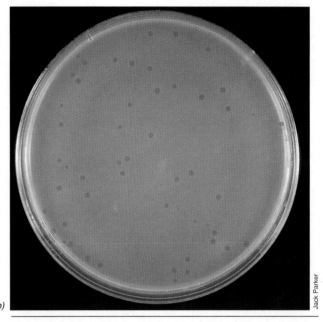

(b)

Jack Parker

Figure 8.6 Quantification of bacterial virus by plaque assay using the agar overlay technique. (a) A dilution of a suspension containing the virus material is mixed in a small amount of melted agar with the sensitive host bacteria, and the mixture poured on the surface of a nutrient agar plate. The host bacteria, which have been spread uniformly throughout the top agar layer, begin to grow, and after overnight incubation form a *lawn* of confluent growth. Each virus particle that attaches to a cell and reproduces may cause cell lysis, and the virus particles released can spread to adjacent cells in the agar, infect them, be reproduced, and again lead to lysis and release. The size of the plaque formed depends on the virus, the host, and conditions of culture. (b) Photograph of a plate showing plaques formed by bacteriophage on a lawn of sensitive bacteria. The plaques shown are about 1–2 mm in diameter.

CONCEPT CHECK   **8.4**

Although it requires only a single virion to initiate an infectious cycle, not all virus particles are equally infectious. One of the most accurate ways of measuring virus infectivity is by the plaque assay. Plaques are clear zones that develop on layers or lawns of host cells, each plaque due to infection by a single virus particle. The virus plaque is analogous to the bacterial colony.

  ✔ *Give a definition of* efficiency of plating.

  ✔ *What is a* plaque-forming unit?

8.5

General Features of Virus Reproduction

The basic problem of virus replication can be simply put: the virus must induce a living host cell to synthesize all the essential components needed to make more virus particles. These components must then be assembled into the proper structure, and the new virions must escape from the cell and infect other cells. The various phases of this replication process in a bacteriophage can be categorized in seven steps (Figure 8.8).

1. **Attachment** (adsorption) of the virion to a susceptible host cell.

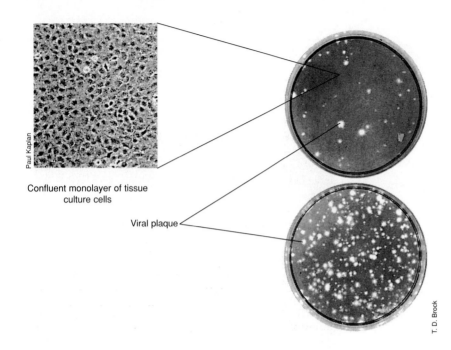

Paul Kaplan

Confluent monolayer of tissue
culture cells

Viral plaque

T. D. Brock

Figure 8.7 Cell cultures in monolayers within Petri plates. Note the presence of plaques where virus-induced cell lysis has occurred. Also shown is a photomicrograph of a cell culture.

2. **Penetration** (injection) of the virion or its nucleic acid into the cell.

3. **Early steps in replication** during which the host cell biosynthetic machinery is altered as a prelude to virus nucleic acid synthesis. Virus-specific enzymes are typically made.

4. **Replication** of the virus nucleic acid.

5. **Synthesis of proteins used as structural subunits** of the virus coat.

6. **Assembly** of structural subunits (and membrane components in enveloped viruses) and **packaging** of nucleic acid into new virus particles.

7. **Release** of mature virions from the cell.

These stages in virus replication are recognized when virus particles infect cells in culture and are illustrated in Figure 8.9, which exhibits what is called a **one-step growth curve.** In the first few minutes after infection the virus is said to undergo an *eclipse.* The virus nucleic acid has become separated from its protein coat, and so even if the infected cell had broken open, the virion no longer exists as an infectious entity. Although virus nucleic acid may be infectious, the infectivity of virus nucleic acid is many times lower than that of whole virions because the machinery for bringing the virus genome into the cell is lacking. Also, outside the virion the nucleic acid is no longer protected from deleterious activities of the environment as it was when it was inside the protein coat.

The eclipse occurs during the early stages of virus replication. This is called the *latent period* because no infectious virions are evident. Finally, maturation begins as the newly synthesized nucleic acid molecules become packaged inside protein coats. During the *maturation* phase, the titer of active virions inside the cell rises dramatically. At the end of maturation, *release* of mature virions occurs, either as a result of cell *lysis* or because of some budding or excretion process. The number of virions released, called the *burst size,* varies with the particular virus and the particular host cell and can range from a few to a few thousand. The timing of this overall virus replication cycle varies from 20–30 min in many bacterial viruses to 8–40 hr in most animal viruses. We now consider each of the steps of the virus multiplication cycle in more detail.

CONCEPT CHECK   **8.5**

The virus life cycle can be divided into seven stages: attachment (adsorption), penetration (injection), early protein synthesis, nucleic acid replication, synthesis of virus protein subunits, assembly of mature virions, and virus release.

8.6

Steps in Virus Multiplication

We will now discuss some of the steps of virus multiplication in more detail. As we have noted, the outcome of a virus infection is the synthesis of viral nucleic acid and viral protein coats. In effect, the virus takes over the biosynthetic machinery of the host and uses it for its own synthesis. A few enzymes needed for virus replication may be present in the virion and may be introduced into the cell during the infection process, but the host supplies

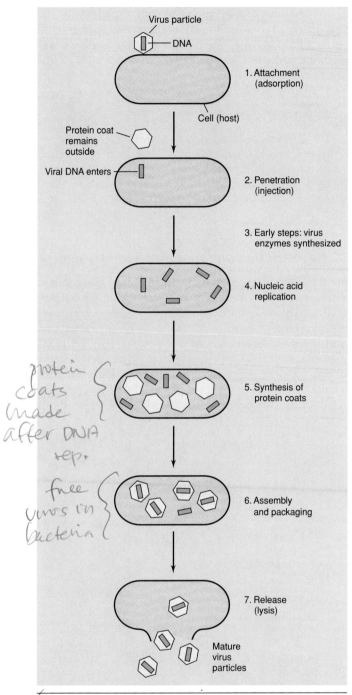

handwritten margin notes:
protein coats made after DNA rep.

free virus in bacteria

Figure 8.8 The replication cycle of a bacterial virus. The general stages of virus replication are indicated.

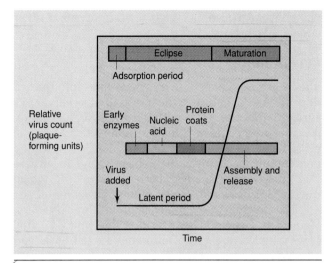

Figure 8.9 The one-step growth curve of virus replication. This graph displays the results of a single round of viral multiplication in a population of cells. Following adsorption, the infectivity of the virus particles disappears, a phenomenon called *eclipse*. This is due to the uncoating of the virus particles. During the *latent period*, replication of viral nucleic acid and protein occurs. The *maturation period* follows, when virus nucleic acid and protein are assembled into mature virus particles. At this time, if the cells are broken up, active virus can be detected. Finally, *release* occurs, either with or without cell lysis. The timing of the one-step growth cycle varies with the virus and host. With many bacterial viruses, the whole cycle may be complete in 30–60 min, whereas with animal viruses 12–24 hr is usually required for a complete cycle. Compare this general picture and color scheme with specific replication events shown for bacteriophage T4 in Figure 8.23.

Attachment

There is a high specificity in the interaction between virus and host. The most common basis for host specificity involves the attachment process. The virus particle itself has one or more proteins on the outside that interact with specific cell surface components called *receptors*. The receptors on the cell surface are normal surface components of the host, such as proteins, polysaccharides, and lipoprotein–polysaccharide complexes, to which the virion attaches. In the absence of the receptor site, the virus cannot adsorb, and hence cannot infect. If the receptor site is altered, the host may become resistant to virus infection. However, mutants of the virus can also arise that are able to adsorb to resistant hosts.

In general, virus receptors carry out normal functions in the cell. For example, in Bacteria some phage receptors are pili or flagella, others are cell envelope components, and others are transport binding proteins. The receptor for influenza virus is a glycoprotein found on red blood cells and on cells of the mucous membrane of susceptible animals, whereas the receptor site of poliovirus is a cell surface lipoprotein. However, many animal and plant viruses do not have specific attachment sites at all, and the virus enters

everything else: energy-generating system, ribosomes, amino acid activating enzymes, transfer RNA (with a few exceptions), and all soluble factors. The virus genome codes for all new proteins. Therefore, the steps in virus multiplication include mechanisms for transporting the viral genome into the cell and ensuring that it is expressed in such a way that new virions will be produced.

passively as a result of phagocytosis or some other endocytotic process.

Penetration

The means by which the virus penetrates into the cell depends on the nature of the host cell, especially on its surface structures. Cells with cell walls, such as bacteria, are infected in a different manner from animal cells, which lack a cell wall. The most complicated penetration mechanisms have been found in viruses that infect bacteria. The bacteriophage T4, which infects *Escherichia coli*, can be used as an example.

The structure of the bacterial virus T4 was shown in Figure 8.5c. The virion has a **head**, within which the viral DNA is folded, and a long, fairly complex **tail**, at the end of which is a series of tail fibers. During the attachment process, the virus particles first attach to cells by means of the tail fibers (Figure 8.10). The ends of the fibers interact specifically with core polysaccharides that are part of the outer layer of the gram-negative cell wall (∞ Section 3.6). These tail fibers then retract, and the core of the tail makes contact with the cell envelope of the bacterium. The action of a lysozyme-like enzyme results in the formation of a small hole. The tail sheath contracts, and the DNA of the virus passes into the cell through a hole in the tip of the tail, the majority of the coat protein remaining outside.

With animal cells, the *whole virion* penetrates the cell, being carried inside by endocytosis (phagocytosis or pinocytosis), an active cellular process. We describe some of these processes in detail later in this chapter.

CONCEPT CHECK   8.6a

The attachment of a virion to a host cell is a highly specific process, involving the interaction of receptors on the host with proteins on the surface of the virus particle. Only after attachment has occurred can the virus or its nucleic acid penetrate the host cell.

Virus restriction and modification by the host

We have already seen that one form of host resistance to virus arises when there is no receptor site on the cell surface to which the virus can attach. Another and more specific kind of host resistance occurs in prokaryotes and involves destruction of the double-stranded DNA genome of a virus after it has been injected. This destruction is brought about by host enzymes that cleave the viral DNA at one or several places, thus preventing its replication. This phenomenon is called *restriction* and is

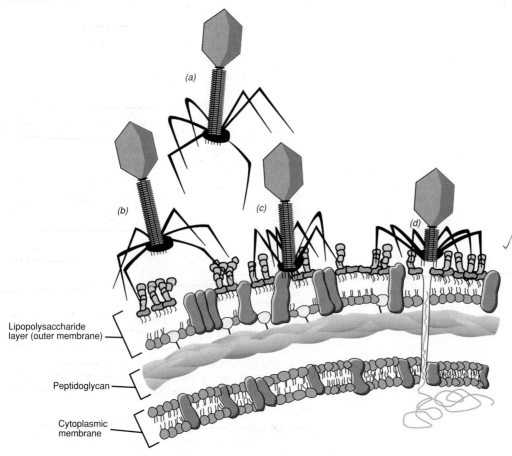

Lipopolysaccharide layer (outer membrane)

Peptidoglycan

Cytoplasmic membrane

Figure 8.10 Attachment of T4 bacteriophage particle to the cell wall of *Escherichia coli* and injection of DNA: (a) Unattached particle. (b) Attachment to the wall by the long tail fibers interacting with core polysaccharide. (c) Contact of cell wall by the tail pins. (d) Contraction of the tail sheath and injection of the DNA. For a detailed description of the gram-negative cell wall see Section 3.6.

part of a general host mechanism to prevent the invasion of foreign nucleic acid. We have already discussed *restriction enzymes* and their action in some detail (∞ Section 6.4) and noted that their cellular role is in defending against foreign DNA. Restriction enzymes are highly specific, attacking only certain sequences (generally four or six base pairs). The host protects its own DNA from the action of restriction enzymes by *modifying* its DNA at the sites where the restriction enzymes act. Modification of host DNA is brought about by methylation of purine or pyrimidine bases (in such a way that their base-pairing properties are not altered).

Some viruses can overcome host restriction mechanisms by modifications of their nucleic acids so that they are no longer subject to enzymatic attack. Two kinds of chemical modifications of viral DNA have been recognized, glucosylation and methylation. For instance, the T-even bacteriophages (T2, T4, and T6) have their DNA glucosylated to varying degrees, and the glucosylation prevents or greatly reduces endonuclease attack. Many other viral nucleic acids have been found to be modified by methylation, but glucosylation has been found only in the T-even bacteriophages. It should be emphasized that modification of viral nucleic acid occurs after replication has occurred and that the modified bases are not copied directly. Other viruses, such as the bacteriophages T3 and T7, avoid restriction by encoding proteins that inhibit the host restriction systems. Some hosts have multiple restriction and methylation systems that help in preventing infection by viruses that can circumvent only one of them.

However, not all restriction systems recognize unmodified DNA. Host restriction systems are also known that restrict only *modified* DNA! Clearly the host containing this enzyme does *not* contain the modification enzyme. However, this host is protected from infection by a virus that was modified during reproduction in its previous host strain.

Hosts also contain other DNA methylases. Some of these methylases may be involved in DNA repair or in gene regulation, but others may offer protection to *host* DNA. This can be during DNA transfer to other cells during genetic recombination (∞ Chapter 9) or because *some viruses themselves encode restriction systems.*

As we discussed in Chapter 6, a knowledge of modification and restriction systems is of considerable practical utility in studying DNA chemistry. We discuss the use of restriction enzymes in genetic engineering in Chapter 10.

CONCEPT CHECK   *8.6b*

The virus nucleic acid is foreign to the host, and the restriction–modification system of the host, which recognizes and destroys foreign DNA, is one means of defense against virus infection.

Production of viral nucleic acid and proteins

New copies of the viral genome must be replicated, and virus-specific proteins must be synthesized in order for virus multiplication to occur. Typically the production of at least some viral proteins begins very early after the viral genome has been taken up by the cell. In order for this to happen, virus-specific messenger RNA must first be made. Exactly how the virus brings about new mRNA synthesis depends on the type of virus and on the structure of its genome. Although many viruses have a double-stranded DNA genome, a great many do not, and these other types of genomes include not only single-stranded DNA but both single- and double-stranded RNA. Furthermore, we have mentioned that some viruses have one type of nucleic acid in the virion but use another as a replicative intermediate. All these "unusual" genomes present problems in understanding virus multiplication because they involve information transfers, such as RNA to RNA and RNA to DNA, that host enzymes do not perform.

The essential features of producing mRNA from double-stranded DNA were discussed in Chapter 6, and it was shown that mRNA represents a complementary copy made by RNA polymerase of one of the two strands of the DNA double helix. Which copy is read into mRNA depends on the location of the appropriate promotor, because the promoter points the direction that the RNA polymerase will follow. In cells uninfected with virus, all mRNA is made on the cell's DNA template, but when viruses are present, the situation is different.

Figure 8.11 shows the nomenclature used to describe different types of nucleic acid based on its information content. Because the nucleic acid sequence of the mRNA can be translated directly into protein, it is by convention considered to be of the *plus* (+) configuration. The sequence of the viral genome nucleic acid is then indicated by a *plus* if it is the same as the mRNA and a *minus* if it is of opposite sense.

These configurations can also be used to describe the virus. For instance, a virus whose genome is single-stranded DNA of the plus configuration is called a *positive-strand DNA virus,* while a virus with an RNA genome of the minus configuration is called a *negative-strand RNA virus.* Figure 8.12 shows how mRNA can be made from different types of genomes. As seen in Figure 8.12, if the virus has double-stranded DNA (ds DNA), then mRNA synthesis can proceed directly as in uninfected cells. However, if the virus has a single-stranded DNA (ss DNA), then it is first converted to ds DNA and the latter serves as the template for mRNA synthesis by the RNA polymerase of the cell.

For RNA viruses, a virus-specific RNA-dependent RNA polymerase is needed because the cell RNA polymerase is DNA-dependent (requires a DNA template) and generally does not copy RNA. Several different types of viruses contain RNA as their genome, and each

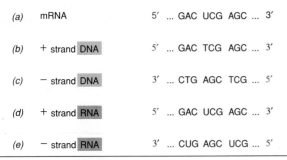

(a) mRNA   5′ ... GAC UCG AGC ... 3′

(b) + strand DNA   5′ ... GAC TCG AGC ... 3′

(c) − strand DNA   3′ ... CTG AGC TCG ... 5′

(d) + strand RNA   5′ ... GAC UCG AGC ... 3′

(e) − strand RNA   3′ ... CUG AGC UCG ... 5′

Figure 8.11 Comparison of sequences of nucleic acids with different configurations. The strands of nucleic acid in viral genomes are called *plus* (+) if they are in the same configuration as mRNA and *minus* (−) if they are in a configuration opposite that of mRNA. (a) The direction and sequence of an mRNA. (b) The sequence of a + strand of a viral DNA genome that could encode that mRNA (using a − strand DNA intermediate). (c) A − strand of a viral DNA genome that could encode the mRNA. (d, e) The + and − strands of RNA genomes that could encode the mRNA. Note that a single-stranded + strand RNA genome has the same sequence and orientation as the mRNA. The 5′-end of each nucleic acid strand is highlighted in red.

requires a different strategy for producing mRNA. The simplest case is the positive-strand RNA viruses in which the single incoming viral RNA strand is the *plus* strand and hence serves directly as mRNA. In addition to the other required proteins, this mRNA encodes the virus-specific RNA polymerase. This polymerase first makes complementary *minus* strands and then uses them as templates to make more plus strands. For negtive-strand RNA viruses (whose virion contains only the minus strand) or double-stranded RNA viruses, the situation is more complicated. In neither case can the incoming RNA serve as mRNA, and therefore mRNA must be synthesized first. However, as mentioned earlier, cells do not typically have an RNA polymerase capable of this. To circumvent this problem, these viruses contain some of this enzyme in their virions, and it is injected into the cell along with the genomic RNA. Therefore, in these cases, the complementary plus strand is synthesized by this RNA-dependent RNA polymerase and used as message.

Retroviruses (causal agents of certain kinds of cancers and acquired immunodeficiency syndrome, AIDS) are RNA viruses that replicate through a DNA intermediate. The process of copying the information found in RNA into DNA is called **reverse transcription,** and thus these viruses require an enzyme called **reverse transcriptase.** (Telomerase is a type of reverse transcriptase; ∞ Section 6.5.) In spite of the fact that the incoming RNA of retroviruses is the plus strand, it is not used as message, and therefore these viruses must carry reverse trancriptase in their virions. After infection, the virion ss RNA is copied to a double-stranded DNA (through an ss DNA intermediate) and the ds DNA then serves as the template for mRNA synthesis (thus, ss RNA → ss

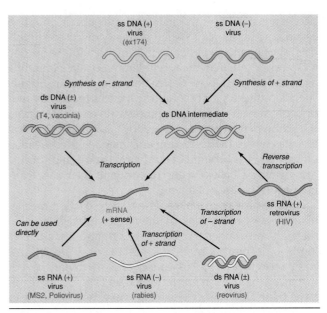

Figure 8.12 Formation of mRNA after infection of cells by viruses of different types. The chemical sense of the mRNA is considered as plus (+). The senses of the various virus nucleic acids are indicated as + if the same as mRNA, as − if opposite, or as ± if double-stranded. Examples are indicated next to the virus nucleic acid. Although examples of viruses containing ss DNA of the − sense are known, none are discussed in the text.

DNA → ds DNA). Retrovirus replication is of unusual complexity and is discussed in Section 8.22.

Viral proteins

Once viral mRNA is made, viral proteins (for example, enzymes and structural subunits) can be synthesized. The proteins synthesized as a result of virus infection can be grouped into two broad categories:

1. Proteins (usually enzymes) synthesized soon after infection, called the **early proteins,** which are necessary for the replication of virus nucleic acid

2. Proteins synthesized later, called the **late proteins,** which include the proteins of the virus coat

Generally, both the time of appearance and the amount of virus proteins are regulated. The early proteins are enzymes that, because they act catalytically, are synthesized in smaller amounts, and the late proteins, often structural, are made in much larger amounts.

Virus infection upsets the regulatory mechanisms of the host because there is a marked overproduction of viral nucleic acid and protein in the infected cell. In some cases, virus infection causes a complete shutdown of host macromolecular synthesis, whereas in other cases host synthesis proceeds concurrently with virus synthesis. In either case, the regulation of virus synthesis is under the control of the virus rather than the host. There are several elements of this control that are similar to the host regulatory mechanisms discussed in Chapter 7, but there are also some uniquely viral regulatory mechanisms. We discuss various regulatory mechanisms when we consider the individual viruses later in this chapter.

CONCEPT CHECK 8.6c

Before replication of viral nucleic acid can occur, new virus proteins are often needed. These are encoded by messenger RNA molecules made from the virus genome. In the case of some RNA viruses, the viral RNA itself acts as mRNA. In other cases, the virus genome serves as a template for the formation of viral mRNA and certain essential enzymes are contained in the virion.

✔ *Give one reason why viruses can infect only specific hosts.*

✔ *What is a positive-strand RNA virus?*

8.7

Overview of Bacterial Viruses

Various kinds of bacterial viruses are illustrated in Figure 8.13. Most of the bacterial viruses that have been studied in detail infect Bacteria of the enteric group, such as *Escherichia coli* and *Salmonella typhimurium*. However, viruses are known that infect a variety of prokaryotes, both Bacteria and Archaea. A few bacterial viruses have lipid envelopes but most do not. However, many bacterial viruses are structurally complex, with head and complex tail structures. As we illustrated in Figure 8.10, the tail is involved in the injection of the nucleic acid into the cell.

We now discuss some of the bacterial viruses for which molecular details of the multiplication process are known. Although these bacterial viruses were first studied as *model systems* for understanding general features of virus multiplication, some of them now serve as convenient tools for *genetic engineering* (discussed in Chapter 10). Thus, the information on bacterial viruses is not only valuable as background for the discussion of animal viruses but also is essential for the material presented in the next two chapters on microbial genetics and genetic engineering.

It should be clear that a great diversity of viruses exists. It should therefore not be surprising that there is also a great diversity in the manner in which virus multiplication occurs. In the present chapter, we are able to present only some of the major types of virus replication patterns and must omit some of the interesting exceptional cases.

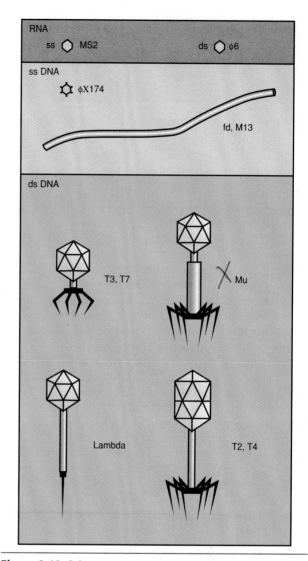

Figure 8.13 Schematic representations of the main types of bacterial viruses. Those discussed in detail are M13, φX174, MS2, T4, lambda, T7, and Mu. Sizes are to approximate scale.

8.8

RNA Bacteriophages

Many viruses have RNA genomes. The best-known bacterial RNA viruses have single-stranded RNA of the plus configuration. Interestingly, the bacterial RNA viruses known in the enteric bacteria group infect only bacterial cells that contain a type of plasmid, called a *conjugative plasmid* (∞ Section 9.9), which allows the bacterial cell to function as a *donor*, or *male*, in a certain type of genetic exchange (the interesting concept of male and female bacteria is discussed in Chapter 9). This restriction to male bacterial cells arises because these viruses infect bacteria by attaching to *pili* (Figure 8.14), which are encoded by the plasmid. Since such pili are absent from

Figure 8.14 Electron micrograph of the pilus of a male bacterial cell of *Escherichia coli* showing virions of a small RNA phage attached to the pilus.

cells that do not carry the plasmid, these RNA viruses are unable to attach to such cells and hence do not initiate infection in these so-called female cells.

The bacterial RNA viruses are all quite small, about 26 nm in size, and they are all icosahedral, with 180 copies of coat protein per virus particle. The complete nucleotide sequences of several RNA phage genomes are known. The genome of the RNA phage MS2, which infects *Escherichia coli*, is 3569 nucleotides long. The RNA strand in the virion acts directly as mRNA on entry into the cell (see Figure 8.12).

The genetic map of MS2 is shown in Figure 8.15*a*, and the flow of events of MS2 multiplication is shown in Figure 8.15*b*. The small genome encodes only four proteins. These are the **maturation protein** (present in the mature virus particle as a single copy), **coat protein, lysis protein** (involved in the lysis process that results in release of mature virus particles), and a subunit of **RNA replicase,** the enzyme that brings about replication of the viral RNA. Interestingly, the RNA replicase is a composite protein, composed partly of the virus-encoded polypeptide and partly of host polypeptides. The host proteins involved in the formation of active viral replicase are part of the cell's normal translational machinery. Thus, the virus appears to employ host proteins that normally have entirely distinct functions and use them to make an active viral replicase.

As noted, the viral RNA is of the plus sense and can thus be translated immediately. After RNA replicase is synthesized, it in turn can synthesize RNA of minus sense using the infecting RNA as template (see Figure 8.12). After minus RNA has been synthesized, more plus RNA is made using this minus RNA as template. The newly made plus RNA strands then serve as messengers for continued virus protein synthesis. The gene for the maturation protein is at the 5'-end of the RNA. Translation of the gene coding for the maturation protein (needed in only one copy per virus particle), occurs only from the nascent form of the plus-strand RNA as the replication process occurs. In this way, the amount of maturation protein needed is limited. The virus RNA is folded into a complex form with extensive secondary structure. Of the four AUG start sites, the most accessible to the translation process is that for the coat protein, and translation begins there very early. The replicase mRNA is also translated early. As coat protein molecules increase in number in the cell, they combine with the RNA around the AUG start site for the replicase protein, effectively turning off synthesis of replicase. The major virus protein synthesized is coat protein, which is needed in the highest amounts.

Another interesting feature of bacteriophage MS2 is that the fourth virus protein, the *lysis* protein, is encoded by a gene that *overlaps* with both the coat protein gene and the replicase gene (see genetic map in Figure 8.15*a*). The phenomenon of **overlapping**

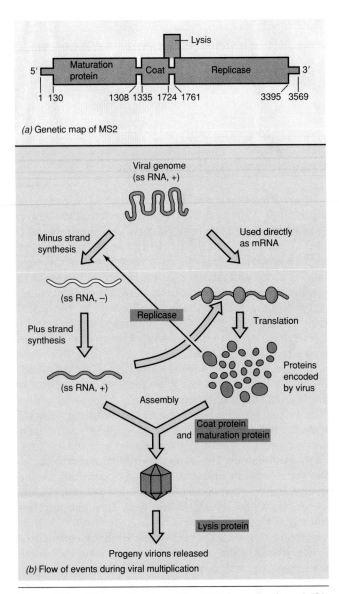

(a) Genetic map of MS2

(b) Flow of events during viral multiplication

Figure 8.15 (a) Genetic map of the RNA bacteriophage MS2. (b) Flow of events during multiplication. The numbers in (a) refer to the nucleotide positions on the RNA.

genes is quite common in very small genomes. It should be noted that although the use of overlapping genes makes possible more efficient use of genetic information, it seriously complicates the evolution process because a mutation in a region of gene overlap may affect two genes simultaneously. The start codon of this lysis gene is not easily accessible to ribosomes because of the secondary structure found in the RNA. When the ribosome terminates synthesis of the coat protein gene, the secondary structure in this region is disrupted, and sometimes this disruption allows a ribosome to begin reading the lysis gene. By restricting the efficiency of translation in this way, premature lysis of the cell is probably avoided. Only after sufficient copies of coat protein are available for the assembly of mature virus particles does lysis commence.

Ultimately, phage assembly takes place, and release of virions from the cell occurs as a result of cell lysis. The features of replication of these simple RNA viruses are themselves fairly simple. The viral RNA itself functions as an mRNA, and regulation occurs primarily by way of controlling access of ribosomes to the appropriate start sites on the viral RNA.

CONCEPT CHECK   **8.8**

A variety of RNA viruses that infect bacteria are known. The small RNA genome of these bacterial viruses acts directly as mRNA and encodes only a few proteins.

✔ *Are these viruses considered positive-strand viruses? Explain.*

✔ *Describe what is meant by overlapping genes.*

8.9
Single-Stranded Icosahedral DNA Bacteriophages

A number of small bacterial viruses have genomes consisting of single-stranded DNA in circular configuration. These viruses are very small, about 25 nm in diameter, and the principal building block of the protein coat is a single protein present in 60 copies (the minimum number of protein subunits possible in an icosahedral virus), to which are attached at the vertices of the icosahedron several other proteins that make up spikelike structures (see Figure 8.13). These small DNA viruses possess only a limited amount of genetic information in their genomes, and the host cell DNA replication machinery is used in the replication of virus DNA.

The genome of phage φX174

The most extensively studied virus of this group is the phage designated φX174, which infects *Escherichia coli*. Its genome consists of a circular single-stranded molecule of 5386 nucleotide residues. The DNA of φX174 was the first DNA to be completely sequenced, a remarkable achievement when it was accomplished by Frederick Sanger and colleagues in 1977. Now, DNA sequencing is a routine procedure (∞ Working with Nucleic Acids: The Tools, Chapter 6). φX174 is also of special interest because it was the first genetic element shown to have *overlapping genes*. In very small viruses such as φX174 there is insufficient DNA to code for all virus-specific proteins unless parts of certain virus nucleotide sequences are read more than once in different reading frames (Figure 8.16).

As seen in the genetic map of φX174, the sequences of genes D and E overlap each other, gene E being contained completely *within* gene D. In addition, the termination codon of gene D overlaps the initiation codon of gene J by one nucleotide. The reading frame of gene E is therefore in a different frame (starting point) from that of gene D. Several other instances of gene overlap occur in the φX174 genome (Figure 8.16). Additionally, a small gene A protein, called A*-protein, is formed by *reinitiation of translation* (not transcription) within the mRNA of gene A, with A protein being read and terminated from the same mRNA reading frame as A-protein but starting at a different codon.

DNA replication by the rolling circle mechanism

The replication process of such a circular single-stranded DNA molecule is of considerable general interest because cellular DNA always replicates in the double-stranded configuration (∞ Section 6.5). The DNA strand in single-stranded DNA phages is of the plus sense. On infection, this DNA strand becomes separated from the protein coat, and entrance into the cell is accompanied by the conversion of this single-stranded DNA to a double-stranded form called **replicative form** (RF) DNA (Figure 8.16*b*). Cell-coded proteins involved in the conversion of viral DNA to RF DNA consist of the enzymes *primase, DNA polymerase, ligase,* and *gyrase*. No virus-encoded poteins are involved in the conversion. The RF DNA is a closed, double-stranded, circular molecule that has extensive supercoiling.

In cells, replication of the lagging strand involves the formation of short *RNA primers* by action of an enzyme called *primase* (∞ Section 6.5). Such RNA primers are made at intervals on the lagging strand and are then removed and replaced with DNA by DNA polymerase (∞ Figure 6.18). Unlike the situation with cellular DNA, however, replication of φX174

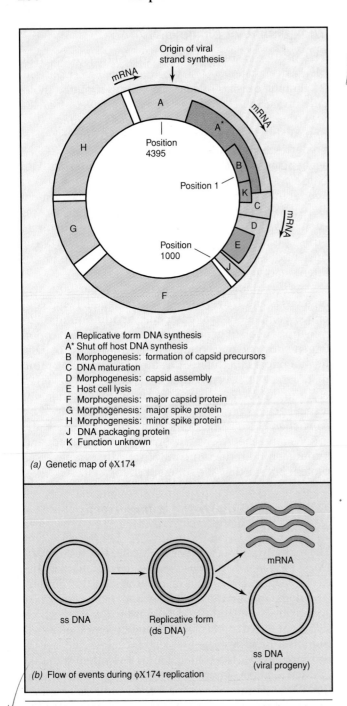

A Replicative form DNA synthesis
A* Shut off host DNA synthesis
B Morphogenesis: formation of capsid precursors
C DNA maturation
D Morphogenesis: capsid assembly
E Host cell lysis
F Morphogenesis: major capsid protein
G Morphogenesis: major spike protein
H Morphogenesis: minor spike protein
J DNA packaging protein
K Function unknown

(a) Genetic map of φX174

(b) Flow of events during φX174 replication

Figure 8.16 Bacteriophage φX174, a single-stranded DNA phage. (a) Genetic map. Note the regions of gene overlap (A/B, K/B, K/C, K/A, A/C, and D/E). Intergenic regions are not colored. Protein A* is formed using only part of the coding sequence of gene A by reinitiation of translation (see text). (b) Flow of events in φX174 multiplication. The production of progeny ss DNA from replicative form ds DNA involves rolling circle replication and is shown in more detail in Figure 8.17.

DNA begins with a single-stranded closed circle. To begin replication of this DNA, primase brings about the synthesis of a short RNA primer at one or more specific initiation sites. DNA is then synthesized by

DNA polymerase III, and the primer is removed and replaced by DNA using DNA polymerase I, exactly as in the case of a lagging strand. This results in the formation of the complete double-stranded RF DNA.

Once the RF is formed, DNA replication occurs by conventional semiconservative replication, involving theta-form intermediates (∞ Figure 6.16) and resulting in the formation of new RF DNA molecules. However, the formation of single-stranded viral genomes involves a different type of replication mechanism called **rolling circle replication** (Figure 8.17). The rolling circle arises because one strand is nicked and the 3'-end of this nick is used to prime synthesis of a new strand. Continued rotation of the circle leads to the synthesis of a linear, single-stranded structure. Note that synthesis is asymmetric because only one of the strands is serving as template. In φX174, synthesis begins when the protein encoded by gene A, called *gene A protein,* cleaves the plus strand of the RF. When the growing viral strand reaches unit length (5386 residues for φX174), gene A protein cleaves and then ligates the two ends of the newly synthesized single strand to give a circular single-stranded DNA.

Many other viruses and some plasmids also use rolling circle replication. We will see that this type of replication can also be used to synthesize double-stranded DNA.

Transcription and translation for φX174

Viral mRNA synthesis is directed by the RF DNA. Synthesis of mRNA begins at several major promoters and terminates at a number of sites (see map, Figure 8.16). The polycistronic mRNA molecules are then translated into the various phase proteins. As we have noted, A-protein and A*-protein are both made from the *same gene,* A*-protein arising as a result of translation from a secondary initiation site internal to the A mRNA. Further, as we have noted, several proteins are made from mRNA transcripts formed from different reading frames from the same DNA sequences (overlapping genes). One can truly be impressed by the effi-

---

CONCEPT CHECK   8.9

The single-stranded DNA genome of the virus φX174 is so small that only the presence of overlapping genes allows it to encode all its essential functions. This virus provided the first example of overlapping genes. The production of progeny viral DNA involves a rolling circle mechanism.

✔ *If the nucleic acid genome of φX174 is in the plus configuration, why can't it be used directly as mRNA?*

✔ *How does the replicative form of the viral nucleic acid differ from the form found in the virion?*

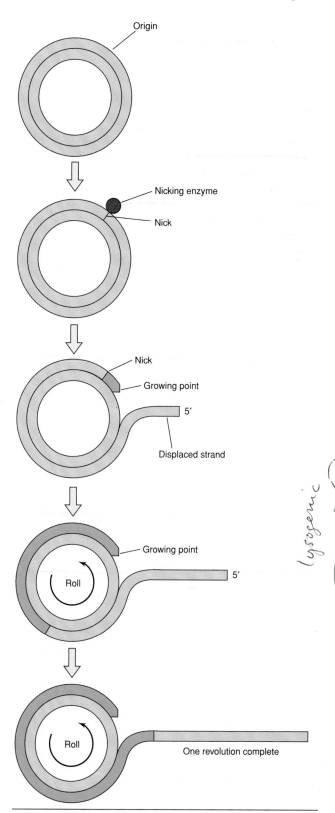

Figure 8.17 Rolling circle replication. Replication begins at the origin by nicking one strand of DNA (in ϕX174, the nicked strand is the *plus* strand, and the gene A protein makes the nick). After one new progeny strand has been synthesized (one revolution of the circle), the gene A protein cleaves the new strand and ligates its two ends.

ciency with which such a small genome as that of ϕX174 can have multiple uses.

Ultimately, assembly of mature virus particles occurs. Release of virions from the cell takes place as a result of cell lysis, which involves the participation of gene E protein.

8.10

Single-Stranded Filamentous DNA Bacteriophages

Quite distinct from ϕX174 are the filamentous DNA phages, which have helical rather than icosahedral symmetry. The most studied member of this group is phage M13, which infects *Escherichia coli*, but related phages include f1 and fd. As with the small RNA bacteriophages, these filamentous DNA phages infect only male cells, entering after attachment to the male-specific pilus. Even though these phages are linear (filamentous) in shape they possess *circular* single-stranded DNA. The DNA is not self-complementary, however, so the two adjacent halves of the molecule that run up and down the virus particle form loops at the ends but exhibit very little if any base pairing. Phage M13 has found extensive use as a cloning vector and DNA sequencing vehicle in genetic engineering (∞ Section 10.4). The virion of M13 is only 6 nm in diameter but is 860 nm long. These filamentous DNA phages have the additional interesting property of being released from the cell *without* killing the host cell. Thus, a cell infected with phage M13 can continue to grow, all the while releasing virus particles. Virus infection causes a slowing of cell growth, but otherwise a cell is able to coexist with its virus. Plaques are thus seen only as areas of reduced cell growth in the bacterial lawn.

Many aspects of DNA replication in filamentous phages are similar to that of ϕX174. The property of release without cell killing occurs by a budding process in which the virus particle is always released from the cell with the end containing the A-protein first (Figure 8.18). There is no accumulation of intracellular virus particles; the assembly of mature virions occurs on the inner surface of the cytoplasmic membrane, and virus assembly is coupled with the budding process.

Several features of these phages make them useful as cloning and DNA sequencing vehicles. First, they have single-stranded DNA, which means that sequencing can be carried out by the Sanger dideoxynucleotide method (∞ Working with Nucleic Acids: The Tools, Chapter 6). Second, as long as infected cells are kept in the growing state, they can be maintained indefinitely with cloned DNA so a continuous source of the cloned DNA is available. Third, there is an intergenic space that does not code for protein and can be replaced by variable amounts of foreign DNA.

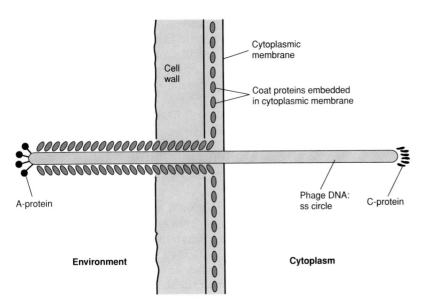

Figure 8.18 Illustration of the manner in which the virion of a filamentous single-stranded phage (such as M13 or fd) leaves an infected cell without lysis. The A-protein passes first through the membrane at a site on the membrane where coat protein molecules have first become embedded. The intracellular circular DNA is coated with dimers of another phage protein, which is displaced by coat protein as the DNA passes through the intact cytoplasmic membrane.

8.11

Double-Stranded DNA Bacteriophages: Lytic Viruses

The first viruses to be studied in any detail were a number of bacteriophages with linear, double-stranded DNA genomes that infect *Escherichia coli* and a number of related bacteria. A group of scientists began studying these viruses as model systems and used them to establish many of the fundamental principles of molecular biology and genetics (see the box, The Phage Group). These phages were given designations of T1, T2, and so on, up to T7. In this section we shall briefly discuss both T4 and T7, beginning with T7, the simpler of the two. Both are typical lytic viruses in that their replication leads to the destruction of the cell.

Replication of bacteriophage T7

Bacteriophage T7 and its close relative T3 are relatively small DNA viruses that infect *Escherichia coli*. (Some strains of *Shigella* and *Pasteurella* are also hosts for phage T7.) The virion has an icosahedral head and a very small tail (see Figure 8.13).

The T7 genome is a linear double-stranded DNA molecule of 39,936 base pairs. The complete genome has been sequenced. About 92% of the DNA of T7 encodes proteins. Gene overlap also occurs in the T7 genome, as do other translational strategies such as

---

LEARNING FROM THE PAST

THE PHAGE GROUP

Historically, the T-even phages provided the study material for early research of the "Phage Group," a group of research workers from various universities and research institutions who spent their summers working together at the Cold Spring Harbor Laboratory on Long Island. The key members of the Phage Group, Max Delbrück, Salvador Luria, and A. D. Hershey, subsequently shared the Nobel Prize for their pioneering work. Among important concepts first uncovered from research on the T-even phages: only the nucleic acid of the virus entered the cell during infection (a discovery that provided key support for the hypothesis that DNA is the genetic material); the existence of genetic recombination in viruses; the phenomenon of restriction and modification (which led to the discovery of the restriction enzymes so important for genetic engineering); the presence in viruses of unique virus-encoded gene functions; the distinction between early and late viral functions. The first ideas of how viruses cause killing of host cells were also developed from research on the T-even phages. Selecting the T-even phages as a model system and concentrating work on them were greatly responsible for the remarkable success of the Phage Group. ■

internal translational reinitiation and internal frame-shifts within certain genes, all apparently to maximize genetic economy.

The genetic map of T7 is shown in Figure 8.19. The order of the genes influences the regulation of virus multiplication. When the virion attaches to the bacterial cell, the DNA is injected in a linear fashion, with the genes at the "left end" of the genetic map always entering the cell first. Several genes at the left end of the DNA are transcribed immediately by the cellular RNA polymerase, using three closely spaced promoters. One of these early proteins inhibits the host restriction system. Note that this protein is synthesized before the entire T7 genome enters the cell. Another one of these early proteins is a viral RNA polymerase, called T7 RNA polymerase. Two other early mRNA molecules code for proteins that stop the action of host RNA polymerase, thus turning off the transcription of the early genes as well as the transcription of host genes. Thus, the host RNA polymerase is used just to transcribe the first few genes and to make the mRNA that codes for the phage-specific RNA polymerase and a few other proteins. The phage-specific RNA polymerase is then involved in the major transcription processes of the phage. This T7 RNA polymerase uses only phage-specific promoters that are distributed along the left-center and center portions of the genome (see Figure 8.19). The T7 RNA polymerase is very specific for these promoters (whose sequence is unrelated to typical *Escherichia coli* promoters) and is also an extremely efficient enzyme. (Genetic engineers have taken advantage of this to fashion genes that can be highly expressed; ∞ Section 10.7.)

DNA replication in T7 begins at a single origin of replication (shown in Figure 8.19) and proceeds *bidirectionally* from this origin (Figure 8.20). Replicating molecules of T7 DNA can be recognized under the electron microscope by their characteristic structures. Because the origin of replication is near the left end, Y-shaped molecules are frequently seen, and earlier in replication, bubble-shaped molecules appear (Figure 8.20). Several virus-encoded proteins are involved in T7 DNA replication, unlike the situation we described for φX174 (see Section 8.9).

A structural feature of the T7 DNA that is important in DNA replication is that there is a *direct terminal repeat* of 160 base pairs at the ends of the molecule. In order to replicate DNA near the 5′-terminus, RNA primer molecules have to be removed before replication is complete. There is thus an unreplicated portion of the T7 DNA at the 5′-terminus of each strand (see lower part of Figure 8.20a). As discussed in Section 6.5, genetic elements with linear DNA genomes have a variety of strategies for solving this problem in DNA replication. The strategy employed by T7 involves the repeated sequence at its ends. The opposite single 3′-strands on two separate DNA molecules, being complementary, can pair with these 5′-strands, forming a DNA molecule twice as long as the original T7 DNA (Figure 8.20b). The unreplicated portions of this end-to-end bimolecular structure are then completed through the action of DNA polymerase and DNA ligase, resulting in a *linear bimolecule* called a *concatamer*. Continued replication and recombination can lead to concatamers of considerable length, but ultimately a phage-encoded endonuclease cuts each concatamer at a specific site, resulting in the formation of virus-sized linear molecules with terminal repeats (Figure 8.20c).

We thus see that T7 has a more complex replication scheme than that seen for the other bacterial viruses discussed earlier. However, it is less complicated than some of the larger bacteriophages, such as T4.

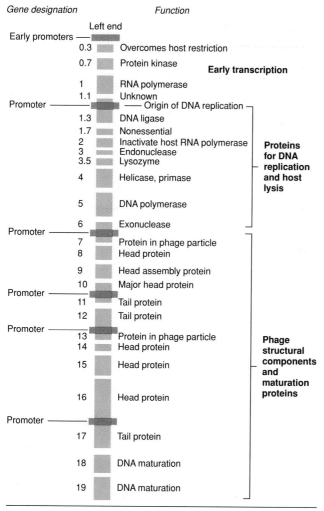

Figure 8.19 Genetic map of phage T7, showing gene numbers, approximate sizes, and functions of the gene products. Transcription from the early promoters involves host RNA polymerase. Transcription from all other promoters involves T7 RNA polymerase. The genes are designated by numbers.

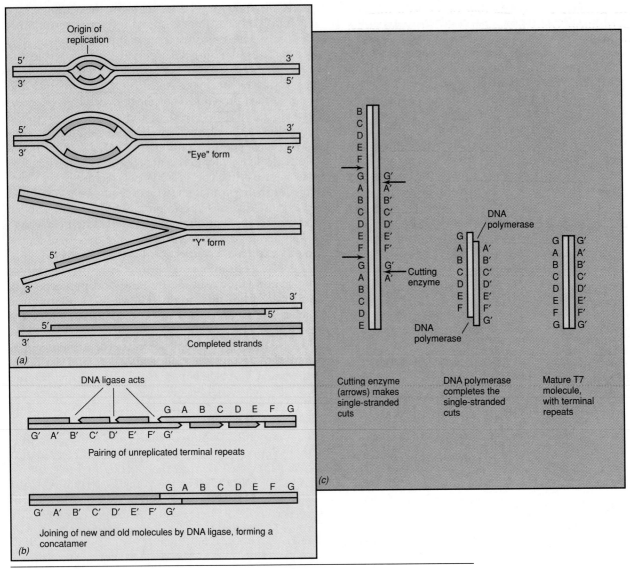

Figure 8.20 Replication of the linear, double-stranded DNA genome of bacteriophage T7. (a) Bidirectional replication of DNA giving rise to intermediate "eye" and "Y" forms. (b) Formation of concatamers by joining DNA molecules at the unreplicated terminal ends. The designation of the genes is arbitrary. (c) Production of mature viral DNA molecules from T7 concatamers by action of cutting enzyme, an endonuclease. Left: The enzyme makes single-stranded cuts of specific sequences (arrows); center: DNA polymerase completes the single-stranded ends; right: the mature T7 molecule with terminal repeats.

Bacteriophage T4 and replication of the T4 genome

One of the most extensively studied groups of DNA viruses is the group called the *T-even phages*, which include the phages T2, T4, and T6. These phages are very closely related and among the largest and most complicated in terms of both structure and manner of replication. We will discuss primarily bacteriophage T4, the representative of this group for which the most information is available.

The virion of phage T4 is structurally complex (see Figure 8.5c). It consists of an icosahedral head that is elongated by the addition of one or two extra bands of proteins, the overall dimensions of the head being 85 × 110 nm. To this head is attached a complex tail consisting of a helical tube (25 × 110 nm) to which are connected a sheath, a connecting "neck" with "collar," and a complex endplate, to which are attached long, jointed tail fibers (see Figure 8.5c). All together, the virus particle has over 25 distinct types of proteins.

The genome of T4, like that of T7, is a double-stranded, linear DNA molecule, but it is quite large, approximately 1.7×10^5 base pairs. The genome encodes over 135 different proteins, and although no known virus encodes its own translational apparatus, T4 does encode several different tRNAs. And like T7, the T4 genome has direct terminal repeats, but they are 3000 to 6000 base pairs. The DNA of T4 has a total length about 650 times longer than the dimension of the head. This means that the DNA must be highly folded and packed very tightly within the head.

Interestingly, the DNA found in T4 is chemically distinct from cell DNA, having a unique base, 5-*hydroxymethylcytosine*, instead of *cytosine* (Figure 8.21). Such a modification makes the DNA resistant to *many* host restriction enzymes. However, the hydroxyl groups of the 5-hydroxymethylcytosine are modified by addition of glucosyl residues. This *glucosylated* DNA is resistant to *virtually all* restriction endonucleases of the host. Thus, this virus-specific DNA modification plays an important role in the ability of the virus to attack a host cell.

The genetic map of T4 is generally represented as a circle, even though the DNA itself is linear. This "genetic circularity" arises because the DNA of the phage exhibits a phenomenon called *circular permutation*. This arises because in different T4 phage particles, the sequence of bases at each end differs (although for a given molecule the same base sequence is repeated at both ends). This structure, a consequence of the way the T4 DNA replicates, results in an appearance of genetic circularity even though the DNA itself is linear.

The process of DNA replication in T4 is similar to that in T7, but in T4 the cutting enzyme that forms virus-sized fragments from concatamers does not recognize specific locations on the long molecule but rather cuts off head-full packages of DNA irrespective of the sequence (Figure 8.22). Because of this and the "extra" room in the head, each virus DNA molecule not only contains repetitious ends but the nucleotide sequences at the ends of different molecules are different as well. Each molecule contains slightly more than one complete copy of the entire genome (Figure 8.22). As shown, the cutting process results in the formation of DNA molecules with permuted sequences at the ends. We have already briefly described the

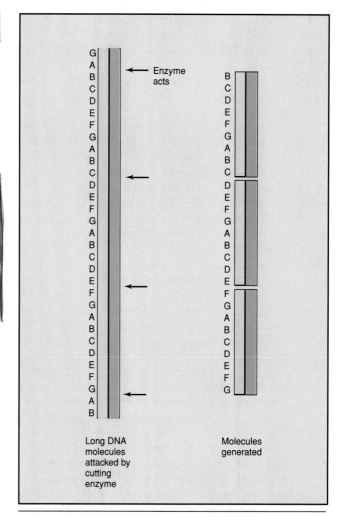

Figure 8.22 Generation in T4 of virus length DNA molecules with permuted sequences by a cutting enzyme, which cuts off constant lengths of DNA irrespective of the sequence. Left: arrows, sites of enzyme attack; right: molecules generated.

infection cycle of T4 in a susceptible cell (see Section 8.7 and Figures 8.8–8.10) and have noted that before DNA synthesis occurs, transcription has begun on the phage genome.

Transcription, translation, and regulation in phage T4

In bacteriophage T4, the details of regulation of replication are more complex than those of T7. Phage T4 is much larger than T7 and has many more genes and phage functions. And, as previously mentioned, the DNA of T4 contains the unusual base 5-hydroxymethylcytosine (Figure 8.21), and some of the OH groups of this base are glucosylated. Thus, enzymes for the synthesis of this unusual base and for its glucosylation must be formed after phage infection, as well as formation of an enzyme that breaks down the normal

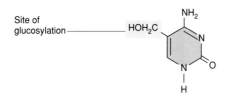

Figure 8.21 The unique base in the DNA of the T-even bacteriophages, 5-hydroxymethylcytosine. The site of glucosylation is shown.

DNA percusor deoxycytidine triphosphate. In addition, T4 encodes a number of enzymes that have functions similar to those of host enzymes in DNA replication but are formed in larger amounts, thus permitting faster synthesis of T4-specific DNA. In all, T4 encodes over 20 new proteins that are synthesized early after infection.

Overall, the T4 genes can be divided into three groups, one encoding early proteins, one middle proteins, and the other, late proteins (Figure 8.23). The **early** and **middle proteins** are the enzymes involved in DNA replication and transcription. The **late proteins** are the head and tail proteins and the enzymes involved in liberating the mature phage particles from the cell.

Unlike T7, T4 does not encode a new phage-specific RNA polymerase. The control of T4 mRNA synthesis involves the production of proteins that sequentially modify the specificity of the host RNA polymerase so that it recognizes different phage promoters. The early promoters are read directly by the host RNA polymerase and involve the function of host *sigma* factor. Phage-specific proteins synthesized from the early genes carry out covalent modifications on the host RNA polymerase α subunits (∞ Section 6.6), and a few phage-encoded proteins also bind to the polymerase. These modifications change the specificity of the polymerase so it now recognizes T4 middle promoters. One of the T4 early proteins, called MotA, apparently recognizes a particular DNA sequence in these promoters. Transcription from the late promoters requires a new T4-encoded sigma factor. Interestingly, it also requires T4 DNA synthesis. Sequential modification of host cell RNA polymerase is used to regulate gene expression by many bacteriophages.

In the case of phage T4, the entire lytic cycle takes about 25 min (Figure 8.23). Assembly of heads and tails occurs independently, DNA is packaged into the assembled head, and the tail and tail fibers are added later. Exit of the virus from the cell occurs as a result of cell lysis (Figure 8.23). The phage codes for a lytic enzyme, the *T4 lysozyme*, which attacks the peptidoglycan of the host cell.

CONCEPT CHECK   **8.11**

The phages T4 and T7 each take over the cell's metabolic machinery, leading to the production of new virions and host cell lysis. Regulation of transcription in T7 involves the production of a virus-specific RNA polymerase, whereas in T4 the host polymerase is modified. The T4 genome is quite large and encodes many enzymes. Both viruses take advantage of repeats at the ends of their genomes to complete DNA replication.

✔ *Describe the strategy employed by T4 and by T7 to ensure that late in infection only virus genes are transcribed.*

✔ *Describe the strategies used by T4 and by T7 to avoid the host's restriction enzymes.*

8.12

Temperate Bacteriophages: Lysogeny and Lambda

Most of the bacterial viruses described previously are called **virulent** viruses, because they usually kill (lyse) the cells they infect. However, many other bacterial

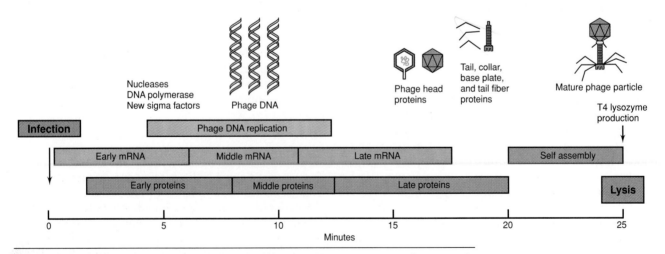

Figure 8.23 Time course of events in phage T4 infection. Following injection of DNA, early and middle mRNA is produced that codes for nucleases, DNA polymerase, new phage-specific sigma factors, and various other proteins involved in DNA replication. Late mRNA codes for structural proteins of the phage virion and for T4 lysozyme, needed to lyse the cell and release new phage particles.

viruses, although also able to kill cells, frequently have more subtle effects. Such viruses are called **temperate.** These viruses can enter into a state called **lysogeny,** where most virus genes are not expressed, and the virus genome is replicated in synchrony with the host chromosome.

Thus, the phage genome is duplicated along with the host material at the time of cell division, being passed from one generation of bacteria to the next. Under certain conditions these bacteria, called **lysogens,** can spontaneously produce virions of the temperate virus. Lysogeny is probably of ecological importance because most bacteria isolated from nature are lysogens for one or more bacteriophages. Lysogeny is not limited to bacteriophages. Many animal viruses set up similar relationships with their hosts.

Overview of the life cycles of a temperate phage

Our discussions of viral reproduction so far in this chapter should make it clear that it is not the presence, or even the replication, of viral DNA that leads to the production of new virions and host cell death. Rather it is *expression* of the viral genome that is deleterious. One could imagine that host cells can harbor virus genomes without harm if the expression of the viral genes can be controlled. This is the situation found in lysogens. However, once this control has been lost, the virus enters the lytic pathway, it produces new virions, and then the host cell lyses. In a culture of lysogens at any one time, only a small fraction of the cells, 0.1–0.0001%, produce virus and lyse, while the majority of the cells neither produce virus nor lyse. Although only rarely do cells of a lysogenic strain actually produce virus, every cell has the potential for virus production. Lysogeny can thus be considered a genetic trait of a bacterial strain.

An overall view of the life cycle of a temperate bacteriophage is shown in Figure 8.24. The temperate virus does not exist in its mature, infectious state inside the cell but rather in a latent form called the **provirus** or **prophage** state. In the example shown in Figure 8.24, the prophage is integrated into the bacterial chromosome. The prophage replicates along with the host cell as long as the genes controlling its lytic pathway are not expressed. Typically this control is maintained by a phage-encoded repressor protein (indicating that at least this gene is being expressed). The virus repressor protein not only controls the lytic genes on the prophage but also prevents the expression of any incoming genomes of the same virus. This results in the lysogens having **immunity** to infection by the same type of virus.

However, if this repressor is inactivated, or if its synthesis is prevented, the prophage is induced (center, Figure 8.24). This induction results in the production of new virions and the lysis of the host cell. In some cases (as we shall see later), induction can be brought about by environmental conditions. If the virus loses the ability to leave the host genome (because of mutation), it becomes a cryptic virus. Interestingly, from sequence studies it has been shown that many bacterial chromosomes contain stretches of DNA that were clearly once part of a virus genome.

It is sometimes possible to eliminate the temperate virus (to "cure" the host) by heavy irradiation or treatment with nitrogen mustards. Among the few survivors may be some cells that have been cured. Presumably the treatment causes the prophage to excise from the host chromosome and be lost during subsequent cell growth. Such a cured strain is no longer immune to the virus and can serve as a suitable host for study of virus replication.

Note that Figure 8.24 shows that infection of a normal cell by a temperate virus can lead to either the lytic pathway or the lysogenic pathway. In this section we shall discuss what factors favor one or the other of pathways during infections by the bacteriophage lambda.

The bacteriophage lambda

One of the best-studied temperate phages is lambda, which infects *Escherichia coli*, and our knowledge of the molecular mechanisms involved in lysogenization and lytic processes in this phage is very advanced. Morphologically, lambda particles look like those of many other bacteriophages (Figure 8.25). The virus particle has an icosahedal head 64 nm in diameter and a tail 150 nm long that has helical symmetry. Attached to the tail is a single 23-nm-long fiber.

The genome of lambda consists of a linear double-stranded DNA molecule, but at the 5'-terminus of each of the single strands is a single-stranded tail 12 nucleotides long. These single-stranded ends are complementary (the ends of the DNA are said to be *cohesive*). Thus, when the two ends of the DNA are free in the host cell, they associate and the genome forms a double-stranded circle. In the circular form the DNA contains 48,502 base pairs, and its complete sequence is known. Figure 8.26 is a representation of the genetic map of lambda after circularization. We discuss the organization and expression of the genes later.

Lambda infection and the lytic pathway

The lambda virion attaches to a specific protein in the cell wall of *Escherichia coli* (a protein involved in the uptake of the sugar maltose) and injects its DNA. The DNA circularizes almost immediately and, if the cell is not a lambda lysogen (and therefore immune), expression of the phage genome begins. The first steps in gene expression are the same whether the final result is lysis or lysogeny.

Production of RNA using host RNA polymerase begins at a few promoters, two of which, called P_L

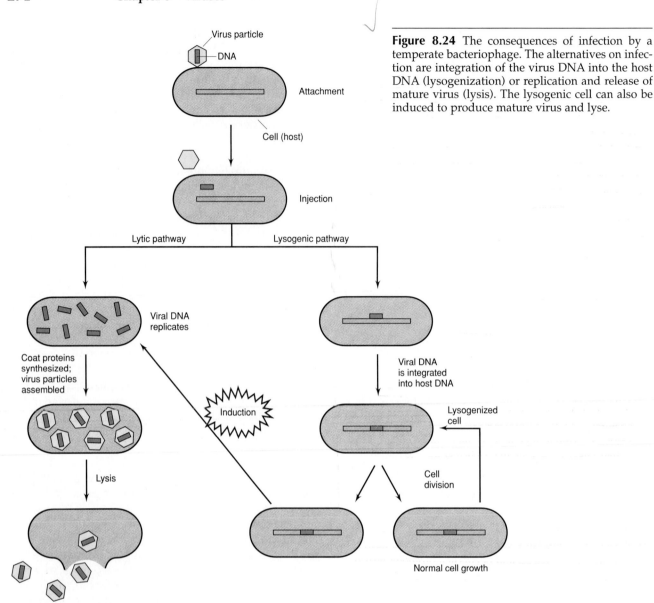

Figure 8.24 The consequences of infection by a temperate bacteriophage. The alternatives on infection are integration of the virus DNA into the host DNA (lysogenization) or replication and release of mature virus (lysis). The lysogenic cell can also be induced to produce mature virus and lyse.

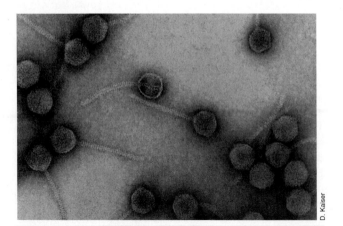

Figure 8.25 Electron micrograph by negative staining of bacteriophage lambda particles. The head of each particle is about 65 nm in diameter.

(promoter left) and P_R (promoter right), are on either side of the regulatory region shown in Figure 8.26. These yield short transcripts that are translated to give the products of the N and the cro genes. Both the proteins encoded by these genes are involved in regulatory events. The Cro protein (the product of the cro gene) participates in the selection between the lytic and lysogenic pathways, and we shall discuss its function later. The N protein is an *antiterminator* protein that allows the RNA polymerase to transcribe past specific terminators (marked on Figure 8.26), making the transcripts from P_L and P_R longer. These longer transcripts can be translated to yield more proteins, including the products of the O, P, cII, and $cIII$ genes. The antiterminator is not completely effective at the terminator before the Q gene, and so only a small amount of the Q protein is made.

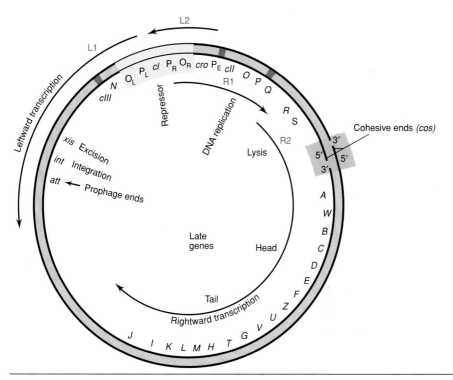

Figure 8.26 Genetic and molecular map of lambda. The genes are designated by letters; *att*, attachment site for phage to host chromosome. Genes of special interest: *cI*, repressor protein; O_R, operator right; P_R, promoter right; O_L, operator left; P_L, promoter left; *cro*, gene for second repressor; N, positive regulator counteracting rho-dependent termination. J through U are genes that encode tail proteins. Genes Z through A encode head proteins. The regulatory region of lambda (shown in yellow) is positioned at the top of this circular map. It is also known as the immunity region and contains the *cI* gene. The site created when the cohesive ends of the lambda genome join is called *cos* (shown in blue). Early transcription in lambda is primarily leftward (counterclockwise) from P_L and rightward (clockwise) from P_R. The main leftward transcript is labeled L1, and the main early rightward transcript is labeled R1. The three transcription terminators affected by the N-protein are shown as gray boxes on the DNA. The late rightward transcript, which encodes head and tail proteins and proteins for lytic function, is labeled R2 and begins at a promoter near the Q gene. The transcript labeled L2 is the positively regulated transcript from P_E that encodes the repressor protein.

The O and P proteins are involved in the initiation of replication of lambda DNA. This early DNA synthesis is bidirectional from a single origin at a site close to the O gene (Figure 8.26) and gives rise to typical theta-like intermediates (∞ Section 6.5). At this stage it is still possible for lambda to switch to a lysogenic cycle. However, let us consider the situation that would result if this switch were not made. It should be remembered that a typical infection by a temperate virus is lytic.

The Q protein is also an antiterminator protein. If its concentration becomes high enough it will allow the transcript from a nearby promoter to synthesize the transcript labeled R2 in Figure 8.26. This transcript is translated to yield the late proteins, all the necessary structural proteins to construct a virion, and the proteins necessary for cell lysis. At the same time the Q protein has built up to this level, the Cro protein (see earlier discussion) has also reached levels where it can block transcription from both P_L and P_R by binding to

both O_L (operator left) and O_R (operator right). Therefore, Cro operates as a repressor protein.

The mechanism of Cro protein action at O_R is diagrammed in more detail in Figure 8.27. Note that there are three similar but nonidentical sites at this operator where the Cro protein can bind. It does so first at site 3, and then site 2, and only when those two sites are filled, at site 1. Note that only when bound to site 1 does it block P_R. Note also that once P_L and P_R are blocked, no more cII or cIII proteins can be synthesized. These proteins are needed to enter the lysogenic pathway (see later), and so when Cro is made in high amounts, lambda is irrevocably committed to the lytic pathway.

The shutoff of these promoters also results in a change of lambda DNA replication. At this stage when the late proteins are being synthesized (and the O and P proteins are not), long, linear concatamers are synthesized by **rolling circle replication** (see Figure 8.17). In this mechanism, replication proceeds in only one

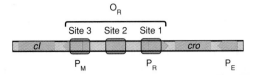

Figure 8.27 Both regulatory proteins Cro and the lambda repressor bind to operator right (O_R) on the lambda genome to carry out regulatory functions. The Cro protein (product of the *cro* gene) binds to the three sites in the order site 3, then site 2, and then site 1. The lambda repressor binds to these sites in the opposite order. The promoter P_R is transcribed immediately on phage entry into the cell. Rightward transcription from this promoter is necessary to produce Cro protein and other downstream genes (see Figure 8.26 for a complete map of the lambda genome). Leftward transcription from either of the promoters P_E or P_M is necessary to synthesize the lambda repressor (product of the *cI* gene). Both these promoters require activation in order to function.

direction and can result in very long chains of replicated DNA. Unlike the replication of ϕX174 DNA (see Figure 8.17), however, rolling circle replication of lambda involves synthesis of both strands (Figure 8.28). This mechanism is efficient in permitting extensive, rapid, relatively uncontrolled DNA replication; thus, it is of value in the later stages of the phage replication cycle when large amounts of DNA are needed to form mature virions. The long concatamers formed are then cut into virus-sized lengths by a DNA-cutting enzyme. In the case of lambda, the cutting enzyme makes staggered breaks at specific sites on the two strands, 12 nucleotides apart, which provide the cohesive ends involved in the cyclization process. These DNA molecules are packaged in phage heads, and then the tails and other proteins are added. The cell is then lysed by the action of phage-encoded proteins.

The lytic pathway of lambda is not much different from those of the other viruses we have discussed earlier. However, the host cell's metabolism is not irreversibly subverted early in the process, ensuring that lysogenization can take place if events favor it.

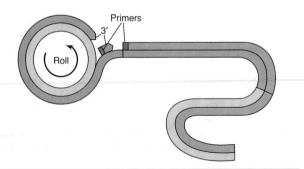

Figure 8.28 A late stage in the rolling circle replication of lambda. Both strands of DNA are being copied at the replicating fork, and two copies of the genome have already been synthesized. Note that this synthesis is *asymmetric* because one of the parental strands continues to serve as a template and the other is used only once.

Lysis or lysogenization?

We have seen that phage genes are controlled in such a way that viral proteins and nucleic acids are made in appropriate amounts and at appropriate times. For many viruses the patterns of expression always proceed in the same programmed manner. However, lambda and other temperate viruses have a *genetic switch* that controls whether the lytic pathway or the lysogenic pathway is followed. So far the steps we have outlined for lambda are those for the lytic pathway. We now consider how the genetic switch can be thrown to lead to lysogeny.

In order to establish lysogeny, two events must happen: the production of all late proteins must be prevented, and a copy of the lambda genome must be integrated into the host chromosome. In order to prevent synthesis of the late proteins, the product of the *cI* gene must be produced. This protein is the **lambda repressor.** If it is synthesized *it will repress the synthesis of all other lambda-encoded proteins*. It is needed to establish lysogeny and to maintain the lysogenic state. The *cI* gene is located between P_L and P_R (see Figure 8.26), but these promoters are oriented in such a way that neither transcribes the *cI* gene. The promoter that can produce mRNA from the *cI* gene during infection is called P_E (promoter establishment) and is located on the map slightly to the right of the *cro* gene but facing the direction opposite that of P_R. Therefore, transcription is in the direction opposite that promoted by P_R (Figures 8.26 and 8.27). Unlike the other promoters we have previously mentioned, P_E must be *activated*. Once it is, lambda repressor protein is synthesized and the lysogenic pathway is followed.

The product of the *cII* gene is an *activator protein* (∞ Section 7.4) that activates promoter P_E (and another promoter required for the production of integrase) (see later). Although the cII protein is made early after infection, it is typically unstable in *Escherichia coli* because it is degraded by a host protease (an enzyme that degrades proteins). If the cII protein is degraded, then there is no possibility that the lysogenic pathway can be chosen. However, this protein can be stabilized by the phage-encoded cIII protein if there is no excess of host protease (or if there is an excess of cIII protein). If the cII protein is stabilized, then it will activate P_E and lambda repressor protein will be made. In a way, this rather complicated process monitors conditions in the host.

Lambda repressor binds to O_L and O_R, as does the Cro protein, but it binds to the sites within these operators in the order opposite that of Cro (see Figure 8.27). That is, it first binds to site 1, turning off P_R (and P_L by a similar mechanism). When this happens, the synthesis of all other lambda proteins is stopped. Without protein Q the late proteins cannot be expressed and lambda cannot enter the lytic pathway.

However, without the cII protein P_E no longer functions. Therefore, if the lysogenic state is to be maintained, there must be another way to transcribe the *cI* gene. Note that in Figure 8.27 there is yet another promoter shown, P_M (promoter maintenance). This promoter is facing toward the *cI* gene (in the same direction as P_E). It is *activated* when lambda repressor binds to site 1 and is repressed only when lambda repressor is bound to all three sites. Therefore, the lambda repressor is both a *repressor* and an *activator* when it binds to site 1, repressing P_R and activating P_M. This type of regulation continues to occur even after lysogenization. Only the lambda repressor is made after lambda is integrated as a prophage.

Integration

Integration of lambda DNA into the host chromosome occurs at a unique site on the *Escherichia coli* genome and is required for lysogeny. Integration occurs by insertion of the virus DNA into the host genome (thus effectively lengthening the host genome by the length of the virus DNA). As illustrated in Figure 8.29, on injection, the cohesive ends of the linear lambda molecule find each other and form a circle, and it is this circular DNA that becomes integrated into the host genome (the site created when these ends join is called *cos*). To estab-

lish lysogeny, genes *cI* and *int* (encoding *integrase*) must be expressed as we discussed. The integration process requires integrase, the product of the *int* gene, which is a site-specific topoisomerase catalyzing recombination of the phage and bacterial attachment sites (labeled *att* in Figures 8.26 and 8.29). The *int* gene has a promoter that, like P_E, is activated by the cII protein.

During cell growth, the lambda repression system prevents expression of the integrated lambda gene except for the gene *cI*, which codes for the lambda repressor. During host DNA replication, the integrated lambda DNA is replicated along with the rest of the host genome and transmitted to progeny cells. When release from repression occurs (see later), the lambda productive cycle occurs. In order to be excised from the chromosome, *excisionase* (the product of the *xis* gene) and the *int* gene product are required.

Lytic growth of lambda after induction

Agents that induce lambda lysogens (cells containing lambda as a prophage) to produce phage are agents that damage DNA, such as ultraviolet irradiation,

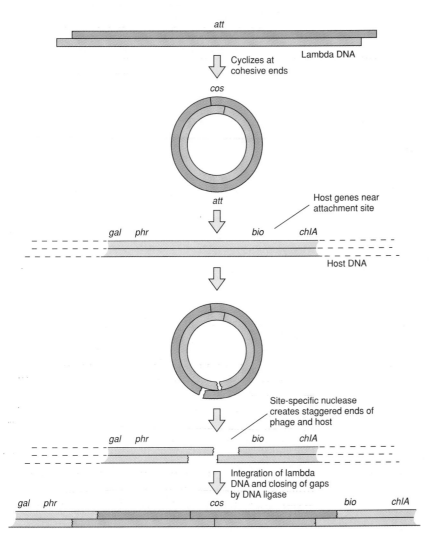

Figure 8.29 Integration of lambda DNA into the host. See the genetic map, Figure 8.26, for details of the gene order. Integration always occurs at a specific site on the host DNA, involving a specific attachment site (*att*) on the phage. Some of the host genes near the attachment site are given. A site-specific enzyme (integrase) is involved, and specific pairing of the complementary ends results in integration of phage DNA.

X-rays, and DNA-damaging chemicals such as the nitrogen mustards. These agents interfere with the function of the lambda repressor. When DNA damage occurs a host defense mechanism called the SOS response (∞ Section 9.3) is brought into play. An array of 10–20 bacterial genes is turned on, some of which help the bacterium survive radiation. However, one result of DNA damage is that a bacterial protein called RecA (normally involved in genetic recombination) is turned into a special kind of protease that participates in the destruction of the lambda repressor. With the lambda repressor destroyed, the inhibition of expression of lambda lytic genes is abolished. We should note that the protease activity of RecA, brought about by DNA damage, normally plays an important role in the cell's response to DNA-damaging agents by participating in the breakdown of a host protein, LexA, which represses a set of host genes involved in DNA repair (∞ Section 9.3). Induction of bacteriophage lambda is thus an indirect consequence of the SOS response.

Once the lambda repressor has been inactivated, the control exerted by this repressor is abolished and new transcriptional events can be initiated. These inevitably lead to lysis because even if the lambda repressor is made, it is inactivated.

Other strategies used by temperate viruses

Lambda provides one of the best-studied examples of how a "decision" is made at the molecular level. It is also widely used in genetic engineering as a vector for carrying recombinant DNA. It has several features that make it an excellent system for genetic engineering. One feature of lambda that makes it of special use for cloning is that there is a long region of DNA, between genes *J* and *att* (Figure 8.26), that does not seem to have any essential functions for replication and can be replaced with foreign DNA. We describe the use of lambda as a cloning vector in Chapter 10. We also describe other uses of lambda as a genetic tool in

Chapter 9. Other types of temperate viruses are known in bacteria, and some have also been widely studied. The virus P1 (which we will also mention in Chapter 9) is a temperate virus that maintains itself in the lysogenic state not as an integrated prophage but replicates as a circular DNA molecule in the cytoplasm, resembling a plasmid. The very interesting temperate phage named Mu is a transposable element. We discuss Mu in the next section.

8.13

A Transposable Phage: Bacteriophage Mu

One of the more interesting bacteriophages is the one called **Mu.** This virus is temperate, like lambda, but has the unusual property of replicating as a transposable element (∞ Section 6.3 and Chapter 9). This phage is called Mu because it is a *mutator* phage, inducing mutations in a host genome into which it becomes integrated. This mutagenic property of Mu arises because the genome of the virus can become inserted into the middle of host genes, causing these genes to become inactive (and hence the host that has become infected with Mu behaves as a mutant). Mu is a useful phage because it can be used to generate a wide variety of bacterial mutants very easily. Also, as we will discuss in Chapter 10, Mu can be used in genetic engineering.

Transposable elements are pieces of DNA that have the ability to move from one site to another as discrete elements. They are found in both prokaryotes and eukaryotes and play important roles in genetic variation (see Section 9.10 for a detailed dicussion of transposition). There are three types of transposable elements: *insertion elements, transposons,* and viruses like Mu (∞ Section 6.3). Mu is a very large transposable element, carrying a number of Mu genes involved in Mu multiplication.

Structure and genetic map of Mu

Structurally, bacteriophage Mu is a large double-stranded DNA virus with an icosahedral head, a helical tail, and six tail fibers (Figure 8.30). The genetic map of Mu is shown in Figure 8.31a. It can be seen that the bulk of the genetic information is involved in the synthesis of the head and tail proteins, but that important genes at each end are involved in replication and immunity. The DNA molecule found within the virion is approximately 39 kilobase pairs long, but only 37.2 kilobase pairs make up the actual Mu genome. This is because both ends of this DNA molecule contain host DNA. At the left end of the Mu DNA are 50–150 base pairs of host DNA, and at the right end are 1–2 kilobase pairs of host DNA. These host DNA sequences are not unique and represent DNA adjacent to the location where Mu was inserted into the genome of its previous host.

CONCEPT CHECK   8.12

Temperate viruses do not always cause the death of the cells they infect. The infected cell sometimes survives because the virus genome becomes a prophage (and replicates with the host chromosome), and the lytic genes of the prophage are kept under the control of a virus-encoded repressor. Sometimes this regulatory system is circumvented and prophage induction occurs, resulting in virus multiplication and lysis of the host cell. Host cells carrying temperate viruses are called lysogens.

✔ *What are the two pathways available to a temperate virus?*

✔ *Describe how a single protein like the lambda repressor can act both as an activator and a repressor.*

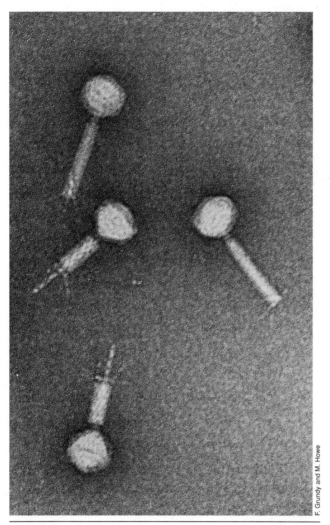

F. Grundy and M. Howe

Figure 8.30 Electron micrograph of virions of bacteriophage Mu, the mutator phage.

When a Mu phage particle is formed, a length of DNA containing the Mu genome just large enough to fill the phage head is cut out of the host, beginning at the left end. The DNA is rolled in until the head is full, but the place at the right end where the DNA is cut varies from one phage particle to another. For that reason, as shown on the genetic map, there is a variable sequence of host DNA at the right-hand end of the phage (right of the *attR* site) that represents the *host* DNA that has become packaged into the phage head. Each virion arising from a single infected cell will have a different amount of host DNA, and the host DNA base sequence in each virion from the same cell will be different.

As shown on the genetic map (Figure 8.31), a specific segment of the Mu genome called G (distinct from the G gene) is invertible, being present either in the orientation designated SU or in the inverted orientation U'S'. The orientation of this segment determines the kind of tail fibers that are made for the phage. Since adsorption to the host cell is controlled by the speci-

ficity of the tail fibers, the host range of Mu is determined by which orientation of this invertible segment is present in the phage. If the G segment is in the orientation designated G⁺, then the phage particle will infect *Escherichia coli* strain K12. If the G segment is in the G⁻ orientation, then the phage particle will infect *E. coli* strain C or several other species of enteric bacteria. The two tail fiber proteins are encoded on opposite strands within this small G segment. Left of the G segment is a promoter that directs transcription into the G segment. In the orientation G⁺, the promoter directing transcription of S and U is active, whereas in the orientation G⁻, a different promoter directs transcription of genes S' and U' on the opposite strand. Regulation involving rearrangement of DNA sequences is known in other viruses as well as both prokaryotic and eukaryotic cells.

Replication of Mu

On infection of a host cell by Mu, the DNA is injected and is protected from host restriction by a modification system in which about 15% of the adenine residues are acetoamidated. In contrast with lambda, integration of Mu DNA into the host genome is essential for both lytic and lysogenic growth. Integration requires the activity of the gene A product, which is a transposase enzyme. At the site where the Mu DNA becomes integrated, a five-base-pair duplication of the host DNA arises at the target site. As shown in Figure 8.31b, this host DNA duplication arises because staggered cuts are made in the host DNA at the point where Mu is inserted, and the resulting single-stranded segments are converted to the double-stranded form as part of the integration process. Duplication of short stretches of host DNA is typical of transposable element insertion (∞ Section 9.10).

Lytic growth of Mu can occur either on initial infection, if the Mu repressor (the product of the *c* gene) is not formed, or by induction of a lysogen. In either case, replication of Mu DNA involves repeated transposition of Mu to multiple sites on the host genome. Initially, transcription of only the early genes of Mu occurs, but after gene C protein, a positive activator of late RNA synthesis, is expressed, the synthesis of the Mu head and tail proteins occurs. Eventually, expression of the lytic function occurs and mature phage particles are released.

8.14
Overview of Animal Viruses

We have discussed in a general way the nature of animal viruses in the first part of this chapter. Now we examine in some detail the structure and molecular biology of a number of important animal viruses.

Viruses will be discussed that illustrate different ways of replicating, and both RNA and DNA viruses

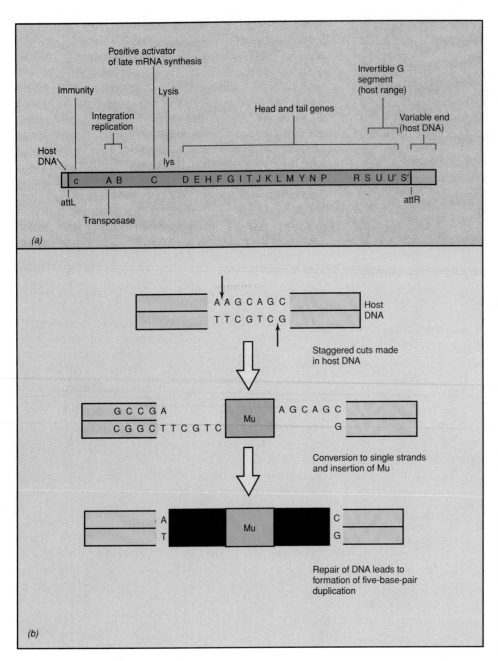

Figure 8.31 Bacteriophage Mu. (a) Genetic map of Mu. See text for details. (Confusingly, there are two G's, the G gene and the invertible G segment. These are different G's. There is also a lowercase c gene, which encodes a repressor, and an uppercase C gene, which encodes an activator protein.) (b) Integration of Mu into the host DNA, showing the generation of a five-base-pair duplication of host DNA.

will be covered. One important group of animal viruses, those called the *retroviruses*, have both an RNA and a DNA phase of replication. Retroviruses are especially interesting not only because of their unusual mode of replication but also because they cause such important diseases as certain *cancers* and *acquired immunodeficiency syndrome* (AIDS).

Before beginning our discussion of the manner of replication of animal viruses, we should remind ourselves of the important differences that exist between eukaryotic and prokaryotic cells. Because virus replication makes use of the biosynthetic machinery of the host, these differences in cellular organization and func-

tion imply differences in the way the viruses themselves replicate.

Differences between prokaryotes and eukaryotes that affect virus multiplication

Prokaryotes do not show compartmentation of the biosynthetic processes. The genome of a bacterium relates directly to the cytoplasm of the cell. Transcription into mRNA can lead directly to translation, and the processes of transcription and translation are not carried out in separate compartments (∞ Figure 6.2a). Animal cells, being eukaryotic, show compartmentation. DNA replication and transcription of the genome into mRNA occur in the nucleus, whereas translation occurs in the cytoplasm (∞ Figure 6.2*b*). This compartmentation has an impact on where viruses replicate. For instance, a DNA virus that uses host polymerases must replicate in the nucleus. Therefore, we can expect differences in replication strategies between viruses that multiply in the nucleus and those that multiply in the cytoplasm.

Furthermore, the transcripts from eukaryotic genes must be processed and transported to the cytoplasm before they can be used as mRNA (∞ Section 6.7). This processing usually involves **splicing** out *introns* as well as adding a **poly-A tail** to the 3′-end and a methylated guanosine triphosphate, called the **cap,** to the 5′-end. The cap is required for binding of the mRNA to the ribosome. The difference between the way ribosomes recognize mRNA allows prokaryotes to use polycistronic mRNA, whereas eukaryotes use monocistronic mRNA (∞ Sections 6.6 and 6.9). All the protein-synthesizing machinery of the eukaryotic cell—the ribosomes, tRNA molecules, and accessory components—is in the cytoplasm, and the mature mRNA associates with the protein-synthesizing apparatus once it leaves the nucleus.

We might also note another important difference between animal and bacterial cells. Bacterial cells have rigid cell walls containing peptidoglycan and associated substances (∞ Section 3.5). Animal cells, on the other hand, lack cell walls. This difference is important for the way in which the virus genome enters and exits the cell. In prokaryotes, the protein coat of the virus remains on the outside of the cell and only the nucleic acid enters. In animal viruses, on the other hand, uptake of the virus often occurs by endocytosis (pinocytosis or phagocytosis), processes that are characteristic of animal cells, and so the whole virus particle enters the cell (Figure 8.32). The separation of animal virus genomes from their protein coats then occurs inside the cell.

Classification of animal viruses

Various types of animal viruses are illustrated in Figure 8.33. Note that the major criteria used in classifying animal viruses are the type of nucleic acid, the presence or absence of an envelope, and, for certain families, the manner of replication. Most of the animal viruses that have been studied in any detail are those that have been amenable to cultivation in cell cultures. Animal viruses are known with either single- or double-stranded DNA or RNA. Some animal viruses are enveloped, others are naked. Size varies greatly, from those large enough to be just visible in the light microscope to those so tiny that they are hard to see even in the electron microscope. In the following sections, we will discuss characteristics and manner of multplication of some of the most important and best-studied animal viruses.

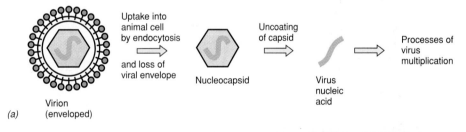

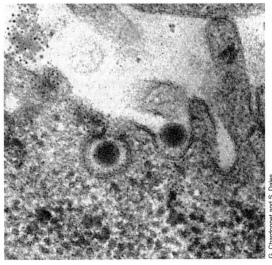

Figure 8.32 Uptake of an enveloped virion by an animal cell. (a) The process by which the viral nucleocapsid is separated from its envelope. (b) Electron micrograph of advenovirus virions entering a cell. Each particle is about 70 nm in diameter.

G. Chardonnet and S. Dales

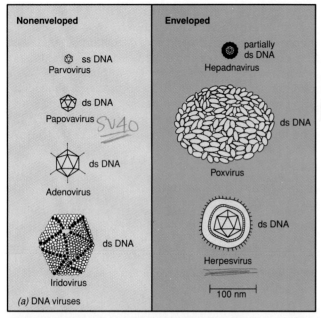

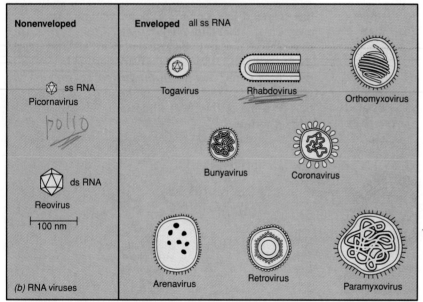

Figure 8.33 The shapes and relative sizes of vertebrate viruses of the major taxonomic groups. The hepadnavirus genome has one complete DNA strand and part of its complement.

Consequences of virus infection in animal cells

Viruses can have varied effects on cells. **Lytic infection** results in the destruction of the host cell (Figure 8.34). However, there are several other possible effects following viral infection of animal cells. In the case of enveloped viruses, release of virions, which occurs by a kind of budding process, may be slow and the host cell may not be lysed. The cell may remain alive and continue to produce virus over a long period of time. Such infections are referred to as **persistent infections** (Figure 8.34). Viruses may also cause **latent infection** of a host. In a latent infection, there is a delay between infection by the virus and the appearance of symptoms. Fever blisters (cold sores), caused by the herpes simplex virus (see Section 8.19), result from a latent viral infection; the symptoms reappear sporadically as the virus emerges from latency. The latent stage in viral infection of an animal cell is generally not due to integation of the viral genome into the genome of the animal cell, as is the case with latent infections by temperate bacteriophages.

Viruses and cancer

A number of animal viruses have the potential to change a cell from a normal one to a cancer or tumor cell (Figure 8.34 and Table 8.2). **Cancer** is a cellular phenomenon of uncontrolled growth. Most cells in a mature animal, although alive, do not divide extensively, apparently because of the presence of growth-

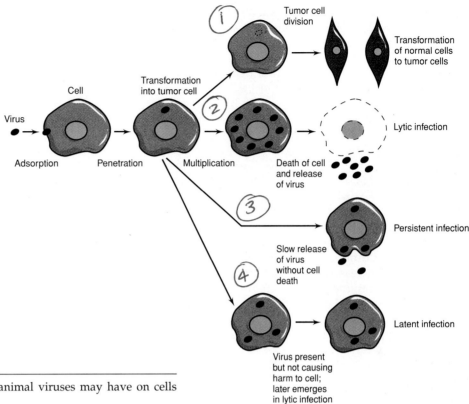

Figure 8.34 Possible effects that animal viruses may have on cells they infect.

inhibiting factors that prevent them from initiating cell division. As noted in Section 8.4, infection by certain types of animal viruses leads to a process called **transformation** during which growth becomes uncontrolled. One of the key differences between normal cells and cancer cells is that the latter have different requirements for growth factors. Rapidly growing cells pile up into accumulations that are visible in culture as **foci of infection**. Because cancerous cells in the animal body have fewer growth requirements, they grow profusely, leading to the formation of large masses of cells, called *tumors*. The term **neoplasm** is often used in the medical literature to describe malignant tumors.

Not all tumors are seriously harmful. The body is able to wall off some tumors so that they do not spread; such noninvasive tumors are said to be **benign**. Other tumors, called **malignant,** invade the body and destroy normal body tissues and organs. In advanced stages of cancer, malignant tumors may develop the ability to spread to other parts of the body and initiate new tumors, a process called **metastasis.**

How does a normal cell become cancerous? The growth and division of normal cells is regulated by at least two types of genes. The first type, called *proto-oncogenes*, promote growth, but they are controlled by the second type, the growth-restraining *tumor suppressor genes*. Changes in either or both types can lead to uncontrolled cell growth and therefore to cancer. The process can be broken down into several stages. In the first step, *initiation*, genetic changes in the cell occur. This step may be induced by certain chemicals, called *carcinogens* (∞ Section 9.3), or by physical stimuli, such as ultraviolet radiation or X-rays. Certain viruses also bring about the genetic change that results in initiation of tumor formation. This initiation event can be the activation of a proto-oncogene into an **oncogene** (a gene that causes a tumor), or the inactivation of a tumor suppressor gene.

TABLE 8.2	Some human cancers that may be caused by viruses		
Cancer	**Virus**	**Family**	**Genome in virion**
Adult T-cell leukemia	Human T-cell leukemia virus (type I)	Retrovirus	RNA
Burkitt's lymphoma	Epstein–Barr virus	Herpes	DNA
Nasopharyngeal carcinoma	Epstein–Barr virus	Herpes	DNA
Hepatocellular carcinoma (liver cancer)	Hepatitis B virus	Hepadna	DNA
Skin and cervical cancers	Papilloma virus	Papova	DNA

Once initiation has occurred, the potentially cancerous cell may remain dormant, but under certain conditions, generally involving some environmental alteration, it may become converted to a tumor cell, a process called **promotion.** Once a cell has been promoted to the cancerous condition, continued cell division can result in the formation of a tumor.

Although the ability of viruses to cause tumors in animals has been proved for many years, the relationship of viruses to cancer in humans has, in most cases, been uncertain. It is difficult to prove the viral origin of a human cancer because of the difficulties of carrying out the necessary experimentation. However, it is now well established that certain specific kinds of human tumors do have a viral origin. A summary of some of the human cancers with definite viral origins is given in Table 8.2. In addition, some viral infections can lead indirectly to an increased risk of cancer, apparently by weakening the immune system's ability to detect and destroy transformed cells. This might be why infection with the retrovirus that causes AIDS increases the risk for developing certain cancers.

CONCEPT CHECK   8.14

Multiplication of animal viruses differs in significant ways from the multiplication of bacterial viruses, because of differences in compartmentation of macromolecule synthesis in eukaryotes as opposed to prokaryotes. Not all infections of animal host cells result in cell lysis or death. In some cases, latent infection occurs, the virus remaining infectious but dormant inside the host and appearing spontaneously at a later time. Some animal viruses cause transformation of host cells to the cancerous state.

✔ *Which macromolecules are synthesized in the nucleus and which are synthesized in the cytoplasm of eukaryotic cells?*

✔ *Contrast the mechanisms by which animal viruses enter cells with those used by bacterial viruses.*

8.15

Positive-Strand RNA Animal Viruses

Just as there are several known positive-strand RNA bacteriophages, there are many different kinds of animal viruses with a positive-strand RNA genome. (The great majority of *plant* viruses are positive-strand RNA viruses. It has been postulated that these small genomes facilitate transfer from cell to cell within the plant.)

An important group of positive-strand RNA animal viruses is the *picornavirus family*, which contains such important viruses infecting humans as the *polioviruses*, the *cold viruses*, and the *hepatitis A virus*. The first animal virus discovered, foot-and-mouth disease virus, is also a picornavirus. These viruses are called *picornaviruses* because they are very small viruses (30 nm in diameter) (*pico* means "small") and contain single-stranded RNA. The virus particle has a simple icosahedral stucture with 60 morphological units per virion, each unit consisting of four distinct proteins (Figure 8.35a).

In poliovirus, the RNA is a linear molecule about 7500 bases in length (see Table 8.1). At the 5'-terminus of the viral RNA is a protein, called the *VPg protein*, that is attached covalently to the RNA. At the 3'-terminus of the RNA is a poly-A tail. RNA molecules that lack the poly-A tail are not infectious, but the VPg protein is not essential for infectivity.

An overview of the manner of multiplication of poliovirus is illustrated in Figure 8.35b. The RNA of the virus acts directly as a messenger RNA. Interestingly, this is so even though the RNA is not capped (∞ Section 6.7). The 5'-end of poliovirus RNA has a long sequence that can fold into several stem-loops (∞ Section 6.6). Somehow these permit binding of the eukaryotic ribosome. The virus RNA is monocistronic but codes for all the proteins of the virus in a single *polyprotein* that is later cleaved into the individual proteins. The coat proteins are encoded by sequences at the 5'-end of the molecule, and the proteins necessary for replication are encoded at the 3'-end. The whole replication process occurs in the cytoplasm.

At initial infection, the virus particle attaches to a specific receptor on the surface of a sensitive cell and enters the cell. Once inside the cell, the virus particle is uncoated, and the free RNA associates with ribosomes. The viral RNA is then translated from a single start codon into a large protein precursor, the polyprotein mentioned earlier, which has a molecular weight of about 240,000. This giant protein then undergoes self-cleavage into about 20 smaller proteins (including cleavage intermediates), among which are the four *structural proteins* of the virus particle, the RNA-linked *VPg protein*, an *RNA polymerase* responsible for synthesis of minus-strand RNA, and at least one *virus-encoded protease*, which carries out the cleavage process. This cleavage process, called *posttranslational cleavage*, occurs in a wide variety of animal viruses as well as in normal cell metabolism in animal cells.

Replication of poliovirus

Replication of viral RNA begins within a short time after infection and is catalyzed by the RNA-dependent RNA polymerase (replicase) made in the process described in the previous paragraph. This replicase transcribes the viral RNA, of plus complementarity, into an RNA molecule of minus complementarity. This minus strand then serves as a template for repeated

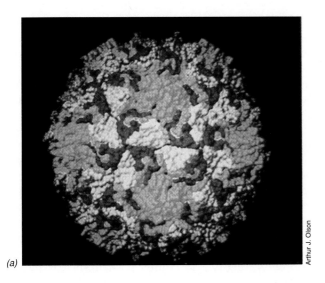

New poliovirus RNA

Replication

Poliovirus RNA serves as mRNA

| Structural genes | Protease genes | RNA polymerase gene |

Translation

A single large protein (polyprotein) is initially produced

Active gene products

Proteases cleave large protein

Structural coat proteins

Proteases

RNA polymerase

(b)

Arthur J. Olson

(a)

Figure 8.35 (a) The poliovirus particle. This is a computer model based on electron diffraction analysis of virus crystals. The various structural proteins are shown in distinct colors. (b) The reproduction of poliovirus. The single-stranded RNA of the virus is translated directly as a messenger RNA, with the production of one large protein molecule. This protein is cleaved, leading to production of the active viral proteins, including the structural coat proteins and the RNA polymerase that brings about replication of the poliovirus RNA. The assembly of intact poliovirus from coat protein molecules and RNA then follows.

transcription of progeny plus strands. Some of the progeny plus strands may again be transcribed into minus strands, and as many as 1000 minus strands may subsequently be present in the cell. From these minus strands, as many as a million plus strands may ultimately be formed. Both the plus and the minus strands become covalently linked to the tiny VPg protein (only 22 amino acids long), and it is thought that VPg serves as a primer for transcription or is added after synthesis. Viral RNA molecules that serve as mRNA lack VPg and are not assembled into mature virus particles.

Once virus multiplication begins, host RNA and protein syntheses are inhibited. Host protein synthesis is inhibited as a result of destruction of a host protein, the cap-binding protein required for translation of capped mRNAs.

At one time, polio was a major infectious disease of humans, but the development of an effective vaccine (∞ Section 20.17) has brought the disease completely under control.

CONCEPT CHECK *8.15*

In small RNA viruses such as polio, the viral RNA acts directly as a single messenger RNA, causing the production of a long polyprotein that is broken down by enzymes into the numerous small proteins necessary for nucleic acid multiplication and virus assembly.

✔ *What is a* cap *and what is its normal function?*

✔ *How can poliovirus RNA be synthesized in the cytoplasm while host RNA must be synthesized in the nucleus?*

8.16

Negative-Strand RNA Viruses

In a number of RNA viruses of animals, the RNA does not serve directly as a messenger but is transcribed into a complement that functions as the mRNA. As discussed in Section 8.6, it is conventional to express the configuration of the mRNA as *plus*, so if the viral genomic RNA is of opposite complementarity, it is called *minus* (see Figure 8.11). This group of viruses is then called minus-strand or *negative-strand RNA viruses*. We discuss here two important negative-strand viruses: rhabdoviruses, including rabies virus, and orthomyxoviruses, including influenza virus. The Ebola virus, a human pathogen responsible for an emerging infectious disease (∞ Section 22.10) is also a negative-strand RNA virus.

Rhabdoviruses

The most important human pathogen that is a negative-stranded RNA virus is the rabies virus, which causes the important disease rabies in animals and humans (∞ Section 23.8). Rabies virus is called a *rhabdovirus*, from *rhabdo* meaning "rod," which refers to the shape of the virus particle. Another rhabdovirus that has been extensively studied is vesicular stomatitis virus (VSV) (Figure 8.36), a virus that causes the disease *vesicular stomatitis* in cattle, pigs, horses, and sometimes humans. Many rhabdoviruses, such as potato yellow dwarf virus, infect both insects and plants and can cause important agricultural problems.

The rhabdoviruses are enveloped viruses, with an extensive and rather complex lipid envelope surrounding the nucleocapsid (Figure 8.36). In animal rhabdoviruses the virus particle is bullet-shaped, about 70 nm in diameter and 175 nm long. The nucleocapsid is helically symmetric and makes up only a small part of the virus particle weight (about 2–3% of the virion is RNA, in contrast to the 30% or more RNA content in nonenveloped RNA viruses). The virion contains several enzymes that are essential for the infection process:

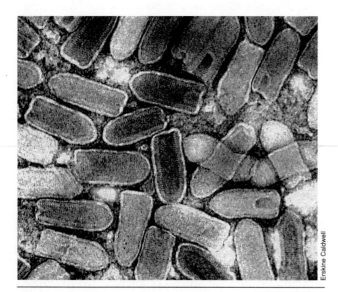

Figure 8.36 Electron micrograph of a rhabdovirus (vesicular stomatitis virus). A particle is about 65 nm in diameter.

RNA-dependent RNA polymerase, RNA methylase, and some capping enzymes. As discussed in Section 8.6, the presence of RNA polymerase is essential because the genome of these negative-strand viruses cannot act as messengers directly, but must first be transcribed into the plus complement, and host enzymes that transcribe RNA into RNA do not exist.

The RNA of the rhabdoviruses is transcribed inside the cytoplasm of the cell into two distinct kinds of RNA (Figure 8.37). The first type of RNA synthesis results in a series of messenger RNAs made from the various genes of the virus (VSV has 5). The second is a plus-strand RNA that is a *copy* of the complete viral genome (the VSV genome is 11,162 nucleotides long). These long plus-strand RNAs then serve as templates for synthesis of the *negative-strand* RNA molecules of the new crop of virions (yet to come). Each mRNA is monocistronic, coding for a single protein. A mechanism exists to ensure that transcription stops at the end of each virus gene and that a series of adenylic acid residues (a poly-A tail) is put on the end of the RNA. Once the mRNA for the virus RNA polymerase is made in this primary transcription process, synthesis of the virus RNA polymerase can begin, leading to the formation of many *plus*-strand RNA molecules, both messengers and full-length genomic (viral) RNA templates.

Translation of viral mRNAs leads to the synthesis of viral coat proteins, and copying of full-length *plus*-strand RNA leads to the formation of full-length *negative*-strand molecules. *Assembly* of an enveloped virus is considerably more complex than assembly of a simple virus particle. Two kinds of coat proteins are formed, *nucleocapsid proteins* and *envelope proteins*. The nucleocapsid is formed first by association of the nucleocapsid protein molecules around the viral RNA. These nucleocapsid protein molecules are synthesized on ribosome complexes in the cytoplasm.

The *envelope proteins* that possess hydrophobic amino acid leader sequences at their amino-terminal ends (∞ Section 6.9) are synthesized on ribosome complexes that are themselves associated with membranes. As these proteins are synthesized, sugar residues are added, leading to the formation of *glycoproteins*. Such glycoproteins, characteristic of membrane-associated proteins, are transported to the cytoplasmic membrane (and the leader sequences are removed), where they replace host membrane proteins. Nucleocapsids then migrate to the areas on the cytoplasmic membrane where these virus-specific glycoproteins exist, recognizing the virus glycoproteins with great specificity. The nucleocapsids then become aligned with the glycoproteins and bud through them, becoming coated by the glycoproteins in the process. The final result is an enveloped virus particle with a nucleocapsid center and a surrounding membrane whose lipid is derived from the host cell but whose membrane proteins are encoded by the virus. The budding

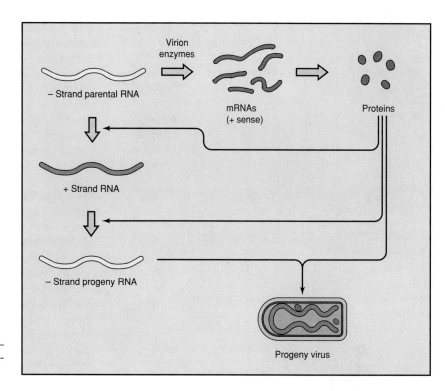

Figure 8.37 Flow of events during multiplication of a negative-strand RNA virus.

process itself does not cause detectable damage to the cell, which may continue to release virus particles in this way for a considerable period of time. (Host damage does occur but is brought about by other unknown factors.)

Influenza and other orthomyxoviruses

Another group of negative-strand viruses of great importance is the group called the *orthomyxoviruses,* which contains the important human virus *influenza.* The term *myxo* refers to the fact that these viruses interact with the *mucus* or *slime* of cell surfaces. In the case of influenza virus, this mucus is at the mucous membrane of the respiratory tract, as these viruses are transmitted primarily by the respiratory route (∞ Section 23.4). The term *ortho* has been added to the influenza virus group to distinguish this group from another group of negative-strand viruses, the *paramyxovirus group.* The paramyxoviruses, which include such important human viruses as those causing mumps and measles, are actually similar in molecular biology to rhabdoviruses. The orthomyxoviruses have been extensively studied over many years, beginning with early work during and after the 1918 pandemic of influenza that caused the deaths of millions of people worldwide (∞ Section 23.4).

The orthomyxoviruses are enveloped viruses in which the viral RNA is present in the virion in a number of separate pieces. The genome of the orthomyxoviruses is thus said to be a **segmented genome.** In the case of influenza A virus, the genome is segmented into *eight* linear single-stranded molecules ranging in size from 890 to 2341 nucleotides. The influenza virus

nucleocapsid is of helical symmetry, about 6–9 nm in diameter and about 60 nm long. This nucleocapsid is embedded in an envelope that has a number of virus-specific proteins as well as lipid derived from the host (Figure 8.38).

Because of the way influenza virus buds as it leaves the cell, the virus has no defined shape and is said to be *polymorphic* (Figure 8.38*a*). There are spikes on the outside of the envelope that interact with the host cell surface. One spike is called a *hemagglutinin* because it causes agglutination of red blood cells. (Agglutination is a process by which cells are caused to clump when they are mixed with an antibody or other protein or polysaccharide molecule that combines specifically with a substance on the cell surface, as described in Sections 20.15 and 21.5.) If the cells undergoing agglutination are red blood cells, then the process is called *hemagglutination.* (*Hema* is the combining form referring to *blood.*) The red blood cell is not the type of host cell the virus normally infects but contains on its surface the same type of membrane component, chemically characterized as *sialic acid,* that the mucous membrane cells of the respiratory tract contain. Thus, the red blood cell is merely a convenient cell type for measurement of agglutination activity. An important feature of the influenza virus hemagglutinin is that antibody directed against this hemagglutinin *prevents* the virus from infecting a cell. Thus, antibody directed against the hemagglutinin *neutralizes* the virus, and this is the mechanism by which immunity to influenza is brought about during the immunization process (∞ Sections 20.17 and 23.4).

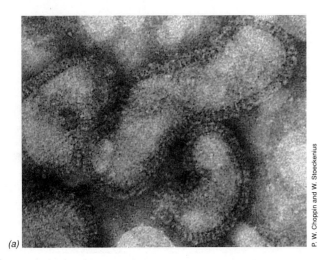

(a)

P. W. Choppin and W. Stoeckenius

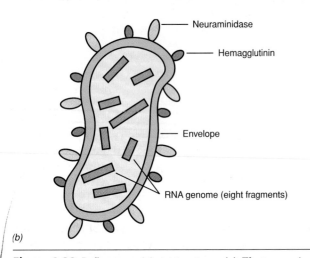

Neuraminidase

Hemagglutinin

Envelope

RNA genome (eight fragments)

(b)

Figure 8.38 Influenza virion structure. (a) Electron micrograph. (b) Diagram, showing some of the components.

A second type of spike on the virus surface is an enzyme called *neuraminidase* (see Figure 8.38). Neuraminidase breaks down the sialic acid component of the cytoplasmic membrane, which is a derivative of neuraminic acid. Neuraminidase appears to function primarily in the virus assembly process, destroying host membrane sialic acid that would otherwise block assembly or become incorporated into the mature virus particle.

In addition to the neuraminidase, the virus possesses two other enzymes, *RNA-dependent RNA polymerase*, which is involved in the conversion of a negative to a positive strand (as already discussed for the rhabdoviruses), and an *RNA endonuclease*, which cuts a primer from capped mRNA precursors.

The virus particle enters via the process of *endocytosis*. Once inside the cytoplasm, the nucleocapsid becomes separated from the envelope and migrates to the nucleus. *Replication* of the viral nucleic acid then occurs in the nucleus. *Uncoating* results in activation of the virus RNA polymerase. The mRNA molecules are then *transcribed* in the nucleus from the virus RNA, using oligonucleotide primers cut from the 5'-ends of newly synthesized capped cellular mRNAs. Thus, the viral mRNAs have 5'-caps. The poly-A tails of the viral mRNAs are added, and the virus mRNA molecules move to the cytoplasm.

Although influenza virus RNA replicates in the nucleus, influenza virus *proteins* are synthesized in the cytoplasm. *Ten* virus proteins are encoded by the *eight* segments of the virus genome (Table 8.3). The six mRNAs transcribed from segments 1 through 6 each encode a single protein, whereas the other two segments (numbers 7 and 8 in Table 8.3) each encode two proteins. This is not done by using true polycistronic

TABLE 8.3	The eight influenza virus RNA segments and the proteins they encode				
Segment	Number of nucleotides	Protein encoded	Size of protein (daltons)	Approximate percentage of virus particle protein	Function of protein
1	2341	P1	96,000	1	Initiation of transcription
2	2300	P2	87,000	1	Cap binding protein
3	2233	P3	85,000	1	Elongation of transcription
4	1765	HA (HA1)	36,000	25	Hemagglutinin spike. HA is converted to HA1 and HA2.
		HA (HA2)	27,000		
5	1565	NP	56,000	30	Structural protein of helical nucleocapsid
6	1413	NA	50,000	4	Neuraminidase
7	1027	M1	27,000	38	Matrix protein
		M2	11,000		Nonstructural protein of unknown function
8	890	NS1	26,000	trace	Nonstructural protein of unknown function
		NS2	12,000	trace	Nonstructural protein of unknown function

mRNA as in prokaryotes because eukaryotic ribosomes typically recognize only the AUG codon closest to the 5'-end of the mRNA as a start codon (∞ Section 6.9). Therefore, they can make only one protein from a given RNA. The original full-length mRNAs transcribed from segments 7 and 8 are each translated to give one protein. In each case, an additional protein is translated from these messages after they have been processed by the host's RNA splicing machinery. Like overlapping genes, this is another example of how RNA viruses make maximum use of their small genome size.

Some of these proteins are involved in virus RNA replication, and others are structural proteins of the virion. The overall strategy of virus RNA synthesis resembles that of the rhabdoviruses, with primary transcription resulting in the formation of *plus*-strand templates for the formation of progeny *minus*-strand molecules. Details of assembly are still uncertain. One possibility is that the nucleocapsid proteins are transported from the cytoplasm to the cell nucleus where *assembly* of the nucleocapsids occurs. The assembled nucleocapsids then migrate to the *cytoplasmic membrane* where the hemagglutinin and neuraminidase are present. The formation of the complete enveloped virus particle occurs by a budding-out process, as was described for the rhabdoviruses.

The segmented genome of the influenza virus has some important practical consequences. Influenza virus and other viruses of this family exhibit a phenomenon called **antigenic shift** in which pieces of the RNA genome from two genetically distinct strains that have infected the same cell become associated. This results in a change in the surface antigens (coat proteins) of the virus, making the virus resistant to antibody that has been formed as a result of an immunization process (∞ Section 23.4). This antigenic shift makes it possible for the newly formed virus to infect hosts that the parent could not have infected. Antigenic shift is thought to bring about major epidemics of influenza.

8.17
Double-Stranded RNA Viruses: Reoviruses

The reoviruses, an important family of animal viruses, have a genome consisting of *double-stranded RNA*. The name *reovirus* is an acronym, derived from the terms *r*espiratory, *e*nteric, and *o*rphan. The term *orphan* was applied because the first viruses of this group to be isolated from humans were not associated with any specific disease syndrome. However, viruses of the reovirus group are known to infect a variety of mammals, as well as insects and plants. The mammalian reoviruses first isolated are now classified in the virus genus *Reovirus*. To date, no specific human syndrome has been associated with this genus, although these viruses are commonly isolated from persons with an array of inconsequential respiratory and/or gastrointestinal illnesses (as well as being isolated from apparently healthy people). A more important genus of reoviruses is the genus *Rotavirus*, a member of which has been implicated in an important diarrheal disease of infants. *Rotavirus* is probably the most common cause of diarrhea in infants from 6 to 24 months of age. Rotaviruses are also known to cause diarrhea in young animals. A reovirus of the genus *Orbivirus* causes Colorado tick fever in humans. Other orbiviruses cause important veterinary diseases such as equine encephalosis and blue-tongue disease in sheep.

As noted, the RNA of the reoviruses is double-stranded. This is the only group of animal viruses with double-stranded RNA; all other RNA virus groups have single-stranded RNA (Figure 8.33). The reovirus particle consists of a nonenveloped nucleocapsid 60–80 nm in diameter, with a *double* shell of icosahedral symmetry (Figure 8.39). Predictably, these double-stranded RNA viruses contain within the virion the virus-encoded enzymes necessary to synthesize RNA.

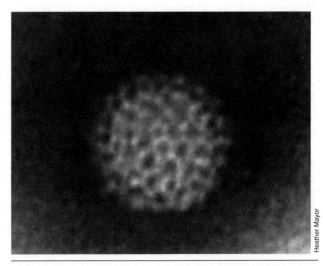

Heather Mayor

Figure 8.39 A reovirus particle. The diameter of the particle is about 70 nm.

CONCEPT CHECK   **8.16**

In a number of RNA viruses, called negative-strand viruses, the virus RNA does not act directly as messenger but is copied into mRNA by an RNA-dependent RNA polymerase present in the virion. An important negative-strand virus is influenza virus.

✔ *Why is it essential that negative-strand viruses carry an enzyme in their virions?*

✔ *What is a segmented genome?*

Present within the virion are *RNA-dependent RNA polymerase, nucleotide phosphohydrolase,* and enzymes that participate in the capping of messenger RNA, *RNA methyltransferase* and *guanyl transferase.*

However, the unique thing about the reoviruses is their *genome,* which consists of 10–12 molecules of double-stranded RNA. Double-stranded RNA is more difficult to unwind than double-stranded DNA, and there are no known RNA helicases (∞ Section 6.5). Replication of the RNA occurs by an asymmetric method. First, one strand is used as the template, and then the single-stranded product serves as a template to form a double helix. This difficulty in unwinding and the susceptibility of single-stranded RNA to cleavage within the cell greatly limits the size of a double-stranded molecule. Segmenting the genome into several molecules seems to be an adaptation to circumvent these problems.

Generally, each molecule of RNA in the genome codes for a single protein, although in a few cases the protein formed is cleaved to give the final product. However, one of the mRNAs produced actually encodes two proteins, and the RNA does not have to be modified to do so. Apparently a ribosome sometimes "misses" the start codon for the first gene in this message and travels on to the start codon of the second gene. As is true for many generalizations in the field of biology (such as "eukaryotic ribosomes initiate at the first AUG codon in mRNA"), continued investigation often uncovers exceptions. These exceptions are usually first found in viruses and often seem related to their small genome size. Replication of reovirus RNA occurs exclusively in the *cytoplasm* of the host. The double-stranded RNA is inactive as mRNA, and the first step in reovirus replication is *transcription,* using the minus strand as a template to make mRNA.

Replication of the reovirus seems to occur within an intracellular equivalent of the viral core, called the *subviral particle,* which remains intact in the cell. Each of the 10 capped, single-stranded plus RNAs is assembled into this double-stranded RNA-synthesizing body. The capped single-stranded plus RNAs act as templates for the synthesis of the progeny minus genomic RNAs, yielding progeny double-stranded viral RNAs. The progeny double-stranded RNAs are further encapsidated, and when enough viral capsid proteins are present, mature virions are assembled.

In the initial infection process, the virion binds to a cellular protein. Once attachment has occurred, the virus enters the cell and is transported into lysosomes. Within the lysosome the outer shell of the virus particle is modified by removal of two proteins and cleavage of another by lysosomal enzymes. This uncoating process activates the viral RNA-dependent RNA polymerase and hence initiates the virus replication process.

8.18
Replication of DNA Viruses of Animals

Most animal viruses with DNA genomes contain double-stranded DNA (one group, the parvoviruses, contains single-stranded DNA). Among these DNA viruses of animals, the four major families are the papovaviruses, the herpesviruses, the pox viruses, and the adenoviruses. Of these, all replicate in the nucleus except for the pox viruses, which have the unique character (for DNA viruses) of replicating in the cytoplasm. In this and the following sections, we discuss the replication of each of these families briefly.

Papovaviruses: SV40

Some viruses of the papovavirus group have the interesting property of inducing tumors in animals. One of these DNA tumor viruses was first isolated from monkeys, and it was thus called *simian virus 40* or SV40. It was one of the first genetic elements to be studied by genetic engineering techniques and has been extensively used as a *vector* for moving genes into eukaryotic cells. (∞ Section 10.4).

The SV40 virion is a simple, nonenveloped particle 45 nm in diameter with an icosahedral head containing 72 protein subunits. There are no enzymes in the virion. In addition to the capsid proteins, however, there are four host-derived *histone proteins* found complexed with the viral DNA. We have mentioned histone proteins during our discussion of chromosome structure (∞ Section 3.14) and have noted that histones play a role in neutralizing the negative charge originating from the phosphates of DNA and aid in packing of the DNA into more compact configurations.

The genome of SV40 consists of one molecule of double-stranded DNA of 5243 base pairs. The DNA is circular (Figure 8.40) and exists in a supercoiled configuration within the virion. The complete base sequence of SV40 has been determined, and the genetic map is known in some detail (Figure 8.41).

The nucleic acid is synthesized in the nucleus, but the proteins are synthesized in the cytoplasm. Final assembly of the virus particle occurs in the nucleus. The replication of these viruses can be divided into two distinct stages, *early* and *late.* During the early stage the *early region* of the viral DNA is transcribed (Figure 8.41). A single RNA molecule, the primary transcript, is made by cellular RNA polymerase, but it is processed into *two species of mRNA,* a large one and a small one. The DNA of SV40 has *introns* that are excised out of the primary RNA transcript. In the cytoplasm, viral mRNA is translated with the formation of two proteins. One of these proteins, the T-antigen, binds to the site on the parental DNA that is the *origin of replication.*

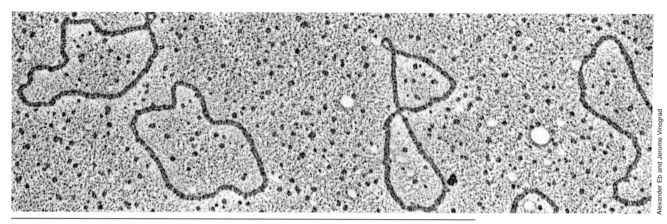

Figure 8.40 Electron micrograph of circular DNA from a tumor virus. The contour length of each circle is about 1.5 μm.

The viral DNA of SV40 is too small to code for its own DNA polymerases; host DNA polymerases are used. Replication occurs in a bidirectional fashion (so-called *theta* replication; ∞ Figure 6.16) from a single replication origin. The process involves the same events that have already been described for host cell DNA replication (∞ Section 6.5): RNA primer synthesis, formation of discontinuous DNA fragments on the lagging strand, gap filling, ligase action, and supercoiling the DNA through the action of gyrase.

Late SV40 mRNA molecules are synthesized using the strand complementary to that used for early mRNA synthesis (see Figure 8.41). Transcription begins at a pro-

moter near the origin of replication. This late RNA is then processed by splicing and polyadenylation to give multiple forms of mRNA corresponding to the three coat proteins. These genes overlap; part of the nucleotide sequence contains information for all three proteins. These late mRNA molecules are transported to the cytoplasm and translated into the viral coat proteins. These proteins are then transported back into the nucleus where *virion assembly* takes place. The mechanism by which virions are released from the cell is not known.

When a virus of the papovavirus group infects a host cell, one of two modes of replication can occur, depending on the type of host cell. In some types of host cells, known as *permissive* cells, virus infection results in the formation of new virions and the lysis of the host cell. In other types of host cells, known as *nonpermissive,* efficient multiplication does not occur, but the virus DNA becomes integrated into some of the host cells, thereby creating new, genetically altered cells. Such cells frequently show loss of growth inhibition and are called *transformed* or tumor cells.

In *nonpermissive hosts,* transformation can take place if the early proteins can be expressed, but the viral DNA cannot be replicated independently. Once the viral DNA has entered the nucleus and transcription of the early genes has taken place, there is a great stimulation of all of the host cell's biosynthetic activities involved in cellular DNA replication and mitosis. However, no replication of viral DNA occurs. Instead, in the transformed cell, the viral DNA becomes stably integrated into the DNA of the host cell (Figure 8.42). In the integrated state, the viral DNA can replicate only as a *cellular gene.* Integration can occur at many sites in the cellular and viral genome. In this integrated form, two viral proteins are made that are essential for the maintenance of a stably integrated viral DNA, but no viral structural proteins are synthesized. Some transformed cells can be converted to cells capable of producing virus, a process that probably involves excision of the viral genome from the host genome.

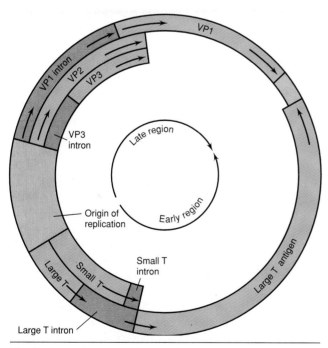

Figure 8.41 Genetic map of SV40. VP1, VP2, and VP3 are the genes coding for the three proteins that make up the coat of SV40. The arrows show the direction of transcription.

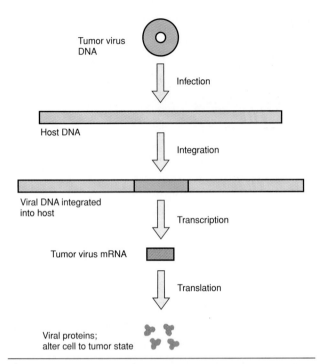

Tumor virus DNA

Infection

Host DNA

Integration

Viral DNA integrated into host

Transcription

Tumor virus mRNA

Translation

Viral proteins; alter cell to tumor state

Figure 8.42 General scheme of molecular events involved in cell transformation by a DNA tumor virus such as SV40. All or portions of viral DNA are incorporated into host cell DNA. The viral genes that encode transforming information are transcribed and processed to viral mRNA molecules, which are transported to the cytoplasm. Here they are translated to form transforming proteins or T-antigens that code for functions that can convert host cells into cancer cells.

A study of the manner of replication of SV40 virus has given some important insights into the manner by which viruses bring about the cancerous state in host cells. However, we note that DNA viruses of other groups can also cause the cancerous transformation. Also, the important family of viruses, the *retroviruses*, are also cancer viruses but have a completely different mode of replication (see Section 8.22).

CONCEPT CHECK   8.18

Most double-stranded DNA animal viruses replicate in the nucleus. Some viruses causing cancer and other conditions in animal cells are double-stranded DNA viruses. In a permissive host, the virus may cause death and lysis, but in a nonpermissive host, it may cause transformation of the infected host cell to a tumor cell.

✔ *Why wouldn't a virus like SV40 be expected to carry enzymes in the virion?*

✔ *How can one transcript yield more than one mRNA?*

8.19

Herpesviruses

The herpesviruses are a large group of double-stranded DNA viruses that cause a wide variety of diseases in humans and animals, including fever blisters (cold sores), venereal herpes, chickenpox, shingles, and infectious mononucleosis; a number of these diseases are discussed in Chapter 23. Some herpesviruses also cause cancer. One of the interesting features of some herpesviruses is their ability to remain *latent* in the body for long periods of time, becoming active only under conditions of stress. Both *herpes simplex*, the virus that causes *fever blisters*, and *varicella-zoster virus*, the cause of *chicken pox* and *shingles*, are able to remain latent in the neurons of the sensory ganglia, from which they are able to emerge to cause infections of the skin.

An important group of herpesviruses are tumorigenic, causing clincial forms of cancer. One herpesvirus that is tumorigenic is the *Epstein–Barr virus*, which causes *Burkitt's lymphoma*, a common tumor among children in Central Africa and New Guinea. Burkitt's lymphoma was among the first human cancers to be linked to virus infection (see Table 8.2).

Molecular biology of herpesviruses

The *herpesvirus particle* is structurally complex, consisting of four distinct morphologic units. In herpes simplex type I, an enveloped virus about 150 nm in diameter (Figure 8.43*a*), the center of the virus is an electron-dense *core* consisting of *double-stranded* DNA. Surrounding this core is the nucleocapsid, of icosahedral symmetry, which consists of 162 capsomeres, each of which is composed of a number of distinct proteins. Outside the nucleocapsid is an amorphous layer that is called the *tegument*, a fibrous structure unique to the herpesviruses. Surrounding the tegument is an *envelope* whose outer surface contains many small *spikes*. A large number of separate proteins are present within the virion, but not all of them have been characterized.

The *genome* of herpes simplex type I virus consists of one large *linear* double-stranded DNA molecule of 152,260 base pairs (about 30 times larger than the SV40 genome). As a further indication of the complexity of herpes simplex, the DNA sequence of this virus indicates that it codes for at least 75 separate proteins.

Infection occurs by attachment of virus particles to specific cell receptors, and, following fusion of the cytoplasmic membrane with the virus envelope, the nucleocapsid is released into the cell. The nucleocapsids are transported to the nucleus, where viral DNA is uncoated. Components of the virus particle inhibit macromolecular synthesis by the host.

There are *three classes of messenger RNAs:* immediate early (also called alpha, α), which codes for five regulatory proteins; delayed early (also called beta, β), which

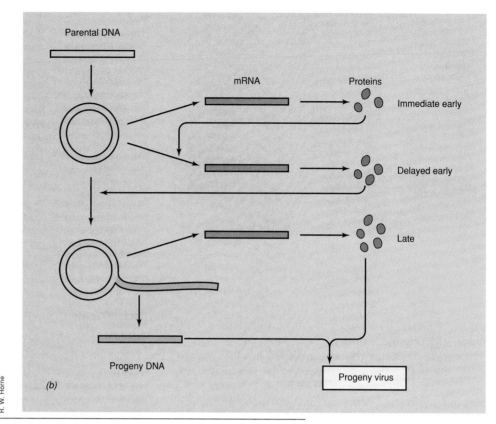

Figure 8.43 Herpesvirus. (a) Electron micrograph of a herpesvirus particle. The diameter of the particle is about 150 nm. (b) Flow of events in multiplication of herpes simplex virus.

codes for DNA replication proteins, including thymidine kinase and DNA polymerase; and late (also called gamma, γ), which codes for the proteins of the virus particle (Figure 8.43*b*). During the *immediate early stage*, about one-third of the viral genome is transcribed by a host cell RNA polymerase. Early mRNA codes for certain positive-acting regulatory proteins that appear to stimulate the synthesis of the delayed early proteins. The second stage, *delayed early*, occurs only after the early proteins have been made. During this stage, about 40% of the viral genome is transcribed. Among the 10 proteins characterized from the delayed early stage are a *DNA polymerase*, enzymes involved in synthesis of *deoxyribonucleotides*, and a *DNA binding protein*. These enzymes are all involved in the process of viral DNA replication.

Herpes viral DNA synthesis itself takes place in the nucleus. After infection, the herpes genome apparently circularizes (remarkably like bacteriophage lambda) (see Section 8.12) and replicates by a rolling circle mechanism (see Figure 8.28). However, three origins seem to be involved. Long concatamers are formed that become processed to virus-length DNA during the assembly process itself (in a manner similar to that described for DNA bacteriophages) (see Sections 8.11 and 8.12).

Viral nucleocapsids are assembled in the nucleus, and acquisition of the virus envelope occurs via a budding process through the *inner membrane of the nucleus.*

Mature virions are subsequently released through the endoplasmic reticulum to the outside of the cell. Thus, the assembly of this enveloped DNA virus differs markedly from that of the enveloped RNA viruses, which were assembled on the *cytoplasmic* membrane instead of the nuclear membrane.

8.20
Pox Viruses

These are the most complex and largest animal viruses known (Figure 8.44) and have some characteristics that approach those of primitive cells. However, the pox viruses, like all viruses, are not able to metabolize and thus depend on the host for the complete machinery of protein synthesis. These viruses are also unique in that they are DNA viruses that replicate in the *cytoplasm*. Thus, a host cell infected with a pox virus exhibits *DNA synthesis outside the nucleus*, something that otherwise occurs only in intracellular organelles such as mitochondria.

General properties of pox viruses

Pox viruses have been important medically as well as historically. *Smallpox* was the first virus to be studied in any detail and was the first virus for which a vac-

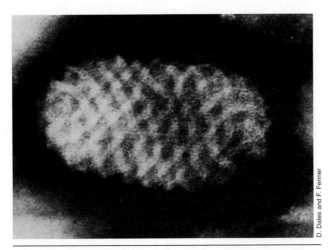

D. Dales and F. Fenner

Figure 8.44 Electron micrograph of a negatively stained vaccinia virus virion. The virion is approximately 400 nm (0.4 μm) long.

cine was developed (described by Edward Jenner in 1798). By diligent application of this vaccine on a worldwide basis, the disease smallpox has been *eradicated*, the first infectious disease to be eliminated in this fashion. Other pox viruses of importance are *cowpox* and *rabbit myxomatosis virus*, an important infectious agent of rabbits and one that was intentionally used in an attempt to control the Australian rabbit population (∞ Section 22.6). Some pox viruses also cause tumors, but these tumors are generally benign.

The pox viruses are very large, so large that they can actually be seen under the light microscope. Most research has been done with *vaccinia virus*, the source of smallpox vaccine. The vaccinia virion is a brick-shaped structure about $400 \times 240 \times 200$ nm in size. The virion is covered on its outer surface with tubules or filaments arranged in a membranelike pattern, although the virus does not have a lipid membrane because the outer envelope consists of protein. Within the virion there are two lateral bodies of unknown composition and a core, the *nucleocapsid,* which contains DNA bounded by a layer of protein subunits.

The pox virus genome consists of double-stranded DNA. The vaccinia virus genome has about 185 kilobase pairs and contains between 150 and 200 genes. The pox DNA is interesting because the two strands of the double helix are cross-linked at the ends as a result of the formation of phosphodiester bonds between adjacent strands (as in the hairpin structure shown in Figure 6.12).

Replication of pox viruses

Vaccinia virions are taken up into cells via a phagocytic process from which the cores are liberated into the cytoplasm. Interestingly, *uncoating* of the virus genome requires the action of a new protein that is synthesized after infection. This protein is encoded by viral DNA,

and the gene specifying this protein is transcribed by an RNA polymerase present *within* the virus particle. In addition to this uncoating gene, a number of other viral genes are transcribed. The primary transcripts are turned into mRNAs by capping and polyadenylation while they are still inside the virus core.

Once the vaccinia DNA is fully uncoated, the formation of *inclusion bodies* within the cytoplasm begins. Within these inclusion bodies, transcription, replication, and encapsidation into progeny virus particles occur. Each infecting virion initiates its own inclusion body, so the number of inclusions depends on the multiplicity of infection. Progeny DNA molecules form a pool from which individual molecules are incorporated into virions. Mature virions accumulate in the cytoplasm. There seems to be no specific release mechanism, and most virions are released only when the infected cell disintegrates.

Pox viruses and recombinant vaccines

Vaccinia virus has been used as a host for genetically altered proteins of other viruses, permitting the construction of genetically engineered vaccines. As we will see in Chapter 20, a vaccine is a substance capable of eliciting an immune response in an animal and serves to protect the animal from future infection with the same agent. Vaccinia virus causes no serious health effects in humans but is highly immunogenic. Molecular cloning methods have been used to express key viral proteins of influenza virus, rabies virus, herpes simplex type I virus, and hepatitis B virus in vaccinia virus virions, and then the latter used as a vaccine (∞ Section 10.13). Thus far very promising results have been achieved in the battle against diseases caused by the viruses previously listed, and research is progressing on using vaccinia virus as a vehicle for cloned viral proteins from the human immunodeficiency virus (HIV). Such an approach, if successful, could result in a safe and effective AIDS vaccine.

A similar vaccine delivery system using adenovirus (see next section) as a vehicle has been developed because, like vaccinia virus, adenoviruses are of little health consequence to humans.

8.21

Adenoviruses

The adenoviruses are a major family of icosahedral DNA-containing viruses that have unique molecular biological properties. The term *adeno* is derived from the Latin for "gland" and refers to the fact that these viruses were first isolated from the tonsils and adenoid glands of humans. Adenoviruses cause generally mild respiratory infections in humans, and a number of such viruses are isolated from apparently healthy individuals.

The genomes of the adenoviruses consist of linear double-stranded DNA of about 36 kilobase pairs. Attached in covalent linkage to the 5′-terminus of the DNA is a protein component essential for infectivity of the DNA. The DNA has inverted terminal repeats of 100–1800 base pairs (this varies with the virus strain). The DNA of the adenoviruses is six to seven times the size of the DNA of SV40.

Replication of the viral DNA occurs in the nucleus (Figure 8.45). After the virus particle has been transported to the nucleus, the core is released and converted to a viral DNA–histone complex. *Early transcription* is carried out by an RNA polymerase of the host, and a number of primary transcripts are made. The transcripts are spliced, capped, and polyadenylated, giving several different mRNAs.

The early proteins are involved in regulation of DNA replication; the later proteins are the virus coat protein. *Viral DNA replication* uses a virus-encoded protein as a primer and another virus-encoded protein

as DNA polymerase. For the replication of a *linear double-stranded* DNA molecule such as that of adenovirus, initiation of replication can begin at either end or at both ends simultaneously (∞ Figure 6.21). In the case of the adenoviruses, replication begins at *either* end, the two strands being replicated asynchronously. The products of a round of replication are double- and single-stranded molecules. The latter then cyclizes by means of the inverted terminal repeats, and a new complementary strand is synthesized beginning from the 5′-end, the products being another double-stranded molecule (Figure 8.45). This mechanism of replication is interesting because it does not involve the formation of discontinuous fragments of DNA on the lagging strand, as occurs in conventional DNA replication.

CONCEPT CHECK   8.21

Most double-stranded DNA animal viruses replicate in the nucleus, although their replication strategies can be quite different. However, the pox viruses replicate in the cytoplasm using enzymes carried in the virion. Some of these viruses are now being used in genetic engineering experiments.

✔ *Except in the case of pox viruses, in what cellular location is the genome of double-stranded DNA viruses replicated? Where does transcription occur?*

✔ *The mRNAs of pox viruses, and all other animal viruses, are translated in what cellular location?*

8.22

Retroviruses

We now discuss one of the most interesting and complex families of animal viruses, the **retroviruses.** The term *retro* means "backward", and the name of this class of virus is derived from the fact that these viruses appear to have a backward mode of nucleic acid replication. The retroviruses are RNA viruses, but they *replicate by means of a DNA intermediate* using the enzyme *reverse transcriptase*. The retroviruses are of interest for a number of other reasons. First, they were the first viruses shown to cause *cancer* and have been studied most extensively for their carcinogenic characteristics. Second, one retrovirus, the one causing *acquired immunodeficiency syndrome (AIDS)* has been known only since the early 1980s but has become a major public health problem. Third, the retrovirus genome can become specifically integrated into the host genome by way of the DNA intermediate, and this integration process is being studied as a means of introducing *foreign* genes into a host, a process called *gene therapy.* Finally, the

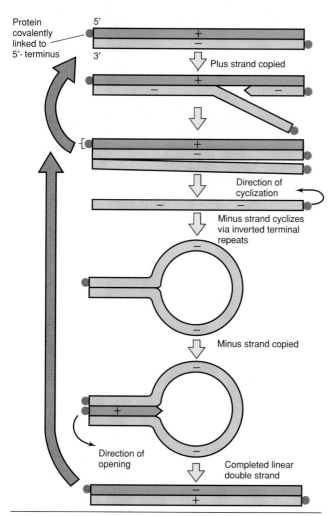

Figure 8.45 Replication of adenovirus DNA. See text for details.

enzyme reverse transcriptase has become a major tool in genetic engineering.

As we will see, the retroviruses have some properties like those of RNA viruses and some like those of DNA viruses. They resemble to a considerable extent movable genetic elements and are sometimes considered to be *escaped cellular transposable elements*. In this respect, the retroviruses resemble bacterial viruses such as Mu (see Section 8.13). We should note that the use of reverse transcriptase is not restricted to the retroviruses because hepatitis B virus (a human virus) and cauliflower mosaic virus (a plant virus) also use reverse transcription in their replication processes. But in contrast to the retroviruses, these latter viruses encapsidate the DNA genome rather than the RNA genome as retroviruses do. Some transposable elements of eukaryotes, called *retrotransposons*, also encode and use reverse transcriptase as part of their replication cycle. In addition, reverse transcriptases capable of producing small multicopy DNA (ms DNA) with an RNA template have been discovered in myxobacteria and *Escherichia coli*. The reverse transcriptase in bacteria is encoded as part of a short genetic element called a *retron*. Although many copies of the ms DNA are made (which contain their RNA template covalently attached), their function is unknown.

The retroviruses are enveloped viruses (Figure 8.46*a*). There are a number of proteins in the virus coat and typically seven internal proteins, four of which are structural and three enzymatic. The enzymatic activities found in the virus particle are *reverse transcriptase*, *DNA endonuclease* (*integrase*), and a *protease*. The virion also contains specific cellular tRNA molecules used in *replication* (see later).

Features of retroviral genomes and replication

The genome of the retrovirus is unique. It consists of *two* identical single-stranded RNA molecules of plus complementarity, each 8.5–9.5 kilobases in length. The 5′-terminus of the RNA is capped and the 3′-terminus is polyadenylated, so the RNA is capable of acting directly as mRNA but is *not* used as such. A genetic map of a typical retrovirus is shown in Figure 8.46*b*. Although there are differences between the genetic maps of different types of retroviruses, all contain the following regions and in the same order: *gag*, encoding internal structural proteins; *pol*, encoding reverse transcriptase and integrase; and *env*, encoding envelope proteins. Some, such as Rous sarcoma virus, carry a fourth gene downstream from *env* that is involved in cellular transformation and cancer (Figure 8.46*c*). The terminal repeats shown on the map play an essential role in the replication process (see later).

The overall process of replication of a retrovirus can be summarized in the following steps (Figure 8.47):

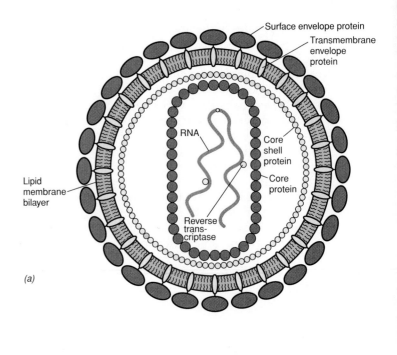

(a)

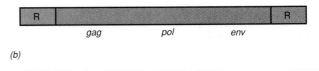

(b)

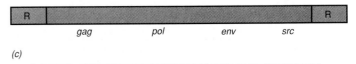

(c)

Figure 8.46 Retrovirus structure and function. (a) Structure of a retrovirus. (b) Genetic map of a typical retrovirus genome. (c) Genetic map of Rous sarcoma virus. Each end of the genomic RNA contains direct repeats (R), and this RNA also has a 5′-cap and a 3′-poly-A tail. See text for more details.

1. **Entrance** into the cell.
2. **Reverse transcription** of *one* of the two RNA genomes into a single-stranded DNA that is subsequently converted to a linear double-stranded DNA by reverse transcriptase.
3. **Integration** of the DNA copy into the host genome.
4. **Transcription** of the viral DNA, leading to the formation of viral mRNAs and progeny viral RNA.
5. **Encapsidation** of the viral RNA into nucleocapsids in the cytoplasm.
6. **Budding** of enveloped virions at the cytoplasmic membrane and release from the cell.

We now discuss some aspects of the retrovirus multiplication process in detail. As we have noted, the first step after the entry of the RNA genome into the cell is reverse transcription: conversion of RNA into a

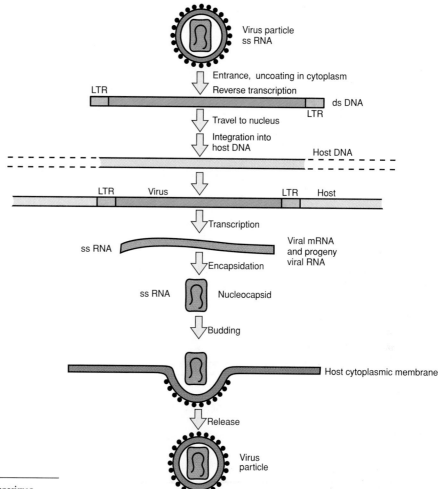

Figure 8.47 Replication process of a retrovirus.

DNA copy using the enzyme reverse transcriptase present in the virion. The DNA formed is a linear double-stranded molecule and is synthesized in the cytoplasm. An outline of the reverse transcription of viral RNA into DNA is given in Figure 8.48.

The enzyme reverse transcriptase is essentially a DNA polymerase, but it actually shows *three* enzymatic activities: (1) synthesis of DNA with an RNA template (reverse transcription), (2) synthesis of DNA with a DNA template, and (3) ribonuclease H activity (an activity that degrades the RNA strand of an RNA:DNA hybrid). Like all DNA polymerases, reverse transcriptase needs a primer for DNA synthesis. The primer for retrovirus reverse transcription is a specific *cellular transfer RNA (tRNA).* The type of tRNA used as primer depends on the virus and is brought into the viron from the previous host cell. In the case of Rous sarcoma virus, the tRNA used is the *tryptophan* tRNA.

Using the tRNA primer, the 100 or so nucleotides at the 5'-terminus of the RNA are reverse-transcribed into DNA. Once transcription reaches the 5'-end of the RNA, the transcription process stops. In order to copy the remaining RNA, which is the bulk of the RNA of the

virus, a different mechanism comes into play. First, terminally redundant RNA sequences at the 5'-end of the molecule are removed by the action of another enzymatic activity of reverse transcriptase, *ribonuclease H.* This leads to the formation of a small, single-stranded DNA that is complementary to the RNA segment at the *other end* of the viral RNA. The small, single-stranded piece of DNA then hybridizes with the other end of the viral RNA molecule, where copying of the viral RNA sequences continues. As summarized in Figure 8.48, continued action of reverse transcriptase and ribonuclease H leads to the formation of a double-stranded DNA molecule with long terminal repeats (LTRs) at each end. These LTRs contain strong promoters of transcription and are involved in the integration process. The *integration* of the viral DNA into the host genome is analogous to the integration of Mu (see Section 8.13) or a bacterial transposon into a bacterial genome. Integration can occur anywhere in the cellular DNA, and once integrated, the element, now called a *provirus,* is a stable genetic element. As a provirus its genetic information may be expressed, or it may remain in a latent state and not be expressed.

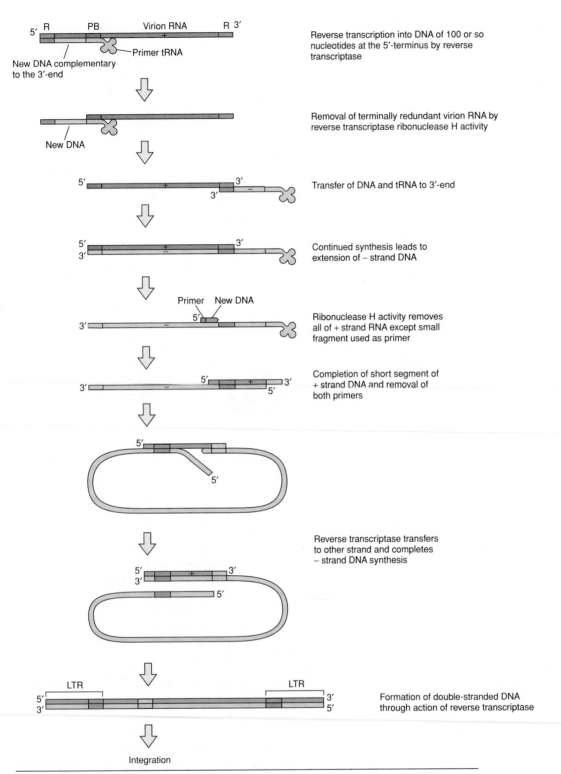

Figure 8.48 Overall steps in the formation of double-stranded DNA from retrovirus single-stranded RNA. The sequences labeled R on the RNA are direct repeats found at either end. The sequence labeled PB is where the primer (tRNA) binds. Note that the process of DNA synthesis has yielded longer direct repeats on the DNA than were originally on the RNA. These are called long terminal repeats (LTRs).

A FOCUS ON . . .

VERY SMALL VIRUSES: ON THE FRINGE

The life cycles of viruses contain a variety of unexpected genome structures and information transfers. Many of these seem exemplified by the hepadnaviruses, such as human hepatitis B virus. The genomes of these fascinating viruses are among the smallest known for a virus, but the virus life cycle is very complex. Like the retroviruses, these viruses utilize reverse transcriptase in their life cycle. In the case of hepadnaviruses the genome in the virion is DNA, but this DNA is replicated using an *RNA* intermediate, the opposite of what occurs in retroviruses.

The genomic DNA of hepadnaviruses is only *partially* double-stranded. One strand is incomplete, and both strands have breaks or gaps, but they are held in a circular form by hydrogen bonding. On entering the cytoplasm a viral polymerase carried in the virion completes the replication of this molecule. (This polymerase is quite a versatile protein; it contains DNA polymerase activity and reverse transcriptase activity and is the protein primer for synthesis of one of the DNA strands.) The figure showing the genetic map indicates the incredible compactness of the genome. Despite the small size (an average of 3200 base pairs), the genome encodes several proteins. Not only do genes overlap, but every base is part of a codon for at least one protein!

Replication of this genome involves transcription by host RNA polymerase (in the nucleus), yielding a transcript with terminal repeats. (The repeats occur because the polymerase proceeds slightly more than once around the circular molecule.) The viral polymerase then copies this into DNA, very much like the replication of retroviruses, but in this case the DNA becomes packaged into new virus particles.

Remarkably, the human hepatitis B virus is a "helper virus" for the *delta agent*. The delta agent is a "sub-virus" and requires hepatitis B virus to supply the proteins necessary for its coat. Therefore, the delta agent is the parasite of a parasite! The delta agent has a circular, negative-strand RNA genome of 1679 bases. Like the viroids (see Section 8.23), it seems to be able to base-pair into a rodlike structure that can then be transcribed by the host RNA polymerase. However, unlike the situation with viroids, the transcript of the delta agent encodes at least one protein, an RNA binding protein carried by the delta agent virion. Incredibly, the RNA is also a ribozyme (∞ Section 6.7) that can cleave itself. This self-cleavage may be involved in mRNA formation.

Very small "viruses" can be very interesting indeed. The hepadnaviruses and the delta agent employ several different strategies for maximizing the information carried in their very small genomes. ■

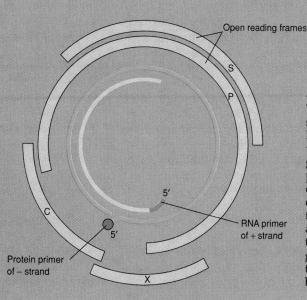

The partially double-stranded genome of human hepatitis B virus is shown in green. Note that the positive strand is not complete. The sizes of the open reading frames C, P, S, and X are also shown. Notice that all these genes overlap and that they cover every base in the genome.

If the promoters in the right LTR are activated, the integrated proviral DNA is transcribed by a cellular RNA polymerase into transcripts that are capped and polyadenylated. These RNA transcripts either may be encapsidated into virus particles or may be processed and translated into virus proteins. Some virus proteins are made initially as a large primary *gag* protein, which is split by proteolytic action into the capsid proteins. Occasionally, read-through past the *gag* region (involving either inserting an amino acid in response to a stop codon or a shift in reading frame by the ribosome) leads to the translation of *pol*, the reverse transcriptase gene. Other proteins are synthesized from spliced transcripts.

When the virus proteins have accumulated in sufficient amounts, *assembly* of the nucleocapsid can occur. This encapsidation process leads to the formation of nucleocapsids, which move to the cytoplasmic membrane for final assembly into the enveloped virus particles.

Not all retroviruses cause cancer, but tumorigenic retroviruses are quite common. Some tumorigenic retroviruses known cause *sarcomas* or *acute leukemia;* they possess high oncogenic potential. Infection with one of these viruses can cause cellular transformation, leading to the formation of a tumor. Why are these viruses tumorigenic? It appears that they possess a transforming gene, or *oncogene,* that encodes a protein that brings about cellular transformation (see Section 8.14). This gene, known in Rous sarcoma virus as the *src* gene (*src* for *sarcoma*) (see Figure 8.46*c*), encodes a phosphoprotein that possesses protein kinase activity. Protein kinases bring about the phosphorylation of proteins, and protein phosphorylation is one mechanism for regulating the activity of proteins (∞ Section 7.7).

Transforming genes analogous to *src* have also been detected in human cancer cells. Interestingly, similar genes have also been detected in *normal* (that is, noncancerous) cells. These cell sequences are the proto-oncogenes (see Section 8.14) and have been found not only in mammalian cells but also in the cells of insects and yeast, suggesting that these sequences are of fundamental importance in the regulation of cell growth. Retroviruses are able to incorporate such normal sequences, which become altered and are abnormally expressed. Retroviruses are thus the agents by which such genes are transferred from cell to cell.

Genetic engineering with retroviruses occurs naturally by the incorporation of oncogene or proto-oncogene sequences. It is also possible to use modified retroviruses to incorporate foreign genes into cells. This is possible because substitution of such foreign genes for essential virus genes can lead to the production of virus particles carrying the foreign gene. Such particles are capable of being integrated into the host genome but are incapable of replicating or causing cancer.

As previously mentioned, one notable retrovirus is HIV, the virus causing AIDS. This virus infects a specific cell type in the human, a kind of T lymphocyte that is vital for proper functioning of the immune system. In later chapters we discuss the medical and immunological aspects of AIDS (∞ Sections 22.4 and 23.7).

Because viruses are not cells but depend on cells for their replication, viral diseases pose serious medical problems; it is frequently difficult to prevent antiviral drugs from doing some damage to host cells. Despite this, certain chemotherapeutic strategies have been devised for use against viral pathogens, including retroviruses. We discuss these in some detail later along with the chemotherapy of other viral diseases (∞ Section 11.11).

CONCEPT CHECK   **8.22**

Retroviruses are enveloped viruses that have complex life cycles since they are RNA viruses that replicate by means of a DNA intermediate. Important retroviruses cause cancer and acquired immunodeficiency syndrome. The retrovirus virion contains an enzyme, reverse transcriptase, that copies the information from its RNA genome into DNA. The DNA becomes integrated into the host chromosome in the manner of a temperate virus. The retrovirus DNA can be transcribed to yield mRNA (and new genomic RNA) or may remain in a latent state.

✔ *What role does reverse transcriptase play in the infection cycle of a retrovirus?*

✔ *How does the life cycle of a temperate bacteriophage differ from that of a retrovirus?*

8.23
Viroids and Prions

So far in this chapter, we have discussed some representative viral groups. Recall that our definition of a virus was a genetic element that subverts the normal cellular process for its own replication and that has an extracellular form. There are a few known entities whose properties are at variance with this definition and which most scientists would not actually consider to be viruses. However, they seem closely related to viruses and are not considered plasmids. Two of the most important of these are *viroids* and *prions*.

Viroids are small, circular, single-stranded RNA molecules that are the smallest known pathogens (ranging from the coconut cadang-cadang viroid, which is 246 nucleotides in size, to citrus exocortis viroid, which has 375 nucleotides). Viroids cause a number of very important crop diseases. The extracellular form of the viroid is naked RNA—there is no capsid of any kind. Even more interestingly, *the RNA molecule contains no protein-encoding genes,* and therefore the viroid is totally dependent on host function for its replication. Although the viroid RNA is a single-stranded circle, there is such considerable secondary structure possible that it resembles a short double-stranded molecule with closed ends (Figure 8.49). The viroid molecule seems to be replicated in the host cell nucleus, and its structure, which somewhat mimics DNA, apparently allows it to be replicated by the host RNA polymerase.

Viroids are sometimes considered "escaped" introns and, like self-splicing introns (∞ Section 6.7), appear to be remnants of an RNA world (see the discussion of the "RNA World" in Section 15.3).

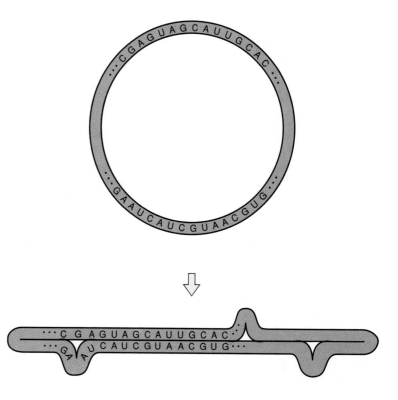

Figure 8.49 Structure of viroids, showing how single-stranded circular RNA can form a seemingly double-stranded structure by intrastrand base pairing.

Prions represent the other extreme from viroids. They have a distinct extracellular form, but the extracellular form seems to be *entirely protein.* It apparently does not contain any nucleic acid, or if it does, the molecule is not long enough to encode the single kind of protein of which the prion is composed. However, the prion protein particle is infectious, and various prions are known to cause a variety of diseases in animals, such as scrapie in sheep, bovine spongiform encephalopathy in cattle (BSE or "mad cow disease"), and kuru and Creutzfeldt–Jakob disease in humans. Interestingly, in 1996, information became available from England that indicated that the prion that causes BSE in cattle might infect humans, resulting in a variant of CJD.

In addition to serious disease, prion infection results in production of more copies of the prion protein. Unless prions violate the central pattern of genetic information flow discussed in Chapter 6, this protein *must* be encoded by nucleic acid. Indeed, it has been discovered that the host cell contains a gene on one of its chromosomes that encodes a protein very similar to the prion protein. The host protein is normally produced and is found mostly in neurons. Apparently, the incoming prion modifies this host protein either during or after synthesis. Therefore, prions do not simply subvert host enzymes but somehow cause a normal host gene to produce more copies of the pathogenic protein itself.

Viroids and prions do more than stretch our definition of a virus. They also demonstrate both the unexpected ways that genetic elements can replicate and the unexpected ways they can subvert the host cells. They are also of interest to us because they cause disease.

CONCEPT CHECK **8.23**

Viroids are circular single-stranded RNA molecules that encode no proteins and are completely dependent on host-encoded enzymes. They are the smallest known pathogens. Unlike viruses, their extracellular form is the same as their intracellular form, and they have no protein coat. Prions have an extracellular form that does contain protein, but it does not contain the nucleic acid that encodes the protein. The gene that encodes the prion protein is found in the host cell, and the prion somehow modifies this protein product.

Material for Review

REVIEW QUESTIONS

1. Define the term *host* as it relates to viruses.

2. Define *virus*. What are the minimal features needed to fit your definition?

3. Figure 8.40 shows circular DNA that has a contour length of 1.5 μm. How many base pairs are there in this DNA molecule (∞ Section 6.2)?

4. Under some conditions, it is possible to obtain nucleic acid–free protein coats (*capsids*) of certain viruses. Under the electron microscope these capsids look very similar to complete virions. What does this fact tell you about the role of the virus nucleic acid in the virus assembly process? Would you expect such virions to be infectious? Why?

5. Write a paragraph describing the events that occur on an agar plate containing a bacterial lawn when a single bacteriophage particle causes the formation of a *bacteriophage plaque*.

6. Describe how a *restriction endonucleae* might play a role in resistance to bacteriophage infection. Why could a restriction endonuclease play such a role whereas a generalized DNase could not?

7. One can divide the replication process of a virus into seven steps. What events are happening in each of these steps?

8. Specifically, why are both the life cycle and the virion of a positive-strand RNA virus likely to be simpler than those of a negative-strand RNA virus?

9. Is rolling circle DNA replication *bidirectional* (∞ Section 6.5)?

10. A bacterium that is missing the outer membrane protein responsible for maltose uptake is *resistant* to lambda infection. A lambda lysogen is *immune* to lambda infection. Describe the functional difference between resistance and immunity. Explain how these conditions are brought about in the examples given.

11. Some RNA bacteriophages are called *male-specific*. Explain.

12. Typically, transfer RNA is used in translation. However, it also plays a role in the replication of retroviral nucleic acid. Explain this role.

13. What is unique about reovirus genomes and what special problems does this introduce for nucleic acid replication?

14. Put together a diagram describing mRNA synthesis and nucleic acid replication for each of the following virus types: RNA tumor virus, reovirus, poliovirus, T4 phage, φX174 phage. For each diagram, be sure to indicate the complementarity of both the virus nucleic acid and its mRNA.

15. Many of the viruses we discussed have *early* genes and *late* genes. What is meant by these two classifications? What types of proteins tend to be encoded by early genes? What type of proteins by late genes? For any three viruses we discussed, describe how expression of the late genes is controlled.

APPLICATION QUESTIONS

1. Can you imagine any advantage for a virus having a metabolically inert extracellular stage rather than having one that is metabolically active?

2. What causes the viral plaques that appear on a bacterial lawn to stop growing larger?

3. Not all proteins are made from the RNA genome of bacteriophage MS2 in the same amounts. Can you explain why? One of the proteins functions very much like a repressor (∞ Section 7.3), but it functions at the translational level. Which protein is it and how does it function?

4. Suppose you want to determine whether a bacterial culture (strain) you are studying is a *lysogen*. You have no information about the genetics of this strain, but you have access to a large number of other strains of the same bacterial species. How would you proceed? Suppose you had no other strains of the same species. Is there any way you might get an idea of lysogenicity?

5. One characteristic of *temperate bacteriophages* is that they cause turbid rather than clear plaques on bacterial lawns. Can you think why this might be? (Remember the process by which a bacterial plaque develops.)

6. There are three lambda genes that when rendered nonfunctional, turn lambda from a temperate to a virulent virus. What are these three genes and how do they normally function?

7. Figure 8.27 shows two promoters, P_R and P_E, on either side of the *cro* gene. Both transcribe through the *cro* gene, but only the transcript from P_R can be translated to yield the Cro protein. Explain.

8. The mechanism of replication of both strands of DNA in some viruses, such as adenoviruses, is continuous. Show how this can be without violating the "rule" learned in Chapter 6 that all DNA synthesis occurs in the overall direction of $5' \rightarrow 3'$.

9. Knowledge of the type of RNA carried in the genome of retroviruses would lead to a prediction that the virion would not carry any enzymes. Explain why one could make this prediction and why in this case it is wrong.

10. Although the RNA in the retrovirus virion is the plus strand and has both caps and tails, it is not used as mRNA. How could its translation be prevented?

11. The promoters for mRNA encoding early proteins in viruses sometimes have a much different sequence than the promoters for mRNA encoding late proteins in the same virus. Explain why this might be true. (*Hint:* What type of RNA polymerase must recognize the "early" promoters?)

12. *Chemotherapeutic agents* are lacking for most virus diseases. From what you know about the stages of virus multiplication, give an explanation of why you think that may be so.

SUPPLEMENTARY READINGS

Boyles, B. A. 1993. *The Biology of Viruses.* Mosby-Year Book. St. Louis, MO. This is an excellent introductory textbook of virology, which is organized around the steps in the virus life cycle.

Dimmock, N. J., and **S. B. Primrose.** 1994. *Introduction to Modern Virology,* 4th edition. Blackwell Science, Cambridge, MA. An introductory text that highlights the basic concepts and principles of virology.

Fields, B. N., D. M. Knipe, and **P. M. Howley** (eds.). 1996. *Fundamental Virology,* 3rd edition. Lippincott-Raven, Philadelphia, PA. A large, detailed treatment of animal and human viruses. Each chapter written by an expert.

Levine, A. J. 1992. *Viruses.* Scientific American Library, New York. An excellent short text dealing with viruses and the history and methodology of virus research.

Prusiner, S. B. (ed.). 1996. *Prions, Prions, Prions.* Springer, New York. A review of recent advances in the study of prions, the infectious agents responsible for a growing list of different diseases.

Ptashne, M. 1992. *A Genetic Switch,* 2nd edition. Blackwell Scientific, Palo Alto, CA. An excellent short book that explains how lambda is regulated and also describes eukaryotic gene regulation.

Watson, J. D., N. H. Hopkins, J. W. Roberts, J. A. Steitz, and **A. M. Weiner.** 1987. *Molecular Biology of the Gene,* 4th edition. Benjamin-Cummings, Menlo Park, CA. Has extensive coverage of how viruses replicate.

 On~line resources for this chapter are on the World Wide Web at: http://www.prenhall.com/~brock (click on the <u>table of contents</u> link and then select Chapter 8).

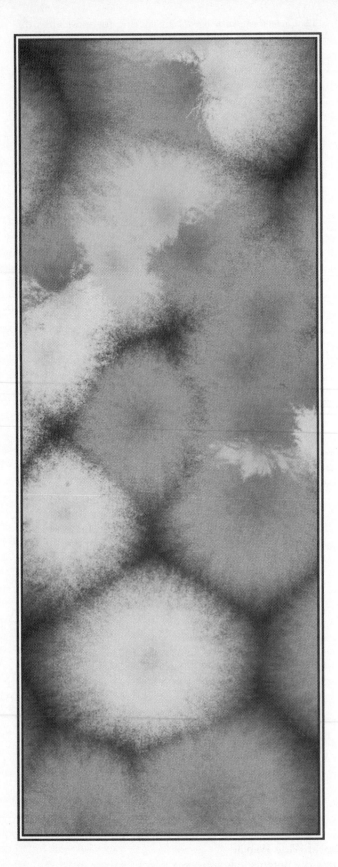

Microbial Genetics

The genetic material of an organism may mutate, giving rise to new forms. The genetic material from one cell can also be transferred to another, allowing recombination to take place. Shown here are various pigmentation mutants of the mold *Aspergillus nidulans*. The wild type (unmutated form) has a green pigment.

MINIGLOSSARY for Chapter 9

Auxotroph an organism that has developed a nutritional requirement through mutation

Conjugation transfer of genes from one prokaryotic cell to another by a mechanism involving cell-to-cell contact and a plasmid

Diploid a eukaryotic cell or organism containing two sets of chromosomes

Electroporation the use of an electric pulse to induce cells to take up free DNA

Gametes in eukaryotic organisms, the haploid germ cells resulting from meiosis

Genetic Map the arrangement of genes on a chromosome

Genotype the precise genetic makeup of an organism

Haploid a cell or organism that has only one set of chromosomes

Mutagens agents that cause mutation

Mutant an organism whose genome carries a mutation

Mutation an inheritable change in the base sequence of the genome of an organism

Phenotype the observable characteristics of an organism

Plasmid an extrachromosomal genetic element that has no extracellular form

Point Mutation a mutation that involves one or only a very few base pairs

Recombination the process by which parts or all of the DNA molecules from two separate sources are exchanged or brought together into a single unit

Selection placing organisms under conditions where the growth of those with a particular genotype will be favored

Transduction transfer of host genes from one cell to another by a virus

Transformation transfer of bacterial genes involving free DNA (but see alternative usage in Chapter 8)

Transposable Element a genetic element that has the ability to move (transpose) from one site on a chromosome to another

Transposon a type of transposable element that carries genes in addition to those involved in transposition

Now that we have introduced the main features of molecular genetics of cells and viruses, we can turn to a discussion of specific aspects of microbial genetics. In this chapter we discuss mutation and explain how genetic material is transferred from one organism to another. Gene transfer can occur in a number of different ways, and if it is accompanied by genetic recombination, it can lead to the formation of new organisms.

Mutation is an inherited change in the base sequence of the nucleic acid comprising the genome of an organism. **Genetic recombination** is the process by which genetic elements contained in two separate genomes are brought together in one unit. Through this mechanism, new combinations of genes can arise even in the absence of mutation. Since the genetic elements brought together may enable the organism to carry out some new function, genetic recombination can result in adaptation to changing environments. Whereas mutation usually brings about only a very small amount of genetic change in a cell, genetic recombination usually involves much larger changes. Entire genes, sets of genes, or even whole chromosomes, are transferred between organisms.

The offspring of eukaryotic organisms that reproduce sexually receive a set of chromosomes from each of their parents (∞ Section 3.14). As a result, offspring are not exactly like either parent; they are *hybrids* and contain a combination of the traits exhibited by each

parent. Prokaryotes do not reproduce sexually and have no exactly analogous process. However, there are mechanisms of genetic exchange in prokaryotes that, although considerably different from those involved in eukaryotic sexual reproduction, allow for both gene transfer and recombination.

Gene transfer and recombination are important research tools that allow analysis of the genetic structure of an organism. They are also of major importance in the construction of new organisms for practical applications, a major activity in the field of genetic engineering. We present the basic principles of microbial genetics in this chapter and then show in the next chapter how these principles apply to research in genetic engineering.

Importance

Microbial genetics is important for a number of reasons:

1. Gene function is at the basis of cell function, and basic research in microbial genetics is necessary to understand how microorganisms function.

2. Microorganisms provide relatively simple systems for studying genetic phenomena and are thus useful tools in attempts to decipher the mechanisms underlying the genetics of all organisms.

3. Microorganisms are used for the isolation and duplication of specific genes from other organisms,

a technique called **molecular cloning** (∞ Chapter 10). In molecular cloning, genes are manipulated and placed in a microorganism where they can be induced to increase in number.

4. Microorganisms produce many substances of value in industry, such as antibiotics, and genetic manipulations can be used to increase yields and improve manufacturing processes. Also, genes of higher organisms that specify the production of particular substances, such as human insulin, can be transferred by molecular cloning into microorganisms and the latter used for the production of these useful substances. The use of genetically modified microorganisms in large-scale industrial processes is an important part of the field of **biotechnology** (∞ Chapter 10).

5. Many diseases are caused by microorganisms, and genetic traits underlie these harmful activities. By understanding the genetics of disease-causing microorganisms, whether cellular or viruses, we can more readily control them and prevent their growth in the body.

6. Some of the types of genetic transfer that occur in prokaryotes, particularly conjugation (see Section 9.8), also play important roles in the spread of genes that confer properties such as resistance to antibiotics. Understanding such processes can help us to determine how genes can be transferred from one organism to another, even from one species to another.

To detect genetic exchange between two organisms, it is necessary to employ *genetic markers* whose transfer can be detected. Genetically altered strains are used for this purpose, the alteration(s) being due to one or more mutations in the DNA of the organism. We begin this chapter on microbial genetics with a consideration of the molecular mechanism of mutation and the properties of mutant microorganisms as a prelude to our discussion of genetic exchange.

9.1
Mutations and Mutants

As previously mentioned, a *mutation* is a heritable change in the base sequence of the nucleic acid genome of an organism. In all cells this nucleic acid is double-stranded DNA (∞ Section 6.1). A strain carrying such a change is called a **mutant.** A mutant by definition differs from its parental strain in **genotype,** the precise sequence of nucleotides in the DNA of a genome. But in addition, the observable properties of the mutant, its **phenotype,** may also be altered relative to the parental strain. It is common to refer to a strain isolated from nature as a *wild-type* strain. Mutant derivatives can be obtained either from wild-type strains or from a strain

derived from the wild type, for example, another mutant.

Depending on the mutation, a mutant may or may not show an altered phenotype from its parent. By convention in microbial genetics, the *genotype* of an organism is designated by three lowercase letters followed by a capital letter (all in italics) indicating the particular gene involved. For example, the *hisC* gene of *Escherichia coli* codes for a protein that could be called the HisC protein. In this case this protein (an enzyme in the biosynthetic pathway of histidine) is usually referred to by the name histidinol-phosphate aminotransferase, which describes it enzymatic activity. However, some proteins, such as the RecA protein (∞ Sections 8.12, 9.3, and 9.5), do not have other names, because enzyme functions can be difficult to describe in a few words. Mutations in the *hisC* gene would be designated as *hisC1*, *hisC2*, and so on, the numbers referring to the order of isolation of the mutant strains.

The *phenotype* of an organism is designated by a capital letter followed by two lowercase letters, with either a plus or minus superscript to indicate the presence or absence of that property. For example, a His$^+$ strain of *E. coli* is capable of making its own histidine whereas a His$^-$ strain is not. Mutations in the *hisC* gene may lead to a His$^-$ phenotype if they eliminate the function of the gene product.

Isolation of mutants

We can distinguish between two kinds of mutations, selectable and nonselectable. An example of a nonselectable mutation is that of loss of color in a pigmented organism (Figure 9.1a). Such colonies usually have neither an advantage nor a disadvantage over the pigmented parent colonies when grown on agar plates (there may be a selective advantage for pigmented organisms in nature, however). This means that the only way we can detect such mutations is to examine large numbers of colonies and look for the "different" ones.

Nonselectable mutants must be found by **screening** a large population of organisms, and the mutant phenotype may not be as easy to recognize as the difference between pigmented and nonpigmented colonies. A *selectable* mutation, on the other hand, confers on the mutant an advantage under certain environmental conditions, so the progeny of the mutant cell are able to outgrow and replace the parent. An example of a selectable mutation is drug resistance: an antibiotic-resistant mutant can grow in the presence of antibiotic concentrations that inhibit or kill the parent (Figure 9.1b). However, the antibiotic-sensitive phenotype cannot be directly selected for by eliminating the antibiotic from the medium. It is relatively easy to detect and isolate selectable mutants by choosing the appropriate envi-

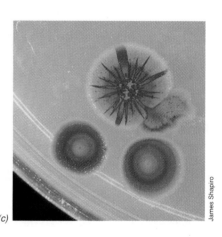

Figure 9.1 Observation of several kinds of mutants. (a) Pigmented mutants and nonpigmented mutants of the fungus *Aspergillus nidulans.* The wild type has a green pigment. The white or colorless mutants make no pigment, whereas the yellow mutants cannot convert the pigment they do make to the normal color. (b) Development of antibiotic-resistant mutants within the inhibition zone of an antibiotic assay disc. (c) Colonies of *Escherichia coli* have been mutated by a derivative of the bacteriophage Mu (∞ Section 8.13) on agar plates that allow detection of cells producing β-galactosidase. The Mu derivative carries the gene for this enzyme, but the enzyme is not produced unless Mu inserts into the *E. coli* chromosome in the proper orientation next to a promoter. Blue colonies produce β-galactosidase, whereas sectored colonies contain some cells where Mu is not inserted in the correct orientation.

Mutants

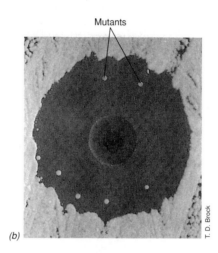

ronmental conditions. Therefore, **selection** is an extremely powerful genetic tool, allowing the isolation of a single mutant from a population containing millions or even billions of parental organisms.

Selection of mutant or recombinant microorganisms is not just of laboratory interest. Every time antibiotics are used to kill pathogenic organisms there is a strong selection for antibiotic resistance.

Virtually any characteristic of a microorganism can be changed through mutation. Nutritional mutants can be detected by the technique of **replica plating** (Figure 9.2a). With the use of sterile velveteen cloth or filter paper, an imprint of colonies from a master plate is made onto an agar plate lacking the nutrient. The colonies of the parental type will grow normally, whereas those of the mutant will not. Thus, the inability of a colony to grow on the replica plate (Figure 9.2b) is a signal that it is a mutant. The colony on the master plate corresponding to the vacant spot on the replica plate (Figure 9.2b) can then be picked, purified, and characterized. A nutritional mutant that has a requirement for a growth factor is often called an **auxotroph,** and the wild-type parent from which the auxotroph was derived is called a **prototroph.** For instance, mutants of *Escherichia coli* with a His⁻ phenotype are said to be *histidine auxotrophs.* Although of great utility, replica plating is a screening process, and it can be laborious to isolate mutants by screening.

An ingenious method widely used to isolate mutants that require amino acids or other growth factors is the **penicillin-selection method.** Ordinarily, mutants that require growth factors are at a disadvantage in competition with the parent cells, and so there is no direct way of isolating them. However, penicillin kills only *growing* cells, and if penicillin is added to a population growing in a medium lacking the growth factor required by the desired mutant, the parent cells will be killed, whereas the nongrowing mutant cells will be unaffected. Thus, after preliminary incubation in the absence of growth factor in a penicillin-containing medium, the population is washed free of penicillin and transferred to plates containing the growth factor. Among the colonies that grow up (including some wild-type cells that have escaped penicillin killing) should be some growth factor mutants. Penicillin selection is a kind of *negative selection;* the selection is not for the mutant but against the parental type.

Some of the most common kinds of mutants and the means by which they are detected are listed in Table 9.1.

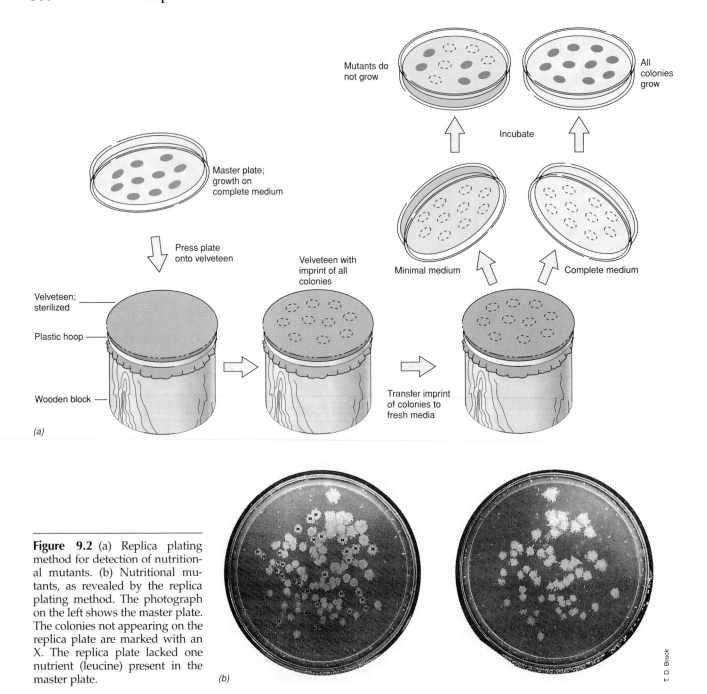

Figure 9.2 (a) Replica plating method for detection of nutritional mutants. (b) Nutritional mutants, as revealed by the replica plating method. The photograph on the left shows the master plate. The colonies not appearing on the replica plate are marked with an X. The replica plate lacked one nutrient (leucine) present in the master plate.

(b)

T. D. Brock

CONCEPT CHECK 9.1

Mutation, a heritable change in DNA, can lead to a change in phenotype. Selectable mutations are those that give the mutant a growth advantage under certain environmental conditions and are especially useful in genetics research.

✔ *Distinguish between* mutation *and* mutant.

✔ *Distinguish between* screening *and* selection.

9.2
Molecular Basis of Mutation

As previously mentioned, mutations arise in cells because of changes in the *base sequence* of an organism's genetic material. In many cases, mutations lead to phenotypic changes in the organism; these changes are mostly harmful, although beneficial changes do occur occasionally.

Mutation can be either spontaneous or induced. **Spontaneous mutations** can occur as a result of the

TABLE 9.1	Kinds of mutants	
Description	**Nature of change**	**Detection of mutant**
Nonmotile	Loss of flagella; nonfunctional flagella	Compact colonies instead of flat, spreading colonies
Noncapsulated	Loss or modification of surface capsule	Small, rough colonies instead of larger, smooth colonies
Rough colony	Loss or change in lipopolysaccharide outer layer	Granular, irregular colonies instead of smooth, glistening colonies
Auxotroph	Loss of enzyme in biosynthetic pathway	Inability to grow on medium lacking the nutrient
Sugar fermentation	Loss of enzyme in degradative pathway	Lack of color change on agar containing sugar and a pH indicator
Drug-resistant	Alteration of permeability to drug or drug target or detoxification of drug	Growth on medium containing a growth-inhibitory concentration of the drug
Virus-resistant	Loss of virus receptor	Growth in presence of large amounts of virus
Temperature-sensitive	Alteration of an essential protein so it is more heat-sensitive	Inability to grow at a temperature normally supporting growth (for example, 40°C) but still growing at a lower temperature (for example, 30°C)
Pigmentless	Loss of enzyme in biosynthetic pathway leading to loss of one or more pigments	Presence of different color or lack of color
Cold-sensitive	Alteration of an essential protein so it is inactivated at low temperature	Inability to grow at a low temperature (for example, 20°C) that normally supports growth

action of natural radiation (cosmic rays, and so on) which alters the structure of bases in the DNA. Spontaneous mutations can also occur during replication, as a result of errors in the pairing of bases, leading to changes in the replicated DNA. In fact, such errors occur at a frequency of about 10^{-7}–10^{-11} per base pair during a single round of replication (a typical gene has about 1000 base pairs). Thus, in a normal, fully grown culture of organisms having approximately 10^8 cells/ml, there are probably a number of different mutants in each milliliter of culture.

Mutations involving one (or a very few) base pairs are sometimes referred to as **point mutations**. Point mutations can result in *base-pair substitutions* in the DNA or in the insertion or deletion of a base pair (called *microinsertions* and *microdeletions*). As is the case with all mutations, the phenotypic change that comes about because of a point mutation depends on exactly where the mutation took place in the gene, what the nucleotide change was, and what product the gene normally encodes.

Base-pair substitutions

If a point mutation occurs within a gene that encodes a protein, any change in the phenotype of the cell is almost certainly the result of a change in the amino acid *sequence* of the protein being produced. Figure 9.3 shows a number of base-pair substitutions that can occur in a short region of DNA within a gene that encodes a protein. The error in the DNA is transcribed into mRNA, and this erroneous mRNA in turn is used as a template and translated into protein. (Because only one strand of the DNA is used as template for the mRNA, an AT base pair does not have the same meaning as a TA base pair.) The

triplet code that directs the insertion of an amino acid via a transfer RNA will thus be incorrect. What are the consequences of base substitutions?

In interpreting the results of mutation, we must first recall that the genetic code is degenerate (∞ Section 6.10). Because of degeneracy, not all mutations in pro-

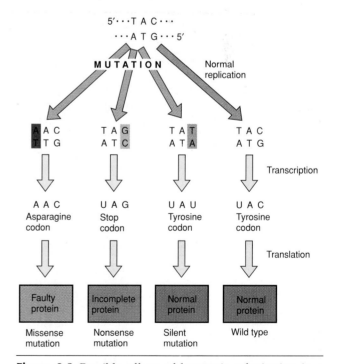

Figure 9.3 Possible effects of base-pair substitution in a gene coding for a protein: three different protein products from changes in the DNA for a single codon.

tein-encoding genes result in changes in protein. This is illustrated in Figure 9.3, which shows several possible results when the DNA that encodes a single *tyrosine codon* undergoes mutation. As seen, a change in the RNA from UAC to UAU would have no apparent effect because UAU is also a tyrosine codon. Mutations that give rise to such changes are called **silent mutations.** Note that silent mutations in coding regions almost always occur in the *third base* of the codon (arginine and leucine can also have silent mutations in the first position). As seen in Table 6.6, changes in the first or second base of the triplet much more often lead to significant changes in the protein. For instance, a single base change from UAC to AAC (Figure 9.3) would result in a change in the protein from tyrosine to asparagine. This is referred to as a **missense mutation** because the chemical "sense" (sequence of amino acids) in the ensuing polypeptide has changed. If the change occurred at a critical point in the polypeptide chain, the protein could be inactive, or of reduced activity. Another possible result of a base pair substitution is the formation of a *stop codon*, which would result in premature termination of translation, leading to an incomplete protein that would almost certainly not be functional (Figure 9.3). Mutations of this type are called **nonsense mutations** because the change is from a codon for an amino acid (sense codon) to a stop codon (nonsense codon).

Thus, not all mutations that cause amino acid substitution necessarily lead to nonfunctional proteins. The outcome depends on where in the polypeptide chain the substitution has occurred and on how it affects the folding and the catalytic activity of the protein. A missense mutation can lead to an enzyme that is temperature-sensitive, and this type of mutation is termed a **temperature-sensitive mutation.** For instance, temperature-sensitive mutants of bacteria are known that function normally at 30°C but cannot grow at 40°C, although the wild-type grows well at both temperatures. Such mutations are also referred to as **conditionally lethal** because the bacteria cannot grow under one condition but can under another. Temperature-sensitive phenotypes often occur because the mutant protein can maintain its correct conformation at the low tempera-

ture but becomes partially unfolded (denatured) at the high temperature.

Frameshift mutations

Because the genetic code is read from one end in consecutive blocks of three bases, any deletion or insertion of a base pair results in a **reading frame shift,** and the translation of the gene is completely upset (Figure 9.4). Partial restoration of gene function can often be accomplished by insertion of another base pair near the one deleted (one kind of suppressor mutation; see later). After correction, depending on the exact amino acids coded by the still faulty region and the region of the protein involved, the protein formed may have some biological activity or even be completely normal.

It is important to remember that microinsertions or microdeletions are frameshift mutations only if they occur in the part of a protein-encoding gene that includes the reading frame. A single base-pair insertion in the promoter of a gene could lead to a dramatic change in the ability of the gene to function, but it would not be a frameshift mutation. (Similarly, base-pair substitutions that are not within the reading frame are not missense or nonsense mutations.)

Back mutations or reversions

Point mutations are reversible. A *revertant* is operationally defined as a strain in which the wild-type phenotype that was lost in the mutant is restored. Revertants can be of two types. In *same-site revertants,* the mutation that restores activity occurs at the same site at which the original mutation occurred. (If the back mutation is not only at the same site but also leads to the wild-type sequence, it is called a *true revertant.*) In *second-site revertants,* the mutation occurs at a different site in the DNA.

Second-site mutations may cause restoration of a wild-type phenotype because of several types of **suppressor mutations** that restore the original phenotype. Suppressor mutations are new mutations that compensate for the effect of the original mutation. Several

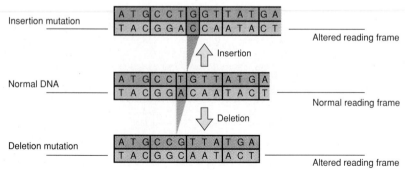

Figure 9.4 Shifts in the reading frame caused by insertion or deletion mutations.

types of suppressor mutations are known: (1) a mutation somewhere else in the same gene can restore enzyme function, such as in a frameshift mutation; (2) a mutation in another gene may restore the wild-type phenotype; and (3) the mutation may result in the production of another enzyme that can replace the mutant one by introducing a metabolic pathway different from that used by the mutant enzyme. In this last type no production of the original enzyme occurs although it does in the other types.

Mutations involving many base pairs

Deletions are mutations in which a region of the DNA has been eliminated. As we have discussed, microdeletions, the removal of one or a very few base pairs (Figure 9.4), are often frameshift mutations and can inactivate a gene. However, deletions can also involve the loss of hundreds or thousands of base pairs. Deletion of a large segment of the DNA results in complete loss of function of any gene that may be involved. Some deletions are so large that they involve several genes (if any of the genes are essential, the mutation will be lethal). Such deletions cannot be restored through further mutations but only through genetic recombination. Indeed, one way in which large deletions are distinguished from point mutations is that the latter are reversible through further mutations, whereas the former are not.

Insertions occur when new bases are added to the DNA. Insertions, like deletions, can involve only a single base or many bases. Microinsertions result from replication errors as do deletions, but larger insertions arise as a result of mistakes that occur during genetic recombination. Insertions inactivate the gene in which they occur. Many insertion mutations are due to the insertion of specific identifiable DNA sequences 700–1400 base pairs in length called **insertion sequences**, a type of transposable element (∞ Section 6.3). The behavior of such insertion sequences is discussed in detail in Section 9.9. Even large insertion mutations can revert by further mutation.

Other types of large-scale mutations exist that also seem to involve rearrangements brought about by mistakes in recombination. These include **translocations,** in which a large section of chromosomal DNA is moved to a new location (and in eukaryotes sometimes to a different chromosome), and **inversions,** in which the orientation of a particular segment of DNA is reversed with respect to the surrounding DNA.

Rates of mutation

There are wide variations in the rates at which various kinds of mutations occur. Some types of mutations occur so very rarely that they are almost impossible to detect, whereas others occur so frequently that they often present difficulties for an experimenter trying to maintain a genetically stable stock culture.

Spontaneous mutations in a gene occur at frequencies of about 10^{-6} per generation (see earlier discussion). This means that there is 1 chance in 1,000,000 that a mutation will arise at some location in a given gene during one cell cycle. Transposition events may occur more frequently, at about 10^{-4}. On the other hand, the occurrence of a nonsense mutation is less frequent, 10^{-6}–10^{-8}, because only a few codons can mutate to nonsense codons. Unless the mutant is selectable, the experimental detection of events of such rarity is difficult and much of the skill of the microbial geneticist is applied to increasing the efficiency of mutation detection. As we will see in the next section, it is possible to significantly increase the rate of mutation by the use of mutagenic treatments.

Mutations in ribonucleic acid (RNA) genomes

Whereas all cells have *DNA* as their genetic material, some viruses have *RNA* genomes (for examples, ∞ Sections 8.8 and 8.15–8.17). These genomes can also mutate. Importantly, the mutation rate in RNA genomes is about *1000-fold higher* than in DNA genomes. Partly, this is because the RNA replicases do not seem to have *proofreading* activities like those of DNA polymerases (∞ Section 6.5). In addition there are many repair systems for DNA that can correct many changes before they become fixed as mutations (see Section 9.3) and there are no comparable RNA repair mechanisms. This very high rate of mutation in RNA viruses is not merely of academic interest. The RNA genomes of viruses that cause disease

---

CONCEPT CHECK   **9.2**

Mutations, which can be either spontaneous or induced, arise because of changes in the base sequence of the nucleic acid of an organism's genome. A point mutation, which is due to a change in a single base pair, can lead to a single amino acid change in a protein or to no change at all, depending on the particular codon involved. In a nonsense mutation, the codon becomes a stop codon and an incomplete protein is made. Deletions and insertions cause more dramatic changes in the DNA, including frameshift mutations, and often result in complete loss of phenotype.

✔ *What does it mean to say that point mutations can spontaneously revert?*

✔ *Do missense mutations occur in genes encoding transfer RNAs (tRNAs)?*

TABLE 9.2 Chemical and physical mutagens and their modes of action

Agent	Action	Result
Base analogs		
5-Bromouracil	Incorporated like T; occasional faulty pairing with G	AT pair → GC pair Occasionally GC → AT
2-Aminopurine	Incorporated like A; faulty pairing with C	AT → GC Occasionally GC → AT
Chemicals reacting with DNA		
Nitrous acid (HNO_2)	Deaminates A and C	AT → GC and GC → AT
Hydroxylamine (NH_2OH)	Reacts with C	GC → AT
Alkylating agents		
Monofunctional (for example, ethyl methane sulfonate)	Put methyl on G; faulty pairing with T	GC → AT
Bifunctional (for example, nitrogen mustards, mitomycin, nitrosoguanidine)	Cross-link DNA strands; faulty region excised by DNase	Both point mutations and deletions
Intercalative dyes (for example, acridines, ethidium bromide)	Insert between two base pairs	Microinsertions and microdeletions
Radiation		
Ultraviolet	Pyrimidine dimer formation	Repair may lead to error or deletion
Ionizing radiation (for example, X-rays)	Free-radical attack on DNA, breaking chain	Repair may lead to error or deletion

can mutate very rapidly, presenting a constantly changing and evolving population of viruses.

9.3
Mutagens

It is now well established that a wide variety of chemical and physical agents can induce mutations. We discuss some of the major categories and their actions here.

Chemical mutagens

An overview of some of the major chemical mutagens and their modes of action is given in Table 9.2. Several classes of chemical mutagens exist. A variety of chemical mutagens are **base analogs,** resembling DNA purine and pyrimidine bases in structure yet showing faulty pairing properties (Figure 9.5). When one of these base analogs is incorporated into DNA, replication may occur normally most of the time, but occa-

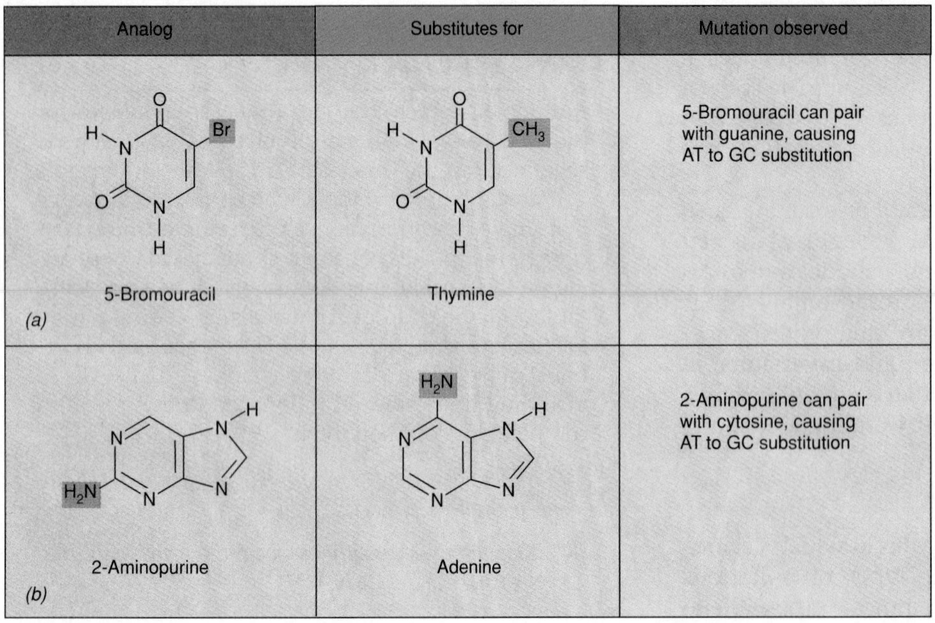

Analog	Substitutes for	Mutation observed
5-Bromouracil	Thymine	5-Bromouracil can pair with guanine, causing AT to GC substitution
2-Aminopurine	Adenine	2-Aminopurine can pair with cytosine, causing AT to GC substitution

Figure 9.5 Structure of two common nucleotide base analogs used to induce mutations, and the normal nucleic acid bases they substitute for.

sional copying errors occur, resulting in incorporation of the wrong base into the copied strand. During subsequent segregation of this strand, the mutation is revealed.

A variety of chemicals react directly on DNA, causing chemical changes in one base or another, which results in faulty pairing or other changes (Table 9.2). *Alkylating agents* such as nitrosoguanidine, for example, are powerful mutagens and generally induce mutations at higher frequency than base analogs. Such chemicals differ in their action from the base analogs in that the chemicals reacting on DNA are able to introduce direct changes even in nonreplicating DNA, whereas the base analogs act only after incorporation during replication. Both base analogs and alkylating agents tend to induce base-pair substitutions (see Section 9.2).

One interesting group of chemicals, the acridines, are planar molecules that act as *intercalating agents.* These mutagens become inserted between two DNA base pairs, thereby pushing them apart. During replication this abnormal conformation can lead to microinsertions or microdeletions in acridine-treated DNA. Thus acridines can induce frameshift mutations (see Section 9.2).

Radiation

Several forms of radiation are highly mutagenic. We can divide mutagenic radiation into two main categories, ionizing and nonionizing (electromagnetic) (Figure 9.6). Although both kinds of radiation are used in microbial genetics, *nonionizing* radiations find the widest use and will be discussed first.

The purine and pyrimidine bases of the nucleic acids absorb ultraviolet (UV) radiation strongly, and the absorption maximum for DNA and RNA is at 260 nm (∞ Working with Nucleic Acids: The Tools, Chapter 6). Proteins also absorb UV but have a peak at 280 nm due to absorption of the aromatic amino acids (tryptophan, phenylalanine, tyrosine). It is now well established that killing of cells by UV radiation is due primarily to its action on DNA, and so UV radiation at 260 nm is most effective as a lethal agent. Although several effects are known, one well-established effect is the induction in DNA of **pyrimidine dimers,** a state in which two adjacent pyrimidine bases become covalently joined so that during replication of the DNA the probability of DNA polymerase inserting an incorrect nucleotide at this position is greatly increased.

The type of UV radiation source most frequently used for mutagenesis is the germicidal lamp, which emits large amounts of UV radiation in the 260-nm region. A dose of UV radiation is used that brings about 90–95% killing of the cell population (∞ Section 11.2), and mutants are then looked for among the survivors. If much higher doses of radiation are used, the number of viable cells will be too low, whereas if lower doses are used, insufficient damage to the DNA will be induced. UV radiation is a very useful tool in isolating mutants of microbial cultures.

Ionizing radiation

Ionizing radiation is a more powerful form of radiation and includes short-wavelength rays such as X-rays, cosmic rays, and gamma rays (Figure 9.6). These radiations cause water and other substances to ionize, and mutagenic effects are brought about indirectly through this ionization. Among the potent chemical species formed by ionizing radiation are chemical free radicals, of which the most important is the hydroxyl radical, OH·. Free radicals react with and inactivate macromolecules in the cell, of which the most important is DNA. DNA is probably no more sensitive to ionizing radiation than other macromolecules, but because each DNA molecule contains only one copy of most genes, inactivation can have a permanent effect. At low doses of ionizing radiation, only a few hits on DNA occur, but at higher doses multiple hits occur, leading to the death of the cell. In contrast to UV radiation, ionizing radiation penetrates readily through glass and other materials. Because of this, ionizing radiation is used frequently to induce mutations in animals and plants (where its penetrating power makes it possible to reach the germ cells of these organisms readily), but because ionizing radiation is more dangerous to use and is less readily available, it finds less use with microorganisms (where penetration with UV is not a problem).

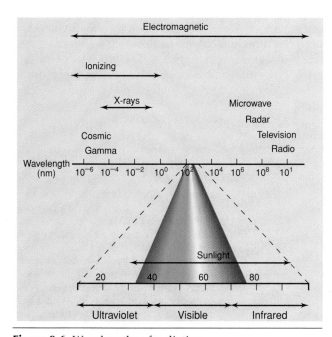

Figure 9.6 Wavelengths of radiation.

Mutations that arise from DNA repair

Recall that a mutation is an *inheritable* change in the genetic material. Therefore, if an error in DNA synthesis can be corrected before the cell divides, there will be no mutation. Furthermore, some DNA damage clearly cannot be replicated and therefore cannot itself be a mutation. For instance, if a DNA molecule contains pyrimidine dimers, it will not be replicated to give two DNA molecules each containing pyrimidine dimers. If such DNA damage cannot be repaired, cells often die. Most cells have a variety of different DNA repair processes to correct mistakes or repair damage. Many of these DNA repair systems do not make mistakes. However, some processes seem to be *error-prone*, and it is the repair process itself that introduces the mutation. Many kinds of mutations arise as a result of faulty repair of damage induced in DNA by some of the various agents just discussed.

Often DNA damage itself can *induce* DNA repair systems. A complex cellular mechanism, called the **SOS regulatory system,** is activated as a result of DNA damage, initiating a number of DNA repair processes. However, in the SOS system, some DNA repair occurs in the absence of template instruction, which results in the creation of many errors, hence many mutations.

In the SOS regulatory system, DNA damage serves as a distress signal to the cell, resulting in the coordinate derepression (induction) of a number of cellular functions involved in DNA repair. The SOS system is normally repressed by a protein called LexA. However, this repressor protein is inactivated by RecA, a protease that is activated as a result of DNA damage (Figure 9.7). Since one of the DNA repair mechanisms of the SOS system is inherently error-prone, many mutations arise. Thus, through the SOS regulatory system, DNA damage by various agents such as chemicals and radiation leads to mutagenesis.

The SOS system senses the presence in the cell of DNA damage, and the repair mechanisms are activated. But once the DNA damage has been repaired, the SOS system is switched off and further mutagenesis ceases. In addition to its effect on cellular mutagenesis, the SOS regulatory system plays a central role in the regulation of temperate virus replication (∞ Section 8.12).

It should be emphasized that not all DNA repair occurs in the absence of template instruction. Cells generally have many DNA repair systems that require template instruction and lead to proper DNA repair. These systems apparently work most of the time but are not sufficient to repair the large amounts of damage done by some of the agents previously mentioned.

Biological mutagens

Mutations can be introduced without the use of chemical or physical agents through the process of *transposon mutagenesis.* We discuss the details of transposon mutagenesis later in this chapter (see Section 9.9) and note here only that if insertion of a transposable element occurs *within* a gene, loss of gene function generally results. For example, bacteriophage Mu, discussed previously (∞ Section 8.13), can serve as a mutagen by disrupting the coding sequence of a gene into which it inserts (see Figure 9.1c). Because transposable elements can enter the chromosome at various locations, transposons are widely used by microbial geneticists as mutagenic agents.

Site-directed mutagenesis

So far, the mutations that we have been discussing have been randomly directed at the genome of the microbial cell. Recombinant DNA technology and the use of synthetic DNA make it possible to induce *specific* mutations in *specific* genes. The procedures for carrying out mutagenesis of specific sites in the genome

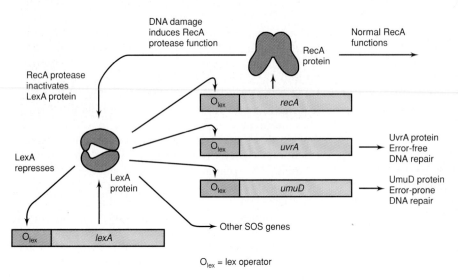

Figure 9.7 Mechanism of the SOS response. DNA damage results in conversion of RecA protein into a protease that cleaves LexA protein. LexA protein normally represses the activities of the *recA* gene and the DNA repair genes *uvrA* and *umuD*. Note, however, that repression is not complete. Some RecA protein is produced even in the presence of LexA protein. With LexA inactivated, these genes become active. As a protease, the RecA protein also cleaves the lambda repressor protein (∞ Section 8.12).

DNA damage induces RecA protease function

Normal RecA functions

RecA protein

RecA protease inactivates LexA protein

O_{lex} *recA*

O_{lex} *uvrA* — UvrA protein Error-free DNA repair

LexA represses

LexA protein

O_{lex} *umuD* — UmuD protein Error-prone DNA repair

Other SOS genes

O_{lex} *lexA*

O_{lex} = lex operator

are called *site-directed mutagenesis* and will be discussed in detail in the next chapter (∞ Section 10.11). Here we briefly describe the overall principle of site-directed mutagenesis.

When the DNA containing a specific gene has been isolated and its sequence determined, it is possible to construct a modified form of this gene in which a specific base or series of bases has been changed. This modified DNA can then be inserted into a recipient cell and *mutants* selected (or screened). These mutants will differ from the wild type by the desired change at a specific site. Site-directed mutagenesis has many uses in microbial genetics and molecular biology and has been especially useful for structure–function studies of enzymes and other proteins (∞ Section 10.11).

CONCEPT CHECK   9.3

Mutagens are chemical, physical, or biological agents that increase the mutation rate. Mutagens can alter DNA in many different ways. However, alterations in DNA are not mutations unless they can be inherited. Some DNA damage can lead to cell death if not repaired.

✔ *How do mutagens work?*

✔ *Differentiate between a mutation and DNA damage.*

9.4

Mutagenesis and Carcinogenesis: The Ames Test

A practical use of mutant bacterial strains has been developed to identify potentially hazardous chemicals in the environment. Because the sensitivity with which selectable mutants can be detected in large populations of bacteria is very high, bacteria can be used as screening agents for the potential mutagenicity of chemicals. This is relevant because it has been found that many mutagenic chemicals are also carcinogenic, capable of causing cancer in animals or humans.

The variety of chemicals, both natural and artificial, that the human population comes into contact with through agricultural and industrial exposure is enormous. There is considerable need for simple tests to ascertain the safety of such compounds. There is good evidence that a large proportion of human cancers have environmental causes, most likely through the agency of various chemicals, making the detection of chemical carcinogens urgent. It does not necessarily follow that because a compound is mutagenic it is also carcinogenic. The correlation, however, is quite high, and the knowledge that a compound is mutagenic in a bacterial system serves as a warning of possible danger. Similarly, the fact

that a compound is not mutagenic in a bacterial system does not mean that it is not carcinogenic, because the bacterial system cannot detect all compounds active in higher animals. The development of bacterial tests for carcinogenic screening was carried out primarily by a group at the University of California in Berkeley under the direction of Bruce Ames, and the mutagenicity test for carcinogens is sometimes called the **Ames test** (Figure 9.8).

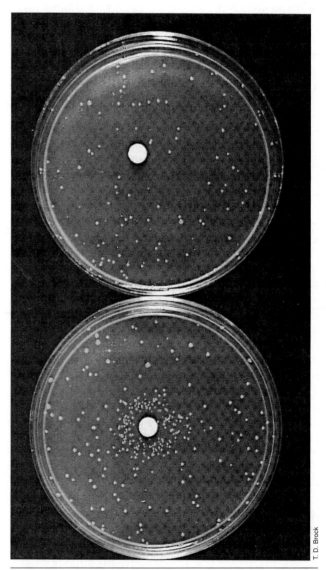

Figure 9.8 The Ames test is used to evaluate the mutagenicity of a chemical. Both plates were inoculated with a culture of a histidine-requiring mutant of *Salmonella typhimurium*. The medium does not contain histidine, so only cells that revert back to wild type can grow. Spontaneous revertants appear on both plates, but the chemical on the filter paper disc in the test plate (bottom) has caused an increase in the mutation rate, as shown by the large number of colonies surrounding the disc. Revertants are not seen very close to the disc because the concentration of the mutagen is so high there that it is lethal.

The standard way to test chemicals for mutagenesis has been to determine if the rate of *back* mutation (reversion) in strains of bacteria that are auxotrophic for some nutrient is increased by the suspected mutagen. It is important that the original mutation be a point mutation so reversion can occur. When cells of such an auxotrophic strain are spread on a medium lacking the required nutrient (for example, an amino acid or vitamin), no growth occurs, and even very large populations of cells can be spread on the plate without formation of visible colonies. However, if back mutants are present, those cells will be able to form colonies. Thus, if 10^8 cells are spread on the surface of a single plate, even as few as 10–20 back mutants (revertants) can be detected by the 10–20 colonies they form. If the back mutation rate has been increased by a chemical mutagen, the number of revertant colonies will also increase. Histidine auxotrophs of *Salmonella typhimurium* (Figure 9.8) and tryptophan auxotrophs of *Escherichia coli* have been the major tools of the Ames test, but a test has also been designed in which the induction of a phage lambda lysogen is used as an assay of DNA damage.

Although the simple testing of chemicals for mutagenesis in bacteria has been carried out for a long time, two elements have been introduced in the Ames test to make it much more powerful. The first of these is the use of strains of bacteria that almost exclusively use error-prone pathways to repair DNA damage (see Section 9.3). The second important element in the Ames test is the use of liver enzyme preparations to convert the chemicals into their active mutagenic (and carcinogenic) forms. It has been well established that many potent carcinogens are not directly carcinogenic or mutagenic but undergo chemical changes in the human body that convert them into active substances. These changes take place primarily in the liver, where enzymes (mixed-function oxygenases) normally involved in detoxification cause formation of epoxides or other activated forms of the compounds, which are then highly reactive with DNA.

In the Ames test, a preparation of enzymes from rat liver is first used to activate the compound. Then the activated complex is taken up on a filter-paper disk, which is placed in the center of a plate on which the proper bacterial strain has been overlaid. After overnight incubation, the mutagenicity of the compound can be detected by looking for a halo of back mutations in the area around the paper disk (Figure 9.8). It is always necessary to carry out this test with several different concentrations of the compound and with appropriate positive and negative controls because compounds vary in their mutagenic activity and are lethal at higher levels. A wide variety of chemicals have been subjected to the Ames test, and it has become one of the most useful prescreens for determining the potential carcinogenicity of a compound.

CONCEPT CHECK   **9.4**

The Ames test employs a sensitive bacterial assay system for detecting chemical mutagens in the environment.

✔ *Why does the Ames test measure the rate of* back *mutation rather than the rate of* forward *mutation?*

✔ *Of what significance is the detection of mutagens to the prevention of cancer?*

9.5
Genetic Recombination

Genetic recombination involves the physical exchange of genetic material between genetic elements. In this section we focus on **general** or **homologous recombination,** which results in genetic exchange between *homologous* DNA sequences from two different sources. Homologous DNA sequences have the same or nearly the same sequence; therefore, base pairing can occur over an extended length of the two DNA molecules.

Homologous recombination is extremely important to all organisms. However, it is also very complex. Even in the bacterium *Escherichia coli* there are at least 25 genes involved. In addition, homologous recombination seems to be of such importance that there are several redundant pathways. Therefore, if one pathway is inhibited or nonfunctional, another may be able to supply necessary functions.

Molecular events in homologous recombination

At the molecular level, recombination has been studied mostly in prokaryotes and viruses. In Bacteria, general recombination involves the participation of a specific protein called the RecA protein, which is specified by the *recA* gene. The RecA protein has been shown to be essential in nearly every homologous recombination pathway. RecA-like proteins have been identified in all prokaryotes examined, including the Archaea. A related protein has also been found in the Eukarya.

An overall molecular mechanism of general recombination is shown in Figure 9.9. The process begins with a *nick* (usually generated by a nuclease) in one of the DNA molecules. This nicked strand must be displaced from the other strand by proteins having helicase activity (∞ Section 6.5). In some pathways specialized enzymes, such as the RecBCD enzyme of *E. coli,* have both nuclease and helicase activities. Single-stranded binding protein (∞ Section 6.5) then binds to the resulting single-stranded segment. Next, the RecA protein binds to the single-stranded fragment, forming a complex that facilitates annealing with a complementary sequence in the adjacent duplex, simultaneously dis-

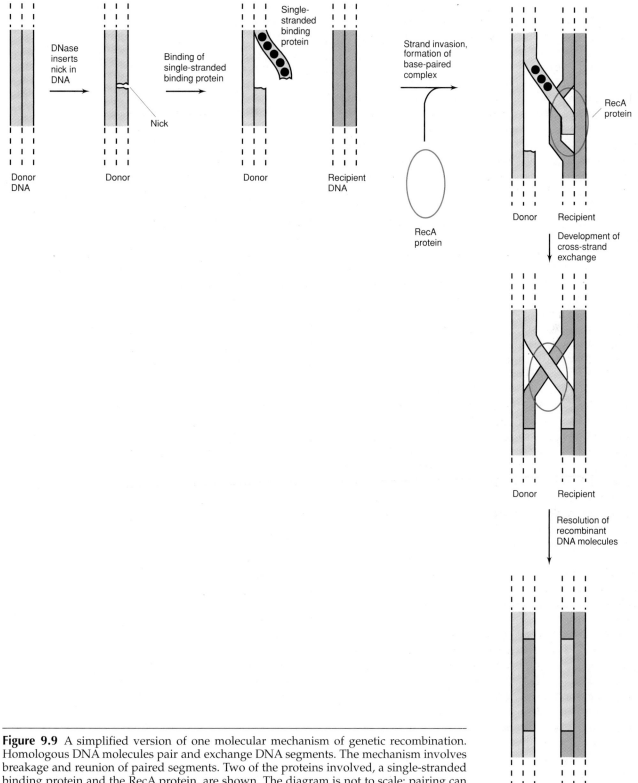

Figure 9.9 A simplified version of one molecular mechanism of genetic recombination. Homologous DNA molecules pair and exchange DNA segments. The mechanism involves breakage and reunion of paired segments. Two of the proteins involved, a single-stranded binding protein and the RecA protein, are shown. The diagram is not to scale: pairing can occur over hundreds or thousands of bases.

placing the resident strand (Figure 9.9). This process is often referred to as *strand invasion*. As noted, recombination involves the *pairing* of DNA molecules over long stretches. Following pairing, *exchange* of homologous

DNA molecules can occur, leading to the formation of recombination intermediates containing extensive **heteroduplex** regions, where each strand has originated from a different chromosome. This process also involves

DNA polymerase and ligase. These regions may be extended by the RecA protein. Finally, the linked molecules are *resolved* by nucleases and DNA ligase to form two recombinant molecules.

Note that this mechanism for the formation of recombinant DNA structures is a completely natural mechanism that occurs extensively within the cell. Whether or not it leads to the formation of new genotypes depends on whether the two molecules undergoing recombination differ genetically in regions outside the region of recombination. Within limits, general recombination can be thought of as occurring at random sites throughout the genome. Thus, the probability of recombination occurring between two genes is proportional to their distance. This fact is useful for genetic mapping. By recombinational analysis it is possible to map the position of genes on chromosomes because the *farther* two genes are apart, the more likely they are to show recombination.

Until the advent of molecular techniques such as restriction enzyme mapping (∞ Section 6.4) and DNA sequencing (∞ Working with Nucleic Acids: The Tools, Chapter 6), recombinational analysis was the only method available for ordering genes on chromosomes and determining the distance between them. It is interesting to note that the order and relative distances as measured by mapping using recombination have been largely confirmed using modern molecular techniques.

For new genotypes to arise as a result of general recombination, it is essential that the two homologous sequences be genetically distinct. This is the case in a diploid eukaryotic cell (see Section 9.11), which has two sets of chromosomes, one from each parent. The two distinct molecules are brought together as the result of sexual reproduction, a process that occurs as part of the regular life cycles of most eukaryotic organisms (see Section 9.11). In prokaryotes, genetically distinct but homologous DNA molecules are brought together in different ways, but the process of genetic recombination is no less important. Recombination can also be critical in the life cycle of some viruses. In Chapter 8 we showed that certain bacteriophages, such as T7 and T4 (∞ Section 8.11), require homologous recombination as a step during DNA replication.

In prokaryotes genetic recombination is observed because fragments of homologous DNA from a donor chromosome are transferred to a recipient cell by one of three processes: (1) **transformation,** which involves donor DNA free in the environment (see Section 9.6), (2) **transduction,** in which the donor DNA transfer is mediated by a virus (see Section 9.7), and (3) **conjugation,** in which the transfer involves cell-to-cell contact and a *conjugative plasmid* in the donor cell (see Section 9.8). These processes are contrasted in Figure 9.10 and will be discussed in detail later in this chapter.

It is *after* the transfer, when the DNA fragment from the host is in the recipient cell, that homologous recombination may occur. Because only a chromosomal fragment is transferred, if recombination does not occur, the fragment will be lost because it cannot replicate independently. Therefore, it is important to remember that in prokaryotes transfer is just the first step in obtaining recombinant organisms.

Detection of recombination

In order to detect physical exchange of DNA segments, the cells resulting from recombination must be phenotypically different from the parents. In crosses involving microorganisms, one must usually use as recipients strains that lack some selectable characteristic that the recombinants will possess. For instance, the recipient may not be able to grow on a particular medium, and genetic recombinants are selected that can. Various kinds of selectable and nonselectable markers (such as drug resistance, nutritional requirements, and so on) were discussed in Section 9.1. The exceedingly great sensitivity of the selection process is shown by the fact that 10^8 or more bacterial cells can be spread on a single plate and, if proper selective conditions are used, no parental colonies will appear, whereas even a few recombinants can form colonies (Figure 9.11). The only requirement is that the *reverse* mutation rate for the selected characteristic be low, because revertants will also form colonies. This problem can often be overcome by using double mutants because it is very unlikely that two back mutations will occur in the same cell. Much of the skill of the bacterial geneticist is exhibited in the choice of proper mutants and selective media for efficient detection of genetic recombination. Because selection is so powerful and because crosses can be made using billions of individual cells, recombinational analysis is a very important tool to the microbial geneticist.

Complementation

When two mutant strains are genetically crossed (mated), homologous recombination can yield a wild-type recombinant unless both of the mutations include changes in exactly the same base pairs. Therefore, if two different Trp⁻ *Escherichia coli* (strains that require the amino acid tryptophan in the medium) are crossed and Trp⁺ recombinants are obtained, it is clear that the mutations in the two strains did not include the same base pairs. However, this experiment cannot detect whether the mutations were in the same gene. This can be determined by a type of experiment called a **complementation test.**

Complementation was first used in *diploid* eukaryotic organisms. Remember that in diploid organisms the cell has *pairs* of chromosomes, one member from each parent (∞ Section 3.14). When the two mutations are present on separate members of a pair (that is, one mutation from each parent), they are said to be in **trans** configuration. On the other hand, if the two

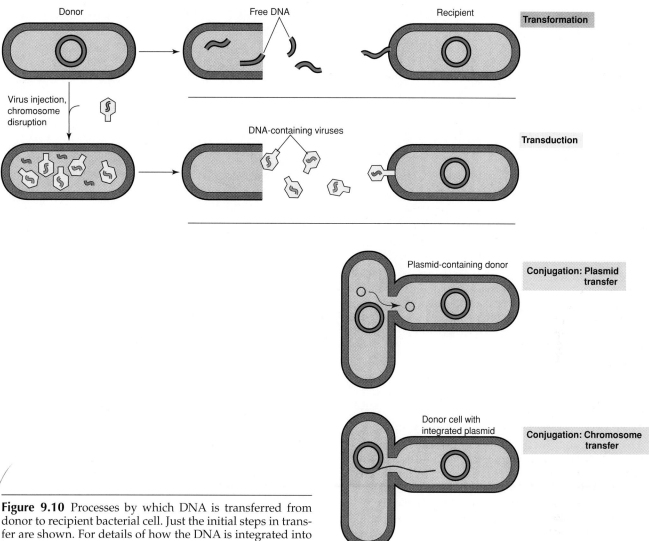

Figure 9.10 Processes by which DNA is transferred from donor to recipient bacterial cell. Just the initial steps in transfer are shown. For details of how the DNA is integrated into the recipient, see text.

mutations are on the *same* chromosome (both from the same parent), then they are said to be in **cis** configuration. True diploidy does not exist in prokaryotes. For complementation tests a diploid state is achieved for a region of the chromosome by transfer of a chromosomal fragment from a donor (see earlier discussion). However, the principle of the test and the nomenclature (*cis* and *trans*) is the same.

Two mutations complement one another (that is, give a wild-type phenotype) when they are present in a diploid state only if they are present in different genes. This is shown diagrammatically in Figure 9.12. Mutations can complement because the genes encode proteins. If each homologous DNA molecule contributes a different required gene, then the cell will have all the enzymes it requires to synthesize tryptophan. Notice that complementation *does not* involve recombination. To do the test, the mutations must be in *trans*. (If one molecule has both mutations, the other is

wild type and should be sufficient itself to confer the wild-type phenotype. If it does not, then one of the mutations must exert its phenotype even in the presence of the wild-type gene. Such a mutation is said to be *dominant*. Therefore, having the mutations in cis serves as a control.)

This type of complementation test, called a *cis–trans test,* is used to define whether two mutations are in the same genetic (functional) unit. The genetic unit defined by the cis–trans test is sometimes called a **cistron** (a term essentially equivalent to a gene). As noted, two mutations in the *same* cistron *cannot* complement each other, and so when complementation is found to exist, this implies that the two mutations lie in *different* cistrons (that is, different genes). The term *cistron* is now rarely used except when describing whether an mRNA has the genetic information from one gene (monocistronic mRNA) or from more than one gene (polycistronic mRNA) (∞ Section 6.6).

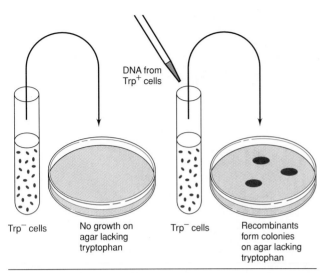

Figure 9.11 How a selective medium can be used to detect rare genetic recombinants among a large population of non-recombinants. On the selective medium only the rare recombinants form colonies. Procedures such as this, which offer high resolution for genetic analyses, can ordinarily be used only with microorganisms.

CONCEPT CHECK   9.5

Homologous recombination arises when closely related sequences from two genetically distinct elements are combined together in the same element. Recombination is an important evolutionary process, and cells have specific mechanisms for ensuring that recombination takes place. In eukaryotes, genetic recombination occurs as a consequence of the sexual cycle. Mechanisms of recombination also occur in prokaryotes but involve DNA transfer during the processes of transformation, transduction, and conjugation.

✔ *What protein, found in all prokaryotes, facilitates the pairing required for homologous recombination?*

✔ *Complementation tests do not involve recombination. Explain.*

9.6

Genetic Transformation

As we have noted, genetic transformation is a process by which free DNA is incorporated into a recipient cell and brings about genetic change. The discovery of genetic transformation in bacteria was one of the outstanding events in biology, as it led to experiments proving without a doubt that DNA is the genetic material (see the box, Origins of Bacterial Genetics). This discovery became the keystone of molecular biology and modern genetics.

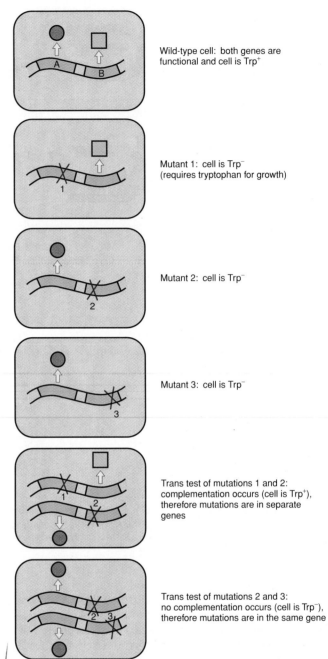

Figure 9.12 Complementation analysis. Mutations 1, 2, and 3 each lead to the same phenotype, a requirement for tryptophan. Complementation analysis indicates that mutations 2 and 3 are in one gene and that mutation 1 is in another. The protein products of both genes (A and B) must be required to synthesize tryptophan.

A number of prokaryotes have been found to be naturally transformable, including certain species of both gram-negative and gram-positive Bacteria and some species of Archaea. However, even within transformable genera, only certain strains or species are transformable. Since the DNA of prokaryotes is present in the cell as a large single molecule, when the cell

ORIGINS OF BACTERIAL GENETICS[a]

 Although genetic recombination in eukaryotes had been known for a long time, the discovery of genetic recombination in bacteria following transformation, transduction, and conjugation has been a relatively recent event. Of the three processes, the discovery of transformation was the most significant as it provided the first direct evidence that DNA is the genetic material. The first evidence of bacterial transformation was obtained by the British scientist Fred Griffith in the late 1920s. Griffith was working with *Streptococcus pneumoniae* (pneumococcus), a bacterium that owes its ability to invade the body in part to the presence of a polysaccharide capsule. Mutants can be isolated that lack this capsule and are thus unable to cause infection; such mutants are called R strains because their colonies appear rough on agar, in contrast to the smooth appearance of capsulated strains. A mouse infected with only a few cells of a smooth (S) strain succumbs in a day or two to pneumococcus infection, whereas even large numbers of R cells do not cause death when injected. Griffith showed that if heat-killed S cells were injected along with living R cells, a fatal infection ensued and the bacteria isolated from the dead mouse were S types. A number of different polysaccharide capsules were known in different pneumococcus S strains, and it was possible to do this experiment with heat-killed S cells from a type different from that from which the R strain was derived. Since the isolated living S cells always had the capsule type of the heat-killed S cells, the R cells had been transformed into a new type, and the process had all the properties of a genetic event. The molecular expla-

nation for the transformation of pneumococcus types was provided by Oswald T. Avery and his associates at Rockefeller Institute in New York in a series of studies carried out during the 1930s, culminating in the now classic paper by Avery, C. M. MacLeod, and M. McCarty in 1944. Avery and his coworkers showed that under certain conditions the transformation process could be carried out in the test tube rather than the mouse and that a cell-free extract of heat-killed cells could induce transformation. In a long series of painstaking biochemical experiments, the active fraction of cell-free extracts was purified and was shown to consist of DNA. The transforming activity of purified DNA preparations was very high, and only very small amounts of material were necessary. Subsequently, others at Rockefeller showed that transformation could occur in pneumococcus not only for capsular characteristics but for other genetic characteristics of the organism as well, such as antibiotic resistance and sugar fermentation.

In 1953, James Watson and Francis Crick announced their model of the structure of DNA, providing a theoretical framework for how DNA could serve as the genetic material. Thus, two types of studies, the bacteriological and biochemical ones of Avery and the physical-chemical ones of Watson and Crick, solidified the concept of DNA as the genetic material. In the subsequent years, this work has led to the whole field of molecular genetics.

Although bacterial transformation resulted from an essentially accidental discovery, bacterial conjugation was initially shown to occur by Joshua Lederberg and E. L. Tatum in 1946 in experiments carefully designed to determine if a sexual process might occur in bacteria. Because it appeared that the process, if present, would be quite rare (no microscopic evidence for bacter-

ial mating had ever been seen, although such evidence can easily be obtained in eukaryotes), Lederberg developed a method involving the use of nutritional mutants of *Escherichia coli*. Fortunately, he isolated these mutants in strain K-12, one of the few wild-type strains now known to contain the F plasmid. The principle was to mix two strains, one requiring biotin and methionine, the other requiring threonine and leucine, and plate the mixture on a minimal medium lacking all four growth factors. Neither parental type could grow on this medium, but any recombinants could, and when about 10^8 cells were plated, a small but significant number of colonies was obtained. Strains with two separate nutritional requirements were employed because it would be unlikely that spontaneous back mutation of both genes would occur in a single cell. Thus, the only explanation for the phenomenon was some sort of genetic recombination. To show that the process required cell-to-cell contact, and hence could not be a type of transformation, it was shown that when culture filtrates or extracts were separated by a sintered glass disc, permeable to macromolecules but not to cells, recombination did not occur. Although initially conjugation appeared to be a very rare event, by the early 1950s a strain of *E. coli* had been isolated by the Italian scientist L. L. Cavalli-Sforza, while he was working in Lederberg's laboratory, that showed a high frequency of recombination. The British physician William Hayes, who independently isolated an Hfr strain, then showed that genetic transfer during mating between Hfr and F$^-$ was a one-way event, with the Hfr serving as donor. The interrupted mating experiment and the demonstration of the circular genetic map of *E. coli* were then carried out by Elie Wollman and Francois Jacob,

[a]For further reading: Brock, T. D. 1990. *The Emergence of Bacterial Genetics.* Cold Spring Harbor Laboratory Press, Cold Spring Harbor, NY.

O. T. Avery

E. L. Tatum (left) and J. Lederberg in 1947

working with Jacques Monod at the Pasteur Institute in Paris. The distinction between Hfr and F⁺ was made by Lederberg, who also showed that F⁺ behaved in an infectious manner. Lederberg coined the term *plasmid* in the 1950s to describe such apparently extrachromosomal genetic elements, although it did not find wide usage until the 1970s when infectious drug resistance became a major medical problem.

Bacterial transduction was discovered by the American scientist Norton Zinder when he was a graduate student working with Lederberg at the University of Wisconsin on genetic recombination in *Salmonella typhimurium.* The original motivation for this work was to show that conjugation occurred in an organism other than *E. coli,* and the techniques involved isolation of mutants and quantification of recombination by observing colony growth on minimal medium. However, although evidence of recombination was obtained, it could be shown that cell-to-cell contact was *not* required. Although this suggested a type of transformation, the process was not affected by DNase, and the gene transfer agent behaved like a bacteriophage. The gene transfer agent could be purified by the same procedures used to purify virus particles, and transduction occurred only with recipient cells that had receptor sites for the virus in question. Further, transducing activity could be eliminated by treatment of a lysate with substances able to adsorb the virus, such as sensitive cells or antibodies. Thus, in all cases, transducing activity and virus activity behaved in similar ways. Zinder and Lederberg coined the word *transduction* to refer to any genetic recombinational process that was only fragmentary and did not involve cell-to-cell contact, intending in this way to encompass processes involving either free DNA (transformation) or phages, but subsequently the word *transduction* has been applied only to virus-mediated genetic transfer. ∎

is gently lysed, the DNA pours out (Figure 9.13). Because of its extreme length (1700 μm in *Bacillus subtilis*), the DNA molecule breaks easily; even after gentle extraction it fragments into 100 or more pieces (*B. subtilis* DNA of 5×10^6 base pairs is converted to fragments of about 15 kilobase pairs). Because the DNA that corresponds to an average gene is about 1000 nucleotides, each of the fragments of purified DNA has about 15 genes. Any cell usually incorporates only one or a few DNA fragments so only a small proportion of the genes of one cell can be transferred to another by a single transformation event.

Competence

A cell that is able to take up a molecule of DNA and be transformed is said to be **competent.** Only certain strains are competent; the ability seems to be an inherited property of the organism. Competence in most naturally transformable bacteria is regulated, and special proteins play a role in the uptake and processing of DNA. These competence-specific proteins may include a membrane-associated DNA binding protein, a cell wall autolysin, and various nucleases. In both *Bacillus subtilis* and *Streptococcus pneumoniae,* induction of competence is dependent on the medium and the growth stage of the culture. In *Bacillus,* about 20% of the cells become competent and stay that way for several hours. However, in *Streptococcus,* 100% of the cells can become competent, but only for a few minutes during the growth cycle.

Uptake of DNA

Bacteria differ in the form in which DNA is taken up. In *Haemophilus,* which is gram-negative, for example, only double-stranded DNA is taken up into the cell despite the fact that only single-stranded segments actually become incorporated into the genome by recombination. In the gram-positive Bacteria *Streptococcus* and *Bacillus,* by contrast, only a single DNA strand is taken up, while the complementary strand is

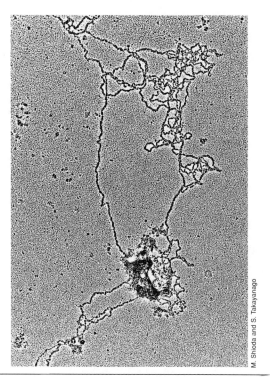

M. Shioda and S. Takayanago

Figure 9.13 The prokaryotic chromosome, as shown in the electron microscope. The circular chromosome is from the hyperthermophile *Sulfolobus*, a member of the Archaea (∞ Section 17.3). See also Figure 9.18.

simultaneously degraded. However, in all cases, double-stranded DNA binds more effectively to the cells.

During the transformation process, competent bacteria first bind DNA reversibly; soon, however, the binding becomes irreversible. Competent cells bind much more DNA than do noncompetent cells—as much as 1000 times more. As we noted earlier, the sizes of the transforming fragments are much smaller than that of the whole genome, and this DNA is further degraded during the uptake process. In *Streptococcus pneumoniae* each cell can bind only about 10 molecules of double-stranded DNA of 15–20 kilobase pairs each. However, as they are taken up, they are converted to single-stranded pieces of about 8 kilobases. The DNA fragments in the mixture compete with each other for uptake, and if excess DNA that does not contain the genetic marker is added, a decrease in the number of transformants occurs. In preparations of transforming DNA, only about 1 out of 100–200 DNA fragments contains the marker being studied. Thus, at high concentrations of DNA, the competition between DNA molecules results in saturation of the system so even under the best conditions it is impossible to transform all the cells in a population for a given genetic marker. The maximum frequency of transformation that has so far been obtained is about 20% of the population; actually the values usually obtained are between 0.1 and 1.0%. The minimum concentration of

DNA yielding detectable transformants is about 0.00001 μg/ml (1×10^{-5} μg/ml), which is so low that it is undetectable chemically.

Interestingly, in *Haemophilus influenzae* there is a requirement that the DNA fragment have a particular 11-base-pair sequence for irreversible binding and uptake to occur. This sequence is found at an unexpectedly high frequency in the *Haemophilus* chromosome. Evidence such as this, and the fact that at least certain bacteria become competent in their natural environment, suggest that transformation is not a laboratory artifact but plays an important role in gene transfer in nature.

Integration of transforming DNA

Transforming DNA is bound at the cell surface by a DNA binding protein, after which either the entire double-stranded fragment is taken up or a nuclease degrades one strand and the other is taken up (Figure 9.14). After uptake, the DNA associates with a competence-specific protein that remains attached to the DNA, presumably preventing it from nuclease attack, until it reaches the chromosome where RecA protein takes over. The DNA is then integrated into the genome of the recipient by recombinational processes (Figure 9.14; see also Figure 9.9). During replication of this heteroduplex DNA, one parental and one recombinant DNA molecule are formed. On segregation at cell division, the latter is present in the transformed cell, which is now genetically altered as compared to the parental type. The preceding discussion pertains to only small pieces of *linear* DNA. The transformation of plasmid DNA generally occurs in the absence of recombination between the plasmid and bacterial chromosome.

Transfection

Bacteria can be transformed with DNA extracted from a *bacterial virus* rather than from another bacterium, a process known as **transfection**. If the DNA is from a lytic bacteriophage, transfection can be measured by the standard phage plaque assay (∞ Section 8.4). Transfection has become a useful tool in studying the mechanism of transformation and recombination because the small size of phage genomes allows for the isolation of a nearly homogeneous population of DNA molecules. By contrast, in conventional transformation, the transforming DNA is generally a random assortment of chromosomal DNA of various lengths and this tends to complicate experiments designed to study the mechanism of transformation.

Artificially induced competence

High efficiency natural transformation is found only in a few bacteria; *Azotobacter*, *Bacillus*, *Streptococcus*, *Haemophilus*, *Neisseria*, and *Thermus*, for example, are easily transformed. Many prokaryotes are trans-

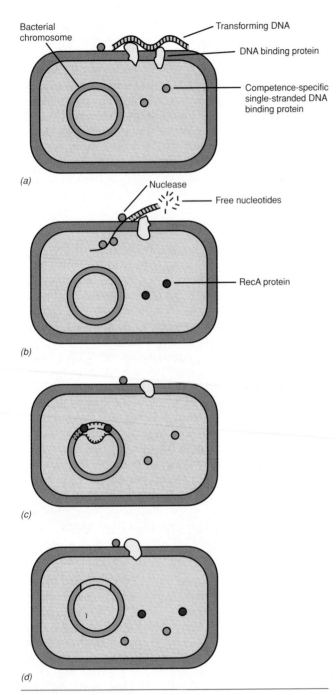

Bacterial chromosome

Transforming DNA

DNA binding protein

Competence-specific single-stranded DNA binding protein

(a)

Nuclease

Free nucleotides

RecA protein

(b)

(c)

(d)

Figure 9.14 Mechanism of DNA transfer by transformation in a gram-positive bacterium. (a) Binding of free DNA by a membrane-bound DNA binding protein. (b) Passage of one of the two strands into the cell while nuclease activity degrades the other strand. (c) The single strand in the cell is bound by specific proteins, and recombination with homologous regions of the bacterial chromosome mediated by RecA protein occurs. (d) Transformed cell.

formed only poorly or not at all under natural conditions. Determination of how to induce competence in such bacteria may involve considerable empirical study, with variation in culture medium, temperature,

and other factors. The nature of the cell surface must be of importance in determining whether a cell can take up DNA, and the presence or absence of intracellular nucleases is also important. To transfer DNA into cells for genetic engineering (∞ Chapter 10), it was necessary to find a way to make *Escherichia coli*, a gram-negative organism, competent. It has been found that when *E. coli* is treated with high concentrations of calcium ions and then stored in the cold, it becomes transformable at low efficiency. With proper procedures it is possible to select *E. coli* transformants for chromosomal genes, but at low frequency. *Escherichia coli* treated in this manner takes up double-stranded DNA, and therefore transformation by plasmid DNA is more efficient because no recombination is required. Why the calcium treatment works is not known, but this procedure also works with some other gram-negative Bacteria. However, such methods of artificially induced competence are rapidly being supplanted by a new method termed *electroporation*.

DNA transfer by electroporation

Small pores are produced in the membranes of cells exposed to pulsed electric fields. When DNA molecules are present outside the cells during the electric pulse, they can then enter the cells through these pores. This process is called **electroporation.** Electroporation requires a sophisticated power supply because the pulses must be carefully controlled and last for only milliseconds. This technique has been used to transport DNA into a large number of different species of prokaryotes, both Archaea and Bacteria. Additionally, electroporation allows an experimenter to transfer a plasmid directly from one cell to another if both are present during electroporation. Therefore, electroporation allows small molecules of DNA to come out of cells as well as to go in! This type of "transformation" eliminates the steps required to isolate the plasmid from the first strain before introducing it into the second.

Transformation (transfection) of eukaryotic cells

Eukaryotic microorganisms and animal and plant cells can take up DNA in a process that resembles bacterial transformation. Because the word *transformation* in mammalian cells is used to describe the conversion of cells to the malignant (tumorous, cancerous) state (∞ Section 8.14), the introduction of DNA into mammalian cells has been called *transfection* (a term with another meaning in bacterial systems; see earlier discussion).

Transfection of cultured animal cells was originally accomplished by precipitating DNA in such a way that the cells would take it up by phagocytosis (∞ Section 20.3) because they do not have cell walls. In yeast, where the introduction of cloned DNA is popular in genetic engineering, transfection at low efficiencies can be mediated by treating cells with enzymes that partially destroy the cell wall, generating spheroplasts, and then adding DNA in the presence of Ca^{2+} and polyethylene glycol (which serves to permeabilize the membrane).

As in the case of prokaryotes, electroporation is becoming widely applied to all types of eukaryotic cells and can be used whether or not the cell wall is removed.

In addition to electroporation, a high velocity microprojectile "gun" has been developed for incorporating DNA into cells. The original **particle gun** operates somewhat like a conventional shotgun. A small steel cylinder containing a gunpowder charge is used to fire nucleic acid–coated particles at the target cells (Figure 9.15). The particles bombard the cell, piercing cell walls and membranes without actually killing the cells. The nucleic acid entering the cells can then recombine with host DNA. The particle gun has been used successfully to transfect yeast, algae, a variety of plant cells, and even mitochondria and chloroplasts. The particle gun is very useful because, unlike electroporation, it can be used on intact tissue such as plant seeds.

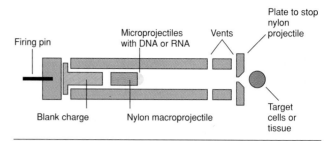

Figure 9.15 Nucleic acid gun for transfection of eukaryotic cells. The inner workings of the gun show how nucleic acids attached to metal pellets are projected at target cells.

CONCEPT CHECK   9.6

Certain prokaryotes exhibit competence, a state in which cells are able to take up free DNA released by other bacteria. This process is called transformation. Relatively few species of prokaryotes can be naturally transformed. However, certain laboratory procedures have been developed that make it possible to introduce DNA into completely unrelated organisms, even eukaryotes. Electroporation involves modification of the cytoplasmic membrane by treatment with an electric field to facilitate DNA uptake.

✔ *The donor bacterial cell in a transformation is probably dead. Explain.*

✔ *Even in naturally transformable cells competency is usually inducible. What does this mean?*

9.7

Transduction

In transduction, DNA is transferred from cell to cell through the agency of viruses. Genetic transfer of host genes by viruses can occur in two ways. In the first, called **generalized transduction,** host DNA derived from virtually any portion of the host genome becomes a part of the DNA of the mature virus particle in place of the virus genome. The second, called **specialized transduction,** occurs only in some temperate viruses; DNA from a specific region of the host chromosome is inte-

grated directly into the virus genome—usually replacing some of the virus genes. The transducing virus particle in both specialized and generalized transduction is usually *defective* as a virus because bacterial genes have replaced some necessary viral genes.

In generalized transduction, if the donor genes do not undergo homologous recombination with the recipient bacterial chromosome, they will be lost. They cannot replicate independently and are not part of a viral genome. In specialized transduction homologous recombination may also occur. However, since the donor bacterial DNA is now actually a part of a temperate phage genome, there are two other possibilities: (1) the DNA may be integrated into the host chromosome during lysogenization, and (2) the DNA may be replicated in the recipient as part of a lytic infection.

Transduction has been found to occur in a variety of prokaryotes, including certain species of the Bacteria: *Desulfovibrio, Escherichia, Pseudomonas, Rhodococcus, Rhodobacter, Salmonella, Staphylococcus,* and *Xanthobacter,* as well as the archaean *Methanobacterium thermoautotrophicum.* Not all phages can transduce, and not all bacteria are transducible; but the phenomenon is sufficiently widespread for us to assume that it plays an important role in genetic transfer in nature.

Generalized transduction

In generalized transduction, virtually any genetic marker can be transferred from donor to recipient. Generalized transduction was first discovered and extensively studied in the bacterium *Salmonella typhimurium* with phage P22 and has also been studied with phage P1 in *Escherichia coli.* An example of how *transducing particles* may be formed is given in Figure 9.16. When the population of sensitive bacteria is infected with a phage, the events of the phage lytic cycle may be initiated. During a lytic infection, the enzymes responsible for packaging viral DNA into the bacteriophage sometimes accidentally package host DNA. The resulting particle is called a *transducing particle.* On lysis of the cell, these particles are released

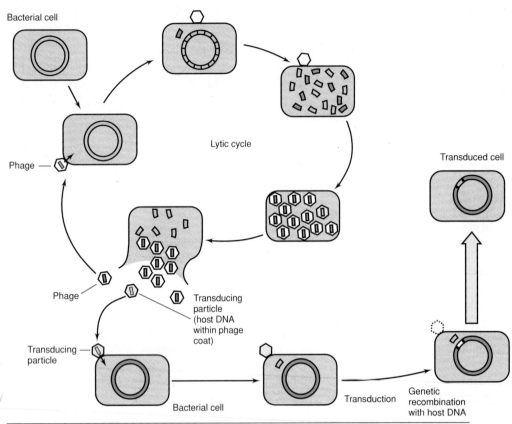

Bacterial cell

Phage

Lytic cycle

Phage

Transducing particle

Transducing particle (host DNA within phage coat)

Bacterial cell

Transduction

Genetic recombination with host DNA

Transduced cell

Figure 9.16 Generalized transduction: one possible mechanism by which virus (phage) particles containing host DNA can be formed.

along with normal virions, and so the lysate contains a mixture of normal virions and transducing particles. Because transducing particles cannot initiate a normal viral infection (they contain no viral DNA), they are said to be *defective*. When this lysate is used to infect a population of recipient cells, most of the cells become infected with normal virus. However, a small proportion of the population receives transducing particles that inject the DNA they received from the previous host bacterium. This DNA can now undergo genetic recombination with the DNA of the new host. Because only a small proportion of the particles in the lysate are of the defective transducing type and each of these contains only a small fragment of donor DNA, the probability of a transducing particle containing a particular gene is quite low, and usually only about 1 cell in 10^6–10^8 is transduced for a given marker.

Phages that form transducing particles can be either temperate or virulent, the main requirements being that they have a DNA packaging mechanism that permits accidental recognition of host DNA and that packaging occurs before the host genome is completely degraded. The detection of transduction is most certain when the multiplicity of phage to host is

low, so a host cell is infected with only a single phage particle; with multiple infection, the cell may be killed by the normal particles.

Specialized transduction

Generalized transduction allows the transfer of DNA from one bacterium to another at a low frequency. However, specialized transduction can allow extremely efficient transfer while also allowing a small region of a bacterial chromosome to be replicated independently of the rest. The example we shall use to discuss specialized transduction was the first to be discovered and involves transduction of the galactose genes by the temperate phage lambda of *Escherichia coli*.

As we discussed (∞ Section 8.12), when a cell is lysogenized by lambda, the phage genome becomes integrated into the host DNA at a specific site. The region in which lambda integrates is immediately adjacent to the cluster of host genes that control the enzymes involved in galactose utilization (∞ Figure 8.29), and the DNA of lambda is inserted into the host DNA at that site. From then on, viral DNA replication is under host control. On induction (for example, by ultraviolet radiation), the viral DNA separates from the host DNA by a process that is the reverse of integration (Figure 9.17). Ordinarily when the lysogenic cell is induced, the lambda DNA is excised as a unit. Under rare conditions, however, the phage genome is excised incorrectly. Some of the adjacent bacterial genes (the galactose cluster) are excised along with

phage DNA. At the same time, some phage genes are left behind. One type of altered phage particle, called **lambda dgal** or λ*dgal* (*dgal* means "defective, galactose"), is defective because of the phage genes lost and does not make mature phage. However, a **helper phage** can provide those functions missing in the defective particles. This "helper" is identical to the

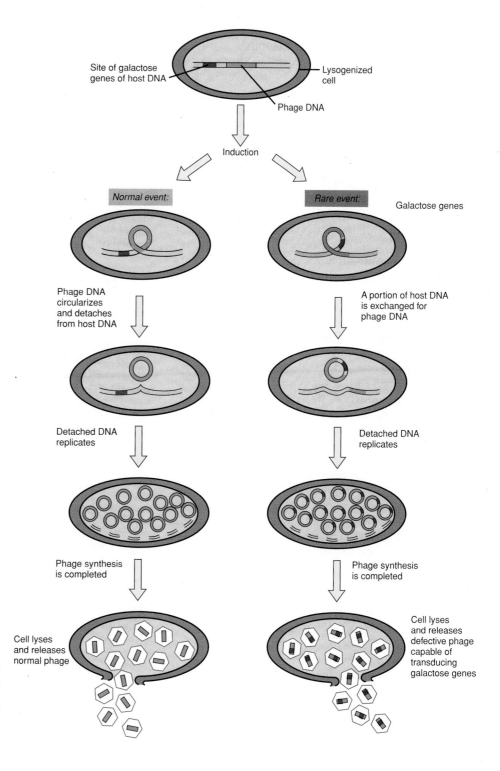

Figure 9.17 Normal lytic events and the production of particles transducing the galactose genes in an *Escherichia coli* cell containing a lambda prophage.

original lambda. Thus, the culture lysate obtained contains a few λ*dgal* particles mixed in with a large number of normal lambda virions.

When a galactose-negative bacterial culture is infected at high multiplicity with such a lysate and *gal*⁺ transductants are selected, many are double lysogens, carrying both lambda and λ*dgal*. (Note then that the bacterium has become a diploid for the *gal* region. Specialized transducing phage can be used in performing complementation tests in bacteria; see Section 9.5.) When such a double lysogen is induced, a lysate is produced containing about equal numbers of lambda and λ*dgal*. Such a lysate can transduce at high efficiency, although only for a restricted group of *gal* genes.

If the phage is to be viable, there is a maximum limit to the amount of phage DNA that can be replaced with host DNA, because sufficient phage DNA must be retained in order to provide the information for production of the phage protein coat and for other phage proteins needed for lysis and lysogenization. However, if a helper phage is used together with the defective phage in a mixed infection, then even less information is needed in the defective phage for transduction. Only the *att* (attachment) region, the *cos* site (cohesive ends, for packaging), and the replication origin of the lambda genome are needed for production of a transducing particle, provided a helper is used (see the genetic map of lambda in Section 8.12).

One important distinction between specialized and generalized transduction is in how the transducing lysate can be formed. In specialized transduction this *must* occur by induction of a lysogen, whereas in generalized transduction it can occur either in this way or by infection of a nonlysogen by the phage, with subsequent phage replication and cell lysis.

Although we have discussed specialized transduction only in the lambda-*gal* system, phage lambda and its relative, φ80, have been widely used to form specialized transducing phages covering many specific regions of the *E. coli* genome. In addition, lambda specialized transducing phage can be constructed by the techniques of genetic engineering to contain genes from any organism (∞ Section 10.3).

Phage conversion

This is a phenomenon analogous in some ways to specialized transduction. When a normal temperate phage (that is, a nondefective one) lysogenizes a cell and its DNA is converted to the prophage state, the lysogen is immune to further infection by the same type of phage. This acquisition of immunity can be considered a change in phenotype. In certain cases other phenotypic alterations can be detected in the lysogenized cell, which seem to be unrelated to the phage immunity system. Such a change, which is brought about through lysogenization by a normal temperate phage, is called **phage conversion.**

Two cases of conversion have been especially well studied. One involves a change in structure of a polysaccharide on the cell surface of *Salmonella anatum* on lysogenization with phage ε[15]. The second involves the conversion of non-toxin-producing strains of *Corynebacterium diphtheriae* to toxin-producing (pathogenic) strains on lysogenization with phage β (we discuss the disease diphtheria in Section 23.2). In these situations the information for production of these new materials is apparently an integral part of the phage genome and hence is automatically and exclusively transferred on infection by the phage and lysogenization.

Lysogeny probably carries a strong selective value for the host cell because it confers resistance to infection by viruses of the same type. Phage conversion seems also to be of considerable evolutionary significance as it results in efficient genetic alteration of host cells. Many bacteria isolated from nature are lysogens. It seems reasonable to conclude, therefore, that lysogeny is the normal state of affairs and may often be essential for survival of the host in nature.

CONCEPT CHECK   **9.7**

Transduction involves transfer of host genes from one bacterium to another by viruses. In generalized transduction, defective virus particles randomly incorporate fragments of the cell DNA; virtually any gene of the donor can be transferred, but the efficiency is low. In specialized transduction, the DNA of a temperate virus excises incorrectly and brings adjacent host gene(s) along with it; only genes close to the integration point of the virus are transduced, but the efficiency may be high.

✔ *What is the important difference between generalized transduction and transformation?*

✔ *In specialized transduction the donor DNA can replicate inside the recipient cell without homologous recombination taking place, but this is not true in generalized transduction. Explain.*

9.8
Plasmids

Before we discuss the third method of genetic transfer, **conjugation,** we must discuss another kind of genetic element called the **plasmid.** Plasmids are genetic elements that replicate independently of the host chromosome (∞ Section 6.3).

Unlike viruses, plasmids do not have an extracellular form and exist inside cells simply as nucleic acid.

However, distinguishing between viruses and plasmids can sometimes present difficulties. As we have discussed, the prophage form of some temperate viruses, such as bacteriophage P1, replicates independently of the host chromosome in a fashion analogous to plasmid replication (∞ Section 8.12).

We noted earlier (∞ Section 6.3) that it is possible to differentiate between a plasmid and a host chromosome in prokaryotes because plasmids do not carry genes required by the host under all conditions. This can be difficult to prove and therefore it may sometimes be difficult to distinguish between chromosomes and very large plasmids in prokaryotes.

In spite of these few difficulties, literally thousands of different types of plasmids are known. Indeed, over 300 different naturally occurring plasmids have been isolated from strains of *Escherichia coli* alone. In this section we shall discuss the properties of a few of them.

Physical nature of plasmids

Almost all of the known plasmids are double-stranded DNA. Most plasmids are circular, but many linear plasmids are also known. Naturally occurring plasmids vary in size from approximately 1 to more than 1000 kilobase pairs. The typical plasmid is a circular double-stranded DNA molecule less than $\frac{1}{20}$ the size of the chromosome (Figure 9.18). Most of the plasmid DNA isolated from cells is in the supercoiled configuration, which is the most compact form within the cell (∞ Figure 6.7). A single break (called a *nick*) in one of the two strands causes the supercoil to convert to an open circular form, and when breaks occur in both strands at the same place, a linear duplex structure is formed.

Isolation of plasmid DNA can generally be readily accomplished by making use of certain physical properties of supercoiled DNA molecules, which can be separated from other types of DNA using an ultracentrifuge. Although chromosomes are also supercoiled inside the cell, isolation of chromosomal DNA almost always leads to breakage of the strands. Plasmid DNA molecules of different sizes can also be readily separated by electrophoresis on agarose gels (∞ Working with Nucleic Acids: The Tools, Chapter 6). Plasmid DNA can be observed under the electron microscope (Figure 9.18).

Replication of plasmids

Most plasmids in gram-negative Bacteria replicate in a manner similar to that already described for the chromosome (∞ Section 6.5). This involves initiation of replication at an origin and bidirectional replication around the circle, giving a *theta* intermediate. However, some plasmids have *unidirectional* replication. Because of the small size of plasmid DNA relative to the chromosome, the whole replication process occurs very quickly, perhaps in $\frac{1}{10}$ or less of the total time of the cell

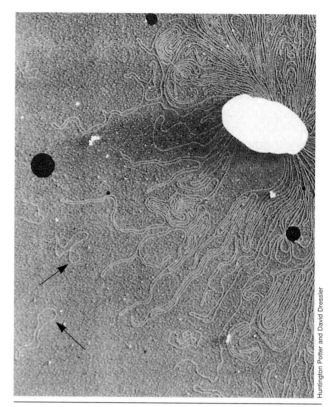

Figure 9.18 The bacterial chromosome and bacterial plasmids, as shown in the electron microscope. The plasmids (arrows) are the circular structures, much smaller than the main chromosomal DNA. The cell (large, white structure) was broken gently so the DNA would remain intact.

division cycle. The enzymes involved in plasmid replication are normal cell enzymes, so the genetic elements within the plasmid itself that control its replication are concerned primarily with control of the timing of the initiation process and with apportionment of the replicated plasmids between daughter cells.

Most plasmids of gram-positive Bacteria replicate by a rolling circle mechanism similar to that used by the phage ϕX174 (∞ Section 8.9 and Figure 8.17). This mechanism gives rise to a single-stranded intermediate, and thus these plasmids are sometimes referred to as *single-stranded DNA plasmids*. Most of the linear plasmids now known replicate using a mechanism involving a protein bound to the 5'-end of each strand that is used in priming DNA synthesis (∞ Figures 6.21 and 8.45). Different plasmids are present in cells in a particular number of plasmid molecules per cell; this is called the *copy number*. Some plasmids are present in the cell in only 1–3 copies, whereas others may be present in as many as 100 copies. Copy number is controlled by genes on the plasmid and by interactions between the host and the plasmid.

Some individual bacterial cells may also contain several different types of plasmids, in some cases more than 10. The ability of two different plasmids to both replicate in the same cell is also controlled by plasmid genes (called *inc*) involved in controlling DNA replication. When a plasmid is transferred into a cell that already carries another plasmid, a common observation is that the second plasmid may not be maintained and is lost during subsequent cell replication. The two plasmids are said to be **incompatible.** A number of incompatibility (Inc) groups have been recognized, the plasmids of one incompatibility group excluding each other but being able to coexist with plasmids from other groups. Plasmids of one incompatibility group are *related* to one another. Therefore, although a bacterial cell may contain different kinds of plasmids, they are not closely related because they must be compatible.

Some plasmids also have the ability to become integrated into the chromosome, and under such conditions their replication comes under control of the chromosome. This situation is remarkably like that found for several viruses whose genomes can become incorporated into the host genome (for example, ∞ Sections 8.12–8.14 and 8.22). Plasmids having the ability to integrate into host chromosomes are called *episomes*. Plasmids can sometimes be eliminated from host cells by various treatments. This process, termed **curing,** apparently results from inhibition of plasmid replication without parallel inhibition of chromosome replication, and as a result of cell division the plasmid is diluted out. Curing may occur spontaneously, but it is greatly increased by use of acridine dyes, which become inserted into DNA, or other treatments that seem to interfere more with plasmid replication than with chromosome replication. Electroporation may also be used to cure a cell of plasmids (see Section 9.6).

We can exemplify many of these characteristics by a very well characterized plasmid called the *F plasmid.* The F plasmid is a circular DNA molecule of 100 kilobase pairs. Cells containing it can be easily cured with acridine orange. Figure 9.19 shows a genetic map of the F plasmid. One region of the plasmid contains genes involved in regulating DNA replication (such as incompatibility, *inc,* and origin of replication, *oriS*). It also contains a number of transposable elements (see Section 9.10) involved in its ability to function as an episome. Last, it has a large region of DNA, the *tra* region, containing genes that permit it to be *tra*nsferred from one cell to another.

Cell-to-cell transfer of plasmids

Because one of the defining characteristics of a plasmid is the lack of a distinct extracellular form, one can imagine that plasmids are confined almost exclusively to transferring only to daughter cells during cell division. Additionally, some prokaryotic cells can take up free DNA from the environment (see Section 9.6), so it is possible that lysis of the host, however it may happen, brings the plasmid in contact with a new host. However, this process occurs naturally in only a few

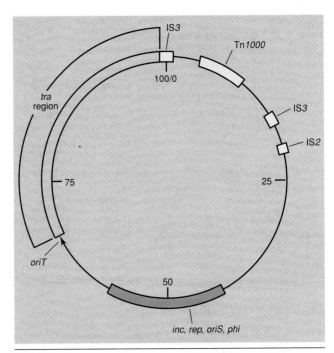

Figure 9.19 Genetic map of the F (fertility) plasmid of *Escherichia coli.* The numbers on the interior show the size of the plasmid in kilobase pairs. The locations of key F genes are shown: *tra,* transfer functions; *oriT,* origin of transfer; *oriS,* origin of replication; *inc,* incompatability group; *rep,* replication functions. The regions shown in yellow on F are transposable elements where integration into identical elements on the bacterial chromosome can occur and lead to the formation of different Hfr strains (see Section 9.9).

bacterial species and is unlikely to account for much cell-to-cell plasmid transfer. The main mechanism of cell-to-cell transfer is **conjugation,** and it is a function encoded by some plasmids themselves. Conjugation is a replicative process, and both cells end up with copies of the plasmid (Figure 9.20).

Plasmids that govern their own transfer by cell-to-cell contact are called **conjugative,** but not all plasmids are conjugative. Transmissability by conjugation is controlled by a set of genes within the plasmid called the *tra region.* The presence of a *tra* region in a plasmid can have another important consequence if the plasmid becomes integrated into the chromosome. In that case, the plasmid can *mobilize* the transfer of chromosomal DNA from one cell to another. Strains of bacteria that transfer large amounts of chromosomal DNA during conjugation are called *Hfr* (high frequency of recombination), and the use of conjugation to transfer host genes is discussed in the next section.

Actually conjugation was discovered because the F plasmid can mobilize chromosomal genes. Therefore, we shall discuss the mechanism of DNA transfer in the next section.

Some conjugative plasmids from *Pseudomonas* are transferrable to a wide variety of other gram-negative

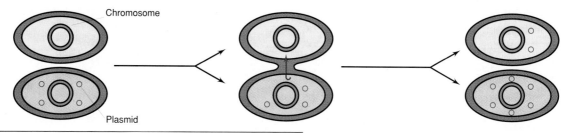

Figure 9.20 Plasmid transfer from cell to cell during conjugation.

Bacteria. Some conjugative plasmids can transfer genetic information between distantly related organisms. Conjugative plasmids have been shown to transfer between gram-negative and gram-positive Bacteria, between Bacteria and plant cells, and between Bacteria and fungi. Even if the plasmid cannot replicate in the new host, the transfer of the DNA itself could have important evolutionary consequences, as well as being involved in pathogenic processes, if it can recombine into the genome of the new host.

Types of plasmids and their biological significance

Clearly all plasmids must carry genes that ensure their own replication. However, for many plasmids we know very little about what other genes they carry. These are called *cryptic plasmids* and were discovered by physical means (such as examining a cell extract using gel electrophoresis.) As we have seen, some plasmids also carry genes necessary for conjugation and they can sometimes be detected biologically, either by the transfer functions themselves or by sensi-

tivities to certain viruses (⚬⚬ Section 8.8). Although plasmids do not carry genes that are essential to the host under all conditions, the presence of plasmids in a cell can have a profound influence on the cell's phenotype. In some cases, plasmids encode properties that we think of as fundamental to the bacterium in question, for example, the ability of *Rhizobium* to interact with plants (⚬⚬ Section 14.23). Some pseudomonad plasmids have been shown to transfer the genetic information for biochemical pathways for the degradation of unusual organic compounds such as camphor, octane, and naphthalene.

However, plasmids can carry a very wide variety of genes. Indeed, the limitation is only that the genes they carry do not interfere either with their own replication or with the survival of the host. Because plasmids can be large and may carry many different genes, it is not always a simple matter to classify a plasmid into a simple phenotypic category. As we shall see, a single plasmid may confer many different phenotypes on its host cell. A summary of plasmids that have been identified and organisms for which evidence of plasmids exists is given in Table 9.3. It is likely that virtu-

TABLE 9.3 Types of plasmids[a]	
Type	**Organisms**
Conjugative plasmids	F plasmid, *Escherichia coli*; pfdm, K, *Pseudomonas*; P, *Vibrio cholerae*; SCP, *Streptomyces*
R plasmids	
Wide variety of antibiotics (⚬⚬ Section 11.13)	Enteric bacteria, *Staphylococcus*
Resistance to mercury, cadmium, nickel, cobalt, zinc, arsenic (⚬⚬ Section 14.18)	*Pseudomonas*
Bacteriocin and antibiotic production	Enteric bacteria; *Clostridium*; *Streptomyces*
Physiological functions	
Lactose, sucrose, urea utilization, nitrogen fixation	Enteric bacteria
Degradation of octane, camphor, naphthalene, salicylate	*Pseudomonas*
Pigment production	*Erwinia*, *Staphylococcus*
Nodulation and symbiotic nitrogen fixation (⚬⚬ Section 14.23)	*Rhizobium*
Virulence plasmids	
Enterotoxin, K antigen, endotoxin (⚬⚬ Sections 19.8–19.11)	*Escherichia coli*
Tumorigenic plasmid (⚬⚬ Section 14.22)	*Agrobacterium tumefaciens*
Adherence to teeth (dextran) (⚬⚬ Section 19.3)	*Streptococcus mutans*
Coagulase, hemolysin, fibrinolysin, enterotoxin (⚬⚬ Sections 19.7 and 19.9)	*Staphylococcus aureus*

[a]Plasmids have been found in most bacterial genera.

ally all prokaryotic groups possess plasmids. In the remainder of this section we will discuss a few of the many phenotypes plasmids may confer on cells.

Resistance plasmids

Among the most widespread and well-studied groups of plasmids are the *resistance plasmids* (*R plasmids*), which confer resistance to antibiotics and various other inhibitors of growth. R plasmids were first discovered in Japan in strains of enteric bacteria that had acquired resistance to a number of antibiotics (multiple resistance) and have since been found in other parts of the world. The emergence of bacteria resistant to several antibiotics is of considerable medical significance and was correlated with the increasing use of antibiotics for the treatment of infectious diseases. Soon after these resistant strains were isolated it was shown that they could transfer resistance to sensitive strains via cell-to-cell contact. This is probably one of the reasons for the rapid rise of multiply resistant strains because it is unlikely that resistance to a number of antibiotics would develop simultaneously by mutation and selection. The infectious nature of the conjugative R plasmids permits rapid spread of the characteristic through populations.

A variety of antibiotic resistance genes can be carried by an R plasmid. In general, these genes encode proteins that either inactivate the antibiotic or affect its uptake into the cell. Plasmid R100, for example, is an 89.3-kilobase-pair plasmid (Figure 9.21) that carries resistance genes for sulfonamides, streptomycin and spectinomycin, fusidic acid, chloramphenicol, and tetracycline. R100 also carries several genes conferring resistance to mercury (∞ Section 14.18). R100 can transfer itself between enteric bacteria of the genera *Escherichia*, *Klebsiella*, *Proteus*, *Salmonella*, and *Shigella* but does not transfer to the nonenteric bacterium *Pseudomonas*. R plasmids with genes for resistance to kanamycin, penicillin, and neomycin are also known. Many drug-resistant elements on R plasmids, such as those on R100, are transposable elements and can be used in transposon mutagenesis (see Section 9.10). Many R plasmids and F plasmids have related *tra* regions, and recombination can occur between F and R plasmids.

Toxins and other virulence characteristics

We will discuss in Chapter 19 the physiological and genetic characteristics of microorganisms that enable them to colonize hosts and set up infections, which can lead to harm. In the present context, we merely note the two major characteristics involved in virulence: (1) the ability of microorganisms to attach to and colonize specific sites in the host; and (2) the formation of substances (toxins, enzymes, and other molecules) that cause damage to the host. It has now

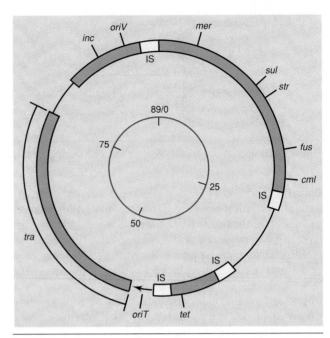

Figure 9.21 Genetic map of the resistance plasmid R100. The inner circle shows the size of the plasmid in kilobase pairs. The outer circle shows the location of major antibiotic resistance genes and other key functions. *inc*, Incompatibility genes; *oriV*, origin of replication site; *oriT*, origin of conjugative transfer; *mer*, mercuric ion resistance; *sul*, sulfonamide resistance; *tet*, tetracycline resistance; *tra*; transfer functions. The locations of insertion sequences (IS) are also shown.

been well established that in several pathogenic bacteria each of these virulence characteristics is carried on plasmids. For example, enteropathogenic strains of *Escherichia coli* are characterized by an ability to colonize the small intestine and to produce a toxin that causes symptoms of diarrhea. Colonization requires the presence of a cell surface protein called the colonization factor antigen (CFA), encoded by a plasmid, which confers on the cells the ability to attach to epithelial cells of the intestine. At least two toxins in enteropathogenic *E. coli* are known to be coded for by a plasmid: the *hemolysin*, which lyses red blood cells, and the *enterotoxin*, which induces extensive secretion of water and salts into the bowel. It is the enterotoxin that is responsible for the induction of diarrhea, as will be discussed in Chapter 19. Since other virulence factors of *E. coli* are carried by chromosomal genes, it is not clear why these toxin genes are carried by plasmids, but the existence of such plasmids raises the question of how widespread plasmid-related virulence is in microorganisms.

In *Staphylococcus aureus*, some virulence-conferring properties are known to be plasmid-linked; *S. aureus* is noteworthy for the variety of enzymes and other extracellular proteins it produces that are involved in its virulence, and the production of *coagulase, hemolysin, fibrinolysin,* and *enterotoxin* is thought to be plasmid-linked.

In addition, the yellow pigment in *S. aureus,* which is probably involved in its ability to resist the destructive action of singlet oxygen in the phagocyte (∞ Section 20.3), is plasmid-linked. *S. aureus* is also a notorious hospital-borne (nosocomial) pathogen, and multiple antibiotic resistance, encoded by plasmids, is common in this species (∞ Section 22.7). Virulence factors from a variety of bacteria are known to be encoded on plasmids, and others are known to be encoded by other types of *mobile genetic elements:* transposons and bacteriophages.

Bacteriocins

Many bacteria produce agents that inhibit or kill closely related species or even different strains of the same species; these agents are called **bacteriocins** to distinguish them from the antibiotics, which have a wider spectrum of activity. Bacteriocins are ribosomally synthesized peptides (although several require extensive posttranslational modification for activity). The structural gene for the bacteriocin and the genes encoding proteins involved with processing and transporting the bacteriocin (and for conferring immunity to its action) are often carried by a plasmid or a transposon. Bacteriocins are named in accordance with the species of organism that produces them. Thus, in *Escherichia coli* we have *colicins,* coded by Col plasmids, *Bacillus subtilis* produces *subtilisin,* and so on.

The Col plasmids of *Escherichia coli* encode various colicins. Colicins released from a producing cell bind to specific receptors on the surface of susceptible cells. The receptors for colicins are generally entities whose normal function is to transport some substance, frequently a growth factor or micronutrient, through the outer membrane (the lipopolysaccharide layer) of the cell. Colicins kill cells by disrupting some critical cell function. Many colicins form channels in the cell membrane that allow potassium ions and protons to leak out, leading to a loss of the cell's energy-forming ability. However, colicin E2 is a DNA endonuclease that can cleave cellular DNA, and colicin E3 is a nuclease that cuts at a specific site in 16S rRNA and inactivates ribosomes. Col plasmids can be either conjugative or nonconjugative.

The bacteriocins or bacteriocin-like agents of gram-positive bacteria are quite different from the colicins but are also often encoded by plasmids, and some even have commercial value. For instance, the lactic acid bacteria produce the bacteriocin Nisin A, which strongly inhibits the growth of a wide range of gram-positive bacteria and is used as a preservative in the food industry.

Engineered plasmids

The techniques of genetic engineering, discussed in the next chapter, have made possible the construction in the laboratory of a limitless number of new, artificial plasmids. Incorporation into artificial plasmids of genes from a wide variety of sources has made possible the transfer of genetic material across virtually any species barrier. It is even possible to synthesize completely new genes and introduce them into plasmids. Such artificial plasmids are useful tools in understanding plasmid structure and function, as well as for the more practical aims of genetic engineering discussed in the next chapter.

We now turn our attention to the details of conjugation and show how certain plasmids can act to mobilize the bacterial chromosome, allowing transfer of chromosomal genes from donor to recipient.

CONCEPT CHECK   9.8

The genetic information that plasmids carry is not essential for cell function under all conditions but may confer selective growth advantage under certain conditions. Examples include antibiotic resistance, enzymes for degradation of unusual organic compounds, and special metabolic pathways. Closely related plasmids cannot both replicate in the same cell. Some plasmids can transfer from cell to cell by a mechanism called conjugation.

✔ *Are two incompatible plasmids likely to be very similar or very different?*

✔ *Conjugative plasmids tend to be large. Why?*

9.9
Conjugation and Chromosome Mobilization

Bacterial conjugation (mating) is a process of genetic transfer that involves cell-to-cell contact. As we discussed previously (see Section 9.8), conjugation is a plasmid-encoded mechanism. A conjugative plasmid uses this mechanism to transfer a copy of itself to a new host. However, sometimes other genetic elements can be *mobilized* during conjugation. These other genetic elements can be other plasmids, or the host chromosome. Indeed, conjugation was discovered because the F plasmid of *Escherichia coli* (see Figure 9.19) can mobilize the host chromosome. Mechanisms of conjugative transfer may differ depending on the plasmid involved, but most plasmids in gram-negative Bacteria seem to employ a mechanism similar to that used by the F plasmid.

Conjugation involves a *donor* cell, which contains a particular type of conjugative plasmid, and a *recipient* cell, which does not. Because conjugation was discovered by performing genetic crosses, the donors were called *males* and the recipients were called *females.* The genes that control conjugation are contained in the *tra* region of the plasmid (see Section 9.8). Many genes in the *tra* region have to do with the synthesis of a surface structure, the **sex pilus** (Figure 9.22). Only donor cells have these pili, explaining why RNA bacteriophages that attach to them are "*male-specific*" (∞ Section 8.8). Different conjugative plasmids may have slightly different *tra* regions, and in some cases the pili are also different and can be distinguished immunologically. The F plasmid and its relatives encode *F pili*.

Pili allow specific pairing to take place between the donor cell and the recipient cell. The pili make specific contact with a receptor on the recipient and then retract, pulling the two cells together so a conjugation bridge can form, through or on which DNA passes from one cell to another. Although the details are not completely understood, it is thought that there is actual membrane fusion between donor and recipient cells and that this in some way triggers DNA transfer.

Mechanism of DNA transfer during conjugation

DNA synthesis is necessary for DNA transfer to occur, and the evidence suggests that one of the DNA strands is derived from the donor cell and the other is newly synthesized in the recipient during the transfer process. A mechanism of DNA synthesis in certain bacteriophages, called **rolling circle replication,** was presented in Figures 8.17 and 8.28. This model best explains DNA transfer during conjugation, and a possible mechanism for this process is outlined in Figure 9.23. The whole series of events is probably triggered by cell-to-cell contact, at which time one strand of the plasmid DNA circle is nicked and one parental strand is transferred. As this transfer occurs, DNA synthesis by the rolling circle mechanism replaces the transferred strand in the donor. A complementary DNA strand is also made in the recipient. The model accounts for the fact that if the DNA of the donor is labeled, some labeled DNA is transferred to the recipient but only a *single* labeled strand is transferred. Therefore, at the end of the process, both donor and recipient possess completely formed plasmids.

The high efficiency of the plasmid DNA transfer process is shown by the fact that under appropriate conditions virtually every recipient cell that pairs acquires a plasmid. When the plasmid genes can be expressed in the recipient, the recipient itself becomes a donor and can transfer the plasmid to other recipients. In this fashion, conjugative plasmids can spread rapidly between populations, behaving like infectious agents. The infectious nature of this phenomenon is of major ecological significance because a few plasmid-positive cells introduced into an appropriate population of recipients can, if they contain genes that confer a selective advantage, convert the whole recipient population into a plasmid-bearing population in a short period of time. The widespread occurrence of drug resistance carried by conjugative plasmids (see Section 9.8) has led to some serious problems for the chemotherapy of infectious disease (∞ Section 11.13).

Chromosome mobilization

The F plasmid of *Escherichia coli* (see Section 9.8) is not only conjugative but also has the special property of being able to mobilize the chromosome so it can be transferred during cell-to-cell contact. The F plasmid is an episome, a plasmid that can integrate into the host chromosome (see Section 9.8). When the F plasmid is integrated into the chromosome, conjugation can lead to transfer of large blocks of chromosomal genes and genetic recombination between donor and recipient can then be very extensive.

Cells possessing an unintegrated F plasmid are called **F⁺**, and strains that can act as recipients for F⁺ (or Hfr, see later) are called **F⁻**. F⁻ cells lack the F plasmid; in general, cells that contain a plasmid are very poor recipients for the same or closely related plas-

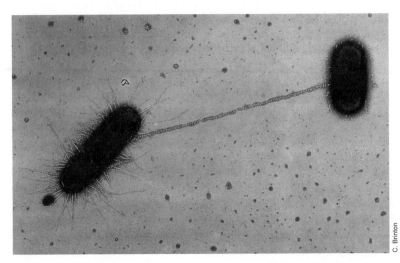

Figure 9.22 Direct contact between two conjugating bacteria is first made via a pilus. The cells are then drawn together for the actual transfer of DNA. Note the F-specific bacteriophages on the pilus (∞ Section 8.8).

C. Brinton

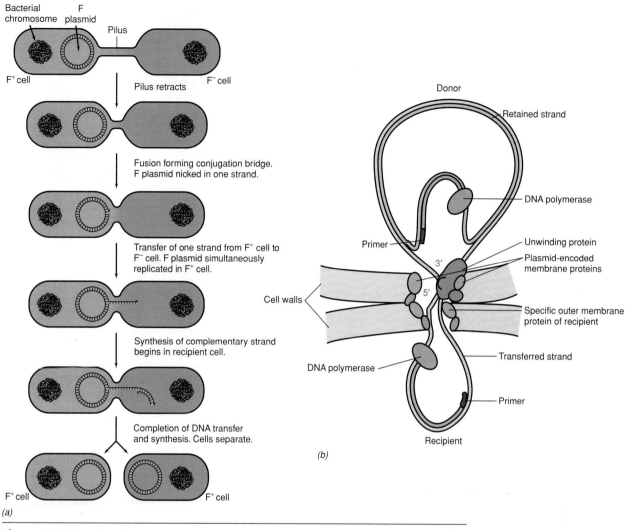

Figure 9.23 Transfer of plasmid DNA by conjugation. (a) In this example, the F plasmid of an F⁺ cell is being transferred to an F⁻ recipient cell. Note the mechanism of rolling circle replication (∞ Figure 8.28). (b) Details of the replication and transfer process.

mids. Bacterial strains that possess a chromosome-integrated F plasmid and show such extensive genetic recombination are called **Hfr** (for high frequency of recombination). Conjugation leads to transfer of the host chromosome in Hfr strains because the plasmid is part of the chromosome. (Plasmid integration is a very simple mechanism of mobilizing other genetic elements.)

We thus see that the presence of the F plasmid results in three distinct alterations in the properties of a cell: (1) ability to synthesize the F pilus, (2) mobilization of DNA for transfer to another cell, and (3) alteration of surface receptors so the cell is no longer able to behave as a recipient in conjugation.

Selection for recombinants formed as the result of mating of an Hfr with an F⁻ strain is accomplished by plating the mating mixture on culture media that allow growth of only the recombinant cells with the desired genotype. For instance, in the experiment shown in Figure 9.24, an Hfr donor that is sensitive to

streptomycin (Str^s) and contains wild-type genes encoding enzymes needed for synthesis of the amino acids threonine and leucine (Thr⁺ and Leu⁺) and for utilization of the energy source lactose (Lac⁺) is mated with a recipient cell that is mutant for these genes but is resistant to streptomycin (Str^r). The selective medium is a minimal medium containing streptomycin so that only recombinant cells can grow. The composition of each selective medium is varied depending on which genotypic characteristics are desired in the recombinant, as shown in Figure 9.24. The frequency of the process is measured by counting the colonies grown on the selective medium.

Formation and behavior of Hfr strains

As noted previously, an F plasmid can become integrated into the chromosome and mobilize it for conjugation. There are several specific sites on the chromo-

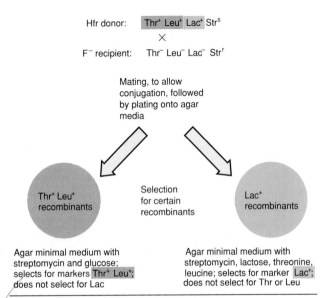

Figure 9.24 Laboratory procedure for the detection of genetic conjugation. Thr, Threonine; Leu, leucine; Lac, lactose; Str, streptomycin. Note that each medium selects for specific classes of recombinants. The controls for the experiment are to plate samples of the donor and the recipient before they are mixed. Neither should be able to grow on the selective media used.

some at which F plasmids can be integrated, and these sites, called IS (for *insertion sequences*) represent regions of homology between chromosome and F plasmid DNA (see Section 9.9 for a discussion of insertion sequences). As seen in Figure 9.25, integration of an F plasmid involves insertion in the chromosome at the specific site. In the particular Hfr shown, the integration site is between the chromosomal genes *pro* and *lac*. The site on the F plasmid (*oriT*, origin of transfer) at which transfer initiates during conjugation is indicated by an arrow. At the time of specific cell pairing, the chromosome is opened (nicked) at the origin and the host genes are inserted into the recipient beginning with the gene downstream from the origin (Figure 9.26). Essentially, the mechanism of transfer is the same as that for just the F plasmid, but here the plasmid is part of the chromosome.

Hfr strains thus arise as a result of integration of the F plasmid into the chromosome. Since a number of distinct insertion sites are present, a number of distinct Hfr strains are possible. A given Hfr strain always donates genes in the same order, beginning with the same position, but Hfr strains of independent origin transfer genes in different sequences. During cell division, the DNA of the Hfr replicates normally, but at the time of pairing with an F⁻ cell, a DNA strand from the Hfr is transferred to the F⁻ cell and replication occurs by the rolling circle process. After transfer, the Hfr strain still remains Hfr because it has retained a copy of the transferred genetic material.

Usually, because of breakage of the DNA strand during transfer, only a *part* of the donor chromosome

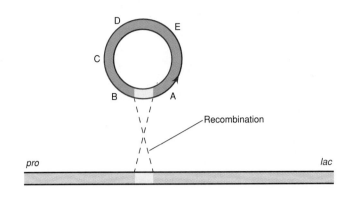

Figure 9.25 Integration of an F plasmid into the chromosome with the formation of an Hfr. The insertion of the F plasmid occurs at a variety of specific sites where IS elements are located, the one here being between chromosomal genes *pro* and *lac*. The letters on the F plasmid represent arbitrary genes. The arrow indicates the origin of transfer, with the arrow as the leading end. The site in the F plasmid at which pairing with the chromosome occurs is between A and B. Thus, in this Hfr *pro* would be the first chromosomal gene to be transferred and *lac* would be among the last.

is transferred. Since only part of the chromosome is transferred, it cannot replicate in the recipient cell. Therefore, donor genes normally cannot be detected unless recombination between the incoming fragment and the recipient chromosome takes place.

Although Hfr strains transmit chromosomal genes at high frequency, they usually do not convert F⁻ cells to F⁺ or Hfr because the entire F plasmid is only rarely transferred. On the other hand, F⁺ cells efficiently convert F⁻ to F⁺ because the entire F plasmid is transferred.

At some insertion sites, the F plasmid is integrated with the origin in one direction, whereas at other sites the origin is in the opposite direction. The direction in which the F plasmid is inserted determines which of the chromosomal genes will be transferred into the recipient first. The manner in which a variety of Hfr strains can arise is illustrated in Figure 9.27. By use of various Hfr strains, it has been possible in *Escherichia coli* to determine the arrangement and orientation of a large number of chromosomal genes, as will be described in Section 9.10.

Transfer of chromosomal genes to the F plasmid

Occasionally integrated F plasmids may be excised from the chromosome, and the possibility exists for the incorporation at that time of *chromosomal* genes into

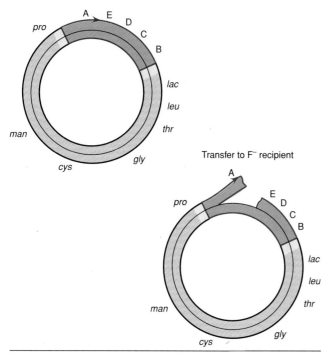

Figure 9.26 Breakage of the Hfr chromosome at the origin and transfer of DNA to the recipient. Replication occurs during transfer, as illustrated in Figure 9.23.

agitated at various times after mixing and the genetic recombinants scored, it is found that the longer the time between pairing and agitation, the greater the number of genes of the Hfr that will appear in the F⁻ recombinant. In addition, gene transfer always occurs in a specific order in a specific Hfr. As shown in Figure 9.28, genes present closer to the origin enter the F⁻ first and are always present in a higher percentage of the recombinants than genes that enter late. In addition to showing that gene transfer from donor to recipient is a sequential process, experiments of this kind provide a method of determining the order of the genes on the bacterial DNA (genetic mapping). The arrangement of gene loci on the chromosome is called a **genetic map** (see Section 9.11).

Genetic recombination between Hfr genes and F⁻ genes in the F⁻ cell requires the presence of enzymes in the recipient cell. This has been shown by the isolation of mutants of F⁻ strains that are unable to form recombinants when mated with Hfr. These mutants are Rec⁻ (recombination minus) and are deficient in the RecA protein because of a mutation in the *recA* gene (see Section 9.5). It is important to remember that recombination is not the same as DNA transfer. Both

the liberated F plasmid. Such F plasmids containing chromosomal genes are called *F'* (F prime) *plasmids.* These F' plasmids differ from normal F plasmids in that they contain identifiable chromosomal genes, and they transfer these genes at high frequency to recipients. F'-mediated transfer resembles specialized transduction in that only a restricted group of chromosomal genes can be transferred. It is often with F' plasmids that complementation tests (see Section 9.5) are done in *Escherichia coli.*

Oriented transfer of Hfr and the phenomenon of interrupted mating

The detection of bacterial conjugation is usually accomplished by use of parental strains having properties that can be selected against on agar plates so that only recombinants can grow (see Figure 9.24). The recipient is usually resistant to an antibiotic and is auxotrophic for one or more nutritional characters. The donor is antibiotic-sensitive but prototrophic for the nutritional characters. With proper agar media, only the recombinants can grow and the large background of nonrecombinants is eliminated.

The oriented transfer of chromosomal genes from Hfr to F⁻ is most clearly shown by a procedure called **interrupted mating.** The mating pairs are rather weakly joined and can be separated by agitation in a mixer or blender. If mixtures of Hfr and F⁻ cells are

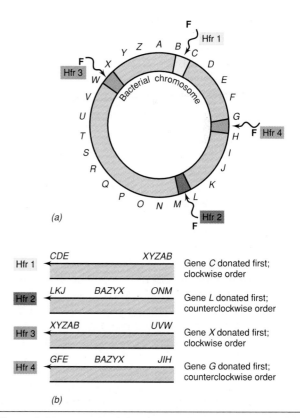

Figure 9.27 Manner of formation of different Hfr strains, which donate genes in different orders and from different origins. The bacterial chromosome is a circle (a) that can open at various insertion sequences, at which F plasmids become inserted. The gene orders are shown in part (b).

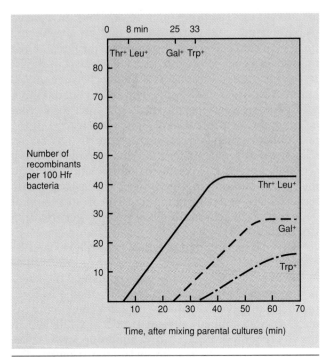

Figure 9.28 Rate of formation of recombinants containing different genes after mixing Hfr and F⁻ bacteria by the process known as interrupted mating. The location of the genes along the Hfr chromosome is shown at the upper left. Note that the genes closest to the origin (0 min) are the first ones detected in the recombinants. The experiment is done by mixing Hfr and F⁻ cells under conditions in which essentially all Hfr cells find mates. The F⁻ recipient was streptomycin-resistant but auxotrophic for the markers being scored. The Hfr donor was streptomycin-sensitive. At various times, samples of the mixture are shaken violently to separate the mating pairs and plated on a selective medium in which only the recombinants can grow and form colonies.

F plasmids and F′ plasmids are transferred normally to a Rec⁻ cell even though recombination does not take place after transfer.

Other conjugation systems

Although we have discussed conjugation almost exclusively as it occurs in *Escherichia coli*, conjugative plasmids have been found in many other gram-negative Bacteria. Indeed, conjugative plasmids of the incompatibility group (see Section 9.8) IncP can be maintained in practically all gram-negative species, and DNA transfer between species and genera can occur. Conjugative plasmids are also known in gram-positive Bacteria (for example, in *Bacteroides*, *Enterococcus*, and *Staphylococcus*). As a rule, the mechanism of conjugation is very similar in the other gram-negative Bacteria to what we have discussed for the F plasmid, whereas conjugation involving gram-positive Bacteria can be quite different.

Some conjugative plasmids mobilize other genetic elements, as we have discussed for the F plasmid, but this is not always the case. There are also elements called *conjugative transposons* that can transfer themselves from the chromosome of a donor host to that of a recipient and can also mobilize other genetic elements. Conjugative transposons have mostly been found in gram-positive Bacteria. They have a very wide host range and can be involved in gene transfer between different bacteria of different genera. The mechanism of conjugation used by these elements is not completely known, but it does seem to involve circular, plasmidlike intermediates. Gene transfer involving cell-to-cell contact has also been described in *Haloferax*, a member of the Archaea, but the mechanism of transfer seems to be much different from that in any other type of prokaryotic conjugation.

CONCEPT CHECK   **9.9**

Conjugation is a mechanism of DNA transfer in prokaryotes that requires cell-to-cell contact. Conjugation is controlled by genes carried by certain plasmids (such as the F plasmid) and typically involves transfer of the plasmid from a donor cell to a recipient cell. However, other genetic elements, including the donor cell chromosome, can sometimes be mobilized and also transferred. Transfer of the host chromosome is rarely complete but can be used to map the order of the genes on the chromosome.

✔ *How are donor and recipient cells brought into contact with each other?*

✔ *In conjugation involving the F plasmid of* Escherichia coli, *how is the host chromosome mobilized?*

9.10

Transposons and Insertion Sequences

We will see in Section 9.11 that the order of genes on a bacterial chromosome can be determined by the methods of gene transfer we have considered, just as genes in eukaryotic chromosomes can be mapped by mating experiments. However, the exact arrangement of the genes along a chromosome is not necessarily permanently fixed; some genes are capable of moving under certain conditions. The process by which a gene moves from one place to another in the genome is called **transposition** and is an important process in evolution and in genetic analysis.

We should emphasize that transposition typically is a *rare* event, occurring at frequencies of 10^{-5}–10^{-7} per generation. Thus, the genes of living organisms are relatively stable.

In addition, not all genes are capable of transposition. Rather, transposition of genes is linked to the presence of special genetic elements called *transposable elements.*

Transposition was originally discovered in corn (maize) and then later in Bacteria owing to the extremely sensitive types of genetic analysis available in these organisms. It has been shown by using DNA hybridization and sequencing techniques (∞ Working with Nucleic Acids: The Tools, Chapter 6) that DNA sequences with the properties of transposable elements are widespread in nature. However, in organisms with poorly characterized genetic systems, it can be difficult to detect transposition (because it is a rare event) and actually prove that a sequence is a transposable element.

Transposable elements

As we discussed earlier (∞ Section 6.3), there are three types of transposable elements in Bacteria: *insertion sequences, transposons,* and some special viruses (such as Mu) (∞ Section 8.13). A brief summary of different types of transposable elements found in both prokaryotes and eukaryotes is given in Table 9.4. In this section we shall confine our discussion primarily to insertion sequences and transposons found in Bacteria. Both these elements have two important features in common: they both carry genes encoding a **transposase,** the enzyme necessary for transposition, and both have short *inverted terminal repeats* at the ends of their DNA (remember that these "ends" are continuous with whatever DNA the element is inserted into). These repeats range in length from about 40 base pairs in simple IS elements to greater than 1000 base pairs in some transposons; each IS has a specific number of base pairs in its terminal repeats. Such inverted terminal repeats are involved in the transposition process (see later). Figure 9.29 shows a genetic map of an insertion element named IS2 and of a transposon named Tn5.

In prokaryotes, insertion sequences are the simplest type and carry no genetic information other than that required for them to move to new locations. Insertion sequences are short segments of DNA, about 1000 nucleotides long, that can become integrated at specific

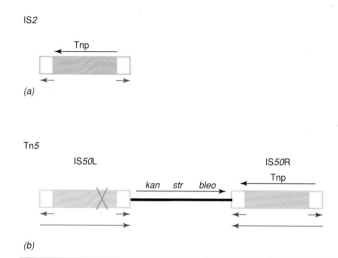

Figure 9.29 Maps of the transposable elements IS2 and Tn5. The red arrows underneath each map indicate the inverted repeats. The arrows above the maps show the direction of transcription of any genes on the elements. *Tnp* is the gene encoding the transposase. The transpose genes of these two elements are not closely related. (a) IS2 is an insertion sequence of 1327 base pairs with inverted repeats of 41 base pairs at its ends. (b) Tn5 is a composite transposon of 5.7 kilobase pairs with the insertion sequences IS50L and IS50R at its left and right ends, respectively. IS50L is not capable of independent transposition because there is a *nonsense mutation* (see Section 9.2) marked by a blue cross in its transposase gene. Otherwise, the two IS50 elements are very nearly identical. Note that these two IS50 elements are inverted with respect to each other. The genes *kan, str,* and *bleo,* confer resistance to the antibiotics kanamycin (and neomycin), streptomycin, and bleomycin. Interestingly, streptomycin resistance is not expressed in *Escherichia coli.*

sites on the genome. Insertion sequences, abbreviated IS are found in both chromosomal and plasmid DNA, as well as in certain bacteriophages. Several distinct IS elements have been characterized, and each is designated by a number identifying its type: IS1, IS2, IS3, and so on. IS elements are scattered about the chromosome, and strains vary in the number and frequency of these elements. For instance, one strain of *Escherichia coli* has five copies of IS2 and five copies of IS3. The F plasmid also carries these insertion sequences (see Figure 9.19), and it is homologous recombination between identical sequences on the plasmid and the chromosome (*not* transposition) that allows the F plasmid to integrate into the bacterial chromosome and mobilize it (see Section 9.9). Some of the Archaea have large numbers of IS elements in their chromosomes.

Transposons are larger than insertion sequences and carry other genes, some of them conferring important properties on the organism carrying them. These often include drug resistance markers and other easily selectable genes. In addition, as we have mentioned (see Section 9.9), there are *conjugative transposons.* These transposons have genes allowing them not only

TABLE 9.4	Transposable elements
Prokaryote	**Eukaryote**
Insertion sequence: IS	Yeast: *sigma*
Transposon: Tn	Yeast: Ty
	Fruit fly: copia, P
	Maize: Ac
Virus: Mu	Retrovirus: Rous sarcoma, human immunodeficiency virus (HIV)

to move from one location on a bacterial genome to another but also to transfer themselves from one bacterium to another.

Some transposons are actually composite structures containing a gene or group of genes lying between two identical insertion sequences. The existence of such *composite transposons* indicates that new transposons probably continue to arise in cells that have insertion sequences.

The mechanism of transposition

As mentioned previously, the inverted repeats found at the ends of transposable elements are essential for transposition. The other essential component is an enzyme called *transposase,* which recognizes these repeats. This enzyme is usually encoded by the transposable element, although some very simple IS elements use an enzyme encoded by another genetic element. The transposase apparently recognizes, cuts, and eventually ligates the DNA during transposition (Figure 9.30).

When a transposable element becomes inserted into another DNA (the target DNA), a short sequence in the target DNA at the site of integration is duplicated. This target DNA sequence was not present in the transposon, but the transposable element has brought about a duplication of this DNA by the insertion process (Figure 9.30*a*). The duplication of the target sequence apparently arises because single-stranded breaks are generated by the transposase (Figure 9.30*b*). The transposon is then attached to the single-stranded ends that have been generated, and repair of the single-strand portions results in the duplication.

Certain transposable elements prefer certain sequences as target sites, but others, including the bacteriophage Mu, can insert themselves almost randomly (for a representation of Mu insertion, see Figure 8.31).

Two mechanisms of transposition are known, called *conservative* and *replicative.* In conservative transposition, such as can occur in the transposon Tn5, the transposable element is excised from one location in the chromosome and becomes reinserted at a second location. The copy number of a conservative transposon therefore remains at one. By contrast, replicative transposons, such as bacteriophage Mu (∞ Section 8.13), are duplicated, and a new copy is inserted at another location. Thus, after the transposition event is completed, *one* copy of the transposing element *remains* at the original site and *another copy* is found at the new site. During this whole transposition process, the source transposon *remains* at its original site; at no time does the source transposon become free in the cell.

Although many of the molecular details of transposition are uncertain, and different transposable elements appear to have different mechanisms, one model for replicative transposition is illustrated in Figure 9.31. As seen, single-strand cuts are made at the ends of the transposon (at the sites of the inverted repeats), and staggered single-strand cuts are made at the target site. The transposon is now joined to the target site via the single-stranded ends, leading to the formation of a composite structure called a *cointegrate.* Replication repair then fills in the single-strand gaps in the target site. This process results in the formation of *direct repeats* in the target site at the ends of the transposon (in addition to the inverted repeats of the transposon). The final event is *resolution* of the cointegrate structure, leading to release of the original transposon and the presence of a new copy of the transposon at the target site. Now that the transposon is present at the new target site, it can also serve as another source of transposition.

It should be emphasized that transposition is essentially a *recombination event,* but one that does not occur between homologous sequences or use the regular genetic recombination system of the cell. It involves the special protein *transposase* rather than the RecA protein that is involved in general recombination. Because this recombination involves a *specific* base sequence, it is called *site-specific recombination* (in contrast to homologous recombination discussed earlier in this chapter).

Mutagenesis with transposable elements

If the insertion site for a transposable element is *within* a gene, insertion of the transposon will result in loss of linear continuity of the gene, leading to mutation (Figure 9.32). Transposons thus provide a facile means of creating mutants throughout the chromosome. The most convenient element for **transposon mutagenesis** is one containing an antibiotic resistance gene. Clones containing the transposon can then be selected by the isolation of antibiotic-resistant colonies. If the antibiotic-resistant clones are selected on rich medium on which all auxotrophs can grow, they can be subsequently screened on minimal medium supplemented with various growth factors to determine if a growth factor is required.

Transposons are also useful for incorporating an auxotrophic gene marker into a wild-type organism. Normally, auxotrophic recombinants cannot be isolated by positive selection, but if the auxotrophic marker to be introduced contains a transposable element with an antibiotic resistance marker, then one can select for antibiotic-resistant clones, a positive selection procedure, and automatically obtain clones that have incorporated the auxotrophic marker.

Two transposons widely used for mutagenesis are Tn5 (see Figure 9.29), which confers neomycin and kanamycin resistance, and Tn10, which contains a marker for tetracycline resistance.

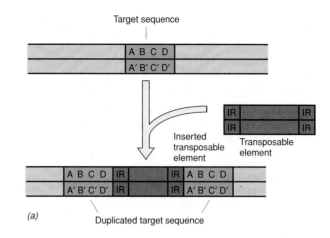

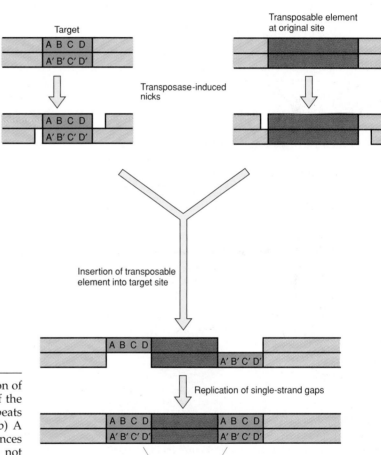

Figure 9.30 The transposition process. (a) Insertion of a transposable element generates a duplication of the target sequence. Note the presence of inverted repeats (IRs) at the ends of the transposable element. (b) A schematic diagram indicating how target sequences might be duplicated. For simplicity, the IRs are not marked.

The bacteriophage Mu (∞ Section 8.13) is also widely used as a *biological mutagen*. Because Mu integrates at a wide variety of host sites, it can be used to induce mutations at many locations. Also, Mu can be used to carry into the cell genes that have been derived from other host cells, a form of *in vivo* genetic engineering. In addition, modified Mu phages have been made artificially in which some of the lytic functions of Mu have been deleted. These phages, called Mini-Mu,

are deleted for significant portions of Mu but have the ends of the phage in normal orientation. Mini-Mu phages are usually defective, unable to form plaques, and their presence must be ascertained by the presence of other genes they carry. One set of Mini-Mu phages containing the β-galactosidase gene of the host (called Mu*d-lac, d* for "defective") can be detected in the integrated state if the *lac* gene is oriented properly in relation to a host promoter. Under these conditions, the

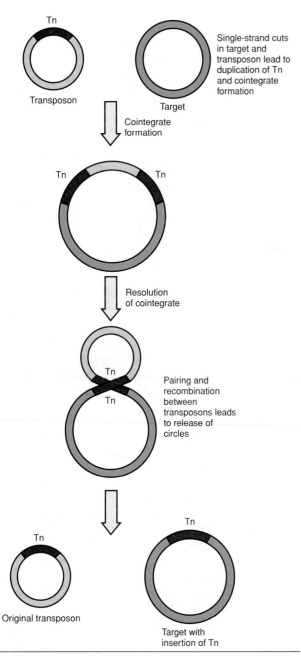

Transposon

Target

Single-strand cuts in target and transposon lead to duplication of Tn and cointegrate formation

Cointegrate formation

Tn   Tn

Resolution of cointegrate

Tn

Tn

Pairing and recombination between transposons leads to release of circles

Tn

Tn

Original transposon

Target with insertion of Tn

Figure 9.31 Replicative transposition. After the formation of single-strand cuts, a cointegrate structure arises by association of the two molecules. After recombination, resolution of the cointegrate structure leads to the release of the original transposon and duplication of the transposon in the target molecule. See Figure 9.30b for an explanation of how the duplication process might occur.

host cell forms the enzyme β-galactosidase, which can be detected in colonies by using a special color indicator (see Figure 9.1c). β-Galactosidase–positive colonies from a β-galactosidase–negative host are thus an indication that Mud-lac infection has occurred.

Invertible DNA and the phenomenon of phase variation

An interesting phenomenon of genetic change that has been recognized in some bacteria involves the **inversion** of a segment of DNA from one orientation to the other. When the segment is oriented in one direction, a particular gene is expressed, whereas when it is oriented in the opposite direction, a different gene is expressed. This "flip-flop" mechanism provides an interesting example of the regulation of gene activity. The process by which gene inversion occurs is another example of *site-specific recombination* (see earlier discussion).

The best-studied case of gene inversion is that called *phase variation,* which has been well studied in bacteria of the genus *Salmonella.* These enteric bacteria are motile by means of peritrichous flagella. As we have noted (∞ Section 3.9), bacterial flagella are composed of a single type of protein. As a result of phase variation, the flagellar protein can be of one of two separate types. Each *Salmonella* cell has two genes, H1 and H2, coding for the two different flagellar proteins, but only one of the two genes is expressed at any one time. Thus, an individual bacterial cell makes either H1-type flagella or H2-type flagella.

The invertible element involved in expression of the flagellar proteins is a 970-base-pair segment in the H2 gene (Figure 9.33). When the invertible segment is in one orientation, the H2 gene is transcribed, but in addition, another gene is transcribed that codes for a protein that represses transcription of gene H1. Thus, when H2 is expressed, H1 is turned off. On the other hand, when the invertible segment is in the opposite orientation, the genes for H2 and the H1 repressor are no longer expressed, so H1 can now be transcribed and expressed.

Invertible segments are also known in other genetic systems. We described an invertible segment involved in host range of bacteriophage Mu (∞ Section 8.13). Regulation by rearrangement also occurs in eukaryotes: we consider the regulation of mating type in yeast in

Figure 9.32 Transposon mutagenesis. The transposon moves into the middle of gene 2. Gene 2 is now disrupted by the transposon and is inactivated. Gene A of the transposon will be expressed in both locations.

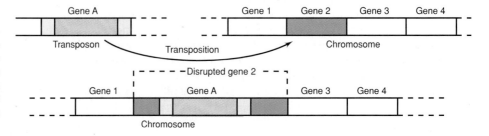

Gene A

Gene 1   Gene 2   Gene 3   Gene 4

Transposon   Transposition   Chromosome

Disrupted gene 2

Gene 1   Gene A   Gene 3   Gene 4

Chromosome

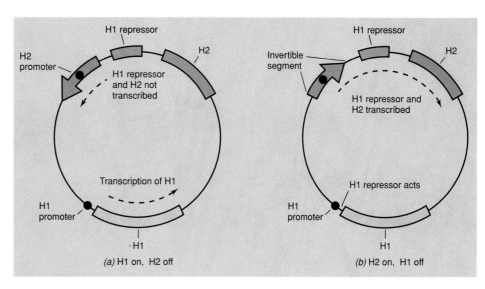

Figure 9.33 Site-specific inversion, the mechanism by which phase variation in *Salmonella* flagella is brought about. The dotted lines show the location and direction of transcription. (a) When the invertible segment is in one orientation, the H2 promoter points away from the H1 repressor and H2 gene; H1 is expressed. (b) In the opposite orientation, H1 repressor is made, turning off H1. At the same time, H2 is expressed. The result is a "flip-flop" between two alternate states.

Section 9.13 and we examine the complex genetic rearrangements involved in the production of antibodies (defense against infection) later (∞ Section 20.12).

9.11
Overview of the Bacterial Genetic Map

The three mechanisms of genetic exchange described in this chapter, transformation, transduction, and conjugation, can be used to map the locations of various genes (actually mutations in genes) on the chromosome. In *Escherichia coli* genes are generally mapped to a particular region of the chromosome using conjugation. Different Hfr strains are used that initiate DNA transfer from different parts of the chromosome, depending on where the F plasmid inserts (see Figure 9.27). By using Hfr strains with origins at different sites, it is possible to map the whole bacterial gene comple-

ment. A circular reference map for *Escherichia coli* strain K-12 is shown in Figure 9.34. The map distances are given in minutes of transfer, with 100 min for the whole chromosome and with "zero time" arbitrarily set as that at which the first genetic transfer (the *threonine* operon) can be detected using the original Hfr strain.

Conjugation experiments do not permit ordering genes that are closely linked to each other. Generalized transduction is used for more fine structure mapping of the *E. coli* chromosome. Bacteriophage P1 can carry fragments of DNA equivalent to about 2 min on the map and has proved very useful for mapping genes. Specialized transducing phages, usually obtained by recombinant DNA techniques (∞ Chapter 10), are also commonly used to study gene organization. The genetic techniques used are dictated by the efficiency with which genetic transfer occurs in the organism to be studied. Transformation is a very inefficient process in *E. coli*, but is more efficient in other organisms and has proved an effective tool for mapping genes in *Bacillus*.

By using a combination of conjugation and transduction, over 1400 genes were positioned on the circular chromosome of *E. coli* (only a few of which are shown in Figure 9.34). Except for genes grouped in operons, there does not seem to be any pattern to the location of specific genes on the *E. coli* genetic map. While some sets of related genes, for example, genes involved in biosynthesis of the amino acid tryptophan (*trp* genes) are tightly clustered, other sets, for example, the genes involved in arginine biosynthesis (*arg* genes), are scattered around the chromosome (Figure 9.34).

Gene clusters and operons

Mapping of the genes that control the enzymes of a single biochemical pathway has shown that these genes are often clustered or closely linked in prokaryotes. The gene clusters for biochemical pathways on the *Escherichia coli* chromosome are shown in Figure

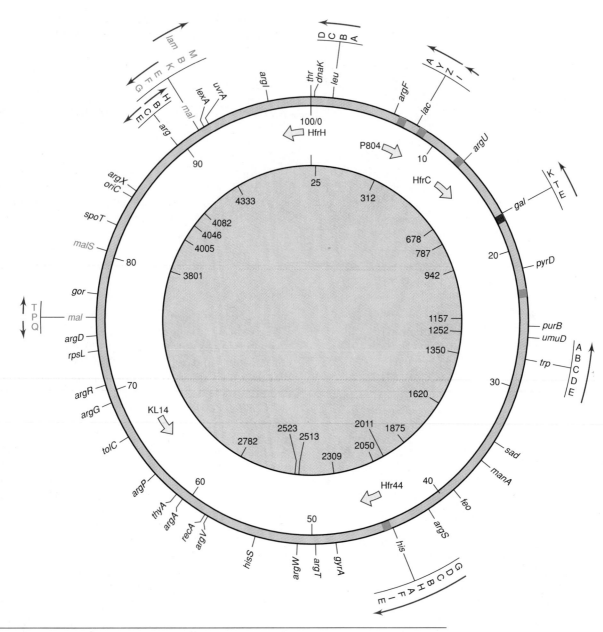

Figure 9.34 Circular linkage map of the chromosome of *Escherichia coli* K-12. On the outer edge of the map, the locations of a few of the mapped genes are indicated. A few operons are also shown, along with the direction in which they are transcribed. Along the inner edge of the map, the numbers from 0 to 100 refer to map position in minutes. The origin of DNA replication is marked *oriC* (84.3 min), and replication proceeds bidirectionally from this point. The inner circle shows the locations, in kilobase pairs, of the sequences recognized by the restriction enzyme *Not*I. Note that 0 min and 0 kilobase pairs are both, by convention, at the *thr* locus. The origins and directions of transfer of a few Hfr strains are also shown (arrows). The positions where five copies of the transposable element IS3 have been located in a particular strain are shown in blue. This element is also found in two copies on the F plasmid and is involved in Hfr formation. The position of the site where the bacteriophage lambda prophage integrates is shown in red. If the prophage were present, it would add an extra 48.5 kilobase pairs (slightly over 1 min) to the map. The genes of the maltose regulon (∞ Section 7.6), which includes several operons, are shown in green. Although most genes in this regulon have an abbreviation beginning with *mal*, note that one of the genes is *lamB*. This gene encodes a membrane protein involved in maltose uptake by the cell, but the protein is also the receptor for bacteriophage lambda. The gene *rpsL* (73 min) encodes a ribosomal protein. The gene was once called *str* because mutations in this gene lead to streptomycin resistance.

9.34; letters are used to indicate the genes for the specific enzymes of the pathway (for instance, the galactose cluster at minute 17, with genes *K, T,* and *E*). It has also been found that all the enzymes of a single gene cluster are often affected simultaneously by induction or repression. In prokaryotes such clustered genes are typically in an **operon** (∞ Sections 6.6 and 7.3) and are transcribed as a single unit.

The transcription of some operons proceeds in one direction along the chromosome, whereas with other operons transcription is in the opposite direction. The direction of mRNA transcription of the operons listed in Figure 9.34 is shown by arrows above the gene clusters. Because transcription always occurs in a $5' \rightarrow 3'$ direction (∞ Section 6.6), this implies that transcription off of either strand of DNA can occur and thus that opersons exist on each of the strands of the DNA double helix. In *E. coli* it appears that approximately equal numbers of operons exist on both strands.

Although many enzymes controlling specific biochemical pathways are linked and their synthesis is regulated by operons, this is not always the case. For some pathways, individual enzymes are encoded by genes at different locations on the chromosome, with the same operator present at each site. Such a collection of genes is termed a *regulon* (∞ Section 7.6), and coordinate regulation may occur because of a shared regulatory mechanism.

In addition to minutes, the map in Figure 9.34 also contains size units in kilobase pairs and shows the location of some restriction enzyme recognition sites. Maps using restriction enzymes (∞ Section 6.4), molecular cloning of genes (∞ Chapter 10), and DNA sequencing are now becoming widely used tools of the geneticist. These techniques have allowed the map of *E. coli* to be nearly completed. Although sequencing of the *E. coli* genome began before large-scale sequencing of other prokaryotic genomes, it was not among the first whose sequence was completed.

Our knowledge of *E. coli* is now very extensive. This knowledge grew from the fact that *E. coli* was the first bacterium studied that had a relatively simple system for mapping genes (see the box, *Escherichia coli,* the Best-Known Prokaryote). However, the advent of recombinant DNA technology has revolutionized the study of gene organization in prokaryotes. Often *physical mapping* using restriction enzymes (∞ Section 6.4) of an entire prokaryotic chromosome is done and then genes are located on the map by hybridization studies (∞ Working with Nucleic Acids: The Tools, Chapter 6). Indeed, several prokaryotic genomes have now been completely sequenced. The first was that of the Bacteria *Haemophilus influenzae.*

This does not mean that genetic studies of *E. coli* are almost over. To understand what a gene does, it is almost always necessary to isolate mutants, map their mutations, and determine their effects on the organism. Knowledge of the sequence of a genome is a powerful tool, but it does not in itself explain how an organism functions.

Organization of prokaryotic chromosomes

As discussed in Chapter 6, most Bacteria have a single circular, double-stranded DNA molecule as their only chromosome. The *Escherichia coli* chromosome of 4700 kilobase pairs seems slightly above average in size, but fewer than 200 different species have been examined. The chromosome of the parasitic bacterium *Mycoplasma genitalium* is very small at 580 kilobase pairs. The complete sequence of the *M. genitalium* genome shows that it contains only 470 protein-encoding genes. Relatively few of these genes encode enzymes involved in metabolic pathways, and therefore this organism requires many metabolic products from its host. In contrast, the free-living bacterium *Myxococcus xanthus* has an unusually large chromosome of 9500 kilobase pairs. The latter chromosome is two-thirds the size of the entire genome of the yeast, which is a eukaryote with 16 different chromosomes (see Section 9.13).

The chromosomes of several Bacteria are being analyzed by genetic and physical techniques, and comparative studies are being done. Bacteria closely related to *E. coli* such as *Salmonella* have very similar arrangements of genes, whereas distantly related bacteria such as *Bacillus* do not. Even the nature and arrangement of genes found in the same operon from two different Bacteria can differ. Even so, the order of genes in some operons is surprisingly conserved, not only in different genera of Bacteria but also in some Archaea. The reason for this is not clear but no doubt reflects some aspect of efficient gene expression.

Although there may be no general rules governing gene location, there do seem to be some constraints. Many Bacteria have a similar organization of genes near the origin of DNA synthesis, and similarities also exist at the terminus of DNA synthesis. Also, in at least some Bacteria, including *E. coli,* it appears that for certain genes or regions it is advantageous for the direction of transcription to be the same as that of the movement of the DNA replication fork. In these cases, an inversion mutation can be very deleterious to the organism.

The pattern of gene location along the chromosome may give clues to the evolution of the chromosome, and it seems clear that in many Bacteria, transposition and other large-scale rearrangements have been important. Good evidence from a number of Bacteria also shows that some groups of genes have been acquired from extrachromosomal genetic elements such as plasmids and bacteriophages.

The genomes of several microorganisms have been completely sequenced and others are being deter-

A FOCUS ON . . .

ESCHERICHIA COLI, THE BEST-KNOWN PROKARYOTE

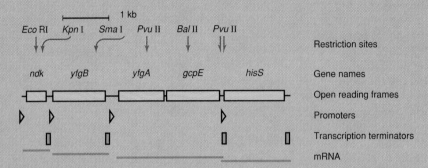

By far, the best-known prokaryotic organism is the intestinal bacterium *Escherichia coli*. Indeed, there are those who say that we know more about the biology of *E. coli* than about any other living organism, even including the human *Homo sapiens*.

Detailed information on the genetics or metabolism of *Escherichia coli* can be found in the two-volume book entitled Escherichia coli *and* Salmonella: *Cellular and Molecular Biology,* by F. C. Neidhardt and his colleagues (see Supplementary Readings), or by going to EcoCyc on the World Wide Web (http://www. ai.sri.com/ecocyc/ecocyc.html). This "web" site is an *E. coli* encyclopedia created by Peter D. Karp of SRI International and Monica D. Riley of the Marine Biological Laboratory (MBL), Woods Hole, MA. Like the book, this site contains a large "knowledge base" of the genes and metabolic pathways of *E. coli.* Kenneth Rudd of the National Library of Medicine contributed much of the genomic data (for other genomic data on the World Wide Web contact the National Center for Biotechnology at http://www.ncbi. nlm.nih.gov/).

The structure and function of *E. coli* is often considered the archetype of *all* living organisms. Why is *E. coli* so well known? There is nothing especially unusual about this organism. It is a run-of-the-mill prokaryote and a common, if minor, inhabitant of the human intestine.

Escherichia coli is so well known primarily because it is easy to work with in the laboratory. Even those who have difficulty with sterile technique and other bacteriological procedures can generally work with *E. coli* without difficulty. It grows rapidly, has simple nutritional requirements, and exhibits a fair bit of interesting biochemistry and physiology. But it was something else that made it so useful in the early days of

molecular biology: "sex." *E. coli* was the first bacterium in which the process of conjugation was discovered, being first recognized in 1946 by Joshua Lederberg and Edward L. Tatum. At the time of its discovery, conjugation seemed very similar to sexual reproduction in eukaryotes. Although we now know that these are quite different, conjugation is an extremely important genetic tool. With conjugation, the possibility of doing real genetics was available. Genetic crosses could be carried out, and genetic properties analyzed. Another valuable property of *E. coli* is its ability to support the growth of a whole range of bacterial viruses, which made it possible to study in detail the nature of viruses and virus multiplication. Thus, with its favorable laboratory properties, its suitability for studies of virology, and the ability to perform genetic crosses, biochemists and molecular biologists were able to probe deep into the nature of life, their work leading ultimately to the sophisticated understanding we now have of molecular biology and virology (Chapters 6–10). The availability of DNA sequence information, coupled with the extensive biochemical and physiological data continues to allow important insights to be developed with this model organism. The figure is a diagram of the genes (and some restriction sites) identified by sequencing the *hisS* gene near 54 minutes on the *E. coli* genetic map (see Figure 9.30). This region of 6 kb con-

tains genes encoding 5 proteins. Only 2 of these genes, *hisS* and *ndk*, were discovered by using the classical techniques of bacterial genetics. These two genes encode essential proteins. The other three genes encode proteins of unknown function and were uncovered only by sequencing. Similar genes have been uncovered by sequencing the genomes of other gram-negative Bacteria. In *Providencia stuartii* the gene homologous to *gcpE* may be part of a two-component regulatory system (Section 7.7).

Escherichia coli was first isolated in 1885 by the German bacteriologist Theodor Escherich, as a normal inhabitant of the intestinal tract. Escherich named the organism *Bacterium coli,* the name reflecting the rod shape of the cell (*Bacterium* means "rod-shaped") and its intestinal habitat (*coli* for "colon"). The genus name *Bacterium* subsequently was changed to *Escherichia* in honor of its discoverer. Although Escherich's strains, and most other strains of *E. coli,* are harmless, some *E. coli* strains are pathogenic, causing diarrhea and urinary tract infections. As we will see in Chapter 19, the pathogenic *E. coli* strains differ in significant ways from the harmless intestinal strains.

It is curious that although *E. coli* is the organism of choice for genetic engineering and biotechnological research (Chapter 10), it has never been a major organism of use in industrial microbiology. The large-scale cultivation of living organisms for industrial purposes (∞ Chapter 12) has involved a quite different group of microorganisms. For complex and not especially revealing reasons, *E. coli* is not nearly as suitable for large-scale cultivation as yeast, *Streptomyces*, and *Bacillus*. Indeed, there were many who worried about the possible use of *E. coli* in industry because of its potential pathogenicity. However, industrialists have "tamed" *E. coli* for industrial purposes, although many other organisms are also used. ∎

mined. As we discussed previously in the case of *Mycoplasma genitalium,* sequence information not only can be used to determine the number and type of genes present, but also can give important clues to the metabolism of the organism. The sequences of the genomes of pathogenic organisms, such as *Neisseria gonorrhoeae* and *Treponema pallidum,* might offer clues on how these organisms cause disease. Computer programs allow for rapid comparisons of sequences between organisms, and these comparisons will vastly improve our understanding of the organization of microbial genomes, the way they are related, and the way they have evolved.

CONCEPT CHECK   9.11

Conjugation, transformation, and transduction have been important methods used to map genes on the chromosomes of Bacteria. These techniques, coupled with restriction enzyme analysis, cloning, and DNA sequencing, have allowed detailed analysis of the bacterial chromosome. Although there are few rules governing gene location, the genes encoding the enzymes for many biochemical pathways are often found tightly linked in operons in prokaryotes. Large-scale sequencing projects have now revealed the complete DNA sequence of the genomes of several prokaryotes.

9.12
Genetics in Eukaryotic Microorganisms

We now turn from a discussion of genetic mechanisms in Bacteria to a consideration of analogous processes in eukaryotes, using yeast as a model system. Eukaryotes can mate during sexual reproduction, and therefore DNA transfer and recombination differ in many ways from that in prokaryotes. The complex nuclear organization of eukaryotes and the existence in each nucleus of a number of chromosomes lead to more regular mechanisms of gene assortment and segregation. Unlike prokaryotes, whose genomes are single DNA molecules, eukaryotic genomes are segmented into a number of chromosomes. While prokaryotic chromosomes are usually circular, eukaryotic chromosomes contain linear DNA molecules.

Typically, eukaryotic cells can be of two types, *haploid* or *diploid* (∞ Section 3.14), depending on the chromosome number. In the *haploid* phase the number of chromosomes per cell is n, and in the *diploid* phase, $2n$ (Figure 9.35). Thus, in the yeast *Saccharomyces cerevisiae* a haploid cell contains 16 chromosomes and the diploid 32. In humans, haploid and diploid cells contain 23 and 46 chromosomes, respectively. Multicellular plants and animals are usually diploid, with the haploid phase present only in the germ cells (sperm and eggs, also called *gametes*), whose life spans are transitory. Eukaryotic microorganisms can be either haploid or diploid. Growing cells of *S. cerevisiae* can be either haploid or diploid (Figure 9.35). In most eukaryotic microorganisms, the diploid phase is transitory.

In diploid cells, two copies of each gene are present, one on each of the two homologous chromosomes. The term *allele* is used to refer to the two alternate forms of the *same gene* present on the two homologous chromosomes. If the allele on one chromosome has a mutation preventing the normal product from being expressed, the allele on the other chromosome can continue to be expressed, and so the effect of the mutation may not be evident. Thus, the expression of one allele may be masked by the other. The gene that is expressed is said to be *dominant* to the other allele, which is said to be *recessive.* Diploidy presents difficulties in genetic research because isolation of mutants is much easier in a haploid cell, where only one form of the gene is present and, therefore, the effect of a mutation can be directly determined.

Meiosis

Mitosis is the process following DNA replication in which chromosomes condense, divide, and are sorted into two identical sets, one for each daughter cell. By contrast, **meiosis** is the process by which the change from the diploid to the haploid state is brought about (∞ Section 3.14). Meiosis involves two divisions. During the first meiotic division, the two sets of homologous chromosomes are segregated into two separate cells, and so the number of chromosomes is reduced from $2n$ to n, yielding haploid cells that are the precursors of the germ cells. The second meiotic division is similar to a mitotic division but involves n chromosomes in each of two daughter cells. The products of meiosis are four haploid gametes. By definition, eggs are formed by females and sperm are

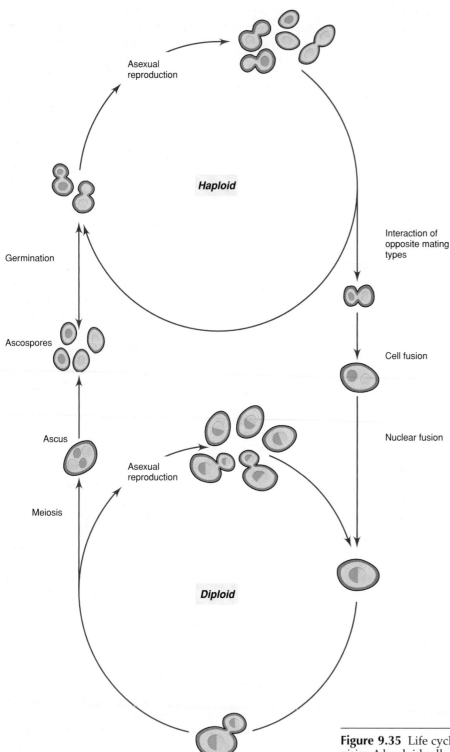

Figure 9.35 Life cycle of a typical yeast, *Saccharomyces cerevisiae*. A haploid cell of *S. cerevisiae* contains 16 chromosomes.

formed by males, but in many microorganisms there is no clear sexual distinction, only different mating types.

If the diploid homologs are genetically identical for a particular allele, the cell is said to be *homozygous*. Under such conditions, the haploid cells formed by

meiosis are also identical for that gene. However, if the two homologs differ, the cell is said to be *heterozygous*. Under such conditions, the four haploid gametes are not identical for that gene.

Once gametes are formed, **mating** of gametes of different type can occur, leading to the formation of diploids. The first diploid cell formed as a result of mating of haploid gametes is called a *zygote*. The diploid cell formed can then become the forerunner of a new population of genetically identical diploid organisms or, as is the case with most eukaryotic microorganisms, meiosis can generate haploid cells that can divide vegetatively.

CONCEPT CHECK   9.12

Eukaryotic organisms can mate and exchange DNA during sexual reproduction. Haploid germ cells formed by meiosis can fuse to form a diploid zygote. Mitosis ensures appropriate segregation of the chromosomes during asexual cell division in eukaryotes.

✔ *If the haploid number of chromosomes of an organism is 10, what is the diploid number?*

✔ *What does it mean to say that an allele is* dominant?

9.13
Yeast Genetics

More is known about the molecular genetics of yeast than about almost any other eukaryote. This is because yeast is easily grown in the laboratory and is an extremely favorable organism for studies of genetic phenomena. The yeast that has been most studied is *Saccharomyces cerevisiae*, the common baker and brewer's yeast. This yeast is not only a useful model system for studies on eukaryotic genetics but is an important organism of commerce, and therefore studies of its genetics can be expected to have practical significance (∞ Section 12.12). An important facet of yeast genetics, discussed in Section 9.14, is the study of mitochondrial inheritance.

A yeast grows as a single cell, and each haploid yeast cell is capable of serving as a gamete. Because yeast is unicellular and can be grown as a haploid, isolation of mutants is straightforward, and a large variety of mutant types are known. By mating mutants, genetic analyses of yeast can be carried out.

The life cycle of a typical yeast was shown in Figure 9.35. Many yeasts have two separate *mating types,* which can be considered analogous to male and female. However, the two mating types of a yeast are alike in structure and can be differentiated only by allowing them to mate. On mating of opposite types, a diploid cell is formed. In many yeasts, this diploid cell is capable of growing vegetatively, leading to the formation of a population of genetically identical, albeit diploid, cells. Under certain conditions, diploid cells of such a population can undergo meiosis and form haploid gametes. Two distinct types of gametes are formed, of opposite mating type. From a single diploid cell, a structure containing four such gametes is formed, two of each mating type. The cell in which the gametes are formed is called an *ascus,* and the cells within the ascus are called *ascospores.*

One important advantage of yeast is that genetic analysis is fairly straightforward. After mating and ascospore formation have occurred, the experimenter can dissect the four ascospores from the ascus, use each ascospore as the forerunner of a separate culture, and analyze the cultures so obtained for phenotypic characters that were in the parent. In this way, it is possible to *map* genes in yeast. Another advantage of yeast is that it can be transformed using exogenous DNA, and plasmids are available for use in genetic engineering.

Haploid cells of *Saccharomyces cerevisiae* have 16 chromosomes, and extensive genetic maps have been prepared for these chromosomes. These chromosomes range in size from 245 to 2200 kilobase pairs. The DNA sequence of all these chromosomes has now been determined, making this yeast the first eukaryote whose genome has been completely sequenced. Such sequence information aids genetic studies in many ways in addition to giving insight into genome evolution. The total haploid genome size of *Saccharomyces cerevisiae* is 12 *mega*base (10^6) pairs. This is only about three times that of *Escherichia coli* (see Section 9.11).

Mating type genetics of yeast

As we have noted, yeast has two mating types, which are indistinguishable except by their behavior in mating. The two mating types of *Saccharomyces cerevisiae* are designated α and a. Cells of type α mate only with cells of type *a*, and whether a cell is α or *a* is itself determined genetically. However, although a yeast cell line generally remains either α or *a*, haploid yeast cells are periodically able to *switch* their mating type from one to the other. (One consequence of this switching is that a pure culture of a single mating type can ultimately form diploids. Note that it is not appropriate to call a yeast cell *bisexual.* Under a given set of conditions, it is either one or the other mating type, never both.)

The switch in mating type from α to *a* and back to α has at its basis the behavior of a mobile genetic element, somewhat reminiscent of transposition or phase variation in *Salmonella* (see Section 9.10). The phenomenon in yeast is illustrated in Figure 9.36. There is a single active genetic locus, called the MAT (for *mating type*) locus, at which either gene α or gene *a* can be inserted. At this active locus, the MAT promoter controls the transcription of whichever mating type gene is present. Thus, if gene α is at that locus, then the cell is mating type α, whereas if gene *a* is at that locus, the cell is mating type *a*. Somewhere else on the yeast genome are copies of both genes, α and *a*, which are

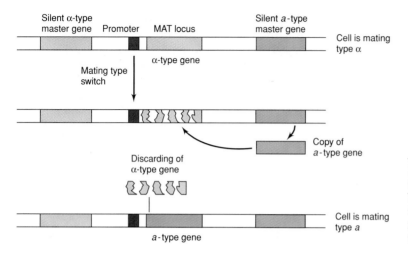

Figure 9.36 The cassette mechanism involved in the switch in yeast from mating type α to a and back again. Whichever "cassette" is inserted at the active locus (reading head) determines the mating type. The process shown is reversible, so type a can revert to type α.

not expressed. These *silent copies* serve as the source of the gene that is inserted when the switch occurs. When the switch occurs, the appropriate gene, α or a, is copied from its silent site and then inserted into the MAT location, *replacing* the gene already present. Thus, the old gene is excised out of the locus and discarded and the new gene is inserted. This mechanism has been called the *cassette mechanism* because each gene can be considered analogous to a tape cassette inserted into a "reading head," the place on the chromosome where active transcription takes place.

What do genes α and a do? At least one function has been identified. Yeast cells undergoing mating excrete *peptide hormones* called α factor and a factor. These hormones bind to cells of opposite mating type and bring about changes in the cell surfaces of these cells so cells of opposite mating type associate and fuse. It seems that α cells have receptors on their surfaces only for a factor, whereas a cells have receptors only for α factor. Once two cells of opposite mating type have associated, a complex series of events is initiated that leads to fusion of these two cells as well as their nuclei, resulting in the formation of a diploid zygote (Figure 9.37).

Yeast plasmids

We have defined plasmids as genetic elements that replicate independently of chromosomal DNA but have no extracellular form (∞ Sections 6.3 and 9.8). Yeast has a number of elements that could be considered plasmids by this strict definition. There are *retrotransposons* that accumulate viruslike particles inside the cell but are never released. In addition yeast also contains double-stranded RNA "viruses" whose particles accumulate inside the cell and are infectious only if yeast cells fuse with each other (which happens in mating). Both these types of elements are much more like viruses than typical plasmids, which are generally circular DNA molecules (with no pro-

tective particle). However, most strains of yeast also contain such a typical plasmid, the so-called 2-μm circle. The 2-μm circle is a 6318-base-pair circular DNA molecule that replicates to high copy number (~30) in cells of *Saccharomyces cerevisiae*. The plasmid contains four protein-encoding genes, all involved in plasmid maintenance. It confers no phenotype on the cell.

Most yeasts contain the 2-μm-circle DNA. It is found within the nucleus packaged into nucleosomes, using histone protein derived from the nucleus. Segregation of the plasmid from mother to daughter cell during mitosis is random, but if the number of 2-μm circles drops to a low value, an increased rate of replication can bring the copy number back up to 30–50 per cell. There is no evidence that this plasmid ever becomes integrated into the nuclear chromosomal DNA.

Although the 2-μm plasmid is a useful model for studying DNA replication in yeast, its greatest value appears to be as a vector for cloning foreign genes into yeast. By use of appropriate treatment (see Section 9.6), it is possible to transform yeast cells using this plasmid, and hence to incorporate genes of interest. The

---

CONCEPT CHECK   **9.13**

The yeast *Saccharomyces cerevisiae* is a eukaryotic microorganism of great commercial value that is also widely used in genetic studies. It is the first eukaryote to have its genome completely sequenced. There are two mating types in yeast, and cells can convert from one to the other. In addition to the nuclear and mitochondrial chromosomes, the cell also has extrachromosomal elements.

✔ *How does mating occur in yeast?*

✔ *Why do you suppose eukaryotic plasmids replicate in the nucleus?*

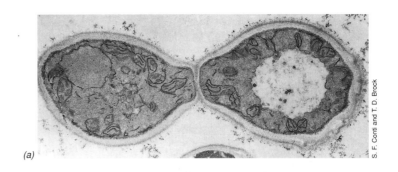

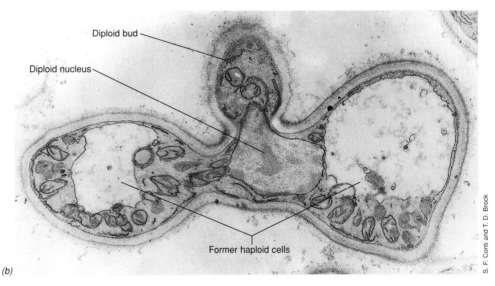

Diploid bud

Diploid nucleus

Former haploid cells

Figure 9.37 Electron micrographs of the mating process in a yeast, *Hansenula wingei*. (a) Two cells have fused at the point of contact and have sent out protuberances toward each other. (b) Late stage of mating. The nuclei of the two cells have fused, and the diploid bud has formed at a right angle to the conjugation tube. This bud eventually separates and becomes the forerunner of a diploid cell line.

(a)

(b)

value of vectors such as this for genetic engineering was discussed in Chapter 10.

We discuss mitochondrial inheritance in yeast in the next section.

9.14
Mitochondrial Genetics

Although most genetic characters are contributed by chromosomal genes, mitochondria and chloroplasts contain separate genetic systems that are required for their own replication. In addition, mitochondria and chloroplasts have their own translation systems, which are needed to synthesize the few proteins these organelles make independently of the nuclear system.

As a eukaryotic cell grows, the number of mitochondria and chloroplasts increases by a process of organellar division—an existing organelle divides and forms two new organelles. Mitochondria and chloroplasts are never made de novo (that is, from scratch); they always arise from preexisting mitochondria and chloroplasts. Thus, if a cell lacks an organelle, its offspring will lack this same type of organelle. Because mitochondria and chloroplasts arise only from preexisting organelles, the genetic characteristics of the

organelle in a new cell are determined by the genetic characteristics of the organelle of its parent.

Mitochondrial inheritance in yeast

One of the most interesting genetic analyses using yeast is that involving *mitochondrial inheritance*. It has been possible to isolate mutant yeast strains in which mutations have occurred in the mitochondrion rather than in the nucleus, and by genetic analysis the inheritance of the mitochondria themselves can be studied. Because inheritance of genetic characteristics via mitochondria occurs outside the nucleus, and outside the process of mitosis and meiosis, it is a form of **cytoplasmic inheritance** (sometimes called *non-Mendelian inheritance* to indicate that it does not follow Mendel's laws.)

An example of the manner in which mitochondrial characteristics are inherited in yeast is shown in Figure 9.38. As seen, when two yeast cells of opposite mating type that also differ in mitochondrial characteristics are mated, the outcome depends on which of the two types of mitochondria replicates most rapidly during subsequent cell divisions. In one class of mitochondrial mutants, called *petite*, large deletions in the mitochondrial DNA have led to abolition of all mitochondrial

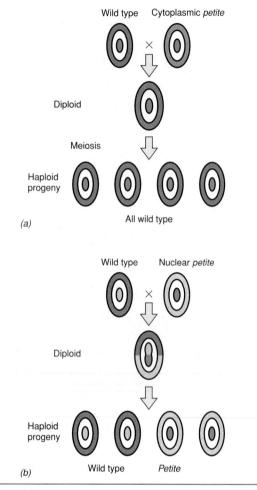

Wild type   Cytoplasmic *petite*

×

Diploid

Meiosis

Haploid
progeny

(a)   All wild type

Wild type   Nuclear *petite*

×

Diploid

Haploid
progeny

(b)   Wild type   Petite

Figure 9.38 Mitochondrial inheritance. Outcome of crosses between petite and wild-type yeast: (a) cytoplasmic petite and (b) nuclear petite.

protein synthesis. Such mitochondria are nonfunctional, and yeast cells containing such mitochondria are unable to carry out respiration, although they are still able to grow anaerobically by fermentation processes. The term *petite* comes from the fact that these mutants produce small colonies on agar both aerobically and anaerobically; wild-type cells form larger colonies aerobically. If a petite yeast cell is mated with a wild-type yeast cell, the diploid zygote will contain both types of mitochondria, but because the mutant mitochondria do not replicate as effectively as the wild-type mitochondria, they lose out in the competition during subsequent cell divisions. Ultimately, the cell lines derived from each of the four ascospores will all be wild type (Figure 9.38*a*). On the other hand, there are also petite mutants of yeast in which the mutation has occurred in a nuclear gene (since most of the proteins of the mitochondria are encoded by nuclear genes rather than mitochondrial

genes). Crosses involving nuclear petites show conventional Mendelian inheritance (Figure 9.38*b*).

Like most mitochondria, those of yeast encode relatively few of the required mitochondrial proteins. Interestingly, most of the yeast mitochondrial DNA does not encode proteins; this noncoding DNA has a very high AT content and its function is unknown. Proteins encoded by yeast mitochondrial DNA include cytochrome *b*, cytochrome *c* oxidase, ATPase, and one ribosome protein. In addition, a number of tRNA molecules are encoded by yeast mitochondrial DNA, as are the ribosomal RNA molecules.

Organellar genomes

The genomes of most mitochondria and chloroplasts exist as circular DNA molecules. Organellar genome sizes vary dramatically among various eukaryotes. Chloroplast genomes fall within a narrow range from 120 to 200 kilobase pairs, but mitochondrial genomes vary much more, from as low as 16 to as high as 2500 kilobase pairs. There is some reason to believe that during evolution, some genes have moved from mitochondrial DNA to nuclear DNA, and vice versa. Oddly, the mitochondria of higher animals have smaller genomes than the mitochondria of many eukaryotic microorganisms. The human mitochondrial genome has been completely sequenced and contains 16,569 base pairs, whereas that of yeast is about five times larger. Plant mitochondrial genomes are often large and complex.

The genetic map of the human mitochondrion is shown in Figure 9.39. By convention, the outer strand is called the *heavy* strand and its complement is called the *light* strand. Both strands are completely transcribed. However, as shown, one strand encodes most of the proteins made in the mitochondrion, as well as a number of tRNAs and the two major species of ribosomal RNA (rRNA). The other strand encodes mainly other tRNAs (Figure 9.39). Genes are tightly packed in the human mitochondrial genome, with genes for tRNAs acting as spacers between sequences coding for protein or rRNA (Figure 9.39). The yeast mitochondrial genome (see earlier) has about the same number of genes as mammalian mitochondria but because it has lots of spacer (noncoding) DNA and several of its genes contain intervening sequences (introns), it is considerably larger. However, like the human mitochondrial genome, most of the genes encoding proteins in the yeast mitochondria reside on just one of the two strands.

Mitochondrial genomes have some surprising molecular features. Whereas prokaryotic or eukaryotic nuclear genomes encode 32 or more different tRNAs, only 22 tRNAs are encoded in most mitochondria. The economy of the mitochondrion in this regard is governed by the fact that only a *single* tRNA is used to encode any amino acid specified by four different

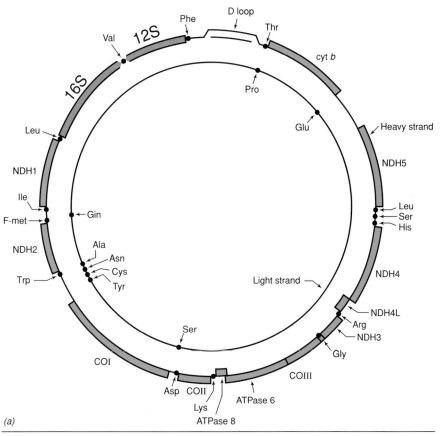

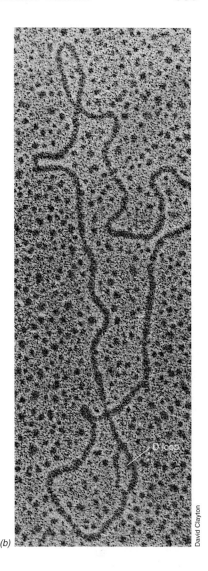

Figure 9.39 (a) Circular genetic map of the human mitochondrion. The complete genome contains 16,569 base pairs. The protein-encoding and ribosomal RNA genes are shown in different colors. Note that in this very small genome, the 16S rRNA corresponds to the prokaryotic 23S rRNA, and the 12S rRNA corresponds to the prokaryotic 16S rRNA (∞ Section 6.9). A dark circle indicates the position of a tRNA gene. The D loop is the site of the origin of replication of the heavy strand. The location of various genes is shown, cyt *b*, cytochrome *b*; NDH 1-5, components of the NADH dehydrogenase complex; COI-III, subunits of the cytochrome oxidase complex; ATPase 6,8, polypeptides of the mitochondrial ATPase complex. The three-letter code for amino acids (∞ Section 2.8) is used to designate a corresponding tRNA. Note that only serine and leucine, each of which has six codons, require more than one tRNA. (b) Electron micrograph of human mitochondrial DNA. Note the circular nature of the DNA. The D loop is shown by an arrow.

codons; this is in contrast to nonmitochondrial systems where the wobble rules (∞ Section 6.10) demand at least two tRNAs for amino acids encoded by four different codons.

Protein synthesis in organelles

Chloroplasts and plant mitochondria use the same genetic code as that used in most prokaryotes and in the cytoplasm of most eukaryotes. This code has been referred to as the *universal code.* However, as discussed earlier (∞ Section 6.10), the mitochondria of other organisms use codes that differ from this universal code and from each other (∞ Table 6.6). This

lack of universality in the code was first discovered in mitochondria, but nonuniversal codes are now also known to be used by some cellular genomes (∞ Section 6.10).

Proteins encoded by the mitochondrial genome include components of the mitochondrial respiratory chain such as cytochromes, dehydrogenases, and ATPases. However, most mitochondrial proteins are not encoded by mitochondrial genes but are transported into the organelle from their sites of synthesis in the cytoplasm. We noted in Section 3.15 that organelles possess rather permeable outer membranes and that transport of macromolecules and other substances that do not normally traverse the cytoplasmic

TABLE 9.5	Some key features of the genetic and translation systems of organelles

System	Features
Genome	Generally small, circular, double-stranded DNA (linear genomes present in the mitochondria of a few species) Histones absent DNA mostly coding and tightly packed in mammalian mitochondria; mostly noncoding with lots of spacer DNA in yeast mitochondria
Genetic code	Some differences from the universal code (∞ Table 6.6)
Translation apparatus	Ribosomes 70S-like; strong relationship in 16S rRNA sequences between organelles and Bacteria Sensitive to antibiotics that affect ribosomes of Bacteria Polypeptides start with formylmethionine (as in Bacteria)

membrane can occur. Of the 300–400 proteins known to reside in mitochondria, only 13 are encoded by mitochondrial genes in the human.

Organellar ribosomes resemble prokaryotic ribosomes in size, although there is some variance, from as small as 50S to as large as 80S (prokaryotic ribosomes are 70S) (∞ Section 6.9). However, an important feature of organellar ribosomes in their sensitivity to antibiotics that affect protein synthesis in Bacteria. Thus, protein synthesis in mitochondria and chloroplasts is inhibited by antibiotics such as chloramphenicol. The phenomenon suggests a close relationship between organelles and Bacteria and is one of several lines of evidence supporting the hypothesis that eukaryotic organelles are derived from *endosymbiotic Bacteria* (∞ Sections 3.15 and 15.4).

Because there are many subtle differences between different mitochondrial genetic codes and because some are identical to the universal code, it is most likely that these changes in the genetic code occurred during the genomic simplification that took place after the endosymbiotic events in evolution (∞ Section 15.4).

Table 9.5 summarizes some of the key features of the genetic and protein synthesis systems of organelles. We should emphasize that unlike mitochondria and chloroplasts, many of the other structures found in eukaryotes do not arise by division of preexisting structures and do not have their own genetic systems. Thus, structures such as lysosomes, Golgi apparatus, flagella, cilia, and so on, all arise de novo and are completely under nuclear control. Only mitochondria and chloroplasts seem to have their own genetic systems, albeit only to a partial extent.

CONCEPT CHECK   9.14

Although mitochondria and chloroplasts are independent genetic elements, they contain only part of the genetic information for their own function and replication, the rest coming from the chromosome. The genomes of these organelles are usually circular DNA molecules, thus resembling prokaryotic chromosomes rather than eukaryotic chromosomes. The molecular evidence supports the hypothesis that mitochondria and chloroplasts evolved in ancient times from endosymbiotic Bacteria.

✔ *What is the function of mitochondria (* ∞ *Section 3.15)?*

✔ *What is* cytoplasmic inheritance?

Material for Review

REVIEW QUESTIONS

1. Write a one-sentence definition of the term *genotype*. Do the same for the term *phenotype*. Does the phenotype of an organism automatically change when a change in genotype occurs? Why or why not? Can phenotype change without a change in genotype? In both cases, give some examples to support your answer.

2. Explain why an *Escherichia coli* strain that is His⁻ is an auxotroph and one that is Lac⁻ is not. (*Hint:* Think about what *E. coli* does with histidine and lactose.)

3. What are silent mutations and why do they occur? From your knowledge of the genetic code, why do you think most silent mutations affect the *third* position of the codon?

4. Microinsertions occur in promoters but are not frameshift mutations. Define the terms *microinsertion, frame shift, mutation,* and *promoter* (∞ Section 6.6). Explain how the statement can be true.

5. Explain how it is possible for a frameshift mutation early in a gene to be corrected by another frameshift mutation farther along the gene.

6. Give an example of one biological, one chemical, and one physical mutagen and describe the mechanism by which each causes a mutation.

7. What is site-specific mutagenesis? How can this procedure target specific genes for mutagenesis?

8. How does homologous recombination differ from site-specific recombination?

9. Why is it difficult in a single experiment using transformation to transfer a large number of genes to a cell?

10. From what you know about cell wall structure of Bacteria (∞ Sections 3.5–3.7), explain the problem a DNA molecule would encounter if the transformation process were to occur.

11. Explain why in generalized transduction one always refers to a transducing *particle* but in specialized transduction one refers to a transducing *virus* (or transducing phage).

12. Explain how it is possible to use the *interrupted mating procedure* to determine the relative order of genes on a bacterial chromosome.

13. Strains of *Escherichia coli* can be Hfr, F⁺, or F⁻. What are the differences between these strains and how would they behave in a mating experiment?

14. Strains that are Hfr can spontaneously become F⁺, and vice versa. Explain.

15. Explain why the insertion of a transposon leads to mutation.

16. The most useful transposons for isolating a variety of bacterial mutants are transposons containing antibiotic-resistance genes. Why are such transposons so useful for this purpose?

17. Compare and contrast the "flip-flop" process for *Salmonella* phase variation and the "cassette" mechanism for yeast mating type switching.

18. Recombination in prokaryotes always involves chromosome fragments, whereas in eukaryotes it involves whole chromosomes. Explain.

19. Describe how certain functions in the yeast cell can be transmitted in non-Mendelian fashion.

APPLICATION QUESTIONS

1. One type of *suppressor mutation* involves a change in tRNA. Draw a diagram with coding sequences and amino acid sequences indicating how this occurs.

2. A constitutive mutant is a strain that continuously makes a protein that in the wild type is inducible. Describe two ways in which a change in a DNA molecule could lead to the development of a constitutive mutant. How could these two types of constitutive mutants be distinguished genetically?

3. In Chapter 6 we saw that it was critical for the ribosome to translocate with great accuracy in order to maintain the proper reading frame. However, sometimes ribosomes make frameshift errors. Compare the impact on the cell of a ribosome periodically making a frameshift error in the mRNA from a particular gene with the impact of a frameshift mutation in the same gene.

4. Although a large number of mutagenic chemicals are known, no chemical is known to induce mutations in a single gene (gene-specific mutagenesis). From what you know about mutagens, explain why it is unlikely that a gene-specific chemical mutagen will be found.

5. Describe the principle behind the Ames test. How is the test run in practice? From your knowledge of how mutants are isolated, why is the back mutation procedure used in the Ames test preferable to a forward mutation procedure?

6. What is the net result of genetic recombination? Why is it that the farther two genes are apart on a chromosome, the more likely they are to show recombination?

7. Suppose you are given the task of developing a genetic transduction system for an industrial organism of interest. You have a large collection of bacterial strains but no phages. Describe the steps you would use to develop such a transduction system. (*Hint:* You must obtain not only transducing phages but also a collection of bacterial mutants to be able to follow genetic markers.)

8. Some retroviruses (∞ Section 8.22) seem to be capable of acting as transducing viruses. Use your knowledge of the life cycle of these viruses to explain whether you think they would more likely participate in generalized or specialized transduction.

9. Design an experiment that would help decide whether a genetic transfer process in a particular bacterium is transformation, transduction, or conjugation. Assume that the following tools are available: appropriate mutants and selective media; DNase (an enzyme that destroys naked DNA); two kinds of filters, one capable of retaining bacteria and bacterial viruses but not free DNA, the other capable of retaining bacteria only; and a glass chamber in which the filters can be inserted to separate the chambers into two compartments. Give the experimental setup and the anticipated results for each of the three gene transfer processes.

10. Insertion sequences transpose (a type of site-specific recombination) in cells that have a defective *recA* gene. However, the formation of an Hfr strain from an F⁺ strain cannot take place in a cell with a defective *recA* gene even though this is an event involving insertion sequences and recombination. Explain how Hfr formation takes place and why the RecA protein is essential.

11. Yeast can exist without mitochondria, but humans cannot. Explain.

SUPPLEMENTARY READINGS

Alberts, B., D. Bray, J. Lewis, M. Raff, K. Roberts, and **J. D. Watson.** 1994. *Molecular Biology of the Cell,* 3rd edition. Garland, New York. Good comparative coverage of prokaryotic and eukaryotic genetics, with extensive treatment of genetic mechanisms in yeast.

Beckwith, J., and **T. J. Silhavy.** 1992. *The Power of Bacterial Genetics: A Literature-based Course.* Cold Spring Harbor Laboratory Press, Cold Spring Harbor, NY. An excellent advanced textbook on prokaryotic genetics containing reprints of the original papers in several key areas as well as commentary and questions.

Berg, P., and **M. Singer.** 1993. *Dealing with Genes: The Language of Heredity.* University Science Books, Mill Valley, CA. An excellent introductory text on modern genetics, stressing molecular biology.

Brock, T. D. 1990. *The Emergence of Bacterial Genetics.* Cold Spring Harbor Laboratory Press, Cold Spring Harbor, NY. A detailed history of the early days of bacterial genetics.

Friedberg, E. C., G. C. Walker, and **W. Siede.** 1994. *DNA Repair and Mutagenesis.* American Society for Microbiology, Washington, DC. This excellent specialized textbook deals with the many ways cells respond to DNA damage and errors in DNA replication.

Hall, M. N., and **P. Linder** (eds.). 1993. *The Early Days of Yeast Genetics.* Cold Spring Harbor Laboratory Press, Cold Spring Harbor, NY. This book contains articles about many of the events and personalities involved during the beginning of the genetic analysis of the yeast *Saccharomyces cerevisiae.*

Maloy, S. R., J. E. Cronan, Jr., and **D. Freifelder.** 1994. *Microbial Genetics,* 2nd edition. This excellent and quite readable textbook covers the genetics of bacteria and bacteriophages.

Miller, J. H. 1992. *A Short Course in Bacterial Genetics: A Laboratory Manual and Handbook for* Escherichia coli *and Related Bacteria.* Cold Spring Harbor Laboratory Press, Cold Spring Harbor, NY. This book is an excellent source for information on the entire range of genetic techniques used in current research in bacterial genetics.

Neidhardt, F. C., R. Curtiss III, J. L. Ingraham, E. C. C. Lin, K. B. Low, B. Magasanik, W. Reznikoff, M. Riley, M. Schaechter, and **H. E. Umbarger** (eds.). 1996. Escherichia coli *and Salmonella. Cellular and Molecular Biology,* 2nd edition. American Society for Microbiology, Washington, DC. This two-volume set contains excellent information on mutations, mapping, and genetic techniques in these important organisms.

Sebald, M. (ed.). 1992. *Genetics and Molecular Biology of Anaerobic Bacteria.* Springer-Verlag, New York. A comprehensive resource book on the genetics and molecular biology of anaerobic prokaryotes, both Bacteria and Archaea.

On~line resources for this chapter are on the World Wide Web at:
http://www.prenhall.com/~brock (click on the table of contents link and then select Chapter 9).

CHAPTER *10*

Genetic Engineering and Biotechnology

Using the techniques of genetic engineering, agricultural crops can be given resistances to herbicides or insect pests.

MINIGLOSSARY for Chapter 10

Biotechnology use of living organisms to carry out defined chemical processes for industrial application

Cloning Vector genetic element into which genes can be recombined and replicated

DNA Fingerprinting use of the techniques of genetic engineering to determine the origin of DNA in a sample of tissue

DNA Library a collection of cloned DNA fragments, which in total contains cloned genes representing the entire genome of an organism; also called a gene library

Expression Vector a cloning vector that contains the necessary regulatory sequences to allow transcription and translation of cloned genes

Gene Disruption use of genetic techniques to inactivate a gene by inserting within it a DNA fragment containing an easily selectable marker. The inserted fragment is called a *cassette*, and the process of insertion, *cassette mutagenesis.*

Gene Therapy replacement or augmentation of a dysfunctional gene for medical purposes

Molecular Cloning isolation and incorporation of a fragment of DNA into a vector where it can be replicated

Nucleic Acid Probe a strand of nucleic acid that can be labeled and used to hybridize to a complementary molecule from a mixture of other nucleic acids

Polymerase Chain Reaction (PCR) a method used to amplify a specific DNA sequence *in vitro* by repeated cycles of synthesis using specific primers and DNA polymerase

Reverse Translation the mental process of using a codon table and the amino acid sequence of a protein to obtain a possible sequence of the mRNA or the gene that encoded the protein

Shuttle Vector a cloning vector that can replicate in two or more dissimilar hosts

Site-Directed Mutagenesis a technique whereby a gene with a specific mutation can be constructed *in vitro*

Synthetic DNA a DNA molecule made by a chemical process in a laboratory

T-DNA the segment of the *Agrobacterium* Ti plasmid that is transferred to plant cells

Ti Plasmid a plasmid in *Agrobacterium* species capable of transferring genes from bacteria to plants

Transgenic Organisms plants or animals that stably pass on cloned DNA that has been inserted into them

---

The concepts of molecular genetics, described in the four preceding chapters, have made possible the development of sophisticated procedures for the isolation, manipulation, and expression of genetic material, a field called **genetic engineering.** Genetic engineering has applications in both basic and applied research. In basic research, genetic engineering techniques are used to study the mechanisms of gene replication and expression in prokaryotes, eukaryotes, and their viruses. Some of the most important basic discoveries of molecular genetics have been made using genetic engineering techniques. For applied research, genetic engineering permits the development of microbial cultures capable of producing valuable products such as human insulin, human growth hormone, interferon, vaccines, and industrial enzymes. Genetic engineering for commercial application, sometimes called **biotechnology,** seems to have limitless potential.

Underlying genetic engineering is the isolation, purification, and replication of specific DNA fragments, a process called **molecular cloning.** Having large amounts of pure DNA allows localization, characterization, and manipulation of genes and their products. With cloned DNA we can determine the *nucleotide sequence* of a gene, from which we can derive, through the genetic code, the *amino acid sequence* of its protein product. The cloned DNA itself can be used as a *probe* to determine the structure of more complex DNA molecules like the human genome. By taking genes from organisms that are difficult or dangerous to work with and moving them into well-characterized, safe microorganisms, valuable biological substances can be produced cheaply and in quantities that were unthinkable until the advent of genetic engineering. By changing the sequence of a cloned gene in a predetermined way, genetic engineers can literally design new, possibly useful biological products that are simply unavailable from natural sources.

In this chapter we discuss the tools and processes involved in creating and purifying desired genes by techniques using *in vitro* recombination, or **recombinant DNA.** Molecular cloning involves creating recombinant DNA and introducing it into a host cell where it will be replicated. We then describe how genetic engineering is used to produce large quantities of desired gene products and how the products themselves can be altered by **site-directed mutagenesis.** We present examples of practical results derived from genetic engineering and conclude the chapter with a review of the principles that underlie biotechnology.

10.1
Molecular Cloning

Molecular cloning is at the base of most genetic engineering procedures. The purpose of molecular cloning (also called *gene cloning*) is to isolate large quantities of specific genes in pure form. While it might be theoretically possible to physically isolate pure DNA fragments with single genes from a restriction enzyme digest of chromosomal DNA (∞ Section 6.4), a little reflection will demonstrate the impracticality of such an approach. Consider that even for a genetically simple organism like *Escherichia coli*, a specific gene represents 1–2 kilobases (kb) out of a genome of about 4700 kb. An average *E. coli* gene is thus less than *0.05%* of the total DNA in the cell. In humans the problem is even worse because the coding regions of average genes are not much bigger than in *E. coli* but the genome is 1000 times larger! In contrast, the DNA of bacteriophage lambda is only 50 kb, and the DNA of some plasmids is less than 5 kb. In these genetic elements, a single average gene constitutes 2–40% of the DNA.

Thus, the basic strategy of molecular cloning is to move the desired gene from a large, complex genome to a small, simple one. Fortunately, our knowledge of DNA chemistry and enzymology allows us to break and join DNA molecules *in vitro*. This process is known as *in vitro* **recombination.** Restriction enzymes, DNA ligase (∞ Sections 6.4 and 6.5), and synthetic DNA (see Section 10.8) are important tools used for *in vitro* recombination.

Molecular cloning can be divided into several steps:

1. Isolation and fragmentation of the source DNA. This can be total genomic DNA from an organism of interest, DNA synthesized from an RNA template by reverse transcriptase (∞ Section 8.22), DNA synthesized by the polymerase chain reaction (see Section 10.9), or even DNA synthesized from nucleotides *in vitro*. If genomic DNA is the source, it is generally cut with restriction enzymes to give a mixture of fragments.

2. Joining the DNA fragments to a **cloning vector** with DNA ligase. The small, independently replicating genetic elements used to replicate genes are known as **cloning vectors.** Cloning vectors are generally designed to allow recombination of foreign DNA at a restriction site that cuts the vector in a way that does not affect its replication. If the source DNA and the vector are cut with the same restriction enzyme, joining can be mediated by annealing of the single-stranded regions called "sticky ends" (∞ Sections 6.4 and 6.11). Blunt ends generated by different restriction enzymes can also be joined, and different sticky ends or blunt ends can be joined by the use of synthetic DNA **linkers** or **adapters.** The

properties of cloning vectors are discussed in Sections 10.2–10.4.

3. Introduction and maintenance of a **host** organism. The recombinant DNA molecule made in a test tube is introduced into a host organism, for example, by DNA transformation (∞ Section 9.6) where it can replicate. Transfer of the DNA into the host usually yields a mixture of clones. Some cells contain the desired cloned gene, whereas other cells contain other clones generated by joining the source DNA to the vector. Such a mixture is known as a **DNA library** or a **gene library** because many different clones can be purified from the mixture, each containing different cloned DNA segments from the source organism.

4. Detection and purification of the desired clone. Often one of the most difficult tasks is finding the right clone in a mixture that may contain thousands of others. Techniques for finding the right clone will be discussed in Section 10.6.

5. Production of large numbers of cells or bacteriophage containing the desired clone for isolation and study of the cloned DNA.

CONCEPT CHECK 10.1

The isolation of large quantities of a specific gene by molecular cloning is usually done using a plasmid or virus as the cloning vector. Restriction enzymes and DNA ligase are used in an *in vitro* recombination procedure to produce the hybrid DNA molecule. Once introduced into a suitable host, the target DNA can be produced in large amounts under the control of the cloning vector.

✔ *What is the purpose of molecular cloning?*

✔ *What are the roles of a cloning vector, restriction enzymes, and DNA ligase in molecular cloning?*

10.2
Plasmids as Cloning Vectors

Plasmids replicate independently of the host chromosome (∞ Section 9.8). In addition to carrying genes required for their own replication, most plasmids are natural vectors because they often carry other genes that confer important properties on their hosts (∞ Section 9.8). Such genes can be acquired by recombination within the host. In genetic engineering, geneticists add genes to a plasmid in a test tube.

Plasmids have very useful properties as cloning vectors. These properties include (1) small size, which makes the DNA easy to isolate and manipulate;

(2) circular DNA, which makes the DNA more stable during chemical isolation; (3) independent origin of replication so plasmid replication in the cell proceeds independently from direct chromosomal control; (4) multiple copy number, so they can be present in the cell in several or numerous copies, making amplification of the DNA possible; (5) the presence of selectable markers such as antibiotic resistance genes, making detection and selection of plasmid-containing clones easier.

Although in the natural environment conjugative plasmids are generally transferred by cell-to-cell contact, plasmid cloning vectors generally have been modified to prevent their transfer conjugatively in order to achieve biological containment. However, transfer in the laboratory can be brought about by transformation or electroporation (∞ Section 9.6). Depending on the host–plasmid system, replication of the plasmid may be under tight cellular control, in which case only a few copies are made, or under relaxed cellular control, in which case a large number of copies are made. Achievement of high copy number is often important in gene cloning, and by proper selection of the host–plasmid system and manipulation of cellular macromolecule synthesis, plasmid copy numbers of several thousand per cell can be obtained.

An example of a suitable cloning plasmid is pBR322, which replicates in *Escherichia coli* (Figure 10.1). Plasmid pBR322 has a number of characteristics that make it suitable as a cloning vehicle:

1. It is relatively small, only 4361 base pairs.

2. It is stably maintained in its host (*Escherichia coli*) in relatively high copy number, 20–30 copies per cell.

3. It can be amplified to a very high number (1000–3000 copies per cell, about 40% of the genome!) by inhibition of protein synthesis by the addition of chloramphenicol.

4. It is easy to isolate in the supercoiled form using a variety of simple techniques (see the box, Working with Nucleic Acids: The Tools, in Chapter 6).

5. A reasonable amount of foreign DNA can be inserted, although inserts of more than 10 kilobases lead to plasmid instability.

6. The complete base sequence of this plasmid is known, making it possible to locate sites where restriction enzymes can act.

7. There are *single* cleavage sites for various restriction enzymes such as *Pst*I, *Sal*I, *Eco*RI, *Hind*III, and *Bam*HI. It is important that only a single recognition site for at least one restriction enzyme is available so treatment with that enzyme opens the plasmid to a full-length linear molecule but does not cut it into pieces.

8. It has a gene conferring ampicillin resistance on the host and another conferring tetracycline resistance. These permit ready selection of hosts containing the plasmid. The sites recognized by some of the restriction enzymes are within one or the other of these resistance genes, facilitating the identification of plasmids carrying cloned DNA (see later).

9. It can be placed into cells easily by transformation.

The use of plasmid pBR322 in gene cloning is shown in Figure 10.2. As seen, the *Bam*HI site is within the gene for tetracycline resistance and the *Pst*I site is within the gene for ampicillin resistance. If a piece of foreign DNA is inserted into one of these sites, the antibiotic resistance conferred by the gene containing this site is lost, a phenomenon called **insertional inactivation**. Insertional inactivation is used to detect the presence of foreign DNA within the plasmid. Thus, when pBR322 is digested with *Bam*HI and linked with foreign DNA, and transformed bacterial clones then isolated, those clones that are both ampicillin-resistant and tetracycline-resistant *lack* the foreign DNA (the plasmid incorporated into these cells represents vector DNA that had recycled without picking up foreign DNA). However, those cells still *resistant* to ampicillin but *sensitive* to tetracycline *contain* the plasmid with inserted foreign DNA. Since ampicillin resistance and tetracycline resistance can be determined independently on agar plates, isolation of bacteria containing the desired clones and elimination of cells not containing the plasmid can readily be accomplished.

The first plasmid cloning vectors used were naturally occurring plasmids. The plasmid pBR322 repre-

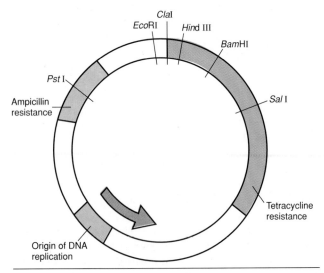

Figure 10.1 The structure of plasmid pBR322, a typical cloning vector, showing the essential features. The arrow indicates the direction of DNA replication from the origin.

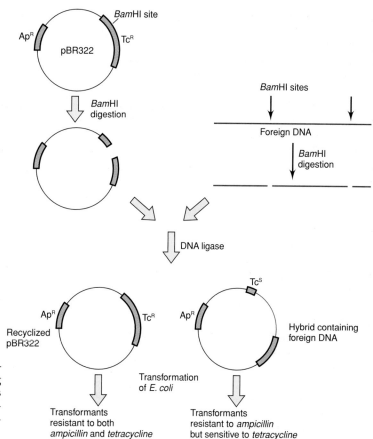

Figure 10.2 The use of plasmid pBR322 as a cloning vector, showing how insertion of foreign DNA causes inactivation of the tetracycline resistance gene, permitting easy identification of transformants containing the cloned DNA fragment.

sents a later generation of cloning vectors, themselves constructed *in vitro*. The tetracycline resistance gene of pBR322 came from one plasmid, the replication origin from another, and the ampicillin resistance gene from the transposon Tn3 (∞ Section 9.10). There are now newer generations of plasmid vectors that have been engineered to have even more useful features and are even simpler to use. These new features almost always include a **polylinker** or **multiple cloning site,** a short segment of DNA with many different restriction sites, each unique to the vector. This polylinker is usually contained in the coding region of a gene where insertional inactivation is very easy to monitor. Such features are also found in bacteriophage vectors, and a specific example is discussed in Section 10.3.

Cloning in plasmids such as pBR322 is a versatile and fairly general procedure of wide use in genetic engineering, particularly when the fragment to be cloned is fairly small. Plasmids are often the best cloning vectors if *expression* of the cloned gene is desired (see Section 10.7). Plasmid vectors based on the 2-μm circle (∞ Section 9.12) are also available for cloning in the yeast *Saccharomyces cerevisiae*.

10.3

Bacteriophage Lambda as a Cloning Vector

We have discussed the fact that during specialized transduction (∞ Section 9.7) some host genes become incorporated into a bacteriophage genome. One phage that is used as a specialized transducing phage is bacteriophage lambda (∞ Section 8.12). During specialized transduction, lambda acts as a vector but the recombination occurs in the cell, not in a test tube. Lambda can also be used as a cloning vector for *in vitro* recombination. It is a particularly useful cloning vector because its molecular genetics is well known, it can hold larger amounts of DNA than most plasmids, and DNA can be efficiently packaged into phage particles *in vitro*. These can be used to infect suitable host cells, and infection is much more efficient than transformation (transfection). Lambda has a complex genetic map (∞ Figure 8.26) and a large number of genes. However, the central third of the lambda genome, between genes *J* and *N*, is not essential and can be replaced with foreign DNA.

Modified lambda phages

Wild-type lambda is not suitable as a cloning vector because it has too many restriction enzyme sites. To avoid this difficulty, modified lambda phages have been constructed that can be used in cloning. In one set of modified lambda phages, the so-called Charon phages, unwanted restriction enzyme sites have been removed by point mutation, deletion, or substitution. In variants that have only a single restriction site, a foreign piece of DNA can be *inserted*, whereas in variants with two sites, the foreign DNA can *replace* a specific segment of the lambda DNA. The latter variants, called **replacement vectors,** are especially useful in cloning large DNA fragments.

Figure 10.3 shows some of the essential features of a wild-type lambda and two of the Charon vectors. Whereas wild-type lambda contains five *Eco*RI sites, Charon 4A contains three and Charon 16 only one. Charon 4A is used as a replacement vector; the two small interior fragments are cut out and discarded during cloning. With Charon 16, the DNA to be cloned is inserted at the single *Eco*RI site. Both Charon 4A and 16 contain deletions (not shown in the figure) that not

only remove some sites found in the wild-type lambda but also make the genome smaller. This allows the cloning of larger DNA fragments.

Both vectors also contain substitution mutations, which are shown in Figure 10.3. One of the substitutions is the gene for β-galactosidase. When the vectors replicate on a lactose-negative (Lac⁻) strain of *Escherichia coli*, β-galactosidase is synthesized from the phage gene and the presence of lactose-positive (Lac⁺) plaques can be detected by using a color indicator agar (see Section 10.4). If a foreign gene is inserted *into* the β-galactosidase gene, the Lac⁺ character is lost. Such Lac⁻ plaques can be readily detected as colorless plaques among a background of colored plaques.

Steps in cloning with lambda

Cloning with lambda replacement vectors involves the following steps (Figure 10.4):

1. Isolation of the vector DNA from phage particles and digestion with the appropriate restriction enzyme.

2. Connection of the two lambda fragments to fragments of foreign DNA using DNA ligase. Conditions are chosen so molecules are formed of a length suitable for packaging into phage particles.

3. Packaging of the DNA by adding cell extracts containing the head and tail proteins and allowing the formation of viable phage particles.

4. Infection of *E. coli* and isolation of phage clones by picking plaques on a host strain.

5. Checking recombinant phage for the presence of the desired foreign DNA sequence using nucleic acid hybridization procedures or observation of genetic properties.

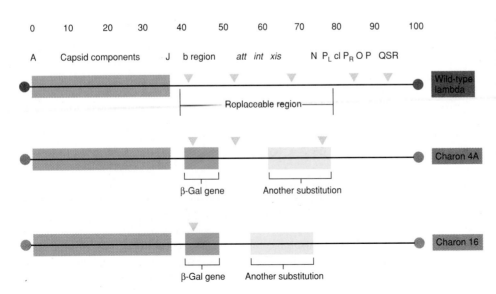

Figure 10.3 Molecular cloning with lambda. Abbreviated genetic map of bacteriophage lambda showing the cohesive ends as circles (∞ Figure 8.26). Charon 4A and 16 are both derivatives of lambda, which have various substitutions and deletions in the nonessential region. One of the substitutions in each case is a gene (β-Gal) that codes for the enzyme β-galactosidase, which permits detection of clones containing this phage. Whereas the wild-type lambda genome is 48.5 kilobase pairs, that for Charon 4A is 45.4 and that for Charon 16 is 41.7 kilobase pairs. The arrows (▼) shown above the maps of each phage indicate the sites recognized by the restriction enzyme *Eco*RI.

Although lambda is a useful cloning vector, there are limits on how much DNA can be inserted. Viability of phage particles is low if the DNA is longer than 105% of normal lambda DNA, and some lambda genes cannot be discarded and still maintain the vector's ability to replicate. Therefore, really large DNA fragments (greater than 20 kb) cannot be efficiently cloned.

Cosmids

A related type of vector that employs specific lambda genes is called a **cosmid**. Cosmids are plasmid vectors containing foreign DNA plus only the *cos* (cohesive end) site from the lambda genome. These *cos* sites are required for packaging DNA into lambda virions. Cosmids are constructed from plasmids containing cloned DNA by ligating the lambda *cos* region to the plasmid DNA. The modified plasmid can then be packaged into lambda virions *in vitro* as described previously, and the phage particles used to transduce *Escherichia coli*. Cosmid construction avoids the necessity of having to transform *E. coli*, which at best is an inefficient process (∞ Section 9.6).

One major advantage of cosmids is that they can be used to clone large fragments of DNA. Therefore, fewer clones are needed to obtain representation of the whole genetic element. This has been especially useful in the cloning of genes from eukaryotic chromosomes, where large amounts of DNA are involved. Another advantage of cosmids is that the DNA can be stored in phage particles instead of in plasmids. Phage particles are much more stable than plasmids, and so the recombinant DNA can be kept for long periods of time (gene banking).

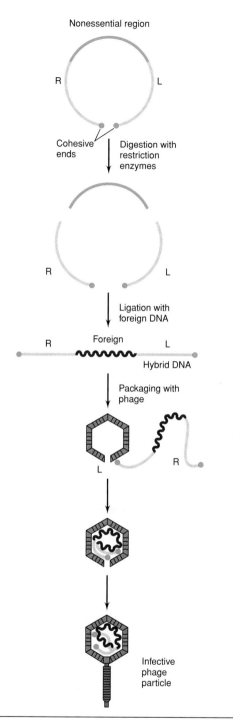

Figure 10.4 The use of bacteriophage lambda as a cloning vector. (See text for details.)

Selection of recombinants is less of a problem with lambda replacement vectors (such as Charon 4A) than with plasmids because (1) the efficiency of transfer of recombinant DNA into the cell by lambda is very high, and (2) lambda fragments that have not received new DNA are too small to be incorporated into phage particles.

CONCEPT CHECK   *10.3*

Bacteriophages such as lambda have been modified to make useful cloning vectors. Larger amounts of foreign DNA can be cloned with lambda than with many plasmids. In addition, the recombinant DNA can be packaged *in vitro* for efficient transfer to a host cell. Plasmid vectors containing the lambda *cos* sites are called cosmids, and they can carry a large fragment of foreign DNA.

✔ *Why is the ability to package recombinant DNA in a test tube useful?*

✔ *What is a* replacement vector?

10.4

Other Vectors

A number of other vectors have been developed that are useful for various purposes in recombinant DNA technology. It is understandable that in the early phases of the development of recombinant DNA techniques, the vectors used were those that had been in most widespread use in other types of genetic research. However, it does not necessarily follow that a genetic system that is useful in basic research automatically provides the best system for the development of a cloning vector. In addition, vectors are sometimes required that have specific applications beyond those needed for simply cloning a DNA fragment. In this section, we discuss briefly some other kinds of vectors and describe some useful modifications.

One important class of newly developed vectors, called **expression vectors,** are used when it is desired to obtain synthesis of the protein coded for by the foreign gene cloned into the vector. We discuss expression vectors in Section 10.7. Related to expression vectors are **secretion vectors,** in which the protein product is not only expressed but also secreted (excreted) from the cell. In these vectors the gene is cloned so the protein expressed carries a *signal sequence* (∞ Section 6.9).

For a number of reasons it is useful to be able to move DNA between completely unrelated organisms. To move DNA between unrelated organisms, a **shuttle vector** is used. A shuttle vector is one that can replicate in two different organisms. Like most specialized vectors, shuttle vectors have themselves been constructed using recombinant DNA techniques. Shuttle vectors have been developed that replicate in both *Escherichia coli* and *Bacillus subtilis*, *E. coli* and yeast, and *E. coli* and mammalian cells, as well as in many other pairs of organisms.

Vectors for DNA sequencing: Bacteriophage M13

M13 is a filamentous phage containing single-stranded DNA and replicates without killing its host (∞ Section 8.10). Mature particles of M13 are released from host cells by a budding process, and it is possible to obtain infected cultures that can provide continuous sources of phage DNA. An important feature of M13 is its single-stranded DNA. In the Sanger procedure for DNA sequencing (∞ Working with Nucleic Acids: The Tools, Chapter 6), single-stranded DNA is needed and DNA cloned into M13 thus provides a ready source of this single-stranded DNA. Also, single-stranded DNA is very useful as a probe for detecting other nucleic acid sequences in transfer procedures such as the Southern blot, and M13 permits ready production of such single-stranded DNA probes.

However, in order to use M13 for cloning, a double-stranded form must be available because restriction enzymes work only on double-stranded DNA. Double-stranded M13 DNA can be obtained from infected cells because M13 replicates in the host as a double-stranded *replicative form* (∞ Section 8.10). Most of the genome of wild-type M13 contains genetic information essential for virus replication.

However, there is a small region called the intergenic sequence that can be used as a cloning site. Variable lengths of foreign DNA, up to about 5 kilobase pairs (kbp), can be cloned without affecting phage viability—as the genome gets larger, the virion gets longer. M13mp18 is a derivative of M13 in which the intergenic region has been modified to facilitate cloning. A map of this vector is shown in Figure 10.5a.

One modification is the insertion of a functional fragment of *lacZ*, the *Escherichia coli* gene that encodes the enzyme β-galactosidase. Therefore, cells infected with M13mp18 can be easily detected by their color on indicator plates (Figure 10.5b). This *lacZ* gene has itself been modified to contain a 54-base-pair DNA fragment called a *polylinker*. The polylinker contains several restriction sites unique in M13 and can therefore be used for cloning. The polylinker is inserted into the beginning of the coding portion of the *lacZ* gene. This small insertion is in-frame, and the extra 18 amino acids do not affect the activity of the enzyme encoded by the gene. However, insertion of additional DNA into the polylinker during cloning inactivates the gene. Phages that contain additional DNA inserts give rise to colorless plaques, and it is therefore very simple to identify clones (Figure 10.5b and c). Similar constructs are used in lambda cloning vectors to allow identification of cells containing cloned DNA.

How then are M13 vectors used in cloning? The replicative double-stranded DNA is isolated from the infected host and treated with a restriction enzyme. The foreign DNA, also double-stranded, is treated with the same restriction enzyme. On ligation, double-stranded M13 molecules are obtained that contain the foreign DNA. When these molecules are introduced into the cell by transformation, they replicate and in time produce mature bacteriophage particles containing single-stranded DNA molecules. Only one strand of DNA is packaged into mature phage. Which of the two foreign strands the mature phage contains depends on the *orientation* in which the strand was inserted. Because foreign DNA can be inserted (in separate phage molecules) in either orientation, *both* strands of the foreign DNA can be cloned.

The single-stranded M13 DNA containing the foreign DNA can then be used in DNA sequencing. Since the base sequence where the foreign DNA is inserted is known (based on the specificity of the restriction enzyme used), it is possible to use a synthetic oligonucleotide complementary to this region as a primer and hence determine the sequence of the whole DNA downstream from this point. In this way, M13 derivatives have proved extremely useful in sequencing foreign DNA, even rather long molecules. (The sequence of the whole bacteriophage lambda DNA, 48,514 bases in length, was determined rather quickly using a gene library constructed in an M13 vector.) A modified system for producing single-stranded DNA has been constructed by placing an M13 origin of replication into a

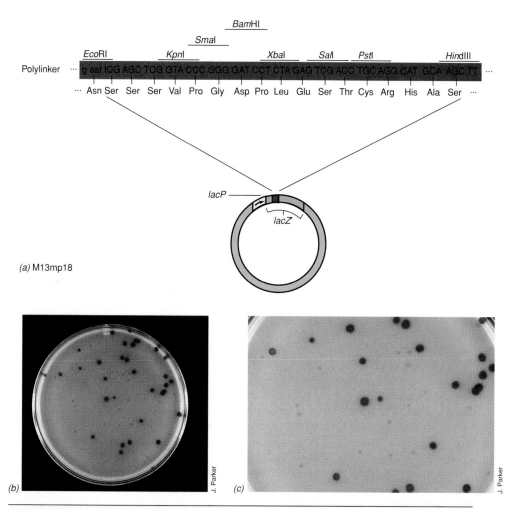

Polylinker

··· g aat TCG AGC TCG GTA CCC GGG GAT CCT CTA GAG TCG ACC TGC AGG CAT GCA AGC TT ···

··· Asn Ser Ser Ser Val Pro Gly Asp Pro Leu Glu Ser Thr Cys Arg His Ala Ser ···

(a) M13mp18

Figure 10.5 (a) A partial map of M13mp18, a derivative of M13 constructed for use as a cloning vector. The vector contains the *lac* promoter and a gene, *lacZ'*, which encodes a functional part of β-galactosidase. At the beginning of this gene is a polylinker that contains several restriction sites but maintains the proper reading frame. The amino acids encoded by the polylinker are shown. Most DNA fragments cloned into the polylinker disrupt the *lacZ'* gene and abolish β-galactosidase activity. (b) A plate with plaques formed by M13mp18 and by clones made using this vector on a lawn of sensitive bacteria plated on a medium containing the chemical 5-bromo-4-chloro-3-indolyl-β-D-galactopyranoside, called X-gal. When β-galactosidase hydrolyzes X-gal, it releases a relatively insoluble blue dye (this same compound was used to make the plate shown in Figure 9.1c). Many plaques on this plate are blue, indicating the presence of vector without cloned DNA. However, many of the plaques are colorless, indicating that foreign DNA has been inserted into the vector and the *lacZ'* gene has been disrupted. (c) An enlargement of a portion of this plate.

plasmid vector. These plasmids can be used for genetic manipulation in the conventional way, and single-stranded DNA can be produced at will by infecting cells containing such a plasmid with an M13 helper phage.

Vectors have also been constructed that are hybrids between a filamentous phage, like M13, and a plasmid. Such vectors are called **phagemids.** They contain both phage and plasmid origins of replication. Normally replication is from the plasmid origin, but when a cell containing a phagemid is infected with a wild-type phage, the phage origin on the vector is used to synthesize single-stranded DNA from the vector (and whatever cloned gene it might carry). This single-

stranded DNA is packaged into virions and can easily be isolated and used for sequencing. Usually, phage-mids can stably carry a larger fragment of cloned DNA than a typical M13-derived vector.

Yeast artificial chromosomes and the human genome project

As mentioned in Sections 10.2 and 10.3, plasmid vectors usually contain less DNA than lambda vectors, and lambda vectors less than cosmid vectors. When making *gene libraries* of bacteria or simple eukaryotes, lambda or cosmid vectors can be used and the entire library will contain at most a few thousand different clones. The size of a gene library (the number of clones) required to contain the complete genome of an organism depends on both the size of the genome and the size of the insert that can be placed in the vector. If the average size of the insert is 20 kb, then it will take many more clones to make a library of a typical mammalian genome (3×10^9 base pairs) than of a bacterial genome (4×10^6 base pairs). Currently, there is a concerted effort in the international genetics community to map, clone, and sequence the entire human haploid genome—the *Human Genome Project*. This is clearly a much larger undertaking than sequencing the prokaryotic chromosomes (∞ Section 9.10). It is useful in such a project to have a cloning vector that can hold very large segments of DNA so the size of the initial gene library can be limited. Such vectors have been developed and are called **yeast artificial chromosomes** (YACs).

These vectors have been designed to replicate in yeast like normal chromosomes, but they have sites where DNA can be inserted. To function like normal eukaryotic chromosomes, YACs must have an *origin of DNA replication, telomeres* at the ends of the chromosome (∞ Section 6.3), and a *centromere* (the section of the chromosome required for segregation during mitosis). They must also contain a cloning site and a gene that can be used for selection after transformation into the host. Figure 10.6 shows a diagram of a YAC vector into which foreign DNA has been cloned. YAC vectors are themselves only about 10 kilobase pairs, but they can have 200–800 kilobase pairs of cloned DNA inserted. After identifying a particular gene or region in the cloned DNA on a YAC, this gene can be *subcloned* into a plasmid or bacteriophage vector for more detailed analysis.

Other eukaryotic vectors

In addition to the YAC vectors described previously, we have mentioned that plasmid vectors are available for use in the yeast *Saccharomyces cerevisiae* (see Section 10.2). The development of vectors for use in eukaryotes, like the same process in prokaryotes, depends on the scientific questions geneticists want to ask or the practical goals of the genetic engineer. YACs are used to clone very large fragments of DNA, and the plasmid vectors, much smaller fragments. Although yeast is an extremely important organism both for genetic studies (∞ Section 9.13) and commercial applications (∞ Section 12.12), it is often important to use other eukaryotes as hosts for cloned DNA. Many cloning vectors have been developed for many different eukaryotes, including plants (see Section 10.14).

Most vectors used in the higher eukaryotes are virus vectors. The DNA virus SV40 (∞ Section 8.18), a virus causing tumors in primates, has been developed as a cloning vector into human tissue culture lines. SV40 virus has double-stranded *circular* DNA, and the entire nucleotide sequence is known. Derivatives of SV40 that do not induce cancer have been developed that permit cloning of mammalian genes, and expression of these genes has been obtained. SV40 or similar mammalian cloning vectors should prove very useful in understanding the events involved in gene expression in these complex organisms.

There are mammalian vectors that utilize *adenovirus* (∞ Section 8.21) and *vaccinia virus* (∞ Section 8.20). Vaccinia virus vectors have been used in the development of new vaccines (see Section 10.13). A variety of eukaryotic expression vectors have also been developed and are essentially of two kinds. One type is designed to produce a particular protein for commercial purposes. Vectors derived from *baculovirus*, a DNA virus that replicates in insect cells, can be used to make large quantities of the products of cloned genes. Other expression vectors are being developed so a cloned gene can be stably maintained and expressed in an organism or tissue, often as an approach to *gene therapy* (see Section 10.15).

The *retroviruses* (∞ Section 8.22) can be used to introduce genes into mammalian cells because these viruses replicate through a DNA form that becomes integrated into the host chromosome.

One further possible approach is the development of human artificial chromosomes (HACs). Unlike YACs, whose original purpose was to carry very large fragments of DNA, HACs are seen as vectors that should be stably maintained in human cells and carry genes that can be expressed in a normal fashion. However, the ability of HACs to carry large amounts

Figure 10.6 Diagram of a yeast artificial chromosome (YAC) containing foreign DNA. The foreign DNA was cloned into the YAC vector at a *Not*I restriction site. The telomeres at the end of the YAC are labeled TEL and the centromere CEN. The origin of replication is labeled ARS (for autonomous replication sequence). For this vector, the gene used for selection is called URA3. The host into which the clone is transformed has a mutation in that gene so that it normally requires uracil for growth (Ura⁻). Host cells containing this YAC become Ura⁺. The diagram is not drawn to scale; the inserted DNA would normally be 200–800 kilobase pairs long and the vector about 10 kilobase pairs.

of cloned DNA could be useful because the presence of some very large introns makes some normal human genes and their regulatory regions very large, over a million base pairs in many cases.

10.5

Hosts for Cloning Vectors

The ideal characteristics of a host for cloned genes are rapid growth, capable of growth in an inexpensive culture medium, not harmful or pathogenic, capable of taking up DNA, and stable in culture. The host must have the appropriate enzymes to allow replication of the vector. The most useful hosts for cloning are microorganisms that grow well and about which a lot of genetic information is available, such as the Bacteria *Escherichia coli* and *Bacillus subtilis* and the yeast *Saccharomyces cerevisiae*. However, many basic scientific questions can be answered only if the cloned DNA can be returned to the species of organism from which it originated. This is particularly true in studies involving gene regulation. Finally, if recombinant DNA itself is to be used therapeutically to treat human disease, the host must be a human being.

Prokaryotic hosts

Although most molecular cloning has been done in *Escherichia coli*, there are perceived disadvantages in using this host. *Escherichia coli* presents dangers for large-scale production of products derived from cloned DNA because it is found in the human intestinal tract and is potentially pathogenic. Also, even nonpathogenic strains produce endotoxins that can contaminate products, an especially bad situation with pharmaceutical injectables. Finally, *E. coli* retains extracellular proteins in the periplasmic space, making isolation and purification potentially difficult. However, modified *E. coli* strains have been developed for which most of these problems have been eliminated. Because of the extensive knowledge of its genetics and biochemistry, *E. coli* remains the organism of choice for most cloning studies.

The gram-positive organism *Bacillus subtilis* can also be used as a host. *Bacillus subtilis* is not potentially pathogenic, does not produce endotoxin, and secretes proteins into the medium. Although the technology for cloning in *B. subtilis* is not nearly as well developed as that for *Escherichia coli*, plasmids and phages suitable for cloning have been developed and transformation is a well-developed procedure in *B. subtilis*. Disadvantages of using *B. subtilis* as a cloning host exist, however. Plasmid instability is a real problem, and it is hard to maintain plasmid replication over many culture transfers. Also, foreign DNA is not well maintained in *B. subtilis* cells and so the cloned DNA is often unexpectedly lost. Adapting a bacterium for use as a host for cloning experiments is not always simple.

Often organisms used as hosts for cloning must have specific genotypes to be effective. For instance, if the vector carries the gene for β-galactosidase, then the host must have a mutation in this gene. Because M13 infects only bacteria with F pili (∞ Sections 8.10 and 9.8), hosts used with M13-derived vectors contain the F plasmid. These types of considerations, and others such as the ability to select for transformants, must be taken into account whether the host is prokaryotic or eukaryotic.

Eukaryotic hosts

Cloning in *eukaryotic microorganisms* has some important uses, especially in understanding the details of gene regulation in eukaryotic systems. The yeast *Saccharomyces cerevisiae* is the best known genetically (∞ Section 9.13) and is being extensively studied as a cloning host. Plasmid vectors, as well as YACs, have been developed for yeast, and transformation using genetically engineered DNA can be accomplished. The ability to clone appropriate genetic material in yeast will advance our understanding of the complex transcription and translation systems of eukaryotes and should provide a better foundation for basic research.

For many purposes, gene cloning in *mammalian cells* would be desirable. Mammalian cell culture systems can be handled in some ways like microbial cultures and find wide use in research on human genetics, cancer, infectious disease, and physiology. In addition, DNA can be introduced into mammalian cells by transfection or electroporation (∞ Section 9.6).

One important advantage of eukaryotic cells as hosts for cloning vectors is that they already possess the complex RNA and posttranslational processing systems involved in the production of gene products in higher organisms, and so these systems do not have to be engineered into the vector as they need to be when production of the desired product is to be carried out in a prokaryote (posttranslational processing, in particular, can create some molecular cloning problems) (see Section 10.10).

A disadvantage of mammalian cells as hosts is that they are expensive and difficult to produce under large-scale conditions and expression levels of cloned

---

CONCEPT CHECK   *10.5*

Like naturally occurring plasmids and viruses, vectors with cloned DNA must be placed in a compatible host in order to replicate. The selection of the host also depends on the nature of the studies to be performed with the cloned DNA. Fast-growing Bacteria like *Escherichia coli* are often used as hosts, but some studies require the use of eukaryotic organisms like the yeast *Saccharomyces cerevisiae* or cultured mammalian cells.

genes are often low. Insect cell lines are simpler to grow, and as we have mentioned, vectors have been developed from an insect DNA virus, the baculovirus.

10.6
Finding the Right Clone

A crucial step in recombinant DNA technology is finding the right clone among the mixture of clones created by the recombinant DNA procedure. The foreign DNA used in the cloning procedure typically contains a large number of genes, only one or a few of which may be the genes of interest. Sections 10.3 and 10.4 discussed how one can select for hosts containing a plasmid vector by selecting for a vector marker, such as antibiotic resistance, so only these cells form colonies. For host cells containing a viral vector, one simply looks for plaques. We also discussed how these colonies or plaques can be screened for vectors that contain foreign DNA inserts by looking for the inactivation of a vector gene, often that for another antibiotic resistance, in the case of a plasmid (see Figure 10.2), or for β-galactosidase (see Figure 10.5). However, one is then left with the biggest challenge: selecting the clone that has the gene of interest. Procedures must be available for examining colonies of bacteria or plaques of infected cells growing on agar plates and detecting those few that contain the gene of interest. It is the purpose of the present section to discuss possible approaches to finding the right clone. We consider first the situation in which the gene is *expressed* (that is, the protein is synthesized) in the cloning host. Then we discuss the situation, rather common, in which the gene is not expressed and we must look for the DNA itself.

If the foreign gene is expressed in the cloning host

If the foreign gene is expressed (that is, the protein product is synthesized) in the cloning host, then procedures can be used that look for the presence of this protein in recombinant colonies. The cloning host itself must *not* produce the protein being studied. If we are looking for clones that express the gene, then we are looking for the rare colonies in which this protein is present. If the protein is one that the cloning host normally produces, then the host used must be defective, that is, mutant, for the gene of interest. Then, when the foreign gene is incorporated, the expression of this foreign gene can be detected by complementation (∞ Section 9.5). If the function is required, a complementing clone can be *selected*, greatly facilitating the process. Clearly, if the host already expressed a protein with the same activity, there will be a large background of this activity against which the protein produced via the foreign gene cannot be detected. If the protein is not normally produced in host bacteria, then the host may be naturally defective.

In some cases these novel activities are expressed and can be detected. A striking example is the cloning of luciferase genes from various types of bioluminescent beetles into *Escherichia coli*; in the dark, the clones glow in various colors depending on the type of luciferase present (Figure 10.7).

Antibody as a method of detecting the protein

If the protein does not have a readily detectable function, then a different approach is needed. It involves the use of an antibody as a reagent that is specific for the protein of interest. We will discuss antibodies and immunology in Chapter 20. For our present purposes, we note that an antibody is a serum protein produced by a mammalian system that combines in a highly specific way with another protein, the *antigen*. In the present case, the protein of interest is the antigen, and this protein is used to produce an antibody in an experimental animal. Since the antibody combines specifically with the antigen, when the antigen is present in one or more colonies on the plate, then the locations of these colonies can be determined by observing the binding of the antibody. Because only a small amount of the protein (antigen) is present in the colonies, only a small amount of antibody is bound, and so a highly sensitive procedure for detecting bound antibody must be available. In practice, this is done using a system involving a radioactive agent or an agent with a specific enzyme attached to it. The radioactivity can be detected by autoradiography using X-ray film. The enzymes used typically convert a colorless substrate into a colored one whose absorbance can be measured very sensitively. Such extremely sensitive techniques

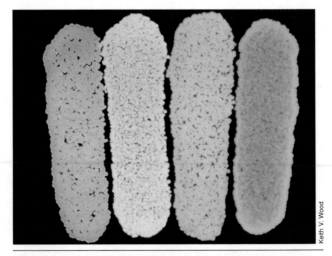

Figure 10.7 Bioluminescence from colony streaks of four strains of *Escherichia coli*, each containing cloned luciferase genes from a different species of click beetle. Each luciferase emits light of a different color. Reprinted with permission from *Science* 244:700–702 (1988), © AAAS.

for detecting antigens are discussed in more detail later (∞ Section 21.8).

Note that this method of detection involves *screening*, not selection, and so thousands of clones must be examined. These can be colonies containing plasmids or

plaques containing viruses that produce the cloned product. The whole procedure, utilizing plasmids and radioactive detection, is outlined in Figure 10.8*a*. As seen, the replica plating procedure (∞ Figure 9.2) is used to make a duplicate of the master plate, but the

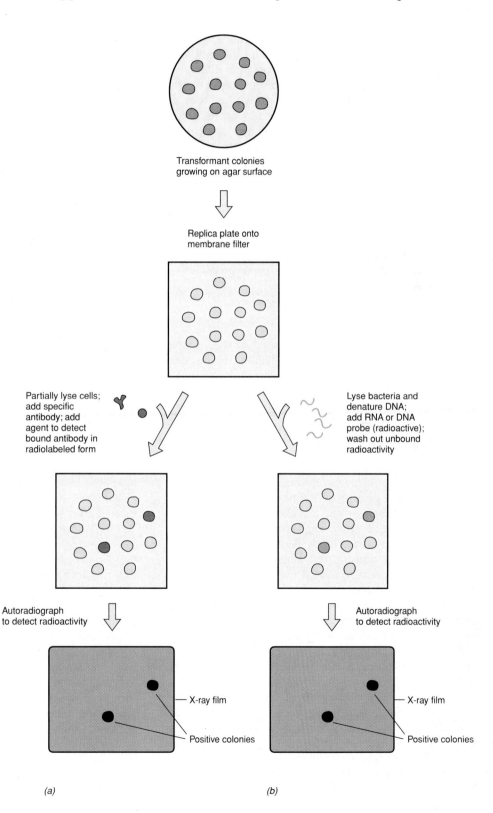

Figure 10.8 Finding the right clone. (a) Method for detecting production of protein by use of specific antibody. (b) Method for detection of recombinant clones by colony hybridization with a radioactive nucleic acid probe.

Transformant colonies growing on agar surface

Replica plate onto membrane filter

Partially lyse cells; add specific antibody; add agent to detect bound antibody in radiolabeled form

Lyse bacteria and denature DNA; add RNA or DNA probe (radioactive); wash out unbound radioactivity

Autoradiograph to detect radioactivity

Autoradiograph to detect radioactivity

X-ray film

Positive colonies

X-ray film

Positive colonies

(a)

(b)

duplication is done onto a membrane filter and all the manipulations are done with this filter. After the duplicate colonies have grown up, they are lysed to release the protein (antigen) of interest. (Screening for expression in phage vectors eliminates this step because the bacteria are already lysed.) The antibody is then added, and the antibody–antigen reaction allowed to proceed. Unbound antibody is then washed off, and a radioactive agent is then added that is specific for the antibody. A piece of X-ray film is placed over the filter and exposed. If a radioactive colony is present, a spot on the X-ray film will be observed after the X-ray film is developed. The location of this spot on the film corresponds to a location on the master plate where a colony is present that produces the protein. This colony can then be picked from the master plate and cultured.

One limitation of this procedure is that an antibody must be available that is *specific* for the protein in question. As we will see in Chapter 20, antibody can be readily produced by injecting the protein (antigen) into an animal, but the protein injected must be pure; otherwise more than one antibody will be formed. Thus, one must have previously purified the protein.

Nucleic acid probes: Searching for the gene itself

Suppose that the gene is not expressed in the cloning host or that no assay or antibody is available for the gene product. How does one detect its presence in colonies? The most general way is to use a **nucleic acid probe** containing a key part of the base sequence of the gene of interest. As we have discussed (∞ Section 6.2), nucleic acid hybridization can be used as a specific means of detecting polynucleotides with specific sequences. Either DNA or RNA can be employed as a probe. The general procedure is to label the nucleic acid probe, often with radioactive phosphate but nonisotopic techniques are increasingly being used, and allow a single-stranded probe to hybridize with single-stranded nucleic acid derived from the cloned DNA. Because of specific complementary base pairing, two single-stranded polynucleotides will hybridize only if they are fairly complementary. By using appropriate hybridization conditions, it is possible to obtain binding of the radioactive probe only to the nucleic acid of interest.

The way in which a nucleic acid probe can be used to detect the presence of recombinant DNA in colonies is shown in Figure 10.8*b*. The procedure, **colony hybridization,** again makes use of replica plating to produce a duplicate of the master plate on a membrane filter. (The same procedure can be carried out with virus vectors by blotting the plaques onto a membrane.) The cells on the filter are lysed in place to release their nucleic acid and to convert the DNA into a single-stranded form and fix it to the filter. This filter is then treated with a radioactive nucleic acid

probe (either RNA or DNA) to allow hybridization, and after removal of unbound radioactive nucleic acid, the filter is subjected to autoradiography. After development, the X-ray film is examined for spots. These correspond to locations on the membrane where the radioactive probe hybridized the DNA from a particular colony. Colonies corresponding to these spots are then picked and studied further. A modification of this procedure, avoiding the use of a radioactive probe, has been developed for clinical microbiology (∞ Section 21.10).

CONCEPT CHECK   **10.6**

Special procedures are needed for detecting the foreign gene in the cloning host. If the gene is expressed, the presence of the foreign protein itself, as detected either by its activity or by reaction with specific antibodies, is evidence that the gene is present. However, if the gene is not expressed, then its presence can be detected by use of a nucleic acid probe.

✔ *Does use of nucleic acid probes depend on gene expression? Explain.*

✔ *Why is it necessary to lyse cells containing plasmids in order to detect the product of the cloned gene?*

10.7
Expression Vectors

For practical applications it is essential that systems be available in which the cloned genes can be *expressed.* Organisms have complex regulatory systems, and many genes are not expressed all the time (∞ Chapter 7). One of the major goals of genetic engineering is the development of vectors in which *high levels* of gene expression can occur. An **expression vector** is a vector that not only can be used to clone the desired gene but also contains the necessary regulatory sequences so expression of the gene is subject to experimental manipulation.

In this section we will emphasize prokaryotic expression vectors. However, expression vectors are also used in eukaryotes. The baculovirus and retrovirus vectors we mentioned (see Section 10.4) are expression vectors, as are the yeast plasmid vectors. Many of the characteristics of eukaryotic and prokaryotic expression vectors are broadly the same (and many expression vectors are also shuttle vectors). However, the details are different because there are differences in gene structure and some aspects of gene expression between prokaryotes and eukaryotes (∞ Section 6.1).

Requirements of a good expression system

Many factors influence the level of expression of a gene, and a vector must be constructed in which all these factors are under control. In addition, a host must be used in which the expression vector is most effective. We summarize the key requirements of a good expression system here.

1. **Number of copies of the gene per cell.** In general, more product is made if many copies of the gene are present. Therefore, if very high levels of expression are desired, high copy number vectors, for example, small plasmids such as pBR322, are valuable. Sometimes for research purposes (or therapeutic purposes with mammalian vectors) it is desirable to have only a single copy of the cloned gene in the cell. For these cases, *integrating vectors* have been developed so the gene can recombine into the host chromosome.

2. **Promoter strength and regulation.** For very high levels of expression, it is essential to produce high levels of mRNA. The promoter region is the site at which binding of RNA polymerase first occurs (∞ Section 6.6), and native promoters in different genes vary considerably in RNA polymerase binding strength. In engineering a practical system, it is important to include a *strong* promoter in the expression vector. For Bacteria, the DNA region around 10 and 35 nucleotides before the start of transcription (called the *−10* and *−35 regions*) (∞ Figure 6.24) is especially important in the promoter. Many *Escherichia coli* genes are controlled by relatively weak promoters, and promoters from eukaryotes and some other prokaryotes function poorly or not at all in *E. coli*. Promoters from *E. coli* that have been used in the construction of expression vectors include *lac* (the *lac* operon promoter), *trp* (the *trp* operon promoter), *tac* (a synthetic hybrid of the −35 region of the *trp* promoter and the −10 region of the *lac* promoter), and lambda P_L (the leftward lambda promoter) (∞ Section 8.12). Note that each of these promoters can be regulated (∞ Sections 7.3, 7.5, and 8.12). In almost all cases it is important to be able to regulate the expression of the cloned gene in order to maximize yield. We shall discuss this in more detail later.

   A novel regulatory system has been created using the bacteriophage T7 promoter and RNA polymerase. When T7 infects *Escherichia coli*, it codes for its own RNA polymerase, which recognizes only T7 promoters, thus effectively shutting down host transcription (∞ Section 8.11). In expression vectors it is possible to place expression of cloned genes under control of a T7 promoter. However, when this is done, it is necessary to engineer into the plasmid the gene for T7 RNA polymerase as well. The latter is placed under control of an easily regulated promoter such as that of lambda or *lac*. Expression of the cloned gene(s) occurs shortly after T7 RNA polymerase transcription has been switched on. Because it recognizes only T7 promoters, T7 RNA polymerase transcribes only the cloned genes; all other host genes remain untranscribed.

3. **Translation initiation.** In order to synthesize protein from an mRNA, it is essential that the ribosomes bind at the correct site and begin reading in the correct frame. In prokaryotes this is accomplished by having a ribosome binding site (Shine–Dalgarno sequence) (∞ Section 6.9) and a nearby start codon on the mRNA. Bacterial ribosome binding sites are not found in eukaryotic genes, and it is thus essential that the bacterial region be present in the cloned gene if high levels of gene expression are to be obtained. Part of the requirement for proper ribosome binding is the necessity for a proper distance between the ribosome binding site and the translation initiation codon. If these sites are too close or too far apart, the gene will be translated at low efficiency.

   In some cases even the initiation codon for the gene to be cloned is part of the expression vector. Because of the way the source DNA is fused into such a vector, three possible reading frames (∞ Section 9.2) can be obtained, only one of which is satisfactory. One approach that can be used if the correct frame is not known is the use of three vectors, each having the restriction site into which new DNA will be inserted positioned such that the insert will be in a different reading frame. The gene fragment is inserted into all three vectors, and the one that gives proper expression is selected by testing.

4. **Codon usage.** There is more than one codon for most of the 20 amino acids (∞ Table 6.6), and some codons are used more frequently than others. Codon usage is partly a function of the concentration of the appropriate tRNA in the cell. A codon frequently used in a mammalian cell may be used less frequently in the organism in which the gene is being cloned. Insertion of the appropriate codon would be difficult because it would have to be changed at all locations in the gene. However, this can be done if necessary by using synthetic DNA and site-directed mutagenesis (∞ Sections 9.3 and 10.11) to create a gene more amenable to the codon usage patterns of the host.

5. **Protein stability.** Some proteins are susceptible to degradation by intracellular proteases and may be destroyed before they can be isolated. Secreted proteins must have the signal sequence attached (∞ Section 6.9) if they are to move through the

cytoplasmic membrane. Some eukaryotic proteins are toxic to the prokaryotic host, and the host for the cloning vector may be killed before a sufficient amount of the product is synthesized. Further engineering of either the host or the vector may be necessary to eliminate these problems.

Often it is advantageous that the protein from the cloned gene be made as a fusion product with a protein encoded by the vector. This not only stabilizes the protein but might simplify purification if the portion encoded by the vector is a protein for which rapid, simple, inexpensive purification techniques are known. Several special fusion vectors are now available. The "cloned protein" is released from the fusion protein after purification by special proteases. Figure 10.9 shows an example of a fusion vector that is also an expression vector.

In some cases the desired protein can also be removed from the fusion protein by chemical means. Fusion systems can also be used for purposes other than achieving increased protein stability. One advantage of making a fusion protein is that the bacterial portion can contain the bacterial sequence coding for the *signal peptide* that enables transport of the protein across the cytoplasmic membrane (∞ Section 6.9), making possible the development of a bacterial system that not only synthesizes the mammalian protein but also actually excretes it.

This list of requirements is by no means exhaustive, but it should give an idea of the challenges that must be met in the construction of an appropriate expression vector. Even with the best-designed vector,

some genes are poorly expressed in a particular cell. In some cases these problems can be rectified by using a mutant host. For instance, some "foreign mRNAs" are degraded very rapidly in wild-type *Escherichia coli* but not in particular mutant strains. In other cases the cloned gene itself must be manipulated. If the cloned gene contains introns (∞ Section 6.7), no protein product will be made if the host is a prokaryote. However, there are methods for circumventing even this problem by creating an intron-free gene (see Section 10.10).

Role of regulatory switches in expression vectors

For maximum production of a protein from a cloned gene, it is usually undesirable to design a vector that permits the gene to be transcribed and translated at all times. There are several reasons for this. As mentioned earlier, some proteins that are of commercial interest are toxic to the bacterial hosts. In addition, some expression systems, such as that involving the T7 promoter, are so powerful that normal host genes cannot be expressed. In either of these cases, it is very desirable that the synthesis of the protein be under the direct control of the experimenter. The ideal situation is to be able to grow the culture containing the expression vector until a large population of cells is obtained, each containing a large copy number of the vector, and then turn on expression in all copies simultaneously by manipulation of a regulatory switch.

We discussed regulatory controls of gene expression in Chapter 7. Recall the major importance of the repressor–operator system in regulating gene tran-

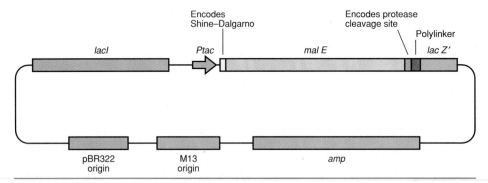

Figure 10.9 An expression vector for fusions. This vector was developed by the New England Biolabs Company. The gene to be cloned is inserted at the polylinker site (see also Figure 10.5) so it is in frame with the *malE* gene, which encodes the maltose binding protein. This insertion inactivates the *lacZ'* gene (see Figure 10.5). The fused gene is under control of the hybrid *tac* promoter *(Ptac)*. The plasmid also contains the *lacI* gene, which encodes the *lac* repressor. Therefore, an inducer must be added to the cells in order to turn on the *tac* promoter. The fusion protein is easily purified by methods involving the affinity of the protein for maltose. Once purified, the two portions of the fusion protein can be separated by a very specific protease (factor Xa). The plasmid contains a gene conferring ampicillin resistance on its host. In addition to the plasmid origin of replication, there is a bacteriophage M13 origin. Therefore, this is a phagemid and can be propagated either as a plasmid or as a phage.

scription (Section 7.3). A strong repressor can completely block the synthesis of the proteins under its control by binding to the operator region. Repressor function can be turned off at the chosen time by adding an inducer, allowing transcription of the genes controlled by the operator.

For the repressor–operator system to work as a regulatory switch for the production of a foreign protein, the expression vector must contain the operator controlled by the repressor to which the cloned gene is fused. This permits proper arrangement of the sequence of genetic elements: promoter–operator–ribosome binding site–structural gene, so efficient transcription and translation can occur. In most cases the operator and promoter correspond to each other (for instance, the *lac* operator is used with the *lac* promoter), but this is not always the case. A vector could easily be constructed to contain a *trp* promoter under the control of a *lac* operator.

For vectors using the *lac* operator, the promoter is switched on by inducers such as lactose or related β-galactosides (Section 7.3). For vectors using the *trp* operator, induction can be brought about by adding a tryptophan analog (such as β-indolacrylic acid) that brings about an apparent tryptophan deficiency. Phasing of cell growth and protein synthesis can thus be achieved by allowing growth to proceed in the absence of inducer until a suitable cell density is achieved, and then adding inducer to bring about synthesis of the desired proteins.

Vectors using bacteriophage lambda promoter P_L (and the corresponding operator, O_L) are controlled by having the lambda repressor protein in the cell (Section 8.12). Typically the lambda repressor is encoded by a mutant gene (carried by the vector or by a prophage in the host) and is temperature-sensitive. By raising the temperature of the culture to the proper value (usually 8–10°C higher than the growth temperature), the lambda repressor is inactivated and transcription from P_L begins.

CONCEPT LINK   *10.7*

Not all cloned genes are expressed at high efficiency in foreign hosts. Expression vectors are special cloning vectors containing various elements necessary for obtaining high levels of gene expression. Regulatory switches are also useful in expression vectors because they can be used by the investigator to turn on expression at the most favorable time in the growth cycle.

✔ *What are five requirements for an efficient prokaryotic expression vector?*

✔ *Discuss the differences in gene structure that may prevent a mammalian gene from being directly expressed in a prokaryote.*

Using such expression systems, one can produce very high levels of foreign proteins in *Escherichia coli*. In many cases the desired protein exceeds 100,000 molecules per cell and makes up over 20% of the protein molecules in a cell.

10.8
Synthetic DNA

Techniques are available for the synthesis of short fragments of DNA of specified base sequence. **Synthetic DNA** is widely used in molecular genetics, especially in genetic engineering but also in basic research. The procedures for synthesis of DNA can be completely automated so an oligonucleotide of 30–35 bases can be easily made in a few hours and oligonucleotides of well over 100 bases in length can be made if necessary. For the synthesis of longer polynucleotides, the oligonucleotide fragments can be joined enzymatically using DNA ligase.

DNA is synthesized in a *solid-phase procedure* in which the first nucleotide in the chain is fastened to an insoluble porous support (such as silica gel with particles about 50 μm in size). The overall procedure, the chemical details of which need not concern us here, is shown in Figure 10.10. Several chemical steps are needed for the addition of each nucleotide. After each

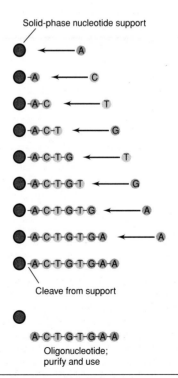

Figure 10.10 Solid-phase procedure for synthesis of a DNA fragment of defined sequence. Chemical synthesis proceeds by adding one nucleotide at a time to the growing chain.

step is completed, the reaction mixtures are flushed out of the solid support and the series of reactions repeated for the addition of the next nucleotide. Once the desired length is achieved, the oligonucleotide is removed from the solid-phase support and purified to eliminate by-products and contaminants.

Synthetic DNA molecules are widely used for various purposes in genetic engineering, for instance, as **probes** to detect, via nucleic acid hybridization, specific DNA sequences (Figure 10.8b). We will describe later (see Section 10.11) how synthetic DNA is used in a procedure called **site-directed mutagenesis** to create mutations at specific locations on the genome. Finally, synthetic DNA is employed extensively as a source of DNA primers for the polymerase chain reaction (see next section).

10.9

Amplifying DNA: The Polymerase Chain Reaction

Conventional molecular cloning methods can be considered *in vivo* DNA-amplifying tools. However, the development of synthetic DNA has spawned a new method for the rapid amplification of DNA *in vitro*, the **polymerase chain reaction (PCR).** The polymerase chain reaction can multiply DNA molecules by up to a billionfold in the test tube, yielding large amounts of specific genes for cloning, sequencing, or mutagenesis purposes. PCR makes use of the enzyme *DNA polymerase*, which copies DNA molecules (∞ Section 6.5).

The PCR technique requires that the nucleotide sequence of a portion of the desired gene be known. This is necessary because short oligonucleotide *primers* complementary to sequences in the gene or genes of interest must be available for PCR to work. The steps in PCR amplification of DNA are as follows. (1) Two oligonucleotide primers flanking the target DNA (Figure 10.11b) are made on an oligonucleotide synthesizer and added in great excess to heat-denatured target DNA (Figure 10.11a). (2) As the mixture cools, the

excess of primers relative to the target DNA ensures that most target strands anneal to a primer and not to each other (Figure 10.11b). (3) DNA polymerase then extends the primers using the target strands as tem-

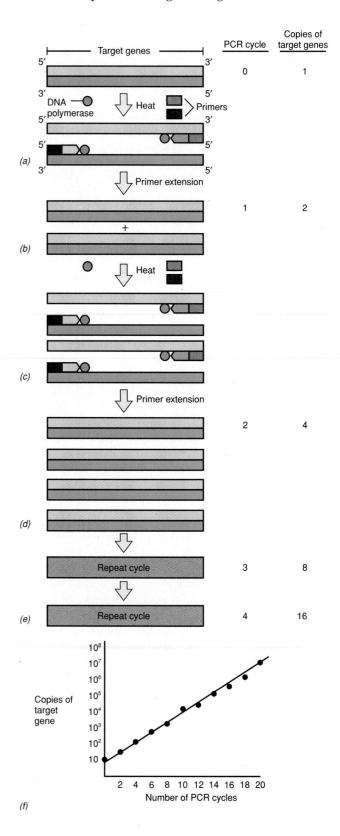

Figure 10.11 The polymerase chain reaction (PCR) for amplifying specific DNA sequences. (a) Target DNA is heated to separate the strands, and a large excess of two oligonucleotide primers, one complementary to the target strand and one to the complementary stand, is added along with DNA polymerase. (b) Following primer annealing, primer extension yields a copy of the original double-stranded DNA. (c) Further heating, primer annealing, and primer extension yields a second double-stranded DNA. (d) The second double-stranded DNA. (e) Two additional PCR cycles yield 8 and 16 copies, respectively, of the original DNA sequence. (f) Effect of running 20 PCR cycles on a DNA preparation originally containing 10 copies of a target gene. Note that the graph is semilogarithmic.

plate (Figure 10.11c). (4) After an appropriate incubation period, the mixture is heated again to separate the strands. The mixture is then cooled to allow the primers to hybridize with complementary regions of newly synthesized DNA, and the whole process is repeated (Figure 10.11e).

Thus, each PCR "cycle" involves the following: (1) heat denaturation of double-stranded target DNA, (2) cooling to allow annealing of specific primers to target DNA, and (3) primer extension by the action of DNA polymerase (Figure 10.11). Note in Figure 10.11 how the extension products of one primer can serve as a template for the other primer in the next cycle. The beauty of the PCR technique lies in the fact that each cycle literally *doubles* the content of the original target DNA. In practice, 20–30 cycles are usually run, yielding a 10^6- to 10^9-fold increase in the target sequence (Figure 10.11f).

PCR at high temperature

The original PCR technique employed *Escherichia coli* DNA polymerase, but because of the high temperatures needed to denature the double-stranded copies of DNA being made, the polymerase itself also was denatured and had to be replenished every cycle. This tended to limit the number of cycles that could be run and was very expensive. This problem was solved by employing a thermostable DNA polymerase isolated from the thermophilic bacterium *Thermus aquaticus*. DNA polymerase from *T. aquaticus*, known as *Taq polymerase,* is stable to 95°C and thus is unaffected by the denaturation step employed in the PCR reaction. The use of *Taq* DNA polymerase also increased the *specificity* of the PCR reaction because the DNA is copied at 72°C rather than 37°C. At high temperatures, nonspecific hybridization of primers to nontarget DNA rarely occurs, thus making the product of *Taq* PCR more homogeneous than that obtained using the *E. coli* enzyme.

One problem with the *Taq* polymerase is that it has no proofreading function (∞ Section 6.5) and consequently makes more mistakes than the *Escherichia coli* enzyme. DNA polymerase from the hyperthermophilic archaean *Pyrococcus furiosus* (growth temperature optimum, 100°C) (∞ Sections 5.8 and 17.3), called *Pfu* polymerase (or "vent polymerase"), is also widely used and is even more thermally stable than *Taq* polymerase. *Pfu* polymerase has proofreading activity, making it a particularly good enzyme when accuracy is crucial.

Because a number of highly repetitive steps are involved in the PCR technique, machines have been developed that can be programmed to run through heating and cooling cycles automatically. Because each cycle requires only about 5 min, the automated procedure allows for large amplifications in only a few hours (by contrast, such amplification by *in vivo* cloning methods would take several days). To supply the demand for thermostable DNA polymerase in the growing PCR and DNA sequencing markets, the genes for these enzymes have been cloned into *E. coli* and produced in large quantities; the cost of the PCR method is now just a fraction of what it was when the technique was first introduced.

Applications of PCR

The polymerase chain reaction has found many practical uses. PCR is extremely valuable in cloning DNA because the gene or genes of interest can be amplified prior to cloning if flanking sequences are known. As a tool for DNA sequencing, PCR yields large amounts of specific DNA sequences to be used as templates. PCR can also be used to produce large amounts of mutated DNA. If made sufficiently long, primers containing one or a few base mismatches will still anneal to the target gene; the introduced mutation can then be amplified by PCR (this is a form of site-directed mutagenesis) (see Section 10.11). The mutated DNA can be used to transform cells, generating mutant derivatives.

Because the primers used do not have to be perfectly complementary, PCR is routinely used in comparative or evolutionary studies to amplify genes from a variety of sources where the DNA has already been cloned (and sequenced) from one organism. In these cases the primers are made to regions of the gene thought to be conserved throughout a wide variety of organisms. Because of the sensitivity of PCR, it has been used to amplify and clone DNA from sources such as mummified human remains and even samples of extinct plants and animals.

PCR can also be used to amplify very small quantities of DNA present in a sample. Using appropriate primers, one can find and identify a single bacterial

---

CONCEPT CHECK   10.9

The polymerase chain reaction, a procedure for amplifying DNA *in vitro*, makes use of a heat-stable DNA polymerase from thermophilic prokaryotes. Heat is used to denature the DNA into two single-stranded molecules, each of which is copied by the polymerase. Beginning with a small oligonucleotide that serves as a primer for the target DNA to be amplified, the polymerase copies the complete DNA to which the primer associates. After a single copy cycle, the newly formed double strands are separated by heat again, and a new round of copying permitted. At each thermal cycle, the amount of target DNA doubles.

✔ *Why does PCR require primers?*

✔ *Why is a primer needed at each "end" of the DNA to be amplified?*

DNA FINGERPRINTING

The techniques used in molecular genetics and genetic engineering are useful not only in determining the evolutionary relationships among organisms (∞ Chapter 15) but also in determining relationships among individuals and in establishing whether a tissue sample, even a very small one, came from a particular individual. These latter techniques are called *DNA fingerprinting.* DNA fingerprinting is made possible both by the technology that allows precise detection and amplification of very small amounts of DNA and by the fact that higher organisms contain repetitive DNA sequences that can exist in different numbers and patterns in the genome.

As discussed (∞ Section 6.3), the genomes of higher eukaryotes contain a very large amount of repetitive DNA. Some of these repeats exist in families of related sequences scattered around the genome, and members of these families have been cloned and sequenced. In order to be useful for identification purposes, a DNA sequence must have a reasonable chance of differing among different groups in a population of organisms. One family of these repetitive sequences was found to vary not simply as to sequence but also as to how many repeats of an individual sequence occurred at a single site on a particular chromosome. These sequences are called *variable number of tandem repeats* (VNTR), and several have been identified.

The use of VNTRs in DNA fingerprinting is illustrated in the accompanying figure. It shows two different alleles (alternative states of the same gene) of a eukaryotic chromosome that differ only in how many copies of the repeated sequence are present. Since the VNTR DNA has been sequenced, it is known which restriction enzyme sites are *not* found in a particular VNTR. Digestion with such an

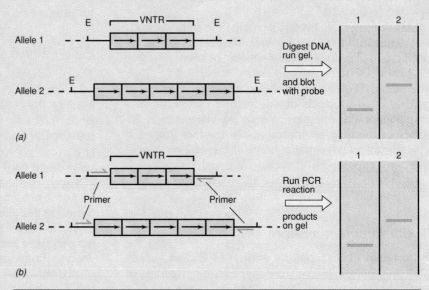

DNA fingerprinting. (a) Two different alleles of a region of a single chromosome. The alleles differ only in the number of repeats in the VNTR. DNA from cells containing these chromosomes can be cut with the restriction enzyme *Eco*RI (which does not cut within the VNTR) and the fragments separated on an agarose gel. The fragments containing the VNTR are then identified after Southern blotting by hybridization with a probe specific to the VNTR. (For simplicity, the figure shows only the result from individuals whose two chromosomes each have the same allele at this site.) (b) The same alleles, but with primers that could be used to amplify the VNTR segments by PCR. The products of the PCR reaction can be loaded directly onto the gel without restriction digestion.

enzyme then releases the complete VNTR intact. When the DNA from two chromosomes with a different number of repeats at this particular locus is digested, the restriction fragments containing this DNA differ in size. (Such a difference is called a *restriction fragment length polymorphism,* or RFLP.) This DNA can be separated by gel electrophoresis, and the VNTR-containing fragments detected by Southern blotting (∞ Working with Nucleic Acids: The Tools, Chapter 6) using a probe made from the cloned VNTR sequence. Having only a single difference in one band is not a very precise way of identifying an individual. Because higher eukaryotes are diploid, there are two copies of the VNTRs, which may or may not be the same. Also, most VNTR sequences are fam-

ilies, and several different loci can be detected with the same probe. In addition, it is possible to probe the digested DNA simultaneously for several different VNTR markers. With the use of these methods, it has been estimated that the probability of identifying a particular individual by comparing two different DNA samples is very high. There is some controversy about exactly how high the probability is with any given protocol, but it is acknowledged that DNA fingerprinting is a very powerful technique.

Notice that the polymerase chain reaction (PCR) does not need to be used in DNA fingerprinting. However, the use of PCR is essential when the amount of DNA in the sample is very small—such as that found in the cells on the root of a single hair. The use of PCR in DNA fingerprinting is also shown in the figure. To use PCR, it is necessary to know the sequences surrounding a particular site that contains a VNTR so primers can be synthesized. However, since PCR amplifies only the DNA between the primers, one does not have to cut the DNA with restriction enzymes before running it on a gel. With enough cycles of amplification, it is also sometimes possible to detect the PCR-generated bands by simply staining the gels rather than using a hybridization probe.

Research labs also use DNA fingerprinting to screen tissue culture cells to determine if the cells are the correct line or from the correct animal. Although it might seem unlikely, tissue culture cells can become contaminated with other cell lines, and there are many fewer tests that can distinguish different species of mammalian cells than those that can differentiate species of bacteria.■

cell in a sample even if large numbers of other species are present. PCR has also been used in conjunction with *DNA fingerprinting,* a powerful technique that permits identification of individuals, or relationships between individuals, from small samples of their DNA (see the box, DNA Fingerprinting).

10.10

Cloning and Expression of Mammalian Genes in Bacteria

So far in this chapter we have described a variety of methods for cloning DNA, finding specific clones, and expressing genes. However, as we mentioned (see Section 10.4), there are sometimes obstacles to obtaining proper expression even using expression vectors. In mammalian genes there are almost always introns (∞ Section 6.3). Indeed, some genes have over 50 introns consisting of tens of thousands of base pairs. Removing introns from genes using the techniques we have described would be difficult. In addition, mammalian genomes are vast, the human haploid genome having 3 billion base pairs. Screening DNA libraries from such a genome for a specific gene would be at best extremely laborious. In some cases very little is known about the gene, making screening almost impossible. In other cases the geneticist may wish to find genes with specific patterns of regulation; for instance, it could be valuable to clone the genes whose products are expressed only in the brain. There are a number of approaches that can be used to circumvent all these problems, and we shall discuss some of them in this section. Not only skill but also intuition and good luck play big roles in a successful outcome. A summary of approaches and procedures is given in Figure 10.12.

Reaching the gene via messenger RNA

One approach to isolating a gene is to get to it through its mRNA. A major advantage of using mRNA is that the noncoding information present in the DNA (introns) has been removed (∞ Section 6.7). The isolated mRNA is used to make complementary DNA (cDNA) by means of reverse transcription (∞ Section 8.22). It is likely that a tissue expressing the gene contains large amounts of the desired mRNA, although except in rare cases this certainly is not the only mRNA produced. In a fortunate situation, where a single mRNA dominates a tissue type, extraction of mRNA from that tissue provides a useful starting point for gene cloning.

In a typical mammalian cell, about 80–85% of the RNA is ribosomal, 10–15% is transfer RNA and other low-molecular-weight RNAs, and 1–5% is messenger RNA. Although low in abundance, the mRNA in a eukaryote is identifiable because of the poly-A tails found at the 3'-end (∞ Section 6.7). In maturing red blood cells, for instance, where virtually the only protein made is the globin portion of hemoglobin, from 50 to 90% of the poly-A-containing cytoplasmic RNA consists of globin mRNA. By passing a poly-A-rich RNA extract over a chromatographic column containing poly-T fragments (linked to a cellulose support), most of the mRNA of the cell can be separated from the other cellular RNA by the specific pairing of A and T bases. Elution of the RNA from the column then gives a preparation greatly enriched in mRNA.

Once the RNA message has been isolated, it is necessary to convert the information to DNA. This is accomplished by use of the enzyme *reverse transcriptase,* which we discussed in Section 8.22. This remarkable enzyme, an essential component of retrovirus replication, copies information from RNA into DNA (Figure 10.13). As we noted, this enzyme requires a primer in order for it to begin working (in retrovirus infection the primer is a tRNA). In the present procedure, an oligo-dT primer is used that is complementary to the poly-A tail of the isolated mRNA. The oligo-dT primer is hybridized with the mRNA, and then reverse transcriptase is allowed to act (Figure 10.13). As seen, the newly synthesized DNA copy has a hairpin loop at its end, which is synthesized because after the enzyme completes copying the mRNA it starts to copy the newly synthesized DNA. This hairpin loop, which is probably an artifact of the test tube reaction, provides a convenient primer for synthesis of the second DNA strand. The resultant double-stranded DNA, with the hairpin loop intact, is then cleaved by a single-strand-specific nuclease to produce the desired

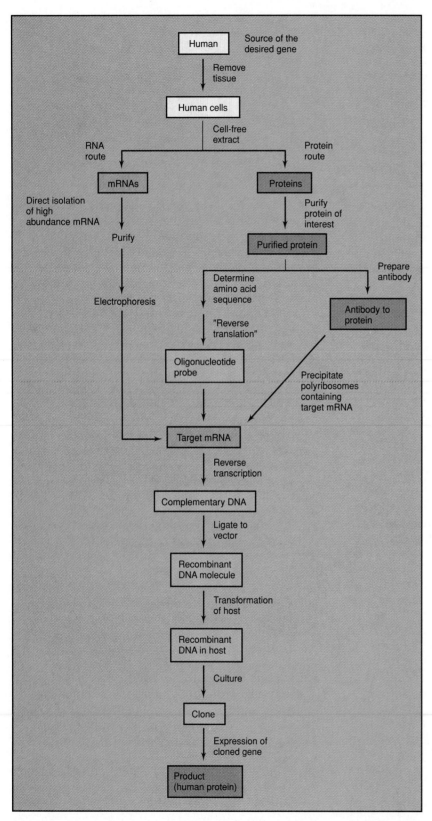

Figure 10.12 Several routes to the isolation and expression of mammalian genes in prokaryotes. The colors indicate whether a step involves DNA (green), RNA (orange), or protein (brown).

double-stranded DNA, one strand of which is complementary to the mRNA. This double-stranded DNA (the gene of interest) can then be inserted into a plasmid or other vector for cloning. The detection of spe-

cific clones makes use of the procedures discussed in Section 10.6.

The cDNA that is then cloned should encode the protein of interest and, therefore, can be considered its

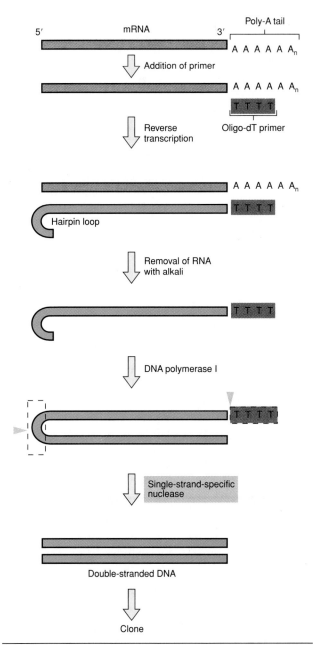

Figure 10.13 Steps in the synthesis of complementary DNA (cDNA) from an isolated mRNA using the retroviral enzyme reverse transcriptase.

"gene." Unlike the "natural" gene on the mammalian chromosome, this one contains no introns. Although there is a start codon, there are no promoters because these are not transcribed and, therefore, their sequence won't be in the mRNA (∞ Section 6.6). The requirements for achieving high levels of expression with genes made in this manner are simply those discussed in the section on expression vectors (see Section 10.7).

One can also make *cDNA libraries* from different tissues when seeking genes whose expression is specific to those tissues. This can be extremely useful because the actual gene (chromosomal DNA) is found in almost all cells but the mRNA is found only in cells actively producing the protein. For instance, the gene encoding the hormone insulin is found throughout the body, but insulin mRNA is found only in certain cells in the pancreas. Therefore, a library made from pancreatic cells is enriched for cDNA corresponding to the insulin gene.

Reaching the gene via the protein

If for some reason, mRNA for the gene of interest cannot be obtained, other approaches are possible. Two such approaches, reverse translation and polyribosome precipitation, will be discussed here.

The most widely used method of cloning low abundance mRNA is to make a **synthetic DNA** that is complementary to part of the mRNA and then use this DNA as a *probe* in a Northern blot procedure (∞ Working with Nucleic Acids: The Tools, Chapter 6) to pull out the mRNA of interest by hybridization. This requires that a partial or complete sequence of the protein be known. Then, from a consideration of the genetic code, the nucleotide sequence of a section of the DNA is deduced, and this piece of DNA is synthesized.

The procedure by which the nucleotide sequence is deduced from the protein sequence is called **reverse translation.** (Note that *reverse translation* is not a cellular process but a mental exercise of the genetic engineer.) The procedure of reverse translation is illustrated in Figure 10.14. From the genetic code, the nucleotide sequence of a section of the DNA is deduced, and this piece of DNA is synthesized. Unfortunately, degeneracy of the genetic code (∞ Section 6.10) somewhat complicates the problem. Most amino acids are encoded by more than one codon, and codon usage varies from organism to organism. The best section of DNA to synthesize is one that corresponds to a part of the protein rich in amino acids specified by only a single codon (methionine, AUG; tryptophan, UGG) or by two codons (for example, phenylalanine, UUU, UUC; tyrosine, UAU, UAC; histidine, CAU, CAC) because this increases the chances that the synthesized DNA will be complementary or nearly complementary to the mRNA of interest. If the complete amino acid sequence of the protein is not known, then the sequence used is generally one at the amino terminus of the protein because it is at the amino terminus that sequencing of the protein begins.

Polyribosome precipitation involves separation from the tissue of *ribosome complexes* that are in the process of synthesizing the desired protein, using an antibody specific for this protein. As we describe in Chapter 20, antibodies are proteins made in response to foreign proteins able to combine with and specifically precipitate them. How is an antibody directed against a *protein* used to detect an *mRNA*?

In the ribosome complex, polypeptides of the protein of interest are still complexed with the protein-

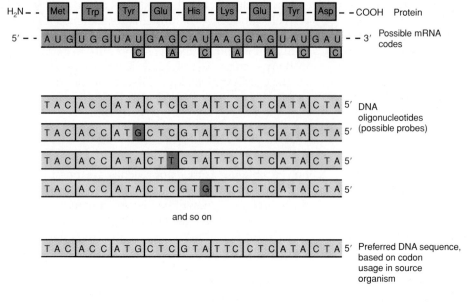

Figure 10.14 Reverse translation: deducing the best sequence of an oligonucleotide probe from the amino acid sequence of the protein. Because of degeneracy, many probes are possible. If codon usage by the same organism is known, then a preferred sequence can be selected. It is not essential that complete accuracy be achieved because a small amount of mismatch can be tolerated.

synthesizing machinery (which contains, among other things, the sought-for mRNA). When the antibody precipitates the protein in the ribosome complex, the mRNA also is precipitated. After isolation, the mRNA can be used to prepare complementary DNA as described in this section.

Synthesis of the complete gene

If the protein is small enough or is of sufficient economic interest to justify a major effort, the complete gene can be synthesized chemically (see Figure 10.10); this requires knowledge of the complete amino acid sequence of the protein. Chemical synthesis not only permits the acquisition of genes that cannot be obtained otherwise but also allows synthesis of *modified genes* that may make new proteins of utility. Tech-

niques for the synthesis of DNA molecules are now well developed, and it is possible to synthesize genes coding for proteins 100–200 amino acid residues in length (300–600 nucleotides). The synthetic approach was used for production of the human hormone insulin in bacteria, as discussed in Section 10.13. We will discuss the use of synthetic DNA in mutagenesis in Section 10.11.

With the use of all these techniques, a large number of different human proteins have been expressed at high yield under the control of bacterial regulatory systems, including human growth hormone, insulin, virus antigens, interferon, and somatostatin (see Section 10.13).

10.11
In Vitro and Site-Directed Mutagenesis

Recombinant DNA technology has opened up a whole new field of mutagenesis. Whereas conventional mutagens (∞ Section 9.3) act at random, by use of synthetic DNA and recombinant DNA techniques it is possible to introduce mutations at *precisely determined sites* on genes. This process is called **site-directed mutagenesis.** Proteins made from strains carrying such mutations can be expected to have properties different from those of the wild-type proteins, properties that may be predicted from a knowledge of protein structure.

Site-directed mutagenesis

The first approaches to site-directed mutagenesis included treating transducing bacteriophage with chemical mutagens and then infecting bacteria and screening for mutants. Such a strategy had the effect of increasing

---

CONCEPT CHECK   **10.10**

To detect and isolate a gene from the large amount of DNA in the mammalian cell, it is often necessary to use special procedures. One approach is to isolate the mRNA from the mammalian cell and make a complementary DNA by use of the enzyme reverse transcriptase. Another approach is to synthesize a nucleic acid probe for the gene in question and use this probe to detect the cloned sequence in bacterial colonies. Under some conditions, the complete mammalian gene can be synthesized chemically and this synthetic gene cloned in bacteria.

✔ *Draw a diagram of how cDNA is made from mRNA.*

✔ *Explain why* reverse translation *is something cells can't do (∞ Section 6.1).*

the mutation rate in a limited region of a genome. New techniques are now available that allow the geneticist much greater specificity; a specific base pair in a gene can be changed to another base pair. Some of the techniques are simple and very powerful.

The basic procedure is to synthesize a short oligodeoxyribonucleotide containing the desired base change and to allow this to pair with a single-stranded DNA containing the gene of interest. Pairing is complete except for the short region of mismatch. Then the short single-stranded fragment of the synthetic oligonucleotide is extended using DNA polymerase, thus copying the rest of the gene. The double-stranded molecule obtained is inserted into a cloning host by transformation, and mutants selected by a procedure already described (∞ Section 9.6). The mutant obtained is then used in production of the modified (mutant) protein.

The whole procedure of **site-directed mutagenesis** is illustrated in Figure 10.15. Several modifications of this technique have been developed to increase the ratio of mutants recovered. As seen, one must begin with the gene of interest cloned into a single-stranded DNA. A widely used vector for site-directed mutagenesis is bacteriophage M13, which, as we have seen (∞ Sections 8.10 and 10.4), has some properties that are useful in recombinant DNA technology. The target DNA is cloned into M13, from which single-stranded DNA can be purified with ease. Because cells infected with this phage remain alive, a ready source of DNA is available.

Total synthesis of mutant genes

Synthetic DNA technology can be used to synthesize a complete gene with a genetic change inserted at the desired location. Although a complete gene of 1000 or so nucleotides would be difficult to synthesize in one piece, it is possible to synthesize smaller portions of the gene and then link these together to produce the final gene. By addition of appropriate sites at the ends of the gene, this gene can then be linked into a vector and cloned. The possibilities of this approach seem virtually limitless, although because of the expense involved, it is essential that one have a carefully considered rationale for the particular sequence. Typically, however, it is more advantageous to synthesize only a small part of a gene and use this to replace the same part of the wild-type gene, a process called *cassette mutagenesis.*

Cassette mutagenesis and gene disruption

Because of the large number of restriction enzymes commercially available and therefore the large number of different DNA sequences that can be cut, it is usually possible to find several different restriction sites in the gene of interest. If sites for the appropriate enzyme are not found in the gene, or at the precise location required, they can also be added by site-directed muta-

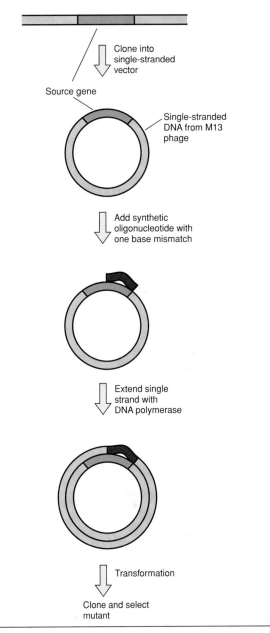

Figure 10.15 Site-directed mutagenesis using short synthetic oligodeoxyribonucleotide fragments.

genesis (see Figure 10.15). If restriction sites are close together, the intervening DNA fragment can be cut out and replaced by a synthetic DNA fragment in which one or more of the bases have been changed. These synthetic fragments are called *cassettes* (or cartridges), and the process is known as **cassette mutagenesis.**

Insertion mutations can also be generated by simply inserting a cassette at a single site. When using cassettes to replace sections of genes, the cassettes are typically the same size as wild-type DNA fragments. However, the cassette used for making insertion mutations can be almost any size and can even be an entire gene. In fact, cassettes that encode proteins that confer a particular antibiotic resistance on the host are commonly used. This type of cassette mutagenesis is used in a process called **gene disruption**. The process of gene disruption is illustrated in Figure 10.16. In this case, a fragment carrying a gene conferring kanamycin resistance, the *kan* cassette, is inserted at a restriction site in a cloned gene. The vector carrying this mutant gene is then linearized by being cut with a different restriction enzyme, and the linear DNA is transformed into the host with kanamycin resistance selected. The linearized plasmid cannot replicate, and so resistant cells likely arise by homologous recombination (∞ Section 9.5) between the mutated gene on the plasmid and the wild-type gene on the chromosome.

The cells have not only gained kanamycin resistance but have also lost the function of the gene in which the *kan* cassette was inserted. Therefore, these mutations are also called "knockout" mutations. This process is similar to searching for insertion mutations made by transposons (∞ Section 9.10), but in this case the geneticist chooses exactly which gene will receive the mutation. Note that gene disruption in haploid organisms yields viable cells only if the disrupted gene is unessential. Gene disruption experiments are now used as one mechanism to find out whether a gene is essential. Methods of obtaining gene disruption have been developed for higher organisms, including mice.

CONCEPT CHECK   *10.11*

Synthetic DNA molecules of desired sequence can be made *in vitro* and used to construct a mutated gene directly or to change specific base pairs within a gene via site-directed mutagenesis. Genes can also be disrupted by inserting DNA fragments, called cassettes, into them. The inserted cassette eliminates the function of the wild-type gene while conferring a new, and usually selectable, phenotype on the cell.

10.12
Practical Applications of Genetic Engineering

We note here some of the main practical applications of recombinant DNA technology and then describe some of the applications in more detail in Sections 10.13–10.15.

Genetic engineering has many commercial and practical applications. A few main areas of interest for commercial development are as follows.

1. **Microbial fermentations.** A number of important products are made industrially using microorgan-

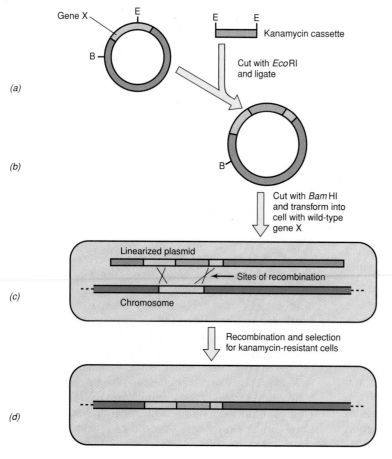

(a)

(b)

(c)

(d)

Figure 10.16 Gene disruption using cassette mutagenesis. (a) A plasmid containing a cloned wild-type copy of gene X is cut with the restriction enzyme *Eco*RI and mixed with a DNA fragment (the kanamycin cassette) that contains a gene capable of conferring kanamycin resistance on a cell and that has been obtained using the same restriction enzyme. The cut plasmid and the cassette are ligated. (b) The product of the ligation is a plasmid that now contains the kanamycin cassette as an insertion mutation within gene X. This new plasmid is now cut with a further restriction enzyme, *Bam*HI, and transformed into a cell containing a wild-type gene X. (c) The transformed cell contains the linearized plasmid with a disrupted gene X and its own chromosome with a wild-type copy of the gene. In some cells, homologous recombination occurs between the wild-type and mutant forms of gene X. (d) Cells that can grow in the presence of kanamycin must have the kanamycin cassette recombined into their chromosome because the linearized plasmid cannot replicate. These cells now have only a single disrupted copy of gene X. This disruption typically abolishes all gene X function.

isms, of which the antibiotics are the most significant (∞ Section 12.6). Genetic engineering procedures can be used to manipulate the antibiotic-producing organism in order to obtain increased yields or produce modified antibiotics.

2. **Virus vaccines.** A vaccine is a material that can induce immunity to an infectious agent (∞ Section 20.17). Frequently, killed virus preparations are used as vaccines, and there is always a potential danger to the patient if the virus has not been completely inactivated. Because the active ingredient in the killed virus vaccine is the protein coat, it is desirable to produce the protein coat separately from the rest of the virus particle. By genetic engineering, viral coat protein genes can be cloned and expressed in bacteria or in nonpathogenic viruses, making possible the development of safe, convenient vaccines (see Section 10.13).

3. **Mammalian proteins.** A number of mammalian proteins are of great medical and commercial interest. Some of these are discussed in Section 10.13. In the case of human proteins, commercial production by direct isolation from tissues or fluids is complicated and expensive, or even impossible. By cloning the gene for a human protein in an appropriate microorganism or a culturable cell, its commercial production is possible.

4. **Transgenic plants and animals.** In addition to the production of valuable *products* by microbial means, genetic engineering promises the advent of genetically altered whole plants and animals. Some of these developments are discussed in Sections 10.14 and 10.15. Such organisms, referred to as *transgenic*, hold great promise for boosting agricultural productivity, altering the nutritional quality of meats and vegetables, and producing certain proteins not readily produced by genetically engineered microorganisms. By introducing cloned DNA into the fertilized eggs of animals, or directly into plant cells grown in tissue culture, it is now possible to grow genetically altered higher organisms.

5. **Environmental biotechnology.** Because of the enormous metabolic diversity of Bacteria (to be discussed in Chapter 13), a large gene pool exists in bacteria from natural habitats. In some cases these genes code for proteins that degrade environmental pollutants. Genes for the biodegradation of many toxic wastes and wastewater pollutants have been shown to exist in natural isolates of bacteria (∞ Section 14.20). Genetic engineering is beginning to tap these resources for the purpose of environmental cleanup. In many cases the gene donors are bacterial strains isolated from contaminated waste sites. Some examples include genes for the biodegradation of chlorinated pesticides,

like 2,4,5-trichlorophenoxyacetic acid (2,4,5-T), chlorobenzenes and related chlorophenolics, naphthalene, toluene, anilines, and various hydrocarbons. The desired genes are isolated from species of *Pseudomonas*, *Alcaligenes*, and a few other Bacteria and then cloned into plasmids. Some plasmids have been constructed containing genes from the biodegradation of several different toxic chemicals.

6. **Gene regulation and gene therapy.** The first use of genetic engineering in the biotechnology industry was primarily for producing useful gene products more easily or for creating transgenic organisms. As we discuss in the following sections, this approach is yielding great benefits. However, much current research in this area involves the creation of new products ("designer drugs") and controlling the expression of specific genes. Much research is currently being directed toward design of antisense RNA (∞ Antisense Nucleic Acid, Chapter 7) and ribozymes (∞ Section 6.7), which regulate the expression of specific genes. (Much research is also directed toward delivering these "drugs" to, and getting them inside, the correct cells and tissues.) An additional area of research showing great promise is *gene therapy*, the use of genetic engineering to treat genetic disease.

Despite the exciting promise of genetic engineering in biotechnology, getting a product to market is an enormous undertaking. Other than the obvious problems of correctly cloning and expressing the gene of interest in a microorganism, generally a bacterium or a yeast, and purifying the desired product, related matters such as clinical trials and governmental approval must be considered. Any microbially synthesized product intended for human use must pass extensive clinical trials. For example, human insulin produced microbially by recombinant DNA technology (see Section 10.13) had to pass strict clinical trials with human volunteers despite the fact that microbially produced insulin was shown to be identical to the protein made in humans. If all goes well in clinical trials, federal approval is usually obtained, in the United States by action of the Food and Drug Administration (FDA), but this can be a time-consuming process. In addition, unexpected problems often arise experimentally; for instance, the vector may be unstable, gene expression may be transient, or gene regulation problems may exist. Problems can also arise during the process of "scaling up" from laboratory to commercial production (∞ Sections 12.3 and 12.4). (This is one reason why biotechnologists should have a good knowledge of basic microbiology!) Finally, some products simply aren't effective when used in an actual commercial or clinical application. Nonetheless, biotechnology shows great promise and we will now consider a few specific areas.

10.13

Production of Mammalian Products and Vaccines by Genetically Engineered Microorganisms

One of the first practical applications of genetic engineering was the use of easily grown bacteria to produce proteins whose genes were from organisms that are more difficult or expensive to grow. Although the special DNA polymerases used in the polymerase chain reaction were originally isolated from thermophilic bacteria, they are now produced in *Escherichia coli* from cloned genes (see Section 10.9). Most restriction enzymes are also produced in *E. coli* from cloned genes. Similarly, many proteins used industrially are now produced from cloned genes, and in some cases, the protein itself has been altered by using site-specific mutagenesis (see Section 10.11) to change the cloned gene.

Many proteins and peptides from mammalian cells have high pharmaceutical value. However, these proteins are usually present in very small amounts in normal tissue, and it is therefore extremely costly to purify them. Another of the first efforts of the biotechnology industry was to use genetic engineering to produce these proteins in microorganisms.

For many early applications, such as the production of insulin, it was known that the product would have great commercial value because of its established therapeutic value in treating a reasonably well-understood disease (in this case diabetes). However, such success is not always guaranteed, even when the protein can be produced and purified. Often this is because the disease process is complex and not well understood or the product has unexpected side effects. Nonetheless, by the mid-1990s biotechnology companies had hundreds of products in clinical trials. There are several different classes of therapeutic protein products that have been produced. These include hormones, interferons, growth factors, and vaccines. Some examples of these and other products are shown in Table 10.1.

Production of insulin

One of the earliest and most dramatic commercial successes was the production of the hormone **insulin.** Many hormones are peptides or small proteins. These molecules are extremely important in controlling mammalian metabolism and have important therapeutic uses. Insulin is a protein produced in the pancreas that is vital for the regulation of carbohydrate metabolism in the body. Diabetes, a disease characterized by insulin deficiency, afflicts millions of people. The standard treatment for diabetes is periodic injections or oral administration of insulin, and because insulins of most mammals are similar in structure, it is

TABLE 10.1	Some therapeutic products made by recombinant DNA techniques[a]
Product	**Function**
Blood proteins	
Erythropoietin	Treats certain types of anemia
Factors VII, VIII, IX	Promote clotting
Tissue plasminogen activator	Dissolves clots
Urokinase	Blood clotting
Human hormones	
Epidermal growth factor	Wound healing
Follicle stimulating hormone	Treatment of reproductive disorders
Human growth hormone	Treatment of dwarfism
Insulin	Treatment of diabetes
Nerve growth factor	Possible treatment of degenerative neurological disorders and stroke
Relaxin	Facilitates childbirth
Immune modulators	
α-Interferon	Antiviral, antitumor agent
β-Interferon	Treatment of multiple sclerosis
Colony stimulating factor	Treatment of infections and cancer
Lysozyme	Anti-inflammatory
Tumor necrosis factor	Antitumor agent, potential treatment of arthritis
Vaccines	
Cytomegalovirus	Prevention of infection
Hepatitis B	Prevention of serum hepatitis
Measles	Prevention of measles
Rabies	Prevention of rabies

[a]Although research is currently progressing in all these areas, not all of the listed products are yet on the market.

possible to treat human diabetes by use of insulin isolated commercially from beef or pork pancreas. However, nonhuman insulin is not as effective as *human insulin,* and the isolation process is expensive and complex. Cloning of a human insulin "gene" in bacteria has hence been carried out.

Producing hormones such as insulin in genetically engineered microorganisms is not simply a matter of cloning a gene (or cDNA) (see Section 10.10) in an expression vector. This is because many of these hormones are only small fragments of the polypeptides encoded by the gene. Insulin in its active form consists of two polypeptides (A and B) connected by disulfide bridges (Figure 10.17a). These two polypeptides are coded by separate parts of a single insulin gene. The insulin gene codes for *preproinsulin,* a longer polypeptide containing a signal sequence (involved in excretion of the protein) (∞ Section 6.9), the A and B polypeptides of the active insulin molecule, and a connecting polypeptide that is absent from mature insulin. *Proinsulin* is formed from preproinsulin, and

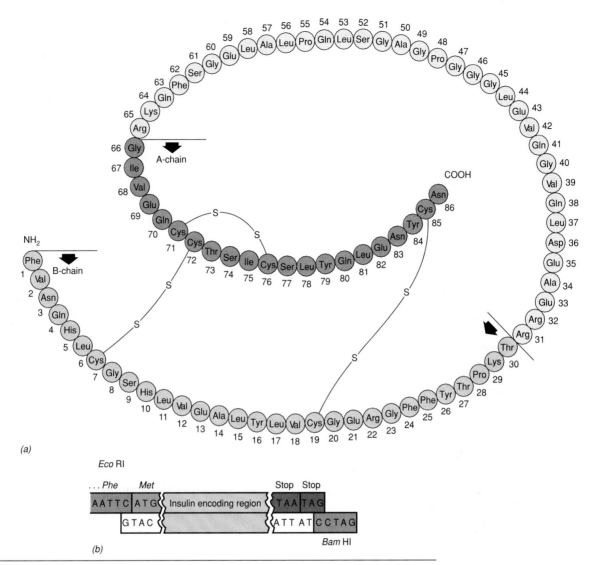

Figure 10.17 Genetic engineering for the production of human insulin in bacteria. (a) Structure of human proinsulin. The peptide shown in yellow must be removed from between the A and B chains in order to make insulin. (b) Chemical synthesis of the insulin gene and suitable linkers, permitting cloning and expression. The synthesized fragments were linked via restriction sites *Eco*RI and *Bam*HI in a plasmid vector in such a way that the insulin chains are formed as a fusion protein (see Section 10.7) with a portion of a gene found on the vector (note that the *Eco*RI site is part of this coding region). The methionine coding sequence was inserted to permit chemical cleavage of the A and B chains from the fused protein made in the bacteria because the reagent cyanogen bromide specifically cleaves at methionine residues and insulin does not contain methionine. Two stop codons were incorporated at the downstream end of the coding sequence.

the conversion of proinsulin to insulin involves enzymatic cleavage of the connecting polypeptide from the A and B chains.

Two approaches have been used to obtain production of human insulin in bacteria: (1) production of proinsulin and conversion to insulin by chemical cleavage, and (2) production of the A and B chains in two separate bacterial cultures, and joining of the two chains chemically to produce insulin. Because the insulin protein is fairly small, it was more convenient

with either approach to synthesize the proper DNA sequence chemically (see Section 10.8) rather than attempt to isolate the insulin gene from human tissue. There are 63 bases encoding the A chain and 90 bases encoding the B chain (Figure 10.17b). In proinsulin there are an additional 105 bases for the peptide connecting the A and B chains. When the polynucleotides were synthesized, suitable restriction enzyme sites were placed at each end so the polynucleotides could be ligated into a plasmid vector. To obtain effective expression, the synthesized genes were inserted downstream from a suitable *Escherichia coli* promoter but in a manner such that the insulin fragment was synthesized as part of a *fusion protein* (see Section 10.7). An important advantage of making the fusion protein is that the fusion product is much more stable in *E. coli* than insulin itself. Finally, a nucleotide triplet coding for methionine was placed at the junction joining the insulin gene to the upstream part of the fusion gene. The reason for this is that the chemical reagent *cyanogen bromide* specifically cleaves polypeptide chains at methionine residues, permitting recovery of the insulin product once the fused protein has been isolated from the bacteria. Insulin itself does not contain methionine and hence is unaffected by cyanogen bromide treatment.

When the proinsulin route is used, the proinsulin isolated from the bacteria via cyanogen bromide treatment is converted to insulin by disulfide bond formation, followed by enzymatic removal of the connecting peptide of proinsulin. Proinsulin naturally folds so the cysteine residues are opposite each other (Figure 10.17a), and chemical treatment then causes the formation of disulfide cross-links. Once this has been accomplished, the connecting peptide can be removed by treatment with the proteases trypsin and carboxypeptidase B, which have no effect on insulin itself.

When insulin is produced by way of the separate A and B peptides, each of the fusion proteins is isolated from a separate bacterial culture and the chains released by cyanogen bromide cleavage. The cleaved chains are then connected by use of chemical treatment that results in disulfide bond formation.

The final product, biosynthetic human insulin, is identical in all respects to insulin purified from the human pancreas and is being marketed commercially. Microbially produced human insulin is less expensive to make and just as effective as porcine or bovine insulin, the major source of insulin for diabetics before the advent of biotechnology.

Recombinant vaccines

Vaccines are suspensions of killed or modified pathogenic microorganisms or specific fractions isolated from the microorganisms that when injected into an animal produce immunity to a particular disease. Often the substance that elicits the immune response is a surface protein, for instance, a virus coat protein. Genetic engineering can be applied in many different ways to the production of vaccines, including those against viral disease and others against bacterial disease. The importance of the recombinant vaccines is the fact that they can replace the killed or inactivated pathogenic organisms used as vaccines.

Genetic engineering has proved successful in the development of some *subunit vaccines*. These vaccines contain only a specific *subunit* of a protein from the pathogenic organism (usually a coat protein), and recombinant DNA techniques are used to produce these subunits in microorganisms. The highly immunogenic coat proteins are purified and used in high dosage to elicit a rapid and high level of immunity with no possibility of transmitting infection. The steps for viral gene cloning are those outlined in the previous sections: fragmentation of viral DNA by restriction enzymes; cloning viral coat protein genes into a suitable vector; providing for proper promoters, reading frame, and ribosome binding sites; and reinsertion and expression of the viral genes in a microorganism. Unfortunately, when *Escherichia coli* is used as the cloning host, the vaccines are often poorly immunogenic and fail to protect animals from subsequent infection with the virus. The problem involves the fact that many key viral coat proteins are posttranslationally modified, generally by the addition of sugar residues (glycosylation), when the virus replicates in its normal host. However, the recombinant proteins produced by *E. coli* or other Bacteria are unglycosylated, and apparently glycosylation is necessary for the proteins to be immunologically active. Therefore, a eukaryotic host is used.

The first recombinant subunit vaccine approved for use in humans was made using yeast. The gene encoding a surface protein from hepatitis B virus was cloned and expressed in yeast. The protein was produced and formed aggregates very similar to those found in patients infected with the virus. These aggregates were purified and used to vaccinate people against hepatitis B virus. Subunit vaccines against a large variety of viruses and pathogenic organisms are also being developed employing genetic engineering. In addition to using yeast, insect cells and even cultured mammalian cells are now being used as hosts to prepare recombinant vaccines. To obtain the correct pattern of glycosylation or other modifications of the protein, it is often important to use a host that is closely related to humans. However, vaccines can also be produced in plants (see Section 10.14).

Many laboratories are working on subunit vaccines against the virus that causes acquired immunodeficiency syndrome (AIDS) (human immunodeficiency virus, or HIV). Because of the serious consequences associated with AIDS, a safe, effective AIDS vaccine would find a huge worldwide market and thus the stakes in this area are high. However, genetic engineering can also be used to develop *recombinant live vaccines*. One way is to isolate deletion mutants of a virus lacking genes that cause disease but which is still infective and still elicits an immune response. Another method is to add genes to a virus so it can confer immunity against viral disease. In this latter category is a live recombinant virus vaccine that offers protection in poultry against both fowlpox (a disease that reduces weight gain and egg production) and Newcastle disease (a viral disease that is often lethal). The fowlpox virus, a typical pox virus (∞ Section 8.20), was first modified to delete genes that cause disease (but not those that elicit immunity). Then immunity-inducing genes from the Newcastle virus were added. This resulted in a *polyvalent vaccine*, in this case a single virus that can confer immunity to two different important diseases.

One vector used to prepare live recombinant vaccines is *vaccinia virus* (∞ Section 8.20). Cloning in vaccinia virus is done using an *Escherichia coli* plasmid containing a fragment of the vaccinia virus thymidine kinase gene (Figure 10.18*a*). An appropriate foreign DNA is inserted into this plasmid, and the recombinant plasmid is mixed with wild-type vaccinia virus DNA (Figure 10.18*b*). If homologous recombination occurs between plasmid DNA and vaccinia genomic DNA (Figure 10.18*c*), recombinant virions can be obtained containing an *inactivated*

thymidine kinase gene. An *active* thymidine kinase leads to growth inhibition by the compound 5-bromodeoxyuridine. Therefore, recombinant vaccinia virions can be selected by allowing viral replication to occur in the presence of this inhibitor (Figure 10.18*d*). Although such recombinant viruses no longer express thymidine kinase, they can still infect human cells and they do express the foreign genes that have been cloned into them. Indeed, some recombinant vaccinia viruses can carry genes from four different viruses!

Vaccinia virus itself is generally not pathogenic for humans (vaccinia virus was originally used as a vaccine against the related virus smallpox). However, vaccinia virus is not completely benign (it causes severe complications in some people), and therefore more research must be done before such vaccines can be used in humans.

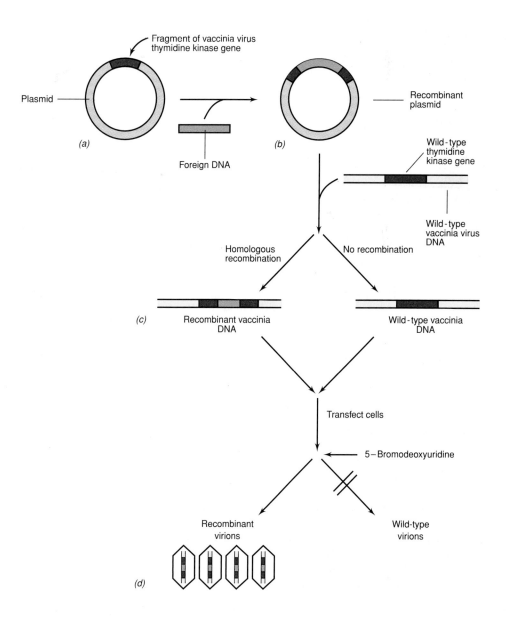

Figure 10.18 Production of recombinant vaccinia virus. (a) Foreign DNA is cloned into a plasmid containing a small piece of the vaccinia virus thymidine kinase gene. (b) Recombinant plasmid formed. (c) The latter is mixed with wild-type vaccinia DNA. If recombination occurs, recombinant vaccinia DNA can be produced. (d) The mixture is used to transfect host cells, and viral replication is allowed to occur in the presence of 5-bromodeoxyuridine, a compound that is toxic to cells having an active thymidine kinase. Only recombinant virions develop under these conditions. If the recombinant vaccinia virions contain genes for other viral coat proteins, these may be expressed.

Genetically engineered vaccines are likely to become increasingly common because (1) they are safer than normal attenuated or killed vaccines, (2) they are more reproducible because their genetic makeup can be carefully monitored, and (3) they can be administered in high doses without fear of side effects.

Vaccines produced using genetic engineering can also usually be made much faster than those produced by more traditional methods. Recombinant vaccines made with cloned influenza virus hemagglutinin genes (∞ Section 8.16) can be made in just 2 or 3 months rather than 6–9 months. This could be of real advantage during an epidemic caused by a new strain of influenza virus (∞ Section 23.4). Recombinant vaccines are also sometimes much less expensive than those produced in other ways, and this in itself can allow for new uses. Bait carrying a recombinant rabies vaccine has been distributed over large tracts of land in Europe and has triggered a dramatic decline in the incidence of rabies among wild foxes. Such a method of vaccination would previously have been too expensive.

One potentially very exciting development is the possible use of the DNA itself as a vaccine. In some cases it appears that if a particular gene is delivered to an animal cell in such a way that it is taken up by the cell, the protein will be produced and the animal will develop immunity. Such an approach would be both safe and inexpensive.

Other proteins and other products

Table 10.1 lists a number of mammalian proteins and products other than insulin and vaccines. These include a number of other hormones and also proteins involved in blood clotting and other blood processes. For example, *tissue plasminogen activator* (TPA) is a protein found in the blood that acts to scavenge and dissolve old blood clots in the final stages of the healing process. The clinical usefulness of TPA is primarily in heart patients or anyone suffering from poor circulation because of excessive clotting tendencies. TPA can be administered following cardiac bypass, transplant, or other open heart surgeries to prevent the development of pulmonary embolisms, which are often life-threatening. Heart disease is a leading cause of death in many developed countries, so microbially produced TPA promises to be in high demand.

In contrast to TPA, blood clotting factors VII, VIII, and IX are critically important for the *formation* of blood clots. Hemophiliacs suffer from a deficiency of one or more clotting factors and can be readily treated with the microbially produced product. Recombinant clotting factors take on added significance when one considers that hemophiliacs have in the past been treated with concentrated clotting factor extracts from pooled human blood, some of which was contaminated with the AIDS virus: this has put hemophiliacs at high risk for contracting AIDS.

A number of proteins have roles as anticancer agents or immune modulators. *Interferons* are a series of proteins made by animal cells in response to viral infection (∞ Section 11.11) or immune activation in the case of one type of interferon.

Even monoclonal antibodies are now produced in microorganisms by genetic engineering (we discuss monoclonal antibodies in some detail in Section 20.14).

Certainly not all the proteins produced using recombinant DNA techniques have therapeutic uses. Many commercial enzymes are produced in this way as well. In addition, even hormones can have other uses. *Bovine somatotropin* produced using these techniques is used to increase milk production in cattle in the United States. Often the "benefits" of genetic engineering can be quite unexpected. *Rennin,* which is used to make cheese, is an animal product. In Great Britain a "vegetarian cheese" containing a recombinant protein produced in a microorganism is being marketed and is finding wide acceptance. The first products made by genetic engineering were primarily the protein products of cloned genes, and there is still a great deal to be done with this approach. Added applications come from being able to use site-directed mutagenesis on the cloned gene so that new products with new attributes can be generated. It must also be remembered that molecules such as antibiotics are synthesized in cells in biochemical pathways using a series of enzymes (proteins). These enzymes can be modified so new antibiotics can be developed.

CONCEPT CHECK   10.13

The first human protein made commercially using engineered bacteria was human insulin, but numerous other hormones and other human proteins are now being produced. Many proteins found in humans that were formerly extremely expensive to produce because they were found in human tissues in only small amounts can now be made in very large amounts from the cloned gene in a suitable expression system. In addition to useful pharmaceuticals such as anticancer agents and immune modulators, even vaccines can be produced using genetic engineering.

✔ *Why is it sometimes important to produce proteins used for therapeutic purposes in a host closely related to humans?*

✔ *Explain why recombinant vaccines might be safer than some vaccines produced by traditional methods.*

10.14

Genetic Engineering in Plant Agriculture

Classical genetic improvement of plants has generally been a slow and difficult task, but recombinant DNA technology promises revolutionary changes. It is possible to use plant tissue culture techniques to select clones of plant cells that have been genetically altered and then, with proper treatments, induce these cell cultures to make whole plants that can be propagated vegetatively or by seeds. Recombinant DNA technology enters into this approach because one can transform plant cells with free DNA by either electroporation or particle gun methods (∞ Section 9.6) or insert foreign DNA into plant cells directly via the bacterium *Agrobacterium tumefaciens.* This organism causes the plant disease *crown gall* by transferring specific genes to the plant (∞ Section 14.22), and plant genetic engineers have used this natural transformation system as a vehicle for the introduction of foreign DNA into plants.

Vectors for cloning in plants

The gram-negative plant pathogen *Agrobacterium tumefaciens* contains a large plasmid called the **Ti plasmid,** which is responsible for its virulence. The plasmid contains genes that mobilize DNA for a transfer to the plant (for details of the disease process and genetic events, see Section 14.22). The segment of the Ti plasmid DNA that is actually transferred to the plant is called **T-DNA.** The sequences at the ends of the T-DNA are essential for transfer, and the DNA to be transferred must be between these ends. One type of vector that has been constructed and is used for the actual transfer of genes to plants is called a *binary vector.* The word *binary* means "consisting of two parts," and a binary vector must be used in conjunction with another plasmid in order for the cloned gene to be transferred to the plant. The vector contains the two ends of the T-DNA on either side of the site used for cloning and an antibiotic resistance marker that can be used in plants. It also contains an origin of replication so that it can replicate in both *A. tumefaciens* and *Escherichia coli* (the latter serves as a host for cloning work), and another antibiotic resistance marker that is expressed in bacteria (Figure 10.19). The DNA to be cloned is inserted into the vector, which is then transformed into *E. coli.* It is then transferred to *A. tumefaciens* (usually by conjugation).

Final transfer to the plant depends on this bacterium also containing a Ti plasmid because the vector does not contain the genes required for T-DNA transfer. The cloned DNA and the kanamycin resistance marker of the vector can be mobilized by the Ti plasmid and transferred into a plant cell (Figure 10.19); note here

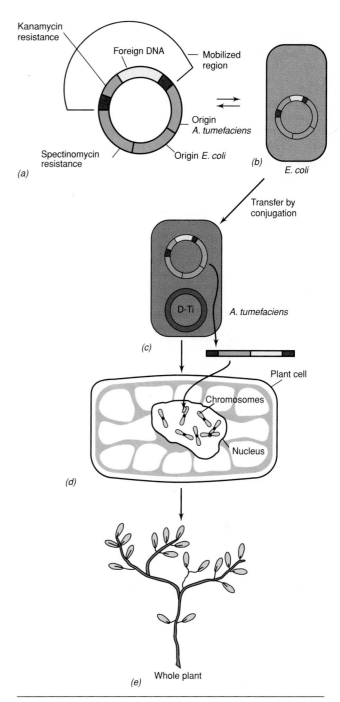

Figure 10.19 Production of transgenic plants using *Agrobacterium tumefaciens.* (a) Generalized plant transfection vector containing ends of T-DNA (in red), foreign DNA (in yellow), origin of replication elements for both *E. coli* and *A. tumefaciens,* and spectinomycin and kanamycin resistance markers. The kanamycin resistance marker can be selected for in plants. (b) The vector can be put into *E. coli* for cloning purposes and then transferred to *A. tumefaciens* by conjugation. (c) The resident Ti plasmid used for transferring the vector to the plant (D-Ti) is itself genetically engineered to remove key pathogenesis genes. (d) However, D-Ti can mobilize the T-DNA region of the vector for transfer to plant cells grown in tissue culture. From the recombinant cell, whole plants can be regenerated.

that the Ti plasmid used for this purpose has been genetically altered in such a way as to prevent the onset of disease in the plant). Following recombination with a host chromosome, the foreign DNA can be expressed to confer new properties on the plant. Many genes are not expressed efficiently in plants unless they are cloned into an expression vector containing a plant promoter. Some of the promoters that have been used in constructing plant expression vectors include some normally found in T-DNA and a promoter from cauliflower mosaic virus, a plant DNA virus (∞ Table 8.1).

With the use of *Agrobacterium tumefaciens,* a number of transgenic plants have been produced. Most successes have come with herbaceous plants (dicots) such as tomato, potato, tobacco, soybean, alfalfa, and cotton, but *A. tumefaciens* has also been used to produce transgenic woody dicots such as walnut and apple. Transgenic crop plants from the grass family (monocots) are difficult to generate using *A. tumefaciens,* but other methods of introducing DNA, such as electroporation or a particle gun (∞ Section 9.6), seem to work well here. In addition, some plant vectors have been developed by modifying plant viruses, but these have not yet proved as useful as the vectors derived from the Ti plasmid.

Applications in plant biotechnology

Major areas targeted for genetic improvement in plants include herbicide, insect, and microbial disease resistance, as well as improved product quality. Some products are already available, and genetically modified plants undergoing large-scale field trials include virus-resistant squash, herbicide-tolerant cotton, and soybeans and oilseed rape with modified oil. More than 1000 different field trials on more than 30 different plant species have been carried out in the last decade. Herbicide resistance can be obtained by genetically engineering the crop plant to no longer respond to the toxic chemical. Many herbicides act by inhibiting a key plant enzyme or protein necessary for growth. For example, the herbicide *glyphosate* kills plants by inhibiting the activity of an enzyme necessary for making aromatic amino acids. Such an herbicide kills both weeds and crop plants and thus must be used as a "preemergence herbicide," that is, before the crop plants emerge from the ground. However, some bacteria contain an enzyme that is naturally resistant to glyphosate. A gene encoding a resistant enzyme from *Agrobacterium* has been cloned, modified for expression in plants, and transferred into crop plants. When sprayed with glyphosate, plants containing the bacterial gene grow as well as unsprayed control plants (Figure 10.20). Soybeans expressing glyphosate resistance have been developed by Monsanto Company and are now available for agricultural production.

Novel means of insect resistance have been genetically introduced into plants. One of the most promis-

Figure 10.20 The photograph shows a portion of a field of soybeans that has been treated with *Roundup,* a glyphosate-based herbicide manufactured by Monsanto. The plants on the right are normal soybeans; those on the left have been genetically engineered to express glyphosate resistance.

ing has been the toxic protein genes of *Bacillus thuringiensis.* This organism produces a crystalline protein (∞ Section 16.26), called *Bt-toxin,* that is toxic to moth and butterfly larvae, and certain strains of *B. thuringiensis* produce additional proteins toxic to beetle and fly larvae and mosquitoes. Biotechnologists are using several different approaches to enhance the use of Bt-toxin for pest control in plants.

One approach is to develop a single Bt-toxin that is effective against many different insects. This can be done because the protein has separate domains for its specificity and its toxic function. The toxic domain is highly conserved in all the various Bt-toxins. Genetic engineers are attempting to make a gene that can encode a Bt-toxin carrying one toxic domain and several different specificity domains. Such a toxin could be applied directly to a number of different plants. Another approach is to transfer the gene to bacteria that normally live (harmlessly) in plant tissue and have the toxin produced by these bacteria in these tissues. An even more effective approach may be to transfer the gene directly into the plant genome. For example, a natural Bt-toxin gene has been cloned into a plasmid vector under control of a chloroplast rRNA promoter and transferred into tobacco plant chloroplasts by particle bombardment (∞ Section 9.6). With this methodology plants were obtained that expressed this protein at levels that were extremely toxic to insect larvae from a number of species (Figure 10.21).

However, there have been reports of insects that have acquired resistance to Bt-toxin. Resistance to insecticides and herbicides is a common problem in agriculture, and the fact that a product has been produced by genetic engineering does not give it any mag-

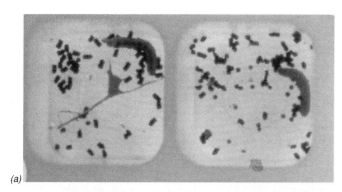

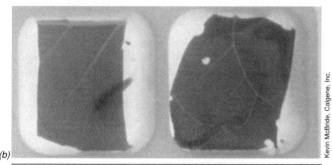

Kevin McBride, Calgene, Inc.

Figure 10.21 Panel (a) shows the results of two different assays to determine the effect of beet armyworm larvae on tobacco leaves from normal plants. Panel (b) shows the results of similar assays but using tobacco leaves taken from transgenic plants that express Bt-toxin in their chloroplasts.

ical properties. This emphasizes that many approaches must be used for pest control, and Bt-toxin is only one of many being developed by biotechnologists.

Genetic engineering has also been used to protect plants from virus infection. For example, it has been discovered that transgenic plants that express the coat protein gene of a virus become resistant to infection by that virus. Although the mechanism of resistance is unknown, the presence of viral coat protein in plant cells apparently interferes with the uncoating of viral particles containing that coat protein, and this interrupts the virus replication cycle.

Not all genetic engineering is directed toward making plants disease-resistant. Genetic engineering can be used in a variety of ways for developing mutant strains of plants with desired characteristics such as delayed spoilage. The *Flavr Savr* tomato developed by the Calgene Corporation produces an antisense RNA (∞ Antisense Nucleic Acid, Chapter 7) that prevents translation of the mRNA from a gene whose protein product normally functions to break down pectin, thereby weakening the structure of the tomato. Therefore, *Flavr Savr* tomatoes do not spoil as rapidly as normal tomatoes and can be harvested at a later stage of maturity.

In addition, transgenic plants can be genetically engineered to produce commercial or pharmaceutical products, as has been done with microorganisms (see

Section 10.13) and animals (see Section 10.15). Crop plants such as tobacco and tomatoes have been engineered to produce a number of different products, such as the human protein interferon. Transgenic plants can be used to produce animal antibodies in quantity (such plant-made antibodies are sometimes called "plantibodies") and even granules of a polyester used to manufacture plastics.

Plant hosts can be useful in producing these types of products because plants typically modify proteins correctly and because crop plants can be efficiently grown and harvested.

Crop plants are also being developed for the production of vaccines. For instance, a recombinant tobacco mosaic virus has been engineered whose coat contains antigens of *Plasmodium vivax*, the organism that causes malaria (∞ Section 23.11). This recombinant virus could be used to develop a malaria vaccine that can be produced in very large amounts at very low cost by harvesting tobacco grown in fields. Another very interesting approach is to produce a vaccine in an edible plant product. Such *edible vaccines* are now under development that could immunize against diseases caused by enteric bacteria, including cholera and diarrhea (∞ Section 23.14).

Many plants have a natural ability to concentrate heavy metals and can be used to remove these contaminants from soil. Genetic engineering can be employed to enhance this ability. When the bacterial gene encoding mercuric reductase was modified and expressed in plants, the plants could grow at otherwise toxic levels of mercury. This enzyme reduces mercuric salts to elemental mercury, which is slowly released into the air. Such plants could be an important tool in removing contamination from soil. However, they would do so by releasing mercury into the air, and this may not be acceptable. This situation illustrates the careful consideration that must be given to the development of any technological solution to a problem. Good science does not necessarily translate directly into commercial viability.

CONCEPT CHECK   *10.14*

Genetic engineering is being employed to make plants resistant to disease, to improve product quality, and to use crop plants as a source of recombinant proteins and even vaccines. One commonly used cloning vector for plants is the Ti plasmid of the bacterium *Agrobacterium tumefaciens*. This plasmid can transfer DNA into plant cells.

✔ *Transfer of DNA by the Ti plasmid most resembles what form of bacterial gene transfer (∞ Chapter 9)?*

✔ *What commercial reasons could there be for producing vaccines or human proteins in a plant?*

10.15

Genetic Engineering in Animal and Human Genetics

This section covers only a few highlights of some of the huge number of uses of genetic engineering in animal and human genetics. Some of these applications have to do with producing products, but more have to do with understanding gene function in mammals and in curing or treating genetic disease.

Transgenic animals

With the use of recombinant DNA technology and microinjection techniques to deliver cloned genes to fertilized eggs, several foreign genes have been expressed in both laboratory research animals and in species important in commercial animal industries. Such animals are called **transgenic animals** and have become increasingly important in basic biomedical research for studying gene regulation and developmental biology. However, many applied aspects of transgenic animals are also of interest.

One approach is to improve the productivity or disease resistance of the animal, as in the case of agricultural plants. However, transgenic animals are also being used to produce proteins of pharmaceutical value—a process some scientists have called "pharming." Transgenic animals may be useful for producing human proteins that require posttranslational modifications for activity, such as certain blood clotting enzymes; many proteins of this type are not produced in an active form by microorganisms. Also, some proteins have been engineered to be secreted into the animals' milk, which can be readily collected and processed. These proteins include α-1-antitrypsin (used to treat lung disease) produced in sheep and tissue plasminogen activator (used to dissolve blood clots) produced in goats. The production of transgenic animals for research and commercial purposes seems likely to continue as an important area in biotechnology.

Human genetics

Conventional genetics, involving genetic crosses or mutagenesis, cannot be done with humans. Therefore, in spite of the obvious interest, our understanding of human genetics has lagged considerably behind our understanding of the genetics of many other organisms. The advent of genetic engineering has quite simply revolutionized studies of the human genome. A detailed discussion of human genetics is beyond the scope of this book, but a few general remarks on the utility of recombinant DNA technology can be made. We have already mentioned the use of PCR for DNA fingerprinting (see the box), and considerable efforts are being made to clone and sequence the entire human genome (see Section 10.4).

However, there are some applications of genetic engineering of the human genome that are directed at treating human disease. A vast number of genetic diseases are known, but except in rare cases, little was known until recently about their molecular bases. By use of recombinant DNA technology, coupled with conventional genetic studies (following family inheritance, and so on), it is possible to localize particular defects to particular chromosomes and to particular locations on chromosomes. With the use of recombinant DNA technology, it is possible to clone the region containing the genetic defect and then to make comparisons between the base sequence in the normal gene and in genetically altered chromosomes. From such studies, even in the absence of knowledge of the enzyme defect, it has been possible to obtain information about the genetic change. Many genes, including those for Huntington's disease, cystic fibrosis, and Duchenne's muscular dystrophy, have been localized with these techniques.

Genetic engineering is employed to provide treatment for some of these diseases using **gene therapy.** In gene therapy, a nonfunctional or dysfunctional gene is augmented or replaced by a functional gene. Not all gene therapy is designed toward treating genetic disease; a considerable effort has been directed toward protocols for treating cancer. Major obstacles to this approach exist in trying to target the correct cells for gene therapy and in successfully transfecting cell lines that will perpetuate the genetic alteration.

The first genetic disease for which an approved gene therapy technique was used is a severe combined immune deficiency caused by the absence of adenosine deaminase (ADA), an enzyme involved in purine metabolism, in bone marrow cells. The procedure involved using a retrovirus as a vector to insert a good copy of the ADA gene into T lymphocytes (cells that are part of the immune system; ∞ Section 20.1) removed from the patient and then placing these "corrected" cells back in the body (the retrovirus also carries a marker gene, resistance to neomycin, so cells carrying the inserted retrovirus can be selected and identified). Since T lymphocytes have a limited life span, it is necessary to repeat the therapy every month or two. Attempts are being made to insert the gene into the stem cells of the bone marrow (which continue to divide) and effect a true cure for the disease.

Several other gene therapy treatments, some using other virus vectors, are currently being tested. Since the first gene therapy experiment with ADA in 1990, there have been no striking practical breakthroughs. Though gene therapy has tremendous practical potential, most applications still remain a distant prospect. Some of the current difficulties are related to the vectors being used. Although transduction using retroviral vectors gives stable integration of the gene, the site of insertion is unpredictable, the amount of cloned

DNA is limited, and expression of the cloned gene is often transient. The vectors also have limited infectivity and are rapidly inactivated in the host. Of even more concern, during vector production replication-competent revertant viruses may arise, which themselves could cause cancer. Many nonretroviral vectors have similar problems. New vectors are being developed to attempt to overcome these problems (see Section 10.7).

It is important to note that in the protocols being tested, the defective gene is not replaced. Rather, the retrovirus (and the good copy of the gene) simply integrates somewhere in the human genome of these cells. Actual gene replacements in germ line cells (cells that give rise to gametes) can be accomplished with some mammals, although the techniques of isolating individual animals with these changes cannot readily be applied to humans. However, attempts to change the germ cells of an individual would raise many ethical and societal questions beyond simple questions of experimental protocols.

10.16

Genetic Engineering as a Microbial Research Tool

Up until now we have mostly presented the principles of genetic engineering in the light of practical goals. However, genetic engineering technology finds many uses in basic research on microorganisms. Molecular cloning and the engineering of new microbial strains provide some of the best ways to understand basic microbial processes such as structure–function relationships, cell growth, enzyme regulation, and microbial ecology. A novel microbial strain can be created that differs from the wild type in a defined manner, and the underlying process can then be studied. In this way, one can observe the importance of a particular gene product to a basic microbial process under precisely controlled conditions. Molecular cloning, restriction mapping, and sequencing also allow geneticists to quickly map and study the genomes of newly identified organisms.

We have discussed how cloned genes can be subjected to site-specific mutagenesis and how gene disruptions or knockout mutations can be introduced into a gene. These mutated genes can then be introduced into the microbial chromosome so the mutant organism can be studied. Genetic engineering also allows the researcher to "tag" a gene so that it is easier to study (Figure 10.22). For instance, if the gene product is difficult to assay or has no known function, it is often very difficult to know under what conditions the cell makes this protein. One can also use this technique to tag the organism itself so that it can be traced in the environment.

In these cases, a gene or its promoter can be fused to a *reporter gene*. Reporter genes are simple to assay, and the gene encoding β-galactosidase (whose activity

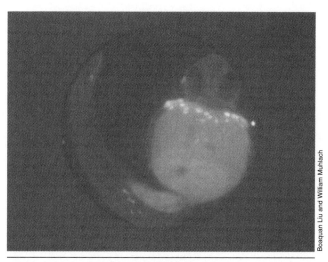

Figure 10.22 The photograph shows a living embryo of a transgenic fish that is expressing the gene encoding the green fluorescent protein under control of a *Xenopus* promoter. The photograph was taken using fluorescent microscopy, and the tissues expressing green fluorescent protein show up as bright green areas. In this organism the gene is being expressed in yolk sac cells.

can be detected on indicator plates and also assayed easily by biochemical means) is commonly used as a reporter (see Figure 10.5). However, there are other choices. For instance, *Escherichia coli* has been engineered in such a way that it produces the luciferase enzyme of the bacterium *Photobacterium* or that of several different luminescent beetles (see Figure 10.7). Because of the presence of this enzyme, the engineered *E. coli* becomes luminescent, and its colonies therefore glow in the dark. One can then detect colonies of the engineered *E. coli* on agar plates by their luminescence against a large background of other colonies. Production and detection of luciferase usually involves more than one gene and accessory factors. Recently a gene encoding a fluorescent protein, called the *green fluorescent protein* (GFP), has been isolated from the jellyfish *Aequorea victoria* and cloned. GFP needs no accessory factors and is already being used as a reporter or tag in a wide variety of organisms (Figure 10.22).

10.17

Summary of Principles at the Basis of Genetic Engineering

We have presented the fundamentals of genetic engineering and have shown how the approaches used have been derived from an understanding of basic concepts of molecular genetics and the "classic" techniques of microbial genetics. We now summarize the principles of genetic engineering by relating current knowledge back to the basic information presented in Chapters 6–9.

The following developments were essential for the development of genetic engineering; their interrelationships are diagrammed in Figure 10.23.

1. **DNA chemistry:** development of procedures for isolation, sequencing, and synthesis of DNA.

2. **DNA enzymology:** discovery of restriction endonucleases, DNA ligases, and DNA polymerases.

3. **DNA replication:** understanding how DNA replication occurs and the importance of DNA vectors capable of independent replication.

4. **Plasmids and conjugation:** discovery of plasmids, determination of the mechanisms by which plasmids replicate, and how some can transfer from cell to cell by conjugation.

5. **Temperate bacteriophage:** understanding how replication and/or integration is controlled in temperate bacteriophages and how specialized transducing phages are formed.

6. **Transformation:** discovery of methods for getting free DNA into cells.

7. **RNA chemistry and enzymology:** understanding how to work with messenger RNA, how eukaryotic mRNA is constructed, and the importance of RNA processing in the formation of mature eukaryotic mRNA.

8. **Reverse transcription:** discovery of the enzyme *reverse transcriptase* in retroviruses and its development as a means for transcribing information from mRNA back into DNA.

9. **Regulation:** understanding the factors involved in the regulation of transcription, including the discovery of promoter sites and operon control.

10. **Translation:** understanding the steps involved in translation, the importance of ribosome binding sites on mRNA, the role of the initiation codon, and the importance of a proper reading frame.

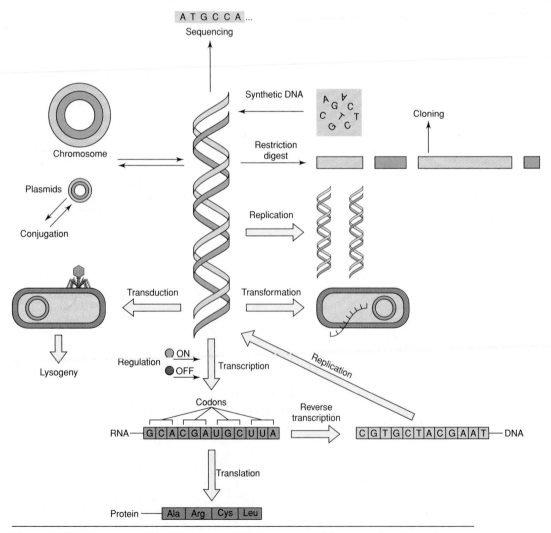

Figure 10.23 Summary of the fundamental processes underlying genetic engineering.

11. **Protein chemistry:** development of methods for isolation, purification, assay, and sequencing of proteins.

12. **Protein excretion and posttranslational modification:** understanding how proteins are built with signal sequences that are removed during or after excretion. Discovery of other kinds of posttranslational modification of proteins.

13. **The genetic code:** elucidation of the genetic code and the determination that it was the same in almost all organisms but that certain codons were less frequently used in some organisms than in others.

CONCEPT CHECK   10.15–10.17

Genetic engineering is providing vast opportunities for improving processes of economic value and in treating disease. Medicine, agriculture, and industry are all finding important uses of cloning and gene expression technology. In addition genetic engineering has revolutionized genetic studies in higher organisms and is a powerful tool in basic research on microbial genetics.

Material for Review

REVIEW QUESTIONS

1. What are the essential features of a cloning vector? What are the characteristics of plasmids that make them especially useful in cloning? Why aren't all plasmids equally useful for cloning?

2. Most of the lambda derivatives used as cloning vectors are virulent. Based on your knowledge of lambda genetics (∞ Section 8.12) and the structure of the lambda Charon vectors, discuss why these viruses are no longer temperate. Why might it be useful to also have a temperate bacteriophage as a cloning vector?

3. The bacteriophage lambda can be an extremely useful cloning vector. However, it cannot be used in *Saccharomyces cerevisiae*. Explain.

4. If *insertional inactivation* is used to detect the presence of an introduced plasmid in a bacterial cell, why is it desirable to have *two* antibiotic resistance markers in the plasmid?

5. Explain why the use of a regulatory switch is desirable for the large-scale production of a protein.

6. How has bacteriophage T7 been used in expressing foreign genes in *Escherichia coli* and what desirable features does this regulatory system possess?

7. The plasmid shown in Figure 10.9 is a cloning vector, an expression vector, a fusion vector, and a phagemid. Define each of these terms and explain how this single plasmid can have all these properties.

8. Describe the basic principles of gene amplification using the polymerase chain reaction (PCR). How have thermophilic bacteria simplified the use of PCR? How many PCR cycles need to be run to get 1000 copies of a specific target sequence (starting from one molecule of double-stranded DNA)?

9. What are the similarities and the differences between *cassette mutagenesis* and *transposon mutagenesis* (∞ Section 9.9)?

10. What is a subunit vaccine and why are subunit vaccines considered a safer way of conferring immunity to viral pathogens than attenuated virus vaccines?

11. What is the Ti plasmid and how has it been of use in genetic engineering?

12. How do transgenic plants and animals differ from plants and animals modified by conventional breeding techniques?

13. What advantages might there be in using a transgenic plant rather than a transgenic animal to produce a protein?

APPLICATION QUESTIONS

1. Suppose you are given the task of constructing a plasmid suitable for molecular cloning in an organism of industrial interest. List the characteristics such a plasmid should have. List the steps you would use to develop such a plasmid.

2. When employing the plasmid vector pUC18, DNA fragments made using the restriction enzyme *Bam*HI inactivates the gene encoding β-galactosidase. When employing pBR322, cloning the same fragments inactivates the gene conferring tetracycline resistance. Both vectors contain a gene for ampicillin resistance that is used to select for bacterial transformants after making the recombinant molecules. Explain why it is much more efficient to use pUC18 rather than pBR322 as a cloning vector. (*Hint:* The increased efficiency has to do with finding the cells that contain vectors with cloned DNA.)

3. The formation of specialized transducing phages inside cells is an example of naturally occurring genetic engineering. Generalized transducing particles (∞ Section 9.7) also carry host DNA, but one wouldn't consider the formation of this particle to be genetic engineering. Why not?

4. Suppose you have just determined the DNA base sequence for an especially strong promoter in *Escherichia coli* and you are interested in incorporating this sequence into an expression vector. Describe the steps you would use. What precautions would be necessary to be sure that this promoter actually works as expected in its new location?

5. Making cDNA libraries from cells undergoing a particular regulatory response can be a powerful way of detecting the genes involved. Even though it is as simple to isolate mRNA from prokaryotes as it is from eukaryotes, this technique isn't used with prokaryotes. Explain. (*Hint:* Think about how cDNA is made.)

6. You have just discovered a protein in mice that may be an effective cure for cancer but it is present only in extremely small amounts. Describe the steps you would use to obtain production of this protein in *Escherichia coli*. If the protein produced in *E. coli* proved to be only marginally effective in clinical trials, what might you do to find a protein that is more effective?

SUPPLEMENTARY READINGS

Glazer, A. N., and **H. Nikaido.** 1995. *Microbial Biotechnology: Fundamentals of Applied Microbiology.* W. H. Freeman, New York. This very fine text covers the fundamentals of biotechnology as well as many specific applications related to microorganisms.

Glick, B. R., and **J. J. Pasternak.** 1994. *Molecular Biotechnology: Principles and Applications of Recombinant DNA.* American Society for Microbiology, Washington, DC. An excellent introductory text covering the basic techniques and applications of genetic engineering.

Mullis, K. B., F. Ferré, and **R. A. Gibbs** (eds.). 1994. *The Polymerase Chain Reaction.* Birkhäuser, Boston. This extremely useful book explains the basic methodology of PCR as well as many new modifications and applications.

Sambrook, J., E. F. Fritsch, and **T. Maniatis.** 1989. *Molecular Cloning: A Laboratory Manual*, 2nd edition. Cold Spring Harbor Laboratory Press, Cold Spring Harbor, NY. A detailed manual in three volumes describing many procedures for cloning and obtaining expression of cloned genes. Brief discussions of the principles behind each procedure.

Watson, J. D., M. Gilman, J. Witkowski, and **M. Zoller.** 1992. *Recombinant DNA*, 2nd edition. Scientific American Books, New York. An excellent introductory text devoted to the principles and applications of genetic engineering.

 On~line resources for this chapter are on the World Wide Web at: http://www.prenhall.com/~brock (click on the <u>table of contents</u> link and then select Chapter 10).

Microbial Growth Control

Steam under pressure attains a temperature above 100°C. An autoclave, such as the one shown here, is a device for sterilizing implements used in microbiology as well as for sterilizing culture media, and works by placing steam under a pressure of 15 psi where it reaches a temperature of 121°C.

MINIGLOSSARY for Chapter 10

Aminoglycosides a group of antibiotics, including streptomycin, containing amino sugars linked by glycosidic bonds

Antibiotic chemical substance produced by a microorganism that kills or inhibits the growth of another microorganism

Antibiotic Resistance the acquired ability of a microorganism to grow in the presence of an antibiotic to which the microorganism is usually sensitive

Antimicrobial Agent a chemical that kills or inhibits the growth of microorganisms

Antiseptic antimicrobial agents that are sufficiently nontoxic to be applied on living tissues

Autoclave a sterilizer that destroys microorganisms with temperature and steam under pressure

β-Lactam Antibiotics a group of antibiotics, including penicillin, that contain the four-membered heterocyclic β-lactam ring

Chemotherapeutic Agent an antimicrobial agent that can be used internally

Cidal lethal or killing

Disinfectant an antimicrobial agent used only on inanimate objects

Growth Factor Analog a chemical agent that is related to and blocks the uptake of a growth factor

Inhibition the reduction of microbial growth because of a decrease in the number of organisms present or alterations in the microbial environment

Lysis loss of cellular integrity with release of cytoplasmic contents

Pasteurization destruction of all disease-producing microorganisms or a reduction in the number of spoilage microorganisms

Semisynthetic Penicillin a natural penicillin that has been chemically altered

Sterilization the killing or removal of all living organisms and their viruses from a growth medium

Tetracycline a class of antibiotics containing the four-membered naphthacene ring

We have now completed our studies of the basic chemical and biological systems used by microorganisms for establishing and maintaining life. Thus far, we have discussed environmental factors that *promote* growth. In this chapter, we will use the principles we have learned, but we will shift our focus and highlight agents and methods used for *control* of microbial growth. In general, control can be effected by limiting microbial growth, the process of **inhibition,** or by destroying the organisms by **sterilization,** the killing or removal of all viable organisms from a growth medium. Agents that destroy or kill bacteria are **bactericidal.** In practice, *sterility* is often not attainable, so in many cases we simply attempt to *inhibit* the rapid growth of organisms. Agents that inhibit bacterial growth are said to be **bacteriostatic.**

Why is microbial growth control necessary? We calculated that a single *Escherichia coli* cell would produce a mass of organisms much greater than the mass of the earth in less than 48 hr under optimal growth conditions (∞ Section 5.3). Of course, our studies of growth in Chapter 5 highlight the absurdity of this example—the growth of a microorganism is limited in the long term by the exhaustion of nutrients. Control of growth, however, is more than simply limiting nutrients. For practical reasons, microbial growth control is often necessary to limit destruction of a valuable "nutrient source." For example, in the food industry, food spoilage creates major economic costs. Therefore,

considerable resources are used to control spoilage. Routinely, microbial control measures such as disinfection and sterilization of food preparation areas and equipment, and pasteurization of consumable liquids are used to inhibit microbial growth and prolong the useful life of perishable materials.

Another major potential nutrient source for microorganisms is humans and other animals. Uncontrolled microbial growth on or in animal tissues causes cell destruction, a process called *infectious disease.* Microbial infectious diseases account for nearly 20 million human deaths each year worldwide (∞ Table 22.1), so the application of microbial control measures to prevent the spread of infectious diseases or to effect cures is very important. In addition to these very practical issues, the study of microbial control agents enhances our understanding of microbial metabolism, spore formation, antibiotic mechanisms, antibiotic resistance, and epidemiology.

This chapter will deal first with physical and chemical means of growth control in the environment. We will then investigate practical applications of growth control by examining methods used for the storage and handling of food. Next, we will explore several classes of compounds, the *growth factor analogs* and the *antibiotics,* which are important for controlling microbial growth in higher animals. Finally, we will study some mechanisms used by microorganisms to defeat growth control attempts. This chapter introduces important concepts for our discussions of the industrial production of antibiotics in Chapter 12 and the clinical use of antibiotics as discussed in Chapter 23.

11.1

Heat Sterilization

We now consider several physical methods of sterilization and growth inhibition including heat, filtration, and radiation. Remember that once a product is sterilized, it remains sterile indefinitely (∞ Pasteur's swan-necked flask, Figure 1.21). Perhaps the most widespread method for controlling microbial growth is heat, which is where we begin our discussion.

Kinetics of heat sterilization

As the temperature rises past the maximum temperature for growth, lethal effects occur. As shown in Figure 11.1, death from heating is an exponential (first-order) function and occurs more rapidly as the temperature is raised. The first-order relationship shown in Figure 11.1 means that the rate of death is proportional at any instant only to the concentration of organisms at that instant; the time taken for a definite fraction (for example, 90%) of the cells to be killed is independent of the initial concentration. These facts have important practical consequences. If we wish to *sterilize* a microbial population, it will take longer at lower temperatures than at higher temperatures. Thus, it is necessary to adjust the time and temperature to achieve sterilization for each specific set of conditions. The nature of the heat is also important: moist heat has better penetrating power than dry heat.

The time required for a 10-fold reduction in the population density at a given temperature, called the **decimal reduction time** or *D,* is the most useful way to characterize heat sterilization. Over the range of temperatures usually used in food sterilization, the relationship between *D* and temperature is essentially exponential. Thus, when the logarithm of *D* is plotted against temperature, a straight line is obtained (Figure 11.2). The slope of the line provides a quantitative measure of the sensitivity of the organism to heat under the conditions employed, and the graph can be used to calculate process times to achieve sterilization, such as in canning operations.

Determination of decimal reduction times is a fairly lengthy procedure because it requires making a number of viable count measurements (∞ Section 5.4). An easier way of characterizing the heat sensitivity of an organism is to determine the **thermal death time,** the *time* at which all cells are killed at a given temperature. This is done simply by heating samples of this suspension for different times, mixing the heated suspensions with culture medium, and incubating. When all cells are killed, no growth is evident in the incubated samples. Thus, the thermal death time depends on the size of the population tested because a longer time is required to kill all cells in a large population than in a small one. When the number of cells is standardized, it is possible to compare the heat sensitivities of different organisms by comparing their thermal death times. When the logarithm of the thermal death time is graphed versus temperature, a straight line similar to that shown in Figure 11.1 is obtained.

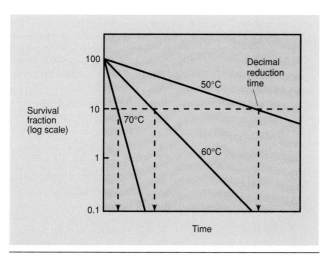

Figure 11.1 The effect of temperature on the viability of a mesophilic bacterium.

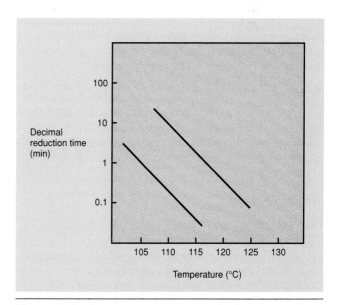

Figure 11.2 The relationship between the temperature and the rate of killing as indicated by the decimal reduction time for two different microorganisms. Data for such a graph are obtained from curves such as those given in Figure 11.1. The upper line in the figure represents data for a very heat-resistant organism.

Spores and heat sterilization

Vegetative cells and bacterial endospores from the same organism vary considerably in heat resistance. For instance, in the autoclave, a temperature of 121°C is normally reached. Under these conditions, endospores may require 4–5 min for a decimal reduction, whereas vegetative cells may require only 0.1–0.5 min at 65°C for a decimal reduction. Thus, practically speaking, heat sterilization must involve procedures for killing endospores.

The nature of the medium in which heating takes place also influences the killing of both vegetative cells and spores. Microbial death is more rapid at acidic pH, and acid foods such as tomatoes, fruits, and pickles are much easier to sterilize than neutral pH foods such as corn and beans. High concentrations of sugars, proteins, and fats decrease heat penetration and usually increase the resistance of organisms to heat, whereas high salt concentrations may either increase or decrease heat resistance, depending on the organism. Dry cells (and spores) are more heat-resistant than moist ones; consequently, heat sterilization of dry objects always requires higher temperatures and longer times than sterilization of moist objects.

Bacterial endospores (∞ Section 3.13) are the most heat-resistant structures known: they are able to survive heat that would rapidly kill vegetative cells of the same species. A major factor in heat resistance is the amount and state of *water* within the endospore. During endospore formation, the protoplasm is reduced to a minimum volume as a result of the accumulation of Ca^{2+} and synthesis of dipicolinic acid, which lead to formation of a gel-like structure. At this stage, the thick cortex forms around the protoplast core. Contraction of the cortex results in a shrunken, dehydrated protoplast. The water content of the protoplast determines the heat resistance of the spore. If endospores have high water content, they have low heat resistance; if they have low water content, they have high heat resistance. Water moves freely in and out of spores, and so it is not the impermeability of the spore coat that excludes water, but the gel-like material in the spore protoplast.

The autoclave

The **autoclave** is a sealed device that allows the entrance of steam under pressure (Figure 11.3). The use of moist heat facilitates killing of all microorganisms, including heat-resistant endospores.

Killing of the heat-resistant endospores requires heating at temperatures above boiling and the use of steam under pressure (Figure 11.3a). The usual procedure is to heat at 1.1 kilograms/square centimeter (kg/cm²) [15 pounds/square inch (lb/in²)] steam pressure, which yields a temperature of 121°C. At 121°C, the time of autoclaving to achieve sterilization

is generally considered to be 10–15 min (Figure 11.3b). If bulky objects are being sterilized, heat transfer to the interior will be slow, and the heating time must be sufficiently long so that the object is at 121°C for 10–15 min. Extended times are also required when large volumes of liquids are being autoclaved because large volumes take longer to reach sterilization temperatures. Note that it is not the *pressure* of the autoclave that kills the microorganisms but the *high temperature* that can be achieved when steam is placed under pressure.

Pasteurization

Pasteurization is a process that reduces the microbial populations in milk and other heat-sensitive foods. It is named for Louis Pasteur, who first used heat for controlling the spoilage of wine (see the box, The Origin of Pasteurization). Pasteurization is not synonymous with sterilization because not all organisms are killed. Originally, pasteurization of milk was used to kill pathogenic bacteria, especially the organisms causing tuberculosis, brucellosis, Q fever, and typhoid fever, but the keeping qualities of milk were also improved following pasteurization. Today, milk rarely comes from cows infected with pathogens and pasteurization is used primarily because it improves the storage life of milk, milk products, and various other beverages and food products (see the box).

Pasteurization of milk is usually achieved by passing the milk through a heat exchanger. Operationally, the milk is fed through tubing that is in contact with a heat source. Careful control of the milk flow rate and the size and temperature of the heat source raises the temperature of the milk to 71°C for 15 sec. The milk is then rapidly cooled. The whole process is aptly called

---

CONCEPT CHECK   11.1

Sterilization is the complete killing of all organisms. The most widely used method for sterilization is the application of heat. The temperature for heat sterilization is selected to eliminate the most heat-resistant organisms in the material, usually bacterial endospores. For routine sterilization, an autoclave is used; this permits application of steam heat under pressure at temperatures above the boiling point of water. Pasteurization rids the material of pathogenic microorganisms and reduces the load of microorganisms to prolong storage life.

✔ *Why is heat an effective sterilizing agent?*

✔ *What steps are necessary to sterilize material that may have bacterial endospores?*

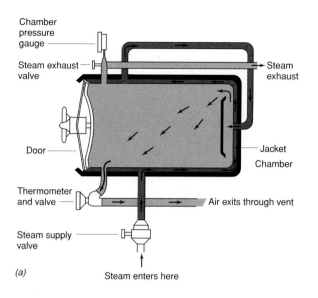

J. Martinko

(a)

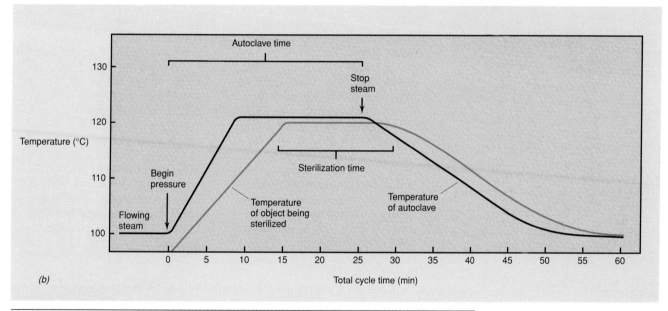

(b)

Figure 11.3 Use of the autoclave for sterilization. (a) The flow of steam through an autoclave. (b) A typical autoclave cycle. Shown is the sterilization of a fairly bulky object. The temperature of the object rises more slowly than the temperature of the autoclave. (c) A modern research autoclave. Note the door and the automatic cycle controls on the right panel.

flash pasteurization. Milk can also be heated in large vats to 63–66°C for 30 min. However, this *bulk pasteurization* method is less satisfactory because the milk heats and cools more slowly and is held at high temperatures for longer times. Flash pasteurization alters the flavor less, kills heat-resistant organisms more effectively, and is normally done on a continuous-flow basis. The flash pasteurization method is more adaptable to large dairy operations, and modern dairies generally employ it, often with even shorter exposure times and higher temperatures.

11.2
Radiation Sterilization

An effective way to sterilize or reduce the microbial burden in almost any substance is through the use of electromagnetic radiation. Microwaves, ultraviolet (UV) radiation, X-rays, gamma rays (γ-rays), and electrons, all forms of electromagnetic radiation (∞ Section 9.3), have the potential to control microbial growth. However, each type of radiation acts through a specific

THE ORIGIN OF PASTEURIZATION

Pasteur's name is forever linked in the public mind with the process of pasteurization. The development of the pasteurization process has been nicely discussed by René Dubos in his book on Pasteur's life.[a] We quote from this book here:

The demonstration that microbes do not generate spontaneously encouraged the development of techniques to destroy them and to prevent or minimize subsequent contamination. Immediately these advances brought about profound technological changes in the preparation and preservation of food products and subsequently in other industrial processes as well. . . .

It was soon discovered that the introduction of microorganisms in biological products can be minimized by an intelligent and rigorous control of the technological operations, but cannot

[a]René Dubos, *Pasteur and Modern Science.* 1988. Science Tech, Madison, WI.

be prevented entirely. The problem therefore was to inhibit the further development of these organisms after they had been introduced into the product. To this end, Pasteur first tried to add a variety of antiseptics, but the results were mediocre and, after much hesitation, he considered the possibility of using heat as a sterilizing agent.

Pasteur's first studies of heat as a preserving agent were carried out with wine. Pasteur had grown up in one of the best wine districts in France, and, as a connoisseur of the beverage, was much disturbed at the thought that heating might alter its flavor and bouquet. He therefore proceeded with very great caution and eventually convinced himself that heating at 55°C would not alter appreciably the bouquet of the wine. . . . These considerations led to the process of partial sterilization, which soon became known the world over under the name of "pasteurization," and which was found applica-

ble to wine, beer, cider, vinegar, milk, and countless other perishable beverages, foods, and organic products.

It was characteristic of Pasteur that he did not remain satisfied with formulating the theoretical basis of heat sterilization, but took an active interest in designing industrial equipment adapted to the heating of fluids in large volumes and at low cost. His treatises on vinegar, wine, and beer are illustrated with drawings and photographs of this type of equipment, and describe in detail the operations involved in the process. The word "pasteurization" is, indeed, a symbol of his scientific life; it recalls the part he played in establishing the theoretical basis of the germ theory, and the phenomenal effort that he devoted to making it useful to his fellow humans. It reminds us also of his well-known statement: "There are no such things as pure and applied science—there are only science, and the application of science."

mechanism. For example, the antimicrobial effects of microwaves are due, at least in part, to thermal effects. UV radiation, normally considered to be between 220 and 300 nm, acts by a different mechanism. UV waves have sufficient energy to cause breaks in DNA, leading to the death of the exposed organism (∞ Section 9.3). This "near-visible" light is useful for disinfecting surfaces, air, and other materials such as water that do not absorb the UV waves. For example, laboratory biological cabinets all come equipped with a "germicidal" UV light to decontaminate the surface after use (Figure 11.4). Ultraviolet radiation cannot penetrate solid, opaque, light-absorbing surfaces, and its usefulness is therefore limited to disinfection of exposed surfaces.

Ionizing radiation

Ionizing radiation is electromagnetic radiation of sufficient energy to produce ions and other reactive molecular species from molecules with which the radiation

Figure 11.4 A biological safety cabinet, shown with an ultraviolet (UV) radiation source (mercury vapor lamp), which is used for decontamination of the inside surfaces.

particles collide. Ionizing radiation produces electrons, e^-, hydroxyl radicals, OH•, and hydride radicals, H•. Each of these reactive molecules is capable of degrading and altering biopolymers such as deoxyribonucleic acid (DNA) and protein. In addition, ionizing radiation can interact directly with DNA, causing breaks in the polymer. The ionization and subsequent degradation of biologically important molecules such as DNA and enzyme proteins leads to the death of irradiated cells. Several radiation sources are potentially useful for sterilization.

The unit of radiation is the *roentgen*, which is a measure of the radiation energy output from a source. The standard for biological applications such as sterilization is the *absorbed radiation dose*. The *absorbed dose* is the rad (100 erg/g), or the gray (1 Gy = 100 rad). Certain microorganisms are much more resistant to radiation than others. Table 11.1 shows the dose of radiation necessary to reduce the numbers of some representative microorganisms or biological functions 10-fold.

In general, microorganisms are much more resistant to ionizing radiation than higher organisms. For example, the amount of energy necessary to reduce the bacterial load by 10-fold is at least 200 Gy. By contrast, the *lethal* dose of radiation for humans is considered to be 10 Gy or less! Note that the figures shown are for a one log reduction in growth of a given organism. The *D10* or *decimal reduction value* gives information similar to the decimal reduction time for heat sterilization (see Section 11.1): the relationship of the survival fraction plotted on a semilogarithmic scale versus the radiation

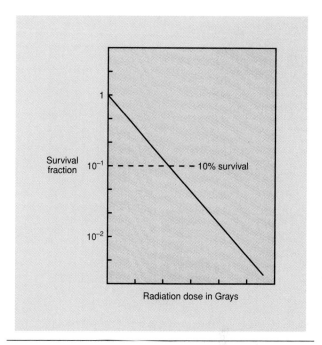

Figure 11.5 Relationship between the survival fraction and the radiation dose. The *D10*, or decimal reduction dose, can be interpolated from the data shown.

dosage in grays is essentially linear (Figure 11.5). A standard killing dose for radiation sterilization can be assigned as 12 *D10* requirements, based on the *D10* dose for destruction of the radioresistant endospores of *Clostridum botulinum*.

Radiation practice

Several sources of ionizing radiation are available, including X-ray machines, cathode ray tubes, and radioactive nuclides. These sources produce either X-rays or gamma rays, both of which have sufficient energy and penetrating power to efficiently inhibit microbial growth in solid and liquid media. The principal sources of commercially useful ionizing radiation are radioactive nuclides that emit γ-rays. The two most commonly used radioisotopes are ^{60}Co and ^{137}Cs, both relatively inexpensive by-products of nuclear fission.

Radiation is currently used for sterilization and decontamination in the medical supplies and food industries. In the United States, the Food and Drug Administration has approved the use of radiation for sterilization of surgical supplies, vaccines, and drugs. In addition, food may be irradiated, although this practice is not yet widespread. However, as shown in Figure 11.6, the practice of sterilizing spices by irradiation has grown enormously in recent years, as the more dangerous alternative, ethylene oxide sterilization (see Section 11.6), has become less common. In addition to sterilization, pasteurization and insect deinfestation

TABLE 11.1	Radiation sensitivity of microorganisms and biological functions	
Species or Function	**Type of Microorganism**	***D10*[a] (Gy)**
Clostridium botulinum	Gram-positive anaerobic sporulating Bacteria	3,300
Clostridium tetani	Gram-positive anaerobic sporulating Bacteria	2,400
Bacillus subtilis	Gram-positive aerobic sporulating Bacteria	600
Salmonella typhimurium	Gram-negative Bacteria	200
Lactobacillus brevis	Gram-positive Bacteria	1,200
Deinococcus radiodurans	Gram-negative radiation-resistant Bacteria	2,200
Aspergillus niger	Mold	500
Saccharomyces cerevisiae	Yeast	500
Foot-and-mouth	Virus	13,000
Coxsackie	Virus	4,500
Enzyme inactivation	—	20,000–50,000
Insect deinfestation	—	1,000–5,000

[a]*D10* is the amount of radiation necessary to reduce the initial population or activity level 10-fold (one logarithm).

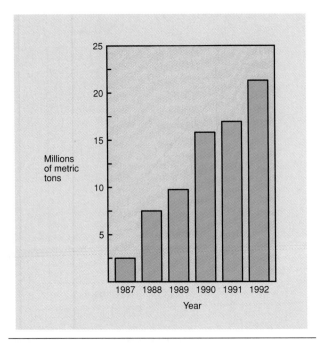

Figure 11.6 Commercial irradiation of spices and seasonings, worldwide.

can be accomplished by carefully adjusting the dose of radiation applied. The use of radiation for all these purposes is an established and accepted technology in many countries but has been slow to gain acceptance in others because of fears of possible radioactive contamination, alteration in nutritional value, production of toxic or carcinogenic products, and production of "off" tastes in irradiated food.

CONCEPT CHECK   **11.2**

Under appropriate conditions, electromagnetic radiation effectively controls microbial growth. Ultraviolet radiation is useful for decontaminating surfaces and materials that do not absorb light, such as air and water. Ionizing radiation is necessary to penetrate solid or light-absorbing materials. Under appropriate conditions, ionizing radiation is very effective for sterilization and decontamination.

✔ *Why is ionizing radiation more effective than ultraviolet radiation for sterilization of food products?*

✔ *What is a common source of ionizing radiation?*

11.3

Filter Sterilization

Although heat is the most common and effective way of sterilizing liquids, it cannot be used for the sterilization of heat-sensitive liquids or gases. An especially

valuable technique for sterilizing such materials is filtration. A filter is a device with pores too small for the passage of microorganisms but large enough to allow the passage of the liquid or gas. The size range of particles involved in sterilization is rather broad. Some of the largest microbial cells are greater than 10 μm in diameter, but at the lower end of the size scale certain bacteria are less than 0.3 μm in diameter. Filters are also frequently used in the field of virology where we deal with even smaller particles—as small as 10 nm. (Historically, filtration was used to demonstrate the existence of virus-sized infectious particles; ∞ The Name *Virus*, Chapter 8.)

Types of filters

There are three major types of filters, which are illustrated in Figure 11.7. One of the oldest types used is the **depth filter**. A depth filter is a fibrous sheet or mat made from a random array of overlapping paper, asbestos, or glass fibers (Figure 11.7*a*). The depth filter traps particles in the tortuous paths created throughout the depth of the structure. Because they are rather porous, depth filters are often used as *prefilters* to remove larger particles from a solution so that clogging does not occur in the final filter sterilization process. They are also used for the filter sterilization of air in industrial processes (∞ Section 12.3).

The most common type of filter for sterilization in the field of microbiology is the **membrane filter** (Figure 11.7*b*). The membrane filter is a tough disc, generally composed of cellulose acetate or cellulose nitrate, which is manufactured in such a way that it contains a large number of tiny holes. The membrane filter differs from the depth filter in that the membrane filter functions more like a sieve, trapping many of the particles on the filter surface. The membranes are open structures with about 80–85% of the filter occupied by space. This openness provides for a relatively high fluid flow rate.

The third type of filter in common use is the **nucleation track (Nucleopore) filter.** These filters are created by treating very thin polycarbonate films (10 μm in thickness) with nuclear radiation and then etching the film with a chemical. The radiation causes localized damage in the film and the etching chemical enlarges these damaged locations into holes. The sizes of the holes can be precisely controlled by the strength of the etching solution and the etching time. A typical Nucleopore filter has very uniform holes arranged almost vertically through the thin film (Figure 11.7*c*). Nucleopore filters are very commonly used in scanning electron microscopy of microorganisms. An organism of interest can be easily removed from a liquid by filtration, with the particles held in a uniform plane on the top of the filter where they can be readily studied in the microscope (Figure 11.8).

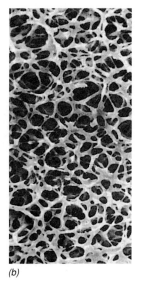
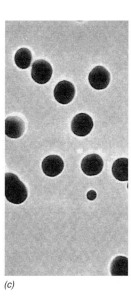

Figure 11.7 The structure of (a) a depth filter, (b) a conventional membrane filter, and (c) a Nucleopore filter.

(a)   *(b)*   *(c)*

Membrane filters for the sterilization of a liquid are illustrated in Figure 11.9. The filter apparatus is generally sterilized separately from the filter, and the apparatus assembled aseptically at the time of filtration. The arrangement shown in Figure 11.9a is suitable for small volumes of liquid. For large-volume sterile filtration, the membrane filter material is arranged in a cartridge and placed in a stainless steel housing. Large-volume filtration of heat-sensitive fluids is very commonly done in the pharmaceutical industry.

Presterilized membrane filter assemblies for sterilization of small to medium volumes are routinely used in most laboratories (Figure 11.9b). Filtration is accomplished by using a syringe, pump or vacuum to force the liquid through the filtration apparatus into a sterile collection vessel.

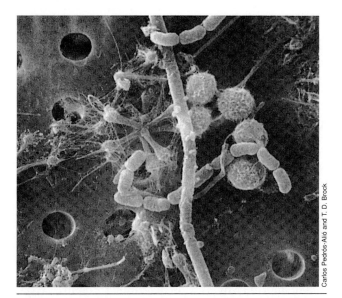

Figure 11.8 Scanning electron micrograph of aquatic bacteria and algae trapped on a Nucleopore filter. The pore size is 5 μm.

CONCEPT CHECK   *11.3*

Filter sterilization involves the *removal* of living microorganisms from liquids. Membrane filters are widely used for sterilization of heat-sensitive liquids in the laboratory.

✔ *Why are depth filters not widely used for sterilization?*

✔ *What advantage does the membrane filter have over the nucleation filter in sterilization?*

11.4

Chemical Growth Control

The growth of microorganisms can also be controlled with chemical agents. An **antimicrobial agent** is a chemical that kills or inhibits the growth of microorganisms. Such a substance may be a synthetic chemical or a natural product. Agents that kill organisms are often called *cidal agents,* with a prefix indicating the kind of organism killed. Thus, we have **bactericidal, fungicidal,** and **viricidal** agents. A bactericidal agent kills bacteria. It may or may not kill other kinds of microorganisms. Agents that do not kill but only inhibit growth are called *static agents,* and we can speak of **bacteriostatic, fungistatic,** and **viristatic** agents.

Antimicrobial agents can vary in their **selective toxicity.** Some act in a rather nonselective manner and have similar effects on all types of cells. Others are far more selective and are more toxic to microorganisms than to animal tissues. Antimicrobial agents with selective toxicity are especially useful as *chemotherapeutic agents* in treating infectious diseases, as they can be used to kill disease-causing microorganisms without harming the host. They will be described later in this chapter.

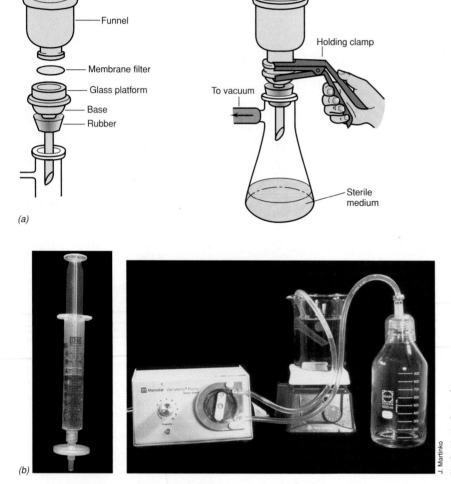

Nonsterile medium added

Funnel

Membrane filter

Glass platform

Holding clamp

To vacuum

Base

Rubber

Sterile medium

(a)

(b)

J. Martinko

Figure 11.9 Membrane filters. (a) Assembly of a reusable membrane filter apparatus. (b) Disposable, presterilized, and assembled membrane filter units. Left: a filter system designed for small volumes. Right: a filter system designed for larger volumes.

Effect of antimicrobial agents on growth

Antimicrobial agents affect growth in a variety of ways, and a study of the action of these agents in relation to the growth curve is important in understanding their modes of action. Three distinct kinds of effects can be observed when an antimicrobial agent is added to an exponentially growing bacterial culture: bacteriostatic, bactericidal, and bacteriolytic. A *bacteriostatic* effect is observed when growth is inhibited, but no killing occurs (Figure 11.10*a*). Bacteriostatic agents are frequently inhibitors of protein synthesis and act by binding to ribosomes. The binding, however, is not tight, and when the concentration of the agent is lowered, the agent becomes free from the ribosome and growth is resumed. The mode of action of protein synthesis inhibitors is discussed in Section 6.9. *Bactericidal* agents kill cells, but lysis or cell rupture does not occur (Figure 11.10*b*). Bactericidal agents are a class of chemical agents that generally bind tightly to their cellular targets and are not removed by dilution. *Bacteriolytic*

agents induce killing by cell lysis, which is observed as a decrease in cell number or in turbidity after the agent is added (Figure 11.10*c*). Bacteriolytic agents include antibiotics that inhibit cell wall synthesis, such as penicillin (∞ Sections 3.7 and 11.9), as well as agents that damage the cytoplasmic membrane.

Measuring antimicrobial activity

Antimicrobial activity is measured by determining the smallest amount of agent needed to inhibit the growth of a test organism, a value called the **minimum inhibitory concentration (MIC).** To determine the MIC, a series of culture tubes is prepared, each tube containing medium with a different concentration of the agent, and all tubes of the series are inoculated. After incubation, the tubes in which growth does *not* occur (indicated by an absence of visible turbidity) are noted, and the MIC is thus determined (Figure 11.11). This simple and effective procedure is often called the *tube dilution technique.* The MIC is not a constant for a

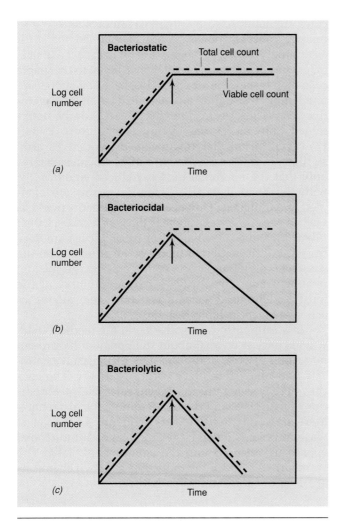

Figure 11.10 (a) Bacteriostatic

Figure 11.10 (b) Bacteriocidal

Figure 11.10 (c) Bacteriolytic

Figure 11.10 Three types of action of antimicrobial agents. At the time indicated by the arrow, a growth-inhibitory concentration was added to the exponentially growing culture. Note the relationships between viable and total cell counts.

given agent, because it is affected by the nature of the test organism used, the inoculum size, the composition of the culture medium, the incubation time, and the conditions of incubation, such as temperature, pH, and aeration. When all conditions are rigorously standardized, it is possible to compare different antimicrobials and determine which is most effective against a given organism or to assess the activity of a single agent against a variety of organisms. This method does not distinguish between a cidal and a static agent because the agent is present in the culture medium throughout the entire incubation period.

Another commonly used procedure for studying antimicrobial action is the **agar diffusion method** (Figure 11.12). A Petri plate containing an agar medium evenly inoculated with the test organism is prepared. Known amounts of the antimicrobial agent are added to filter paper discs, which are then placed on

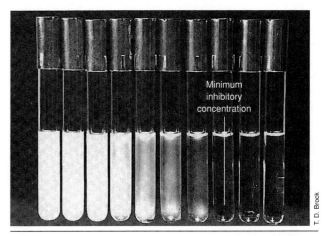

Figure 11.11 Antibiotic assay by tube dilution, permitting detection of the *minimum inhibitory concentration* (MIC). A series of increasing concentrations of antibiotic is prepared in the culture medium. Each tube is inoculated, and incubation is allowed to proceed. Growth (turbidity) occurs in those tubes with antibiotic concentrations below the MIC.

Figure 11.12 Agar diffusion method for assaying antibiotic activity.

the surface of the agar. During incubation, the agent diffuses from the filter paper into the agar; the further it gets from the filter paper, the smaller the concentration of the agent. At some distance from the disc, the MIC is reached. Past this point growth occurs, but closer to the disc growth is absent. A **zone of inhibition** is thus created; the diameter of the zone is proportional to the amount of antimicrobial agent added to the disc and to the overall effectiveness of the agent. This method is routinely used to test for antibiotic sensitivity in pathogens (∞ Section 21.3).

CONCEPT CHECK   **11.4**

Chemicals are often used to control microbial growth. Chemicals that kill organisms are called cidal agents; those that inhibit growth are called static agents. The value of a chemical agent is assessed by determining the minimum concentration necessary to kill or inhibit growth and by determining whether it exhibits selective toxicity.

✔ *With regard to antibacterial agents, what is meant by* selective toxicity?

✔ *Describe the* minimum inhibitory concentration *of a bactericidal agent.*

11.5
Disinfectants and Antiseptics

Disinfectants are chemicals that kill microorganisms and are used on inanimate objects. **Antiseptics,** on the other hand, are chemical agents that kill or inhibit growth of microorganisms and that are sufficiently nontoxic to be applied to living tissues (Table 11.2). Chemical antimicrobial agents, which are frequently referred to as **germicides,** have wide use in situations where it is impractical to use heat for sterilization. For example, hospitals must sterilize heat-sensitive materials, such as thermometers, lensed instruments, polyethylene tubing and catheters, and inhalation and anesthesia equipment. For these applications, *cold sterilization* is commonly used. Cold sterilization is performed in enclosed devices that resemble autoclaves but employ a chemical agent such as ethylene oxide, formaldehyde, or hydrogen peroxide. In the food industry, floors, walls, and surfaces of equipment must often be treated with germicides to reduce the load of microorganisms. In addition, drinking water is commonly treated with chlorine to eliminate any potentially harmful organisms. All these examples use chemical agents to eliminate microorganisms.

The determination of germicide efficacy in practical cases is often very difficult because many germicides are neutralized by organic materials and so germicidal concentrations are not maintained for sufficient time. Further, pathogens are often encased in particles, and penetration of a chemical agent to the viable cells may be slow or absent. Also, bacterial spores are much more resistant to any germicide than are vegetative cells. Thus, germicide effectiveness can be determined only under the intended conditions of use. However, germicidal treatments do not necessarily sterilize. In most cases, the use of germicides ensures only that the microbial load is reduced significantly, with the hope that pathogenic organisms are completely eliminated. However, bacterial endospores as well as vegetative cells such as those of *Mycobacterium tuberculosis,* the causal agent of tuberculosis, are very resistant to the action of many germicides, and so the complete elimination of pathogens by germicidal treatment may not always occur. A summary of the most widely used germicides and their modes of action is given in Table 11.2.

In addition to their wide use in the medical field, where concern is great about the potential hazards of infectious wastes, antiseptic and disinfectant chemicals find many important uses in industry. The concern here is generally to prevent microbial deterioration of materials, and the use of antimicrobial agents is therefore quite extensive. This frequently leads to toxic waste problems when large amounts of antimicrobial agents are released into the environment. Table 11.3 summarizes some of the industries in which chemicals are used to control microbial growth.

CONCEPT CHECK   **11.5**

Disinfectants are chemical compounds used to decontaminate or sterilize inanimate objects. Antiseptics can be used to decontaminate living tissues. These compounds are used in many commercial, health care, and industrial applications.

✔ *Distinguish between a* disinfectant *and an* antiseptic.

✔ *What disinfectants are routinely used for sterilization of water?*

11.6
Microbial Growth Control in Food

In the previous sections in this chapter, we presented several important physical and chemical methods for controlling microbial growth. In this section, we will discuss some of the practical applications of these methods as they relate to the storage and preservation of food. Microbial growth destroys vast quantities of food, causing not only considerable economic prob-

TABLE 11.2   Antiseptics and disinfectants

Agent	Use in health-related fields	Mode of action
Antiseptics		
Organic mercurials	Skin	Combines with —SH groups of proteins
Silver nitrate	Eyes of newborn to prevent blindness due to infection by *Neisseria gonorrhoeae*	Protein precipitant
Iodine solution	Skin	Iodinates tyrosine residues of proteins; oxidizing agent
Alcohol (70% ethanol in water)	Skin	Lipid solvent and protein denaturant
Bisphenols (hexachlorophene)	Soaps, lotions, body deodorants	Disrupts cell membrane
Cationic detergents (quaternary ammonium compounds)	Soaps, lotions	Interact with phospholipids of membrane
Hydrogen peroxide (3% solution)	Skin	Oxidizing agent
Disinfectants		
Mercuric dichloride	Tables, bench tops, floors	Combines with —SH groups
Copper sulfate	Algicide in swimming pools, water supplies	Protein precipitant
Iodine solution	Medical instruments	Iodinates tyrosine residues
Chlorine gas	Purification of water supplies	Oxidizing agent
Chlorine compounds	Dairy, food industry equipment, water supplies	Oxidizing agent
Phenolic compounds	Surfaces	Protein denaturant
Cationic detergents (quaternary ammonium compounds)	Medical instruments; food, dairy equipment	Interact with phospholipids
Ethylene oxide (gas)	Temperature-sensitive laboratory materials such as plastics	Alkylating agent
Ozone	Drinking water	Strong oxidizing agent

lems but also considerable loss of important nutrients. In some cases, consumption of contaminated food causes serious food infections or food poisonings (∞ Section 23.13). In a few cases, methods of microbial growth control that inhibit microbial growth but do not harm the host are peculiar to food preservation.

Food spoilage

Food spoilage is defined as any change in the visual appearance, smell, or taste of a food product that makes it unacceptable to the consumer. From a health standpoint, spoiled food is not necessarily food that is unsafe to eat. However, unpalatable food will not be purchased by the average consumer, and thus preventing food spoilage is a major goal of the food industry. Foods are attacked by microorganisms in a variety of ways that are generally harmful to the quality of the food. Foods are organic and hence provide adequate nutrients for the growth of a wide variety of chemoorganotrophic bacteria. What determines whether a food can support the growth of a microorganism? The physical and chemical characteristics of the food and how it is stored determine its degree of susceptibility

TABLE 11.3   Industries that use chemicals to control microbial growth

Industry	Chemicals	Use
Paper	Organic mercurials, phenolics	To prevent microbial growth during manufacture
Leather	Heavy metals, phenolics	Antimicrobial agents are present in the final product
Plastic	Cationic detergents	To prevent growth of bacteria on aqueous dispersions of plastics
Textile	Heavy metals, phenolics	To prevent microbial deterioration of fabrics exposed in the environment such as awnings, tents
Wood	Phenolics	To prevent deterioration of wooden structures
Metal working	Cationic detergents	To prevent growth of bacteria in aqueous cutting emulsions
Petroleum	Mercurics, phenolics, cationic detergents	To prevent growth of bacteria during recovery and storage of petroleum and petroleum products
Air conditioning	Chlorine, phenolics	To prevent growth of bacteria (for example, *Legionella*) in cooling towers
Electrical power	Chlorine	To prevent growth of bacteria in condensors and cooling towers
Nuclear	Chlorine	To prevent growth of radiation-resistant bacteria in nuclear reactors

to microbial attack. Foods can be classified into three major categories: (1) *highly perishable foods* such as meats, fish, poultry, eggs, milk, and most fruits and vegetables; (2) *semiperishable foods* such as potatoes, some apples, and nuts; and (3) *stable* or *nonperishable foods* such as sugar, flour, rice, and dry beans. How do these three categories differ? To a great extent, perishability is a function of *moisture content*, which is related, as we saw in Section 5.10, to water activity, a_w. Stable foods have *low* water activity and can generally be stored for considerable lengths of time without deterioration. Perishable and semiperishable foods are those with *high* water activity. These foods must be stored under conditions that slow or stop microbial growth.

Fresh foods are spoiled by a variety of both bacteria and fungi, but each type of fresh food is typically attacked only by particular microorganisms (Table 11.4). This is because the chemical properties of foods vary widely, and different foods are colonized by the indigenous spoilage organisms that are best able to utilize the nutrients available. For example, enteric bacteria are rarely implicated in fruit or vegetable spoilage but can spoil meat products because their habitat is the gut of warm-blooded animals: they can easily contaminate the meat when the animal is slaughtered. Likewise, lactic acid bacteria, the most common microorganisms in dairy products, are the major spoilage organisms of milk and milk products. *Pseudomonas* species inhabit both soil and the animal body and are thus widely involved in the spoilage of various fresh foods (Table 11.4).

Microbial growth in foods follows the standard pattern for a bacterial growth curve (∞ Figure 5.4). The lag phase may be of variable duration in a food, depending on the contaminating organism and its previous growth history. The rate of growth during the exponential phase depends on the temperature,

the available nutrients, and other conditions of growth. The time required for the population density to reach a significant level in a given food product depends on both the initial inoculum and the rate of growth during the exponential phase. However, it is only when the microbial population density reaches a substantial level that harmful effects are usually observed. Indeed, throughout much of the exponential phase of growth, population densities may be low enough that no perceptible effect can be observed, but because of the nature of exponential growth (∞ Figure 5.2), it is only the *last* doubling or two that leads to problems. Thus, for much of the period of microbial growth in a food, the observer is unaware of impending problems.

Food preservation

Besides moisture, one of the most crucial factors affecting microbial growth in food is **temperature** (∞ Section 5.7). In general, the *lower* the storage temperature, the *less* rapid the spoilage rate, although, as we have seen, psychrophilic microorganisms grow well at refrigerator temperatures. Therefore, storage of perishable food products for long periods of time is possible only at temperatures below freezing. Freezing alters the physical structure of many food products and therefore cannot be universally used, but it is widely and successfully used for the preservation of meats and many vegetables and fruits. Freezers providing a temperature of $-20°C$ are most commonly used. At $-20°C$, storage for weeks or months is possible, but some microbial growth may still occur, usually in pockets of liquid water trapped within the frozen mass. Also, nonmicrobial chemical changes in the food may continue. For very long-term storage, temperatures lower than $-20°C$ are necessary, such as $-80°C$ (dry ice temperature), but maintenance at such low

TABLE 11.4	Microbial spoilage of fresh food[a]	
Food product	**Type of microorganism**	**Common spoilage organisms**
Fruits and vegetables	Bacteria	*Erwinia, Pseudomonas, Corynebacterium* (mainly vegetable pathogens; rarely spoil fruit)
	Fungi	*Aspergillus, Botrytis, Geotrichium, Rhizopus, Penicillium, Cladosporium, Alternaria, Phytophora,* various yeasts
Fresh meat, poultry, and seafood	Bacteria	*Acinetobacter, Aeromonas, Pseudomonas, Micrococcus, Achromobacter, Flavobacterium, Proteus, Salmonella, Escherichia*
	Fungi	*Cladosporium, Mucor, Rhizopus, Penicillium, Geotrichium, Sporotrichium, Candida, Torula, Rhodotorula*
Milk	Bacteria	*Streptococcus, Leuconostoc, Lactococcus, Lactobacillus, Pseudomonas, Proteus*
High sugar foods	Bacteria	*Clostridium, Bacillus, Flavobacterium*
	Fungi	*Saccharomyces, Torula, Penicillium*

[a]Several other organisms not listed here have also been isolated, but the organisms listed are the most commonly observed spoilage agents of the fresh foods indicated.

temperatures is expensive and consequently is not used for routine food storage.

Another major factor affecting microbial growth in food is **pH** or **acidity.** Foods vary widely in pH, although most are neutral or acidic. As we have seen (∞ Section 5.9), microorganisms differ in their ability to grow under acidic conditions: most food spoilage bacteria do not grow at pH values below 5. Therefore, acid is often used in food preservation, in the process called *pickling*. Foods commonly pickled include cucumbers (sweet, sour, and dill pickles), cabbage (sauerkraut), and some meats and fruits. The food can be made acid either by addition of vinegar or by allowing acidity to develop directly in the food through microbial action, in which case the product is called a *fermented food*. The microorganisms involved in food fermentations are acid-tolerant bacteria, the lactic acid bacteria, the acetic acid bacteria, and the propionic acid bacteria. But even these bacteria cannot grow below about pH 4, so the food fermentation is self-limiting. Vinegar, frequently added to food to lower the pH, is essentially dilute acetic acid. Vinegar itself is a product of the action of acetic acid bacteria; its industrial production is discussed in Section 12.10.

Because microorganisms do not grow at low water activities, microbial growth can be controlled by lowering the water activity of the product by drying or by adding salt. Sun-drying is the least expensive way of drying foods if the climate is right. Some foods can be successfully dried with artificial heat, but heating generally results in deterioration of quality. The least damaging way of drying foods is freeze-drying (lyophilization), but this is quite expensive and can be justified only if the food has a high economic value. Milk, meats, fish, vegetables, fruits, and eggs are all commonly preserved by some form of drying.

A number of foods are preserved by addition of salt or sugar to lower water activity. Foods preserved by addition of sugar are mainly fruits (jams, jellies, and preserves). Salted products are primarily meats and fish. Sausage and ham are preserved by salt, although these products vary in water activity depending on how much salt is added and how much the meat has been dried. Several famous sausages such as *landjaeger* can be stored indefinitely in the absence of refrigeration but are very salty.

Canning

Canning is a process in which a food is sealed and heated so as to kill all living organisms, or at least to ensure that there will be no growth of residual organisms in the can. Canning is hence a type of heat sterilization, and the principles already presented in Section 11.1 apply. When the can is properly sealed and heated, the food should remain stable and unspoiled indefinitely, even when stored in the absence of refrigeration. Home-canned foods are usually prepared in glass containers, whereas commercial products most often come in tin-coated steel cans. In any canning process, the *seal* on the can or jar is the most critical part. The heating process itself is done by submersion in water, generally under pressure.

The temperature–time relationships for canning depend on the type of food, its pH, the size of the container, and the consistency or bulkiness of the food. Because heat must penetrate completely to the center of the food within the can, heating times must be longer for large cans or very viscous foods. Acid foods can often be canned effectively by heating just to boiling, 100°C, whereas nonacid foods must be heated to autoclave temperatures. The canning process may not *sterilize* the food. However, the numbers of organisms are reduced greatly during the heating process, and the product is probably sterile, but if the initial load of organisms in the food is high, then not every cell may be killed. Heating times long enough to guarantee absolute sterility of every can would change the food so greatly that it would likely be unpalatable and nutritionally altered.

The environment inside a can is anoxic, and microbial growth in a canned food frequently is the result of fermentative organisms that produce extensive amounts of gas. This can result in pressure buildup inside the can, resulting in bulges or, in severe cases, even an explosion of the can (Figure 11.13). Because many of the anoxic bacteria that grow in canned foods are powerful toxin producers (∞ Section 23.13), food from a visibly altered can should never be eaten.

Chemical food preservation

Although chemicals should never be used in place of careful food sanitation, there are a few chemical antimicrobial agents that are used commercially to control microbial growth in foods. These are classified by the U.S. Food and Drug Administration as "generally recognized as safe" and find wide application in the food industry. These are summarized in Table 11.5. Many of these chemicals, like sodium propionate, have been used for many years with no evidence of human toxicity. Others, like nitrites, ethylene or propylene oxides, or antibiotics, are more controversial food supplements because of evidence that these compounds are detrimental to human health.

Nitrites can react at acid pH with secondary amines in the body to form *nitrosamines,* a class of potentially carcinogenic chemical compounds. Ethylene and propylene oxides are alkylating agents and therefore mutagenic (∞ Section 9.3), and there is some suggestion that these compounds are also carcinogenic.

Because of lengthy and costly testing programs for any new chemical proposed as a food preservative

(a)   (b)   (c)   (d)

Figure 11.13 Changes in cans as a result of microbial spoilage. (a) Normal can; note that the top of the can is indented due to negative pressure (vacuum) inside. (b) Slight swell resulting from minimal gas production. Note that the lid is slightly raised. (c) Severe swell due to extensive gas production. Note the great deformation of the can. (d) The can shown in (c) was dropped and the gas pressure resulted in a violent explosion. Note that the lid has been torn apart.

TABLE 11.5	Chemical food preservatives
Chemical	**Foods**
Sodium or calcium propionate	Bread
Sodium benzoate	Carbonated beverages, fruit juices, pickles, margarine, preserves
Sorbic acid	Citrus products, cheese, pickles, salads
Sulfur dioxide, sulfites, bisulfites	Dried fruits and vegetables; wine
Formaldehyde (from food-smoking process)	Meat, fish
Ethylene and propylene oxides	Spices, dried fruits, nuts
Sodium nitrite	Smoked ham, bacon

today, it is unlikely that new compounds will be added to the list in Table 11.5 in the near future. An alternative to chemical preservatives, preservation by ionizing radiation, was discussed in Section 11.2.

CONCEPT CHECK   11.6

Food microbiology deals with methods for keeping microorganisms from growing in food during processing and storage. Foods vary considerably in their sensitivity to microbial growth, acid foods being much more resistant to spoilage than those of neutral or alkaline pH. Microbial growth in foods can be controlled by use of heat, refrigeration, or chemical agents, but only a few chemical agents are approved for direct addition to foods.

✔ *What advantages does heat have in food preservation? What disadvantages?*

✔ *Why is salt an effective food preservative?*

11.7
Growth Factor Analogs

The preceding discussion dealt with chemical agents used to inhibit microbial growth *outside* the human body. Most of the chemicals mentioned were too toxic to be used in the body, although antiseptics can be used on the skin. For control of infectious disease, agents that can be used internally are essential. Such agents are called **chemotherapeutic agents,** and they have played major roles in modern medicine (see the box, Microbiology and "Magic Bullets"). The key requirement of a successful chemotherapeutic agent is *selective toxicity,* the ability to inhibit bacteria or other microorganisms without affecting the body (see the box).

Sulfa drugs

In Section 4.2 we discussed growth factors and defined them as specific chemical substances *required* in the medium because the organism cannot synthesize them. A substance that is related to a growth factor but blocks utilization of the growth factor is known as a **growth factor analog.** Growth factor analogs are usually structurally similar to the growth factors in question but are sufficiently different that they cannot duplicate the function of the growth factor in the cell. The first of these to be discovered were the *sulfa drugs,* the first modern chemotherapeutic agents to specifically inhibit the growth of bacteria (see the box); they have been highly successful in the treatment of certain diseases. The simplest sulfa drug is **sulfanilamide** (Figure 11.14*a*). Sulfanilamide acts as an analog of *p*-aminobenzoic acid (Figure 11.14*b*), which is itself a part of the vitamin folic acid (Figure 11.14*c*). Sulfanilamide acts by blocking the synthesis of folic acid. Sulfanilamide is active in Bacteria but not in

LEARNING FROM THE PAST

MICROBIOLOGY AND "MAGIC BULLETS"

 The development of chemotherapeutic agents has had a greater impact on clinical medicine than any other discovery. Although a variety of natural chemical agents had been used earlier, the real advances in work with chemotherapeutic agents began with the German scientist Paul Ehrlich. In the early 1900s, Ehrlich developed the concept of selective toxicity. He began his work by studying the staining of microorganisms and observed that some dyes stained microorganisms but not animal tissue. He assumed that if a dye did not stain a tissue, the dye molecules were unable to combine with the cell constituents. He then reasoned that if such a dye had toxic properties, it should not affect the animal cells because it could not combine with them, but it should attack the microbial cells. In an infected animal, chemicals of this sort should behave like "magic bullets," striking the pathogen but missing the host. Ehrlich proceeded to test large numbers of chemicals for selectivity and discovered the first chemotherapeutic agents, of which Salvarsan, an arsenic-containing drug for the cure of syphilis, was the most famous (see the figure).

HCl·H₂N   NH₂·HCl

HO— As═As —OH

Salvarsan

However, no chemical agents were discovered that affected the vast majority of infectious agents until the 1930s, when Gerhard Domagk discovered the sulfa drugs. The discovery of the sulfas came about through the large-scale screening of chemicals for activity in infectious diseases in experimental animals. Domagk, at the Bayer Chemical Company in Germany, tested a large variety of synthetic organic chemicals, mainly dyes, for their ability to cure streptococcal infections in mice. The first active compound was Prontosil, which was active in mice but had no activity against streptococci grown in the test tube. Domagk discovered that in the animal body, Prontosil broke down to sulfanilamide, which was the actual active agent. It was possible to embark on a program of synthesis based on the sulfanilamide structure, which yielded a large number of active drugs. D. D. Woods in England then showed that *p*-aminobenzoic acid specifically counteracted the inhibitory action of sulfanilamide, and he also showed that streptococci required *p*-aminobenzoic acid for growth. This led to the concept of the *growth factor analog*, which enabled chemists to pursue the synthesis of a wide variety of chemotherapeutic agents.

Despite the successes of the sulfa drugs, most infectious diseases were still not under chemical control. It took the discovery of the first antibiotic, penicillin, by Alexander Fleming, a Scottish physician engaged in research at St. Mary's Hospital in London, to point investigators in the right direction. Fleming's first paper on penicillin, published in 1929, begins as follows:

While working with staphylococcus variants a number of culture plates were set aside on the laboratory bench and examined from time to time. In the examination these plates were necessarily exposed to the air and they became contaminated with various micro-organisms. It was noticed that around a large colony of contaminating mould the staphylococcus colonies became transparent and were obviously undergoing lysis. Subcultures of this mould were made and experiments conducted with a view of ascertaining something of the properties of the bacteriolytic substance which had evidently been formed in the mould culture and which had diffused into the surrounding medium.

Fleming characterized the product, and since it was produced by a fungus of the genus *Penicillium*, gave it the name *penicillin*. His work, however, did not include a process for large-scale production nor did it show that penicillin was effective in the treatment of infectious disease. This was done by a group of British scientists at Oxford University, headed by Howard Florey in 1939, motivated in part by the impending World War II and the knowledge that infectious disease was the leading cause of death among soldiers on the battlefield. Florey and his colleagues developed methods for the analysis and testing of penicillin and for its production in large quantities. They then proceeded to test penicillin against bacterial infections in humans. Penicillin was dramatically effective in controlling staphylococcal and pneumococcal infections and was also more effective for streptococcal infections than the sulfa drugs. With the effectiveness of penicillin demonstrated and the war in Europe becoming more intense, Florey brought cultures of the penicillin-producing fungus to the United States in 1941. He persuaded the U.S. government to create a large-scale research program, which led to a joint effort of the pharmaceutical industry, the U.S. Department of Agriculture at its laboratory in Peoria, Illinois, and several universities. By the end of World War II, penicillin was available in large amounts, for civilian as well as military use. As soon as the war was over, pharmaceutical companies entered into commercial production of penicillin on a competitive basis and began to look for other antibiotics. Success was quick and dramatic, and the impact on medicine has been close to phenomenal. Infant and child mortality have been greatly reduced, and many diseases that formerly had high fatality rates are now no more than medical curiosities. ∎

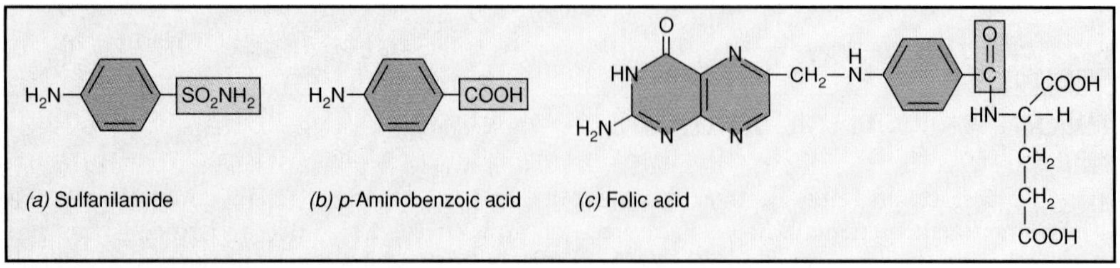

Figure 11.14 (a) The simplest sulfa drug, sulfanilamide. (b) Sulfanilamide is an analog of *p*-aminobenzoic acid, which itself is part of (c) the growth factor folic acid.

higher animals because Bacteria synthesize their own folic acid, whereas higher animals obtain folic acid from their diet.

Other growth factor analogs

The concept that a chemical substance can act as a competitive inhibitor of an essential biosynthetic pathway has had far-reaching effects on chemotherapeutic research. Analogs are now known for various vitamins, amino acids, purines, pyrimidines, and other compounds. A few examples are given in Figure 11.15. In these examples, the analog has been formed by addition of a fluorine or a bromine atom. Fluorine is a relatively small atom and does not alter the overall shape of the molecule, but it changes the chemical properties sufficiently so that the compound does not act normally in cell metabolism. Fluorouracil resembles the nucleic acid base uracil; bromouracil resembles another base, thymine (chemicals that resemble bases, such as bromouracil, are also used as mutagens) (∞ Section 9.3).

Growth factor analogs are also useful for treatment of viral and fungal infections. We will discuss a number of these compounds and investigate the mechanisms of their selective toxicity in Sections 11.11 and 11.12.

CONCEPT CHECK   *11.7*

Growth factor analogs are powerful metabolic inhibitors. As a general class of compounds, they can be taken internally, display bacteriostatic properties, and exhibit selective toxicity.

✔ *There are several features that distinguish growth factor analogs from antiseptics and disinfectants. Identify at least two of them.*

✔ *What is a competitive inhibitor?*

11.8
Antibiotics

Antibiotics are chemical substances produced by certain microorganisms that inhibit or kill other microorganisms. Antibiotics constitute a special class of chemo-

therapeutic agents, distinguished from growth factor analogs because they are natural products (products of microbial activity) rather than synthetic chemicals (products of human activity). Antibiotics constitute one of the most important classes of substances produced by large-sale microbial processes. The industrial production of antibiotics and the methods used to discover new ones will be investigated in Chapter 12. In this section we present a broad overview of the antibiotics. In the next section, we discuss the structure and function of a specific group of antibiotics, the β-lactam antibiotics, which are produced by fungi. We conclude our

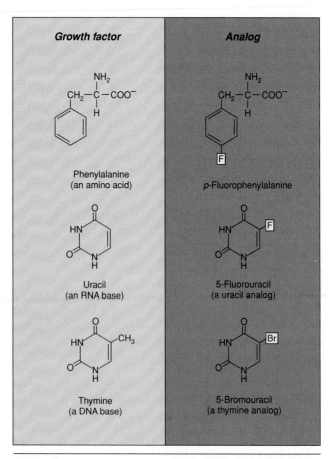

Figure 11.15 Growth factors and structurally similar analogs.

consideration of antibacterial agents with a discussion of some important antibiotics made by prokaryotes of the genus *Streptomyces*.

Targets of antibiotics

A very large number of antibiotics have been discovered, but probably less than 1% of them have been of practical value in medicine. Those that have been useful have had a dramatic impact on the treatment of infectious diseases. Further, some antibiotics can be made more effective by chemical modification; these are said to be *semisynthetic*.

The sensitivity of microorganisms to antibiotics and other chemotherapeutic agents varies (Figure 11.16). Gram-positive Bacteria are usually more sensitive to antibiotics than are gram-negative Bacteria, although some antibiotics act only on gram-negative Bacteria. An antibiotic that acts on both gram-positive and gram-negative Bacteria is called a **broad-spectrum antibiotic.** In general, a broad-spectrum antibiotic finds wider medical usage than a *narrow-spectrum antibiotic,* which acts on only a single group of organisms. A narrow-spectrum antibiotic may, however, be quite valuable for the control of microorganisms that fail to respond to other antibiotics. Some antibiotics have an extremely limited spectrum of action, being effective for only one bacterial species.

Antibiotics and other chemotherapeutic agents can be grouped based on chemical structure (Figure 11.17) or on mode of action (Figure 11.18). In Bacteria, the important targets of antibiotic action are the cell wall, the cytoplasmic membrane, and the biosynthetic processes of protein and nucleic acid synthesis. Some chemotherapeutic agents, such as the sulfa drugs, which are growth factor analogs (see Section 11.6), work because they mimic important growth factors needed in cell metabolism (see the box).

We begin our study of the antibiotics with the β-lactam group, which includes the penicillins and related compounds of major clinical significance.

> *CONCEPT CHECK*   *11.8*
>
> Antibiotics are a chemically diverse group of static or cidal compounds produced by microorganisms. They can act by several diverse mechanisms to disrupt microbial metabolism. Most known antibiotics have no useful applications.
>
> ✔ *Distinguish* antibiotics *from* growth factor analogs.
>
> ✔ *What is meant by a* broad-spectrum antibiotic?

11.9

β-Lactam Antibiotics: Penicillins and Cephalosporins

One of the most important groups of antibiotics, both historically and medically, is the β-lactam group. The β-lactam antibiotics include the penicillins, cephalosporins, and cephamycins, all medically useful antibiotics. These antibiotics are called β-lactams because they contain the β-lactam ring system (Figure 11.19, page 418).

Types of penicillin

The first β-lactam antibiotic discovered, **penicillin G** (Figure 11.19), is active primarily against gram-positive Bacteria. Its action is restricted to gram-positive Bacteria primarily because gram-negative Bacteria are impermeable to the antibiotic. As a result of extensive research, a vast number of new penicillins have been discovered, some of which are quite effective against gram-negative Bacteria. One of the most significant developments in the antibiotic field over the past several decades has been the discovery and development of these new penicillins.

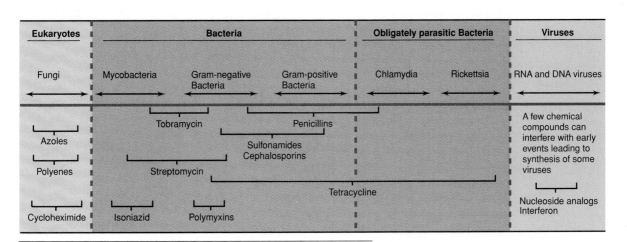

Figure 11.16 Ranges of action of selected chemotherapeutic agents.

Antibiotic classification	Subclassification	Example	Representative structure
I. Carbohydrate-containing compounds	Pure sugars Aminoglycosides Orthosomycins N-Glycosides C-Glycosides Glycolipids	Nojirimycin Streptomycin Everninomicin Streptothricin Vancomycin Moenomycin	
II. Macrocyclic lactones	Macrolide antibiotics Polyene antibiotics Ansamycins Macrotetrolides	Erythromycin Candicidin Rifampin Tetranactin	
III. Quinones and related compounds	Tetracyclines Anthracyclines Naphthoquinones Benzoquinones	Tetracycline Adriamycin Actinorhodin Mitomycin	
IV. Amino acid and peptide analogs	Amino acid derivatives β-Lactam antibiotics Peptide antibiotics Chromopeptides Depsipeptides Chelate-forming peptides	Cycloserine Penicillin, ceftriaxone Bacitracin Actinomycin Valinomycin Bleomycin	
V. Heterocyclic compounds containing nitrogen	Nucleoside antibiotics	Polyoxins	
VI. Heterocyclic compounds containing oxygen	Polyether antibiotics	Monensin	
VII. Alicyclic derivatives	Cycloalkane derivatives Steroid antibiotics	Cycloheximide Fusidic acid	
VIII. Aromatic compounds	Benzene derivatives Condensed aromatics Aromatic ether	Chloramphenicol Griseofulvin Novobiocin	
IX. Aliphatic compounds	Compounds containing phosphorus	Fosfomycin	
X. Quinolone compound	4-Quinolone Fluoro-4-quinolones	Nalidixic acid Norfloxacin	
XI. Oxazolidinone	Cyclic lactone	2-Oxazolidinone	

Rifampin

Streptomycin

Mitomycin C

Ceftriaxone

Polyoxin B

Monensin

Cycloheximide

Griseofulvin

Fosfomycin

Nalidixic acid

2-Oxazolidinone

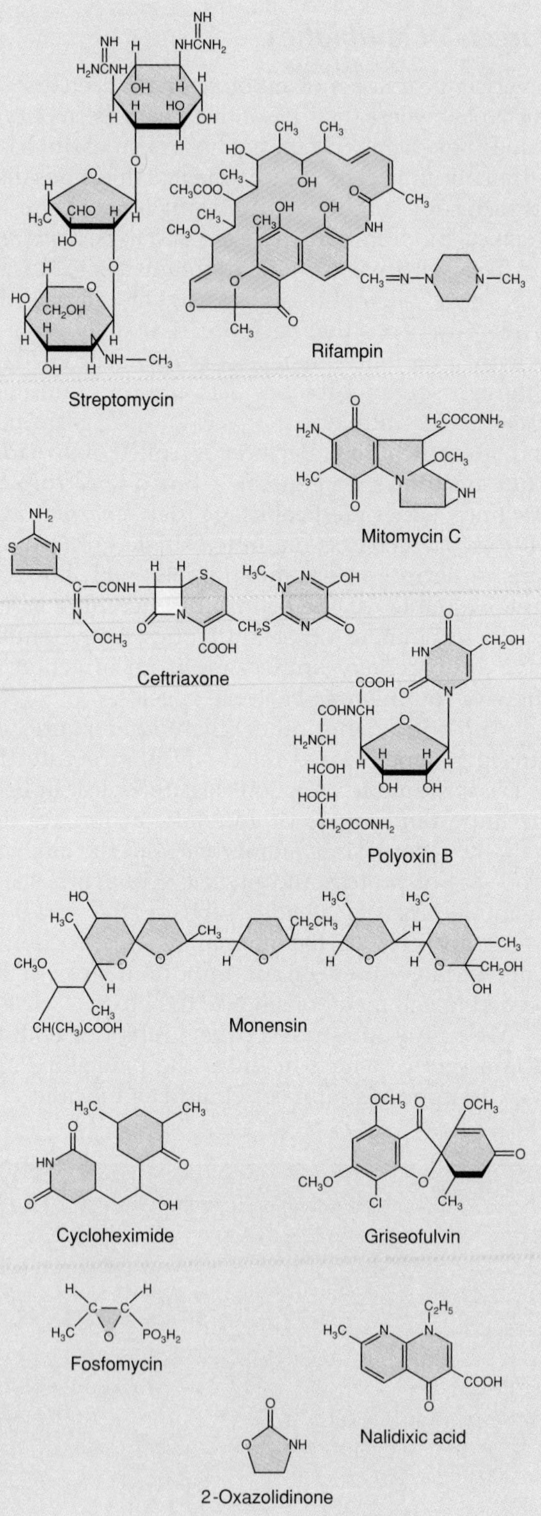

Figure 11.17 Classification of antibiotics according to chemical structure. A representative example is shown for each group.

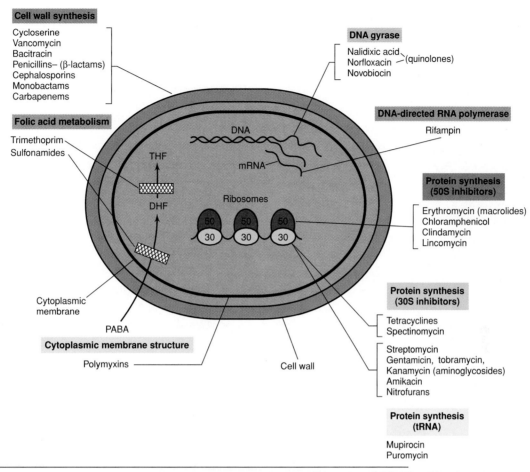

Figure 11.18 Mode of action of major antibacterial antibiotics and growth factor analogs. THF, Tetrahydrofolate; DHF, dihydrofolate; mRNA, messenger RNA; tRNA, transfer RNA.

Figure 11.19 shows the complex structures of some of these new penicillins. Modifications of the basic penicillin G structure by chemical synthesis methods, as in the semisynthetic penicillins shown, significantly change the properties of the resulting antibiotics. For example, ampicillin and carbenicillin are semisynthetic penicillins that have a broader spectrum of antibiotic activity, which includes some gram-negative Bacteria. The structural differences in the *N*-acyl groups of these semisynthetic penicillins allow them to be transported inside the gram-negative cell envelope (∞ Section 3.6) where they inhibit cell wall synthesis. Note also that penicillin G is sensitive to β-lactamase, an enzyme produced by a number of penicillin-resistant Bacteria (see Section 11.13). The semisynthetic penicillins oxacillin and methicillin are useful because they are β-lactamase resistant.

Mechanisms of action

The β-lactam antibiotics are potent inhibitors of cell wall synthesis. As we discussed in Sections 3.5–3.7, an important feature of cell wall synthesis is the transpep-tidation reaction, which results in the cross-linking of two glycan-linked peptide chains (∞ Figures 3.29 and 3.38). The enzymes that accomplish this task, the transpeptidases, are also capable of binding to penicillin or other antibiotics with the β-lactam ring. Thus, these transpeptidases are known as *penicillin binding proteins* (PBPs). The PBPs bind very tightly to penicillin and can no longer catalyze the transpeptidase reaction. The cell wall continues to be formed but is no longer cross-linked and becomes progressively weaker as the peptidoglycan backbone is laid down. In addition, the antibiotic–PBP complex stimulates the release of autolysins that digest the existing cell wall. The result is a weakened, eventually degraded cell wall. Under normal circumstances, the osmotic pressure differences inside the cell, as compared to outside, lyse the cell. By contrast, vancomycin, a peptide analog (Figure 11.17), does not bind to PBPs but acts directly on the terminal D-alanyl-D-alanine peptide on the peptidoglycan precursors (∞ Figure 3.28), blocking the transpeptidase reaction. Because the cell wall and its synthesis mechanisms are unique to Bacteria, the β-lactam antibiotics have very high specificity and are not toxic to host cells.

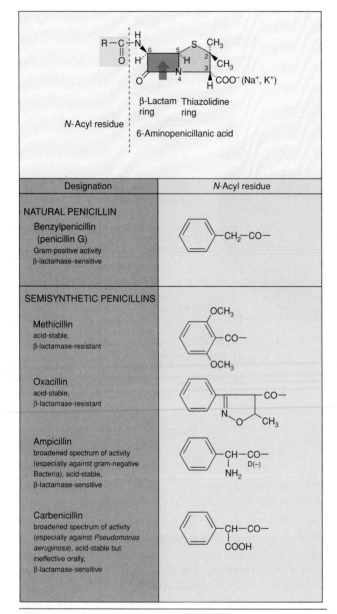

Figure 11.19 The structures of some important penicillins. The red arrow (top panel) is the site of action for most β-lactamases.

Cephalosporins

The cephalosporins are another group of clinically important antibiotics that contain the β-lactam ring. They differ structurally from the penicillins because they have a six-member dihydrothiazine ring instead of the five-member thiazolidine ring. The cephalosporins have the same mode of action as the penicillins. That is, they bind irreversibly to the PBPs and prevent the cross-linking of peptidoglycan. Cephalosporins generally have a broader spectrum of antibiotic activity than the penicillins and are often more resistant to the action of

enzymes that destroy β-lactam rings, the β-lactamases. For example, ceftriaxone (Figure 11.17), a widely used cephalosporin, is highly resistant to the β-lactamases and has now supplanted penicillin as the drug of choice for treatment of infections due to *Neisseria gonorrhoeae* (see Sections 11.13 and 23.6 and Figure 11.25) because many *N. gonorrhoeae* strains have developed β-lactamases that cleave the β-lactam rings of penicillin.

CONCEPT CHECK   **11.9**

The β-lactam compounds are the most important clinical antibiotics. This group includes the penicillins and the cephalosporins. These antibiotics are specific for the cell wall synthesis enzymes of Bacteria and, as a group, have a very broad spectrum of activity.

✔ *Draw the common structure of the β-lactam antibiotics.*

✔ *How do the β-lactam antibiotics function?*

11.10
Antibiotics from Prokaryotes

Many antibiotics active against prokaryotes are also produced by prokaryotes. These include the aminoglycosides, the macrolides, and the tetracyclines. Many of these antibiotics have major clinical applications, and thus we discuss their general properties here.

Aminoglycoside antibiotics

Aminoglycoside antibiotics contain amino sugars bonded by glycosidic linkage (∞ Section 2.5) to other amino sugars. A number of clinically useful antibiotics are aminoglycosides, including *streptomycin* (Figure 11.17; ∞ Section 12.6, and Figure 12.13) and its relatives, *kanamycin* (Figure 11.20), *gentamicin*, and *neomycin*. The aminoglycoside antibiotics are used clinically against gram-negative Bacteria. Streptomycin has also been used extensively in the treatment of tuberculosis. Historically, the use of streptomycin for tuberculosis treatment was a major medical advance, as it was the first antibiotic capable of controlling this infectious disease. However, none of the aminoglycoside antibiotics are widely used today. Streptomycin has been supplanted by several synthetic chemicals for tuberculosis treatment because streptomycin causes several serious side effects and bacterial resistance readily develops. The use of aminoglycosides for treatment of gram-negative infections has decreased since the development of the semisynthetic penicillins (see ear-

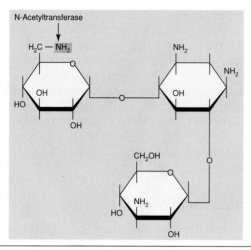

Figure 11.20 Structure of kanamycin, an aminoglycoside antibiotic. The amino sugars are in yellow. The site of modification by an *N*-acetyltransferase, encoded by a resistance plasmid, is indicated.

lier discussion) and the tetracyclines (see later). The aminoglycoside antibiotics are now considered reserve antibiotics used primarily when other antibiotics fail. The aminoglycosides act by inhibiting protein synthesis at the level of the 30S subunit of the ribosome (Figure 11.18).

Macrolide antibiotics

Macrolide antibiotics contain large lactone rings connected to sugar moieties (Figure 11.21). Variations in both the macrolide ring and the sugar moieties are known, and so a large variety of macrolide antibiotics exist. The best-known macrolide antibiotic is *erythromycin*, but other macrolides include *oleandomycin*, *spiramycin*, and *tylosin*. Erythromycin acts as a protein synthesis inhibitor at the level of the 50S subunit of the ribosome (Figure 11.18). Erythromycin is commonly used clinically in place of penicillin in patients allergic to penicillin or other β-lactam antibi-

otics. Erythromycin has been particularly valuable in treating cases of legionellosis (∞ Section 23.2) because of the exquisite sensitivity of the causative agent, the bacterium *Legionella pneumophila*, to this antibiotic.

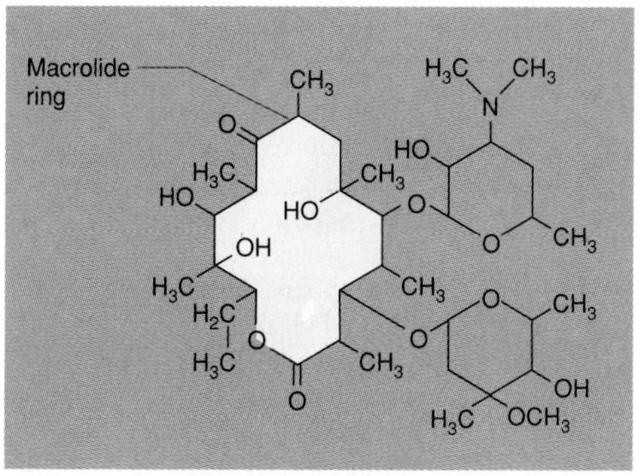

Tetracyclines

The tetracyclines are an important group of antibiotics that find widespread medical use. They were some of the first so-called *broad-spectrum* antibiotics, inhibiting almost all gram-positive and gram-negative Bacteria. The basic structure of the tetracyclines consists of a naphthacene ring system (Figure 11.22). To this ring is added any of several constituents. *Chlortetracycline*, for instance, has a chlorine atom, whereas *oxytetracycline* has an additional hydroxyl (OH) group and no chlorine (Figure 11.22). All three of these antibiotics are produced microbiologically, but there are also semisynthetic tetracyclines on the market, into which other constituents have been inserted chemically into the naphthacene ring system. Like erythromycin and the aminoglycoside antibiotics, tetracycline is a protein synthesis inhibitor. It interferes with 30S ribosomal subunit function (see Figure 11.18).

The tetracyclines and the β-lactam antibiotics are the two most important groups of antibiotics in the medical field. The tetracyclines also find use in veterinary medicine, and in some countries are also used as nutritional supplements for poultry and swine. However, such nonmedical uses of medically important antibiotics are now discouraged because of the potential danger of development of antibiotic resistance (see Section 11.13).

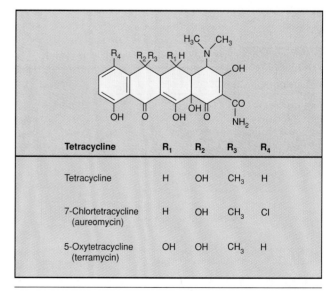

Tetracycline	R₁	R₂	R₃	R₄
Tetracycline	H	OH	CH₃	H
7-Chlortetracycline (aureomycin)	H	OH	CH₃	Cl
5-Oxytetracycline (terramycin)	OH	OH	CH₃	H

Figure 11.22 Structure of tetracycline and important derivatives.

Figure 11.21 Structure of erythromycin, a typical macrolide antibiotic.

CONCEPT CHECK   **11.10**

The aminoglycosides, macrolides, and tetracycline antibiotics are structurally complex molecules produced by prokaryotes and are active against other prokaryotes. Erythromycin and the various tetracyclines are used widely in clinical medicine.

✔ *What are the biological sources of the aminoglycoside, tetracycline, and macrolide antibiotics?*

✔ *What is the mechanism of action for each of these classes of antibiotics?*

11.11

Viral Control

In Sections 11.7–11.10, we examined a variety of control agents that are effective against Bacteria. One of the key features for every agent we discussed was selective toxicity. For example, in Section 11.9 we saw that since the β-lactam antibiotics inhibit cell wall synthesis, they are selectively toxic for Bacteria and there is no toxicity to the animal host cell, which lacks cell walls, the target structure. We now begin discussions of two classes of infectious agents, the viruses and the fungi, both of which have a great deal of their metabolic machinery in common with the animal host. Viruses actually use the host cell machinery to perform their metabolic functions (∞ Chapter 8). Therefore, most attempts at chemical control of viruses result in toxicity for the host. However, as we shall see, several agents are more toxic for the virus than the host and there are a few agents produced by the host that specifically target viruses: from detailed knowledge of

the molecular events involved in viral replication, some clues have finally emerged concerning control of viral infection. The fungi are eukaryotic microorganisms, and so treatment is again complicated by potential host toxicity, but several agents are now known that target fungus-specific features of structures such as membranes and cell walls. First, we shall discuss the control of viruses. We will examine two approaches to viral therapy that have been studied in considerable detail: chemical inhibition of growth and replication, and interferon action.

Chemical inhibition

Since a virus depends on its host cell for most aspects of virus replication, it is difficult to inhibit virus multiplication without affecting the host cell. Because of this, the spectacular medical successes with antibacterial agents have not been followed by similar successes in the search for specific antiviral agents. A few antiviral compounds are effective in controlling virus infections in laboratory situations (Table 11.6), and certain of these have been used in restricted clinical cases; but no substance has yet been found with more than limited practical use.

One interesting inhibitor listed in Table 11.6 is *rifamycin*, which is an inhibitor of RNA polymerase in Bacteria but not in eukaryotes or Archaea (∞ Sections 6.6 and 15.8). The RNA polymerase of vaccinia and other poxviruses is also inhibited by rifamycin. Thus, this antibiotic specifically inhibits the replication of poxviruses, although it has no effect on a wide range of other viruses affecting animal cells.

Another interesting chemical, *azidothymidine* (AZT), inhibits retroviruses such as the human immunodeficiency virus (HIV), the causative agent of AIDS (∞ Section 23.7). Azidothymidine is chemically related to

---

TABLE 11.6   **Stages of virus replication at which chemical inhibition of virus action is known**

Stage of replication	Chemical	Virus
Free virus	Kethoxal	Influenza virus
Absorption	None known	—
Entry of nucleic acid (uncoating)	Amantadine	Influenza virus
	Carbobenzoxypeptides	Measles
	3-Methylisoxazole compounds	Rhinoviruses (cold viruses)
Nucleic acid replication	Benzimidazole, guanidine	Poliovirus
	5-Fluorodeoxyuridine (FUDR)	Herpesvirus
	5-Iododeoxyuridine (IUDR)	Herpesvirus
	Acyclovir	Herpesvirus, *Varicella zoster*
	Rifamycin	Vaccinia virus
	Azidothymidine (AZT)	Retrovirus (HIV)
	Dideoxyinosine (ddI)	Retrovirus (HIV)
	Stavudine (d4T)	Retrovirus (HIV)
	Zalcytobine (ddC)	Retrovirus (HIV)
	Lamivudine (3TC)	Retrovirus (HIV)
Maturation (or late protein synthesis)	Isatin-thiosemicarbazone	Smallpox virus
Release	None known	—

thymidine but is a dideoxy derivative, lacking the 3'-hydroxyl (thus analogous to the dideoxynucleotides used in the Sanger DNA sequencing technique— ∞ Working with Nucleic Acids: The Tools, Chapter 6). AZT inhibits multiplication of retroviruses by blocking the synthesis of the DNA intermediate (reverse transcription) and is used to inhibit multiplication of HIV. A number of other nucleotide analogs that have similar modes of action have been developed for the treatment of HIV, as shown in Table 11.6. In virtually all cases, these drugs exhibit some level of toxicity and many lose their antiviral potency with time (∞ Section 23.7). However, the search for new anti-HIV drugs has greatly accelerated research in this area.

Interferon

Interferons are antiviral substances produced by many animal cells in response to infection by certain viruses. They are low-molecular-weight proteins (17,000 MW) that prevent viral multiplication in normal cells. There are three molecular types, IFN-α, produced by leukocytes, IFN-β, produced by fibroblasts, and IFN-γ, produced by immune cells known as lymphocytes (∞ Section 20.8). All three types are effective viral inhibitors. They were first discovered in the course of studies on virus interference, a phenomenon whereby infection with one virus interferes with subsequent infection with another virus, hence the name *interferon*. Interferons are formed in response to live virus, viral nucleic acids, and also to virus inactivated by radiation. Interferon is produced in larger amounts by cells infected with viruses of low virulence, whereas little is produced against highly virulent viruses. Apparently highly virulent viruses inhibit cell protein synthesis before any interferon can be produced. Interferon is also induced by a variety of double-stranded RNA molecules, either natural or synthetic. Since double-stranded RNA does not exist in uninfected cells but exists as the replicative form in RNA virus-infected cells, double-stranded RNA may serve as a signal of virus infection in the animal cell and brings into action the interferon-producing system.

Interferons are not virus-specific but *host*-specific. Interferon produced by a member of one species recognizes specific receptors only on cells of the same species. Therefore, interferon produced by one type of animal (for example, chicken) in response to influenza virus inhibits multiplication of other viruses in the same species but has no effect on the multiplication of influenza virus in other animal species. Interferon has no effect on uninfected cells; it seems to inhibit viral synthesis specifically. It acts by preventing RNA synthesis directed by virus, thus inhibiting synthesis of virus-specific proteins.

Interferons have been of interest as possible antiviral agents and possibly also as anticancer agents. Their use as therapeutic agents was long hindered by the difficulty and expense of producing large quantities, but genetic engineering techniques (∞ Section 10.10) have now made the production of interferon possible on a commercial scale, and clinical trials are under way.

CONCEPT CHECK   **11.11**

Viruses use host metabolic machinery for multiplication. Therefore, it is difficult to use chemotherapy for inhibition of virus growth, although a few chemical agents inhibit virus-specific functions. Animal cells produce antiviral protein substances called interferons that affect the virus multiplication process. Gene technology has now made interferons available to test clinical effectiveness.

✔ *Why are there no antiviral antibiotics?*

✔ *What are some concerns about effective antiviral drugs?*

11.12
Fungal Control

Like the viruses, fungi pose special problems for successful chemotherapy. Since fungi are members of the Eukarya, much of their cellular machinery is the same as in higher animals and humans, and so chemotherapeutic agents that affect metabolic pathways in fungi often affect corresponding pathways in host cells. This results in drug toxicity in higher animals. Thus, many antifungal drugs can be used only for topical (surface) applications. However, some drugs are selectively toxic for fungi. Drugs for fungal treatment are becoming increasingly important as fungal infections in immunosuppressed individuals become more prevalent (∞ Sections 23.7 and 23.16). We will look in detail at the selective toxicity of several classes of chemicals that are effective against fungi.

Ergosterol inhibitors

Two major groups of antifungal compounds work by interacting with ergosterol or inhibiting its synthesis. In most fungi, ergosterol replaces the cholesterol component found in higher eukaryotic cell membranes (∞ Section 3.3). The first group includes the *polyenes*, a group of antibiotics produced by *Streptomyces* species. Polyenes bind to ergosterol, which disrupts membrane function, eventually causing membrane permeability and cell death (Figure 11.23). A second major group of antifungal compounds includes the *azoles* and the *allylamines*, agents that selectively inhibit ergosterol biosynthesis and therefore have broad antifungal activity. Treatment with azoles results in the inability to produce

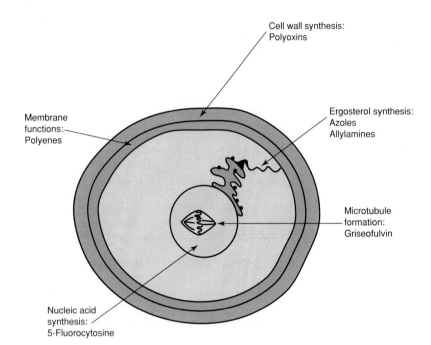

Figure 11.23 Sites of action of some antifungal chemotherapeutic agents. Because fungi are eukaryotic cells, antibacterial antibiotics do not affect them.

a normal membrane, leading to membrane damage and alteration of critical membrane activities such as nutrient transport. Allylamines also inhibit ergosterol biosynthesis but are useful only topically because they are not readily taken up and utilized by animal cells and tissues.

Other antifungal agents

A number of other antifungal drugs interfere with fungus-specific structures and functions. For example, most fungal cell walls contain *chitin,* a polymer of *N*-acetylglucosamine found only in fungi and insects (∞ Table 13.9). Several drugs such as the *polyoxins* inhibit cell wall synthesis by interfering with chitin biosynthesis. However, no cell wall targeting drugs are currently used clinically. Other drugs inhibit folate biosynthesis, interfere with DNA topology during replication, or, like *griseofulvin,* disrupt microtubule aggregation during mitosis (∞ Section 3.14). The nucleic acid analog *5-fluorocytosine* is an effective nucleic acid synthesis inhibitor in fungi. Some very effective antifungal drugs also have other biological applications. For example, *vincristine, vinblastin,* and *toxol* are effective antifungal agents and have known anticancer properties.

Unfortunately, the use of antifungal drugs has predictably resulted in the emergence of populations of resistant fungi and the emergence of "new" fungal pathogens. For example, *Candida* species, which are normally not pathogenic, now produce disease in individuals who have been treated with antifungal drugs. These drug-resistant *Candida* pathogens are not treatable by employing any of the currently used antifungal agents.

As the use of chemotherapeutic agents, both antibacterial and antifungal, increases, the possibilities for opportunistic fungal infections and the corresponding need for specific fungal control agents will increase.

CONCEPT CHECK   **11.12**

Antifungal agents fall into a wide variety of chemical categories. As with viruses, selective toxicity is hard to achieve, but there are some effective chemotherapeutic agents. Treatment of fungal infection is now an important human health issue.

- ✔ *Why are there very few clinically effective antifungal antibiotics?*
- ✔ *What factors are contributing to the apparent rise in fungal infections?*

11.13
Antibiotic Resistance

We have discussed the major antibiotics and their action in a number of situations. We now come to the concept of **antibiotic resistance,** the acquired ability of an organism to resist the effects of an antibiotic to which it is normally susceptible. First, we should note that resistance genes are probably acquired through a process of genetic exchange from the antibiotic producers: in order to protect themselves from the antibiotics they produce, these organisms have developed genetic mechanisms to neu-tralize or destroy their own antibiotics. The existence of these genes means that, under the right circumstances, resistance can be transferred to other organisms. First, we will look at some of the mechanisms of antibiotic resistance. Then we will explore the consequences of resistance in practical situations.

Resistance mechanisms

Not all antibiotics act against all microorganisms. Some microorganisms are naturally resistant to some antibi-otics. Antibiotic resistance can be an inherent property of a microorganism, or it can be acquired. There are sev-eral reasons why microorganisms may have an inherent resistance to an antibiotic. (1) The organism may lack the structure an antibiotic inhibits. For instance, some bacteria, such as mycoplasmas, lack a typical bacterial cell wall and are resistant to penicillins. (2) The organ-ism may be impermeable to the antibiotic. For example, gram-negative Bacteria are impermeable to penicillin G. (3) The organism may be able to alter the antibiotic to an inactive form. Many staphylococci contain β-lactamases that cleave the β-lactam ring of most penicillins (Figure 11.19). (4) The organism may modify the *target* of the antibiotic. (5) By genetic change, alteration may occur in a metabolic pathway that the antimicrobial agent blocks. Thus, the organism develops a resistant bio-chemical pathway. (6) The organism may be able to pump out an antibiotic entering the cell (efflux).

We give some specific examples of bacterial resis-tance to antibiotics in Table 11.7. Also, as discussed in Section 9.8, antibiotic resistance can be genetically encoded by the microorganism at either the chromoso-mal or the plasmid level on so-called *resistance plasmids (R factors);* specific types of resistance typically have a genetic basis in one location or the other (Table 11.7). Because of the development of antibiotic resistance, testing of bacteria isolated from clinical material for antibiotic sensitivity must be carried out. Details of the sensitivity testing of clinical isolates are described in Section 21.3.

Mechanism of resistance mediated by R plasmids

In the *laboratory* antibiotic-resistant cells are often iso-lated from cultures that were predominantly antibi-otic-sensitive. The resistance of these isolates is usually due to mutations in *chromosomal* genes. On the other hand, the majority of drug resistant bacteria isolated from *patients* contain the drug resistance genes on R plasmids. The mechanism of R plasmid resistance is different from that of chromosomal resistance. In most cases, antibiotic resistance mediated by chromosomal genes arises because of a modification of the *target* of antibiotic action (for example, a ribosome).

By contrast, R plasmid resistance is in most cases due to the presence in the R plasmid of genes encoding new enzymes that *inactivate* the drug (Figure 11.24) or genes that encode enzymes that either prevent uptake of the drug or actively pump it out. For instance, a number of antibiotics are known that have similar chemical structures containing aminoglycoside units.

TABLE 11.7   Mechanisms of bacterial resistance to antibiotics

Resistance mechanism	Antibiotic example	Genetic basis of resistance	Mechanism present in:
Reduced permeability	Penicillins	Chromosomal	*Pseudomonas aeruginosa* Enteric Bacteria
Inactivation of antibiotic (for example, penicillinase; modifying enzymes methylases, acetylases, and phosphorylases; and others)	Penicillins	Plasmid and chromosomal	*Staphylococcus aureus* Enteric Bacteria *Neisseria gonorrhoeae*
	Chloramphenicol	Plasmid and chromosomal	*Staphylococcus aureus* Enteric Bacteria
	Aminoglycosides	Plasmid	*Staphylococcus aureus*
Alteration of target (for example, RNA polymerase, rifamycin; ribosome, erythromycin, and streptomycin; DNA gyrase, quinolones)	Erythromycin Rifamycin Streptomycin Norfloxacin	Chromosomal	*Staphylococcus aureus* Enteric Bacteria Enteric Bacteria Enteric Bacteria *Staphylococcus aureus*
Development of resistant biochemical pathway	Sulfonamides	Chromosomal	Enteric Bacteria *Staphylococcus aureus*
Efflux	Tetracyclines	Plasmid	Enteric Bacteria
	Chloramphenicol	Chromosomal	*Staphylococcus aureus* *Bacillus subtilis*

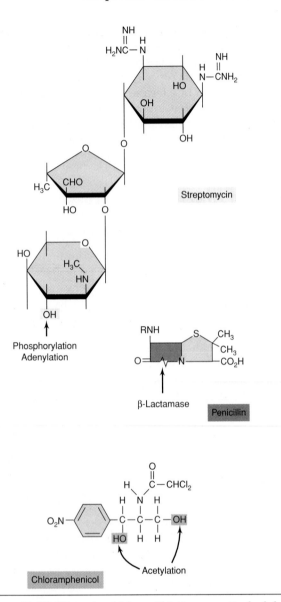

Streptomycin

Phosphorylation
Adenylation

β-Lactamase

Penicillin

Chloramphenicol

Acetylation

Figure 11.24 Sites at which antibiotics are attacked by enzymes encoded by R plasmid genes. In aminoglycoside antibiotics with a free amino group, inactivation is also by *N*-acetylation (see Figure 11.20).

Among the aminoglycoside antibiotics are streptomycin, neomycin, kanamycin, and spectinomycin. Strains carrying R plasmids conferring resistance contain enzymes that chemically modify the antibiotics either by phosphorylation, acetylation, or adenylylation. The modified drug then lacks antibiotic activity (Figures 11.19 and 11.20). In the case of the penicillins, R plasmid resistance is due to the formation of penicillinase (β-lactamase), which splits the β-lactam ring, thus destroying the molecule. Chloramphenicol resistance mediated by an R plasmid arises because of the presence of an enzyme that acetylates the antibiotic. Thus, the fact that R plasmids can confer multiple antibiotic resistance does not imply that the mode of action of the R plasmid genes is similar. The presence of multiple

antibiotic resistance is due to the fact that a single R plasmid contains a number of genes encoding different antibiotic inactivating enzymes.

Origin of resistance plasmids

Although specific evidence for the origin of multiple drug resistance R plasmids is not available, a number of lines of circumstantial evidence suggest that plasmids with R plasmid-type character existed before the antibiotic era. The widespread use of antibiotics provided selective conditions for the spread of R plasmids with one or more antibiotic resistance genes (see the box, Nonmedical Uses of Antibiotics). Indeed, a strain of *Escherichia coli* that was freeze-dried in 1946 contained a plasmid with genes conferring resistance to tetracycline and streptomycin, even though neither of these antibiotics were used clinically until several years later. Also, strains carrying R plasmid genes for resistance to semisynthetic penicillins were shown to exist before the semisynthetic penicillins had been synthesized. Of perhaps even more ecological significance, R plasmids conferring antibiotic resistance have been detected in some nonpathogenic gram-negative soil Bacteria. In the soil, such resistance may confer selective advantage because major antibiotic-producing organisms (*Streptomyces, Penicillium*) are also normal soil organisms. Thus, it seems that R plasmids are not a recent phenomenon but existed in the natural bacterial population before the antibiotic era. Later, the widespread use of antibiotics provided selective conditions for the rapid spread of these R plasmids. R plasmids are thus a predictable outcome of natural selection. They pose significant limits for the long-term use of any antibiotic as an effective chemotherapeutic agent.

Spread of antibiotic resistance

Inappropriate, extensive use of antibiotics is leading to the rapid development of antibiotic resistance in disease-causing microorganisms. The discovery and clinical use of the many known antibiotics has been paralleled by the emergence of bacteria that resist their action. There are numerous examples of the overuse of antibiotics and the concomitant development of resistance. Figure 11.25*a* shows a correlation between the number of tons of antibiotics used and the percentage of bacteria resistant to each antibiotic. The resistant organisms were isolated from patients with diarrheal disease. In general, high levels of antibiotic use resulted in high levels of resistance.

There are many examples of diseases in which the drug prescribed for treatment has changed because of increased resistance of the microorganism causing the disease. A classic example is the development of resistance to penicillin in *Neisseria gonorrhoeae*, the bacterium that causes gonorrhea (Figure 11.25*b*). Penicillin is no longer a useful antibiotic for treatment of gonor-

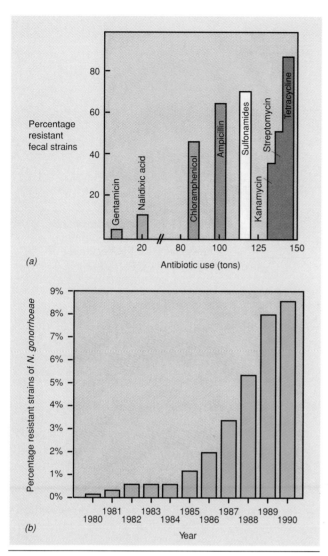

Figure 11.25 The emergence of antibiotic-resistant bacteria. (a) Relationship between antibiotic use and the percentage of bacteria isolated from diarrheal patients resistant to the antibiotic. Those antibiotics that have been used in the largest amounts, as indicated by the amount of antibiotic produced commercially, are those for which antibiotic-resistant strains are most frequent. (b) Percentage of reported cases of gonorrhea caused by antibiotic-resistant strains. The actual number of reported antibiotic-resistant cases in 1985 was 9000. This number rose to 59,000 in 1990. Greater than 95% of the reported antibiotic-resistant cases are due to penicillinase-producing strains of *Neisseria gonorrhoeae*. (Source: Centers for Disease Control, Atlanta, GA).

rhea because a large percentage of the clinical isolates produce β-lactamase and are resistant. Virtually all resistant strains have developed since 1980. The current drug of choice is ceftriaxone, but new treatment modalities are recommended nearly every year, simply to limit the effects of rapidly emerging resistance genes (∞ Section 23.6). Thus, resistant microorganisms appear to be *selected* by the presence of the antibiotic in the environment.

Surveys have shown that antibiotics are used far more often than is necessary. As a result, almost all pathogenic microorganisms have developed resistance to some antibiotics since widespread use of antimicrobial chemotherapy began in the 1950s (Figure 11.26). Penicillin and sulfa drugs, the first widely used chemotherapeutic agents, are not as widely used today because many pathogens have acquired some resistance to them. Even the organisms that are still uniformly sensitive to penicillin, such as *Streptococcus pyogenes* (the bacterium that causes strep throat, scarlet fever, and rheumatic fever) (∞ Section 23.2), now need significantly more penicillin for successful treatment than a decade ago. Other indiscriminant, nonessential uses of antibiotics may have worsened this situation. For example, antibiotics are used in agriculture both as growth-promoting substances in animal feeds and as prophylactics (to prevent the occurrence of disease rather than to treat an existing one). Several recent food poisoning outbreaks have been blamed on the use of antibiotics in animal feeds. By overloading various environments with antibiotics, rapid development of drug resistance may result. Resistance can be minimized if drugs are used only for serious diseases and are given in sufficiently high doses so that the microbial population level is reduced before mutants have a chance to appear. Resistance can also be minimized by combining two unrelated chemotherapeutic agents because it is likely that a mutant strain resistant to one antibiotic will still be sensitive to the other. However, with the increasing prevalence of resistance plasmids (so-called R factors), in pathogenic bacteria (∞ Section 9.8), multiple antibiotic therapy is proving less attractive as a clinically useful strategem. These R factors are capable of transferring multiple drug resistance to all plasmid-susceptible Bacteria.

There is, however, some encouraging information concerning antibiotic resistance. Several reports suggest that if the use of a particular antibiotic is stopped, the resistance to that antibiotic will be reversed over time. Although many of these reports are from limited environments such as hospitals, at least one report suggests this phenomenon is occurring on a nationwide scale (see the box, Reversing Antibiotic Resistance). This information implies that resistance is reversible and that the efficacy of some antibiotics may be reestablished by long-term monitoring and prudent use.

Overcoming drug resistance

Given sufficient drug exposure and time, resistance can develop to any antimicrobial drug. Prudent, conservative use of existing drugs is an important way to prolong their useful clinical life. However, the long-term solution to microbial drug resistance is to develop new antimicrobial drugs. Two strategies are used to find new agents: either analogs of existing compounds can be created, or entirely new classes of drugs can be developed.

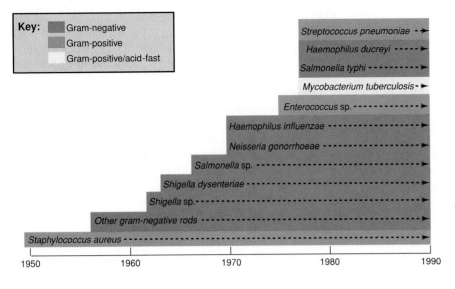

Figure 11.26 Appearance of antibiotic resistance in several human pathogens since the beginning of antibiotic therapy.

NONMEDICAL USES OF ANTIBIOTICS

A major nonmedical use of antibiotics in the United States is addition to animal feed. The addition of low levels of antibiotics to animal feeds stimulates animal growth, shortening the period required to get the animal to market. For example, addition of 25 milligrams (mg) of penicillin per pound of chicken feed saves 2 billion lb (900 million kg) of feed each year because of more rapid weight gains and feeding efficiency. The antibiotics probably act by inhibiting organisms responsible for low grade infections and by reducing intestinal epithelial inflammation. Studies with germ-free animals have confirmed this idea. The growth of germ-free animals is not accelerated by antibiotic-supplemented feed. In addition, the intestinal wall of the normal animal is much thicker than that of the germ-free animal, probably because of low level inflammation caused by the normal bacterial flora. The lessening of inflammation in the gut of animals fed low levels of antibiotics probably promotes nutrient uptake and could account for the more efficient utilization of feed observed.

The problem with low levels of antibiotics in animal feeds is that an antibiotic-resistant microflora is selected by the constant exposure to antibiotics. The use of antibiotics in animal feed therefore expands the gene pool of antibiotic resistance in nature. Because some of the animal gut flora also inhabit the human gut, the transmission of resistant flora from animals to humans is a real possibility. Indeed, studies on antibiotic resistance in human gut flora have shown that many strains of human enteric bacteria are multiply resistant, especially among those who work in animal husbandry.

Molecular studies of resistant strains of *Salmonella* isolated from poultry have shown that the resistance resides on conjugative plasmids or transposons (Sections 9.8 and 9.10). Resistance genes are rapidly transferred between different species and even between different genera. Resistant organisms can then infect humans through contaminated meat or by contact with live animals.

Unfortunately, long-term studies on animals previously fed antibiotics and then put on antibiotic-free rations have shown that antibiotic-resistant bacteria are not quickly lost from the gut. The resistance genes may have integrated with stable plasmids in the gut flora, and, in the absence of counterselective forces, these resistance determinants have been maintained and will remain a part of the gut flora for some time, even if supplementation of feeds with antibiotics were to stop. Nonmedical use of antibiotics has therefore reinforced, albeit painfully, a simple lesson in microbial ecology: the environment selects the best adapted species.

Although continued use of clinically useful antibiotics in animal feeds will undoubtedly widen the dissemination of resistance genes, it is not clear that halting this single practice will effectively solve the problem. Continued veterinary use of antibiotics may by itself maintain resistant animal microflora. However, in hope of reducing the spread of antibiotic resistance in Europe, most European countries have banned the use of antibiotics in animal feeds. In the United States large amounts of antibiotics continue to be used in the cattle, poultry, and swine industries. ■

REVERSING ANTIBIOTIC RESISTANCE

 The widespread use of antibiotics has increased the number of pathogenic microorganisms that display antibiotic resistance. As a result, many antibiotics have lost effectiveness and some, including many penicillins, are no longer useful for treating certain infections. However, there are some indications that the process of selecting antibiotic-resistant organisms is reversible. In the 1980s, Hungary was extremely dependent on penicillin for treatment of ear and sinus infections, very common childhood diseases. The antibiotic was cheap, readily available, and widely used. However, the causative agent in most cases of these diseases, *Streptococcus pneumoniae*, became resistant to penicillin—in the early 1980s, 50% of the diagnosed cases were caused by penicillin-resistant pneumococci. As a result, simple infections became very difficult to treat and routine childhood illnesses became long, painful, debilitating diseases. However, the Hungarian National Institute of Public Health, through 23 national microbiology laboratories, was able to carefully monitor this trend. Eventually, physicians became convinced that penicillin was no longer useful because of their own clinical observations and the records and educational efforts of the institute. As a result, physicians began to prescribe non-β-lactam antibiotics to treat these diseases. From 1983 to 1992, the consumption of penicillin decreased by one-half. Along with the decrease in penicillin use, the levels of penicillin-resistant pneumococci surprisingly decreased from the high of 50% to 34%, presumably because there was no longer any selection for penicillin-resistant organisms. Thus, by shifting drug use away from penicillin, Hungarian physicians were able to stop and to actually reverse a trend toward complete penicillin resistance in pneumococci. If the levels of penicillin-resistant organisms continue to fall, penicillin may again become a useful chemotherapeutic agent in Hungary. This encouraging example suggests that prudent use, careful monitoring, and patient and physician education may reverse the effects of previous overuse of some antibiotics. If the same principles of education and prudent use of antibiotics are applied *before* resistance becomes a widespread problem, then the effectiveness of antibiotics can be maintained indefinitely. ∎

The production of new *analogs* of existing antimicrobial compounds is generally straightforward and often productive, largely because new compounds that are structural mimics of older ones have a predictable site of action (many studies of drug analogs begin at the computer, where new drugs are rapidly "created" and even tested for binding and toxicity in the computer environment) (⚭ Section 12.5). In many cases, parameters such as solubility and affinity can be changed by altering chemical structure without altering the site of drug action. The new compound may actually be more potent than the parent compound, and, because resistance is based on structural recognition, the new compound may not be recognized by resistance factors. With this basic strategy in mind, new β-lactam antibiotics (see Section 11.9) and new tetracycline-related compounds (see Section 11.10) are routinely synthesized and tested. New analogs of vancomycin (Figure 11.17) are 100 times as potent as the original compound and are capable of killing even vancomycin-resistant organisms. However, based on previous experiences with the parent compounds, resistance will eventually develop and render the altered compounds less useful over time.

New *classes* of antimicrobial compounds are much more difficult to identify. They must be isolated from natural sources or synthetically designed to interact with specific microbial structures and then screened for efficacy and toxicity (⚭ Sections 12.5 and 23.4). To be effective, new antibacterial compounds must work at novel sites in bacterial metabolism and biosynthesis, or be structurally dissimilar to existing compounds, thus avoiding existing resistance factors. The *oxazolidinones* (Figure 11.17), a novel class of experimental synthetic chemicals, fit this description. Soluble forms are highly effective against gram-positive pathogens, including many pathogens that exhibit resistance to other drugs. Oxazolidinones are structurally unlike other antibiotics and are thought to inhibit protein synthesis in a unique way, probably at a very early stage such as the mRNA-tRNA-30S ribosomal interaction (⚭ Section 6.9). In the near future, members of this new class of structurally and functionally unique compounds may become the antimicrobial agents of choice to treat drug-resistant pathogens.

CONCEPT CHECK   11.13

An important side effect of the use of antibiotics is the development of resistance by the targeted microorganisms. In many cases this can be attributed to selection of existing resistance genes. A major selection factor for resistance genes is the indiscriminate use of antibiotics. New antimicrobial compounds are constantly being developed to deal with drug-resistant organisms.

✔ *Is antibiotic resistance a naturally occurring phenomenon?*

✔ *What is one way resistance genes for antibiotics can be controlled?*

Material for Review

REVIEW QUESTIONS

1. Why is the decimal reduction time (*D*) important in heat sterilization? How would the presence of bacterial endospores affect *D*?

2. Describe the effects of lethal irradiation at the molecular level.

3. What are the principal advantages of using membrane filters instead of depth filters?

4. Describe the procedure for obtaining the minimum inhibitory concentration (MIC) for a chemical that is bactericidal for *Escherichia coli*.

5. Contrast the action of disinfectants and antiseptics. Why can't disinfectants normally be used on living tissue?

6. Compare the growth of food spoilage organisms on food and in an artificial culture medium. What specific factors influence the growth rate?

7. Growth factor analogs are generally distinguished from antibiotics by a single important criterion. Explain.

8. Most antibiotics are made by only certain groups of organisms. Is this statement true? What groups of organisms make antibiotics?

9. Describe the mechanism of action that characterizes a β-lactam antibiotic.

10. Distinguish between the mode of action of at least three of the protein synthesis-inhibiting antibiotics.

11. Why do antiviral drugs generally exhibit host toxicity?

12. Define some of the targets for selective toxicity of chemotherapeutic agents in fungi.

13. What is the ultimate origin of bacterial resistance genes?

APPLICATION QUESTIONS

1. Describe in a graph the experimental results you would expect for the decimal reduction time of a very heat-sensitive organism. How would this graph be affected if the vegetative cells were heat-sensitive but the organism formed heat-resistant endospores?

2. What are some potential drawbacks to the use of radiation in food preservation? Do you think these drawbacks could be manifested as health hazards? Why or why not? How would you test for radiation-damaged and radiation-contaminated food?

3. Filtration is an acceptable means of pasteurization for some liquids. Design a filtration system for pasteurization of a heat-sensitive liquid. Why might filtration be desirable over a heat pasteurization system?

4. Design an experiment to distinguish between a cidal and a static agent. Can you use the minimum inhibitory concentration (MIC) test in your experiments? Explain.

5. What tests would you perform to decide whether a chemical agent could be used as an antiseptic? As a disinfectant? Some chemicals might serve both purposes. Describe the properties of such a chemical and given an example.

6. Although growth factor analogs may inhibit microbial metabolism, only a few of the agents are practically useful. Many potential agents, and some that are in wide use, such as azidothymidine (see Table 11.6), exhibit significant host cell toxicity. Describe a growth factor that is effective and has low toxicity for host cells. Why might the toxicity be low for the agent you chose? Also describe a growth factor analog that is effective against an infectious disease but also exhibits toxicity for host cells. A toxic agent might still be used in certain situations to treat infectious diseases. Explain.

7. Track the contamination of canned meat with *Escherichia coli* from the source through the canning process, to the consumer. In the absence of gas formation, what evidence might be available to the consumer to indicate spoilage? Suggest alterations to the canning process, based on your knowledge of the biology of *E. coli* and the canning process, that would prevent future contaminations.

8. We estimated that less than 1% of all known antibiotics have any practical value for either research or clinical use. Indicate why this might be so. Do you think it is important to expand and continue searches for new antibiotics? What alternatives to antibiotic treatments are, or could be, available for the treatment of human disease?

9. Although the β-lactam antibiotics demonstrate clear selective toxicity for Bacteria, many groups of Bacteria are innately resistant to their effects. Without invoking bacterial resistance genes, indicate why gram-negative Bacteria are resistant to the effects of most, but not all, β-lactam antibiotics. Further explain why some β-lactam antibiotics are useful against these organisms.

10. What potential advantages might the aminoglycosides, macrolides, and tetracyclines have over penicillin G for chemotherapy? Explain.

11. List the features of an ideal antiviral drug, especially with regard to selective toxicity. Do such drugs exist? What factors might limit use of such a drug?

12. Like viruses, fungi present special chemotherapeutic problems but generally can be targeted more readily than viruses. Explain the problems inherent in chemotherapy of both groups and explain whether or not you agree with the preceding statement. Give specific examples and suggest at least one group of chemotherapeutic agents that might target both types of infectious agents.

13. Explain the genetic basis of acquired resistance to β-lactam antibiotics in *Staphylococcus aureus*. Can you reverse the resistance phenomenon? Design a set of experiments to test your answer.

SUPPLEMENTARY READINGS

Block, S. S. (ed.). 1991. *Disinfection, Sterilization, and Preservation,* 4th edition. Lea and Febiger, Philadelphia. A standard reference on chemical disinfection, with separate chapters on each of the different groups of agents.

Brody, T. M., J. Larner, K. P. Minneman, and **H. Nev.** 1994. *Human Pharmacology–Molecular to Clinical,* 2nd edition. Mosby Year Book, St. Louis. An up-to-date source for information on drug actions and mechanisms.

Bryan, L. E. (ed.). 1989. *Microbial Resistance to Drugs.* Springer-Verlag, Berlin. An excellent source of information on the mechanisms of antimicrobial agent resistance and methods for testing clinical isolates for resistance.

Conte, J. E., Jr. 1995. *Manual of Antibiotics and Infectious Diseases,* 8th edition. Williams and Wilkins, Baltimore. A compendium of clinical antibiotics and strategies for their use.

Hardman, J. G., and **L. E. Limbird** (eds.). 1995. *The Pharmacological Basis of Therapeutics,* 9th edition. Pergamon Press, New York. A comprehensive reference source covering all aspects of chemotherapy.

Hugo, W. B., and **A. D. Russell.** 1992. *Pharmaceutical Microbiology,* 5th edition. Blackwell Scientific, Oxford, England. Detailed treatment of antibiotic production, and microbial contamination problems in large-scale production processes.

Page, M. I. (ed.). 1992. *The Chemistry of β-Lactams.* Chapman & Hall, New York. An authoritative review of all aspects of β-lactam antibiotics including mode of action, structural features, and microbial resistance.

Pratt, W. B., and **P. Taylor** (eds.). 1990. *Principles of Drug Action.* Churchill Livingstone, New York. An advanced treatment of drug mechanisms and interactions.

 On~line resources for this chapter are on the World Wide Web at: http://www.prenhall.com/~brock (click on the <u>table of contents</u> link and then select Chapter 11).

CHAPTER *12*

Industrial Microbiology

The production of alcoholic beverages, such as the white and red wines shown here, involve the large-scale use of microorganisms and is just one of many areas of industrial microbiology.

MINIGLOSSARY for Chapter 12

Aminoglycosides a group of antibiotics including streptomycin, containing amino sugars linked by glycosidic bonds

β-Lactam Antibiotics a group of antibiotics including penicillin, containing the four-membered heterocyclic β-lactam ring

Bioconversion the use of microorganisms to carry out a chemical reaction that is more costly or not feasible nonbiologically

Brewing the manufacture of alcoholic beverages such as beer from the fermentation of malted grains

Commodity Chemicals chemicals such as ethanol that have low monetary value and are thus sold primarily in bulk

Distilled Beverage a beverage containing alcohol concentrated by distillation

Fermentation in an industrial context, any large-scale microbial process whether carried out aerobically or anaerobically

Fermentor the tank in which an industrial fermentation is carried out

Immobilized Enzyme an enzyme attached to a solid support over which substrate is passed and converted to product

Primary Metabolite a metabolite excreted during the growth phase

Scale-up conversion of an industrial process from a small laboratory setup to a large commercial fermentation

Secondary Metabolite a metabolite excreted at the end of the primary growth phase and into the stationary phase

Secondary Treatment in sewage treatment, either the aerobic or anoxic decomposition of sewage following the removal of nondegradable objects by primary treatment

Semisynthetic Penicillin a penicillin produced using components derived from both microbial fermentation and chemical syntheses

Single-Cell Protein protein derived from microbial cells for use as food or a food supplement

Tetracyclines a class of antibiotics containing the four-membered naphthacene ring

Wine a product of the alcoholic fermentation of fruit juices, usually grape juice, by yeast

Industrial microbiology is the discipline that uses microorganisms, usually grown on a large scale, to produce valuable commercial products or carry out important chemical transformations. Industrial microbiology originated with alcoholic fermentation processes, such as those for making beer and wine. Subsequently, microbial processes were developed for the production of pharmaceutical agents (such as antibiotics), food additives (such as amino acids), enzymes, and chemicals such as butanol and citric acid. All these industrial microbiological processes are enhancements of metabolic reactions that microorganisms were already capable of carrying out, with the goal in most cases of simply overproducing the product of interest. In research on such processes, the industrial microbiologist's main task is to modify the organism, usually by traditional genetic methods, or to modify the process so that the *highest yield* of the desired product is obtained.

But now, in addition to traditional industrial microbiology, a new era has unfolded—that of *microbial biotechnology*. In biotechnology, methods for gene manipulation have given rise to new microbial products, most of which are not naturally produced by microorganisms. Microbial production arises from the fact that the microorganisms have been genetically engineered to make the desired product. In biotechnology, the microbiologist employs genetic engineering methods to actually *create new organisms* containing new complements of genes that enable them to make novel products, usually of high commercial value.

In the present chapter, we focus on traditional industrial microbiological processes, emphasizing the unique problems and requirements of large-scale industrial cultivation of microorganisms. We discuss a number of large-scale microbial processes and concentrate on the key industrial products of antibiotics, foods, food supplements, and alcohol, with emphasis on those aspects of these processes that are of special interest to the industrial microbiologist. We conclude with coverage of wastewater treatment, an important, if not glamorous, large-scale use of microorganisms. The biotechnological aspects of industrial microbiology were covered in Chapter 10, where the discussion focused on the gene manipulations necessary to produce "biotech drugs" and related products and on the nature of some of the products themselves.

12.1

Industrial Microorganisms and Products

Not all microorganisms find industrial use. Whereas microorganisms isolated from nature exhibit cell growth as their main physiological property, industrial microorganisms are organisms that have been selected carefully so that they manufacture one or more specific *products*. Even if the industrial microorganism is one that has been isolated by traditional techniques, it becomes a highly "modified" organism before it enters large-scale industries. To a great extent, industrial microorganisms are metabolic specialists capable of producing particular metabolites specifically and to high yield. In order to achieve this high metabolic specialization, industrial strains are genetically altered by mutation or recombination. Minor metabolic pathways are usually repressed or eliminated, and metabolic imbalance frequently is present. Industrial microorganisms may show many altered cellular and biochemical properties. Although industrial strains may grow satisfactorily under the highly specialized conditions of the industrial fermentor, they may show poor growth properties in competitive environments in nature.

Origin of industrial strains

The ultimate source of all strains of industrial microorganisms is the natural environment. But through the years, as large-scale microbial processes have been perfected, a number of industrial strains have been deposited in *culture collections*. When a new industrial process is patented, the applicant for the patent is required to deposit a strain capable of carrying out the process in a recognized culture collection. There are a number of culture collections that serve as repositories of microbial cultures (Table 12.1). Although these culture collections can serve as ready sources of cultures, it should be understood that most industrial companies are reluctant to deposit their *best* cultures in culture collections. In addition to cultures of microorganisms, many culture collections also have collections of various plasmids, cloned genes, and vectors for use in genetic engineering animal cell lines for growth of animal viruses, and hybridomas for producing monoclonal antibodies (∞ Section 20.14).

Strain improvement

As we have noted, the initial source of an industrial microorganism is the natural environment, but the original isolate is greatly modified in the laboratory. As a result of this modification, progressive improvement in the yield of a product can be anticipated. The most dramatic example of such progressive improvement is that of penicillin, the antibiotic produced by the fungus *Penicillium chrysogenum*. When penicillin was first produced on a large scale (∞ Microbiology and "Magic Bullets," Chapter 11), yields of 1–10 μg/ml were obtained. Over the years, as a result of strain improvement coupled with changes in the medium and growth conditions, the yield of penicillin has been increased to about 50,000 μg/ml! It is of interest that all of this 50,000-fold increase in yield was obtained by mutation and selection; no genetic engineering manipulations were involved. The introduction of new genetic techniques has led to further, albeit much more modest, yield increases.

Properties of a useful industrial microorganism

A microorganism suitable for industrial use must produce the substance of interest, but there is much more than that. The organism must be available in pure culture, must be genetically stable, and must grow in large-scale culture. It must also be possible to maintain cultures of the organism for a long period of time in the laboratory and in the industrial plant. The culture should preferably produce spores or some other reproductive cell form so the organism can be easily inoculated into large fermentors.

An important characteristic is that the industrial organism grow rapidly and produce the desired product in a relatively short period of time. The organism must also be able to grow in a relatively inexpensive liquid culture medium obtainable in bulk quantities.

TABLE 12.1	**Culture collections that supply cultures of industrial microorganisms**[a]	

Abbreviation	Name	Location
ATCC	American Type Culture Collection	Rockville, MD, United States
CBS	Centraalbureau voor Schimmelculturen	Baarn, The Netherlands
CCM	Czechoslovak Collection of Microorganisms	J. E. Purkyne University, Brno, Czech Republic
CDDA	Canadian Department of Agriculture	Ottawa, Canada
CMI	Commonwealth Mycological Institute	Kew, United Kingdom
DSM	Deutsche Sammlung von Mikroorganismen und Zellkulturen GmbH	Braunschweig, Germany
FAT	Faculty of Agriculture, Tokyo University	Tokyo, Japan
IAM	Institute of Applied Microbiology	University of Tokyo, Japan
NCIB	National Collection of Industrial Bacteria	Aberdeen, Scotland
NCTC	National Collection of Type Cultures	London, United Kingdom
NRRL	Northern Regional Research Laboratory	Peoria, IL, United States
PCC	Pasteur Culture Collection	Paris, France

[a]Listed here are just a few of the general culture collections. Many universities and research laboratories maintain collections of specific microbial groups.

Many industrial microbiological processes use waste carbon from other industries as major or supplemental ingredients for large-scale culture media. These include *corn steep liquor* (a product of the corn wet milling industry that is rich in nitrogen and growth factors), *whey* (a waste liquid of the dairy industry containing lactose and minerals), and other industrial waste materials having high organic carbon contents [utilization of waste carbon materials in this way also helps to solve disposal problems created by the high biological oxygen demand (BOD) of organic industrial wastes] (see Section 12.15). In addition, an industrial microorganism should not be harmful to humans or economically important animals or plants. Because of the large population size in the industrial fermentor and the virtual impossibility of avoiding contamination of the environment outside the fermentor, a pathogen would present potentially disastrous problems.

Another important requisite of an industrial microorganism is that it be possible to remove the microbial cells from the culture medium relatively easily. In the laboratory, cells are removed from culture media primarily by centrifugation, but centrifugation may be difficult or expensive on a large scale. The most favorable industrial organisms are those of large cell size because larger cells settle rapidly from a culture or can be easily filtered out with relatively inexpensive filter materials. Fungi, yeasts, and filamentous bacteria are preferred. Unicellular bacteria, because of their small size, are difficult to separate from a culture fluid.

Finally, an industrial microorganism should be amenable to genetic manipulation. In industrial microbiology, increased yields have been obtained genetically primarily by means of mutation and selection. It is also desirable for the industrial organism to be capable of genetic recombination, either by a sexual or by some sort of parasexual process. Genetic recombination permits the incorporation in a single genome of genetic traits from more than one organism. However, many industrial strains have been greatly improved by mutation and selection without use of any genetic recombination.

Industrial products

Microbial products of industrial interest are of several major types (Figure 12.1). These include the microbial cells themselves, for example, yeast cultivated for food, baking, or brewing (see Figure 12.24), and substances produced by cells. Examples of the latter include enzymes such as glucose isomerase, pharmacologically active agents such as antibiotics, steroids, and alkaloids, specialty chemicals and food additives such as the currently popular aspartame food and drink sweetener, and commodity chemicals, such as ethanol. A summary of some important industrial products, many of which will be discussed in more detail later, is given in Figure 12.1.

CONCEPT CHECK 12.1

An industrial microorganism must produce the product of interest in high yield, grow rapidly on inexpensive culture media available in bulk quantities, be amenable to genetic manipulation, and, if possible, be nonpathogenic. Industrial products are many and include both cells and substances made by cells.

✔ *Why should industrial microorganisms be genetically manipulable?*

✔ *List three important products of industrial microbiology.*

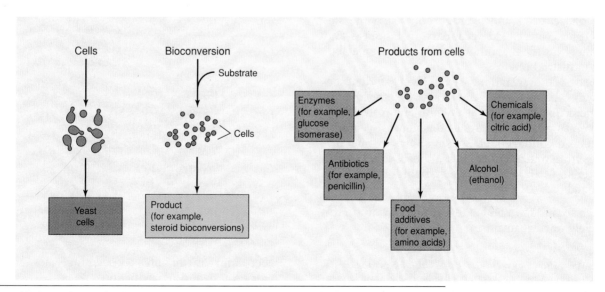

Figure 12.1 Products of industrial microbiology. The products may be the cells themselves or products made from cells. In the case of bioconversion, cells are used to chemically convert a specific substance from one form to another.

12.2
Growth and Product Formation in Industrial Processes

In Section 5.3, we discussed the microbial growth process and described the various stages: lag, log, and stationary phase. In the present section, we discuss the microbial growth process as it occurs in an *industrial* process. We are concerned here primarily with those processes in which a microbial metabolite is the desired product. There are two basic types of microbial metabolites: primary and secondary. A *primary metabolite* is one that is formed during the primary growth phase of the microorganism, whereas a *secondary metabolite* is one that is formed near the end of the growth phase, frequently at, near, or in the stationary phase of growth. The contrast between a primary metabolite and a secondary metabolite is illustrated in Figure 12.2.

Primary microbial metabolites

A typical microbial process in which the product is formed during the primary growth phase is *alcohol (ethanol) fermentation*.* Ethanol is a product of anoxic metabolism of yeast and certain bacteria (∞ Section 4.10) and is formed as part of energy metabolism. Because growth can occur only if energy production can occur, ethanol formation takes place in parallel with growth. A typical alcohol fermentation, showing the formation of microbial cells, ethanol, and sugar utilization, is illustrated in Figure 12.3*a*.

Secondary microbial metabolites

A more complex type of microbial industrial process is one in which the desired product is not produced during the primary growth phase but instead in the *stationary* phase. Metabolites produced during the stationary phase are called **secondary metabolites** and are some of the most common and important metabolites of industrial interest.

The best-known and most extensively studied secondary metabolites are the antibiotics, and the kinetics of the penicillin process are shown in Figure 12.3*b*.

Whereas primary metabolism is generally similar in all cells, secondary metabolism shows distinct differences from one organism to another. The following characteristics of secondary metabolites have been recognized:

1. Each secondary metabolite is formed by only a relatively few organisms.

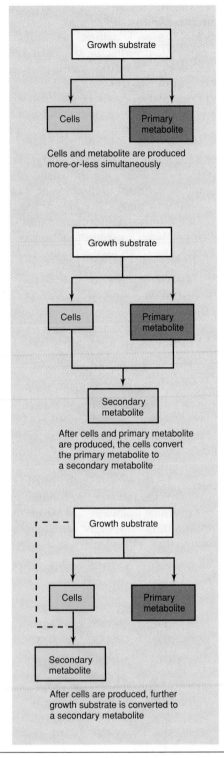

Figure 12.2 Contrast between primary and secondary metabolites.

2. Secondary metabolites are seemingly not essential for growth and reproduction.

3. The formation of secondary metabolites is extremely dependent on growth conditions, especially on the

---

*In industrial microbiology, the term *fermentation* refers to *any* large-scale microbial process, whether or not it is biochemically a fermentation. In fact, most industrial fermentations are aerobic. The *tank* in which the industrial fermentation is carried out is called a fermentor; the microorganism involved is the fermenter.

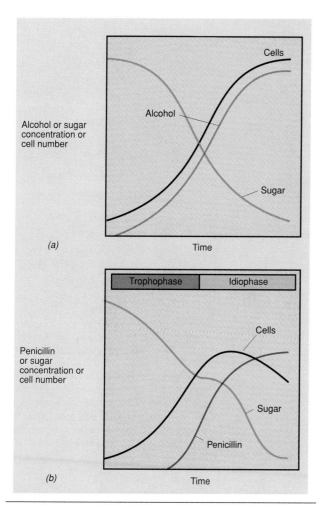

Figure 12.3 Contrasts between primary and secondary metabolism. (a) Primary metabolism: the formation of alcohol from sugar by yeast. (b) Secondary metabolism: the formation of penicillin from sugar by the fungus *Penicillium chrysogenum*, showing the separation of the growth phase (trophophase) and the production phase (idiophase). Note that in (b) most of the product is produced *after* growth has entered the stationary phase.

composition of the medium. Repression of secondary metabolite formation frequently occurs.

4. Secondary metabolites are often produced as a group of closely related structures. For instance, a single strain of a species of *Streptomyces* has been found to produce 32 related but different anthracycline antibiotics.

5. It is often possible to get dramatic *overproduction* of secondary metabolites, whereas primary metabolites, linked as they are to primary metabolism, usually cannot be overproduced in such a dramatic manner.

Trophophase and idiophase

In secondary metabolism, the two distinct phases of metabolism are called the *trophophase* and the *idiophase* (Figure 12.3*b*). The trophophase is the *growth* phase (*tropho* is a prefix meaning "growth"), whereas the metabolite production phase is the idiophase. If we are dealing with a secondary metabolite, then we should ensure that appropriate conditions are provided during the trophophase for excellent growth and that conditions are properly altered at the appropriate time to obtain excellent product formation.

In secondary metabolism, the product in question may be derived not from the primary growth substrate but from a product that itself was formed from the primary growth substrate. Thus, the secondary metabolite is generally produced from several intermediate products that accumulate, either in the culture medium or in the cells, during primary metabolism.

One characteristic of secondary metabolites is that the enzymes involved in production of the secondary metabolite are regulated separately from the enzymes of primary metabolism. In some cases, specific *inducers of secondary metabolite* production have been identified. For instance, a specific inducer has been identified for *streptomycin* production, a compound called *Factor A* (see Section 12.6).

Relationship between primary and secondary metabolism

Most secondary metabolites are complex organic molecules that require a large number of specific enzymatic reactions for synthesis. For instance, it is known that at least 72 separate enzymatic steps are involved in synthesis of the antibiotic *tetracycline* and over 25 steps in the synthesis of *erythromycin*, none of which are reactions occurring during primary metabolism. The metabolic pathways of these secondary metabolites do arise out of primary metabolism, however, because the starting materials for secondary metabolism come from the major biosynthetic pathways. This is summarized in Figure 12.4, which shows the interrelationship of the main primary metabolic pathway for aromatic amino acid synthesis with the secondary metabolic pathways for a variety of antibiotics. As can be seen, many structurally complex secondary metabolites originate from structurally quite similar precursors (Figure 12.4).

CONCEPT CHECK   12.2

Primary and secondary metabolites are produced during active cell growth and near the onset of stationary phase, respectively. Many economically valuable microbial products are secondary metabolites.

✔ *Is penicillin a primary or a secondary metabolite? Why?*

✔ *What type of metabolite, primary or secondary, can be more easily overproduced? Why?*

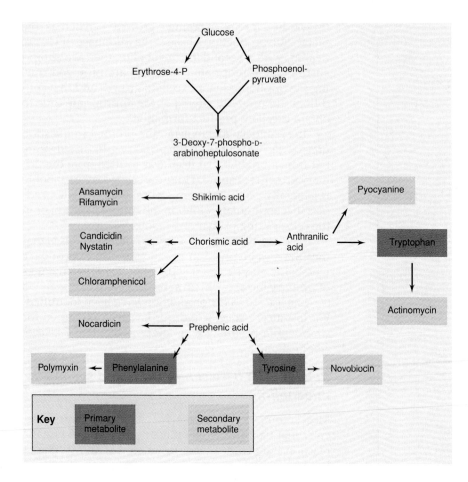

Glucose

Erythrose-4-P

Phosphoenol-
pyruvate

3-Deoxy-7-phospho-D-
arabinoheptulosonate

Ansamycin
Rifamycin ← Shikimic acid

Pyocyanine

Candicidin
Nystatin ← ← Chorismic acid → Anthranilic
acid → Tryptophan

Chloramphenicol

Actinomycin

Nocardicin ← Prephenic acid

Polymyxin ← Phenylalanine   Tyrosine → Novobiocin

Key   Primary
metabolite   Secondary
metabolite

Figure 12.4 Relationship of the primary metabolic pathway for the synthesis of aromatic amino acids (∞ Section 4.19) and formation of a variety of secondary metabolites containing aromatic rings. Note that this is a composite scheme of processes occurring in a variety of microorganisms: no one organism produces all these secondary metabolites.

12.3

Characteristics of Large-Scale Fermentations

The vessel in which the industrial process is carried out is called a *fermentor*. Fermentors can vary in size from the small 5- to 10-liter laboratory scale (Figure 12.5a) to the enormous 500,000-liter industrial scale. The size of the fermentor used depends on the process and how it is operated. Processes operated in batch mode require larger fermentors than processes operated continuously or semicontinuously. A summary of fermentor sizes for some common microbial fermentations is given in Table 12.2.

TABLE 12.2	Fermentor sizes for various industrial processes
Size of fermentor (liters)	**Product**
1–20,000	Diagnostic enzymes, substances for molecular biology
40–80,000	Some enzymes, antibiotics
100–150,000	Penicillin, aminoglycoside antibiotics, proteases, amylases, steroid transformations, amino acids
200,000–500,000	Amino acids (glutamic acid)

Industrial fermentors can be divided into two major classes, those for *anaerobic* processes and those for *aerobic* processes. Anaerobic fermentors require little special equipment except for removal of heat generated during the fermentation, whereas aerobic fermentors require much more elaborate equipment to ensure that mixing and adequate aeration are achieved. Because most industrial fermentations are aerobic, the present discussion will be confined to aerobic fermentors.

Construction of an aerobic fermentor

Large-scale industrial fermentors are almost always constructed of stainless steel. Such a fermentor is essentially a large cylinder, closed at the top and bottom, into which various pipes and valves have been fitted (Figure 12.5b). Because sterilization of the culture medium and removal of heat are vital for successful operation, the fermentor generally has an external *cooling jacket* through which steam or cooling water can be run. For very large fermentors, insufficient heat transfer occurs through the jacket, and so *internal coils* must be provided through which either steam or cooling water can be piped.

An important part of the fermentor is the aeration system. With large-scale equipment, transfer of oxygen from the gas to the liquid is a very difficult

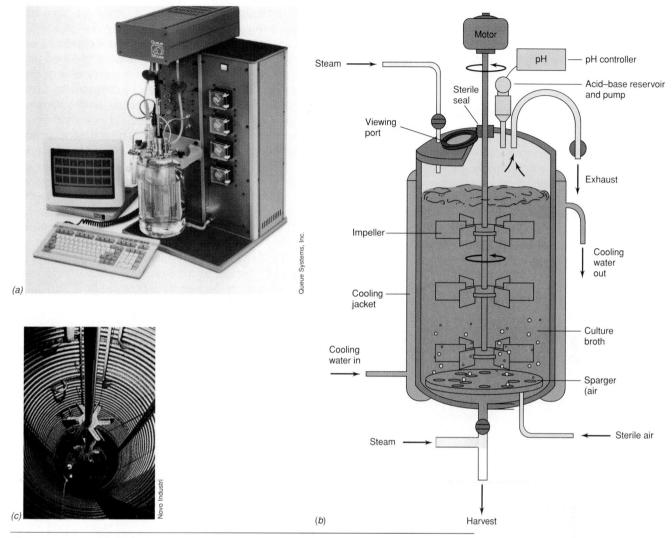

Figure 12.5 An industrial fermentor. (a) Photograph of a small research model. (b) Diagram of a fermentor, illustrating construction and facilities for aeration and process control. (c) Photograph of the inside of a large fermentor, showing the impeller and internal heating and cooling coils.

process and elaborate precautions must be taken to ensure proper aeration. Oxygen is poorly soluble in water, and in a fermentor with a high microbial population density, there is a tremendous oxygen demand by the culture. Two separate installations are used to ensure adequate aeration: an aeration device, called a *sparger,* and a stirring device called an *impeller* (Figure 12.5*b*). The sparger is a device, often a series of holes in a metal ring or a nozzle, through which filter-sterilized air can be passed into the fermentor under high pressure. The air enters the fermentor as a series of tiny bubbles from which the oxygen passes by diffusion into the liquid. A good sparging system should ensure that the bubbles are of very small size so diffusion of oxygen from the bubble into the liquid can occur readily.

In small fermentors use of a sparger alone may be sufficient to ensure adequate aeration, but in indus-

trial-size fermentors, *stirring* of the fermentor with an impeller is essential (Figure 12.5*c*). Stirring accomplishes two things: it mixes the gas bubbles through the liquid; and it mixes the organism through the liquid, thus ensuring uniform access of the microbial cells to the nutrients. One of the most common stirrers is a flat blade or flat disc, fastened to a center shaft, which is rotated at a high rate by a shaft attached to a motor. In order to ensure the most effective mixing by the impeller, *baffles* are generally installed vertically along the inside diameter of the fermentor. As it is stirred by the impeller, the liquid passes the baffles and is broken up into smaller patches. Fluid dynamics in fermentors is extremely complex but is important for the industrial microbiologist to understand because effective design and operation of a fermentor depend on adequate mixing.

The shaft that drives the impeller is attached to the motor by way of a drive shaft that must penetrate into the fermentor from outside. Because of the need to maintain sterility, it is vital that the seal connecting the drive shaft to the motor be arranged so that contaminants cannot pass through (see Figure 12.5*b*). A typical large-scale fermentor installation is shown in Figure 12.6.

Process control and monitoring

Any microbial process must be monitored to ensure that it is proceeding properly, but it is especially important that industrial fermentors be monitored carefully because there is such a major expense involved. In most cases, it is necessary not only to measure growth and product formation but also to *control* the process by altering environmental parameters as the process proceeds. Environmental factors that are frequently monitored include temperature, oxygen concentration, pH, cell mass, and product concentration.

(a)

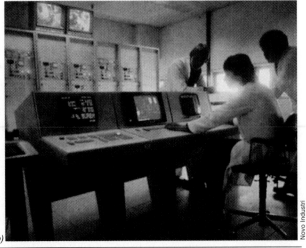

(b)

Figure 12.6 (a) A large industrial fermentation plant. Only the tops of the fermentors, which can be several stories high, are visible. (b) Computer control room for a large fermentation plant.

Computers play important roles in fermentor process control (Figure 12.6*b*). During the growth and product formation process in a large-scale fermentation, it is essential, if the fermentation is to be operated properly, that data be obtained on the process as it actually takes place. For instance, it may be desirable to change one of the environmental parameters as the fermentation progresses, or to feed a nutrient at a rate that exactly balances growth. The computer can be used to process the data on-line and then respond as directed by a program in regard to when and how much nutrient to add. In this way, nutrient is added when needed and not before, thus avoiding potential diversion of nutrient from the desired product into unwanted products.

Computers can also be used to *model* fermentation processes. Using a mathematical model, one can test the effect of various parameters on growth and product yield quickly and interactively and then make modifications in the parameters to see how they affect the process. In this way, many variations in the fermentation can be studied inexpensively at the computer terminal, rather than expensively at the pilot plant stage or in the industrial plant (see next section).

CONCEPT CHECK   **12.3**

Large-scale industrial fermentations present several engineering problems. In aerobic processes, ensuring adequate oxygen availability in a large industrial tank is a difficult problem, requiring installations for stirring and aeration. The microbial process must be continuously monitored in order to ensure satisfactory yields of the desired product.

✔ *What types of devices are used to ensure proper aeration in a large-scale fermentation?*

✔ *How are computers used to monitor large-scale fermentations?*

12.4

Fermentation Scale-up

One of the most important and complicated aspects of industrial microbiology is the transfer of a process from small-scale laboratory equipment to large-scale commercial equipment, a procedure called **scale-up**. An understanding of the problems of scale-up is extremely important because rarely does a microbial process behave the same way in large-scale fermentors as in small-scale laboratory equipment (Figure 12.7).

Why does a microbial process differ between large-scale and small-scale equipment? Mixing and aeration are much easier to accomplish in the small laboratory flask than in the large industrial fermentor. As the *size* of the equipment is increased, the *surface-to-*

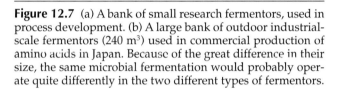

Figure 12.7 (a) A bank of small research fermentors, used in process development. (b) A large bank of outdoor industrial-scale fermentors (240 m³) used in commercial production of amino acids in Japan. Because of the great difference in their size, the same microbial fermentation would probably operate quite differently in the two different types of fermentors.

volume ratio changes (the principle of surface-to-volume ratio was explained in Figure 3.2); the large fermentor has much more volume for a given surface area. Because gas transfer and mixing depend more on surface exposed than on fermentor volume, it is obviously more difficult to mix a large tank than a small flask. Oxygen transfer especially is much more difficult to obtain in a large fermentor, again because of the different surface-to-volume ratio. Because most industrial fermentations are aerobic, effective oxygen transfer is essential. With the rich culture media used in industrial processes, a high biomass is obtained, leading to a high oxygen demand. If aeration is reduced, even for a short period, the culture may experience partial anoxic conditions, with serious consequences in terms of product yield. Scale-up of an industrial process is the task of the *biochemical engineer,* who is familiar with gas transfer, fluid dynamics, mixing, and thermodynamics.

The scale-up process

In transferring an industrial process from the laboratory to the commercial fermentor, several stages can be envisioned. (1) Experiments in the *laboratory flask* which are generally the first indication that a process of commercial interest is possible. (2) The *laboratory fermentor,* a small-scale fermentor, generally of glass and generally 5- to 10-liter size, in which the first efforts at scale-up are made (Figure 12.7*a*). In the laboratory fermentor, it is possible to test variations in medium, tem-

perature, pH, and so on, inexpensively because little cost is involved for either equipment or culture medium. (3) The *pilot plant stage,* usually carried out in equipment 300–3000 liters in size. Here, the conditions more closely approach the commercial scale; however, cost is not yet a major factor. In the pilot plant fermentor, careful instrumentation and computer control are desirable so the conditions most similar to those in the laboratory fermentor can be obtained. (4) The *commercial fermentor* itself, generally 10,000–500,000 liters (Figures 12.6*a* and 12.7*b*).

It is generally found in scale-up studies on aerobic fermentations that the oxygen transfer rate in the fermentor is best kept *constant* as the size of the fermentor is increased. Thus, if an oxygen transfer rate of 200 millimoles (mmol) O_2 liter^{-1}hr^{-1} is required to obtain optimal yield in a small fermentor, then stirring and aeration in the large fermentor should be adjusted to ensure this same rate. This requires more rapid stirring as well as a higher pressure of the inlet air as the culture becomes more dense. Because stirring is a mechanical process that can be monitored in terms of *power,* one approach to scale-up is to maintain constant power to the fermentor when going from small- to large-scale equipment.

We now consider the industrial production of microbial products, beginning with the antibiotics. Antibiotic production is a huge industry worldwide and one where many important principles of large-scale microbial cultures were first developed.

Scale-up is the process of gradually converting a useful industrial fermentation from laboratory scale to production scale. Aeration is a particularly critical aspect to monitor during scale-up studies.

✔ *What are the differences in size among a typical laboratory fermentor, a pilot plant fermentor, and a commercial fermentor?*

12.5

Antibiotics: Isolation and Characterization

Of the microbial products manufactured commercially, probably the most important are the antibiotics. As we discussed in Chapter 11, antibiotics are chemical substances produced by microorganisms that kill or inhibit the growth of other microorganisms. The development of antibiotics as agents for treatment of infectious disease has probably had more impact on the practice of medicine than any other single development.

Antibiotics are products of secondary metabolism. Although their yields are relatively low in most industrial fermentations, because of their high therapeutic activity and consequently high economic value, they can be produced commercially by microbial fermentation. Many antibiotics can be synthesized chemically, but because of the chemical complexity of the antibiotics and the great expense attendant on chemical synthesis, rarely is it possible for chemical synthesis to compete with microbial fermentation.

Commercially useful antibiotics are produced primarily by filamentous fungi and by Bacteria of the actinomycete group. A listing of the most important antibiotics produced by large-scale industrial fermentation is given in Table 12.3. Frequently, a number of chemically related antibiotics exist, and so *families* of antibiotics are known. Antibiotics can hence be classified according to their chemical structure, as was shown in Figure 11.17.

Search for new antibiotics

Over 8000 antibiotic substances are known, and several hundred antibiotics are discovered yearly. Are there more antibiotics waiting to be discovered? Almost certainly yes, because most of the microorganisms that have been examined for their ability to produce antibiotics are members of a few genera, such as *Streptomyces, Penicillium,* and *Bacillus.* Many antibiotic researchers believe that a vast number of new antibiotics will be discovered if other groups of microorganisms are examined. It also seems likely that genetic engineering techniques will permit the artificial construction of new antibiotics and that computer modeling will eventually replace traditional screening methods for new drugs. To some extent this is already occurring, as new more powerful computer methods are being used for *drug design.* The approach here is to use a computer to model interactions between a drug's target (for example, a specific protein) and a modified version of a known drug. The computer can then be used to screen various hypothetical (not yet existing) drugs and actually predict their efficacy. Particularly promising drugs identified in this way can then be synthesized by chemical or biological modifications of existing drugs and tested for clinical effectiveness.

However, the main way in which new antibiotics were discovered in the past was by *screening.* In the screening approach, a large number of isolates of possible antibiotic-producing microorganisms are obtained from nature in pure culture (Figure 12.8*a*), and these isolates are then tested for antibiotic production by seeing

TABLE 12.3	Some antibiotics produced commercially	
Antibiotic	**Producing microorganism**	**Type of microorganism**
Bacitracin	*Bacillus licheniformis*	Endospore-forming bacterium
Cephalosporin	*Cephalosporium* sp.	Fungus
Chloramphenicol	Chemical synthesis (formerly produced microbially by *Streptomyces venezuelae*)	
Cycloheximide	*Streptomyces griseus*	Actinomycete
Cycloserine	*Streptomyces orchidaceus*	Actinomycete
Erythromycin	*Streptomyces erythreus*	Actinomycete
Griseofulvin	*Penicillium griseofulvin*	Fungus
Kanamycin	*Streptomyces kanamyceticus*	Actinomycete
Lincomycin	*Streptomyces lincolnensis*	Actinomycete
Neomycin	*Streptomyces fradiae*	Actinomycete
Nystatin	*Streptomyces noursei*	Actinomycete
Penicillin	*Penicillium chrysogenum*	Fungus
Polymyxin B	*Bacillus polymyxa*	Endospore-forming bacterium
Streptomycin	*Streptomyces griseus*	Actinomycete
Tetracycline	*Streptomyces rimosus*	Actinomycete

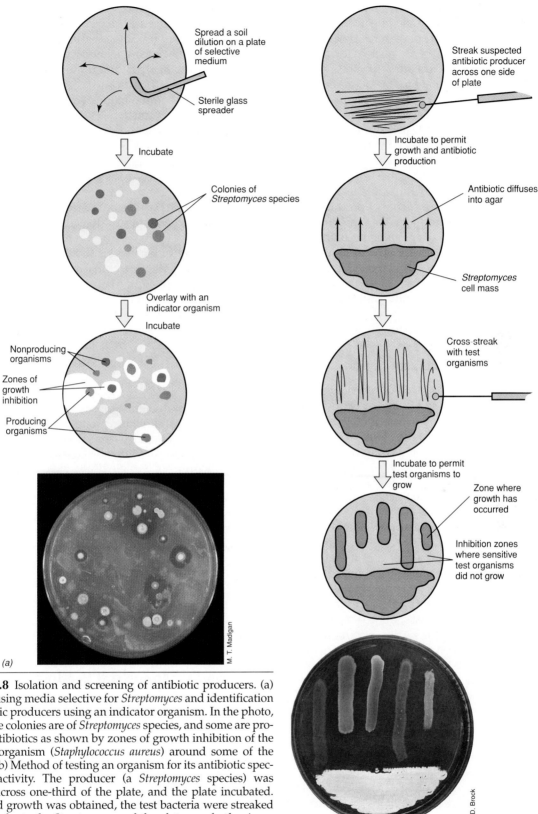

Figure 12.8 Isolation and screening of antibiotic producers. (a) Isolation using media selective for *Streptomyces* and identification of antibiotic producers using an indicator organism. In the photo, most of the colonies are of *Streptomyces* species, and some are producing antibiotics as shown by zones of growth inhibition of the indicator organism (*Staphylococcus aureus*) around some of the colonies. (b) Method of testing an organism for its antibiotic spectrum of activity. The producer (a *Streptomyces* species) was streaked across one-third of the plate, and the plate incubated. After good growth was obtained, the test bacteria were streaked perpendicular to the *Streptomyces* and the plate was further incubated. The failure of several organisms to grow near the mass growth of *Streptomyces* indicates that the *Streptomyces* produced an antibiotic active against these bacteria. Test organisms (left to right): *Escherichia coli, Bacillus subtilis, Staphylococcus aureus, Klebsiella pneumoniae, Mycobacterium smegmatis.*

whether they produce any diffusible materials that are inhibitory to the growth of test bacteria. The test bacteria used are selected from a variety of bacterial types but are chosen to be representative of or related to bacterial pathogens. The classical procedure for testing new microbial isolates for antibiotic production is the cross-streak method, first used by Fleming in his pioneering studies on penicillin (∞ Microbiology and "Magic Bullets," Chapter 11 and Figure 12.8b). Those isolates that show evidence of antibiotic production are then studied further to determine if the antibiotics they produce are new. In most screening programs, most of the isolates obtained produce *known* antibiotics, so the industrial microbiologist must quickly identify producers of known antibiotics and discard them. Once an organism producing a *new* antibiotic is discovered, the antibiotic is produced in sufficient amounts for structural analyses and then tested for toxicity and therapeutic activity in infected animals. Most new antibiotics *fail* these animal tests, but a few pass them successfully. Ultimately, only a very few of these new antibiotics prove to be medically useful and are produced commercially.

Steps toward commercial production

An antibiotic that is to be produced commercially must first be produced successfully in large-scale industrial fermentors. We have discussed in general the problems of scale-up earlier in this chapter. One of the most important tasks thereafter is the development of efficient purification methods. Because of the relatively small amounts of antibiotic present in the fermentation liquid, elaborate methods for extraction and purification of the antibiotic are necessary (Figure 12.9). If the antibiotic is soluble in an organic solvent that is immiscible in water, it may be relatively simple to purify the antibiotic by extracting it into a small volume of the solvent, thus concentrating the antibiotic. If the antibiotic is not solvent-soluble, then it must be removed from the fermentation liquid by adsorption, ion exchange, or chemical precipitation. In all cases, the goal is to obtain a crystalline product of high purity, although some antibiotics do not crystallize readily and are difficult to purify. A related problem is that cultures often produce other end products, including other antibiotics, and it is essential to end up with a product consisting of only a single chemical compound. The purification chemist may be required to develop methods for eliminating undesirable by-products, but in some cases it may be necessary for the microbiologist to find strains that do not produce such undesirable chemicals.

Rarely do antibiotic-producing strains just isolated from nature produce the desired antibiotic at sufficiently high concentration that commercial production can begin immediately. One of the main tasks of the industrial microbiologist is thus to isolate new *high-yielding strains*. The industrial microbiologist has made significant contributions to the antibiotic industry by developing high-yielding processes. As we noted earlier, the yield of penicillin has been greatly increased by strain selection and appropriate medium development. Strain selection involves mutagenesis of the initial culture, plating of mutant types, and testing of these mutants for antibiotic production. In most cases, mutants produce *less* antibiotic than the parent, so only rarely is a higher yielding strain obtained.

In recent years, the development of genetic engineering techniques has greatly improved the procedures for seeking high-yielding strains. The technique of *gene amplification* makes it possible to place additional copies of genes of interest into a cell by means of a vector such as a plasmid. Alterations in regulatory processes also may permit increased yields. However, one difficulty with using genetic procedures for increasing antibiotic yield is that the biosynthetic pathways for the synthesis of most antibiotics involve large numbers of steps with many genes (see Section 12.6), and it is not clear which genes should be altered or increased in number to increase yields. Thus, it is critical that the rate-limiting step in a given biochemical pathway be identified by basic research.

CONCEPT CHECK   *12.5*

The industrial production of antibiotics begins with screening for antibiotic producers. Once new producers are identified, purification and chemical analyses of the antimicrobial agent are made. If the new antibiotic is biologically active *in vivo*, the industrial microbiologist may seek high-yielding strains or may genetically modify the wild-type isolate to increase yields to levels acceptable for commercial development.

✔ *What is the natural habitat of most antibiotic-producing microorganisms?*

✔ *What is meant by the word* screening *in the context of finding new antibiotics?*

12.6

Antibiotics: Industrial Production

Once an antibiotic has been structurally characterized, has been proven medically effective in tests on experimental animals and sufficiently nontoxic, and, finally, has passed clinical trials (this sequence of events can take several years in actual practice), it is ready to be produced commercially and marketed. For antibiotics like penicillin, tetracycline, and streptomycin, these hurdles were passed long ago; today, literally tons of these antibiotics are produced for medical and veterinary use. We focus here on the industrial production of these three antibiotics as examples of antibiotic production in general.

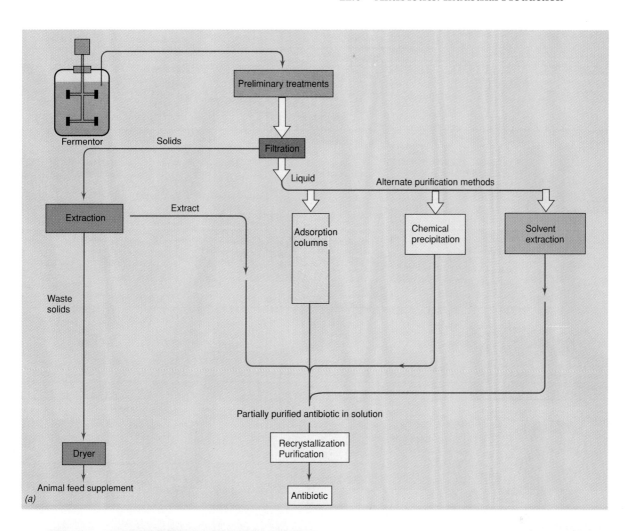

Figure 12.9 Purification of an antibiotic. (a) Overall process of extraction and purification. (b) Installation for the solvent extraction of an antibiotic from fermentation broth.

β-Lactam antibiotics: Penicillin and its relatives

The basic structure and mode of action of β-lactam antibiotics was discussed in Section 11.8. A variety of useful antibiotics have the basic β-lactam ring, characteristic of the penicillins, and the structures of these and the producing organisms are shown in Figure 12.10. Note that several prokaryotes as well as the more familiar eukaryotes (molds) like *Penicillium* and *Aspergillus*, produce β-lactam antibiotics (Figure 12.10). Among penicillins, several different types are produced and the final product may actually be a combination of substances produced biologically and chemically.

The basic structure of the penicillins is *6-aminopenicillanic acid* (6-APA), which consists of a thiazolidine ring with a condensed β-lactam ring (Figures 12.10 and 12.11). The 6-APA carries a variable side chain in position 6. If the penicillin fermentation is carried out without addition of side-chain precursors, the **natural penicillins,** such as benzylpenicillin (penicillin G), are produced (Figure 12.11). The fermentation can be better

Basic structures		Antibiotics	Most important producing species
Penam 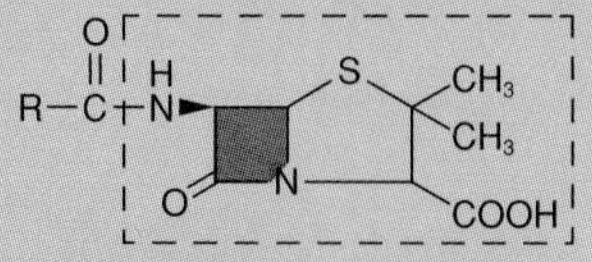		Penicillins (dashed lines outline 6-aminopenicillanic acid)	*Penicillium chrysogenum* *Aspergillus nidulans* *Cephalosporium acremonium* *Streptomyces clavuligerus*
Ceph-3-em 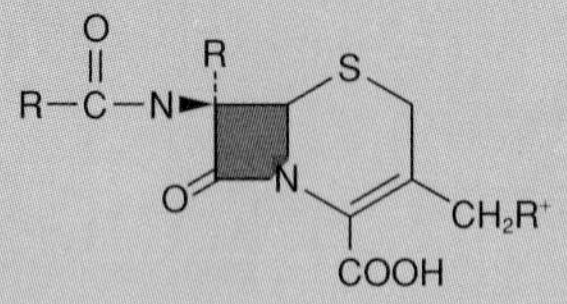		Cephalosporins	*Cephalosporium acremonium* *Nocardia lactamdurans* *Streptomyces clavuligerus*
Clavam 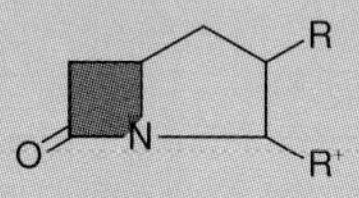		Clavulanic acids	*Streptomyces clavuligerus*
Carbapenem 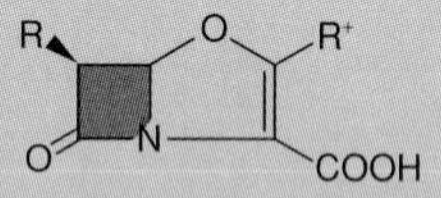		Thienamycins Olivanic acids Epithienamycins	*Streptomyces cattleya* *Streptomyces olivaceus* *Streptomyces flavogriseus*
Monolactam 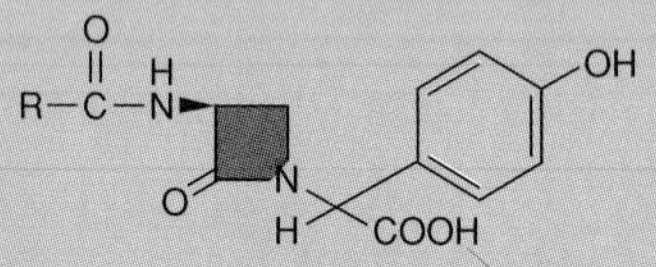		Nocardicins	*Nocardia uniformis* subsp. *tsuyamanesis*
		Monobactams	*Gluconobacter* sp. *Chromobacterium violaceum* *Agrobacterium radiobacter* *Pseudomonas acidophila* *Pseudomonas mesoacidophila* *Flexibacter* sp. *Acetobacter* sp.

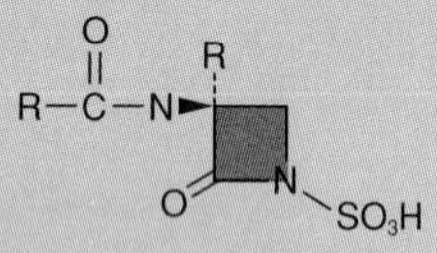

Figure 12.10 The basic structures of the naturally occurring β-lactam antibiotics and the major producing organisms. The positions where chemical substitutions can occur are indicated by R. The β-lactam ring is shown in red.

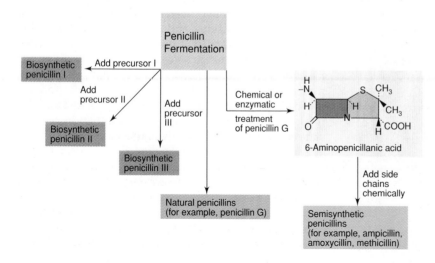

6-Aminopenicillanic acid

Figure 12.11 Industrial production of penicillins. The normal fermentation leads to the natural penicillins. If specific precursors are added during the fermentation, various biosynthetic penicillins are formed. Semisynthetic penicillins are produced by chemically adding a specific side chain to the 6-aminopenicillanic acid nucleus.

controlled by adding to the broth a *side-chain precursor* so only one desired penicillin is produced. The product formed under these conditions is referred to as a **biosynthetic penicillin** (Figure 12.11). However, in order to produce the most useful penicillins, those with activity against *gram-negative Bacteria,* a combined fermentation and chemical approach is used that leads to the production of **semisynthetic penicillins** (Figure 12.11). In this case, microbially produced benzylpenicillin is split either chemically or enzymatically to yield 6-APA and the latter chemically modified by the addition of a side chain (Figure 12.11). Semisynthetic penicillins have many significant clinical advantages in terms of their spectrum of activity and the fact that many of them, for example, ampicillin, can be taken orally and thus do not require injection. For these reasons, semisynthetic penicillins make up the bulk of the penicillin market today.

Production methods for β-lactam antibiotics

Penicillin G is produced using a submerged fermentation process in 40,000- to 200,000-liter fermentors. Penicillin production is a highly aerobic process, and efficient aeration is necessary. Penicillin is a typical secondary metabolite, as was illustrated in Figure 12.3*b*. During the growth phase (trophophase), very little penicillin is produced, but once the carbon source has been nearly exhausted (idiophase), the penicillin production phase begins (Figure 12.12). By feeding with various culture medium components, the production phase can be extended for several days (Figure 12.12).

A major ingredient of most penicillin production media is **corn steep liquor.** This substance contains the nitrogen source as well as other growth factors. The carbon source is generally *lactose* (Figure 12.12). The side chain of penicillin G is the phenylacyl moiety, and yields of penicillin G are markedly increased if phenylacetic acid is fed as a precursor. Penicillin G is excreted into the medium, and after the cells are removed by filtration, the pH of the medium is lowered and the antibiotic extracted from the filtered broth with amyl or butyl acetate. After concentration into the solvent, the antibiotic is back-extracted into an alkaline aqueous medium, concentrated further, and crystallized. Highly purified penicillin can be readily obtained in this way.

Cephalosporins are β-lactam antibiotics containing a dihydrothiazine instead of a thiazolidine ring system (Figure 12.10). Cephalosporins were first discovered as products of the fungus *Cephalosporium acremonium,* but a number of other fungi also produce antibiotics with this ring system. In addition, a number of semisynthetic cephalosporins are produced. Cephalosporins are valued clinically not only because of their low toxicity but also because they are broad-spectrum antibiotics (∞ Section 11.9).

Intensive screening for new β-lactam antibiotics has led to the development of compounds whose struc-

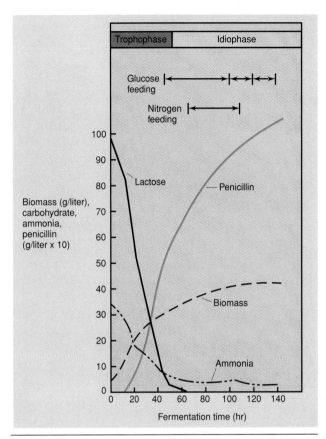

Figure 12.12 Kinetics of the penicillin fermentation with *Penicillium chrysogenum.* Note how the production of penicillin occurs as cells are entering stationary phase and most of the carbon and nitrogen is exhausted. Nutrient "feedings" keep penicillin production high.

tures are different from those of both the penicillins and the cephalosporins. Included in this category are *nocardicin, clavulanic acid,* and *thienamycin* (Figure 12.10). Clavulanic acid is of particular interest because, although it is not very effective as an antibiotic by itself, it inhibits the activity of β-lactamases. These enzymes are produced by certain bacteria and function to destroy β-lactam antibiotics, rendering them ineffective in treating disease (∞ Section 11.13). Thus, when used in combination with β-lactamase-sensitive penicillins and cephalosporins, clavulanic acid causes a distinct increase in the activity of these antibiotics.

Production of aminoglycosides

The structures and modes of action of the aminoglycoside and tetracycline antibiotics were discussed in Section 11.9. We focus here on their industrial production. The production of streptomycin by the bacterium *Streptomyces griseus* is an extremely complex process (Figure 12.13). However, one of the most interesting features of streptomycin synthesis is its *regulation.* As seen in Figure 12.13, the three parts of the strepto-

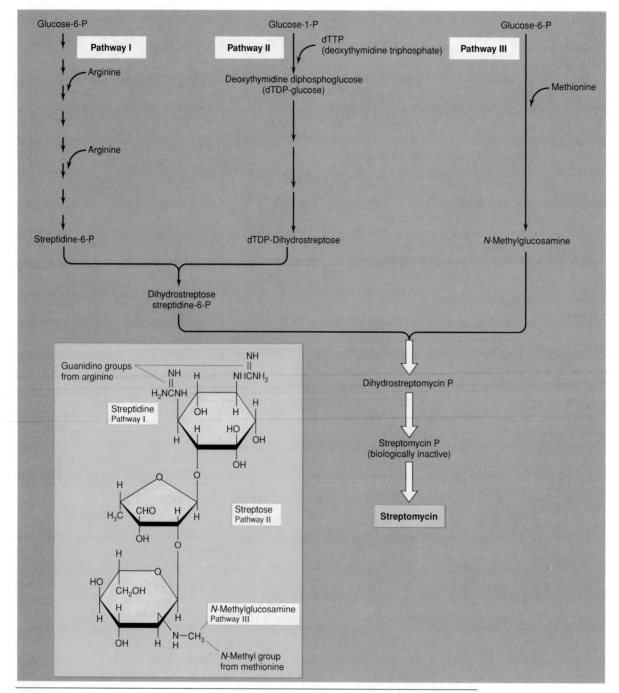

Figure 12.13 Streptomycin biosynthesis in *Streptomyces griseus*. The three pathways leading to the streptomycin precursors streptidine, streptose, and *N*-methylglucosamine are shown on top. The structure of streptomycin is shown at the left.

mycin molecule are synthesized by separate biochemical pathways and the subunits brought together at the end. The final intermediate in the pathway, streptomycin P, is biologically inactive but becomes active following removal of the phosphate (Figure 12.13).

Streptomycin is synthesized as a typical secondary metabolite. One aspect of the regulation involves the production of an inducer called Factor A (Figure 12.14).

Factor A is not chemically related to streptomycin but is instead involved in carbohydrate metabolism. Key enzymes in streptomycin biosynthesis are not synthesized until the Factor A concentration builds up, thus explaining how this substance might function as a trigger of secondary metabolism. During the growth phase, Factor A is excreted and gradually builds up in the medium. However, only when Factor A concentra-

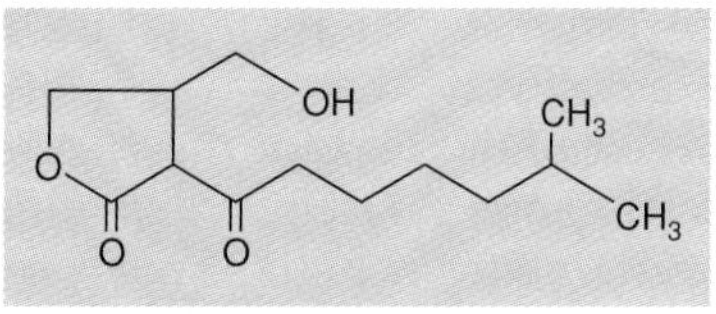

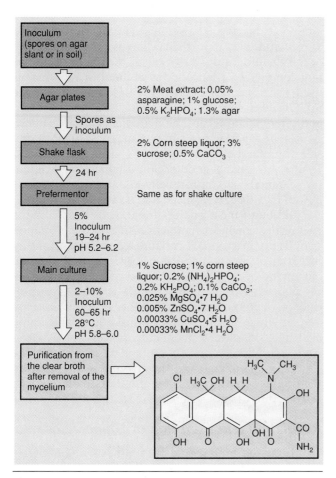

Figure 12.14 Structure of Factor A from *Streptomyces griseus*.

tion reaches a critical level does it induce synthesis of the key streptomycin biosynthesis enzymes. Factor A is not itself a precursor of streptomycin, as shown by its structure (Figure 12.14) and by the fact that the addition of a tiny amount of pure Factor A, *1 μg*, to cultures of a Factor A-negative mutant can lead to the production of over *1 g* of streptomycin. Although other regulatory processes are important in streptomycin biosynthesis, Factor A is a major controlling feature.

Production of tetracyclines

The biosynthesis of a tetracycline involves a large number of enzymatic steps. In the case of chlortetracycline (Figure 12.15), as many as 72 intermediate products may

be involved, most of which are known in only a general way. Studies on the genetics of *Streptomyces aureofaciens*, the producer of chlortetracycline, have shown that over 300 genes are involved! With such a large number of genes, regulation of biosynthesis of this antibiotic is obviously quite complex. However, a few regulatory signals are known and production schemes are well worked out. For example, repression of chlortetracycline synthesis by both glucose and phosphate is known to occur. Phosphate repression is especially significant, and so the medium used in commercial production contains low phosphate concentrations. A production scheme for chlortetracycline is shown in Figure 12.15. Note that as in penicillin production, corn steep liquor is used in the large-scale production of chlortetracycline and the use of glucose avoided; chlortetracycline production is typically run with sucrose as the carbon source. Glucose is avoided because it causes catabolite repression (∞ Section 7.6) of antibiotic production.

CONCEPT CHECK   *12.6*

Major antibiotics of clinical significance include the β-lactam antibiotics penicillin and cephalosporin, and the aminoglycoside and tetracycline antibiotics. All these antibiotics are typical secondary metabolites, and their industrial production is well worked out despite the fact that the biochemistry and genetics of their biosynthesis are only partially understood.

✔ *What chemical structure is common to all penicillins?*

✔ *In terms of penicillin production, what is meant by the term semisynthetic?*

✔ *Why is clavulanic acid of importance to clinical medicine?*

12.7
Vitamins and Amino Acids

Vitamins and amino acids are growth factors that are often used pharmaceutically or are added to foods. Several important vitamins and amino acids are produced commercially by microbial processes.

Vitamins

Vitamins are used as supplements for human food and animal feeds. Production of vitamins is second only to that of antibiotics in terms of total sales of pharmaceuticals—nearly $1 billion per year. Most vitamins are made commercially by chemical synthesis. However, a few are too complicated to be synthesized inexpensively but can be made by microbial fermentation.

Figure 12.15 Production scheme for chlortetracycline with *Streptomyces aureofaciens*. The structure of chlortetracycline is shown on the bottom right.

Vitamin B_{12} and riboflavin are the most important of this class of vitamins.

Vitamin B_{12} (Figure 12.16a) is synthesized in nature exclusively by microorganisms. As a coenzyme, vitamin B_{12} plays an important role in animal biochemistry in various intramolecular rearrangements in which a hydrogen atom on one carbon atom and a substituent on the adjacent carbon atom exchange places. In humans, a major deficiency of vitamin B_{12} leads to a severe condition called *pernicious anemia,* characterized by low production of red blood cells and nervous system disorders. The requirements of animals for vitamin B_{12} are satisfied by food intake or by absorption of the vitamin produced in the gut of the animal by intestinal microorganisms. Plants do not produce or use vitamin B_{12}.

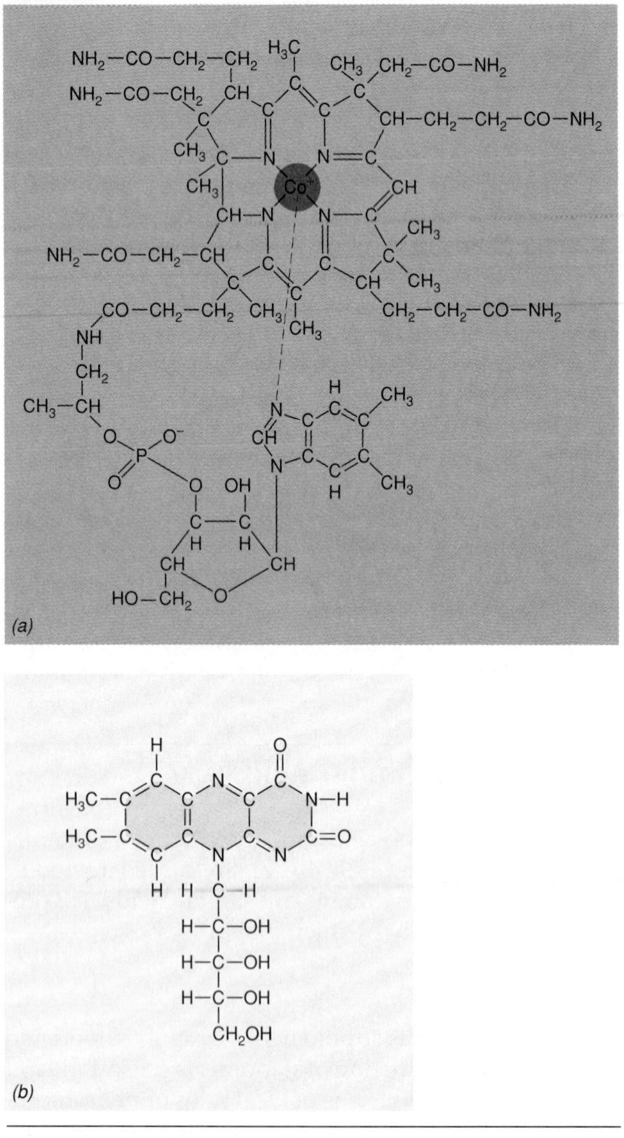

(a)

(b)

Figure 12.16 Vitamins produced by microorganisms on an industrial scale. (a) Vitamin B_{12}. Shown is the structure of cobalamin; note the central cobalt atom. The coenzyme form of vitamin B_{12} contains a deoxyadenosyl group attached to Co above the plane of the ring. (b) Riboflavin (vitamin B_2).

For industrial production of vitamin B_{12}, microbial strains are employed that have been specifically selected for their high yields of the vitamin. Members of the bacterial genus *Propionibacterium* give yields of the vitamin ranging from 19 to 23 mg/l in a two-stage process, and another bacterium, *Pseudomonas denitrificans,* produces 60 mg/l in a one-stage process that uses sugarbeet molasses as the carbon source. Vitamin B_{12} contains cobalt as an essential part of its structure (Figure 12.16a), and yields of the vitamin are greatly increased by addition of cobalt to the culture medium.

Riboflavin (Figure 12.16b) is the parent compound of the flavins, FAD and FMN, coenzymes that play important roles in enzymes involved in oxidation–reduction reactions in virtually all organisms (∞ Section 4.12). Riboflavin is synthesized by many microorganisms, including bacteria, yeasts, and fungi. The fungus *Ashbya gossypii* produces a huge amount of this vitamin (up to 7 g/l) and is therefore used for most of the microbial production processes. In spite of this good yield, there is great economic competition between this microbiological process and chemical synthesis.

Amino acids

Amino acids have extensive uses in the food industry, as feed additives, in medicine, and as starting materials in the chemical industry (Table 12.4). The most important commercial amino acid is **glutamic acid,** which is used as a flavor enhancer [monosodium glutamate (MSG)]. Two other important amino acids, **aspartic acid** and **phenylalanine,** are the ingredients of the artificial sweetener **aspartame,** an important constituent of diet soft drinks and other foods sold as sugar-free products. **Lysine,** an essential amino acid for humans and certain farm animals, is commercially produced by the bacterium *Brevibacterium flavum* for use as a food additive.

Although most of the amino acids can be made chemically, chemical synthesis results in the formation of optically inactive D, L mixtures. If the biochemically important L form (∞ Section 2.9) is desired, then an enzymatic or microbiological method of manufacturing is needed (see Figure 12.7b). Microbiological production of amino acids can be either by *direct fermentation,* in which the microorganism produces the amino acid in a standard fermentation process, or by *enzymatic synthesis,* in which the microorganism is the source of an enzyme and the enzyme is then used in the production process. The present discussion will be restricted to direct fermentation processes.

Regulation and excretion of amino acids

We discussed amino acid biosynthesis in Chapter 4 (∞ Section 4.19), and the regulation of enzymatic pathways by both feedback inhibition and repression in Chapter 7 (∞ Sections 7.1 and 7.3). Because amino acids are used by microorganisms as building blocks of proteins, strict cellular regulation of their produc-

TABLE 12.4 Amino acids used in the food industry[a]

Amino acid[b]	Annual production worldwide (metric tons)	Uses	Purpose
L-Glutamate (monosodium glutamate, MSG)	370,000	Various foods	Flavor enhancer; meat tenderizer
L-Aspartate and alanine	5,000	Fruit juices	"Round off" taste
Glycine	6,000	Sweetened foods	Improve flavor; starting point for organic syntheses
L-Cysteine	700	Bread	Improves quality
		Fruit juices	Antioxidant
L-Tryptophan + L-Histidine	400	Various foods, dried milk	Antioxidant, prevents rancidity; nutritive additive
Aspartame (made from L-phenylalanine + L-aspartic acid)	7,000	Soft drinks	Low calorie sweetener
L-Lysine	70,000	Bread (Japan), feed additives	Nutritive additive
DL-Methionine	70,000	Soy products, feed additives	Nutritive additive

[a]Data from Glazer, A. N., and H. Mikaido. 1995. *Microbial Biotechnology.* W. H. Freeman, New York.
[b]The structures of these amino acids are shown in Figure 2.12.

tion generally occurs. However, for the industrial production of an amino acid, ways to circumvent these regulatory mechanisms are necessary in order to obtain an *overproducing strain* capable of producing the amino acid economically. Such overproducing strains have been developed, and we discuss here a typical example, *Brevibacterium flavum,* used in the production of the amino acid *lysine* (Figure 12.17).

The production of lysine in *Brevibacterium flavum* is biochemically controlled at the level of the enzyme aspartokinase, in that excess lysine feedback inhibits activity of this enzyme (Figure 12.17a) (the general phenomenon of feedback inhibition was described in Section 6.1). However, overproduction of lysine can be obtained by isolating mutants of *B. flavum* in which aspartokinase is no longer subject to feedback inhibition. This is done by isolating mutants resistant to the lysine analog S-aminoethylcysteine (AEC), which binds to the allosteric site of aspartokinase and shuts down activity of the enzyme (Figure 12.17b). AEC-resistant mutants, which are easily obtained by positive selection, produce a mutant form of aspartokinase with an allosteric site that no longer recognizes AEC *or* lysine, and thus feedback inhibition by lysine is greatly reduced. Such mutants of *B. flavum* can produce over 60 g of lysine per liter in industrial scale reactors (see Figure 12.7b).

Another factor important in the commercial production of amino acids is *excretion.* In general, organisms do not excrete essential metabolites such as amino acids. By arranging for excretion, high concentrations that might cause feedback inhibition or repression even in resistant mutants usually do not occur inside the cells. Excretion is stimulated in various ways, and an interesting procedure for obtaining high excretion of the amino acid *glutamic acid* has been worked out. Production and excretion of excess glutamic acid is dependent on cell permeability. The industrial organism producing glu-

tamic acid, *Corynebacterium glutamicum,* requires the vitamin biotin, an essential cofactor in fatty acid biosynthesis (∞ Section 4.21). Deficiency in biotin leads to membrane damage (as a result of poor phospholipid production), and under these conditions intracellular glutamic acid is excreted. The medium used for commercial glutamic acid production thus contains sufficient biotin to obtain good growth, after which biotin deficiency is imposed and glutamic acid is excreted.

CONCEPT CHECK 12.7

Microbial products used in foods and as food supplements include vitamins and amino acids. Vitamins produced microbially include vitamin B_{12} and riboflavin, whereas the most important amino acids produced commercially are glutamic acid, aspartic acid, phenylalanine, and lysine. High yields of amino acids are obtained by modifying regulatory signals that control synthesis of the particular amino acid such that overproduction occurs.

✔ *What amino acid is commercially produced in the greatest amounts?*

✔ *How can overcoming feedback inhibition improve the yield of an amino acid?*

12.8

Microbial Bioconversion

One of the most far-reaching discoveries in industrial microbiology was the understanding that microorganisms can be used to carry out specific chemical reactions beyond the capabilities of organic chemistry. The use of microorganisms for this purpose is called **bio-**

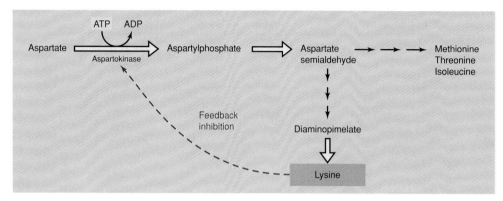

(a)

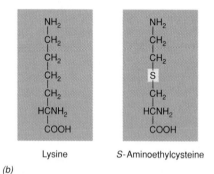

Lysine   S-Aminoethylcysteine

(b)

Figure 12.17 Industrial production of lysine using *Brevibacterium flavum*. (a) Biochemical pathway leading from aspartate to lysine; note that lysine can feedback-inhibit (∞ Section 7.1) activity of the enzyme aspartokinase, leading to cessation of lysine production. (b) Structure of lysine and the lysine analog *S*-aminoethylcysteine (AEC). AEC normally inhibits growth, but AEC-resistant mutants of *B. flavum* have an altered allosteric site on their aspartokinase and grow and overproduce lysine because feedback inhibition no longer occurs.

conversion or **biotransformation** and involves growth of the organism in large fermentors, followed by the addition at an appropriate time of the chemical to be converted. Following a further incubation period during which the chemical is acted on by the organism, the fermentation broth is extracted and the desired product purified. Although in principle bioconversion may be used for a wide variety of processes, its major practical use has been in the production of certain steroid hormones (Figure 12.18).

We discussed the role of sterols in eukaryotic membranes in Section 3.3. Steroids, which are derivatives of sterols, are important hormones in animals that regulate various metabolic processes. Some steroids are also used as drugs in human medicine. Members of one group, the *adrenal cortical steroids*, reduce inflammation and hence are effective in controlling the symptoms of arthritis and allergy. Members of another group, the estrogens and androgenic steroids, are involved in human fertility, and some of them can be used in the control of fertility. Steroids can be obtained by complete chemical synthesis, but this is a complicated and expensive process. Certain key steps in chemical synthesis can be carried out more efficiently by microorganisms, and commercial production of steroids usually has at least one microbial step.

Cortisone and hydrocortisone

In the production of hydrocortisone and cortisone, steroids used to reduce swelling and itching from minor skin irritations, the fungus *Rhizopus nigricans* carries out

a key stereospecific hydroxylation of a cortisone precursor (Figure 12.18). Most steroid bioconversions involve hydroxylations of this type, and a variety of different fungi are used industrially to carry out one or another specific hydroxylation. Steroid production is currently a big business, as worldwide sales of the four major steroids, hydrocortisone, cortisone, prednisone and prednisolone, amount to over 800 tons/year.

CONCEPT CHECK   **12.8**

Microbial bioconversion employs microorganisms to carry out a specific step or steps in an otherwise strictly chemical synthesis.

✔ *Give an example of a microbial bioconversion. Why is this bioconversion necessary?*

12.9

Enzymes

Each organism produces a large variety of enzymes, most of which are made in only small amounts and are involved in cellular processes. However, certain enzymes are produced in much larger amounts by some organisms, and instead of being held within the cell, they are excreted into the medium. Extracellular enzymes are usually capable of digesting insoluble nutrient materials such as cellulose, protein, and starch, the products of digestion then being transported into

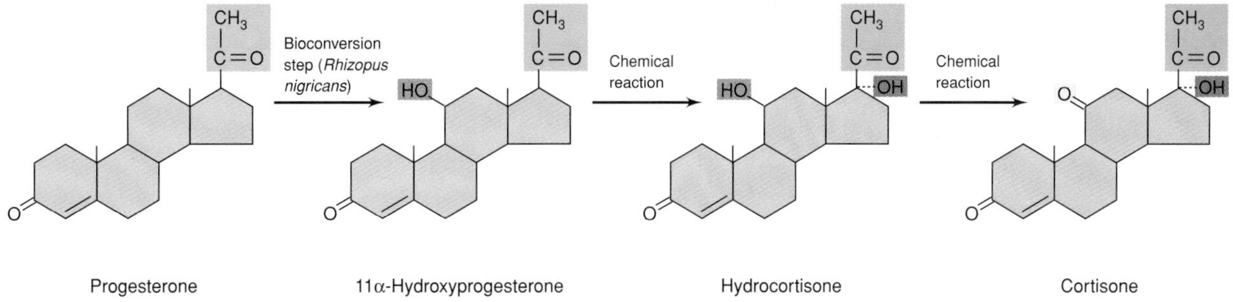

Figure 12.18 Cortisone production using a microorganism. The first reaction is a typical microbial bioconversion, the formation of 11α-hydroxyprogesterone from progesterone. This highly specific oxidation, carried out by the fungus *Rhizopus nigricans*, bypasses a difficult chemical synthesis. All the other steps, from progesterone to the steroid hormone cortisone, are performed chemically.

the cell where they are used as nutrients for growth. Some of these extracellular enzymes are used in the food, dairy, pharmaceutical, and textile industries and are produced in large amounts by microbial synthesis (Table 12.5). They are especially useful because they often act on single chemical functional groups, they easily distinguish between similar functional groups on a single molecule, and in many cases, they catalyze reactions in a stereospecific manner producing only one of two possible enantiomers (for example, a D-sugar or an L-amino acid; ∞ Section 2.9).

Enzymes are produced commercially from both fungi and bacteria. The production process is usually aerobic, and culture media similar to those used in antibiotic fermentations are employed. The enzyme itself is generally formed in only small amounts during the active growth phase but accumulates in large amounts during the stationary phase of growth. As we have seen (∞ Section 7.3), induced enzymes are produced only when an appropriate inducer is present in the medium. The potential for the production of useful enzymes has improved markedly in recent years because of the increased ease with which genes can be manipulated.

Proteases and amylases

The microbial enzymes produced in the largest amounts on an industrial basis are the bacterial proteases, used as additives in laundry detergents. Most laundry detergents today contain enzymes, chiefly proteases but also amylases, lipases, reductases, and others. Many of these enzymes are isolated from alkaliphilic bacteria (∞ Section 5.9), mainly species of *Bacillus* like *Bacillus licheniformis* (Table 12.5). These enzymes, which have pH optima between 9 and 10, remain active at the alkaline pH of laundry detergent solutions.

Other important enzymes manufactured commercially are amylases and glucoamylases, which are used in the production of glucose from starch. The glucose so produced can then be attacked by glucose isomerase to produce fructose (which is sweeter than either glucose or sucrose), resulting in the final production of a high fruc-

tose sweetener from corn, wheat, or potato starch. The use of this process in the food industry has been increasing, especially in the production of soft drinks.

Three reactions, each catalyzed by a separate microbial enzyme, operate in sequence in the conversion of cornstarch into the final product called **high fructose corn syrup.**

1. The enzyme **α-amylase** brings about the initial attack on the starch polysaccharide, shortening the chain and reducing the viscosity of the polymer. This is called the *thinning reaction.*

2. The enzyme **glucoamylase** produces glucose monomers from the shortened polysaccharides, a process called *saccharification.*

3. The enzyme **glucose isomerase** brings about the final conversion of glucose to fructose, a process called *isomerization.*

All three enzymes are produced industrially by microbial fermentation. The end product of this series of reactions is a syrup containing about equal amounts of glucose and fructose, which can be added directly to soft drinks and other food products, thereby greatly increasing their sweetness.

Extremozymes: Enzymes from prokaryotes that inhabit extreme environments

In Chapter 5 we considered aspects of microbial growth at high temperature and discovered that some prokaryotes, called *hyperthermophiles,* grow optimally at very high temperatures including, in some cases, above the boiling point of water. Hyperthermophiles are able to grow at such high temperatures because they produce heat-stable macromolecules (∞ Section 5.8) including enzymes, some of which catalyze the reactions shown in Table 12.5 but do so at very high temperatures. The term *extremozyme* has been coined to refer to enzymes that function at extremely high temperature (or enzymes that function optimally under any environmental extreme; for example, in the cold, in very high salt, or at very acid

TABLE 12.5	Microbial enzymes and their applications		
Enzyme	**Source**	**Application**	**Industry**
Amylase (starch-digesting)	Fungi	Bread	Baking
	Bacteria	Starch coatings	Paper
	Fungi	Syrup and glucose manufacture	Food
	Bacteria	Cold-swelling laundry starch	Starch
	Fungi	Digestive aid	Pharmaceutical
	Bacteria	Removal of coatings (desizing)	Textile
	Bacteria	Removal of stains; detergents	Laundry
Protease (protein-digesting)	Fungi	Bread	Baking
	Bacteria	Spot removal	Dry cleaning
	Bacteria	Meat tenderizing	Meat
	Bacteria	Wound cleansing	Medicine
	Bacteria	Desizing	Textile
	Bacteria	Household detergent	Laundry
Invertase (sucrose-digesting)	Yeast	Soft-center candies	Candy
Glucose oxidase	Fungi	Glucose removal, oxygen removal	Food
		Test paper for diabetes	Pharmaceutical
Glucose isomerase	Bacteria	High fructose corn syrup	Soft drink
Pectinase	Fungi	Pressing, clarification	Wine, fruit juice
Rennin	Fungi	Coagulation of milk	Cheese
Cellulase	Bacteria	Fabric softening, brightening; detergent	Laundry
Lipase	Fungi	Breaks down fat	Dairy, laundry
Lactase	Fungi	Breaks down lactose to glucose and galactose	Dairy, health foods
DNA polymerase	Bacteria	DNA replication in polymerase chain	Biological research;
	Archaea	reaction (PCR) technique (∞ Section 10.9)	forensics

or alkaline pH), and the organisms that produce them *extremophiles,* to indicate that they are organisms that grow best under conditions unsuitable for most microorganisms.

Because many industrial processes operate best at high temperatures, extremozymes from hyperthermophiles are becoming increasingly attractive as biocatalysts for the industrial applications shown in Table 12.5 and also for many research applications that require enzymes. Besides the *Taq* and *Pfu* DNA polymerases for use in the polymerase chain reaction (PCR) described in Sections 5.8 and 10.9, extremely thermostable proteases, amylases, cellulases, pullulanases (Figure 12.19), and xylanases have been isolated and characterized from various hyperthermophiles. The pullulanase from the hyperthermophile *Thermococcus litoralis,* which is related to the organism *Pyrococcus woesei* (Figure 12.19), is catalytically most active at a temperature of 118°C and ferredoxin (an iron–sulfur protein involved in electron transfer reactions; ∞ Section 4.12) from *Pyrococcus furiosus* is active at similar temperatures and does not denature until 140°C! Such temperature-resistant biocatalysts as well as extremozymes that are cold-active (from psychrophiles), active in the presence of high salt (from halophiles) or active at high or low pH (from alkaliphiles and acidophiles, respectively) will undoubtedly find more industrial applications in the coming years in situations that call for biocatalysis under extreme conditions. Indeed, the great specificity of enzymes and their ability to distinguish between

chiral isomers make those that also function at environmental extremes particularly important to the chemical industry.

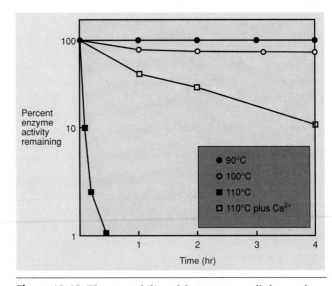

Figure 12.19 Thermostability of the enzyme pullulanase from the hyperthermophile *Pyrococcus woesei.* Purified pullulanase was incubated at the temperatures shown and then assayed for enzyme activity. The enzyme is strongly activated by Ca^{2+}. Pullulanase converts starch to small oligosaccharides and could find industrial application in high temperature starch-digesting processes. The growth temperature optimum of *P. woesei* is near 100°C. Data from Rudiger, A., P. L. Jorgensen, and G. Anthranikian. 1995. *Appl. Environ. Microbiol. 61:*567–575.

Immobilized enzymes

For use in industrial processes, it is frequently desirable to convert soluble enzymes into some sort of immobilized state. Immobilization not only makes it easier to carry out the enzymatic reaction under large-scale conditions but also generally stabilizes the enzyme to denaturation. There are three basic approaches to enzyme immobilization (Figure 12.20):

1. **Cross-linkage (polymerization)** of enzyme molecules. Linkage of enzyme molecules with each other is usually done by chemical reaction with a bifunctional cross-linking agent such as glutaraldehyde. Cross-linking of enzymes involves the chemical reaction of amino groups of the enzyme protein with glutaraldehyde (Figure 12.20b). If the reaction is carried out properly, the enzyme molecules can be linked in such a way that most enzymatic activity is maintained.

2. **Bonding** of the enzyme to a carrier. The bonding can be through adsorption, ionic bonding, or covalent bonding. Carriers used include modified celluloses, activated carbon, clay minerals, aluminum oxide, and glass beads (Figure 12.20a).

3. **Enzyme inclusion,** which involves incorporation of the enzyme into a *semipermeable membrane.* Enzymes can be enclosed in microcapsules, gels, semipermeable polymer membranes, or fibrous polymers such as cellulose acetate (Figure 12.20a).

Each of these methods has advantages and disadvantages, and the procedure used depends on the enzyme and on the particular industrial application.

Immobilized cells

In some cases it is not necessary to use purified enzyme. Rather, enzyme-rich *cells* can themselves be immobilized and the industrial process operated continuously. An example is the immobilization of glucose isomerase–containing cells of *Bacillus coagulans* in the production of high fructose corn syrup. The glucose syrup is passed through columns containing the immobilized cells and fructose syrup is produced. Much smaller installations are generally necessary for such continuous-flow processes.

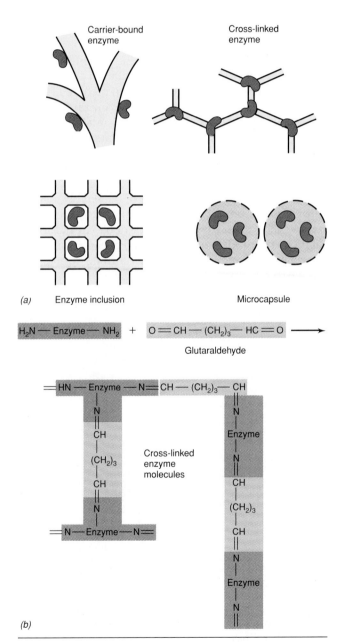

Figure 12.20 Immobilized enzymes. (a) Procedures for the immobilization of enzymes. (b) Procedure for cross-linking with glutaraldehyde.

CONCEPT CHECK   *12.9*

Microorganisms are ideal for the large-scale production of enzymes. Many enzymes are used in the laundry industry to remove stains from clothing, and thermostable enzymes have many advantages in these markets. The production of high fructose corn syrup involves the participation of three microbial enzymes, of which glucose isomerase (which converts glucose to the sweeter sugar fructose) is the most important. When an enzyme is used in a large-scale process, it is usually desirable to immobilize it by chemically bonding it to an inert substrate.

✔ *How are enzymes of use in the laundry industry?*

✔ *What reaction does the enzyme glucose isomerase carry out?*

✔ *What is an* extremozyme?

✔ *How is the chemical glutaraldehyde used in the enzyme industry?*

12.10

Vinegar

Vinegar is the product resulting from the conversion of ethyl alcohol to acetic acid by **acetic acid bacteria,** members of the genera *Acetobacter* and *Gluconobacter.* The word *vinegar* is English but is derived from the French word *vinaigre,* meaning, literally, "sour wine." Vinegar can be produced from any alcoholic substance, although the usual starting material is wine or alcoholic apple juice (cider). Vinegar can also be produced from a mixture of pure alcohol in water, in which case it is called *distilled vinegar,* the term *distilled* referring to the alcohol from which the product is made rather than the vinegar itself. Vinegar is used as a flavoring ingredient in salads and other foods, and because of its acidity, it is also used in pickling. Meats and vegetables properly pickled in vinegar can be stored unrefrigerated for years.

The aerobic *acetic acid bacteria* are an interesting group of bacteria (∞ Section 16.17); however, do not confuse these aerobic acetic acid bacteria with the anaerobic *homoacetogenic* bacteria (∞ Section 16.9). The aerobic acetic acid bacteria differ from most other aerobes in that they do not oxidize their energy sources completely to CO_2 and water (Figure 12.21). Thus, when provided with ethyl alcohol as electron donor, they oxidize it to only acetic acid, which accumulates in the medium. Acetic acid bacteria are quite acid-tolerant and are not killed by the acidity that they produce. There is a high oxygen demand during growth, and the main problem in the production of vinegar is to ensure sufficient aeration of the medium.

Vinegar production

There are three different processes for the production of vinegar. The **open-vat** or **Orleans method** was the original process and is still used in France where it was developed. Wine is placed in shallow vats with considerable exposure to the air, and the acetic acid bacte-

ria develop as a slimy layer on the top of the liquid. This process is not very efficient because the only place that the bacteria come in contact with both the air and the substrate is at the surface. The second process is the **trickle method,** in which the contact between the bacteria, air, and substrate is increased by trickling the alcoholic liquid over beechwood twigs or wood shavings packed loosely in a vat or column while a stream of air enters at the bottom and passes upward. The bacteria grow on the surface of the wood shavings and thus are maximally exposed both to air and liquid. The vat is called a vinegar generator (Figure 12.22), and the whole process is operated in a continuous fashion. The life of the wood shavings in a vinegar generator is long, from 5 to 30 years, depending on the kind of alcoholic liquid used in the process.

The third vinegar process is the **bubble method.** This is basically a submerged fermentation process such as already described for antibiotic production. Efficient aeration is even more important with vinegar than with antibiotics, and special highly efficient aeration systems have been devised. The process is operated in a continuous fashion: alcoholic liquid is added at a rate just sufficient to balance removal of vinegar. The efficiency of the process is high, and 90–98% of the alcohol is converted to acid. One disadvantage of the bubble method is that the product must undergo more filtering to remove the bacteria, whereas in the open-vat and trickle methods the product is virtually free of bacteria because the cells are bound in the slimy layer in the former and adhere to the wood chips in the latter.

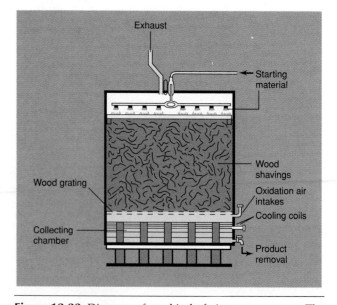

Figure 12.22 Diagram of one kind of vinegar generator. The alcoholic juice is allowed to trickle through the wood shavings, and air is passed up through the shavings from the bottom. Acetic acid bacteria develop on the wood shavings and convert alcohol to acetic acid. The acetic acid solution accumulates in the collecting chamber and is removed periodically. The process can be run semicontinuously.

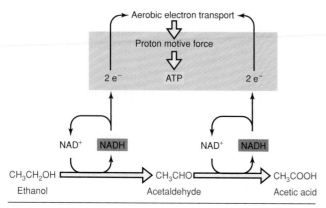

Figure 12.21 Oxidation of ethanol to acetic acid, the key process in the production of vinegar.

Although acetic acid can be easily made chemically from alcohol, the microbial product, vinegar, is a distinctive material, the flavor being due in part to other substances present in the starting material. For this reason, the fermentation process has not been supplanted by a chemical process.

CONCEPT CHECK   12.10

The active ingredient in vinegar is acetic acid, which is produced by an acetic acid bacterium oxidizing an alcohol-containing fruit juice. Adequate aeration is the most important consideration in ensuring a successful vinegar process.

✔ *Write a balanced reaction for the microbial production of vinegar.*

✔ *Why does vinegar produced by the trickle method have a more distinctive taste than vinegar produced by the bubble method?*

✔ *Why is O_2 necessary in vinegar production?*

12.11

Citric Acid and Other Organic Compounds

Many organic chemicals are produced by microorganisms in sufficient yields that they can be manufactured commercially by fermentation. *Citric acid*, used widely in foods and beverages, *itaconic acid*, used in the manufacture of acrylic resins, and *gluconic acid*, used in the form of calcium gluconate to treat calcium deficiencies in humans and industrially as a washing and softening agent, are produced by fungi. *Sorbose*, which is produced when *Acetobacter* oxidizes sorbitol, is used in the manufacture of *ascorbic acid*, vitamin C. (In fact, this sorbitol–sorbose reaction is the only biological step in the otherwise entirely nonbiological chemical synthesis of ascorbic acid.) *Gibberellin*, a plant growth hormone used to stimulate growth of plants, is produced by a fungus. *Dihydroxyacetone*, produced by allowing *Acetobacter* to oxidize glycerol, is used as a suntanning agent. *Dextran*, a gum employed as a blood plasma extender and as a biochemical reagent, and *lactic acid*, used in the food industry to acidify foods and beverages, are produced by lactic acid bacteria. *Acetone* and *butanol* can be produced in fermentations by *Clostridium acetobutylicum* but are now prepared mainly from petroleum by strictly chemical synthesis.

Citric acid

Citric acid is produced microbiologically by a fermentation using the mold *Aspergillus niger*. Although citric acid is normally considered in connection with the citric acid cycle (∞ Section 4.14), in certain organisms

such as *A. niger*, excretion of large amounts of citric acid can be obtained. The fermentation is carried out aerobically in large fermentors, and a key requirement for high citric acid yield is that the medium be *iron-deficient* because citric acid is overproduced by the fungus as a chelator to scavenge iron (Figure 12.23*a*). Therefore, the medium used for citric acid production is treated to remove most of the iron, and the fermentors are made of stainless steel to prevent leaching of iron from the fermentor walls at the low pH values generated by citric acid accumulation.

The media used for citric acid production have been highly perfected over the many years that the commercial process has been under way. A variety of starting materials can be used as carbohydrate sources: starch from potatoes, starch hydrolysates, glucose syrup from saccharified starch, sucrose (Figure 12.23*b*), sugarcane syrup, sugarcane molasses, and sugar beet molasses. If starch is used, amylases formed by the producing fungus or added to the fermentation broth (see Table 12.5) hydrolyze the starch to sugars. The sugars are catabolized through the glycolytic pathway (∞ Section 4.10) and enter the citric acid cycle where citrate production occurs.

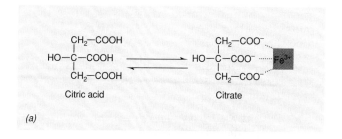

(a)

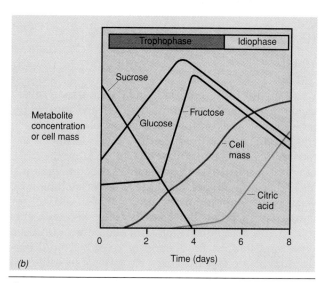

(b)

Figure 12.23 Citric acid fermentation. (a) Structure of citric acid. Note how the ionized form, citrate, contains three carboxylic acid groups, which can chelate ferric iron (Fe^{3+}). (b) Kinetics of citric acid fermentation. See text for further details.

Although both surface and submerged processes for citric acid production have been developed, most citric acid today is produced by submerged processes in large fermentors. Because *Aspergillus niger* is a strict aerobe, it is crucial to this fermentation to make sure that the culture stays properly aerated. Citric acid is produced in this way as a typical secondary metabolite. In the trophophase of the citric acid fermentation, part of the added sugar is used for the production of fungal mycelium and part is converted through respiration to CO_2. In the idiophase, the rest of the sugar is converted to citric acid, and during this phase there is a minimal loss through respiration (Figure 12.23b).

Historically, the development of a submerged process for citric acid was of great importance because it was the first *aerobic* industrial fermentation. The technology for manufacturing aerobic fermentors was perfected with the citric acid process. This technology was then applied to penicillin and the other important antibiotic fermentations. Thus, we owe some of our current success with large-scale production of antibiotics to the pioneering work done on citric acid fermentation.

CONCEPT CHECK   12.11

A number of organic chemicals are produced commercially by use of microorganisms, of which the most important economically is citric acid, produced by certain fungi.

✔ *Why is citric acid produced by* Aspergillus niger *considered a secondary metabolite (see Figure 12.23b)?*

✔ *What is the relationship between iron and citric acid production by A. niger?*

12.12
Yeast

Yeasts are the most important and the most extensively used microorganisms in industry. They are cultured for the cells themselves, for cell components, and for the end products they produce during the alcoholic fermentation (Table 12.6). Yeast cells are used in the manufacture of bread and also as sources of food, vitamins, and other growth factors. Large-scale fermentation by yeast is responsible for the production of alcohol for industrial purposes, but yeast is better known for its role in the manufacture of alcoholic beverages: beer, wine, and liquors. Production of yeast cells and production of alcohol by yeast are two quite different processes industrially, in that the first process requires the presence of oxygen for maximum production of cell material and hence is an *aerobic* process, whereas the alcoholic fermentation is *anaerobic* and takes place only

TABLE 12.6	Industrial uses of yeast and yeast products

Production of yeast cells
  Baker's yeast, for bread making
  Dried food yeast, for food supplements
  Dried feed yeast, for animal feeds

Yeast products
  Yeast extract, for culture media
  B vitamins, vitamin D
  Enzymes for food industry; invertase, galactosidase
  Biochemicals for research; ATP, NAD^+, RNA

Fermentation products from yeast
  Ethanol, for industrial alcohol
  Glycerol

Beverage alcohol
  Beer
  Wine

Distilled beverages
  Whiskey
  Brandy
  Vodka
  Rum

in the absence of oxygen. However, the same or similar species of yeasts are used in virtually all industrial processes. The yeast *Saccharomyces cerevisiae* was derived from wild yeast used in ancient times for the manufacture of wine and beer. The yeasts currently used are descendants of early *S. cerevisiae*. Because they have been cultivated in laboratories for such a long time, there has been ample opportunity for selection of strains according to particular desirable properties. In addition, it is possible to genetically alter yeasts in the laboratory, using genetic exchange methods to produce new strains that contain desirable qualities from two separate parent strains (∞ Section 9.13). By the techniques of genetic engineering, it is now also possible to improve strains by direct intervention.

Yeast production

Bakers use yeast as a leavening agent in the rising of the dough prior to baking. A secondary contribution of yeast to bread is its flavor. In the leavening process, the yeast is mixed with the moist dough in the presence of a small amount of sugar. The yeast converts the sugar to alcohol and CO_2, and the gaseous CO_2 expands, causing the dough to rise. When the bread is baked, the heat drives off the CO_2 (and incidentally, the alcohol) and holes are left within the bread mass, thus giving bread its characteristic light texture. That yeast contributes more to bread than CO_2 is shown by the fact that dough raised with baking powder, a chemical source of CO_2, produces a product quite different from that produced by dough raised by yeast. Only the latter bears the name *bread*.

Yeast for baking or nutritional purposes is cultured in large aerated fermentors in a medium containing molasses as a major ingredient. Molasses, a byproduct of sugar refining from beets or cane, still contains large amounts of sugar that serve as the source of carbon and energy. Molasses also contains minerals, vitamins, and amino acids used by the yeast. To make a complete medium for yeast growth, phosphoric acid (a phosphorus source) and ammonium sulfate (a source of nitrogen and sulfur) are added.

Fermentation vessels for yeast production range from 40,000 to 200,000 liters. Beginning with the pure stock culture, several intermediate stages are needed to scale up the inoculum to a size sufficient to inoculate the final stage (Figure 12.24a). Fermentors and accessory equipment are made of stainless steel and are sterilized by high pressure steam. The actual operation of the fermentor requires special control to obtain the maximum amount of yeast. It is undesirable to add all the molasses to the tank at once because this results in a sugar excess and the yeast ferments some

of this surplus sugar to alcohol plus CO_2 rather than turning it into yeast cells. Therefore, only a small amount of the molasses is added initially, and then as the yeast grows and consumes this sugar, more is added.

At the end of the growth period, the yeast cells are recovered from the broth by centrifugation. The cells are usually washed by dilution with water and recentrifuged until they are light in color. Baker's yeast is marketed in two ways, either as compressed cakes or as a dry powder. *Compressed yeast* cakes (Figure 12.24b) are made by mixing the centrifuged yeast with emulsifying agents, starch, and other additives that give it a suitable consistency and reasonable shelf life, and the product is then formed into cubes or blocks of various sizes for domestic or commercial use. A yeast cake contains about 70% moisture and about 2×10^{10} cells/g. Compressed yeast must be stored in the refrigerator so its activity is maintained. Yeast marketed in the dry state for baking is usually called *active dry yeast* (Figure 12.24b). The washed yeast is mixed with additives and

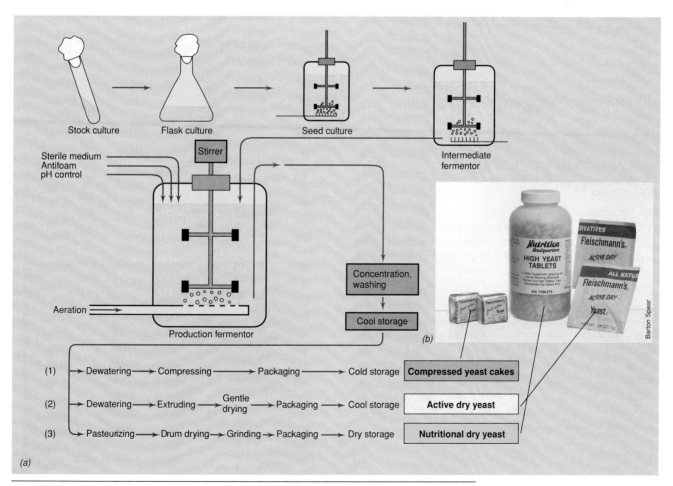

Figure 12.24 Industrial production of yeast cells. (a) Stages in production. (b) Photograph of common yeast products: yeast cake; package of active dry yeast; bottle of nutritional yeast.

dried under vacuum at 25–45°C for a 6-hr period until its moisture is reduced to about 8%. It is then packed in airtight containers such as fiber drums, cartons, or multiwall bags, sometimes under a nitrogen atmosphere to promote long shelf life. Active dry yeast does not exhibit as great a leavening action as compressed fresh yeast but has a much longer shelf life. *Nutritional yeast*, marketed as a food supplement (Figure 12.24*b*), is heat-killed and usually dried. Yeast cells are rich in B vitamins and in protein, except for sulfur-containing amino acids. Yeast is added to wheat or corn flour to increase the nutritional value of these foods and is also sold in pelleted form as a health food (Figure 12.24*b*).

CONCEPT CHECK   **12.12**

Yeast cells are grown for use in the baking and food industry. Commercial yeast is produced in large-scale aerated fermentors using molasses as the main carbon and energy source.

✔ *Write a balanced chemical reaction that accounts for the action of yeast in bread making.*

✔ *Why is it important when growing yeast for cells to maintain oxic conditions in the fermentor?*

12.13

Alcohol and Alcoholic Beverages

Ethyl alcohol (ethanol, C_2H_5OH) has been produced on a large scale for centuries. Early production was associated with alcoholic beverages, but today, besides the enormous alcoholic beverage industry, alcohol is produced in large quantities as an industrial solvent and gasoline additive. We begin our discussion here with the production of alcoholic beverages.

The use of yeast in the production of alcoholic beverages is an ancient process. Most fruit juices undergo a natural fermentation caused by wild yeasts that are present on the fruit. From these natural fermentations, yeasts have been selected for more controlled production, and today alcoholic beverage production is a large industry worldwide. The most important alcoholic beverages are *wine*, produced by the fermentation of fruit juice; *beer*, produced by the fermentation of malted grains; and *distilled beverages*, produced by concentrating alcohol from a fermentation by distillation. The biochemistry of alcohol fermentation by yeast was discussed in Section 4.10.

Wine

Wine is a product of the alcoholic fermentation by yeast of fruit juices or other materials that are high in sugar. Most wine is made from grapes, and unless otherwise specified, the word *wine* refers to the product resulting from the fermentation of grape juice. Wine

manufacture occurs in parts of the world where grapes can be most economically grown. The greatest wine-producing countries, in order of decreasing volume of production, are Italy, France, Spain, Algeria, Argentina, Portugal, and the United States. Wine manufacture originated in Egypt and Mesopotamia well before 2000 B.C. and spread from there throughout the Mediterranean region, which is still the largest wine-producing area in the world. Other parts of the world where wine is extensively produced often have a climate similar to that of the Mediterranean, for example, California (Figure 12.25), Chile, South Africa, and

(a)

(b)

(c)

Figure 12.25 Commercial wine making. (a) Equipment for transporting grapes to the winery for crushing. (b) Large tanks where the main wine fermentation takes place. (c) Barrels where the aging process takes place.

Australia. There are a great number of different wines, and their quality and character vary considerably. *Dry wines* are wines in which the sugars of the juice are practically all fermented, whereas in *sweet wines,* some of the sugar is left or additional sugar is added after the fermentation. A *fortified wine* is one to which brandy or some other alcoholic spirit is added after the fermentation; sherry and port are the best-known fortified wines. A *sparkling wine,* such as champagne, is one in which considerable carbon dioxide is present, arising from a final fermentation by the yeast directly in the bottle.

The yeasts involved in wine fermentation are of two types: the so-called wild yeasts, which are present on the grapes as they are taken from the field and are transferred to the juice, and the cultivated wine yeast, *Saccharomyces ellipsoideus,* which is added to the juice to begin the fermentation. One important distinction between wild yeasts and the cultivated wine yeast is their alcohol tolerance. Most wild yeasts can tolerate only about 4% alcohol, and when the alcohol concentration reaches this point, the fermentation stops. The best wine yeasts can tolerate up to 14% alcohol before they stop growing, although above about 10% alcohol growth can be very slow. In unfortified wine, the final alcoholic content reached is determined partly by the alcohol tolerance of the yeast and partly by the amount of sugar present in the juice. The alcohol content of most unfortified wines ranges from 8 to 14%. Fortified wines such as sherry have an alcohol content as high as 20%, but this is achieved by adding distilled spirits such as brandy. In addition to the lower alcohol content produced, wild yeasts do not produce some of the flavor components considered desirable in the final product, and hence the presence and growth of wild yeasts during fermentation is unwanted.

Wine production

The production of wine begins in the early fall with the harvesting of grapes. The grapes are crushed by machine, and the juice, called *must*, is squeezed out. Depending on the grapes used and on how the must is prepared, either white or red wine may be produced (Figure 12.26). A white wine is made either from white grapes or from the juice of red grapes from which the skins, containing the red coloring matter, have been removed. In the making of red wine, the *pomace* (skins, seeds, and pieces of stem) is left in during the fermentation. In addition to the color difference, red wine has a stronger flavor than white because of the presence of larger amounts of chemicals called *tannins,* which are extracted into the juice from the grape skins during the fermentation.

It is the practice in many wineries to kill the wild yeasts present in the must by adding sulfur dioxide (listed on the bottle as "sulfites") at a level of about 100 parts per million (ppm). The cultivated wine yeast is resistant to this concentration of sulfur dioxide and is

added as a starter culture from a pure culture grown on sterilized or pasteurized grape juice. During the initial stages, air is present in the liquid and rapid aerobic growth of the yeast occurs; then, as the air is used up, anoxic conditions develop and alcohol production begins. The fermentation may be carried out in vats of various sizes, from 50-gallon (gal) casks to 55,000-gal tanks made of oak, cement, stone, or glass-lined metal (see Figure 12.25b). Temperature control during the fermentation is important because heat produced during metabolism potentially raises the temperature above the point where yeast can function. Temperatures must be kept below 29°C, and the finest wines are produced at lower temperatures, from 21 to 24°C. Temperature control is achieved by using jacketed tanks through which cold water is circulated. The fermentor must be constructed so that the large amount of carbon dioxide produced during the fermentation can escape but air cannot enter; this is often accomplished by fitting the tank with a special one-way valve.

With a red wine, after 3–5 days of fermentation, sufficient tannin and color have been extracted from the pomace and the wine is drawn off for further fermentation in a new tank, usually for another week or two. The next step is called *racking;* the wine is separated from the sediment (called *lees*), which contains the yeast and organic precipitate, and then stored at lower temperature for aging, flavor development, and further clarification. The final clarification may be hastened by the addition of materials called fining agents, such as casein, tannin, or bentonite clay, or the wine may be filtered through diatomaceous earth, asbestos, or membrane filters. The wine is then bottled and either stored for further aging or sold. Red wine is usually aged for several years or more after bottling (see Figure 12.25c), but white wine is usually sold without much aging. During the aging process, complex chemical changes occur, including reduction of bitter components, resulting in improvement in flavor and odor, or *bouquet.*

Brewing

The manufacture of alcoholic beverages made from malted grains is called *brewing.* Typical malt beverages include beer, ale, porter, and stout. *Malt* is prepared from germinated barley seeds, and it contains natural enzymes that digest the starch of grains and convert it to sugar. Since brewing yeasts are unable to digest starch, the malting process is essential for the preparation of a fermentable material from cereal grains. Malted beverages are made in many parts of the world but are most common in areas with cooler climates where cereal grains grow well and where wine grapes grow poorly.

The fermentable liquid from which beer and ale are made is prepared by a process called *mashing.* The grain of the mash may consist only of malt, or other grains such as corn, rice, or wheat may be added. The

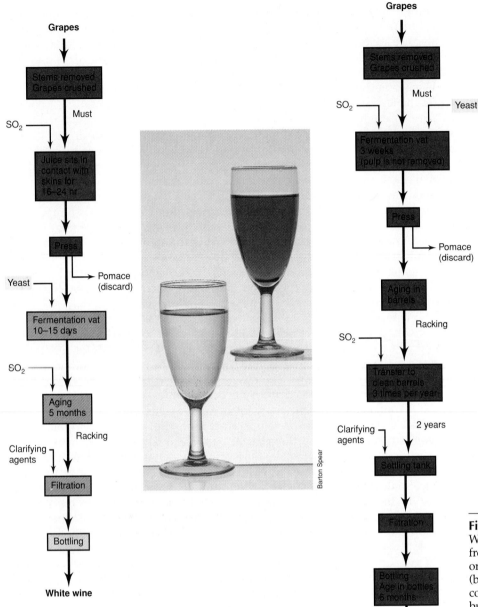

Figure 12.26 Production of wine. (a) White wine. White wines can vary from nearly colorless to straw-colored depending on the grapes used. (b) Red wine. Red wines can vary in color from a faint red to a deep, rich burgundy. The photograph shows a glass of Chenin Blanc, a typical white wine (left), and a glass of a light red wine (right).

mixture of ingredients in the mash is cooked and allowed to steep in a large mash tub at warm temperatures. There are a number of different methods of mashing, involving heating at different temperatures for various lengths of time; the particular combination of temperature and time used considerably influences the character of the final product. During the heating period, enzymes from the malt cause digestion of the starches and liberate sugars, which are fermented by the yeast. Proteins and amino acids are also liberated into the liquid, as are other nutrient ingredients necessary for the growth of yeast.

After cooking, the aqueous extract, called *wort*, is separated by filtration from the husks and other grain residues of the mash. *Hops*, an herb derived from the female flowers of the hops plant, is added to the wort at this stage. Hops is a flavoring ingredient, but it also has antimicrobial properties, which probably help to prevent contamination in the subsequent fermentation. The wort is then boiled for several hours, usually in large copper kettles (Figure 12.27a,b), during which time desired ingredients are extracted from the hops, proteins present in the wort that are undesirable from the point of view of beer stability are coagulated and removed, and the wort is sterilized. Heating is accomplished either by passing steam through a jacketed kettle or by direct heating of the kettle from below by fire. Then the wort is filtered again, cooled, and transferred to the fermentation vessel.

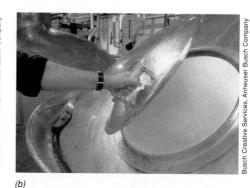

Figure 12.27 Brewing beer in a commercial brewery. (a, b) The copper brew kettle is where the wort is mixed with hops and then boiled. From the brew kettle the liquid is passed to large fermentation tanks where yeast ferments glucose to ethanol plus CO_2. (c) The fermented liquid is then stored for several weeks at low temperature in lagering tanks where settling of particulate matter including yeast cells occurs. (d) The beer is then filtered and placed in storage tanks from which it is packaged into kegs, bottles, or cans.

Brewery yeast strains are of two major types: top-fermenting and bottom-fermenting. The main distinction between the two is that **top-fermenting yeasts** remain uniformly distributed in the fermenting wort and are carried to the top by the CO_2 gas generated during the fermentation, whereas **bottom yeasts** settle to the bottom. Top yeasts are used in the brewing of ales, and bottom yeasts are used to make lager beers. Bottom yeasts are usually given the species designation *Saccharomyces carlsbergensis*, and top yeasts are called *Saccharomyces cerevisiae*. Fermentation by top yeasts usually occurs at higher temperatures (14–23°C) than that by bottom yeasts (6–12°C) and is accomplished in a shorter period of time (5–7 days for top fermentation versus 8–14 days for bottom fermentation). After completion of lager beer fermentation by bottom yeasts, the beer is pumped off into large tanks where it is stored at a cold temperature (about −1°C) for several weeks (in German, *lager* means "to store") (Figure 12.27c). Lager beer is the most widely manufactured type of beer and is made by large breweries in the United States, Germany, Scandinavia, the Netherlands, and the Czech Republic. Following the lagering process, the beer is filtered and placed in storage tanks (Figure 12.27d) from which packaging occurs. Top-fermented ale is almost exclusively a product of England and certain former British colonies. After its fermentation, the clarified ale is stored at a higher temperature (4–8°C), which assists in development of the characteristic ale flavor.

For more details on the brewing process, refer to the box, Home Brew.

Distilled alcoholic beverages

Distilled alcoholic beverages are made by heating a fermented liquid at a high temperature that volatilizes most of the alcohol. The alcohol is then condensed and collected, a process called *distilling*. A product much higher in alcohol content can be obtained by this process than is possible by direct fermentation. Virtually any alcoholic liquid can be distilled, and each yields a characteristic distilled beverage. The distillation of malt brews yields *whiskey*, distilled wine yields *brandy*, distillation of fermented molasses yields *rum*, distillation of fermented grain or potatoes yields *vodka*, and distillation of grain and juniper berries yields *gin* (Figure 12.28).

The distillate contains not only alcohol but also other volatile products arising either from the yeast fermentation or from the mash itself. Some of these other products are desirable flavor ingredients, whereas others are undesirable substances called *fusel oils*. To eliminate the latter, the distilled product is almost always aged, usually in wood barrels. During the aging process, fusel oils are removed and desirable new flavor ingredients develop. The fresh distillate is usually colorless, whereas the aged product is often brown or yellow (Figure 12.28). The character of the final product is partly determined by the manner and length of aging (aging times of 10 years or more are not uncommon for some distilled spirits), and the whole process of manufacturing distilled alcoholic beverages is highly complex. To a great extent, the process is carried out by traditional methods that have been found to yield a particular product, rather than by scientifically proven methods.

HOME BREW[a]

The skills of the brewer can be applied by anyone who is willing to learn aseptic technique and the first principles of microbiology. The amateur brewer can make many kinds of beer, from English bitters and India pale ale to German bock and Russian Imperial stout. The necessary equipment and supplies can be purchased from a local beer and winemakers shop (the Home Wine and Beer Trade Association, 604 N. Miller Road, Valrico, FL 33594, can supply the address of a nearby shop).

The brewing process can be divided into three basic stages: making the wort, carrying out the fermentation, and bottling and aging. The character of the brew depends on many factors: the proportion of malt, sugar, hops, and grain; the kind of yeast; the temperature and duration of the fermentation; and how the aging process is carried out. The instructions provided here are for a simple and relatively foolproof beer (so-called single-stage fermentation).

The fermentor itself consists of a 20-liter (5-gal) glass jar or carboy that can be fitted with a tightly fitting closure. In order to have a good quality beer, it is essential that *everything* be sterilized that comes into contact with the wort. This includes the fermentor, tubing, stirring spoon, and bottles. The best procedure is to use a sterilizing rinse consisting of 50–60 ml of liquid bleach in 20 liters of water. Soak the items for 15 min and then rinse lightly with hot water or air-dry.

1. **Making the wort.** In commercial brewing, the wort is made by extracting fermentable sugars and yeast nutrients from malt, sugar, and hops. The process is complex and relatively difficult to carry out satisfactorily. Many home brewers make their own

wort from malt, but a reasonably satisfactory beer can be made with hop-flavored malt extract purchased ready-made. Malt extracts come in a variety of flavors and colors, and the kind of beer depends on the type of malt extract used. A simple recipe for making the wort uses 5–6 lb (2.25–2.75 kg) of hop-flavored malt extract and 20 liters of water. The malt extract and 6 liters of water are brought to a boil for 15 min in an enamel or stainless steel container (aluminum heating kettles must be avoided because of the inhibiting action of metals leached from aluminum containers) (see photo *a*). The hot wort is then poured into 14–15 liters of clean, cold water that has already been added to the fermentor. After the temperature has dropped below 30°C the yeast can be added to initiate the fermentation.

2. **Carrying out the fermentation.** The process by which yeast is added to the wort is called *pitching*. Brewer's yeast can be purchased as active dry yeast from the home brew supplier. Different yeasts are available for producing different kinds of beer. If the brewing is carried out in the summer time when the temperature is higher, a yeast suitable for a high-temperature fermentation should be used. Add two packs of fresh beer yeast to the cooled wort and cover the fermentor with a rubber stopper into which a plastic hose has been inserted. The hose is directed into a bucket containing water. During the initial 2–3 days of the fermentation, large amounts of CO_2 will be given off, which will exit through the hose. The water trap is to prevent wild yeasts or bacteria from the air from getting back into the fermentor. After about 3 days, the activity will diminish as the fermentable sugars are used up. At

this time, the rubber stopper and hose are replaced with an inexpensive fermentation lock. The fermentation lock (see photo *b*), which can be purchased at the home brew store, prevents contamination while permitting the small amount of gas still being produced to escape. Allow the beer to ferment for 7–10 days at 10–15°C or higher. The fermentation should begin within 24 hr after pitching the yeast. If it does not, the yeast used may not have been active, or the temperature too high when pitching was carried out.

3. **Bottling and aging.** The fermentation should be allowed to proceed for the full 7–10 days, even if the vigorous fermentation action ceases earlier. Most of the yeast should have settled to the bottom of the fermentor. Carefully siphon the beer off the yeast layer, allowing it to run into glass beer bottles. The bottles themselves should have been sanitized first. Take care that the yeast at the bottom of the fermentor is not stirred up and leave the yeast-rich liquid at the bottom. The bottles used should accept standard crown caps, and new, clean caps should be used (see photo *c*). Before capping, add $\frac{3}{4}$ teaspoon of corn sugar syrup to each bottle. Be certain not to add more than $\frac{3}{4}$ teaspoon of syrup because if excess sugar is added, the buildup of carbon dioxide in the bottles may cause them to burst. Once the bottles are capped, turn each one upside down once to mix the sugar syrup and then allow the beer to age upright at room temperature for at least 7–10 days. If another large container is available, a better way of adding the sugar is to siphon the beer into this second container, add the proper amount of sugar for the whole brew, dissolve, and then siphon

[a]Primary reference source: Burch, B. 1992. *Brewing Quality Beers—The Home Brewer's Essential Guidebook*, 2nd edition. Joby Books, Fulton, CA.

into the bottles. After this aging period, the beer may be stored at a cooler temperature.

All homemade beer has a natural yeast sediment in the bottom of the bottles. The beer will improve if it is allowed to age for several weeks. Aging tends to make beer smoother. Using the same basic production equipment, several different types of beer can be made, each with its own distinctive taste and character (the reference listed here contains a number of beer recipes). Dark beers, which generally contain more alcohol than lighter beers, require more malt for their production and are usually brewed from a combination of different malts such as ones obtained from darker varieties of grain or ones that have been roasted to carmelize the sugars and yield a darker color. A typical American style light lager (see photo *d*, left) contains about 3.5% alcohol (by volume) whereas a Munich style dark (see photo *d*, right) contains 4.25% alcohol and bock beers contain about 5% alcohol.

The trend toward "individuality" in beer can be attested to not only by the growing number of home brewers but also by the fact that major brewers in the United States are feeling more and more competition from new, usually very small, breweries called *microbreweries*. Although total production by a microbrewery may pale by comparison to that of a major brewer, the products themselves often have their own distinctive character and local appeal. Part of these differences probably has to do with the smaller scale on which the brewing takes place but also undoubtedly has to do with the use of different sources of ingredients, yeast strains, brewing times, and other aspects of the brewing process described in this chapter. ■

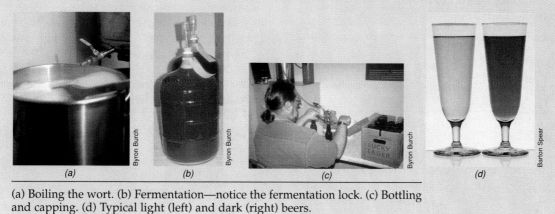

(a) Boiling the wort. (b) Fermentation—notice the fermentation lock. (c) Bottling and capping. (d) Typical light (left) and dark (right) beers.

Figure 12.28 Typical distilled spirits. These alcoholic beverages contain not only alcohol but also distinctive volatile flavoring agents obtained from the fermented substrate. Aging in wood casks yields the distinctive amber or yellow color of certain distilled spirits. Top row (left to right) gin, vodka. Bottom row (left to right) dark rum, brandy, whiskey.

Commodity ethanol

Production of ethanol as a commodity chemical is a major industrial process, and today close to a billion gallons of alcohol are produced yearly in the United States, primarily from the fermentation of cornstarch. This ethanol is used as an industrial solvent and also for the production of gasohol, a lead-free fuel containing 10% ethanol in gasoline. The combustion of gasohol produces lower amounts of carbon monoxide and nitrogen oxides than pure gasoline; thus gasohol is marketed as a cleaner burning fuel, and its use is encouraged in major cities where automobile pollution is extensive. If automobile engines are modified to burn it, they can be run on pure ethanol, and this has been done in certain countries such as Brazil where sugarcane (as a fermentable substrate) is plentiful but oil is scarce.

Various yeasts have been used in commodity ethanol production, including species of *Saccharomyces*, *Kluyveromyces*, and *Candida*, but most ethanol in the United States is produced by *Saccharomyces* in the reaction

$$C_6H_{12}O_6 \rightarrow 2\ C_2H_5OH + 2\ CO_2$$

(this occurs by glycolysis) (∞ Section 4.10).

If 100% of the glucose were fermented to products, alcohol would compose 51% by weight of the original glucose. Although for technical reasons this level of alcohol production has not been achieved, commodity ethanol production has been refined to the point where ethanol distilled from the fermentation broth is obtained at 90–95% of the theoretical yield.

CONCEPT CHECK   12.13

Alcoholic beverages are produced by yeast from the fermentation of sugar to ethyl alcohol and CO_2. Wine is produced from grape juice, beer from malted grain, and distilled spirits from the distillation of fermented solutions. Commodity alcohol is used as a gasoline additive and industrial solvent.

✔ *How does wine differ from beer in terms of the fermentable substrate used?*

✔ *What are the major differences between a beer and an ale?*

12.14

Food from Microorganisms

Microorganisms can be grown to produce food for humans, and we will discuss here the production of microorganisms as food (so-called *single-cell protein*) and mushrooms. In recent years, there has been considerable interest in the expanded production of microorganisms as food, especially in parts of the world where conventional sources of food are in short supply. Perhaps the most important potential use of microorganisms is not as a complete diet for humans but as a *protein supplement*. It is usually protein that is in shortest supply in food, and it is in the production of protein that microorganisms are perhaps the most successful. In many cases, microbial cells contain greater than 50% protein, and in at least some species this is complete protein; that is, it contains sufficient amounts of all the amino acids essential to humans. The protein produced by microorganisms as food has been called *single-cell protein* to distinguish it from the protein produced by multicellular animals and plants. The only organism presently used as a source of single-cell protein is yeast (∞ Section 12.12), as previously mentioned, but algae, bacteria, and fungi have also been considered and in some cases have been marketed as supplements for animal feeds and as health foods (especially algae and cyanobacteria).

Mushrooms

Several kinds of *fungi* are sources of human food, of which the most important are the mushrooms. Mushrooms are a group of filamentous fungi that form large, complicated structures called **fruiting bodies** (Figure 12.29). The fruiting body is commonly called the *mushroom* and is formed through the association of a large number of individual hyphae to form a mycelium (Figure 12.29).

During most of its existence, the mushroom fungus lives as a simple mycelium, growing in soil, leaf litter, or decaying logs. However, when environmental conditions are favorable, the fruiting body develops, beginning first as a small button-shaped structure underground and then expanding into the full-grown fruiting body that we see above ground (Figure 12.29). The nutrients for growth come from organic matter in the soil and are taken up by the hyphal filaments, which, like the roots of a plant, feed the growing fruiting body. Sexual spores, called **basidiospores,** are formed, borne on the underside of the fruiting body on flat plates called **gills** (Figure 12.29). The spore is the agent of dispersal of mushrooms and is carried away by the wind. If it alights in a favorable place, it will germinate and initiate the growth of new hyphae, mycelium, and fruiting body.

The mushroom commercially available in most parts of the world is *Agaricus bisporus,* and it is generally cultivated on mushroom farms. The organism is grown in special beds, usually in buildings where temperature and humidity are carefully controlled (Figure 12.30*a*). Since light is not necessary, mushrooms may even be grown in basements of homes or in caves. Beds are prepared by mixing soil with a material very rich in organic matter, such as horse manure, and the beds are then inoculated with mushroom *spawn*. The spawn is actually a pure culture of the mushroom fungus that has been grown in large bottles on an organic-rich medium. In the bed, the mycelium grows and spreads through the substrate, and after several weeks it is ready for the next step, the induction of mushroom formation. This is accomplished by adding to the surface of the bed a layer of soil called *casing soil*. The appearance of mushrooms on the surface of the bed is called a *flush* (Figure 12.30*b*), and when flushing occurs, the mushrooms must be collected immediately while still fresh. After collection they are packaged and kept cool until brought to market. Several flushes take place on a single bed, and after the last flush the bed must be cleaned out and the process begun again.

Another widely cultured mushroom is **shiitake,** *Lentinus edulus.* The most widely cultivated mushroom in the Far East, shiitake is now finding expanding demand in North America. Shiitake is a cellulose-digesting fungus that grows well on hardwood trees and is cultivated on small logs (Figure 12.30*c*). The logs are soaked in water to hydrate them and then inoculated by inserting plugs of spawn into small holes drilled in them. The fungus grows through the log, and after about a year forms a flush of fruiting bodies (see Figure 12.30*d*). Shiitake has the advantage that it can be cultivated on waste or scrap wood. Some people find it to be much tastier than *Agaricus bisporus.*

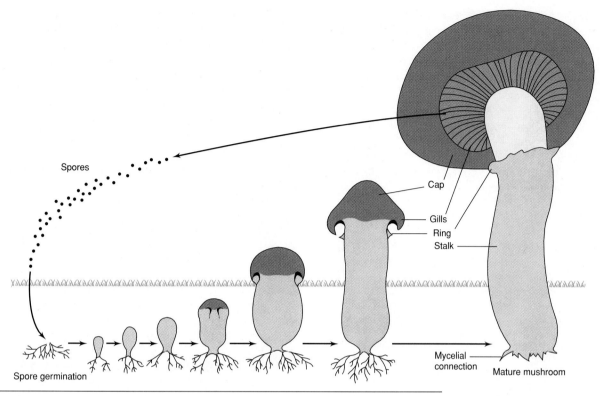

Figure 12.29 Mushroom life cycle, showing how the fruiting body develops from underground hyphae.

Although mushrooms make flavorful food, their digestibility and nutritional value are not very high. They are low in protein and deficient in certain essential amino acids; they are also not exceptionally rich in vitamins. Mushrooms and filamentous fungi are definitely inferior to yeast as food sources, although they serve as valuable flavoring ingredients.

CONCEPT CHECK   **12.14**

Many microorganisms are high in protein and could become sources of protein for humans, but only yeast has thus far been utilized on a large scale. The most important food produced from a microorganism is the mushroom, which is produced not for its protein but for its flavor.

✔ *What is single-cell protein?*

✔ *Why are mushrooms considered microorganisms?*

12.15
Sewage and Wastewater Microbiology

Although production of a valuable commercial product is not the goal of sewage and wastewater treatment, the process itself is clearly a large-scale use of microorganisms and can be considered a type of bioconversion: wastewaters enter a treatment plant and, following microbial treatment, water suitable for release to rivers and streams or to drinking water purification facilities is produced.

Wastewaters are materials derived from domestic sewage or industrial effluents, which for reasons of public health and for recreational, economic, and aesthetic considerations, cannot be disposed of merely by discarding them untreated into lakes or streams. Wastewaters contain both inorganic and organic components, and microorganisms play a particularly important role in removing organic compounds. However, proper wastewater treatment also results in the elimination of pathogenic microorganisms, thus preventing these organisms from getting into rivers or other water supply sources.

About 15,000 wastewater treatment facilities exist in the United States. The vast majority of them are fairly small, treating 1 million gal or less of wastewater per day. However, collectively, these plants treat nearly 40 *billion* gal of wastewater every day. Wastewater plants are usually constructed to handle both domestic and industrial wastes. Domestic wastewaters are made up of sewage, "gray water" (the water resulting from washing, bathing, and cooking), and wastewater from food processing. Industrial wastewaters include those from the petrochemical, pesticide, food, plastics, and

Figure 12.30 Commercial mushroom production. (a) An installation for *Agaricus bisporus*, the common commercial mushroom of the Western world. (b) Close-up of a mushroom flush. (c) Shiitake, *Lentinus edulus*, the most common commercial mushroom of the Far East but finding increasing production in the West. A large Japanese installation where the mushroom is cultivated on hardwood logs. (d) Close-up of the shiitake mushroom.

pharmaceutical industries, and from metallurgical industries, such as electroplating.

Many industrial wastes contain toxic substances and must be pretreated before they can be released for wastewater treatment. Pretreatment is generally a mechanical process in which debris that could clog equipment in the wastewater treatment plant is removed. However, certain wastewaters are pretreated biologically to remove highly poisonous substances such as cyanide and high levels of heavy metals. These substances can be converted to less toxic forms by pretreating them with specific microorganisms capable of oxidizing, precipitating, or volatilizing the toxic components.

We now consider the workings of a typical wastewater treatment facility using a domestic sewage treatment facility as an example of the processes involved.

Levels of sewage treatment

Sewage treatment is generally a multistep process employing both physical and biological treatment steps (Figure 12.31). **Primary treatment** of sewage consists only of physical separations. Sewage entering the treatment plant is passed through a series of grates and screens that remove large objects, and then the effluent is left to settle for a number of hours to allow suspended solids to sediment.

Because of the high nutrient loads that remain in sewage effluent following primary treatment, municipalities that treat sewage no further than the primary stage suffer from extremely polluted water when the sewage is dumped into adjacent waterways. This is why the majority of sewage plants employ **secondary treatment** processes to reduce the organic load of the sewage to acceptable levels before releasing it to natural waterways. Secondary treatment is intimately tied to microbiological processes as described in the following sections.

Tertiary treatment is the most complete method of treating sewage but has not been widely adopted because it is so expensive. Tertiary treatment is a physicochemical process employing precipitation, filtration, and chlorination to sharply reduce the levels of inorganic nutrients, especially phosphate and nitrate, from the final effluent. Wastewater receiving

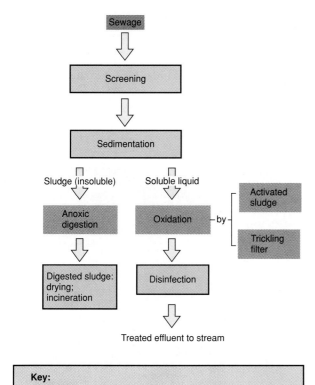

Figure 12.31 An overview of sewage treatment processes.

itself is carried out in large enclosed tanks called **sludge digestors** or **bioreactors** (Figure 12.32*a* and *b*), and requires the collective activities of many different types of microorganisms; the reactions are summarized in Figure 12.32*c*. The macromolecular components must first be digested by polysaccharidases, proteases, and lipases into soluble components. The latter are fermented to a mixture of fatty acids, H_2, and CO_2, and the fatty acids further fermented to acetate, CO_2 plus H_2. These are all substrates for methanogenic bacteria (∞ Section 17.2), which are capable of carrying out the reactions $CH_3COOH \rightarrow CH_4 + CO_2$ and $4 H_2 + CO_2 \rightarrow CH_4 + 2 H_2O$ (Figure 12.32*c*). Thus, major products of anoxic sewage treatment are CH_4 (methane) and CO_2. The methane is collected and either burned off or used as fuel to drive electric generators used to heat and power the treatment plant.

Aerobic secondary treatment process

Several kinds of aerobic decomposition processes are used in sewage treatment, but the trickling filter and activated sludge methods are the most common. A **trickling filter** (Figure 12.33*a*) is a bed of crushed rocks, about 2 m thick, on top of which the wastewater is sprayed. The liquid slowly passes through the bed, the organic matter adsorbs to the rocks, and microbial growth takes place. The complete mineralization of organic matter to carbon dioxide, ammonia, nitrate, sulfate, and phosphate occurs.

The most common aerobic treatment system is the **activated sludge** process. Here, the wastewater to be treated is mixed and aerated in a large tank (Figure 12.33*b–d*). Slime-forming bacteria (primarily a species called *Zoogloea ramigera*) grow and form flocs (**zoogloeas**) (Figure 12.34), and these flocs form the substratum to which protozoa and small animals attach. Occasionally, filamentous bacteria and fungi are also present. The basic process of oxidation is similar to that in a trickling filter. The effluent containing the flocs is pumped into a holding tank or clarifier where the flocs settle. Some of the floc material is then returned to the aerator to serve as inoculum, and the rest is sent to the sludge digestor. The residence time in an activated sludge tank is generally 5–10 hr, too short for complete oxidation of organic matter. The main process occurring during this short time is *adsorption of soluble organic matter* to the floc and incorporation of some of the soluble material into microbial cell material. The BOD of the liquid is thus considerably reduced by this process (75–90%), but the overall BOD (liquid plus solids) is only slightly reduced because most of the adsorbed organic matter still resides in the floc. The main process of BOD reduction thus occurs in the anoxic

proper tertiary treatment is unable to support extensive microbial growth. We focus our discussion here on secondary treatment processes, for it is these that have the most microbial components and are the most widely practiced in sewage treatment facilities today.

Anoxic secondary treatment processes

Anoxic sewage treatment involves a complex series of digestive and fermentative reactions carried out by a host of different bacterial species. The efficiency of a treatment process is expressed in terms of the percentage decrease in the biochemical oxygen demand (BOD), a measure of the amount of dissolved oxygen consumed by microorganisms for the oxidation of organic and inorganic matter; the higher the level of oxidizable organic and inorganic materials in the wastewater, the higher the BOD. A well-operated wastewater treatment plant can remove 95% or greater of the initial BOD.

Anoxic decomposition is usually employed in the treatment of materials that have much insoluble organic matter, such as fiber and cellulose, or of concentrated industrial wastes. The degradation process

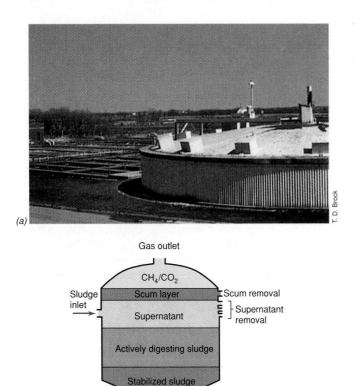

(a)

(b)

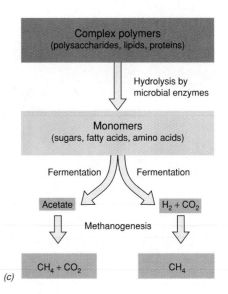

Figure 12.32 (a) Anoxic sludge digestor. Only the top of the tank is shown; the remainder is underground. (b) Inner workings of a sludge digestor. (c) Major microbial processes occurring during sludge digestion. Note how methane (CH_4) is the major product of anaerobic biodegradation.

Gas outlet

CH_4/CO_2

Sludge inlet

Scum layer — Scum removal

Supernatant — Supernatant removal

Actively digesting sludge

Stabilized sludge

Sludge outlet

(c)

Complex polymers
(polysaccharides, lipids, proteins)

Hydrolysis by microbial enzymes

Monomers
(sugars, fatty acids, amino acids)

Fermentation   Fermentation

Acetate $H_2 + CO_2$

Methanogenesis

$CH_4 + CO_2$ CH_4

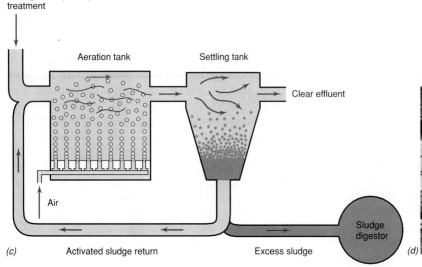

(a)   (b)

(c)

Sewage from primary treatment

Aeration tank   Settling tank

Clear effluent

Air

Sludge digestor

Activated sludge return   Excess sludge

(d)

Figure 12.33 Aerobic sewage treatment processes. (a) Trickling filter. (b–d) Activated sludge process. (b) Aeration tank of an activated sludge installation in a metropolitan sewage treatment plant. (c) Inner workings of an activated sludge installation. (d) A small-scale activated sludge operation used to process dairy waste.

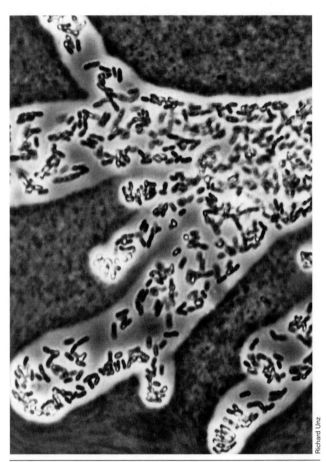

Figure 12.34 Photomicrograph of a floc formed by *Zoogloea ramigera*, the characteristic organism of the activated-sludge process. Note the large number of small, rod-shaped bacteria surrounded by a polysaccharide slime and the characteristic fingerlike projections of the floc. Negative stain using india ink.

sludge digestor to which the flocs are transferred and degraded.

Water purification

Wastewaters treated as previously described are generally of such quality that they can be discharged into rivers and streams. However, such water is not *potable*, that is, suitable for drinking. Potable water requires further treatment to remove potentially pathogenic microorganisms, decrease turbidity, eliminate taste and odor, and reduce nuisance chemicals such as iron and manganese.

A typical drinking water treatment installation for a large city is shown in Figure 12.35. Water is first pumped from the source, in this case a river, to **sedimentation basins** where sand, gravel and other particles contributing to turbidity settle out. From here the

water travels to a **coagulation basin** where chemicals containing aluminum and iron are added to form a floc that traps microorganisms, absorbs organic matter and sediment, and removes them from the water. After coagulation, the clarified water is filtered to remove the remaining suspended particles and microorganisms. This is usually done by passing the water through thick layers of sand, which, when combined with previous purification steps, removes greater than 99% of the bacteria present in the original untreated water.

Chlorination is the most common method of ensuring microbiological safety in a water supply. In sufficient doses it causes the death of most microorganisms within 30 min. In addition, chlorine reacts with organic compounds, oxidizing and effectively neutralizing them. Therefore, since most taste- and odor-producing compounds are organic in nature, chlorine treatment also improves water taste and smell. Chlorine is added to water either from a concentrated solution of sodium or calcium hypochlorite or as a gas from pressurized tanks. The latter method is used most commonly in large water treatment plants (Figure 12.35), as it is most amenable to automatic control.

When chlorine reacts with organic materials, it is consumed. Therefore, if a water supply is high in organic materials, sufficient chlorine must be added so there is a residual amount left to react with the microorganisms after all reactions with organic materials have occurred. The water plant operator performs chlorine analyses on the treated water to determine the residual level of chlorine. A chlorine residual of 0.2–0.6 μg/ml is an average level suitable for most water supplies. After chlorine treatment, the now-potable water is pumped to storage tanks from which it flows by gravity to the consumer.

CONCEPT CHECK   12.15

Sewage treatment processes are industrial-scale microbial culture systems in which the organic materials of the sewage are converted to CO_2, CH_4, and inorganic nutrients. Two kinds of sewage treatment processes are used: anoxic, in which organic materials are converted principally to methane and carbon dioxide; and aerobic, in which organic materials are converted to microbial cells and carbon dioxide. Properly treated wastewater must be further purified before it is suitable for drinking.

✔ *What is biological oxygen demand (BOD)? Why is its reduction necessary in sewage treatment?*

✔ *Why is chlorine added to water used for drinking purposes? What does chlorine do?*

Figure 12.35 Water purification plant. Aerial view of water treatment plant in Louisville, Kentucky. The arrows indicate direction of flow of water through the plant.

Material for Review

REVIEW QUESTIONS

1. In what ways do industrial microorganisms differ from conventional microorganisms? In what ways are they similar?

2. Describe some of the techniques that can be used to improve strains of industrial microorganisms.

3. List three major types of industrial products that can be obtained with microorganisms and give two examples of each.

4. Give an example of a *commodity chemical* produced by a microorganism and describe briefly the process by which this chemical is manufactured.

5. Compare and contrast *primary* and *secondary metabolites* and give an example of each. List at least two molecular explanations for why some metabolites are secondary rather than primary.

6. Define *trophophase* and *idiophase*.

7. How does an industrial fermentor differ from a laboratory culture vessel? How does a fermentor differ from a fermenter?

8. Discuss the problems of scale-up from the viewpoints of *aeration, sterilization,* and *process control.* Why is sterility so much more important in an industrial fermentor than in a laboratory fermentor?

9. List three examples of *antibiotics* that are important industrially. For each of these antibiotics, list the producing organisms, the general chemical structure, and the mode of action.

10. Why are the β-lactam antibiotics so important medically? Compare and contrast the production of *natural, biosynthetic,* and *semisynthetic* β-lactam antibiotics.

11. Describe briefly the unique aspect of the regulation of streptomycin biosynthesis.

12. Addition of what metal to the fermentation medium can markedly improve production of vitamin B_{12}?

13. What unusual characteristics must an organism have if it is to overproduce and excrete an amino acid such as *glutamic acid*?

14. Define *microbial bioconversion* and give an example. Explain why the chemical reactions involved in microbial bioconversions are preferably carried out microbially rather than chemically.

15. List three different kinds of enzymes that are produced commercially. For each enzyme, list the organism used in commercial production, the action of the enzyme, and how the enzyme is used in commerce.

16. Describe the stages involved in the production of *high fructose syrup* and explain the role of an enzyme in each step. How is high fructose syrup used in the food industry?

17. Why is it desirable to *immobilize* enzymes? Give examples of two different immobilization procedures and describe how each is carried out.

18. Give two reasons why stainless steel fermentors are used in the industrial production of citric acid.

19. Why are yeasts of such great industrial importance?

20. In what way is the manufacture of *beer* similar to the manufacture of *wine*? In what ways do these two processes differ? How does the production of *distilled alcoholic beverages* differ from that of beer and wine?

21. What part of the mushroom is actually consumed as food? What is contained within this structure?

22. Why is sewage that has received only primary treatment both a serious health hazard and a source of environmental pollution?

23. Compare and contrast aerobic and anoxic means of *secondary* sewage treatment. What types of compounds are best degraded by each method? What are the major products of the treatment process in each case?

APPLICATION QUESTIONS

1. You have just isolated a strain of the yeast *Saccharomyces cerevisiae* that grows over twice as fast as any known strain of *S. cerevisiae* and may thus be useful in ethanol production. However, the yeast ferments glucose to a mixture of ethanol and other products and the yield of ethanol is only about 50% that of industrial yeast strains. Recalling what you have learned in this chapter about industrial microorganisms and also your knowledge of microbial physiology (Chapter 4), genetics (Chapter 9), and genetic engineering (Chapter 10), outline a plan for converting your new yeast strain to a strain useful for the production of commodity ethanol.

2. As a researcher in a pharmaceutical company you are assigned the task of finding and developing an antibiotic effective against a new bacterial pathogen. Outline a complete plan for this process, starting from isolation of the low-yield producing organism to high yield industrial production of the new antibiotic.

3. A partially consumed bottle of an "organic" (containing no preservatives) red wine is recapped and stored under refrigeration for 2 months. On tasting the wine again, you notice a distinct bitter taste, making the wine undrinkable. Using information presented in his chapter and the principles of microbial growth control discussed in Chapter 11, describe (a) what microbe-mediated process occurred in the wine, and (b) two ways in which this process could have been prevented.

4. You wish to produce high yields of the amino acid phenylalanine for use in production of the sweetener *aspartame*. The overproducing organism you wish to use is not subject to feedback inhibition by phenylalanine but is subject to typical repression of phenylalanine biosynthesis enzymes by excess phenylalanine. Applying the principles of enzyme regulation studied in Chapter 7 and microbial genetics in Chapter 9, describe two classes of mutants you could isolate that would overcome this problem and detail the genetic lesions each would have.

SUPPLEMENTARY READINGS

Betina, V. 1994. *Bioactive Secondary Metabolites.* Elsevier Science, Amsterdam. A detailed treatment of major secondary metabolites of industrial importance.

Bitton, G. 1994. *Wastewater Microbiology.* Wiley-Liss, New York. A thorough treatment of microbial processes in wastewater bioconversions. Also contains a readable account of engineering aspects for nonexperts.

Burch, B. 1992. *Brewing Quality Beers—The Home Brewer's Essential Guidebook,* 2nd edition. Joby Books, Fulton, CA. A step-by-step tour through the home brewing process.

Crueger, W., and **A. Crueger.** 1990. *Biotechnology: A Textbook of Industrial Microbiology,* 2nd edition. English edition edited by Thomas D. Brock. Sinauer Associates, Sunderland, MA. A concise textbook of industrial microbiology that emphasizes the economically important microbial processes. Has excellent chapters on the various large-scale processes and a good chapter on fermentor design and scale-up.

Doran, P. M. 1995. *Bioprocessing Engineering Principles.* Academic Press, San Diego, CA. A book on engineering principles essential for microbiologists interested in large-scale fermentations.

Glazer, A. N., and **H. Nikaido.** 1995. *Microbial Biotechnology: Fundamentals of Applied Microbiology.* W. H. Freeman, New York. A textbook of industrial microbiology and biotechnology. Although the emphasis is on genetically engineered systems, a fair bit of coverage is devoted to traditional industrial fermentations.

Godfrey, T., and **S. West** (eds.). 1996. *Industrial Enzymology–The Application of Enzymes in Industry,* 2nd edition. Stockton, New York. A major reference source on the use of enzymes in various industries. Packed with reference data.

Hershberger, C. L., S. W. Queener, and **G. Hegeman** (eds.). 1989. *Genetics and Molecular Biology of Industrial Microorganisms.* American Society for Microbiology, Washington, DC. Detailed reviews on the genetics of biosynthesis of most major antibiotics and on the molecular biology of industrially useful Bacteria.

Hugo, W. B., and **A. D. Russell.** 1992. *Pharmaceutical Microbiology,* 5th edition. Blackwell Scientific, Oxford, England. Detailed treatment of antibiotic production and microbial contamination problems in large-scale production processes.

Leatham, G. F., and **M. E. Himmel** (eds.). 1991. *Enzymes in Biomass Conversion.* American Chemical Society, Washington, DC. A detailed treatment of the major enzymes in industrial use and their production.

Page, M. I. (ed.). 1992. *The Chemistry of β-Lactams.* Chapman & Hall, New York. An authoritative review of all aspects of β-lactam antibiotics including mode of action, structural features, and microbial resistance.

Rose, A. H., and **J. S. Harrison** (eds.). 1989. *Metabolism and Physiology of Yeasts.* Academic Press, New York. An advanced treatise on yeast physiology.

Sikyta, B. 1995. *Techniques in Applied Microbiology—Progress in Industrial Microbiology,* Vol. 31. Elsevier Science, Amsterdam. Covers many of the topics in this chapter.

Vining, L. C., and **C. Stuttard** (eds.). 1994. *Genetics and Biochemistry of Antibiotic Production.* Butterworth-Heinemann, Stoneham, MA. A treatise on molecular aspects of antibiotic production.

On~line resources for this chapter are on the World Wide Web at:
http://www.prenhall.com/~brock (click on the table of contents link and then select Chapter 12).

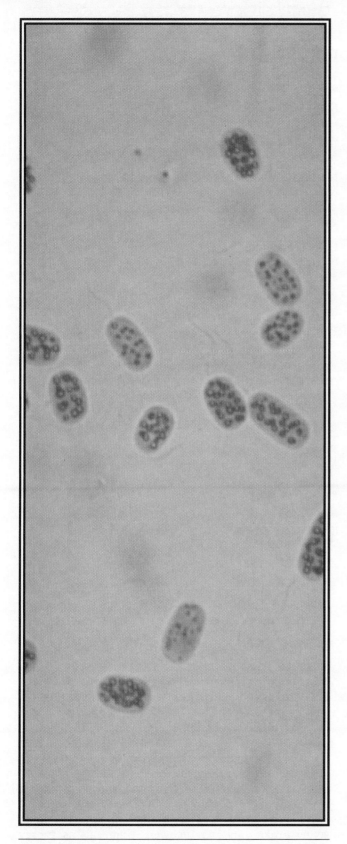

The purple phototrophic bacterium *Chromatium* has been used as a model system for studying green plant photosynthesis.

CHAPTER 13

Metabolic Diversity among the Microorganisms

MINIGLOSSARY for Chapter 13

Anaerobic Respiration respiration in which some substance such as SO_4^{2-} or NO_3^- serves as terminal electron acceptor instead of O_2

Anoxic oxygen-free

Anoxygenic Photosynthesis photosynthesis in which O_2 is not produced

Bacteriochlorophyll the chlorophyll pigment of anoxygenic phototrophs

Calvin Cycle the biochemical route of CO_2 fixation in many autotrophic organisms

Carotenoids hydrophobic accessory pigments present along with chlorophyll in photosynthetic membranes

Chemolithotroph a microorganism capable of oxidizing inorganic compounds as energy sources

Chlorophyll the light-sensitive, Mg-containing porphyrin of photosynthetic organisms that initiates the process of photophosphorylation

Denitrification anaerobic respiration in which NO_3^- is reduced to gaseous nitrogen compounds, primarily N_2

Disproportionation splitting of a compound into two new compounds, one more oxidized and one more reduced than the original compound

Fermentation anaerobic catabolism of an organic compound in which the compound serves as both an electron donor and an electron acceptor and in which ATP is produced by substrate-level phosphorylation

Hydrogenase an enzyme, widely distributed in anaerobic microorganisms, capable of taking up or evolving H_2

Methanogenesis the biological production of methane (CH_4)

Mixotrophic a nutritional state in which an inorganic compound serves as energy source (electron donor) and organic compounds serve as carbon source

Monooxygenase an enzyme that catalyzes the incorporation of one atom of O_2 into a substrate while the other atom is reduced to H_2O

Nitrification the microbial conversion of NH_3 to NO_3^-

Nitrogenase an enzyme capable of reducing N_2 to NH_3 in the process of nitrogen fixation

Nitrogen Fixation the biological reduction of N_2 to NH_3 by nitrogenase

Oxygenic Photosynthesis photosynthesis carried out by cyanobacteria and green plants in which O_2 is evolved

Photophosphorylation the production of ATP in photosynthesis

Phototroph an organism capable of using light as an energy source

Reaction Center a photosynthetic complex containing chlorophyll (or bacteriochlorophyll) and several other components within which occurs the initial electron transfer reactions of photosynthetic electron flow

Recalcitrant resistant to microbial attack

Reversed Electron Transport the energy-dependent movement of electrons against the thermodynamic gradient to form a strong reductant from a weaker electron donor

Syntrophy a process whereby two or more microorganisms cooperate to degrade a substance neither can degrade alone

---

Up until now, we have discussed microorganisms that can be said to have a "conventional" metabolism. These are organisms that use organic compounds as energy sources, primarily via aerobic (respiratory) metabolism. In Chapter 4, we considered briefly alternate modes of energy metabolism and indicated that there are interesting types of metabolism that we would return to later in this book. We discuss in this chapter some of these alternate microbial "lifestyles."

The diversity of organisms and metabolisms we are about to consider are some of the most widespread and significant on Earth. They play major roles in the functioning of the whole biosphere and are of great importance in agriculture and other aspects of human affairs, as we will discuss in Chapter 14.

13.1
Energy-Yielding Metabolism

Microorganisms show an impressive diversity of metabolic processes. Figure 13.1 reviews the basic types of energy metabolism and lists some of the terms that have been used. As seen, energy-yielding metabolism can be divided into two broad categories, that in which the energy source is a *chemical* compound, and that in which it is *light*. Many microorganisms resemble animals and humans, using *organic* chemicals as energy sources. We call these organisms **chemoorganotrophs**. An interesting collection of microorganisms, exclusively prokaryotic, use *inorganic* chemicals as energy sources. These organisms are called **chemolithotrophs**.

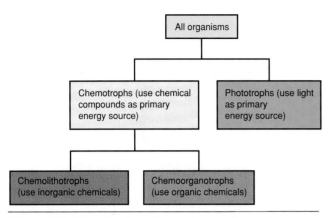

Figure 13.1 Classification of organisms in terms of energy.

TABLE 13.1	Reduction potentials of some biochemically important redox pairs[a]
Redox pair	E_0' **(V)**
SO_4^{2-}/HSO_3^-	−0.52
$2\,H^+/H_2$	−0.41
$S_2O_3^{2-}/HS^- + HSO_3^-$	−0.40
Ferredoxin ox/red	−0.39
$NAD^+/NADH$ (or $NADP^+/NADPH$)	−0.32
Cytochrome c_3 ox/red	−0.29
CO_2/acetate	−0.29
S^0/HS^-	−0.27
CO_2/CH_4	−0.24
SO_4^{2-}/HS^-	−0.22
Pyruvate⁻/lactate⁻	−0.19
HSO_3^-/HS^-	−0.11
Fumarate²⁻/succinate²⁻	+0.033
Cytochrome b ox/red	+0.035
Ubiquinone ox/red	+0.115
Cytochrome c_1 ox/red	+0.23
NO_2^-/NO	+0.36
Cytochrome a_3 ox/red	+0.385
NO_3^-/NO_2^-	+0.43
Fe^{3+}/Fe^{2+}	+0.77
O_2/H_2O	+0.82
N_2O/N_2	+1.36

[a]See Appendix 1 for details and Table A1.1 for a more complete list of reduction potentials.

Another major class of living organisms, called **phototrophs,** use *light* as an energy source. Phototrophs include green plants and also many microorganisms, both prokaryotic and eukaryotic. We will see that there are two distinct types of photosynthesis, that typified by green plants but also found in both eukaryotic and certain prokaryotic microorganisms, and that found exclusively in special groups of bacteria, the purple and green bacteria.

Organisms that use inorganic chemicals or light as energy sources are frequently able to grow in the complete absence of organic materials, using *carbon dioxide* as their sole source of carbon. The term **autotroph** (meaning literally, "self-feeding") is applied to organisms able to obtain all the carbon they need from inorganic sources. Note that autotrophy does not refer to the *energy* source used but to the *carbon* source. Autotrophs are of great importance in the functioning of the biosphere because they are able to bring about the synthesis of organic matter from inorganic (nonliving) sources; this process is called *primary production*. Because humans and other animals require *organic* carbon and are thus nutritionally called **heterotrophs,** the life of the biosphere itself is dependent on the activities of such autotrophic primary producers.

It might be useful at this point to review briefly the concept of oxidation–reduction, especially the terms *oxidation, reduction, electron acceptor, electron donor,* and *electron transport.* We use these terms in this chapter and assume that the student is familiar with them from Section 4.6. A summary of reduction potentials of some redox pairs important in the present chapter is given in Table 13.1 (full details can be found in Appendix 1).

In addition to alternate styles of energy metabolism, we also discuss in this chapter a number of specialized aspects of catabolism that we passed over lightly earlier. For instance, our discussion of respiratory metabolism centered primarily on O_2 as an electron acceptor, yet there are a number of other electron acceptors that can participate in the electron transport process in particular groups of microorganisms. The process in which an electron acceptor other than molecular oxygen is used is called **anaerobic respiration** and will be one of the major features of metabolic diversity considered here.

Moreover, our discussion of organic energy metabolism dealt primarily with the utilization of glucose and related sugars, but there are many other organic compounds that can be energy sources for chemoorganotrophs. The utilization of these "unusual" organic compounds is not only of biochemical interest but of applied interest as well, as we will see when we discuss topics like petroleum microbiology, and biodegradation in the next chapter.

Finally, although the major thrust of this chapter is *carbon* metabolism, our discussion would not be complete without a consideration of one of the important metabolic reactions of microorganisms that involves *nitrogen.* This is the utilization of nitrogen gas, N_2, as a source of nitrogen for biosynthesis, a process called **nitrogen fixation.** This important process contributes to the recycling of nitrogen into living matter and is important not only agriculturally but also in the total function of the biosphere. It is a process unique to prokaryotes.

13.2

Photosynthesis

One of the most important biological processes on Earth is **photosynthesis,** the conversion of light energy to chemical energy. Organisms that can carry out photosynthesis are called *phototrophs.* Most phototrophic organisms are also *autotrophs,* capable of growing with CO_2 as sole carbon source. Energy from light is thus used in the *reduction* of CO_2 to organic compounds. The ability to photosynthesize is dependent on the presence of light-sensitive pigments, the *chlorophylls,* found in plants, algae, and some bacteria. Light reaches phototrophic organisms in distinct units of energy called *quanta.* Absorption of light quanta by chlorophyll pigments begins the process of photosynthetic energy conversion.

Light and dark reactions

The growth of a photoautotroph can be characterized by two distinct sets of reactions: the **light reactions,** in which light energy is converted to chemical energy, and the **dark reactions,** in which this chemical energy is used to reduce CO_2 to organic compounds. For autotrophic growth, energy is supplied in the form of adenosine triphosphate (ATP), while electrons for the reduction of CO_2 come from NADPH. The latter is produced by the reduction of $NADP^+$ by electrons originating from various electron donors to be discussed later.

The light reactions bring about the conversion of *light* energy to *chemical* energy in the form of ATP. Purple and green bacteria use light primarily to form ATP; they produce NADPH from electron donors present in their environment, such as H_2S and organic com-

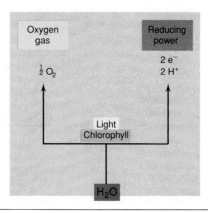

Figure 13.2 Production of O_2 and reducing power from H_2O by oxygenic phototrophs.

pounds. Green plants, algae, and cyanobacteria, however, do not generally use H_2S or organic compounds as electron donors. Instead, they obtain electrons for $NADP^+$ reduction by oxidizing water molecules, producing O_2 as a by-product (Figure 13.2). The reduction of $NADP^+$ to NADPH by these organisms is therefore a *light-mediated* event. Because molecular oxygen, O_2, is produced, the process of photosynthesis in these organisms is called **oxygenic** photosynthesis. In contrast, the purple and green bacteria do not produce oxygen; their process is called **anoxygenic** photosynthesis.

The dark reactions of photosynthesis are so named because unlike the light reactions, where light is essential, the dark reactions themselves do not require light in order to be carried out. However, the dark reactions do require the *products* of the light reactions, ATP and reducing power. Thus, the dark reactions can be considered "light-dependent" in the sense that without the ATP and reducing power generated by the light reactions, they could not occur. Thus, the dark reactions can occur in either the light or the dark, provided that the necessary components (ATP and reducing power) are available.

13.3

Role of Chlorophyll and Bacteriochlorophyll in Photosynthesis

Photosynthesis occurs only in organisms that possess some type of **chlorophyll.** Chlorophyll is a porphyrin, as are the cytochromes (∞ Section 4.12), but unlike the cytochromes, chlorophyll contains a *magnesium* atom instead of an iron atom at the center of the porphyrin ring. Chlorophyll also contains specific substituents bonded to the porphyrin ring, as well as a hydrophobic alcohol molecule. Because of this alcohol side chain, chlorophyll associates with lipid and hydrophobic proteins of photosynthetic membranes.

The structure of chlorophyll *a,* the principal chlorophyll of higher plants, most algae, and the cyanobacteria, is shown in Figure 13.3. Chlorophyll *a* is green in color because it *absorbs* red and blue light preferentially and *transmits* green light. We discussed the electromagnetic spectrum in Section 9.3 (∞ Figure 9.6). The spectral properties of any pigment can best be expressed by its *absorption spectrum,* which indicates the degree to which the pigment absorbs light of different wavelengths. The absorption spectrum of cells containing chlorophyll *a* shows strong absorption of red light (maximum absorption at a wavelength of 680 nm) and blue light (maximum at 430 nm) (Figure 13.4a).

There are a number of chemically different chlorophylls that are distinguished by their different absorption spectra. Chlorophyll *b,* for instance, absorbs max-

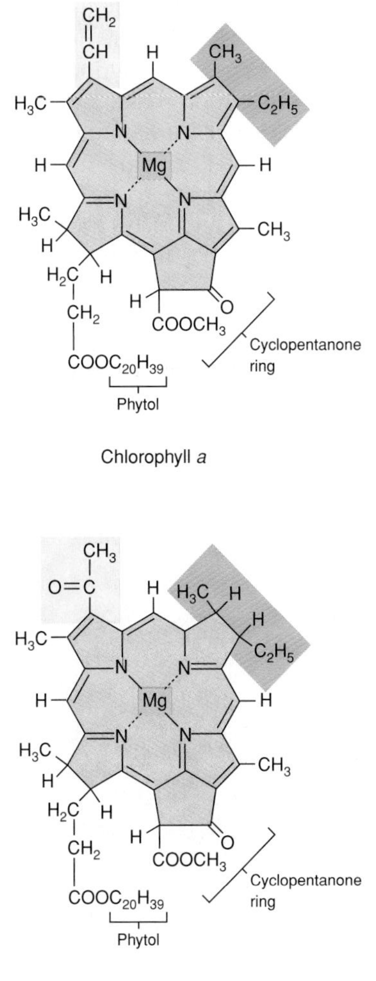

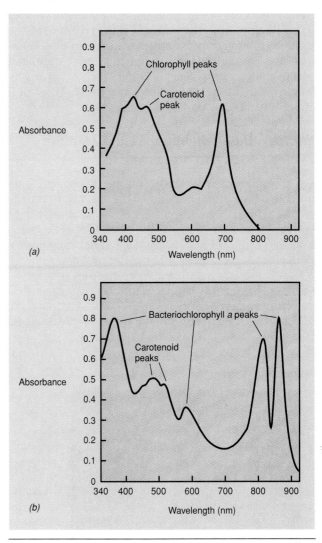

Figure 13.4 (a) Absorption spectrum of cells of the green alga *Chlamydomonas.* The peaks at 680 and 430 nm are due to chlorophyll *a;* the peak at 480 nm is due to carotenoids. (b) Absorption spectrum of cells of the phototrophic purple bacterium *Rhodopseudomonas palustris.* Peaks at 870, 800, 590, and 360 nm are due to bacteriochlorophyll *a* (870 = light-harvesting I; 800 = light-harvesting II); peaks at 525 and 475 nm are due to carotenoids.

Figure 13.3 Structures of chlorophyll *a* and bacteriochlorophyll *a,* both of which are magnesium tetrapyrroles. The two molecules are identical except for those portions contrasted in yellow and green. The central Mg atom is shown in blue.

imally at 660 nm rather than at 680 nm. Many plants have more than one chlorophyll, but the most common are chlorophylls *a* and *b*. Among prokaryotes, the cyanobacteria have chlorophyll *a*, but the purple and green bacteria have chlorophylls of slightly different structure, called **bacteriochlorophyll.** Bacteriochlorophyll *a* (Figure 13.3) from purple phototrophic bacteria absorbs maximally at about 800 and 850–880 nm (Figure 13.4*b*); other bacteriochlorophylls absorb maximally at 720–790 nm, and one type of bacteriochlorophyll absorbs at 1020 nm (∞ Figure 16.3).

Why do organisms have several kinds of chlorophylls absorbing light at different wavelengths? One reason appears to be to make it possible to use more of the energy of the electromagnetic spectrum. Only light energy that is *absorbed* can be used to make energy, and so by having more than one chlorophyll, more of the incident light energy becomes available to the organism. By having different pigments, two unrelated organisms can coexist in a habitat, each using wavelengths of light that the other is not using. Thus, pigment diversity has ecological significance.

Photosynthetic membranes and reaction center versus antenna pigments

Just where are the chlorophyll pigments located inside the cell? These pigments, and all the other components of the light-gathering apparatus, are associated with special membrane systems, the **photosynthetic membranes.** The location of the photosynthetic membranes within the cell differs between prokaryotic and eukaryotic microorganisms. In eukaryotes, photosynthesis is associated with special intracellular organelles, the **chloroplasts** (∞ Section 3.15 and Figures 3.76, 3.77, and 13.5*a*). The chlorophyll pigments are attached

to sheetlike (lamellar) membrane structures of the chloroplast (Figure 13.5*b*). These photosynthetic membrane systems are called **thylakoids;** stacks of thylakoids are called *grana* (Figure 13.5*b*). The thylakoids are so arranged that the chloroplast is divided into two regions, the matrix space that surrounds the thylakoids, and the inner space within the thylakoid array (Figure 13.5*b*). This arrangement makes possible the development of a light-driven proton motive force which can be used to synthesize ATP, as will be described in the next section.

Within the thylakoid membrane, the chlorophyll molecules are associated in complexes consisting of about 200–300 molecules. Only relatively few of these chlorophyll molecules participate *directly* in the conversion of light energy to ATP. These special chlorophyll molecules are referred to as **reaction center chlorophyll** and receive energy by transfer from the more numerous **light-harvesting** (or **antenna**) chlorophyll molecules (Figure 13.6). The chlorophyll molecules are bound to proteins that precisely control their orientation in the membrane so energy absorbed by one chlorophyll molecule can be efficiently transferred to another.

In prokaryotes, chloroplasts are not present and photosynthetic pigments are integrated into internal membrane systems that arise from (1) invagination of the cytoplasmic membrane (purple bacteria) (see for example, Figures 16.5*a* and *b*), (2) the cytoplasmic membrane itself (heliobacteria) (∞ Figure 16.5*d*), or (3) in both the cytoplasmic membrane and specialized nonunit membrane-enclosed structures called *chlorosomes* (green bacteria) (∞ Figure 16.5*c*). In chlorosomes, antenna bacteriochlorophyll molecules are not bound to proteins as in purple bacteria but still function to absorb light energy and transfer it to reaction center bacteriochlorophyll present in the cytoplasmic membrane.

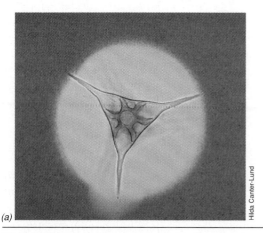
(a)

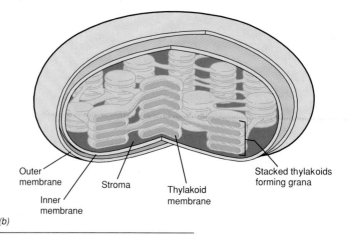
(b)

Outer membrane   Stroma   Thylakoid membrane   Stacked thylakoids forming grana
Inner membrane

Figure 13.5 The chloroplast. (a) Photomicrograph of an algal cell showing chloroplasts. (b) Details of chloroplast structure, showing how the convolutions of the thylakoid membranes define an inner space called the stroma, and form membrane stacks called grana.

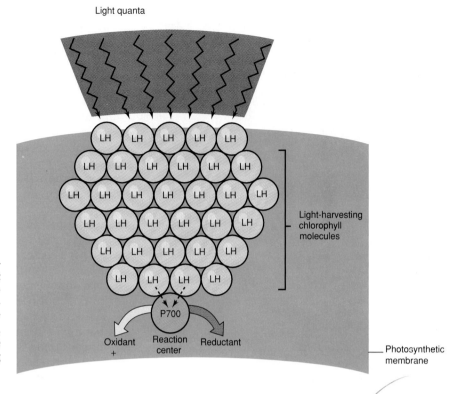

Light quanta

LH LH LH LH LH
LH LH LH LH LH LH
LH LH LH LH LH LH LH
LH LH LH LH LH LH
LH LH LH LH LH
LH LH LH LH

Light-harvesting
chlorophyll
molecules

P700

Oxidant
+

Reaction
center

Reductant

Photosynthetic
membrane

Figure 13.6 The photosynthetic unit and its associated reaction center in photosynthetic membranes of oxygenic phototrophs. Light energy, absorbed by light-harvesting chlorophyll molecules, travels to the reaction center where the actual charge separation occurs leading to ATP synthesis.

Antenna chlorophyll molecules make possible a dramatic increase in the rate at which photosynthesis can be carried out. At the low light intensities often prevailing in nature, reaction centers can be excited only about once per second, which would not be sufficient to carry out a significant photosynthetic process. The additional antenna chlorophyll molecules permit collection of light energy at a much more rapid rate. Since reaction center chlorophyll absorbs light energy over only a very narrow range of the spectrum, antenna pigments also perform the additional function of spreading out the spectral range available for use.

CONCEPT CHECK   *13.3*

The central light-harvesting pigment of photosynthesis is chlorophyll (or bacteriochlorophyll). Chlorophylls are located in membrane systems called photosynthetic membranes, where the light reactions of photosynthesis are carried out. Most chlorophyll molecules are antenna molecules and function only to harvest light energy and transfer it on to special molecules called reaction center chlorophylls.

✔ *Why is it necessary for chlorophyll pigments to be located in membranes?*

✔ *What is the difference between* antenna *and* reaction center *chlorophyll molecules? Which are more abundant in the cell and why?*

13.4
Anoxygenic Photosynthesis

The process of light-mediated ATP synthesis in all phototrophic organisms involves electron transport through a sequence of electron carriers. These electron carriers are arranged in the photosynthetic membrane in series from those with electronegative to those with more electropositive potentials. Conceptually, the process of photosynthetic electron flow resembles that of respiratory electron flow (∞ Sections 4.12 and 4.13). In fact, in phototrophic bacteria capable of aerobic (dark) growth, many of the same electron transport components are present in the membranes of cells grown in either the light (anoxic) or the dark (aerobic). We now consider the structure of the photosynthetic apparatus in anoxygenic phototrophs and the details of photosynthetic electron flow in purple bacteria, where much is known concerning the molecular events of photosynthesis.

Structure of the purple bacterial photosynthetic apparatus

The photosynthetic apparatus of purple phototrophic bacteria is contained in intracytoplasmic membrane systems of various morphologies. Membrane vesicles (chromatophores) (Figure 13.7) are a commonly observed membrane type. The photosynthetic apparatus consists of four membrane-bound pigment–protein complexes plus an ATPase complex that allows ATP

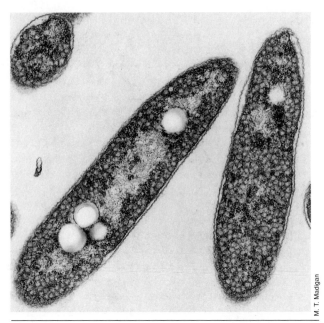

Figure 13.7 Chromatophores. Section through a cell of the phototrophic purple bacterium *Rhodobacter capsulatus* containing an abundance of vesicular photosynthetic membranes. The vesicles arise by invagination of the cytoplasmic membrane. The clear areas are regions in the cell in which the reserve polymer, poly-β-hydroxybutyrate (∞ Section 3.11), was stored.

synthesis at the expense of a proton gradient (∞ Section 4.13). Three of the four complexes specific to photosynthesis are the *reaction center, light-harvesting I,* and *light-harvesting II* components. The fourth complex of the photosynthetic apparatus, the *cytochrome bc₁ complex,* is common to both respiratory and photosynthetic electron flow. We discussed the structure and function of the cytochrome *bc₁* complex in Section 4.13.

In purple bacteria, the light-harvesting complexes (*antenna complexes*) contain two distinct spectral forms of bacteriochlorophyll *a* called B870 (light-harvesting I) and B800–850 (light-harvesting II) (see Figure 13.4*b*). The numbers refer to the wavelengths of radiation that are most strongly absorbed by these forms of bacteriochlorophyll in the cell. As previously described for oxygenic phototrophs (see Figure 13.6), the function of the antenna complexes is to absorb radiation and funnel the energy to the reaction center.

The purple bacterial photosynthetic reaction center has been crystallized and its structure determined to atomic resolution by X-ray diffraction (Figure 13.8*a*). Reaction centers of purple bacteria contain three polypeptides, designated the L, M, and H subunits. These proteins are firmly embedded in the photosynthetic membrane and traverse the membrane several times (Figure 13.8*b*). The L, M, and H polypeptides bind the reaction center photochemical complex, which consists of two molecules of bacteriochlorophyll *a* called the *special pair*, two additional bacteriochlorophyll *a* molecules whose function is unknown, two molecules of *bacteriopheophytin* (bacteriochlorophyll *a* minus its magnesium atom), two molecules of quinone, and two molecules of a carotenoid pigment (accessory photosynthetic pigment) (see

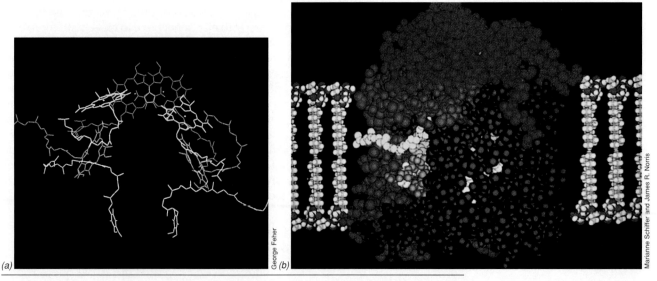

Figure 13.8 Structure of the reaction center of purple phototrophic bacteria. (a) Arrangement of components in the reaction center. The "special pair" of bacteriochlorophyll molecules are overlapping and shown in red, and molecules of quinone are in dark yellow and point downward in the figure. The accessory bacteriochlorophylls are in lighter yellow near the special pair, and the bacteriopheophytin molecules are shown in blue. (b) Molecular model of the protein structure of the reaction center. The pigments discussed in (a) are bound to membranes by three reaction center proteins called protein H (blue), protein M (red), and protein L (green). The reaction center pigment–protein complex is integrated into the lipid bilayer.

Section 13.6; Figure 13.8). All components of the reaction center are integrated in such a way that they can interact in very fast electron transfer reactions that, as we will see, ultimately result in ATP production.

Photosynthetic electron flow

It should be recalled that the photosynthetic reaction center is surrounded by light-harvesting antenna bacteriochlorophyll *a* molecules that function to funnel light energy to the reaction center (see earlier and Figure 13.8). Light energy is transferred from the antenna to the reaction center in packets called *excitons*, mobile electronic states that migrate through the antenna to the reaction center at high efficiency. Photosynthesis begins when exciton energy strikes the special pair of bacteriochlorophyll *a* molecules (Figure 13.8*a*). The absorption of energy excites the special pair, converting it to a good electron donor with a sufficiently low E_0' to reduce a very electronegative acceptor molecule. This represents work done on the system by light energy.

Before excitation, the bacterial reaction center, which is referred to as *P870*, has an E_0' of about +0.5

V; after excitation it has a potential of about −1.0 V (Figure 13.9). The excited electron within P870 proceeds to reduce a molecule of bacteriopheophytin within the reaction center (Figures 13.8*b* and 13.9). This transition takes place incredibly fast, taking about three-trillionths of a second (3×10^{-12} sec) to occur. Once reduced, bacteriopheophytin *a* reduces several intermediate quinone molecules, with the electron eventually reducing a quinone in the "quinone pool" within the membrane. This transition is also very fast, taking less than one-billionth of a second (Figures 13.9 and 13.10). Relative to what has happened in the reaction center, further electron transport reactions occur rather slowly, on the order of microseconds to milliseconds. From the quinone, electrons are transported in the membrane through a series of iron–sulfur proteins and cytochromes (Figures 13.9 and 13.10), eventually returning to the reaction center. Key electron transport proteins include cytochrome bc_1 and cytochrome c_2 (Figure 13.9). Cytochrome c_2 is a periplasmic cytochrome and serves as an electron shuttle between the membrane-bound bc_1 complex and the reaction center (∞ Section 4.13 and Figures 13.9 and 13.10).

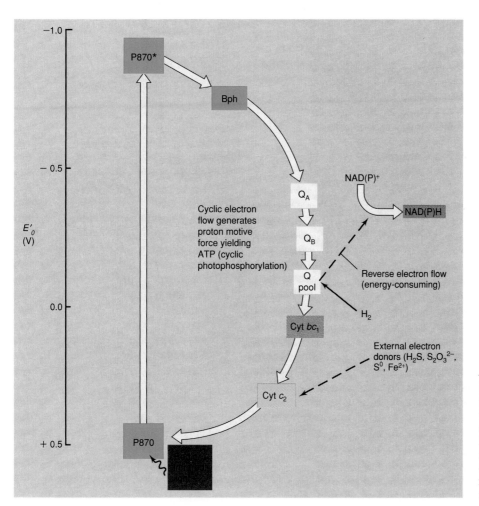

Figure 13.9 General scheme of electron flow in anoxygenic photosynthesis in a purple bacterium. Only a single light reaction occurs. RC, Reaction center; Bchl, bacteriochlorophyll; Bph, bacteriopheophytin; Q_A, Q_B, intermediate quinones; Q pool, quinone pool in membrane; Cyt, cytochrome.

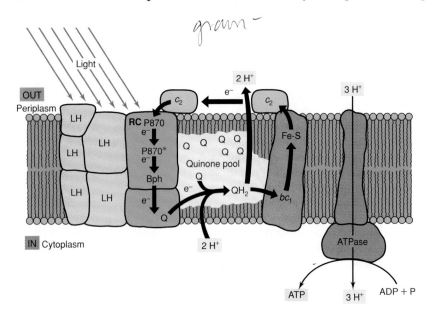

gram–

Figure 13.10 Arrangement of protein complexes in the photosynthetic membrane of a purple phototrophic bacterium. The light-generated proton gradient is used in the synthesis of ATP by the ATP synthase (ATPase). LH, Light-harvesting bacteriochlorophyll complexes; RC, reaction center; Bchl, bacteriochlorophyll; Bph, bacteriopheophytin; Q, quinone; FeS, iron–sulfur protein; bc_1, cytochrome bc_1 complex; c_2, cytochrome c_2. For description of the functioning of the cytochrome bc_1 complex and the Q cycle, the details of which are not shown here, see Figure 4.19.

Photophosphorylation

Synthesis of ATP during photosynthetic electron flow occurs as a result of the formation of a *proton motive force* generated by proton extrusion during electron transport and the activity of ATPases in coupling the dissipation of the proton motive force to ATP formation (see Section 4.13 for a discussion of the proton motive force). The reaction series is completed when cytochrome c_2 donates an electron to the special pair bacteriochlorophylls (Figure 13.9), returning these molecules to their original ground state potential ($E_0' = +0.5$ V). The reaction center is then capable of absorbing new energy and repeating the process. This method of making ATP is called **cyclic photophosphorylation** because electrons are repeatedly moved around a closed circle. Cyclic photophosphorylation resembles respiration in that electron flow through the membrane establishes a proton motive force. However, unlike respiration, in cyclic photophosphorylation *there is no net input or consumption of electrons;* electrons simply travel a closed route.

The spatial relationships of the electron transport components in the bacterial photosynthetic membrane are illustrated in Figure 13.10. Note that as in respiratory electron flow (∞ Section 4.13), the cytochrome bc_1 complex interacts with the quinone pool during photosynthetic electron flow to allow functioning of the Q cycle (∞ Figure 4.19), which is a major means of establishing the proton motive force used to drive ATP synthesis (Figure 13.10). Thus, the major function of light in photosynthesis in purple bacteria is to generate a strongly electronegative electron donor that can transfer electrons through the reaction center to quinones; the remaining reactions leading to ATP synthesis have a strong parallel in respiratory electron flow.

Genetics of bacterial photosynthesis

Purple phototrophic bacteria are gram-negative prokaryotes (∞ Section 16.1), and certain species are readily amenable to genetic manipulation. Species of the genus *Rhodobacter*, especially *R. capsulatus* and *R. sphaeroides*, have been the main subjects of genetic research in bacterial photosynthesis. In *R. capsulatus*, most genes involved in photosynthesis are clustered in several operons that span a 45- to 50-kb region of the chromosome that has been called the **photosynthetic gene cluster** (Figure 13.11). Genes in the photosynthetic gene cluster encode proteins involved in bacteriochlorophyll biosynthesis (*bch* genes) and in carotenoid biosynthesis (*crt* genes) and polypeptides that bind pigment molecules in the reaction center and light-harvesting complexes (*puf* and *puh* genes) (Figure 13.11).

As can be imagined, efficient synthesis of bacteriochlorophyll, carotenoids, and specific pigment binding proteins in phototrophic bacteria must be a highly coordinated process. When new photosynthetic complexes are synthesized, the correct proportions of each of the components of the complex must be available within the cell for final assembly. Biochemical and genetic analyses of photosynthesis in *Rhodobacter capsulatus* have shown that coordinate expression of photosynthetic components does indeed occur because the operons are arranged to form **superoperons.** Instead of terminating at the end of one operon, transcripts of pigment biosynthesis operons (*bch* and *crt*) (Figure 13.11) extend through the promoters and structural genes encoding the polypeptides of the photosynthetic complexes, yielding large transcripts encoding many proteins. Photosynthesis superoperons thus allow for transcription of many functionally related genes whose products interact and form the photosynthetic

Figure 13.11 Map of the photosynthetic gene cluster of the purple phototrophic bacterium, *Rhodobacter capsulatus*. Genes are arranged in superoperons where transcripts of pigment biosynthesis operons extend through to include transcription of polypeptides of the photosynthetic complexes. The *bch* genes, which code for bacteriochlorophyll synthesis proteins, are shown in green, while *crt* genes, which code for proteins that synthesize carotenoids, are shown in red. Genes encoding reaction center polypeptides (*puh* and *puf* genes) are shown in blue, and genes encoding light-harvesting I polypeptides (B870 complex) (*puf* genes) are shown in yellow. Genes shown by diagonal lines are of unknown function. Not all genes have been given letter designations. Arrows indicate direction of transcription.

complexes that eventually integrate into the membrane. The master regulatory signal governing transcription of the photosynthetic gene cluster in these organisms is O_2. Molecular oxygen represses pigment synthesis such that photosynthesis in anoxygenic phototrophs occurs only under *anoxic* conditions. Oxygen controls gene expression through interaction with a regulatory protein that prevents transcription of the photosynthetic gene cluster until conditions become anoxic.

Anoxygenic photosynthesis shares a number of important features with the oxygenic process of green plants, and genetic analysis of photosynthesis in purple bacteria has thus contributed enormously to our understanding of basic photosynthetic processes, especially as regards light-mediated ATP production. The remarkable nutritional versatility of purple bacteria (∞ Section 16.1) has made it possible to isolate mutants unable to photosynthesize (such mutants can still grow by respiration or fermentation in darkness). Nonphotosynthetic mutants of *Rhodobacter capsulatus* and *R. sphaeroides* in particular have been used to identify and genetically manipulate the key genes involved in photosynthesis (Figure 13.11).

Autotrophy and anoxygenic photosynthesis: reverse electron flow

As we have noted, purple and green bacteria do *not* produce O_2 during photosynthesis. Their photosynthesis is thus said to be *anoxygenic*. The reactions described previously have led to the conversion of light energy to high energy phosphate bonds of ATP. However, if an anoxygenic phototroph is going to grow with CO_2 as its sole or major carbon source, formation of ATP is not enough. Reducing power (NADPH) must also be made so CO_2 can be reduced to the level of cell material.

The ultimate source of electrons for anoxygenic phototrophs is some reduced substance from the environment; unlike oxygenic phototrophs, the source of reducing power in anoxygenic phototrophs is *not* water. Examples of electron donors used in anoxy-

genic photosynthesis are reduced sulfur compounds (for example, H_2S, elemental sulfur, thiosulfate), H_2, or organic compounds (for example, succinate, malate, butyrate). A few purple phototrophic bacteria can even use ferrous iron (Fe^{2+}) as photosynthetic electron donor, producing Fe^{3+} as the oxidized product (see Section 13.11). During autotrophic growth, anoxygenic phototrophs oxidize the external electron donor and reduce $NADP^+$ to NADPH. For instance, during growth on H_2S, elemental sulfur is produced:

$$6\,CO_2 + 12\,H_2S \rightarrow C_6H_{12}O_6 + 6\,H_2O + 12\,S^0$$

The sulfur formed is deposited either outside the cells in phototrophic green bacteria (Figure 13.12*a*) or inside the cells in most of the purple bacteria (Figure 13.12*b*).

How are electrons transferred from reduced substances to $NADP^+$? At least two ways are known for using reduced substances to produce NADPH. The first involves direct transfer from a highly reduced compound to $NADP^+$. The best example of an electron donor capable of direct transfer to $NADP^+$ is H_2. The $E_0{'}$ of H_2 is sufficiently low (-0.42 V) to reduce $NADP^+$ (-0.32 V) directly, provided the organism has the enzyme *hydrogenase* to oxidize H_2. However, many anoxygenic phototrophs can grow autotrophically using electron donors such as sulfide or thiosulfate, whose reduction potential is *more electropositive* than that of the $NADP^+/NADPH$ couple. Under these conditions, electrons from the quinone pool are forced backward (against the electropotential gradient) by the membrane potential, eventually reducing $NADP^+$ to NADPH (Figure 13.9). This energy-consuming process, called **reversed electron transport,** is necessitated by the need for a low potential reductant for CO_2 fixation, coupled with the reality of having to use electron donors of higher reduction potential (Figure 13.9). It is because electrons travel in a thermodynamically *un*favorable direction during reverse electron flow that energy, in this case the membrane potential, is required to form NADPH from $NADP^+$. Reversed electron transport is also a key feature of the synthesis of reducing power in chemolithotrophic prokaryotes (see Section 13.13).

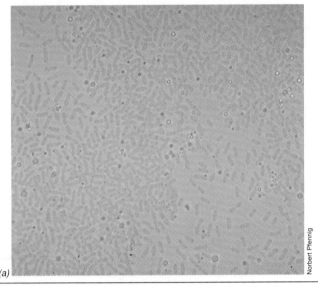

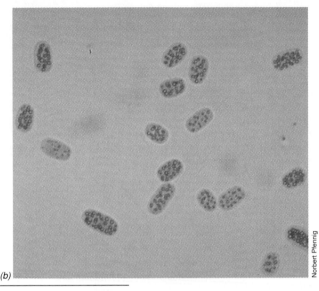

Figure 13.12 Photomicrographs of phototrophic bacteria taken by bright-field microscopy. (a) Green bacterium: *Chlorobium limicola*. The refractile bodies are sulfur granules deposited *outside* the cell. (b) Purple bacterium: *Chromatium okenii*. Notice the sulfur granules deposited *inside* the cell.

CONCEPT CHECK   13.4

A complex series of electron transport reactions occur in the photosynthetic reaction center of anoxygenic phototrophs, which result in the formation of a proton motive force and the synthesis of ATP. The reducing power for CO_2 fixation comes from reductants present in the environment and usually requires reverse electron transport.

✔ *How does photophosphorylation compare with electron transport phosphorylation in respiration?*

✔ *What is the value of having photosynthesis genes arranged in* superoperons?

✔ *What is* reverse electron flow *and why is it necessary?*

13.5

Oxygenic Photosynthesis

Electron flow in oxygenic phototrophs involves two distinct, but interconnected, photochemical reactions. Oxygenic phototrophs use light to generate both ATP *and* NADPH, the electrons for the latter arising from the splitting of *water* into oxygen and electrons (see Figure 13.2). The two systems of light reactions are called *photosystem I* and *photosystem II*, each photosystem having a spectrally distinct form of reaction center chlorophyll *a*. Photosystem I chlorophyll, called P700, absorbs light at long wavelengths (far red light), whereas photosystem II chlorophyll, called P680, absorbs at shorter wavelengths (near red light). Like

anoxygenic photosynthesis, oxygenic photochemical reactions occur in membranes. In eukaryotic cells, these membranes are found in the *chloroplast*, whereas in cyanobacteria, photosynthetic membranes are arranged in stacks within the cytoplasm. In both groups of phototrophs the membranes are arranged in a similar way and the two forms of chlorophyll *a* are attached to specific proteins in the membrane and interact as shown in Figure 13.13.

Electron flow in oxygenic photosynthesis

The path of electron flow in oxygenic phototrophs roughly resembles the letter Z turned on its side, and scientists studying oxygenic photosynthesis have come to refer to the electron flow of oxygenic phototrophs as the "Z" scheme. We should first note that the reduction potential of the P680 chlorophyll *a* molecule in photosystem II is very electropositive, slightly more positive than that of the O_2/H_2O couple (see Table 13.1). This facilitates the first step in oxygenic electron flow, the *splitting* of water into oxygen and hydrogen atoms (Figure 13.13), a thermodynamically unfavorable reaction. An electron from water is donated to the oxidized P680 molecule following the absorption of a quantum of light near 680 nm. Light energy converts P680 into a moderately strong reductant, capable of reducing an intermediary molecule of E_0' about −0.5 V. The nature of this molecule is uncertain, but it is likely a pheophytin *a* molecule (chlorophyll *a* without the magnesium atom). From here the electron travels through several membrane carriers including quinones, cytochromes, and a copper-containing protein called **plastocyanin;** the latter donates electrons to photosystem I. The electron is accepted by the reaction center chlorophyll of photo-

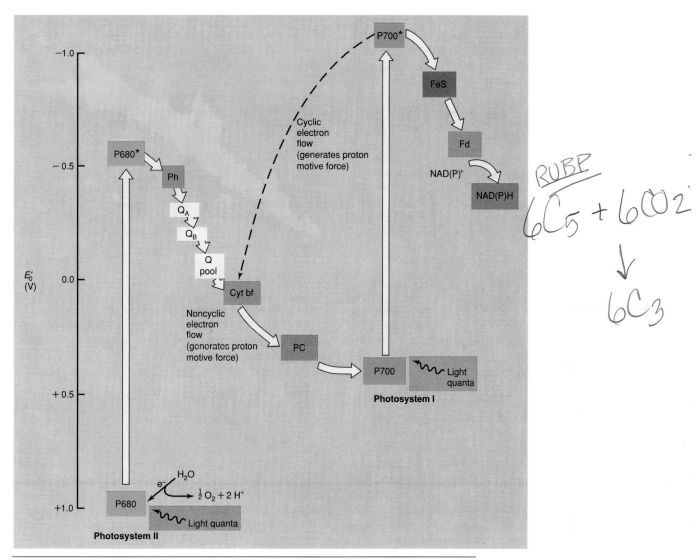

Figure 13.13 Electron flow in oxygenic (green plant) photosynthesis, the "Z" scheme. Two photosystems (PS) are involved, PS I and PS II. Ph, Pheophytin; Q, quinone; Cyt, cytochrome; PC, plastocyanin; FeS, nonheme iron–sulfur protein; Fd, ferredoxin. P680 and P700 are the reaction center chlorophylls of PS II and PS I, respectively. Compare with Figure 13.9.

system I, P700, which has previously absorbed light quanta and donated electrons to the primary acceptor of photosystem I; this acceptor has a very electronegative E_0', about -1.1 V. As in photosystem II, the primary acceptor of electrons from photosystem I has not been positively identified but is thought to be a free radical form of chlorophyll a. At any rate, the acceptor in photosystem I, once reduced, is at a reduction potential sufficiently negative to reduce the iron–sulfur protein **ferredoxin** ($E_0' = -0.39$ V), which then reduces NADP$^+$ to NADPH plus H$^+$ (Figure 13.13).

ATP synthesis in oxygenic photosynthesis

Besides the net synthesis of reducing power (that is, NADPH), other important events take place while electrons flow in the membrane from one photosystem

to another. During transfer of an electron from the acceptor in photosystem II to the reaction center chlorophyll molecule in photosystem I, electron transport occurs in a thermodynamically favorable (negative-to-positive) direction. This generates a proton motive force from which ATP can be produced. This type of ATP generation has been called *noncyclic* photophosphorylation because electrons whose transport results in ATP formation do not cycle back to reduce the oxidized P680: they are ultimately used in the reduction of NADP$^+$. (Contrast cyclic photophosphorylation shown in Figure 13.9 with noncyclic photophosphorylation shown in Figure 13.13.) When sufficient reducing power is present, ATP can also be produced in oxygenic phototrophs by *cyclic* photophosphorylation involving only photosystem I (Figure 13.13.) This occurs when the primary acceptor of photosystem I, instead of reducing ferredoxin (and hence $NADP^+$), returns the electron to the P700 molecule via membrane-bound cytochromes *bf*. This flow creates a membrane potential and synthesis of additional ATP (see dashed line in Figure 13.13).

Anoxygenic photosynthesis in oxygenic phototrophs

Photosystems I and II normally function together in the oxygenic process. However, under certain conditions many algae and some cyanobacteria are able to carry out cyclic photophosphorylation using *only* photosystem I, obtaining reducing power from sources other than water. This is, in effect, photosynthesizing *anoxygenically,* as purple and green bacteria do. This alteration requires the presence of anoxic conditions as well as a reducing substance, such as H_2 or H_2S. Under these conditions the electrons for CO_2 reduction come not from water but from the reducing substance. In algae, H_2 is generally the reductant, and following a period of adaptation to anoxic conditions, the enzyme hydrogenase is made and is used to oxidize H_2, which reduces $NADP^+$ to NADPH directly.

A number of cyanobacteria can use H_2S as an electron donor for anoxygenic photosynthesis. When H_2S is used, it is oxidized to elemental sulfur (S^0), and sulfur granules similar to those produced by green sulfur bacteria are deposited outside the cells (see Figure 13.12*a*). The filamentous cyanobacterium *Oscillatoria limnetica* lives in sulfide-rich saline ponds where it carries out anoxygenic photosynthesis along with photosynthetic green and purple bacteria and produces sulfur as an oxidation product of sulfide (Figure 13.14). In cultures of *O. limnetica*, electron flow from photosystem II is strongly inhibited by H_2S, thus necessitating anoxygenic photosynthesis if the organism is to survive.

From an evolutionary point of view, the existence of cyclic photophosphorylation indicates a close relationship between green plant and bacterial photosyn-

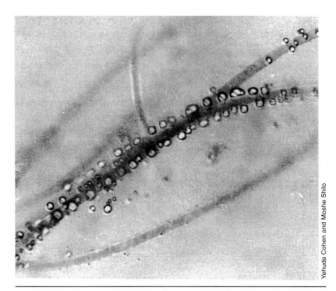

Figure 13.14 Cells of the filamentous cyanobacterium *Oscillatoria limnetica* grown anaerobically on sulfide as photosynthetic electron donor. Note the globules of elemental sulfur, the oxidation product of sulfide, formed *outside* the cells.

thesis. Although organisms such as *Oscillatoria limnetica* that carry out oxygenic photosynthesis have acquired photosystem II, and hence the ability to split H_2O, they still retain the ability under certain conditions to use photosystem I alone.

13.6

Additional Aspects of Photosynthetic Light Reactions: Carotenoids and Phycobilins

Although chlorophyll or bacteriochlorophyll is obligatory for photosynthesis, phototrophic organisms have other pigments involved in the capture of light energy. These include the **carotenoids** and the **phycobilins.** They function as *accessory* pigments and primarily play a photoprotective role (carotenoids) or serve as light-harvesting pigments (phycobilins). We consider each of these groups of pigments now.

Carotenoids

The most widespread accessory pigments are the **carotenoids,** which are always found in phototrophic organisms. Carotenoids are water-insoluble pigments firmly embedded in the membrane; the structure of a typical carotenoid is shown in Figure 13.15. Carotenoids have long hydrocarbon chains with alternating C—C and C=C bonds, an arrangement called a *conjugated* double-bond system. As a rule, carotenoids are yellow, red, brown, or green in color (∞ Figure 16.4*b*) and absorb light in the blue region of the spectrum (see

---

CONCEPT CHECK   **13.5**

In oxygenic photosynthesis, water is used as the source of electrons and oxygen is produced. Electron transport in oxygenic photosynthesis follows the Z scheme in which two separate light reactions are involved, photosystems I and II. Photosystem I resembles the system in anoxygenic photosynthesis. Photosystem II is responsible for splitting H_2O to yield $\frac{1}{2}O_2 + 2 e^- + 2 H^+$.

✔ *Why is the term* noncyclic *electron flow used in reference to oxygenic photosynthesis?*

✔ *What is a major difference between the two reaction center chlorophyll molecules in photosystems I and II?*

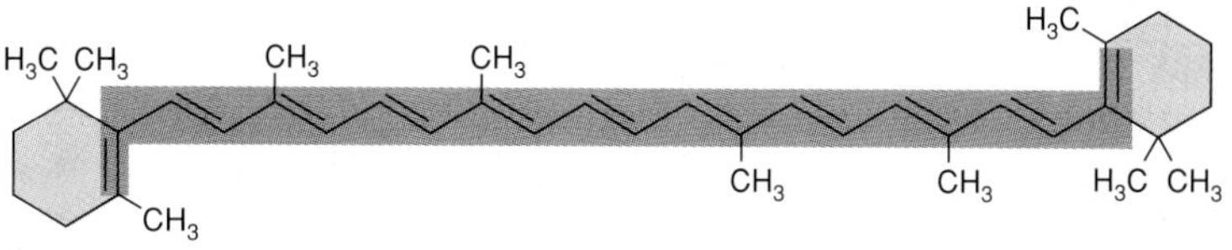

Figure 13.15 Structure of β-carotene, a typical carotenoid. The conjugated double-bond system is highlighted in orange.

Figure 13.4). Carotenoids are closely associated with chlorophyll in the photosynthetic membrane but do not function directly in photophosphorylation reactions. They can, however, transfer energy to the reaction center, and this transferred energy may be used in photophosphorylation in the same way as light energy captured directly by chlorophyll.

Another function of the carotenoids is as photoprotective agents. Bright light can often be harmful to cells in that it causes various photooxidation reactions that can lead to the production of singlet oxygen and destruction of chlorophyll and of the photosynthetic apparatus itself. Carotenoids quench singlet oxygen and absorb much of this harmful light, thus providing a shield for the light-sensitive chlorophyll. Because phototrophic organisms must by their very nature live in the light, the photoprotective role of carotenoids is of obvious advantage. However, some nonphototrophic organisms produce carotenoids as well. For example, many airborne bacteria produce carotenoid pigments to protect them from the lethal action of toxic oxygen molecules (∞ Section 5.11) produced by photochemical reactions. Exposing a nutrient medium plate to air and dust particles often reveals brightly pigmented colonies of carotenoid-containing bacteria. Carotenoids are also found in extremely halophilic bacteria such as *Halobacterium* (∞ Section 17.1). Although this organism is not able to carry out chlorophyll-based photosynthesis, it does employ a type of light-driven ATP synthesis as described in Section 17.1.

Phycobilins and phycobilisomes

Cyanobacteria and red algae contain **phycobiliproteins,** which are the main light-harvesting pigments of these organisms. Phycobiliproteins are red or blue in color and consist of open-chain tetrapyrroles coupled to proteins (Figure 13.16a). The red pigment, called *phycoerythrin,* absorbs light most strongly at wavelengths around 550 nm, whereas the blue pigment, *phycocyanin* (Figure 13.16a), absorbs most strongly at 620 nm (Figure 13.17). A third pigment, called *allophycocyanin* absorbs at about 650 nm.

Phycobiliproteins occur as high-molecular-weight aggregates, called **phycobilisomes,** attached to the photosynthetic membranes (Figure 13.16b). Phycobilisomes are constructed in such a way that the allophycocyanin molecules make physical contact with the photosynthetic membrane and are surrounded by molecules of phycocyanin and phycoerythrin. The latter pigments absorb shorter (higher energy) wavelengths of light and transfer the energy to allophycocyanin, which is closely linked to the reaction center chlorophyll and transfers energy to this site. The phycobilisome thus yields very efficient energy transfer from the biliprotein complex to chlorophyll *a,* which allows for growth of cyanobacteria at fairly low light intensities. Indeed, phycobilisome content *increases* in cells of cyanobacteria as light intensity *decreases,* such that phycobilisome-rich cells are those grown at the *lowest* light intensities.

Figure 13.16 Phycobilins and phycobilisomes. (a) A typical phycobilin. This compound is an open-chain tetrapyrrole derived biosynthetically from a closed porphyrin ring by loss of one carbon atom as carbon monoxide. The structure shown is the prosthetic group of phycocyanin, a proteinaceous pigment found in cyanobacteria and red algae. (b) Electron micrograph of a thin section of the cyanobacterium *Synechocystis* sp. Note the darkly staining ball-like phycobilisomes attached to the lamellar membranes.

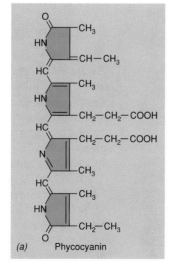

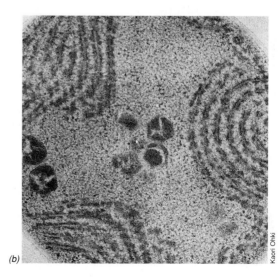

The light-gathering function of accessory pigments like carotenoids and phycobilins is of obvious advantage to the organism. Light from the sun is distributed over the whole visible range, yet chlorophylls absorb well in only part of this spectrum. By having accessory pigments, the organism is able to capture more of the available light (Figure 13.17).

CONCEPT CHECK   13.6

Accessory pigments such as carotenoids and phycobilins can absorb light and transfer the energy to reaction center chlorophyll, thus broadening the wavelengths of light usable in photosynthesis. Carotenoids also play an important photoprotective role in preventing photooxidative damage to the cell.

✔ *What are the functions of carotenoids in cells?*

✔ *How does the structure of a phycobilin compare with that of a chlorophyll?*

13.7

Autotrophic CO_2 Fixation: The Calvin Cycle

In the first part of this chapter, we discussed the *light* reactions of phototrophic organisms. We now consider the so-called *dark* reactions, the reactions by which autotrophs, both phototrophic and chemolithotrophic, convert CO_2 to organic matter.

Although all organisms require small amounts of their carbon for cellular biosynthesis to come from

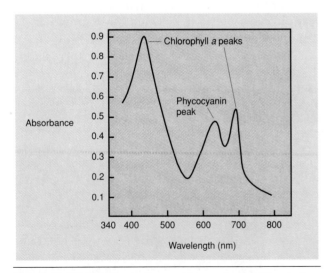

Figure 13.17 The absorption spectrum of a cyanobacterium that has a phycobiliprotein (phycocyanin) as an accessory pigment. Note how the presence of phycocyanin broadens the wavelengths of usable light energy (between 600 and 700 nm). Compare with Figure 13.4.

CO_2, autotrophs can obtain *all* their carbon from CO_2. The overall process is called *CO_2 fixation,* and all the reactions of CO_2 fixation can occur in complete darkness, using ATP and reducing power (NADPH) generated either during the light reactions of photosynthesis or during oxidation of inorganic compounds (see Section 13.8). Many autotrophs have a special pathway for CO_2 reduction, the **Calvin cycle,** also referred to as the *reductive pentose cycle.* This cycle is summarized later, but first let us consider some of the key individual reactions.

Key enzymes of the Calvin cycle

The first step in CO_2 reduction in the Calvin cycle is the reaction catalyzed by the enzyme *ribulose bisphosphate carboxylase* (RubisCO), which involves a reaction between CO_2 and ribulose bisphosphate (Figure 13.18*a*) leading to the formation of two molecules of *3-phosphoglyceric acid* (PGA), one of which contains the carbon atom from CO_2. PGA constitutes the first identifiable intermediate in the CO_2 reductive process. The carbon atom in PGA derived from CO_2 is still at the same oxidation level as it was in CO_2, and the next two steps involve reduction of PGA to the oxidation level of carbohydrate (Figure 13.18*b*). In these steps, *both* ATP and NADPH are required: the former is involved in the phosphorylation reaction that activates the carboxyl group, the latter in the reduction itself. [Note that this reaction is the reversal of a key glycolytic step (∞ Figure 4.12) except that NADPH is involved instead of NADH.] The average carbon atom in glyceraldehyde phosphate is now at the reduction level of carbohydrate (CH_2O), but only one of the carbon atoms of glyceraldehyde phosphate has been derived from CO_2, the other two having arisen from the ribulose bisphosphate. However, this is only the first part of the Calvin cycle. Most of the remaining reactions involve rearrangements to regenerate ribulose bisphosphate molecules, with some of the recently fixed CO_2 going to new cell synthesis.

Stoichiometry of the Calvin cycle

The overall scheme of the Calvin cycle is shown in Figure 13.19. The series of reactions leading to the synthesis of ribulose bisphosphate involve a number of sugar rearrangements. Through the action of enzymes that rearrange pentose phosphate compounds and enzymes of the glycolytic pathways, glyceraldehyde 3-phosphate is converted to ribulose 5-phosphate and subsequently to ribulose bisphosphate.

It is best to consider reactions of the Calvin cycle based on the incorporation of 6 molecules of CO_2. For RubisCO to incorporate 6 molecules of CO_2, 6 ribulose bisphosphate molecules are required as acceptor molecules (Figure 13.19). This yields 12 molecules of 3-phosphoglyceric acid (a total of 36 carbon atoms). These 12

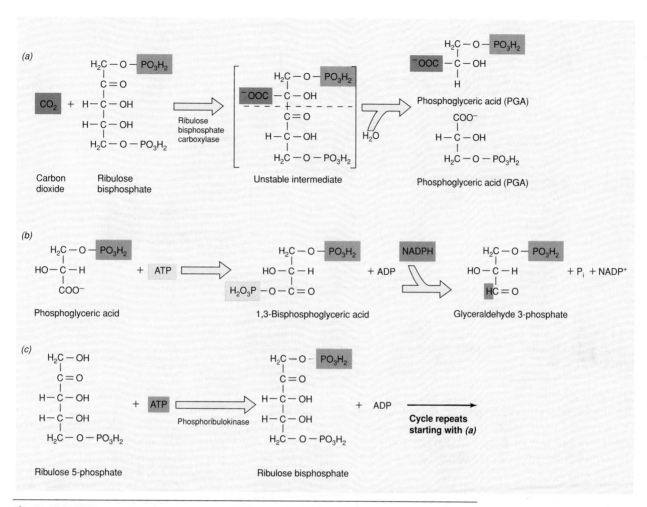

Figure 13.18 Key enzyme reactions of the Calvin cycle. (a) Reaction of the enzyme ribulose bisphosphate carboxylase. (b) Steps in the conversion of 3-phosphoglyceric acid (PGA) to glyceraldehyde 3-phosphate. Note that both ATP and NADPH are required. (c) Conversion of ribulose 5-phosphate to ribulose bisphosphate by the enzyme phosphoribulokinase.

molecules serve as carbon skeletons to form 6 *new* molecules of ribulose bisphosphate (a total of 30 carbon atoms), and 1 molecule of hexose for cell biosynthesis. A complex series of rearrangements involving C$_3$, C$_4$, C$_5$, C$_6$, and C$_7$ intermediates finally yields 6 molecules of ribulose 5-phosphate, from which the 6 ribulose bisphosphates are generated. The final step in the regeneration of ribulose bisphosphate is the phosphorylation of ribulose 5-phosphate with ATP by the enzyme *phosphoribulokinase* (Figure 13.18c). This enzyme is another catalyst unique to the Calvin cycle.

Let us consider now the *overall* stoichiometry for conversion of 6 molecules of CO$_2$ into 1 molecule of fructose 6-phosphate (Figure 13.19). Twelve molecules each of ATP and NADPH are required for the reduction of 12 molecules of phosphoglyceric acid (PGA) to glyceraldehyde phosphate, and 6 ATP molecules are required for conversion of ribulose phosphate to ribulose bisphosphate. Thus, *12 NADPH and 18 ATP are required to synthesize 1 hexose molecule from CO$_2$.* Hexose molecules can

be converted to *storage polymers* such as glycogen, starch, or poly-β-hydroxyalkanoates (∞ Section 3.11) during periods when ATP and NADPH are abundant and then can be used in other periods, such as in darkness, as sources of carbon and energy.

Alternatives to the Calvin cycle

The key enzymes of the Calvin cycle, *ribulose bisphosphate carboxylase* and *phosphoribulokinase,* are unique to autotrophs that fix CO$_2$ via the Calvin cycle. These enzymes have been found in virtually all phototrophic organisms examined—plants, algae, and bacteria. They are also found in many chemolithotrophic Bacteria, such as the sulfur, iron, and nitrifying bacteria (see later). The enzyme ribulose bisphosphate carboxylase is sometimes stored inside the cell in large aggregates called *carboxysomes* (∞ Section 16.5 and Figure 16.29)

However, several groups of autotrophs, including the phototrophic green sulfur bacteria, *Chloroflexus,* the

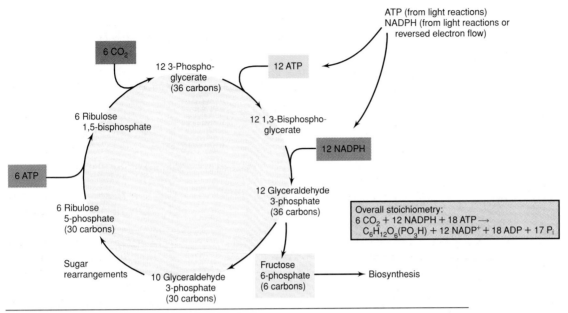

Figure 13.19 The Calvin cycle. For each *six* molecules of CO_2 incorporated, *one* fructose 6-phosphate is produced. Use the color coding here to follow the biochemical reactions occurring in Figure 13.18.

methanogenic Archaea, certain sulfate-reducing bacteria, and the homoacetogenic bacteria, do not use the Calvin cycle for CO_2 fixation. In these groups, Calvin cycle enzymes are not present. We discuss the pathways by which these groups of prokaryotes fix CO_2 in Chapters 16 and 17.

CONCEPT CHECK   *13.7*

The fixation of CO_2 by most phototrophic and other autotrophic organisms occurs via the Calvin cycle, in which the enzyme ribulose bisphosphate carboxylase (RubisCO) plays a key role. The Calvin cycle is an energy-demanding process in which NADPH and ATP are used to reduce CO_2 to the oxidation state of cell material.

✔ *What reaction does the enzyme ribulose bisphosphate carboxylase carry out?*

✔ *Why is reducing power needed for autotrophic growth?*

✔ *Of what importance is the enzyme phosphoribulokinase?*

13.8

Chemolithotrophy: Energy from the Oxidation of Inorganic Electron Donors

Organisms that obtain energy from the oxidation of inorganic compounds are called **chemolithotrophs.** Most chemolithotrophic bacteria are also able to obtain

all their carbon from CO_2, and they are therefore also autotrophs. As we have noted, two components are needed for growth on CO_2 as sole carbon source: energy in the form of ATP and reducing power. In chemolithotrophs, ATP generation is in principle similar to that in chemoorganotrophs except that the electron donor is *inorganic* rather than organic. Thus, ATP synthesis is coupled to *oxidation* of the electron donor. Reducing power in chemolithotrophs is obtained either directly from the inorganic compound, if it has a sufficiently low reduction potential, or by *reverse electron transport reactions,* as discussed for phototrophic bacteria in Section 13.4.

There are many sources of inorganic electron donors for chemolithotrophs. These include geological, biological, and anthropogenic sources. Volcanic activity is a major source of reduced sulfur compounds, as is biological sulfate reduction. Agricultural and mining operations add inorganic electron donors to the environment, as does the burning of fossil fuels and the input of industrial wastes. The ecological success (∞ Chapter 14) and metabolic diversity (this chapter) of chemolithotrophs indicates that sources and supplies of inorganic electron donors in nature are abundant.

Energetics of chemolithotrophy

A review of reduction potentials listed in Table 13.1 reveals that a number of inorganic compounds can provide sufficient energy for ATP synthesis when O_2 is used as electron acceptor. Recall from Chapter 4 that the further apart two half reactions are, the greater the amount of energy released. For instance, the difference in reduction potential between the H^+/H_2 couple and the $\frac{1}{2} O_2/H_2O$ couple is -1.23 V, which is equivalent to a free-energy yield of -237 kJ/mol (see Appendix 1 for calculations). On the other hand, the potential difference between the H^+/H_2 couple and the NO_3^-/NO_2^- couple is less, -0.84 V, equivalent to a free-energy yield of -163 kJ/mol. This is still quite sufficient for the production of ATP (the high energy phosphate bond of ATP has a free energy of about -31.8 kJ/mol). However, a similar calculation will show that there is insufficient energy available from the oxidation of H_2S using CO_2 as electron acceptor.

From such energy calculations, it is possible to predict the kinds of chemolithotrophs that might be found in nature. Since organisms obey the laws of thermodynamics, only reactions that are thermodynamically favorable are potential energy-yielding reactions. However, just because a reaction is theoretically possible does not mean that it actually occurs. There may be ecological reasons why a particular reaction does not take place. Table 13.2 summarizes energy yields for some reactions known to be carried out by chemolithotrophic microorganisms. We discuss some of these processes briefly in the rest of this chapter, and the organisms involved are considered in more detail in Chapters 16 and 17. We also examine ecological aspects of chemolithotrophy in Chapter 14.

CONCEPT CHECK   13.8

A number of specialized prokaryotes, called chemolithotrophs, are able to oxidize inorganic chemicals as their sole sources of energy and reducing power. Most chemolithotrophs are also able to grow autotrophically.

✔ *For what two purposes is a given inorganic compound used by a chemolithotroph?*

✔ *Why does the oxidation of H_2 yield more energy with O_2 as electron acceptor than with SO_4^{2-} as electron acceptor?*

13.9
Hydrogen-Oxidizing Bacteria

Hydrogen, H_2, is a common product of microbial metabolism, and a number of chemolithotrophs are able to use it as an energy source. A wide variety of H_2-oxidizing bacteria are known, differing in the electron *acceptor* they use. Those H_2-oxidizing bacteria that use electron acceptors other than oxygen carry out a type of anaerobic respiration and are discussed later in this chapter (see Sections 13.14–13.18). We discuss here the *aerobic* H_2-oxidizing species, which are the types most commonly referred to when the term *hydrogen bacteria* is used.

Most hydrogen bacteria are facultative chemolithotrophs, capable of growing either as conventional chemoorganotrophs on organic compounds or chemolithotrophically by H_2 oxidation. Taxonomically and phylogenetically, the hydrogen bacteria belong to various groups of Bacteria (∞ Section 16.6). When growing autotrophically, hydrogen bacteria show the following stoichiometry:

$$6 H_2 + 2 O_2 + CO_2 \rightarrow (CH_2O) + 5 H_2O$$

where the formulation (CH_2O) signifies cell material. The actual generation of ATP comes from the oxidation of H_2 by O_2:

$$H_2 + \frac{1}{2} O_2 \rightarrow H_2O$$

We discussed the free-energy change occurring in the preceding reaction earlier in this chapter. The enzyme **hydrogenase** catalyzes the initial oxidation of H_2, the electrons being transferred to a quinone acceptor. Electrons introduced into the electron transport chain from reduced quinone undergo electron transport reactions leading to the formation of a proton motive force, and ATP formation occurs via the membrane-bound ATPase. Electrons for CO_2 reduction can come directly from the activity of hydrogenase or from

---

TABLE 13.2 Energy yields from the oxidation of various inorganic electron donors[a]

Reaction	Type of chemolithotroph	E_0' of couple (V)	$\Delta G^{0\prime}$ (kJ/reaction)	$\Delta G^{0\prime}$ (kJ/2 e$^-$)
$H_2 + \frac{1}{2} O_2 \rightarrow H_2O$	Hydrogen bacteria	-0.42	-237.2	-237.2
$HS^- + H^+ + \frac{1}{2} O_2 \rightarrow S^0 + H_2O$	Sulfur bacteria	-0.27	-209.4	-209.4
$S^0 + 1\frac{1}{2} O_2 + H_2O \rightarrow SO_4^{2-} + 2 H^+$	Sulfur bacteria	-0.25	-587.1	-195.7
$NH_4^+ + 1\frac{1}{2} O_2 \rightarrow NO_2^- + 2 H^+ + H_2O$	Nitrifying bacteria	0[b]	-274.7	-137.4
$NO_2^- + \frac{1}{2} O_2 \rightarrow NO_3^-$	Nitrifying bacteria	$+0.43$	-75.8	-75.8
$Fe^{2+} + H^+ + \frac{1}{4} O_2 \rightarrow Fe^{3+} + \frac{1}{2} H_2O$	Iron bacteria	$+0.77$	-31	-62

[a]Data calculated from values in Appendix 1; values for Fe^{2+} are for pH 2, and others are for pH 7. At pH 7 the Fe^{3+}/Fe^{2+} couple is about $+0.2$ V.
[b]E_0' of the NH_3/NH_2OH couple.

the quinone pool via reverse electron transport reactions (see Section 13.4). In some hydrogen bacteria a separate hydrogenase exists that can reduce $NADP^+$ to NADPH for biosynthetic purposes (∞ Section 16.6). However, in most aerobic hydrogen bacteria, electrons must be pumped backward to reduce $NADP^+$.

When growing autotrophically, hydrogen bacteria fix CO_2 by the Calvin cycle. When organic compounds are present, synthesis of the Calvin cycle enzymes is repressed and enzymes involved in utilization of the organic compound induced. Regulation of chemolithotrophic metabolism in hydrogen bacteria is very precise and specific, with aspects of both carbon metabolism and energy metabolism playing important roles in gene expression.

Other H_2 bacteria

Many bacteria incapable of autotrophy can still oxidize H_2 as a sole energy source. For instance, some organisms can use H_2 as an energy source but cannot grow on CO_2 as sole carbon source. To obtain growth on H_2, therefore, these organisms must be given an organic compound as carbon source. Organisms that use an inorganic compound as energy source and an organic compound as carbon source are sometimes called *mixotrophs*. Sulfate-reducing bacteria, the homo-acetogenic bacteria, and most methanogens also use H_2 as electron donor, as discussed later in this chapter, but these organisms are all obligately anaerobic prokaryotes; many phototrophic purple bacteria can grow as H_2 bacteria under aerobic conditions in darkness.

The most common sulfur compounds used as energy sources are hydrogen sulfide (H_2S), elemental sulfur (S^0), and thiosulfate ($S_2O_3^{2-}$). The final product of sulfur oxidation in most cases is sulfate (SO_4^{2-}), and the total number of electrons involved between H_2S (oxidation state, -2) and sulfate (oxidation state, $+6$) is eight (see Table 13.5 for a summary of sulfur oxidation states). Less energy is available when one of the intermediate sulfur oxidation states is used:

$$H_2S + 2\,O_2 \rightarrow SO_4^{2-} + 2\,H^+$$
$$-798.2 \text{ kJ/reaction}$$

$$HS^- + \tfrac{1}{2}\,O_2 + H^+ \rightarrow S^0 + H_2O$$
$$-209.4 \text{ kJ/reaction}$$

$$S^0 + H_2O + 1\tfrac{1}{2}\,O_2 \rightarrow SO_4^{2-} + 2\,H^+$$
$$-587.1 \text{ kJ/reaction}$$

$$S_2O_3^{2-} + H_2O + 2\,O_2 \rightarrow 2\,SO_4^{2-} + 2\,H^+$$
$$-822.6 \text{ kJ/reaction}$$
$$(-411 \text{ kJ/S atom oxidized})$$

The oxidation of the most reduced sulfur compound, H_2S, occurs in stages, and the first oxidation step results in the formation of elemental sulfur, S^0. Some H_2S-oxidizing bacteria deposit the elemental sulfur formed inside the cell (Figure 13.20a). The sulfur deposited as a result of the initial oxidation is an energy reserve, and when the supply of H_2S has been depleted, additional energy can be obtained from the oxidation of sulfur to sulfate.

CONCEPT CHECK   13.9

The hydrogen bacteria use hydrogen gas (H_2) as energy source and fix CO_2 into cell carbon. Hydrogen bacteria are facultative chemolithotrophs also capable of using organic compounds when they are present.

✔ *What does the enzyme* hydrogenase *do?*

13.10

Sulfur Bacteria

Many reduced sulfur compounds can be used as electron donors by a variety of colorless sulfur bacteria [called "colorless" to distinguish them from the bacteriochlorophyll-containing (pigmented) green and purple sulfur bacteria discussed earlier in this chapter]. Indeed, the whole concept of chemolithotrophy emerged from studies of the sulfur bacteria as the great Russian microbiologist Winogradsky first proposed the idea of chemolithotrophy from studies of these organisms (see the box, Winogradsky's Legacy).

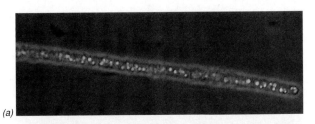
(a)

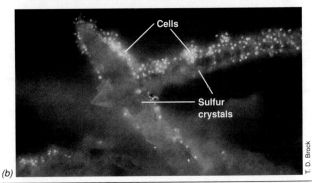
(b)

Cells

Sulfur crystals

T. D. Brock

Figure 13.20 Sulfur bacteria. (a) Deposition of internal sulfur granules by *Beggiatoa*. **(b)** Attachment of the sulfur-oxidizing archaean *Sulfolobus acidocaldarius* to a crystal of elemental sulfur. Visualized by fluorescence microscopy after staining the cells with the dye acridine orange. The sulfur crystal does not fluoresce.

LEARNING FROM THE PAST

WINOGRADSKY'S LEGACY

The discovery of autotrophy in chemolithotrophic bacteria was of major significance in the advance of our understanding of cell physiology because it showed that CO_2 could be converted to organic carbon without the intervention of chlorophyll. Previously, it had been thought that only green plants converted CO_2 to organic form. The idea of chemolithotrophic autotrophy was first developed by the great Russian microbiologist Sergei Winogradsky. Winogradsky studied sulfur bacteria because certain colorless sulfur bacteria (*Beggiatoa, Thiothrix*) are very large (see the figure) and hence are easy to investigate even in the absence of pure cultures. Springs with waters rich in H_2S are fairly common around the world, and Winogradsky studied several such springs in the Bernese Oberland district of Switzerland. In the outflow channels of sulfur springs, vast populations of *Beggiatoa* and *Thiothrix* develop, and suitable material for microscopic and physiological studies could be obtained by merely lifting up the white filamentous masses. Pure cultures were not needed for many studies. As Winogradsky noted, "This study of demipurity would be poor in an ordinary culture but is sufficient for a culture under the microscope, since it is possible to observe the development from day to day, almost from hour to hour, and see easily the presence of contaminants." Winogradsky first showed that the colorless sulfur bacteria were present only in water containing H_2S. As the water flowed away from the source, the H_2S gradually dissipated, and sulfur bacteria were no longer present. This suggested to him that their development was dependent on the presence of H_2S. Winogradsky then showed that by starving *Beggiatoa* filaments for a while, they lost their sulfur granules; he found, however, that the granules were rapidly restored if a small amount of H_2S was added (see the figure). He thus concluded that H_2S was being oxidized to elemental sulfur. But what happened to the sulfur granules when the filaments were starved of H_2S? Winogradsky showed by some clever microchemical tests that when the sulfur granules disappeared, sulfate appeared in the medium. Thus, he formulated the idea that *Beggiatoa* (and by inference other colorless sulfur bacteria) oxidize H_2S to elemental sulfur and subsequently to sulfate. Because they seemed to require H_2S for development in the springs, he postulated that this oxidation was the principal source of energy for these organisms.

Studies on *Beggiatoa* provided the first evidence that an organism could oxidize an *inorganic* substance as a possible energy source, and this was the origin of the concept of chemolithotrophy. However, the sulfur bacteria proved difficult to work with, primarily because there are a number of spontaneous chemical changes in sulfur compounds that can also occur and confuse the study. Winogradsky thus turned to a study of the nitrifying bacteria, and it was with this group that he clearly was able to show that autotrophic fixation of CO_2 was coupled to the oxidation of an inorganic compound in the complete absence of light and chlorophyll. The process of nitrification had been known before Winogradsky's work from studies on the fate of sewage when added to soil. Two French soil scientists, T. Schloesing and A. Mutz, had shown that the process was due to living organisms. Winogradsky proceeded to isolate some of the bacteria using completely mineral media in which CO_2 was the sole carbon source and ammonia was the sole electron donor. Because ammonia is chemically stable, it was easy to show that the oxidation of ammonia to nitrite, and subsequently to nitrate, is a strictly bacterial process. As no organic materials were present in the medium, it was also possible to show that organic matter (the bacterial cell material) was formed only from CO_2. When the ammonia was left out of the medium, no growth occurred. Careful chemical analyses showed that the amount of organic matter formed by the bacteria was proportional to the amount of ammonia or nitrite they oxidized. Winogradsky concluded, "This [process] is contradictory to that fundamental doctrine of physiology which states that a complete synthesis of organic matter cannot take place in nature except through chlorophyll-containing plants by the action of light." His basic conclusion has been confirmed by a large number of subsequent studies. At least in one way, however, autotrophy in most chemolithotrophs and phototrophs is similar in that in both processes the pathway of CO_2 fixation follows the same biochemical steps (the Calvin cycle) involving the enzyme ribulose bisphosphate carboxylase (see Section 13.7). ∎

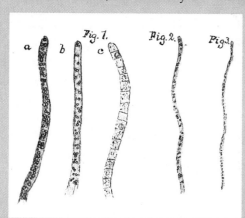

Drawings made by Winogradsky of *Beggiatoa* and translation (from the French) of the legend accompanying these figures. "Fig. 1. The tip of a filament of *Beggiatoa alba:* (a) in sulfurous [sulfide-containing] water, (b) after 24 hr in water nearly depleted in H_2S, (c) after 48 hr in water without H_2S [note depletion of sulfur globules with time]. Fig. 2. The tip of a filament of *Beggiatoa media.* Fig. 3. The tip of a filament of *Beggiatoa minima.*" From Winogradsky, S. 1949. *Microbiologie du Sol.* Masson, Paris.

When elemental sulfur is provided externally as an electron donor, the organism must grow attached to the sulfur particle because of the extreme insolubility of elemental sulfur (Figure 13.20*b*). By adhering to the particle, the organism can efficiently obtain the atoms of sulfur needed. This is thought to occur through the action of membrane or periplasmic proteins that solubilize the sulfur, probably by reduction of S^0 to HS^-, from which it is transported into the cell and enters chemolithotrophic metabolism.

Note that in the sulfur oxidation reactions shown here one of the products is H^+. Production of protons results in a lowering of the pH, and one result of the oxidation of reduced sulfur compounds is the acidification of the medium. The acid formed by the sulfur bacteria is *sulfuric acid*, H_2SO_4, and sulfur bacteria are often able to bring about a marked reduction in the pH of the medium. Some sulfur bacteria have been found to lower the pH to values less than 1!

Energetics and carbon metabolism in sulfur chemolithotrophs

The electron transport system of a typical sulfur-oxidizing bacterium is illustrated in Figure 13.21*a*. As shown, electrons from reduced sulfur compounds enter the chain at various points (depending on their reduction potentials) and are transported to molecular oxygen, the whole process generating a *proton motive force* that leads to ATP synthesis by membrane-bound ATPases. Electrons for CO_2 reduction come from reverse electron transport reactions (see Section 13.4) eventually yielding NADPH (Figure 13.21*a*). Note in Figure 13.21 how the relatively high reduction potentials of compounds like $S_2O_3^{2-}$ and S^0 result in a rather

abbreviated electron transport chain for energy purposes and a rather long route to the reduction of NADPH; the latter is an energy-dependent series of reactions. These biochemical limitations are in part responsible for the relatively poor growth yields of most sulfur chemolithotrophs.

When growing autotrophically, the sulfur bacteria use the Calvin cycle to fix CO_2. However, some sulfur bacteria are also able to grow using organic compounds. A few sulfur bacteria can grow anaerobically on reduced sulfur compounds, utilizing nitrate (NO_3^-) as electron acceptor. Under these latter conditions, a type of anaerobic respiration is carried out (see Section 13.14). Several sulfur-oxidizing bacteria appear to grow only mixotrophically, using H_2S as energy donor and an organic compound as carbon source. In this category is *Beggiatoa*, an organism widespread in marine and fresh waters and also in sulfur springs (see the box).

CONCEPT CHECK   13.10

The sulfur bacteria are chemolithotrophs that can use reduced sulfur compounds such as H_2S and S^0 as electron donors. Most sulfur bacteria fix CO_2 via the Calvin cycle with reducing power coming from reverse electron flow.

✔ *How do sulfur chemolithotrophs obtain carbon for cell growth?*

✔ *How many electrons are available from the oxidation of H_2S to SO_4^{2-}?*

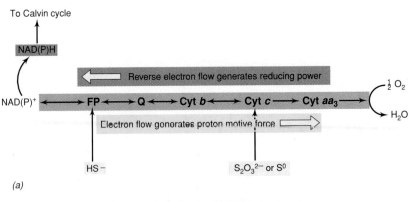

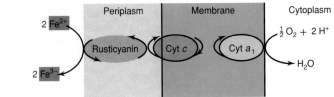

Figure 13.21 Electron flow in *Thiobacillus* grown on (a) sulfur, thiosulfate, sulfide, or (b) ferrous iron. In (a) note the direction of electron flow in ATP synthesis versus reverse electron flow. FP, Flavoprotein; Q, quinone; Cyt, cytochrome.

13.11

Iron-Oxidizing Bacteria

The aerobic oxidation of iron from the ferrous (Fe^{2+}) to the ferric (Fe^{3+}) state is an energy-yielding reaction for a few bacteria. Only a small amount of energy is available from this oxidation (see Table 13.2), and for this reason the iron bacteria must oxidize large amounts of iron in order to grow. Ferric iron forms very insoluble ferric hydroxide [$Fe(OH)_3$] precipitates in water (Figure 13.22a). Many iron-oxidizing bacteria also oxidize sulfur and are thus obligate acidophiles. This is in part because at neutral pH ferrous iron rapidly oxidizes nonbiologically to the ferric state and is thus stable only under *anoxic* conditions. At acid pH, however, ferrous iron is stable to chemical oxidation. Since spontaneous oxidation of ferrous iron in oxic environments

(a)

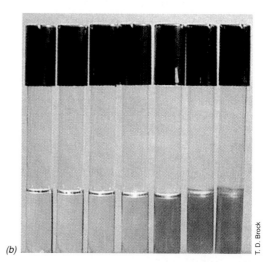

(b)

(c)

Figure 13.22 Iron-oxidizing bacteria. (a) Acid mine drainage, showing the confluence of a normal river and a creek draining a coal-mining area. The acidic creek is very high in ferrous iron. At low pH values, ferrous iron does not oxidize spontaneously in air, but *Thiobacillus ferrooxidans* carries out the oxidation. Insoluble ferric hydroxide and complex ferric salts precipitate, forming the precipitate called "yellow boy" by coal miners. (b) Cultures of *Thiobacillus ferrooxidans*. Shown is a dilution series, with no growth in the tube on the left and increasing amounts of growth from left to right. Growth is evident from the production of Fe^{3+}, which readily forms $Fe(OH)_3$, leading to the yellow-orange color. (c) Extensive development of insoluble ferric hydroxide in a small pool draining a bog in Iceland. Iron deposits such as this are widespread in cooler parts of the world and are modern counterparts of the extensive bog iron deposits of earlier geological eras. These ancient deposits are now the sources of much commercially mined iron ore. In the water-saturated bog soil, facultatively aerobic bacteria reduce ferric iron to the more soluble ferrous state. The ferrous iron leaches into the drainage area surrounding the bog, where oxidation occurs, either spontaneously or through the agency of iron-oxidizing bacteria (*Gallionella* and *Leptothrix*), and the insoluble ferric hydroxide deposit is formed.

at neutral pH is so rapid, significant amounts of ferrous iron do not accumulate under these conditions. This explains why most iron-oxidizing bacteria are obligately acidophilic.

The best-known iron-oxidizing bacterium is *Thiobacillus ferrooxidans,* which is able to grow autotrophically using either ferrous iron (Figure 13.22b) or reduced sulfur compounds as electron donors. The electron transport system was illustrated in Figure 13.21b. This organism is very common in acid-polluted environments such as coal-mining dumps (Figure 13.22a), and we discuss its role in acid-mine pollution and mineral oxidation in Sections 14.16 and 14.17. Another iron-oxidizing prokaryote is the archaean *Sulfolobus,* which lives in hot, acid springs at temperatures up to the boiling point of water (see Figure 13.20b).

Energy from ferrous iron

The bioenergetics of iron oxidation by *Thiobacillus ferrooxidans* is of biochemical interest because of the very electropositive reduction potential of the Fe^{3+}/Fe^{2+} couple (+0.77 V at pH 3). The respiratory chain of *T. ferrooxidans* contains cytochromes of the c and a_1 types and a copper-containing protein called *rusticyanin.* Because the reduction potential of the Fe^{3+}/Fe^{2+} couple is so high, the route of electron transport to oxygen ($\frac{1}{2} O_2/H_2O$, $E_0' = +0.82$ V) is very short. Electrons from the oxidation of Fe^{2+} cannot reduce NAD^+, FAD, and many of the other components of the electron transport chain. How, then, do acidophilic iron-oxidizing bacteria make ATP?

Biochemical studies of electron flow in *Thiobacillus ferrooxidans* suggest that acidophilic iron bacteria take advantage of the preexisting proton gradients of their environment for energy-generating purposes. Remember that the cytoplasm of any organism, including that of acidophiles, must remain near neutrality; in *T. ferrooxidans* the internal pH is about 6. However, the pH of the environment of *T. ferrooxidans* is much lower, near pH 2. This pH difference across the cytoplasmic membrane represents a natural proton motive force that can play a role in ATP synthesis. However, to retain a neutral intracellular pH, protons entering the cell through the proton translocating ATPase (driving the phosphorylation of ADP in the process) must be consumed. It is here that the oxidation of Fe^{2+} plays an important role.

The oxidation of Fe^{2+} to Fe^{3+} ($2 Fe^{2+} + \frac{1}{2} O_2 + 2 H^+ \rightarrow 2 Fe^{3+} + H_2O$) is a *proton-consuming* reaction. Experimental evidence suggests that the reaction $\frac{1}{2} O_2 + 2 H^+ \rightarrow H_2O$ occurs on the *inner* face of the cytoplasmic membrane, whereas the reaction $Fe^{2+} \rightarrow Fe^{3+}$ occurs near the *outer* face of the membrane. Electrons from Fe^{2+} are accepted in the periplasmic space by rusticyanin (Figure 13.21b); as befits its periplasmic location, rusticyanin is acid-stable and functions optimally at pH 2. Electrons are donated from rusticyanin to an unusually high potential membrane-bound cytochrome c which subsequently transfers electrons to cytochrome a_1, the terminal oxidase (Figure 13.21b). Cytochrome a_1 donates electrons to $\frac{1}{2} O_2$ with the two protons required to form water coming from the cytoplasm. An influx of protons via the ATPase replenishes the proton supply, and as long as Fe^{2+} remains available, the natural proton motive force across the *Thiobacillus ferrooxidans* membrane can continue to drive ATP synthesis.

As we discussed earlier, autotrophic CO_2 fixation requires the presence of both ATP and reducing power. Reducing power in the form of NADH or NADPH is formed by the process of *reverse electron transport* considered earlier in this chapter, using energy from the membrane potential to reverse electron flow and form NADH or NADPH. Because of the high potential of the Fe^{3+}/Fe^{2+} couple, this process takes a lot of energy, and it is therefore understandable that the cell yields of iron-oxidizing bacteria are quite low. In an environment where these organisms are living, their presence is signaled not by the formation of much cell material but by the presence of massive amounts of ferric iron precipitation (Figure 13.22a and b). We consider the important ecological roles of the iron-oxidizing bacteria in Section 14.16, and their taxonomy is discussed in Section 16.5.

Iron oxidation at neutral pH

In addition to the acidophilic iron-oxidizing bacteria just mentioned, there are other iron bacteria that live at near-*neutral* pH and are commonly found in environments where ferrous iron is moving from anoxic to oxic conditions (see Figure 13.22c for a typical site for these bacteria). As we have discussed, at neutral pH, ferrous iron is not stable in the presence of oxygen and is rapidly oxidized to the ferric (insoluble) state. Because of this, the only neutral pH environments where ferrous iron is present are *interfaces* between anoxic and oxic conditions. *Gallionella ferruginea, Sphaerotilus natans,* and *Leptothrix ochracea* live at such interfaces and are generally seen mixed in with the characteristic deposits they form (Figure 13.23). Only in *Gallionella* has true autotrophy been documented. Cultures of *G. ferruginea* grow on CO_2 as sole carbon source via the Calvin cycle. Autotrophy in other neutral pH iron bacteria is still unproven. We discuss the taxonomy of these interesting organisms in Section 16.10.

In addition to ferrous iron, the chemically related metal *manganese* is also oxidized by a few bacteria at neutral pH. The most common manganese-oxidizing bacterium is *Leptothrix discophora,* which is discussed

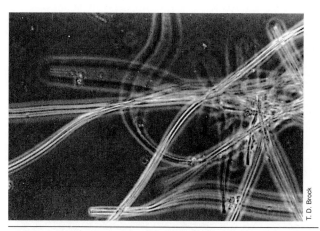

Figure 13.23 Phase contrast photomicrograph of empty iron-encrusted sheaths of *Sphaerotilus* collected from seepage at the edge of a small swamp.

in Section 16.14. We discuss the process of manganese oxidation itself in Section 14.16.

Ferrous iron oxidation by anoxygenic phototrophs

Ferrous iron can be oxidized under *anoxic* conditions by certain anoxygenic phototrophic bacteria (Figure 13.24). The ferrous iron is used in this case as an electron donor for photoautotrophic growth. At neutral pH, the Fe^{3+}/Fe^{2+} couple is much less electropositive than at pH 2, about $+0.2$ V, and thus electrons from Fe^{2+} can reduce cytochrome c in the photosystem of purple bacteria (see Section 13.4 for a discussion of anoxygenic photosynthesis). Anoxically, Fe^{2+} is stable at neutral pH where these organisms flourish. In culture, iron-oxidizing anoxygenic phototrophs are supplied with Fe^{2+} in the form of $FeCO_3$ and the following reaction is observed:

$$FeCO_3 + 10\ H_2O$$
$$\rightarrow 4\ Fe(OH)_3 + (CH_2O) + 3\ HCO_3^- + 3\ H^+$$

with (CH_2O) representing new cell material (Figure 13.24*a*). The organisms involved, which are species of purple bacteria (Figure 13.24*b*), can also use FeS; under these conditions both Fe^{2+} and S^{2-} are oxidized as electron donors.

The discovery of Fe^{2+}-oxidizing phototrophs has important implications for both understanding the evolution of photosynthesis and explaining the large depositions of ferric iron found in ancient sediments. Such ferric iron was previously thought to have been formed from the oxidation of Fe^{2+} by O_2 produced by oxygenic phototrophs (∞ Section 15.1). However, because of the age of these sediments, it is more likely that the ferric iron was formed by anoxygenic phototrophs oxidizing Fe^{2+} in anoxic environments.

CONCEPT CHECK   *13.11*

The iron bacteria are chemolithotrophs able to use ferrous iron (Fe^{2+}) as sole energy source. Most iron bacteria grow only at acid pH and are often associated with acid pollution from mineral and coal mining. Some phototrophic purple bacteria can oxidize Fe^{2+} to Fe^{3+} anoxically.

✔ *Why does most iron oxidation in nature occur under* acidic *conditions?*

✔ *Why is only a very small amount of energy available from the oxidation of Fe^{2+} to Fe^{3+}?*

✔ *How can Fe^{2+} be oxidized anoxically?*

13.12
Ammonium and Nitrite-Oxidizing Bacteria

The most common *inorganic nitrogen compounds* used as electron donors are ammonia (NH_3) and nitrite (NO_2^-), which are oxidized aerobically by the **nitrifying bacteria.** The nitrifying bacteria are widely distributed in soil and water. One group of organisms (*Nitrosomonas* is one genus) oxidizes ammonia to nitrite, and another group (*Nitrobacter*) oxidizes nitrite to nitrate; the complete oxidation of ammonia to nitrate, an eight-electron transfer (see Table 13.4 in Section 13.15), is thus carried out by members of these two groups of organisms acting in sequence. Nitrifying bacteria are widespread in soil, and their significance in soil fertility and in the nitrogen cycle is discussed in Section 14.14; their morphology, taxonomy, and physiology, is considered in Section 16.4. We deal here primarily with energetic principles.

Energy-yielding reactions in nitrifying bacteria

The electrons from nitrogen compounds enter an electron transport chain, and electron flow establishes a membrane potential and proton motive force linked to ATP synthesis. However, because of the reduction potential of their electron donors, nitrifying bacteria are faced with bioenergetic problems similar to those of the sulfur chemolithotrophs. The E_0' of the NH_2OH/NH_3 couple, the first step in the oxidation of NH_3, is about 0 V. The E_0' of the NO_3^-/NO_2^- couple is very high, about $+0.43$ V. These relatively high reduction potentials mean that nitrifying bacteria must donate electrons to their electron transport chains at rather late steps in the overall process. This effectively limits the amount of ATP that can be produced from each pair of electrons introduced.

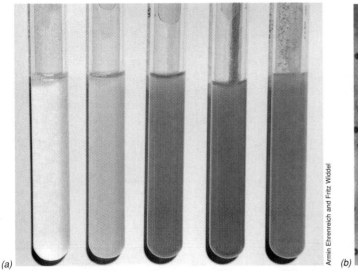

(a)

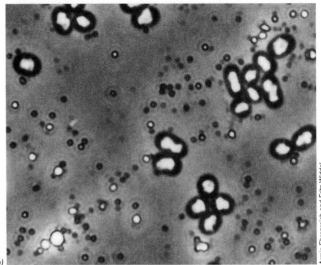

(b)

Figure 13.24 Ferrous iron oxidation by anoxygenic phototrophic bacteria. (a) Fe²⁺ oxidation in anoxic tube cultures. Left to right: Sterile medium, inoculated medium, growth to increasing cell densities. The brown-red color is due mainly to $Fe(OH)_3$ precipitate. (b) Phase contrast photomicrograph of an iron-oxidizing purple bacterium. The bright refractile areas within cells are gas vesicles. The granules outside the cells are iron precipitates. This organism is phylogenetically related to the purple sulfur bacterium *Chromatium* (∞ Section 16.1).

Several key enzymes are involved in oxidizing reduced nitrogen compounds. In ammonia-oxidizing bacteria, NH_3 is oxidized by *ammonia monooxygenase* (see Section 13.24 for a discussion of monooxygenase enzymes) that produces NH_2OH and H_2O (Figure 13.25). *Hydroxylamine oxidoreductase* then oxidizes NH_2OH to NO_2^-, removing *four* electrons in the process. Ammonia monooxygenase is an integral membrane protein, whereas hydroxylamine oxidoreductase is periplasmic (Figure 13.25). In the reaction carried out by ammonia monooxygenase,

$$NH_3 + O_2 + 2\,H^+ + 2\,e^- \rightarrow NH_2OH + H_2O$$

there is a need for two exogenously supplied electrons to reduce one atom of dioxygen to water. These electrons originate from the oxidation of hydroxylamine and are supplied to ammonia monooxygenase from hydroxylamine oxidoreductase via cytochrome *c* and ubiquinone (Figure 13.25). Thus, for every four electrons generated from the oxidation of NH_3 to NO_2^-, only two actually reach the terminal oxidase (cytochromes aa_3, Figure 13.25).

Nitrite-oxidizing bacteria employ the enzyme *nitrite oxidase* to oxidize nitrite to nitrate, with electrons traveling a very short electron transport chain (because of the high potential of the NO_3^-/NO_2^- couple) to the terminal oxidase (Figure 13.26). Cytochromes of the *a* and *c* types are present in the electron transport chain of nitrite oxidizers, and generation of a proton motive force (which ultimately drives ATP synthesis) occurs

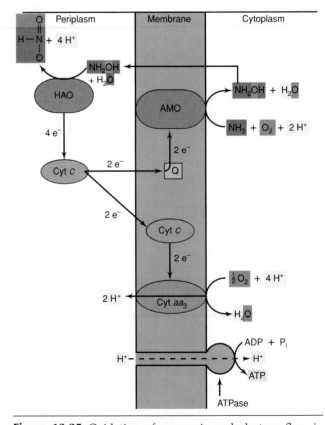

Figure 13.25 Oxidation of ammonia and electron flow in ammonia-oxidizing bacteria. The reactants and the products of this reaction series are highlighted. The cytochrome *c* (cyt *c*) in the periplasm is a different form of cyt *c* than that in the membrane. AMO, Ammonia monooxygenase; HAO, hydroxylamine oxidoreductase; Q, ubiquinone.

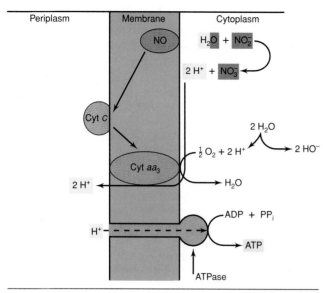

Figure 13.26 Oxidation of nitrite to nitrate by nitrifying bacteria. The reactants and products of this reaction series are highlighted. NO, Nitrite oxidase.

through the action of cytochromes aa_3 (Figure 13.26). And like the situation with iron oxidation (see Section 13.11), only small amounts of energy are available. Thus, growth yields of nitrifying bacteria are relatively low.

Anoxic ammonia oxidation: Anammox

Although classical nitrifying bacteria such as *Nitrosomonas* and *Nitrobacter* are strictly aerobic bacteria, at least when growing as chemolithotrophs on ammonia or nitrite, ammonia can also be oxidized under *anoxic* conditions. Anoxic wastewater or sludge supplemented with nitrate to stimulate denitrification ($NO_3^- \rightarrow N_2$) (see Section 13.15) also oxidizes ammonia to N_2. This reaction is nitrate-dependent and shows the following stoichiometry:

$$5 NH_4^+ + 3 NO_3^- \rightarrow 4 N_2 + 9 H_2O + 2 H^+$$
$$(\Delta G^{0\prime} = -1483 \text{ kJ/reaction})$$

This reaction, referred to as the *anammox reaction* (for *anoxic ammonia oxidation*) is highly exergonic, releasing an abundant amount of energy, which is presumably linked to energy conservation in the microorganisms responsible. The microbiology of annamox is not yet understood, but the process is not thought to be due to anoxic metabolic activities of known nitrifying bacteria. Interestingly, however, discovery of anammox has disproven the long-held assumption that ammonia is stable under anoxic conditions and can be oxidized only by the aerobic nitrifying bacteria.

Carbon metabolism in nitrifying bacteria

Like sulfur- and iron-oxidizing chemolithotrophs, aerobic nitrifying bacteria employ the Calvin cycle for CO_2 fixation, and the ATP requirements of this process (18 ATP are required for every glucose molecule produced) place additional burdens on an already inefficient energy-generating system. The energetic constraints are particularly severe for nitrite oxidizers, and it is perhaps for this reason that most nitrite oxidizers can also grow chemoorganotrophically on glucose and certain other organic substrates (∞ Section 16.4).

CONCEPT CHECK   *13.12*

Ammonia (NH_3) and nitrite (NO_2^-) can be used as electron donors by the nitrifying bacteria. The ammonia-oxidizing bacteria produce nitrite, which is then oxidized by the nitrite-oxidizing bacteria to nitrate (NO_3^-). Anoxic NH_3 oxidation is coupled to a denitrification process.

✔ *What is the inorganic electron donor for* Nitrosomonas? *For* Nitrobacter?

✔ *What are the substrates for the enzyme* ammonia monooxygenase?

✔ *What do nitrifying bacteria use as a carbon source?*

✔ *What is the anammox reaction?*

13.13

Summary of Autotrophy and ATP Production among Chemolithotrophs

As we have noted, when growing autotrophically, chemolithotrophs must also produce reducing power in addition to ATP if they are to grow with CO_2 as sole carbon source. Although NAD^+ can be reduced to NADH directly by H_2, all the other inorganic electron donors have reduction potentials more electropositive than that of NADH (Table 13.1). If the reduction potential of an electron donor is more electropositive than that of NADH, there is no way in which its oxidation can be directly coupled to the reduction of NAD^+ to NADH. Thus, NAD^+ must be reduced by energy-driven reversed electron transport as already discussed in Sections 13.4 and 13.11.

ATP production by chemolithotrophs

As noted in the discussions of the individual groups of chemolithotrophs, these bacteria all contain electron transport chains consisting of cytochromes, quinones, and many of the major components described in aerobic chemoorganotrophic organisms (∞ Sections 4.12

and 4.13). However, with the exception of H_2, no inorganic reductant has a sufficiently low E_0' to reduce NAD^+ directly, and so electrons feed into the electron transport chain at a point where the reduction potential of the intermediate is more electropositive than that of the $NAD^+/NADH$ couple. In most cases the electron acceptor turns out to be a cytochrome or an enzyme that feeds electrons into a cytochrome (see Figures 13.21, 13.25, and 13.26). As we noted, the situation is an extreme one for the iron-oxidizing and the nitrite-oxidizing bacteria because the reduction potential of their electron donors is already very high. This directly affects growth yields.

Table 13.2 summarized theoretical energy yields from the aerobic oxidation of various inorganic electron donors. It can be seen that the energetics of hydrogen and sulfur oxidation are more favorable than those of nitrite or iron oxidation because the amount of energy released is proportional to the *difference* in reduction potential between the electron donor and electron acceptor couples and the differences are greater in the former two oxidations. Hence, per mole of substrate oxidized, H_2 and S^0 oxidizers typically produce more biomass than NO_2^- or Fe^{2+} oxidizers (Figure 13.27). However, owing to the added burden of autotrophy, most chemolithotrophs synthesize only small amounts of cell material while oxidizing huge amounts of substrate. Chemoorganotrophs, by contrast, do not have as severe a problem (Figure 13.27) because their substrates are already preformed organic compounds that can be diverted into biosynthetic intermediates at less cost to the cell in terms of ATP and reducing power. From an ecological standpoint, however, it should be remem-

bered that chemolithotrophs are utilizing electron donors not catabolized by chemoorganotrophic organisms. Hence, chemolithotrophs survive in nature without severe nutritional competition from the bulk of the microbial world.

Comparison of energy and carbon metabolism

A summary of energy and carbon metabolism and the terminology used to describe different nutritional groups of organisms is indicated in Table 13.3. The ability of certain chemolithotrophs to assimilate organic compounds as carbon sources while using inorganic oxidations for energy (Table 13.3) makes possible the growth of these organisms as chemolithotrophic heterotrophs. This type of growth is referred to as **mixotrophic.** Some chemolithotrophs grow best under mixotrophic conditions because they are free of the energy-demanding process of autotrophic CO_2 fixation. At least one sulfur bacterium, *Beggiatoa*, seems to require mixotrophic conditions to use reduced sulfur compounds as electron donors. Most strains of *Beggiatoa* are unable to grow on a completely inorganic medium with reduced sulfur compounds and CO_2 but use reduced sulfur compounds as electron donors when acetate or some other suitable organic compound is present. *Beggiatoa* also grows on acetate as sole source of carbon and energy in the absence of reduced sulfur compounds. Mixotrophy allows an organism the best of both chemolithotrophic and chemoorganotrophic worlds. However, growth studies in mixed culture have shown that most mixotrophs compete poorly with strict chemolithotrophs or strict chemoorganotrophs when the mixotroph is forced to grow under one or the other contrasting nutritional regime. Thus, most mixotrophs are successful in nature only when *both* inorganic and organic compounds are present simultaneously.

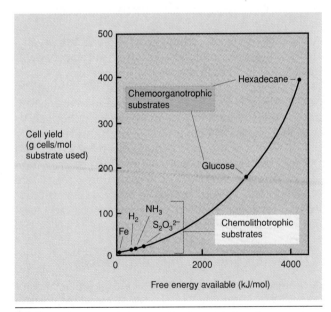

Figure 13.27 Comparison of cell yield (data from various microorganisms) on electron donors with different free energies of reaction. In all cases, the electron acceptor is O_2.

CONCEPT CHECK   13.13

The chemolithotrophic bacteria produce ATP via electron transport processes in which a proton motive force is generated from the oxidation of an inorganic compound. Reducing power for CO_2 fixation usually comes from reverse electron transport reactions, resulting in the formation of NAD(P)H. Because the energy yield from chemolithotrophic metabolism is often low, chemolithotrophs must oxidize a large amount of energy source in order to produce a significant amount of cell material.

✔ *What major energetic burdens do chemolithotrophs face that are not problems for chemoorganotrophs?*

✔ *What is a mixotroph?*

TABLE 13.3 Summary of energy and carbon metabolism and nutritional groups of organisms

Carbon source	Energy source		
	Light	Inorganic chemicals	Organic chemicals
Inorganic (CO_2, HCO_3^-, CO_3^{2-})	Phototrophic eukaryotes; cyanobacteria (photoautotrophs, H_2O is electron donor)	Chemolithotrophic autotrophs: hydrogen, sulfur, iron, and nitrifying bacteria	None known
	Purple and green bacteria (photoautotrophs, H_2S or H_2 is electron donor)		
Organic	Purple and green bacteria using organic carbon sources (photoheterotrophs)	Mixotrophs	Chemoorganotrophs (most prokaryotes and all nonphototrophic eukaryotes)

13.14

Anaerobic Respiration

We examined in some detail in Chapter 4 the process of *aerobic* respiration. As we noted, molecular oxygen (O_2) functions as an external electron acceptor, accepting electrons from electron carriers such as NADH by way of an electron transport chain. However, we noted in Chapter 4 that a variety of other electron acceptors can be used instead of O_2, in which case the process is called *anaerobic* respiration. We now consider some of the anaerobic respiration processes that have been recognized.

In most cases, the energy source(s) used by organisms carrying out anaerobic respiration are *organic* compounds, but several chemolithotrophic organisms can also carry out this process. In addition, although most of the electron acceptors to be discussed are *inorganic*, several organic compounds can also serve as electron acceptors. Most organisms carrying out anaerobic respiration are prokaryotes, and the chemical transformations they carry out during their energy generation process are frequently of great importance ecologically or industrially, as we will discuss in Chapter 14.

The bacteria carrying out anaerobic respiration generally possess electron transport systems containing cytochromes, quinones, iron–sulfur proteins and other typical electron transport proteins. Their respiratory systems are thus analogous to those of conventional aerobes. In some cases, such as with the denitrifying bacteria, the anaerobic respiration process competes in the same organism with an aerobic one. In such cases, if O_2 is present, aerobic respiration is usually favored, and when O_2 is depleted from the environment, the alternate electron acceptor is reduced. Other organisms carrying out anaerobic respiration are *obligate* anaerobes and are unable to use O_2.

Alternative electron acceptors and the electron tower

The energy released from the oxidation of an electron donor using O_2 as electron acceptor is higher than if the same compound is oxidized with an alternate electron acceptor. These energy differences are apparent if the reduction potentials listed in Table 13.1 are examined. As discussed at the beginning of this chapter, and in detail in Appendix 1, theoretical energy yields can be calculated using differences in reduction potential between the electron donor and the electron acceptor. Because the O_2/H_2O couple is the most oxidizing, more energy is available when O_2 is used than when another electron acceptor is used. As noted in Table 13.1, other electron acceptors that are near O_2 are Fe^{3+}, NO_3^-, and NO_2^-. Farther up the scale are S^0, CO_2, and SO_4^{2-}. A summary of the most common types of anaerobic respiration is given in Figure 13.28.

Assimilative and dissimilative metabolism

Inorganic compounds such as NO_3^-, SO_4^{2-}, and CO_2 are reduced by many organisms as sources of nitrogen, sulfur, and carbon, respectively. The end products of such reductions are primarily amino groups (—NH_2), sulfhydryl groups (—SH), and organic carbon compounds (at the oxidation level of carbohydrate or lower), respectively. We briefly examined the *nutrition* of microorganisms in Sections 4.2 and 4.3 and noted that all organisms need sources of N, S, and C for growth. When an inorganic compound such as NO_3^-, SO_4^{2-}, or CO_2 is reduced for use as a nutrient source, it is said to be *assimilated*, and the reduction process is called *assimilative* metabolism. We emphasize here that assimilative metabolism of NO_3^-, SO_4^{2-}, and CO_2 is quite different from the use of these compounds as electron acceptors for *energy* metabolism. To distinguish these two kinds of reduction processes, the use of these compounds as elec-

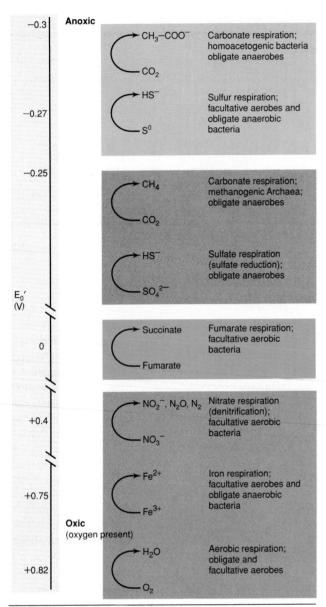

Figure 13.28 Various kinds of anaerobic respirations. The couples are arranged in order from those with the most electronegative E_0' (top) to those with the most electropositive E_0' (bottom).

and higher plants), whereas only a restricted variety of organisms carry out dissimilative metabolism (almost all are prokaryotes). We now discuss the main types of dissimilative metabolism.

CONCEPT CHECK   **13.14**

Although oxygen (O_2) is the most widely used electron acceptor in energy-yielding metabolism, a number of other compounds can be used as electron acceptors. This process of anaerobic respiration is less energy-efficient but makes it possible for respiration to occur in environments where oxygen is absent.

✔ *What is anaerobic respiration?*

✔ *With H_2 as electron donor, why might the reduction of NO_3^- be a more favorable reaction than the reduction of S^0?*

13.15

Nitrate Reduction and the Denitrification Process

Inorganic nitrogen compounds are some of the most common electron acceptors in anaerobic respiration. A summary of the various inorganic nitrogen species with their oxidation states is given in Table 13.4. The most widespread inorganic nitrogen species in nature are ammonia and nitrate, both of which are formed in the atmosphere by inorganic chemical processes, and nitrogen gas, N_2, also an atmospheric gas, which is the most stable form of nitrogen in nature. We discuss *nitrogen fixation*, the utilization of N_2 as a nitrogen source, later in this chapter.

One of the most common alternative electron acceptors is nitrate, NO_3^-, which is converted to more reduced forms of nitrogen, N_2O, NO, and N_2. Because these products of nitrate reduction are all gaseous, they can easily be lost from the environment, and because of this the process is called **denitrification**.

tron acceptors in energy metabolism is called *dissimilative* metabolism.

Assimilative and dissimilative metabolism differ markedly. In assimilative metabolism, only enough of the compound (NO_3^-, SO_4^{2-}, or CO_2) is reduced to satisfy the needs of the nutrient for growth. The reduced atoms are eventually converted to cell material in the form of macromolecules. In dissimilative metabolism, a comparatively large amount of the electron acceptor is reduced, and the reduced product is *excreted* into the environment. Many organisms carry out assimilative metabolism of compounds such as NO_3^-, SO_4^{2-}, and CO_2 (for example, many Bacteria, Archaea, fungi, algae,

TABLE 13.4	Oxidation states of key nitrogen compounds
Compound	**Oxidation state**
Organic N (R—NH$_2$)	−3
Ammonia (NH$_3$)	−3
Nitrogen gas (N$_2$)	0
Nitrous oxide (N$_2$O)	+1 (average per N)
Nitrogen oxide (NO)	+2
Nitrite (NO$_2^-$)	+3
Nitrogen dioxide (NO$_2$)	+4
Nitrate (NO$_3^-$)	+5

Assimilative nitrate reduction, in which nitrate is reduced to the oxidation level of ammonia for use as a nitrogen source for growth, and *dissimilative nitrate reduction,* in which nitrate is used as an alternative electron acceptor in energy generation, are contrasted in Figure 13.29. Under most conditions, the end product of dissimilative nitrate reduction is N_2 or N_2O. The process is the main means by which gaseous N_2 is formed biologically, and because N_2 is much less readily available to organisms than nitrate as a source of nitrogen, for agricultural purposes at least, denitrification is a detrimental process. For sewage treatment (∞ Section 12.15), however, denitrification is beneficial because it converts NO_3^- to N_2, effectively decreasing the amount of available nitrogen in the sewage treatment effluent that can stimulate algal growth.

Biochemistry of dissimilative nitrate reduction

The enzyme involved in the first step of nitrate reduction, *nitrate reductase,* is a molybdenum-containing enzyme. In general, assimilative nitrate reductases are soluble proteins that are ammonia-repressed, whereas dissimilative nitrate reductases are membrane-bound proteins whose synthesis is repressed by O_2 and thus they are synthesized only under anoxic conditions.

Thus, in most bacteria that have been examined, the process of denitrification is strictly an *anoxic* process, whereas assimilative nitrate reduction can occur quite well under fully aerobic conditions. Assimilative nitrate reduction occurs in all plants and most fungi, as well as in many prokaryotes, whereas dissimilative nitrate reduction is restricted to prokaryotes, although a wide diversity of such organisms can carry out this process.

In all cases, the first product of nitrate reduction is nitrite, NO_2^-, and another enzyme, *nitrite reductase,* is responsible for the next step. In the dissimilative process, two routes are possible, one to ammonia and the other to N_2. The route to ammonia is carried out by a fairly large number of bacteria but is of less practical significance. There are also some bacteria that do not reduce nitrate but do reduce nitrite to ammonia. This may be a detoxification mechanism because nitrite can be toxic under acidic conditions (nitrous acid is an effective mutagen) (∞ Table 9.2). The pathway to nitrogen gas proceeds via two intermediate gaseous forms of nitrogen, nitric oxide (NO) and nitrous oxide (N_2O). Several organisms are known that produce only N_2O during the denitrification process, whereas other organisms produce N_2 as the gaseous product. Because of their global significance, the formation of gaseous nitrogen compounds by denitrifying bacteria has been under considerable study (∞ Section 14.14).

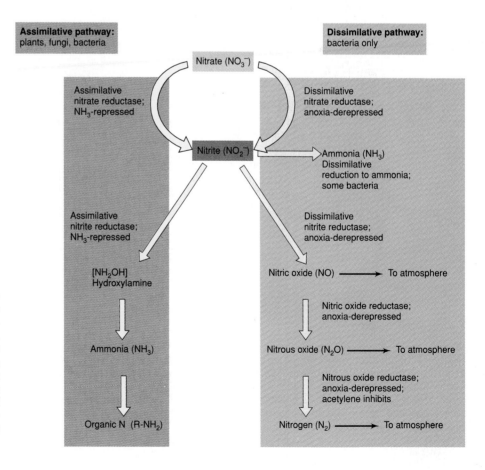

Figure 13.29 Comparison of assimilative and dissimilative processes for the reduction of nitrate. *Assimilative* processes lead to the incorporation of the nitrogen atom of nitrate into organic nitrogen compounds in the cell. *Dissimilative* processes lead to the reduction of nitrate to various inorganic nitrogen compounds during anaerobic respiration.

The biochemistry of dissimilative nitrate reduction has been studied in considerable detail in *Escherichia coli,* an organism that is able to carry out only the first step of the process, the reduction of NO_3^- to NO_2^-. The enzyme nitrate reductase is membrane-bound and accepts electrons from a cytochrome *b* when O_2 is *absent* from the environment. A comparison of the electron transport chains in aerobic metabolism and nitrate respiration in *E. coli* is given in Figure 13.30. As seen, because of the reduction potential of the NO_3^-/NO_2^- couple, only two potential proton-translocating processes occur instead of the three that can occur in aerobic respiration. This is consistent with the lower reduction potential of the NO_3^-/NO_2^- couple as compared to the O_2/H_2O couple (see Table 13.1).

Isolation of denitrifying bacteria

Enrichment culture of denitrifying bacteria is straightforward. A defined culture medium is used in which potassium nitrate is added as an electron acceptor and anoxic conditions maintained. An energy source (electron donor) must be added that is not readily fermentable (see Section 13.19) so fermentative organisms will not be selected. Suitable energy sources include ethanol, acetate, succinate, and benzoate. Because most denitrifying organisms are facultative aerobes, great care to establish anoxic conditions is not necessary. A simple glass-stoppered incubation vessel is sufficient. The residual oxygen present in the culture medium is quickly used up, and then the population switches to anaerobic respiration. Once good growth is obtained, the culture should become quite turbid and bubbles may appear under the glass stopper; the bubbles consist of N_2 as well as CO_2 derived from oxidation of the energy source. The most common bacteria enriched in this way are *Pseudomonas* species, such as *Pseudomonas fluorescens, Pseudomonas aeruginosa,* and related pseudomonads.

CONCEPT CHECK   13.15

Nitrate is a commonly used electron acceptor in anaerobic respiration. Its utilization involves participation of the enzyme nitrate reductase, a molybdenum-containing enzyme capable of reducing nitrate to nitrite. Many bacteria that use nitrate in anaerobic respiration eventually produce nitrogen gas (N_2), a process called denitrification.

✔ *Why is more energy released when aerobic respiration occurs instead of denitrification?*

✔ *Where is the dissimilative nitrate reductase found in the cell? What metal(s) does it contain?*

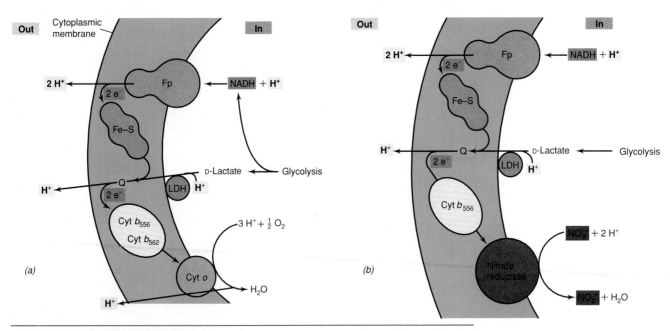

Figure 13.30 Electron transport processes in *Escherichia coli* when (a) O_2 or (b) NO_3^- is used as an electron acceptor. Fp, Flavoprotein; Q, ubiquinone; LDH, lactate dehydrogenase. Under high [O_2] conditions, the sequence of carriers is cyt $b_{562} \rightarrow$ cyt $o \rightarrow O_2$. However, under low [O_2] conditions, the sequence is cyt $b_{558} \rightarrow$ cyt $d \rightarrow O_2$. Note how more protons are translocated aerobically during electron transport reactions than with nitrate as electron acceptor.

13.16
Sulfate Reduction

Several inorganic sulfur compounds are important electron acceptors in anaerobic respiration. A summary of the oxidation states of the key sulfur compounds is given in Table 13.5. Sulfate, the most oxidized form of sulfur, is one of the major anions in seawater and is used by the *sulfate-reducing bacteria,* a group that is widely distributed in nature. The end product of sulfate reduction is H_2S, an important natural product that participates in many biogeochemical processes (∞ Section 14.15). Again, as with nitrogen, it is important to distinguish between assimilative and dissimilative sulfate reduction. Many organisms, including higher plants, algae, fungi, and most prokaryotes, use sulfate as a sulfur source for biosynthesis. But the ability to utilize sulfate as an *electron acceptor* for energy-generating processes involves a large-scale reduction of SO_4^{2-} and is restricted to the sulfate-reducing bacteria. In assimilative sulfate reduction, the H_2S formed is immediately converted into organic sulfur in the form of amino acids, and so on, but in dissimilative sulfate reduction, the H_2S is excreted.

As the reduction potential in Table 13.1 shows, sulfate is a much less favorable electron acceptor than either O_2 or NO_3^-. However, sufficient energy to make ATP is available when an electron donor that yields NADH or FADH is used. Because of the less favorable energetics, growth yields are lower for an organism growing on SO_4^{2-} than for one growing on O_2 or NO_3^-.

TABLE 13.5	**Sulfur compounds and electron donors for sulfate reduction**

Compound	Oxidation state
Oxidation states of key sulfur compounds	
Organic S (R—SH)	−2
Sulfide (H_2S)	−2
Elemental sulfur (S^0)	0
Thiosulfate ($S_2O_3^{2-}$)	+2 (average per S)
Tetrathionate ($S_4O_6^{2-}$)	+2.5 (average per S)
Sulfur dioxide (SO_2)	+4
Sulfite (SO_3^{2-})	+4
Sulfur trioxide (SO_3)	+6
Sulfate (SO_4^{2-})	+6
Some electron donors used for sulfate reduction	
H_2	Acetate
Lactate	Propionate
Pyruvate	Butyrate
Ethanol and other alcohols	Long-chain fatty acids
Fumarate	Benzoate
Malate	Indole
Choline	Hexadecane

A list of some of the electron donors used by sulfate-reducing bacteria is given in Table 13.5. The first three compounds listed, H_2, lactate, and pyruvate, are used by a wide variety of sulfate-reducing bacteria; the others have more restricted use. However, a large variety of morphological and physiological types of sulfate-reducing bacteria are known; their characteristics and taxonomy are discussed in Section 16.8.

Biochemistry and energetics of sulfate reduction

The reduction of SO_4^{2-} to hydrogen sulfide, an eight-electron reduction, proceeds through a number of intermediate stages. The sulfate ion is stable and cannot be used without first being activated. Sulfate is activated by means of ATP. The enzyme *ATP sulfurylase* catalyzes the attachment of the sulfate ion to a phosphate of ATP, leading to the formation of **adenosine phosphosulfate (APS)** as shown in Figure 13.31. In dissimilative sulfate reduction, the sulfate moiety of APS is reduced directly to sulfite (SO_3^{2-}) with the release of AMP. In assimilative reduction, another P is added to APS to form **phosphoadenosine phosphosulfate (PAPS)** (Figure 13.31*b*), and only then is the sulfate moiety reduced. In both cases, the first product of sulfate reduction is *sulfite,* SO_3^{2-}. Once SO_3^{2-} is formed, the subsequent reductions proceed readily. Several organisms unable to carry out dissimilative sulfate (SO_4^{2-}) reduction are able to carry out dissimilative sulfite (SO_3^{2-}) reduction, presumably because although they can convert sulfite to H_2S, they lack the APS system and thus are unable to reduce sulfate to sulfite.

The sulfate-reducing bacteria carry out a cytochrome-based electron transport process, the electrons from the energy source being transferred to the sulfate ion in APS and to sulfite. The cytochrome of the sulfate-reducing bacteria is a very electronegative cytochrome *c* called *cytochrome c_3*. This cytochrome is not found in organisms using other electron acceptors. Other electron carriers in the electron transport chain of the sulfate-reducing bacteria include ferredoxin and flavodoxin. Type II sulfate reducers (species capable of degrading acetate and other fatty acids) (∞ Section 16.8) also contain a cytochrome of the *b* type which is presumably involved in the electron transport chain of these species; cytochrome *b* is absent from species of sulfate-reducing bacteria that do not degrade fatty acids. The electron transport system is shown in Figure 13.32.

In electron transport in sulfate-reducing bacteria, hydrogen, H_2, either directly from the environment or generated from certain organic electron donors including lactate (Figure 13.32), transfers electrons to the enzyme *hydrogenase,* which is situated in the periplasm in close association with cytochrome c_3. Because of the spatial arrangement of the electron transport compo-

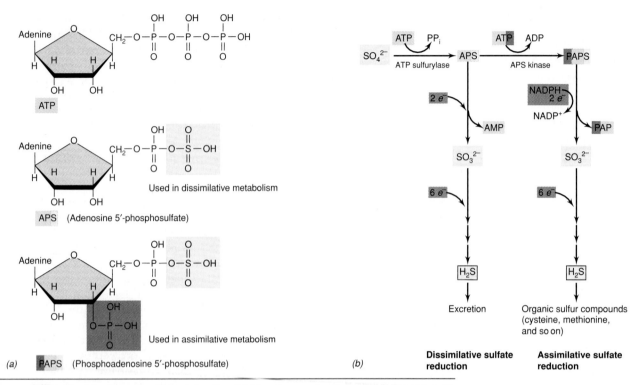

Figure 13.31 Biochemistry of sulfate reduction. (a) Two forms of *active sulfate,* adenosine 5'-phosphosulfate (APS) and phosphoadenosine 5'-phosphosulfate (PAPS). (b) Schemes of assimilative and dissimilative sulfate reduction.

nents in the membrane, when the H atoms of H_2 are oxidized, the protons (H^+) remain *outside* the membrane, whereas the electrons are transferred *across* the membrane. In this way, a proton motive force is set up that can be used for the synthesis of ATP. In the cytoplasm, the electrons are used in the reduction of APS and sulfite.

When sulfate-reducing bacteria grow on H_2/SO_4^{2-}, they are growing chemolithotrophically as H_2 bacteria (see Section 13.9). Some species can even grow autotrophically under these conditions, using CO_2 as

sole source of carbon. However, most sulfate-reducing bacteria are chemoorganotrophs and use various organic compounds as electron donors (some of which were listed in Table 13.5).

Growth of sulfate-reducing bacteria on acetate

Many sulfate-reducing bacteria are capable of growth on *acetate* as sole energy source; most such organisms are of marine origin (∞ Section 16.8). These organ-

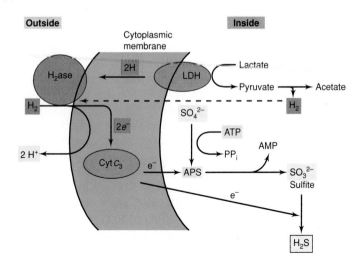

Figure 13.32 Electron transport and generation of a proton motive force in a sulfate-reducing bacterium. In addition to external hydrogen (H_2) arising from the metabolism of fermentative anaerobes, this compound can also originate from the catabolism of organic compounds, such as lactate and pyruvate.

isms oxidize acetate completely to CO_2 and reduce sulfate to sulfite:

$$Acetate^- + SO_4^{2-} + 3\ H^+ \rightarrow 2\ CO_2 + H_2S + 2\ H_2O$$
$$\Delta G^{0'} = -57.5\ kJ/reaction$$

Although the citric acid cycle (∞ Section 4.14) is the major means by which acetate is oxidized in most organisms, it does not function as such in acetate-oxidizing sulfate reducers. Instead, two biochemical mechanisms for acetate oxidation by sulfate-reducing bacteria are known, a modified citric acid cycle (CAC) and the acetyl-CoA pathway.

Organisms like *Desulfobacter* use the modified citric cycle for acetate oxidation. This series of reactions involves most of the enzymes of the citric acid cycle. *Desulfobacter* contains an enzyme that first activates acetate by reaction with succinyl-CoA to yield succinate and acetyl-CoA; the latter then enters the CAC and is oxidized to CO_2 (Figure 13.33). As previously mentioned, because the SO_4^{2-}/SO_3^{2-} couple has such a low reduction potential (see Table 13.1), energy is actually *required* to drive this first step in sulfate reduction (see Figure 13.31). However, this energy requirement creates a special problem for acetate-oxidizing sulfate reducers because the ATP yield from the oxidation of acetate via the CAC is about that required for activation of sulfate to form APS (and hence sulfite) (see Figure 13.31). However, a novel enzyme present in *Desulfobacter* allows for net ATP synthesis (and thus growth) at the expense

of acetate. *Desulfobacter* uses *citrate lyase*, an enzyme that can couple to ATP synthesis by *substrate-level phosphorylation* from the conversion of acetyl-CoA (via acetyl-P) to acetate during the production of citrate (Figure 13.33). This additional ATP makes growth on acetate possible.

Most acetate-oxidizing, sulfate-reducing bacteria do not use this modified CAC but instead employ the acetyl-CoA pathway for acetate oxidation. This pathway allows for oxidation of acetate to CO_2 via a quite different series of reactions from that of the CAC, employing the key enzyme *carbon monoxide dehydrogenase*. However, because the acetyl-CoA pathway was first discovered in *homoacetogenic* bacteria (wherein it is used in acetate *formation*, as discussed in Section 16.9) and its presence only later discovered in sulfate-reducing bacteria, we reserve discussion of this pathway for later (∞ Sections 16.8, 16.9, and 17.2).

Sulfite, thiosulfate, and S^0 disproportionation

Certain sulfate-reducing bacteria are capable of a unique form of energy metabolism called *disproportionation*, using sulfur compounds of intermediate oxidation state. The term disproportionation refers to the splitting of a compound into two new compounds, one of which is *more oxidized* and one of which is *more reduced* than the original substrate. In the present discussion, we describe the disproportionation of thiosulfate ($S_2O_3^{2-}$), sulfite (SO_3^{2-}), and sulfur (S^0).

Desulfovibrio sulfodismutans can disproportionate sulfur compounds as follows:

$$S_2O_3^{2-} + H_2O \rightarrow SO_4^{2-} + H_2S$$
$$\Delta G^{0'} = -21.9\ kJ/reaction$$

Note that one sulfur atom of $S_2O_3^{2-}$ becomes more oxidized (forming SO_4^{2-}) and the other more reduced (forming H_2S). Another disproportionation involves sulfite:

$$4\ SO_3^{2-} + 2\ H^+ \rightarrow 3\ SO_4^{2-} + H_2S$$
$$\Delta G^{0'} = -235.6\ kJ/reaction$$

In these reactions electrons from either $S_2O_3^{2-}$ or SO_3^{2-} enter the electron transport chain and eventually reduce other molecules of $S_2O_3^{2-}$ or SO_3^{2-}, respectively, to H_2S.

Elemental sulfur (S^0) can also be disproportionated, in this case, to H_2S and SO_4^{2-}:

$$4\ S^0 + 4\ H_2O \rightarrow 3\ H_2S + SO_4^{2-} + 2\ H^+$$
$$\Delta G^{0'} = +48\ kJ$$

As written, this is an energetically unfavorable reaction. However, if the H_2S that is formed is oxidized back to S^0 by chemical reaction with Mn^{4+} as

$$H_2S + MnO_2 \rightarrow S^0 + Mn^{2+} + 2\ OH^-$$
$$\Delta G^{0'} = -92\ kJ$$

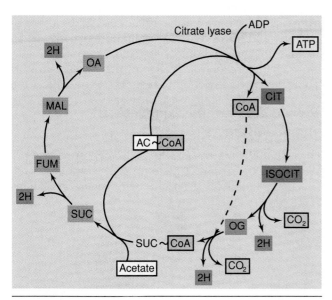

Figure 13.33 Mechanism of acetate oxidation to CO_2 in sulfate-reducing bacteria that use the citric acid cycle for acetate oxidation. Note how coenzyme A is recycled in this reaction series through interaction of succinyl-CoA and acetate and note the use of citrate lyase to form citrate. See text for details. OA, Oxalacetate; MAL, malate; FUM, fumarate; SUC, succinate; SUC~CoA, succinyl~coenzyme A; OG, α-ketoglutarate; ISOCIT, isocitrate; CIT, citrate; AC~CoA, acetyl~coenzyme A. C_4 Compounds are shown in orange, C_5 in purple, and C_6 in red.

the sum of these two reactions:

$$3 \, S^0 + 4 \, H_2O + MnO_2$$
$$\rightarrow 2 \, H_2S + SO_4^{2-} + 2 \, H^+ + Mn^{2+} + 2 \, OH^-$$
$$(\Delta G^{0\prime} = -44 \text{ kJ})$$

yields sufficient energy to support growth of sulfur-disproportionating bacteria. Thus, unlike sulfite- or thiosulfate-disproportionating bacteria, sulfur-disproportionating bacteria require an electron acceptor such as Mn^{4+} to drive the energetics of their reaction.

Although the enzymology of sulfur disproportionation has not been worked out, it is likely that proton motive force formation is central to the bioenergetics of these energy-yielding reactions. In addition, however, in the case of SO_3^{2-} oxidation, substrate-level phosphorylation may increase the ATP yield as SO_3^{2-} is oxidized to SO_4^{2-} via reversal of the APS reductase system (see Figure 13.31).

CONCEPT CHECK   **13.16**

The sulfate-reducing bacteria are able to reduce sulfate to hydrogen sulfide. Because the reduction of sulfate requires first its activation by a reaction with ATP, the sulfate reduction process is not very efficient energetically. Electron donors for sulfate reduction include hydrogen gas and a number of organic compounds, of which acetate and lactate are two of the most important. Disproportionation of sulfur compounds is an additional energy-yielding strategy for certain members of this group.

✔ *Define the following: S^0, SO_4^{2-}, SO_3^{2-}, $S_2O_3^{2-}$, H_2S.*

✔ *How is sulfate converted to sulfite?*

✔ *Why is H_2 of importance to sulfate-reducing bacteria?*

✔ *Give an example of disproportionation.*

13.17

Carbon Dioxide as an Electron Acceptor: Methanogenesis and Acetogenesis

Carbon dioxide, CO_2, is common in nature and is a major product of the energy metabolism of chemoorganotrophs. Several prokaryotic groups are able to use CO_2 as an electron acceptor in anaerobic respiration. The most important CO_2-reducing prokaryotes are the **methanogens,** a major group of Archaea. Many of these organisms utilize H_2 as the electron donor (energy source); the overall reaction is

$$4 \, H_2 + H^+ + HCO_3^- \rightarrow CH_4 + 3 \, H_2O$$
$$\Delta G^{0\prime} = -135.6 \text{ kJ/reaction}$$

When growing on H_2/CO_2, the methanogens are chemolithotrophic autotrophs (Figure 13.34). However, the process by which CO_2 is fixed is not the Calvin cycle of conventional autotrophs but instead the acetyl-CoA pathway (∞ Sections 16.9 and 17.2). We discuss the ecology of methanogenesis in Section 14.12 and the methanogens themselves in Section 17.2.

Another group of CO_2-reducing bacteria is the **homoacetogens,** which produce *acetate* rather than CH_4 from CO_2 and H_2 (Figure 13.34). The overall reaction of homoacetogenesis is

$$4 \, H_2 + H^+ + 2 \, HCO_3^- \rightarrow CH_3COO^- + 4 \, H_2O$$
$$\Delta G^{0\prime} = -104.6 \text{ kJ/reaction}$$

The homoacetogenic bacteria are not a defined taxonomic group because they include such widely differing organisms as the gram-positive spore-forming bacterium *Clostridium aceticum* and the gram-negative non-spore-forming *Acetobacterium woodii.* Homoacetogenic bacteria are discussed in detail in Section 16.9.

When comparing the two preceding equations, it is seen that somewhat more energy is released when methane is formed from CO_2 reduction than when acetate is formed, in line with the differences in reduction potential of the two couples involved (Figure 13.28). When growing on $H_2 + CO_2$, both methanogens and homoacetogens are autotrophs (they are also chemolithotrophs). Both groups are also obligate anaerobes. Note, however, that methanogens are not restricted to $H_2 + CO_2$; many methanogens grow and form methane from methanol, formic acid, and acetate, as discussed in Sections 14.12 and 17.2. In addition, as discussed in Section 16.9, most homoacetogenic bacteria can also grow chemoorganotrophically by the fermentation of sugars and a variety of other organic compounds.

Bioenergetics of methanogenesis and acetogenesis

How do methanogens and homoacetogens make ATP? We reserve the details of these processes for our discussion of the homoacetogens and methanogens in Sections

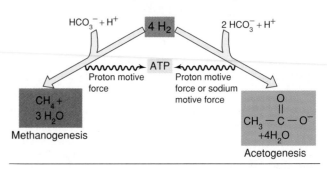

Figure 13.34 The contrasting processes of methanogenesis and acetogenesis. The free energy released in each reaction as drawn is: (a) methanogenesis, -136 kJ, and (b) acetogenesis, -104.6 kJ.

16.9 and 17.2, respectively, and consider here only general aspects of their bioenergetics. In methanogens, ATP synthesis is linked to a proton motive force and proton-translocating ATPase; the terminal step in the reduction of CO_2 to CH_4 (the CH_3-to-CH_4 conversion) establishes a proton motive force whose energy is captured by membrane-bound ATPases (Figure 13.34). In homoacetogens, two bioenergetic patterns have been found. Certain homoacetogens, such as *Acetobacterium woodii*, require Na^+ for growth. These species link the synthesis of acetate to formation of a *sodium motive force* (a Na^+ gradient across the membrane) that is dissipated by a Na^+-translocating ATPase generating ATP (Figure 13.34). Other homoacetogens, such as *Clostridium thermoaceticum*, do not require Na^+ and generate a *proton motive force* during acetogenesis that functions to drive ATP synthesis (Figure 13.34). However, in addition to ion gradients, *substrate-level* phosphorylation is possible during acetogenesis, as the final step in the formation of acetate involves the reaction acetyl ~ P + ADP → acetate + ATP. The biochemistry of the acetyl-CoA pathway is covered in Section 16.9.

CONCEPT CHECK   13.17

Methanogens are able to use carbon dioxide (CO_2) as an electron acceptor in a reaction in which H_2 is electron donor. The methanogens produce methane (CH_4), but another group, the homoacetogens, convert H_2 plus CO_2 into acetate.

✔ *Draw the chemical structures of the products of methanogenesis and acetogenesis.*

✔ *With H_2 as electron donor, why is more energy released in the formation of methane than in the formation of acetate?*

✔ *How is ATP produced during methanogenesis and acetogenesis?*

13.18
Other Electron Acceptors for Anaerobic Respiration

In addition to nitrate, sulfate, and carbon dioxide, a variety of other compounds, both organic and inorganic, are used by one or another group of bacteria in anaerobic respiration. We discuss the most important in this section and summarize the reactions involved in Figure 13.35.

Ferric iron reduction

Ferric (Fe^{3+}) iron can be reduced to the ferrous (Fe^{2+}) state by a variety of microorganisms, and because Fe^{3+} is abundant in many microbial habitats, its reduction

can be a major form of anaerobic respiration. Ferric iron can function as an electron acceptor for energy metabolism in certain fungi and in a wide variety of both chemoorganotrophic and chemolithotrophic bacteria. The reduction potential of the Fe^{3+}/Fe^{2+} couple is very electropositive ($E_0' = 0.77$ V at pH 2, $E_0' = 0.2$ V at pH 7), and because of this, Fe^{3+} reduction can be coupled to the oxidation of a wide variety of both organic and inorganic electron donors. Various organic compounds including aromatic compounds can be oxidized anaerobically by ferric iron reducers with electrons presumably traveling through electron transport chains that terminate in a ferric iron reductase system. Such electron flow establishes a proton gradient that can be used to generate ATP. Most research on the energetics of ferric iron reduction has been done in the gram-negative bacterium *Shewanella putrefaciens*, in which Fe^{3+}-dependent anaerobic growth occurs with various organic electron donors.

Ferric iron is one of the most common metals present in soils and rocks, and its reduction leads to the production of ferrous iron, a more soluble form of iron. Bacterial iron reduction can thus lead to solubilization of iron, an important geochemical process. As illustrated in Figure 13.22c, one type of iron deposit, called *bog iron*, is formed as a result of the activities of ferric iron-reducing microorganisms.

Reduction of manganese and other inorganic substances

The metal manganese has a number of oxidation states, of which Mn^{4+} and Mn^{2+} are the most stable and biologically relevant. Anoxic reduction of Mn^{4+} to Mn^{2+} is carried out by a variety of microorganisms, mostly chemoorganotrophs. However, in most cases it has been difficult to show that Mn^{4+} reduction is energetically beneficial because chemical reduction of Mn^{4+} also occurs. However, in *Shewanella putrefaciens* and a few other bacteria, anoxic growth on acetate and several other nonfermentable carbon sources occurs with Mn^{4+} as electron acceptor. This indicates that Mn^{4+}-supported anaerobic respiration is possible, presumably at the expense of a proton motive force formed during electron transport from an electron donor to Mn^{4+} as electron acceptor. The reduction potential of the Mn^{4+}/Mn^{2+} couple is extremely high (Figure 13.35); thus, several compounds should be able to donate electrons to Mn^{4+} reduction.

Other inorganic substances can function as electron acceptors for anaerobic respiration. These include selenium and arsenic compounds (Figure 13.35). Although usually not present in large amounts in natural systems, arsenic and selenium compounds are occasional pollutants and can support anoxic growth of various bacteria. The reduction of SeO_4^{2-} to SeO_3^{2-} and eventually to Se^0 (metallic selenium) is an important

Acceptor	Reaction	E'_0 of couple (V)	Product
Manganic ion	$Mn^{4+} \xrightarrow{2e^-} Mn^{2+}$	+0.798	Manganous ion
Ferric ion	$Fe^{3+} \xrightarrow{e^-} Fe^{2+}$	+0.77	Ferrous ion
Selenate	$^-O-\overset{O}{\underset{O}{\overset{\|}{\underset{\|}{Se}}}}-O^- \xrightarrow[2H^+]{2e^-} \overset{O^-}{\underset{O^-}{\overset{\|}{Se}}}=O+H_2O$	+0.475	Selenite
Dimethyl sulfoxide (DMSO)	$H_3C-\overset{\|}{\underset{O}{S}}-CH_3 \xrightarrow[2H^+]{2e^-} (CH_3)_2S+H_2O$	+0.16	Dimethyl sulfide (DMS)
Arsenate	$^-O-\overset{O^-}{\underset{O^-}{\overset{\|}{As}}}=O \xrightarrow[2H^+]{2e^-} \overset{O^-}{\underset{O^-}{As}}-O^-+H_2O$	+0.139	Arsenite
Trimethylamine-N-oxide (TMAO)	$H_3C-\overset{CH_3}{\underset{O}{\overset{\|}{N}}}-CH_3 \xrightarrow[2H^+]{2e^-} (CH_3)_3N+H_2O$	+0.13	Trimethylamine (TMA)
Fumarate	$^-O-\overset{O}{\overset{\|}{C}}-\overset{H}{\underset{H}{C}}=C-\overset{O}{\overset{\|}{C}}-O^- \xrightarrow[2H^+]{2e^-} {}^-O-\overset{O}{\overset{\|}{C}}-CH_2-CH_2-\overset{O}{\overset{\|}{C}}-O^-$	+0.03	Succinate
Glycine	$H_3N^+-CH_2-\overset{O}{\overset{\|}{C}}-O^- \xrightarrow[2H^+]{2e^-} CH_3-\overset{O}{\overset{\|}{C}}-O^-+NH_4^+$	−0.01	Acetate + ammonia

Figure 13.35 Some alternative electron acceptors for anaerobic respirations.

method of selenium removal from water and has been used as a means of cleaning up (bioremediation) (∞ Section 14.20) of selenium-contaminated soils. Most bacteria capable of selenate or arsenate reduction can also use several other electron acceptors, such as Fe^{3+}, Mn^{4+}, and organic compounds, and in most cases show a facultatively aerobic form of metabolism.

Organic electron acceptors

Several organic compounds can participate in anaerobic respirations (Figure 13.35). Note that many organic compounds also function as electron acceptors in conventional chemoorganotrophic metabolism. However, in these cases, for example, during fermentation (∞ Section 4.10), the organic electron acceptor is produced *internally* as part of the metabolic process and then reduced internally. In the present case, we are discussing organic compounds that function as electron acceptors when added *externally*.

Several common electron acceptors for anaerobic respiration are listed in Figure 13.35. Of those listed, the compound that has been most extensively studied is **fumarate,** which is reduced to **succinate.** An examination of the *citric acid cycle* in Figure 4.21 will indicate that fumarate and succinate are important intermediates. Fumarate's role as an electron acceptor for anaer-

obic respiration derives from the fact that the fumarate–succinate couple has a reduction potential near 0 V (see Figure 13.28), which allows coupling of fumarate reduction to NADH oxidation. The energy yield is sufficient for the synthesis of one ATP. Bacteria able to use fumarate as an electron acceptor include *Wolinella succinogenes* (which can grow on H_2 as sole energy source using fumarate as electron acceptor), *Desulfovibrio gigas* (a sulfate-reducing bacterium that can also grow under non-sulfate-reducing conditions), some clostridia, *Escherichia coli*, and *Proteus rettgeri*. Another bacterium, *Streptococcus faecalis*, can use fumarate as an electron acceptor but does not couple this to electron transport phosphorylation. In the latter case, fumarate merely serves in the reoxidation of NADH that had been formed during glycolysis.

The compound **trimethylamine oxide** shown in Figure 13.35 is an interesting electron acceptor. Trimethylamine oxide (TMAO) is an important osmotic solute in marine fish, where it serves in these animals as a means of excreting excess nitrogen, but a variety of bacteria are able to reduce TMAO to trimethylamine (TMA). TMA has a strong odor and flavor, and some of the odor that frequently occurs in spoiled marine fish is due to TMA produced by bacterial action. A variety of facultatively aerobic bacteria are able to utilize TMAO as an alternate electron acceptor. In addition, several phototrophic purple bacteria (∞ Section 16.1) are able to use TMAO as an electron acceptor for anoxic metabolism in darkness. A compound analogous to TMAO is **dimethyl sulfoxide** (DMSO), which is reduced by a variety of bacteria to dimethyl sulfide (DMS). DMSO is a common natural product and is found in both marine and freshwater environments. DMS has a strong, pungent odor, and bacterial reduction of DMSO to DMS is signaled by the presence of the characteristic odor of DMS. A variety of bacteria, including *Campylobacter*, *Escherichia*, and many purple bacteria, are able to use DMSO as an electron acceptor in energy generation (see Section 14.15 for further discussion of DMSO metabolism).

The reduction potentials of the TMAO/TMA and DMSO/DMS couples are similar, near +0.15 V, which means that any electron transport chain that ends with TMAO or DMSO reduction must be rather brief. In most instances of TMAO and DMSO reduction, cytochromes of the *b* type (with reduction potentials near 0 V) have been identified as terminal oxidases.

Glycine is an amino acid but can also function as an electron acceptor for certain obligately anaerobic bacteria. In certain clostridia, glycine is reduced to acetate (Figure 13.35) with electrons derived from oxidation of a second amino acid. This metabolism of an amino acid *pair* is called a *Stickland reaction*, and specific examples of this process are given in Section 16.26.

CONCEPT CHECK   13.18

Besides inorganic nitrogen and sulfur compounds or CO_2, a variety of other substances, both organic and inorganic, can function as electron acceptors for anaerobic respiration. These include in particular Fe^{3+}, Mn^{4+}, and fumarate.

✔ *With H_2 as electron donor why is reduction of Fe^{3+} a much more favorable reaction than reduction of fumarate?*

✔ *What is a Stickland reaction?*

13.19

Fermentations: Energetic and Redox Considerations

Because oxygen is not highly soluble (9.6 mg/l distilled water in equilibrium with air at 25°C), many environments easily become anoxic. In such environments, decomposition of organic materials occurs anaerobically. If adequate supplies of the electron acceptors previously considered are not available in such anoxic environments, much of the carbon will be catabolized by fermentation. We discussed the overall process of fermentation in Sections 4.9 and 4.10 and showed that it was an internally balanced oxidation–reduction process in which carbon from the same external organic compound was partially oxidized and partially reduced (Figure 13.36).

There are two problems an organism faces if it is to catabolize organic compounds in energy-yielding metabolism: (1) conserving some of the energy released as ATP, and (2) disposing of electrons removed from the electron donor. In fermentation, ATP synthesis generally occurs by way of *substrate-level phosphorylation*, a mechanism by which high energy phosphate bonds from organic intermediates of the fermentation are transferred to ADP (∞ Section 4.9). The second problem, that of redox balance, is solved by production and excretion by the organism of *fermentation products* generated from the original substrate (Figure 13.36). We now consider these basic principles of fermentation in more detail and highlight the enormous diversity of microbial fermentations known.

High energy compounds and substrate-level phosphorylation

Energy can be obtained by substrate-level phosphorylation in many different ways. However, central to the mechanism of ATP synthesis is the production of one or another *high energy compound*. These are generally organic compounds containing a phosphate group or

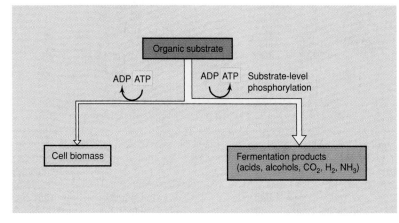

Figure 13.36 Overall process of fermentation. Note how in a typical fermentation, most of the carbon is excreted as a partially reduced end product of energy metabolism and only a small amount is used in biosynthesis.

a coenzyme-A molecule, the hydrolysis of which is highly exergonic. A list of the major high energy intermediates is given in Table 13.6. This list is not complete, but it includes most recognized high energy intermediates known to be formed during biochemical processes. Because most of the compounds listed in Table 13.6 can couple directly to ATP synthesis (-31.8 kJ/mol), if an organism can form one or another of these compounds during fermentative metabolism, it can make ATP. Substrate-level phosphorylation is a more direct way of making ATP than via a proton motive force but requires that the energy source couple directly to a high energy intermediate.

Pathways for the anaerobic breakdown of various fermentable substances to high energy intermediates are summarized in Figure 13.37. It should be noted that this figure is organized by the high energy compounds listed in Table 13.6 and that either one of these compounds, or a related derivative, is generated in each case and leads to ATP synthesis. Thus, Figure 13.37 and Table 13.6 should be examined together.

| TABLE 13.6 | Energy-rich compounds involved in substrate-level phosphorylation[a] |

Compound	Free energy of hydrolysis, $\Delta G^{0'}$ (kJ/mol)
Acetyl-CoA	-35.7
Propionyl-CoA	-35.6
Butyryl-CoA	-35.6
Succinyl-CoA	-35.1
Acetylphosphate	-44.8
Butyrylphosphate	-44.8
1,3-Bisphosphoglycerate	-51.9
Carbamyl phosphate	-39.3
Phosphoenolpyruvate	-51.6
Adenosine-phosphosulfate (APS)	-88
N^{10}-formyltetrahydrofolate	-23.4
Energy of hydrolysis of ATP (ATP→ ADP + P$_i$)	-31.8

[a]Data from Thauer, R. K., K. Jungermann, and K. Decker. 1977. Energy conservation in chemotrophic anaerobic bacteria. *Bacteriol. Rev.* 41:100–180.

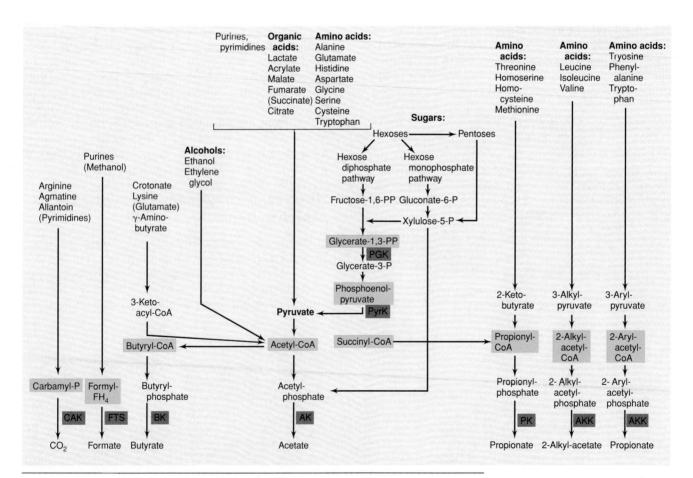

Figure 13.37 Major routes of the anaerobic breakdown of various fermentable substances. The sites of substrate-level phosphorylation are shown by the abbreviations. CAK, Carbamyl phosphate kinase; FTS, formyltetrahydrofolate synthetase; AK, acetate kinase; PK, propionate kinase; BK, butyrate kinase; AKK, alkyl (aryl) acetate kinase; PGK, phosphoglycerate kinase; PyrK, pyruvate kinase. High energy CoA derivatives and other key high energy compounds are highlighted in blue while enzymes are in red. Refer back to Table 13.6.

Energy yields of fermentative organisms

How much ATP can be produced by a fermentative organism? As we have seen, glucose fermenters produce 2–3 ATPs/mol of glucose fermented in glycolysis (∞ Figure 4.12). This is about the maximum amount of ATP produced by fermentation; many other substrates provide less energy. The potential energy released from a particular fermentation can be calculated from the balanced reaction and from the free energy values given in Appendix 1. For instance, the fermentation of glucose to ethanol and CO_2 has a theoretical energy yield of -235 kJ/mol, enough to produce about 7 ATPs. However, only 2 ATPs are actually produced, which implies that the organism operates at considerably less than 100% efficiency (see Section 4.15 for a discussion of energy efficiency).

Oxidation–reduction balance

In any fermentation reaction, there must be a *balance* between oxidation and reduction. The total number of electrons in the products on the right side of the equation must balance the number in the substrates on the left side of the equation. When fermentations are studied experimentally in the laboratory, it is conventional to calculate a *fermentation balance* to make certain that no products are missed. The fermentation balance can also be calculated theoretically from the oxidation states of the substrates and products (see Appendix 1 for the procedure for calculating oxidation states).

In a number of fermentations, electron balance is maintained by the production of molecular hydrogen, H_2. In H_2 production, protons (H^+) derived from water serve as electron acceptors. Production of H_2 is generally associated with the presence in the organism of an iron–sulfur protein called *ferredoxin*, a very electronegative electron carrier. The transfer of electrons from ferredoxin to H^+ is catalyzed by the enzyme **hydrogenase**, as illustrated in Figure 13.38. We have already discussed the enzyme hydrogenase earlier in this chapter in reference to the *utilization* of hydrogen by sulfate-reducing bacteria and the aerobic hydrogen bacteria. In the present case, hydrogenase is involved in the *production* of hydrogen.

The energetics of hydrogen production are actually somewhat unfavorable, and so most fermentative organisms produce only a relatively small amount of hydrogen along with other fermentation products. Hydrogen production thus functions primarily to maintain redox balance. If hydrogen production is prevented, for instance, then the oxidation–reduction balance of the other fermentation products will be shifted toward *more reduced* products. Thus, many fermentative organisms that make H_2 produce both ethanol and acetate. Since ethanol is more reduced than acetate, its formation is favored when hydrogen production is inhibited.

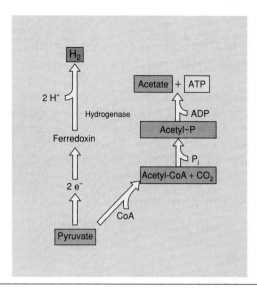

Figure 13.38 Production of molecular hydrogen from pyruvate. Note how production of acetate leads to ATP synthesis.

Numerous anaerobic bacteria produce *acetate* as one of the products of fermentation. The production of acetate is energetically advantageous because it allows the organism to make ATP by substrate-level phosphorylation. The key intermediate generated in acetate production is acetyl-CoA (see Table 13.6), a high energy intermediate. Acetyl-CoA can be converted to acetyl phosphate (also listed in Table 13.6), and the high energy phosphate group of acetyl phosphate subsequently transferred to adenosine diphosphate (ADP) by acetate kinase, yielding ATP. One of the main substrates that is converted to acetyl-CoA is pyruvate, a major product of glycolysis. However, the conversion of pyruvate to acetyl-CoA is an oxidation reaction (Figure 13.38) because pyruvate is more reduced than acetyl-CoA. The excess electrons generated must be used either to make a more reduced end product or in the production of H_2 as discussed previously.

CONCEPT CHECK   13.19

In the absence of an external electron acceptor, organic compounds can be catabolized only by fermentation. Only certain compounds are fermentable, and a requirement for most fermentations is that an energy-rich organic intermediate be formed that can yield ATP by substrate-level phosphorylation. Redox balance must also be achieved in fermentations, and H_2 production is one means of disposing of excess electrons.

✔ *What is substrate-level phosphorylation?*

✔ *Why is acetate formation in fermentation energetically beneficial?*

/13.20

Fermentations: Diversity, Syntrophy, and Role in Anoxic Decomposition

Fermentations can be classified either in terms of the substrate fermented or in terms of the fermentation products formed. Many of the specific fermentation reactions of bacteria will be discussed when the individual groups are discussed in Chapters 16 and 17. Here we present an overview of common fermentations.

Diversity of fermentations

Table 13.7 summarizes some of the main types of fermentations as classified on the basis of *products formed.* Note some of the broad categories, such as alcohol, lactic acid, propionic acid, mixed acid, butyric acid, and homoacetic acid. A number of fermentations are classified on the basis of the *substrate fermented* rather than the fermentation product. For instance, many of the spore-forming anaerobic bacteria (genus *Clostridium*) ferment *amino acids* with the production of acetate, lactate, ammonia, and H_2. Other *Clostridium* species, such

as *C. acidi-urici* and *C. purinolyticum*, ferment *purines* such as xanthine or adenine with the formation of acetate, formate, CO_2, and ammonia. Still other anaerobes ferment *aromatic* compounds. As an example, the bacterium *Pelobacter acidigallici* ferments the aromatic compound *phloroglucinol* (1,3,5-benzenetriol, $C_6H_6O_3$) via the following overall pathway:

$$\text{Phloroglucinol } (C_6H_6O_3) + 3\,H_2O \rightarrow 3 \text{ acetate}^- + 3\,H^+$$
$$\Delta G^{0\prime} = -142.5 \text{ kJ/reaction}$$

Many unusual fermentations are carried out by only a very restricted group of anaerobes and, in some cases, by only a single known bacterium. Some examples are listed in Table 13.8. Many of these bacteria can be considered metabolic specialists, having evolved biochemical capabilities to catabolize a substrate or substrates not catabolized by other bacteria. However, as for the substances listed in Table 13.7, successful fermentation of these more unusual substrates requires that the organism be able to produce a high energy intermediate, usually a coenzyme-A derivative of the type listed in Table 13.7, during the fermentation in order to recover the energy released as ATP.

TABLE 13.7 Examples of common bacterial fermentations and some of the organisms carrying them out

Type	Overall reaction[a]	Organisms
Alcoholic fermentation	Hexoses → 2 Ethanol + 2 CO_2	Yeast *Zymomonas*
Homolactic fermentation	Hexose → 2 Lactate$^-$	*Streptococcus* Some *Lactobacillus*
Heterolactic fermentation	Hexose → Lactate$^-$ + Ethanol + CO_2	*Leuconostoc* Some *Lactobacillus*
Propionic acid	Lactate$^-$ → Propionate$^-$ + Acetate$^-$ + CO_2	*Propionibacterium* *Clostridium propionicum*
Mixed acid	Hexoses → Ethanol + 2,3-Butanediol + Succinate^{2-} + Lactate$^-$ + Acetate$^-$ + Formate$^-$ + H_2 + CO_2	Enteric bacteria *Escherichia* *Salmonella* *Shigella* *Klebsiella* *Enterobacter*
Butyric acid	Hexoses → Butyrate$^-$ + Acetate$^-$ + H_2 + CO_2	*Clostridium butyricum*
Butanol	Hexoses → Butanol + Acetate$^-$ + Acetone + Ethanol + H_2 + CO_2	*Clostridium acetobutylicum*
Caproate	Ethanol + Acetate$^-$ + CO_2 → Caproate$^-$ + Butyrate$^-$ + H_2	*Clostridium kluyveri* *Clostridium aceticum*
Homoacetogenic	Fructose → 3 Acetate$^-$ + 3 H^- 4 H_2 + 2 CO_2 + H^+ → Acetate$^-$ + 2 H_2O	*Acetobacterium*
Methanogenic	Acetate$^-$ + H_2O → CH_4 + HCO_3^-	*Methanothrix* *Methanosarcina*

[a]Reactions are intended as an overview of the process and are not necessarily balanced.

TABLE 13.8	Some unusual bacterial fermentations	
Type	**Overall balanced reaction**	**Organisms**
Acetylene	$2\ C_2H_2 + 3\ H_2O \rightarrow$ ethanol $+$ acetate$^-$ $+$ H$^+$	*Pelobacter acetylenicus*
Glycerol	4 Glycerol $+ 2\ HCO_3^- \rightarrow 7$ acetate$^-$ $+ 5$ H$^+$ $+ 4\ H_2O$	*Acetobacterium* spp.
Resorcinol (an aromatic compound)	$2\ C_6H_4(OH)_2 + 6\ H_2O \rightarrow 4$ acetate$^-$ $+$ butyrate$^-$ $+ 5$ H$^+$	*Clostridium* spp.
Cinnamate (an aromatic compound)	$2\ C_9H_7O_2 + 2\ H_2O \rightarrow C_9H_9O_2 +$ benzoate$^-$ $+$ acetate$^-$	*Acetivibrio multivorans*
Phloroglucinol (an aromatic compound)	$C_6H_6O_3 + 3\ H_2O \rightarrow 3$ acetate$^-$ $+ 3$ H$^+$	*Pelobacter massiliensis* *Pelobacter acidigallici*
Putrescine	$10\ C_4H_{14}N_2 + 26\ H_2O \rightarrow 6$ acetate$^-$ $+ 7$ butyrate$^-$ $+ 10\ NH_4^+ + 16\ H_2 + 13$ H$^+$	Unclassified gram-positive nonsporing anaerobes
Citrate	Citrate^{3-} $+ 2\ H_2O \rightarrow$ formate$^-$ $+ 2$ acetate$^-$ $+ HCO_3^-$ $+$ H$^+$	*Bacteroides* sp.
Glyoxylate	4 Glyoxylate$^-$ $+ 3$ H$^+$ $+ 3\ H_2O \rightarrow 6\ CO_2 + 5\ H_2 +$ glycolate$^-$	Unclassified gram-negative bacterium
Succinate	Succinate^{2-} $+ H_2O \rightarrow$ propionate$^-$ $+ HCO_3^-$	*Propionigenium modestum*
Oxalate	Oxalate^{2-} $+ H_2O \rightarrow$ formate$^-$ $+ HCO_3^-$	*Oxalobacter formigenes*
Malonate	Malonate^{2-} $+ H_2O \rightarrow$ acetate$^-$ $+ HCO_3^-$	*Malonomonas rubra* *Sporomusa malonica*

Fermentations without substrate-level phosphorylation: Decarboxylations of organic acids

With certain substrates there is insufficient energy released to couple to the synthesis of ATP directly by substrate-level phosphorylation, yet these compounds support fermentative growth of an organism. In these cases, catabolism of the substrate is linked to ion pumps that establish a proton or sodium gradient across the membrane. Examples of this include the fermentations by *Propionigenium modestum* and *Oxalobacter formigenes;* both of these organisms couple the fermentation of dicarboxylic acids to membrane-bound energy-linked ion pumps. *Propionigenium modestum* carries out the following reaction:

$$\text{Succinate}^{2-} + H_2O \rightarrow \text{propionate}^- + HCO_3^-$$
$$\Delta G^{0\prime} = -20.5 \text{ kJ/reaction}$$

This overall reaction yields insufficient free energy to couple to ATP synthesis directly by substrate-level phosphorylation, but nevertheless it serves as the sole energy-yielding reaction for growth of the organism. This is possible because the decarboxylation of succinate (via methylmalonyl-CoA and its membrane-bound decarboxylase) by *Propionigenium modestum* is coupled to the export of Na$^+$ across the cytoplasmic membrane (Figure 13.39a). A Na$^+$-translocating ATPase in the membrane of *P. modestum* employs this Na$^+$ gradient to drive ATP synthesis (Figure 13.39a). *Oxalobacter formigenes* carries out the fermentation of oxalate:

$$\text{Oxalate}^{2-} + H_2O \rightarrow \text{formate}^- + HCO_3^-$$
$$\Delta G^{0\prime} = -26.7 \text{ kJ/reaction}$$

At neutral pH, oxalate exists in the ionized form as oxalate^{2-}, and its decarboxylation to formate$^-$ consumes one proton. The subsequent export of formate from the cell then builds a proton motive force that can be coupled to ATP synthesis by a proton-translocating ATPase in the membrane (Figure 13.39b).

The interesting and unique aspect of the metabolism of both *Propionigenium modestum* and *Oxalobacter formigenes* is the fact that ATP synthesis occurs without substrate-level phosphorylation *or* electron transport occurring; however, chemiosmotic ATP formation still occurs as a result of a Na$^+$/H$^+$ pump linked to decarboxylation of organic acids. The lesson to be learned from these fermentations is clear: any chemical reaction that yields less than the 31.8 kJ required to make one ATP (see Table 13.6) or that appears unable to couple to a substrate-level phosphorylation, cannot be automatically ruled out as a potential growth-supporting reaction for a bacterium. If the reaction can be coupled to an ion gradient, ATP production (and subsequently growth) remains a possibility. Therefore, because the influx of approximately 3 H$^+$ (or 3 Na$^+$) is required to drive ATP formation by a membrane-associated ATPase, a reaction must yield at least the energy required to pump a single H$^+$ or Na$^+$ ion to the outside of the cell membrane to be theoretically capable of supporting growth.

Methane as the final product of anoxic decomposition processes

Many of the products of fermentative metabolism listed in Tables 13.7 and 13.8 are themselves energy sources for other fermentative organisms. It is reasonable that organisms might exist that are able to use the

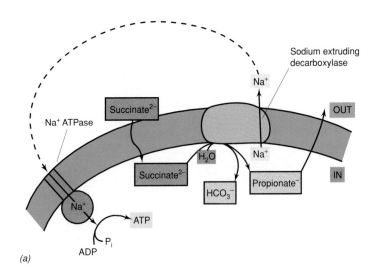

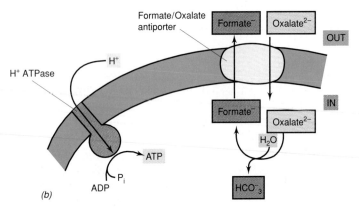

Figure 13.39 The unique fermentations of succinate and oxalate. (a) Succinate fermentation by *Propionigenium modestum*. A sodium-translocating ATPase produces ATP; sodium export is linked to succinate decarboxylation. (b) Oxalate fermentation by *Oxalobacter formigenes*. Oxalate import and formate export by a formate–oxalate antiporter consume protons. ATP synthesis is linked to a proton-driven ATPase. All substrates and products of a given reaction are shown in contrasting colors.

fermentation products of other organisms because all these anaerobes are likely to be found together in environments where fermentation is taking place. For example, succinate, lactate, and ethanol, produced from the fermentation of sugars, can themselves be fermented by other organisms. Fermentation of these "fermentation products" leads ultimately to the formation of acetate, H_2, and CO_2, substrates for the methanogenic Archaea. However, two fermentation products listed in Tables 13.7 and 13.8 cannot be further fermented: CO_2 and CH_4, the most oxidized and the most reduced forms of carbon. Thus, the ultimate products of anoxic decomposition are CH_4 and CO_2.

Coupled fermentation reactions: Syntrophy

Even if a single fermentation reaction does not provide favorable energetics, this reaction may still be carried out by an organism and support its growth if the *product* of the reaction, especially H_2, can be used by another organism in an energetically favorable reaction. We are dealing here with the phenomenon of **syntrophy** (mutual feeding), where two organisms do something together that neither can do separately. One of the best-studied syntrophic systems in fermentative

metabolism is the fermentation of ethanol to acetate and methane by two organisms, an ethanol fermenter and a methanogen. This coupled scheme is illustrated in Figure 13.40. As seen, the ethanol fermenter produces hydrogen and acetate, but this reaction has an unfavorable (positive) standard free-energy balance.

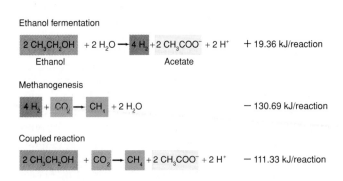

Figure 13.40 Fermentation of ethanol to methane and acetate by the coupling of an energetically unfavorable reaction (ethanol fermentation) with an energetically favorable reaction (methanogenesis). The overall energy yield is thus greater than -100 kJ/reaction. This is an example of *syntrophy* based on *interspecies hydrogen transfer*.

However, the H_2 produced by the ethanol fermenter is consumed by the methanogen in an energetically favorable reaction. When the energies of these two reactions are summed, the overall reaction is favorable energetically. The reaction summarized here is an example of a phenomenon called **interspecies hydrogen transfer,** which is discussed in more detail in Section 14.12. Note that the methanogen is unable by itself to utilize ethanol so the ethanol fermenter and the methanogen both benefit from this relationship.

In *aerobic* organisms, syntrophic relationships are less important than in anaerobic food chains. Aerobic organisms can frequently degrade fairly complex molecules completely to CO_2 and H_2O without obligate relationships with other organisms. This may have to do with the fact that the energetics of catabolism of complex materials with O_2 as electron acceptor is much more favorable than fermentative catabolism of the same substrates. However, when conditions become anoxic and alternative electron acceptors scarce, fermentative catabolism predominates but in many cases requires syntrophic interactions; we will see several examples of these relationships later (∞ Section 14.12).

Limits of anoxic catabolism

Two groups of natural organic materials appear to be refractory to fermentative breakdown: lignin and hydrocarbons. **Lignin** is a complex aromatic polymer of phenylpropane building blocks held together by C—C and C—O—C (ether) linkages. Lignin is a major component of wood and is responsible for conferring rigidity on the cellulosic walls of woody plants. Lignin appears to be stable to anoxic degradation and hence does not decompose in anoxic habitats. *Coal* is an organic material that is ultimately derived from woody plants. Coal would not exist today if lignin did not stabilize woody material to anoxic decay.

The other major natural materials that are refractory to fermentation are the long-chain **aliphatic hydrocarbons,** such as hexadecane ($C_{16}H_{34}$) and octadecane ($C_{18}H_{38}$). These hydrocarbons, which are major constituents of petroleum, either are produced directly by plants and animals or are derived chemically by reduction from long-chain fatty acids. Protected from decay by anoxic conditions, these hydrocarbons are stable in nature and become the basis of petroleum reserves. However, although unable to serve as substrates for fermentation, saturated hydrocarbons can be slowly degraded anaerobically by certain sulfate-reducing bacteria. *Aromatic* hydrocarbons can be degraded by a variety of anaerobic bacteria including sulfate reducers and denitrifiers. But as in the case of aliphatic hydrocarbons, *fermentation* of aromatic hydrocarbons by pure cultures has not been reported although degradation by microbial consortia does occur.

13.21

Hexose, Pentose, and Polysaccharide Utilization

We complete our discussion of chemoorganotrophic metabolism with consideration of a few special aspects of the catabolism of organic compounds, especially the use of polymeric substances that must first be hydrolyzed to monomeric units before energy-generating mechanisms can be employed. We begin with the microbial degradation of polysaccharides.

Hexose and polysaccharide utilization

Sugars with six carbon atoms, called **hexoses,** are the most important electron donors for many chemoorganotrophs and are also important structural components of microbial cell walls, capsules, slimes, and storage products. The most common hexose sources in nature are listed in Table 13.9, from which it can be seen that most are polysaccharides, although a few are disaccharides. Cellulose and starch are two of the most important natural polysaccharides.

Although both starch and cellulose are composed of glucose units, they are connected differently (Table 13.9), and this profoundly affects their properties. Cellulose is much more insoluble than starch and is usually less rapidly digested. Cellulose forms long fibrils, and organisms that digest cellulose are often found closely associated with them (Figure 13.41). Many fungi are able to digest cellulose, and these are mainly responsible for decomposition of plant materials on the forest floor. Among bacteria, however, cellulose digestion is restricted to relatively few groups, of which the gliding bacteria such as *Sporocytophaga* and *Cytophaga* (Figures 13.41 and 13.42), clostridia, and

TABLE 13.9   Naturally occurring polysaccharides yielding hexose and pentose sugars[a]

Substance	Composition	Sources	Catabolic enzymes
Cellulose	Glucose polymer (β-1,4-)	Plants (leaves, stems)	Cellulases (β,1-4-glucanases)
Starch	Glucose polymer (α-1,4-)	Plants (leaves, seeds)	Amylase
Glycogen	Glucose polymer (α-1,4- and α-1,6-)	Animals (muscle)	Amylase, phosphorylase
Laminarin	Glucose polymer (β-1,3-)	Marine algae (Phaeophyta)	β-1,3-Glucanase (laminarinase)
Paramylon	Glucose polymer (β-1,3-)	Algae (Euglenophyta and Xanthophyta)	β-1,3-Glucanase
Agar	Galactose and galacturonic acid polymer	Marine algae (Rhodophyta)	Agarase
Chitin	N-Acetylglucosamine polymer (β-1,4-)	Fungi (cell walls) Insects (exoskeletons)	Chitinase
Pectin	Galacturonic acid polymer (from galactose)	Plants (leaves, seeds)	Pectinase (polygalacturonase)
Dextran	Glucose polymer	Capsules or slime layers of bacteria	Dextranase
Xylan	Heteropolymer of xylose and other sugars (β-1,4- and α-1,2 or α-1,3 side groups)	Plants	Xylanases
Sucrose	Glucose–fructose disaccharide	Plants (fruits, vegetables)	Invertase
Lactose	Glucose–galactose disaccharide	Milk	β-Galactosidase

[a]Each of these is subject to degradation by microorganisms.

actinomycetes are among the most common. Anoxic digestion of cellulose is carried out by a few *Clostridium* species, which are common in lake sediments, animal intestinal tracts, and systems for anaerobic sewage digestion. Cellulose digestion is also a major process in the rumen of ruminant animals where *Fibrobacter* and *Ruminococcus* species actively degrade cellulose (∞ Section 14.13).

Starch is digestible by many fungi and bacteria; this is illustrated for a laboratory culture in Figure 13.43. Starch-digesting enzymes, called *amylases,* are of

considerable practical utility in many industrial situations where starch must be digested, such as the textile, laundry, paper, and food industries, and fungi and bacteria are the commercial sources of these enzymes (∞ Section 12.9).

All the polysaccharides occurring extracellularly and utilized as substrates are broken down to monomeric units by hydrolysis. In contrast, the polysaccharides formed within cells as storage products are broken down not by hydrolysis but by **phosphorolysis**. This process, involving the addition of *inor-*

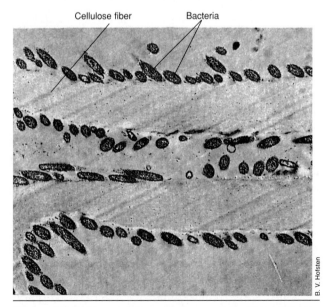

Figure 13.41 Transmission electron micrograph showing attachment of cellulose-digesting bacteria, *Sporocytophaga myxococcoides,* to cellulose fibers. Cells are about 0.5 μm in diameter.

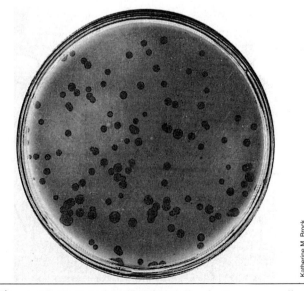

Figure 13.42 *Cytophaga hutchinsonii* colonies on a cellulose–agar plate. Clear areas are where cellulose has been digested.

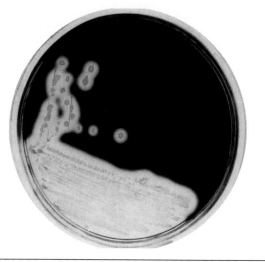

Figure 13.43 Demonstration of hydrolysis of starch by colonies of *Bacillus subtilis*. After incubation, the plate was flooded with Lugol's iodine solution. Where starch hydrolysis occurred, the characteristic purple color of the starch–iodine complex is absent. Hydrolysis of starch occurs at some distance from the bacterial colonies because of the production of extracellular amylase, which diffuses into the surrounding medium.

ganic phosphate, results in the formation of hexose phosphate rather than the free hexose and may be summarized as follows for the degradation of starch, an α-1,4 polymer of glucose:

$$(C_6H_{12}O_6)_n + P_i \rightarrow (C_6H_{12}O_6)_{n-1} + \text{glucose 1-phosphate}$$

Because glucose 1-phosphate can be easily converted to glucose 6-phosphate, a key intermediate in glycolysis (∞ Figure 4.12), and no ATP is required to form it, phosphorolysis represents a net energy savings to the cell.

Disaccharides

Many microorganisms can use *disaccharides* for growth (Table 13.9). *Lactose* utilization by microorganisms is of considerable economic importance because milk-souring organisms produce lactic acid from lactose. *Sucrose,* the common disaccharide of higher plants, is usually first hydrolyzed to its component monosaccharides (glucose and fructose) by the enzyme *invertase,* and the monomers are then metabolized by normal pathways. *Cellobiose,* β-1,4-diglucose and a major product of cellulose digestion, is also readily degraded by a variety of bacteria that cannot degrade the cellulose polymer itself.

The microbial polysaccharide *dextran* is synthesized by some bacteria using the enzyme *dextransucrase* and sucrose as starting material:

$$n \text{ sucrose} \rightarrow \underset{\text{dextran}}{(\text{glucose})_n} + n \text{ fructose}$$

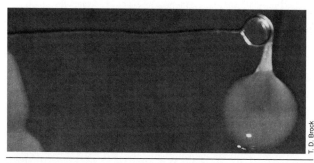

Figure 13.44 Slimy colony formed by the dextran-producing bacterium, *Leuconostoc mesenteroides,* growing on a sucrose-containing medium. When the same organism is grown on glucose, the colonies are small and not slimy.

Dextran is formed in this way by the bacterium *Leuconostoc mesenteroides* and a few others, and the polymer formed accumulates around the cells as a massive slime or capsule (Figure 13.44). Because sucrose is required for dextran formation, no dextran is formed when the bacterium is cultured on a medium with glucose or fructose. In nature, when cells containing dextran or other polysaccharide capsules die, these materials once again become available for attack by fer- mentative or other chemoorganotrophic microorganisms.

CONCEPT CHECK   *13.21*

Polysaccharides are abundant in nature and can be broken down, usually by phosphorolysis, into hexose or pentose monomers and used as energy sources. Starch and cellulose are common polysaccharides.

✔ *What is* phosphorolysis?

✔ *What disaccharides are common in nature?*

13.22
Organic Acid Metabolism

A variety of organic acids can be utilized by microorganisms as carbon sources and electron donors. The acids of the citric acid cycle, such as *citrate, malate, fumarate,* and *succinate,* are common natural products formed by plants and are also fermentation products of microorganisms. Because the citric acid cycle has major *biosynthetic* (∞ Section 4.19) as well as *energetic* (∞ Section 4.15) functions, the complete cycle or major portions of it are nearly universal in microorganisms. Thus, it is not surprising that many microorganisms are able to utilize these acids as electron donors and carbon sources. Aerobic utilization of four-, five-, and six-carbon acids can be accomplished by means of enzymes of the citric acid cycle, with ATP formation by oxidative phosphorylation.

Anaerobic utilization of organic acids usually involves conversion to pyruvate followed by formation of acetate via acetyl phosphate with consequent ATP production by substrate-level phosphorylation (see Section 13.19).

Glyoxylate cycle

Utilization of two- or three-carbon acids as carbon sources cannot occur by means of the citric acid cycle alone. This cycle can continue to operate only if the acceptor molecule, the four-carbon acid *oxalacetate*, is regenerated at each turn of the cycle; any removal of carbon compounds for biosynthetic reactions would prevent completion of the cycle. When acetate is utilized, the oxalacetate needed to continue the cycle is produced through the **glyoxylate cycle** (Figure 13.45), so called because glyoxylate is a key intermediate. This cycle is composed of most of the citric acid cycle reactions plus two additional enzymes: *isocitrate lyase*, which splits isocitrate to succinate and glyoxylate, and *malate synthase*, which converts glyoxylate and acetyl-CoA to malate.

Biosynthesis through the glyoxylate cycle occurs as follows. The splitting of isocitrate into succinate and glyoxylate allows the succinate molecule (or another citric acid cycle intermediate derived from it) to be drawn off for biosynthesis because glyoxylate combines with acetyl-CoA to yield malate. Malate can be converted to oxalacetate to maintain the cyclic nature of the citric acid cycle despite the fact that a C_4 intermediate (succinate) has been drawn off. The succinate molecule can be used directly in the production of porphyrins, be oxidized to oxalacetate and serve as a carbon skeleton for C_4 amino acids, or be converted (via oxalacetate and phosphoenolpyruvate) to glucose.

Pyruvate and C_3 utilization

Three-carbon compounds such as pyruvate or compounds converted to pyruvate (for example, lactate or carbohydrates) also cannot be utilized as energy sources through the citric acid cycle alone. Because some of the citric acid cycle intermediates are used for biosynthesis, the oxalacetate needed to keep the cycle going is synthesized from pyruvate or phosphoenolpyruvate by the addition of a carbon atom from CO_2. In some organisms this step is catalyzed by the enzyme *pyruvate carboxylase*:

Pyruvate + ATP + CO_2 → oxalacetate + ADP + P_i

whereas in others it is catalyzed by *phosphoenolpyruvate carboxylase*:

Phosphoenolpyruvate + CO_2 → oxalacetate + P_i

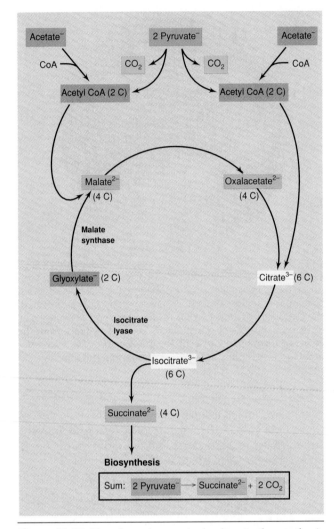

Figure 13.45 The glyoxylate cycle, leading to the synthesis of oxalacetate from acetate. Two unique enzymes, isocitrate lyase and malate synthase, operate with a majority of the citric acid cycle reactions. In addition to growth on pyruvate, the glyoxylate cycle can also operate during growth on acetate. All compounds containing the same number of carbons are shown in a single color.

These reactions replace oxalacetate that is lost when intermediates of the citric acid cycle are removed for use in biosynthesis, and the cycle can continue to function.

CONCEPT CHECK   **13.22**

Organic acids are frequently metabolized through the citric acid cycle or through the glyoxylate cycle. Isocitrate lyase and malate synthase are the key enzymes of the glyoxylate cycle.

✔ *Why is the glyoxylate cycle necessary for growth on acetate but not on succinate?*

13.23

Lipids as Microbial Nutrients

Lipids are abundant in nature. The cytoplasmic membranes of all cells contain lipids, and many microorganisms as well as macroorganisms produce lipid storage materials. These substances are all biodegradable and can be excellent substrates for microbial energy-yielding metabolism.

Fat and phospholipid hydrolysis

Fats are esters of glycerol and fatty acids (∞ Section 2.6). Microorganisms utilize fats only after hydrolysis of the ester bond, and extracellular enzymes called **lipases** are responsible for the reaction (Figure 13.46). The end result is formation of glycerol and free fatty acids (Figures 13.46 and 13.47). Lipases are not highly specific and attack fats containing fatty acids of various chain lengths. Phospholipids are hydrolyzed by specific enzymes called *phospholipases,* given different letter designations depending on which ester bond they cleave (Figure 13.47). Phospholipases A and B

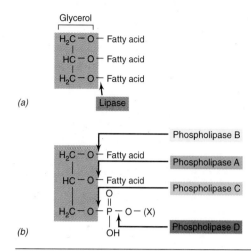

(a)

(b)

Figure 13.47 (a) Action of lipase on a fat. (b) Phospholipase action on phospholipid. The sites of action of the four distinct phospholipases A, B, C, and D are shown. X refers to a number of small organic molecules that may be at this position in different phospholipids. Compare this diagram to the more complete figure of a phospholipid in Figure 2.7.

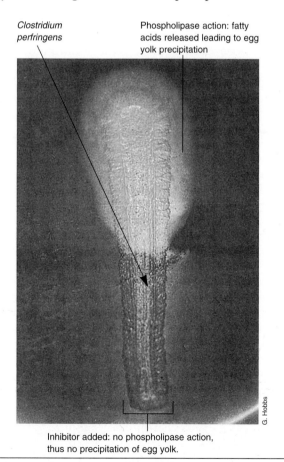

Figure 13.46 Action of phospholipase around streak of *Clostridium perfringens* growing on an agar medium containing egg yolk. On half of the plate, an inhibitor of phospholipase was added, preventing action of the enzyme.

cleave fatty acid esters and thus resemble the lipases described earlier, but phospholipases C and D cleave phosphate ester linkages and hence are quite different types of enzymes. The result of lipase action is the release of free fatty acids and glycerol, and all these substances can be attacked both anaerobically as well as aerobically by various chemoorganotrophic microorganisms.

Fatty acid oxidation

Fatty acids are oxidized by a process called *beta oxidation,* in which two carbons of the fatty acid are split off at a time (Figure 13.48). In eukaryotes the enzymes are in the mitochondria, whereas in prokaryotes they are cytoplasmic. The fatty acid is first activated with coenzyme A; oxidation results in the release of *acetyl-CoA* and the formation of a fatty acid shorter by two carbons (Figure 13.48). The process of beta oxidation is then repeated, and another acetyl-CoA molecule is released. Two separate dehydrogenation reactions occur. In the first, electrons are transferred to flavin-adenine dinucleotide (FAD), whereas in the second they are transferred to NAD^+. Most fatty acids have an even number of carbon atoms, and complete oxidation yields only acetyl-CoA. The acetyl-CoA formed is then oxidized by way of the citric acid cycle or is converted to hexose and other cell constituents via the glyoxylate cycle. Fatty acids are good electron donors. For example, the anaerobic oxidation of the 16-carbon fatty acid palmitic acid results in the net synthesis of 129 ATP molecules from electron transport phosphorylation from electrons generated dur-

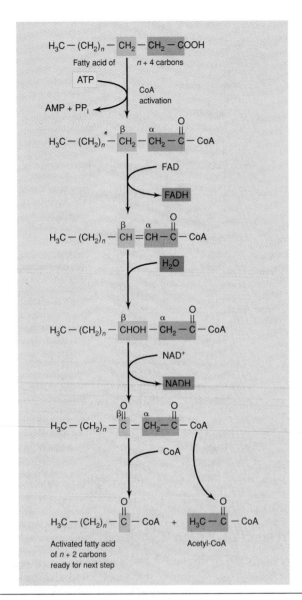

Figure 13.48 Mechanism of beta oxidation of a fatty acid, which leads to successive formation of two-carbon fragments of acetyl-CoA.

ing the formation of acetyl-CoA from beta oxidations and from oxidation of the acetyl-CoA units themselves through the citric acid cycle (∞ Section 4.15).

CONCEPT CHECK   **13.23**

Fats are metabolized by hydrolysis by lipases or phospholipases to free fatty acids. The latter are oxidized by beta oxidation to acetyl-CoA units, which are subsequently oxidized to CO_2 by the citric acid cycle.

✔ *What are the functions of phospholipases?*

✔ *What is meant by the term β-oxidation?*

13.24

Molecular Oxygen (O_2) as a Reactant in Biochemical Processes

We have discussed in some detail the role of O_2 as an *electron acceptor* in energy-generating reactions. Although this is by far the most important role of O_2 in cellular metabolism, O_2 plays an interesting and important role as a direct reactant in certain types of anabolic and catabolic processes.

Oxygenases are enzymes that catalyze the incorporation of oxygen from O_2 into organic compounds. There are two kinds of oxygenases: *dioxygenases*, which catalyze the incorporation of *both* atoms of O_2 into the molecule; and *monooxygenases*, which catalyze the transfer of *only one* of the two O_2 atoms to an organic compound as a hydroxyl (OH) group, with the second atom of O_2 being reduced to water, H_2O. Because monooxygenases catalyze the formation of hydroxyl groups (OH) in organic compounds, they are sometimes called *hydroxylases.* In most monooxygenases, the electron donor is NADH or NADPH, although the direct coupling to O_2 is through a flavin that is reduced by the NADH or NADPH donor. In the case of ammonia monooxygenase discussed previously (see Section 13.12), the electron donor is cytochrome *c*.

There are several types of reactions in living organisms that require O_2 as a reactant. One of the best examples is the involvement of O_2 in sterol biosynthesis. The formation of the fused sterol ring system (∞ Figure 3.18) requires the participation of molecular oxygen. Such a reaction can obviously not take place under anoxic conditions so organisms that grow anaerobically must either dispense with this reaction or obtain the required substance (sterol) preformed from their environment. The requirement of O_2 as a reactant in biosynthesis is of considerable evolutionary significance, as molecular O_2 was originally absent from the atmosphere of the earth when life evolved and became available only after the evolution of cyanobacteria, the first phototrophic organisms to produce O_2 (∞ Chapter 15). The role of O_2 in hydrocarbon utilization is discussed below.

CONCEPT CHECK   **13.24**

In addition to its role as an electron acceptor, oxygen (O_2) is also a chemical reactant in certain biochemical processes. Enzymes called oxygenases introduce O_2 into a biochemical compound.

✔ *How do monooxygenases differ in function from dioxygenases?*

13.25

Hydrocarbon Transformations

Hydrocarbons are organic compounds containing only carbon and hydrogen and are highly insoluble in water. Low-molecular-weight hydrocarbons are gases, whereas those of higher molecular weight are liquids or solids at room temperature. Some hydrocarbons are aliphatic compounds, a class of carbon compounds in which the carbon atoms are joined in open chains. There is a tremendous variation among aliphatic hydrocarbons in chain length, degree of branching, and number of double bonds. Another important group of hydrocarbons contains the aromatic ring and can be viewed as derivatives of benzene.

Aliphatic hydrocarbons

Only relatively few kinds of microorganisms (for example, *Nocardia, Pseudomonas, Mycobacterium,* and certain yeasts and molds) can utilize hydrocarbons for growth. For the most part, utilization of saturated aliphatic hydrocarbons is an *aerobic* process: in the absence of O_2, saturated hydrocarbons are virtually unaffected by microorganisms (a novel sulfate-reducing bacterium is an exception) (see Section 13.20).

The initial oxidation step of saturated aliphatic hydrocarbons involves molecular oxygen (O_2) as a reactant, and one of the atoms of the oxygen molecule is incorporated into the oxidized hydrocarbon. This reaction is carried out by a monooxygenase (see Section 13.24), and a typical reaction sequence is that shown in Figure 13.49. The end product of the reaction sequence is acetyl-CoA. However, the initial oxidation is not at the terminal carbon in all cases. Oxidation may sometimes occur at the second carbon, and then quite different subsequent reactions occur. *Unsaturated* aliphatic hydrocarbons containing a terminal double bond are not refractory to anoxic decomposition and can be oxidized by certain sulfate-reducing and other anaerobic bacteria.

Aromatic hydrocarbons

Many aromatic hydrocarbons can be used as electron donors aerobically by microorganisms, of which bacteria of the genus *Pseudomonas* have been the best studied. It has been demonstrated that the metabolism of these compounds, some of which are quite complex, frequently has as its initial stage the formation of either of two molecules, *protocatechuate* or *catechol*, as shown in Figure 13.50a.

These single-ring compounds are referred to as *starting substrates* because oxidative catabolism proceeds only after the complex aromatic molecules have been converted to these more simple forms.

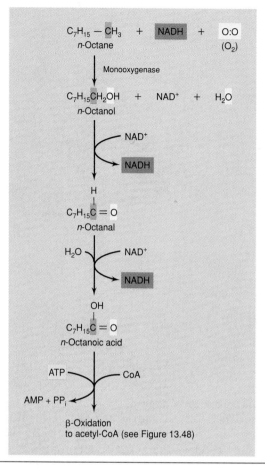

Figure 13.49 Steps in oxidation of an aliphatic hydrocarbon, the first of which is catalyzed by a monooxygenase.

Protocatechuate and catechol may then be further degraded to compounds that can enter the citric acid cycle: succinate, acetyl-CoA, pyruvate. Several steps in the catabolism of aromatic hydrocarbons usually require oxygenases. Figures 13.50b and c show two different oxygenase-catalyzed reactions, one using a monooxygenase and the other a dioxygenase.

Aromatic compounds can also be degraded anaerobically and quite readily if they already contain an atom of oxygen. Mixed bacterial cultures have been shown to degrade benzoate and other substituted phenolic compounds, yielding CH_4 and CO_2 as final products. Substituted phenolic compounds are also degraded by certain denitrifying, phototrophic, ferric iron-reducing, and sulfate-reducing bacteria. The anoxic catabolism of aromatic compounds proceeds by *reductive* rather than oxidative ring cleavage (Figure 13.51). This is reasonable because anaerobically, oxygen cannot be involved in the oxidation of aromatic compounds. Anoxic catabolism involves *ring reduction* followed by *ring cleavage* to yield a straight-chain fatty acid or dicarboxylic acid. These intermediates can generally be converted to acetyl-CoA and used for both biosynthetic and energy-yielding purposes. Benzoate

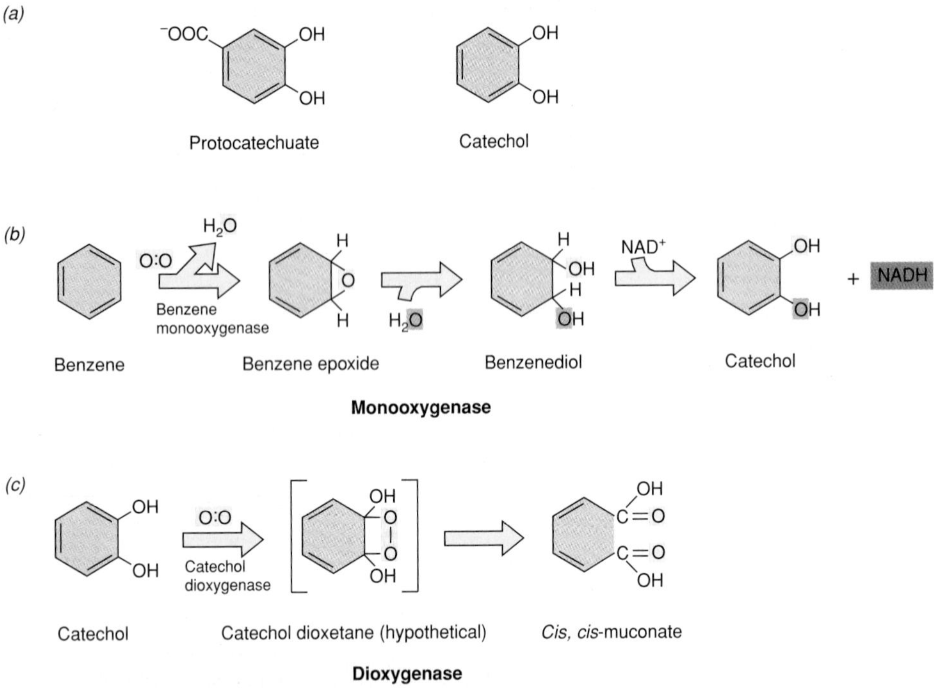

(a)

Protocatechuate

Catechol

(b)

Benzene

Benzene epoxide

Benzenediol

Catechol

Monooxygenase

(c)

Catechol

Catechol dioxetane (hypothetical)

Cis, cis-muconate

Dioxygenase

Figure 13.50 Roles of oxygenases in catabolism of aromatic compounds. (a) Protocatechuate and catechol, two common oxidation products of aromatic compounds. (b) Hydroxylation of benzene to catechol by a monooxygenase in which NADH is an electron donor. (c) Cleavage of catechol to *cis,cis*-muconate by a dioxygenase. Reactant oxygen atoms are shown in color in both reactions to demonstrate the different mechanisms.

and benzoate derivatives are common natural products and are readily degraded anaerobically.

Evidence for the anoxic degradation of benzene and toluene, aromatic compounds lacking an oxygen atom, has also been obtained. Catabolism of benzene occurs by anaerobic microbial consortia leading to methanogenesis, but toluene oxidation to CO_2 can occur in pure culture. In the latter instance, growth on toluene is supported by anaerobic respiration coupled to the reduction of ferric iron or nitrate. The biochemistry of anaerobic toluene oxidation is unknown.

13.26
Nitrogen Metabolism

We have discussed nitrogen nutrition in general terms in Sections 4.19 and 4.20. Here we discuss some of the special biochemistry and genetics of nitrogen utilization, with emphasis on the process of **nitrogen fixation.** Nitrogen can be obtained by microorganisms either from inorganic or organic forms. The most common

inorganic nitrogen sources are nitrate and ammonia, but other inorganic sources used by certain microorganisms include cyanide (CN^-), cyanate (OCN^-), thiocyanate (SCN^-), cyanamide (NCN^{2-}), nitrite (NO_2^-), and hydroxylamine (NH_2OH). As discussed in the next section, free nitrogen gas (N_2) is also used by a variety of bacteria.

We discussed the general biochemistry of the utilization of organic nitrogen in Section 4.19. There we

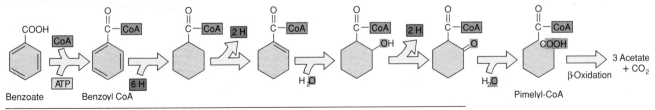

Benzoate

Benzoyl CoA

Pimelyl-CoA

3 Acetate + CO_2

β-Oxidation

Figure 13.51 Anoxic degradation of benzoate by reductive-ring cleavage. Note that all intermediates of the pathway are bound to coenzyme A.

described enzymes such as *transaminases* and *amino acid dehydrogenases*, which catalyze the reductive assimilation of ammonia from the environment. Another very important means of incorporating ammonia is via the enzyme **glutamine synthetase.** As shown in Figure 13.52, this enzyme catalyzes the addition of ammonia to glutamate to yield *glutamine.* Then in a second reaction, the enzyme **glutamate synthase** [formally known as glutamate:oxoglutarate aminotransferase (GOGAT)] catalyzes the transfer of the amide group of glutamine to α-ketoglutarate, forming two molecules of *glutamate* (Figure 13.52). When these two reactions are summed, the net reaction is the conversion of one molecule of α-ketoglutarate and one molecule of ammonia to glutamate, with the consumption of one high energy phosphate bond from ATP. In addition to its role in ammonia assimilation, glutamine also functions as an amino donor for a variety of reactions, for example, in the synthesis of the purine ring (∞ Section 4.20).

Control of ammonia assimilation via glutamine synthetase

A rather interesting metabolic control is exerted on the activity of the enzyme glutamine synthetase. Glutamine synthetase exists in two forms, an active form and an inactive form; the latter form occurs when adenyl residues from ATP are attached to the molecule, a process called *adenylylation.* Adenylylation of glutamine synthetase is promoted by high concentrations of ammonia, leading to inactivation of the enzyme. However, since glutamate dehydrogenase functions when ammonia concentrations are high, ammonia assimilation is shifted from the glutamine pathway to the glutamate pathway. When ammonia concentrations fall, glutamine synthetase becomes deadenylylated and becomes active again. Control via adenylylation thus involves a *covalent modification* of the enzyme rather than

merely a change in its conformation as was described for enzyme inhibition in Section 7.1. The presumed function of adenylylation is to prevent the wasteful consumption of ATP by glutamine synthetase when ammonia levels are sufficiently high for the operation of amino acid dehydrogenases; the latter do not require ATP for their function. We discussed the utilization of nitrate and the whole process of assimilative nitrate reduction earlier in this chapter (see Section 13.15).

13.27
Nitrogen Fixation

The utilization of nitrogen gas (N_2) as a source of nitrogen is called *nitrogen fixation* and is a property of only certain prokaryotes. An abbreviated list of nitrogen-fixing organisms is given in Table 13.10, from which it can be seen that a variety of prokaryotes, both anaerobic and aerobic, fix nitrogen. In addition, there are some bacteria, called *symbiotic,* that fix nitrogen only in association with certain plants. As far as is currently known, no eukaryotic organisms fix nitrogen. Symbiotic nitrogen fixation will be discussed in Section 14.23.

Nitrogenase

In the fixation process, N_2 is *reduced* to ammonium and the ammonium converted to organic form. The reduction process is catalyzed by the enzyme complex **nitrogenase,** which consists of two separate proteins called *dinitrogenase* and *dinitrogenase reductase.* Both components contain iron, and dinitrogenase contains molybdenum as well. The iron and molybdenum in dinitrogenase are contained in a cofactor known as *FeMo-co* (Figure 13.53), and the actual reduction of N_2 occurs on this iron–molybdenum center. The composition of FeMo-co is $MoFe_7S_9$ homocitrate (Figure 13.53), and FeMo-co is present in two copies per molecule of nitrogenase.

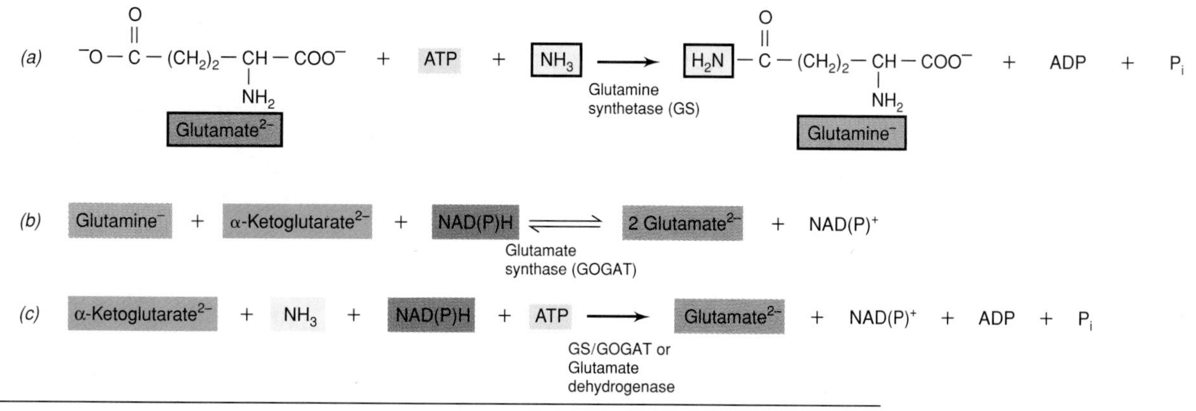

Figure 13.52 Role of glutamine synthetase in the assimilation of ammonia. (a) Glutamine synthetase reaction. (b) Glutamate synthase (GOGAT) reaction. (c) Sum of the two reactions. Reaction (c) can also be carried out by the enzyme glutamate dehydrogenase in the absence of ATP but only at high ammonia concentrations. Glutamine synthetase and glutamate synthase operate at low ammonia concentrations.

TABLE 13.10   Some nitrogen-fixing organisms

Free-living aerobes		
Chemo-organotrophs	Phototrophs	Chemo-lithotrophs
Bacteria: *Aztobacter* spp. *Klebsiella*[a] *Beijerinckia* *Bacillus polymyxa* *Mycobacterium flavum* *Azospirillum lipoferum* *Citrobacter freundii* *Acetobacter diazotrophicus* *Methylomonas* *Methylococcus*	Cyanobacteria (various, but not all)	*Alcaligenes* *Thiobacillus* (some species)

Free-living anaerobes		
Chemo-organotrophs	Phototrophs	Chemo-lithotrophs
Bacteria: *Clostridium* spp. *Desulfovibrio* *Desulfotomaculum*	Bacteria: *Chromatium* *Thiocapsa* *Chlorobium* *Rhodospirillum* *Rhodopseudomonas* *Rhodomicrobium* *Rhodopila* *Rhodobacter* *Heliobacterium* *Heliobacillus* *Heliophilum*	Archaea: *Methanosarcina* *Methanococcus*

Symbiotic	
Leguminous plants	Nonleguminous plants
Soybeans, peas, clover, locust, and so on, in association with a bacterium of the genus *Rhizobium, Bradyrhizobium,* or *Azorhizobium*	*Alnus, Myrica, Ceanothus,* *Comptonia, Casuarina;* in association with actinomycetes of the genus *Frankia*

[a] N_2 fixation occurs only under anoxic conditions.

Some nitrogen-fixing bacteria can synthesize non-molybdenum nitrogenases under certain growth conditions, and these so-called *alternative nitrogenases* do not contain molybdenum but instead contain either vanadium (and iron) or iron only. Cofactors similar to FeMo-co are present in both alternative nitrogenases as well, FeVa-co in the vanadium nitrogenase and an iron–sulfur cluster resembling FeMo-co and FeVa-co but lacking both Mo and Va, in the iron nitrogenase. Alternative nitrogenases are not synthesized when sufficient molybdenum is present, as the molybdenum nitrogenase is generally the main nitrogenase in the cell. Alternative nitrogenases presumably serve as a backup mechanism to ensure that N_2 fixation can still

occur when molybdenum is limiting in the habitat (∞ Section 16.16).

Owing to the stability of the N≡N triple bond (which has a dissociation energy of 940 kJ compared with 493 kJ for O_2), N_2 is extremely inert and its activation is a very energy-demanding process. Six electrons must be transferred to reduce N_2 to 2 NH_3, and several intermediate steps might be visualized; it is thought that the three successive reduction steps occur directly on nitrogenase with no free intermediates accumulating (Figure 13.54*a*). Nitrogen fixation is highly reductive in nature, and the process is inhibited by oxygen because both dinitrogenase and dinitrogenase reductase are rapidly and irreversibly inactivated by O_2 (even when isolated from *aerobic* nitrogen fixers). In aerobic bacteria, N_2 fixation occurs in the presence of O_2 in whole cells but not in purified enzyme preparations, and nitrogenase in such organisms is protected from O_2 inactivation either by rapid removal of O_2 by respiration, the production of O_2-retarding slime layers, or by compartmentalization of nitrogenase in a special type of cell (the heterocyst) (∞ Section 16.2). In addition, although N_2 fixation does not occur in oxic cell extracts, in aerobic nitrogen fixers like *Azotobacter*, nitrogenase is protected from oxygen inactivation by complexing with a specific protein; this has been referred to as *conformational protection*.

Electron flow in nitrogen fixation

The sequence of electron transfer in nitrogenase is as follows: electron donor → dinitrogenase reductase → dinitrogenase → N_2 → 2 NH_3. The electrons for nitrogen reduction are transferred to dinitrogenase reductase from ferredoxin or flavodoxin, low potential iron–sulfur proteins. In *Clostridium pasteurianum*, ferredoxin is the electron donor and is reduced by phosphoroclastic splitting of pyruvate to acetyl-CoA + CO_2. In addition to reduced ferredoxin, ATP is required for N_2 fixation. The ATP requirement for nitrogen fixation is very high, about 4–5 ATPs hydrolyzed for each 2 e^- transferred. ATP is required to lower the reduction potential of nitrogenase sufficiently so that N_2 may be reduced. The reduction potential of dinitrogenase reductase is −0.30 V, and this is lowered to −0.40 V when the electrons are transferred to the enzyme and the enzyme hydrolyzes ATP (Figure 13.54). This complex then combines with dinitrogenase, and the latter becomes reduced. Reduced dinitrogenase then reduces N_2 to NH_3, with the actual reduction occurring at the FeMo-co center. Although only *six* electrons are necessary to reduce N_2 to 2 NH_3, *eight* electrons are actually consumed in the process, *two* electrons being lost as hydrogen (H_2), for each mole of N_2 reduced (Figure 13.54*a*). The reason for this apparent wastage is not known, but evidence is strong that H_2 evolution is an intimate part of the reaction mechanism of nitrogenase.

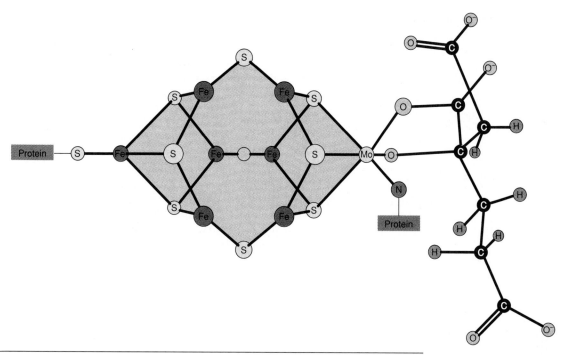

Figure 13.53 Structure of FeMo-co, the iron–molybdenum cofactor from nitrogenase. On the left is the Fe_7S_8 cube that binds to molybdenum along with oxygen atoms from homocitrate (right, all oxygen atoms shown in green) and N and S atoms from dinitrogenase. Two molecules of FeMo-co are present per molecule of nitrogenase.

Genetics and regulation of nitrogen fixation

The genes for dinitrogenase and dinitrogenase reductase in *Klebsiella pneumoniae*, a well-studied N_2 fixer, are part of a complex regulon (a large network of operons) called the *nif regulon* (Figure 13.54*b*); the *K. pneumoniae nif* regulon spans 24 kb of DNA and contains 20 genes arranged in several transcriptional units (Figure 13.54). In addition to nitrogenase structural genes, the genes for FeMo-co, genes controlling the electron transport proteins, and a number of regulatory genes are also present in the *nif* regulon. Dinitrogenase is a complex protein made up of two subunits, α (product of the *nifD* gene) and β (product of the *nifK* gene), each of which is present in two copies. Dinitrogenase reductase is a protein dimer consisting of two identical subunits, the product of *nifH*. FeMo-co is synthesized through the participation of several genes, including *nifN, V, Z, W, E,* and *B,* as well as *Q,* which controls a product involved in molybdenum processing. The *nifA* gene encodes a positive regulatory protein that serves to activate transcription of other *nif* genes.

Nitrogenase is subject to strict regulatory controls. Nitrogen fixation is blocked by O_2 and by fixed nitrogen, including NH_3, NO_3^-, and certain amino acids. A major part of this regulation is at the level of transcription. The various transcriptional units of the *nif* regulon are shown in Figure 13.54*b*. While transcrip-

tion of the *nif* structural genes is *activated* by the NifA protein (positive regulation), this activation is eliminated by the NifL protein under certain conditions.

The ammonia produced by nitrogenase does not repress enzyme synthesis because as soon as it is made, it is incorporated into organic form and used in biosynthesis. But when ammonia is in excess (as in environments high in ammonia), nitrogenase synthesis is quickly repressed. This prevents the wastage of ATP by not making a product already present in ample amounts. In certain nitrogen-fixing bacteria, nitrogenase *activity* is also regulated by ammonia, a phenomenon called the ammonia "switch-off" effect. In this case, excess ammonia causes a covalent modification of dinitrogenase reductase, which results in a loss of enzyme activity. When ammonia again becomes limiting, this modified protein is converted back to the active form and N_2 fixation resumes. Ammonia switch-off is thus a rapid and reversible method of controlling ATP consumption by nitrogenase.

Nitrogenase has been purified from a large number of nitrogen-fixing organisms and in all cases has been shown to be a two-protein complex. It is of considerable evolutionary interest that within molybdenum nitrogenases, dinitrogenase from one organism usually functions with dinitrogenase reductase from another organism. This can be interpreted to mean that the structures of the nitrogenase components have not changed markedly during evolution, suggesting that

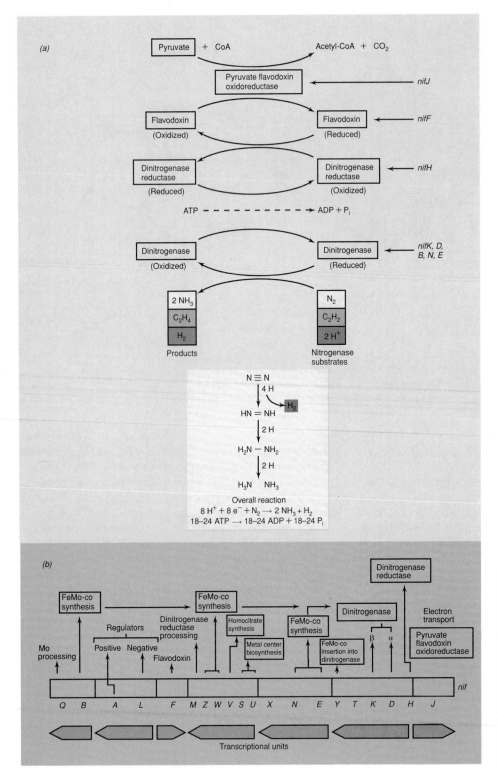

Figure 13.54 The nitrogenase system. (a) Steps in nitrogen fixation: reduction of N_2 to 2 NH_3. (b) The genetic structure of the *nif* regulon in *Klebsiella pneumoniae*, the best-studied nitrogen-fixing organism. The functions of some of the genes are uncertain. Several other genes suspected of being *nif* genes are known but not shown in this figure. The mRNA transcripts (transcriptional units) are shown below the genes; arrows indicate the direction of transcription.

the molecular requirements for N_2 reduction are fairly specific. Indeed, studies of the "HDK" gene cluster have confirmed this, in that molecular probes containing cloned *nifHDK* genes from one nitrogen-fixing bacterium hybridize to DNA from virtually all N_2 fixers but not to DNA from non-N_2-fixing bacteria.

Assaying nitrogenase: Acetylene reduction

Nitrogenase is not entirely specific for N_2, as it also reduces cyanide (CN^-), acetylene ($HC{\equiv}CH$), and several other triply bonded compounds. The reduction of acetylene by nitrogenase is only a two-electron

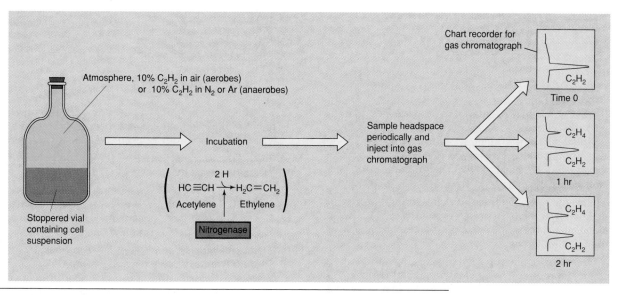

Figure 13.55 The acetylene reduction assay for nitrogenase activity. The results show no C_2H_4 when the experiment begins (time 0), but increasing production of C_2H_4 as the assay proceeds. Note how as C_2H_4 is produced, C_2H_2 is consumed. If the vial contained an enzyme extract, conditions would be anoxic, even if the nitrogenase were from an aerobic bacterium.

process, and *ethylene* ($H_2C{=}CH_2$) is produced. The reduction of acetylene probably serves no useful purpose to the cell, but it does provide the experimenter with a simple and rapid way of measuring the activity of nitrogen-fixing systems because it is fairly easy to measure the reduction of acetylene to ethylene by gas chromatography (Figure 13.55). This technique is now widely used to detect nitrogen fixation in unknown systems. Previously, it was not easy to prove that an organism fixed N_2; indeed, many claims for nitrogen fixation in microorganisms were shown to be erroneous. The growth of an organism in a medium to which no nitrogen compounds have been added does not mean that the organism is fixing nitrogen from the air, because traces of fixed nitrogen compounds often occur as contaminants in the ingredients of culture media or enter the media in gaseous form or as dust particles.

Definitive proof of N_2 fixation is obtained using an isotope of nitrogen, ^{15}N, as a tracer. (^{15}N is not a radioisotope but a stable isotope. It is detected with a mass spectrometer.) The gas phase of a culture is enriched with ^{15}N, and after incubation, the cells and medium are digested, the ammonia produced being distilled off and assayed for its ^{15}N content. If there has been a significant production of ^{15}N-labeled NH_3, it is proof of nitrogen fixation. However, the acetylene reduction method is a more rapid and sensitive, albeit indirect, way of measuring nitrogen fixation. The sample, which may be soil, water, a culture, or a cell extract, is incubated with acetylene, and the gas

phase of the reaction mixture is later analyzed by gas chromatography for production of ethylene (Figure 13.55). This method is far simpler and faster than other methods and can easily be adapted for field use in ecological studies of N_2-fixing bacteria directly in their habitats.

CONCEPT CHECK 13.27

Nitrogen fixation, the reduction of N_2 to NH_3, involves a complex enzyme system called nitrogenase, which consists of dinitrogenase and dinitrogenase reductase, metal-containing enzymes found only in certain prokaryotic organisms. Most nitrogenases contain molybdenum or vanadium and iron as metal cofactors, and the process of nitrogen fixation is highly energy-demanding. Nitrogenase and associated regulatory proteins are encoded by the *nif* regulon. Certain artificial substrates that are structurally similar to N_2, such as acetylene and cyanide, are also reduced by nitrogenase.

✔ *Write a balanced equation for the reaction carried out by the enzyme nitrogenase.*

✔ *What is FeMo-co?*

✔ *What metals are necessary for nitrogen fixation?*

✔ *What chemical and physical factors affect the function of nitrogenase?*

✔ *How is C_2H_2 useful for studies of nitrogen fixation?*

Material for Review

REVIEW QUESTIONS

1. In what nutritional class would you place an organism that uses *glucose* as sole carbon and energy source? *Elemental sulfur* as an energy source? How would we refer to the latter organism if it grew with CO_2 as sole carbon source? What is the energy source for *phototrophic* organisms?

2. What is the role of light in the photosynthetic process of green and purple bacteria? Of cyanobacteria? Compare and contrast the photosynthesis process in these two groups of prokaryotes.

3. What are the functions of light-harvesting and reaction center chlorophylls? Why would a mutant incapable of making light-harvesting chlorophylls (such mutants can be readily isolated in the laboratory) probably not be a successful competitor in nature?

4. Where are the photosynthetic pigments located in a purple bacterium? A cyanobacterium? A green alga? Considering the function of chlorophyll pigments, why can't they be located elsewhere in the cell, for example, in the cytoplasm or in the cell wall?

5. How does light result in ATP production in an *anoxygenic phototroph*? In what ways are photosynthetic and respiratory electron flow similar? In what ways do they differ?

6. How is reducing power made for autotrophic growth in a purple bacterium? In a cyanobacterium?

7. How does the reduction potential of chlorophyll *a* in photosystem I and photosystem II differ? Why must the reduction potential of photosystem II chlorophyll be so highly electropositive?

8. What is the major function of carotenoids and phycobilins in phototrophic microorganisms?

9. What two enzymes are unique to organisms that carry out the Calvin cycle? What reactions do these enzymes carry out? What would be the consequences if a mutant arose that lacked either of these enzymes?

10. For conversion of 6 molecules of CO_2 into 1 fructose molecule, 18 molecules of ATP are required. Where in the Calvin cycle reactions are these ATPs consumed?

11. Compare and contrast the utilization of H_2S by a purple phototrophic bacterium and by a colorless sulfur bacterium like *Beggiatoa*. What role does H_2S play in the metabolism of each organism?

12. Discuss why the growth yield (grams of cells per mole of substrate) of *Thiobacillus ferrooxidans* is considerably greater when the organism is growing aerobically on elemental sulfur than on ferrous iron as electron donor (assume the organism is growing autotrophically in both cases).

13. What is a mixotroph? Why does an organism capable of mixotrophy frequently grow better mixotrophically than chemolithotrophically or even chemoorganotrophically?

14. In *Escherichia coli* synthesis of the enzyme *nitrate reductase* is repressed by oxygen. On the basis of bioenergetic arguments, why do you think this repression phenomenon might have evolved?

15. Discuss at least three major differences between *assimilative* and *dissimilative* nitrate reduction.

16. Define the term *substrate-level phosphorylation*. How does it differ from oxidative phosphorylation? Assuming an organism is facultative, what basic nutritional conditions dictate whether the organism obtains energy from substrate-level rather than oxidative phosphorylation?

17. Although many different compounds are theoretically fermentable, in order to support a fermentative process, most organic compounds must be eventually converted to one of a relatively small group of molecules. What are these molecules and why must they be produced?

18. To a culture of *Escherichia coli* growing fermentatively you add 1 g/l of $NaNO_3$. Would you expect the growth *yield* of the culture to increase or decrease? Why?

19. How can fermentations occur in the absence of substrate-level phosphorylation?

20. Why have hydrocarbons accumulated in large reservoirs on Earth despite the fact that they are readily degradable microbiologically under certain conditions?

21. How are xenobiotic compounds defined? Give an example of a compound you think qualifies as a xenobiotic.

22. Compare and contrast the conversion of cellulose and intracellular starch to glucose units. What enzymes are involved and which process is the more energy-efficient?

23. How do *monooxygenases* differ from *dioxygenases* in the reactions they catalyze?

24. How do ammonia levels in a cell control the activities of the enzyme glutamine synthetase? How does such control benefit the cell? Although ammonia controls the *activity* of glutamine synthetase, this is not a typical "feedback" inhibition phenomenon. Explain.

25. Write out the reaction catalyzed by the enzyme *nitrogenase*. How many electrons are required in this reaction? How many are actually used? Explain.

26. What is the function of ATP in the nitrogen fixation process? How does ATP affect dinitrogenase reductase?

27. What metals are found in nitrogenase? What is the evidence that nitrogenase is a highly conserved protein in an evolutionary sense?

APPLICATION QUESTIONS

1. Compare and contrast the absorption spectrum of chlorophyll *a* and bacteriochlorophyll *a*. What wavelengths are preferentially absorbed by each pigment and how do the absorption properties of these molecules compare with the regions of the spectrum visible to our eye? Why are most plants green in color?

2. The growth rate of the phototrophic purple bacterium *Rhodobacter* is about twice as fast when the organism is grown phototrophically in a medium containing malate as carbon source as when it is grown with CO_2 as carbon source (with H_2 as electron donor). Discuss the reasons why this is true and list the nutritional class we would place *Rhodobacter* in when growing under each of the two different conditions.

3. Discuss the nature of the evidence obtained from studies on the photosynthetic process of certain cyanobacteria that supports the hypothesis that these organisms evolved from anoxygenic phototrophs.

4. Although physiologically distinct, chemolithotrophs and chemoorganotrophs share a number of features with respect to the production of ATP. Discuss these common features along with reasons why the growth yield (grams of cells per mole of substrate) of a chemoorganotroph respiring glucose is so much higher than for a chemolithotroph respiring sulfur.

5. Why is the following statement, if taken literally, incorrect? "Anaerobic respiration is simply a process where an alternative electron acceptor is substituted for O_2 in the respiratory process."

6. Although dextran is a glucose polymer, glucose cannot be used to make dextran. Explain. How is dextran synthesis important in oral hygiene (∞ Section 19.3)?

7. *Pseudomonas fluorescens* can grow on benzoate aerobically, whereas the phototrophic bacterium *Rhodopseudomonas palustris* can grow on benzoate anaerobically. Compare and contrast the metabolism of benzoate by these two species, focusing on the following considerations: requirement for oxygenases, initial reactions leading to the opening of the ring, and product(s) formed that can feed into central metabolic pathways.

SUPPLEMENTARY READINGS

Balows, A., H. G. Trüper, M. Dworkin, W. Harder, and **K.-H. Schleifer** (eds.). 1992. *The Prokaryotes: A Handbook on the Biology of Bacteria, Ecophysiology, Isolation, Identification, and Applications,* 2nd edition. Springer-Verlag, New York. A complete reference source on all groups of bacteria. Excellent chapters on phototrophs, chemolithotrophs, methylotrophs, nitrogen fixers, and various fermentative anaerobes. In many cases the chapters on the different groups of organisms are prefaced by an overview chapter on the particular metabolic process characteristic of the group.

Barton, L. L. (ed.). 1995. *Sulfate-Reducing Bacteria.* Plenum, New York. A summary of current information on the biology of those interesting organisms.

Blankenship, R. E., M. T. Madigan, and **C. E. Bauer** (eds.). 1995. *Anoxygenic Photosynthetic Bacteria.* Kluwer Academic, Dordrecht, The Netherlands. The most up-to-date compendium on all aspects of the biology of anoxygenic phototrophic bacteria.

Krulwich, T. A. (ed.). 1990. *Bacterial Energetics.* Academic Press, New York. Detailed coverage of bioenergetic principles in a wide variety of metabolic types of prokaryotes.

Moat, A. G., and **J. W. Foster.** 1995. *Microbial Physiology.* John Wiley, New York. An undergraduate-level textbook of microbial physiology with the focus on prokaryotes.

Nicholls, D. G., and **S. J. Ferguson** (eds.). 1992. *Bioenergetics 2.* Academic Press, Orlando, FL. An advanced treatment of bioenergetic principles focusing on the generation of a proton motive force by various means and recovery of membrane energy as ATP.

Schlegel, H. G., and **B. Bowien** (eds.). 1989. *Autotrophic Bacteria.* Science-Tech Publishers, Madison, WI. A series of chapters written by experts on all phases of the microbiology, physiology, and biochemistry of autotrophic bacteria.

Sebald, M. (ed.). 1992. *Genetics and Molecular Biology of Anaerobic Bacteria.* Springer-Verlag, New York. Genetics and other molecular information on a wide variety of anaerobic bacteria carrying out reactions that are described in this chapter.

Silver, S., A. M. Chakrabarty, B. Iglewski, and **S. Kaplan** (eds.). 1990. *Pseudomonas: Biotransformations, Pathogenesis, and Evolving Biotechnology.* American Society for Microbiology, Washington, DC. A compendium of papers on a very metabolically diverse group of bacteria.

White, D. 1995. *The Physiology and Biochemistry of Prokaryotes.* Oxford University Press, New York. An excellent textbook of bacterial physiology, it discusses in detail the physiology of most of the groups considered in this chapter.

Zehnder, A. J. B. (ed.). 1988. *Biology of Anaerobic Bacteria.* John Wiley, New York. A collection of review chapters covering all major groups of anaerobes and anoxic processes.

 On~line resources for this chapter are on the World Wide Web at:
http://www.prenhall.com/~brock (click on the <u>table of contents</u> link and then select Chapter 13).

CHAPTER *14*

Microbial Ecology

Microbial interactions with plants and animals are common in nature, and in many cases, the interactions are symbiotic, such as in these nitrogen-fixing nodules on the roots of a soybean plant.

MINIGLOSSARY for Chapter 14

Acid Mine Drainage acidic water containing H_2SO_4 derived from the microbial oxidation of iron sulfide minerals

Anoxic an oxygen-free environment, usually also highly reducing (low E_0')

Bacteroid morphologically misshapen *Rhizobium* cells inside a leguminous plant root nodule; can fix N_2

Barophilic an organism that grows best when placed under a pressure greater than 1 atm

Barotolerant an organism that can grow under elevated pressures but that grows best at atmospheric pressure

Biofilm colonies of microbial cells encased in slime and attached to a surface

Biogeochemistry study of biologically mediated chemical transformations

Black Smoker an extremely hot (250–350°C) deep-sea hot spring emitting both hot water and various minerals

Cometabolism metabolism of a compound in the presence of a second organic compound, which is used as the primary energy source

Ecosystem a community of organisms and their natural environment

Enrichment Culture a means of obtaining cultures of microorganisms from a natural environment by using highly selective culture methods

Guild a population of metabolically related microorganisms

Hydrothermal Vent a deep-sea warm or hot spring

Infection Thread in the formation of root nodules, a cellulosic tube through which *Rhizobium* cells can travel to reach and infect root cells

Interspecies Hydrogen Transfer the production and subsequent consumption of H_2 by different groups of microorganisms that interact closely during anaerobic catabolism

Leaching solubilization and removal of metals from an ore by microbial attack

Lichen a fungus and an alga (or cyanobacterium) living in symbiotic association

Microbial Plastics biodegradable polymeric materials obtained from microorganisms that have properties similar to those of synthetic plastics

Microenvironment the immediate environmental surroundings of a microbial cell or group of cells

Mycorrhiza a symbiotic association between a fungus and the roots of a plant

Oxic an oxygen-containing environment frequently possessing a high E_0'

Primary Producer an organism that uses light to synthesize new organic material from CO_2

Pyrite a common iron-containing ore, FeS_2

Reductive Dechlorination removal of Cl as Cl^- from an organic compound by reducing the carbon atom from C—Cl to C—H

Rhizosphere the region immediately adjacent to plant roots

Root Nodule a tumorlike growth on plant roots that contains symbiotic nitrogen-fixing bacteria

Rumen the first vessel in the multichambered stomach of ruminant animals in which cellulose digestion occurs

Ti Plasmid a conjugative plasmid present in the bacterium *Agrobacterium tumefaciens* that can transfer genes into plants

Xenobiotic a totally synthetic product not naturally occurring in nature

U p to this point we have mainly considered microorganisms as laboratory entities. In this chapter we examine microorganisms in soil, water, and other environments and discuss how they act to change the chemical and physical properties of their environments. The term *environment* refers to everything surrounding a living organism: the chemical, physical, and biological factors and forces that act on a living organism. From an ecological perspective, microorganisms are part of organismal communities called *ecosystems.* Each organism in an ecosystem interacts with its surroundings and in some cases greatly modifies the characteristics of the ecosystem in the process. This is particularly true of microorganisms,

where significant chemical changes can occur because of their metabolic activities. In a microbial ecosystem individual cells grow to form *populations* (Figure 14.1a). Metabolically related populations constitute groupings called *guilds*, and sets of guilds conducting complementary physiological processes interact to form microbial *communities* (Figure 14.1b). Microbial communities then interact with communities of macroorganisms to define the entire ecosystem.

Energy enters ecosystems in the form of sunlight, organic carbon, or reduced inorganic substances. Light is used by phototrophic organisms (∞ Sections 13.1–13.6, 16.1, and 16.2) to synthesize new organic matter (Figure 14.1b). The latter contains not only carbon but

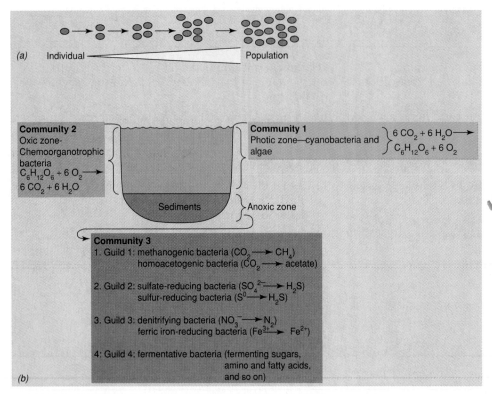

(a) Individual ——————————— Population

Community 2
Oxic zone-
Chemoorganotrophic
bacteria
$C_6H_{12}O_6 + 6\,O_2 \longrightarrow$
$6\,CO_2 + 6\,H_2O$

Community 1
Photic zone—cyanobacteria and
algae

$6\,CO_2 + 6\,H_2O \longrightarrow$
$C_6H_{12}O_6 + 6\,O_2$

Sediments

Anoxic zone

Community 3
1. Guild 1: methanogenic bacteria ($CO_2 \longrightarrow CH_4$)
   homoacetogenic bacteria ($CO_2 \longrightarrow$ acetate)

2. Guild 2: sulfate-reducing bacteria ($SO_4^{2-} \longrightarrow H_2S$)
   sulfur-reducing bacteria ($S^0 \longrightarrow H_2S$)

3. Guild 3: denitrifying bacteria ($NO_3^- \longrightarrow N_2$)
   ferric iron-reducing bacteria ($Fe^{3+} \longrightarrow Fe^{2+}$)

4: Guild 4: fermentative bacteria (fermenting sugars,
   amino and fatty acids,
   and so on)

(b)

Figure 14.1 Populations, guilds, and communities—an example of microbial community structure in a lake ecosystem. (a) Microbial guilds consist of populations of cells of various species that arise from cell division. (b) A lake ecosystem. For simplicity, only three microbial communities are depicted: phototrophic, aerobic chemoorganotrophic, and anaerobic. In the anaerobe community, examples of guild structure are given.

also nitrogen, sulfur, phosphorus, iron, and a host of other elements. This newly synthesized organic material, along with organic matter that enters the ecosystem from the outside (*allochthonous* organic matter) and reduced inorganic substances, drives the metabolic activities of chemoorganotrophic and chemolithotrophic organisms, whose metabolic diversity was discussed in the previous chapter. For the key elements of living systems, a *biogeochemical cycle* can be defined in which the element undergoes changes in oxidation as it moves through the ecosystem (Figure 14.1*b*). Microorganisms are intimately involved in biogeochemical cycling and in many instances are the only biological agents capable of regenerating forms of the elements usable by other organisms, particularly plants. We will discuss many biogeochemical cycles in this chapter.

The science of microbial ecology has two broad objectives: (1) to understand the *biodiversity* of microorganisms in nature and how different guilds interact in microbial communities, and (2) to measure the *activities* of microorganisms in nature and monitor their effects on ecosystems. Unfortunately, despite everything we discussed in Chapter 13 about metabolic diversity, we know relatively little about the diversity of *organisms* that carry out these reactions. In many cases only one or a few organisms able to carry out a particular metabolic transformation are known from the study of pure cultures; indeed, most microbiologists agree that there are many microorganisms left to be discovered and this is a major goal of microbial ecology. However, measurements of microbial *activities* in nature have allowed us to assess the

metabolic functioning of microbial communities even if we know little about the actual organisms that compose them. The occurrence of a specific biochemical transformation is good evidence that a corresponding microbial guild is present and metabolically active, and such information is often useful in designing procedures for attempting to isolate the responsible organisms.

In this chapter we discuss modern methods of assessing and tracking microbial communities in their natural habitats and consider how the activities of microorganisms in nature play important roles in biogeochemical cycling. We also examine a number of specialized microbial habitats where much is understood about the role of microorganisms in the ecosystem. We begin our treatment of microbial ecology with

---

CONCEPT CHECK

Ecology is the study of organisms in their natural environments. Microbial communities consist of various guilds of metabolically related organisms. Microorganisms play major roles in energy transformations and biogeochemical processes. Knowledge of both microbial diversity and microbial activity is necessary in order to fully characterize a natural microbial community.

✔ *How does a microbial* guild *differ from a microbial* community?

✔ *What is a* biogeochemical cycle?

a consideration of microbial habitats because the habitat is where the microorganism actually lives and carries out its important functions.

14.1

Microorganisms in Nature

The natural habitats of microorganisms are exceedingly diverse. Any habitat suitable for the growth of higher organisms can also support growth of microorganisms. But in addition, there are many habitats where, because of some physical or chemical extreme, higher organisms are absent yet microorganisms exist and occasionally even flourish. Microorganisms inhabit the surfaces of higher organisms and in some cases actually live *within* plants and animals. Microorganisms frequently reach large numbers in such habitats and may benefit the plant or animal in a nutritionally significant way. On the other hand, as we will see in Chapters 19–23, some microorganisms are pathogenic and bring harm to the host. We focus now on the microbial habitat from the standpoint of the microorganism and emphasize the heterogeneous and rapidly changing nature of typical microbial habitats.

The microorganism and the microenvironment

As in laboratory culture, the growth of microorganisms in nature depends on the *resources* (nutrients) available and on the growth *conditions*. Differences in the type and quantity of different resources and the physicochemical conditions (temperature, pH, water availability, light, oxygen) (∞ Chapter 5) of a habitat define the *niche* for each particular microorganism. Ecological theory states that for every organism there exists at least one niche, the *prime* niche, in which that organism is most successful. The organism may also inhabit other niches, but in these it is less ecologically successful than in its prime niche. Countless microbial niches exist on Earth and are in part responsible for the great metabolic diversity (∞ Chapter 13) and biodiversity (∞ Chapters 16–18) of microorganisms we see today.

Because microorganisms are so small, their habitats are also small. A microbiologist must therefore learn to "think small" when considering the microorganism in its environment. For example, for a typical 3-μm rod-shaped bacterium, a distance of 3 mm in its habitat is the same that a human experiences over a distance of 2 km! And across that 3-mm distance chemical and physical gradients might exist that could greatly affect the organism. Thus, we must be more precise in our characterization of a microorganism's habitat, and microbial ecologists use the term *microenvironment* to describe where a microorganism actually lives and metabolizes within its habitat. In a 3-mm particle of soil, for exam-

ple, several different microenvironments could exist, differing chemically and physically in many ways. This can be visualized by considering the distribution of an important microbial nutrient like oxygen in a soil particle. Using microelectrodes (a technique to be described in Section 14.5) it is possible to measure oxygen concentrations throughout small soil particles. As shown in data from an actual experiment in Figure 14.2, soil particles are not homogeneous in terms of their oxygen content. The outer zones of a small soil particle may be fully oxic, whereas the center, only a very short distance away, can remain completely anoxic (Figure 14.2). This finding shows that different niches can exist across a very small spatial dimension and explains how various physiological types of microorganisms could coexist in such a soil particle. Anaerobic organisms could be active near the center of the particle shown in Figure 14.2, microaerophiles could be active further out, and obligately aerobic organisms could metabolize in the outer 2–3 mm of the particle; facultatively aerobic bacteria could be distributed throughout the particle.

Physicochemical conditions in the microenvironment can change rapidly in terms of both time and space. Because the oxygen concentrations shown in Figure 14.2 represent only "instantaneous" measurements, oxygen measurements taken following a period of microbial respiration or after an increase in soil water content could show a drastically different gradient of oxygen across the microenvironment. It can thus be said

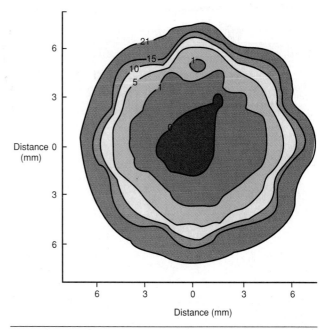

Figure 14.2 Contour map of O_2 concentrations in a soil particle. The axes show the dimensions of the particle. The numbers on the contours are O_2 concentrations (in percent; air is 21% O_2). In terms of oxygen relationships for microorganisms, each zone can be considered a different microenvironment.

that microenvironments are *heterogeneous* and that conditions in a given microenvironment can change rapidly. Thus, because of microenvironments, high microbial diversity is possible in a relatively small physical area.

Surfaces and biofilms

Surfaces are often of considerable importance as microbial habitats because nutrients can adsorb to them; in the microenvironment of a surface, nutrient levels may be much higher than they are in the bulk solution. This phenomenon can greatly affect the rate of microbial metabolism. On surfaces microbial numbers and activity are usually much greater than in free water because of adsorption effects. Microscope slides can serve as experimental surfaces on which organisms can attach and grow. When a slide is immersed in a microbial habitat, left for a period of time, and then retrieved and examined by microscopy, the importance of the surface to microbial development is apparent (Figure 14.3). Microcolonies readily develop on such surfaces much as they do on natural surfaces in nature. Microscopic examination of immersed microscope slides can actually be used as a technique to measure growth rates of attached organisms in nature.

A surface may itself also be a nutrient, such as a particle of organic matter, where attached microorganisms catabolize organic or inorganic nutrients directly from the surface of the particle. Dead plant material, for example, is rapidly colonized by microorganisms in soil, and simple staining techniques can detect microbial populations attached to the solid surface (Figure 14.4).

Studies of microbial colonization of surfaces have shown that most microorganisms grow on surfaces enclosed in **biofilms**. These are encased microcolonies of bacterial cells attached to a surface by way of adhesive polysaccharides excreted by the cells. Biofilms trap nutrients for growth of the enclosed microbial population and help prevent detachment of cells on surfaces

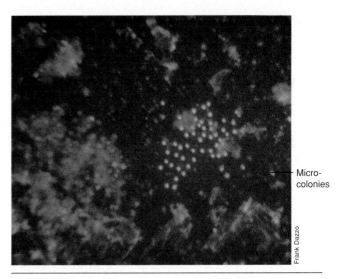

Micro-
colonies

Figure 14.4 Fluorescence photomicrograph of a natural microbial community colonizing plant roots in soil. Note microcolony development. The preparation has been stained with acridine orange.

in flowing systems. Biofilms have significant implications in human medicine and commerce. In the human body, bacterial cells within a biofilm are made unavailable for attack by the immune system. This fact thus complicates the use of introduced artificial surfaces such as medical implants, which can serve as sites of development of biofilms containing pathogenic microorganisms. Biofilms are also important in oral hygiene; dental plaque, a typical biofilm, contains acid-producing bacteria responsible for dental caries (∞ Section 19.3). In industrial situations, biofilms can slow the flow of water or oil through pipelines, accelerate the corrosion of the pipes themselves, and initiate degradation of submerged objects such as offshore oil rigs, boats, and shoreline installations.

Nutrient levels and growth rates

Nutrients (or in ecological terms, resources) often enter an ecosystem in intermittent fashion. A large pulse of nutrients—for example, an input of leaf litter or the carcass of a dead fish or animal—may be followed by a period of severe nutrient deprivation. Microorganisms in nature are thus often faced with a "feast-or-famine" type of existence, and many microorganisms have evolved biochemical systems for production of storage polymers that serve as reserve material. Reserve polymers store excess nutrients present under favorable growth conditions for use during periods of nutrient deprivation. Examples of reserve materials are poly-β-hydroxybutyrate and other alkanoates, polysaccharides, polyphosphate, and so on (see Section 3.10 for a discussion of reserve materials).

Extended periods of exponential growth in nature are rare. Growth more often occurs in spurts, linked

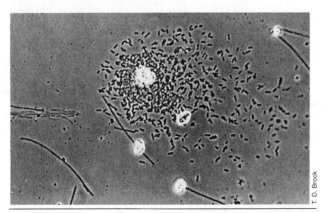

Figure 14.3 Bacterial microcolonies developing on a microscope slide immersed in a small river. The bright particles are mineral matter. The short, rod-shaped cells are about 3 μm long.

closely to the availability of nutrients. Because physicochemical conditions in nature are rarely optimal all at the same time, growth rates of microorganisms in nature are generally well below the maximum growth rates recorded in the laboratory. For instance, the generation time of *Escherichia coli* in the intestinal tract is about 12 hr (two doublings per day), whereas in pure culture it grows much faster, with a minimum generation time of 20 min under the best of conditions. Estimates of the growth rate of certain soil bacteria have shown that they grow in nature at less than 1% of the maximal growth rate measured in the laboratory. On average, these slow growth rates are a reflection of the fact that (1) nutrients (resources) are frequently in low supply, (2) the distribution of nutrients throughout the microbial habitat is not uniform, and (3) except for rare instances, microorganisms do not grow in pure culture in natural environments and thus must deal with the competitive effects of other microorganisms, a situation not encountered in growth in pure culture.

Microbial competition and cooperation

Competition among microorganisms for available resources may be intense. In the simplest case, the outcome of a competitive interaction depends on rates of nutrient uptake, inherent metabolic rates, and, ultimately, growth rates. However, in some cases of microbial competition, one organism may inhibit growth or metabolism of other organisms. This may occur by excretion of a specific inhibitor such as an antibiotic or because of the physiological activities of an organism producing a toxic product, like acid from the fermentation of sugars.

Instead of competing for the same nutrient, some microorganisms work together to carry out a particular transformation that neither organism can carry out alone. These types of microbial interactions, called *syntrophy*, are crucial to the competitive success of certain anaerobic bacteria, as will be described in Section 14.12. Syntrophic relations generally require that the two or more organisms involved in the process share the same microenvironment because the product of the metabolism of one organism must be easily accessible to the second. Metabolic cooperation is also seen in the activities of groups of organisms that carry out *complementary* metabolisms. For example, in Chapter 13 we discussed metabolic transformations that involved two distinct groups of organisms, such as those of the *nitrosifying* and the *nitrifying* bacteria, which combine to oxidize NH_3 to NO_3^- although neither group is capable of doing this alone (∞ Section 13.12). Or consider the activities of sulfate-reducing and sulfide-oxidizing bacteria; in this example the product of one organism (H_2S from the reduction of SO_4^{2-}) is the substrate for the other ($H_2S + \frac{1}{2} O_2 \rightarrow S^0 + H_2O$) (∞ Sections 13.10 and 13.16). Such types of cooperative interactions are common in microbial habitats.

CONCEPT CHECK   *14.1*

Microorganisms are very small, and their natural environments are likewise small. The microenvironment is the place in which the microorganism actually lives. Microorganisms in nature often live a feast-or-famine existence such that only the best-adapted species survive in a given niche. Cooperation among microorganisms is also important in many microbial interrelationships.

✔ *What aspects define the niche of a particular microorganism?*

✔ *Why can many different physiological groups of organisms live in a single habitat?*

✔ *What is a* biofilm?

14.2
Methods in Microbial Ecology

As previously mentioned, microbial ecology focuses on two major issues: (1) *biodiversity*, including the isolation, identification, and quantification of microorganisms in various habitats, and (2) *microbial activity*, that is, what are microorganisms *doing* in their habitats. Although both types of studies are important, studies designed to measure microbial activities *in situ* have the advantage of allowing the microbial ecologist to make perturbations in a sample of the environment and measure what effects such chemical or physical changes have on microbial activity. However, biodiversity studies have a role in microbial ecology as well and go hand in hand with activity measurements to generate a more complete picture of the microbial ecology of a given habitat.

We begin here with a consideration of methods for assessing biodiversity through enrichment and isolation, and then consider nonculture methods of identification and enumeration based on fluorescent antibodies and nucleic acid probes. We go from there to a consideration of microbial activity and describe the key methods that have been developed for measuring microbial activities directly in natural environments.

14.3
Enrichment and Isolation Methods

Rarely does a natural environment contain only a single type of microorganism. In most cases, a microbial community exists, and it is particularly challenging for the microbiologist to devise methods and procedures that permit the isolation and culture of organisms of interest. The most common approach to this goal is the **enrichment culture technique.** In this method, a medium and a set of incubation conditions are used that are *selective* for the desired organism and are counter-selective for the undesired organisms. We recall here the concept of *resources* and *conditions* of the microbial niche (see Section 14.1): the enrichment culture strategy is to duplicate as closely as possible resources and conditions in the niche and then probe for potential inhabitants of such a niche. An overview of some successful enrichment culture procedures is given in Table 14.1.

Successful enrichment requires that an appropriate *inoculum* source be used. Thus, we begin an enrichment culture protocol by going to the appropriate habitat (see Table 14.1) and sampling to obtain an enrichment inoculum. Enrichment cultures are frequently established by placing the inoculum directly into a highly selective medium; many common prokaryotes can be isolated this way. For example, by pasteurizing the inoculum to kill vegetative cells, endospore-forming organisms can be isolated on various enrichment media. The nitrogen-fixing bacterium *Azotobacter* can be easily isolated aerobically in a mannitol mineral salts medium lacking fixed forms of nitrogen (see the box, Rise of General Microbiology), and so on. Literally hundreds of different enrichment approaches have been tried in the past, and some of the more dependable ones are listed in Table 14.1. Both the culture medium *and* the general incubation conditions are listed because in devising enrichment culture protocols, one must optimize both resources and conditions in order to have the best chance at obtaining the organism of interest.

The Winogradsky column

For isolation of purple and green phototrophic bacteria and other anaerobes, the **Winogradsky column** has traditionally been used. Named for the famous Russian microbiologist Sergei Winogradsky (∞ Winogradsky's Legacy, Chapter 13), the column was devised by him in the 1880s to study soil microorganisms. The column is a miniature anaerobic ecosystem that can be an excellent long-term source of all types of prokaryotes involved in nutrient cycling.

A Winogradsky column can be prepared by filling a large glass cylinder about one-third full with organic-rich, preferably sulfide-containing, mud (Figure 14.5, page 541). Carbon substrates are first mixed into the mud; organic additions that have been successfully used in the past include hay, shredded newsprint, sawdust, shredded leaves or roots, ground meat, hard boiled eggs, and even dead animals! The mud is also supplemented with $CaCO_3$ and $CaSO_4$ as a buffer and as a source of sulfate, respectively. The mud is packed tightly in the container, care being taken to avoid entrapping air (Figure 14.5). The mud is then covered with lake, pond, or ditch water, and the top of the

cylinder covered with aluminum foil. The cylinder is placed in a north window so as to receive adequate (but not excessive) sunlight and left to develop for a period of weeks (in the Southern Hemisphere use a south window).

In a typical Winogradsky column a mixture of many different types of organisms develops. Algae and cyanobacteria appear quickly in the upper portions of the water column and by producing O_2 help to keep this zone oxic. Fermentative decomposition processes in the mud quickly lead to the production of organic acids, alcohols, and H_2, suitable substrates for sulfate-reducing bacteria. As a result of the production of sulfide, purple and green patches appear on the outer layers of the mud exposed to light, the purple patches, consisting of purple sulfur bacteria, frequently developing in the upper layers, and the green patches, consisting of green sulfur bacteria, in the lower layers (this occurs because of differences in sulfide tolerance between green and purple bacteria) (∞ Section 16.1). At the mud–water interface the water is frequently quite turbid and may be colored as a result of the growth of purple sulfur and purple nonsulfur bacteria (Figure 14.5). Sampling of the column for phototrophic bacteria is performed by inserting a long, thin pipet into the column and removing some colored mud or water. These can be used to inoculate enrichment media.

Winogradsky columns have been used to enrich for a variety of prokaryotes, both aerobes and anaerobes. The great advantage of a column, besides the ready availability of inocula for different enrichment cultures, is that it can be spiked with a particular compound whose degradation one wishes to study and then allowed to select from the inoculum for an organism or organisms that can degrade it. In addition, because the Winogradsky column more nearly resembles the natural environment than do culture media, a variety of each physiological type of microorganism is more frequently observed in Winogradsky columns than in enrichments where a liquid culture medium is inoculated directly with a natural sample. In the latter approach, rapidly growing species frequently rise to the forefront, leaving the slower growing species behind.

From enrichments to pure cultures

The objective of an enrichment culture study is usually to obtain a pure culture. Pure cultures can be obtained in many ways, but the most frequently employed means are the streak plate, the agar shake, and liquid dilution methods. For organisms that grow well on agar plates, the **streak plate** is the method of choice (∞ Section 1.7). By repeated picking and restreaking of a well-isolated colony, one usually obtains a pure culture that can then be transferred to a liquid medium. With proper incubation it is possible to purify both

RISE OF GENERAL MICROBIOLOGY

The field of general (primarily nonmedical) microbiology owes a great debt to the Delft School of Microbiology, which was initiated by Martinus Beijerinck and continued by A. J. Kluyver and C. B. van Niel. Beijerinck, one of the world's greatest microbiologists (among other things, he was the first to characterize viruses), first devised the enrichment culture technique and used this technique in the isolation and characterization of a wide variety of bacteria. Subsequently, Kluyver and van Niel used the enrichment culture technique to isolate phototrophic and chemolithotrophic bacteria and to show the fascinating physiological diversity among the bacteria. Another important figure from the Dutch school was L. G. M. Baas-Becking, who carried out the first calculations of the energetics of phototrophic and chemolithotrophic bacteria and emphasized the importance of these organisms in geochemical processes. Van Niel subsequently came to the United States and was responsible for the training of a number of general bacteriologists who carried on the Delft tradition, including the late R. Y. Stanier and M. Doudoroff, and Robert E. Hungate, now retired from the University of California at Davis. In addition, a number of scientists visited van Niel's laboratory at Pacific Grove, California, in the 1950s and 1960s and learned his methods and enrichment approaches. In recent years, the "Delft tradition" has been carried on in Germany by Norbert Pfennig, Bernhard Schink, Karl Stetter, and Fritz Widdel, and by many others in Europe, the countries of the former Soviet Union, and the United States.

An excellent example of the application of the enrichment culture technique can be found in the isolation of *Azotobacter*, a nitrogen-fixing bacterium discovered by Beijerinck in 1901 (see the figure and Section 16.16). Beijerinck was interested in knowing whether any aerobic bacteria capable of fixing nitrogen existed in the soil. The only other nitrogen-fixing organism known at that time, *Clostridium pasteurianum*, was an anaerobe discovered several years earlier by Sergei Winogradsky. To look for an aerobic nitrogen fixer, Beijerinck added a small amount of soil to an Erlenmeyer flask that contained a thin layer of mineral salts medium devoid of ammonia, nitrate, or any other form of fixed nitrogen and that contained mannitol as carbon source. Within 3 days a thin film developed on the surface of the liquid and the liquid became quite turbid. Beijerinck observed large, rod-shaped cells that appeared quite distinct from the spore-containing rods of *C. pasteurianum*. Beijerinck streaked agar plates containing phosphate and mannitol with the turbid liquid, incubated aerobically, and within 48 hr obtained large, slimy colonies typical of *Azotobacter* (see the figure). Pure cultures were obtained by picking and restreaking well-isolated colonies a number of times. Beijerinck assumed his new organism was using N_2 from air as its source of cell nitrogen and later proved this by showing total nitrogen increases in pure cultures of *Azotobacter* grown in the absence of fixed nitrogen.

The selective pressure of the enrichment approach used here should be evident. By omitting from his medium any source of combined nitrogen and incubating aerobically, Beijerinck placed constraints on the microbial population. Any organism that developed had to be able to both fix its own nitrogen and tolerate the presence of molecular oxygen. Beijerinck noted that if he placed too much liquid in his flasks or employed a more readily fermentable organic substrate, such as glucose or sucrose, in place of mannitol, his enrichment frequently turned anoxic and favored the growth of *Clostridium* rather than *Azotobacter*. The addition of ammonia or nitrate to the original enrichment never resulted in the isolation of *Azotobacter* but only in a variety of non-nitrogen-fixing bacteria instead (see the figure). Hence Beijerinck showed that *Azotobacter* has a strong selective advantage over other soil bacteria under a specific set of nutritional conditions and in the process defined the key physiological properties of *Azotobacter*: aerobiosis and nitrogen fixation. Beijerinck also demonstrated that the composition of the growth medium as well as the incubation conditions employed were of paramount importance in the proper development of an enrichment culture.

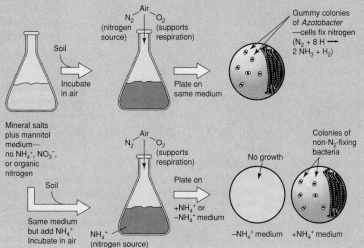

TABLE 14.1   Enrichment culture methods for prokaryotes[a]

Light-phototrophic bacteria: main C source, CO_2

Aerobic incubation	Organisms enriched	Inoculum
N_2 as nitrogen source	Cyanobacteria	Pond or lake water; sulfide-rich muds; stagnant water; raw sewage; moist, decomposing leaf litter; moist soil exposed to light; pasteurized soil (heliobacteria)
NO_3^- as nitrogen source, 55°C	Thermophilic cyanobacteria	Hot spring microbial mat
Anaerobic incubation		
H_2 or organic acids; N_2 as sole nitrogen source	Nonsulfur purple bacteria, heliobacteria	
H_2S as electron donor	Purple and green sulfur bacteria	

Dark-chemolithotrophic bacteria: main C source, CO_2 (medium must lack organic C)

Aerobic incubation

Electron donor	Electron acceptor	Organisms enriched	Inoculum
NH_4^+	O_2	Nitrosifying bacteria (*Nitrosomonas*)	Soil, mud; sewage effluent
NO_2^-	O_2	Nitrifyng bacteria (*Nitrobacter*)	
H_2	O_2	Hydrogen bacteria (various genera)	
H_2S, S^0, $S_2O_3^{2-}$	O_2	*Thiobacillus* spp.	
Fe^{2+}, low pH	O_2	*Thiobacillus ferrooxidans*	

Anaerobic incubation

			Inoculum
S^0, $S_2O_3^{2-}$	NO_3^-	*Thiobacillus denitrificans*	
H_2	NO_3^- + yeast extract	*Paracoccus denitrificans*	Mud, lake sediments, soil

Dark-chemoorganotrophic bacteria and methanogens: main C source, organic compounds

Aerobic incubation: respiration

Electron donor and nitrogen source	Electron acceptor	Organisms enriched	Inoculum
Lactate + NH_4^+	O_2	*Pseudomonas fluorescens*	Soil, mud; lake sediments; decaying vegetation; pasteurize inoculum (80°C for 15 min) for all *Bacillus* enrichments
Benzoate + NH_4^+	O_2	*Pseudomonas fluorescens*	
Starch + NH_4^+	O_2	*Bacillus polymyxa*, other *Bacillus* spp.	
Ethanol (4%) + 1% yeast extract, pH 6.0	O_2	*Acetobacter*, *Gluconobacter*	
Urea (5%) + 1% yeast extract	O_2	*Sporosarcina ureae*	
Hydrocarbons (e.g., mineral oil) + NH_4^+	O_2	*Mycobacterium*, *Nocardia*	
Cellulose + NH_4^+	O_2	*Cytophaga*, *Sporocytophaga*	
Mannitol or benzoate, N_2 as N source	O_2	*Azotobacter*	

Anaerobic incubation: anaerobic respiration

Main ingredients	Electron acceptor	Organisms enriched	Inoculum
Organic acids	KNO_3 (0.2%)	*Pseudomonas* (denitrifying species)	Soil, mud; lake sediments
Yeast extract	KNO_3 (1%)	*Bacillus* (denitrifying species)	
Organic acids	Na_2SO_4	*Desulfovibrio*, *Desulfotomaculum*	
Acetate, propionate, butyrate	Na_2SO_4	Type II sulfate reducers	As above; or sewage digestor sludge; rumen contents
Acetate, ethanol	S^0	*Desulfuromonas*	

TABLE 14.1   (continued)

Anaerobic incubation: anaerobic respiration *(continued)*

Main ingredients	Electron acceptor	Organisms enriched	Inoculum
H$_2$	Na$_2$CO$_3$	Methanogens (chemolithotrophic species only), homoacetogens	Mud, sediments, sewage sludge
CH$_3$OH	Na$_2$CO$_3$	*Methanosarcina barkeri*	
CH$_3$NH$_2$	KNO$_3$	*Hyphomicrobium*	

Anaerobic incubation: fermentation

Electron donor and nitrogen source	Electron acceptor	Organisms enriched	Inoculum
Glutamate or histidine	No exogenous electron acceptors added	*Clostridium tetanomorphum*	Mud, lake sediments; rotting plant material; dairy products (lactic and propionic acid bacteria); rumen or intestinal contents (enteric bacteria); sewage sludge
Starch + NH$_4^+$	None	*Clostridium* spp.	
Starch, N$_2$ as N source	None	*Clostridium pasteurianum*	
Lactate + yeast extract	None	*Veillonella* spp.	
Glucose or lactose + NH$_4^+$	None	*Escherichia, Enterobacter,* other fermentative organisms	
Glucose + yeast extract (pH 5)	None	Lactic acid bacteria (*Lactobacillus*)	
Lactate + yeast extract	None	Propionic acid bacteria	
Succinate + NaCl	None	*Propionigenium*	

^{a}All media must contain an assortment of mineral salts including N, P, S, Mg^{2+}, Mn^{2+}, Fe^{2+}, Ca^{2+}, and other trace elements.

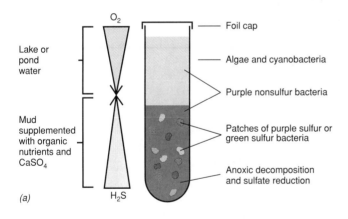

(a)

(b)

Figure 14.5 The Winogradsky column. (a) Schematic view of a typical column. The column is placed so as to receive subdued sunlight. Chemoorganotrophic bacteria grow throughout the column, aerobes and microaerophiles in the upper regions, anaerobes in the zones containing H$_2$S. Anoxic decomposition leading to sulfate reduction creates the gradient of H$_2$S. Green and purple sulfur bacteria stratify according to their tolerance for H$_2$S. (b) Photo of Winogradsky columns that have remained anoxic up to the top where blooms of three different phototrophic bacteria have occurred in the mud and up into the water column. Left to right: *Thiospirillum jenense, Chromatium okenii,* and *Chlorobium limicola.*

Figure 14.6 Agar shake technique for isolation of anaerobic bacteria in pure culture. A dilution series was established from right to left, eventually yielding well-isolated colonies. The tubes are sealed with a sterile mixture of paraffin and mineral oil to maintain anaerobiosis.

aerobes and anaerobes on agar plates via the streak plate method.

The **agar shake tube method** (Figure 14.6) involves the dilution of a mixed culture in tubes of molten agar, resulting in colonies embedded *in* the agar rather than on the surface of a plate. The shake tube method has been found useful for purifying particular types of microorganisms (for example, phototrophic sulfur bacteria and sulfate-reducing bacteria). Purification can be obtained by successively diluting a cell suspension in tubes of liquid medium until a dilution is reached in which well-isolated colonies are obtained (Figure 14.6). By repeating this procedure using the highest dilution showing growth as inoculum for a new set of dilutions, it is possible in most casts to eventually obtain pure cultures.

Once a putative pure culture has been obtained, it is essential to check its purity. This is usually done through a combination of microscopy and checking for growth in a variety of culture media that favors growth of contaminants but not the organism of interest.

CONCEPT CHECK   14.3

The enrichment culture technique is a means of selecting from natural samples microorganisms that are already present and are capable of carrying out specific reactions. The Winogradsky column is a type of enrichment culture that is suitable for obtaining phototrophic bacteria and other anaerobes. Once a vigorous enrichment culture has been obtained, a pure culture can usually be obtained by use of conventional bacteriological procedures.

✔ *What things are important to the success of an enrichment culture?*

✔ *How does the agar shake tube method differ from conventional streaking to obtain colonies on plates?*

14.4

Identification and Quantification: Nucleic Acid Probes, Fluorescent Antibodies, and Viable Counts

Methods that have been devised for counting microorganisms in pure culture can easily be adapted for use on natural samples. For example, dilution and plate counts (∞ Section 5.4) can be used to enumerate bacteria from water or soil samples. However, depending on the culture medium and incubation conditions chosen, only a fraction of the total microbial community can ever be measured by viable counting methods. In certain instances, highly selective media may be available for counting a specific subset of the total community, and such counts are usually more reliable estimates than are total counts.

In some cases the microorganisms of interest in an ecosystem may not yet be culturable. If this is the case, methods are needed to at least detect that they are present. Several fairly specific methods have been developed for identification and enumeration, including methods based on the specificity of antibodies and nucleic acid sequences.

Staining and fluorescent antibodies

Microorganisms can be identified and enumerated by direct microscopic examination of the habitat. However, special procedures are needed to make microorganisms visible in opaque habitats. In soil, for instance, organisms can be stained with any of a variety of different fluorescent dyes specific for living cells (see Figure 14.4). Acridine orange is such a dye; this dye stains deoxyribonucleic acid (DNA) and ribonucleic acid (RNA). Direct staining techniques using acridine orange are widely used for estimating total microbial numbers in soil and water.

Fluorescent antibody staining is a method for identifying a single species (or even a single strain of a given species) in soil samples; in certain applications the technique can be used as a method of quantification as well. We discuss the theory of fluorescent antibodies in Section 21.7. The great specificity of antibodies prepared against cell surface constituents of a particular organism can be exploited to identify that organism in a complex habitat such as soil (Figure 14.7). Fluorescent antibodies are therefore most useful for tracking a single microbial species in soil or other habitats.

Nucleic acid probes

A very powerful approach to the identification and quantification of microorganisms in nature is the use of *nucleic acid probes*. As discussed in the box, Working

T. D. Brock

Figure 14.7 Visualization of bacterial microcolonies (bacterial cells appear as greenish-yellow dots) on the surface of soil particles by use of the fluorescent antibody technique. Cells are about 1 μm in diameter.

with Nucleic Acids: The Tools, in Chapter 6, a nucleic acid probe is a short piece of DNA or RNA *complementary* in base sequence to part of a gene to which the probe can hybridize. Some probes that have been used employ 16S rRNA sequences as tools for differentiating microorganisms in natural environments. We discuss in Chapter 15 how 16S ribosomal RNA (rRNA) sequencing can be used to identify the primary phylogenetic domain (Bacteria, Archaea, Eukarya) to which a given microorganism belongs. And, because the number of available 16S rRNA sequences has increased greatly since the technique was first introduced, highly specific probes capable of differentiating organisms *within* a phylogenetic domain can now be constructed. By using a collection of such probes to test natural samples, it is possible to identify different species in a single sample (see Figure 14.8).

Two types of labeling systems have been used in phylogenetic probes, *radioisotopes* and *fluorescent dyes*. The radioisotopic method employs ^{35}S- or ^{32}P-labeled probes added to formaldehyde-fixed cells immobilized on glass fiber filters. Before treatment with a probe, the dried cells are treated in such a way that they are made permeable, to allow the probe to penetrate and find a ribosome to bind to. Labeling of individual cells is then visualized by autoradiography: the exposure of a photosensitive emulsion by the radioactivity in the ribosomes to which the probe has bound (Figure 14.8*a*).

An alternative detection system employs fluorescent dyes. The dyes are chemically attached to the probe, and the cells visualized by fluorescent microscopy (Figure 14.8*b*). Using fluorescent microscopy and probes labeled with various dyes that fluoresce at different wavelengths, it is possible to treat a sample with several different probes at once and then identify and quantify different cell types in the sam-

ple by use of the appropriate fluorescence filter (Figure 14.8*b* and *c*).

Study of microbial communities using probes

Study of the biodiversity of microbial communities can be done using ribosomal RNA probes. The result obtained is a phylogenetic picture of the species present in the community. In brief, the methods involve extracting either the bulk DNA or RNA from representatives of the entire microbial community and then, using specific nucleic acid probes, obtaining a set of ribosomal RNA clones (Figure 14.9*a*). The latter are generated by either polymerase chain reaction (PCR) amplification of ribosomal RNA genes or indirectly by reverse transcriptase-mediated production of cDNA complementary to ribosomal RNA (Figure 14.9*a*).

Following sequencing of the ribosomal RNA gene, phylogenetic trees showing the species composition of the microbial community can be constructed (see Chapter 15 for a discussion of these techniques). Applications of these methods to actual microbial community analyses have yielded interesting results. In almost all cases phylogenetic analyses of microbial communities have shown them to contain phylogenetically distinct organisms that had not been previously cultured (Figure 14.9*b*). This supports the hypothesis that biodiversity of microbial communities is much greater than the isolated organisms obtained thus far would suggest and indicates that much work lies ahead for the enrichment culture microbiologist.

One can also use probe technology for *tracking* purposes (for example, monitoring the colonization success of a genetically engineered microorganism released into a new habitat) or for *identifying particular genes* (for example, genes that encode proteins involved in the biodegradation of a toxic compound) in organisms in the environment. In either case, increases or decreases in "gene dosage" of the specific gene or genes to which a probe reacts can be used to monitor the success of the organism(s) that carry them. In addition, using probe methods, experimental perturbations of an environment can be made and their effect on the rise or decline of specific microbial populations (or specific gene dosage) assessed.

The future of nucleic acid probes in microbial ecology is very bright, especially for use where culture methods have not yet been successful or where the organisms of interest are present in numbers below the sensitivity of activity measurements. However, a major limitation of all the methods described thus far is that the data obtained are limited to the detection, enumeration, or phylogenetic positioning of the organisms only. In some types of ecological studies, this may be the main goal. However, in many microbial ecology studies, knowledge of microbial *activity* as well as biodiversity is needed to conclude whether the microor-

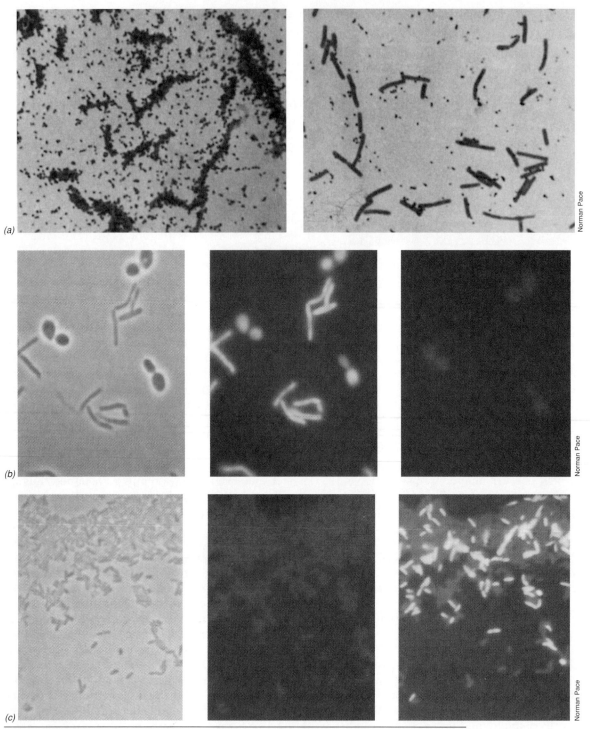

Figure 14.8 Nucleic acid probe methods for identifying microorganisms in natural samples. (a) Microautoradiographs of single cells of *Bacillus megaterium* hybridized with ribosomal RNA oligonucleotide probes complementary to specific sequences in the 16S rRNA of Bacteria (left) and to the 18S rRNA of Eukarya (right). Note the extensive number of silver grains around cells that have hybridized to the Bacterial probe. (b) Fluorescently labeled ribosomal RNA probes. Left, phase contrast photomicrograph of *B. megaterium* and the yeast *Saccharomyces cerevisiae* (no probes present). Center, same field, cells stained with universal rRNA probe. Right, same field, cells stained with eukaryal probe (only cells of *S. cerevisiae* react). (c) Differentiation of closely related gram-negative Bacteria. Left, phase micrograph of mixture of *Proteus vulgaris* and a related bacterium isolated from wasps. Center, same field stained with the bacterial probe. Right, same field stained with a probe specific for the bacterium from wasps. Cells of *B. megaterium* are about 1.5 μm in diameter, *S. cerevisiae* about 6 μm in diameter, and *P. vulgaris* about 0.8 μm in diameter.

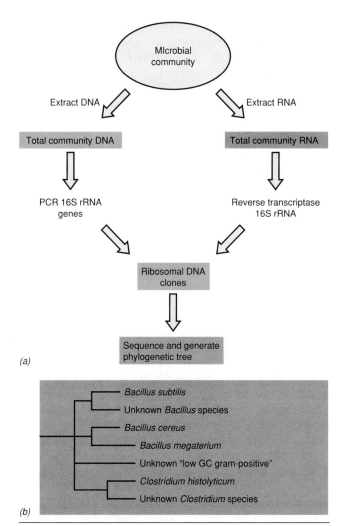

Figure 14.9 Steps in biodiversity analysis of microbial community using phylogenetic nucleic acid probes. (a) Ribosomal DNA clones are obtained in either of two ways and then sequenced. (b) Hypothetical example of a phylogenetic tree that might be generated if polymerase chain reaction (PCR) or reverse transcriptase primers were used specific for the "low GC gram-positive" lineage of the phylogenetic domain Bacteria, which includes species of *Bacillus* and *Clostridium* (∞ Section 16.26). Note that some of the sequences do not match any previously sequenced species and therefore represent previously unrecognized (unknown) species. See Sections 15.5–15.8 for more details concerning microbial phylogeny and phylogenetic trees.

ganisms present are of real ecological significance. We now discuss some methods for measuring microbial activities in nature.

14.5

Microbial Activity Measurements: Radioisotopes and Microelectrodes

Several methods have been used to measure microbial activity. Activity measurements are generally estimates of microbial *guild processes*, where the collective activities of several metabolically related populations are as-

sessed by measuring a single type of transformation. An important feature of many of these methods is *sensitivity*; activity measurements must be sensitive enough to detect whether a given process has occurred. High sensitivity also allows for short-term incubations if a natural sample must be removed from its habitat to make the activity measurement. We begin our discussion with a consideration of radioisotopic methods, a highly sensitive measure of microbial activities.

Radioisotopes

Although in many situations direct chemical measurement of microbial transformations using nonradioactive methods is satisfactory, **radioisotopes** are very useful in measuring specific microbial processes at high sensitivity and also in obtaining information on the turnover rates of chemical species in nature. For instance, if photoautotrophy is to be measured, the light-dependent uptake of $^{14}CO_2$ into microbial cells can be measured; if sulfate reduction is of interest, the rate of conversion of $^{35}SO_4^{2-}$ to $H_2^{35}S$ can be assessed (Figure 14.10). Methanogenesis in natural environments can be studied by measuring the conversion of $^{14}CO_2$ to $^{14}CH_4$ in the presence of a suitable reductant such as H_2, or by the conversion of methyl compounds such as ^{14}C-labeled methanol or acetate (Figure 14.10c) to $^{14}CH_4$. Chemoorganotrophic activity can be measured by following incorporation of ^{14}C-organic compounds. For example, the uptake of ^{14}C-glucose or ^{14}C-amino acids by chemoorganotrophs in lake water can be studied by filtering the bacterial cells from suspension following an incubation period with the isotope. Alternatively, one might measure the extent of $^{14}CO_2$ production from the added radioisotope. In all these studies if actual *rates* of a given chemical process are to be determined, it is necessary to know the amount of *nonradioactive* as well as radioactive substrate available to the population in order for a *specific activity* of the substrate to be determined; from this, total rates of conversion can be calculated.

Isotope methods are widely used to evaluate the activity of microorganisms in nature. However, because

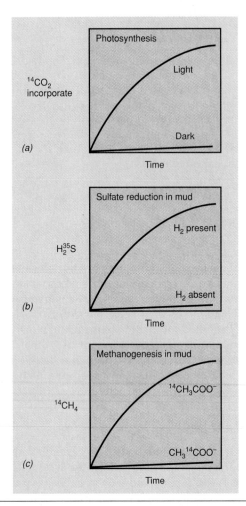

(a)

Photosynthesis

$^{14}CO_2$ incorporate

Light

Dark

Time

(b)

Sulfate reduction in mud

$H_2^{35}S$

H_2 present

H_2 absent

Time

(c)

Methanogenesis in mud

$^{14}CH_4$

$^{14}CH_3COO^-$

$CH_3^{14}COO^-$

Time

Figure 14.10 Use of radioisotopes to measure microbial activity in nature. (a) Photosynthesis measured in natural seawater with $^{14}CO_2$. (b) Sulfate reduction in mud measured with $^{35}SO_4^{2-}$. (c) Methanogenesis measured in mud with acetate labeled in either the methyl ($^{14}CH_3COO^-$) or the carboxyl ($CH_3^{14}COO^-$) carbon.

there is always the possibility that some transformation of a labeled compound might be due to a strictly chemical, rather than a microbial, process it is essential when using isotopes to employ proper controls. The key control necessary is the *killed cell control*. It is absolutely essential to show that the transformation being measured in nature is prevented by microbicidal agents or heat treatments known to block microbial action or kill the organisms. Formalin at a final concentration of 4% is frequently used as a chemical sterilant in microbial ecology studies.

It is possible to combine the use of radioisotopes with the study of the activities of *single* microbial cells using *microautoradiography*. We discussed this technique in connection with the use of nucleic acid probes in microbial ecology in Section 14.4 (see Figure 14.8a). Here we consider other uses of this method. Autoradiography can be employed to measure the uptake of any substrate available in radioactive form,

including nucleotides, amino acids, and many other organic compounds. If autoradiography is used in conjunction with an identification method, such as immunofluorescence or phylogenetic staining (see Figure 14.8), then information about both the activity and the taxonomy or phylogeny of a microorganism can be obtained simultaneously.

Microelectrodes

We discussed earlier the fact that the environments of microorganisms are very small; that is, they are *microenvironments*. Microbial ecologists have used small glass electrodes, referred to as **microelectrodes,** to study the activity of microorganisms in their microenvironments (we discussed the use of O_2 microelectrodes in Section 14.1) (see Figure 14.2). Although limited to measurements of chemical species that can be detected with an electrode, several types of electrodes have been used in field studies. Currently, the most popular microelectrodes are those that measure pH, oxygen, N_2O, and sulfide. In addition to chemical microelectrodes, spectroradiometers are available that have microelectrode-sized probes for measuring light absorption. A probe can be inserted into an environment containing various phototrophic microbial species, like a microbial mat (see later), and the absorption properties of the organisms determined; the latter can help in identifying the types of phototrophic organisms present (∞ Sections 13.1–13.6 and 16.1).

As the name implies, microelectrodes are very small, the tips of the electrode ranging in diameter from 2 to 100 μm; O_2 microelectrodes 2–3 μm in diameter can be routinely made (Figure 14.11a). The electrodes are carefully inserted into the habitat using a micromanipulator, a device that allows for precise movement of the electrode through distances of a millimeter or less (Figure 14.11b). Microelectrodes have been used extensively in the study of chemical transformations and photosynthesis in *microbial mats*. The latter are layered microbial communities usually containing cyanobacteria in the uppermost layer, anoxygenic phototrophic bacteria in subsequent layers (until the mat becomes light-limited), and chemoorganotrophic, especially sulfate-reducing, bacteria in the lower layers (Figure 14.12). Microbial mats are found in a variety of environments, especially in hot springs (Figure 14.12a) and shallow marine basins and are dynamic systems where photosynthesis occurring in the upper layers of the mat is balanced by decomposition from below.

With the use of O_2 microelectrodes, oxygen concentrations in microbial mats (Figure 14.12b) or soil particles (Figure 14.2) can be sensitively measured over extremely fine intervals. With the use of a micromanipulator, electrodes can be immersed through a microbial habitat and readings taken every 0.05–0.1 mm (50–100 μm) (Figures 14.11b and 14.12). With a bank of microelectrodes, each sensitive to a different chemical, simulta-

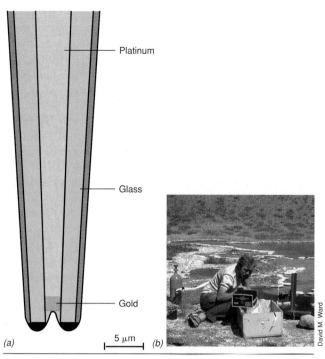

Figure 14.11 Microelectrodes. (a) Schematic drawing of an oxygen microelectrode. Note the scale of the electrode. (b) Photo of microelectrodes being used in a microbial mat (see Figure 14.12).

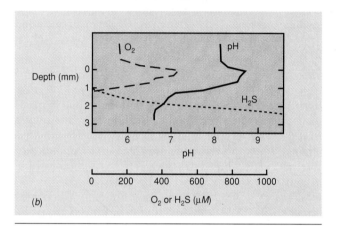

Figure 14.12 Microbial mats and the use of microelectrodes to study them. (a) Photograph of a core taken through the kind of hot spring microbial mat used in the experiment shown in part (b). Upper layer (dark green) contains cyanobacteria, beneath which are several layers of anoxygenic phototrophic bacteria (orange and yellow). The whole thickness of the mat is about 2 cm. (b) Oxygen, sulfide, and pH microprofiles in a hot spring microbial mat. Note the millimeter scale on the ordinate.

neous measurements of a number of microbial transformations can be made at one time (Figure 14.12b). Oxygen and sulfide measurements are often taken together because gradients of both chemical species form in many microbial environments as a result of photosynthesis and sulfate reduction, respectively (Figure 14.12b) Note also in Fig. 14.12b how the H_2S concentration diminishes near the oxic zone. This is a result of H_2S oxidation by phototrophic bacteria and chemolithotrophic sulfur bacteria. Thus, the concentration of H_2S at any point in the mat is a function of both H_2S production and H_2S consumption, complementary metabolisms (see Section 14.1), and an H_2S measurement is the net result of both activities at a given location in the mat.

CONCEPT CHECK   *14.5*

The activity of microorganisms in natural samples can be assessed very sensitively using radioisotopes or microelectrodes. In most cases these measurements are of the net activity of a microbial guild or guilds rather than of a population of a single species.

✔ *Why are radioisotopes so useful in measuring microbial activities?*

✔ *What is a* microelectrode?

14.6

Microbial Activity Measurements: Stable Isotopes

We learned in Chapter 2 that different isotopes of most elements exist and that although certain isotopes are unstable and break down (as a result of radioactive decay), others are stable and simply contain a different number of neutrons than the major isotopic form. Such isotopes, known as *stable isotopes*, can be used to study various microbial transformations.

The two chemical elements that have proven most useful for stable isotope studies in microbial ecology are carbon and sulfur. Carbon exists in nature primarily as ^{12}C. However, a small amount of carbon is found as ^{13}C. Likewise, sulfur exists primarily as ^{32}S, although some sulfur is found as ^{34}S. All these are stable isotopes. The reason stable isotope measurements are

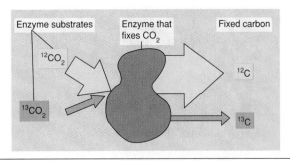

Figure 14.13 Mechanism of isotopic fractionation using carbon as an example. Although the ratio of natural abundance of $^{12}CO_2$ to $^{13}CO_2$ is about 19:1, enzymes that fix CO_2 preferentially fix the lighter isotope (^{12}C). This results in fixed carbon being enriched in ^{12}C and depleted in ^{13}C relative to the starting substrate. The degree of ^{13}C depletion is calculated as an isotopic fractionation (see legend to Figure 14.14 for calculation).

useful in microbial ecology is that most biochemical reactions involving carbon or sulfur favor the *lighter* isotope. That is, the heavier isotope is discriminated against when the elements are acted on biochemically (Figure 14.13). Thus, when CO_2 is fed to a phototrophic organism, the cellular carbon becomes enriched in ^{12}C whereas the CO_2 remaining in the medium becomes enriched in ^{13}C. Likewise, sulfide produced from the bacterial reduction of sulfate is much "lighter" than sulfide of strictly geochemical origin. This phenomenon is known as *isotopic fractionation*.

Use of isotopic fractionation in microbial ecology

How can differences in isotopic composition be used to assess microbial activities? The isotopic composition of a sample contains a record of its past biological activity (Figures 14.14 and 14.15). In the case of carbon, it is easy to see that plant material and petroleum (which is derived from plant material) have similar isotopic compositions; carbon from both sources is isotopically lighter than the standard because it was fixed by a pathway that discriminated against $^{13}CO_2$ (Figure 14.14). Methane of biological origin can be extremely light, indicating that CO_2-reducing methanogens dramatically discriminate between isotopic forms of CO_2 (see Sections 14.12 and 17.2 for discussion of the biochemistry of methanogenesis). Marine carbonates, on the other hand, are clearly of geological origin because they are isotopically much "heavier" than biogenic carbon (Figure 14.14). This large carbon reservoir has obviously not yet passed through phototrophic organisms. Because of the differences in the proportion of ^{12}C and ^{13}C in carbon of biological and geological origin, the $^{13}C/^{12}C$ isotopic ratio of various geological strata has been used to detect the onset of living (in this case, autotrophic) processes in ancient rocks. Interestingly, organic carbon in rocks as old as 3.5 billion years shows some isotopic lightness (Figure 14.14), suggesting that autotrophy might have evolved very early in the diversification of living organisms.

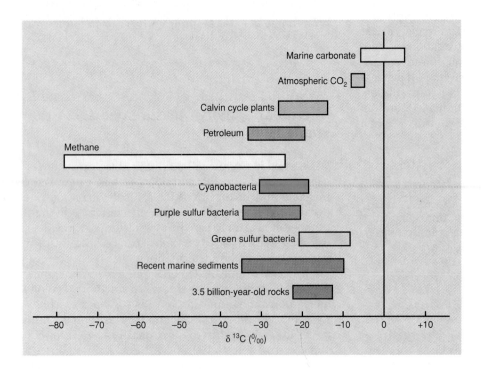

Figure 14.14 Carbon isotopic compositions of various substances. The values were calculated using the formula

$$\frac{(^{13}C/^{12}C \text{ sample}) - (^{13}C/^{12}C \text{ standard})}{(^{13}C/^{12}C \text{ standard})} \times 1000$$

The standard is a belemnite sample from the PeeDee rock formation.

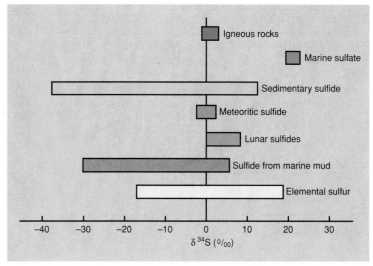

Figure 14.15 Summary of the isotope geochemistry of sulfur, indicating the range of values for ^{34}S and ^{32}S in various sulfur-containing substances. The values were calculated using the formula

$$\frac{(^{34}S/^{32}S\ \text{sample}) - (^{34}S/^{32}S\ \text{standard})}{(^{34}S/^{32}S\ \text{standard})} \times 1000$$

The standard is an iron sulfide mineral from the Canyon Diablo meteorite. Note that sulfide and sulfur of biogenic origin tend to be depleted in ^{34}S (enriched in ^{32}S).

The key role of sulfate-reducing bacteria in the formation of sulfur deposits is known from studies on the fractionation of ^{34}S/^{32}S (Figure 14.15). As compared to a meteoritic sulfur standard, sedimentary sulfide is highly enriched in ^{32}S. Nonbiogenic sulfide (as, for instance, in igneous rocks from volcanic deposits) does not show this bias toward the lighter isotope (Figure 14.15). Also, biological oxidation of sulfide to sulfur or sulfate, either aerobically or anaerobically, shows a preference for the lighter isotope, although the fractionation is not nearly as great as that occurring during sulfate reduction (Table 14.2).

Isotope fractionation studies thus have several uses in microbial ecology. Sulfur isotopes have been used to distinguish between biogenic and abiogenic ores (iron sulfides) and elemental sulfur deposits. Knowledge of the fractionations observed in sulfate reduction and subsequent sulfide oxidation by phototrophic or chemolithotrophic bacteria allows monitoring of important links in the sulfur cycle. As previously mentioned, carbon isotopic analyses have been used to distinguish biogenic from abiogenic organic matter, and oxygen analyses (using ^{18}O/^{16}O measurements of rocks of various ages) have been used to trace the earth's transition from an anoxic to an oxic environment (the earth's molecular oxygen originated from oxygenic photosynthesis by cyanobacteria) (∞ Section 15.1).

Stable isotope analyses have also been used as evidence for the lack of living processes on the moon. For example, the data in Figure 14.15 show that the sulfide isotopic composition of lunar rocks closely approximates that of meteoritic sulfide and not that typical of biogenic sulfide. Stable isotope analyses can also be used to track microbial and higher organism food chains. Because the isotopic composition of a given animal should approximate the isotopic composition of its major food source, it has been possible to trace the flow of carbon through an ecosystem in ways previously not possible.

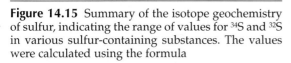	Selective utilization of ^{32}S over ^{34}S (isotope fractionation) in certain parts of the microbiological sulfur cycle			
Process		**Starting substance**	**End product**	**Isotope fractionation**[a]
Dissimilatory sulfate reduction (*Desulfovibrio*)		SO_4^{2-}	HS^-	−46.0
Dissimilatory sulfite reduction (*Desulfovibrio*)		SO_3^{2-}	HS^-	−14.3
Assimilatory sulfate reduction (*Escherichia coli*)		SO_4^{2-}	Organic S	−2.8
Putrefaction and desulfurylation (*Proteus*)		Organic S	HS^-	−5.1
Chemolithotrophic sulfide oxidation (*Thiobacillus*)		HS^-	S^0	−2.5
		HS^-	SO_4^{2-}	−18.0
Phototrophic sulfide oxidation (*Chromatium*)		HS^-	S^0	−10.0
		HS^-	SO_4^{2-}	0

[a]See legend to Figure 14.15 for calculation.

CONCEPT CHECK   **14.6**

Isotopic fractionation can yield information on the biological origin of various substances. Fraction-ation is a result of the activity of certain enzymes that can discriminate, usually against the heavier form of an element, when acting on their substrates.

✔ *How can the $^{12}C/^{13}C$ composition of a substance tell us anything about its possible biological origin?*

✔ *Why are lunar sulfides isotopically heavy?*

14.7

Aquatic Habitats

Typical aquatic environments are the oceans, estuaries, salt marshes, lakes, ponds, rivers, and springs. Aquatic environments differ considerably in chemical and physi-cal properties, and it is not surprising that their microbial species compositions also differ. The predominant pho-totrophic organisms in most aquatic environments are microorganisms; in oxic areas cyanobacteria and algae prevail, and in anoxic areas anoxygenic phototrophic bac-teria are preponderant. Algae floating or suspended freely in the water are called **phytoplankton;** those attached to the bottom or sides are called **benthic algae.** Because these phototrophic organisms utilize energy from light in the initial production of organic matter, they are called **primary producers.** In the final analysis, the biological activity of an aquatic ecosystem is dependent on the rate of primary production by the phototrophic organisms.

The activities of primary producers are in turn affected by both resources and conditions. Open oceans are very low in primary productivity, whereas inshore ocean areas are high, with some lakes and springs being highest of all. The open ocean is relatively infer-tile because the inorganic nutrients needed for phyto-plankton growth are present only in low concentra-tions. The more fertile inshore ocean areas, on the other hand, receive extensive nutrient enrichment from rivers and other polluted water inputs (Figure 14.16). The amount of economically important crops such as fish or shellfish is determined ultimately by the rate of primary production; lakes and inshore ocean areas are high in primary production (Figure 14.16) and thus are the richest sources of fish and shellfish.

Oxygen relationships in lakes and rivers

We discussed oxygen requirements and anaerobiosis in Section 5.11, the production of oxygen via photosynthe-sis in Section 13.5, and oxygen in microenvironments in Section 14.1. Although oxygen is one of the most plenti-ful gases in the atmosphere (~21% of air), it has limited solubility in water, and in a large water mass its exchange with the atmosphere is slow. Significant photosynthetic production of oxygen occurs only in the *surface layers* of a

Figure 14.16 Distribution of chlorophyll in the western North Atlantic Ocean as recorded by satellite. The east coast of the United States from mid-Florida to northern Maine is shown. Near the center of the photograph is Chesapeake Bay; the Great Lakes are at the upper left. Areas rich in phy-toplankton are shown in red (>1 mg chlorophyll/m³); blue and purple areas have lower chlorophyll concentrations (<0.01 mg/m³). Note the high primary productivity of coastal areas and the Great Lakes.

lake or ocean, where light is available. Organic matter that is not consumed in these surface layers sinks to the depths and is decomposed by facultative microorgan-isms, using oxygen dissolved in the water. In lakes, once the oxygen is consumed, the deep layers become anoxic; here strictly aerobic organisms such as higher plants and animals cannot grow, and the bottom layers have a species composition restricted to anaerobic bacteria and a few kinds of microaerophilic animals. In addition, there is a conversion from a respiratory to a fermentative metabolism, with important consequences for the carbon cycle and other nutrient cycles (see Sections 14.14–14.20 for a consideration of this).

Whether or not a body of water becomes depleted of oxygen depends on several factors. If organic matter is sparse, as it is in unproductive lakes or in the open ocean, there may be insufficient substrate available for chemoorganotrophs to consume all the oxygen. Also important is how rapidly the water from the depths exchanges with surface water. Where strong currents or turbulence occurs, the water mass may be well mixed, and consequently oxygen may be transferred to the deeper layers. In many bodies of water in temperate climates, however, the water mass becomes *stratified* during the summer, with the warmer and less dense surface layers, called the *epilimnion,* separated from the

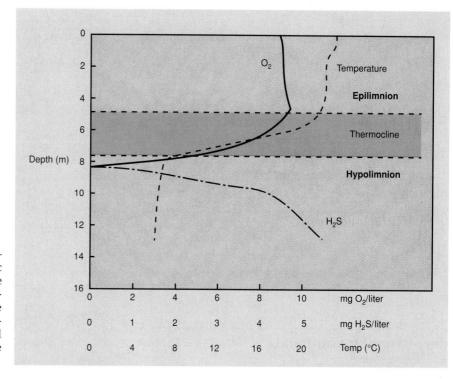

Figure 14.17 Development of anoxic conditions in the depths of a temperate climate lake as a result of summer stratification. The colder bottom waters are more dense and contain H₂S from bacterial sulfate reduction. The zone of rapid temperature change is referred to as the thermocline.

colder and denser bottom layers (the *hypolimnion*) (Figure 14.17). After stratification sets in, usually in early summer, the bottom layers become anoxic (Figure 14.17). In the late fall and early winter, the surface waters become colder and heavier than the bottom layers, and the water "turns over," leading to a reaeration of the bottom. Most lakes in temperate climates thus show an annual cycle in which the bottom layers of water pass from oxic to anoxic and back to oxic.

Rivers

The oxygen relations in a river are of particular interest, especially in regions where rivers receive much organic matter in the form of sewage and industrial pollution. Even though a river may be well mixed because of rapid water flow and turbulence, large amounts of added organic matter can lead to a marked oxygen deficit. This is illustrated in Figure 14.18. As the water moves away from a sewage outfall, organic matter is gradually consumed and the oxygen content returns to normal. Oxygen depletion in a body of water is undesirable because aquatic animals require O₂ and die under even very temporary anoxic conditions. Further, conversion to anoxia results in the production by anaerobic bacteria of odoriferous compounds (for example, amines, H₂S, mercaptans, fatty acids), some of which are also toxic to higher organisms.

Biochemical oxygen demand

Sanitary engineers term the oxygen-consuming property of a body of water its **biochemical oxygen demand** (BOD). The BOD is determined by taking a sample of water, aerating it well, placing it in a sealed bottle, incubating for a standard period of time (usually 5 days at 20°C), and determining the residual oxygen in the water at the end of incubation. A BOD determination thus gives a measure of the amount of organic material in the water that could be oxidized by microorganisms. As a river recovers from contamination with an organic pollutant, the drop in BOD is accompanied by a corresponding increase in dissolved oxygen (Figure 14.18).

We thus see that in a water body, the oxygen and carbon cycles are interrelated, with the concentrations of organic carbon and oxygen in general being inversely related. This is particularly evident in *anoxic* environments, which are frequently rich in organic carbon (see Sections 14.11–14.13).

CONCEPT CHECK   *14.7*

In aquatic ecosystems phototrophic microorganisms are usually the main primary producers. Most of the organic matter produced is consumed by bacteria, which can lead to depletion of oxygen in the environment. BOD is a measure of the oxygen-consuming properties of a water sample.

✔ *What is a primary producer?*

✔ *In a lake, where is the epilimnion and where is the hypolimnion?*

✔ *Will addition of organic matter to a water sample increase or decrease its BOD?*

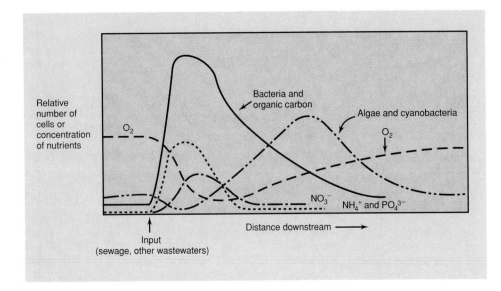

Figure 14.18 Effect of input of sewage or other organic-rich wastewaters into a river. Increase in heterotrophic bacterial numbers and decrease in O_2 levels occur immediately. If NH_4^+ is present in the input, for example, from sewage, NH_4^+ is oxidized to NO_3^- by nitrifying bacteria. The rise in numbers of algae and cyanobacteria is primarily a response to inorganic nutrients, especially PO_4^{3-}.

14.8
Terrestrial Environments

In the consideration of terrestrial environments, our attention inevitably turns to *soil* and *plants* because it is within the soil and on or near plants that many of the key processes occur that influence the functioning of the ecosystem. The process of soil development involves complex interactions among the parent material (rock, sand, glacial drift, and so on), topography, climate, and living organisms. Soils can be divided into two broad groups—**mineral soils** and **organic soils**—depending on whether they derive initially from the weathering of rock and other inorganic material or from sedimentation in bogs and marshes, respectively. Our discussion will concentrate on mineral soils, the predominant soil in most areas.

Soil formation

Soils form as a result of combined physical, chemical, and biological processes. An examination of almost any exposed rock reveals the presence of algae, lichens, or mosses. These organisms are able to remain dormant on the dry rock and then grow when moisture is present. They are phototrophic and produce organic matter, which supports the growth of chemoorganotrophic bacteria and fungi. The numbers of chemoorganotrophs increase directly with the degree of plant cover of the rocks. Carbon dioxide produced during respiration by chemoorganotrophs is converted to carbonic acid ($CO_2 + H_2O \rightleftharpoons H_2CO_3$), which is an important agent in the dissolution of rocks, especially those composed of limestone. Many chemoorganotrophs also excrete organic acids, which further promote the dissolution of rock into smaller particles. Freezing and thawing and other physical processes lead to the develop-

ment of cracks in the rocks. In these crevices, a raw soil forms in which pioneering higher plants can develop. The plant roots penetrate farther into crevices and increase the fragmentation of the rock, and their excretions promote the development of a **rhizosphere** (the soil that surrounds plant roots) microflora. When the plants die, their remains are added to the soil and become nutrients for an even more extensive microbial development. Minerals are further rendered soluble, and as water percolates, it carries some of these chemical substances deeper. As weathering proceeds, the soil increases in depth, thus permitting the development of larger plants and trees. Soil animals become established and play an important role in keeping the upper layers of the soil mixed and aerated. Eventually the movement of materials downward results in the formation of layers, and a typical *soil profile* becomes evident (Figure 14.19). The rate of development of a typical soil profile depends on climatic and other factors, but it is usually very slow, taking hundreds of years.

Soil as a microbial habitat

The most extensive microbial growth takes place on the *surfaces* of soil particles, usually within the rhizosphere (see Figures 14.4, 14.7 and 14.20). As pointed out in Section 14.1, even a small soil aggregate can have many differing microenvironments (compare Figures 14.2 and 14.20), and thus several different types of microorganisms may be present. To examine soil particles directly for microorganisms, fluorescence microscopes are often used, the organisms in the soil being stained with a dye that fluoresces (see Figure 14.4). To observe a *specific* microorganism in a soil particle, **fluorescent-antibody staining** (see Figure 14.7) can be used. Microorganisms can also be visualized on such opaque surfaces as soil by means of the **scanning electron microscope** (Figure 14.21). The scanning electron

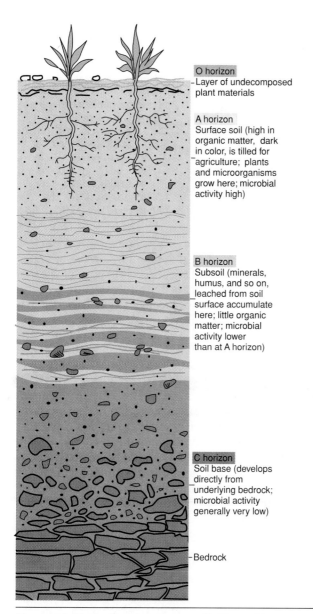

Figure 14.19 Profile of a mature soil. The soil horizons are soil zones as defined by soil scientists.

O horizon
Layer of undecomposed plant materials

A horizon
Surface soil (high in organic matter, dark in color, is tilled for agriculture; plants and microorganisms grow here; microbial activity high)

B horizon
Subsoil (minerals, humus, and so on, leached from soil surface accumulate here; little organic matter; microbial activity lower than at A horizon)

C horizon
Soil base (develops directly from underlying bedrock; microbial activity generally very low)

Bedrock

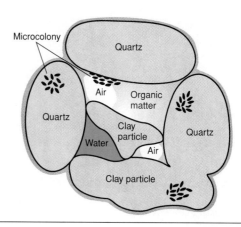

Figure 14.20 A soil aggregate composed of mineral and organic components, showing the localization of soil microorganisms. Very few microorganisms are found free in the soil solution; most of them occur as microcolonies attached to the soil particles.

microscope gives excellent information on the morphology of soil bacteria and can also be used to enumerate cells on soil particle surfaces.

One of the major factors affecting microbial activity in soil is the availability of *water*, and we have previously discussed the importance of water availability to microbial growth (∞ Section 5.10). Water is a highly variable component of soil, its presence depending on soil composition, rainfall, drainage, and plant cover. Water is held in the soil in two ways, by adsorption onto surfaces or as free water existing in thin sheets or films between soil particles. The water present in soils has a variety of material dissolved in it, the whole mixture being referred to as the *soil solution*. In well-drained soils, air penetrates readily and oxygen concentrations

can be high. In waterlogged soils, however, the only oxygen present is that dissolved in the water, and this is soon consumed by microorganisms. Such soils quickly become anoxic, showing profound changes in their biological properties. We discussed oxygen relationships in soil in Section 14.1 and Figure 14.2.

The *nutrient status* (resources) of a soil is the other major factor affecting microbial activity. The greatest microbial activity is in the organic-rich surface layers, especially in and around the rhizosphere. The numbers and activity of soil microorganisms depend to a great extent on the balance of nutrients present. In some soils carbon is not the limiting nutrient, but instead the availability of *inorganic* nutrients such as phosphorus and nitrogen limit microbial productivity.

Deep subsurface microbiology

Interest in the chemistry of groundwater and the potential leaching of pollutants and their transfer in groundwater aquifers had led to consideration of the role microorganisms play in the *deep subsurface* terrestrial environment. The deep soil subsurface, which can extend for several hundred meters below the soil surface, is not a biological wasteland. A variety of microorganisms, primarily bacteria, are present in most deep underground soils. In samples collected aseptically from bore holes drilled down to 300 m, a diverse array of bacteria have been found including anaerobes such as sulfate-reducing bacteria, methanogens, and homoacetogens, and various aerobes and facultative aerobes. Microorganisms in the deep subsurface presumably have access to nutrients because groundwater flows through their habitats, but activity measurements suggest that metabolic rates of these bacteria are rather low in their natural habitats (however, see the box, Microbial Life Deep Underground). Compared to microorganisms in the upper layers of soil, the biogeo-

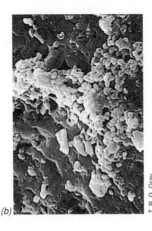

Figure 14.21 Visualization of microorganisms on the surface of soil particles by use of the scanning electron microscope. (a) Rod-shaped bacteria. (b) Actinomycete spores. The cells in (a) and the spores in (b) are about 1–2 μm wide. (c) Fungus hyphae. The fungal hyphae are about 4 μm wide.

chemical significance of deep subsurface microorganisms may thus be minimal. However, there is evidence that the metabolic activities of these buried microorganisms may over very long periods be responsible for some mineralization of organic compounds and release of products into the groundwater. The potential for *in situ* bioremediation (see Section 14.20) of toxic substances leached from soil into groundwater (for example, benzenes and agricultural chemicals) by deep subsurface microorganisms is of particular current interest.

CONCEPT CHECK   14.8

The soil is a complex habitat with numerous microenvironments and niches. Microorganisms are present in the soil primarily attached to soil particles. The most important factor influencing microbial activity in surface soil is the availability of water, whereas in deep soil (the subsurface environment) nutrient availability plays a major role.

✔ *What is the difference between* mineral *soils and* organic *soils?*

✔ *What factors govern the extent and type of microbial activity in soils?*

14.9
Deep-Sea Microbiology

The oceans cover over three-fourths of the earth's surface, and marine microbiology is consequently an important area of study. However, because open ocean waters are relatively poor in nutrients (see Section 14.7), microbial activity in these areas is generally not extensive. Nevertheless, marine microorganisms are of interest for several reasons, including their ability to grow at low-nutrient concentrations and cold temperatures and, in microorganisms inhabiting the deep sea, to withstand enormous hydrostatic pressures. We focus in this section on microbial life in the deep sea.

What is the deep sea? Visible light penetrates no further than about 300 m in open ocean waters; this upper region is referred to as the **photic zone.** Beneath the photic zone, down to a depth of about 1000 m, considerable biological activity still occurs as a result of the action of animals and chemoorganotrophic microorganisms. Water at depths greater than 1000 m is, by comparison, relatively biologically inactive and has come to be known as the "deep sea." Greater than 75% of all ocean water is in the deep sea, primarily at depths between 1000 and 6000 m. Organisms that inhabit the deep sea are faced with three major environmental extremes: low temperature, high pressure, and low nutrient levels. Below depths of about 100 m ocean water stays a constant 2–3°C. We discussed the responses of microorganisms to changes in temperature in Section 5.7. As would be expected, bacteria isolated from depths below 100 m are *psychrophilic.* Some are *extreme* psychrophiles, growing only in a narrow range near the *in situ* temperature (see later). Deep-sea microorganisms must also be able to withstand the enormous hydrostatic pressures associated with great depths. Pressure increases by *1 atm* for every *10 m* depth. Thus, an organism growing at a depth of 5000 m must be able to withstand pressures of 500 atm. We focus on this interesting aspect of life in the deep sea here.

Barotolerant and barophilic bacteria

Do deep-sea bacteria simply *tolerate* high pressure (that is, are they *barotolerant*) or are they actually *dependent* on pressure (that is, are they *barophilic*)? Studies of various deep-sea bacteria have shown that both patterns exist and that the distribution of barotolerant and barophilic bacteria is basically a function of depth. Organisms isolated from depths down to about 4000 m and tested for growth or metabolic activity as a function of pressure are **barotolerant;** higher metabolic rates are observed at 1 atm than at 400 atm, although growth rates at the two pressures are about the same (Figure 14.22). However, barotolerant isolates do not grow at pressures above 500 atm. By contrast, cultures derived from samples taken at greater depths, 5000–6000 m, are **barophilic,** growing optimally at pressures of about 400 atm (Figure 14.22). Note that although barophiles grow best under pressure, they retain the ability to grow at 1 atm (Figure 14.22).

A FOCUS ON . . .

MICROBIAL LIFE DEEP UNDERGROUND

Microbiologists studying the deep terrestrial subsurface have found viable prokaryotes at depths of several thousand meters below the surface, existing in different physical and chemical environments. How are these microorganisms making a living? Initial findings indicated that these buried microorganisms were very slow growing chemoorganotrophic bacteria surviving by the slow catabolism of organic carbon deposited within the sediments. However, studies on the microbial ecology of deep basalt aquifers[a] have shown that sluggish chemoorganotrophs are not the only prokaryotes that live deep underground.

Basalts are iron-rich volcanic rocks that are essentially devoid of organic matter. In certain basalts up to 1500 m deep from the Columbia River Basin (Washington, USA), large numbers of anaerobic, *chemolithotrophic* bacteria have been found (see the photograph below), including sulfate-reducing bacteria, methanogens, and homoacetogens.[a] These

organisms were shown to be metabolically active by carbon-stable isotope analyses of methane (CH_4) present in the rocks and surrounding groundwaters; measurements showed a strong enrichment in the lighter isotope of carbon (^{12}C), typical of biological methanogenesis (see Section 14.6).

A common metabolic link among these anaerobes is a thirst for H_2, an excellent electron donor for their various energy-yielding metabolisms [∞ Sections 13.16 and 13.17 and (b) below]. Hydrogen is a common product of the anoxic decomposition of organic matter (∞ Sections 13.19 and 13.20). But if basalts contain very little organic material, where does the H_2 come from to support metabolism of the H_2 consumers? Interestingly, H_2 in the Columbia basalts apparently originates from the *strictly chemical* interaction of water with iron minerals in the rocks [see proposed reaction in (b) below]. Such reactions are known from inorganic chemistry, and in laboratory studies in which crushed Columbia basalt was mixed with sterile

water under anoxic conditions, rapid H_2 evolution occurred.[a] H_2 was also detected *in situ* in groundwater percolating through the basalts.

From analyses of these experimental results it was hypothesized that H_2 formed in deep underground basalts is indeed the electron donor that supports growth of the substantial populations of anaerobic prokaryotes found there. If this is true, these organisms would be living a strictly *geochemical* existence because their electron acceptor (CO_2 in the case of methanogens and homoacetogens and SO_4^{2-} in the case of sulfate-reducing bacteria) and electron donor, H_2, are all derived from *inorganic* materials. These buried chemolithotrophs are also novel because of their total independence of photosynthetic primary production, which generates the molecular oxygen and/or organic matter required by virtually all surface-dwelling organisms.

How widespread this unique microbial lifestyle is awaits further research. But as our understanding of microbial life deep underground increases, there is a good chance that other surprises lie in store as well. ■

[a]Stevens, T. O., and J. P. McKinley. 1995. Lithoautotrophic microbial ecosystems in deep basalt aquifers. *Science 270*:450–454.

(a)   Todd O. Stevens

(a) Laser confocal photomicrograph of a microbial film attached to the surface of basalt chips. Green is reflected light from the basalt surface while the red color is from Nile-red stained bacterial cells. The cells in the film were grown on H_2 from basalt as depicted in the bottom reaction of (b).

Methanogenesis:	$4 H_2 + CO_2 \longrightarrow CH_4 + 2 H_2O$
Acetogenesis:	$4 H_2 + 2 HCO_3^- + H^+ \longrightarrow CH_3COO^- + 4 H_2O$
Sulfate reduction:	$4 H_2 + SO_4^{2-} + H^+ \longrightarrow HS^- + 4 H_2O$
Proposed inorganic H_2 production:	$FeO + H_2O \longrightarrow H_2 + FeO_{3/2}$

(b)

(b) Key metabolic reactions of anaerobic prokaryotes growing in anoxic deep basalt aquifers.

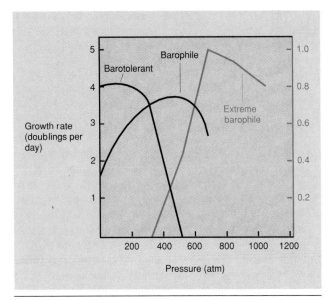

Figure 14.22 Growth of barotolerant, barophilic, and extremely barophilic bacteria. The extreme barophile was isolated from the Mariana Trench (10,500 m). Note the much slower growth rate (at any pressure) of the extreme barophile (right ordinate) as compared to the barotolerant and barophilic bacteria (left ordinate). Note also the inability of the extreme barophile to grow at low pressures.

Samples from even deeper water (10,000 m) have yielded **extreme (obligate) barophiles.** One strain studied in detail grew fastest at a pressure of 700–800 atm and grew nearly as well at 1035 atm, the pressure it was experiencing in its natural habitat (see Figure 14.22). The unique aspect of this extremely barophilic isolate was that it not only *tolerated* pressure but actually *required* pressure for growth; the isolate would not grow at pressures of less than about 400 atm. Interestingly, however, this extreme barophile was not killed by decompression because it could tolerate moderate periods of decompression; however, viability was lost when the culture was left for several hours in a decompressed state.

Barotolerant and barophilic bacteria are also cold-loving, that is, psychrophilic. This property appears to be more prevalent among extremely barophilic isolates. The extreme barophile described in Figure 14.22 was found to be sensitive to temperature; the optimal growth temperature was determined to be the environmental temperature of 2°C, and temperatures above 10°C significantly reduced viability.

Physiology of barophiles

Pressure is known to affect cellular physiology and biochemistry. It has been established that increased pressure decreases the binding capacity of enzymes for their substrates. Thus, the enzymes of extreme barophiles must be folded in such a way as to minimize these pressure-related effects. Other potential pressure-sensitive targets include protein synthesis and membrane phenomena such as transport. An organism grown under high pressure has an increase in the proportion of unsaturated fatty acids in its cytoplasmic membrane. This change is presumably of adaptive significance because it makes the membrane less likely to gel at high pressures. The rather slow growth rates of extreme barophiles (see Figure 14.22) are probably due to a combination of pressure effects on cellular biochemistry and to the fact that these organisms grow only at low temperatures, where reaction rates are decreased considerably to begin with.

Molecular genetic tools have given new insight into the physiology of barophilism. In barophiles capable of growth up to 500–600 atm, it has been shown that growth at high pressure is accompanied by changes in the protein composition of the cell wall outer membrane. In one barophile studied in detail, a specific outer membrane protein called the OmpH protein (*o*uter *m*embrane *p*rotein H) is synthesized in cells grown at high pressures but not in cells grown at 1 atm pressure. The OmpH protein is a type of *porin.* Porins are structural proteins that form channels for the diffusion of organic molecules through the outer membrane and into the periplasm (∞ Section 3.6 and Figure 3.34). Presumably, the porin present in cells of the barophile grown at low pressures cannot function properly at high pressure, and thus a new type of porin molecule must be synthesized.

The *ompH* gene (which encodes the OmpH protein) from this barophile has been cloned into *Escherichia coli.* Studies of gene expression of the *ompH* gene in this experimental system have shown that pressure affects its transcription; messenger RNA (mRNA) coding for *ompH* is expressed in cells of *E. coli* grown at 200 atm but not in cells grown at 1 atm (*E. coli* can grow slowly at high pressure). Further, sequencing of the *ompH* gene has shown that the OmpH protein is related but clearly distinct from the porin produced at 1 atm.

Studies of the *ompH* system thus show that pressure can affect gene expression in barophilic bacteria. How this happens is unclear but could involve the activities of pressure-sensitive repressor proteins or pressure-dependent gene activators. However, it appears that relatively few proteins are controlled by

---

CONCEPT CHECK   **14.9**

The deep sea is a cold, dark habitat where high hydrostatic pressure and low nutrient availability prevails. Barophiles grow best under pressure, and extreme barophiles, obtained from the greatest depths, require high pressures for growth.

✔ *What is a* barophilic *bacterium?*

✔ *What molecular adaptations occur in barophiles that allow them to grow optimally under pressure?*

pressure in barophiles because many proteins seem to be the same in cells grown at both high and low pressure; cell wall and related structural proteins and transport proteins seem to be the major variable components. Thus, pressure acts selectively to turn on or off the transcription of specific genes coding for proteins needed for growth at high pressure. Control of most genes in barophilic bacteria is presumably affected by environmental and nutritional factors in the same manner as it is in nonbarophilic bacteria (∞ Chapter 7).

14.10

Hydrothermal Vents

The general conception of the deep sea as a remote, low temperature, high pressure environment capable of supporting only the slow growth of barotolerant and barophilic bacteria is generally correct, but there are some amazing exceptions. A number of dense, thriving *invertebrate* communities supported by the activities of microorganisms exist clustered about thermal springs in deep waters throughout the world. Geophysical measurements have identified thermal springs in several locations on both the Atlantic and Pacific Ocean

floors. Geologically, these springs are associated with *ocean floor spreading centers*, regions where hot basalt and magma very near the sea floor cause the floor to slowly drift apart. Seawater seeping into these cracked regions mixes with hot minerals and is emitted from the springs (Figure 14.23); because of their unique properties these springs have come to be known as **hydrothermal vents.**

Two major types of vents have been found. *Warm vents* emit hydrothermal fluid at temperatures of 6–23°C (into seawater at 2°C). *Hot vents*, usually referred to as "black smokers" because the mineral-rich hot water forms a dark cloud of precipitated material on mixing with seawater, emit hydrothermal fluid at 270–380°C (Figure 14.23). The flow rates of these two vent types are also characteristic; warm vents emit fluid at 0.5–2 cm/sec, whereas hot vents have higher flow rates, about 1–2 m/sec.

Animals living at thermal vents

Using small pressurized submarines, it is possible to study the organisms associated with the thermal vents. Thriving invertebrate communities have been found, with tube worms over 2 m in length and large numbers of giant clams and mussels (Figure 14.24).

Figure 14.23 Schematic diagram showing the morphology and major chemical species occurring at warm vents and black smokers. At warm vents, the hot hydrothermal fluid is cooled by cold (2–3°C) seawater permeating the sediments. In black smokers, hot hydrothermal fluid reaches the sea floor directly. Warm vents and black smokers are typically found at about 2000-meter depth. Such depths can be explored by humans in small submersibles such as *Alvin,* used by researchers at the Woods Hole Marine Laboratory and Oceanographic Institution, Woods Hole, MA.

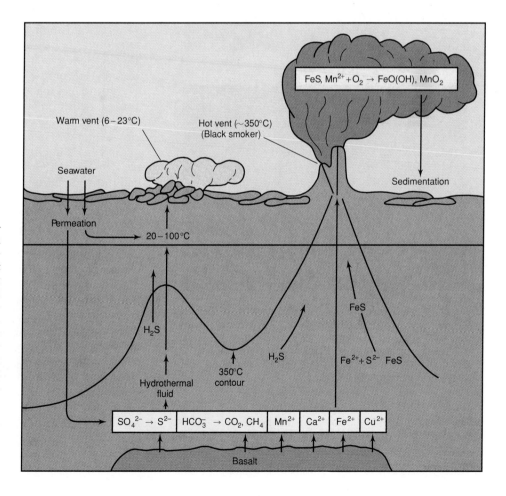

(a)

Dudley Foster, Woods Hole Oceanographic Institution

(b)

James Agviar, Woods Hole Oceanographic Institution

(c)

Carl Wirsen, Woods Hole Oceanographic Institution

Figure 14.24 Invertebrates from habitats near deep-sea thermal vents. (a) Tube worms (family *Pogonophora*), showing the sheath (white) and plume (red) of the worm bodies. (b) Close-up photograph showing worm plume. (c) Mussel bed in vicinity of warm vent. Note yellow deposition of elemental sulfur in and around mussels.

Considering that other locations in the deep sea are so biologically unproductive, how do dense animal communities exist in the absence of phototrophic primary producers? What is the nature of the energy source(s)?

Chemical analyses of hydrothermal fluid show large amounts of reduced inorganic materials, including H_2S, Mn^{2+}, H_2, and CO. Some vents contain little H_2S but have high levels of NH_4^+. Organic matter is not present in the fluid emitted from any of the hydrothermal vents thus far examined. From studies of vent chemistry and associated microbial processes, it is clear that the animals are dependent on the activities of chemolithotrophic bacteria (∞ Sections 13.8–13.13), which grow at the expense of inorganic energy sources emitted from the vents. Carbon dioxide, abundant in seawater as CO_3^{2-} and HCO_3^-, is fixed into organic carbon by the chemolithotrophs, and the latter form the base of an extremely short food chain for hydrothermal vent animals. We now discuss these communities in detail.

Microorganisms in hydrothermal vents

Large numbers of sulfur-oxidizing chemolithotrophs such as *Thiobacillus*, *Thiomicrospira*, *Thiothrix*, and *Beggiatoa* (∞ Section 13.10) are present in and around the vents. Samples collected from near the vents have yielded cultures of these organisms, and *in situ* experiments have shown fixation of CO_2 and oxidation of H_2S and $S_2O_3^{2-}$ by natural populations of these bacteria. Other vents yield nitrifying bacteria, hydrogen-oxidizing bacteria, iron- and manganese-oxidizing bacteria, and methylotrophic bacteria, the latter presumably growing on the methane and carbon monoxide (CO) emitted from the vents (see Chapter 13 and Section 16.7 for discussion of some of these physiological groups). Table 14.3 summarizes the electron donors and electron acceptors for chemolithotrophs suspected of playing a role in hydrothermal vent ecology. However, there is no direct evidence that the animals of the vents directly *eat* these chemolithotrophic bacteria. Instead, a different role of chemolithotrophs in animal nutrition occurs.

Nutrition of animals living near hydrothermal vents

Perhaps the most exciting discovery is that certain chemolithotrophs live directly in association with animals of the thermal vents. The 2-m-long tube worms (see Figure 14.24) lack a mouth, gut, or anus but contain a modified gastrointestinal tract consisting primarily of spongy tissue called the **trophosome**. Making up about 50% of the weight of the worm, trophosome tissue is loaded with sulfur granules, and microscopy of trophosome tissue shows large numbers of prokaryotic cells (Figure 14.25), an average of 3.7×10^9 cells/g of trophosome tissue. The large spherical cells observed in the trophosome are structurally similar to the marine sulfur-oxidizing bacterium *Thiovulum*. Trophosome tissue (containing symbiotic sulfur bacteria) also shows activity of the enzyme *rhodanese*, an

TABLE 14.3   Chemolithotrophic prokaryotes of potential significance to hydrothermal vent primary productivity[a]

Chemolithotroph	Electron donor	Electron acceptor
Sulfur-oxidizing bacteria	HS^-, S^0, $S_2O_3^{2-}$	O_2, NO_3^-
Nitrifying bacteria	NH_4^+, NO_2^-	O_2
Sulfate-reducing bacteria	H_2	S^0, SO_4^{2-}
Methanogenic Archaea	H_2	CO_2
Hydrogen-oxidizing bacteria	H_2	O_2, NO_3^-
Iron and manganese-oxidizing bacteria	Fe^{2+}, Mn^{2+}	O_2
Methylotrophic bacteria	CH_4, CO	O_2

[a]See Sections 13.8–13.13 for a discussion of chemolithotrophs.

enzyme capable of disproportionating $S_2O_3^{2-}$ to S^0 and SO_3^{2-}. Also present are enzymes of the *Calvin cycle*, the pathway by which most autotrophic organisms fix CO_2 into cellular material (∞ Section 13.7).

The chemolithotrophic bacteria supply the worm with its nourishment, the animal living off the excretory products and dead cells of its chemolithotrophic symbionts. The bright red plume (Figure 14.24*b*) is rich in blood vessels and serves as a trap for O_2 and H_2S (see later) for transport to chemolithotrophs in the trophosome. Similar conclusions can be reached concerning the nutrition of the giant clams and mussels (see Figure 14.24*c*) present around the vents because sulfur-oxidizing bacterial communities are found in the gill tissues of these animals as well. Use of nucleic acid sequencing techniques (∞ Working with Nucleic Acids: The Tools, Chapter 6 and Sections 15.5 and 15.7) has shown that each vent animal harbors only one major species of bacterial symbiont and that the species of symbiont varies among the different animal types.

Further study of the tube worms has shown that these animals contain unusual soluble hemoglobins that bind H_2S as well as O_2 and transport both substrates to the trophosome where they are released to the bacterial symbiont; trapping and transporting sulfide are necessary to prevent the H_2S from poisoning the animal. Furthermore, stable isotope analyses (see Section 14.6) of the elemental sulfur found within the bacterial symbionts have shown the $^{34}S/^{32}S$ isotope composition to be the same as the sulfide emitted from the vent. This ratio is distinctly different from that of seawater sulfate and serves as proof that geothermal sulfide is entering the worm.

A link between animal nutrition and other physiological groups of chemolithotrophs (for example, H_2 oxidizers and nitrifying bacteria) has been suggested.

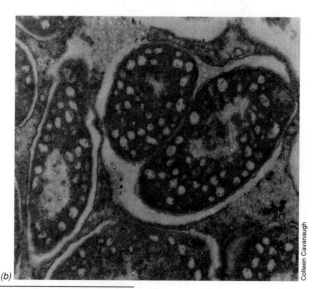

Figure 14.25 Chemolithotrophic sulfur-oxidizing bacteria associated with the trophosome tissue of tube worms from hydrothermal vents. (a) Scanning electron microscopy of trophosome tissue showing spherical chemolithotrophic sulfur-oxidizing bacteria. Cells are 3–5 μm in diameter. Reprinted with permission from *Science 213*:340–342 (1981), © AAAS. (b) Transmission electron micrograph of bacteria in sectioned trophosome tissue. The cells are frequently enclosed in pairs by an outer membrane of unknown origin.

Methanotrophic symbionts have been shown to play a nutritional role for animals living in symbiotic association with giant clams near natural gas seeps at relatively shallow depths in the Gulf of Mexico. Although not truly autotrophs (CH_4 is an organic compound), these symbionts support growth of the animal, in this case by the oxidation of CH_4 as an energy source.

Other chemolithotrophs, iron, H_2, and manganese oxidizers, for example, are probably not animal symbionts but instead exist as free-living bacteria growing at the expense of reduced substances emitted from the vents. Nevertheless, these chemolithotrophs probably contribute to overall primary productivity in the vent ecosystem.

Black smokers

The great depths of the deep sea create huge hydrostatic pressures that affect the physical properties of water. At a depth of 2600 m, water does not boil until it reaches a temperature of about 450°C. At certain vent sites superheated (but not boiling) hydrothermal fluid is emitted at temperatures of 270–380°C (see Figure 14.23) and could theoretically be a habitat for hyperthermophilic bacteria (∞ Section 5.8). The hydrothermal fluid emitted from black smokers contains abundant metal sulfides, especially iron sulfides, and cools quickly as it emerges into cold seawater. The precipitated metal sulfides form a tower referred to as a "chimney" about the source (Figure 14.26). Although it is very doubtful that prokaryotes actually live in the hot (>250°C) hydrothermal fluid (∞ Section 17.5), good evidence exists for the presence of thermophilic or hyperthermophilic bacteria of various types in the seawater or hydrothermal fluid *gradient* that forms as the hot water blends with cold ocean water. For example, the walls of smoker chimneys are teaming with hyperthermophilic prokaryotes such as *Methanopyrus*, a methanogenic archaean that oxidizes H_2 and is of great evolutionary interest (∞ Chapter 17).

In efforts to define the upper temperature limits for life, attempts have been made to detect bacterial growth in the outflows of black smokers at various temperatures. By fitting a vent with a titanium "cap" (Figure 14.27) from which glass microscope slides and other surfaces for microbial colonization can be suspended in the emerging hot water, evidence for colonization and growth of prokaryotes has been obtained at temperatures above 125°C to about 140°C. Similar surfaces exposed to outflows at 200°C or higher showed no microbial attachment or growth. Although hyperthermophiles from water 125°C or higher have not yet been cultured, the formation of microcolonies on artificial surfaces at these temperatures is similar to those observed at lower temperatures (see Figure 14.3) and is good evidence that microbial growth is occurring here. These results suggest that the upper temperature limit for microbial life is probably under 150°C. We discuss some of the reasons for this upper temperature limit in Section 17.5.

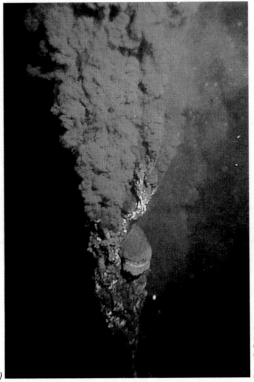

(a)

Robert D. Ballard

(b)

Dudley Foster, Woods Hole Oceanographic Institution

Figure 14.26 Black smokers emitting sulfide- and mineral-rich water at temperatures of 350°C. (a) The chimney is quite large, about 1 m in length. (b) The chimney is much smaller. Note the scientific equipment near the smoker in (b), indicating the relatively small size of the chimney.

Norman Pace

Figure 14.27 Methodology for studying extremely hyperthermophilic prokaryotes in black smokers. The titanium cap placed over the vent serves as a support for microscope slides submerged to regions of the vent of known temperature. With the vent cap system, evidence for organisms growing at temperatures above 125°C has been obtained. The vent cap shown covers a 160°C hydrothermal vent located at a depth of 2000 m in the Guayamas Basin (Gulf of California).

CONCEPT CHECK 14.10

Hydrothermal vents are specialized environments in the deep sea where thermal (volcanic) activity leads to the presence of fluids containing large amounts of inorganic energy sources that can be used by chemolithotrophic bacteria. The chemolithotrophic bacteria fix CO_2 autotrophically into organic carbon, some of which is then used by the deep-sea animals. The deep-sea hydrothermal vents are thus habitats where the primary producers are chemolithotrophic rather than phototrophic.

✔ *How does a* warm hydrothermal vent *differ from a* black smoker, *both chemically and physically?*

✔ *How do giant tube worms receive their nutrition?*

✔ *Why aren't phototrophic microorganisms important in hydrothermal vent food chains?*

14.11

Carbon Cycle

On a global basis, carbon is cycled through all the earth's major carbon reservoirs: the atmosphere, the land, the oceans and other aquatic environments, and sediments and rocks (Figure 14.28). The largest carbon reservoir is present in the sediments and rocks of the earth's crust, but the turnover time is so long that flux out of this compartment is relatively insignificant on a human scale. From the viewpoint of living organisms, a large amount of organic carbon is found in land plants. This represents the carbon of forests and grasslands and constitutes the major site of photosynthetic CO_2 fixation. However, more carbon is present in dead organic material, called *humus*, than in living organisms.

Humus is a complex mixture of organic materials. It is derived partly from the protoplasmic constituents of soil microorganisms that have resisted decomposition and partly from resistant plant material. Some humic substances are fairly stable, with a global turnover time of about 40 years, although certain other humic components decompose much more rapidly than this.

The most rapid means of global transfer of carbon is via the CO_2 of the atmosphere. Carbon dioxide is removed from the atmosphere primarily by photosynthesis of land plants and is returned to the atmosphere by respiration of animals and chemoorganotrophic microorganisms. An analysis of the various processes suggests that the single most important contribution of CO_2 to the atmosphere is via microbial decomposition of dead organic material, including humus.

Importance of photosynthesis in the carbon cycle

The only major way in which new organic carbon is synthesized on Earth is via photosynthesis. Phototrophic organisms are therefore at the basis of the carbon cycle (Figure 14.28). We discussed in Sections 13.4 and 13.5 the processes of anoxygenic and oxygenic photosynthesis. The bulk of the photosynthesis on Earth is carried out by oxygenic phototrophs, and so the carbon cycle and oxygen cycle on Earth are intimately intertwined.

Phototrophic organisms are found in nature almost exclusively in habitats where light is available. Thus, the deep sea and other permanently dark habitats are devoid of phototrophs. Phototrophic organisms can be divided into two major groups: higher plants and microorganisms. Higher plants are the dominant phototrophic organisms of terrestrial environments, whereas phototrophic microorganisms are the most abundant photosynthesizers of aquatic environments.

The redox cycle for carbon is shown in Figure 14.29. We begin with photosynthesis. The overall equation for oxygenic photosynthesis is

$$CO_2 + H_2O \xrightarrow{\text{light}} (CH_2O) + O_2$$

where (CH_2O) represents organic matter at the oxidation state of cell material such as polysaccharides (the main form in which photosynthesized organic matter is stored in the cell). Phototrophic organisms also carry out respiration, both in the light and the dark. The overall equation for respiration is the reverse of the preceding equation:

$$(CH_2O) + O_2 \xrightarrow{\text{light or dark}} CO_2 + H_2O$$

where (CH_2O) again represents storage polysaccharides. If an organism is to grow (that is, increase in cell number or mass) phototrophically, then the rate of photosynthesis must exceed the rate of respiration. If this occurs, then

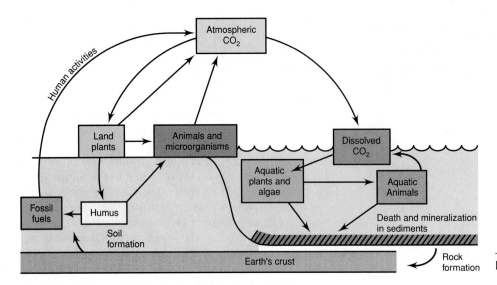

Figure 14.28 The carbon cycle.

some of the carbon fixed from CO_2 into polysaccharide can become the starting material for biosynthesis. The whole carbon cycle is built on a net positive balance of the rate of photosynthesis over the rate of respiration.

Decomposition

Photosynthetically fixed carbon is eventually degraded by various organisms and two major oxidation states of carbon result: methane (CH_4) and carbon dioxide (CO_2) (Figure 14.29). These two gaseous products are formed from the activities of methanogens (CH_4) or from vari-

ous chemoorganotrophs via fermentation, anaerobic respiration, or aerobic respiration (CO_2). When methane is transported to oxic environments, it is oxidized to CO_2 by methanotrophic bacteria (∞ Section 16.7). Hence all carbon eventually returns to CO_2 from which autotrophic metabolism once again begins the cycle.

The balance between the oxidative and reductive portions of the carbon cycle is critical; the products of metabolism of some organisms are the substrates for others. Any significant changes in levels of gaseous forms of carbon may have serious global consequences (as we are already experiencing from the increasing

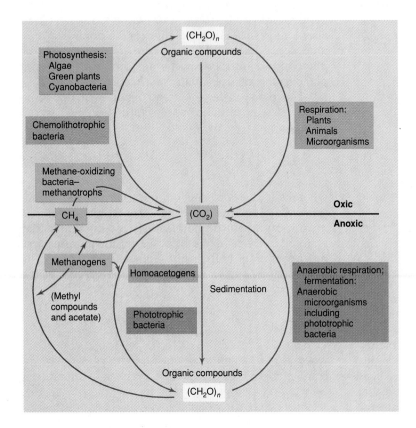

Figure 14.29 Redox cycle for carbon; note in particular the contrasts between autotrophic and heterotrophic processes. Photosynthesis in oxic habitats is mainly oxygenic, whereas in anoxic environments it is mainly anoxygenic. Under anoxic conditions, besides homoacetogens and methanogens, certain sulfate-reducing bacteria are also autotrophic.

CO_2 levels in the atmosphere caused by deforestation and the burning of fossil fuels). We continue our treatment of the carbon cycle in the next two sections with detailed discussions of methanogenesis and carbon cycling in the rumen of ruminant animals.

14.12

Methanogenesis and Syntrophy

Although methane (CH_4) is a relatively minor component of the global carbon cycle, it is of great importance in many localized situations. In addition, it is of considerable microbiological interest because it is a product of anoxic microbial metabolism. Methane production is carried out by a group of Archaea, the *methanogens*, which are obligate anaerobes. We discussed methane formation in Section 13.17 and will discuss the organisms themselves in Section 17.2. Most methanogens use CO_2 as their terminal electron acceptor in anaerobic respiration, reducing it to methane; the electron donor used in this process is generally hydrogen, H_2. The overall reaction of **methanogenesis** in this pathway is as follows:

$$4 H_2 + CO_2 \rightarrow CH_4 + 2 H_2O$$
$$\Delta G^{0\prime} = -130.7 \text{ kJ/reaction}$$

which shows that CO_2 reduction to methane is an eight-electron process (four H_2 molecules). A few other substrates can be converted to methane, including methanol, CH_3OH; formate, $HCOO^-$; methyl mercaptan, CH_3SH; acetate, CH_3COO^-; and methylamines (∞ Section 17.2). It is of interest that despite the widespread production of methane, very few carbon compounds serve as direct precursors of methanogenesis. Thus, methanogenesis is dependent on the production of these few carbon compounds by other organisms from complex organic matter.

Anoxic decomposition, interspecies hydrogen transfer, and syntrophy

We mentioned these concepts briefly in Section 13.20 during our discussion of fermentative metabolism. Here we present the significance for the whole problem of the anoxic carbon cycle. High-molecular-weight sub-

stances such as polysaccharides, proteins, and fats, are converted to CH_4 by the cooperative interaction of several physiological groups of prokaryotes. In many anoxic environments the immediate precursors of CH_4 are H_2 and CO_2, these substrates being generated by the activities of fermentative anaerobes. For the conversion of a typical polysaccharide such as cellulose to methane, as many as five major physiological groups of prokaryotes may be involved in the overall process (Figure 14.30 and Table 14.4). *Cellulolytic bacteria* cleave the high-molecular-weight cellulose molecule into cellobiose (glucose–glucose) and into free glucose. Glucose is then fermented by *primary fermenters* to a variety of fermentation products, acetate, propionate, butyrate, succinate, alcohols, H_2, and CO_2 being the major ones observed. Any H_2 produced in primary fermentative processes is immediately consumed by methanogens, homoacetogens, or sulfate-reducing bacteria (in environments con-

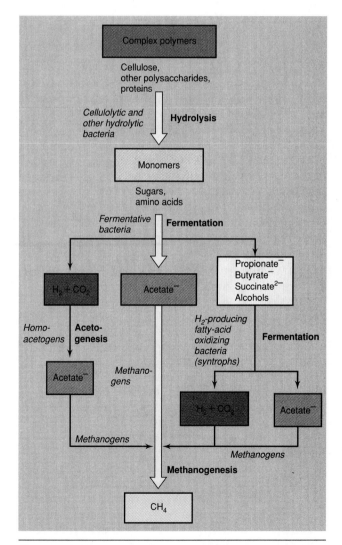

Figure 14.30 Overall process of anoxic decomposition, showing the manner in which various groups of fermentative anaerobes act together in the conversion of complex organic materials ultimately to methane (CH_4) and CO_2.

TABLE 14.4 Major reactions occurring in the anoxic conversion of organic compounds to methane[a]

Reaction type	Reaction	Free-energy change (kJ/reaction)	
		$\Delta G^{0'b}$	ΔG^{c}
Fermentation of glucose to acetate, H_2, and CO_2	Glucose + 4 $H_2O \rightarrow$ 2 acetate$^-$ + 2 HCO_3^- + 4 H^+ + 4 H_2	−207	−319
Fermentation of glucose to butyrate, CO_2, and H_2	Glucose + 2 $H_2O \rightarrow$ butyrate$^-$ + 2 HCO_3^- + 2 H_2 + 3 H^+	−135	−284
Fermentation of butyrate to acetate and H_2	Butyrate$^-$ + 2 $H_2O \rightarrow$ 2 acetate$^-$ + H^+ + 2 H_2	+48.2	−17.6
Fermentation of propionate to acetate, CO_2, and H_2	Propionate$^-$ + 3 $H_2O \rightarrow$ acetate$^-$ + HCO_3^- + H^+ + H_2	+76.2	−5.5
Fermentation of benzoate to acetate, CO_2 and H_2	Benzoate$^-$ + 6 $H_2O \rightarrow$ 3 acetate$^-$ + 2 H^+ + CO_2 + 3 H_2	+49.5	−18
Methanogenesis from H_2 + CO_2	4 H_2 + HCO_3^- + $H^+ \rightarrow CH_4$ + 3 H_2O	−136	−3.2
Methanogenesis from acetate	Acetate$^-$ + $H_2O \rightarrow CH_4$ + HCO_3^-	−31	−24.7
Acetogenesis from H_2 + CO_2	4 H_2 + 2 HCO_3^- + $H^+ \rightarrow$ acetate$^-$ + 2 H_2O	−105	−7.1

[a]Data adapted from Zinder, S. 1984. Microbiology of anaerobic conversion of organic wastes to methane: Recent developments. *Am. Soc. Microbiol. News* 50:294–298.

[b]Standard conditions: solutes, 1 M; gases, 1 atm.

[c]Concentrations of reactants in typical anoxic freshwater ecosystem: fatty acids, 1 mM; HCO_3^-, 20 mM; glucose, 10 μM; CH_4, 0.6 atm; H_2, 10^{-4} atm.

taining significant levels of sulfate). In addition, acetate can be converted to methane by certain methanogens.

Key organisms in the conversion of complex organic materials to methane are *secondary fermenters*, especially the *H_2-producing fatty acid-oxidizing bacteria*. These organisms use fatty acids or alcohols as energy sources but grow poorly or not at all on these substrates in pure culture. However, in association with a *H_2-consuming* organism (such as a methanogen or a sulfate-reducing bacterium), the H_2-producing bacteria grow luxuriantly. As explained later, H_2 consumption by a second organism is critically important to the growth of H_2-producing fatty acid-oxidizing bacteria. Examples of H_2 producers include *Syntrophomonas* and *Syntrophobacter*. The genus names of these organisms reflect their dependence on *syntrophic* relationships with other bacteria (in this case H_2-consuming bacteria) for growth (∞ Figure 13.40), and the organisms are collectively referred to simply as *syntrophs* (see Figure 14.31). **Syntrophy** means, literally, "eating together." *Syntrophomonas wolfei* oxidizes C_4 to C_8 fatty acids yielding acetate, CO_2 (if the fatty acid contained an odd number of carbon atoms), and H_2 (from the reduction of protons) (Table 14.4). Other species of *Syntrophomonas* use fatty acids up to C_{18}, including some unsaturated fatty acids (Figure 14.31b). *Syntrophobacter wolinii* specializes in propionate oxidation and generates acetate, CO_2, and H_2 (Table 14.4). When written with all reactants at standard conditions (solutes, 1 *M*; gases, 1 atm), these fatty acid conversions yield free-energy changes that are *positive* (Table 14.4). That is, the $\Delta G^{0'}$ associated with these conversions are such that the reactions do *not* occur with the release of free energy.

Energetics of syntrophy

Without energy release, how can these reactions support growth of the syntrophic fatty acid-oxidizing bacteria? If growth of the syntrophs occurs when H_2 is removed, the anaerobic oxidation of fatty acids to H_2 must somehow be coupled to ATP production. This implies that the presence or absence of H_2 affects the energetics of the reaction. A brief review of the principles of free energy given in Appendix 1 indicates that the *actual* concentration of reactants and products in a given reaction can drastically change the available free energy of a reaction (ΔG^0 is calculated on the basis of *standard* conditions, whereas ΔG is calculated on the basis of actual concentrations that are present). In a natural situation, reactants and products of fatty acid oxidation, acetate and especially H_2, are consumed by methanogenic Archaea (or sulfate-reducing bacteria); measurements of H_2 in actively methanogenic ecosystems are generally below 10^{-4} atm.

As shown in Table 14.4, if the concentration of H_2 is kept very low by constant removal, the free-energy change associated with the oxidation of fatty acids to acetate plus H_2 by H_2-producing syntrophs becomes *negative* in sign, indicating that energy is released. Although the mechanism of ATP production by syntrophic fatty acid oxidizers is not clear, it probably involves the conversion of acetyl-CoA (generated by β-oxidation of the fatty acid) to acetate (Figure 14.31a). Moreover, electron transport-mediated ATP production is likely under certain circumstances because *Syntrophomonas* can be grown in pure culture on certain unsaturated fatty acids. Crotonate, for example, sup-

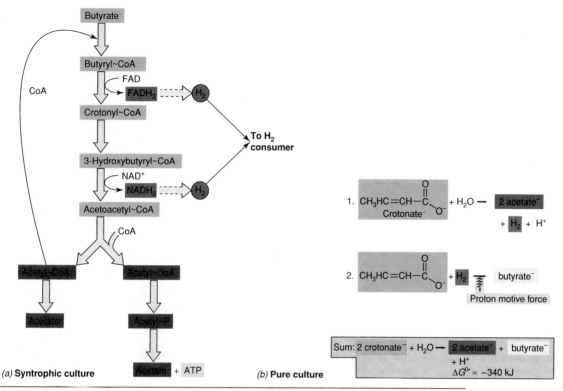

Figure 14.31 Energetics of growth of *Syntrophomonas wolfei* in syntrophic culture (a) and in pure culture (b). In syntrophic culture growth is dependent on the presence of a H_2-consuming organism such as a methanogen. H_2 production probably involves proton motive force-driven reverse electron flow because the E_0' of FAD/FADH$_2$ or NAD$^+$/NADH is more electropositive than that of 2 H$^+$/H$_2$. In pure culture, energy production is linked to crotonate reduction to butyrate.

ports growth of *Syntrophomonas,* with some of the crotonate being oxidized to acetate and some reduced to butyrate (Figure 14.31*b*). It is likely that crotonate reduction by *Syntrophomonas* is coupled to formation of a proton motive force and ATP synthesis as in other types of anaerobic respirations employing organic electron acceptors, such as fumarate reduction to succinate (∞ Section 13.18). However, in the syntrophic situation, *Syntrophomonas* must produce H$_2$ from more electropositive precursors such as FADH$_2$ and NADH$_2$ (Figure 14.31*a*), suggesting that part of the ATP produced by substrate-level phosphorylation in this organism may be needed to drive reverse electron transport reactions (∞ Sections 13.4 and 13.13) to yield H$_2$.

The term **interspecies hydrogen transfer** describes the interdependent series of reactions involved in the anaerobic conversion of complex polymers to methane (∞ also Section 13.20). With the exception of the cellulolytic microflora, all the remaining microbial groups are in some way dependent on each other, with interspecies transfer of hydrogen generally terminating at the methane "sink." In most anoxic ecosystems, the rate-limiting step in methanogenesis from organic compounds is not the terminal step of methane formation but instead the steps involved in the production of acetate and H$_2$ by syntrophs. Growth rates of syntrophic

fatty acid oxidizers are generally very slow. As soon as H$_2$ is formed during their fermentations, it is quickly consumed by a methanogen, a homoacetogen, or a sulfate reducer. Indeed, many syntrophs tend to form flocs, pellets, or other types of aggregates containing one or more H$_2$-consuming organisms such that both organisms—H$_2$ producer and H$_2$ consumer—are in close association for effective H$_2$ transfer.

Methanogenic habitats

Despite the obligate anaerobiosis and specialized metabolism of methanogens, they are quite widespread on Earth. Although high levels of methanogenesis occur only in anoxic environments, such as swamps and marshes, or in the rumen (see Section 14.13), the process also occurs in habitats that normally might be considered oxic, such as forest and grassland soils. In such habitats methanogenesis occurs in anoxic microenvironments, for example, in the midst of soil crumbs (see Figure 14.2). An overview of the rates of methanogenesis in different kinds of habitats is given in Table 14.5. It should be noted that biogenic production of methane by the methanogenic Archaea exceeds the production rate from gas wells and other abiogenic sources. Eructation by ruminants (see Section 14.13) and CH$_4$ released from

TABLE 14.5	Estimates of CH_4 released into the atmosphere[a]	
Source	**CH_4 emission (10^{12} g/year)**	
Biogenic		
Ruminants	80–100	
Termites	25–150	
Paddy fields	70–120	
Natural wetlands	120–200	
Landfills	5–70	
Oceans and lakes	1–20	
Tundra	1–5	
Abiogenic		
Coal mining	10–35	
Natural gas flaring and venting	10–30	
Industrial and pipeline losses	15–45	
Biomass burning	10–40	
Methane hydrates	2–4	
Volcanoes	0.5	
Automobiles	0.5	
Total	349–820	
Total biogenic	302–665	81–86% of total
Total abiogenic	48–155	13–19% of total

[a]Data adapted from estimates of Tyler in Tyler, S. C. 1991. The global methane budget, pp. 7–58, in E. J. Rogers and W. B. Whitman (eds.), *Microbial Production and Consumption of Greenhouse Gases: Methane, Nitrogen Oxides, and Halomethanes*, American Society for Microbiology, Washington, DC.

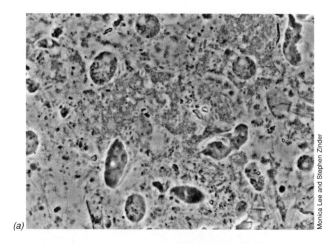

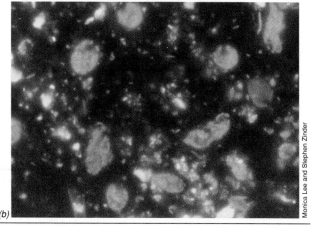

Figure 14.32 Microorganisms from the hindgut of the termite *Zootermopsis angusticolis*. A single microscope field was photographed by two different methods. (a) Phase contrast. (b) Epifluorescence, showing color typical of methanogens due to the high content of the fluorescent coenzyme F_{420}. The methanogens are inside cells of the protozoan *Tricercomitis* sp. Plant particles fluoresce yellow. The average diameter of a protozoan cell is 15–20 μm.

wetlands are the largest sources of biogenic methane. Methanogens are also found in the mammalian intestinal tract, in the guts of wood-eating insects, and in most other anoxic habitats (Table 14.5).

Methanogens have also been found living as endosymbionts of certain protozoa. Several types of protozoa, including free-living aquatic amebas and flagellates found in the insect gut, have been shown to harbor methanogens. In termites, for example, methanogens are present within cells of several small trichomonal protozoa inhabiting the termite hindgut (Figure 14.32). Methanogenic symbionts of protozoa resemble rod-shaped species of the genus *Methanobacterium* or *Methanobrevibacter* (∞ Section 17.2), but their exact relationship to other methanogens is unclear. In the termite hindgut, endosymbiotic methanogens are thought to benefit their protozoan hosts by consuming H_2 generated from glucose fermentation by cellulolytic protozoans. As shown in Table 14.5, termites can be a major source of biogenic CH_4.

Methanogenesis in the oceans

Methanogenesis is more extensive in freshwater and terrestrial environments than in the oceans. The reason for this appears to be that marine waters and sediments contain rather high levels of sulfate, and sulfate-reducing bacteria (see Sections 14.16 and 16.8), which are abundant in marine sediments, effectively compete with the methanogenic population for available acetate and H_2:

$$4 H_2 + SO_4^{2-} \rightarrow H_2S + 2 H_2O + 2 OH^-$$
$$\Delta G^{0\prime} = -155.13 \text{ kJ/reaction}$$

$$CH_3COO^- + SO_4^{2-} \rightarrow 2 HCO_3^- + HS^-$$
$$\Delta G^{0\prime} = -47.6 \text{ kJ/reaction}$$

The biochemical basis for the success of sulfate-reducing bacteria in scavenging H_2 lies in the increased *affinity* sulfate-reducing bacteria have for H_2 as compared to methanogens. When H_2 levels get below about 1 μM, as they often do in sulfate-rich environments, methanogens are no longer able to compete because their H_2 uptake systems function less efficiently at such low H_2 concentrations. Sulfate reducers, on the other hand, can grow at these low pressures of H_2 (assuming that sulfate is present as an electron acceptor), effectively preventing H_2-mediated methanogenesis. Sulfate reduction can also be a significant process in fresh water, but because the sulfate concentration of fresh water is so low, sulfate is rapidly depleted at the surface of anoxic sediments; thus, throughout the bulk of the sediment, methanogenesis is the major process consuming H_2. Acetate utilization is also more efficient in sulfate-reducing bacteria. The affinity for acetate of some sulfate reducers is over 10 times that of methanogens.

Because of consumption of H_2 and acetate by sulfate reducers, the major precursors of methane in marine environments are methylated substrates, such as methylamines and methanol, which are poorly used by sulfate reducers. Trimethylamine, a major excretory product of marine animals, is readily converted to CH_4 by the methanogens *Methanosarcina* and *Methanococcus*. Thus, it is of interest that methanogenesis in marine sediments is not supported by H_2 and acetate, the major methanogenic substrates in other methanogenic ecosystems.

CONCEPT CHECK **14.12**

Under anoxic conditions, organic matter is degraded principally to CH_4 and CO_2. Much CH_4 is formed from the reduction of CO_2 by H_2 supplied by H_2-producing syntrophic bacteria that depend on H_2 consumption to balance their energetics. On a global basis, biogenic CH_4 is a much larger source than abiogenic CH_4.

✔ *What is it about the metabolism of butyrate by* Syntrophomonas wolfei *that makes it dependent on syntrophy?*

✔ *What kinds of organisms can grow in coculture with* Syntrophomonas?

✔ *Why is methanogenesis from H_2 not an abundant process in ocean sediments?*

14.13

Rumen Microbial Ecosystem

Ruminants are herbivorous mammals that possess a special organ, the **rumen,** within which the digestion of cellulose and other plant polysaccharides occurs through the activity of special microbial populations. Some of the most important domestic animals, the cow, sheep, and goat, are ruminants. Because the human food economy depends to a great extent on these animals, rumen microbiology is of considerable economic significance.

Rumen anatomy and action

The bulk of the organic matter in terrestrial plants is present in insoluble polysaccharides, of which *cellulose* is the most important. Mammals, and indeed almost all animals, lack the enzymes necessary to digest cel-

lulose, but all mammals that subsist primarily on grasses and leafy plants can metabolize cellulose by making use of microorganisms as digestive agents. Unique features of the rumen as a site of cellulose digestion are its relatively large size (100–150 liters in a cow, 6 liters in a sheep) and its position in the alimentary tract as the organ where ingested food goes first. The high constant temperature (30°C), constant pH (6.5), and anoxic nature of the rumen are also important factors in overall rumen function. The rumen operates in a more-or-less continuous fashion and in some ways can be considered analogous to a microbial chemostat (∞ Section 5.5).

The relationship of the rumen to other parts of the ruminant digestive system is shown in Figure 14.33a. Food enters the rumen mixed with saliva containing bicarbonate and is churned in a rotary motion during which the microbial fermentation occurs. This peristaltic action grinds the cellulose into a fine suspension, which assists in microbial attachment. The food mass then passes gradually into the reticulum where it is formed into small clumps called cuds, which are regurgitated into the mouth where they are chewed again. The now finely divided solids, well mixed with saliva, are swallowed again, but this time the material passes to the omasum, finally ending in the abomasum, an organ more like a true (acidic) stomach. Here chemical digestive processes begin that continue in the small and large intestine.

Microbial fermentation in the rumen

Food remains in the rumen about 9–12 hr. During this period cellulolytic bacteria and cellulolytic protozoa hydrolyze cellulose to the disaccharide cellobiose and to free glucose units. The released glucose then undergoes a bacterial fermentation with the production of **volatile fatty acids** (VFAs), primarily *acetic, propionic,* and *butyric,* and the gases *carbon dioxide* and *methane* (Figure 14.33b). The fatty acids pass through the rumen wall into the bloodstream and are oxidized by the animal as its main source of energy. In addition to their digestive functions, rumen microorganisms synthesize amino acids and vitamins that are the main source of these essential nutrients for the animal. The rumen contents consist of enormous numbers of microbial cells (10^{10}–10^{11} bacteria/ml rumen fluid) plus partially digested plant materials; these proceed through the gastrointestinal tract of the animal where they undergo further digestive processes similar to those of nonruminants. Many microbial cells formed in the rumen are digested in the gastrointestinal tract and serve as a major source of proteins and vitamins for the animal. Since many of the microorganisms of the rumen are able to grow on urea as a sole nitrogen source, it is often supplied in cattle feed in order to promote microbial protein synthesis. The bulk of this protein ends up in the animal itself. A ruminant is thus

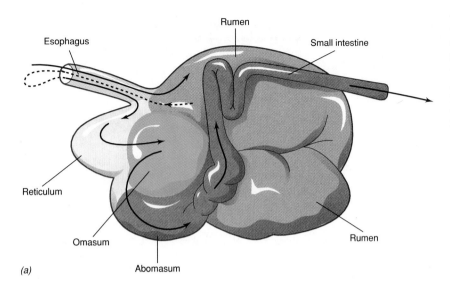

(a)

Figure 14.33 The rumen. (a) Schematic diagram of the rumen and gastrointestinal system of a cow. Food travels from the esophagus to the rumen and is then regurgitated and travels to the reticulum, omasum, abomasum, and intestines, in that order. See text for details. (b) Biochemical reactions in the rumen. The major substrate, glucose, and end products are highlighted; dashed lines indicate minor pathways. Approximate steady state rumen levels of volatile fatty acids (VFAs) are acetate, 60 mM; propionate, 20 mM; butyrate, 10 mM.

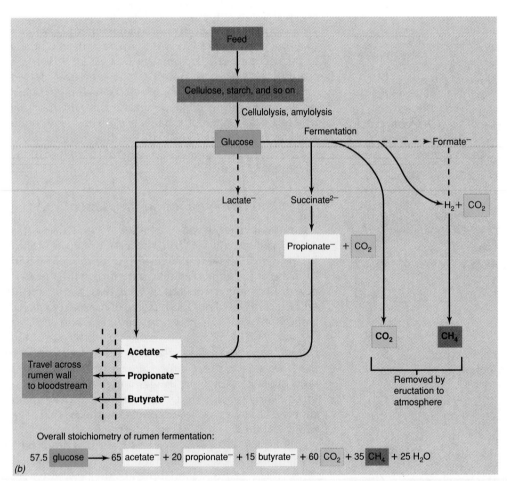

Overall stoichiometry of rumen fermentation:

57.5 glucose $\longrightarrow$ 65 acetate$^-$ + 20 propionate$^-$ + 15 butyrate$^-$ + 60 CO_2 + 35 CH_4 + 25 H_2O

(b)

nutritionally superior to a nonruminant when subsisting on foods that are deficient in protein, such as grasses.

Rumen bacteria

The biochemical reactions occurring in the rumen are complex and involve the combined activities of a variety of microorganisms. Because the reduction potential of the rumen is −0.4 V (the O_2 concentration at this highly reducing potential is 10^{-22} M), anaerobic bacterial naturally dominate. Furthermore, because the conversion of cellulose to CO_2 and CH_4 involves a multistep microbial food chain, a variety of anaerobes can be expected (Table 14.6).

Several different rumen bacteria hydrolyze polymers such as cellulose to sugars and ferment the sugars

TABLE 14.6   Characteristics of some rumen bacteria

Organism	Gram stain	Shape	Motility	Fermentation products	DNA (mol % GC)
Cellulose decomposers					
Fibrobacter succinogenes[a]	Negative	Rod	−	Succinate, acetate, formate	45–51
Butyrivibrio fibrisolvens[a]	Negative	Curved rod	+	Acetate, formate, lactate, butyrate, H_2, CO_2	41
Ruminococcus albus[a]	Positive	Coccus	−	Acetate, formate, H_2, CO_2	43–46
Clostridium lochheadii	Positive	Rod (spores)	+	Acetate, formate, butyrate, H_2, CO_2	—
Starch decomposers					
Bacteroides ruminicola	Negative	Rod	−	Formate, acetate, succinate	40–42
Ruminobacter amylophilus	Negative	Rod	−	Formate, acetate, succinate	49
Selenomonas ruminantium	Negative	Curved rod	+	Acetate, propionate, lactate	49
Succinomonas amylolytica	Negative	Oval	+	Acetate, propionate, succinate	—
Streptococcus bovis	Positive	Coccus	−	Lactate	37–39
Lactate decomposers					
Selenomonas lactilytica	Negative	Curved rod	+	Acetate, succinate	50
Megasphaera elsdenii	Positive	Coccus	−	Acetate, propionate, butyrate, valerate, caproate, H_2, CO_2	54
Pectin decomposer					
Lachnospira multiparus	Positive	Curved rod	+	Acetate, formate, lactate, H_2, CO_2	—
Methanogens					
Methanobrevibacter ruminantium	Positive	Rod	−	CH_4 (from H_2 + CO_2 or formate)	31
Methanomicrobium mobile	Negative	Rod	+	CH_4 (from H_2 + CO_2 or formate)	49

[a]These species also degrade xylan, a major plant cell wall polysaccharide (∞ Table 13.9).

to volatile fatty acids. *Fibrobacter succinogenes* and *Ruminococcus albus* are the two most abundant cellulolytic rumen anaerobes. Although both organisms produce cellulases, *Fibrobacter*, a gram-negative bacterium, employs a periplasmic cellulase (see Section 3.6 for a discussion of the periplasm) to break down cellulose; thus, the organism must remain attached to the cellulose fibril while digesting it. *Ruminococcus*, on the other hand, produces an extremely large (2×10^6 kD) cellulase that is excreted into the rumen contents where it degrades cellulose outside the bacterial cell proper. However, the end result is the same in both cases: free glucose is made available for fermentative anaerobes. If a ruminant is gradually switched from cellulose to a diet high in starch (grain, for instance), then starch-digesting bacteria such as *Ruminobacter amylophilus* and *Succinomonas amylolytica* develop; on a low starch diet these organisms are in a minority. If an animal is fed legume hay, which is high in pectin, then the pectin-digesting bacterium *Lachnospira multiparus* is a common member of the rumen flora.

Some of the fermentation products of the saccharolytic rumen microflora are used as energy sources by other rumen bacteria. Thus, *succinate* is converted to *propionate* and *CO_2* (Figure 14.33*b*), and *lactate* is fermented to *acetic* and other acids by *Selenomonas* and *Megasphaera* (Table 14.6). A number of rumen bacteria produce ethanol as a fermentation product when grown in pure culture, yet ethanol rarely accumulates in the rumen because it can be fermented to acetate + H_2. Hydrogen

produced in the rumen by fermentative processes never accumulates because it is quickly used to reduce CO_2 to CH_4 by methanogens. Another source of H_2 and CO_2 for methanogens is *formate* (Figure 14.33*b*). Large amounts of CH_4 and CO_2 accumulate in the rumen, the average gas composition being about 65% CO_2 and 35% CH_4. These gases leave the ruminant during eructation (belching). Acetate is not converted to CH_4 in the rumen because the retention time is too short for development of acetoclastic methanogens, which typically grow very slowly (∞Section 17.2). In addition, syntrophic fatty acid-degrading bacteria are not abundant in the rumen because of the short retention time and because the ruminant itself is a major sink for fatty acids.

Rumen protozoa and fungi

In addition to prokaryotes, the rumen has a characteristic protozoal fauna (about 10^6/ml), composed almost exclusively of ciliates (see Section 18.4 for a discussion of ciliated protozoa). Many of these protozoa are obligate anaerobes, a property that is rare among eukaryotes. Although protozoa are not essential for the rumen fermentation, they definitely contribute to the overall process. They are able to hydrolyze cellulose and starch and ferment glucose with the production of the same organic acids formed by the bacteria. Rumen protozoa ingest rumen bacteria and are thought to play a role in controlling rumen bacterial densities.

Anaerobic fungi also inhabit the rumen and are known to play a role in ruminal digestive processes. Rumen fungi are generally species that alternate between a flagellated and a thallus form, and studies with pure cultures show that they can ferment cellulose to VFAs. Rumen fungi play an important role in the degradation of other plant polysaccharides as well, including a partial degradation of lignin (the strengthening agent in the cell walls of woody plants), hemicellulose, and pectins.

Dynamics of the rumen ecosystem

A major feature of the rumen ecosystem is its *constancy*. Studies on various ruminant species in different parts of the world show that most animals contain the same major rumen bacterial species, with the proportions of each species varying somewhat with diet. In addition, the nature and proportions of the volatile fatty acids produced and the levels of rumen CO_2 and CH_4 are relatively constant among different ruminant species.

Occasionally, changes in the microbial composition of the rumen cause illness or even death of the animal. For example, if a cow is changed abruptly from forage to a completely grain diet, an explosive growth of *Streptococcus bovis* is observed in the rumen; the normal level of *S. bovis*, about 10^7 cells/ml, quickly expands to over 10^{10} cells/ml. This occurs because *S. bovis* grows rapidly on starch and grain contains high levels of starch, whereas grasses contain mainly cellulose. Being a lactic acid bacterium, *S. bovis* produces large amounts of lactate from the fermentation of starch and this acidifies the rumen (a condition called *acidosis*), killing off the normal rumen flora. Severe acidosis can cause death of the animal. To avoid acidosis, animals are switched from forage rations to grain *gradually* over a period of a few days. A slow introduction of starch selects for volatile fatty acid-producing starch degraders instead of *S. bovis*, and thus normal rumen biochemical processes are not disrupted.

Other animals

The familiar ruminants are cows and sheep. However, goats, camels, buffalo, deer, reindeer, caribou, and elk are also ruminants. There is even some evidence that baleen whales have a rumenlike fermentation. Baleen whales contain a multichambered stomach consisting of a forestomach similar to the rumen. Samples of forestomach material from gray and bowhead whales show abundant volatile fatty acid production in proportions typical of the volatile fatty acids observed in the rumen of cattle or sheep. The diet of baleen whales is primarily chitinous invertebrates, small fish, and kelp. It is thought that *N*-acetylglucosamine, the major

monomeric unit of chitin, is the primary energy source for the forestomach microbial fermentation of baleen whales.

At least one *bird* has been shown to have a foregut fermentation that resembles that of the rumen. The *hoatzin*, a tropical bird, is one of the only obligate folivorous (leaf-eating) birds known. Probably because of its restricted diet—cellulose is its sole carbon source—the hoatzin has evolved a rumenlike forestomach to allow for cellulose digestion. In the hoatzin, a structure called the *crop* functions as the major digestive organ. The pH of the crop is near neutrality, and the organ contains high bacterial numbers and a volatile fatty acid content similar to that found in the rumen. After digestion in the crop, food travels through the esophagus and the proventriculus (an acidic organ analogous to a true stomach) to the small intestine.

Horses and rabbits are also herbivorous mammals, but they are not ruminants. Instead, these animals have only one stomach but use an organ called the **cecum,** a small digestive organ located posterior to the large intestine (just before the anus), as their cellulolytic fermentation vessel. The cecum contains a cellulolytic microflora, and digestion of cellulose occurs there. The precise microflora of the cecum is not well understood, but the species involved are not thought to be the same as those in the rumen. Nutritionally, ruminants have an advantage over horses and rabbits in that the cellulolytic microflora of the ruminant eventually passes through a true (acidic) stomach and as such is a protein source for the animal. In horses and rabbits, the cellulolytic microflora is passed out of the animal in the feces. To recover some of this lost protein, rabbits frequently practice coprophagy (eating of the feces), which releases microbial protein into the stomach.

CONCEPT CHECK   14.13

Ruminants are animals that have a special digestive organ, the rumen, that is a unique ecosystem in which anaerobic microorganisms digest insoluble feed materials such as cellulose and starch. Bacteria, protozoa, and fungi of the rumen produce volatile fatty acids that are used by the ruminant. In addition to their role in the digestive process, rumen microorganisms synthesize vitamins and amino acids used by the ruminant.

✔ *What physical and chemical conditions prevail in the rumen?*

✔ *What are* VFAs *and of what value are they to the ruminant?*

✔ *Why is the metabolism of* Streptococcus bovis *of special concern to ruminant nutrition?*

14.14

Biogeochemical Cycles: Nitrogen

The element nitrogen, N, a key constituent of protoplasm, exists in a number of oxidation states (∞ Table 13.4). We discussed two major processes of microbial nitrogen transformation in Chapter 13: denitrification in Section 13.15 and nitrification in Section 13.12. These and several other nitrogen transformations are summarized in the redox cycle shown in Figure 14.34.

Several of the key redox reactions of nitrogen are carried out in nature almost exclusively by microorganisms, and so microbial involvement in the nitrogen cycle is of great importance. Thermodynamically, nitrogen gas, N_2, is the most stable form of nitrogen, and it is to this form that nitrogen reverts under equilibrium conditions. This explains the fact that a major reservoir for nitrogen on Earth is the atmosphere. This is in contrast to carbon, for which the atmosphere is a relatively minor reservoir (CO_2, CH_4). The high energy necessary to break the $N{\equiv}N$ bond of molecular nitrogen (∞ Section 13.27) means that the reduction of N_2 is an energy-demanding process. Only a relatively small number of organisms are able to utilize N_2, in the process called **nitrogen fixation;** thus, the recycling of nitrogen on Earth involves to a great extent the more easily available forms, ammonia and nitrate. However, because N_2 constitutes by far the greatest reservoir of nitrogen available to living organisms, the ability to utilize N_2 is of great ecological importance. In many environments, productivity is limited by the short supply of combined nitrogen compounds, putting a premium on biological nitrogen fixation.

The global nitrogen cycle is given in Figure 14.35. Transfer of nitrogen in and out of the atmosphere is to a great extent as N_2, with a smaller amount of transfer as nitrous oxides, N_2O and NO, and as gaseous ammonia, NH_3. Transfer between terrestrial and aquatic compartments is primarily as organic nitrogen, ammonium ion, and nitrate ion.

Nitrogen fixation

We examined the biochemistry and microbiology of nitrogen fixation ($N_2 + 8\,H^+ + 8\,e^- \rightarrow 2\,NH_3 + H_2$) in Section 13.27, and we discuss symbiotic nitrogen fixation by legumes in Section 14.23. Nitrogen fixation can also occur chemically in the atmosphere, to a small extent, via lightning discharges, and a certain amount of nitrogen fixation occurs in the industrial production of nitrogen fertilizers (labeled as industrial fixation in Figure 14.35). Some nitrogen fixation also occurs during artificial combustion processes because air contains 78% N_2 by weight and burning in air inevitably involves high temperature combustion of some N_2 (to nitrogen oxides and ultimately to nitrate). However, as can be calculated from the fluxes given in Figure 14.35, about 85% of nitrogen fixation on Earth is of *biological* origin. As can also be calculated from Figure 14.35, about 60% of biological nitrogen fixation occurs on land, and the other 40% in the oceans.

Denitrification

We discussed the role of nitrate as an alternative electron acceptor in Section 13.15. Assimilatory nitrate reduction, in which nitrate is reduced to the oxidation level of ammonia for use as a nitrogen source for growth, and dissimilatory nitrate reduction, in which nitrate is used as an alternative electron acceptor in energy generation, were contrasted in Figure 13.29.

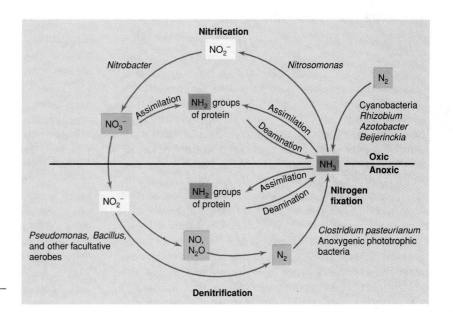

Figure 14.34 Redox cycle for nitrogen.

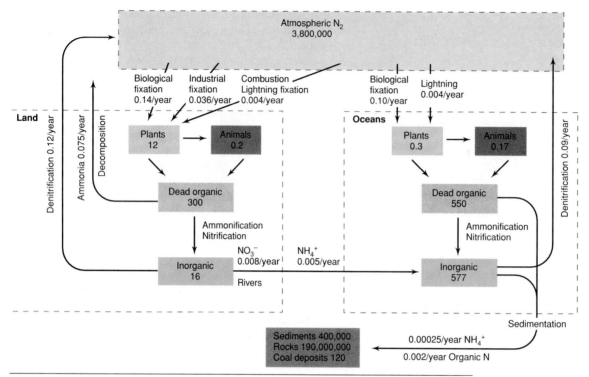

Figure 14.35 The global nitrogen cycle. Minor compartments and fluxes are not given. All numbers should be multiplied by 10^{15} to give grams of nitrogen (global scale).

Under most conditions, the end product of dissimilatory nitrate reduction is N_2 or N_2O, and the conversion of nitrate to gaseous nitrogen compounds is called **denitrification**. This process is the main means by which gaseous N_2 is formed biologically, and because N_2 is much less utilizable by organisms than nitrate as a source of nitrogen, denitrification is a detrimental process because it removes fixed nitrogen from the environment.

Ammonia fluxes and nitrification

Ammonia is produced during the decomposition of organic nitrogen compounds **(ammonification)** and exists at neutral pH as the ammonium ion (NH_4^+). Under anoxic conditions, ammonia is stable (but see Section 13.12), and it is in this form that nitrogen predominates in most anoxic sediments. In soils, much of the ammonia released by aerobic decomposition is rapidly recycled and converted to amino acids in plants. Because ammonia is volatile, some loss can occur from soils (especially highly alkaline soils) by vaporization, and major losses of ammonia to the atmosphere occur in areas of dense animal populations (for example, cattle feedlots). On a global basis, ammonia constitutes only about 15% of the nitrogen released to the atmosphere, the majority of the rest being in the form of N_2 or N_2O (from denitrification).

In oxic environments ammonia can be oxidized to nitrogen oxides and nitrate, but ammonia is a rather stable compound and strong oxidizing agents or catalysts are usually needed for the chemical reaction. However, a specialized group of bacteria, the *nitrifying bacteria*, are biological catalysts, oxidizing ammonia to nitrate in a process called **nitrification** (∞ Section 13.12).

Nitrification is a major *aerobic* process and occurs readily in well-drained soils at neutral pH; it is inhibited by anoxic conditions or in highly acidic soils; however, nitrification can occur anoxically if high levels of nitrate are present (∞ Section 13.12). If materials high in protein, such as manure or sewage, are added to soils, the rate of nitrification is increased. Although nitrate is readily assimilated by plants, it is very water-soluble and is rapidly leached from soils receiving high rainfall. Consequently, nitrification is not beneficial in agricultural practice. Ammonia, on the other hand, is cationic and consequently is strongly adsorbed to negatively charged clay minerals.

Anhydrous ammonia is used extensively as a nitrogen fertilizer, and chemicals are commonly added to the fertilizer to inhibit the nitrification process. One of the most common inhibitors of nitrification is a substituted pyridine compound called *nitrapyrin* (2-chloro-6-trichloromethylpyridine). Nitrapyrin specifically inhibits the first step in nitrification, the oxidation of NH_3 to NO_2^- (∞ Section 13.12), thus effectively inhibiting both steps in the nitrification process. The addition of nitrification inhibitors has served to greatly increase the efficiency of fertilization and helps to prevent pollution of waterways from nitrate leached from fertilized soils.

14.15
Biogeochemical Cycles: Sulfur

Sulfur transformations are even more complex than those of nitrogen because of the variety of oxidation states of sulfur and the fact that some transformations occur at significant rates *chemically* as well as biologically. We discussed the processes of sulfate reduction and chemolithotrophic sulfur oxidation in Sections 13.10 and 13.16. The redox cycle for sulfur and the involvement of microorganisms in sulfur transformations are given in Figure 14.36. Although a number of oxidation states are possible, only three form significant amounts of sulfur in nature, −2 (sulfhydryl, R—SH, and sulfide, HS^-), 0 (elemental sulfur, S^0), and +6 (sulfate, SO_4^{2-}). The bulk of the sulfur of the earth is found in sediments and rocks in the form of sulfate minerals (primarily gypsum, $CaSO_4$) and sulfide minerals (primarily pyrite, FeS_2), although the oceans constitute the most significant reservoir of sulfur for the biosphere (in the form of inorganic sulfate). The global transport cycle

for sulfur is given in Figure 14.37, and some of the components of this cycle are discussed later.

Hydrogen sulfide and sulfate reduction

A major volatile sulfur gas is *hydrogen sulfide.* As we saw in Section 13.16, this substance is formed primarily by the bacterial reduction of sulfate:

$$SO_4^{2-} + 8\,e^- + 8\,H^+ \rightleftarrows H_2S + 2\,H_2O + 2\,OH^-$$

The form in which sulfide is present in an environment depends on pH due to the following equilibria:

$$H_2S \underset{\text{Low pH}}{\rightleftarrows} HS^- \underset{\text{Neutral pH}}{\rightleftarrows} S^{2-}_{\text{High pH}}$$

At high pH, the dominant form is sulfide, S^{2-}. At neutral pH, HS^- predominates, and below pH 6, H_2S, a gaseous product, is the major species. HS^- and S^{2-} are very water-soluble, but H_2S is not and readily volatilizes. Even at neutral pH, some volatilization of H_2S from HS^- can occur because there is an equilibrium between HS^- and H_2S, and as volatilization occurs, the reaction is pulled toward H_2S.

A wide variety of organisms can use sulfate as a sulfur source and carry out *assimilative* sulfate reduction, converting the HS^- formed to organic sulfur, R—SH (∞Figure 13.31*b*). HS^- is ultimately reformed from the decomposition of this organic sulfur by putrefaction and desulfurylation (Figure 14.36), and this is a significant source of HS^- in fresh water. In the marine environment, because of the vast amount of sulfate present, dissimilatory sulfate reduction is the main source of HS^-. Dissimilatory sulfate reduction, in which sulfate is an electron acceptor, is carried out by a variety of bacteria, collectively referred to as the *sulfate-reducing bacteria* (∞ Sections 13.16 and 16.8), which are all obligate anaerobes. It should be empha-

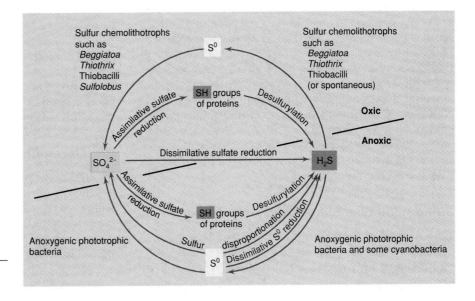

Figure 14.36 Redox cycle for sulfur.

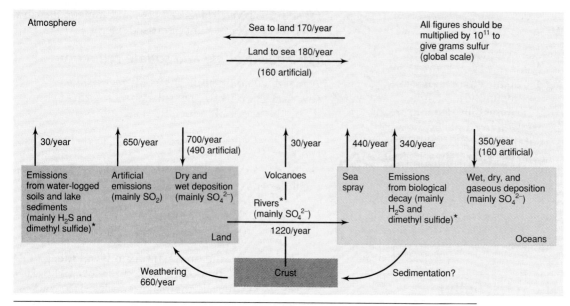

Figure 14.37 The global sulfur cycle. Artificial emissions are derived from human activities. An asterisk indicates a process that is partially or solely due to microbial action.

sized that the sulfate anion is very stable chemically, and its reduction does not occur spontaneously in nature under normal environmental conditions.

As discussed in Section 14.12, there is a competition in nature between methanogenic and sulfate-reducing bacteria for available electron donors, especially H_2 and acetate, and as long as sulfate is present, the sulfate-reducing bacteria are favored. Because of the necessity of organic electron donors (or molecular H_2, which is itself derived from the fermentation of organic compounds), sulfate reduction occurs extensively only where significant amounts of organic matter are present. In many marine sediments, the rate of sulfate reduction is carbon-limited, and the rate can be greatly increased by addition of organic matter. This is of considerable importance for marine pollution because disposal of sewage, sewage sludge, and garbage in the sea can lead to marked increases in organic matter in the sediments. Since HS^- is a toxic substance to many organisms, formation of HS^- by sulfate reduction is potentially detrimental. Sulfide is toxic because it combines with the iron of cytochromes and other essential iron-containing compounds in the cell. One common detoxification mechanism for sulfide in the environment is combination with iron, leading to formation of the insoluble FeS. The black color of many sediments where sulfate reduction is taking place is due to the accumulation of FeS.

Sulfide and elemental sulfur oxidation

Under oxic conditions, sulfide (HS^-) rapidly oxidizes spontaneously at neutral pH (∞ Section 13.10). Sulfur-oxidizing bacteria are also able to catalyze the

oxidation of sulfide, but because of the rapid spontaneous reaction, bacterial oxidation of sulfide occurs only in areas in which H_2S rising from anoxic areas meets O_2 descending from oxic areas. If light is available, anoxic oxidation of HS^- can also occur, catalyzed by the phototrophic sulfur bacteria (∞ Sections 13.4 and 16.1), but this occurs only in restricted areas, usually in lakes, where sufficient light can penetrate to anoxic zones.

Elemental sulfur, S^0, is chemically stable in most environments in the presence of oxygen but is readily oxidized by sulfur-oxidizing bacteria. Although a number of sulfur-oxidizing bacteria are known, members of the genus *Thiobacillus* (∞ Section 13.10) are most commonly involved in elemental sulfur oxidation. Elemental sulfur is very insoluble, and the bacteria that oxidize it attach firmly to the sulfur crystals (∞ Figure 13.20). Oxidation of elemental sulfur results in the formation of sulfate and hydrogen ions, and sulfur oxidation characteristically results in a *lowering* of the pH. Elemental sulfur is sometimes added to alkaline soils to effect a lowering of the pH, reliance being placed on the ubiquitous thiobacilli to carry out the acidification process.

Organic sulfur compounds

In addition to the *inorganic* forms of sulfur whose biogeochemistry was just discussed, a vast array of *organic* sulfur compounds are also synthesized by living organisms, and these enter into biogeochemical sulfur cycling as well. The most abundant organic sulfur compound in nature is dimethyl sulfide ($H_3C-S-CH_3$). It is produced primarily in marine environments as a degradation product of dimethylsulfonium propionate, a major osmoregulatory solute in marine algae. Dimethylsulfonium propionate can be used as a carbon and energy source by microorganisms and is catabolized to dimethyl sulfide and acrylate; the latter compound, a derivative of the fatty acid propionate, is used to support growth.

Microbial production of dimethyl sulfide in nature is dramatic, about 45 million tons being produced annually. Dimethyl sulfide released to the atmosphere undergoes photochemical oxidation to methane sulfonic acid ($CH_3SO_3^-$), SO_2, and SO_4^{2-}, but dimethyl sulfide produced in anoxic habitats can be used microbiologically as a substrate for methanogenesis (yielding CH_4 and H_2S), as an electron donor for photosynthetic CO_2 fixation in phototrophic purple bacteria [yielding dimethyl sulfoxide (DMSO)], and as an electron donor in energy metabolism in certain chemoorganotrophs and chemolithotrophs (also yielding DMSO). Anaerobically, DMSO can be an electron acceptor for anaerobic respiration (∞ Section 13.18), once again yielding dimethyl sulfide. Many other organic sulfur compounds impact on the global sulfur cycle, including methanethiol (CH_3SH), dimethyl disulfide ($H_3C—S—S—CH_3$), and carbon disulfide (CS_2), but on a global basis, dimethyl sulfide production and consumption are quantitatively the most significant.

CONCEPT CHECK   14.15

Bacteria play major roles in both the oxidative and reductive sides of the sulfur cycle. Sulfur- and sulfide-oxidizing bacteria, which are often chemolithotrophs, produce sulfate, and sulfate-reducing bacteria use sulfate as electron acceptor in anaerobic respiration and produce hydrogen sulfide. Because sulfide is toxic and also reacts with various metals, sulfate reduction is an important biogeochemical process. Dimethyl sulfide is the major organic sulfur compound of ecological significance in nature.

✔ *How many electrons are required to reduce SO_4^{2-} to H_2S?*

✔ *Why is acid generated from the bacterial oxidation of sulfur?*

✔ *What organic sulfur compound is most abundant in nature?*

14.16
Biogeochemical Cycles: Iron

Iron is one of the most abundant elements in the earth's crust but is a relatively minor component in aquatic systems because of its relative insolubility in water.

Iron exists in two oxidation states, ferrous (+II) and ferric (+III). We discussed iron oxidation and reduction in Sections 13.11 and 13.18, respectively. The form in which iron is found in nature is greatly influenced by pH and oxygen. Because of the high reduction potential of the Fe^{3+}/Fe^{2+} couple, +0.77 V, the only electron acceptor able to oxidize ferrous iron is oxygen, O_2. At neutral pH, ferrous iron oxidizes spontaneously in air to ferric iron, which forms highly insoluble precipitates of ferric hydroxide and ferric oxides. Thus, at neutral pH, the only way that iron is maintained in solution is by chelation with organic materials.

Bacterial iron reduction and oxidation

The bacterial reduction of ferric iron to the ferrous state is a major means by which iron is solubilized in nature. As we noted in Section 13.18, a number of organisms can use ferric iron as an electron acceptor. In addition to the bacterially catalyzed reduction, if hydrogen sulfide is present, as it is in many anoxic environments (see Section 14.15), ferric iron is also reduced chemically to FeS (ferrous sulfide). Thus, there are complex interactions in many environments between the iron and sulfur cycles.

Ferric iron reduction is very common in waterlogged soils, bogs, and anoxic lake sediments. Movement of iron-rich groundwater from anoxic bogs or waterlogged soils can result in the transport of considerable amounts of ferrous iron. Once this iron-laden water reaches oxic regions, the ferrous iron is quickly oxidized spontaneously and ferric compounds precipitate, leading to the formation of a brown deposit (∞ Figure 13.22c). The overall reaction of ferrous iron oxidation is as follows:

$$Fe^{2+} + \tfrac{1}{4} O_2 + H^+ \rightarrow Fe^{3+} + \tfrac{1}{2} H_2O$$

$$Fe^{3+} + 3\ H_2O \rightarrow \underset{\text{precipitate}}{Fe(OH)_3} + 3H^+$$

Sum:

$$Fe^{2+} + \tfrac{1}{4} O_2 + 2\tfrac{1}{2} H_2O \rightarrow Fe(OH)_3 + 2\ H^+$$

Note that although the initial oxidation of ferrous iron consumes hydrogen ions and thus leads to a rise in pH, the hydrolysis of Fe^{3+} and formation of $Fe(OH)_3$ consumes hydroxyl ions (and thus produces H^+) and leads to acidification of the medium. This is one way in which iron oxidation leads to the formation of acidic conditions in the environment.

Although ferric iron forms very insoluble hydroxides, some ferric iron can be kept in solution in natural waters by forming complexes with organic materials. If an organism is present that can oxidize the organic compound, then the iron present will precipitate. This is probably a major mechanism of iron precipitation in

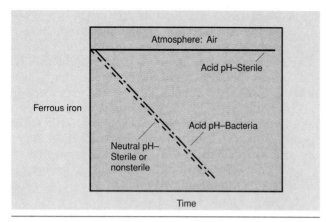

Figure 14.38 Oxidation of ferrous iron as a function of pH and the presence of the bacterium *Thiobacillus ferrooxidans*.

many neutral pH environments. In addition, at neutral pH, organisms such as *Gallionella* (∞ Figure 16.44) and *Leptothrix* (∞ Figure 16.66) contribute to the oxidation of ferrous iron, presumably in energy-yielding reactions. Also at neutral pH in *anoxic* environments receiving light, anoxygenic phototrophic bacteria can oxidize Fe^{2+} to Fe^{3+}; this process was discussed in Section 13.11.

Ferrous iron oxidation at acid pH

At low pH, chemical oxidation of ferrous iron to the ferric state is slow (∞ Section 13.11). However, the acidophilic chemolithotroph *Thiobacillus ferrooxidans* is able to catalyze the oxidation (Figure 14.38); *T. ferrooxidans* oxidizes ferrous iron as its primary energy-gen-

erating process. Because very little energy is generated in the oxidation of ferrous to ferric iron (∞ Section 13.11), these bacteria must oxidize large amounts of iron in order to grow, and consequently even a small number of cells can be responsible for precipitating a large amount of iron. This iron-oxidizing bacterium, which is a strict acidophile, is very common in acid mine drainages and in acid springs and is probably responsible for most of the ferric iron precipitated at acid pH values.

Thiobacillus ferrooxidans lives in environments in which sulfuric acid is the dominant acid and large amounts of sulfate are present. Under these conditions, ferric iron does not precipitate as the hydroxide but as a complex sulfate mineral called *jarosite* [usually found as $HFe_3(SO_4)_2(OH)_6$]. Jarosite is a yellowish or brownish precipitate and is responsible for one of the manifestations of acid mine drainage, an unsightly yellow stain called "yellow boy" by U.S. miners (∞ Figures 13.22*a* and 14.41).

Pyrite oxidation

One of the most common forms of iron and sulfur in nature is *pyrite*, which has the overall formula FeS_2. Pyrite is formed from the reaction of sulfur with ferrous sulfide (FeS) to form a highly insoluble crystalline structure, and pyrite is very common in bituminous coals and in many ore bodies (Figure 14.39). The bacterial oxidation of pyrite is of great significance in the development of acidic conditions in mines and mine drainages. Additionally, oxidation of pyrite by bacteria is of considerable importance in the process called

(a)   *(b)*   *(c)*

Figure 14.39 Pyrite-rich microbial habitats in bituminous coal- and copper-mining environments. (a) Bituminous coal-mining operation in a strip mine. The shovel is removing the soil (overburden) to reach the coal seam. (b) The coal seam. Removal of the bituminous coal exposes the environment to air. The pyrite associated with the coal is colonized with iron-oxidizing bacteria. (c) Copper ore deposit rich in pyrite. Mining of the copper ore results in exposure of the formation to air.

microbial leaching of ores (see later). The oxidation of pyrite is a combination of chemically and bacterially catalyzed reactions. Two electron acceptors for this process can function: molecular oxygen (O_2) and ferric ions (Fe^{3+}). However, ferric ions are present only when the solution is acidic, at pH values below about 2.5. At pH values above 2.5, ferric ion reacts with water to form the insoluble ferric hydroxide. When pyrite is first exposed, as in a mining operation, a slow chemical reaction with molecular oxygen occurs as shown in the following reaction:

$$FeS_2 + 3\tfrac{1}{2}\,O_2 + H_2O \rightarrow Fe^{2+} + 2\,SO_4^{2-} + 2\,H^+$$

This reaction, called the *initiator reaction*, leads to the development of acidic conditions under which the ferrous iron formed is relatively stable in the presence of oxygen. However, *Thiobacillus ferrooxidans* catalyzes the oxidation of ferrous to ferric ions. The ferric ions formed under these acidic conditions, being soluble, can readily react spontaneously with more pyrite to oxidize the pyrite to ferrous ions plus sulfate ions:

$$FeS_2 + 14\,Fe^{3+} + 8\,H_2O \rightarrow 15\,Fe^{2+} + 2\,SO_4^{2-} + 16\,H^+$$

The ferrous ions formed are again oxidized to ferric ions by the bacteria, and these ferric ions again react with more pyrite. Thus, there is a progressive, rapidly increasing rate at which pyrite is oxidized, called the *propagation cycle*, as illustrated in Figure 14.40. Under natural conditions some of the ferrous iron generated by the bacteria leaches away, being carried by groundwater into surrounding streams. However, because oxygen is present in the aerated drainage, bacterial oxidation of the ferrous iron takes place in these outflows and an insoluble ferric precipitate is formed.

Acid mine drainage

Bacterial oxidation of sulfide minerals is the major factor in the formation of **acid mine drainage,** a common environmental problem in coal-mining regions (Figure 14.41; ∞ also Figure 13.22a). Acid mine drainage occurs because of the attack by *Thiobacillus ferrooxidans* on pyrite, following the steps outlined above.

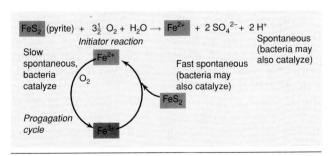

Figure 14.40 Role of iron-oxidizing bacteria in oxidation of the mineral pyrite.

T. D. Brock

Figure 14.41 Acid mine drainage from a bituminous coal region. Note the yellowish-red color due to precipitated iron oxides. See also Figure 13.22.

Not all coal seams contain iron sulfide; thus, acid mine drainage does not occur in all coal-mining regions. Where acid mine drainage does occur, however, it is often a very serious problem. Mixing of acidic mine waters with natural waters in rivers and lakes causes serious degradation in the quality of the natural water because both the acid and the dissolved metals are toxic to aquatic life (∞ Figures 13.22a and 14.41). In addition, such polluted waters are unsuitable for human consumption and industrial use.

We outlined in Figure 14.40 the steps in the bacterial oxidation of pyrite, and we noted that certain of the steps occur chemically but that the rate-limiting step, the oxidation of ferrous to ferric iron, occurs at acid pH only in the presence of the bacterium. The breakdown of pyrite leads ultimately to the formation of sulfuric acid and ferrous iron, and pH values can be as low as pH 2. The acid formed attacks other minerals in the rock associated with the coal and pyrite, causing breakdown of the whole rock fabric. A major rock-forming element, aluminum, is soluble only at low pH, and often several grams of Al^{3+}, which can be toxic to aquatic organisms, are present per liter.

The requirement for O_2 in the oxidation of ferrous to ferric iron helps to explain how acid mine drainage develops. As long as the coal is unmined, oxidation of pyrite cannot occur because neither air nor the bacteria can reach it. When the coal seam is exposed, it quickly becomes contaminated with *Thiobacillus ferrooxidans*, and O_2 is introduced, making oxidation of pyrite possible. The acid formed can then leach into the surrounding streams (Figure 14.41).

CONCEPT CHECK   14.16

Iron exists in nature primarily in two oxidation states, ferrous (Fe^{2+}) and ferric (Fe^{3+}), and bacterial transformation of these cations is of great geological and ecological importance. Bacterial ferric iron reduction occurs in anoxic environments and results in the mobilization of iron from swamps, bogs, and other iron-rich aquatic habitats. Bacterial oxidation of ferrous iron occurs significantly only at low pH and is very common in coal-mining regions, where it results in a type of pollution called acid mine drainage.

✔ *What form (ferrous or ferric) is iron in the mineral Fe(OH)$_3$? FeS? How is Fe(OH)$_3$ formed?*

✔ *Why does Fe^{2+} oxidation under oxic conditions occur mainly at acidic pH?*

14.17

Microbial Leaching

We consider here a situation in which acid production and metal solubility by acidophilic bacteria play a beneficial role in mining. Sulfide forms highly insoluble minerals with many metals, and many ores used as sources of these metals are sulfides. If the concentration of metal in the ore is low, it may not be economically feasible to concentrate the mineral by conventional chemical means. Under these conditions, **microbial leaching** is frequently practiced. Microbial leaching is especially useful for *copper* ores because copper sulfate, formed during oxidation of the copper sulfide ores, is very water-soluble. We have noted that sulfide itself, HS^-, oxidizes spontaneously in air. Most metal sulfides also oxidize spontaneously, but the rate is very much slower than that of free sulfide. Bacteria such as *Thiobacillus ferrooxidans* are able to catalyze a much faster rate of oxidation of the sulfide minerals, thus aiding in solubilization of the metal. The relative rate of oxidation of a copper mineral in the presence and absence of bacteria is illustrated in Figure 14.42. The susceptibility to oxidation also varies among minerals, and those minerals that are most readily oxidized are most amenable to microbial leaching. Thus, iron and copper sulfide ores such as pyrrhotite (FeS) and covellite (CuS) are readily leached, whereas lead and molybdenum ores are much less so.

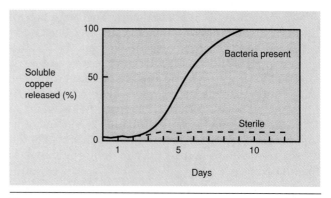

Figure 14.42 Effect of the bacterium *Thiobacillus ferrooxidans* on the leaching of copper from the mineral covellite. The leaching was done in a laboratory column, and the acid leach solution contained inorganic nutrients necessary for development of the bacterium. The leaching activity was monitored by assaying for soluble copper in the leach solution at the bottom of the column. The leach solution was continuously recirculated, maintaining an essentially closed system.

Leaching process

In the microbial leaching process, low grade ore is dumped in a large pile (the leach dump) and a dilute sulfuric acid solution (pH about 2) is percolated down through the pile (Figure 14.43*a*). The liquid coming out of the bottom of the pile (Figure 14.43*b*), rich in the mineral, is collected and transported to a precipitation plant (Figure 14.43*c*) where the metal is reprecipitated and purified. The liquid is then pumped back to the top of the pile and the cycle is repeated. As needed, more acid is added to maintain the low pH.

There are several mechanisms by which the bacteria can catalyze oxidation of the sulfide minerals. To illustrate this, examples will be used of the oxidation of two copper minerals, chalcocite, Cu_2S, in which copper has a valence of $+1$, and covellite, CuS, in which copper has a valence of $+2$. As illustrated in Figure 14.44, *Thiobacillus ferrooxidans* is able to oxidize Cu^+ in chalcocite (Cu_2S) to Cu^{2+}, thus removing some of the copper in the soluble form, Cu^{2+}, and forming the mineral covellite. Note that in this reaction there is no change in the valence of sulfide, the bacteria utilizing the reaction Cu^+ to Cu^{2+} as a source of energy. This is analogous to the oxidation by the same bacterium of Fe^{2+} to Fe^{3+}. Covellite can then be oxidized releasing sulfate and soluble Cu^{2+} (Figure 14.44).

A second mechanism, and probably the most important in most mining operations, involves an indirect oxidation of the copper ore with *ferric* ions formed by the bacterial oxidation of ferrous ions (Figure 14.44). In almost any ore, pyrite is present, and the oxidation of this pyrite (see Section 14.16) leads to the formation of ferric iron. Ferric iron is a very good oxidant for sulfide minerals, and reaction of CuS with ferric iron results in solubilization of the copper and the formation of fer-

(a)

(b)

(c)

Figure 14.43 The leaching of low-grade copper ores using bacteria. (a) A typical leaching dump. The low-grade ore has been dumped in a large pile. Pipes distribute the acidic leach water over the surface of the pile. The acidic water slowly percolates through the pile and exits at the bottom. (b) Effluent from a copper leaching dump. The acidic water is very rich in dissolved copper. (c) Recovery of dissolved copper by passage of the copper-rich water over metallic iron in a long flume. (d) A small pile of recovered copper metal removed from the flume, ready for further purification.

(d)   Sprinkling of acid leach liquor on copper ore: Fe^{3+} and H_2SO_4

rous iron. In the presence of O_2, at the acid pH values involved, *Thiobacillus ferrooxidans* reoxidizes the ferrous iron back to the ferric form so it can oxidize more copper sulfide. Thus, the process is kept going indirectly by the oxidation of Fe^{2+} to Fe^{3+} by the bacterium.

Another source of iron in leaching operations is at the precipitation plant used in recovery of the soluble copper from the leaching solution (Figure 14.43c and d). Scrap iron, Fe^0, is used to recover copper from the leach liquid by the reaction shown in the lower part of Figure 14.44, and this results in the formation of considerable Fe^{2+}. In most leaching operations, the Fe^{2+}-rich liquid remaining after the copper is removed is conducted to an oxidation pond, where *Thiobacillus ferrooxidans* proliferates and forms Fe^{3+}. Acid is added to the pond to keep the pH low, thus keeping the Fe^{3+} in solution, and this ferric-rich liquid is then pumped to the top of the pile and the Fe^{3+} is available to oxidize more sulfide mineral.

Because of the huge dimensions of copper leach dumps, penetration of oxygen from air is poor, and the interior of these piles usually becomes anoxic. Although most of the reactions written in Figure 14.44 require molecular O_2, because *Thiobacillus ferrooxidans* can use Fe^{3+} as an electron *acceptor* in the absence of O_2, the oxidation reactions can also proceed anaerobically; the large amounts of Fe^{3+} added to the leach solution from scrap oxidized iron drive the process forward, even under anoxic conditions.

CONCEPT CHECK   **14.17**

Oxidation of copper ores by bacteria can lead to the solubilization of copper, a process called microbial leaching. Leaching is important in the recovery of copper and uranium from low grade ores. Bacterial oxidation of iron in the iron sulfide mineral pyrite is also an important part of the microbial leaching process because the ferric iron produced is itself an oxidant of ores.

✔ *How is CuS oxidized under* anoxic *conditions?*

✔ *Why is it important to keep the leach liquor acidic in the copper ore leaching process?*

14.18

Biogeochemical Cycles: Trace Metals and Mercury

Trace elements are elements that are present in low concentrations in rocks, soils, waters, and the atmosphere. Some trace elements (for example, cobalt, copper, zinc, nickel, molybdenum) are nutrients (∞ Section 4.2), but a number of trace elements in high concentrations are actually toxic to organisms. Of these toxic elements,

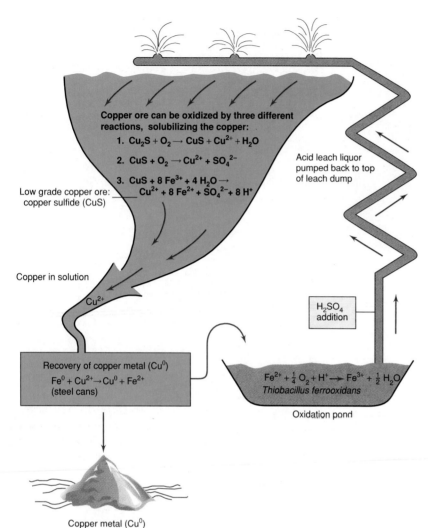

Figure 14.44 Arrangement of a leaching pile and reactions involved in the microbial leaching of copper sulfide minerals to yield Cu^0 (copper metal). Reaction 1 is primarily bacterial while Reaction 2 occurs both biologically and chemically. Reaction 3 is strictly chemical, but is probably the most important reaction in copper-leaching processes.

several are sufficiently volatile that they exhibit significant atmospheric transport, and hence are of some environmental concern. These include mercury, lead, arsenic, cadmium, and selenium. Many of these trace elements undergo redox reactions catalyzed by microorganisms, and several are also converted to organic form via microbial action. Because of environmental concern and significant microbial involvement, we focus our discussion on the biogeochemistry of the element mercury.

Global cycling of mercury and methylmercury

Although mercury is present in extremely low concentrations in most natural environments, averaging about 1 nanogram (ng)/liter, it is a widely used industrial product and is the active component of many pesticides that have been introduced into the environment. Because of its unusual ability to be concentrated in living tissues and its high toxicity, mercury is of considerable environmental importance. The mining of

mercury ores and the burning of fossil fuels release about 40,000 *tons* of mercury into the environment each year; an even greater amount is released by geochemical processes. Other anthropogenic sources of mercury include the electronics industry, especially battery and wiring production, the chemical industry, and the burning of municipal wastes.

The major form of mercury in the atmosphere is elemental mercury (Hg^0), which is volatile and is oxidized to mercuric ion (Hg^{2+}) photochemically; most of the mercury entering aquatic environments is thus Hg^{2+} (Figure 14.45). Mercuric ion readily adsorbs to particulate matter and can be metabolized from there by microorganisms. The major microbial reaction observed is the *methylation* of mercury, yielding methylmercury, CH_3Hg^+ (Figure 14.45). Methylmercury is soluble and can be concentrated in the aquatic food chain, primarily in fish, or further methylated by microorganisms to yield the volatile compound called dimethylmercury, $CH_3—Hg—CH_3$. Metabolically, methylation of mercury occurs by donation of methyl groups from $CH_3—B_{12}$.

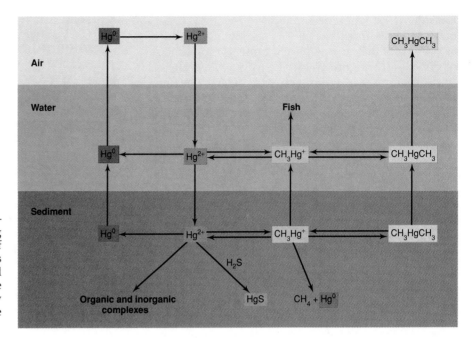

Figure 14.45 Biogeochemical cycling of mercury. The major reservoirs of mercury are in water and in sediments where it can be concentrated in animal tissues or precipitated out as HgS. The various forms of mercury commonly found in aquatic environments are each shown in a different color.

Both methylmercury and dimethylmercury bond to proteins and tend to accumulate in animal tissues, especially muscle. Methylmercury is about 100 times more toxic than Hg^0 or Hg^{2+} and can be concentrated over 1 million times in fish, where it is a potent neurotoxin, eventually causing death. Methylmercury is thus a major environmental toxin, and its accumulation seems to be a particular problem in freshwater lakes where enhanced levels of methylmercury have been observed in fish caught for human consumption. Mercury can also cause liver and kidney damage in humans and other animals.

Several other mercury transformations occur on a global scale, including reactions involving sulfate-reducing bacteria ($H_2S + Hg^{2+} \rightarrow HgS$) and methanogens ($CH_3Hg^+ \rightarrow CH_4 + Hg^0$) (see Figure 14.45). The solubility of HgS is very low, so in anoxic sulfate-reducing sediments, most mercury is found as HgS. But on aeration, oxidation of HgS can occur, primarily by thiobacilli, leading to the formation of Hg^{2+} and eventually methylmercury.

Mercuric resistance

At sufficiently high concentrations, Hg^{2+} and CH_3Hg^+ can be toxic not only to higher organisms but also to microorganisms. Several methods of detoxifying toxic mercury species exist. An NADPH-linked enzyme called *mercuric reductase* transfers two electrons to Hg^{2+}, reducing it to Hg^0. The Hg^0 produced in this reaction is volatile but is essentially nontoxic to humans and microorganisms, compared with Hg^{2+}. Bacterial conversion of Hg^{2+} to Hg^0 then allows more CH_3Hg^+ to be converted to Hg^{2+}.

Mercury resistance has been intensively studied in the gram-negative bacterium *Pseudomonas aeruginosa*, where genes for mercury resistance reside on a plasmid. These genes, called *mer* genes, are arranged in an operon and are under control of the regulatory protein MerR (the product of *merR*) (Figure 14.46a). Interestingly, MerR

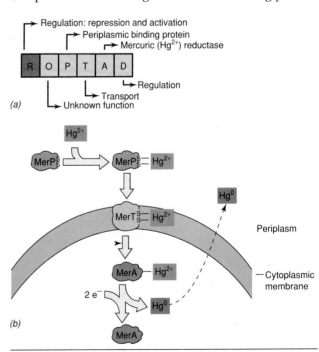

Figure 14.46 Mechanism of Hg^{2+} reduction to Hg^0 in *Pseudomonas aeruginosa*. (a) The *mer* operon. MerR can function as either a repressor (absence of Hg^{2+}) or transcriptional activator (presence of Hg^{2+}). (b) Transport and reduction of Hg^{2+}. Hg^{2+} is bound by cysteine residues in both MerP and MerT.

functions as both a repressor and an activator (∞ Section 7.4). In the absence of Hg^{2+}, MerR binds to the operator region of the *mer* operon and presents transcription of *mer TPCAD* (Figure 14.46a). However, when Hg^{2+} is present, it forms a complex with MerR, which then functions as an *activator* of transcription of the *mer* operon. The mercuric reductase, mentioned previously, is the product of the *merA* gene. MerD, the product of *merD*, also plays a regulatory role, whereas *merP* encodes a periplasmic Hg^{2+} binding protein (Figure 14.46b). This protein, MerP, binds Hg^{2+} and transfers it to a membrane protein MerT (the product of *merT*), which transports Hg^{2+} into the cell for reduction by mercuric reductase (Figure 14.46b). The final result is reduction of Hg^{2+} to Hg^0, which is volatile and is released from the cell (Figure 14.46b).

Resistance to other heavy metals

A variety of plasmids (∞ Section 9.8) isolated from both gram-positive and gram-negative Bacteria have been found to encode resistance to the effects of heavy metals. Certain antibiotic resistance plasmids also have genes for resistance to mercury and arsenic. Other plasmids encode only heavy-metal resistances. A large plasmid isolated from *Staphylococcus aureus* has been found to encode resistance to mercury, cadmium, arsenate, and arscnite. The mechanism of resistance to any specific metal varies. For example, arsenate and cadmium resistances are due to the action of enzymes that immediately pump out any arsenate or cadmium ions incorporated, thus preventing the metals from denaturing proteins.

Studies on nickel- and cobalt-resistant bacteria have shown that in most cases the resistance genes are plasmid-borne; resistance to both metals on a single plasmid is typical. Enrichment culture studies have shown that nickel-resistant bacteria are uncommon in soils and other environments in which this metal is absent in significant amounts. Bacteria highly resistant to nickel or other metals are most common in wastewaters of the metal processing industry or in mining operations where heavy metals are leached out along with iron or copper ores.

CONCEPT CHECK   **14.18**

Trace metals such as mercury and arsenic are often mobilized by bacteria capable of oxidizing them. A major toxic form of mercury is methylmercury. The ability of bacteria to resist the toxicity of heavy metals is often due to the presence of specific plasmids that encode enzymes capable of detoxifying the metals.

✔ *What form of mercury is most toxic to organisms?*

✔ *How is mercury detoxified by bacteria?*

---

for Mon

14.19
Petroleum Biodegradation

Microbial decomposition of petroleum and petroleum products is of considerable economic and environmental importance. Because petroleum is a rich source of organic matter and the hydrocarbons within it are readily attacked aerobically by a variety of microorganisms, it is not surprising that when petroleum is brought into contact with air and moisture, it is subject to microbial attack. Under some circumstances such as in bulk storage tanks, microbial growth is not desirable. However, in other situations, such as in oil spills, microbial utilization of oil is desirable and may even be promoted by the addition of inorganic nutrients. The term *bioremediation* refers to the cleanup of oil or other pollutants by microorganisms, and in recent years the importance of bioremediation in oil spills has been amply demonstrated in several major crude oil spills in the marine environment.

Hydrocarbon decomposition

Hydrocarbon-oxidizing bacteria and fungi are the main agents responsible for decomposition of oil and oil products. A wide variety of bacteria, several molds and yeasts, and certain cyanobacteria and green algae have been shown to be able to oxidize hydrocarbons. Bacteria and yeasts, however, appear to be the prevalent hydrocarbon degraders in aquatic ecosystems. Small-scale oil pollution of aquatic and terrestrial ecosystems from human as well as natural activities is very common, and hence it is not surprising that a diverse microbial community exists capable of using hydrocarbons as an electron donor. Methane, the simplest hydrocarbon, is degraded by only a specialized group of bacteria, the *methanotrophic* bacteria (∞ Section 16.7), and these organisms do not oxidize higher hydrocarbons.

Hydrocarbon-oxidizing microorganisms develop rapidly on oil films and slicks. However, as we saw in Section 13.25, aliphatic hydrocarbons are not fermentable. Thus, significant aliphatic hydrocarbon oxidation occurs only in the presence of O_2; if the oil is carried into anoxic sediments, it will decompose very slowly and may remain in place for many years (natural oil deposits in anoxic environments are millions of years old). Even in oxic environments, hydrocarbon-oxidizing microorganisms can act only if other environmental conditions, such as temperature, pH, and inorganic nutrients, are adequate. Because oil is insoluble in water and is less dense, it floats to the surface and forms slicks. Hydrocarbon-oxidizing bacteria are able to attach to insoluble oil droplets and can often be seen there in large numbers (Figure 14.47). The action of these bacteria eventually leads to decomposition of the oil and dispersal of the slick. A wide variety of microor-

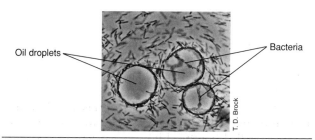

Figure 14.47 Hydrocarbon-oxidizing bacteria in association with oil droplets. The bacteria are concentrated in large numbers at the oil–water interface but are not within the droplet.

ganisms are capable of petroleum biodegradation, including pseudomonads, various corynebacteria and mycobacteria, and even some yeasts.

Microorganisms participate in oil spill cleanups by oxidizing the oil to CO_2. In large spills, volatile hydrocarbon fractions evaporate quickly, leaving longer-chain aliphatic and aromatic components for cleanup crews or microorganisms to tackle. In oil spills where careful bioremediational studies have been performed, it has been shown that hydrocarbon-oxidizing bacteria increase in number 10^3–10^6 times shortly after the oil spill. Experiments using radioisotopic hydrocarbons as tracers or O_2 uptake as a measure of heterotrophic activity have shown that under ideal conditions up to 80% of the nonvolatile components are oxidized by bacteria within 6 months to a year of the spill. Certain fractions, such as branched-chain and polycyclic hydrocarbons, however, remain in the environment much longer. Spilled oil that travels to the sediments is only slowly degraded and can have a significant long-term impact on fisheries and related activities that depend on unpolluted waters for productive yields.

Large oil spills are not common, but when they do occur, such as the 11-million-gal spill from the super-

tanker *Exxon Valdez* that ran aground near Prince William Sound, Alaska, in March 1989, the input of oil to the environment can be ecologically devastating (Figure 14.48a) and cleanup costs staggering (estimated at $1.5 billion).

Addition of inorganic nutrients such as phosphorus and nitrogen to oil spill areas can increase bioremediation rates significantly. In the *Exxon Valdez* spill, for example, accelerated bioremediation of oil washed up on beaches was observed following spraying of the beaches with a mixture of inorganic nutrients (Figure 14.48b). For most large-scale oil spills, however, a combination of both human and microbial efforts is needed to effect a satisfactory cleanup in a reasonable amount of time.

Interfaces where oil and water meet often occur on a massive scale. It is virtually impossible to keep moisture from bulk storage tanks; it accumulates as a layer of water beneath the petroleum. Gasoline storage tanks (Figure 14.49) are thus potential habitats for hydrocarbon-oxidizing microorganisms, which can accumulate and grow at the oil–water interface. Gasoline generally has additional chemicals (added to aid combustion and inhibit corrosion in engines) that

Figure 14.48 Environmental consequences of large oil spills in the marine environment and the effect of bioremediation. (a) A contaminated beach along the coast of Alaska containing oil from the *Exxon Valdez* spill of 1989. (b) The center rectangular plot was treated with inorganic nutrients to stimulate bioremediation of spilled oil by microorganisms, whereas areas to the left and right were untreated.

Figure 14.49 Bulk fuel storage tanks, where massive microbial growth may occur at oil–water interfaces.

inhibit the growth of microorganisms, but not all fuels have such additives.

Petroleum production

Although microbial degradation of petroleum can be quite extensive, microbial *production* of hydrocarbons also occurs, particularly in certain green algae. For example, in the colonial alga *Botryococcus braunii*, growth of the alga is accompanied by the excretion of long-chain hydrocarbons (C_{30} to C_{36}) that have the consistency of oil (Figure 14.50). In *B. braunii* about 30% of the cell dry weight is petroleum, and there has been interest in using this and other oil-producing algae as renewable sources of petroleum. There is even evidence that oil in certain types of oil shale originated from green algae like *B. braunii* that grew in lake beds in ancient times.

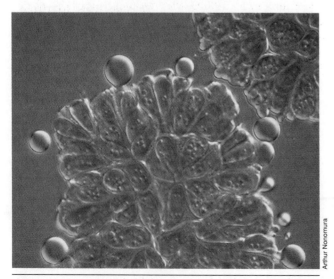

Figure 14.50 Photomicrograph by Nomarski interference contrast of cells of the green alga *Botryococcus braunii*. Note oil droplets, produced and excreted by this alga, at the margin of the cells.

CONCEPT CHECK   14.19

Hydrocarbons are generally stable in anoxic environments but are subject to microbial attack in oxic habitats. Addition of inorganic nutrients is important for bioremediation of spilled oil. Some algae can produce hydrocarbons.

✔ *What is* bioremediation?

✔ *Why might inorganic nutrients stimulate oil degradation?*

14.20 ~~for Mon.~~

Biodegradation of Xenobiotics

Xenobiotics are chemically synthesized compounds that have never existed naturally. Thus, organisms capable of using them may not exist in nature. However, a number of xenobiotic compounds are subject to microbial attack, and we examine this process here.

Pesticides

Some of the most widely distributed xenobiotics are the **pesticides,** which are common components of toxic wastes. Over 1000 pesticides have been marketed for chemical pest control purposes. These include primarily *herbicides, insecticides,* and *fungicides.* Pesticides are of a wide variety of chemical types, such as chlorophenoxyalkyl carboxylic acids, substituted ureas, nitrophenols, triazines, phenylcarbamates, organochlorines and organophosphates, and others (Figure 14.51). Some of these substances are suitable as carbon sources and electron donors for certain soil microorganisms, whereas others are not. If a substance can be attacked by microorganisms, it will eventually disappear from the soil. Such degradation in the soil is usually desirable because toxic accumulations of the compound are avoided. However, even closely related compounds may differ remarkably in their degradability, as is shown for the relative persistence rates of a number of herbicides in Table 14.7. However, these figures are only approximate because a variety of environmental factors, such as temperature, pH, aeration, and organic matter content of the soil, influence decomposition. Some of the chlorinated insecticides are so recalcitrant that they have persisted for over 10 years. Disappearance of a pesticide from an ecosystem does not necessarily mean that it was degraded by microorganisms because pesticide loss can also occur by volatilization, leaching, or spontaneous chemical breakdown.

The organisms that are able to metabolize pesticides and herbicides are fairly diverse, including genera of both bacteria and fungi. Some pesticides can be both carbon and energy sources and are oxidized com-

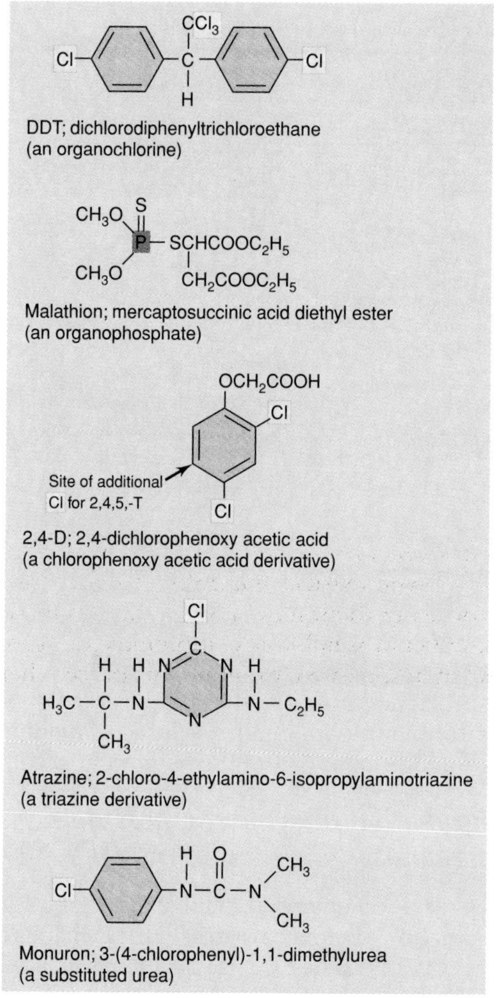

Figure 14.51 Some xenobiotic compounds. Although none of these compounds exist naturally, various microorganisms exist (or have been developed experimentally) that will break them down.

pletely to CO_2. However, other compounds are much more recalcitrant and are attacked only slightly or not at all, although they may often be degraded either partially or totally provided some other organic material is present as primary energy source, a phenomenon called **cometabolism.** However, when the breakdown is only partial, the microbial degradation product of a pesticide may sometimes be even more toxic than the original compound.

Reductive dechlorination

Significant chlorinated pesticide degradation occurs in anoxic environments. In these cases, anoxic biodegradation is linked to *reductive dechlorination* of the molecule, the dechlorinated derivative being much less toxic than the original chlorinated molecule. For example, a sulfate-reducing bacterium implicated in this process, *Desulfomonile*, reduces 3-chlorobenzoate (used

TABLE 14.7 Persistence of herbicides and insecticides in soils

Substance	Time for 75–100% disappearance
Chlorinated insecticides	
DDT [1,1,1-trichloro-2,2-bis-(p-chlorophenyl)ethane]	4 years
Aldrin	3 years
Chlordane	5 years
Heptachlor	2 years
Lindane (hexachlorocyclohexane)	3 years
Organophosphate insecticides	
Diazinon	12 weeks
Malathion	1 week
Parathion	1 week
Herbicides	
2,4-D (2,4-dichlorophenoxyacetic acid)	4 weeks
2,4,5-T (2,4,5-trichlorophenoxyacetic acid)	20 weeks
Dalapin	8 weeks
Atrazine	40 weeks
Simazine	48 weeks
Propazine	1.5 years

as a model compound for studies of chlorinated pesticide degradation) to benzoate and Cl^-:

$$C_7H_4O_2Cl^- + 2\,H \rightarrow C_7H_5O_2^- + HCl$$

3-Chlorobenzoate   Benzoate

Electrons for this reduction can come from acetate, formate, or H_2, and the reductive reaction is linked to establishment of a proton gradient across the cytoplasmic membrane, which *Desulfomonile* uses to drive adenosine triphosphate (ATP) synthesis. Thus, the reductive dechlorination of 3-chlorobenzoate is a type of anaerobic respiration (∞ Section 13.14) (Table 14.8). Anoxic reductive dehalogenation can also result in methanogenesis, either in microbial consortia or in the case of certain compounds like chloroform, by pure cultures of methanogens.

Direct evidence for reductive dechlorinations has been obtained in the case of dichloroethylene, trichloroethylene, tetrachloroethylene (perchloroethylene), chloroform, dichloromethane, and certain brominated and fluorinated compounds. These toxic compounds, some of which (particularly trichloroethylene) are suspected of being carcinogenic, are widely used as industrial solvents and degreasing agents and are among the most frequently detected groundwater contaminants in the United States. Besides *Desulfomonile*, a variety of genera of bacteria are known to reductively dechlorinate, and at least one genus, *Dehalobacter,* can use *only* chlorinated compounds as anaerobic electron acceptors (Table 14.8).

TABLE 14.8   Characteristics of major genera of bacteria capable of reductive dechlorination

Property	Genus			
	Dehalobacter	*Desulfomonile*	*Desulfitobacterium*	*Dehalospirillum*
Electron donors	H_2	H_2, formate, pyruvate, lactate, benzoate	H_2, formate, pyruvate, lactate	H_2, formate, pyruvate, lactate, ethanol, glycerol
Electron acceptors	Trichloroethylene, tetrachloroethylene	Metachlorobenzoates, SO_4^{2-}, SO_3^{2-}, $S_2O_3^{2-}$	Orthochlorophenols, NO_3^-, fumarate, SO_3^{2-} $S_2O_3^{2-}$, S^0	Trichloroethylene, tetrachloroethylene, NO_3^-, fumarate
Other properties[a]	Capable of autotrophic growth; no cytochromes	Contains cytochrome c_3; requires organic carbon source	Fast grower (g = 3.5 hr); requires organic carbon source	Contains b- and c-type cytochromes; grows very fast (g = 2.5 hr); requires organic carbon source
Phylogeny[b]	Related to low GC gram-positive bacteria	Related to delta Proteobacteria	Related to low GC gram-positive bacteria	Related to epsilon Proteobacteria

[a]All organisms are obligate anaerobes.
[b]See Chapter 15 for detailed discussion of prokaryotic phylogeny and Table 16.1 for the phylogenetic characterization of Proteobacteria.

Aerobic dechlorination of chlorinated organic compounds also occurs (see Figure 14.52), probably by different biochemical mechanisms than for anaerobic biodegradation, but reductive dechlorination is of particular environmental interest because of the rapidity with which anoxic conditions can develop in polluted microbial habitats in nature.

Biodegradation of xenobiotics and microbial evolution

The existence of organisms able to metabolize xenobiotics is of considerable evolutionary interest because these compounds are completely new to the earth in the past 50 years or so. Observations on the rapidity with which organisms metabolizing new compounds arise can give us some idea of the rates of microbial evolution in general. In Section 11.13 we discussed the evolution in past decades of plasmids conferring resistance to antibiotics. The evolution of pesticide-degrading bacteria seems to be a similar case. For example, enrichment cultures (see Section 14.3) yield bacteria capable of degrading 2,4,5-trichlorophenoxyacetic acid (2,4,5-T) and other recalcitrant pesticides. These organisms appear to be common *Pseudomonas* species that are now capable of growing on these pesticides as sole sources of carbon and energy (Figure 14.52). In a study of 2,4,5-T and 2,4-dichlorophenoxyacetic acid (2,4-D) biodegradation, it has been shown that portions of plasmids that code for 2,4-D biodegradation are "recruited" to form new plasmids conferring the ability to degrade 2,4,5-T. Because this can happen relatively quickly, it can be concluded that if biodegradation of a particular xenobiotic compound is possible, evolutionary events will move rather rapidly to establish microorganisms with new genetic properties to allow for breakdown of the compound. This assumes that a sufficient amount of the compound is present in the environment to maintain a selective advantage for biodegradation potential in the new population.

Biodegradation of synthetic polymers and the landfill crisis

A major area of environmental concern besides the biodegradation of toxic wastes like pesticides is the disposal of solid wastes. Currently, the plastics industry produces nearly 75 *billion* lb of plastics per year, 40% of which is discarded in landfills. Solid wastes, which beside plastics include packaging materials, polyurethane, polystyrenes, and other polymeric substances, are rapidly filling up available landfill sites. The chemical structures of the building blocks of some of these synthetic polymers are shown in Figure 14.53a.

It is now recognized that many synthetic polymers are highly recalcitrant to microbial degradation and remain essentially unaltered for decades in landfills and other refuse dumps. This problem has fueled the search for *biodegradable* alternatives to the common synthetic polymers now in use. Some improvements have been made in this area, including photodegradable, starch-linked, and "microbial" plastics (see later). Photobiodegradable plastics consist of material whose polymeric structure is altered by exposure to ultraviolet radiation (from sunlight), generating modified polymers amenable to microbial attack. Other biodegradable plastic bags have been made that incorporate starch to hold together short fragments of a biodegradable polymer. This design accelerates biodegradation because starch-digesting bacteria in soil attack the starch, releasing polymer fragments that are degraded by other microorganisms.

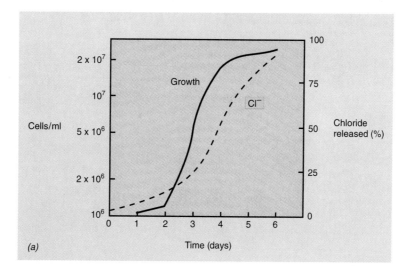

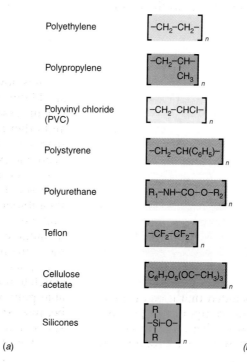

Figure 14.52 Biodegradation of the herbicide 2,4,5-T. (a) Growth of *Burkholderia* (formerly *Pseudomonas*) *cepacia* on 2,4,5-T as sole source of carbon and energy. The strain was enriched from nature using a chemostat to keep the concentration of herbicide low. Growth here is aerobic on 1.5 g/l of 2,4,5-T. The release of chloride from the molecule is indicative of biodegradation. (b) Pathway of aerobic 2,4,5-T biodegradation. Note the steps in which Cl⁻ is released. The final products, succinate and acetate, are catabolized in the citric acid cycle. For the mechanism of action of a dioxygenase, see Figure 13.50c.

Figure 14.53 Synthetic and microbial polymers. (a) Monomeric structure of a number of common synthetic polymers. (b) A brand of shampoo marketed in Europe and packaged in a bottle made of "bacterial plastic." The bottle consists of a copolymer of poly-β-hydroxybutyrate and poly-β-hydroxyvalerate (marketed under the trade name *Biopol*; see the box). Because this material is a natural product, the bottle readily degrades both aerobically and anaerobically.

Polyethylene	$\left[-CH_2-CH_2-\right]_n$
Polypropylene	$\left[\begin{matrix}-CH_2-CH- \\ \quad\quad\; CH_3\end{matrix}\right]_n$
Polyvinyl chloride (PVC)	$\left[-CH_2-CHCl-\right]_n$
Polystyrene	$\left[-CH_2-CH(C_6H_5)-\right]_n$
Polyurethane	$\left[R_1-NH-CO-O-R_2\right]_n$
Teflon	$\left[-CF_2-CF_2-\right]_n$
Cellulose acetate	$\left[C_6H_7O_5(OC-CH_3)_3\right]$
Silicones	$\left[\begin{matrix}R \\ -Si-O- \\ R\end{matrix}\right]_n$

(a)

(b)

BIODEGRADABLE POLYMERS FROM BACTERIA

Although humans in developed countries have grown accustomed to life in a "plastic society," it is now clear that municipal landfills cannot continue to be filled up with synthetic (xenobiotic) plastics. Because no significant biodegradation of many of these materials has occurred in landfills over relatively long periods of time, naturally occurring polymeric materials have been sought as substitutes for xenobiotic plastics. Because any substance of biological origin can be degraded by one or another microorganism, the widespread use of biologically produced plastics could solve major waste problems.

For several reasons, poly-β-hydroxyalkanoates (PHAs) appear to be the most ideal candidates for synthetic plastic substituents (see Figure 14.53b). PHAs are microbially produced storage polymers that have many of the general properties of synthetic plastics and can be synthesized by cells in various chemical forms. Depending on the length of the side chain in the monomeric units of the PHA polymer (a property that can be adjusted by modifying the composition of the growth medium or by genetically modifying the producing bacterium), PHAs of varying melting points, crystallinity, flexibility, and tensile strength can be obtained. Various bacteria have been tested for use as catalysts for com-

mercial production of PHAs, but as previously discussed in Chapter 10, an industrial fermentation is commercially feasible only if the organism can be made to overproduce the product and can be grown on a large scale using inexpensive carbon sources such as glucose or ethanol as chemical feedstocks. Thus far, only a relatively few bacterial species fit these criteria.

At a manufacturing plant in Billingham, England, the British chemical giant Imperial Chemical Industries (ICI) is producing PHAs and marketing the material as a packaging polymer under the trade name *Biopol.* ICI uses the bacterium *Alcaligenes eutrophus* to produce PHAs using glucose as feedstock. With *A. eutrophus,* ICI obtains yields of a poly-β-hydroxybutyrate/poly-β-hydroxyvalerate (PHB/PHV) copolymer (see below) of greater than 80% of cell dry weight (see also Figure 3.59). The polymer is extracted from the cells and milled to a powder or pellet form for use in the production of injection-molded articles such as the shampoo bottle

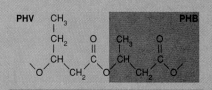

PHV/PHB copolymer

shown in Figure 14.53b. Because PHA production by ICI is on a relatively small scale right now, about 600 tons/year, the current cost of a PHB/PHV bottle is several times that of a synthetic plastic bottle. However, plant expansion to a capacity 20 times this size could put PHAs on a cost-competitive basis with petroleum-based plastics, especially in markets for environmentally conscious consumers. Moreover, because the metabolic diversity of bacteria is so substantial (Chapter 13), novel polymers probably await discovery, and thus far several variations on the basic PHB or PHV pattern have already been discovered. Commercial production of these polymers could boost the future market share of bacterial plastics, especially in the specialty plastics markets where materials of specific physical properties are often needed.

It is of interest that the field of polymer science, which until recently has been dominated by chemical and physical scientists, may now be strengthened by microbiologists who offer microbial solutions to two of the most serious problems in the synthetic polymer industry today: (1) the ecological problem surrounding the recalcitrance of many synthetic disposable plastics, and (2) the production of novel polymers possessing unusual physical properties not obtainable by strictly chemical syntheses. ∎

Bacterial plastics

Bacterial plastics is a term used to describe an exciting new area of research where naturally synthesized bacterial polymers such as the lipid storage material poly-β-hydroxybutyrate (PHB) (Section 3.11) are being used as raw materials for plastic-based packaging materials (Figure 14.53b) (see the box, Biodegradable Polymers from Bacteria). Research in PHB production has shown that the precise chemical nature (and hence chemical properties) of the polymer produced by a bacterium can be controlled by varying the substrates used to grow the organism.

For example, in certain bacteria acetate and butyrate lead to poly-β-hydroxybutyrate (C_4) production, whereas caproate (a C_6 fatty acid) leads to a polymer containing C_6 units and valerate (a C_5 fatty acid) yields a polymer containing C_5 units. *Copolymers,* containing alternating repeats of various monomeric units, can also be synthesized (see the box). Bacterial plastics are particularly appealing because they have been shown to undergo rapid biodegradation under both aerobic and anoxic conditions. Thus, as substitutes for synthetic polymers, increased use of bacterial plastics could significantly reduce the need for new landfill space.

Many chemically synthesized compounds such as insecticides, herbicides, and plastics (all called xenobiotics) are completely foreign to the environments in which they are used and can persist because microorganisms capable of degrading these xenobiotics may not naturally occur. Although appropriate microorganisms may eventually evolve, it is desirable to use materials that are readily biodegradable and thus readily degraded by indigenous microorganisms.

✔ *What is* reductive dechlorination *and in what types of environments does it occur?*

✔ *What advantages would* biopolymers *have over synthetic polymers?*

14.21

Plant–Microorganism Interactions

As microbial habitats, plants are clearly vastly different from animals. Compared with warm-blooded animals, plants vary greatly in temperature, both diurnally and throughout the year, and compared with the complex circulatory system of animals, the internal communication system of the plant is only poorly developed, and so transfer of microorganisms within the plant is relatively inefficient. The aboveground parts of the plant, especially the leaves and stems, are subjected to frequent drying, and for this reason many plants have developed waxy coatings that retain moisture and keep out microorganisms. The roots, on the other hand, exist in an environment in which moisture is less variable and nutrient concentrations are higher. For this reason, the roots of plants are a main area of microbial action.

The **rhizosphere** is the region immediately outside the root; it is a zone where microbial activity is usually high. The bacterial count is almost always higher in the rhizosphere than it is in regions of the soil devoid of roots, often many times higher. This is because roots excrete significant amounts of sugars, amino acids, hormones, and vitamins, which promote such an extensive growth of bacteria and fungi that these organisms often form microcolonies on the root surface. The **phyllosphere** is the surface of the plant leaf, and under conditions of high humidity, as in wet forests in tropical and temperate zones, the microbial flora of leaves may be quite high. Many of the bacteria on leaves fix nitrogen (∞ Sections 13.27 and 14.23), and nitrogen fixation presumably aids these organisms in growing with the predominantly carbohydrate nutrients provided by leaves. We proceed now to consider two highly developed associations between plants and microorganisms, lichens and mycorrhizae.

Lichens

Lichens are leafy or encrusting growths that are widespread in nature and are often found growing on bare rocks, tree trunks, house roofs, and surfaces of bare soils (Figure 14.54). The lichen plant consists of two organisms, a fungus and an alga. However, little specificity resides in the relationship, as a given fungus can establish the lichen symbiosis with several different algae, and vice versa. The alga is phototrophic and is able to produce organic matter, which is then used for nutrition of the fungus. Because the fungus is unable to carry out photosynthesis, its ability to live in nature is dependent on the activity of its algal partner. Lichens are usually found in environments where other organisms do not grow, and their success in colonizing such environments is due to the mutual interrelationships between the alga and fungus partners.

Lichens consist of a tight association of many fungal cells within which the algal cells are embedded (Figure 14.55). The shape of the lichen is determined primarily by the fungal partner, and a wide variety of fungi are able to form lichen associations. The diversity of algal types is much smaller, and many different kinds of lichens may have the same algal component. Some lichens contain cyanobacteria instead of algae as the phototrophic component. The algae or cyanobacteria are usually present in defined layers or clumps within the lichen structure.

The fungus clearly benefits from associating with the alga, but how does the alga benefit? The fungus provides a firm anchor within which the alga can grow protected from erosion by rain or wind. In addition, the fungus facilitates the uptake of water and absorbs from the rock or other substrate on which the lichen is living the inorganic nutrients essential for the growth of the alga. *Lichen acids,* complex organic compounds excreted by the fungus, promote the dissolution and chelation of nutrients. Another role of the fungus is to protect the alga from drying; most of the habitats in which lichens live are dry (rock, bare soil, roof tops) (see Figure 14.54), and fungi are in general much better able to tolerate dry conditions than are algae.

Most lichens grow extremely slowly—a 2-cm lichen observed on the surface of a rock may actually be several years old. Measurements of lichen growth vary from 1 mm or less per year to over 3 cm/year, depending on the organisms composing the symbiosis, the amount of rainfall and sunlight received, and general weather conditions. Although lichens live in nature under rather harsh conditions, they are extremely sensitive to air pollution and quickly disappear in areas experiencing heavy air pollution. One reason for this sensitivity is that they absorb and concentrate materials from rainwater and air and have no means for excreting them so lethal concentrations of compounds such as SO_2 are easily reached.

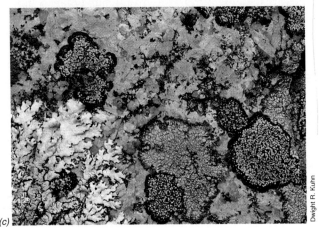

Figure 14.54 Lichens. (a) A lichen growing on a branch of a dead tree. (b) Lichens coating the surface of a large rock. (c) Several different lichens growing on the surface of a rock.

Mycorrhizae

Mycorrhiza literally means "root fungus" and refers to the symbiotic association that exists between plant roots and fungi. Probably the roots of the majority of terrestrial plants are mycorrhizal. There are two gen-

eral classes of mycorrhizae: **ectomycorrhizae,** in which fungal cells form an extensive sheath around the outside of the root with only little penetration into the root tissue itself, and **ericoid mycorrhizae,** in which the fungal mycelium is embedded in the root tissue. The present discussion will deal only with ectomycorrhizae.

Ectomycorrhizae are found mainly in forest trees, especially conifers, beeches, and oaks, and are most highly developed in temperate forests. In a forest, almost every root of every tree is mycorrhizal. The root system of a mycorrhizal tree is composed of both long and short roots. The short roots, which are characteristically dichotomously branched (Figure 14.56), show the typical fungal sheath, whereas long roots are usually uninfected. Most mycorrhizal fungi do not attack cellulose and leaf litter but instead use simple carbohydrates for growth and usually have one or more vi-

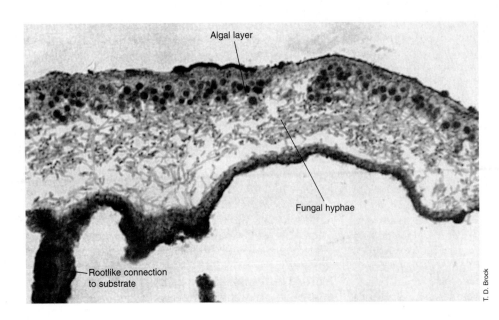

Algal layer

Fungal hyphae

Rootlike connection to substrate

Figure 14.55 Photomicrograph of a cross section through a lichen.

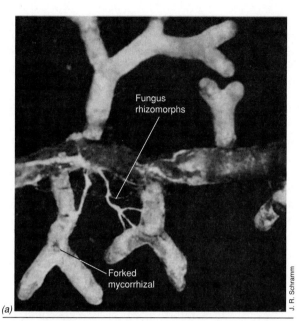

(a)

(b)

Figure 14.56 Mycorrhizae. (a) Typical mycorrhizal root of the pine, *Pinus rigida*, with rhizomorphs of the fungus *Thelophora terrestris*. (b) Cross section of a seedling of *Pinus contorta* (lodgepole pine), showing extensive development of mycorrhizae. The seedling extends about 4 cm above the soil surface.

tamin requirements; they obtain their nutrients from root secretions. The mycorrhizal fungi are never found in nature except in association with roots and hence can be considered obligate symbionts. These fungi produce plant growth substances that induce morphological alterations in the roots, causing characteristically short dichotomously branched mycorrhizal roots to be formed. Despite the close relationship between fungus and root, there is little species specificity involved; a single species of pine can form mycorrhizae with over 40 species of fungi.

The beneficial effect on the plant of the mycorrhizal fungus is best observed in poor soils, where trees that are mycorrhizal thrive but nonmycorrhizal ones do not (Figure 14.57). When trees are planted in

prairie soils, which ordinarily lack a suitable fungal inoculum, trees that were artificially inoculated at the time of planting grow much more rapidly than uninoculated trees (Figure 14.57). It is well established that the mycorrhizal plant is able to absorb nutrients from its environment more efficiently than does a nonmycorrhizal one. This improved nutrient absorption is probably due to the greater surface area provided by the fungal mycelium (Figure 14.56).

14.22

Agrobacterium and Plant Interactions: Crown Gall and Hairy Root

The genus *Agrobacterium* comprises organisms that cause the formation of tumorous growths on a wide variety of plants. The two species most widely studied are *A. tumefaciens*, which cause *crown gall*, and *A. rhizogenes*, which causes *hairy root*. Although plants often form benign accumulations of tissue, called a **callus,** when wounded, the growth induced by *A. tumefaciens* (Figure 14.58) is different in that the callus shows uncontrolled growth. It thus resembles tumor growth in animals, and considerable research on crown gall has been carried out with the idea that it may provide a model for how malignant growths occur in humans. *Agrobacterium rhizogenes* causes a malignant mass of roots to emanate from the infected wound site, creating

Figure 14.57 Six-month-old seedlings of Monterey pine (*Pinus radiata*) growing in prairie soil: left, nonmycorrhizal; right, mycorrhizal.

Figure 14.58 Photograph of tumor on a tobacco plant caused by crown gall bacteria of the genus *Agrobacterium*.

the condition known as *hairy root.* Interestingly, once induced, these tumors continue to grow in the absence of *Agrobacterium* cells. Thus, once *Agrobacterium* has brought about the induction of the tumorous condition, its presence is no longer necessary.

An overview of the events in crown gall formation is given in Figure 14.59. It is now well established that a large plasmid called the *Ti* (*tumor induction*) *plasmid* (Figure 14.60) must be present in the *Agrobacterium* cells if they are to induce tumor formation. In *Agrobacterium rhizogenes*, a similar plasmid called the *Ri plasmid* is necessary for induction of hairy root. However, the best-studied system is that of the *A. tumefaciens*–crown gall disease so we focus on that here. Following infection, a part of the Ti plasmid, called the *transfer DNA* (T-DNA), is integrated into the plant's genome. T-DNA carries the genes for tumor formation and also for the production of a number of modified amino acids called **opines.** *Octopine* [N^2-(1,3-dicarboxyethyl)-L-arginine] and *nopaline* [N^2-(1,3-dicarboxypropyl)-L-arginine] are the two most common opines. Opines are produced by plant cells transformed by T-DNA and are a source of carbon and nitrogen for *Agrobacterium* cells (Figure 14.59). The Ti plasmid is also used in genetic engineering (∞ Section 10.14 and see later).

Recognition of Agrobacterium *by the plant*

To initiate the tumorous state, cells of *Agrobacterium* must first attach to a wound site on the plant. The recognition of *Agrobacterium* by plant tissue involves complementary receptor molecules on the surfaces of the bacterial and plant cells. It is thought that the plant receptor molecule is a type of *pectin* (a complex poly-saccharide) and that the bacterial receptor is a type of polysaccharide containing β-glucans, embedded in the cell wall lipopolysaccharide.

Studies with nontumorigenic mutants of *Agrobacterium tumefaciens* have shown clearly that most functions necessary for attachment of the bacterium to plant are borne on the bacterial chromosome. Presumably β-glucans modify the bacterial cell wall in a way that facilitates binding of the bacterial cell to the plant cell. In addition to β-glucan, attachment of *A. tumefaciens* to certain plant types, notably to carrot or tobacco tissue, is mediated by bacterial production of *cellulose* microfibrils. Following recognition and initial binding [mediated by lipopolysaccharide (LPS) and β-glucan], the rapid synthesis of cellulose microfibrils anchors the inoculum to the wound site and literally entraps the bacterial cells, forming large bacterial aggregates on the plant cell surface. Although helpful in mediating colonization and rapid growth of agrobacteria at the site of infection, studies with mutants of *A. tumefaciens* incapable of synthesizing cellulose indicate that cellulose production is not necessary for induction of the tumorous state.

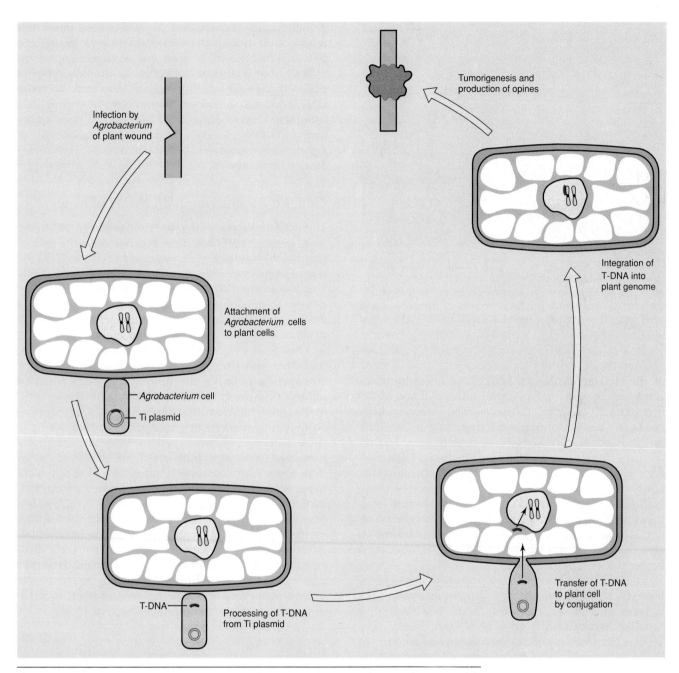

Figure 14.59 Overview of events of crown gall disease following infection of a susceptible plant by *Agrobacterium tumefaciens*.

Plasmid transfer and tumorigenesis: vir *genes and T-DNA*

The structure of the Ti plasmid is given in Figure 14.60. Note that although a number of genes are needed for infectivity, only a smaller portion is needed for tumorigenesis. Before the onset of disease symptoms in a plant, Ti plasmid DNA must be transferred from the bacterium to the plant. Only a small portion of the Ti plasmid, a region called the T-DNA (Figures 14.59 and 14.60), is actually transferred to the plant. The T-DNA

contains oncogenes that direct events leading to tumorigenesis. The *vir* genes that reside on the Ti plasmid code for proteins that are essential for T-DNA transfer (Figure 14.60). *Vir* gene expression is induced by plant signal molecules synthesized by wounded plant tissues. Some inducers that have been identified include the phenolic compounds acetosyringone, *p*-hydroxybenzoic acid, and vanillin.

The *vir* genes are the key to T-DNA transfer. The *virA* gene codes for a protein kinase that interacts with signal molecules and then phosphorylates the product

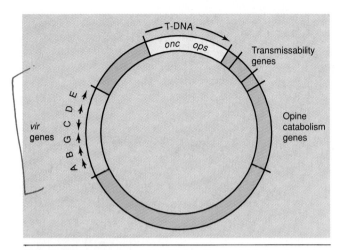

Figure 14.60 Structure of the Ti plasmid of *Agrobacterium tumefaciens*. T-DNA is the region actually transferred to the plant. *vir*, Virulence genes; *onc*, oncogenes; *ops*, opine synthesis genes. Arrows indicate the direction of transcription of each gene. The entire Ti plasmid is about 200 kb of DNA, and the T-DNA about 20 kb.

of the *virG* gene (Figure 14.61). The latter becomes activated by the phosphorylation event and functions to activate other *vir* genes. The product of the *virD* gene has endonuclease activity and nicks DNA in the Ti plasmid in a region adjacent to the T-DNA (Figures 14.60 and 14.61). The product of the *virE* gene is a single-stranded DNA binding protein that binds the *single strand* of T-DNA generated from endonuclease activity and transports this small fragment of DNA into the plant cell. The *virB* gene product is located in the bacterial membrane and mediates transfer of the single strand of DNA between bacterium and plant.

T-DNA transfer occurs in a process that resembles bacterial conjugation, the *virB* protein facilitating the transfer of T-DNA across the membrane through a porelike structure (Figures 14.59 and 14.61). The T-DNA becomes inserted into the nuclear genome of the plant (apparently plant cell organellar genomes are not infected) where integration can occur at any number of sites where specific inverted or direct tandem repeats are present. The oncogenes of the Ti plasmid (Figure 14.60) encode enzymes involved in plant hormone production and at least one key enzyme of opine biosynthesis. Expression of these oncogenes leads to tumor formation.

The Ri plasmid involved in hairy root disease also contains oncogenes. In this case, the oncogenes confer increased auxin responsiveness to the plant cells, which leads to overproduction of root tissue, resulting in the symptoms of the disease. The Ri plasmid also codes for several opine biosynthetic enzymes.

Despite differences in tumor morphology, the similar nature of the molecular events involved in crown gall and hairy root disease suggest a close molecular relationship among the two diseases and among the two infecting plasmids. From the standpoint of microbiology both diseases involve a unique type of plant–microorganism interaction in which bacterial DNA serves to transform plant cells. When this was recognized, the usefulness of this natural plant transformation system as a vector for the introduction of genetically engineered DNA into plants became immediately obvious.

Genetic engineering with the Ti plasmid

The Ti and Ri plasmid systems have revolutionized plant genetics and have done much to open the field of plant biotechnology. We discussed in Section 10.14 the use of *Agrobacterium* to genetically transform plants and investigated some of the applications of this technology to genetic engineering. Here we briefly consider a novel use of the *Agrobacterium tumefaciens* Ti plasmid system in plant molecular biology.

Besides using *Agrobacterium* to transfer genes for herbicide and drought resistance into plants or for attempts to improve the nutritional composition of plants (∞ Section 10.14), agrobacteria have been used as vectors for introducing reporter genes in plants as easily visible markers of gene expression. For example, *A. tumefaciens* has been used to introduce the firefly luciferase gene into tobacco plants, yielding plants that are bioluminescent (Figure 14.62). Used as a reporter and linked to other key genes of interest, luciferase is used as a visible marker of gene expression, as a genetic marker in classical plant breeding, and as a general tool for studying plant molecular biology. Experience has shown that plants are fairly difficult to transform by other means. Thus, the Ti and Ri plasmids have opened up plant biology to the power of molecular genetics, and the field of plant biotechnology is rapidly expanding (∞ Section 10.14).

CONCEPT CHECK   **14.22**

The crown gall bacterium *Agrobacterium* enters into a unique relationship with higher plants. A plasmid in the bacterium (the Ti plasmid) is able to transfer part of itself to the genome of the plant, in this way bringing about the production of the crown gall disease. The crown gall plasmid has also found extensive use in the genetic engineering of crop plants.

✔ *What are* opines *and why are they produced?*

✔ *How do the* vir *genes differ from* T-DNA *in the Ti plasmid?*

✔ *How has an understanding of crown gall disease benefited the area of plant molecular biology?*

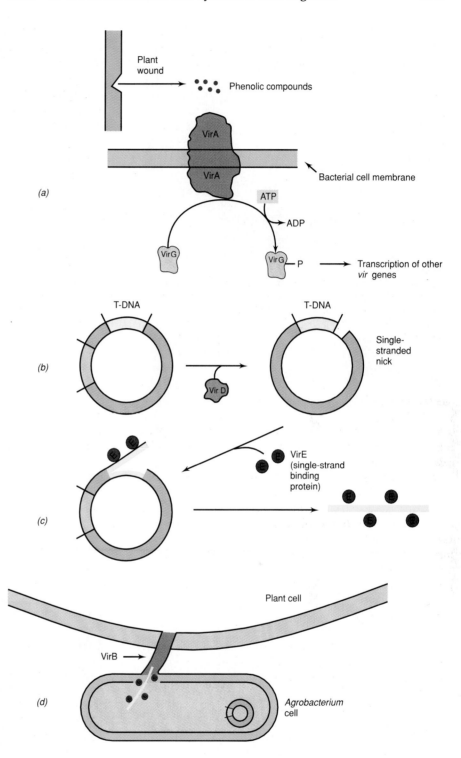

Figure 14.61 Mechanism of transfer of T-DNA to the plant cell by *Agrobacterium tumefaciens*. (a) Levels of VirA protein increase dramatically upon stimulation with plant phenolic inducer molecules. VirA activates VirG by phosphorylation, and VirG activates transcription of other *vir* genes. (b) VirD is an endonuclease. (c) VirE is a single-stranded binding protein. (d) VirB acts as a conjugation bridge between *Agrobacterium* and the plant cell.

14.23

Root Nodule Bacteria and Symbiosis with Legumes

One of the most interesting and important plant bacterial interactions is that between leguminous plants and bacteria of the genera *Rhizobium*, *Bradyrhizobium*, and *Azorhizobium*. Legumes are a large group that includes such economically important plants as soybeans, clover, alfalfa, beans, and peas and are defined as plants that bear seeds in pods. *Rhizobium*, *Bradyrhizobium*, and *Azorhizobium* are gram-negative motile rods. Infection of the roots of a leguminous plant with the appropriate species of *Rhizobium* or *Bradyrhizobium* leads to the formation of **root nodules** (Figure 14.63) that are able to

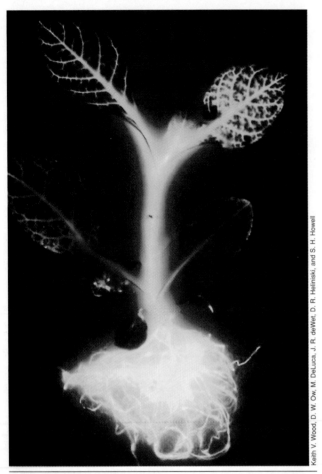

Figure 14.62 Leaf from a tobacco plant, *Nicotiana tabacum*, that has been genetically engineered by insertion of the luciferase gene from the firefly, *Photinus pyralis*. Note that the leaf is luminescent as the result of expression of the luciferase gene. Reprinted with permission from *Science* 234:856–859 (1986), © AAAS.

convert gaseous nitrogen to combined nitrogen, a process called *nitrogen fixation* (Section 13.27); *Azorhizobium* forms stem nodules (see later). Nitrogen fixation by the legume–*Rhizobium* symbiosis is of considerable agricultural importance, as it leads to very significant increases in combined nitrogen in the soil. Because nitrogen deficiencies often occur in unfertilized bare soils, nodulated legumes are at a selective advantage under such conditions and can grow well in areas where other plants cannot (Figure 14.64).

Leghemoglobin and cross-inoculation groups

Under normal conditions, neither legume nor *Rhizobium* alone is able to fix nitrogen; yet the interaction between the two leads to the development of nitrogen-fixing ability. In pure culture, the *Rhizobium* is able to fix N_2 alone when grown under strictly controlled *microaerophilic* conditions. Apparently *Rhizobium* needs some O_2 to generate energy for N_2 fixation, yet its

Figure 14.63 Soybean root nodules. The nodules develop by infection with *Bradyrhizobium japonicum*.

nitrogenase (like those of other nitrogen-fixing organisms) (Section 13.27) is inactivated by O_2. In the nodule, precise O_2 levels are controlled by the O_2 binding protein **leghemoglobin**. This is a red, iron-containing protein that is always found in healthy N_2-fixing nodules (Figure 14.65). Neither plant nor *Rhizobium* alone synthesizes leghemoglobin, but formation is thought to be induced through the interaction of these two organisms. Leghemoglobin functions as an "oxygen buffer" cycling between the oxidized (Fe^{3+}) and reduced (Fe^{2+}) forms to keep free O_2 levels within the nodule at a low

Figure 14.64 A field of unnodulated (left) and nodulated soybean plants growing in nitrogen-poor soil.

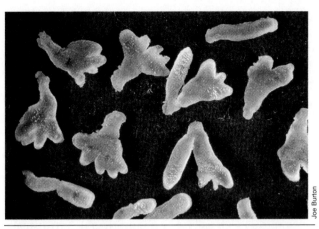

Joe Burton

Figure 14.65 Sections of root nodules from the legume *Coronilla varia,* showing the reddish pigment leghemoglobin.

but constant level. The ratio of leghemoglobin-bound O_2 to free O_2 in the root nodule is on the order of 10,000:1.

About 90% of all leguminous plant species are capable of becoming nodulated. However, there is a marked specificity between species of legume and strains of *Rhizobium*. A single *Rhizobium* strain is generally able to infect certain species of legumes and not others. A group of *Rhizobium* strains able to infect a group of related legumes is called a *cross-inoculation group.* The major rhizobial cross-inoculation groups are listed in Table 14.9. Even if a *Rhizobium* strain is able to infect a certain legume, it is not always able to bring about the production of nitrogen-fixing nodules. If the strain is *ineffective,* the nodules formed will be small, greenish-white, and incapable of fixing nitrogen; if the strain is *effective,* on the other hand, the nodule will be large, reddish (Figure 14.65), and nitrogen-fixing. Effectiveness is determined by genes in the bacterium (see the section, Genetics of nodule formation: *nod* genes).

TABLE 14.9	Major cross-inoculation groups of leguminous plants

Plant	Nodulated by
Pea	*Rhizobium leguminosarum* biovar *viciae*[a]
Bean	*Rhizobium leguminosarum* biovar *phaseoli*[a]
Bean	*Rhizobium tropici*
Bean	*Rhizobium etli*
Clover	*Rhizobium leguminosarum* biovar *trifolii*[a]
Alfalfa	*Rhizobium meliloti*
Soybean	*Bradyrhizobium japonicum*
Soybean	*Bradyrhizobium elkanii*
Soybean	*Rhizobium fredii*
Sesbania rostrata (a tropical legume)	*Azorhizobium caulinodans*

[a]Several varieties (biovars) of *Rhizobium leguminosarum* exist, each capable of nodulating a different legume.

Stages in nodule formation

The stages in the infection and development of root nodules are now fairly well understood (Figure 14.66). They include

1. **Recognition** of the correct partner on the part of both plant and bacterium and **attachment** of the bacterium to root hairs.
2. **Invasion** of the root hair by the bacterial formation of an infection thread.
3. **Travel** to the main root via the infection thread.
4. Formation of deformed bacterial cells, **bacteroids,** within the plant cells and development of the nitrogen-fixing state.
5. Continued plant and bacterial division and formation of the mature **root nodule.**

We now explore some of these stages in nodule formation in more detail.

The roots of leguminous plants secrete a variety of organic materials that stimulate the growth of a rhizosphere microflora. This stimulation is not restricted to the rhizobia but occurs with a variety of rhizosphere bacteria. If there are rhizobia in the soil, they grow in the rhizosphere and build up to high population densities. Attachment of bacterium to plant in the legume–*Rhizobium* symbiosis is the first step in the formation of nodules. A specific adhesion protein called *rhicadhesin* is present on the surfaces of all species of *Rhizobium* and *Bradyrhizobium*. Rhicadhesin is a calcium binding protein and may function by binding calcium complexes on the root hair surface. Other substances, such as carbohydrate-containing proteins called *lectins,* also play some role in plant–bacterium attachment. Lectins have been identified on both root hair tips and on the surface of *Rhizobium* cells, but the interaction of lectins in the binding–recognition process is thought to be less important than that of rhicadhesin.

Initial penetration of *Rhizobium* cells into the root hair is via the root hair tip. Following binding, the root hair curls as a result of the action of substances excreted by the bacterium called *Nod factors* (see later) and the bacteria enter the root hair and induce formation by the plant of a cellulosic tube, called the **infection thread,** which spreads down the root hair. Root cells adjacent to the root hairs subsequently become infected by rhizobia, and Nod factors stimulate plant cell division, eventually leading to formation of the nodule (Figures 14.63, 14.65, 14.67, and 14.68).

The bacteria multiply rapidly within the plant cells and are transformed into swollen, misshapen, and branched forms called **bacteroids.** Bacteroids become surrounded singly or in small groups by portions of the plant cell membrane, called the *peribacteroid membrane.* Only after the formation of bacteroids does nitrogen fixation begin. (Effective nitrogen-fixing

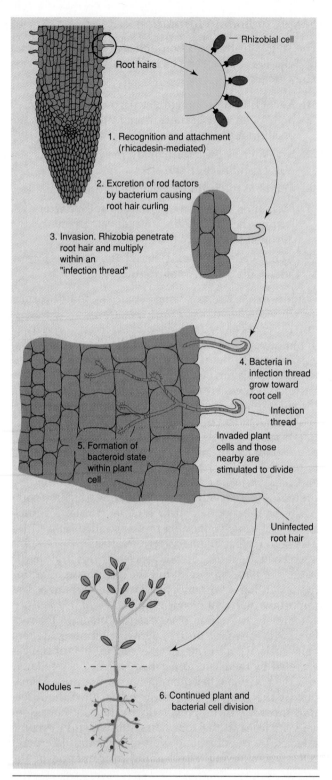

Root hairs

— Rhizobial cell

1. Recognition and attachment (rhicadesin-mediated)

2. Excretion of rod factors by bacterium causing root hair curling

3. Invasion. Rhizobia penetrate root hair and multiply within an "infection thread"

4. Bacteria in infection thread grow toward root cell

— Infection thread

Invaded plant cells and those nearby are stimulated to divide

5. Formation of bacteroid state within plant cell

Uninfected root hair

Nodules —

6. Continued plant and bacterial cell division

Figure 14.66 Steps in the formation of a root nodule in a legume infected by *Rhizobium*.

nodules can be detected by acetylene reduction) (∞ Section 13.27 and Figure 13.55). When the plant dies the nodule deteriorates, releasing bacteria into the soil. The bacteroid forms are incapable of division, but there are always a small number of dormant rod-shaped cells present. These now proliferate, using some of the products of the deteriorating nodule as nutrients, and the bacteria can initiate the infection in other roots or maintain a free-living existence in the soil.

Biochemistry of nitrogen fixation in nodules

As discussed in Section 13.27, nitrogen fixation involves the activity of the enzyme **nitrogenase,** a large two-component protein containing iron and molybdenum. Nitrogenase in root nodules has characteristics similar to the enzyme from free-living N_2-fixing bacteria, including O_2 sensitivity and the ability to reduce acetylene as well as N_2 (∞ Figure 13.55). Nitrogenase is localized within the bacteroids themselves and is not released into the plant cytosol.

Bacteroids are totally dependent on the plant for supplying them with energy sources for N_2 fixation. The major organic compounds transported across the peribacteroid membrane and into the bacteroid proper are citric acid cycle intermediates, in particular the C_4 acids *succinate, malate,* and *fumarate* (Figure 14.69). These are used as electron donors for ATP production and, following conversion to pyruvate, as the ultimate source of electrons for the reduction of N_2.

The first stable product of N_2 fixation is *ammonia*, and several lines of evidence suggest that assimilation of ammonia into organic nitrogen compounds in the root nodule is carried out primarily by the plant. Although bacteroids can assimilate some ammonia into organic form, the levels of ammonia-assimilatory enzymes in bacteroids are quite low. By contrast, the ammonia-assimilating enzyme *glutamine synthetase* (∞ Figure 13.52) is present in high levels in the plant cell cytoplasm. Hence, ammonia transported from the bacteroid to the plant cell can be assimilated by the plant as the amino acid *glutamine*. Besides glutamine, other nitrogenous compounds, in particular other amino acid amides, such as asparagine and 4-methylene glutamine, and the ureides *allantoin* and *allantoic acid*, are synthesized by the plant and subsequently transported to plant tissues (see Figure 14.69).

Genetics of nodule formation: Nod *genes*

Genes directing specific steps in nodulation of a legume by a strain of *Rhizobium* have been called *nod genes* (Figure 14.70). Many *nod* genes from different *Rhizobium* species are highly conserved and are borne on large plasmids called *Sym plasmids*. In addition to *nod* genes, which direct specific nodulation events, Sym plasmids contain *specificity genes*, which restrict a strain of *Rhizobium* to a particular host plant. Indeed, cross-inoculation group specificity can be transferred across species of rhizobia by simply transferring its Sym plasmid. For example, when the Sym plasmid of *Rhizobium leguminosarum* biovar *viciae* (whose host is

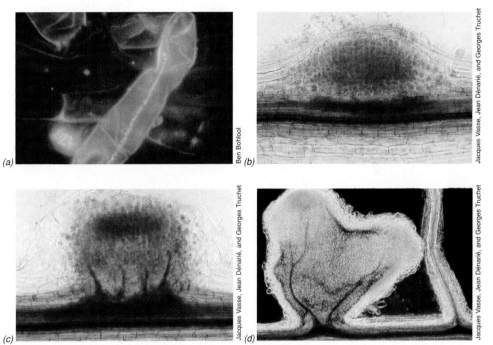

Figure 14.67 The infection thread and formation of root nodules. (a) An infection thread formed by cells of *Rhizobium leguminosarum* biovar *trifolii* formed on a root hair of white clover *(Trifolium repens)*. The infection thread consists of a cellulosic tube through which bacteria move to root cells. (b–d) Nodules from alfalfa roots infected with cells of *Rhizobium meliloti* shown at different stages of development. Cells of both *R. leguminosarum* biovar *trifolii* and *Rhizobium meliloti* are about 2 μm in length. Photos b–d reprinted with permission from *Nature 351:* 670–673 (1991), © Macmillan Magazines Ltd.

the pea) is transferred to *Rhizobium leguminosarum* biovar *trifolii* (whose host is clover), cells of the latter species effectively nodulate pea.

In the Sym plasmid of *Rhizobium leguminosarum* biovar *viciae, nod* genes are located between two clusters of genes for nitrogen fixation, the *nif* genes (in this species and in certain other *Rhizobium* species, *nif* genes are plasmid-borne). The arrangement of *nod* genes in the *R. leguminosarum* Sym plasmid is shown in Figure 14.70. Ten *nod* genes have been identified in this species. The entire *nod* region has been sequenced, and the function of many *nod* proteins is known. The *nodABC* genes are common to all species of *Rhizobium* and are involved in the production of chitinlike molecules, called *Nod factors,* which induce root hair curling and trigger cortical plant cell division, eventually leading to formation of the nodule (Figure 14.67). Chem-

ically, Nod factors consist of a backbone of *N*-acetylglucosamine to which various substituents are linked (Figure 14.71). Host specificity is determined by the precise structure of the Nod factor produced by a given species of *Rhizobium*. Thus, besides the common *nodABC* genes, whose products synthesize the nod backbone, each species of *Rhizobium* contains certain unique *nod* genes responsible for introducing chemical variations on the basic Nod factor backbone. That Nod factors alone are responsible for root hair curling and initiation of the nodule has been shown in experiments in which purified Nod factors have been added to root tissue; such tissue proceeds to form a nodule, even in the absence of *Rhizobium* cells.

In *Rhizobium leguminosarum* biovar *viciae nodD* encodes a regulatory protein, NodD, which controls transcription of other *nod* genes (Figure 14.70). NodF is

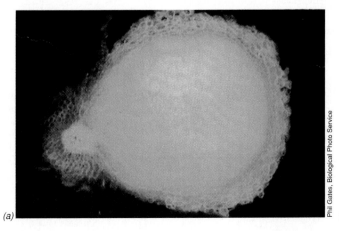

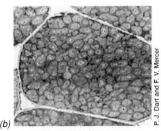

Figure 14.68 (a) Cross section through a legume root nodule, as seen by fluorescence microscopy. The darkly stained region contains plant cells filled with bacteria. (b) Electron micrograph of a thin section through a single bacteria-filled cell of a subterranean clover nodule. Cells of *Rhizobium trifolii* are about 2 μm long.

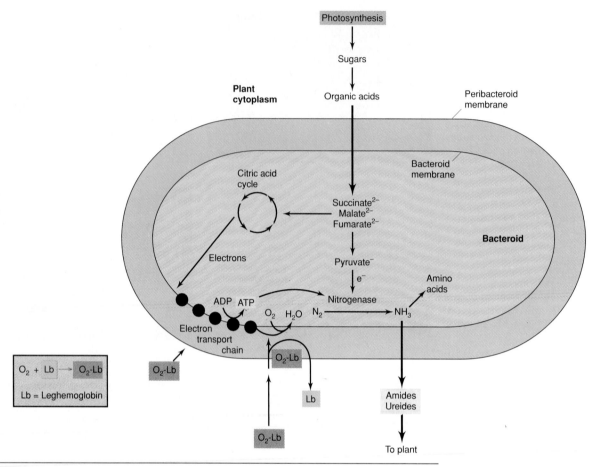

Figure 14.69 Schematic diagram of major metabolic reactions and nutrient exchanges occurring in the bacteroid.

involved in transport of substances needed for effective nodulation, NodE and NodL are involved in host range (cross-inoculation groups), NodM is a glucosamine synthase involved in Nod factor synthesis, and NodI and NodJ are membrane proteins that function to export Nod factors from the bacterial cells.

NodD is a member of a family of regulatory proteins that function to activate transcription of other genes by bending DNA at the promoter; this enhances binding of RNA polymerase. Thus, NodD can be considered a type of *positive* regulatory element (∞ Section

7.4). It is thought that NodD binds to regions upstream of *nod* structural gene operons (regions that have been called *nod boxes*) and, after interacting with inducer molecules, bend DNA in this region, which promotes transcription. Several inducer molecules have been identified. In most cases these are plant *flavonoids*, complex organic molecules that are widespread plant products (Figure 14.72*a* and *b*). Flavonoids have many functions in plants, including growth regulation and attraction of pollinating animals. However, leguminous plants are unusual because, unlike those of other

Figure 14.70 Organization of the *nod* gene cluster on the Sym plasmid of *Rhizobium leguminosarum* biovar *viciae*. The product of *nodD* controls transcription of other *nod* genes. The *nod* boxes are highlighted in red, and the arrows indicate the direction of transcription of *nod* genes. See text for further details.

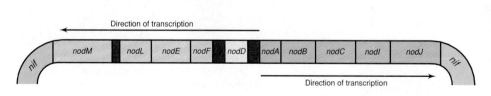

(b)

Species	R_1	R_2
Rhizobium meliloti	C16:2 or C16:3	SO_4^{2-}
Rhizobium leguminosarum biovar *viciae*	C18:1 or C18:4	H or Ac

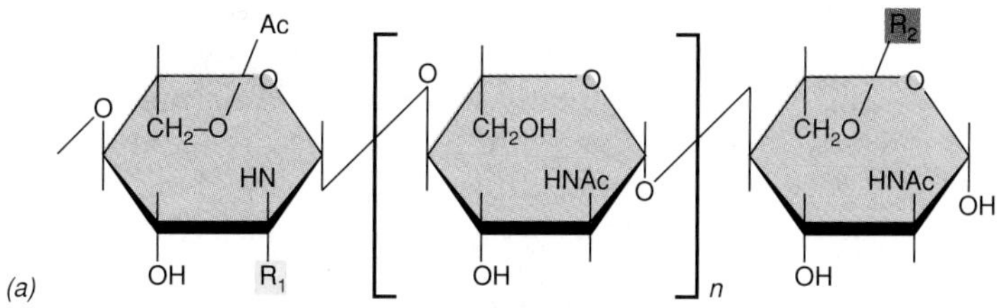

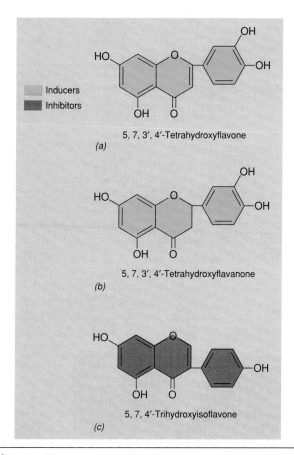

Figure 14.71 Nod factors. (a) The general structure of the Nod factor produced by *Rhizobium meliloti* and *Rhizobium leguminosarum* biovar *viciae* and (b) in the table the structural differences (R_1, R_2) that define the precise Nod factor of each species. The central hexose unit can repeat up to three times. C16:2, Palmitic acid with two double bonds; C16:3, palmitic acid with three double bonds; C18:1, oleic acid with one double bond; C18:4, oleic acid with four double bonds; Ac, acetyl.

plants, their *roots* secrete large amounts of flavonoids, presumably to trigger *nod* gene expression in nearby rhizobial cells in the soil. Interestingly, some flavonoids that are structurally very closely related to *nodD* inducers (Figure 14.72*c*) strongly *inhibit* induction of *nod* genes in certain rhizobial species, suggesting that part of the specificity observed between plant and bacterium in the *Rhizobium*–legume symbiosis could lie in the chemical nature of the flavonoids excreted by a particular plant.

Genetic cooperativity in the Rhizobium*–legume symbiosis*

Other genetically directed functions crucial to the *Rhizobium*–legume symbiosis have also been identified. As previously noted, the important O_2 binding protein in the root nodule, *leghemoglobin,* is genetically encoded in part by both the plant and the bacterium. Nitrogen fixation itself is clearly a process genetically directed by the bacterium; *nif* genes reside either on Sym plasmids or, in some *Rhizobium* species, on the chromosome itself. By contrast, plant lectin and flavonoid synthesis clearly is a genetic property of the plant.

The gene for the enzyme hydrogenase, the *hup* gene, is bacterial and is frequently encoded on the Sym plasmid. The role of hydrogenase in rhizobia is to incorporate H_2 evolved by the activity of nitrogenase (∞ Section 13.27). Screening of wild-type strains of various species of *Rhizobium* have shown that only about 20% of all *Rhizobium leguminosarum* and *Bradyrhizobium japonicum* strains contain *hup* genes and produce hydrogenase, and that this important enzyme is totally absent from field strains of many other *Rhizobium* species. Yield studies have shown that strains of rhizobia containing hydrogenase increase total plant nitrogen levels significantly over strains that lack hydrogenase activity because some of the otherwise wasted H_2 (from the activity of nitrogenase) is recycled. This finding has stimulated efforts to genetically engineer *hup* genes into all strains of *Rhizobium* used for field inoculation.

Figure 14.72 Structure of flavonoid molecules that are (a, b) inducers of *nod* gene expression, and (c) inhibitors of *nod* gene expression in *Rhizobium leguminosarum* biovar *viciae.* Note the similarities in the structures of all three molecules. The common name of the structure shown in (a) is *luteolin* and it is a flavone derivative. The structure in (b) is called *eriodictyol* and is chemically a flavanone. The structure in (c) is called *genistein* and it is an isoflavone derivative.

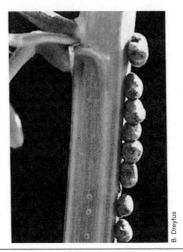

Figure 14.73 Stem nodules caused by stem-nodulating *Rhizobium*. The photograph shows the stem of the tropical legume *Sesbania rostrata*. On the left side of the stem are uninoculated sites, on the right identical sites inoculated with stem-nodulating rhizobia.

Stem-nodulating rhizobia

Although most leguminous plants form nitrogen-fixing nodules on their *roots*, a few legume species bear nodules on their *stems*. Stem-nodulated leguminous plants are quite widespread in tropical regions where soils are often nitrogen-deficient because of leaching and intense biological activity. The major experimental system here has been the tropical legume *Sesbania*, which is nodulated by the bacterium *Azorhizobium caulinodans* (Figure 14.73). Stem nodules usually form in the submerged portion of the stems or on portions of the stem just above the water level (Figure 14.73). From studies carried out thus far, the general sequence of events in formation of stem nodules strongly resembles that of root nodules. An infection thread is formed, and bacteroid formation predates the N$_2$-fixing state.

One curious finding is that some stem-nodulating rhizobia produce bacteriochlorophyll *a* and thus may

have the potential to carry out anoxygenic photosynthesis (∞ Section 13.4). Bacteriochlorophyll-containing rhizobia are apparently widespread in nature, particularly in association with tropical legumes, and in these species, light could drive at least part of the energy-demanding process of N$_2$ fixation in the bacterium.

Nonlegume nitrogen-fixing symbioses and other associations

In addition to the legume–*Rhizobium* relationship, nitrogen-fixing symbioses occur in a variety of non-leguminous plants, involving microorganisms other than rhizobia. Nitrogen-fixing cyanobacteria form symbioses with a variety of plants. The water fern *Azolla* contains a species of heterocystous N$_2$-fixing cyanobacteria called *Anabaena azollae* within small pores of its fronds (Figure 14.74). *Azolla* has been used for centuries to enrich rice paddies with fixed nitrogen. Before planting rice, the farmer allows the surface of the rice paddy to become densely covered with the fern. As the rice plants grow, they eventually crowd out the *Azolla–Anabaena* mixture, leading to death of the fern and release of fixed nitrogen, which is assimilated by the rice plants. By repeating this process each growing season, the farmer can obtain high yields of rice without the need for nitrogenous fertilizers.

The alder tree (genus *Alnus*) has nitrogen-fixing root nodules (Figure 14.75) that harbor a filamentous, streptomycete-like, nitrogen-fixing organism called *Frankia*. Although when assayed in cell extracts the nitrogenase of *Frankia* is sensitive to molecular oxygen, like intact cells of *Azotobacter* (∞ Section 16.16), intact cells of *Frankia* fix N$_2$ at full oxygen tensions. This is because *Frankia* protects its nitrogenase by localizing it in terminal swellings on the cells called *vesicles* (Figure 14.75b). The vesicles contain thick walls that act as a barrier to O$_2$ diffusion, thus maintaining the O$_2$ tension within vesicles at levels compatible with nitrogenase activity. In this regard, *Frankia* vesicles resemble the heterocysts produced by some filamentous cyanobacteria as localized sites of N$_2$ fixation (see Figure 14.74b and Section 16.2).

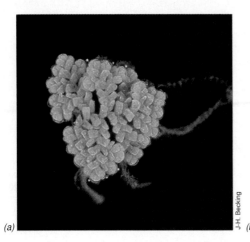

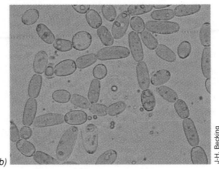

Figure 14.74 *Azolla–Anabaena* symbiosis. (a) Intact association showing a single plant of *Azolla pinnata*. The diameter of the plant is approximately 1 cm. (b) Cyanobacterial symbiont *Anabaena azollae* as observed in crushed leaves of *A. pinnata*. Single cells of *Anabaena azollae* are about 5 μm wide. Note the oval-shaped *heterocysts* (lighter color), the site of nitrogen fixation in the cyanobacterium.

(a)   (b)

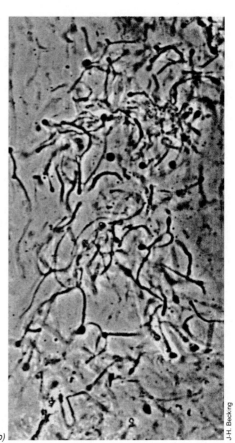

Figure 14.75 *Frankia* nodules and *Frankia* cells. (a) Root nodules of the common alder *Alnus glutinosa*. (b) *Frankia* culture purified from nodules of *Comptonia peregrina*. Note vesicles (spherical structures) on the tips of hyphal filaments. *(a)*   *(b)*

Alder is a characteristic pioneer tree able to colonize bare soils at nutrient-poor sites, and this is likely due to its ability to enter into a symbiotic nitrogen-fixing relationship with *Frankia*. A number of other small woody plants are nodulated by *Frankia*. This root nodule symbiosis has been reported in at least eight families of plants, many of which show no evolutionary relationships to one another. This suggests that the nodulation process in the *Frankia* symbiosis is more of a generalized phenomenon than the highly specific process observed in the *Rhizobium*–legume symbiosis and this holds promise for experimental attempts to expand the *Frankia* symbiosis to agriculturally important plants.

Azospirillum lipoferum is a N₂-fixing bacterium that lives in a rather casual association with roots of tropical grasses. It is found in the rhizosphere, where it grows on products excreted from the roots. It also has the ability to grow around the roots of cultivated grasses, such as corn (*Zea mays*), and inoculation of corn with *A. lipoferum* may lead to small increases (about 10%) in growth yield of the plant. In sugarcane, the organism *Acetobacter diazotrophicus* grows in plant vascular tissue and fixes substantial amounts of N₂, which benefits the plant. Cultures of *A. diazotrophicus* require high sucrose concentrations, suggesting that this plant–bacterium association may be more highly developed than the *Azospirillum* association. It is likely

that many casual to tight associations exist between bacteria and plants, although only in the *Rhizobium–Frankia* nodule has direct evidence for a symbiotic relationship been found.

CONCEPT CHECK   14.23

One of the most widespread and important plant–microbial relationships is that between legumes and certain nitrogen-fixing bacteria. The bacteria induce the formation of root nodules within which the nitrogen-fixing process occurs. The plant provides the organic energy source needed by the root nodule bacteria, and the bacteria provide fixed nitrogen for the growth of the plant. The root nodule bacteria play an important agricultural role because many important crop plants are legumes (alfalfa, clover, soybeans, peas, and so on).

✔ *What is a* leguminous *plant?*

✔ *What is* leghemoglobin *and what is its function?*

✔ *What is a* bacteroid *and what occurs within it?*

✔ *What are the major similarities and differences between* Rhizobium *and* Frankia?

Material for Review

REVIEW QUESTIONS

1. Explain why both obligately *anaerobic* and obligately *aerobic* bacteria can be isolated from the same soil sample.

2. What is the basis of the *enrichment culture technique?* Why is an enrichment medium usually suitable for the enrichment of only a certain group or groups of bacteria?

3. What is the principle of the Winogradsky column and what types of organisms does it serve to enrich? How might a Winogradsky column be used to study the breakdown of a xenobiotic compound?

4. Compare and contrast the use of fluorescent dyes and fluorescent antibodies for use in enumerating bacteria in natural environments. What advantages and limitations do each of these methods have?

5. Can nucleic acid probes in microbial ecology be as sensitive as culturing methods? If so, how? What advantages do nucleic acid methods have over culture methods? What disadvantages?

6. What are the major advantages of *radioisotopic methods* in the study of microbial ecology? What type of controls (discuss at least two) would you include in a radioisotopic experiment to show $^{14}CO_2$ incorporation by phototrophic bacteria or to show $^{35}SO_4^{2-}$ reduction by sulfate-reducing bacteria?

7. What information is obtained from knowledge of the biochemical oxygen demand (BOD) of a water source? Which should have a higher BOD, drinking water or raw sewage? Why?

8. What is the difference between *barotolerant* and *barophilic* bacteria? Between these two groups and *extreme barophiles?* What properties do barotolerant, barophilic, and extremely barophilic bacteria have in common?

9. How can organisms such as *Syntrophobacter* and *Syntrophomonas* grow when their metabolism is based on thermodynamically unfavorable reactions? How are these organisms grown in laboratory culture?

10. Why is sulfate reduction the main form of anaerobic respiration in marine environments, whereas methanogenesis dominates in fresh waters? Does any methanogenesis occur in the marine environment? If so, how?

11. What is the *rumen* and how do the digestive processes operate in the ruminant digestive tract? What are the major benefits and the disadvantages of a rumen system? How does a cecal animal compare with a ruminant?

12. Why can urea or ammonia be a nitrogen source for ruminants but not for humans?

13. Compare and contrast the processes of *nitrification* and *denitrification* in terms of the organisms involved, the environmental conditions that favor each process, and the changes in nutrient availability that accompany each process.

14. What organisms are involved in cycling sulfur compounds anoxically? If sulfur chemolithotrophs had never evolved, would there be a problem in the microbial cycling of sulfur compounds? What *organic* sulfur compounds are of interest in nature?

15. Why are all iron-oxidizing chemolithotrophs obligate aerobes and why are most iron oxidizers acidophilic?

16. Explain how spontaneous chemical reactions can acidify a coal seam and how both chemical reactions and *Thiobacillus ferrooxidans* continue the production of acid thereafter.

17. How is *Thiobacillus ferrooxidans* useful in the mining of copper ores? What crucial step in the indirect oxidation of copper ores is carried out by *T. ferrooxidans?* How is copper recovered from copper solutions produced by leaching?

18. How is mercury detoxified by the *mer* system?

19. What physical and chemical conditions are necessary for the rapid microbial degradation of oil in aquatic environments? Design an experiment that would allow you to test what conditions optimized the oil oxidation process.

20. Why is bioremediation a more attractive method of cleaning up oil spills or other toxic wastes than more conventional methods of attacking these problems?

21. What are *xenobiotic compounds* and why might microorganisms have difficulty catabolizing them?

22. How is the organism *Desulfomonile* of benefit in solving environmental pollution problems?

23. Describe the chemical properties of a "microbial plastic" now in commercial production.

24. Compare and contrast the production of a plant tumor by *Agrobacterium tumefaciens* and a root nodule by a *Rhizobium* species. In what ways are these structures similar? In what ways are they different? Of what importance are plasmids to the development of both structures?

25. What genetic information resides on the Ti plasmid? Which gene(s) could be deleted from the Ti plasmid without affecting tumorigenesis?

26. What ecological advantages do leguminous plants have over nonlegumes?

27. What substances of plant origin are found in root nodules of legumes that are required for the rhizobial symbiont to fix nitrogen?

28. Describe the steps in the development of root nodules on a leguminous plant. What is the nature of the recognition between plant and bacterium? How does this compare with recognition in the *Agrobacterium*–plant system?

29. How does nitrogen fixed by *Rhizobium* become part of the plant's proteins?

30. Describe the *nod* operons of a typical *Rhizobium* species such as *R. leguminosarum.* Explain how *nod* genes are regulated.

APPLICATION QUESTIONS

1. Compare and contrast a lake ecosystem with a hydrothermal vent ecosystem. How does energy enter each ecosystem? What is the basis of primary production in each ecosystem? What nutritional classes of organisms exist in each ecosystem and how do they feed themselves?

2. How do stains like acridine orange differ from phylogenetic stains in assessing microbial diversity of a habitat?

3. Why is the enrichment for nitrifying bacteria as described in Table 14.1? Why are each of the resources and conditions necessary?

4. Design an experiment for measuring the activity of sulfur-oxidizing bacteria in soil. How would you prove that your activity measurement was due to biological activity?

5. You wish to identify in soil samples any organism capable of autotrophic growth using the Calvin cycle. Keeping in mind that such autotrophs contain a unique enzyme, ribulose bisphosphate carboxylase, design a procedure that could be used to identify such organisms in nature.

6. ^{14}C-Labeled cellulose is added to a vial containing a small amount of sewage sludge and sealed under anaerobic conditions. A few hours later ^{14}CH$_4$ appears in the vial. Discuss what has happened to yield such a result.

7. Compare and contrast the microbiological steps involved in the conversion of cellulose to methane in the rumen as compared to cellulose conversion in lake sediments. What organisms are involved in each ecosystem and why?

8. Suppose you have discovered a new animal that consumes only grass in its diet. You suspect it to be a ruminant and have available a specimen for anatomical inspection. If this animal is a ruminant, describe the position and basic components of the digestive tract you would expect to find and any key microorganisms and substances you might look for.

SUPPLEMENTARY READINGS

Ahmadjian, V. 1993. *The Lichen Symbiosis.* John Wiley, New York. The most complete reference to experimental studies on lichens.

Akermans, A. D. L., J. D. van Elsas, and **F. J. Bruizn** (eds.). 1995. *Molecular Microbial Ecology Manual.* Kluwer Academic Publishers, Norwell, MA. A detailed guide to laboratory and field study of microorgansims using molecular approaches.

Andrews, J. H. 1991. *Comparative Ecology of Microorganisms and Macroorganisms.* Springer-Verlag, New York. A comparative view of basic ecological principles as they pertain to organisms of dramatically different size.

Atlas, R. M., and **R. Bartha.** 1993. *Microbial Ecology—Fundamentals and Applications.* Benjamin-Cummings, Redwood City, CA. An elementary textbook of microbial ecology.

Chaudhry, G. R. (ed.). 1995. *Biological Degradation and Bioremediation of Toxic Chemicals.* Timber Press, Portland, OR. An up-to-date treatment of various aspects of toxic waste bioconversions.

Ford, T. E. (ed.). 1993. *Aquatic Microbiology: An Ecological Approach.* Blackwell Scientific, Cambridge, MA. A series of chapters on various aspects of microorganisms and their activities in aquatic ecosystems.

Gaylarde, C. C., and **H. A. Videla** (eds.). 1995. *Bioextraction and Biodeterioration of Metals.* Cambridge University Press, New York. Reviews of microbial corrosion processes and the use of microorganisms in microbial mining.

Hinchee R. E, B. C. Alleman, R. E. Hoeppel, and **R. N. Miller** (eds.). 1994. *Hydrocarbon Bioremediation.* Lewis Publishers, Boca Raton, FL. A multiauthored volume on various issues concerning the bioremediation of spilled petroleum. Many case history studies.

Hobson, P. N. (ed.). 1988. *The Rumen Microbial Ecosystem.* Elsevier Applied Science, London. A detailed treatment of the organisms and microbial processes of the rumen.

Humphris, S. E., R. A. Zierenberg, L. S. Mullineaux, and **R. E. Thomson.** (eds.). 1995. *Seafloor Hydrothermal Systems—Physical, Biological, and Geological Interactions.* American Geophysical Union, Washington, DC. A monograph on the biology and geochemistry of deep-sea hydrothermal vents.

Lederberg, J. (ed.). 1992. *Encyclopedia of Microbiology.* Academic Press, San Diego. Several good chapters on activities of microorganisms in nature.

Ratledge, C. (ed.). 1994. *Biochemistry of Microbial Degradation.* Kluwer Academic, Dordrecht, The Netherlands. A detailed treatment of the microbiology of oil, halogenated hydrocarbons, polymers, and various other environmental pollutants.

Rogers, J. E., and **W. B. Whitman** (eds.). 1991. *Microbial Production and Consumption of Greenhouse Gases.* American Society for Microbiology, Washington, DC. A consideration of production and sinks for methane, nitrogen oxides, and halogenated gases in natural ecosystems.

Saltzman, E. S., and **W. J. Cooper** (eds.). 1989. *Biogenic Sulfur in the Environment.* American Chemical Society, Washington, DC. An ACS series publication focusing on biological aspects of the sulfur cycle.

Spain, J. C. (ed.). 1995. *Biodegradation of Nitroaromatic Compounds.* Plenum Press, New York. Scientific accounts of the biodegradation of many toxic aromatic compounds, in particular, 2, 4, 6–Trinitrotoluene (TNT).

Varma, A., and **B. Hock.** (eds.). 1995. *Mycorrhiza—Structure, Function, Molecular Biology and Biotechnology.* Springer, New York. Discussion of the most recent aspects of these interesting root fungi symbioses.

Young, L. Y., and **C. E. Cerniglia** (eds.). 1995. *Microbial Transformation and Degradation of Toxic Organic Chemicals.* Wiley-Liss, New York. A thorough treatment of a number of toxic waste problems using microbial solutions.

 On~line resources for this chapter are on the World Wide Web at: http://www.prenhall.com/~brock (click on the table of contents link and then select Chapter 14).

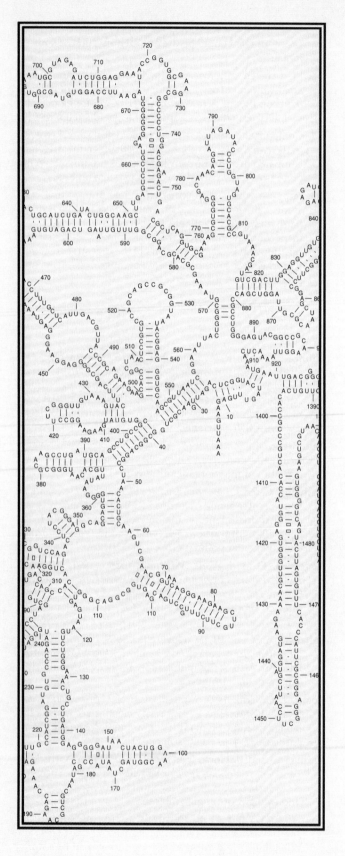

Microbial Evolution, Systematics, and Taxonomy

Evolutionary relationships between microorganisms can be deduced from the sequence of nucleotides in their ribosomal RNAs, such as that of the 16S rRNA of *Escherichia coli* shown here. See Figure 15.8c for the complete sequence.

MINIGLOSSARY for Chapter 15

Archaea a group of phylogenetically related prokaryotes distinct from Bacteria

Bacteria a group of phylogenetically related prokaryotes distinct from Archaea

Domain in a taxonomic sense, the highest level of biological classification (see alternative usage in Chapter 20)

Endosymbiosis a theory stating that the mitochondrion and chloroplast were originally free-living Bacteria that established stable residence in primitive eukaryotic cells, eventually yielding the modern eukaryotic cell

Eukarya all eukaryotic cells: algae, protozoa, fungi, slime molds, plant and animal cells

Evolutionary Distance in phylogenetic trees, the sum of the physical distance on a tree separating organisms; this distance is inversely proportional to evolutionary relatedness

Family in biological classification, an intermediate level of taxonomic hierarchy. Contains several genera, each of which consists of one or more species

GC Base Ratio in DNA (or RNA) from any organism, the percentage of the total nucleic acid that consists of guanine and cytosine bases

Genus a collection of different species, each sharing one or more (usually several) major properties

Phylogeny the evolutionary history of organisms

Proteobacteria a large group of phylogenetically related gram-negative Bacteria

RNA Life a life form lacking DNA and protein that may have existed on early Earth and in which RNA served both a genetic coding and a catalytic function

Signature Sequence short oligonucleotides of defined sequence in 16S or 18S rRNA characteristic of specific organisms or a group of phylogenetically related organisms

16S rRNA a large polynucleotide (~1500 bases) that functions as part of the small subunit of the ribosome of prokaryotes and from whose sequence evolutionary information can be obtained; eukaryotic counterpart, 18S rRNA

Species in microbiology, a collection of strains that all share the same major properties but differ in one or more significant properties from other collections of strains; two prokaryotic species generally show differences in 16S rRNA sequence of 3% or more

Stromatolites laminated microbial mats, typically built from layers of filamentous and other microorganisms, which can become fossilized

Taxonomy the science of identification, classification, and nomenclature of organisms

A recurrent theme throughout this book has been the enormous diversity of microorganisms on Earth. In Chapter 13 we considered the diversity of metabolic activities of microorganisms. In Chapter 14 we discussed how these activities impact on the environment and other organisms. In Chapters 16–18 we will cover the biology of the organisms themselves. In this chapter we consider how microbial diversity arose and how the evolutionary history of present-day microorganisms can be determined. We also discuss the properties of early Earth on which life arose.

Like the evolution of any living system, microbial evolution occurred by natural selection. Over time, mutation and genetic recombination led to more fit microorganisms, better suited to life in particular environments. And as new microbial habitats arose on Earth, they were eventually colonized by microorganisms genetically capable of existing in them. Through this interaction of genetics and environment, the great diversity of habitats that exist on Earth eventually selected for the great microbial diversity we see today.

With this in mind, let us start our discussion of microbial evolution at the beginning—Earth before life was present—and see where microbial diversity has taken us today.

15.1
Evolution of Earth and Earliest Life Forms

Origin of Earth

Earth is about 4.6 billion years old as determined by radiodating measurements. Our solar system is thought to have been formed when a large, very hot star exploded, generating a new star (our sun) and the other components of our galaxy. Although no rocks dating to this period have yet been discovered on Earth, rocks dating back to nearly 4 *billion* years ago have been found in several locations on Earth. The oldest rocks discovered thus far are those of the Isua Formation in Greenland, which date to about 3.8 billion years ago. The Isua rocks are of three types: sedimentary, volcanic, and carbonate. The sedimentary composition of the Isua rocks is of particular evolutionary interest because from our understanding of how modern sedimentary rocks are formed, the presence of sedimentary rocks 3.8 billion years old strongly suggests that *liquid water* was present at that time. The presence of liquid water in turn implies that conditions on Earth at that time were likely to be compatible with life as we know it. Other rocks of ancient orgin include the Warrawoona series, Towers Formation, and Early Archaean Apex Basalt in Western Australia and the Swaziland series in southern Africa; all of these rock formations are about 3.5 billion years old.

Evidence for microbial life on early Earth

Fossil evidence of microbial life exists in rocks 3.6 billion years old and younger. Most of the microfossils in the oldest rocks resemble simple rod-shaped bacteria. But in certain rocks of this age, stromatolitic microfossils containing morphologically diverse prokaryotes are present in abundance. *Stromatolites* are fossilized microbial mats consisting of layers of primarily filamentous prokaryotes containing trapped sediment (Figure 15.1*a* and *b*). We discussed some characteristics of microbial mats in Section 14.5 (∞ Figure 14.12*a*). By comparison with the microbial composition of modern stromatolites growing in shallow marine basins (Figure 15.1*c–e*) and in hot springs in various locations in the world (∞Figures 14.12*a* and 15.1*f*), it is assumed that ancient stromatolites consisted of filamentous phototrophic bacteria. Although modern stromatolites are frequently composed of filamentous *cyanobacteria* (oxygenic phototrophs) this would not have been the case in the oldest stromatolites. Because Earth was still anoxic at that time, stromatolites dating from 3 billion years or older were probably made exclusively by *anoxygenic* phototrophs (purple and green bacteria) rather than O_2-evolving cyanobacteria.

In rocks younger than about 2 billion years old, the morphological diversity of fossil microorganisms is considerably greater than that of 3.5-billion-year-old rocks. Figure 15.2 shows some photomicrographs of

(a)

(b)

(c)

(d)

(e)

(f)

Figure 15.1 Ancient and modern stromatolites. (a) The oldest known stromatolite, found in a rock about 3.5 billion years old, from the Warrawoona Group in Western Australia. Shown is a vertical section through a laminated, hemispheroidal structure, which has been preserved in the rock. Scale, 10 cm. (b) Stromatolites of conical shape from 1.6-billion-year-old dolomite rock of the McArthur basin of the Northern Territory of Australia. (c) Modern stromatolites in a warm marine bay, Shark Bay, Western Australia. (d) Another view of large modern stromatolites from Shark Bay. Note the resemblance to the ancient stromatolites shown in (b). (e) Underwater photograph of modern stromatolites growing in Shark bay. The diver indicates the scale. Shown are large columns formed by a complex community of diatoms, cyanobacteria, and green algae, to which are attached various macroscopic algae. (f) Modern stromatolites composed of thermophilic cyanobacteria growing in a thermal pool in Yellowstone National Park.

thin sections of rocks containing structures remarkably similar to modern filamentous bacteria. Structures such as those shown in Figure 15.2a are morphologically similar to filamentous green phototrophic bacteria and cyanobacteria (∞ Sections 16.1 and 16.2) and provide strong evidence that prokaryotes as a group had evolved an impressive morphological diversity long before the advent of modern eukaryotic cells.

Conditions on early Earth

The atmosphere of early Earth was devoid of significant amounts of O_2 and hence constituted a *reducing* environment. Besides H_2O, a variety of gases were present, the most abundant being CH_4, CO_2, N_2, and NH_3. In addition, trace amounts of CO and H_2 existed, as well as considerable amounts of sulfide, as a mixture

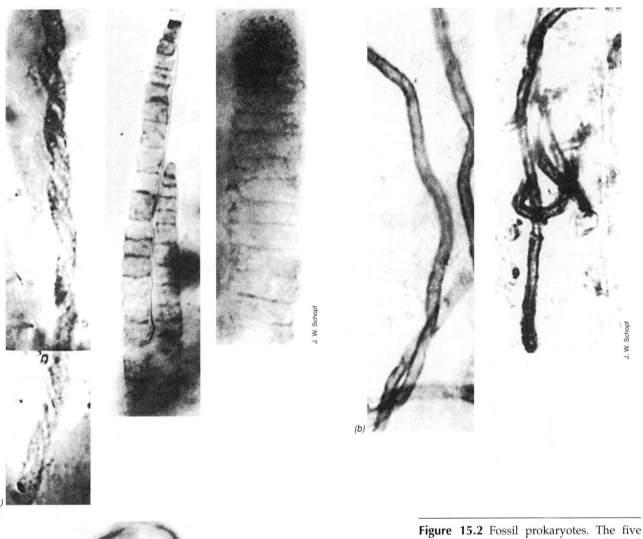

(a)

(b)

J. W. Schopf

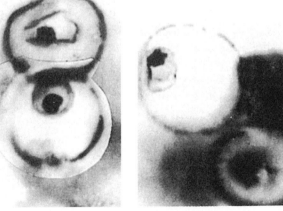

(c)

J. W. Schopf

Figure 15.2 Fossil prokaryotes. The five photographs in (a) (magnification, 2000×) and (b) (magnification, 920×) show fossil prokaryotic microorganisms found in the Bitter Springs Formation, a rock formation in central Australia about 1 billion years old. These forms bear a striking resemblance to modern cyanobacteria, anoxygenic phototrophs, or filamentous sulfur chemolithotrophs. The two photographs in (c) (magnification, 2000×) show microfossils possibly of a eukaryotic alga. The cellular structure is remarkably similar to that of certain modern green algae, such as *Chlorella* sp. These are from the same rock formation as are the prokaryotic organisms.

of H$_2$S and FeS. It is also likely that a considerable amount of hydrogen cyanide, HCN, was produced on early Earth when NH$_3$ and CH$_4$ reacted chemically to yield HCN. Geochemical estimates of the temperature of early Earth also suggest that it was a much hotter planet than it is today. For the first half billion or so years of its existence it is likely that the surface of Earth was greater than 100°C; thus free water probably did not exist on early Earth but accumulated only later as Earth cooled. How fast Earth cooled is unknown, but it is likely that living organisms first appeared at a time when Earth was much hotter than it is now. Thus, early life forms must have been quite heat-tolerant and resembled in this respect the hyperthermophilic prokaryotes that inhabit thermal environments today (∞ Sections 5.8 and 17.3).

Origin of life

It is now well established that the synthesis of biologically important molecules can occur if reducing atmospheres containing the aforementioned gases are subjected to intense energy sources. Of the energy sources available on primitive Earth, the most important was probably ultraviolet (UV) radiation from the sun, but lightning discharges, radioactivity, and thermal energy from volcanic activity were also available. If gaseous mixtures resembling those thought to be present on primitive Earth are irradiated with UV or subjected to electric discharges in the laboratory, a wide variety of biochemically important molecules can be made, such as sugars, amino acids, purines, pyrimidines, various nucleotides, thioesters, and fatty acids. It has also been shown that under prebiological conditions some of these biochemical building blocks could have polymerized, leading to the formation of polypeptides, polynucleotides, and the like. We can therefore imagine that on primitive Earth a rich mixture of organic compounds eventually accumulated, but in the absence of living organisms these compounds would have been stable (that is, not consumed) and should have persisted for countless years. Thus, with time, there should have been an extensive accumulation of organic materials, setting the stage for biological evolution.

A major difficulty with most hypotheses of prebiotic synthesis surrounds how *macromolecules* could have originated from monomeric constituents spontaneously in an aqueous environment. From a chemical standpoint, nucleic acids and proteins are polymerized via *dehydration* reactions; thus, it is difficult to conceive of how, in the absence of enzymes, macromolecules could have arisen in an aqueous setting. To avoid this problem, it has been hypothesized that relatively anhydrous *exposed surfaces* such as clays, pyrite, or basaltic glasses functioned as supports for prebiotic polymer-

ization reactions. Such surfaces would have provided a stable, relatively dry environment for the synthesis and accumulation of macromolecules into organic films from which a primitive cellular structure could have emerged. Pyrite (FeS$_2$) in particular is favored for this process because of the crucial role it may have played in early energy-generating systems (see next section).

CONCEPT CHECK   *15.1*

Earth is thought to be 4.6 billion years old; the first evidence of microbial life emerges in rocks about 3.6 billion years old. Early Earth was anoxic and much hotter than at present. The first biochemical compounds were made by abiotic syntheses that set the stage for the origin of life.

✔ *How old is Earth? How old are the oldest known microfossils?*

✔ *How did the atmosphere of early Earth compare with that of Earth today?*

✔ *How were early biochemicals formed?*

15.2
Primitive Organisms and Metabolic Strategies

What was the first living organism like? This is something we will probably never know. However, even very primitive organisms must have possessed the following: (1) **metabolism,** that is, the ability to accumulate, convert, and transform nutrients and energy; and (2) a **hereditary mechanism,** the ability to replicate and transfer its properties to its offspring. Both of these features as we currently understand them require the development of a *cellular* structure. Cell-like structures probably arose through the spontaneous aggregation of lipid and protein molecules to form membranous structures within which were trapped polynucleotides, polypeptides, and other substances. This step may have occurred countless times on early Earth to no avail, but just once the proper set of constituents could have become associated, and a primitive organism arose. This primordial organism would certainly have been structurally simple (that is, resembling a *prokaryotic* cell) and would have depended on very simple energy-generating and self-replicating mechanisms.

Although no precise time period can be given for when primitive, self-replicating entities first arose, the fact that prokaryotic microfossils are found in rocks nearly 3.6 billion years old (see Figure 15.1), implies that the earliest life forms probably date from between 3.6 and 4 billion years ago. Remarkably, this means that after early Earth cooled to the point at which liquid water was present (thought to be about 4.0–4.2 billion years ago), life arose relatively rapidly, perhaps in as little as 200–400 million years.

Metabolism in primitive organisms

Because of the reducing conditions prevailing on early Earth, primitive organisms must have been able to carry out some form of *anaerobic* metabolism. However, from our consideration of metabolic diversity, this constraint rules out very little; early organisms could have used chemoorganotrophic, chemolithotrophic, or phototrophic metabolism because all these energy-generating processes can occur anoxically (∞ Chapter 13).

It is unlikely that bioenergetic mechanisms in primitive cells employed multiple steps, common in catabolic pathways in modern organisms (∞ Chapters 4 and 13). Instead, energy-yielding reactions were likely quite simple, perhaps requiring as few as one or two proteins. A potential mechanism that fits this scenario well is one in which ferrous iron, such as ferrous carbonate or ferrous sulfide (compounds that were abundant on early Earth; see Section 15.1), reacts with hydrogen sulfide to form pyrite and molecular hydrogen:

$$FeCO_3 + 2\,H_2S \rightarrow FeS_2 + H_2 + H_2O + CO_2$$
$$\Delta G^{0\prime} = -61.7 \text{ kJ/reaction}$$

$$FeS + H_2S \rightarrow FeS_2 + H_2$$
$$\Delta G^{0\prime} = -41.9 \text{ kJ/reaction}$$

Both of the preceding reactions are exergonic and yield sufficient free energy to couple to the formation of adenosine triphosphate (ATP) or other energy-trapping molecules such as thioesters (∞ Sections 4.8 and 13.9) that may have been used by early cells. Interestingly, both of the reactions shown also yield H_2, and it is possible that H_2 produced from these reactions could have been an electron donor for reduction of elemental sulfur (S^0) to H_2S (Figure 15.3). Energy coupling in this chemolithotrophic reaction could have occurred by separating H_2 into electrons and protons across a simple cytoplasmic membrane, thereby establishing a proton motive force to drive a primitive ATPase (Figure 15.3). Interestingly, most hyperthermophilic Archaea, which are probably the closest extant relatives of Earth's earliest organisms (see Section 15.7), are able to reduce S^0 with H_2 and form H_2S; most can also produce pyrite when supplied with Fe^{2+} (∞ Section 17.3). Perhaps these reactions are modern remnants of primitive metabolic schemes employed on early Earth by the first living organisms (∞ Microbial Life Deep Underground, Chapter 14, for another example).

Carbon to make cell material of primitive organisms could have come from organic carbon com-

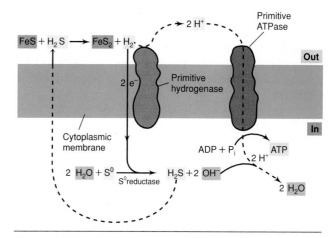

Figure 15.3 A hypothetical energy-generating scheme for primitive cells. Formation of pyrite leads to H_2 production and S^0 reduction, which fuels a primitive ATPase. Note how H_2S plays only a catalytic role; the net substrates would be FeS and S^0. Also note how few different proteins would be required. The $\Delta G^{0\prime}$ of the reaction $FeS + H_2S \rightarrow FeS_2 + H_2 = -41.9$ kJ.

pounds present in the environment from prebiotic syntheses. Autotrophy, the use of CO_2 as carbon source, is also a possibility, but it is hard to imagine a simple means by which CO_2 could have been converted to organic intermediates because modern pathways of CO_2 fixation are highly complex and require many enzymes (∞ Sections 13.7, 16.1, and 16.9).

In addition to these chemolithotrophic energy-yielding reactions, the earliest cells could also have obtained energy by chemoorganotrophic mechanisms, most likely simple fermentations. But even a relatively "simple" fermentative pathway such as glycolysis (∞ Section 4.10) requires many individually catalyzed reactions. Thus, simple chemolithotrophic reactions seem more likely as the basis of early metabolism. Photosynthesis is also a possibility but seems less likely than chemolithotrophic or chemoorganotrophic reactions for the earliest cells because of the biochemical requirements and complexity of photosynthetic reactions (∞ Sections 13.2–13.6). The attractive feature of the pyrite-based energy scheme just discussed is that it would have required very few enzymes, perhaps only a primitive hydrogenase and an ATPase to trap energy released from the reaction (Figure 15.3).

Regardless of the energy sources for life on early Earth, the organisms present were undoubtedly biochemically very simple, in the sense of possessing very few enzymes. Nutritional requirements would likely have been complex. No cytochrome system would have been present, and no flagella or other special morphological properties would have been necessary. To make survival in osmotically varying environments possible, a cell wall probably evolved fairly early. The earliest organisms would probably have required a variety of growth factors and other complex organic nutrients. However, as time went on, mutation and selection would have yielded new organisms with greater biosynthetic capacities, better adapted to the changing chemical environment.

Further metabolic evolution and photosynthesis

Another highlight in metabolic evolution would have been the evolution of the first porphyrins because this, along with refinements in membrane structure over time, could have led to the construction of cytochromes and other tetrapyrroles and electron transport chains able to carry on electron transport phosphorylation (Figure 15.4). This could have opened up many new possibilities for anaerobic respiration (∞ Sections 13.14–13.18).

The evolution of porphyrins would also have allowed development of light-sensitive magnesium tetrapyrroles, the bacteriochlorophylls, and thus photosynthesis (Figure 15.4). Following this major event, a great explosion of life could have occurred because of the availability of the enormous amount of energy from the radiation of the sun. The first phototrophic organisms were no doubt *anoxygenic*, probably using light only for ATP synthesis and employing reduced compounds from its environment—such as H_2S—as sources of reducing power. Such an organism may have resembled one of the modern-day purple or green sulfur bacteria (∞ Section 16.1).

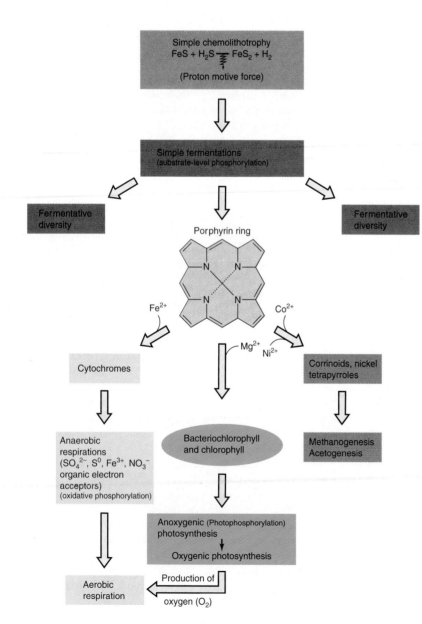

Figure 15.4 Possible scenario for the evolution of energy-generating mechanisms in prokaryotes. Because photosynthesis and virtually all forms of anaerobic and aerobic respiration require cytochromes or related tetrapyrroles, it is likely that evolution of the porphyrin ring was a milestone in metabolic evolution. Oxygenic photosynthesis would have first evolved in cyanobacteria about 2.5 billion years ago. For a description of the energetic principles of FeS-based chemolithotrophy, see Figure 15.3.

A monumental step in microbial evolution occurred with development of the second light reaction of photosynthesis (∞ Section 13.5), making it possible to use the plentiful supply of H_2O as an electron donor. Because both ATP and reduced pyridine nucleotides (NADH, NADPH) could now be made photosynthetically, light energy could be used more efficiently. Such organisms would have considerable competitive advantage over other phototrophic organisms. The first of this type was prokaryotic, similar to cyanobacteria. The evolution of oxygenic photosynthesis also had enormous consequences for the environment of Earth because as O_2 gradually accumulated, the atmosphere changed from an anoxic to an oxic one (Figure 15.5). With O_2 now available as an electron acceptor, aerobic organisms evolved; these were able to obtain much more energy from the oxidation of organic compounds than anaerobes could (∞ Chapter 4). More energy was made available, and higher population densities could develop, increasing the chances for the evolution of new types of organisms and metabolic schemes. There is good evidence from the fossil record that, at about the time that Earth's atmosphere became highly oxidizing, there was an enormous burst in the rate of evolution, leading to the appearance of eukaryotic microorganisms with organelles and from them to metazoa (multicelled organisms) and eventually to higher animals and plants (see Figure 15.5 and Section 15.7).

The ozone shield

Another major consequence of the appearance of O_2 was the formation of ozone (O_3), a substance that provides a barrier preventing the intense ultraviolet radiation of the sun from reaching Earth. When O_2 is subject to short-wavelength ultraviolet radiation, it is converted to O_3, which strongly absorbs wavelengths

---

CONCEPT CHECK   **15.2**

The first organisms must have employed a simple strategy to obtain energy. Primitive metabolism was anaerobic and likely chemolithotrophic, exploiting the abundant sources of FeS and H_2S present. Fermentations and anaerobic respiration appeared later along with anoxygenic followed by oxygenic photosynthesis. The latter led to development of an oxic environment and to great bursts of biological evolution.

✔ *How could energy have been obtained from FeS + H_2S by early cells?*

✔ *Give at least two reasons why it is unlikely that oxygenic photosynthesis was the first energy-generating mechanism on Earth.*

✔ *Of what importance was evolution of the first porphyrins?*

Figure 15.5 Major landmarks in biological evolution. The positions of the stages on the time scale are approximate. Note how the oxygenation of the atmosphere due to cyanobacterial metabolism was a gradual process, occurring over a period of about a billion years. Also note that for the bulk of Earth's history, only microbial life forms existed.

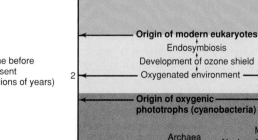

up to 300 nm. Until an ozone shield developed, evolution could have continued only in environments protected from direct radiation from the sun, such as under rocks or in the oceans. After the photosynthetic production of O₂ and development of an ozone shield, organisms could have ranged generally over the surface of Earth, permitting evolution of a greater diversity of living organisms. A summary of the steps that could have occurred in biological evolution is shown in Figure 15.5.

15.3
Primitive Organisms and Molecular Coding

Besides the generation of energy, early cells would have had to have some mechanism for simple genetic coding in order to replicate themselves with fidelity. How did this occur? It is tempting to extrapolate backward from the present and postulate that early organisms were much like modern cells but contained only a very few genes [made of deoxyribonucleic acid (DNA)] and had very limited transcriptional and translational abilities. But even something like this would have been very complex, much more complex than what the first self-replicating entities might have been. What would these creatures have been like? Following the discovery that certain types of ribonucleic acid (RNA) are catalytic (see discussion of ribozymes in Section 6.7),

many scientists now believe that the earliest life forms probably lacked DNA altogether, contained just a very few, if any, proteins, and consisted primarily of *RNA* (Figure 15.6)—simply put, this was the age of *RNA life.*

RNA life

If an "RNA world" consisting of self-replicating RNA molecules actually existed eons ago, it would have been during a period that predated cellular life. In the early RNA world, RNA molecules would have functioned simply to replicate themselves and would have carried out the minimal number of catalytic reactions necessary for this purpose (Figure 15.6). This period of RNA life could have evolved into the first *cellular* life forms when self-replicating RNAs became enclosed within lipoprotein vesicles (Figure 15.6). This could have occurred countless times until a mechanism evolved for the coordinated replication of the RNA along with the lipoprotein vesicle. Although still lacking DNA and proteins, this life form would otherwise have resembled a modern cell, and as such life forms became more widespread, natural selection led to their further evolutionary development.

How would RNA life forms have accomplished all the reactions we associate with living systems? Without proteins, biochemistry in RNA organisms would by necessity have been very simple compared with that in modern cells. RNA has poor catalytic specificity, and in modern cells those few reactions

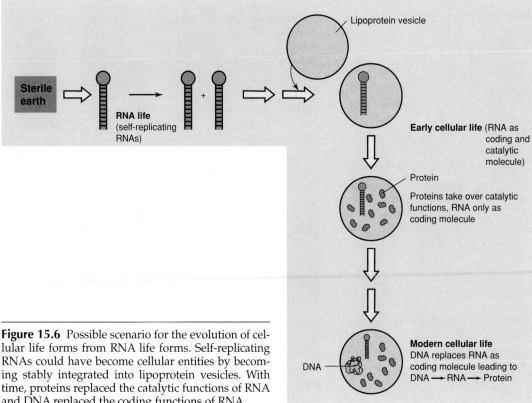

Figure 15.6 Possible scenario for the evolution of cellular life forms from RNA life forms. Self-replicating RNAs could have become cellular entities by becoming stably integrated into lipoprotein vesicles. With time, proteins replaced the catalytic functions of RNA and DNA replaced the coding functions of RNA.

still catalyzed by RNA are biochemically simple compared with typical reactions catalyzed by protein enzymes (∞ Section 6.7). Thus, before life diversified significantly, a few types of RNA may have been able to catalyze the relatively few reactions required. However, as living organisms became biochemically more complex, an evolutionary push to *proteins* as the major biocatalysts probably occurred. It is likely that proteins appeared gradually in cells, perhaps at first complexed with RNA, and as evolution selected for more rapid and precise catalysts, RNA was eventually replaced by protein as major cellular enzymes (Figure 15.6).

The modern cell: DNA → RNA → Protein

The establishment of DNA as the genome of the cell may have resulted from the need to store genetic information in a more stable form. By storing all the genetic information in one place in the cell and processing only what was needed under a specific set of growth conditions (that is, *regulation of gene expression*), cells would have saved energy, which would have increased their competitive fitness. However, an additional reason for the origin of DNA as the genetic material may have been the highly error-prone nature of enzymes that copy RNA. For unknown reasons, enzymes that copy RNA are inherently less precise than DNA polymerases. Maintaining RNA as the genetic material and relying on inherently error-prone copying systems for its replication would likely have been incompatible with increasing cellular complexity and the precise genetic demands this would entail. Evolution would therefore have favored transfer of genetic storage to a form that had inherently high replicative fidelity, like DNA, and this would have eventually eliminated the less precise RNA life forms (Figure 15.6).

Somewhere in the early stages of microbial evolution, the three-part system—DNA, RNA, and protein—became fixed in cellular life as the best solution to biological information flow. That this system was an evolutionary success can be attested to by the fact that all extant cells contain all three kinds of informational macromolecules. Thus, although modern life forms employ DNA and proteins as essential parts of cellular function, early life forms may have accomplished all this with only RNA, employing a much simpler system of information storage and a rather nonspecific and inefficient means of biochemical catalysis.

15.4
Eukaryotes and Organelles

From comparative nucleic acid sequencing studies we now know that the three main lines of descent, *Bacteria*, *Archaea*, and *Eukarya* (*nuclear*), were established relatively early in cellular evolution (see Section 15.7). However, the eukaryotic cell we know today undoubtedly differs structurally from primitive eukaryotic cells. The nuclear line of descent (primitive eukaryotic cells) originally consisted of structurally quite simple cells, resembling modern-day prokaryotic cells in lacking mitochondria, chloroplasts, and a membrane-enclosed nucleus. Modern eukaryotic cells, with their distinctive cellular organelles, were the result of endosymbiotic events (see later) that took place billions of years after divergence of the nuclear line of descent from the universal ancestor (see Figures 15.7 and 15.12 and Section 15.7).

Origin of the nucleus

It seems likely that the eukaryotic nucleus and mitotic apparatus arose as a necessity for ensuring the replication and orderly partitioning of DNA once the genome size had increased to the point where replication as one molecule (as in prokaryotes) was no longer feasible. Origin of the nucleus made it possible for eukaryotic cells to manage the huge genomes needed to encode the genetic blueprint of multicellular organisms and also made possible the recombination of genomes through sexual reproduction (∞ Sections 3.14 and 9.12–9.14).

The widespread occurrence of plasmids in bacteria suggests that even in prokaryotes there are evolutionary advantages for segregation of genetic information into more than one DNA molecule. It is thus possible to imagine how separate chromosomes might have arisen in a primitive eukaryotic cell from plasmidlike structures and have become segregated within the cell into a membrane-enclosed nucleus. Probably spindle fibers and the mitotic apparatus would also have had to evolve at the same time. There is no obvious reason why this primitive eukaryote would have needed other typical eukaryotic organelles, and these could have arisen later. Indeed, even today there are eukaryotes known that lack organelles (see Figures 15.7 and 18.1 and Section 15.7).

CONCEPT CHECK   15.3

Primitive organisms may have used RNA in both a genetic and an enzymatic capacity, but because catalytic RNA is not very efficient, evolution of protein catalysts would have led to a marked improvement in cellular efficiency. DNA may have arisen because it provides a more stable form of genetic information than RNA and one that can be copied more accurately.

✔ *What evidence supports an era of "RNA life"?*

✔ *Why would RNA life forms not have survived to the present?*

Endosymbiosis

Strong evidence now exists that the modern eukaryotic cell evolved in steps through incorporation into cells from the nuclear line of descent of chemoorganotrophic and phototrophc symbionts. This theory, referred to as the **endosymbiotic theory** of eukaryotic evolution, has through the years gathered increasing experimental support (∞ Sections 3.15 and 9.13). The theory postulates that an aerobic bacterium established residency within the cytoplasm of a primitive eukaryote and supplied the larger cell with energy in exchange for a stable, protected environment and a ready supply of nutrients (Figure 15.7). This aerobic bacterium would represent the forerunner of the present mitochondrion.

In similar fashion, the endosymbiotic uptake of an oxygenic phototrophic prokaryote would have made the primitive eukaryote photosynthetic and no longer dependent on organic compounds for energy production. The phototrophic endosymbiont would then be considered the forerunner of the chloroplast (Figure 15.7). Following the acquisition of prokaryotic endosymbionts, eukaryotic cells underwent an explosion in biological diversity. The period from about 1.5 billion years ago to the present saw the rise of the metazoa and diversification of this highly successful group, culminating in the structurally complex higher plants and animals (see Figure 15.12 and Section 15.7). Studies on the nucleic acids and ribosomes of eukaryotic cells and their organelles have built an impressive case for the

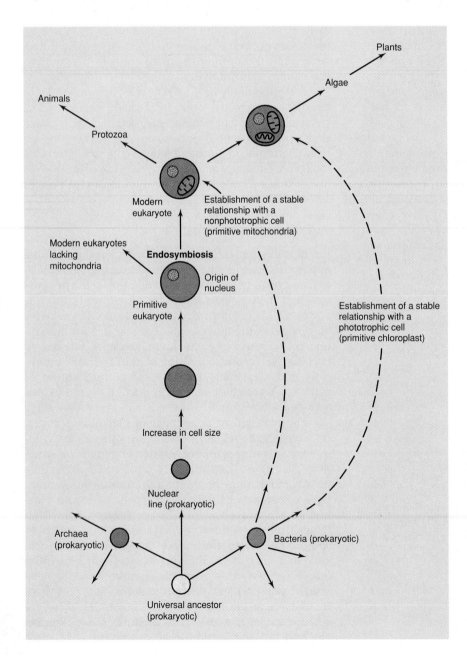

Figure 15.7 Origin of modern eukaryotes by endosymbiotic events. Note how organelles originated from Bacteria rather than Archaea. Endosymbiosis was unlikely to have been a one-time event and probably occurred in various types of cells of the nuclear line of descent. Note, however, how some primitive eukaryotes either never underwent endosymbiotic events or permanently lost their symbionts, but that they otherwise maintained the basic properties of eukaryotic cells. Extant examples of such eukaryotes, all microbial, are known today (∞ Figure 18.1).

theory of endosymbiosis. Briefly, these include the fact that the chloroplast and mitochondrion contain ribosomes of the prokaryotic type (and whose function is inhibited by antibiotics that inhibit ribosome function in Bacteria), contain small amounts of DNA arranged in a covalently closed, circular fashion typical of prokaryotes (∞ Section 3.15), and finally, show ribosomal RNA (rRNA) sequences (see Section 15.6) characteristic of certain Bacteria.

However, to evolve to the modern eukaryotic stage, primitive eukaryotic cells had to abandon several advantageous features of prokaryotic cells such as a haploid genome (and the concomitant ability to adapt rapidly to new environments) and structural simplicity. Also, the greater complexity of the eukaryotic cell meant that it would face difficulty in adapting to life in extreme environments, such as thermal areas, where the prokaryote is today preeminent. For these reasons the evolution of the eukaryote did not toll the death bell for the prokaryotes. All three cell lines, Archaea, Bacteria, and Eukarya, continued to evolve, carving out their ecological niches and serving as the forerunners of the various species we know today.

Biological evolution and geological time scales

In terms of geological time, the period from the origin of the first metazoa to the present represents only one-sixth of the total time that life has existed on Earth. Put another way, five-sixths of Earth's history was restricted to *microbial* life, the bulk of this period to *prokaryotes only* (see Figure 15.5). However, because metazoa have left a considerable and highly diverse fossil record, our understanding of biological evolution from the time of the metazoa is much greater than our knowledge of evolutionary relationships among prokaryotes. Fortunately, this has changed with the advent of molecular methods for discerning evolutionary relationships. The objective of most of the remainder of this chapter is to present in some detail the

---

CONCEPT CHECK   **15.4**

The eukaryotic nucleus and mitotic apparatus probably arose as a necessity for ensuring the orderly partitioning of DNA in large-genome organisms. Mitochondria and chloroplasts, the principal organelles of eukaryotes, probably arose from prokaryotes that became symbiotic within eukaryotic cells. This process has been called endosymbiosis.

✔ *Why did the eukaryotic nucleus evolve?*

✔ *What is the endosymbiotic theory and what evidence supports it?*

---

modern approaches to an understanding of microbial evolution based on molecular biology and genetics. From these methods a totally new evolutionary picture of life on Earth has emerged.

15.5
Evolutionary Chronometers

It is now clear that certain cellular macromolecules are evolutionary chronometers—actual measures of evolutionary change. From studies on the sequence of monomers in certain informational macromolecules, it has been shown that the *evolutionary distance* between two species can be measured by differences in the nucleotide or amino acid sequence of *homologous* macromolecules from the two species. This is so because the number of sequence differences in a molecule is proportional to the number of stable mutational changes fixed in the DNA encoding that molecule in species that diverged from a common ancestor. As different mutations become fixed in different populations, biological evolution results.

Choosing the right chronometer

In order to determine true evolutionary relationships between species, it is essential that the correct molecules be chosen for sequencing studies. This is important for several reasons. First, the molecule should be *universally distributed* across the group chosen for study. Second, it must be *functionally homologous* in each organism; phylogenetic comparisons must start with molecules of *identical* function. One cannot compare the amino acid sequences of enzymes that carry out different reactions, or the nucleotide sequences of nucleic acids of different function, because functionally unrelated molecules would not be expected to show sequence similarities. Third, it is crucial in sequence comparisons to be able to properly *align* the two molecules in order to identify regions of sequence homology and sequence heterogeneity. Finally, the sequence of the molecule chosen should change at a rate commensurate with the evolutionary distance measured. And in fact the broader the phylogenetic distance being measured, the *slower* must be the rate at which the sequence changes. A molecule that has undergone too many sequence changes is essentially useless as a tool for determining evolutionary relationships because regions of common sequence would eventually be lost.

Ribosomal RNAs as evolutionary chronometers

Because of the likely antiquity of the protein-synthesizing process and for several other reasons, ribosomal RNAs turn out to be excellent molecules for discerning evolutionary relationships among living organisms.

Ribosomal RNAs are ancient molecules, functionally constant, universally distributed, and moderately well conserved across broad phylogenetic distances. Also, because the *number* of different possible sequences of large molecules such as ribosomal RNAs is so large, similarity in two sequences always indicates *some* phylogenetic relationship. However, it is the *degree* of similarity in ribosomal RNA sequences between two organisms that indicates their relative evolutionary relatedness. From comparative sequence analyses, molecular genealogies can be constructed leading to phylogenetic trees that show the true evolutionary position of organisms relative to one another (see Figure 15.12).

Recall the structure of the ribosome (Figure 15.8). There are three ribosomal RNA molecules, which in prokaryotes have sizes of 5S, 16S, and 23S. The large bacterial rRNAs, 16S (Figure 15.8*c*) and 23S rRNA (approximately 1500 and 2900 nucleotides, respectively) contain several regions of highly conserved sequence useful for obtaining proper sequence alignments, yet contain sufficient sequence variability in other regions of the molecule to serve as excellent phylogenetic chronometers.

The 5S rRNA has also been used for phylogenetic measurements, but its small size (~120 nucleotides) limits the information obtainable from this molecule. Because 16S RNA is more experimentally manageable than 23S RNA, it has been used extensively to develop the phylogeny of both prokaryotes and eukaryotes (using the 18S rRNA counterpart of prokaryotic 16S rRNA). The database of rRNA sequences now numbers over 10,000 and can be accessed on the World Wide Web (http://rdpwww.life.uiuc.edu/). Use of 16S rRNA as a phylogenetic tool was pioneered in the early 1970s by Carl Woese at the University of Illinois, and the method is now widely used.

15.6
Ribosomal RNA Sequences and Evolution

The protocol for using RNA as a molecular chronometer of evolutionary relationships involves growing organisms and extracting their ribosomal RNA, determining the sequence of the 16S (prokaryotic) or 18S (eukaryotic) ribosomal RNA, entering the sequence data into a computer and choosing known sequences from which to compare the new sequence(s), and finally, allowing the computer to generate phylogenetic analyses, usually as phylogenetic trees. Although growing a pure culture of the organism to be analyzed is obviously a prerequisite for study of its phylogenetic position, *cultures* of organisms are not necessary in order to do some phylogenetic studies. Microbial ecologists have turned to ribosomal RNA sequencing methods to determine the phylogeny of organisms in microbial communities by extracting ribosomal RNA from cells in natural samples and performing phylogenetic analysis directly on these (∞ Section 14.4). However, our discussion here will proceed on the assumption that a pure culture of the organism to be analyzed is available.

Sequence methodology

Ribosomal RNAs are now relatively easy to analyze because they can be directly sequenced from crude cell extracts using reverse transcriptase and the dideoxy sequencing method (∞ Working with Nucleic Acids: The Tools, Chapter 6). A typical rRNA sequencing experiment involves phenol extraction of a relatively small (0.3–0.5 g) amount of cells to remove protein and other nonnucleic acid material and release the RNA. Total RNA is then precipitated with alcohol and salt, and then a small DNA oligonucleotide primer of about 15–20 nucleotides in length *complementary* in base sequence to some highly conserved region of the 16S rRNA molecule is added to the mixture (Figure 15.9). The enzyme reverse transcriptase (∞ box, Chapter 6, and Section 8.22) is then added along with a ^{32}P-labeled deoxyadenosine triphosphate and the other unlabeled deoxyribonucleotides, and then the mixture is divided into four identical portions. To each portion, a small amount of a different 2′,3′-*di*deoxynucleotide is added. Reverse transcriptase reads the 16S rRNA template and begins making a DNA copy terminated at various spots by the incorporation of *di*deoxynucleotides (Sanger sequencing method; ∞ Working with Nucleic Acids: The Tools, Chapter 6). The fragments are then separated by electrophoresis. From knowledge of the cDNA sequence obtained by this dideoxy sequencing, the sequence of the original 16S rRNA can be deduced.

Although direct sequencing of ribosomal RNA is still used, newer methods are beginning to supplant this approach. Specifically, the polymerase chain reaction (PCR) technique (∞ Section 10.9) is being used to amplify rRNA *genes* (the DNA that encodes the 16S rRNA) using synthetically produced primers complementary to conserved sequences in rRNA as PCR templates (Figure 15.10). PCR amplification of the DNA encoding rRNA requires less cell material than direct rRNA sequencing and is more rapid and convenient for large-scale studies than the direct method. The amplified DNA is then sequenced directly using the dideoxy sequencing method (Figure 15.10).

Phylogenetic trees from RNA sequences

A major method of generating phylogenetic trees from rRNA sequences is the so-called *distance-matrix method* (Figure 15.11, page 621). Two rRNA sequences are aligned, and an **evolutionary distance** (E_D) is calcu-

Figure 15.8 Ribosomal RNA. (a) Electron micrograph of 70S ribosomes from the bacterium *Escherichia coli*. (b) Parts of the ribosome; 5S, 16S, and 23S refer to different forms of RNA in the small subunit of the ribosome. (c) Primary and secondary structure of 16S ribosomal RNA (rRNA). This is the 16S rRNA from *Escherichia coli* (Bacteria); 16S rRNA from Archaea has general similarities in secondary structure (folding) but numerous differences in primary structure (sequence). The counterpart to 16S rRNA in eukaryotes is 18S rRNA present in cytoplasmic ribosomes.

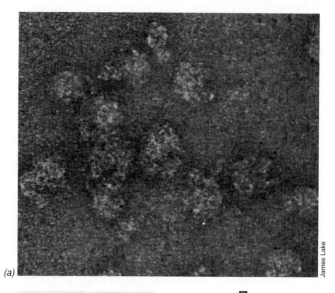

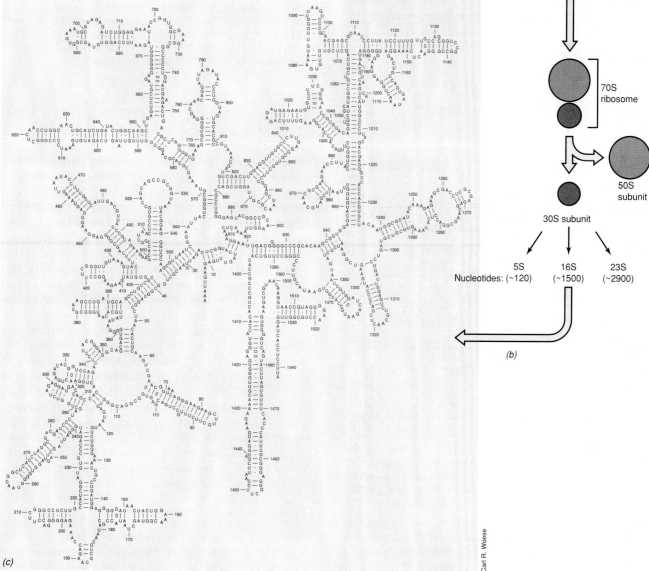

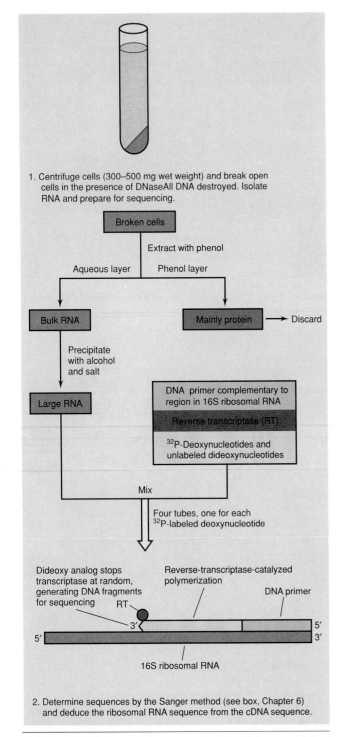

1. Centrifuge cells (300–500 mg wet weight) and break open cells in the presence of DNaseAll DNA destroyed. Isolate RNA and prepare for sequencing.

Broken cells

Extract with phenol

Aqueous layer   Phenol layer

Bulk RNA   **Mainly protein** → Discard

Precipitate with alcohol and salt

Large RNA

DNA primer complementary to region in 16S ribosomal RNA

Reverse transcriptase (RT)

^{32}P-Deoxynucleotides and unlabeled dideoxynucleotides

Mix

Four tubes, one for each ^{32}P-labeled deoxynucleotide

Dideoxy analog stops transcriptase at random, generating DNA fragments for sequencing   RT

Reverse-transcriptase-catalyzed polymerization

DNA primer

16S ribosomal RNA

2. Determine sequences by the Sanger method (see box, Chapter 6) and deduce the ribosomal RNA sequence from the cDNA sequence.

Figure 15.9 Steps in sequencing 16S rRNA from a microbial culture using the retroviral enzyme reverse transcriptase and the dideoxy sequencing (Sanger) method. Although ^{32}P is radioactive, other nonradioactive methods are available for sequencing (∞ Working with Nucleic Acids: The Tools, Chapter 6).

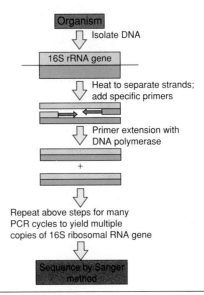

Organism

Isolate DNA

16S rRNA gene

Heat to separate strands; add specific primers

Primer extension with DNA polymerase

+

Repeat above steps for many PCR cycles to yield multiple copies of 16S ribosomal RNA gene

Sequence by Sanger method

Figure 15.10 Ribosomal RNA sequencing using the polymerase chain reaction (PCR). The 16S rRNA gene is amplified and then sequenced by the Sanger method. The primers added are complementary to conserved sequences in one of the domains of 16S rRNA (see Figure 15.8c).

lated by recording (with a computer) the number of positions in the sequence at which the two *differ*. A statistical correction factor is then used to account for the possibility that multiple changes might have occurred

that would lead back to the same sequence. A matrix of evolutionary distances generated from sequence comparisons is then analyzed by a computer algorithm designed to produce phylogenetic trees from E_D measurements. In effect, this amounts to examining all possible branching arrangements for the set of distances compared, and then arranging the branch lengths for each branching arrangement to optimally fit the data. The E_D separating any two organisms is directly proportional to the *total length* of the branches separating them. The principles of phylogenetic tree formation are shown in Figure 15.11 with a sample tree constructed from sequence data from four hypothetical organisms.

Different formats of phylogenetic trees can be generated depending on the computer program used and the number of organisms to be analyzed. For examining only a few major groups, fanlike trees (for example, see Figure 15.12) are most useful. However, when large groups of organisms are to be compared, dendrograms are used. Dendrograms yield highly branched, more dichotomous-appearing trees. Both types of trees are shown on the endpapers of this book where the universal tree is in fanlike format and the detailed trees are shown as dendrograms.

Signature sequences

Computer analysis of ribosomal RNA sequences has revealed **signature sequences,** short oligonucleotides unique to a certain group or groups of organisms. Oligonucleotide signatures defining each of the three

Organism   Sequence

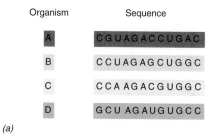

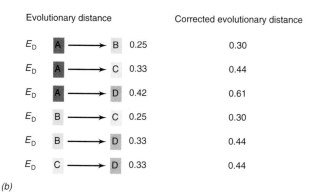

(a)

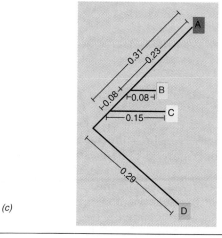

(b)

Phylogenetic tree (computer-generated best fit to corrected evolutionary distances)

(c)

Figure 15.11 Preparing a phylogenetic distance-matrix tree from 16S ribosomal RNA sequences. For illustrative purposes, only short sequences are shown, but it should be considered that the sequences in (a) are representative of the entire 16S rRNA. The evolutionary distance (E_D) in (b) is calculated as the percentage of *nonhomologous* sequence between the RNAs of any total two organisms. The corrected E_D is a statistical correction necessary to account for either back mutations to the original genotype or additional forward mutations at the same site that could have occurred. The tree (c) is ultimately generated by computer analysis of the data to give the best fit. The length of the branches separating any two organisms is proportional to the evolutionary distance between them.

primary domains (Table 15.1) have been identified. Other signatures defining the major taxa within each domain have also been detected. The "signatures" are generally found in defined regions of the 16S rRNA molecule but are revealed only when computer scans

of two aligned sequences are performed. The exclusivity of signature sequences makes them particularly useful for placing unknown organisms in the correct major phylogenetic group. In addition, signature information is useful for constructing genus and species-specific *nucleic acid probes,* which are used extensively for identification purposes in microbial ecology (∞ Section 14.4) and in clinical diagnostics (∞ Section 21.10).

In a few cases *single base* signatures are known (Table 15.1). For example, at position 675 in 16S rRNA from all Bacteria an adenine (A) is found (Figure 15.8c), whereas this base is never found at this same position in the 16S rRNA of species of Archaea and only rarely in species of Eukarya (Table 15.1). Single base signatures can be used to instantly place an organism in the correct domain for further sequence analysis.

We now proceed to examine what ribosomal RNA sequencing has taught us about the evolution of microorganisms.

CONCEPT CHECK   *15.6*

Comparisons of sequences of ribosomal RNA can be used to determine the relationships between organisms. Phylogenetic trees based on ribosomal RNA have now been prepared for all the major prokaryotic and eukaryotic groups.

✔ *What is the function of* reverse transcriptase *in ribosomal RNA sequencing?*

✔ *What is an* evolutionary distance?

✔ *What is a* signature sequence?

15.7

Microbial Phylogeny as Revealed by Ribosomal RNA Sequencing

Molecular sequencing has revealed a previously unsuspected phylogeny of living organisms, a phylogeny quite different from previous ones based primarily on phenotypic relationships. Biologists have often grouped the living world into five kingdoms, only one of which is prokaryotic, based on structural similarities between organisms. Molecular phylogeny, on the other hand, has revealed that the five kingdoms do not represent five major evolutionary lines. Instead, cellular life on Earth has evolved along *three* major lineages, two of which are exclusively microbial and composed only of prokaryotic cells. The third line constitutes the eukaryotic lineage (Figure 15.12). The two prokaryotic lines are the *Bacteria* and the *Archaea.* The eukaryotic line is called the *Eukarya* (Figure 15.12). These terms define the three

| TABLE 15.1 | Signature sequences from 16S or 18S rRNA defining the three domains of living organisms |

Oligonucleotide signatures[a]	Approximate position[b]	Occurrence among[c]		
		Archaea	Bacteria	Eukarya
CACYYG	315	0	>95	0
CYAAYUNYG	510	0	>95	0
AAACUCAAA	910	3	100	0
AAACUUAAAG	910	100	0	100
NUUAAUUCG	960	0	>95	0
YUYAAUUG	960	100	<1	100
CAACCYYCR	1110	0	>95	0
UUCCCG	1380	0	>95	0
UCCCUG	1380	>95	0	100
CUCCUUG	1390	>95	0	0
UACACACCG	1400	0	>99	100
CACACACCG	1400	100	0	0

Single-position signatures	Position[b]	Occurrence among[c]		
		Archaea	Bacteria	Eukarya
U	549	98	0	0
A	675	0	100	2
U	880	0	2	100

[a]Y, Any pyrimidine; R, any purine; N, any purine or pyrimidine.
[b]Refer to Figure 15.8c for numbering scheme of 16S rRNA.
[c]Occurrence refers to percentage of organisms examined in any domain.

domains of life, the domain being the highest of biological taxons. Thus plants, animals, fungi, and protists are all kingdoms within the domain Eukarya.

The universal tree of life

The universal phylogenetic tree (Figure 15.12) shows the relative evolutionary positions of major groups of living organisms. Interestingly, the tree shows that the Eukarya are not of recent origin but instead are as ancient as either of the prokaryotic lineages. Thus, before endosymbiotic events led to the modern eukaryotic cell (that is, with membrane-enclosed organelles) (see Section 15.4 and Figure 15.7), organelle-less Eukarya inhabited Earth (and remnants exist to this day, see later), and these organisms shared common ancestry with the other two evolutionary lines (Figure 15.12).

The *root* of the universal tree (Figure 15.12) represents a point in evolutionary time when all extant life on Earth shared a common ancestor. The root of the tree has been determined by rRNA sequencing and by related macromolecular sequencing methods, but it clearly indicates that initial evolution from the universal ancestor was at first in *two* directions, the Bacteria versus the Archaea–Eukarya line. Later, the Archaea and Eukarya diverged to yield two major lineages. Archaea and Eukarya are therefore phylogenetically more related to each other than either group is to the Bacteria.

Notice also how the universal phylogenetic tree shows the Archaea branching off the tree at a point closest to the root (Figure 15.12). This leads to the conclusion that Archaea remain, even today, as the *most primitive* (least evolved) of organisms in the three domains, whereas the Eukarya, now the farthest away from the universal ancestor, are the *least primitive*, that is, the most evolved (Figure 15.12). The latter suggests that the eukaryotic lineage evolved some important characteristic or set of characteristics that allowed for rapid, albeit late, evolutionary diversity. Perhaps this evolutionary event was something as simple as a larger cell size; this would have facilitated the large amounts of DNA needed for the development of metazoan organisms and produced a cell sufficiently large to accept prokaryotic endosymbionts. Placement of the Archaea nearest the universal ancestor is supported by the fact that many Archaea inhabit extreme environments, such as those of high temperature, low pH, high salinity, and so on (∞ Chapter 17); these environmental extremes, especially that of high temperature, reflect the environmental conditions under which life originated (see Section 15.1). Thus, members of the Archaea may well be evolutionary relics of Earth's earliest life forms.

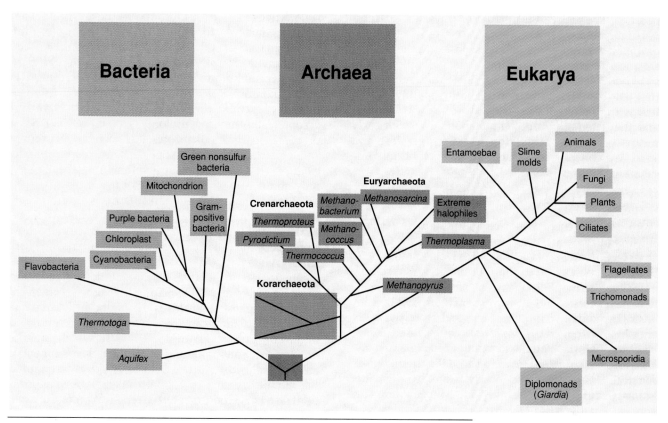

Figure 15.12 Rooted universal phylogenetic tree as determined from comparative ribosomal RNA sequencing. The data support the separation of three domains, two of which (Bacteria and Archaea) contain only prokaryotic representatives. The location highlighted in red is the root of the tree, which represents the position of the universal ancestor of all cells.

A point that deserves emphasis here is that *none* of the organisms living today are *primitive*. All extant life forms are *modern* organisms, well adapted to, and successful in, their ecological niches. Certain of these organisms may indeed be phenotypically similar to primitive organisms and may represent stems of the evolutionary tree that have changed little for millions if not billions of years (∞ Section 17.6); in this respect they are evolutionarily *related* to primitive organisms, but they are not themselves primitive.

Organelles

The overall phylogenetic picture presented in Figure 15.12 also tells us about other evolutionary events. For example, it is clear that mitochondria and chloroplasts arose from endosymbiotic Bacteria that established stable relationships, perhaps several different times, with cells from the nuclear line of descent (see also Figure 15.7). Molecular sequencing suggests that mitochondria arose from a small group of organisms that includes the modern prokaryotes *Agrobacterium, Rhizobium,* and the rickettsias. The latter organisms belong to a larger group of Bacteria called the purple bacteria (organisms such as

Rhodopseudomonas and *Rhodobacter*). (This whole group is called the *Proteobacteria.*) It is interesting to note that the mitochondrion, itself an intracellular symbiont, is specifically related to organisms (*Agrobacterium, Rhizobium,* and the rickettsias) that are capable of an intracellular existence (∞ Sections 14.22, 14.23, and 16.22). The other eukaryotic endosymbiont, the chloroplast, originated from Bacteria as well. The universal evolutionary tree in Figure 15.12 also shows that the chloroplast and cyanobacteria *shared* a common ancestor. Thus, molecular sequencing supports the endosymbiotic theory (Figure 15.7) and has independently shown that the evolutionary roots of the mitochondrion and the chloroplast are within the Bacteria.

Bacteria

In the domain Bacteria at least 12 distinct phylogenetic lineages (kingdoms) exist, and several key ones are shown in the universal tree in Figure 15.12. Some lineages were previously recognized on a phenotypic basis such as morphology and physiology; good examples here are the spirochetes and the cyanobacteria, respectively. However, most of the kingdoms are composed of species with a mixture of physiologies, morphologies, and other phenotypic properties. The branch of the Bacteria tree that emerges closest to the universal ancestor contains the genus *Aquifex* (Figure 15.12), a hyperthermophilic, chemolithotrophic prokaryote that oxidizes H_2 (∞ Section 17.3); this is fully consistent with the hypothesis that early Earth was very hot and that chemolithotrophic metabolism was an early means of energy generation (see Sections 15.1 and 15.2). We consider the properties of Bacteria in some detail in Chapter 16.

Archaea

From a phylogenetic perspective, the domain Archaea consists of three major groups (kingdoms), the Crenarchaeota, Euryarchaeota, and Korarchaeota (Figure 15.12). Branching very close to the root of the universal tree are hyperthermophilic members of the Crenarchaeota like *Thermococcus* and *Pyrodictium* (Figure 15.12). They are followed by two groups of Euryarchaeota, the methanogens *Methanococcus–Methanobacterium* and *Methanosarcina*, and the extreme halophiles *Halobacterium* and *Halococcus; Thermoplasma*, an acidophilic, thermophilic cell wall-less bacterium is loosely related to this latter group (Figure 15.12). Another methanogen branches very close to the root of the tree. This is *Methanopyrus*, a methanogen capable of growth up to 110°C! Note that this hyperthermophilic organism is also a chemolithotroph, growing at the expense of H_2 oxidation at very high temperature (∞ Section 17.3). *Methanopyrus* is another good example of the type of organism that could have inhabited early Earth (see Sections 15.1 and 15.2).

Finally in the phylogenetic tree of Archaea we see a very early lineage, the kingdom Korarchaeota. Interestingly, this lineage has been identified only from sequences of 16S rRNA obtained by cloning the corresponding genes from cells in natural samples (∞ Section 14.4); cultures of these organisms have yet to be obtained. However, from what we can deduce about the Korarchaeota using this method it is clear that representatives of this kingdom are, like Crenarchaeota, hyperthermophilic, and probably share some of the interesting physiological properties of Crenarchaeota as well (discussed in Section 17.3). We consider the properties of Archaea in more detail in Chapter 17.

Eukarya

Phylogenetic trees of species in the domain Eukarya are generated from sequence data of *18S* rRNA, the functional equivalent of 16S rRNA in eukaryotic cytoplasmic ribosomes. Inspection of the Eukarya tree suggests that evolution in this lineage was not a continuous process but instead occurred in major epochs. Early eukaryotes were probably similar to present-day microsporidia and diplomonads. These organisms are all obligate parasites that live in association with rep-

resentatives of various groups of eukaryotes, from microorganisms to humans [for example, the pathogen *Giardia* (∞ Section 23.14) is a member of the diplomonad group]. And interestingly, although they contain a membrane-enclosed nucleus, microsporidia and diplomonads lack mitochondria; in this connection they resemble the type of cell that might have first accepted stable endosymbionts (see Figure 15.7). Most of these "early" eukaryotes also contain extremely small genomes for eukaryotic cells; microsporidia, for example, have genomes of about six megabase pairs, only about 20% larger than that of *E. coli*.

Later branches on the Eukarya tree go from microbial forms to multicellular plants and animals (the metazoa), culminating with the largest and structurally most complex of the Eukarya, the plants and animals (Figure 15.12). When the fossil record is compared with the phylogenetic tree of Eukarya, rapid evolutionary radiation can be dated to about 1.5 billion years ago. Geochemical evidence suggests that this is the period in Earth's history in which significant oxygen levels had accumulated in the atmosphere (Figure 15.5). It is thus likely that the onset of oxic conditions and subsequent development of an ozone shield (which would have greatly expanded the number of surface habitats available for colonization) also triggered the rapid diversification of Eukarya. We consider the major groups of microbial Eukarya and their basic biology in Chapter 18.

CONCEPT CHECK   15.7

The major conclusion from phylogenies derived from comparisons of ribosomal RNA sequences is that life on Earth evolved along three major lines, called domains, all derived from a common (universal) ancestor. Two of the lines, Bacteria and Archaea, remained prokaryotic, whereas the third line, Eukarya, evolved into the modern eukaryotic cell.

✔ *Which domains contain only prokaryotic cells?*

✔ *How does the universal tree support the idea that early Earth was very hot?*

✔ *How does the universal tree support the hypothesis presented in Figure 15.7?*

15.8
Characteristics of the Primary Domains

Although the primary domains—Bacteria, Archaea, and Eukarya—are defined on the basis of comparative ribosomal RNA sequencing, subsequent studies have shown that each domain has associated with it a number of *phenotypic* properties. Some of these characteristics are unique to one domain, whereas others are found in only two out of the three domains. We present here an overview of major phenotypic traits of phylogenetic value.

Cell walls

Virtually all Bacteria have cell walls containing *peptidoglycan* (∞ Section 3.5). The only known exceptions are members of the *Planctomyces–Pirella* group (∞ Section 16.10), whose cells are composed of protein, and the *Chlamydia–Mycoplasma* groups, which lack cell walls altogether (∞ Sections 16.23 and 16.27). Peptidoglycan can thus be considered a "signature" molecule for species of Bacteria (when assaying for peptidoglycan, it is *muramic acid* that is actually detected because this is unique to peptidoglycan).

Eukarya and Archaea lack peptidoglycan. In eukaryotes, if cell walls are present, they are usually made of cellulose or chitin (∞ Sections 18.1 and 18.2). In Archaea, as previously discussed (∞ Section 3.5), various cell wall types exist, from the peptidoglycan analog pseudopeptidoglycan to walls made of polysaccharide, protein, or glycoprotein. Thus, great diversity is observed in the chemistry of microbial cell walls, but it is the presence or absence of peptidoglycan that is most useful in assessing phylogeny.

Lipids

The chemical nature of membrane lipids is perhaps the most useful of all nongenetic criteria for differentiating Archaea from Bacteria. Bacteria and eukaryotes synthesize membrane lipids with a backbone consisting of fatty acids hooked in *ester* linkage to a molecule of glycerol (∞ Figure 2.7). Although the nature of the fatty acid can be highly variable, the key point is that the chemical linkage to glycerol is an **ester link.** By contrast, archaeal lipids consist of **ether-linked** molecules (Figure 15.13). In ester-linked lipids, the fatty acids are straight-chain (linear) molecules, whereas in Archaea, long-chain, branched hydrocarbons, either of the phytanyl or biphytanyl type, are present and are bonded by ether linkage to glycerol molecules (Figure 15.13).

In addition to the differences in chemical *linkage* between the hydrocarbon and alcohol portions of archaeal and bacterial and eukaryotic lipids, the chirality of the glycerol moiety differs in the lipid of representatives of the three domains. The central carbon atom of the glycerol molecule is stereoisomerically of the R form in Bacteria and Eukarya and of the L form in Archaea (see Section 2.9 for more discussion about stereoisomerism and life).

RNA polymerase

Transcription is carried out by DNA-dependent RNA polymerases in all oganisms; DNA is the template, and RNA is the product (∞ Section 6.6). Cells of Bacteria

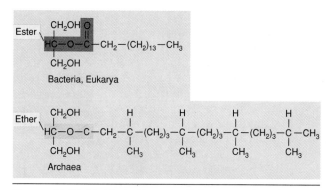

Figure 15.13 Lipids in Bacteria, Eukarya, and Archaea. In Bacteria and Eukarya, lipid side chains are fatty acids (palmitic acid is shown) bonded by *ester* linkages to glycerol. In Archaea, the side chains are branched hydrocarbon (phytanyl, C_{20}, is shown) bonded by *ether* linkages to glycerol. Phytanyl is synthesized from isoprene (∞ Figure 3.19c).

contain a single type of RNA polymerase of rather simple quaternary structure. This is the classic RNA polymerase containing *four* polypeptides, α, β, β', and one of a variety of different σ factors, combined in a ratio of 2:1:1:1, respectively, in the active polymerase (Figure 15.14) (∞ also Section 6.6).

Archaeal RNA polymerases are of several types and are structurally more complex than those of Bacteria. The RNA polymerases of methanogens and halophiles contain *eight* polypeptides, five large ones and three smaller ones (Figure 15.14). Hyperthermophilic Archaea contain an even more complex RNA polymerase consisting of at least *10* distinct polypeptides (Figure 15.14). The major RNA polymerase of eukaryotes (there are several, only one of which makes messenger RNA) contains 10–12 polypeptides, and the relative sizes of the peptides coincide most closely with those from the hyperthermophilic Archaea (Figure 15.14). Two other RNA polymerases are known in eukaryotes, and each specializes in transcribing certain regions of the genome, one for rRNA and one for transfer RNA (tRNA). In terms of phylogenetic signatures, the $\alpha_2\beta\beta'\sigma$ polymerase of Bacteria is highly diagnostic whereas the remaining polymerases are too complex to be phylogenetically definite. It should also be noted that the antibiotic rifamycin, which functions by specifically interfering with the beta subunit of RNA polymerase, inhibits only Bacteria because Archaea and Eukarya lack this form of RNA polymerase.

Features of protein synthesis

Because of differences in ribosomal RNA sequences and several protein synthesis factors, it is not surprising that certain aspects of the protein-synthesizing machinery differ in representatives of the three domains. Although ribosomes of Archaea and Bacteria are the same size (70S, as compared with the 80S ribosomes in the cytoplasm of eukaryotes), several steps in archaeal protein synthesis more strongly resemble

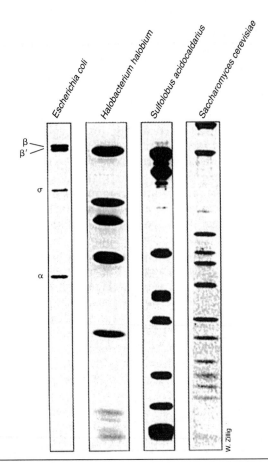

β
β'

σ

α

W. Zillig

Escherichia coli

Halobacterium halobium

Sulfolobus acidocaldarius

Saccharomyces cerevisiae

Figure 15.14 RNA polymerases from representatives of the three domains: *Escherichia coli* (Bacteria), *Halobacterium halobium* (Euryarchaeota, Archaea), *Sulfolobus acidocaldarius* (Crenarchaeota, Archaea), and *Saccharomyces cerevisiae* (Eukarya). The purified RNA polymerase proteins have been denatured and separated by electrophoresis on a polyacrylamide gel. The largest subunits are on the top, and the smallest subunits are on the bottom. Only members of the Bacteria *E. coli* contain the simple (four-polypeptide) RNA polymerase.

otic) protein synthesis. The sensitivity of representatives of the three domains to various protein synthesis inhibitors is shown in Table 15.2 (∞ also Section 6.9), where various antibiotics are grouped according to their modes of action in blocking protein synthesis in various domains of organisms. Knowledge of the mode of action of specific antibiotics and the organisms whose growth they inhibit (Table 15.2), has yielded valuable information about the evolution of the translational apparatus.

Functional similarities in the translational machinery of Archaea and Eukarya have been revealed from measurements of *in vitro* protein synthesis by *hybrid ribosomes* made up of ribosomal components from organisms of different domains. Specifically, hybrid ribosomes composed of the large subunit (50S) of ribosomes from *Sulfolobus* (Archaea) and the small subunit (40S) of ribosomes from yeast (Eukarya) have been shown to be functional *in vitro* and to actually produce polypeptides. However, if large ribosomal subunits from *Escherichia coli* (Bacteria) pair with small ribosomal subunits from yeast, no protein-synthesizing activity occurs. These results suggest that ribosomal proteins from cells of Archaea and Eukarya are more similar to each other than they are to ribosomal proteins from Bacteria. Such results reinforce conclusions about the relationships of various domains to one another (Figure 15.12).

Other features defining the domains

A number of other more subtle differences can be listed that assist in delineating organisms at the domain level. For example, the number of modified nucleotides present in eukaryotic 18S rRNA (the functional equivalent of prokaryotic 16S rRNA) is considerable. Depending on the eukaryote, 25–35 or more nucleotides are chemically modified. A similar num-

those in eukaryotes than in Bacteria. Recall that translation always begins at a unique codon, the so-called *start codon*. In Bacteria this start codon (AUG) calls for the incorpoation of an initiator tRNA containing a modified methionine residue, *formyl*methionine (∞Section 6.9). By contrast, in eukaryotes and in Archaea, the initiator tRNA carries an *unmodified* methionine.

The exotoxin produced by *Corynebacterium diphtheriae* is a potent inhibitor of eukaryotic protein synthesis because it ADP-ribosylates (adds ADP to) an elongation factor required to translocate the ribosome along the mRNA; the modified elongation factor is inactive (∞Section 19.8). Diphtheria toxin also inhibits protein synthesis in Archaea but does not affect this process in Bacteria.

Most antibiotics that specifically affect protein synthesis in Bacteria do not affect archaeal (or eukary-

TABLE 15.2   Sensitivity of representatives of the three domains to various protein synthesis inhibitors[a]

| Antibiotics | Mode of action | Archaea | | Bacteria | Eukarya |
		Methanobacterium	*Sulfolobus*	*Escherichia coli*	*Saccharomyces cerevisiae*
Fusidic acid, sparsomycin	Inhibits elongation steps	+	−	+	+
Anisomycin, narciclasine	Inhibits peptidyl transfer	+	−	−	+
Cycloheximide	Blocks initiation	−	−	−	+
Erythromycin, streptomycin, chloramphenicol	Increases error frequencies and other effects	−	−	+	−
Virginiamycin, pulvomycin	Inhibits elongation steps	+	−	+	−
Neomycin, puromycin	Causes premature termination	+	+	+	+
Rifamycin	Inhibits RNA polymerase	−	−	+	−

[a]A + indicates that protein synthesis (and growth) is inhibited. Data adapted from Bock, A., J. H. Hummel, and G. Schmid. 1985. Evolution of translation, pp. 73–90, in K.-H. Schleifer and E. Stackebrandt (eds.), *Evolution of Prokaryotes*, Academic Press, London.

ber of modified nucleotides are found in the 16S rRNA of hyperthermophilic Archaea. By contrast, Bacteria and the methanogen–halophile– *Thermopolasma* archaeal line, contain only 4–8 modified residues in their 16S rRNA. The modified base dihydrouracil (Figure 15.15) is present in the tRNAs of all Bacteria and eukaryotes but is absent from Archaea, except for one group of methanogens.

Table 15.3 summarizes some major points that define the three domains. When examining this table, it should be understood that some of the features listed will only be present in certain representatives of a given domain.

15.9

Taxonomy, Nomenclature, and *Bergey's Manual*

Taxonomy is the science of classification and consists of two major subdisciplines, *identification* and *nomenclature*. It is important to distinguish between bacterial *taxonomy* and the main topic of this chapter up to this point, bacterial *phylogeny*, for the terms really mean different things. Bacterial taxonomy has traditionally relied on *phenotypic* analyses as the basis of classifica-

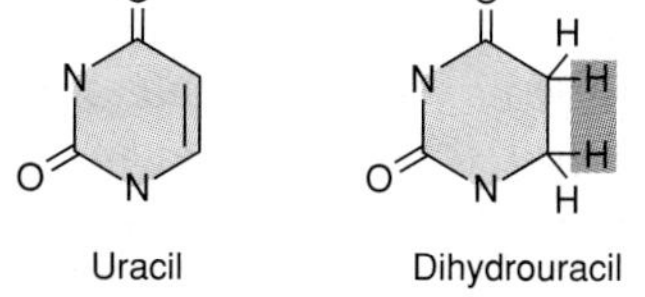

Figure 15.15 Comparison of the structures of uracil and dihydrouracil. Dihydrouracil is found in the RNAs of Bacteria and Eukarya but not in the RNAs of Archaea (with one exception).

tion. By contrast, because bacteria are so small and contain relatively few structural clues to their evolutionary roots, phylogenetic relationships between prokaryotes have emerged only from the *genotypic* analyses discussed in the previous sections. Phenotypic analyses have traditionally played an important role in bacterial identification and classification, especially in practical situations where identification may be an end in itself, for example, in clinical diagnostic microbiology. However, even in an area such as diagnostic microbiology where traditional microbiological methods have ruled for decades, genotypic methods (for example, nucleic acid probes) are taking over because of the sensitivity, rapidity, and specificity of these identification methods (∞ Chapter 21).

Classification and the species concept

Because microbiologists need to identify and classify bacteria for a variety of practical reasons, bacterial taxonomy is still an important discipline. In microbiology, the basic taxonomic unit is the **species**. A species can be operationally defined as a collection of similar strains that differ sufficiently from other groups of strains to warrant recognition as a basic taxonomic unit. However, a prokaryotic species can be more precisely defined on the basis of rRNA sequences. Although not yet universally accepted in microbiology, it has been proposed (with considerable experimental support) that two prokaryotes whose 16S rRNA sequences are greater than 97% identical are likely to be of the same species.

A species is usually defined from the characterization of several strains or clones. The use of the word *clone* in this sense can be taken to mean a population of *genetically identical* cells derived from a single cell. The species concept is important because it gives the col-

TABLE 15.3   Summary of major differentiating features among Bacteria, Archaea, and Eukarya

Characteristic	Bacteria	Archaea	Eukarya
Prokaryotic cell structure	Yes	Yes	No
DNA present in covalently closed and circular form	Yes	Yes	No
Membrane-enclosed nucleus	Absent	Absent	Present
Cell wall	Muramic acid present	Muramic acid absent	Muramic acid absent
Membrane lipids	Ester-linked	Ether-linked	Ester-linked
Ribosomes	70S	70S	80S
Initiator tRNA	Formylmethionine	Methionine	Methionine
Introns in tRNA genes	Yes	Yes	Yes
Introns in most genes	No	No	Yes
Operons	Yes	Yes	No
Capping and poly-A tailing of mRNA	No	No	Yes
Plasmids	Yes	Yes	Rare
Ribosome sensitivity to diphtheria toxin	No	Yes	Yes
RNA polymerases (see Figure 15.14)	One (4 subunits)	Several (8–12 subunits each)	Three (12–14 subunits each)
Sensitivity to chloramphenicol, streptomycin, and kanamycin	Yes	No	No
Methanogenesis	No	Yes	No
Reduction of S^0 to H_2S	Yes	Yes	No
Nitrification	Yes	No	No
Denitrification	Yes	Yes	No
Nitrogen fixation	Yes	Yes	No
Chlorophyll-based photosynthesis	Yes	No	Yes
Chemolithotrophy (Fe, S, H_2)	Yes	Yes	No
Gas vesicles	Yes	Yes	No
Synthesis of carbon storage granules composed of poly-β-hydroxyalkanoate	Yes	Yes	No

lected strains formal taxonomic identity. Groups of species are collected into **genera** (singular, **genus**). By analogy to the species, a genus can be defined as a collection of different species, each sharing some major property or properties that define the genus but differing from one another by the presence or absence of other (usually less significant) characteristics. Two prokaryotes of different genera will show greater variation in 16S rRNA sequence than will two species of a single genus, although no guidelines for distinguishing genera on the basis of molecular sequencing have yet been proposed. Groups of genera are collected into **families,** families into **orders,** orders into **divisions,** and so on up to the highest level taxon, the **domain** (Table 15.4). However, the *family* is the highest level taxon used routinely in taxonomic studies of prokaryotes.

In the identification of an unknown organism, it is essential that the organism satisfy all the taxonomic criteria of ranks *above* its species designation. Thus, in the example given in Table 15.4, all species of the genus *Chromatium* must be rod-shaped purple sulfur gram-negative Bacteria. The converse, however, is not true; not all members of the domain Bacteria are phototrophic purple bacteria. In other words, as one *descends* the taxonomic hierarchy from the level of domain to that of species, the criteria used become less general and more specific (Table 15.4).

Nomenclature and formal taxonomic standing

Following the **binomial system** of nomenclature, all bacteria are given genus and species names. The genus name of an organism is usually abbreviated to a single (capital) letter; the species name is never abbreviated. Thus, *Escherichia coli* is usually written *E. coli*. The genus and species names are either Latin or Greek derivations of some descriptive property appropriate for the species and are set in print in *italics*. For example, several species of the genus *Bacillus* have been described, including *Bacillus (B.) subtilis*, *B. cereus*, *B. stearothermophilus*, and *B. acidocaldarius*. The species names mean "slender," "waxen," "heat-loving," and "acid-thermal," respectively, and in each case refer to key morphologial, physiological, or ecological traits characteristic of each organism. The rules for bacterial nomenclature are fixed in a publication called *The International Code of Nomenclature of Bacteria*, which spells out rules for naming newly isolated organisms as new genera or new species. The code governs nomenclatural policy for all prokaryotes, both Bacteria and Archaea.

When a new organism is isolated and thought to be unique, a decision must be made as to whether it is sufficiently different from other species to be described as a new species, or perhaps even sufficiently different

TABLE 15.4 Taxonomic hierarchy for the purple sulfur bacterium *Chromatium warmingii*

Taxonomic division	Name	Properties	Confirmed by
Domain	Bacteria	Prokaryotic cells; ribosomal RNA sequences typical of purple bacterial lineage (see Figure 15.12)	Microscopy; 16S ribosomal RNA sequencing; presence of unique biomarkers, for example, peptidoglycan
Division	Gracilicutes	Gram-negative bacteria	Gram-staining, microscopy
Order	Rhodospirillales	Phototrophic purple bacteria	Characterizing pigments (∞ Figures 16.3 and 16.4)
Family	Chromatiaceae	Purple sulfur bacteria	Ability to oxidize H_2S and store S^0 within cells; observe culture microscopically for presence of S^0 (see photo)
Genus	*Chromatium*	Rod-shaped purple sulfur bacteria	Microscopy (see photo)
Species	*warmingii*	Cells 3.5–4.0 μm × 5–11 μm; store sulfur mainly in poles of cell (see photo)	Measure cells in microscope using a micrometer; look for position of S^0 globules in cells (see photo)

Photograph of cells of *Chromatium warmingii*:

Norbert Pfennig

from all described genera to warrant description as a new genus (in which a species is automatically created). In order to achieve formal taxonomic standing as a new genus or species, a description of the isolate and the proposed name is published and a pure culture of the organism is deposited in an approved culture collection, usually the American Type Culture Collection (ATCC) or the Deutsche Sammlung von Mikroorganismen und Zellkulturen (DSM, German Collection for Microorganisms). (See Section 12.1 for a discussion of microbial culture collections.) The deposited strain serves as the *type* strain of the new species or genus and species and remains as the standard by which other strains thought to be the same can be compared.

If the description of the new organism is published in a journal *other than* the *International Journal of Systematic Bacteriology* (IJSB), the official publication of record for the taxonomy and classification of microorganisms, a copy of the published paper must be submitted to this journal and the name validated before it is formally accepted as a new microbiological taxon. Periodically, the IJSB publishes an approved list of bacterial names, which formalizes any new names and paves the way for their inclusion in *Bergey's Manual*, a major taxonomic treatment of prokaryotes (see later).

Culture collections preserve the deposited culture, generally by freezing or freeze-drying. This practice is very different from the botanical or zoological approach. These disciplines employ preserved (dead) specimens (either dried herbarium material or chemically fixed animal specimens) as the basis for comparison with proposed new species. Microbiologists rely on a *living type strain*, and this approach allows for more detailed and reproducible comparisons, especially at the molecular level.

Conventional bacterial taxonomy

There are many ways in which to group prokaryotes. In conventional bacterial taxonomy, a variety of characteristics of different strains or species are measured and these traits are then used to group the organisms. Characteristics of taxonomic value that are widely used include morphology, Gram reaction, nutritional classification (phototroph, chemoorganotroph, chemolithotroph), cell wall chemistry, presence of cell inclusions and storage products, capsule chemistry, pigments, nutritional requirements, ability to use various carbon, nitrogen, and sulfur sources, fermentation products, gaseous needs, temperature and pH requirements (and tolerances), antibiotic sensitivity, pathogenicity, symbiotic relationships, immunological characteristics, and habitat.

To identify an organism a series of criteria are used that proceed from the general to the specific. Examples of this are shown in Table 15.4 and Figure 15.16. Using a dichotomous key approach and testing several phenotypic properties, it is possible to eliminate from consideration more and more organisms until the organism in question is positively identified (Figure 15.16). These types of identification methods are widespread in clinical microbiology and other applications where microbial isolates need to be routinely identified. However, today, many of these growth-dependent methods are being supplanted by molecular methods employing nucleic acid or immunological probes (∞ Chapter 21). These have the advantage that they can be made extremely specific and are by their very nature highly sensitive. This combination is ideal for the rapid identification of a given microorganism and in most cases does not even require that a culture be obtained (∞ Chapter 21).

Molecular taxonomy: GC ratios

Nucleic acid analyses have become part of conventional taxonomy, but the techniques involved differ from the nucleic acid *sequencing* methods discussed in Sections 15.5 and 15.6. Determination of the **guanine plus cytosine base composition** of the DNA of an organism is required in order to name it as a new taxon:

DNA base ratios, expressed as mole percent (mol %) GC,

$$\frac{G + C}{A + T + G + C} \times 100\%$$

vary over a wide range (Figure 15.17). GC ratios as low as 20% and as high as 78% are known among prokaryotes, a somewhat broader range than for eukaryotes (Figure 15.17).

Base compositions of DNA have been determined (for methods see the box, Working with Nucleic Acids: The Tools, Chapter 6) for a wide variety of organisms, and several correlations can be observed. (1) Organisms with highly similar phenotypes often but not

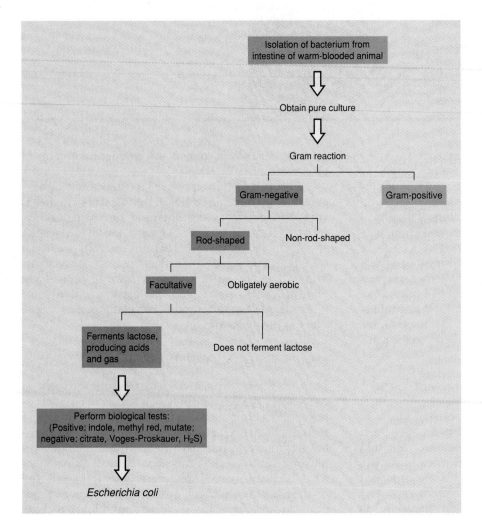

Figure 15.16 Example of methods that would be used for identification of a newly isolated enteric bacterium, using conventional microbiological methods (the example given shows the procedures that would be used for identifying *Escherichia coli*). Note that most of the analyses here require that the organisms be grown in pure culture and that solely phenotypic criteria be used in the identification. For a description of biochemical tests, see Chapter 21.

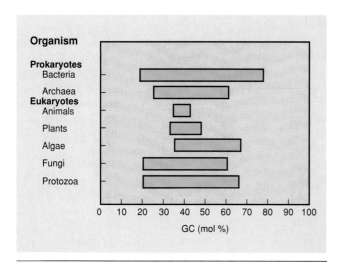

Figure 15.17 Ranges of DNA base composition of various organisms.

always possess similar DNA base ratios. (2) If two organisms thought to be closely related by phenotypic criteria are found to have widely different base ratios, closer examination usually indicates that the organisms are not so closely related as was supposed. (3) Two organisms can have identical base ratios and yet be quite unrelated (both taxonomically and phylogenetically) because a variety of *sequences* is possible with DNA of a given base composition.

Molecular taxonomy: DNA:DNA hybridization

GC base ratio determinations yield information on the percentage of each nucleotide present in an organism's DNA but give absolutely no information on the *sequence* of nucleotides in the DNA. Sequences are critical because if two organisms have many similar or identical genes, they have many of the same nucleotide sequences in their DNAs. Such DNAs would be expected to *hybridize* to one another in proportion to the similarities in their sequences. Genomic hybridization can address this issue and is a useful method in bacterial taxonomy, especially for resolving relationships at the species level and below.

We discussed the theory and methodology of nucleic acid hybridization in Chapter 6 (∞ Working with Nucleic Acids: The Tools). In an actual genomic comparison experiment, DNA isolated from one organism is made radioactive with ^{32}P or ^{3}H, sheared to a relatively small size, heated to denature, and mixed with an excess of unlabeled DNA prepared in the same way from a second organism (Figure 15.18). The DNA mixture is then cooled to allow it to reanneal and double-stranded DNA is separated from any unhybridized single-stranded DNA. Following this, the amount of radioactivity in the hybridized DNA is determined and

compared with the control, which is taken as 100% (Figure 15.18).

There is no uniform convention on how much genomic homology is necessary to assign two organisms to the same taxonomic rank. However, homology values of above 60% are considered evidence for two different strains being of the same species, and values above 20% are generally taken as evidence that two organisms probably reside in the same genus. DNA from unrelated genera usually show only background hybridization, about 1–5% (Figure 15.18). Genomic hybridization is thus a useful tool in taxonomic studies of closely related bacteria but is useless as a method for determining relationships at higher taxonomic rank. In addition, DNA:DNA hybridization does not reveal the phylogenetic history of an organism; only molecular sequencing can accomplish this (see Sections 15.5–15.7). Despite these limitations, genomic hybridization experiments are useful in many taxonomic studies, and newer methods for nonradioactive labeling (∞ Working with Nucleic Acids: The Tools, Chapter 6) and processing of the DNA have made the actual hybridizations much easier and quicker to do than when the technique was first introduced many years ago.

Bergey's Manual *and* The Prokaryotes

Bergey's Manual of Systematic Bacteriology is a compendium of classical and molecular information on all recognized species of prokaryotes and contains a number of dichotomous keys that are useful for identification purposes. Although still a taxonomic (rather than phylogenetic) treatise, the latest edition of *Bergey's Manual* has

---

CONCEPT CHECK   15.9

Although phylogenetic analyses allow for the grouping of organisms based on evolutionary relationship, taxonomy is the science of identification and naming of microorganisms, based primarily on phenotypic properties. The species concept applies to prokaryotes just as it does to eukaryotes, and a similar taxonomic hierarchy exists, with the domain as the highest level taxon. Bacterial taxonomy requires testing numerous properties of a given isolate and yields a positive identification in most cases if a sufficient number of properties have been examined. *Bergey's Manual* is a major taxonomic compilation of Bacteria and Archaea.

✔ *How does the science of* taxonomy *differ from that of* phylogeny?

✔ *What is a* bacterial species?

✔ *Define the term* GC ratio.

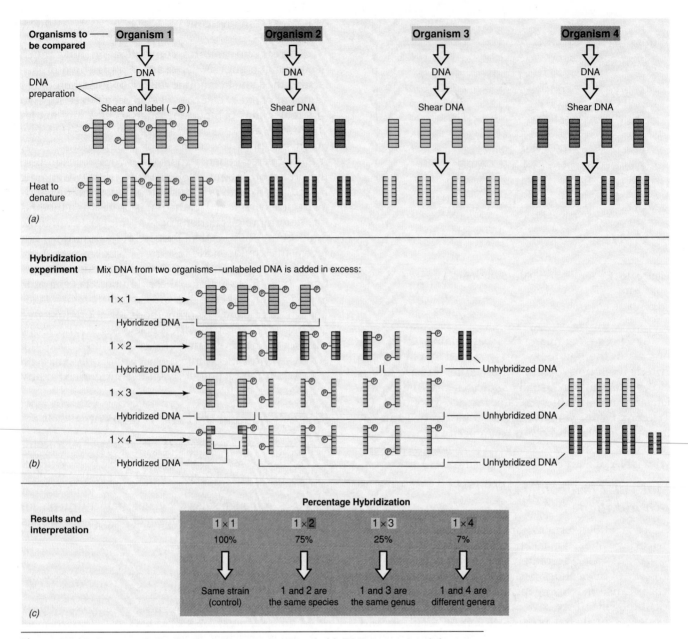

Figure 15.18 Genomic hybridization as a taxonomic tool. (a) DNA is prepared from test organisms. One of the DNAs is labeled (shown here as radioactive phosphate in the DNA of Organism 1). (b) Actual hybridization experiment. All combinations are tried and excess unlabeled DNA is added in each experiment to prevent labeled DNA from reannealing with itself. Following hybridization, hybridized DNA is separated from unhybridized DNA before measuring radioactivity in the hybridized DNA only. (c) Results. Radioactivity in the control is taken as the 100% hybridization value.

incorporated some molecular sequencing information in the descriptions of various bacterial groups. *Bergey's Manual* consists of four volumes. Volume I covers gram-negative Bacteria of medical or industrial significance; volume II, gram-positive Bacteria of medical or industrial significance; volume III, the remaining gram-negative Bacteria, the Archaea and cyanobacteria; and volume IV, the gram-positive, filamentous spore-forming Bacteria (Actinomycetes). The genera described in *Bergey's Manual* are given in Appendix 3 of this book.

Although *Bergey's Manual* recognizes the contribution of molecular phylogeny to bacterial evolution, *Bergey's Manual* retains many classical taxonomic groupings because it is not yet clear how molecular sequencing studies will affect the nomenclatural aspects of taxonomy. A more extensive coverage of prokaryotic taxonomy with a stronger emphasis on phylogeny is the four-volume treatise *The Prokaryotes*, 2nd edition. This work of more than 4100 pages is the most complete reference source on prokaryotes available today.

Material for Review

REVIEW QUESTIONS

1. What is the evidence that prokaryotic life forms were present on Earth billions of years before eukaryotic life forms?

2. What major features would primitive organisms have had to have in order to replicate copies of themselves. Why?

3. Why was the evolution of porphyrin rings of possible importance to prokaryotic evolution and diversification?

4. Why was the evolution of cyanobacteria of such importance to the further evolution of life on Earth?

5. What properties of RNA could have made possible an era of RNA life? If RNA life forms ever existed, why is there no trace of them today?

6. What might have been the advantages of abandoning the era of RNA life for cellular life based on DNA, RNA, and protein?

7. Why are macromolecules like nucleic acids or proteins excellent phylogenetic markers, whereas polysaccharides and lipids are not? (You may want to review the material in Chapter 2 before answering this question.)

8. Why are ribosomal RNAs better molecules for phylogenetic studies than proteins like ferredoxin, cytochromes, or specific enzymes?

9. What are signature sequences and of what phylogenetic value are they? How are signature sequences discerned?

10. Describe the methods involved in obtaining 16S rRNA sequences. How has the polymerase chain reaction benefited molecular phylogeny?

11. What major evolutionary finding has emerged from the study of ribosomal RNA sequences? How did this modify the classic view of evolution? How has this discovery changed our thinking on the origin of eukaryotic organisms?

12. What major lesson has microbiology learned from RNA sequencing concerning the use of phenotypic criteria in establishing evolutionary relationships?

13. What major physiological and biochemical properties do Archaea share with Eukarya? With Bacteria?

14. Examine the following bacterial name: *Pseudomonas aeruginosa*. What part of this name is the *species* name. What is the other name? In reference to taxonomic hierarchy, which of the two names might have several other names listed *under* it?

15. What major phenotypic properties are used to group organisms in classical bacterial taxonomy? Which, if any, of these properties have phylogenetic predictive value?

16. Why aren't GC base ratios useful for making phylogenetic determinations? In what situations are GC base ratios of use in taxonomic studies?

APPLICATION QUESTIONS

1. Why is it highly unlikely that life could originate today as it did billions of years ago?

2. Carefully review Figure 4.12 and compare it with Figure 15.3. Then defend the following statement: Chemolithotrophic metabolism was more likely an earlier form of metabolism than chemoorganotrophic metabolism.

3. Imagine that you are debating someone who is arguing against the theory of endosymbiosis. List five forms of evidence you would use to convince your opponent that endosymbiosis did occur. (You may wish to review Section 3.15 before writing your answer.)

4. On the basis of the following sequences, calculate an evolutionary distance between these three organisms and predict which two of the three are most closely related.

> Organism 1: AGGUACGUUA
> Organism 2: UGCCACGGUU
> Organism 3: AGGUACGGUA

Draw a phylogenetic tree that shows the approximate evolutionary relationships of these three organisms.

5. How has the discovery of retroviruses helped advance methodology in molecular phylogeny? (You may want to review the basic properties of retroviruses in Section 8.22 before answering this question.)

6. Determine the GC ratio of the following stretch of DNA:

> TAAGCCTGCAAGCTTAGCTA
> ATTCGGACGTTCGAATCGAT

7. What reference resource in your library would you check for a taxonomic treatment of prokaryotes? For a more phylogenetic approach to bacterial diversity? Does your library have these resources? If so, examine the table of contents of each and describe why these two reference sources are considered to take either a *taxonomic* or a *phylogenetic* approach.

SUPPLEMENTARY READINGS

Allsopp, D., R. R. Colwell, and D. L. Hawksworth (eds.). 1995. *Microbial Diversity and Ecosystem Function.* CAB International, Oxon, UK. A collection of chapters on relating microbial diversity to the available methods for assessing it.

Balows, A., H. G. Trüper, M. Dworkin, W. Harder, and K. H. Schleifer (eds.). 1992. *The Prokaryotes,* 2nd edition. Springer-Verlag, New York. The most up-to-date treatment of the biology of the prokaryotes that blends both taxonomic and phylogenetic information. Highly recommended for beginning a detailed study of any prokaryotic group.

Bengtson, S. (ed.). 1994. *Early Life on Earth*. Columbia University Press, New York. Forty-three chapters of coverage on early Earth, early organisms, and early ecosystems. Highly recommended.

Gesteland, R. F., and **J. F. Atkins** (eds.). 1993. *The RNA World*. Cold Spring Harbor Laboratory Press, Plainview, NY. A compilation of theories and experimental support for the idea of a period of RNA life.

Goodfellow, M., and **A. G. O'Donnell** (eds.). 1993. *Handbook of New Bacterial Systematics*. Academic Press, Orlando, FL. A comprehensive treatment of modern bacterial systematics including traditional as well as modern molecular methods.

Holt, J. G. (editor-in-chief). *Bergey's Manual of Systematic Bacteriology*, Vol. I, 1984; vol. II, 1986; vols. III and IV, 1989. Williams and Wilkins, Baltimore. The major taxonomic reference for the prokaryotes.

Holt, J. G. (editor-in-chief). 1994. *Bergey's Manual of Determinative Bacteriology*, 9th edition. Williams and Wilkins, Baltimore. Focuses on descriptions of genera with only brief listings of species traits. A manual suitable for identifying bacteria.

Li, W. H., and **D. Graw.** 1991. *Fundamentals of Molecular Evolution*. Sinauer Associates, Sunderland, MA. A textbook of molecular sequencing explaining many of the theoretical concepts behind going from sequences to evolutionary principles.

Logan, N. A. 1994. *Bacterial Systematics*. Blackwell Scientific Publications, Oxford, England. A concise overview of bacterial systematics with an emphasis on molecular aspects of classification including ribosomal RNA sequencing and other comparative methods.

Schopf, J. W., and **C. Klein.** (eds.). 1992. *The Proterozoic Biosphere—A Multidisciplinary Study*. Cambridge University Press, New York. A detailed treatment of the paleobiology of early Earth including geology, biogeochemistry, origin of life, microbial mats, and related topics.

Sneath, P. H. A. 1992. *International Code of Nomenclature of Bacteria*. American Society for Microbiology, Washington, DC.

Woese, C. R. 1987. Bacterial evolution. *Microbiol. Rev.* 51:221–271. An excellent summary of what ribosomal RNA sequencing has taught us about the evolution of both prokaryotes and eukaryotes. Highly recommended as a starting point for a detailed study of molecular evolution.

On~line resources for this chapter are on the World Wide Web at: http://www.prenhall.com/~brock (click on the <u>table of contents</u> link and then select Chapter 15).

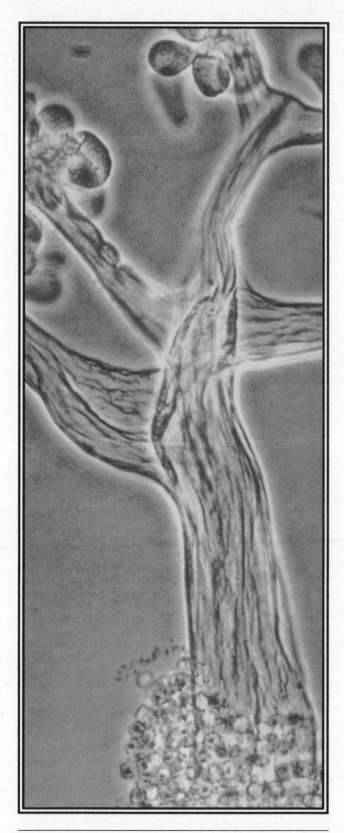

The diversity of prokaryotes is enormous and includes organisms that undergo complex life cycles, such as that of *Chondromyces crocatus*, shown here.

Prokaryotic Diversity: Bacteria

MINIGLOSSARY for Chapter 16

Acid–Alcohol Fastness (Acid Fastness) a property of *Mycobacterium* species in which cells stained with the dye basic fuchsin resist decolorization with acidic alcohol

Carboxysomes polyhedral cellular inclusions of crystalline ribulose bisphosphate carboxylase (RubisCO), the key enzyme of the Calvin cycle

Chemolithotrophs organisms able to oxidize inorganic compounds as energy sources

Chlorosomes cigar-shaped structures bounded by a nonunit membrane and containing the light-harvesting bacteriochlorophyll (*c, d,* or *e*) in green bacteria and *Chloroflexus*

Cyanobacteria prokaryotic oxygenic phototrophs that contain chlorophyll *a* and phycobilins but not chlorophyll *b*

Enteric Bacteria a large group of gram-negative rod-shaped Bacteria characterized by a facultatively aerobic metabolism

Green Bacteria anoxygenic phototrophs containing chlorosomes and bacteriochlorophyll *c, c_s, d,* or *e* as light-harvesting chlorophyll

Heliobacteria anoxygenic phototrophs containing bacteriochlorophyll *g*

Heterocyst a differentiated cyanobacterial cell that carries out nitrogen fixation but not oxygenic photosynthesis

Heterofermentative in reference to lactic acid bacteria, capable of making more than one fermentation product

Homoacetogen an obligately anaerobic bacterium that produces only acetate from fermentation of sugars or from CO_2 reduction with H_2

Homofermentative in reference to lactic acid bacteria, producing only lactic acid as a fermentation product

Methanotroph an organism capable of oxidizing methane (CH_4)

Methylotroph an organism capable of oxidizing organic compounds that do not contain carbon–carbon bonds; if able to oxidize CH_4, also a methanotroph

Nitrifying Bacteria chemolithotrophs capable of carrying out the transformation $NH_3 \rightarrow NO_2^-$ or $NO_2^- \rightarrow NO_3^-$

Nonsulfur Purple Bacteria a group of phototrophic prokaryotes containing bacteriochlorophylls *a* or *b* that grow best as photoheterotrophs and have a relatively low tolerance for H_2S

Prochlorophyte a prokaryotic oxygenic phototroph that contains chlorophylls *a* and *b* but lacks phycobilins

Prostheca an extrusion of cytoplasm often forming a distinct appendage, bounded by the cell wall

Pseudomonad member of the genus *Pseudomonas*, a large group of gram-negative, obligately respiratory (never fermentative) Bacteria

Purple Sulfur Bacteria a group of phototrophic prokaryotes containing bacteriochlorophylls *a* or *b* and characterized by the ability to oxidize H_2S and store elemental sulfur inside the cells (or in the genus *Ectothiorhodospira*, outside the cell)

Spirochete a slender, tightly coiled gram-negative prokaryote characterized by possession of axial filaments used for motility

Stickland Reaction fermentation of an amino acid pair in which one amino acid serves as an electron donor and a second serves as an electron acceptor

Sulfate-Reducing Bacteria a large group of anaerobic Bacteria that respire anaerobically with SO_4^{2-} as elec-

I n the preceding chapter we stressed the evolutionary relationships among microorganisms. In this and the next chapters we describe the major groups of prokaryotes in the domains Bacteria and Archaea. We begin this and Chapter 17 with a phylogenetic overview of the major evolutionary lineages in each domain, illustrating this with a "close-up" phylogenetic tree in each case. We follow this with descriptions of key groups of prokaryotes within that domain. Additional information on all the organisms described here can be found in *Bergey's Manual* and in *The Prokaryotes* (2nd edition) (∞ Section 15.9), and the student seeking more detailed information on any group should refer to these sources.

Phylogenetic overview of Bacteria

Twelve major lineages (kingdoms) of Bacteria can be resolved by ribosomal ribonucleic acid (rRNA) sequencing (Figure 16.1). The most ancient of these is the *Aquifex–Hydrogenobacter* group, hyperthermophilic chemolithotrophs that oxidize H_2 or reduced sulfur compounds. *Aquifex* grows at up to 95°C and is the closest known relative of the universal ancestor of all Bacteria (∞ Section 17.3). Next in line is *Thermotoga* (Figure 16.1), also a hyperthermophile but an anaerobic fermentative (chemoorganotrophic) organism rather than a chemolithotroph (∞ Section 17.3). Following up the evolutionary ladder from *Thermotoga* are the "green nonsulfur bac-

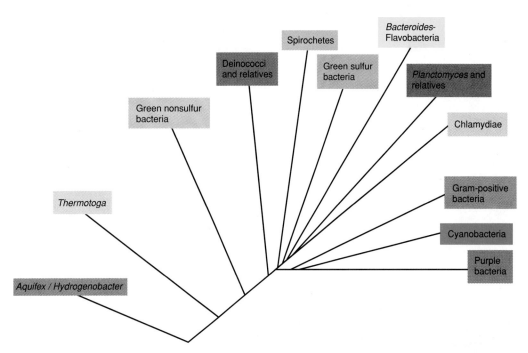

Figure 16.1 Detailed phylogenetic tree of the major lineages (kingdoms) of Bacteria based on 16S ribosomal RNA sequence comparisons. The relative positions of branches on this detailed tree differ slightly (for statistical reasons) from those shown in the universal tree (∞ Figure 15.12), but the branch lengths on the tree remain proportional to the corrected evolutionary distances calculated between any two groups. The purple bacteria are also called the Proteobacteria. See Section 17.3 for a discussion of *Aquifex* and *Hydrogenobacter*.

teria," sometimes known as the *Chloroflexus* group. This group contains the phototrophic species *Chloroflexus* as well as two nonphototrophic genera, *Herpetosiphon* and *Thermomicrobium*. All these organisms are thermophiles but not hyperthermophiles, suggesting that this lineage may have evolved at a time when the earth was still quite warm but not extremely hot.

Continuing through the domain Bacteria past the green nonsulfur bacteria, thermophily and hyperthermophily is lost as an obligatory trait, and a number of lineages of Bacteria fan out (Figure 16.1). These are the deinococci and relatives, which include the highly radiation-resistant organisms *Deinococcus* (see Section 16.24) and the thermophile *Thermus aquaticus* (see the back cover of this book), the morphologically unique spirochetes (see Section 16.12), and the phototrophic green sulfur bacteria (see Section 16.1). Following these lineages is the *Bacteroides* group, which incudes a mixture of physiological types such as the strictly anaerobic *Bacteroides* and the strictly aerobic *Sporocytophaga*, and a variety of related aerobic rod-shaped gliding bacteria such as *Cytophaga* and *Flexibacter* (see Section 16.13). The *Planctomyces–Pirella* group (Figure 16.1) contains organisms that reproduce by budding and lack peptidoglycan in their cell walls. They are primarily aquatic, are aerobic in metabolism, and require very dilute media for cultivation (see Section 16.10). The *Chlamydia* group contains obligately intracellular parasites, many of which cause disease in humans and other animals (see Section 16.23).

The remaining three evolutionary lineages of Bacteria will constitute the majority of our discussion in this chapter. They include the gram-positive bacteria, the cyanobacteria, and the purple bacteria. Each of

these is a large group containing many genera and species and are groups about which much phenotypic information was known long before their evolutionary stature was appreciated. The gram-positive bacteria include both rods and cocci and can be separated into two subgroups, called *low GC* and *high GC*. The "GC" designation refers to the fact that members of a particular subgroup tend to have DNA GC base ratios (∞ Section 15.9) either well above (high GC) or well below (low GC) 50%. The gram-positive bacteria are a large heterogeneous group and are discussed in detail in Sections 16.24–16.32. The cyanobacteria are oxygenic phototrophic organisms with evolutionary roots near those of the gram-positive bacteria; these organisms are covered in Section 16.2.

The final lineage on the Bacteria tree is the "purple bacteria," also called the *Proteobacteria* (Figure 16.1). This group is the largest and most physiologically diverse of all Bacteria (see Sections 16.1, 16.4–16.8, 16.11, and 16.13–16.22). The group comprises five subdivisions: *alpha, beta, gamma, delta,* and *epsilon* (Table 16.1). Many organisms grouped into the purple bacteria are indeed phototrophic (this is how the group originally got its name), and it has thus been proposed that the common ancestor of the purple bacterial group was itself phototrophic. However, many other "purple" bacteria are nonphototrophic. The physiological diversity characteristic of the purple bacterial group probably arose over time by exchange of photosynthetic capacity for other energy-yielding strategies (such as chemoorganotrophy and chemolithotrophy) better fitted to organisms evolving to colonize new ecological niches. How this may have happened is

TABLE 16.1	Major genera that can be grouped in the purple bacteria (Proteobacteria)
Group	**Genera**
Alpha group	*Rhodospirillum*[a], *Rhodopseudomonas*[a], *Rhodobacter*[a], *Rhodomicrobium*[a], *Rhodovulum*[a], *Rhodopila*[a], *Rhizobium*, *Nitrobacter*, *Agrobacterium*, *Aquaspirillum*, *Hyphomicrobium*, *Acetobacter*, *Gluconobacter*, *Beijerinckia*, *Paracoccus*, *Pseudomonas* (some species)
Beta group	*Rhodocyclus*[a], *Rhodoferax*[a], *Rubrivivax*[a], *Spirillum*, *Nitrosomonas*, *Sphaerotilus*, *Thiobacillus*, *Alcaligenes*, *Pseudomonas*, *Bordetella*, *Neisseria*, *Zymomonas*
Gamma group	*Chromatium*[a], *Thiospirillum*[a], and other purple sulfur bacteria[a], *Beggiatoa*, *Leucothrix*, *Escherichia* and other enteric bacteria, *Legionella*, *Azotobacter*, fluorescent *Pseudomonas* species, *Vibrio*
Delta group	*Myxococcus*, *Bdellovibrio*, *Desulfovibrio* and other sulfate-reducing bacteria, *Desulfuromonas*
Epsilon group	*Thiovulum*, *Wolinella*, *Campylobacter*, *Helicobacter*

[a]Phototrophic representatives.

containing polar stacks of lamellar membranes; phylogenetically, *Rhodopseudomonas* is very closely related to *Nitrobacter*, a budding *chemolithotrophic* bacterium that also contains polar stacks of lamellar membranes (Table 16.1). Among gamma purple bacteria there is a close phylogenetic relationship between the *phototrophic* organism *Chromatium* and the *chemolithotrophic* organism *Beggiatoa* (Table 16.1). Both organisms oxidize H_2S and store elemental sulfur within their cells (compare Figures 16.7a and b and 16.58) but for quite different physiological reasons that will be discussed later (see Sections 16.1 and 16.13). We will see many parallels of this sort as we consider the various groups of purple bacteria.

With this introduction to the phylogeny of the domain Bacteria, let us proceed to a description of some representative groups. We begin with organisms able to carry out anoxygenic photosynthesis (∞ Sections 13.1–13.4), which includes organisms from four of the phylogenetic lineages we have just discussed (Figure 16.1): the purple bacteria, green sulfur bacteria, green nonsulfur bacteria, and gram-positive bacteria.

16.1
Purple and Green (Anoxygenic Phototrophic) Bacteria

Bacteria able to use light as an energy source comprise a large and heterogeneous group of organisms, grouped together primarily because they possess one or more pigments called *chlorophylls* and are able to carry out light-mediated generation of adenosine triphosphate (ATP), a process called **photophosphorylation.** Two major groups are recognized, the **purple**

depicted in Figure 16.2. This hypothesis accounts nicely for the finding of close phylogenetic relationships between various purple bacteria whose physiologies are quite different. For example, the alpha purple bacterium *Rhodopseudomonas* is a budding *phototrophic* bacterium

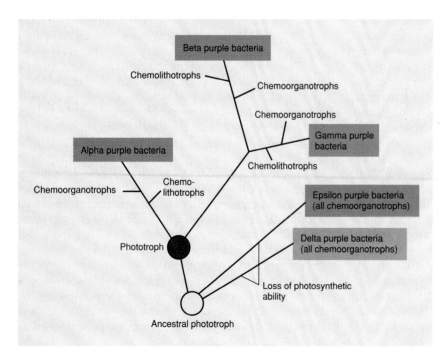

Figure 16.2 Evolutionary divergence of the purple Bacteria (Proteobacteria) to give rise to nonphototrophic relatives.

and green bacteria as one group, and the **cyanobacteria** as the other group. The basic distinction between the purple and green bacteria and the cyanobacteria is based on the photopigments and overall photosynthetic process. Cyanobacteria are *oxygenic* phototrophs, employing chlorophyll *a* and two photosystems in their photosynthetic process. Purple and green bacteria are *anoxygenic* phototrophs, employing bacteriochlorophyll (of several different types) but only *one* photosystem in their photosynthetic process.

The biochemistry of photosynthesis was discussed in some detail in Chapter 13. Green plant photosynthesis, which is also exhibited by the cyanobacteria, involves two light reactions, and H_2O serves as the electron donor. As a result of the photolysis of water, O_2 is produced. By contrast, during photosynthesis by anoxygenic phototrophs water is not photolysed, and O_2 is not produced. Because the purple and green bacteria are unable to photolyse water, they must obtain their reducing power for CO_2 fixation from a reduced substance in their environment. This can be an organic compound, a reduced sulfur compound, or H_2.

Anoxygenic phototrophs can grow phototrophically only under *anaerobic* conditions because pigment synthesis in these organisms is repressed by O_2. Cyanobacteria, on the other hand, develop readily under aerobic conditions with no reduced compounds present. Many anoxygenic phototrophs can grow autotrophically with CO_2 as carbon source and a reduced sulfur compound or H_2 as electron donor under anaerobic conditions in the light. Under microaerobic conditions, these same electron donors support chemolithotrophic growth of many purple bacteria. Phototrophic growth is also possible with light as energy source and an *organic* compound as a carbon source (photoheterotrophy); under such conditions, CO_2 is usually only a minor source of cell carbon. Some purple and green bacteria can also grow chemoorganotrophically in the dark using an organic compound as electron donor, but under these conditions a suitable electron acceptor is necessary.

Bacteriochlorophylls

We compared a typical bacteriochlorophyll with algal chlorophyll in Figure 13.3. A number of bacteriochlorophylls exist, differing in substituents on various parts of the porphyrin ring, and these are outlined in Figure 16.3. The various modifications lead to changes in the absorption spectra of the bacteriochlorophylls so an organism containing a certain bacteriochlorophyll is best able to utilize light of particular wavelengths. The ecological significance of the ability to utilize different wavelengths of light was discussed in Section 13.5, and it is likely that this selective absorption provides an evolutionary pressure for the development of organisms with various chlorophylls. The long-wavelength absorption maxima of the *bacteriochlorophylls*

(Bchl) are the most characteristic, and these are given in Figure 16.3 as measured in the living cell (*in vivo*) and in solvent extract. Although the *in vivo* absorption maximum is of most significance ecologically, from the viewpoint of characterizing the various purple and green bacteria taxonomically, the absorption spectrum in solvent extract is most convenient because its measurement is easier. Thus, to characterize the bacteriochlorophyll of a new isolate of a purple or green bacterium, a cell extract in methanol or other organic solvent is made and the absorption spectrum determined. From the long-wavelength maximum listed in Figure 16.3, the identification of the bacteriochlorophyll can be made.

Classification

The purple and green bacteria are a diverse group morphologically and phylogenetically (∞ Section 15.7). Morphologically, cocci, rods, vibrios, spirals, budding, and gliding species are known. It thus seems likely that the ability to grow phototrophically has developed in a wide variety of bacteria and that the only unifying thread among the whole group is the ability to carry out photophosphorylation. Anoxygenic phototrophic bacteria have been classified into three major groups: purple bacteria, green bacteria, and heliobacteria. The major properties of each group are as follows:

Group	Bacteriochlorophylls	Photosynthetic membrane systems
Purple bacteria	Bchl *a*, Bchl *b*	Lamellae, tubes, or vesicles continuous with cytoplasmic membrane (see Figure 16.5*a* and *b*)
Green bacteria	Bchl *c*, Bchl *d*, or Bchl *e*, plus small amounts of Bchl *a*	Chlorosomes, attached to but not continuous with cytoplasmic membrane (see Figure 16.5*c*)
Helio-bacteria	Bchl *g*	Cytoplasmic membrane only (see Figure 16.5*d*)

Anoxygenic phototrophic bacteria also produce **carotenoid pigments** (Figure 16.4), and the carotenoids of the purple bacteria generally differ from those of the green bacteria and heliobacteria. Carotenoid pigments are responsible for the purple color of the purple bacteria, and mutants lacking carotenoids are blue-green in color, reflecting the actual color of Bchl *a* (see Figure 16.4). In fact, purple phototrophic bacteria are frequently not purple but brown, pink, brown-red, or purple-violet, depending on their carotenoid pigments (see Figure 16.4*b*). Also, many of the "green" bacteria are actually brown in color, owing to their complement of carotenoids (Figure 16.4). Thus, color is not a good criterion for use in identifying isolates as either green bacteria or purple bacteria.

Pigment	R$_1$	R$_2$	R$_3$	R$_4$	R$_5$	R$_6$	R$_7$	Infrared absorption maxima (nm) In vivo	Extract (methanol)
Bacteriochlorophyll *a* (purple bacteria)	—C(=O)—CH$_3$	—CH$_3$ [b]	—CH$_2$—CH$_3$	—CH$_3$	—C(=O)—O—CH$_3$	P/Gg[a]	—H	805 830–890	771
Bacteriochlorophyll *b* (purple bacteria)	—C(=O)—CH$_3$	—CH$_3$ [c]	=C—CH$_3$ (H)	—CH$_3$	—C(=O)—O—CH$_3$	P	—H	835–850 1020–1040	794
Bacteriochlorophyll *c* (green sulfur bacteria)	—C(H)(OH)—CH$_3$	—CH$_3$	—C$_2$H$_5$ / —C$_3$H$_7$[d] / —C$_4$H$_9$	—C$_2$H$_5$ / —CH$_3$	—H	F	—CH$_3$	745–755	660–669
Bacteriochlorophyll *c$_s$* (green nonsulfur bacteria)	—C(H)(OH)—CH$_3$	—CH$_3$	—C$_2$H$_5$	—CH$_3$	—H	S	—CH$_3$	740	667
Bacteriochlorophyll *d* (green sulfur bacteria)	—C(H)(OH)—CH$_3$	CH$_3$	—C$_2$H$_5$ / —C$_3$H$_7$ / —C$_4$H$_9$	—C$_2$H$_5$ / —CH$_3$	—H	F	—H	705–740	654
Bacteriochlorophyll *e* (green sulfur bacteria)	—C(H)(OH)—CH$_3$	—C(=O)—H	—C$_2$H$_5$ / —C$_3$H$_7$ / —C$_4$H$_9$	—C$_2$H$_5$	—H	F	—CH$_3$	719–726	646
Bacteriochlorophyll *g* (heliobacteria)	—C(H)=CH$_2$	—CH$_3$ [b]	—C$_2$H$_5$	—CH$_3$	—C(=O)—O—CH$_3$	F	—H	670, 788	765

[a]P, Phytyl ester (C$_{20}$H$_{39}$O—); F, farnesyl ester (C$_{15}$H$_{25}$O—); Gg, geranylgeraniol ester (C$_{10}$H$_{17}$O—); S, stearyl alcohol (C$_{18}$H$_{37}$O—).
[b]No double bond between C$_3$ and C$_4$; additional H atoms are in positions C$_3$ and C$_4$.
[c]No double bond between C$_3$ and C$_4$; an additional H atom is in position C$_3$.
[d]Bacteriochlorophylls *c, d,* and *e* consist of isomeric mixtures with the different substituents on R$_3$ as shown.

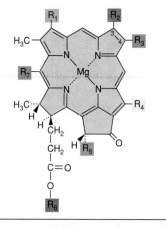

Figure 16.3 Structure of all known bacteriochlorophylls. The different substituents present in the positions R$_1$ to R$_7$ are given in the accompanying table.

Photosynthetic membrane systems

A major difference between the green and purple bacteria is in the nature of the photosynthetic membrane system. In the purple bacteria, the photosynthetic pigments are part of an elaborate internal membrane system, connected to and produced from the cytoplasmic membrane; the membrane often occupies much of the cell interior. In some cases, the membrane system is an array of flat sheets called **lamellae** (Figure 16.5*a*), whereas in others it consists of round tubes referred to as **vesicles** (Figure 16.5*b*). The membrane content of the cell varies with pigment content, which is itself affected by light intensity and the presence of O$_2$. When cells are grown aerobically, synthesis of bacteriochlorophyll is repressed, and the organisms may be

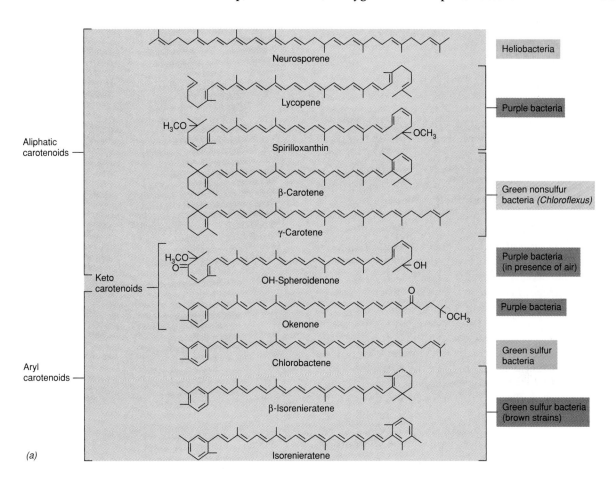

(a)

Group	Names	Color of organisms
1	Lycopene, rhodopin, spirilloxanthin, neurosporene	Orange-brown brownish-red, pink, purple-red, green
2	Spheroidene, hydroxyspheroidene, spheroidenone, hydroxyspheroidenone, spirilloxanthin	Red (aerobic) Brownish red to purple (anaerobic)
3	Okenone, methoxylated keto carotenoids (*Rhodopila globiformis* only)	Purple-red
4	Lycopenal, lycopenol, rhodopin, rhodopinal, rhodopinol	Purple-violet
5	Chlorobactene, hydroxychlorobactene, β-isorenieratene, isorenieratene	Green (chlorobactene) Brown (isorenieratene)
6	β-carotene, γ-carotene	Orange-green

(b)

Based on Schmidt, K. 1978, pp. 729–750, and Liaaen-Jensen, S. 1978, pp. 233–247, in R. K. Clayton and W. R. Sistrom (eds.), *The Photosynthetic Bacteria*, Plenum Press, New York.

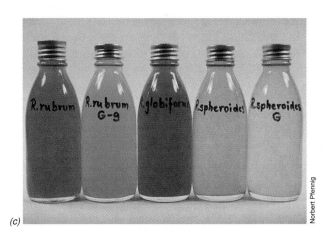

(c)

Figure 16.4 (a) Structures of major carotenoids of phototrophic bacteria. (b) This table lists a few representative carotenoids of the various groups. (c) Photograph of mass cultures of phototrophic bacteria showing the color of species with various carotenoid pigments. The blue culture is a carotenoid-less mutant derivative of *Rhodospirillum rubrum* showing how bacteriochlorophyll *a* is actually *blue* in color. The bottle on the far right (*R. sphaeroides* strain G) lacks one of the carotenoids of the wild type and thus is more green in color.

virtually devoid of photopigments as well as the internal membrane systems. Consequently, phototrophic growth is possible only under *anaerobic* conditions, where bacteriochlorophyll synthesis can occur.

Superimposed on this O_2 effect is an effect of light intensity. Even under anaerobic conditions, when synthesis of the photosynthetic apparatus is not repressed, the level of the photopigments and internal membranes is affected by light intensity. At *high* light intensity, the synthesis of the photosynthetic apparatus is inhibited, whereas when cells are grown at *low* light intensity the bacteriochlorophyll content is high and the cells are packed with membranes. (This increase in cell pigment content at low light intensities allows the organism to better utilize the available light.) Usually carotenoid pigment synthesis is coordinately regulated with the bacteriochlorophyll content (∞ Figure 13.11).

In the green bacteria, the photosynthetic apparatus is structurally quite different, consisting of a series of cylindrically shaped structures called **chlorosomes** underlying and attached to the cytoplasmic membrane (Figure 16.5c). These vesicles are enclosed within a thin membrane that does not have the usual bilayered appearance (it is referred to as a *nonunit* membrane). Bacteriochlorophylls *c*, *c$_s$*, *d*, or *e* (depending on the species) are present inside the chlorosomes, while most of the Bchl *a* and components of the photosynthetic electron transport chain are located in the cytoplasmic membrane.

In heliobacteria, a group of strictly anaerobic phototrophs that are green in color but not phylogenetically related to any other green bacteria, neither internal membranes, typical of purple bacteria, nor chlorosomes, typical of green bacteria, are observed. Bacteriochlorophyll in heliobacteria is associated with the cytoplasmic membrane (Figure 16.5d). High resolution electron microscopy of thin sections of heliobacteria have shown the membrane to be studded with particles that may be the site of bacteriochlorophyll:protein complexes in these organisms.

Aerobic anoxygenic phototrophs

An interesting group of bacteria from marine environments produce bacteriochlorophyll only under *aerobic* conditions. These aerobic phototrophs resemble nonsulfur purple bacteria, but they do not grow or produce bacteriochlorophyll anaerobically. Significant pigment levels are observed only when cells are grown aerobically in either the light or the dark. The function of bacteriochlorophyll in the aerobic phototrophs is presumably to carry out photophosphorylation, and it is of interest that O_2 regulates pigment synthesis in these organisms in a manner opposite that of typical purple bacteria. The majority of aerobic phototrophic bacteria have been placed in the genus *Erythrobacter,* but a few isolates have been classified within the genus *Pseudomonas.*

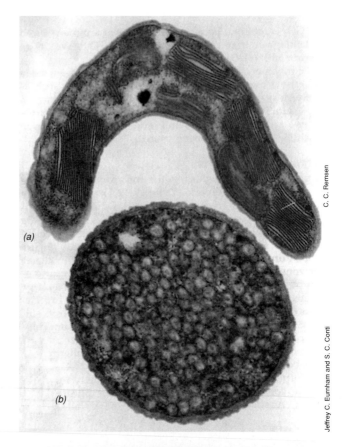
(a)

(b)

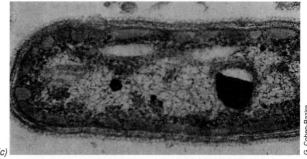
(c)

(d)

Figure 16.5 Membrane systems of phototrophic bacteria as revealed by the electron microscope. (a) Purple phototrophic bacterium, *Ectothiorhodospira mobilis,* showing the photosynthetic membranes in flat sheets (lamellae). (b) *Chromatium* sp., strain D, another purple phototrophic bacterium, showing the membranes as individual vesicles. (c) Green phototrophic bacterium, *Pelodictyon* sp., showing location of chlorosomes. (d) *Heliobacterium chlorum,* showing the lack of internal photosynthetic membranes.

Enrichment culture

The purple and green sulfur bacteria are generally found in *anoxic* zones of aquatic habitats, often where H_2S accumulates. These organisms can be enriched by duplicating the habitat in the laboratory (Table 16.2). A basal mineral salts medium is used, to which bicarbonate is added as a source of CO_2. Because many phototrophic bacteria require vitamin B_{12}, this vitamin is usually added. The optimum pH range is 6–8, and the optimum temperature for mesophilic species is about 30°C. Incubation is anoxic, and a small amount of sodium sulfide (0.05–0.1% $Na_2S \cdot 9\ H_2O$) is added as photosynthetic electron donor. Selection of appropriate light conditions is important. Light intensities should not be too high because these bacteria usually live in deep areas of lakes or other anoxic habitats where light is low. Intensities between 200 and 1000 lux (lx) are adequate (about the intensity 1–2 feet (ft) away from a 40-Watt (W) incandescent light bulb). The quality of light is also important. Purple and green bacteria use radiation of the near infrared, 700–1000 nm, which can be most easily obtained in the laboratory by use of conventional tungsten light bulbs. Fluorescent bulbs are usually deficient in infrared radiation and are not as satisfactory, especially for purple bacteria. Purple bacteria that contain bacteriochlorophyll *b* can be enriched by using infrared filters that pass wavelengths greater than 900 nm because bacteriochlorophyll *b* has an absorption maximum at 1020 nm. After inoculation, culture tubes or bottles are incubated for several weeks and examined periodically for signs of visible growth. Enrichment cultures should appear pigmented, and microscopic examination of positive cultures should reveal organisms resembling purple and green bacteria. Pure cultures can be obtained from positive enrichments by shake tube methods (∞ Section 14.3 and Figure 14.6), care being taken to keep cultures anaerobic throughout. However, the purple and green bacteria are not as sensitive to oxygen as some other anaerobes, and so extreme precautions to maintain anaerobic conditions are usually not necessary. A widely used medium for the culture of purple and green sulfur bacteria is given in Table 16.2.

For *nonsulfur* purple bacteria, the sulfide concentration of the medium should be reduced to a very low level, 0.005–0.01% $Na_2S \cdot 9\ H_2O$ (or sulfide can be eliminated altogether), and an *organic* compound added to provide a carbon source and/or electron donor. Because many nonsulfur purple bacteria have multiple growth-factor requirements, usually one or more B vitamins, addition of 0.01% yeast extract as a source of growth factors is recommended. The organic substrate used should not be widely fermentable (to avoid competition in the enrichment from fermentative organisms); acetate, ethanol, benzoate, isopropanol, and butyrate or dicarboxylic acids such as succinate are ideal.

Heliobacteria can be isolated by taking advantage of a major property apparently shared by all representatives. All known heliobacteria grow well at temperatures up to 42°C, and most species produce endospores. This temperature is restrictive for most other phototrophic bacteria. Hence, to enrich heliobacteria, anaerobic enrichments containing dilute yeast extract and lacking sulfide are inoculated with soil (heliobacteria are very resistant to drying, and thus dry soil often yields strains of these organisms) pasteurized to eliminate purple bacteria and incubated in the light at 40–42°C.

TABLE 16.2	Pfennig's medium for the culture of phototrophic sulfur bacteria[a]
Solution	**Medium**
Solution 1	0.83 g $CaCl_2 \cdot 2\ H_2O$ in 2.5 liters H_2O. For marine organisms, add NaCl, 130 g.
Solution 2	H_2O, 67 ml; KH_2PO_4, 1 g; NH_4Cl, 1 g; $MgCl_2 \cdot 2\ H_2O$, 1 g; KCl, 1 g; 3 ml vitamin B_{12} solution (2 mg/100 ml H_2O); 30 ml trace element solution (1 liter H_2O; ethylenediaminetetraacetic acid, 500 mg; $FeSO_4 \cdot 7\ H_2O$, 200 mg; $ZnSO_4 \cdot 7\ H_2O$, 10 mg; $MnCl_2 \cdot 4\ H_2O$, 3 mg; H_3BO_3, 30 mg; $CoCl_2 \cdot 6\ H_2O$, 20 mg; $CuCl_2 \cdot 2\ H_2O$, 1 mg; $NiCl_2 \cdot 6\ H_2O$, 2 mg; $Na_2MoO_4 \cdot 2\ H_2O$, 3 mg, pH 3).
Solution 3	Na_2CO_3, 3 g; H_2O, 900 ml. Autoclave in a container suitable for gassing aseptically with CO_2.
Solution 4	$Na_2S \cdot 9\ H_2O$, 3 g; H_2O, 200 ml. Autoclave in flask containing a Teflon-covered magnetic stirring rod.

[a]Dispense the bulk of solution 1 in 67-ml aliquots in 30 screw-capped bottles of 100-ml capacity and autoclave with screw caps loosely on. Autoclave the other solutions in bulk. After autoclaving, cool all solutions rapidly by placing the bottles in a cold water bath (to prevent lengthy exposure to air). Solution 3 is then gassed with CO_2 gas until it is saturated (about 30 min; pH drops to 6.2). Add this to cooled solution 2 and aseptically place 33 ml of this mixture in each bottle containing solution 1. Solution 4 is partially neutralized by adding dropwise (while stirring on a magnetic stirrer) 1.5 ml of sterile 2 *M* H_2SO_4. Add 5-ml portions of solution 4 to each bottle, fill the bottles completely with remaining solution 1, and tightly close. The final pH should be between 6.7 and 7.2. Store overnight to consume residual oxygen before using. (For some organisms, the sulfide concentration should be reduced. Use 2.5 ml instead of 5 ml of solution 4 per bottle.) It may be necessary to "feed" the cultures with sulfide (solution 4) from time to time, as the sulfide is used up during growth. To do this, remove 2.5 or 5 ml of liquid from a bottle and refill with an equal amount of sterile, neutralized solution 4.

Originally described by Pfennig, N. 1965. Anreicherungskulturen für rote und grüne Schwefelbakterien. *Zentrabl. Bakteriol. Parasitenkd. Infektionskr. Hyg. Abt. 1, Suppl. 1*;179. For an English version, see van Niel, C. B. 1971. Techniques for the enrichment, isolation and maintenance of photosynthetic bacteria. *Meth. Enzymol.* 23:3–28.

It should be noted that many purple and green bacteria are incapable of assimilatory sulfate reduction, and so they must be given a source of reduced sulfur. The sulfide added to the medium not only uses up the remaining O_2 and serves as an electron donor but also provides this source of reduced sulfur. If an electron donor such as H_2 or an organic compound is to be used, and sulfide must for some reason be avoided, then it is necessary to add another source of reduced sulfur, such as methionine or cysteine (if yeast extract is added, it contains reduced sulfur as well as growth factors).

Nonsulfur purple bacteria

These bacteria have been called *nonsulfur* because it was originally thought that they were unable to use sulfide as an electron donor for the reduction of CO_2 to cell material. However, sulfide can be used by most species provided the concentration is maintained at a low level. It appears that levels of sulfide utilized well by green or purple *sulfur* bacteria are toxic to most *nonsulfur* purple bacteria. Some nonsulfur purple bacteria can also grow anaerobically in the dark using fermentative metabolism, and most can grow aerobically in darkness by respiration. Under the latter conditions, the electron donor can be an organic compound, or in some species even an inorganic compound such as H_2. Nonsulfur purple bacteria are thus among the most energetically versatile of all prokaryotes. However, because of their great photoheterotrophic abilities, these organisms are generally enriched and cultured under phototrophic conditions with an organic compound as carbon source. Most members of this group also require vitamins, and so yeast extract or some other source of vitamins is usually provided.

The morphological diversity of this group is typical of that of other purple and green bacteria (Table 16.3 and Figure 16.6), and it is clearly a heterogeneous group as it contains both polarly and peritrichously flagellated genera, the latter growing by budding.

Some nonsulfur purple bacteria have the ability to utilize methanol as sole carbon source for phototrophic growth. When growing anaerobically with methanol, some CO_2 fixation is necessary, and the following stoichiometry is observed:

$$2 CH_3OH + CO_2 \rightarrow 3 (CH_2O) + H_2O$$
Cell material

CO_2 is required because methanol is at a more reduced oxidation state than cell material, and the CO_2 serves as an electron sink (electron acceptor).

Enrichments for nonsulfur purple bacteria can be made highly selective by omitting fixed nitrogen sources, such as ammonia or nitrate, from the medium and substituting an ample supply of gaseous nitrogen, N_2. Most nonsulfur purple bacteria are active N_2 fixers and grow well in a medium in which N_2 is the sole nitrogen source.

Purple sulfur bacteria

Purple bacteria that deposit sulfur and oxidize it to sulfate are morphologically diverse (Table 16.4). The cell is usually larger than that of green bacteria and in sulfide-rich environments may be packed with sulfur granules (Figure 16.7a and b; see also Figure 16.15), although in the smaller-celled genera the sulfur granules may not be so obvious (Figure 16.7c). Purple sulfur bacteria are commonly found in anoxic zones of lakes as well as in sulfur springs; because of their conspicuous purple color, they are easily visible as large blooms or masses (see Figures 16.15, 16.16, and 16.18). The genus *Ectothiorhodospira* is of interest because it deposits sulfur externally and is also halophilic, growing at sodium chloride concentrations approaching saturation, often at very high pH. It is found in saline

---

TABLE 16.3   Genera and characteristics of nonsulfur purple bacteria

Characteristics	Genus	Number of species	16S rRNA group[a]	DNA (mol % GC)
Spirilla, polarly flagellated	*Rhodospirillum*	9	Alpha purple	62–68
Rods, polarly flagellated; divide by budding	*Rhodopseudomonas*	9	Alpha purple	64–72
Rods; divide by binary fission	*Rhodobacter*	6	Alpha purple	62–71
Ovoid to rod-shaped cells	*Rhodovulum*	4	Alpha purple	64–68
Ovals, peritrichously flagellated; growth by budding and hypha formation	*Rhodomicrobium*	1	Alpha purple	61–63
Large spheres, acidophilic (pH 5 optimum)	*Rhodopila*	1	Alpha purple	66
Ring-shaped or spirilla	*Rhodocyclus*	2	Beta purple	64–66
Curved rods	*Rubrivivax*	1	Beta purple	70–72
Curved rods	*Rhodoferax*	1	Beta purple	59–60

[a] ∞ Section 15.7.

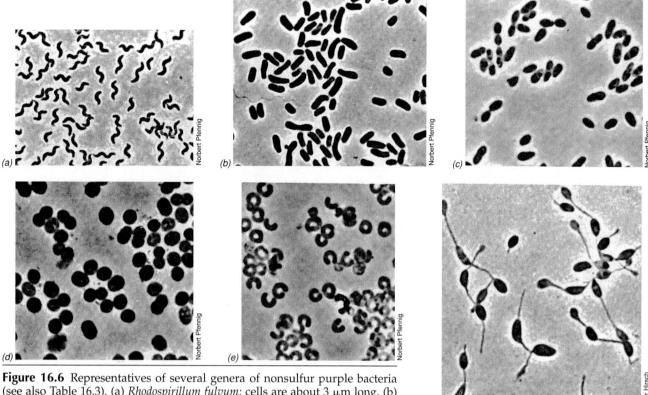

Figure 16.6 Representatives of several genera of nonsulfur purple bacteria (see also Table 16.3). (a) *Rhodospirillum fulvum;* cells are about 3 μm long. (b) *Rhodopseudomonas acidophila;* cells are about 4 μm long. (c) *Rhodobacter sphaeroides;* cells are about 1.5 μm wide. (d) *Rhodopila globiformis;* cells are about 1.6 μm wide. (e) *Rhodocyclus purpureus;* cells are about 0.7 μm in diameter. (f) *Rhodomicrobium vannielii;* cells are about 1.2 μm wide.

lakes, soda lakes, salterns, and other bodies of water high in salt (∞ Section 17.1).

Purple sulfur bacteria have a limited ability to utilize organic compounds as carbon sources for phototrophic growth. Acetate and pyruvate are utilized by all species, and some species use a few other organic compounds. A few purple sulfur bacteria grow chemolithotrophically in darkness with thiosulfate as electron donor, and *Thiocapsa* grows chemoorganotrophically on acetate.

Green sulfur and green nonsulfur bacteria

Green bacteria are morphologically quite diverse, including nonmotile rods, spirals, and spheres (green sulfur bacteria) and motile filamentous gliding forms (green nonsulfur or *Chloroflexus* group) (see Figure 16.8 and Table 16.5). Green bacteria are also very phylogenetically diverse, as the green sulfur bacteria and *Chloroflexus* represent two distinct lineages of Bacteria (∞ Section 15.7). Several other green bacteria have complex appendages called **prosthecae,** and in this connection resemble the budding and/or appendaged bacteria (see Section 16.10).

The green sulfur bacteria that live planktonically in lakes generally possess gas vesicles, whereas the species that live in the outflow of sulfur and hot springs, or in

other benthic habitats, are not gas-vesiculate. Members of one genus, *Pelodictyon,* consist of rods that undergo branching, and because the rods remain attached, a three-dimensional network is formed (Figure 16.8b).

Green sulfur bacteria are strictly anaerobic and obligately phototrophic, being unable to carry out respiratory metabolism in the dark. Most green sulfur bacteria can assimilate simple organic substances for phototrophic growth, provided that a reduced sulfur compound is present as a sulfur source (because they are incapable of assimilatory sulfate reduction). Organic compounds used by these species include acetate, propionate, pyruvate, and lactate. *Chloroflexus* is much more versatile than green sulfur bacteria, being able to grow chemoorganotrophically in the dark under aerobic conditions, as well as phototrophically on a wide variety of sugars, amino acids, and organic acids, or photoautotrophically with H_2S or H_2 and CO_2. Although capable of oxidizing sulfide, *Chloroflexus* grows best as a photoheterotroph, and thus, in analogy to the nutritional situation in purple nonsulfur bacteria, it has been given the designation *green nonsulfur bacterium*.

Green sulfur bacteria share some photochemical characteristics with heliobacteria (see later discussion of the latter group). The chain of electron carriers between the primary electron acceptor and the bacteriochloro-

TABLE 16.4   Genera and characteristics of purple sulfur bacteria[a]

Characteristics	Genus	Number of species	DNA (mol % GC)
Sulfur deposited externally:			
Spirilla, polar flagella	*Ectothiorhodospira*	6	62–67
Spirilla, extreme halophiles	*Halorhodospira*	3	50–69
Sulfur deposited internally:			
Do not contain gas vesicles			
Ovals or rods, polar flagella	*Chromatium*	12	48–70
Spheres, diplococci, tetrads, nonmotile	*Thiocapsa*	3	63–70
Spheres or ovals, polar flagella	*Thiocystis*	2	61–68
Large spirilla, polar flagella	*Thiospirillum*	1	45
Small spirilla	*Thiorhodovibrio*	1	61–62
Rod- to spindle-shaped cells, 1.5–1.7 μm wide and 16–32 μm long	*Rhabdochromatium*	1	60
Contain gas vesicles			
Irregular spheres, ovals, nonmotile	*Amoebobacter*	4	63–65
Rods	*Lamprobacter*	1	64
Spheres, ovals, polar flagella	*Lamprocystis*	1	64
Rods, nonmotile; forming irregular network	*Thiodictyon*	2	65–66
Spheres, nonmotile; forming flat sheets of tetrads	*Thiopedia*	1	62–64

[a]From a phylogenetic standpoint, all are members of the gamma subdivision of the purple Bacteria (Proteobacteria) (∞ Section 15.7 and Table 16.1).

phyll of green bacteria and heliobacteria is quite similar. Of particular interest is the fact that the primary electron acceptor in green bacteria and heliobacteria is poised at a reduction potential of about −0.5 V. This is much more reducing than the primary electron acceptor of purple bacteria, which has a reduction potential of about −0.15 V (the primary electron acceptor in *Chloroflexus* closely resembles that of purple bacteria). The significance of

the low potential primary acceptor in green sulfur bacteria and heliobacteria lies in the fact that such an acceptor, once reduced, is sufficiently negative to reduce NAD^+ directly; this alleviates the need for reverse electron flow as a means of reducing NAD^+ (Figure 16.9). Reverse electron flow (∞ Section 13.4) is the mechanism by which purple bacteria and *Chloroflexus* produce NADH (Figure 16.9). Thus, although they employ

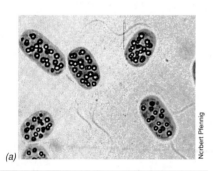
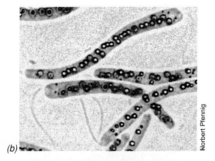
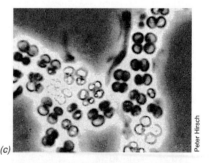
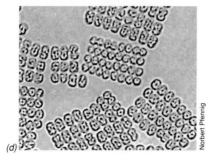

Figure 16.7 Bright-field photomicrographs of purple sulfur bacteria. (a) *Chromatium okenii*; cells are about 5 μm wide. Note the globules of elemental sulfur inside the cells. (b) *Thiospirillum jenense*, a very large, polarly flagellated spiral; cells are abut 30 μm long. Note the sulfur globules. (c) *Thiocapsa*; cells are about 2 μm wide. (d) *Thiopedia rosea*; cells are about 1.5 μm wide. (e) Scanning electron micrograph of a sheet of 16 cells of *Thiopedia rosea* showing the major division planes.

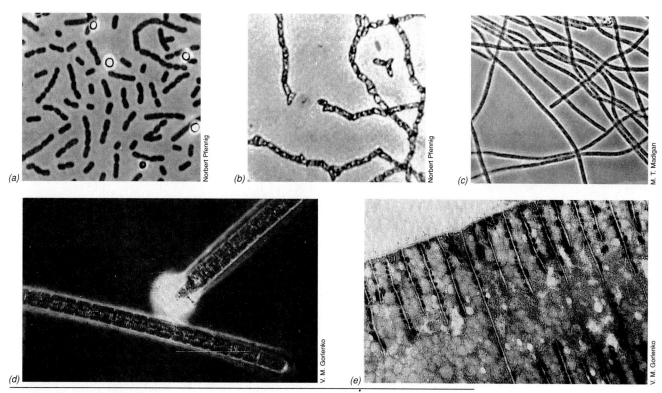

Figure 16.8 Green sulfur and green nonsulfur phototrophic bacteria. (a) *Chlorobium limicola;* cells are about 0.8 μm wide. Note the sulfur granules deposited *extra*cellularly. (b) *Pelodictyon clathratiforme,* a bacterium forming a three-dimensional network; cells are about 0.8 μm wide. (c) *Chloroflexus aurantiacus,* a filamentous gliding bacterium; cells are about 1 μm wide. (d) *Oscillochloris,* a large filamentous, gliding green bacterium; cells are about 5 μm wide. Phase contrast. The brightly contrasting material is the holdfast. (e) Electron micrograph of *Oscillochloris.* The chlorosomes in this preparation are darkly stained.

TABLE 16.5	Genera and characteristics of green phototrophic bacteria and heliobacteria			
Characteristics	Genus	Number of species	16S rRNA group[a]	DNA (mol % GC)
No gas vesicles:				
Straight or curved rods, nonmotile	*Chlorobium*	6	Green sulfur	49–58
Spheres and ovals, nonmotile, forming prosthecae (appendages)	*Prosthecochloris*	2	Green sulfur	50–56
Filamentous, gliding	*Chloroflexus*	2	Green nonsulfur	53–55
	Heliothrix	1	Green nonsulfur	—
Filamentous, gliding, large diameter (2–5 μm)	*Oscillochloris*	1	Green nonsulfur	59
Contain gas vesicles:				
Branching nonmotile rods, in loose irregular network	*Pelodictyon*	4	Green sulfur	48–58
Spheres with prosthecae	*Ancalochloris*	1	Green sulfur	—
Rods, gliding	*Chloroherpeton*	1	Green sulfur	45–48
Filamentous, gliding, large diameter (2–2.5 μm)	*Chloronema*	1	—	—
Heliobacteria:				
Rod-shaped, motile by gliding or by polar flagella	*Heliobacterium*	3	Gram-positive	52–55
Rod-shaped, motile by peritrichous flagella	*Heliobacillus*	1	Gram-positive	50
Rod-shaped, motile, cells attached in bundles and move as a unit	*Heliophilum*	1	Gram-positive	51

[a] ∞ Section 15.7. Dash indicates organisms not available in pure culture.

different bacteriochlorophylls in their photosynthetic reactions (see Figure 16.3), green sulfur bacteria and heliobacteria are remarkably similar in terms of primary photochemical events.

Heliobacteria

The heliobacteria are a phylogenetically separate group of anoxygenic phototrophic bacteria that contain a structurally distinct form of bacteriochlorophyll, bacteriochlorophyll *g* (see Figures 16.3 and 16.11 and Table 16.5). Bacteriochlorophyll *a* is absent. These bacteria are both physiologically and phylogenetically unique from the other phototrophic bacteria. The group consists of three genera, *Heliobacterium*, *Heliophilum*, and *Heliobacillus*. *Heliobacterium* (Figure 16.5*d*) contains gliding rod and motile spirilla species, while *Heliobacillus* (Figure 16.10) is an actively motile rod. *Heliophilum* forms bundles of cells that are motile as a unit. Although physiologically similar to nonsulfur purple bacteria, heliobacteria are strictly anaerobic phototrophs and are unable to grow by respiratory means as is typical of nonsulfur purple bacteria. In addition, autotrophic growth of heliobacteria has not been achieved.

Several species of *Heliobacterium* produce endospores. These structures contain dipicolinic acid and elevated Ca²⁺ levels, typical of the endospores of *Bacillus* or *Clostridium* (∞ Section 3.13). Heliobacteria

seem most abundant in tropical soils, especially in rice soils where extremes of heat and alternating flooding and drying might favor sporulating phototrophs. Interestingly, phylogenetic studies of heliobacteria show them to be closely related to clostridia. Thus, endospore formation by heliobacteria is explainable in terms of their evolutionary roots.

Figure 16.11 shows the structure of bacteriochlorophyll *g* and that of chlorophyll *a*. Unlike all other bacteriochlorophylls, but like chlorophyll *a*, bacteriochlorophyll *g* contains a vinyl (H₂C═CH₂) group on ring I of the tetrapyrrole molecule (Figures 16.3 and 16.11). The only structural differences between bacteriochlorophyll *g* and chlorophyll *a* therefore lie in ring II of the tetrapyrrole. Like other bacteriochlorophylls, ring II of bacteriochlorophyll *g* is *reduced*, unlike that of chlorophyll *a*. However, if cells of *Heliobacterium* (which are originally brown-green in color) are exposed to air and light, a bond in ring II of bacteriochlorophyll *g* becomes oxidized, and cultures of the organism slowly turn emerald green as a result of the conversion of *bacteriochlorophyll g* to *chlorophyll a* (Figure 16.11). This transition is apparently irreversible and leads to loss of cell viability.

Detailed studies of heliobacterial photosynthetic reaction centers have shown that a modified form of chlorophyll *a* called *hydroxychlorophyll a* is present in normal heliobacterial membranes and that it plays a role in

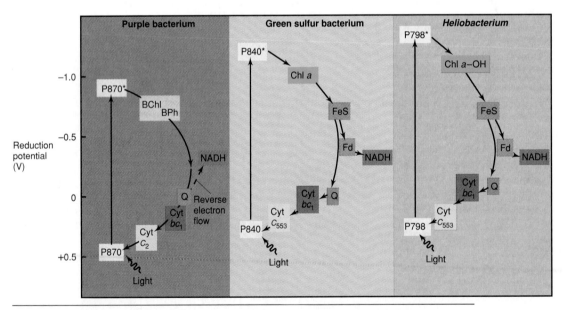

Figure 16.9 A comparison of electron flow in green sulfur bacteria, heliobacteria, and purple bacteria. Note how reverse electron flow in purple bacteria is necessitated by the fact that the primary acceptor (quinone, Q) is more positive in potential than the NAD⁺/NADH couple. In green and heliobacteria NADH production is light-driven. Bchl, Bacteriochlorophyll; BPh, bacteriopheophytin. P870 and P840 are reaction centers of purple and green bacteria, respectively, and consist of Bchl *a*. The reaction center of heliobacteria (P798) contains Bchl *g*. The reaction center of *Chloroflexus* is similar to that of purple bacteria. Note the presence of forms of chlorophyll *a* in the reaction centers of green bacteria and heliobacteria.

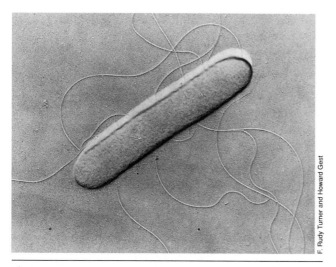

Figure 16.10 Electron micrograph of a representative species of heliobacteria, *Heliobacillus mobilis*. Cells are about 1 μm wide. Notice the peritrichous flagellation.

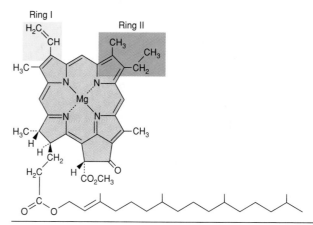

Figure 16.11 Comparison of the structures of bacteriochlorophyll *g* (top) and chlorophyll *a* (bottom). Note the differences on ring II of the tetrapyrrole but the similarities on ring I and elsewhere. No other bacteriochlorophyll contains a vinyl group (H_2C=CH—) on ring I. Note also that the esterifying alcohols are different on the two pigments; in bacteriochlorophyll *g* it is farnesyl, and in chlorophyll *a* it is phytol.

photosynthetic electron flow (Figure 16.9). Thus, although clearly *anoxygenic* phototrophs, heliobacteria contain a slightly modified form of the major pigment of *oxygenic* phototrophs. The phylogenetic standing of heliobacteria may be important in the pigment complement of these organisms because the gram-positive roots of these organisms are also shared by the cyanobacteria, probably the first oxygenic phototrophs to have evolved (∞ Section 15.2).

Physiology of phototrophic growth

In anoxygenic phototrophs light is used primarily in the generation of ATP and is not generally involved in the generation of reducing power as it is in organisms exhibiting oxygenic photosynthesis. Exceptions to this rule exist with green sulfur bacteria and heliobacteria, as discussed earlier (see Figure 16.9). However, whether or not a source of reducing power is actually needed depends on the carbon source supplied. If CO_2 is the sole carbon source, then reducing power is needed, and this can come from a reduced sulfur compound or H_2. Reducing power for CO_2 fixation may also come from an organic compound, but if an organic compound is supplied (photoheterotrophy), the situation is more complex. This is because the organic compound is a carbon source itself so CO_2 need not necessarily be reduced. Whether or not CO_2 is reduced when an organic compound is added depends at least in part on the oxidation state of the organic compound. Compounds such as acetate, glucose, and pyruvate are at about the oxidation level of cell material and can thus be assimilated directly as carbon sources with no requirement for either oxidation or reduction. Fatty acids longer than acetate (for example, propionate, butyrate, caprylate) are more reduced than cell mate-

rial, and some means of disposing of excess electrons is necessary, such as the reduction of CO_2. Thus, the amount of CO_2 fixed by a purple or green bacterium growing with an organic compound depends on the oxidation state of the compound and whether or not inorganic electron donors such as sulfide are also present. Many purple and green bacteria can grow phototrophically using H_2 as sole electron donor, with CO_2 as carbon source. These organisms have a *hydrogenase* for activating H_2 for CO_2 reduction. Most species also fix N_2 and contain a typical nitrogenase system for this purpose (∞ Section 13.27).

Autotrophy in green bacteria

Although the Calvin cycle (∞ Section 13.7) is used for CO_2 fixation in phototrophic purple bacteria, it is not the mechanism by which CO_2 is fixed in green sulfur bacteria or in *Chloroflexus*. In *Chlorobium*, CO_2 fixation occurs by a reversal of steps in the citric acid cycle, a pathway referred to as the **reverse citric acid cycle** (Figure 16.12a). *Chlorobium* contains two ferredoxin-linked enzymes, which catalyze a reductive fixation of CO_2 into intermediates of the citric acid cycle (Figure 16.12a). The two ferredoxin-linked reactions involve the carboxylation of succinyl-CoA to α-ketoglutarate and the carboxylation of acetyl-CoA to pyruvate. Most of the other reactions in the reverse citric acid cycle in *Chlorobium* are enzymes of the cycle working in reverse of the normal oxidative direction of the cycle.

One exception is *citrate lyase,* an ATP-dependent enzyme that cleaves citrate into acetyl-CoA and oxalacetate in green sulfur bacteria. In the oxidative direction, citrate is produced from these same components by the enzyme *citrate synthase.*

Chloroflexus grows autotrophically with either H_2 or H_2S as electron donor. However, neither the Calvin cycle not the reverse citric acid cycle operates in this organism. Instead, two molecules of CO_2 are converted to glyoxylate by a unique autotrophic pathway, the **hydroxypropionate pathway** (Figure 16.12b). This

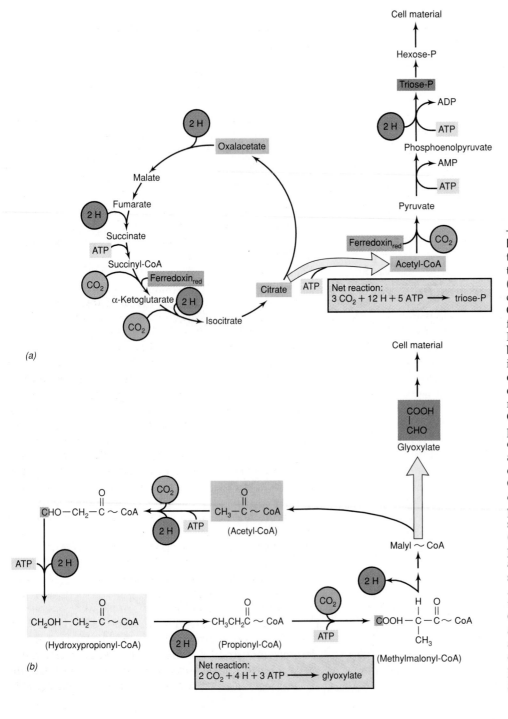

Figure 16.12 Unique autotrophic pathways in phototrophic green bacteria. (a) The reverse citric acid cycle is the mechanism of CO_2 fixation in the green sulfur bacterium *Chlorobium.* Ferredoxin$_{red}$ indicates carboxylation reactions requiring reduced ferredoxin (2 H each). Starting from oxalacetate, each turn of the cycle results in three molecules of CO_2 being incorporated and pyruvate as the product. The cleavage of citrate regenerates the C_4 acceptor oxalacetate and produces acetyl-CoA for biosynthesis. The conversion of pyruvate to phosphoenolpyruvate consumes two ~P equivalents. (b) The hydroxypropionate pathway is the means of autotrophy in the green nonsulfur bacterium *Chloroflexus.* Acetyl-CoA is carboxylated twice to yield methylmalonyl-CoA. This intermediate is rearranged to yield acetyl-CoA and glyoxylate. The latter is converted to cell material probably through a serine or glycine intermediate.

pathway leads to the synthesis of hydroxypropionate as a key intermediate. It is of evolutionary interest that *Chloroflexus*, a phylogenetically ancient member of the Bacteria (see Figure 16.1), has a mechanism for autotrophy unknown in any other organism. This suggests that the hydroxypropionate pathway may have been one of the earliest attempts at autotrophy by evolving Bacteria.

Sulfur metabolism in the purple and green bacteria

Most of the purple and green bacteria are able to oxidize reduced sulfur compounds under anaerobic conditions, with the formation of sulfate. The most common reduced sulfur compounds used are *sulfide* and *thiosulfate*. Elemental sulfur is frequently formed during the oxidation of sulfide or thiosulfate and is deposited either inside or outside the cells. Elemental sulfur deposited inside the cells (see Figure 16.7) is readily available as a further source of electrons. The overall pathway of oxidation of reduced sulfur compounds in purple and green bacteria is shown in Figure 16.13. As seen, either sulfide or thiosulfate is oxidized first to sulfite (SO_3^{2-}) and the enzymes adenosylphosphosulfate reductase (APS reductase) and adenosine diphosphate (ADP) sulfurylase catalyze the oxidation of sulfite to sulfate. Note that a substrate-level phosphorylation occurs at this step, providing for the synthesis of a high energy phosphate bond (in ADP).

Elemental sulfur is not an obligatory intermediate between sulfide and sulfate but merely a side product.

Elemental sulfur is a *storage* product, formed when sulfide concentrations in the environment are high. Indeed, under limiting sulfide levels, sulfide is oxidized directly to sulfate without formation of elemental sulfur. The formation of elemental sulfur as a storage product has been most clearly shown in *Chromatium* where it is deposited *inside* the cells (Figure 16.7a; also ∞ Figure 13.12b). Intracellular elemental sulfur in *Chromatium* is an electron donor for phototrophic growth when sulfide is absent. In green bacteria, sulfur is formed *outside* the cells but often remains attached to the outer surface of the cell and can still be used as an electron donor (see Figure 16.8a).

Symbiotic associations and mixed culture interactions

A two-membered system in which each organism does something for the benefit of the other has been called a **consortium**. An association of two organisms consisting of a large, colorless central bacterium, which is polarly flagellated, surrounded by 12–24 smaller, ovoid- to rod-shaped green sulfur bacteria arranged in rows has been called *Chlorochromatium aggregatum* (Figure 16.14a). Electron micrographs of thin sections of this association clearly show the chlorosomes of the green bacterial partner and the intimate relationship that exists between the two components of *C. aggregatum* (Figure 16.14b). Cell division in the *C. aggregatum* symbiosis is synchronous, suggesting that the two cell types have some means of communicating with each other.

The genus and species name *Chlorochromatium aggregatum* are invalid in formal taxonomy because they refer not to a single organism but to an association of two or more organisms, yet the association is seen quite commonly in lakes and in muds and is probably a rather specific one. The green organism in this association has been cultured and resembles *Chlorobium limicola*, but the large, colorless central organism has not been cultured. A similar association in which the colored organism is a brown-pigmented *Chlorobium* has been called *Pelochromatium roseum*. The symbiosis presumably functions to allow the normally nonmotile green bacterium to move up and down in the water column in response to gradients of light and sulfide.

A clue to the possible role of the colorless central organism comes from the common observation that the sulfide needed by purple and green bacteria can be derived from sulfate- or sulfur-reducing bacteria associated with them. Sulfate- and sulfur-reducing bacteria use organic compounds such as ethanol, lactate, formate, or fatty acids as electron donors and reduce sulfate or sulfur to sulfide. The sulfide produced can then be used by associated phototrophic bacteria, which in the light oxidize the sulfide back to sulfate. If some of the organic matter produced by the phototrophic green bacterium is used by the sulfate reducer, a self-

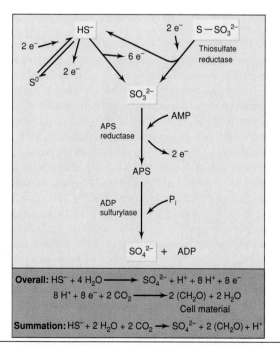

Figure 16.13 Pathways of oxidation of reduced sulfur compounds in phototrophic bacteria. Note that a substrate-level phosphorylation occurs via the enzyme ADP sulfurylase.

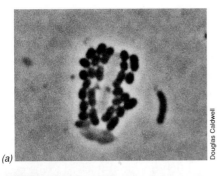

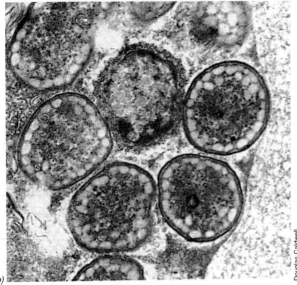

Figure 16.14 (a) Phase contrast micrograph and (b) transmission electron micrograph of the green bacterial consortium *Chlorochromatium aggregatum*. In (a) the nonphototrophic central organism is much lighter in color than the pigmented phototrophic bacteria. Note the chlorosomes (arrows) in (b). The entire consortium is about 3 × 6 μm.

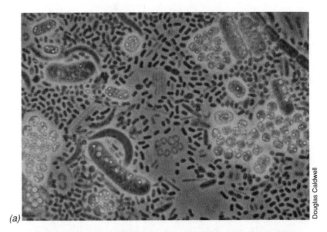

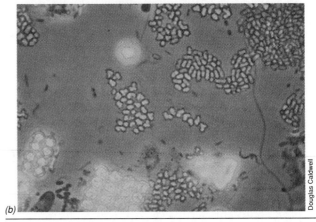

Figure 16.15 (a) Phase contrast photomicrographs of layers containing purple sulfur bacteria and (b) green sulfur bacteria from the water column of a small, stratified lake in Michigan. The purple bacteria include large-celled *Chromatium* species and *Thiocystis*. The green bacteria are predominantly cells of *Ancalochloris* filled with gas vesicles.

feeding system can develop, driven by light energy. This interrelationship may also control the motility of the consortium.

Ecology

Masses of purple and green bacteria are frequently found in the depths of certain kinds of lakes where stable conditions for growth occur. We discussed the development of thermal stratification in lakes in Chapter 14 (∞ Figure 14.17). After stratification occurs, stable anoxic conditions may continue in the deep waters throughout the summer season. If there is a sufficient supply of H_2S and the lake water is sufficiently clear so light penetrates to the anoxic zone, a massive layer of purple or green bacteria can develop (Figure 16.15). This layer, often hidden from the view of the observer on the surface, can be studied by sampling water at various depths (Figure 16.16). The bac-

teria often form a distinct layer just at the depth where H_2S is first present (Figure 16.17). The most favorable lakes for development of these bacteria are those called **meromictic** lakes, which are permanently stratified because of the presence of denser (usually saline) water in the bottom, or **holomictic** lakes, which stratify seasonally because of thermal differences. In meromictic lakes, the bloom of purple or green bacteria may be present throughout the year. Often the photosynthetic activity of these bacteria is sufficiently great that it is an important source of organic matter to the lake ecosystem. In some cases, more photosynthesis may occur in the bacterial layer than in the surface algal layer. In general, blooms of purple and green bacteria are more common in small lakes and ponds than in large bodies of water, mainly because the stratification of large bodies of water is more affected by wind.

One of the easiest places to observe purple and green bacteria in nature is in sulfur springs, where massive blooms are often present a few centimeters

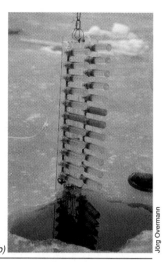

Figure 16.16 Vertical stratification of purple sulfur bacteria in a Canadian lake. (a) Sample of water from 7 m in Lake Mahoney, British Columbia. The major organism is *Amoebobacter purpureus*. (b) A syringe sampling device that can collect water at intervals. Note concentration of *A. purpureus* cells at the 7-m depth.

below the surface of the water (Figure 16.18). Along the seacoast, blooms also occur in warm, shallow pools of seawater not connected to the open ocean, where the activities of sulfate-reducing bacteria lead to production of large amounts of H_2S.

Hot springs are habitats for a variety of purple and green bacteria. *Chloroflexus* forms thick mats (frequently in association with cyanobacteria) in hot springs at temperatures from 40 to 70°C (∞ Figure 14.12*a* for photograph of a microbial mat). *Chlorobium* mats are occasionally found in high sulfide acidic hot springs, generally at temperatures below 50°C. One species of the purple sulfur bacterium *Chromatium*, *Chromatium tepidum*, grows in high sulfide springs at temperatures up to 60°C. A filamentous, phototrophic gliding bacterium that resembles *Chloroflexus* but lacks chlorosomes and bacteriochlorophyll c_s grows in certain hot springs in association with *Chloroflexus*. This organism, named *Heliothrix*, is unusual among phototrophic bacteria because it requires extremely high light intensities for growth. Although *Heliothrix* contains only bacteriochlorophyll *a* and thus should technically be considered a purple bacterium, the morphological similarity between *Heliothrix* and *Chloroflexus*, the ability to glide, and the fact that the two organisms have nearly identical carotenoids have been used to provisionally classify *Heliothrix* in with the other gliding green bacteria. Nucleic acid sequencing of ribosomal RNA (rRNA) (∞ Sections 15.6 and 15.7) from *Heliothrix* and *Chloroflexus* supports this classification.

Mats of phototrophic bacteria also form in non-thermal environments. Cyanobacterial-phototrophic bac-

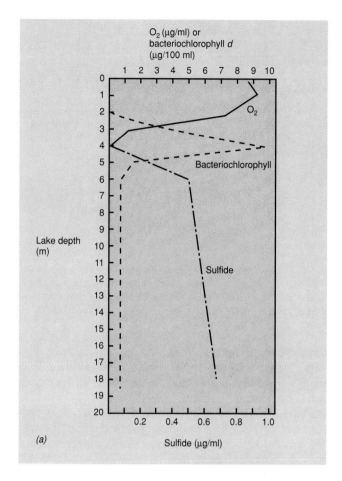

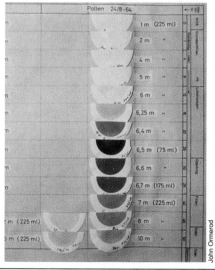

Figure 16.17 Vertical distribution of *Chlorobium* in stratified lakes. (a) Lake Mary, Wisconsin. The phototrophic bacterium forms a layer just at the top of the anoxic zone. The bacterial population size is quantified by measuring bacteriochlorophyll *d* concentration. (b) Stratified Norwegian lake. Series of membrane filters through which were passed water samples taken at varying depths. At 6.4–6.6 m a heavy cyanobacterial bloom is present, and at 6.7–8 m the green sulfur bacterium *Pelodictyon luteolum* reaches its highest population density.

Figure 16.18 Massive accumulation of purple sulfur bacteria. *Thiopedia roseopersicinia*, in a spring in Madison, Wisconsin. The bacteria grow near the bottom of the spring pool and float to the top (by virtue of their gas vesicles) when disturbed. The green color is from cells of the eukaryotic alga *Spirogyra*.

terial mats are common in salt marsh and marine intertidal regions (Figure 16.19). The top layer of the mat usually contains filamentous cyanobacteria, and phototrophic purple (generally purple sulfur) bacteria lie underneath (Figure 16.19). Some marine mats also contain *Chloroflexus*. The lower layers of these mat communities are typically black, as a result of active sulfate reduction, which generates the sulfide necessary for development of the purple bacteria.

Purple and green bacteria make up one of the most diverse groups of bacteria known and are of interest for a wide variety of reasons. They have provided extremely useful systems for studying fundamental aspects of photosynthesis (∞ Chapter 13) and are of great evolutionary significance. Their ecological roles may also be important, and their extreme diversity challenges the bacterial taxonomist.

Figure 16.19 Cross section through a bacterial mat in the Sippewisset salt marsh near Woods Hole, Massachusetts. The layers are as follows: top, cyanobacteria; pink, phototrophic purple sulfur bacteria attached to sand grains; black, sulfate-reducing bacteria; peach, bacteriochlorophyll *b*-containing cells of *Thiocapsa pfennigii*.

16.2
Cyanobacteria

The **cyanobacteria** comprise a large and heterogeneous group of phototrophic Bacteria. Cyanobacteria differ in fundamental ways from purple and green bacteria, most notably in the fact that they are *oxygenic* phototrophs. Cyanobacteria represent one of the major phylogenetic lines of Bacteria and show a distant relationship to gram-positive Bacteria (see Figure 16.1).

Structure and classification

The morphological diversity of the cyanobacteria is considerable (Figure 16.20). Both unicellular and filamentous forms are known, and considerable variation within these morphological types occurs. *Bergey's Manual* has divided the cyanobacteria into five morphological groups: unicellular dividing by binary fission (see Figure 16.20*a*); unicellular dividing by multiple fission (colonial) (see Figure 16.20*b*); filamentous containing differentiated cells called heterocysts that function in nitrogen fixation (see Figures 16.20*d* and 16.22); filamentous nonheterocystous forms (see Figure 16.20*c*); and branching filamentous types (see Figure 16.20*e*). Table 16.6 lists the genera currently recognized in each group. Cyanobacterial cells range in size from those of typical bacteria (0.5–1 μm in diameter) to cells as large as 60 μm in diameter (in the species *Oscillatoria princeps*).

The cyanobacteria differ in fatty acid composition from all other prokaryotes. Other Bacteria contain almost exclusively saturated and monounsaturated fatty acids (one double bond), but the cyanobacteria frequently contain unsaturated fatty acids with two or more double bonds.

The fine structure of the cell wall of some cyanobacteria is similar to that of gram-negative Bacteria, and peptidoglycan can be detected in the walls. Many cyanobacteria produce extensive mucilaginous envelopes, or sheaths, that bind groups of cells or filaments together (see, for example, Figure 16.20*a*). The photosynthetic lamellar membrane system is often complex and multilayered (∞ Figure 13.16*b*), although in some of the simpler cyanobacteria the lamellae are regularly arranged in concentric circles around the periphery of the cytoplasm (Figure 16.21). Cyanobacteria have only one form of chlorophyll, chlorophyll *a*, and all of them also have characteristic biliprotein pigments, **phycobilins** (∞ Figure 13.16), which function as accessory pigments in photosynthesis. One class of phycobilins, *phycocyanins*, are blue, absorbing light maximally at about 625 nm (∞ Figure 13.17), and together with the green chlorophyll *a* are responsible for the blue-green color of the bacteria. However, some cyanobacteria produce *phycoerythrin*, a red phycobilin absorbing light maximally at

TABLE 16.6 Genera and grouping of cyanobacteria

Group	Genera
Group I—Unicellular: single cells or cell aggregates	*Gloeothece* (Figure 16.20a), *Gloeobacter, Synechococcus, Cyanothece, Gloeocapsa, Synechocystis, Chamaesiphon*
Group II—Pleurocapsalean: reproduce by formation of small spherical cells called baeocytes produced through multiple fission	*Dermocarpa* (Figure 16.20b), *Xenococcus, Dermocarpella, Pleurocapsa, Myxosarcina, Chroococcidiopsis*
Group III—Oscillatorian: filamentous cells that divide by binary fission in a single plane	*Oscillatoria* (Figure 16.20c), *Spirulina, Arthrospira, Lyngbya, Microcoleus, Pseudanabaena*
Group IV—Nostocalean: filamentous cells that produce heterocysts	*Anabaena* (Figure 16.20d), *Nostoc, Calothrix, Nodularia, Cylinodrosperum, Scytonema*
Group V—Branching: cells divide to form branches	*Fischerella* (Figure 16.20e), *Stigonema, Chlorogloeopsis, Hapalosiphon*

about 550 nm, and species possessing this pigment are red or brown in color. Even more confusing, the eukaryotic red algae (Rhodophyta) (∞ Section 18.1) are red because of phycoerythrin, but some species have phycocyanin instead and are blue-green.

Structural variations: gas vesicles and heterocysts

Among the cytoplasmic structures seen in many cyanobacteria are **gas vesicles** (∞ Section 3.12), which are especially common in species that live in open waters (planktonic species). Their function is to provide the organism with flotation (∞ Figure 3.62) so it may remain where there is most light. Some cyanobacteria form **heterocysts,** which are rounded, seemingly more-or-less empty cells, usually distributed regularly along a filament or at one end of a filament (Figure 16.22). Heterocysts arise from differentiation of vegetative cells and are the sole sites of *nitrogen fixation* in heterocystous cyanobacteria. In *Anabaena,* a well-studied heterocystous cyanobacterium, complex gene rearrangements occur within the heterocyst to yield a continuous cluster of *nif* genes that can be expressed as a unit (∞ Section 13.27).

Heterocysts have intercellular connections with adjacent vegetative cells, and there is mutual exchange of materials between these cells, with products of photosynthesis moving from vegetative cells to heterocysts and products of nitrogen fixation moving from heterocysts to vegetative cells. Heterocysts are low in phycobilin pigments and *lack* photosystem II, the oxy-

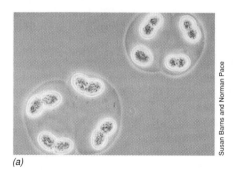
(a)

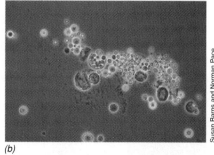
(b)

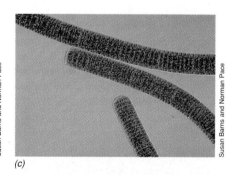
(c)

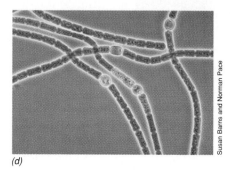
(d)

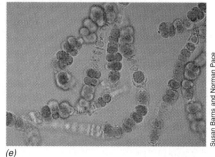
(e)

Figure 16.20 Morphological diversity among the cyanobacteria: the five major morphological types of cyanobacteria. (a) Unicellular, *Gloeothece,* phase contrast; a single cell measures 5–6 μm in diameter; (b) colonial, *Dermocarpa,* phase contrast; (c) filamentous, *Oscillatoria,* brightfield; a single cell measures about 15 μm wide; (d) filamentous heterocystous, *Anabaena,* phase contrast; a single cell measures about 5 μm wide; (e) filamentous branching, *Fischerella,* bright-field.

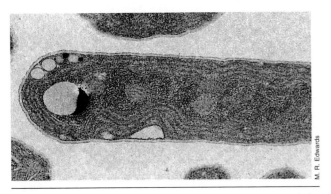

Figure 16.21 Electron micrograph of a thin section of the cyanobacterium *Synechococcus lividus*. A cell is about 5 μm in diameter.

gen-evolving photosystem. They are also surrounded by a thickened cell wall containing large amounts of glycolipid, which serves to slow the diffusion of O_2 into the cell. Because of the oxygen lability of the enzyme nitrogenase (∞ Section 13.27), it seems likely that the heterocyst, by maintaining an anoxic environment, stabilizes the nitrogen-fixing system in organisms that are not only aerobic but also oxygen-producing. Indeed, some nonheterocystous filamentous cyanobacteria produce nitrogenase and fix nitrogen in normal vegetative cells if they are grown anaerobically. However, a few unicellular cyanobacteria of the sheath-forming *Gloeothece* type (Figure 16.20a) do not produce heterocysts but nevertheless fix nitrogen under oxic conditions. Some marine *Oscillatoria* also fix nitrogen without heterocysts and produce a series of cells in the center of the filament that lack photosystem II activity; nitrogen fixation only occurs in this non-O_2-producing region.

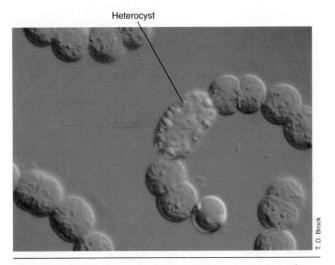

Heterocyst

Figure 16.22 Heterocysts in the cyanobacterium *Anabaena* sp. Heterocysts are the sole site of nitrogen fixation in heterocystous cyanobacteria.

Cyanophycin and other structures

A structure called **cyanophycin** can be seen in electron micrographs of many cyanobacteria. This structure is a simple polymer of aspartic acid, with each aspartate residue containing an arginine molecule:

$$\text{Asp—Asp—Asp—Asp—Asp—}$$
$$\quad|\qquad|\qquad|\qquad|\qquad|$$
$$\text{Arg}\quad\text{Arg}\quad\text{Arg}\quad\text{Arg}\quad\text{Arg}$$

and can constitute up to 10% of the cell mass. This copolymer is a nitrogen storage product in many cyanobacteria, and when nitrogen in the environment becomes deficient, this polymer is broken down and used. The phycobilin pigments can also constitute a major portion of the cell mass, up to 10%, and likewise serve as a nitrogen storage material, being broken down under nitrogen starvation. Because of this, nitrogen-starved cyanobacteria often appear green instead of blue-green in color. Cyanophycin is also an energy reserve in cyanobacteria. Arginine, derived from cyanophycin, can be hydrolyzed to yield ornithine, with the production of ATP through the action of the enzyme *arginine dihydrolase* with carbamyl phosphate (∞ Section 13.19) occurring as an intermediate:

Arginine + ADP + P_i + H_2O →
$$\text{Ornithine} + 2\,NH_3 + CO_2 + \text{ATP}$$

Arginine dihydrolase is present in many cyanobacteria and may function there as a source of ATP for maintenance purposes during dark periods.

Many, but by no means all, cyanobacteria exhibit gliding motility; flagella have never been found. Gliding occurs only when the cell or filament is in contact with a solid surface or with another cell or filament. In some cyanobacteria gliding is not a simple translational movement but is accompanied by rotations, reversals, and flexings of filaments. Most gliding forms exhibit directed movement in response to light (phototaxis); it is usually positive, although negative movement from bright light may also occur. Chemotaxis (∞ Section 3.10) may occur as well. One species of marine cyanobacteria is capable of swimming motility, but the mechanism of motility, which does not involve flagella, is unknown.

Among the filamentous cyanobacteria, fragmentation of the filaments often occurs by formation of **hormogonia** (Figure 16.23a and b), which break away from the filaments and glide off. In some species resting spores or **akinetes** (Figure 16.23c) are formed, which protect the organism during periods of darkness, drying, or freezing. These are cells with thickened outer walls; they germinate through the breakdown of the outer wall and outgrowth of a new vegetative filament. However, even the vegetative cells of many cyanobacteria are relatively resistant to drying or low temperatures.

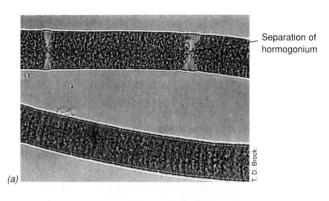

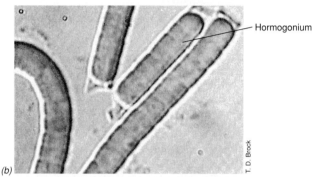

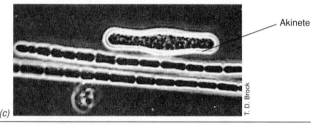

Figure 16.23 Structural differentiation in filamentous cyano-bacteria. (a) Initial stage of hormogonium formation in *Oscillatoria*. Notice the empty spaces where the hormogonium is separating from the filament. (b) Hormogonium of a smaller *Oscillatoria* species. Notice that the cells at both ends are rounded. Nomarski interference contrast microscopy. (c) Akinete (resting spore) of *Anabaena* sp. by phase contrast.

Physiology

The nutrition of cyanobacteria is simple. Vitamins are not required, and nitrate or ammonia is used as nitrogen source. Nitrogen-fixing species are also common. Most species tested are obligate phototrophs, being unable to grow in the dark on organic compounds. However, some cyanobacteria can assimilate simple organic compounds such as glucose and acetate if light is present. Apparently they are unable to make ATP by oxidation of organic compounds, but if ATP is provided by means of photophosphorylation, organic compounds can be utilized as carbon sources. Some species, mainly filamentous forms, can grow in the dark on glucose or other sugars, using the organic material as both carbon and energy source. As we dis-

cussed in Section 13.5, a number of cyanobacteria can carry out anoxygenic photosynthesis using only photosystem I when sulfide is present in the environment.

Several metabolic products of cyanobacteria are of considerable practical importance. Many cyanobacteria produce potent neurotoxins, and during water blooms when massive accumulations of cyanobacteria may develop, animals ingesting such water may succumb rapidly. Fortunately, the massive accumulations needed to cause death do not occur extensively, although subclinical manifestations of cyanobacterial water blooms may be a common, but unobserved, occurrence. Many cyanobacteria are also responsible for the production of earthy odors and flavors in fresh waters, and if such waters are used as drinking water sources, aesthetic problems may arise. The compound produced is **geosmin** (trans-1,10-dimethyl-trans-9-decalol). This substance is also produced by many actinomycetes (see the discussion in Section 16.32) and is responsible for the distinctive "earthy" odor of soil.

Ecology and evolution

Cyanobacteria are widely distributed in nature in terrestrial, freshwater, and marine habitats. In general they are more tolerant to environmental extremes than are algae and are often the dominant or sole phototrophic organisms in hot springs (∞ Table 5.2), saline lakes, and other extreme environments. Many members are found on the surfaces of rocks or soil and occasionally even within rocks themselves. In desert soils subject to intense sunlight, cyanobacteria often form extensive crusts over the surface, remaining dormant during most of the year and growing during the brief winter and spring rains. In shallow marine bays, where relatively warm seawater temperatures exist, cyanobacterial mats of considerable thickness may form. Freshwater lakes, especially those that are fairly rich in nutrients, may develop blooms of cyanobacteria (∞ Figures 1.13*a* and 3.62). A few cyanobacteria are symbionts of liverworts, ferns, and cycads; a number are found as the phototrophic component of lichens. In the case of the water fern *Azolla* (∞ Section 14.23), it has been shown that the cyanobacterial endophyte (a species of *Anabaena*) fixes nitrogen that becomes available to the plant.

Base compositions of DNA of a variety of cyanobacteria have been determined. Those of the unicellular forms vary from 35 to 71% GC, a range so wide as to suggest that this group contains many members with little relationship to each other. On the other hand, the values for the heterocyst formers vary much less, from 39 to 47% GC. Phylogenetically, cyanobacteria group along morphological lines in most cases. Filamentous heterocystous and nonheterocystous species form distinct groups, as do the branching forms. However, unicellular cyanobacteria are phylogenetically highly diverse, as representatives show evolutionary relationships to several different groups and show little phylogenetic relationship to one another.

The evolutionary significance of cyanobacteria was discussed in Section 15.2, and it is nearly certain that these organisms were the first oxygen-evolving phototrophic organisms and were responsible for the conversion of the atmosphere of the earth from anoxic to oxic. Microfossil evidence of cyanobacteria-like organisms over 3 billion years old is good (∞ Section 15.1), and there is evidence that cyanobacteria occupied vast areas of the earth in those ancient times. Although quantitatively much less significant today, cyanobacteria can be the dominant phototrophic members of certain communities, such as microbial mats (∞ Section 14.5).

16.3
Prochlorophytes

Prochlorophytes are prokaryotic phototrophs that contain chlorophyll *a* and *b* but do *not* contain phycobilins. Prochlorophytes therefore resemble both cyanobacteria (because they are prokaryotic and produce chlorophyll *a*) and the plant chloroplast (because they contain chlorophyll *b* instead of phycobilins).

Prochloron

Prochloron was the first prochlorophyte discovered. It is found in nature as a symbiont of marine invertebrates (didemnid ascidians), and all studies of the organism to date have relied on material collected from natural samples. Cells of *Prochloron* expressed from the cavities of didemnid tissue are roughly spherical in morphology (Figure 16.24), 8–10 μm in diameter. Electron micrographs of thin sections (Figure 16.24) show that *Prochloron* has an extensive thylakoid membrane system similar to that observed in the chloroplast (∞ Figure 3.77). Evidence that *Prochloron* is phylogenetically a member of the Bacteria is the presence of muramic acid in the cell walls, indicating that peptidoglycan is present.

The ratio of chlorophyll *a*:chlorophyll *b* in cells of *Prochloron* is about 4–7:1, somewhat higher than the value of 1–2:1 typical of green algae. The carotenoids of *Prochloron* are similar to those of cyanobacteria, predominantly β-carotene and zeaxanthin. The GC base ratio of different samples of *Prochloron* isolated from different ascidians varies from 31 to 41%, indicating a fair bit of genetic heterogeneity. Different species of *Prochloron* probably exist, but confirmation of this must await laboratory culture and study of pure strains.

Other prochlorophytes

Another prochlorophyte has been discovered that can be grown in pure culture. This prochlorophyte is filamentous (Figure 16.25) and grows as one of the dominant phototrophic bacteria in several shallow Dutch lakes. This filamentous prochlorophyte has been given the genus name *Prochlorothrix*. Like *Prochloron*, *Prochlorothrix* is prokaryotic and contains chlorophylls *a* and *b* and lacks phycobilins. The ratio of chlorophyll *a* to chlorophyll *b* is somewhat higher in *Prochlorothrix* (about 8:1 to 9:1) than in *Prochloron*, and the thylakoid membranes are much less developed than in *Prochloron* (compare Figures 16.24 and 16.25*b*).

A novel type of prochlorophyte has been found in the deep euphotic zone of the open oceans. These phototrophs are extremely small cocci, measuring less than 1 μm in diameter, and have been given the genus name *Prochlorococcus*. Like those of other prochlorophytes, cells of *Prochlorococcus* are prokaryotic and contain chlorophyll *b*. However, these phototrophs lack true chlorophyll *a* and produce instead a modified form of chlorophyll *a* called *divinyl chlorophyll a*. Cells of *Prochlorococcus* also contain α- (instead of β-) carotene, a pigment previously unknown in prokaryotes. Interestingly, the chlorophyll *a/b* ratio of *Prochlorococcus* is near 1, similar to that of the chloroplasts of marine green algae. Because their numbers in the oceans are relatively large (10^4–10^5 cells/ml), prochlorophytes, like *Prochlorococcus*, may have considerable ecological significance as primary producers in open ocean waters.

Prochlorophytes and evolution

Based on our discussion of endosymbiosis (∞ Section 15.4), the evolutionary significance of prochlorophytes should be apparent. Until the discovery of prochlorophytes, it was always assumed that the chloroplast originated from endosymbiotic association of *cyanobacteria* with a primitive eukaryotic cell. However, this hypothesis has never been scientifically satisfying for

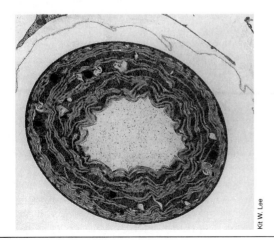

Figure 16.24 Electron micrograph of the prochlorophyte *Prochloron*. Note the extensive intracytoplasmic membranes (thylakoids). Cells are about 10 μm in diameter.

Kit W. Lee

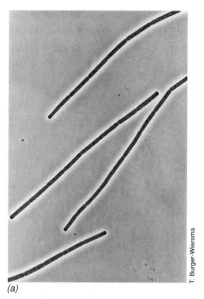 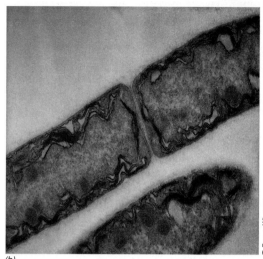

Figure 16.25 Phase and electron micrographs of the filamentous prochlorophyte *Prochlorothrix.* (a) Phase contrast. (b) Electron micrograph of thin section showing arrangement of membranes. The diameter of cells is about 1–2 μm.

(a)

(b)

at least one major reason: how did the plant chloroplast evolve the pigment complement it has today if it originated from a *cyanobacterial* endosymbiont that contained *phycobilins* instead of chlorophyll *b*? The hypothesis that *prochlorophytes* instead of cyanobacteria were the ancestors of the green plant chloroplast eliminates this major point of contention. However, comparative sequencing of 16S rRNA (∞ Section 15.7) does not show *Prochloron, Prochlorococcus,* or *Prochlorothrix* to be the immediate ancestor of the green plant chloroplast. Instead, prochlorophytes, cyanobacteria, and the plant chloroplast all *share* a common ancestor. Thus, although prochlorophytes have arisen from the same evolutionary roots as the chloroplast and are phenotypically highly related to chloroplasts, it appears that the closest extant free-living ancestor of chloroplasts has thus far not been found.

16.4
Chemolithotrophs: Nitrifying Bacteria

We discussed in Chapter 13 the conceptual basis of chemolithotrophy. Various chemolithotrophic bacteria are known, but they are all physiologically united by their ability to utilize *inorganic* electron donors as energy sources. Most chemolithotrophs are also capable of autotrophic growth and in this way share a major physiological trait with phototrophic bacteria and cyanobacteria. The best-studied chemolithotrophs are those capable of oxidizing reduced sulfur and nitrogen compounds, and the hydrogen-oxidizing bacteria, and we focus on these groups here and in the next two sections.

Bacteria able to grow chemolithotrophically at the expense of reduced inorganic nitrogen compounds are called **nitrifying bacteria.** Several genera are recognized on the basis of morphology and the particular steps in the oxidation sequences that they carry out (Table 16.7). No chemolithotroph is known that will carry out the complete oxidation of ammonia to nitrate; thus, **nitrification** of ammonia in nature results from the sequential action of two separate groups of organisms, the **ammonia-oxidizing bacteria,** the **nitrosifyers** (Figure 16.26), and the **nitrite-oxidizing bacteria,** the true **nitrifying** (nitrate-producing) bacteria (Figure 16.27). Some chemoorganotrophic bacteria and fungi oxidize ammonia completely to nitrate, but the rate of the process is much less than that accomplished by the chemolithotrophic nitrifying bacteria and may not be significant ecologically. Historically, the nitrifying bacteria were the first organisms to be shown to grow chemolithotrophically; Winogradsky showed that they were able to produce organic matter and cell mass when provided with CO_2 as sole carbon source (∞ Winogradsky's Legacy, Chapter 13).

TABLE 16.7   Characteristics of the nitrifying bacteria[a]

Characteristics	Genus	DNA (mol % GC)	Habitats
Oxidize ammonia:			
Gram-negative short to long rods, motile (polar flagella) or nonmotile; peripheral membrane systems	Nitrosomonas	45–53	Soil, sewage, freshwater, marine
Large cocci, motile; vesicular or peripheral membranes	Nitrosococcus	49–50	Freshwater, marine
Spirals, motile (peritrichous flagella); no obvious membrane system	Nitrosospira	54	Soil
Pleomorphic, lobular, compartmented cells; motile (peritrichous flagella)	Nitrosolobus	54	Soil
Slender, curved rods	Nitrosovibrio	54	Soil
Oxidize nitrite:			
Short rods, reproduce by budding, occasionally motile (single subterminal flagellum); membrane system arranged as a polar cap	Nitrobacter	59–62	Soil, freshwater, marine
Long, slender rods, nonmotile; no obvious membrane system	Nitrospina	58	Marine
Large cocci, motile (one or two subterminal flagella); membrane system randomly arranged in tubes	Nitrococcus	61	Marine
Helical to vibrioid-shaped cells, nonmotile; no internal membranes	Nitrospira	50	Marine

[a]Phylogenetically, all nitrifying bacteria thus far examined are either α or β purple Bacteria (∞ Section 15.7 and Table 16.1).

Many of the nitrifying bacteria have remarkably complex internal membrane systems (see Figures 16.26 and 16.27), although not all genera have such membranes.

Biochemistry of nitrification

Molecular oxygen is required for ammonia oxidation, the initial step involving a *monooxygenase*, which uses NADH as electron donor. The first product of ammonia oxidation is *hydroxylamine*, NH₂OH, and no energy is generated in this step (energy is actually used up via the oxidation of NADH) (Figure 16.28). Hydroxylamine is then oxidized to nitrite, and ATP formation occurs at this step via electron transport phosphorylation through a cytochrome system (Figure 16.28; ∞ also Figure 13.25).

There are some interesting similarities between the ammonia-oxidizing bacteria and the methane-oxidizing bacteria. As noted in Section 16.7, methane-oxidizing bacteria generally oxidize ammonia to nitrite, and ammonia inhibits methane oxidation. In a similar fashion, certain ammonia oxidizers are capable of oxidizing methane and can incorporate significant amounts of carbon originating from methane into cell material. However, methane does *not* serve as sole carbon and electron donor for growth of ammonia oxidizers. Nitrite oxidizers do not oxidize methane. It appears that the two substrates methane and ammonia have some structural similarities, and so an enzyme that recognizes one can also combine with, and be inhibited by, the other. Another similarity between ammonia oxidizers and methane oxidizers is that both groups generally possess extensive internal membrane systems (compare Figure 16.26 with Figure 16.33).

Only a single step is involved in the oxidation of nitrite to nitrate by the nitrite-oxidizing bacteria (Figure 16.28; ∞ also Figure 13.26). This reaction is carried out by a *nitrite oxidase* system, the electrons

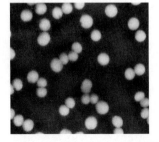
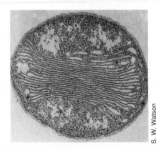

Figure 16.26 Phase contrast photomicrograph (left) and electron micrograph (right) of the nitrosifying bacterium *Nitrosococcus oceanus.* A single cell is about 2 μm in diameter.

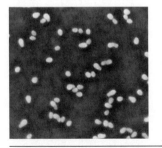
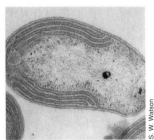

Figure 16.27 Phase contrast photomicrograph (left) and electron micrograph (right) of the nitrifying bacterium *Nitrobacter winogradskyi.* A cell is about 0.7 μm in diameter.

Nitrosifying bacteria

1. $NH_3 + O_2 + 2\ e^- + 2\ H^+ \longrightarrow NH_2OH + H_2O$

2. $NH_2OH + H_2O + \frac{1}{2}\ O_2 \longrightarrow NO_2^- + 2\ H_2O + H^+$

Sum: $NH_3 + 1\frac{1}{2}\ O_2 \longrightarrow NO_2^- + H^+ + H_2O$

$\Delta G^{0'} = -287$ kJ/reaction

Nitrifying bacteria

$NO_2^- + \frac{1}{2}\ O_2 \longrightarrow NO_3^-$

$\Delta G^{0'} = -76$ kJ/reaction

Figure 16.28 Reactions involved in the oxidation of inorganic nitrogen compounds by chemolithotrophic nitrifying bacteria (∞ also Figures 13.25 and 13.26).

being transported to O_2 via cytochromes, with ATP being generated by electron transport phosphorylation. The energy available from the oxidation of nitrite to nitrate is only 76 kJ/mol, which is sufficient for the formation of two ATPs. However, careful measurements of molar growth yields suggest that only *one* ATP is produced per each NO_2^- oxidized to NO_3^-. As we discussed in Section 13.8, the generation of reducing power (NADPH) for the reduction of CO_2 to organic compounds comes from ATP-driven *reversed electron transport* reactions because the NO_3^-/NO_2^- reduction potential is too high to reduce $NADP^+$ directly.

Ecology

The nitrifying bacteria are widespread in soil and water. They can be expected to be present in highest numbers in habitats where considerable amounts of ammonia are present, such as sites where extensive protein decomposition occurs (ammonification). Nitrifying bacteria develop especially well in lakes and streams that receive inputs of untreated (or even treated) sewage because sewage effluents are generally high in ammonia. Because O_2 is required for ammonia oxidation by nitrosifying bacteria (however, ∞ Section 13.12), ammonia tends to accumulate in anoxic habitats, and in stratified lakes nitrifying bacteria may develop especially well at the thermocline, where both ammonia and O_2 are present. Nitrification results in acidification of the habitat, owing to the buildup of nitric acid (the situation is analogous to the buildup of sulfuric acid through the activities of sulfur-oxidizing bacteria) (∞ Section 14.15). Since nitrous acid can form at acid pH values as a result of ammonia oxidation, the accumulation of this toxic (and mutagenic) (∞ Table 9.2) agent in acidic environments can result in inhibition of further nitrification. In general, nitrification is much more extensive in neutral and alkaline than in acidic habitats.

Enrichment and culture

Enrichment cultures of nitrifying bacteria are readily obtained by using selective media containing ammonia or nitrite as electron donor and bicarbonate as sole carbon source. Because of the inefficiency of growth of these organisms, visible turbidity may not develop even after extensive nitrification has occurred, and so the best

means of monitoring growth is to assay for production of nitrite (with ammonia as electron donor) or disappearance of nitrite (with nitrite as electron donor). After 1 or 2 weeks of incubation, chemical assays reveal whether a successful enrichment has been obtained, and attempts can be made to obtain pure cultures by streaking on agar plates. Because many common chemoorganotrophs present in the enrichment grow rapidly on the traces of organic matter present in the medium, purification of nitrifying bacteria must be done by repeated picking and streaking, followed by testing to be certain that chemoorganotrophic contaminants are no longer present. In addition, many nitrifying bacteria, especially the ammonia oxidizers, appear to be inhibited by the traces of organic material present in most agar preparations. Culture of these organisms on solid media sometimes requires the use of extensively washed, high purity agar or the completely *inorganic* solidifying agent **silica gel.**

Most of the nitrifying bacteria are obligate chemolithotrophs. *Nitrobacter* is an exception, however, and is able to grow, although slowly, on acetate or pyruvate as sole carbon and energy source. None of these bacteria require growth factors. Although the group is somewhat heterogeneous morphologically, it seems to be more homogeneous than the sulfur-oxidizing or phototrophic bacteria, as shown by the fairly narrow range of deoxyribonucleic acid (DNA) base compositions (Table 16.7) and the similar biochemical properties. Phylogenetically, nitrifying bacteria are members of the purple bacteria (∞ Section 15.7 and Table 16.1).

16.5

Chemolithotrophs: Sulfur- and Iron-Oxidizing Bacteria

The ability to grow chemolithotrophically on reduced sulfur compounds is a property of a diverse group of microorganisms. However, only six genera, *Thiobacillus, Thiosphaera, Thiomicrospira, Thermothrix, Beggiatoa,* and *Sulfolobus* have been consistently cultured, and so our discussion here is restricted to these genera. With the exception of *Sulfolobus*, which is an archaean and is discussed in Section 17.3, the remaining sulfur chemolithotrophs are Bacteria and phylogenetically fall into the purple bacteria group (see Table 16.1). Two broad ecological classes of sulfur-oxidizing bacteria can be discerned, those living at neutral pH and those living at acid pH. Many of the forms living at acid pH also have the ability to grow chemolithotrophically using ferrous iron as electron donor. We discussed the biogeochemistry of these acidophilic sulfur- and iron-oxidizing bacteria in Sections 13.10 and 13.11.

Thiobacillus

The genus *Thiobacillus* contains those gram-negative, polarly flagellated rods that are able to derive their energy from the oxidation of elemental sulfur, sulfides, and thiosulfate (Table 16.8). Some pseudomonads have been confused with members of the genus *Thiobacillus*. Morphologically, *Thiobacillus* (Figure 16.29) is similar to *Pseudomonas*, but the two differ in that *Thiobacillus* can grow chemolithotrophically using reduced sulfur compounds. A few thiobacilli can also grow chemoorganotrophically with organic electron donors, and under such conditions resemble pseudomonads. Further, a few pseudomonads (both marine and freshwater) can oxidize thiosulfate to tetrathionate:

$$2\ S\text{—}SO_3^{2-} \rightarrow\ ^-O_3S\text{—}S\text{—}S\text{—}SO_3^- + 2\ e^-$$

and although they cannot grow solely from the energy obtained in this reaction, they do obtain some slight growth advantage from this process when growing on organic compounds (mixotrophy) (∞ Section 13.13).

The biochemical steps in the oxidation of various sulfur compounds are summarized in Figure 16.30. Oxidation of sulfide and sulfur involves first the reaction of these substances with sulfhydryl groups of the cell, such as glutathione with formation of a sulfide–sulfhydryl complex (Figure 16.30). The sulfide ion is then oxidized to sulfite (SO_3^{2-}) by the enzyme *sulfide oxidase*. There are two ways in which sulfite can be oxidized to produce high energy phosphate bonds. In one, sulfite is oxidized to sulfate by a cytochrome-linked sulfite oxidase, with the formation of ATP via electron transport phosphorylation. This pathway is universally present in thiobacilli. In the second, sulfite reacts with adenosine monophosphate (AMP); two electrons are removed, and adenosine phosphosulfate (APS) (∞ Section 13.16) is formed. The electrons removed in either case are transferred to O_2 via the cytochrome system, leading to the formation of high energy phosphate bonds through electron transport phosphorylation. In addition, a substrate-level phosphorylation occurs, APS reacting with P_i and being converted to ADP and sulfate. With the enzyme adenylate kinase, two ADP can be converted to one ATP and one AMP. Thus, oxidation of two sulfite ions via this system produces three ATP, two via electron transport phosphorylation and one via substrate-level phosphorylation. The significance of the APS pathway in thiobacilli in general is unclear, however, because it has been found in only a few *Thiobacillus* species.

Thiosulfate ($S_2O_3^{2-}$) is split into sulfite and sulfur (Figure 16.30). The sulfite is oxidized to sulfate with production of ATP, and the other sulfur atom is converted to insoluble elemental sulfur. Thus, when they oxidize thiosulfate, the thiobacilli produce elemental sulfur, but when they oxidize sulfides they do not. The elemental sulfur produced can itself be oxidized later when the thiosulfate supply is exhausted. If thiosulfate is low, elemental sulfur does not accumulate, probably being oxidized as soon as it is formed.

TABLE 16.8 Physiological characteristics of sulfur-oxidizing chemolithotrophic prokaryotes[a]

Genus and/or species	Inorganic electron donor	Range of pH for growth	DNA (mol % GC)
Thiobacillus species growing poorly in organic media:			
T. thioparus	H_2S, sulfides, S^0, $S_2O_3^{2-}$	6–8	61–66
T. denitrificans[b]	H_2S, S^0, $S_2O_3^{2-}$	6–8	63–68
T. neapolitanus	S^0, $S_2O_3^{2-}$	6–8	52–56
T. thiooxidans	S^0	2–4	51–53
T. ferrooxidans	S^0, metal sulfides, Fe^{2+}	2–4	55–65
Thiobacillus species growing well in organic media:			
T. novellus	$S_2O_3^{2-}$	6–8	66–68
T. intermedius	$S_2O_3^{2-}$	3–7	64
Filamentous sulfur chemolithotrophs:			
Beggiatoa	H_2S, $S_2O_3^{2-}$	6–8	37–43
Thiothrix	H_2S	6–8	—
Thioploca[b]	H_2S, S^0	—	—
Other genera:			
Thiomicrospira[c]	$S_2O_3^{2-}$, H_2S	6–8	36–44
Thiosphaera	H_2S, $S_2O_3^{2-}$, H_2	6–8	66
Thermothrix[b]	H_2S, $S_2O_3^{2-}$, SO_3^-	6.5–7.5	—
Thiovulum	H_2S, S^0	6–8	—
Sulfolobus[d]	H_2S, S^0	1–5	37
Acidianus[d]	S^0	1–5	31

[a]Phylogenetically, all sulfur-oxidizing bacteria are α or β purple Bacteria (∞ Section 15.7 and Table 16.1).

[b]Facultative aerobes; use NO_3^- as electron acceptor anaerobically.

[c]One of its species is capable of using NO_3^- anaerobically.

[d]Hyperthermophilic Archaea; ∞ Section 17.3.

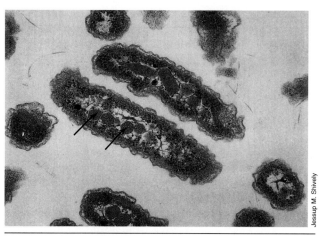

Jessup M. Shively

Figure 16.29 Transmission electron micrograph of cells of the chemolithotrophic sulfur oxidizer *Thiobacillus neapolitanus*. A single cell is about 0.4 μm in diameter. Note the polyhedral bodies (carboxysomes) distributed throughout the cell (arrows). See text for details.

Enrichment and ecology

Enrichment cultures of thiobacilli are easy to prepare. Sulfur or thiosulfate is added to a basal salts medium with NH_4^+ as a nitrogen source and bicarbonate as a carbon source, and the medium is then inoculated with a sample of soil or mud. After aerobic incubation at room temperature for a few days, the liquid should appear turbid owing to the growth of thiobacilli. If thiosulfate is used, droplets of amorphous sulfur will also be present. If elemental sulfur is used, many of the bacteria may be attached to the insoluble sulfur crystals, and the edges of such crystals should be examined for the presence of bacteria. The bacteria also attach to the crystals of metal sulfides such as PbS, HgS, or CuS if these are used as energy sources. From thiosulfate enrichment cultures, pure cultures can be obtained by streaking onto a solidified medium of the same composition. However, pure cultures are fairly difficult to obtain because growth is usually slow and chemoorganotrophic contaminants grow on small amounts of organic matter released by the thiobacilli. Another problem in obtaining pure cultures of thiobacilli that grow best at neutral pH, such as *Thiobacillus thioparus*, is that sulfur oxidation leads to sulfuric acid production and a drop in pH, resulting in the death of the culture. Thus, highly buffered media and frequent transfers of the culture are necessary. On the other hand, isolation of the acidophilic *Thiobacillus thiooxidans* is relatively easy because this organism is resistant to the acid it produces and most chemoorganotrophic contaminants cannot grow at low pH values (2–5) where *T. thiooxidans* thrives.

Many isolates of acidophilic thiobacilli can also oxidize ferrous iron. The geochemical aspects of iron oxidation were discussed in Sections 14.16 and 14.17. At acid pH, ferrous iron is not readily oxidized spontaneously, and the acid-tolerant thiobacilli are the main agents in nature for the oxidation of ferrous iron in acidic environments. Those isolates that oxidize both iron and sulfur compounds are currently classified as the species *Thiobacillus ferrooxidans*, the species *T. thiooxidans* being restricted to those acid-tolerant thiobacilli that cannot oxidize iron.

Thiosphaera is a coccoid-shaped relative of *Thiobacillus* that is capable of growing chemolithotrophically on H_2 as well as reduced sulfur compounds and can also grow chemoorganotrophically.

Thiomicrospira *and* Thermothrix

Members of the genera *Thiomicrospira* and *Thermothrix* are thiosulfate-oxidizing bacteria that grow at neutral pH. *Thiomicrospira*, as its name implies, is a tiny, spiral-shaped bacterium. In fact, this organism is so small that it can be selectively enriched by inoculating a thiosulfate or hydrogen sulfide-containing medium with the filtrate remaining from filtering a mud slurry through a membrane filter with a pore size of just 0.22 μm. Such a filter would retain organisms the size of *Thiobacillus*. The genus *Thiomicrospira* consists of two species; both are obligate chemolithotrophs, and one species can grow anaerobically with nitrate as electron acceptor. *Thermothrix* is a filamentous bacterium that inhabits hot sulfur springs having a neutral or slightly acidic pH. Cultures of *Thermothrix* grow aerobically or anaerobically (with nitrate as electron acceptor) at

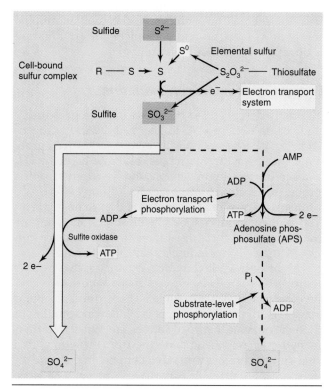

Figure 16.30 Steps in the oxidation of different compounds by thiobacilli. The sulfite oxidase pathway accounts for the majority of sulfite oxidized.

temperatures between 55 and 85°C (optimum at 70°C). *Thermothrix* is capable of chemoorganotrophic growth and utilizes a variety of organic compounds, including glucose, acetate, and amino acids as electron donors.

Carbon metabolism and carboxysomes

Several carbon nutritional classes of chemolithotrophs are known, from pure autotrophs to pure heterotrophs. The sulfur chemolithotrophs discussed thus far, such as *Thiobacillus,* are generally capable of *autotrophic* growth with CO_2 as sole carbon source. However, certain sulfur chemolithotrophs are incapable of autotrophic growth. Organisms such as most species of *Beggiatoa,* certain pseudomonads, and one species of *Thiobacillus, T. perometabolis,* fall into this category. Most of the remaining chemolithotrophs are true autotrophs and presumably grow autotrophically in nature.

Studies on CO_2 fixation in *Thiobacillus* have shown that reactions of the Calvin cycle (∞ Section 13.7) are responsible for CO_2 fixation in this chemolithotroph. Thin sections of cells of various thiobacilli have revealed polyhedral cell inclusions scattered throughout the cell (see Figure 16.29). The inclusions are about 100 nm in diameter (Figure 16.31) and are surrounded by a thin, nonunit membrane. The polyhedral bodies have been referred to as *carboxysomes* and have been shown to consist of molecules of the enzyme *ribulose 1,5-bisphosphate carboxylase* (RubisCO, the key enzyme of the Calvin cycle) in crystalline form.

Although the reason for carboxysome formation is not clear, they may increase the amount of ribulose bisphosphate carboxylase in the cell. Because the enzyme is in an insoluble form, the carboxysome is an effective mechanism for concentrating enzyme in the cell without affecting osmolarity. Carboxysomes have been observed in nitrifying bacteria, various sulfur chemolithotrophs, and the cyanobacteria, but not in chemolithotrophic bacteria that require organic compounds as a carbon source or in phototrophic purple bacteria, many of which can grow chemoorganotrophically. This suggests that carboxysomes are unique to organisms that have specialized in the autotrophic way of life.

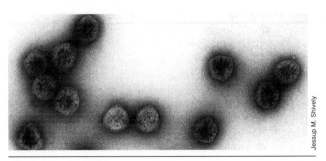

Figure 16.31 Polyhedral bodies (carboxysomes) purified from the chemolithotrophic sulfur oxidizer *Thiobacillus neapolitanus.* The bodies are about 100 nm in diameter.

16.6
Chemolithotrophs: Hydrogen-Oxidizing Bacteria

A wide variety of bacteria are capable of growing with H_2 as sole electron donor and O_2 as electron acceptor using the "knallgas" reaction, the reduction of O_2 with H_2:

$$2 H_2 + O_2 \rightarrow 2 H_2O$$

Many, but not all, of these organisms can also grow autotrophically (using reactions of the Calvin cycle to incorporate CO_2) and are grouped together here as the chemolithotrophic *hydrogen-oxidizing bacteria*. Both gram-positive and gram-negative hydrogen bacteria are known, with the best-studied representatives classified in the genera *Pseudomonas, Paracoccus,* and *Alcaligenes* (Table 16.9). All hydrogen-oxidizing bacteria contain one or more *hydrogenase* enzymes that function to bind H_2 and use it either to produce ATP or for reducing power for autotrophic growth.

Almost all hydrogen bacteria are *facultative chemolithotrophs,* meaning that they can also grow chemoorganotrophically with organic compounds as energy sources. This is a major distinction between hydrogen chemolithotrophs and many sulfur chemolithotrophs or nitrifying bacteria; most representatives of the latter two groups are *obligate chemolithotrophs*—growth does not occur in the absence of the inorganic energy source. By contrast, hydrogen chemolithotrophs can switch between chemolithotrophic and chemoorganotrophic modes of metabolism and presumably do so in nature as nutritional conditions warrant.

Energetics of hydrogen oxidation

We briefly described the energetics of hydrogen-oxidizing bacteria in Section 13.9. The majority of hydrogen-oxidizing bacteria are obligate aerobes; a few can grow anaerobically with nitrate as electron acceptor. H_2 is taken up by hydrogenase, and electrons shunted through an electron transport chain leading to the establishment of a membrane potential and a proton gradient; ATP is then produced by chemiosmotic mechanisms (see Section 4.13 for a discussion of chemiosmosis and related topics). In *Alcaligenes eutrophus* (Figure 16.32), one of the best-studied hydrogen bacteria, two distinct hydrogenases are present, one membrane-bound and the other cytoplasmic (soluble). The membrane-bound enzyme is involved in energetics. Following binding of H_2 to the enzyme, hydrogen atoms are transferred from the hydrogenase to a quinone and from there through a series of cytochromes to O_2. Cytochromes of the *c* and aa_3 types are present in *A. eutrophus*. In other organisms, such as *Paracoccus denitrificans,* a similar sequence is observed, but *b*-type cytochromes are also present.

TABLE 16.9   Differential characteristics of species of hydrogen-oxidizing bacteria[a]

Genus and/or species	Denitri-fication	Growth on fructose	Motility	DNA (mol% GC)	Other characteristics
Gram-negative					
Acidovorax facilus	−	+	+	64	Membrane-bound hydrogenase
Alcaligenes eutrophus	+	+	+	66	Membrane-bound and cytoplasmic hydrogenases
Alcaligenes xylosoxidans	−	+	+	—	Membrane-bound and cytoplasmic hydrogenases
Aquaspirillum autotrophicum	−	−	+	61	Only membrane-bound hydrogenase present
Pseudomonas carboxydovorans	−	−	+	60	Only membrane-bound hydrogenase present; also oxidizes CO
Hydrogenophaga flava	−	+	+	67	Colonies are bright yellow
Seliberia carboxydohydrogena	−	?	+	58	Also oxidizes CO
Paracoccus denitrificans	+	+	−	66	Only membrane-bound hydrogenase present; strong denitrifier
Aquifex pyrophilus	+	−	+	65	Hyperthermophile, grows microaerophilically or anaerobically (with NO_3^-), obligate chemolithotroph; also uses S^0 or $S_2O_3^{2-}$ as electron donor
Hydrogenobacter thermophilus	−	−	−	37–46	As for *Aquifex*, but obligate aerobe (microaerophile)
Gram-positive					
Bacillus schlegelii	−	−	+	66	Produces endospores; thermophile; also uses CO or $S_2O_3^{2-}$ as electron donor
Arthrobacter sp.	−	+	−	70	Only membrane-bound hydrogenase present
Mycobacterium gordonae	−	?	−	—	Acid-fast; colonies yellow to orange

[a]Phylogenetically, most H_2-oxidizing bacteria are α or β purple Bacteria (see Table 16.1), except for the gram-positive organisms; *Aquifex* and *Hydrogenobacter* represent one of the major lineages of Bacteria (∞ Section 15.7).

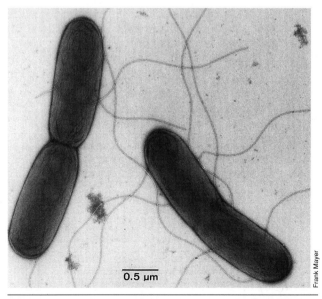

Figure 16.32 Transmission electron micrograph of negatively stained cells of the hydrogen-oxidizing chemolithotroph *Alcaligenes eutrophus.*

The second hydrogenase of *Alcaligenes eutrophus* is involved in the production of *reducing power* for autotrophic growth. This soluble hydrogenase takes up H_2 and reduces NAD^+ to NADH directly. NADH can be converted to NADPH (by enzymes called *transhydrogenases*) for use as a reductant in the Calvin cycle, but many hydrogen chemolithotrophs use NADH itself, instead of NADPH, as the reductant in Calvin cycle reactions. Because the reduction potential of H_2 is so low (−0.42 V), it can be used to reduce NAD^+ directly in hydrogen bacteria like *A. eutrophus* that contain two hydrogenases. Thus, in *A. eutrophus*, ATP-dependent reverse electron flow reactions (∞ Section 13.8) are *not* required. Because of this, growth rates and cell yields of organisms like *A. eutrophus* are typically much higher than in chemolithotrophic bacteria that must use energy to make NAD(P)H. Many hydrogen bacteria contain only a membrane-bound hydrogenase (see Table 16.9), however, and this enzyme does not couple to the reduction of NAD^+. In these organisms reverse electron flow operates as a means of generating NADH.

Physiology and isolation of hydrogen bacteria

Most hydrogen bacteria prefer microaerobic conditions when growing chemolithotrophically on H_2 because hydrogenases are oxygen-sensitive enzymes. Typically, oxygen levels of about 10% (half the level in air) support best growth. *Nickel* must be present in the medium for growth of hydrogen bacteria because virtually all hydrogenases contain Ni^{2+} as a metal cofactor. A few hydrogen bacteria also fix molecular N_2, and when growing on N_2, the organisms are quite oxygen-sensitive because the enzyme nitrogenase needed for the reduction of molecular nitrogen (∞ Section 13.27) is an oxygen-sensitive enzyme. Unlike nitrifying bacteria or methanotrophs, hydrogen-oxidizing bacteria generally lack internal membranes; apparently the cytoplasmic membrane of hydrogen bacteria can integrate sufficient levels of hydrogenase to maintain high rates of H_2 oxidation.

All species of H_2 bacteria will incorporate organic compounds if they are added to the medium. In some hydrogen bacteria organic compounds such as glucose and acetate strongly repress the synthesis of both Calvin cycle enzymes and hydrogenase. Some hydrogen bacteria also grow on carbon monoxide, CO, as energy source, with electrons from the oxidation of CO to CO_2 entering the electron transport chain to drive ATP synthesis. CO-oxidizing bacteria, which are known as *carboxydotrophic* bacteria, grow autotrophically using Calvin cycle reactions to fix the CO_2 generated from the oxidation of CO. CO is oxidized to CO_2 by the enzyme *carbon monoxide dehydrogenase*, which is a molybdenum-containing enzyme. The molybdenum in CO dehydrogenase is bound to a small cofactor consisting of a multiringed structure called a *pterin*, similar to the situation in the enzyme nitrate reductase (∞ Section 13.15).

Hydrogen-oxidizing bacteria are fairly easy to isolate. A small amount of mineral salts medium containing trace metals (especially Ni^{2+} and Fe^{2+}) is inoculated with soil or water and incubated in a large, sealed flask containing a head space of 5% O_2, 10% CO_2, and 85% H_2. When the liquid becomes turbid, plates of the same medium are streaked and incubated in a glass jar containing the same gas mixture. Pure cultures are easily obtained by picking and restreaking colonies several times (one must exercise care in mixing gas phases, however—mixtures of O_2 and H_2 are potentially explosive).

Ecology of hydrogen bacteria and carboxydotrophic bacteria

The ecological importance of aerobic hydrogen-oxidizing bacteria is difficult to assess. This is because in natural environments there is great competition for H_2 among various anaerobic bacteria. Methanogens, homoacetogens, sulfate-reducing bacteria, phototrophic bacteria, and certain denitrifying bacteria all compete for H_2 produced from fermentation. And because hydrogen levels in actively methanogenic or sulfide-rich ecosystems are frequently very low, it is doubtful whether large amounts of H_2 ever reach oxic habitats from sources in anoxic environments. It is perhaps for this reason that most aerobic hydrogen bacteria also grow chemoorganotrophically; a "backup" system is required if the primary energy source is not always available.

CO consumption by carboxydotrophic bacteria is a very significant ecological process. Although much CO is generated from various human and other sources, CO levels in air have not risen significantly over many years. Microbial CO consumption is probably the reason why. Under anoxic conditions CO is oxidized by certain phototrophic bacteria, methanogens, and sulfate reducers. However, the most significant releases of CO (primarily from automobile exhaust, incomplete combustion of fossil fuels, and the catabolism of lignin) occur in oxic environments. Thus, carboxydotrophic bacteria in the upper layers of soil probably represent the most significant sink for CO in nature. Carboxydotrophic bacteria include both gram-negative and gram-positive types. Some of the best-studied carboxydobacteria include *Pseudomonas carboxydovorans*, *Bacillus schlegelii*, and *Alcaligenes carboxydus* (Table 16.9). At least one carboxydobacterium can grow on CO anaerobically with nitrate as electron acceptor, but this does not seem to be a widespread property of the group. Like the hydrogen bacteria, all isolates of carboxydotrophic bacteria, with the exception of a thermophilic *Streptomyces* species, grow chemoorganotrophically on organic substrates as well as on CO.

> **CONCEPT CHECK** 16.6
>
> Chemolithotrophs are prokaryotes that can oxidize inorganic electron donors and in many cases use CO_2 as sole carbon source.
>
> ✔ Compare and contrast the nitrifying bacteria with the sulfur, iron, and hydrogen bacteria in terms of inorganic electron donors used, carbon sources, E_0' of electron donors (∞ Chapter 13), and habitats.
>
> ✔ What major pathway is present for assimilation of CO_2 in many chemolithotrophs?

16.7
Methanotrophs and Methylotrophs

Methane, CH_4, is found extensively in nature. It is produced in anoxic environments by methanogenic bacteria (∞ Section 14.12) and is a major gas of anoxic muds, marshes, anoxic zones of lakes, the rumen, and the mammalian intestinal tract. Methane is the major constituent of natural gas and is also present in many coal formations. It is a relatively stable molecule; but a variety of bacteria, the **methanotrophs,** oxidize it readily, utilizing methane and a few other one-carbon compounds as electron donors for energy generation and as sole sources of carbon. These bacteria are all aerobes and are widespread in nature in soil and water. They are also of a diversity of morphological types, seemingly related only in their ability to oxidize methane. There has been considerable interest in methane-utilizing bacteria in recent years because of the possibility of using this simple and widely available energy source to produce bacterial protein as a food or feed supplement (single-cell protein).

C_1 metabolism

In addition to methane, a number of other one-carbon compounds are known to be utilized by microorganisms. A list of these compounds is given in Table 16.10. From a biochemical viewpoint, these compounds share a key characteristic: they contain no carbon–carbon bonds. Thus, all carbon–carbon bonds of the cell must be synthesized de novo. Organisms that can grow using only one-carbon compounds are generally called **methylotrophs.** Many, but not all, methylotrophs are also methanotrophs. From the viewpoint of carbon assimilation, methanotrophs and methylotrophs have something in common with autotrophs (∞ Chapter 13), which also use a carbon compound lacking a carbon–carbon bond, CO_2. The two groups differ, however, in that the methylotrophs require a carbon compound *more reduced* than CO_2.

It is important to distinguish between methylotrophs and methanotrophs. A wide variety of bacteria are known that can grow on methanol, methylamine, or formate, but not methane, and these bacteria are members of various genera of chemoorganotrophs: *Hyphomicrobium, Pseudomonas, Bacillus,* and *Vibrio.* In contrast, methanotrophs are unique in that they can grow not only on some of the more oxidized one-carbon compounds but also on methane. The methane-oxidizing bacteria possess a specific enzyme system, *methane monooxygenase,* for the introduction of an oxygen atom into the methane molecule, leading to the formation of methanol. It should be noted that although the methanotrophs can also oxidize more oxidized one-carbon compounds such as methanol and formate, initial isolation from nature requires the use of methane as a sole electron donor because if one of these other one-carbon compounds is used in initial enrichment, a nonmethanotrophic methylotroph will almost certainly be isolated. All methanotrophs appear to be obligate C_1 utilizers, unable to utilize compounds with carbon–carbon bonds. By contrast, many nonmethanotrophic methylotrophs are able to utilize organic acids, ethanol, and sugars.

Methane-oxidizing bacteria are also unique among prokaryotes in possessing relatively large amounts of **sterols.** As we noted in Section 3.3, sterols are found in eukaryotes as a functional part of the membrane system but are absent from most prokaryotes. In methanotrophs, sterols may be an essential part of the complex internal membrane system (see later) involved in methane oxidation.

Classification

An overview of the classification of methanotrophs is given in Table 16.11. These bacteria were initially distinguished on the basis of morphology and formation of resting stages, but it was then found that they could

---

TABLE 16.10   Substrates used by methylotrophic bacteria[a]

Substrates used for growth	Substrates oxidized but not used for growth (cometabolism)
Methane, CH_4	Ammonium, NH_4^+
Methanol, CH_3OH	Ethylene, $H_2C{=}CH_2$
Methylamine, CH_3NH_2	Chloromethane, CH_3Cl
Dimethylamine, $(CH_3)_2NH$	Bromomethane, CH_3Br
Trimethylamine, $(CH_3)_3N$	Higher hydrocarbons (ethane, propane)
Tetramethylammonium, $(CH_3)_4N^+$	
Trimethylamine N-oxide, $(CH_3)_3NO$	
Trimethylsulfonium, $(CH_3)_3S^+$	
Formate, $HCOO^-$	
Formamide, $HCONH_2$	
Carbon monoxide, CO	
Dimethyl ether, $(CH_3)_2O$	
Dimethyl carbonate, $CH_3OCOOCH_3$	
Dimethyl sulfoxide, $(CH_3)_2SO$	
Dimethylsulfide, $(CH_3)_2S$	

[a]A single isolate does not use all of the above, but at least one methylotrophic bacterium has been reported to oxidize each of the listed compounds.

TABLE 16.11   Some characteristics of methanotrophic bacteria

Organism	Morphology	16S rRNA group[a]	Resting stage	Internal membranes[b]	Citric acid cycle[c]	Carbon assimilation pathway[d]	N₂ fixa-tion	DNA (mol % GC)
Methylomonas	Rod	Gamma purple bacteria	Cystlike body	I	Incomplete	Ribulose monophosphate	No	50–54
Methylomicrobium	Rod	Gamma purple bacteria	None	I	Incomplete	Ribulose monophosphate	No	49–60
Methylobacter	Coccus to ellipsoid	Gamma purple bacteria	Cystlike body	I	Incomplete	Ribulose monophosphate	No	50–54
Methylococcus	Coccus	Gamma purple bacteria	Cystlike body	I	Incomplete	Ribulose monophosphate	Yes	62–64
Methylosinus	Rod or vibrioid	Alpha purple bacteria	Exospore	II	Complete	Serine	Yes	63
Methylocystis	Rod	Alpha purple bacteria	Exospore	II	Complete	Serine	Yes	63

[a] ∞ Section 15.7 and Table 16.1.

[b] Internal membranes: Type I, bundles of disc-shaped vesicles distributed throughout the organism; Type II, paired membranes running along the periphery of the cell. See Figure 16.33.

[c] Organisms with an incomplete citric acid cycle lack the enzyme α-ketoglutarate dehydrogenase.

[d] See Figures 16.34 and 16.35. Unlike other methylotrophs, *Methylococcus* species contain Calvin cycle enzymes.

be divided into two major groups depending on their internal cell structure and carbon assimilation pathway. *Type I* organisms assimilate one-carbon compounds via a unique pathway, the **ribulose monophosphate cycle,** whereas *Type II* organisms assimilate C_1 intermediates via the **serine pathway.** The requirement for O_2 as a reactant in the initial oxidation of methane explains why all methanotrophs are obligate aerobes, whereas some organisms using methanol as electron donor can grow anaerobically (with nitrate or sulfate as electron acceptor).

Both groups of methanotrophs contain extensive internal membrane systems, which appear to be related to their methane-oxidizing ability. Type I bacteria are characterized by internal membranes arranged as bundles of disc-shaped vesicles distributed throughout the cell (Figure 16.33*b*), whereas Type II bacteria possess paired membranes running along the periphery of the cell (Figure 16.33*a*). Type I methanotrophs are also characterized by a lack of a complete citric acid cycle (the enzyme *α-ketoglutarate dehydrogenase* is absent), whereas Type II organisms possess a complete cycle. Absence of a complete citric acid cycle greatly diminishes the ability of an organism to grow chemoorganotrophically. If these reactions cannot be run as a cycle, NADH cannot be generated from reactions of the cycle, thus preventing growth at the expense of organic compounds metabolized through the citric acid cycle.

Biochemistry of methane oxidation

Two aspects of the biochemistry of methanotrophs are of interest, the manner of oxidation of methane to CO_2 (and how this is coupled to ATP synthesis) and the manner in which one carbon compounds are assimilated into cell material. The overall pathway of methane oxidation involves stepwise, two-electron oxidations:

$$CH_4 \xrightarrow[-126 \text{ kJ}]{} CH_3OH \xrightarrow[-193 \text{ kJ}]{} HCHO \xrightarrow[-214 \text{ kJ}]{}$$

Methane   Methanol   Formaldehyde

$$HCOO^- \xrightarrow[-239 \text{ kJ}]{} HCO_3^-$$

Formate   Bicarbonate

The initial step in the oxidation of methane involves an enzyme called **methane monooxygenase.** As we discussed in Section 13.23, oxygenase enzymes catalyze the incorporation of oxygen from O_2 into carbon compounds and seem to be widely involved in the metabolism of hydrocarbons. Oxygenases require a source of reducing power, usually NADH, but in *Methylosinus*, where the process has been most thoroughly studied, electrons come not from NADH but from cytochrome *c*. This cytochrome *c* is involved in recycling of electrons from methanol dehydrogenase (and possibly from formaldehyde dehydrogenase). No ATP synthesis occurs during the first step, the oxidation of methane to methanol, and this is consistent with the fact that growth yields of methanotrophs are the same whether methane or methanol is used as substrate. Thus, although considerable energy is potentially available in the oxidation of CH_4 to CH_3OH (about 126 kJ/mol), this energy is not available to the organism, apparently because no biochemical mechanism is available for conserving the energy of methane oxidation.

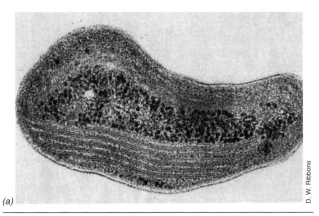

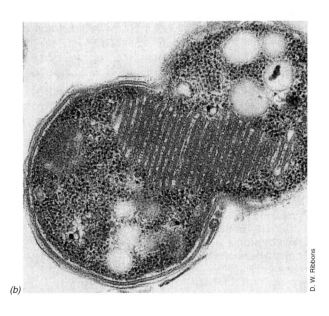

Figure 16.33 Electron micrographs of methanotrophs. (a) A *Methylosinus* species, illustrating a Type II membrane system. Cells are about 0.6 μm in diameter. (b) *Methylococcus capsulatus*, illustrating a Type I membrane system. Cells are about 1 μm in diameter.

Biochemistry of one-carbon assimilation

As noted earlier, all Type I methanotrophs possess the ribulose monophosphate pathway for carbon assimilation, whereas Type II methanotrophs have the serine pathway.

The serine pathway is outlined in Figure 16.34. It is present not only in Type II methanotrophs but is also the pathway for C_1 incorporation in facultative methylotrophs (*Hyphomicrobium* and *Pseudomonas*, for example). In this pathway, a two-carbon unit, acetyl-CoA, is synthesized from one molecule of formaldehyde and one molecule of CO_2. The pathway requires the introduction of reducing power and energy in the form of *two* molecules each of NADH and ATP for each acetyl-CoA synthesized.

The ribulose monophosphate pathway, present in Type I methanotrophs, is outlined in Figure 16.35. It is more efficient than the serine pathway in that *all* the carbon atoms for cell material are derived from formaldehyde, and because formaldehyde is at the same oxidation level as cell material, no reducing power is needed. The ribulose monophosphate pathway requires the introduction of energy in the form of *one* molecule of ATP for each molecule of 3-phosphoglyceric acid synthesized (Figure 16.35). Consistent with the lower energy requirements of the ribulose monophosphate pathway, the cell yield of Type I organisms from a given amount of methane or methanol is *higher* than the cell yield of Type II organisms.

Ecology and isolation

Methanotrophs are widespread in aquatic and terrestrial environments, being found wherever stable sources of methane are present. Methane produced in the anoxic regions of lakes rises through the water column, and methanotrophs are often concentrated in a narrow band at the thermocline, where methane from the anoxic zone meets oxygen from the oxic zone. Although methanotrophs are obligate aerobes, they are sensitive to O_2 at normal concentrations and prefer microaerobic habitats for development. One reason for this O_2 sensitivity may be that many aquatic methanotrophs simultaneously fix N_2, and nitrogenase is O_2-sensitive, and so optimal development occurs where O_2 concentrations are reduced. Methane-oxidizing bacteria play a small but probably important role in the carbon cycle, converting methane derived from anoxic decomposition back into cell material (and CO_2).

The initial enrichment of methanotrophs is relatively easy, and all that is needed is a mineral salts medium over which an atmosphere of 80% methane and 20% air is maintained. Once good growth is obtained, purification is carried out by streaking on mineral salts agar plates, which are incubated in a jar with the methane–air mixture. Colonies appearing on the plates are of two types, common chemoorganotrophs growing on traces of organic matter in the medium, which appear in 1–2 days, and methanotrophs, which appear after about a week. The colonies of many methanotrophs are pink in color. Colonies of methanotrophs should be picked when small, and purification by continued picking and restreaking is essential. It should be emphasized that cultures of methanotrophs are often contaminated with chemoorganotrophs, or with methanol oxidizers, so careful attention should be paid to the purity of an isolate before any detailed studies are undertaken.

Methanotrophs are able to oxidize ammonia, although they cannot grow chemolithotrophically using ammonia as sole electron donor. Methane monooxygenase also functions as an ammonia oxygenase, and a competitive interaction between the two substrates exists (see Section 16.4). For this reason, ammonia is generally

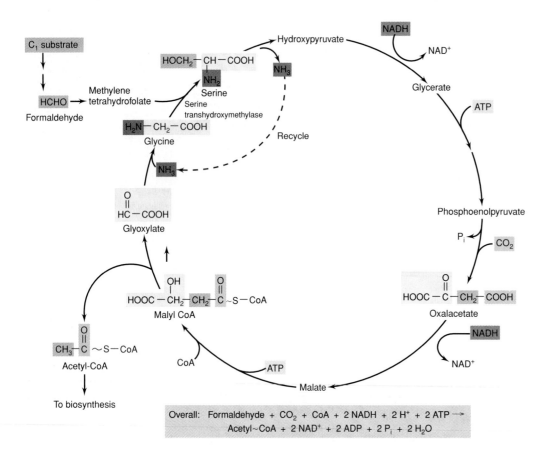

Figure 16.34 The serine pathway for the assimilation of C_1 units into cell material by Type II methylotrophic bacteria. The product of the pathway, acetyl-CoA, is used as the starting point for making new cell material. The key enzyme of the pathway is serine transhydroxymethylase.

toxic to methanotrophs, and the preferred nitrogen source is nitrate. It has been speculated that methanotrophic bacteria could have arisen from the nitrosifying bacteria via genetic changes causing the conversion of ammonia oxidase to methane oxidase. The fact that both groups of bacteria have elaborate internal membrane systems (see Section 16.4) supports such a theory.

Methanotrophic symbionts of animals

A symbiotic association between methanotrophic bacteria and marine mussels and certain types of marine sponges is known to occur. Mussels live in the vicinity of hydrocarbon seeps where methane is released in substantial amounts. Intact mussels as well as isolated mussel gill tissue consume methane at high rates in the presence of O_2. In the gill tissue of the mussel, coccoid-shaped bacteria are present in high numbers (Figure 16.36a). The bacterial symbionts contain stacks of intracytoplasmic membranes (Figure 16.36b) typical of Type I methanotrophs. The symbionts are found in vacuoles within animal cells near the gill surface, which probably ensures an effective gaseous exchange with seawater. The oxidation of methane by mussel gill tissue is strictly O_2-dependent and totally inhibited by acetylene, a known poison of biological methane oxidation. This is consistent with the hypothesis that

methane oxidation is carried out by the methanotrophic symbiont.

Stable isotope studies of mussel and sponge tissue support the concept that methane serves as the major food source for the animal. Using $^{13}C/^{12}C$ analyses (∞ Section 14.6), it was shown that animal tissues had $^{13}C/^{12}C$ isotopic compositions nearly identical to that of

---

CONCEPT CHECK   **16.7**

Methylotrophs can grow on various C_1-type compounds. Some methylotrophs are also methanotrophs (can grow on CH_4), but many methylotrophs cannot use CH_4. Two different routes of C_1 incorporation are used by methylotrophs, both pathways incorporating much of the carbon at the level of formaldehyde.

✔ *In what ways are the serine and ribulose monophosphate pathways similar? In what ways do they differ?*

✔ *Compare and contrast Type I with Type II methanotrophs in regard to their phylogeny, carbon assimilation pathways, and internal membranes.*

✔ *Draw the reaction catalyzed by the enzyme* methane monooxygenase.

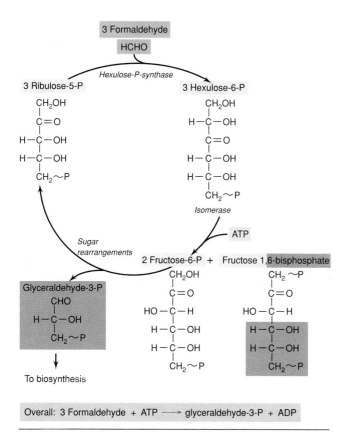

Figure 16.35 The ribulose monophosphate pathway for assimilation of one-carbon compounds, as found in Type I methylotrophic bacteria. The complete name of the hexulose sugar is D-erythro-L-glycero-3-hexulose 6-phosphate. Three formaldehydes are needed to carry the cycle to completion, with the net result being one molecule of glyceraldehyde-3-P. Regeneration of ribulose 5-phosphate and formation of phosphoglyceraldehyde occur via a series of pentose phosphate reactions.

the methane being consumed from their environment. Presumably methane assimilated by the methanotrophs is distributed throughout the animals by the excretion of carbon compounds by the methanotrophs. The methanotrophic symbiosis is therefore conceptually similar to the symbiosis established between sulfide-oxidizing chemolithotrophs and hydrothermal vent tube worms and giant clams discussed in Section 14.10. Animal–bacteria symbioses, such as the methanotrophic mussel/sponge symbiosis and the sulfide-oxidizing vent animal symbioses, thus show that prokaryotic cells can occasionally constitute the basis of a one-step food chain.

16.8
Sulfate- and Sulfur-Reducing Bacteria

Sulfate is used as a terminal electron acceptor under anoxic conditions by a heterogeneous assemblage of bacteria that utilize organic acids, fatty acids, alcohols, and H_2 as electron donors. Although morphologically

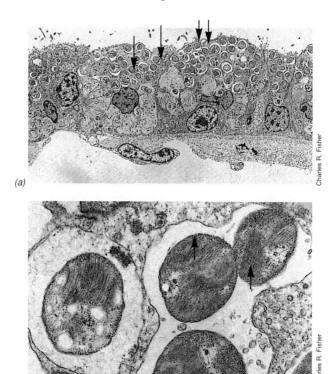

Figure 16.36 Methanotrophic symbionts of marine mussels. (a) Electron micrograph of a thin section at low magnification of gill tissue of a marine mussel living near hydrocarbon seeps in the Gulf of Mexico. Note the symbiotic methanotrophs (arrows) in the tissues. (b) High magnification view of gill tissue showing Type I methanotrophs. Note membrane bundles (arrows). The methanotrophs are about 1 μm in diameter.

diverse, the **sulfate-reducing bacteria** can be considered a physiologically unified group in the same manner as the phototrophic or methanotrophic bacteria. Eighteen genera of dissimilatory sulfate-reducing bacteria are currently recognized and can be placed in two broad physiological subgroups as outlined in Table 16.12. The genera in group I, such as *Desulfovibrio* (Figure 16.37a), *Desulfomonas*, *Desulfotomaculum*, and *Desulfobulbus* (Figure 16.37c), utilize lactate, pyruvate, ethanol, or certain fatty acids as carbon and energy sources, reducing sulfate to hydrogen sulfide. The genera in group II, such as *Desulfobacter* (Figure 16.37d), *Desulfococcus*, *Desulfosarcina* (Figure 16.37e), and *Desulfonema* (Figure 16.37b), specialize in the oxidation of fatty acids, particularly *acetate*, reducing sulfate to sulfide. The sulfate-reducing bacteria are all obligate anaerobes, and strict anoxic techniques must be used in their cultivation.

Sulfate-reducing bacteria are widespread in aquatic and terrestrial environments that become anoxic as a result of microbial decomposition processes. The best-known genus is *Desulfovibrio* (Figure 16.37a), which is common in aquatic habitats or waterlogged soils containing abundant organic material and sufficient levels of sulfate. *Desulfotomaculum* consists of endospore-

TABLE 16.12   Characteristics of sulfate- and sulfur-reducing bacteria[a]

Genus	Characteristics	DNA (mol % GC)
Group I sulfate reducers: Nonacetate oxidizers		
Desulfovibrio	Polarly flagellated, curved rods, no spores; gram-negative; contain desulfoviridin; twelve species, one thermophilic	46–61
Desulfomicrobium	Motile rods, no spores; gram-negative; desulfoviridin absent; two species	52–57
Desulfobotulus	Vibrios; gram-negative; motile; desulfoviridin absent; one species	53
Desulfotomaculum	Straight or curved rods; motile by peritrichous or polar flagellation; gram-negative; desulfoviridin absent; produce endospores; four species, one thermophilic; one species capable of utilizing acetate as energy source	37–46
Desulfomonile	Rod; capable of reductive dechlorination of 3-chlorobenzoate to benzoate (∞ Section 14.20)	49
Desulfobacula	Oval to coccoid cells, marine; can oxidize various aromatic compounds including the aromatic hydrocarbon toluene, to CO_2; one species	42
Archaeoglobus	Archaean; hyperthermophile, temperature optimum, 83°C; contains some unique coenzymes of methanogenic bacteria, makes small amount of methane during growth; H_2, formate, glucose, lactate, and pyruvate are electron donors, SO_4^{2-}, $S_2O_3^{2-}$, or SO_3^{2-}, electron acceptors; two species (∞ Section 17.3)	41–46
Desulfobulbus	Ovoid or lemon-shaped cells; no spores; gram-negative; desulfoviridin absent; if motile, by single polar flagellum; utilizes propionate as electron donor with acetate + CO_2 as products; three species	59–60
Thermodesulfobacterium	Small, gram-negative rods; desulfoviridin present; thermophilic, optimum growth at 70°C	34
Group II sulfate reducers: Acetate oxidizers		
Desulfobacter	Rods; no spores, gram-negative; desulfoviridin absent; if motile, by single polar flagellum; utilizes only acetate as electron donor and oxidizes it to CO_2 via the citric acid cycle; four species	45–46
Desulfobacterium	Rods, some with gas vesicles, marine; capable of autotrophic growth via the acetyl-CoA pathway; three species	41–59
Desulfococcus	Spherical cells; nonmotile; gram-negative; desulfoviridin present, no spores; utilizes C_1 to C_{14} fatty acids as electron donor with complete oxidation to CO_2; capable of autotrophic growth via the acetyl-CoA pathway; two species	57
Desulfonema	Large, filamentous gliding bacteria; gram-positive, no spores; desulfoviridin present or absent; utilizes C_2 to C_{12} fatty acids as electron donor with complete oxidation to CO_2; capable of autotrophic growth via the acetyl-CoA pathway (H_2 as electron donor); two species	35–42
Desulfosarcina	Cells in packets (sarcina arrangement); gram-negative; no spores; desulfoviridin absent; utilizes C_2 to C_{14} fatty acids as electron donor with complete oxidation to CO_2; capable of autotrophic growth via the acetyl-CoA pathway (H_2 as electron donor); one species	51
Desulfoarculus	Vibrios; gram-negative; motile; desulfoviridin absent; utilizes only C_1 to C_{18} fatty acids as electron donor	66
Desulfacinum	Cocci to oval-shaped cells; gram-negative; utilizes C_1 to C_{18} fatty acids, very nutritionally diverse, capable of autotrophic growth; thermophile	64
Desulforhabdus	Rods; no spores; gram-negative; nonmotile; utilizes fatty acids with complete oxidation to CO_2	52
Thermodesulforhabdus	Gram-negative motile rods; thermophilic; uses fatty acids up to C_{18}	51
Dissimilatory sulfur reducers		
Desulfuromonas	Straight rods, single lateral flagellum; no spores; gram-negative; does not reduce sulfate; acetate, succinate, ethanol, or propanol used as electron donor; obligate anaerobe; four species	50–63
Desulfurella	Motile short rods; gram-negative; requires acetate; thermophilic	31
Campylobacter	Curved, vibrio-shaped rods; polar flagella; gram-negative; no spores; unable to reduce sulfate but can reduce sulfur, sulfite, thiosulfate, nitrate, or fumarate anaerobically with acetate or a variety of other carbon or electron donor sources; facultative aerobe	40–42
Hyperthermophilic Archaea (∞ Section 17.3) Several genera are dissimilatory sulfur reducers		

[a]Phylogenetically, most sulfate- and sulfur-reducing bacteria are delta purple Bacteria (∞ Section 15.7 and Table 16.1).

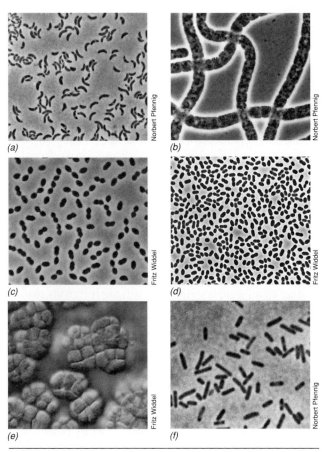

Figure 16.37 Phase contrast photomicrographs of (a–e) representative sulfate-reducing and (f) sulfur-reducing bacteria. (a) *Desulfovibrio desulfuricans;* cell diameter about 0.7 μm. (b) *Desulfonema limicola;* cell diameter 3 μm. (c) *Desulfobulbus propionicus;* cell diameter about 1.2 μm. (d) *Desulfobacter postgatei;* cell diameter about 1.5 μm. (e) *Desulfosarcina variabilis* (interference contrast microscopy); cell diameter about 1.25 μm. (f) *Desulfuromonas acetoxidans;* cell diameter about 0.6 μm.

By contrast, species of the sulfur-reducing organism *Desulfuromonas* (see later) are very closely related to one another but are not closely related (in an evolutionary sense) to sulfate-reducing bacteria.

Sulfur reduction

A variety of bacteria are known that can reduce elemental sulfur to sulfide but are unable to reduce sulfate to sulfide. These organisms are referred to as **dissimilatory sulfur-reducing bacteria.** Members of the genus *Desulfuromonas* (Figure 16.37f) can grow anaerobically by coupling the *oxidation* of substrates such as acetate or ethanol to the *reduction* of elemental sulfur to hydrogen sulfide. However, the ability to reduce elemental sulfur, as well as other sulfur compounds such as thiosulfate, sulfite, or dimethyl sulfoxide (DMSO), is a widespread property of a variety of chemoorganotrophic, generally facultatively aerobic bacteria (for example, *Proteus, Campylobacter, Pseudomonas,* and *Salmonella*). *Desulfuromonas* differs from the latter in that it is an obligate anaerobe and utilizes only sulfur as an electron acceptor (see Table 16.12). In addition, certain sulfate-reducing bacteria are capable of substituting sulfur for sulfate as an electron acceptor for anaerobic growth.

Studies of sulfur-rich thermal environments have led to the discovery of several hyperthermophilic dissimilatory sulfur-reducing Archaea (∞ Section 17.3). Many of these organisms have growth temperature optima near 90°C or higher, and these organisms represent a distinct branch of dissimilatory sulfur reducers, sharing little with organisms like *Desulfuromonas* except the ability to use sulfur as an electron acceptor. One extremely thermophilic archaean, *Archaeoglobus,* has been shown to reduce sulfate but is incapable of reducing elemental sulfur (∞ Section 17.3).

Physiology

The range of electron donors used by sulfate-reducing bacteria is broad. H_2, lactate, and pyruvate are almost universally used, and many group I species utilize malate, formate, and certain primary alcohols (for example, methanol, ethanol, propanol, and butanol). Some strains of *Desulfotomaculum* utilize glucose, but this is rather rare among sulfate reducers in general. Group I sulfate reducers oxidize their energy source to the level of acetate and excrete this fatty acid as an end product. Group II organisms differ from those in group I by their ability to oxidize fatty acids, lactate, succinate, and even benzoate in some cases, all the way to CO_2. *Desulfosarcina, Desulfonema, Desulfococcus, Desulfobacterium, Desulfotomaculum,* and certain species of *Desulfovibrio* (see Figure 16.37), are unique in their ability to grow chemolithotrophically with H_2 as electron donor, sulfate as electron acceptor, and CO_2 as sole carbon source (that is, autotrophic growth).

forming rods found primarily in soil, and one species is thermophilic. Growth and reduction of sulfate by *Desulfotomaculum* in certain canned foods leads to a type of spoilage called *sulfide stinker.* The remaining genera of sulfate reducers are indigenous to anoxic freshwater or marine environments; *Desulfomonas* can also be isolated from the mammalian intestine.

The sulfate- and sulfur-reducing bacteria represent a rather broad phylogenetic group. By 16S rRNA sequence analysis the endospore-forming sulfate reducer *Desulfotomaculum* groups, as expected, with the *Clostridium* subdivision of the gram-positive Bacteria (∞ Sections 15.7 and 16.26). The remaining sulfate reducers align most closely with other gram-negative Bacteria in the delta subdivision of the purple bacterial group (see Table 16.1). The sulfate reducers *Desulfovibrio, Desulfobacter,* and so on, are phylogenetically related to the bacterial predator *Bdellovibrio* (see Section 16.11) and the gliding myxobacteria (see Section 16.13).

In addition to using sulfate as an electron acceptor, many sulfate-reducing bacteria grow using *nitrate* (NO_3^-) as an electron acceptor, reducing NO_3^- to ammonia, NH_3, or can use certain organic compounds for energy generation by fermentative pathways in the complete absence of sulfate or other terminal electron acceptors. The most common fermentable compound is *pyruvate*, which is converted via the phosphoroclastic reaction to acetate, CO_2, and H_2 (∞ Figure 13.38). With lactate or ethanol, insufficient energy is available from fermentation, so sulfate is required. The value of sulfate as an electron acceptor can be readily demonstrated by comparing growth yields on pyruvate with and without sulfate. In the presence of sulfate, the growth yield is higher because of the larger amount of energy available when pyruvate utilization is coupled with sulfate reduction.

Certain sulfate-reducing bacteria can fix N_2, but this property is not uniformly distributed across the group. Species of *Desulfovibrio* and *Desulfobacter* are most likely to fix N_2.

Biochemistry

The enzymology of sulfate reduction was discussed in Section 13.16. Recall that the first step in dissimilatory sulfate reduction is the formation of *adenosine phosphosulfate* (APS) from ATP and sulfate; the enzyme APS reductase then catalyzes the reduction of the sulfate moiety to sulfite (∞ Figure 13.31). Sulfite is then reduced to sulfide in a series of steps about which little is known biochemically. A sulfite reductase has been identified, but it is not clear whether this enzyme serves to reduce all or only some of the potential intermediates between sulfite and sulfide.

Sulfate-reducing bacteria contain a number of electron transfer proteins including cytochromes of the c_3 and b types (∞ Section 13.16), ferredoxins, flavodoxins, and hydrogenases, although the actual function of many of the components (other than hydrogenase and cytochrome c_3) remains obscure. Because many sulfate reducers can use H_2 as an energy source, however, it is clear that the reduction of sulfate is linked to ATP synthesis via electron transport and chemiosmotic mechanisms (∞ Section 13.16 and Figure 13.32). The utilization of organic compounds as electron donors presumably involves the transfer of electrons through the same electron transport chain, generating a membrane potential.

Growth yield studies with sulfate-reducing bacteria have been used to determine the stoichiometry between ATP production and sulfate reduction. Studies with *Desulfovibrio* suggest that one (net) ATP is produced per sulfate reduced to sulfide and that three (net) ATPs are produced per sulfite reduced to sulfide. These values are in line with what is known concerning the conversion of sulfate to sulfite. During this conversion, two high energy phosphate bond equivalents are consumed because ATP and sulfate are the initial reactants and AMP and sulfite are the final products (∞ Section 13.16).

Certain sulfate reducers grow completely autotrophically with CO_2 as sole carbon source, H_2 as electron donor, and sulfate as electron acceptor. As discussed in Section 16.9 in connection with the homoacetogenic bacteria, autotrophic sulfate reducers use the *acetyl-CoA pathway* for fixation of CO_2 into cell material (see Table 16.12). In addition, most group II sulfate reducers (those capable of oxidizing acetate to CO_2) *reverse* the steps of the acetyl-CoA pathway to yield CO_2 from acetate and do not employ the more common citric acid cycle for this purpose.

Isolation and ecology

The enrichment of *Desulfovibrio* is relatively easy on an anoxic lactate–sulfate medium to which ferrous iron is added. A reducing agent such as thioglycolate or ascorbate is also added to achieve a lower E_0'. The sulfide formed from sulfate reduction combines with the ferrous iron to form black, insoluble ferrous sulfide. This blackening not only indicates sulfate reduction, but the iron also ties up and detoxifies the sulfide, making possible growth to higher cell yields. The conventional procedure is to set up liquid enrichments, and after some growth has occurred as evidenced by blackening of the medium, purification is accomplished by streaking onto a tube coated on the inside surface with a thin layer of agar (called roll tubes) or on Petri plates in an anoxic glove box. Although the sulfate reducers are obligate anaerobes, they are not as rapidly inactivated by oxygen as methanogens (∞ Section 17.2), and so streaking on plates can be done in air, provided a reducing agent is present in the medium and plates are quickly incubated in an anoxic environment.

Alternatively, agar shake tubes can be used for purification purposes. In the *shake tube method* a small amount of liquid from the original enrichment is added to a tube of molten agar growth medium, mixed thoroughly, and sequentially diluted through a series of molten agar tubes (∞ Section 14.3 and Figure 14.6). On solidification, individual cells distributed throughout the agar form colonies that can be removed aseptically, and the whole process is repeated until pure cultures are obtained. Colonies of sulfate-reducing bacteria are recognized by the black deposit of ferrous sulfide and purified by further streaking.

16.9

Homoacetogenic Bacteria

Homoacetogenic bacteria are obligate anaerobes that utilize CO_2 as a terminal electron acceptor, producing acetate as the sole product of anaerobic respiration; we discussed the overall formation of acetate by these organisms in Section 13.17. Electrons for the reduction of CO_2 to acetate can come from H_2, a variety of C_1 compounds, sugars, organic acids, alcohols, amino acids, and certain nitrogen bases. Many homoacetogens can also reduce NO_3^- and $S_2O_3^{2-}$; however, CO_2 reduction is probably the major reaction of ecological significance.

The major unifying thread among homoacetogens is the pathway of CO_2 reduction. Homoacetogens convert CO_2 to acetate by the **acetyl-CoA pathway** (see later), and in many homoacetogens autotrophic growth via the acetyl-CoA pathway also occurs. The acetyl-CoA pathway is also known as the Ljungdahl–Wood pathway in honor of its discovers, Lars Ljungdahl and Harland Wood. A list of the major organisms that produce acetate or oxidize acetate via the acetyl-CoA pathway is given in Table 16.13. Organisms such as *Acetobacterium woodii* and *Clostridium aceticum* can grow either chemoorganotrophically or chemolithotrophically by carrying out a homoacetic acid fermentation of sugars or through the reduction of CO_2 to acetate with H_2 as electron donor, respectively. The stoichiometries observed are as follows:

$$C_6H_{12}O_6 \rightarrow 3 \ CH_3COOH \qquad (1)$$

$$2 \ CO_2 + 4 \ H_2 \rightarrow CH_3COOH + 2 \ H_2O \qquad (2)$$

Homoacetogens ferment glucose via the glycolytic pathway converting glucose to two molecules of pyruvate and two molecules of NADH (the equivalent of

4 H). From this point, two molecules of acetate are produced as follows:

$$2 \ \text{pyruvate} \rightarrow 2 \ \text{acetate} + 2 \ CO_2 + 4 \ H \qquad (3)$$

The third acetate of the homoacetate fermentation comes from the reduction of the two molecules of CO_2 generated in reaction (3), using the four electrons generated from glycolysis *plus* the four electrons produced during the oxidation of two pyruvates to two acetates [reaction (3)]. Starting from pyruvate, then, the overall production of acetate can be written as

$$2 \ \text{pyruvate} + 4 \ H \rightarrow 3 \ \text{acetate}$$

All homoacetogenic bacteria that produce and excrete acetate in *energy metabolism* are gram-positive, and many are classified in the genus *Clostridium*. A few other gram-positive and many different gram-negative bacteria utilize the acetyl-CoA pathway for *autotrophic* purposes, reducing CO_2 to acetate, which then serves as a source of cell carbon. The acetyl-CoA pathway functions in autotrophic growth for those sulfate reducers capable of $H_2 + CO_2 + SO_4^{2-}$-mediated autotrophy (see Section 16.8) and is also used by the methanogenic bacteria, most of which can grow autotrophically on $H_2 + CO_2$ (∞ Section 17.2 and Table 16.13). By contrast, certain bacteria employ the reactions of the acetyl-CoA pathway primarily in the *reverse* direction as a means of *oxidizing* acetate. These include acetotrophic methanogens (∞ Section 17.2) and most type II (acetate-oxidizing) sulfate reducers (see Section 16.8).

Reactions of the acetyl-CoA pathway

We will consider the acetyl-CoA pathway here from the standpoint of organisms using H_2 as electron donor and CO_2 as electron acceptor for energy metabolism and autotrophic growth. Unlike other autotrophic pathways, such as the Calvin cycle (∞ Section 13.7) and the reverse citric acid cycle (∞ Section 16.1), the acetyl-CoA pathway of CO_2 fixation is *not* a cycle (Figure 16.38). Instead it involves the direct reduction of CO_2 to acetyl-CoA; one molecule of CO_2 is reduced to the methyl group of acetate, and the other molecule of CO_2 is reduced to the carbonyl group. The acetyl-CoA pathway has thus far been found only in certain obligate anaerobes, and it requires H_2 as an electron donor. A key enzyme of the acetyl-CoA pathway is *carbon monoxide (CO) dehydrogenase*. CO dehydrogenase is a complex enzyme that contains the metals Ni, Zn, and Fe as metal cofactors. CO dehydrogenase catalyzes the following reaction:

$$CO_2 + H_2 \rightleftharpoons CO + H_2O$$

The importance of CO dehydrogenase in the acetyl-CoA pathway is that it catalyzes the reduction of CO_2

TABLE 16.13	Organisms employing the acetyl-CoA pathway of CO_2 fixation

I.   Acetate synthesis the result of energy metabolism
    Acetoanaerobium noterae
    Acetobacterium woodii
    Acetobacterium wieringae
    Acetogenium kivui
    Acetitomaculum ruminis
    Clostridium aceticum
    Clostridium thermoaceticum
    Clostridium formicoaceticum
    Desulfotomaculum orientis
    Sporomusa paucivorans
    Eubacterium limosum (also produces butyrate)

II. Acetate synthesis in autotrophic metabolism
    Autotrophic homoacetogenic bacteria
    Autotrophic methanogens (∞ Section 17.2)
    Autotrophic sulfate-reducing bacteria
        (see Section 16.8)

III. Acetate oxidation in energy metabolism
    Reaction: Acetate + 2 $H_2O \rightarrow 2 \ CO_2$ + 8 H
        Group II sulfate reducers (other than *Desulfobacter*)
    Reaction: Acetate $\rightarrow CO_2 + CH_4$
        Acetotrophic methanogens (*Methanosarcina*, *Methanothrix*)

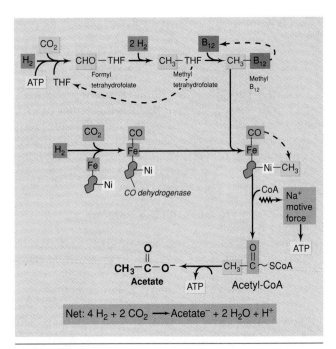

Net: $4 H_2 + 2 CO_2 \longrightarrow Acetate^- + 2 H_2O + H^+$

Figure 16.38 Reactions of the acetyl-CoA pathway, the mechanism of autotrophy in homoacetogenic, sulfate-reducing, and methanogenic bacteria. THF, Tetrahydrofolate; B_{12}, vitamin B_{12} in an enzyme-bound intermediate. CO is bound to an Fe atom in CO dehydrogenase, and the CH_3 group to an organic nickel compound in CO dehydrogenase. Note how formation of acetate powers a Na^+ pump that is used to drive ATP synthesis.

to CO, the CO ending up in the *carbonyl* position of acetate. The reactions of the acetyl-CoA pathway are shown in Figure 16.38. The methyl group of acetate originates from the reduction of CO_2 by a series of enzymatic reactions involving the coenzyme *tetrahydrofolate*. The latter is a common C_1 carrier in all organisms but plays a special role in homoacetogenic organisms. CO_2 is first reduced to formate by the enzyme formate dehydrogenase, and then the formate is converted to formyl tetrahydrofolate (Figure 16.38). The addition of two more pairs of electrons yields *methyl tetrahydrofolate*. The methyl group is then transferred from methyl tetrahydrofolate to an enzyme containing vitamin B_{12} as cofactor (Figure 16.38). In the final synthesis of acetate the CH_3 group is combined with CO in CO dehydrogenase with the CH_3 being attached to an atom of nickel and the CO to an atom of iron within the enzyme (Figure 16.38). Coenzyme A is added at this point and CO dehydrogenase then catalyzes formation of the final product, acetyl-CoA (Figure 16.38).

An interesting homoacetogen capable of methylotrophic growth has also been described. This organism, *Sporomusa paucivorans*, converts methanol into acetate, presumably by oxidizing methanol to CO_2 + 6 H and then using the acetyl-CoA pathway to reduce CO_2 to acetate. However, because methyl groups are

directly available from CH_3OH, *Sporomusa* probably puts methyl groups in at the terminal part of the tetrahydrofolate branch of the acetyl-CoA pathway (Figure 16.38) and joins the methyl group to CO, produced from the reduction of CO_2 by the enzyme CO dehydrogenase.

Because homoacetogens can grow at the expense of reactions of the acetyl-CoA pathway, this reaction sequence must be an overall energy-releasing one. One potential site of ATP synthesis in the reduction of CO_2 to acetate is the terminal step, the conversion of acetyl-CoA to acetate and ATP (via acetyl-P) (∞ Section 13.19). However, additional energy-linked steps occur because a Na^+ gradient is established across the cytoplasmic membrane during growth of homoacetogens. This gradient drives ATP synthesis through a Na^+-driven ATPase as discussed previously for the succinate fermenter, *Propionigenium* (∞ Section 13.20).

CONCEPT CHECK   16.9

Sulfate- and sulfur-reducing bacteria are obligately anaerobic bacteria that use SO_4^{2-} or S^0, respectively, as electron acceptor. Homoacetogenic bacteria are also obligate anaerobes but use CO_2 instead of sulfur compounds as electron acceptor, producing acetate as a reduced product.

✔ *Compare and contrast sulfate-, sulfur-, and CO_2-reducing homoacetogens in terms of potential electron donors and acceptors and carbon source.*

✔ *What pathway is used for CO_2 incorporation in autotrophic sulfate-reducing bacteria?*

✔ *What is the major physiological difference between* Desulfovibrio *and* Desulfobacter? *Between* Desulfobacter *and* Desulfococcus?

16.10
Budding and Appendaged (Prosthecate) Bacteria

This large and rather heterogeneous group contains bacteria that form various kinds of cytoplasmic extrusions: *stalks, hyphae,* or *appendages* (Table 16.14). Extrusions of these kinds, which are smaller in diameter than the mature cell, contain cytoplasm, and are bounded by the cell wall, are called **prosthecae** (singular, **prostheca**) (Figure 16.39). Of considerable interest in this group of bacteria is that cell division often occurs as a result of unequal cell growth. In contrast to cell division in the typical bacterium, which occurs by *binary fission* and results in the formation of two equivalent cells (Figure 16.40), cell division in the stalked and budding bacteria involves the formation of a new daughter cell with the mother cell retaining its identity after the cell division

TABLE 16.14   Characteristics of stalked, appendaged (prosthecate), and budding bacteria[a]

Characteristics	Genus	DNA (mol % GC)
Stalked bacteria:		
Stalk an extension of the cytoplasm and involved in cell division	*Caulobacter*	62–67
Stalked, fusiform-shaped cells	*Prosthecobacter*	54–60
Stalked, but stalk an excretory product not containing cytoplasm:		
Stalk depositing iron, cell vibrioid	*Gallionella*	55
Laterally excreted gelatinous stalk not depositing iron	*Nevskia*	60
Pear-shaped or globular cells with long stalk	*Planctomyces*	50
Appendaged (prosthecate) bacteria:		
Single or double prosthecae	*Asticcacaulis*	55–61
Multiple prosthecae		
Short prosthecae, multiply by fission, some with gas vesicles	*Prosthecomicrobium*	64–70
Flat, star-shaped cells, some with gas vesicles	*Stella*	69–74
Long prosthecae, multiply by budding, some with gas vesicles	*Ancalomicrobium*	70–71
Phototrophic	*Prosthechochloris*	50–56
With gas vesicles	*Ancalochloris*	—
Budding bacteria:		
Phototrophic, produce hyphae	*Rhodomicrobium*	61–63
Phototrophic, budding without hyphae	*Rhodopseudomonas*	64–72
Chemoorganotrophic, pear-shaped, stalks lacking	*Pirella*	57
Chemoorganotrophic, rod-shaped cells	*Blastobacter*	59–66
Chemoorganotrophic, buds on tips of slender hyphae		
Single hyphae from parent cell	*Hyphomicrobium*	59–65
Multiple hyphae from parent cell	*Pedomicrobium*	62–67

[a]Phylogenetically, *Planctomyces* and *Pirella* form a distinct line of Bacteria; most other budding or stalked organisms are members of the gamma or alpha subdivisions of the purple Bacteria (Proteobacteria) (∞ Section 15.7).

process is completed (Figure 16.40). The genera from Table 16.14 that show this unequal cell division process are indicated in Figure 16.40.

The critical difference between these bacteria and conventional bacteria is not the formation of buds or stalks but the formation of a new cell wall from a *single point* (polar growth) rather than throughout the whole cell (intercalary growth). Several genera not normally considered to be budding bacteria show polar growth without differentiation of cell size (Figure 16.40). An important consequence of **polar growth** is that internal structures, such as membrane complexes, are not involved in the cell division process, thus permitting the formation of more complex internal structures than in cells undergoing intercalary growth. Several additional consequences of polar growth include the following. *Aging* of the mother cell occurs; because only a certain number of buds can be formed, cells are mortal (rather than immortal as in intercalary growth); cell division may be asymmetric; the daughter cell at division is immature and must form internal or budding structures before it can divide; organisms showing polar growth have a potential for morphogenetic evolution not possible in cells with intercalary growth. Thus, some of the most complex morphogenetic processes in the prokaryotes are found in the budding and stalked bacteria.

Evolution and ecology

Although the majority of budding and/or appendaged bacteria phylogenetically group with the purple Bacteria group, specifically the alpha purple Bacteria (see Table 16.1), the genera *Planctomyces* and *Pirella* do not. These two organisms are related to each other but are phylogenetically unrelated to any other group of Bacteria. *Planctomyces* and *Pirella* are also unusual because their cell walls consist primarily of protein. The walls of *Planctomyces* and *Pirella* contain large amounts of cysteine (cross-linked as cystine) and proline but lack peptidoglycan. As would be expected of organisms lacking peptidoglycan, *Planctomyces* and *Pirella* are totally resistant to the antibiotics penicillin, cephalosporin, and cycloserine, all drugs that affect peptidoglycan synthesis.

Most budding and/or appendaged bacteria are aquatic; in nature many live attached to surfaces, their stalks or appendages serving as attachment sites. Many of the prosthecate forms are free-floating, and it is thought that their appendages serve as absorptive organs, making possible more efficient growth in the nutritionally dilute aquatic environment. Calculations of surface-to-volume ratios (∞ Section 3.2) of appendaged bacteria suggest that a major function of bacterial appendages may be to *increase* the surface-to-volume ratio; such structures increase the cell's surface area and

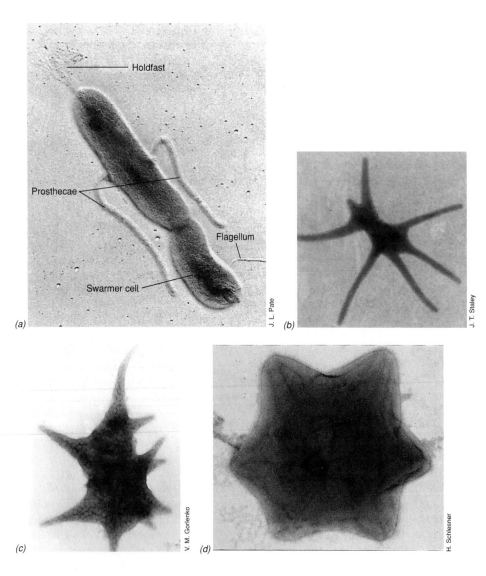

(a) Holdfast
Prosthecae
Flagellum
Swarmer cell
J. L. Pate
(b) J. T. Staley
(c) V. M. Gorlenko
(d) H. Schlesner

Figure 16.39 Prosthecate bacteria. (a) Electron micrograph of a shadow-cast preparation of *Asticcacaulis biprosthecum,* illustrating the location and arrangement of the prosthecae. Cells are about 0.6 μm wide. Note also the holdfast material and the swarmer cell in the process of differentiation. (b) Electron micrograph of a negatively stained preparation of a cell of the prosthecate bacterium *Ancalomicrobium adetum.* The appendages are cellular (prosthecae) because they are bounded by the cell wall and contain cytoplasm and are about 0.2 μm in diameter. (c) Electron micrograph of a whole cell of a prosthecate phototrophic green bacterium, *Ancalochloris perfilievii;* cells are about 0.7 μm in diameter. The structures within the cell are gas vesicles. (d) Electron micrograph of the star-shaped bacterium *Stella.* Cells are about 0.8 μm in diameter.

thus confer a competitive advantage for survival in dilute environments. Many of the free-floating forms have gas vesicles, presumably an adaptation to the planktonic existence.

Stalked bacteria

The stalked bacteria comprise a group of gram-negative, polarly flagellated rods that possess a **stalk,** a structure by which they attach to solid substrates. Most members of this group are classified in the genus *Caulobacter.* Stalked bacteria are frequently seen in aquatic environments attached to particulate matter, plant materials, or other microorganisms; generally they are found attached to microscope slides that have been immersed in lake or pond water for a few days. When many *Caulobacter* cells are present in the suspension, groups of stalked cells are seen attached, exhibiting the formation of *rosettes* (Figure 16.41*a*). Electron microscopic studies reveal that the stalk is not an excretion product but an outgrowth of the cell because it contains cytoplasm surrounded by cell wall

and cytoplasmic membrane (Figure 16.41*b* and *c*). The *holdfast* by which the stalk attaches the cell to a solid substrate is at the tip of the stalk, and, once attached, the cell usually remains permanently fixed. Because the stalk is cytoplasmic, it is also a prostheca. A stalk that is *not* a prostheca can be seen in an electron micrograph of the budding bacterium *Planctomyces* (Figure 16.42). In this organism the stalk contains no cytoplasm or cell wall and is probably proteinaceous in nature. The stalk functions to anchor the cell to surfaces via a holdfast located at the tip of the stalk.

Caulobacter

The *Caulobacter* cell division cycle (Figure 16.43) is of special interest because it involves a process of *unequal binary fission.* Cell division occurs by elongation of the cell followed by fission, a single flagellum forming at the pole opposite the stalk. The flagellated cell so formed, called a *swarmer,* separates from the nonflagellated mother cell, swims around, and settles down on a new surface, forming a new stalk at the flagellated

Products of cell division are equal:

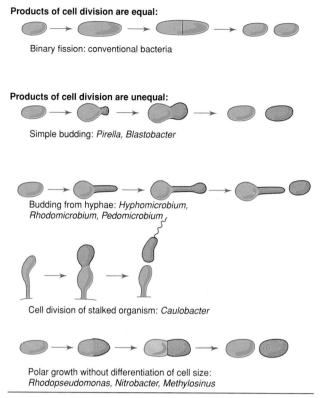

Binary fission: conventional bacteria

Products of cell division are unequal:

Simple budding: *Pirella, Blastobacter*

Budding from hyphae: *Hyphomicrobium,*
Rhodomicrobium, Pedomicrobium

Cell division of stalked organism: *Caulobacter*

Polar growth without differentiation of cell size:
Rhodopseudomonas, Nitrobacter, Methylosinus

Figure 16.40 Contrast between cell division in conventional bacteria and in budding and stalked bacteria.

pole; the flagellum then disappears (Figure 16.43). Stalk formation is a necessary precursor of cell division and is coordinated with DNA synthesis (Figure 16.43). The time span for the division of the stalked cell is thus shorter than the time span for division of the swarmer cell, owing to the requirement that a swarmer (flagellated) cell must synthesize a stalk before it divides (Figure 16.43). The cell division cycle in *Caulobacter* is thus more complex than simple binary fission because the stalked and swarmer cells have polar differentiation and the cells themselves are structurally different.

Caulobacters are chemoorganotrophic aerobes; they usually have one or several vitamin requirements but are able to grow on a variety of organic carbon compounds as sole sources of carbon and energy. Various amino acids are used as nitrogen sources. The enrichment culture for caulobacters makes use of the fact that they occur quite commonly in the organic film that develops at the surface of an undisturbed liquid. If pond, lake, or seawater is mixed with a small amount of organic material such as 0.01% peptone and incubated at 20–25°C for 2–3 days, a surface film consisting of bacteria, fungi, and protozoa develops, and in this microbial film caulobacters are common. A sample of the surface film is then streaked on an agar medium containing 0.05% peptone, and after 3–4 days the plates are examined under a dissecting microscope for the presence of microcolonies typical of *Caulobacter*. These colonies are then picked and streaked on fresh medium containing a higher concentration of organic matter (for example, 0.5% peptone + 0.1% yeast extract), and the resulting colonies are examined microscopically for stalked bacteria. The ability to grow in dilute media is a common property of organisms that attach to solid substrates; in the case of stalked organisms such as *Caulobacter* the stalk itself may also function as an absorptive organ, making it possible for the organism to acquire larger amounts of the restricted supply of organic nutrients than can organisms lacking these appendages.

A stalked organism sometimes classified with the caulobacters is *Gallionella*, which forms a twisted stalk containing ferric hydroxide (Figure 16.44). However, the stalk of *Gallionella* is not an integral part of the cell but is *excreted* from the cell surface. It contains an organic matrix on which the ferric hydroxide accumulates. *Gallionella* is frequently found in the waters draining bogs, iron springs, and other habitats where ferrous iron is present, usually in association with sheathed bacteria such as *Sphaerotilus*. In very acidic waters containing iron, *Gallionella* is not present, and acid-tolerant thiobacilli replace it.

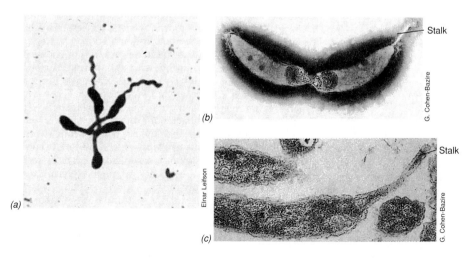

Figure 16.41 (a) A *Caulobacter* rosette. A single cell is about 0.5 μm wide. The five cells are attached by their stalks (prosthecae). Two of the cells have divided, and the daughter cells have formed flagella. (b, c) Electron micrographs of *Caulobacter* cells. (b) Negatively stained preparation of a cell in division. (c) A thin section. Notice that cytoplasmic constituents are present in the stalk region.

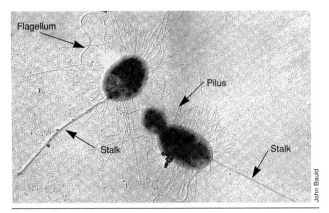

Figure 16.42 An electron micrograph of a metal-shadowed preparation of *Planctomyces maris*. A single cell is about 1–1.5 μm long. Note the fibrillar nature of the stalk. Pili are also abundant. Note also the flagella (curly appendages) on each cell and the bud that is developing from the nonstalked pole of one cell.

Budding bacteria

The two best-studied budding bacteria are *Hyphomicrobium*, which is chemoorganotrophic, and *Rhodomicrobium*, which is phototrophic; both organisms release buds from the ends of long, thin hyphae. The process of reproduction in a budding bacterium is illustrated in Figure 16.45. The mother cell, which is often attached by its base to a solid substrate, forms a thin outgrowth that lengthens to become a hypha, and at the end of the hypha a bud forms. This bud enlarges, forms a flagellum, breaks loose from the mother cell, and swims away. Later, the daughter cell loses its flagellum and after a period of maturation forms a hypha and buds. Further buds can also form at the hyphal tip of the mother cell. Many variations on this cycle are possible. In some cases the daughter cell does not break away from the mother cell but forms a hypha from its other pole. Complex arrays of cells connected by hyphae are frequently seen (Figure 16.46). In some cases a bud begins to form directly from the mother cell without the intervening formation of hypha, whereas in other cases a single cell forms hyphae from each end (Figure 16.46). The hypha is a direct cellular extension of the mother cell (Figure 16.47), containing cell wall, cytoplasmic membrane, ribosomes, and occasionally DNA.

Chromosomal replication events during the budding cycle are of interest (Figure 16.45). The DNA located in the mother cell replicates, and then once the bud has formed, a copy of the circular chromosome is

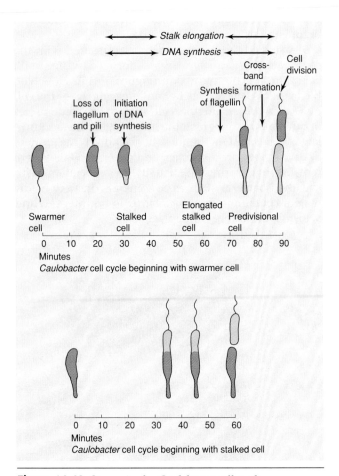

Figure 16.43 Stages in the *Caulobacter* cell cycle.

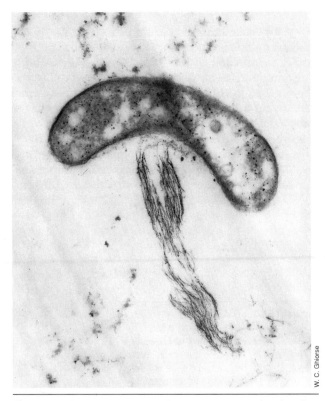

Figure 16.44 Transmission electron micrograph of a thin section of the iron bacterium *Gallionella ferruginea*. Cells are about 0.6 μm wide. Note the twisted stalk of ferric hydroxide emanating from the center of the cell.

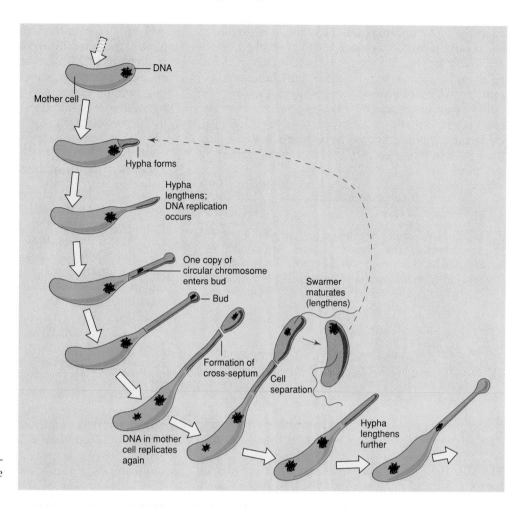

Figure 16.45 Stages in the *Hyphomicrobium* cell cycle.

moved down the length of the hypha and into the bud. A cross-septum then forms, separating the still developing bud from the hypha and mother cell.

Physiology and ecology

Hyphomicrobium is a methylotrophic bacterium (see Section 16.7). Preferred carbon sources are *one-carbon* compounds such as methanol, methylamine, formalde-

hyde, and formate. Growth on acetate, ethanol, or higher aliphatic compounds is usually slow, and growth is poor on sugars or most amino acids. Urea, amides, ammonia, nitrite, and nitrate can be utilized as nitrogen sources; no vitamins are required. *Hyphomicrobium* is widespread in freshwater, marine, and terrestrial habitats. Initial enrichment cultures can be prepared using a mineral–salts medium lacking organic carbon and nitrogen, to which a sample of natural material is added. After several weeks of incubation, the surface film that develops is streaked out on

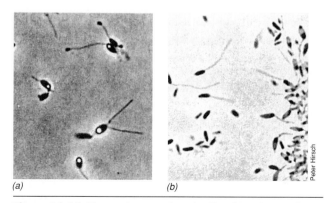

Figure 16.46 Photomicrographs of cells of *Hyphomicrobium*. Cells are about 0.7 μm wide. By phase contrast, showing typical fields. Notice the long hyphae and the occasional budding cell.

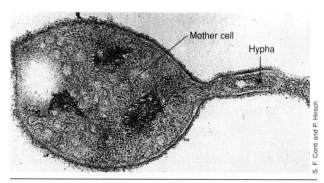

Figure 16.47 Electron micrograph of a thin section of a single *Hyphomicrobium* cell. The hypha is about 0.2 μm wide.

agar medium containing methylamine or methanol as a sole carbon source. Colonies are then checked microscopically for the characteristic *Hyphomicrobium* cellular morphology. A fairly specific enrichment procedure for *Hyphomicrobium* uses methanol as electron donor with nitrate as electron acceptor under anoxic conditions. Virtually the only denitrifying organism using methanol is *Hyphomicrobium,* and so this procedure selects this organism out of a wide variety of environments.

CONCEPT CHECK   16.10

Budding and appendaged bacteria divide by polar growth rather than binary fission. A variety of structurally distinct appendages are produced by these bacteria to facilitate attachment or nutrient absorption in dilute environments.

✔ *How does the stalk of* Planctomyces *differ from the stalk of* Caulobacter?

✔ *How does* budding *division differ from* binary *fission?*

16.11

Spirilla

The **spirilla** are gram-negative, motile, spiral-shaped bacteria with a wide variety of physiological attributes. Some of the key taxonomic criteria used are cell shape, size, kind of polar flagellation (single or multiple), relation to oxygen (obligately aerobic, micro-

aerophilic, facultative), relationship to plants (as symbionts or plant pathogens) or animals (as pathogens), fermentative ability, and certain other physiological characteristics (nitrogen-fixing ability, halophilic nature, thermophilic nature). Phylogenetically, spirilla constitute a line of purple Bacteria. The genera to be covered in the present discussion are given in Table 16.15.

Spirillum, Aquaspirillum, Oceanospirillum, *and* Azospirillum

These are helically curved rods, which are motile by means of polar flagella (usually tufts at both poles) (see Figure 16.48). The number of turns in the helix may vary from less than one complete turn (in which case the organism looks like a vibrio) (see Section 16.19) to many turns. Spirilla with many turns can superficially resemble spirochetes (see Section 16.12) but differ in that they do not have an outer sheath and axial filaments but instead contain typical bacterial flagella. Some spirilla are very large bacteria and were seen by early microscopists. It is likely that van Leeuwenhoek first described *Spirillum* in the 1670s, and the genus was first created by the protozoologist Ehrenberg in 1832. The organism seen by these workers is now called *Spirillum volutans* and is a rather large bacterium (Figure 16.48*a* and *b*). A phototrophic organism resembling *S. volutans* is *Thiospirillum* (see Figure 16.7*b*). *Spirillum volutans* is microaerobic, requiring O_2, but is inhibited by O_2 at normal levels. The simplest way of achieving microaerobic growth of *S. volutans* is to use a medium such as nutrient broth in a closed vessel with a large head (gas) space, with an atmosphere of N_2, and to then inject sufficient

---

TABLE 16.15   Characteristics of the genera of spiral-shaped bacteria[a]

Genus	16S rRNA group[b]	Characteristics	DNA (mol % GC)
Spirillum	Beta	Cell diameter 1.7 μm; microaerophilic; freshwater	36–38
Aquaspirillum	Alpha or beta	Cell diameter 0.2–1.5 μm; aerobic; freshwater	49–66
Oceanospirillum	Gamma	Cell diameter 0.3–1.2 μm; aerobic; marine (require 3% NaCl)	42–51
Azospirillum	Alpha	Cell diameter 1 μm; microaerophilic; soil and rhizosphere; fixes N_2	68–70
Herbaspirillum	—	Cell diameter 0.6–0.7 μm; microaerophilic; soil and rhizosphere; fixes N_2	66–67
Camplylobacter	Epsilon	Cell diameter 0.2–0.8 μm; microaerophilic to anaerobic; pathogenic or commensal in humans and animals; single polar flagellum	30–38
Helicobacter	Epsilon	Cell diameter 0.5–1 μm; tuft of polar flagella; associated with pyloric ulcers in humans	—
Bdellovibrio	Delta	Cell diameter 0.25–0.4 μm; aerobic; predatory on other bacteria; single polar sheathed flagellum	33–52
Spirosoma	Bacteroides–Flavobacterium group	Cell diameter 0.5 μm; curved rods forming rings; nonmotile; aerobic; sometimes gas-vesiculate	66–69

[a]All are gram-negative and respiratory but never fermentative.

[b]All genera except for *Spirosoma* are members of the purple Bacteria (Proteobacteria) (∞ Section 15.7 and Table 16.1).

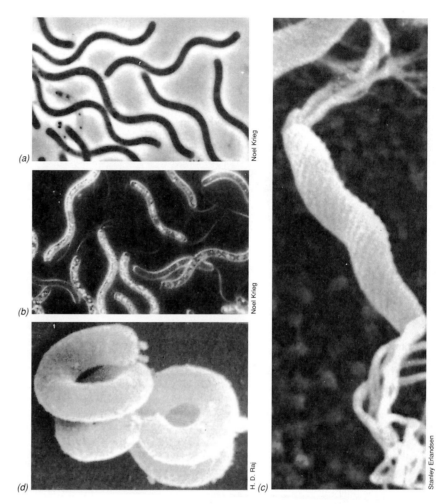

Figure 16.48 (a) Photomicrograph by phase contrast of *Spirillum volutans,* a large spirillum. Cells are about 1.6 by 20–50 μm. (b) *Spirillum volutans,* by dark-field microscopy, showing flagellar bundles and volutin (polyphosphate) granules. (c) Scanning electron micrograph of an intestinal spirillum. Note the polar flagellar tufts and the spiral structure of the cell surface. (d) Scanning electron micrograph of cells of *Spirosoma linguale.* Cells are about 0.5 μm in diameter.

O_2 from a hypodermic syringe to occupy 1–5% of the head space volume. Because *S. volutans* is dependent on O_2 for growth, high cell yields cannot be expected in this way because the O_2 is quickly used up. For cultivation for physiological studies, a continuous stream of 1% O_2 in N_2 can be passed through the culture vessel. Another characteristic of *S. volutans* is the formation of prominent granules (volutin granules) consisting of polyphosphate (see Figure 16.48*b* and Section 3.11).

Azospirillum lipoferum is a nitrogen-fixing organism, which was originally described and named *Spirillum lipoferum* by Beijerinck in 1922. It has become of considerable interest in recent years because this bacterium has been found to enter into a loose symbiotic relationship with tropical grasses and grain crops (∞ Section 14.23).

The genus *Spirillum* includes only a single species, *S. volutans,* characterized by its microaerophilic character, large size, and formation of volutin granules. The small-diameter spirilla (which are not microaerophilic) have been separated into two genera, *Aquaspirillum* and *Oceanospirillum,* the former for freshwater forms and the latter for those living in seawater and requiring

NaCl for growth (Table 16.15). At least 17 species of *Aquaspirillum* have been described and 9 species of *Oceanospirillum,* the various species being separated on physiological grounds. These organisms undoubtedly play an important role in the recycling of organic matter in aquatic environments.

Highly motile microaerophilic spirilla have been isolated from freshwater habitats. These organisms demonstrate a dramatic directed movement in a magnetic field referred to as **magnetotaxis.** In an artificial magnetic field magnetotactic spirilla quickly orient their long axis along the north–south magnetic moment of the field. Within the cells, chains of 5–40 magnetic particles consisting of Fe_3O_4 called **magnetosomes** (∞ Section 3.11) are present (Figure 16.49), and these function as internal magnets that orient the cells along a specific magnetic field. Magnetotactic bacteria can have one of two magnetic polarities depending on the orientation of magnetosomes within the cell. Cells in the Northern Hemisphere have the north-seeking pole of their magnetosomes forward with respect to their flagella and thus move in a northward direction. Cells in the Southern Hemisphere have the opposite polarity and move southward. Although the ecological

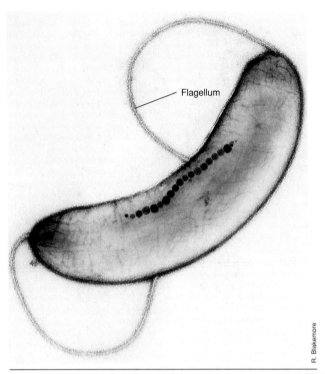

Flagellum

R. Blakemore

Figure 16.49 Negatively stained electron micrograph of a magnetotactic spirillum, *Aquaspirillum magnetotacticum*. A cell measures 0.3×2 µm. This bacterium contains particles of Fe_3O_4 (magnetite) called magnetosomes arranged in a chain; the particles align the cell along geomagnetic lines. The organism was isolated from a water treatment plant in Durham, New Hampshire.

role of bacterial magnets is unclear, it has been suggested that the ability to orient in a magnetic field is of selective advantage in directing these microaerophilic organisms downward toward less oxic zones near the sediments.

Bdellovibrio

These small vibroid organisms have the unusual property of preying on other bacteria, using as nutrients the cytoplasmic constituents of their hosts. These bacterial predators are small, highly motile cells that stick to the surfaces of their prey cells. Because of the latter property, they have been given the name *Bdellovibrio* (*bdello-* is a combining form meaning "leech"). Other predatory bacteria have been isolated and given such names as *Vampirococcus*. However, it appears that *Bdellovibrio* has a unique mode of attack and development intraperiplasmically. After attachment of a *Bdellovibrio* cell to its prey, the predator penetrates through the prey wall and replicates in the space between the prey wall and membrane (the periplasmic space), eventually forming a spherical structure called a **bdelloplast**. The stages of attachment and penetration are shown in electron micrographs in Figure 16.50 and diagrammatically in Figure 16.51. A wide variety of gram-negative Bacteria can be attacked by a single *Bdellovibrio* species; gram-positive cells are not attacked.

As originally isolated, *Bdellovibrio* cells grow only on living prey, but it is possible to isolate mutants that are prey-independent and are able to grow on complex organic media such as yeast extract–peptone. These strains, like the wild-type strain, are unable to utilize sugars as electron donors but are proteolytic and can oxidize the amino acids liberated by protein digestion. Prey-dependent revertants can be reisolated from the prey-independent mutants by introducing a host strain.

Bdellovibrio is an obligate aerobe, obtaining its energy from the oxidation of amino acids and acetate (via the citric acid cycle). It apparently is unable to utilize sugars as electron donors. In addition, *Bdellovibrio*

Figure 16.50 Stages of attachment and penetration of a prey cell by *Bdellovibrio*. A *Bdellovibrio* cell measures about 0.3 µm in diameter. (a) Electron micrograph of a shadowed whole-cell preparation showing *Bdellovibrio bacteriovorus* attacking *Pseudomonas*. (b, c) Electron micrographs of thin sections of *Bdellovibrio* attacking *Escherichia coli*; (b) early penetration; (c) complete penetration. The *Bdellovibrio* cell is enclosed in a membranous infolding of the prey cell (the bdelloplast) and replicates in the periplasmic space between the wall and the membrane.

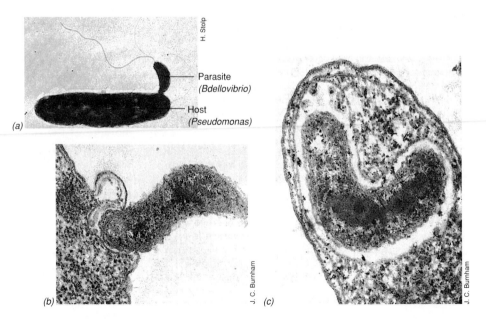

H. Stolp

Parasite (*Bdellovibrio*)

Host (*Pseudomonas*)

(a)

(b)   J. C. Burnham

(c)   J. C. Burnham

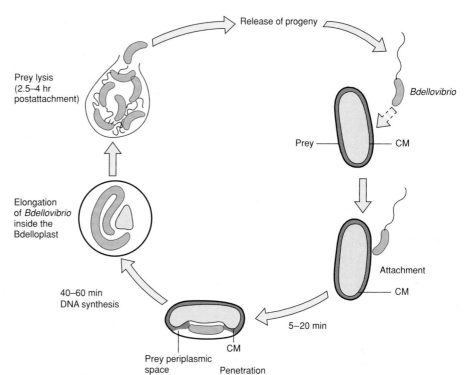

Release of progeny

Bdellovibrio

Prey — CM

Attachment

CM

5–20 min

CM

Prey periplasmic
space   Penetration

40–60 min
DNA synthesis

Elongation
of *Bdellovibrio*
inside the
Bdelloplast

Prey lysis
(2.5–4 hr
postattachment)

Figure 16.51 Developmental cycle of the bacterial predator *Bdellovibrio bacteriovorus*. Following primary contact between a highly motile *Bdellovibrio* cell and a gram-negative bacterium, attachment and penetration into the prey periplasmic space occurs. Once inside, *Bdellovibrio* elongates and within 4 hours progeny cells are released. The number of progeny cells released varies with the size of the prey bacterium. For example, 5–6 bdellovibrios are released from each infected *Escherichia coli* cell, and 20–30 for *Spirillum* sp. CM, Prey cytoplasmic membrane.

assimilates nucleoside phosphates, fatty acids, and even whole proteins in some cases directly from its host without first breaking them down. Thus it is clear that the predatory mode of existence has involved the development in *Bdellovibrio* of interesting and unusual biochemical processes.

Phylogenetically, bdellovibrios fall into the delta group of purple Bacteria (see Table 16.1). Taxonomically, three species of *Bdellovibrio* are recognized as shown in Table 16.16. Nucleic acid studies in bdellovibrios have shown them to be a rather heterogeneous group. Both GC base ratio and DNA:DNA hybridization analyses suggest that bdellovibrios are genetically diverse (Table 16.16) even though they share a unique lifestyle. In addition to being predators themselves, bdellovibrios,

like most bacteria, are subject to attack by bacteriophages. Nicknamed *bdellophages*, most phages that plaque on lawns of *Bdellovibrio* cells are strictly lytic, single-stranded DNA phages.

Members of the genus *Bdellovibrio* are widespread in soil and water, including the marine environment. Their detection and isolation require methods reminiscent of those used in the study of bacterial viruses (∞ Section 8.4). Prey bacteria are spread on the surface of an agar plate to form a lawn, and the surface is inoculated with a small amount of soil suspension that has been filtered through a membrane filter; the latter retains most bacteria but allows the small *Bdellovibrio* cells to pass. On incubation of the agar plate, plaques analogous to those produced by bacteriophages are formed at locations

TABLE 16.16   Genetic relationships among species of *Bdellovibrio*[a]

Species	DNA (mol % GC)	Genome size (kilobase pairs)	Genomic DNA hybridization to		
			B. bacteriovorus	**B. stolpii**	**B. starrii**
Bdellovibrio					
B. bacteriovorus	50	2000	100	28	26
B. stolpii	42	2270	28	100	16
B. starrii	44	2575	23	16	100
Escherichia (for comparison)					
E. coli	51	4700	—	—	—

[a]Data from Torrella, F., R. Guerrero, and R. J. Seidler. 1978. Further taxonomic characterization of the genus *Bdellovibrio*. *Can. J. Microbiol.* 24:1387–1394, and Ruby, E. G. 1992. The genus *Bdellovibrio*, pp. 3400–3415, in A. Balows, et al., *The Prokaryotes*, 2nd edition. Springer-Verlag, New York. See Section 15.9 for a discussion of DNA:DNA hybridization.

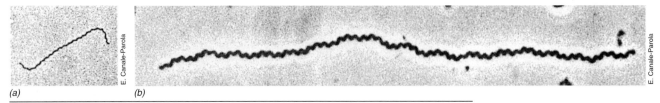

(a)   *(b)*

Figure 16.52 Two spirochetes at the same magnification, showing the wide size range in the group. (a) *Spirochaeta stenostrepta*, by phase contrast microscopy. A single cell is 0.25 μm in diameter. (b) *Spirochaeta plicatilis*. A single cell is 0.75 μm in diameter and can be up to 250 μm (0.25 mm) in length.

where *Bdellovibrio* cells are growing. However, unlike phage plaques, which continue to enlarge only as long as the bacterial host is growing, *Bdellovibrio* plaques continue to enlarge even after the prey has stopped growing, resulting in large plaques on the agar surface. Pure cultures of *Bdellovibrio* can then be isolated from these plaques. *Bdellovibrio* cultures have been obtained from a wide variety of soils and are thus common members of the soil population.

Spirosoma

Members of the genus *Spirosoma* are ring-shaped, non-motile, chemoorganotrophic bacteria (Figure 16.48d). They resemble very tightly curved vibrios and are widely distributed in aquatic environments. A phototrophic counterpart to *Spirosoma* exists in the genus *Rhodocyclus* (see Section 16.1 and Figure 16.6e).

16.12
Spirochetes

Spirochetes are bacteria with a unique morphology and mechanism of motility. They are widespread in aquatic environments and in the bodies of animals. Some of them cause diseases of animals and humans, of which the most important is *syphilis*, caused by *Treponema pallidum*. The spirochete cell is typically slen-

der, flexuous, helical (coiled) in shape, and often rather long (Figure 16.52). The "protoplasmic cylinder," consisting of the regions enclosed by the cytoplasmic membrane and the cell wall, constitutes the major portion of the spirochetal cell. Fibrils, referred to as **axial fibrils** or **axial filaments,** are attached to the cell poles and wrapped around the coiled protoplasmic cylinder (Figure 16.53). Both the axial fibrils and the protoplasmic cylinder are surrounded by a three-layered membrane called the *outer sheath* or *outer cell envelope* (Figure 16.53). The outer sheath and axial fibrils are usually not visible by light microscopy but are observable in negatively stained preparations or thin sections examined by electron microscopy (see, for example, Figure 16.54).

Motility of spirochetes

From 2 to more than 100 axial fibrils are present per cell, depending on the type of spirochete. The ultrastructure and the chemical composition of axial fibrils are similar to those of bacterial flagella (∞ Section 3.9). As in typical bacterial flagella (∞ Figure 3.49), basal hooks and paired discs are present at the insertion end. The shaft of each fibril is composed of a core surrounded by an "axial fibril sheath," and so the spirochete axial fibrils are in a sense analogous to sheathed flagella.

The axial fibrils play a significant role in spirochete motility. Each fibril is anchored at one end and

Figure 16.53 (a) Electron micrograph of a negatively stained preparation of *Spirochaeta zuelzerae*, showing the position of the axial fibril. A single cell is about 0.3 μm in diameter. (b) Cross section of a spirochete cell, showing the arrangement of the protoplasmic cylinder, axial fibrils, and external sheath and the manner in which the rotation of the rigid axial fibril can generate rotation of the protoplasmic cylinder and (in the opposite direction) rotation of the external sheath. If the sheath is free, the cell will rotate about its longitudinal axis and move along it. If the sheath is in contact with a solid surface, the cell will creep forward. See text for details.

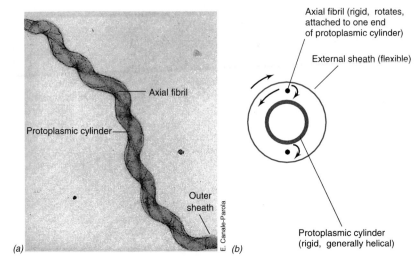

Axial fibril

Protoplasmic cylinder

Outer sheath

(a)

Axial fibril (rigid, rotates, attached to one end of protoplasmic cylinder)

External sheath (flexible)

Protoplasmic cylinder (rigid, generally helical)

(b)

extends for approximately two-thirds of the length of the cell. The axial fibrils rotate rigidly, as do bacterial flagella (∞ Section 3.9). Because the protoplasmic cylinder is also rigid, whereas the outer sheath is flexible when both axial fibrils rotate in the same direction, the protoplasmic cylinder rotates in the *opposite* direction, as illustrated in Figure 16.53*b*. If the sheath is not in contact with a surface, it also rotates (see arrow in Figure 16.53*b*). This simple mechanism is all that is needed to generate the wide variety of motions exhibited by spirochetes. When the protoplasmic cylinder is helical (as in most spirochetes), then forward motion is generated when the sheath is moving through a liquid or semisolid medium by the circumferential slip of the helix through the medium. If the sheath is in contact along its length with a solid surface, the protoplasmic cylinder may not be able to rotate, and so the roll of the sheath causes the cell to slide in a direction nearly parallel to the axis of the helix, generating a "creeping" motility. In free liquid, many spirochetes show flexing or lashing motions due to torque exerted at the ends of the protoplasmic cylinder by the twisting axial fibrils. Thus, despite superficial differences, spirochetes have fundamentally the same motility mechanism as other bacteria, namely, the rotation of rigid flagellar fibrils attached in the cell membrane via a basal hook.

Classification

Spirochetes are classified into six genera primarily on the basis of habitat, pathogenicity, and morphological and physiological characteristics. Table 16.17 lists the major genera and their characteristics. Molecular sequencing studies of 16S rRNA from various spirochetes show that spirochetes form a phylogenetic cluster. In addition, it has been found that most spirochetes are naturally resistant to the antibiotic rifampicin, an indication that their RNA polymerases may differ from those of other bacteria (∞ Section 6.6).

Spirochaeta *and* Cristispira

The genus *Spirochaeta* includes free-living, anaerobic, and facultatively aerobic spirochetes. These organisms are common in aquatic environments, such as the water and mud of rivers, ponds, lakes, and oceans. One species of the genus *Spirochaeta* is *S. plicatilis* (Figure 16.52*b*), a fairly large spirochete found in freshwater and marine H_2S-containing habitats and is

TABLE 16.17	**Genera of spirochetes and their characteristics**[a]						
Genus	Dimensions (μm)	Number of species recognized	General characteristics	Number of axial fibrils	DNA (mol % GC)	Habitat	Diseases
Cristispira	30–150 × 0.5–3.0	1	3–10 complete coils; bundle of axial fibrils visible by phase contrast microscopy	>100	—	Digestive tract of molluscs; has not been cultured	None known
Spirochaeta	5–250 × 0.2–0.75	7	Anaerobic or facultatively aerobic; tightly or loosely coiled	2–40	50–65	Aquatic, free-living, freshwater and marine	None known
Treponema	5–15 × 0.1–0.4	13	Microaerophilic or anaerobic; coil amplitude up to 0.5 μm	2–15	25–53	Commensal or parasitic in humans, other animals	Syphilis, yaws, swine dysentery, pinta
Borrelia	8–30 × 0.2–0.5	15	Anaerobic; 5–7 coils of approx. 1 μm amplitude	Unknown	46	Humans and other mammals, arthropods	Relapsing fever, Lyme disease, ovine and bovine borreliosis
Leptospira	6–20 × 0.1	2	Aerobic, tightly coiled, with bent or hooked ends; requires long-chain fatty acids	2	33–43	Free-living or parasitic in humans, other mammals	Leptospirosis
Leptonema	6–20 × 0.1	1	Aerobic; does not require long-chain fatty acids	—	54	Free-living	None known

[a]Phylogenetically, spirochetes form a distinct line of Bacteria (∞ Section 15.7).

probably anaerobic. The axial fibrils of *S. plicatilis* are arranged in a bundle that winds around the coiled protoplasmic cylinder. From 18 to 20 axial fibrils are inserted at each pole of this spirochete. Another species, *Spirochaeta stenostrepta*, has been cultured and is shown in Figure 16.52a. It is an obligate anaerobe commonly found in H_2S-rich, black muds. It ferments sugars via the glycolytic pathway to ethanol, acetate, lactate, CO_2, and H_2. The species *Spirochaeta aurantia* is an orange-pigmented facultative aerobe, fermenting sugars via the glycolytic pathway under anaerobic conditions and oxidizing sugars aerobically mainly to CO_2 and acetate.

The genus *Cristispira* (Figure 16.54) contains organisms with a unique distribution, being found in nature primarily in the *crystalline style* of certain molluscs, such as clams and oysters. The crystalline style is a flexible, semisolid rod seated in a sac and rotated against a hard surface of the digestive tract, thereby mixing with and grinding the small particles of food. Being large spirochetes, the cristispiras can readily be seen microscopically within the style as they rapidly rotate forward and backward in corkscrew fashion. *Cristispira* may occur in both freshwater and marine molluscs, but not all species of molluscs possess them. Unfortunately, *Cristispira* has not been cultured, and so the physiological reason for its restriction to this unique habitat is not known. There is no evidence that *Cristispira* is harmful to its host; in fact, the organism may be more common in healthy than in diseased molluscs.

Treponema

Anaerobic, host-associated spirochetes that are commensals or parasites of humans and animals are placed in the genus *Treponema*. *Treponema pallidum*, the causal agent of syphilis (∞ Section 23.6), is the best-known species of *Treponema*. It differs in morphology from other spirochetes; the cell is not helical but has a flat wave form. Furthermore, electron microscopy does not show the presence of an outer sheath surrounding both the axial fibrils and the protoplasmic cylinder. Apparently the axial fibrils of *T. pallidum* lie on the outside of the organism. The *T. pallidum* cell is remarkably thin, measuring approximately 0.2 μm in diameter. Living cells are clearly visible in the dark-field microscope or after staining with

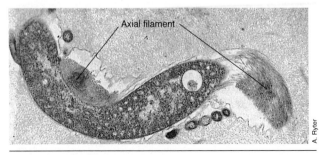

Figure 16.54 Electron micrograph of a thin section of *Cristispira*, a very large spirochete. A cell measures about 2 μm in diameter. Notice the numerous fibrils in the axial filament.

fluorescent antibody; dark-field microscopy has long been used to examine exudates from suspected syphilitic lesions (∞ Figure 23.25). In nature *T. pallidum* is restricted to humans, although artificial infections have been established in rabbits and monkeys. Although never grown in laboratory culture, it has been established from animal studies that virulent *T. pallidum* cells (purified from infected rabbits) contain a cytochrome system and are in fact microaerophiles. This property further separates *T. pallidum* from the other species of *Treponema*, which, as noted, are obligate anaerobes. Meaningful taxonomic studies on this species will have to await cultivation away from its host.

Treponema pallidum is quite sensitive to increased temperature, being rapidly killed by exposure to 41.5–42°C. The heat sensitivity of *T. pallidum* is also reflected in the fact that the organism becomes most easily established in cooler sites of the body, such as the male genital organs, although once established in other areas of the body it multiplies there. The organism is rapidly killed by drying, and this at least partially explains why transmission of *T. pallidum* between persons is only by direct contact, usually sexual intercourse.

Other species of the genus *Treponema* are common commensal organisms in the oral cavity of humans and can generally be seen in material scraped from between the teeth and from the narrow space between the gums and the teeth. Various oral spirochetes can be cultivated anaerobically in complex media containing serum. Three species, *Treponema denticola*, *T. macrodentium*, and *T. oralis*, have been described, differing in morphology and physiological characteristics. *Treponema denticola* ferments amino acids such as cysteine and serine, forming acetate as the major fermentation acid as well as CO_2, NH_3, and H_2S. This spirochete can also ferment glucose, but in media containing both glucose and amino acids, the amino acids are used preferentially. The true relationship between *T. pallidum* and the remaining members of the genus *Treponema* may be a distant one because the GC base ratio of *T. pallidum* is about 53% whereas other species of this genus cluster between 38 and 40% or 25 and 26%.

Spirochetes are also found in the rumen. *Treponema saccharophilum* (Figure 16.55) is a large, *pectinolytic* spirochete found in the bovine rumen. *Treponema saccharophilum* is an obligately anaerobic bacterium that ferments pectin, starch, inulin, and other plant polysaccharides. This and other spirochetes may play an important role in the conversion of plant polysaccharides to volatile fatty acids, usable as energy sources by the ruminant (∞ Section 14.13).

Leptospira *and* Leptonema

The genera *Leptospira* and *Leptonema* contain strictly *aerobic* spirochetes that use long-chain fatty acids (for example, oleic acid) as electron donor and carbon

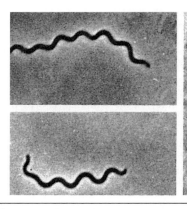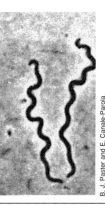

B. J. Paster and E. Canale-Parola

Figure 16.55 Phase contrast photomicrographs of *Treponema saccharophilum,* a large pectinolytic spirochete from the bovine rumen. A cell measures about 0.4 μm in diameter. Left, Regularly coiled cells; right, irregularly coiled cells.

sources. With few exceptions, these are the only substrates utilized by leptospiras for growth. The leptospira cell is thin, finely coiled, and usually bent at each end into a semicircular hook. At present, several species are recognized in this group, some free-living and some parasitic. Two major species are *Leptospira interrogans* (parasitic) and *L. biflexa* (free-living). Strains of *L. interrogans* are parasitic for humans and animals. Different strains are distinguished serologically by agglutination tests, and a large number of serotypes have been recognized. Rodents are the natural hosts of most leptospiras, although dogs and pigs are also important carriers of certain strains. In humans the most common leptospiral syndrome is *Weil's disease;* in this disorder the organism usually localizes in the kidney. Leptospiras ordinarily enter the body through the mucous membranes or through breaks in the skin. After a transient multiplication in various parts of the body the organism localizes in the kidney and liver, causing nephritis and jaundice. The organism passes out of the body in the urine and infection of another individual is most commonly by contact with infected urine. Therapy with penicillin, streptomycin, or the tetracyclines is possible but may require extended courses to eliminate the organism from the kidney; this is probably because of the slow growth and protected location of the leptospiras. Domestic animals are vaccinated against **leptospirosis** with a killed virulent strain; dogs are usually immunized routinely with a combined distemper–leptospira– hepatitis vaccine. In humans, prevention is effected primarily by elimination of the disease from animals. The serotype that infects dogs, *L. interrogans* serotype *canicola,* does not ordinarily infect humans, but the strain attacking rodents, *L. interrogans* serotype *icterohaemorrhagiae,* does; hence elimination of rats from human habitation is of considerable aid in preventing the organism from reaching human beings.

Borrelia

The majority of species in the genus *Borrelia* are animal or human pathogens. *Borrelia recurrentis* is the causative agent of **relapsing fever** in humans and is transmitted via an insect vector, usually by the human body louse. Relapsing fever is characterized by a high fever and generalized muscular pain, which lasts for 3–7 days followed by a recovery period of 7–9 days. Left untreated, the fever returns in two to three more cycles (hence the name relapsing fever) and causes death in up to 40% of those infected. Fortunately, the organism is quite sensitive to tetracycline, and if the disease is correctly diagnosed, treatment is straightforward. Other borrelia are of veterinary importance causing diseases in cattle, sheep, horses, and birds. In most of these diseases the organism is transmitted by ticks. *Borrelia burgdorferi* is the causative agent of the tick-borne disease called *Lyme disease,* which infects humans and other animals. Lyme disease is discussed in Section 23.10. *Borrelia burgdorferi* is also of interest because it is as yet one of the only known prokaryotes with a *linear* (as opposed to a *circular*) chromosome (∞ Section 6.3).

CONCEPT CHECK 16.12

Spirilla and spirochetes are both spiral-shaped bacteria, but spirochetes are much more tightly coiled and are motile by the action of special types of flagella called axial fibrils. Spirilla and spirochetes are phylogenetically distinct Bacteria.

✔ *Compare and contrast spirilla with spirochetes as to morphology, phylogenetic position, habitats, and pathogenicity.*

✔ *Among spirilla, what is unique about the genus* Bdellovibrio?

16.13

Gliding Bacteria

A variety of bacteria exhibit gliding motility. These organisms have no flagella but are able to move when in contact with surfaces. All **gliding bacteria** are gramnegative. One group of gliding bacteria, the fruiting myxobacteria, possesses the interesting property of forming multicellular structures of complex morphology called **fruiting bodies.** Table 16.18 gives a brief outline of some of the genera of gliding bacteria.

The mechanism of gliding by gliding bacteria appears complex, and it is likely that more than one mechanism is responsible. Gliding is most apparent when cells are on a solid surface. Some gliding bacteria rotate along their long axis while moving, whereas others seem to keep only one side of the cell in contact with the surface while moving. Gliding motility is

TABLE 16.18   Characteristics of some genera of chemoorganotrophic gliding bacteria

Characteristics	Genus	16S rRNA group[a]	DNA (mol % GC)
Rods and nonseptate filaments:			
Unicellular, rod-shaped; many digest cellulose, chitin, or agar	*Cytophaga* (no microcysts)	*Bacteroides–Flavobacterium*	30–40
	Sporocytophaga (microcysts formed)	*Bacteroides–Flavobacterium*	36
Inhabit human oral cavity; facultative aerobes	*Capnocytophaga*	*Bacteroides–Flavobacterium*	33–41
Helical or spiral-shaped	*Saprospira*	*Bacteroides–Flavobacterium*	35–48
Filamentous	*Microscilla, Flexibacter*		37–47
Can lyse a variety of microorganisms	*Lysobacter*	Gamma purple Bacteria	65–70
Septate filaments:			
Filamentous, chemoorganotrophic or chemolithotrophic, producing S^0 granules from H_2S	*Beggiatoa*	Gamma purple Bacteria	37–51
Filamentous, chemoorganotrophic; life cycle involving gonidia and rosette formation	*Leucothrix*	Gamma purple Bacteria	46–51
Filamentous, chemolithotrophic sulfur oxidizer; life cycle like *Leucothrix*	*Thiothrix*	Gamma purple Bacteria	52
Cells in chains or short filaments; or chemoorganotrophic; occurs in oral cavity or digestive tract of humans and other animals	*Simonsiella, Alysiella*	Beta purple Bacteria	41–55
Filamentous	*Vitreoscilla*	Beta purple Bacteria	44–45
Rods, forming fruiting bodies:			
Unicellular, rod-shaped; life cycle involving aggregation, fruiting-body formation, and myxospore formation	Fruiting myxobacteria; *Archangium, Chondromyces, Myxococcus, Polyangium,* and so on (see Table 16.19)	Delta purple Bacteria	67–71

[a] ∞ Section 15.7 and Table 16.1.

generally much slower than flagellar motility, but absolute rates of gliding are somewhat dependent on cell length; filamentous gliding bacteria usually move much faster than unicellular gliding bacteria. In *Cytophaga*, there is evidence that small, rotating particles, presumably made of protein, lie between the cytoplasmic membrane and the gram-negative outer membrane. Acting like miniature ball bearings, these particles rotate, presumably at the expense of a membrane potential or perhaps ATP directly, and this rotary motion slides the cell along the solid surface. By contrast, in *Myxococcus*, gliding seems to occur as a result of the secretion of a chemical surfactant that affects surface tension forces between the cell and the solid surface and is apparently sufficient to slide the cell along the surface.

Cytophaga *and related genera*

Organisms of the genus *Cytophaga* are long, slender rods, often with pointed ends, that move by gliding. Many digest cellulose, agar, or chitin. They are widespread in the soil and water, often being present in great abundance. The cellulose decomposers can be easily isolated by placing small crumbs of soil on pieces of cellulose filter paper laid on the surface of mineral agar. The bacteria attach to and digest the cellulose fibers, forming transparent spreading colonies that are usually yellow or orange in color. Microscopic examination reveals the bacteria aligned on the surface of the cellulose fibrils. The cytophagas do not produce soluble, extracellular, cellulose-digesting enzymes (cellulases); the enzymes probably remain attached to the cell envelope, accounting for the fact that the cells must adhere to cellulose fibrils in order to digest them.

Organisms of the genus *Sporocytophaga* are similar to *Cytophaga* in morphology and physiology but form resting spherical structures called *microcysts,* similar to those produced by some fruiting myxobacteria (see below), although they are produced without formation of fruiting bodies. Despite its ability to form microcysts, *Sporocytophaga* is not related to the fruiting myxobacteria; its DNA base composition is similar to that of *Cytophaga* but far removed from those of the fruiting myxobacteria (Table 16.18). In pure culture, *Cytophaga* can be cultured on agar containing embedded cellulose fibers, the presence of the organism being indicated by the clearing that occurs as the cellulose is digested (∞ Figure 13.42). *Lysobacter,* a gliding rod-shaped bacterium, has the ability to lyse both bacteria and fungi through the action of an array of proteases, chitinases, and other lytic enzymes excreted into the medium.

Beggiatoa

Organisms of this genus are morphologically similar to filamentous cyanobacteria (see Section 16.2). The filaments of *Beggiatoa* are usually quite large in diameter and long, consisting of many short cells attached end to end (Figure 16.56*a*). In addition to moving by gliding, they can flex and twist so that many filaments may become intertwined to form a complex tuft. *Beggiatoa* is found in nature primarily in habitats rich in H_2S, such as sulfur springs, decaying seaweed beds, mud layers of lakes, and waters polluted with sewage, and in these habitats the filaments of *Beggiatoa* are usually filled with sulfur granules (Figures 16.57 and 16.58). *Beggiatoa* are also common inhabitants of hydrothermal vents (∞ Section 14.10). It was with *Beggiatoa* that Winogradsky first demonstrated that a living organism could oxidize H_2S to S^0 and then to SO_4^{2-}, leading him to formulate the concept of chemolithotrophy (∞ Winogradsky's

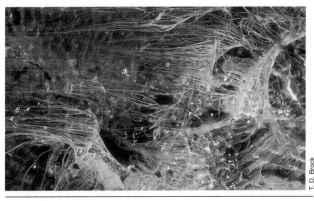

Figure 16.57 Filamentous sulfur-oxidizing bacteria in a small stream. The filamentous cells twist together to form thick streamers, and the white color is due to the abundant elemental sulfur content of the cells.

Legacy, Chapter 13). However, most pure cultures of *Beggiatoa* so far isolated grow best *chemoorganotrophically* on compounds such as acetate, succinate, and glucose, and when H_2S is provided as an electron donor, they still require organic substances for growth. On the other hand, some marine strains of *Beggiatoa* have been isolated and shown to be true chemolithotrophic autotrophs, but this does not seem to be common.

An interesting habitat of *Beggiatoa* is the rhizosphere of plants (rice, cattails, and other swamp plants) living in flooded, and hence anoxic, soils. Such plants pump oxygen down into their roots so a sharply defined boundary develops at the root surface between O_2 on the root and H_2S in the soil. *Beggiatoa* (and probably other sulfur bacteria) develops at this boundary, and it has been suggested that *Beggiatoa* plays a beneficial role for the plant by oxidizing and thus detoxifying hydrogen sulfide. The growth of *Beggiatoa* is greatly stimulated by the addition to culture media of the enzyme *catalase* (which converts hydrogen peroxide into water and oxygen), and because plant roots contain catalase, it has been suggested that

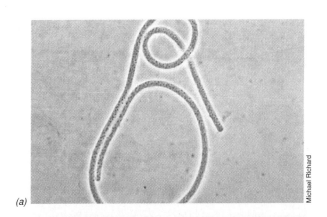

(a)

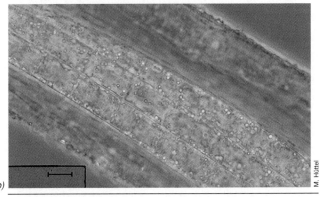

(b)

Figure 16.56 *Beggiatoa* and *Thioploca*. (a) Phase contrast photomicrograph of a *Beggiatoa* species. A cell is about 2 μm in diameter. (b) Cells of a marine *Thioploca* species. Cells contain sulfur granules and are about 40–50 μm wide.

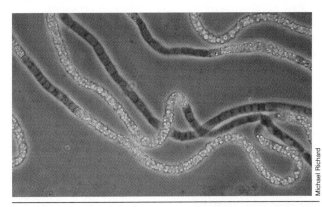

Figure 16.58 Phase contrast photomicrograph of a filamentous sulfur-oxidizing bacterium isolated from a sewage treatment plant. Note the abundant elemental sulfur granules in some of the cells.

the plant promotes the growth of *Beggiatoa* in its rhizo-sphere via catalase production, thus leading to the development of a loose mutualistic relationship between the plant and the bacterium.

Beggiatoa and related filamentous bacteria (Figure 16.58) can cause major settling problems in sewage treatment facilities and in industrial waste lagoons such as from canning, paper pulping, brewing, and milling. These problems are generally referred to as *bulking* and occur when filamentous bacteria overgrow the normal flora of the waste system, producing a loose detrital floc instead of the normal and more easily settling tight floc. If bulking occurs, the wastewater remains improperly treated because the effluent discharged is still high in biological oxygen demand (BOD); in sewage treatment, for example, bulking occurs when *Beggiatoa* or other fil-amentous bacteria replace *Zoogloea* in the activated sludge process (∞ Section 12.15).

Another gliding sulfur-oxidizing bacterium is *Thioploca*. *Thioploca* is a very large, filamentous sulfur-oxidizing chemolithotroph that forms cell bundles sur-rounded by a common sheath (Figure 16.56*b*). Thick mats of a marine *Thioploca* species have been found on the ocean floor off the coast of Chile and Peru. Studies on the ecology of these organisms have shown that they carry out the anoxic oxidation of H_2S coupled to the reduction of nitrate (NO_3^-) presumably to N_2 (de-nitrification) (∞ Section 14.14). Interestingly, it has been shown that cells of *Thioploca* can accumulate huge amounts of nitrate intracellularly and that this nitrate can then support extended periods of anaerobic respiration with H_2S as electron donor. It is postulated

that these marine *Thioploca* mats fix substantial amounts of CO_2 and also play a major role in sulfur and nitrogen cycling. Similar mats consisting primar-ily of *Beggiatoa* are found near hydrothermal vents (∞ Section 14.10), but the connection with nitrate res-piration in these cases is not as well established.

Leucothrix *and* Thiothrix

These two genera are related in cell structure and life cycle. *Thiothrix*, a chemolithotroph that oxidizes H_2S, grows only as a mixotroph (requiring both H_2S and an organic compound). *Leucothrix* is the chemoorgan-otrophic counterpart of *Thiothrix*, and because its mem-bers have been more amenable to cultivation, the details of its life cycle and physiology are fairly well established. *Leucothrix* is a filamentous organism that has been found in nature only in marine environments, where it grows most commonly as an epiphyte on marine algae. *Leucothrix* filaments are 2–5 μm in diam-eter and may reach lengths of 0.1–0.5 cm. The filaments have clearly visible cross-walls, and cell division is not restricted to either end but occurs throughout the length of the filament. The free filaments never glide (thus distinguishing them from *Beggiatoa*), although they occasionally wave back and forth in a jerky fash-ion. Under environmental conditions unfavorable to rapid growth, individual cells of the filaments become round and form ovoid structures called *gonidia,* which are released individually, often from the tips of the fila-ments (Figure 16.59*a*). The gonidia are able to glide in a jerky manner when they come into contact with a solid

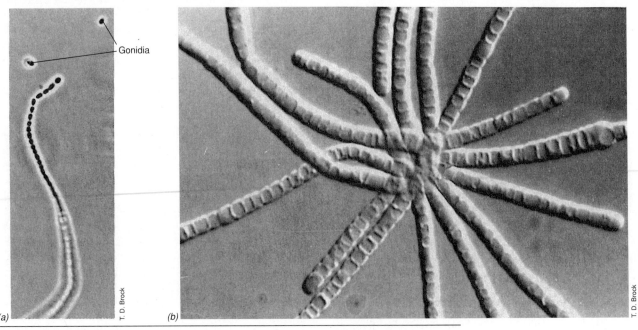

Gonidia

(a)   T. D. Brock   *(b)*   T. D. Brock

Figure 16.59 *Leucothrix mucor.* A cell is about 2 μm in diameter. (a) Filaments, showing multicellular nature and release of gonidia. (b) Rosette composed of several multicellular filaments. Nomarski interference contrast microscopy.

surface. They settle down on solid surfaces, synthesize a holdfast, and through growth and successive cell divisions form new filaments. Presumably in nature the gonidia are elements of dispersal, enabling the organism to spread to other areas. If there are high concentrations of gonidia, individual cells may aggregate, probably because of mutual attraction; they then synthesize a holdfast that causes their ends to adhere in a rosette, and new filaments grow out (Figure 16.59b). Rosette formation is found in both *Leucothrix* and *Thiothrix* and is an important means of distinguishing these organisms from many other filamentous bacteria.

Fruiting myxobacteria

The fruiting myxobacteria exhibit the most complex behavioral patterns and life cycles of all known prokaryotic organisms. In keeping with this complexity, the chromosome size of some myxobacteria can be rather large. *Myxococcus xanthus,* for example, has a single circular chromosome of 9500 kilobase pairs, *twice* as large as that of *Escherichia coli.* Indeed, this is two-thirds the size of the entire yeast genome, which is contained on 16 chromosomes (∞ Section 9.12). The vegetative cells of the fruiting myxobacteria are simple, nonflagellated, gram-negative rods that glide across surfaces and obtain their nutrients primarily by causing the lysis of other bacteria. Under appropriate conditions, a swarm of vegetative cells aggregate and construct *fruiting bodies*, within which some of the cells become converted to resting structures called **myxospores.** (A myxospore is a resting cell contained in a fruiting body.) It is the ability to form complex fruiting bodies that distinguishes the myxobacteria from all other prokaryotes. Because the vegetative cells of fruiting myxobacteria look like those of nonfruiting gliding bacteria, it is only through observation of the fruiting bodies that these organisms can be identified (see Table 16.19).

The fruiting bodies of the myxobacteria vary from simple globular masses of myxospores in loose slime to complex forms with a fruiting-body wall and a stalk (Figure 16.60). The fruiting bodies are often strikingly colored (Table 16.19). Occasionally they can be seen with a hand lens or dissecting microscope on pieces of decaying wood or plant material. Fruiting bodies of myxobacteria often develop on dung pellets (for example, those of the sheep or rabbit) after they have been incubated for a few days in a moist chamber. Although the vegetative cells are common in soils, the fruiting bodies themselves are less common. An effective means of isolating fruiting myxobacteria is to prepare Petri plates of water agar (1.5% agar in distilled water with no added nutrients) on which is spread a heavy suspension of any of several bacteria that the myxobacteria can lyse and use as a source of nutrients (for example, *Micrococcus luteus* or *Escherichia coli*). In the center of the plate a small amount of soil, decaying bark, or other natural material is placed.

Myxobacteria in the inoculum lyse the bacterial cells and use their liberated products as nutrients; as they grow, they swarm out across the plate from the inoculum site. After several days to a week, the plates are examined under a dissecting microscope for myxobacterial swarms or fruiting bodies, and pure cultures are obtained by transfer of cells from the fruiting bodies or from the edge of the swarm to organic media.

Life cycle of a fruiting myxobacterium

The life cycle of a typical fruiting myxobacterium is shown in Figure 16.61. The vegetative cells are typical gram-negative rods and do not reveal in their fine structure any clue as to their gliding motility or to their ability to aggregate and form fruiting structures. The vegetative cells of many strains grow poorly or not at all when first dispersed in liquid medium but can often be adapted to growth in liquid by making several passages through shaken liquid growth medium. A vegetative cell usually excretes slime, and as it moves across a solid surface it leaves a slime trail behind (Figure 16.62a, page 696). This trail is preferentially used by other cells in the swarm so that often a characteristic radiating pattern is soon created, with cells migrating along slime trails (Figure 16.62b). The fruiting body ultimately formed (Figure 16.62c) is a complex structure formed by the differentiation of cells in the stalk region and in the myxospore-bearing head. A wide variety of gram-positive and gram-negative Bacteria, as well as fungi, yeasts, and algae, can be used as food sources. A few fruiting myxobacteria can also use cellulose. Many myxobacteria can be grown in the laboratory on media containing peptone or casein hydrolysate, which provides organic nutrients in the form of amino acids or small peptides. The organisms are typical aerobes with a complete citric acid cycle and cytochrome system.

Fruiting-body formation does not occur so long as adequate nutrients for vegetative growth are present, but on exhaustion of amino acids, the vegetative swarms begin to fruit. Cells aggregate, possibly through a chemotactic response, with the cells migrating toward each other and forming mounds or heaps (Figure 16.63a, page 696). A single fruiting body may have 10^9 or more cells. As the cell mounds become higher, the differentiation of the fruiting body into stalk and head begins (Figure 16.63b and c). Figure 16.63d clearly illustrates the differentiation of the fruiting body into stalk and head. The stalk is composed of slime, within which a few cells may be trapped. The majority of the cells accumulate in the fruiting-body head and undergo differentiation into *myxospores* (Figure 16.64, page 697, and Table 16.19). And, in some genera, the myxospores are enclosed in large walled structures called **cysts.** Compared to the vegetative cell, the myxospore is more resistant to drying, sonic vibra-

Typical fruiting bodies of selected myxobacteria

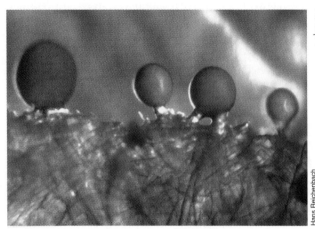

Myxococcus fulvus, about 125 μm high

Myxococcus stipitatus, about 170 μm high

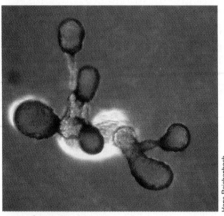

Mellitangium erectum, about 50 μm high

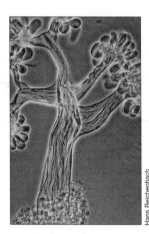

Chondromyces crocatus, about 560 μm high

Stigmatella aurantiaca, about 150 μm high

TABLE 16.19	Classification of the fruiting myxobacteria[a]		
Characteristics		**Genus**	**DNA (mol % GC)**
Vegetative cells tapered			
Spherical or oval myxospores, fruiting bodies usually soft and slimy without well-defined sporangia or stalks		*Myxococcus*	68–71
Tough, cartilaginous, ridged fruiting bodies		*Corallococcus*	—
Rod-shaped myxospores:			
Myxospores not contained in sporangia, fruiting bodies without stalks		*Archangium*	67–68
Myxospores embedded in slime envelope:			
Fruiting bodies without stalks		*Cystobacter*	68
Stalked fruiting bodies, single sporangia		*Melittangium*	—
Stalked fruiting bodies, multiple sporangia		*Stigmatella*	68–69
Fruiting bodies are dark-brown clusters consisting of tiny spherical or disclike sporangia with an outer wall		*Angiococcus*	—
Vegetative cells not tapered (blunt, rounded ends); myxospores resemble vegetative cells; sporangia always produced:			
Fruiting bodies without stalks; myxospores rod-shaped		*Polyangium*	69
Fruiting bodies without stalks; myxospores oval; highly cellulolytic		*Sorangium*	—
Fruiting bodies without stalks; myxospores coccoid		*Nannocystis*	70–72
Large, solitary yellow fruiting bodies with netlike surface		*Haploangium*	—
Stalked fruiting bodies		*Chondromyces*	69–70

[a]Phylogenetically, those species examined fall into the delta subdivision of the purple Bacteria (Proteobacteria, see Table 16.1).

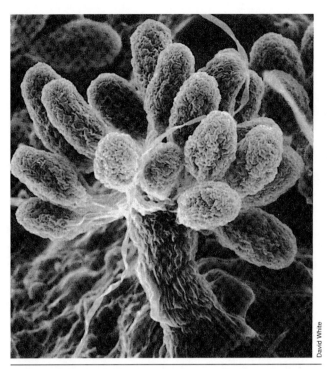

Figure 16.60 Scanning electron micrograph of a fruiting body of the gliding myxobacterium *Stigmatella aurantiaca* growing on a piece of wood. Note the individual cells visible in each fruiting structure.

tion, UV radiation, and heat, but the degree of heat resistance is much less than that of the bacterial endospore. It seems likely that the main function of encysted myxospores is to enable the organism to survive desiccation during dispersal or during drying of the habitat. The myxospore eventually germinates by a localized rupture of the capsule, with the growth and emergence of a typical vegetative rod.

Myxobacteria are usually colored by carotenoid pigments (see Table 16.19), and the main pigments are carotenoid glycosides. Pigment formation is promoted by light, and at least one function of the pigment is photoprotection. Since in nature the myxobacteria usually form fruiting bodies in the light, the presence of these photoprotective pigments is understandable. In the genus *Stigmatella*, light greatly stimulates fruiting-body formation, and it is thought that light catalyzes production of a pheromone that initiates the aggregation step. The fruiting myxobacteria are classified primarily on morphological grounds using characteristics of the vegetative cells, the myxospores, and fruiting-body structure (Table 16.19). Phylogenetically, gliding myxobacteria belong to the purple Bacteria group (∞ Section 15.7).

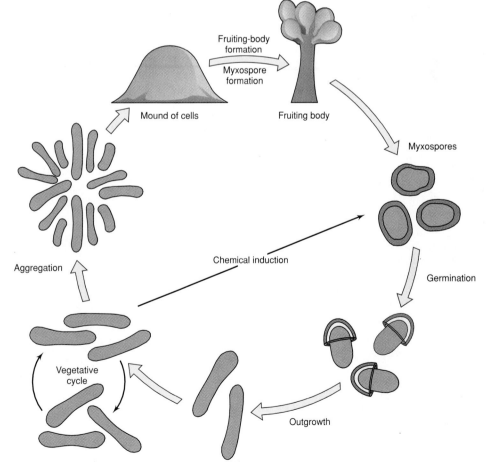

Figure 16.61 Life cycle of *Myxococcus xanthus*. Aggregation serves to assemble vegetative cells for fruiting-body formation. Vegetative cells undergo morphogenesis to resting cells called myxospores. The latter germinate under favorable nutritional and physical conditions to yield vegetative cells. Vegetative cells can be converted directly to myxospores without fruiting-body formation by certain chemical inducers, notably high concentrations of glycerol. See photograph of *Myxococcus* fruiting bodies in Table 16.19.

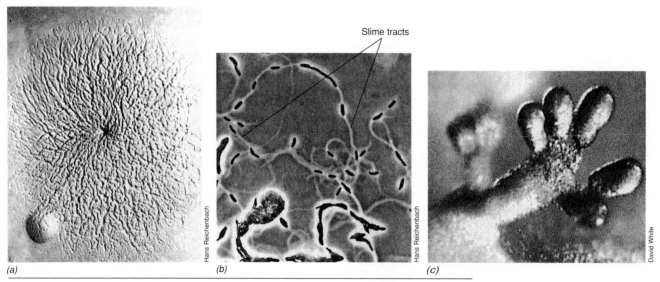

Figure 16.62 (a) Photomicrograph of a swarming colony (9-mm diameter) of *Myxococcus xanthus* on agar. (b) Single cells of *Myxococcus fulvus* from an actively gliding culture, showing the characteristic slime tracks on the agar. (c) Scanning electron micrograph of a fruiting body of *Stigmatella aurantiaca*.

16.14

Sheathed Bacteria

Sheathed bacteria are filamentous organisms with a unique life cycle involving formation of flagellated swarmer cells within a long tube or sheath. Under certain (generally unfavorable) conditions, the swarmer cells move out and become dispersed to new environments, leaving behind the empty sheath. Under favorable conditions, vegetative growth occurs within the filament, leading to the formation of long, cell-packed sheaths. Sheathed bacteria are common in freshwater habitats that are rich in organic matter, such as polluted streams, and trickling filters and activated sludge plants (∞ Section 12.15), being found primarily in flowing waters. In

habitats where reduced iron or manganese compounds are present, the sheaths may become coated with ferric hydroxide or manganese oxide (see, for example, Figure 13.23). Iron precipitation is probably due to chemical reactions, but some sheathed bacteria have the ability to oxidize manganous ions to manganese oxide. Three genera are currently recognized: *Sphaerotilus*, in which manganese oxidation does not occur, *Leptothrix*, whose members do oxidize Mn^{2+}, and *Crenothrix*, which contains a multilayered sheath and whose unicellular forms are nonmotile. A single species of *Sphaerotilus* is recognized, *S. natans*, but several species of *Leptothrix* have been discerned, distinguished primarily by size, flagellation of swarmers, and some other morphological characteristics. Most of our discussion will concern *S. natans*, the organism that has been most extensively studied.

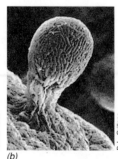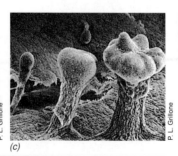

Figure 16.63 Scanning electron micrographs of fruiting-body formation in *Chondromyces crocatus*. (a) Early stage, showing aggregation and mound formation. (b) Initial stage of stalk formation. Slime formation in the head has not yet begun, and so the cells of which the head is composed are still visible. (c) Three stages in head formation. Note that the diameter of the stalk also increases. (d) Mature fruiting bodies. The entire fruiting structure is about 700 μm in height (see Table 16.19).

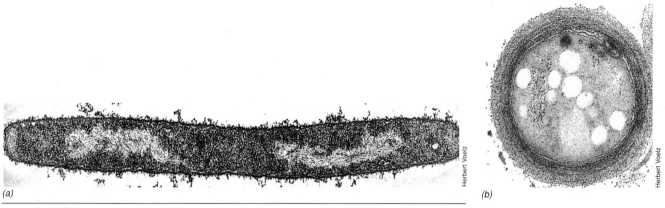

Figure 16.64 (a) Electron micrograph of a thin section of a vegetative cell of *Myxococcus xanthus.* A cell measures about 0.75 μm wide. (b) Myxospore of *M. xanthus,* showing the multilayered outer wall. Myxospores measure about 2 μm in diameter.

(a)   (b)

Sphaerotilus

The *Sphaerotilus* filament is composed of a chain of rod-shaped cells with rounded ends enclosed in a closely fitting sheath. This thin, transparent sheath is difficult to see when it is filled with cells, but when it is partially empty, the sheath can easily be seen by phase contrast microscopy (Figure 16.65a) or by staining. The cells within the sheath divide by binary fission (Figure 16.65b), and the new cells pushed out at the end synthesize new sheath material. Thus, the sheath is always formed at the *tips* of the filaments. Individual cells are 1–2 μm wide and 3–8 μm long and stain gram-negatively. Eventually cells are liberated from the sheaths, probably when the nutrient supply is low. These free cells are actively motile, the flagella being arranged lophotrichously (in a bundle at one pole) (Figure 16.65c). Probably the flagella are synthesized before the cells leave the sheath and, if so, may even aid in their liberation. It is thought that the swarmer cells then migrate, settle down, and begin to grow, each swarmer being the forerunner of a new filament. The sheath, which is devoid of muramic acid or other components of the peptidoglycan cell wall, is a protein–polysaccharide–lipid complex, possibly analogous to the capsules formed by many gram-negative Bacteria but differing in that it forms a linear structure.

Sphaerotilus cultures are nutritionally versatile, able to use a wide variety of simple organic compounds as carbon and energy sources, with inorganic nitrogen sources. Many strains require vitamin B_{12}, a substance frequently needed by aquatic microorganisms. Befitting its habitat in flowing waters, *Sphaerotilus* is an obligate aerobe.

Sphaerotilus blooms often occur in the fall of the year in streams and brooks when leaf litter causes a temporary increase in the organic content of the water. Its filaments are the main component of a microbial complex that sanitary engineers call "sewage fungus," which is the funguslike filamentous slime found on the rocks in streams receiving sewage pollution. In activated sludge

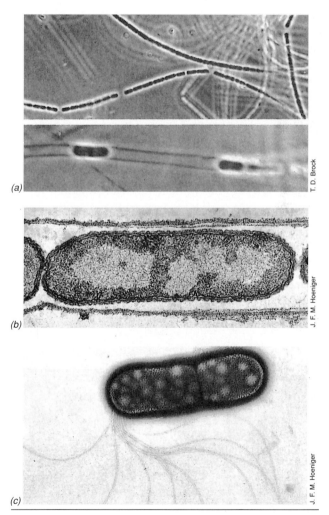

(a)

(b)

(c)

Figure 16.65 *Sphaerotilus natans.* A single cell is about 2 μm wide. (a) Phase contrast photomicrographs of material collected from a polluted stream. Active growth stage (above) and swarmer cells leaving the sheath. (b) Electron micrograph of a thin section through a filament. (c) Electron micrograph of a negatively stained swarmer cell. Notice the polar flagellar tuft.

plants (∞ Section 12.15), *Sphaerotilus* growth, like that of *Beggiatoa* (see Section 16.13), is often responsible for the detrimental condition called "bulking." The tangled masses of *Sphaerotilus* filaments so increase the bulk of the sludge that it does not settle properly, thus presenting difficulties in sludge clarification.

However, the ability of *Sphaerotilus* and *Leptothrix* to cause precipitation of iron oxides on their sheaths is well established. Such iron-encrusted sheaths are frequently seen in iron-rich waters (Figure 16.66). The process whereby iron deposition occurs is as follows: in iron-rich waters, ferrous iron is often held in solution as a chelate with organic materials such as humic and tannic acids. The sheathed bacteria can take up these soluble chelates, oxidize the organic compound, liberating the ferrous ions, which then oxidize spontaneously and become precipitated, generally in the region of the sheath (see Figure 16.66).

In the case of Mn^{2+}, specific oxidation by *Leptothrix*, but not by *Sphaerotilus*, is known to occur. (Oxidation of Mn^{2+} is not unique to *Leptothrix*, as a number of other bacteria, as well as fungi and yeasts, are able to carry out this process) (∞ Section 13.18). At pH values below 8, Mn^{2+} does not oxidize spontaneously, and so a significant biological oxidation is possible. However, even with Mn^{2+}, there is no evidence that *Leptothrix* obtains *energy* from the process, either chemolithotrophically or mixotrophically. Although *Leptothrix* produces an enzyme that catalyzes Mn^{2+} oxidation, it is unclear what benefit the organism derives from this oxidation process.

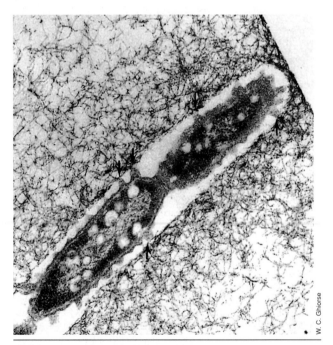

Figure 16.66 Transmission electron micrograph of a thin section of *Leptothrix* sp. in a sample from a ferromanganese film in a swamp in Ithaca, New York. A single cell measures about 0.9 μm in diameter. Note the protuberances of the cell envelope that contact the sheath (arrows).

CONCEPT CHECK   **16.14**

Gliding bacteria include the myxobacteria and relatives, many of which undergo a life cycle producing a fruiting body containing sporelike cells. Sheathed bacteria produce a tubelike outer covering, the sheath, that encloses individual cells.

✔ *What is a* myxospore *and how is it produced?*

✔ *Why is* Beggiatoa *of historical interest in terms of energy metabolism?*

16.15

Pseudomonads

All the genera in this group are straight or slightly curved rods with *polar* flagella (Figure 16.67). Pseudomonads are a major group of chemoorganotrophic aerobic gram-negative rods that never show a fermentative metabolism. (Fermentative organisms with polar flagella are generally classified in the genus *Aeromonas* or *Vibrio*, as indicated in Section 16.19.) The important genera are *Pseudomonas*, *Commamonas*, and *Burkholderia*, discussed in some detail here. Other genera include *Xanthomonas*, primarily a plant pathogen that is responsible for a number of necrotic plant lesions and that is characterized by its yellow-colored pigments; *Zoogloea*, characterized by its formation of an extracellular fibrillar polymer, which causes the cells to aggregate into distinctive flocs (this organism is a dominant component of activated sludge) (∞ Section 12.15), and *Gluconobacter*, characterized by its incomplete oxidation of sugars or alcohols to acids, such as the oxidation of glucose to gluconic acid or ethanol to acetic acid (this organism is discussed briefly with the other acetic acid bacteria in Section 16.17). Phylogenetically, the various genera of pseudomonads scatter within the purple Bacteria (see Table 16.1 and Section 15.7). Presumably the pseudomonads are derived from ancestral phototrophic bacteria that dispensed with the property of photosynthesis in evolving to colonize habitats in which the ability to carry out anoxygenic photosynthesis was not a significant advantage (for example, in soil and on the surfaces of plants and animals) (see Figure 16.2).

Characteristics

The distinguishing characteristics of the pseudomonad group are given in Table 16.20. Also listed in this table are the minimal characteristics needed to identify an organism as a pseudomonad. Key identifying characteristics are the absence of gas formation from glucose, and the positive oxidase test, both of which help to distinguish pseudomonads from enteric bacteria (Section 16.20).

The species of the genus *Pseudomonas* are defined on the basis of various physiological characteristics, as

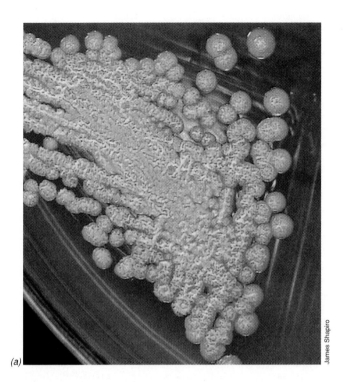

(a)

(b)

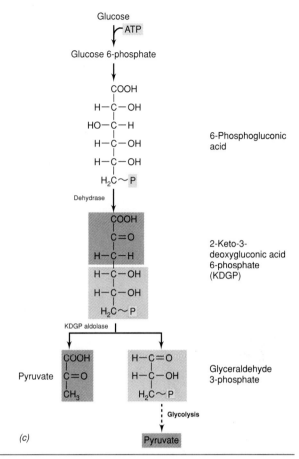

(c)

Figure 16.67 Typical pseudomonad colony and cell morphology and a biochemical pathway common in pseudomonads. (a) Photograph of colonies of *Pseudomonas cepacia* on an agar plate. (b) Shadow-cast preparation of a *Pseudomonas* species. The cell measures about 1 μm in diameter. (c) The Entner–Doudoroff pathway, the major means of glucose catabolism in pseudomonads.

TABLE 16.20	Genera *Pseudomonas, Commamonas,* and *Burkholderia*

General characteristics:
Straight or curved rods but not vibrioid; size 0.5–1.0 μm by 1.5–4.0 μm; no spores; gram-negative; polar flagella: single or multiple; no sheaths, appendages, or buds; respiratory metabolism, never fermentative, although may produce small amounts of acid from glucose aerobically; use low-molecular-weight organic compounds, not polymers; some are chemolithotrophic, using H_2 or CO as sole electron donor; some can use nitrate as electron acceptor anaerobically; some can use arginine as energy source anaerobically

Minimal characteristics for identification:
Gram-negative, straight or slightly curved; no spores; motile (always); polar flagella (flagellar stain); oxidative-fermentative medium with glucose: tube open, acid produced; tube sealed, acid not produced; gas not produced from glucose (distinguishes them easily from enteric bacteria and *Aeromonas*); oxidase, almost always positive (enterics are oxidase-negative); catalase always positive; photosynthetic pigments absent (distinguishes them from nonsulfur purple bacteria); indole-negative; methyl red-negative; Voges–Proskauer-negative (for discussion of many of these biochemical tests, see Section 21.2)

outlined in Tables 16.21 and 16.22. Some species (for example, *Pseudomonas aeruginosa*) are quite homogeneous, and so all isolates fit into a very narrow range of distribution of characteristics, whereas other species are much more heterogeneous. The taxonomy of the genus *Pseudomonas* has been clarified by DNA hybridization studies (⚬⚬ Section 15.9), and the subgroups given in Table 16.21 are supported by 16S rRNA sequencing.

Pseudomonads have very simple nutritional requirements and grow chemoorganotrophically at neutral pH and at temperatures in the mesophilic range. One of the striking properties of the pseudomonads is the wide variety of organic compounds used as carbon sources and as electron donors for energy generation. Some species utilize over *100* different compounds, and only a few species utilize fewer than 20. As an example of this versatility, a single strain of *Pseudomonas aeruginosa* can make use of many different sugars, fatty acids, dicarboxylic acids, tricarboxylic acids, alcohols, polyalcohols, glycols, aromatic compounds, amino acids, and amines, plus miscellaneous organic compounds not fit-

TABLE 16.21 Characteristics of subgroups and species of the genera *Pseudomonas, Commamonas,* and *Burkholderia*

Group	16s rRNA group[a]	Characteristics	DNA (mol % GC)
Fluorescent subgroup	Gamma	Most produce water-soluble, yellow-green fluorescent pigments; do not form poly-β-hydroxybutyrate; single DNA homology group	
Pseudomonas aeruginosa		Pyocyanin production, growth at up to 43°C, single polar flagellum, capable of denitrification	67
Pseudomonas fluorescens		Does not produce pyocyanin or grow at 43°C; tuft of polar flagella	59–61
Pseudomonas putida		Similar to *P. fluorescens* but does not liquefy gelatin and does grow on benzylamine	60–63
Pseudomonas syringae		Lacks arginine dihydrolase, oxidase-negative, pathogenic to plants	58–60
Acidovorans subgroup	Beta	Nonpigmented, form poly-β-hydroxybutyrate, tuft of polar flagella, do not use carbohydrates; single DNA homology group	
Commamonas acidovorans		Uses muconic acid as sole carbon source and electron donor	67
Commamonas testosteroni		Uses testosterone as sole carbon source	62
Pseudomallei-cepacia subgroup	Beta	No fluorescent pigments, tuft of polar flagella, forms poly-β-hydroxybutyrate; single DNA homology group	62
Burkholderia cepacia		Extreme nutritional versatility; some strains pathogenic to plants	67
Pseudomonas pseudomallei		Causes melioidosis in animals; nutritionally versatile	69
Pseudomonas mallei		Causes glanders in animals; nonmotile; nutritionally restricted	69
Diminuta-vesicularis subgroup	Alpha	Single flagellum of very short wavelength, require vitamins (pantothenate, biotin, B_{12})	
Pseudomonas diminuta		Nonpigmented, does not use sugars	66–67
Pseudomonas vesicularis		Carotenoid pigment, uses sugars	66
Miscellaneous species	—		
Pseudomonas solanacearum		Plant pathogen	66–68
Pseudomonas saccharophila		Grows chemolithotrophically with H_2, digests starch	69
Pseudomonas maltophilia		Requires methionine, does not use NO_3^- as N source, oxidase-negative	67

[a]All pseudomonads are members of the purple Bacteria (Proteobacteria) (∞ Section 15.7 and Table 16.1).

ting into any of the preceding categories. On the other hand, pseudomonads are generally unable to break down polymers into their component monomers. Nutritionally versatile pseudomonads typically contain numerous inducible operons (∞ Section 7.3) because the catabolism of unusual organic substrates often requires the activity of several different enzymes. The pseudomonads are ecologically important organisms in soil and water and are probably responsible for the degradation of many soluble compounds derived from the breakdown of plant and animal materials in oxic habitats.

A few pseudomonads are pathogenic (Table 16.22). Among the fluorescent pseudomonads, the species *Pseudomonas aeruginosa* is frequently associated with infections of the urinary and respiratory tracts in humans. *Pseudomonas aeruginosa* infections are also common in patients receiving treatment for severe burns or other traumatic skin damage. *Pseudomonas aeruginosa* is not an obligate parasite, however, because it can be readily isolated from soil, and as a denitrifier (∞ Section 14.14), it plays an important role in the nitrogen cycle in nature. As a pathogen it appears to be primarily an opportunist, initiating infections in individuals whose resistance is low. In addition to urinary infections, it can also cause systemic infections, usually in individuals who have experienced extensive skin damage. The organism is naturally resistant to many of the widely used antibiotics, so chemotherapy is often difficult. Resistance is frequently due to a *resistance transfer plasmid (R plasmid)* (∞ Sections 9.8 and 11.13), which is a plasmid carrying genes coding for detoxification of various antibiotics. *Pseudomonas aeruginosa* is commonly found in the hospital environment and can easily infect patients receiving treatment for

TABLE 16.22	Pathogenic pseudomonads
Species	**Relationship to disease**
Animal pathogens	
P. aeruginosa	Opportunistic pathogen, especially in hospitals; in patients with metabolic, hematologic, and malignant diseases; hospital-acquired infections from catheterizations, tracheostomies, lumbar punctures, and intravenous infusions; in patients given prolonged treatment with immunosuppressive agents, corticosteroids, antibiotics and radiation; may contaminate surgical wounds, abscesses, burns, ear infections, lungs of patients treated with antibiotics; primarily a soil organism
P. fluorescens	Rarely pathogenic, as does not grow well at 37°C; may grow in and contaminate blood and blood products under refrigeration
P. maltophilia	A ubiquitous, free-living organism that is a common nosocomial pathogen
B. cepacia	Causes onion bulb rot; has also been isolated from humans and from environmental sources of medical importance
P. pseudomallei	Causes melioidosis, a disease endemic in animals and humans in Southeast Asia
P. mallei	Causes glanders, a disease of horses that is occasionally transmitted to humans
P. stutzeri	Often isolated from humans and environmental sources; may live saprophytically in the body
Plant pathogens	
P. solanacearum	Causes wilts of many cultivated plants (for example, potato, tomato, tobacco, peanut)
P. syringae	Attacks foliage, causing chlorosis and necrotic lesions on leaves; rarely found free in soil
P. marginalis	Causes soft rot of various plants; active pectinolytic species
Xanthomonas	Causes necrotic lesions on foliage, stems, fruits; also causes wilts and tissue rots; rarely found free in soil

other illnesses (see Section 22.7 for a discussion of hospital-acquired infections). Polymyxin, an antibiotic not ordinarily used in human therapy because of its toxicity, is effective against *P. aeruginosa* and can be used with caution.

Many pseudomonads, as well as a variety of other gram-negative Bacteria, metabolize glucose via the Entner–Doudoroff pathway (Figure 16.67). Two key enzymes of the Entner–Doudoroff pathway are *6-phosphogluconate dehydrase* and *ketodeoxyglucosephosphate aldolase* (Figure 16.67c). A survey for the presence of these enzymes in a wide variety of bacteria has shown that they are absent from all gram-positive Bacteria (except a few *Nocardia* isolates) and are generally present in bacteria of the genera *Pseudomonas*, *Rhizobium*, and *Agrobacterium*, as well as in some isolates of several other genera of gram-negative Bacteria.

Phytopathogens

Certain species of *Pseudomonas* and *Burkholderia* and the genus *Xanthomonas* are well-known plant pathogens (phytopathogens) (see Table 16.22). In many cases these organisms are so highly adapted to the plant environment that they can rarely be isolated from other habitats, including soil. Phytopathogens frequently inhabit nonhost plants (where disease symptoms are not apparent) and from there become transmitted to host plants and initiate infection. Disease symptoms vary considerably depending on the particular phytopathogen and host plant and are generally due to the release by the bacterium of plant toxins, lytic enzymes, plant growth factors, and other substances that destroy or distort plant tissue. In many cases the disease symptoms are highly diagnostic of the type of phytopathogen and are actually used in the taxonomy of phytopathogenic pseudomonads. Thus, *Pseudomonas syringae* is frequently isolated from leaves showing chlorotic lesions, whereas *P. marginalis* is a typical "soft-rot" pathogen, infecting stems and shoots but rarely leaves.

16.16

Free-Living Aerobic Nitrogen-Fixing Bacteria

A variety of organisms that inhabit primarily the soil are capable of fixing N_2 *aerobically* (Table 16.23). The genus *Azotobacter* comprises large, gram-negative, obligately aerobic rods capable of fixing N_2 nonsymbiotically (Figure 16.68). The first member of this genus was discovered by the Dutch microbiologist M. W. Beijerinck early in the twentieth century, using an enrichment culture technique with a medium devoid of a combined nitrogen source (∞ Rise of General Microbiology, Chapter 14). Although capable of growth on N_2, *Azotobacter* grows more rapidly on NH_3; indeed, adding NH_3 actually *represses* nitrogen fixation (∞ Section 13.27). Much work has been done in seeking to evaluate the role of *Azotobacter* in nitrogen fixation in nature, especially in comparison to the anaerobic organism *Clostridium pasteurianum* and the symbiotic organisms of the genus *Rhizobium*. *Azotobacter* is also of interest because it has the highest respiratory rate (measured as the rate of O_2 uptake) of any living organism. In addition to its ecological and physiological importance, *Azotobacter* is of interest because of its ability to form an un-

TABLE 16.23	Genera of free-living aerobic nitrogen-fixing bacteria[a]		
Genus	**Number of species**	**Characteristics**	**DNA (mol % GC)**
Azotobacter	6	Large rod; produces cysts; primarily found in neutral to alkaline soils	63–67
Azomonas	3	Large rod; no cysts; primarily aquatic	52–59
Azospirillum	4	Microaerophilic rod; associates with plants	69–71
Beijerinckia	4	Pear-shaped rod with large lipid bodies at each end; produces extensive slime; inhabits acidic soils	54–59
Derxia	1	Rods; form coarse, wrinkled colonies	69–73

[a]All species examined are members of the purple Bacteria, primarily the alpha and gamma subdivisions (see Table 16.1).

usual resting structure called a *cyst* (Figure 16.68*b*). Phylogenetically, *Azotobacter* belongs to the gamma group of purple Bacteria (∞ Section 15.7 and Table 16.1).

Azotobacter *and nitrogenases*

Azotobacter cells are rather large for prokaryotes, many isolates being almost the size of yeasts, with diameters of 2–4 μm or more. Pleomorphism is common, and a variety of cell shapes and sizes have been described. Some strains are motile by peritrichous flagella. On carbohydrate-containing media, extensive capsules or slime layers are produced by free-living N_2-fixing bacteria (Figure 16.69). *Azotobacter* is able to grow on a wide variety of carbohydrates, alcohols, and organic acids. The metabolism of carbon compounds is strictly oxidative, and acids or other fermentation products are rarely produced. All members fix nitrogen, but growth also occurs on simple forms of combined nitrogen: ammonia, urea, and nitrate.

Despite the fact that *Azotobacter* is an obligate aerobe, its nitrogenase is as O_2-sensitive as all other nitrogenases (∞ Section 13.27). It is thought that the high respiratory rate of *Azotobacter* (mentioned earlier) has something to do with protection of nitrogenase from O_2. The intracellular O_2 concentration is kept low

enough by metabolism that inactivation of nitrogenase does not occur.

The remaining genera of free-living N_2 fixers include *Azomonas*, a genus of large, rod-shaped bacteria that resemble *Azotobacter* except that they do not produce cysts and are primarily aquatic, and *Beijerinckia* and *Derxia* (Figure 16.70), two genera that grow well in acidic soils.

Like bacterial endospores, *Azotobacter* cysts (Figure 16.68) show negligible endogenous respiration and are resistant to desiccation, mechanical disintegration, and ultraviolet and ionizing radiation. In contrast to endospores, however, they are *not* especially heat-resistant, and they are not completely dormant because they

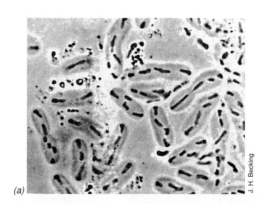

(a)

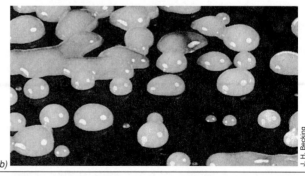

(b)

Figure 16.69 Examples of slime production by free-living N_2-fixing bacteria. (a) Cells of *Derxia gummosa* encased in slime. Cells are about 1–1.2 μm wide. (b) Colonies of *Beijerinckia* species growing on a carbohydrate-containing medium. Note the raised, glistening appearance of the colonies due to abundant capsular slime.

(a)

(b)

Figure 16.68 *Azotobacter vinelandii:* (a) vegetative cells and (b) cysts by phase contrast microscopy. A cell measures about 2 μm in diameter, and a cyst about 3 μm.

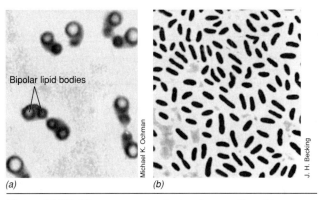

Bipolar lipid bodies

(a)   (b)

Michael K. Ochman

J. H. Becking

Figure 16.70 Phase-contrast photomicrographs of two genera of acid-tolerant, free-living N_2-fixing bacteria. (a) *Beijerinckia indica*. The cells are roughly pear-shaped, about 0.8 μm in diameter, and contain a large globule of poly-β-hydroxybutyrate at each end. (b) *Derxia gummosa*.

rapidly oxidize exogenous energy sources. The carbon source of the medium greatly influences the extent of cyst formation, butanol being especially favorable; compounds related to butanol, such as β-hydroxybutyrate, also promote cyst formation.

The species *Azotobacter chroococcum* was the first N_2-fixing bacterium shown capable of growth on N_2 in the complete *absence* of molybdenum. When cells of *A. chroococcum* are placed in a mineral medium lacking ammonia and molybdenum but containing the metal vanadium, an alternative nitrogenase containing *vanadium* in place of *molybdenum* is produced and functions to support growth of the organism on N_2. Like the molybdenum enzyme (∞ Section 13.27), vanadium nitrogenase consists of *two* proteins, one of which contains iron, the second iron and vanadium, and reduces N_2 to NH_3, H^+ to H_2, and C_2H_2 to C_2H_4 (∞ Figure 13.55). However, the rates of substrate reduction are considerably *lower* by the vanadium nitrogenase than by the molybdenum enzyme, indicating that the vanadium enzyme is a less effective catalyst. Interestingly, the vanadium nitrogenase also reduces C_2H_2 to ethane, C_2H_6. Although only about 3% of the acetylene reduced goes to ethane, the ability of the vanadium nitrogenase to catalyze the reduction of C_2H_2 to C_2H_6 can serve as a test for detecting N_2-fixing bacteria capable of synthesizing this alternative nitrogenase. *Azotobacter* can also produce a nitrogenase containing only *iron*. This nitrogenase is produced if both molybdenum and vanadium are absent, but it is a very poor N_2-reducing catalyst. The iron-only nitrogenase also produces some C_2H_6.

Vanadium nitrogenase is not the major nitrogenase of *Azotobacter chroococcum* because addition of molybdenum to the culture medium represses synthesis of both the vanadium and iron-only nitrogenases, and instead the conventional nitrogenase is formed. Vanadium nitrogenases are not universal among N_2-fixing bacteria. However, it is known that vanadium stimu-

lates N_2 fixation in a number of organisms including various species of *Azotobacter*, some cyanobacteria and phototrophic bacteria, and *Clostridium pasteurianum*. Alternative nitrogenases are probably "backup" systems for fixing N_2 in molybdenum-deficient habitats.

16.17
Acetic Acid Bacteria

As originally defined, the *acetic acid bacteria* comprised a group of gram-negative, aerobic, motile rods that carried out *incomplete* oxidation of alcohols, leading to the accumulation of organic acids as end products. With *ethanol* as a substrate, *acetic acid* is produced; hence the derivation of the common name for these bacteria. Another property is the relatively high tolerance to acidic conditions, most strains being able to grow at pH values lower than 5. This acid tolerance should of course be essential for an organism producing large amounts of acid. The acetic acid bacteria are a heterogeneous assemblage, comprising both peritrichously and polarly flagellated organisms. The *polarly* flagellated organisms are related to the pseudomonads, differing mainly in their acid tolerance and their inability to carry out a complete oxidation of alcohols. These organisms are now classified in the genus *Gluconobacter* (Table 16.24). All acetic acid bacteria group phylogenetically with the purple Bacteria (see Table 16.1).

The genus *Acetobacter* comprises the *peritrichously* flagellated organisms. In addition to flagellation, *Acetobacter* differs from *Gluconobacter* in being able to further oxidize the acetic acid it forms to CO_2. This difference in ability to oxidize acetic acid is related to the presence of a complete citric acid cycle. *Gluconobacter*, which *lacks* a complete citric acid cycle. is unable to oxidize acetic acid, whereas *Acetobacter*, which has all enzymes of the cycle, can oxidize it. Because of these differences in oxidative potential, gluconobacters are sometimes called *underoxidizers*, and acetobacters, *overoxidizers*.

The acetic acid bacteria are frequently found in association with alcoholic juices and probably evolved from pseudomonads inhabiting sugar-rich flowers and fruits where a yeast-mediated alcoholic fermentation is common. Acetic acid bacteria can often be isolated from an alcoholic fruit juice such as cider or wine. Colonies of acetic acid bacteria can be recognized on $CaCO_3$–agar plates containing ethanol, the acetic acid produced causing a dissolution and clearing of the insoluble $CaCO_3$ (Figure 16.71). Cultures of acetic acid bacteria are used in commercial production of vinegar (∞ Section 12.10). Acetate produced by acetic acid bacteria and by homoacetogens (see Section 16.9) is also used (in the form of the calcium salt) as a road deicer.

In addition to ethanol, these organisms carry out an incomplete oxidation of such organic compounds

TABLE 16.24   Differentiation of *Acetobacter*, *Gluconobacter*, and *Pseudomonas*

Characteristic	Acetobacter	Gluconobacter	Pseudomonas
Flagellation	Peritrichous	Polar	Polar
Growth at pH 4.5	+	+	−
Oxidation of ethanol to acetic acid at pH 4.5	+	+	−
Complete citric acid cycle	Present	Absent	Present
DNA (mol % GC)	53–65	56–64	58–70
Number of species	7	3	28
16S rRNA group[a]	Alpha purple Bacteria	Alpha purple Bacteria	Alpha, beta, or gamma purple Bacteria

[a] ∞ Section 15.7 and Table 16.1.

as higher alcohols and sugars. For instance, glucose is oxidized only to gluconic acid, galactose to galactonic acid, arabinose to arabonic acid, and so on. This property of underoxidation is exploited in the manufacture of ascorbic acid (vitamin C). Ascorbic acid can be formed from sorbose, but sorbose is difficult to synthesize chemically. It is, however, conveniently obtainable microbiologically from acetic acid bacteria, which oxidize sorbitol (a readily available sugar alcohol) only to sorbose, a process called *bioconversion* (∞ Section 12.8). The use of acetic acid bacteria makes the manufacture of ascorbic acid economically feasible.

Another interesting property of some acetic acid bacteria is their ability to synthesize *cellulose*. The cellulose formed does not differ significantly from that of plant cellulose, but instead of being a part of the cell wall, the bacterial cellulose is formed as a matrix outside the wall and the bacteria become embedded in the tangled mass of cellulose microfibrils. When these species of acetic acid bacteria grow in an unshaken vessel, they form a surface pellicle of cellulose in which the bacteria develop. Since these bacteria are

obligate aerobes, the ability to form such a pellicle may be a means by which the organisms are assured of remaining at the surface of the liquid where oxygen is readily available.

CONCEPT CHECK   16.17

Pseudomonads include many gram-negative chemoorganotrophic aerobic rods; many N₂-fixing species are phylogenetically closely related and can reduce N₂ to NH₃ in the process of nitrogen fixation. The acetic acid bacteria are also phylogenetically related to pseudomonads and are characterized by an ability to oxidize ethanol to acetate aerobically.

✔ *Compare and contrast the pseudomonads,* Azotobacter, *and the acetic acid bacteria in terms of O_2 requirements, electron donors, pathogenicity, and habitats.*

✔ *Compare and contrast the organism* Acetobacter *with the organism* Acetobacterium *(see Section 16.9) in as many ways as you can think of.*

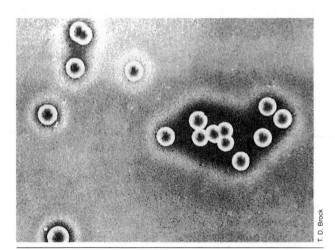

Figure 16.71 Photograph of colonies of *Acetobacter aceti* on calcium carbonate agar containing ethanol as energy source. Note the clearing around the colonies due to the dissolution of calcium carbonate by the acetic acid produced by the bacteria.

16.18
Zymomonas and *Chromobacterium*

The genera *Zymomonas* and *Chromobacterium* are grouped together as facultatively aerobic gram-negative rods of uncertain taxonomic affiliation. Phylogenetically, both organisms are members of the beta purple Bacteria group (∞ Section 15.7 and Table 16.1). The best-known *Chromobacterium* species is *Chromobacterium violaceum*, a bright purple pigmented organism (Figure 16.72) found in soil and water and occasionally in pyogenic infections of humans and other animals. *Chromobacterium violaceum* and a few other chromobacteria produce the purple pigment *violacein* (Figure 16.73), a water-insoluble pigment that has antibiotic properties and is produced only in media containing the amino acid tryptophan. *Chromobacterium* is a facultative aerobe, growing fermentatively on sugars and aerobically on a variety of carbon sources.

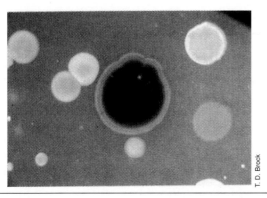

Figure 16.72 A large colony of *Chromobacterium violaceum* growing among other colonies on an agar plate.

The genus *Zymomonas* consists of large, gram-negative rods (2–6 μm long and 1–1.4 μm wide), which carry out a vigorous fermentation of sugars to ethanol. Although not all strains are motile, if motility occurs, it is by lophotrichous flagella. *Zymomonas* is a common organism involved in alcoholic fermentation of various plant saps, and in many tropical areas of South and Central America, Africa, and Asia, it occupies a position similar to that of *Saccharomyces cerevisiae* (yeast) in North America and Europe. *Zymomonas* is involved in the alcoholic fermentation of agave in Mexico, and palm sap in many tropical areas. It also carries out an alcoholic fermentation of sugarcane juice and honey. Although *Zymomonas* is rarely the sole organism involved in these alcoholic fermentations, it is often the dominant organism and is probably responsible for the production of most of the ethanol (the desired product) in these beverages. *Zymomonas* is also responsible for spoilage of fruit juices such as apple cider and perry. It also may be a constituent of the bacterial flora of spoiled beer and may be responsible for the production in beer of an unpleasant odor of rotten apples (resulting from traces of acetaldehyde and H_2S).

Zymomonas is distinguished from *Pseudomonas* by its fermentative metabolism, microaerophilic to anaerobic nature, oxidase negativity, and other molecular taxonomic characteristics. It resembles most closely the acetic acid bacteria, specifically *Gluconobacter*

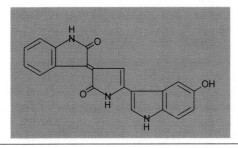

Figure 16.73 Structure of the pigment violacein, produced by various species of the genus *Chromobacterium*.

because of its polar flagellation, and it is often found in nature associated with the acetic acid bacteria. This is of interest because *Zymomonas* ferments glucose to ethanol, whereas the acetic acid bacteria oxidize ethanol to acetic acid. Thus, the acetic acid bacteria may depend on the activity of *Zymomonas* for the production of their growth substrate, ethanol. Like the acetic acid bacteria, *Zymomonas* is quite tolerant of low pH. Unlike yeast, which ferments glucose to ethanol via the Embden–Meyerhof (glycolytic) pathway, *Zymomonas* employs the Entner–Doudoroff pathway. This pathway is active in many pseudomonads as a means of catabolizing glucose (see Figure 16.67c).

Zymomonas is of interest to the ethanol industry because it shows higher rates of glucose uptake and ethanol production and gives a higher yield of ethanol than many yeasts. *Zymomonas* is also rather tolerant of high ethanol concentrations (up to 10%) but is not quite as tolerant as some of the best yeast strains, which can grow to 12–15% ethanol. However, the fact that *Zymomonas* is a *gram-negative* bacterium and can thus be readily manipulated genetically makes it an attractive candidate for use by ethanol production industries.

16.19
Vibrio and Related Genera

The *Vibrio* group contains gram-negative, facultatively aerobic rods and curved rods that possess a fermentative metabolism. Most of the members of the *Vibrio* group are polarly flagellated, although some are peritrichously flagellated. One key difference between the *Vibrio* group and enteric bacteria is that members of the former are oxidase-*positive* (∞ Table 21.3), whereas members of the latter are oxidase-*negative*. Although *Pseudomonas* is also polarly flagellated and oxidase-positive, it is *not* fermentative and hence can be separated from the vibrios by simple sugar fermentation tests.

Most vibrios and related bacteria are aquatic, found either in freshwater or marine habitats, although one important organism, *Vibrio cholerae*, is pathogenic for humans. The group contains four genera, *Vibrio, Aeromonas, Photobacterium,* and *Plesiomonas,* and the major characteristics of each are given in Table 16.25.

Vibrio cholerae is the specific cause of the disease *cholera* in humans (∞ Sections 19.9 and 23.14); the organism does not normally infect other hosts. Cholera is one of the most common infectious human diseases in underdeveloped countries and one that has had a long history. The organism is transmitted almost exclusively via water, and studies on its distribution in the nineteenth century played a major role in demonstrating the importance of water purification in urban areas (∞ Snow on Cholera, Chapter 22). We discuss the pathogenesis of *V. cholerae* in Section 19.9. *Vibrio cholerae* is capable of good growth at a pH of over 9,

TABLE 16.25 Distinguishing characteristics of genera of the family Vibrionaceae and related genera[a]

Characteristic	Vibrio	Aeromonas	Photobacterium	Plesiomonas
Morphology	Straight or curved rods, single-sheathed polar flagellum	Straight rods, single polar flagellum	Straight rods, 1–3 polar flagella	Straight rods with round ends, 2–5 polar flagella
Sodium required for growth	+	−	+	−
Gas production	−	+	+	−
Sensitivity to vibriostat (2,4-diamino-6,7-diisopropyl pteridine)	+	−	+	+
Fermentation of mannitol	+	+	−	−
Luminescence	+ or −	−	+	−
DNA (mol % GC)	38–51	57–63	40–44	51
Pigment	None	None or brown	None	None
Cause disease in humans	+	+	−	+

[a]All are gram-negative rods, straight or curved, facultative aerobes with a fermentative metabolism (O/F medium) but oxidase-positive (∞ Section 21.2). They are predominantly aquatic organisms and are phylogenetically members of the gamma purple Bacteria (see Table 16.1).

and this characteristic is frequently employed in the selective isolation and identification of this organism.

Vibrio parahemolyticus is a marine organism. It is a major cause of gastroenteritis in Japan (where raw fish is widely consumed) and has also been implicated in outbreaks of gastroenteritis in other parts of the world, including the United States. The organism can be frequently isolated from seawater or from shellfish and crustaceans, and its primary habitat is probably marine animals, with human infection being a secondary development (∞ Section 23.14).

Luminescent bacteria

A number of gram-negative, polarly flagellated rods possess the interesting property of emitting light (luminescence). Most of these bacteria have been classified in the genus *Photobacterium* (Table 16.25; see Figure 16.74), but a few *Vibrio* isolates are also luminescent (Figure 16.74a). Most **luminescent bacteria** are marine forms, usually found associated with fish. Some fish possess a special organ in which luminescent bacteria grow (Figure 16.74c–f). Other luminescent marine bacteria live saprophytically on dead fish. A good way of isolating luminescent bacteria is to incubate a dead marine fish for 1 or 2 days at 10–20°C; the luminescent bacterial colonies that usually appear on the surface of the fish can be easily seen and isolated (Figure 16.74a and b). (To see luminescence readily, one should observe the material in a completely dark room after the eyes have become adapted to the dark.)

Although *Photobacterium* isolates are facultative aerobes, they are luminescent only when O_2 is present. Several components are needed for bacterial luminescence: the enzyme **luciferase** and a long-chain aliphatic aldehyde (for example, *dodecanal*); flavin mononucleotide (FMN) and O_2 are also involved. The primary electron donor is NADH, and the electrons pass through FMN to the luciferase. The reaction can be expressed as

$$FMNH_2 + O_2 + RCHO \xrightarrow{\text{Luciferase}} FMN + RCOOH + H_2O + Light$$

The light-generating system constitutes a bypass route for shunting electrons from $FMNH_2$ to O_2, without involving other electron carriers such as quinones and cytochromes.

Regulation and genetics of bioluminescence

The enzyme luciferase shows a unique kind of regulatory synthesis called **autoinduction.** The luminous bacteria produce a specific substance, the *autoinducer,* which accumulates in the culture medium during growth, and when the amount of this substance reaches a critical level, induction of the enzyme occurs. The autoinducer in *Vibrio fischeri* has been identified as N-β-ketocaproylhomoserine lactone. Thus, cultures of luminous bacteria at low cell density are not luminous but only become luminous when growth reaches a sufficiently high density that the autoinducer can accumulate and function, a mechanism referred to as "quorum sensing" for the density-dependent nature of the phenomenon. Because of autoinduction, it is obvious that free-living luminescent bacteria in seawater are not luminous because the autoinducer cannot accumulate, and luminescence develops only when conditions are favorable for the development of high population densities. Although it is not clear why luminescence is density-dependent in free-living bacte-

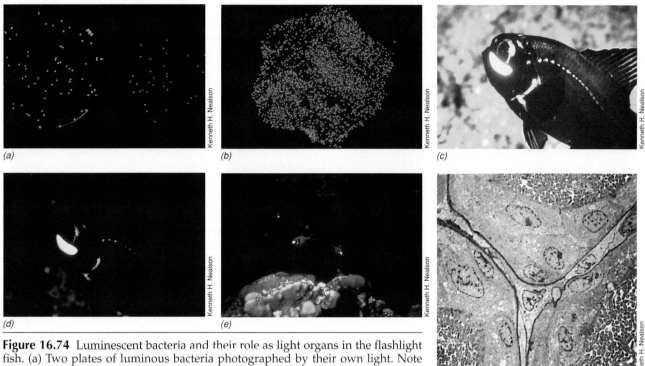

Figure 16.74 Luminescent bacteria and their role as light organs in the flashlight fish. (a) Two plates of luminous bacteria photographed by their own light. Note the different colors. Left, *Vibrio fischeri* strain MJ-1, blue light, and right, *V. fischeri* strain Y-1, green light. (b) Colonies of *Photobacterium phosphoreum* photographed by their own light. (c) The flashlight fish *Photoblepharon palpebratus*; the bright area is the light organ containing luminescent bacteria. (d) Same fish photographed by its own light. (e) Underwater photograph taken at night of *P. palpebratus* in coral reefs in the Gulf of Eilat. (f) Electron micrograph of a thin section through the light-emitting organ of *P. palpebratus*, showing the dense array of luminescent bacteria.

ria, in symbiotic strains of luminescent bacteria (see Figure 16.74), the rationale for density-dependent luminescence is clear: luminescence develops only when sufficiently high population densities are reached in the light organ of the fish to allow a visible flash of light.

Much new information about bioluminescence has emerged from studies on the genetics of this process. Several *lux* operons have been identified in luminescent *Vibrio* species, and the key structural genes have been cloned and sequenced. The *luxA* and *luxB* genes encode the α and β subunits, respectively, of bacterial luciferase. The *luxC, luxD,* and *luxE* genes encode polypeptides that function in the bioluminescence reaction and in the generation and activation of fatty acids for the luminescence system. Light-emitting strains of *Escherichia coli* have been constructed by inserting *lux* genes cloned from a *Vibrio* species into cells of *E. coli* and then placing their expression under the control of specific and easily manipulable *E. coli* promoters. This construction accelerated the pace of research on the regulatory aspects of bioluminescence because many *E. coli* genetic tools can be employed to probe aspects of *lux* gene control. In addition, cloned *lux* genes have stimulated biotechnological exploitation of bioluminescence in clinical diagnostics and other biomedical fields (∞ Figure 10.7).

16.20
Facultatively Aerobic Gram-Negative Rods

We focus here on the **enteric bacteria,** which comprise a relatively homogeneous phylogenetic group within the gamma purple Bacteria (∞ Section 15.7 and Table 16.1) and are characterized phenotypically as follows: gram-negative, nonsporulating rods, nonmotile or motile by *peritrichous* flagella (Figure 16.75), facultative aerobes, oxidase-*negative* with relatively simple nutritional requirements, fermenting sugars to a variety of end products. The phenotypic characteristics used to separate the enteric bacteria from other bacteria of similar morphology and physiology are given in Table 16.26.

Among the enteric bacteria are many strains pathogenic to humans, animals, or plants as well as other strains of industrial importance. Probably more is known about *Escherichia coli* than about any other bacterial species (∞ *Escherichia coli,* the Best-Known Prokaryote, Chapter 9).

Because of the medical importance of the enteric bacteria, an extremely large number of isolates have been studied and characterized, and a fair number of distinct genera have been defined. Despite the fact that

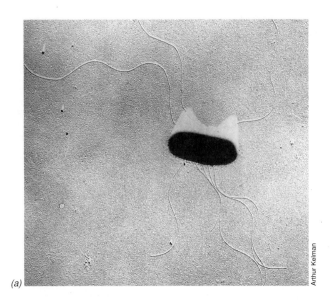

(a)

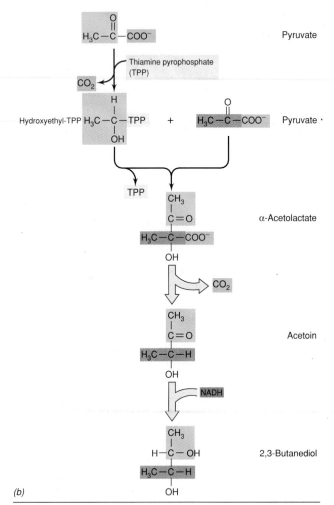

(b)

Figure 16.75 (a) Electron micrograph of a shadow-cast preparation of cells of the butanediol-producing enteric bacterium *Erwinia carotovora*. Cells are about 0.8 μm wide. Note the peritrichously arranged flagella. (b) Biochemical pathway for formation of butanediol from two molecules of pyruvate by butanediol fermenters.

TABLE 16.26	Defining characteristics of the enteric bacteria

General characteristics:
Gram-negative straight rods; motile by peritrichous flagella, or nonmotile; nonsporulating; facultative aerobes, producing acid from glucose; sodium neither required nor stimulatory; catalase-positive; oxidase-negative; usually reduce nitrate to nitrite (not to N_2); 16S rRNA of gamma section of purple Bacteria (Proteobacteria, see Table 16.1)

Key tests to distinguish enteric bacteria from other bacteria of similar morphology:
Oxidase test, enterics always negative—separates enterics from oxidase-positive bacteria of genera *Pseudomonas, Aeromonas, Vibrio, Alcaligenes, Achromobacter, Flavobacterium, Cardiobacterium*, which may have similar morphology; nitrate reduced only to nitrite, (assay for nitrite after growth)—distinguishes enteric bacteria from bacteria that reduce nitrate to N_2 (gas formation detected), such as *Pseudomonas* and many other oxidase-positive bacteria; ability to ferment glucose—distinguishes enterics from obligately aerobic bacteria

there is marked genetic relatedness among many of the enteric bacteria, as shown by DNA homologies and genetic recombination, separate genera are maintained, largely for practical reasons. Since these organisms are frequently cultured from diseased states (∞ Sections 19.4, 23.13, and 23.14), some means of identifying them is necessary.

Fermentation patterns in enteric bacteria

One of the key taxonomic characteristics separating the various genera of enteric bacteria is the type and proportion of *fermentation products* produced by anaerobic fermentation of glucose. Two broad patterns are recognized, the **mixed-acid fermentation** and the **2,3-butanediol fermentation** (Figure 16.76). In mixed-acid fermentation, *three* acids are formed in significant amounts—acetic, lactic, and succinic; ethanol, CO_2, and H_2 are also formed, but *not* butanediol. In butanediol fermentation, *smaller* amounts of acids are formed, and butanediol, ethanol, CO_2, and H_2 are the main products. As a result of a mixed-acid fermentation *equal* amounts of CO_2 and H_2 are produced, whereas with a butanediol fermentation *more* CO_2 than H_2 is produced. This is because mixed-acid fermenters produce CO_2 only from formic acid by means of the enzyme system *formic hydrogen lyase*:

$$HCOOH \rightarrow H_2 + CO_2$$

and this reaction results in equal amounts of CO_2 and H_2. The butanediol fermenters also produce CO_2 and H_2 from formic acid, but they produce additional CO_2 during the reactions that lead to butanediol (Figure 16.75b).

A variety of diagnostic tests and differential media are used to separate the various genera in the two broad

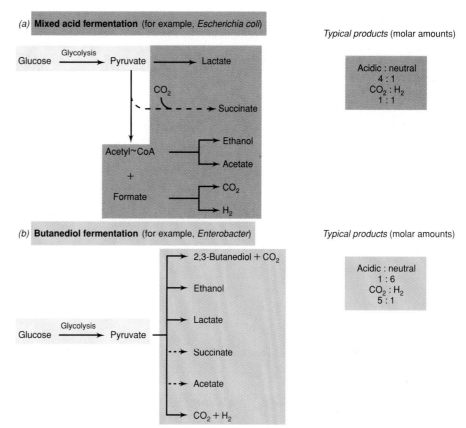

Typical products (molar amounts)

Acidic : neutral
4 : 1
CO_2 : H_2
1 : 1

Typical products (molar amounts)

Acidic : neutral
1 : 6
CO_2 : H_2
5 : 1

Figure 16.76 Distinction between mixed acid and butanediol fermentation in enteric bacteria. The bold arrows indicate reactions leading to major products. Dashed arrows indicate minor products.

groups of enteric bacteria, and these are listed in Table 16.27. On the basis of these and other tests, the genera can be defined, as outlined in Tables 16.28 and 16.29.

Positive identification of enteric bacteria often presents considerable difficulty because of the large num-ber of strains that have been characterized. Almost any small set of diagnostic characteristics fails to provide clear-cut distinctions between genera if a large number of strains are tested because exceptional strains are iso-lated. In advanced work in clinical laboratories, iden-

TABLE 16.27 **Key differential media and tests for classifying enteric bacteria[a]**

Media and tests	Composition and procedure
Triple sugar iron (TSI) agar	Butt reactions: acid (yellow), gas Slant reactions: acid (yellow), alkaline (red) H_2S reaction: blackening of butt (in the absence of acid butt reaction)
Urea medium	Urea agar with phenol red indicator; look for alkaline reaction due to urease action on urea, liberating ammonia
Citrate medium	Growth on citrate as sole energy source
Indole test	Peptone broth with high tryptophan content; assay for indole
Voges–Proskauer (VP) test	Glucose-containing broth; assay for acetoin
Methyl red (MR) test	Buffered glucose–peptone broth; measure pH drop with methyl red indicator
KCN medium	Broth with 75 μg/ml potassium cyanide; look for growth
Phenylalanine agar	Contains 0.1% phenylalanine; after growth, look for phenylpyruvic acid production (indicative of phenylalanine deaminase) by adding ferric chloride solution and looking for green color formation
β-Galactosidase	Medium containing the lactose analog, o-nitrophenylgalactoside (ONPG); positive strains cleave this to yield yellow product
Utilization of carbon sources	Mannitol, tartrate, mucate, acetate, dulcitol, sorbitol, adonitol, inositol

[a]Several of these tests are depicted in color in Figure 21.7.

TABLE 16.28 Key diagnostic reactions used to separate the various genera of enteric bacteria[a]

Genus	H₂S (TSI)	Urease	VP	Indole	Motility	Gas from glucose	β-Galac-tosidase
Escherichia	−	−	−	+	+ or −	+	+
Enterobacter	−	−	+	−	+	+	+
Shigella	−	−	−	+ or −	−	−	+ or −
Edwardsiella	+	−	−	+	+	+	−
Salmonella	+	−	−	−	+	+	+ or −
Klebsiella	−	+	+ or −	−	−	+	+
Arizona	+	−	−	−	+	+	+
Citrobacter	+ or −	−	−	−	+	+	+
Proteus	+ or −	+	−	+ or −	+	+ or −	−
Providencia	−	−	−	+	+	−	−
Yersinia	−	+	−	−	+[b]	−	+
Hafnia	−	−	+	−	+	+	+ or −

Genus	KCN	Citrate	Mucate	Phenyl-methyl red	Tartrate	Alanine deaminase	DNA (mol % GC)
Escherichia	−	−	+	+	+	−	48–52
Enterobacter	+	+	+	−	−	−	52–60
Shigella	−	−	−	+	−	−	50
Edwardsiella	−	−	−	+ or −	−	−	53–59
Salmonella	−	+ or −	+ or −	+	+ or −	−	50–53
Klebsiella	+	+	+	−	+ or −	−	53–58
Arizona	−	+	+ or −	+	−	−	50
Citrobacter	+ or −	+	+	+	+	−	50–52
Proteus	+	+ or −	−	+	+	+	38–41
Providencia	+	+	−	+	+	+	39–42
Yersinia	−	−	−	+	−	−	46–50
Hafnia	+	+	−	+	−	−	48–49

[a]See Tables 16.26 and 16.27 for the procedures for these diagnostic reactions. Tartrate: utilization as carbon source; mucate: fermentation.
[b]Motile when grown at room temperature; nonmotile at 37°C.

tification is now based on computer analysis of a large number of diagnostic tests carried out using miniaturized rapid diagnostic media kits and immunological and nucleic acid probes (∞ Figure 21.1), with consideration being given for variable reactions of exceptional strains. Thus, the separation of genera given in Tables 16.28 and 16.29 must be considered approximate, for it is always possible to isolate a strain that does not possess one or another characteristic normally considered positive for the genus as a whole. With these limitations in mind, an even more simplified separation of the key genera is found in Figure

16.77. This key permits a quick decision on the likely *genus* in which to place a new isolate.

Escherichia

Members of the genus *Escherichia* are almost universal inhabitants of the intestinal tract of humans and warm-blooded animals, although they are by no means the dominant organisms in these habitats. *Escherichia* may play a nutritional role in the intestinal tract by synthesizing vitamins, particularly vitamin K. As a facultative aerobe, this organism probably also

TABLE 16.29 Key diagnostic reactions used to separate the various genera of 2,3-butanediol producers

Genus	Ornithine decarboxyl-ase	Gelatin hydrolysis	Temperature optimum (°C)	Pigmentation	Motility	Lactose	DNase	Sorbitol	DNA (mol % GC)
Klebsiella	−	−	37–40	None	−	+	−	+	53–58
Enterobacter	+	Slow	37–40	Yellow (or none)	+	+	−	+	52–60
Serratia	+	+	37–40	Red (or none)	+	−	+	−	52–60
Erwinia	−	+ or −	27–30	Yellow (or none)	+	+ or −	−	+	50–58
Hafnia	+	−	35	None	+	−	−	−	48–49

Diagnostic test	Go to number
1 MR +; VP −	
(mixed-acid fermenters)	2
MR −; VP +	
(butanediol producers)	7
2 Urease +	*Proteus*
Urease −	3
3 H₂S (TSI) +	4
H₂S (TSI) −	6
4 KCN +	*Citrobacter*
KCN −	5
5 Indole +; citrate −	*Edwardsiella*
Indole −; citrate +	*Salmonella*
6 Gas from glucose	*Escherichia*
No gas from glucose	*Shigella*
7 Nonmotile; ornithine −	*Klebsiella*
Motile; ornithine +	8
8 Gelatin +; DNAse +	*Serratia*
	(red pigment)
Gelatin slow; DNAse −	*Enterobacter*

Key
■ Mixed-acid fermenters
▨ Butanediol producers

Figure 16.77 A simplified key to the main genera of enteric bacteria. Only the most common genera are given. See text for precautions in the use of this key. Diagnostic tests for use with this figure are given in Table 16.27. Other characteristics of the genera are given in Tables 16.28 and 16.29. Color coding is as in Figure 16.76.

helps consume oxygen, thus rendering the large intestine anoxic. Wild-type *Escherichia* strains rarely show any growth-factor requirements and are able to grow on a wide variety of carbon and energy sources such as sugars, amino acids, organic acids, and so on. Some strains of *Escherichia* are pathogenic. The latter strains of *Escherichia* have been implicated in diarrhea in infants, occasionally occurring in epidemic proportions in children's nurseries or obstetric wards, and *Escherichia* may also cause urinary tract infections in older persons or in those whose resistance has been lowered by surgical treatment or by exposure to ionizing radiation. Enteropathogenic strains of *E. coli* are becoming more frequently implicated in dysentery-like infections and generalized fevers (∞ Sections 19.6, 19.9 and 23.13). As noted, these strains form *K antigen*, permitting attachment and colonization of the small intestine, and *enterotoxin*, responsible for the symptoms of diarrhea.

Shigella

The shigellas are genetically very closely related to *Escherichia;* in fact they are so similar that they are able to undergo genetic recombination with each other and are susceptible to some of the same bacteriophages. Tests for DNA homology show strains of *Shigella* having 70 to nearly 100% homology with *Escherichia coli.* In contrast to *Escherichia,* however, *Shigella* is commonly pathogenic to humans, causing a rather severe gas-

troenteritis usually called *bacillary dysentery. Shigella dysenteriae* is transmitted by food and waterborne routes and is capable of invading intestinal epithelial cells. Once established, it produces both an endotoxin and a neurotoxin that exhibits enterotoxic effects.

Salmonella

Salmonella and *Escherichia* are quite closely related; the two genera have about 45–50% of their DNA sequences in common. However, in contrast to *Escherichia,* members of the genus *Salmonella* are usually pathogenic, either to humans or to other warm-blooded animals. In humans the most common diseases caused by salmonellas are *typhoid fever* and *gastroenteritis* (∞ Sections 23.13 and 23.14). The salmonellas are characterized immunologically on the basis of three cell surface antigens, the O, or cell wall (somatic) antigen; the H, or flagellar, antigen; and the Vi (outer polysaccharide layer) antigen, found primarily in strains of *Salmonella* causing typhoid fever. The O antigens are complex lipopolysaccharides that are part of the endotoxin structure of these organisms (∞ Section 19.10). We discussed the chemical structure of lipopolysaccharides in Section 3.6 (∞ Figures 3.33 and 3.34). The genus *Salmonella* contains over 1000 distinct types having different antigenic specificities in their O antigens. Additional antigenic subdivisions are based on the antigenic specificities of the flagellar (H) antigens. There is little or no correlation between the antigenic type of a *Salmonella* and the disease symptoms elicited, but antigenic typing permits tracing a single strain involved in an epidemic.

Proteus

The genus *Proteus* is characterized by rapid motility and by production of the enzyme *urease.* By DNA homology it shows only a distant relationship to *Escherichia coli. Proteus* is a frequent cause of urinary tract infections in people and rarely may cause enteritis. Because of its active urea-splitting ability, *Proteus* has been implicated in several kidney infections (∞ Section 19.5). The species of *Proteus* probably do not form a homogeneous group, as is indicated by the fact that DNA base compositions vary over a fairly wide range (39–50% GC). Because of the rapid motility of *Proteus* cells, colonies growing on agar plates often exhibit a characteristic **swarming** phenomenon. Cells at the edge of the growing colony are more rapidly motile than those in the center of the colony; the former move a short distance away from the colony in a mass and then undergo a reduction in motility, settle down, and divide, forming a new crop of motile cells that again swarm. As a result, the mature colony appears as a series of concentric rings, with higher concentrations of cells alternating with lower concentrations.

Although all enteric bacteria can use nitrate as alternate electron acceptor anaerobically, *Proteus* has the additional ability to use several *sulfur* compounds as electron acceptors for anaerobic growth: thiosulfate, tetrathionate, and dimethyl sulfoxide.

Yersinia

The genus *Yersinia* consists of three species, *Y. pestis*, the causal agent of the ancient and dreaded disease bubonic plague (∞ Section 23.12); *Y. pseudotuberculosis*, the causal agent of a tuberculosis-like disease of the lymph nodes in animals (and rarely in humans); and *Y. enterocolitica*, the causal agent of an intestinal infection (and also occasionally systemic infections) in humans and animals. Although the latter two species are rarely involved in fatal infections, *Y. pestis* was responsible for the so-called Black Death that ravaged Europe during the fourteenth century and killed over *one-fourth* of the European population. Other pandemics of bubonic plague occurred in subsequent centuries, causing great suffering and countless thousands of deaths. The differentiation of *Yersinia* from other mixed-acid fermenters is given in Table 16.28. In recent years, the prevalence of *Y. enterocolitica* as a waterborne and foodborne pathogen has been recognized (∞ Sections 23.13 and 23.14), and now that methods are available for its rapid isolation and recognition, it is being detected with increasing frequency.

Butanediol fermenters

The butanediol fermenters are genetically more closely related to each other than to the mixed-acid fermenters, a finding that is in agreement with the observed physiological differences. Their DNA base composition is higher, 53–58% GC, and genetic recombination does not occur between the two groups. However, it is possible to transfer plasmids from a mixed-acid fermenter to a butanediol fermenter; the plasmid replicates in the latter but does not become integrated into the chromosome. A current classification of this group is outlined in Table 16.29.

One species of *Klebsiella*, *K. pneumoniae*, occasionally causes pneumonia in humans, but klebsiellas are most commonly found in soil and water. Most *Klebsiella* strains fix N_2 (∞ Section 13.27), a property not found among other enteric bacteria. The genus *Serratia* is also a butanediol producer and forms a series of red pyrrole-containing pigments called **prodigiosins**. This pigment, a linear tripyrrole, is of interest because it contains the pyrrole ring also found in the pigments involved in energy transfer: porphyrins, chlorophylls, and phycobilins. There is no evidence that prodigiosin plays any role in energy transfer, however, and its exact function is unknown. Species of *Serratia* can be isolated from water and soil as well as from the gut of various insects and vertebrates and occasionally from the intestines of humans.

> ### CONCEPT CHECK   16.20
>
> *Zymomonas* is obligately anaerobic, while vibrios and enteric bacteria are facultative aerobes. From a phylogenetic standpoint, all these organisms are purple bacteria.
>
> ✔ *Compare and contrast* Zymomonas *with* Vibrio *and* Escherichia *in terms of metabolism and pathogenicity.*
>
> ✔ *What major features differentiate* Escherichia *from* Enterobacter?

16.21
Neisseria and Other Gram-Negative Cocci

This group comprises a diverse collection of organisms, related by Gram stain, morphology, lack of motility, nonfermentative aerobic metabolism, and similar DNA base composition. The five genera *Neisseria*, *Moraxella*, *Kingella*, *Psychrobacter*, and *Acinetobacter* are distinguished as outlined in Table 16.30. A major distinction is made on the basis of oxidase reaction. The oxidase-positive organisms (such as most isolates of *Neisseria gonorrhoeae*) (∞ Figure 21.5b) are unusually sensitive to penicillin, being inhibited by 1 μg/ml, differing in this way from most other gram-negative Bacteria, including *Acinetobacter*.

In the genus *Neisseria*, the cells are always cocci (∞ Figure 23.24a), whereas the cells of the other genera are rod-shaped, becoming coccoid in the stationary phase of growth. This had led to designation of these organisms as **coccobacilli.** Organisms of the genera *Neisseria*, *Kingella*, and *Moraxella* are commonly isolated from animals, and some of them are pathogenic, whereas organisms of the genus *Acinetobacter* are common soil and water organisms, although they are occasionally found as parasites of certain animals and have been implicated in nosocomial infections. Some strains of *Moraxella* and *Acinetobacter* possess the interesting property of **twitching motility,** exhibited as brief translocative movements or "jumps" covering distances of about 1–5 μm. We discuss the clinical microbiology of *Neisseria gonorrhoeae* in Section 21.1 and the disease gonorrhea in Section 23.6.

16.22
Rickettsias

The rickettsias are gram-negative, coccoid or rod-shaped cells in the size range of 0.3–0.7 μm wide and 1–2 μm long. They are, with one exception, *obligate intracellular parasites* (Figure 16.78a) and have not yet been cultivated in the absence of host cells. Rickettsias are the causative agents of such diseases as typhus fever, Rocky Mountain spotted fever, and Q fever (∞ Section 23.9). Electron

TABLE 16.30   Characteristics of the genera of Gram-negative cocci[a]

Characteristics	Genus	Number of species	DNA (mol % GC)
I. Oxidase-positive, penicillin-sensitive:			
Cocci; complex nutrition, utilize carbohydrates, obligate aerobes	*Neisseria*	12	49–55
Rods or cocci; generally no growth-factor requirements, generally do not	*Moraxella, Branhamella*	10	40–47
utilize carbohydrates; do not contain flagella, but some species exhibit	*Kingella*	2	47–55
"twitching" motility; many are commensals or pathogens of animals	*Psychrobacter*	1	44–46
II. Oxidase-negative, penicillin-resistant: some strains can utilize a restricted range of sugars, and some exhibit "twitching" motility	*Acinetobacter*	13	38–47

[a]Phylogenetically, those species examined are members of the beta purple Bacteria (∞ Section 15.7 and Table 16.1).

micrographs of thin sections of rickettsias show cells with a normal bacterial morphology (Figure 16.78*b*); both cell wall and cell membrane are visible. The cell wall contains muramic acid and diaminopimelic acid. Both RNA and DNA are present, and the DNA is in the normal double-stranded form, with a GC content varying from 29 to 33% in various species of the genus *Rickettsia* and 43% for *Coxiella burnetii* (the causal agent of Q fever). The rickettsias divide by normal binary fission, with doubling times of about 8 hr. The penetra-

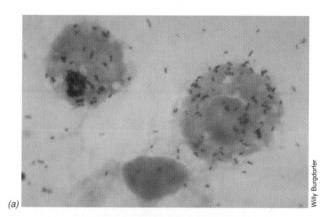

(a)

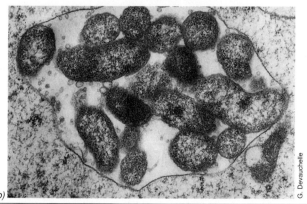

(b)

Figure 16.78 Rickettsias growing within host cells. (a) *Rickettsia rickettsii* in tunica vaginalis cells of the vole, *Microtus pennsylvanicus.* Cells are about 0.3 μm in diameter. (b) Electron micrograph of cells of *Rickettsiella popilliae* within a blood cell of its host, the beetle *Melolontha melolontha.* Notice that the bacteria are growing within a vacuole within the host cell.

tion of a host cell by a rickettsial cell is an active process, requiring both host and parasite to be alive and metabolically active. Once inside the phagocytic cell, the bacteria multiply primarily in the cytoplasm and continue replicating until the host cell is loaded with parasites (see Figures 16.78 and 23.37), at which time the host cell bursts and liberates the bacteria into the surrounding fluid.

Metabolism and pathogenesis

Much attention has been directed to the metabolic activities and biochemical pathways of rickettsias in an attempt to explain why they are obligate intracellular parasites. Many rickettsias possess a highly distinctive energy metabolism, being able to oxidize only glutamate or glutamine and being unable to oxidize glucose, glucose 6-phosphate, or organic acids. However, *Coxiella burnetii* is able to utilize both glucose and pyruvate as electron donors. Rickettsias possess a respiratory chain complete with cytochromes and are able to carry out electron transport phosphorylation, using NADH as electron donor. They are also able to synthesize at least some of the small molecules needed for macromolecular synthesis and growth, and they obtain the rest of their nutrients from the host cell. There is some suggestion that the host also provides some key coenzymes, such as NAD^+ and coenzyme A. A summary of the biochemical properties of rickettsias is given in Table 16.31.

Rickettsias do not survive long outside their hosts, and this may explain why they must be transmitted from animal to animal by arthropod vectors. When the arthropod obtains a blood meal from an infected vertebrate, rickettsias present in the blood are inoculated directly into the arthropod, where they penetrate to the epithelial cells of the gastrointestinal tract, multiply, and appear later in the feces. When the arthropod feeds on an uninfected individual, it then transmits the rickettsias either directly with its mouthparts or by contaminating the bite with its feces. However, the causal agent of Q fever, *Coxiella burnetii* (∞ Section 23.9), can also be transmitted to the respiratory system by aerosols. *Coxiella burnetii* is the most resistant of the

TABLE 16.31   Major properties of rickettsias and comparison of rickettsias with chlamydias and viruses

Property	Rickettsias	Chlamydias	Viruses
Structural			
Nucleic acid	RNA and DNA	RNA and DNA	Either RNA or DNA, never both
Ribosomes	Present	Present	Absent
Cell wall	Muramic acid, diaminopimelic acid (DPA) present	Muramic acid, DPA not present	No wall
Structural integrity during multiplication	Maintained	Maintained	Lost
Metabolic capacities			
Macromolecular synthesis	Carried out	Carried out	Only with use of host machinery
ATP-generating system	Present	Absent	Absent
Capable of oxidizing glutamate	Yes	No	No
Sensitivity to antibacterial antibiotics	Sensitive	Sensitive (except for penicillin)	Resistant

rickettsias to physical damage, probably because it produces an endospore-like form, and this explains its ability to survive in air. *Rochalimaea* is an atypical rickettsia because it can be grown in culture and is thus not an obligate intracellular parasite. In addition, when growing in tissue culture, cells of *Rochalimaea* grow on the *outside surface* of the eukaryotic host cells rather than within the cytoplasm or the nucleus. *Rochalimaea quintana* is the causative agent of *trench fever*, a disease that decimated troops in World War I. A summary of the major properties of species of *Rickettsia*, *Rochalimaea*, and *Coxiella* is given in Table 16.32.

Phylogeny of rickettsias

Despite their intracellular existence, the phylogenetic relationships of the rickettsias have been determined.

By 16S rRNA sequencing, the rickettsias clearly group within the purple Bacteria (∞ Section 15.7), more specifically with the plant pathogen *Agrobacterium*. Although at first this may appear puzzling, it should be remembered that *Agrobacterium*, like the rickettsias, has evolved close associations with eukaryotic cells, being responsible for the plant disease crown gall (∞ Section 14.22). The 16S rRNA sequences of rickettsias and *Agrobacterium tumefaciens* are about 95% homologous; those of *Escherichia coli* or *Bacillus subtilis* and the rickettsias are less than 80% homologous. The evolutionary relationships between rickettsias and intracellular symbionts such as *A. tumefaciens* therefore suggest that rickettsias evolved from plant-associated bacteria. Perhaps the rickettsias were originally *plant* pathogens that evolved to be associated with animals following transfer from plants to animals by insect vehicles.

TABLE 16.32   Characteristics of representative *Rickettsia*, *Rochalimaea*, and *Coxiella* species[a]

Species	Rickettsial group	Alternate host	Cellular location	DNA (mol % GC)	DNA hybridization to *R. rickettsii* DNA (%)[b]
Rickettsia					
R. rickettsii	Spotted fever	Tick	Cytoplasm and nucleus	32–33	100
R. prowazekii	Typhus	Louse	Cytoplasm	29–30	53
R. typhi	Typhus	Flea	Cytoplasm	29–30	36
Rochalimaea					
R. quintana	Trench fever	Louse	Epicellular	39	30
R. vinsonii	—	Vole	Epicellular	39	30
Coxiella					
C. burnetii	Q fever	Tick	Vacuoles	43	—

[a]Phylogenetically, those species examined fall within the alpha purple Bacteria (∞ Section 15.7 and Table 16.1).
[b]For discussion of DNA:DNA hybridization, see Section 15.9.

16.23

Chlamydias

Organisms of the genus *Chlamydia* probably represent a further stage in degenerate evolution from that discussed above for the rickettsias because the chlamydias are obligate parasites in which there has been an even greater loss of metabolic function. Three species of *Chlamydia* are recognized (Table 16.33): *C. psittaci*, the causative agent of the disease *psittacosis*; *C. trachomatis*, the causative agent of *trachoma* and a variety of other human diseases; and *C. pneumoniae*, the cause of a variety of respiratory syndromes (Table 16.34). **Psittacosis** is an epidemic disease of birds that is occasionally transmitted to humans and causes pneumonia-like symptoms. **Trachoma** is a debilitating disease of the eye characterized by vascularization and scarring of the cornea. Trachoma is the leading cause of blindness in humans. Other strains of *C. trachomatis* infect the genitourinary tract, and chlamydial infections are one of the leading sexually transmitted diseases today (∞ Section 23.6). A comparison of the properties of *C. psittaci*, *C. trachomatis*, and *C. pneumoniae* is shown in Table 16.33.

Besides being disease entities, the chlamydias are intriguing because of the biological and evolutionary problems they pose. Biochemical studies show that the chlamydias have gram-negative cell walls (although peptidoglycan is absent), and they have both DNA and RNA. Electron microscopy of thin sections of infected cells shows forms that clearly are undergoing binary fission (Figure 16.79). The biosynthetic capacities of the chlamydias are even more restricted than those of the rickettsias. This raises the interesting question of the limits to which evolutionary loss of function can be pushed while independence of macromolecular function is still retained. Phylogenetically, chlamydias form a major branch of Bacteria, totally unrelated to rickettsias or other gram-negative Bacteria (∞ Section 15.7).

Life cycle of Chlamydia

The life cycle of a typical member of the genus *Chlamydia* is shown in Figure 16.80. Two cellular types are seen in a typical life cycle: a small, dense cell, called an **elementary body,** which is relatively resistant to drying and is the means of *dispersal* of the agent, and a larger, less dense cell, called a **reticulate body,** which divides by binary fission and is the *vegetative* form. Elementary bodies are nonmultiplying cells specialized for transmission, whereas reticulate bodies are noninfectious forms that specialize in intracellular multiplication. Unlike the rickettsias just discussed, the chlamydias are not transmitted by arthropods but are primarily *airborne* invaders of the respiratory system—hence the significance of resistance to drying of elementary bodies. When a virus infects a cell, it loses its structural integrity and liberates nucleic acid. When an elementary body enters a cell, however, although it changes form, it remains a structural unit and enlarges and begins to undergo binary fission. A reticulate body is seen in Figure 16.79. After a number of divisions, the vegetative cells are converted to elementary bodies that are released when the host cell disintegrates and can then infect other cells. Generation times of 2–3 hr have been reported, which are considerably faster than those found for the rickettsias.

Molecular and metabolic properties

As we noted, chlamydia cells have a chemical composition similar to that of other Bacteria. Both RNA and DNA are present. The DNA content of a chlamydia corresponds to about twice that of vaccinia virus and one-eighth to one-fifth that of *Escherichia coli;* genome sizes range between about 500 and 1000 kilobase pairs. At least some of the RNA is in the form of ribosomes, and, like those of other prokaryotes, the ribosomes are 70S particles composed of one 50S and one 30S unit (∞ Sections 6.9 and 15.5).

TABLE 16.33	Differential characteristics of species of the genus *Chlamydia*[a]		
Characteristic	**C. trachomatis**	**C. psittaci**	**C. pneumoniae**
Hosts	Humans	Birds, lower mammals, occasionally humans	Humans
Usual site of infection	Mucous membrane	Multiple sites	Respiratory mucosa
Human-to-human transmission	Common	Rare	Probable
Mol % GC	42–45	39–43	40
Percent homology to *C. trachomatis* DNA by DNA:DNA hybridization[b]	100	10	10
DNA, kilobase pairs/genome (*Escherichia coli* = 4700)	1000	550	~1000

[a]Phylogenetically, species of *Chlamydia* form a distinct line of Bacteria (∞ Section 15.7 and Figure 16.1).
[b]For discussion of DNA:DNA hybridization, see Section 15.9.

TABLE 16.34 Chlamydial diseases

Human diseases caused by *C. trachomatis*
Trachoma
Inclusion conjunctivitis
Otitis media
Infant pneumonia
Nongonococcal urethritis (males)
Urethral inflammation (females)
Lymphogranuloma venereum
Cervicitis

Human diseases caused by *C. pneumoniae*
Respiratory syndromes probably transmitted from person to person

Human diseases caused by *C. psittaci*
Psittacosis

Diseases caused by *C. psittaci* in nonhuman hosts
Avian chlamydiosis (parrots, parakeets, pigeons, turkeys, geese, other birds)
Seminal vesiculitis (sheep, cattle)
Pneumonia (kittens, lambs, calves, piglets, foals)
Conjunctivitis (lambs, calves, piglets, cats)
Synovial tissue arthritis (lambs, calves, foals, piglets)

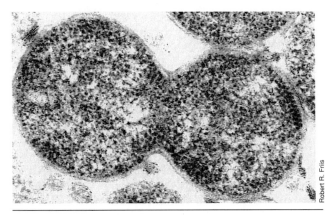

Figure 16.79 Electron micrograph of a thin section of a dividing cell (reticulate body) of *Chlamydia psittaci*, a member of the psittacosis group, within a mouse tissue culture cell. A single chlamydial cell is about 1 μm in diameter.

The metabolic properties of chlamydias purified from infected cells have been studied by methods similar to those used with the rickettsias. The biosynthetic capacities of the chlamydias are much more limited than are those of the rickettsias (see Table 16.31). Although macromolecular syntheses occur in the chlamydias, no energy-generating system is present and the cells are thus "energy parasites" on their host cells. From the limited biosynthetic and catabolic capacities of chlamydias it is easy to see why they are obligate parasites. The chlamydias probably have the simplest biochemical abilities of all cellular organisms.

CONCEPT CHECK   **16.23**

Rickettsia and *Chlamydia* species share a common property of being obligate intracellular parasites. They are deficient in many metabolic functions and cause a variety of diseases.

✔ *Compare and contrast* Neisseria *with* Rickettsia *and* Chlamydia *in terms of ability to be grown in artificial media, major diseases produced, and phylogenetic position.*

16.24

Gram-Positive Bacteria: Cocci

With the exception of *Deinococcus*, all gram-positive Bacteria form a phylogenetically coherent group. However, two major subgroups of gram-positive Bacteria emerge from ribosomal RNA sequencing studies:

the *Clostridium* subgroup, consisting of the endospore formers, the lactic acid bacteria, and most gram-positive cocci (these organisms generally have a fairly *low* mole percent GC in their DNA) and the Actinomycetes subgroup, which essentially consists of genera of actinomycetes, most of which have a rather *high* GC content, and the genus *Propionibacterium*. We begin here with the gram-positive cocci, common organisms to the beginning student in microbiology.

The gram-positive cocci (Figure 16.81) include bacteria with widely differing physiological characteristics. The major genera of gram-positive cocci and some differentiating characteristics are given in Table 16.35, page 719. The genus *Streptococcus* is not considered here but instead is discussed in Section 16.25 along with the other lactic acid bacteria.

Staphylococcus *and* Micrococcus

Staphylococcus and *Micrococcus* (Figure 16.82, page 719) are both aerobic organisms with a typical respiratory metabolism. They are catalase-positive, and this test permits their distinction from *Streptococcus* and some other genera of gram-positive cocci.

As we discussed in Section 5.10, the gram-positive cocci are relatively resistant to reduced water potential and tolerate drying and high salt fairly well. Their ability to grow in media with high salt provides a simple means for isolation. If an inoculum is spread on an agar plate with a fairly rich medium containing about 7.5% NaCl and the plate incubated aerobically, gram-positive cocci often form the predominant colonies. Often, these organisms are pigmented, and this provides an additional aid in selecting gram-positive cocci.

The two genera *Micrococcus* and *Staphylococcus* can easily be separated based on the oxidation–fermentation (O/F) (∞ Table 21.3) test. *Micrococcus* is an obligate aerobe and produces acid from glucose only aerobically, whereas *Staphylococcus* is a facultative aerobe

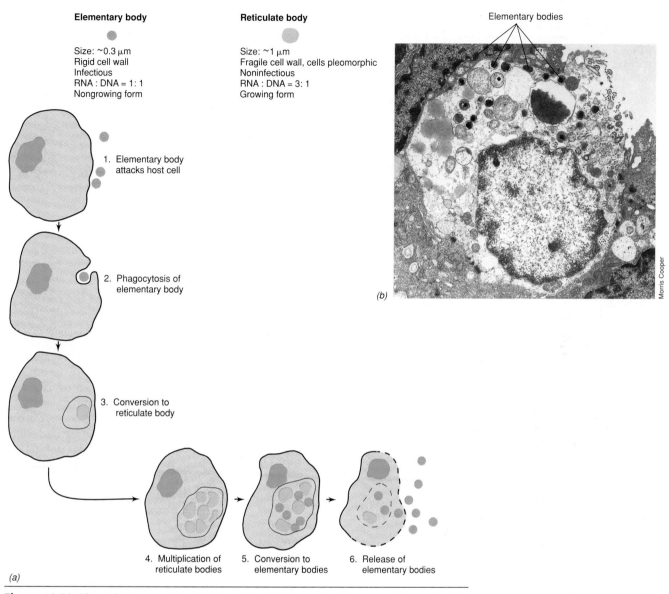

Elementary body

Size: ~0.3 μm
Rigid cell wall
Infectious
RNA : DNA = 1 : 1
Nongrowing form

Reticulate body

Size: ~1 μm
Fragile cell wall, cells pleomorphic
Noninfectious
RNA : DNA = 3 : 1
Growing form

Elementary bodies

1. Elementary body attacks host cell

2. Phagocytosis of elementary body

3. Conversion to reticulate body

4. Multiplication of reticulate bodies

5. Conversion to elementary bodies

6. Release of elementary bodies

(a)

(b)

Morris Cooper

Figure 16.80 The infectious cycle of chlamydia. (a) Schematic diagram of the cycle; the whole cycle takes about 48 hr. (b) Human chlamydial infection. An infected fallopian tube is bursting, releasing mature elementary bodies.

and produces acid from glucose both aerobically and anaerobically. Their DNA base compositions are also widely different: *Micrococcus* species have very high GC ratios, whereas those of *Staphylococcus* species are rather low (Table 16.35).

Staphylococci are common parasites of humans and animals and occasionally cause serious infections. In humans, two major forms are recognized, *Staphylococcus epidermidis,* a nonpigmented, nonpathogenic form usually found on the skin or mucous membranes, and *Staphylococcus aureus,* a yellow pigmented form that is most commonly associated with pathological conditions, including boils, pimples, pneumonia, osteomyelitis, meningitis, and arthritis. We list the exotoxins of *S. aureus* in Table 19.4. One of the significant

exotoxins is **coagulase,** an enzymelike factor that causes fibrin to coagulate and form a clot. Strains of *S. aureus* are generally coagulase-*positive,* whereas *S. epidermidis* is coagulase-*negative.* We discuss the possible role of the yellow carotenoid pigment of *S. aureus* in resistance to phagocytosis in Section 20.3, and staphylococcal food poisoning is discussed in Section 23.13.

Sarcina

The genus *Sarcina* contains two species of bacteria that divide in three perpendicular planes to yield packets of eight cells or more (Figure 16.81a). *Sarcina* are obligate anaerobes and are extremely acid-tolerant, being able to ferment sugars and grow down to pH 2. Cells

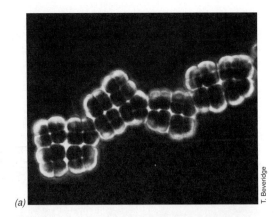

(a)

T. Beveridge

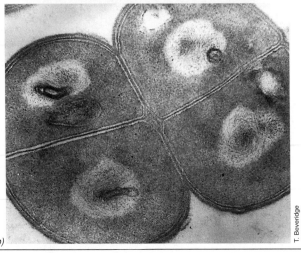

(b)

T. Beveridge

Figure 16.81 (a) Phase contrast photomicrograph of cells of a typical gram-positive coccus *Sarcina* sp. A single cell is about 2 μm in diameter. (b) Electron micrograph of a thin section.

of one species, *Sarcina ventriculi,* contain a thick fibrous layer of cellulose surrounding the cell wall (Figure 16.81b). The cellulose layers of adjacent cells become attached, and this functions as a cementing material to hold together packets of *S. ventriculi* cells. *Sarcina* can be isolated from soil, mud, feces, and stomach contents. Because of its extreme acid tolerance, *S. ventriculi* is one of the few bacteria that can actually grow in the stomach of humans and other monogastric animals. Rapid growth of *S. ventriculi* is observed in the stomach of humans suffering from certain gastrointestinal pathological conditions (such as pyloric ulcerations) that retard the flow of food to the intestine.

Deinococcus

The genus *Deinococcus* is an unusual genus of gram-positive cocci. Besides their unique phylogenetic stature (∞ Section 15.7), species of *Deinococcus* differ from other gram-positive cocci in a number of interesting chemical and physiological properties. The cell walls of deinococci are structurally complex and consist of several layers, including an *outer membrane* layer

(Figure 16.83) normally present only in gram-negative Bacteria (∞ Section 3.6). However, the outer membrane of *Deinococcus* is chemically unique and does not contain heptoses and lipid A typical of that of gram-negative Bacteria.

Most deinococci are bright red or pink in color because of the variety of carotenoids found in these organisms, and many strains are highly resistant to ultraviolet radiation and to desiccation. Resistance to radiation can be used to advantage in isolating deinococci. These remarkable organisms can be isolated from soil, ground meat, dust, and filtered air following exposure of the sample to intense ultraviolet (or even gamma) radiation and plating on a rich medium containing tryptone and yeast extract. Because many strains of *Deinococcus radiodurans* are even more resistant to radiation than bacterial endospores, treatment of a sample with strong doses of radiation effectively sterilizes the sample of organisms other than *D. radiodurans,* making isolation of deinococci relatively straightforward. For example, *D. radiodurans* cells can survive exposure to up to 30,000 Gy of ionizing radiation (1 Gy = 100 rad), sufficient to literally shatter the organism's chromosome into hundreds of fragments (by contrast, a human can be killed by exposure to less than 5 Gy). A powerful DNA repair machinery exists in *Deinococcus* cells (see later) and it is able to repair the organism's chromosome even from a fragmented state.

In addition to impressive radiation resistance, *Deinococcus radiodurans* is resistant to the mutagenic effects of many highly mutagenic chemicals. Studies of the mutability of *D. radiodurans* have shown it to be highly efficient in repairing damaged DNA. Several different DNA repair enzymes exist in *D. radiodurans* for repairing breaks in single- or double-stranded DNA and for excising and repairing thymine dimers formed by the action of ultraviolet light. These efficient repair mechanisms have actually hindered studies of the molecular genetics of *D. radiodurans.* Because it is difficult to induce stable mutations in the DNA of *D. radiodurans,* it has been nearly impossible to obtain *mutants* of this organism for genetic study! The only chemical mutagens that seem to work on *D. radiodurans* are agents like nitrosoguanidine, which tend to induce *deletions* in DNA; deletions are apparently not repaired as efficiently as point mutations in this organism. Consistent with the fact that *D. radiodurans* is an extremely radiation-resistant bacterium, strains of this organism have been isolated from near atomic reactors and other potentially lethal radiation sources.

16.25
Lactic Acid Bacteria

The lactic acid bacteria are gram-positive, usually nonmotile, nonsporulating Bacteria that produce lactic acid as a major or sole product of fermentative metab-

TABLE 16.35 Distinguishing features of gram-positive cocci[a]

Genus	Motility	Arrangement of cells	Growth by fermentation	DNA (mol % GC)	16S rRNA group[a]	Other characteristics
Micrococcus	−	Clusters, tetrads	−	66–73	High GC	Strict aerobe
Staphylococcus	−	Clusters, pairs	+	30–39	Low GC	Only genus to contain teichoic acid in cell wall
Stomatococcus	−	Clusters, pairs	+	56–60	—	Only genus containing a capsule
Planococcus	+	Pairs, tetrads	−	39–52	Low GC	Primarily marine
Sarcina	−	Cuboidal packets of eight or more cells	+	28–31	Low GC	Extremely acid-tolerant; cellulose in cell wall
Deinococcus	−	Pairs, tetrads	−	62–70	Deinococcus–Thermus	Phylogenetically unique (see Figure 16.1)
Ruminococcus	+	Pairs, chains	+	39–46	Low GC	Obligate anaerobe; inhabits rumen, cecum, and large intestine of many animals
Peptococcus	−	Clusters, pairs	+	50–51	Low GC	Obligate anaerobe; ferments peptone but not sugars
Peptostreptococcus	−	Clumps, short chains	+	28–37	Low GC	Obligate anaerobe; ferments peptone; common member of human normal flora, skin, intestine, vagina; also isolated from vaginal and purulent discharges

[a]With the exception of *Deinococcus*, all organisms listed belong to the high GC or low GC gram-positive Bacteria (∞ Section 15.7). Cocci of the lactic acid group are considered in Table 16.36.

olism. Members of this group lack porphyrins and cytochromes, do not carry out electron transport phosphorylation, and hence obtain energy only by *substrate-level phosphorylation.* All lactic acid bacteria grow anaerobically. Unlike many anaerobes, however, most lactic acid bacteria are not sensitive to O_2 and can grow in its presence as well as in its absence; thus they are **aerotolerant anaerobes.** Some strains are able to take up O_2 through the mediation of flavoprotein oxidase systems, producing H_2O_2, although most strains lack catalase and most dispose of H_2O_2 via alternative enzymes referred to as *peroxidases* (∞ Section 5.11). No ATP is formed in the flavoprotein oxidase reaction, but the oxidase system can be used for reoxidation of NADH generated during fermentation. Most lactic acid bacteria obtain energy only from the metabolism of sugars and related fermentable compounds and hence are usually restricted to habitats in which sugars are present. They usually have only limited biosynthetic ability, and their complex nutritional requirements include needs for amino acids, vitamins, purines, and pyrimidines.

Homo- and heterofermentation

One important difference between subgroups of the lactic acid bacteria lies in the nature of the products formed during the fermentation of sugars. One group, called **homofermentative,** produces virtually a single fermentation product, *lactic acid,* whereas the other group, called **heterofermentative,** produces other products, mainly *ethanol* and CO_2 as well as lactate. Abbreviated pathways for the fermentation of glucose by a homo- and a heterofermentative organism are shown in Figure 16.84. The differences observed in the fermentation products are determined by the presence or absence of the enzyme **aldolase,** one of the key enzymes in *glycolysis* (∞ Figure 4.12). The heterofermenters, lacking aldolase, cannot break down fructose bisphosphate to triose phosphate. Instead, they oxidize glucose 6-phosphate to 6-phosphogluconate and then decarboxylate this to pentose phosphate, which is broken down to triose phosphate and acetylphosphate by means of the enzyme **phosphoketolase** (Figure 16.84).

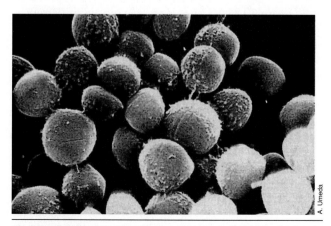

Figure 16.82 Scanning electron micrograph of typical *Staphylococcus,* showing the irregular arrangement of the cell clusters. Individual cells are about 0.8 μm in diameter.

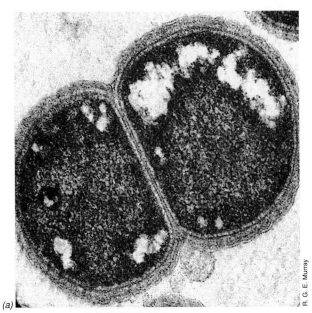

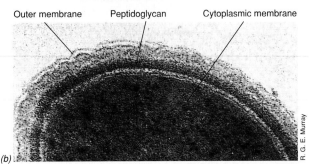

Outer membrane   Peptidoglycan   Cytoplasmic membrane

Figure 16.83 The radiation-resistant coccus *Deinococcus radiodurans*. An individual cell is about 2.5 µm in diameter. (a) Transmission electron micrograph of *D. radiodurans*. Note the outer membrane layer. (b) High magnification micrograph of wall layer.

In heterofermenters, triose phosphate is converted ultimately to lactic acid with the production of 1 mol of ATP, while the acetylphosphate accepts electrons from the NADH generated during the production of pentose phosphate and is thereby converted to ethanol *without* yielding ATP. Because of this, heterofermenters produce only *1 mol* of ATP from glucose instead of the 2 mol produced by homofermenters. This difference in ATP yield from glucose is reflected in the fact that homofermenters produce twice as much cell mass as heterofermenters from the same amount of glucose. Because the heterofermenters decarboxylate 6-phosphogluconate, they produce CO_2 as a fermentation product, whereas the homofermenters produce little or no CO_2; therefore one simple way of detecting a heterofermenter is to observe for production of CO_2 in laboratory cultures. At the enzyme level, heterofermenters are characterized by the lack of aldolase and the presence of phosphoketolase. Many strains of heterofermenters can use O_2 as an electron acceptor with a reduced flavoprotein serving as electron donor. In this reaction, half of the NADH generated from the oxidation of glucose to ribose is transferred to a flavin and on to O_2. Acetylphosphate can then be converted to acetate instead of being reduced to ethanol, and an additional ATP is synthesized.

The various genera of lactic acid bacteria have been defined on the basis of cell morphology, DNA base composition, and type of fermentative metabolism, as is shown in Table 16.36. Members of the genera *Streptococcus*, *Enterococcus*, *Lactococcus*, *Leuconostoc*, and *Pediococcus* have fairly similar DNA base ratio compositions; in addition, there is very little variation from strain to strain. The genus *Lactobacillus*, on the other hand, has members with widely diverse DNA compositions and hence does not constitute a homogeneous group.

Streptococcus *and other cocci*

The genus *Streptococcus* (Figure 16.85) contains a wide variety of species with quite distinct habitats, whose activities are of considerable practical importance to humans. Some members are pathogenic to people and animals (∞ Section 23.2). As producers of lactic acid, certain streptococci play important roles in the production of buttermilk, silage, and other fermented products (∞ Section 11.6). To distinguish generally nonpathogenic streptococci from human pathogenic species, the genus *Streptococcus* has been split into three genera. The genus *Lactococcus* contains those streptococci of dairy significance, whereas the genus *Enterococcus* includes streptococci that are primarily of fecal origin.

Organisms remaining in the genus *Streptococcus* have been further divided into two groups of related species on the basis of characteristics enumerated in Table 16.37, page 723. Hemolysis on blood agar is of considerable importance in the subdivision of the genus. Colonies of those strains producing streptolysin O or S are surrounded by a large zone of complete red blood cell hemolysis, a condition called **β hemolysis** (∞ Figure 19.16a). On the other hand, many streptococci and the lactococci and enterococci do not produce hemolysins but instead cause the formation of a greenish or brownish zone around colonies on blood agar, which is due not to true hemolysis but to discoloration and loss of potassium from the red cells. This type of reaction has classically been referred to as **α hemolysis.** Streptococci and related cocci are also divided into *immunological* groups based on the presence of specific carbohydrate antigens. These antigenic groups (or **Lancefield groups** as they are commonly known, named for Rebecca Lancefield, a pioneer in *Streptococcus* taxonomy), are designated by letters; A through O are currently recognized. Those β-hemolytic streptococci found in human beings usually contain the group A antigen, which is a cell wall polymer containing

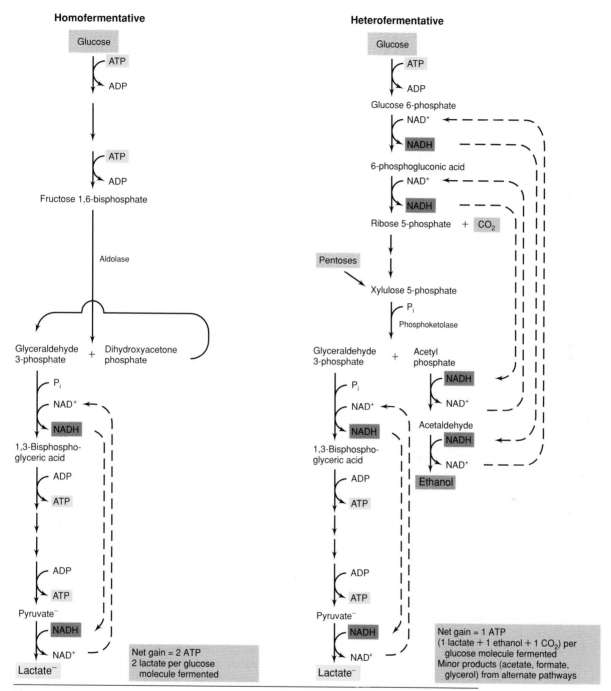

Figure 16.84 The fermentation of glucose in homofermentative and heterofermentative lactic acid bacteria. Note that no ATP is made in reactions leading to ethanol formation.

N-acetylglucosamine and rhamnose. Enterococci contain the group D antigen, a glycerol teichoic acid containing glucose side chains. Group B streptococci are usually found in association with animals, are a cause of mastitis in cows, and have also been implicated in human infections. Lactococci are of antigen group N and are not pathogenic.

Placed in the genus *Leuconostoc* are cocci that are morphologically similar to streptococci but are *hetero-*

fermentative. Strains of *Leuconostoc* also produce the flavoring ingredients diacetyl and acetoin by breakdown of citrate and have been used as starter cultures in dairy fermentations, but their place has now been taken by *Streptococcus diacetilactis*. Some strains of *Leuconostoc* produce large amounts of dextran polysaccharides (α-1,6-glucan) when cultured on sucrose (∞ Figure 13.44). Dextrans produced by *Leuconostoc* have found some medical use as plasma extenders in

TABLE 16.36 Differentiation of the principal genera of lactic acid bacteria[a]

Genus	Cell form and arrangement	Fermentation	DNA (mol % GC)
Streptococcus	Cocci in chains	Homofermentative	34–46
Leuconostoc	Cocci in chains	Heterofermentative	38–41
Pediococcus	Cocci in tetrads	Homofermentative	34–42
Lactobacillus	(1) Rods, usually in chains	Homofermentative	32–53
	(2) Rods, usually in chains	Heterofermentative	34–53
Enterococcus	Cocci in chains	Homofermentative	38–40
Lactococcus	Cocci in chains	Homofermentative	38–41

[a]Phylogenetically, all organisms are members of the low GC gram-positive subdivision of the gram-positive Bacteria (◦◦ Section 15.7).

blood transfusions. Other strains of *Leuconostoc* produce fructose polymers called *levans*.

Lactobacillus

Lactobacilli are typically rod-shaped, varying from long and slender to short, bent rods (Figure 16.86). Most species are homofermentative, but some are heterofermentative. The genus has been divided into three major subgroups, and over 70 species are recognized (Table 16.38).

Lactobacilli are often found in dairy products, and some strains are used in the preparation of fermented milk products. For instance, *Lactobacillus delbrueckii* (Figure 16.86c) is used in the preparation of yogurt, *L. acidophilus* (Figure 16.86a) in the production of acidophilus milk, and other species in the production of sauerkraut, silage, and pickles (◦◦ Section 11.6). The lactobacilli are usually more resistant to acidic conditions than are the other lactic acid bacteria, being able to grow well at pH values around 4–5. Because of this, they can be selectively isolated from natural materials by use of carbohydrate-containing media of acid pH, such as tomato juice–peptone agar. The acid resistance of the lactobacilli enables them to continue growing during natural lactic fermentations when the pH value has dropped too low for other lactic acid bacteria to grow, and the lactobacilli are therefore responsible for the final stages of most lactic acid fermentations. The lactobacilli are rarely if ever pathogenic.

16.26
Endospore-Forming Gram-Positive Rods and Cocci

The structure and heat resistance of the bacterial endospore along with the process of spore formation itself was discussed in Section 3.13. Several genera of endospore-forming bacteria have been recognized, distinguished on the basis of morphology, relationship to O_2, and energy metabolism (Table 16.39). The two genera most frequently studied are *Bacillus*, the species of which are aerobic or facultatively aerobic, and *Clostridium*, which contains the strictly anaerobic species. The genera *Bacillus* and *Clostridium* consist of gram-positive or gram-variable rods, which are usually motile, possessing peritrichous flagella. A major property of taxonomic value for distinguishing between species of *Bacillus* and *Clostridium* is the shape and position of endospores. These properties vary considerably among different species of endospore formers (Figure 16.87, page 725) and are useful starting points for beginning a taxonomic study of the group. Most endospore-forming bacteria produce one spore per cell. Sporulation is thus a means of *survival*, not reproduction. *Anaerobacter polyendosporus* is unusual in this connection because up to five endospores can be produced per cell. Members of the genus *Bacillus* produce the enzymes catalase and superoxide dismutase. Clostridia do not produce catalase and produce only low levels of superoxide dismutase, and it is thought that one reason they are obligately anaerobic is that they have no way of getting rid

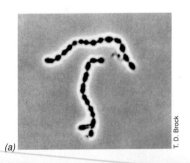
(a)
T. D. Brock

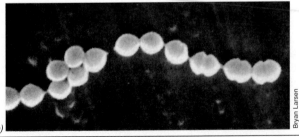
(b)
Bryan Larsen

Figure 16.85 Phase contrast (a) and scanning electron (b) micrographs of *Streptococcus* species. (a) *Streptococcus lactis*. (b) *Streptococcus* sp. Cells in both cases are 0.5–1 μm in diameter.

TABLE 16.37 Differential characteristics of streptococci, lactococci, and enterococci

Group	Antigenic (Lancefield) groups	Representative species	Type of hemolysis on blood agar	Good growth at 10°C	Good growth at 45°C	Survive 60°C for 30 min	Growth in Milk with 0.1% methylene blue	Growth in Broth with 40% bile	Habitat
Streptococci									
Pyogenes subgroup	A,B,C,F,G	*Streptococcus pyogenes*	Lysis (β)	−	−	−	−	−	Respiratory tract, systemic
Viridans subgroup	—	*Streptococcus mutans*	Greening (α)	−	+	−	−	−	Mouth, intestine
Enterococci	D	*Enterococcus faecalis*	Lysis (β), greening (α), or none	+	+	+	+	+	Intestine, vagina, plants
Lactococci	N	*Lactococcus lactis*	None	+	−	+	+	+	Plants, dairy products

of the toxic H_2O_2 and O_2^- produced from molecular oxygen (∞ Section 5.11).

Molecular taxonomic studies of the genus *Bacillus* have shown that it is a heterogeneous group and can

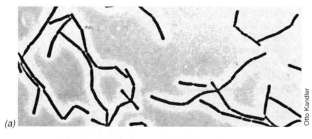

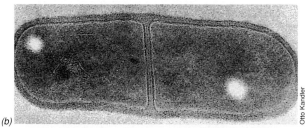

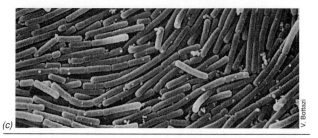

Figure 16.86 Phase contrast and electron micrographs of *Lactobacillus* species. (a) *Lactobacillus acidophilus.* Cells are about 0.75 μm wide. (b) *Lactobacillus brevis,* transmission electron micrograph. Cells measure about 0.8 × 2 μm. (c) *Lactobacillus delbrueckii,* scanning electron micrograph. Cells are about 0.7 μm in diameter.

hardly be considered an assemblage of closely related organisms. The DNA base composition of *Bacillus* species varies from about 30% to nearly 70% GC, suggesting considerable genetic heterogeneity. Even though they are not closely related genetically, all the endospore-forming bacteria are *ecologically* related because they are found in nature primarily in soil. Even those species that are pathogenic to humans or animals are primarily saprophytic soil organisms and infect hosts only incidentally. Spore formation should be advantageous for a soil microorganism because the soil is a highly variable environment. Although at some times nutrient supply is in excess, at other times it is deficient. Soil temperatures can be quite high in summer, especially at the surface. Thus, a heat-resistant dormant structure should offer considerable survival value in nature. On the other hand, the ability to germinate and grow quickly when nutrients become available is also of value as it enables the organism to capitalize on a transitory food supply.

Bacillus

Members of the genus *Bacillus* are easy to isolate from soil or dust particles in air and are among the most common organisms to appear when soil samples are streaked on agar plates containing various nutrient media. Spore formers can be selectively isolated from soil, food, or other material by exposing the sample to 80°C for 10 min, a treatment that effectively destroys vegetative cells while many of the spores present remain viable. When such pasteurized samples are streaked on plates and incubated aerobically, the colonies that develop are almost exclusively of the genus *Bacillus*. Bacilli usually grow well on synthetic media containing sugars, organic acids, alcohols, and so on, as sole carbon sources and ammonium as the sole nitrogen source; a few isolates have vitamin requirements. Many bacilli produce extracellular *hydrolytic enzymes* that break down

TABLE 16.38   Characteristics of subgroups in the genus *Lactobacillus*

Characteristics	Representative species	DNA (mol % GC)
Homofermentative:		
Lactic acid the major product (>85% from glucose)		
No gas from glucose; aldolase present		
(1) Grow at 45°C but not at 15°C, long rods; glycerol teichoic acid	*L. delbrueckii*	49–51
	L. acidophilus	34–37
(2) Grow at 15°C, variable growth at 45°C; short rods and coryneforms; ribitol and glycerol teichoic acids; can produce more oxidized fermentation products if O_2 is present	*L. casei, L. plantarum*	45–46
	L. curvatus	42–44
Heterofermentative:		
Produce about 50% lactic acid from glucose; produce CO_2 and ethanol; aldolase absent; phosphoketolase present; long and short rods; glycerol teichoic acid	*L. fermentum*	52–54
	L. brevis, L. buchneri	44–47
	L. kefir	41–42

polysaccharides, nucleic acids, and lipids, permitting the organisms to use these products as carbon sources and electron donors. Many bacilli produce antibiotics, of which bacitracin, polymyxin, tyrocidin, gramicidin, and circulin are examples. In most cases, antibiotic production seems to be related to the sporulation process, the antibiotic being released when the culture enters the stationary phase of growth and after it is committed to sporulation. An outline of the subdivision of the genus *Bacillus* is given in Table 16.40.

Bacterial insecticides

Several *Bacillus* species, most notably *B. popilliae* and *B. thuringiensis*, produce insect larvicides. *Bacillus popilliae* causes a fatal disease called *milky disease* in Japanese beetle larvae and larvae of closely related beetles of the family Scarabaeidae. *Bacillus thuringiensis* causes a fatal disease of larvae of many different groups of insects, although individual strains are specific as to host affected. Strains exist that are specific for lepidopterans, such as the silkworm, the cabbage worm, the tent caterpillar, and the gypsy moth. Some strains kill dipterans such as mosquitoes and black

flies. Others kill coleopterans such as Colorado potato beetles. Strains of *B. thuringiensis* have also been discovered that are toxic to Japanese beetles.

The disease caused by *Bacillus popilliae* is a septicemia, whereas the disease caused by *B. thuringiensis* is essentially an intoxication. Both of these insect pathogens form a crystalline protein during sporulation called the *parasporal body*, which is deposited within the sporangium but outside the spore proper (Figure 16.88). In the case of *B. thuringiensis*, the crystal (parasporal body) protein is a protoxin that is ultimately responsible for the fatal insect disease. The function of the crystal protein in *B. popilliae* is uncertain. Purified crystals do not cause the disease. Also, a closely related *Bacillus* species, *B. lentimorbus*, causes essentially the same disease as *B. popilliae*, but it does not produce parasporal crystals.

The crystal protoxin protein of *Bacillus thuringiensis* is converted to a toxin by proteolytic cleavage in the larval gut. The toxin binds to intestinal epithelial cells and induces pore formation that causes leakage of the host cells followed by lysis. Death is probably due to starvation. Thus, the disease is essentially an intoxication. Genes encoding crystal proteins from several *B. thuringiensis* strains have been isolated. Separate

TABLE 16.39   Genera of endospore-forming bacteria[a]

Characteristics	Genus	DNA (mol % GC)
Rods		
Aerobic or facultative, catalase produced	*Bacillus*	32–69
Microaerophilic, no catalase; homofermentative lactic acid producer	*Sporolactobacillus*	46–47
Anaerobic:		
Sulfate-reducing	*Desulfotomaculum*	38–50
Does not reduce sulfate, fermentative	*Clostridium* (see Figure 16.87)	21–54
Gram-negative; can grow as homoacetogen on $H_2 + CO_2$	*Sporomusa*	41–49
Halophile, isolated from the Dead Sea	*Sporohalobacter*	31
Produces up to five spores per cell; fixes N_2	*Anaerobacter*	29
Phototrophic	*Heliobacterium, Heliophilum*	50–58
Syntrophic, degrades fatty acids but only in coculture with a H_2-utilizing bacterium	*Syntrophospora*	37
Cocci (usually arranged in tetrads or packets), aerobic	*Sporosarcina* (see Figure 16.91)	40–41

[a]Phylogenetically, all organisms are members of the low GC subdivision of the gram-positive Bacteria (∞ Section 15.7).

Figure 16.87 Phase contrast photomicrographs of various *Clostridium* species, showing the different locations of the endospore. (a) *Clostridium cadaveris*, terminal spores. Cells are about 0.9 μm wide. (b) *Clostridium sporogenes*, subterminal spores. Cells are about 1 μm wide. (c) *Clostridium bifermentans*, central spores. Cells are about 1.2 μm wide.

(a) 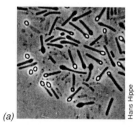   *(b)* 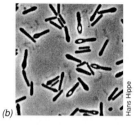   *(c)*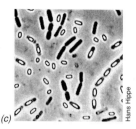

domains of the toxin molecule have been identified that are responsible for toxicity and binding to host cells. Genetic engineering is being used to modify host specificity of the crystal proteins. The genes for the *B. thuringiensis* crystal protein (known commercially as "Bt-toxin") have been introduced into plants to render the plants "naturally" resistant to insects. This strategy has been shown to be effective in controlled situations, and a variety of Bt-toxins are being developed by genetic engineering (∞ Section 10.14).

Spore (and parasporal crystal) preparations derived from insect pathogens are widely used as biological insecticides. Interest in such biological control is currently high because of declining use of chemical insecticides for human food crops.

Clostridium

The clostridia lack a cytochrome system and a mechanism for electron transport phosphorylation, and hence they obtain ATP *only* by substrate-level phosphorylation. A wide variety of anaerobic energy-yielding mechanisms are known in the clostridia (fermentative diversity is discussed in Section 13.20); indeed, the separation of the genus into subgroups is based primarily on these properties and on the nature of the electron donors used (Table 16.41).

A number of clostridia ferment sugars, producing as a major end product *butyric acid*. Some of these also produce *acetone* and *butanol,* and at one time acetone–butanol fermentation by clostridia was of great industrial importance as it was the main commercial source of these products. Today, however, the chemical synthesis of acetone and butanol from petroleum products has mostly replaced the microbiological process. Some clostridia of the acetone–butanol type fix N_2; the most vigorous N_2 fixer is *Clostridium pasteurianum,* which probably is responsible for most anaerobic nitrogen fixation in the soil. One group of clostridia ferments cellulose with the formation of acids and alcohols, and these are the main organisms decomposing cellulose anaerobically in soil. There is considerable industrial interest in the production of ethanol (an automotive fuel additive) by the clostridial fermentation of *cellulose,* and genetic studies are under way to increase the yield of ethanol and reduce the formation of acidic fermentation products, the goal being to use waste cellulose as a motor fuel.

TABLE 16.40   Characteristics of representative species of the genus *Bacillus*

Characteristics	Species	Spore position	DNA (mol % GC)
I. Spores oval or cylindrical, facultative aerobes, casein and starch hydrolyzed; sporangia not swollen, spore wall thin			
Thermophiles and acidophiles	*B. coagulans*	Central or terminal	47
	B. acidocaldarius	Terminal	60
Mesophiles	*B. licheniformis*	Central	46
	B. cereus	Central	35
	B. anthracis	Central	33
	B. megaterium	Central	37
	B. subtilis	Central	43
Insect pathogen	*B. thuringiensis*	Central	34
Sporangia distinctly swollen, spore wall thick			
Thermophile	*B. stearothermophilus*	Terminal	52
Mesophiles	*B. polymyxa*	Terminal	44
	B. macerans	Terminal	52
	B. circulans	Central or terminal	35
Insect pathogens	*B. larvae*	Central or terminal	—
	B. popilliae	Central	41
II. Spores spherical, obligate aerobes, casein and starch not hydrolyzed			
Sporangia swollen	*B. sphaericus*	Terminal	37
Sporangia not swollen	*B. pasteurii*	Terminal	38

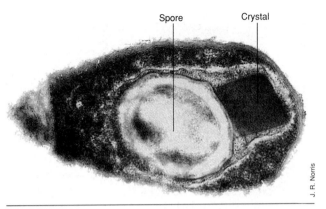

Spore   Crystal

J. R. Norris

Figure 16.88 Formation of the toxic parasporal crystal in the insect pathogen *Bacillus thuringiensis*. Electron micrograph of a thin section.

The biochemical steps in the formation of butyric acid and butanol from sugars are well understood (Figure 16.89). Glucose is converted to pyruvate via the Embden–Meyerhof pathway, and pyruvate is split to acetyl-CoA, CO_2, and hydrogen (reduced ferredoxin) by the phosphoroclastic reaction (∞ Section 13.20 and Figure 13.38). Acetyl-CoA is then reduced to fermentation products using the NADH derived from glycolytic reactions. The proportions of the various products are influenced by the duration and the con-

ditions of the fermentation. During the early stages, butyric and acetic acids are the predominant products, but as the pH of the medium drops, synthesis of acids ceases and the neutral products acetone and butanol begin to accumulate. If the medium is kept alkaline with $CaCO_3$, very little of the neutral products are formed and the fermentation products consist of about three parts butyric and one part acetic acid. Studies of the enzymology of solvent production by clostridia indicate that acid production in some way derepresses synthesis of enzymes involved in acetone and butanol synthesis.

Another group of clostridia obtain their energy by fermenting *amino acids*. Some strains do not ferment single amino acids, but only amino acid *pairs*. In this situation one functions as the electron *donor* and is *oxidized*, whereas the other acts as the electron *acceptor* and is *reduced*. The type of coupled decomposition is known as the **Stickland reaction.** For instance, *Clostridium sporogenes* catabolizes a mixture of glycine and alanine, as outlined in Figure 16.90.

Various amino acids that can function as either electron donors or acceptors in Stickland reactions are listed in Table 16.42. The products of Stickland oxidation are always NH_3, CO_2, and a carboxylic acid with one *less* carbon atom than the amino acid that is oxidized (Figure 16.90).

TABLE 16.41   Characteristics of some groups of the genus *Clostridium*

Key characteristics	Other characteristics	Species	DNA (mol % GC)
I. Ferment carbohydrates			
Ferment cellulose	Fermentation products: acetate, lactate, succinate, ethanol, CO_2, H_2	*C. cellobioparum*	28
		C. thermocellum	38–39
Ferment sugars, starch, and pectin	Fermentation products: acetone, butanol, ethanol, isopropanol, butyrate, acetate, propionate, succinate, CO_2, H_2; some fix N_2	*C. butyricum*	27–28
		C. acetobutylicum	28–29
		C. pasteurianum	26–28
		C. perfringens	24–27
		C. thermosulfurogenes	33
Ferment sugars primarily to acetic acid	Total synthesis of acetate from CO_2; cytochromes present in some species	*C. aceticum*	33
		C. thermoaceticum	54
		C. formicoaceticum	34
Ferments only pentoses or methylpentoses	Ring-shaped cells form left-handed, helical chains; fermentation products: acetate, propionate, *n*-propanol, CO_2, H_2	*C. methylpentosum*	46
II. Ferment proteins or amino acids	Fermentation products: acetate, other fatty acids, NH_3, CO_2, sometimes H_2; some also ferment sugars to butyrate and acetate; may produce exotoxins	*C. sporogenes*	26
		C. tetani	25–26
		C. botulinum	26–28
		C. tetanomorphum	25–28
	Ferments three-carbon compounds to propionate, acetate, and CO_2	*C. propionicum*	35
III. Ferments carbohydrates or amino acids	Fermentation products from glucose: acetate, formate, small amounts of isobutyrate and isovalerate	*C. bifermentans*	27
IV. Purine fermenters	Ferments uric acid and other purines, forming acetate, CO_2, NH_3	*C. acidurici*	27–30
V. Ethanol fermentation to fatty acids	Produces butyrate, caproate, and H_2; requires acetate as electron acceptor; does not use sugars, amino acids, or purines	*C. kluyveri*	30

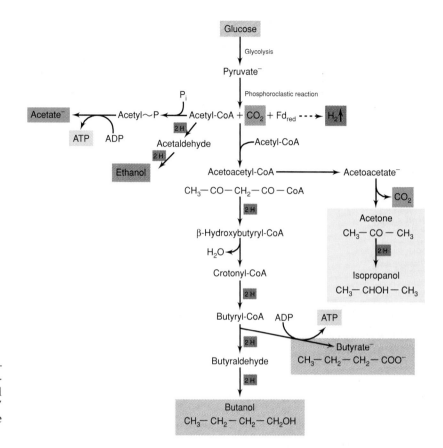

Figure 16.89 Pathway of formation of fermentation products from the butyric acid group of clostridia. The designation "2 H" represents two electrons from one molecule of NADH.

Some amino acids can be fermented singly, rather than in a Stickland-type reaction. These are alanine, cysteine, glutamate, glycine, histidine, serine, and threonine. The products are generally acetate, butyrate, CO_2, and H_2. It is usually found that each group of clostridia is specific in the kinds of substances it can ferment; usually either sugars or amino acids are utilized, although some strains ferment both. Many of the products of amino acid fermentation by clostridia are foul-smelling substances, and the odor that results from putrefaction is a result mainly of clostridial action. In addition to butyric acid, other odoriferous compounds produced are isobutyric acid, isovaleric acid, caproic acid, hydrogen sulfide, methylmercaptan (from sulfur amino acids), cadaverine (from lysine), putrescine (from ornithine), and ammonia.

The main habitat of clostridia is the soil, where they live primarily in anoxic "pockets," made anoxic primarily by facultative organisms metabolizing various organic compounds present. In addition, a number of clostridia have adapted to the anoxic environment of the mammalian intestinal tract. Also, as is discussed in Section 19.8, several clostridia that live primarily in soil are capable of causing disease in humans under specialized conditions. Botulism is caused by *Clostridium botulinum*, tetanus by *C. tetani*, and gas gangrene by *C. perfringens* and a number of other clostridia, both sugar and amino acid fermenters. These pathogenic clostridia seem in no way unusual metabolically but are distinct in that they pro-

duce specific toxins or, in the case of those causing gas gangrene, a group of toxins (∞ Section 19.8 and Table 19.4). Many gas gangrene clostridia also cause diseases in domestic animals, and botulism occurs in sheep and ducks and a variety of other animals. An unsolved ecological problem is what role these toxins play in the natural habitat of the organism.

Sporosarcina

The genus *Sporosarcina* is unique among endospore formers because cells are *cocci* instead of rods. *Sporosarcina* consists of spherical to oval cells that divide in two or three perpendicular planes to form tetrads or packets of eight or more cells (Figure 16.91). The organism is motile and strictly aerobic. Two species of *Sporosarcina* are known, *S. ureae* and *S. halophila*. The latter species is of marine origin and differs from *S. ureae* primarily in its requirement for sodium ions for growth. The endospores of *Sporosarcina* species are highly refractile, are centrally located (Figure 16.91), and contain dipicolinic acid, typical of endospores from other genera (∞ Section 3.13).

Sporosarcina ureae can easily be enriched from soil by plating dilutions of a pasteurized soil sample on nutrient agar supplemented with 8% urea and incubating in air. Most soil bacteria are strongly inhibited by as little as 2% urea. However, *S. ureae* actively decomposes urea to CO_2 and NH_3 and in so doing can

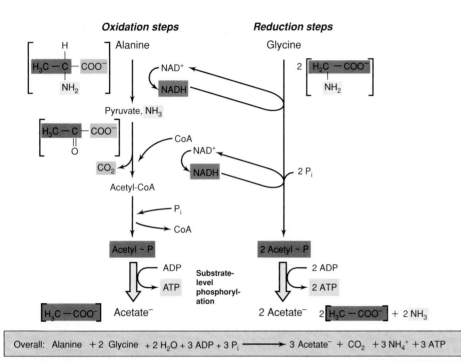

Figure 16.90 Coupled oxidation–reduction reaction (Stickland reaction) between alanine and glycine in *Clostridium sporogenes*. The structures of the key substrates, intermediates, and products are shown (in brackets) to allow the chemistry of the reaction to be followed.

dramatically raise the pH of unbuffered media (*S. ureae* is remarkably alkaline-tolerant and grows in media up to pH 10–11). *Sporosarcina ureae* is common in soils, and studies on its distribution suggest that numbers of *S. ureae* are greatest in soils that receive inputs of urine (a source of urea), such as soils in which animals periodically urinate. Since many soil organisms are quite urea-sensitive, these results suggest that *S. ureae* is ecologically important as a major urea degrader in nature.

TABLE 16.42	Amino acids participating in coupled fermentations (Stickland reaction)
Amino acids oxidized:	**Amino acids reduced:**
Alanine	Glycine
Leucine	Proline
Isoleucine	Hydroxyproline
Valine	Tryptophan
Histidine	Arginine

clostridia. The mycoplasmas are organisms without cell walls that do not revert to walled organisms. They are probably the smallest organisms capable of autonomous growth and are of special evolutionary interest because of their extremely simple cell structure and small genomes.

CONCEPT CHECK   16.26

Gram-positive bacteria constitute a major phylogenetic line of Bacteria and include both nonsporing and endospore-forming groups.

✔ *Compare and contrast* Micrococcus, Lactobacillus, *and* Clostridium *in terms of metabolism, pathogenicity, sporulation, and habitats.*

✔ *Although a gram-positive coccus, why is the organism* Deinococcus *distinct among gram-positive bacteria?*

16.27

Mycoplasmas

Although seeming at first to be out of place in a discussion of gram-positive Bacteria because their lack of cell walls yields no reaction in the Gram stain, phylogenetically mycoplasmas belong with the gram-positive Bacteria because they have close evolutionary ties to the

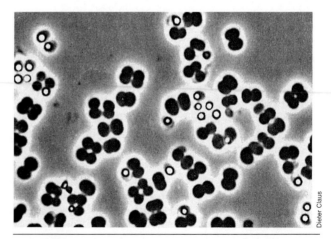

Figure 16.91 Phase contrast photomicrograph of cells of *Sporosarcina ureae*. A single cell is about 2 μm wide. Note bright refractile endospores. Most cell packets contain 8 cells.

Properties of mycoplasmas

The lack of cell walls in the mycoplasmas observed by electron microscopy has been reinforced by chemical analysis, the latter showing that the key wall components, muramic acid and diaminopimelic acid, are missing. In Chapter 3 we discussed protoplasts and showed how these structures can be formed when cell wall-digesting enzymes act on cells that are in an osmotically protected medium, and that when the osmotic stabilizer is removed, protoplasts take up water, swell, and burst (∞ Figure 3.32). Mycoplasmas resemble protoplasts in their lack of a cell wall, but they are more resistant to osmotic lysis and are able to survive conditions under which protoplasts lyse. This ability to resist osmotic lysis is at least partially determined by the nature of the mycoplasma cytoplasmic membrane, which is more stable than that of other prokaryotes. In one group of mycoplasmas, the membrane contains **sterols** that seem to be responsible for stability, whereas in other mycoplasmas, carotenoids or other compounds may be involved instead. Those mycoplasmas possessing sterols in their membrane do not synthesize them but require them preformed in the culture medium. This sterol requirement is a basis for separating the mycoplasmas into two groups (Table 16.43). *Thermoplasma* is phylogenetically unrelated to all other mycoplasmas because it is an archaean. It is thermophilic and acidophilic, growing optimally at 55°C and pH 2; the properties of *Thermoplasma* are considered in more detail in Section 17.4.

Certain mycoplasmas contain compounds called **lipoglycans** (see Table 16.43). Lipoglycans are long-chain heteropolysaccharides covalently linked to membrane lipids and embedded in the cytoplasmic membrane of many mycoplasmas. Lipoglycans resemble the lipopolysaccharides (LPSs) of gram-negative Bacteria (∞ Section 3.6) except that they lack the lipid A backbone and the phosphate typical of bacterial LPSs. Lipoglycans also function to help stabilize the membrane and have also been identified as facilitating attachment of mycoplasmas to cell surface receptors of animal cells. Like LPSs, lipoglycans stimulate antibody production when injected into experimental animals.

Growth of mycoplasmas

Mycoplasma cells are small and they are highly pleomorphic, a consequence of their lack of rigidity. A single culture may exhibit small coccoid elements, larger, swollen forms, and filamentous forms of variable lengths, often highly branched (Figure 16.92). It is from the production of filamentous, funguslike forms that the name *Mycoplasma* (*myco* means "fungus") derives. A common growth form is seen in cultures that divide by budding; division occurs with the cells remaining either directly attached or connected by thin hyphae (Figure 16.93).

TABLE 16.43 Major characteristics of mycoplasmas[a]

Genus	Number of recognized species	Properties	DNA (mol % GC)	Genome size (kilobase pairs)	Presence of lipoglycans
Require sterols					
Mycoplasma	87	Many pathogenic; require sterols; facultative aerobes (see Figure 16.92)	23–41	580–1100	+
Anaeroplasma	4	May or may not require sterols; obligate anaerobes; degrade starch, producing acetic, lactic, and formic acids plus ethanol and CO_2; inhibited by thallium acetate; found in the bovine and ovine rumen	29–33	1500	+
Spiroplasma	11	Spiral to corkscrew-shaped cells; associated with various phytopathogenic (plant disease) conditions (see Figure 16.95)	25–31	1500	–
Ureaplasma	5	Coccoid cells; occasional clusters and short chains; growth optimal at pH 6; strong urease reaction; associated with certain urinary tract infections in humans; inhibited by thallium acetate	27–30	750	–
Do not require sterols					
Acholeplasma	11	Facultative aerobes	27–36	1500	+
Asteroleplasma	1	Obligate anaerobe; isolated from the bovine or ovine rumen	40	1500	+
Thermoplasma	1	Thermophilic, acidophilic archaean found in heated coal refuse sites; no sterol requirement; DNA contains histones (∞ Section 17.4)	46	1500	+

[a]Phylogenetically, all mycoplasmas (except for *Thermoplasma*) are members of the low GC subdivision of the gram-positive Bacteria (∞ Section 15.7).

The small coccoid elements (0.2–0.3 μm in size) are the smallest mycoplasma units capable of independent growth. Because of flexibility due to lack of a cell wall, mycoplasma cells pass through filters with pore sizes smaller than the true diameter of the cells, and this has led to erroneous estimates of the minimum cell size capable of growth. Cellular elements of diameters close to 0.1 μm exist in mycoplasma cultures, but these are not capable of growth. Even so, the minimum reproductive unit of 0.2–0.3 μm probably represents the smallest *free-living* cell. Additionally, the genome size of mycoplasmas is also smaller than that of most prokaryotes, between 500 and 1100 kilobase pairs of DNA, which is comparable to that of the obligately parasitic chlamydia and rickettsia (see Sections 16.22 and 16.23) and about one-fifth to one-fourth that of *Escherichia coli*. The genome of at least one *Mycoplasma* species, *M. genitalium* contains 580 kilobase pairs and has been completely sequenced (∞ Section 9.11).

The mode of growth of mycoplasmas differs in liquid and agar cultures. On agar, there is a tendency for the organisms to grow so that they become embedded in the medium, and the fibrous nature of the agar gel seems to affect the division process, perhaps by promoting separation of units from the growing mass. Colonies of mycoplasmas on agar exhibit a characteristic "fried-egg" appearance because of the formation of a dense central core, which penetrates downward into the agar, surrounded by a circular spreading area that is lighter in color (Figure 16.94). Growth of mycoplasmas is not inhibited by penicillin, cycloserine, or other antibiotics that inhibit cell wall synthesis, but the organisms are as sensitive as other Bacteria to antibiotics that act on targets other than the cell wall. Use is made of the natural penicillin resistance of mycoplasmas in preparing selective media for their isolation from natural materials. The culture media used for the growth of most mycoplasmas have usually been quite complex. Growth is poor or absent even in complex yeast extract–peptone–beef heart

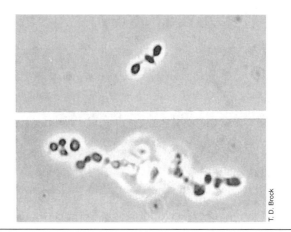

Figure 16.93 Photomicrograph by phase contrast microscopy of a mycoplasma culture, showing an irregular cell arrangement.

infusion media unless fresh serum or ascitic fluid is added. The main constituents provided by these two adjuncts are unsaturated fatty acids and sterols. Some mycoplasmas can be cultivated on relatively simple media, however, and defined media have been developed for some strains. Most mycoplasmas use carbohydrates as energy sources and require a range of vitamins, amino acids, purines, and pyrimidines as growth factors. The energy metabolism of mycoplasmas is not unique. Some species are oxidative, possessing a cytochrome system and making ATP by electron transport phosphorylation. Other species resemble the lactic acid bacteria in being strictly fermentative, producing energy by substrate-level phosphorylation and yielding lactic acid as the final product of sugar fermentation. Members of the genus *Anaeroplasma* are obligate anaerobes that ferment glucose or starch to a variety of acidic products.

Spiroplasma

The genus *Spiroplasma* consists of pleomorphic cells, spherical or slightly ovoid, which are often helical or spiral in shape (Figure 16.95). Although they lack a cell wall

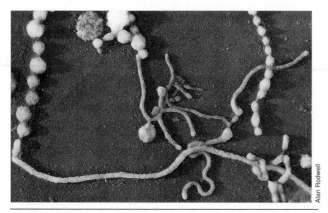

Figure 16.92 Electron micrograph of a metal-shadowed preparation of *Mycoplasma mycoides*. Note the coccoid and hyphalike elements. The average diameter of cells in chains is about 0.5 μm.

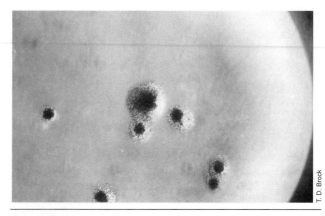

Figure 16.94 Typical "fried egg" appearance of mycoplasma colonies on agar. The colonies are about 0.5 mm in diameter.

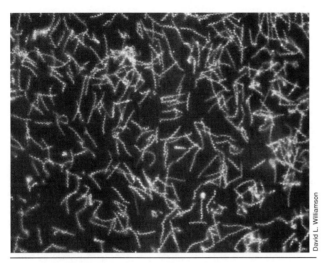

David L. Williamson

Figure 16.95 Dark-field micrograph of the "sex ratio" spiroplasma removed from the hemolymph of the fly *Drosophila pseudoobscura.* Female flies infected with the sex ratio spiroplasma bear only female progeny. Individual spiroplasma cells are about 0.15 μm in diameter.

and flagella, they are motile by means of a rotary (screw) motion or a slow undulation. Intracellular fibrils that are thought to play a role in motility have been demonstrated. The organism has been isolated from ticks, the hemolymph and gut of insects, vascular plant fluids and insects that feed on fluids, and the surfaces of flowers and other plant parts. *Spiroplasma citri* has been isolated from the leaves of citrus plants, where it causes a disease called *citrus stubborn disease* and from corn plants suffering from *corn stunt disease*. A number of other mycoplasma-like bodies have been detected in diseased plants by electron microscopy, which indicates that a large group of plant-associated mycoplasmas may exist. Four species of *Spiroplasma* are recognized that cause a variety of animal diseases such as *honeybee spiroplasmosis, suckling mouse cataract disease,* and *lethargy disease* of the beetle *Melolontha*.

CONCEPT CHECK   **16.27**

The mycoplasma group are organisms that lack cell walls and contain a very small genome. Many species require sterols to strengthen their membranes, and several are pathogenic for humans and other animals.

✔ *Where do the mycoplasmas group phylogenetically?*

✔ *Compare and contrast the genus* Mycoplasma *with the genus* Acholeplasma *in terms of growth requirements, genome size, and metabolism.*

✔ *Compare and contrast* Mycoplasma *with* Thermoplasma *in terms of metabolism, phylogenetic position, habitat, and pathogenicity.*

16.28
High GC Gram-Positive Bacteria: "Actinomycetes"

An extremely large variety of bacteria fall under this heading, as evidenced by the entire volume (Volume IV) of *Bergey's Manual,* which is devoted to the filamentous actinomycetes, and major portions of Volume II devoted to the rod-shaped relatives of this group. Phylogenetically, the actinomycetes form a subdivision of gram-positive Bacteria, distinct from the endospore formers and gram-positive cocci and asporogenous rods. Despite great morphological variability, the actinomycetes form a tight phylogenetic unit and most representatives have mole percent GC ratios in the 60s and 70s. There are considerable difficulties in drawing clear-cut distinctions between various genera of actinomycetes. Most of these organisms show a few common features: they are gram-positive, rod-shaped to filamentous, aerobic, and generally nonmotile in the vegetative phase (although motile stages are known). A continuum exists from simple rod-shaped organisms to rod-shaped organisms that occasionally grow in a filamentous manner, to strictly filamentous forms. Table 16.44 provides an overview of this group. In the following sections, we discuss some of the more interesting and important genera.

16.29
Coryneform Bacteria

The coryneform bacteria are gram-positive, aerobic, nonmotile, rod-shaped organisms with the characteristic of forming irregular-shaped, club-shaped, or V-shaped cell arrangements during normal growth. V-shaped cell groups arise as a result of a snapping movement that occurs just after cell division (called postfission snapping movement or, simply, *snapping division*) (Figure 16.96). **Snapping division** has been shown to occur in one species because the cell wall consists of two layers; only the inner layer participates in cross-wall formation, and so after the cross-wall is formed, the two daughter cells remain attached by the outer layer of the cell wall. Localized rupture of this outer layer on one side results in a bending of the two cells away from the ruptured side (Figure 16.97) and thus development of V-shaped forms.

The main genera of coryneform bacteria are *Corynebacterium* and *Arthrobacter*. The genus *Corynebacterium* consists of an extremely diverse group of bacteria, including animal and plant pathogens as well as saprophytes. The genus *Arthrobacter*, consisting primarily of soil organisms, is distinguished from *Corynebacterium* on the basis of a cycle of development in *Arthrobacter* involving conversion from rod to sphere and back to rod again (Figure 16.98, page 734).

TABLE 16.44   Actinomycetes and related genera (all gram-positive)[a]

Major groups	DNA (mol % GC)
Cornyeform group of bacteria: rods, often club-shaped, morphologically variable; not acid-fast or filamentous; snapping cell division	
Corynebacterium: irregularly staining segments, sometimes granules; club-shaped swellings frequent; animal and plant pathogens, also soil saprophytes	51–65
Arthrobacter: coccus–rod morphogenesis; soil organisms	59–70
Cellulomonas: coryneform morphology; cellulose digested; facultative aerobe	71–73
Kurthia: rods with rounded ends occurring in chains; coccoid later	36–38
Brevibacterium: coccus–rod morphogenesis; cheese, skin	60–67
Propionic acid bacteria: anaerobic to aerotolerant; rods or filaments, branching	
Propionibacterium: nonmotile; anaerobic to aerotolerant; produce propionic acid and acetic acid; dairy products (Swiss cheese); skin, may be pathogenic	53–68
Eubacterium: obligate anaerobes; produce mixture of organic acids, including butyric, acetic, formic, and lactic; intestine, infections of soft tissue, soil; may be pathogenic; probably the predominant member of the intestinal flora	26–48
Obligate anaerobes	
Bifidobacterium: smooth microcolony, no filaments; coryneform cells common; found in intestinal tract of breast-fed infants	55–67
Acetobacterium: homoacetogen; sediments and sewage	39–43
Butyrivibrio: curved rods; rumen	36–42
Thermoanaerobacter: rods, thermophilic, found in hot springs	37–39
Actinomycetes: filamentous, often branching; highly diverse	
Group I. Actinomycetes: not acid–alcohol-fast; facultatively aerobic; mycelium not formed; branching filaments may be produced; rod, coccoid, or coryneform cells	
Actinomyces: anaerobic to facultatively aerobic; filamentous microcolony, but filaments transitory and fragment into coryneform cells; may be pathogenic for humans or animals; found in oral cavity	57–69
Other genera: *Arachnia, Bacterionema, Rothia, Agromyces*	
Group II. Mycobacteria: acid–alcohol-fast, filaments transitory	
Mycobacterium: pathogens, saprophytes; obligate aerobes; lipid content of cells and cell walls high; waxes, mycolic acids; simple nutrition; growth slow; tuberculosis, leprosy, granulomas, avian tuberculosis; also soil organisms; hydrocarbon oxidizers	62–70
Group III. Nitrogen-fixing actinomycetes: nitrogen-fixing symbionts of plants; true mycelium produced	
Frankia: forms nodules of two types on various plant roots; probably microaerophilic; grows slowly; fixes N_2	67–72
Group IV. Actinoplanes: true mycelium produced; spores formed, borne inside sporangia	
Actinoplanes, Streptosporangium	69–71
Group V. Dermatophilus group: mycelial filaments divide transversely, and in at least two longitudinal planes, to form masses of motile, coccoid elements; aerial mycelium absent; occasionally responsible for epidermal infections	
Dermatophilus, Geodermatophilus	56–75
Group VI. Nocardias: mycelial filaments commonly fragment to form coccoid or elongate elements; aerial spores occasionally produced; sometimes acid–alcohol-fast	
Nocardia: common soil organisms; obligate aerobes; many hydrocarbon utilizers	61–72
Rhodococcus: soil saprophytes, also common in gut of various insects; utilize hydrocarbons	59–69
Group VII. Streptomycetes: mycelium remains intact, abundant aerial mycelium and long spore chains	
Streptomyces: Nearly 500 recognized species, many produce antibiotics	69–75
Other genera (differentiated morphologically): *Streptoverticillium, Sporichthya, Microcellobosporia, Kitasatoa, Chainia*	67–73
Group VIII. Micromonosporas group: mycelium remains intact; spores formed singly, in pairs, or short chains; several thermophilic; saprophytes found in soil, rotting plant debris; one species produces endospores	
Micromonospora, Thermoactinomyces, Thermomonospora	54–79

[a]Phylogenetically, all species (except for *Acetobacterium, Butyrivibrio,* and *Thermoanaerobacter*) fall into the high GC subdivision of the gram-positive Bacteria (∞ Section 15.7).

However, some corynebacteria are pleomorphic and form coccoid elements during growth, and so the distinction between the two genera on the basis of life cycle is not absolute. The *Corynebacterium* cell frequently has a swollen end, so it has a club-shaped appearance (hence the name of the genus: *koryne* is the Greek word for "club"), whereas *Arthrobacter* is less commonly club-shaped.

Organisms of the genus *Arthrobacter* are among the most common of all soil bacteria. They are remarkably resistant to desiccation and starvation, despite the fact that they do not form spores or other resting cells. Arthrobacters are a heterogeneous group that have considerable nutritional versatility, and strains have been isolated that decompose herbicides, caffeine, nicotine, phenol, and other unusual organic compounds.

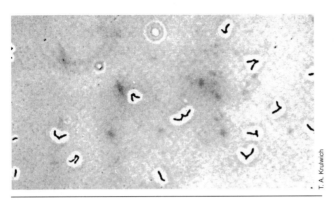

Figure 16.96 Photomicrograph of characteristic V-shaped cell groups in *Arthrobacter crystallopoietes,* resulting from snapping division. Cells are about 0.9 μm in diameter.

16.30
Propionic Acid Bacteria

The propionic acid bacteria (genus *Propionibacterium*) were first discovered as inhabitants of Swiss (Emmentaler) cheese, where their fermentative production of CO_2 produces the characteristic holes; the presence of propionic acid is at least partly responsible for the unique flavor of the cheese. Although this acid is produced by some other bacteria, its production in large amounts by the propionic acid bacteria is a distinguishing characteristic of the genus. The bacteria in

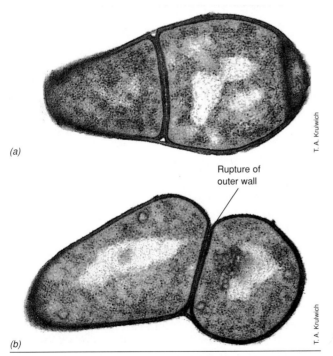

Figure 16.97 Electron micrograph of cell division in *Arthrobacter crystallopoietes,* illustrating how snapping division and V-shaped cell groups arise. (a) Before rupture of the outer cell wall layer. (b) After rupture of the outer layer on one side. Cells are 0.9–1 μm in diameter.

this group are gram-positive, pleomorphic, nonsporulating rods, nonmotile and anaerobic. They ferment lactic acid, carbohydrates, and polyhydroxy alcohols, producing propionic acid, succinic acid, acetic acid, and CO_2. Their nutritional requirements are complex, and they usually grow rather slowly.

The enzymatic reactions leading from glucose to propionic acid are of interest (Figure 16.99). The initial catabolism of glucose to pyruvate follows the Embden–Meyerhof pathway as in the lactic acid bacteria, but the NADH formed is reoxidized as one part of a cycle in which *propionic acid* is formed. Pyruvate accepts a carboxyl group from methylmalonyl-CoA by a transcarboxylase reaction, leading to the formation of oxalacetate and propionyl-CoA. The latter substance reacts with succinate in a step catalyzed by a CoA transferase, producing succinyl-CoA and propionate. The succinyl-CoA is then isomerized to methylmalonyl-CoA, and the cycle is complete (Figure 16.99). Reoxidation of NADH occurs in the steps between oxalacetate and succinate, and the oxidation–reduction balance is restored.

Fermentation of lactate to propionate

Most propionic acid bacteria also ferment lactate with the production of propionate, acetate, and CO_2. The anaerobic fermentation of lactic acid to propionate is of interest because lactic acid itself is an end product of fermentation for many bacteria (see Section 16.25). The propionic acid bacteria are thus able to obtain energy anaerobically from a substance that other bacteria have produced by carrying out a "secondary fermentation" of a primary fermentation product (lactate).

It is the fermentation of lactate to propionate that is important in Swiss cheese manufacture. The starter culture consists of a mixture of homofermentative streptococci and lactobacilli, plus propionic acid bacteria. The initial fermentation of lactose to lactic acid during formation of the curd is carried out by the homofermentative organisms. After the curd (protein and fat) has been drained, the propionic acid bacteria develop rapidly and reach numbers of 10^8–10^9 per gram by the time the cheese is 2 months old. Swiss cheese "eyes" (holes) are formed by the accumulation of CO_2, the gas diffusing through the curd and gathering at weak points. In the fermentation, lactate is oxidized to pyruvate from which it is converted to propionate as shown in Figure 16.99.

Fermentation of succinate by *Propionigenium*

A second bacterium, quite unrelated to *Propionibacterium*, also produces propionic acid fermentatively. *Propionigenium* is a gram-negative, strictly anaerobic bacterium that ferments *succinate* to propionate and CO_2:

$$\text{Succinate}^{2-} + H_2O \rightarrow \text{propionate}^- + HCO_3^-$$
$$\Delta G^{0\prime} = -20.5 \text{ kJ/reaction}$$

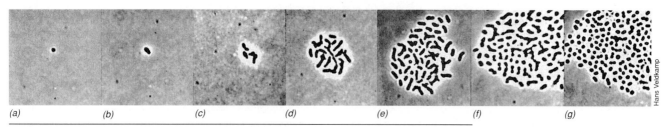

(a)   (b)   (c)   (d)   (e)   (f)   (g)

Hans Veldkamp

Figure 16.98 Stages in the life cycle of *Arthrobacter globiformis* as observed in slide culture: (a) Single coccoid element; (b–e) conversion to rod and growth of a microcolony consisting predominantly of rods; (f–g) conversion of rods to coccoid forms. Cells are about 0.9 μm in diameter.

We discussed the bioenergetics of succinate fermentation by *Propionigenium* in Section 16.19, pointing out that this fermentation is unusual because of the role Na$^+$ plays in establishing an ion gradient that ulti-

mately drives ATP synthesis. *Propionigenium* also grows on fumarate, malate, aspartate, oxalacetate, and pyruvate as sole energy sources but does not ferment sugars or carry out anaerobic respiration linked to nitrate, sulfate, or other potential electron acceptors. In nature, *Propionigenium* presumably specializes in succinate fermentation, receiving a steady supply of substrate from other fermentations or from organisms that produce succinate during anaerobic respiration with fumarate as electron acceptor.

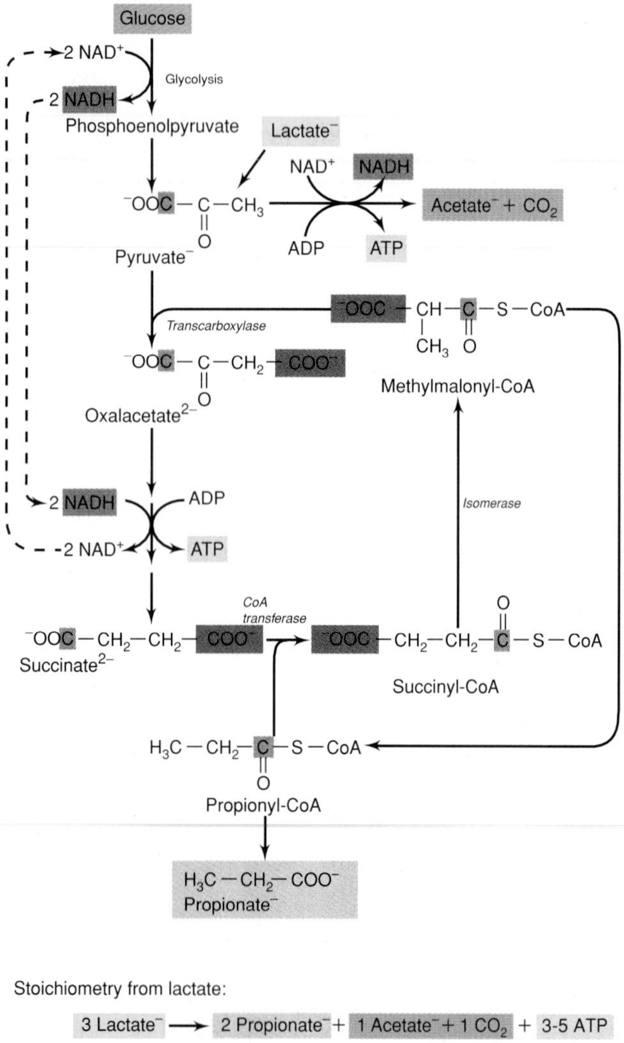

16.31

Mycobacterium

The genus *Mycobacterium* consists of rod-shaped organisms, which at some stage of their growth cycle possess the distinctive staining property called **acid–alcohol fastness**. This property is due to the presence on the surface of the mycobacterial cell of unique lipid components called **mycolic acids** and is found only in the genus *Mycobacterium*. First discovered by Robert Koch during his pioneering investigations on tuberculosis (∞ Section 1.9), this unique staining property permitted the identification of the organism in tuberculous lesions; it has subsequently proved to be of great taxonomic use in defining the genus *Mycobacterium*.

Acid–alcohol fastness (Ziehl–Neelsen stain)

A mixture of the dye basic fuchsin and phenol is used in the acid–alcohol staining procedure, the stain being driven into the cells by slow heating of the microscope slide to the steaming point for 2–3 min. The role of the phenol is to enhance penetration of the fuchsin into the lipids. After washing in distilled water, the preparation is decolorized with acid–alcohol (3% HCl in 95% ethanol); this procedure removes the fuchsin dye from other organisms, but it is retained by the mycobacteria. After another wash in water, a final counterstain of methylene blue is used. Acid–alcohol-fast organisms on the final preparation appear *red*, whereas the background and non-acid–alcohol-fast organisms appear *blue*.

As noted, the key component necessary for acid–alcohol fastness is a unique lipid fraction of mycobacterial cells called *mycolic acid*. Mycolic acid is

Figure 16.99 The formation of propionic acid by *Propionibacterium*. Either lactate, produced by the fermentative activities of other bacteria, or glucose can be fermented in the propionate fermentation. ATP synthesis is associated with electron transport reactions occurring during the formation of succinate and by substrate level phosphorylation in the production of acetate.

Stoichiometry from lactate:

3 Lactate$^-$ ⟶ 2 Propionate$^-$ + 1 Acetate$^-$ + 1 CO$_2$ + 3-5 ATP

actually a group of complex branched-chain hydroxy lipids with the overall structure shown in Figure 16.100a. The carboxylic acid group of the mycolic acid must be free (unesterified), and it reacts on a one-to-one basis with the fuchsin dye (Figure 16.100b). The mycolic acid is complexed to the peptidoglycan of the mycobacterial wall, and this complex somehow prevents approach of the acid–alcohol solvent during the decolorization step. It was also first demonstrated by Koch that disruption of cellular integrity destroys the acid–alcohol-fast property; thus cellular integrity is a necessary prerequisite of this property.

Mycobacteria are not readily stained by the Gram method because of the high surface lipid content, but if the lipoidal portion of the cell is removed with alkaline ethanol (1% KOH in absolute ethanol), the intact cell remaining is non-acid–alcohol-fast but instead is gram-positive. However, if the lipoidal portion is not removed, then the cells are resistant to decolorization by the Gram procedure even when stained with crystal violet alone (in the absence of iodine), whereas iodine is essential for the conventional Gram-staining procedure (∞ Figure 3.6). Mycobacterium can thus be considered a true gram-positive bacterium, and phylogenetic studies have confirmed this—Mycobacterium groups with other high GC actinomycetes by 16S ribosomal RNA sequencing (∞ Section 15.7).

Characteristics of mycobacteria

Mycobacteria are rather pleomorphic and may undergo branching or filamentous growth. However, in contrast to those of the actinomycetes, filaments of the mycobacteria become fragmented into rods or coccoid elements on slight disturbance; a true mycelium is not formed. In general, mycobacteria can be separated into two major groups, *slow growers* and *fast growers* (Table 16.45).

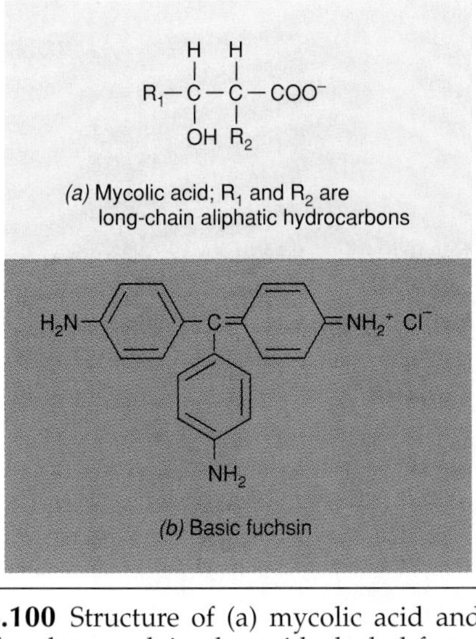

(a) Mycolic acid; R_1 and R_2 are long-chain aliphatic hydrocarbons

(b) Basic fuchsin

Figure 16.100 Structure of (a) mycolic acid and (b) basic fuchsin, the dye used in the acid–alcohol-fast stain. The fuchsin dye probably combines with the mycolic acid via ionic bonds between COO^- and NH_2^+.

Mycobacterium tuberculosis is a typical slow grower, and visible colonies are produced from dilute inoculum only after days to weeks of incubation. (The reason Koch was successful in first isolating *M. tuberculosis* was that he waited long enough after inoculating media; ∞ Section 1.9.) When growing on solid media, mycobacteria generally form tight, compact, often wrinkled colonies, the organisms piling up in a mass rather than spreading out over the surface of the agar (Figure 16.101a). This formation is probably due to the high lipid content and hydrophobic nature of the cell surface. The characteristic slow growth of most mycobacteria is probably also due, at least in part, to the *hydrophobic* character of the cell surface, which renders the cells strongly impermeable to nutrients; species having less lipid grow considerably more rapidly.

For the most part, mycobacteria have relatively simple nutritional requirements. Growth often occurs in simple mineral salts medium with ammonium as nitrogen source and glycerol or acetate as sole carbon source and electron donor incubated in air. Growth of *Mycobacterium tuberculosis* is stimulated by lipids and fatty acids, and egg yolk (a good source of lipids) is often added to culture media to achieve more luxuriant growth. A glycerol–whole egg medium (Lowenstein–Jensen medium) is often used in primary isolation of *M. tuberculosis* from pathological materials. Perhaps because of the high lipid content of its cell walls, *M. tuberculosis* is able to resist such chemical agents as alkali and phenol for considerable periods of time, and this property is used in the selective isolation of the organism from sputum and other materials that are grossly contaminated. The sputum is first treated with 1 N NaOH for 30 min and then neutralized and streaked onto an isolation medium.

A characteristic of many mycobacteria is their ability to form yellow carotenoid pigments (Figure 16.101c). Based on pigmentation, the mycobacteria can be classified into three groups: nonpigmented (including *Mycobacterium tuberculosis*, *M. bovis*); forming pigment only when cultured in the light, a property called **photochromogenesis** (including *M. kansasii*, *M. marinum*); and forming pigment even when cultured in the dark, a property called **scotochromogenesis** (including *M. gordonae*, *M. paraffinicum*). The property of photochromogenesis is of some interest and has been extensively studied. This property is not unique to mycobacteria, as it also occurs in a number of fungi. Photoinduction of carotenoid formation involves short-wavelength (blue) light and occurs only in the presence of O_2. The evidence indicates that the critical event in photoinduction is a light-catalyzed oxidation event, and it appears that one of the early enzymes in carotenoid biosynthesis is photoinduced. As with other carotenoid-containing bacteria, it has been suggested that carotenoids protect mycobacteria against oxidative damage involving singlet oxygen (∞ Section 5.11).

The cell walls of mycobacteria contain a peptidoglycan that is covalently bound to an arabinose–galactose–mycolic acid polymer, and it is this lipid–polysaccharide–peptidoglycan complex that confers the

TABLE 16.45 Some characteristics of representative mycobacteria

Species	Growth in 5% NaCl	Nitrate reduction	Growth at 45°C	Human pathogen	Pigmentation
Slow-growing species					
Mycobacterium tuberculosis	–	+	–	+	None
Mycobacterium avium	–	–	–	+	Old colonies pigmented (see Figure 16.101c)
Mycobacterium bovis	–	–	+	+	None
Mycobacterium kansasii	–	+	–	+	Photochromogenic
Fast-growing species					
Mycobacterium smegmatis	+	+	+	–	None
Mycobacterium phlei	+	+	+	–	Pigmented
Mycobacterium chelonae	+	–	–	+	None
Mycobacterium parafortuitum	+	+	–	–	Photochromogenic

hydrophobic character on the mycobacterial cell surface. In addition to this lipid component, mycobacteria form a wide variety of other lipids, providing lipid chemists with a fascinating amount of material.

The virulence of *Mycobacterium tuberculosis* cultures has been correlated with the formation of long, cordlike structures (Figure 16.101*b*) on agar or in liquid medium, due to side-to-side aggregation and intertwining of long chains of bacteria. Growth in cords reflects the presence on the cell surface of a characteristic lipid, the **cord factor,** which is a glycolipid (Figure 16.102). The pathogenesis of the disease tuberculosis is discussed in detail in Section 23.3.

16.32
Filamentous Actinomycetes

The actinomycetes are a large group of filamentous Bacteria, usually gram-positive, that form branching filaments. As a result of successful growth and branch-

ing, a ramifying network of filaments is formed, called a *mycelium* (Figure 16.103). Although it is of bacterial dimensions, the mycelium is in some ways analogous to the mycelium formed by the filamentous fungi. Most actinomycetes form spores; the manner of spore formation varies and is used in separating subgroups, as outlined in Table 16.44. The DNA base compositions of most members of the actinomycetes fall within the range of 63–78% GC. Organisms at the upper end of this range have the highest GC percentage of any bacteria known. Phylogenetically, the filamentous actinomycetes form a coherent group; thus, the mycelial spore-forming habit is of both phylogenetic as well as taxonomic importance. In the present discussion we concentrate on the genus *Streptomyces*.

Streptomyces

Streptomyces is a genus represented by a large number of species and varieties. Over 500 species of *Streptomyces* are recognized by *Bergey's Manual*, although GC base

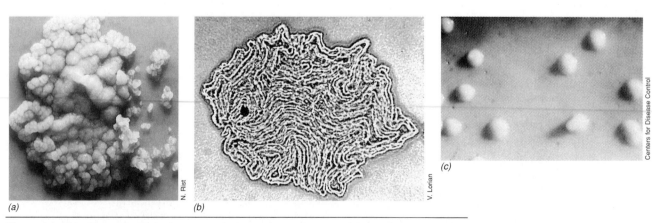

(a)   (b)   (c)

Figure 16.101 Characteristic colony morphology of mycobacteria. (a) *Mycobacterium tuberculosis*, showing the compact, wrinkled appearance of the colony. The colony is about 7 mm in diameter. (b) A colony of virulent *M. tuberculosis* at an early stage, showing the characteristic cordlike growth. Individual cells are about 0.5 μm in diameter. (See also the historic drawings of *M. tuberculosis* cells made by Robert Koch, in Figure 1.23). (c) Colonies of *Mycobacterium avium* from a strain of this organism isolated as an opportunistic pathogen from an AIDS patient.

ratios cluster tightly between 69 and 73 mol %. *Streptomyces* filaments are usually 0.5–1.0 μm in diameter, are of indefinite length, and often lack cross-walls in the vegetative phase. Growth occurs at the tips of the filaments and is often accompanied by branching so that the vegetative phase consists of a complex, tightly woven matrix, resulting in a compact, convoluted colony. As the colony ages, characteristic aerial filaments called *sporophores* are formed, which project above the surface of the colony and give rise to spores (Figure 16.104). *Streptomyces* spores, usually called **conidia,** are not related in any way to the endospores of *Bacillus* and *Clostridium* because the streptomycete spores are produced simply by the formation of cross-walls in the multinucleate sporophores followed by separation of the individual cells directly into spores (Figure 16.105). The surface of the conidial wall often has convoluted projections, the nature of which is characteristic of each species. Differences in shape and arrangement of aerial filaments and spore-bearing structures of various species are among the fundamental features used in separating the *Streptomyces* groups (Figure 16.106). The conidia and sporophores are often pigmented and contribute a characteristic color to the mature colony; in addition, pigments sometimes are produced by the substrate mycelium and contribute to the final color of the colony (Figure 16.107*a*). The dusty appearance of the mature colony, its compact nature, and its color make detection of *Streptomyces* colonies on agar plates relatively easy (Figure 16.107*b*).

Ecology and isolation of Streptomyces

Although a few streptomycetes can be found in aquatic habitats, they are primarily *soil* organisms. In fact, the characteristic earthy odor of soil is caused by the production of a series of streptomycete metabolites called **geosmins.** These substances are sesquiterpenoid compounds, unsaturated ring compounds of carbon, oxygen, and hydrogen. The geosmin first dis-

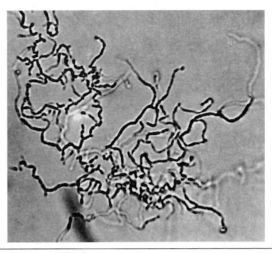

Figure 16.103 A young colony of an actinomycete of the genus *Nocardia,* showing typical filamentous cellular structure (mycelium). Each filament is about 0.8–1 μm in diameter.

covered has the chemical name trans-1,10-dimethyl-trans-9-decalol. Geosmins are also produced by some cyanobacteria (see Section 16.2).

Alkaline and neutral soils are more favorable for the development of *Streptomyces* than are acid soils. Higher numbers of *Streptomyces* are usually found in well-drained soils (such as sandy loams, or soils covering

(a)

(b)

Figure 16.104 Photomicrographs of several spore-bearing structures of actinomycetes. (a) *Streptomyces,* a monoverticillate type. (b) *Streptomyces,* a spiral type. Filaments are about 0.8 μm wide in both cases.

Figure 16.102 Structure of cord factor, a mycobacterial glycolipid: 6,6′-dimycolyltrehalose. The two identical long-chain dialcohol groups are shown in purple.

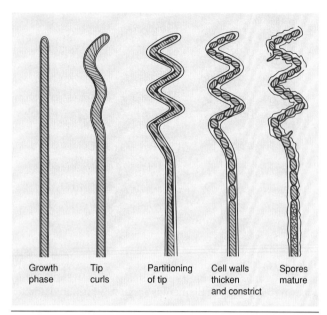

Figure 16.105 Diagram of stages in the conversion of a streptomycete's aerial hypha into spores (conidia).

limestone), and there is some evidence to suggest that *Streptomyces* require a lower water potential for growth than many other soil bacteria. Isolation of *Streptomyces* from soil is relatively easy: a suspension of soil in sterile water is diluted and spread on selective agar medium, and the plates are incubated at 25°C (∞ Figure 12.8). Media often selective for *Streptomyces* contain the usual assortment of inorganic salts to which starch, asparagine, or calcium malate is added as a carbon source and undigested casein or potassium nitrate as nitrogen source. After incubation for 5–7 days in air, the plates are examined for the presence of the characteristic *Streptomyces* colonies (Figure 16.107b), and spores of interesting colonies can be streaked and pure cultures isolated.

Nutritionally, the streptomycetes are quite versatile. Growth-factor requirements are rare, and a wide variety of carbon sources, such as sugars, alcohols, organic acids, amino acids, and some aromatic compounds, can be utilized. Most isolates produce extracellular hydrolytic enzymes that permit utilization of polysaccharides (starch, cellulose, hemicellulose), proteins, and fats, and some strains can use hydrocarbons, lignin, tannin, or even rubber. *Streptomyces* can often be obtained by spreading a soil dilution on an agar medium containing polymers such as casein and starch (Figure 16.107b). A single isolate may be able to break down over 50 distinct carbon sources. Streptomycetes are strict aerobes whose growth in liquid culture is usually markedly stimulated by forced aeration. Sporulation usually does not take place in liquid culture but only when the organism is growing on the surface of agar or another solid substrate; it can occur, however, when organisms form a pellicle on the surface of an unshaken liquid culture.

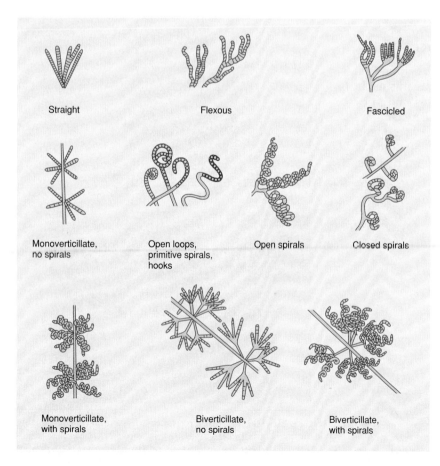

Figure 16.106 Various types of spore-bearing structures in the streptomycetes.

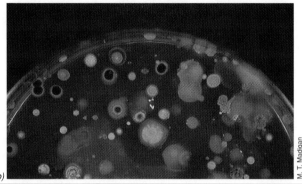

Figure 16.107 *Streptomyces.* (a) Typical appearance of a streptomycete growing on agar slants. Varying degrees of pigmentation are shown on different culture media. The coloration results both from the production of soluble pigments that diffuse into the agar and from the production of pigmented spores. (b) Colonies of *Streptomyces* and other soil bacteria derived from spreading a soil dilution on a casein–starch agar plate. The *Streptomyces* colonies are of various colors (several black *Streptomyces* colonies are in the foreground) but can easily be identified by their opaque, rough, nonspreading morphology.

Antibiotics of Streptomyces

Perhaps the most striking property of the streptomycetes is the extent to which they produce **antibiotics** (Table 16.46). Evidence for antibiotic production is often seen on the agar plates used in the initial isolation of *Streptomyces:* adjacent colonies of other bacteria show zones of inhibition (Figure 16.108; see also Figure 12.8). In some studies close to 50% of all *Streptomyces* isolated have proved to be antibiotic producers. Because of the great economic and medical importance of many streptomycete antibiotics, an enormous amount of work has been done on these producers. Over 500 distinct antibiotic substances have been shown to be produced by streptomycetes, and a large number of these have been studied chemically. Some organisms produce more than one antibiotic, and often the several kinds produced by one organism are not even chemically related. The same antibiotic may be formed by different species found in widely scattered parts of the world. A change in nutrition of the organism may result in a change in the nature of the antibiotic produced. The organisms are usually resistant to their own antibiotics, but they may be sensitive to antibiotics produced by other streptomycetes.

The genetics of antibiotic production is now an active area of research and was briefly discussed in Section 12.6. Interestingly, several *Streptomyces* sp. have been found to contain large *linear* plasmids of over 500 kilobases in length and some of these contain genes for antibiotic biosynthesis. Linear plasmids have occasionally been found in other bacteria but are apparently widespread among *Streptomyces* species.

More than 50 streptomycete antibiotics have found practical application in human and veterinary medicine, agriculture, and industry. Some of the more common antibiotics of *Streptomyces* origin are listed in Table 16.46. They are grouped into classes based on the chemical structure of the parent molecule. The search for new streptomycete antibiotics continues because many infectious diseases are still not adequately controlled by existing antibiotics. Also, the development of antibiotic-resistant strains requires the continual discovery of new agents. We discussed the antibiotic industry in general

TABLE 16.46 Some common antibiotics synthesized by species of *Streptomyces*

Chemical class	Common name	Produced by	Active against[a]
Aminoglycosides	Streptomycin	*S. griseus*	Most gram-negative Bacteria
	Spectinomycin	*Streptomyces* spp.	*M. tuberculosis*, penicillinase-producing *N. gonorrhoeae*
	Neomycin	*S. fradiae*	Broad spectrum, usually used in topical applications because of toxicity
Tetracyclines	Tetracycline	*S. aureofaciens*	Broad spectrum, gram-positive and gram-negative Bacteria, rickettsias and chlamydias, *Mycoplasma*
	Chlortetracycline	*S. aureofaciens*	As for tetracycline
Macrolides	Erythromycin	*S. erythreus*	Most gram-positive Bacteria, frequently used in place of penicillin, *Legionella*
	Clindamycin	*S. lincolnensis*	Effective against obligate anaerobes, especially *Bacteroides fragilis*
Polyenes	Nystatin	*S. noursei*	Fungi, especially *Candida* infections
	Amphocetin B	*S. nodosus*	Fungi
None	Chloramphenicol	*S. venezuelae*	Broad spectrum; drug of choice for typhoid fever

[a]Most antibiotics are effective against several different Bacteria. The entries in this column refer to the common clinical application of a given antibiotic. The structures and mode of action of many of these antibiotics are discussed in Sections 11.8–11.10.

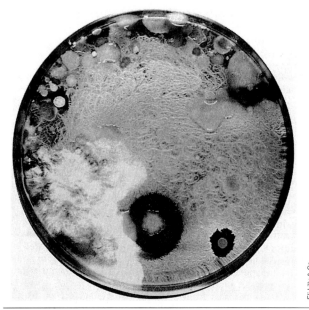

Eli Lilly & Co.

Figure 16.108 Antibiotic action of soil microorganisms on a crowded plate. The smaller colonies surrounded by inhibition zones are streptomycetes; the larger, spreading colonies are *Bacillus* species.

and the role of *Streptomyces* in the commercial production of antibiotics in Sections 12.5 and 12.6. Ironically, despite the extensive work on antibiotic-producing streptomycetes and the fact that the antibiotic industry is

a multibillion dollar enterprise, the ecology of *Streptomyces* remains poorly understood. The ecological rationale for why antibiotics are produced is not clear. However, one hypothesis for why *Streptomyces* species produce antibiotics is that antibiotic production, which is linked to sporulation (a process itself triggered by nutrient depletion), might be a mechanism to inhibit the growth of other organisms competing with differentiating *Streptomyces* cells for limiting nutrients. Whatever the ecological reasoning for antibiotic production by *Streptomyces* sp. and other antibiotic producers, humans have reaped the benefits of this process and it has revolutionized clinical medicine.

CONCEPT CHECK   16.32

The "high GC" gram-positive bacteria are a large, phylogenetically related group of aerobic, primarily soil-borne Bacteria. Many representatives produce antibiotics, and a few species are pathogenic.

✔ *Which of the high GC gram-positive Bacteria are acid-fast? How would you determine whether an organism is acid-fast?*

✔ *Why are members of the genus* Streptomyces *of such great economic and medical importance?*

✔ *What is the main habitat of most high GC gram-positive Bacteria? Name at least two organisms that live in habitats distinctly different from this.*

Material for Review

SUPPLEMENTARY READINGS

Balows, A., H. G. Trüper, M. Dworkin, W. Harder, and **K.-H. Schleifer** (eds.). 1992. *The Prokaryotes,* 2nd edition. Published in four volumes. Springer-Verlag, New York. The most complete reference on the characteristics of bacteria. Also includes formulation of various media and isolation procedures for virtually every prokaryotic group known. Well illustrated.

Barton, L. L. (ed.). 1995. *Sulfate-Reducing Bacteria.* Plenum Press, New York. A collection of chapters covering all aspects of sulfate-reducing bacteria.

Blankenship, R. E., M. T. Madigan, and **C. E. Bauer** (eds.). 1995. *Anoxygenic Photosynthetic Bacteria.* Kluwer Academic, Dordrecht, Netherlands. The most complete source of information on anoxygenic phototrophs. Over 1300 pages. Highly recommended.

Drake H. L. (ed.). 1994. *Acetogenesis.* Chapman and Hall, New York. A collection of chapters covering all aspects of homoacetogenic bacteria.

Holt, J. G. (editor-in-chief). *Bergey's Manual of Systematic Bacteriology.* Williams and Wilkins, Baltimore. Vol. I, 1984. Gram-negative Bacteria of medical or industrial

importance. Vol. II, 1986. Gram-positive Bacteria of medical or industrial importance. Vol. III, 1989. Other Gram-negative Bacteria, cyanobacteria, Archaea. Vol. IV, 1989. Other Gram-positive Bacteria. This manual is a recognized authority on bacterial taxonomy. A good place to begin a literature survey on a specific bacterial group.

Holt, J. G. (editor-in-chief). 1994. *Bergey's Manual of Determinative Bacteriology.* Williams and Wilkins, Baltimore. A condensed version of the systematic manual with emphasis on genera.

Lederberg, J. (ed.). 1992. *Encyclopedia of Microbiology.* Academic Press, San Diego, CA. A multivolume collection of articles on microorganisms and their habitats, metabolism, activities in nature, and human significance.

Schlegel, H. G., and **B. Bowien** (eds.). 1989. *Autotrophic Bacteria.* Science Tech, Madison, WI, and Springer-Verlag, Heidelberg. An excellent treatment of all autotrophic bacteria. Also has a nice historical survey.

On~line resources for this chapter are on the World Wide Web at: http://www.prenhall.com/~brock (click on the table of contents link and then select Chapter 16).

Prokaryotic Diversity: Archaea

Many Archaea live in extreme environments, such as the hot, acidic, iron-rich spring shown here, which is home to the hyperthermophilic acidophile, *Sulfolobus acidocaldarius.*

MINIGLOSSARY for Chapter 17

Acetotrophic acetate-consuming. When used to describe a methanogen, an organism capable of splitting acetate into CH_4 and CO_2

Acetyl-CoA (Ljungdahl–Wood) Pathway a pathway of autotrophic CO_2 fixation widespread in obligate anaerobes including methanogens, homoacetogens, and sulfate-reducing bacteria

Bacteriorhodopsin a membrane protein containing retinal, a pigment produced by certain extreme halophiles and capable of light-mediated proton motive force formation

Coenzyme M a coenzyme involved in the terminal step of methanogenesis

Corrinoid a porphyrin-like compound containing cobalt and involved in methyl transfers in methanogenesis

Crenarchaeota a kingdom of Archaea that contains primarily hyperthermophilic prokaryotes

Euryarchaeota a kingdom of Archaea that contains primarily methanogens, the extreme halophiles, and *Thermoplasma*

Extreme Halophile an organism whose growth is dependent on large concentrations (generally >10%) of NaCl

Factor$_{420}$ (F$_{420}$) an electron donor involved in certain steps in the reduction of CO_2 to CH_4

Hyperthermophile a prokaryote with a growth temperature optimum of 80°C or greater

Korarchaeota a kingdom of hyperthermophilic Archaea identified from 16S rRNA sequences obtained from a terrestrial hot spring but not yet represented by laboratory cultures

Methanogen a methane-producing prokaryote

Phytanyl a branched-chain hydrocarbon containing 20 carbon atoms and commonly found in the lipids of Archaea

Solfatara a hot, sulfur-rich, generally acidic environment commonly inhabited by hyperthermophilic Archaea

---

We now consider members of the domain Archaea. We emphasized in Chapter 15 the profound evolutionary differences between Bacteria and Archaea—despite the fact that they are prokaryotes, by molecular sequencing criteria Archaea are no more closely related to Bacteria than they are to Eukarya. In fact, the universal phylogenetic tree (∞ Figure 15.12) shows that Archaea are actually more closely related to Eukarya than to Bacteria.

In Chapter 3 we covered aspects of the unique lipids, cell membranes, and cell walls of Archaea, and in Chapter 9 the strong parallels between the genetics of Archaea and Bacteria were discussed. In this chapter we consider the organisms themselves and focus on their physiology and ecology. We begin with overviews of the evolutionary relationships *within* the domain Archaea and the general metabolic patterns observed.

Phylogenetic overview of Archaea

The detailed phylogenetic tree of Archaea is shown in Figure 17.1. As previously discussed (∞ Section 15.7), closest to the root of the tree is the Korarchaeota kingdom, a group of as yet uncultured, hyperthermophilic Archaea. Also quite close to the root are the hyperthermophiles, including the methanogen *Methanopyrus* and several sulfur-reducing hyperthermophiles (Crenarchaeota, Figure 17.1). In the latter organisms, sulfur functions as an electron acceptor for anaerobic respiration (see Section 17.3). Going up the tree, several phylogenetically distinct groups of methanogens

branch off, including the *Methanococcus* group, the *Methanobacterium* group, and the *Methanosarcina–Methanospirillum* group. The extreme halophiles (for example, *Halobacterium–Halococcus*) are a group unto themselves, as is the acidophilic, thermophilic cell-wall-less prokaryote *Thermoplasma* (Figure 17.1). Collectively, these groups make up the kingdom Euryarchaeota. We thus see considerable phylogenetic diversity within the domain Archaea, and we will also see enormous physiological diversity as we make our way through this chapter.

Energy metabolism and central metabolic pathways in Archaea

Several Archaea are chemoorganotrophic and thus use organic compounds as energy sources for growth. Catabolism of glucose in extremely halophilic and hyperthermophilic Archaea proceeds via slight modifications of the Entner–Doudoroff (E-D) pathway (∞ Section 16.15 for a description of this pathway in Bacteria). By contrast, gluconeogenesis, the *production* of glucose from noncarbohydrate precursors (∞ Section 4.18), proceeds in Archaea by reversal of steps in the Embden–Meyerhof pathway (glycolysis) (∞ Section 4.10), a reaction series almost universally distributed among Bacteria.

Oxidation of acetate to CO_2 in Archaea proceeds through the citric acid cycle (∞ Section 4.14) or some slight variation of this reaction series, or by the acetyl-CoA (Ljungdahl–Wood) pathway (∞ Section 16.9). Little is known concerning biosynthesis of amino acids

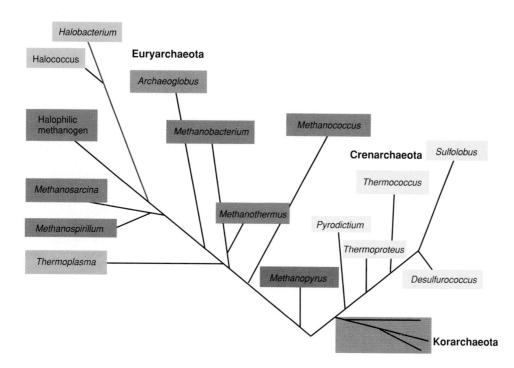

Figure 17.1 Detailed phylogenetic tree of the Archaea based on 16S ribosomal RNA sequence comparisons. See note in Figure 16.1 for comparisons of detailed tree with the universal tree (∞ Figure 15.12).

and other macromolecular precursors in Archaea, but presumably key monomers are produced from the central biosynthetic intermediates discussed previously for Bacteria (∞ Section 4.19 and Figure 4.25). Electron transport chains including cytochromes of the *a*, *b*, and *c* types exist in extreme halophiles, and *a*-type cytochromes are present in certain hyperthermophiles. Employing these and other electron carriers, chemoorganotrophic metabolism in most Archaea probably proceeds by introduction of electrons from organic electron donors, probably at the level of NADH, into an electron transport chain, leading to the reduction of O_2, S^0, or some other electron acceptor, with concurrent establishment of a proton motive force that drives adenosine triphosphate (ATP) synthesis through membrane-bound ATPases (∞ Section 4.13 for a description of ATPase function). Chemolithotrophy is also well established in the Archaea, with H_2 being a common electron donor (see Section 17.3).

Autotrophy is widespread among Archaea and occurs by several different means. In methanogens, and presumably in most chemolithotrophic hyperthermophiles, CO_2 is incorporated via the acetyl-CoA pathway or some modification thereof (∞ Section 16.9 for a description of the acetyl-CoA pathway). In the hyperthermophiles *Aquifex* (which is actually a member of the Bacteria) (∞ Section 15.7) and *Thermoproteus*, however, CO_2 fixation occurs via the reverse citric acid cycle, a reaction series previously described as the autotrophic pathway in the green sulfur bacteria (∞ Section 16.1). Although extreme halophiles are all chemoorganotrophs, autotrophic pathways also exist, as various species of extreme halophiles fix significant amounts of CO_2 via the Calvin cycle, the

most widespread autotrophic pathway in Bacteria and eukaryotes (∞ Section 13.7).

We thus see that many of the major catabolic and anabolic sequences in the Archaea are familiar ones from our study of these processes in various Bacteria. Methanogenesis is a major exception and will be covered in detail in this chapter. With this in mind, we now begin our study of the basic biology of the major groups of Archaea.

17.1
Extremely Halophilic Archaea

We begin our discussion with a consideration of the extreme halophiles. **Extremely halophilic Archaea** are a diverse group of prokaryotes that inhabit highly saline environments such as solar salt evaporation ponds and natural salt lakes, or artificial saline habitats such as the surfaces of heavily salted foods like certain fish and meats. Such habitats are often called *hypersaline*. The term *extreme halophile* is used to indicate not only that these organisms are halophilic, but also that their requirement for salt is *very high*, in some cases near that of saturation. A generally accepted definition of an extreme halophile is that the organism requires at least 1.5 *M* (about 9%) NaCl for growth, but most species require 2–4 *M* NaCl (12–23%) for optimal growth. Virtually all extreme halophiles can grow at 5.5 *M* NaCl (32%, the limit of saturation for NaCl), although some species grow only very slowly at this salinity.

We begin with a consideration of the habitats of these unusual prokaryotes and proceed to discuss the taxonomy and physiology of the group and the unique mechanism of salt tolerance. Finally, we consider a unique mechanism of light-driven ATP synthesis that does not involve chlorophyll or bacteriochlorophyll.

Saline environments

Salty habitats are common throughout the world, but *extremely* saline habitats are rather rare. Most extremely saline environments are in hot, dry areas of the world, and such climatic conditions encourage evaporation and further concentration of the salts. Salt lakes can vary considerably in ionic composition. The predominant ions in a saline lake depend to a major extent on the surrounding topography, geology, and general climatic conditions. Great Salt Lake in Utah (USA) (Figure 17.2a), for example, is essentially concentrated seawater because the relative proportions of the various ions are those of seawater, although the overall concentration of ions is much higher. Sodium is the predominant cation in Great Salt Lake, whereas chloride is the predominant anion; significant levels of sulfate are also present at a slightly alkaline pH (Table 17.1). By contrast, another very saline basin, the Dead Sea, is relatively low in sodium but contains high levels of magnesium because of the abundance of magnesium minerals in the surrounding rocks (Table 17.1). The water chemistry of soda lakes resembles that of saline lakes such as Great Salt Lake, but because high levels of *carbonate* minerals are present in the surrounding rocks, the pH of soda lakes is quite high; pH values of 10–12 are not uncommon in these environments (Table 17.1).

Despite what may seem like rather harsh conditions, salt lakes can be highly productive ecosystems. Archaea are not the only microorganisms found. The eukaryotic alga *Dunaliella* is the major, if not sole, oxygenic phototroph in most salt lakes. In highly alkaline soda lakes where *Dunaliella* is absent, anoxygenic phototrophic purple bacteria of the genus *Ectothiorhodospira* (∞ Section 16.1) predominate. Because sulfate is generally abundant in salt lakes (Table 17.1), significant sulfate reduction can occur, leading to formation of the sulfide needed for development of purple bacteria. At the high salt concentrations found in saline lakes, no higher plants can develop, although in addition to microorganisms a few types of animals are capable of living (brine shrimp, brine flies). Organic matter originating from primary production by oxygenic or anoxygenic phototrophs then sets the stage for development of the extremely halophilic Archaea, all of which are chemoorganotrophic. In addition, a few extremely halophilic anaerobic chemoorganotrophic Bacteria, such as *Haloanaerobium* and *Halobacteroides*, are present in such environments.

Figure 17.2 Hypersaline habitats. (a) Great Salt Lake, Utah, a hypersaline lake in which the ratio of ions is similar to that in seawater but in which absolute concentrations of ions are about 10 times that of seawater. (b) Aerial view near San Francisco Bay, California, of a series of seawater evaporating ponds where solar salt is prepared. The red-purple color is predominantly due to bacterioruberins and bacteriorhodopsin in cells of *Halobacterium* and β-carotene in *Dunaliella salina* cells.

TABLE 17.1   Ionic composition of some highly saline environments

Ion	Concentration (g/l)			
	Great Salt Lake	Dead Sea	Typical soda lake	Seawater (for comparison)
Na^+	105	40.1	142	10.6
K^+	6.7	7.7	2.3	0.38
Mg^{2+}	11	44	<0.1	1.27
Ca^{2+}	0.3	17.2	<0.1	0.40
Cl^-	181	225	155	18.9
Br^-	0.2	5.3	—	0.065
SO_4^{2-}	27	0.5	23	2.65
HCO_3^- or CO_3^{2-}	0.7	0.2	67	0.14
pH	7.7	6.1	11	8.1

Marine salterns are also habitats for extremely halophilic prokaryotes. Marine salterns are small basins filled with seawater that are left to evaporate, yielding NaCl and other salts of commercial value (Figure 17.2b). As salterns approach the minimum salinity limits for extreme halophiles, the waters turn a reddish purple in color indicative of the massive growth (called a *bloom*) of halophilic Archaea (the red coloration apparent in Figure 17.2b comes from carotenoids and other pigments discussed later). Extreme halophiles have also been found in high salt foods such as certain sausages, marine fish, and salt pork. Other than causing some aesthetic problems, growth of extreme halophiles in foods is of little consequence because no extreme halophile has been shown to cause foodborne illness.

Taxonomy of extremely halophilic Archaea

Table 17.2 lists the currently recognized species of extremely halophilic Archaea. 16S ribosomal ribonucleic acid (rRNA) sequencing and other studies have defined eight genera of extreme halophiles: *Halobacterium*,

TABLE 17.2   Taxonomy of some extremely halophilic Archaea

Genus	Morphology	DNA (mol % GC)	Habitat and comments
Halobacterium	Rods		
H. salinarum		66–71	Isolated from salted fish, hides, hypersaline lakes; related organisms, probably all the same species are *H. halobium* and *H. cutirubrum*
Halorubrum	Rods		
H. sodomense		68	Dead Sea; requires high Mg^{2+}
H. saccharovorum		71	Salterns; uses sugars
H. lacusprofundi	Pleomorphic rods	65–66	From an Antarctic lake; grows at 4°C
H. trapanicum		64	From salt works, Trapani, Italy
Halobaculum	Rods		
H. gomorrense		70	Dead Sea; requires high Mg^{2+}, optimal NaCl about 1.5 M
Haloferax	Flattened disc or		
H. volcanii	cup-shaped	63–66	Dead Sea; requires high Mg^{2+}
H. mediterranei		60–62	Salterns; uses starch
H. gibbonsii		62	Marine salterns in Spain
H. denitrificans		64	Baja California saltern; capable of dissimilative nitrate reduction
Haloarcula	Irregular discs,		
H. vallismortis	triangles, rectangles	65	Death Valley, CA
H. hispanica		63	Marine salterns in Spain
Halococcus	Cocci		
H. morrhuae		60–66	Salted fish
H. saccharolyticus		59	Salterns; uses sugars
Natronobacterium	Rods		
N. gregoryi		65	All species from highly saline soda lakes; optimum pH for growth, 9.5
N. magadii		63	
N. pharaonis		64	
Natronococcus	Cocci		
N. occultus		64	Highly saline soda lakes

Halococcus, Haloferax, Haloarcula, Halorubrum, Halobaculum, Natronobacterium, and *Natronococcus.* The extremely halophilic Archaea are frequently referred to collectively as "halobacteria," because the genus *Halobacterium* was the first in this group to be described and is still the best-studied representative of the group. Halobacteria lack peptidoglycan in their cell walls and contain ether-linked lipids and archaean-type RNA polymerases; they also are insensitive to most antibiotics and possess the other general attributes of representatives of this domain (∞ Section 15.8). *Natronobacterium* and *Natronococcus* differ from other extreme halophiles in being extremely *alkaliphilic* as well as halophilic. As befits their soda lake habitat (see Table 17.1), growth of natronobacteria is optimal at very low Mg^{2+} concentrations and high pH (9–11). Natronobacteria also contain unusual diether lipids not found in other extreme halophiles and cluster tightly as a phylogenetic group.

All halophilic Archaea stain gram *negatively,* reproduce by binary fission, and do not form resting stages or spores. Most halobacteria are nonmotile, but a few strains are weakly motile by lophotrichous flagella. The genomic organization of *Halobacterium* and *Halococcus* is highly unusual in that large plasmids containing up to 25–30% of the total cellular DNA are frequently present and the GC base ratio of these plasmids (57–60% GC) is significantly different from that of chromosomal DNA (66–68% GC). Plasmids from extreme halophiles are among the largest naturally occurring plasmids known. In addition to the large amount of nonchromosomal DNA, the *Halobacterium* genome also contains considerable amounts of highly repetitive DNA, the function of which is unknown.

Physiology of extremely halophilic Archaea

All extremely halophilic Archaea are chemoorganotrophs, and most species are obligate *aerobes.* Most halobacteria use amino acids or organic acids as energy sources and require a number of growth factors (mainly vitamins) for optimal growth. A few *Halobacterium* species oxidize carbohydrates, but this ability is relatively rare. Electron transport chains containing cytochromes *a, b,* and *c* are present in *Halobacterium,* and energy is conserved during aerobic growth via a proton motive force arising from membrane-mediated chemiosmotic events. Some strains of *Halobacterium* have been shown to grow anaerobically. Anoxic growth at the expense of sugar fermentation and by anaerobic respiration (∞ Section 13.14) linked to the reduction of nitrate or fumarate has been demonstrated in certain strains.

All extremely halophilic Archaea require large amounts of sodium for growth. In the case of *Halobacterium,* where detailed salinity studies have been performed, the requirement for Na^+ cannot be satisfied by replacement with another ion, even with the chemically related ion K^+. We learned in Section 5.10 that certain microorganisms can withstand the osmotic forces that accompany life in a high solute environment by accumulating organic compounds *intracellularly;* the latter compounds are referred to as **compatible solutes.** These compounds counteract the tendency of the cell to become dehydrated under conditions of high osmotic strength by placing the cell in positive water balance with its surroundings (∞ Section 5.10). Ironically, however, although *Halobacterium* thrives only in an osmotically stressful environment, it produces no *organic* compatible solute. Instead, cells of *Halobacterium* pump large amounts of K^+ from the environment into the cell such that the concentration of K^+ *inside* the cell is considerably greater than the concentration of Na^+ *outside* the cell; thus, *Halobacterium* employs an *inorganic ion* as its compatible solute and in this way remains in positive water balance (Table 17.3).

The cell wall of *Halobacterium* is stabilized by sodium ions; in low sodium environments the cell wall breaks down. In electron micrographs of thin sections of *Halobacterium* (Figure 17.3), the organism appears similar to gram-negative Bacteria in many respects, but the cell wall is quite different (compare Figure 17.3 with Figure 3.26). Na^+ binds to the outer surface of the *Halobacterium* wall and is absolutely essential for maintaining cellular integrity; when insufficient Na^+ is present, the cell wall breaks apart and the cell lyses. Peptidoglycan is absent in the cell wall of *Halobacterium* (as it is in the walls of all other Archaea) (∞ Section 3.5), and instead the cell wall is composed of *glycoprotein.* This protein has an exceptionally high content of the *acidic* (negatively charged) amino acids aspartate and glutamate. The negative charges contributed by the carboxyl groups of these amino acids in the cell wall glycoprotein are shielded by Na^+; when Na^+ is diluted away, the negatively charged parts of the proteins actively repel each other, leading to cell lysis.

Cytoplasmic proteins of *Halobacterium* are also highly acidic, but studies on several halobacterial enzymes have shown that K^+, not Na^+, is required for activity. This, of course, is not surprising when it is recalled that K^+ is the predominate internal cation in cells of *Halobacterium* (Table 17.3). Besides a high acidic amino acid composition, halobacterial cytoplasmic

TABLE 17.3	Concentration of ions in cells of *Halobacterium salinarum*	
Ion	**Concentration in medium (*M*)**	**Concentration in cells (*M*)**
Na^+	3.3	0.8
K^+	0.05	5.3
Mg^{2+}	0.13	0.12
Cl^-	3.3	3.3

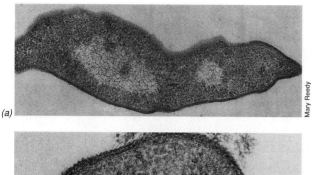

Mary Reedy

Mary Reedy

Figure 17.3 Electron micrographs of thin sections of the extreme halophile *Halobacterium salinarum*. A cell is about 0.8 μm in diameter. (a) Longitudinal section. (b) High magnification electron micrograph showing the regular structure of the cell wall.

proteins typically contain very low levels of *hydrophobic* amino acids. This phenomenon probably represents an evolutionary adaptation to the highly ionic cytoplasm of *Halobacterium*; in such an environment highly polar proteins tend to remain in solution, whereas nonpolar molecules would tend to cluster and perhaps lose activity. The ribosomes of *Halobacterium* also require high K⁺ levels for stability (ribosomes of nonhalophiles have no K⁺ requirement). It thus appears that the extremely halophilic Archaea are highly adapted, both internally and externally, to life in a highly ionic environment. Cellular components exposed to the external environment require high Na⁺ for stability, whereas internal components require high K⁺. In no other group of prokaryotes do we find this unique requirement for specific cations in such high amounts.

Bacteriorhodopsin and light-mediated ATP synthesis

Certain species of extreme halophiles have the additional interesting property of showing a *light-mediated* synthesis of ATP that does not involve chlorophyll pigments. As discussed in Section 13.4, one of the key aspects of photosynthesis is *photophosphorylation*, the synthesis of ATP as a consequence of the light-dependent generation of a proton motive force. We have seen the highly pigmented nature of extremely halophilic Archaea in Figure 17.2. Pigmentation is due to red- and orange-colored carotenoids, primarily C₅₀ carotenoids called *bacterioruberins*, and also to inducible pigments involved in energy metabolism. Although lacking chlorophylls or bacteriochlorophylls, under conditions of low aeration, *Halobacterium salinarum* and certain other extreme halophiles synthesize and insert a protein called

bacteriorhodopsin into their membranes. Bacteriorhodopsin was named because of its structural and functional similarity to the visual pigment of the eye, *rhodopsin*. Conjugated to bacteriorhodopsin is a molecule of *retinal*, a carotenoid-like molecule that can absorb light and catalyze the transfer of protons across the cytoplasmic membrane (Figure 17.4). Because of its retinal content, bacteriorhodopsin is purple in color, and cells of *Halobacterium* switched from growth under conditions of high aeration to oxygen-limiting conditions gradually change from an orange or red color to a more reddish-purple color because of the insertion of bacteriorhodopsin into the cytoplasmic membrane. Isolated purple membranes of *Halobacterium* contain about 25% lipid and 75% protein, and the purple membrane component is inserted at random on the surface of the cytoplasmic membrane; apparently bacteriorhodopsin is wedged into preexisting cytoplasmic membranes during the switch to low O₂.

Bacteriorhodopsin absorbs light strongly in the green region of the spectrum at about 570 nm. The retinal chromophore of bacteriorhodopsin, which normally exists in an *all-trans* configuration, is excited and temporarily converted to the *cis* form following the absorption of light (Figure 17.4). This transformation results in the transfer of protons to the *outside* surface of the membrane. The retinal molecule then relaxes and returns to its more stable all-trans isomer in the dark following the uptake of a proton from the cytoplasm, thus completing the cycle (Figure 17.4). As protons accumulate on the outer surface of the membrane, the proton motive force (∞ Section 4.13) increases until the membrane is sufficiently "charged" to drive ATP synthesis through action of the membrane-bound ATPase (Figure 17.4).

Light-mediated ATP production in *Halobacterium salinarum* has been shown to support slow growth of this organism anaerobically under nutritional conditions in which other energy-generating reactions do not occur, and light has been shown to maintain the viability of anoxic cultures of *Halobacterium* incubated in the absence of organic energy sources. The light-stimulated proton pump of *H. salinarum* also functions to pump Na⁺ out of the cell by action of a Na⁺/H⁺ antiport system (∞ Section 3.4) and to drive the uptake of a variety of nutrients, including the K⁺ needed for osmotic balance. The uptake of amino acids by *H. salinarum* has been shown to be indirectly driven by light because the transport of amino acids occurs with Na⁺ uptake by an amino acid–Na⁺ symporter (∞ Section 3.4). Continued uptake depends on the removal of Na⁺ via the (light-driven) Na⁺/H⁺ antiporter.

A separate light-driven pump called **halorhodopsin** is present in halobacteria to pump Cl⁻ into the cell as an anion for K⁺. Like bacteriorhodopsin, halorhodopsin also contains retinal, and Cl⁻ binds to this retinal (instead of H⁺ as in bacteriorhodopsin, see Figure 17.4) and is transported across the membrane from outside to inside.

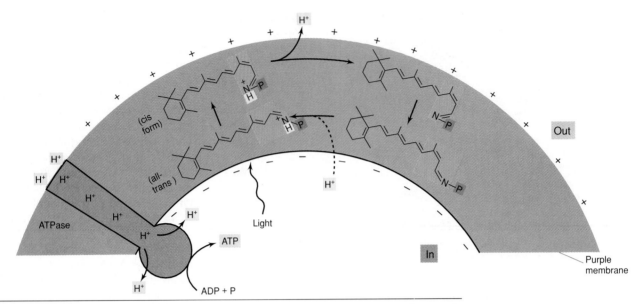

Figure 17.4 Model of the light-mediated bacteriorhodopsin proton pump in the purple membrane of *Halobacterium*. The P stands for the protein to which the chromophore retinal is attached. Out and In designate opposite sides of the cytoplasmic membrane.

Evolution of extreme halophiles

Although they share little with them from a *phenotypic* standpoint, the extreme halophiles are phylogenetically quite closely related to certain methanogens (see Figure 17.1). This at first seems odd because most extreme halophiles are obligate aerobes, whereas methanogenic bacteria are all obligate anaerobes (see later). However, more detailed studies of halophilic Archaea have shown them to be remarkably versatile, including many species capable of anoxic growth by fermentation or anaerobic respiration. Indeed, extremely halophilic and alkaliphilic methanogens are known (see Table 17.5), and so

the connection between the two groups may not be as unusual as first thought. We will return to the theme of the evolution of Archaea in Section 17.6 after we have considered the remaining groups of Archaea.

17.2
Methane-Producing Archaea: Methanogens

We described the overall process of methanogenesis in Section 13.17 and the ecology of methanogenesis in Sections 14.12 and 14.13 and have noted that the biological production of methane is carried out by a unique group of prokaryotes, the **methanogens.** Methane formation occurs only under strictly *anoxic* conditions. Thus, methanogenesis is restricted to habitats that are anoxic.

Substrates and energy yields

At least 10 substrates have been shown to be converted to methane by one or another methanogen. Carbon dioxide, CO_2, is a nearly universal substrate for methanogens, the needed electrons usually being derived from H_2. When growing on $H_2 + CO_2$, the methanogens are *autotrophic*, with CO_2 serving as both carbon source and electron acceptor. In addition to CO_2, however, a variety of other compounds can be converted to methane by certain methanogenic species; these substrates are listed in Table 17.4. As described in Section 13.17, methane formation from $H_2 + CO_2$ can be viewed as a type of *anaerobic respiration* in which CO_2 is the electron acceptor. However, biochemical studies on methanogenesis indicate that a conventional electron

TABLE 17.4	Substrates converted to methane by various methanogenic Archaea

CO$_2$-type substrates
Carbon dioxide, CO$_2$ (with electrons derived from H$_2$, certain alcohols, or pyruvate)
Formate, HCOO$^-$
Carbon monoxide, CO

Methyl substrates
Methanol, CH$_3$OH
Methylamine, CH$_3$NH$_3^+$
Dimethylamine, (CH$_3$)$_2$NH$_2^+$
Trimethylamine, (CH$_3$)$_3$NH$^+$
Methylmercaptan, CH$_3$SH
Dimethylsulfide, (CH$_3$)$_2$S

Acetotrophic substrate
Acetate, CH$_3$COO$^-$

transport system involving cytochromes and quinones is absent from methanogens grown on H$_2$ + CO$_2$. Nevertheless, the electron carriers involved in the reduction of CO$_2$ to methane are well understood. This process requires several specific coenzymes unique to methanogens that function as cofactors for enzymes that sequentially reduce C$_1$ intermediates starting with CO$_2$ and yielding CH$_4$ (see Figure 17.10).

Three *classes* of methanogenic substrates are known (see Table 17.4) and all release free energy suitable for ATP synthesis. The first class involves the use of *CO$_2$-type substrates:*

$$CO_2 + 4\,H_2 \rightarrow CH_4 + 2\,H_2O$$
$$\Delta G^{0'} = -131 \text{ kJ/reaction}$$

$$4\,HCOO^- + 4\,H^+ \rightarrow CH_4 + 3\,CO_2 + 2\,H_2O$$
$$\Delta G^{0'} = -145 \text{ kJ/reaction}$$

$$4\,CO + 2\,H_2O \rightarrow CH_4 + 3\,CO_2$$
$$\Delta G^{0'} = -210 \text{ kJ/reaction}$$

The second class of reaction involves reduction of the *methyl group* of methyl-containing compounds to methane. In the case of methanol or methylamine, the overall reaction to methane has the following stoichiometry:

$$4\,CH_3OH \rightarrow 3\,CH_4 + CO_2 + 2\,H_2O$$
$$\Delta G^{0'} = -319 \text{ kJ/reaction}$$

$$4\,CH_3NH_3Cl + 2\,H_2O \rightarrow 3\,CH_4 + CO_2 + 4\,NH_4Cl$$
$$\Delta G^{0'} = -230 \text{ kJ/reaction}$$

In these reactions, some molecules of the substrate function as an electron *donor* and are oxidized to CO$_2$, whereas other molecules are reduced and are thus as electron *acceptors*. During growth on methyl compounds the reducing power for methanogenesis can also come from H$_2$. In fact, one species of methanogen, *Methanosphaera stadtmaniae*, can grow on methanol *only* in the presence of H$_2$. In this case the stoichiometry is:

$$CH_3OH + H_2 \rightarrow CH_4 + H_2O$$
$$\Delta G^{0'} = -113 \text{ kJ/reaction}$$

A few methanogens grow on alcohols other than methanol. *Methanospirillum* and *Methanogenium*, for example, grow with 2-propanol as electron donor, four molecules of 2-propanol being converted to four molecules of acetone along with reduction of CO$_2$ to CH$_4$. Ethanol, 1-propanol, and 1-butanol are also oxidized by certain strains, the products being acetate, propionate, or butyrate, and CH$_4$ (from the reduction of CO$_2$). Pyruvate can also function as electron donor for CO$_2$ reduction to CH$_4$ by certain *Methanococcus* species.

The final methanogenic reaction is **acetotrophic,** the cleavage of acetate to CH$_4$ plus CO$_2$:

$$CH_3COO^- + H_2O \rightarrow CH_4 + HCO_3^-$$
$$\Delta G^{0'} = -31 \text{ kJ/reaction}$$

Only two genera of methanogens, *Methanosarcina* and *Methanothrix* (*Methanosaeta*), have species that are acetotrophic. The conversion of acetate to methane appears to be a very significant ecological process, especially in sewage digestors and in freshwater anoxic environments where competition for acetate between sulfate-reducing bacteria and methanogenic bacteria is not extensive (∞ Section 14.12).

Diversity and physiology of methanogenic Archaea

A variety of morphological types of methanogenic bacteria have been isolated, and studies of their physiology and molecular properties have served to classify methanogens into seven major groups containing a total of 17 genera (Table 17.5). Short and long rods, cocci in various cell arrangements, plate-shaped cells, and filamentous methanogens are all known (Figures 17.5–17.7). Both gram-positive and gram-negative methanogens are known (see Table 17.5); thus the Gram stain is of little use in classifying these organisms. The current taxonomy of methanogens relies primarily on molecular methods, in particular on 16S rRNA sequence comparisons (∞ Section 15.7 for a discussion of molecular phylogeny) and, to a lesser degree, on immunological methods. The creation of seven major groups (Table 17.5) has emerged from extensive 16S ribosomal RNA sequence analyses.

When growing autotrophically, CO$_2$ is the carbon source for methanogens (see the section, Autotrophy in methanogens). However, growth of virtually all methanogens is stimulated by acetate, and the growth of some species is also stimulated by certain amino acids. For laboratory culture, some methanogens require complex additions such as yeast extract or casein digests, and some rumen methanogens require a mixture of branched-chain fatty acids (∞ Section 14.13).

TABLE 17.5   Characteristics of methanogenic Archaea

Genus	Morphology	Gram reaction	Number of species	Substrates for methanogenesis	DNA (mol % GC)
Group I					
Methanobacterium	Long rods	+ or −	8	$H_2 + CO_2$, formate	29–61
Methanobrevibacter	Short rods	+	3	$H_2 + CO_2$, formate	27–31
Methanosphaera	Cocci	+	1	Methanol + H_2 (both needed)	26
Group II					
Methanothermus	Rods	+	2	$H_2 + CO_2$; can also reduce S^0	33
Group III					
Methanococcus	Irregular cocci	−	5	$H_2 + CO_2$, pyruvate + CO_2, formate	29–34
Group IV					
Methanomicrobium	Short rods	−	2	$H_2 + CO_2$, formate	45–49
Methanogenium	Irregular cocci	−	3	$H_2 + CO_2$, formate	51–61
Methanospirillum	Spirilla	−	1	$H_2 + CO_2$, formate	46–50
Methanoplanus	Plate-shaped cells—occurring as thin plates with sharp edges	−	2	$H_2 + CO_2$, formate	38–47
Group V					
Methanosarcina	Large irregular cocci in packets	+	6	$H_2 + CO_2$, methanol, methylamines, acetate	41–43
Methanolobus	Irregular cocci in aggregates	−	5	Methanol, methylamines	38–42
Methanoculleus	Irregular cocci	−	4	$H_2 + CO_2$, alcohols, formate	54–62
Methanohalobium	Irregular cocci	−	1	Methanol, methylamines; halophilic	44
Methanococcoides	Irregular cocci	−	2	Methanol, methylamines	42
Methanohalophilus	Irregular cocci	−	3	Methanol, methylamines, methyl sulfides; halophile	41
Methanothrix (*Methanosaeta*)	Long rods to filaments	−	3	Acetate	52–61
Group VI					
Methanopyrus	Rods in chains	+	1	$H_2 + CO_2$; hyperthermophile, growth at 110°C	60
Group VII					
Methanocorpusculum	Irregular cocci	−	3	$H_2 + CO_2$, formate, alcohols	48–52

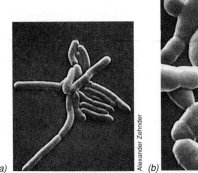
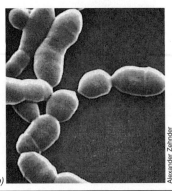

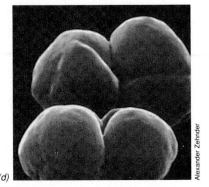

Figure 17.5 Scanning electron micrographs of cells of methanogenic Archaea, showing the considerable morphological diversity. (a) *Methanobrevibacter ruminantium.* A cell is about 0.7 μm in diameter. (b) *Methanobacterium* strain AZ. A cell is about 1 μm in diameter. (c) *Methanospirillum hungatii.* A cell is about 0.4 μm in diameter. (d) *Methanosarcina barkeri.* A cell is about 1.7 μm wide.

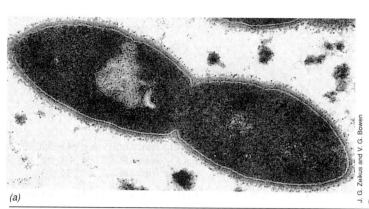
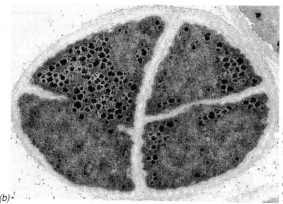

Figure 17.6 Transmission electron micrographs of thin sections of methanogenic Archaea. (a) *Methanobrevibacter ruminantium*. A cell is 0.7 μm in diameter. (b) *Methanosarcina barkeri*, showing the thick cell wall and the manner of cell segmentation and cross-wall formation. A cell is 1.7 μm in diameter.

All methanogens use NH_4^+ as a nitrogen source, and a few species are known to fix molecular nitrogen (N_2 fixation) (∞ Section 13.27). The trace metal *nickel* is required by all methanogens; it is a component of an important methanogenic coenzyme, $Factor_{430}$, and is also present in the enzymes *hydrogenase* and *carbon monoxide dehydrogenase* (discussion follows). Iron and cobalt are also important trace metals for methanogens and, along with nickel, have been the only trace metals shown to be absolutely required for growth of these organisms.

Unique methanogenic coenzymes

A number of coenzymes have been found in methanogens that are unique to this group of prokaryotes (Figure 17.8). These coenzymes play important roles in the biochemistry of methanogenesis. Because of their importance in the energy-yielding pathway, many of these coenzymes are present at far *higher* levels in methanogens than many of the common coenzymes such as NAD^+ or FMN are in other prokaryotes. We detail here the structure and function of each of these coenzymes as a prelude to our discussion of the biochemistry of methanogenesis.

Coenzymes as C_1 carriers in methanogenesis

Methanofuran is a low-molecular-weight coenzyme that interacts in the *first* step of methanogenesis from CO_2 (see Figure 17.10). Methanofuran consists of a molecule of phenol, two glutamic acid molecules, an unusual long-chain dicarboxylic fatty acid, and a furan ring (Figure 17.8c). CO_2 is reduced to the formyl level and bound by the amino side chain of the furan in the initial step of methanogenesis and is subsequently transferred to a second coenzyme in later steps of the pathway (see Figure 17.10).

Methanopterin is a methanogenic coenzyme containing a substituted pterin ring (Figure 17.8d). Methanopterin exhibits a bright blue fluorescence following absorbance at 342 nm. Structurally, methanopterin resembles the vitamin *folic acid* (∞ Figure 11.14c) and is a C_1 carrier during the reduction of CO_2 to CH_4 (the nitrogen atoms highlighted in Figure 17.8d are the atoms to which the C_1 intermediate binds). Methanopterin carries the C_1 unit during the majority of reductive steps in the methanogenic pathway, from the formyl (—CHO) level to the methyl (—CH_3) level (see Figure 17.10). *In vivo*, the reduced form of methanopterin, *tetrahydromethanopterin*, is the active form of the coenzyme.

Coenzyme M is involved in the *final step* in methane formation. Coenzyme M (Figure 17.8e), a very simple structure, has the chemical name 2-mercaptoethanesulfonic acid. The coenzyme is the carrier of the *methyl* group that is reduced to methane by the F_{430}–methyl reductase enzyme complex in the final step of methanogenesis:

$$CH_3—S—CoM + 2\ H \rightarrow HS—CoM + CH_4$$

Despite the structural simplicity of coenzyme M, it is highly specific in the methyl reductase reaction. A number of closely related analogs of coenzyme M have been found to be inactive in methanogenesis *in vitro*. The rumen methanogen, *Methanobrevibacter ruminantium*, has the interesting property of requiring coenzyme M as a growth factor. In this way coenzyme M can be considered a *vitamin*. Some methanogenic bacteria excrete coenzyme M, and this is apparently the source of coenzyme M for *M. ruminantium* in the rumen (∞ Section 14.13). Coenzyme M is so active as a vitamin for *M. ruminantium* that the organism shows a growth response at concentrations as low as 5 nanomolar (n*M*) ($5 \times 10^{-9}\ M$).

A potent inhibitory analog of coenzyme M is *bromo*ethanesulfonic acid, $Br—CH_2—CH_2—SO_3H$. This compound causes 50% inhibition of methyl reductase

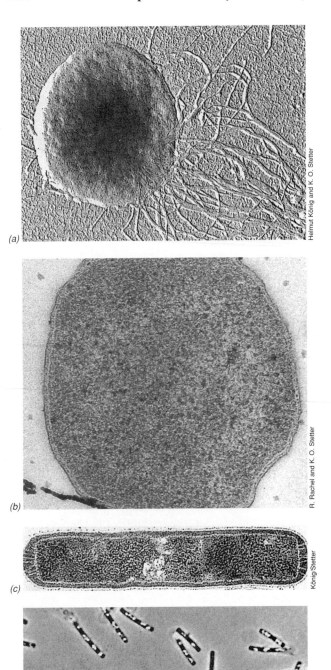

(a) Helmut König and K. O. Stetter

(b) R. Rachel and K. O. Stetter

(c) König/Stetter

(d) Stephen Zinder

Figure 17.7 Hyperthermophilic and thermophilic methanogens. (a) *Methanococcus jannaschii* (temperature optimum, 85°C), shadowed preparation electron micrograph. A cell is about 1 μm in diameter. (b) *Methanococcus igneus* temperature optimum, 88°C), thin section. A cell is about 1 μm in diameter. (c) *Methanothermus fervidus* (temperature optimum, 88°C), thin-sectioned electron micrograph. A cell is about 0.4 μm in diameter. (d) *Methanothrix* (*Methanosaeta*) *thermophila* (temperature optimum, 60°C), phase contrast micrograph. A cell is about 1 μm in diameter. The refractile bodies inside the cells are gas vesicles.

activity at a concentration of 10^{-6} M, and it also inhibits growth of methanogenic bacteria. Because coenzyme M is restricted to methanogenic bacteria, the bromo analog can be used experimentally to specifically inhibit methanogenesis in natural environments in ecological studies of the anoxic degradation of organic matter to methane.

Although not a C_1 carrier, **coenzyme F_{430}** is a yellow, soluble, nickel-containing tetrapyrrole (Figure 17.8*b*) and, like CoM, plays an intimate role in the *terminal* step of methanogenesis as part of the methyl reductase system. Coenzyme F_{430} absorbs light strongly at 430 nm but, unlike F_{420}, does not fluoresce. The nickel requirement for growth of methanogens reflects the abundance of F_{430} in the cells because most of the nickel in cells of methanogens is associated with the F_{430}–methyl reductase system.

Coenzymes involved in redox reactions

Coenzyme F_{420} and the coenzyme 7-mercaptoheptanoylthreonine phosphate (HS-HTP) are electron donors in methanogenesis. Coenzyme F_{420} is a flavin derivative, structurally resembling the common flavin coenzyme FMN (∞ Figure 4.14). The structure of F_{420} (Figure 17.8*a*) resembles that of FMN, but F_{420} lacks one of the nitrogen atoms of FMN in its middle ring and also lacks the methyl groups found on the benzene ring typical of true flavins (∞ Figure 4.14). F_{420} is a *two*-electron carrier of low reduction potential ($E_0' = -0.37$ V). Coenzyme F_{420} interacts with a number of different enzymes in methanogens including hydrogenase and NADP$^+$ reductase. F_{420} also plays a role in methanogenesis as the electron donor in at least one of the steps of CO_2 reduction (see Figure 17.10). The oxidized form of F_{420} absorbs light at 420 nm and fluoresces blue-green (Figure 17.9); on reduction, the coenzyme becomes colorless. The fluorescence of F_{420} is a useful tool for preliminary identification of an organism as a methanogen (Figure 17.9). Although important in methanogenesis, F_{420} may have other biological functions because cofactors similar to F_{420} have been detected in sulfate-reducing Archaea and in low levels in various Bacteria, including *Streptomyces* and some cyanobacteria.

HS-HTP (7-mercaptoheptanoyl threonine phosphate) is the final unique coenzyme of the methanogens to be considered. Like coenzyme M, this cofactor is involved in the *terminal* step of methanogenesis catalyzed by the methyl reductase system. As shown in Figure 17.8*f*, the structure of HS-HTP is rather simple. It is a phosphorylated derivative of the amino acid threonine containing a fatty acid side chain with a terminal SH group. HS-HTP resembles the vitamin *pantothenic acid* (part of acetyl-CoA) (∞ Figure 4.22) and serves as an electron *donor* to the methyl reductase system. HS-HTP reduction of coenzyme M (discussed later) is at the core of energy conservation in methanogenesis.

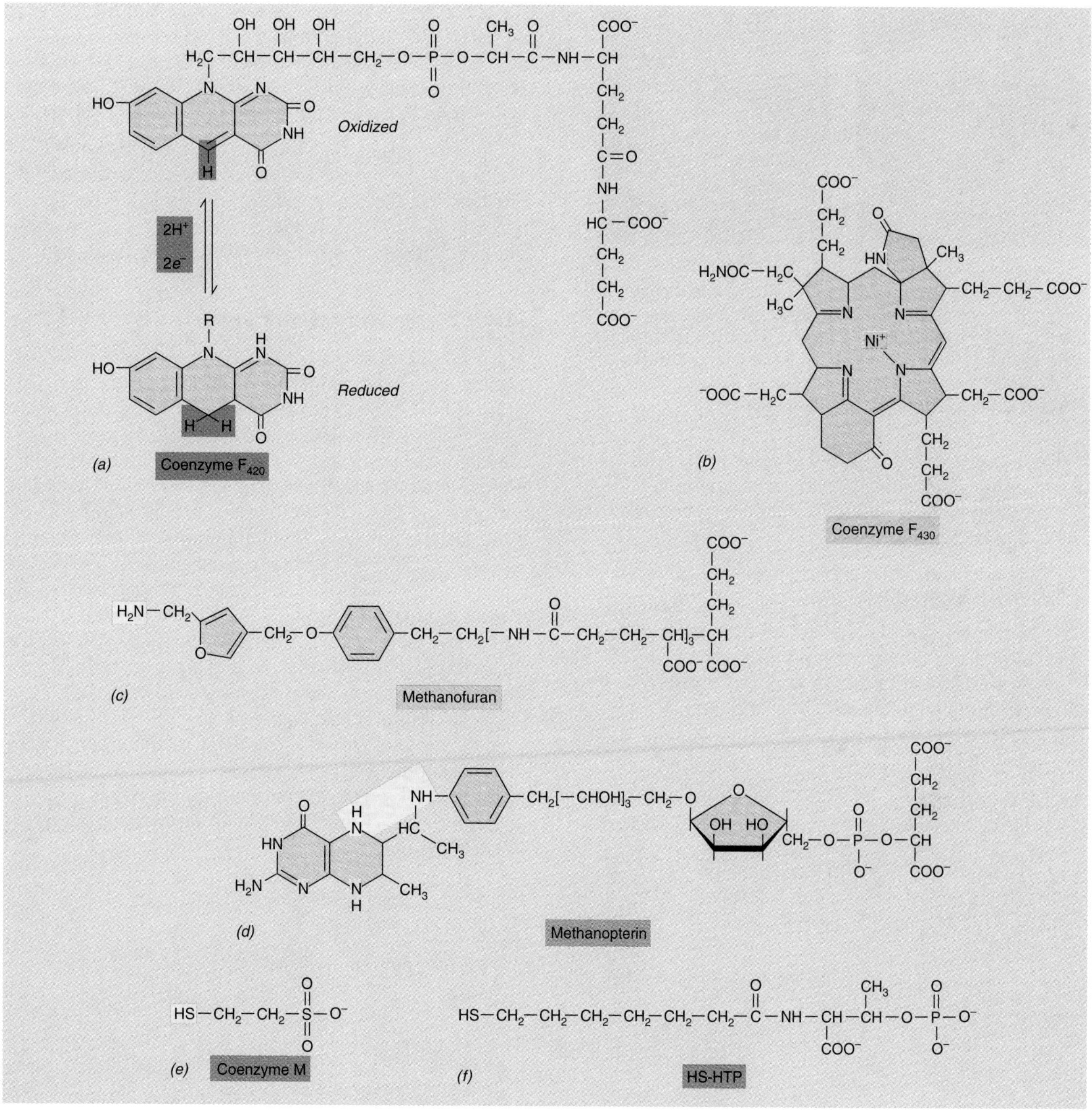

Figure 17.8 Coenzymes unique to methanogenic Archaea. The atoms shaded in brown or yellow are the sites of oxidation–reduction reactions (F_{420}—brown) or the position to which the C_1 moiety is attached during the reduction of CO_2 to CH_4 (methanofuran, methanopterin, and coenzyme M—yellow). The colors used to highlight a particular coenzyme itself (HS-HTP is orange, for example) are used throughout in Figures 17.10–17.13 and can be used to follow the reactions in each figure.

Biochemistry of CO_2 reduction to CH_4

Now that we have discussed the major coenzymes involved in methanogenesis from CO_2 plus H_2, we consider how these molecules interact with their specific enzymes in the conversion of CO_2 to CH_4. The reduction of CO_2 to CH_4 is generally H_2-dependent, but formate, carbon monoxide, and even elemental iron (Fe^0) can

serve as electron donors for methanogenesis. In the latter case, Fe^0 is oxidized to Fe^{2+}, with the electrons released combining with protons to form H_2, which is the immediate electron donor in methanogenesis. And as previously mentioned, in a few methanogens even certain simple *organic* compounds such as alcohols can supply the electrons for CO_2 reduction. For example, 2-propanol can be oxidized to acetone, yielding electrons for meth-

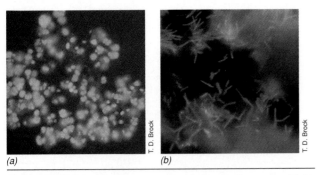

(a)   *(b)*

Figure 17.9 (a) Autofluorescence of the methanogen *Methanosarcina barkeri* due to the presence of the unique electron carrier F_{420}. A single cell is about 1.7 μm in diameter. The organisms were visualized with blue light in a fluorescence microscope. (b) F_{420} fluorescence in the methanogen *Methanobacterium formicicum*. A single cell is about 0.6 μm in diameter.

anogenesis in some species. But in general, the production of CH_4 from CO_2 is driven by molecular hydrogen (H_2).

The steps in CO_2 reduction, shown in Figure 17.10, are summarized as follows:

1. CO_2 is activated by methanofuran and subsequently reduced to the *formyl* level.

2. The formyl group is transferred from methanofuran to tetrahydromethanopterin (MP in Figure 17.10) and subsequently dehydrated and reduced in two separate steps to the *methylene* and *methyl* levels.

3. The methyl group is transferred from methanopterin to coenzyme M.

4. Methyl-coenzyme M is reduced to *methane* by the methyl reductase system in which F_{430} and HS—

HTP are involved. The electron donor for this reaction is HS—HTP, and the product of the reaction, besides CH_4, is a disulfide of CoM and HTP (CoM—S—S—HTP). Free CoM and HS—HTP are regenerated by reduction with H_2.

In the steps of methanogenesis shown in Figure 17.10, reduced F_{420} is the electron donor in the reduction of the methenyl to the methylene group, but the nature of the electron donors in the initial step of CO_2 reduction and in the step between methylene and methyl is unknown.

Autotrophy in methanogens

Carbon dioxide is converted to organic form in methanogenic Archaea through the reactions of the **acetyl-CoA (Ljungdahl–Wood) pathway** used also by homoacetogens and sulfate-reducing bacteria. (See Section 16.9 for detailed discussion of this pathway.) However, unlike the other anaerobes that use this pathway, methanogens growing on H_2 + CO_2 integrate their biosynthetic and bioenergetic pathways because common intermediates are shared. This is possible because both the acetyl-CoA pathway and the methanogenic pathway lead to the production of CH_3 groups. As shown in Figure 17.11, autotrophically grown methanogens lack that part of the acetyl-CoA pathway leading to the production of methyl groups via tetrahydrofolate intermediates and instead obtain methyl groups for the production of acetate for biosynthesis from the methanogenic pathway (compare Figure 17.11 with Figure 16.38). Specifically, methyl tetrahydromethanopterin donates methyl groups to a corrinoid-containing enzyme (see later) to

Figure 17.10 Pathway of methanogenesis from CO_2. MF, Methanofuran; MP, tetrahydromethanopterin; CoM, coenzyme M; F_{420}, coenzyme F_{420}; F_{430}, coenzyme F_{430}; HS-HTP, 7-mercaptoheptanoylthreonine phosphate. The carbon atom reduced is shown in yellow, and the source of electrons are highlighted in brown. See Figure 17.8 for the structures of the coenzymes and the text for discussion of the reversible Na⁺ pump.

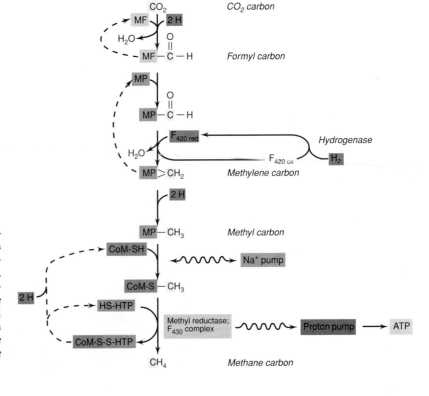

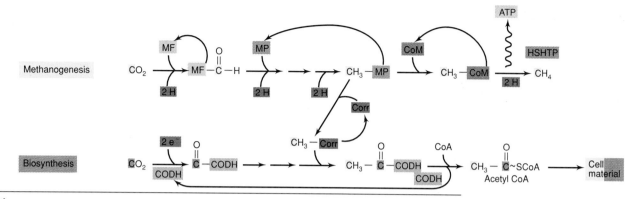

Figure 17.11 How autotrophic methanogens combine aspects of biosynthesis and bioenergetics. Abbreviations and color coding are as in Figure 17.8. CODH, Carbon monoxide dehydrogenase; Corr, C_1-carrying corrinoid protein. Note how half of the acetyl-CoA molecule produced comes from reactions leading to methanogenesis.

yield CH_3–corrinoid (Figure 17.11). The CH_3 group is then transferred to carbon monoxide dehydrogenase (which has previously reduced CO_2 to the level of CO), to eventually yield acetyl-CoA (Figure 17.11). Because relative to methanogenesis only a small amount of CO_2 is incorporated into cell material, this small drain on the methanogenic pathway is of little consequence, and the merging of the two pathways probably effects an energy savings to the organism because synthesis of additional enzymes is not necessary in order to make the CH_3 group of acetate for biosynthesis.

Methanogenesis from methyl compounds and acetate

Reactions of the acetyl-CoA pathway discussed earlier are intimately involved in the production of methane from methyl compounds and from acetate. Methyl compounds such as methanol are catabolized by donating methyl groups to a corrinoid protein to form CH_3–corrinoid (Figure 17.12a). Corrinoids are the parent structures of such compounds as vitamin B_{12}, which contain a porphyrin-like corrin ring with a central cobalt atom. The CH_3–corrinoid complex donates the methyl group to CoM to give CH_3—CoM, from which methane is obtained by reduction with electrons derived from oxidation of other molecules of methanol to CO_2 (Figure 17.12a). Carbon for biosynthesis originates by formation of CO from the oxidation of a methyl group from methanol and the combination of CO and a methyl group by carbon monoxide dehydrogenase to yield acetate (Figure 17.12a).

Growth of acetotrophic (acetate-degrading) methanogens is also tied to reactions of the acetyl-CoA pathway. In acetotrophic methanogens acetate is used directly for biosynthesis. For energy purposes, acetate is also the energy source. Acetate is activated to acetyl-CoA, which can interact with carbon monoxide dehydrogenase, following which the methyl group of acetate is transferred to the corrinoid enzyme of the

acetyl-CoA pathway to yield CH_3–corrinoid (Figure 17.12b). From here the methyl group is transferred to tetrahydromethanopterin and then to coenzyme M to yield CH_3—CoM. The latter is then reduced to CH_4 using electrons generated from the oxidation of CO to CO_2 by CO dehydrogenase (Figure 17.12b). Energy conservation occurs in acetotrophy as a result of a proton motive force formed during the methyl reductase step, just as it does for $H_2 + CO_2$ or methylotrophically grown methanogens (see below).

Energetics of methanogenesis

Under standard conditions, the free-energy change of the reduction of CO_2 to CH_4 with H_2 is −131 kJ/mol. However, concentrations of H_2 in methanogenic habitats are usually quite low, no higher than 10 micromolar (μM), and because of the influence of concentration of reactants on free-energy change (see Appendix 1), the free energy for the reaction forming CH_4 from $H_2 + CO_2$ by methanogens in their natural habitat is probably much lower, only about −30 kJ. Thus, probably only one ATP is formed during CO_2 reduction to CH_4, and this agrees with molar growth yield data obtained for methanogens growing on $H_2 + CO_2$.

The *terminal step* of methanogenesis, in which the methyl reductase enzyme complex reduces CH_3—CoM to CH_4 (Figure 17.10), is the energy conservation step. The interaction of HS—HTP with CH_3—CoM in this terminal step forms CH_4 and CoM—S—S—HTP (see Figure 17.10). The latter is then reduced with electrons derived from reduced F_{420} (or possibly from H_2 itself) to yield CoM—SH and HS—HTP (Figure 17.13). This reduction, carried out by the enzyme *heterodisulfide reductase*, is exergonic and is associated with the extrusion of protons across the membrane, creating a proton motive force (Figure 17.13). Dissipation of the proton gradient by a membrane-integrated proton-translocating ATPase (see Section 4.13 for a discussion of ATPases) drives ATP synthesis during methanogenesis in the

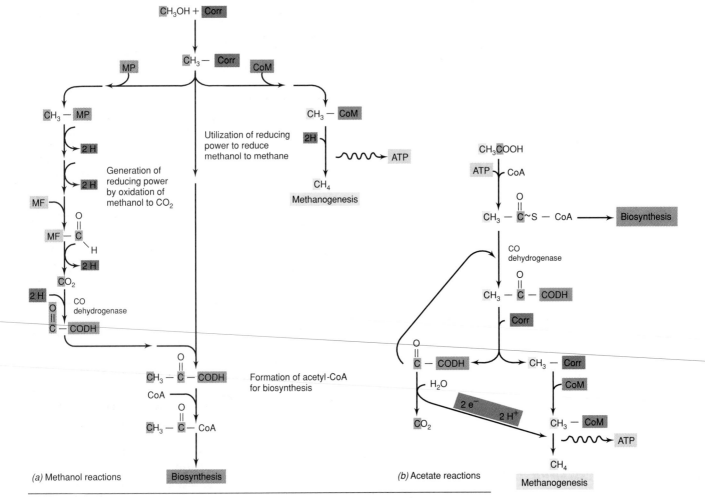

Figure 17.12 Utilization of reactions of the acetyl-CoA pathway during growth on methanol (a) or acetate (b) by methanogenic Archaea. For growth on methanol, most methanol carbon is converted to CH_4, while a smaller amount is converted to either CO_2 or, via formation of acetyl-CoA, is assimilated into cell material. Abbreviations and color coding are as in Figures 17.8, 17.10, and 17.11.

same way that this process occurs in other forms of respiratory metabolism.

Growth of methanogens on methyl compounds is also linked to the heterodisulfide reductase proton pump, but an additional factor is involved. In the absence of H_2, methanogenesis from methylated compounds requires that some of the substrate be oxidized to generate the electrons needed for methyl reduction to methane. This occurs at the expense of a membrane-bound *sodium pump*, which establishes a sodium gradient across the cytoplasmic membrane to drive the oxidation of methyl groups. The sodium pump is linked to the interconversion of CH_3–tetrahydromethanopterin and CH_3—CoM during methanogenesis; production of CH_3—CoM establishes the sodium gradient, while production of CH_3–tetrahydromethanopterin from CH_3—CoM, an endergonic reaction, dissipates the gradient (see Figure 17.10).

CONCEPT CHECK   **17.2**

Methanogenic Archaea are strictly anaerobic prokaryotes that produce natural gas (CH_4) as the result of their energy metabolism. Several unique coenzymes are involved in methanogenesis, and it is the terminal step of this process that is linked to ATP synthesis. Parts of the acetyl-CoA pathway are important in biosynthesis and the production of methane from methyl substrates or acetate.

✔ *What is* coenzyme M *and how does it participate in methanogenesis?*

✔ *Draw the sequence of events in the reduction of CO_2 to CH_4. In what order do the various unique coenzymes participate in this process?*

✔ *How is ATP synthesized in methanogens?*

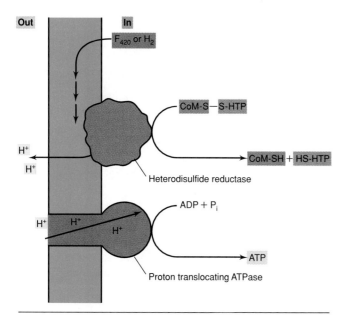

Figure 17.13 Bioenergetics of methanogenesis. ATP synthesis linked to a proton motive force established during the terminal step of methanogenesis. In some methanogens, the electron transport steps shown by arrows from F_{420} to the heterodisulfide reductase are mediated by b- and c-type cytochromes. Refer to Figure 17.8 for structures of the coenzymes involved.

Further oxidative steps in the conversion of methyl groups to CO_2 to yield electrons for methyl reduction to methane proceed by reversal of the enzymatic steps leading to CH_4 formation from CO_2 (Figure 17.10) and show no energy requirement. However, energy from the Na^+ gradient is also required to drive the carboxylation of methanofuran in methanogenesis (see Figure 17.10). This reaction is sufficiently endergonic that an energetic push (the Na^+ gradient) is needed to drive the reaction. Thus, in methanogens we see two types of ion pumps in play: a typical proton pump used to drive adenosine triphosphate (ATP) synthesis and a reversible Na^+ pump that can function to drive methyl group oxidation and methanofuran carboxylation.

17.3
Hyperthermophilic Archaea

A phylogenetically distinct kingdom of Archaea (Crenarchaeota) consists of organisms unified by their extremely thermophilic nature. This branch, referred to as the **hyperthermophiles,** contains representatives that are the most heat-loving of all known prokaryotes (∞ Sections 5.7 and 5.8 for a discussion of thermophily). Several hyperthermophiles are capable of growth at temperatures above the normal boiling point of water, and all have temperature optima *above* 80°C! We begin with a brief overview of the group and proceed to a more detailed discussion of the major genera that have been characterized.

Overview of hyperthermophilic Archaea

Most hyperthermophilic Archaea have been isolated from geothermally heated soils or waters containing elemental sulfur and sulfides, and most species metabolize sulfur in some way. Elemental sulfur is formed from geothermal H_2S either by the spontaneous oxidation of H_2S with O_2 or by reaction of H_2S with SO_2 (the latter is a common component of volcanic gases). In terrestrial environments, sulfur-rich springs, mud pots, and soils may have temperatures up to 100°C and are generally mildly to extremely acidic owing to production of sulfuric acid, H_2SO_4, from the biological oxidation of H_2S and S^0 (∞ Section 14.15). Such hot, sulfur-rich environments are called *solfataras,* and solfataric fields are found throughout the world (Figure 17.14a). Extensive solfataras are found in Italy, Iceland, New Zealand, and Yellowstone National Park in Wyoming. Depending on the surrounding geology, solfataric environments may be either slightly alkaline to mildly acidic, pH 5–8, or extremely acidic, with pH values below 1 not uncommon, and are generally anoxic. Hyperthermophiles have been obtained from both types of environments, but the majority of these organisms inhabit neutral or mildly acidic habitats. In addition to these natural habitats, hyperthermophilic Archaea also thrive within artificial thermal habitats, in particular the boiling outflows of geothermal power plants.

With only a few exceptions, hyperthermophiles are *obligate anaerobes.* Their energy-yielding metabolism is either chemoorganotrophic or chemolithotrophic. The requirement for sulfur, nearly universal across the group, is based on the need for an electron *acceptor* to carry out anaerobic respiration or an electron *donor* for chemolithotrophic metabolism. Elemental sulfur (S^0) is reduced to H_2S using electrons derived from the oxidation of organic compounds or from H_2. The energy-yielding reactions of some hyperthermophiles are shown in Table 17.6. Chemoorganotrophs, such as those of the genera *Thermococcus* and *Thermoproteus,* oxidize a variety of organic compounds, in particular small peptides, glucose, and starch anaerobically in the presence of S^0 as electron acceptor. *Sulfolobus,* on the other hand, utilizes a variety of organic compounds (as well as S^0) as electron donor with O_2 as electron acceptor.

Many hyperthermophilic Archaea can grow *chemolithotrophically* with H_2 as energy source (Table 17.6). *Pyrodictium,* for example, grows strictly anaerobically in a mineral-salts medium supplemented with H_2 and S^0 at temperatures up to 110°C. Cultures are usually grown in sealed glass or stainless steel vessels to which H_2 is added under pressure. Growth of *Pyrodictium* is stimulated by the addition of organic compounds, but the latter cannot replace H_2. *Acidianus* and *Sulfolobus* can grow aerobically with H_2 as electron donor. Other chemolithotrophic growth modes among hyperthermophiles include the oxidation of elemental sulfur and

Figure 17.14 Habitats of hyperthermophilic Archaea. (a) A typical solfatara in Yellowstone National Park. Steam rich in hydrogen sulfide rises to the surface of the earth. Because of the heat and acidity, higher forms of life do not develop. (b) Sulfur-rich hot spring, a habitat containing dense populations of *Sulfolobus*. (c) A typical boiling spring of neutral pH in Yellowstone Park; Imperial Geyser. (d) An iron-rich geothermal spring, another *Sulfolobus* habitat.

ferrous iron (aerobically) by various *Sulfolobus* species and the oxidation of H_2 or Fe^{2+} coupled to NO_3^- reduction (producing NO_2^- and eventually N_2 or NH_4^+) (see Table 17.6). Thus, a variety of respiratory processes can be carried out by hyperthermophilic Archaea, but in many cases elemental sulfur plays a key role, as either an electron donor or an electron acceptor.

Hyperthermophiles from volcanic habitats

As mentioned previously, volcanic habitats can have temperatures as high as 100°C and are thus suitable for hyperthermophilic Archaea. The first such organism dis-

covered, *Sulfolobus*, grows in sulfur-rich hot acid springs (Figure 17.14*b*) at temperatures up to 90°C and at pH values of 1–5*. *Sulfolobus* (Figure 17.15*a*) is an obligate aerobe capable of oxidizing H_2S or S^0 to H_2SO_4 and fixing CO_2 as carbon source. *Sulfolobus* can also grow chemoorganotrophically. Cells of *Sulfolobus* are generally spherical but form distinct lobes (Figure 17.15*a*). Cells adhere tightly to sulfur crystals where they can be visualized microscopically by use of fluorescent dyes (∞ Figure 13.20*b*). Besides an active *aerobic* metabolism, *Sulfolobus* can also reduce Fe^{3+} to Fe^{2+} (but not grow) *anaerobically*. The ability of *Sulfolobus* to oxidize Fe^{2+} to Fe^{3+} aerobically (Figure 17.14*c*), however, has been used quite successfully in the high temperature leaching of iron and copper ores (∞ Section 14.17).

A facultative aerobe resembling *Sulfolobus* is also present in acidic solfataric springs. This organism, named *Acidianus* (Figure 17.15*b*), differs from *Sulfolobus* primarily by virtue of its ability to grow *anaerobically*.

---

*Historical note: *Sulfolobus* was first discovered by Thomas Brock and colleagues in 1970 and formally described in 1972. The discovery of *Sulfolobus*, along with the previously isolated *Thermus aquaticus* (source of the extremely thermostable *Taq* DNA polymerase, see back cover of this book), is generally credited with launching the field of hyperthermophilic microbiology. Thomas Brock was the senior author of the first seven editions of this book. In the 1980s to the present, Karl Stetter and colleagues in Germany have greatly expanded the field of hyperthermophilic microbiology with the discovery of many new genera and species.

	TABLE 17.6	Energy-yielding reactions of hyperthermophilic Archaea	

Nutritional class	Energy-yielding reaction	Example
Chemoorgano-trophic	Organic compound + $S^0 \rightarrow H_2S + CO_2$	*Thermoproteus, Thermococcus, Desulfurococcus, Thermofilum, Pyrococcus*
	Organic compound + $SO_4^{2-} \rightarrow H_2S + CO_2$	*Archaeoglobus*
	Organic compound + $O_2 \rightarrow H_2O + CO_2$	*Sulfolobus*
	Organic compound $\rightarrow CO_2$ + fatty acids	*Staphylothermus*
	Organic compound $\rightarrow CO_2 + H_2$	*Pyrococcus*
Chemolitho-trophic	$H_2 + S^0 \rightarrow H_2S$	*Acidianus, Pyrodictium, Thermoproteus*
	$H_2 + NO_3^- \rightarrow NO_2^- + H_2O$	*Pyrobaculum, Stygiolobus, Aquifex,*
	(NO_2^- reduced to N_2 by some species)	*Pyrodictium, Thermoproteus*
	$4 H_2 + NO_3^- + 2 H^+ \rightarrow NH_4^+ + 3 H_2O$	*Pyrolobus*
	$2 H_2 + O_2 \rightarrow 2 H_2O$	*Acidianus, Sulfolobus, Pyrobaculum, Aquifex*[a]
	$2 S^0 + 3 O_2 + 2 H_2O \rightarrow 2 H_2SO_4$	*Sulfolobus, Acidianus*
	$4 FeS_2 + 15 O_2 + 2 H_2O \rightarrow 2 Fe_2(SO_4)_3 + 2 H_2SO_4$	*Sulfolobus*
	$10 FeCO_3 + 2 NO_3^- + 24 H_2O \rightarrow 10 Fe(OH)_3 + N_2 + 10 HCO_3^- + 8 H^+$	*Ferroglobus*
	$4 H_2 + SO_4^{2-} + 2 H^+ \rightarrow 4 H_2O + H_2S$	*Archaeoglobus*
	$4 H_2 + CO_2 \rightarrow CH_4 + 2 H_2O$	*Methanopyrus, Methanococcus*

[a]Member of the Bacteria.

Remarkably, *Acidianus* is able to use S^0 in both its aerobic and anoxic metabolism. Under *aerobic* conditions the organism uses S^0 as an electron *donor*, oxidizing S^0 to H_2SO_4. Anaerobically, *Acidianus* uses S^0 as an electron *acceptor* (with H_2 as electron *donor*) forming H_2S as the reduced product. Thus, the metabolic fate of S^0 in cultures of *Acidianus* depends on the presence of O_2 and/or an electron donor.

Like *Sulfolobus*, *Acidianus* is roughly spherical in shape (Figure 17.15*b*). It grows at temperatures from about 65°C up to a maximum of 95°C, with an optimum of about 90°C. Another property shared by *Sulfolobus*

and *Acidianus* is an unusually low GC base ratio. The DNA of *Sulfolobus* is about 38% GC, whereas that of *Acidianus* is even lower, about 31%; many other hyperthermophiles have DNA of low GC content as well (see Table 17.7). These low GC base ratios are intriguing when one considers the hyperthermophilic nature of these organisms; how do they prevent their DNA from melting? In the test tube, DNA of 30–40% GC content would melt almost instantly at 90°C. Obviously hyperthermophiles have evolved protective mechanisms to prevent DNA melting *in vivo* and we discuss these in Section 17.5.

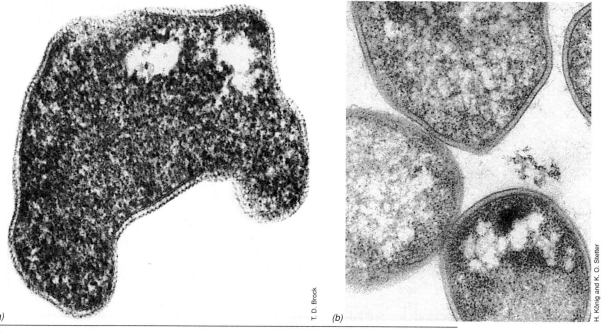

(a) T. D. Brock

(b) H. König and K. O. Stetter

Figure 17.15 Acidophilic hyperthermophilic Archaea. (a) *Sulfolobus acidocaldarius.* Electron micrograph of a thin section. (b) *Acidianus infernus.* Electron micrograph of a thin section. Cells of both organisms vary from 0.8 to 2 μm in diameter.

TABLE 17.7   Properties of hyperthermophilic Archaea

Genus	Morphology	Number of species	DNA (mol % GC)	Temperature (°C) Minimum	Temperature (°C) Optimum	Temperature (°C) Maximum	Optimum pH
Terrestrial volcanic isolates							
Sulfolobus	Lobed sphere	4	37	55	75–85	87	2–3
Acidianus	Sphere	2	31	65	85–90	95	2
Thermoproteus	Rod	2	56	60	88	96	6
Thermofilum	Rod	2	57	70	88	95	5.5
Desulfurococcus	Sphere	2	51	70	85	95	6
Desulfurolobus	Lobed sphere	1	32	65	80	87	2.5
Pyrobaculum	Rod	2	46	74	100	102	6
Methanothermus	Rods in clusters	2	33	65	83–88	97	6–7
Stygiolobus	Lobed sphere	1	38	57	80	89	3
Submarine volcanic isolates							
Pyrodictium	Disc-shaped with attached filaments	3	62	82	105	110	6
Pyrococcus	Sphere	2	38	70	100	106	6–8
Pyrolobus	Lobed cocci	1	53	90	105	113	5.5
Thermodiscus	Disc-shaped	1	49	75	90	98	5.5
Staphylothermus	Spheres in clumps	1	35	65	92	98	6–7
Thermococcus	Sphere	3	38–57	70	88	98	6–7
Methanopyrus	Rods	1	60	85	100	110	6.5
Methanococcus	Irregular cocci	3	31	45	88	91	6
Archaeoglobus	Cocci	2	46	64	83	95	7
Thermotoga[a]	Rods with a covering (see Figure 17.23a)	4	46	55	80	90	7
Aquifex[a]	Rods	2	40	67	85	95	6

[a]Member of the Bacteria.

The genera *Thermoproteus* and *Thermofilum* consist of *rod-shaped* cells that inhabit neutral or slightly acidic hot springs. Cells of *Thermoproteus* are stiff rods about 0.5 μm in diameter and are highly variable in length, ranging from short cells of 1–2 μm up to filaments 70–80 μm in length (Figure 17.16a). Filaments of *Thermofilum* are thinner, some 0.17–0.35 μm in width with filament lengths ranging up to 100 μm (Figure 17.16b). Both *Thermoproteus* and *Thermofilum* are strict anaerobes that carry out a S^0-based anaerobic respiration. Unlike most hyperthermophiles, the oxygen sensitivity of *Thermoproteus* and *Thermofilum* is extreme, comparable to that of the methanogens (see Section 17.2); thus, strict precautions must be taken in their culture. Most *Thermoproteus* isolates can grow chemolithotrophically on H_2 or chemoorganotrophically on complex carbon substrates such as yeast extract, small peptides, starch, glucose, ethanol, malate, fumarate, or formate. *Thermofilum* can be grown in either mixed culture with *Thermoproteus* or in pure culture only by the addition of a highly polar lipid fraction isolated from cells of *Thermoproteus*; in nature the two organisms probably coexist in close association. Both *Thermoproteus* and *Thermofilum* have similar GC base ratios (56–58% GC) but are phylogenetically distinct by nucleic acid hybridization analyses.

Desulfurococcus (Figure 17.17) is a spherical, obligately anaerobic, S^0-respiring organism. *Desulfurococcus* grows best at neutral pH and 80–90°C. The major features that differentiate *Desulfurococcus* from other hyperthermophilic Archaea are *genotypic* rather than phenotypic. DNA from *Desulfurococcus* contains 51% GC and hybridizes at only low levels with nucleic acid from *Thermoproteus*.

Hyperthermophiles from submarine volcanic areas

We now turn our attention to *submarine* volcanic habitats where a phylogenetically distinct set of hyperthermophilic Archaea exist. Although these underwater microbial habitats are generally quite shallow, the pressure of even a few meters of water can raise the boiling point of water sufficiently to select for organisms capable of growth above 100°C.

Geothermally heated sea floors exist in various locations around the world, and from a series of such habitats in the Mediterranean several genera of hyperthermophilic Archaea have been isolated. In submarine solfatara fields located in 2–10 m of water off the coast of Vulcano, Italy, the sea floor consists of sandy

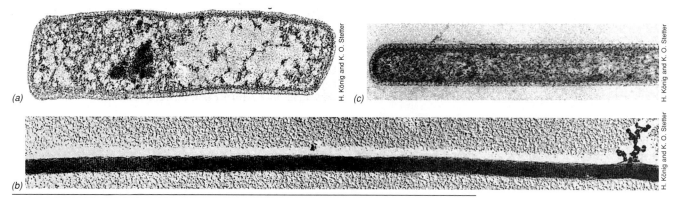

Figure 17.16 Rod-shaped hyperthermophilic Archaea from terrestrial volcanic habitats. (a) *Thermoproteus neutrophilus.* Electron micrograph of a thin section. A cell is about 0.5 μm in diameter. (b) *Thermofilum librum.* A cell is only about 0.25 μm in diameter. Electron micrograph of shadowed cells. (c) *Thermofilum librum.* Electron micrograph of a thin section.

sediments with cracks and holes from which geothermally heated (but not yet boiling) water is emitted at temperatures up to 103°C. From such waters several genera of hyperthermophiles have been isolated. But in addition, various Archaea have also been isolated from near very hot deep-sea hydrothermal vents (black smokers, ∞ Section 14.10) where hot water is emitted at much higher temperatures.

Pyrodictium *and* Thermodiscus

Pyrodictium is one of the most fascinating of all submarine volcanic hyperthermophiles because its growth temperature optimum, 105°C, is *above* the boiling point of water (at 1 atm). Cells of *Pyrodictium* are irregularly disc- and dish-shaped (Figure 17.18) and grow in culture as a moldlike layer on sulfur crystals suspended in

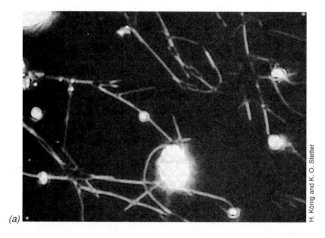

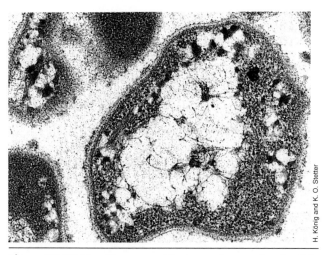

Figure 17.17 *Desulfurococcus.* Electron micrograph of a thin section of *D. saccharovorans.* A cell is about 0.7 μm in diameter.

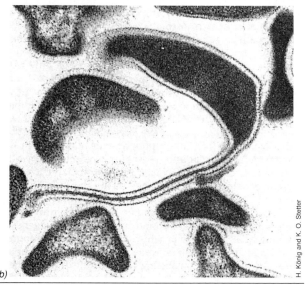

Figure 17.18 *Pyrodictium occultum,* a prokaryote with a growth temperature optimum of 105°C. (a) Dark-field micrograph. (b) Electron micrograph of a thin section. A cell is highly variable in diameter, from 0.3 to 2.5 μm.

the medium. The cell mass consists of a huge network of fibers to which individual cells are attached (Figure 17.18). The fibers are hollow and consist of proteinaceous subunits arranged in a fashion similar to that of the flagellin protein of the bacterial flagellum (∞ Section 3.9). The filaments of *Pyrodictium* do not function in motility but apparently serve to *attach* the cells to a solid substratum. *Pyrodictium* is a strict anaerobe that grows chemolithotrophically at neutral pH on H_2 with S^0 as electron acceptor. Growth occurs between 82° and 110°C and is stimulated by the addition of organic compounds. The cell envelope of *Pyrodictium* consists of glycoprotein and lacks peptidoglycan (as do all archaeal cell walls) (∞ Section 3.5). Phylogenetically, *Pyrodictium* shows affinities to *Sulfolobus*. However, because of the very large number of posttranscriptionally modified bases in the ribosomal RNAs of *Pyrodictium,* the exact phylogenetic position of this organism is unclear. Deoxyribonucleic acid (DNA) from *Pyrodictium* shows a substantially higher base ratio (62% GC) than that of any hyperthermophilic archaean from terrestrial habitats. *Pyrolobus* is found in environments similar to those of *Pyrodictium* but grows to even higher temperatures, 113°C (Table 17.7), and has a unique metabolism in which NO_3^- is reduced to NH_4^+ with H_2 as electron donor (Table 17.6).

Thermodiscus resembles *Pyrodictium* morphologically but is *chemoorganotrophic* rather than chemolithotrophic and is unable to grow above 100°C. *Thermodiscus* is an obligate chemoorganotroph, with growth occurring on complex organic mixtures with S^0 as electron acceptor at temperatures between 75 and 98°C (optimum at about 90°C). The DNA of *Thermodiscus* contains a significantly lower GC base ratio (49%) than that of *Pyrodictium*, but analyses of 16S rRNA sequences show that *Thermodiscus* is specifically related to *Pyrodictium*.

Pyrobaculum (Figure 17.19) is a physiologically unique archaean. It is capable of both aerobic respiration and denitrification ($NO_3^- \rightarrow N_2$), using either organic or inorganic electron donors at temperatures up to 103°C. H_2 as well as various complex nutrients support growth of *Pyrobaculum*, but sugars are not used. Elemental sulfur (S^0), the main electron acceptor of many hyperthermophilic Archaea, is not used by *Pyrobaculum*, and for unknown reasons, S^0 actually inhibits its growth. Interestingly, however, from a

phylogenetic perspective, *Pyrobaculum* is most clearly related to *Thermoproteus*, an obligately anaerobic sulfur-reducing hyperthermophile. *Pyrobaculum* is thus a good example of how difficult it is to draw phylogenetic conclusions from knowledge of metabolic capacities, a point previously made in our discussion of bacterial phylogeny (∞ Section 15.8).

Thermococcus, Pyrococcus, *and* Staphylothermus

Thermococcus is a spherical hyperthermophilic archaean indigenous to anoxic submarine thermal waters in various locations throughout the world. The spherical cells contain a tuft of polar flagella and are thus highly motile (Figure 17.20a). *Thermococcus* is an obligately anaerobic chemoorganotroph that grows on proteins and other complex organic mixtures (including some sugars) with S^0 as electron acceptor. It is not as thermotolerant as *Pyrodictium,* the optimum growth temperature being only 88°C, with a temperature range of 70–95°C.

An organism morphologically similar to *Thermococcus* is *Pyrococcus* (Figure 17.20b). *Pyrococcus* (the Latin derivation literally means "fireball") differs from *Thermococcus* primarily by its significantly different GC ratio (see Table 17.7) and its higher temperature requirements; *Pyrococcus* grows at between 70 and 106°C with an optimum of 100°C. However, metabolically *Thermococcus* and *Pyrococcus* are quite similar: proteins, starch, or maltose are oxidized as energy sources and S^0 is reduced to H_2S. Comparisons of 16S rRNA sequences (see Figure 17.1) show *Thermococcus* and *Pyrococcus* to be specifically related, despite their significantly different GC base ratios and optimal growth temperatures (see Table 17.7).

The genus *Staphylothermus* consists of spherical cells about 1μm in diameter that form aggregates of up to 100 cells, resembling those of *Staphylococcus* (Figure 17.21). *Staphylothermus* is a strictly anaerobic hyper-

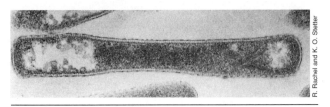

Figure 17.19 *Pyrobaculum aerophilum.* Cells measure 0.5 × 3.5 μm. *Pyrobaculum aerophilum* can grow by aerobic respiration or by denitrification with a 100°C growth temperature optimum.

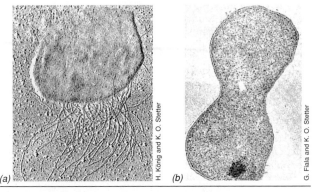

Figure 17.20 Spherical hyperthermophilic Archaea from submarine volcanic areas. (a) *Thermococcus celer.* Electron micrograph of shadowed cells. (b) Dividing cell of *Pyrococcus furiosus.* Electron micrograph of thin section. Cells of both organisms are about 0.8 μm in diameter.

thermophile growing optimally at 92°C and capable of growth between 65 and 98°C. Although S⁰ is required for growth, oxidation of complex organic compounds is not tightly coupled to S^0 reduction in *Staphylothermus* as it is in most other hyperthermophiles; fatty acids such as acetate and isovalerate are produced during growth by this organism, suggesting that a fermentative metabolism is occurring. The GC base ratio of *Staphylothermus* is quite low, about 35%, and 16S rRNA sequence studies have not shown this organism to be closely related to any other hyperthermophile.

Isolates of *Staphylothermus* have been obtained from both shallow marine hydrothermal vents and from samples taken from a "black smoker" at a depth of 2500 m (∞ Section 14.10 for a discussion of black smokers). Since the temperature of the black smoker yielding the *Staphylothermus* isolate was about 330°C, it is likely that the organism originated from thermally heated seawater and not from the superheated spring water itself (see Section 17.5). Nevertheless, *Staphylothermus* may be widely distributed near hot hydrothermal vents and may be an ecologically important microbial consumer in these environments.

Table 17.7 summarizes the major properties of both submarine and terrestrial hyperthermophilic Archaea.

Archaeoglobus

In the preceding discussion we emphasized that many hyperthermophilic Archaea share a major metabolic theme: the use of *elemental sulfur (S⁰)* as an electron acceptor for anoxic growth. Interestingly, however, all hyperthermophilic Archaea discussed thus far are unable to use *sulfate* as an electron acceptor. *Archaeoglobus*, however, is a true sulfate-reducing hyperthermophile. This organism, isolated from hot sediments from marine hydrothermal vents, couples the oxidation of H_2, lactate, pyruvate, glucose, or complex organic mixtures to the reduction of sulfate to sulfide (sulfite and thiosulfate are also reduced as electron acceptors). Cells of *Archaeo-*

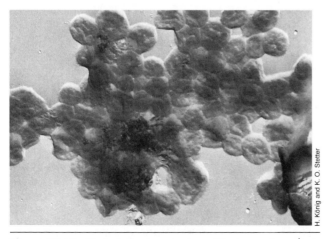

Figure 17.21 *Staphylothermus marinus.* Electron micrograph of shadowed cells. A single cell is about 1 μm in diameter.

globus are irregular motile spheres (Figure 17.22*a*), and cultures of this organism grow at between 64 and 92°C with an optimum of 83°C. *Archaeoglobus* clearly belongs to the Archaea as adjudged from ribosomal RNA analyses, its lack of peptidoglycan, and its pattern of antibiotic resistance (∞ Section 15.8). However, one of the most remarkable aspects of this organism (besides its ability to reduce sulfate) is that it contains certain coenzymes previously found only in *methanogenic* Archaea. Specifically, coenzyme F_{420} and methanopterin, coenzymes intimately involved in biochemical reactions leading to the production of methane (see Section 17.2), are also present in cells of *Archaeoglobus*. In fact, *Archaeoglobus* actually produces small amounts of methane during growth as a sulfate-reducing organism!

Archaeoglobus thus shares some metabolic features with methanogens, despite its major phenotypic property of reducing sulfate. Supporting this is the finding of a specific relationship between *Archaeoglobus* and certain methanogens by molecular sequencing methods (see Figure 17.1). Thus, sulfate reduction might have been a transitional type of metabolism important in the metabolic diversification of Archaea from sulfur-respiring phenotypes to methanogens and extreme halophiles.

Methanopyrus

A methanogenic prokaryote capable of growth at over 100°C is also known. *Methanopyrus*, a gram-positive rod-shaped methanogen (Figure 17.22*b*) has been isolated from sediments near submarine hydrothermal vents (∞ Section 14.10 for a discussion of thermal vents). We saw the unique phylogenetic position of *Methanopyrus* on the archaeal tree in Figure 17.1; it is the most ancient (least derived) of all known hyperthermophilic Archaea and shares phenotypic properties with both the hyperthermophiles (growth temperature maximum, 110°C) and the methanogens (4 H_2 + $CO_2 \rightarrow CH_4$ + 2 H_2O). *Methanopyrus* produces methane only from H_2 + CO_2 and grows rapidly (generation time less than 1 hr) at its temperature optimum of 100°C. Unlike most methanogens, however, cells of *Methanopyrus* have large amounts of the glycolytic derivative, cyclic 2,3-diphosphoglycerate, dissolved in the cytoplasm. This compound, present at more than 1 M concentration in *Methanopyrus* and other hyperthermophilic methanogens such as *Methanothermus*, is thought to function as a thermostabilizing agent to prevent denaturation of enzymes and DNA inside the cell.

The discovery of *Methanopyrus* may explain the origin of hydrocarbon-like materials in hot oceanic sediments previously thought to be too hot to support biogenic methanogenesis. In addition, at the depth at which *Methanopyrus* was found, approximately 2000 m, water remains liquid at temperatures up to 350°C, suggesting that other hyperthermophilic methanogens may exist capable of growth at 110°C or at perhaps even higher temperatures.

(a)

R. Rachel and K. O. Stetter

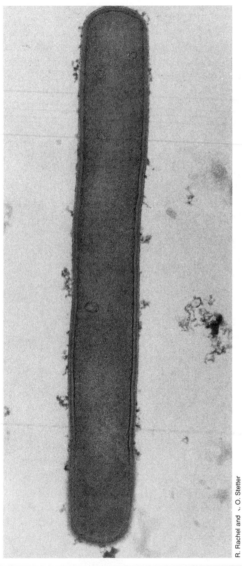

(b)

R. Rachel and ... O. Stetter

Figure 17.22 *Archaeoglobus* and *Methanopyrus*. (a) *Archaeoglobus lithotrophicus*, cells measure 1 × 2 μm, shadowed preparation. *Archaeoglobus* is the only known sulfate-reducing hyperthermophile; it also produces small amounts of methane. (b) *Methanopyrus kandleri*—the most thermophilic of all known methanogens (upper limit, 110°C). Thin section of cells that measure 0.5 × 2–14 μm.

Aquifex *and* Thermotoga

Although these are not Archaea, *Aquifex* and *Thermotoga* are hyperthermophilic Bacteria that otherwise strongly resemble hyperthermophilic Archaea. *Thermotoga* (Figure 17.23a) grows at up to 90°C (optimum 80°C) and comprises both marine and continental hot spring species. *Aquifex* (Figure 17.23b) is a submarine volcanic hot spring bacterium and is more thermophilic than *Thermotoga,* growing at up to 95°C with an optimum of 85°C. *Thermotoga* and *Aquifex* have quite contrasting metabolisms. *Thermotoga* is exclusively chemoorganotrophic, growing anaerobically on various sugars or complex protein digests. *Aquifex,* by contrast, is an obligate chemolithotroph; it grows microaerobically or anaerobically with only H_2, S^0, or $S_2O_3^{2-}$ as electron donor and O_2 or NO_3^- as electron acceptor. Phylogenetically, *Thermotoga* and *Aquifex* branch closest to the root of the universal tree (∞ Figure 16.1) and thus represent the most ancient of all lineages in the Bacteria domain. *Aquifex* branches closest to the root (Figure 16.1), and its H_2-oxidizing metabolism and thermophily fit nicely with scenarios for early Earth conditions and metabolic systems (∞ Sections 15.1 and 15.2).

CONCEPT CHECK   17.3

Hyperthermophilic Archaea are prokaryotes whose growth temperature optimum is above 80°C, and some hyperthermophiles have temperature optima above 100°C. Both terrestrial volcanic and submarine volcanic habitats contain hyperthermophiles, and many require sulfur (S^0) as an electron acceptor for anaerobic respiration.

✔ *What is unique about* Pyrodictium?

✔ *What electron donor is commonly used by chemolithotrophic hyperthermophiles?*

✔ *Compare and contrast* Methanopyrus *with* Pyrodictium *in terms of morphology, energy generation, and temperature requirements.*

17.4

Thermoplasma: A Cell-Wall-less Archaean

Thermoplasma acidophilum is a cell-wall-less prokaryote that in this respect resembles the mycoplasmas. However, phylogenetically, *Thermoplasma* is a member of the Archaea. *Thermoplasma* (Figure 17.24a) is an acidophilic, aerobic chemoorganotroph and is also thermophilic. *Thermoplasma* grows optimally at 55°C and pH 2 in complex media. With one exception, all strains of *Thermoplasma* have been obtained from self-heating coal refuse piles (Figure 17.25). Coal refuse contains coal fragments, pyrite, and other organic materials extracted from coal, and when dumped into piles in coal-mining operations, tends to self-heat by spontaneous combustion (Figure

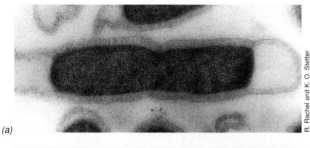

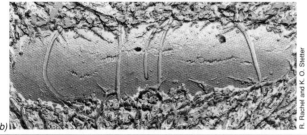

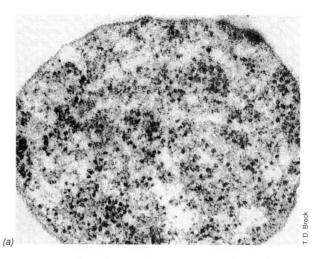

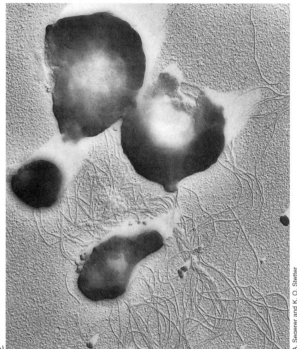

Figure 17.23 Hyperthermophilic Bacteria. (a) *Thermotoga maritima*—temperature optimum, 80°C. Note the outer covering on the cell (the "toga"). (b) *Aquifex pyrophilus*—temperature optimum, 85°C. Cells of *Thermotoga* (thin section) measure 0.6 × 3.5 μm; cells of *Aquifex* (freeze-fracture micrograph) measure 0.5 × 2.5 μm.

17.25). This sets the stage for growth of *Thermoplasma*, which apparently metabolizes organic compounds leached from the hot coal refuse. Because coal refuse piles are transitory environments, other potential habitats for *Thermoplasma* have been sought. Although acid hot springs appear to be ideal habitats for this organism, only one isolate of *Thermoplasma* has ever been obtained from such an environment, suggesting that coal refuse piles, or perhaps coal itself, are the primary habitat of *Thermoplasma*. A second species of *Thermoplasma*, *T. volcanium* (Figure 17.24b), has been isolated from solfatara fields throughout the world. It is genetically distinct from *T. acidophilum* but resembles the latter species in many phenotypic properties. *Thermoplasma volcanium* is also highly motile, and cells show multiple flagella (Figure 17.24b).

To survive the osmotic stresses of life without a cell wall and to withstand the dual environmental extremes of low pH and high temperature, *Thermoplasma* has evolved a cell membrane of chemically unique structure. The membrane contains lipopolysaccharide (referred to as *lipoglycan* in mycoplasmas) (∞ Section 16.27) consisting of a *tetraether* lipid with mannose and glucose units (Figure 17.26). This molecule constitutes a major fraction of the total lipid composition of *Thermoplasma*. The membrane also contains glycoproteins but not sterols. Together these and other molecules render the *Thermoplasma* membrane stable to hot acid conditions.

Thermoplasma *genome*

The genome of *Thermoplasma* is of interest. Like other mycoplasmas (∞ Section 16.27), *Thermoplasma* contains an extremely small genome. The DNA of *Thermoplasma*

Figure 17.24 *Thermoplasma* species. (a) *Thermoplasma acidophilum*, an acidophilic, thermophilic mycoplasma-like archaean. Electron micrograph of thin section. The diameter of cells is highly variable from 0.2 to 5 μm. The cell shown is about 1 μm in diameter. (b) Shadowed preparation of cells of *Thermoplasma volcanium* isolated from hot springs. Cells are 1–2 μm in diameter. Notice abundant flagella.

(about 1100 kilobase pairs) has a GC content of 46% and is surrounded by a highly basic DNA binding protein that organizes the DNA into globular particles resembling the nucleosomes of eukaryotic cells (see Section 3.14 for a discussion of the arrangement of DNA in eukaryotes). This protein strongly resembles the basic histone proteins of eukaryotic cells, and comparisons of amino acid sequences between the *Thermoplasma* protein and eukaryotic nuclear histones show significant sequence homology. Phylogenetically, *Thermoplasma* groups with the Euryarchaeota (see Figure 17.1).

(a)

(b)

Figure 17.25 A typical self-heating coal refuse pile, habitat of *Thermoplasma*. (a) Spontaneous heat production can ignite nearby vegetation. (b) Photo of a large hot refuse pile.

CONCEPT CHECK   **17.4**

Thermoplasma is a cell-wall-less archaean that is thermophilic and extremely acidophilic. *Thermoplasma* inhabits acidic thermally heated coal refuse piles and certain acidic hot springs.

✔ *Among archaeans, what is unique about* Thermoplasma?

✔ *How does* Thermoplasma *strengthen its membrane to survive its harsh living conditions?*

17.5

Limits of Microbial Existence: Temperature

In the case of the hyperthermophilic Archaea, we see growth at temperatures far higher than those supporting growth of any other prokaryotes. What is the nature of this extreme heat tolerance and what are the temperature limits beyond which life is impossible? As we discussed in Sections 5.7 and 5.8, the *macromolecules* of thermophiles and hyperthermophiles are stable to heat, and in the case of proteins, this stability resides in the unique folding of the molecules; the latter, of course, ultimately depends on the *sequence* of amino acids in the protein. But is this all there is to it? No. Life at extremely high temperatures requires a number of molecular adaptations, some of which are just beginning to be understood.

Protein stability in hyperthermophiles may be assisted by the accumulation of solutes, such as the cyclic 2,3-diphosphoglycerate found in large amounts in cells of *Methanopyrus* and other thermophilic methanogens (see Section 17.3). In addition, special proteins may be necessary for growth at the highest temperatures. For example, in *Pyrodictium* (Figure 17.27), cells grown at 110°C produce 80% of their total protein as a single protein that has two enzymatic activities. Besides possessing ATPase activity, this protein functions as a molecular chaperonin, stabilizing other cellular proteins by refolding them as they begin to denature near the upper temperature limits for growth (the function of chaperonin proteins was discussed in Section 7.6). At 100°C (near the optimum for growth of *Pyrodictium*), very little of this chaperonin protein is made, suggesting that only at very extreme temperatures close to the upper limit do the otherwise thermally stable proteins of this organism begin to denature.

Opening and closing of the DNA helix in hyperthermophiles is also an interesting question. How does DNA in hyperthermophiles keep from denaturing? How is the helix opened up to allow exposure of the strands for transcription? This problem has been studied most extensively in *Methanothermus*, a hyperthermophilic methanogen. A DNA binding protein has been found in *Methanothermus* cells that is closely related to the histone proteins of eukaryotic cells. This protein binds tightly to DNA and probably functions to open up the helix for transcription. It is not known how DNA is prevented from melting, especially in

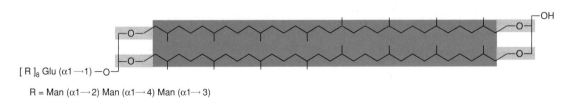

[R]₈ Glu (α1 → 1) — O

R = Man (α1→ 2) Man (α1→ 4) Man (α1→ 3)

Figure 17.26 Structure of the tetraether lipoglycan of *Thermoplasma acidophilum*. Glu, Glucose; Man, mannose. Note the ether linkages (shown in green) and compare with Figures 3.19 and 15.13.

very low GC hyperthermophiles (such as *Pyrococcus*, whose DNA of 38% GC should quickly melt at its optimum growth temperature of 100°C). However, protection may be afforded by a combination of high cytoplasmic solute concentrations (the melting temperature of DNA increases as the solute concentration increases) such as the cyclic 2, 3-diphosphoglycerate mentioned earlier, or by the activity of specific DNA binding proteins that somehow prevent DNA from melting, perhaps by folding the DNA into a conformation consistent with thermal stability.

Besides the stability of *macromolecules*, however, a second consideration is important in governing the upper temperature limits for growth of hyperthermophilic bacteria: the thermal lability of *monomers*. At temperatures above 100°C, a number of biologically significant monomers show some degree of heat lability. And, no matter how stable the macromolecules are, life is impossible if the basic building blocks themselves are unstable. The thermal stability of biomolecules therefore may dictate the upper temperature for life, and at temperatures as low as 120°C some important biomolecules are destroyed. For example, molecules such as ATP and NAD$^+$ hydrolyze quite rapidly at high temperature; the half-life of both of these molecules is less than 30 min at 120°C. However, temperatures up to 113°C are clearly still compatible with life because at least one hyperthermophile, *Pyrolobus*, grows at this temperature.

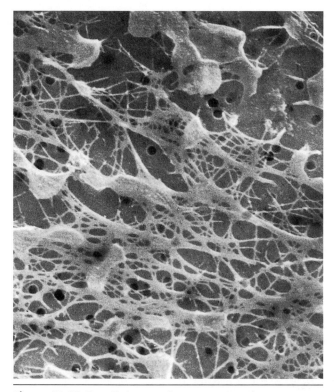

Figure 17.27 *Pyrodictium occultum*, scanning electron micrograph. Cells of this hyperthermophile grow optimally at 105°C and maximally at 110°C and have been studied as a model of macromolecular stability at high temperatures. Cells are enmeshed in a sticky glycoprotein matrix that binds them together.

The upper temperature limits for life

Because life depends on liquid water, microbial habitats in excess of 100°C are limited to environments where cells are under *pressure*, such as the sea floor. Indeed, all hyperthermophilic Archaea capable of growth at 100°C or higher seem restricted to these environments. However, another possible habitat is the deep subsurface of the earth where heat from the earth's interior coupled with immense pressures creates conditions suitable for growth of hyperthermophiles. Although such organisms have not been extensively sought out, various lines of evidence, primarily geochemical, suggest that hyperthermophiles may exist kilometers below the earth's crust.

As discussed in Section 14.10, extremely hot (but not boiling) water is emitted from deep-sea thermal vents (black smokers; ∞ Figure 14.26) at temperatures of 250–350°C. Will prokaryotes capable of growth at 250–350°C ever be isolated? Probably not. Life in such environments seems impossible. Although hyperthermophiles have been isolated from very near black smoker emissions, presumably existing in the gradient of hot water formed as the hydrothermal fluid mixes with 2°C seawater (see, for example, the discussion of *Staphylothermus* and Figure 17.21), at 250°C or higher, macromolecules and simple organic molecules such as amino acids and nucleotides are spontaneously hydrolyzed so quickly that life *as we know it* could not occur at such temperatures. Hence, although the upper temperature for life has not yet been defined, the discovery of hyperthermophilic Archaea has shown that this limit is not lower than 113°C and is probably a little higher. Laboratory experiments on the heat stability of biomolecules suggest that living processes could be maintained at temperatures as high as 140–150°C but that above this temperature organisms would probably not be able to overcome the heat lability of the important biomolecules of life. Indeed, some evidence for prokaryotes growing at 125–140°C was previously discussed (∞ Section 14.10 and Figure 14.27), but chemical principles tell us that if organisms do grow at these temperatures, they must be living near the limits of microbial existence.

17.6
Archaea: Earliest Life Forms?

Now that we have considered the various groups of Archaea, what can we say about their role in microbial evolution? Although metabolically diverse organisms, a common theme running through the Archaea is *adaptation to environmental extremes*. High salt, high temperature, low pH, strictly anoxic conditions—these environmental obstacles have all been overcome by one or another species of Archaea. Recalling our description of the geochemical conditions prevalent on Earth at the time that life arose (∞ Section 15.2), do any particular groups of Archaea stand out as candidates for the earliest life forms? Yes, the hyperthermophiles. Recall that the surface of early Earth was likely much *hotter* than it is today, perhaps as high as 100°C or higher, when living organisms first appeared. This implies that the earliest life forms had to have been hyperthermophilic, perhaps not unlike present-day hyperthermophilic Archaea.

Current phylogenetic evidence deduced from comparison of 16S rRNA sequences (∞ Chapter 15) suggests that Archaea have evolved *slower* than either Bacteria or the eukaryotes. This is especially true of hyperthermophilic Archaea (see Figure 17.1). It is not known why Archaea are the slowest evolving of the three domains, but it may be related to their inhabiting extreme environments. It is not hard to imagine that organisms living in thermal environments must maintain those genes that specify phenotypic characteristics critical to life at high temperatures; these genes cannot be significantly changed during evolution if the organism is to maintain itself in these environments. Thus, organisms like the hyperthermophilic Archaea are likely to have been among the earliest life forms. The phenotypic properties of this group, thermophily, and anaerobic chemoorganotrophic or chemolithotrophic metabolism, agree well with the phenotype of primitive organisms predicted from a consideration of early Earth geochemical conditions (∞ Section 15.2).

Perhaps H_2-oxidizing sulfur-reducing hyperthermophiles or a methanogenic organism like *Methanopyrus* (see Figure 17.22b) were among the earliest types of cells on Earth. From here, it is possible that the methanogenic subline of Archaea diversified, perhaps via transitional forms like *Archaeoglobus*. Finally, extreme halophiles, the most derived (least ancient) of all Archaea, arose from methanogenic ancestors (see Figure 17.1).

In closing this chapter, it is interesting to note how often H_2 *metabolism* enters into the evolutionary picture of Archaea (Table 17.6); indeed, much speculation on early metabolic strategies have emphasized the likely importance of H_2 as an electron donor (see, for example, Figure 15.3). In this connection it will be especially important to obtain cultures of Korarchaeota species since, from a phylogenetic perspective, they are positioned curiously close to the root of the archaeal tree (see Figure 17.1). These and other hyperthermophilic Archaea undoubtedly hold many secrets about early life on planet Earth.

CONCEPT CHECK   17.6

The most thermophilic of all known organisms can grow at up to 113°C; the upper temperature limit for life is probably above this, but not much. The instability of small molecules at high temperatures is probably what limits the upper temperature for life.

✔ *Why is it unlikely that hyperthermophiles capable of growth at 250°C exist?*

✔ *What attributes of hyperthermophiles make them excellent candidates for models of early Earth life forms?*

Material for Review

SUPPLEMENTARY READINGS

Balows, A., H. G. Trüper, M. Dworkin, W. Harder, and **K.-H. Schleifer** (eds.). 1992. *The Prokaryotes,* 2nd edition. Springer-Verlag, New York. This large compendium on the biology of bacteria has several chapters devoted to Archaea.

Danson, M. J., D. W. Hough, and **G. G. Lunt.** 1992. *Archaebacteria: Biochemistry and Biotechnology.* Biochemical Society Symposium, No. 58. Portland Press, Colchester, UK. A volume devoted to the biochemistry and potential biotechnological applications of Archaea. Balanced coverage of extreme halophiles, methanogens, and hyperthermophiles.

Ferry, J. G. (ed.). 1993. *Methanogenesis: Ecology, Physiology, Biochemistry, and Genetics.* Chapman and Hall, New York. An excellent treatment of all aspects of methanogenesis and methanogens.

Kates, M., D. J. Kushner, and **A. T. Matheson** (eds.). 1993. *The Biochemistry of Archaea (Archaebacteria).* Elsevier, Amsterdam. A comprehensive volume on the biology of the archaeans.

Kristjansson, J. K. (ed.). 1992. *Thermophilic Bacteria.* CRC Press, Boca Raton, FL. Focuses on thermophilic and hyperthermophilic Bacteria (rather than Archaea) but has good coverage of a number of areas relating to life at high temperatures.

Robb, F. T., A. R. Place, K. R. Sowers, H. J. Schreier, H. Das Sarma, and **E. M. Fleischman** (eds.). 1995. *Archaea: A Laboratory Manual.* Cold Spring Harbor Laboratory Press, Plainview, NY. A three-volume handbook on growth and laboratory manipulation of various Archaea.

Sharp, R., and **R. Williams** (eds.). 1995. *Thermus Species.* Plenum Press, New York. Although a member of the Bacteria rather than the Archaea, *Thermus* is a thermophilic prokaryote about which much biochemical information is known, and is well presented in this volume.

Staley, J. T. (ed.). 1989. *Bergey's Manual of Systematic Bacteriology,* Vol. III. Williams and Wilkins, Baltimore, MD. This volume of *Bergey's Manual* includes the Archaea.

Woese, C. R., and **Wolfe, R. S.** (eds.). 1985. *The Bacteria—A Treatise on Structure and Function,* Vol. VIII, *Archaebacteria.* Academic Press, New York. Although slightly dated, still a well-illustrated series of reviews on the basic biology of Archaea.

 On~line resources for this chapter are on the World Wide Web at: http://www.prenhall.com/~brock (click on the <u>table of contents</u> link and then select Chapter 17).

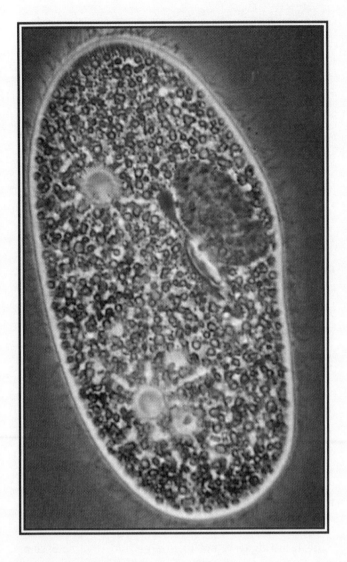

Eukarya: Eukaryotic Microorganisms

Eukaryotic microorganisms, like those of the protozoan *Paramecium*, shown here, are not only larger and structurally more complex than prokaryotes, they also encompass an evolutionary lineage distinct from that of Bacteria and Archaea.

MINIGLOSSARY for Chapter 18

Algae phototrophic eukaryotic microorganisms

Ameboid Movement a type of motility in which cytoplasmic streaming moves the organism forward

Chitin a polymer of *N*-acetylglucosamine commonly found in the cell walls of algae and fungi

Chloroplast the photosynthetic organelle of eukaryotic phototrophs

Ciliates a group of protozoa characterized by rapid motility driven by numerous short appendages called cilia

Conidia asexual spores of fungi

Eukarya all eukaryotic organisms

Flagellates a group of protozoa characterized by motility driven by the whiplike action of one or more long, thin appendages called flagella

Fungi nonphotosynthetic eukaryotic microorganisms that contain rigid cell walls

Mitochondrion the respiratory organelle of eukaryotic organisms

Molds filamentous fungi

Mushrooms filamentous fungi that produce large, often edible structures called fruiting bodies

Phagocytosis a mechanism for ingesting particulate food in which a portion of the cell membrane surrounds the particle and brings it into the cell

Protozoa unicellular eukaryotic microorganisms that lack cell walls

Slime Molds nonphototrophic eukaryotic microorganisms that lack cell walls and that aggregate to form fruiting structures (cellular slime molds) or masses of protoplasm (acellular slime molds)

Sporozoa nonmotile parasitic protozoa

Yeasts unicellular fungi

We complete our tour of the microbial world with a consideration of eukaryotic microorganisms. We have seen in various sections of this book how eukaryotes differ from Bacteria and Archaea in many fundamental ways, including cell size and internal structure (Chapter 3), genetic properties (Chapters 6–9), and evolutionary history (Chapter 15). We now consider eukaryotic microorganisms themselves and the major properties that are used to separate them into groups. We introduce the material from a phylogenetic perspective and then proceed to discuss the traditional groups of microbial Eukarya, the algae, fungi (molds and yeasts), slime molds, and protozoa.

Phylogenetic overview of Eukarya

We discussed the phylogeny of Eukarya in the context of the universal tree of life in Figure 15.12. Figure 18.1 shows details of the Eukarya branch of this tree. As mentioned in Section 15.7, the obligately parasitic diplomonads and microsporidia are the most phylogenetically ancient of the eukaryotes known (Figure 18.1). Going up the tree from them we find the flagellates, which include some algae. Following them are the slime molds, the brown algae–diatom group, and then in a burst of evolutionary radiation, the remaining microbial eukaryotes and the plants and animals (Figure 18.1). From the universal phylogenetic tree (∞ Figure 15.12) we can also see that the Eukarya as a domain are the most derived of all living organisms, having evolved away from the universal ancestor of life further than either prokaryotic domain, the Bacteria or the Archaea. This, of course, makes sense in

light of the structural (∞ Chapter 3) and genetic (∞ Chapters 6–10) complexity we have already seen in the Eukarya and will continue to see in the present chapter. We begin our discussion here with *phototrophic* microbial Eukarya: the algae.

18.1
Algae

The term **algae** refers to a large and diverse assemblage of eukaryotic organisms that contain *chlorophyll* and carry out oxygenic photosynthesis (Figure 18.2). Algae should not be confused with *cyanobacteria*, which are also oxygenic phototrophs but which are Bacteria and thus evolutionarily quite distinct from algae (∞ Section 16.2 and Figure 16.1). Although most algae are of microscopic size and hence are clearly microorganisms, a number of forms are macroscopic, some seaweeds growing to over 100 ft in length.

Algae are either unicellular (Figure 18.2*a* and *c*) or colonial, the latter occurring as aggregates of cells (Figure 18.2*b*). When the cells are arranged end to end, the alga is said to be filamentous (Figure 18.2*d*). Among the filamentous forms, both unbranched filaments and more intricate branched filaments occur. Most algae contain chlorophyll and are thus green in color. However, a few kinds of common algae are not green but appear brown or red because in addition to chlorophyll, other pigments such as carotenoids are present that mask the green color. Algal cells contain one or more **chloroplasts,** membranous structures that house the photosynthetic

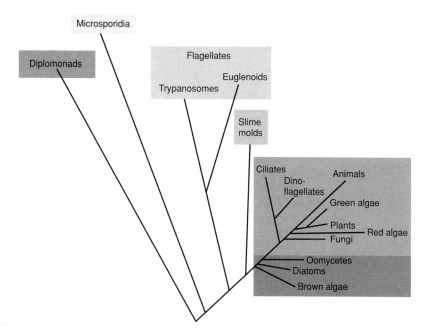

Figure 18.1 Phylogenetic tree of Eukarya based on 18S ribosomal RNA sequence comparisons. Cells of microsporidia and diplomonads are phylogenetically most ancient of known Eukarya and contain a nucleus but lack mitochondria. The Oomycetes are covered in this chapter with the fungi (see Table 18.2), although they are phylogenetically distinct from most fungi. See note in Figure 15.12 for comparisons of detailed tree with the universal tree.

pigments. Chloroplasts can often be recognized microscopically within algal cells by their distinct green color (Figure 18.2). We discussed the general structure and properties of chloroplasts in Section 3.15.

Using ribosomal RNA sequencing as a phylogenetic tool (∞ Section 15.7), the algae can be seen to constitute a phylogenetically heterogeneous group. Although green algae (Chlorophyta, Table 18.1) and to

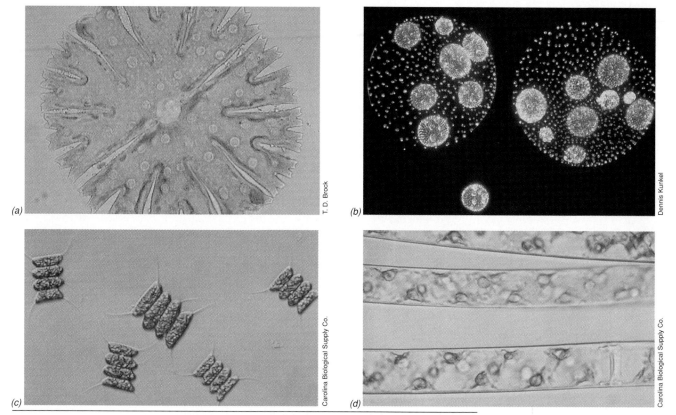

Figure 18.2 Light micrographs of representative green algae. (a) *Micrasterias*. A single cell. (b) *Volvox* colony, containing a large number of cells. (c) *Scenedesmus*. A packet of four cells. (d) *Spirogyra*. A filamentous alga. Note the green spiral-shaped chloroplasts.

a lesser extent red algae (Rhodophyta, Table 18.1) are quite closely related to green plants, other algal groups such as the brown algae and diatoms are more ancient (Figure 18.1). Even less derived are the euglenoids, such as the alga *Euglena* (see Figure 18.3*a*). Euglenoids show a phylogenetic relationship to flagellated protozoa (Figure 18.1), and in fact, cells of *Euglena* can spontaneously lose their chloroplasts and exist as completely heterotrophic organisms (see Section 18.4).

Pigments and reserve polymers

Several characteristics are used to classify algae, including the nature of the chlorophyll(s) present, the carbon reserve polymers produced, the cell wall structure, and the type of motility. All algae contain chlorophyll *a*. Some, however, also contain other chlorophylls that differ in minor ways from chlorophyll *a*. The presence of these additional chlorophylls is characteristic of particular algal groups. The distribution of chlorophylls and other photosynthetic pigments in algae is summarized in Table 18.1.

All algae carry out oxygen-evolving (oxygenic) photosynthesis, using H_2O as an electron donor. In addition, some algae can use H_2 to carry out photosynthesis without yielding oxygen. Many algae are obligate phototrophs and are thus unable to grow in darkness on organic carbon compounds. However, some algae can grow chemoorganotrophically and catabolize simple sugars or organic acids in the dark. One of the organic compounds most widely used by algae is acetate, which can be used as a sole carbon and energy source by many flagellates and chlorophytes. In addition, some algae can assimilate simple organic compounds in the light (photoheterotrophy; ∞ Section 16.1 and Table 13.3) but cannot grow on them as sole energy sources. One of the key characteristics used in the classification of algal groups is the nature of the **reserve polymer** synthesized as a result of photosynthesis. Algae of the division Chlorophyta produce starch (α-1,4-glucose) in a form very similar to that of higher plants. By contrast, algae of other groups produce a variety of reserve substances, some polymeric and some as free monomers, and the major ones are listed in Table 18.1.

Cell walls of algae

Algae show considerable diversity in the structure and chemistry of their cell walls. In many cases the cell wall is composed of a network of cellulose fibrils, but it is

TABLE 18.1	Properties of major groups of algae						
Algal group	Common name	Morphology	Pigments	Typical representative	Carbon reserve materials	Cell wall	Major habitats
Chlorophyta	Green algae	Unicellular to leafy	Chlorophylls *a* and *b*	*Chlamydomonas*	Starch (α-1,4-glucan), sucrose	Cellulose	Freshwater, soils, a few marine
Euglenophyta[a]	Euglenoids	Unicellular, flagellated	Chlorophylls *a* and *b*	*Euglena*	Paramylon (β-1,2-glucan)	No wall present	Freshwater, a few marine
Chrysophyta	Golden-brown algae, diatoms	Unicellular	Chlorophylls *a*, *c*, and *e*	*Navicula*	Lipids	Many have two overlapping components made of silica	Freshwater, marine, soil
Phaeophyta	Brown algae	Filamentous to leafy, occasionally massive and plantlike	Chlorophylls *a* and *c*, xanthophylls	*Laminaria*	Laminarin (β-1,3-glucan), mannitol	Cellulose	Marine
Pyrrophyta	Dinoflagellates	Unicellular flagellated	Chlorophylls *a* and *c*	*Gonyaulax*	Starch (α-1,4-glucan)	Cellulose	Freshwater, marine
Rhodophyta	Red algae	Unicellular, filamentous to leafy	Chlorophylls *a* and *d*, phycocyanin, phycoerythrin	*Polysiphonia*	Floridean starch (α-1,4- and α-1,6-glucan), fluoridoside (glycerolgalactoside)	Cellulose	Marine

[a]This group is also considered with the protozoa (see Section 18.4).

usually modified by the addition of other polysaccharides such as pectin (highly hydrated polygalacturonic acid containing small amounts of the hexose rhamnose), xylans, mannans, alginic acids, or fucinic acid. In some algae, the wall is additionally strengthened by the deposition of calcium carbonate; these forms are often called "calcareous" or "coralline" (corallike) algae. Sometimes chitin, a polymer of *N*-acetylglucosamine, is also present in the cell wall. In euglenoids a cell wall is absent. In diatoms (Figure 18.3*b*) the cell wall is composed of *silica*, to which protein and polysaccharide are added. Even after the diatom dies and the organic materials have disappeared, the external structure remains, showing that the siliceous component is indeed responsible for the rigidity of the cell. Because of the extreme resistance to decay of these diatom frustules, they remain intact for long periods of time and constitute some of the best algal fossils ever found. From this excellent fossil record, it is known that diatoms first appeared on Earth about 200 million years ago. Diatomaceous earth, an industrial filtering agent, is composed of fossilized diatom cells. This material is mined in large quantities from ancient sea beds, emphasizing the large numbers of these organisms found in prehistoric seas.

Algal cell walls are freely permeable to low-molecular-weight constituents such as water, ions, gases, and other nutrients. Their cell walls are essentially impermeable, however, to larger molecules or to macromolecules. Algal cell walls contain pores 3–5 nm wide, which are sufficiently small to pass molecules of a molecular weight of about 15,000 or less. Thus, although animal cells can eat particulate matter, a process called *phagocytosis* (see Section 18.4), phagocytic activities are impossible in algae; particles large enough to be phagocytized never reach the cytoplasmic membrane because they are unable to penetrate the cell wall.

Motility and ecology of algae

A number of algae are motile, usually because of flagella; cilia (∞ Sections 3.9 and 18.4) do not occur in algae. Simple flagellate forms, such as *Euglena* (see Figure 18.3*a*), usually have a single polar flagellum, whereas flagellated representatives of the Chlorophyta

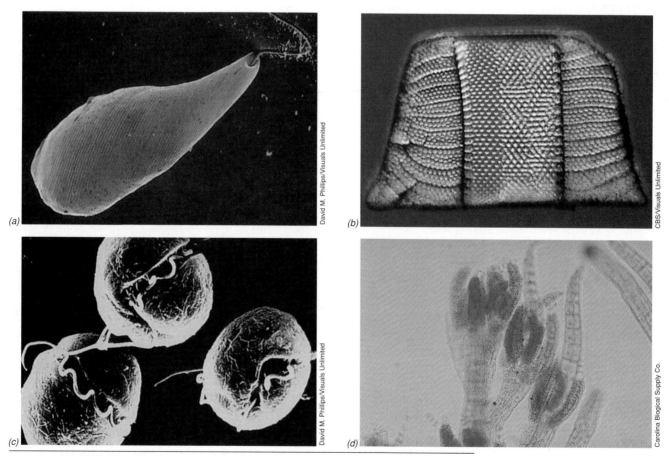

Figure 18.3 Scanning electron (a–c) and light (d) micrographs of algae other than Chlorophyta. (a) *Euglena*, a member of the Euglenophyta. (b) *Isthmix*, a diatom (Chrysophyta). (c) *Gonyaulax*, a dinoflagellate (Pyrrophyta). (d) *Polysiphonia*. This marine red alga (Rhodophyta) grows attached to the surfaces of various marine plants.

have either two or four polar flagella. Dinoflagellates (Figure 18.3c) have two flagella of different lengths and with different points of insertion into the cell. The transverse flagellum is attached laterally, whereas the longitudinal flagellum originates from the lateral groove of the cell and extends lengthwise. In many cases, algae are nonmotile in the vegetative state and form motile gametes only during sexual reproduction. Gliding motility, widespread among filamentous cyanobacteria (∞ Section 16.2), is present in only one group of eukaryotic algae, the diatoms.

Algae abound in nature in aquatic habitats, both freshwater and marine. Algae are also found in moist soils and artificial aquatic habitats like fish tanks and swimming pools, and in temporary pools of water formed from rainwater runoff. Algae are also common in moist soils while a few species thrive in dry and even extremely dry soils where the water potential (∞ Section 5.10) can be very low. Algae are often the dominant or sole phototrophic microorganisms in acidic habitats as well; below pH 4–5 cyanobacteria are absent and several algal species flourish.

CONCEPT CHECK   18.1

Algae are phototrophic Eukarya that contain photosynthetic pigments within a structure called the chloroplast.

✔ *How do* algae *differ from* cyanobacteria?

✔ *What is the primary habitat of algae?*

18.2

Fungi

In contrast to the algae, the fungi lack chlorophyll. Fungi can be differentiated from prokaryotes by the fact that fungal cells are usually much larger and contain a nucleus, vacuoles, and mitochondria, typical of eukaryotic cells. Although the fungi are a large and diverse group of eukaryotic microorganisms, three groups of fungi have major practical importance: the **molds, yeasts,** and **mushrooms.**

The habitats of fungi are quite diverse. Some are aquatic, living primarily in fresh water, and a few marine fungi are also known. Most fungi, however, have terrestrial habitats, in soil or on dead plant matter, and these types often play crucial roles in the mineralization of organic carbon in nature. A large number of fungi are parasites of terrestrial plants. Indeed, fungi cause the majority of economically significant diseases of crop plants (see Table 18.2). A few fungi are parasitic on animals, including humans, although in general fungi are less significant as animal pathogens than are bacteria and viruses (∞ Section 23.16 for a discussion of pathogenic fungi).

Like algae, fungi also contain rigid cell walls. Fungal cell walls resemble plant cell walls architecturally, but not chemically. Although cellulose is present in the walls of certain fungi, many fungi have noncellulosic walls. **Chitin** is a common constituent of fungal cell walls. It is laid down in microfibrillar bundles like cellulose; other glucans such as mannans, galactosans, and chitosans replace chitin in some fungal cell walls. Fungal cell walls are generally 80–90% polysaccharide, with proteins, lipids, polyphosphates and inorganic ions making up the wall-cementing matrix. An understanding of fungal cell wall chemistry is important because of the extensive biotechnological uses of fungi (∞ Chapter 12) and because the chemical nature of the fungal cell wall has been useful in classifying fungi for research and industrial purposes.

All fungi are chemoorganotrophs. Lacking chlorophyll, they of course cannot photosynthesize, and the group also lacks chemolithotrophic forms. When compared to bacteria, the fungi in general have fairly simple nutritional requirements, and their metabolic and biosynthetic processes are not particularly diverse or unusual. In addition, unlike the algae, which are phylogenetically diverse (see Section 18.1), fungi, with the exception of the Oomycetes, form a phylogenetically very coherent group in the tree of Eukarya (Figure 18.1). It is only in their morphological properties and in their sexual life cycles that the fungi exhibit considerable diversity; hence, it is on the basis of these characteristics that the fungi are currently classified. We present an overview of fungal classification in Table 18.2.

Molds

The molds are *filamentous* fungi. They are widespread in nature and are commonly seen on stale bread, cheese, or fruit. Each filament grows mainly at the tip, by extension of the terminal cell (Figure 18.4). A single filament is called a *hypha* (plural, *hyphae*). Hyphae usually grow together across a surface and form compact tufts, collectively called a *mycelium,* which can be seen easily without a microscope. The mycelium arises because the individual hyphae form branches as they grow, and these branches intertwine, resulting in a compact mat. In most cases, the vegetative cell of a fungal hypha contains more than one nucleus—often hundreds of nuclei are present. Thus, a typical hypha is a nucleated tube containing cytoplasm (referred to as *coenocytic*). Usually there is extensive cytoplasmic movement within a hypha, generally in a direction toward the hyphal tip, and the older portions of the hypha usually become vacuolated and virtually devoid of cytoplasm. Even if a hypha has cross-walls, cytoplasmic movement is often not prevented, as there is usually a pore in the center of the septum through which nuclei and cytoplasmic particles can move.

From the fungal mycelium, other hyphal branches may reach up into the air above the surface, and on

TABLE 18.2   Classification and major properties of fungi[a]

Group	Common name	Hyphae	Typical representatives	Type of sexual spore	Habitats	Common diseases
Ascomycetes	Sac fungi	Septate	*Neurospora, Saccharomyces, Morchella* (morels)	Ascospore	Soil, decaying plant material	Dutch elm, chestnut blight, ergot, rots
Basidiomycetes	Club fungi, mushrooms	Septate	*Amanita* (poisonous mushroom), *Agaricus* (edible mushroom)	Basidiospore	Soil, decaying plant material	Black stem, wheat rust, corn smut
Zygomycetes	Bread molds	Coenocytic	*Mucor, Rhizopus* (common bread mold)	Zygospore	Soil, decaying plant material	Food spoilage; rarely involved in parasitic disease
Oomycetes	Water molds	Coenocytic	*Allomyces*	Oospore	Aquatic	Potato blight, certain fish diseases
Deuteromycetes	Fungi imperfecti	Septate	*Penicillium, Aspergillus, Candida*	None	Soil, decaying plant material, surfaces of animal bodies	Plant wilt, infections of animals such as ringworm, athlete's foot, and other dermatomycoses, surface or systemic infections (*Candida*)

[a]With the exception of the Oomycetes, which are phylogenetically distinct, the other groups of fungi are closely related (see Figure 18.1).

these aerial branches spores called *conidia* are formed (Figure 18.4*a*). Conidia are *asexual* spores, often highly pigmented (∞ Figure 9.1*a*) and resistant to drying, and function in the dispersal of the fungus to new habitats.

When conidia form, the white color of the mycelium changes, taking on the color of the conidia, which may be black, blue-green, red, yellow, or brown. The presence of these spores gives the mycelial mat a rather

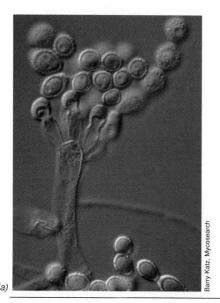
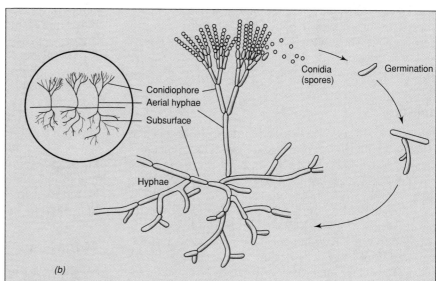

(a)
(b)

Barry Katz, Mycosearch

Figure 18.4 Mold structure and growth. (a) Photomicrograph of a typical mold. Conidia are seen as the spherical structures at the ends of aerial hyphae. (b) Diagram of a mold life cycle.

dusty appearance. Conidia are considered *asexual* spores because no sexual reproduction is involved in their formation. Because these spores are so numerous and spread so easily through the air, molds are common laboratory contaminants. Airborne mold spores are also often responsible for allergies (∞ Section 20.16).

Some molds also produce *sexual* spores, formed as a result of sexual reproduction (Table 18.2). The latter occur from the fusion either of unicellular gametes or of specialized hyphae called *gametangia*. Alternatively, sexual spores can originate from the fusion of two haploid cells to yield a diploid cell, which then undergoes meiosis and mitosis to yield individual spores. Depending on the group to which a particular fungus belongs (see Table 18.2), different types of sexual spores are produced. Spores formed within an enclosed sac *(ascus)* are called *ascospores*, and those produced on the ends of a club-shaped structure *(basidium)* are *basidiospores* (Table 18.2). Sexual spores of fungi are usually resistant to drying, heating, freezing, and some chemical agents. However, fungal sexual spores are not as resistant to heat as bacterial endospores (∞ Section 3.13). Either an asexual or a sexual spore of a fungus can germinate and develop into a new hypha and mycelium.

A major ecological activity of many fungi, especially members of the Basidiomycetes (see Table 18.2), is the decomposition of wood, paper, cloth, and other products derived from natural sources. Basidiomycetes that attack these products are able to utilize cellulose or lignin from the product as carbon and energy sources. Lignin is a complex polymer in which the building blocks are phenolic compounds. It is an important constituent of woody plants, and in association with cellulose it confers rigidity on them. The decomposition of lignin in nature occurs almost exclusively through the action of certain Basidiomycetes called *wood-rotting fungi*. Two types of wood rots are known: *brown rot*, in which the cellulose is attacked preferentially and the lignin left unchanged, and *white rot*, in which both cellulose and lignin are decomposed. The white rot fungi are of considerable ecological interest because they play such an important role in decomposing woody material in forests.

Yeasts

The yeasts are *unicellular* fungi, and most of them are classified with the Ascomycetes. Yeast cells are usually spherical, oval, or cylindrical, and cell division generally takes place by budding (Figure 18.5). In the budding process, a new cell forms as a small outgrowth of the old cell; the bud gradually enlarges and then separates (Figures 18.4 and 18.5; see also Figure 18.6). Yeasts usually do not form filaments or a mycelium, and the population of yeast cells remains a collection of single cells. However, some yeasts can form a filamentous phase that in some cases is essential for expression of certain characteristics. For example, in the yeast *Candida*

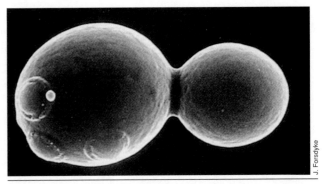

Figure 18.5 Scanning electron micrograph of the common baker's and brewer's yeast *Saccharomyces cerevisiae*. Note the budding division and previous bud scars.

albicans, a pathogenic yeast causing vaginal infections, thrush, lung infections, and systemic tissue damage in acquired immunodeficiency syndrome (AIDS) patients (∞ Figure 23.30e), the filamentous phase is essential for pathogenicity. Even the baker's yeast *Saccharomyces cerevisiae* has been shown to be capable of forming a filamentous phase under certain conditions.

Yeast cells are much larger than bacterial cells and can be distinguished microscopically from bacteria by their size and by the obvious presence of internal cell structures, such as the nucleus (Figure 18.6). Some yeasts exhibit sexual reproduction by a process called *mating*, in which two yeast cells fuse. Within the fused cell, called a *zygote*, ascospores are eventually formed. We discussed the sexual cycle of a typical yeast, *Saccharomyces*, including the important property of *mating types*, in Section 9.13.

Yeasts usually flourish in habitats where sugars are present, such as fruits, flowers, and the bark of trees. A number of yeast species live symbiotically with animals, especially insects, and a few species are pathogenic for animals and humans (∞ Section 23.16). The most important commercial yeasts are the baker's and brewer's yeasts, which are members of the genus *Saccharomyces*. The original habitats of these yeasts were undoubtedly fruits and fruit juices, but the commercial yeasts of today are probably quite different from wild strains because they have been greatly improved through the years by careful selection and genetic manipulation by industrial microbiologists. Baker's and brewer's yeasts are probably the best known scientifically of all fungi because they are easily manipulable eukaryotic cells, and they are thus excellent models for the study of many important problems in eukaryotic biology (∞ Section 9.13).

Mushrooms

Mushrooms are filamentous fungi that typically form large structures called *fruiting bodies*, the edible part of the mushroom (Figure 18.7). Many mushrooms live as

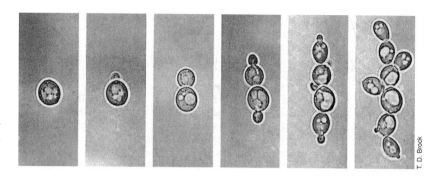

Figure 18.6 Growth by budding division in *Saccharomyces cerevisiae.* Note the pronounced nucleus.

mycorrhizae (∞ Section 14.21), whereas others live on dead organic matter in the soil or on the trunks of trees. Sexual spores called *basidiospores* are produced by mushrooms (Figure 18.7) and are dispersed through the air and initiate mycelial growth on favorable substrates. The resulting haploid mycelium may grow extensively, but it does not form a fruiting body, the latter occurring only after the fusion of haploid mycelia. Once fusion has occurred, each cell contains two nuclei and is called *dikaryotic*. Dikaryotic hyphae grow together and form small buttonlike structures, which constitute the mushroom primordia. Buttons may remain underground for long periods until favorable conditions, usually heavy rains, stimulate enlargement and development of mature fruiting bodies. This expansion can occur rapidly, within a few hours or days, and is due primarily to the uptake of water. If many fruiting bodies mature simultaneously, a so-called *flush* is produced (∞ Figure 12.30) to yield a large number of mushrooms in a given locality, which may have been devoid of these fruiting bodies the previous day.

Because many mushrooms are edible, and in some cases sought as delicacies, the large-scale production of mushrooms is an important industrial process. We therefore considered other aspects of

---

CONCEPT CHECK   18.2

Fungi include the molds and yeasts and differ from algae in that they lack chlorophyll. Many economically valuable microorganisms are fungi.

✔ *How do* molds *differ from* yeasts?

✔ *What is* chitin *and what is its function?*

✔ *How do* mushrooms *most clearly differ from other filamentous fungi?*

✔ *How do* conidia *differ from* ascospores?

Figure 18.7 Mushrooms. (a) *Amanita*, a highly poisonous mushroom. (b) Gills on the underside of the mushroom contain the spore-bearing basidia. (c) Scanning electron micrograph of basidiospores released from mushroom basidia.

mushroom biology, particularly their growth as food products, in Section 12.14.

18.3

Slime Molds

Slime molds are nonphototrophic eukaryotic microorganisms that have phenotypic similarity to both fungi and protozoa. From a phylogenetic perspective, however, slime molds are more ancient than fungi and some protozoa (such as the ciliates), but more derived than flagellated protozoa and their evolutionary predecessors (Figure 18.1).

The slime molds can be divided into two groups, the *cellular slime molds*, whose vegetative forms are composed of single amebalike cells, and the *acellular slime molds*, whose vegetative forms are naked masses of protoplasm of indefinite size and shape called *plasmodia*. Slime molds live primarily on decaying plant matter, such as leaf litter, logs, and soil (Figure 18.8). Their food consists mainly of other microorganisms, especially bacteria, which they ingest by phagocytosis. One of the easiest ways of detecting and isolating slime molds is to bring small pieces of rotting logs into the laboratory and place them in moist chambers; the amebas or plasmodia proliferate (Figure 18.8), migrate over the surface of the wood, and eventually form fruiting bodies, which can then be observed under a dissecting microscope.

Cellular slime molds: Dictyostelium

Dictyostelium discoideum, a cellular slime mold, undergoes a remarkable life cycle in which vegetative cells aggregate, migrate as a cell mass, and eventually produce fruiting bodies in which cells differentiate and form spores (Figures 18.9 and 18.10). As cells of *Dictyostelium* become starved, they aggregate and form a *pseudoplasmodium*, a structure in which the cells lose

their individuality but do not fuse (Figures 18.9 and 18.10). This aggregation is triggered by the production of two compounds, *cyclic adenosine monophosphate* (cAMP) and a specific glycoprotein, both of which function as chemotactic agents (we discuss the involvement of cAMP in various regulatory systems in prokaryotes in Section 7.6). Those cells that are the first to produce these compounds serve as centers for the attraction of other vegetative cells, leading to aggregating masses of cells, which come together and form a slimy migrating mass referred to as a *slug* (Figure 18.9*d*).

Fruiting-body formation begins when the slug ceases to migrate and becomes vertically oriented (Figures 18.9 and 18.10). The fruiting body then becomes differentiated into a stalk and a head; cells in the forward end of the slug become stalk cells, and those in the posterior end become spores. Cells that form stalk cells begin to secrete cellulose, which provides the rigidity of the stalk. Cells from the rear of the slug swarm up the stalk to the tip and form the head. Most of these posterior cells differentiate into spores. On maturation of the head the spores are released and dispersed. Each spore germinates and becomes a vegetative ameba.

The cycle of fruiting-body and spore formation in *Dictyostelium* is an *asexual* process. However *sexual* spores called *macrocysts* are also produced by this organism and by other cellular slime molds. Macrocysts are structures of multicellular origin that develop from simple aggregates of amebas, which, when mature, are enclosed in a thick cellulose wall. As a result of the conjugation of two amebas, a single large ameba develops near the center of the cell mass and becomes actively phagocytic. This phagocytic cell continues to enlarge and engulf amebas until eventually all surrounding cells have been engulfed. At this stage a cellulose wall develops around the now greatly enlarged ameba and the mature macrocyst is formed. During a period of dormancy, the diploid nucleus undergoes meiosis, haploid nuclei are formed, and by progressive nuclear divisions and cytoplasmic cleavage, a new generation of

(a)   Kerry D. Givens   *(b)*   Carolina Biological Supply Co.

Figure 18.8 Slime molds. Plasmodia of acellular slime molds (a) growing on a decaying log and (b) growing on an agar surface.

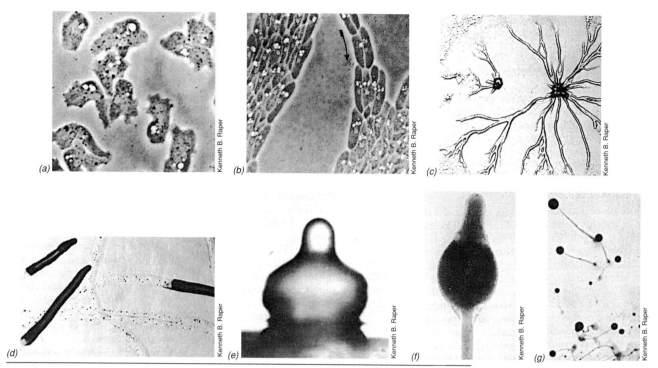

Figure 18.9 Photomicrographs of various stages in the life cycle of the cellular slime mold *Dictyostelium discoideum*. (a) Amebas in preaggregation stage. Note irregular shape and lack of orientation. (b) Aggregating amebas. Notice the regular shape and orientation. The cells are moving in streams in one direction. (c) Low power view of aggregating amebas. (d) Migrating pseudoplasmodia (slugs) moving on an agar surface and leaving trails of slime in their wake. (e, f) Early stage of fruiting body. (g) Mature fruiting bodies. See Figure 18.10 for sizes of these structures.

vegetative amebas is formed. The macrocyst then germinates and releases amebas that reinitiate vegetative growth.

Acellular slime molds

In the vegetative phase, acellular slime molds exist as a mass of protoplasm of indefinite extent, which might be compared to a giant ameba (Figure 18.8). This structure is actively motile by *ameboid motion*, the plasmodium flowing over the surface of the substratum, engulfing food particles as it moves. Ameboid movement is the result of cytoplasmic streaming. Cytoplasm flows forward because the tip of the plasmodium is less contracted and viscous, and thus cytoplasm takes the path of least resistance. Cytoplasmic streaming is facilitated by filaments of a protein called *actin*, which exists in a

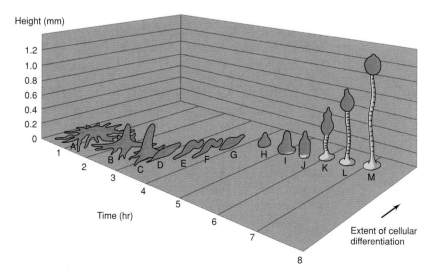

Figure 18.10 Stages in fruiting-body formation in the cellular slime mold *Dictyostelium discoideum*. (A–C) Aggregation of amebas; (D–G) migration of the slug formed from aggregated ameba; (H–L) culmination and formation of the fruiting body; (M) mature fruiting body.

thin layer just beneath the cytoplasmic membrane of eukaryotic cells. In acellular slime molds, cytoplasmic streaming occurs in definite strands, each surrounded by the thin cytoplasmic membrane (Figure 18.8b). The strands of protoplasm do not retain their individuality indefinitely but fuse occasionally into larger masses, which then separate again into smaller strands. Slime mold plasmodia are often brilliantly colored and frequently can be seen spreading across the surface of a piece of wood that has been kept moist (Figure 18.8).

The acellular slime mold plasmodium is diploid. From it, two kinds of differentiated structures are produced, *sporangia* and *sclerotia*. Sporangia arise as part of a sexual cycle; their production is initiated by aggregation of the protoplasmic mass into a compact unit and culminates in secretion by the plasmodium of a stalk, usually devoid of protoplasm, and in the formation of a complex head within which spores develop. Soon after the formation of the head, nuclei in this structure undergo meiosis, and these haploid nuclei then become forerunners of the spores. The protoplasmic mass in the head collects around the haploid nuclei, and a thick cell wall is formed around each. The spores thus formed are discharged when the noncellular part of the fruiting body disintegrates. These resting spores are relatively resistant to drying and other unfavorable conditions and are able to remain dormant for long periods of time. Under favorable conditions, germination occurs. Each spore forms one to four swarm cells that eventually produce more swarm cells by binary fission. Eventually swarm cells conjugate and are converted to a diploid ameboid cell. This cell then grows and undergoes successive nuclear divisions, producing the new diploid plasmodium.

Under starvation conditions, some acellular slime molds form another kind of resting structure, the sclerotium, an irregular, hard, dry mass of protoplasm, resistant to drying and desiccation. No sexual process is involved in this conversion, but the resistant sclerotium, having enabled the organism to overwinter, can form a new plasmodium when conditions are favorable.

CONCEPT CHECK   18.3

Slime molds are nonphototrophic motile cells. Cellular slime molds aggregate to form fruiting bodies from which spores are eventually released. Acellular slime molds are masses of motile protoplasm that form less highly differentiated structures than in cellular slime molds.

✔ *In what ways are the slime molds similar to fungi and in what way are they similar to protozoa?*

✔ *What is a* macrocyst *and what types of slime molds produce them?*

18.4
Protozoa

Protozoa are unicellular eukaryotic microorganisms that lack cell walls (Figure 18.11). They are generally colorless and motile. **Protozoa** are distinguished from prokaryotes by their usually greater size and eukaryotic nature, from algae by their lack of chlorophyll, from yeasts and other fungi by their motility and absence of a cell wall, and from the slime molds by their lack of fruiting-body formation. Protozoa are also phylogenetically distinct from all of the aforementioned groups, existing in several lineages on the Eukarya tree (Figure 18.1).

Protozoa usually obtain food by ingesting other organisms or organic particles. Protozoa are found in a variety of freshwater and marine habitats; a large number are parasitic in other animals, including humans, and some are found growing in soil or in aerial habitats, such as on the surface of trees.

Protozoa feed by ingesting particulate or macromolecular materials. The uptake of macromolecules in solution occurs by a process called pinocytosis. Fluid droplets are sucked into a channel formed by the invagination of the cell membrane, and when portions of this channel are pinched off, the fluid becomes enclosed within a membrane-enclosed vacuole. Most protozoa are also able to ingest particulate material by **phagocytosis,** a process of surrounding a food particle with a portion of their flexible cell membrane to engulf the particle and bring it into the cell. Some protozoa can literally swallow particulate matter (such as bacterial cells) by operation of a special structure called a gullet (see Figure 18.15).

As is appropriate for organisms that "catch" their own food, most protozoa are motile. Indeed, their mechanisms of motility are key characteristics used to divide them into taxonomic groups (Table 18.3). Protozoa that move by ameboid motion are called Sarcodina; those using flagella, the Mastigophora; and those using cilia, the Ciliophora. The Sporozoa, a fourth group, are generally nonmotile and are all parasitic for higher animals.

Mastigophora: The Flagellates

Members of this protozoal group are motile by the action of flagella (Figure 18.11c and Figure 18.12). Although many flagellated protozoa are free-living organisms, a number are parasitic in, or pathogenic for, animals, including humans. The most important pathogenic Mastigophora are the *trypanosomes*. These organisms cause a number of serious diseases in humans and vertebrate animals, including the feared disease *African sleeping sickness*. In *Trypanosoma*, the genus infecting humans, the protozoa are rather small, about 20 μm in length, and are thin, crescent-shaped organisms. They

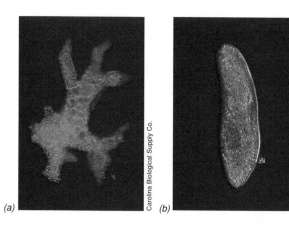
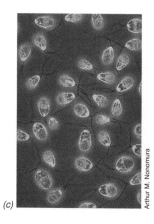
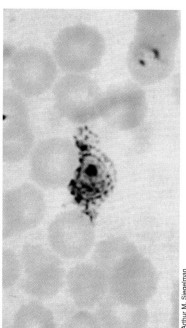

Figure 18.11 Typical protozoa. (a) *Amoeba.* (b) A typical ciliate, *Paramecium.* (c) A flagellate, *Dunaliella* (this flagellate contains chloroplasts and thus can also be considered an alga). (d) *Plasmodium vivax,* a sporozoan, growing in red blood cells.

have a single flagellum that originates in a basal body and folds back laterally across the cell where it is enclosed by a flap of surface membrane (Figure 18.12). Both the flagellum and the membrane participate in propelling the organism, making effective movement possible even in blood, which is rather viscous. *Trypanosoma gambiense* is the species that causes the chronic and usually fatal African sleeping sickness. In humans, the parasite lives and grows primarily in the bloodstream, but in the later stages of the disease, invasion of the central nervous system occurs, causing an inflammation of the brain and spinal cord that is responsible for the characteristic neurological symptoms of the disease. The parasite is transmitted from host to host by the tsetse fly,

Glossina sp., a bloodsucking fly found only in certain parts of Africa. The parasite proliferates in the intestinal tract of the fly and invades the insect's salivary glands and mouthparts, from which it can be transferred to a new human host following a single fly bite.

We also described *phototrophic* flagellates in Section 18.1. These are the *euglenoids,* flagellates that contain chloroplasts, which allow for photosynthetic growth. However, in darkness, cells of *Euglena* (Figure 18.3a) can survive and grow as completely heterotrophic organisms, as can cells in which the chloroplast has been lost. Many euglenoids are known and they are all aquatic, inhabiting primarily fresh waters. Unlike other flagellated protozoa, the euglenoids are nonpathogenic.

TABLE 18.3 Characteristics of the major groups of protozoa

Group	Common name	Typical representatives	Habitats	Common diseases
Mastigophora	Flagellates	*Trypanosoma, Giardia, Leishmania*	Freshwater; parasites of animals	African sleeping sickness, giardiasis, leishmaniasis
Euglenoids[a]	Phototrophic flagellates	*Euglena*	Freshwater; some marine	None known
Sarcodina	Amebas	*Amoeba, Entamoeba*	Freshwater and marine; animal parasites	Amebic dysentery (amebiasis)
Ciliophora	Ciliates	*Balantidium, Paramecium*	Freshwater and marine; animal parasites; rumen	Dysentery
Sporozoa	Sporozoans	*Plasmodium, Toxoplasma*	Primarily animal parasites; insects (vectors for parasitic diseases)	Malaria, toxoplasmosis

[a]This group is also considered with the algae (see Section 18.1 and Table 18.1).

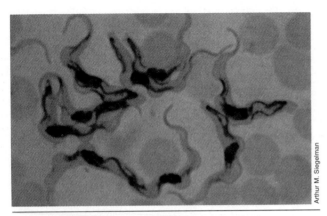

Figure 18.12 Photomicrograph of the flagellated protozoan *Trypanosoma gambiense,* the causative agent of African sleeping sickness, from a blood smear.

Sarcodina: The Amebas

Among the sarcodines are organisms such as *Amoeba,* which are always naked in the vegetative phase (Figure 18.11*a*), and the foraminifera, amebas that secrete a shell during vegetative growth (see Figure 18.13). A wide variety of naked amebas are parasites of humans and other vertebrates, and their usual habitat is the oral cavity or the intestinal tract. They move in these habitats by *ameboid movement* (Figure 18.14), a mechanism previously discussed for the acellular slime molds (see Section 18.3). *Entamoeba histolytica* (∞ Figure 23.53) is a good example of a parasitic ameba. In many cases infection causes no obvious symptoms, but in some individuals it produces ulceration of the intestinal tract, which results in a diarrheal condition called *amebic dysentery* (amebiasis). The organism is transmitted from person to person in the cyst form by fecal contamination of water and food. We discuss the etiology and pathogenesis of amebic dysentery in Section 23.14.

Shelled sarcodines present a variety of interesting morphological forms. The best-known of the shelled

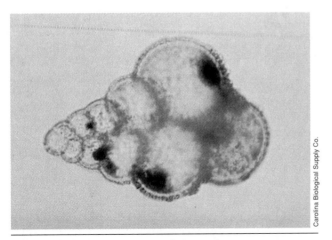

Figure 18.13 Shelled amebas: foraminifera.

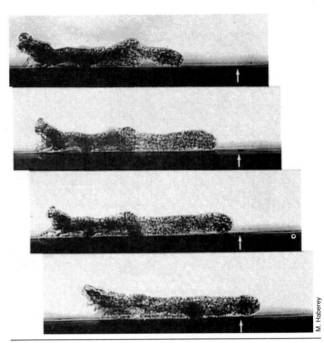

Figure 18.14 Side view of a moving ameba, *Amoeba proteus,* taken from a film, the time interval between frames being 2 sec. The arrows point to a fixed spot on the surface. A single cell is about 80 μm in diameter.

forms are the *foraminifera.* Foraminifera are exclusively marine organisms, living primarily in coastal waters. The shells, called *tests,* of different species show distinctive characteristics and are often quite ornate (Figure 18.13). Tests are usually made of calcium carbonate. The cell is not firmly attached to the test, and the ameba cell may extend partway out of the shell during feeding. However, because of the weight of the test, the cell usually sinks to the bottom, and it is thought that the organisms feed on particulate deposits in the sediments, primarily bacteria and detritus. The shells of foraminifera are relatively resistant to decay and hence readily become fossilized (the White Cliffs of Dover, England, are composed to a great extent of foraminiferal shells). Because of the excellent fossil record these organisms leave, we have a better idea of their distribution through geological time than for virtually any other protozoa.

Ciliophora: The Ciliates

Ciliates are those protozoa that, in some stage of their life cycle, possess cilia (Figure 18.15). They are also unique among protozoa in having *two kinds* of nuclei: the *micronucleus,* which is concerned only with inheritance and sexual reproduction; and the *macronucleus,* which is involved only in the production of messenger ribonucleic acid (mRNA) for various aspects of cell growth and function. Probably the best-known and most widely distributed of the ciliates are those of the genus *Paramecium* (Figure 18.15), which will be used

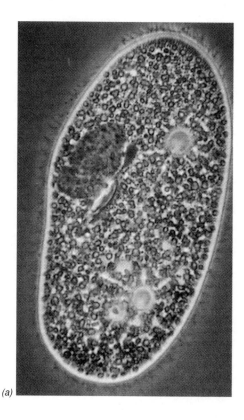

(a)

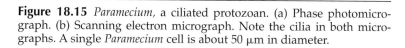

Lines of cilia

Mouth

Yeast cell
(for scale)

Sydney Tamm

(b)

Figure 18.15 *Paramecium,* a ciliated protozoan. (a) Phase photomicrograph. (b) Scanning electron micrograph. Note the cilia in both micrographs. A single *Paramecium* cell is about 50 μm in diameter.

here as an example of the group. Most ciliates obtain their food by ingesting particulate materials through a distinct oral region or mouth connected to an underlying gullet (Figure 18.15b). Once inside, the food particles are carried down the gullet and into the cytoplasm where they are enclosed in a food vacuole, a structure into which digestive enzymes are secreted. In addition to cilia, which function in motility, many ciliates have *trichocysts,* which are long, thin filaments of a contractile nature, anchored beneath the surface of the outer cell layer. These enable the protozoa to attach to a surface and can aid in defense by indicating to the cell that it is being attacked by a predator. In the case of predatory ciliates, such as *Didinium,* trichocysts hold onto and paralyze the prey as a prelude to ingestion.

The two kinds of nuclei in the ciliates set this group aside from all other protozoa. The micronucleus is diploid, but the macronucleus is polyploid, containing from about 40 to over 500 times as much deoxyribonucleic acid (DNA) as does the micronucleus. The micronucleus plays no direct role in vegetative growth and cell division because strains can be developed that

lack micronuclei but continue to divide normally. However, if the macronucleus is removed, the cell quickly dies. Sexual reproduction in ciliates involves conjugation between two cells of opposite mating type. Conjugation involves exchange and fusion of micronuclei with the net result being the formation of exconjugants that are hybrid for certain genes of the two strains.

Many *Paramecium* species (as well as many other protozoa) contain endosymbiotic bacteria living in the cytoplasm or the macronucleus. In some cases, evidence exists that these endosymbionts play a nutritional role for the host, synthesizing vitamins or other growth factors that would otherwise have to be obtained from the environment. In the case of protozoa existing in the termite gut, we previously saw how endosymbiotic methanogens remove H_2 produced from fermentation, yielding CH_4, which is released to the atmosphere (∞Section 14.12 and Figure 14.32).

Although a few ciliates are parasitic for animals, this mode of existence is less extensively developed in the ciliates than it is in other groups of protozoa. The species *Balantidium coli* (Figure 18.16) is primarily a parasite of domestic animals, but occasionally it infects the intestinal tract of humans, producing intestinal dysentery symptoms similar to those caused by *Entamoeba histolytica*. In addition, there is usually a characteristic fauna of *obligately anaerobic* ciliates in the rumen, the forestomach of ruminant animals (∞ Section 14.13); these protozoa are thought to play a beneficial role in the digestive and fermentative processes that occur there.

Sporozoa

The Sporozoa comprise a large group of protozoa, all of which are obligate parasites. They are characterized by a *lack* of motile adult stages and by a nutritional mode of life in which food is generally not ingested but instead is absorbed in soluble form through the outer wall, such as

occurs in prokaryotes and in fungi. Although the name *Sporozoa* implies the formation of spores, these organisms do not form true resting spores, like those of bacteria, algae, and fungi, but instead produce analogous structures called *sporozoites*, which are involved in transmission to a new host. Numerous kinds of vertebrates and invertebrates are hosts for Sporozoa, and in some cases an alternation of hosts takes place, with some stages of the life cycle occurring in one host and some in another. The most important members of the sporozoa are the coccidia, usually parasites of birds, and the plasmodia (malaria parasites) (Figure 18.11*d*), which infect birds and mammals, including humans. Because malaria is a major disease of humans, especially in developing countries, we devote a considerable discussion to this disease and the properties of malarial parasites in Section 23.11.

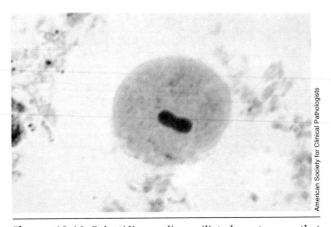

American Society for Clinical Pathologists

Figure 18.16 *Balantidium coli*, a ciliated protozoan that causes a dysentery-like disease in humans. The dark blue stained structure is the macronucleus.

CONCEPT CHECK   **18.4**

Protozoa are unicellular microbial Eukarya that lack cell walls and are usually motile by various means. Many protozoa are pathogenic to humans and other animals.

✔ *List at least two major ways in which the protozoan* Paramecium *differs from the protozoan* Trypanosoma.

✔ *How do the Sporozoa differ from all other protozoa?*

✔ *Why is the alga* Euglena *of interest to protozoologists (scientists who study protozoa)?*

Material for Review

SUPPLEMENTARY READINGS

Alexopoulous, C. J., C. W. Mims, and **M. Blackwell.** 1995. *Introductory Mycology,* 4th edition. John Wiley, New York. An excellent general textbook of fungal biology written by experts in the field.

Carile, M. J., and **S. Watkinson.** 1994. *The Fungi.* Academic Press, San Diego, CA. A comprehensive introduction to mycology—the study of molds and yeasts.

Carter-Lund, H., and **J. W. G. Lund.** 1995. *Freshwater Algae: Their Microscopic World Explored.* Biopress (Lubrecht and Gramer, Ltd), Forestburgh, NY. A beautifully illustrated introduction to phycology (the study of algae) with numerous color photomicrographs to illustrate various groups of algae.

Dix, N. J. and **J. Webster.** 1995. *Fungal Ecology.* Chapman and Hall, New York. A well-illustrated treatment of the ecology of fungi including their habitats and activities in nature.

Moore-Landecker, E. 1990. *Fundamentals of the Fungi,* 3rd edition. Prentice-Hall, Englewood Cliffs, NJ. A general textbook of fungi.

Sleigh, M. A. 1992. *Protozoa and Other Protists.* Cambridge University Press, New York. An introduction to protozoology.

Van Den Hoek, C. D. Mann, and **H. M. Jahns.** 1996. *Algae— An Introduction to Phycology.* Cambridge University Press, New York. An excellent textbook of phycology: coverage of all groups.

On~line resources for this chapter are on the World Wide Web at:
http://www.prenhall.com/~brock (click on the table of contents link and then select Chapter 18).

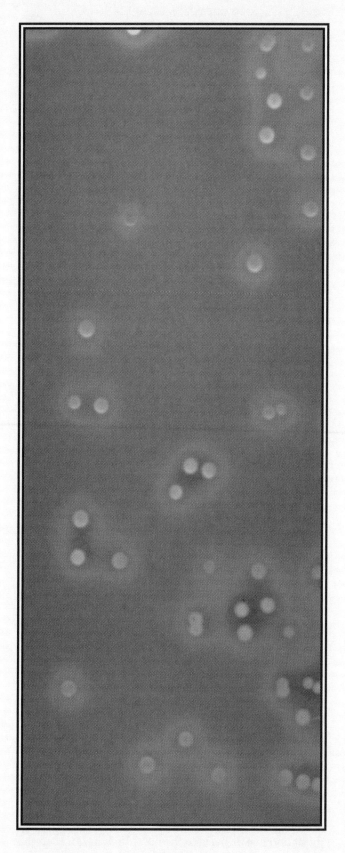

CHAPTER *19*

Host–Parasite Relationships

Zones of hemolysis surrounding colonies of *Streptococcus pyogenes*, a respiratory pathogen commonly found in the upper respiratory tract. *S. pyogenes* combines the virulence factors of invasiveness and toxicity to cause disease.

MINIGLOSSARY for Chapter 19

Adherence a property of bacteria that allows them to stick to surfaces

Attenuation decrease or loss of virulence

Colonization multiplication of a pathogen after it has gained access to host tissues

Dental Caries tooth decay resulting from bacterial infection

Dental Plaque bacterial cells encased in a matrix of extracellular polymers and salivary products, found on the teeth

Disease injury to the host that impairs host function

Endotoxin the lipopolysaccharide portion of the cell wall of certain gram-negative Bacteria, which acts as a toxin when solubilized

Enterotoxin a protein released by an organism as it grows (see exotoxin) that has its effect on the small intestine

Exotoxin a protein released by an organism as it grows that has toxic effects on the host

Glycocalyx bacterial polysaccharides important for attachment of bacterial cells to host structures and to one another

Host an organism that harbors a parasite

Infection growth of organisms in the host

Inflammation host response to injury or infection, characterized by redness, swelling, heat, and pain

Invasiveness pathogenicity caused by the ability of a pathogen to enter the body and spread

Leukocytes nucleated cells found in the blood (white blood cells)

Lower Respiratory Tract trachea, bronchi, and lungs

Normal Flora microorganisms that are usually found associated with healthy body tissue

Parasite an organism that grows in or on a host

Pathogen a parasite that does harm to a host

Pathogenicity the ability of a parasite to inflict damage on the host

Toxigenicity pathogenicity caused by toxins produced by a pathogen

Upper Respiratory Tract the nasopharynx, oral cavity, and throat

Virulence the degree of pathogenicity produced by a pathogen

This chapter begins a major section of the book that deals with the roles of microorganisms in infectious disease. Infectious disease is the most important applied aspect of microbiology and is the area in which most microbiologists work. Up until now, we have surveyed the microbial world and have acquired a great deal of knowledge about its inhabitants. We now understand microbial growth requirements and, to some extent, how to control microbial growth using various physical, chemical, and biochemical means. From this point on, our discussions will be more applied: we will concentrate on the interactions of microorganisms with the animal body, a process that sometimes leads to disease.

We first outline the basic principles of microbial growth on and in the animal body. These principles apply to all potentially harmful microorganisms, but we place particular emphasis on the Bacteria. Second, we examine some of the major mechanisms that invading organisms use to displace normal flora and damage the host. Finally, we investigate the general physical, chemical, and physiological means used by the host to suppress or destroy the microbial invaders. In the next chapter, we will discuss the immune response, a group of disease-specific processes used by the animal body to counteract infectious diseases. Then in Chapter 21 we will discuss the use of techniques for the laboratory

diagnosis of infectious disease agents and for the selection of proper chemotherapeutic procedures. This is followed in Chapter 22 by a discussion of *epidemiology*, the spread of infectious disease within populations, and how epidemiological information can aid in disease control. Finally, in Chapter 23, we will examine a particular group of diseases caused by bacteria, viruses, fungi, and protozoa, with particular emphasis on the causative organisms and the methods used to interrupt or prevent pathogenesis.

The animal body is in continual contact with microorganisms. Literally billions of cells are present in and on the human body, and most play beneficial, sometimes essential, roles in the overall health of the person. These organisms are collectively referred to as the **normal flora** and are species that have developed an intimate relationship with certain tissues of the animal body.

However, some microorganisms do not have beneficial effects on the host. For example, a **parasite** is an organism that lives on or in a second organism, called the **host.** In some cases, the parasite has little or no harmful effect on the host and its presence may be inapparent. In other cases, however, the parasite brings about damage or harm to the host. Such harmful organisms are called **pathogens.** The relationship between host and parasite is dynamic because each modifies the activities and functions of the other. The outcome of the host–parasite relationship depends on the **pathogenicity** of the parasite, that is, on the ability of the parasite to inflict damage on the host, and on the resistance or susceptibility of the host to the parasite. The term **virulence** is a quantitative term used to indicate the degree of pathogenicity of the parasite and is usually expressed as the dose or cell number that will elicit a pathological response within a given time period. However, neither the virulence of the parasite nor the resistance of the host are constant factors: each varies under the influence of external factors or as a result of the host–parasite relationship itself.

Infection refers to the growth of microorganisms in the host. *Infection is not synonymous with disease* because infection does not always lead to injury of the host, even if the pathogen is potentially virulent. In a diseased state, the host is harmed in some way, whereas infection refers to any situation in which a microorganism is established and growing in a host, whether or not the host is harmed.

The ability to cause infectious disease is one of the most dramatic properties of microorganisms. Understanding of the physiological and biochemical basis of infectious disease has led to therapeutic and preventive measures that have had a far-reaching influence on medicine and human affairs. We begin this chapter by considering the normal flora of the healthy human adult. By understanding the microbial ecology of the human body we will be in a better position to appreciate the competitive forces that govern the success or failure of a potential pathogen in initiating disease. We end this chapter by discussing general mechanisms that the host uses to limit parasitism and prevent permanent tissue damage.

CONCEPT CHECK

Organisms that grow in or on other organisms are parasites. Infection is the process by which a parasite becomes established and grows in its host. If the parasite causes harm, it is a pathogen. Most pathogens are rejected by a variety of host defense mechanisms.

✔ *Are all* parasites *also* pathogens?

✔ *Are all* pathogens *also* parasites?

✔ *Distinguish between* infection *and* disease.

19.1

Microbial Interactions with Higher Organisms

Animal bodies provide favorable environments for the growth of many microorganisms. Animals are rich in organic nutrients and growth factors required by chemoorganotrophs, they provide relatively constant conditions of pH and osmotic pressure, and warm-blooded animals have highly constant temperatures. However, the animal body should not be considered one uniform microbial environment. Each region or organ differs chemically and physically from other regions and thus provides a selective environment where certain microorganisms are favored over others. The skin, respiratory tract, gastrointestinal tract, and so on, each provide a wide variety of chemical and physical conditions in which different microorganisms can grow selectively. Animals also possess a variety of defense mechanisms that act in concert to prevent or inhibit microbial invasion and growth. The microorganisms that ultimately colonize the host successfully are those that have developed ways of circumventing these defense mechanisms.

Infections frequently begin at sites in the animal body called *mucous membranes*. Mucous membranes are found throughout the body including the mouth, pharynx, esophagus, and the urinary, respiratory, and gastrointestinal tracts. Mucous membranes consist of single or multiple layers of *epithelial cells,* tightly packed cells that exist in direct contact with the external environment. Mucous membranes are frequently coated with a protective layer of mucus, primarily glycoproteins, which serves to protect epithelial cells. When bacteria contact host tissues at mucous membranes, they may associate either loosely or firmly. If they associate loosely with the mucosal surface, they are usually swept away by physical processes, but they may also attach specifically to the epithelial surface as a result of specific cell–cell recognition between pathogen and host. From there, actual tissue infection may follow. When this occurs, the mucosal barrier is breached, allowing the pathogen to invade deeper tissues (Figure 19.1).

Microorganisms are almost always found in those regions of the body exposed to the outside world, such as the skin, oral cavity, respiratory tract, intestinal tract, and genitourinary tract. They are not normally found in the organs, blood, and lymph systems of the body; if microorganisms are found in significant quantities in any of these latter areas, it is usually indicative of a disease state.

Table 19.1 shows some of the major types of microorganisms normally found in association with body surfaces. As we shall see, the most visible exposed body surface, the skin (2 m²), has a number of normal microbial

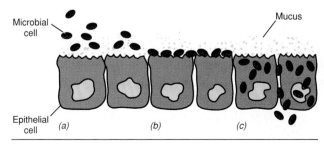

Figure 19.1 Bacterial interactions with mucous membranes. (a) Loose association. (b) Adhesion. (c) Invasion into submucosal cells.

TABLE 19.1 Representative microorganisms in the normal flora of humans

Anatomical site	Organism[a]
Skin	*Staphylococcus, Corynebacterium, Acinetobacter, Pityrosporum* (yeast), *Propionibacterium*
Mouth	*Streptococcus, Lactobacillus, Fusobacterium, Veillonella, Corynebacterium, Neisseria, Actinomyces*
Respiratory tract	*Streptococcus, Staphylococcus, Corynebacterium, Neisseria*
Gastrointestinal tract	*Lactobacillus, Streptococcus, Bacteroides, Bifidobacterium, Eubacterium, Peptococcus, Peptostreptococcus, Ruminococcus, Clostridium, Escherichia, Klebsiella, Proteus, Enterococcus*
Urogenital tract	*Escherichia, Klebsiella, Proteus, Neisseria, Lactobacillus* (vagina of mature females)

[a]Many of the genera listed also contain human pathogens.

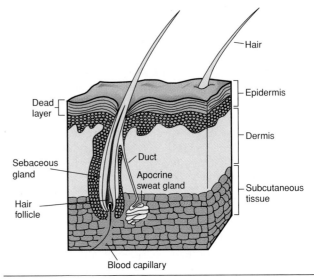

Figure 19.2 Anatomy of the human skin. Microorganisms are associated primarily with the sweat ducts and the hair follicles.

inhabitants. However, mucosal surfaces have an even larger variety of associated microorganisms. This is due in part to the sheltered, moist, hospitable environment of the various mucosal surfaces and also to the huge overall surface area of the mucosae (400 m²). For example, the specialized function of a mucosal organ such as the small intestine requires a large specialized surface area for nutrient transport, and this surface also serves as a site for microbial growth. We shall now examine these normal microbial interactions in greater detail.

CONCEPT CHECK   19.1

Animal bodies are favorable environments for the growth of many microorganisms, including pathogens. The initial colonization by a pathogen is on the surface of the host, often on the mucous membranes.

✔ *Are pathogens more likely to grow on skin or in the throat? Why?*

✔ *Why might one area of the body be more suitable for microbial growth than another?*

19.2

Normal Flora of the Skin

An average human adult has about 2 m² of skin surface that can vary greatly in chemical composition and moisture content. Figure 19.2 shows the anatomy of the skin and regions in which bacteria may live. The skin surface (epidermis) is not a favorable place for microbial growth, as it is subject to periodic drying. Only in certain areas of the body, such as the scalp, face, ears, underarm regions, genitourinary and anal regions, and palms and interdigital spaces of the toes, are moisture conditions on the skin surface sufficiently high to support resident microbial populations.

Most skin microorganisms are associated directly or indirectly with the sweat glands, of which there are several kinds. The **eccrine glands** are not associated with hair follicles and are rather unevenly distributed over the body, with denser concentrations on the palms, finger pads, and soles of the feet. They are the main glands responsible for perspiration. Eccrine glands seem to be relatively devoid of microorganisms, perhaps because of the extensive flow of fluid, because when the flow of an eccrine gland is blocked, bacterial invasion and multiplication do occur. The **apocrine glands** are more restricted in their distribution, being confined mainly to the underarm and genital regions, the nipples, and the umbilicus. They are inactive in childhood and become fully functional only at puberty. Bacterial populations on the surface of the skin in these warm, humid places are relatively high, in contrast to the situation on the smooth surface skin. *Underarm odor develops as a result of bacterial activity on the secretions of the apocrines*; aseptically collected apocrine secretion is odorless but develops odor on inoculation with bacteria. Each hair follicle is associated with a **sebaceous gland**, which secretes a lubricant fluid. Hair follicles provide an attractive habitat for microorganisms; a variety of aerobic and anaerobic bacteria and fungi inhabit these regions, mostly within the area just below the surface of the skin. The secretions of the skin glands are rich in microbial nutrients. Urea, amino acids, salts, lactic acid, and lipids are present in considerable amounts. The pH of human secretions is almost always acidic, the usual range being between pH 4 and 6.

The microorganisms of the normal flora of the skin are either *transients* or *residents*. The skin as an external organ is continually being inoculated with transients, virtually all of which are unable to multiply and usually die. Resident organisms are able to multiply, not merely survive, on the skin. The normal flora of the skin consists primarily of gram-positive Bacteria restricted to a few groups (Table 19.1). These include several species of *Staphylococcus* and a variety of both aerobic and anaerobic corynebacteria. Of the latter, *Propionibacterium acnes* is ordinarily a harmless resident but can incite or contribute to the condition known as *acne*. Gram-negative Bacteria are almost always minor constituents of the normal flora, even though such intestinal organisms as *Escherichia coli* are being continually inoculated onto the surface of the skin by fecal contamination. *Acinetobacter* is an exception and is one of the few gram-negative Bacteria commonly found on skin. It is thought that the lack of colonization of gram-negative Bacteria on the skin is due to their inability to compete with gram-positive organisms that are better adapted to the dry conditions of the skin; if the latter are eliminated by antibiotic treatment, the gram-negative Bacteria can flourish. Yeasts are uncommon on the skin surface, but the lipophilic yeast *Pityrosporum ovalis* is occasionally found on the scalp.

Although the resident microflora remains more-or-less constant, various factors can affect the nature and extent of the normal flora: (1) The weather may cause an increase in temperature and humidity, which increases the density of the skin microflora. (2) Age has an effect, and young children have a more varied microflora and carry more gram-negative Bacteria and potential pathogens than adults. (3) Personal hygienic habits influence the resident microflora, and unclean individuals usually have higher microbial population densities on their skin. Organisms that cannot survive on the skin generally succumb from either the skin's low moisture content or low pH (due to organic acid content).

CONCEPT CHECK 19.2

The skin is a dry, acidic environment that is not conducive to the growth of most microorganisms. However, moist areas, especially around sweat glands, are colonized by gram-positive Bacteria and other members of the skin normal flora.

✔ *How large is the surface area of the skin?*

✔ *Describe the properties of microorganisms that grow well on the skin.*

19.3
Normal Flora of the Oral Cavity

The oral cavity is one of the more complex and heterogeneous microbial habitats in the body. This cavity includes the teeth and tongue and the central space that they fill. Although saliva is the most pervasive source of microbial nutrients in the oral cavity, it is not an especially good microbial culture medium. Saliva contains about 0.5% dissolved solids, about half of which are inorganic (mostly chloride, bicarbonate, phosphate, sodium, calcium, potassium, and trace elements); the predominant organic constituents of saliva are proteins, such as salivary enzymes, mucoproteins, and some serum proteins. Small amounts of carbohydrates, urea, ammonia, amino acids, and vitamins are also present. A number of antibacterial substances have been identified in saliva, of which the most important are the enzymes *lysozyme* and *lactoperoxidase*. Lysozyme is an enzyme that cleaves glycosidic linkages in peptidoglycan in the bacterial cell wall, leading to weakening of the wall and cell lysis (∞ Section 3.5). Lactoperoxidase, an enzyme present in both milk and saliva, kills bacteria by a reaction involving chloride ions and H_2O_2, in which singlet oxygen is probably generated (∞ Sections 5.11 and 20.3). The pH of saliva is controlled primarily by a bicarbonate buffering system ($H_2CO_3 \rightleftharpoons H^+ + HCO_3^-$) and varies between 5.7 and 7.0, with a mean pH near 6.7. The composition of saliva varies, and even within the same individual, variations due to physiological and emotional factors are seen. Despite the activity of antibacterial substances, the presence of food particles and epithelial debris makes the oral cavity a very favorable microbial habitat.

The teeth and dental plaque

The tooth consists of a mineral matrix of calcium phosphate crystals (enamel), within which the living tissue of the tooth (dentin and pulp) is present (Figure 19.3). The teeth influence the nature of the microbial flora. Bacteria found in the mouth during the first year of life (when teeth are absent) are predominantly aerotolerant anaerobes such as streptococci and lactobacilli, but a variety of other bacteria, including some aerobes, occur in small numbers. When the teeth appear, there is a pronounced shift in the balance of the microflora toward anaerobes, and a variety of bacteria specifically adapted for growth on surfaces and in crevices of the teeth develop.

The bacterial colonization of smooth tooth surfaces occurs as a result of firm attachment of single bacterial cells, followed by growth in the form of microcolonies. Beginning with a freshly cleaned tooth surface, the first event is the formation of a thin organic film several micrometers thick as a result of the attachment of acidic glycoproteins from the saliva. This film provides a

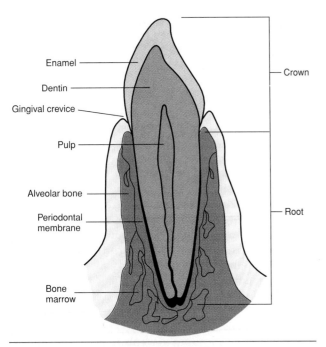

Figure 19.3 Section through a tooth showing the surrounding tissues that anchor the tooth in the gum.

firmer attachment site for the colonization and growth of bacterial microcolonies (Figure 19.4). The colonization of this glycoprotein film is highly specific, and only a few species of *Streptococcus* (primarily *S. sanguis*, *S. sobrinus*, *S. mutans*, and *S. mitis*) are involved. As a result of extensive growth of these organisms, a thick bacterial zone, called **plaque**, is formed (Figures 19.5 and 19.6). If plaque continues to form, filamentous bacteria, usually *Fusobacterium* species, begin to grow. The filamentous bacteria are embedded in the matrix formed by the streptococci, and extend perpendicular to the tooth surface, making an ever-thicker bacterial layer. Associated with the filamentous bacteria, in addition to the streptococci, are spirochetes such as *Borrelia* species (∞ Section 16.12), gram-positive rods, and gram-negative cocci. In heavy plaque, filamentous organisms such as anaerobic *Actinomyces* species may predominate.

The anaerobic nature of the flora may seem surprising, considering that the mouth has good accessibility to oxygen. It is likely that anoxia develops through the action of facultative bacteria growing aerobically on organic materials on the tooth because the dense matrix of the plaque decreases oxygen diffusion onto the tooth surface. The microbial populations of the dental plaque exist in a microenvironment partly of their own making and maintain themselves in the face of wide variations in the macroenvironment of the oral cavity.

Dental caries

As dental plaque accumulates and acid products are formed, **dental caries** (tooth decay) is the usual result. Thus, tooth decay is an infectious disease caused by

Figure 19.4 (a) Bacterial microcolonies growing on a model tooth surface inserted into the mouth for 6 hr. (b) Higher magnification of the preparation in (a). Note the diverse morphology of the organisms present and the slime material (arrows) holding the organisms together.

microorganisms. The role of the oral flora in tooth decay has now been well established through studies on germ-free animals. The smooth surfaces of the teeth that are exposed to frequent cleaning by the tongue, cheek, saliva, or toothbrush or to the abrasive action of food mastication are relatively resistant to dental caries. The tooth surfaces in crevices, where food particles can be retained, are the sites where tooth decay predominates. Thus, the shape of the teeth is important. For instance, dogs are highly resistant to tooth decay because the shape of their teeth does not favor retention of food. Diets high in sugars are especially cariogenic because lactic acid bacteria ferment the sugars to lactic acid, which causes decalcification of the enamel (see Figure 19.3) of the tooth. Once breakdown of the hard tissue has begun, proteolysis of the matrix of the tooth enamel occurs through the action of proteolytic enzymes released by bacteria. Microorganisms penetrate further into the decomposing matrix, but the later stages of the process may be exceedingly slow

Day 1   1436 mm²

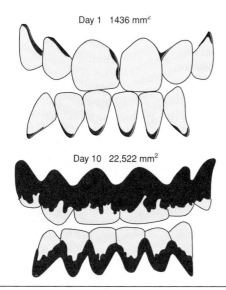

Day 10   22,522 mm²

Figure 19.5 Distribution of dental plaque, as revealed by use of a disclosing agent, on brushed (top) and unbrushed (bottom) teeth. The numbers give the total area of dental plaque.

and are often highly complex. The structure of the calcified tissue also plays an important role in the extent of dental caries. Incorporation of fluoride into the calcium phosphate crystal matrix makes the matrix more resistant to decalcification by acid. Thus, fluorides are used in drinking water and dentifrices to aid in controlling tooth decay.

Two organisms that have been implicated in dental caries are *Streptococcus sobrinus* and *Streptococcus mutans*, both lactic acid–producing bacteria. *S. sobrinus* is able to colonize smooth tooth surfaces because of its specific affinity for salivary glycoproteins (Figure 19.6), and this organism is probably the primary organism involved in decay of smooth surfaces. *S. mutans* is found predominantly in crevices and small fissures, and its ability to attach to tooth surfaces is the result of its ability to produce a dextran polysaccharide that is strongly adhesive (Figure 19.7). *S. mutans* produces dextran only when sucrose is present, by means of the enzyme *dextransucrase*:

$$n \text{ Sucrose} \rightarrow \underset{\text{Dextran}}{(\text{glucose})_n} + n \text{ fructose}$$

Sucrose is common table sugar prevalent in the diet of most Western Europeans and North Americans. Its ability to act as a substrate for dextransucrase is one reason that sucrose is highly cariogenic.

Susceptibility to tooth decay varies greatly among individuals and is affected by inherent traits in the individual as well as by diet and other extraneous factors. Studies of the distribution of oral streptococci have shown a direct correlation between the presence of *S. mutans*, and to a lesser degree *S. sobrinus*, in humans

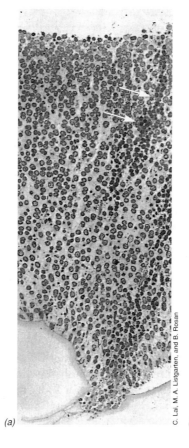

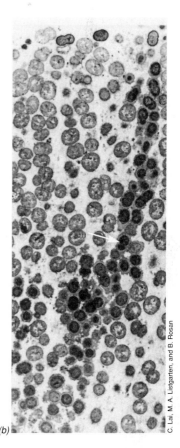

Figure 19.6 Electron micrographs of thin sections of dental plaque. Bottom is the base of the plaque; top is the portion exposed to the oral cavity. (a) Low power electron micrograph. Organisms are predominantly streptococci. The species *Streptococcus sobrinus* has been labeled by an antibody-microchemical technique, and these cells appear darker than the rest. They are seen as two distinct chains (arrows). The total thickness of the plaque layer shown is about 50 μm. (b) Higher power electron micrograph showing the region with *S. sobrinus* cells (dark, arrow). Note the extensive glycocalyx (see Section 19.6) surrounding the *S. sobrinus* cells.

(a)

(b)

C. Lai, M. A. Listgarten, and B. Rosan

C. Lai, M. A. Listgarten, and B. Rosan

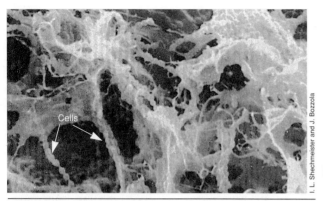

Figure 19.7 Scanning electron micrograph of the cariogenic bacterium *Streptococcus mutans*. The sticky dextran material can be seen as masses of filamentous particles. Individual cells are about 1 μm in diameter.

and the extent of dental caries. In the United States and Western Europe, for example, 80–90% of all people have their teeth colonized by *S. mutans* and dental caries is a nearly universal phenomenon. By contrast, dental caries do not occur in Tanzanian children, presumably because of dietary factors, and *S. mutans* is absent from the plaque of these individuals.

Although tooth decay is an infectious disease, we tend to place it in a different category from other infectious diseases. However, microorganisms in the mouth can also cause other infections such as periodontal disease, gingivitis, and infections of the tooth pulp (abscesses).

CONCEPT CHECK   **19.3**

Bacteria can grow on tooth surfaces in thick layers called plaque. The microorganisms in plaque produce adherent substances that encourage further colonization. Acid-producing organisms in plaque damage tooth surfaces, and dental caries form.

✔ *How do anaerobic microorganisms become established in the mouth?*

✔ *Is dental caries an infectious disease? Give at least one reason for your answer.*

19.4
Normal Flora of the Gastrointestinal Tract

The general anatomy of the gastrointestinal tract is shown in Figure 19.8. The human gastrointestinal tract, the site of food digestion, consists of the stomach, small intestine, and large intestine. The pH of stomach fluids is low, about pH 2. The stomach can thus be viewed as a microbiological barrier against

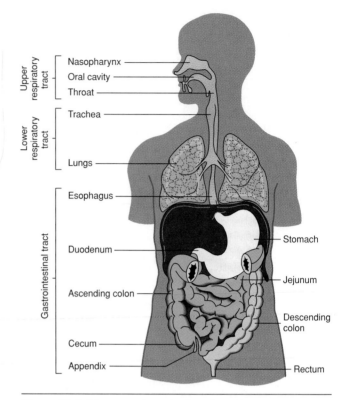

Figure 19.8 The gastrointestinal and respiratory tracts.

entry of foreign bacteria into the intestinal tract. Although the bacterial count of the stomach contents is generally low, the walls of the stomach are often heavily colonized with bacteria. These are primarily acid-tolerant lactobacilli and streptococci and can be seen in large numbers in histological sections or by scanning electron microscopy of the stomach epithelium (Figure 19.9). These bacteria appear very early after birth and are well established by the first week.

The intestinal tract

The intestinal tract consists of the *small intestine* and the *large intestine*, each of which is further subdivided into different anatomical structures.

The **small intestine** is separated into two parts, the *duodenum* and the *ileum*. The former, adjacent to the stomach, is fairly acidic and resembles the stomach in its microbial flora. From the duodenum to the ileum the pH gradually becomes less acidic and bacterial numbers increase. In the lower ileum, bacteria are found in the intestinal cavity (the lumen), mixed with digestive material. Cell numbers of 10^5–10^7 per gram are common.

In the **large intestine,** bacteria are present in enormous numbers, so much so that this region can be viewed as a specialized fermentation vessel. Many bacteria live within the lumen itself, probably using as nutrients some products of the digestion of food (Table 19.1). Facultative aerobes, such as *Escherichia coli*, are

Figure 19.9 Light microscopy and scanning electron microscopy of the stomach microflora of a mouse. (a) Section through a Gram-stained preparation of the stomach wall of a 14-day-old mouse, showing extensive development of lactic acid bacteria (*Lactobacillus* sp.) in association with the epithelial layer. (b) Bacterial growth on the surface of the keratinized epithelium of the stomach of an adult mouse as observed by scanning electron microscopy. Individual cells are 3–4 μm in length. *(a)*

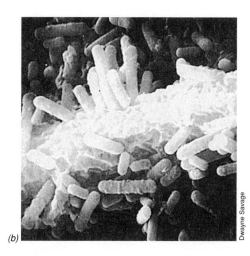

present but in smaller numbers than other bacteria; total counts of facultative aerobes are generally less than 10^7 per gram of intestinal contents. The activities of facultative aerobes consume any oxygen present, making the environment of the large intestine strictly anaerobic and favorable for the profuse growth of obligate anaerobes. Many of these anaerobes are long, thin, gram-negative rods with tapering ends (called *fusiform*) and are attached end-on to small indentations in the intestinal wall (Figure 19.10). Other obligate anaerobes include species of *Clostridium* and *Bacteroides.* The total number of obligate anaerobes is enormous. Counts of 10^{10}–10^{11} cells/g of intestinal contents are not uncommon, with various species of *Bacteroides* accounting for the majority of intestinal obligate anaerobes. In addition, *Enterococcus faecalis* is almost always present in significant numbers.

The intestinal flora of the newborn becomes established early. In breast-fed human infants the flora is often fairly simple, consisting largely of *Bifidobacterium* sp. As the infant ages and its diet becomes more complex, the composition of the intestinal flora also becomes more complex, ultimately approaching that of the adult.

The normal flora of the gastrointestinal tract varies among species. For example, in guinea pigs lactobacilli make up 80% of the intestinal flora, whereas the same organisms are only minor components of human gastrointestinal flora. The gut flora in humans can also vary qualitatively depending on the diet. Persons who consume a considerable amount of meat show higher numbers of *Bacteroides* and lower numbers of coliforms and lactic acid bacteria than those on a vegetable diet. An overview of microorganisms of the gastrointestinal tract is given in Figure 19.11.

The intestinal flora has a profound influence on the animal, carrying out a wide variety of metabolic reactions (Table 19.2). Not all microorganisms carry out these reactions, and changes in the intestinal flora due to diet or disease may thus affect the animal. Of special note in Table 19.2 are the roles of the intestinal flora in modifying the bile acids. Bile acids are steroids produced in the liver and secreted into the intestine via the gallbladder. Their role is to promote emulsification of fats in the diet so the fats can be effectively digested. Intestinal microorganisms cause a variety of transformations of these bile acids so the materials excreted in the feces are quite different from the original bile acids. Other products of microbial fermentation are the odor-producing substances listed in Table 19.2. Composition of the intestinal

Figure 19.10 Scanning electron micrographs of the microbial community on the surface of the columnar epithelium in the mouse ileum. (a) An overview at low magnification. Note the long, filamentous *fusiform* bacteria lying on the surface. (b) Higher magnification, showing several filaments attached at a single depression. Note that the attachment is at the end of the filaments only. Individual cells are 10–15 μm in length. *(a)*

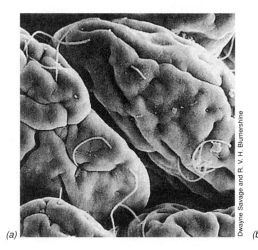

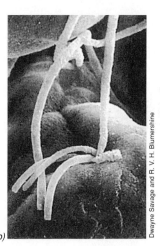

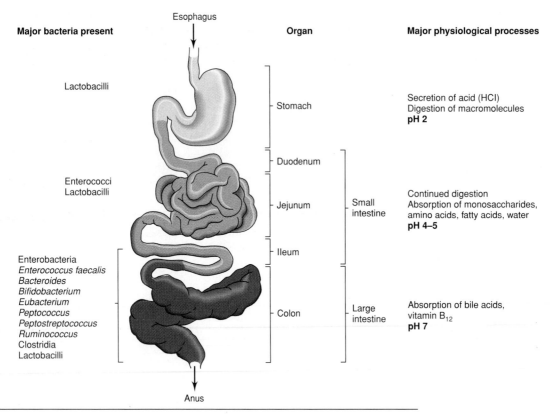

Figure 19.11 The human gastrointestinal tract showing functions and the distribution of microorganisms.

microflora as well as diet influences the amount of gas and the amount of odoriferous materials present.

During the passage of food through the gastrointestinal tract, water is withdrawn from the digested material, and it gradually becomes more concentrated and is converted to feces. Bacteria make up about one-third of the weight of fecal matter. Organisms living in the lumen of the large intestine are continuously being displaced downward by the flow of material, and if bacterial numbers are to be maintained, those bacteria

that are lost must be replaced by new growth. Thus, the large intestine resembles a chemostat (∞ Section 5.5). The time needed for passage of material through the complete gastrointestinal tract is about 24 hr in humans; the growth rate of bacteria in the lumen is one to two doublings per day.

When an antibiotic is given orally, it may inhibit the growth of the normal flora as well as pathogens; continued movement of the intestinal contents then leads to loss of the preexisting bacteria and virtual sterilization of the intestinal tract. In the absence of the normal flora, the environmental conditions of the large intestine change, and opportunistic microorganisms such as antibiotic-resistant *Staphylococcus*, *Proteus*, or the yeast *Candida albicans* may become established. These organisms usually do not grow in the intestinal tract because they cannot compete with the normal flora. Occasionally, establishment of these opportunistic pathogens can lead to a harmful alteration in digestive function or even to disease. After antibiotic therapy stops, the normal flora eventually becomes reestablished, but often only after a considerable period.

Intestinal gas

The gas produced within the intestines, called *flatus*, is the result of the action of fermentative and methanogenic microorganisms. Some foods can be metabolized

TABLE 19.2	Biochemical/metabolic contributions of intestinal microorganisms
Vitamin synthesis	Product: thiamine, riboflavin, pyridoxine, B_{12}, K
Gas production	Product: CO_2, CH_4, H_2
Odor production	Product: H_2S, NH_3, amines, indole, skatole, butyric acid
Organic acid production	Product: acetic, propionic, butyric acids
Glycosidase reactions	Enzyme: β-glucuronidase, β-galactosidase, β-glucosidase, α-glucosidase, α-galactosidase
Steroid metabolism	Process: esterification, dehydroxylation, oxidation, reduction, inversion

by fermentative bacteria in the intestines, resulting in the production of hydrogen (H_2) and carbon dioxide (CO_2). Methanogens (∞ Section 17.2) are found in the intestines of over one-third of normal adults. The methanogens convert to H_2 and CO_2 produced by the other intestinal microorganisms to methane (CH_4). Normal adults expel several hundred milliliters of gas, of which about half is N_2 from swallowed air, from the intestines every day.

CONCEPT CHECK   19.4

The stomach is very acidic and is a barrier to most microbial growth. The intestinal tract is slightly acid to neutral and supports a large, diverse population of microorganisms in a variety of nutritional and environmental conditions.

✔ *Why does* Lactobacillus *inhabit the* stomach?

✔ *What particular conditions favor the survival of* Bacteroides *in the* large intestine?

19.5
Normal Flora of Other Body Regions

Each individual mucous membrane supports the growth of a specialized group of microorganisms. These organisms are part of the normal local environment and are characteristic of healthy tissue. In many cases, potentially pathogenic microorganisms cannot colonize mucous membranes because of the presence of the normal resident population of microorganisms. In this section, we discuss two such mucosal environments and their resident microorganisms.

Respiratory tract

The anatomy of the respiratory tract was shown in Figure 19.8 (see also Figure 23.2). In the **upper respiratory tract** (nasopharynx, oral cavity, and throat) microorganisms live primarily in areas bathed with the secretions of the mucous membranes. Bacteria enter the upper respiratory tract from the air during breathing, but most of them are trapped in the nasal passages and expelled again with the nasal secretions. The resident organisms most commonly found are staphylococci, streptococci, diphtheroid bacilli, and gram-negative cocci. Potentially harmful bacteria, such as *Staphylococcus aureus, Streptococcus pneumoniae, Streptococcus pyogenes,* and *Corynebacterium diphtheriae* are often part of the normal flora of the nasopharynx of healthy individuals (Table 19.1). These individuals are *carriers* of the pathogens but do not normally acquire disease because the other resident microorganisms compete successfully for resources and limit pathogen growth. The local immune system (∞ Section

20.5) is particularly active at mucosal surfaces and may also inhibit the growth of pathogens.

The **lower respiratory tract** (trachea, bronchi, and lungs) is essentially sterile, in spite of the large numbers of organisms potentially able to reach this region during breathing. Dust particles, which are fairly large, settle out in the upper respiratory tract. As the air passes into the lower respiratory tract, its rate of flow decreases markedly, and organisms settle onto the walls of the passages. The walls of the entire respiratory tract are lined with ciliated epithelium, and the cilia, beating upward, push bacteria and other particulate matter toward the upper respiratory tract where they are then expelled in the saliva and nasal secretions. Only particles smaller than about 10 μm in diameter are able to reach the lungs.

Urogenital tract

The main anatomical features of the male and female urogenital tracts are shown in Figure 19.12*a*. In both male and female, the bladder itself is usually sterile, but the epithelial cells lining the urethra are colonized by facultatively aerobic gram-negative rods and cocci (Table 19.1). These organisms, including *Escherichia coli* and *Proteus mirabilis* and others, can occasionally become *opportunistic pathogens*. These organisms are normally present in the body or in the local environment, but they are not pathogenic under normal circumstances. Changes in the body, such as local pH changes or decreases in immune system functions (∞ Section 20.16), allow the organisms to multiply and become pathogenic. Such organisms frequently cause urinary tract infections, especially in women.

The vagina of the adult female generally is weakly acidic and contains significant amounts of the polysaccharide glycogen. *Lactobacillus acidophilus* ferments glycogen and produces acid. It is usually present in the vagina and may be responsible for the acidity (Figure 19.12*b*). Other organisms—yeasts (*Torulopsis* and *Candida* species), streptococci, and *E. coli*—may also be present. Before puberty, the female vagina is alkaline and does

---

CONCEPT CHECK   19.5

The presence and maintenance of a population of normal nonpathogenic microorganisms in the respiratory and urogenital tracts is essential for normal organ function and often prevents the colonization of pathogens.

✔ *Pathogens are sometimes found in the normal flora of the upper respiratory tract. Why do they not cause disease in some cases?*

✔ *Why is* Lactobacillus *found in the urogenital tract of normal adult women?*

---

competitive exclusion

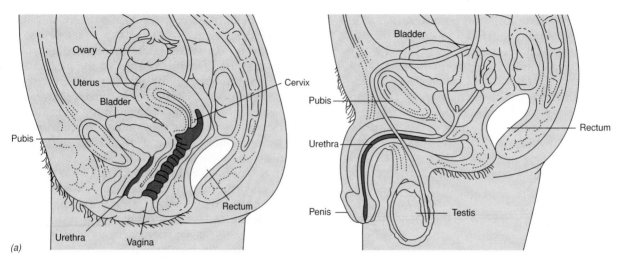

(a)

(b)

John Durham

Figure 19.12 (a) The genitourinary tracts of the human female and male, showing regions (color) where microorganisms often grow. (b) Gram stain of *Lactobacillus acidophilus,* the predominant organism in the vagina of women. Individual rods are 3–4 µm in length.

not produce glycogen, *L. acidophilus* is absent, and the flora consists predominantly of staphylococci, streptococci, diphtheroids, and *E. coli.* After menopause, glycogen disappears, the pH rises, and the flora again resembles that found before puberty.

19.6
Entry of the Pathogen into the Host

We now start discussion of mechanisms used by pathogens to alter host function. The steps of *pathogenesis,* the progression of a disease state, include entry, colonization and growth, and the use of several strategies to establish *virulence,* the relative ability of a pathogen to cause disease in the host (Figure 19.13). We will start our discussion by considering the factors responsible for entry of a pathogen into a host.

A pathogen must usually gain access to host tissues and multiply before damage can be done. In most cases, this requires that the organism penetrate the skin, mucous membranes, or intestinal epithelium, surfaces that normally act as microbial barriers. Passage through the skin into subcutaneous layers almost always occurs through wounds; in rare instances pathogens penetrate through the unbroken skin.

Specific adherence

Most microbial infections begin on the mucous membranes of the respiratory, alimentary, or genitourinary tract. There is considerable evidence that bacteria or viruses able to initiate infection can adhere specifically to epithelial cells (Figure 19.14). The evidence for specificity is of several types. First, there is *tissue specificity.* An infecting microorganism does not adhere to all epithelial cells equally but selectively adheres to cells in the particular region of the body where it normally gains entrance. For example, *Neisseria gonorrhoeae,* the causative agent of the sexually transmitted disease gonorrhea, adheres much more strongly to urogenital epithelia than to other tissues. Second, there is *host specificity.* A bacterial strain that normally infects humans adheres more strongly to the appropriate human epithelial cells than to similar cells in another animal (for example, the rat), whereas a strain that specifically colonizes the rat adheres more firmly to rat cells than to human cells.

Many bacteria possess specific surface macromolecules that bind to complementary receptor molecules on the surfaces of certain animal cells, thus promoting specific and firm adherence. Certain of these macromolecules are polysaccharide in nature and form a sticky meshwork of fibers called the bacterial **glycocalyx** (Figure 19.14*b*). The glycocalyx is important not only in attaching bacterial cells to host cell surfaces but also in adherence between bacterial cells (Figure 19.14*b*). In addition, fimbriae (∞ Section 3.11) may be important in

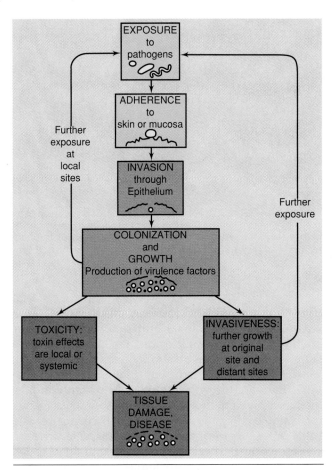

Figure 19.13 Microorganisms and pathogenesis. The presence of microorganisms on the host does not always lead to disease.

the attachment process. For instance, the fimbriae of *N. gonorrhoeae* play a key role in the attachment of this organism to urogenital epithelium, and fimbriated strains of *Escherichia coli* (Figure 19.15) are much more frequent causes of urinary tract infections than strains lack-

ing fimbriae. Among the best-characterized fimbriae are the so-called *type I fimbriae* of enteric bacteria (*Escherichia, Klebsiella, Salmonella, Shigella*). Type I fimbriae are 0.2–1 μm in length and 7 nm wide and are uniformly distributed on the surface of cells. Type I fimbriae function in attachment by binding mannose residues of specific host cell glycoproteins to initiate the attachment event.

Evidence of the specific interaction between mucosal epithelium and the pathogen comes from studies of diarrhea caused by *Escherichia coli*. Most strains of *E. coli* are nonpathogenic and are part of the normal flora of the *large* intestine. A few strains (only a handful of the 160 different *E. coli* serotypes known) are *enteropathogenic*, possessing the ability to colonize the *small* intestine and initiate diarrhea. Such strains possess specific surface structures, the **colonization factor antigens (CFA),** which are fimbrial proteins involved in specific attachment to intestinal mucosa. Thus, two kinds of *E. coli* can be recognized: pathogenic strains, which are able to adhere to the mucosal surface of the small intestine and cause disease symptoms, and "normal" *E. coli* strains, which are unable to adhere to the small intestine or produce enterotoxin (see Section 19.9). The normal *E. coli* strains grow in the large intestine (cecum and colon), and often enter into a long-lasting symbiotic relationship with the mammalian host. A summary of major factors in microbial adherence is given in Table 19.3.

Invasion

A few microorganisms are pathogenic solely because of the toxins they produce. These organisms do not need to gain access to host tissues, and we will discuss them separately (see Sections 19.8 and 19.9). However, most pathogens penetrate the epithelium to initiate pathogenicity, a process called *invasion*. At the point of entry, usually at small breaks or lesions in the skin or in mucosal surfaces, growth is often established in the

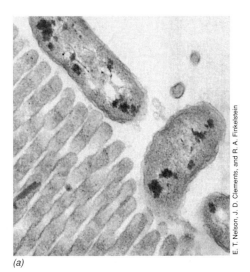

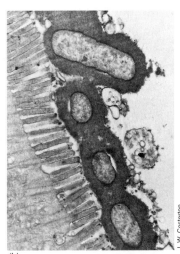

Figure 19.14 Adherence of pathogens to animal tissues. (a) Transmission electron micrograph of a thin section of *Vibrio cholerae* adhering to the brush border of rabbit villi. Note the absence of the outer layer (glycocalyx). (b) Enteropathogenic *Escherichia coli* in a fatal model infection in the newborn calf. The bacterial cells are attached to the brush border of calf villi via an extensive glycocalyx. The rods are about 0.5 μm in diameter.

submucosa. Growth may also be established on intact mucosal surfaces, especially if the normal flora is altered or eliminated, for example, by antimicrobial chemotherapy. Pathogens may then more readily colonize the tissue and begin the invasion process. Pathogen growth may also be established at sites distant from the original point of entry. Access to distant, usually interior, sites is through the blood or lymphatic circulatory system (∞ Section 20.1).

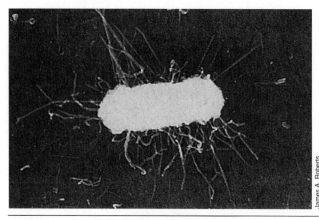

Figure 19.15 Shadow-cast electron micrograph of the bacterium *Escherichia coli,* showing type P fimbriae. Type P fimbriae resemble type I fimbriae but are somewhat longer. The cell shown is about 0.5 μm wide.

CONCEPT CHECK   19.6

Pathogens may first gain access to host tissues by adherence to specific host molecules, usually on mucosal surfaces. Invasion starts at the site of adherence and may spread throughout the host via the circulatory systems.

✔ *Distinguish between* adherence *and* invasion.

✔ *Why is invasion usually necessary to establish pathogenicity?*

19.7

Colonization and Growth

If a pathogen gains access to tissues, it must multiply, a process called *colonization.* Colonization requires that the pathogen bind to specific tissue surface receptors and overcome any nonspecific or immune host defenses (see Section 19.12 and Chapter 20). The initial inoculum is rarely sufficient to cause damage; a pathogen must *grow* within host tissues in order to produce a disease. If the pathogen is to grow, it must find appropriate nutrients and environmental conditions in the host. Temperature, pH, and reduction potential are environmental factors that affect pathogen growth, but the availability of microbial nutrients in host tissues is most important. Although a vertebrate host might seem to be

a nutritional paradise for microorganisms, not all nutrients are plentiful. Soluble nutrients such as sugars, amino acids, and organic acids may often be in short supply, and organisms able to utilize complex nutrient sources such as glycogen may be favored. Not all vitamins and other growth factors are necessarily in adequate supply in all tissues at all times. *Brucella abortus,* for example, can grow slowly in most tissues of infected cattle but grows very rapidly in the placenta, where it causes abortion. This specificity is due to the elevated concentration of erythritol found in the placenta, a nutrient that greatly stimulates growth of *B. abortus* (see Table 19.6).

Trace elements may also be in short supply and can influence establishment of the pathogen. For example, considerable evidence exists for the influence of **iron** on microbial growth. Specific proteins called *transferrin* and *lactoferrin,* present in animals, bind iron tightly and transfer it through the body. Such is the affinity of these proteins for iron that microbial iron deficiency may be common; indeed, administration of

---

TABLE 19.3   Major adherence factors used to facilitate attachment of microbial pathogens to host tissues[a]

Factor	Example
Sticky outer capsule (glycocalyx)	Enterotoxic *Escherichia coli* (ETEC)—glycocalyx promotes adherence to the brush border of intestinal villi
	Dental caries—binding to tooth surface by *Streptococcus mutans*
Adherence proteins	M-protein on surface of *Streptococcus pyogenes* binds receptor on respiratory mucosa
Lipoteichoic acid	Along with M-protein of *Streptococcus pyogenes*—facilitates binding to respiratory mucosal receptor
Fimbriae (pili)	Gonorrhea—pili on *Neisseria gonorrhoeae* facilitate binding to urogenital epithelium
	Salmonellosis—type I fimbriae facilitate binding to small intestinal epithelium
	Enterotoxic *Escherichia coli*—colonization factor antigens (CFAs), which are fimbrial, facilitate binding to small intestinal epithelium

[a]For the most part, receptor sites on host tissues are glycoproteins or complex lipids such as gangliosides or globosides.

a soluble iron salt to an infected animal may greatly increase the virulence of some pathogens. As we noted in Section 4.2, many bacteria produce iron-chelating compounds (siderophores), which help them to obtain iron from the environment. Some iron chelators isolated from pathogenic bacteria are so efficient that they can actually remove iron from animal iron-binding proteins. For example, a siderophore called *aerobactin*, produced by certain strains of *Escherichia coli* and encoded by the Col V plasmid (∞ Section 9.8), readily removes iron bound to transferrin.

Localization in the body

After initial entry, the organism often remains localized and multiplies, producing a small focus of infection such as the boil, carbuncle, or pimple that commonly arises from *Staphylococcus* infections of the skin. Alternatively, the organisms may pass through the lymphatic vessels and be deposited in lymph nodes. If an organism reaches the blood, it will be distributed to distant parts of the body, usually concentrating in the liver or spleen. Spread of the pathogen through the blood and lymph systems can result in a generalized (systemic) infection of the body, with the organism growing in a variety of tissues. If extensive bacterial growth in tissues occurs, some of the organisms are usually shed into the bloodstream in large numbers, a condition called **bacteremia.** Generalized infections of this type almost always start as a localized infection at a specific organ site but fortunately are quite rare.

Virulence factors

A number of pathogen-produced extracellular proteins aid in the establishment and maintenance of disease. These proteins, which are mostly enzymes, are called *virulence factors.* For example, streptococci, staphylococci, pneumococci, and certain clostridia produce **hyaluronidase** (Table 19.4), an enzyme that promotes spreading of organisms in tissues by breaking down hyaluronic acid, a polysaccharide that functions in the body as a tissue cement. Production of this enzyme enables these organisms to spread from an initial focus. Streptococci and staphylococci also produce a vast array of proteases, nucleases, and lipases that serve to depolymerize host proteins, nucleic acids, and fats, respectively. Clostridia that cause gas gangrene produce **collagenase,** or κ-toxin (Table 19.4), which breaks down the collagen network supporting the tissues; the resulting dissolution of tissue is a factor in enabling these organisms to spread through the body. Fibrin clots are often formed by the host in a region of microbial invasion and wall off the organism, preventing its spread through the body. Some organisms are able to produce fibrinolytic enzymes to dissolve these clots and make further invasion possible. One such fibrinolytic substance, produced by *Streptococcus pyo-*genes, is known as **streptokinase** (Table 19.4). On the other hand, some organisms produce enzymes that actually promote fibrin clotting, which causes localization of the organism rather than its spread. The best-studied fibrin-clotting enzyme is **coagulase** (Table 19.4), produced by pathogenic *Staphylococcus aureus*, which causes the fibrin material to be deposited on the cocci and may offer them protection from attack by host cells. The fibrin matrix produced as a result of coagulase activity probably accounts for localization of many staphylococcal infections in boils and pimples (∞ Figure 23.6).

Various pathogens produce proteins that are able to act on the animal cytoplasmic membrane, causing cell lysis and hence cell death. The action of these toxins is most easily detected with red blood cells (erythrocytes), hence they are often called **hemolysins** (Table 19.4); in probably all cases, however, they also work on cells other than erythrocytes. The production of such toxins is most readily demonstrated in the laboratory by streaking the organism on a *blood agar plate.* During growth of the colonies, some of the hemolysin is released and lyses the surrounding red blood cells, typically clearing a zone of hemolysis (Figure 19.16*a*). Some hemolysins are enzymes that attack the phospholipid of the host cytoplasmic membrane. Because the phospholipid lecithin (phosphatidylcholine) is often used as a substrate, these enzymes are called **lecithinases** or **phospholipases** (Figure 19.16*b*). Since the cytoplasmic membranes of all organisms, both prokaryotes and eukaryotes, contain phospholipids, hemolysins that are phospholipases sometimes destroy bacterial as well as animal cytoplasmic membranes. Some hemolysins are not phospholipases, however. Streptolysin O, a hemolysin produced by streptococci, affects the sterols of the host cytoplasmic membrane, and its action is neutralized by addition of cholesterol or other sterols. **Leukocidins** (Table 19.4) are lytic agents capable of lysing white blood cells and hence serve to decrease host resistance (∞ Section 20.1).

CONCEPT CHECK   19.7

The disease process requires that a pathogen gain access to host-provided nutrients, followed by colonization and growth in substantial numbers in host tissue. A number of pathogen-produced extracellular virulence factors are designed to protect the pathogen from host defenses or to provide increased access to nutrients.

✔ *Why is colonization necessary for the success of most pathogens?*

✔ *Why do bacterial enzymes attack structural components of host cells?*

TABLE 19.4	Exotoxins and extracellular virulence factors produced by certain bacteria pathogenic for humans		
Organism	**Disease**	**Toxin or factor**	**Action**
Clostridium botulinum	Botulism	Neurotoxin	Flaccid paralysis (see Figure 19.18a)
Clostridium tetani	Tetanus	Neurotoxin	Spastic paralysis (see Figure 19.18b)
Clostridium perfringens	Gas gangrene, food poisoning	α-Toxin	Hemolysis (lecithinase, see Figure 19.16b)
		β-Toxin	Hemolysis
		γ-Toxin	Hemolysis
		δ-Toxin	Hemolysis
		θ-Toxin	Hemolysis (cardiotoxin)
		κ-Toxin	Collagenase
		λ-Toxin	Protease
		Enterotoxin	Alters permeability of intestinal epithelium
Corynebacterium diphtheriae	Diphtheria	Diphtheria toxin	Inhibits protein synthesis in eukaryotes and in Archaea (see Figure 19.17)
Staphylococcus aureus	Pyogenic (pus-forming) infections (boils, and so on), respiratory infections, food poisoning, toxic shock syndrome, scalded skin syndrome	α-Toxin	Hemolysis
		Toxic shock syndrome toxin	Systemic shock
		Exfoliating toxins A and B	Peeling of skin, shock
		Leukocidin	Destroys leukocytes
		β-Toxin	Hemolysis
		γ-Toxin	Kills cells
		δ-Toxin	Hemolysis, leukolysis
		Enterotoxins A, B, C, D, and E	Induce vomiting, diarrhea, shock
		Coagulase	Induces fibrin clotting
Streptococcus pyogenes	Pyogenic infections, tonsillitis, scarlet fever	Streptolysin O	Hemolysin
		Streptolysin S	Hemolysin
		Erythrogenic toxin	Causes scarlet fever rash
		Streptokinase	Dissolves fibrin clots
		Hyaluronidase	Dissolves hyaluronic acid in connective tissue
Vibrio cholerae	Cholera	Enterotoxin	Induces fluid loss from intestinal cells (see Figure 19.19)
Escherichia coli (enteropathogenic strains only)	Gastroenteritis	Enterotoxin	Induces fluid loss from intestinal cells
Bacillus cereus	Food poisoning	Enterotoxin	Induces fluid loss from intestinal cells
Shigella dysenteriae	Bacterial dysentery	Neurotoxin	Paralysis, hemorrhage
Yersinia pestis	Plague	Plague toxin	Kills cells
Bordetella pertussis	Whooping cough	Whooping cough (Pertussis) toxin	Kills cells
Pseudomonas aeruginosa	Various *P. aeruginosa* infections	Exotoxin A	Kills cells

19.8

Exotoxins

The ways in which pathogens bring about damage to the host are diverse. Only rarely are symptoms of a disease due simply to the presence of large numbers of microorganisms. Although a large mass of cells can block vessels or heart valves or clog the air passages of the lungs, in many cases pathogens produce *toxins* that are responsible for host damage.

Toxins released extracellularly as the organism grows are called **exotoxins.** These toxins may travel from a focus of infection to distant parts of the body

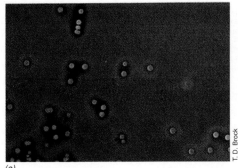

Figure 19.16 (a) Zones of hemolysis around colonies of *Streptococcus pyogenes* growing on a blood agar plate. (b) Action of lecithinase, a phospholipase, around colonies of *Clostridium perfringens,* growing on an agar medium containing egg yolk.

(a)   *(b)*

and cause damage in regions far removed from the site of microbial growth. Table 19.4 provides a summary of the properties and actions of some of the best-known exotoxins.

Diphtheria toxin

The toxin produced by *Corynebacterium diphtheriae,* the causal agent of diphtheria, was the first exotoxin to be discovered. It differs markedly in its action on different animal species; rats and mice are relatively resistant, whereas humans, rabbits, guinea pigs, and birds are susceptible. Diphtheria toxin is very potent; only a single molecule is required to kill a single cell. The toxin binds irreversibly to the cell, and within a few hours the cell loses its ability to synthesize protein because the toxin interferes with protein synthesis by blocking transfer of an amino acid from a transfer ribonucleic acid (tRNA) to the growing peptide chain. The toxin specifically inactivates one of the elongation factors (elongation factor 2) involved in growth of the polypeptide chain (Figure 19.17) by catalyzing the attachment of the adenosine diphosphate (ADP) ribose moiety of NAD^+ to the elongation protein. The elongation protein

is ADP-ribosylated at a single amino acid residue, a modified histidine molecule called *diphthamide;* following ADP-ribosylation, the activity of elongation factor 2 drops dramatically and protein synthesis stops.

Diphtheria toxin is formed by strains of *C. diphtheriae* that are lysogenized by a bacteriophage called phage β, and the toxin production is encoded in the phage genome. Nontoxigenic and hence nonpathogenic strains of *C. diphtheriae* can be converted to pathogenic strains by infection with the β phage (the process of phage conversion) (∞ Section 9.7).

The toxin as excreted by *C. diphtheriae* cells is a single polypeptide of 62,000 molecular weight containing 535 amino acids. Following binding to the host cell, the polypeptide is cleaved by a protease into two fragments. Fragment A (193 amino acids) enters the cell and disrupts protein synthesis, whereas the remaining piece, Fragment B (342 amino acids), is discarded. Before cleavage, Fragment B promotes specific binding of the toxin to the host cell, and following cleavage it assists in the entry of Fragment A into the host cytoplasm.

A factor in toxin production is the concentration of *iron* present in the environment. In media containing sufficient iron for optimal growth, no toxin is produced.

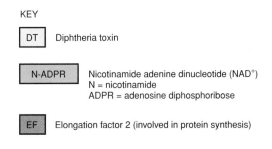

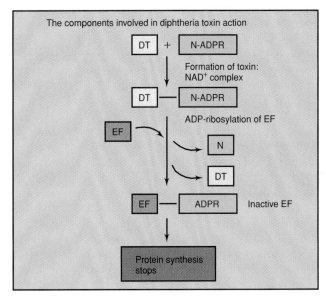

Figure 19.17 Catalysis by diphtheria toxin of attachment of the adenyldiphosphoribosyl (ADPR) portion of NAD^+ to elongation factor 2, leading to inhibition of protein synthesis.

When the iron concentration is reduced to growth-limiting levels, toxin production occurs. The role of iron is to bind to a regulatory protein in *C. diphtheriae* (that is, act as a negative control element) (∞ Section 7.3). The iron-binding protein then combines with a control region of the DNA of β phage and prevents expression of the diphtheria toxin gene. When iron is absent, the regulatory protein does not act, and toxin synthesis can occur. The disease diphtheria is discussed in Section 23.2. This strategy for toxin production may also be found in other microorganisms. For example, exotoxin A of *Pseudomonas aeruginosa* (Table 19.4) has an action quite similar to that of diphtheria toxin, also transferring the ADP-ribosyl portion of NAD^+ to elongation factor 2.

Tetanus and botulinum toxins

These toxins are produced by two species of obligately anaerobic bacteria, *Clostridium tetani* and *C. botulinum*, which are normal soil organisms that occasionally become involved in disease situations in animals. *C. tetani* grows in the body in deep wound punctures that become anaerobic, and although *C. tetani* does not invade the body from the initial site of infection, the toxin it produces can spread and cause severe neurological symptoms that can result in death. *C. botulinum* rarely grows directly in the body, but it does grow and produce toxin in improperly preserved foods. Ingestion of toxin-containing food results in neurological disease and death.

However, in *infant botulism*, infection of the intestinal tract from a *C. botulinum*-containing food product such as raw honey results in chronic infection by the toxin-producing organism. Elaboration of the toxin then causes the disease. *C. botulinum* infection is rare in normal adults because the normal flora and the immune response are developed and prevent colonization of the intestinal tract by the pathogen.

Tetanus toxin is a protein of molecular weight 150,000, containing two polypeptides. On entry into the central nervous system, this toxin becomes fixed to nerve synapses, binding specifically to a ganglioside lipid. This binding blocks the release of glycine, a factor that induces relaxation of the muscles. Thus, the toxin allows constant firing of the motor neurons and continual contraction. This is very different from the normal muscle action pattern (Figure 19.18). Usually, two neurons innervate each muscle fiber. One neuron transmits activation (contraction) signals from the central nervous system. The other neuron transmits inhibition (relaxation) signals. Muscles throughout the body are arranged in opposing pairs. Thus, when one muscle of each pair has received an activation signal and is contracted, the other has received an inhibition signal and is relaxed. However, if tetanus toxin is bound to the inhibitory motor neurons as described above, it blocks the inhibitory signal, resulting in the simultaneous contraction of both of the paired muscles.

The outcome is a spastic, twitching paralysis, with both muscles contracted and opposing one another at the same time. If the muscles of the mouth are involved, the prolonged spasm restricts the mouth's movement, resulting in the condition known as *lockjaw*. If the respiratory muscles are involved, death may be due to asphyxiation.

Botulinum toxin is a series of seven related toxins that are the most poisonous substances known. One milligram of pure botulinum toxin is enough to kill more than 1 million guinea pigs. Of the seven distinct botulinum toxins described, at least two of these are encoded on lysogenic bacteriophages specific for *Clostridium botulinum*. The major toxin is a protein of about 150,000 molecular weight, which readily forms complexes with nontoxic botulinum proteins to give an active form of the toxin of almost 10^6 molecular weight. Toxicity occurs because the toxin binds to presynaptic membranes at the nerve–muscle junction, blocking the release of acetylcholine. Because transmission of the nerve impulse to the muscle is by means of acetylcholine action, muscle contraction is inhibited, causing a flaccid paralysis. The fatality rate from botulism poisoning can approach 100% but can be significantly reduced by quick administration of an antitoxin antibody (∞ Section 20.17) and by use of an artificial respirator to prevent respiratory failure. Death in cases of botulism is usually due to respiratory or cardiac failure. The mode of action of tetanus and botulinum toxins is contrasted in Figure 19.18.

CONCEPT CHECK   19.8

The most potent biological toxins are the exotoxins produced by pathogens. Each exotoxin acts on specific host cells or molecules.

✔ *What key features are shared by all exotoxins?*

✔ *What factors make botulinum toxin so lethal?*

19.9

Enterotoxins

Enterotoxins are exotoxins that act on the *small* intestine, generally causing massive secretion of fluid into the intestinal lumen, leading to the symptoms of diarrhea. Enterotoxins are produced by a variety of bacteria, including the food-poisoning organisms *Staphylococcus aureus*, *Clostridium perfringens*, and *Bacillus cereus*, and the intestinal pathogens *Vibrio cholerae*, *Escherichia coli*, and *Salmonella enteritidis*. The *E. coli* enterotoxin is plasmid-encoded. It is likely that this plasmid also encodes synthesis of the specific surface antigens that are essential for attachment of enteropathogenic *E. coli* to intestinal epithelial cells (see Section 19.6).

PERMANENT RELAXATION | PERMANENT CONTRACTION

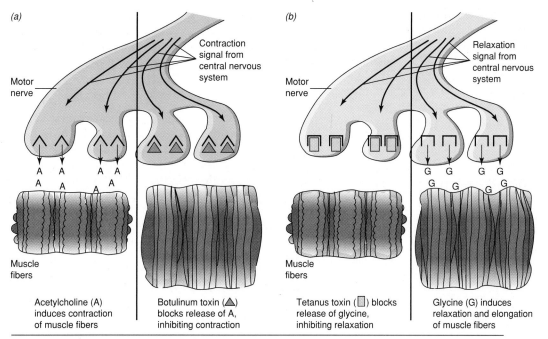

Figure 19.18 Actions of neurotoxins on the motor end plate at the muscle fiber–motor nerve junction. (a) Botulinum toxin from *Clostridium botulinum*. This toxin results in irreversible relaxation and flaccid paralysis in the affected muscles. (b) Tetanus toxin from *Clostridium tetani*. This toxin results in irreversible contraction and spastic paralysis in the affected muscles.

Cholera toxin

The enterotoxin produced by *Vibrio cholerae*, the causal agent of cholera, is the best understood. Cholera toxin is a protein consisting of three polypeptides, the A_1, A_2, and B polypeptide chains of 82,000 total molecular weight. Chains A_1 and A_2 are covalently connected by a disulfide bridge to make a dimer called subunit A, and this is loosely associated with a variable number of B chains. The B subunit contains the binding site by which the cholera toxin combines specifically with the ganglioside GM1 (a complex glycolipid) in the epithelial cytoplasmic membrane (Figure 19.19*a*), but the B subunit itself does not cause an alteration in membrane permeability. Rather, the toxic action is in the A_1 chain, which activates the cellular enzyme *adenyl cyclase,* causing the conversion of adenosine triphosphate (ATP) to cyclic adenosine monophosphate (cAMP).

As we discussed in Section 7.6, cyclic AMP is a specific mediator of a variety of regulatory systems in cells. In mammals, cyclic AMP is involved in the action of a variety of hormones, as well as in synaptic transmission in the nervous system, and in inflammatory and immune reactions of tissues, including allergies. Although the A_1 subunit of cholera toxin is responsible for activation of adenyl cyclase, A_1 must first be activated by a cellular enzyme that requires NAD^+ and ATP. In the action of cholera enterotoxin, the increased cyclic AMP levels bring about the active secretion of chloride and bicarbonate ions from the mucosal cells

into the intestinal lumen. This change in ionic balance leads to the secretion of large amounts of water into the lumen (Figure 19.19*b*). In the acute phase of cholera, the rate of water loss into the small intestine is greater than reabsorption of water by the large intestine, and so massive net fluid loss occurs. Cholera victims generally die from extreme dehydration, and the best treatment for the disease is the oral administration of electrolyte solutions containing solutes (Figure 19.19*c*) to replace the lost fluid and ions.

At the molecular level, cholera enterotoxin has a mode of action (formation of cyclic AMP) identical to that of some normal mammalian hormones, and it has been suggested that cholera toxin may represent an ancestral hormone. Because cholera enterotoxin activates adenyl cyclase in a variety of cells and tissues, pathological manifestations of cholera toxin are related more to the specific site at which it binds, the epithelial cells of the small intestine, than to toxin activation of adenyl cyclase. Indeed, purified B subunits devoid of adenyl cyclase activity can actually *prevent* the action of cholera enterotoxin, if they are administered first, because they bind to the specific cholera receptors on the mucosal cells and block the binding of the complete toxin.

Genetic studies of cholera toxin have shown that the cholera enterotoxin is encoded by two genes, *ctxA* and *ctxB*. Expression of *ctxA* and *ctxB* is controlled by a positive regulatory element, a protein encoded by the *toxR* gene. The *toxR* gene product is a transmembrane protein

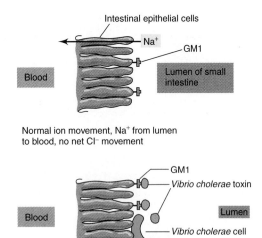

Normal ion movement, Na⁺ from lumen
to blood, no net Cl⁻ movement

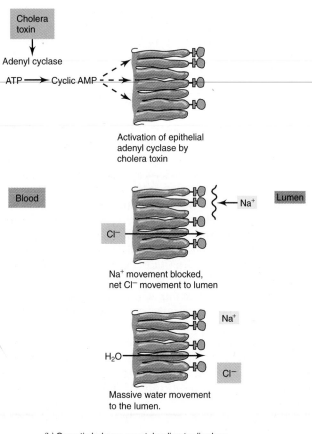

(a) Bacterial colonization of the small intestine. Production
and binding of cholera toxin to GM1.

Activation of epithelial
adenyl cyclase by
cholera toxin

Na⁺ movement blocked,
net Cl⁻ movement to lumen

Massive water movement
to the lumen.

(b) Osmotic balance upset, leading to diarrhea.

(c) **Treatment:** oral solution for cholera therapy (ingredients in g/l):
Glucose, 20; NaCl, 4.2; NaHCO₃, 4.0; KCl, 1.8

Figure 19.19 Action of cholera enterotoxin.

that controls not only cholera toxin production but also several other important virulence factors, such as outer membrane proteins and pili required for successful colonization of *Vibrio cholerae* in the small intestine.

Other enterotoxins

There is good evidence that the enterotoxins produced by enteropathogenic *Escherichia* and *Salmonella* have modes of action similar to that of cholera toxin, and antibody against cholera enterotoxin also inactivates these other enterotoxins, suggesting a similar structure. The sequence of the cholera toxin genes *ctxA and B* further supports this relationship: cholera toxin genes show greater than 75% sequence homology with the genes encoding the heat-labile enterotoxin produced by enteropathogenic *Escherichia coli*. Also, the active component of *Escherichia* enterotoxin is activated by a cellular enzyme system requiring ATP and NAD⁺. As discussed in Section 9.8, *Escherichia* enterotoxin is controlled by a conjugative plasmid, but the enterotoxin gene of *Vibrio cholerae* is chromosomal, although transmissible by conjugation. However, the enterotoxins produced by the food-poisoning bacteria (*Staphylococcus aureus*, *Clostridium perfringens*, *Bacillus cereus*) may be quite different in their modes of action (see Section 20.16 for a discussion of *S. aureus* toxic mechanisms) because their action is at least partly systemic and cannot be explained by alterations in intestinal permeability alone.

CONCEPT CHECK   **19.9**

Enterotoxins are exotoxins that act specifically on the small intestine, causing changes in intestinal permeability that lead to diarrhea. Many food-poisoning microorganisms produce enterotoxins.

✔ *Are all enterotoxins also exotoxins?*

✔ *Do all enterotoxins share a common mechanism of action?*

19.10

Endotoxins

Gram-negative Bacteria produce lipopolysaccharides as part of the outer layer of their cell walls (∞ Figures 3.33 and 3.34), which under many conditions are toxic. These are called **endotoxins** because they are generally cell-bound and released in large amounts only when cells lyse. In most cases, *endotoxin* can be equated with lipopolysaccharide toxin. Endotoxins have been studied primarily in the genera *Escherichia*, *Shigella*, and especially *Salmonella*. The major differences between exotoxins and endotoxins are listed in Table 19.5.

TABLE 19.5	Basic properties of exotoxins and endotoxins	
Property	**Exotoxins**	**Endotoxins**
Chemical properties	Proteins, excreted by certain gram-positive or gram-negative Bacteria; generally heat-labile	Lipopolysaccharide–lipoprotein complexes (see Figures 3.33 and 3.34); released on cell lysis as part of the outer membrane of gram-negative Bacteria; extremely heat-stable
Mode of action; symptoms	Specific; either cytotoxin, enterotoxin, or neurotoxin with defined specific action on cells or tissues	General; fever, diarrhea, vomiting
Toxicity	Highly toxic, often fatal	Weakly toxic, rarely fatal
Immunogenicity	Highly immunogenic; stimulate the production of neutralizing antibody (antitoxin)	Relatively poor immunogen; immune response not sufficient to neutralize toxin
Toxoid potential	Treatment of toxin with formaldehyde will destroy toxicity, but treated toxin (toxoid) remains immunogenic	None
Fever potential	Do not produce fever in host	Pyrogenic, often produce fever in host

Endotoxin structure and function

When injected into an animal, endotoxins cause a variety of physiological effects. *Fever* is an almost universal symptom because endotoxin stimulates host cells to release proteins called *endogenous pyrogens,* which affect the temperature-controlling center of the brain. In addition, however, the animal may develop diarrhea, experience a rapid decrease in lymphocyte, leukocyte, and platelet numbers, and enter into a generalized inflammatory state. Large doses of endotoxin can cause death, primarily through hemorrhagic shock and tissue necrosis. However, the toxicity of endotoxins is much *lower* than that of exotoxins. For instance, in the mouse the amount of endotoxin required to kill 50% of a population of test animals (the so-called LD_{50}) is 200–400 µg per mouse, whereas the LD_{50} for botulinum toxin is about 25 picograms (pg) per mouse, about 10 million times less! (A picogram is 10^{-12} g or 10^{-6} µg.)

The overall structure of lipopolysaccharide (LPS) was diagrammed in Figure 3.34. Lipopolysaccharide consists of lipid A, a core polysaccharide, which in *Salmonella* is the same for many species, consisting of ketodeoxyoctonate, seven-carbon sugars (heptoses), glucose, galactose, and *N*-acetylglucosamine, and the *O-polysaccharide,* a highly variable molecule that usually contains galactose, glucose, rhamnose, and mannose and generally contains one or more unusual dideoxy sugars such as abequose, colitose, paratose, or tyvelose. The sugars of the *O-polysaccharide* are connected in four- to five-sugar sequences (often branched), which then repeat to form the complete molecule (∞ Figures 3.33 and 3.34). Lipid A is not a normal glycerol lipid, but instead the fatty acids are connected by ester linkage to *N*-acetylglucosamine. Fatty acids frequently found in the lipid include β-hydroxymyristic, lauric, myristic, and palmitic acids.

Purification of lipopolysaccharide fractions has shown that it is the *lipid A complex* that is responsible for toxicity and that the polysaccharide acts mainly to render the lipid water-soluble. However, animal studies have shown that the entire endotoxin complex, which contains both polysaccharide and lipid, is required to obtain a response.

Limulus *assay for endotoxin*

Because endotoxins are fever inducers, pharmaceuticals such as antibiotics and intravenous solutions must be endotoxin-free. An endotoxin assay of very high sensitivity has been developed using lysates of amebocytes from the horseshoe crab, *Limulus polyphemus.* Although the mechanism of this assay is not understood, endotoxin specifically causes lysis of amebocytes (Figure 19.20). In a commercial assay, amebocyte extracts are mixed with the solution to be tested. If endotoxin is present, the amebocyte extract gels and precipitates, causing a marked change in turbidity. This reaction can be measured quantitatively with a spectrophotometer. A measurable reaction can be obtained with as little as 10–20 pg/ml of lipopolysaccharide. Apparently the active component of the *Limulus* extract reacts with the lipid component of lipopolysaccharide. The *Limulus* assay has been used to detect the presence of minute quantities of endotoxin in serum, cerebrospinal fluid, drinking water, and fluids used for injection.

The *Limulus* test is so sensitive that considerable care must be taken to avoid contamination of the equipment, solutions, and reagents with the gram-negative Bacteria in the laboratory and clinical environment, for example, as contaminants in the distilled water. In clinical work, detection of endotoxin by the *Limulus* assay in serum or cerebrospinal fluid is presumptive evidence of gram-negative infection of these body fluids.

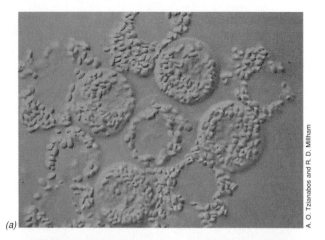

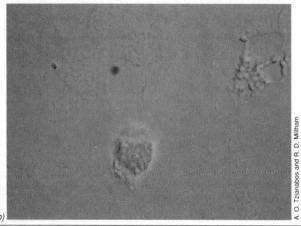

A. O. Tzianabos and R. D. Millham

A. O. Tzianabos and R. D. Millham

Figure 19.20 Photomicrographs of *Limulus* amebocytes. (a) Normal amebocytes. (b) Amebocytes following exposure to bacterial lipopolysaccharide. Treatment with lipopolysaccharide causes degranulation of the cells, and this response can be used as an assay for lipopolysaccharide content.

CONCEPT CHECK   **19.10**

Endotoxins are toxic outer cell wall components derived from gram-negative Bacteria. Host fever is a symptom of endotoxin action.

✔ *Why do gram-positive Bacteria not produce endotoxins?*

✔ *Are endotoxins generally as potent as exotoxins? Why or why not?*

19.11
Virulence

Virulence is the relative ability of a parasite to cause disease. In the last five sections, we described several specific virulence factors, all of which dealt with the ability of a pathogen either to *invade* a host or to cause damage

by producing *toxins*. In this section, we will deal with specific examples of particularly virulent organisms and we will apply our knowledge of virulence factors to explain the virulence of these organisms.

Both *invasiveness* and *toxigenicity* are quantitative properties and may vary over a wide range from very high to very low. An organism that is only weakly invasive may still be virulent if it is highly toxigenic. A good example of this is the organism *Clostridium tetani*. The cells of this organism rarely leave the wound where they were first deposited; yet they are able to bring about death of the host because they produce the potent tetanus exotoxin, which can move to distant parts of the body and initiate paralysis. On the other hand, a weakly toxigenic organism may still be able to produce disease if it is highly invasive. *Streptococcus pneumoniae* is not known to produce any toxin but is able to cause extensive damage and even death because it is highly invasive, being able to grow in lung tissues in enormous numbers and initiate host responses that lead to disturbance of lung function. These two organisms exemplify the extremes of invasiveness and toxigenicity; most pathogens fall somewhere between these two extremes.

We have discussed several virulence factors used by pathogens, including the toxins (see Sections 19.8–19.10). Several other virulence factors have also been identified in pathogenesis. In *Salmonella*, for example, a genus in which genetic studies can be readily done, a variety of virulence factors are known. Toxin production contributes to the virulence of *Salmonella* sp., and at least three toxins are produced: enterotoxin, endotoxin, and *cytotoxin*. Cytotoxin acts by inhibiting host cell protein synthesis, and because it is associated with the cell surface, it may also be involved in *adherence*, which allows *Salmonella* to bind to epithelial cells. Other factors involved in adherence are the cell surface polysaccharide O antigen (∞ Figure 3.33) and the flagellar Vi antigen. Fimbriae may also enhance adherence. *Invasion factors* include the O and Vi antigens. These invasion factors are important because they prevent killing by *phagocytes*, a group of white blood cells that normally ingest and kill bacteria (∞ Section 20.3). *Salmonella* is thought to establish infections through *intracellular parasitism*, the practice of residing in host cells, eventually growing and destroying those cells, and spreading to other cells. A plasmid-borne virulence factor is responsible for intracellular persistence and spread in most species of *Salmonella*. Thus, *Salmonella*, and probably most other pathogens, use several virulence factors simultaneously to initiate infection. Figure 19.21 summarizes the known virulence factors in *Salmonella*.

The virulence of a pathogen can be estimated from experimental studies of the LD_{50}. Highly virulent pathogens frequently show little difference in the number of cells required to kill 100% of the population as compared to the number required to kill 50% of the population. This is illustrated in Figure 19.22 for exper-

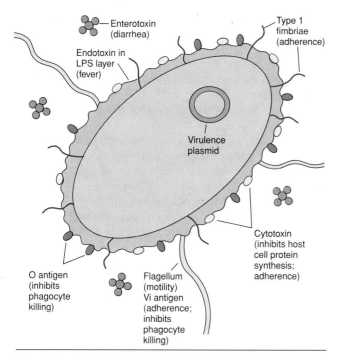

Figure 19.21 Summary of virulence factors important in *Salmonella* pathogenesis. See text for discussion.

imental *Streptococcus* and *Salmonella* infections in mice. Only a few cells of *Streptococcus pneumoniae* are required to establish a fatal infection in mice. Fewer than 100 cells per mouse are necessary to kill every member of a test population once the virulence of a particular strain has been established. In fact, the LD_{50} for this organism is hard to ascertain because so few organisms are needed to produce a lethal infection. By contrast, the LD_{50} for

Salmonella typhimurium, also a mouse pathogen but a much less virulent one, is much higher than for *S. pneumoniae,* and the number of cells required to kill 100% of the population is much higher than the LD_{50}.

When pathogens are kept in laboratory culture and not passed through animals for long periods, their virulence is often decreased or even completely lost. Such organisms are said to be **attenuated.** Attenuation probably occurs because nonvirulent mutants may grow faster and, through successive transfers to fresh media, such mutants are selectively favored. Attenuation often occurs more readily when culture conditions are not optimal for the species. If an attenuated culture is reinoculated into an animal, virulent organisms are sometimes reisolated, but in many cases loss of virulence is permanent. Attenuated strains find frequent use in the production of vaccines, especially viral vaccines (∞ Section 20.17). Measles and mumps vaccines, for example, are composed of attenuated viruses, as is the rabies vaccine given to domesticated animals.

CONCEPT CHECK   **19.11**

Virulence is determined by the invasiveness and toxigenicity of a pathogen. In most pathogens, a number of factors contribute to virulence. Attenuation is loss of virulence.

✔ *Why is* Streptococcus pneumoniae *highly virulent even though it produces no toxins?*

✔ *Suggest a method for producing an attenuated pathogen.*

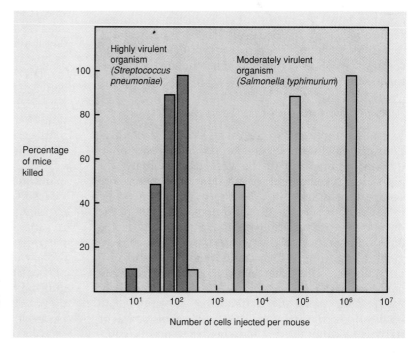

Figure 19.22 An example of differences in microbial virulence as shown by the number of cells of *Streptococcus pneumoniae* or *Salmonella typhimurium* required to kill a mouse population.

19.12

Nonspecific Host Defenses

Many of the mechanisms responsible for the suppression of pathogens are innate "resistance factors." These resistance factors can be divided into two categories: *specific* host defenses, which are directed against individual species or strains of pathogens, and *nonspecific* host defenses, directed against a variety of pathogens. In this chapter we consider the major *nonspecific* host defenses that have been identified as important in preserving the healthy state of the host. In the following chapter we consider *specific* host defenses—the immune response.

Natural host resistance

The ability of a particular pathogen to cause disease in different animal species is highly variable. In *rabies,* for instance, death usually occurs in all species of mammals once symptoms of the disease develop. Nevertheless, certain animal species are much more susceptible to rabies than others. Raccoons and skunks, for example, are extremely susceptible to rabies infection as compared to opossums, which are rarely linked to cases of rabies in wild animals. *Anthrax* infects a variety of animals and causes disease symptoms varying from mild pustules in humans to a fatal blood poisoning in cattle. However, birds are totally resistant to anthrax. Finally, diseases of warm-blooded animals are rarely transmitted to cold-blooded species, and vice versa. Why should this be so?

Resistance to certain diseases and susceptibility to others is an innate property of a given species and is governed by complex and interdependent factors. Differences in physiology and nutrition as well as anatomical differences are important, as is variation in tissue surface receptors, as discussed later. The net result is that different animal species, even very closely related species, may show completely different susceptibilities to the same disease agent.

Age, stress, and diet

Age is an important factor in susceptibility to infectious disease. Infectious diseases are more common in the very young and in the aged. In the infant, for example, development of an intestinal microflora occurs quite quickly, but the normal flora of a young infant is not the same as that of the adult. Before the development of an adult flora, and especially in the days immediately following birth, pathogens have a greater opportunity to become established and produce disease. Thus, diarrhea caused by enteropathogenic strains of *Escherichia coli* (∞ Section 23.13) or *Pseudomonas aeruginosa* is frequently encountered in infants under the age of 1 year. These organisms can be transmitted from the mother where they may be causing no ill effects because they have established a stable residence as part of the mother's flora. The undeveloped state of the infant's microflora provides poor competition for pathogenic species. As we discussed previously, infant botulism is encountered only in very young infants because establishment of the intestinal normal flora in older children precludes intestinal infection with *Clostridium botulinum*, which causes the disease (see Section 19.8).

In individuals over the age of 65, infectious diseases are much more common than in younger adults. For example, the elderly are much more susceptible to respiratory infections, particularly influenza (∞ Section 23.4), probably because of a declining ability to make an effective immune response to respiratory pathogens. In addition, anatomical changes associated with age may also encourage infection. Enlargement of the prostate gland, a common condition in men over the age of 50, frequently leads to a decreased urine flow. This, in turn, allows pathogens to colonize the male urinary tract (Figure 19.12) more readily, leading to an increase in these infections in elderly men.

Stress can predispose a normally healthy individual to disease. In studies with rats and mice, fatigue, exertion, poor diet, dehydration, or drastic climatic changes, all sources of physiological stress, increase the incidence and severity of infectious diseases. For example, rats subjected to intense physical activity for long periods of time show a higher mortality rate from experimental *Salmonella* infections than well-rested control animals. The interaction of hormones that are produced under stress with the immune system may play a role in stress-mediated disease. Hormonal balances change dramatically when an animal is placed under stressful conditions. The hormone *cortisone,* for example, is produced at much higher levels in times of stress than during calm periods, and this hormone is an effective anti-inflammatory agent. Suppression of inflammation removes one of the normal defenses against disease (see Section 19.13).

Diet plays a role in host resistance. The correlation between famine and infectious disease has been known for centuries. Protein shortages may alter the composition of the normal flora, thus allowing opportunistic pathogens a better chance to multiply. For example, cholera is much more prevalent in malnourished individuals than in well-nourished ones. On the other hand, the number of *Vibrio cholerae* required to cause infection is drastically reduced when the *V. cholerae* is ingested in food, presumably because the food neutralizes the stomach acids that would normally destroy the pathogen (∞ Section 23.14). Overeating may be harmful as well. Studies on clostridial diseases of sheep, in particular bloats caused by excessive gas accumulation, indicate that constant overeating affects the composition of the normal flora, leading to massive growth of bacterial species normally present in low numbers.

Not eating a particular substance needed by a pathogen may serve to prevent disease. The best example here is the effect sucrose has on the development of dental caries. As explained in Section 19.3, absence of sucrose from the diet (along with good oral hygiene) virtually eliminates tooth decay. In the absence of sucrose, the highly cariogenic bacteria *Streptococcus mutans* and *S. sobrinus* are unable to synthesize the gummy outer surface polysaccharide needed to keep the bacterial cells attached to the teeth.

Anatomical defenses

The structural integrity of tissue surfaces poses a barrier to penetration by microorganisms. In the skin and mucosal tissues potential pathogens must not only bind to tissue surfaces but also grow at these sites before traveling elsewhere in the body. Intact surfaces form an effective barrier to colonization, but microbial access to damaged surfaces is more easily obtained. Resistance to colonization and invasion is due to the production of host defense substances and to various mechanical actions that disrupt colonization. A summary of the major anatomical defenses is shown in Figure 19.23.

The **skin** is an effective barrier to the penetration of microorganisms. *Sebaceous glands* in the skin (Figure 19.2) secrete fatty acids and lactic acid, which lower skin pH and inhibit colonization of pathogenic bacteria. Microorganisms inhaled through the *nose* or *mouth* are removed by the action of ciliated epithelial cells in the mucous surfaces of the nasopharynx and tracheal regions. Cilia push bacterial cells upward until they are caught in oral secretions and either are expectorated or are swallowed and killed in the stomach. Potential pathogens entering the host via the oral route must first survive the *acidity* of the stomach (which is about pH 2) and then successfully compete with the increasingly abundant resident microflora present in the small intestine (which is about pH 5) and finally in the large intestine (pH 6–7). The latter organ contains bacterial numbers of about 10^{10} per gram of intestinal contents in a normal adult (see Section 19.4).

In a healthy adult, the kidney and the surface of the eye are constantly bathed with secretions containing *lysozyme* that markedly reduce microbial populations. Extracellular fluids such as blood plasma also contain bactericidal substances. For example, blood proteins called β-lysins bind and destroy microbial cells. β-Lysins are basic proteins that act by disrupting the bacterial cytoplasmic membrane, leading to leakage of cytoplasmic constituents and cell death.

However effective these defenses may be, damage to physical barriers and changes in other nonspecific defenses can quickly lead to growth of the pathogen and initiation of disease.

Tissue specificity

Most pathogens must first establish themselves at the site of infection. If the site is not compatible with their nutritional and environmental needs, the organisms cannot multiply. Thus, if *Clostridium tetani* were ingested, it would not bring about tetanus because the pathogen is killed by the acidity of the stomach. If, on the other hand, *C. tetani* cells were introduced into a deep wound, the organism would grow in the anaerobic zones created by localized tissue destruction and produce tetanus toxin (see Section 19.8). By contrast, enteric bacteria such as *Salmonella* and *Shigella* do not cause wound infections but successfully colonize the intestinal tract. Table 19.6 summarizes a number of examples of tissue specificity.

The compromised host

The term *compromised host* refers to hosts in which one or more resistance mechanisms are malfunctioning and in which the probability of infection is therefore increased.

Hospital patients are often compromised hosts. Many hospital procedures such as catheterization, hypodermic injection, spinal puncture, and biopsy can also introduce pathogens into the patient. Surgical procedures expose highly susceptible parts of the body to sources of contamination. The stress of surgery also diminishes the resistance of the patient to infection. Finally, in organ transplant procedures (∞ Section 20.7), drugs are used that suppress the immune system to prevent rejection of the transplant. Immunosuppressive drugs greatly increase susceptibility to infection. Thus, many hospital patients with noninfectious primary ailments (for example, cancer and heart disease) die of microbial infection because they are compromised hosts (∞ Section 22.7).

Compromised hosts exist even outside the hospital. Smoking, excess consumption of alcohol, intravenous drug usage, lack of sleep, poor nutrition, and infection itself are condition that compromise a host.

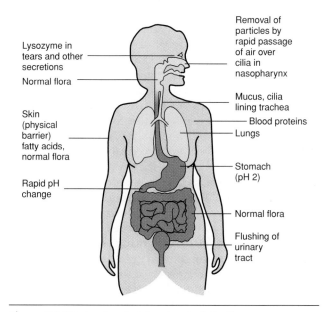

Figure 19.23 Anatomical barriers to infection.

Lysozyme in tears and other secretions

Normal flora

Skin (physical barrier) fatty acids, normal flora

Rapid pH change

Removal of particles by rapid passage of air over cilia in nasopharynx

Mucus, cilia lining trachea

Blood proteins

Lungs

Stomach (pH 2)

Normal flora

Flushing of urinary tract

TABLE 19.6	Tissue specificity as a factor in infectious disease	
Disease	Tissue infected	Organism
Diphtheria	Throat epithelium	*Corynebacterium diphtheriae*
Gonorrhea	Urogenital epithelium	*Neisseria gonorrhoeae*
Cholera	Small intestine epithelium	*Vibrio cholerae*
Pyelonephritis	Kidney medulla	*Proteus* sp.
Dental caries	Oral epithelium	*Streptococcus mutans, S. sobrinus, S. sanguis, S. mitis*
Spontaneous abortion (cattle)	Placenta	*Brucella abortus*
Acquired immunodeficiency syndrome (AIDS)	T helper lymphocytes	Human immunodeficiency virus (HIV)
Malaria	Blood (erythrocytes)	*Plasmodium* sp.

For instance, the virus causing acquired immunodeficiency syndrome (AIDS) destroys one type of cell involved in the immune response (T helper cells). Therefore, AIDS patients are unable to mount effective resistance to infection; death is generally due to some infectious agent (∞ Sections 22.4 and 23.7).

Finally, there are certain genetic conditions that may compromise the host, such as genetic diseases that eliminate important parts of the immune system. Individuals with such conditions frequently die at an early age, not from the genetic condition itself but from microbial infection.

CONCEPT CHECK 19.12

Nonspecific physical, anatomical, and chemical barriers prevent colonization of the host by most pathogens. Breakdown in these defenses results in a compromised host who is more susceptible to infection.

✔ *How can diet influence host resistance to a pathogen?*

✔ *How might smoking compromise an otherwise healthy host?*

19.13
Inflammation and Fever

Inflammation is a general nonspecific reaction to foreign particles and other noxious stimuli such as toxins and pathogens. The characteristic inflammatory response results in redness, swelling, pain, and heat, all localized at the site where the host contacted the noxious stimuli (∞ Section 23.3 and Figure 20.32). The mediators of inflammation include a group of proteins called *cytokines* (∞ Section 20.8), which are produced by white blood cells or *leukocytes* (∞ Section 20.3). Leukocytes are also involved in pathogen-specific responses to noxious stimuli in the immune response, which we will discuss in

Chapter 20. The most important outcome of the inflammatory response is the immediate localization of the noxious agent, often by the production of a fibrin clot at the site of inflammation.

Inflammation is one of the most important and ubiquitous aspects of host defense against invading microorganisms. However, inflammation is also an important aspect of microbial pathogenesis because the inflammatory response elicited by an invading microorganism can result in considerable host damage.

Fever

The healthy human body maintains a surprisingly constant temperature. Over an average 24-hr period, body temperature varies over the narrow range of 1–1.5°C. However, individuals vary in their "normal" temperatures, and although 37°C is considered the standard normal temperature, the actual normal temperature in some individuals may be as low as 36°C or as high as 38°C. Also, body temperature varies with the amount of physical activity and can be as much as 2°C below normal in sleep and as much as 4°C above normal during strenuous exercise.

Fever is defined as an abnormal increase in body temperature. Although fever can be caused by noninfectious disease, most fevers are caused by infection. At least one reason why fever occurs during many infections is that certain products of pathogenic organisms are *pyrogenic* (fever-inducing). The most well-studied pyrogenic agents are the endotoxins of gram-negative Bacteria (see Section 19.10). However, many organisms that do not produce endotoxins are able to cause fever on infection. In these organisms, proteins known as *endogenous pyrogens* are released when leukocytes destroy them (∞ Section 20.3). Slight temperature increases benefit the host by accelerating phagocytic and antibody responses, while strong fevers of 40°C (104°F) or greater may benefit the pathogen if host tissues are further damaged.

Three kinds of characteristic fever patterns have been recognized in infectious disease. (1) *Continuous fever* is that condition in which the body temperature remains elevated over a whole 24-hr period and the total range of variation in temperature is less than 1°C. Continuous fever is seen in *typhoid fever* (∞ Section 23.13) and *typhus fever* (∞ Section 23.9). (2) A *remittent fever* is one in which the body temperature is abnormal over the whole of a 24-hr period and the daily range shows swings greater than 1°C. This occurs in some *pyogenic infections* (∞ Section 23.2) and in *tuberculosis* (∞ Section 23.3). (3) An *intermittent fever* is one in which the temperature is normal for part of the day and then rises above normal. Most infectious diseases elicit some intermittent fever, and the condition is a diagnostic characteristic of malaria (∞ Section 23.11), a protozoan infection. *Relapsing fever,* caused by various *Borrelia* species (∞ Sections 23.9 and 16.12) is an intermittent fever in which the temperature remains normal for a long period of time, followed by a new attack of fever. This is characteristic of an incomplete recovery from an infectious disease, the fever arising when the infection periodically reestablishes itself.

CONCEPT CHECK   19.13

Inflammation and fever are nonspecific responses to noxious stimuli such as pathogens. These host responses can result in accelerated isolation and destruction of the pathogen.

✔ *Describe the chief symptoms of inflammation.*

✔ *Describe the three types of fever.*

Material for Review

REVIEW QUESTIONS

1. Distinguish between a parasite and a pathogen. Distinguish between infection and disease.

2. Which parts of the human body are normally heavily colonized with microorganisms? Which body parts are normally devoid of microorganisms?

3. Distinguish between the resident and transient microorganisms at a body site. How could you distinguish between resident and transient microorganisms experimentally?

4. Why are members of the genus *Streptococcus* instrumental in forming dental caries?

5. What region of the gastrointestinal tract has the highest concentrations of bacteria? What region has the lowest concentrations? Why?

6. Describe the relationship between *Lactobacillus acidophilus* and glycogen in the vaginal tract.

7. Give two examples of adherence factors important for pathogen attachment. At least one example should not be a protein.

8. What do hyaluronidase, collagenase, streptokinase, and coagulase have in common? What is the mode of action of each in promoting disease?

9. Define and contrast exotoxin, enterotoxin, and endotoxin. Give two examples of each and the name of an organism producing each.

10. For each of the exotoxins listed below, describe (i) the producing organism, (ii) the mode of action in the host, (iii) its role in pathogenicity, and (iv) how its effects can be counteracted. (a) Diphtheria toxin, (b) tetanus toxin, (c) botulinum toxin, (d) cholera toxin.

11. Give an example of a microorganism that is pathogenic almost solely because of its toxin-producing ability. Give an example of a microorganism that is pathogenic almost solely because of its invasive characteristics.

12. How do temperature and pH work to limit bacterial infections? What organisms might be susceptible to either of these agents?

13. Distinguish between a continuous fever and an intermittent fever. Which type most commonly occurs in infectious diseases?

APPLICATION QUESTIONS

1. Describe experiments to demonstrate the effects of mucus in protection against bacterial colonization.

2. What steps are involved in the formation of dental plaque? Describe and discuss experiments that demonstrate the buildup of plaque on toothlike surfaces and discuss experiments designed to illustrate biological methods for removal of plaque.

3. Obligately anaerobic bacteria are very common in the large intestine, yet they are able to grow there only if facultatively aerobic bacteria are also present. Explain. How could you test the validity of your answer in the laboratory?

4. Certain antibiotics, even antibiotics whose mode of action is bacteriostatic instead of bactericidal, sterilize

the intestinal tract. How could a bacteriostatic antibiotic bring about this result?

5. Design an experiment to demonstrate the likely route of infection of a urinary tract pathogen in a catheterized patient.

6. Describe how enteropathogenic strains of *Escherichia coli* differ from normal strains of *E. coli*. Include a discussion of structural and ecological variables.

7. Although mutants incapable of producing exotoxins are relatively easy to isolate, mutants incapable of producing endotoxin are much harder to isolate. From what you know of the structure and function of these types of toxins, explain the differences in mutant recovery.

8. Should fever always be treated? Give reasons for your answer based on your knowledge of the importance of the inflammatory response in limiting the spread of infection.

SUPPLEMENTARY READINGS

Ayoub, E. M., G. H. Cassell, W. C. Branche, Jr., and **T. J. Henry** (eds.). 1990. *Microbial Determinants of Virulence and Host Response.* American Society for Microbiology, Washington, DC. Coverage of the molecular biology of factors involved in bacterial and fungal pathogenicity.

Brooks, G. F., J. S. Butel, and **L. N. Ornston.** 1995. *Jawetz, Melnick and Adelberg's Medical Microbiology,* 20th edition. Appleton and Lange, Norwalk, CT. A complete, short medical text covering many aspects of medical microbiology. Very current.

Friedman, H., T. W. Klein, M. Nakano, and **A. Nowotny** (eds.). 1990. *Endotoxin.* Plenum Press, New York. An advanced treatise on endotoxin structure and biological activity.

Henry, J. B. 1991. *Clinical Diagnosis and Management by Laboratory Methods,* 18th edition. W. B. Saunders, Philadelphia. A standard technical medical reference source.

Howard, B. J., J. F. Keiser, T. F. Smith, A. S. Weissfeld, and **R. C. Tilton.** 1994. *Clinical and Pathogenic Microbiology,* 2nd edition. C. V. Mosby, St. Louis. An excellent source of up-to-date diagnostic and metabolic information on pathogenic bacteria.

Iglewski, B., and **V. L. Clark** (eds.). 1990. *Molecular Basis of Bacterial Pathogenesis.* Academic Press, New York. A multiauthored volume on mechanisms of bacterial pathogenesis, especially in gram-negative Bacteria.

Koneman, E. W., S. D. Allen, W. M. Janda, P. C. Schreckenberger, and **W. C. Winn, Jr.** 1992. *Color Atlas and Textbook of Diagnostic Microbiology,* 4th edition. J. B. Lippincott, Philadelphia. A diagnostic, medical reference atlas with excellent illustrations.

Kreier, J. P., and **R. F. Mortensen.** 1990. *Infection, Resistance, and Immunity.* Harper & Row, New York. An elementary textbook of host–parasite relationships and immune mechanisms.

Murray, P. R., E. J. Baron, M. A. Pfaller, F. C. Tenover, and **R. H. Yolken.** 1995. *Manual of Clinical Microbiology,* 6th edition. American Society for Microbiology, Washington, DC. A very basic, thorough treatment of medical microbiology, including the science behind clinical information. An indispensable reference source for the clinical microbiologist.

Schaechter, M., G. Medoff, and **D. Schlessinger** (eds.). 1993. *Mechanisms of Microbial Disease,* 2nd edition. Williams and Wilkins, Baltimore, MD. This medical school textbook has a good section on host–parasite relationships.

Wachsmuth, I. K., P. A. Blake, and **O. Olsvik** (eds.). 1994. *Vibrio cholerae and Cholera: Molecular to Global Perspectives.* American Society for Microbiology, Washington, DC. A very modern, thorough treatment of an ancient epidemic human disease.

On~line resources for this chapter are on the World Wide Web at: http://www.prenhall.com/~brock (click on the table of contents link and then select Chapter 19).

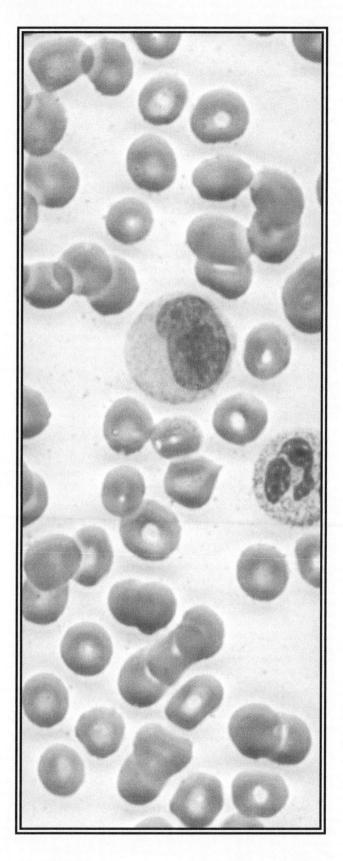

CHAPTER *20*

Concepts of Immunology

The immune system is composed of several different cell types that interact to elicit an immune response. Shown is a photomicrograph of human blood showing red blood cells and two phagocytic cells (with violet-stained nuclei). Phagocytes contact, ingest, and destroy foreign cells and cell products and also initiate the immune response.

MINIGLOSSARY for Chapter 20

Antibody a soluble protein, produced by B cells, that interacts with antigen; also called immunoglobulin

Antigen a molecule capable of interacting with specific components of the immune system

Antigenic Determinant that portion of an antigen that is reactive with a specific antibody or T cell receptor; also called an epitope

Antigen-Presenting Cell (APC) any cell that functions primarily to present antigen to a T cell

Autoantibody an antibody that reacts to self antigens

B Cell a lymphocyte that produces immunoglobulin

Cell-Mediated Immunity (CMI) immunity resulting from direct interaction with antigen-specific T cells

Class I MHC Proteins antigen-presenting molecules found on all nucleated vertebrate cells

Class II MHC Proteins antigen-presenting molecules found primarily on macrophages, B cells, and other antigen-presenting cells

Clonal Selection a theory that each B or T cell, when stimulated by antigen, produces copies of itself

Complement a series of proteins that react in a sequential manner with antibody–antigen complexes to amplify or potentiate their activity

Cytokine a soluble immune response modulator produced by leukocytes

Domain a region of a protein having a distinct function

Hapten a low-molecular-weight substance that combines with specific antibodies but that is incapable of eliciting an immune response by itself

Humoral Immunity immunity resulting from direct interaction with antibodies

Hybridoma an artificially fused product of two unrelated cells that exhibits properties of both cells; used to produce monoclonal antibodies

Hypersensitivity an immune response leading to damage to host tissues, sometimes referred to as allergies

Immunity the ability of an organism to resist infection

Immunodeficient having a dysfunctional or completely nonfunctional immune system

Immunogen a molecule capable of eliciting an immune response

Immunoglobulin (Ig) a soluble protein, produced by B cells, that interacts with antigens; also called antibody

Immunological Memory ability to rapidly produce large quantities of specific immune cells or antibodies after subsequent exposure to a previously encountered antigen

Leukocytes nucleated cells found in the blood (white blood cells)

Lymph a fluid similar to blood but which lacks red blood cells and travels through a separate circulatory system (the lymphatic system) containing lymph nodes, which filter out particulate materials such as bacterial cells

Lymphocytes a subset of nucleated cells found in the blood that are involved in the immune response

Macrophage a type of large leukocyte that has phagocytic properties

Major Histocompatibility Complex (MHC) a genetic complex responsible for encoding several cell surface proteins important in antigen presentation

Monoclonal Antibody an antibody that is the product of a single B cell clone

Natural Killer (NK) Cell a specialized lymphocyte that recognizes and destroys foreign cells or infected host cells in a nonspecific manner

Neutrophil a circulating phagocyte also known as a polymorphonuclear leukocyte (PMN)

Plasma the liquid portion of the blood with cells removed and clotting proteins deactivated

Polyclonal Antiserum a serum containing antibodies derived from many different B cell clones, as occurs in a normal immune response

Polymorphonuclear Leukocyte (PMN) a type of leukocyte exhibiting phagocytic properties, a granular cytoplasm, and a multilobed nucleus. A neutrophil

Primary Antibody Response antibodies made on first exposure to antigen; mostly of the class IgM

Secondary Antibody Response antibody made on second (or any subsequent) exposure to antigen; mostly of the class IgG

Serology the study of antigen–antibody reactions *in vitro*

Serum the liquid portion of the blood with clotting proteins and cells removed

Specificity the ability of the immune response to interact with individual antigens

T Cell a lymphocyte responsible for antigen-specific cellular interactions

T Cell Receptor (TCR) antigen-specific receptor protein on the surface of T cells

Tolerance inability to make an immune response to specific antigens

Vaccination inoculation of a host with inactive or weakened pathogens or pathogen products to stimulate protective immunity

In the previous chapter, we discussed some general mechanisms that the human body uses to prevent colonization and infection by microorganisms. In this chapter, we will examine the events that take place when physical and chemical host defense factors are not sufficient to destroy invading microorganisms.

We will first consider the cells and mechanisms responsible for **immunity,** the ability of an organism to resist infection. Next, we will examine *nonspecific immunity,* the general ability of certain cells to resist most pathogenic viruses, bacteria, and fungi. Nonspecific immunity is the first line of defense after pathogens break through the anatomical barriers discussed in Chapter 19. Nonspecific reactions act as triggers for the *specific immune response,* a highly sophisticated mechanism found in vertebrates for developing immunity to individual pathogens. A specific immune response is elicited by a wide variety of individual molecules that are foreign to the vertebrate host. These foreign molecules, collectively known as **immunogens,** are usually macromolecular components of the pathogens, such as surface proteins. Foreign molecules are known as **antigens** when they are recognized by the immune system.

Antigens are passed along to antigen-specific cells, the *T lymphocytes* or *T cells,* by a process called *antigen presentation.* The T cells can interact directly with the antigen (*cell-mediated immunity*) or they can pass information on to a second group of antigen-specific cells, the *B lymphocytes* or *B cells,* which then make proteins called *immunoglobulins* or *antibodies.* The antibodies, generally found as soluble proteins in serum or body secretions, react with the antigen and help destroy or neutralize it. Antibody-mediated immunity is known as *humoral immunity.*

The specific immune response has three major characteristics: specificity, memory, and tolerance. First, the **specificity** of the antigen–antibody or antigen–T cell interaction is unlike the other host resistance mechanisms we have discussed previously. Anatomical and nonspecific host responses develop immediately against virtually any invading microorganism, even those the host has never before encountered. In the specific immune response, however, each new microorganism must interact with the immune system *before* a response occurs. In most cases, no specific immune response can be detected for several days after the first contact with the pathogen. However, when the immune response occurs, it is directed solely toward that particular microorganism.

Second, once the immune system produces a specific type of antibody or activated T cell, challenge by further exposure to the same microorganism results in the rapid production of large amounts of the same antibody or large numbers of antigen-reactive T cells. The specific immune effectors, either T cells or antibodies, then interact with the invader and destroy it. This capacity for responding to challenge after additional exposure to a pathogen is known as **immunological memory.** Memory allows the host to resist reinfection by specific pathogens that have been previously encountered. We take advantage of this principle by employing the procedure of **vaccination,** the practice of inoculating the host with inactive or weakened pathogens to artificially stimulate immunity and actively enhance specific protection against individual pathogens.

Finally, **tolerance,** the inability to make an immune response to certain antigens, occurs because macromolecules in the host are also potential antigens. Tolerance, like the immune response, is learned by antigen exposure. However, in this case, the immune systems learns to *not* recognize the host antigens. Host molecules would be damaged if they were recognized by specific antibodies and activated T cells. Through tolerance, the host immune response distinguishes between foreign macromolecules (nonself and potentially dangerous) and host macromolecules (self and not dangerous) and interacts appropriately with them.

Figure 20.1 is an overview of the humoral and cell-mediated aspects of specific immunity. We will first discuss the cells and molecules of the immune system and the molecular basis for antigen recognition. Then we will explain how the individual components of the immune system interact to produce the specific immunity. Next, we will look at instances where this process breaks down and the host acquires diseases resulting from inappropriate immune responses. Finally, we will examine specific instances where we can use the immune response to stimulate protection against a pathogen and avoid a serious infection or disease.

20.1
Cells and Organs of the Immune System

Both nonspecific and specific immunity result from the actions of cells found circulating in the blood and lymphatic systems. In this section, we will discuss some of the important cells involved in immunity and explore their origin and functions in the body. Many of the substances and cells involved in the immune response are found in the *blood* and *lymph,* two major body fluids that directly or indirectly interact with every major organ system.

Blood and lymph

Blood consists of cellular and noncellular components. Changes in blood properties and constituents are sensitive reflections of body changes, including disease states. Blood also contains many of the cells and molecules involved in the immune response. Because blood can be easily and safely obtained from patients, it is a valuable source of material for clinical analytical pro-

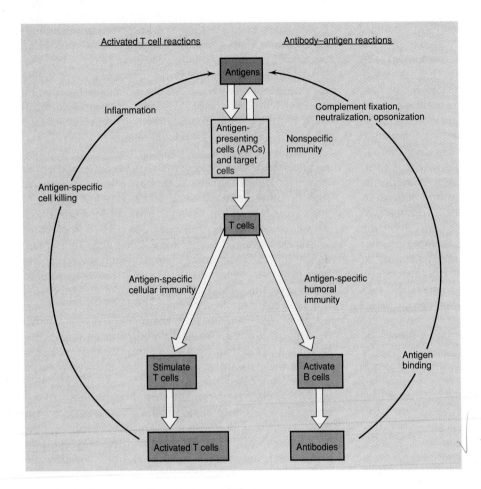

Figure 20.1 Overview of the immune response.

cedures, including most immune response assays. The most numerous cells in human blood are *erythrocytes* (red blood cells), which are nonnucleated cells that function to carry oxygen from the lungs to the tissues (Table 20.1). White blood cells, or *leukocytes,* include a variety of phagocytic cells such as *monocytes,* as well as cells called *lymphocytes,* which are involved in antibody production and cell-mediated immunity. Erythrocytes outnumber leukocytes by roughly a factor of 1000. Platelets are small cell-like constituents that lack a nucleus and play an important role in preventing leakage of blood from damaged blood vessels.

TABLE 20.1	Major formed elements in normal human blood

Cell type	Cells per milliliter
Erythrocytes	$4.2–6.2 \times 10^9$
Leukocytes	$4.5–11 \times 10^6$
Lymphocytes	$1.0–4.8 \times 10^6$
Monocytes	Up to 8.0×10^5
Platelets	$1.5–4.0 \times 10^8$

Source: Henry, J. B. 1991. *Clinical Diagnosis and Management by Laboratory Methods,* 18th edition. W. B. Saunders, Philadelphia.

Platelets clump together to form a temporary plug in a damaged vessel until a permanent clot forms through the action of various clotting agents, some of which are released from the platelets themselves. **Lymph** is a fluid similar to blood but lacks red blood cells.

All the blood and lymph cells and elements have a common origin. As shown in Figure 20.2, common stem cells in the bone marrow are the progenitors of all the mature cells. Stem cells differentiate to produce mature cells largely because of the influence of a group of soluble cell proteins known as **cytokines** (see Section 20.8).

When cells and platelets are removed from blood, the remaining fluid is called *plasma.* An important component of plasma is the protein fibrinogen, which undergoes a complex set of reactions during the formation of a fibrin clot. Clotting can be prevented by the addition of an anticoagulant such as potassium oxalate, potassium citrate, or heparin. Plasma is stable only when such an anticoagulant is added. If no anticoagulant is added, whole blood or plasma quickly forms a clot. The fluid left behind is called *serum.* Serum consists of proteins and the other noncellular components of plasma, except for fibrin. Since serum contains a high concentration of antibody proteins, it is widely used in immunological investigations (∞ Section 21.4).

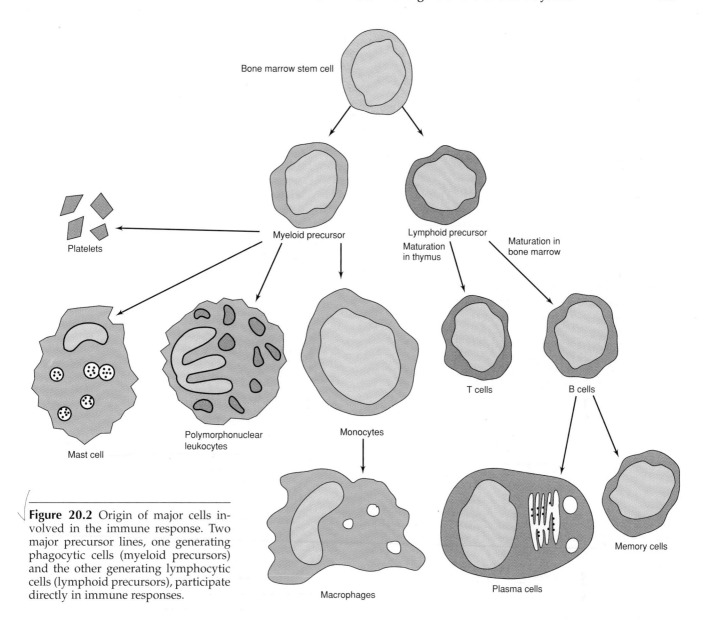

Figure 20.2 Origin of major cells involved in the immune response. Two major precursor lines, one generating phagocytic cells (myeloid precursors) and the other generating lymphocytic cells (lymphoid precursors), participate directly in immune responses.

Blood is pumped by the heart through a network of arteries and capillaries to various parts of the body and is returned through the veins (Figure 20.3). The circulatory system carries not only nutrients (including O_2) but also the components of the blood involved in host resistance to infection. Figure 20.3 shows the capillary beds, a site where leukocytes may pass to and from the blood into the *lymphatic system,* a separate circulatory system through which lymph circulates.

Lymph drains from extravascular tissues into lymphatic capillaries and then into **lymph nodes** (Figure 20.3*d*) found at various locations throughout the lymph system. Lymph nodes filter out microorganisms and other foreign particulate materials. The spleen serves an analogous function in the blood circulatory system. As a result of this filtration activity, lymph nodes may become sites of infection because organisms collected by the filtering mechanisms may proliferate if they are

not destroyed. The lymph nodes are also the sites of most immune responses. Lymph eventually flows back into the circulatory system via the thoracic lymph duct.

CONCEPT CHECK   **20.1**

All the cells involved in immunity originate from a common stem cell. The blood and lymph circulation systems include cellular and noncellular elements that are important for physiological functions, including the cells and organs of the immune system. A variety of leukocytes participate in immune responses.

✔ *What cells are most prevalent in the blood?*

✔ *Describe the circulation of a leukocyte from the blood to the lymph and back to the blood.*

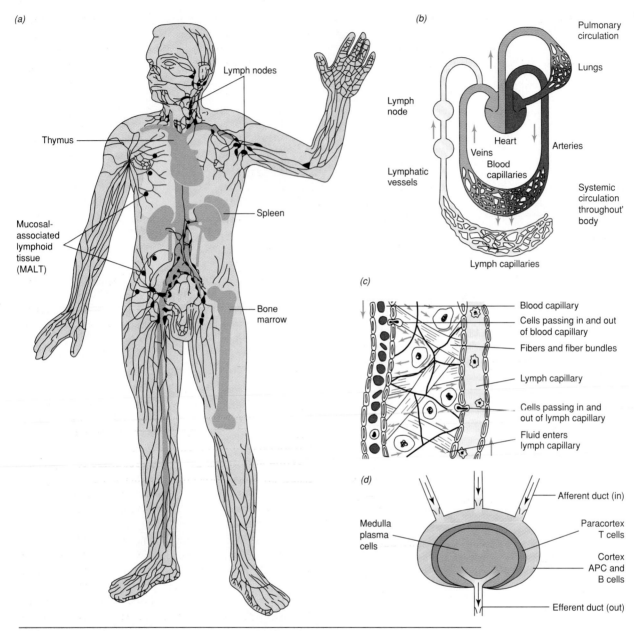

(a)

Lymph nodes

Thymus

Mucosal-
associated
lymphoid
tissue
(MALT)

Spleen

Bone
marrow

(b)

Pulmonary
circulation

Lungs

Lymph
node

Heart

Veins

Arteries

Lymphatic
vessels

Blood
capillaries

Systemic
circulation
throughout'
body

Lymph capillaries

(c)

Blood capillary

Cells passing in and out
of blood capillary

Fibers and fiber bundles

Lymph capillary

Cells passing in and
out of lymph capillary

Fluid enters
lymph capillary

(d)

Afferent duct (in)

Medulla
plasma
cells

Paracortex
T cells

Cortex
APC and
B cells

Efferent duct (out)

Figure 20.3 The blood and lymph systems. (a) Overall view of the major lymph systems, showing the locations of major organs. (b) Diagrammatic relationship between the lymph and blood systems. Blood flows from the veins to the heart, then to the lungs where it becomes oxygenated, and then through the arteries to the tissues. (c) Connection between the blood and lymph systems is shown microscopically. Both blood and lymph capillaries are closed vessels, but cells and fluids can pass from one vessel to another by a process known as extravasation. (d) A lymph node. The diagram depicts major anatomical areas and the immune cells present.

Leukocytes, or white blood cells, are nucleated cells in the blood and the lymph. There are several distinct kinds of leukocytes (Table 20.1), but all participate in nonspecific or specific immune functions (see Sections 20.3 and 20.4). Specialized white blood cells called **macrophages** are found in abundance in the lymph nodes and carry out the lymph filtering action,

as will be described later (see Section 20.3). **Lympho-cytes,** another special type of cell found within the lymphatic system (see Section 20.4), are involved in the specific immune response. They also concentrate in the lymph nodes and spleen and interact with the macrophages. Lymphocytes and some other leukocytes can travel throughout the body and pass freely from blood to interstitial spaces to lymph and back, a process called *extravasation*. (Figure 20.3*c*). Later we will examine these leukocyte types and some of the ways in which each functional leukocyte contributes to the overall immune response.

20.2
Immunogens and Antigens

Immunogens are substances that, when administered to an animal in the appropriate manner, induce an immune response. The immune response may involve antibody production, the activation of specific immunologically competent cells (called *activated T cells*), or both. **Antigens** are substances that react with either antibodies or antigen-specific receptors known as **T cell receptors (TCRs)** that are found on T cells. Most antigens are also immunogens. However, some substances recognized by immune systems are not true immunogens. For example, **haptens** are low-molecular-weight substances that combine with specific antibody molecules but do not by themselves induce antibody formation. Haptens include such molecules as sugars, amino acids, and small polymers.

An enormous variety of macromolecules that are foreign to the host can act as immunogens under appropriate conditions. These include complex macromolecules, among them virtually all proteins and lipoproteins, many polysaccharides, some nucleic acids, and certain teichoic acids. One important requirement is that the molecules must be of fairly high molecular weight, usually greater than 10,000. However, the antibody or TCR does not interact with the antigenic macromolecule as a whole but only against distinct portions of the molecule that are called its **antigenic determinants** or **epitopes** (Figure 20.4). Chemically, antigenic determinants include sugars, amino acid side chains, organic acids and bases, hydrocar-bons, and aromatic groups. Thus, the haptens mentioned previously are actually examples of individual antigenic determinants. Antibodies are formed most readily to determinants that project from the foreign molecule or to terminal residues of a polymer chain. In proteins, for example, the majority of antibodies react with accessible surface determinants. A region of as few as four or five amino acids can define an antigenic determinant on a protein. Also, the surface of a protein can and frequently does have many overlapping antigenic determinants. A cell or virus is a mosaic of proteins, polysaccharides, and other macromolecules, each of which is a potential antigen. Each antigen of the cell is also a mosaic of side chains and residues, each of which is a potential antigenic determinant. The antigenic components of a typical microorganism are thus extremely complex.

In general, the specificity of antibodies is comparable to that of enzymes, which are also able to distinguish between closely related substances. For instance, antibodies can distinguish between the sugars glucose and galactose, which differ only in the orientation of a hydroxyl group. However, specificity is not absolute, and an antibody may react, at least to some extent, with other epitopes. The antigen that induced the antibody is called the **homologous antigen,** and other antigens that react with the antibody are called **heterologous antigens.** The interaction between an antibody and a heterologous antigen is called a *cross-reaction.*

While antibodies generally recognize epitopes expressed on macromolecular surfaces, TCRs recognize determinants only after the macromolecules have been partially degraded. This degraded or "processed" antigen is then presented to T cells on the surface of specialized antigen-presenting cells (APCs) or target cells (see Section 20.7 and Figure 20.16). Since antigen processing and presentation normally destroy the conformational structure of an antigen, T cell epitopes generally consist of sequential linear portions of macromolecules rather than the conformational epitopes recognized by antibodies.

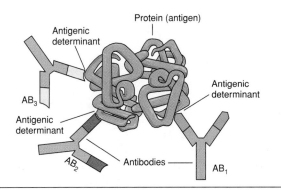

Figure 20.4 Antigens and antigenic determinants for antibodies. Antigens may contain several different antigenic determinants, each capable of reacting with a specific antibody.

CONCEPT CHECK   *20.2*

Immunogens are foreign molecules of sufficient size, complexity, and accessibility that are recognized by the immune system. Epitopes are the individual molecular sites at which immune molecules interact with antigens. Antibody epitopes are native antigen structures, whereas T cell epitopes are generally linear.

✔ *Distinguish among* immunogens, antigens, *and* haptens.

✔ *Why do* antibody *and* TCR *epitopes differ?*

20.3

Nonspecific Immunity: Phagocytes and Phagocytosis

On rare occasions, pathogens break through the host physical and chemical defense mechanisms described in Chapter 19 (∞ Section 19.12). The pathogen can then access host tissues and begin to colonize and infect the host (∞ Section 19.7). At this point, the immune system must become mobilized.

The starting point for immunity, whether the final effect is specific or nonspecific, cellular or humoral, is contact of a cell with the pathogen or immunogenic protein. The cell type involved in this initial contact is a *phagocyte* (literally, a cell that eats). The primary function of a phagocyte is to engulf and destroy pathogens and digest their remains. In this process, immunogens are often generated and presented to antigen-specific immune cells (see Section 20.7). In this section, we will examine some of the important phagocytic cells and discuss their cytolytic and antigen-processing functions.

Phagocytes

Some of the leukocytes found in blood are phagocytes, and phagocytes are also found in various tissues and fluids of the body. Phagocytes are usually motile and move by ameboid action. Most have granular inclusions called *lysosomes*, which contain bactericidal substances such as hydrogen peroxide, lysozyme, proteases, phosphatases, nucleases, and lipases. Attracted to microorganisms by chemotactic esterases and agents such as complement proteins (see Section 20.13), phagocytes work best when they can trap a pathogen on a surface such as a blood vessel wall or a fibrin clot. After adhering to the cell, the phagocyte's cytoplasmic membrane invaginates and engulfs the foreign cell. The entire complex pinches off and eventually fuses with the lysosomes, forming a new inclusion, a *phagolysosome.* The toxic substances and enzymes inside the phagolysosome are usually capable of killing and digesting the engulfed microorganism.

One group of phagocytes, the **neutrophils,** or **polymorphonuclear leukocytes** (sometimes abbreviated PMNs), are actively motile cells containing large numbers of lysosomes (Figure 20.5a). PMNs are short-lived cells (2–3 days) that are found predominantly in the bloodstream and bone marrow but may appear in large numbers at sites of active infection. They can move rapidly, up to 40 μm/min, and are attracted chemotactically to bacteria and cellular components, especially by immune mechanisms (see Section 20.13). In general, large numbers of PMNs in the blood or at a site of inflammation indicate an active infection.

Macrophages and **monocytes** are the other major groups of phagocytic cells. Macrophages are large cells capable of ingesting and destroying most pathogens and

antigens as well as cooperating with lymphocytes in the production of specific immunity. Monocytes are circulating cells that differentiate to become macrophages (Figures 20.5a and 20.6). Thus, the term *macrophage* is generally used to describe phagocytes that are fixed to tissue surfaces, and the term *monocyte* describes the circulating precursor. Macrophages are up to 10 times larger than monocytes and are abundant in lymphoid tissue and spleen, whereas monocytes circulate in the blood and lymph. Macrophages are important *antigen-presenting cells* (APCs); they can present partially degraded foreign antigens to antigen-specific T cells, the first step in antibody production (see Section 20.11). This specialized feature of macrophages makes them a very important component of antigen-specific immunity, and we will examine their role as APCs in more detail in Section 20.7.

During the process of phagocytosis, phagocytes convert from aerobic respiration to anaerobic metabo-

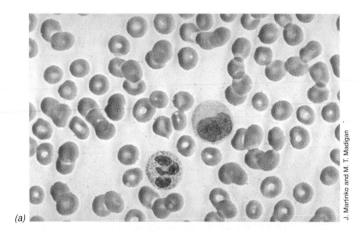

(a)

J. Martinko and M. T. Madigan

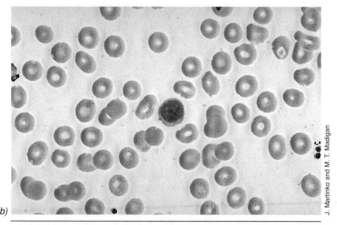

(b)

J. Martinko and M. T. Madigan

Figure 20.5 Major immune cell types. (a) Phagocytic cells. The nucleated cell in the lower left center is a neutrophil (PMN), characterized by a segmented nucleus and granulated cytoplasm. The nucleated cell to the right and slightly above the PMN is a monocyte. Phagocytes are 12–15 μm in diameter. The nonnucleated red blood cells are about 6 μm in diameter. (b) The nucleated cell is a circulating lymphocyte. The lymphocyte has almost no visible cytoplasm and is smaller than the phagocytes, about 10 μm in diameter.

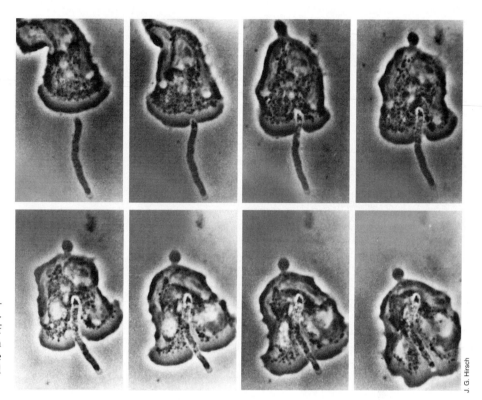

J. G. Hirsch

Figure 20.6 Phagocytosis: engulfment and digestion of a chain of *Bacillus megaterium* cells by a human macrophage, observed by phase contrast microscopy. The bacterial chain is about 18–20 μm long.

lism. Anaerobic glycolysis results in the formation of lactic acid and a consequent drop in pH. This lowered pH is partly responsible for the death of the microbial cell because the hydrolytic lysosomal enzymes all have acid pH optima. Finally, the initial act of phagocytosis conditions the phagocyte so that it is more efficient—a cell that has recently phagocytized can take up bacteria about 10 times more effectively than a cell that has not.

Oxygen-dependent phagocytic killing

As we discussed in Section 5.11, various biochemical reactions can lead to the formation of toxic oxygen-containing compounds including hydrogen peroxide (H_2O_2), superoxide anions (O_2^-), hydroxyl radicals (OH•), and singlet oxygen ($^1O_2$). Phagocytic cells make use of toxic forms of oxygen to kill ingested bacterial cells. Superoxide, formed by the reduction of O_2 by NADPH oxidase, reacts at the acid pH of the phagocyte to yield singlet oxygen and hydrogen peroxide (Figure 20.7). The phagocytic enzyme myeloperoxidase forms hypochlorous acid (HOCl) from chloride ions and H_2O_2, and the HOCl reacts with a second molecule of H_2O_2 to yield additional singlet oxygen. The combined action of these oxygen-dependent phagocyte enzymes forms sufficient levels of toxic oxygen compounds to kill ingested bacterial cells by oxidizing key cellular constituents. These reactions occur within the phagocytic cell itself, which is not damaged by the toxic oxygen products. The action of phagocytic cells in oxygen-mediated killing is summarized in Figure 20.7.

Phagocyte failure

In some cases, pathogens have developed mechanisms for neutralizing the effects of toxic phagocyte products, for killing the phagocyte, or for avoiding phagocytosis. For example, *Staphylococcus aureus* (∞ Section 23.2) produces pigmented compounds called *carotenoids*, which quench singlet oxygen and prevent killing (∞ Section 5.11). Intracellular pathogens such as *Mycobacterium leprae* (leprosy bacillus) and *Mycobacterium tuberculosis* (tuberculosis bacillus) grow and persist within phagocytic cells (∞ Section 23.3). They apparently use cell wall-associated phenolic glycolipids (∞ Section 16.31) to scavenge toxic oxygen compounds. These glycolipids are highly effective in removing hydroxyl radicals and superoxide anions, the most damaging of the toxic oxygen species produced by phagocytic cells.

Other intracellular pathogens produce proteins called *leukocidins*, which destroy phagocytes. In such cases, the pathogen is not killed when ingested but instead kills the phagocyte and is then released. *Streptococcus pyogenes* and *Staphylococcus aureus* are the major leukocidin producers. Destroyed phagocytes make up much of the material of *pus*; organisms that produce leukocidins are therefore usually *pyogenic* (pus-forming) and cause localized infections resulting in boils or abscesses (∞ Sections 19.7 and 23.2).

Another important microbial defense against phagocytosis is the bacterial capsule (∞ Section 3.11). Capsulated bacteria are often highly resistant to phagocytosis, apparently because the capsule somehow pre-

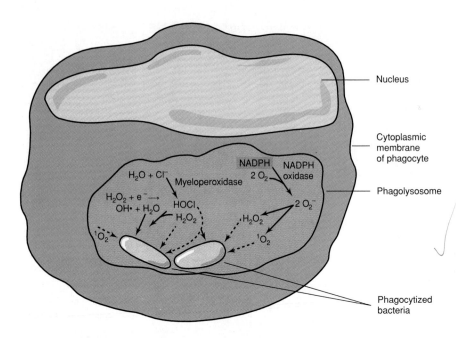

Figure 20.7 Action of phagocyte enzymes in generating toxic oxygen species. These include hydrogen peroxide (H_2O_2), the hydroxyl radical ($OH\bullet$), hypochlorous acid ($HOCl$), the superoxide anion (O_2^-), and singlet oxygen ($^1O_2$).

vents adherence of the phagocyte to the bacterial cell. The clearest case of the importance of a capsule that prevents phagocytosis is that of *Streptococcus pneumoniae*. If only a few cells of a capsulated strain of this species are injected into a mouse, an infection is initiated that leads to death in a few days (∞ Figure 19.22). On the other hand, even large numbers of a noncapsulated mutant strain are completely avirulent. Surface components other than capsules can also inhibit phagocytosis. For instance, pathogenic *Streptococcus pyogenes* produces a specific substance, the M-protein, on both the cell surface and fimbriae (∞ Section 23.2). M-protein apparently alters the surface properties of the bacterial cell in such a way that phagocytes cannot act.

Antibodies to capsules or other cell surface molecules often reverse the protective effect of these bacterial defense mechanisms and enhance phagocytosis, a process known as *opsonization* (see Section 20.13). In

the following sections, we will examine the specific immune response as it is directed against individual structure like the capsule substances.

20.4
Specific Immunity: Lymphocytes

As we have discussed, one of the functions of phagocytes is to present antigens to other cells, the antigen-specific leukocytes known as **lymphocytes.** Lymphocytes are one of the most prevalent mammalian cell types (the average adult human has about 10^{12} lymphocytes) (Figure 20.5*b*). Both B lymphocytes and T lymphocytes are involved in antigen-specific immune responses and are derived from stem cells in the bone marrow. The differentiation of stem cells into mature lymphocytes is determined by the organ in which precursor lymphocytes become established (Figure 20.2). B cells mature in the bone marrow in mammals, but develop in a special organ called the *Bursa of Fabricius* in birds (hence, the designation "B"). T cells mature in the thymus (thus, the designation "T"). Because of their role in the initial development and maturation of B and T cells, the bone marrow, bursa, and thymus are called *primary lymphoid organs.* After maturation, B and T cells are dispersed throughout the body via the blood and lymph (Figure 20.3). Mature T and B cells come to reside in the lymph nodes, spleen or *mucosa-associated lymphoid tissue (MALT)* (Figure 20.3*a*), which are collectively known as *secondary lymphoid organs.* The spleen and lymph nodes are positioned in the blood and lymph and act as filters where the macrophages are situated to trap antigens that pass through. The MALT interacts with antigens that are found on mucosal surfaces. The B and T cells in these organs can then produce an immune response.

CONCEPT CHECK   20.3

The phagocytes are the first cells involved in protection against invading pathogens after physical and chemical barriers have been breached. They are capable of destroying most, but not all, pathogens. In many cases, phagocytes also process and present molecules to antigen-specific leukocytes. Many pathogens enhance their survival with mechanisms for inhibiting phagocytosis and killing.

✔ *Describe separately the functions of PMNs and macrophages.*

✔ *What oxygen-dependent mechanisms are used by phagocytes to kill pathogens?*

B lymphocytes

B cells are responsible for antigen interaction, antibody production, and immune memory. B lymphocytes are readily distinguished from T lymphocytes by the presence of antibody molecules on their surface. These surface antibodies are copies of the single type of antibody that a given B cell will produce later in its development. Surface antibodies on B cells recognize antigen in its native configuration, usually on the pathogen surface (see Section 20.11). B cells are concentrated in the cortex of the lymph nodes where they can contact antigens. After antigen exposure, B cells divide into memory cells or plasma cells. The memory cells are long-lived and may remain in the cortical area for years. On restimulation with antigen, memory cells quickly proliferate to produce more memory cells and plasma cells. The differentiated, antibody-producing plasma cells live for only several days. They are found in the medulla where the antibodies can drain directly into an efferent lymph vessel (Figure 20.3d).

T lymphocytes

The situation with T cells is more complex. All T cells have antigen-specific T cell receptors (TCRs) on their surface and interact specifically with antigen. In the case of TCRs, the antigen is always "presented" by another cell, often a macrophage. Because this antigen is always in a processed form, degraded by the lysosomal proteases, the determinant is always a short, linear peptide derived from the intact immunogen (see Sections 20.2 and 20.7).

Several functionally distinct subsets of T cells have been identified. Two major subpopulations are distinguished from each other by the presence of specific cell surface proteins called CD4 or CD8: mature T cells carry only one of these proteins (Figure 20.8). The CD4 population is subdivided into two functional subsets. One of these is the *T helper* or T_H subset. T_H cells stimulate B lymphocytes to produce large amounts of immunoglobulin. In most cases, little if any antibody is made by B cells without T_H interaction (see Section 20.11). T_H cells also interact with other T cells and stimulate them to become effector cells. The *T delayed-type hypersensitivity*

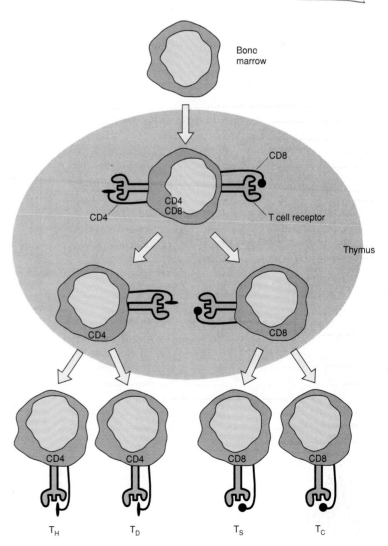

Figure 20.8 T cell subsets. T cell subsets develop in the thymus from bone marrow precursor cells. T_H, T helper cells; T_D, T delayed-type hypersensitivity cells; T_S, T suppressor cells; T_C, T cytotoxic cells.

or T_D *cell* also has CD4. T_D cells participate in cell-mediated reactions but do not interact with B cells; T_D cells are responsible for recruiting and activating nonspecific effector cells such as phagocytes. The second major T cell subpopulation, the CD8 cells, also have at least two functional subsets. The first of these consists of the *T cytotoxic* or T_C *cells*, which directly kill cells having antigen on their surface. The other CD8 subpopulation consists of *T suppressor* or T_s *cells*, which regulate the immune response by suppressing the action of immune cells such as B cells. Table 20.2 compares B and T lymphocytes with respect to development, surface antigens, and functions.

CONCEPT CHECK   20.4

Lymphocytes are antigen-specific leukocytes. B lymphocytes have immunoglobulin antigen receptors on their surface, whereas T cells have antigen-specific T cell receptors. T cells, especially T_H cells, contact antigen through the action of the nonspecific phagocytes. T cells are divided into a number of subsets based on their surface proteins and functional characteristics.

✔ *What are the chief differences between B cells and T cells?*

✔ *Differentiate between T_H cells and T_D cells. Are their surface proteins different?*

20.5

Immunoglobulins (Antibodies)

The next three sections of this chapter discuss the antigen-specific molecules of the immune system. We begin with immunoglobulins because we understand

their structure and function in detail. They serve as the model for an antigen-specific receptor molecule. *Immunoglobulins (antibodies)* are protein molecules that are able to combine with antigenic determinants. They are found in the serum and in other body fluids such as gastric secretions and milk. Serum containing antigen-specific antibody is often called **antiserum.** Immunoglobulins (Ig's) can be separated into five major classes on the basis of their physical, chemical, and immunological properties: **IgG, IgA, IgM, IgD,** and **IgE** (Table 20.3). Immunoglobulin class IgG has been further divided into four immunologically distinct subclasses called IgG_1, IgG_2, IgG_3, and IgG_4. These subclasses are genetically, structurally, and functionally different from one another. Antibody molecules specific for a given antigenic determinant can be found in each of the several classes, even in a single immunized individual. On initial immunization, the first immunoglobulin to appear is IgM, a pentameric immunoglobulin with a molecular weight of about 970,000; IgG appears later. In most individuals about 80% of the serum immunoglobulins are IgG proteins, and these have therefore been studied extensively.

Immunoglobulin structure

Immunoglobulin G is the most common circulating antibody, and thus we will discuss its structure in some detail. Immunoglobulin G has a molecular weight of about 146,000 and is composed of four polypeptide chains (Figure 20.9). Both intrachain and interchain disulfide (S—S) bridges are present. The two light (short) chains are identical in amino acid sequence, as are the two heavy (longer) chains. The molecule as a whole is thus symmetric. Each light chain consists of about 212 amino acids, and each heavy chain consists of about 450 amino acids.

When an IgG molecule is treated with the proteolytic enzyme *papain*, it breaks into several fragments

---

TABLE 20.2 Comparison of B and T lymphocytes	
T cells	**B cells**
Origin: bone marrow	Origin: bone marrow
Maturation: thymus	Maturation: bone marrow
Long-lived: months to years	Long-lived: years (memory cells); short-lived: days (plasma cells)
Mobile	Relatively immobile (stationary)
T cell antigen receptor (TCR) on surface	Immunoglobulin antigen receptor on surface
CD4 or CD8 on surface	Complement receptors on surface
CD3 on surface	
Restricted antigenic specificity	Restricted antigenic specificity
Proliferate on antigenic stimulation	Proliferate on antigenic stimulation into plasma cells and memory cells
Produce cytokines	Synthesize immunoglobulin (antibody)
Show delayed hypersensitivity (T_D cells)	May serve as an antigen-presenting cell (APC)
Help in immunoglobulin production by B cells (T_H cells)	
Perform as killer T lymphocytes in cell-mediated immunity (T_C cells)	
Control immune response (T_s cells)	

(Figure 20.9c). Two fragments contain the complete light chain plus the amino-terminal half of the heavy chain. These portions combine with antigen and are called *Fab* fragments (*f*ragment of *a*ntigen *b*inding). The fragment containing the carboxy-terminal half of both heavy chains, called *Fc* (*f*ragment *c*rystallizable), does *not* combine with antigen. Therefore, each antibody molecule of the IgG class contains *two antigen combining sites* (and is thus *bivalent*). This bivalency is of considerable importance in understanding the manner in which some antigen–antibody reactions occur (see Section 20.15). The antigen binding site is in the amino-terminal portion of both the heavy and the light chains (Figure 20.9). Immunoglobulins also contain small amounts of complex carbohydrates consisting mainly of hexose and hexosamine, which are attached to portions of the heavy chain; the carbohydrate is not involved in the antigen binding site.

Although the view of the IgG molecule shown in Figure 20.9 is adequate for conveying the general structure of this molecule, immunoglobulins are very complex proteins. They are twisted and folded in their final conformation and assume a complex three-dimensional structure. This is illustrated in Figure 20.10 with a computer-generated model and an electron micrograph of a single IgG molecule. Although the basic Y-shaped structure is apparent in both views of the IgG molecule shown in Figure 20.10, the twisting and folding characteristic of the secondary and tertiary structure of proteins is readily apparent in Figure 20.10a. These higher order structures of the polypeptide chains combine to form an intact immunoglobulin. In the process, the heavy- and light-chain interaction produces a unique binding site on each antibody molecule. This binding site is ultimately responsible for the specificity of antigen–antibody reactions.

Light chains of IgG

Each IgG light chain contains two amino acid domains, the *variable domain* and the *constant domain*. The sequences of amino acids in a major portion of the light chains of immunoglobulins of the class IgG are frequently identical, even in IgGs directed against completely different antigenic determinants. This is because the amino acid sequence in the carboxy-terminal half of the light chain constitutes one of two specific and constant sequences, referred to as the *lambda* (λ) sequence and the *kappa* (κ) sequence. One IgG molecule has either two λ chains or two κ chains but never one of each (Figure 20.9a). By contrast, light-chain *variable* regions, located in the amino-terminal half of the light chain, always differ in amino acid sequence from one IgG molecule to the next unless both molecules are produced by the same cell or clone of cells.

Heavy chains of IgG

Each IgG heavy chain contains four amino acid regions, one variable and three constant regions (referred to as *variable* and *constant domains*, respectively). Each domain is approximately 110 amino acids in length. Analogous to the situation that exists in the light chain, all immunoglobulins of the IgG class have a portion of their heavy chain (the carboxy-terminal three domains) in which the amino acid sequence is identical (C_H1, C_H2, and C_H3) (Figure 20.9) from one IgG molecule to another. In addition, each heavy chain has a region in the amino-terminal domain (antigen binding site V_H) (Figure

TABLE 20.3   Properties of human immunoglobulins

Class designation	Molecular weight	Proportion of total antibody (%)	Concentration in serum (mg/ml)	Antigen binding sites	Properties	Distribution
IgG	146,000	80	13	2	Major circulating antibody; four subclasses exist: IgG_1, IgG_2, IgG_3, IgG_4; binds complement weakly	Extracellular fluid; blood and lymph; crosses placenta
IgM	970,000 (pentamer)	6	1.5	10	First antibody to appear after immunization; binds complement strongly	Blood and lymph; B lymphocyte surfaces (as monomer)
IgA	160,00 385,000 (secretory form)	13	3 0.05	2 4	Major secretory antibody	Secretions (saliva, colostrum, serum), cellular and blood fluids; exists as a monomer in serum and as a dimer in secretions
IgD	184,000	1	0.03	2	Minor circulating antibody; heat-labile	Blood and lymph; B lymphocyte surfaces
IgE	188,000	0.002	0.00005	2	Involved in allergic reactions: contains mast cell binding fragment	Blood and lymph only

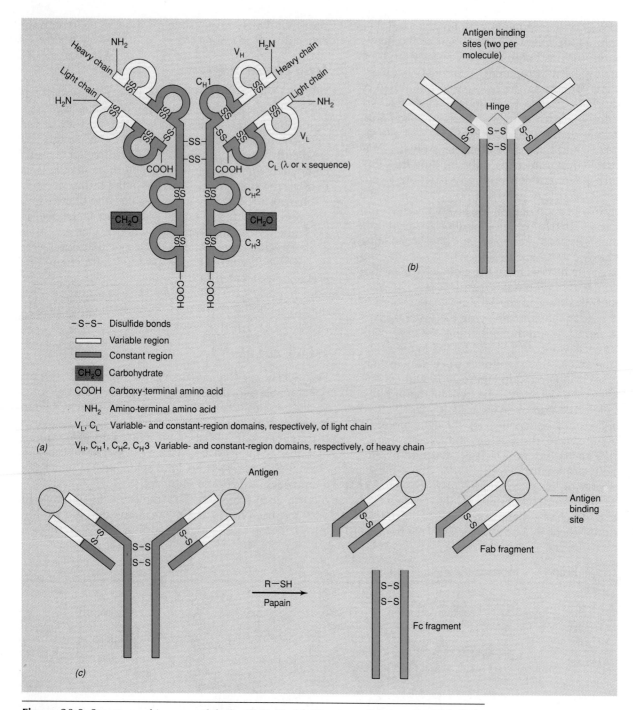

- –S–S– Disulfide bonds
- ▭ Variable region
- ▬ Constant region
- CH₂O Carbohydrate
- COOH Carboxy-terminal amino acid
- NH₂ Amino-terminal amino acid
- V_L, C_L Variable- and constant-region domains, respectively, of light chain
- V_H, C_H1, C_H2, C_H3 Variable- and constant-region domains, respectively, of heavy chain

Figure 20.9 Structure of immunoglobulin G (IgG). (a) Structure showing disulfide linkages within and between chains. (b) Alternative structural diagram, which omits the intrachain disulfide bonds and carbohydrates to simplify the diagram. (c) Effect of papain treatment on immunoglobulin structure. R—SH represents one of several organic thiols that react with selected S—S bonds of the immunoglobulin molecule.

20.9) where considerable amino acid sequence variation occurs from one IgG to the next. The great specificity of a given antibody molecule for a particular antigen lies in the unique three-dimensional structure of the antigen binding site, formed by a combination of the amino acid sequences in the variable regions of the heavy and light chains (Figure 20.10).

Other classes of immunoglobulins

How do immunoglobulins of the other classes differ from IgG? The heavy-chain *constant* region of a given immunoglobulin molecule defines its class and can have one of five amino acid sequences: gamma, alpha, mu, delta, or epsilon. These sequences constitute the

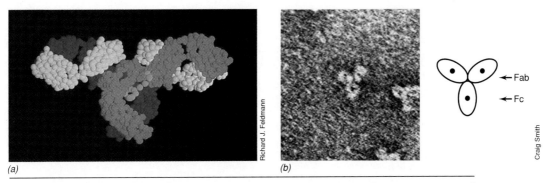

(a)   (b)

Richard J. Feldmann

Craig Smith

Figure 20.10 Three-dimensional views of immunoglobulin IgG. (a) Computer-generated model of IgG; the heavy chains are shown in red and dark blue, and the light chains in green and light blue. The light blue near the Fc portion of the two heavy chains represents attached carbohydrate. (b) Electron micrograph of a single molecule of IgG.

carboxy-terminal three-fourths of the heavy chains of immunoglobulins of the class IgG, IgA, IgM, IgD, and IgE, respectively (Figure 20.9*a*). Each antibody of the IgM class, for example, contains amino acids in its heavy-chain constant region that constitute the mu sequence. If two immunoglobulins of *different classes* react with the same antigenic determinant, then the variable regions of their heavy and light chains can be identical, but their class-determining sequences, specific to their *heavy* chains, would be different. It is not unusual in a typical immune response to observe the production of antibodies of two different classes to the same antigenic determinant.

The structure of **immunoglobulin M** (IgM) is shown in Figure 20.11. It is usually found as an aggregate of five immunoglobulin molecules attached as shown in Figure 20.11 by short peptides called *J chains*. IgM accounts for 5–10% of the total serum immunoglobulins. Each heavy chain of IgM contains an extra constant-region domain (C_H4), and IgM has a very high carbohydrate content. IgM is the first class of immunoglobulin made in a typical immune response to a bacterial infection, but immunoglobulins of this class are generally of low affinity. Antigen binding strength is enhanced to some degree, however, by the high *valency* of the pentameric IgM molecule; 10 binding sites are available for interaction with antigen (Table 20.3 and Figure 20.11). The term *avidity* is used to describe the *strength of binding* by multivalent antigen binding molecules; thus, IgM is said to be of *low* affinity but *high* avidity.

Immunoglobulin A (IgA) is present in body secretions. It is the dominant antibody in all fluids bathing organs and tissues in contact with the outside world. IgA is present in saliva, tears, breast milk and colostrum, gastrointestinal secretions, and mucus secretions of the respiratory and genitourinary tracts. The mucosal surfaces of the human body total about 400 m². All these mucosal surfaces are associated with the MALT lymph nodes (Figure 20.3) that secrete IgA. As a result, the total amount of *secretory* IgA produced

by the body is higher than the amount of *serum* IgG. IgA is also present in the second highest concentration in serum (Table 20.3). The secretory form of IgA has an altered molecular structure, consisting of a dimeric immunoglobulin attached to a protein high in carbohydrate, called the *secretory piece*, and a J chain peptide (Figure 20.12). These proteins help hold the dimeric immunoglobulin molecule together and aid in the transport of IgA across membranes and into secretions.

Immunoglobulin E (IgE) is found in serum in extremely small amounts (in an average human about 1 of every 50,000 serum immunoglobulin molecules is

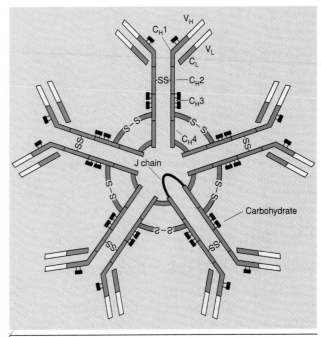

Figure 20.11 Structure of IgM, a large immunoglobulin with five molecules (a pentamer). Note that each heavy chain has four rather than three constant regions and that the five molecules are themselves held together by disulfide bonds. Also note that 10 antigen binding sites are available.

IgE). Despite its low concentration, it is important because immediate-type hypersensitivities (allergies) (see Section 20.16) are mediated by IgE. The molecular weight of an IgE molecule is significantly higher than most other immunoglobulins (Table 20.3) because, like IgM, it contains an additional constant region. This additional constant region functions to bind IgE to mast cell surfaces (see Figure 20.32), an important prerequisite for certain allergic reactions.

Immunoglobulin D (IgD) is also present in low concentrations, and its function in the overall immune response is unclear. IgD is abundant on the surfaces of antibody-producing B cells and may play a role along with monomeric IgM in binding antigen. Antigen-bound surface antibody serves as a signal to the B lymphocyte to begin antibody production.

CONCEPT CHECK   20.5

Immunoglobulins (antibodies) are proteins consisting of four chains, two heavy and two light. The antigen binding site is found at the juxtaposition of variable regions of heavy and light chains. Each different class of immunoglobulin has different structural and functional characteristics.

✔ *What immunoglobulin domains are involved in antigen binding?*

✔ *What structural characteristics differentiate Ig classes?*

20.6

T Cell Receptors

As we have seen, T cells play a variety of complex roles in the overall immune response. Although T cells do not produce antibody, they do recognize antigen, and this recognition process is due to antigen-specific receptor molecules located on T cell surfaces called **T cell receptors (TCRs).** T cell receptors have antigen specificity, but unlike most immunoglobulins, they are integrated into the T cell membrane. Both CD4 and CD8 lymphocytes have TCRs on their cell surface. Thus, all T cells are equipped to recognize specific antigens with their TCRs. How is this accomplished at the molecular level?

Structure of the T cell antigen receptor

T cells must be structurally diverse. For example, a different TCR must be available to distinguish each virus or virus strain that infects the body. Thus, TCRs must have at least as many different antigen binding sites as antibodies. Although TCRs are not antibody molecules, they resemble antibody molecules in many ways. Indeed, TCRs and antibodies have much in common and in evolutionary terms are clearly related molecules (see Section 20.12).

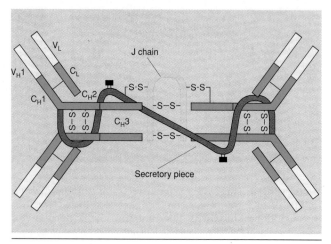

Figure 20.12 Structure of human secretory immunoglobulin A (IgA). Although serum IgA is monomeric, the secretory form of IgA is a dimeric molecule.

The TCR consists of two disulfide-linked peptides, called alpha (α) and (β). The alpha chain is of about 40,000 molecular weight, and the beta chain is of about 43,000 molecular weight. The α-β heterodimer is found on 85% of mature T cells. The remaining T cells have analogous polypeptide chains designated gamma (γ) and delta (δ). Both types of TCRs contain regions of highly variable amino acid sequences. These *variable* domains (which are the amino-terminal portion of each of the two polypeptide chains) combine to form the actual antigen binding site, much as a combination of the heavy- and light-chain variable regions forms the antigen binding site in immunoglobulins. Each TCR polypeptide also contains one constant domain in which the amino acid sequence is invariant from chain to chain within a corresponding type. Thus, all α chain constant-region sequences from any TCR are the same. The variable and constant domains are roughly equivalent in size to the immunoglobulin domains (see Section 20.5). A comparison of a TCR with an analogous antibody molecule is shown in Figure 20.13. Note that the TCR is an integral membrane protein, with both chains spanning the membrane.

Although the TCR is structurally quite similar to immunoglobulin, there are very significant differences

---

CONCEPT CHECK   20.6

T cell receptors are antigen-specific proteins found on the surface of T cells. They are structurally and evolutionarily related to immunoglobulins. T cells recognize antigen only in the context of other proteins found on cell surfaces.

✔ *Where are TCRs found?*

✔ *How do TCRs differ from serum immunoglobulins in structure? How are they similar?*

gen-presenting molecules and interact with both the antigen and the TCR. Thus, MHC proteins are a *third* set of antigen binding molecules and play an integral role in the immune response. We focus first on the structure and genetics of the MHC proteins in humans. Then we will investigate the function of the MHC molecules.

Structure of human major histocompatibility complex proteins

The MHC genes encode two distinct types of proteins known as *class I* and *class II*. The entire MHC has several genes that encode class I proteins and additional genes that encode class II proteins. These class I and class II molecules are cell surface proteins and are intimately involved in immune recognition events. Class I MHC proteins are found on the surfaces of *all* nucleated cells. Class II MHC proteins are found only on the surface of B lymphocytes, macrophages, and other antigen-presenting cells (APCs). The reasons for this differential distribution will become apparent when we discuss the function of these molecules.

Class I MHC proteins consist of two polypeptides (Figure 20.14), one, an alpha (α) chain of 45,000 molecular weight, is encoded in the MHC gene region. The

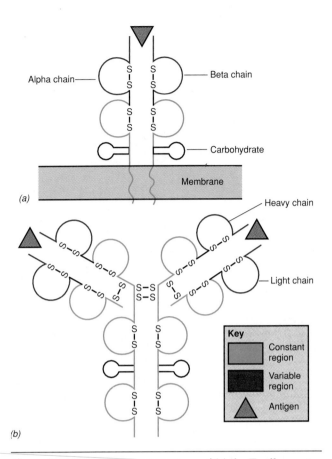

Figure 20.13 Structural comparison of (a) the T cell receptor with (b) an immunoglobulin. Note the presence of variable regions.

in the way that it binds antigen. As we shall discuss in detail in the following section, TCR can bind antigen effectively only if it is presented on the surface of another cell. Finally, the antigen must be presented in direct association with a self protein, which is expressed on the presenting cell surface.

20.7
Histocompatibility Proteins

Antibodies recognize antigens in solution. However, although they have much in common structurally with antibodies, T cell receptors (TCRs) can recognize only an antigen that is bound to a set of *self* proteins found on the surface of normal cells. These proteins are encoded by a genetic region, present in all vertebrates, called the *major histocompatibility complex* (MHC). MHC proteins are produced by a number of genes in this complex and are collectively called *human leukocyte antigens* or HLAs. MHC molecules were first discovered as the major target molecules for transplantation rejection; if tissues from one animal, a donor, are immunologically rejected when transplanted to another animal, a recipient, then their MHC proteins are different. We now know that MHC proteins function as anti-

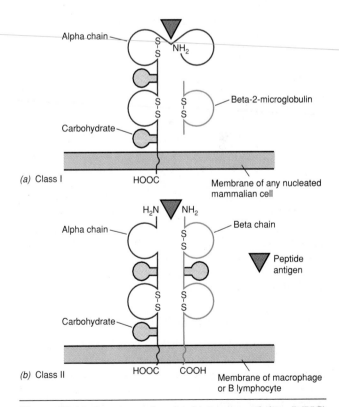

Figure 20.14 Structure of major histocompatibility (MHC) proteins. (a) Class I type. (b) Class II type. Note that class I molecules are present on the surfaces of all nucleated cells and that class II molecules are present only on certain specialized APC cells.

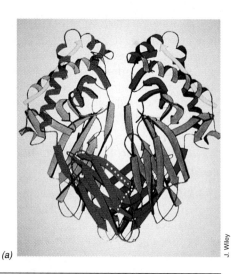

(a)

J. Wiley

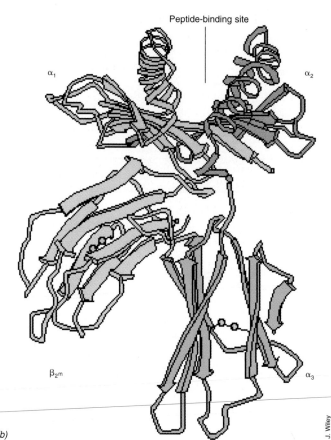

Peptide-binding site

α_1

α_2

β_2m

α_3

(b)

J. Wiley

Figure 20.15 Three-dimensional structures of major histo-compatibility complex (MHC) proteins. (a) The class II protein in its dimeric form. The yellow arrows indicate the peptide binding positions. Reprinted with permission from *Nature 364:* 33 (1993), © Macmillan Magazines. (b) The class I protein. (c) A space-filling model of a class I protein with a bound peptide as seen from above. The peptide is shown as a stick structure embedded in the space-filling model of the class I protein.

other class I polypeptide, a 12,000-molecular-weight protein called *β-2 microglobulin (β₂m)*, is encoded by a non-MHC gene. The MHC-encoded polypeptide is a glycoprotein firmly anchored in the cell membrane; β_2m is noncovalently linked to class I (Figure 20.14a). Class II MHC proteins consist of two noncovalently linked gly-cosylated polypeptides, called α and β, of 33,000 and 28,000 molecular weight, respectively. Like class I mole-cules, these polypeptides are embedded in the cytoplas-mic membrane and project outward from the cell sur-face (Figure 20.14b). Class II molecules usually associate in *pairs*. Apparently the paired structure is much more efficient for binding to TCRs. Figure 20.15a shows a three-dimensional structure for the MHC class II pair.

Detailed studies of the crystal structure of class I and class II MHC molecules have revealed a distinc-tive shape that suggests how these proteins interact with antigens and the cells of the immune system. The α chain of a class I molecule forms three separate extracellular domains. Between two of these domains, a large groove exists, and it is within this groove that the MHC molecule binds foreign antigens. The α and β chains of the class II protein interact to form a simi-lar groove (Figure 20.15). As we will see, the binding of a foreign antigen in the groove of an MHC protein and the interaction of this foreign antigen–MHC complex with a TCR is the trigger for the immune response.

MHC proteins are not structurally identical *within* a given species. Different individuals often show subtle differences in the amino acid sequence of their MHC molecules. These limited sequence variations are called *polymorphisms*. In humans, for example, there may be 100

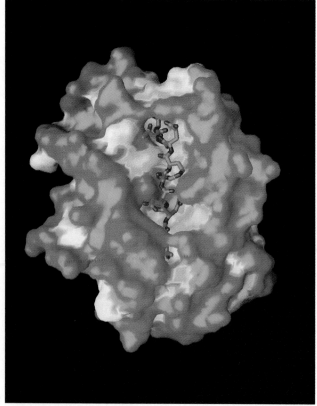

(c)

Aideen C. M. Young

different MHC alleles, or polymorphisms, for each gene position, but each individual has only two of these alleles and only two proteins are expressed; one allele is of paternal origin, and one is of maternal origin. In all, there are three gene positions (loci) for class I genes and three loci for class II genes. Thus, an individual has only six class I and six class II proteins of different sequences. In addition, major structural changes in MHC proteins always exist *among* different animal species. These differences, both within and among species, are the major reason why tissues transplanted from one individual or one species to another seldom match, are recognized as nonself, and are usually rejected.

Functions of MHC proteins

MHC proteins serve as molecular reference points that permit T cells to identify foreign antigens. In a normal animal, T cells, through their TCRs, constantly interact with proteins or other potential antigens. It is critical that they be able to discriminate self from nonself antigens. The T cell, through its TCR, binds to MHC molecules and can then recognize foreign antigens embedded in the MHC structure; a T cell cannot recognize a foreign antigen unless it is presented in the context of an MHC protein.

How does this happen? As we discussed previously, when a foreign antigen is taken up by a host cell, the cell "processes" or degrades it. This processed antigen then becomes embedded, or bound, to the MHC protein, and the complex is passed through the cytoplasmic membrane and expressed on the surface of the cell. Two distinct antigen-processing schemes are known, one for class I antigen presentation and one for class II antigen presentation (Figure 20.16). In the class I scheme (Figure 20.16*a*), antigens that are manufactured by host degradation reactions are bound by class I proteins in the endoplasmic reticulum. The actual processed peptide is about 10 amino acids in length. This method of antigen contact is very important in virus infections, where the host cell manufactures viral proteins. Degraded viral peptides, which are nonself, then complex with class I proteins and move to the cell surface where they are recognized by peptide-specific T cells through the antigen- and MHC-specific TCRs with the aid of the CD8 molecule. The T cells, in turn, activate immune response effectors and destroy the virus-containing cell, which has now become a *target cell* for the immune response.

In effect, the MHC molecules act as a *platform* on which the foreign antigen is bound (Figure 20.15). For example, viral infection of an animal cell leads to the embedding of viral antigens in class I molecules on the infected cell's surface (Figure 20.16*a*). T_C cells are constantly exposed to the entire cell population. These T_C cells have MHC–peptide specificity for nonself peptides. Normal, healthy cells all express class I proteins on their

surface, but the class I molecules contain self peptides, which are not recognized by the T cells. However, the T cells *do* recognize the virus-infected cell because it exhibits the nonself viral antigen bound by the self class I MHC molecule (Figure 20.16*a*). Thus, the TCR on the surface of the T cells interacts with both antigen-specific (nonself) and MHC molecule-specific (self) sites.

A second antigen presentation scheme involves the class II molecule (Figure 20.16*b*). In this case, class II molecules, complete with a self peptide called *Ii,* or invariant chain, line the cell vacuoles (lysosomes) (see Section 20.3) that degrade antigens phagocytized by APCs. When the phagosome containing the foreign antigen fuses with the lysosome forming a *phagolysosome,* the antigens are digested by proteolytic enzymes along with the Ii. The foreign peptides, generally about 11 to 15 amino acids in length (slightly larger than class I binding peptides), are then bound by the newly opened class II antigen binding site (Figure 20.15*a*), and the whole complex is eventually expressed on the external cytoplasmic membrane where it is presented to the T_H cells. The T_H cells, through the TCR and the CD4 coreceptor, then recognize the class II MHC–foreign peptide complex on the surface of the APCs. The T_H cell is activated by contact with foreign antigen and secretes molecules that stimulate antibody production by specific B cell clones.

Finally, MHC proteins obviously do not recognize every antigenic peptide with an individual MHC molecule as TCRs and antibodies do. Because of the limited individual variation in the MHC proteins (only two per locus and only three loci each for class I and class II), a different mechanism for binding large numbers of different peptides must be used. The peptides bound by a single MHC protein share common structural patterns, or **motifs,** and each different MHC molecule can bind to a different motif. For example, one class I protein binds all peptides having tyrosine at position 2 and leucine at position 7. Thus, this single MHC protein is able to present every peptide with the amino acid sequence X-tyrosine-X-X-X-X-leucine-X-X-X, where X

CONCEPT CHECK   *20.7*

Proteins encoded by the MHC are molecular reference markers that permit T cells to interact with antigen. Found on the surfaces of APCs and target cells, MHC proteins embed processed antigen and present it to T cells. This is the only way T cells can recognize nonself antigens and initiate the specific immune response.

✔ *On what cells are class I and class II MHC proteins found?*

✔ *What are the differences between antigens presented by class I MHC and those presented by class II MHC?*

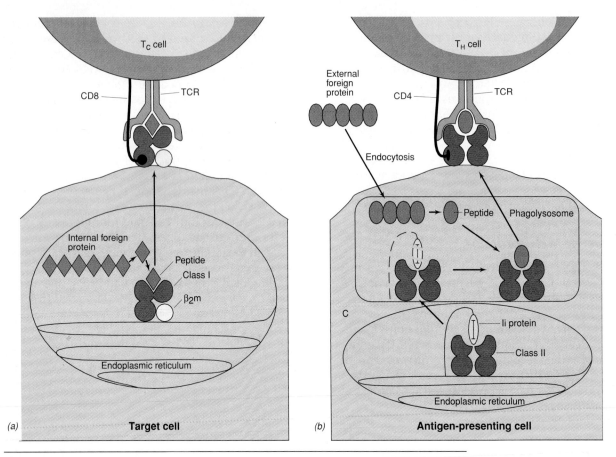

Figure 20.16 Antigen presentation by class I and class II major histocompatibility complex (MHC) proteins. (a) In the class I antigen presentation pathway, the class I proteins are made and assembled in the endoplasmic reticulum along with other proteins, including viral and tumor antigens. Some of these proteins are digested in the endoplasmic reticulum. Digested peptides then bind to class I and are expressed on the cell surface where they interact with T cell receptors (TCRs) on the surface of T_C cells. The CD8 coreceptor on the T_C cell also engages the class I MHC, resulting in a stronger complex. The T_C cells then release cytokines and cytotoxins, proteins that kill the target cell. Any nucleated cell can act as a target cell for class I. (b) In the class II antigen presentation pathway, class II proteins are produced in the endoplasmic reticulum and are assembled with a blocking protein, Ii (invariant chain), which prevents class II from complexing with other peptides made in the endoplasmic reticulum of antigen-presenting cells (APCs). Class II then goes to the phagolysosome where the Ii and foreign proteins, imported from outside the cell (by endocytosis), are digested. The class II protein then binds to the processed foreign peptides, and the complex is expressed on the cell surface where it interacts with TCRs and the CD4 coreceptor on T_H cells. The T_H cells then release cytokines, which act on B cells to cause antibody production. The APCs include certain phagocytes and B cells.

is any amino acid. In this way, each MHC protein can present a large number of different peptides with a limited number of peptide binding sites.

20.8
Cytokines

Section 20.1 described how the cells of the immune system develop from a variety of leukocyte cell types (Figure 20.3). In order to accomplish individual tasks,

the cells in the immune system must communicate, and one method for doing this is through a number of soluble proteins known as **cytokines.** Cytokines are a group of soluble proteins that regulate cellular functions. Specialized cytokines produced by lymphocytes are sometimes known as *lymphokines.* In general, cytokines are secreted from one cell and bind to corresponding specific receptors on a target cell. Some cytokines bind to receptors on the cell that produced them. Thus, these cytokines have *autocrine* (self-stimulation) abilities. The receptors responsible for signal transduction

(∞ Section 7.7) send information inside the cell to either increase or decrease metabolic activity such as protein synthesis and cell division. These signals can ultimately result in differentiation and clonal proliferation of leukocytes. Many of the cytokines are designated *interleukins* (ILs) because they are molecules that mediate interactions between leukocytes. Table 20.4 lists some of the immunologically important cytokines, their producer cells, their target cells, and their biological effects. In all, there are nearly 40 known cytokines, most of which are produced by either T_H cells or monocytes and macrophages. Cytokines are usually small proteins of less than 30,000 molecular weight, and most belong to one of four distinct families, as defined by their protein structure. As examples of cytokine function, we now explore the action of two cytokines that are essential for the specific immune response.

IL-1 and IL-2

As we have seen, macrophages are responsible for antigen uptake, processing, and presentation (see Sections 20.3 and 20.7). In addition, they secrete a potent cytokine known as IL-1, which acts on several different cell types (Table 20.4 and Figure 20.17). IL-1 is a key component of the immune response because T_H cells are one of its main targets. Because of the prox-

imity of macrophages and T_H cells in the lymph nodes, the T_H cells that are activated are generally those that are nearby. Thus, the T_H cells in direct contact with macrophages when specific antigen is being presented are the most likely to become activated. IL-1 binding by IL-1 receptors (IL-1R) on the T_H cell acts as an activation signal and causes the T_H cell to divide, producing clonal copies (Figure 20.17). During this activation stage, the stimulated T_H cells begin to make IL-2, which in turn stimulates other T_H and T_C precursors to form active T_C cells. IL-2 secreted by T_H cells acts with other cytokines such as IL-4 (see below) to stimulate antigen-activated B cells to proliferate and produce antibody-forming plasma cells. The end result is stimulation of several cells through the actions of IL-1 and IL-2, resulting in both humoral and cell-mediated immune responses.

Other cytokines

IL-1 and IL-2 have very specific effects on specific cell types. Table 20.4 shows the action of other cytokines. Many of these proteins affect cells involved in specific immunity. For example, IL-4 and IL-6 act primarily on B cells. However, several cytokines affect nonimmune cells and serve as important modulators of nonspecific host responses. For example, interferon (IFN-α and IFN-γ) is

---

TABLE 20.4   Properties of some major cytokines

Cytokine[a]	Producer cells	Target cells	Effect
IL-1	Monocytes, Macrophages, B cells	T_H B	Activation Maturation, expansion
IL-2	T_H	B T_C, T_H	Proliferation, of activated cells Induce proliferation
IL-4	T_H	B (antigen-primed) B (activated)	Activation Proliferation, class switching
IL-6	T_H Monocytes Macrophages	Proliferating B cells Plasma cells Myeloid stem cells	Differentiation to plasma cells Antibody secretion Differentiation
IL-10	T_H	Macrophages	Suppress IL-1 production
IL-12	Macrophages, B cells	T_C T_H	Differentiation to T_C Proliferation
IFN-α	Leukocytes	Normal cells	Antiviral
IFN-γ	T_H, T_C, NK	Macrophages Normal cells	Activation Antiviral
GM-CSF	Macrophages, T cells	Myeloid stem cells	Differentiation to granulocytes, monocytes
TNF-α	Macrophages, NK	Tumor cells	Cytotoxic
TNF-β	T, B	Tumor cells	Cytotoxic

[a]IL, Interleukin; IFN, interferon; GM-CSF, granulocyte, monocyte colony stimulating factor; TNF, tumor necrosis factor.

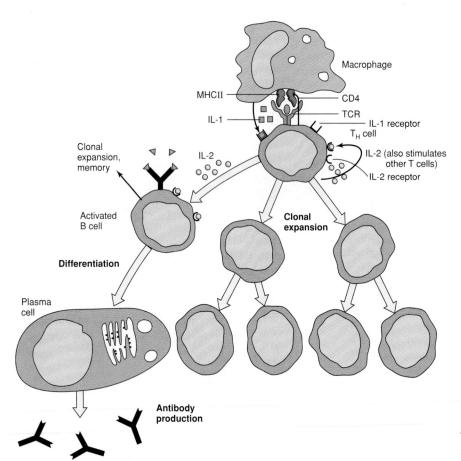

Figure 20.17 Some major effects of IL-1 and IL-2 cytokines.

produced by leukocytes and inhibits viral replication in virtually any cell in the body. Tumor necrosis factors (TNF-α and TNF-β) are capable of destroying a variety of tumors provided that the TNF-producing cells have access to the tumor. Interferons and TNFs appear to have no target cell specificity, but because they are produced by T cells, they augment the effects of immune cells.

Cytokines are important mediators of a variety of immune functions. Some cytokines have broader activities, interacting with a variety of normal cells. However, the unifying feature of these heterogeneous proteins is that they are all produced by leukocytes.

CONCEPT CHECK   20.8

Cytokines are soluble mediators that regulate interactions between leukocytes. Several, such as IL-1 and IL-2, act only on leukocytes and are critical components in the generation of a specific immune response. Others, such as IFNs and TNFs, are produced by leukocytes and act on a wide variety of cells.

✔ *What cells make IL-2? What cells respond to IL-2?*

✔ *Compare the target cells for IL-1, IL-4, IL-12, IFN-γ, and TNF-β.*

20.9
Cell-Mediated Immunity

The terms **cell-mediated immunity (CMI)** and **cellular immunity** are used to describe any immune response in which antigen-specific cells of the immune system are directly involved but in which antibody production or activity is of minor importance.

Cellular immunity differs from antibody-mediated immunity in that the immune response cannot be transferred from animal to animal by antibodies or antiserum but only by lymphocytes removed from the blood. The lymphocytes that function in the transfer of cellular immunity are activated T cells.

The events that occur when an animal is exposed to an antigen are summarized in Figure 20.18. We now turn our attention to the cytotoxic T cells, the natural killer cells, and the delayed-type hypersensitivity T cells and discuss the role each plays in cellular immunity.

Cytotoxic T lymphocytes

Cytotoxic T cells (T$_C$ cells) (see Section 20.4) are involved in the destruction of foreign cells. As previously mentioned in connection with the major histocompatibility

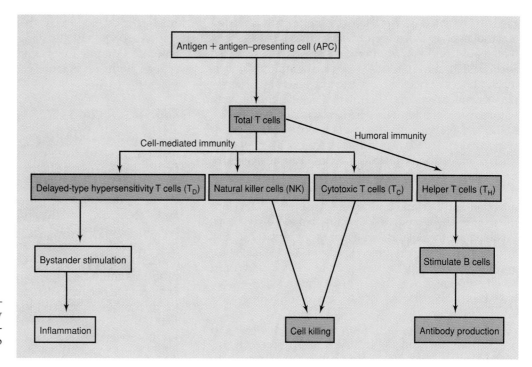

Figure 20.18 Overview of cell-mediated immunity and its relationship to humoral immunity.

complex (MHC) (see Section 20.7), T_C cells recognize foreign antigens embedded in MHC class I molecules. Any cell carrying a foreign antigen, such as those introduced by incompatible tissue grafts or cells harboring viruses, can be lysed by T_C cells (recognition of viral antigens by T_C cells was discussed in Section 20.6 in connection with the T cell receptor). If tissue grafts are made between two different species, or even between two individuals of the same species with different histocompatibility antigens, T_C cells respond by killing the foreign tissue, resulting in rejection of the graft. In human organ grafts, tissue cross-matching is done to ensure that the major histocompatibility antigens are identical (or nearly so) in donor and recipient. An identical or nearly identical tissue match is desirable for successful skin grafting or organ transplants.

Contact between a T_C cell and the target cell is required for lysis. The contact is initiated by the TCR: Ag-MHC complex (Figure 20.16a). On contact with the target cell, granules in the T cell are drawn to the contact site and the contents of the granules are released. The granules contain *perforin* monomers, which enter the membrane of the target cell, polymerize, and form a pore. The cellular contents leak out and the target cell dies, but the T cell remains unaffected. Interestingly, in contrast to the nonspecific nature of activated macrophage killing, T_C cells are effective only at lysing the specific target cells containing the foreign antigen; killing of other neighboring cells occurs at much lower frequencies. Presumably, the concentration of perforin that escapes from the cell contact area is not sufficient to damage bystander cells.

Natural killer cells

Killer cells are an additional class of lymphocytes distinct from cytotoxic T cells that play a role in destroying foreign cells. *Natural killer* (NK) cells are neither T cells nor B cells. Their numbers are not enhanced, nor do they exhibit memory after stimulation. Nevertheless, NK cells resemble T_C cells in their ability to kill foreign cells. For example, NK cells also use perforin to lyse their targets. However, NK cells differ from T_C cells in that they are able to kill in the *absence* of stimulation by a specific antigen. NK cells are capable of destroying malignant and virus-infected cells *in vitro* without previous exposure or contact with the foreign antigen. The molecular target for NK cells seems to be the *lack* of appropriate MHC class I proteins. NK cells recognize normal cells and their class I proteins through a set of special class I receptors. Binding of these receptors to class I *deactivates* the lytic mechanism, but in the absence of binding, the lytic mechanism is active and the NK cell lyses the unrecognized target. As mentioned above, the main targets of NK cells are tumor cells and virus infected cells, both of which often have reduced or altered class I expression.

T_D cells and macrophage activation

Macrophages play a central role as **antigen-presenting cells (APCs)** in both antibody-mediated and cell-mediated immunity. As illustrated in Figure 20.16, macrophages bind, process, and present antigen to T cells. As phagocytic cells, however, macrophages also take up and kill certain foreign cells by themselves, and this ability can be stimulated by T cells. A key property of activated macrophages is that they can kill intracellular bacteria that would normally multiply. As we have noted (see Section 20.3), some bacteria survive and multiply within macrophages, whereas most bacteria taken into macrophages are killed and digested. Bacteria multiplying within macrophages include *Mycobacterium tuberculosis*, *M. leprae* (causal agents of tuberculosis and leprosy, respectively), *Listeria monocytogenes* (causal agent of listeriosis), and various *Brucella* species (causal agents of undulant fever and infectious abortion). Animals given a moderate dose of *M. tuberculosis* are able to overcome the infection and become immune because of the development of a T cell-mediated immune response. The cells involved are the delayed-type hypersensitivity (T_D) cells. Surprisingly, such immunized animals also phagocytize and kill unrelated organisms such as *Listeria,* and it can be shown that macrophages in the immunized animal have been activated so that they more readily kill the secondary invader.

Macrophages not only kill foreign pathogens but are also involved in the destruction of foreign mammalian cells. This shows up in the development of transplantation immunity, and is a major problem in the transplantation of organs and tissues from one person to another. Macrophages also target tumor cells in some cases. Tumor cells contain some specific antigens not seen on normal cells, and tumors function like self-inflicted transplants. There is considerable evidence that tumor cells are normally recognized as foreign and are destroyed primarily by macrophages, which are recruited by cytokines from activated T cells.

CONCEPT CHECK   *20.9*

Cell-mediated immunity involves a variety of T cells, which are activated by interaction with MHC–antigen complexes. Activated T cells may interact directly to kill target cells, or they may facilitate involvement of other cellular effector cells, particularly macrophages, with cytokines.

✔ *What antigens do T_C cells recognize?*

✔ *How do NK cells recognize targets?*

✔ *Why are macrophages sometimes involved in the response to specific antigens?*

20.10
Clonal Selection and Immune Tolerance

We have now developed an overview of the immune response, especially in relation to effector molecules and mechanisms. In this section, we examine the mech-

anisms used to focus this very powerful, specific response on potentially dangerous nonself antigens. We will also examine the mechanisms used to avoid interactions with the equally complex but nonthreatening self antigens.

Clonal selection

The **clonal selection theory** states that each antigen-reactive B cell or T cell has only a single type, or specificity, of antigen-specific receptor on its surface. When stimulated by a specific antigen, each cell is capable of dividing, making a copy of itself. The antigen-driven B and T cells continue to divide, whereas cells that are not antigen-stimulated do not divide. The key feature of clonal selection is that each cell divides and makes an exact copy of itself. Thus, one antigen-reactive cell produces multiple copies, or *clones*, of itself after antigen contact. Because each precursor cell creates its own unique antigen receptor, the resulting clones produce many copies of the same receptor (Figure 20.19).

Because of the infinite variety of antigens available, a large number of antigen-reactive cells must be available in the body, and each cell is capable of expanding into an antigen-reactive clone. However, antigen-reactive cells must avoid interactions and subsequent immune reactions with *self* antigens in the host. How does this occur? The clonal selection theory again provides an answer. Ultimately, the immune response must develop the ability to discriminate between foreign invaders (nonself and potentially dangerous) and host (self and not dangerous) tissue by deleting or inactivating self-reactive clones. This capacity for antigen-specific immune *un*responsiveness to self antigens is known as **tolerance** and occurs in both T and B cells. We consider the normal development of tolerance now. In Section 20.16, we will discuss the failure of tolerance and the subsequent development of autoimmunity.

Immune tolerance

The development of T cell tolerance occurs in the thymus, a walnut-sized organ that lies on top of the heart. The thymus has a central role in the maturation and development of T cells and is therefore a primary lymphoid organ (Figure 20.3). In the first T cell maturation stage, called *positive selection,* lymphocytes that will become T cells leave the bone marrow and enter the thymus from the lymphatic ducts (Figure 20.20). After entry, some of these immature T cells interact, using their newly developed T cell receptors (TCRs) to bind to self major histocompatibility complex (MHC) molecules on the thymus. The T cells that do not bind MHC molecules are then programmed to die, a process called *apoptosis,* whereas the T cells that bind thymic MHC proteins continue to mature and proliferate. Thus, positive selection *retains* T cells that *can* recognize self MHC

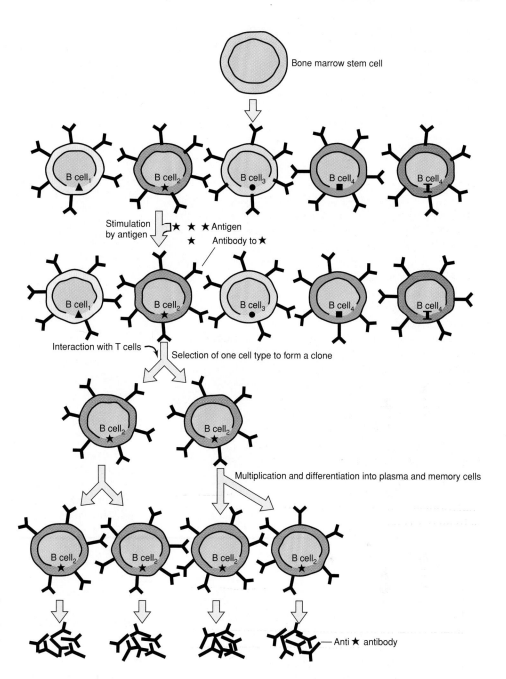

Figure 20.19 Clonal selection. The expansion of a particular antibody-producing cell line after stimulation with antigen.

proteins and *deletes* those that *cannot* recognize self MHC proteins.

The second stage of T cell development is termed *negative selection*. After proliferation, positively selected T cells continue to react with MHC molecules, which are complexed with antigen (see Section 20.7). The antigens in the thymus are mostly of self origin. T cells that react with the self antigens in the thymus are potentially dangerous because they can lead to destruction of self tissue. Therefore, these *autoreactive* T cells must be eliminated. The autoreactive T cells bind very tightly to thymus tissues and cannot leave; they remain bound to the thymus

and eventually die. T cells that interact with nonself antigens (plus MHC), however, do not bind to the thymus as tightly, presumably because no nonself antigens are available to make a tight bond. These T cells do not die, but leave the thymus and migrate to the spleen and lymph nodes where they can contact foreign antigens presented by B lymphocytes and other antigen-presenting cells (APCs) (Figure 20.20).

This mechanism for inducing tolerance is called *clonal deletion*; precursors of T cell clones that are not useful, or even harmful, are deleted during development. There is also evidence for another tolerance

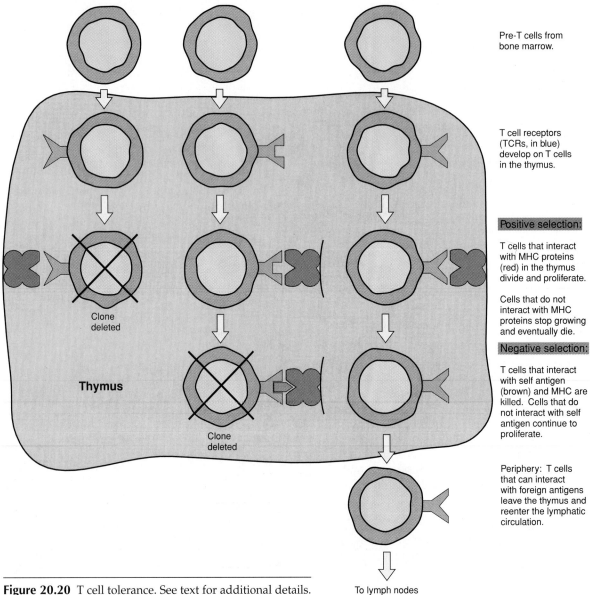

Pre-T cells from bone marrow.

T cell receptors (TCRs, in blue) develop on T cells in the thymus.

Positive selection:

T cells that interact with MHC proteins (red) in the thymus divide and proliferate.

Cells that do not interact with MHC proteins stop growing and eventually die.

Negative selection:

T cells that interact with self antigen (brown) and MHC are killed. Cells that do not interact with self antigen continue to proliferate.

Periphery: T cells that can interact with foreign antigens leave the thymus and reenter the lymphatic circulation.

Clone deleted

Thymus

Clone deleted

To lymph nodes

Figure 20.20 T cell tolerance. See text for additional details.

mechanism. In some cases, self-reactive T cell clones somehow avoid clonal deletion but become unresponsive to self antigen through interactions in the secondary lymphoid organs. This mechanism is known as *clonal anergy* or *clonal paralysis*. However, these clones remain in circulation and, in certain individuals, can be reactivated later in life. These reactivated self-reactive clones are often associated with autoimmune diseases (see Section 20.16).

The acquisition of immune tolerance in B cells is also necessary because antibodies produced by self-reactive B cells (**autoantibodies**) may damage host tissue (see Section 20.16). The mechanisms for achieving B cell tolerance parallel those already seen in T cell tolerance.

B cells also undergo clonal deletion; many self-reactive B cells are eliminated during development. In addition, clonal anergy also plays a role; cells that are self-reactive do not develop when exposed to antigen. Several explanations are possible for induction of B cell clonal anergy. It can result when high concentrations of self antigens are always present. These antigens constantly occupy all the reactive B cell surface receptors, making the B cell unable to react with antigen in a normal manner. B cell anergy can also result if antigen-reactive T cells are not available (because they were eliminated during T cell maturation) to help B cells in the production of antibodies, as we will discuss in the next section.

20.11

Mechanism of Immunoglobulin Formation

Antibody formation is a complex process that we will divide into two separate sections. In this section, we will define the nature of a typical antibody response, especially with regard to the cell interactions necessary to produce effective immunity. In the next section, we will examine the unique genetic mechanisms used to generate the incredible antibody diversity necessary for antigen-specific host defense.

The production of immunoglobulins in response to stimulation by antigen involves T cells, B cells, and antigen-presenting cells (APCs such as macrophages) and the intimate interaction of the various cell surface molecules discussed previously. As we have described, APCs act nonspecifically to ingest foreign antigens. Once the antigen is processed, the APC presents it to T_H cell–B cell pairs. Following recognition, the T_H cell then releases cytokines such as IL-2, IL-4, and IL-6 (see Section 20.8) that stimulate B cells to begin antibody synthesis. Although the initial interaction between antigen and the APC may be nonspecific, all further steps in the process of antibody production are *highly* specific; receptor molecules on the surfaces of both T cells (T cell receptors) (see Section 20.6) and B cells (antibody molecules) (see Section 20.5) are highly specific for that antigen.

The genetic control of antibody production is also a highly complex process. An animal may be capable of making over 1 billion structurally distinct antibody molecules. Although at first this seems like an enormous demand on an organism's genetic coding potential, we will see that only a relatively small number of genes is required to encode this immense antibody diversity because of a phenomenon known as **gene rearrangement**. During development of lymphocytes in the bone marrow, gene rearrangements and deletions occur in B cells to eventually yield two complete transcriptional units, one of which codes for the synthesis of a specific heavy chain and one for a light chain of the antibody molecule. The number of possible gene rearrangements, even of a relatively small number of genes, is sufficient to account for the diversity of immunoglobulin molecules (see Section 20.12). Similar rearrangements occur during T cell development and result in the diversity of T cell receptors observed. We now detail the steps in antibody production, beginning with the injection of an antigen and ending with the production of a specific antibody that will react with that antigen.

Exposure to antigen

In considering the mechanism of antibody formation, we must first explore what happens to the antigen in the whole animal body. Antigens are carried to all parts of the body by the blood and lymph systems, which were illustrated in Figure 20.3. The main sites of antigen localization in the body are the lymph nodes, the spleen, and the liver. However, antibodies are formed in only the spleen and lymph nodes; the liver is not involved. If the antigen is injected intravenously, the spleen is the site of greatest antibody formation, whereas subcutaneous, intradermal, and intraperitoneal administration of antigens lead to antibody formation in lymph nodes. Fragments of lymph node or spleen from immunized animals can continue to produce antibody when placed in tissue culture or when transferred into other, nonimmunized, animals.

Following the primary or initial introduction of an antigen, there is a lapse of time (latent period) before specific antibody appears in the blood, followed by a gradual increase in antibody *titer* (that is, concentration) and then a slow fall. This reaction to a single antigen exposure is called the **primary antibody response** (Figure 20.21). When a second exposure to antigen is made some days or weeks later, the titer rises rapidly to a maximum of 10–100 times above the level achieved following the first exposure. This large rise in antibody titer is referred to as the **secondary antibody response** (Figure 20.21). With time, the titer slowly drops again, but later exposures to the same antigen can bring it back up. The secondary response is the basis for the vaccination procedure known as a "booster shot" (for example, the yearly rabies shot given to domestic animals) to maintain high levels of circulating antibody specific for a certain antigen.

Cell interactions

How do B cells, T_H cells, and APCs cooperate to produce immunoglobulins? The APCs for antibody production include macrophages, dendritic cells, and B cells. Macrophages are capable of processing and presenting antigen to T cells in a nonspecific manner. *Dendritic cells* are another group of phagocytic cells that, along with macrophages, are capable of presenting antigen in a nonspecific manner. They also have class II proteins on

booster

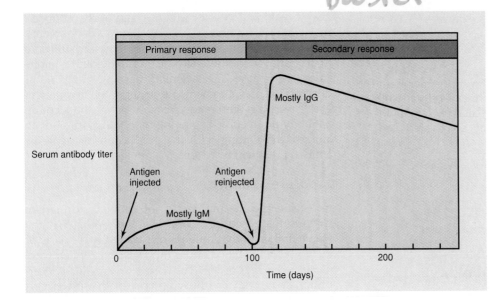

Primary response   Secondary response

Serum antibody titer

Antigen injected

Antigen reinjected

Mostly IgG

Mostly IgM

0   100   200

Time (days)

Figure 20.21 Primary and secondary antibody responses. The secondary response may be 10- to 100-fold higher than the primary response.

their surface and are found in the lymph node cortex (see Figure 20.3*c*). However, the B cell is probably the most efficient APC (Figure 20.22). B cell APCs, through their cell surface immunoglobulin antigen receptors, are capable of binding antigen and internalizing it by phagocytosis. The antigen is then processed (digested) in the phagolysosomes (Figure 20.16*b*), and the resulting peptides are put back on the surface, bound to class II major histocompatibility complex (MHC) proteins (Figure 20.22). Macrophage and dendritic cell APCs work essentially the same way except that they have no antigen-specific receptor like immunoglobulin. However, macrophages and particularly dendrocytes are sticky and efficiently adhere to and phacocytize foreign antigens. The antigens are then processed and presented by class II antigens on the surface, as is the case for B cell APCs. Next, the APC presents the antigen–MHC complex to the T cell via the T cell receptor (TCR). The class II MHC molecules on the APCs interact specifically with the CD4 molecule on T_H cells (Figure 20.16*b*), focusing the antigen presentation to the T_H cells. The T_H cell, through the action of released cytokines (see Section 20.8), then triggers the B cell to divide and form many copies of itself, resulting in a B cell clone. This clone then produces only one specificity of antibody molecule.

Further differentiation of the activated B cell clone then occurs, resulting in the formation of large antibody-secreting cells called **plasma cells** and special B cells called **memory cells** (Figure 20.22). Plasma cells are relatively short-lived (less than 1 week) but excrete large amounts of antibody during this period. Memory cells, in contrast, are very long-lived cells, and on reexposure to the initial stimulating antigen, they

quickly transform into plasma cells and begin secreting antibody. This accounts for *immunological memory*, also known as the secondary response, and results in the rapid and more abundant antibody production observed after repeated antigenic stimulation (Figure 20.21). A secondary response usually involves a change in antibody class, often from IgM to IgG. This is known as *class switching* and is mediated by IL-4 produced by the T_H cell that activates the memory B cell (see Table 20.4).

Certain antigens can stimulate low level antibody production in the *absence* of previous T cell interactions (these are the so-called *T-independent antigens*). Most T-independent antigens are large polymeric molecules with repeating antigenic determinants (for example, polysaccharides). The immunoglobulins produced to T-independent antigens are usually of the IgM class and are of low affinity. In addition, B cells that respond to T-independent antigens do not have immunological memory.

The preceding discussion describes the general principles of antibody production. Each antigen (strictly speaking, each antigenic determinant) catalyzes, through the action of APCs and T_H helper cells, the growth of a *different* B cell line, which is capable of producing antibodies that react specifically with that antigen. In this fashion, the normal animal can respond to perhaps as many as 1 billion distinct antigens by developing a specific B cell clone in response to stimulation by antigen.

We turn now to the genetic orchestration of these immune events and investigate the molecular diversity of antibodies and T cell receptors at the deoxyribonucleic acid (DNA) level.

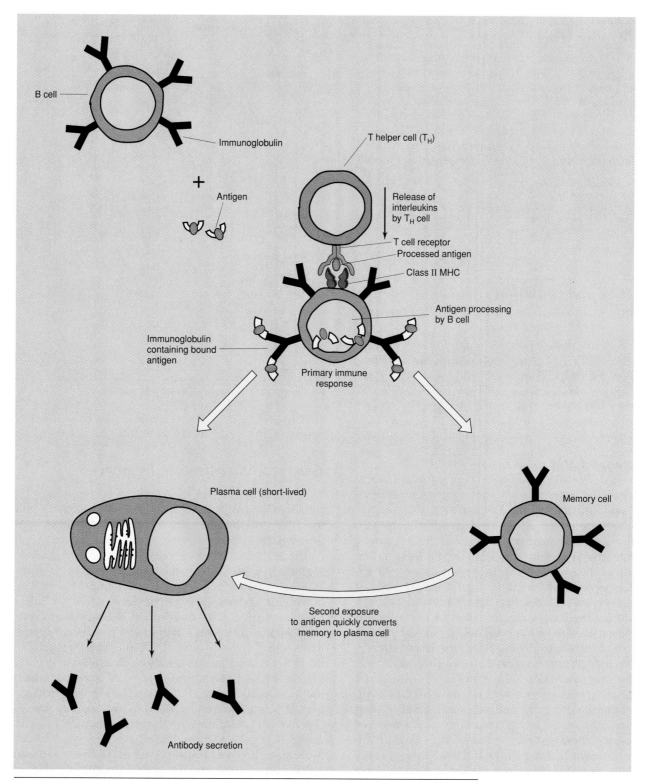

Figure 20.22 T cell–B cell interaction and antibody production. The B cell functions as an antigen-presenting cell (APC) and concentrates antigen with specific immunoglobulin receptors. After processing, antigen is presented to the T_H cell by a class II MHC molecule. The T_H cell in turn signals the same B cell to proliferate and form plasma cells (antibody producers) or memory cells. After subsequent antigen exposure, memory cells quickly convert to plasma cells.

CONCEPT CHECK   **20.11**

Antibody production is initiated by antigen contact, through an APC, with an antigen-specific T_H cell. The T_H cell, in turn, signals an antigen-specific B cell to produce antibody. The antigen acts as a selective agent, causing proliferation of only those clones of cells that react with that particular antigen. Clones remain active for years as memory cells and can be immediately stimulated after reexposure to antigen.

✔ *What APCs are involved in antigen presentation?*

✔ *How do T_H cells activate antigen-specific B cells?*

20.12

Genetics and Evolution of Immunoglobulins and T Cell Receptors

If one B lymphocyte produces immunoglobulins of a single specificity, one gene should encode the two identical light chains and one gene should encode two identical heavy chains. However, this is not the case. A single light or heavy chain is actually encoded by several genes that undergo a complex series of rearrangements as B cells mature. How does this series of events occur?

Immunoglobulin gene organization

To understand the genetics of immunoglobulin synthesis, it is necessary to examine the structure of immunoglobulins in more detail. As illustrated in Figure 20.9, variation in amino acid sequence exists in the *variable region* of different immunoglobulins. Further, amino acid variability is especially apparent in several so-called **hypervariable regions** (Figure 20.23). It is at these hypervariable sites that combination with antigen actually occurs. Each variable region in the light and heavy chains has three hypervariable regions. A portion of the third hypervariable region on the heavy chain is encoded by a distinct gene called the D (diversity) gene, with the first two hypervariable sites being encoded by the variable-region gene itself (Figure 20.24). In addition, at the site of joining between the diversity region and the constant-region genes, there is a stretch of nucleotides, about 40 bases in length, called the J (joining) region that is encoded by a distinct gene (*J gene*). Thus, the third hypervariable region is encoded by a portion of the V gene and the entire D and J genes. Finally, the class-defining constant region of the immunoglobulin molecule is encoded by its own gene, the C (constant) gene. Light chains are encoded by their own variable-region genes, joining-region genes, and constant-region genes but do not contain diversity regions (Figure 20.24).

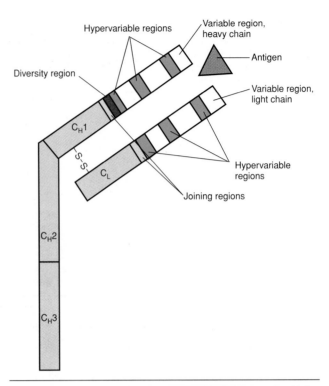

Figure 20.23 Detailed structure of variable regions of immunoglobulin light and heavy chains. Only one half of a typical immunoglobulin molecule is shown. C_H and C_L are constant regions of heavy and light chains, respectively.

Gene rearrangements in lymphocytes

The origin of cells of the immune system was presented in Figure 20.2. All the genetic diversity required to make antibodies exists in each lymphocyte as it develops from the stem cells of the bone marrow. Each *immature* B cell contains about 150 light-chain variable-region genes and 5 distinct joining sequences, whereas 100–200 variable-region genes, 4 joining sequences, and about 50 diversity region genes exist for the heavy chains (Figure 20.24). In addition, 5 heavy-chain constant-region genes and 2 light-chain constant-region genes are present. The V, J, D, and C genes are not located adjacent to one another but are separated by noncoding sequences (introns) typical of gene arrangements in eukaryotes (∞ Section 6.1). During maturation of B lymphocytes, genetic recombination in each B cell occurs, resulting in the construction of an *active heavy-chain gene* and an *active light-chain gene*, which are transcribed and translated to make the heavy and light chains of the immunoglobulin molecule (Figure 20.24). *Randomly selected* V, D, and J segments are fused in each B cell by enzymes that delete all intervening DNA. The active gene (containing an intervening sequence between the VDJ gene segment and the constant-region gene segment) is transcribed, and the resulting primary ribonucleic acid (RNA) transcript is spliced to yield the final messenger RNA (mRNA) (Figure 20.24).

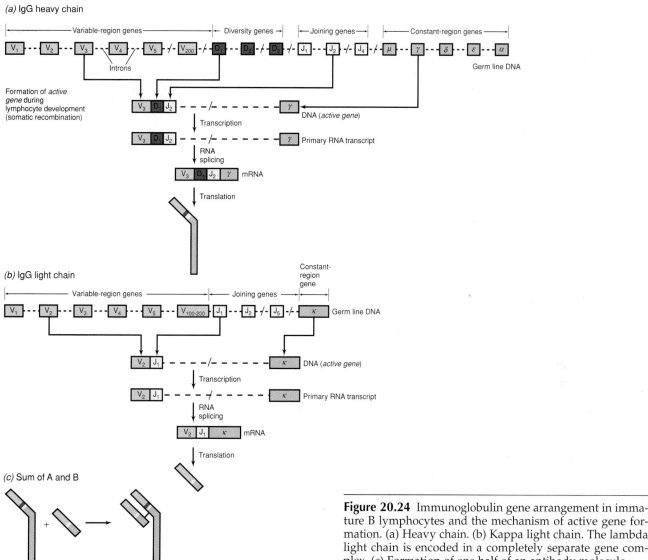

Figure 20.24 Immunoglobulin gene arrangement in immature B lymphocytes and the mechanism of active gene formation. (a) Heavy chain. (b) Kappa light chain. The lambda light chain is encoded in a completely separate gene complex. (c) Formation of one half of an antibody molecule.

The final light and heavy gene complement of a given B cell is largely a matter of *chance rearrangement*. The number of possible gene combinations is such that as many as *1 billion* different antibody molecules can be made. However, for each B cell, only a single protein-producing rearrangement occurs for both the heavy- and light-chain genes. Therefore, each B cell produces a single immunoglobulin. In other words, each B cell makes copies of a single immunoglobulin with identical variable regions.

Additional antibody diversity occurs as the result of mutations arising during formation of the active heavy-chain gene. The DNA joining mechanism appears to be rather imprecise and frequently varies the site of VDJ fusions by a few nucleotides. This, of course, is sufficient to change an amino acid or two, and this apparently sloppy mechanism leads to even greater antibody diversity. Finally, after B cells have been exposed to antigen,

mutational events often occur. These antigen-directed somatic mutation events add further to immunoglobulin diversity (see the box, Catalytic Antibodies, Enzymes, and Natural Selection).

Thus, an enormous number of antibodies can be encoded by a relatively small amount of DNA. The discovery of gene rearrangements in lymphocytes has had a marked impact on our understanding of eukaryotic genetics. It had always been assumed that every cell of a multicellular organism was genetically identical (barring any somatic cell mutation) and that the differences between cell types were a function of differential gene expression. This clearly is not the case with B lymphocytes because several million (if not billion) genetically distinct lymphocytes, products of *somatic cell* recombination, exist in vertebrates. Lymphocytes constitute the only eukaryotic cells in which this type of gene rearrangement is know to occur.

CATALYTIC ANTIBODIES, ENZYMES, AND NATURAL SELECTION

The clonal selection theory (see Section 20.10) states that interaction with antigen is the driving force that stimulates precommitted, antigen-specific B cells to expand and produce antibodies, which then react with the original antigen. Immunologists believe that clonal selection is a critical feature of the immune response. However, other factors help to shape the final strength of the antibody–antigen interaction, even after the original clone has been selected by antigen. These factors are collectively called *somatic* mechanisms because they do not affect the original germ-line genes, but rather act on the rearranged Ig genes in mature B cells. Thus, each clone, even if derived from the same set of germ-line genes, may undergo separate somatic changes that will produce a different antibody.

As a result of the somatic mechanisms, a second exposure to the immunizing antigen usually results in a change in the predominant antibody class produced—usually IgM to IgG (class switching)—*and* an increase in the strength of the antibody–antigen interaction (affinity). This *affinity maturation* is one of the factors responsible for the dramatically stronger secondary response in immunity. How does this happen? Perhaps the best evidence comes from experimental studies on antibodies that function as enzymes, known as *abzymes*. Abzymes are antibodies that are raised to enzyme substrates and analogs. Abzymes perform catalytic reactions on bound substrates, resulting in covalent modification of the substrate and formation of a product. The reaction mechanisms involved in antibody-mediated catalysis seem to be identical to the mechanisms involved in traditional enzyme–substrate interactions (Section 4.5). Unfortunately,

most abzymes have very low substrate affinity, and do not efficiently convert substrate to product. However, when abzyme-producing B-cell clones are reexposed to antigen, the substrate affinity for some of the antibodies frequently increases (see the figure). In one study, an abzyme produced by a clone isolated after repeated immunizations bound the substrate more than 10,000 times stronger than the original antibody.

Studies of the clone that evolved into a better abzyme producer revealed several basic principles. First, nine amino acid alterations were observed in the high affinity abzyme when compared to the abzyme from the original clone. These alterations correlated with increased substrate binding affinity, and resulted from mutations in the heavy- and light-chain V region genes, mostly in the hypervariable regions. Surprisingly, the mutations did not involve the amino acids that directly interacted with substrate; the mutations occurred in adjacent areas, which influenced the overall shape of the binding site.

Protein biochemists are very excited about these results because they provide a simple way to design highly efficient catalytic proteins. If useful catalytic antibodies can be developed simply through repeated immunizations with substrate analogs; the possibilities for practical applications are limitless. Through careful selection of the appropriate substrate antigen, B cells can be induced to make abzymes that will react with virtually any imaginable structure. Finally, this system provides an excellent experimental model for investigating the natural selection and evolution of enzymes using the substrate antigen as a well-defined selective agent. ∎

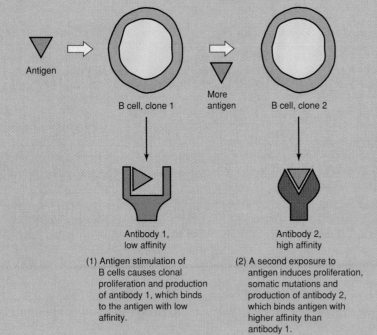

(1) Antigen stimulation of B cells causes clonal proliferation and production of antibody 1, which binds to the antigen with low affinity.

(2) A second exposure to antigen induces proliferation, somatic mutations and production of antibody 2, which binds antigen with higher affinity than antibody 1.

The selection of a certain B cell as a forerunner of an immunoglobulin-producing clone is, to a major degree, a function of the antigenic history of the animal. From the diverse pool of mature B lymphocytes previously discussed, a specific B cell is stimulated by antigen to expand and form a clone of B cells, all of identical genetic makeup (see Figure 20.19). The clonal selection theory predicts that many B cell types will remain forever "silent" within the animal's body because of a lack of exposure to a corresponding antigen. Presumably, this large pool of antibody diversity remains available for future selection if new immunogens are encountered.

Genetics of T cell receptors

We discussed the detailed structure of the T cell receptor (TCR) in Section 20.6. Recall that TCRs recognize antigen fragments embedded in major histocompatibility complex (MHC) molecules on the cell surface. Therefore, TCRs must be able to recognize both conserved regions of the MHC molecule and a specific antigenic determinant. How is this accounted for in genetic terms?

Like immunoglobulins, TCRs contain constant and variable regions of amino acid sequences. Therefore, it is not surprising that, as for immunoglobulins, a number of genes encode the variable regions of TCRs. In the mouse, the α chain has about 100 variable-region genes and 50 joining segments, whereas the β chain has 25 variable genes, 2 diversity genes, and 12 joining segments. In addition, several small additions of from 1 to 6 nucleotides can be inserted between variable, diversity, and joining gene segments in TCRs. Thus, the number of possible sequence combinations is enormous, probably on the order of 10^{15}, which is even greater than the number of immunoglobulin combinations.

From the structural and genetic similarities between immunoglobulins and TCRs, it is easy to imagine that they are evolutionarily related molecules, and we consider this possibility now.

Evolution of molecules of the immune system

From an evolutionary standpoint, the three categories of immunologically specific molecules, MHC proteins, TCRs, and immunoglobulins, share a number of common structural features and undoubtedly represent closely related molecules. Molecular cloning and sequencing of genes coding for immunoglobulins and TCRs strongly suggest that a *family* of genes has evolved from a primordial gene that originally encoded very simple, and presumably much less specific, receptor molecules. This can be visualized by comparing the structures of the molecules discussed thus far in this chapter (Figure 20.25).

Because the molecules of the immune system are obviously related in an evolutionary sense, the term gene *superfamily* has been proposed to describe the genes that code for these molecules. This *immunoglobulin gene superfamily* presumably arose from evolutionary selection on genes coding for primitive immune molecules in ancient organisms, and this genetic progression is apparent today when the immune systems of various animals are compared. For example, a comparison of immunelike reactions in vertebrates and invertebrates suggests that only portions of the gene superfamily exist in invertebrates. Invertebrates are capable of eliciting simple nonspecific defense responses, such as secreting antibacterial substances and carrying out phagocytosis. Horseshoe crabs, for example, secrete lymphokines and proteins with antibacterial activity similar to that of antibodies and also have cells capable of phagocytosis (∞ Section 19.10). However, invertebrates do not show T cell- or

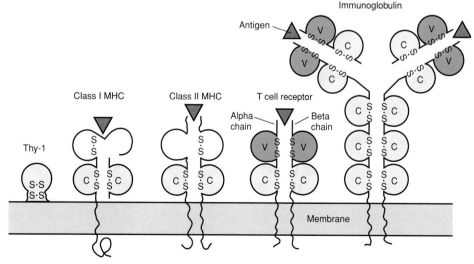

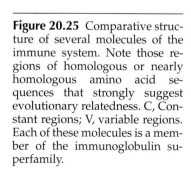

Figure 20.25 Comparative structure of several molecules of the immune system. Note those regions of homologous or nearly homologous amino acid sequences that strongly suggest evolutionary relatedness. C, Constant regions; V, variable regions. Each of these molecules is a member of the immunoglobulin superfamily.

B cell-mediated immune responses. Although some higher invertebrates can actually reject grafted foreign tissues, suggesting that MHC-like molecules may be present on their cell surfaces, it is only in the vertebrates that we see the highly developed MHC-dependent immune functions and the complex genetic systems required to encode the many molecules involved.

CONCEPT CHECK   **20.12**

Diversity in immunoglobulin and TCR genes is generated by a process of gene shuffling. The resulting diversity is much higher than expected for the number of genes involved.

✔ *Describe the gene organization of immunoglobulin heavy-chain genomic DNA.*

✔ *What molecules are evolutionarily related to immunoglobulins?*

20.13

The Complement System

Complement acts with specific antigen–antibody complexes to bring about reactions that would not otherwise occur. Complement is an important effector mechanism and, as we shall see, it enhances the effects of antibody–antigen interactions. These reactions are described in Table 20.6.

Complement is composed of a number of enzymes found in serum that interact in a sequential, ordered fashion with bacterial cells or other foreign material, causing lysis or leakage of cellular constituents as a result of damage to the cell membrane. These enzymes, found at comparable levels in all individuals, are normally inactive but become active when an antibody–antigen reaction occurs. In fact, a major function of antibody is to recognize invading cells and activate the complement system for attack. There is considerable economy in an arrangement such as this because a wide variety of antibodies, each specific for a single antigen, can recruit the complement enzymatic machinery; thus, the body does not need separate enzymes to attack each invading agent.

Some reactions in which complement participates include (1) bacterial lysis, especially in gram-negative Bacteria, when specific antibody combines with antigen on bacterial cells in the presence of complement; (2) microbial killing, even in the absence of lysis; and (3) phagocytosis, which may not occur during infection if the invading microorganism possesses a capsule or other surface structure that prevents the phagocyte from acting (see Section 20.3). When specific antibody combines with the cell in the presence of complement, the cell is changed

in such a way that phagocytosis can occur. This process in which antibody plus complement renders a cell more susceptible to phagocytosis is called **opsonization.**

Activation of the complement system

Complement is a system of 11 proteins, designated C1, C2, C3, and so on. Activation of complement occurs only with antibodies of the IgG and IgM classes (see Table 20.3). When such antibodies combine with their respective antigens, especially on cell surfaces, the antibodies assume new conformations in their constant-region domains, which allow them to fix (attach to) the ever-present complement proteins (Figure 20.26). The complement proteins act in a cascade fashion, activation of one component resulting in activation of the next. In summary, the steps are (1) binding of antibody to antigen (initiation); (2) recognition of antigen–antibody complex by C1; (3) C4-C2 binding to an adjacent membrane site; (4) activation of C3; (5) formation of the C5-C6-C7 complex at another membrane site; and (6) formation of the C8-C9 complex at the same site, causing cell lysis.

As seen in Figure 20.26, complement reactions at the C3 level result in chemotactic attraction of phagocytes to invading agents and to phagocytosis following opsonization. Reaction at C5 also leads to leukocyte attraction. The terminal series of reactions from C5 through C9 results in cell lysis and death. Lysis results from destruction of the cytoplasmic membrane, leading to the formation of holes through which cytoplasm can leak (Figure 20.27).

Complement is necessary in the bactericidal and lytic actions of antibodies against many gram-negative Bacteria. (Interestingly, gram-positive Bacteria are not killed by specific antibody, in either the presence or absence of complement, although gram-positive Bacteria are opsonized.) Death of gram-negative Bacteria involves antibodies against antigens on the surface of the cell; complement perhaps brings about an actual change in the cell surface, possibly by an enzymelike reaction, after antibodies have prepared the way. No cytocidal or lytic effect is seen when cells and complement are mixed alone, but if antibody has bound to cells first, death or lysis occurs rapidly after complement is added. The presence of antibodies is absolutely required to fix complement.

As noted, complement is necessary for opsonization, the promotion of phagocytosis by antibodies against bacterial capsules. However, opsonization does not require the whole complement pathway but only the proteins through C3. Opsonization occurs because the binding of C3 causes the cells to adhere to phagocytes that have C3 receptors on their surfaces. It is presumably the C3 receptors that are involved in adherence of the C3-coated bacterial cell to the phagocyte.

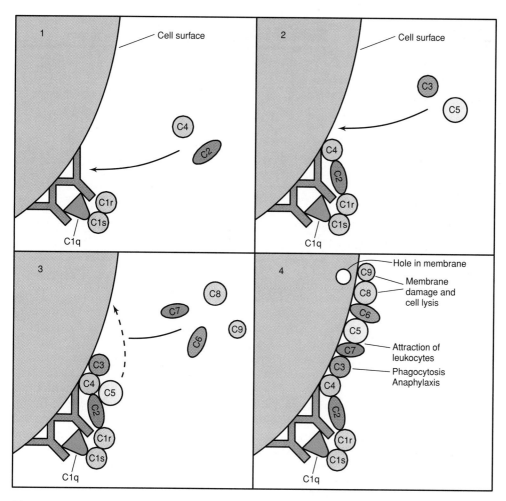

(a)

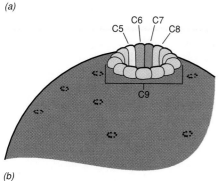

(b)

Figure 20.26 The complement system. (a) The sequence, orientation, and activity of the various components as they interact to lyse a cell. Panel 1: Binding of the antibody recognition unit C1q and other C1 proteins; panel 2: the C4-C2 complex; panel 3: the C4-C2-C3-C5 complex—after activation the C5 unit travels to an adjacent membrane site; panel 4, binding of C6, C7, C8, and C9, resulting in membrane damage. (b) Three-dimensional view of the hole formed by complement components C5 through C9.

CONCEPT CHECK   *20.13*

The complement protein cascade is a nonspecific mechanism for cell destruction and opsonization. It is triggered by specific immune antibody interactions and is a critical component of host defense.

✔ *What classes of immunoglobulin fix complement?*

✔ *What complement components are necessary to complete cell lysis?*

20.14
Polyclonal and Monoclonal Antibodies

In the whole animal, a typical immune response results in the production of a broad spectrum of immunoglobulin molecules of various classes, affinities, and specificities for the determinants present on the antigen. The immunoglobulins directed toward a particular determinant represent only a portion of the total antibody pool. Each immunoglobulin is produced by a single clone

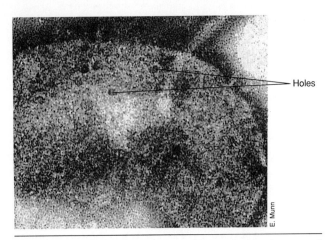

E. Munn

Figure 20.27 Electron micrograph of a negatively stained preparation of *Salmonella paratyphi,* showing holes created in the cell envelope as a result of reaction involving cell envelope antigens, specific antibody, and complement.

TABLE 20.5	Characteristics of monoclonal and polyclonal antibody production

Polyclonal	Monoclonal
Contains many antibodies recognizing many determinants on an antigen	Contains a single antibody recognizing only a single determinant
Various classes of antibodies are present (IgG, IgM, and so on)	Single class of antibody produced
To make a specific antibody, a highly purified antigen is necessary	Can make a specific antibody using an impure antigen
Reproducibility and standardization difficult	Highly reproducible

that has proliferated from antigen stimulation; an antiserum containing such a mixture of antibodies is thus referred to as a **polyclonal antiserum.** However, it is possible to produce antibodies of only a *single* specificity, derived from a single B cell clone. A variety of such monospecific antibodies, called **monoclonal antibodies,** have been generated for use in research and clinical medicine. Table 20.5 compares the properties of antibodies prepared against an antigen in the usual way—by preparing a polyclonal antiserum—with monoclonal antibodies.

Hybridomas and monoclonal antibodies

Each antigen-activated B cell becomes a clone that produces a monoclonal antibody—it is genetically programmed to do so (see Section 20.12). But we consider here how to isolate and grow a single B cell clone for experimental production of monoclonal antibodies. Normal lymphocytes cannot be grown and maintained in cell culture. However, B lymphocytes can be fused with myeloma (tumor) cells to form hybrid cell lines that will grow in culture and retain the ability to produce antibodies. This technique of B cell–myeloma cell fusion is known as the **hybridoma technique** and is summarized in Figure 20.28.

First, a mouse is immunized with the antigen of interest and a period of weeks allowed for specific B cell clones to proliferate and begin producing antibody by the normal sequence of events. Then, spleen tissue, rich in B lymphocytes, is removed from the mouse and the B cells are fused with myeloma cells (Figure 20.28). Although true fusions represent only a small fraction of the total cell population remaining in the mixture, addition of the compounds hypoxanthine, aminopterin, and thymidine to the medium (the so-called HAT medium)

strongly selects for *fused* hybrids. This occurs because unfused myeloma cells are unable to use the metabolites hypoxanthine and thymidine to bypass a metabolic block caused by aminopterin, a cell poison. By contrast, fused hybrid cells are able to use hypoxanthine and thymidine. The hybrids grow normally in HAT medium because they received the genetic information for this metabolic pathway from the normal cell fusion partner. Unfused B cells die off in a week or two because they are unable to grow in culture for more than a few cell divisions. Following fusion, the clones of interest must be identified; that is, if individual fused cells are placed in the wells of microtiter plates and allowed to grow and produce antibody, which cells produce the antibodies of interest?

A variety of assay techniques (see Sections 20.15 and 21.8) can be used to identify clones producing monoclonal antibodies. From a typical fusion, several distinct clones are isolated, each making a monoclonal antibody to a different determinant on the antigen. Once the clones of interest are identified, they can be grown in cell culture or they can be injected into mice and be perpetuated as a mouse myeloma tumor. As a mouse tumor line, hybridomas secrete large amounts of monoclonal antibody (in mice over 10 mg of pure monoclonal antibody can be obtained per milliliter of mouse peritoneal fluid). Specific hybridomas of interest can also be frozen. The hybridoma can then be carefully thawed and injected back into mice, or grown in tissue culture when more of a specific monoclonal antibody is needed.

Monoclonal antibodies are extremely useful reagents for research and medical science. For research purposes, monoclonal antibodies directed against specific cell markers, for example, the CD4 and CD8 lymphocyte markers, can be used to identify and separate mixtures of T cells from one another. We will examine methods for utilizing the discriminatory abilities of monoclonal antibodies in Chapter 21.

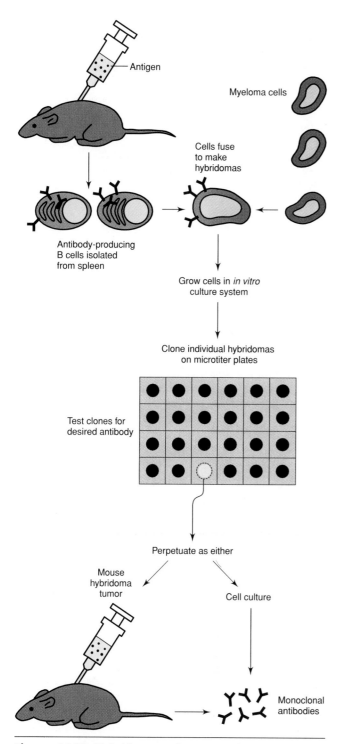

Figure 20.28 Hybridoma technique and production of monoclonal antibodies.

Other research applications of monoclonal antibodies include studies on the active site of enzymes by observing the effect on an enzyme's activity when treated with monoclonal antibodies to specific determinants on the enzyme. Monoclonal antibodies are also useful in the immunological typing of bacteria and in the identification of cells containing foreign surface antigens (for example, a virus-infected cell). Monoclonal antibodies have also been used in genetic engineering for identifying and measuring levels of gene products not detectable by other methods.

Monoclonal antibodies and medicine

Of considerable importance are the applications of monoclonal antibodies to clinical diagnostics and medical therapeutics. We discuss the use of monoclonal antibodies in clinical diagnostics in Section 21.4. In therapeutics, perhaps the boldest application of monoclonal antibodies may be in the detection and treatment of human malignancies. Malignant cells contain a variety of surface antigens not expressed on the surfaces of normal cells. Tumor antigens include differentiation antigens expressed during fetal development but not in the normal adult, and unique, tumor-specific cell surface antigens (the latter were originally detected by monoclonal antibodies prepared against random tumor cell surface antigens). Because monoclonal antibodies prepared against cancer-related antigenic determinants should be able to distinguish between normal and malignant cells, monoclonal antibodies serve as vehicles for delivering toxins to malignant cells. Such tumor-specific antibodies, covalently linked to toxins, are now undergoing clinical testing. The specificity of the monoclonal antibody treatment can greatly improve cancer chemotherapy and avoid damage to normal cells by the toxic chemical and radiation treatments used in conventional cancer chemotherapy.

Monoclonal antibodies have also been prepared that can distinguish human transplantation antigens, thus improving the specificity of the tissue matching process critical for successful transplantations. Monoclonal antibodies also show great promise for increasing the specificity of a variety of conventional clinical tests including blood typing, rheumatoid factor determination, and even pregnancy determination; the latter test employs monoclonal antibodies prepared against the specific hormones associated with pregnancy (see the box, Over-the-Counter Immunodiagnostic Kits). Other monoclonals are being developed for *in vivo* diagnostics. For example, monoclonal antibodies have been used experimentally to detect exposed myosin in heart muscle; myosin is a major heart muscle protein that normally becomes detectable only following damage to the heart muscle. A myosin-specific monoclonal antibody would therefore be useful for diagnosing the extent of heart damage following a heart attack.

As with any clinical reagent, the greater the specificity, the more useful it is in clinical diagnostics. Monoclonal antibodies now offer the physician and

OVER-THE-COUNTER IMMUNODIAGNOSTIC KITS

A number of manufacturers now market immunodiagnostic kits for use by the general public. For example, virtually every pharmacy now carries several brands of pregnancy tests and many also stock kits to determine the onset of ovulation in women.

These kits are all based on principles involving detection of a hormone excreted in the urine. All pregnancy tests detect the presence of human chorionic gonadotropin (HCG). When the ovum is fertilized, it produces HCG, which functions to maintain the corpus luteum and sustain pregnancy. As pregnancy continues, the new embryo produces and releases increasing amounts of HCG. The hormone, by this time being produced in massive amounts, is released into the bloodstream, removed by the kidneys, and excreted in the urine. Normal menstruation, which occurs about 10–14 days after ovulation if the ovum is not fertilized, does not take place, and at this time the level of HCG present in the urine of the pregnant woman is high enough to be detected easily by these tests. Thus, detectable HCG in the urine means that the individual is pregnant. The tests are advertised to be sensitive enough to detect pregnancy on the first day of a missed menstrual period, that is, 10–14 days *after* fertilization, but *not* on the first day of a pregnancy.

The key to marketing these kits is their simplicity. Since most people using the tests are not trained in standard laboratory procedures, the test methods must be straightforward and the results must be easy to interpret. Typically, the patient must first provide a urine sample. In the easiest test, the urine sample is simply applied to a test strip. In others, the first step involves pouring the sample through a filter or immersing a "dipstick" in the urine, followed by exposing the filter or dipstick to a second solution. The results are interpreted as a simple color change: a positive test turns the strip, dipstick, or filter from white to pink (see figures) within 30 min or less. All tests share certain principles. At some point, the HCG in the urine is bound to an HCG-specific monoclonal antibody that is immobilized on a solid support—the dipstick, test strip, or filter. Next, a second HCG-specific monoclonal antibody, provided in a solution for the dipstick or filter test, and already soaked onto the test strip for the one-step test, reacts with the immobilized antibody–HCG complex. This second antibody is chemically linked to colloidal gold particles. If this second antibody reacts with the HCG, it becomes immobilized and the whole complex appears colored.

These kinds of tests, all modifications of sol particle immunoassay (SPIA) tests, are also extensively used in doctor's offices to help diagnose diseases. For example, a number of available tests are specific for cell surface antigens on *Streptococcus pyogenes*, which causes strep throat. Formerly, the only way to positively identify a streptococcal infection was to culture and identify the organism, a process that involved skilled personnel and a laboratory. Because of the requirement for incubation of the culture, results were not available for at least 16 hr. Today, results are available in minutes, using material from a throat swab as the antigen source. Several test kits are also available for detecting serum antibodies that react with self proteins. These kits are useful in diagnosing autoimmune diseases such as rheumatoid arthritis.

These tests have several advantages. They are simple to perform, require no technical expertise, are relatively inexpensive, and are reasonably accurate. They also have a long shelf life, and no special storage conditions are required. However, there are disadvantages. They are qualitative tests, which means that they can be used only when the outcome is absolute. For example, one is either pregnant or not pregnant. However, quantitative measures of antigen levels might be desirable in some cases, such as in the tests that identify autoimmune antibodies. In addition, many manufacturers do not provide adequate positive and negative controls to ensure that the test works properly. Thus, results should always be interpreted with caution, and negative or positive results may be in error because of improper test procedures or faulty test reagents. ■

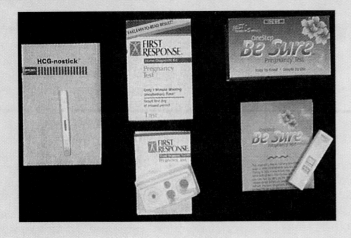

clinical microbiologist the ultimate in immunological specificity. (See the box, Urine Testing for Drug Abuse, for a description of the use of monoclonal antibodies for detecting illicit drugs.)

20.15

Antigen–Antibody Reactions

Antigen–antibody reactions are most easily studied *in vitro* using preparations of antigens and antisera. The study of antigen–antibody reactions *in vitro* is called **serology** and is especially important in clinical diagnostic microbiology. A variety of serological reactions can be observed (Table 20.6), depending on the specific properties of the antigen and antibody and on the conditions chosen for reaction. Serology has many applications, only a few of which will be discussed here. Further practical applications of these basic reactions will be discussed in detail in the next chapter.

The principle of all antigen–antibody reactions lies in the *specific* combination of determinants on the antigen with the *variable* region of the antibody molecule (Figure 20.29). All antigen–antibody reactions rely on antigen binding as a required first step. The combining site of an antibody molecule measures about 2×3 nm and is capable of binding to a small number of amino acids, about 10–15, of a protein antigen. Recall that the recognition of antigen by antibody is ultimately governed by the secondary, tertiary, and quarternary struc-

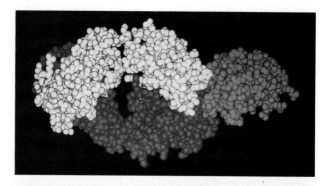

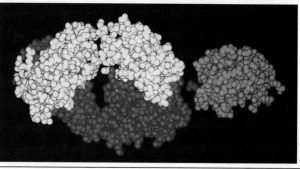

Figure 20.29 Structure of the combining site of an antigen and immunoglobulin. The antigen (lysozyme) is in green. The variable region of the immunoglobulin heavy chain is shown in blue, and the light chain in yellow. The amino acid shown in red is a glutamine residue of lysozyme. The glutamine residue fits into a pocket on the immunoglobulin molecule, but the overall antigen–antibody recognition involves contacts made between several other amino acids on both the immunoglobulin and antigen as well. Reprinted with permission from *Science 233:* 747 (1986) © AAAS.

ture of the polypeptide chains in the antibody molecule. The folding itself is dictated by the primary structure—the amino acid sequence—which in turn is controlled by the genetic constitution of the B cell. The end result of this complex process is a specific antibody molecule that combines with a specific antigenic determinant as shown in Figure 20.30. This combination can block or distort the antigen sufficiently to reduce or totally eliminate its biological activity.

Some typical antigen–antibody reactions are illustrated in Figure 20.30.

TABLE 20.6	Types of antigen–antibody reactions	
Location of antigen	**Accessory factors required**	**Reaction observed**
Soluble	None	Precipitation
On cell or inert particle	None	Agglutination
Flagellum	None	Immobilization or agglutination
On bacterial cell	Complement	Lysis
On bacterial cell	Complement	Killing
On erythrocyte	Complement	Hemolysis
Toxin	None	Neutralization
Virus	None	Neutralization
On bacterial cell	Phagocyte, complement	Phagocytosis (opsonization)

URINE TESTING FOR DRUG ABUSE

Many government agencies and private employers carry out drug testing programs in an effort to control drug abuse in the workplace. If a person uses a drug, metabolites of the drug are excreted into the urine. Testing urine for the presence of such metabolites thus permits detection of drug use. Common drugs tests include those for cannabinoids (found in marijuana), cocaine, phencyclidine, amphetamines, propoxyphene, benzodiazepine, opiates, steroids, and barbiturates.

Because of the minute quantities of drugs or drug metabolites that occur in the urine, extremely sensitive detection methods are needed, but these methods must also be very specific. Immunological procedures are among the most sensitive and specific methods known for testing urine. Two types of immunoassays are usually employed for drug testing, radioimmunoassay (RIA) and enzyme-linked immunosorbent assay (ELISA). For a drug immunoassay, an antibody must first be prepared against the drug or drug metabolite to be assayed. The antibody thus specifically recognizes an antigenic determinant of the drug. The assay is based on the principle of *competition* between a labeled antigen present in the drug testing system and an unlabeled antigen (the drug or drug metabolite in the urine) for binding sites on the specific antibody. In both the RIA and ELISA tests, the *greater* the amount of drug in the urine, the *less* of the labeled drug will be bound.

In RIA drug testing, known amounts of radiolabeled drug are added to a urine sample together with known amounts of an antibody specific for the drug being tested. The presence of the drug is measured by determining the radioactivity of the antibody bound to a solid substrate. As the concentration of drug in the urine goes up, the radioactivity bound goes down. Concentrations of drug as low as 1–5 μg/ml can be detected in 1–5 hr. The RIA test is very suitable for large-scale screening programs because automatic pipetting and counting procedures can be used.

In the ELISA drug testing procedure, the antigen (drug) is covalently linked to an enzyme, commonly glucose-6-phosphate dehydrogenase, whose activity can be detected by a simple and sensitive colorimetric assay. If the enzyme, via its linked drug ligand, becomes associated with the antibody, it loses its enzymatic activity, but free enzyme can react with the substrate. Urine is mixed with a reagent consisting of antibody to the drug, the glucose-6-phosphate dehydrogenase–drug derivative, and glucose 6-phosphate. If drug is present in the urine, it competes with the enzyme–drug derivative for specific antibody and more enzyme is then free to react with its substrate. Thus, the more drug present in the subject's urine, the more intense the color. The ELISA procedure has the advantage that the analysis time is short and the color change can be detected simply. However, the ELISA procedure is sometimes less sensitive than the RIA procedure. The ELISA procedure is especially valuable where a small number of samples are to be analyzed at the site where the urine

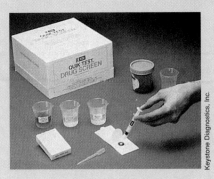

Keystone Diagnostics, Inc.

is collected (for instance, away from a laboratory). In some cases, the reagents have been incorporated into paper strips, providing means for urine testing that can be used even by those lacking laboratory experience (see the photo). For large-scale testing, robotic devices have the capability of processing up to 18,000 urine samples per hour with each sample identified by a bar code identification sticker.

Either polyclonal or monoclonal antibodies can be used for urine testing. Monoclonal antibodies have the advantage of defined specificity but may not necessarily have as high an affinity for the drug as a selected polyclonal antibody. The selection of an antibody depends on the specificity and sensitivity needed, the cost of the product, and the ease of use.

One problem with the use of any of these immunological methods for urine testing is cross-reaction with other chemicals that might be present in the urine. For instance, the pain killer ibuprofen cross-reacts with marijuana metabolites in some tests, and the antihistamine drug diphenhydramine can cross-react in the ELISA test for methadone, a controlled narcotic. Because of such false-positive reactions, positive tests must be confirmed by independent tests, such as the use of gas chromatography or high performance liquid chromatography. Highly sensitive combined mass spectrometer–gas chromatographic methods are also available and are valuable for identifying exotic drugs or drugs that elicit a poor immune response.

Immunological tests for drugs in the urine have provided a new and potentially widespread technique for controlling the use of illegal drugs in the workplace. Urine drug testing is another example of the power and utility of immunology in modern society. ∎

Neutralization

Neutralization of microbial toxins by specific antibody can occur when toxin molecules and antibody molecules directed against the toxin combine in such a way that the active portion of the toxin is blocked (Figure 20.30*a*). Neutralization reactions of this type are known for a wide variety of exotoxins, including most of those listed in Table 19.4. An antiserum containing an antibody that neutralizes a toxin is referred to as an **antitoxin.** Reactions of a similar nature also occur between viruses and their specific antibodies. For instance, anti-

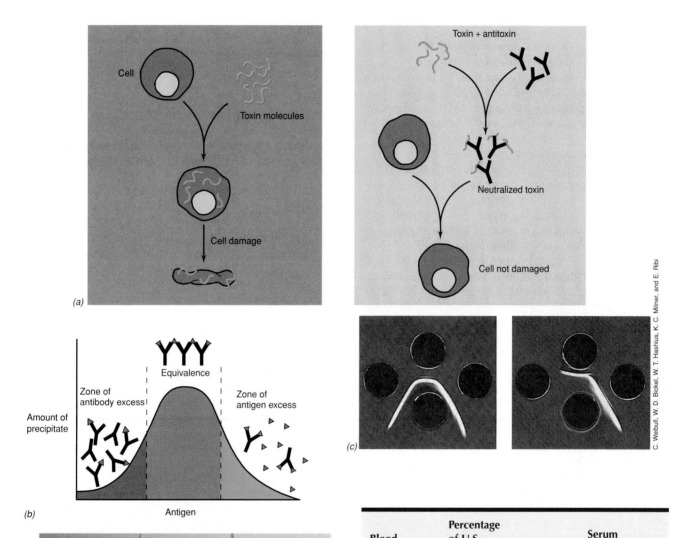

Figure 20.30 Some antigen–antibody reactions. (a) Neutralization of toxin by antitoxin. (b) Precipitation reaction between soluble antigen and antibody. The graph shows the extent of precipitation as a function of antigen and antibody concentration. (c) Precipitation in an agar gel, a process known as immunodiffusion. Wells in agar labeled S contained antibodies to cells of *Proteus mirabilis*. Wells labeled A, B, and C contained extracts of *P. mirabilis* cells. (Left) a line of identity is observed. (Right) Antigen E does not react and antigen F shows a line of partial identity. (d) Agglutination. Typing of human blood in the ABO system. (Left) Typical agglutination pattern. (Right) Table of expected results with different blood types.

Blood type	Percentage of U.S. population	Serum	
		Anti A	Anti B
Type O	47	No aggl.	No aggl.
Type A	42	Aggl.	No aggl.
Type B	8	No aggl.	Aggl.
Type AB	3	Aggl.	Aggl.

bodies directed against the protein coats of viruses may prevent the adsorption of the viruses to host cells. This is one means by which a viral vaccine can act. If the vaccine stimulates antibodies that react to free virus, the virus is unable to attach to and enter host cells and cannot initiate infection.

Precipitation

Since antibody molecules generally have two combining sites (that is, they are bivalent) (see Figure 20.9), it is possible for each site to combine with a separate antigen molecule, and if the antigen also has more than one available epitope, a **precipitate** may develop consisting of aggregates of antibody and antigen molecules (Figure 20.30b). Because they are easily observed *in vitro*, precipitation reactions are very useful serological tests, especially in the quantitative measurement of antibody concentrations. Precipitation occurs maximally only when there are optimal proportions of the two reacting substances because when either reactant occurs in excess, the formation of large antigen–antibody aggregates is not possible (Figure 20.30b).

Precipitation reactions carried out in agar gels, referred to as *immunodiffusion*, are useful in the study of the specificity of antigen–antibody reactions. Both antigen and antibody diffuse outward from separate wells cut in the agar gel, and precipitation bands form in the region where antibody and antigen meet in optimal proportions (Figure 20.30c). The shapes of the precipitation bands and the distance from the wells are characteristic for the reacting substances, and it is possible to determine whether two antigens reacting against antibodies in an antiserum are identical by observing the bands formed when the two antigens are placed in adjacent wells near the antiserum well. For example, if two antigens in adjacent wells are identical, they will form a fused precipitin band. This is referred to as a *line of identity*. If, on the other hand, adjacent wells contain one antigen in common but one well contains a second antigen, a *line of partial identity* will form (Figure 20.30c). The small protruding precipitin line (representing a reaction between the antiserum and the second antigen) is referred to as a *spur*. Immunodiffusion is often used as a tool in biochemical research to assess the relatedness of proteins obtained from different sources. Unfortunately, the readily visible precipitation reactions are not very sensitive. Microgram quantities of antibody are necessary to visualize a precipitate (∞ Table 21.6). Therefore, precipitation reactions are not widely used.

Agglutination

If an antigen is not in solution but is present on the surface of a cell or other particle, an antigen–antibody reaction can lead to clumping of the particles, called **agglutination.**

Soluble antigens can be adsorbed to or coupled chemically to cells or other particulate structures such as latex beads, and they can then be detected by agglutination reactions, the cell or particle serving only as an inert carrier. This greatly increases the ability to detect the presence of antibodies against soluble antigens because agglutination is much more sensitive than precipitation.

Agglutination of red blood cells is referred to as *hemagglutination*. Red blood cells contain a variety of antigens, and people differ in the antigens present on their red blood cells. The classification of blood antigens is called *blood typing* and is a good example of an agglutination reaction (Figure 20.30d). The principle of blood typing is that antibodies, specific for particular erythrocyte surface antigens, cause red blood cells to visibly clump. The human red cell antigens most commonly tested for are called A, B, and D. Humans also make antibodies to some nonself blood group antigens. For example, type A humans have antibodies to group B antigens, while type B individuals have antibodies to group A antigens. Type AB individuals have neither A nor B antibodies, whereas type O individuals have anti-A and anti-B activity. Antibodies against A and B antigens are called *natural antibodies* because they appear to be produced by most individuals in response to ubiquitous antigen sources such as enteric Bacteria.

The reason for blood typing before blood transfusion is to prevent the red blood cell destruction that would occur if blood containing a particular antigen were transfused into a person having an antibody against this antigen. If agglutination occurred, the clumps of cells could lodge in blood vessels or arteries and block the flow of blood, causing serious illness or death; if the antibodies caused hemolysis through the action of complement, anemia could result.

The presence or absence of the A and B antigens or antibodies is the basis for the ABO blood grouping assay (Figure 20.3d). A small drop of blood is placed on a microscope slide containing antiserum prepared against either the A or B antigens of human erythrocytes. These commercially available high titer antisera are obtained from human donors who have been immunized to these antigens by natural or artificial means. The agglutination reaction observed in the ABO blood grouping test is rapid and easy to interpret (Figure 20.30d); hence this simple assay is used routinely to group human blood. At least 15 different blood group systems have been described in humans, the ABO system being one of the most important.

The agglutination reaction is about 100-fold more sensitive than precipitation, but many clinical and diagnostic immunoassays are based on techniques that are even more sensitive (∞ Table 21.6). We will examine a number of these sensitive techniques in the context of applied immunology in Chapter 21.

20.16

Immune Diseases

Immune diseases occur when the immune system reacts in a way that results in host damage. In this section, we first deal with **hypersensitivity,** the overreaction of the cell-mediated immune response to cause *delayed-type hypersensitivity* or the antibody-mediated immune response to cause *immediate hypersensitivity.* We next focus on the increasingly important **autoimmune diseases,** which result from a breakdown of tolerance (see Section 20.10) and inappropriate reactions to self antigens. Finally, we look at a group of toxins called **superantigens,** which are produced by bacteria and result in massive immune cell stimulation and host damage.

Delayed-type hypersensitivity

This form of hypersensitivity is cell-mediated and involves the activities of a special subset of T cells called T delayed-type hypersensitivity (T_D) cells, with symptoms appearing several hours to a few days after exposure to the eliciting agent. Delayed-type hypersensitivity reactions occur against invasion by certain microorganisms and as a result of skin contact with certain sensitizing chemicals. The latter phenomenon, known as **contact dermatitis,** is responsible for many of the common allergic skin disorders in humans, including those to poison ivy, cosmetics, and certain drugs or chemicals. Shortly after exposure to the agent, the skin feels itchy at the site of contact, and within several hours reddening and swelling appear, indicative of a general inflammatory response. Localized tissue destruction occurs as a result of the activities of immune cells.

A good example of a delayed-type hypersensitivity reaction is the development of immunity to the causal agent of tuberculosis, *Mycobacterium tuberculosis* (Figure 20.31). This cellular immune response was first discovered by Robert Koch during his classic work on tuber-culosis and has been widely studied. Antigens derived from the bacterium, when injected subcutaneously into an animal previously immunized with the same antigen, elicit a characteristic skin reaction that develops fully only after a period of 24–48 hr. (In contrast, skin reactions to antibody-mediated responses as seen in conventional allergic reactions discussed later, develop almost immediately after antigen injection.) In the region of the injected antigen, T_D cells become stimulated by the antigen and release cytokines, which attract large numbers of macrophages. The macrophages are responsible for the ingestion and digestion of the invading antigen. The characteristic skin reaction seen at the site of injection is a result of an inflammatory response arising as a result of the release of cytokines by activated T cells. This skin response serves as the basis for the **tuberculin test** for determining prior exposure to *M. tuberculosis* (Figure 20.31).

A number of microbial infections elicit delayed-type hypersensitivity reactions. In addition to tuberculosis, these include leprosy, brucellosis, psittacosis (all caused by Bacteria), mumps (caused by a virus), and coccidioidomycosis, histoplasmosis, and blastomycosis (caused by fungi). In all these cellular immune reactions, characteristic skin reactions are elicited after injection of antigens derived from the pathogens, and these skin reactions can be used to diagnose prior exposure to the pathogen.

Immediate-type hypersensitivity

Another form of hypersensitivity is known in which an immune response occurs more quickly than in the delayed-type response. This form of hypersensitivity is referred to as **immediate-type hypersensitivity** and is mediated by *antibodies* instead of activated T cells. Immediate-type hypersensitivities commonly involve

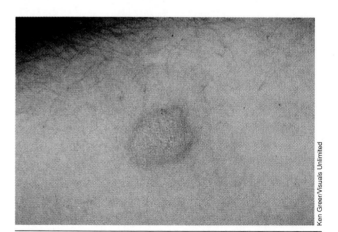

Figure 20.31 Cell-mediated immunity. A positive tuberculin test, typical for delayed hypersensitivity, and the result of the action of T_D effector cells. The raised area of inflammation is 1.5 cm in diameter.

allergies and, depending on the individual and the antigen, may cause mild or extremely severe, even life-threatening reactions, a process known as *anaphylaxis*. Antigens that cause these hypersensitivities are known as *allergens*.

From 10 to 20% of the human population suffer from immediate-type hypersensitivities, involving allergic (*anaphylactic*) reactions to specific allergens such as pollens, animal dander, and a variety of other agents (Table 20.7). In a typical anaphylactic reaction, an allergen elicits (on first exposure) the production of immunoglobulins of the class IgE. Instead of circulating like immunoglobulins of the class IgG or IgM, IgE molecules tend to become attached via a cell-binding constant domain (see Section 20.5) to the surfaces of mast cells and basophils (Figure 20.32). **Mast cells** are nonmotile connective tissue cells found adjacent to capillaries throughout the body, and **basophils** are motile white blood cells (leukocytes) that make up about 1% of the total leukocyte population. On subsequent exposure to antigen, the cell-bound IgE molecules bind the antigen. This triggers the release of several allergic mediators from mast cells and basophils. An antigen must bridge at least two IgE molecules on the cell to initiate release of these active substances.

The primary chemical mediators released from mast cells and basophils are **histamine** and **serotonin** (both are modified amino acids). Several other anaphylactic mediators have been characterized as small peptides. The release of histamine and serotonin causes

TABLE 20.7	Common immediate-type hypersensitivity allergens

Pollen and fungal spores (hay fever)
Insect venoms (bee sting)
Penicillin and other drugs
Certain foods
Animal dander
Mites in house dust

dilation of blood vessels and contraction of smooth muscle, which initiate the typical symptoms of anaphylaxis. These symptoms include, among others, difficulty in breathing, flushed skin, copious mucus production, sneezing, and itchy, watery eyes. In general, the symptoms are relatively short-lived, but once initially sensitized by an allergen, an individual can respond repeatedly on subsequent exposure to the antigen. Depending on the individual, the magnitude of the anaphylactic reactions may vary from mild symptoms (or none), to such severe symptoms that the individual goes into **anaphylactic shock.** In humans, anaphylactic shock is characterized by severe respiratory distress, capillary dilation (causing a sharp drop in blood pressure), and flushing and itching of the skin. If severe cases of anaphylactic shock are not treated immediately with large doses of adrenalin to counter smooth muscle contraction and promote breathing, death can occur.

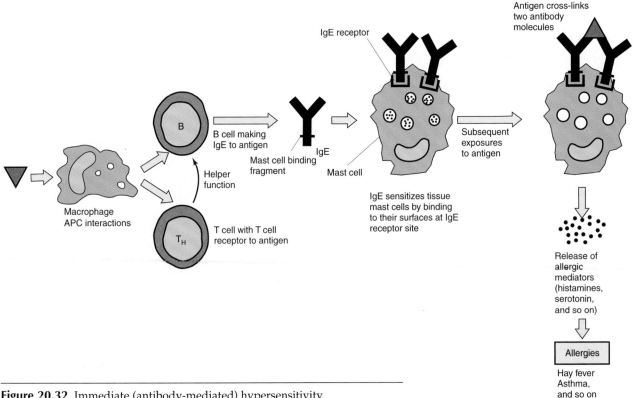

Figure 20.32 Immediate (antibody-mediated) hypersensitivity.

Autoimmune diseases

We saw in Section 20.10 how T and B cells destined to react with self antigens are eliminated or anergized during the process of lymphocyte maturation. However, in some individuals, the anergized T and B lymphocytes become reactivated. This resurgence of self-reactive clones leads to immunological disorders referred to as *autoimmune diseases*. Most are slow, progressive disorders in which the function of some specific organ or set of tissues becomes increasingly poorer with time (Table 20.8). Depending on the specific disorder, autoimmunity may involve autoantibodies or a cellular immune response to self constituents. Certain autoimmune diseases are highly organ-specific. For example, in *Hashimoto's disease*, autoantibodies are made against thyroglobulin, the major iodine-containing protein in the thyroid. In *juvenile diabetes* (insulin-dependent diabetes mellitus), autoantibodies against the insulin-producing cells, the islets of Langerhans, are observed. Such antibodies may cause the disease by destroying the cells, or they may be the result of antigens released from islet cells damaged by other mechanisms. *Systemic lupus erythematosis* (SLE) involves a large-scale production of autoantibodies against many self constituents, including DNA. This disease and others like it are induced by circulating antigen–antibody complexes that may deposit in several different body tissues, such as the kidney and the spleen. Complement fixation and the resulting lytic and inflammatory responses cause local but often severe cell damage at the site of the complex deposition. Organ-specific autoimmune diseases are sometimes more easily controlled clinically because the product of organ function, such as thyroxin in hypothyroidism or insulin in diabetes, can often be supplied in pure form from another source. More generalized syndromes such as SLE can be controlled only by immunosuppressive therapy, but this approach is not without risk because of the increased chance of opportunistic infections.

Evidence is accumulating that heredity has an important influence on the incidence, type, and severity of autoimmune diseases. An inherited tendency to develop certain autoimmune diseases is known to exist; many autoimmune diseases correlate strongly with the presence or absence of certain major histocompatibility complex (MHC) antigens (see Section 20.7). Studies of model autoimmune diseases in mice support such a genetic link, but the conditions necessary for developing autoimmunity are also known to be dependent on other factors, including hormone levels and the presence of infectious agents such as pathogenic bacteria and viruses.

Superantigens

In Sections 19.8 and 19.9, we discussed the effects of bacterial exotoxins and enterotoxins on the host. As we noted, not all exotoxins or even enterotoxins work by the method described for cholera toxin (∞ Figure 19.19). In fact, a completely different mechanism of action is used by a family of exotoxins known as **superantigens.** These toxins are produced by the staphylococci and streptococci. They include *toxic shock syndrome toxin, staphylococcal enterotoxin,* and *exfoliating toxin* (∞ Table 19.4), and all have a toxic mechanism dependent on the immune system. These anti-

TABLE 20.8   Some autoimmune diseases of humans

Disease	Organ or area affected	Mechanism
Juvenile diabetes (insulin-dependent diabetes mellitus)	Pancreas	Autoantibodies against surface and cytoplasmic antigens of islets of Langerhans
Myasthenia gravis	Skeletal muscle	Autoantibodies against acetylcholine receptors on skeletal muscle
Goodpasture's syndrome	Kidney	Autoantibodies against basement membrane of kidney glomeruli
Rheumatoid arthritis	Cartilage	Autoantibodies against self IgG antibodies, which form complexes deposited in joint tissue, leading to inflammation and cartilage destruction
Hashimoto's disease (hypothyroidism)	Thyroid	Autoantibodies to thyroid surface antigens
Male infertility (some cases)	Sperm cells	Autoantibodies agglutinate host sperm cells
Pernicious anemia	Intrinsic factor	Autoantibodies prevent absorption of vitamin B_{12}
Systemic lupus erythematosis	DNA, cardiolipin, nucleoprotein, blood clotting factors	Massive autoantibody response to various cellular constituents
Addison's disease	Adrenal glands	Autoantibodies to adrenal cell antigens
Allergic encephalomyelitis	Brain	Cell-mediated response against brain tissue
Multiple sclerosis	Brain	Cell-mediated and autoantibody response against central nervous system

gens all bind to the β chain of the T cell receptor (see Section 20.6) at a site that is outside the normal antigen binding site. In some cases, this binding may stimulate greater than 10% of the normal T cells in the affected individual. A normal antigen, even a very potent one, stimulates less than 0.1% of the total T cell population. Such massive stimulation results in simultaneous participation of all these T cells in an overwhelming cell-mediated response characterized by systemic inflammatory effects, sometimes resulting in generalized shock (such as toxic shock syndrome). This general method for stimulating the host immune system to induce host damage may be a common pathogenic strategy for the staphylococci and streptococci.

CONCEPT CHECK   20.16

Allergy results when foreign antigens stimulate cellular or humoral immunity and effector cells react so vigorously that host tissue is damaged by the resulting inflammation. Autoimmunity results when the immune response reacts or cross-reacts with self antigens.

✔ *Distinguish between* hay fever *and* poison ivy *allergies.*

✔ *Why is insulin-dependent diabetes mellitus considered an autoimmune disease?*

20.17
Immunity to Infectious Diseases

The major role of the immune response in the body is to protect the animal from the consequences of infection. The importance of antibodies in disease resistance is shown dramatically in individuals with the genetic disorder **agammaglobulinemia,** in whom antibodies are not produced because their B cells are defective. Such individuals are unusually sensitive to infectious diseases, especially those involving bacterial infections, and in the days before antibiotic therapy, few of them survived infancy. The general lack of an antibody response is also observed in those suffering from acquired immunodeficiency syndrome (AIDS). However, in this case the problem is not due to defective B cells. Instead, AIDS patients suffer from a virtually total cessation of CD4 (primarily T_H) cell activities (∞ Section 23.7). The crucial importance of T cells in the production of immunity is clearly evident in AIDS patients: the inability to mount an immune response leads to the eventual death of AIDS patients from infectious diseases (∞ Section 22.4).

The purposeful artificial induction of specific immunity to infectious diseases provided one of the first real triumphs of the scientific method in medicine and was one of the outstanding contributions of microbiology to the treatment and prevention of infectious diseases. An animal or human may acquire immunity to a disease in several ways.

(1) The individual may acquire infection and develop immunity. This is **natural active immunity** because the immunization was a natural outcome of infection and the infected individual produced the immune response. (2) The individual may be given injections of an antigen known to induce formation of antibodies, a type of immunity known as **artificial active immunity** because the individual in question produced the antibodies. (3) Alternatively, the individual may receive injections of an antiserum derived from another individual who has previously formed antibodies against the antigen in question. This is called **artificial passive immunity** because the individual receiving the antibodies played no active part in the antibody-producing process. (4) Finally, **natural passive immunity** also occurs. For several months after birth, newborns have maternal IgG antibodies in their blood. These antibodies, acquired through the placenta before birth, provide valuable disease protection while the immune system of the newborn is maturing. Active and passive immunity are contrasted in Table 20.9.

An important distinction between active and passive immunity is that in *active* immunity the immunized individual is fundamentally changed because the individual is able to continue to make the antibody in question and will exhibit a secondary or booster response if exposed to another dose of the same antigen (see Figure 20.21). Active immunity often remains throughout life. A *passively* immunized individual never has more antibodies than it received in the initial injection, and these antibodies gradually disappear from the body; moreover, a later inoculation with the antigen does not elicit a booster response. Active immunity is usually used as a *prophylactic* measure, to protect a person against future attack by a pathogen. Passive immunity is usually *therapeutic,* designed to cure a person who is presently suffering from the disease. For example, tetanus toxoid (see the following section) actively immunizes an individual against future encounters with *Clostridium tetani* exotoxin, whereas tetanus antiserum (antitoxin) (see later) is administered to passively immunize an individual suspected of coming in contact with *C. tetani* exotoxin via growth of the organism in a penetrating wound.

Vaccination

The material used in inducing active immunity, the antigen or mixture of antigens, is known as a **vaccine.** However, to induce active immunity to toxin-caused dis-

TABLE 20.9	Comparison of active and passive immunity	
Active immunity	**Passive immunity**	
Exposure to antigen; immunity achieved by injecting antigen	No exposure to antigen; immunity achieved by injecting antibodies to antigen	
Antibodies made by individual achieving immunity	Antibodies made in a secondary host	
Immune system activated to antigen; immunological memory in effect	No immune system activation; no immunological memory	
Antibody titer can remain high through subsequent boosters	Antibody titer progressively decays	
Immune state develops over a period of weeks	Immune state develops immediately	

eases, the toxin itself is not injected. Many exotoxins can be modified chemically so they retain their antigenicity but are no longer toxic. Such a modified exotoxin is called a **toxoid.** Toxoids are usually not such efficient antigens as the original exotoxin, but they can be given safely and in high doses. When immunization against whole microorganisms is necessary, such as for endotoxin-producing organisms, the microorganism is usually killed by an agent such as formaldehyde, phenol, or heat, and the dead cells are then injected. Endotoxin-caused diseases for which vaccines are made routinely are whooping cough and typhoid. Formaldehyde treatment is also used to inactivate viruses in preparing some vaccines, such as the Salk polio vaccine.

Immunization with live cells or virus is usually more effective than with dead or inactivated material. Often it is possible to isolate a mutant strain of a pathogen that has lost its virulence but still retains the immunizing antigens; strains of this type are called **attenuated strains** (∞ Section 19.11).

A summary of vaccines available for use in humans is given in Table 20.10.

Vaccination practices

Infants possess antibodies derived from their mothers and hence are relatively immune to infectious disease during the first 6 months of life. However, it is desirable to immunize infants for key infectious diseases as soon as possible so that their own *active* immunity can replace the *passive* immunity received from the mother. However, infants have a rather poorly developed ability to form antibodies, and so immunization is not begun until a few months after birth. As discussed in Section 20.11, a single injection of antigen does not lead to a high antibody titer; it is desirable therefore to use a series of injections so that a *high titer* of antibody is developed. A typical vaccination schedule for children from birth to school age is given in Table 20.11.

The importance of immunization procedures in controlling infectious diseases is well established. On introduction of a specific immunization procedure into a population, the incidence of the disease often

drops markedly (Figure 20.33). The degree of immunity obtained by vaccination varies greatly, depending on the individual and on the quality and quantity of the vaccine. However, lifelong immunity is rarely achieved by means of a single injection, or even a series of injections, and the population of immune cells induced by immunization gradually disappears from the body. One way in which antigenic stimulation occurs even in the absence of immunization is by nonsymptomatic or minor infections. A natural infection results in a rapid booster response, leading to both a further increase in activated antibody-producing cells and to production of antibody. It is not known how long immunity can last in the complete absence of antigenic stimulation, the immune period varying with different antigens. However, active immunity to certain vaccines such as tetanus toxoid can last many years, and in some cases, a lifetime.

Immunization procedures are not only beneficial to the individual but are effective public health procedures because disease spreads poorly through a population in which a large proportion of the individuals are immune (∞ Section 22.6).

Passive immunity

The material used in inducing passive immunity—the serum containing antibodies—is known as a *serum,* an *antiserum,* or an *antitoxin* (the last applies to a serum containing antibodies directed against a toxin). Antisera are obtained either from large-sized immunized animals, such as horses, or from humans who have high antibody **titers** (quantities). These individuals are said to be **hyperimmune.** The antiserum or antitoxin is standardized to contain a known antibody titer; a sufficient number of units of antiserum must be inoculated to neutralize any antigen that might be present in the body. Sometimes the immunoglobulin fraction of pooled human serum is used as a source of antibodies. It contains a wide variety of antibodies that normal people have formed through the years by artificial or natural exposure to various antigens. Pooled sera are used when hyperimmune antisera are not available, but in recent years the routine use of

TABLE 20.10   Available vaccines for infectious diseases in humans

Disease	Type of vaccine used
Bacterial diseases	
Diphtheria	Toxoid
Tetanus	Toxoid
Pertussis	Killed bacteria (*Bordetella pertussis*)
Typhoid fever	Killed bacteria (*Salmonella typhi*)
Paratyphoid fever	Killed bacteria (*Salmonella paratyphi*)
Cholera	Killed cells or cell extract (*Vibrio cholerae*)
Plague	Killed cells or cell extract (*Yersinia pestis*)
Tuberculosis	Attenuated strain of *Mycobacterium tuberculosis* (BCG)
Meningitis	Purified polysaccharide from *Neisseria meningitidis*
Bacterial pneumonia	Purified polysaccharide from *Streptococcus pneumoniae*
Typhus fever	Killed bacteria (*Rickettsia prowazekii*)
Haemophilus influenzae meningitis	Conjugated vaccine (polysaccharide of *Haemophilus influenzae* conjugated to protein)
Viral diseases	
Yellow fever	Attenuated virus
Measles	Attenuated virus
Mumps	Attenuated virus
Rubella	Attenuated virus
Polio	Attenuated virus (Sabin) or inactivated virus (Salk)
Influenza	Inactivated virus
Rabies	Inactivated virus (human) or attenuated virus (dogs and other animals)
Hepatitis A	Recombinant DNA vaccine
Hepatitis B	Recombinant DNA vaccine or inactivated virus
Varicella (chickenpox)	Attenuated virus

pooled sera is declining because of the threat of contamination [for example, with the human immunodeficiency (HIV) virus, the cause of AIDS.]

The most common use of passive immunization is in the prevention of infectious hepatitis (resulting from hepatitis A virus) (∞ Section 23.13). Pooled human immunoglobulin, often referred to as *gamma globulin,* contains fairly high titers of antibody against hepatitis A virus because infection with hepatitis A virus is widespread in the population. Travelers to areas where the incidence of infectious hepatitis is high, such as tropical Africa and North Africa, the Middle East, Asia, and parts of South America, may be given

prophylactic doses of pooled human gamma globulin. A single dose of 0.02 ml/kg body weight should protect for up to 2 months, but for more prolonged exposures, doses at repeated intervals should be given. Pooled human gamma globulin may also be of value in the *therapy* of infectious hepatitis if given early in the incubation period.

Alternate vaccine strategies

Most vaccines currently in use are produced from whole organisms or toxoids, as described earlier in this section. However, there are several other methods for

TABLE 20.11   Immunizations recommended for children, United States

Vaccine	At birth	1–2 months	2 months	4 months	6 months	12 months	15 months	6–18 months	4–6 years
DTP (diphtheria, tetanus, pertussis)			X	X	X		X		X
Polio (oral, trivalent)			X	X			X		X
MMR (measles, mumps, rubella)							X		X
HbCV (*Haemophilus influenzae* conjugate)									
Option 1			X	X	X		X		
Option 2			X	X		X			
HBv (hepatitis B virus)									
Option 1	X	X						X	
Option 2		X		X				X	

Source: Centers for Disease Control, Atlanta, GA.

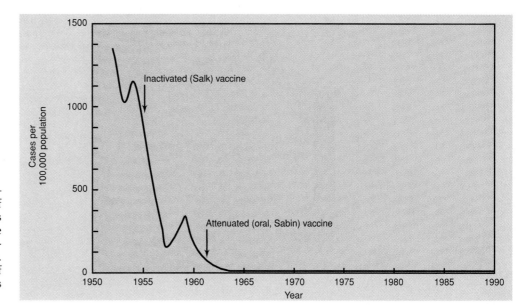

Figure 20.33 Cases of polio in the United States 1950–1990, showing the consequences of the introduction of polio vaccine. There have been no cases of polio in the United States since 1990.

producing antigens suitable for vaccination purposes. Vaccines made by some of these methods are replacing more traditional vaccines.

The simplest alternate approach to vaccine development is the use of *synthetic peptides.* To make a vaccine, a peptide can be synthesized that corresponds to a known epitope on an infectious agent. For example, the structure of the protein antigen responsible for immunity to foot-and-mouth virus, an important animal pathogen, is known. A synthetic peptide of 20 amino acids constituting the antigenic portion of the protein has been made and attached to suitable carrier molecules. This synthetic vaccine evokes an excellent neutralizing antibody response to foot-and-mouth virus. However, as a general method, this approach has one major problem: the entire antigenic structure of the protein must be known to make an effective vaccine. Although this condition has been met for foot-and-mouth virus, very few pathogens have such a well-defined antigenic profile.

Sophisticated molecular biology techniques can also be used to make vaccines. As shown in Figure 10.18, genes that encode antigens from virtually any virus can be cloned into the vaccinia virus genome and expressed. The new *genetically engineered* vaccinia virus can then be used directly to induce immunity to the product of the cloned gene. Such vaccines for hepatitis B surface antigen and chickenpox are now in use. This method depends on the availability of the cloned gene that encodes the antigen and also on the ability of the vaccinia virus to express the cloned gene as an antigenic protein. The use of recombinant DNA methods to develop vaccines was discussed in Section 10.13.

Finally, a novel experimental approach to vaccine production has resulted from our knowledge of anti-

body structure. The foreign antigen binding site of an antibody molecule can itself act as an antigen to elicit a new antibody response, even in the animal that generated the first antibody. The variable region of an antibody molecule is referred to as its *idiotype.* Because they are structurally unique (recall there are about 10^9 different variable regions), idiotypes can serve as *antigens* in the same animal that produced them, yielding *anti-idiotypic* antibodies (Figure 20.34). If the variable region of an anti-idiotypic antibody is *complementary* to the idiotype, this region should structurally resemble the original antigen itself. Anti-idiotypic antibodies appear to be a normal part of the immune response and because they bind to idiotype, may play a role in curbing overproduction of specific idiotypes.

Idiotypes and anti-idiotypes have practical implications in vaccine immunology. Because the variable region of an anti-idiotypic antibody mimics the original antigen, there is considerable interest in the use of these antibodies as *antigen-free vaccines.* How does this work? The anti-idiotypic antibody does not cause disease but can serve as an antigen to elicit protective antibodies in an animal. Vaccines of this type are developed by isolating a monoclonal antibody (which contains a single idiotype) reactive against a specific pathogenic microorganism and then using this antibody as an *antigen* to produce an anti-idiotypic antibody (Figure 20.34). The latter, when properly purified, can then be employed as a vaccine against the pathogen. Anti-idiotype vaccines to *Escherichia coli,* rabies virus, hepatitis virus, and *Trypanosoma cruzi,* a parasite, have been effective in experimental animal models. If antigen-free vaccines can be developed for a number of human diseases, genetic engineering might also enter the pic-

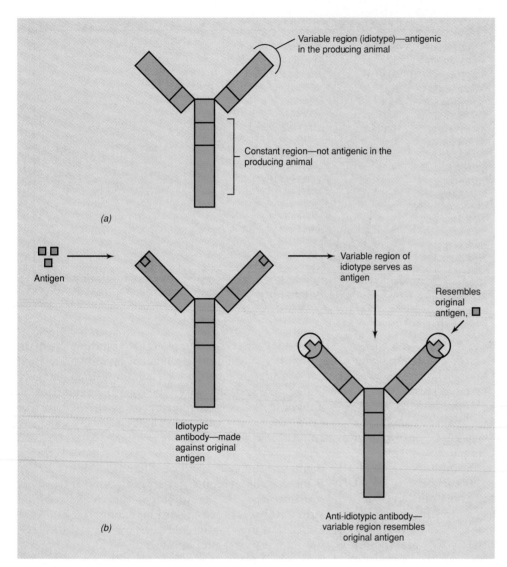

(a)

Variable region (idiotype)—antigenic in the producing animal

Constant region—not antigenic in the producing animal

Antigen

Variable region of idiotype serves as antigen

Resembles original antigen,

Idiotypic antibody—made against original antigen

Anti-idiotypic antibody—variable region resembles original antigen

(b)

Figure 20.34 Generation of idiotypic and anti-idiotypic antibodies during an immune response. (a) Structure of an immunoglobulin showing portions of the molecule that are immunogenic in the producing animal. (b) Generation of anti-idiotypic antibodies. Note how the *variable* region of the anti-idiotypic antibody resembles the original antigen to which the idiotypic antibody was formed.

ture because cloned genes coding for anti-idiotypic antibodies (obtained by standard genetic engineering methods) (∞ Section 10.13) would allow antigen-free vaccines to be produced in microorganisms.

Antigen-free vaccines based on anti-idiotypic antibodies represent a novel and safe means of disease prophylaxis. This therapeutic alternative has emerged from basic knowledge of the structure of antibody molecules and the nature of the immune response.

CONCEPT CHECK   20.17

Immunity to infectious disease can be either passive or active, natural or artificial. Vaccination, a form of active artificial immunity, is widely employed in preventing and alleviating infectious diseases.

✔ *What is an example of natural, passive immunity?*

✔ *What alternative strategies exist to immunization with live vaccines?*

Material for Review

REVIEW QUESTIONS

1. What is the origin of the cells involved in the immune response? Where do B cells and T cells mature?

2. What substances provoke immune responses? What substances do not? What properties are necessary for a substance to provoke an immune response?

3. Explain how phagocytes engulf and kill microorganisms, with particular attention to oxygen-dependent mechanisms.

4. Differentiate between T and B lymphocytes. Differentiate between T cell subsets. *Hint:* Define each cell type or subset by surface markers.

5. Describe the basic protein structure of an IgG molecule. How do molecules of the other major Ig classes differ?

6. Describe the basic structure of the T cell receptor (TCR). Why do TCRs need specificity comparable to the immunoglobulins?

7. Describe the basic structure of class I and class II major histocompatibility complex (MHC) proteins. In what fundamental ways do they differ?

8. Describe the action of IL-2 on cells in the immune system. How and where is it produced? What cells does it affect?

9. What important protective functions are ascribed to cell-mediated immunity? How do T_C cells kill targets?

10. How does positive and negative selection of the T cell repertoire result in tolerance? Why does tolerance sometimes break down?

11. Describe the various antigen-presenting cells (APCs) that may be involved in antibody formation. What do they have in common?

12. Explain how up to 10^9 immunoglobulin binding sites are generated from the ~300 genes encoding portions of the variable region.

13. Describe the complement cascade. Is the order of protein interaction important? Why or why not?

14. Describe two cells that would be used to generate a hybridoma. How could you select for a fused hybrid cell?

15. How does precipitation differ from agglutination? What do these two basic reactions have in common?

16. Define the differences between immediate and delayed-type hypersensitivity in terms of immune effectors, target tissues, antigens, and clinical outcome.

17. List the diseases for which you have been vaccinated. List the diseases for which you have naturally acquired immunity.

APPLICATION QUESTIONS

1. Trace the path of a stem cell that becomes a macrophage, a B lymphocyte, or a T lymphocyte. What environmental factors induce stem cells to become one of these end cells?

2. All immunogens are antigens. However, all antigens are not immunogens. Explain these statements, using relevant examples.

3. Phagocytes are intensely involved with antigen processing and presentation, as are B cells. Explain the differences in the antigens that may be processed by each cell type.

4. T lymphocytes interact with a variety of cells. Describe each interaction at the level of the cell surface receptors and cytokines involved. What do all these reactions have in common? How do they differ?

5. Antibodies of the IgA class are probably more prevalent than those of the IgG class. What advantage does this have for the host?

6. How does TCR interaction with antigen differ from that of immunoglobulin (Ig) interaction? What fundamental difference does this make in the kinds of antigens recognized by the TCR and Ig?

7. Why, in your opinion, do MHC proteins not need the level of diversity found in TCRs and Igs? Is there diversity in MHC at any level? Explain the value of diversity for each of the proteins listed.

8. Some cytokines have very general functions, such as cell killing. Based on your knowledge of cytokine signaling, how might this occur?

9. Cell-mediated immunity has often been implicated in tumor surveillance and defense. Explain why this might be so and contrast this mechanism with an antibody-mediated tumor surveillance system.

10. Why might tolerance break down and what potential outcomes would result? *Hint:* Think about cross-reactions, especially with regard to pathogens.

11. In certain situations, B cells are activated in the absence of T cells. Why might T cells not be able to "help" B cells? *Hint:* Think about the chemical nature of the antigens.

12. Why do TCRs need a diversity level comparable to that of Igs? Trace the evolutionary development of each of these molecules back to a theoretical precursor molecule. What features would the precursor have had in common with both TCRs and Igs?

13. Complement is regarded as the most important humoral defense mechanism. Do you agree with this statement? Explain your answer. Predict what might happen to individuals who lack complement.

14. Predict the protective ability of a passively administered monoclonal antibody against a viral surface protein. Would you expect these monoclonal antibodies to be effective over a long time period? *Hint:* Think about antibiotics and resistance.

15. How can the sensitivity (that is, the amount of antigen detected) be increased in antigen–antibody reactions? *Hint:* What do you know about the sensitivity of the precipitation and agglutination reactions?

16. Immune diseases are becoming much more prevalent, especially in people over 50 years of age. Why is this happening?

17. Many infectious diseases have no effective vaccines. Pick several of these diseases (for example, AIDS, malaria, the common cold) and explain why current vaccine methods have not been effective. Prepare some alternate strategies for vaccination against the diseases you have chosen.

SUPPLEMENTARY READINGS

Abbas, A. K., A. H. Lichtman, and **J. S. Pober.** 1994. *Cellular and Molecular Immunology,* 2nd edition. W. B. Saunders, Philadelphia. A very readable, basic-science approach to immunology.

Honjo, T., F. W. Alt, and **T. H. Rabbitts** (eds.). 1989. *Immunoglobulin Genes.* Academic Press, New York. A classic advanced-level treatment of immunoglobulin gene structure and rearrangements.

Janeway, C. A., and **P. Travers.** 1996. *Immunobiology,* 2nd edition. Garland Publishing, New York. A very complete up-to-date treatment of immunology.

Kuby, J. 1994. *Immunology.* A well-organized, readable basic treatment of immunology.

Roitt, I. M. 1994. *Essential Immunology,* 8th edition. Blackwell Scientific, Oxford. Probably the best short treatment of immunology, with emphasis on the general processes of immune functions rather than molecular details.

Roitt, I. M., J. Brostoff, and **D. K. Male.** 1996. *Immunology,* 4th edition. C. V. Mosby, St. Louis. A well-illustrated textbook of immunology for advanced undergraduates.

Stites, D. P., A. I. Terr, and **T. G. Parslow.** 1994. *Basic and Clinical Immunology,* 8th edition. Appleton and Lange, Norwalk, CT. A very thorough, up-to-date text oriented toward medical professionals.

Tizard, I. R. 1995. *Immunology: An Introduction,* 4th edition. Saunders College Publishing, New York. A good introductory immunology textbook.

On~line resources for this chapter are on the World Wide Web at: http://www.prenhall.com/~brock (click on the table of contents link and then select Chapter 20).

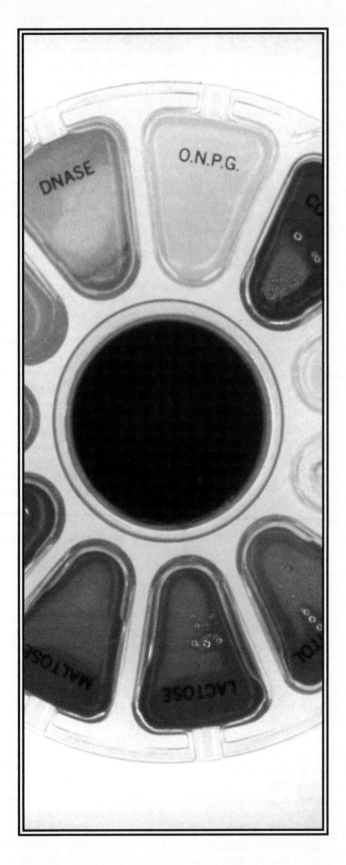

Clinical and Diagnostic Microbiology and Immunology

The miniaturization and automation of clinical culture methods permits rapid identification of clinical pathogens. Shown here is a miniaturized test system for observing carbohydrate utilization by organisms isolated from clinical samples. All of the tests are inoculated simultaneously from a single specimen.

MINIGLOSSARY for Chapter 21

Agglutination reaction between antibody and particle-bound antigen, resulting in visible clumping of the particles

Bacteremia the presence of bacteria in the blood

Complement Fixation the consumption of complement by an antibody–antigen reaction

ELISA *enzyme-linked immunosorbent assay*

Fluorescent Antibody covalent modification of an antibody molecule with a fluorescent dye; the dye makes the antibody visible under fluorescent light

Gonococcus *Neisseria gonorrhoeae,* the gram-negative diplococcus that causes gonorrhea

Immunoblot (Western Blot) electrophoresis of proteins followed by transfer to a membrane and detection by addition of specific antibodies

Nucleic Acid Probe in clinical microbiology, a short oligonucleotide of unique sequence used as a hybridization probe for identifying pathogens (see Chapter 6 for general usage)

Precipitation reaction between antibody and a soluble antigen resulting in a visbible, insoluble complex

RIA *radioimmunoassay*

Septicemia blood infection

Titer in an immunological context, the quantity of antibody present in a solution

---

The most important activity of the microbiologist in medicine is to isolate and identify the agents that cause infectious disease. This major area of microbiology is called **clinical** or **diagnostic microbiology.** There is increasing awareness of the importance of precise identification of the pathogen for proper treatment of infectious disease, and new sophisticated methods are being continually developed. Clinical laboratories are generally able to isolate, identify, and determine the antibiotic sensitivity of most routinely encountered pathogenic bacteria within 48 hr of sampling. However, recent advances in rapid diagnostic methods have made it possible to identify some pathogens in minutes and antibiotic susceptibility patterns in hours. Diagnostic methods based on immunological and molecular biology methods make it possible to identify many pathogens without culturing the organism at all. This is particularly important for the diagnosis of viral and protozoal infections, diseases that are typically difficult to identify because of the difficulty of culturing the agent. The clinical microbiologist works with and advises the physician in matters relating to the diagnosis and treatment of infectious diseases.

21.1
Isolation of Pathogens from Clinical Specimens

The physician, following clinical examination of the patient, may suspect that an infectious disease is present. Samples of infected tissues or fluids are then collected for microbiological, immunological, and molecular biological analyses (Figure 21.1). Depending on the kind of infection, materials collected may include blood, urine, feces, sputum, cerebrospinal fluid, or pus. A sterile swab may be passed across a suspected infected area (Figure 21.2). The swab is then streaked over the surface of an agar plate or placed directly in a liquid culture medium. In some cases, small pieces of living tissue may be aseptically removed (biopsy) for culture. Table 21.1 summarizes current recommendations for culture of organisms isolated from typical clinical specimens.

If clinically relevant organisms are to be isolated and a correct diagnosis made, care must be taken in obtaining samples of clinical specimens. The physician must ensure that the specimen is removed from the *actual site of the infection.* Recovery or detection of pathogens may not be possible if insufficient inoculum is available. The sample must also be taken under aseptic conditions so that contamination is avoided. Care must also be taken to ensure that metabolic requirements for certain organisms, such as anoxic conditions, are maintained. Once taken, the sample is analyzed as soon as possible. If it cannot be analyzed immediately, it is usually refrigerated to slow down deterioration. In the rest of this section, we describe some of the most common microbiological procedures used to obtain and culture microorganisms in the clinical laboratory.

Blood cultures

Bacteremia means the presence of bacteria in the blood (∞ Section 19.7). Bacteria are normally cleared from the bloodstream rapidly. Therefore, bacteremia is uncommon in healthy individuals, and the presence of bacteria in the blood is generally indicative of systemic infection. The most common pathogens found in blood include *Pseudomonas aeruginosa,* enteric bacteria, especially *Escherichia coli* and *Klebsiella pneumoniae,* and the gram-positive cocci *Staphylococcus aureus* and *Streptococcus pyogenes.* The classic type of blood infection is **septicemia,** resulting from a virulent organism entering the blood from a focus of infection, multiplying, and traveling to various body tissues to initiate

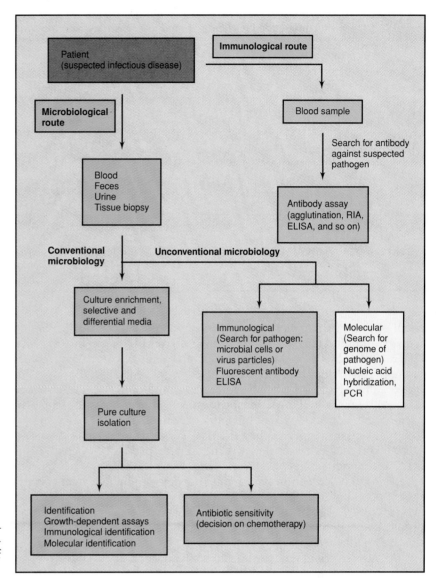

Figure 21.1 Clinical and diagnostic methods used for isolation and identification of infectious pathogens.

new infections. Septicemia is indicated by the presence of severe systemic symptoms, usually with fever and chills, followed by prostration. In many disease situations, culture of the blood provides the only immediate way of isolating and identifying the causal agent, and diagnosis therefore depends on careful and proper blood culture.

The standard blood culture procedure is to remove 10 ml of blood aseptically from a vein and inject it into a blood culture bottle containing an anticoagulant and

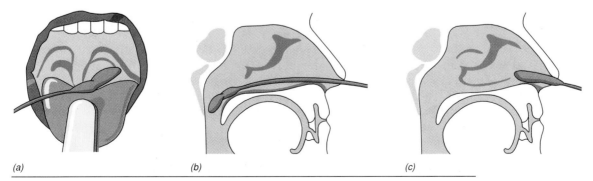

(a)   *(b)*   *(c)*

Figure 21.2 Methods for obtaining specimens from the upper respiratory tract. (a) Throat swab. (b) Nasopharyngeal swab passed through the nose. (c) Swabbing the inside of the nose.

TABLE 21.1   Recommended enrichment media for primary isolation purposes in a clinical microbiology laboratory[a]

Specimen	Media[b]				
	Blood agar	Enteric agar	CA	TM	Anaerobic agar
Fluids: chest, abdomen, pericardium	+	+	+	−	+
Feces: rectal swabs	+	+	+	−	−
Surgical tissue biopsies: lung, lymph nodes	+	+	−	−	+
Throat: swabs, sputum, tonsil, nasopharynx	+	+	+	−	−
Genitourinary swabs: urethra, vagina, cervix	+	+	+	+	−
Urine	+	+	−	−	−
Blood	+	−	−	−	+
Swabs: wounds, abscesses, exudates	+	+	+	−	+

[a]Data from Murray, P. R., E. J. Baron, M. A. Pfaller, F. C. Tenover, and R. H. Yolken. 1995. *Manual of Clinical Microbiology,* 6th edition. American Society for Microbiology, Washington, DC.

[b]Blood agar, 5% whole sheep blood added to trypticase soy agar; enteric agar, either eosin–methylene blue (EMB) agar or MacConkey agar; CA, chocolate (heated blood) agar; TM, Thayer–Martin agar; anaerobic agar, thioglycolate-containing blood agar or supplemented thioglycolate agar incubated anaerobically.

an all-purpose culture medium. Two cultures are set up with one bottle being incubated aerobically and one anaerobically. Media used are all relatively rich, containing protein digests and other complex ingredients. Blood culture bottles are incubated at 35°C and examined daily for up to 7 days. Most clinically significant bacteria are recovered within this period. Some blood isolation systems employ a chemical that lyses red and white blood cells, releasing potential intracellular pathogens that might otherwise be overlooked. Microorganisms in blood cultures are commonly detected by visual inspection (turbidity), microscopic examination, and subculture. Automated blood culture systems detect growth by continuously monitoring carbon dioxide production and turbidity.

Because a certain amount of skin contamination is unavoidable during initial drawing of the blood, a contamination rate of 2–3% can be expected. Contamination may be indicated if certain organisms commonly found on the skin are isolated, such as *Staphylococcus epidermidis,* coryneform bacteria, or propionibacteria, although even these organisms can occasionally cause infection of the wall of the heart (subacute bacterial endocarditis). Thus, considerable microbiological and clinical experience is necessary when interpreting blood cultures.

Urine cultures

Urinary tract infections are very common, and because the causal agents are often identical or similar to bacteria of the normal flora (for example, *Escherichia coli*), considerable care must be taken in the bacteriological analysis of urine. Since urine supports extensive bacterial growth under many conditions, fairly high cell numbers are often found in urinary infection. In most cases, the infection occurs as a result of an organism ascending the urethra from the outside. Occasionally,

even the bladder may become infected. Urinary tract infections are the most common form of *nosocomial* (hospital-acquired) infection (∞ Section 22.7).

Significant urinary infection generally results in bacterial counts of 10^5 or more organisms per milliliter of a clean-voided midstream specimen, whereas in the absence of infection, contamination of the urine from the external genitalia (almost unavoidable to some extent) results in less than 10^3 organisms per milliliter. The most common urinary tract pathogens are members of the enteric bacteria, with *E. coli* accounting for about 90% of the cases. Other urinary tract pathogens include *Klebsiella, Enterobacter, Proteus, Pseudomonas, Staphylococcus saprophyticus,* and *Enterococcus faecalis. Neisseria gonorrhoeae,* the causal agent of gonorrhea, does not grow in the urine itself, but in the urethral epithelium, and must be diagnosed by different methods (see later).

Direct microscopic examination of urine may be used to indicate *bacturia,* the presence of abnormal numbers of bacteria in the urine. However, because nearly all urine contains some level of bacterial growth, significant bacturia is most commonly monitored by using a variety of commercially available dipstick tests. For example, one dipstick test monitors the reduction of nitrate by detecting the reduction product, nitrite. A positive test is indicated by a color change on the dipstick (Figure 21.3). Since nitrite production occurs only when significant numbers ($>10^5$ per milliliter) of enteric organisms are present, the method is a virtually instantaneous check for urinary tract infections. Other dipstick tests for urinary tract infections, often used in conjunction with nitrate reduction, detect esterase (produced by leukocytes) (∞ Section 20.3) and peroxidase (produced by a variety of bacteria) (∞ Sections 5.11 and 13.24). A positive dipstick test is then followed by a urine culture.

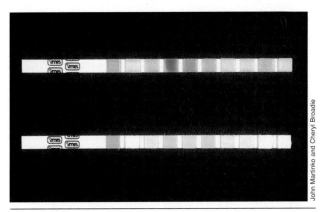

Figure 21.3 Urinalysis dipstick test. A control strip is shown underneath the test strip. From left to right, the strip measures abnormal levels of glucose, bilirubin, ketones, specific gravity, blood, pH, protein, urobilinogen, nitrite, and leukocytes (esterase) in a urine sample. Abnormal readings for esterase (trace positive, far right) and nitrite (strong positive, second from right) indicate bacturia. Subsequent culture of this sample indicated the presence of *Escherichia coli.*

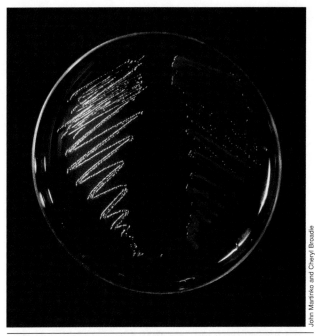

Figure 21.4 An eosin–methylene blue (EMB) agar plate showing a lactose fermenter, *Escherichia coli* (left), and a non-lactose fermenter, *Pseudomonas aeruginosa* (right). Note the green metallic sheen of the *E. coli* colonies.

To culture potential urinary tract pathogens, two media are used: blood agar as a nonselective general medium, and a medium selective for enteric bacteria, such as MacConkey or eosin–methylene blue agar (EMB) (see Section 21.2 and Figure 21.4). These specialized media permit the initial differentiation of lactose fermenters from nonfermenters, and the growth of gram-positive organisms such as *Staphylococcus* spp. (common skin contaminants) is inhibited. Organisms isolated can be identified, and antibiotic susceptibility tests performed. Experienced clinical microbiologists may make a tentative identification of an isolate by observing the color and morphology of colonies of the suspected pathogen grown on various selective media as described in Table 21.2. Such an identification must be followed up with more detailed analyses, but clinical microbiologists use this information in conjunction with more detailed test results, discussed throughout the remainder of this chapter, to make a positive identification.

Finally, if no bacterial growth is obtained in spite of persistent urinary tract infection symptoms, a clinician may request direct cultures for a number of fastidious organisms, especially *Neisseria gonorrhoeae, Chlamydia trachomatis, Branhamella* spp., mycoplasma, or several anaerobic organisms (∞ Section 23.6).

Fecal cultures

Proper collection and preservation of feces is important in the isolation of intestinal pathogens. During storage, the pH of feces drops, and thus an extended delay between sampling and sample processing must be avoided. This is especially critical for the isolation of *Shigella* and *Salmonella* species, both of which are rather sensitive to acid pH. Samples, collected from feces

freshly voided into a sterile plastic cup, are placed in a vial containing phosphate buffer for transport to the lab. If a patient has a bloody or pus-containing stool, this material is always sampled; such discharges generally contain a large number of the organisms of interest. In the case of suspected foodborne or waterborne infections, fecal samples should be inoculated into a variety of selective media (see Section 21.2) for the isolation of specific bacteria or characterization of intestinal parasites. Positive identifications are made by the techniques described in later sections.

Wounds and abscesses

Infections associated with traumatic injuries such as animal or human bites, burns, cuts, or the penetration of foreign objects, must be carefully sampled in order to recover the relevant pathogen. This is because wound infections and abscesses are frequently contaminated with members of the normal flora. Swab samples of such lesions are frequently misleading. The best sampling method is to aspirate purulent (pus-containing) lesions with a sterile syringe and needle following disinfection of the skin surface with 70% ethyl or isopropyl alcohol. Internal purulent discharges are usually sampled by biopsy or from tissues removed in surgery.

A variety of pathogens can be associated with wound infections, and because some of these are anaerobes, samples should be transported from the collection site under anaerobic conditions. A common pathogen

TABLE 21.2   Colony characteristics of frequently isolated gram-negative rods cultured on various clinically useful media[a]

Organism	Agar media[b]			
	EMB	MC	SS	BS
Escherichia coli	Dark center with greenish metallic sheen (see Figure 21.4)	Red or pink	Red to pink	Mostly inhibited
Enterobacter	Similar to *E. coli*, but colonies are larger	Red or pink	White or beige	Mucoid colonies with silver sheen
Klebsiella	Large, mucoid, brownish	Pink	Red to pink	Mostly inhibited
Proteus	Translucent, colorless	Transparent, colorless	Black center, clear periphery	Green
Pseudomonas	Translucent, colorless to gold (see Figure 21.4)	Transparent, colorless	Mostly inhibited	No growth
Salmonella	Translucent, colorless to gold	Translucent, colorless	Opaque	Black to dark green
Shigella	Translucent, colorless to gold	Transparent, colorless	Opaque	Brown or inhibited

[a]Adapted from Murray, P. R., E. J. Baron, M. A. Pfaller, F. C. Tenover, and R. H. Yolken. 1995. *Manual of Clinical Microbiology,* 6th edition. American Society for Microbiology, Washington, DC.
[b]BS, Bismuth sulfite agar; EMB, eosin–methylene blue agar; MC, MacConkey agar; SS, *Salmonella–Shigella* agar.

associated with purulent discharges is *Staphylococcus aureus*, but enteric bacteria, *Pseudomonas aeruginosa*, and the anaerobes *Bacteroides* and *Clostridium* species are also commonly encountered. The major isolation media are blood agar, several selective media for enteric bacteria (Tables 21.1 and 21.2), and blood agar containing additional supplements and reducing agents for obligate anaerobes. Smears from such specimens should also be examined directly by microscopy.

Genital specimens and the laboratory diagnosis of gonorrhea

In males, a purulent urethral discharge is the classic symptom of the sexually transmitted disease gonorrhea (∞ Section 23.6). If no discharge is present, a suitable sample can be obtained using a sterile narrow-diameter cotton swab that is inserted into the anterior urethra, left in place a few seconds to absorb any exudate, and then removed for culture of *Neisseria gonorrhoeae*, the causative agent of gonorrhea. Alternatively, a sample of the first early morning urine of an infected individual usually contains viable cells of *N. gonorrhoeae*. In females suspected of having gonorrhea or other genital infections, samples are usually obtained by swab from the cervix and the urethra.

Gonorrhea is one of the most common infectious diseases in adults, and clinical microbiological procedures are central to its diagnosis. *N. gonorrhoeae* (referred

to clinically as the *gonococcus*) colonizes mucosal surfaces of the urethra, uterine cervix, anal canal, throat, and conjunctiva. The organism is quite sensitive to drying and therefore is transmitted almost exclusively by direct person-to-person contact, usually by sexual intercourse. The major goal of public health measures to control gonorrhea involves identification of asymptomatic carriers, and this requires microbiological analysis.

Because the gonococcus is a gram-negative coccus, usually observed as diplococci, and similar organisms are not very common in the normal flora of the urogenital tract, direct microscopy of Gram-stained material is of value. For example, observation of gram-negative diplococci in a urethral discharge or in a vaginal or cervical smear is presumptive evidence for gonorrhea. In acute gonorrhea, microscopy usually reveals phagocytized gram-negative diplococci in the polymorphonuclear leukocytes (∞ Section 20.3), with virtually no other organisms present (Figure 21.5a).

Cultural procedures have a higher degree of sensitivity than microscopic analyses. Most media for the culture of *N. gonorrhoeae* contain heated blood or hemoglobin (referred to as *chocolate agar* because of its deep brown appearance), the heating causing the formation of a precipitated material, which is quite effective in absorbing toxic products present in the agar and other media constituents. A second primary isolation medium, called *Thayer–Martin agar,* also is used for isolation of *N. gonorrhoeae* (Figure 21.5b). This medium incorporates the antibiotics vancomycin, nystatin, and

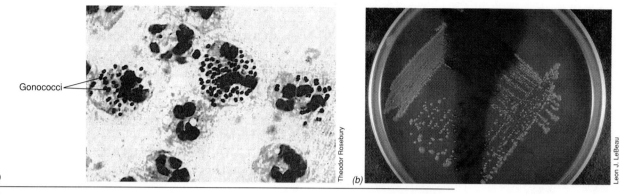

Gonococci

(a)

Theodor Rosebury

(b)

Leon J. LeBeau

Figure 21.5 (a) Photomicrograph of *Neisseria gonorrhoeae* within human polymorphonu-clear leukocytes from a cervical smear. Note how many cells are in pairs as diplococci (arrows). (b) Colonies of *N. gonorrhoeae* growing on Thayer–Martin agar. The plate has been stained in the middle with a reagent that turns colonies blue if cells contain cytochrome *c* (the oxidase test).

colistin, to which most clinical isolates of *N. gonorrhoeae* are naturally resistant.

After streaking, the plates must be incubated in a humid environment in an atmosphere containing 3–7% CO_2 (CO_2 is required for growth of gonococci). The plates are examined after 24 and 48 hr, and portions of colonies should be immediately tested by the oxidase test because all *Neisseria* are oxidase-positive (see Section 21.2). Oxidase-positive gram-negative diplococci grow-ing on chocolate agar can be presumed to be gonococci if the inoculum was derived from genitourinary sources, but definitive identification requires determination of carbohydrate utilization patterns or immunological or nucleic acid probe tests (see Sections 21.4–21.10).

A rapid test employing chromogenic substrates has been developed for differentiating *N. gonorrhoeae* from other species of *Neisseria*. With the use of colonies on plates, various species can be differentiated by the substrate color reaction obtained following incubation of the test medium with bacterial cells. The test is designed to detect the presence of specific enzymes present in one species of *Neisseria* but absent in the others. The enzymes act on substrates that yield col-ored products. A simple, sensitive, highly specific nucleic acid probe test (see Section 21.10) has also been developed for identifying *N. gonorrhoeae*.

Culture of anaerobes

Obligately anaerobic bacteria are common causes of infection and are completely missed in clinical diagnosis unless special precautions are taken for their isolation and culture. We have discussed anaerobes in general in Section 5.11, and we noted that many anaerobes are extremely susceptible to oxygen. Because of this, speci-men collection, handling, and processing require special attention if an obligate anaerobe may be involved. There are several habitats in the body (for example, the oral cavity and the intestinal tract) (∞ Sections 19.3 and 19.4) that are generally anoxic and in which obligately anaerobic bacteria can be found as part of the normal flora. However, other parts of the body can become anoxic as a result of tissue injury or trauma, which results in reduction of blood supply to the injured site. These anaerobic sites are then available for colonization by obligate anaerobes. In general, pathogenic anaerobic bacteria are part of the normal flora and are only oppor-tunistic pathogens, although two important pathogenic anaerobes, *Clostridium tetani* (causal agent of tetanus) and *C. perfringens* (causal agent of gas gangrene and one type of food poisoning), both endospore-forming Bac-teria, are predominantly soil organisms.

With anaerobic culture, the microbiologist is pre-sented not only with the usual problems of obtaining and maintaining an uncontaminated specimen but also with ensuring that the specimen not come in con-tact with air. Samples, collected by suction or biopsy, must be immediately placed in a tube containing oxy-gen-free gas, preferably containing a small amount of a dilute salts solution with a reducing agent such as thioglycolate and the redox indicator resazurin. This dye is colorless when reduced and becomes pink when oxidized, thus quickly indicating any oxygen contam-ination of the specimen. If a proper anaerobic trans-port tube is not available, the syringe itself can be used to transport the specimen, the needle being inserted into a sterile rubber stopper so that no air is drawn into the syringe.

For anaerobic incubation, agar plates are placed in a sealed jar, which is made anoxic by either replacing the atmosphere in the jar with an oxygen-free gas mix-ture (a mixture of N_2 and CO_2 is frequently employed) or by adding some compound to the enclosed vessel that removes O_2 from the atmosphere. For example, as shown in Figure 21.6, H_2 is generated and, in the pres-ence of a suitable catalyst, usually palladium, the H_2 is combined with free O_2 to form H_2O, thus removing the contaminating oxygen. Alternative means for providing

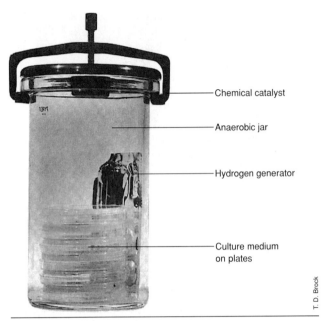

Figure 21.6 Sealed jar for incubating cultures under anoxic conditions.

anaerobic conditions include the use of culture media containing reducing agents and the use of anaerobic glove boxes. The latter are large gas-impermeable bags filled with an oxygen-free gas such as nitrogen or hydrogen that are fitted with an airlock for inserting and removing cultures (∞ Figure 5.22). The advantage of an anaerobic glove box is that manipulations can be done as one would normally perform them on a laboratory bench. However, because of their expense, anaerobic glove boxes are not employed extensively in clinical laboratories but are in widespread use in research laboratories that specialize in anaerobic microorganisms.

In general, media for anaerobes do not differ greatly from those used for aerobes, except that they are generally richer in organic constituents, and con-

---

CONCEPT CHECK   21.1

Culture of the suspected pathogen is the most reliable way to identify an organism that causes a disease. For successful microbial culture, the growth needs of the organisms must be met. This requires knowledge of bacterial physiology and nutrition. A variety of rapid tests that do not require microbial culture are also being developed and used to identify pathogens.

✔ *Why are urine cultures almost always positive for bacterial growth?*

✔ *Describe the specialized methods and precautions necessary for successful isolation of anaerobic pathogens.*

---

tain reducing agents (usually cysteine or thioglycolate) and a redox indicator such as resazurin. Once positive cultures have been obtained, they must be characterized and identified, to be certain that the isolate is not a member of the normal flora.

21.2
Growth-Dependent Identification Methods

If the inoculation of a primary medium results in bacterial growth, the clinical microbiologist must identify the organism or organisms present. Identification of a clinical isolate can frequently be made using a variety of growth-dependent assays. We discuss some of these methods here.

Growth on selective and differential media

On the basis of growth characteristics in primary isolation media, a clinical microbiologist subcultures an unknown pathogen on perhaps several of the dozens of available, diagnostically useful culture media. In many large hospitals and clinics, media are purchased from commercial sources, which ensures quality control and reliable testing in different clinical settings (Figure 21.7). Many of these media are available in miniaturized kits containing a number of different media in separate wells, all of which can be inoculated at one time (Figure 21.7d and e).

The battery of media employed are selective, differential, or both. A *selective medium* is one to which compounds have been added to selectively inhibit the growth of certain microorganisms but not others. A *differential medium* is one to which some sort of indicator, usually a dye, has been added, which allows the clinician to differentiate between various chemical reactions carried out during growth. Eosin–methylene blue (EMB) agar, for example, is a widely used selective *and* differential medium. EMB agar is used for the isolation of gram-negative enteric Bacteria. The methylene blue is present to inhibit gram-positive Bacteria; although the mechanism is unclear, small amounts of this dye effectively inhibit the growth of most gram-positive Bacteria. Eosin is a dye that responds to changes in pH, going from colorless to black under acidic conditions. EMB agar medium contains lactose and sucrose, but not glucose, as energy sources. Lactose-fermenting (generally enteric) bacteria, such as *Escherichia coli, Klebsiella,* and *Enterobacter,* acidify the medium and the colonies appear black with a greenish sheen. Colonies of lactose nonfermenters, such as *Salmonella, Shigella,* and *Pseudomonas,* are translucent or pink (Figure 21.4).

In the battery of tests performed to help identify an organism, many different biochemical reactions can be

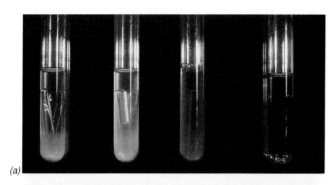

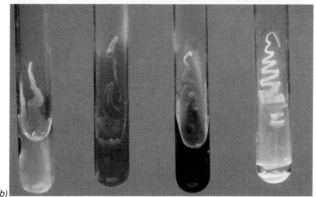

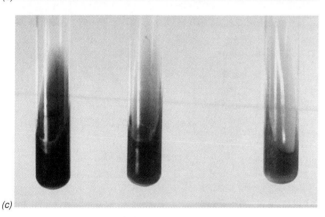

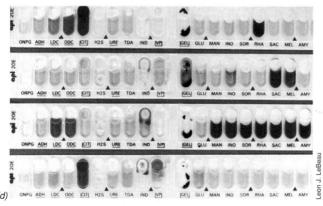

(e)

Leon J. LeBeau

Figure 21.7 Growth-dependent diagnostic methods used for the identification of clinical isolates by color changes in various diagnostic media. (a) Use of a differential medium to assess sugar fermentation. Acid production is indicated by color change of the pH-indicating dye added to the liquid medium. If gas production occurs, a bubble appears in the inverted vial in each tube. From left to right: acid, acid and gas, negative, uninoculated. (b) A conventional diagnostic test for enteric bacteria in a medium called *triple sugar iron (TSI) agar*. The medium is inoculated both on the surface of the slant and by stabbing into the butt. The medium contains a small amount of glucose and a large amount of lactose and sucrose. Organisms able to ferment only the glucose cause acid formation only in the butt, whereas lactose or sucrose-fermenting organisms also cause acid formation in the top. Gas formation is indicated by the breaking up of the agar in the butt. Hydrogen sulfide formation (either from protein degradation or from reduction of thiosulfate in the medium) is indicated by a blackening due to reaction of the H_2S with ferrous iron in the medium. From left to right: fermentation of glucose only; no reaction; hydrogen sulfide formation; fermentation of glucose and another sugar. (c) Measurement of citrate utilization by *Salmonella* on Simmons citrate agar. The change in pH causes a change in the color of the indicating dye. From left to right: positive, negative, uninoculated. (d) Media kits used for the rapid identification of clinical isolates. The principle is the same as in (a), but the whole arrangement has been miniaturized so that a number of tests can be run at the same time. Four separate strips, each with a separate culture, are shown. (e) Another arrangement of a miniaturized test kit. This one defines sugar utilization in nonfermentative organisms.

measured. The most important tests are summarized in Table 21.3. These tests measure the presence or absence of *enzymes* involved in catabolism of the substrate or substrates added to the differential medium. Fer-

mentation of sugars is measured by incorporating pH indicator dyes that change color on acidification (Figure 21.7*a*). Production of hydrogen gas and/or carbon dioxide during sugar fermentation is assayed by observing gas production either in gas collection vials or in agar (Figure 21.7*a* and *b*). Hydrogen sulfide production is indicated following growth in a medium containing ferric iron. If sulfide is produced, ferric iron complexes with H_2S to form a black precipitate of iron

TABLE 21.3   Important clinical diagnostic tests for bacteria

Test	Principle	Procedure	Most common use
Carbohydrate fermentation	Acid and/or gas produced during fermentative growth with sugars or sugar alcohols	Broth medium with carbohydrate and phenol red as pH indicator; inverted tube for gas	Enteric bacteria differentiation (also several other genera or species separations with some individual sugars) (Figure 21.7)
Catalase	Enzyme decomposes hydrogen peroxide, H_2O_2	Add drop of H_2O_2 to dense culture and look for bubbles (O_2) (Figure 5.25)	*Bacillus* (+) from *Clostridium* (−); *Streptococcus* (−) from *Micrococcus–Staphylococcus* (+)
Citrate utilization	Utilization of citrate as sole carbon source, results in alkalinization of medium	Citrate medium with bromthymol blue as pH indicator. Look for intense blue color (alkaline pH)	*Klebsiella–Enterobacter* (+) from *Escherichia* (−), *Edwardsiella* (−) from *Salmonella* (+) (Figure 21.7)
Coagulase	Enzyme causes clotting of blood plasma	Mix dense liquid suspension of bacteria with plasma, incubate, and look for fibrin clot	*Staphylococcus aureus* (+) from *S. epidermidis* (−)
Decarboxylases (lysine, ornithine, arginine)	Decarboxylation of amino acid releases CO_2 and amine	Medium enriched with amino acids. Bromcresol purple pH indicator becomes purple (alkaline pH) if there is enzyme action	Aid in determining bacterial group among the enteric bacteria
β-Galactosidase (ONPG) test	Orthonitrophenyl-β-galactoside (ONPG) is an artificial substrate for the enzyme. When hydrolyzed, nitrophenol (yellow) is formed	Incubate heavy suspension of lysed culture with ONPG. Look for yellow color	*Citrobacter* and *Arizona* (+) from *Salmonella* (−). Identifying some *Shigella* and *Pseudomonas* species
Gelatin liquefaction	Many proteases hydrolyze gelatin and destroy the gel	Incubate in broth with 12% gelatin. Cool to check for gel formation. If gelatin is hydrolyzed, tube remains liquid on cooling	To aid in identification of *Serratia, Pseudomonas, Flavobacterium, Clostridium*
Hydrogen sulfide (H_2S) production	H_2S produced by breakdown of sulfur amino acids or reduction of thiosulfate	H_2S detected in iron-rich medium from formation of black ferrous sulfide (many variants: Kliger's iron agar and triple sugar iron agar, also detect carbohydrate fermentation)	In enteric bacteria, to aid in identifying *Salmonella, Arizona, Edwardsiella*, and *Proteus* (Figure 21.7)
Indole test	Tryptophan from proteins converted to indole	Detect indole in culture medium with dimethyl-aminobenzaldehyde (red color)	To distinguish *Escherichia* (+) from *Klebsiella* (−) and *Enterobacter* (−); *Edwardsiella* (+) from *Salmonella* (−)
Methyl red test	Mixed-acid fermenters produce sufficient acid to lower pH below 4.3	Glucose-broth medium. Add methyl red indicator to a sample after incubation	To differentiate *Escherichia* (+, culture red) from *Enterobacter* and *Klebsiella* (usually −, culture yellow)
Nitrate reduction	Nitrate as alternate electron acceptor, reduced to NO_2^- or N_2	Broth with nitrate. After incubation, detect nitrite with α-naphthylamine-sulfanilic acid (red color). If negative, confirm that NO_3^- still present by adding zinc dust to reduce NO_3^- to NO_2^-. If no color after zinc, then $NO_3^- \rightarrow N_2$	To aid in identification of enteric bacteria (usually +)

TABLE 21.3   (continued)

Test	Principle	Procedure	Most common use
Oxidase test	Cytochrome *c* oxidizes artificial electron acceptor: tetramethyl (or dimethyl)-*p*-phenylenediamine	Broth or agar. Oxidase-positive colonies on agar can be detected by flooding plate with reagent and looking for blue or brown colonies	To separate *Neisseria* and *Moraxella* (+) from *Acinetobacter* (−). To separate enteric bacteria (all −) from pseudo-monads (+). To aid in identification of *Aeromonas* (+)
Oxidation–fermentation (O/F) test	Some organisms produce acid only when growing aerobically	Acid production in top part of sugar-containing culture tube; soft agar used to restrict mixing during incubation	To differentiate *Micrococcus* (acid produced aerobically only) from *Staphylococcus* (acid produced anaerobically). To characterize *Pseudomonas* (aerobic acid production) from enteric bacteria (acid produced anaerobically)
Phenylalanine deaminase test	Deamination produces phenylpyruvic acid, which is detected in a colorimetric test	Medium enriched in phenylalanine. After growth, add ferric chloride reagent and look for green color	To characterize the genus *Proteus* and the *Providencia* group
Starch hydrolysis	Iodine-iodide gives blue color with starch	Grow organism on plate containing starch. Flood plate with Gram's iodine and look for clear zones around colonies	To identify typical starch hydrolyzers such as *Bacillus* spp.
Urease test	Urea ($H_2N—CO—NH_2$) split to $2\ NH_3 + CO_2$	Medium with 2% urea and phenol red indicator. Ammonia release raises pH, intense pink-red color	To distinguish *Klebsiella* (+) from *Escherichia* (−). To distinguish *Proteus* (+) from *Providencia* (−)
Voges–Proskauer test	Acetoin produced from sugar fermentation	Chemical test for acetoin using α-naphthol	To separate *Klebsiella* and *Enterobacter* (+) from *Escherichia* (−). To characterize members of genus *Bacillus*

sulfide (Figure 21.7*b*). Utilization of citric acid, a six-carbon acid containing three carboxylic acid groups, is accompanied by a pH rise, and a specific dye incorporated into this test medium changes color as conditions become alkaline (Figure 21.7*c*). Hundreds of differential tests have been developed for clinical use, but only about 20 are used routinely (Figure 21.7*d*).

The typical reaction patterns for large numbers of strains of various pathogens have been published, and in the modern clinical microbiology laboratory, all this information is stored in a computer. The results of the differential tests on an unknown pathogen are entered, and the computer makes the best match by comparing the characteristics of the unknown with the species in the data bank. For many organisms, as few as three or four key tests are all that are required to make an unambiguous identification. In cases of a dubious match, however, more sophisticated identification procedures may be called for, especially if the chemotherapy regimens are different for several pathogens with similar growth characteristics.

Clinical diagnosis

Many companies market their own versions of growth-dependent rapid identification systems (Figure 21.7*d* and *e*). Such systems are frequently designed for use in identifying enteric bacteria because enterics are frequently implicated in routine urinary tract and intestinal infections (see Section 21.1).

Other growth-dependent rapid identification kits are available for other bacterial groups or even for single bacterial species. For example, commercial kits containing a battery of tests have been developed for *Staphylococcus aureus*, *Streptococcus pyogenes*, *Neisseria gonorrhoeae*, *Haemophilus influenzae*, and *Mycobacterium tuberculosis*. Other kits are available for identification of the pathogenic fungi *Candida albicans* and *Cryptococcus neoformans* (∞ Section 23.16).

The decision to use a specific diagnostic test is usually made by the clinical microbiologist. This individual takes into consideration the nature of the clinical specimen, basic characteristics (especially the Gram stain) of pure cultures obtained, and previous experience with similar cases. For example, an enteric identification kit would be useless in identifying a gram-positive coccus isolated from an abscess. Instead, a *Staphylococcus aureus* or *Streptococcus pyogenes* kit would be used to make a positive identification.

CONCEPT CHECK   **21.2**

Traditional methods for identifying pathogens depend on observing metabolic changes induced as a result of growth. These growth-dependent methods provide rapid and reasonably accurate means of diagnosing many infectious diseases.

✔ *Distinguish between* selective *and* differential *identification methods. Give an example of a medium used for each purpose.*

✔ *What parameters would a clinical microbiologist use to prescribe a specific diagnostic test kit for identification of an infectious agent?*

21.3

Testing Cultures for Antibiotic Sensitivity

In medical practice, microbial cultures are isolated from diseased patients to confirm diagnoses and to aid in decisions on therapy. Determination of the sensitivity of microbial isolates to antimicrobial agents is one of the most important tasks of the clinical microbiologist.

We discussed the principles for the measurement of antimicrobial activity in Chapter 11. The sensitivity of a culture can be most easily determined by an agar diffusion method or by using a tube dilution technique to determine the *minimum inhibitory concentration* (MIC) of an agent that is necessary to inhibit growth (∞ Section 11.4). Food and Drug Administration (FDA) regulations now control the procedures used for sensitivity testing in the United States, and similar regulations exist in other countries. A recommended agar diffusion procedure is called the *Kirby–Bauer method*, named after the workers who developed it (Figure 21.8). A plate of suitable culture medium is inoculated by spreading a sample of culture evenly across the agar surface. Filter paper discs containing known concentrations of different antimicrobial agents are then placed on the plate. The concentration of each agent on the disc is specified, and after incubation, the presence and size of inhibition zones around the discs of the different agents are noted. Table 21.4 presents typical zone sizes for several antibiotics. Zones observed on the plate are measured and compared to standard data to determine if the isolate is truly sensitive to a given antibiotic.

The MIC procedure for antibiotic sensitivity testing involves an *antibiotic dilution assay*, either in culture tubes (∞ Figure 11.11) or in the wells of a microtiter plate (Figure 21.8e). A series of twofold dilutions of each antibiotic are made in the wells, and then all wells are inoculated with a standard amount of the same test organism. After incubation, the inhibition of growth by the various antibiotics can be observed by measuring turbidity. Sensitivity is usually expressed as the *highest dilution* (lowest concentration) of antibiotic that completely inhibits growth. The dilution assay, because it can be performed in microtiter plates, is readily automated.

Because of the widespread occurrence of antibiotic resistance (∞ Section 11.13), an antibiotic sensitivity test is essential for pathogens isolated from each patient. Data such as those in Table 21.4 are useful to the physician in choosing the best antibiotic for a specific bacterial infection. Fortunately, many potentially serious pathogens are susceptible to a number of different antibiotics, and this allows the physician considerable latitude in the course of treatment. However, some pathogens, for example, *Pseudomonas aeruginosa,* are sensitive to very few drugs. Other pathogens, such as some encountered in hospital environments, have developed antibiotic resistance (∞ Sections 22.7 and 11.13). Thus, antibiotic sensitivity testing for these organisms is absolutely essential for effective chemotherapy. Using the drug sensitivity information gathered in this fashion, the clinical microbiologist generates periodic reports to the physician and pharmacist. These reports, called *antibiograms*, indicate the sensitivity of clinically isolated organisms to the antibiotics in current use. This report is particularly valuable for tracking the emergence of antibiotic-resistant strains of pathogens in facilities such as hospitals and nursing homes.

CONCEPT CHECK   **21.3**

Antibiotics are in wide use for the treatment of infectious diseases. Pathogens must be tested for sensitivity to individual antibiotics *before treatment* to ensure appropriate chemotherapy.

✔ *Describe the Kirby–Bauer technique. What does it indicate?*

✔ *Why is antibiotic sensitivity testing important for the clinical microbiologist, the physician, and the patient?*

21.4

Immunodiagnostics

In this section, we will apply the principles of immunity to the diagnosis of infectious diseases. First, we will briefly review the immune response to pathogens. Next, we will observe the immune response in a normal individual. Finally, we will examine immunologi-

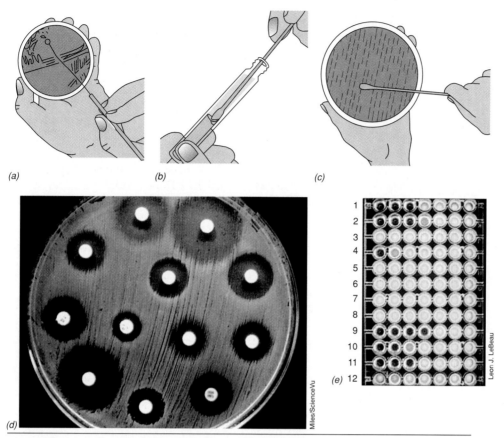

(a)   (b)   (c)

(e)

(d)

Miles/ScienceVu

Leon J. LeBeau

Figure 21.8 Antibiotic sensitivity testing. (a–d) the Kirby–Bauer procedure for determining the sensitivity of an organism to antibiotics. (a) A colony is picked from an agar plate. It is inoculated into a tube of liquid culture medium and allowed to grow to a specified density. (b) A swab is dipped in the liquid culture. (c) The swab is streaked evenly over a plate of sterile agar medium. (d) Discs containing known amounts of different antibiotics are placed on the plate. After incubation, inhibition zones are observed. The susceptibility of the organism is determined by reference to a chart of zone sizes (Table 21.4). (e) Antibiotic sensitivity determined by the dilution method. The organism is *Pseudomonas aeruginosa*. Each row has a different antibiotic. The use of the microtiter plate enables automation of these tests. The end point is read as the well with the lowest concentration of antibiotic that shows no evidence of bacterial growth. The highest concentration of antibiotic is in the well at the left; serial dilutions are made in the wells to the right. For example, in rows 1 and 2, the end point is the third well. In row 3, the antibiotic is ineffective at the concentrations tested, since there is bacterial growth in all the wells. In row 4, the end point is in the first well. The lowest concentration of antibiotic that completely inhibits bacterial growth defines the minimum inhibitory concentration (MIC) for that agent (∞ Section 11.4).

cal reagents that are useful for diagnostic applications. In the following sections, we will examine specific applications of these reagents.

Immunity to infection: Overview and review

The immune response was discussed in Chapter 20. A summary of the major aspects of immunity is shown in Figure 20.1. The body responds to pathogens in a three-step process. For a pathogen that the body has never before encountered, the pathogen must first be recognized. This is usually accomplished by a group of cells called phagocytes (∞ Section 20.3), which constitute

the first line of defense against any pathogen that gains access to body tissues. Fortunately, phagocytes ingest and destroy most pathogens (a process called *phagocytosis*). Phagocytosis is *nonspecific*, and the target may be any foreign substance, including the pathogens and their components.

In the second phase of immunity, the phagocytes present pathogen-derived *antigens* (proteins obtained from the destroyed pathogen) to antigen-specific immune lymphocytes known as T cells (∞ Section 20.7). Some T cells known as T helper (T_H) cells do not act directly on the pathogen but recruit and stimulate (help) another group of antigen-specific cells known as B cells.

TABLE 21.4   Zone sizes for some antimicrobial disc susceptibility tests

Antibiotic	Amount on disc	Inhibition zone diameter (mm)[a]		
		Resistant	Intermediate	Sensitive
Ampicillin[b]	10 µg	11 or less	12–13	14 or more
Ampicillin[c]	10 µg	28 or less	—	29 or more
Cephoxitin	30 µg	14 or less	15–17	18 or more
Cephalothin	30 µg	14 or less	15–17	18 or more
Chloramphenicol	30 µg	12 or less	13–17	18 or more
Clindamycin	2 µg	14 or less	15–16	17 or more
Erythromycin	15 µg	13 or less	14–17	18 or more
Gentamicin	10 µg	12 or less	13–14	15 or more
Kanamycin	30 µg	13 or less	14–17	18 or more
Methicillin[c]	5 µg	9 or less	10–13	14 or more
Neomycin	30 µg	12 or less	13–16	17 or more
Nitrofurantoin	300 µg	14 or less	15–16	17 or more
Penicillin G[d]	10 units	28 or less	—	29 or more
Penicillin G[e]	10 units	11 or less	12–21	22 or more
Polymyxin B	300 units	8 or less	9–11	12 or more
Streptomycin	10 µg	11 or less	12–14	15 or more
Tetracycline	30 µg	14 or less	15–18	19 or more
Trimethoprim-sulfamethoxazole	1.25/23.75 µg	10 or less	11–15	16 or more
Tobramycin	10 µg	12 or less	13–14	15 or more

[a]See Figure 21.8d for an illustration of a typical test.

[b]For gram-negative organisms and enterococci.

[c]For staphylococci and highly penicillin-sensitive organisms.

[d]For staphylococci.

[e]For organisms other than staphylococci. Includes some organisms, such as enterococci and some gram-negative rods, that may cause some systemic infections treatable with high doses of penicillin G.

The B cells then respond by producing soluble, antigen-specific binding proteins known as *antibodies* (∞ Sections 20.5 and 20.11). A *primary antibody response* generally occurs within 5 days, but antibodies do not reach peak quantities for several weeks. The antibody proteins, because they are antigen-specific, and thus pathogen-specific, are critical components of the immune response.

The antibodies interact specifically with the antigen on target cells, but cannot kill the cells. A group of nonspecific enzyme proteins, known collectively as complement (∞ Section 20.13), may attach to antibodies bound to the pathogen and lyse all cells with attached antibody. For example, antibodies specific for cell surface proteins of *Salmonella* spp. interact only *Salmonella*: complement causes lysis of the antibody-sensitized *Salmonella* cell, but not of an *Escherichia coli* cell that is not antibody-sensitized. Thus, the immune response is *specific* for individual antigens, by virtue of specific antibodies, but may be mediated or enhanced through nonspecific mechanisms such as complement.

In many cases, antibody-mediated immunity is not an effective mechanism for controlling the spread of infection. Some infectious agents parasitize the body from *within* cells. For example, animal viruses reproduce using host cell systems and, therefore, spend a large portion of their life cycle within the host cells (∞ Section 8.14). Likewise, bacteria such as *Mycobacterium tuberculosis*, the causative agent of tuberculo-sis, take up residence preferentially within phagocytes (∞ Sections 20.3 and 23.3). Because antibodies are geared to recognize the free pathogen in the blood or at mucosal cell surfaces, the infected host cells must be identified and destroyed by other means, usually involving the cell-to-cell interactions of the *cell-mediated immunity*. Fortunately, the internal pathogens all produce antigens that are in turn presented on the surface of infected target cells. A T cytotoxic cell (T_C) recognizes the antigen (∞ Section 20.7) and acts directly on the infected target cell by secreting cytolytic proteins called *perforins*, which destroy the infected cell (∞ Section 20.9).

No useful specific immunity exists before exposure to antigen, but after the first antigen exposure, specific immune T and B cells are present and some level of detectable circulating antibody may persist for years. More importantly, the cells that are capable of making antibody are now present in large numbers; a second antigen stimulation through reinfection generates a very rapid and very strong immune response, which peaks within several days, often at a level several orders of magnitude higher than the primary response (∞ Section 20.11). This *secondary antibody response* quickly targets and destroys the pathogen. Thus, the immune response has *memory*. Memory is characterized by a rapid rise in antibody *titer*, or quantity, and we will now use this principle to track infections.

Antibody titers and the diagnosis of infectious disease

In the diagnosis of an infectious disease, isolation of the pathogen is not always possible. One alternative is to measure antibody titer to a suspected pathogen. As we discussed earlier, if an individual is infected with a suspected pathogen, the antibody titer to that pathogen should be elevated. Antibody titer can be measured by agglutination, precipitation, enzyme-linked immunosorbent assay (ELISA), immobilization methods, or radioimmunoassay (RIA), depending on the situation. The general procedure is to set up a series of dilutions of serum (usually twofold dilutions: 1:2, 1:4, 1:8, 1:16, 1:32, and so on) and to determine the *highest* dilution at which the antigen–antibody reaction occurs.

A *single* measure of antibody titer does not indicate active infection. Many antibodies remain at high titer for long times after infection; to establish that an acute illness is due to a particular pathogen, it is essential to show a *rise* in antibody titer in successive samples of serum from the same patient. Frequently, the antibody titer is low during the acute stage of the infection and rises during convalescence (Figure 21.9). Such a rise in antibody titer is the best indication that the illness is due to the suspected agent and is also useful in diagnosis of infectious diseases of a rather chronic nature, such as typhoid fever and brucellosis. In some cases, however, the mere presence of antibody may be sufficient to indicate infection. This is the case for a pathogen that is quite rare in a population, and so the presence of antibody is sufficient to indicate that the individual has experienced an infection. A relevant example here is acquired immunodeficiency syndrome (AIDS). As will be discussed in Section 21.8, an extremely sensitive and highly reliable ELISA test is now available for detecting antibodies to human immunodeficiency virus (HIV), the AIDS virus. After infection with HIV, an antibody response occurs but, unlike the case in most diseases where antibody titers *increase* in the later stages of the disease, the loss of T helper cell function (∞ Section 23.7) actually causes HIV-specific antibody titers to *decrease* in the later stages of AIDS. Nevertheless, the exquisite sensitivity of ELISA allows detection of even very low antibody titers, and the HIV-ELISA is used to routinely screen blood samples for evidence of HIV infection.

Unfortunately, not all infections result in formation of systemic antibody. If a pathogen is extremely localized, there may be little induction of an immunological response and no rise in antibody titer even if the pathogen is proliferating profusely at its site of infection. A good example is the disease gonorrhea. Infection with *Neisseria gonorrhoeae*, the causative agent of gonorrhea, does not elicit a systemic immune response, and thus reinfection of a cured individual is not uncommon (see Sections 21.1 and 23.6). In other cases, the presence of antibody in the serum may have been due to vaccination. In fact, measurement of the rise in antibody titer following vaccination is one of the best ways of determining that the vaccination is effective.

Some of the most common clinical immunological procedures are outlined in Table 21.5.

Figure 21.9 The course of infection in a typical untreated typhoid fever patient. Measurement of body temperature provides a measure of the course of clinical symptoms. The antibody titer was measured by determining the highest dilution (twofold series) causing agglutination of a test strain of *Salmonella typhi*. Presence of viable bacteria in blood, feces, and urine was determined from periodic cultures. Note that the pathogen clears from the blood as the antibody titer rises, and clearance from feces and urine requires longer time. Body temperature gradually drops to normal as the antibody titer rises. The data given do not represent a single patient but are a composite of the picture seen in large numbers of patients.

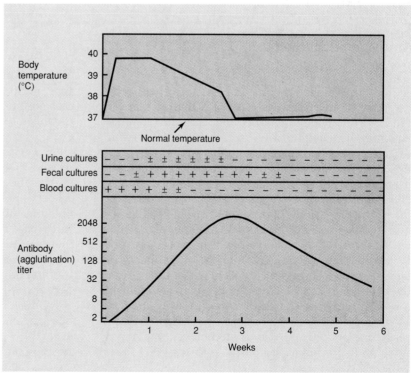

TABLE 21.5   Some clinical immunological procedures for identification of infectious agents

Pathogen or disease	Antigen	Serological procedure[a]
Streptococcus (group A)	Streptolysin O (exotoxin)	Neutralization of hemolysis
	DNase (extracellular protein)	Neutralization of enzyme
Neisseria meningitidis	Capsular polysaccharide	Passive hemagglutination (N. meningitidis polysaccharide adsorbed to red cells)
	N. meningitidis cells	Indirect fluorescent antibody
Salmonella	O or H antigen	Agglutination (Widal test)
		ELISA
Vibrio cholerae	O antigen	Agglutination
		Bactericidal test (in presence of complement)
		ELISA
Borrelia burgdorferi (Lyme disease)	Flagellin	ELISA
	Surface proteins	Immunoblot
		Bactericidal test (∞ Section 23.10)
Brucella	Cell wall antigen	Agglutination
		ELISA
Corynebacterium diphtheriae	Toxin	Skin test (Schick test)
Mycobacterium tuberculosis	Tuberculin (partially purified bacterial proteins, PPD)	Skin test (tuberculin test)
		ELISA
Syphilis (Treponema pallidum)	Cardiolipin-lecithin-cholesterol	Flocculation [Venereal Disease Research Laboratory (VDRL) test]
	T. pallidum antigens	ELISA
	T. pallidum cells	Indirect fluorescent antibody (FTA)
Rickettsial diseases (Q fever, typhus, Rocky Mountain spotted fever)	Killed rickettsial cells	Complement-based assay or cell agglutination tests
		ELISA
Influenza virus	Influenza virus suspensions	Complement-based assay
	Nasopharynx cells containing influenza virus	Immunofluorescence
AIDS	Human immunodeficiency virus (HIV)	ELISA
		Immunoblot
Pneumocystis carinii	P. carinii cells	Immunofluorescence

[a]Except for the skin tests and the immunofluorescent tests, the serum of the patient is assayed for antibody against the specific antigen by the methods shown.

Monoclonal antibodies for immunodiagnostics

As we discussed in Section 20.14, the normal antibody response to an antigen is *polyclonal*; that is, many B cells are stimulated to produce antibodies to a complex antigen. The resulting antiserum consists of a mixture of different antibodies. Although this antibody population can give adequate immune protection to the host, it is usually not specific for a single, defined antigen and is not precisely reproducible because it was produced in a single individual animal at a single time. For immunodiagnostic procedures, these types of antibodies, while they may be very potent, are extremely hard to standardize. *Monoclonal* antibodies, on the other hand, are products of clones of single cells, and because the cell clones can be stored and grown indefinitely, reproducibility and standardization are easily attained. Monoclonal antibody technology, therefore, has supplanted standard polyclonal techniques for many immunodiagnostic applications.

A monoclonal antibody is generally highly specific for a *single* antigenic determinant and hence is very useful in immunodiagnostics. For example, fluorescent antibodies (see Section 21.7) against *Chlamydia trachomatis* membrane proteins can be used to detect this organism in host tissues (*C. trachomatis* causes a variety of sexually transmitted diseases as well as trachoma, a serious eye disease) (∞ Section 23.6). These monoclonal antibodies react with *C. trachomatis* but are so specific they fail to react with even the closely related species *C. psittaci*. *C. trachomatis* is also an obligate intracellular parasite and is not easily cultured because it is dependent on the host cell to complete its life cycle. Use of the fluorescent anti-*C. trachomatis* monoclonal antibody on cervical scrapings or urethral or vaginal exudates makes positive identification of chlamydial infections almost routine.

Other monoclonal antibodies have been developed against outer membrane proteins of *Neisseria gonorrhoeae*. These probes are not only monospecific, reacting only with *N. gonorrhoeae*, but are also capable of differentiating *strains* of *N. gonorrhoeae*. The use of fluorescent monoclonal antibodies therefore eliminates much of the cross-reactivity problem observed when polyclonal sera are used.

With presently available technology, it is possible to generate monoclonal antibodies that react with only a certain bacterial species or even with only a certain *strain* of a species. In addition, viral antigens can be detected with the appropriate monoclonals. For example, both fluorescent and enzyme-conjugated monoclonal antibodies (subsequently assayed by ELISA) (see Section 21.9) have been developed for the diagnosis of herpes infections and the typing of herpes virus obtained from clinical specimens. Hence, monoclonal antibodies are useful for broad screening purposes as well as highly detailed analyses. Monoclonal antibodies have also been widely used in noninfectious disease diagnoses as well (for example, see Figure 21.15).

CONCEPT CHECK   **21.4**

An immune response is a natural outcome of infection. A specific immune response to a pathogen can be used as a diagnostic aid. Monoclonal antibodies are widely used for immunodiagnostic applications.

✔ *How might a patient develop an antibody titer to an organism?*

✔ *Why does the antibody titer to an organism rise during convalescence?*

✔ *What advantages do* monoclonal *antibodies have over* polyclonal *antibodies in immunodiagnostic tests?*

21.5

Agglutination

Agglutination is due to the binding of a particulate antigen by antibody. Agglutination reactions were discussed in Section 20.15, with the well-known ABO blood grouping reaction serving as a prime example of direct agglutination. However, many other agglutination reactions are used for the detection of antigens or antibodies associated with certain disease. Although not as sensitive a test as some other immunoassays (Table 21.6), agglutination remains useful in clinical diagnostics as an inexpensive, highly specific, rapid immunoassay.

Coated-particle agglutination

The agglutination of antigen-coated or antibody-coated latex beads by complementary antibody or antigen from a patient is a typical method of rapid diagnosis. Small

(0.8-μm) latex beads coated with a specific antigen or antibody are mixed with patient serum on a microscope slide and incubated for a short period. If the antibody complementary to the molecule bound to the bead surface is present in the patient's serum, the milky-white latex suspension will be visibly clumped, indicative of the agglutination reaction. Latex agglutination is also used to detect bacterial surface antigens by mixing a small amount of a bacterial colony with antibody-coated latex beads. For example, a commercially available suspension of latex beads containing antibodies to protein A and clumping factor, two molecules found exclusively on the surface of *Staphylococcus aureus*, is virtually 100% accurate in identifying clinical isolates of *S. aureus*. Unlike traditional tests for *S. aureus*, many of which are growth-dependent assays, identification of *S. aureus* by the latex bead assay takes only 30 seconds (Figure 21.10). Other latex bead agglutination assays have been developed to identify *Streptococcus pyogenes*, *Neisseria gonorrhoeae*, *Haemophilus influenzae*, *Campylobacter* spp. and the yeasts *Cryptococcus neoformans* and *Candida albicans*.

A very widely employed latex agglutination assay is that used for detecting specific serum antibodies for *rheumatoid factor*, an antibody directed against the body's own immunoglobulins and associated with the autoimmune disease *rheumatoid arthritis* (∞ Section 20.16). Latex beads coated with human immunoglobulin are mixed with whole blood or serum, and agglutination scored versus positive and negative control sera run in parallel. Latex bead assays are simple and specific. In addition, the inexpensive nature of the assays makes them suitable for large-scale screening purposes; the widespread use of the rheumatoid test is a good example of this. Because they require no expensive equipment or particular expertise, they are in wide use in virtually all clinical settings.

Some agglutination assays use a suspension of activated charcoal as the carrier. For example, a rapid diagnostic test for detection of the virus *Herpes simplex*, fre-

---

TABLE 21.6   **Sensitivity of immunodiagnostic assays**

Assay	Sensitivity (μg antibody/ml)[a]
Precipitin reaction	
In fluids	24–160
In gels (double immunodiffusion)	24–160
Agglutination reactions	
Direct	0.4
Passive	0.08
Radioimmunoassay (RIA)	0.0008–0.008
Enzyme-linked immunosorbent assay (ELISA)	0.0008–0.008
Immunofluorescence	8.0

[a]The smallest amount of antibody necessary to give a positive reaction in the presence of antigen.

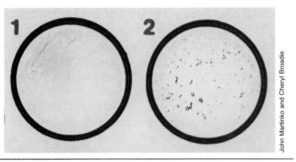

Figure 21.10 Latex bead agglutination test for *Staphylococcus aureus*. Panel 1 shows a negative control. Note the uniform pink color of the suspended latex beads coated with antibodies to protein A and clumping factor, two antigens found exclusively on the surface of *S. aureus* cells. Panel 2 shows the same suspension after a loopful of material from a bacterial colony was mixed into the suspension. The bright red clumps indicate a positive agglutination reaction took place and indicates that the colony is *S. aureus*.

quently associated with oral fever blisters or genital sores (∞ Section 23.6), employs anti-*H. simplex* virus antibodies adsorbed to small particles of activated charcoal. Cotton swabs used to sample suspected herpes lesions are placed in a buffer solution, and samples of the buffer solution, possibly now containing virus, are used to test for charcoal agglutination. A positive test is indicated by visible clumping of the charcoal into large, black aggregates. Because of the specificity of the antiserum used (and here, naturally, monoclonal antibodies are ideal), complicating cross-reactions with related pathogens are not a problem. Like latex beads, charcoal agglutination tests can be rapid and cost-effective diagnostic tools.

Coated-particle tests are *passive* agglutination reactions and are up to five times more sensitive than the direct agglutination tests we discussed in Section 20.15 (Table 21.6).

CONCEPT CHECK   21.5

A number of clinically useful agglutination tests are available. These tests are rapid, relatively sensitive, and inexpensive methods for identifying a variety of pathogens.

✔ *Distinguish between* direct *and* passive *agglutination. Which tests are more sensitive?*

✔ *What advantages do agglutination tests have over other immunoassays? What disadvantages?*

21.6
Immunoelectron Microscopy

Antibodies to which heavy metals have been chemically conjugated can be used to locate antigens in cells by electron microscopy. This is possible because heavy metals scatter the electron beam of the electron microscope. This technique, called *immunoelectron microscopy,* is used primarily in research where there is a need to determine where a specific antigen (usually a protein) is localized in a particular region of the cell (Figure 21.11). Cells, following chemical fixation and other preparations necessary for observation by the electron microscope, are treated with antibodies covalently conjugated to a heavy metal, usually gold or platinum. The electron-dense metals scatter electrons, and thus the presence of bound antibody can be detected by dense black spots in photographs of the preparation.

In immunoelectron microscopy, although the cell is dead and chemically fixed, most protein antigens retain sufficient native structure and antibodies still react with little nonspecific cross-reaction. Immunoelectron microscopy has been used extensively to pinpoint the location of enzymes in cells, especially those suspected to be associated with the cytoplasmic membrane or some other internal structure (Figure 21.11).

Although immunoelectron microscopy can be used for identifying pathogens such as human immunodefi-

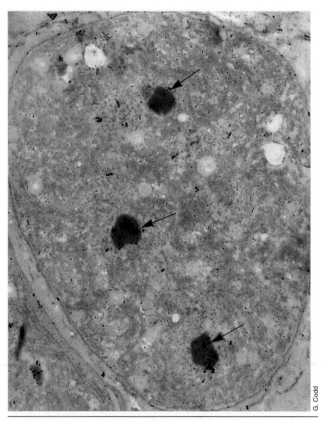

Figure 21.11 Immunoelectron microscopy. Antibodies made in rabbits to the enzyme ribulose-1,5-bisphosphate carboxylase from the cyanobacterium *Chlorogloeopsis fritschii* were added to thin sections of *C. fritschii* and the preparation treated with goat anti-rabbit IgG conjugated to 20-nm colloidal gold particles. The concentration of the particles around large inclusions called carboxysomes (arrows) indicate that these are sites of large amounts of the enzyme.

ciency virus (HIV) in cells (∞ Figure 23.32), the time, expense, expertise, and specialized equipment involved make it impractical for diagnostic procedures in all but the most specialized clinical research settings.

21.7

Fluorescent Antibodies

In addition to heavy metals, there are a number of other chemical methods for covalently modifying antibodies that make them readily detectable. In this section, we will discuss the use of antibodies chemically modified with fluorescent dyes. This procedure makes it possible to detect reactions of antibodies with single cells. Virtually all well-equipped clinical laboratories make extensive use of fluorescent antibodies for clinical diagnostic procedures.

Fluorescent methods

Antibody molecules can be made fluorescent by covalently attaching them to fluorescent organic compounds such as rhodamine B, which fluoresces red, or fluorescein isothiocyanate, which fluoresces yellow-green. This does not alter the specificity of the antibody but makes it possible to detect the antibody bound to cell or tissue surface antigens by use of the fluorescence microscope (Figure 21.12). Cells to which fluorescent antibodies have bound emit a bright fluorescent color, usually red or yellow-green, depending on the dye used. Fluorescent antibodies have been of

considerable utility in diagnostic microbiology because they permit the study of immunological reactions on single cells. The fluorescent antibody technique is also very useful in microbial ecology as one of the few methods for directly identifying microbial cells in natural environments.

Two distinct fluorescent antibody procedures, the **direct** and the **indirect** staining methods, are used (Figure 21.13). In the direct method, the antibody

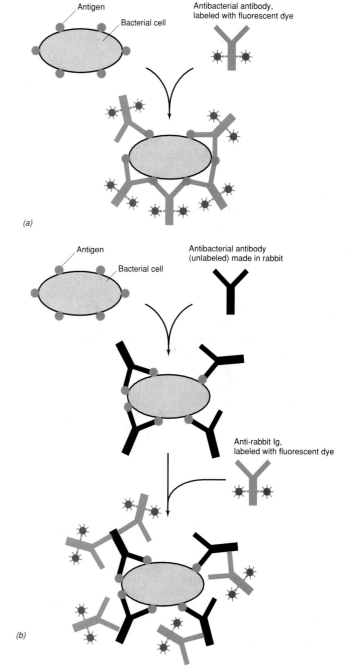

(a)

(b)

Figure 21.13 Methods of using fluorescent antibodies to detect bacterial surface antigens. (a) Direct staining method. (b) Indirect staining method.

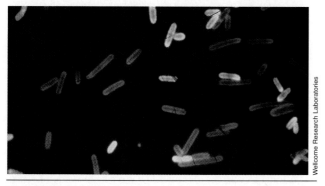

Figure 21.12 Fluorescent antibody reactions. Cells of *Clostridium septicum* were teated with antibody conjugated with fluorescein isothiocyanate, which fluoresces yellow-green. Cells of *Clostridium chauvei* were stained with antibody conjugated with rhodamine B, which fluoresces red.

against the organism itself is fluorescent. In the indirect method, the presence of a nonfluorescent antibody on the surface of the cell is detected by the use of a fluorescent antibody directed against the nonfluorescent antibody (Figure 21.13). This is done by immunizing one animal species, for example, a goat, with antibodies from a second species, for example, a rabbit, and then conjugating the fluorescent dye to the goat antibodies. The fluorescent goat anti-rabbit antibodies can then be used to detect the presence of rabbit immunoglobulin previously bound to cells. One of the advantages of the indirect staining method is that it eliminates the need to make a fluorescent antibody for each antigen of interest.

Clinical applications

In a typical clinical test using fluorescent antibodies, a smear of material containing a suspected pathogen is allowed to react with a specific fluorescent antibody and observed with a fluorescent microscope. If the pathogen contains surface antigens against which the fluorescent antiserum was prepared (that is, the suspected pathogen is identical to or immunologically closely related to the cells used to generate the antibodies), the cells will fluoresce (Figure 21.14). Organisms immunologically unrelated to the control organism generally do not react or react only weakly.

Fluorescent antibodies can also be applied directly to infected host tissues, permitting diagnosis long before primary isolation techniques yield a suspected organism. For example, in diagnosing legionellosis (∞ Section 23.2) a positive diagnosis can be made by staining biopsied lung tissue with fluorescent antibod-

ies prepared against cell walls of *Legionella pneumophila*, the causative agent of legionellosis (Figure 21.14a). Likewise, a fluorescent antibody against the capsule of *Bacillus anthracis* can be used in the microscopic diagnosis for anthrax. Fluorescent antibody reactions can also be used in diagnosis of viral infection (Figure 21.14b) and in a variety of noninfectious diseases. For example, in identifying cell types expressing a particular antigen, such as malignant cells, fluorescent antibodies may be very valuable in following the course of the disease (Figure 21.15).

Fluorescent antibodies can also be used to separate mixtures of cells into relatively pure populations or to define the numbers of certain cell types in complex mixtures such as blood. Fluorescent-labeled monoclonal antibodies directed against the CD4 and CD8 surface antigens of T lymphocytes (∞ Section 20.4) are routinely used to identify and enumerate these cells in the blood leukocyte population (Figure 21.16). For

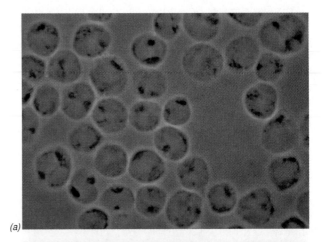
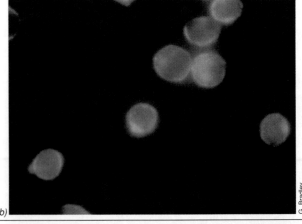

(a)

(b)

G. Bradley

Figure 21.15 Use of fluorescent antibodies in noninfectious disease diagnostics. (a) Human leukemic cells, some of which are sensitive to a toxic anticancer drug and some of which are not, appear indistinguishable. (b) When the cells in (a) are treated with a fluorescent monoclonal antibody that binds specifically to a protein found only on the surface of drug-resistant cells, the latter fluoresce whereas drug-sensitive cells do not.

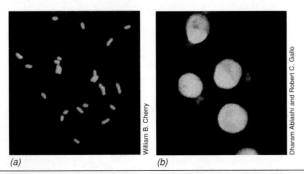

William B. Cherry

Dharam Ablashi and Robert C. Gallo

(a)   (b)

Figure 21.14 Examples of the use of fluorescent antibodies in clinical microbiology. (a) Immunofluorescent stained cells of *Legionella pneumophila*, the cause of legionellosis. The individual organisms are 2–5 μm in length. (b) Detection of virus-infected cells by immunofluorescence. Human B lymphotrophic virus (HBLV)-infected spleen cells were incubated with serum containing antibodies to HBLV from a patient with a lymphoproliferative disorder. Cells were then treated with fluorescein isothiocyanate-conjugated anti-human IgG antibodies. HBLV-infected cells fluoresce bright yellow. Cells in the background did not react with the patient's serum. Individual cells are about 10–15 μm in diameter.

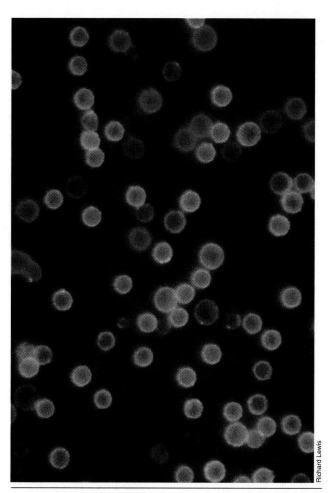

Richard Lewis

Figure 21.16 T lymphocytes stained with fluorescent-tagged monoclonal antibodies to specific surface markers. Yellow-green cells are cytotoxic (CD8) T cells; red cells are T helper (CD4) cells. The different-colored cells can be separated from one another by a fluorescence-activated cell sorter to yield enriched populations of different cell types. Reprinted with permission from *Science 239:* Cover (Feb. 12, 1988), © AAAS.

example, the definition of acquired immunodeficiency syndrome (AIDS) includes a reduction in CD4 cells. In addition, the CD4:CD8 ratio changes during the progression of AIDS. Thus, by defining the CD4 and CD8 numbers, the clinician can identify the reduction in CD4 cells and, with successive assays over time, can follow the progress of the disease (∞ Section 23.7).

Fluorescing cells can be visualized, counted, and separated with an instrument called a *fluorescence spectrometer*, often referred to as a fluorescence-activated cell sorter (FACS). The FACS uses a laser beam to activate the fluorescent molecules (in this case, fluorescent antibody bound to cells), placing a charge on the labeled cells. Following laser exposure, an electric field is applied to the cell mixture. The fluorescing and non-fluorescing cells are then deflected to opposite ends of the electric field where each cell population is counted and deposited in a tube. The use of several antibodies, each labeled with a different fluorescent dye, can result

in the simultaneous identification of several cell markers. A typical application used for identifying CD3 and CD4 positive T cells in normal and AIDS patients is shown in Figure 21.17. FACS analysis is also useful for research applications. For example, immunologists routinely use FACS methods to separate complex mixtures of immune cells. They can then study the properties of the highly enriched cell populations.

Under appropriate conditions, fluorescent antibodies yield rapid, highly specific, useful information about a variety of clinical conditions. However, immuno-diagnoses using fluorescent antibody techniques are not without their pitfalls. Nonspecific staining can be a problem because of surface antigens that may *cross-react* between various bacterial species, some of which may be members of the normal flora. This is a major problem among enteric bacteria, where antigens derived from lipopolysaccharides are frequently sufficiently similar among species to cause binding or partial binding of the fluorescent probe. The clinical microbiologist must therefore be careful to perform controls using nonspecific sera and confirm all positive immunofluorescent findings by other immunological or microbiological tests.

CONCEPT CHECK   21.7

Fluorescent antibodies can be employed for quick, accurate identification of pathogens and other antigenic substances in tissue samples and other complex environments. Through use of cell sorting, fluorescent antibodies can be used for quantitative enumeration of a variety of cell types, including pathogens.

✔ *Are fluorescent antibodies more sensitive for detecting antigens than normal antibodies?*

✔ *How are fluorescent antibodies used to identify specific cells in complex mixtures like blood?*

21.8

Enzyme-Linked Immunosorbent Assay and Radioimmunoassay

The specificity of antibodies is such that the limiting factor in most of the immunological reactions discussed thus far is not *specificity* but *sensitivity* (Table 21.6). Because of their exquisite sensitivity, radioimmunoassay (RIA) and enzyme-linked immunosorbent assay (ELISA) are two widely used immunological techniques. These methods employ radioisotopes and enzymes, respectively, to detect antibody molecules. Because radioactivity and the products of certain enzymatic reactions can be measured in very small amounts, the attachment of radioactive or enzyme ligands to anti-

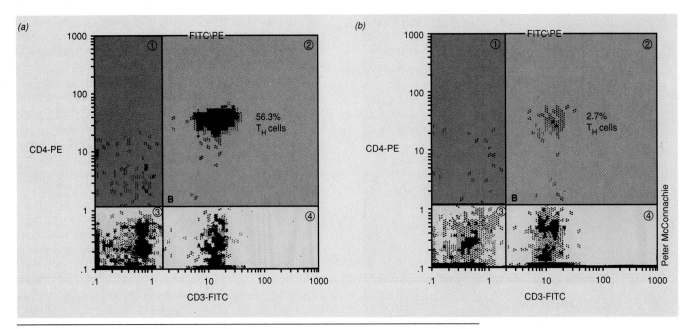

Figure 21.17 CD3 and CD4 cell enumeration from a healthy human (a) and from a human with acquired immunodeficiency syndrome (AIDS) (b) using a fluorescence-activated cell sorter (FACS). Peripheral blood cells were simultaneously labeled with monoclonal antibody to CD4 conjugated to phycoerythrin (PE) and with monoclonal antibody to CD3 conjugated to fluorescein isothiocyanate (FITC). CD3 is found on all T cells. CD4 is found on T helper (T_H) cells only. Quadrant 3 shows cells that were stained with neither antibody. Quadrant 1 shows cells stained with only anti-CD4. Quadrant 4 shows cells stained with only anti-CD3. Quadrant 2 shows cells stained with both anti-CD3 and anti-CD4. (a) Results from a healthy human. In this case, 56.3% of the T cells were T_H cells. Thus, quadrant 2 shows a dense staining pattern. (b) Results from a patient with clinical AIDS. In this case, only 2.7% of the total T cells are T_H cells. This is indicated by the very light staining pattern in quadrant 2. [Original data from Peter McConnachie used with permission.]

body molecules serves to decrease the amount of antigen–antibody complex required to detect a reaction. This increased sensitivity has been extremely helpful in clinical diagnostics and research and has opened the door to the development of a variety of new immunological tests, previously impossible because the methods available were not sufficiently sensitive (∞ Urine Testing for Drug Abuse, and Over-the-Counter Immunodiagnostic Kits, Chapter 20).

ELISA

The covalent attachment of enzymes to antibody molecules creates an immunological tool possessing both high specificity and high sensitivity. The technique, called **ELISA** (for *e*nzyme-*l*inked *i*mmuno-*s*orbent *a*ssay), makes use of antibodies to which enzymes have been covalently bound such that the enzyme's catalytic properties and the antibody's specificity are unaltered. Typical linked enzymes include peroxidase, alkaline phosphatase, and β-galactosidase, all of which catalyze reactions whose products are colored and can be measured in very low amounts.

Two basic ELISA methodologies have been developed, one for detecting antigen (*direct* ELISA) and the other for detecting antibodies (*indirect* ELISA). For detecting antigens such as virus particles from a blood or fecal sample, the direct ELISA method is used. In this procedure the antigen is "trapped" between two layers of antibodies (Figure 21.18). Thus, this method is sometimes called the "sandwich ELISA." The specimen is added to the wells of a microtiter plate (see the box, The Microtiter Plate and Immunoassays) previously coated with antibodies specific for the antigen to be detected. If the antigen (virus particle) is present in the sample, it will be trapped by the antigen binding sites on the antibodies. After washing unbound material away, a second antibody containing a conjugated enzyme is added. The second antibody is also specific for the antigen, and so it binds to any remaining exposed determinants. Following a wash, the enzyme activity of the bound material in each microtiter well is determined by adding the substrate of the enzyme. The color formed is proportional to the amount of antigen present (Figure 21.18).

To detect *antibodies* in human serum, an indirect ELISA is employed. An indirect ELISA test is widely used to detect antibodies to human immunodeficiency virus (HIV), and we will discuss this test in detail because the principles involved are applicable to all indirect ELISA tests.

The HIV-ELISA

The causative agent of AIDS, the human immunodeficiency virus (HIV) (∞ Section 23.7), is transmitted by bodily fluids including blood. Rapid, efficient, cost-effective screening tools are needed to test blood samples to ensure that HIV is not being inadvertently transmitted during blood transfusions or through the transfer of blood products. An ELISA test is used for the routine screening of blood for signs of exposure to HIV (and hence possible AIDS).

The HIV-ELISA test is an *indirect* ELISA designed to measure *antibodies* to HIV present in serum. Initial infection with HIV leads to the production of antibod-

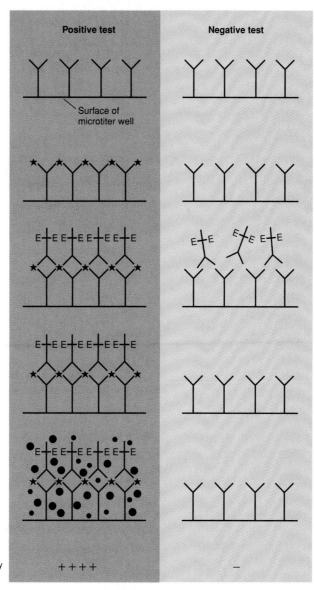

Procedure

1. Antibodies (Y) to virus (★) bound to wells of microtiter plate

2. Add patient sample (feces, secretions, serum, and so on) suspected of containing virus particles or virus antigens and wash wells with buffer

3. Add antivirus antibody containing conjugated enzyme
   (E┼E)

4. Wash with buffer

5. Add substrate for enzyme and measure amount of colored product (●). Colored product observed is proportional to amount of antigen.

Color intensity   + + + +   −

Positive test   **Negative test**

Surface of microtiter well

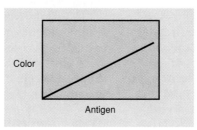

Color / Antigen

Figure 21.18 Detection of viruses by a direct ELISA test.

THE MICROTITER PLATE AND IMMUNOASSAYS

A number of immunoassays have been developed requiring that either the antigen or the antibody be bound to a solid support. A variety of solid phase carriers such as latex beads or plastic tubes have been used, but the plastic, disposable *microtiter plate*

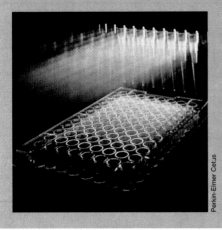

Perkin-Elmer Cetus

has been the most successful solid support in modern immunoassays.

Microtiter plates are small, plastic trays containing a number of wells. The plates are made of polystyrene or polypropylene, both of which are transparent plastics. Proteins, either antigens or antibodies, adsorb to the plastic surface of the microtiter plate as a result of interactions between hydrophobic regions of the protein and the nonpolar plastic surface. Once bound, the proteins are not easily removed by washing or other manipulations, and thus bound proteins can be treated with various reagents and washed repeatedly without being removed from the plate surface. The standard microtiter plate contains 96 wells, which allows for several replicates of each sample to be run. For quantitation, several dilutions are prepared. Although only small amounts of pro-

tein can be adsorbed to each well, the sensitivity of ELISAs and radioimmunoassays are such that only small amounts of antigen or antibody are needed anyway.

Automatic machines are available for adding reagents and preparing dilutions. Depending on whether the microtiter plate assay is an ELISA or a radioimmunoassay, special machines are available to read the absorbance of a colored product (ELISA) or amount of radioactivity (radioimmunoassay) in each of the wells of the plate. Such automation allows these immunoassays to be performed both rapidly and routinely at relatively low cost. The microtiter plate has therefore helped to revolutionize research and diagnostic immunoassays. ELISAs in particular have become part of the everyday activities of the clinical laboratory. ∎

ies to several HIV antigens, in particular those of the HIV envelope. These antibodies can be detected by the HIV-ELISA test (Figure 21.19).

To carry out an HIV-ELISA test, microtiter plates are first coated with a disrupted preparation of HIV particles; about 200 ng of disrupted HIV is required in each well. Following a brief incubation period to ensure binding of the antigens to the surface of the microtiter wells, a diluted serum sample is added and the mixture incubated to allow HIV-specific antibodies to bind to HIV antigens. To detect the presence of antigen–antibody complexes, a second antibody is then added. This second antibody is an enzyme-conjugated anti-human IgG preparation. Following a brief incubation period with the second antibody and a washing step to remove any unbound second antibody, the enzyme activity is assayed (the anti-human IgG antibodies bind to any HIV-specific IgG antibodies previously bound to the HIV antigen preparation). A color is obtained in the enzyme assay in proportion to the amount of anti-human IgG antibody bound (Figure 21.19). The binding of the second antibody is an indication that antibodies from the patient's serum recognized the HIV antigens, the patient has antibodies to HIV, and the patient has been exposed to HIV. Control sera (known to be HIV-

negative) are assayed in parallel with any samples to measure the extent of background absorbance in the assay.

The HIV-ELISA test is a rapid, highly sensitive, specific method for detecting exposure to HIV. Since ELISAs in general are highly adaptable to mass screening and automation, the HIV-ELISA test is used as a standard screening method for blood. However, this test method can give erroneous results under certain circumstances.

For example, the test occasionally gives false positive results. Because a number of factors can contribute to these results, none of which are related to exposure to HIV, all positive HIV-ELISA tests *must* be confirmed by another independent test, usually the Western blot (immunoblot) test (see Section 21.9). A positive HIV Western blot test after a positive HIV-ELISA test is considered proof of HIV infection.

A final drawback to the HIV-ELISA test is the possibility of obtaining false negative results. As we learned in Section 20.11, it takes the immune system some time to develop an effective antibody response with a detectable antibody titer. In the case of HIV infection, this lag time is estimated to be 6 weeks to a year. Therefore, individuals who have been recently infected with HIV may not yet be producing detect-

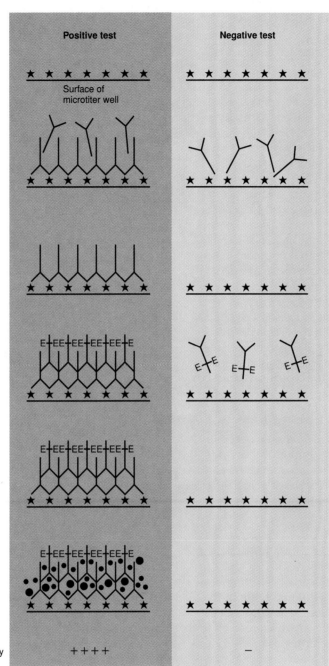

Procedure

1. Coat microtiter wells with antigen preparation from disrupted HIV particles (★)

   Surface of microtiter well

2. Add patient serum sample. HIV-specific antibodies bind to HIV antigen

3. Wash with buffer

4. Add human anti-IgG antibodies conjugated to enzyme (E ┼ E)

5. Wash with buffer

6. Add substrate for enzyme and measure amount of colored product (●). Colored product observed is proportional to the antibody concentration.

   Color intensity

Positive test

++++

Negative test

−

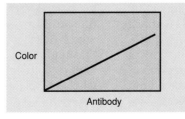

Color

Antibody

Figure 21.19 Indirect ELISA test for detecting antibodies to human immunodeficiency virus (HIV), the causal agent of acquired immunodeficiency syndrome (AIDS).

able amounts of antibody when they are tested. Another reason for a false negative result in the HIV-ELISA test is the total destruction of the immune system seen in advanced cases of AIDS; if no immune cells are left in the body, no antibodies can be made and the ELISA test is not useful. However, at this stage of disease, a clinical diagnosis is possible and the ELISA test is useful only as a confirmatory indicator.

Other ELISA tests of clinical importance

Besides the ELISA test for HIV, literally hundreds of clinically useful ELISAs have been developed. Some of these are direct ELISAs for detecting antigens. Direct ELISAs for detecting bacterial toxins such as cholera toxin, enteropathogenic *Escherichia coli* toxin, and *Staphylococcus aureus* enterotoxin have been developed. Viruses currently detected using direct ELISA techniques include rotavirus, hepatitis viruses, rubella virus, bunyavirus, measles and mumps viruses, and parainfluenza virus.

Indirect ELISAs have been developed for detecting antibodies to a variety of clinically important bacteria. Although not meant to be a complete list, ELISAs for detecting serum antibodies to *Salmonella* (gastrointestinal diseases), *Yersinia* (plague), *Brucella* (brucellosis), a variety of rickettsias (Rocky Mountain spotted fever, typhus, Q fever), *Vibrio cholerae* (cholera), *Mycobacterium tuberculosis* (tuberculosis), *Mycobacterium leprae* (leprosy), *Legionella pneumophila* (legionellosis), *Borelia burgdorferi* (Lyme disease), and *Treponema pallidum* (syphilis) have been developed. ELISAs have also been developed for detecting antibodies to *Candida* (yeast) and antibodies to a variety of parasites, including those causing amebiasis, Chagas' disease, schistosomiasis, toxoplasmosis, and malaria.

The speed, low cost, lack of radioactive waste, and long shelf life make ELISA tests particularly attractive for many laboratories. But it is the extreme *sensitivity* of ELISAs that really make them important immunodiagnostic tools. New ELISA tests are marketed each year, and many of them are rapidly replacing older, less-sensitive methods such as agglutination.

Radioimmunoassay

Radioimmunoassay (RIA) employs radioisotopes instead of enzymes as antibody conjugates. The isotope iodine-125 is the most commonly used detecting system, as proteins can be readily iodinated without disrupting their specificity. RIA is used clinically to measure rare serum proteins such as human growth hormone, glucagon, vasopressin, testosterone, and insulin present in humans in extremely small amounts (Figure 21.20) and also in some urine tests for drug abuse (∞ Urine Testing for Drug Abuse, Chapter 20). In most cases a *direct* RIA is employed. The direct assay is a two-step procedure. First, radioactive antigen-specific antibodies are added to a series of microtiter wells containing known concentrations of pure antigen (such as a hormone), which is first bound to the wells. The radioactivity in each of these standard wells is then measured. Next, the antigen sample from a patient is allowed to bind to another well, and radioactive antibodies are added and measured, as before. The amount of radioactivity bound by the patient sample is then compared to a standard plot generated from the binding data obtained using the pure antigen, and the concentration of antigen in the patient serum is interpolated from the standard plot (Figure 21.20).

RIA has the same sensitivity range as ELISA and can also be performed very rapidly. However, the instruments used to detect radioactivity are quite specialized and expensive. RIA generates a considerable amount of radioactive waste, and the radioactive decay time (half life) of the radioisotopes used for detection may limit the useful life of the test kit. As a result, RIA is often used only when ELISA is not sufficiently accurate or sensitive. For example, RIA is often more useful than ELISA for detecting serum protein levels (as described earlier) because some serum components may inhibit ELISA enzyme–substrate reactions. Thus, for certain applications, each test system has clear advantages over the other.

CONCEPT CHECK   **21.8**

ELISA and RIA methods are the most sensitive known immunoassay techniques. Both involve linking a detection system, either an enzyme or a radioactive molecule, to an antibody or antigen, enhancing sensitivity. ELISA and RIA are used for clinical and research work; tests have been designed to detect either antibody or antigen in a vast number of applications.

✔ *Why are ELISA and RIA techniques more sensitive than standard immunoassays such as precipitation and agglutination?*

✔ *What hazards are associated with radioimmunoassays?*

21.9
Immunoblot Procedures

Antibodies can be also used in clinical diagnostics to identify individual specific *proteins* associated with specific pathogens. The procedure employs three techniques discussed previously: (1) the separation of proteins on polyacrylamide gels, (2) the transfer (blotting) of proteins from gels to nitrocellulose paper (∞ Working with Nucleic Acids: The Tools, Chapter 6), and (3) identification of the proteins by specific antibodies. Protein blotting and the subsequent identification of the proteins by specific antibodies is sometimes called the "Western" blot technique to distinguish it from the (DNA) "Southern" blot technique.

The immunoblot is a very sensitive method for detecting specific proteins in complex mixtures. In the first step of an immunoblot, a protein mixture is subjected to electrophoresis on a polyacrylamide gel. This separates the proteins into several distinct bands, each of

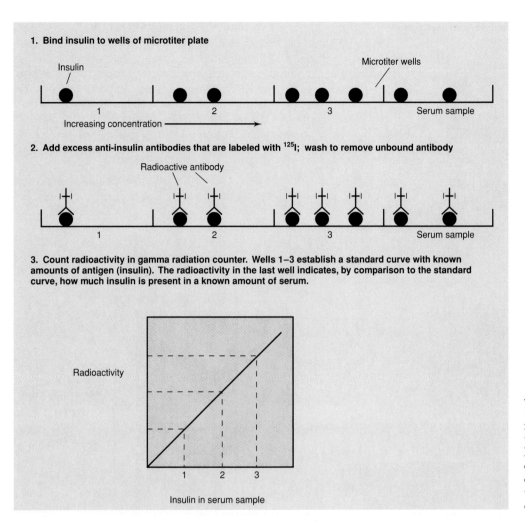

1. Bind insulin to wells of microtiter plate

Insulin

Microtiter wells

1   2   3   Serum sample

Increasing concentration ⟶

2. Add excess anti-insulin antibodies that are labeled with ^{125}I; wash to remove unbound antibody

Radioactive antibody

1   2   3   Serum sample

3. Count radioactivity in gamma radiation counter. Wells 1–3 establish a standard curve with known amounts of antigen (insulin). The radioactivity in the last well indicates, by comparison to the standard curve, how much insulin is present in a known amount of serum.

Radioactivity

1   2   3

Insulin in serum sample

Figure 21.20 Radioimmunoassay (RIA). Using RIA to detect insulin levels in human serum. Following establishment of a standard curve, the insulin concentration in a serum sample can be estimated.

which represents a single protein of specific molecular weight (Figure 21.21). The proteins are then transferred to nitrocellulose paper by an electrophoretic transfer process that forces the proteins out of the gel and onto the paper. At this point, antibodies raised against a protein or group of proteins from a pathogen are added to the nitrocellulose blot. Following a short incubation period to allow the antibodies to bind, a radioactive marker that binds antigen–antibody complexes is added. The most common radioactive marker used is *Staphylococcus* protein A iodinated with radioactive iodine, ^{125}I. Protein A has a strong affinity for antigen–antibody complexes and binds firmly to them. Once the radioactive marker has bound, its vertical position on the blot can be detected by exposing the nitrocellulose blot to X-ray film; the gamma rays emitted by the ^{125}I expose the film only in the region where the radioactive antibody has bound to antigen–antibody complexes (Figure 21.21).

For many clinical applications, immunoblots employ enzyme-linked immunosorbent assay (ELISA) technology (see Section 21.8) for detection of bound antigen–antibody complexes. Following treatment of

the blotted proteins with specific antibody, the paper is washed and then treated with a second antibody, which binds to the first. For example, if antibodies from a human were used in the first step, then the second antibody could be a rabbit anti-human antibody. Covalently attached to this second antibody is an enzyme. The original antigen–antibody complexes are visualized when the enzyme is assayed because the product of the enzyme reaction leaves a colored product on the nitrocellulose filter at any spot where rabbit antibodies are bound to the human antibodies. By comparing the location of the color bands on the nitrocellulose paper with the position of colored bands from control samples, a protein associated with a given pathogen can be positively identified.

The immunoblot procedure can be used to detect either antigen (*direct* evidence for pathogen presence) or antibody (*indirect* evidence for pathogen exposure). Thus, this very sensitive, extremely accurate method is analogous to the direct and indirect ELISA procedures detailed in Section 21.8. We now examine a widely used immunoblot test designed to identify exposure to HIV.

1. Denature proteins by boiling in detergent

2. Subject mixture to electrophoresis; proteins separate by molecular weight

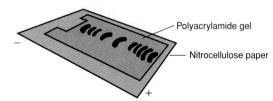

Polyacrylamide gel

Nitrocellulose paper

3. Blot the separated proteins from the gel to nitrocellulose paper

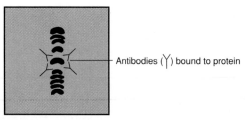

Antibodies ( ) bound to protein

4. Treat nitrocellulose paper containing blotted proteins with antibodies; each antibody recognizes and binds to a specific protein

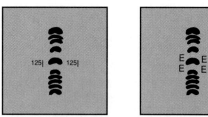

125I   125I

E   E
E   E

5. Add marker to bind to antigen:antibody complexes, either (left) radioactive *Staphylococcus* protein A–125I, or (right) antibody containing conjugated enzyme

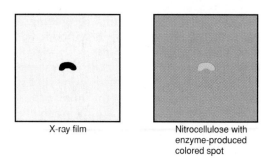

X-ray film

Nitrocellulose with enzyme-produced colored spot

(a)

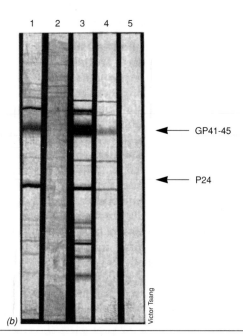

1   2   3   4   5

GP41-45

P24

Victor Tsang

(b)

Figure 21.21 The Western blot (immunoblot) and its use in the diagnosis of human immunodeficiency virus (HIV) infection. (a) Protocol for an immunoblot. (b) Developed HIV immunoblot. The proteins P24 and GP41-45 are coat proteins of the virus and are diagnostic for HIV. Lane 1, Positive control serum (from known AIDS patients); lane 2, negative control serum (from healthy volunteer); lane 3, strong positive from patient sample; lane 4, weak positive from patient sample; lane 5, reagent blank to check for background binding.

The HIV immunoblot

Immunoblots have had a significant clinical impact on the diagnosis and confirmation of cases of AIDS. Because an immunoblot is more laborious, time-consuming, and costly than the ELISA test, HIV-ELISA tests have been widely used for screening purposes. However, the HIV-ELISA test occasionally yields false positive results. Thus, an immunoblot is virtually always used to confirm positive ELISA results.

Like the HIV-ELISA, the HIV immunoblot is designed to detect the presence of *antibodies* to HIV in a serum sample. To perform the immunoblot, a purified preparation of HIV is treated with the detergent sodium dodecyl sulfate (SDS), which solubilizes HIV proteins and also renders the virus inactive. HIV proteins are then resolved by polyacrylamide gel electrophoresis. The HIV proteins are then blotted from the gel onto sheets of nitrocellulose paper (Figure 21.21). At least seven major HIV proteins are resolved by electrophoresis, and two of them, designated P24 and GP41-45, are used as specific diagnostic proteins in the AIDS immunoblot. Protein P24 is the HIV core protein, and proteins GP41-45 are HIV coat proteins (∞ Section 8.22).

Following blotting of the proteins, the nitrocellulose strips are incubated with a serum sample previously identified as HIV-positive (a positive control) by HIV-ELISA. If the sample is truly HIV-positive, antibodies against HIV proteins will be present and will bind to the HIV proteins separated on the nitrocellulose paper (Figure 21.21). To detect whether antibodies from the serum sample have bound to HIV antigens, a *detecting antibody*, anti-human IgG conjugated to the enzyme peroxidase, is added to the strips. If detecting antibody binds, the activity of the conjugated enzyme will form a brown band on the strip at the site of antibody binding after addition of substrate. The serum from the HIV-ELISA-positive patient is assayed in parallel with the positive control serum. The patient can be confirmed as HIV-positive if the position of the bands in the patient and the positive control sera are identical; negative control sera are also analyzed in parallel and must show no bands (Figure 21.21).

Although the intensity of the bands obtained in the HIV immunoblot varies somewhat from sample to sample (Figure 21.21*b*), the interpretation of an immunoblot is generally unequivocal, and thus the test is valuable for confirming HIV-ELISA positives and eliminating HIV-ELISA false positives. To make the HIV immunoblot clinically accessible, nitrocellulose strips containing inactivated HIV antigens (previously separated by electrophoresis) are available commercially. Separate strips can be incubated directly with the patient and control serum samples and subsequently treated with the detecting antibody. In addition to standardizing the assay, this eliminates the need for clinical microbiology laboratories to produce HIV as a source of antigen for the HIV immunoblot.

The immunoblot technique is considered the definitive test for serodiagnosis of HIV infection and is virtually always used to confirm positive HIV-ELISA tests. This technique is also used to confirm the specificity of screening tests for the Lyme disease antibody (see Table 21.5). However, because of the expense, technical requirements, and time involved, immunoblot tests are not likely to supplant the rapid, low cost ELISA methods for general screening purposes.

CONCEPT CHECK   *21.9*

Immunoblot procedures can be used to detect antibodies to specific antigens or to detect the presence of the antigens themselves. The antigens are electrophoresed, transferred (blotted) to a filter, and exposed to antibody. Immune complexes are visualized through the use of enzyme-labeled or radioactive second antibodies. Immunoblots are extremely sensitive *and* accurate, but procedures are complex and time-consuming.

✔ *What advantage does the immunoblot have over immunoassays such as ELISA and RIA?*

✔ *What alternative labeling methods are used for immunoblot detection systems?*

21.10
Nucleic Acid Probes in Clinical Diagnostics

The emergence of molecular biology has given rise to new molecular tools that are rapidly being adapted to the field of diagnostic microbiology. *DNA diagnostics*, as this area has come to be known, is revolutionizing the whole approach to identifying and monitoring infectious diseases, genetic and malignant diseases, and other medical conditions such as coronary artery disease and diabetes. This new approach uses *genotypic* rather than *phenotypic* factors to identify specific pathogens. The power of DNA diagnostics is a consequence of two facts: (1) nucleic acids can be rapidly and sensitively measured, and (2) the *sequence* of nucleotides in a given DNA molecule is so specific that hybridization analyses can be used for reliable clinical diagnoses.

Automation is helping to make DNA analysis virtually routine in the clinical setting. Automated DNA extractors, polymerase chain reaction (PCR) machines (∞ Section 10.9), DNA sequencers, and pulse field gel electrophoresis equipment (for separating large DNA segments such as whole chromosomes) (∞ Working with Nucleic Acids: The Tools, Chapter 6) are all available for use in the diagnosis of diseases. The **nucleic acid probe** is now a major molecular tool in clinical laboratories.

Nucleic acid probes

One of the most powerful analytical tools available to clinical microbiologists is *nucleic acid hybridization*. Instead of detecting a whole organism or its products (for example, antigens), hybridization detects the presence or absence of *specific DNA sequences* associated with a specific organism. To identify a microorganism through

DNA analysis, the clinical microbiologist must have available a *nucleic acid probe* to that microorganism, a *single strand* of DNA containing sequences unique to the organism. The probe may be up to several kilobases in length, but many synthetic oligonucleotides consist of 20 bases or less and are still highly specific. If a microorganism in a clinical specimen contains DNA sequences complementary to the probe, the two sequences can hybridize (following appropriate sample preparation to yield single-stranded DNA from the microorganism), forming a *double-stranded* molecule (Figure 21.22). To detect that a reaction has occurred, the probe is labeled with a *reporter molecule,* a radioisotope, an enzyme, or a fluorescent compound that can be measured in small amounts following hybridization. Depending on the reporter used (radioisotopes are the most sensitive), as little as 0.25 μg of DNA per sample can be detected.

Nucleic acid probes offer many advantages over clinical immunological assays. Nucleic acids are much more stable than proteins to high temperatures, high pH, organic solvents, and other chemicals. This means that a clinical sample can be treated in a relatively harsh manner so as to destroy most interfering material, leaving behind the nucleic acid. In addition, nucleic acid probes are more defined entities than antibodies; the composition of a probe can be accurately checked by sequence analysis, and new molecules of the short probes can be produced in DNA synthesizers whenever necessary.

Nucleic acid probes are also very sensitive. With current technology it is possible to detect less than 1 μg of nucleic acid per sample. This translates into about 10^6 bacterial cells or virus particles. Although probes used in this fashion are not as sensitive as direct culture (where as few as 1–10 cells per sample can be detected), probe methods can be useful in situations where culture of the organism is difficult or even impossible.

Perhaps one of the most promising methods for clinical diagnostics involves the use of sequence-specific probes for polymerase chain reaction amplification of DNA or ribonucleic acid (RNA) from specific pathogens. As we discussed in Section 10.9, PCR uses two sequence-specific oligonucleotides to amplify target DNA. A DNA amplification of a millionfold or more increases the probe sensitivity and theoretically makes this procedure capable of detecting DNA from a single bacterial cell. For example, probes for a pathogen might be used to examine DNA derived from suspected infected tissue, even in the absence of an observ-

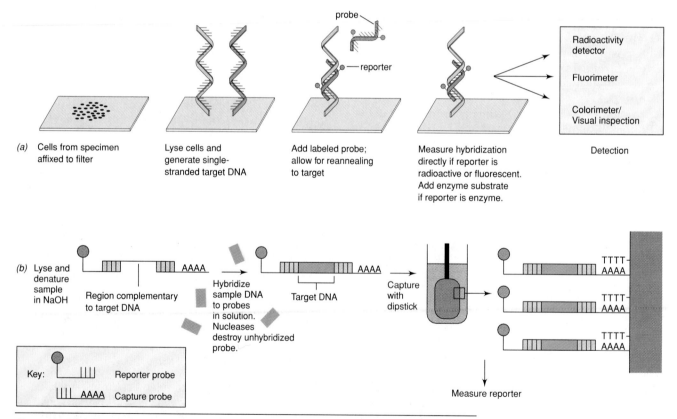

(a) Cells from specimen affixed to filter

Lyse cells and generate single-stranded target DNA

Add labeled probe; allow for reannealing to target

Measure hybridization directly if reporter is radioactive or fluorescent. Add enzyme substrate if reporter is enzyme.

probe
reporter

Radioactivity detector
Fluorimeter
Colorimeter/Visual inspection

Detection

(b) Lyse and denature sample in NaOH

Region complementary to target DNA

Hybridize sample DNA to probes in solution. Nucleases destroy unhybridized probe.

Target DNA

Capture with dipstick

Measure reporter

Key: Reporter probe
        AAAA   Capture probe

Figure 21.22 Nucleic acid probe methodology in clinical diagnostics. (a) Membrane filter assay. The detecting system (reporter) can be a radioisotope, a fluorescent dye, or an enzyme. (b) Dipstick assay. In the dipstick assay a dual reporter or capture probe is used. The capture probe contains a poly-dA tail that hybridizes to a poly-dT oligonucleotide affixed to the dipstick.

able, culturable pathogen. These methods are particularly useful for identifying viral and intracellular infections. The presence of the appropriate amplified gene segment (Figure 21.23) confirms the presence of the pathogen. Several of the specific organisms for which either hybridization or PCR methods are in use are listed in Table 21.7. We now consider several specific examples of nucleic acid probes and discuss some applications in more detail.

Probes in the clinical laboratory

In most clinical probe assays, colonies from plates or pieces of infected tissue are treated with strong alkali, usually NaOH, to lyse the cells and partially denature the DNA, forming single-stranded molecules (Figure 21.22). This mixture is then affixed to a filter or left in solution (for dipstick assays, see later), and the labeled probe added. Hybridization is allowed to occur at a temperature at which considerable sequence homology between target DNA and probe DNA is necessary to form a stable duplex (the actual temperature used in a given probe assay is governed by the length and nucleic acid composition of the probe and target DNA). Following a wash to remove any unhybridized probe DNA, the extent of hybridization that occurred is measured using the reporter molecule attached to the probe. Depending on how the probe was labeled, this involves measurement of radioactivity, enzyme activity, or fluorescence from an attached dye.

Nucleic acid probes have been marketed for the identification of several major microbial pathogens and are in widespread use for the detection of *Neisseria gonorrhoeae* and *Chlamydia trachomatis* (see Table 21.7 and Section 23.6). However, in addition to their clinical usefulness, probes are finding widespread application in food industries and in food regulatory agencies. Probe detection systems can be used to routinely monitor foods for their content of important pathogens such as *Salmonella* and *Staphylococcus.* In probe assays of food, an enrichment period is usually employed to allow low numbers of cells in the food to multiply to a sufficient number to be detectable by the probe. However, the use of PCR gene amplification techniques eliminates the need for the enrichment period.

Probes designed for use in the food industry employ probe dipsticks to remove hybridized DNA from solution. Two component probes are used here, one serving as a *reporter probe* and the other as a *capture probe* (Figure 21.22). Following hybridization of the reporter or capture probe to target DNA, the dipstick, which contains a sequence complementary to the capture probe (usually poly dT to capture poly dA on the probe) (see Figure 21.22b) is inserted into the hybridization solution, and it traps hybridized DNA for removal and measurement.

Probes for detecting certain cancer viruses are also being developed. For example, a probe is now available to detect DNA sequences unique to human papilloma viruses. These viruses sometimes cause skin and cervical cancer in humans (∞ Section 8.14), and a specific group of papilloma viruses causes genital warts (∞ Section 23.6). In women, an increased incidence of genital papilloma virus infection is associated with an increased risk of cervical cancer. The DNA probe developed for papilloma viruses can be used to search by hybridization for papilloma virus sequences in tissues removed during a cervical exam. Early detection and treatment of papilloma virus infections decreases the risk of cervical cancer.

The development of new nucleic acid probes is a major activity in pharmaceutical, biotechnology, and clinical diagnostic companies. Although developing probes for diseases for which no probe yet exists is a top priority, a major goal of probe research and development is to make existing probes even more specific and to continue to simplify the procedures necessary for implementing probe-based assays in the clinical setting.

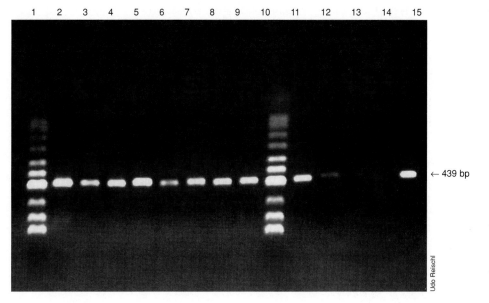

Figure 21.23 Polymerase chain reaction (PCR) analysis of patient sputum for *Mycobacterium tuberculosis* in the diagnosis of tuberculosis. Sputum samples from patients were used as a source of DNA. Amplification was initiated with a primer pair, which produced the indicated 439-base pair product when a pure culture of *M. tuberculosis* was used as the DNA source (lane 15). Lanes 2–9, 11 and 12 are from sputums positive for *M. tuberculosis* (lane 12 is only a weak positive). Lanes 13 and 14 are from *M. tuberculosis*-negative sputum samples. Lanes 1 and 10 are molecular weight reference markers.

← 439 bp

TABLE 21.7   Pathogens that can be identified with nucleic acid probes

Pathogen	Disease
Bacteria	
Campylobacter spp.	Food infections
Chlamydia trachomatis	Venereal syndromes; trachoma
Escherichia coli (enteropathogenic strains)	Gastrointestinal disease
Haemophilus influenzae	Infectious meningitis
Legionella pneumophila	Pneumonia
Listeria monocytogenes	Listeriosis
Mycobacterium avium	Tuberculosis
Mycobacterium tuberculosis	Tuberculosis
Mycoplasma hominis	Urinary tract infection; pelvic inflammatory disease
Mycoplasma pneumoniae	Pneumonia
Neisseria gonorrhoeae	Gonorrhea
Salmonella spp.	Gastrointestinal disease
Shigella spp.	Gastrointestinal disease
Staphylococcus aureus	Purulent discharges (boils, blisters, pus-forming skin infections)
Streptococcus pyogenes	Scarlet, rheumatic fever; "strep throat"
Fungi	
Blastomyces dermatitidis	Blastomycosis
Coccidioides immitis	Coccidioidomycosis
Histoplasma capsulatum	Histoplasmosis
Viruses	
Cytomegalovirus	Congenital viral infections
Epstein–Barr virus	Burkitt's lymphoma; mononucleosis
Hepatitis viruses A, B, C, delta	Hepatitis
Herpes virus (types I and II)	Cold sores; genital herpes
Human immunodeficiency virus (HIV)	Acquired immunodeficiency syndrome (AIDS)
Human papilloma virus	Genital warts; cervical cancer
Protozoa	
Leishmania donovanii	Leishmaniasis
Plasmodium spp.	Malaria
Pneumocystis carinii	Pneumonia
Trypanosoma spp.	Trypanosomiasis

A major impediment to more extensive use of nucleic acid probes is the high cost per test, both for special reagents and for trained personnel. However, as automated methods and new tests become available, these problems will be overcome. Another problem is the lack of specific probes for many pathogens. However, recent advances in our understanding of bacterial phylogenetics based on 16S ribosomal RNA (rRNA) sequencing (∞ Section 15.7) now allow new and more specific nucleic acid probes to be constructed.

For example, *ribotyping* is a new method based on DNA probes that recognize conserved RNA operon genes. Ribotyping is essentially a Southern blot analysis in which strains are characterized for restriction fragment length polymorphisms (RFLPs; ∞ DNA Fingerprinting, Chapter 10) of their individual ribosomal genes. Within a species, and particularly within a strain, the DNA sequences and restriction digest patterns of genes encoding 16S rRNA, 25S rRNA and tRNAs are highly conserved and serve as a molecular fingerprint for that organism.

Since *all* organisms have ribosomal genes, this technique is universally applicable and is finding wide acceptance as a clinical and phylogenetic tool.

CONCEPT CHECK   *21.10*

Nucleic acid hybridization is a powerful tool for disease diagnosis. A nucleic acid sequence specific for the virus or microbial pathogen must be available to design a probe. In addition to their use in clinical diagnosis, probes are also finding wide application in food and environmental monitoring.

✔ *What advantage does nucleic acid hybridization have over standard culture methods for identification of microorganisms? What disadvantages?*

✔ *Cite an example where identification of a pathogen can be made with a nucleic acid probe that cannot be made with standard culture techniques.*

21.11
Diagnostic Virology

Because of their unique characteristics as cellular parasites, identification and diagnosis of pathogenic viruses pose significantly different problems from the bacterial pathogens on which we have focused (∞ Chapter 8). For example, viruses cannot be directly cultured on artificial media, but they can be grown in host cells. Thus, viral growth-dependent assays are extremely complex. However, immunodiagnostic methods (see Sections 21.4–21.9) and nucleic acid hybridization methods (see Section 21.10) are widely used for viral identification, using the same principles applied to bacterial identification schemes. Finally, electron microscopy is often used for direct examination of viral specimens. In this section, we will concentrate on specific methods useful for viral identification, especially for the pathogenic viruses we will discuss in Chapter 23.

First, laboratory cultivation of viruses from clinical materials is more difficult, time-consuming, and specialized than the cultivation of most bacterial pathogens. This is because viruses grow only in living cells. We discussed the use of cell cultures for the growth of viruses in Section 8.3, and such cultures are commonly used in diagnostic virology. A common cell line is a human diploid fibroblast culture called WI-38, which grows rapidly and reproducibly in cell culture medium. Another cell line sometimes used is HeLa, a cell culture derived initially from a human cancer. This cell line has been maintained *in vitro* for so many successive transfers that it has greatly changed its character (for instance, it is no longer diploid). In addition to these two cell lines, which can be maintained in the laboratory indefinitely, cultures are also made from Rhesus monkey kidneys. Monkey kidney cell lines are called *primary* because they are not maintained by successive transfer in the laboratory. Primary monkey kidney cells support growth of a number of pathogenic viruses and are therefore of value in initial isolation of unknown viruses, but are not routinely used in most clinical laboratories. Because of the technical expertise and expense involved (for example, laboratories that do primary culture must maintain or purchase Rhesus monkeys), most diagnostic virology laboratories are located at specialized government facilities or major clinical research institutions.

Although the best diagnostic technique for most viral infections is actual isolation, growth, and identification of the virus, this is not practical in most clinical settings. Instead, several immunological tests and nucleic acid probes for viruses have been developed. Most of the immunological tests are either direct enzyme-linked immunosorbent assays (ELISAs) that detect viral particles (see Section 21.8 and Figure 21.18) or fluorescent antibody methods in which antibodies made against viral antigens are used to detect cells con-

taining viruses (see Section 21.7 and Figure 21.14*b*). Viruses currently detectable by direct ELISA include rotavirus, hepatitis viruses, rubella virus, bunyavirus, measles virus, mumps virus, and parainfluenza virus. Agglutination tests are also available in which antiviral antibodies conjugated to latex beads or activated charcoal (see Section 21.5) are used to test for agglutination of viral antigens released from lysed tissue samples. Nucleic acid probes for detection and identification of clinically significant viruses are also becoming a major diagnostic tool in clinical virology (see Section 21.10).

Electron microscopy

In addition to the tools described previously, diagnostic virology can be done by electron microscopy. Because many viruses have distinctive morphologies (∞ Chapter 8), their presence in clinical specimens can often be detected by observing the sample under the electron microscope (Figure 21.24). In most specimens the virus particles must first be concentrated and separated from human tissues, and a variety of techniques, generally employing centrifugation and filtration, are used to obtain a sample enriched in virus particles. Although not as reliable as immunological or nucleic acid probe methods, the observation of virus particles of a specific morphology in a particular type of human tissue can often serve as presumptive evidence for a disease. Treating the sample with antibodies prepared against particular viruses can be used to increase the sensitivity and specificity of this method; such antibodies may cause the viral particles to agglutinate, and this makes them easier to distinguish from cellular

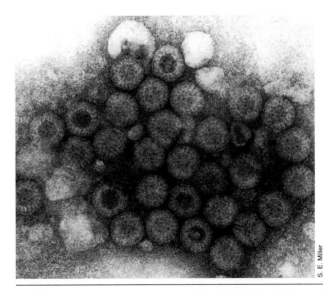

Figure 21.24 Electron microscopic observation of clinical specimens to detect viruses. Human rotavirus from a fecal sample. The distinct spherical nature of the virus, coupled with the source, is a highly diagnostic criterion. Each rotavirus particle is approximately 75 nm in diameter.

debris under the electron microscope. Viruses can also be visualized following treatment of specimens with antiviral antibodies containing conjugated heavy metals (see Section 21.6). The electron microscope can thus serve as a provisional means of identifying viruses directly from patient specimens and offers the major advantage of speed. With the use of negative staining techniques (Figure 21.24), results from electron microscopic analysis can be available 20 minutes after collection of the specimen.

A summary of a few laboratory procedures used in diagnostic virology is given in Table 21.8. Most of these procedures are used only under special circumstances. For routine virus infections, diagnoses are made by assessing symptoms or by immunological or other indirect means. For example, testing for human immunodeficiency virus (HIV) infection involves the ELISA test and the immunoblot test. Both these methods detect the presence of *antibodies* to HIV, not HIV itself; as is the case for most viruses, there is currently no *routine* method used to detect the presence of HIV directly.

CONCEPT CHECK   *21.11*

Virus culture can be accomplished only in susceptible tissue or organs. Therefore, most diagnostic techniques for viral identification are not growth-dependent but routinely rely on immunoassays and nucleic acid hybridization techniques. Electron microscopy techniques are useful for direct observation of virus in host samples.

✔ *Why must viruses be grown in tissue or organ culture and not on artificial, inert media?*

✔ *How can individual pathogenic viruses be identified?*

21.12
Safety in the Clinical Laboratory

By their very nature, clinical laboratories are areas in which potentially dangerous biological specimens must be handled on a routine basis. Hence, a defined protocol for handling clinical samples must be established to avoid laboratory accidents. In the United States, every clinical and research institution that deals with human or primate tissue is required by law to have an occupational exposure control plan in place for the handling of all bloodborne pathogens. This law was specifically designed to protect workers from infection by hepatitis B virus (HBV) (∞ Section 23.13) and human immunodeficiency virus (HIV) (∞ Section 23.7) but effectively protects workers from infection by virtually all pathogens because of the stringent precautions.

Studies of laboratory-associated infections have indicated that most such infections do not result from known exposures or accidents but instead from routine handling of patient specimens. The two most common causes of laboratory accidents are ignorance and carelessness. Infectious aerosols, generated during processing of the specimen, are the most likely cause of laboratory infections. In attempts to minimize the exposure of clinicians to infectious agents and to thereby reduce the number of nonaccident-associated laboratory infections, well-run clinical laboratories stress the safety rules outlined here. These rules, if applied stringently, ensure the prevention of pathogen spread and meet the legal requirements of the United States law.

1. Laboratories handling hazardous materials must restrict access to laboratory and support personnel. These individuals must have knowledge of the bio-

---

TABLE 21.8   Some laboratory procedures used in diagnostic virology[a]

Condition	Possible viral cause	Sample source	Inoculation procedure
Upper respiratory infection	Rhinovirus Coronavirus Adenovirus	Nasopharyngeal or tracheal fluid (aspirate)	Human fibroblast culture
Pneumonia	Influenza	Nasopharyngeal fluid or swab	Human fibroblast cultures or embryonated eggs
Measles	Measles virus	Nasopharyngeal fluid or swab	Monkey kidney cells
Vesicular rash	Herpes simplex	Vesicular fluid by aspiration	Human fibroblast culture
Diarrhea	Rotavirus (infants) Norwalk agent (adults)	Feces or rectal swab	Look for characteristic virus particles with the electron microscope (Figure 21.24)
Nonbacterial meningitis	Enterovirus Mumps Herpes simplex	Spinal fluid	Human fibroblast or monkey kidney cultures

[a]Immunological methods and nucleic acid probe methods are also widely used in the diagnosis of viral infections (see Sections 21.4–21.10).

logical risks involved in the clinical laboratory and act accordingly.

2. Effective procedures for decontaminating infectious materials or wastes, including specimens, syringes and needles, inoculated media, bacterial cultures, tissue cultures, experimental animals, glassware, instruments, and surfaces must be in place and be practiced without compromise. A 5.25% (full strength) chlorine bleach solution or other approved disinfectant is recommended for decontaminating spilled infectious material. All potentially infectious waste must be burned in a certified incinerator or handled by a licensed waste handler.

3. Personnel working with hazardous infectious agents or vaccines (for example, rabies, polio, or diphtheria-pertussis-tetanus vaccines) must be properly vaccinated against the agent. Persons working with human or primate tissue must be vaccinated against HBV.

4. All clinical specimens should be considered potentially infectious and handled in the appropriate manner. This is especially important for preventing laboratory-acquired hepatitis because of the relative frequency with which hepatitis viruses are present in clinical specimens.

5. All pipetting must be done with automatic pipetting devices (not by mouth), and devices such as syringes, needles, and clinical centrifuges must always be used with proper biological containment equipment.

6. Animals should be handled only by trained laboratory personnel, and anesthetics or tranquilizers should be used to avoid injury to both personnel and animals.

7. Laboratory personnel must wear laboratory coats or gowns, sealed shoes, rubber gloves, masks, eye protection, respiratory devices when needed, and other barrier protection as deemed appropriate by the level of exposure and the severity of the potential infection. These barrier devices must also be properly stored and decontaminated after use. Laboratory personnel must also practice good personal hygiene with respect to hand washing.

Eating and drinking, applying lip balm, or wearing contact lenses is never permitted in the clinical laboratory.

8. Because of the special risks associated with AIDS, all clinical specimens should be treated as if they contain HIV (which they might). Latex or vinyl gloves should be worn whenever handling specimens of *any* kind. Masks must be worn any time there is a possibility of generating an aerosol during specimen preparation. Needles must not be resheathed, bent, or broken; they should be placed in a labeled container designated expressly for this purpose that can be sealed and autoclaved before disposal.

These safety rules should be the norm for all clinical laboratories. Specialized clinical laboratories may have additional rules to ensure a safe work environment. For example, if laboratory personnel handle extremely hazardous airborne pathogens (such as the causative agent of tuberculosis, *Mycobacterium tuberculosis*) on a routine basis, the laboratory should be fitted with special features, such as negatively pressurized rooms, biological safety cabinets (∞ Figure 11.4), and air filters, to prevent accidental release of the pathogen from the laboratory. In the final analysis, however, it is the attitude of the personnel that makes the laboratory a safe or an unsafe place to work. Any clinical laboratory is a potentially hazardous place for untrained personnel or those unwilling to take the necessary steps to prevent laboratory-acquired infection.

CONCEPT CHECK   21.12

Safety in the clinical laboratory requires effective training, planning, and care to prevent the infection of laboratory workers with pathogens. Materials such as inoculated culture media, needles, and patient specimens require specific precautions for safe handling.

✔ *What are the major precautions necessary to prevent spread of a bloodborne pathogen to laboratory personnel?*

Material for Review

REVIEW QUESTIONS

1. Describe the standard procedure for obtaining and culturing a blood sample for bacteria.

2. Why is the *number* of bacterial cells in urine, rather than simply the *presence* of bacteria in urine, of significance? What organism is responsible for most urinary tract infections? Why?

3. Describe the procedures used for culturing anaerobic microorganisms. Why is it important to process all clinical specimens quickly? What special procedures and precautions are necessary for the isolation and culturing of anaerobes?

4. Differentiate between *selective* and *differential* media. Is eosin–methylene blue (EMB) agar selective or differential? How and why is it used in a clinical laboratory?

5. Describe the Kirby–Bauer test for antibiotic sensitivity. Why should potential pathogens from patient isolates be tested by this method?

6. Why does the antibody titer *rise* after infection? Why is it necessary to draw two serum samples to monitor infections?

7. How are fluorescent antibodies used for the diagnosis of viral diseases?

8. Agglutination tests are significantly more sensitive than precipitation tests. Why might this be the case?

9. Likewise, radioimmunoassay (RIA) and enzyme-linked immunosorbent assay (ELISA) tests are extremely sensitive, as compared to agglutination. Why?

10. What advantages do *monoclonal* antibodies have over *polyclonal* antibody preparations, especially with regard to standardization of antibody preparations?

11. Why is the immunoblot (Western blot) procedure used to confirm positive human immunodeficiency virus (HIV)-ELISA results?

12. What information is essential for the design of a pathogen-specific nucleotide probe? Where can one obtain such information? Is this information available for all pathogens?

13. What is a primary cell line? Why do some animal viruses grow in primary cell lines but not in cell lines such as HeLa cells?

14. How are most laboratory-associated infections contracted? What actions can be taken to prevent them?

APPLICATION QUESTIONS

1. From a blood culture, you obtain a culture positive for *Staphylococcus aureus*. Interpret and explain the results. Is it likely that the patient has a *S. aureus* bacteremia? Why or why not?

2. Describe the microscopic and cultural evidence that would support a diagnosis of gonorrhea. Why is Thayer–Martin agar a "better" medium than chocolate agar for the isolation of *Neisseria gonorrhoeae*?

3. Compare and contrast the changes in color due to pH-sensitive dyes in tests for carbohydrate fermentation and citrate utilization. Is the same dye used in both tests? Why or why not?

4. Why should it not be a common medical practice to treat an infectious disease with antibiotics before isolating the suspected pathogen? What further steps should be taken before antibiotic therapy is initiated? Why are these steps seldom taken outside a hospital environment?

5. What are the advantages of rapid identification systems such as agglutination tests as compared to standard clinical diagnostic procedures? Also discuss the potential disadvantages of rapid, non-culture-based tests.

6. Design a fluorescent antibody assay for confirming an initial diagnosis of "strep throat" (*Streptococcus pyogenes* is the causative agent of strep throat). Discuss all aspects of the assay, including preparation of antisera, necessary controls, and clinical interpretation.

7. Design an ELISA test for detecting hepatitis A virus in feces. Likewise, design an *indirect* hepatitis A virus test for detection of exposure. Would either test require anti-human IgG antibodies? Why or why not?

8. What are the major advantages of using DNA probes in diagnostic microbiology? Discuss at least four aspects of probe technology that benefit clinical medicine. Where can you find information to design polymerase chain reaction (PCR) assay probes for the hepatitis A virus in Question 7? Remember, the probes must be sequence-specific for the virus.

9. As a new professional in a clinical laboratory, you are assigned the task of formalizing the laboratory safety requirements to prevent infectious diseases. Explain how you would monitor and enforce the recommendations outlined in Section 21.12.

SUPPLEMENTARY READINGS

Baron, E. J., and **S. M. Finegold.** 1994. *Bailey and Scott's Diagnostic Microbiology,* 9th edition. C. V. Mosby, St. Louis. A good source of clinical microbiological methods and diagnostic and identification procedures.

Brooks, G. F., J. S. Butel, L. N. Ornston, E. Jawetz, J. L. Melnick, and **E. A. Adelberg.** 1995. *Jawetz, Melnick and Adelberg's Medical Microbiology,* 20th edition. Appleton and Lange, Norwalk, CT. A brief, up-to-date treatment of infectious diseases and clinical information.

Hames, B. D., and **S. J. Higgins.** 1995. *Gene Probes: A Practical Appoach.* Oxford University Press, New York. A comprehensive source listing methods of preparation and application for a variety of gene probes.

Henry, J. B. (ed.) 1991. *Clinical Diagnosis and Management by Laboratory Methods,* 18th edition. W. B. Saunders, Philadelphia. A comprehensive review of clinical methods.

Joklik, W. K., H. P. Willet, D. B. Amos, and **C. M. Wilfert.** 1992. *Zinsser Microbiology,* 20th edition. Appleton and Lange, East Norwalk, CT. A comprehensive overview of medical microbiology, including immunological and scientific principles.

Koneman, E. W., S. D. Allen, V. R. Dowell, Jr., W. M. Janda, H. M. Sommers, and **W. C. Winn, Jr.** 1992. *Color Atlas and Textbook of Diagnostic Microbiology,* 4th edition. J. B. Lippincott, Philadelphia. A well-illustrated primary reference on clinical microbiology.

Macario, A. J. L., and **E. Conway de Macario** (eds.). 1990. *Gene Probes for Bacteria.* Academic Press, Orlando, FL. A comprehensive listing of the clinical applications of nucleic acid probes for pathogenic bacteria.

Murray, P. R., E. J. Baron, M. A. Pfaller, F. C. Tenover, and **R. H. Yolken.** 1995. *Manual of Clinical Microbiology,* 6th edition. American Society for Microbiology, Washington, DC. The standard reference work for clinical microbiologists. Includes discussions of the science behind clinical applications.

Murray, P. R., G. G. Kobayashi, M. A. Pfaller, and **K. S. Rosenthal.** 1994. *Medical Microbiology,* 2nd edition, C. V. Mosby, St. Louis. An excellent medical text covering immunology as well as microbiology.

Rayburn, S. R. 1990. *The Foundations for Laboratory Safety: A Guide for the Biomedical Laboratory.* Springer-Verlag, New York. An up-to-date and well written textbook on laboratory safety. Excellent treatment of the hazards of biological agents.

Stine, G. J. 1996. *AIDS Update 1996.* Prentice-Hall, Englewood Cliffs, NJ. An excellent monograph disccussing various aspects of AIDS biology, pathology, and societal issues. Updated yearly.

 On~line resources for this chapter are on the World Wide Web at: http://www.prenhall.com/~brock (click on the table of contents link and then select Chapter 21).

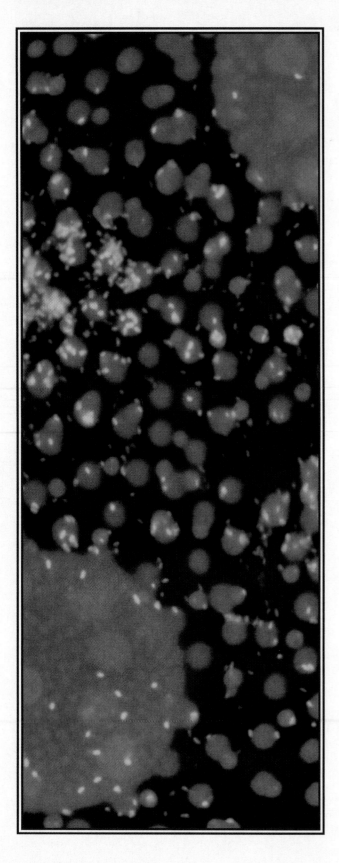

Epidemiology and Public Health Microbiology

Many human pathogens exist as normal inhabitants of the environment. Shown here are cells of *Vibrio cholerae* (green) attached to the surface of *Volvox* (red), a freshwater alga. The *V. cholerae* organism causes cholera. This isolate was found in a water source in Bangladesh. The ability of these organisms to exist outside the human host makes their eradication as pathogens nearly impossible.

MINIGLOSSARY for Chapter 22

Acute short-term infection usually characterized by dramatic onset and rapid recovery

Carrier subclinically infected individuals who may spread a disease

Chronic long-term infection

Common-Source Epidemic an epidemic resulting from infection of a large number of people from a single contaminated source

Emerging Infections infectious diseases whose incidence has increased in the past 20 years or whose incidence threatens to increase in the near future

Endemic disease constantly present, usually in low numbers

Epidemic the occurrence of a disease in unusually high numbers in a localized region

Epidemiology the study of the occurrence, distribution, and control of infectious diseases

Fomites inanimate objects that, when contaminated with a viable pathogen, can transfer the pathogen to a host

Herd Immunity resistance of a group to a pathogen as a result of the immunity of a large portion of the group

Host-to-Host Epidemic an epidemic resulting from person-to-person contact, characterized by a gradual rise and fall in numbers of cases

Incidence the number of cases of disease in a population

Morbidity incidence of illness in a population

Mortality incidence of death in a population

Nosocomial Infection hospital-acquired infection

Outbreak the occurrence of a large number of cases of a disease in a short period of time

Pandemic a worldwide epidemic

Prevalence the proportion or percentage of individuals in the population having a disease

Public Health the health of the population as a whole

Quarantine the practice of restricting the movement of individuals with highly contagious serious infections to prevent spread of the disease

Reservoir sites in which viable infectious agents remain and from which infection of individuals may occur

Resurgent Infections infectious diseases, thought to be under control, that reemerge

Surveillance observation, recognition, and reporting of diseases as they occur

Vector a living agent that transfers a pathogen (note alternative usage in Chapter 8)

Vehicle nonliving source of pathogens that infect large numbers of individuals; common vehicles are food and water

Zoonosis a disease that occurs primarily in animals but can be transmitted to humans

---

In Chapters 19 and 20 we considered the general principles of how microorganisms cause infectious disease and how the host responds to microbial onslaught. In Chapter 21, we discussed the methods for diagnosis of infectious diseases. In this chapter, we consider how a pathogen spreads from an infected individual to others in a population. Thus, we are dealing here with *public health*. In the next chapter we consider the diseases themselves. The principles put forward in this chapter are vital for controlling the spread of infectious disease.

One measure of our success in the control of infectious disease was shown by the data presented in Figure 1.18, which compared the present causes of death in the United States with those at the beginning of the twentieth century. Many microbial diseases are no longer the threat to public health they once were in developed countries. However, we will discuss a number of infectious diseases that are emerging as important public health problems, even in developed countries. In developing countries, which include about 75% of the world's population, infectious diseases are still a major problem. Worldwide, infectious diseases account for nearly 40% of the total of 50 million annual estimated deaths. Table 22.1 shows the most prevalent causes of death. Note that, for many of these diseases (for example, measles and whooping cough), effective vaccines are manufactured (∞ Section 20.17) but are often not available or are not used outside of developed countries. Clearly, infectious diseases will remain an important public health problem throughout the world. The current acquired immunodeficiency syndrome (AIDS) epidemic, which has apparently spread worldwide in 20 years or less, is only one example of the devastating consequences of an infectious disease in a global theater. Eradication or even effective control of infectious diseases must involve scientific, economic, political, and educational solutions, and ultimately, global cooperation.

TABLE 22.1	Infectious diseases: The leading human killers

Cause of death	Estimated yearly deaths[a]	Infectious agents
Acute respiratory infections	6,900,000	Bacteria, viruses, protozoa, fungi
Diarrheal diseases	4,200,000	Bacteria, viruses
Tuberculosis	3,300,000	Bacteria
Acquired immunodeficiency syndrome (AIDS)	1,000,000–2,000,000	Virus
Malaria	1,000,000–2,000,000	Protozoa
Hepatitis	1,000,000–2,000,000	Viruses
Measles	220,000	Virus
Meningitis, bacterial	200,000	Bacteria
Schistosomiasis	200,000	Parasitic worm
Pertussis (whooping cough)	100,000	Bacterium
Amebiasis	40,000–100,000	Protozoa
Hookworm	50,000–60,000	Parasitic worm
Rabies	35,000	Virus
Yellow fever	30,000	Virus
African trypanosomiasis (sleeping sickness)	20,000 or more	Protozoan

[a]Data represent estimates for yearly worldwide total deaths each year from each disease. There are approximately 50 million deaths per year, worldwide, from all causes. Over 20 million deaths are caused by infectious disease each year.

Source: World Health Organization.

22.1

The Science of Epidemiology

The most visible aspect of microbial disease is the actual diseased individual. However, individuals do not live alone, and when we consider infectious diseases in populations, some new factors arise. The study of the occurrence, distribution, and control of disease in populations is the field of **epidemiology.**

To continue existing in nature, a pathogen must be able to grow and reproduce. For this reason, an important aspect of the epidemiology of any disease is a consideration of the natural history of the pathogen. In many cases the pathogen cannot grow outside the host, and if the host dies, the pathogen will also die. Pathogens that kill the host before they are transmitted to a new host would thus become extinct. This raises the question of why pathogens occasionally kill their hosts. Actually, a well-adapted parasite lives in synchrony with its host, taking only what it needs for existence and causing only a minimum of harm. However, serious host damage often occurs when new varieties of pathogens arise for which the host has not developed resistance, or when the resistance of the host changes because of the factors discussed in Chapter 19. Pathogens are selective forces in the evolution of the host, just as hosts are selective forces in the evolution of pathogens. When equilibrium between host and pathogen exists, both coexist in a stable relationship.

The epidemiologist traces the spread of a disease to identify its origin and mode of transmission. The epidemiologist relies heavily on data obtained from clinical studies, disease reporting surveys, insurance questionnaires, and interviews with patients to define common factors that constitute a disease. The science of epidemiology has been referred to as "medical ecology" because the study of a disease in populations is really a study of a disease in its natural environment. This is in contrast to the clinical or laboratory study of disease, where the focus is on treating the individual patient. Knowledge of both the clinical aspects and ecological aspects of a given disease are important if public health measures to control diseases are to be effective.

> **CONCEPT CHECK**   *22.1*
>
> Epidemiology follows the spread of disease in populations. For infectious disease, the epidemiologist develops methods for the control of infectious disease by defining the interactions of the pathogen in the host population.
>
> ✔ *How does an epidemiologist differ from a microbiologist?*
>
> ✔ *What data do epidemiologists acquire for infectious diseases?*

22.2

Epidemiological Terminology

A number of terms having specific meanings are used by the epidemiologist to describe patterns of disease. The **prevalence** of a disease in a population is defined as the proportion (or percentage) of diseased individuals in a population at any one time. The **incidence** of a disease is the *number* of diseased individuals in a population at risk. A disease is said to be **epidemic** when it occurs in an unusually high number of individuals in a community at the same time; a **pandemic** is a widely distributed epidemic (Figure 22.1). By contrast, an **endemic** disease is one that is constantly present, usually at low incidence, in a population. In an endemic disease, the pathogen may not be highly virulent, or the majority of the individuals may be immune, and so the incidence of disease is low. However, as long as an endemic situation lasts, a few individuals remain who serve as reservoirs of infection.

Sporadic cases of a disease occur when individual cases are recorded in geographically separated areas, implying that the incidents are not related. A disease **outbreak,** on the other hand, occurs when a number of cases are observed, usually in a relatively short period of time, in an area previously experiencing only sporadic cases of the disease. Finally, *subclinical infection* is used to describe diseased individuals who show no or only mild symptoms. Subclinically infected individuals are frequently identified as **carriers** of a particular disease, because even though they themselves show few or perhaps no symptoms, they may still be actively carrying and shedding the pathogenic agent (see Section 22.3).

Mortality and morbidity

In practice, the incidence and prevalence of disease is determined from statistics of illness and death. From these data a picture of the public health in a population can be obtained. The population under consideration could range in size from the total global population of humans down to the population of a localized region of a country or district. Public health varies from region to region, as well as with time; thus, assessment of public health at a given moment provides only a snapshot of the situation. By continuing to examine health statistics over many years, it is possible to assess the value of various public health policies that may influence the incidence of disease.

Mortality expresses the incidence of *death* in the population. Infectious diseases were the major causes of death in 1900 in developed countries (∞ Figure 1.18), whereas currently they are of much less significance; noninfectious diseases such as heart disease and cancer are of greater importance. However, the current situation could rapidly change if a breakdown in public health measures were to occur and, in fact, does not mirror the worldwide situation. In developing countries, infectious diseases are still the major killers (Table 22.1).

Morbidity refers to the incidence of *disease* in populations and includes both fatal and nonfatal diseases. Clearly, morbidity statistics define the health of the population more precisely than mortality statistics because many diseases that affect health in important ways have only a low mortality. The major causes of illness are quite different from the major causes of death. Major illnesses are acute respiratory diseases (the common cold, for instance) and acute digestive system conditions, which are generally due to infectious agents.

Disease progression

In terms of clinical symptoms, the course of a typical disease can be divided into stages:

1. *Infection:* the organism begins to grow in the host.
2. *Incubation period:* the time between infection and the appearance of disease symptoms. Some diseases, like influenza (∞ Section 23.4), have short incubation periods, measured in days; others, like AIDS, have longer ones, measured in years (∞ Section 23.7). The incubation period for a given disease is determined by inoculum size, virulence of the pathogen, resistance of the host, and distance of the site of entrance from the focus of infection (∞ Sections 19.6 and 19.7). At the end of incubation, the first symptoms, such as headache and a feeling of illness, appear.

Figure 22.1 Classification of disease by incidence. Each dot represents several cases of a particular disease.

(a) Endemic disease   *(b)* Epidemic disease   *(c)* Pandemic disease

3. *Acute period:* the disease is at its height, with overt symptoms such as fever and chills.

4. *Decline period:* disease symptoms are subsiding, the temperature falls, usually following a period of intense sweating, and a feeling of well-being develops. The decline may be rapid (within 1 day), in which case it is said to occur by *crisis,* or it may be slower, extending over several days, in which case it is said to be by *lysis.*

5. *Convalescent period:* the patient regains strength and returns to normal.

During the later stages of the infection cycle, the immune mechanisms of the host become increasingly important, and in most cases complete recovery from the disease requires and results in active immunity.

CONCEPT CHECK   **22.2**

An endemic disease is constantly present in a population in low numbers. In epidemics, an unusually high incidence of disease occurs. An infection may cause morbidity (disease) or mortality (death) in a population. A disease follows a predictable clinical pattern in the host.

✔ *Distinguish between* morbidity *and* mortality.

✔ *What is the normal course of an infectious disease?*

22.3
Disease Reservoirs

Reservoirs are sites in which viable infectious agents remain alive and from which infection of individuals may occur. Reservoirs may be either animate or inanimate. Table 22.2 lists some common human diseases and their reservoirs. Some pathogens are primarily saprophytic (living on dead matter) and only incidentally infect and cause disease. For example, *Clostridium tetani* (the causal agent of tetanus) normally inhabits the soil. Infection of animals by this organism is an accidental event; infection of a host is not essential for its continued existence and even if there were no susceptible hosts, *C. tetani* would still survive in nature.

However, many pathogens have other living organisms as their only reservoirs. In these cases, the reservoir is an essential component of the natural life cycle of the infectious agent. Some infections occur only in humans, and maintenance of the cycle involves person-to-person transmission. This type of pathogen cycle is common for viral and bacterial respiratory diseases, sexually transmitted diseases, staphylococcal and streptococcal infections, diphtheria, typhoid fever, and mumps.

Zoonosis

A number of infectious diseases that occur in humans also occur in animals. A disease that occurs primarily in animals but is occasionally transmitted to humans is called a **zoonosis.** Because public health measures for animal populations are much less developed than for humans, the infection rate for many diseases is much higher in animals, and animal-to-animal transmission is the rule. However, occasionally transmission is from animal to human. It is less likely for transmission to also occur from person to person in such diseases. Thus, maintenance of the pathogen in nature depends on animal-to-animal transfer. However, control of a zoonosis in the human population in no way eliminates it as a public health problem. Indeed, more effective human control can generally be achieved through elimination of the disease in the animal reservoir. Marked success has been achieved in the control of two diseases that were often transferred to humans from domestic animals, bovine tuberculosis and brucellosis. Control was achieved primarily by identifying and destroying infected animals. Pasteurization of milk was also of considerable importance in the prevention of the spread of bovine tuberculosis to humans because milk was the main vehicle of transmission.

Certain infectious diseases have more complex cycles, involving an obligate transfer from animal to human to animal. These are due to organisms with complex life cycles like metazoans (for example, tapeworms) or protozoa (for example, malaria, ∞ Section 23.11). In such cases, control of the disease in the population can be either through control in humans or in the alternate animal host.

Carriers

A carrier is an infected individual with no obvious signs of clinical disease. Carriers are potential sources of infection for others and are important for understanding the spread of disease. Carriers may be individuals in the incubation period of the disease, in which case the carrier state precedes the development of actual symptoms. Carriers of this sort are prime sources of infectious agents for respiratory infections because they are not yet aware of their infection and so are not taking any precautions against infecting others. Such persons are **acute carriers** because the carrier state lasts for only a short time. Also significant from the public health standpoint are chronic carriers, who may remain infected for long periods of time. **Chronic carriers** may be either individuals who had a clinical disease and recovered, or they may have a subclinical infection that has remained inapparent. These individuals may be perfectly healthy, but they harbor and spread viable pathogens (see the box, The Tragic Case of Typhoid Mary).

TABLE 22.2   Epidemic diseases: Agents, sources, reservoirs and control

Disease	Causative agent[a]	Infection sources	Reservoirs	Control measures
Common-source epidemics[b]				
Anthrax	*Bacillus anthracis* (B)	Milk or meat from infected animals	Cattle, swine, goats, sheep, horses	Destruction of infected animals
Bacillary dysentery	*Shigella dysenteriae* (B)	Fecal contamination of food and water	Humans	Detection and control of carriers; oversight of food handlers; decontamination of water supplies
Botulism	*Clostridium botulinum* (B)	Soil-contaminated food	Soil	Proper preservation of food
Brucellosis	*Brucella melitensis* (B)	Milk or meat from infected animals	Cattle, swine, goats, sheep, horses	Pasteurization of milk; control of infection in animals
Cholera	*Vibrio cholerae* (B)	Fecal contamination of food and water	Humans	Decontamination of public water sources; vaccination
Giardiasis	*Giardia* spp. (P)	Fecal contamination of water	Wild mammals	Decontamination of public water sources
Hepatitis	Hepatitis A, B, C, D, E (V)	Infected humans	Humans	Decontamination of contaminated fluids and fomites, vaccination if available (A and B only)
Paratyphoid	*Salmonella paratyphi* (B)	Fecal contamination of food and water	Humans	Decontamination of public water sources; oversight of food handlers; vaccination
Typhoid fever	*Salmonella typhi* (B)	Fecal contamination of food and water	Humans	Decontamination of public water sources; oversight of food handlers; pasteurization of milk; vaccination
Host-to-host epidemics				
Respiratory diseases				
Diphtheria	*Corynebacterium diphtheriae* (B)	Human cases and carriers; infected food and fomites	Humans	Vaccination; quarantine of infected individuals
Hantavirus pulmonary syndrome	Hantavirus (V)	Inhalation of contaminated fecal material	Rodents	Control rodent population and exposure
Meningicoccal meningitis	*Neisseria meningitidis* (B)	Human cases and carriers	Humans	Exposure treated with sulfadiazine for susceptible strains
Pneumococcal pneumonia	*Streptococcus pneumoniae* (B)	Human carriers	Humans	Antibiotic treatment; isolation of cases for period of communicability
Tuberculosis	*Mycobacterium tuberculosis* (B)	Sputum from human cases; contaminated milk	Humans, cattle	Treatment with isoniazid; pasteurization of milk
Whooping cough	*Bordetella pertussis* (B)	Human cases	Humans	Vaccination; case isolation
German measles	Rubella virus (V)	Human cases	Humans	Vaccination; avoid contact between infected individuals and pregnant women
Influenza	Influenza virus (V)	Human cases	Humans, animals	Vaccination (recommended only in certain cases)
Measles	Measles virus (V)	Human cases	Humans	Vaccination

TABLE 22.2   (continued)

Disease	Causative agent[a]	Infection sources	Reservoirs	Control measures
Sexually transmitted diseases[c]				
Acquired immuno-deficiency syndrome (AIDS)	Human immuno-deficiency virus (HIV)	Infected body fluids, especially blood and semen	Humans	Treatment with viral replication inhibitors (not curative)
Chlamydia	*Chlamydia trachomatis* (B)	Urethral, vaginal, and anal secretions	Humans	Testing for organism during routine pelvic examinations; chemotherapy of carriers and potential contacts; case tracing and treatment
Gonorrhea	*Neisseria gonorrhoeae* (B)	Urethral and vaginal secretions	Humans	Chemotherapy of carriers and potential contacts; case tracing and treatment
Syphilis	*Treponema pallidum* (B)	Infected exudate or blood	Humans	Identification by serological tests; antibiotic treatment of seropositive individuals
Trichomoniasis	*Trichomonas vaginalis* (P)	Urethral, vaginal, prostate secretions	Humans	Chemotherapy of infected individuals and contacts
Vector-borne diseases				
Epidemic thyphus	*Rickettsia prowazekii* (B)	Bite by infected louse	Humans, lice	Control louse population
Lyme disease	*Borrelia burgdorferi* (B)	Bite from infected tick	Rodents, deer, ticks	Avoid tick exposure; treat infected individuals with antibiotics
Malaria	*Plasmodium* spp. (P)	Bite from *Anopheles* mosquito	Humans, mosquito	Control mosquito population; treat infected humans with antimalarial drugs
Plague	*Yersinia pestis* (B)	Bite by flea	Wild rodents	Control rodent populations
Rocky Mountain spotted fever	*Rickettsia rickettsii* (B)	Bite by infected tick	Ticks, rabbits, mice	Avoid tick exposure; treat infected individuals with antibiotics
Direct-contact diseases				
Psittacosis	*Chlamydia psittaci* (B)	Contact with birds or bird excrement	Wild and domestic birds	Avoid contact with birds; treat infected individuals with antibiotics
Rabies	Rabies virus (V)	Bite by carnivores	Wild and domestic carnivores	Avoid animal bites; vaccination of animal handlers and exposed individuals
Tularemia	*Franciscella tularensis* (B)	Contact with rabbits	Rabbits	Avoid contact with rabbits; treat infected individuals with antibiotics

[a]B, Bacteria; V, virus; P, protozoan.
[b]Some common-source diseases can also be spread from host to host.
[c]Sexually transmitted diseases can also be controlled by effective use of condoms and by sexual abstinence.

Carriers can be identified by routine surveys of populations using cultural, radiological (chest X-ray), or immunological techniques. In general, carriers are sought only among groups of individuals who may be sources of infection for the public at large, such as food handlers and health care workers.

Diseases in which carriers are important for the spread of infection include hepatitis (∞ Section 23.13), tuberculosis (∞ Section 23.3), and typhoid fever (see the box, The Tragic Case of Typhoid Mary). Surveys of food handlers and health care workers are sometimes made to detect inapparent cases of these infections.

CONCEPT CHECK   22.3

To understand how diseases develop, the pathogen reservoir must be known. Some pathogens exist in soil, water, or animals. Other pathogens are restricted to humans and are maintained solely by person-to-person contact. An understanding of disease carriers is critical for controlling disease.

✔ *What is a disease reservoir?*

✔ *Distinguish between* acute *and* chronic *carriers.*

LEARNING FROM THE PAST

THE TRAGIC CASE OF TYPHOID MARY

The classic example of a chronic carrier was the woman known as "Typhoid Mary," a cook in New York City and Long Island in the early part of the twentieth century. Typhoid Mary (her real name was Mary Mallon) was employed in a number of households and institutions, and as a cook she was in a central position to infect large numbers of people. Extensive epidemiological investigation of a number of typhoid outbreaks by Dr. George Soper revealed that Mary was the likely source of contamination. When her feces were examined bacteriologically, she had very high numbers of the typhoid bacterium, *Salmonella typhi*. She remained a carrier for many years, probably because her gallbladder was infected, and organisms were continuously being excreted from there into her intestine. Public health authorities offered to remove her gallbladder, but she refused the operation, and to prevent her from continuing to serve as a source of infection, she was imprisoned. After almost 3 years in prison, she was released on the pledge that she would not cook or handle food for others and that she was to report to the health department every 3 months. She promptly disappeared, changed her name, and cooked in hotels, restaurants, and sanitariums, leaving behind a wake of typhoid fever. After 5 years she was captured as a result of the investigation of an epidemic at a New York hospital. She was again arrested and imprisoned and remained in custody on North Brother Island in the East River of New York City for 23 years. She died in 1938, 32 years after epidemiologists had first discovered she was a chronic typhoid carrier. ∎

22.4

Epidemiology of AIDS: An Example of How Epidemiological Research Is Done

Cases of acquired immunodeficiency syndrome (AIDS), the virus-mediated infectious disease that severely cripples the body's immune system (∞ Section 23.7), were first reported in the United States in 1981. Since then, the number of new AIDS cases in the United States has risen dramatically nearly every year (Figure 22.2). The extraordinary rise for 1993–1994 was probably due to a change in the definition of AIDS (∞ A Definition of AIDS, Chapter 23) but *about 50,000 new cases of AIDS will be diagnosed every year in the* United States for the foreseeable future. *Worldwide, 30 to 40 million individuals will be infected with human immunodeficiency virus* (HIV) (∞ Section 23.7) *by the year 2000. At least 1 million people now die each year from AIDS.* A vast majority of the new infections and deaths will occur in developing countries.

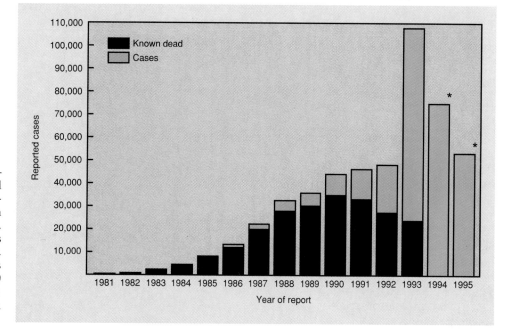

Figure 22.2 Number of total cases of acquired immunodeficiency syndrome (AIDS) in the United States since 1981. The numbers of known deaths are also shown. Cumulatively, there were about 500,000 cases of AIDS and about 300,000 deaths due to AIDS from 1981 to 1995. *Data for deaths in these years are incomplete.

Tracking an epidemic

Initial case control studies suggested an unusually high AIDS prevalence among homosexual men and intravenous drug abusers. This in turn strongly implicated a transmissible agent, presumably transferred during sexual activity or by contaminated needles. The finding that individuals requiring blood or blood products were also at high risk strengthened the case for a transmissible agent (Figure 22.3).

Soon after the discovery of HIV, laboratory tests were developed to detect antibodies to the virus in serum (∞ Sections 21.8 and 21.9). This made possible more extensive surveys of the incidence of HIV infection in different populations and also served as a screening method to ensure that new cases of AIDS were not transmitted by blood transfusions. Such tests revealed a fourth group of individuals at high risk for AIDS—the children of mothers who are themselves at high risk for AIDS.

The pattern illustrated in Figure 22.3 is typical of an agent transmissible by sexual activity or by blood, and the association of AIDS cases with *specific* groups was an important epidemiological finding. The identification of certain well-defined high risk groups implied that AIDS was *not* transmitted from person to person by casual contact, such as the respiratory route, or by contaminated

food or water. Instead, epidemiological findings pointed clearly to bodily fluids, primarily blood and semen, as the major vehicles for transmission of HIV.

In the United States AIDS has affected mainly homosexual men (Figure 22.3a), but intravenous drug users represent an increasing proportion of AIDS cases. Subsequent epidemiological surveys have shown that homosexual men with multiple sexual partners are more likely to contract AIDS than monogamous homosexual males. In addition, the fastest growing category for new AIDS cases is the heterosexual contact group. This group is already a major risk group among women (Figure 22.3b). This undoubtedly reflects the increased probability of contacting an HIV-infected individual when engaging in sexual activity with multiple partners.

The incidence of AIDS in hemophiliac transfusion recipients has been reduced greatly in recent years (Figure 22.3). This is due not only to screening of the blood supply but also because many blood clotting factors needed by hemophiliacs can withstand a heat treatment sufficient to inactivate HIV. Pediatric AIDS cases are still a major concern. In 1993, there were 958 new cases among this group. HIV can be transmitted to the fetus by infected mothers and probably also in mother's milk. All infants born to HIV-infected mothers have maternally derived antibodies to HIV in their

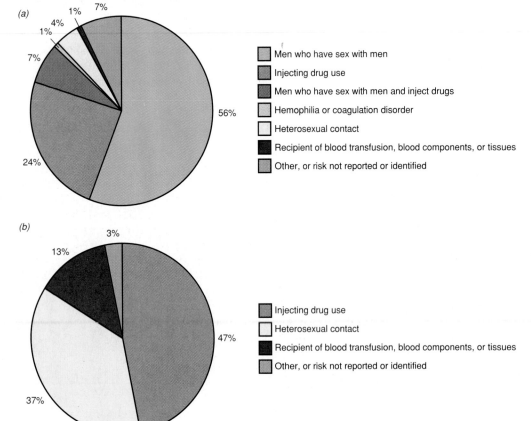

(a)

Men who have sex with men

Injecting drug use

Men who have sex with men and inject drugs

Hemophilia or coagulation disorder

Heterosexual contact

Recipient of blood transfusion, blood components, or tissues

Other, or risk not reported or identified

(b)

Injecting drug use

Heterosexual contact

Recipient of blood transfusion, blood components, or tissues

Other, or risk not reported or identified

Figure 22.3 Distribution of AIDS cases by risk group and sex in the United States, 1993. The total number of cases reported was 105,989. (a) AIDS in men. $N = 89,165$. (b) AIDS in women. $N = 16,824$.

blood, but a clear diagnosis of AIDS in infants must wait a year or more after birth because about 70% of infants showing maternal HIV antibodies at birth do not go on to develop HIV infection.

Epidemiological studies of AIDS in Africa, where the disease is thought to have originated, have clearly shown that transmission of AIDS is not linked to particular sexual practices, such as homosexuality, but instead to person-to-person transfer of HIV-infected fluids. In Africa heterosexual transmission of AIDS seems the norm, with about equal numbers of men and women infected. Unfortunately, reliable global statistics on AIDS are not available because of differences in reporting in various countries. Current estimates are about 10 million AIDS cases worldwide with up to 40 million infected people who do not (yet) show clinical symptoms; that is, they are carriers. The identification of high risk groups through epidemiological studies led to the development of health education campaigns to inform the public of how AIDS is transmitted and what activities constitute high risk behavior (∞ Sexual Activity and AIDS, Chapter 23). Because no cure for AIDS is yet available, public health education offers the most effective approach to the control of AIDS and is the major weapon for preventing the spread of infection. We discuss the pathology and therapy of AIDS in Section 23.7.

CONCEPT CHECK 22.4

AIDS is one of the newest and most studied pandemics. Current information suggests that AIDS will continue to be a major public health problem, especially in developing countries, although we now know a great deal about its pathology and spread.

✔ Describe the major risk factors for acquiring AIDS.

✔ Estimate the total number of individuals in the United States who now have AIDS and make a prediction for this number in the year 2000.

22.5

Infectious Disease Transmission

Epidemiologists follow the incidence of a disease by correlating geographical, seasonal, and age group distribution of a disease with possible modes of transmission. A disease limited to a restricted geographical location may suggest a particular vector; malaria, for example, is transmitted by a mosquito found mainly in tropical regions. A marked seasonality to a disease is often indicative of certain modes of transmission, such as in the case of measles and chickenpox, where the number of cases jumps sharply when children enter school and come in close contact (∞ Figure 23.21). The age group distribution of a disease can also be an important epidemiological statistic, frequently suggesting or eliminating particular routes of transmission.

Different pathogens have different modes of transmission, which are usually related to the habitats of the organisms in the body. For instance, respiratory pathogens are generally airborne, whereas intestinal pathogens are spread by food or water. If the pathogen is to survive, it must undergo transmission from one host to another. Even environmental factors may play a role in survival of the pathogen, and such variables as weather conditions may influence exposure to a pathogen. For example, aseptic meningitis, caused by any of a group of more than 60 different enteroviruses, is most prevalent in the summer (Figure 22.4) and is thought to be spread via fecal contamination of water sources such as public swimming areas.

Thus pathogens generally are associated with specific features or mechanisms that permit or ensure transmittal. Transmission involves three stages: (1) escape from the host, (2) travel, and (3) entry into a new host. We give here a brief overview of transmission mechanisms. Several of these will be discussed in detail for certain diseases in the next chapter.

Direct host-to-host transmission

Host-to-host transmission occurs whenever an infected host transmits the disease to a susceptible host. Transmission by the respiratory route and by direct contact is very common. Transmission by infectious droplets is the most frequent means by which upper respiratory infections such as the common cold and influenza are propagated. However, some pathogens are so sensitive to environmental influences that they are unable to survive for significant periods of time away from the host and must be transmitted from host to host by direct contact. The best examples of pathogens transmitted in this way are those responsible for sexually transmitted diseases, such as *Treponema pallidum* (syphilis) and *Neisseria gonorrhoeae* (gonorrhea). These agents are extremely sensitive to drying and do not survive away from the body, even for a few moments. Intimate person-to-person contact, such as kissing or sexual intercourse, provides a direct means for the transmission of such pathogens. However, such intimate transfer can occur only if the viable pathogen is present on the transmitting person at the body site that comes in direct contact with that of the recipient. Thus, pathogens causing sexually transmitted diseases live in genitalia, the mouth, or the anus because these are the sites involved in sexual contact.

Direct contact is also involved in the transmittal of skin pathogens, such as staphylococci (boils and pimples) and fungi (ringworm). However, these pathogens are relatively resistant to environmental influ-

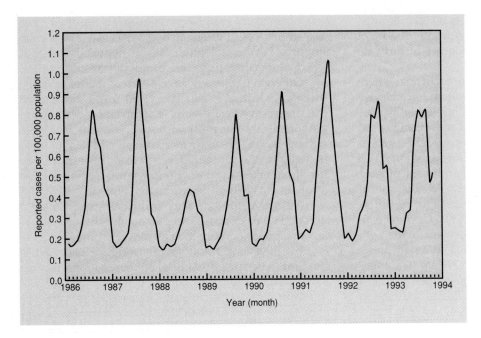

Figure 22.4 The incidence of aseptic (viral) meningitis in the United States. Note the marked summer rise and annual cyclical nature of disease incidence.

ences such as drying, and intimate person-to-person contact is not the only means of transmission. Many respiratory pathogens are also transmitted by direct means because they are spread by droplets resulting from sneezing or coughing. However, many of these droplets do not remain airborne for long. Transmission, therefore, requires close, although not necessarily intimate, person-to-person contact.

Indirect host-to-host transmission

Indirect transmission can occur by either living or inanimate means. Living agents transmitting pathogens are called **vectors;** they are generally arthropods (for example, insects, mites, or fleas) or vertebrates (for example, dogs, rodents). Arthropod vectors may not be hosts for the disease but simply carry the agent from one host to another. Large numbers of arthropods obtain nourishment by biting, and if the pathogen is present in the blood, the arthropod vector may ingest the pathogen and transmit it when biting another individual. In some cases, the pathogen actually replicates in the arthropod, which is then considered an alternate host. Such replication leads to a buildup of the inoculum, increasing the probability that a subsequent bite will lead to infection.

Inanimate agents such as bedding, toys, books, and surgical instruments can also transmit disease. These inanimate objects are collectively referred to as **fomites.** Food and water are referred to as disease **vehicles.** Fomites can also be disease vehicles, but major epidemics originating from a single source are usually traced to food or water because these are actively consumed in large amounts by a number of individuals in a population.

Epidemics

Two major types of epidemics can be distinguished: common-source and host-to-host. These two types are contrasted in Figure 22.5. A **common-source epidemic** arises as the result of infection (or intoxication) of a large number of people from a contaminated common source, such as food or water. Usually such contamination occurs because of a malfunction in the sanitation of a central distribution system. Foodborne and waterborne diseases are primarily *intestinal* diseases; the pathogen leaves the body in fecal material, contaminates food or water via improper sanitary procedures, and then enters the intestinal tract of the recipient during ingestion. Because foodborne and waterborne diseases are some of those that are most amenable to control by public health measures, we shall discuss them in some detail in Chapter 23 (also see the boxes, Snow on Cholera, and The Tragic Case of Typhoid Mary in this chapter). The disease incidence for a common-source outbreak is characterized by a rapid rise to a peak because a large number of individuals succumb within a relatively brief period of time (Figure 22.5). The common-source outbreak also declines rapidly, although the decline is less rapid than the rise. Cases continue to be reported for a period of time approximately equal to the duration of one incubation period of the disease.

In a **host-to-host epidemic,** the disease incidence shows a relatively slow, progressive rise (Figure 22.5) and a gradual decline. Cases continue to be reported over a period of time equivalent to several incubation periods of the disease. The epidemic may have been initiated by the introduction of a single infected individual into a susceptible population, and this individual has infected one or more people in the population. The pathogen then repli-

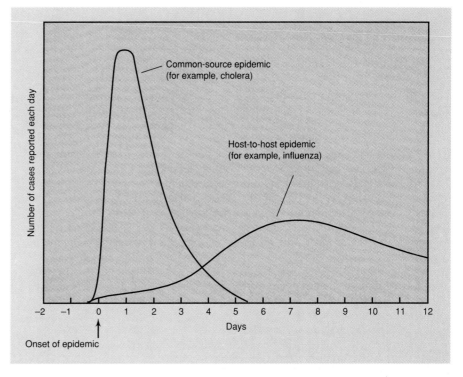

Figure 22.5 Origins of epidemics. The shape of the epidemic curve helps to distinguish the likely origin. In a common-source epidemic, such as from contaminated food or water, the curve is characterized by a sharp rise to a peak, with a rapid decline, which is less abrupt than the rise. Cases continue to be reported for a period approximately equal to the duration of one incubation period of the disease. In a host-to-host epidemic, the curve is characterized by a relatively slow, progressive rise, and the cases continue to be reported over a period equivalent to several incubation periods of the disease.

cated in the susceptible individuals, reached a communicable stage, and was transferred to others, where it replicated and again became communicable. Table 22.2 summarizes some of the key epidemiological features of some major epidemic diseases observed today.

CONCEPT CHECK   *22.5*

A pathogen can be transmitted directly from one host to another, or indirectly by means of another living agent called a vector. Pathogens can also be transmitted by inanimate objects (fomites) and common vehicles such as food and water. Epidemics may be of common-source or host-to-host origin. By observing disease incidence over time, the type of epidemic can be determined.

✔ *Compare a* common-source *epidemic to a* host-to-host *epidemic. Cite at least one example of each.*

✔ *Suggest one method for halting the spread of a common-source epidemic and a host-to-host epidemic.*

22.6
The Host Community

The colonization of a susceptible, unimmunized host by a parasite may first lead to an explosive infection and an epidemic. As the host population develops resistance, however, the spread of the parasite is checked, and eventually a *balance* is reached in which host and parasite are

in equilibrium. A subsequent genetic change in the parasite could lead to the formation of a more virulent form, which would then initiate another explosive epidemic until the host again responds and another balance is reached. In effect, the host and parasite are affecting each other's evolution; that is, the host and parasite are *co-evolving*.

Coevolution of a host and a parasite

An excellent experimental example of coevolution of host and parasite occurred when a virus was intentionally introduced for purposes of population control in the wild rabbits of Australia.

Wild rabbits were introduced into Australia from Europe in 1859 and quickly spread until they were overrunning large parts of the continent. Myxoma virus was discovered in South American rabbits, which are a different species from the European rabbits in Australia. In South America the virus and its hosts are apparently in equilibrium, and the virus causes only a minor disease. However, this same virus is extremely virulent in the European rabbit and almost always causes a fatal infection. The virus is spread from rabbit to rabbit by mosquitoes and other biting insects, and is capable of rapid spread in areas with appropriate insect vectors.

Myxoma virus was introduced into Australian rabbits in 1950 to control the rabbit population. Within several months, the virus was well established in the population and spread over an area in Australia as large as all of Western Europe. The disease showed a marked seasonal pattern, rising to a peak in the sum-

SNOW ON CHOLERA

 The importance of drinking water as a vehicle for the spread of cholera was first shown in 1855 by British physician John Snow, who at that time had no knowledge of the bacterial causation of the disease. Snow's study is one of the great classics of epidemiology and serves as a model for how a careful study can lead to clear and meaningful conclusions.

In London, the water supplies to different parts of the city were from different sources and were transmitted in different ways. In a large area south of the Thames River, across the river from Westminster Abbey and the Parliament Building, the water was supplied to houses by two competing private water companies, the Southwark and Vauxhall Company, and the Lambeth Company. It was the water of the former company that was the major vehicle for the transmission of cholera. When Snow began to suspect the water supply of the Southwark and Vauxhall Company, he made a careful survey of the residence of every cholera death in this district and determined which company supplied the water to that residence. In some parts of the area served by these two companies, each had a monopoly, but in a fairly large area the two companies competed directly, each having run independent water pipes along the various streets. Houses had the option of connecting with either supply, and the distribution of houses between the two companies was random. The clear-cut results of Snow's survey were completely convincing, even to those skeptical about the importance of polluted water in the transmission of cholera: in the first seven weeks of the epidemic, there were 315 deaths per 10,000 houses supplied by the Southwark and Vauxhall Company, and only 37 per 10,000 houses supplied by the Lambeth Company. In the rest of London, there were 59 deaths per 10,000 houses, showing that those supplied by the Lambeth Company had fewer deaths than the general population. In the districts where each company had exclusive rights, it could of course be argued that it was not the water, but some other factor (soil, air, general layout of houses, and so on), that might have been responsible for the differences in disease incidence, but in the districts where the two companies competed, all of these other factors were the same, yet the incidence was high for those supplied with Southwark and Vauxhall water and low for those supplied with Lambeth water. Snow attempted to relate these differences in disease incidence to the sources of the waters used by the two companies. Since he suspected that the excrement and evacuations from cholera patients were highly infectious, he considered that sewage contamination of the water supply might exist. In those days, sewage treatment did not exist and raw sewage was dumped directly into the Thames River. The Southwark and Vauxhall Company obtained its water supply from the Thames right in the heart of London, where sewage contamination could occur, while the Lambeth Company obtained its water from a point on the river considerably above the city, and hence was relatively free of pollution. It was this difference in source that accounted for the difference in disease incidence. In Snow's words:

As there is no difference whatever, either in the houses or the people receiving the supply of the two Water Companies, or in any of the physical conditions with which they are surrounded, it is obvious that no experiment could have been devised which would more thoroughly test the effect of water supply on the progress of cholera than this.... The experiment, too, was on the grandest scale. No fewer than three hundred thousand people of both sexes, of every age and occupation, and of every rank and station, from gentlefolk down to the very poor, were divided into two groups without their choice, and, in most cases, without their knowledge; one group being supplied with water containing the sewage of London, and, amongst it, whatever might have come from cholera patients, the other group having water quite free from such impurity. ∎

---

mer when the mosquito vectors were present and declining in the winter. The epidemiology of myxoma virus was studied as a model of a virus-induced epidemic by Australian scientists. Virus was isolated from wild rabbits, and the isolated strains were characterized for virulence with laboratory rabbits. At the same time, baby rabbits were removed from their dens before infection could occur and reared in the laboratory. Then these wild rabbits were challenged with standard virulent strains of myxoma virus to determine their susceptibility. The results of this large-scale model study are shown in Figure 22.6.

During the first year of the epidemic, over 95% of the infected rabbits died. However, within 6 years both the virus and the rabbit population had changed. Over this interval, rabbit mortality dropped to about 84%, and the virus isolated was of decreased virulence. In addition, changes were noted in the resistance of the

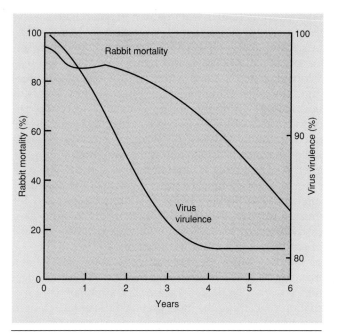

Figure 22.6 Changes in virulence of myxoma virus and in susceptibility of the Australian rabbit during the years after the virus was introduced into Australia in 1950. Virus virulence is given as the average mortality in standard laboratory rabbits for virus recovered from the field each year. Rabbit susceptibility was determined by removing young rabbits from their dens and challenging with a virus strain of moderately high virulence, which killed 90–95% of normal laboratory rabbits.

rabbit. In parts of Australia where the virus was first introduced, the remaining rabbit population had been subjected to selective pressure by the virus for several years. As seen in Figure 22.6, within years the resistance of the rabbits had increased dramatically. This resistance was due to innate changes in the rabbit population and not to immunological responses, for the rabbits tested had been removed from their mothers at birth and had never been in contact with the virus. Their resistance was due to some genetic change in the animal that made it less susceptible to myxoma virus.

As a result of the introduction of myxoma virus, the Australian rabbit population was controlled, but the genetic changes in virus and host prevented complete eradication of the rabbit from Australia. A steady-state rabbit population of about 20% of that present before the introduction of myxoma virus was observed (Figure 22.6). The virus was thus a major factor in population control but did not totally eliminate rabbits because of coevolutionary events in both host and parasite. The Australian experiment reveals how quickly an equilibrium is reached between host and parasite. The manner in which the malaria parasite has affected biochemical evolution in humans is another

example of host–parasite coevolution and will be discussed in Section 23.11.

Herd immunity

An analysis of the immune state of a group is of great importance in understanding the role of immunity in the development of epidemics. **Herd immunity** is the resistance of a group to invasion and spread of an infectious agent resulting from immunity of a high proportion of the members of the group. If the proportion of immune individuals is sufficiently great, then the whole population will be protected. The fraction of resistant individuals necessary to prevent an epidemic is higher for a highly virulent agent or one with a long period of infectivity, and lower for a mildly virulent agent or one with a brief period of infectivity.

The proportion of the population that must be immune to prevent infection of the rest of the population can be estimated from data on poliovirus immunization in the United States. From epidemiological studies of the incidence of polio in large populations, it appears that if a population is 70% immunized, polio will be essentially absent in the population. Clearly, these immunized individuals protect the rest of the population. For a highly infectious disease such as influenza, the proportion of immune individuals necessary to confer herd immunity is higher, about 90–95%. A value of about 70% has also been estimated for diphtheria, but further study of several small diphtheria outbreaks has shown that in densely settled areas a much higher proportion must be immunized to prevent development of an epidemic. Apparently, in dense populations, person-to-person transmission can occur even if the agent is not highly infectious. In the case of diphtheria, an additional complication arises because immunized persons can still harbor the pathogen (inapparent infection) and thus act as chronic carriers, serving as potential sources of infection.

Cycles of disease

The concepts of propagated epidemics and herd immunity can also explain why certain diseases occur in *cycles*. A good example of a cyclical disease is chickenpox, which occurs in a high proportion of school children. Because the chickenpox virus (∞ Section 8.19) is transmitted by the respiratory route, its infectivity is high in crowded situations such as schools. On entry into school at age 5, most children are susceptible, so that on the introduction of virus into the school, an explosive propagated epidemic results. Virtually every individual becomes infected and develops immunity, and as the immune population builds up, the epidemic dies down. Chickenpox shows an annual cycle (∞ Figure 23.21) probably because a new group of nonimmune children arrives each year; the phasing of the epidemic is related to the time of the year at which school begins after the summer vacation.

CONCEPT CHECK   **22.6**

Hosts and pathogens coevolve with time and arrive at a steady state that favors the continued survival of both. With herd immunity, a large fraction of a population is immune to a given disease, and it is difficult for the disease to spread. Disease cycles occur when a large, recurring, nonimmune population such as children entering school is exposed to a pathogen.

✔ *Explain coevolution of host and pathogen. Cite a specific example.*

✔ *How does herd immunity work to prevent a nonimmune individual from acquiring a disease?*

22.7

Hospital-Acquired (Nosocomial) Infections

A hospital may not only be a place where sick people get well but may also be a place where sick people get sicker. Cross-infection from patient to patient or from hospital personnel to patients presents a constant hazard. Hospital infections are often called *nosocomial infections* (*nosocomium* is the Latin word for "hospital") and occur in about 5% of all patients admitted. In certain clinical services, such as intensive care units, up to 10% of the patients acquire a nosocomial infection. In all, there are about 2 million nosocomial infections each year in the United States, leading directly or indirectly to 80,000 deaths. Hospital infections are partly due to the prevalence of diseased patients but are often due to the presence of pathogenic microorganisms that are selected for and maintained within the hospital environment. Most nosocomial infections are endemic rather than epidemic. These infections result from organisms already in the hospital environment. Even multiple-drug-resistant organisms are often spread from host to host as normal flora. Therefore, virtually all the important nosocomial pathogens are normal flora in either patients or hospital staff.

The hospital environment

Hospitals are special environments. Infectious diseases are spread easily and rapidly in hospital environments for several reasons. (1) Many patients have weakened resistance to infectious disease because of their illness (compromised hosts) (∞ Section 19.12). (2) Hospitals treat patients suffering from infectious disease, and these patients may be reservoirs of highly virulent pathogens. (3) The crowding of patients in rooms and wards increases the chance of cross-infection. (4) Hos-

pital personnel move from patient to patient, increasing the probability of transfer of pathogens. (5) Many hospital procedures, such as catheterization, hypodermic injection, spinal puncture, and removal of tissue samples (biopsy) or fluids, carry with them the risk of introducing pathogens to the patient. (6) In maternity wards of hospitals, newborn infants are unusually susceptible to certain kinds of infection because they lack well-developed immune systems. (7) Surgical procedures are a major hazard because not only are internal organs exposed to sources of contamination but the stress of surgery often diminishes the resistance of the patient to infection (∞ Section 19.12). (8) Many drugs used for immunosuppression (for instance, in organ transplant procedures) increase susceptibility to infection. (9) Use of antibiotics to control infection carries with it the risk of selecting antibiotic-resistant organisms, which then may not be easily controlled if they cause further infection (∞ Section 11.13). Figure 22.7 summarizes information concerning the most prevalent hospital-acquired infections.

Hospital pathogens

A relatively limited number of organisms cause the majority of hospital infections. *Escherichia coli,* presumably introduced from the normal flora, is the most common cause of urinary tract infections in hospitals, but other gram-negative bacteria and *Pseudomonas aeruginosa* (see later) are often implicated as well. *Enterococcus* is also a common urinary tract pathogen; the yeast *Candida* is also encountered (Figure 22.7).

One of the most important and widespread hospital pathogens is *Staphylococcus aureus*. It is most commonly associated with blood (septicemia), surgical (wound), and lower respiratory tract infections and is a particular problem in infections acquired by newborns in the hospital (Figure 22.7). Certain strains of unusual virulence have been widely associated with hospital infections. Although only the coagulase-positive strains of *S. aureus* were normally considered as pathogens in the past, a number of other strains of *Staphylococcus* spp. (most of which are coagulase-negative) are now collectively the most common cause of hospital-acquired septicemia (an acute host response due to the presence of organisms in the blood) (∞ Section 23.2) and are also very prominent as agents of wound infections (Figure 22.7). The habitat of these staphylococci is the upper respiratory tract, usually the nasal passages, and they often become established as normal flora in hospital personnel. In healthy personnel the organism may cause no disease, but these symptomless carriers may be a source of infection for patients. Because staphylococci are resistant to drying, they survive for long periods on dust particles and other fomites and can subsequently infect patients. Because of the potential seriousness of infection with hospital

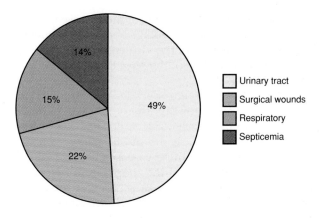

Pathogen	Urinary tract infection (%)	Surgical wound (%)	Respiratory tract (%)	Septicemia (%)
Escherichia coli	26	10	6	6
Enterococcus	16	3	2	8
Pseudomonas aeruginosa	12	8	17	4
Staphylococcus aureus	2	17	16	16
Staphylococcus spp.[a]	4	12	2	27
Candida spp.	9	2	5	8
Streptococcus spp.	0	3	1	4
Other gram-negative organisms	20	18	26	12
All other organisms	11	27	25	15

[a]Coagulase-negative staphylococci.

Figure 22.7 Major sites and pathogen distribution for nosocomial infections.

staphylococci, careful adherence to proper hospital sanitation procedures is necessary.

Pseudomonas aeruginosa is important in causing infections of the lower respiratory and urinary tracts (Figure 22.7). It is also an important cause of infections in burn patients who have lost their primary barrier to skin infection. *P. aeruginosa* exhibits one of the most significant features complicating the treatment of nosocomial infections, *antibiotic resistance.* Isolates of this bacterium from patients with hospital infections are commonly resistant to many antibiotics. A somewhat lower degree of resistance has been noted among *Staphylococcus aureus* isolates (with occasional highly resistant strains not uncommon), whereas *Escherichia coli* isolates generally remain sensitive to antibiotics. Antibiotic-resistant pathogenic bacteria in hospitals generally contain plasmids encoding *multiple antibiotic resistance* (∞ Section 11.13).

CONCEPT CHECK   22.7

Hospital patients often have impaired immune systems or are unusually sensitive to infectious disease. Certain organisms that are ordinarily not important pathogens in normal populations cause serious disease in hospitals.

✔ *Why are hospital patients more susceptible than normal individuals to pathogens?*

✔ *Why are antibiotic-resistant staphylococci a major problem in hospital settings?*

22.8
Public Health Measures for the Control of Epidemics

An understanding of the epidemiology of an infectious disease makes it possible to develop methods for control of the disease. **Public health** refers to the health of the population as a whole and to the activities of public health authorities in the control of disease. However, the incidence of many infectious diseases has dropped dramatically, especially in developed countries, over the past 100 years not because of public health efforts but because of general increases in the well-being of the population. Better nutrition, less crowded living quarters, and lighter work loads have probably done as much as public health measures to control diseases such as tuberculosis, primarily by reducing the risk factors related to disease (∞ Section 19.12). However, diseases such as typhoid fever, diphtheria, brucellosis, and poliomyelitis owe their low incidence to active and specific public health measures.

Overall public health depends on application of control measures to prevent the spread of infectious disease, and we discuss these here.

Controls directed against the reservoir

If the disease occurs primarily in *domestic animals*, then infection of humans can be prevented if the disease is eliminated from the infected animal population. Immunization procedures or destruction of infected animals may be used to eliminate the disease in animals. These procedures have been quite effective in eliminating brucellosis and bovine tuberculosis from humans. Not incidentally, the health of the domestic animal population is also enhanced, with likely economic benefits to the farmer.

When the reservoir is a *wild animal*, then eradication is much more difficult. **Rabies** is a disease that occurs in both wild and domestic animals but is transmitted to domestic animals primarily by wild animals. Thus *control* of rabies can be achieved by immunization of domestic animals, although this will never lead to complete *eradication* of the disease. The majority of rabies cases are in wild rather than domestic animals, at least in the United States (∞ Figure 23.36). There-fore, eradication of rabies would require the immu-nization or destruction of all wild animal reservoirs, which includes such diverse species as raccoons, bats, skunks, and foxes. Although oral rabies immunization is practical and recommended for rabies control in lim-ited animal populations, its efficacy is untested in large, diverse animal populations.

If the reservoir is an *insect* (such as a mosquito in the case of malaria), effective control of the disease can be accomplished by eliminating the reservoir with chemical insecticides or other lethal agents. However, such use must be balanced with environmental con-cerns about the use of toxic or carcinogenic chemicals—in some cases the elimination of one public health prob-lem only creates another. For example, the insecticide dichlorodiphenyl trichloroethane (DDT) (∞ Figure 14.51) is very effective against mosquitoes and is cred-ited with eradicating yellow fever and malaria in North America. However, its use is currently banned in the United States because of environmental concerns. DDT is still widely used in many developing countries to control mosquito-borne diseases.

When *humans* are the reservoir (for example, AIDS), then control and eradication are much more difficult, especially if there are asymptomatic carriers.

Controls directed against transmission of the pathogen

If the organism is transmitted via food or water, then public health procedures can be instituted either to prevent contamination of these vehicles or to destroy the pathogen in the vehicle. Water purification meth-ods (∞ Section 23.15) have been responsible for dra-matic reductions in the incidence of typhoid fever, and the pasteurization of milk has helped in the control of bovine tuberculosis in humans. Food protection laws have been devised that greatly decrease the probabil-ity of transmission of a number of enteric pathogens to humans (∞ Section 23.13). Transmission of respira-tory pathogens is much more difficult to prevent. Attempts at chemical disinfection of air have been unsuccessful. In Japan, many individuals wear face masks when they have upper respiratory infections to prevent transmission to others, but such methods, although effective, are voluntary, and would be diffi-cult to institute as public health measures.

Vaccination

Immunization of the host has been the prime means by which smallpox, diphtheria, tetanus, pertussis (whoop-ing cough), and poliomyelitis have been controlled. As we discussed in Section 22.6, 100% immunization is

not necessary in order to control the disease in a pop-ulation, although the percentage needed to ensure dis-ease control varies with the virulence of the pathogen and with the condition of the population (for example, crowding).

Over the past several decades, the proportion of children vaccinated for diphtheria, tetanus, pertussis, measles, mumps, rubella, and polio has been decreas-ing, apparently because the public has become less fearful of contracting these diseases because of their very low incidence in the population. However, because none of these diseases has been eradicated from the United States (indeed, the reservoir of tetanus is the soil, and so it will never be eradicated), and with a decrease in the proportion of individuals immunized, the protection afforded by herd immu-nity (see Section 22.6) can be overcome, and these infectious diseases could reappear in epidemic form. For example, Table 22.3 shows the vaccination rate for measles in selected countries in the Americas. In the United States, nearly 30,000 cases of measles were reported in 1990 (∞ Section 23.4), but renewed efforts to increase vaccination levels in preschool chil-dren have reversed this alarming trend (∞ Figure 23.19). Presumably, as the percentage of the immu-nized preschool population approached 70% in the United States, the benefits of herd immunity disap-peared but have now been reestablished as the result

TABLE 22.3   Infants immunized against measles in the Americas (1990)[a]

Country	Immunized (%)
Panama	99
Chile	98
Dominican Republic	96
Argentina	94
Cuba	94
Honduras	91
Bahamas	86
Costa Rica	85
Colombia	82
Uruguay	82
Belize	81
Nicaragua	81
Brazil	77
Paraguay	77
El Salvador	75
Jamaica	74
United States	**70**
Guatemala	68
Mexico	66
Peru	64
Venezuela	64
Ecuador	62
Bolivia	53
Haiti	31

[a]Data are for children less than 2 years of age.

of an aggressive immunization program. In countries such as Haiti, measles remains a significant cause of morbidity and mortality because of inadequate vaccination standards.

Many *adults* are inadequately immunized to a variety of infectious agents, either because they received low titer vaccines when they were children or because their immunity has gradually disappeared with age. In the United States, up to 80% of adults may lack solid immunity to important childhood diseases. When these so-called childhood diseases occur in adults, they can have severe effects. If a woman contracts rubella (a viral disease) (∞ Section 23.4) during pregnancy, the unborn child can be seriously impaired. Measles and polio are also much more serious diseases in adults than in children.

All adults are advised to review their immunization status, checking their medical records (if available) to ascertain dates of vaccinations. *Tetanus* vaccinations should be renewed at least every 10 years. Surveys of adult populations have shown that more than 10% of adults under the age of 40 and over 50% of those over 60 are not protected. *Measles* immunity in adults also needs to be reviewed. People born before 1957 probably had measles as children and are immune. Those born after 1956 may have been vaccinated, but the effectiveness of early vaccines was variable and solid immunity may not be present, especially if the vaccine was given before 1 year of age. Revaccination for polio is not recommended for adults unless they are traveling to countries in Africa and Asia where polio is still prevalent.

Vaccination practices and procedures have been discussed in Section 20.17, and those for particular infections will be discussed in Chapter 23.

Quarantine

Quarantine involves restricting the movement of individuals with active infections to prevent spread of disease to other members of the population. The time limit of quarantine is the longest period of communicability of the given disease. Quarantine must be done in such a manner that the infected individual cannot contact individuals who have not been exposed. Quarantine is not as severe a measure as strict isolation, which is used for unusually infectious diseases in hospital situations.

By international agreement, six diseases are considered quarantinable: smallpox, cholera, plague, yellow fever, typhoid fever, and relapsing fever. Although smallpox has been eliminated from the world, quarantine for the other five diseases is still mandated. Each of them is considered a highly serious, particularly communicable disease. Thus, it is essential to quarantine an infected individual for the period of communicability.

Surveillance

Surveillance is the observation, recognition, and reporting of diseases as they occur. The diseases that are under surveillance in the United States are listed in Table 22.4. Note that several of the epidemic diseases listed in Table 22.2 are not on the surveillance list. Several diseases like influenza are, however, surveyed through regional laboratories that identify *index cases*—those cases that exhibit new syndromes, characteristics, or pathogens indicating high potential for new epidemics.

CONCEPT CHECK   22.8

Food and water purity regulations, vector control, vaccination, quarantine, and disease surveillance are public health measures that play a major role in reduction of disease incidence.

✔ *Compare public measures for controlling infectious disease caused by insect reservoirs and by human carriers.*

✔ *What public health methods can be used to halt the spread of an epidemic disease once it has begun?*

22.9
Global Health Considerations

The United States is typical of countries where public health protection is highly developed. Other countries with similar characteristics include Japan, Australia, New Zealand, Israel, and the European countries. However, only about one-quarter of the nearly 6 billion people in the world live in these developed countries. In quite another category as far as infectious disease is concerned are the developing countries, a category that includes most of the countries in Africa, Central and South America, and Asia. In these countries, infectious diseases are still major causes of death (Figure 22.8).

Infectious disease in developing countries

There is a sharp contrast in the degree of importance of infectious diseases as causes of death in developing versus developed nations. In developing regions of the world, infectious diseases account for about 40% of deaths, whereas infectious diseases account for about 4% of deaths in developed regions (Figure 22.8). Diseases that were leading causes of death in the United States nearly a century ago, such as tuberculosis and gastroenteritis, are still leading causes of death in developing countries today (see Figures 22.8 and 1.18). Furthermore, the majority of deaths due to infectious disease in developing regions occur among infants and

TABLE 22.4   Reportable infectious diseases in the United States

Diseases caused by Bacteria	Diseases caused by Bacteria (cont.)
Anthrax	Syphilis
Botulism	Tetanus
Brucellosis	Toxic shock syndrome
Chancroid	Tuberculosis
Chlamydia	Tularemia
Cholera	Typhoid fever
Diphtheria	**Diseases caused by viruses**
Escherichia coli 0157:H7	Acquired immunodeficiency syndrome
Gonorrhea	(AIDS)
Granuloma inguinale	Aseptic meningitis
Haemophilus influenzae	Hepatitis
Hansen disease (leprosy)	Measles
Legionellosis	Mumps
Leptospirosis	Poliomyelitis, paralytic
Lyme disease	Rabies, animal
Lymphogranuloma venereum	Rabies, human
Meningococcal infections	Rubella
Murine typhus fever	Varicella (chicken pox)
Pertussis	**Diseases caused by protozoa**
Plague	Amebiasis
Psittacosis	Malaria
Rheumatic fever	**Disease caused by a helminth**
Rocky Mountain spotted fever	Trichinosis
Salmonellosis	
Shigellosis	

children. Thus, the average age of individuals dying as a result of infectious diseases in developing versus developed countries is also dramatically different.

The distinct differences in the health status of people in different regions of the world are due in part to general nutritional deficiency in individuals in developing countries and to a lower overall standard of living. As discussed in Section 19.12, factors such as physical stress and diet play important roles in the ability of the host to ward off infection. Thus, it is not surprising that in developing countries death from infection is about 10-fold more likely. In addition, the generally lower levels of public health protection and lack of economic resources for implementing widespread vaccination and food and water purity programs in developing countries make infection more likely in the first place. Statistics on disease in developed countries show that control of many diseases is possible. However, statistics on the worldwide incidence of disease show that infectious disease remains an important public health problem.

Travel to endemic areas

The high incidence of disease in many parts of the world is also a concern for people traveling to such areas. It is possible to be immunized against many of the diseases that are endemic in foreign countries. Some typical recommendations for immunization for those traveling abroad are shown in Table 22.5. Many foreign countries currently require immunization certificates for yellow fever, but most other immunizations are recommended only for people who are expected to be at high risk. There is also risk in many parts of the world of exposure to diseases for which there is no effective vaccine available. These include amebiasis, dengue fever, encephalitis, giardiasis, malaria, and typhus. Travelers are advised to take reasonable precautions such as avoiding insect and animal bites, drinking only water that has been properly treated and eating food properly stored and prepared, and undergoing chemotherapeutic programs when exposure is suspected.

CONCEPT CHECK   22.9

Infectious diseases account for over one-third of all deaths, worldwide. Control measures such as adequate immunizations are important for maintaining health, especially when traveling in developing countries.

✔ *Contrast the morbidity and mortality due to infectious diseases in developing and developed countries.*

✔ *List a series of infectious diseases for which you have not been immunized and with which you could come into contact next year.*

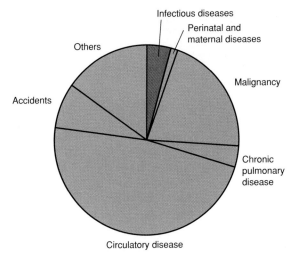

(a) Developed countries

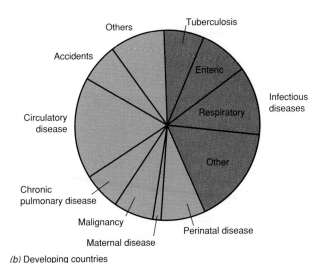

(b) Developing countries

Figure 22.8 Leading causes of death in developed and developing countries. Infectious diseases are shown in pink, and noninfectious diseases are shown in blue. (a) Developed countries. Of approximately 11.5 million deaths per year, about 500,000 are attributed to infectious disease, with nearly all deaths in this category due to pneumonia. (b) Developing countries. Of approximately 38.5 million deaths per year, about 17.5 million are attributed to infectious disease. Major infectious disease categories are shown.

22.10

Emerging and Resurgent Infectious Diseases

Infectious diseases are *global* health problems, and the scope and focus of these problems are constantly changing. In this section, we will examine some recent changes in patterns of infectious disease outbreaks, the reasons for the changing patterns, and the methods used by epidemiologists to identify and deal with new threats to public health.

The worldwide distribution of diseases can change dramatically and rapidly. Alterations in the pathogen, the environment, or the host population can contribute to the rapid spread of new diseases, with potential for high morbidity and mortality among infected individuals. We refer to diseases that suddenly become prevalent as *emerging* diseases. Emerging infections are not limited to "new" diseases but also include *resurgence* of diseases thought to be controlled, especially as antibiotics become less effective and public health systems fail. Some of the most recent, dramatic examples of emerging and resurgent disease are shown in Figure 22.9 on a global scale.

The phenomenon of suddenly emerging diseases of epidemic proportions is not new. Some of the diseases that suddenly emerged into prominence in the past were syphilis (caused by *Treponema pallidum*) (∞ Section 23.6) and plague (caused by *Yersinia pestis*) (∞ Section 23.12). In the Middle Ages, up to one-third of all living humans were killed by the plague epidemics that swept Europe, Asia, and Africa. More recently, influenza (∞ Section 23.4) became a major public health threat in the early part of the twentieth century. In the 1980s, legionellosis (caused by *Legionella pneumophila*) (∞ Section 23.2), acquired immunodeficiency syndrome (AIDS) (∞ Section 23.7), and Lyme disease (∞ Section 23.10) became major epidemic diseases.

Emergence factors

Some factors responsible for the emergence of new pathogens are (1) human demographics and behavior, (2) technology and industry, (3) economic development and land use, (4) international travel and commerce, (5) microbial adaptation and change, (6) breakdown of public health measures, and (7) abnormal natural occurrences that upset the usual host–pathogen balance.

The *demographics* of human populations have changed dramatically in the last two centuries. In 1800, less than 2% of the world's population lived in urban areas. By contrast, today nearly one-half of the world's population lives in cities. The numbers, sizes, and population densities of modern urban centers make disease transmission much easier. For example, dengue fever (Figure 22.9 and Table 22.6) is now recognized as a serious hemorrhagic disease in tropical cities, largely because of the spread of dengue virus by the mosquito *Aedes aegypti*. The disease now spreads as an epidemic in tropical urban areas. Prior to 1950, dengue fever was rare, presumably because the virus was not easily spread among a more dispersed, smaller population.

Human behavior, especially in large population centers, also contributes to disease spread. For example, sexual promiscuity and the use of injectable drugs, centered mainly in large urban areas, have been a

TABLE 22.5   Immunizations required or recommended for travel to developing countries[a]

Disease	Destination	Recommendation
Cholera	Many central African nations, India, Pakistan, South Korea, Albania, Malta, endemic areas in South America	*Vaccination recommended* if entering from or continuing to endemic areas
Yellow fever	Tropical and subtropical countries, worldwide	*Vaccination often required* for entry; or if entering from or continuing to endemic areas
Plague	Mostly rural mountainous and upland areas of Africa, Asia, and South America	*Vaccination recommended* if direct contact with rodents is anticipated
Infectious hepatitis (A)	Specific tropical areas and many developing countries	*Vaccination recommended*
Serum hepatitis (B)	Africa, Indochina, eastern and southern Europe, countries in the former Soviet Union, Central and South America	*Vaccination recommended*
Typhoid fever	Many African, Asian, Central and South American countries	*Vaccination recommended*

[a]*Current Health Information for International Travelers,* U.S. Department of Health and Human Services.

Vaccinations are also recommended for diphtheria, pertussis, tetanus, polio, measles, mumps, and rubella. Most U.S. citizens are already immunized through normal immunization practices.

major contributing factor to the spread of AIDS and hepatitis (Table 22.6; ∞ Sections 23.6 and 23.7).

Although *technological advances* and *industrial development* have had a generally positive impact on living standards worldwide, in some cases these advances have contributed to the spread of diseases. For example, one of the chief technological advances of the twentieth century has been in the health care area. However, as we noted in Section 22.7, the health care environment, especially in hospitals, has resulted in an explosive increase in nosocomial infections. For example, during the 1980s there was a threefold rise in hospital-associated bacteremias in the United States (see Section 22.7 and Figure 22.7). Antibiotic resistance in microorganisms is another negative outcome of modern health care practices; vancomycin-resistant

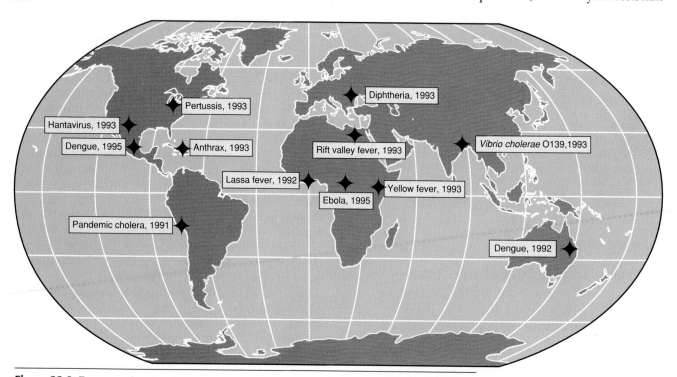

Figure 22.9 Recent outbreaks of emerging and resurgent infectious diseases on a global scale.

TABLE 22.6   **Some emerging and resurgent infectious diseases**

Agent	Disease and symptoms	Mode of transmission	Cause(s) of emergence
Bacteria, Rickettsias, and Chlamydias			
Borrelia burgdorferi	Lyme disease: rash, fever, neurological and cardiac abnormalities, arthritis	Bite of infective *Ixodes* tick	Increase in deer and human populations in wooded areas
Campylobacter jejuni	Campylobacter enteritis: abdominal pain, diarrhea, fever	Ingestion of contaminated food, water, or milk; fecal-oral spread from infected person or animal	Increased recognition; consumption of undercooked poultry
Chlamydia trachomatis	Trachoma, genital infections, conjunctivitis, infant pneumonia	Sexual intercourse	Increased sexual activity; changes in sanitation
Escherichia coli O157:H7	Hemorrhagic colitis; thrombocytopenia; hemolytic uremic syndrome	Ingestion of contaminated food, especially undercooked beef and raw milk	Development of a new pathogen
Haemophilus influenzae biogroup *aegyptus*	Brazilian purpuric fever: purulent conjunctivitis, fever, vomiting	Discharges of infected persons; flies are suspected vectors	Possible increase in virulence due to mutation
Helicobacter pylori	Gastritis, peptic ulcers, possibly stomach cancer	Contaminated food or water, especially unpasteurized milk; contact with infected pets	Increased recognition
Legionella pneumophila	Legionnaires' disease: malaise, myalgia, fever, headache, respiratory illness	Air-cooling systems, water supplies	Recognition in an epidemic situation
Mycobacterium tuberculosis	Tuberculosis: cough, weight loss, lung lesions; infection can spread to other organ systems	Sputum droplets (exhaled through a cough or sneeze) of a person with active disease	Immunosuppression, immunodeficiency
Staphylococcus aureus	Abscesses, pneumonia, endocarditis, toxic shock	Contact with the organism in a purulent lesion or on the hands	Recognition in an epidemic situation; possibly mutation
Streptococcus pyogenes	Scarlet fever, rheumatic fever, toxic shock	Direct contact with infected persons or carriers; ingestion of contaminated foods	Change in virulence of the bacteria; possibly mutation
Vibrio cholerae	Cholera: severe diarrhea, rapid dehydration	Water contaminated with the feces of infected persons; food exposed to contaminated water	Poor sanitation and hygiene; possibly introduced via bilge water from cargo ships
Viruses			
Dengue	Hemorrhagic fever	Bite of an infected mosquito (primarily *Aedes aegypti*)	Poor mosquito control; increased urbanization in tropics; increased air travel
Filoviruses (Marburg, Ebola)	Fulminant, high mortality, hemorrhagic fever	Direct contact with infected blood, organs, secretions, and semen	Unknown; in Europe and the United States, virus-infected monkeys shipped from developing countries via air
Hantaviruses	Abdominal pain, vomiting, hemorrhagic fever	Inhalation of aerosolized rodent urine and feces	Human intrusion into virus ecological niche
Hepatitis B	Nausea, vomiting, jaundice; chronic infection leads to hepatocellular carcinoma and cirrhosis	Contact with saliva, semen, blood, or vaginal fluids of an infected person; mode of transmission to children not known	Probably increased sexual activity and intravenous drug abuse; transfusion (before 1978)
Hepatitis C	Nausea, vomiting, jaundice; chronic infection leads to hepatocellular carcinoma and cirrhosis	Exposure (percutaneous) to contaminated blood or plasma; sexual transmission	Recognition through molecular virology applications; blood transfusion practices, especially in Japan
Hepatitis E	Fever, abdominal pain, jaundice	Contaminated water	Newly recognized
Human immuno-deficiency viruses: HIV-1 and HIV-2	HIV disease, including AIDS: severe immune system dysfunction, opportunistic infections	Sexual contact with or exposure to blood or tissues of an infected person; vertical transmission	Urbanization; changes in lifestyle or mores; increased intravenous drug use; international travel; medical technology (transfusions and transplants)

TABLE 22.6   (continued)

Agent	Disease and symptoms	Mode of transmission	Cause(s) of emergence
Viruses (cont.)			
Human papillomavirus	Skin and mucous membrane lesions (often, warts); strongly linked to cancer of the cervix and penis	Direct contact (sexual contact or contact with contaminated surfaces)	Newly recognized; perhaps changes in sexual lifestyle
Human T-cell lymphotrophic viruses (HTLV-I and HTLV-II)	Leukemias and lymphomas	Vertical transmission through blood or breast milk; exposure to contaminated blood products; sexual transmission	Increased intravenous drug abuse; medical technology (transfusion and transplantation)
Influenza pandemic	Fever, headache, cough, pneumonia	Airborne; especially in crowded, enclosed spaces	Animal–human virus reassortment; antigenic shift
Lassa	Fever, headache, sore throat, nausea	Contact with urine or feces of infected rodents	Urbanization and conditions favoring infestation by rodents
Measles	Fever, conjunctivitis, cough, red blotchy rash	Airborne; direct contact with respiratory secretions of infected persons	Deterioration of public health infrastructure supporting immunization
Norwalk and Norwalk-like agents	Gastroenteritis; epidemic diarrhea	Most likely fecal-oral; vehicles may include drinking and swimming water, and uncooked foods	Increased recognition
Rabies	Acute viral encephalomyelitis	Bite of a rabid animal	Introduction of infected host reservoir to new areas
Rift Valley	Febrile illness	Bite of an infective mosquito	Importation of infected mosquitoes and/or animals; development (dams, irrigation)
Rotavirus	Enteritis: diarrhea, vomiting, dehydration, and low grade fever	Primarily fecal-oral; fecal-respiratory transmission can also occur	Increased recognition
Venezuelan equine encephalitis	Encephalitis	Bite of an infective mosquito	Movement of mosquitoes and hosts (horses)
Yellow fever	Fever, headache, muscle pain, nausea, vomiting	Bite of an infective mosquito (*Aedes aegypti*)	Lack of effective mosquito control and widespread vaccination; urbanization in tropics; increased air travel
Protozoa and Fungi			
Candida	Candidiasis: fungal infections of the gastrointestinal tract, vagina, and oral cavity	Endogenous flora; contact with secretions or excretions from infected persons	Immunosuppression; medical management (catheters); antibiotic use
Cryptococcus	Meningitis; sometimes infections of the lungs, kidneys, prostate, liver	Inhalation	Immunosuppression
Cryptosporidium	Cryptosporidiosis: infection of epithelial cells in the gastrointestinal and respiratory tracts	Fecal-oral, person to person, waterborne	Development near watershed areas; immunosuppression
Giardia lamblia	Giardiasis: infection of the upper small intestine, diarrhea, bloating	Ingestion of fecally contaminated food or water	Inadequate control in some water supply systems; immunosuppression; international travel
Microsporidia	Gastrointestinal illness, diarrhea; wasting in immunosuppressed persons	Unknown; probably ingestion of fecally contaminated food or water	Immunosuppression; recognition
Plasmodium	Malaria	Bite of an infective *Anopheles* mosquito	Urbanization; changing parasite biology; environmental changes; drug resistance; air travel
Pneumocystis carinii	Acute pneumonia	Unknown; possibly reactivation of latent infection	Immunosuppression
Toxoplasma gondii	Toxoplasmosis: fever, lymphadenopathy, lymphocytosis	Exposure to feces of cats carrying the protozoan; sometimes foodborne	Immunosuppression; increase in cats as pets

enterococci and multiple-drug-resistant *Streptococcus pneumoniae* have become important emerging diseases, especially in developed countries.

Transportation, bulk processing, and central distribution methods have become an important factor for quality assurance and economy in the food industry. However, these same factors can increase the potential for common-source epidemics when sanitation measures fail. For example, a single meat processing plant spread *Escherichia coli* O157:H7 (Table 22.6) to at least 500 individuals in four states in the United States. Finally, the food source, ground beef, was recalled and the epidemic was curtailed, but not before several people died (∞ Section 23.13).

Economic development and changes in land use also have potential implications for promoting disease spread. For example, Rift Valley fever, a mosquito-borne viral infection, has been on the increase since completion of the Aswan High Dam in Egypt in 1970. The dam created 2 million acres of flooded land, which dramatically increased mosquito breeding grounds at the edge of the new reservoir. The first major epidemic of Rift Valley fever occurred in Egypt in 1977 when an estimated 200,000 people became ill and 598 died. Several epidemic outbreaks have occurred since then including a major outbreak in 1993 (Figure 22.9), and the disease has become endemic near the reservoir.

Lyme disease, the most common vector-borne disease in the United States, is probably on the rise because of *changes in land use*. Reforestation and the concomitant increase in the numbers of deer (the natural host for the disease-producing *Borrelia burgdorferi*) have resulted in greater numbers of infected ticks, the arthropod vector (∞ Section 23.10). In addition, increasing numbers of people are building homes and pursuing recreational activities in and near forests, resulting in increased contact between the infected ticks and humans and, consequently, increased disease.

International travel and commerce can also affect the spread of pathogens. For example, filoviruses (Filoviridae), a group of ribonucleic acid (RNA) viruses, (∞ Section 8.16), cause fevers culminating in hemorrhagic disease in infected hosts. These diseases, because of their viral origin, are not treatable. They generally have a mortality rate higher than 20%. Most outbreaks of these diseases have been restricted to equatorial central Africa, where the still-unidentified natural hosts and vectors undoubtedly live (Figure 22.9). Travel of potential hosts to or from endemic areas is usually implicated in disease transmission. For example, one of these viruses was imported into Marburg, Germany, with a shipment of African green monkeys, a species used for laboratory work. The virus quickly spread from the primate vector to some of the human handlers. Twenty-five people were initially infected, and six more developed disease as a result of contact with the human cases. Seven people died in this outbreak of what came to be known as the Marburg virus. Another shipment of laboratory monkeys brought a filovirus to the United States. At least four individuals who worked with the imported monkeys were infected with what is now called the Reston virus (named for Reston, Virginia, the site of the outbreak). The Reston virus was highly contagious and spread through the monkeys, presumably by a respiratory route. However, only four humans were infected and none developed clinical disease. Fortunately, this virus did not cause significant human disease. These two filoviruses are closely related to the Ebola virus (Table 22.6 and Figure 22.9). Recent Ebola outbreaks in central Africa, characterized by mortality rates of greater than 50%, have again underscored the existence of highly virulent human pathogens for which there is little or no immunity. These pathogens could potentially be disseminated via air travel throughout the world in a matter of days. A single agent that combines the highly contagious respiratory transmission route of the Reston virus and the high mortality rate of the Ebola virus could start a major pandemic that could devastate population centers worldwide in a matter of weeks.

Microbial adaptation and change also contribute to pathogen emergence. For example, nearly all RNA viruses, including influenza and human immunodeficiency virus (HIV), undergo genetic mutations. Hepatitis B virus, a deoxyribonucleic acid (DNA) virus known for rapidly mutating, also uses reverse transcriptase to replicate (∞ Section 8.14). These viruses lack correction mechanisms for replication steps, and so they incorporate genomic mutations at an extremely high rate compared to most DNA viruses. RNA viruses are considered to be major epidemiological problems because of their constantly changing genomes.

Bacteria also have genetic mechanisms that enhance virulence and promote emergence of new epidemics. One group of virulence-enhancing mechanisms are the mobile genetic elements: bacteriophages, plasmids, and transposons (∞ Sections 8.7, 9.8, and 9.10, respectively). Table 22.7 shows some representative virulence factors that are carried on these mobile genetic elements and contribute to pathogen emergence.

Antibiotic resistance is also a major factor in bacterial pathogen resurgence (∞ Section 11.13). Drug resistance is also a factor for virus emergence. Although several drugs are effective against certain viral diseases (∞ Section 11.11), resistance to these drugs is very common, especially among the RNA viruses. For example, most strains of HIV develop resistance to azidothymidine very rapidly unless it is used in combination with other drugs (∞ Section 23.7).

A breakdown of public health measures is sometimes responsible for the emergence or resurgence of diseases. For instance, cholera (caused by *Vibrio cholerae*) (see Figure 22.9 and Section 23.14) can be adequately controlled, even in endemic areas, by providing proper sanitation, especially for water sources. However, contaminated municipal water supplies in Peru led to a

TABLE 22.7   Virulence factors encoded by bacteriophages, plasmids, and transposons

Genetic element	Organism	Virulence factors
Bacteriophage	*Streptococcus pyogenes*	Erythrogenic toxin
	Escherichia coli	Shiga-like toxin
	Staphylococcus aureus	Enterotoxins A, D, E, staphylokinase, toxic shock syndrome toxin-1 (TSST-1)
	Clostridium botulinum	Neurotoxins C, D, E
	Corynebacterium diphtheriae	Diphtheria toxin
Plasmid	*Escherichia coli*	Enterotoxins, pili colonization factor, hemolysin, urease, serum resistance factor, adherence factors, cell invasion factors
	Bacillus anthracis	Edema factor, lethal factor, protective antigen, poly-D-glutamic acid capsule
	Yersinia pestis	Coagulase, fibrinolysin, murine toxin
Transposon	*Escherichia coli*	Heat-stable enterotoxins, aerobactin siderophores, hemolysin and pili operons
	Shigella dysenteriae	Shiga toxin
	Vibrio cholerae	Cholera toxin

major cholera pandemic, involving nearly 400,000 people by 1991, with almost 4000 deaths. In another case, the municipal water supply of Milwaukee, Wisconsin, was contaminated with the chlorine-resistant protozoan *Cryptosporidium* in 1993. The contamination resulted in 370,000 cases of intestinal disease, 4000 of which required hospitalization. More effective treatment procedures including enhanced filtration systems were required to rid the water supply of the pathogen.

Inadequate public vaccination programs are an important potential reason for the resurgence of some previously controlled diseases. For example, recent outbreaks of diphtheria (caused by *Corynebacterium diphtheriae*) (see Figure 22.9 and Section 23.2) in the former Soviet Union are the result of inadequate immunization of susceptible children resulting from the breakdown of the formerly centralized public health infrastructure. Pertussis, another vaccine-preventable childhood respiratory disease (caused by *Bordetella pertussis*) (see Figure 22.9 and Section 23.2), has increased recently in the United States because of inadequate immunization and record keeping. As we mentioned in Section 22.8, the incidence of measles was also on the rise in the United States owing to a lack of effective, timely vaccination programs.

Finally, *abnormal natural occurrences* such as rapid environmental changes sometimes upset the usual host–pathogen balance. For example, hantavirus is a well-known human pathogen that occurs in many rodent populations, even in laboratory animals. Over the last decade, several isolated cases of hantavirus infection have occurred in laboratory animal handlers. However, a number of lethal cases of hantavirus infection were reported in 1993 in the American Southwest and were linked to exposure to wild animal droppings. Abundant rainfall and a long growing season, coupled with a mild winter, caused a tremendous increase in the number of mice in 1993. Virtually everyone who acquired the hantavirus infection had

been exposed to rodents or their droppings. Thus, increased human contact with the larger-than-normal mouse population resulted in propagation and transfer of a deadly virus to a large number of human hosts, all because of abnormally mild weather conditions.

Addressing emerging diseases

Many of the emerging diseases we have discussed are absent from the official notifiable disease list for the United States (Table 22.4). How then do public health officials define and deal with emerging diseases to prevent major epidemics? The key features for addressing emerging diseases are *recognition* of the disease and *intervention* to prevent spread of the disease.

The first step in disease *recognition* is surveillance. *Epidemic* diseases that exhibit particular *clinical syndromes* warrant intensive public health surveillance. These syndromes are (1) acute respiratory diseases, (2) encephalitis and aseptic meningitis, (3) hemorrhagic fever, (4) acute diarrhea, (5) clusterings of high fever cases, (6) unusual clusterings of any disease or deaths, and (7) resistance to common drugs or treatment. Thus, new diseases are recognized because of their epidemic incidence, clusterings, and syndromes. As the prevalence and pathology of an emerging disease are recognized, it is added to the notifiable disease list. For example, AIDS was recognized as a disease in 1981 and was added to the notifiable disease list in 1984. Lyme disease was first recognized as a separate clinical disease in the 1980s and added to the notifiable disease list in 1991. Likewise, outbreaks of gastrointestinal disease due to enteropathogenic *Escherchia coli* O157:H7 have been increasing in recent years, and the strain was added to the notifiable disease list in 1995.

Intervention to prevent spread of emerging infections must be a public health response involving a variety of methods. General strategies such as strengthening the public health system and supporting research and training are useful, but disease-specific intervention is the key to controlling individual outbreaks. Public health methods such as vector control and quarantine were discussed in Section 22.8. In addition, intervention must include drug and vaccine development (∞ Sections 12.5 and 20.17) to prevent and treat specific diseases. Finally, a number of the emerging diseases are propagated in nonhuman hosts, or vectors. We must identify the alternate hosts and vectors and develop means to intervene in the life cycle of the pathogen to prevent disease propagation.

In the following chapter, we will examine a number of infectious diseases, including several emerging and resurgent diseases. We will define their individual effects on the host and identify specific intervention strategies.

CONCEPT CHECK   22.10

Emerging and resurgent diseases are of major global concern. Changes in host, vector, or pathogen conditions, whether natural or artificial, can result in conditions that encourage the explosive emergence of certain infectious diseases. Global surveillance and intervention programs must be maintained and enhanced to prevent major new epidemics and pandemics.

✔ *What factors are important in the emergence or resurgence of potential pathogens?*

✔ *Indicate general and specific methods that would be useful for dealing with perceived and actual emerging infectious diseases.*

Material for Review

REVIEW QUESTIONS

1. What are the most common causes of death due to infectious diseases throughout the world?

2. Describe the stages involved in a typical infectious disease in which the host recovers.

3. Explain the difference between a chronic carrier and an acute carrier of an infectious disease.

4. Identify the major risk factors for acquiring human immunodeficiency virus (HIV) infection. Does this indicate a common-source or a host-to-host epidemic?

5. Give examples of host-to-host transmission of disease via direct contact. Also give examples of indirect host-to-host transmission of disease.

6. Some diseases produce high mortality on introduction to a susceptible population, but after time these diseases usually generate much lower mortality. Explain this phenomenon.

7. Hospitals are particularly hazardous environments for the spread of infectious diseases. Review the reasons for the enhanced spread of infection in hospitals.

8. Many factors that can control the spread of infection are important considerations for public health personnel. Describe the major methods used to control the spread of infectious diseases.

9. Compare the role of infectious diseases on mortality in developed and developing countries.

10. Review the major reasons for the emergence of new infectious diseases. Review the major methods available for controlling the emergence of new infectious diseases.

APPLICATION QUESTIONS

1. How would an epidemiologist acquire data concerning a potential common-source epidemic? What resources are currently at the epidemiologist's disposal and what resources must be enhanced to better define serious infectious disease outbreaks?

2. If an infectious disease causes high *mortality*, then *morbidity* may be quite low. On the other hand, diseases characterized by high morbidity often induce very low mortality. Explain these statements and present examples to support your explanation. Can you identify any infectious diseases that do not fit these generalizations?

3. Smallpox, a disease that was limited to humans, was eradicated. Plague, a disease with a zoonotic reservoir in rodents (Table 22.2) will never be eradicated. Explain this statement and why you agree or don't agree with the possibility of eradicating plague. Could you eradicate plague in limited environments (that is, individual cities)? Why or why not?

4. Acquired immunodeficiency syndrome (AIDS) transmission is considered to be person to person. How did epidemiologists determine this fact? AIDS is a candidate for a disease that can be eliminated because it is propagated by person-to-person contact and there are no known animal reservoirs. Design a program for eliminating AIDS in a developed country and in a developing country. How would these programs differ from one

another? What factors would work against the success of your program, both in terms of human behavior and in terms of the AIDS disease itself? Why are the numbers of HIV-infected and AIDS patients continuing to grow?

5. Transmission of many epidemic diseases is host to host, whereas other epidemics are spread via a common source. Some epidemics can be transmitted by both routes. Explain how this might happen, using specific infectious agents (at least one bacterium and one virus) as examples. Is one disease category more likely to show this pattern? Why or why not?

6. What is the overall advantage of pathogen–host coevolution in terms of species survival? Is it beneficial to the pathogen to cause high mortality in the host? Why or why not? What diseases in Table 22.2 have caused high mortality? Compare their reservoir or host to diseases with low mortality.

7. Why are diseases due to antibiotic-resistant organisms of a given species more common in hospital environments than in the general population? What special precautions must one take when diagnosing and treating infectious diseases in a hospital setting?

8. As a public health official, you are faced with a common-source epidemic and you believe the source is the municipal water supply. How would you use your limited resources to stop the epidemic? Do not focus on treatment of the disease unless you, as a public health officer, believe that treatment will stop disease spread. List the steps you would take in priority order. Do the same exercise for a host-to-host epidemic for which there are available vaccines and chemotherapeutic agents.

9. Travel to developing countries involves a certain amount of exposure to infectious diseases. What general precautions should you take before, during, and after visits to developing countries? Where can you obtain information on the infectious disease status in a specific foreign country? When you return from a foreign country, are you a disease risk to your family or your associates? Explain.

10. Although many factors may be involved in the emergence of an infectious disease, some diseases develop to the pandemic stage while others never get beyond localized epidemics. Examples of this are HIV and Ebola virus infections, which were both identified in the last 20 years. What factors do these viruses share that led to their emergence? What factors cause them to be quite different in terms of their spread?

SUPPLEMENTARY READINGS

Brachman, P. S. 1991. *Bacterial Infections of Humans: Epidemiology and Control.* Plenum Press, New York. The epidemiology of bacterial diseases.

Brock, T. D. (ed.). 1989. *Microorganisms: From Smallpox to Lyme Disease.* W. H. Freeman, New York. A collection of fascinating articles on infectious disease from *Scientific American* magazine.

Brooks, G. F., J. S. Butel, and **L. N. Ornston.** 1995. *Jawetz, Melnick and Adelberg's Medical Microbiology,* 20th edition. Appleton and Lange, Norwalk, CT. A complete, short medical text covering many aspects of host–parasite interactions.

Centers for Disease Control. *Morbidity and Mortality Weekly Report.* Atlanta, GA. Available from Massachusetts Medical Society, Waltham, MA. Issued weekly. This publication, available in most large libraries, gives an instantaneous view of epidemiological problems in the United States. Lists the incidence of all major infectious diseases by geographic region and gives updates on particular disease problems. An *Annual Summary* is published yearly providing an overview of the incidence of past and present infectious diseases. A good source of graphs, tables, and statistical analyses of disease trends.

Fenner, F. 1983. Biological control as exemplified by smallpox eradication and myxomatosis. *Proc. Roy. Soc. London Ser. B 218:*259–285. An excellent historical account of the control of the Australian rabbit population and the global eradication of smallpox.

Howard, B. J. (ed.). 1994. *Clinical and Pathogenic Microbiology,* 2nd edition. Mosby–Year Book, St. Louis. Excellent treatments of nosocomial infections.

Lederberg, J., R. E. Shope, and **S. C. Oaks, Jr.** (eds.). 1992. *Emerging Infections: Microbial Threats to Health in the United States.* National Academy Press, Washington, DC. A chilling account of emerging infectious diseases worldwide.

Morse, S. S. (ed.). 1993. *Emerging Viruses.* Oxford University Press, New York. A readable collection of articles by investigators who have encountered emerging viral pathogens. Fascinating!

Roizman, B. (ed.). 1995. *Infectious Diseases in an Age of Change: The Impact of Human Ecology and Behavior on Disease Transmission.* National Academy Press, Washington, DC. Current information about the epidemiology and risk factors of emerging and resurgent infectious diseases.

Snow, J. 1936. *Snow on Cholera.* A reprint of two papers by John Snow, M.D. Commonwealth Fund, New York. A reprinting of Snow's 1855 study of cholera, the first epidemiological investigation, and a fascinating detective story. Highly recommended.

Stine, G. J. 1996. *AIDS Update 1996.* Prentice-Hall, Englewood Cliffs, NJ. An excellent monograph describing the current status of the AIDS epidemic. Updated annually.

Walker, D. H. (ed.). 1993. *Global Infectious Diseases: Prevention, Control, and Eradication.* Springer-Verlag, New York. Review of the epidemiology, ecology, pathogenesis, and immunity of several infectious diseases of worldwide significance.

Wilson, M. E. 1991. *A World Guide to Infections: Disease, Distribution, Diagnosis.* Oxford University Press, New York. Epidemiological considerations of immigration and travel.

World Health Organization. 1992. *Global Estimates for Health Situation Assessment and Projections 1992.* WHO, Geneva, Switzerland. An excellent source of epidemiological data from worldwide sources, including information concerning the most prevalent infectious diseases and predictions for future disease problems.

On~line resources for this chapter are on the World Wide Web at: http://www.prenhall.com/~brock (click on the table of contents link and then select Chapter 22).

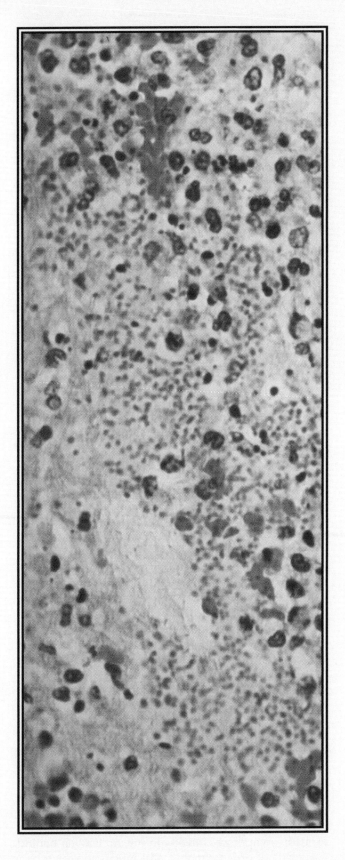

Major Microbial Diseases

Many organisms are capable of causing disease in compromised hosts. Pictured here is a section of a small intestine from an AIDS patient. The tissue shows invasion and infection by *Mycobacterium* spp., the red-stained cells. Members of the *Mycobacterium* genus are common, life-threatening opportunistic pathogens in AIDS patients.

MINIGLOSSARY for Chapter 23

Antigenic Drift minor changes in antigens due to gene mutation in influenza virus

Antigenic Shift major changes in antigens due to gene reassortment in influenza virus

Chlorination a highly effective disinfectant procedure for drinking water using chlorine gas or other chlorine-containing compounds

Coliform a large group of gram-negative, facultative Bacteria

Congenital Syphilis syphilis contracted by an infant from its mother during birth

Food Infection infection resulting from ingestion of contaminated food

Food Poisoning disease resulting from ingestion of a bacterial exotoxin

Lyme Disease a tick-transmitted disease caused by the spirochete *Borrelia burgdorferi*

Mycoses infections caused by fungi

Rheumatic Fever an inflammatory autoimmune disease triggered by an immune response to infection by *Streptococcus pyogenes*

Rickettsias obligate intracellular parasites that cause a variety of diseases including typhus and Rocky Mountain spotted fever

Scarlet Fever characteristic reddish rash resulting from an exotoxin produced by *Streptococcus pyogenes*

Sexually Transmitted Disease (STD) a disease that is usually transmitted by sexual contact

Sickle-Cell Anemia a genetic trait that confers resistance to malaria but causes a reduction in the efficiency of red blood cells

Toxic Shock Syndrome acute shock resulting from a host response to an exotoxin produced by *Staphylococcus aureus*

Tuberculin Test a skin test for previous infection with *Mycobacterium tuberculosis*

Within the microbial world only a relatively few species are pathogens. The majority of microorganisms carry out essential activities in nature, and many are closely associated with plants or animals in stable, beneficial relations (∞ Chapter 14). However, as we have seen, pathogenic organisms can have profound effects on animals and humans. In this chapter we consider the major human diseases and group together those microbial diseases that affect a specific human organ system. We will continually make reference to *modes of transmission* in our discussion of diseases because the pathology of a disease is best considered in light of its ecology. For example, streptococcal sore throat and influenza are diseases whose etiological agents are completely different, one a bacterium and the other a virus. Yet, both diseases are transmitted from person to person primarily by an airborne route. Hence, if we group diseases by their modes of transmission, what appears to be a long list of unrelated diseases becomes a short list of ecologically related diseases.

23.1

Airborne Transmission of Pathogens

Air is not a suitable medium for the growth of microorganisms; organisms found in air are derived from soil, water, plants, animals, people, or other sources. In outdoor air, soil organisms predominate. Microbial numbers indoors are considerably higher than those out-doors, and the organisms are mostly those commonly found in the human respiratory tract.

Windblown dust carries with it significant microbial populations that can travel long distances. Most of these organisms survive poorly in air, and so effective transmittal to a suitable habitat (another human) occurs only over short distances. However, certain human pathogens (*Staphylococcus*, *Streptococcus*) survive under dry conditions fairly well and may remain alive in dust for long periods of time. Gram-positive Bacteria are in general more resistant to drying than gram-negative Bacteria because of their thicker, more rigid cell wall. Spore-forming Bacteria are extremely resistant to drying but are not generally passed from human to human in the spore form. However, airborne *Clostridium* spp. (∞ Section 16.26) are derived from soil, are spore formers, and are occasional pathogens, although they are seldom respiratory pathogens.

An enormous number of droplets of moisture are expelled during sneezing (Figure 23.1), and a considerable number are expelled during coughing or even merely talking. Each infectious droplet has a size of about 10 μm and contains one or two bacteria. The speed of the droplet movement is about 100 m/sec (more than 200 mi/hr) in a sneeze and about 16–48 m/sec during coughing or loud talking. The number of bacteria in a single sneeze varies from 10,000 to 100,000. Because of the small size of the droplets, the moisture evaporates quickly in the air, leaving behind a nucleus of organic matter and mucus to which bacterial cells are attached.

Figure 23.1 High speed photograph of an unstifled sneeze.

Respiratory infection

The average human breathes several million cubic feet of air in a lifetime, much of it containing microorganism-laden dust, which is a potential source of inoculum for upper respiratory infections caused by streptococci and staphylococci. The speed at which air moves through the respiratory tract varies, and in the lower respiratory tract the rate is quite slow. As the air slows down, particles in it stop moving and settle. The larger particles settle first and the smaller ones later, and in the tiny bronchioles of the lung, only particles smaller than 3 μm are present (Figure 23.2). Different organisms reach different levels in the tract, thus accounting for the differences in the kinds of infections that occur in the upper and lower respiratory tracts.

CONCEPT CHECK   **23.1**

Many respiratory pathogens are gram-positive Bacteria that are very resistant to drying and can be transmitted by air. Other less hardy organisms can be transferred in respiratory aerosols such as sneezes.

✔ *What physical features of gram-positive organisms allow them to survive for long periods in the air?*

✔ *Why are certain infections usually found in the upper respiratory tract and not in the lower respiratory tract?*

23.2
Respiratory Infections: Bacterial

A variety of bacterial pathogens affect the respiratory tract and because their mode of transmission is air, they are predominantly gram-positive Bacteria. Because secondary problems associated with an initial bacterial respiratory infection can often be quite serious, it is impor-

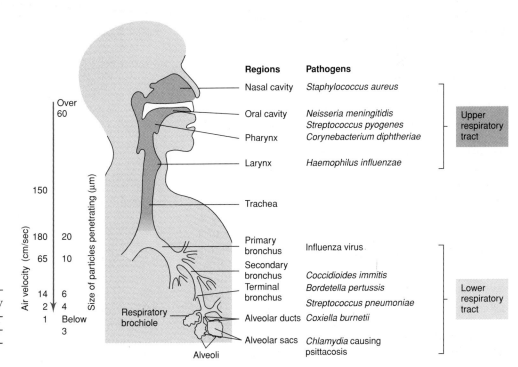

Figure 23.2 The respiratory system of humans and locations at which various organisms generally initiate infections.

tant to swiftly diagnose and treat these infections to prevent damage to host tissues. Fortunately, most respiratory bacterial pathogens respond readily to antibiotic therapy and many can also be controlled by immunization programs. Nevertheless, bacterial respiratory infections are still rather common, and we begin here with a consideration of two of the most common bacterial respiratory pathogens, *Streptococcus pyogenes* and *Streptococcus pneumoniae*.

Streptococcal diseases

Streptococci are gram-positive cocci that typically grow in elongated chains (∞ Section 16.25). Several species of streptococci are potent human pathogens. Of particular importance are *Streptococcus pyogenes* and *S. pneumoniae*.

Streptococcus pyogenes and *S. pneumoniae* are frequently isolated from the upper respiratory tract of healthy adults. Although numbers of *S. pyogenes* and *S. pneumoniae* are usually low here, if the host's defenses are weakened or a new, highly virulent strain is introduced, acute streptococcal bacterial infections are possible. *S. pyogenes* is the cause of streptococcal pharyngitis, so-called strep throat (the pharynx is the tube that connects the oral cavity to the larynx and the esophagus) (Figure 23.2). Most isolates from clinical cases of strep throat produce a toxin that lyses red blood cells, a condition called *β-hemolysis* (∞ Figure 19.16). Streptococcal pharyngitis is characterized by intense inflammation of the mucous membranes of the throat, a mild fever, and a general feeling of malaise. *Streptococcus pyogenes* can also cause related infections of the inner ear (otitis media), the tonsils (tonsillitis), the mammary glands (mastitis), and infections of the superficial layers of the skin, a condition referred to as impetigo (however, most cases of impetigo are caused by *Staphylococcus aureus*) (Figure 23.3).

About half of the clinical cases of severe sore throat turn out to be due to *Streptococcus pyogenes*, the remainder being of viral origin. An accurate, prompt diagnosis is important because if the sore throat is due to a virus, the application of antibacterial antibiotics will be useless, whereas if the sore throat is due to *S. pyogenes*, immediate antibiotic therapy is indicated. This is because occasional cases of streptococcal sore throat can lead to more serious streptococcal syndromes such as scarlet fever, rheumatic fever, and acute glomerulonephritis.

Certain strains of *Streptococcus pyogenes* carry a lysogenic bacteriophage that encodes production of a potent exotoxin, responsible for most of the symptoms of *scarlet fever*. The erythrogenic toxin causes a pink-red rash to develop (Figure 23.4) and also acts to damage small blood vessels and initiate fever. The condition is acute and easily treated with antibiotics.

Occasionally, *Streptococcus pyogenes* may cause fulminant systemic infections, often marked by necrotizing fasciitis, a rapid and progressive infection of subcutaneous tissue. These infections are responsible for

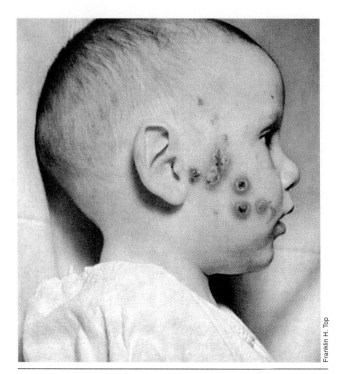

Figure 23.3 Typical lesions of impetigo, commonly caused by *Streptococcus pyogenes* or *Staphylococcus aureus*.

the dramatic, but fortunately rare, reports of "flesh-eating bacteria." In these cases, exotoxins A and B and the surface M-protein act as superantigens (∞ Section 20.16) that recruit massive numbers of T cells. The T cells then secrete cytokines, which activate large num-

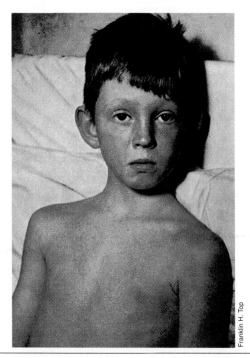

Figure 23.4 The typical rash of scarlet fever, resulting from the action of the erythrogenic toxin produced by *Streptococcus pyogenes*.

bers of effector cells, resulting in massive systemic inflammation, tissue destruction, and death in up to 30% of the cases.

Untreated or insufficiently treated cases of *Streptococcus pyogenes* infection may lead to severe *delayed sequelae*, or follow-up diseases. **Rheumatic fever,** one of these delayed sequelae, is caused by strains of *S. pyogenes* containing cell surface antigens that are similar to certain human cell surface antigens. When an immune response to the invading pathogen is made, the antibodies produced cross-react with host tissues, in particular those of the heart, joints, and kidneys. This results in significant tissue destruction. In essence, rheumatic fever is a type of autoimmune disease—antibodies react with self constituents (∞ Section 20.16). Damage may be permanent and is often accelerated by later infections.

Another potential delayed sequela of *Streptococcus pyogenes* infection is *acute glomerulonephritis*, a painful disease of the kidney. This is an immune complex disease (∞ Section 20.16) resulting from the formation of streptococcal antigen–antibody complexes in the bloodstream during the recovery phase of a streptococcal infection. These immune complexes lodge in the *glomeruli*, or filtration membranes of the kidney, causing inflammation of the kidney (*nephritis*) accompanied by severe kidney pain. Within several days, these complexes are usually dissolved and the patient quickly returns to normal. However, as in the case of rheumatic fever, timely, adequate antibiotic treatment of the initial *S. pyogenes* infection prevents the disease from occurring.

The other major pathogenic streptococcal species, *Streptococcus pneumoniae,* causes lung infections that often develop as secondary infections to other respiratory disorders. A characteristic of *S. pneumoniae* is that cells are typically present in pairs (or short chains) and are surrounded by a large capsule (Figure 23.5). The capsule enables the cells to resist phagocytosis; capsulated strains of *S. pneumoniae* are very invasive. Cells invade alveolar tissues (lower respiratory tract) of the lung and elicit a strong host inflammatory response. Reduced lung function can result from accumulation of phagocytic cells and fluid, and the *S. pneumoniae* cells can spread from the focus of infection as a bacteremia, sometimes resulting in bone infections, inner ear infections, and endocarditis. Pneumococcal pneumonia is a serious infection, untreated cases having a mortality rate of about 30%. At this time, most strains of *S. pneumoniae* and *S. pyogenes* respond dramatically to penicillin therapy. However, although penicillin remains the clinical drug of choice, an increasing incidence of penicillin-resistant streptococci has been reported in recent years, and erythromycin is the next best therapeutic agent.

The potential for serious host damage following a streptococcal sore throat has encouraged the development of clinical methods for rapid identification of *Streptococcus pyogenes*. At least two immunological

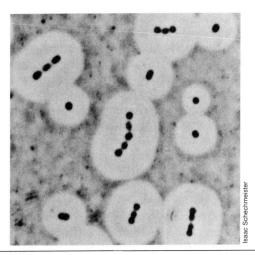

Figure 23.5 India ink negatively stained preparation of cells of *Streptococcus pneumoniae*. Note the extensive capsule surrounding the cells.

techniques have been developed thus far that can detect surface proteins unique to *S. pyogenes* by latex bead agglutination or by fluorescent antibody staining (∞ Sections 21.5 and 21.7) directly from a patient throat swab. In addition, a nucleic acid probe test (∞ Section 21.10) is available for detecting deoxyribonucleic acid (DNA) sequences unique to *S. pyogenes* in host tissues. Such procedures allow the confident initiation of immediate antibiotic therapy in order to avoid complications such as rheumatic fever. However, the definitive diagnostic confirmation of infection by pathogenic streptococci is a positive culture from the throat (*S. pyogenes*) or the sputum (*S. pneumoniae*).

CONCEPT CHECK   23.2a

Two respiratory diseases caused by streptococci are streptococcal sore throat and pneumococcal pneumonia. Under certain conditions, simple *Streptococcus pyogenes* infections can develop into more serious conditions such as scarlet fever and rheumatic fever. Pneumonia caused by *Streptococcus pnemoniae* is always a serious disease.

✔ *Why does* Streptococcus pyogenes *infection occasionally cause rheumatic fever?*

✔ *What is the primary virulence factor for* Streptococcus pneumoniae?

Staphylococcus

The genus *Staphylococcus* contains common pathogens of humans and animals, including some that are occasionally serious pathogens. Staphylococci are gram-positive cocci that divide in several planes to form irregular clumps (∞ Section 16.24 and Figure 16.82).

Staphylococci are relatively resistant to drying and hence can be readily dispersed in dust particles through the air. In humans, two species are important: *Staphylococcus epidermidis*, a nonpigmented, nonpathogenic form usually found on the skin or mucous membranes, and *S. aureus*, a yellow-pigmented form associated with pathological conditions, including boils, pimples, and impetigo (Figures 23.3 and 23.6), pneumonia, osteomyelitis, carditis, meningitis, and arthritis.

Those strains of *Staphylococcus aureus* most frequently causing human disease produce a number of extracellular enzymes or toxins (∞ Sections 19.8 and 19.9). At least four different *hemolysins* have been recognized, a single strain often being capable of producing more than one hemolysin. The production of these is responsible for the hemolysis seen around colonies on blood agar plates. *S. aureus* is also capable of producing an *enterotoxin*, commonly associated with foodborne illness (∞ Sections 19.9, 20.16, and 23.13).

Another substance produced by *Staphylococcus aureus* is *coagulase*, an enzymelike factor that causes fibrin to coagulate and form a clot (∞ Section 19.8). The production of coagulase is generally associated with pathogenicity. It seems likely that clotting induced by coagulase results in the accumulation of fibrin around the bacterial cells and makes them resistant to phagocytosis (Figure 23.6). In addition, the formation of such fibrin clots results in isolation of the infected area, making it difficult for host defense agents to come into contact with the bacteria. Most *S. aureus* strains also produce *leukocidin,* which causes the destruction of leukocytes, allowing the *S. aureus* cells to escape phagocytosis unharmed. Production of leukocidin in skin lesions

such as boils and pimples results in much cell destruction and is one of the factors responsible for pus formation (Figure 23.6). Other extracellular factors produced by some strains of *S. aureus* include proteolytic enzymes, hyaluronidase, fibrinolysin, lipase, ribonuclease, and deoxyribonuclease.

The most common habitat of *Staphylococcus aureus* is the upper respiratory tract, especially the nose and throat, as well as the surface of the skin. Many healthy people are carriers of this organism. Most infants become infected during the first week of life from the mother or from another close human contact. In most cases, these strains do not cause disease. Serious staphylococcal infections occur only when the resistance of the host is low because of hormonal changes, debilitating illness, wounds to the skin's surface, or treatment with steroids or other anti-inflammatory drugs. Hospital epidemics have occurred in recent years (∞ Section 22.7), which have usually involved antibiotic-resistant strains. Extensive use of antibiotics has resulted in the natural selection of resistant strains of *Staphylococcus aureus*. Hospital epidemics with these antibiotic-resistant staphylococci often occur in patients whose resistance to infection is lowered due to other diseases, surgical procedures, or drug therapy. These patients acquire staphylococcal cells from hospital personnel, who are often normal carriers of antibiotic-resistant strains. Control of such hospital epidemics requires careful attention to the maintenance of asepsis.

Certain strains of *Staphylococcus aureus* have been implicated as the agents responsible for **toxic shock syndrome (TSS),** a severe result of staphylococcal infec-

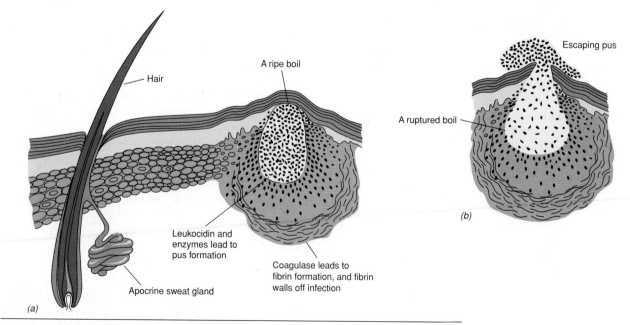

Figure 23.6 The structure of a boil. (a) Staphylococci initiate a localized infection of the skin and become walled off by coagulated blood and fibrin through the action of coagulase. (b) The rupture of the boil releases pus and bacteria.

tion characterized by high fever, rash, vomiting, diarrhea, and occasionally death. Toxic shock has occurred in menstruating women and has been associated with use of tampons. In addition, several cases of toxic shock have been reported in both men and women from staphylococcal infections following surgery. In females, blood and mucus in the vagina become colonized by hemolytic *S. aureus* from the skin, and the presence of a tampon concentrates this material, creating ideal microbial growth conditions.

The symptoms of TSS result indirectly from an exotoxin called *toxic shock toxin*. This toxin is a superantigen (∞ Section 20.16). The exotoxin is released by the growing staphylococci, causing the massive T cell reaction and inflammatory response characteristic of superantigen reactions. Since 1981, the incidence of TSS has declined sharply, primarily because of changes in the absorbent materials used in tampons, proper use of superabsorbent tampons, and public awareness of TSS symptoms. However, there is increasing evidence that TSS may also be caused by different exotoxins and Bacteria, including *Streptococcus pyogenes* (see earlier discussion).

Staphylococcal enterotoxin A, which causes the most prevalent form of food poisoning in the United States, is also a superantigen. Presumably, after ingestion of toxin-contaminated food, the toxin stimulates T cells localized along the intestine, resulting in a massive T cell response, release of mediators, and increased permeability of the intestine. The final outcome is the severe but short-lived diarrhea and vomiting associated with this type of food poisoning (see Section 23.13).

Finally, appropriate antibiotic therapy for *Staphylococcus aureus* infections is a problem. Although some community-acquired infections are treatable with penicillin, nosocomial (hospital-acquired) infections due to *S. aureus* are often drug-resistant (∞ Section 22.7). Therefore, disease-producing isolates of *S. aureus* must be individually checked for antibiotic sensitivity.

CONCEPT CHECK   23.2b

Although staphylococci are usually harmless inhabitants of the upper respiratory tract and skin, several serious diseases can result from infection, including some caused by staphylococcal toxins that act as superantigens.

✔ *Distinguish between an* infection *and an* intoxication.

✔ *Describe the action of a superantigen.*

Diphtheria

Corynebacterium diphtheriae, the causative agent of diphtheria, is a gram-positive irregular rod (∞ Section 16.29). *Corynebacterium diphtheriae* enters the body via the respiratory route with cells lodging in the throat

and tonsils. Although limited information is available concerning the mechanism of adherence of *C. diphtheriae* to these tissues, the organism produces a neuraminidase capable of splitting *N*-acetylneuraminic acid (a component of glycoproteins found on animal cell surfaces), and this may enhance the invasion process. The inflammatory response of throat tissues to *C. diphtheriae* infection results in formation of a characteristic lesion called a *pseudomembrane* (Figure 23.7), which consists of damaged host cells and cells of *C. diphtheriae*. As described in Section 19.8, certain strains of *C. diphtheriae* are lysogenized by bacteriophage β, and these strains produce a powerful exotoxin, the *diphtheria toxin*. Diphtheria toxin inhibits eukaryotic protein synthesis and thus kills cells (∞ Sections 15.8 and 19.8).

The pseudomembrane that forms in diphtheria may block the passage of air, and death from diphtheria is usually due to a combination of the effects of partial suffocation and tissue destruction by exotoxin. Although diphtheria was once a major childhood disease, it is now rarely encountered because an effective vaccine is available. This vaccine is made by treating the diphtheria exotoxin with formalin to yield an immunogenic, yet nontoxic, toxoid (see Section 20.17 for a general discussion of toxoids).

Diphtheria toxoid is part of the *DTP* (*d*iphtheria, *t*etanus, *p*ertussis) vaccine, administered several times in the first year of life (∞ Section 20.17). A patient diagnosed as having diphtheria by culture of *Corynebacterium diphtheriae* from a pseudomembrane in the throat is usually treated simultaneously with antibiotics and diphtheria antitoxin (an antitoxin contains neutralizing antibodies formed in another animal) (see Section 20.17 for a discussion of antitoxins). Penicillin, erythromycin, or gentamicin is generally effective for diphtheria therapy. Early administration of both antibiotics and antitoxin is necessary for effective control of the disease.

CONCEPT CHECK   23.2c

Diphtheria is caused by the gram-positive bacterium *Corynebacterium diphtheriae*. A standard early childhood vaccine (DTP) is very effective in preventing this very serious respiratory disease.

✔ *Is the pathogenesis of diphtheria due to infection?*

Legionellosis (Legionnaires' disease)

This disease, caused by the organism *Legionella pneumophila*, derives its name from the fact that it was first recognized as a disease entity from an outbreak of pneumonia occurring during a convention of the American Legion in the summer of 1976. *Legionella* is a thin gram-negative rod (Figure 23.8) with complex nutritional requirements, including an unusually high iron require-

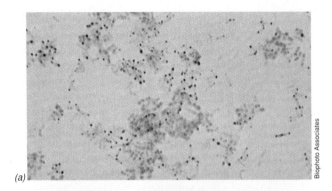

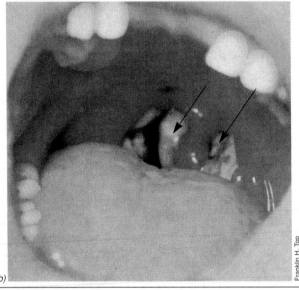

Figure 23.7 Diphtheria. (a) Cells of *Corynebacterium diphtheriae* stained to show metachromatic (polyphosphate) granules. (b) Pseudomembrane (arrows) in an active case of diphtheria caused by the bacterium *C. diphtheriae.*

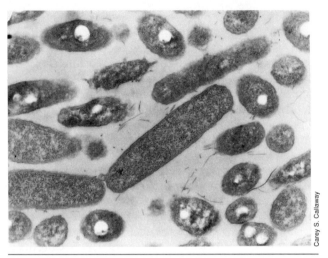

Figure 23.8 Transmission electron micrograph of *Legionella pneumophila.* Cells are approximately 0.6 μm in diameter.

ment, and is immunologically distinct from any other pathogen associated with respiratory infections. *Legionella* can be detected by immunofluorescence techniques (∞ Section 21.7 and Figure 21.14*a*) and can be isolated from many terrestrial and aquatic habitats as well as from patients suffering from legionellosis. Epidemiological studies have shown that *Legionella* is a common inhabitant of cooling towers of air conditioning units and that infectious *Legionella* aerosols from such sources can spread the organisms to humans. Curiously, although apparently spread via an airborne route, no evidence of direct person-to-person transmission of *Legionella* has been obtained. Consistent with these findings is the fact that cases of legionellosis tend to peak in midsummer to late summer months when air conditioners are most extensively used. This is in contrast to an airborne disease such as chickenpox, which is spread from person to person and peaks in the winter months (see Figure 23.21) when people are more frequently indoors and in close contact. More recently, a trend toward late fall–early winter peaks of Legionnaires' dis-

ease has been observed, and if this continues, it may suggest alternative vehicles for transmission of the disease. Overall, the incidence of legionellosis has been steadily increasing (Figure 23.9). However, up to 90% of actual cases may go unreported.

Legionella infections may be totally asymptomatic and occasionally result in only mild symptoms such as headache and fever. The majority of cases of *Legionella* pneumonia are in elderly individuals whose resistance has been previously compromised. In addition, certain serotypes of *Legionella* (eight are known) are more strongly associated with the pneumonic form of the illness than others. Prior to the onset of pneumonia, intestinal disorders are common, followed by high fever, chills, and muscle aches. These symptoms precede the dry cough and chest and abdominal pains typical of legionellosis. Death, if it occurs, is usually due to respiratory failure. Clinical detection of *Legionella pneumophila* is now straightforward because of fluorescent antibodies and other highly specific immunological probes (∞ Figure 21.14*a*). *Legionella pneumophila* is sensitive to the antibiotics rifampicin and erythromycin, and intravenous administration of erythromycin is the treatment of choice in most cases.

CONCEPT CHECK   *23.2d*

Legionellosis incidence is on the rise. This may reflect increased familiarity and precision in diagnosing the disease, but *Legionella* is becoming a more prevalent respiratory pathogen in artificial microenvironments like air-conditioned buildings.

✔ *Classify* Legionella pneumophila *as a primary respiratory pathogen or an opportunistic pathogen. Give reasons for your answer.*

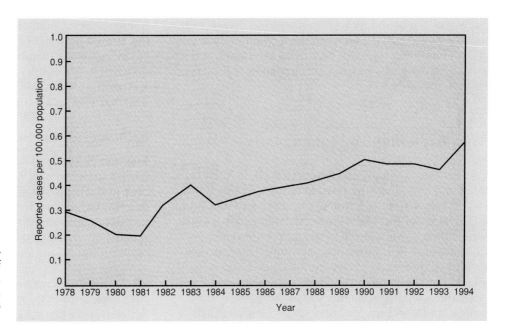

Figure 23.9 Annual incidence of legionellosis in the United States. Note the general upward trend. In 1994 the total number of cases of legionellosis was 1615.

Whooping cough

Whooping cough (pertussis) is an acute, highly infectious respiratory disease generally observed in children under 1 year of age. Whooping cough is caused by a small, gram-negative, strictly aerobic coccobacillus, *Bordetella pertussis*. The organism attaches to cells of the upper respiratory tract by producing a specific adherence factor called *filamentous hemagglutinin antigen*, which recognizes a complementary molecule on the surface of host cells. Once attached, *B. pertussis* grows and produces pertussis exotoxin that induces synthesis of cyclic adenosine monophosphate (cyclic AMP) (∞ Figure 7.19), which is at least partially responsible for the events that lead to host tissue damage. *Bordetella pertussis* also produces an endotoxin, which also may induce some of the symptoms of whooping cough. Clinically, whooping cough is characterized by a recurrent, violent cough that usually lasts up to 6 weeks. The spasmodic coughing gives the disease its name, for a whooping sound results from the patient inhaling in deep breaths to obtain sufficient air.

A vaccine consisting of killed whole cells of *Bordetella pertussis* is part of the routinely administered DTP vaccine (∞ Section 20.17). This vaccine, while normally very effective, must be given to susceptible individuals, usually children, at appropriate intervals beginning soon after birth (∞ Section 20.17). In the United States, up to 50% of children who acquire whooping cough have not been properly immunized. The threat of this very communicable disease remains high, as illustrated by epidemic outbreaks in several locations in the United States (∞ Section 22.10) and in the former Soviet Union, presumably as a result of recent breakdowns in public health programs.

Because of undesirable side effects of pertussis vaccine, including local swelling and redness, fever, and occasional more serious problems such as encephalitis and convulsions, a "second generation" pertussis vaccine, containing purified cell fractions of *Bordetella pertussis* rather than whole cells, is now licensed for use in the United States and is already in wide use in Japan.

Diagnosis of whooping cough can be made by fluorescent antibody staining of throat smears or by actually culturing the organism. For culture of *Bordetella pertussis*, the "cough-plate" method is used. The patient is asked to cough directly into a blood–glycerol–potato extract agar plate (although not selective, this medium supports good recovery of *B. pertussis*). Alternatively, throat and nose swabs (if streaked immediately after sampling) can be used. β-Hemolytic colonies containing small gram-negative coccobacilli are tested for *B. pertussis* by a latex bead agglutination test or are stained with an anti-*B. pertussis* fluorescent antibody for positive identification (∞ Sections 21.5 and 21.7). Cultures of *B. pertussis* are killed by ampicillin, tetracycline, and erythromycin, although antibiotics alone do not seem to be sufficient to kill the pathogen *in vivo*. Because a

---

CONCEPT CHECK   *23.2e*

There has been a disturbing increase in the number of annual cases of whooping cough in the last decade. From 1975 to 1982, there was an average of less than 2000 cases per year, but that number has risen to more than 4000 cases per year at present. Up to 60% of preschool children in some cities are inadequately immunized, creating a major potential public health threat.

✔ *What measures can be taken to decrease the current incidence of whooping cough in a population?*

patient with whooping cough remains infectious for up to 2 weeks following commencement of antibiotic therapy, the immune response may be as important, if not more so, than antibiotics, in the elimination of *B. pertussis* from the body.

23.3

Mycobacterium and Tuberculosis

The famous German microbiologist Robert Koch isolated and described the causative agent of tuberculosis, *Mycobacterium tuberculosis,* in 1882 (see the box, Discoverers of the Main Bacterial Pathogens). At one time, tuberculosis was the single most important infectious disease of humans and accounted for one-seventh of all deaths. At present in the United States, over 20,000 new cases of tuberculosis are diagnosed each year and about 2000 deaths occur. Worldwide, tuberculosis still accounts for more than 3 million deaths per year, and up to one-third of the world's population have been infected with *M. tuberculosis* (∞ Table 22.1).

In recent years, many of the new tuberculosis cases in the United States result at least in part from the elevated incidence of tuberculosis in acquired immunodeficiency syndrome (AIDS) patients. After influenza and pneumonia, tuberculosis is the leading cause of death by infectious disease in the United States.

Pathology of tuberculosis

The microbiology of *Mycobacterium tuberculosis* is discussed in Section 16.31. The interaction of the human host and *M. tuberculosis* is extremely complex, being determined in part by the virulence of the strain but probably more importantly by the specific and nonspecific resistance of the host. Cell-mediated immunity plays an important role in the development of disease symptoms. It is convenient to distinguish between two kinds of human tuberculosis infections: *primary* and *postprimary* (or reinfection). Primary infection is the first infection that an individual acquires and usually results from inhalation of droplets containing viable bacteria from an individual with an active pulmonary

---

LEARNING FROM THE PAST

DISCOVERERS OF THE MAIN BACTERIAL PATHOGENS

The history of the discovery of the microbial role in infectious disease was described briefly in Chapter 1. Once the concept of specific microbial disease agents was clarified and the procedures for culture of microorganisms developed, it was a relatively simple procedure to isolate a large number of microbial pathogens. The two decades after the formulation of Koch's postulates were indeed fruitful for medical microbiology. The rapid development of this field is indicated by the accompanying table, which lists the main bacterial pathogens isolated during the "golden age of bacteriology." ∎

Year	Disease	Organism	Discoverer
1877	Anthrax	*Bacillus anthracis*	Koch, R.
1878	Suppuration	*Staphylococcus*	Koch, R.
1879	Gonorrhea	*Neisseria gonorrhoeae*	Neisser, A. L. S.
1880	Typhoid fever	*Salmonella typhi*	Eberth, C. J.
1881	Suppuration	*Streptococcus*	Ogston, A.
1882	Tuberculosis	*Mycobacterium tuberculosis*	Koch, R.
1883	Cholera	*Vibrio cholerae*	Koch, R.
1883	Diphtheria	*Corynebacterium diphtheriae*	Klebs, T. A. E.
1884	Tetanus	*Clostridium tetani*	Nicolaier, A.
1885	Diarrhea	*Escherichia coli*	Escherich, T.
1886	Pneumonia	*Streptococcus pneumoniae*	Fraenkel, A.
1887	Meningitis	*Neisseria meningitidis*	Weichselbaum, A.
1888	Food poisoning	*Salmonella enteritidis*	Gaertner, A. A. H.
1892	Gas gangrene	*Clostridium perfringens*	Welch, W. H.
1894	Plague	*Yersinia pestis*	Kitasato, S., Yersin, A. J. E. (independently)
1896	Botulism	*Clostridium botulinum*	van Ermengem, E. M. P.
1898	Dysentery	*Shigella dysenteriae*	Shiga, K.
1900	Paratyphoid	*Salmonella paratyphi*	Schottmüller, H.
1903	Syphilis	*Treponema pallidum*	Schaudinn, F. R., and Hoffman, E.
1906	Whooping cough	*Bordetella pertussis*	Bordet, J., and Gengou, O.

infection. Dust particles that have become contaminated from sputum of tubercular individuals are another source of primary infection. The bacteria settle in the lungs and grow. A delayed-type hypersensitivity reaction (∞ Sections 20.9 and 20.16) results in the formation of aggregates of activated macrophages, called *tubercles*, characteristic of tuberculosis. However, the bacteria are often able to survive and grow to some extent within the macrophages. In a few individuals with low resistance, the bacteria are not effectively controlled, and an acute pulmonary infection occurs, which can lead to the extensive destruction of lung tissue, the spread of the bacteria to other parts of the body, and death.

In most cases of tuberculosis, however, acute infection does not occur, and the infection remains localized and is usually inapparent; later it subsides. But this initial infection hypersensitizes the individual to the bacteria or their products and consequently alters the response of the individual to subsequent *M. tuberculosis* exposures. A diagnostic test, called the **tuberculin test,** can be used to measure this hypersensitivity. When *tuberculin,* a protein fraction extracted from *Mycobacterium tuberculosis,* is injected intradermally into a hypersensitive individual, it elicits a localized immune reaction within 1–3 days at the site of injection. The reaction is characterized by *induration* (hardening) and *edema* (swelling) (∞ Figure 20.31). An individual exhibiting this reaction is said to be *tuberculin-positive,* and many healthy adults give positive reactions as a result of previous inapparent infections. A positive tuberculin test does not indicate active disease but only that the individual has been exposed to the organism at some time.

It is in tuberculin-positive individuals that the postprimary type of tuberculosis infection can occur. When renewed pulmonary infections occur in tuberculin-positive individuals, they are usually chronic infections that involve destruction of lung tissue, followed by partial healing and a slow spread of the lesions within the lungs. Spots of destroyed tissue may be revealed by X-ray examination (Figure 23.10), but viable bacteria are found in the sputum only in individuals with extensive tissue destruction. In many cases, symptoms in tuberculin-positive individuals are a result of reactivation and growth of bacteria that have remained alive and dormant in the lungs for long periods of time. Malnutrition, overcrowding, stress, and hormonal imbalance often are factors predisposing an individual to reinfection.

Control and treatment

Individuals who have active cases of tuberculosis may spread the disease simply by coughing on uninfected individuals. Because tuberculosis is so highly contagious, the United States Occupational Safety and Health Administration has stringent requirements for the protection of health care workers who are responsible for

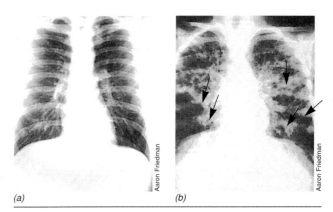

(a)   *(b)*

Figure 23.10 X-ray photographs. (a) Normal chest X-ray. The faint white lines are arteries and other blood vessels. The heart is visible as a white bulge in the lower right quadrant. (b) An advanced case of pulmonary tuberculosis; white patches (arrows) indicate areas of disease. These patches, or tubercles, may contain live *Mycobacterium tuberculosis*. Lung tissue and function in these lesions is permanently destroyed.

tuberculosis patient care. For example, patients with infectious tuberculosis must be hospitalized in negative-pressure rooms. In addition, health care workers who have patient contact must be provided with personally fitted face masks with high energy particulate air (HEPA) filters. These special filters prevent the passage of *Mycobacterium tuberculosis* in sputum or on dust particles.

Chemotherapy of tuberculosis has been a major factor in control of the disease. The initial success in chemotherapy occurred with the introduction of streptomycin, but the real revolution in tuberculosis treatment came with the discovery of isonicotinic acid hydrazide (isoniazid or INH) (Figure 23.11), a nicotinamide derivative virtually specific for mycobacteria. This agent is not only effective and free from toxicity but is also inexpensive and readily absorbed when given orally. Although the mode of action of INH is not completely understood, it apparently affects in some way the synthesis of mycolic acid by *Mycobacterium* (mycolic acid is a complex lipid that complexes with the peptidoglycan of the mycobacterial cell wall) (∞ Section 16.31). INH may act by mimicking the activity of a structurally related molecule, nicotinamide (Figure 23.11), becoming incorporated in place of nicotinamide and thus inactivating enzymes requiring this compound for activity. Treatment of

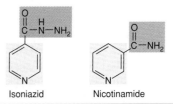

Isoniazid   Nicotinamide

Figure 23.11 Structure of isoniazid (isonicotinic acid hydrazide), an effective chemotherapeutic agent for tuberculosis. Note the structural similarity to nicotinamide.

mycobacteria with very small amounts of INH (as little as 5 picomoles (pmol) per 10^9 cells) results in complete inhibition of mycolic acid synthesis, and continued incubation results in a complete loss of outer membrane areas of the cell, a loss of cellular integrity, and death. Following treatment with INH, mycobacteria lose their acid–alcohol fastness, in keeping with the role of mycolic acid in this staining property (∞ Section 16.31). However, mycobacterial resistance to INH and other drugs is increasing at an alarming rate, especially in AIDS patients (see Section 23.7).

The development of drug resistance may be encouraged by patients themselves because many patients do not complete their therapy. The typical course of INH treatment is 12 months, and patients must take daily medication throughout this time to eradicate the tubercle bacilli. Failure to complete the entire prescribed treatment may result in resurgence of the infection, and the resurgent organisms often are resistant to the original treatment drug. In certain populations, such as in hospitals and nursing homes, patients are routinely treated with several drugs simultaneously because of the emergence and prevalence of multiple-drug-resistant tuberculosis.

Other pathogenic mycobacteria

Mycobacterium bovis is also pathogenic for humans as well as other animals. A common pathogen of dairy cattle, *M. bovis* enters humans via the intestinal tract, typically from the ingestion of raw milk. After a localized intestinal infection, the organism eventually spreads to the respiratory tract and initiates the classic symptoms of tuberculosis. The question as to whether *M. bovis* is really a different organism from *Mycobacterium tuberculosis* is unclear because of the nearly 100% homology observed in hybridization of the two organisms' DNA. Pasteurization of milk and elimination of diseased cattle have essentially eradicated bovine-to-human transmission of tuberculosis.

Mycobacterium leprae is the causative agent of the ancient and dreaded disease *leprosy* (Hansen's disease). Unfortunately, *M. leprae* has never been grown on artificial media. It can be grown in mice, but the typical human symptoms of leprosy are not observed. The only experimental animal that has been successfully used is the armadillo. The symptoms of leprosy are the characteristic folded, bulblike lesions on the body, especially on the face and extremities (Figure 23.12) due to growth of *M. leprae* cells in the skin. In severe cases the disfiguring lesions lead to destruction of peripheral nerves and loss of motor function. Leprosy can be treated with the drug dapsone (4,4'-sulfonylbisbenzeneamine). As in the case of tuberculosis, extended drug therapy is required to effect a cure.

Despite ancient myths, leprosy is not a highly contagious disease. Little is known of the pathogenicity of *M. leprae* or even of its mode of transmission, although

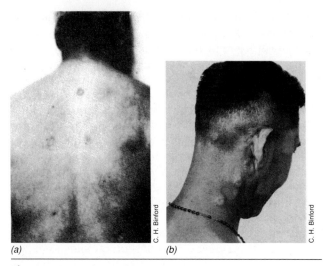

Figure 23.12 Leprosy lesions on the skin due to infection with *Mycobacterium leprae*.

it seems likely that it is transmitted by direct contact. The bacterium grows within macrophages, causing an intracellular infection that can result in an enormous population of bacteria within the skin. In many areas of the world the incidence of leprosy is very low, although in ancient times it was apparently much more common, perhaps due to crowding and poor sanitation. In the United States, fewer than 300 cases of leprosy are diagnosed per year. In other areas, such as tropical areas, the incidence is much higher; leprosy remains a medical problem for about 15 million people worldwide.

CONCEPT CHECK   **23.3**

Tuberculosis is one of the most prevalent single diseases in the world. Its incidence is on the increase in developed countries, in part because of the emergence of drug-resistant pathogens. The pathology of tuberculosis and leprosy is largely the result of delayed-type hypersensitivity reactions.

✔ *Why is* Mycobacterium tuberculosis *such a widespread respiratory pathogen?*

✔ *What factors aid in controlling the spread of tuberculosis?*

23.4

Respiratory Infections: Viral

As we discussed (∞ Section 11.11), viruses, by their very nature, are less easily controlled by chemotherapeutic means than bacteria or other microorganisms. Since the growth of viruses is intimately tied to host cell functions, it is difficult to specifically attack viruses with chemotherapeutic agents without causing at least some harm to host cells as well. Not surprisingly, therefore, the most prevalent infectious diseases today are of viral etiology (Figure 23.13). On the other hand, many viral diseases are acute, self-limiting infections that are rarely fatal in normal healthy adults. In addition, serious viral diseases such as smallpox and rabies have been effectively controlled by immunization. We begin here by describing the two most common viral infections, the common cold and influenza, and proceed to discuss measles, mumps, and chickenpox; these viral diseases are all transmitted in infectious droplets by an airborne route.

The common cold

The common cold is one of the most prevalent diseases of children and adults. The symptoms include rhinitis (inflammation of the nasal region, especially the mucous membranes), nasal obstruction, watery nasal discharges, and a general feeling of malaise. Rhinoviruses [single-stranded ribonucleic acid (RNA)] viruses of the picornavirus group) (see Figure 23.14a and Section 8.15), are the dominant etiological agents of the common cold. Over 100 different serotypes of rhinoviruses are known, and hence immunity to the common cold via vaccination or previous exposure is not to be expected. Another group of single-stranded RNA viruses, the coronaviruses (Figure 23.14b), are responsible for about 15% of all colds in adults. A variety of other viruses including adenoviruses, coxsackie viruses, respiratory syncytial virus, and orthomyxoviruses, are responsible for about 10% of common colds.

Although airborne transmission of the common cold virus is suspected of being a major means of spreading the infection, transmission experiments with human volunteers suggest that direct contact and/or fomite contact is also an important means of transmission. In fact, one effective experimental method for the prevention of rhinovirus spread is through the use of disposable tissues impregnated

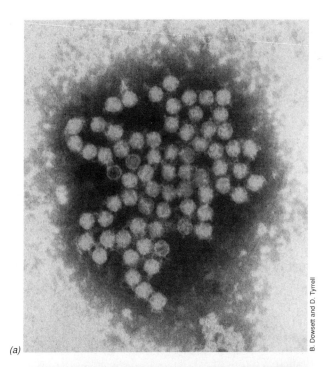

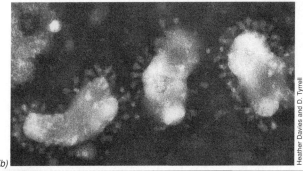

Figure 23.14 Electron micrographs of some common cold viruses. (a) Human rhinovirus. (b) Human coronavirus. Each rhinovirus virion is about 30 nm in diameter. Each coronavirus virion is about 60 nm in diameter.

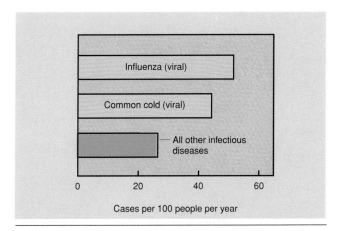

Figure 23.13 Viruses are the leading causes of acute illness in the United States. The data are typical of recent years.

with antiviral disinfectant agents. Most antiviral drugs are ineffective, but a pyrazidine derivative (Figure 23.15a) has proven promising for preventing colds in volunteers. In addition, new experimental antiviral drugs are being designed based on information derived from three-dimensional structures. For example, the antirhinovirus drug WIN 52084 (Figure 23.15b) binds to the virus and disrupts a cellular binding site, thus preventing infection. Interferon-α, a cytokine (∞ Section 20.8), is also effective in preventing the onset of colds. Monoclonal antibodies (∞ Section 20.14) offer the possibility of binding and blocking cell receptor sites on the virus. Thus, there are several experimental possibilities for cold prevention and treatment, although none are widely accepted as effective and safe. The accepted treatment for colds is to treat the symptoms, especially nasal discharges, with a variety of antihistamine and decongestant drugs.

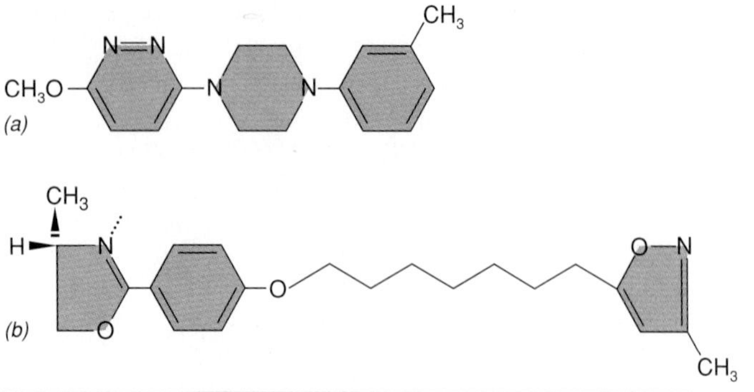

Figure 23.15 Experimental antirhinovirus drugs. (a) The structure of 3-methoxy-6-[4-(3-methylphenyl)]-1-piperazinyl. (b) The structure of WIN 52084, a receptor-blocking drug.

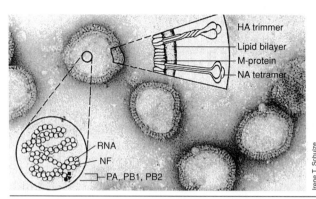

Figure 23.16 Electron micrograph of the influenza virus, showing the location of the major viral coat proteins and the nucleic acid. Each virion is about 100 nm in diameter. HA, Hemagglutinin (three copies make up the HA coat spike); NA, neuraminidase (four copies make up the NA coat spike); M, coat protein; NP, nucleoprotein; PA, PB1, PB2, other internal proteins, some of which may have enzymatic functions.

Influenza

Influenza is caused by an RNA virus of the orthomyxovirus group (∞ Section 8.16). Influenza virus is an enveloped virus, the RNA genome being surrounded by an envelope made up of protein, a lipid bilayer, and external glycoproteins (see Figures 23.16 and 8.38 and Table 8.3). Human influenza virus exists in nature only in humans. It is transmitted from person to person through the air, primarily in droplets expelled during coughing and sneezing. The virus infects the mucous membranes of the upper respiratory tract and occasionally invades the lungs. Symptoms include a low grade fever for 3–7 days, chills, fatigue, headache, and general aching (see the box, Is It a Cold or Is It the Flu?) Recovery is usually spontaneous and rapid. Most of the serious consequences of influenza infection do not occur because of the viral infection but because bacterial invaders may be able to develop as secondary infections

in persons whose resistance has been lowered. Especially in infants and elderly people, influenza is often followed by bacterial pneumonia; death, if it occurs, is usually due to the bacterial infection.

Influenza often occurs in pandemics. Early pandemics, of which the one in 1918 is the most famous, occurred before knowledge was sufficiently advanced to make careful analysis possible, but the 1957 pandemic of the so-called Asiatic flu provided an opportunity for careful study of how a worldwide epidemic develops (Figure 23.17). The epidemic probably arose when a virulent mutant virus strain that differed from all previous strains in antigenicity appeared in the population. Since immunity to this strain was not present, the virus was able to advance rapidly

---

A FOCUS ON . . .

IS IT A COLD OR IS IT THE FLU?

The symptoms of a common cold and symptoms of "the flu" (influenza) often seem similar, but the two diseases are distinct and caused by quite different viruses. A typical common cold caused by a rhinovirus is associated with nasal discharges, cough, chills, and perhaps a sore throat. Influenza, caused by an orthomyxovirus, is generally associated with a different set of symptoms. Although either condition may make one feel miserable for a period, colds are usually of shorter duration and the symptoms are milder. The following can serve as a guideline for determining whether you have "caught a cold" or "caught the flu": ∎

Symptoms	Common cold	Influenza
Fever	Rare	Common (39–40°C); sudden onset
Headache	Rare	Common
General malaise	Slight	Common; often quite severe; can last several weeks
Nasal discharge	Common and abundant	Less common; usually not abundant
Sore throat	Common	Much less common
Vomiting and/or diarrhea	Rare	Common

throughout the world. It first appeared in the interior of China in late February 1957 and by early April had been brought to Hong Kong by refugees. It spread from Hong Kong along air and naval routes and was apparently transferred to San Diego, California, by naval ships. In May, an outbreak occurred in Newport, Rhode Island, on a naval vessel. Other outbreaks occurred in various parts of the United States. Peak incidence occurred in the last 2 weeks of October, during which time 22 million new cases developed. Afterward, there was a progressive decline.

The genetic material of influenza virus, single-stranded RNA, is arranged in a highly unusual manner. As discussed in Section 8.16, the influenza virus genome is *segmented*, with genes found on each of eight distinct fragments of its single-stranded RNA (∞ Figure 8.38 and Table 8.3). Such an arrangement allows the rapid and constant reassortment of genes with genes from a different strain of influenza virus because more than one strain of influenza virus can infect a cell at one time. This allows reassorted viruses to arise at frequent intervals and is probably responsible for the generally unsuccessful attempts to completely control influenza by vaccination. This reassortment of genes in different strains of influenza virus usually manifests itself in the phenomenon called *antigenic shift*. Antigenic shift refers to modifications in the protein coat of the virions, especially to two proteins important in the attachment and eventual release of virus from host cells, hemagglutinin and neuraminidase, respectively (∞ Figures 8.38 and 23.16). Immunity to

influenza in humans is largely dependent on the production of secretory antibody (IgA) (∞ Section 20.5), especially to antigenic determinants of the hemagglutinin and neuraminidase proteins.

Once a strain of influenza virus has passed through the population, a majority of the people are immune to that strain and it is impossible for a strain of similar antigenic type to cause an epidemic for about 3 years. During this time, the hemagglutinin and neuraminidase antigens exhibit frequent minor antigenic variation because of genetic mutations that result in the change of one or more amino acids. This phenomenon is known as *antigenic drift* and is responsible for the recurrence of minor epidemics of influenza in a 2- to 3-year cycle. There is some evidence that in 1918 (the year of the unusually serious *pandemic*, or worldwide epidemic), the strain may have originated from a related virus that infects swine (swine flu) and that the 1957 strain may have arisen from a similar animal reservoir (perhaps a wild animal) somewhere in Asia. Although a vaccine can be prepared to any strain, the large number of strains and the phenomena of antigenic shift and antigenic drift make it difficult to prevent influenza epidemics.

Control of influenza epidemics can be carried out by vaccination, but this is complicated by the fact that so many different virus strains may cause the disease. Obviously, the vaccine must be derived from the strain causing the epidemic. Vaccines prepared from several different virus strains can be mixed, producing what is called a *polyvalent* vaccine. When new strains develop, vaccines are not immediately available, but through

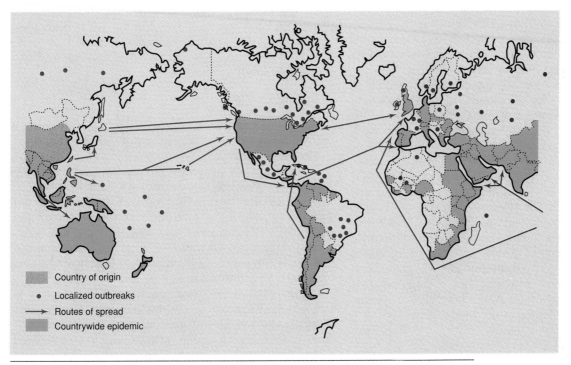

Figure 23.17 Route of spread of a major influenza epidemic, the Asian flu pandemic of 1957.

careful worldwide surveillance (∞ Section 22.10), it is usually possible to obtain samples of the major emerging strains of influenza virus *before* the disease reaches epidemic proportions. Vaccines can thus be produced prior to a potential epidemic. The strategy used in many countries is to recommend influenza vaccination only for those most likely to succumb to severe or fatal secondary illnesses, such as the aged and those suffering from chronic debilitating diseases, and to health care workers. The duration of effective artificial immunity for influenza is usually only for a few years, and it is strain-specific. Therefore, revaccination is necessary when a new epidemic occurs.

Influenza may also be controlled by use of the chemicals *amantadine* and *ramantadine*. These drugs have been used as chemoprophylactic agents to prevent the spread of influenza to those at high risk. They may also help to shorten the course of infection. The treatment of influenza symptoms with aspirin is not recommended, as there is evidence of a link between aspirin treatment of influenza and Reye's syndrome (a rare but occasionally fatal affliction involving the central nervous system) in children.

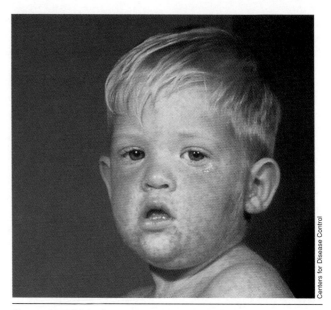

Figure 23.18 Typical rash associated with measles in children.

Measles

Measles (rubeola) virus causes an acute highly infectious childhood disease characterized by nasal discharges, redness of the eyes, cough, and fever. The measles virus is a paramyxovirus (∞ Section 8.16) and enters the body in the nose and throat by airborne transmission and quickly leads to systemic viremia. As the disease progresses, the fever and cough intensify and a rash appears (Figure 23.18); in most cases measles lasts a total of 7–10 days. Circulating antibodies to measles virus are measurable about 5 days after initiation of infection, and the activities of both serum antibodies and cytotoxic T lymphocytes (∞ Sections 20.5 and 20.9) combine to eliminate the virus from the system. As a consequence of measles, a variety of complications may occur, including inner ear infection, pneumonia, and, in rare cases, measles encephalomyelitis. Encephalomyelitis can cause neurological disorders and a form of epilepsy and has a mortality rate of nearly 20%.

Although once a common childhood illness, measles generally occurs nowadays in rather isolated outbreaks because of widespread vaccination programs begun in the mid-1960s (Figure 23.19a). In the United States, most school systems require proof of measles vaccination before allowing children to enroll because of the potential complications and highly infectious nature of measles. Active immunization is done with the MMR (measles, mumps, and rubella) vaccine (∞ Tables 20.10 and 20.11). A childhood case of measles generally confers lifelong immunity to reinfection.

Mumps

Mumps is caused by a different paramyxovirus than that causing measles but shares with the disease measles a highly infectious character. Mumps is

spread by airborne droplets, and the disease is characterized by inflammation of the salivary glands leading to swelling of the jaws and neck (Figure 23.20). The virus spreads through the bloodstream and may infect other organs including the brain, testes, and pancreas. The host immune response produces antibodies to mumps virus surface proteins, and this generally leads to a quick recovery. An attenuated mumps vaccine is highly effective in preventing the disease (Figure 23.19b). Hence, like measles, the incidence of mumps has also been greatly reduced in the last two decades, with mumps epidemics usually restricted to those individuals who did not receive the MMR vaccine during childhood.

Rubella

Rubella (*German measles*) is caused by a positive-strand RNA virus of the togavirus group (∞ Section 8.15). The symptoms of the disease resemble those of measles but are generally milder. German measles is less contagious than true measles, and thus a good proportion of the population has never been infected. However, during the first 3 months of pregnancy rubella virus can infect the fetus by placental transmission and cause a host of serious fetal abnormalities. Rubella can cause stillbirth, or deafness, heart and eye defects, and brain damage in live births. Thus, it is important that pregnant women not be exposed to, vaccinated against, or contract German measles during this period. For this reason, routine childhood vaccination against German measles should be practiced. An attenuated virus vaccine is administered with attenuated measles and mumps viruses in the MMR vaccine mentioned previously (see Figure 23.19c and Tables 20.10 and 20.11).

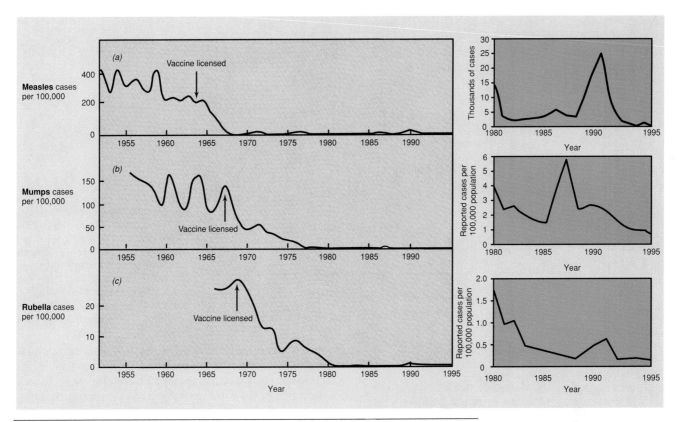

Figure 23.19 The effect of vaccines on the incidence in the United States of the major childhood viral diseases now controlled by the MMR (measles, mumps, rubella) vaccine. (a) Measles. (b) Mumps. (c) Rubella. The insets show a more detailed picture of these diseases in the past decade.

Chickenpox and shingles

Chickenpox (varicella) is a common childhood disease caused by a herpes virus (∞ Section 8.19). Chickenpox is highly contagious and is transmitted by infectious droplets, especially when susceptible individuals

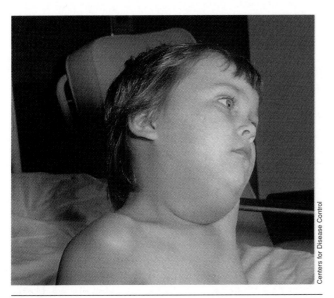

Figure 23.20 Typical glandular swelling associated with mumps.

are in close contact. The incidence of chickenpox shows a disease cycle typical of a respiratory infection (Figure 23.21). In schoolchildren, for example, close confinement during the winter months leads to the spread of chickenpox by airborne secretions from infected classmates and through contact with contaminated fomites. The virus enters the respiratory tract, multiplies, and is quickly disseminated via the bloodstream, resulting in a systemic papular rash that quickly heals, rarely leaving disfiguring marks (Figure 23.22). For unknown reasons, early chickenpox vaccines have been poorly immunogenic, but a successful and highly protective attenuated virus vaccine (∞ Table 20.10) has been marketed in Japan and a similar vaccine is available in the United States.

The chickenpox virus can remain dormant in nerve cells for years with no apparent symptoms. The virus occasionally migrates from this reservoir to the skin surface, causing a painful skin eruption referred to as **shingles** (zoster). Shingles most commonly strikes immunosuppressed individuals or the elderly. Studies with human volunteers suggest that T cells are important in destroying the virus. The prophylactic use of human hyperimmune globulin prepared against the virus is useful for preventing the onset of symptoms of shingles. Such therapy is advised only for patients where secondary infections occasionally associated with shingles, such as pneumonia or encephalitis, may be life-threatening.

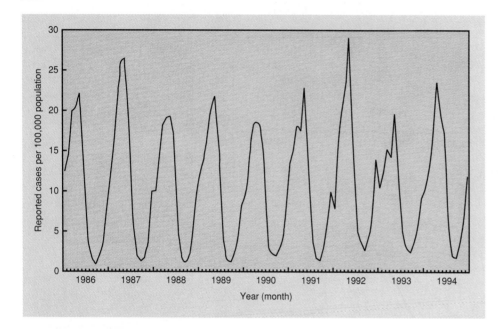

Figure 23.21 Reported incidence of chickenpox (varicella) by month in the United States, 1986–1994. Note the early spring seasonal incidence, typical of a disease transmitted by a respiratory airborne route. In 1994 there were 151,219 cases reported, which is estimated to be 4–5% of the total number of cases.

CONCEPT CHECK   23.4

Most respiratory infections are caused by viruses. The most prevalent is the common cold, and the most serious is influenza, which occurs in cyclical epidemics. These diseases all share a common mode of transmission, the airborne route.

✔ *List the viral respiratory diseases for which an effective vaccine is not available.*

✔ *Should antibiotics be used for treatment of viral diseases? Why or why not?*

23.5

Why Are Respiratory Infections So Common?

Respiratory infections have several common features. They are generally transmitted by *airborne droplets*. The infectious agents occur in finely dispersed form in air, on dust, or on the surfaces of fomites, and the diseases typically show marked seasonality. In addition, because they are easily transmitted by normal human activities, respiratory infections are the most common infectious diseases.

Respiratory infections are frequently difficult to control for several reasons. First, many respiratory infections are caused by viruses; antibiotics, and indeed most other therapeutic agents, are completely ineffective in controlling viral diseases. Second, most respiratory diseases are acute, showing a rapid onset of symptoms and a rapid recovery. Although this may seem a desirable outcome, in many ways it isn't. Patients suffering from the common cold, for example, are fre-

quently communicable before symptoms begin; chickenpox can spread in explosive fashion among schoolchildren for the same reason. Third, effective vaccines are not available for a number of respiratory diseases. This can be due to unusual biological properties of the pathogen. For example, the antigenic drift and genomic rearrangements of the influenza virus generate new strains (from an immunological standpoint) every few years. Also, unusually large numbers of genetic variants of a particular pathogen (for example, the more than 100 serotypes of rhinovirus, the agent responsible for most common colds) makes vaccine production very difficult. In addition, vaccines against certain respiratory pathogens, for example against *Streptococcus pyogenes*, might actually harm the host because of cross-reacting antigens on bacterial and host cell surfaces (∞ Section 20.16). Fourth, many respiratory infections are transmitted from healthy carriers to susceptible hosts. This is very common in the case of streptococcal

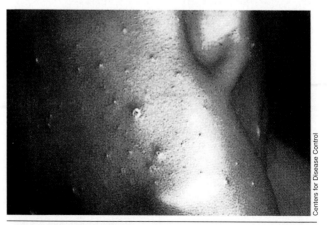

Figure 23.22 Mild papular rash associated with the disease chickenpox.

infections because about 20% of the adult population carry, apparently without ill effect, strains of *Streptococcus pyogenes* in their nose and throat.

Respiratory diseases therefore represent a difficult challenge for public health officials. Because many respiratory diseases cause discomfort but are not life-threatening, the tendency for many individuals suffering a respiratory illness is to simply "wait it out" while carrying on normal everyday activities. Unfortunately, such individuals are often unaware that they are actively infecting others. Because of this and the other problems discussed above, no practical means of control exist for many respiratory infections. It indeed appears that pathogens transmitted by the respiratory route have evolved a highly successful strategy for maintaining themselves in nature.

CONCEPT CHECK   **23.5**

Acute respiratory diseases spread in a predictable seasonal pattern based on normal human activities. Most are very difficult to control or treat.

✔ *Give an example of seasonal incidence and inability to control the spread of a viral respiratory pathogen.*

✔ *Give another example of a viral respiration pathogen that has been controlled. How was control achieved?*

23.6

Sexually Transmitted Diseases

Various sexually transmitted diseases (STDs) are caused by bacteria, viruses, and protozoa. Table 23.1 summarizes the major sexually transmitted diseases seen today. STDs are of major medical importance worldwide.

Control of STDs presents are unusually difficult public health problem for several reasons. First, nearly one-third of all cases involve teenagers. Sexual activity in this age group is on the rise, and sexually active young people today are more likely to have more than one sex partner, further complicating the disease picture. Second, many STDs initially cause no symptoms, or the symptoms that develop may be confused with the symptoms of other diseases not of a sexually transmitted nature. Third, the social stigma attached to STDs tends to inhibit some individuals from seeking prompt medical care; this is especially true of young people.

On the other hand, there are many reasons to seek medical treatment of STDs. First of all, virtually all sexually transmitted diseases are curable [acquired immunodeficiency syndrome (AIDS) being a notable exception]. As shown in Table 23.1, antibiotic and chemotherapeutic agents are highly effective in treating *most* STDs. Second, the long-term health consequences of many STDs can be quite serious. For example, certain STDs can cause secondary health problems

such as pelvic inflammatory disease (a major cause of infertility), cervical cancer, and heart and nerve damage. Pregnant women infected with a sexually transmitted pathogen can pass the agent on to the fetus, resulting in birth defects or even stillbirths. Third, diagnosis and treatment of one STD often reveals the presence of a second, inapparent infection that can still be transmitted after treatment for the first (for example, inapparent chlamydial infections are often diagnosed in individuals treated for gonorrhea).

Despite the fact that most sexually transmitted diseases can be controlled, the incidence of many of these diseases is still quite high; sexually transmitted diseases are obviously a social as well as a medical problem (Figure 23.23). We begin our discussion here with the disease gonorrhea because, despite the value of antibiotics in treating this disease, the prevalence of inapparent infections and the use of birth control pills have made gonorrhea the most widespread of the reportable sexually transmitted diseases.

Gonorrhea

Gonorrhea is one of the most widespread human diseases, and in spite of the availability of excellent treatment it is still a common disease (Figure 23.23). The disease symptoms of gonorrhea are quite different in the male and female. In the female the symptoms are usually a mild vaginitis that is difficult to distinguish from vaginal infections caused by other organisms, and the infection may easily go unnoticed; in the male, however, the organism causes a painful infection of the urethral canal (∞ Figure 19.12). The causative agent of gonorrhea, *Neisseria gonorrhoeae*, is killed quite rapidly by drying, sunlight, and ultraviolet light; this extreme sensitivity probably explains in part the sexually transmitted nature of the disease, the organism being transmitted from person to person only by intimate direct contact. In addition to gonorrhea, the organism also causes eye infections in the newborn and adult. Infants born of infected mothers may acquire eye infections during birth. Therefore, prophylactic treatment of the eyes of all newborns with silver nitrate or an ointment containing penicillin is generally mandatory and has helped to control infection in infants. Complications arising from untreated gonorrhea include pelvic inflammatory disease and damage to heart valves and joint tissues.

We discussed the clinical microbiology of *Neisseria gonorrhoeae* in Section 21.1, and the general bacteriology of the genus *Neisseria* is described in Section 16.21. The pathogen enters the body by way of the mucous membranes of the genitourinary tract (Figure 23.24a), being transmitted during sexual intercourse. Treatment of the infection with penicillin has been successful in the past, with a single injection usually resulting in elimination of the organism and a complete cure. For many years, all isolates of *N. gonorrhoeae* were found to be sensitive to

TABLE 23.1   Summary of some sexually transmitted diseases and treatment guidelines

Disease	Causative organisms[a]	Recommended treatment[b]
Gonorrhea	*Neisseria gonorrhoeae* (B)	Ceftriaxone plus doxycycline
Syphilis	*Treponema pallidum* (B)	Penicillin
Chlamydia trachomatis infections	*Chlamydia trachomatis* (B)	Doxycycline
Nongonococcal urethritis	*C. trachomatis* (B) or *Ureaplasma urealyticum* (B) or *Trichomonous vaginalis* (P)	Doxycycline or tetracycline
Lymphogranuloma venereum	*C. trachomatis* (B)	Doxycycline
Chancroid	*Haemophilus ducreyi* (B)	Erythromycin
Genital herpes	Herpes simplex type 2 (V)	No known cure; symptoms can be controlled with topical application of acyclovir (see Figure 23.29).
Genital warts	Papilloma virus (certain strains)	No known cure; warts can be removed surgically, chemically, or by cryotherapy.
Trichomoniasis	*T. vaginalis* (P)	Metronidazole
Acquired immunodeficiency syndrome (AIDS)	Human immunodeficiency virus (HIV)	No known cure; nucleotide base analogs clinically useful in some treatments (see Table 23.2).
Pelvic inflammatory disease	*N. gonorrhoeae* (B) or *C. trachomatis* (B)	Cefoxitin plus doxycycline
Vulvovaginal candidiasis	*Candida albicans* (F)	Miconazole nitrate

[a]B, Bacterium; V, virus; P, protozoan; F, fungus.
[b]Recommendations of the U.S. Department of Health and Human Services, Public Health Service.

penicillin or ampicillin (a semisynthetic penicillin), and sensitivity testing was not necessary. However, strains of *N. gonorrhoeae* resistant to penicillin are now widespread, and this resistance is due to a plasmid-encoded penicillinase. The incidence of penicillinase-producing *N. gonorrhoeae* has increased dramatically since the discovery of these strains in the mid-1970s (Figure 23.24*b*). In the United States, more than 8% of all clinical isolates are penicillinase-producing. Fortunately, however, the majority of penicillinase-producing strains respond to alternative antibiotic therapy, with a single dose of ceftriaxone, followed by doxycycline for 7 days. Doxycycline, a tetracycline derivative, is given because it is also an antichlamydial agent (Table 23.1), and nearly 50% of gonorrhea patients are also infected with the harder-to-diagnose *Chlamydia trachomatis* organism.

Despite the ease with which gonorrhea can be cured, the incidence of gonococcus infection remains relatively high. The reasons for this are threefold. (1)

Acquired immunity does not exist; hence repeated reinfection is possible (whether this is due to lack of local immunity or to the fact that at least 16 distinct serotypes of *Neisseria gonorrhoeae* have been isolated is not understood). (2) The use of oral contraceptives alters the local mucosal environment in favor of the pathogen. Oral contraceptives induce the body to mimic pregnancy, which results, among other things, in a lack of glycogen production in the vagina and a raising of the vaginal pH. Lactic acid bacteria, normally found in the adult vagina (∞ Section 19.5) fail to develop under such circumstances, and this allows *N. gonorrhoeae* transmitted from an infected partner to colonize more easily than in an acidic vagina containing lactobacilli. (3) Symptoms in the female are so mild that the disease may be unrecognized, and a promiscuous infected female can serve as a reservoir for the infection of many males. The disease can be controlled if the sexual contacts of infected persons are quickly identified

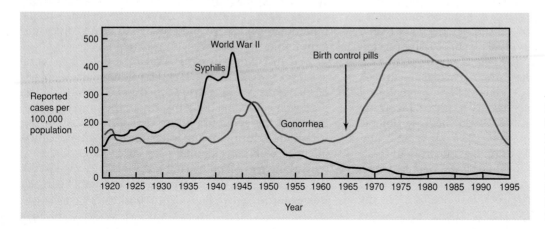

Figure 23.23 Sexually transmitted diseases: reported cases of gonorrhea and syphilis (primary and secondary only) per 100,000 population in the United States. After 1940, military cases are excluded.

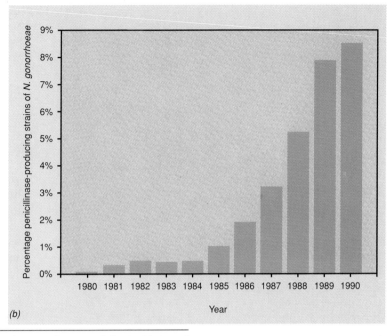

Figure 23.24 The causative agent of gonorrhea, *Neisseria gonorrhoeae*, and the incidence of penicillinase production in this organism. (a) Scanning electron micrograph of the microvilli of human fallopian tube mucosa, showing how cells of *N. gonorrhoeae* attach to the surfaces of epithelial cells. Note the distinct diplococcus morphology. Cells of *N. gonorrhoeae* are about 0.8 μm in diameter. (b) Reported penicillinase-producing *N. gonorrhoeae* (PPNG) in the United States. Note the rapid rise in resistance.

and treated, but it is often difficult to obtain this information and even more difficult to arrange treatment.

Syphilis

The sexually transmitted disease syphilis is potentially much more serious than gonorrhea, but because of differences in pathobiology, the incidence of syphilis in the United States has always been much lower (Figure 23.23). Syphilis is caused by a spirochete, *Treponema pallidum,* an organism that has been very difficult to cultivate (Figure 23.25). We mentioned the clinical immunology and diagnostic methods for syphilis in Table 21.5, and the biology of the spirochetes is discussed in Section 16.12.

Syphilis exhibits variable symptoms. The organism does not pass through unbroken skin, and initial infection most probably takes place through tiny breaks in the epidermal layer. In the male, initial infection is usually on the penis; in the female it is most often in the vagina, cervix, or perineal region. In about 10% of cases, infection is extragenital, usually in the oral region. During pregnancy, the organism can be transmitted from an infected woman to the fetus; the disease acquired in this way by an infant is called **congenital syphilis.** *Treponema pallidum* multiplies at the initial site of entry, and a characteristic *primary* lesion known as a **chancre** (Figure 23.26) is formed within 2 weeks to 2 months. Dark-field microscopy of the exudate from syphilitic chancres often reveals the actively motile spirochetes (Figure 23.25*a*). In most cases the chancre

heals spontaneously and the organisms disappear from the site. Some cells, however, spread from the initial site to various parts of the body, such as the mucous membranes, the eyes, joints, bones, or central nervous system, and extensive multiplication occurs. A hypersensitivity reaction to the treponeme then often takes place, revealed by the development of a generalized skin rash; this rash is the key symptom of the *secondary* stage of the disease. At this stage the patient's condition may be highly infectious, but eventually the organism disappears from secondary lesions and infectiousness ceases.

The subsequent course of the disease in the absence of treatment is highly variable. About one-fourth of infected individuals undergo a spontaneous cure, and another one-fourth do not exhibit any further symptoms, although the infection may persist. In about half of the patients the disease enters the *tertiary* stage, with symptoms ranging from relatively mild infections of the skin and bone to serious or fatal infections of the cardiovascular system or central nervous system. Involvement of the nervous system is the most serious phase of the illness because generalized paralysis or other severe neurological damage may result. In the tertiary stage very few organisms are present, and most of the symptoms probably result from delayed hypersensitivity reactions (∞ Sections 20.9 and 20.16) to the spirochetes.

Penicillin is highly effective in syphilis therapy, and the early stages of the disease can usually be controlled by a series of injections over a period of 1–2 weeks. In the secondary and tertiary stages, treatment

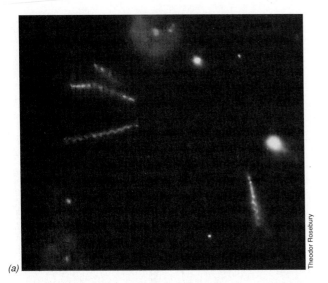

Theodor Rosebury

(a)

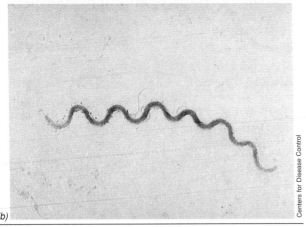

Centers for Disease Control

(b)

Figure 23.25 The spirochete of syphilis, *Treponema pallidum.* (a) Dark-field microscopy of an exudate. *Treponema pallidum* cells measure 0.15 μm in width and 10–15 μm in length. (b) Shadow-cast electron micrograph of a cell of *T. pallidum.* Note the axial fibrils, typical of spirochetes.

must extend for longer periods of time. However, even though syphilis is easily diagnosed and effectively treated, the number of individuals in the United States with infectious primary and secondary syphilis has been *rising* at a steady rate during the last decade. The disease is now concentrated in sexually promiscuous heterosexuals, individuals of low socioeconomic status, and illegal drug users.

Chlamydial infections

Now a *reportable* disease, as are gonorrhea and syphilis (∞ Table 22.4), a host of sexually transmitted syndromes can be ascribed to the obligate intracellular bacterium *Chlamydia trachomatis* (Figure 23.27 and Section 16.23). The total incidence of sexually transmitted *C. trachomatis* infections probably greatly outnumbers the incidence of gonorrhea. There may be up to 3 million new *C. trachomatis* sexually transmitted infections every year, making this organism the most prevalent cause of venereal disease. *C. trachomatis* also causes a serious disease of the eye called *trachoma* (∞ Section 16.23), but the strains responsible for venereal infections consist of a group of *C. trachomatis* serotypes distinct from those causing trachoma. Chlamydial infections may also be transmitted congenitally to the newborn from contamination in the birth canal, causing newborn conjunctivitis and pneumonia.

Chlamydial nongonococcal urethritis (NGU) is one of the most frequently observed sexually transmitted diseases today. *Chlamydia trachomatis* causes urethritis in males and urethritis, cervicitis, and pelvic inflammatory disease in females. In both the male and female, inapparent chlamydial infections are common, and this undoubtedly accounts for the prevalence of chlamydial infections in sexually active individuals. In a small percentage of cases, NGU can lead to serious complications, including testicular swelling and prostate inflammation in men and pelvic inflammatory diseases and fallopian tube damage in women. In the latter instance, cells of *C. trachomatis* attach to microvilli of fallopian tube cells, enter, multiply, and eventually lyse the cells (Figure 23.27b). Untreated NGU can cause infertility.

Chlamydia NGU is relatively difficult to diagnose by traditional isolation and identification methods. To

Figure 23.26 Primary syphilis lesions. (a) Chancre on lip. (b) Several chancres on penis.

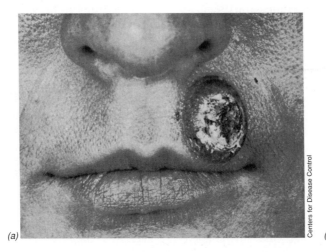

(a)

Centers for Disease Control

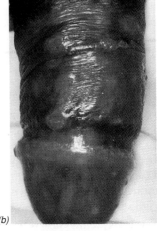

(b)

S. Olansky and L. W. Shaffer

Figure 23.27 Cells of *Chlamydia trachomatis* (arrows) attached to human fallopian tube tissues. (a) Cells attached to microvilli of fallopian tube. (b) Damaged fallopian tube containing a cell of *C. trachomatis* (arrow) in the lesion.

expedite diagnoses, a variety of immunological tests have been developed for identifying *Chlamydia trachomatis* from a vaginal or pelvic swab or from discharges. These clinical tests include fluorescent monoclonal antibodies and various enzyme-linked immunosorbent assay (ELISA) tests for detecting specific *C. trachomatis* antigens. If a chlamydial infection is suspected, treatment is initiated with doxycycline or tetracycline; penicillin is ineffective against *C. trachomatis* because the organisms lack peptidoglycan, the target of penicillin (∞ Section 16.23 and Table 23.1).

Chlamydial NGU is frequently observed as a secondary event following gonorrhea infection. If both *Neisseria gonorrhoeae* and *Chlamydia trachomatis* are transmitted to a new host in a single event, treatment of gonorrhea with ceftriaxone is usually successful but does not eliminate the chlamydia. Although cured of gonorrhea, such patients are still infected with chlamydia and eventually experience an apparent recurrence of gonorrhea that is instead a case of NGU. Thus, patients undergoing gonorrhea therapy should also be tested for chlamydial infections as well, because a latent chlamydial infection requires additional chemotherapy, usually with doxycycline, a tetracycline derivative.

Lymphogranuloma venereum is a sexually transmitted disease also caused by a specific strain of *Chlamydia trachomatis*. The disease, which occurs most frequently in males, consists of a swelling of the lymph nodes in and about the groin. From the infected lymph nodes, chlamydial cells may travel to the rectum and cause a painful inflammation of rectal tissues called *proctitis*. Because of the potential for regional lymph node damage and the complications of proctitis, lymphogranuloma venereum is considered to be one of the most serious sexually transmitted chlamydial syndromes.

Herpes

We discussed the molecular biology of herpesviruses in Section 8.19. Herpes simplex viruses are responsible for fever blisters and cold sores but can also cause genital infections. There are two main types of herpesviruses and many serotypes of each main type. **Herpesvirus type 1** (HV1) is generally associated with cold sores and fever blisters in and around the mouth and lips (Figure 23.28*a*). The incubation period of HV1 infections is short (3–5 days), and the lesions heal without treatment in 2–3 weeks. Relapses of HV1 infections are relatively common, and it is thought that the virus is spread primarily via contact with infectious lesions. Latent herpes infections are apparently quite common, with the virus persisting in low numbers in nerve tissue. Recurrent acute herpes infections are due to a periodic triggering of virus activity by unknown causes.

Herpesvirus type 2 (HV2) infections are associated primarily with the anogenital region, where the virus causes painful blisters on the penis of males or on the cervix, vulva, or vagina of females (Figure 23.28*b*). HV2 infections are transmitted by direct sexual contact, and the disease is most easily transmitted during the active blister stage rather than during periods of inapparent (presumably latent) infection.

Genital herpes infections are presently incurable, although a limited number of drugs have been successful in controlling the infectious blister stages. The guanine analog **acyclovir** (Figure 23.29), in ointment form, is particularly effective in limiting the shed of active virus from blisters and promoting the healing of blistering lesions. Acyclovir acts by specifically interfering with herpesvirus DNA polymerase, inhibiting viral DNA replication.

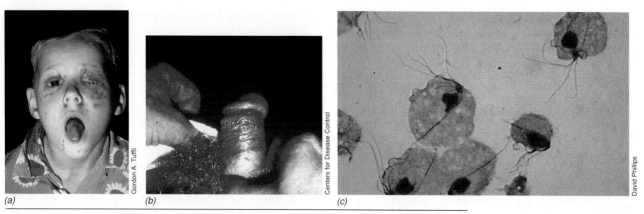

Figure 23.28 Nonbacterial sexually transmitted pathogens: herpesvirus and *Trichomonas*. (a) A severe case of herpes blisters on the face due to infection with herpesvirus type 1. (b) Genital herpes due to infection on the penis with herpesvirus type 2. (c) Cells of the flagellated protozoan *Trichomonas vaginalis*.

The overall health implications of genital herpes infections are not yet understood. Oral herpes is quite common and apparently has no harmful effects on the host beyond the oral blisters. However, epidemiological studies have shown a significant correlation between *genital* herpes infections and cervical cancer in females. In addition, herpesvirus type 2 can be transmitted to a newborn at birth by contact with herpetic lesions in the birth canal. The disease in the newborn varies from latent infections with no apparent damage to systemic disease that can result in brain damage or death. To avoid passing herpes infections to newborns, delivery by caesarean section is advised for pregnant women with genital herpes infections.

Trichomoniasis

Nongonococcal urethritis may also be caused by infections with the protozoan *Trichomonas vaginalis* (Figure 23.28c). Although many protozoa produce resting cells called *cysts*, *T. vaginalis* does not produce cysts. Thus transmission must be from person to person, generally by sexual intercourse. However, cells of *T. vaginalis* can survive a few hours outside the host, provided they do not dry out. Thus, transmission of *T. vaginalis* by contaminated toilet seats, sauna benches, and paper towels occasionally occurs. *Trichomonas vaginalis* infects the vagina in women, the prostate and seminal vesicles of men, and the urethra of both males and females.

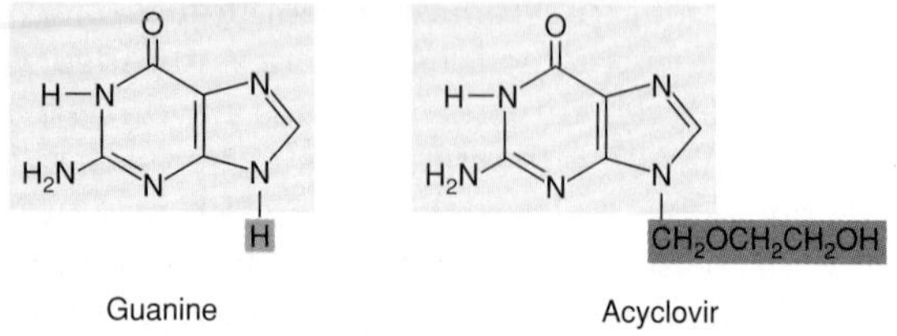

Many cases of trichomoniasis are totally asymptomatic. In fact, asymptomatic cases are the rule in males. In women trichomoniasis is characterized by a vaginal discharge, vaginitis, and painful urination. The organism is commonly found in females, surveys indicating from 25 to 50% of sexually active women being infected; only about 5% of men are infected. The male partner of an infected female should be examined for *Trichomonas vaginalis* and treated if necessary because promiscuous asymptomatic males can serve as reservoirs, transmitting the infection to several females. Trichomoniasis is diagnosed by preparing and microscopically examining a wet mount of fluid discharged from the patient for the motile protozoa. The antiprotozoal drug *metronidazole* is particularly effective in treating trichomoniasis (Table 23.1).

Ecology of sexually transmitted diseases

Thus far we have seen that a variety of microorganisms can cause sexually transmitted diseases. What properties do these organisms share in common that limit their distribution to the human genitourinary tract and their mode of transmission to sexual activity? Unlike respiratory infections where large numbers of infectious particles may be expelled by an individual, sexually transmitted pathogens are generally *not* shed in large numbers other than during sexual activity. Consequently, transmission is limited to physical contact, generally during sexual intercourse. This is very clearly shown by the fact that spread of venereal diseases is controlled very effectively by sexual abstinence (no fluid exchange) or by the use of barriers such as condoms that stop the normal exchange of body fluids during sexual activity. In addition, many sexually transmitted pathogens are very sensitive to drying. Their habitat, the human genitourinary tract, is generally a moist environment. Thus, these organisms colonize moist niches and have apparently lost the ability to survive outside the animal host.

Guanine

Acyclovir

Figure 23.29 Structure of guanine and the guanine analog acyclovir. Acyclovir has been used therapeutically to control genital herpes (HV2) blisters.

We now consider AIDS the most serious of all sexually transmitted diseases. We have seen that this disease has other modes of transmission as well (∞ Section 22.4), but unlike the diseases we have discussed thus far, AIDS is of unusual health concern because the disease is invariably fatal.

23.7
Acquired Immunodeficiency Syndrome

Acquired immunodeficiency syndrome (AIDS) has received considerable attention since it was recognized as a disease in 1981. More than 500,000 cases of AIDS have been reported since then in the United States alone, and more than 300,000 people have died (∞ Section 22.4 and Figure 22.2). Worldwide, the outlook is even more serious, with more than 10 million people already infected with the human immunodeficiency virus (HIV), the causative agent of AIDS, which now seems to be a universally lethal disease. These numbers will continue to rise dramatically unless effective treatment or prevention methods are discovered. We have already discussed the epidemiology of AIDS (∞ Section 22.4), including these grim predictions, and the clinical diagnostic methods for identifying HIV infection (∞ Sections 21.8 and 21.9). In this section, we will concentrate on the pathogenesis of AIDS.

AIDS was first suspected of being a disease of the immune system when a startling increase in the number of so-called opportunistic infections was observed (see the box, A Definition of AIDS). *Opportunistic infections* are defined as infections rarely observed in humans with normal immune responses (∞ Section 19.12).

The most common AIDS-associated opportunistic infections include pneumonia caused by the protozoan *Pneumocystis carinii* (Figure 23.30a) and other protozoal infections such as cryptosporidiosis, caused by *Cryptosporidium* species (Figure 23.30b), and toxoplasmosis, caused by *Toxoplasma gondii* (Figure 23.30c), systemic yeast infections due to *Cryptococcus neoformans* (Figure 23.30d), *Candida albicans* (Figure 20.30e), and *Histoplasma capsulatum* (Figure 23.30f), viral infections due to herpes simplex (Figure 8.43a) or cytomegalovirus, tuberculosis, and other mycobacterial infections (Figure 23.30g), and enteric helminthic infections due to *Strongyloides stercoralis* (Figure 23.30h). However, *Pneumocystis* pneumonia is by far the most common opportunistic disease encountered, being observed at some time in nearly two-thirds of all cases of AIDS.

Besides the opportunistic infections associated with many AIDS cases, a rare form of cancer called

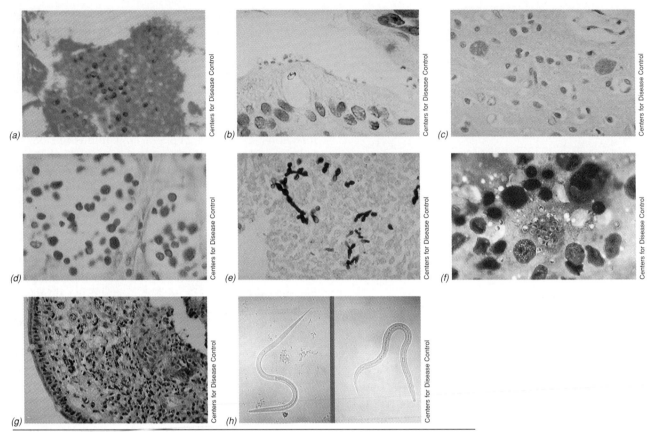

Figure 23.30 Opportunistic pathogens associated with cases of acquired immunodeficiency syndrome (AIDS). (a) *Pneumocystis carinii*, from patient with pulmonary pneumocystosis. (b) *Cryptosporidium* sp., from biopsy of small intestine. (c) *Toxoplasma gondii*, from brain tissue of patient with toxoplasmosis. (d) *Cryptococcus neoformans*, from liver tissue of patient with cryptococcosis. (e) *Candida albicans*, from heart tissue of patient with systemic *Candida* infection. (f) *Histoplasma capsulatum*, from liver tissue of patient with histoplasmosis. (g) Mycobacterial infection of small bowel, acid-fast stain. (h) *Strongyloides stercoralis*, filariform larvae.

Kaposi's sarcoma is also observed in many AIDS patients. Kaposi's sarcoma is a cancer of the cells lining blood vessel walls and is diagnosed by the characteristic purplish patches it leaves on the surface of the skin (Figure 23.31). The incidence of Kaposi's sarcoma is much higher in homosexual men suffering from AIDS than in the other high risk groups identified (∞ Figure 22.3), but the cause of Kaposi's sarcoma is not known.

Human immunodeficiency virus

The disease AIDS is caused by human immunodeficiency virus. HIV is a retrovirus (∞ Section 8.22) containing 9749 nucleotides in each of its two identical single-stranded RNA genomes. Using the enzyme *reverse transcriptase*, which is present in the intact virion, HIV forms a complementary single-stranded DNA molecule using RNA as a template and converts the complementary DNA (cDNA) formed into double-stranded DNA, which can enter the host cell genome. The target host cell of HIV is the CD4 class of T lymphocyte (T helper cells) (∞ Sections 20.4 and 20.9), and HIV infection prevents normal division processes in these cells. Thus, immune functions are compro-

mised in those infected with HIV, and this leads to development of opportunistic infections. We now consider the AIDS disease process itself and see how HIV effectively dismantles the immune system.

HIV: T lymphocyte interactions

HIV has the specific capacity to infect the CD4 class of T lymphocytes; infected CD4 lymphocytes produce high numbers of HIV particles 90–120 nm in diameter (Figure 23.32). Recall that CD4 lymphocytes include the T helper (T_H) and T delayed-type hypersensitivity (T_D) subsets (∞ Sections 20.4 and 20.9). CD4 cells respond to antigens presented in association with class II major histocompatibility (MHC) antigens on the surface of macrophages and B cells (∞ Section 20.7). Once activated by antigen, normal T_H cells secrete cytokines that, together with contact-mediated signals, stimulate particular clones of B lymphocytes to multiply and secrete antibody specific for the antigen (∞ Section 20.11).

The specificity of HIV for CD4 T cells is due to the fact that the CD4 molecule acts as a *cell surface receptor* for HIV. CD4 normally plays a role as a co-receptor molecule on cell surfaces by binding to the class II MHC

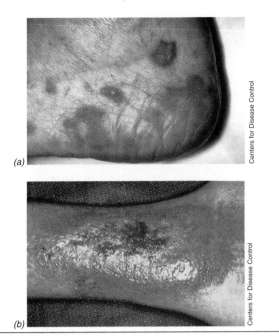

(a)

Centers for Disease Control

(b)

Centers for Disease Control

Figure 23.31 Kaposi's sarcoma lesions as they appear on (a) the heel and lateral foot, and (b) the distal leg and ankle.

molecule on the antigen-presenting cell (∞ Figure 20.16*b*). However, a great deal of complementarity exists between CD4 and coat proteins of HIV. Binding of CD4 by the coat proteins acts as a mechanism for entrance of HIV into the T lymphocyte. CD4 interacts with a major envelope coat protein of HIV, called *gp120* (Figure 23.33). This interaction causes the membrane of HIV and the host cytoplasmic membrane to fuse with

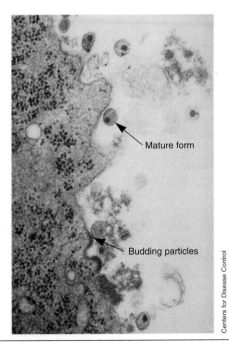

Mature form

Budding particles

Centers for Disease Control

Figure 23.32 Transmission electron micrograph of a thin section of a lymphocyte releasing human immunodeficiency virus (HIV). Cells were from a hemophiliac patient who developed AIDS. HIV particles are 90–120 nm in diameter.

release of the nucleocapsid (nucleic acid and reverse transcriptase) of HIV into the cell (Figure 23.33).

Besides this one class of T lymphocytes, several other cell types in the human body have the CD4 molecule on their surface in small amounts and can also be infected with HIV. Monocytes and macrophages express low levels of CD4 and can be infected, apparently without ill effect, but serve as a reservoir of HIV to spread the infection to T lymphocytes. A small percentage of B lymphocytes can also be infected. Certain human brain cells and intestinal cells as well as a variety of cultured cell lines derived from human brain or bowel tissues can also be infected with HIV, although many of these cells express the CD4 molecule on their surfaces in only trace amounts.

In cases of clinical AIDS, lymphocytes bearing the CD4 molecule (mainly T_H cells) are greatly reduced in number. However, unlike lytic animal viruses, HIV usually does not immediately kill and lyse its host cell. Following reverse transcription to produce DNA from the RNA genome, the viral cDNA integrates into host chromosomal DNA and exists as a provirus. The cell may show no outward sign of infection, and HIV DNA can remain in a latent state for long periods. Eventually, however, productive virus synthesis occurs and new HIV particles are produced and released from the cell. T cells producing HIV no longer divide and eventually die.

Accelerated destruction of CD4 cells occurs following the processing of HIV antigens by infected T cells. Such cells embed molecules of gp120 from HIV particles on their cell surfaces. The embedded gp120 protein on the infected cells then sticks to uninfected T cells by binding to the CD4 molecule. Eventually, numerous cells of each type fuse to produce multinucleate giant cells called *syncytia*. One HIV-infected T cell may eventually bind and fuse with up to 50 uninfected T cells. Shortly after syncytia formation occurs, the resulting cells lose immune function and die.

The end result of HIV infection is that CD4 cells progressively decline in number. This has serious health consequences. In a normal human, CD4 cells constitute about 70% of the total T cell pool; in AIDS patients, the number of CD4 cells steadily decreases, and by the time opportunistic infections set in, CD4 cells may be almost absent (∞ Figures 21.17 and 23.34). As CD4 cells decline in number, there is a concomitant loss in the cytokines they produce. This leads to a gradual reduction in all types of lymphocytes, effectively shutting down the immune system in those suffering from clinical AIDS. This loss of both humoral and cellular immune function is readily apparent in the opportunistic infections observed. Systemic infections by fungi and mycobacteria (Figure 23.30) point to a loss in cellular immunity (∞ Section 20.9). Other opportunistic infections, such as the various viral and bacterial infections associated with AIDS, indicate the loss of humoral immunity; decline in antibody production is due to the loss of T_H cells necessary to stimulate antibody production by B cells (∞ Section 20.11).

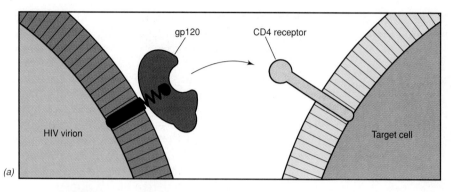

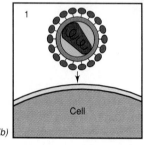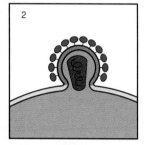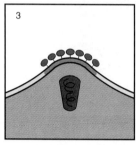

Figure 23.33 Process of infection with human immunodeficiency virus (HIV). (a) Uptake of HIV by CD4 cells by interaction of HIV protein gp120 and the CD4 molecule on host cells. (b) Fusion of the HIV envelope with the host cell envelope and entry of the nucleocapsid.

The overall picture of typical AIDS progression indicates that during the clinical latency period, a very active infectious process is proceeding. First, there is an intense immune response to HIV: about 1 billion virions are destroyed each day. However, this means that HIV is replicating at a very high rate, and this replication results in the corresponding destruction of about 100 million CD4 T cells each day. Eventually, the immune response is simply overwhelmed, and the T cells are finally completely destroyed, crippling the immune response and allowing the emergence of opportunistic infections. The example in Figure 23.34 documents T cell destruction over a typical time course.

Diagnosis and treatment of AIDS

The prognosis of an HIV-infected individual is not encouraging. Opportunistic pathogens or malignancies (Figures 23.30 and 23.31) eventually kill most AIDS patients. Long-term studies of AIDS patients indicate that the average person infected with HIV progresses through several stages of decreasing immune function, with CD4 cells dropping from a normal range of 600–1000/mm^3 of blood to near zero over a period of 5–7 years (Figure 23.34). Although the *rate* of decline in immune function varies from one HIV-infected individual to another for reasons that are not yet clear, virtually every person infected with HIV will eventually contract AIDS, usually within 10 years of initial infection, and will eventually die from AIDS.

No *cure* for HIV infection is known, although research is very intensive in the areas of vaccine production and chemotherapy. Several drugs have been identified as helpful in delaying symptoms of AIDS and in some cases in prolonging the life of those infected with HIV (Table 23.2). The most promising drugs discovered thus far are chemicals that inhibit the activity of *reverse*

transcriptase, the enzyme that converts the genetic information (which resides in single-stranded RNA) of retroviruses, such as HIV, into a complementary DNA copy (∞ Section 8.22). For example, the nucleotide base analog **azidothymidine** (AZT) (Figure 23.35) is an effective inhibitor of HIV replication because it closely resembles thymidine but lacks the correct attachment point for the next nucleotide in the chain and thus serves as a *DNA chain terminator*. The dideoxynucleotide analogs (Table 23.2) also function as chain terminators (recall that dideoxynucleotides were used to interrupt nucleic acid synthesis in DNA sequencing as well) (∞ Working with Nucleic Acids: The Tools, Chapter 6). Other AIDS drugs have uncertain modes of action but have simply been observed to improve the clinical picture of AIDS in experimental situations (Table 23.2).

Resistance to the nucleoside analog drugs by HIV is now a major problem. Based on the tremendous numbers of virions produced, HIV replicates every 30 hours. Thus, a normal mutation rate allows resistant strains of HIV to occur within several weeks. As a result, virtually everyone treated with single chemotherapeutic agents, usually the nucleoside analogs (Table 23.2), develops drug-resistant HIV strains. Currently, drugs are given in pairs to avoid this problem. For example, AZT given in combination with 3TC has been shown to significantly reduce the viral burden in most patients over long periods of time without producing mutant HIV resistant to both drugs. Such combination therapies are now the method of choice. Unfortunately, the most effective drug strategies currently involve nucleoside analogs, and many patients also suffer toxicity from them. However, a new class of drugs are now experimentally available. These drugs are known as protease inhibitors (Table 23.2) because they inhibit the HIV protease that is responsible for digesting the HIV-encoded polypeptide into functional proteins (∞ Section 8.22). This protease is an attractive target for

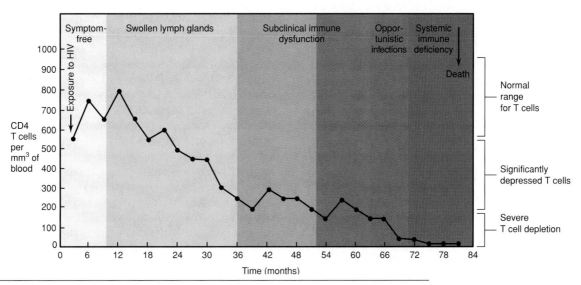

Figure 23.34 Decline in CD4 T lymphocytes in blood of a young male AIDS patient showing the typical progressive loss in CD4 cells and immune function.

chemotherapeutic agents because it does not resemble any human proteins. Thus, toxic side effects due to modifications of host functions are unlikely. These drugs significantly reduce viral numbers and, *in vitro*, they act synergistically with nucleoside analogs to inhibit HIV replication. Protease inhibitors may be the next generation of effective HIV chemotherapeutic agents.

AIDS vaccines

The genetic variability of HIV has thus far hampered the development of an AIDS vaccine. One vaccine strategy is to make antibodies to the envelope protein,

gp120, and use these antibodies to block CD4–gp120 interactions (Figure 23.33*a*) and thus block infection. However, this approach has not been successful thus far because the gene encoding gp120 mutates frequently, forming antigenic variants of the protein that are not recognized by antibodies made to a different form. The most impressive results from clinical immunization trials have emerged from *subunit vaccines* (∞ Section 10.13), where genes for several HIV envelope proteins have been engineered into vaccinia virus or adenovirus particles. Using these harmless viruses as expression vectors and vehicles for delivery of HIV antigens, several subunit vaccines have been

TABLE 23.2 Some chemotherapeutic strategies for AIDS	
Drug[a]	**Mechanism of action and comments**
Azidothymidine (AZT, ZDV or Zidovudine)	Reverse transcriptase inhibitor (chain terminator); available as prescription drug; increases survival time and reduces incidence of opportunistic infection in AIDS patients; toxic to bone marrow cells.
Dideoxycytidine (ddC) Dideoxyinosine (ddI or Didanosine) Stavudine (d4T) Lamivudine (3TC)	Reverse transcriptase inhibitors (chain terminators); alternatives to AZT; may be used in conjunction with AZT in multiple-drug-treatment protocols; may have fewer side effects than AZT in some patients.
Phosphonoformate	Reverse transcriptase inhibitor; also active against some viral opportunistic pathogens
Dextran sulfate	Blocks binding of HIV to CD4; little or no toxicity.
Interferon-α	Reduces HIV budding from infected cells; strong antitumor agent against Kaposi's sarcoma.
Ampligen	Interferon inducer; nontoxic.
Soluble CD4	Inhibits binding of HIV to T cells; genetically engineered form of CD4.
Phosphorothioate oligodeoxynucleotides complementary to HIV genome sequences	Serves as "antisense" strand of DNA (∞ Antisense Nucleic Acid, Chapter 7). Binds HIV RNA and, because of the unique chemistry of antisense DNA nucleotides, prevents viral protein synthesis.
Protease inhibitors (a variety of compounds)	Experimental drugs designed to bind to HIV protease, inhibiting processing of viral polypeptide and virus maturation.

[a]The only drugs approved for clinical use by the United States Food and Drug Administration are AZT, ddC, ddI, d4T, and 3TC. The other drugs are experimental.

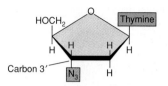

Figure 23.35 Structure of azidothymidine (AZT), a drug used for AIDS therapy. Because AZT lacks a hydroxyl group on the 3′ carbon, replication ceases when it is incorporated into the growing DNA chain. AZT is also known as Zidovudine (ZDV).

shown to elicit a potent humoral and cellular immune response to HIV. Clinical trials of subunit vaccines are under way.

Other potential vaccines include *killed intact HIV*. These inactivated vaccines are restricted to use in HIV-infected individuals because inactivation procedures may not kill 100% of the HIV; it would be unethical to expose uninfected individuals to even a small risk of HIV infection. Next, some laboratories are exploring the possibilities of producing *live attenuated vaccines*. This vaccine strategy is bolstered by the finding that individuals infected with HIV-2, a related virus that causes a very mild form of AIDS with a very long latent period, protects carriers from infection with HIV-1, the strain responsible for severe AIDS. However, there are serious potential risks. For example, integrated virus could cause cancer, mutations might reactivate virulence, and so on. Another approach uses *anti-idiotypic antibodies*. In this approach, antibodies to CD4 are used as an antigen. Antibodies raised to the idiotype (binding site) should resemble the molecular configuration of CD4 (∞ Section 20.17). In sufficient quantity, anti-idiotypic antibodies could then bind HIV particles by gp120–CD4 type interactions. A related approach is to use the CD4 protein itself as an AIDS antagonist. Referred to as "soluble CD4" (and a product of recombinant DNA technology), this protein could interact with HIV and block its attachment to intact CD4 T cells (Table 23.2).

No effective AIDS vaccine is currently available. Additionally, an AIDS vaccine would most probably not be useful in *treating* most patients that already have the disease because these individuals already have a debilitated immune system or are dealing with large amounts of virus. Most of these individuals lack significant immune function and thus would not respond to a vaccine. Thus, despite considerable advances in our molecular understanding of HIV and in our clinical understanding of the AIDS disease process, public education about AIDS and avoidance of high risk behavior are still the major tools to combat AIDS today (see the box, Sexual Activity and AIDS).

Detection of HIV infection

Exposure to HIV can be diagnosed by immunological means. Both radioimmunoassay (RIA) and enzyme-linked immunosorbent assay (ELISA) tests (∞ Section 21.8) have been developed for screening blood samples to detect anti-HIV antibodies. The ELISA test has

proven particularly valuable for large-scale screening of donated blood to prevent transfusion-associated HIV. Statistics have shown that about 0.25% (2–3 per thousand) of all blood donated by volunteer donors in the United States tests HIV-positive in the ELISA assay. A positive HIV-ELISA test must be confirmed by a second procedure called immunoblotting (Western blotting), a technique that combines the analytical tools of protein purification and immunology (∞ Section 21.9).

A number of other rapid tests are being developed and marketed to identify individuals who are infected with HIV. One test uses a single drop of patient blood and a single reagent. The reagent is a bioengineered antibody in which one binding site is directed to a red blood cell antigen, while the other site is directed to the gp41 HIV surface antigen. In a positive test, the bifunctional antibody cross-links the red blood cells to the HIV, resulting in a visible agglutination (∞ Section 21.5). In another test, saliva is used as a source of secretory antibody to HIV. The saliva is expelled onto a cartridge containing immobilized HIV antigens. A second antibody, reactive with the bound antibody and conjugated to an enzyme, is then added. After addition of the enzyme substrate, a positive reaction shows a colored product, as in the ELISA methods (∞ Section 21.8). The rapid tests are designed to provide maximum convenience, speed (*minutes* instead of the hours or days required for ELISA or immunoblot), extended shelf life, portability, and ease of application and interpretation. However, in general, the rapid tests are not as sensitive or accurate as the standard HIV-ELISA and HIV-immunoblot tests (∞ Sections 21.8 and 21.9).

All tests, no matter how sensitive or accurate, can fail to detect HIV-positive individuals who have recently acquired the virus but have not yet made an antibody response, which may be 6 weeks to a year after exposure to HIV. In spite of this drawback, these tests ensure the general safety of the blood supply, and statistics indicate

---

CONCEPT CHECK   23.7

AIDS is one of the most prevalent infectious diseases in the human population. Concern about HIV infection is particularly great because virtually all infected individuals eventually progress to AIDS, which has a 100% fatality rate. HIV first destroys a central part of the immune system, and opportunistic pathogens then kill the host. There is still no effective vaccine for HIV. Likewise, the few drugs that inhibit HIV multiplication are also toxic to the host. The only prevention for the spread of HIV infection is through total avoidance of risky behavior such as intravenous drug use (needle sharing) and unsafe sexual practices.

✔ *Review the definition of AIDS. What diagnostic features are shared by all AIDS patients?*

✔ *What is the current treatment for AIDS? Is it effective? What are the side effects?*

Sexual Activity and AIDS

Sexual promiscuity has always been associated with sexually transmitted diseases, but the acquired immunodeficiency syndrome (AIDS) epidemic, discussed in this chapter and elsewhere in this book, has focused attention on the dangers of multiple sex partners and on the high risk associated with certain sex practices. AIDS, caused by the human immunodeficiency virus (HIV), is only one type of sexually transmitted disease. Others include gonorrhea, syphilis, herpes simplex, nonspecific urethritis (caused by *Chlamydia*), protozoal vaginitis (caused by *Trichomonas vaginalis*), fungal vaginitis (caused by *Candida albicans*), and venereal warts (caused by the human papilloma virus). Some of these sexually transmitted diseases have been associated with human society for all of recorded history. The unique aspect of AIDS is that it is almost uniformly fatal. There are neither drugs to cure AIDS nor vaccines to prevent it, and it is unlikely that highly effective drugs or vaccines will be available in the near future. We do not at this time know the extent of the AIDS epidemic because the long latent period means that many people now infected (and perhaps infectious) have not yet exhibited symptoms. Most public health officials believe that over the next 10 years there will be a massive increase in the incidence of the disease worldwide, as latent cases develop into full-blown AIDS.

Because AIDS is linked to certain sex practices, prevention means avoidance of these sex practices. The United States Surgeon General has issued a report that makes specific recommendations that individuals can follow if they wish to reduce the likelihood of AIDS infection. Among the recommendations are

1. Avoid mouth contact with penis, vagina, or rectum.
2. Avoid all sexual activities that could cause cuts or tears in the linings of the rectum, vagina, or penis.
3. Avoid sexual activities with individuals from high risk groups. These include prostitutes (both male and female), homosexual and bisexual individuals, and intravenous drug users.
4. If a person has had sex with a member of one of the high risk groups, a blood test should be done to determine if infection with HIV has occurred. If the test is positive, then it is essential that sexual partners of an HIV-positive individual be protected by use of a condom during sexual intercourse.

It is important to emphasize that AIDS is *not* just a disease of male homosexuals. In certain cultures, AIDS is as common in women as in men. The disease is linked to promiscuous sexual activities and other activities that involve exchange of body fluids, which include not only male homosexuality but also female prostitution and intravenous drug use.

Is it possible, then, to have sex without incurring the risk of AIDS? Certain sex practices are inherently much safer than others. Safe sex practices include dry kissing (mouths closed), mutual masturbation (in the absence of breaks in the skin), and intercourse protected by a condom. Dangerous sex practices include wet kissing (mouths open), masturbation where breaks in the skin occur, oral sex (either male or female), and unprotected sexual intercourse (either vaginal or anal). The U.S. Surgeon General has recommended that if the health status of the partner is unknown, a condom be used for all sex practices in which exchange of body fluids occurs.

The AIDS epidemic has focused new attention on the condom (see photo). Condoms have always played two roles in sexual activity: disease protection and prevention of pregnancy. Although the best way to avoid AIDS is to avoid dangerous sex practices, if sexual intercourse is to be carried out with an individual whose infection status is unknown, then a latex condom should be used. The U.S. Surgeon General strongly recommends the use of condoms for all extramarital sexual activity. In certain countries, advertising campaigns to promote the use of condoms are widespread.

Moralistic statements alone (prescriptions for monogamy, abstinence, avoidance of sexual activity outside of matrimony), will *not* control the AIDS epidemic. Epidemiological studies on all previously known sexually transmitted diseases have shown that fear of disease is not, by itself, sufficient to prevent sexual activities that put an individual at risk for a sexually transmitted disease. The sex drive in some individuals is so strong that it will suppress the fear of disease, even a disease like AIDS. Every individual must therefore take the responsibility for protecting himself or herself from this widespread and extremely dangerous infectious disease.

For more information on prevention of AIDS, see the *Surgeon General's Report on Acquired Immune Deficiency Syndrome*, U.S. Department of Health and Human Services. For more information on protection against AIDS, the Public Health Service has established a toll-free telephone number, called the PHS AIDS Hotline. The number to call is 800-342-2437. The CDC National AIDS Clearinghouse can be contacted at http://www.cdc-nac.org. ∎

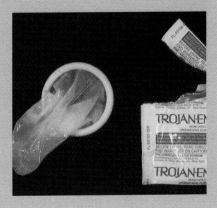

that the risk of contracting HIV through contaminated blood or blood products is now very low. Sexual promiscuity and group intravenous drug use are the major routes of HIV infection today (∞ Figure 22.3).

23.8
Animal-Transmitted Diseases: Rabies

Animals can contract a number of infectious diseases, some of which can be passed to humans. Through the use of effective vaccination practices and good veterinary care, domestic animal populations are generally maintained in good overall health. But when a new pathogen, or a new strain of a pathogen currently under control, is introduced into the domestic animal population, infections can spread quickly, leading to substantial economic losses. This situation is different in the wild animal population. Wild animals cannot be routinely vaccinated and do not receive veterinary care. Thus, animal diseases (zoonoses) cycle through wild animal populations on a periodic basis.

Most zoonoses are of little serious consequence to human health. Rabies is an exception. Rabies is one of the best examples of a disease that occurs primarily in animals but under certain conditions occurs in humans. The major reservoir of rabies in the United States is in wild animals, primarily carnivores. However, a significant number of cases are still seen in domestic animals (Figure 23.36). Surprisingly, rabies is still a major disease in humans as well: worldwide, approximately 35,000 people die every year from this disease, primarily in developing countries (∞ Table 22.1). Rabies is caused by a single-stranded RNA virus of the rhabdovirus family (negative-strand viruses) (∞ Section 8.16) that attacks the central nervous system of most warm-blooded animals, almost invariably leading to death if not treated. The virus enters the body through a bite wound from a rabid animal. The virus multiplies at the site of inoculation and then travels to the central nervous system. The incubation period for the onset of symptoms is highly variable, depending on the size, location, and depth of the wound and the actual number of viral particles transmitted in the bite. In dogs, the incubation period averages 10–14 days. In humans, up to 9 months may elapse before the onset of rabies symptoms. The virus proliferates in the brain (especially in the thalamus and hypothalamus), leading to fever, excitation, dilation of the pupils, excessive salivation, and anxiety. A fear of swallowing (hydrophobia) develops from uncontrollable spasms of the throat muscles; death eventually results from respiratory paralysis.

The first rabies vaccine was developed by Louis Pasteur. Today, highly effective rabies vaccines for both animals and humans are available, and an average of less than one case of human rabies is reported in the United States each year. Because of the long incubation period, a rabies vaccine can even be used in therapy; treatment of an infected human before the onset of symptoms is almost always successful in preventing rabies. An unprovoked animal that bites a human is generally held 10 days for observation, even if no signs of rabies are apparent. Treatment of the human with rabies vaccine or rabies immune serum (antirabies virus antibodies obtained from a hyperimmune individual) is generally not undertaken. However, if the animal is rabid, the patient will be passively immunized with rabies immune globulin (injected at both the site of the bite and intramuscularly) and also will be vaccinated with inactivated rabies virus, preferably virus grown in human cell lines.

A "second generation" rabies vaccine is now under development. This vaccine is prepared by recombinant DNA techniques involving the cloning and expression in *Escherichia coli* of highly immunogenic rabies virus surface glycoproteins. Subunit vaccines (∞ Section

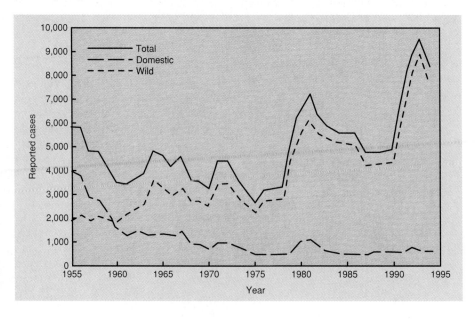

Figure 23.36 Reported rabies cases in wild and domestic animals in the United States. Most cases of rabies occur in wild animal populations. During the time shown, there has not been a single year with more than 10 human rabies cases.

TABLE 23.3	Guidelines for treating possible human exposure to rabies virus

Unprovoked bite by a domestic animal

Animal suspected of rabies
1. Sacrifice animal and test for rabies.
2. Begin treatment of human immediately.[a]

Animal not suspected of rabies
1. Hold for 10 days. If no symptoms, do not treat human.
2. If symptoms develop, treat human immediately.[a]

Bite by wild carnivore (for example, skunk, bat, fox, raccoon, coyote)

Regard animal as rabid.
1. Sacrifice animal and test for rabies.
2. Begin treatment of human immediately.[a]

Bite by wild rodent, squirrel, livestock, rabbit

Consult local or state public health officials about possible recent cases of rabies transmitted by these animals (these animals rarely transmit rabies). If no reports, do not treat human.

[a]All bites should be thoroughly cleansed with soap and water. Treatment is generally a combination of rabies immunoglobulin and human diploid cell rabies vaccine (five injections intramuscularly).

10.13) are also in development. In these, rabies virus genes encoding coat proteins are expressed in vaccinia virus. Such vaccines would be a completely safe means of controlling rabies in both animals and humans. A summary of guidelines for treating possible human exposure to rabies is shown in Table 23.3.

Rabies has been effectively controlled in domestic animal populations by vaccination, but it is still a problem in the United States because of the apparently large wild animal reservoir of rabies virus (Figure 23.36). All dogs and cats should be vaccinated beginning at 3 months of age, and booster inoculations given yearly. Under normal circumstances, humans are not vaccinated for rabies. However, vaccination is recommended for persons at high risk, such as veterinarians and others who have routine contact with susceptible species. Because of the highly fatal nature of the disease and the effective treatment available, any unprovoked animal bite must be taken seriously. A wild animal suspected of being rabid should be captured, sacrificed, and examined immediately for evidence of rabies in order to expedite treatment of those exposed. Characteristic virus inclusion bodies in the cytoplasm of nerve cells, called *Negri bodies,* are taken as confirmation of rabies. Fluorescent antibody preparations (∞ Section 21.7) that recognize rabies-infected brain tissue are also very useful for confirming a diagnosis of rabies.

CONCEPT CHECK   23.8

Rabies occurs primarily in wild animals but can be transmitted to domestic animals and from them to humans. Vaccination of dogs and cats is of central importance for the control of rabies. In developing countries, rabies is still an important human disease.

✔ *Why is rabies vaccine not given routinely to humans in the United States?*

✔ *How can rabies be treated in humans?*

23.9

Insect- and Tick-Transmitted Diseases: Rickettsias

The rickettsias are small bacteria that have a strictly intracellular existence in vertebrates, usually in mammals, and are also associated at some point in their natural cycle with blood-sucking arthropods such as fleas, lice, or ticks. We discussed the biology of rickettsias in Section 16.22. Rickettsias cause a variety of diseases in humans and animals, of which the most important are typhus fever, Rocky Mountain spotted fever, scrub typhus (tsutsu-gamushi disease), and Q fever. Rickettsias take their name from Howard Ricketts, a scientist at the University of Chicago who first provided evidence for their existence and who died from infection with the rickettsia that causes typhus fever, *Rickettsia prowazekii.* Rickettsias have not been cultured in artificial media but can be cultured in laboratory animals, lice, mammalian tissue culture cells, and the yolk sac of chick embryos. In animals, growth takes place primarily in phagocytic cells (∞ Section 20.3).

Typhus fever (epidemic typhus)

Typhus fever is caused by *Rickettsia prowazekii.* Epidemic typhus is transmitted from host to host by the common body or head louse. Typhus can be a serious disease. During World War I, an epidemic of typhus spread throughout eastern Europe and claimed almost 3 million lives. Typhus has frequently been a problem among military troops during wartime. Cells of *R. prowazekii* are introduced through the skin when the puncture caused by the louse bite becomes contaminated with louse feces, the major source of rickettsial cells. During an incubation period of 1–3 weeks, the organism multiples inside cells lining the small blood vessels. Symptoms of typhus (fever, headache, and general body weakness) then begin to appear. Five to nine days later a characteristic *rash* is observed in the armpits and generally spreads throughout the body *except* for the face, palms of the hands, and soles of the feet. Complications from untreated typhus may develop involving damage to the central nervous system, lungs, kidneys, and heart. Several drugs are effective against rickettsias. Tetracycline and chloramphenicol are most commonly used to control *R. prowazekii*.

Murine typhus, caused by *Rickettsia typhi,* is much less common than epidemic typhus. The disease occurs primarily in rodents and is transmitted to humans by feces from the rat flea. Symptoms are similar to those of epidemic typhus, including the formation of a rash. Unlike epidemic typhus, however, murine typhus is rarely fatal, although kidney problems are observed in untreated cases. Prevention of murine typhus depends on effective rat control, and thus the disease is more common in areas with poor public health standards.

Rocky Mountain spotted fever

Rocky Mountain spotted fever was first recognized in the western United States around the turn of the twentieth century but is actually more common today in the southeastern United States. Rocky Mountain spotted fever is caused by *Rickettsia rickettsii* and is transmitted to humans by various species of ticks, most commonly the dog and wood ticks (Figure 23.37). Humans acquire the pathogen from tick fecal matter, which is injected into the body during a bite, or by rubbing infectious material into the skin by scratching. Cells of *R. rickettsii*, unlike other rickettsias, grow within the nucleus of the host cell as well as in host cell cytoplasm (Figure 23.37*a* and *b*). Following an incubation period of 3–12 days, an abrupt onset of symptoms occurs, including fever and a severe headache. Three to five days later, a rash is observed that is present on the palms of the hands and soles of the feet (Figure 23.37*c*) (the location of rickettsial rashes is of some diagnostic value). Gastrointestinal problems such as diarrhea and vomiting are usually observed as well, and the clinical symptoms of Rocky Mountain spotted fever may exist for over 2 weeks if the disease is left untreated. Tetracycline or chloramphenicol generally promotes a prompt recovery from Rocky Mountain spotted fever if administered early in the course of the infection.

Q fever

Q fever is a pneumonia-like infection caused by an obligate intracellular parasite, *Coxiella burnetii,* related to the rickettsias (∞ Section 16.22). Although not transmitted to *humans* directly by an insect bite, the agent of Q fever is transmitted to *animals* by insect bites, and various arthropod species serve as a reservoir of infection (the "Q" stands for *query,* because when cases of the disease were first observed, no pathogenic agent could be clearly

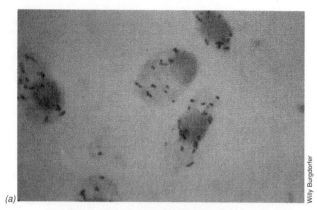
(a)

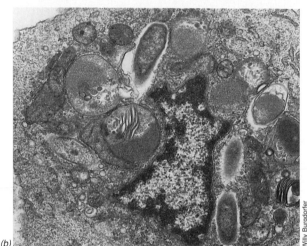
(b)

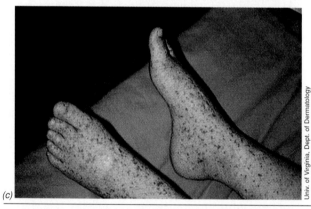
(c)

Figure 23.37 *Rickettsia rickettsii,* the causative agent of Rocky Mountain spotted fever. (a) Cells of *R. rickettsii,* growing in the cytoplasm and nucleus of tick hemocytes. Individual cells are about 0.4 μm in diameter. (b) Cells of *R. rickettsii* in a granular hemocyte of an infected wood tick, *Dermacentor andersoni.* Transmission electron micrograph. (c) Rash of the disease on the feet.

implicated). Domestic animals generally have inapparent infections, but may shed large quantities of *C. burnetii* in their urine, feces, milk, and other body fluids.

Cattle ticks are frequently implicated in transmitting the infection to dairy herds, where the disease Q fever can be a considerable occupational hazard to farmers and others involved in animal husbandry. Outbreaks of Q fever in humans have usually been traced to improperly pasteurized milk. With the recognition of Q fever as a potential milk pathogen, dairy processing facilities have modified their pasteurization protocol to ensure that pasteurized milk is *Coxiella*-free. Because *Coxiella burnetii* is a thermotolerant organism, the "holding method" of milk pasteurization, originally set by federal law at 60°C for 20 min, was raised to 62.8°C for 30 min.

Symptoms of Q fever are quite variable. They include an influenza-like illness, prolonged fever, headache and chills, chest pains, and pneumonia. Although Q fever is rarely fatal, complications arising from a *Coxiella burnetii* infection may be severe. In particular, endocarditis (inflammation of the lining of heart and heart valves) is associated with many cases. Q fever endocarditis frequently occurs months or even years after the primary infection, the organisms remaining dormant in liver cells during the interim. Damage to heart valves as a result of Q fever endocarditis may necessitate heart valve replacement surgery in later life.

An obligate intracellular parasite, *Coxiella burnetii* is difficult to isolate and culture. However, a diagnosis of Q fever can readily be made by immunological tests designed to measure host antibodies to the organism (an indirect enzyme-linked immunosorbent assay (ELISA) test has made the diagnosis of Q fever almost routine). *C. burnetii* infections respond quite dramatically to the antibiotic tetracycline, and therapy is usually begun quickly in any suspected human case of Q fever in order to prevent heart damage. Finally, Q fever is one of the infectious diseases that has been studied as a possible agent for biological warfare.

Diagnosis and control of rickettsial diseases

In the past, rickettsial infections have been difficult to diagnose because the characteristic rash associated with many rickettsial diseases may be mistaken for measles, scarlet fever, or adverse drug reactions. Clinical confirmation of rickettsial diseases has now been greatly aided by the introduction of specific immunological reagents. These include polyclonal or monoclonal antibodies that detect rickettsial surface antigens by immunofluorescence, latex bead agglutination assays, and ELISA analyses (∞ Sections 21.5, 21.7, and 21.8). Control of most rickettsial diseases requires control of the vectors: lice, fleas, and ticks. For humans traveling in wooded or grassy areas, the use of insect repellants on the exposed extremities usually prevents tick attachment. Firmly attached ticks should be removed gently with forceps, care being taken to remove all the mouth parts. A solvent such as gasoline or ethanol applied to a tick with a saturated swab usually expedites removal. Although a vaccine is available

for the prevention of typhus, the few cases reported do not warrant its general administration. No vaccines are currently available for the prevention of Rocky Mountain spotted fever or Q fever. However, an experimental Q fever vaccine, using a mixed *Coxiella burnetii* antigen preparation, is now undergoing clinical trials in human volunteers.

CONCEPT CHECK   **23.9**

Rickettsias are obligate intracellular parasitic Bacteria that are transmitted by insect or tick bite. Most rickettsial infections are readily controlled by antibiotic therapy.

✔ What are the arthropod vectors for typhus, Rocky Mountain spotted fever, and Q fever?

✔ What are the normal mammalian hosts for these same diseases?

23.10

Tick-Transmitted Diseases: Lyme Disease

Lyme disease is a tick-borne disease that affects humans and other animals. Lyme disease was named for Lyme, Connecticut, where cases were first recognized, and has rapidly become the most prevalent tick-borne disease in the United States. Lyme disease is caused by a spirochete, *Borrelia burgdorferi* (Figure 23.38), which is spread primarily by the deer tick, *Ixodes dammini*, but can also be spread by the common dog (wood) tick and other types of ticks (Figure 23.39). The ticks that carry *B. burgdorferi* cells feed on the blood of birds, domesticated animals, various wild animals, and occasionally humans; deer and the white-footed field mouse are prime hosts of the tick in the northeastern portions of the United States. However, in other parts of the country, different species of rodents and ticks are involved in the transmission of Lyme disease. Lyme disease has now also been identified in Europe and Asia, again with its life cycle in different rodent reservoirs and different species of the *Ixodes* genus of tick. The deer tick and other members of the genus are much smaller than many other types of ticks and thus are easy to overlook (Figure 23.39). Unlike the case with other tick-borne diseases, a very high percentage (up to 50% in certain regions of the Northeast) of deer ticks carry *B. burgdorferi* cells. Thus, extended contact with a vector gives a high probability of disease transmission.

Although most cases of Lyme disease have been reported from the northeastern and upper midwestern United States, cases have been observed in nearly every state, and Lyme disease is rapidly spreading west and south. For example, in California, Lyme disease is becoming more prevalent; different species of ticks

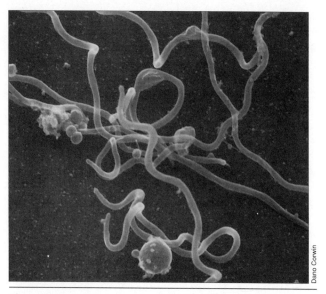

Dario Corwin

Figure 23.38 Electron micrograph of the Lyme spirochete, *Borrelia burgdorferi*. The diameter of a single cell is approximately 0.4 μm.

and mammals (in this case, the dusky-footed woodrat) have been implicated as nonhuman reservoirs. Figure 23.40 shows the recent spread of Lyme disease across the continental United States and the alarming and rapid rise in the total number of cases.

Transmission and pathogenesis of Lyme disease

Cells of *Borrelia burgdorferi* are transmitted to humans while the tick is obtaining a blood meal (Figure 23.41). A systemic infection develops leading to the main symptoms of Lyme disease, which include an acute

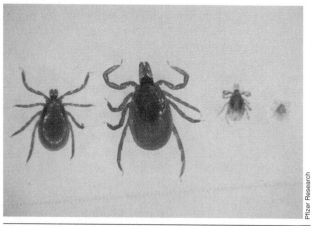

Pfizer Research

Figure 23.39 Deer ticks (*Ixodes dammini*), the major vectors of Lyme disease. Left to right, male and female adult ticks, nymph, and larval forms. The length of an adult female is about 3 mm. All forms feed on humans and are capable of transmitting *Borrelia burgdorferi*.

headache, backache, chills, and fatigue. In about 75% of all cases, a large rash is observed at the site of the tick bite (Figure 23.41). If a correct diagnosis is made at this point, Lyme disease is easily treatable with tetracycline or penicillin. However, if not treated properly, Lyme disease may progress to a chronic stage, causing a crippling numbness of the limbs and severe exhaustion. Treatment at this stage generally requires intravenous antibiotics. The drug ceftriaxone, a highly active β-lactam antibiotic, is used to treat chronic Lyme disease because it is one of the few antibiotics that can cross the blood–brain barrier and can thus attack spirochetes residing in the central nervous system. If no treatment is obtained, cells of *B. burgdorferi* infecting the central nervous system may lie dormant for long periods before eliciting a variety of additional chronic symptoms, including visual disturbances, facial paralysis, and seizures. There is also some indication that Lyme disease can cause miscarriages and stillbirths. Finally, up to 60% of untreated individuals will develop arthritis.

Because the disease is so new—the causative agent was first identified in 1982—little is known about the pathogenesis of Lyme disease. No toxins or other virulence factors have yet been identified. In many respects the latent symptoms of Lyme disease resemble those of syphilis, caused by a different spirochete, *Treponema pallidum* (see Section 23.6). Indeed, some of the neurological symptoms of Lyme disease resemble those of chronic syphilis. However, unlike syphilis, Lyme disease has not been reported to be spread by sexual intercourse or other types of human contact. Small numbers of *Borrelia burgdorferi* cells are shed in the urine of infected individuals, and there is some indication that Lyme disease can spread through domestic animal populations, particularly cattle, by infected urine.

Detection, treatment, and prevention of Lyme disease

Serological tests have been developed for detection of antibodies to *Borrelia burgdorferi*. Antibodies appear about 4–6 weeks after infection and can be detected by an indirect enzyme-linked immunosorbent assay (ELISA) or a fluorescent antibody assay (∞ Sections 21.7 and 21.8). However, because these serological tests recognize antigens on all species of *Borrelia*, the most definitive test for Lyme disease is a Western blot (∞ Section 21.9). Unfortunately, the immune response to Lyme disease is frequently not strong (as is also true of syphilis) and antibodies against the pathogen are difficult to detect in many cases of the disease. New diagnostic tools are thus being developed, including a nucleic acid probe assay that employs the polymerase chain reaction (∞ Section 21.10) to increase the number of *Borrelia burgdorferi*-specific DNA sequences in clinical specimens to a

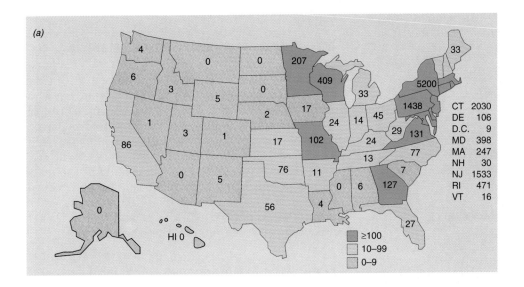

CT	2030
DE	106
D.C.	9
MD	398
MA	247
NH	30
NJ	1533
RI	471
VT	16

≥100
10–99
0–9

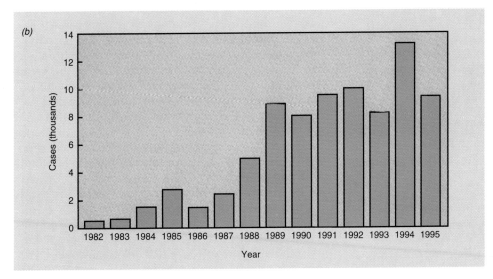

Figure 23.40 Incidence of Lyme disease in the United States. (a) Geographical incidence for 1994. Lyme disease is spreading south and west from its original focus in the Northeast. (b) Total number of reported cases of Lyme disease by year. Data are from the Centers for Disease Control, Atlanta, GA.

detectable level. Other assays involve immunological methods for detecting *B. burgdorferi* antigens in urine. However, the usefulness of all these assays has not been proven. In general, the surface antigens of *B. burgdorferi* are not well defined, and many cross reactions occur with related organisms such as *Treponema pallidum* (the causative agent of syphilis) and *Rickettsia rickettsii* (the causative agent of Rocky Mountain spotted fever). As a result, the serology-based Gunderson Lyme test is one of the most accurate indicators of *B. burgdorferi* exposure. Patient serum and complement (∞ Section 20.13) are incubated with *B. burgdorferi* cells. If antigen is bound by the patient serum, the *B. burgdorferi* cell membrane is damaged and the cell contents begin to escape. A dye, acridine orange, stains escaping DNA and confirms the presence of antibodies to *B. burgdorferi* in the serum, indicating that the patient has been exposed to the pathogen.

Early detection and treatment of Lyme disease generally afford complete recovery without further symptoms. In cases where a typical Lyme rash is observed surrounding a tick bite (Figure 23.41b), antibiotic therapy is usually initiated without waiting for an antibody titer to develop. If no rash is apparent but Lyme disease is still suspected, a Lyme ELISA test or the Gunderson Lyme test is done after an appropriate interval.

Prevention of Lyme disease requires proper precautions to prevent tick attachment. In tick-infested areas like woods, tall grass, and brush, it is advisable to wear protective clothing such as shoes, long pants, and a long-sleeved shirt with a snug collar and cuffs. Tucking the pants into tight-fitting socks worn with boots forms an effective barrier to tick attachment. After spending time in a tick-infested environment, individuals should check themselves carefully for ticks and gently remove any attached ticks (including

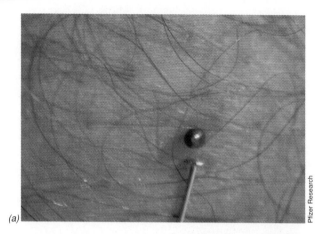

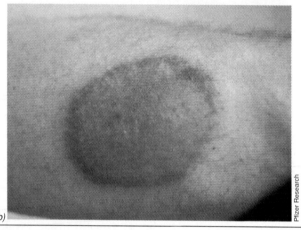

Figure 23.41 (a) Deer tick obtaining a blood meal from a human. (b) Characteristic circular rash associated with Lyme disease. The rash typically starts at the site of the bite and grows in a circular fashion over a period of several days. This rash is about 5 cm in diameter.

the head). Insect repellants containing diethyl-*m*-toluamide (DEET) are very effective if applied to both skin and clothing. Finally, several experimental vaccines are promising. In one case, vaccination of the white-footed mouse with *Borrelia* cell-wall antigens results in protection of the mouse and also transfers bactericidal immunity, presumably by means of antibodies in the blood meal, to the tick. If this early study is confirmed, this vaccine could virtually eliminate the pathogen.

CONCEPT CHECK   23.10

Lyme disease is now the most prevalent arthropod-borne disease in the United States. It is transferred from several mammalian host vectors to humans via ticks. Prevention and treatment of Lyme disease are simple, but proper diagnosis is a major problem.

✔ *What are the primary symptoms of Lyme disease?*

✔ *What antibiotics can be used to treat Lyme disease?*

23.11
Insect-Transmitted Diseases: Malaria

Malaria is a disease caused by a protozoan, a member of the Sporozoa group. We discussed sporozoa as a group in Section 18.4. The malaria parasite is one of the most important human pathogens and has played an extremely significant role in the development and spread of human culture. Indeed, as we will see, malaria has even affected human evolution. The magnitude of the malaria problem can be appreciated when it is considered that over 100 *million* cases of malaria are estimated worldwide and that at least 1 million people die from the disease each year.

Ecology and pathogenesis of malaria

Four species of sporozoa infect humans, of which the most widespread is *Plasmodium vivax*. This parasite carries out part of its life cycle in humans, and part in the mosquito vector, which spreads the parasite from person to person. Only female mosquitoes of the genus *Anopheles* are involved, and because these inhabit primarily warmer parts of the world, malaria occurs predominantly in the tropics and subtropics. Malaria did not exist in the northern regions of North America prior to settlement by Europeans but was a major problem in the South, where appropriate breeding grounds for the mosquito existed. The disease is associated with swampy low-lying areas, and the name *malaria* is derived from the Italian words for "bad air."

The life cycle of the malaria parasite is complex (Figure 23.42). First, the human host is infected by plasmodial **sporozoites,** small, elongated cells produced in the mosquito, which localize in the salivary gland of the insect. The sporozoites replicate in the liver, where they become transformed into a stage referred to as a *schizont,* which subsequently enlarges and segments into a number of small cells called **merozoites;** these cells are liberated from the liver into the bloodstream. Some of the merozoites then infect red blood cells (erythrocytes). The cycle in erythrocytes proceeds as in the liver and usually repeats at regular intervals of 48 hr in the case of *Plasmodium vivax.* It is during this period that the characteristic symptoms of malaria occur, characterized by a fever of up to 40°C (104°F) followed by chills. The chills occur when a new generation of *P. vivax* cells is liberated from erythrocytes. Vomiting and severe headache may accompany the fever–chill cycles, but asymptomatic periods generally alternate with periods in which the characteristic symptoms are present. Because of the loss of red blood cells, malaria generally causes anemia and some enlargement of the spleen as well.

Not all protozoal cells liberated from red blood cells are able to infect other erythrocytes; those that cannot,

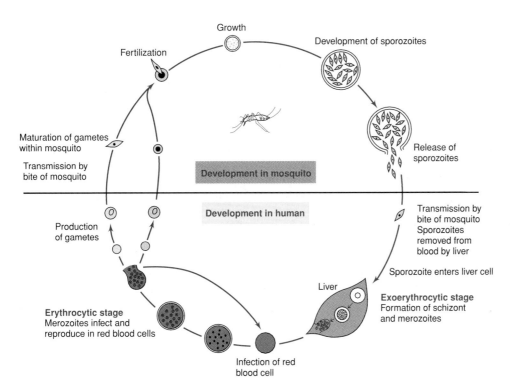

Growth

Fertilization

Development of sporozoites

Maturation of gametes
within mosquito

Transmission by
bite of mosquito

Release of
sporozoites

Development in mosquito

Development in human

Production
of gametes

Transmission by
bite of mosquito
Sporozoites
removed from
blood by liver

Sporozoite enters liver cell

Liver

Exoerythrocytic stage
Formation of schizont
and merozoites

Erythrocytic stage
Merozoites infect and
reproduce in red blood cells

Infection of red
blood cell

Figure 23.42 Life cycle of the malaria parasite, *Plasmodium vivax.*

called *gametocytes,* are infective only for the mosquito. These gametocytes are ingested when another *Anopheles* mosquito bites the infected person and they mature within the mosquito into *gametes.* Two gametes fuse, and a zygote is formed; the zygote migrates by ameboid motility to the outer wall of the insect's intestine where it enlarges and forms a number of sporozoites. These are released and some reach the salivary gland of the mosquito, from where they can be inoculated into another person; the cycle then begins again.

Conclusive evidence for the diagnosis of malaria in humans is obtained by examining blood smears for the presence of infected erythrocytes (Figure 23.43). *Chloroquine* is the drug of choice for treating parasites *within* red blood cells, but this quinine derivative does not kill stages of the malarial parasite residing *outside* erythrocytes. The related drug *primaquine* effectively eliminates the latter, and treatment of malarial patients with chloroquine and primaquine together effects a complete cure. However, because recurrences of malaria many years after a primary infection are not uncommon, small numbers of sporozoites apparently survive in the liver, protected from the effects of quinine drugs, and release merozoites months or years later to reinitiate the disease.

Plasmodium vivax has been very difficult to grow in culture, and because of its specificity for humans, experimental infection has also been difficult. For these reasons, most of the experimental work on malarial parasites has been done with species of *Plasmodium* that infect birds or rats, and it is with these species that most of the studies on the development of new drugs

have been carried out. However, scientists have been able to grow *Plasmodium falciparum,* another human pathogen, in laboratory culture, and this organism is the current model for experimental malarial research.

Eradication of malaria

Although quinine derivatives are effective in treating human cases of malaria, because of the obligatory alternation of hosts, *control* of malaria can be best effected

Infected cells

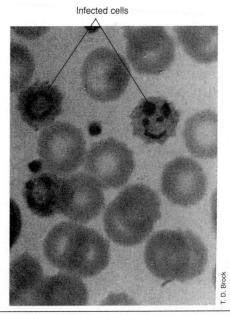

Figure 23.43 *Plasmodium vivax,* the causative agent of malaria, growing inside human red blood cells.

by elimination of the *Anopheles* mosquito. Two approaches to mosquito control are possible: (1) elimination of habitat by drainage of swamps and similar breeding areas, and (2) elimination of the mosquito by insecticides. Both approaches have been extensively used. The marked drop in incidence of malaria as a result of mosquito eradication programs in the United States is shown in Figure 23.44. During the 1930s, about 33,000 miles of ditches were dug in 16 southern states, removing 544,000 acres of mosquito breeding area. Millions of gallons of oil were also spread on swamps to reduce the oxygen supply to mosquito larvae. With the discovery of the insecticide dichlorodiphenyltrichloroethane (DDT) (∞ Figure 14.51), chemical control of both larvae and adult mosquitoes was possible. During World War II, the Public Health Service organized an Office of Mosquito Control in War Areas, and because many U.S. military bases were in the southern states, this organization carried out an extensive eradication program in the United States as well as overseas. In 1946, a year in which there were 48,610 cases of malaria in the United States, Congress established a 5-year malaria eradication program. In endemic areas, the program involved drug prophylaxis and treatment regimens for individuals, along with DDT treatment of mosquito infestations. By 1953 there were only 1310 malaria cases. In 1935 there were about 4000 deaths from malaria; in 1952 there were only 25 deaths. Thus, the overall public health threat from malaria in the United States is now minimal. However, *endemic* malaria has occurred in recent years, albeit in very low numbers, as far north as New York City.

In other parts of the world, eradication has been much slower, but the same control measures are used. Despite environmental problems (∞ Section 14.20),

DDT is still an effective agent for mosquito control and has been used for the control of malaria in developing countries.

Malaria and biochemical evolution of humans

Malaria has undoubtedly been endemic in Africa for thousands of years. In West Africans, resistance to malaria caused by *Plasmodium falciparum* is associated with the presence in their red blood cells of an altered blood protein, hemoglobin S, which differs from normal hemoglobin A at only a single amino acid in each of the two identical subunits of the molecule. In hemoglobin S, the neutral amino acid *valine* is substituted for the acidic amino acid *glutamic acid* in hemoglobin A. Red blood cells containing hemoglobin S have reduced affinity for oxygen, and the malaria parasite, having a highly aerobic metabolism, cannot grow as well in these red blood cells as it can in normal ones. However, individuals who are heterozygous for the hemoglobin S trait are less able to survive at high altitudes, where oxygen pressures are lower, but in tropical lowland Africa this disadvantage is not manifested. Individuals who are homozygous for hemoglobin S have a marked survival disadvantage, and most do not live past childhood. This is known as the sickle-cell anemia trait. In West Africans, resistance to another malarial parasite, *Plasmodium vivax*, is associated with the presence of another abnormal hemoglobin, hemoglobin E. In certain Mediterranean regions where malaria is endemic, resistance to *P. falciparum* is associated with a deficiency in the red blood cells of the enzyme glucose-6-phosphate dehydrogenase (GPD). The faulty GPD leads to higher levels of oxidants, which damage parasite membranes. In many Mediterranean

Figure 23.44 The incidence of malaria in the United States. Note the dramatic decrease from 1945 onward, principally as a result of mosquito eradication programs. The ordinate is logarithmic, and increases after 1960 are actually very small. There are currently less than 1500 cases of malaria per year in the United States. Prior to 1947, there were at least 48,000 cases per year.

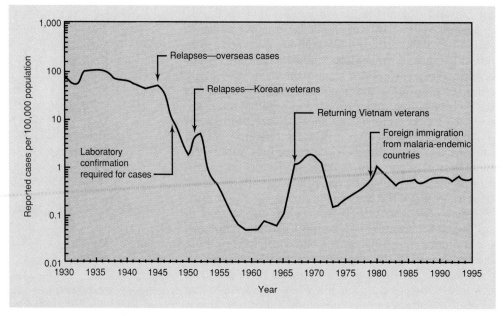

populations, a diverse group of genetic abnormalities affect hemoglobin production and efficiency. These are known collectively as the thalassemias, and all involve some aspect of hemoglobin synthesis, expression, or function. Like hemoglobins S and E and the GPD traits, the thalassemias are genetic mutations that confer a level of resistance to malaria infections and thus are selected in the population in spite of the fact that the carriers all have red blood cell defects.

Another case in which the malaria parasite influences biochemical evolution involves the major histocompatibility complex (MHC) and the immune system (∞ Section 20.7). As discussed previously, the MHC class I and class II proteins present antigens to T cells for initiation of an immune response. In malaria-prone equatorial West Africa, individuals are very likely to have one particular MHC class I gene and one particular set of class II genes, selected from a total of perhaps 100 different genes and gene sets. As a result, these particular selected MHC genes are more common in the West African population than in any other human population group. Individuals who express these genes have as much resistance to severe fatal malaria infections as those with the hemoglobin S trait. These particular MHC proteins are exceptionally good antigen-presenting molecules for certain malarial antigens and confer protective resistance to *Plasmodium* sp. infection as a result of the powerful immune response they help initiate.

Like the hemoglobin variants, the parasite acts as a strong selection agent for individual genes important for host survival: individuals with these MHC genes have a measurable survival advantage and are more likely to reproduce and pass the resistance-conferring genes to their progeny. The malaria parasite has thus been a selection factor in human evolution. Other microbial parasites have also probably promoted evolutionary changes in their hosts, but in no case do we have such clear evidence as for malaria.

CONCEPT CHECK   *23.11*

Malaria is a widespread, mosquito-borne infectious disease occurring mainly in tropical and subtropical portions of the world. It is a major cause of morbidity and mortality in developing countries and has been responsible for the evolution of several resistance genes. The disease is preventable with a combination of public health and chemotherapy measures.

✔ *How can malaria be prevented?*

✔ *Review genetic mechanisms responsible for malaria resistance. Why are anti-malarial genes not found in all humans?*

23.12
Insect-Transmitted Diseases: Plague

Pandemic occurrences of **plague** have been directly responsible for more human deaths than any other infectious disease other than malaria. Plague is caused by a gram-negative, facultatively aerobic rod called *Yersinia pestis* (∞ Section 16.20). Plague is a natural disease of domestic and wild rodents, but rats are the primary disease reservoir. Most infected rats die soon after symptoms appear, but a low proportion develop a chronic infection and can serve as a source of virulent *Y. pestis*. The majority of cases of human plague in the United States occur in the southwestern states, where the disease is endemic among wild rodents (sylvatic plague) (Figure 23.45).

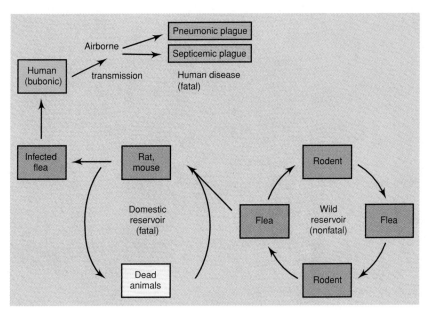

Figure 23.45 The ecology of plague. Plague in most animals is generally a mild infection. Plague in rats and humans is frequently fatal.

Plague is transmitted by the rat flea (*Xenopsylla cheopis*), which ingests *Yersinia pestis* cells by sucking blood from an infected animal. Cells multiply in the flea's intestine and can be transmitted to a healthy animal in the next bite. As the disease spreads, rat mortality becomes so great that infected fleas seek new hosts, including humans. Once in humans, cells of *Y. pestis* usually travel to the lymph nodes, where they cause the formation of swollen areas referred to as *buboes*. For this reason the disease is frequently referred to as **bubonic plague** (Figure 23.46*a*). The buboes become filled with *Y. pestis*, but the distinct capsule on cells of *Y. pestis* prevents them from being phagocytized (∞ Section 20.3). Secondary buboes form in peripheral lymph nodes, and cells eventually enter the bloodstream, causing a generalized septicemia. Multiple hemorrhages produce dark splotches on the skin (Figure 23.46*b*). If not treated prior to the septicemic stage, the symptoms of plague, including extreme lymph node pain, prostration, shock, and delirium, usually cause death within 3–5 days.

The pathogenesis of plague is not clearly understood, but it is known that cells of *Yersinia pestis* produce a number of antigenically distinct molecules, including toxins, that undoubtedly contribute to the disease process. The V and W antigens of *Y. pestis* cell walls are protein–lipoprotein complexes that serve to prevent phagocytosis. Other envelope proteins are also present. An exotoxin called *murine toxin,* because of its extreme toxicity for mice, is produced by all virulent strains of *Y. pestis.* Murine toxin is a respiratory inhibitor that blocks mitochondrial electron transport reactions at the point of coenzyme Q (∞ Section 4.12). Although it is not clear that murine toxin is involved in the pathogenesis of human plague (because murine toxin is highly toxic for certain animal species but not for others), it produces systemic shock, liver damage, and respiratory distress in mice. These symptoms are similar to those in human cases of plague. *Y. pestis* also produces a highly immunogenic *endotoxin* that may also play a role in the disease process.

Pneumonic plague occurs when cells of *Yersinia pestis* are either inhaled directly or reach the lungs during bubonic plague (Figure 23.46*c*). Symptoms are usually absent until the last day or two of the disease when large amounts of bloody sputum are emitted. Untreated cases rarely survive more than 2 days. Pneumonic plague, as one might expect, is a highly contagious dis-

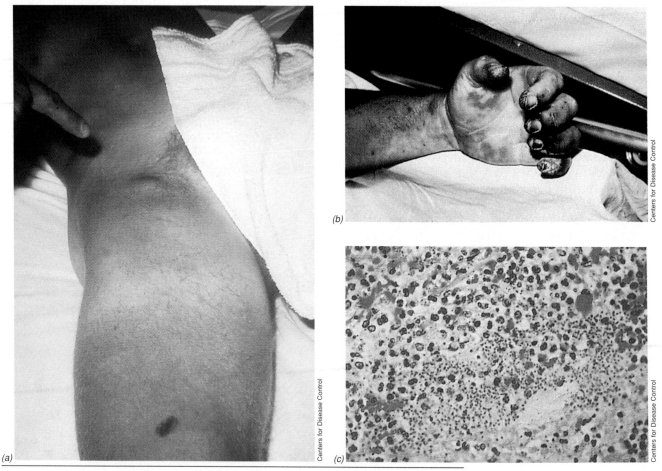

(a)

(b)

(c)

Centers for Disease Control

Figure 23.46 Plague in humans. (a) Bubo formed in the groin. (b) Gangrene and sloughing of skin in hand. (c) *Yersinia pestis,* the causative agent of plague. Cells are seen as very small blue cells from the lung tissue of a pneumonic plague victim.

ease and can spread rapidly via the respiratory route if infected individuals are not immediately quarantined. **Septicemic plague** involves the rapid spread of *Y. pestis* throughout the body via the bloodstream without the formation of buboes and usually causes death before a diagnosis can be made.

Plague can be successfully treated if swiftly diagnosed. Although *Yersinia pestis* is naturally resistant to penicillin, most strains are sensitive to streptomycin, chloramphenicol, or the tetracyclines. If treatment is started promptly, mortality from bubonic plague can be reduced to as few as 1–5% of those infected. Pneumonic and septicemic plague can also be treated, but these forms progress so rapidly that antibiotic therapy in the latter stages of the disease is usually too late. Although potentially a devastating disease, an average of fewer than 20 cases of human plague are reported each year in the United States (Figure 23.47). Worldwide, there are fewer than 1500 confirmed cases and fewer than 300 deaths per year. This is undoubtedly due to improved public health practices and the overall control of rat populations.

CONCEPT CHECK   **23.12**

Plague is largely confined to individuals who come into contact with rodent populations that are endemic reservoirs for *Yersinia pestis*. A disseminated systemic infection often leads to rapid death, but localized infections are treatable with antibiotics.

✔ *Distinguish among* bubonic, septicemic, *and* pneumonic *plague*.

✔ *What are the insect reservoir, natural host, and treatment for plague?*

23.13
Foodborne Diseases

There are two categories of foodborne disease, *food poisoning,* caused by toxins produced by microorganisms, and *food infection,* caused by growth of the microorganisms in the human body after the contaminated food has been eaten.

In addition to passive transfer of pathogens in food, active *growth* of a pathogen may also occur in foods (for example, because of improper storage), leading to marked increases in the microbial load. We discussed some aspects of heat sterilization, pasteurization, and food microbiology in Sections 11.1 and 11.6. A summary of major food poisoning and food infection agents is given in Table 23.4. We begin our discussion here with the common food poisonings, *staphylococcal* and *perfringens,* and proceed to a discussion of the most serious of all poisonings, *botulism.* We then turn our attention to common food infections including salmonellosis and similar ailments, *Campylobacter* infection, hepatitis, and traveler's diarrhea.

Staphylococcal food poisoning

The most common food poisoning is caused by the gram-positive coccus *Staphylococcus aureus.* This organism produces several *enterotoxins* (∞ Section 19.9) that are released into the surrounding medium or food; if food containing the toxin is ingested, severe reactions are observed within 1–6 hr, including nausea with vomiting and diarrhea. Six types of *S. aureus* enterotoxin have been identified, A, B, C_1, C_2, D, and E. Enterotoxin A is most frequently associated with outbreaks of staphylococcal food poisoning. The mecha-

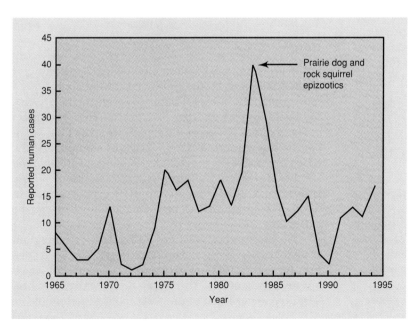

Figure 23.47 Cases of human plague in the United States in recent years.

TABLE 23.4 Major food-poisoning and food infection bacteria

Organism	Products involved
Salmonella spp.	Poultry, other meats; milk and cream, eggs
Staphylococcus aureus	Meat dishes, desserts
Clostridium perfringens	Cooked and reheated meats and meat products
Vibrio parahaemolyticus	Sea foods
Bacillus cereus	Rice and other starchy foods
Clostridium botulinum	Home-canned vegetables (especially beans and corn), smoked fish
Campylobacter jejuni	Poultry, milk
Yersinia enterocolitica	Pork, milk
Escherichia coli O157:H7	Meat

nism of action of enterotoxin A is that of a superantigen (∞ Sections 20.16 and 23.2) and involves systemic stimulation of large numbers of T cells. *S. aureus* enterotoxin A is a small single peptide of 30,000 molecular weight that is encoded by a chromosomal gene. Cloning and sequencing of this gene, the *entA* gene, and of several other *S. aureus* enterotoxin genes, show that this family of toxins is genetically related. Although the *entA* gene is chromosomally located, the B- and C-type *S. aureus* enterotoxins may be plasmid- or transposon-encoded, or alternatively encoded by a lysogenic bacteriophage. We discussed the importance of accessory genetic elements such as plasmids and bacteriophages as vectors for toxin production in Sections 19.8 and 19.9.

The kinds of foods most commonly involved in *Staphylococcus aureus* food poisoning are custard- and cream-filled baked goods, poultry, meat and meat products, gravies, egg and meat salads, puddings, and creamy salad dressings. If such foods are kept refrigerated after preparation, they remain relatively safe, as *Staphylococcus* is unable to grow at low temperatures. However, foods of this type are often not refrigerated, and are frequently kept in warm kitchens or outdoors at summer picnics. Under these conditions, *Staphylococcus*, which might have entered the food from a food handler during preparation, grows and produces enterotoxin. Many of the foods involved in staphylococcal food poisoning are not cooked again before eating, but even if they are, this toxin is relatively heat-stable and may remain active. Staphylococcal food poisoning can be prevented by careful sanitation methods or by storage of the food at low temperatures to prevent bacterial growth, and by the discarding of foods stored for any period of hours above 4°C (refrigerator temperature).

Perfringens food poisoning

Clostridium perfringens is a major cause of food poisoning in the United States. *Clostridium perfringens* produces an enterotoxin (∞ Table 19.4) that elicits diarrhea and intestinal cramps, but nausea and vomiting in only one-third of those infected. The symptoms last for about 24 hr, and fatalities are rare. The disease results from the ingestion of a large dose (>10^8 cells) of *C. perfringens* and is most frequently associated with the consumption of tainted meat or meat products.

Clostridium perfringens is quite common in a variety of cooked and uncooked foods, especially meat, poultry, and fish, and in soil and sewage. The organism is present naturally in low numbers in the human gut but apparently does not reach large numbers in the presence of the other competing intestinal microorganisms. Large doses of *C. perfringens* are most likely to be obtained from meat dishes cooked in bulk lots (heat penetration in these situations is often slow and insufficient) and then left at 20–40°C for short periods. Spores of *C. perfringens* germinate and grow quickly in the meat. Upon consumption of the contaminated meat, sporulation begins in the intestine and toxin is produced, altering the permeability of the intestinal epithelium and leading to gastrointestinal symptoms. The onset of perfringens food poisoning begins about 7–15 hr after consumption of the contaminated food. Diagnosis of perfringens food poisoning is made by isolation of *C. perfringens* from the gut or, more reliably, by a direct enzyme-linked immunosorbent assay (ELISA) to detect *C. perfringens* enterotoxin in feces (∞ Section 21.8).

Botulism

Botulism is the most severe type of food poisoning; it is often fatal and occurs following the consumption of food containing the exotoxin produced by the anaerobic bacterium *Clostridium botulinum*. This bacterium normally lives in soil or water, but its spores may contaminate raw foods before harvest or slaughter. If the foods are properly processed so that the *C. botulinum* spores are killed, no problem arises; but if viable spores are present, they may initiate growth and even a small amount of the resultant neurotoxin can be extremely poisonous.

We discussed the nature and action of botulinum toxin in Section 19.8 (∞ Figure 19.18). At least seven distinct types of botulinum toxin are known, most of which are toxic to humans. The toxins are destroyed by heat (80°C for 10 min), and so properly cooked food should be harmless, even if it originally contained toxin. Most cases of botulism occur as a result of eating foods that are not cooked after processing (Figure 23.48*a*). Canned vegetables and beans are often used without cooking in making cold salads. Similarly, smoked fish and meat and most vacuum-packed sliced meats are often eaten directly, without heating. If these products contain the botulinum toxin, then ingestion of even a small amount will result in this severe and highly dangerous type of food poisoning. In the United States, the disease is fortunately quite rare. In Japan cases of botulism are often linked to the consumption of *sushi,* a raw fish preparation.

Infant botulism occurs when spores of *Clostridium botulinum* are ingested, often from raw honey (Figure 23.48*b*). If the infant's normal flora is not well developed or if the infant is undergoing antibiotic therapy, the spores may germinate and *C. botulinum* cells may grow and release toxin. Most cases of infant botulism occur between the first week of life and 2 months of age; infant botulism is rare in children older than 6 months.

Salmonellosis

Although sometimes called food poisoning, gastrointestinal disease due to foodborne *Salmonella* is more

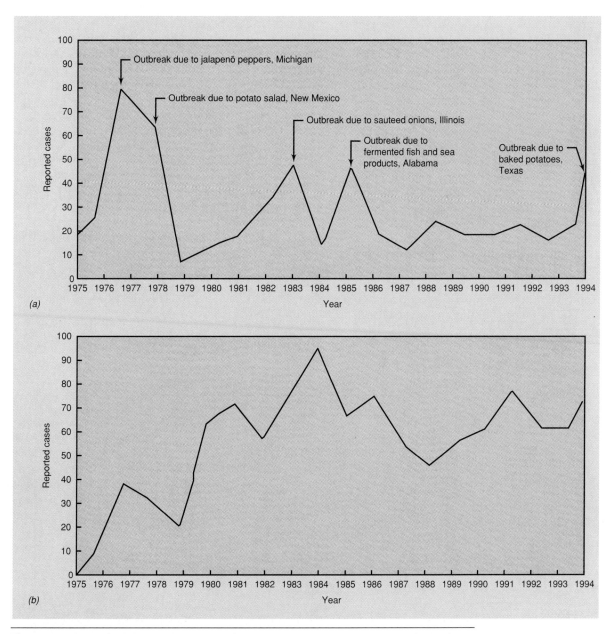

Figure 23.48 Cases of botulism in the United States in recent years. (a) Foodborne botulism. In years with higher numbers of cases, major outbreaks that account for the increase are indicated. (b) Infant botulism.

aptly called *Salmonella* **food infection** because symptoms arise only after the pathogen grows in the intestine (hence symptoms can begin several *days* after eating a contaminated food). The symptoms of salmonellosis include the sudden onset of headache, chills, vomiting, and diarrhea, followed by a fever that lasts a few days. Diagnosis is from symptoms and by culture of the organism from feces. Virtually all species of *Salmonella* are pathogenic for humans: one, *S. typhi,* causes the serious human disease typhoid fever, and a small number of other species cause foodborne gastroenteritis. *Salmonella typhimurium* is the most common cause of salmonellosis in humans.

The ultimate sources of the foodborne salmonellas are humans and warm-blooded animals. The organism reaches most food by contamination from food handlers (∞ The Tragic Case of Typhoid Mary, Chapter 22). For foods such as eggs or meat, the animal that produced the food may be the source of contamination. The foods most commonly implicated are meats and meat products (such as meat pies, sausage, and cured meats), poultry, eggs, and milk and milk products. *Salmonella* food infections are often traced to products made with *uncooked* eggs such as custards, cream cakes, meringues, pies, and eggnog. Properly cooked foods are safe if consumed immediately or if stored immediately at 4°C or less. However, cooked or canned foods that become contaminated by an infected food handler can support the growth of *Salmonella* if they are held for long periods of time, especially without refrigeration. *Salmonella* infection is more common in summer than in winter, probably because warm environmental conditions are more favorable for growth of microorganisms in foods. The incidence of salmonellosis has been steadily and significantly increasing over the last several decades, as shown in Figure 23.49.

Escherichia coli

Several strains of *Escherichia coli* have recently emerged as potent foodborne pathogens. They are characterized by their ability to produce potent enterotoxins (∞ Section 19.9) and thus are designated *enterotoxic E. coli,* or ETEC strains. One particular strain, *E. coli* O157:H7, causes at least 20,000 cases of food infection and 250 deaths each year. The disease produced is characterized by bloody diarrhea and is therefore further classified as an *enterohemorrhagic E. coli* (EHEC). It is also a leading cause of kidney failure in children. *E. coli* O157:H7 produces potent toxins related to the *Shigella* toxins, which we will discuss in the following section.

The most common source of this infection is contaminated uncooked or undercooked meat. For example, several major outbreaks have involved infected ground beef. In at least one instance, a regional distribution center was involved and the infected product caused disease in four western states. Another outbreak involved processed, cured, but uncooked, beef. The major source of contamination seems to be the source of the beef, and contamination probably originated from strains in slaughtered beef carcasses.

Diagnosis of infection by *Escherichia coli* O157:H7 involves culture from the feces and identification of the O and H antigens and toxins by serology. Subtyping of strains is also done using molecular methods such as restriction fragment length polymorphism (RFLP) and pulse field gel electrophoresis (∞ DNA Fingerprinting, Chapter 10, and Working with Nucleic Acids: The Tools, Chapter 6). This is now a nationally reportable infectious disease (∞ Table 22.4). The most effective way to prevent infection with *E. coli* O157:H7 is to make sure that meat is cooked thoroughly, which

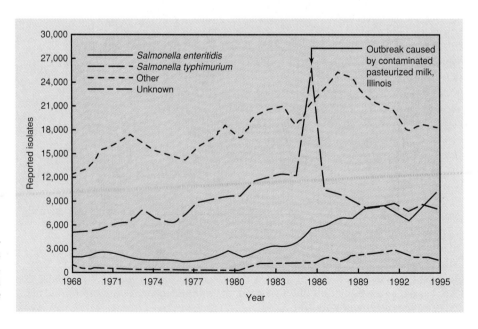

Figure 23.49 The number of cases of salmonellosis in the United States, by year and species of isolate. Data are from the Centers for Disease Control, Atlanta, GA.

means that it should appear gray or brown and juices should be clear.

Traveler's diarrhea is another extremely common enteric infection in North Americans and Europeans traveling to developing countries. The primary causal agents are the enteropathogenic *Escherichia coli*, although *Salmonella* and *Shigella* are sometimes implicated. Several studies have been done on groups of U.S. citizens traveling in Mexico. Such studies have shown that the infection rate in travelers is often greater than 50% and that the prime vehicles are foods, such as uncooked vegetables (for example, lettuce in salads) and water. The very high infection rate in travelers is due to contamination of local public water supplies. The local population is usually immune to the infecting strains, undoubtedly because they have lived with the agent for a long period of time. Secretory antibodies present in the bowel may prevent successful colonization of the pathogen in local residents, but when the organism colonizes the intestine of a nonimmune person, it finds a hospitable environment. Also, stomach acidity, so often a barrier to intestinal infection, may not be able to act if only small amounts of liquid are consumed (as, for instance, the melting ice of a cocktail) because small amounts of liquid induce rapid emptying of the stomach and hence pass through so quickly that stomach acidity may have no effect on an enteric pathogen present in the liquid.

Campylobacter

Campylobacter is a gram-negative, curved rod that grows at reduced oxygen tensions, that is, as a microaerophile (∞ Section 16.11). Two major species are recognized, *Campylobacter jejuni* and *C. fetus,* and these species probably account for the majority of cases of bacterial diarrhea in children. *Campylobacter fetus* is also of economic importance because it is a major cause of sterility and spontaneous abortion in cattle and sheep. The symptoms of *Campylobacter* infection include a high fever (usually greater than 104°F or 40°C), nausea, abdominal cramps, and a watery, frequently bloody, stool. Diagnosis requires isolation of the organism from stool samples and identification by growth-dependent tests or immunological assays. Because of the frequency with which *C. jejuni* infections are observed in infants, a variety of selective media and highly specific immunological methods have been developed for positive identification of this organism.

Campylobacter is transmitted to humans via contaminated food, most frequently in poultry, pork, raw clams, and other shellfish, or by a water route in surface waters not subjected to chlorination. Poultry is a major reservoir of *C. jejuni,* and virtually all chicken and turkey carcasses contain this organism (Table 23.4). Beef, on the other hand, is rarely a vehicle for *Campylobacter*. Proper washing of uncooked poultry (and any utensils coming in contact with uncooked poultry) and thorough cooking of the meat eliminate the possibility of *Campylobacter* infection. *Campylobacter* species also infect domestic animals such as dogs, causing a milder form of diarrhea than that observed in humans. Infant cases of *Campylobacter* infection are frequently traced to infected domestic animals, especially dogs.

Helicobacter pylori *and peptic ulcers*

Helicobacter pylori is a gram-negative spiral-shaped bacterium that is related to *Campylobacter* (∞ Section 16.11). *H. pylori,* first identified in human intestinal biopsies in 1983, is associated with gastritis and peptic ulcers. Up to 80% of ulcer patients have concomitant *H. pylori* infection, and about 50% of all adults, especially in developing countries, are chronically infected. Although the mode of transmission has not been established, there is no known nonhuman reservoir of *H. pylori*. In addition, infection occurs in high incidence in certain families, and the overall incidence in the population increases with age. These factors suggest a host-to-host type of transmission (∞ Sections 22.3 and 22.5). However, infections sometimes occur in epidemic clusters, suggesting that a common source such as food or water may also be involved (∞ Section 22.5).

Helicobacter pylori is not invasive, but colonizes the gastric mucosal surfaces, where a host response causes inflammation. Researchers speculate that severe and untreated inflammation leads to tissue destruction and ulceration. Antibodies to *H. pylori* are usually present in infected individuals, but do not prevent colonization. Therefore, individuals who acquire *H. pylori* tend to have chronic, long-term infections.

More evidence for a causal association between *Helicobacter pylori* and ulcers comes from treatments for the disease. Long-term treatment of most ulcers with antacid preparations has never been uniformly successful, and most patients relapse within one year. However, by treating ulcers as an infectious disease, permanent cures are often possible. Treatment usually consists of a combination of drugs including metranidazole, a second antibiotic such as tetracycline, and a bismuth-containing antacid preparation. The combination treatment abolishes the *H. pylori* infection and seems to cure the ulcers on a long-term basis. In spite of all of this anecdotal information, the causal relationship between *H. pylori* infection and ulcers has not been unequivocally established. The current battery of circumstantial evidence, however, points more and more to *H. pylori* as a major cause of ulcers.

Hepatitis

Infectious hepatitis (hepatitis A) is a virus-mediated inflammation of the liver caused by a picornavirus (positive, single-strand RNA virus) (∞ Section 8.15).

The virus is transmitted primarily through fecal contamination of water, food, or milk. **Hepatitis A** infection can be subclinical in mild cases or can lead to severe liver damage in chronic infections. The type A virus spreads from the intestine via the bloodstream to the liver and usually results in jaundice, a yellowing of the skin and eyes, and a browning of the urine as a result of stimulation of bile pigment production by infected liver cells. An immune response is initiated against the hepatitis A virus, and this eventually brings the condition under control. However, in severe cases, permanent loss of a portion of liver function can occur.

The most significant food vehicles for type A hepatitis are shellfish (oysters and clams) harvested from waters polluted with human feces. As filter feeders, shellfish living in such environments tend to concentrate the hepatitis virus. Only *raw* shellfish are a problem because hepatitis A virus is destroyed by heating. Therefore, the best means of controlling hepatitis transmission include sound sanitary practices, especially sewage treatment, the prevention of fecal contamination of food products by infected food handlers, and avoiding the consumption of uncooked shellfish.

Serum hepatitis (hepatitis B) is caused by a *DNA-containing* hepatitis virus transmitted primarily by infected blood or blood products (∞ Section 21.12). The type B virus can also be spread by maternal transmission *in utero* and from one infected individual to another by sexual intercourse. Serum hepatitis frequently results in more severe liver damage than infectious hepatitis, leading to death in up to 10% of those infected; hepatitis A rarely causes death. Long-term infection with type B virus is also associated with increased risk of liver cancer. Those at high risk for serum hepatitis include individuals who require frequent blood transfusions, dialysis patients, health care workers, and intravenous drug abusers. As in the case of infectious hepatitis, no specific therapy for serum hepatitis is currently available. However, because of the seriousness of the infection, a vaccine against serum hepatitis is now available. Use of the serum hepatitis vaccine is recommended for those in the high risk groups already mentioned, as well as for health personnel who come in frequent contact with blood or blood products. A variety of immunological tests, especially ELISAs (∞ Section 21.8), have been devised for the diagnosis of infectious and serum hepatitis by either antibody detection or direct viral detection methods.

A third type of hepatitis virus, unrelated to type A or type B and referred to as type C hepatitis virus, is also known. Although not yet reliably grown in tissue culture, the type C virus is widespread in the human population and is now the most common hepatitis virus encountered in cases of hepatitis mediated by blood transfusion. Like type B, type C hepatitis virus is spread primarily by contaminated blood and possibly also by sexual transmission. Although less likely to cause serious liver damage than type A or type B hepatitis viruses, the type C virus can cause chronic ailments such as cirrhosis. In addition, *at least* two other poorly characterized strains, hepatitis D and hepatitis E, are now known (∞ Table 22.6).

Altogether, the various forms of hepatitis account for over 40,000 illnesses per year (Figure 23.50). Control of these infections is based primarily on hygiene and cleanliness measures, but vaccines are also important. As mentioned above, there is now an effective vaccine in widespread use for hepatitis B and a vaccine is also available for hepatitis A (∞ Section 20.17), but none is yet available for hepatitis C, D, or E.

Assessing microbial content of foods

All fresh foods have some viable microorganisms present. The purpose of assay methods is to detect evidence of abnormal microbial growth in foods or to detect the presence of specific organisms of public health concern, such as *Salmonella*, *Staphylococcus*, and *Clostridium botulinum*. We discussed in Section 21.10 the use of nucleic acid probes for the detection of specific foodborne pathogens, and these methods are finding increasing use. For cultural studies of nonliquid food products, preliminary treatment is usually required to suspend microorganisms embedded or entrapped within the food in a liquid medium. The most suitable method for treatment is high speed blending. Examination of the food should be done as soon after sampling as possible, and if examination cannot begin within 1 hr of sampling, the food should be refrigerated. A frozen food should be thawed in its original container in a refrigerator and examined as soon as possible after thawing is complete. For *Salmonella*, several selective media are available (∞ Section 21.2), and tests for its presence are most commonly done on animal food products, such as raw meat, poultry, eggs, and powdered milk, because *Salmonella* from production animals is the usual source of food contamination. For staphylococcal counts, a high-salt medium (either sodium chloride or lithium chloride at a final concentration of 7.5%) is used. Of the

---

CONCEPT CHECK   23.13

Many microbial diseases are foodborne. Some foodborne diseases are food poisonings in which the causal agent produces an exotoxin or enterotoxin. This toxin is eaten with the food and causes the food poisoning symptoms. Other foodborne diseases are food infections, in which the food is the agent by which the pathogen is transmitted to the host.

✔ *What are some common food infections? Intoxications?*

✔ *How can you treat a food infection? An intoxication?*

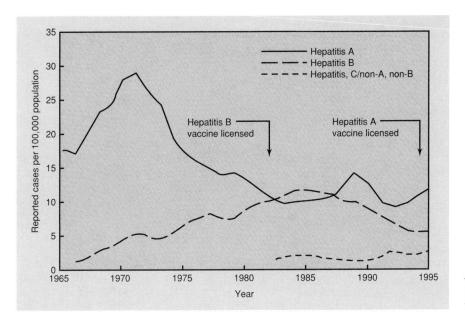

Figure 23.50 Incidence of hepatitis in the United States in recent years.

organisms present in foods, staphylococci are the only common ones tolerant of such levels of salt. Because *Staphylococcus aureus* is responsible for one of the most common types of food poisoning, staphylococcal counts are of considerable importance. We discussed food preservation in Section 11.6.

23.14
Waterborne Diseases

Human pathogens that are transmitted by water include bacteria, viruses, and protozoa (Table 23.5). Organisms transmitted by water usually grow in the intestinal tract and leave the body in the feces. Thus, they are *infections*. Fecal pollution of water supplies may then occur, and if the water is not properly treated, the pathogens enter a new host when the water is consumed. Because water is consumed in large quantities, it may be infectious even if it contains only a small number of pathogenic organisms. We discussed wastewater treatment and drinking water purification in Section 12.15.

Pathogenic organisms transmitted by water

Probably the most important pathogenic *bacteria* transmitted by the water route are *Salmonella typhi*, the organism causing typhoid fever, and *Vibrio cholerae*, the organism causing cholera. Although the causal agent of typhoid fever may also be transmitted by contaminated food (see Section 23.13) and by direct contact from infected people, the most common and serious means of transmission is the water route. Typhoid fever has been virtually eliminated in many parts of

the world, primarily as a result of the development of effective water treatment methods. However, a breakdown in water purification methods, contamination of water during floods, earthquakes, and other disasters, or cross-contamination of water pipes from leaking sewer lines occasionally results in epidemics of typhoid fever.

The causal agent of cholera is also usually transmitted by the water route. However, during the recent cholera epidemic in the Americas, consumption of raw shellfish and raw vegetables was also implicated as a means of cholera spread. Presumably, the vegetables were washed in contaminated water and the shellfish beds were contaminated by raw sewage. At one time, cholera was common in Europe and North America, but the disease has been virtually eliminated from these areas by effective water purification. The disease is still common in Asia and in certain parts of Central and South America Travelers to these ares are advised to be vaccinated for cholera. Both *Vibrio cholerae* and *Salmonella typhi* are eliminated from sewage during proper sewage treatment and hence do not enter water courses receiving treated sewage effluent. More frequent than typhoid, but generally less serious, is salmonellosis caused by species of *Salmonella* other than *S. typhi* (see Section 23.13). As seen in Table 23.5, the largest number of *cases* of waterborne bacterial disease in the United States have been due to *Salmonella* spp. infections, except for a single, massive outbreak due to *Cryptosporidium* (see later). *Salmonella* spp. also account for the largest number of waterborne disease *outbreaks*.

Enteric bacteria are effectively eliminated from water during water purification and so they should never be present in properly treated drinking water.

TABLE 23.5　Waterborne disease outbreaks due to microorganisms[a]

Disease	Agent	Outbreaks[b] (%)	Cases[c] (%)
Bacteria			
Typhoid fever	*Salmonella typhi*	10	0.1
Shigellosis	*Shigella* spp.	9	2.6
Salmonellosis	*Salmonella paratyphi* and other *Salmonella* species	3	3.5
Gastroenteritis	*Escherichia coli*	0.3	0.7
	Campylobacter spp.	0.3	0.7
Viruses			
Infectious hepatitis	Hepatitis A virus	11	0.5
Diarrhea	Norwalk virus	1.5	0.6
Protozoa			
Giardiasis	*Giardia lamblia*	7	3.8
Cryptosporidiosis[d]	*Cryptosporidium parvum*	0.2	71
Unknown etiology			
Gastroenteritis		57	16.7

[a]Compiled from data provided by the Centers for Disease Control, Atlanta, GA.

[b]Of more than 650 outbreaks in recent decades.

[c]Of 520,000 cases over the same period.

[d]A single outbreak of cryptosporidiosis in 1993 caused illness in 370,000 individuals from Milwaukee, Wisconsin (USA). This is the largest single recorded outbreak of a waterborne disease in history.

Most outbreaks of waterborne disease in the United States are due to breakdowns in treatment systems or are a result of postcontamination in pipelines. This latter problem can be controlled by maintaining a detectable level of free chlorine in pipelines.

Viruses transmitted by the water route include poliovirus and other viruses of the enterovirus group, as well as hepatitis A (Table 23.5). Poliovirus has several modes of transmittal, and transmission by water may be of serious concern in some areas. Poliovirus, however, has now been eliminated from the Americas.

Because viruses are acellular, they are more stable in the environment and are not as easily killed as bacteria. However, both poliovirus and infectious hepatitis virus are eliminated from water by proper treatment practices, and the maintenance of 0.6 parts per million (ppm) free chlorine in a water supply generally ensures its safety.

Cholera

We discussed the action of cholera enterotoxin in Section 19.9. The disease cholera is caused by *Vibrio cholerae*, a gram-negative, curved rod transmitted almost exclusively via contaminated water. Cholera enterotoxin catalyzes a life-threatening diarrhea that can result in dehydration and death unless the patient is given fluid and electrolyte therapy. The disease has swept the world in seven major pandemics, the most recent of which began in South America in 1991. Since then, more than 900,000 cases of cholera and more than 8000 deaths have been reported. Today, the disease is threatening to take

hold in developed countries (for instance, cases have been reported along the Gulf coast of the United States) and is very common in developing countries in the Americas, especially in areas where sewage treatment is either not practiced or poorly performed.

Two major biotypes of *Vibrio cholerae* have been recognized, the *classic* and the *El Tor* types. Each biotype has two major serotypes. The classic strain, such as the *V. cholerae* strain first isolated by Robert Koch in 1883, was more prevalent in cholera outbreaks before 1960, whereas the El Tor strain has been more frequently observed since that time. Following ingestion of a substantial inoculum, the *V. cholerae* cells take up residence in the *small* intestine. Studies with human volunteers have shown that the acidity of the stomach is responsible for the large inoculum needed to initiate cholera. Although the ingestion of 10^8–10^9 cholera vibrios is generally required to cause cholera, human volunteers given bicarbonate to neutralize gastric acidity developed the disease when only 10^4 cells were administered. Far lower cell numbers are required to initiate infection if administered with food. Cholera vibrios attach firmly to small intestinal epithelium and grow and release enterotoxin. The enterotoxin causes huge fluid losses, 20 liters per day not being uncommon in a fulminant case of cholera. If untreated, the mortality rate can be as high as 60%. Intravenous or oral liquid replacement therapy is the major means of treatment (∞ Section 19.9). Streptomycin or tetracycline may shorten the course of cholera, but antibiotics are of little benefit without simultaneous fluid replacement.

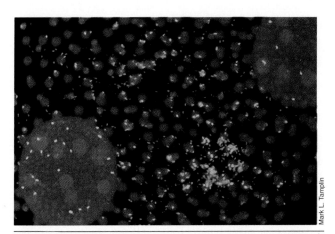

Figure 23.51 *Vibrio cholerae* attached to the surface of *Volvox,* a freshwater alga. The isolate was from a cholera-endemic area in Bangladesh. The green *V. cholerae* cells are stained with a monoclonal antibody to bacterial cell surface proteins. The red color is due to the fluorescence of chlorophyll *a* in algal cells.

Control of cholera depends primarily on satisfactory sanitation measures, particularly in the treatment of sewage and the purification of drinking water. *Vibrio cholerae* organisms can adhere to normal flora in fresh water and can survive for long periods of time (Figure 23.51). In the last 10 years sporadic outbreaks of cholera (less than 300 total cases) have been reported in the United States, and some evidence exists that raw shellfish may be an alternate vehicle, as mentioned previously. Cholera is heavily endemic in India, Pakistan, Bangladesh, and the Americas, with occasional epidemics flourishing from time to time. Although most of these cases are due to water contaminated with human feces, cholera is also spread within households in these regions by fecally contaminated food.

Giardiasis

Giardiasis is an acute form of gastroenteritis caused by the protozoan parasite *Giardia lamblia* (Figure 23.52). *G. lamblia* is a flagellated protozoan that is transmitted to humans primarily by contaminated water, although foodborne and even sexual transmission of giardiasis has been documented (∞ Section 18.4). The protozoal cells, called trophozoites (Figure 23.52*a*) produce a resting stage called a cyst (Figure 23.52*b*), and this is the primary form transmitted by water. Cysts germinate in the gastrointestinal tract and bring about the symptoms of giardiasis: an explosive, foul-smelling, watery diarrhea and intestinal cramps, flatulence, nausea, and malaise. The foul-smelling nature of the diarrhea and the absence of blood or mucus in the stool are diagnostically helpful in distinguishing giardiasis from diarrhea of bacterial or viral origin. The drugs quinacrine and metronidazole are useful in treating the disease.

Between 1965 and 1981, 53 waterborne outbreaks of giardiasis affecting over 20,000 people were reported in the United States. Most outbreaks occurred in undeveloped or mountainous regions where surface water sources are used for drinking purposes. *Giardia* cysts are fairly resistant to chlorine, and many outbreaks have been associated with water systems using only chlorination as a means of water purification. Water subjected to proper sedimentation, filtration, and chlorination is generally free of *Giardia* cysts. Many cases of giardiasis have been associated with untreated drinking water in wilderness areas. Studies of wild animals have indicated that beavers and muskrats are major carriers of *Giardia* and may transmit cells or cysts to water supplies. As a safety precaution, all water consumed from rivers and streams, for example, during a camping or hiking trip, should be filtered, treated with iodine or chlorine, or boiled. Boiling is the preferred method of rendering water microbially safe.

Cryptosporidiosis

The largest single outbreak of a waterborne disease ever recorded took place in Milwaukee, Wisconsin, in the spring of 1993 (Table 23.5). About 370,000 people developed a diarrheal illness that was traced to the municipal water supply. Apparently, spring rains and runoff from

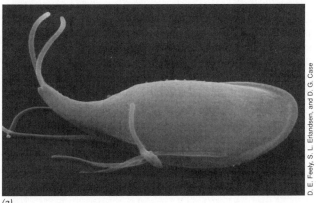

(a)

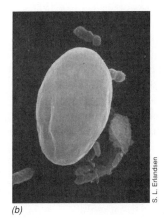

(b)

Figure 23.52 Scanning electron micrographs of the parasite *Giardia.* (a) Motile trophozoite. (b) Cyst.

surrounding farmland had overburdened the water supply system, leading to contamination by the protozoan *Cryptosporidium parvum*. The protozoan is a significant intestinal pathogen in dairy cattle, a likely source of this outbreak.

Cryptosporidiosis is characterized by severe diarrhea, but in normal individuals this is self-limiting and most people recover without incident within 2 weeks. However, individuals with impaired immunity such as acquired immunodeficiency syndrome (AIDS) patients or the very young or old can develop serious complications. In this case, about 4000 people required hospital care and several died from complications of the disease, including severe dehydration.

Cryptosporidium is highly resistant to chlorine (up to 14 times as resistant as the chlorine-resistant *Giardia*), and therefore methods for removing it rely on sedimentation and filtration. The Milwaukee outbreak points out the fragility of water purification systems, the need for constant water monitoring and surveillance, and the consequences of failure in a large supply system.

Amebiasis

A number of different amebas inhabit the tissues of humans and other vertebrates, usually in the oral cavity or intestinal tract. Some of these are pathogenic. We discussed the general properties of ameboid protozoa in Section 18.4. *Entamoeba histolytica*, the causative agent of amebiasis, is a common pathogenic protozoan transmitted to humans primarily by contaminated water and occasionally by the foodborne route (Figure 23.53). *E. histolytica* is an *anaerobic* ameba, the trophozoites lacking mitochondria. Like *Giardia,* the trophozoites of *Entamoeba* produce cysts. Cyst germination occurs in the intestine, and cells grow both on and in intestinal mucosal cells. Continued growth leads to ulceration of intestinal mucosa, causing diarrhea and severe intestinal cramps. Diarrhea is replaced by a condition referred to as *dysentery,* characterized by the passage of intestinal exudates, blood, and mucus. If not treated, trophozoites of *E. histolytica* can migrate to the liver, lung, and brain. Growth in these tissues can cause abscesses and other tissue damage.

Amebiasis can be treated with the drugs metronidazole and chloroquine, but amebicidal drugs are not always effective. Spontaneous cures do occur, implying that the host immune system plays some role in ending the infection. However, protective immunity is not afforded by primary infection because reinfection is not uncommon. Amebiasis is rather easily diagnosed by examining stool samples for the morphologically distinct cysts of *Entamoeba histolytica*. The disease occurs at very low incidence in regions that practice adequate sewage treatment. Ineffective sewage treatment and use of untreated surface waters for drinking purposes are the usual scenarios for cases of amebiasis.

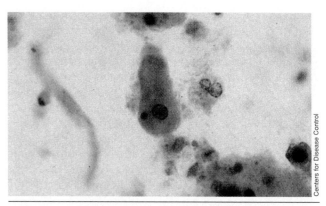

Figure 23.53 Trophozoites of *Entamoeba histolytica,* the causative agent of amebiasis. The small red structures are red blood cells.

CONCEPT CHECK   23.14

Waterborne diseases are a very significant source of morbidity, especially in developing countries. A variety of etiological agents are involved. Recent massive outbreaks in developed countries and the newest cholera pandemic emphasize the need for water sanitation.

✔ *What are the common sources of waterborne infection due to bacteria?*

✔ *Is cholera an infection or is the disease due to a toxin?*

23.15
Public Health and Water Quality

As we have just discussed, a number of important diseases are waterborne. Even water that looks clear and pure may be sufficiently contaminated with pathogenic microorganisms to be a health hazard. Some means are necessary to ensure that drinking water is safe. Unfortunately, it is not usually practical to examine water directly for the various pathogenic organisms that may be present. As stated earlier, a wide variety of organisms may be present, including bacteria, viruses, and protozoa. To check each drinking water supply for each of these agents would be difficult and time-consuming. In practice, *indicator organisms* are used instead. These are organisms, usually associated with the intestinal tract, whose presence in water indicates that the water has received contamination of an intestinal origin. The most widely used indicator is the **coliform group** of organisms. This group is defined in water bacteriology as all the aerobic and facultatively aerobic, gram-negative, non-spore-forming, rod-shaped Bacteria that ferment lactose with gas formation within 48 hr at 35°C. This is an operational rather than a taxonomic definition, and the coliform group includes a variety of organisms, mostly of intestinal origin. In practice, the coliform organisms are almost always members of the enteric bacterial group (∞ Section 16.20). The coliform group includes the organism *Escherichia coli*, a common intestinal organism, and the organism *Klebsiella pneumoniae*, a less common intestinal inhabitant. The definition also currently includes organisms of the species *Enterobacter aerogenes* not generally associated with the intestine.

The coliform group of organisms are suitable as indicators because they are common inhabitants of the intestinal tract, both of humans and warm-blooded animals, and are present in the intestinal tract in large numbers. When excreted into water, the coliforms eventually die, but they do not die at a faster rate than the pathogenic bacteria *Salmonella* and *Shigella*, and both the coliforms and the pathogens behave similarly during water purification processes. Thus, it is likely that if coliforms are found in a water sample, the water has received fecal contamination and may be unsafe for drinking purposes. Finally, the coliform group includes organisms derived not only from humans but also from other warm-blooded animals. Because many of the pathogens (for example, *Salmonella, Leptospira*) found in warm-blooded animals also infect humans, an indicator of both human and animal pollution is desirable.

The coliform test

There are two types of procedures that are used for the coliform test. These are the **most-probable-number** (MPN) procedure and the **membrane filter** (MF) procedure. The MPN procedure employs liquid culture medium in test tubes, the samples of drinking water being added to the tubes of media. In the more common MF procedure, the sample of drinking water is passed through a sterile membrane filter, which removes the bacteria (∞ Figure 11.9), and the filter is then placed on a culture medium for incubation. When using the membrane filter method with drinking water, at least 100 ml of water should be filtered, although in clean water systems, even larger volumes can be filtered. After filtration of a known volume of water, the filter is placed on the surface of a plate of eosin–methylene blue (EMB) culture medium, which is highly selective for coliform organisms (∞ Section 21.2 and Figure 21.4). The coliform colonies (Figure 23.54) are counted, and from this value the number of coliforms in the original water sample can be determined. In well-regulated water supply systems, coliform tests are always negative. If the coliform tests are not uniformly negative, a breakdown in the system has occurred (such as in chlorination) or in the distribution network (pipelines).

Drinking water standards in the United States are specified under the Safe Drinking Water Act, which provides a framework for the development of drinking water standards by the Environmental Protection Agency (EPA). Current standards prescribe that when the MF technique is used, 100-ml samples must be filtered and the number of coliform bacteria shall not exceed any of the following: (1) 1 per 100 ml as the arithmetic mean of all samples examined per month; (2) 4 per 100 ml in more than one sample when less than 20 are examined per month; or (3) 4 per 100 ml in more than 5% of the samples when 20 or more are examined per month. Water utilities report their results to the EPA, and if they do not meet the prescribed standards, they must notify the public and take steps to correct the problem. Many smaller communities and even large cities sometimes fail to meet the standards.

Public health significance of drinking water purification

Today the incidence of waterborne disease in developed countries is so low that it is difficult to appreciate the significance of treatment practices and drinking water standards. Most intestinal infection today is not due to transmission by the water route but via food (see Section 23.13). It was not always so. Until the twentieth century, effective water treatment practices did not exist, and there were no bacteriological methods for evaluating the health significance of polluted drinking water (∞ Snow on Cholera, Chapter 22). The first coliform counting procedures were introduced about 1905. Until then, water purification, if practiced at all, was primarily for aesthetic

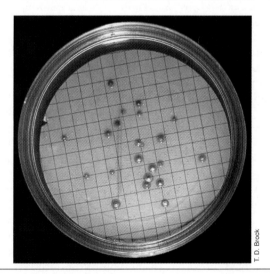

Figure 23.54 Coliform colonies growing on a membrane filter. A drinking water sample has been passed through the filter. The filter was placed on eosin–methylene blue (EMB) media that is both selective and differential for lactose-fermenting bacteria (coliforms) (∞ Section 21.2). The dark color of the colonies is characteristic of coliforms. A count of the number of colonies gives a measure of the coliform count of the original water sample.

purposes, to remove turbidity. Actually, turbidity removal by filtration provides a significant decrease in the microbial load of water, and so filtration did play a part in providing safer drinking water. But filtration alone was of only partial value because many organisms passed through the filters. It was the discovery of chlorine as an extremely efficient water disinfectant, in about 1910, that had major impact. Chlorine is so effective and so inexpensive that its use spread widely; chlorination was of major significance in reducing the incidence of waterborne disease. However, the effectiveness of chlorination would not have been realized, and the necessary doses could not have been determined, if standard methods for assessing the coliform content of drinking water had not been developed. Thus, engineering and microbiology moved forward together.

The significance of filtration and chlorination for ensuring the safety of drinking water cannot be overemphasized. Figure 23.55 illustrates the dramatic drop in incidence of typhoid fever in a major American city after these two purification procedures were introduced. Similar results were obtained in other major cities. The dramatic improvement in the health of the American people in the early decades of the twentieth century was due to a large extent to the establishment of satisfactory water purification procedures.

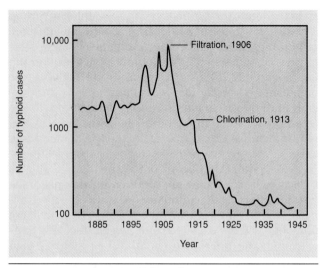

Figure 23.55 The dramatic effect of water purification on the incidence of waterborne disease. The graph shows the incidence of typhoid fever in Philadelphia during the early part of the twentieth century. Note the marked reduction in incidence of the disease after the introduction of filtration and chlorination.

CONCEPT CHECK   23.15

Assessment of possible contamination of drinking water sources is accomplished by counting coliform bacteria. Strict microbiological standards make this method a reliable indicator of fecal contamination. The introduction of water purification measures is probably the single most important public health measure ever devised.

✔ *Is the coliform test an assay for pathogens? Is it an indicator of water quality? Why or why not?*

✔ *Why is filtration an important component in water purification?*

23.16
Pathogenic Fungi

The pathogens discussed in this chapter are bacteria, viruses, and protozoa. However, some human diseases are caused by fungi. Discussion of these diseases has been saved until last because the fungi normally occur in nature as free-living saprophytes and are primarily opportunistic pathogens. Thus, they are found in some form in nearly every ecological environment, usually growing on nonliving organic materials. The fungi include the eukaryotic organisms we commonly refer to as *yeasts,* which normally grow as single cells

(Figure 23.56*a*), and *molds,* which grow in branching chains called *hyphae* (Figure 23.56*b*). The taxonomy and biological diversity of these organisms were discussed in Section 18.2. Fortunately, most fungi are harmless to humans. Only about 50 species cause human disease, and the overall incidence of serious fungal infections is rather low, although certain superficial fungal infections are quite common.

Fungi cause disease through three major mechanisms. First, some fungi cause immune responses that can result in allergic (hypersensitivity) reactions following exposure to specific fungal antigens (∞ Section 20.16). Exposure to fungi, whether growing on the host or in the environment, may cause allergic symptoms on reexposure. For example, *Aspergillus* spp., a common saprophyte often found in nature as a leaf mold, is a potent and common allergen, often causing asthma and other hypersensitivity reactions. As we shall see later, *Aspergillus* also has other mechanisms for producing disease.

A second fungal disease-producing mechanism involves the production and action of *mycotoxins,* a large, diverse group of fungal exotoxins. The best-known examples of mycotoxins are produced by *Aspergillus flavus,* an organism that commonly grows on improperly stored food such as grain. The toxins produced by *A. flavus* are known as *aflatoxins* (Figure 23.57). Aflatoxins are highly toxic and induce tumors in some animals, especially in birds that feed on contaminated grain. Their direct role in human disease is not well defined.

The third fungal disease-producing mechanism is through infection. The growth of a fungus on or in the body is called a *mycosis* (plural, *mycoses*). Mycoses can

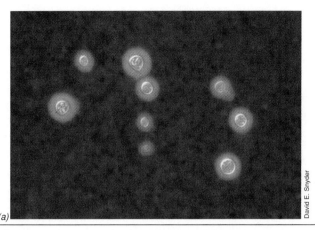
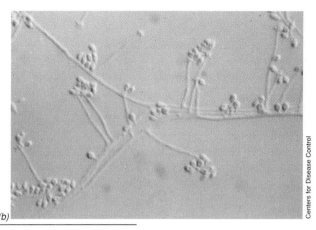

Figure 23.56 Typical forms of pathogenic fungi. (a) Yeast form of encapsulated *Cryptococcus neoformans*, stained with india ink. The cell itself ranges from 4 to 20 μm. (b) *Sporothrix schenckii*, showing the branching, or hyphae, characteristic of the mold form of fungi. The round basidiospores are about 2 μm in diameter.

range in severity from relatively innocuous, superficial diseases to serious, life-threatening diseases.

Mycoses

Mycoses are subdivided into three categories. The first of these is the *superficial mycoses*. These diseases involve colonization of the skin, hair, or nails and infect only the surface layers (Figure 23.58a). Table 23.6 lists some of the common superficial fungi. In general, these diseases are relatively benign and self-limiting. Some, such as *Trichophyton* infections of the feet (athlete's foot), are quite common. Spread is by personal contact with an infected person or by contact with contaminated surfaces such as bathtubs or shower stalls or other contaminated shared articles such as towels or bed linens. Treatment for severe cases is with topical application of miconazole nitrate or griseofulvin. Griseofulvin is also administered orally. After entering the bloodstream, it passes to the skin where it can inhibit fungal growth.

The *subcutaneous mycoses* involve deeper layers of skin (Figure 23.58b) and a different group of organisms (Table 23.6). One disease in this category is *sporotrichosis*, an occupational hazard of agricultural workers, miners, and other workers who come into contact with the soil. The causative organism is found as a ubiqui-

tous saprophyte on wood and in soil. The lesions, usually initiated by fungal infection of a small wound or abrasion, resemble the ulcers and chancres seen in ulcerating diseases such as syphilis (compare Figures 23.58b and 23.26). The causative organism, *Sporothrix schenckii*, can readily be isolated from the lesion and cultured *in vitro*. Treatment is with oral potassium iodide or oral ketoconazole.

The *systemic mycoses* involve fungal growth in internal organs of the body. These are subclassified as *primary* or *secondary* infections. A primary infection is one resulting directly from the fungal pathogen in otherwise normal, healthy individuals. A secondary infection involves infection with the pathogen only in hosts with a predisposing condition such as antibiotic therapy or immunosuppression. In the United States, the most widespread primary fungal infections are *histoplasmosis*, caused by *Histoplasma capsulatum*, and *coccidioidomycosis* (San Joaquin Valley fever), caused by *Coccidioides immitis*. Both of these organisms normally live in soil. These are both respiratory diseases in which the host becomes infected by breathing in airborne spores, which germinate and grow in the lungs. Histoplasmosis is primarily a disease of rural areas in the midwestern United States, especially in the Ohio and Mississippi river valleys. Most cases are mild and can be mistaken for more common respiratory infections. San Joaquin Valley fever is generally restricted to the desert regions of the southwestern United States. The fungus lives in desert soils, and the spores are disseminated on dry, windblown particles that are inhaled. In some areas in the southwestern United States, as many as 80% of the inhabitants may be infected, although most individuals suffer no apparent ill effects.

A number of fungal infections, including histoplasmosis and coccidioidomycosis, are especially serious and common in individuals whose immune systems have been impaired, either by acquired immunodefi-

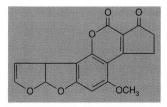

Figure 23.57 Structure of aflatoxin B1. This toxin is one of a group of related compounds produced by *Aspergillus flavus*.

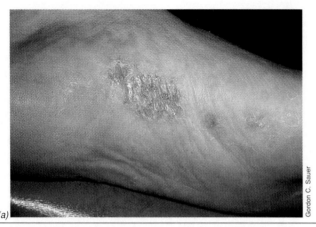
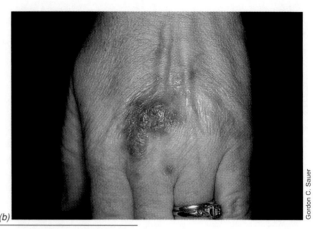

Figure 23.58 Fungal infections. (a) Superficial mycosis of the foot (athlete's foot) due to infection with *Trichophyton rubrum*. (b) Sporotrichosis, a subcutaneous infection due to *Sporothrix schenckii*.

ciency syndrome (AIDS) (see Section 23.7) or by immunosuppressive drugs. These are secondary fungal diseases because normal individuals either do not get the disease or generally have a less severe form. Examples of other fungal organisms involved as secondary pathogens are given in Table 23.6. These fungi are known as *opportunistic* pathogens because of their particular ability to cause serious infections only in individuals with impaired defense mechanisms (in particular AIDS patients) (see Section 23.7, the box, A Definition of AIDS, and Figure 23.30).

Effective chemotherapy against systemic fungal infections is very difficult (∞ Section 11.12). Most antibiotics that inhibit fungi also harm other eukaryotic organisms, including the human host. One of the most effective antibiotics, amphotericin B, is widely used to treat systemic fungal infections of humans, but serious side effects such as kidney toxicity may occur.

CONCEPT CHECK   **23.16**

Fungal diseases are often difficult to control because of a lack of suitable fungus-specific antibiotics. Serious systemic infections due to the normally nonpathogenic *Candida albicans* and other fungi have been routinely seen in AIDS patients as opportunistic pathogens. The three classes of fungal infections (mycoses) include the superficial, the subcutaneous, and the systemic mycoses.

✔ *Distinguish a superficial mycosis from a systemic mycosis.*

✔ *Why are mycoses so difficult to treat with antibiotics?*

TABLE 23.6	Some pathogenic fungi and the diseases they cause	
Disease	**Causal organism**	**Main disease foci**
Superficial mycoses (dermatomycoses)		
Ringworm	*Microsporum*	Scalp of children
Favus	*Trichophyton*	Scalp
Athlete's foot	*Epidermophyton, Trichophyton*	Between toes, skin
Jock itch	*Trichophyton, Epidermophyton*	Genital region
Subcutaneous mycoses		
Sporotrichosis	*Sporothrix schenckii*	Arms, hands
Chromoblastomycosis	Several fungal genera	Legs, feet
Systemic mycoses		
Cryptococcosis[a]	*Cryptococcus neoformans*	Lungs, meninges
Coccidioidomycosis[a]	*Coccidioides immitis*	Lungs
Histoplasmosis[a]	*Histoplasma capsulatum*	Lungs
Blastomycosis	*Blastomyces dermatitidis*	Lungs, skin
Candidiasis[a]	*Candida albicans*	Oral cavity, intestinal tract

[a]Considered opportunistic pathogens and have been implicated in the pathogenesis of AIDS.

Material for Review

REVIEW QUESTIONS

1. Why do many gram-positive Bacteria cause respiratory diseases?

2. What are the typical symptoms of a staphylococcal skin infection and what properties of *Staphylococcus aureus* are responsible for these symptoms?

3. Describe the process of infection by *Mycobacterium tuberculosis*. Does infection always lead to active disease? Why or why not?

4. Why are colds and influenza such common respiratory diseases? Discuss at least five reasons for their high incidence.

5. Compare and contrast the diseases *measles, mumps,* and *rubella.* Include in your discussion a description of the pathogen, major symptoms encountered, and any potentially serious consequences of these infections. Why is it essential that females be vaccinated against rubella at an early age?

6. Discuss at least three reasons why the incidence of gonorrhea rose dramatically in the mid-1960s, whereas the incidence of syphilis did not rise.

7. For the sexually transmitted diseases listed below, describe the methods for treating each infection. In each case, is the treatment an effective cure? Why or why not? Why is trichomoniasis not treatable with antibiotics such as penicillin and tetracycline?

   (a) Gonorrhea, (b) syphilis, (c) chlamydial infections, (d) genital herpes, and (e) trichomoniasis.

8. Describe how human immunodeficiency virus (HIV) effectively shuts down most aspects of both humoral immunity and cell-mediated immunity. Are there any effectors of the immune system that remain functional in cases of acquired immunodeficiency syndrome (AIDS)?

9. Describe the sequence of events you would take if a child received an unprovoked bite from a stray dog.

10. Discuss at least three common properties of the disease agents and review the disease process for the following diseases: Rocky Mountain spotted fever, typhus, Q fever, and Lyme disease.

11. Discuss the following as they relate to Lyme disease: causative agent, mode of transmission, symptoms, therapy, prevention, and diagnosis.

12. The classic symptoms of malaria, fever followed by chills, are related to activities of the pathogen *Plasmodium vivax* or *P. falciparum.* Describe the growth stages of this pathogen in the human host and relate them to the classic symptoms. Why might a person of Western European descent be more susceptible to malaria than a person of West African descent?

13. Contrast *food infection* and *food poisoning.* Give an example of each and indicate how each disease could be effectively controlled.

14. Compare and contrast the toxins responsible for cholera, botulism, and staphylococcal food poisoning. What is the mechanism of action of each toxin? In each case, is *infection* with the causative organism necessary for the initiation of disease symptoms?

15. Describe the steps used to carry out a coliform test on a sample of drinking water. What are the acceptable limits for coliforms in drinking water?

16. What are the three categories of fungal infections? Why are fungal infections particularly difficult to treat?

APPLICATION QUESTIONS

1. Most sneezes generate an *aerosol* of mucous droplets containing microorganisms. Why are infections due to sneeze transmission more likely to be *upper* rather than *lower* respiratory tract infections?

2. Explain why *rheumatic fever* is usually considered a type of *autoimmune* disease.

3. Why does the disease *tuberculosis* often lead to a permanent reduction in lung capacity, whereas most other respiratory diseases cause only temporary respiratory problems?

4. Discuss the molecular biological and immunological bases of *antigenic shift* and the reasons why this phenomenon effectively thwarts control measures for influenza. Why do cycles of influenza occur about every 3–5 years? Next, compare antigenic *shift* to antigenic *drift.* Which mechanism is more important for the influenza virus? Which causes the greatest antigenic change? Which creates the biggest problems for vaccine developers? Why?

5. Despite the ease with which gonorrhea can be diagnosed and cured, its incidence remains very high. Why?

6. As the director of your dormitory's public health advisory group, you are charged to present information on sexually transmitted diseases (STDs), including prevention, symptoms, notification, and treatment. Outline your information program for three separate STDs. Will your program for each disease overlap? For each of the diseases, justify the need for reporting an occurrence of the disease to sexual partners of the infected individual.

7. What is the current basis for treatment of AIDS? Does it work? Why or why not? Describe the various vaccine possibilities for AIDS. How might each vaccine type actually induce immunity to the disease? Is passive immunity of potential use for treatment of AIDS? Why or why not? What risks may be involved?

8. From what has been described about the reservoir of *Legionella pneumophila,* explain why this disease shows

an atypical disease cycle as compared with other respiratory diseases.

9. What similarities does Lyme disease have to the diseases Rocky Mountain spotted fever and typhus? To syphilis? How might each disease be controlled at the vector level? Is this possible for all, or any, of these diseases? Has control been accomplished for any of these diseases? Explain.

10. Write a scenario for a typical food poisoning outbreak due to *Staphylococcus*. What conditions might lead to the development of the toxic food, how might the outbreak arise, and how should patients be treated?

11. Compare and contrast *Clostridium perfringens* and *C. botulinum* food poisoning. Discuss similarities and differences between the organisms involved, the role of infec-

tion in the disease, the toxins produced, and the symptoms observed.

12. Why is the incidence of *Salmonella* infection higher in the human population in the summer than in the winter? What measures can be used to reduce this trend?

13. Why was the development of the coliform test an important step in the development of effective methods for drinking water purification? Describe a molecular biology-based test of your own design that could replace the coliform test. What organisms would you try to identify?

14. With regard to infections, why are fungi often secondary opportunistic pathogens in immunologically compromised individuals? What particular problems, especially in terms of therapy, do fungi pose for the clinician?

SUPPLEMENTARY READINGS

Baker, H. F., and **R. M. Ridley.** 1996. *Prion Diseases.* Humana Press, Totowa, NJ. The newest hypotheses and findings concerning this new and frightening category of latent infectious diseases.

Baron, S. 1992. *Medical Microbiology.* Addison-Wesley, Menlo Park, CA. An advanced-level medical microbiology text; very complete and up-to-date.

Brock, T. D. (ed.). 1989. *Microorganisms: From Smallpox to Lyme Disease.* W. H. Freeman, New York. A collection of short articles on infectious diseases taken from the pages of *Scientific American.*

Brooks, G. F., J. S. Butel, and **L. N. Ornston.** 1995. *Jawetz, Melnick and Adelberg's Medical Microbiology,* 20th edition. Appleton and Lange, Norwalk, CT. A complete, short medical text covering many aspects of medical microbiology. Very current.

Conte, J. E., Jr. 1995. *Manual of Antibiotics and Infectious Diseases,* 8th edition. Williams and Wilkins, Baltimore, MD. An invaluable reference source for clinical treatment and diagnostic information about infectious diseases.

Holmes, K. K., P. Mardh, P. F. Sparling, and **P. J. Wiener** (eds.). 1990. *Sexually Transmitted Diseases,* 2nd edition. McGraw-Hill, New York. A comprehensive, authoritative treatment of sexually transmitted diseases.

Howard, B. J., J. K. Reiser, T. F. Smith, A. S. Weissfeld, and **R. C. Tilton.** 1994. *Clinical and Pathogenic Microbiology,* 2nd edition. Mosby and Sons, St. Louis. An up-to-date, complete medical reference.

Morbidity and Mortality Weekly Report. Massachusetts Medical Society, Waltham, MA. The weekly report of the Centers for Disease Control. Regular issues and supplements are invaluable sources of information concerning infectious diseases in the United States.

Murray, P. R., E. J. Baron, M. A. Pfaller, F. C. Tenover, and **R. H. Yolken.** 1995. *Manual of Clinical Microbiology,* 6th edition. American Society for Microbiology, Washington, DC. A standard reference work for clinical microbiologists. Includes discussions of the science behind clinical applications.

Sauer, G. C. 1991. *Manual of Skin Diseases,* 6th edition. J. B. Lippincott, Philadelphia. An excellent source of information on fungal and bacterial diseases of the skin.

Schaechter, M., G. Medoff, and **B. Eisenstein.** 1993. *Mechanisms of Microbial Disease,* 2nd edition. Williams and Wilkins, Baltimore, MD. A medical school textbook that takes a case history approach to medical microbiology. Very readable.

Stine, G. J. 1996. *AIDS Update 1996.* Prentice-Hall, Englewood Cliffs, NJ. An excellent monograph describing the current status of the AIDS epidemic. Updated annually.

World Health Organization. 1996. *World Health Report 1996.* WHO, Geneva. A survey of infectious diseases throughout the entire world. An invaluable source for global information.

 On~line resources for this chapter are on the World Wide Web at: http://www.prenhall.com/~brock (click on the table of contents link and then select Chapter 23).

APPENDIX 1 *Energy Calculations in Microbial Bioenergetics*

The information in Appendix 1 is intended to help students calculate changes in free energy accompanying chemical reactions carried out by microorganisms. It begins with definitions of the terms required to make such calculations and proceeds to show how knowledge of redox state, atomic and charge balance, and other factors are necessary to calculate free-energy problems successfully.

Definitions

1. ΔG^0 = standard free-energy change of the reaction at 1 atm pressure and 1 M concentrations; ΔG = free-energy change under the conditions specified; $\Delta G^{0'}$ = free-energy change under standard conditions at pH 7.

2. Calculation of ΔG^0 for a chemical reaction from the free energy of formation, G_f^0, of products and reactants:

$$\Delta G^0 = \Sigma\, \Delta G_f^0 \text{ (products)} - \Sigma\, \Delta G_f^0 \text{ (reactants)}$$

   That is, sum the ΔG_f^0 of products, sum the ΔG_f^0 of reactants, and subtract the latter from the former.

3. For energy-yielding reactions involving H^+, converting from standard conditions (pH 0) to biochemical conditions (pH 7):

$$\Delta G^{0'} = \Delta G^0 + m\, \Delta G_f'(H^+)$$

   where m is the net number of protons in the reaction (m is negative when more protons are consumed than formed) and $\Delta G_f'(H^+)$ is the free energy of formation of a proton at pH 7 = -39.87 kJ at 25°C.

4. Effect of concentrations on ΔG: with soluble substrates, the concentration ratios of products formed to exogenous substrates used are generally equal to or greater than 10^{-2} at the beginning of growth and equal to or less than 10^{-2} at the end of growth. From the relation between ΔG and the equilibrium constant (see item 8), it can be calculated that ΔG for the free-energy yield in practical situations differs from the free-energy yield under standard conditions by at most 11.7 kJ, a rather small amount, and so for a first approximation, standard free-energy yields can be used in most situations. However, with H_2 as a product, H_2-consuming bacteria present may keep the concentration of H_2 so low that the free-energy yield is significantly affected. Thus, in the fermentation of ethanol to acetate and H_2 ($C_2H_5OH + H_2O \rightarrow C_2H_3O_2^- + 2\,H_2 + H^+$), the $\Delta G^{0'}$ at 1 atm H_2 is $+9.68$ kJ, but at 10^{-4} atm H_2 it is -36.03 kJ. With H_2-consuming bacteria present, therefore, the ethanol fermentation becomes useful. (See also item 9.)

5. Reduction potentials: by convention, electrode equations are written in the direction, oxidant + $ne^- \rightarrow$ reductant (that is, as reductions), where n is the number

of electrons transferred. The standard potential (E_0) of the hydrogen electrode, $2\,H^+ + 2\,e^- \rightarrow H_2$ is set by definition at 0.0 V at 1.0 atm pressure of H_2 gas and 1.0 M H^+, at 25°C. E_0' is the standard reduction potential at pH 7. See also Table A1.2.

6. Relation of free energy to reduction potential:

$$\Delta G^{0'} = -nF\, \Delta E_0'$$

   where n is the number of electrons transferred, F is the Faraday constant (96.48 kJ/V), and $\Delta E_0'$ is the E_0' of the electron-accepting couple minus the E_0' of the electron-donating couple.

7. Equilibrium constant, K. For the generalized reaction $aA + bB \rightleftharpoons cC + dD$,

$$K = \frac{[C]^c[D]^d}{[A]^a[B]^b}$$

   where A, B, C, and D represent reactants and products; a, b, c, and d represent number of molecules of each; and brackets indicate concentrations. This is true only when the chemical system is in equilibrium.

8. Relation of equilibrium constant, K, to free-energy change. At constant temperature and pressure,

$$\Delta G = \Delta G^0 + RT \ln K$$

   where R is a constant (8.29 J/mol/K) and T is the absolute temperature (K).

9. Two substances can react in a redox reaction even if the standard potentials are unfavorable, provided that the concentrations are appropriate.

Assume that normally the reduced form of A would donate electrons to the oxidized form of B. However, if the concentration of the reduced form of A were low and the concentration of the reduced form of B were high, it would be possible for the reduced form of B to donate electrons to the oxidized form of A. Thus, the reaction would proceed in the direction opposite that predicted from standard potentials. A practical example of this is the utilization of H^+ as an electron acceptor to produce H_2. Normally, H_2 production in fermentative bacteria is not extensive because H^+ is a poor electron acceptor; the E_0' of the 2 H^+/H_2 pair is -0.42 V. However, if the concentration of H_2 is kept low by continually removing it (a process done by methanogenic prokaryotes, which use $H_2 + CO_2$ to produce methane, CH_4, or by many other anaerobes capable of consuming H_2 anaerobically), the potential will be more positive and then H^+ will serve as a suitable electron acceptor.

Oxidation State or Number

1. The oxidation state of an element in an elementary substance (for example, H_2, O_2) is zero.

2. The oxidation state of the ion of an element is equal to its charge (for example, $Na^+ = +1$, $Fe^{3+} = +3$, $O^{2-} = -2$).

3. The sum of oxidation numbers of all atoms in a neutral molecule is zero. Thus, H_2O is neutral because it has two H at +1 each and one O at −2.

4. In an ion, the sum of oxidation numbers of all atoms is equal to the charge on that ion. Thus, in the OH^- ion, $O(-2) + H(+1) = -1$.

5. In compounds, the oxidation state of O is virtually always −2, and that of H is +1.

6. In simple carbon compounds, the oxidation state of C can be calculated by adding up the H and O atoms present and using the oxidation states of these elements as given in item 5, because in a neutral compound the sum of all oxidation numbers must be zero. Thus, the oxidation state of carbon in methane, CH_4, is −4 (4 H at +1 each = +4); in carbon dioxide, CO_2, the oxidation state of carbon is + 4 (2 O at −2 each = −4).

7. In organic compounds with more than one C atom, it may not be possible to assign a specific oxidation number to each C atom, but it is still useful to calculate the oxidation state of the compound as a whole. The same conventions are used. Thus, the oxidation state of carbon in glucose, $C_6H_{12}O_6$, is zero (12 H at +1 = 12; 6 O at −2 = −12) and the oxidation state of carbon in ethanol, C_2H_6O, is −4 (6 H at +1 = +6; one O at −2).

8. In all oxidation–reduction reactions there is a balance between the oxidized and reduced products. To calculate an oxidation–reduction balance, the number of molecules of each product is multiplied by its oxidation state. For instance, in calculating the oxidation–reduction balance for the alcoholic fermentation, there are two molecules of ethanol at −4 = −8 and two molecules of CO_2 at +4 = +8 so the net balance is zero. When constructing model reactions, it is useful to calculate redox balances to be certain that the reaction is possible.

Calculating Free-Energy Yields for Hypothetical Reactions

Energy yields can be calculated either from free energies of formation of the reactants and products or from differences in reduction potentials of electron-donating and electron-accepting partial reactions.

Calculations from free energy

Free energies of formation are given in Table A1.1. The procedure to use for calculating energy yields of reactions follows.

1. *Balancing reactions*   In all cases, it is essential to ascertain that the coupled oxidation–reduction reaction is balanced. Balancing involves three things: (*a*) the total number of each kind of atom must be identical on both sides of the equation; (*b*) there must be an ionic balance so that when positive and negative ions are added up on the right side of the equation, the total ionic charge (whether positive, negative, or neutral) exactly balances the ionic charge on the left side of the equation; and (*c*) there must be an oxidation–reduction balance so that all the electrons removed from one substance must be transferred to another substance. In general, when constructing balanced reactions, one proceeds in the reverse of the three steps just listed. Usually, if steps (*c*) and (*b*) have been properly handled, step (*a*) becomes correct automatically.

2. *Examples*   (*a*) What is the balanced reaction for the oxidation of H_2S to SO_4^{2-} with O_2? First, decide how many electrons are involved in the oxidation of H_2S to SO_4^{2-}. This can be most easily calculated from the oxidation states of the compounds, using the rules given previously. Because H has an oxidation state of +1, the oxidation state of S in H_2S is −2. Because O has an oxidation state of −2, the oxidation state of S in SO_4^{2-} is +6 (because it is an ion, using the rules given in items 4 and 5 of the previous section). Thus, the oxidation of H_2S to SO_4^{2-} involves an eight-electron transfer (from −2 to +6). Because each O atom can accept two electrons (the oxidation state of O in O_2 is zero, but in H_2O is −2), this means that two molecules of molecular oxygen, O_2, are required to provide sufficient electron-accepting capacity. Thus, at this point, we know that the reaction requires 1 H_2S and 2 O_2 on the left side of the equation, and 1 SO_4^{2-} on the right side. To achieve an ionic balance, we must have two positive charges on the right side of the equation to balance the two negative charges of SO_4^{2-}. Thus, 2 H^+ must be added to the right side of the equation, making the overall reaction

$$H_2S + 2\,O_2 \rightarrow SO_4^{2-} + 2\,H^+$$

By inspection, it can be seen that this equation is also balanced in terms of the total number of atoms of each kind on each side of the equation.

(*b*) What is the balanced reaction for the oxidation of H_2S to SO_4^{2-} with Fe^{3+} as electron acceptor? We have just ascertained that the oxidation of H_2S to SO_4^{2-} is an eight-electron transfer. Because the reduction of Fe^{3+} to Fe^{2+} is only a one-electron transfer, 8 Fe^{3+} will be required. At this point, the reaction looks like

$$H_2S + 8\,Fe^{3+} \rightarrow 8\,Fe^{2+} + SO_4^{2-} \qquad \text{(not balanced)}$$

We note that the ionic balance is incorrect. We have 24 positive charges on the left and 14 positive charges on the right (16+ from Fe, 2− from sulfate). To equalize the charges, we add 10 H^+ on the right. Now our equation looks like

$$H_2S + 8\,Fe^{3+} \rightarrow 8\,Fe^{2+} + 10\,H^+ + SO_4^{2-}$$
$$\text{(not balanced)}$$

To provide the necessary hydrogen for the H^+ and oxygen for the sulfate, we add 4 H_2O to the left and find that the equation is now balanced:

$$H_2S + 4\ H_2O + 8\ Fe^{3+} \rightarrow 8\ Fe^{2+} + 10\ H^+ + SO_4^{2-}$$
(balanced)

In general, in microbiological reactions, ionic balance can be achieved by adding H^+ or OH^- to the left or right side of the equation, and because all reactions take place

in an aqueous medium, H_2O molecules can be added where needed. Whether H^+ or OH^- is added generally depends on whether the reaction is taking place under acid or alkaline conditions.

3. *Calculation of energy yield for balanced equations from free energies of formation* Once an equation has been balanced, the free-energy yield can be calculated by inserting the values for the free energy of formation of each reactant and product from Table A1.1 and using the formula in item 2 of the first section of this appendix.

TABLE A1.1 **Free energies of formation, G^0_f, for some substances (kJ/mol)^a**

Carbon compound	Metal	Nonmetal	Nitrogen compound
CO, −137.34	Cu$^+$, +50.28	H$_2$, 0	N$_2$, 0
CO$_2$, −394.4	Cu^{2+}, +64.94	H$^+$, 0 at pH 0	NO, +86.57
CH$_4$, −50.75	CuS, −49.02	−5.69 per pH unit	NO$_2$, +51.95
H$_2$CO$_3$, −623.16	Fe^{2+}, −78.87	O$_2$, 0	NO$_2^-$, −37.2
HCO$_3^-$, −586.85	Fe^{3+}, −4.6	OH$^-$, −157.3 at pH 14	NO$_3^-$, −111.34
CO$_3^{2-}$, −527.90	FeCO$_3$, −673.23	−198.76 at pH 7	NH$_3$, −26.57
Acetate, −369.41	FeS$_2$, −150.84	−237.57 at pH 0	NH$_4^+$, −79.37
Alanine, −371.54	FeSO$_4$, −829.62	H$_2$O, −237.17	N$_2$O, +104.18
Aspartate, −700.4	PbS, −92.59	H$_2$O$_2$, −134.1	
Benzoic acid, −245.6	Mn^{2+}, −227.93	PO$_4^{3-}$, −1026.55	
Butyrate, −352.63	Mn^{3+}, −82.12	Se^0, 0	
Caproate, −335.96	MnO$_4^{2-}$, −506.57	H$_2$Se, −77.09	
Citrate, −1168.34	MnO$_2$, −456.71	SeO$_4^{2-}$, −439.95	
Crotonate, −277.4	MnSO$_4$, −955.32	S^0, 0	
Cysteine, −339.8	HgS, −49.02	SO$_3^{2-}$, −486.6	
Ethanol, −181.75	MoS$_2$, −225.42	SO$_4^{2-}$, −744.6	
Formaldehyde, −130.54	ZnS, −198.60	S$_2$O$_3^{2-}$, −513.4	
Formate, −351.04		H$_2$S, −27.87	
Fructose, −915.38		HS$^-$, +12.05	
Fumarate, −604.21		S^{2-}, +85.8	
Gluconate, −1128.3			
Glucose, −917.22			
Glutamate, −699.6			
Glycerate, −658.1			
Glycerol, −488.52			
Glycine, −314.96			
Glycolate, −530.95			
Guanine, +46.99			
Lactate, −517.81			
Lactose, −1515.24			
Malate, −845.08			
Mannitol, −942.61			
Methanol, −175.39			
Oxalate, −674.04			
Phenol, −47.6			
n-Propanol, −175.81			
Propionate, −361.08			
Pyruvate, −474.63			
Ribose, −757.3			
Succinate, −690.23			
Sucrose, −370.90			
Urea, −203.76			
Valerate, −344.34			

^{a}Values for free energy of formation of various compounds can be found in Dean, J. A. 1973. *Lange's Handbook of Chemistry*, 11th edition. McGraw-Hill, New York; Garrels, R. M., and C. L. Christ. 1965. *Solutions, Minerals, and Equilibria*. Harper and Row, New York; Burton, K. 1957. In Krebs, H. A., and H. L. Kornberg. Energy transformations in living matter, *Ergebnisse der Physiologie* (appendix). Springer-Verlag, Berlin; and Thauer, R. K., K. Jungermann, and H. Decker. 1977. Energy conservation in anaerobic chemotrophic bacteria. *Bacteriol. Rev.* 41:100–180.

For instance, for the equation

$$H_2S + 2\,O_2 \rightarrow SO_4^{2-} + 2\,H^+$$

$$G'_f \text{ values} \rightarrow (-27.87) + (0)\quad (-744.6) + 2\,(-39.83)$$

$$\text{(assuming pH 7)}$$

$$\Delta G^{0\prime} = -796.39 \text{ kJ/mol}$$

The values on the right are summed and subtracted from the values on the left, taking care to ensure that the signs are correct. From the data in Table A1.1, a wide variety of free-energy yields for reactions of microbiological interest can be calculated.

Calculation of free-energy yield from reduction potential

Reduction potentials of some important redox pairs are given in Table A1.2. The amount of energy that can be released from two half reactions can be calculated from the differences in reduction potentials of the two reactions and from the number of electrons transferred. The further apart the two half reactions are, and the greater the number of electrons, the more energy released. The conversion of potential difference to free energy is given by the formula $\Delta G^0 = -nF\,\Delta E_0'$, where n is the number of electrons, F is the Faraday constant (96.48 kJ/V), and $\Delta E_0'$ is the difference in potentials. Thus, the 2 H^+/H_2 couple has a potential of -0.41 and the $\tfrac{1}{2}O_2/H_2O$ pair has a potential of $+0.82$, and so the potential difference is 1.23, which (because two electrons are involved) is equivalent to a free-energy yield (ΔG^0) of -237.34 kJ/mol. On the other hand, the potential difference between the 2 H^+/H_2 and the NO_3^-/NO_2^- reactions is less, 0.84 V, which is equivalent to a free-energy yield of -162.08 kJ/mol.

Because many biochemical reactions are two-electron transfers, it is often useful to give energy yields for two-electron reactions, even if more electrons are involved. Thus, the SO_4^{2-}/H_2 redox pair involves eight electrons, and complete reduction of SO_4^{2-} with H_2 requires 4 H_2 (equivalent to eight electrons). From the reduction potential difference between 2 H^+/H_2 and SO_4^{2-}/H_2S (0.19 V), a free-energy yield of -146.64 kJ is calculated, or -36.66 kJ per two electrons. By convention, reduction potentials are given for conditions in which equal concentrations of oxidized and reduced forms are present. In actual practice, the concentrations of these two forms may be quite different. As discussed earlier in this appendix (item 9, first section), it is possible to couple half reactions even if the potential difference is unfavorable, providing the concentrations of the reacting species are appropriate.

TABLE A1.2	**Microbiologically important reduction potentials**[a]

Redox pair	E_0' (V)
SO_4^{2-}/HSO_3^-	-0.52
$CO_2/\text{formate}^-$	-0.43
$2\,H^+/H_2$	-0.41
$S_2O_3^{2-}/HS^- + HSO_3^-$	-0.40
Ferredoxin ox/red	-0.39
Flavodoxin ox/red[b]	-0.37
$NAD^+/NADH$	-0.32
Cytochrome c_3 ox/red	-0.29
$CO_2/\text{acetate}^-$	-0.29
S^0/HS^-	-0.27
CO_2/CH_4	-0.24
FAD/FADH	-0.22
SO_4^{2-}/HS^-	-0.217
Acetaldehyde/ethanol	-0.197
Pyruvate$^-$/lactate$^-$	-0.19
FMN/FMNH	-0.19
Dihydroxyacetone phosphate/glycerolphosphate	-0.19
$HSO_3^-/S_3O_6^{2-}$	-0.17
Flavodoxin ox/red[b]	-0.12
HSO_3^-/HS^-	-0.116
Menaquinone ox/red	-0.075
APS/AMP + HSO_3^-	-0.060
Rubredoxin ox/red	-0.057
Acrylyl-CoA/propionyl-CoA	-0.015
Glycine/acetate$^-$ + NH_4^+	-0.010
$S_4O_6^{2-}/S_2O_3^{2-}$	$+0.024$
Fumarate^{2-}/succinate^{2-}	$+0.033$
Cytochrome b ox/red	$+0.035$
Ubiquinone ox/red	$+0.113$
AsO_4^{3-}/AsO_3^{3-}	$+0.139$
Dimethyl sulfoxide (DMSO)/dimethylsulfide (DMS)	$+0.16$
$Fe(OH)_3 + HCO_3^-/FeCO_3$	$+0.20$
$S_3O_6^{2-}/S_2O_3^{2-} + HSO_3^-$	$+0.225$
Cytochrome c_1 ox/red	$+0.23$
NO_2^-/NO	$+0.36$
Cytochrome a_3 ox/red	$+0.385$
NO_3^-/NO_2^-	$+0.43$
SeO_4^{2-}/SeO_3^{2-}	$+0.475$
Fe^{3+}/Fe^{2+}	$+0.77$
Mn^{4+}/Mn^{2+}	$+0.798$
O_2/H_2O	$+0.82$
NO/N_2O	$+1.18$
N_2O/N_2	$+1.36$

[a]Data from Thauer, R. K., K. Jungermann, and K. Decker, 1977. Energy conservation in anaerobic chemotrophic bacteria. *Bacteriol. Rev.* 41:100–180.

[b]Separate potentials are given for each electron transfer in this potentially two-electron transfer.

APPENDIX *2* *Advanced Mathematics of Microbial Growth and Chemostat Operation*

Exponential Growth

Although analyses of microbial growth can be done graphically or algebraically, as discussed in Chapter 5, for some purposes it is necessary to use a differential equation to express the quantitative relationships of growth:

$$\frac{dX}{dt} = \mu X \qquad (1)$$

where X may be cell number or some specific cellular component such as protein and μ is the *instantaneous growth-rate constant*. When this equation is integrated, we obtain a form that reflects the activities of a typical microbial population in batch culture:

$$\ln X = \ln X_0 + \mu(t) \qquad (2)$$

where ln refers to the natural logarithm (logarithm base *e*), X_0 is cell number at time 0, X is cell number at time t, and t is elapsed time during which growth is measured. This equation fits experimental data from the exponential phase of bacterial cultures very well, such as those in Figure 5.2. Taking the antilogarithm of each side gives

$$X = X_0 e^{\mu t} \qquad (3)$$

This equation is useful because it allows prediction of population density at a future time from the present value and μ, the growth-rate constant. As discussed in Section 5.2, an important constant parameter for an exponentially growing population is the doubling (or generation) time. Doubling of the population has occurred when $X/X_0 = 2$. Rearranging and substituting this value into Equation (3) gives

$$2 = e^{\mu(t_{gen})} \qquad (4)$$

Taking the natural logarithm of each side and rearranging gives

$$\mu = \frac{\ln 2}{t_{gen}} = \frac{0.693}{t_{gen}} \qquad (5)$$

The generation time, t_{gen}, may be used to define another growth parameter, k, as follows:

$$k = \frac{1}{t_{gen}} \qquad (6)$$

where k is the growth-rate constant for a batch culture. Combining Equations (5) and (6) shows that the two growth-rate constants, μ and k, are related:

$$\mu = 0.693k \qquad (7)$$

It is important to understand that μ and k are both reflections of the same growth process of an exponentially increasing population. The difference between them may be seen in their derivation: μ is the *instantaneous rate constant*, and k is an *average* value for the population over a finite period of time. (See Table A2.1 for calculation of k values from experimental results.) This distinction is more than a mathematical point. As was emphasized in Chapter 5, microbial growth studies must deal with *population* phenomena, not the activities of individual cells. The constant k reflects this averaging assumption. However, the constant μ, being instantaneous, is a closer approximation of the rate at which individual activities are occurring. Further, the instantaneous constant μ allows us to consider bacterial growth dynamics in a theoretical framework separate from the traditional batch culture.

Mathematical relationships of chemostats

An especially important application of the instantaneous growth-rate constant, μ, is the chemostat, a culture device (∞ Section 5.5) in which population size and growth rate may be maintained at constant values of the experimenter's choosing over a wide range of values. As we saw in Section 5.5, the rate of bacterial growth is a function of nutrient concentration. This function represents a saturation process that may be described by the equation

---

TABLE A2.1   Calculation of *k* values

Bacterial population densities are expressed in scientific notation with powers of 10, and so Equation (3) may be converted to terms of logarithm base 10 and k substituted for the instantaneous constant μ:

$$k = \frac{\log_{10} X_t - \log_{10} X_0}{0.301t}$$

Example 1:
$X_0 = 1000\ (= 10^3)\ \log_{10}$ of $1000 = 3$
$X_t = 100{,}000\ (= 10^5)\ \log_{10}$ of $100{,}000 = 5$
$t = 4$ hr

$$k = \frac{5 - 3}{(0.301)4} = \frac{2}{1.204}$$

$k = 1.66$ doublings/hr
$t_{gen} = 0.60$ hr (36 min) for population to double

Example 2:
$X_0 = 1000\ (= 10^3)\ \log_{10}$ of $1000 = 3$
$X_t = 100{,}000{,}000\ (= 10^8)\ \log_{10}$ of $10^8 = 8$
$t = 120$ hr

$$k = \frac{8 - 3}{(0.301)120} = \frac{5}{36.12}$$

$k = 0.138$ doubling/hr
$t_{gen} = 7.2$ hr (430 min) for population to double

$$\mu = \mu_{max} \frac{S}{K_s + S} \tag{8}$$

where μ_{max} is the growth rate at nutrient saturation, S is the concentration of nutrient,* and K_s is a saturation constant that is numerically equal to the nutrient concentration at which $\mu = \frac{1}{2}\mu_{max}$. Equation (8) is formally equivalent to the Michaelis–Menten equation used in enzyme kinetic analyses. Estimates of the unknown parameters μ_{max} and K_s can be made by plotting $1/\mu$ on the y axis versus $1/s$ on the x axis. A straight line will be obtained that intercepts the y axis, and the value at the intercept is $1/\mu_{max}$. The slope of the line is K_s/μ_{max}, and because μ_{max} is known, K_s can then be calculated. Values for K_s as a rule are very low. A few examples are 2.1×10^{-5} M for glucose (*Escherichia coli*), 1.34×10^{-6} M for oxygen (*Candida utilis*), 3.5×10^{-8} M for phosphate (*Spirillum* sp.), and 4.7×10^{-13} M for thiamine (*Cryptococcus albidus*).

The key to the operation of the chemostat is that, in a steady-state population, the nutrient concentration is very low, usually very near 0. Therefore, each drop of fresh medium entering the chemostat supplies nutrient, which is consumed almost instantaneously by the population and the experimenter has direct control over the growth rate: if nutrient is added more rapidly, μ increases; if it is added more slowly, μ decreases. In a chemostat the rate at which medium enters is equal to the rate at which spent medium and cells exit through the overflow. This rate is called the *dilution rate, D*, and is defined as

$$D = \frac{F}{V} \tag{9}$$

where F is the flow rate (in units of volume per time, usually ml/hr) and V is chemostat volume (ml). Therefore, D has units of time^{-1} (usually hr^{-1}) as does μ. In fact, at steady state, $\mu = D$. That μ must equal D may be seen by consideration of Equations (8) and (9) and the preceding discussion. When S is low, μ is a direct function of S. But because nutrient is instantly consumed on addition, the value of S at any moment is controlled by the rate of medium addition, that is, the dilution rate D. A further crucial consequence of the relation $\mu = D$ is that a chemostat is a *self-regulating* system within broad limits of D. Consider a steady-state chemostat functioning with $\mu = D$. If μ were to increase momentarily because of some variation in the culture, S would necessarily decline, as the increased growth caused increased nutrient consumption. The lower value of S would in turn cause μ to decrease back to its stable level. Conversely, if μ were to decrease momentarily, then S would increase as "extra" nutrient were unconsumed. This rise in S would lead to higher values of μ until steady state was again reached. This biological feedback system functions well in practice, as it does in theory. Chemostat populations may be maintained at constant growth rates (μ) for long periods of time: months or even years.

---

*It is usual to speak of a *limiting nutrient,* that is, some component of the growth medium (often the electron donor) that, if increased in concentration, causes an increase in growth rate. One must be careful with this terminology, however, for, as Section 5.5 (and Figure 5.10) showed, a saturation level is usually observed for each nutrient. At saturation values, the particular nutrient under consideration no longer limits growth and some other component or factor becomes limiting.

There is another important aspect of chemostat studies, and that is *population density.* The relation between cell growth and nutrient consumption may be defined as

$$Y = \frac{dX}{dS} \tag{10}$$

where Y is the yield constant for the particular organism on the particular medium and dX/dS is the amount of cell increase per unit of substrate consumed. The yield constant is in units of

$$Y = \frac{\text{weight of bacteria formed}}{\text{weight of substrate consumed}}$$

and is a measure of the efficiency with which cells convert nutrient to more cell material. The composition of the medium in the reservoir is central to the successful steady-state operation of the chemostat. All components in this medium are present in excess except one, the growth-limiting substrate. The symbol for the concentration of that substrate in the reservoir is S_R, whereas that for the same substrate in the culture vessel is S. Taking into account Equations (1), (8), and (10), the following considerations can be made.

When a chemostat is inoculated with a small number of bacteria and the dilution rate does not exceed a certain critical value (D_c), the number of organisms starts to increase. The increase is given by

$$\frac{dX}{dt} = \text{growth} - \text{output or } \frac{dX}{dt} = \mu X - DX \tag{11}$$

because initially $\mu > D$, dX/dt is positive. However, as the population density increases, the concentration of the growth-limiting substrate decreases, causing a decrease in μ [Equation (8)]. As discussed earlier, if D is kept constant, a steady state will be reached in which $\mu = D$ and $dX/dt = 0$.

Steady states are possible only when the dilution rate does not exceed a critical value D_c. This value depends on the concentration of the growth-limiting substrate in the reservoir (S_R):

$$D_c = \mu_{max} \left(\frac{S_R}{K_s + S_R} \right) \tag{12}$$

Thus, when $S_R >> K_s$ (which is usually the case), steady states can be obtained by growth rates close to μ_{max}. The change in S is given by

$$\frac{dS}{dt} = \text{input} - \text{output} - \text{consumption}$$

Because consumption is growth divided by yield constant [see Equation (10)], we have

$$\frac{dS}{dt} = S_R D - SD - \frac{\mu X}{Y} \tag{13}$$

When a steady state is reached, $dS/dt = 0$.

From Equations (11) and (13), the steady-state values of organism concentration ($\overline{X}$) and growth-limiting substrate ($\overline{S}$) can be calculated:

$$\overline{X} = Y(S_R - S) \tag{14}$$

$$\overline{S} = K_s \left(\frac{D}{\mu_{max} - D} \right) \tag{15}$$

These equations have the constants S_R, Y, K_s, and μ_{max}. Once their values are known, the steady-state values of $\overline{X}$ and $\overline{S}$ can be predicted for any dilution rate. As can be seen from Equation (15), the steady-state substrate concentration ($\overline{S}$) depends on the dilution rate (D) applied and is independent of the substrate concentration in the reservoir (S_R). Thus, D determines $\overline{S}$ and $\overline{S}$ determines $\overline{X}$. A specific example of the use of the equations to predict chemostat behavior can be obtained by study of Figure 5.11. In that figure, the constants are $\mu_{max} = 1.0 \text{ hr}^{-1}$, $Y = 0.5$, $K_s = 0.2$ g/l, $S_R = 10$ g/l. With these constants, the curves in Figure 5.11 can be generated.

Experimental Uses of the Chemostat

One of the major theoretical advantages of a chemostat is that this device allows the experimenter to control growth rate (μ) and population density ($\overline{X}$) independently of each other. Over rather wide ranges (see earlier discussion and Section 5.5), any desired value of μ can be obtained by alteration of D. In practice this means changing the flow rate because volume is usually fixed [Equation (9)]. Similarly, the population density $\overline{X}$ may be determined by varying nutrient concentration in the reservoir [Equation (14)]. This independent control of these two crucial growth parameters is not possible with conventional batch cultures.

A practical advantage to the chemostat is that a population may be maintained in a desired growth condition (μ and $\overline{X}$) for long periods of time. Therefore, experiments can be planned in detail and performed whenever most convenient. Comparable studies with batch cultures are essentially impossible because specific growth conditions are constantly changing.

Mathematics of the Stationary Phase

The differential equation [Equation (1)] expressing the exponential growth phase can be modified to include the trend toward the stationary phase, which occurs at high population density. A second term is added to the equation that expresses the maximum population attainable for that organism under the environmental conditions specified, here called X_m. The equation then assumes a form often called the *logistics equation:*

$$\frac{dX}{dt} = \mu X - \frac{\mu}{X_m}X^2 \tag{16}$$

The second term in this equation essentially expresses self-crowding effects, such as nutrient depletion or inhibitor buildup. In integrated form, this equation graphs as a typical microbial growth curve, showing exponential and stationary phases. The length of the exponential phase is determined by the size of X_m. When X is small, the second term has little effect, whereas as X approaches X_m, the growth rate (dX/dt) approaches zero. The logistics equation is widely used by ecologists studying higher organisms to express the growth rate and carrying capacity of a habitat for animals and plants. The equation is equally applicable to microbial situations.

Mathematics of Disinfection

An equation can also be developed expressing the death rate of unicellular organisms on treatment with a disinfectant or other lethal agent. If X_d represents the number of dead cells at time t, then $X_0 - X_d$ represents the survivors. The death rate is proportional to the survivors:

$$\frac{dX}{dt} = k(X_0 - X_d) \tag{17}$$

On integration, this equation gives a typical first-order relationship:

$$kt = \ln\frac{X_0}{X_0 - X_d} \tag{18}$$

and when the logarithm of $X_0 - X_d$ (that is, the number of survivors) is plotted against time, a straight line is obtained. This equation is useful when determining the length of time necessary to allow a disinfectant to completely sterilize something (for example, the contact time for chlorine gas in a water purification plant).

APPENDIX *3 Bergey's Classification of Prokaryotes*

Bergey's Manual of Systematic Bacteriology is a recognized authority on bacterial taxonomy.* The *Manual* is divided into four volumes. Each volume contains several sections, and each section contains a number of related genera. In brief, the contents of each volume are as follows:

---

*Besides *Bergey's Manual of Systematic Bacteriology,* a single-volume version, *Bergey's Manual of Determinative Bacteriology* (1994) and a four-volume treatise called *The Prokaryotes,* 2nd edition (edited by A. Balows, H. G. Trüper, M. Dworkin, W. Harder, and K.-H. Schleifer, 1992) are available and should be consulted for more recent developments in bacterial classification.

Volume I **1984. Gram-negative Bacteria of medical and commercial importance:** spirochetes, spiral and curved Bacteria, gram-negative aerobic and facultatively aerobic rods, gram-negative obligate anaerobes, gram-negative aerobic and anaerobic cocci, sulfate- and sulfur-reducing Bacteria, rickettsias and chlamydias, mycoplasmas.

Volume II **1986. Gram-positive Bacteria of medical and commercial importance:** gram-positive cocci, gram-positive endospore-forming and nonsporing rods, mycobacteria, nonfilamentous actinomycetes.

Volume III   1989. Remaining gram-negative Bacteria and the Archaea: phototrophic, gliding, sheathed, budding, and appendaged Bacteria, cyanobacteria, chemolithotrophic Bacteria; methanogens, extreme halophiles, hyperthermophiles, *Thermoplasma*, and other Archaea.

Volume IV   1989. Filamentous actinomycetes and related bacteria.

The detailed list of genera in *Bergey's Manual of Systematic Bacteriology* follows. Names given in quotation marks refer to organisms for which the nomenclature is not yet clear.

Volume I

SECTION 1
The Spirochetes
Order I: Spirochaetales
  Family I: Spirochaetaceae
    Genus I: *Spirochaeta*
    Genus II: *Cristispira*
    Genus III: *Treponema*
    Genus IV: *Borrelia*
  Family II: Leptospiraceae
    Genus I: *Leptospira*
Other Organisms
  Hindgut Spirochetes of Termites and *Cryptocercus punctulatus*

SECTION 2
Aerobic/Microaerophilic, Motile, Helical/Vibrioid Gram-Negative Bacteria
  Genus: *Aquaspirillum*
  Genus: *Spirillum*
  Genus: *Azospirillum*
  Genus: *Oceanospirillum*
  Genus: *Campylobacter*
  Genus: *Bdellovibrio*
  Genus: *Vampirovibrio*

SECTION 3
Nonmotile (or Rarely Motile), Gram-Negative Curved Bacteria
  Family I: Spirosomaceae
    Genus I: *Spirosoma*
    Genus II: *Runella*
    Genus III: *Flectobacillus*
Other Genera
  Genus: *Microcyclus*
  Genus: *Meniscus*
  Genus: *Brachyarcus*
  Genus: *Pelosigma*

SECTION 4
Gram-Negative Aerobic Rods and Cocci
  Family I: Pseudomonadaceae
    Genus I: *Pseudomonas*
    Genus II: *Xanthomonas*
    Genus III: *Frateuria*
    Genus IV: *Zoogloea*
  Family II: Azotobacteraceae
    Genus I: *Azotobacter*
    Genus II: *Azomonas*
  Family III: Rhizobiaceae
    Genus I: *Rhizobium*
    Genus II: *Bradyrhizobium*
    Genus III: *Agrobacterium*
    Genus IV: *Phyllobacterium*
  Family IV: Methylococcaceae
    Genus I: *Methylococcus*
    Genus II: *Methylomonas*
  Family V: Halobacteriaceae
    Genus I: *Halobacterium*
    Genus II: *Halococcus*
  Family VI: Acetobacteraceae

  Genus I: *Acetobacter*
  Genus II: *Gluconobacter*
Family VII: Legionellaceae
  Genus I: *Legionella*
Family VIII: Neisseriaceae
  Genus I: *Neisseria*
  Genus II: *Moraxella*
  Genus III: *Acinetobacter*
  Genus IV: *Kingella*
Other Genera
  Genus: *Beijerinckia*
  Genus: *Derxia*
  Genus: *Xanthobacter*
  Genus: *Thermus*
  Genus: *Thermomicrobium*
  Genus: *Halomonas*
  Genus: *Alteromonas*
  Genus: *Flavobacterium*
  Genus: *Alcaligenes*
  Genus: *Serpens*
  Genus: *Janthinobacterium*
  Genus: *Brucella*
  Genus: *Bordetella*
  Genus: *Francisella*
  Genus: *Paracoccus*
  Genus: *Lampropedia*

SECTION 5
Facultatively Anaerobic Gram-Negative Rods
  Family I: Enterobacteriaceae
    Genus I: *Escherichia*
    Genus II: *Shigella*
    Genus III: *Salmonella*
    Genus IV: *Citrobacter*
    Genus V: *Klebsiella*
    Genus VI: *Enterobacter*
    Genus VII: *Erwinia*
    Genus VIII: *Serratia*
    Genus IX: *Hafnia*
    Genus X: *Edwardsiella*
    Genus XI: *Proteus*
    Genus XII: *Providencia*
    Genus XIII: *Morganella*
    Genus XIV: *Yersinia*
Other Genera of the Family Enterobacteriaceae
    Genus: *Obesumbacterium*
    Genus: *Xenorhabdus*
    Genus: *Kluyvera*
    Genus: *Rahnella*
    Genus: *Cedecea*
    Genus: *Tatumella*
  Family II: Vibrionaceae
    Genus I: *Vibrio*
    Genus II: *Photobacterium*
    Genus III: *Aeromonas*
    Genus IV: *Plesiomonas*
  Family III: Pasteurellaceae
    Genus I: *Pasteurella*
    Genus II: *Haemophilus*
    Genus III: *Actinobacillus*

Other Genera
  Genus: *Zymomonas*
  Genus: *Chromobacterium*
  Genus: *Cardiobacterium*
  Genus: *Calymmatobacterium*
  Genus: *Gardnerella*
  Genus: *Eikenella*
  Genus: *Streptobacillus*

SECTION 6
Anaerobic Gram-Negative Straight, Curved, and Helical Rods
  Family I: Bacteroidaceae
    Genus I: *Bacteroides*
    Genus II: *Fusobacterium*
    Genus III: *Leptotrichia*
    Genus IV: *Butyrivibrio*
    Genus V: *Succinimonas*
    Genus VI: *Succinivibrio*
    Genus VII: *Anaerobiospirillum*
    Genus VIII: *Wolinella*
    Genus IX: *Selenomonas*
    Genus X: *Anaerovibrio*
    Genus XI: *Pectinatus*
    Genus XII: *Acetivibrio*
    Genus XIII: *Lachnospira*

SECTION 7
Dissimilatory Sulfate- and Sulfur-Reducing Bacteria
  Genus: *Desulfuromonas*
  Genus: *Desulfovibrio*
  Genus: *Desulfomonas*
  Genus: *Desulfococcus*
  Genus: *Desulfobacter*
  Genus: *Desulfobulbus*
  Genus: *Desulfosarcina*

SECTION 8
Anaerobic Gram-Negative Cocci
  Family I: Veillonellaceae
    Genus I: *Veillonella*
    Genus II: *Acidaminococcus*
    Genus III: *Megasphaera*

SECTION 9
The Rickettsias and Chlamydias
Order I: Rickettsiales
  Family I: Rickettsiaceae
  Tribe I: *Rickettsieae*
    Genus I: *Rickettsia*
    Genus II: *Rochalimaea*
    Genus III: *Coxiella*
  Tribe II: *Ehrlichieae*
    Genus IV: *Ehrlichia*
    Genus V: *Cowdria*
    Genus VI: *Neorickettsia*
  Tribe III: *Wolbachieae*
    Genus VII: *Wolbachia*
    Genus VIII: *Rickettsiella*
  Family II: Bartonellaceae
    Genus I: *Bartonella*

Genus II: *Grahamella*
Family III: Anaplasmataceae
  Genus I: *Anaplasma*
  Genus II: *Aegyptianella*
  Genus III: *Haemobartonella*
  Genus IV: *Eperythrozoon*
Order II: Chlamydiales
  Family I: Chlamydiaceae
    Genus I: *Chlamydia*

SECTION 10
The Mycoplasmas
Order I: Mycoplasmatales
  Family I: Mycoplasmataceae
    Genus I: *Mycoplasma*
    Genus II: *Ureaplasma*
  Family II: Acholeplasmataceae
    Genus I: *Acholeplasma*
  Family III: Spiroplasmataceae
    Genus I: *Spiroplasma*
  Other Genera
    Genus: *Anaeroplasma*
    Genus: *Thermoplasma*
  Mycoplasma-like Organisms of Plants
  and Invertebrates

SECTION 11
Endosymbionts
A: Endosymbionts of Protozoa
  Endosymbionts of ciliates
  Endosymbionts of flagellates
  Endosymbionts of amebas
  Taxa of endosymbionts:
    Genus I: *Holospora*
    Genus II: *Caedibacter*
    Genus III: *Pseudocaedibacter*
    Genus IV: *Lyticum*
    Genus V: *Tectibacter*
B: Endosymbionts of Insects
  Blood-sucking insects
  Plant sap-sucking insects
  Cellulose and stored grain feeders
  Insects feeding on complex diets
  Taxon of endosymbionts:
    Genus: *Blattabacterium*
C: Endosymbionts of Fungi and
Invertebrates other than Arthropods
  Fungi
  Sponges
  Coelenterates
  Helminthes
  Annelids
  Marine worms and molluscs

Volume II
SECTION 12
Gram-Positive Cocci
  Family I: Micrococcaceae
    Genus I: *Micrococcus*
    Genus II: *Stomatococcus*
    Genus III: *Planococcus*
    Genus IV: *Staphylococcus*
  Family II: Deinococcaceae
    Genus I: *Deinococcus*
  Other Genera
    Genus: *Streptococcus*
      Pyogenic hemolytic streptococci
      Oral streptococci
      Enterococci
      Lactic acid streptococci
      Anaerobic streptococci
      Other streptococci

Genus: *Leuconostoc*
Genus: *Pediococcus*
Genus: *Aerococcus*
Genus: *Gemella*
Genus: *Peptococcus*
Genus: *Peptostreptococcus*
Genus: *Ruminococcus*
Genus: *Coprococcus*
Genus: *Sarcina*

SECTION 13
Endospore-Forming Gram-Positive
Rods and Cocci
  Genus: *Bacillus*
  Genus: *Sporolactobacillus*
  Genus: *Clostridium*
  Genus: *Desulfotomaculum*
  Genus: *Sporosarcina*
  Genus: *Oscillospira*

SECTION 14
Regular, Nonsporing, Gram-Positive
Rods
  Genus: *Lactobacillus*
  Genus: *Listeria*
  Genus: *Erysipelothrix*
  Genus: *Brochothrix*
  Genus: *Renibacterium*
  Genus: *Kurthia*
  Genus: *Caryophanon*

SECTION 15
Irregular, Nonsporing, Gram-Positive
Rods
  Genus: *Corynebacterium*
  Plant Pathogenic Species of
  Corynebacterium
  Genus: *Gardnerella*
  Genus: *Arcanobacterium*
  Genus: *Arthrobacter*
  Genus: *Brevibacterium*
  Genus: *Curtobacterium*
  Genus: *Caseobacter*
  Genus: *Microbacterium*
  Genus: *Aureobacterium*
  Genus: *Cellulomonas*
  Genus: *Agromyces*
  Genus: *Arachnia*
  Genus: *Rothia*
  Genus: *Propionibacterium*
  Genus: *Eubacterium*
  Genus: *Acetobacterium*
  Genus: *Lachnospira*
  Genus: *Butyrivibrio*
  Genus: *Thermoanaerobacter*
  Genus: *Actinomyces*
  Genus: *Bifidobacterium*

SECTION 16
The Mycobacteria
  Family: Mycobacteriaceae
    Genus: *Mycobacterium*

SECTION 17
Nocardioforms
  Genus: *Nocardia*
  Genus: *Rhodococcus*
  Genus: *Nocardioides*
  Genus: *Pseudonocardia*
  Genus: *Oerskovia*
  Genus: *Saccharopolyspora*
  Genus: *Micropolyspora*

Genus: *Promicromonospora*
Genus: *Intrasporangium*

Volume III
SECTION 18
Anoxygenic Phototrophic Bacteria
Purple Bacteria
  Family I: Chromatiaceae
    Genus I: *Chromatium*
    Genus II: *Thiocystis*
    Genus III: *Thiospirillum*
    Genus IV: *Thiocapsa*
    Genus V: *Lamprobacter*
    Genus VI: *Lamprocystis*
    Genus VII: *Thiodictyon*
    Genus VIII: *Amoebobacter*
    Genus IX: *Thiopedia*
  Family II: Ectothiorhodospiraceae
    Genus: *Ectothiorhodospira*
  Purple Nonsulfur Bacteria
    Genus: *Rhodospirillum*
    Genus: *Rhodopila*
    Genus: *Rhodobacter*
    Genus: *Rhodopseudomonas*
    Genus: *Rhodomicrobium*
    Genus: *Rhodocyclus*
Green Bacteria
  Green Sulfur Bacteria
    Genus: *Chlorobium*
    Genus: *Prosthecochloris*
    Genus: *Pelodictyon*
    Genus: *Ancalochloris*
    Genus: *Chloroherpeton*
  Symbiotic Consortia
  Multicellular, Filamentous, Green
  Bacteria
    Genus: *Chloroflexus*
    Genus: *Heliothrix*
    Genus: *"Oscillochloris"*
    Genus: *Chloronema*
  Genera Incertae Sedis
    Genus: *Heliobacterium*
    Genus: *Erythrobacter*

SECTION 19
Oxygenic Photosynthetic Bacteria
Group I: Cyanobacteria
Subsection I: Order Chroococcales
    Genus I: *Chamaesiphon*
    Genus II: *Gloeobacter*
    Genus III: *Gloeothece*
Subsection II: Order Pleurocapsales
    Genus I: *Dermocarpa*
    Genus II: *Xenococcus*
    Genus III: *Dermocarpella*
    Genus IV: *Myxosarcina*
    Genus V: *Chroococcidiopsis*
Subsection III: Order Oscillatoriales
    Genus I: *Spirulina*
    Genus II: *Arthrospira*
    Genus III: *Oscillatoria*
    Genus IV: *Lyngbya*
    Genus V: *Pseudanabaena*
    Genus VI: *Starria*
    Genus VII: *Crinalium*
    Genus VIII: *Microcoleus*
Subsection IV: Order Nostocales
  Family I: Nostocaceae
    Genus I: *Anabaena*
    Genus II: *Aphanizomenon*
    Genus III: *Nodularia*
    Genus IV: *Cylindrospermum*
    Genus V: *Nostoc*

Family II: Scytonemataceae
  Genus I: *Scytonema*
Family III: Rivulariaceae
  Genus I: *Calothrix*
Subsection V: Order Stigonematales
  Genus I: *Chlorogloeopsis*
  Genus II: *Fischerella*
  Genus III: *Stigonema*
  Genus IV: *Geitleria*
Group II: Order Prochlorales
Family I: Prochloraceae
  Genus: *Prochloron*
Other taxa:
  Genus: *"Prochlorothrix"*

SECTION 20
**Aerobic Chemolithotrophic
Bacteria and Associated Organisms**
A: Nitrifying Bacteria
  Family: Nitrobacteraceae
    Genus I: *Nitrobacter*
    Genus II: *Nitrospina*
    Genus III: *Nitrococcus*
    Genus IV: *Nitrospira*
    Genus V: *Nitrosomonas*
    Genus VI: *Nitrosococcus*
    Genus VII: *Nitrosospira*
    Genus VIII: *Nitrosolobus*
    Genus IX: *Nitrosovibrio*
B: Colorless Sulfur Bacteria
  Genus: *Thiobacterium*
  Genus: *Macromonas*
  Genus: *Thiospira*
  Genus: *Thiovulum*
  Genus: *Thiobacillus*
  Genus: *Thiomicrospira*
  Genus: *Thiosphaera*
  Genus: *Acidiphilium*
  Genus: *Thermothrix*
C: Obligate Hydrogen Oxidizers
  Genus: *Hydrogenobacter*
D: Iron and Manganese Oxidizing and/or
Depositing Bacteria
  Family: Siderocapsaceae
    Genus I: *Siderocapsa*
    Genus II: *Naumanniella*
    Genus III: *Siderococcus*
    Genus IV: *Ochrobium*
E: Magnetotactic Bacteria
  Genus: *Aquaspirillum (A. magneto
    tacticum)*
  Genus: *Bilophococcus*

SECTION 21
Budding and/or Appendaged Bacteria
A: Prosthecate Bacteria
  1: Budding Bacteria
    Genus: *Hyphomicrobium*
    Genus: *Hyphomonas*
    Genus: *Pedomicrobium*
    Genus: *Ancalomicrobium*
    Genus: *Prosthecomicrobium*
    Genus: *Stella*
    Genus: *Labrys*
  2: Nonbudding Bacteria
    Genus: *Caulobacter*
    Genus: *Asticcacaulis*
    Genus: *Prosthecobacter*
B: Nonprosthecate Bacteria
  1: Budding Bacteria
    Genus: *Planctomyces*
    Genus: *"Isosphaera"*
    Genus: *Blastobacter*

Genus: *Angulomicrobium*
Genus: *Gemmiger*
Genus: *Ensifer*
2: Nonbudding Bacteria
  Genus: *Gallionella*
  Genus: *Nevskia*
C: Morphologically Unusual Budding
Bacteria (involved in iron and manganese
deposition)
  Genus: *Seliberia*
  Genus: *Metallogenium*
  Genus: *Caulococcus*
  Genus: *Kuznezovia*
  Genus: *Thiodendron*
D: Others
  Spinate bacteria

SECTION 22
Sheathed Bacteria
  Genus: *Sphaerotilus*
  Genus: *Leptothrix*
  Genus: *Haliscominobacter*
  Genus: *"Lieskeella"*
  Genus: *"Phragmidiothrix"*
  Genus: *Crenothrix*
  Genus: *"Clonothrix"*

SECTION 23
**Nonphotosynthetic, Nonfruiting,
Gliding Bacteria**
Order I: Cytophagales
  Family: Cytophagaceae
    Genus I: *Cytophaga*
    Genus II: *Sporocytophaga*
    Genus III: *Capnocytophaga*
    Genus IV: *Flexithrix*
  Other genera
    Genus: *Flexibacter*
    Genus: *Chitinophaga*
    Genus: *Microscilla*
Order II: Lysobacterales
  Family: Lysobacteriaceae
    Genus: *Lysobacter*
Order III: Beggiatoales
  Family: Beggiatoaceae
    Genus I: *Beggiatoa*
    Genus II: *Thioploca*
    Genus III: *"Thiospirillopsis"*
    Genus IV: *Thiothrix*
  Others
  Family I: Simonsiellaceae
    Genus I: *Simonsiella*
    Genus II: *Alysiella*
  Family II: *"Pelonemataceae"*
    Genus I: *"Pelonema"*
    Genus II: *"Peloploca"*
    Genus III: *"Achroonema"*
    Genus IV: *"Desmanthus"*
  Other genera
    Genus: *Herpetosiphon*
    Genus: *Toxothrix*
    Genus: *Leucothrix*
    Genus: *Vitreoscilla*
    Genus: *Desulfonema*
    Genus: *Agitococcus*
    Genus: *Achromatium*

SECTION 24
Gliding, Fruiting Bacteria
Order: Myxococcales
  Family I: Myxococcaceae
    Genus: *Myxococcus*

Family II: Archangiaceae
  Genus: *Archangium*
Family III: Cystobacteraceae
  Genus I: *Cystobacter*
  Genus II: *Melittangium*
  Genus III: *Stigmatella*
Family IV: Polyangiaceae
  Genus I: *Polyangium*
  Genus II: *Nannocystis*
  Genus III: *Chondromyces*

SECTION 25
Archaeobacteria*
**Group I: Methanogenic
Archaeobacteria**
Order I: Methanobacteriales
  Family I: Methanobacteriaceae
    Genus I: *Methanobacterium*
    Genus II: *Methanobrevibacter*
  Family II: Methanothermaceae
    Genus: *Methanothermus*
Order II: Methanococcales
  Family: Methanococcaceae
    Genus: *Methanococcus*
Order III: Methanomicrobiales
  Family I: Methanomicrobiaceae
    Genus I: *Methanomicrobium*
    Genus II: *Methanogenium*
    Genus III: *Methanospirillum*
  Family II: Methanosarcinaceae
    Genus I: *Methanosarcina*
    Genus II: *Methanococcoides*
    Genus III: *Methanolobus*
    Genus IV: *Methanothrix*
  Other taxa
  Family: Methanoplanaceae
    Genus: *Methanoplanus*
  Others
    Genus: *Methanosphaera*
**Group II: Archaeobacterial
Sulfate Reducers**
Order: Archaeoglobales
  Family: *"Archaeoglobaceae"*
    Genus: *Archaeoglobus*
**Group III: Extremely Halophilic
Archaeobacteria**
Order: Halobacteriales
  Family: Halobacteriaceae
    Genus I: *Halococcus*
    Genus II: *Halobacterium*
    Genus III: *Haloferax*
    Genus IV: *Haloarcula*
    Genus V: *Natronococcus*
    Genus VI: *Natronobacterium*
**Group IV: Cell-Wall-less
Archaeobacteria**
  Genus: *Thermoplasma*
**Group V: Extremely Thermophilic
S⁰ Metabolizers**
Order I: Thermoproteales
  Family I: Thermoproteaceae
    Genus I: *Thermoproteus*
    Genus II: *Thermophilum*
  Family II: Desulfurococcaceae
    Genus I: *Desulfurococcus*
    Genus II: *Thermococcus*
    Genus III: *Thermodiscus*
    Genus IV: *Pyrodictium*

*An older term for the Archaea.

Order II: Thermococcales
　Family: Thermococcaceae
　　Genus I: *Thermococcus*
　　Genus II: *Pyrococcus*
Order III: Sulfolobales
　Family: Sulfolobaceae
　　Genus: *Sulfolobus*
　　Genus: *Acidianus*

Volume IV

SECTION 26
Nocardioform Actinomycetes
　Genus: *Nocardia*
　Genus: *Rhodococcus*
　Genus: *Nocardioides*
　Genus: *Pseudonocardia*
　Genus: *Oerskovia*
　Genus: *Saccharopolyspora*
　Genus: *Faenia (Micropolyspora)*
　Genus: *Promicromonospora*
　Genus: *Intrasporangium*
　Genus: *Actinopolyspora*
　Genus: *Saccharomonospora*
　Genus: *Amycolatopsis*
　Genus: *Amycolata*

SECTION 27
Actinomycetes with Multilocular Sporangia
　Genus: *Geodermatophilus*
　Genus: *Dermatophilus*
　Genus: *Frankia*

SECTION 28
Actinoplanetes
　Genus: *Actinoplanes*
　Genus: *Ampullariella*
　Genus: *Pilimelia*
　Genus: *Dactylosporangium*
　Genus: *Micromonospora*

SECTION 29
Streptomycetes and Related Genera
　Genus: *Streptomyces*
　Genus: *Streptoverticillium*
　Genus: *Kineosporia*
　Genus: *Sporichthya*

SECTION 30
Maduromycetes
　Genus: *Actinomadura*
　Genus: *Microbispora*

　Genus: *Microtetraspora*
　Genus: *Planobispora*
　Genus: *Planomonospora*
　Genus: *Spirillospora*
　Genus: *Streptosporangium*

SECTION 31
Thermomonospora and Related Genera
　Genus: *Thermomonospora*
　Genus: *Actinosynnema*
　Genus: *Nocardiopsis*
　Genus: *Streptoalloteichus*

SECTION 32
Thermoactinomycetes
　Genus: *Thermoactinomyces*

SECTION 33
Other Genera
　Genus: *Glycomyces*
　Genus: *Kibdelosporangium*
　Genus: *Kitasatosporia*
　Genus: *Saccharothrix*

Glossary

Only the major terms and concepts are included. If a term is not here, consult the index.

Abscess A localized infection characterized by production of pus.

Acetotrophic Splitting of acetate into CH_4 plus CO_2 by certain methanogens.

Acetyl-CoA (Ljungdahl–Wood) pathway A pathway of autotrophic CO_2 fixation widespread in obligate anaerobes including methanogens, homoacetogens, and sulfate-reducing bacteria.

Acetylene reduction assay Method of measuring activity of nitrogenase by substituting acetylene for the natural substrate of the enzyme, N_2. Acetylene is reduced to ethylene or ethane.

Acid–alcohol fastness A staining property of *Mycobacterium* species where cells stained with hot carbolfuschin do not decolorize with acid–alcohol.

Acid mine drainage Acidic water containing H_2SO_4 derived from the microbial oxidation of iron sulfide minerals.

Acidophile Organism that grows best at acidic pH values.

Activation energy Energy needed to make substrate molecules more reactive; enzymes function by lowering activation energy.

Activator protein A regulatory protein that binds to specific sites on DNA and stimulates transcription; involved in positive control.

Active immunity An immune state achieved by self-production of antibodies. Compare with *Passive immunity*.

Active site The portion of an enzyme that is directly involved in binding substrate(s).

Active transport The energy-dependent process of transporting substances into or out of the cell in which the transported substances are chemically unchanged.

Acute In reference to infections, short-term, usually characterized by dramatic onset and rapid recovery.

Adherence A property of bacteria that allows them to stick to host surfaces.

Aerobe An organism that grows in the presence of O_2; may be facultative, obligate, or microaerobic.

Aerosol Suspension of particles in airborne water droplets.

Aerotolerant Of an anaerobe, not being inhibited by O_2.

Agglutination Reaction between antibody and particle-bound antigen resulting in clumping of the particles.

Algae Phototrophic eukaryotic microorganisms.

Alkaliphile An organism that grows best at high pH.

Allergy A harmful immune reaction, usually caused by a foreign antigen in food, pollen, or chemicals; immediate-type or delayed-type hypersensitivity.

Allosteric enzyme An enzyme that contains two combining sites, the active site (where the substrate binds) and the allosteric site (where an effector molecule binds).

Ameboid movement A type of motility in which cytoplasmic streaming moves the organism forward.

Aminoacyl-tRNA synthetase An enzyme that catalyzes the attachment of the correct amino acid to the correct tRNA.

Aminoglycoside An antibiotic such as streptomycin that consists of amino sugars linked by glycosidic bonds.

Anabolism The biochemical processes involved in the synthesis of cell constituents from simpler molecules, usually requiring energy.

Anaerobe An organism that grows in the absence of O_2. Some may even be killed by O_2.

Anaerobic respiration Use of an electron acceptor other than O_2 in an electron transport-based oxidation and leading to a proton motive force.

Anaphylatoxins The C3a and C5a fractions of complement that act to mimic some of the reactions of anaphylaxis.

Anaphylaxis (anaphylactic shock) A violent allergic reaction caused by an antigen–antibody reaction.

Anoxic Absence of oxygen. Usually used in reference to a microbial habitat.

Anoxygenic photosynthesis Use of light energy to synthesize ATP by cyclic photophosphorylation without O_2 production.

Antibiotic A chemical agent produced by one organism that is harmful to other organisms.

Antibiotic resistance The acquired ability of a microorganism to grow in the presence of an antibiotic to which the microorganism is usually sensitive.

Antibody A protein present in serum or other body fluid that combines specifically with antigen. An immunoglobulin.

Anticodon A sequence of three bases in transfer RNA that base-pairs with a codon in messenger RNA during protein synthesis.

Antigen A substance that interacts with a T cell receptor or an immunoglobulin.

Antigen-presenting cell (APC) A cell that processes and presents antigen to T lymphocytes.

Antigenic determinant The portion of an antigen that interacts with an immunoglobulin or T cell receptor. Also called an *epitope*.

Antigenic drift In influenza virus, minor changes in viral proteins (antigens) due to gene mutation.

Antigenic shift In influenza virus, major changes in viral proteins (antigens) due to gene reassortment.

Antimicrobial Harmful to microorganisms by either killing or inhibiting growth.

Antimicrobial agent A chemical that kills or inhibits the growth of microorganisms.

Antiparallel In reference to double-stranded DNA, one strand runs $5' \rightarrow 3'$, the other $3' \rightarrow 5'$.

Antiseptic An agent that kills or inhibits microbial growth but is not harmful to human tissue.

Antiserum A serum containing antibodies.

Antitoxin An antibody that specifically interacts with and neutralizes a toxin.

Archaea A phylogenetic domain of prokaryotes consisting of the methanogens, most extreme halophiles and hyper-thermophiles, and *Thermoplasma*.

Aseptic technique Manipulation of sterile instruments or culture media in such a way as to maintain sterility.

Atomic weight The total mass of protons and neutrons in an atom.

ATP Adenosine triphosphate, the principal energy carrier of the cell.

Attenuation Selection of nonvirulent strains of a pathogen still capable of immunizing. Also, a mechanism for controlling gene expression. Typically transcription is terminated after initiation but before a full-length mRNA is produced.

Autoantibody An antibody that reacts to self antigens.

Autoclave A sterilizer that destroys microorganisms by high temperature using steam under pressure.

Autoimmunity Immune reactions of a host against its own self antigens.

Autolysis The lysis of a cell brought about by the activity of the cell itself.

Autoradiography Detection of radioactivity in a sample, for example, a cell or gel, by placing it in contact with a photographic film.

Autotroph Organism able to utilize CO_2 as a sole source of carbon.

Auxotroph An organism that has developed a nutritional requirement through mutation. Contrast with a *Prototroph.*

B lymphocyte A cell of the immune system that differentiates into an immunoglobulin-producing cell.

Bacteremia The transient appearance of bacteria in the blood.

Bacteria All prokaryotes that are not members of the domain Archaea.

Bactericidal Capable of killing bacteria.

Bacteriocins Agents produced by certain bacteria that inhibit or kill closely related species.

Bacteriophage A virus that infects prokaryotic cells.

Bacteriorhodopsin A protein containing retinal that is found in the membranes of certain extremely halophilic Archaea and that is involved in light-mediated ATP synthesis.

Bacteriostatic Capable of inhibiting bacterial growth without killing.

Bacteroid A swollen, deformed *Rhizobium* cell found in the root nodule; capable of nitrogen fixation.

Barophile An organism that lives optimally at high hydrostatic pressure.

Barotolerant An organism able to tolerate high hydrostatic pressure, although growing better at 1 atm.

Base composition In reference to nucleic acids, the proportion of the total bases consisting of guanine plus cytosine or thymine plus adenine base pairs. Usually expressed as a guanine + cytosine (G + C) value for example, 60% G + C.

Batch culture A closed-system microbial culture of fixed volume.

Beta-lactam An antibiotic such as penicillin that contains the four-membered heterocyclic beta-lactam ring.

Bioconversion In industrial microbiology, use of microorganisms to convert a substance to a chemically modified form.

Biofilm Microbial colonies encased in an adhesive, usually polysaccharide material, and attached to a surface.

Biogeochemistry Study of microbially mediated chemical transformations of geochemical interest, for example, nitrogen or sulfur cycling.

Bioremediation Use of microorganisms to remove or detoxify toxic or unwanted chemicals in an environment.

Biosynthesis The production of needed cellular constituents from other (usually simpler) molecules.

Biotechnology The use of living organisms to carry out defined chemical processes for industrial application.

Black smoker A thermal vent emitting very hot (270–380°C) water and minerals.

Brewing The manufacture of alcoholic beverages such as beer from the fermentation of malted grains.

Calvin cycle The biochemical route of CO_2 fixation in many autotrophic organisms.

Capsid The protein coat of a virus.

Capsomere An individual protein subunit of the virus capsid.

Capsule A compact layer of polysaccharide exterior to the cell wall in some bacteria. See also *Glycocalyx* and *Slime layer.*

Carboxysomes Polyhedral cellular inclusions of crystalline ribulose bisphosphate carboxylase (RubisCO), the key enzyme of the Calvin cycle.

Carcinogen A substance that causes the initiation of tumor formation. Frequently a mutagen.

Carrier An individual that harbors infectious organisms but does not show symptoms of disease.

Catabolism The biochemical processes involved in the breakdown of organic or inorganic compounds, usually leading to the production of energy.

Catabolite repression Repression of a variety of unrelated enzymes when cells are grown in a medium containing glucose.

Catalysis Increase in rate of a chemical reaction.

Catalyst A substance that promotes a chemical reaction without itself being changed in the end.

CD4 cells T helper cells. They are targets for HIV infection.

Cell The fundamental unit of life.

Cell-mediated immunity An immune response generated by the activities of non-antibody-producing cells such as T cells. Compare with *Humoral immunity.*

Chemiosmosis The use of ion gradients, especially proton gradients, across membranes to generate ATP. See *Proton motive force.*

Chemolithotroph An organism obtaining its energy from the oxidation of inorganic compounds.

Chemoorganotroph An organism obtaining its energy from the oxidation of organic compounds.

Chemostat A continuous culture device controlled by the concentration of limiting nutrient and dilution rate.

Chemotaxis Movement toward or away from a chemical.

Chemotherapeutic agent An antimicrobial agent that can be used internally.

Chemotherapy Treatment of infectious disease with chemicals or antibiotics.

Chlorination A highly effective disinfectant procedure for drinking water using chlorine gas or other chlorine-containing compounds as disinfectant.

Chlorophyll and bacteriochlorophyll Pigments of phototrophic organisms consisting of light-sensitive magnesium tetrapyrroles.

Chloroplast The chlorophyll-containing organelle of phototrophic eukaryotes.

Chlorosomes Cigar-shaped structures enclosed by a nonunit membrane and containing the light-harvesting bacteriochlorophyll (c, c_s, d, or e) in green sulfur bacteria and in *Chloroflexus.*

Chromogenic Producing color; a chromogenic colony is a pigmented colony.

Chromosome A genetic element carrying genes essential to cellular function. Prokaryotes typically have a single chromosome consisting of a circular DNA molecule. Eukaryotes typically have several chromosomes, each containing a linear DNA molecule.

Chronic Long-term.

Cidal Lethal or killing.

Cilium Short, filamentous structure that beats with many others to make a cell move.

Citric acid cycle A cyclical series of reactions resulting in the conversion of acetate to CO_2 and NADH. Also called the *Tricarboxylic acid cycle* or the *Kreb's cycle.*

Class I MHC proteins Antigen-presenting molecules found on all nucleated vertebrate cells.

Class II MHC proteins Antigen-presenting molecules found primarily on macrophages and B lymphocytes in vertebrates.

Clonal selection A theory that each B or T lymphocyte, when stimulated by antigen, divides to form a clone of itself.

Clone A population of cells all descended from a single cell; a number of copies of a DNA fragment obtained by allowing an inserted DNA fragment to be replicated by a phage or plasmid.

Cloning vectors Genetic elements into which genes can be recombined and replicated.

Coccoid Sphere-shaped.

Coccus A spherical bacterium.

Codon A sequence of three bases in messenger RNA that encodes a specific amino acid.

Coenzyme A low-molecular-weight molecule that participates in an enzymatic reaction by accepting and donating electrons or functional groups. Examples: NAD⁺, FAD.

Coliforms Gram-negative, nonsporing, facultative rods that ferment lactose with gas formation within 48 hr at 35°C.

Colonization Multiplication of a microorganism after it has attached to host tissues or other surfaces.

Colony A macroscopically visible population of cells growing on solid medium, arising from a single cell.

Cometabolism The metabolic transformation of a substance while a second substance serves as primary energy or carbon source.

Commodity chemicals Chemicals such as ethanol that have low monetary value and are thus sold primarily in bulk.

Common-source epidemic An epidemic resulting from infection of a large number of people from a single contaminated source.

Compatible solutes Organic compounds that serve as cytoplasmic solutes to balance water relations for cells growing in environments of high salt or sugar.

Competence Ability to take up DNA and become genetically transformed.

Complement A complex of proteins in the blood serum that interacts sequentially with specific antigen–antibody complexes.

Complement fixation The consumption of complement by an antibody–antigen reaction.

Complementary Nucleic acid sequences that can base-pair with each other.

Complex media Culture media whose precise chemical composition is unknown. Also called *undefined media.*

Concatamer A DNA molecule consisting of two or more separate molecules linked end to end to form a long, linear structure.

Congenital syphilis Syphilis contracted by an infant from its mother during birth.

Conjugation Transfer of genes from one prokaryotic cell to another by a mechanism involving cell-to-cell contact.

Consensus sequence A nucleic acid sequence in which the base present in a given position is that base most commonly found when many experimentally determined sequences are compared.

Consortium A two- (or more) membered bacterial culture (or natural assemblage) in which each organism benefits from the others.

Contagious Transmissible.

Cortex The region inside the spore coat of an endospore, around the core.

Covalent bond A nonionic chemical bond formed by a sharing of electrons between two atoms.

Crista Inner membrane in a mitochondrion, site of respiration.

Culture A particular strain or kind of organism growing in a laboratory medium.

Culture medium An aqueous solution of various nutrients suitable for the growth of microorganisms.

Cutaneous Relating to the skin.

Cyanobacteria Prokaryotic oxygenic phototrophs containing chlorophyll *a* and phycobilins.

Cyst A resting stage formed by some bacteria and protozoa in which the whole cell is surrounded by a protective layer; not the same as a spore.

Cytochrome Iron-containing porphyrin complexed with proteins, which functions as an electron carrier in the electron transport system.

Cytokine A soluble immune response modulator produced by cells other than lymphocytes, usually phagocytic cells.

Cytoplasm Cellular contents inside the cytoplasmic membrane, excluding the nucleus.

Cytoplasmic membrane The permeability barrier of the cell, separating the cytoplasm from the environment.

Defined media Culture media whose exact chemical composition is known. Compare with *Complex media.*

Degeneracy In relation to the genetic code, the fact that more than one codon can code for the same amino acid.

Deletion Removal of a portion of a gene.

Denaturation Irreversible destruction of a macromolecule, as for example the destruction of a protein by heat.

Denitrification Conversion of nitrate into nitrogen gases under anoxic conditions.

Dental caries Tooth decay resulting from bacterial infection.

Dental plaque Bacterial cells encased in a matrix of extracellular polymers, found on the teeth.

Deoxyribonucleic acid (DNA) A polymer of nucleotides connected via a phosphate–deoxyribose sugar backbone; the genetic material of the cell.

Desiccation Drying.

Dideoxynucleotide A nucleotide lacking the 3'-hydroxyl group on the deoxyribose sugar. Used in the Sanger method of DNA sequencing.

Differentiation The modification of a cell in terms of structure and/or function occurring during the course of development.

Diploid In eukaryotes, an organism or cell with two chromosome complements, one derived from each haploid gamete.

Disease Injury to the host that impairs host function.

Disinfectant An agent that kills microorganisms but may also be harmful to human tissue.

Disproportionation The splitting of a chemical compound into two new compounds, one more oxidized and one more reduced than the original compound.

DNA fingerprinting Use of genetic engineering to determine the origin of DNA in a sample of tissue.

DNA library A collection of cloned DNA fragments that in total contain genes from the entire genome of an organism; also called a *gene library.*

DNA polymerase An enzyme that synthesizes a new strand of DNA in the 5′ → 3′ direction using an antiparallel DNA strand as a template.

Domain The highest level of biological classification. The three domains of biological organisms are the Bacteria, the

Archaea, and the Eukarya. Also used to describe a region of a protein having a distinct function.

Doubling time The time needed for a population to double. See also *Generation time*.

Downstream position Refers to nucleic acid sequences on the 3′ side of a given site on the DNA or RNA molecule. Compare with *Upstream position*.

Ecology Study of the interrelationships between organisms and their environments.

Ecosystem A community of organisms and their natural environment.

Electron acceptor A substance that accepts electrons during an oxidation–reduction reaction.

Electron donor A compound that donates electrons in an oxidation–reduction reaction.

Electron transport phosphorylation Synthesis of ATP involving a membrane-associated electron transport chain and the creation of a proton motive force. Also called *Oxidative phosphorylation*. See also *Chemiosmosis*.

Electrophoresis Separation of charged molecules in an electric field.

Electroporation The use of an electric pulse to enable cells to take up DNA.

ELISA Enzyme-linked immunosorbent assay. An immunoassay that uses specific antibodies to detect antigens or antibodies in body fluids. The antibody-containing complexes are visualized through enzyme coupled to the antibody. Addition of substrate to the enzyme–antibody–antigen complex results in a colored product.

Emerging infection An infectious disease that has increased in incidence in the last 20 years or threatens to increase in incidence in the future.

Endemic A disease that is constantly present in low numbers in a population. Compare with *Epidemic*.

Endergonic reaction A chemical reaction requiring input of energy to proceed.

Endocytosis A process in which a particle such as a virus is taken intact into an animal cell. Phagocytosis and pinocytosis are two kinds of endocytosis.

Endoplasmic reticulum An extensive array of internal membranes in eukaryotes.

Endospore A differentiated cell formed within the cells of certain gram-positive bacteria that is extremely resistant to heat as well as to other harmful agents.

Endosymbiosis The hypothesis that mitochondria and chloroplasts are the descendants of ancient prokaryotic organisms from the domain Bacteria.

Endotoxin A toxin not released from the cell; bound to the cell surface or intracellular. Compare with *Exotoxin*.

Enrichment culture Use of selective culture media and incubation conditions to isolate microorganisms directly from natural samples.

Enteric Intestinal.

Enterotoxin A toxin affecting the intestine.

Entropy A measure of the degree of disorder in a system; entropy always increases in a closed system.

Enzyme A catalyst, usually composed of protein, that promotes specific reactions or groups of reactions.

Epidemic A disease occurring in an unusually high number of individuals in a population at the same time. Compare with *Endemic*.

Epidemiology The study of the incidence and prevalence of disease in populations.

Epitope Antigenic determinant.

Eukarya The phylogenetic domain containing all eukaryotic organisms.

Eukaryote A cell or organism having a unit membrane-enclosed (true) nucleus and usually other organelles.

Evolution Change in a line of descent over time leading to the production of new species or varieties within a species.

Evolutionary distance In phylogenetic trees, the sum of the physical distance on a tree separating organisms; this distance is inversely proportional to evolutionary relatedness.

Exergonic reaction A chemical reaction that proceeds with the liberation of energy.

Exons The coding sequences in a split gene. Contrast with *Introns*, the intervening noncoding regions.

Exotoxin A toxin released extracellularly. Compare with *Endotoxin*.

Exponential growth Growth of a microorganism where the cell number doubles within a fixed time period.

Exponential phase A period during the growth cycle of a population in which growth increases at an exponential rate.

Expression The ability of a gene to function within a cell in such a way that the gene product is formed.

Expression vector A cloning vector that contains the necessary regulatory sequences allowing transcription and translation of a cloned gene or genes.

Extreme halophile An organism whose growth is dependent on large amounts (generally >10%) of NaCl.

Facultative A qualifying adjective indicating that an organism is able to grow in either the presence or absence of an environmental factor (for example, "facultative aerobe").

Feedback inhibition A decrease in the activity of the first enzyme of a pathway caused by the final product of the pathway.

Fermentation Catabolic reactions producing ATP in which organic compounds serve as both primary electron donor and ultimate electron acceptor and ATP is produced by substrate-level phosphorylation.

Fermentation (industrial) A large-scale microbial process.

Fermenter An organism that carries out the process of fermentation.

Fermentor A large growth vessel used to culture microorganisms on a large scale frequently for the production of some commercially valuable product.

Ferredoxin An electron carrier of low reduction potential; small protein containing iron–sulfur clusters.

Fever A rise of body temperature above normal.

Filamentous In the form of very long rods, many times longer than wide.

Fimbria (plural fimbriae) Short, filamentous structure on a bacterial cell; although flagella-like in structure, generally present in many copies and not involved in motility. Plays a role in adherence to surfaces and in the formation of pellicles. See also *Pilus*.

Flagellum (plural flagella) A thin, filamentous organ of motility in prokaryotes that functions by rotating.

Flavoprotein A protein containing a derivative of riboflavin, which functions as electron carrier in the electron transport system.

Fluorescent Having the ability to emit light of a certain wavelength when activated by light of another wavelength.

Fluorescent antibody Immunoglobulin molecule that has been coupled with a fluorescent molecule so that it exhibits fluorescence.

Fomites Inanimate objects that, when contaminated with a viable pathogen, can transfer the pathogen to a host.

Food infection Microbial infection resulting from ingestion of contaminated food.

Food poisoning Disease resulting from ingestion of food contaminated with a toxin produced by a microorganism.

Frame shift   Because the genetic code is read three bases at a time, if reading begins at either the second or third base of a codon, a faulty product usually results.

Free energy   Energy available to do useful work.

Fruiting body   A macroscopic reproductive structure produced by some fungi (for example, mushrooms) and some Bacteria (for example, myxobacteria). Fruiting bodies are distinct in size, shape, and coloration for each species.

Fungi   Nonphototrophic eukaryotic microorganisms that contain rigid cell walls.

Fusion protein   The result of translation of two or more genes joined such that they retain their correct reading frames but make a single protein.

G + C base ratio   In DNA (or RNA) from any organism, the percentage of the total nucleic acid that consists of guanine plus cytosine bases.

Gametes   In eukaryotes, the haploid germ cells that result from meiosis.

Gas vesicle   A gas-filled structure made of protein that confers ability to float.

Gel   An inert polymer, usually made of agarose or polyacrylamide, used for separating macromolecules such as nucleic acids and proteins by electrophoresis.

Gene   A unit of heredity; a segment of DNA specifying a particular protein or polypeptide chain, a tRNA or an rRNA.

Gene cloning   See *Molecular cloning.*

Gene disruption   Use of genetic techniques to inactivate a gene by inserting within it a DNA fragment containing an easily selectable marker. The inserted fragment is called a *cassette,* and the process of insertion, *cassette mutagenesis.*

Gene library   A collection of cloned DNA fragments that contains all the genetic information for a particular organism.

Gene therapy   Replacement or augmentation of a dysfunctional gene for medical purposes.

Generation time   Time needed for a population to double. See also *Doubling time.*

Genetic engineering   The use of *in vitro* techniques in the isolation, manipulation, recombination, and expression of DNA.

Genetic map   The arrangement of genes on a chromosome.

Genetics   Heredity and variation of organisms.

Genome   The complete set of genes present in an organism.

Genotype   The precise genetic constitution of an organism. Compare with *Phenotype.*

Genus   A taxonomic group of related species.

Germicide   A substance that inhibits or kills microorganisms.

Glycocalyx   General term for polysaccharide components outside the bacterial cell wall. See also *Capsule* and *Slime layer.*

Glycolysis   Reactions of the Embden–Meyerhof pathway in which glucose is oxidized to pyruvate.

Glycosidic bond   A type of covalent bond that links sugar units together in a polysaccharide.

Gonococcus   *Neisseria gonorrhoeae,* the gram-negative diplococcus that causes the disease gonorrhea.

Gram-negative cell   A prokaryotic cell whose cell wall contains relatively little peptidoglycan but has an outer membrane composed of lipopolysaccharide, lipoprotein, and other complex macromolecules.

Gram-positive cell   A prokaryotic cell whose cell wall consists chiefly of peptidoglycan and lacks the outer membrane of gram-negative cells.

Growth   In microbiology, an increase in cell number.

Growth-factor analog   A chemical agent that is related to and blocks the uptake or utilization of a growth factor.

Growth rate   The rate at which growth occurs, usually expressed as the generation time.

Guild   A group of metabolically related organisms.

Habitat   The location in nature where an organism resides.

Halophile   An organism requiring salt (NaCl) for growth.

Halotolerant   Capable of growing in the presence of NaCl but not requiring it.

Haploid   An organism or cell containing only one set of chromosomes.

Hapten   A low-molecular-weight substance not inducing antibody formation but able to combine with a specific antibody.

Helix   A spiral structure in a macromolecule that contains a repeating pattern.

Hemagglutination   Agglutination of red blood cells.

Hemolysins   Bacterial toxins capable of lysing red blood cells.

Hemolysis   Lysis of red blood cells.

Herd immunity   Resistance of a group to a pathogen as a result of the immunity of a large proportion of the group to that pathogen.

Heterocyst   A differentiated cyanobacterial cell that carries out nitrogen fixation.

Heteroduplex   A double-stranded DNA in which one strand is from one source and the other strand is from another, usually related, source.

Heterofermentation   Fermentation of glucose or another sugar to a mixture of reduced products.

Heterotroph   Chemoorganotroph.

Homoacetogens   Bacteria that produce acetate as the sole product of sugar fermentation or from $H_2 + CO_2$.

Homofermentation   Fermentation of glucose or other sugar leading to a single product, lactic acid.

Homologous antigen   An antigen that reacts with the antibody it has induced.

Host   An organism capable of supporting the growth of a virus or other parasite.

Host-to-host epidemic   An epidemic resulting from host-to-host contact, characterized by a gradual rise and fall in disease incidence.

Humoral immunity   An immune response involving antibodies.

Hybridization   The natural formation or artificial construction of a duplex nucleic acid molecule by complementary base pairing between two nucleic acid strands derived from different sources.

Hybridoma   The fusion of an immortal cell with a lymphocyte to produce an immortal lymphocyte.

Hydrogen bond   A weak chemical bond between a hydrogen atom and a second, more electronegative element, usually an oxygen or nitrogen atom.

Hydrolysis   Breakdown of a polymer into smaller units, usually monomers, by addition of water; digestion.

Hydrophobic interactions   Attractive forces between molecules due to the close positioning of nonhydrophilic portions of the two molecules.

Hydrothermal vents   Warm or hot water-emitting springs associated with crustal spreading centers on the sea floor.

Hypersensitivity   An immune reaction, usually harmful to the animal, caused either by antigen–antibody reactions or cellular immune processes (see *Allergy*).

Hyperthermophile   A prokaryote having a growth temperature optimum of 80°C or higher.

Icosahedron   A geometrical shape occurring in many virus particles, with 20 triangular faces and 12 corners.

Immobilized enzyme   An enzyme attached to a solid support over which substrate is passed and converted to product.

Immune   Able to resist infectious disease.

Immunity   The ability of an organism to resist infection.

Immunization Induction of specific immunity by injecting antigens, antibodies, or immune cells.

Immunoblot (Western blot) Detection of proteins immobilized on a filter by complementary reaction with specific antibody. Compare with *Southern* and *Northern blot*.

Immunodeficiency Having a dysfunctional or completely nonfunctional immune system.

Immunogen An antigen that can induce the production of an immune response.

Immunoglobulin Antibody

Immunological memory The ability to rapidly produce large quantities of specific immune cells following reexposure to a previously encountered antigen.

In vitro In glass, away from the living organism.

In vivo In the body, in a living organism.

Incidence In reference to disease transmission, the number of cases of the disease in a specific subset of the population.

Induced enzyme An enzyme subject to induction.

Induction The process by which an enzyme is synthesized in response to the presence of an external substance, the inducer.

Infection Growth of an organism within the body.

Infection thread In the formation of root nodules, a cellulosic tube through which *Rhizobium* cells can travel to reach and infect root cells.

Inflammation Characteristic reaction to foreign particles and noxious stimuli, resulting in redness, swelling, heat, and pain.

Inhibition Prevention of growth or function.

Inoculum Material used to initiate a microbial culture.

Insertion A genetic phenomenon in which a piece of DNA is inserted into the middle of a gene.

Insertion sequence (IS elements) The simplest type of transposable element. Has only genes involved in transposition.

Integration The process by which a DNA molecule becomes incorporated into another genome.

Interferon A protein produced by cells as a result of virus infection that interferes with virus replication.

Interspecies hydrogen transfer The process by which organic matter is degraded by the interaction of several groups of microorganisms in which H_2 production and H_2 consumption are closely coupled.

Introns The intervening noncoding sequences in a split gene. Contrasted with *Exons*, the coding sequences.

Invasiveness The degree to which an organism is able to spread through the body from a focus of infection.

Ionophore A compound that can cause the leakage of ions across membranes.

Isotopes Different forms of the same element containing the same number of protons and electrons but differing in the number of neutrons.

Joule (J) A unit of energy equal to 10^7 ergs; 1000 Joules equal 1 kilojoule (kJ).

Kilobase (kb) A 1000-base fragment of nucleic acid. A *kilobase pair* is a fragment containing 1000 base pairs.

Kinase An enzyme that adds a phosphoryl group to a compound.

Lag phase The period after inoculation of a population before growth begins.

Latent virus A virus present in a cell, yet not causing any detectable effect.

Leaching Removal of valuable metals from ores by microbial action.

Leukocidin A substance able to destroy phagocytes.

Leukocyte A white blood cell.

Lichen A fungus and an alga (or a cyanobacterium) living in symbiotic association.

Lipid Water-insoluble organic molecules important in structure of the cytoplasmic membrane and (in some organisms) the cell wall. See also *Phospholipid*.

Lipopolysaccharide (LPS) Complex lipid structure containing unusual sugars and fatty acids found in many gram-negative Bacteria and constituting the chemical structure of the outer layer.

Lophotrichous Having a tuft of polar flagella.

Lower respiratory tract Trachea, bronchi, and lungs.

Luminescence Production of light.

Lymph A clear, yellowish fluid found in the lymphatic vessels that carries various white (but not red) blood cells.

Lymphocyte A white blood cell involved in antibody formation or cellular immune responses.

Lysin An antibody that induces lysis.

Lysis Rupture of a cell, resulting in a loss of cell contents.

Lysogen A prokaryote containing a prophage. See also *Temperate virus*.

Lysosome A cell organelle containing digestive enzymes.

Macromolecule A large molecule (polymer) formed by the connection of a number of small molecules (monomers).

Macrophage Large, noncirculating phagocytic cells involved in both phagocytosis and the antibody production process.

Magnetosomes Small particles of Fe_3O_4 present in cells that exhibit magnetotaxis.

Magnetotaxis Directed movement of bacterial cells by a magnetic field.

Major histocompatability complex (MHC) A cluster of genes encoding cell surface proteins important for antigen presentation to T cells.

Malignant In reference to a tumor, an infiltrating metastasizing growth no longer under normal growth control.

Mast cells Tissue cells adjoining blood vessels throughout the body that contain granules with inflammatory mediators.

Medium (plural media) In microbiology, the nutrient solution(s) used to grow microorganisms.

Meiosis In eukaryotes, reduction division, the process by which the change from diploid to haploid occurs.

Membrane Any thin sheet or layer. See especially *Cytoplasmic membrane*.

Memory cell A differentiated B lymphocyte capable of rapid conversion to an antibody-producing plasma cell on subsequent stimulation with antigen.

Mesophile Organism living in the temperature range near that of warm-blooded animals, and usually showing a growth temperature optimum between 25 and 40°C.

Messenger RNA (mRNA) An RNA molecule transcribed from DNA that contains the genetic information necessary to encode a particular protein.

Metabolism All biochemical reactions in a cell, both anabolic and catabolic.

Methanogen A methane-producing prokaryote; member of the Archaea.

Methanogenesis The biological production of methane (CH_4).

Methanotroph An organism capable of oxidizing methane.

Methylotroph An organism capable of oxidizing organic compounds that do not contain carbon–carbon bonds; if able to oxidize CH_4, also a methanotroph.

Microaerophilic Requiring O_2 but at a level lower than atmospheric.

Microenvironment The immediate physical and chemical surroundings of a microorganism.

Micrometer One-millionth of a meter, or 10^{-6} m (abbreviated μm), the unit used for measuring microorganisms.

Microorganism A microscopic organism consisting of a single cell or cell cluster including the viruses.

Microtubules Tubes that are the structural entity for eukaryotic flagella, have a role in maintaining cell shape, and function as mitotic spindle fibers.

Minus (negative)-strand nucleic acid An RNA or DNA strand that has the opposite sense of (would be complementary to) the mRNA of a virus.

Mitochondrion Eukaryotic organelle responsible for the processes of respiration and electron transport phosphorylation.

Mitosis A highly ordered process by which the nucleus divides in eukaryotes.

Mixotroph An organism able to assimilate organic compounds as carbon sources while using inorganic compounds as electron donors.

Molds Filamentous fungi.

Molecular chaperone A protein that helps other proteins fold or refold properly.

Molecular cloning Isolation and incorporation of a fragment of DNA into a vector where it can be replicated.

Molecule Two or more atoms chemically bonded to one another.

Monoclonal antibody An antibody produced from a single clone of cells. This antibody has uniform structure and specificity.

Monocytes Circulating white blood cells that contain many lysosomes and can differentiate into macrophages.

Monomer A building block of a polymer.

Monotrichous Having a single polar flagellum.

Morbidity Incidence of disease in a population, including both fatal and nonfatal cases.

Mortality Incidence of death in a population.

Motility The property of movement of a cell under its own power.

Mushrooms Filamentous fungi that produce large, often edible structures called fruiting bodies.

Mutagen An agent that induces mutation, such as radiation or certain chemicals.

Mutant A strain differing from its parent because of mutation.

Mutation An inheritable change in the base sequence of the genome of an organism.

Mycorrhiza A symbiotic association between a fungus and the roots of a plant.

Mycoses Infections caused by fungi.

Myeloma A malignant tumor of a plasma cell (antibody-producing cell).

Natural killer (NK) cell A specialized lymphocyte that recognizes and destroys foreign cells or infected host cells in a nonspecific manner.

Neutrophil A polymorphonuclear leukocyte.

Neutrophile An organism that grows best around pH 7.

Nitrification The microbiological conversion of ammonia to nitrate.

Nitrogen fixation Reduction of nitrogen gas to ammonia by the enzyme nitrogenase.

Nodule A tumorlike structure produced by the roots of symbiotic nitrogen-fixing plants. Contains the nitrogen-fixing microbial component of the symbiosis.

Nonpolar Possessing hydrophobic (water-repelling) characteristics and not easily dissolved in water.

Nonsense mutation A mutation that changes a sense codon into one that does not code for an amino acid.

Normal flora Microorganisms that are usually found associated with healthy body tissue.

Northern blot Hybridization of a single strand of nucleic acid (DNA or RNA) to RNA fragments immobilized on a filter. Compare with *Southern* and *Western blot*.

Nosocomial infection Hospital-acquired infection.

Nucleic acid A polymer of nucleotides. See *Deoxyribonucleic acid* and *Ribonucleic acid*.

Nucleic acid probe A strand of nucleic acid that can be labeled and used to hybridize to a complementary molecule from a mixture of other nucleic acids. In clinical microbiology, short oligonucleotides of unique sequences used as hybridization probes for identifying pathogens.

Nucleoid The aggregated mass of DNA that makes up the chromosome of prokaryotic cells.

Nucleoside A nucleotide minus phosphate.

Nucleotide A monomeric unit of nucleic acid, consisting of a sugar, a phosphate, and a nitrogenous base.

Nucleus A membrane-enclosed structure in eukaryotes containing the genetic material (DNA) organized in chromosomes.

Nutrient A substance taken by a cell from its environment and used in catabolic or anabolic reactions.

Obligate A qualifying adjective referring to an environmental factor always required for growth (for example, "obligate anaerobe").

Oligonucleotide A short nucleic acid molecule, either obtained from an organism or synthesized chemically.

Oligotrophic Describing a habitat in which nutrients are in low supply.

Oncogene A gene whose expression causes formation of a tumor.

Open reading frame (ORF) The entire length of a DNA molecule that starts with a start codon and ends with a stop codon.

Operator A specific region of the DNA at the initial end of a gene, where the repressor protein binds and blocks mRNA synthesis.

Operon A cluster of genes whose expression is controlled by a single operator. Typical of prokaryotic cells.

Opsonization Promotion of phagocytosis by a specific antibody in combination with complement.

Organelle A membrane-enclosed structure found in eukaryotic cells.

Osmosis Diffusion of water through a membrane from a region of low solute concentration to one of higher concentration.

Outbreak The occurrence of a large number of cases of a disease in a short period of time.

Oxic Containing oxygen; aerobic. Usually used in reference to a microbial habitat.

Oxidation A process by which a compound gives up electrons (or H atoms) and becomes oxidized.

Oxidation–reduction (redox) reaction A pair of reactions in which one compound becomes oxidized while another becomes reduced and takes up the electrons released in the oxidation reaction.

Oxidative (electron transport) phosphorylation The non-phototrophic production of ATP at the expense of a proton motive force formed by electron transport.

Oxygenic photosynthesis Use of light energy to synthesize ATP and NADPH by noncyclic photophosphorylation with the production of O_2 from water.

Palindrome A nucleotide sequence on a DNA molecule in which the same sequence is found on each strand but in the opposite direction.

Pandemic   A worldwide epidemic.

Parasite   An organism able to live on and cause damage to another organism.

Passive immunity   Immunity resulting from transfer of antibodies or immune cells from an immune to a nonimmune individual.

Pasteurization   A process using mild heat (usually 80°C for 15 min) to reduce the microbial level in heat-sensitive materials.

Pathogen   An organism able to inflict damage on a host it infects.

Pathogenicity   The ability of a parasite to inflict damage on the host.

Peptide bond   A type of covalent bond joining amino acids in a polypeptide.

Peptidoglycan   The rigid layer of the cell walls of Bacteria, a thin sheet composed of *N*-acetylglucosamine, *N*-acetylmuramic acid, and a few amino acids. Also called *murein*.

Periplasmic space   The area between the cytoplasmic membrane and the cell wall in gram-negative Bacteria.

Peritrichous flagellation   Condition of having flagella attached to many places on the cell surface.

Phage   See *Bacteriophage*.

Phagemid   A cloning vector that can replicate either as a plasmid or as a bacteriophage.

Phagocyte   A body cell able to ingest and digest foreign particles.

Phagocytosis   Ingestion of particulate material such as bacteria by protozoa and phagocytic cells of higher organisms.

Phenotype   The observable characteristics of an organism. Compare with *Genotype*.

Phosphodiester bond   A type of covalent bond linking nucleotides together in a polynucleotide.

Phospholipid   Lipids containing a substituted phosphate group and two fatty acid chains on a glycerol backbone.

Photoautotroph   An organism able to use light as its sole source of energy and CO_2 as its sole carbon source.

Photoheterotroph   An organism using light as a source of energy and organic materials as carbon source.

Photophosphorylation   Synthesis of high energy phosphate bonds as ATP, using light energy.

Photosynthesis   The use of light energy to drive the incorporation of CO_2 into cell material. See also *Anoxygenic photosynthesis* and *Oxygenic photosynthesis*.

Phototaxis   Movement toward light.

Phototroph   An organism that obtains energy from light.

Phylogeny   The ordering of species into higher taxa and the construction of evolutionary trees based on evolutionary (natural) relationships.

Phytanyl   A branched-chain hydrocarbon containing 20 carbon atoms, commonly found in the lipids of Archaea.

Pilus   A fimbria-like structure that is present on fertile cells, both Hfr and F$^+$, and is involved in DNA transfer during conjugation. Sometimes called a *sex pilus*. See also *Fimbria*.

Pinocytosis   In eukaryotes, phagocytosis of soluble molecules.

Plaque   A zone of lysis or cell inhibition caused by virus infection on a lawn of cells.

Plasma   The noncellular portion of blood.

Plasma cell   A large, differentiated, short-lived B lymphocyte specializing in abundant (but short-term) antibody production.

Plasmid   An extrachromosomal genetic element that is not essential for growth and has no extracellular form.

Platelet   A noncellular disc-shaped structure containing protoplasm found in large numbers in blood and functioning in the blood clotting process.

Plus-strand nucleic acid   An RNA or DNA strand that has the same sense as the mRNA of a virus.

Point mutation   A mutation that involves one or only a very few base pairs.

Polar   Possessing hydrophilic characteristics and generally water-soluble.

Polar flagellation   Condition of having flagella attached at one end or both ends of the cell.

Poly-β-hydroxybutyrate (PHB)   A common storage material of prokaryotic cells consisting of a polymer of β-hydroxybutyrate (PHB) or other β-alkanoic acids (PHA).

Polyclonal antiserum   A mixture of antibodies to a variety of antigens or to a variety of determinants on a single antigen.

Polymer   A large molecule formed by polymerization of monomeric units.

Polymerase chain reaction (PCR)   A method used to amplify a specific DNA sequence *in vitro* by repeated cycles of synthesis using specific primers and DNA polymerase.

Polymorphonuclear leukocyte (PMN)   Motile white blood cells containing many lysosomes and specializing in phagocytosis. Characterized by a distinct segmented nucleus. Also, a neutrophil.

Polynucleotide   A polymer of nucleotides bonded to one another by phosphodiester bonds.

Polypeptide   Several amino acids linked together by peptide bonds.

Polysaccharide   A long chain of monosaccharides (sugars) linked by glycosidic bonds.

Porins   Protein channels in the lipopolysaccharide layer of gram-negative Bacteria through which small to medium-sized molecules can flow.

Porters   Membrane proteins that function to transport substances into and out of the cell.

Precipitation   A reaction between antibody and soluble antigen resulting in visible antibody–antigen complexes.

Prevalence   The *proportion* of individuals in a population having a disease.

Pribnow box   The consensus sequence TATAAT located approximately 10 base pairs upstream from the transcriptional start site. A binding site for RNA polymerase.

Primary antibody response   Antibodies made on first exposure to antigen; mostly of the class IgM.

Primary metabolite   A metabolite excreted during the growth phase.

Primary producer   An organism that uses light to synthesize new organic material from CO_2.

Primary structure   In an informational macromolecule, such as a polypeptide or a nucleic acid, the *precise sequence* of monomeric units.

Primary transcript   An unprocessed RNA molecule that is the direct product of transcription.

Primer   A molecule (usually a polynucleotide) to which DNA polymerase can attach the first nucleotide during DNA replication.

Prion   An infectious agent whose extracellular form may contain no nucleic acid.

Probe   See *Nucleic acid probe*.

Prochlorophyte   A prokaryotic oxygenic phototroph that contains chlorophylls *a* and *b* but lacks phycobilins.

Prokaryote   A cell or organism lacking a nucleus and other membrane-enclosed organelles, usually having its DNA in a single circular molecule.

Promoter   The site on DNA where the RNA polymerase binds and begins transcription.

Prophage   The state of the genome of a temperate virus when it is replicating in synchrony with that of the host, typically integrated into the host genome.

Prophylactic   Treatment, usually immunological or chemotherapeutic, designed to protect an individual from a future attack by a pathogen.

Prostheca   A cytoplasmic extrusion from a cell such as a bud, hypha, or stalk.

Prosthetic group   The tightly bound, nonprotein portion of an enzyme; not the same as a *Coenzyme*.

Protein   A polymeric molecule consisting of one or more polypeptides.

Proton motive force   An energized state of a membrane created by expulsion of protons usually occurring through action of an electron transport chain. See also *Chemiosmosis*.

Protoplasm   The complete cellular contents, cytoplasmic membrane, cytoplasm, and nucleus/nucleoid.

Protoplast   A cell from which the wall has been removed.

Prototroph   The parent from which an auxotrophic mutant has been derived. Contrast with *Auxotroph*.

Protozoa   Unicellular eukaryotic microorganisms that lack cell walls.

Provirus   See *Prophage*.

Psychrophile   An organism able to grow at low temperatures and showing a growth temperature optimum of <15°C.

Psychrotolerant   Able to grow at low temperature but having a growth temperature optimum of >15°C.

Public health   The health of the population as a whole.

Pure culture   A culture containing a single kind of microorganism.

Pyogenic   Pus-forming; causing abscesses.

Pyrite   A common iron ore, FeS_2.

Pyrogenic   Fever-inducing.

Quarantine   The limitation on the freedom of movement of an individual to prevent spread of a disease to other members of a population.

Quaternary structure   In proteins, the number and arrangement of individual polypeptides in the final protein molecule.

Radioimmunoassay   An immunological assay employing radioactive antibody or antigen for the detection of antigen or antibody binding.

Radioisotope   An isotope of an element that undergoes spontaneous decay with the release of radioactive particles.

Reaction center   A photosynthetic complex containing chlorophyll (or bacteriochlorophyll) and other components, within which occurs the initial electron transfer reactions of photophosphorylation.

Reading-frame shift   See *Frame shift*.

Recalcitrant   Resistant to microbial attack.

Recombinant DNA   A DNA molecule containing DNA originating from two or more sources.

Recombination   Process by which genetic elements in two separate genomes are brought together in one unit.

Redox   See *Oxidation–reduction reaction*.

Reduction   A process by which a compound accepts electrons to become reduced.

Reduction potential (E_0')   The inherent tendency, measured in volts, of a compound to act as an electron donor or an electron acceptor.

Reductive dechlorination   Removal of Cl as Cl^- from an organic compound by reducing the carbon atom from C—Cl to C—H.

Regulation   Processes that control the rates of synthesis of proteins. Induction and repression are examples of regulation.

Regulon   A set of operons that are all controlled by the same regulatory protein (repressor or activator).

Replacement vector   A cloning vector, such as a bacteriophage, in which some of the DNA of the vector can be replaced with foreign DNA.

Replication   Synthesis of DNA using DNA as a template.

Repression   The process by which the synthesis of an enzyme is inhibited by the presence of an external substance, the repressor.

Repressor protein   A regulatory protein that binds to specific sites on DNA and blocks transcription; involved in negative control.

Reservoir   In epidemiology, the organism or environment that normally harbors a pathogen.

Respiration   Catabolic reactions producing ATP in which either organic or inorganic compounds are primary electron donors and organic or inorganic compounds are ultimate electron acceptors.

Response regulator protein   One of the members of a two-component system; a regulatory protein that is phosphorylated by a sensor protein (see *Sensor protein*).

Restriction endonucleases (restriction enzymes)   Enzymes that recognize and cleave specific DNA sequences, generating either blunt or single-stranded (sticky) ends.

Resurgent infection   An infectious disease thought to be under control that reemerges in high incidence.

Retrovirus   A virus containing single-stranded RNA as its genetic material, which produces a complementary DNA by action of the enzyme reverse transcriptase.

Reverse electron transport   The energy-dependent movement of electrons against the thermodynamic gradient to form a strong electron donor from a weaker electron donor.

Reverse transcription   The process of copying information found in RNA into DNA.

Reverse translation   The mental process of using a codon table and the amino acid sequence of a protein to obtain a possible sequence of the mRNA or the gene that encoded the protein.

Rheumatic fever   An inflammatory autoimmune disease triggered by an immune response to infection by *Streptococcus pyogenes*.

Rhizosphere   The region immediately adjacent to plant roots.

Ribonucleic acid (RNA)   A polymer of nucleotides connected via a phosphate–ribose backbone; involved in protein synthesis.

Ribosomal RNA (rRNA)   Type of RNA found in the ribosome; some rRNAs participate actively in the process of protein synthesis.

Ribosome   A cytoplasmic particle composed of ribosomal RNA and protein, which is part of the protein-synthesizing machinery of the cell.

Ribozyme   An RNA molecule that can catalyze a chemical reaction.

Rickettsias   Obligate intracellular parasites that cause a variety of diseases, including typhus and Rocky Mountain spotted fever.

RNA life   A hypothetical ancient life form lacking DNA and protein that may have existed on early Earth and in which RNA had both a genetic coding and a catalytic function.

RNA polymerase   An enzyme that synthesizes RNA in the $5' \rightarrow 3'$ direction using an antiparallel DNA strand as a template.

RNA processing   The conversion of a precursor RNA to its mature form.

Root nodule   A tumorlike growth on certain plant roots that contains symbiotic nitrogen-fixing bacteria.

Rumen   The forestomach of ruminant animals in which cellulose digestion occurs.

S-layer   A paracrystalline outer wall layer composed of protein or glycoprotein and found in many prokaryotes.

Scale-up   Conversion of an industrial process from a small laboratory setup to a large commercial fermentation.

Scarlet fever   Disease characterized by high fever and a reddish skin rash resulting from an exotoxin produced by cells of *Streptococcus pyogenes.*

Secondary antibody response   Antibody made on second (subsequent) exposure to antigen; mostly of the class IgG.

Secondary metabolite   A product excreted by a microorganism near the end of the growth phase or during the stationary phase.

Secondary structure   The initial pattern of folding of a polypeptide or a polynucleotide, usually the result of hydrogen bonding.

Secondary treatment   In sewage treatment, either the aerobic or anaerobic decomposition of sewage following the removal of nondegradable objects by primary treatment.

Secretion vector   A DNA vector in which the protein product is both expressed in and secreted (excreted) from the cell.

Selection   Placing organisms under conditions where the growth of those with a particular genotype will be favored.

Semiconservative replication   DNA synthesis yielding new double helices, each consisting of one parental and one progeny strand.

Semisynthetic penicillin   A natural penicillin that has been chemically altered.

Sensor protein   One of the members of a two-component system; a kinase found in the cell membrane that phosphorylates itself in response to an external signal and then passes the phosphoryl group to a response regulator protein (see *Response regulator protein*).

Septicemia   Infection of the bloodstream by microorganisms.

Serology   The study of antigen–antibody reactions *in vitro.*

Serum   Fluid portion of blood remaining after the blood cells and materials responsible for clotting are removed.

Sexually transmitted disease (STD)   A disease whose usual means of transmission is by sexual contact.

Shine–Dalgarno sequence   A short stretch of nucleotides on a prokaryotic mRNA molecule upstream of the translational start site that binds to ribosomal RNA and thereby brings the ribosome to the initiation codon on the mRNA.

Shuttle vector   A cloning vector that can replicate in two different organisms; used for moving DNA between unrelated organisms.

Sickle-cell anemia   A genetic trait that confers resistance to malaria but causes a reduction in the number and efficiency of red blood cells.

Siderophore   An iron chelator that can bind iron present at very low concentrations.

Signal sequence   A short stretch of amino acids found at the beginning of proteins that are typically excreted from the cell. The signal sequence is usually rich in hydrophobic amino acids, which helps transport the entire polypeptide through the membrane.

Signature sequence   Short oligonucleotides of unique sequence found in 16S ribosomal RNA of a particular group of prokaryotes.

Single-cell protein   Protein derived from microbial cells for use as food or a food supplement.

Site-directed mutagenesis   a technique whereby a gene with a specific mutation can be constructed *in vitro.*

16S rRNA   A large polynucleotide (~1500 bases) that functions as a part of the small subunit of the ribosome of prokaryotes and from whose sequence evolutionary relationships can be obtained; eukaryotic counterpart, 18S rRNA.

Slime layer   A diffuse layer of polysaccharide exterior to the cell wall in some prokaryotes. See also *Capsule* and *Glycocalyx.*

Slime molds   Nonphototrophic eukaryotic microorganisms lacking cell walls, which aggregate to form fruiting structures (cellular slime molds) or simply masses of protoplasm (acellular slime molds).

Solfatara   A hot, sulfur-rich, generally acidic environment, commonly inhabited by hyperthermophilic Archaea.

Southern blot   Hybridization of a single strand of nucleic acid (DNA or RNA) to DNA fragments immobilized on a filter. Compare with *Northern* and *Western blot.*

Species   Of prokaryotes, a collection of closely related (>97% 16S rRNA sequence homology) strains sufficiently different from all other strains to be recognized as a distinct unit.

Specificity   The ability of the immune response to interact with individual antigens.

Spheroplast   A spherical, osmotically sensitive cell derived from a bacterium by loss of some but not all of the rigid wall layer. If all the rigid wall layer has been completely lost, the structure is called a *Protoplast.*

Spontaneous generation   The hypothesis that living organisms can originate from nonliving matter.

Spore   A general term for resistant resting structures formed by many prokaryotes and fungi.

Sporozoa   Nonmotile parasitic protozoa.

Stalk   An elongate structure, either cellular or excreted, that anchors a cell to a surface.

Static   Inhibitory.

Stationary phase   The period during the growth cycle of a population in which growth ceases.

Stereoisomers   Mirror image forms of two molecules having the same molecular and structural formulas.

Sterile   Free of living organisms and viruses.

Sterilization   Treatment resulting in death of all living organisms and viruses in a material.

Strain   A population of cells all descended from a single cell; a clone.

Stromatolites   Laminated microbial mats, typically built from layers of filamentous and other microorganisms which can become fossilized.

Substrate   The molecule undergoing reaction with an enzyme.

Substrate-level phosphorylation   Synthesis of high energy phosphate bonds through reaction of inorganic phosphate with an activated organic substrate.

Supercoil   Highly twisted form of circular DNA.

Superoxide anion (O_2^-)   A derivative of O_2 capable of oxidative destruction of cell components.

Suppressor   A mutation that restores a wild-type phenotype without altering the original mutation, usually arising by mutation in another gene.

Surveillance   Observation, recognition, and reporting of diseases as they occur.

Symbiosis   A relationship between two organisms.

Synthetic DNA   A DNA molecule that has been made by a chemical process in a laboratory.

Syntrophy   A nutritional situation in which two or more organisms combine their metabolic capabilities to catabolize a substance not capable of being catabolized by either one alone.

Systemic   Not localized in the body; an infection disseminated widely through the body is said to be systemic.

T cell receptor   The antigen-specific receptor on the surface of T lymphocytes.

T-DNA   The segment of the *Agrobacterium* Ti plasmid that is transferred to plant cells.

T lymphocyte   A type of immune cell responsible for antigen-specific cell-mediated immune responses plus stimulation of differentiation of antibody-producing (B) lymphocytes during the humoral immune response.

Taxis   Movement toward or away from a stimulus.

Taxonomy   The study of scientific classification and nomenclature.

Temperate virus   A virus that on infection of a host does not necessarily cause lysis but whose genome may replicate in synchrony with that of the host. See *Lysogen.*

Tertiary structure   The final folded structure of a polypeptide that has previously attained secondary structure.

Tetracycline   A member of a class of antibiotics containing the four-membered naphthacene ring.

Thermocline   Zone of water in a stratified lake in which temperature and oxygen concentration drop precipitously with depth.

Thermophile   An organism with a growth temperature optimum between 45 and 80°C.

Ti plasmid   A conjugative plasmid present in the bacterium *Agrobacterium tumefaciens* that can transfer genes into plants.

Titer   Measure of antibody quantity.

Tolerance   In reference to immunology, the acquisition of nonresponsiveness to a molecule normally recognized by the immune system.

Toxic shock syndrome   Acute shock resulting from host response to an exotoxin produced by *Staphylococcus aureus.*

Toxigenicity   The degree to which an organism is able to elicit toxic symptoms.

Toxin   A microbial substance able to induce host damage.

Toxoid   A toxin modified so that it is no longer toxic but is still able to induce antibody formation.

Transcription   Synthesis of an RNA molecule complementary to one of the two strands of a double-stranded DNA molecule.

Transduction   Transfer of host genes from one cell to another by a virus.

Transfection   The transformation of a prokaryotic cell by DNA or RNA from a virus. Used also to describe the process of genetic transformation in eukaryotic cells.

Transfer RNA (tRNA)   A type of RNA that carries amino acids to the ribosome during translation; contains the anticodon.

Transformation   Transfer of genetic information via free DNA. Also, a process, sometimes initiated by infection with certain viruses, whereby a normal animal cell becomes a cancer cell.

Transgenic organisms   Plants or animals that stably pass on cloned DNA that has been inserted into them.

Translation   The synthesis of protein using the genetic information in a messenger RNA as a template.

Transpeptidation   The formation of peptide bonds between the short peptides present in the cell wall polymer, peptidoglycan.

Transposable element   A genetic element that has the ability to move (transpose) from one site on a chromosome to another.

Transposon   A type of transposable element that, in addition to genes involved in transposition, carries other genes; often genes conferring selectable phenotypes such as antibiotic resistance.

Transposon mutagenesis   Insertion of a transposon into a gene; this inactivates the host gene, leading to a mutant phenotype, and also confers the phenotype associated with the transposon gene.

Tuberculin test   A test for previous infection with *Mycobacterium tuberculosis* characterized by an inflammatory cell-mediated immune response.

Two-component system   A regulatory system containing a sensor protein and a response regulator protein (see *Sensor protein* and *Response regulator protein*).

Upper respiratory tract   The nasopharynx, oral cavity, and throat.

Upstream position   Refers to nucleic acid sequences on the 5′ side of a given site on a DNA or RNA molecule. Compare with *Downstream position.*

Vaccination   Inoculation of a host with inactive or weakened pathogens or pathogen products to stimulate protective immunity.

Vaccine   Material used to induce specific protective immunity to a pathogen.

Vacuole   A small space in a cell that contains fluid and is surrounded by a membrane. In contrast to a vesicle, a vacuole is not rigid.

Vector   An agent, usually an insect or other animal, able to carry pathogens from one host to another. Also, a genetic element able to incorporate DNA and cause it to be replicated in another cell.

Vehicle   Nonliving source of pathogens that infect large numbers of individuals; common vehicles are food and water.

Viable   Alive; able to reproduce.

Viable count   Measurement of the concentration of live cells in a microbial population.

Virion   A virus particle; the virus nucleic acid surrounded by a protein coat and in some cases other material.

Viroid   A small RNA molecule with viruslike properties.

Virulence   Degree of pathogenicity of a parasite.

Virus   A genetic element containing either DNA or RNA that replicates in cells but is characterized by having an extracellular state.

Water activity (a_w)   An expression of the relative availability of water in a substance. Pure water has an a_w of 1.000.

Western blot   See *Immunoblot.*

Wild type   A strain of microorganism isolated from nature. The usual or native form of a gene or organism.

Wobble   In reference to reading the genetic code, the concept that nonstandard base pairing is allowed between the anticodon and the third position of the codon.

Xenobiotic   A completely synthetic chemical compound not naturally occurring on Earth.

Xerophile   An organism adapted to growth at very low water potentials.

Yeasts   Unicellular fungi.

Zoonoses   Diseases primarily of animals that are occasionally transmitted to humans.

Zygote   In eukaryotes, the single diploid cell resulting from the union of two haploid gametes.

Index